"It's not for the textbook online," I replied. "It's for a website that supports the textbook with all sorts of study materials. Using those, in addition to taking good lecture notes and thoroughly reading the textbook, has made a real difference in my understanding of the topics."

How I Learned Intro. Biology

Then I explained that for years I had wanted to be a better student. "When I bought the Freeman textbook," I continued, "I saw the Companion Website URL and the access code for it. I went to the site and found an Online Study Guide that offers key summary points and review questions for each section of each chapter. Web tutorials on the website, as well as on the student CD, use animations to help me understand dynamic and challenging concepts in biology.

"There are also PDF files of book figures that I can print out before class, so I don't have to spend half the lecture trying to redraw them as I take notes. That allows me to pay closer attention to the lecture and take better notes than ever before. Together, these online learning tools—along with my lecture notes and a close reading of our unusually well written textbook—have given me the confidence in my abilities that I have lacked for years.

"I have been doing quite well on the quizzes, and I can't wait to take the first exam next week. And as Bronson will tell you, I did OK in high school, but I've never had a special understanding of biology or science."

Try It *Before* the First Exam

You, too, can use the Companion Website to help you learn introductory biology. Forget the idea that you had to be a high school valedictorian to do well in this course. Success in college is all about being well organized and knowing how to make the most of your limited study time. Whether you are an average student or one of *People* magazine's "50 Most Promising Freshmen," you will be interested in using Freeman's Companion Website.

With Apologies to John Caples

This is a parody of one of the most successful ads of all time: "They Laughed When I Sat Down at the Piano, But When I Started to Play!" created in 1925 by advertising genius John Caples. We wanted to try something different to draw your attention to the Freeman Companion Website, because too many students learn about textbook websites only after they are disappointed by early exam results. We hope that this ad will lead you to visit the Companion Website.

ONLIN

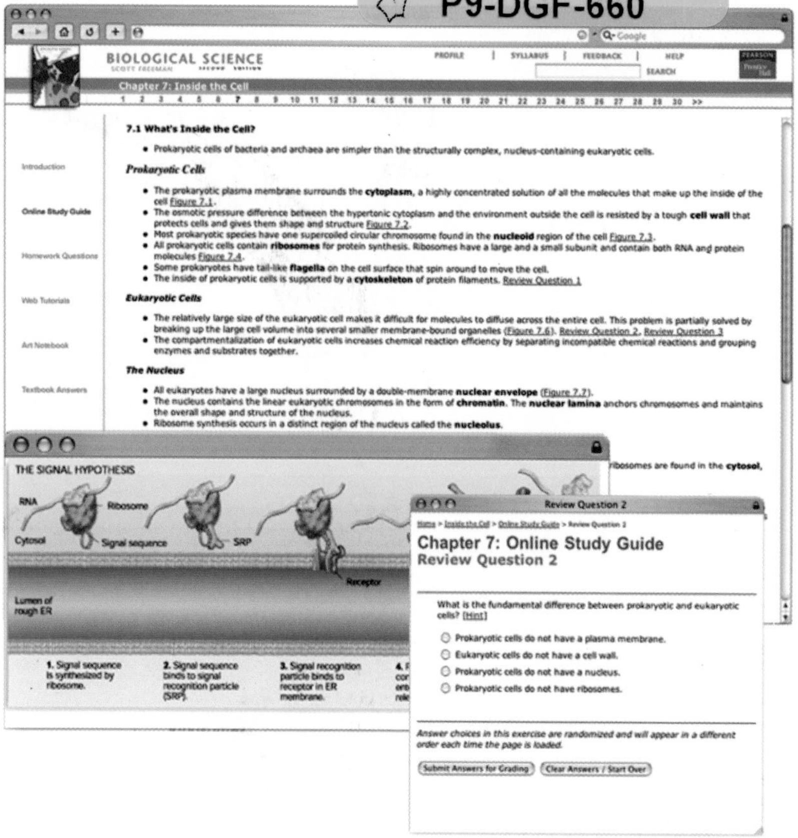

WEB TUTORIALS

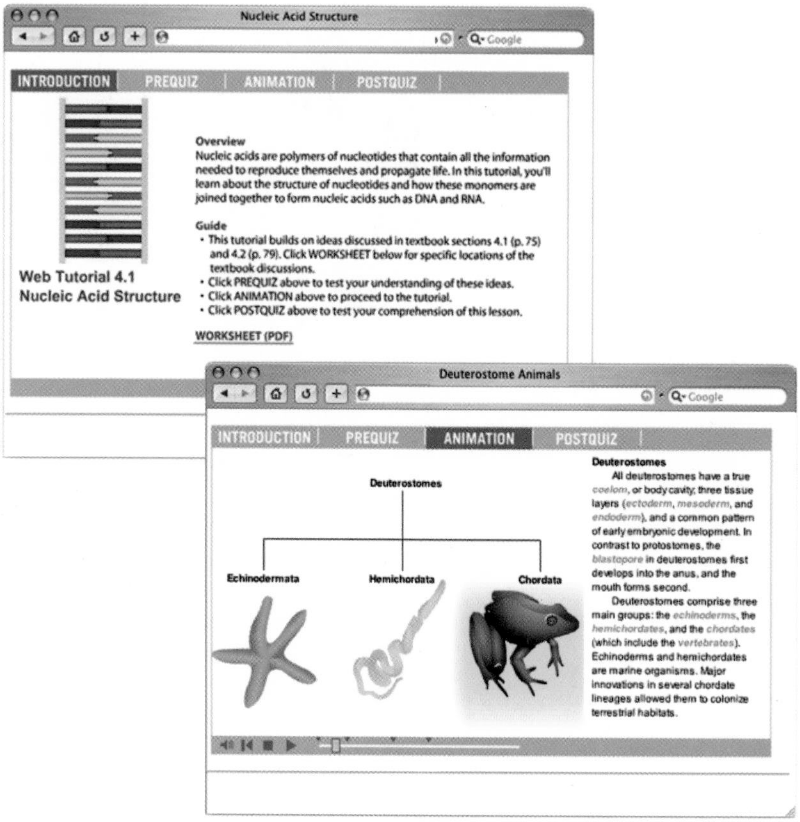

Go to www.prenhall.com/freeman/biology and enter the access code on the card that comes with a new copy of Freeman's *Biological Science*.

Classification of Life

This appendix presents a taxonomic classification for the major extant groups of organisms discussed in this text; not all phyla are included. The classification presented here is based on the three-domain system, which assigns the two major groups of prokaryotes, bacteria and archaea, to separate domains (with eukaryotes making up the third domain).

Various alternative classification schemes are discussed in Unit Five of the text. The taxonomic turmoil includes debates about the number and boundaries of kingdoms and about the alignment of the Linnaean classification hierarchy with the findings of modern cladistic analysis. In this review, asterisks (*) indicate currently recognized phyla thought by some systematists to be paraphyletic.

DOMAIN BACTERIA

- **Proteobacteria**
- **Chlamydia**
- **Spirochetes**
- **Gram-positive Bacteria**
- **Cyanobacteria**

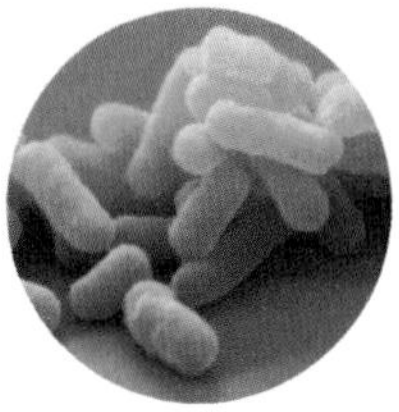

DOMAIN ARCHAEA

- **Korarchaeota**
- **Euryarchaeota**
- **Crenarchaeota**
- **Nanoarchaeota**

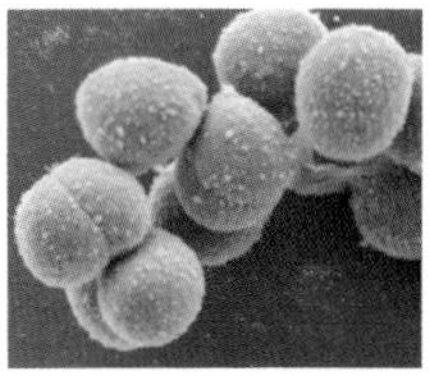

DOMAIN EUKARYA

In the phylogenetic hypothesis we present in Chapter 28, major clades of eukaryotes are grouped together in the five "supergroups" listed below in blue type. Formerly, all the eukaryotes generally called protists were assigned to a single kingdom, Protista. However, advances in systematics have made it clear that Protista is in fact polyphyletic: Some protists are more closely related to plants, fungi, or animals than they are to other protists. As a result, the kingdom Protista has been abandoned.

Excavata

- Diplomonadida (diplomonads)
- Parabasala (parabasalids)
- Euglenozoa (euglenozoans)
  - Kinetoplastida (kinetoplastids)
  - Euglenophyta (euglenids)

Chromalveolata

- Alveolata (alveolates)
  - Dinoflagellata (dinoflagellates)
  - Apicomplexa (apicomplexans)
  - Ciliophora (ciliates)
- Stramenopila (stramenopiles)
  - Bacillariophyta (diatoms)
  - Chrysophyta (golden algae)
  - Phaeophyta (brown algae)
  - Oomycota (water moulds)

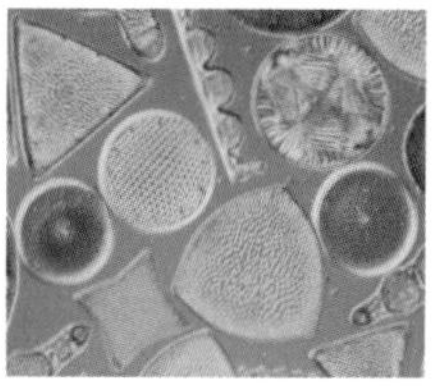

Archaeplastida

- Rhodophyta (red algae)
- Chlorophyta (green algae: chlorophytes)
- Charophyta (green algae: charophytes)
- Plantae

Phylum Hepatophyta (liverworts) Phylum Bryophyta (mosses) Phylum Anthocerophyta (hornworts)	Nonvascular plants (bryophytes)	
Phylum Lycophyta (lycophytes) Phylum Pterophyta (ferns, horsetails, whisk ferns)	Seedless vascular plants	
Phylum Ginkgophyta (ginkgo) Phylum Cycadophyta (cycads) Phylum Gnetophyta (gnetophytes) Phylum Coniferophyta (conifers)	Gymnosperms	Seed plants
Phylum Anthophyta (flowering plants)	Angiosperms	Seed plants

Rhizaria

- Radiolaria (radiolarians)
- Foraminifera (forams)
- Cercozoa (cercozoans)

Periodic Table of the Elements

Atomic number (number of protons) → 6
Element symbol → C
12.01 ← Atomic mass (number of protons plus number of neutrons averaged over all isotopes)

Metals | Metalloids | Nonmetals

Representative elements

Groups: Elements in a vertical column have the same number of electrons in their valence (outer) shell and thus have similar chemical properties.

Periods: Each horizontal row contains elements with the same total number of electron shells. Across each period, elements are ordered by increasing atomic number.

Alkali metals (Group 1) — Alkaline earth metals (Group 2) — Transition elements (Groups 3–12) — Halogens (Group 17) — Noble gases (Group 18)

Increasing electronegativity → (across a period); Increasing electronegativity ↑ (up a group)

Period number	1 Group 1A	2 Group 2A	3 3B	4 4B	5 5B	6 6B	7 7B	8 8B	9 8B	10 8B	11 1B	12 2B	13 Group 3A	14 Group 4A	15 Group 5A	16 Group 6A	17 Group 7A	18 Group 8A
1	1 H 1.008																	2 He 4.003
2	3 Li 6.941	4 Be 9.012											5 B 10.81	6 C 12.01	7 N 14.01	8 O 16.00	9 F 19.00	10 Ne 20.18
3	11 Na 22.99	12 Mg 24.31											13 Al 26.98	14 Si 28.09	15 P 30.97	16 S 32.07	17 Cl 35.45	18 Ar 39.95
4	19 K 39.10	20 Ca 40.08	21 Sc 44.96	22 Ti 47.87	23 V 50.94	24 Cr 52.00	25 Mn 54.94	26 Fe 55.85	27 Co 58.93	28 Ni 58.69	29 Cu 63.55	30 Zn 65.41	31 Ga 69.72	32 Ge 72.64	33 As 74.92	34 Se 78.96	35 Br 79.90	36 Kr 83.80
5	37 Rb 85.47	38 Sr 87.62	39 Y 88.91	40 Zr 91.22	41 Nb 92.91	42 Mo 95.94	43 Tc (98)	44 Ru 101.1	45 Rh 102.9	46 Pd 106.4	47 Ag 107.9	48 Cd 112.4	49 In 114.8	50 Sn 118.7	51 Sb 121.8	52 Te 127.6	53 I 126.9	54 Xe 131.3
6	55 Cs 132.9	56 Ba 137.3	57* La 138.9	72 Hf 178.5	73 Ta 180.9	74 W 183.8	75 Re 186.2	76 Os 190.2	77 Ir 192.2	78 Pt 195.1	79 Au 197.0	80 Hg 200.6	81 Tl 204.4	82 Pb 207.2	83 Bi 209.0	84 Po (209)	85 At (210)	86 Rn (222)
7	87 Fr (223)	88 Ra (226)	89† Ac (227)	104 Rf (261)	105 Db (262)	106 Sg (266)	107 Bh (264)	108 Hs (269)	109 Mt (268)	110 Ds (271)	111 Rg (272)	112 Cn (285)	113 — (284)	114 Fl (289)	115 — (288)	116 Lv (293)	117 — (294?)	118 — (294)

*Lanthanides	58 Ce 140.1	59 Pr 140.9	60 Nd 144.2	61 Pm (145)	62 Sm 150.4	63 Eu 152.0	64 Gd 157.3	65 Tb 158.9	66 Dy 162.5	67 Ho 164.9	68 Er 167.3	69 Tm 168.9	70 Yb 173.0	71 Lu 175.0
†Actinides	90 Th 232.0	91 Pa 231.0	92 U 238.0	93 Np (237)	94 Pu (244)	95 Am (243)	96 Cm (247)	97 Bk (247)	98 Cf (251)	99 Es 252	100 Fm 257	101 Md 258	102 No 259	103 Lr 260

Name (Symbol)	Atomic Number	Name (Symbol)	Atomic Number	Name (Symbol)	Atomic Number	Name (Symbol)	Atomic Number	Name (Symbol)	Atomic Number
Actinium (Ac)	89	Copernicium (Cn)	112	Iodine (I)	53	Osmium (Os)	76	Silicon (Si)	14
Aluminum (Al)	13	Copper (Cu)	29	Iridium (Ir)	77	Oxygen (O)	8	Silver (Ag)	47
Americium (Am)	95	Curium (Cm)	96	Iron (Fe)	26	Palladium (Pd)	46	Sodium (Na)	11
Antimony (Sb)	51	Darmstadtium (Ds)	110	Krypton (Kr)	36	Phosphorus (P)	15	Strontium (Sr)	38
Argon (Ar)	18	Dubnium (Db)	105	Lanthanum (La)	57	Platinum (Pt)	78	Sulphur (S)	16
Arsenic (As)	33	Dysprosium (Dy)	66	Lawrencium (Lr)	103	Plutonium (Pu)	94	Tantalum (Ta)	73
Astatine (At)	85	Einsteinium (Es)	99	Lead (Pb)	82	Polonium (Po)	84	Technetium (Tc)	43
Barium (Ba)	56	Erbium (Er)	68	Lithium (Li)	3	Potassium (K)	19	Tellurium (Te)	52
Berkelium (Bk)	97	Europium (Eu)	63	Livermorium (Lv)	116	Praseodymium (Pr)	59	Terbium (Tb)	65
Beryllium (Be)	4	Fermium (Fm)	100	Lutetium (Lu)	71	Promethium (Pm)	61	Thallium (Tl)	81
Bismuth (Bi)	83	Flerovium (Fl)	114	Magnesium (Mg)	12	Protactinium (Pa)	91	Thorium (Th)	90
Bohrium (Bh)	107	Fluorine (F)	9	Manganese (Mn)	25	Radium (Ra)	88	Thulium (Tm)	69
Boron (B)	5	Francium (Fr)	87	Meitnerium (Mt)	109	Radon (Rn)	86	Tin (Sn)	50
Bromine (Br)	35	Gadolinium (Gd)	64	Mendelevium (Md)	101	Rhenium (Re)	75	Titanium (Ti)	22
Cadmium (Cd)	48	Gallium (Ga)	31	Mercury (Hg)	80	Rhodium (Rh)	45	Tungsten (W)	74
Calcium (Ca)	20	Germanium (Ge)	32	Molybdenum (Mo)	42	Roentgenium (Rg)	111	Uranium (U)	92
Californium (Cf)	98	Gold (Au)	79	Neodymium (Nd)	60	Rubidium (Rb)	37	Vanadium (V)	23
Carbon (C)	6	Hafnium (Hf)	72	Neon (Ne)	10	Ruthenium (Ru)	44	Xenon (Xe)	54
Cerium (Ce)	58	Hassium (Hs)	108	Neptunium (Np)	93	Rutherfordium (Rf)	104	Ytterbium (Yb)	70
Cesium (Cs)	55	Helium (He)	2	Nickel (Ni)	28	Samarium (Sm)	62	Yttrium (Y)	39
Chlorine (Cl)	17	Holmium (Ho)	67	Niobium (Nb)	41	Scandium (Sc)	21	Zinc (Zn)	30
Chromium (Cr)	24	Hydrogen (H)	1	Nitrogen (N)	7	Seaborgium (Sg)	106	Zirconium (Zr)	40
Cobalt (Co)	27	Indium (In)	49	Nobelium (No)	102	Selenium (Se)	34		

BIOLOGICAL SCIENCE

SECOND EDITION

SCOTT FREEMAN

University of Washington

CONTRIBUTORS

Healy Hamilton
California Academy of Science, Center for Biodiversity

Sara Hoot
University of Wisconsin, Milwaukee

Greg Podgorski
Utah State University

James M. Ryan
Hobart and William Smith Colleges

Sally Sommers Smith
Boston University

Carol Trent
Western Washington University

Charles Walcott
Cornell University

D. Scott Weigle
University of Washington

Upper Saddle River, New Jersey 07458

PEARSON Prentice Hall

Library of Congress Cataloging-in-Publication Data

Freeman, Scott
Biological science / Scott Freeman.— 2nd ed.
p. cm.
ISBN 0-13-140941-7 (hardcover)
1. Biology. I. Title.

QH308.2.F73 2005
570—dc22
2004027312

Publisher: Sheri L. Snavely
Senior Development Editor: Karen Karlin
Production Editor: Donna Young
Project Manager: Karen Horton
Senior Media Editor: Patrick Shriner
Manager of Electronic Composition: Allyson Graesser
Electronic Production Specialists/Electronic Page Makeup: Karen Stephens, Vicki Croghan, Julita Nazario, Richard Foster, Jim Sullivan
Project Art Director: Kenny Beck
Editor in Chief: John Challice
Editor in Chief of Development: Carol Trueheart
Senior Marketing Manager: Shari Meffert
Marketing Manager: Andrew Gilfillan
Director of Science Marketing: Linda Taft MacKinnon
Executive Managing Editor: Kathleen Schiaparelli
Director of Creative Services: Paul Belfanti
Managing Editor, Audio/Video Assets: Patricia Burns
Manufacturing Buyer: Alan Fischer
Editorial Assistants: Nancy Bauer, Lisa Tarabokjia
Marketing Assistant: Laura Rath
Assistant Managing Editor, Science Media: Nicole Jackson
Assistant Managing Editor, Science Supplements: Becca Richter
Media Production Editor: Aaron Reid
Copy Editor: Chris Thillen
Proofreader: Brian I. Baker
Cover and Interior Designer: Joseph Sengotta
AV Editor: Connie Long
Illustrators: Pearson Artworks, Quade Paul, Imagineering
National Sales Director for Key Markets: David Theisen
Photo Researcher: Yvonne Gerin
Director, Image Resource Center: Melinda Reo
Manager, Rights and Permissions: Zina Arabia
Manager, Visual Research: Beth Boyd-Brenzel
Manager, Cover Visual Research and Permissions: Karen Sanatar
Image Permission Coordinator: LaShonda Morris
Cover photo: © Joseph Scheer
Other image credits appear in the backmatter.

PEARSON Prentice Hall

Pearson Prentice Hall
Pearson Education, Inc.
Upper Saddle River, NJ 07458

Printed in the United States of America
10 9 8 7 6 5 4 3 2

ISBN 0-13-140941-7 (Student Edition)
ISBN 0-13-141050-4 (Instructor's Edition)
ISBN 0-13-150293-X (Volume 1)
ISBN 0-13-150295-6 (Volume 2)
ISBN 0-13-150296-4 (Volume 3)

Pearson Education LTD., *London*
Pearson Education Australia PTY, Limited, *Sydney*
Pearson Education Singapore, Pte. Ltd
Pearson Education North Asia Ltd, *Hong Kong*
Pearson Education Canada, Ltd., *Toronto*
Pearson Educación de Mexico, S.A. de C.V.
Pearson Education—Japan, *Tokyo*
Pearson Education Malaysia, Pte. Ltd

Brief Contents

About the Author

Scott Freeman received his Ph.D. in Zoology from the University of Washington and was subsequently awarded an Albert Sloan Postdoctoral Fellowship in Molecular Evolution at Princeton University. His research publications explore a range of topics, including the behavioral ecology of nest parasitism and the molecular systematics of the blackbird family. Scott teaches the majors' general biology course as a Lecturer at the University of Washington. He assisted in the groundbreaking and influential redesign of the course, which emphasizes an inquiry-based approach and the logic of experimental design. With Jon Herron, Scott is co-author of the standard-setting *Evolutionary Analysis*, which over 50,000 students have used to explore evolution with the same spirit of inquiry. He is currently conducting research on how active learning and peer teaching techniques affect student learning.

CONTRIBUTORS

The author and publisher are grateful to the large number of teachers and experts in all fields of biology who shared their experiences and offered advice at every stage of planning and writing the second edition. In particular, we'd like to thank the following contributors for lending their expertise and providing new ideas and material for this revision.

Healy Hamilton
California Academy of Science, Center for Biodiversity

Sara Hoot
University of Wisconsin, Milwaukee

Greg Podgorski
Utah State University

James M. Ryan
Hobart and William Smith Colleges

Sally Sommers Smith
Boston University

Carol Trent
Western Washington University

Charles Walcott
Cornell University

D. Scott Weigle
University of Washington

ILLUSTRATOR

Kim Quillin combines training and experience in biology and art to create effective and scientifically accurate visual representations of biological principles. She received her B.A. in Biology at Oberlin College and her Ph.D. in Integrative Biology from the University of California, Berkeley, and has taught undergraduate biology at both schools. Students and instructors alike have praised Kim's illustration programs for *Biology: A Guide to the Natural World*, by David Krogh, and *Biology: Science for Life*, by Colleen Belk and Virginia Borden, for their success at clearly conveying complex biological ideas in a visually appealing manner.

Preface

Cultural evolution can be defined as a change in the frequency of ideas and practices over time. Introductory biology courses and *Biological Science* are textbook examples.

The courses we design for our majors are changing in response to selection pressure from two sources: the knowledge explosion in biology and dramatic advances in research on how introductory students learn. The knowledge explosion has made it less and less viable to teach an introductory biology course that emphasizes the memorization of facts. At the same time, research on student learning has shown that introductory students struggle to differentiate key unifying concepts from supporting details and that the greatest gains in understanding occur when students have to apply the facts and concepts they are learning to new situations. In a recent article, Handelsman et al.[1] noted that "There is mounting evidence that supplementing or replacing lectures with active learning strategies and engaging students in discovery and scientific process improves learning and knowledge retention."

Instead of being satisfied with memorization, instructors are training students to use facts. The goal is to have students demonstrate a mastery of content and concepts by applying them in new contexts.

The second edition of *Biological Science* is designed to make the transition to active, higher-level learning easier for both professors and students. Every sentence and figure in the text has been revised with that goal in mind.

Ease of Use

To make the transition to inquiry-based active learning easier, I made two major changes to the second edition: I increased the amount of content coverage to give you more flexibility in the topics you emphasize, and I added study aids to help students with the task of stepping up to a college-level biology course.

Increased Content Coverage

Compared with the first edition of *Biological Science*, this book contains much more content. Recommendations from well over 500 instructors guided decisions on which topics and terms to add. The goal was to provide students with more core coverage and vocabulary, and thereby provide instructors with more flexibility in designing a syllabus and better support for organizing lectures and labs. Experiments still play a central role in this edition, but I trimmed the overall number so that the remaining experiments could be developed more thoroughly and with a clearer focus on the concept they illustrate. Throughout, the text retains its commitment to presenting topics in the context of questions, hypotheses, tests, and conclusions. Facts are tools for understanding—not ends in themselves.

New Study Tools

As introductory biology instructors, one of our most important jobs is to help our students become better students. As my colleague Mary Pat Wenderoth says, "We need to help them learn how to learn." The students in our courses are novices in biology. Like novices in any field, they have a difficult time distinguishing important points from unimportant points. They also struggle with self-diagnosis—to recognize that they do not understand something well. To help students get better at studying biology, and to take some of the burden for doing so off you, this edition offers several new features:

- **Key Concepts** are listed at the start of the chapter and then revisited in the Summary of Key Concepts. Each chapter's "big ideas" are laid out at the start, developed in detail, and then summarized.
- **Check Your Understanding** boxes appear at the ends of key sections within each chapter. These features briefly summarize one or two fundamental points and then present two to three tasks that students should be able to complete in order to demonstrate a mastery of the material. These boxes are checkpoints—a way for students to make sure that they understand what is going on before they move ahead.
- **Diversity Boxes** serve as the capstone for each of the chapters on biodiversity (Chapters 27–34). Their goal is to present a focused summary of features in key lineages. The detailed information about each group is tied to (1) where it occurs on the tree of life and (2) how and why the featured lineage diversified. Instead of swamping students with details during a traditional "march through lineages," the diversity boxes present selected information in a well-developed conceptual context.

The Forest and the Trees: Helping Students Synthesize and Unify

In addition to coping with an enormous amount of content in this course, instructors have to manage its diversity. In *Biological Science*, the emphasis on inquiry and experimentation provides a unifying theme from biochemistry through ecosystem ecology. In addition, the text highlights the fundamental how and why

[1]Handelsman, J., D. Ebert-May, R. Beichner, P. Bruns, A. Chang, R. DeHaan, J. Gentile, S. Lauffer, J. Stewart, S. M. Tilghman, and W. B. Wood. April 23, 2004. Scientific teaching. *Science* 304(5670): 521–522.

questions of biology. How does this event or process occur at the molecular level? In an evolutionary context, why does it exist?

Most chapters include at least one case history of an analysis done at the molecular level. Natural selection is introduced by exploring the evolution of antibiotic resistance via point mutations in the RNA polymerase gene of *Mycobacterium tuberculosis*. A box in the chapter on behavior features research on alleles that influence fruit-fly foraging behavior.

Similarly, evolutionary analyses do not begin or end with the unit on evolution. Concepts such as adaptation, homology, natural selection, and phylogenetic thinking are found in virtually every chapter. Unit 1, for example, presents traditional content in biochemistry—ranging from covalent bonding to the structure and function of macromolecules—in the context of chemical evolution and the origin of life. Meiosis is analyzed in terms of its consequences for generating genetic variation and hypotheses to explain the evolution of sex.

The overriding idea is that molecular and evolutionary analyses can help unify introductory biology courses, just as molecular tools and evolutionary questions are helping to unify many formerly disparate research fields within biology.

Supporting Visual Learners

Clear, attractive, and extensive graphics are critical to our success in the classroom. The second edition offers a major improvement in the visual presentation of the material. Kim Quillin has revised virtually every figure in the book to increase clarity, accuracy, and visual appeal and to tighten the focus on the central teaching point. Compared with the first edition, this book has 350 additional diagrams and 325 additional photographs.

To support active learning and conceptual understanding, the figures contain several important features:

- **Caption Questions and Exercises** challenge students to critically examine the information in the figure—not just absorb it.
- **Experiment Boxes** offer a standardized design to help students see how biologists answer questions by posing hypotheses and testing predictions, and to give students practice with interpreting data. In some experiment boxes, space is left blank for the null hypothesis, predicted outcomes, or conclusion. Students are challenged to fill them in.
- **Figure Pointers** act like your hand at the whiteboard so that students can easily find a figure's central teaching point.

Throughout the revision, the goal was to build an art program that supports the book's focus on thinking like a biologist. Color is used judiciously to distinguish important points from supporting details and general context. Layouts flow from top to bottom or left to right, and extensive labeling lets students work through each figure in a step-by-step manner. The overall look and feel of the art is clean, clear, accessible, and inviting.

Serving a Community of Teachers

As instructors, we have at least four major texts available that are essentially well organized, well written, and beautifully illustrated encyclopedias of the life sciences. *Biological Science* is different. By de-emphasizing the encyclopedic approach to learning biology and focusing more on the questions and experimental tools that make the science come alive, my aim is to offer a book that is more readable, attractive, and contemporary than traditional texts. Learning concepts well enough to apply them to new examples and data sets may be more challenging for some students than simply memorizing facts, but also it is more compelling. By motivating the presentation with questions and then using facts as tools to find answers, students of biology may come to think and feel more like the people who actually do biology.

Thank you for your devotion to biology, for your commitment to teaching, and for considering *Biological Science*.

Scott Freeman
University of Washington

Acknowledgments

Contributors

First and foremost, I'd like to acknowledge the second edition contributors for lending their expertise and for providing new ideas and material for this revision. Their commitment to scholarship and their passion for teaching resonated throughout the contributed chapters and had an enormous impact on the published version. In writing an introductory text, it is challenging to appeal to biologists of all specialties yet focus on what a student needs. The contributors were invaluable. They made the material more accurate and teachable and gave me much-needed partners in this endeavor.

Healy Hamilton, *California Academy of Science, Center for Biodiversity*
Sara Hoot, *University of Wisconsin, Milwaukee*
Greg Podgorski, *Utah State University*
James M. Ryan, *Hobart and William Smith Colleges*
Sally Sommers Smith, *Boston University*
Carol Trent, *Western Washington University*
Charles Walcott, *Cornell University*
D. Scott Weigle, *University of Washington*

Focus Group Participants

I have been fortunate to be the beneficiary of advice and inspiration from biology instructors attending a series of workshops at Sundance, Utah, since the inception of this book. The first and second editions are influenced by the experiences and wisdom of these visionaries. The focus group attendees read through chapters and helped me make countless critical decisions about content that should be added or deleted or handled differently.

Michel Bellini, *University of Illinois, Urbana-Champaign*
Peter Berget, *Carnegie Mellon University*
Jack Burk, *University of California, Fullerton*
Ruth Buskirk, *University of Texas, Austin*
Patrick Carter, *Washington State University*
Thomas Christianson, *University of Chicago*
Jim Colbert, *Iowa State University*
William Collins, *Stony Brook University*
Mark Decker, *University of Minnesota*
Kathryn L. Edwards, *Kenyon College*
Jeffrey Feder, *University of Notre Dame*
Ross Feldberg, *Tufts University*
Michael Gaines, *University of Miami*
Miriam Golbert, *College of the Canyons*
Harry Greene, *Cornell University*
Judith Heady, *University of Michigan*
Jean Heitz, *University of Wisconsin*
Carole Kelley, *Cabrillo College*
Kevin Kelley, *California State University, Long Beach*
Stephen Kelso, *University of Illinois, Chicago*
Judith Kjelstrom, *University of California, Davis*
Jeff Klahn, *University of Iowa*
Loren Knapp, *University of South Carolina*
Karen Koster, *University of South Dakota*
Dan Krane, *Wright State University*
Mary Rose Lamb, *University of Puget Sound*
Andrea Lloyd, *Middlebury College*
Jim Manser, *Harvey Mudd College*
Mike Meighan, *University of California, Berkeley*
Robert Newman, *University of North Dakota*
Harry Nickla, *Creighton University*
Shawn Nordell, *Saint Louis University*
Julie Palmer, *University of Texas, Austin*
Marc Perkins, *Orange Coast College*
Randall Phillis, *University of Massachusetts, Amherst*
Carol Reiss, *Brown University*
Amanda Schivell, *University of Washington*
Tom Sharkey, *University of Wisconsin*
Fred Singer, *Radford University*
Sally Sommers Smith, *Boston University*
Lori Stevens, *University of Vermont*
Briana Timmerman, *University of South Carolina*
Carol Trent, *Western Washington University*
Barbara Wakimoto, *University of Washington*
Charles Walcott, *Cornell University*
D. Scott Weigle, *University of Washington*
John Whitmarsh, *University of Illinois*
Susan Whittemore, *Keene College*
David Wilson, *Parkland College*
Dan Wivagg, *Baylor University*

Media and Supplements Contributors

The media and support materials that accompany the second edition were created by a team of talented and dedicated introductory biology instructors who brought an extraordinarily high level of creativity, experience, and ability to their respective projects. Our goal was to provide an innovative and tightly focused support package that addresses the unique challenges facing instructors and students in introductory biology today. I thank the instructors who attended workshops in which the critical roles of assessment and media in this course were carefully considered and discussed and that lead to guidelines inspiring the creation of content throughout the textbook's support package.

Media Contributors

Jennifer Butler, *Willamette University*
Carol Chihara, *University of San Francisco*
Cheryl Ingram-Smith, *Clemson University*
James M. Ryan, *Hobart and William Smith Colleges*
Eric Stavney, *DeVry University*
Mark D. Decker, *University of Minnesota*

Supplements Contributors

Marc Albrecht, *University of Nebraska, Kearney*
Brian Bagatto, *University of Akron*
Jay Brewster, *Pepperdine University*
Warren Burggren, *University of North Texas*
Sharon Eversman, *Montana State University*
Michelle Fay
Michael Gaines, *University of Miami*
Vanessa Handley, *University of California, Berkeley*
Harry Nickla, *Creighton University*
Laurel Hester, *University of South Carolina*
Christopher Keller, *Minot State University*
Marc Perkins, *Orange Coast College*
Randall Phillis, *University of Massachusetts, Amherst*
Greg Podgorski, *Utah State University*
Susan Rouse, *Brenau University*
Elena Shpak, *University of Washington*
Sally Sommers Smith, *Boston University*
Ellen Smith
Briana Timmerman, *University of South Carolina*
David Wilson, *Parkland College*
Cindy Wedig, *University of Texas, Pan American*

The Book Team

The production team for this edition brought a high level of experience and expertise to bear and was characterized by a single-minded focus on quality. The tenacity, work rate, and attention to detail of Senior Development Editor Karen Karlin were both instrumental and exemplary. Production Editor Donna Young held the reins with an expert's touch as a large team moved forward at high speed. Illustrator Kim Quillin acted as lead on the figure program and is a talent that comes along once in a generation. In addition to coordinating what is probably the largest review program in the history of textbook publishing, Project Manager Karen Horton managed focus groups, contributors, and media and supplements authors. Senior Media Editor Patrick Shriner worked tirelessly to revise and improve the quality of the media program—never losing sight of the fact that media must solve problems for professors and students. Photo Researcher Yvonne Gerin again provided superb photo research and was particularly effective at contacting scientists for images of research results that are not available from stock agencies. Formatters Karen Stephens, Vicki Croghan, and Julita Nazario worked patiently under relentless deadlines to create pages for the book and ensure that the layout works for students. Research Scientist Kathleen Hunt, of the University of Washington, is responsible for the dramatically improved Glossary. The entire team was assembled and inspired by Publisher Sheri Snavely, who has been the driving force behind this project since its inception. Her commitment to innovative biology publishing and devotion to meeting the needs of instructors are the reasons that this book exists.

The art program was executed by the talented crews at Pearson Artworks, Quade Paul, and Imagineering; a special note of thanks goes to Managing Editor Patricia Burns and AV Editor Connie Long. I'm particularly grateful to Lee Wilcox, of the University of Wisconsin, who worked with Kim Quillin to upgrade the photo and art program dramatically for all of the plant chapters, and to Robin Manasse, of RMBlueStudios, who worked with Kim to improve the illustrations in the anatomy and physiology unit. Project Art Director Kenny Beck managed multiple rounds of revision on the design and cover and added key creative input. Designer Joseph Sengotta is responsible for creating a clear and accessible text design and a striking cover.

A textbook can help students and professors only if it ends up on their desks. The marketing and sales efforts for this edition are directed by Director of Science Marketing Linda Taft MacKinnon, who was instrumental in making the first edition the most successful launch in the history of majors' biology. Sincere thanks to Senior Marketing Manager Shari Meffert and to Marketing Manager Andrew Gilfillan for their work on the thoughtful preview booklet and their continued tireless efforts in the field on behalf of *Biological Science*. I extend a special thank you to Director of ESM Sales Programs Meghan O'Donnell and to the Sales Directors—especially Don O'Neal, Rebecca Bersagel, Kate Brousseau, Brian Buckley, Megan Donnelly, Meghan Duffy, Christine Henry, Tom Johnson, and Michelle Renda—for their input, travel, and commitment to this effort. I'm also deeply grateful to Dave Theisen, National Sales Director for Key Markets, for his tactical skill and devotion to this book.

Finally, I thank my students at the University of Washington for inspiration, Barb Radin for invaluable help with cataloging and organizing reviewer comments, and Ben and Peter Freeman for love and support.

This book has two dedications. The first is to the memory of Bill Keeton and Neil Campbell, whose books inspired two generations of introductory biology students. As teachers and authors, they are the giants whose shoulders I try to stand on. The second is to my wife, Susan. After 24 years together, I have one thing to say: I am the luckiest man alive.

Reviewers

The review program for the second edition was even more rigorous than that for the first edition. The chapters were reviewed three times as they moved through the revision process. Reviewers included star instructors who addressed issues such as level, pacing, accuracy, and student comprehension. Other reviewers were experts in particular fields who focused primarily on making sure that the details are correct and that chapters are authoritative and current. In addition, all 55 chapters underwent a fourth and final review for accuracy just prior to publication. To a person, our reviewers supplied exemplary attention to detail, expertise, and empathy for students. I am deeply indebted to all of the colleagues who reviewed chapters of *Biological Science*; it is not possible to overstate how crucial these individuals are to the success of this book. Their effort reflects a deep commitment to excellence in teaching and a profound belief in the importance of introductory courses for training the next generation of professionals.

Julie Aires, *Florida Community College at Jacksonville*
Marc Albrecht, *University of Nebraska, Kearney*
Terry C. Allison, *University of Texas, Pan American*
Jorge E. Arriagada, *St. Cloud State University*
David Asai, *Harvey Mudd College*
David Asch, *Youngstown State University*
Karl Aufderheide, *Texas A & M University*
Christopher Austin, *Louisiana State University*
Ellen Baker, *Santa Monica College*
Christopher Beck, *Emory University*
Robert Beckman, *North Carolina State University*
Patricia Bedinger, *Colorado State University*
Peter Bednekoff, *Eastern Michigan University*
Michel Bellini, *University of Illinois*
Carl Bergstrom, *University of Washington*
John Bishop, *Washington State University, Vancouver*
Meredith Blackwell, *Louisiana State University*
Andrew Blaustein, *Oregon State University*
Dona F. Boggs, Eastern Michigan *University*
Barry Bowman, *University of California, Santa Cruz*
Jerry Brand, *University of Texas, Austin*
Angela Brown, *University of Idaho*
Albert Burchsted, *College of Staten Island*
Warren Burggren, *University of North Texas*
John Burr, *University of Texas, Dallas*
Scott Burt, *Truman University*
David Byres, *Florida Community College at Jacksonville*
Jeff Carmichael, *University of North Dakota*
Patrick Carter, *Washington State University*
David Champlin, *University of Southern Maine*
Jung Choi, *Georgia Institute of Technology*
Thomas Christianson, *University of Chicago*
Cynthia Church, *Metropolitan State College*
Vitaly Citovsky, *Stony Brook University*
Michael Clancy, *Boston University*

Anne B. Clark, *Binghamton University*
Jim Colbert, *Iowa State University*
Jerry L. Cook, *Sam Houston State University*
Scott Cooper, *University of Wisconsin, LaCrosse*
Erica Corbett, *Southeastern Oklahoma State University*
David Craig, *Williamette University*
Sarah Cunningham, *University of California, Berkeley*
Elizabeth Dahlhoff, *Santa Clara University*
David Dalton, *Reed College*
Sandra Davis, *University of Louisiana, Monroe*
Neta Dean, *Stony Brook University*
Lynda Delph, *Indiana University*
Charles F. Delwiche, *University of Maryland*
Jean DeSaix, *University of North Carolina, Chapel Hill*
Kathryn Dodd, *University of Texas, Pan American*
John Downing, *Iowa State University*
Marvin Druger, *Syracuse University*
Ernest F. Dubrul, *University of Toledo*
John Dudley, *University of Illlinois Champaign-Urbana*
Charles Duggins, Jr., *University of South Carolina*
William Eckberg, *Howard University*
Jean Everett, *College of Charleston*
Sharon Eversman, *Montana State University*
Stephanie Fabritius, *Southwestern University*
Scott Fay, *University of California, Berkeley*
Ross Feldberg, *Tufts University*
Siobhan Fennessy, *Kenyon College*
Anne Findley, *University of Louisiana, Monroe*
Jon Fischer, *Saint Louis University*
Teresa G. Fischer, *Indian River Community College*
Robert Fogel, *University of Michigan, Ann Arbor*
Don Fontes, *Community College of Rhode Island*
Larry J. Forney, *University of Idaho*
Irwin Forseth, *University of Maryland*
Marty Fox, *Edinboro University*
Krista Frankenberry, *Marshall University*
Michael Gaines, *University of Miami*
Gary Galbreath, *Northwestern University*
George Gilchrist, *College of William and Mary*
John Godwin, *North Carolina State University*
Miriam Golbert, *College of the Canyons*
Walter M. Goldberg, *Florida International University*
Sara V. Good-Avila, *Acadia University*
John Graham, *Bowling Green State University*
Eileen Gregory, *Rollins College*
Patricia A. Grove, *College of Mount Saint Vincent*
Alan Gubanich, *University of Nevada*
Cary Guffey, *Our Lady of the Lake University*
Bill Hamilton, *Washington and Lee University*
Samuel Hammer, *Boston University*
Vanessa Handley, *University of California, Berkeley*
Alice Harmon, *University of Florida*
Carla Hass, *Pennsylvania State University*
Stephen Hauschka, *University of Washington*
Judith Heady, *University of Michigan, Dearborn*
Albert Herrera, *University of Southern California*
Karen Hicks, *Kenyon College*
Jay Hirsch, *University of Virginia*
William Hoese, *California State University, Fullerton*
Mark Holbrook, *University of Iowa*
John Hoogland, *University of Maryland*
Sara Hoot, *University of Wisconsin, Milwaukee*
Margaret Horton, *University of North Carolina, Greensboro*
Kelly Howe, *University of New Mexico*
Lawrence E. Hurd, *Washington and Lee University*
Erin Irish, *University of Iowa*
Donald Jackson, *Brown University*
Rebecca Jann, *Queens University of Charlotte*
Lee Johnson, *Ohio State University*
Walter S. Judd, *University of Florida*
Thomas C. Kane, *University of Cincinnati*
Elizabeth A. Kellogg, *University of Missouri, St. Louis*
Stephen Kelso, *University of Illinois, Chicago*
Chris Kennedy, *Simon Fraser University*
Gwendolyn Kinnebrew, *John Carroll University*
John Kiss, *Miami University of Ohio*
Jeff Klahn, *University of Iowa*
Helen Koepfer, *Queens College*
John La Claire, *University of Texas at Austin*
Mary Rose Lamb, *University of Puget Sound*
Thomas Lehman, *Morgan Community College*
Paula Lemons, *Duke University*
Lynn Lewis, *Mary Washington College*
Andi Lloyd, *Middlebury College*
Paula Lovett, *University of Maryland*
John Lugthart, *Dalton State College*
Robert D. Lynch, *University of Massachusetts, Lowell*
Richard Malkin, *University of California, Berkeley*
Charles Mallery, *University of Miami*
Paul Manos, *Duke University*
James Manser, *Harvey Mudd College*
Diane Marshall, *University of New Mexico*
Steven L. Matzner, *Augustana College*
Michael Mazurkiewicz, *University of Southern Maine*
Richard E. McCarty, *John Hopkins University*
Kelly McLaughlin, *Tufts University*
Mona Mehdy, *University of Texas*
Eli Meier, *Simbiotic, Inc.*
Michael Meighan, *University of California, Berkeley*
Madeline Mignone, *Dominican College*
Molly Morris, *Ohio University*
Dale Mueller, *Texas A & M University*
Leann Naughton, *University of Wyoming*
Jacalyn S. Newman, *University of Pittsburgh*
Robert Newman, *University of North Dakota*
Karen Bushaw Newton, *American University*
Harry Nickla, *Creighton University*
Shawn Nordell, *Saint Louis University*
Amanda Norvell, *The College of New Jersey*
Deborah O'Dell, *Mary Washington College*
John Olsen, *Rhodes College*
John Osterman, *University of Nebraska*
Norman Pace, *University of Colorado, Boulder*
Julie Palmer, *University of Texas, Austin*

Matthew B. Parks, *University of Idaho*
C. O. Patterson, *Texas A & M University*
Andrew Pease, *Villa Julie College*
Craig L. Peebles, *University of Pittsburgh*
Curtis Pehl, *University of California, Berkeley*
Marc Perkins, *Orange Coast College*
Gary Peterson, *South Dakota State University*
Patricia Phelps, *Austin Community College*
Randall Phyllis, *University of Massachusetts*
Greg Podgorski, *Utah State University*
Frank Polanowski, *Elizabethtown College*
Donald Potts, *University of California, Santa Cruz*
F. Harvey Pough, *Rochester Institute of Technology*
Jerry Purcell, *San Antonio College*
Jonathan Reed, *Community College of Southern Nevada*
Stuart Reichler, *University of Texas, Austin*
Robin Richardson, *Winona State University*
Jared Rifkin, *Queens College*
Bruce Riley, *Texas A & M University*
John Romeo, *University of South Florida*
Peter Russell, *Reed College*
James M. Ryan, *Hobart and William Smith Colleges*
Brody Sandel, *University of California, Berkeley*
Amanda Schivell, *University of Washington*
Brian G. Scholtens, *College of Charleston*
Susan Schreier, *Villa Julie College*
Robert Seagull, *Hofstra University*
Eli Seigel, *Tufts University*
Marty Shankland, *University of Texas, Austin*
Thomas B. Shea, *University of Massachusetts, Lowell*
Allen Shearn, *Johns Hopkins University*
Tim Sherman, *University of South Alabama*
Richard M. Showman, *University of South Carolina*
Michele Shuster, *New Mexico State University*
Amanda Simcox, *Ohio State University*
Anne Simon, *University of Maryland*
Sally Sommers Smith, *Boston University*
William Smith, *Wake Forest University*
James Sniezek, *Montgomery College*
Phillip Sokolove, *University of Maryland, Baltimore County*
James Staples, *University of Western Ontario*
Eric Stavney, *University of Washington*
David Steen, *Andrews University*
William Stein, *State University of New York, Binghamton*
Lori Stevens, *University of Vermont*
Charles Stinemetz, *Rhodes College*
Sarah H. Swain, *Middle Tennessee State University*
Kevin Teather, *University of Prince Edward Island*
Ethan J. Temeles, *Amherst College*
Joshua Tewksbury, *University of Washington*
Briana Timmerman, *University of South Carolina*
Albert Torzill, *George Mason University*
Carol Trent, *Western Washington University*
John True, *Stony Brook University*
Nancy Trun, *Duquesne University*
Stephen Turnbull, *University of New Brunswick*
Elizabeth Van Volkenburgh, *University of Washington*
Neal J. Voelz, *St. Cloud State University*
Susan Waaland, *University of Washington*
Charles Walcott, *Cornell University*
Margaret Wallace, *John Jay College of Criminal Justice*
Jennifer Warner, *University of North Carolina, Charlotte*
Cindy Wedig, *University of Texas, Pan American*
D. Scott Weigle, *University of Washington*
Elizabeth A. Weiss, *University of Texas, Austin*
Barry Welch, *San Antonio College*
Sue Simon Westendorf, *Ohio University*
Susan Whittemore, *Keene College*
David Wilkes, *Harvey Mudd College*
Judy Williams, *Southeastern Oklahoma State University*
David Wilson, *Parkland College*
Eric Winkler, *Blinn College*
Bob Winning, *Eastern Michigan University*
Dan Wivagg, *Baylor University*
C. B. Wolfe, *University of North Carolina, Charlotte*
Lorne Wolfe, *Georgia Southern University*
Denise Woodward, *Pennsylvania State University*
Shawn Wright, *Albuquerque TVI*
Richard P. Wunderlin, *University of South Florida*
Peter H. Wyckoff, *University of Minnesota, Morris Campus*
Todd Christian Yetter, *Cumberland College*
Stephen Yezerinac, *Reed College*
Donna Young, *University of Winnipeg*
Gregory Zimmerman, *Lake Superior State University*

Print and Media Resources for Instructors and Students

For the Instructor

Lecture Presentation Tools

Instructor Resource Center on CD/DVD

As the demands of teaching introductory biology continue to grow, it becomes increasingly important for instructors that the right textbook be accompanied by the right tools to aid them during every stage—in lecture preparation and presentation, in assessment, and in the lab, as well as in the overall management of the course. The Freeman *Biological Science* Instructor Resource Center, provided in both CD and DVD formats, offers a rich suite of electronic tools designed to help instructors make the most of their limited time. Features include:

- As JPEG files, the entire textbook illustration program: all line drawings with labels individually enhanced for optimal projection results (as well as unlabeled versions), all tables, and all photos, as well as additional photos not found in the textbook
- The entire textbook illustration program pre-loaded into comprehensive PowerPoint presentations for each chapter
- A second set of PowerPoint presentations consisting of a thorough lecture outline for each chapter, augmented by key text illustrations
- An impressive series of concise instructor animations, Web Tutorials, and video clips adding depth and visual clarity to the most important topics and dynamic processes described in the text
- These same illustrative animations, Web Tutorials, and video clips pre-loaded into PowerPoint presentations for each chapter
- PowerPoint presentations containing a comprehensive set of in-class CRS questions for each chapter
- In Word files, a complete set of the assessment materials and study questions and answers from the Test Bank for Assessment, the textbook's in-chapter questions, and the student media practice questions, as well as files containing the entire Instructor's Guide
- Finally, to help instructors keep track of all that is available in this media package, a printable Media Integration Guide consisting of a PDF file that lists the media offerings for each chapter

Transparency Package and Instructor Resource Kit

Transparencies are an effective way to reinforce your lecture presentation visually. Every illustration from the text is available on four-color transparency acetates. The transparency set is three-hole-punched and organized in manila folders, which are stored in an Instructor Resource Kit file box along with the printed lecture tools from the Instructor's Resource Guide. Labels and images have been enlarged and modified to ensure optimal readability in a large lecture hall.

Instructor's Guide

Susan Rouse (Brenau University) with Brian Bagatto (University of Akron)

The Instructor's Guide includes traditional instructor support tools—lecture outlines, answers to end-of-chapter questions, vocabulary lists—as well as innovative activities to help motivate students and reinforce their understanding of the material. All of the content in the Instructor's Guide is available electronically on the Instructor Resource Center on CD/DVD.

Assessment Tools

Test Bank for Assessment and TestGen EQ

Edited by Harry Nickla (Creighton University) and Marc Perkins (Orange Coast College)
Contributors: Marc Albrecht (University of Nebraska, Kearney), Brian Bagatto, (University of Akron), Warren Burggren (University of North Texas), Sharon Eversman (Montana State University), Michelle Fay, Michael Gaines (University of Miami), Vanessa Handley (University of California, Berkeley) Laurel Hester (University of South Carolina), Christopher Keller (Minot State University), Randall Phillis (University of Massachusetts, Amherst), Greg Podgorski (Utah State University), Elena Shpak (University of Washington), Sally Sommers Smith (Boston University), Briana Timmerman (University of South Carolina), David Wilson (Parkland College), and Cindy Wedig (University of Texas, Pan American)

Our assessment materials were guided by a team of faculty who participated in an assessment workshop. The group developed guidelines for writing effective test questions for students and set our standards for quality and accuracy. The Test Bank features questions that:

- are based on the major biological themes and processes covered in each chapter rather than the minutiae

WEB TUTORIAL

- require critical thinking or application/integration of preexisting knowledge, not rote memorization and recall
- are biologically accurate, with only one clearly correct answer choice
- are relatively resistant to test-taking strategy
- have been peer reviewed and student tested

To make this test bank more useful, we have made a few significant changes compared with the typical test bank, including:

- *In-class questions*: For each chapter, we feature three to five questions that could be used in class to help motivate discussion and formative assessment during class sessions.
- *Literature questions*: Questions based on data or ideas contained in a relevant peer-reviewed journal article. These questions extrapolate from material covered in the chapter.

Laboratory Support

Symbiosis

We have assembled a set of inquiry-based experiments to complement the approach of *Biological Science*. Symbiosis, the state-of-the-art Pearson Custom Publishing service, allows you to choose from hundreds of other experiments and/or include your own. The result is a well-integrated lab manual customized to your needs.

Course Management Tools

Prentice Hall's **OneKey** online course management content offers instructors using *Biological Science* everything they need to plan and administer their course. It includes the best teaching and learning resources all in one place. Organized by textbook chapter, the compiled resources help you save time and help your students reinforce and apply what they have learned in class. All resources from the Instructor Resource Center on CD/DVD and the Companion Website are included. OneKey is available for institutions using **WebCT** or **Blackboard** and is also available in Prentice Hall's nationally hosted **Course Compass** course management system. If desired, WebCT and Blackboard cartridges containing only the Test Bank for Assessment are available for download. Visit http://cms.prenhall.com for details.

For The Student

Companion Website (www.prenhall.com/freeman/biology)

Student study time is an increasingly valuable resource in introductory biology courses, and the greatest challenge many students face is making sure they use it wisely. The Companion Website has been designed to allow students using *Biological Science* to zero in quickly on the chapter sections and topics where they need review or further explanation. Its **Online Study Guide** offers concise lesson summaries punctuated with key illustrations, as well as probing review questions that offer hints and feedback on correct and incorrect responses. **Web Tutorials** offer students the opportunity to visualize complex topics and dynamic processes. The media's strict adherence to both the principles and specific lessons of the textbook means that study time is not being wasted. Tabs like the one in the margin of this page alert students that a Web Tutorial exists on a topic related to the coverage on that page.

Additional features of the Companion Website include an **Art Notebook**, hundreds of section-specific **Weblinks** to aid in online research; and **Textbook Answers** for all end-of-chapter and figure-caption questions.

Accelerator CD-ROM

The Accelerator CD-ROM that accompanies all student editions of *Biological Science,* Second Edition, offers the same **Web Tutorials** that are found on the Companion Website but in a stand-alone CD format.

Student Study Guide

Edited by Warren Burggren (University of North Texas)
Contributors: Jay Brewster (Pepperdine University), Laurel Hester (University of South Carolina), and Brian Bagatto (University of Akron)

The Student Study Guide helps students focus on the fundamentals chapter by chapter. Each chapter presents a breakdown of key biological concepts, difficult topics, and quizzes. In addition, the Study Guide features four introductory, stand-alone chapters: Introduction to Experimentation and Research in the Biological Sciences, Presenting Biological Data, Understanding Patterns in Biology and Improving Study Techniques, and Reading and Writing to Understand Biology.

Contents

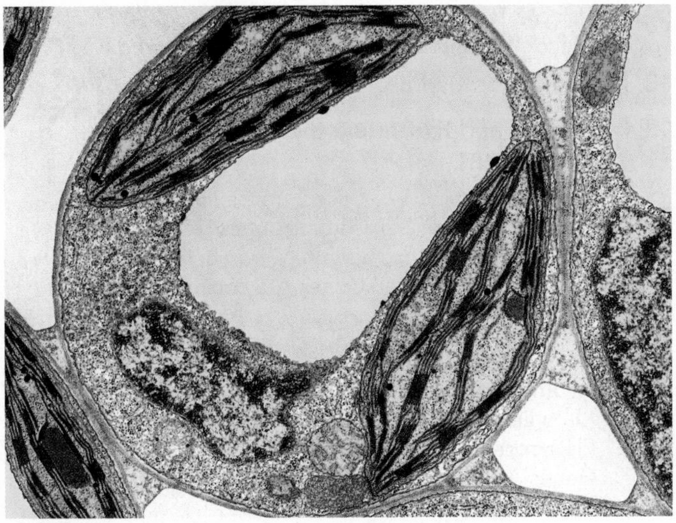

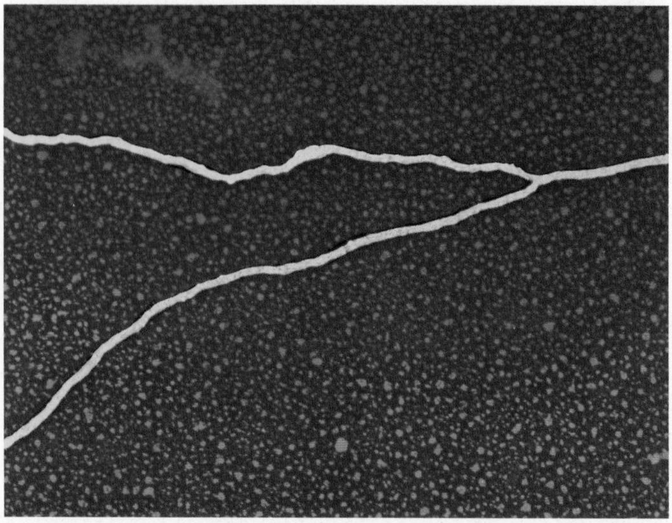

Biology and the Tree of Life

1

A biologist measuring an Atlantic puffin chick on an island off the coast of Norway. The data he's collecting will help him test hypotheses about factors that influence the growth and survival of seabirds.

KEY CONCEPTS

- Biological science was founded with the development of (1) the cell theory, which proposes that all organisms are made of cells and that all cells come from preexisting cells, and (2) the theory of evolution by natural selection, which maintains that the characteristics of species change through time—primarily because individuals with certain heritable traits produce more offspring than do individuals without those traits.
- A phylogenetic tree is a graphical representation of the evolutionary relationships among species. Phylogenies can be estimated by analyzing similarities and differences in traits. Species that share many traits are closely related and are placed close to each other on the tree of life.
- Biologists ask questions, generate hypotheses to answer them, and design experiments that test the predictions made by competing hypotheses.

In essence, biology is a search for ideas and observations that unify the incredible diversity of life. Chapter 1 is an introduction to this search. Its goals are to introduce the amazing variety of life-forms alive today, consider some fundamental traits shared by all organisms, and explore how biologists go about answering questions about life. Appreciating the diversity of life, understanding its underlying unity, and learning how to think like a biologist are themes that will resonate throughout this book.

We begin by examining two of the greatest unifying ideas in all of science: the cell theory and the theory of evolution by natural selection. When these concepts emerged in the mid-1800s, they revolutionized the way that biologists understand the world. The cell theory proposed that all organisms are made of cells and that all cells come from preexisting cells. The theory of evolution by natural selection maintained that species have changed through time and that all species are related to one another through common ancestry. The theory of evolution by natural selection established that bacteria, mushrooms, roses, and robins are all part of a family tree, similar to the genealogies that connect individual people.

A **theory** is an explanation for a very general class of phenomena or observations. The cell theory and the theory of evolution provided a foundation for the development of modern biology because they focused on two of the most general questions possible: How are organisms structured? Where did they come from? Let's begin by tackling the first of these two questions.

1.1 The Cell Theory

The initial conceptual breakthrough in biology—the cell theory—emerged after some 200 years of work. In 1665 Robert Hooke used a crude **microscope** to examine the structure of cork (a bark tissue) from an oak tree. The instrument magnified objects to just 30 times (30×) their normal size, but it allowed Hooke to see something extraordinary. In the cork he observed small, pore-like compartments that were invisible to the naked eye (**Figure 1.1a**). These structures came to be called cells.

Soon after Hooke published his results, Anton van Leeuwenhoek succeeded in developing much more powerful microscopes, some capable of magnifications up to 300×. With these instruments, Leeuwenhoek inspected samples of pond water and made the first observations of single-celled organisms like the *Paramecium* in **Figure 1.1b**. He also observed and described the structure of human blood cells and sperm cells.

In the 1670s a researcher who was studying the leaves and stems of plants with a microscope concluded that these large, complex structures are composed of many individual cells. By the early 1800s enough data had been accumulated for a biologist to claim that *all* organisms consist of cells. This conclusion was a classic example of inductive reasoning. Scientists made a broad generalization only after making thousands of supporting observations.

(a) The first view of cells: Robert Hooke's drawing from 1665

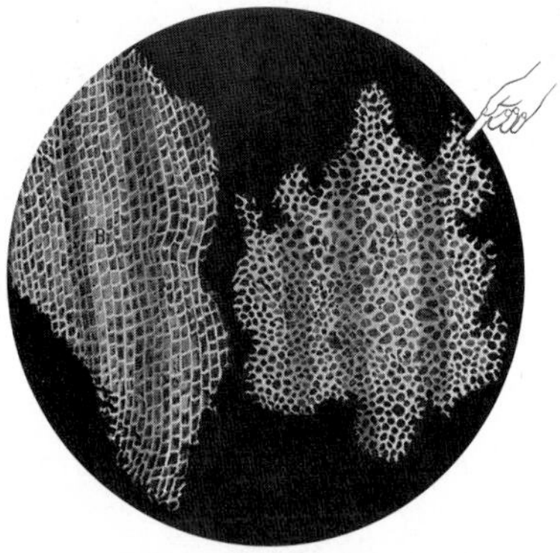
The pore-like compartments are cork cells from oak bark

(b) Anton van Leeuwenhoek was the first to view single-celled "animalcules" in pond water.

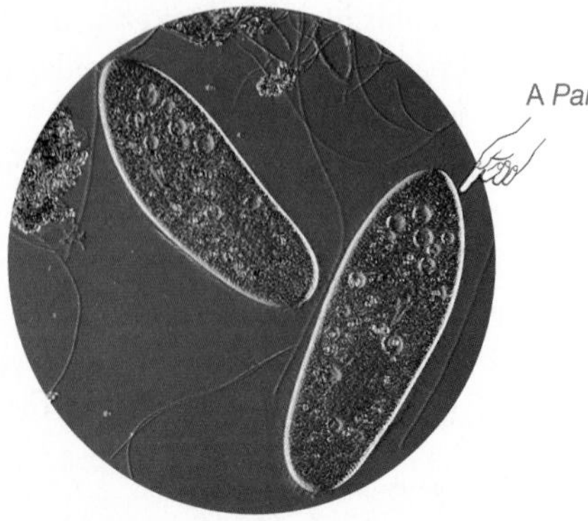
A *Paramecium*

FIGURE 1.1
The Discovery of Cells

Are *All* Organisms Made of Cells?

The smallest organisms known today are bacteria that are barely 200 nanometers wide, or 200 *billionths* of a meter. (See the Appendix at the back of this book to review the metric system and its prefixes.) Lining up bacteria end to end would take 5000 of these organisms to span a millimeter—the distance between the smallest hash marks on a metric ruler. In contrast, sequoia trees can be over 100 meters tall—the equivalent of a 20-story building. Bacteria and sequoias are composed of the same fundamental building block, however—the cell. Bacteria are unicellular (one-celled) organisms; sequoias are multicellular (many-celled) organisms.

Biologists have become increasingly dazzled by the diversity and complexity of cells as advances in microscopy have made it possible to examine cells at higher magnifications. The basic conclusion made in the 1800s is intact, however: As far as is known, all organisms are made of cells. Today, a **cell** is defined as a highly organized compartment that is bounded by a thin, flexible structure called a plasma membrane and that contains concentrated chemicals in an aqueous (watery) solution. The chemical reactions that sustain life take place inside cells. Most cells are also capable of reproducing by dividing—in effect, by making a copy of themselves.

The realization that all organisms are made of cells was fundamentally important, but it formed only the first part of the cell theory. In addition to understanding what organisms are made of, scientists wanted to understand how cells come to be.

Where Do Cells Come From?

Most scientific theories have two components: The first describes a pattern in the natural world, and the second identifies a mechanism or process that is responsible for creating that pattern. The early workers had articulated the pattern component of the cell theory. In 1858 Rudolph Virchow added a process component by stating that all cells arise from preexisting cells. The complete **cell theory**, then, can be stated as follows: All organisms are made of cells, and all cells come from preexisting cells.

This claim was a direct challenge to the prevailing explanation, called **spontaneous generation.** At the time, most biologists believed that organisms arise spontaneously under certain conditions. For example, the bacteria and fungi that spoil foods such as milk and wine were thought to appear in these nutrient-rich media of their own accord—meaning they spring to life from nonliving materials. Spontaneous generation was a **hypothesis:** a proposed explanation. The all-cells-from-cells hypothesis, in contrast, maintained that cells do not spring to life spontaneously but are produced only when preexisting cells grow and divide.

Soon after the all-cells-from-cells hypothesis appeared in print, Louis Pasteur set out to test its predictions experimentally. Pasteur wanted to determine whether microorganisms could arise spontaneously in a nutrient broth or whether they appear only when a broth is exposed to a source of preexisting cells.

To address the question, he created two treatment groups: a broth that was not exposed to a source of preexisting cells and a broth that was. The spontaneous generation hypothesis predicted that cells would appear in both treatments. The all-cells-from-cells hypothesis predicted that cells would appear only in the treatment exposed to a source of preexisting cells.

Figure 1.2 shows Pasteur's experimental setup. Note that the two treatments are identical in every respect but one. Both used glass flasks filled with the same amount of the same nutrient broth. Both were boiled for the same amount of time to kill any existing organisms such as bacteria or fungi. But because the flask pictured in Figure 1.2a had a straight neck, it was exposed to preexisting cells after sterilization by the heat treatment. These preexisting cells are the bacteria and fungi that cling to dust particles in the air. They could drop into the nutrient broth because the neck of the flask was straight. In contrast, the flask drawn in Figure 1.2b had a long swan neck. Pasteur knew that water would condense in the crook of the swan neck after the boiling treatment and that this pool of water would trap any bacteria or fungi that entered on dust particles. Thus, the swan-necked flask was isolated from any source of preexisting cells even though it was still open to the air. The experimental setup was effective because there was only one difference between the two treatments and because that difference was the factor being tested—in this case, a broth's exposure to preexisting cells.

The result? As Figure 1.2 shows, the treatment exposed to preexisting cells quickly filled with bacteria and fungi. This treatment was important because it showed that the heat sterilization step had not altered the nutrient broth's capacity to support growth. But the treatment in the swan-necked flask remained sterile. Even when the broth was left standing for months, no organisms appeared in it.

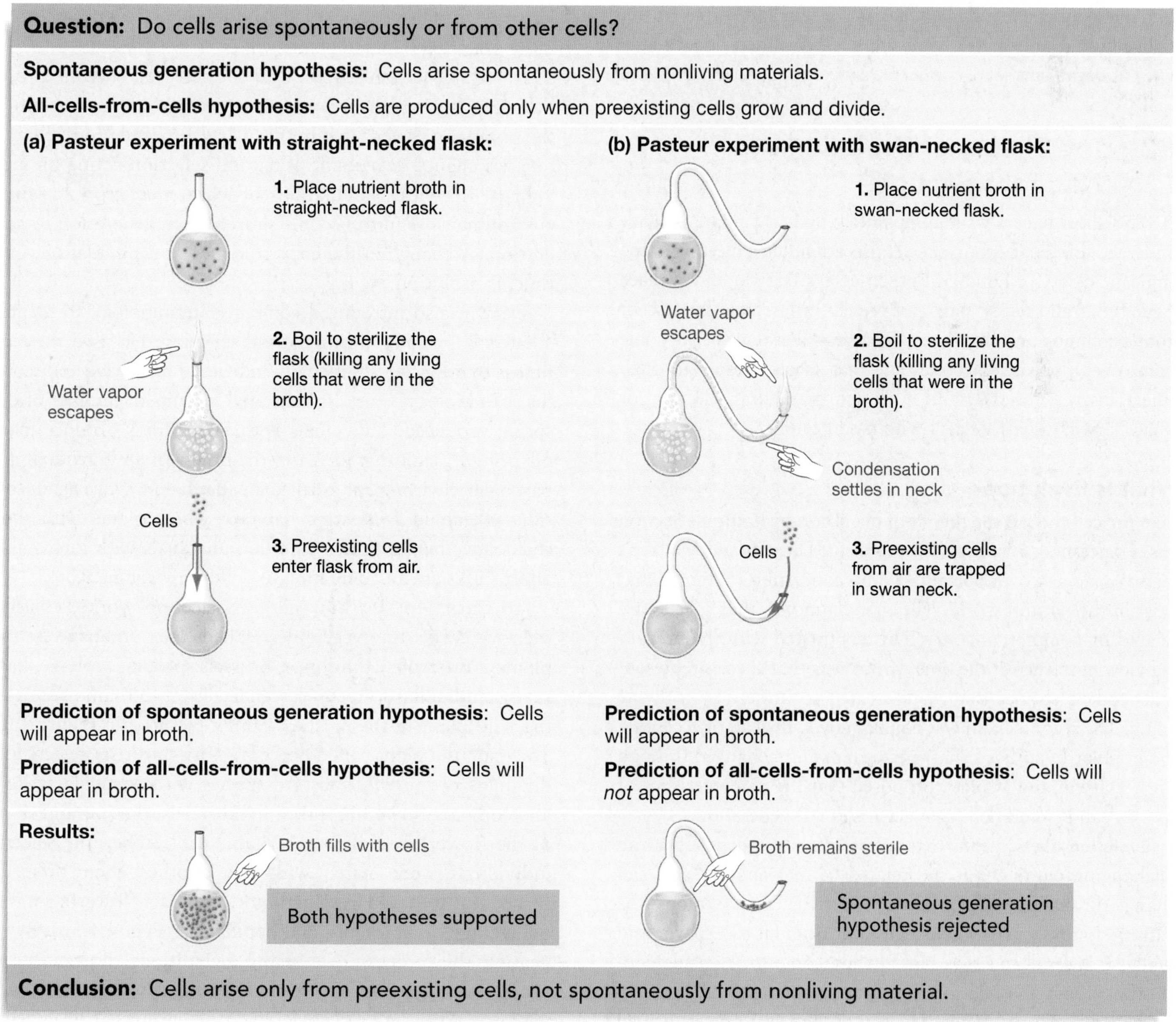
Question: Do cells arise spontaneously or from other cells?
Spontaneous generation hypothesis: Cells arise spontaneously from nonliving materials.
All-cells-from-cells hypothesis: Cells are produced only when preexisting cells grow and divide.
(a) Pasteur experiment with straight-necked flask:
1. Place nutrient broth in straight-necked flask.
Water vapor escapes
2. Boil to sterilize the flask (killing any living cells that were in the broth).
Cells
3. Preexisting cells enter flask from air.
(b) Pasteur experiment with swan-necked flask:
1. Place nutrient broth in swan-necked flask.
Water vapor escapes
2. Boil to sterilize the flask (killing any living cells that were in the broth).
Condensation settles in neck
Cells
3. Preexisting cells from air are trapped in swan neck.
Prediction of spontaneous generation hypothesis: Cells will appear in broth.
Prediction of all-cells-from-cells hypothesis: Cells will appear in broth.
Prediction of spontaneous generation hypothesis: Cells will appear in broth.
Prediction of all-cells-from-cells hypothesis: Cells will not appear in broth.
Results:
Broth fills with cells
Both hypotheses supported
Broth remains sterile
Spontaneous generation hypothesis rejected
Conclusion: Cells arise only from preexisting cells, not spontaneously from nonliving material.

FIGURE 1.2 The Spontaneous Generation Hypothesis Was Tested Experimentally

Because Pasteur's data were in direct opposition to the predictions made by the spontaneous generation hypothesis, the results persuaded most biologists that the all-cells-from-cells hypothesis was correct.

The success of the cell theory's process component had an important implication: If all cells come from preexisting cells, then all individuals in a population of single-celled organisms must be related by common ancestry. Similarly, in a multicellular individual such as you, all of the cells present are descended from preexisting cells, tracing back to a fertilized egg. A fertilized egg is a cell created by the fusion of sperm and egg—cells that formed in individuals of the previous generation. In this way, all of the cells in unicellular and multicellular organisms are connected by common ancestry.

The second great founding idea in biology is similar, in spirit, to the cell theory. It also happened to be published the same year as the all-cells-from-cells hypothesis. This was the realization, made independently by Charles Darwin and Alfred Russel Wallace, that all *species*—meaning all distinct, identifiable types of organisms—are connected by common ancestry.

1.2 The Theory of Evolution by Natural Selection

In 1858 short papers written separately by Darwin and Wallace were read to a small group of scientists attending a meeting of the Linnean Society of London. In their essays Darwin and Wallace argued that all species, past and present, are related by descent from a common ancestor. Their hypothesis was that species come from other, preexisting species and that species change through time. This was the theory of evolution by natural selection, or what Darwin called "descent with modification."

WEB TUTORIAL 1.1 Artificial Selection

What Is Evolution?

Like the cell theory, the theory of evolution by natural selection has a pattern and a process component. Darwin and Wallace's theory made two important claims concerning patterns that exist in the natural world. The first claim was that species are related by common ancestry. This contrasted with the prevailing view in science at the time, which was that species represent independent entities that were created separately by a divine being. The second claim was equally novel. Instead of accepting the popular hypothesis that species remain unchanged through time, Darwin and Wallace proposed that the characteristics of species can be modified from generation to generation.

Evolution, then, means that species are not independent and unchanging entities, but are related to one another and can change through time. This part of the theory of evolution—the pattern component—was actually not original to Darwin and Wallace. Several scientists had already come to the same conclusions about the relationships among species. The great insight by Darwin and Wallace was in proposing a process, called **natural selection**, that explained *how* evolution occurs.

What Is Natural Selection?

Natural selection occurs whenever two conditions are met. The first condition is that individuals within a population vary in characteristics that are heritable. A **population** is defined as a group of individuals of the same species living in the same area at the same time. Darwin and Wallace had studied natural populations long enough to realize that variation among individuals is almost universal. In wheat, for example, some individuals are taller than others. **Heritable** traits are characteristics that can be passed on to offspring. As a result of work by wheat breeders, Darwin and Wallace knew that short parents tend to have short offspring. Subsequent research has shown that heritable variation exists in most traits and populations. The second condition of natural selection is that in a particular environment, certain versions of these heritable traits help individuals survive better or reproduce more than do other versions. For example, if tall wheat plants are easily blown down by wind, then in windy environments shorter plants will tend to survive better and leave more offspring than tall plants will.

If certain heritable traits lead to increased success in producing offspring, then those traits become more common in the population over time. In this way, the population's characteristics change as a result of natural selection acting on individuals. In wheat, populations of wheat that grow in windy environments tend to become shorter from generation to generation. A change in the characteristics of a population, over time, is evolution.

Darwin also introduced some new terminology to identify what is happening during natural selection. He used the term **fitness** to mean the ability of an individual to survive and reproduce. In biology, fitness is measured in units of number of offspring produced. Individuals with high fitness produce many offspring. A trait that increases the fitness of an individual in a particular environment is called an **adaptation**. Once again consider wheat: In windswept habitats, wheat plants with short stalks have higher fitness than do individuals with long stalks. Short stalks are an adaptation to windy environments.

To clarify further how natural selection works, consider the origin of the vegetables called the "cabbage family plants." Broccoli, cauliflower, Brussels sprouts, cabbage, kale, savoy, and collard greens descended from the same species—the wild plant in the mustard family pictured in **Figure 1.3a**. To create the plant called broccoli, horticulturists selected individuals of the wild mustard species with particularly large and compact flowering stalks. In mustards, the size and shape of the flowering stalk is a heritable trait. When the selected individuals were mated with one another, their offspring turned out to have larger and more compact flowering stalks, on average, than the original population (**Figure 1.3b**). By repeating this process over many generations, horticulturists produced a population with extraordinarily large and compact flowering stalks. The derived population has been artificially selected for the size and shape of the flowering stalk;

Spindly
flowering
stalk

(a) Wild *Brassica oleracea*: spindly flowering stalks

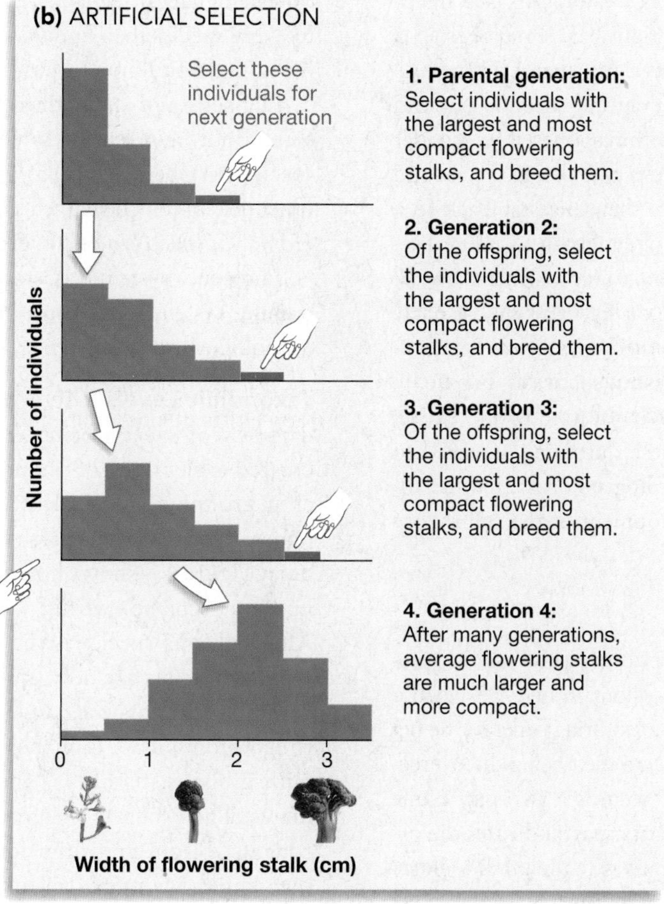
(b) ARTIFICIAL SELECTION
Select these individuals for next generation
1. Parental generation: Select individuals with the largest and most compact flowering stalks, and breed them.
2. Generation 2: Of the offspring, select the individuals with the largest and most compact flowering stalks, and breed them.
3. Generation 3: Of the offspring, select the individuals with the largest and most compact flowering stalks, and breed them.
4. Generation 4: After many generations, average flowering stalks are much larger and more compact.
Number of individuals
0
1
2
3
Width of flowering stalk (cm)
These charts, called histograms, are charts that illustrate how the width of the flowering stalk changed through time in response to selection

Large,
compact
flowering
stalks

(c) Broccoli: extremely large, compact flowering stalks

FIGURE 1.3 Artificial Selection Can Produce Dramatic Changes in Organisms

as **Figure 1.3c** shows, it barely resembles the ancestral form. Note that during this process, the size and shape of the flowering stalk in each individual plant did not change within its lifetime—the change occurred in the characteristics of the population over time.

Darwin pointed out that natural selection changes the characteristics of a wild population over time, just as the deliberate manipulation of "artificial selection" changes the characteristics of a domesticated population over time. But no horticulturist is involved in the case of natural selection. Natural selection occurs naturally, simply because certain individuals in wild populations have heritable traits that allow them to leave more offspring than do individuals without those traits. Evolution, or change in the population over time, is the outcome of this process.

Since Darwin and Wallace published, biologists have succeeded in measuring hundreds of examples of natural selection in wild populations and have accumulated a massive body of evidence documenting that species have changed through time.

Together, the cell theory and the theory of evolution provided the young science of biology with two central, unifying ideas:

1. The cell is the fundamental structural unit in all organisms.
2. All species are related by common ancestry and have changed over time in response to natural selection.

✓CHECK YOUR UNDERSTANDING

Natural selection occurs when heritable variation in certain traits leads to improved success in reproduction. Because individuals with these traits produce many offspring with the same traits, these traits increase in frequency and evolution occurs. Evolution is simply a change in the characteristics of a population over time. Although these ideas appear simple, they are often misunderstood. Research has shown that some biology students think that evolution is progressive, meaning that species always get larger, more complex, or "better" in some sense. In addition, it is common for students to think that *individuals* as well as populations change when natural selection occurs, or that individuals with high levels of fitness are stronger or bigger or "more dominant." None of these ideas are correct. Using the example of selection on height of wheat stalks, you should be able to explain why each of these three common misconceptions is wrong.

1.3 The Tree of Life

In Section 1.2 we focused on how individual populations change through time in response to natural selection. But over the past several decades, biologists have also documented dozens of cases in which natural selection has caused populations of one species to diverge and form new species. This divergence process is called **speciation.** In several instances, biologists are documenting the formation of new species right before our eyes.

Research on speciation supports a claim that Darwin and Wallace made over a century ago—that natural selection can lead to change *between* species as well as within species. The broader conclusions are that all species come from preexisting species and that all species, past and present, trace their ancestry back to a single common ancestor. If the theory of evolution by natural selection is valid, biologists should be able to reconstruct a **tree of life**—a family tree of organisms. If life on Earth arose just once, then such a diagram would describe the genealogical relationships among species with a single, ancestral species at its base.

Has this task been accomplished? If the tree of life exists, what does it look like? To answer these questions, we need to step back in time and review how biologists organized the diversity of organisms *before* the development of the cell theory and the theory of evolution.

Linnaean Taxonomy

In science, the effort to name and classify organisms is called **taxonomy.** This branch of biology began to flourish in 1735 when a botanist named Carolus Linnaeus set out to bring order to the bewildering diversity of organisms that were then being discovered.

The building block of Linnaeus' system is a two-part name unique to each type of organism. The first part indicates the organism's **genus** (plural: **genera**). A genus is made up of a closely related group of species. For example, Linnaeus put humans in the genus *Homo*. Although humans are the only living species in this genus, several extinct organisms, all of which walked upright and made extensive use of tools, were later also assigned to *Homo*. The second term in the two-part name identifies the organism's species. In Section 1.1 we defined a species as a distinct, identifiable type of organism. More formally, a **species** is made up of individuals that regularly breed together or have characteristics that are distinct from those of other species. Linnaeus gave humans the specific name *sapiens*.

An organism's genus and species designation is called its scientific name or Latin name. Scientific names are always italicized. Genus names are always capitalized, but species names are not: for instance, *Homo sapiens*. Scientific names are based on Latin or Greek word roots or on "Latinized" words from other languages (see **Box 1.1**). Linnaeus gave a scientific name to every species then known. (He also latinized his own name—from Karl von Linné to Carolus Linnaeus.)

Linnaeus also maintained that different types of organisms should not be given the same genus and species names. Other species may be assigned to the genus *Homo*, and members of other genera may be named *sapiens*, but only humans are named *Homo sapiens*. As a result, each scientific name is unique.

Linnaeus' system has stood the test of time. His two-part naming system, or **binomial nomenclature,** is still the standard in biological science.

Taxonomic Levels To organize and classify the tremendous diversity of species being discovered in the 1700s, Linnaeus created a hierarchy of taxonomic groups: From the most specific grouping to the least specific, the levels are **species, genus, family, order, class, phylum** (plural: **phyla**), and **kingdom. Figure 1.4** shows how this nested, or hierarchical, classification scheme works, using humans as an example. Although our species is the sole living member of the genus *Homo*, humans are now grouped with the orangutan, gorilla, common chimpanzee, and pygmy chimpanzee in a family called Hominidae. Linnaeus grouped members of this family with gibbons, monkeys, and lemurs in an order called Primates. The Primates are grouped in the class Mammalia with rodents, bison, and other organisms that have fur and produce milk. Mammals, in turn, join other animals with structures called notochords in the phylum Chordata, and all other animals in the kingdom Animalia. Each of these named groups—primates, mammals, or *Homo sapiens*—can be referred to as a **taxon** (plural: **taxa**). The essence of Linnaeus' system is that lower-level taxa are nested within higher-level taxa.

Aspects of this hierarchical scheme are still in use. As biological science matured, however, several problems with Linnaeus' original proposal emerged.

How Many Kingdoms Are There? Linnaeus proposed that species could be organized into two kingdoms: plants and animals. According to Linnaeus, organisms that do not move and that produce their own food are plants; organisms that move and acquire food by eating other organisms are animals.

BOX 1.1 Scientific Names and Terms

Scientific names and terms are often based on Latin or Greek word roots that are descriptive. For example, *Homo sapiens* is derived from the Latin *homo* for "man" and *sapiens* for "wise" or "knowing." The yeast that bakers use to produce bread and that brewers use to brew beer is called *Saccharomyces cerevisiae*. The Greek root *saccharo* means "sugar," and *myces* refers to a fungus. *Saccharomyces* is aptly named "sugar fungus" because yeast is a fungus and because the domesticated strains of yeast used in commercial baking and brewing are often fed sugar. The specific name of this organism, *cerevisiae*, is Latin for *beer*. Loosely translated, then, the scientific name of brewer's yeast means "sugar fungus for beer."

Most biologists find it extremely helpful to memorize some of the common Latin and Greek roots. To aid you in this process, new terms in this text are often accompanied by a reference to their Latin or Greek word roots in parentheses.

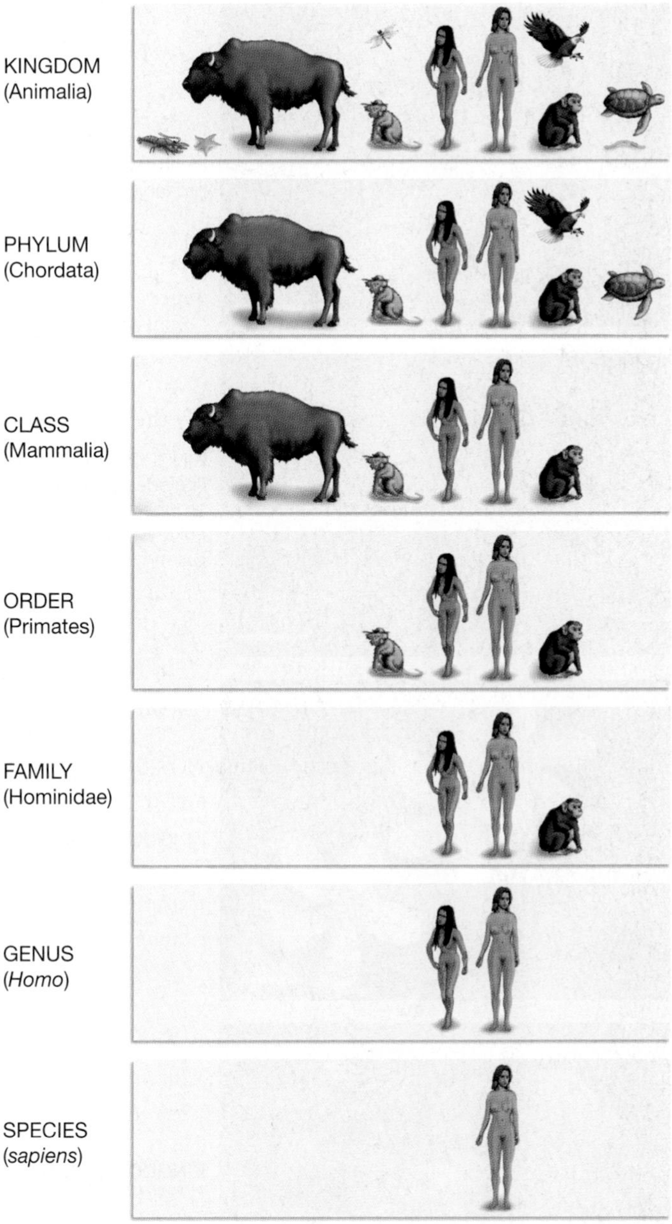
KINGDOM
(Animalia)
PHYLUM
(Chordata)
CLASS
(Mammalia)
ORDER
(Primates)
FAMILY
(Hominidae)
GENUS
(*Homo*)
SPECIES
(*sapiens*)

FIGURE 1.4 Linnaeus Defined Taxonomic Levels
In the Linnaean system, each animal species is placed in a taxonomic hierarchy with seven levels. Lower levels are nested within higher levels. Linnaeus proposed that these levels reflected the natural order of organisms.

Not all organisms fall neatly into these categories, however. Molds, mushrooms, and other fungi survive by absorbing nutrients from dead or living plants and animals. Even though they do not make their own food, they were placed in the kingdom Plantae because they do not move. The tiny, single-celled organisms called bacteria also presented problems. Some bacteria can move, and many can make their own food. Initially they, too, were thought to be plants. Eventually, though, it became clear that the two-kingdom system was simply inadequate.

Further, the development of the theory of evolution suggested a new goal for taxonomy. As evidence for evolution mounted, biologists concentrated on understanding how classification systems such as the one invented by Linnaeus could be modified to reflect the genealogical relationships among organisms—not just how similar they are in appearance or other broad characteristics. The goal of taxonomy became an attempt to reflect **phylogeny** (meaning "tribe-source")—the true historical relationships among types of organisms.

Linnaeus proposed the two-kingdom system because he thought it reflected a fundamental pattern in the natural world. His two-kingdom system was a hypothesis—a proposed explanation for an observed phenomenon. In this case, the pheno-menon that Linnaeus and other biologists were trying to explain was the relationship among the diverse forms of life on Earth.

For example, Linnaeus proposed that the fundamental division in organisms was between plants and animals. But when advances in microscopy allowed biologists to study the contents of individual cells in detail, a different fundamental division emerged and his hypothesis was rejected. In plants, animals, and the organisms that taxonomists call protists, cells contain a prominent component called a nucleus (**Figure 1.5a**). But in bacteria, cells lack this kernel-like structure (**Figure 1.5b**). Organisms with a nucleus are called **eukaryotes** ("true-kernel"); organisms without a nucleus are called **prokaryotes** ("before-kernel"). The vast majority of prokaryotes are unicellular ("one-celled"), many eukaryotes; are multicellular ("many-celled").

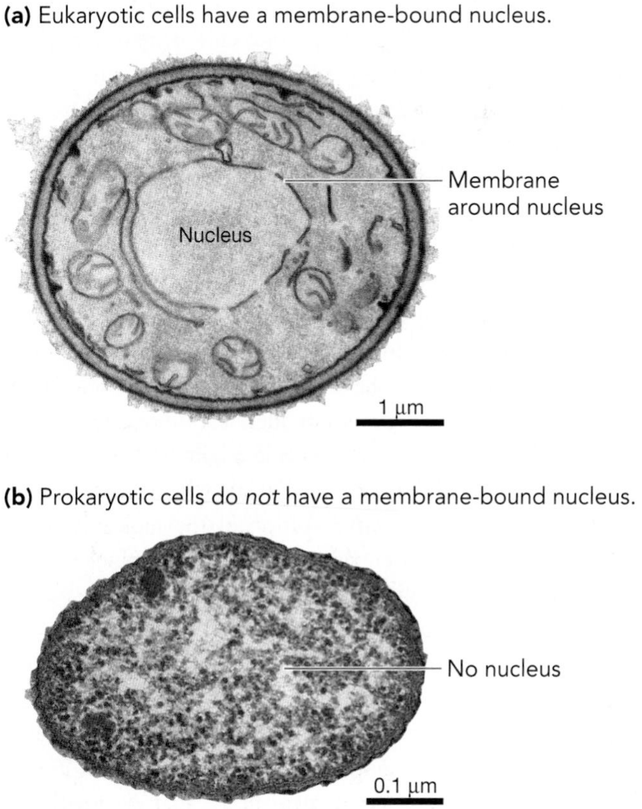
(a) Eukaryotic cells have a membrane-bound nucleus.
Membrane around nucleus
Nucleus
1 μm
(b) Prokaryotic cells do *not* have a membrane-bound nucleus.
No nucleus
0.1 μm

FIGURE 1.5 Eukaryotes and Prokaryotes
Cross sections of **(a)** a eukaryotic and **(b)** a prokaryotic cell.
EXERCISE Study the scale bars; then draw two ovals that accurately represent the relative sizes of a eukaryotic cell and a prokaryotic cell.

When data began to conflict with Linnaeus' original scheme, biologists proposed alternative hypotheses. In the late 1960s one researcher suggested that a system of five kingdoms best reflects the patterns observed in nature. This five-kingdom system is depicted in **Figure 1.6**. Although the scheme has been widely used, it represents just one proposal out of many. Other biologists proposed that organisms are organized into three, four, six, or eight kingdoms. But it was still not clear which of these schemes, if any, accurately described the phylogeny of organisms.

About the time that the five-kingdom proposal was published, however, Carl Woese (pronounced "woes") and colleagues began working on the problem from a radically different angle. Instead of assigning organisms to kingdoms on the basis of characteristics such as the presence of a nucleus or the ability to move or to manufacture food, these researchers attempted to understand the relationships among organisms by analyzing their chemical components.

5 µm
KINGDOM MONERA
(includes all
prokaryotes)
20 µm
KINGDOM PROTISTA
(includes several
groups of unicellular
eukaryotes)
3 cm
KINGDOM PLANTAE
1 cm
KINGDOM FUNGI
1 mm
KINGDOM ANIMALIA

FIGURE 1.6 The Five-Kingdom Scheme
For decades, most biologists accepted the hypothesis that organisms naturally fall into the five kingdoms illustrated here. QUESTION How many times bigger is a fruit fly than one of the prokaryotic cells pictured here?

Using Molecules to Understand the Tree of Life

Woese and his co-workers had an explicit goal: to estimate where major branches occurred on the tree of life by analyzing the molecular components of cells. To accomplish this goal they needed to study a molecule that is found in all organisms. The molecule they selected is called small subunit rRNA. It is an essential part of the machinery that all cells use to grow and reproduce.

Although rRNA is a large and complex molecule, its underlying structure is simple. The rRNA molecule is made up of sequences of four smaller chemical components called ribonucleotides. Ribonucleotides are symbolized by the letters A, U, C, and G. In rRNA, ribonucleotides are connected to one another linearly, like boxcars of a freight train (**Figure 1.7**).

Why might rRNA be useful for understanding the relationships among organisms? The answer is that the ribonucleotide sequence in rRNA is a trait, similar to the height of wheat stalks or the size of flowering stalks of broccoli, that can change during the course of evolution. Although rRNA performs the same function in all organisms, the sequence of ribonucleotide building blocks in this molecule is not identical among species. In land plants, for example, the molecule might start with the sequence A-U-A-U-C-G-A-G. In green algae, which are closely related to land plants, the same section of the molecule might contain A-U-A-U-*G*-G-A-G. But in brown algae, which are not closely related to green algae or to land plants, the same part of the molecule might consist of A-*A*-A-U-*G*-G-A-G.

The research program that Woese and co-workers pursued was based on a simple premise: If the theory of evolution is correct, then rRNA sequences should be very similar in closely related organisms but less similar in organisms that are less closely related. The rRNA sequences of two species that diverged from each other long ago should be quite different, while rRNA sequences from two species that diverged from each other more recently should be much more alike.

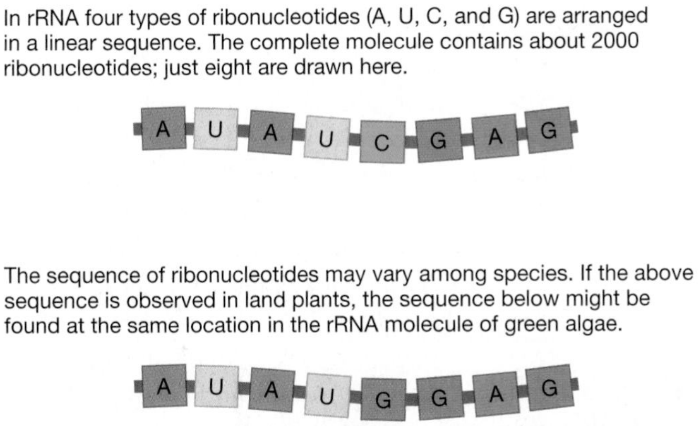

FIGURE 1.7 RNA Molecules Are Made Up of Smaller Molecules
The four smaller molecules that make up an rRNA molecule are symbolized A, U, C, and G. The sequence of A, U, C, and G subunits in rRNA varies among species. **QUESTION** Suppose that in the same portion of rRNA, molds and other fungi have the sequence A-U-A-U-G-G-A-C. According to these data, are fungi more closely related to green algae or to land plants? Explain your logic.

To put this insight to work, the researchers determined the sequence of ribonucleotides in the rRNA of a wide array of species. Then they considered what the similarities and differences in the sequences implied about relationships among the species. The goal was to produce a diagram that described the phylogeny of the organisms in the study. A diagram that depicts evolutionary history in this way is called a **phylogenetic tree.** Just as a family tree shows relationships among individuals, a phylogenetic tree shows relationships among species. On a phylogenetic tree, branches that are close to one another represent species that are closely related; branches that are farther apart represent species that are more distantly related.

The rRNA Tree To construct a phylogenetic tree, researchers use a computer to find the arrangement of branches that is most consistent with the similarities and differences observed in the data—in this case, in the sequences of ribonucleotides observed in rRNA. The tree produced by comparing these sequences is shown in **Figure 1.8**. Because this tree includes species from many different kingdoms and phyla, it is often called the universal tree, or the tree of life.

The tree of life implied by rRNA data astonished biologists. According to data from this molecule:

- The fundamental division in organisms is not between plants and animals or even between prokaryotes and eukaryotes. Rather, *three* major groups occur: (1) the Bacteria; (2) another group of prokaryotic, single-celled organisms called

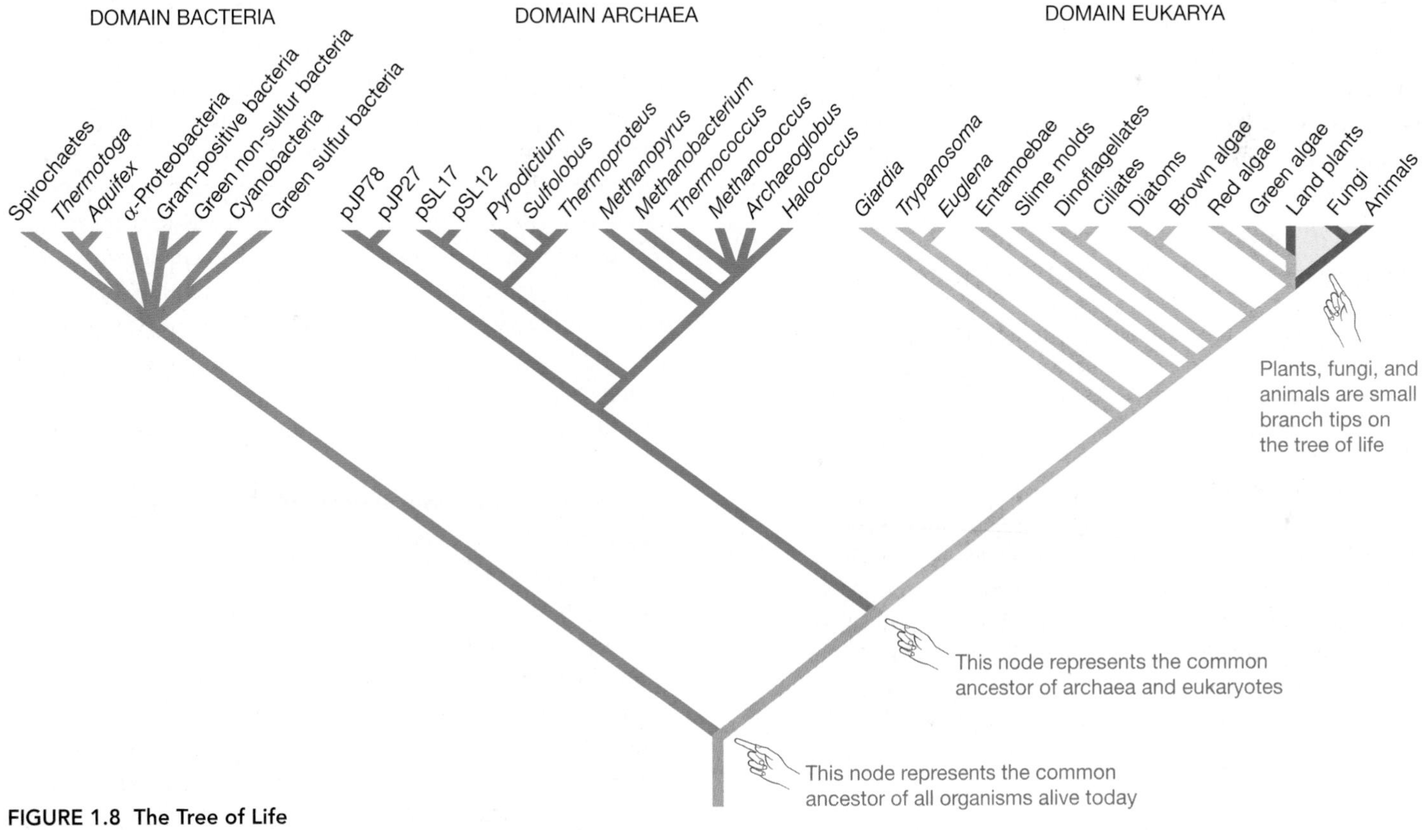

FIGURE 1.8 The Tree of Life
"Universal tree" estimated from rRNA sequence data. The three domains of life revealed by the analysis are labeled. Common names are given for most lineages in the domains Bacteria and Eukarya. Genus names are given for members of the domain Archaea, because most of these organisms have no common names. Archaean species with labels such as "pSL17" have not yet been given scientific names. **EXERCISE** Circle the branches and tips that represent prokaryotes.

the Archaea; and (3) the eukaryotes. To accommodate this new perspective on the diversity of organisms, Woese created a new taxonomic level called the **domain.** As Figure 1.8 indicates, the three domains of life are now called the Bacteria, Archaea, and Eukarya.

- Some of the kingdoms that had been defined earlier do not reflect how evolution actually occurred. For example, recall that Linnaeus grouped the multicellular eukaryotes known as fungi with plants. But the rRNA data indicate that fungi are much more closely related to animals than they are to plants.
- Bacteria and Archaea are much more diverse than anyone had imagined. If the differences among animals, fungi, and plants warrant placing them in separate kingdoms, then dozens of kingdoms exist among the prokaryotes.

The Tree of Life Is a Work in Progress The rRNA tree was a hypothesis that inspired a flurry of research. Biologists in laboratories all over the world tested the conclusions by determining the sequences of other molecules found in cells and by reanalyzing older data in light of the new findings. In general, these studies have confirmed the major features of the tree in Figure 1.8. The discovery of the Archaea and the placement of lineages such as the fungi qualify as exciting breakthroughs in our understanding of life's diversity.

Work on the tree of life continues at a furious pace, however, and the location of certain branches on the tree is hotly debated. As databases expand and as techniques for analyzing data improve, the shape of the tree of life presented in Figure 1.8 will undoubtedly change.

✓CHECK YOUR UNDERSTANDING

A phylogenetic tree is analogous to a family tree. A phylogenetic tree shows the evolutionary relationships among species, just as a family tree shows the genetic relationships among individuals.

To infer where species belong on a phylogenetic tree, biologists examine the characteristics of the species involved. Closely related species should have similar characteristics, while less closely related species should be less similar. A characteristic that has been used extensively in phylogenetic analyses is the sequence of subunits in a molecule called rRNA, which is found in all species. You should be able to examine the sequences given below and draw a phylogenetic tree showing the relationships among Species A, B, and C that these data imply.

Species A:	A	A	C	T	A	G	C	G	C	G	A	T
Species B:	A	A	C	T	A	G	C	G	C	C	A	T
Species C:	T	T	C	T	A	G	C	G	G	T	A	T

1.4 Doing Biology

This chapter has introduced some of the great ideas in biology. The development of the cell theory and the theory of evolution by natural selection provided cornerstones when the science was young; the tree of life is a relatively recent insight that has revolutionized the way researchers understand the diversity of life on Earth.

These theories are considered great because they explain fundamental aspects of nature, and because they have consistently been shown to be correct. They are considered correct because they have withstood extensive testing. How do biologists test ideas about the way the natural world works? The answer is that they test the predictions made by alternative hypotheses, often by setting up carefully designed experiments. To illustrate how this approach works, let's consider two questions currently being addressed by researchers.

Why Do Giraffes Have Long Necks? An Introduction to Hypothesis Testing

If you were asked why giraffes have long necks, you might say that long necks enable giraffes to reach food that is unavailable to other mammals. This hypothesis is expressed in African folktales and has traditionally been accepted by many biologists. The food competition hypothesis is so plausible, in fact, that for decades no one thought to test it. Recently, however, Robert Simmons and Lue Scheepers assembled data suggesting that the food competition hypothesis is only part of the story. Their analysis supports an alternative hypothesis—that long necks allow giraffes to use their heads as effective weapons for battering their opponents.

How did biologists test the food competition hypothesis? What data support their alternative explanation? Before we answer these questions, it's important to recognize that hypothesis testing is a two-step process. The first step is to state the hypothesis as precisely as possible and list the predictions it makes. The second step is to design an observational or experimental study that is capable of testing those predictions. If the predictions are accurate, then the hypothesis is supported. If the predictions are not met, then researchers do further tests, modify the original hypothesis, or search for alternative explanations.

The Food Competition Hypothesis: Predictions and Tests Stated precisely, the food competition hypothesis claims that giraffes compete for food with other species of mammals. When food is scarce, as it is during the dry season, giraffes with longer necks can reach food that is unavailable to other species and to giraffes with shorter necks. As a result, the longest-necked individuals in a giraffe population survive better and produce more young than do shorter-necked individuals, and average neck length of the population increases with each generation. To use the terms introduced earlier in this chapter, long necks are adaptations that increase the fitness of individual giraffes during competition for food. This type of natural selection has gone on so long that the population has become extremely long necked.

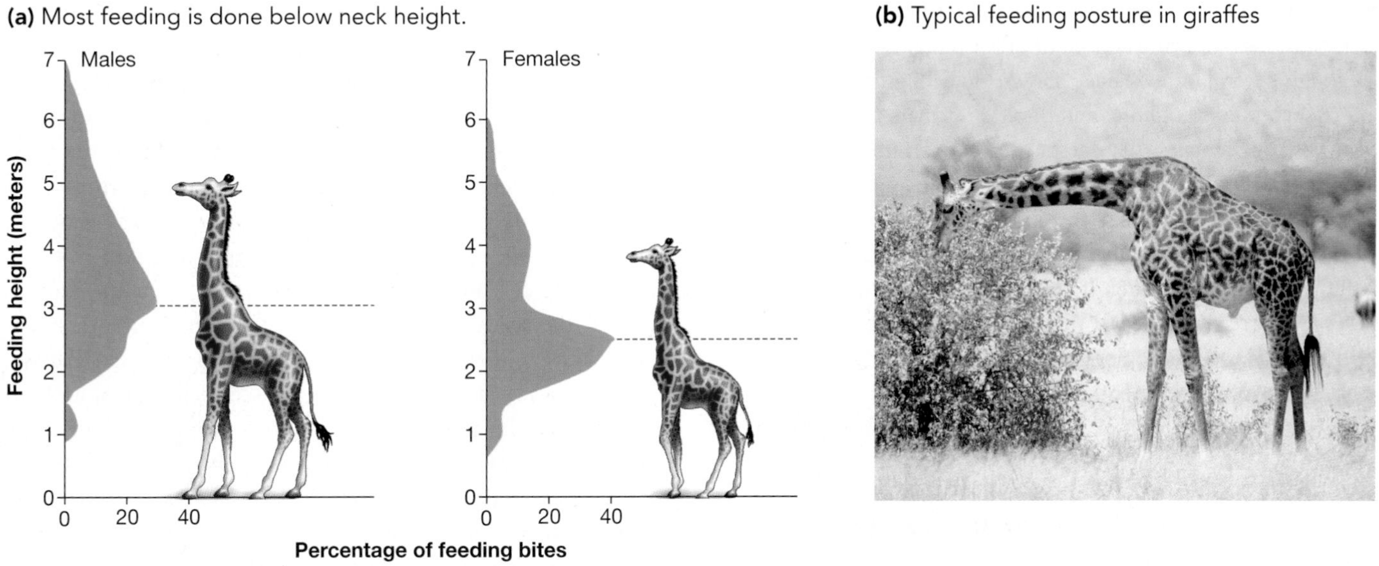
(a) Most feeding is done below neck height.
Males
Females
Feeding height (meters)
7
6
5
4
3
2
1
0
0 20 40
Percentage of feeding bites
(b) Typical feeding posture in giraffes

FIGURE 1.9 Giraffes Do Not Usually Extend Their Necks to Feed
(a) The proportion of bites that male and female giraffes of average size take at different heights. (b) Although it is common to see photos of giraffes straining to reach leaves high in trees, giraffes usually feed at the heights shown here.

The food competition hypothesis makes several explicit *predictions*. A prediction is something that can be measured and that must be correct if a hypothesis is valid. For example, the food competition hypothesis predicts that (1) neck length is variable among giraffes; (2) neck length in giraffes is heritable; and (3) giraffes feed high in trees, especially during the dry season, when food is scarce and the threat of starvation is high.

The first prediction is clearly correct. Studies in zoos and natural populations confirm that neck length is variable among individuals.

The researchers were unable to test the second prediction, however, because they studied giraffes in a natural population and were unable to do breeding experiments. As a result, they simply had to accept this prediction as an assumption. In general, though, biologists prefer to test every assumption behind a hypothesis.

What about the prediction regarding feeding high in trees? According to Simmons and Scheepers, this is where the food competition hypothesis breaks down. Consider, for example, data collected by a different research team about the amount of time that giraffes spend feeding in vegetation of different heights. **Figure 1.9a** shows that in a population from Kenya, both male and female giraffes spend most of their feeding time eating vegetation that averages just 60 percent of their full height. Studies on other populations of giraffes, during both the wet and dry seasons, are consistent with these data. Giraffes usually feed with their necks bent (**Figure 1.9b**).

These data cast doubt on the food competition hypothesis, because one of its predictions does not appear to hold. Biologists have not abandoned this hypothesis completely, though, because feeding high in trees may be particularly valuable during extreme droughts, when a giraffe's ability to reach leaves far above the ground could mean the difference between life and death. Still, Simmons and Scheepers have offered an alternative explanation for why giraffes have long necks. The new hypothesis is based on the mating system and social behavior of giraffes.

The Sexual Competition Hypothesis: Predictions and Tests

Giraffes have an unusual social system. Breeding occurs year round rather than seasonally. To determine when females are coming into estrus (or "heat") and are thus receptive to mating, the males nuzzle the rumps of females. In response, the females urinate into the males' mouths. The males then tip their heads back and pull their lips to and fro, as if tasting the liquid. Biologists who have witnessed this behavior have proposed that the males taste the females' urine to detect whether estrus has begun.

Once a female giraffe enters estrus, males fight among themselves for the opportunity to mate. Combat is spectacular. The bulls stand next to one another, swing their necks, and strike thunderous blows with their heads. Researchers have seen males knocked unconscious for 20 minutes after being hit and have cataloged numerous instances in which the loser died. Giraffes are the only animals that fight in this way.

These observations inspired a new explanation for why giraffes have long necks. The sexual competition hypothesis is based on the idea that longer-necked giraffes are able to strike harder blows during combat than can shorter-necked giraffes. In engineering terms, longer necks provide a longer moment arm. A long moment arm increases the force of the impact. (Think about the type of sledge hammer you'd use to bash down a concrete wall—one with a short handle or one with a long handle?) Thus, longer-necked males should win more fights and, as a result, father more offspring than do shorter-necked males. If neck length in giraffes is inherited, then the average neck length in the population should increase over time. Under the sexual

WEB TUTORIAL 1.2
Introduction to Experimental Design

competition hypothesis, long necks are adaptations that increase the fitness of males during competition for females.

Although several studies have shown that long-necked males are more successful in fighting and that the winners of fights gain access to estrous females, the question of why giraffes have long necks is not closed. With the data collected to date, most biologists would probably concede that the food competition hypothesis needs further testing and refinement and that the sexual selection hypothesis appears promising. It could also be true that both hypotheses are correct. But clearly, more work needs to be done.

In many cases in biological science, "more work" involves experimentation. Experimenting on giraffes is difficult. But in the case study considered next, biologists were able to test an interesting hypothesis experimentally.

Why Are Chili Peppers Hot? An Introduction to Experimental Design

Experiments are a powerful scientific tool because they allow researchers to test the effect of a single, well-defined factor on a particular phenomenon. Experiments that test the effect of neck length on food and sexual competition in giraffes have yet to be done. Instead, as an example of how experiments are designed, let's consider a different phenomenon: Why do chili peppers taste so spicy?

The jalapeño, anaheim, and cayenne peppers used in cooking descended from a wild shrub that is native to the deserts of the American Southwest. As **Figure 1.10a** shows, wild chilies produce fleshy fruits with seeds inside, just like their domesticated descendants. In both wild chilies and the cultivated varieties, the "heat" or pungent flavor of the fruit and seeds is due to a molecule called capsaicin. In humans and other mammals, capsaicin binds to heat-sensitive cells in the tongue and mouth. In response to this binding, signals are sent to the brain that produce the sensation of burning. Similar signals would be transmitted if you drank boiling water. Asking why chilies are hot, then, is the same as asking why chilies contain capsaicin.

Josh Tewksbury and Gary Nabhan proposed that the presence of capsaicin is an adaptation that protects chili fruits from being eaten by animals that destroy the seeds inside. To understand this hypothesis, it's important to realize that the seeds inside a fruit have one of two fates when the fruit is eaten. If the seeds are destroyed in the animal's mouth or digestive system, then they never germinate (sprout). In this case, "seed predation" has occurred. But if seeds can travel undamaged through the animal, then they are eventually "planted" in a new location along with a valuable supply of fertilizer. In this case, seeds are dispersed. Here's the key idea: Natural selection should favor fruits that taste bad to animal species that act as seed predators. But these same fruits should not deter species that act as seed dispersers. This proposal is called the *directed dispersal hypothesis*.

Does capsaicin deter seed predators, as the directed dispersal hypothesis predicts? To answer this question, the researchers captured some cactus mice (**Figure 1.10b**) and birds called curve-billed thrashers (**Figure 1.10c**). These species are among the most important fruit- and seed-eating animals in habitats where chilies grow. Based on earlier observations, the biologists predicted that cactus mice destroy chili seeds but that curve-billed thrashers disperse them effectively.

To test the directed dispersal hypothesis, the biologists offered both cactus mice and curve-billed thrashers three kinds of fruit: hackberries, chilies from a strain of plant that can't synthesize capsaicin, and pungent chilies that have lots of cap-

(a) Chilies produce fruits that contain seeds.

(b) Cactus mice are seed eaters.

(c) Curve-billed thrashers eat chili fruits.

FIGURE 1.10 Chilies ... and Chili Eaters?
(a) Chili fruits are hot because they contain a molecule called capsaicin. **(b)** Cactus mice and **(c)** curve-billed thrashers are common in habitats where chilies grow.

saicin. The non-pungent chilies are about the same size and color as normal chilies and have similar nutritional value. The hackberries don't look anything like chilies, however, and contain no capsaicin. The three fruits were present in equal amounts. For each animal tested, the researchers recorded the percentage of hackberry, non-pungent chili, and pungent chili that was eaten during a specific time interval. Then they calculated the average amount of each fruit that was eaten by five test individuals from each species.

The directed dispersal hypothesis predicts that seed dispersers will eat the pungent chilies readily but that seed predators won't. Recall that a prediction specifies what we should observe if a hypothesis is correct. Good scientific hypotheses make testable predictions—predictions that can be supported or rejected by collecting and analyzing data. If the directed dispersal hypothesis is wrong, however, then there shouldn't be any difference in what various animals eat. This latter possibility is called a **null hypothesis.** A null hypothesis specifies what we should observe when the hypothesis being tested doesn't hold. These predictions are listed in **Figure 1.11.**

Do the predictions of the directed dispersal hypothesis hold? To answer this question, look at the results plotted in Figure 1.11. View the data for each type of fruit to see if different amounts were eaten by different types of predators. Then fill in the following table:

Type of fruit	Did the two predators eat about the same amount or very different amounts of this fruit?	If the amount eaten was very different, comment on the nature of the difference.
Hackberry		
Non-pungent chilies (no capsaicin)		
Pungent chilies (lots of capsaicin)		

Based on your analysis of the data, decide whether the results support the directed dispersal hypothesis or the null hypothesis. Use the conclusion stated in Figure 1.11 to check your answer.

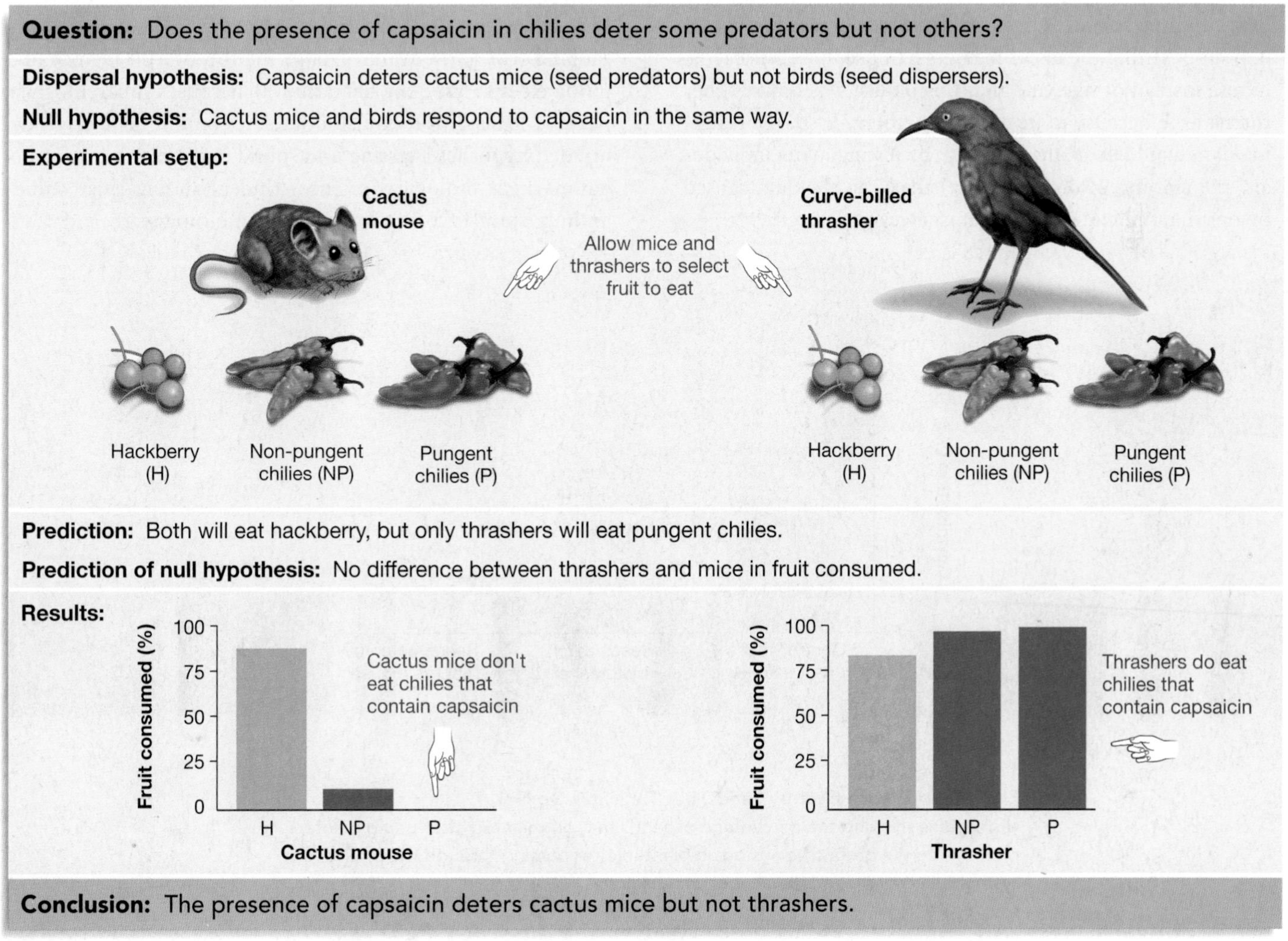
Question: Does the presence of capsaicin in chilies deter some predators but not others?
Dispersal hypothesis: Capsaicin deters cactus mice (seed predators) but not birds (seed dispersers).
Null hypothesis: Cactus mice and birds respond to capsaicin in the same way.
Experimental setup:
Cactus mouse
Allow mice and thrashers to select fruit to eat
Curve-billed thrasher
Hackberry (H)
Non-pungent chilies (NP)
Pungent chilies (P)
Hackberry (H)
Non-pungent chilies (NP)
Pungent chilies (P)
Prediction: Both will eat hackberry, but only thrashers will eat pungent chilies.
Prediction of null hypothesis: No difference between thrashers and mice in fruit consumed.
Results:
Fruit consumed (%)
100
75
50
25
0
H
NP
P
Cactus mouse
Cactus mice don't eat chilies that contain capsaicin
Fruit consumed (%)
100
75
50
25
0
H
NP
P
Thrasher
Thrashers do eat chilies that contain capsaicin
Conclusion: The presence of capsaicin deters cactus mice but not thrashers.

FIGURE 1.11 An Experimental Test: Does Capsaicin Deter Some Fruit Eaters?

In relation to designing effective experiments, this study illustrates several important points:

- *It is critical to include control groups.* A **control** checks for factors, other than the one being tested, that might influence the experiment's outcome. For example, if hackberries had not been included as a control, it would have been possible to claim that the cactus mice in the experiment didn't eat pungent chilies simply because they weren't hungry. But the not-hungry hypothesis can be rejected because all of the animals ate hackberries.
- *The experimental conditions must be carefully controlled.* The investigators used the same feeding choice setup, the same time interval, and the same definitions of predator response in each test. Controlling all of the variables except one—the types of fruits presented—is crucial because it eliminates alternative explanations for the results. For example, what types of problems could arise if the cactus mice were given less time to eat than the thrashers, or if the test animals were always presented with hackberries first and pungent chilies last?
- *Repeating the test is essential.* It is almost universally true that larger sample sizes in experiments are better. For example, suppose that the experimenters had used just one cactus mouse instead of five, and that this mouse was unlike other cactus mice because it ate almost anything. If so, the resulting data would be badly distorted. By testing many individuals, the amount of distortion or "noise" in the data caused by unusual individuals or circumstances is reduced.

To test the assumption that cactus mice are seed predators and curve-billed thrashers are seed dispersers, the researchers did a follow-up experiment. They fed fruits of the non-pungent chili to each type of predator. When the seeds had passed through the animals' digestive systems and were excreted, the researchers collected and planted the seeds—along with 14 uneaten seeds. Planting uneaten seeds served as a control treatment, because it tested the hypothesis that the seeds were viable and would germinate if they were not eaten. **Figure 1.12** shows the percentage of seeds that germinated. The data indicate that seeds pass through curve-billed thrashers unharmed but are destroyed when eaten by cactus mice.

Based on the outcomes of these two experiments, the researchers concluded that curve-billed thrashers are efficient seed dispersers and are not deterred by capsaicin. The cactus mice, in contrast, refuse to eat chilies. If they ate chilies, the mice would kill the seeds. These are exactly the results predicted by the directed dispersal hypothesis. The biologists concluded that the presence of capsaicin in chilies is an adaptation that keeps their seeds from being destroyed by mice. In habitats that contain cactus mice, the production of capsaicin increases the fitness of individual chili plants.

These experiments are a taste of things to come. In this text you will encounter hypotheses and experiments on questions ranging from how water gets to the top of 100-meter-tall sequoia trees to why the bacterium that causes tuberculosis has become resistant to antibiotics. A commitment to tough-minded hypothesis testing and sound experimental design is a hallmark of biological science. Understanding their value is an important first step in becoming a biologist.

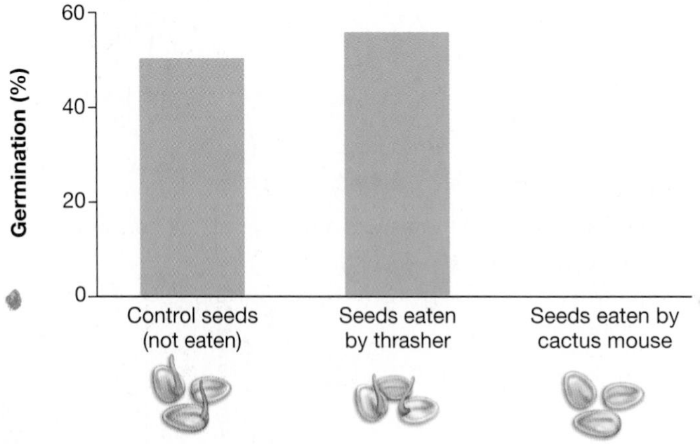

FIGURE 1.12 Mice Destroy Seeds, but Thrashers Do Not
Histograms showing the percentage of seeds that germinated after being planted directly into the ground versus being planted after passing through the digestive system of a curve-billed thrasher or cactus mouse. QUESTION Seven thrashers and five cactus mice were used in this experiment. Why wan't just one individual of each species used?

ESSAY Where Do Humans Fit on the Tree of Life?

Given the vast diversity of organisms that make up the tree of life, where do humans fit in? Most of the major branches on the tree diagrammed in Figure 1.8 represent unicellular organisms. Familiar multicellular species such as grasses and mushrooms are part of the twigs labeled "Plants" and "Fungi," respectively. At the tree's tip, the label "Animals" identifies the kingdom that appeared most recently among the eukaryotes, about 800 million years ago. Humans are found on this branch, along with millions of other species ranging from sponges and corals to insects and fish.

Our species is a tiny new twig on an enormous and ancient tree of life.

Just how recently did our species appear? In Chapter 2 we examine a technique called radiometric dating that allows geologists to estimate the age of rocks containing bones, teeth, shells, and other traces of organisms that lived in the past. According to this technique, the first traces left by our species appear about 100,000 years ago. To clarify just how recent this is in the sweep of Earth history, consider the calendar in **Figure 1.13**. The 12 months shown are scaled to represent the 4.6 billion years of Earth history. At this scale, 7 seconds make up a millennium; each hour denotes a span of 525,000 years; and each day represents an interval of 12.6 million years. Note that the first unicellular organisms appear in late March and the first multicellular life in early October. Hominids (members of the family Hominidae) walk upright for the first time on mid-afternoon of New Year's Eve. *Homo sapiens* appears about 15 minutes before the stroke of midnight.

The message of these analyses is clear: Our species is a tiny new twig on an enormous and ancient tree of life.

Given that our species evolved so recently, some other interesting questions arise. For example, a typical species of mammal lasts about 2.5 million years in the fossil record. In your opinion, how likely is it that our species will continue to leave fossil evidence for another 2.4 million years? What factors might contribute to the long-term survival of *Homo sapiens*? What factors might contribute to our demise?

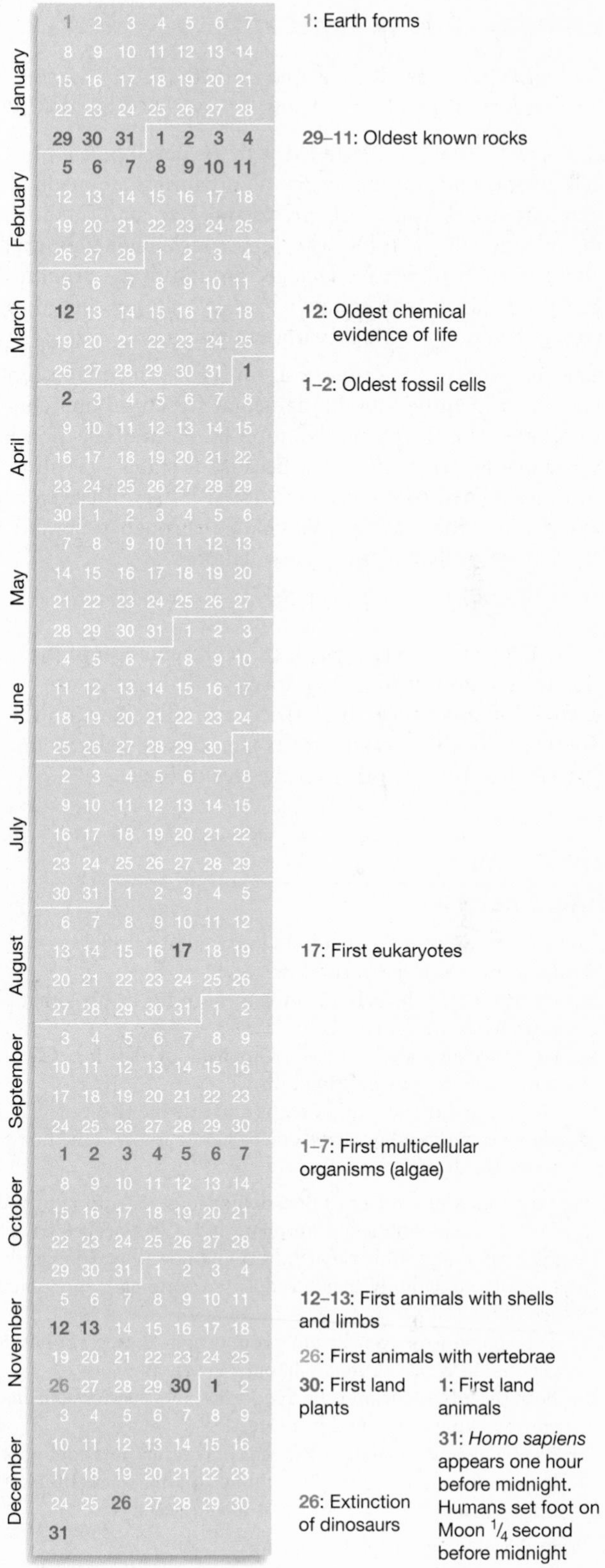
January
February
March
April
May
June
July
August
September
October
November
December
1: Earth forms
29–11: Oldest known rocks
12: Oldest chemical evidence of life
1–2: Oldest fossil cells
17: First eukaryotes
1–7: First multicellular organisms (algae)
12–13: First animals with shells and limbs
26: First animals with vertebrae
30: First land plants
1: First land animals
31: *Homo sapiens* appears one hour before midnight. Humans set foot on Moon 1/4 second before midnight
26: Extinction of dinosaurs

FIGURE 1.13 Earth's History as a Calendar

1 day = 12.6 million years
1 second = 143 years

QUESTION If the length of Earth's history (4.6 billion years) were to correspond to the length of a football field (100 yards), where on the football field does the first fossil evidence of humans (100,000 years ago) appear?

CHAPTER REVIEW

Summary of Key Concepts

Biologists have been discovering traits that unify the spectacular diversity of living organisms for over two hundred years.

- **Biological science was founded with the development of (1) the cell theory, which proposes that all organisms are made of cells and that all cells come from preexisting cells, and (2) the theory of evolution by natural selection, which maintains that the characteristics of species change through time—primarily because individuals with certain heritable traits produce more offspring than do individuals without those traits.**

The cell theory is an important unifying principle in biology, because it identified the fundamental structural unit common to all life. The theory of evolution by natural selection is another key unifying principle, because it states that all organisms are related by common ancestry. It also offered a robust explanation for why species change through time and why they are so well adapted to their habitats.

Web Tutorial 1.1 Artificial Selection

- **A phylogenetic tree is a graphical representation of the evolutionary relationships among species. Phylogenies can be estimated by analyzing similarities and differences in traits. Species that share many traits are closely related and are placed close to each other on the tree of life.**

The cell theory and the theory of evolution predict that all organisms are part of a genealogy of species, and that all species trace their ancestry back to a single common ancestor. To reconstruct this phylogeny, biologists have analyzed the sequence of components in a molecule called rRNA, which is found in all cells. A tree of life, based on similarities and differences in these sequences, has recently been constructed. According to the information contained in rRNA, the tree of life has three major lineages: the Bacteria, Archaea, and Eukarya. Analyses of other molecules and traits have largely supported the conclusions made from rRNA data.

- **Biologists ask questions, generate hypotheses to answer them, and design experiments that test the predictions made by competing hypotheses.**

Another unifying theme in biology is a commitment to hypothesis testing and to sound experimental design. Analyses of neck length in giraffes and the capsaicin found in chilies are case studies in the value of testing alternative hypotheses and conducting experiments. Biology is a hypothesis-driven, experimental science.

Web Tutorial 1.2 Introduction to Experimental Design

Questions

Content Review

1. Anton van Leeuwenhoek made an important contribution to the development of the cell theory. How?
   a. He articulated the pattern component of the theory—that all organisms are made of cells.
   b. He articulated the process component of the theory—that all cells come from preexisting cells.
   c. He invented the first microscope and saw the first cell.
   d. He invented more powerful microscopes and was the first to describe the diversity of cells.
2. Suppose that a proponent of the spontaneous generation hypothesis claimed that cells would appear in Pasteur's swan-necked flask eventually. According to this view, Pasteur did not allow enough time to pass before concluding that life does not originate spontaneously. Which of the following is the best response?
   a. The spontaneous generation proponent is correct: Spontaneous generation would probably happen eventually.
   b. Both the all-cells-from-cells hypothesis and the spontaneous generation hypothesis could be correct.
   c. If spontaneous generation happens only rarely, it is not important.
   d. If spontaneous generation did not occur after weeks or months, it is not reasonable to claim that it would occur later.
3. What does the term *evolution* mean?
   a. The strongest individuals produce the most offspring.
   b. The characteristics of an individual change through the course of its life, in response to natural selection.
   c. The characteristics of populations change through time.
   d. The characteristics of species become more complex over time.
4. What does it mean to say that a characteristic of an organism is heritable?
   a. The characteristic evolves.
   b. The characteristic can be passed on to offspring.
   c. The characteristic is advantageous to the organism.
   d. The characteristic does not vary in the population.
5. In biology, what does the term *fitness* mean?
   a. how well trained and muscular an individual is, relative to others in the same population
   b. how slim an individual is, relative to others in the same population
   c. how long a particular individual lives
   d. the ability to survive and reproduce
6. Could *both* the food competition hypothesis and the sexual selection hypothesis explain why giraffes have long necks? Why or why not?
   a. No. In science, only one hypothesis can be correct.
   b. No. Observations have shown that the food competition hypothesis cannot be correct.
   c. Yes. Long necks could be advantageous for more than one reason.
   d. Yes. All giraffes have been shown to feed at the highest possible height and fight for mates.

Conceptual Review

1. The Greek roots of the term *taxonomy* can be translated as "arranging rules." Explain why these roots were an appropriate choice for this term.
2. It was once thought that the deepest split among life-forms was between two groups: prokaryotes and eukaryotes. Draw and label a phylogenetic tree that represents this hypothesis. Then draw and label a phylogenetic tree that shows the actual relationships among the three domains of organisms.
3. Why was it important for Linnaeus to establish the rule that only one type of organism can have a particular genus and species name?
4. What does it mean to say that an organism is adapted to a particular habitat?
5. Compare and contrast natural selection with the process that led to the divergence of a wild mustard plant into cabbage, broccoli, and Brussels sprouts.
6. The following two statements explain the logic behind the use of molecular sequence data to estimate evolutionary relationships:

   "If the theory of evolution is true, then rRNA sequences should be very similar in closely related organisms but less similar in organisms that are less closely related."

   "On a phylogenetic tree, branches that are close to one another represent species that are closely related; branches that are farther apart represent species that are more distantly related."

   Is the logic of these statements sound? Why or why not?

Group Discussion Problems

1. A scientific theory is a set of propositions that defines and explains some aspect of the world. This definition contrasts sharply with the everyday usage of the word *theory*, which often carries meanings such as "speculation" or "guess." Explain the difference between the two definitions, using the cell theory and the theory of evolution by natural selection as examples.
2. Turn back to the tree of life shown in Figure 1.8. Note that Bacteria and Archaea are prokaryotes, while Eukarya are eukaryotes. On the simplified tree below, draw an arrow that points to the branch where the structure called the nucleus originated. Explain your reasoning.

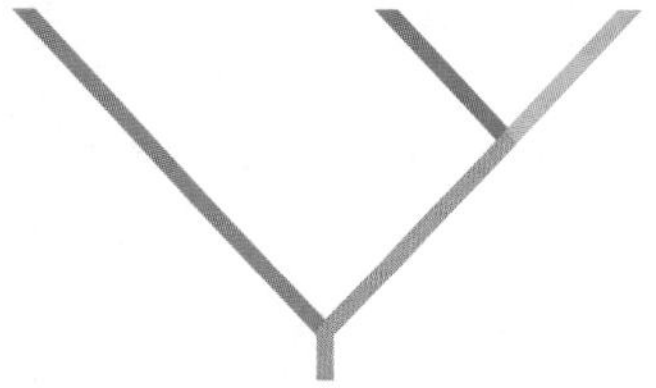

3. The proponents of the cell theory could not "prove" that it was correct in the sense of providing incontrovertible evidence that all organisms are made up of cells. They could state only that all organisms examined to date were made of cells. Why was it reasonable for them to conclude that the theory was valid?
4. How do the tree of life and the taxonomic categories created by Linnaeus (kingdom, phylum, class, order, family, genus, and species) relate to one another?

Answers to Multiple-Choice Questions 1. d; 2. d; 3. c; 4. b; 5. d; 6. c

www.prenhall.com/freeman is your resource for the following: Web Tutorials; Online Quizzes and other Online Study Guide materials; Answers to Conceptual Review Questions; Solutions to Group Discussion Problems; Answers to Figure Caption Questions and Exercises; and Additional Readings and Research.

UNIT 1

The Origin and Early Evolution of Life

When life began, Earth's surface and oceans were roiled by volcanoes; the atmosphere lacked oxygen. This volcanic eruption in Hawaii can only hint at conditions that prevailed when the first organisms evolved.

2 The Atoms and Molecules of Ancient Earth

KEY CONCEPTS

- Molecules form when atoms bond to each other. Chemical bonds are based on electron sharing. The degree of electron sharing varies from nonpolar covalent bonds, to polar covalent bonds, to ionic bonds.
- Chemical reactions tend to be spontaneous if they lead to lower potential energy and higher entropy (more disorder). An input of energy is required for nonspontaneous reactions to occur.
- During chemical evolution, (1) energy in sunlight was converted to chemical energy and (2) carbon atoms were reduced.
- Water is a small, highly polar molecule. As a result, it is an extremely efficient solvent and has a high capacity for absorbing energy.

This chapter explores the earliest events in Earth's history, from the formation of the planet through the development of the first oceans and atmosphere.

Chapter 1 introduced experiments on the hypothesis of spontaneous generation, which tested the idea that life arises from nonliving materials. This work helped build a consensus that spontaneous generation does not occur. But for life to exist now, spontaneous generation must have occurred at least once, early in Earth's history. How did life begin? This simple query has been called "the mother of all questions."

In this chapter we examine a hypothesis, called chemical evolution, that attempts to answer this question. In its simplest sense, the term *evolution* means change through time. **Chemical evolution** is the proposition that early in Earth's history, simple chemical compounds in the atmosphere and ocean combined to form larger, more complex substances. As a result, the chemistry of the oceans and atmosphere changed over time. The hypothesis maintains that product chemicals then reacted with one another to produce even more complicated compounds and that continued chemical evolution eventually led to the origin of life. More specifically, the hypothesis is that one of these complex compounds was able to make a copy of itself, or *self-replicate*. As this molecule multiplied, the process of evolution by natural selection took over—there was a switch from chemical evolution to biological evolution. Eventually, a self-replicating molecule became surrounded by a membrane, and cellular life began.

At first glance, the chemical evolution hypothesis seems wildly implausible. But is it? What evidence do biologists have that chemical evolution occurred? Let's start with the first steps—the formation of Earth and the chemical reactions that could have gotten chemical evolution started.

2.1 The Ancient Earth

Advocates of the chemical evolution hypothesis contend that when this process took place, Earth was a radically different place than it is today. To understand whether chemical evolution is plausible, then, it is important to understand what these claims about the nature of the early Earth are, and what evidence researchers use to support their argument. Let's take a closer look.

Studying the Formation of Planets

Astronomers and geologists pursue two strategies to study how Earth formed: They perform computer simulations to model conditions that led to the development of Earth, and they analyze how planets are being created elsewhere in the universe.

In biology and the other sciences, computer simulation is an important tool for analyzing processes that are difficult to study directly. Researchers write a computer program that specifies a set of starting conditions and the properties of the entities involved. The computer then records the outcome of interactions among these entities as the simulation proceeds.

For example, consider the results of simulations based on the condensation theory of planet formation. This theory maintains that our solar system originated when many billions of dust-sized particles accumulated around the forming Sun. To test this idea, researchers have created mathematical models that describe how small particles behave in orbit around the Sun. The properties of the particles are based on the types of atoms found in the solar system and their relative abundance. Then these equations are incorporated into computer simulations of planetary formation. In the initial stages of these simulations, dust and gas particles start colliding. Particles frequently stick together after the collisions, due to electrical and gravitational attraction. As the simulation continues to run, the collision-and-sticking events begin producing small bodies of rock called planetesimals. Collisions between these rocks eventually create larger bodies, called proto-planets, whose gravity attracts still more planetesimals. In this process of accumulation, called accretion, planet-sized bodies form.

Recently, these types of computer simulations have been augmented by direct studies of planet formation. Prior to 1992, astronomers had no firm evidence that planets existed anywhere outside our solar system. But numerous stars that appear to be surrounded by orbiting planets or planetary systems have since been identified. At least some of these planet-sized bodies are located in the middle of enormous clouds of orbiting dust and rock. These observations are consistent with the condensation theory, which predicts that particularly young planets should still be surrounded by dust and planetesimals.

Based on the success of these simulations and observational studies, the condensation theory is now widely accepted as the most plausible explanation for the formation of Earth. This theory also makes important predictions about the nature of the young Earth. Because immense heat would have been generated by the larger impacts during the "early accretion" phase of planet formation, most scientists conclude that Earth was molten initially. As Earth began to cool enough to form a crust of solid rock, two important events would have occurred. First, water would have rained out of the cooling atmosphere to form the first ocean. Second, volcanoes would have proliferated, as molten rock from below punctured the developing crust.

Several lines of evidence argue that heavy bombardment from space continued, however. For example, most astronomers believe that a particularly spectacular blow—a collision with a proto-planet the size of Mars—knocked Earth into its present tilt and blasted debris into space that formed the Moon. The extensive cratering on the Moon's surface suggests that Earth was also bombarded long after this giant impact.

When Did Chemical Evolution Take Place?

Geologists and astronomers estimate when these events occurred using a procedure called **radiometric** ("ray-measure") **dating**. To understand this technique, it is essential to recognize that all **atoms** share the same basic structure.

Atomic Components and Isotopes Extremely small particles called **electrons** orbit an atomic **nucleus** made up of larger particles called **protons** and **neutrons** (**Figure 2.1**). Protons have a positive electric charge, neutrons are electrically neutral, and

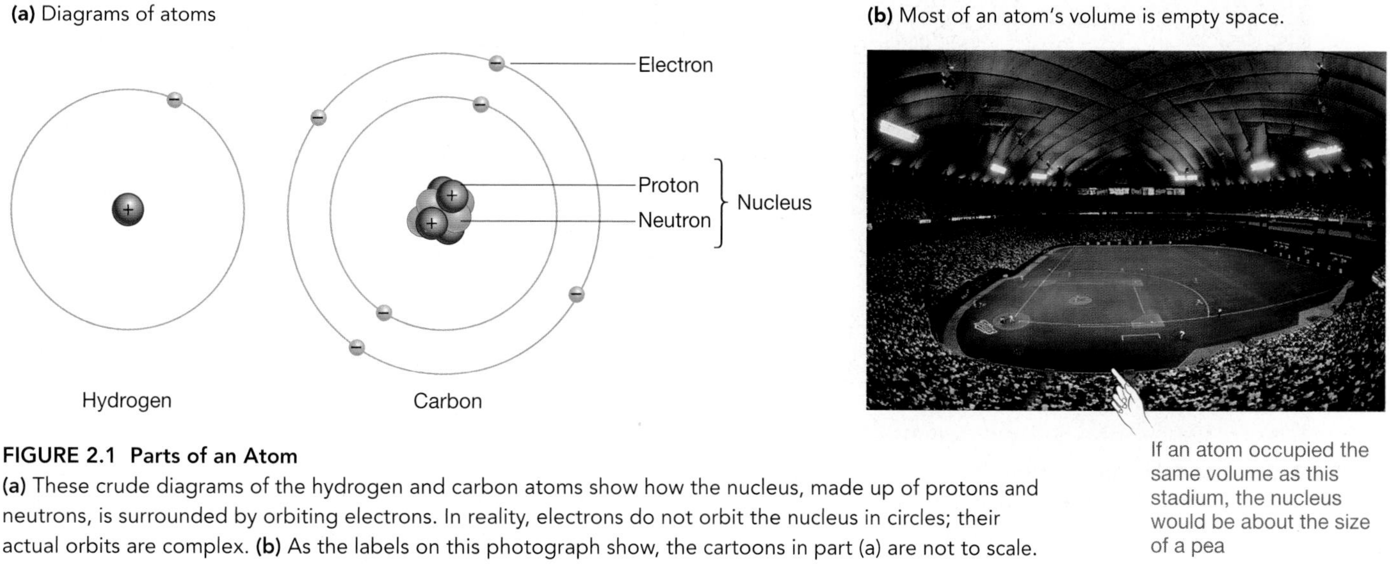
(a) Diagrams of atoms
Electron
Proton
Neutron
Nucleus
Hydrogen
Carbon
(b) Most of an atom's volume is empty space.
If an atom occupied the same volume as this stadium, the nucleus would be about the size of a pea

FIGURE 2.1 Parts of an Atom
(a) These crude diagrams of the hydrogen and carbon atoms show how the nucleus, made up of protons and neutrons, is surrounded by orbiting electrons. In reality, electrons do not orbit the nucleus in circles; their actual orbits are complex. **(b)** As the labels on this photograph show, the cartoons in part (a) are not to scale.

electrons have a negative electric charge. Opposite charges attract; like charges repel. When the number of protons and the number of electrons in an atom (or molecule) are the same, the charges balance and the atom is electrically neutral.

Each of the **elements** contains a characteristic number of protons. The number of neutrons present can vary, however. Forms of an element with different numbers of neutrons are known as **isotopes** ("equal-places"). For example, all atoms of the element uranium have 92 protons. But naturally occurring isotopes of uranium can have either 143 or 146 neutrons, giving them a total of 235 or 238 protons and neutrons, respectively. The sum of the protons and neutrons in an atom is called its **mass number**. Although the masses of protons, neutrons, and electrons can be measured in grams, the numbers involved are so small that chemists and physicists prefer to use a special unit called the **atomic mass unit (amu)**, or the dalton. The masses of protons and neutrons are virtually identical and are routinely rounded to 1 amu. A carbon atom that contains 6 protons and 6 neutrons has a mass of 12 amu and a mass number of 12. Uranium occurs in two isotopes with different mass numbers, which represent their masses in amu. They are symbolized ^{235}U and ^{238}U (or U-235 and U-238, respectively).

Radioactive Decay One of the most striking observations about atomic structure is that certain numerical combinations of protons and neutrons make the nucleus that contains them inherently unstable. That is, these nuclei break down. Atoms of ^{235}U and ^{238}U, for example, are not stable (**Figure 2.2**). Like other **radioactive isotopes**, their nuclei tend to change spontaneously by emitting radiation or a particle. These changes are called **radioactive decay**. Most types of radioactive decay result in the formation of a different isotope or element. The original isotope is called the parent, and the resulting isotope or element is called the daughter.

FIGURE 2.2 Some Isotopes Are Unstable
A technician loads pellets of the radioactive isotope ^{235}U into a long rod. The isotope will be used as fuel to generate electricity in a nuclear reactor.

Radioactive decay has an important feature: The proportion of atoms that decay in a sample of an isotope over a given time period is constant. More specifically, the proportion of the sample that decays per unit of time does not change as a function of either the isotope's age or the quantity found in a particular sample. The rate at which decay occurs varies widely from isotope to isotope, however, and is usually reported in terms of a unit called a half-life. A **half-life** is the time it takes for half of the parent isotope to decay to a particular daughter isotope (**Figure 2.3**). The half-lives of many long-lived isotopes, such as ^{235}U and ^{238}U, have been measured. The half-life of ^{235}U decaying to lead-207 (^{207}Pb), for example, is 713 million years.

How Old Is the Earth? To use radiometric dating to determine when a rock formed, geologists must estimate two quantities:

1. *The relative amounts of parent and daughter isotopes in a rock sample.* This can be measured directly, using instruments that identify the isotopes in the sample and record their abundance.

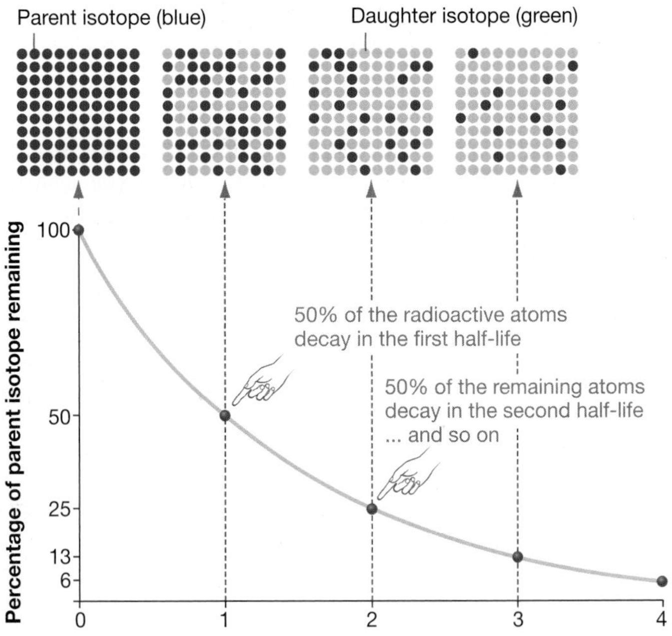

FIGURE 2.3 Radioactive Decay
A graph showing how a sample of radioactive atoms decays with time. When a sample forms, only the parent isotope exists. After the time represented by one half-life passes, the parent and daughter isotopes are present in equal proportions; after two half-lives, the ratio of parent to daughter isotopes is 1:3; after three half-lives, it is 1:7. **QUESTION** What percentages of the original parent isotope and daughter isotope exist after four half-lives? Five? Six? **EXERCISE** Suppose that a rock contains 3 percent parent isotope and 97 percent daughter isotope, and that the half-life of the parent isotope is 10 million years. How old do you estimate this rock to be?

2. *The relative amounts of parent and daughter isotopes that existed when the sample formed.* This ratio is estimated from information about how the rock formed. For example, when molten rock containing uranium atoms cools, the uranium forms crystals that do not contain lead. Based on this observation, geologists assume that ^{235}U-containing rocks that formed during Earth's early accretion phase started out with crystals containing ^{235}U but no ^{207}Pb. Over time (that is, as half-lives passed), the crystals accumulated more and more ^{207}Pb until they reached their present quantities of ^{235}U and ^{207}Pb.

Once the ratio of parent to daughter isotopes has been determined in a sample, geologists can calculate the number of half-lives that have passed and multiply that number by the length of a half-life of the parent isotope to estimate the rock's age.

To determine the age of Earth, geologists would need to find a uranium-bearing rock that formed during the planet's early accretion phase. Unfortunately, this is impossible—the planet was molten when it formed, so no rocks existed. As an alternative, researchers assume that all components of our solar system formed at about the same time. The uranium:lead ratios in meteorites, for example, indicate that meteorites formed 4.58 billion years ago, or 4.58 Ga. (The phrase "billion years ago" is often abbreviated as Ga, for "giga-years ago.") The timeline presented in **Figure 2.4a** starts with these data, which coincide with the start of Earth's formation. The ages of the oldest Moon rocks indicate that the gigantic impact responsible for the Moon's creation occurred at 4.51 Ga. Because the most ancient rock crystals found thus far on Earth date to about 4.40 Ga, it is clear that some solid crust had begun to form by this time. Indeed, most astronomers agree that Earth's formation was largely complete by 4.47 Ga. Based on the ages of the most recent craters of the Moon, it is estimated that the era of heavy bombardment—when both Earth and Moon were regularly pummeled by large objects from space—had ended by about 3.9 billion years ago.

When did the oceans originate? The oldest sedimentary rocks on Earth date to 3.85 Ga. Because sedimentary rocks form from deposits of sand, mud, or other particles carried by water, geologists are confident that an ocean existed by this time. Other evidence suggests that oceans may have been present as early as 4.28 Ga. But the heat generated when Earth was smashed by asteroids and other objects would have repeatedly vaporized most or all of this early ocean.

When Did Life Begin? If chemical evolution took place, it occurred after Earth formed but before life began. To estimate when the origin of life took place, biologists examine fossils—traces of living organisms left in rocks. The most ancient fossils that have been found to date are in rocks that formed about 3.85 Ga. As **Figure 2.4b** shows, these fossils consist of tiny grains of carbon. Although the grains do not contain cells or other direct evidence of life, a recent analysis suggests that the carbon atoms present were once part of organisms. Specifically, the isotope ^{12}C is much more abundant in the grains than expected, given the normal abundance of ^{12}C relative to heavier carbon isotopes. This is significant because living organisms that take up carbon in the form of CO_2 use ^{12}C much more readily than they use the heavier isotopes of carbon. Researchers hypothesize that even though heat and pressure have eliminated most traces of the original cells, the carbon isotope "signature" of life remains.

If these carbon grains actually do represent the remains of organisms, they suggest that life started just after the end of massive bombardment. The current consensus is that chemical evolution occurred during a fairly narrow window of time, perhaps 300 million years or less. Almost as soon as conditions on Earth stabilized enough for life to begin, it did.

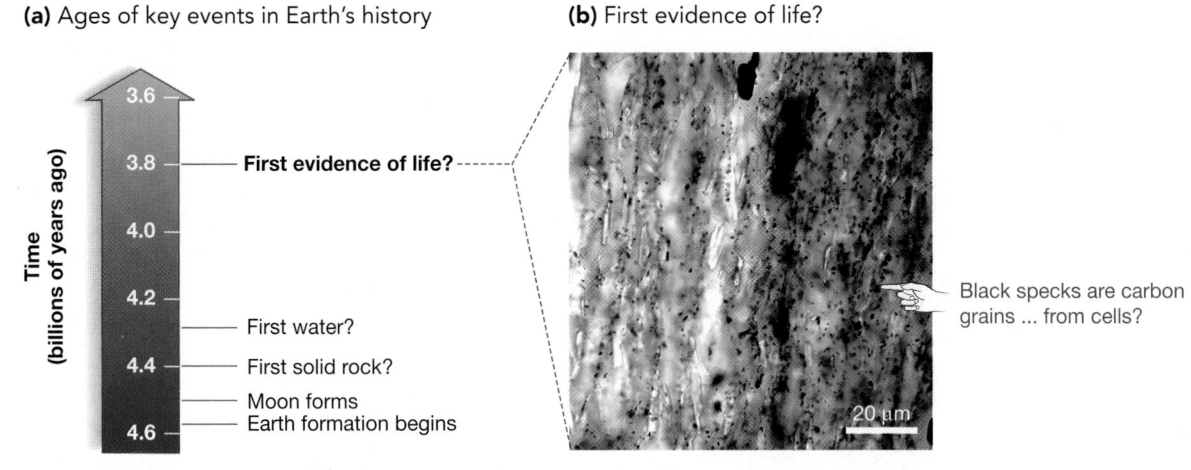
(a) Ages of key events in Earth's history
(b) First evidence of life?
Time (billions of years ago)
3.6
3.8
4.0
4.2
4.4
4.6
First evidence of life?
First water?
First solid rock?
Moon forms
Earth formation begins
Black specks are carbon grains ... from cells?
20 μm

FIGURE 2.4 Early Events in the History of Earth
(a) Radiometric dating has assigned ages to key events in early Earth's history. The dates with question marks are still under investigation. **(b)** The black material is carbon that was found inside a 3.8-billion-year-old rock. The ratio of ^{13}C to ^{12}C inside is identical to the ratio observed in today's organisms, but very different from the ratios observed in the nonliving environment (air, water, and soil). Researchers hypothesize that this carbon is the remains of some of Earth's earliest organisms.

2.2 The Building Blocks of Chemical Evolution

Just four types of atoms—hydrogen, carbon, nitrogen, and oxygen—make up 96 percent of all matter found in organisms today. Many of the compounds found in living cells contain thousands, or even millions, of these atoms. But early in Earth's history, it is likely that these elements existed only in simple substances such as water and carbon dioxide, which contain just three atoms apiece.

The chemical evolution hypothesis maintains that simple compounds in the ancient atmosphere and ocean combined to form the larger, more complex substances found in living cells. To understand how this process could have begun, we need to consider the following questions:

- What is the physical structure of hydrogen, carbon, nitrogen, oxygen, and other atoms found in living cells?
- How do these atoms combine to form simple molecules such as water and carbon dioxide—the building blocks of chemical evolution?

What Atoms Are Found in Organisms?

Figure 2.5 shows one way that chemists summarize the structure of atoms. Notice that each atom has a symbol, such as the C shown for carbon, along with a superscript and a subscript. The subscript is the element's **atomic number**, meaning the number of protons in its nucleus. The superscript is the element's mass number (Section 2.1), which is the sum of the number of protons and neutrons in the nucleus. The mass number given for each element is that of the most common isotope of the element. For example, the most abundant isotope of carbon, ^{12}C, has six protons and six neutrons in its nucleus.

If an element is uncharged, its protons are matched by an equal number of electrons. Electrons carry a negative electric charge but have very little mass compared with protons. To understand how the atoms involved in chemical evolution behave, it is critical to understand how electrons are arranged around the nucleus.

Electrons move around atomic nuclei in specific regions called **orbitals**. Each electron orbital has a distinctive shape, and each orbital can hold up to two electrons. Orbitals, in turn, are grouped into levels called **electron shells**. These are numbered 1, 2, 3, and so on, to indicate their relative distance from the nucleus. The electrons of an atom fill the innermost shells first, before filling outer shells.

To understand how the structures of atoms differ, take a moment to study **Figure 2.6**. This chart, which is a subset of the periodic table of the elements, highlights those atoms that are most abundant in living cells. The diagrams in each box of the table show how electrons are distributed in the shells of carbon and other key elements. The outermost shell of an atom is called the atom's **valence shell**, and the electrons found in that shell are referred to as **valence electrons**. Two observations are important:

1. In each of the highlighted elements, the outermost electron shell is not full. The highlighted elements have at least one unpaired valence electron.
2. The number of unpaired electrons in the valence shell varies among elements. Carbon has four unpaired electrons in its outermost shell; hydrogen has one. The number of unpaired electrons found in an atom is called its **valence.** Carbon's valence is four; hydrogen's valence is one.

These observations are significant because an atom is most stable when its valence shell is filled. One way that shells can be filled is through the formation of **chemical bonds**—strong attractions that bind atoms together.

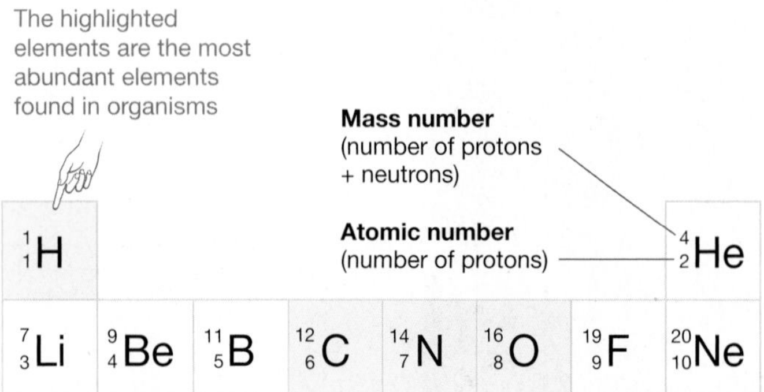

FIGURE 2.5 Chemical Symbols of Atoms Found in Organisms A segment of the periodic table, highlighting just the elements commonly found in organisms.

How Does Covalent Bonding Hold Molecules Together?

To understand how atoms can become more stable by bonding to another atom, consider hydrogen. The hydrogen atom has just one electron, which resides in a shell that can hold two electrons. The hydrogen atom is most stable when this shell is full. When two atoms of hydrogen approach each other, the two electrons present become shared by the two nuclei (**Figure 2.7**). Both atoms now have a completely filled shell, and they have formed a chemical bond. Together, the bonded hydrogen atoms are more stable than the two individual hydrogen atoms. The shared electrons "glue" the atoms together in a **covalent bond**. Substances that are held together by covalent bonds are called **molecules**. In this case of two hydrogen atoms, the bonded atoms form a single molecule of hydrogen, written as H–H or H_2.

To visualize what is happening when a covalent bond forms, imagine two children who want the same toy. If both children grasp the toy tightly, through it they become attached to each

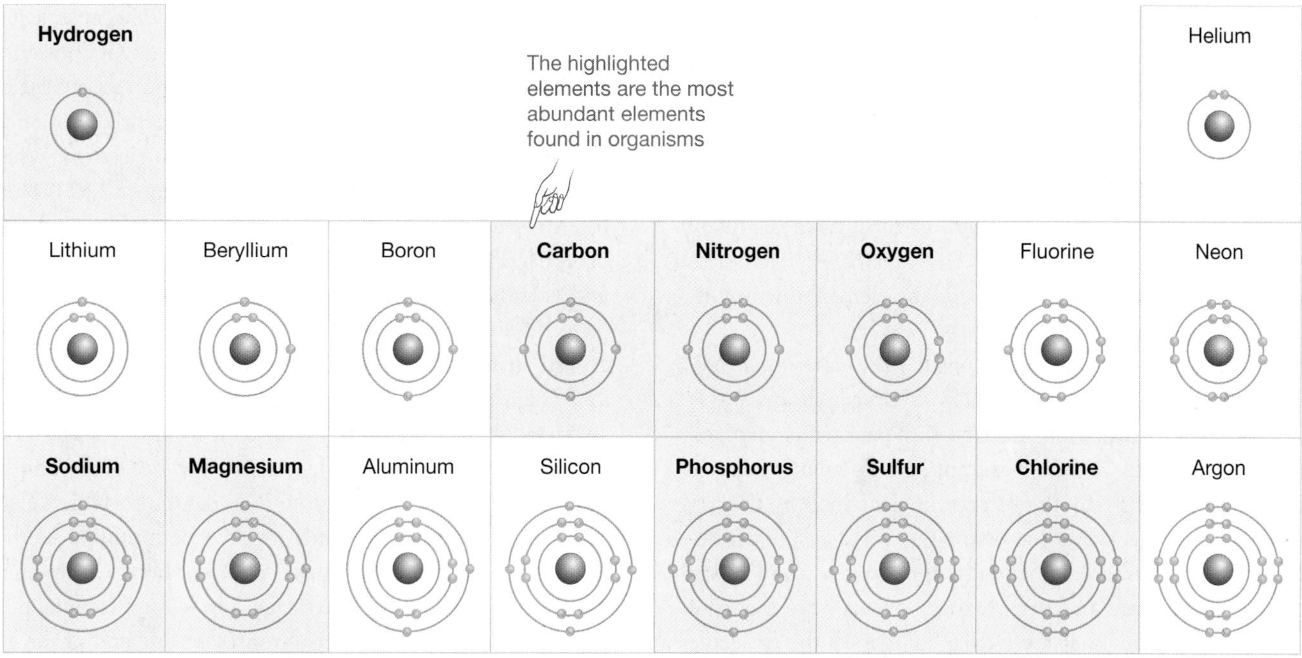

FIGURE 2.6 The Structure of Atoms Found in Organisms
The atomic nucleus is represented by a closed sphere. The first (inner), second, and third electron shells are shown as rings. The dots on the rings represent electrons. Electrons are drawn in pairs if they occupy filled orbitals within the same electron shell; they are drawn singly if they occupy unfilled orbitals. EXERCISE For each highlighted atom, write down the number of bonds it is capable of forming based on the number of unpaired electrons in its outermost electron shell.

other. For as long as they hold onto the toy, neither child can leave the other's side.

Another way to think about covalent bonding is in terms of electrical attraction and repulsion. As two hydrogen atoms move closer together, their positively charged nuclei repel each other and their negatively charged electrons repel each other. But each proton attracts both electrons, and each electron attracts both protons. Covalent bonds form when the attractive forces overcome the repulsive forces. This is the case when hydrogen atoms interact to form the hydrogen molecule (H_2).

To summarize, covalent bonding is based on electron sharing. In a hydrogen molecule, the electrons involved in the covalent bond are shared between the two nuclei. This situation is illustrated in **Figure 2.8a**. The covalent bond between hydrogen atoms is represented by a dash, and the electrons are drawn as dots that are between the two nuclei. An equal sharing of electrons, as in Figure 2.8a, results in a **nonpolar covalent bond**—a covalent bond that is symmetrical.

It's important to note, though, that the electrons participating in covalent bonds are not always shared equally between the atoms involved. Asymmetric sharing of electrons results in **polar covalent bonds.** Some atoms hold electrons in covalent

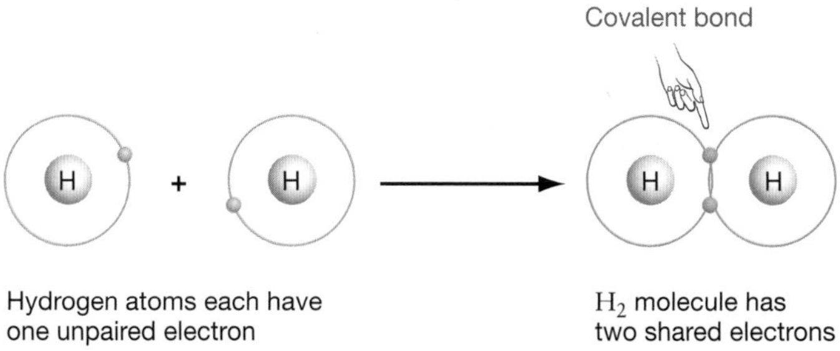

FIGURE 2.7 Covalent Bonds Result from Electron Sharing
When two hydrogen atoms come into contact, their electrons are attracted to the positive charge in each nucleus. As a result, their orbitals overlap, the electrons are shared by each nucleus, and a covalent bond forms.

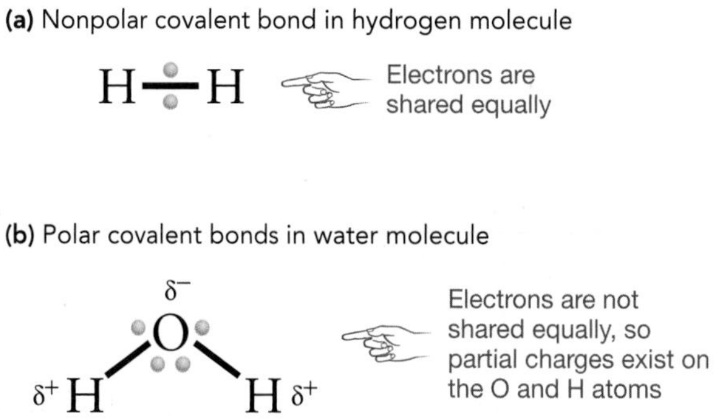

FIGURE 2.8 Electron Sharing and Bond Polarity
(a) In a hydrogen molecule, electrons are shared equally by the two hydrogen nuclei. **(b)** In a water molecule, oxygen's high electronegativity causes that atom to hold electrons more tightly than do the hydrogen atoms. As a result, different parts of the molecule have partial charges. The delta (δ) symbols in the polar covalent bonds refer to these partial positive and negative charges.

bonds much more tightly than do other atoms, so the degree of sharing varies by element. Chemists call this property an atom's **electronegativity.** Oxygen is among the most electronegative of all elements: It attracts covalently bonded electrons more strongly than does any other atom commonly found in organisms. Nitrogen's electronegativity is somewhat lower than oxygen's. Carbon and hydrogen, in turn, have relatively low and approximately equal electronegativities. The electronegativities of the four most abundant elements in organisms are related as follows: $O \gg N > C \cong H$.

To put these ideas into practice, consider the water molecule. Water consists of oxygen bonded to two hydrogen atoms and is written H_2O. As **Figure 2.8b** illustrates, the electrons involved in the covalent bonds in water are not shared equally but are held much more tightly by the oxygen nucleus than by the hydrogen nuclei. Hence, water has two polar covalent bonds. Because electrons are shared unequally, the oxygen atom has a partial negative charge and the hydrogen atoms have a partial positive charge. Notice that the partial charges on the molecule are symbolized by the lowercase Greek letter delta, δ.

How Does Ionic Bonding Hold Ionic Compounds Together?

Ionic bonds are similar in principle to covalent bonds, but instead of sharing electrons between two atoms, the electrons in ionic bonds are completely transferred from one atom to the other. The electron transfer occurs because it gives the resulting atoms a full outermost shell. Sodium atoms (Na), for example, tend to lose an electron, leaving them with a full second shell. This is a much more stable arrangement, energetically, than having a lone electron in their third shell (**Figure 2.9a**). The atom that results has a net electric charge of +1, because it has one more proton than it has electrons. An atom or molecule that carries a charge is called an **ion.** The sodium ion is written Na^+ and, like other positively charged ions, is called a **cation.** Chlorine atoms (Cl), in contrast, tend to gain an electron. When this occurs, the atom's outermost shell is full (**Figure 2.9b**). The atom has a net charge of −1, because it has one more electron than protons. This negatively charged ion, or **anion,** is written Cl^- and is called chlor*ide*. When sodium and chlorine combine to form table salt (sodium chloride, NaCl), the atoms pack into a crystal structure consisting of sodium cations and chloride anions (**Figure 2.9c**). The electrical attraction between the ions is so strong that salt crystals are difficult to break apart.

This discussion of covalent and ionic bonding supports an important general observation: The degree to which electrons are shared in chemical bonds forms a continuum. As the left-hand side of **Figure 2.10** shows, covalent bonds between atoms with exactly the same electronegativity—for example, between the atoms of hydrogen in H_2—represent one end of the continuum. The electrons in these nonpolar bonds are shared equally. In the middle of the continuum are bonds where one atom is much more electronegative than the other. In these asymmetric bonds, substantial partial charges exist on each of the atoms. These types of polar covalent bonds occur when a highly electronegative atom such as oxygen or nitrogen is bound to an atom with a lower affinity for electrons, such as carbon or hydrogen. Water (H_2O) and ammonia (NH_3) contain polar covalent bonds. At the right-hand side of the continuum are molecules made up of atoms with extreme differences in their electronegativities. In this case, electrons are transferred rather than shared, the atoms have full charges, and the bonding is

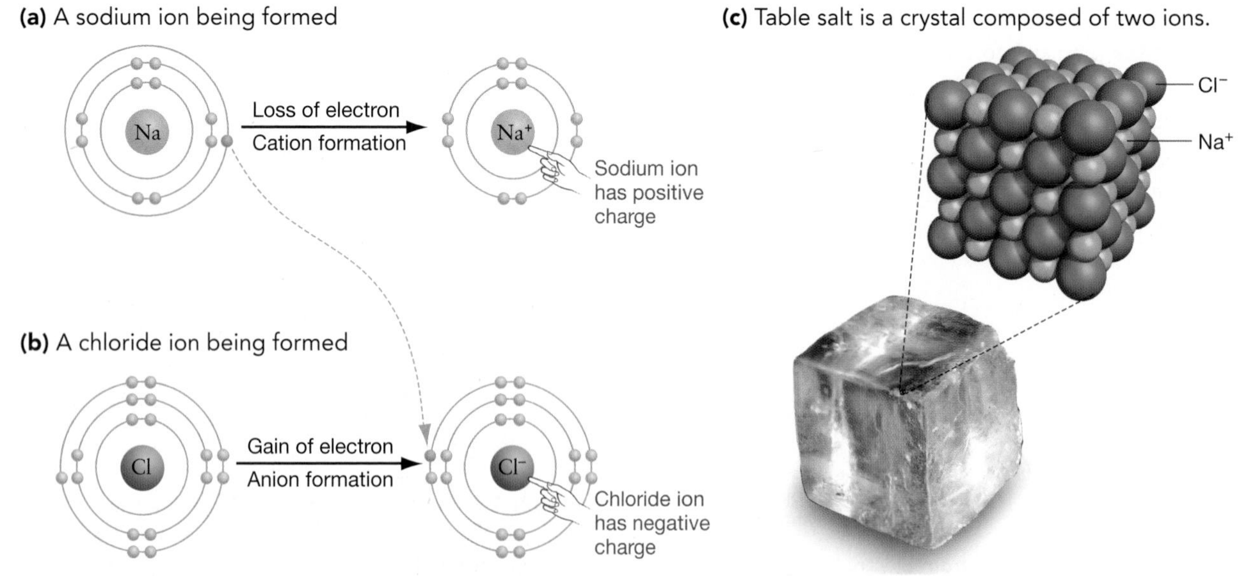

FIGURE 2.9 Ion Formation and Ionic Bonding
(a) When an atom of sodium (Na) loses an electron, it forms a sodium ion (Na^+). The sodium ion is stable because it has a full valence shell. **(b)** When an atom of chlorine (Cl) gains an electron, it forms a chloride ion (Cl^-), which also has a full valence shell. **(c)** In table salt (NaCl), sodium and chloride ions pack into a crystal structure held together by electrical attraction.

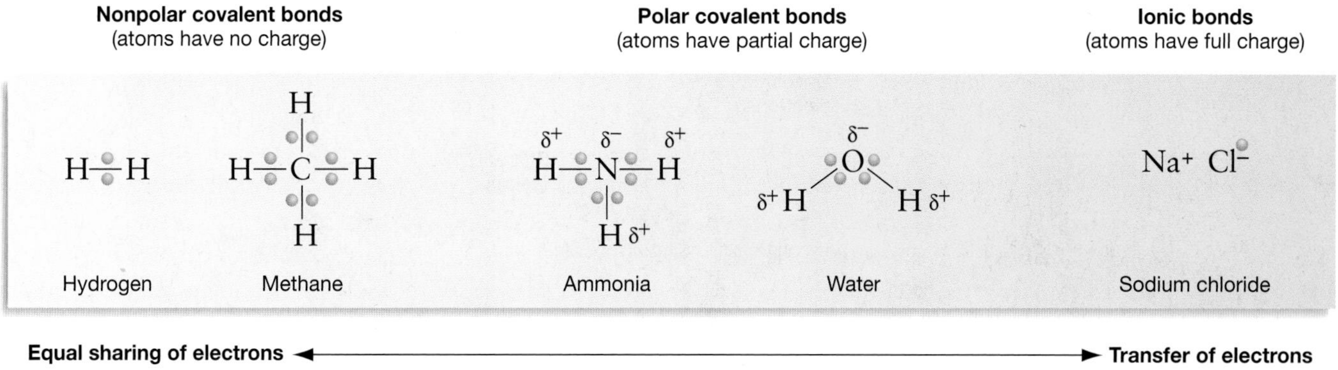

Nonpolar covalent bonds
(atoms have no charge)
Polar covalent bonds
(atoms have partial charge)
Ionic bonds
(atoms have full charge)
Hydrogen
Methane
Ammonia
Water
Sodium chloride
Na+ Cl−
Equal sharing of electrons
Transfer of electrons

FIGURE 2.10 The Electron-Sharing Continuum
The degree of electron sharing in chemical bonds can be thought of as a continuum, from equal sharing in nonpolar covalent bonds to no sharing in ionic bonds.

ionic. Common table salt, NaCl, is a familiar example. In biology, ionic bonding is rare. The chemical bonds that occur in biological molecules are on the left-hand side of this continuum.

Some Simple Molecules Formed from H, C, N, and O

Look again at Figure 2.6 and count the number of unpaired electrons in the valence shells of carbon, nitrogen, and oxygen atoms: Carbon has four, nitrogen has three, and oxygen has two. Each unpaired electron can make up half of a covalent bond. As a result, a carbon atom can form a total of four covalent bonds; nitrogen can form three; and oxygen, two. When each of the four unpaired electrons of a carbon atom covalently bonds with a hydrogen atom, the molecule that results is written CH_4 and is called methane (**Figure 2.11a**). This is the most common molecule found in natural gas. When a nitrogen atom's three unpaired electrons bond with three hydrogen atoms, the result is NH_3, or ammonia. Similarly, an atom of oxygen can form covalent bonds with two atoms of hydrogen, resulting in a water molecule (H_2O).

Double and Triple Bonds In addition to forming multiple single bonds, atoms with more than one unpaired electron in the valence shell can sometimes produce double or triple bonds. **Figure 2.11b** shows how carbon forms double bonds with oxygen atoms to produce the molecule called carbon dioxide (CO_2). Triple bonds occur when three pairs of electrons are shared. **Figure 2.11c** shows the structure of molecular nitrogen (N_2), which forms when two nitrogen atoms establish a triple bond.

Bond Angles and the Shape of Molecules Proponents of the chemical evolution hypothesis argue that simple molecules such as CH_4, NH_3, H_2O, CO_2, and N_2 were important components of Earth's ancient atmosphere and ocean. Their reasoning is based on the observation that these molecules are found in volcanic gases on Earth and in the atmospheres of nearby planets. If so, then such molecules provided the building

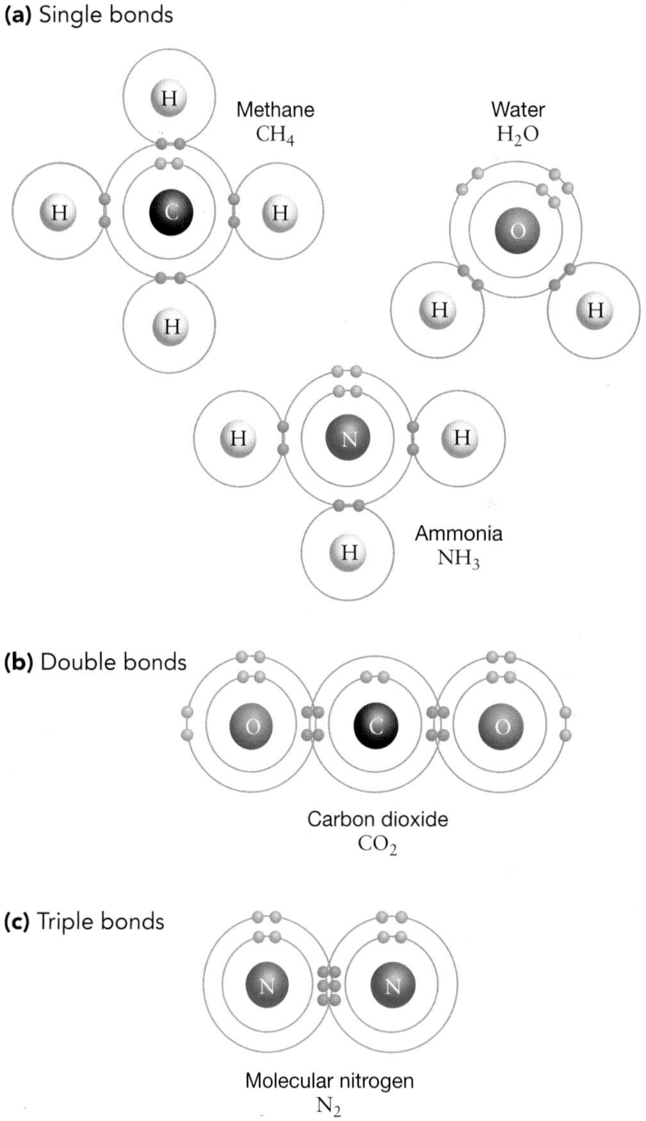

(a) Single bonds
Methane
CH4
Water
H2O
Ammonia
NH3
(b) Double bonds
Carbon dioxide
CO2
(c) Triple bonds
Molecular nitrogen
N2

FIGURE 2.11 Unpaired Electrons in the Valence Shell Participate in Covalent Bonds
Covalent bonding is based on sharing of electrons in the outermost shell. Covalent bonds can be **(a)** single, **(b)** double, or **(c)** triple.
QUESTION How do the electron configurations of H, C, O, and N in these molecules compare with their configurations shown in Figure 2.6?

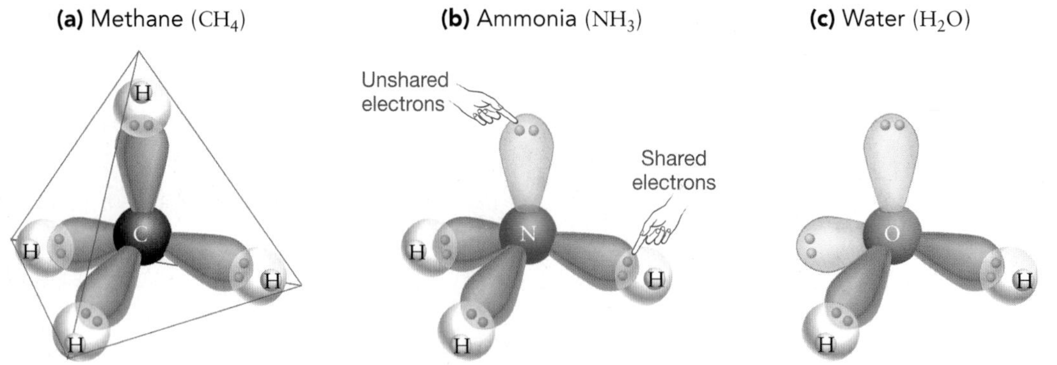
(a) Methane (CH_4)
(b) Ammonia (NH_3)
(c) Water (H_2O)
Unshared electrons
Shared electrons
H
C
N
O

FIGURE 2.12 The Geometry of Methane, Ammonia, and Water
Diagrams showing the shapes of the orbitals in the outermost shells of **(a)** methane, **(b)** ammonia, and **(c)** water. The dots represent electrons. Even though some of the same orbitals are involved in bonding, the overall shape of each molecule is radically different. EXERCISE Label which molecule is bent and planar, which forms a tetrahedron (having four identical faces), and which forms a pyramid. In parts (b) and (c), draw lines to indicate the overall shape of ammonia and water, similar to those drawn for methane.

blocks for chemical evolution. Before analyzing how these molecules could have combined to form more complex compounds, however, we need to consider their shape and how scientists represent their composition and shape. This is a crucial point. The overall shape of a molecule often dictates its behavior and function.

The shape of a simple molecule is governed by the geometry of its bonds. Bond angles, in turn, are determined by electrical repulsion that occurs between pairs of electrons. For example, **Figure 2.12a** shows the shape of methane (CH_4). The four pairs of electrons in carbon's outer shell form a tetrahedron when bonded to hydrogen because repulsion among the shared electron pairs orients them at an angle of 109.5° to one another. Ammonia (NH_3), in contrast, is shaped like a pyramid. As **Figure 2.12b** shows, the four orbitals in nitrogen's outermost shell form a tetrahedron, just as they do in the carbon atom in methane. But because one of the four orbitals in the nitrogen atom's outer shell is already filled with a pair of electrons, only the three orbitals with unpaired electrons bond with hydrogen. In this case, the geometry of the orbitals and the geometry of the molecule are different. In the oxygen atom, two of the four orbitals present in the outermost shell are filled with electron pairs. When hydrogen atoms bond to the remaining two electrons to form water (H_2O), the resulting molecule is bent and two-dimensional, or planar (**Figure 2.12c**).

Representing Molecules Molecules can be represented in a variety of ways. The simplest representation is a **molecular formula**, which indicates the numbers and types of atoms in a molecule. Water has the molecular formula H_2O; methane has the molecular formula CH_4 (**Figure 2.13a**).

Molecular formulas are a compact way of stating a molecule's composition, but they contain no information about how the molecule is put together. **Structural formulas** indicate which atoms are bonded together. In a structural formula, single, double, and triple bonds are represented by single, double, and triple dashes, respectively. **Figure 2.13b** gives structural formulas for molecules that were important components of ancient Earth's atmosphere and ocean.

The limitation of structural formulas is that they are two-dimensional, while molecules are three-dimensional. More complex diagrams, such as *ball-and-stick models* and *space-filling models*, are based on the three-dimensional structure of molecules. **Figures 2.13c** and **2.13d** show how ball-and-stick and space-filling models indicate a molecule's geometry as well as the relative sizes of the atoms involved.

Quantifying the Concentration of Key Molecules Scientists quantify the number of molecules present in a sample by using a unit called the **mole**. A mole refers to the number 6.022×10^{23} (just as the unit called the *dozen* refers to the number 12). The mole is a useful unit because the mass of one mole of any molecule is the same as its molecular weight expressed in grams. **Molecular weight** is the sum of the mass numbers of all of the atoms in a molecule. For example, summing the mass numbers of two atoms of hydrogen and one atom of oxygen gives $1 + 1 + 16$, or 18, which is water's molecular weight. It follows that if you weighed a sample of 18 grams of water, it would contain 6.022×10^{23} water molecules, or one mole of water molecules.

Many important biological substances are normally dissolved in liquid. A **solution** is a homogenous (uniform) mixture of one or more substances dissolved in a liquid. When substances are dissolved in a liquid, their concentration is expressed in terms of molarity (symbolized by "M"). **Molarity** is the number of moles of the substance present per liter of solution. A 1-molar solution of carbon dioxide gas (CO_2) in water, for instance, means that 1 mole of CO_2—that is, 6.022×10^{23} CO_2 molecules—is contained in 1 liter of water.

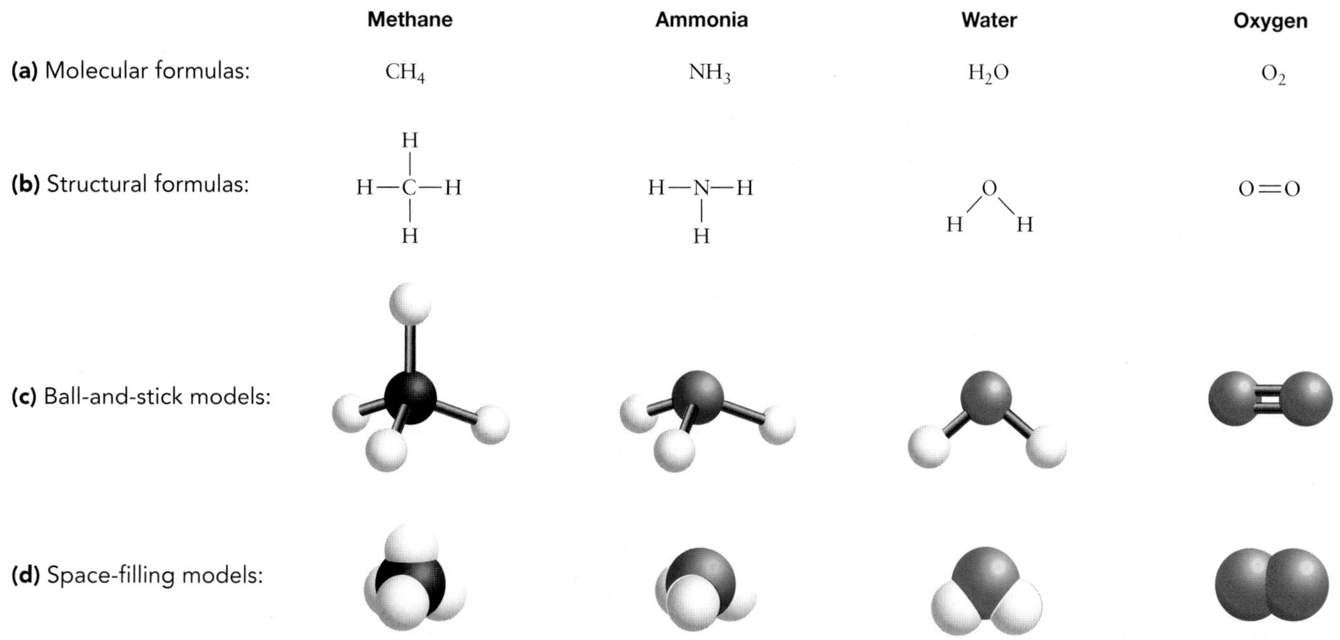

Methane
Ammonia
Water
Oxygen
(a) Molecular formulas:
CH_4
NH_3
H_2O
O_2
(b) Structural formulas:
(c) Ball-and-stick models:
(d) Space-filling models:

FIGURE 2.13 Molecules Can Be Represented Several Ways
Each method of representing a molecule has particular advantages. **(a)** Molecular formulas are compact and indicate the number and identity of atoms involved. **(b)** Structural formulas show which atoms are bonded to one another. **(c)** Ball-and-stick models take up more space than do structural formulas but include information about bond geometry. **(d)** Space-filling models are not as easy to read as ball-and-stick models but more accurately depict the spatial relationships between atoms.

✓ CHECK YOUR UNDERSTANDING

Covalent bonds are based on electron sharing, while ionic bonds are based on electrical attraction between ions with opposite charges. Covalent bonds can be polar or nonpolar, depending on whether the electronegativities of the two atoms involved are the same or different. You should be able to (1) draw the structural formulas of methane (CH_4) and ammonia (NH_3) and add dots to indicate the relative locations of the covalently bonded electrons, and (2) draw the electron shells around sodium ions (Na^+) and chloride ions (Cl^-) and explain why table salt (NaCl) exists.

2.3 Chemical Reactions, Chemical Evolution, and Chemical Energy

Proponents of the chemical evolution hypothesis contend that simple molecules present in the atmosphere and ocean of ancient Earth participated in chemical reactions that produced larger, more complex molecules. A **chemical reaction** is an event in which one substance is combined with others or broken down into another substance. According to one hypothesis, the early atmosphere was made up of gases ejected from volcanoes (**Figure 2.14a**). Carbon dioxide, water, and nitrogen are the dominant gases ejected from volcanoes today; a small amount

(a) Volcanoes eject an array of gases, including water vapor.

(b) Liquid water forms when warm water vapor cools.

FIGURE 2.14 What Molecules Were Present at the Start of Chemical Evolution?
(a) When volcanoes erupt, they eject enormous quantities of dust, N_2, CO_2, and water into the atmosphere. **(b)** When water vapor cools, condensation occurs and liquid water forms. This process was responsible for the formation of the world's first lakes and oceans.

of molecular hydrogen (H_2) and methane (CH_4) may also be present. But when these molecules are placed in a glass tube together and allowed to interact, very little happens. Water vapor condenses to liquid water as the mixture cools (**Figure 2.14b**), but the simple molecules do not suddenly link together to create large, complex substances like those found in living cells. Instead, their bonds remain intact. How, then, did chemical evolution occur?

To answer this question we must explore two topics: (1) how chemical reactions occur, and (2) how conditions on ancient Earth made certain reactions possible.

How Do Chemical Reactions Happen?

Chemical reactions are written in a format similar to mathematical equations, with the initial, or **reactant**, atoms or molecules shown on the left and the resulting reaction **product**(s) shown on the right. For example, the most common reaction in the mix of gases and water that emerge from volcanoes is

$$CO_2(g) + H_2O(l) \rightleftharpoons H_2CO_3(aq)$$

This expression indicates that carbon dioxide (CO_2) reacts with water (H_2O), forming carbonic acid (H_2CO_3). The state of each reactant and product is indicated as gas (*g*), liquid (*l*), in aqueous solution (*aq*), or solid (*s*).

Note that the previous expression is balanced; that is, one carbon, three oxygen, and two hydrogen atoms are present on each side of the expression. Note also that the expression contains a double arrow, meaning that the reaction is reversible. When the forward and reverse reactions proceed at the same rate, the quantities of reactants and products remain constant, although not necessarily equal. A dynamic but stable state such as this is termed a **chemical equilibrium.** A chemical equilibrium can be disturbed by changing the quantity (*concentration*) of reactants or products. For example, adding CO_2 to the mixture would drive the reaction to the right, creating more H_2CO_3 until the equilibrium proportions of reactants and products are reestablished. Adding H_2CO_3 instead would drive the reaction to the left. Removing CO_2 would also drive the reaction to the left; removing H_2CO_3 would drive it to the right.

A chemical equilibrium can also be altered by changes in temperature. For example, the water molecules in this set of interacting elements, or **system**, would be present as a combination of liquid water and water vapor:

$$H_2O(l) \rightleftharpoons H_2O(g)$$

If liquid water molecules absorb enough heat, they transform to the gaseous state. This is called an **endothermic** ("within heating") process because heat is absorbed during the process. In contrast, the transformation of water vapor to liquid water releases heat and is called **exothermic** ("outside heating"). Raising the temperature of this system drives the equilibrium to the right; cooling the system drives it to the left. In endothermic and exothermic processes, however, the system must be *isolated*, meaning one with no external source of energy. The components of the system can exchange energy (such as heat), but not matter, with their surroundings. If water vapor is lost to the environment, the system is not isolated and the equilibrium may not proceed as described earlier.

In relation to chemical evolution, though, these reactions and changes of state are not particularly interesting. Carbonic acid is not an important intermediate in the formation of more complex molecules. According to models developed by a series of researchers, however, interesting things do begin to happen when larger amounts of energy are added to the system.

What Is Energy?

Energy can be defined as the capacity to do work or to supply heat. This capacity exists in one of two ways—as a stored potential or as an active motion.

Stored energy is called **potential energy.** An object gains or loses its ability to store energy as a consequence of its position. An electron that resides in an outer electron shell will, if the opportunity arises, fall into a lower electron shell closer to the positive charges on the protons in the nucleus. As a result, an electron in an outer electron shell has more potential energy than does an electron in an inner shell (**Figure 2.15**).

Kinetic energy is the energy of motion. Molecules have kinetic energy because they are constantly in motion. This form of kinetic energy—the kinetic energy of molecular motion—is called **thermal energy.** The **temperature** of an object is a measure of how much thermal energy its molecules possess. If an object has a low temperature, its molecules are moving slowly (we perceive this as "cold"). If an object has a high temperature, its molecules are moving rapidly (we perceive this as "hot"). When two objects with different temperatures come into contact, thermal energy is transferred between them. We call this transferred energy **heat.**

There are many forms of potential and kinetic energies, and energy can change from one form into another. To drive this

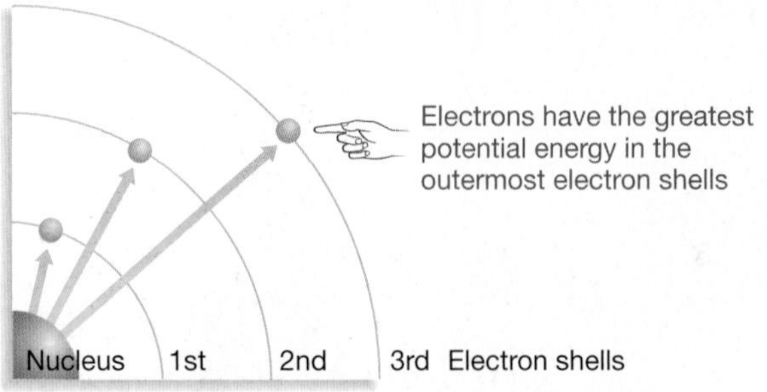

Electrons have the greatest potential energy in the outermost electron shells
Nucleus
1st
2nd
3rd Electron shells

FIGURE 2.15 Potential Energy as a Function of Electron Shells
Electrons in outer shells have more potential energy than do electrons in inner shells, because negative charges in outer shells are farther from the positive charges in the nucleus. Each shell represents a distinct level of potential energy.

point home, consider a water molecule sitting at the top of a waterfall, as in **Figure 2.16a**. This molecule has potential energy as a result of its position. If the molecule passes over the waterfall, its potential energy is converted to the kinetic energy of motion. When the molecule has reached the rocks below, it has experienced a change in potential energy because it has changed position. Panel 3 in Figure 2.16a shows that this change in potential energy is transformed into an equal amount of energy in other forms: mechanical energy, which tends to break up the rocks; heat (thermal energy), which raises the temperature of the rocks and the water itself; and sound.

An electron in an outer electron shell is analogous to the water molecule at the top of a waterfall (**Figure 2.16b**). If the electron falls to a lower shell, its potential energy is converted to the kinetic energy of motion. After the electron occupies the lower electron shell, it experiences a change in potential energy. As panel 3 in Figure 2.16b shows, the change in potential energy is transformed into an equal amount of energy in other forms—usually thermal energy, but sometimes light.

These examples illustrate the **first law of thermodynamics**, which states that energy is conserved. Energy cannot be created or destroyed, but only transferred and transformed.

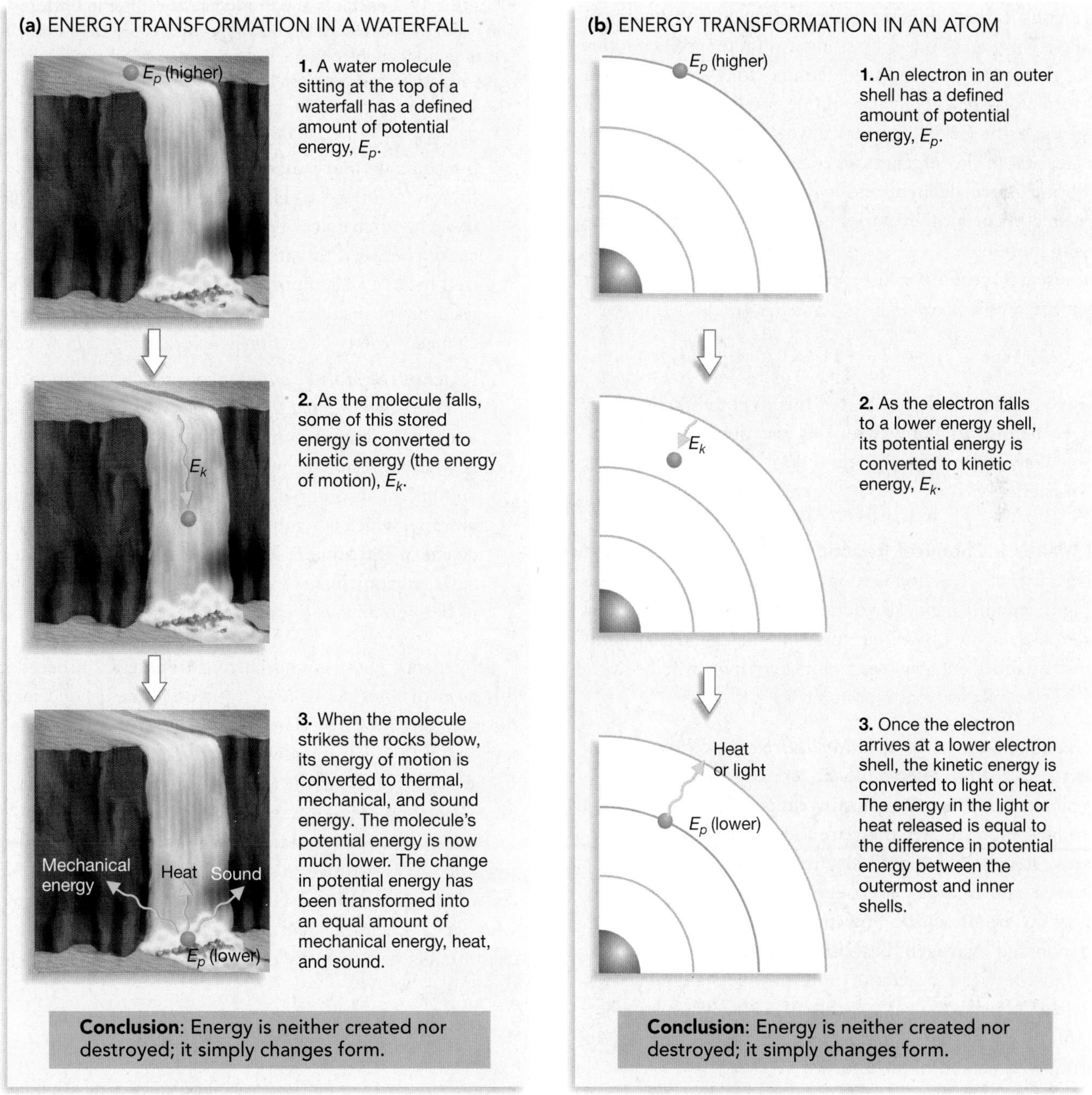

FIGURE 2.16 Energy Transformations
During an energy transformation, the total amount of energy in the system remains constant.

If the Earth, as it cooled from a molten state, was being bombarded with large objects from space, then the simple molecules present in the ancient Earth's atmosphere and ocean would have been exposed to massive amounts of thermal energy from volcanic eruptions and asteroid impacts, in addition to high-energy ultraviolet radiation from the Sun. How would the application of large amounts of energy affect the course of chemical evolution?

Chemical Evolution: A Model System

To assess the impact of energy inputs on the simple molecules present in the early oceans and atmosphere, researchers have constructed computer models to simulate the reactions that can occur among carbon dioxide, water, nitrogen, and hydrogen molecules. The goal of one such study was quite specific: In this case the researchers wanted to determine whether a molecule called formaldehyde (H_2CO) could be produced. Along with hydrogen cyanide (HCN), formaldehyde is a key intermediate in the creation of the larger, more complex molecules found in cells. Forming formaldehyde and hydrogen cyanide is the critical first step in chemical evolution—a trigger that could set the process in motion.

The research group began by proposing that the following reaction could take place:

$$CO_2(g) + 2\,H_2(g) \longrightarrow H_2CO(g) + H_2O(g)$$

Before we explore how they tested this hypothesis, however, it will be helpful to know why this reaction doesn't occur spontaneously—that is, why doesn't it occur *without* an input of energy?

What Makes a Chemical Reaction Spontaneous? When chemists say that a reaction is spontaneous, they have a precise meaning in mind: Chemical reactions are spontaneous if they proceed on their own, without any continuous external influence such as added energy. Two factors determine whether a reaction is spontaneous or nonspontaneous:

1. *Reactions tend to be spontaneous if the products have lower potential energy than the reactants.* Reaction products have lower potential energy if their electrons are held more tightly than are the electrons of the reactants. Recall that highly electronegative atoms such as oxygen and nitrogen hold electrons much more tightly than do atoms with a lower electronegativity, such as carbon and hydrogen. Because the difference in potential energy between reactants and products is given off as heat, the reaction is exothermic. For example, when natural gas burns, methane reacts with oxygen gas to produce carbon dioxide and water:

$$CH_4(g) + 2\,O_2(g) \longrightarrow CO_2(g) + 2\,H_2O(g)$$

FIGURE 2.17 Reactants and Products May Differ in Entropy
The chemical reaction called burning results in a system that is much less ordered than the original system. Stated another way, the reaction results in an increase in entropy.

The electrons involved in the C–O and H–O bonds of carbon dioxide and water are held much more tightly than they were in the C–H and O–O bonds of methane and oxygen. In chemical reactions, the difference in potential energy between the products and the reactants is symbolized by ΔH. (The uppercase Greek letter Δ, delta, is often used in chemical and mathematical notation to represent change.) When a reaction is exothermic, ΔH is negative.

2. *Reactions tend to be spontaneous when the product molecules are less ordered than the reactant molecules.* Wood is a highly ordered structure. But when wood burns, the molecules that result are much less ordered (**Figure 2.17**). The amount of disorder in a group of molecules is called its **entropy**, which is symbolized by S. When the products of a chemical reaction are less ordered than the reactant molecules, entropy increases and ΔS is positive. Reactions tend to be spontaneous if they increase entropy.

In general, physical and chemical processes proceed in the direction that results in lower potential energy and increased disorder (**Figure 2.18**). The **second law of thermodynamics**, in fact, states that entropy always increases in an isolated system. In the case of burning, the reaction is exothermic *and* results in higher entropy—less-ordered products.

Because reactions tend to be spontaneous when ΔH is negative and ΔS is positive, it's necessary to assess the *combined* contributions of changes in heat and disorder to determine whether a chemical reaction is spontaneous. To do this, chemists define a quantity called the **Gibbs free-energy change**, symbolized by ΔG:

$$\Delta G = \Delta H - T\Delta S$$

Here T stands for temperature measured on the Kelvin scale. The $T\Delta S$ term simply means that entropy becomes more im-

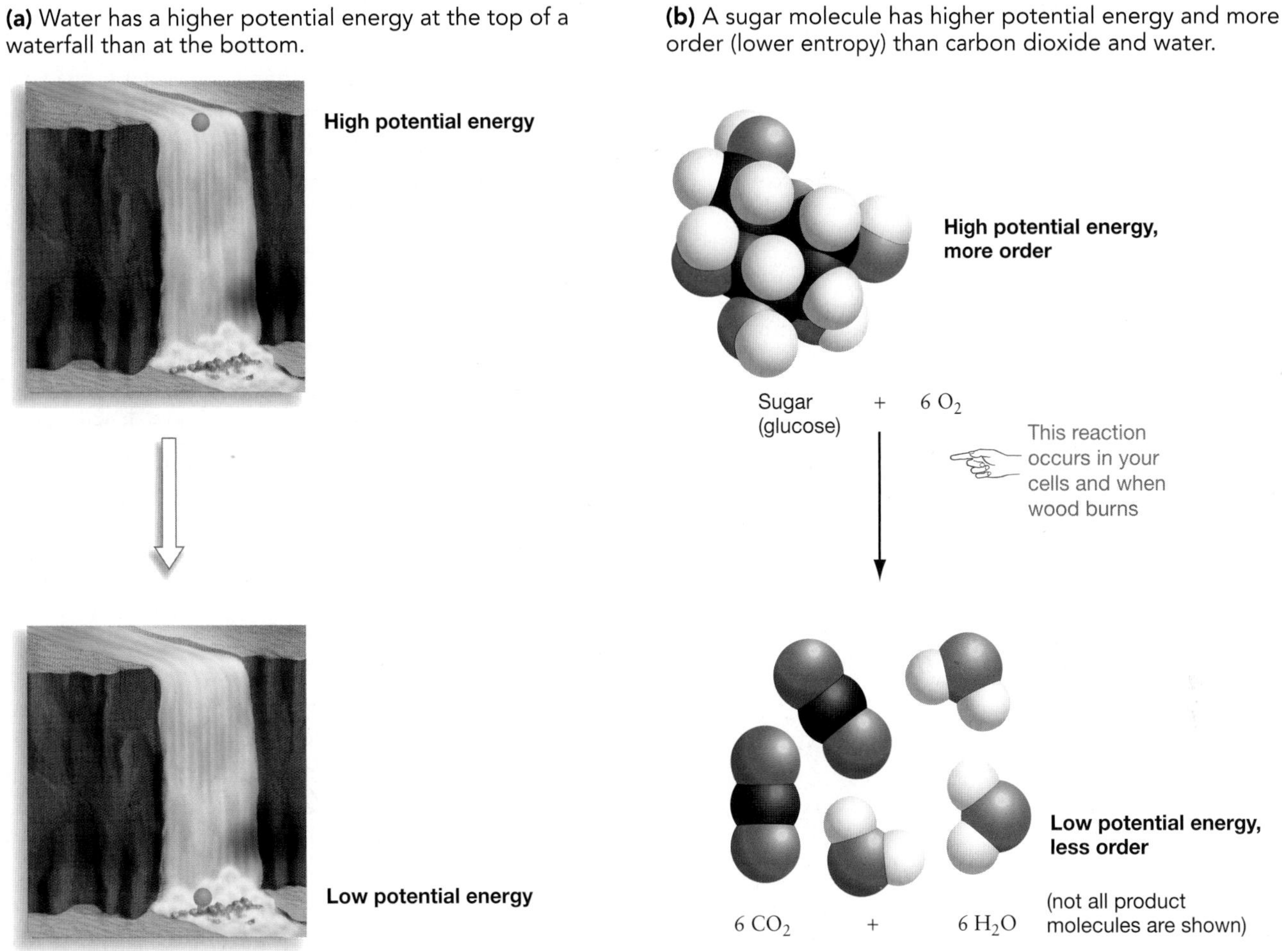
(a) Water has a higher potential energy at the top of a waterfall than at the bottom.
High potential energy
Low potential energy
(b) A sugar molecule has higher potential energy and more order (lower entropy) than carbon dioxide and water.
High potential energy, more order
Sugar (glucose) + 6 O2
This reaction occurs in your cells and when wood burns
Low potential energy, less order
6 CO2 + 6 H2O
(not all product molecules are shown)

FIGURE 2.18 Spontaneous Processes Result in Lower Potential Energy or Increased Disorder, or Both
Examples of spontaneous processes include **(a)** a water molecule going over a waterfall, and **(b)** a sugar molecule undergoing a chemical reaction that forms smaller molecules.

portant in determining free-energy change as the temperature of the molecules increases.

Chemical reactions are spontaneous when ΔG is less than zero. Such reactions are said to be **exergonic.** Reactions are nonspontaneous when ΔG is greater than zero. Such reactions are termed **endergonic.** When ΔG is zero, reactions are at equilibrium. The reaction between carbon dioxide and hydrogen gas that forms formaldehyde and water is endergonic. It is nonspontaneous because it is endothermic and because it results in a decrease in entropy. For the reaction to occur, a large input of energy is required.

Energy Inputs and the Start of Chemical Evolution To explore how carbon dioxide (CO_2) and hydrogen gas (H_2) could have reacted to form formaldehyde (H_2CO) and trigger chemical evolution, a research group constructed a computer model of the ancient atmosphere. The model consisted of a list of all possible chemical reactions that can occur among CO_2, H_2O, N_2, NH_3, CH_4, and H_2 molecules. In addition to the spontaneous reactions, they included reactions that occur when these molecules are struck by sunlight. This was crucial because sunlight represents a source of energy.

The sunlight that strikes Earth is made up of packets of light energy called photons. The amount of light energy contained in a photon can vary widely. Today, most of the higher-energy photons in sunlight never reach Earth's lower atmosphere. Instead, they are absorbed by a molecule called ozone (O_3) in the upper atmosphere. But if Earth's early atmosphere was filled with volcanic gases, it is extremely unlikely that appreciable quantities of ozone existed. As a result, we can infer that when chemical evolution was occurring, large quantities of high-energy photons bombarded the planet.

Why was the energy in photons important? Recall that the atoms in hydrogen and carbon dioxide have full outermost shells. As a result, these molecules are largely unreactive. But energy from photons can break molecules apart by knocking electrons away from the outer shells of atoms. The atoms that result, called **free radicals,** have unpaired electrons and are

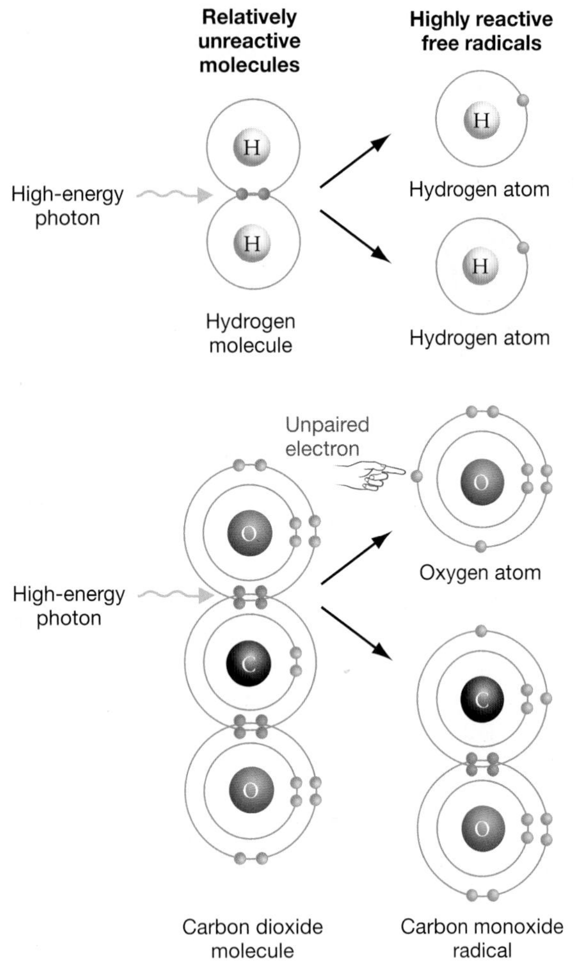
Relatively unreactive molecules
Highly reactive free radicals
High-energy photon
H
H
Hydrogen atom
Hydrogen atom
Hydrogen molecule
Unpaired electron
O
C
Oxygen atom
High-energy photon
Carbon dioxide molecule
Carbon monoxide radical

FIGURE 2.19 Free Radicals
When a high-energy photon strikes a hydrogen or carbon dioxide molecule, free radicals can be created. Formation of free radicals is thought to be responsible for some key reactions in chemical evolution.

extremely unstable (**Figure 2.19**). To mimic the conditions on early Earth more accurately, the computer model included several reactions that produce highly reactive free radicals.

To understand which of the long list of possible reactions would actually occur, and to estimate how much formaldehyde could have been produced in the ancient atmosphere, the researchers needed to consider the effects of two additional factors: temperature and concentration.

The Roles of Temperature and Concentration in Chemical Reactions Even if a chemical reaction occurs spontaneously, it may not happen quickly. For most reactions to proceed, one chemical bond has to break and another one has to form. For this to happen, the substances involved must collide in a specific orientation that brings the electrons involved near each other.

The number of collisions occurring among the substances in a mixture depends on the temperature and the concentrations of the reactants. When the concentration of reactants is high, more collisions occur and reactions proceed more quickly. When their temperature is high, reactants move faster and collide more frequently. Higher concentrations and higher temperatures tend to speed up chemical reactions.

To model the behavior of simple molecules in the ancient atmosphere, then, the researchers needed to specify both the concentration of each molecule and the temperature. Then they were able to assign a rate to each of the reactions listed in their model based on the actual reaction rates observed in experiments conducted at controlled temperatures and concentrations.

Their result? They calculated that, under temperature and concentration conditions accepted as reasonable approximations of early Earth conditions by most atmospheric scientists, appreciable quantities of formaldehyde would have been produced. Using a similar model, other researchers have shown that significant amounts of hydrogen cyanide (HCN) could also have been produced in the ancient atmosphere. According to this research, large quantities of the critical intermediates in chemical evolution would have formed in the ancient atmosphere.

How Did Chemical Energy Change during Chemical Evolution?

The initial products of chemical evolution are important, for a simple reason: They have more potential energy than do the reactant molecules. When formaldehyde is produced, an increase in potential energy occurs because the electrons that bond CO_2 and H_2 together are held more tightly than they are in H_2CO or H_2O. This form of potential energy—the potential energy stored in chemical bonds—is called **chemical energy.**

This observation gets right to the heart of chemical evolution: The energy in sunlight was converted to chemical energy—meaning potential energy in chemical bonds. This energy transformation explains how chemical evolution was possible. When small, simple molecules absorb energy, chemical reactions can occur that transform the external energy into potential energy stored in chemical bonds. More specifically, the energy in sunlight was converted to chemical energy in the form of formaldehyde and hydrogen cyanide. The complete reaction that results in the formation of formaldehyde is written as

$$CO_2(g) + 2\,H_2(g) + \text{sunlight} \longrightarrow H_2CO(g) + H_2O(g)$$

This reaction is balanced in terms of the atoms *and* the energy involved. An increase in chemical energy occurred that made the production of larger, more complex molecules possible.

✓CHECK YOUR UNDERSTANDING

Chemical reactions tend to be spontaneous if they lead to lower potential energy and higher entropy (more disorder). The combined effects of potential energy and entropy changes are summarized in the equation for the Gibbs free-energy change. You should be able to (1) write out the Gibbs equation and define each of the components, and (2) explain why potential energy might decrease as the result of a chemical reaction and why entropy might increase.

2.4 The Composition of the Early Atmosphere: Redox Reactions and the Importance of Carbon

In the models of the early atmosphere we just reviewed, chemical evolution did not begin until energy in the form of sunlight was added to the system. This conclusion makes sense at a very basic level. Energy is the capacity to do work, and it seems logical that building larger, more complex molecules requires work to be done. More specifically, the reactions involved in chemical evolution are endergonic, so inputs of energy were required.

It's important to recognize, though, that the start of chemical evolution also depended on the types of building-block molecules that were present in the atmosphere. In models, key compounds such as formaldehyde (H_2CO) and hydrogen cyanide (HCN) are produced only if molecular hydrogen (H_2), ammonia (NH_3), and methane (CH_4) are present in the atmosphere. No matter what type of energy is added to the model systems, chemical evolution does not take place unless these molecules are present. What evidence do biologists have that these molecules existed in the ancient atmosphere?

The earliest hypotheses for the composition of Earth's first atmosphere, developed in the 1920s and 1940s, proposed that H_2, NH_3, and CH_4 were abundant. Investigators based this claim on the idea that intense gravity and lack of volcanic activity had kept the atmospheres of Jupiter and Saturn unchanged since the founding of the solar system. If so, then ancient Earth's atmosphere was similar to the current atmosphere of Jupiter and Saturn. But in 1951 other researchers began arguing that volcanic gases such as CO_2, N_2, and H_2O dominated Earth's original atmosphere. Who is correct?

This controversy is difficult to resolve, because there is no direct evidence about the composition of the ancient atmosphere. The current consensus is based on recent models of the volcanic gases produced as Earth's crust formed. These emissions were undoubtedly very different from the volcanic gases produced today. These models support the hypothesis that small but significant amounts of H_2, NH_3, and CH_4 were in the early atmosphere, along with abundant CO_2, N_2, and H_2O.

Why are these particular molecules essential to chemical evolution? They trigger the most important chemical reactions in biology: reduction-oxidation, or redox, reactions.

WEB TUTORIAL 2.1 Redox Reactions

What Is a Redox Reaction?

Reduction-oxidation reactions, or redox reactions, are a class of chemical reactions that involve the loss or gain of an electron. In a redox reaction, the atom that loses one or more electrons is said to be **oxidized**, and the atom that gains one or more electrons is said to be **reduced**. To help keep these terms straight, chemists use the mnemonic "LEO the lion goes GER"—Loss of Electrons is Oxidation; Gain of Electrons is Reduction. (An alternative is OIL RIG—Oxidation Is Loss; Reduction Is Gain.) Oxidation events are always coupled with a reduction; if one atom loses an electron, another has to gain it. Stated another way, a reactant that acts as an **electron donor** is always paired with a reactant that acts as an **electron acceptor**.

The gain or loss of an electron can be relative, however. That is, during a redox reaction, electrons can be transferred completely from one atom to another, or the electrons can simply shift their positions in covalent bonds. For example, consider the burning of methane diagrammed in **Figure 2.20**. The dots in the illustration represent the electrons involved in covalent bonds. Compare the position of the electrons in the reactant, methane, with their position in the product, carbon dioxide. Note that the electrons have moved farther from the carbon nucleus in the carbon dioxide product. This means that carbon has been oxidized. It has "lost" electrons. The change occurred because the carbon and hydrogen in CH_4 share electrons equally, while the carbon and oxygen in CO_2 do not. In carbon dioxide, the high electronegativity of oxygen pulls electrons away from the carbon atom. Now compare the position of the electrons in the oxygen reactant and their position in the product, water. These electrons have moved closer to the oxygen in the water molecule, meaning that the oxygen has been reduced. In this reaction, oxygen has "gained" electrons.

These shifts in electron position change the amount of chemical energy in the reactants and products. When methane burns, electrons are held more tightly in the product molecules than in

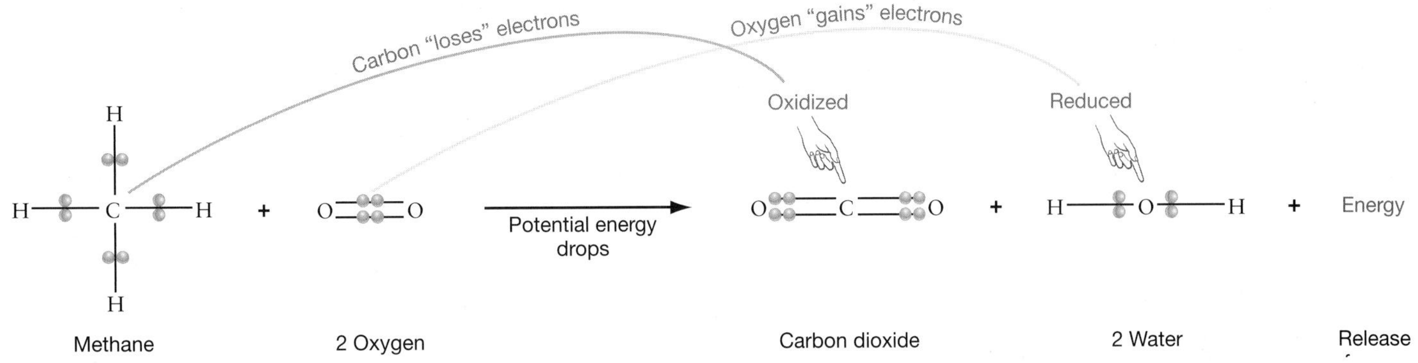

FIGURE 2.20 Redox Reactions Involve a Transfer of Electrons
Equation of the burning of methane. Burning is a redox reaction. Lines between atoms indicate covalent bonds; dots represent the relative positions of the electrons involved in those bonds. EXERCISE Circle the most electronegative atoms in this reaction.

BOX 2.1 Some Other Approaches to Understanding Redox Reactions

Redox reactions are absolutely fundamental to life, so it is critical to gain a solid understanding of redox chemistry. To further your understanding of how these reactions work, here are additional descriptions of redox dynamics:

- The electrons and bonds involved in redox reactions can be likened to a seesaw. In this analogy, an electron pair that is being shared equally by two atoms is like a seesaw balanced on its fulcrum.

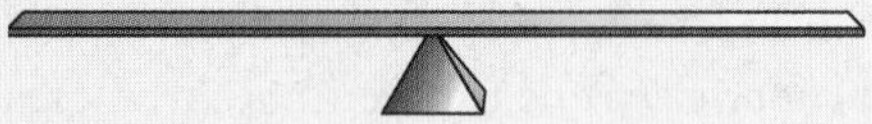

Because this state is inherently unstable, it has high potential energy. A redox reaction occurs, and an electron is transferred toward a more electronegative atom, like a seesaw with one end lowered to the ground.

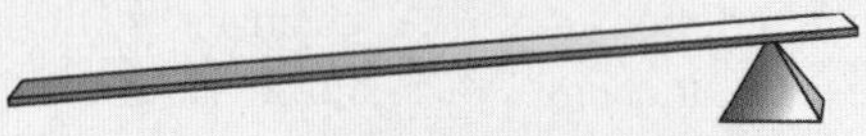

In this state, the seesaw is stable—meaning that it has much lower potential energy. In a redox reaction, the difference in potential energy between the two states is released as heat or transferred to another atom.

- During the redox reactions that occur in cells, electrons (e^-) are often transferred from an atom in one molecule to an atom in a different molecule. When this occurs, the electron is usually accompanied by a proton (H^+). As a result, the molecule that contains the reduced atom gains a hydrogen (H) atom. In many redox reactions in biology, understanding where oxidation and reduction have occurred becomes a matter of following hydrogen atoms. In many biological systems, reduction means "adding H's" and oxidation means "removing H's."

the reactant molecules, meaning their potential energy has dropped. As a result, the reaction is exothermic. **Box 2.1** introduces other ways to think about redox reactions.

Now consider **Figure 2.21**, which illustrates one of the key reactions in chemical evolution. The covalently bonded electrons in the hydrogen molecules (H_2) are shared equally and have relatively high potential energy. As a result, H_2 is a relatively unstable molecule and can act as an electron donor. The reactant, carbon dioxide (CO_2), in contrast, has covalently bonded electrons that are tightly held by oxygen. Consequently, CO_2 is a relatively stable molecule and can act as an electron acceptor. The interaction between H_2 and CO_2 is a redox reaction that couples an electron donor and an electron acceptor. Using Figure 2.20 as a guide, you should be able to add the electron positions for each of the bonds involved in this reaction. Then you should be able to determine whether carbon is reduced or oxidized as a result of this reaction. Based on this analysis, the following claim should make sense: The key step in launching chemical evolution was the reduction of carbon.

What Happens When Carbon Is Reduced?

Life has been called a carbon-based phenomenon, and with good reason. With the exception of water, almost all of the molecules found in organisms contain this atom.

Carbon is so important in biology because it is the most versatile atom on Earth. Due to its four valence electrons, it can form a large number of covalent bonds. With different combinations of single and double bonds, an almost limitless array of molecular shapes is possible. You have already examined the tetrahedral structure of methane and the linear shape of carbon dioxide. When molecules contain more than one carbon atom, these atoms can be bonded to one another in long chains, as in the component of gasoline called octane (C_8H_{18}; **Figure 2.22a**), or in a ring, as in the sugar glucose ($C_6H_{12}O_6$; **Figure 2.22b**). Molecules that contain carbon are called **organic molecules.** Other types of molecules are referred to as *inorganic compounds*.

Linking Carbon Atoms Together The formation of carbon-carbon bonds was an important event in chemical evolution. It represented a crucial step toward the production of the types of

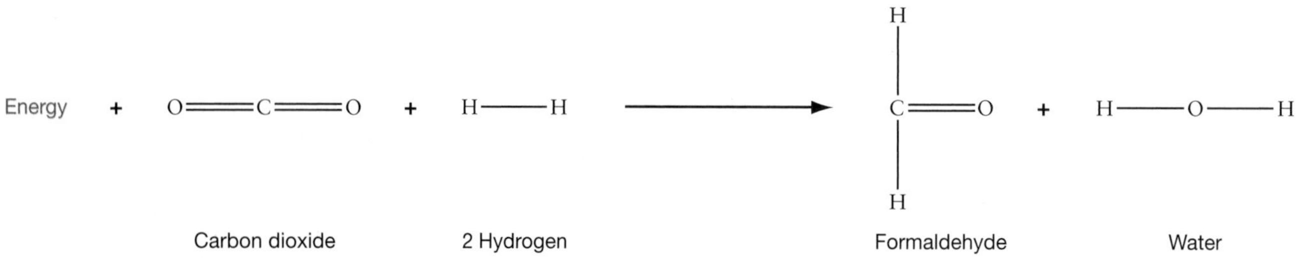
Energy
Carbon dioxide
2 Hydrogen
Formaldehyde
Water

FIGURE 2.21 Tracking Electron Transfer during a Redox Reaction
A reaction that occurred during chemical evolution. EXERCISE Using Figure 2.20 as a guide, add dots to each covalent bond to show the relative positions of the electrons involved. Is the carbon atom reduced or oxidized in this reaction? Label the reactant molecules that act as the electron donor and the electron acceptor.

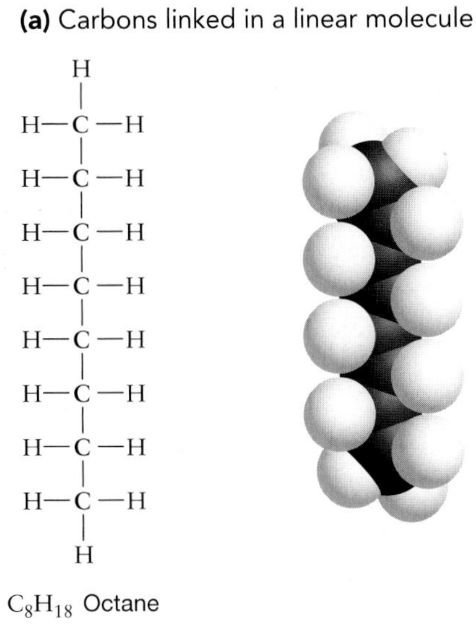
(a) Carbons linked in a linear molecule
C_8H_{18} Octane

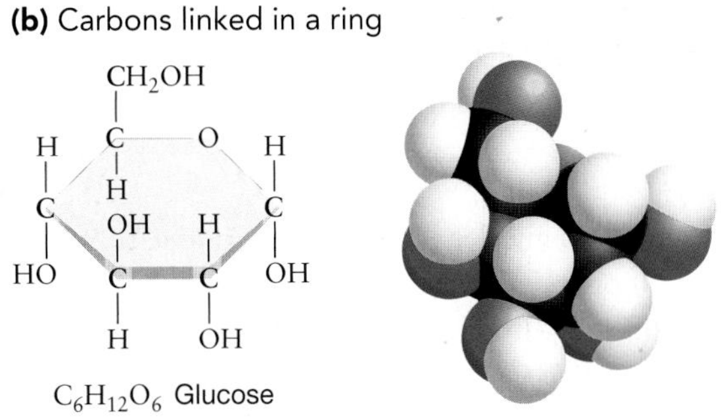
(b) Carbons linked in a ring
$C_6H_{12}O_6$ Glucose

FIGURE 2.22 The Shapes of Carbon-Containing Molecules
(a) Octane is one of the primary ingredients in gasoline. It is a linear molecule with carbon atoms that are highly reduced. **(b)** Glucose is a sugar that can form the ring-like structure illustrated here.

molecules found in living organisms. Once compounds with reduced carbon atoms such as formaldehyde and hydrogen cyanide had formed, continued chemical evolution could occur by the addition of heat alone. For example, when molecules of formaldehyde are heated, they react with one another to form a molecule called acetaldehyde. Acetaldehyde contains a carbon-carbon bond. With continued heating, reactions between formaldehyde and acetaldehyde molecules can produce the larger carbon-containing compounds called sugars.

In sum, advocates of the chemical evolution hypothesis propose that two sources of energy—the potential energy in inorganic compounds such as H_2 and the energy in sunlight—made the production of reduced carbon-containing compounds possible. Subsequently, the potential energy in these carbon-containing molecules made the production of the first complex organic compounds possible. **Figure 2.23** details these steps. In studying the figure, be sure to note two key messages: (1) Chemical evolution got under way because molecules with high potential energy were

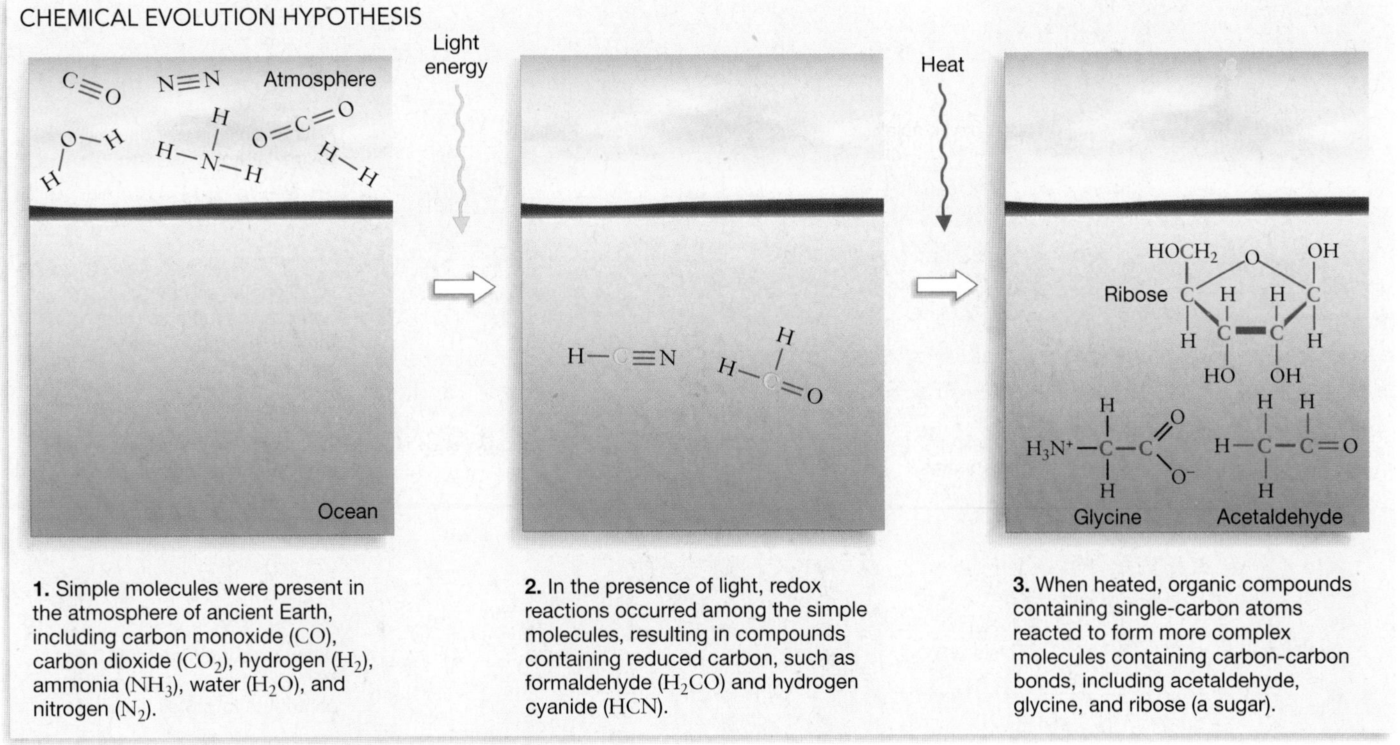
CHEMICAL EVOLUTION HYPOTHESIS
Atmosphere
Light energy
Heat
Ocean
Ribose
Glycine
Acetaldehyde
1. Simple molecules were present in the atmosphere of ancient Earth, including carbon monoxide (CO), carbon dioxide (CO_2), hydrogen (H_2), ammonia (NH_3), water (H_2O), and nitrogen (N_2).
2. In the presence of light, redox reactions occurred among the simple molecules, resulting in compounds containing reduced carbon, such as formaldehyde (H_2CO) and hydrogen cyanide (HCN).
3. When heated, organic compounds containing single-carbon atoms reacted to form more complex molecules containing carbon-carbon bonds, including acetaldehyde, glycine, and ribose (a sugar).

FIGURE 2.23 The Start of Chemical Evolution—an Overview
Chemical evolution is a process whereby simple molecules containing C, H, O, and N react to form molecules with reduced carbon atoms, which then react to form organic compounds with carbon-carbon bonds. The process is triggered by an energy source such as sunlight or the heat released in a volcanic eruption.

available to act as electron donors and because abundant sources of outside energy were available to trigger endergonic reactions; and (2) some of these molecules produced by chemical evolution are found in organisms living today.

Functional Groups Once chemical evolution was under way, a wide variety of small organic molecules would have accumulated in the ancient Earth. Although the molecular and structural formulas of small organic compounds are enormously variable, their chemical functions are reasonably predictable. In general, the carbon atoms in an organic molecule furnish a skeleton that gives the molecule its shape. The chemical behavior of the compound, however, is dictated by groups of H, N, or O atoms that are bonded to one of the carbon atoms in a specific way. These H-, N-, and O-containing groups of atoms are called **functional groups.** The composition and properties of six prominent functional groups recognized by organic chemists are summarized in **Table 2.1.**

TABLE 2.1 Six Functional Groups Commonly Attached to Carbon Atoms

EXERCISE Add notes to the table, predicting whether each functional group is polar or nonpolar, based on the electronegativities of the atoms involved.

Functional Group	*Formula	Family of Molecules	Example	Properties of Functional Group
Amino	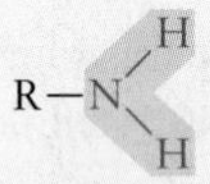	Amines	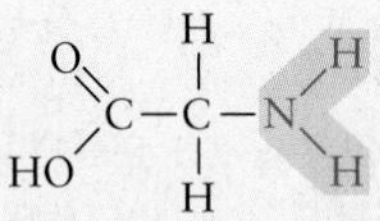 Glycine (an amino acid)	Acts as a base—tends to attract a proton to form 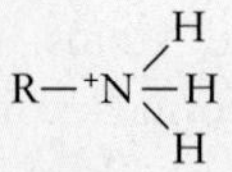
Carbonyl	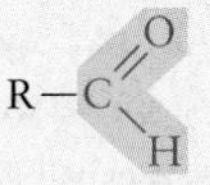	Aldehydes	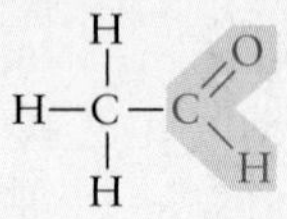 Acetaldehyde	Aldehydes, especially, react with compounds of form HR_2 to produce larger molecules with form 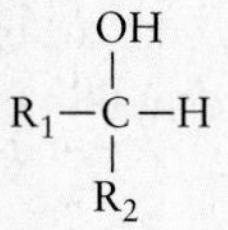
	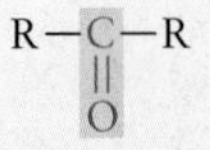	Ketones	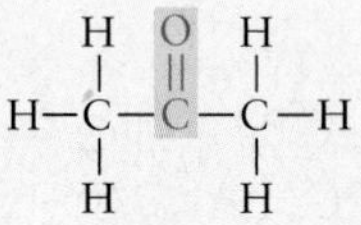 Acetone	
Carboxyl	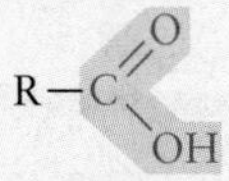	Carboxylic acids	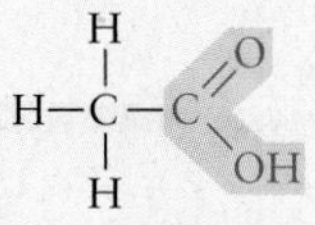 Acetic acid	Acts as an acid—tends to lose a proton to form 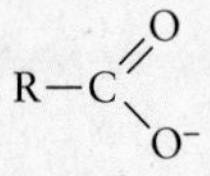
Hydroxyl	R—OH	Alcohols	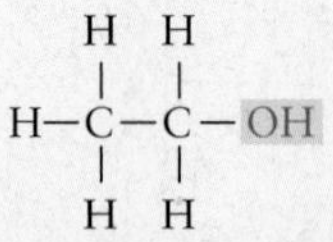 Ethanol	Highly polar, so makes compounds more soluble through hydrogen bonding with water
Phosphate	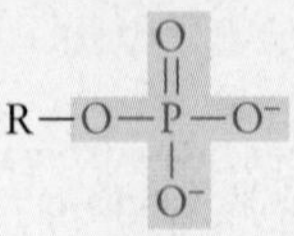	Organic phosphates	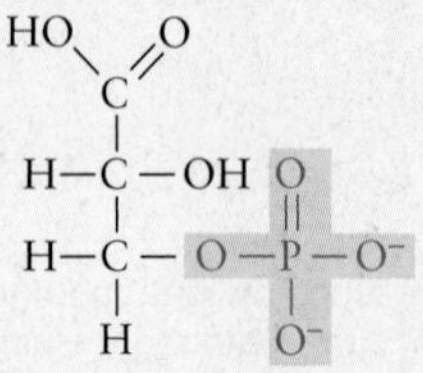 3-Phosphoglyceric acid	When several phosphate groups are linked together, breaking O–P bonds between them releases large amounts of energy
Sulfhydryl	R—SH	Thiols	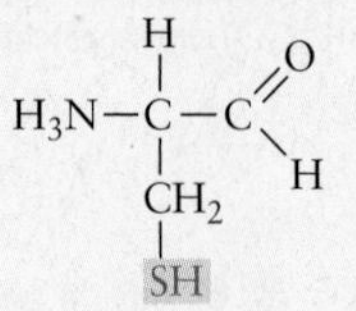 Cysteine	When present in proteins, can form disulfide (S–S) bonds that contribute to protein structure

*In these structural formulas, "R" stands for the rest of the molecule.

The *carbonyl group*, for example, is found on aldehyde molecules such as formaldehyde and acetaldehyde. This functional group is the site of the reaction that links these molecules into larger, more complex compounds. The other functional groups listed in Table 2.1 confer equally distinctive properties, ranging from a propensity to participate in acid-base reactions to increasing the solubility of the compound in water.

Reactions among organic molecules launched the next phase of chemical evolution: the formation of the larger and more complex molecules found in living cells. Advocates of the chemical evolution hypothesis propose that these reactions occurred in water rather than in the atmosphere. Why?

2.5 The Early Oceans and the Properties of Water

Although it is accurate to say that life is based on carbon, it is even more meaningful to say that it is based on water. In a typical living cell, over 75 percent of the volume consists of this molecule. Nearly 70 percent of your body, by weight, is water (**Figure 2.24**). Virtually all researchers agree that most of the important steps in chemical evolution, including the origin of life itself, occurred in water. This hypothesis is logical because, as a dissolving agent, or **solvent**, water can dissolve more types of substances than any other molecule known. The hypothesis is also important because chemical reactions depend on direct, physical interaction between the reactants and because substances are most likely to collide when they are dissolved. Thus, life is based on water primarily because of water's solvent properties.

The formation of Earth's first ocean was a turning point in chemical evolution. It gave chemical evolution a place to happen.

Why Is Water Such an Efficient Solvent?

To understand why water is such an effective solvent, recall that the molecule contains two hydrogen atoms bonded to an oxygen atom. Then recall from Section 2.2 that oxygen and hydrogen differ in their ability to attract the electrons involved in covalent bonds, or their electronegativity. Because oxygen is among the most electronegative of all elements, it attracts covalently bonded electrons much more strongly than does the hydrogen nucleus. As a result, the electrons that participate in the bonds are not shared equally. The covalent bonds in water are polar.

If water were a linear molecule like carbon dioxide, the polarity of its covalent bonds would not matter very much. But recall from Section 2.2 that the water molecule is bent. The simple fact that water is a bent molecule with polar covalent bonds has an enormously important consequence: The molecule as a whole is **polar**, meaning that the overall distribution of charge is asymmetrical. As **Figure 2.25a** shows, the side of the molecule containing the oxygen atom is slightly more negative, and the side with the hydrogen atoms is slightly more positive. As before, the partial charges on the molecule are symbolized by the lowercase Greek letter delta, δ.

Figure 2.25b illustrates how the polarity of water affects its interactions with other substances in solution. When two liquid water molecules approach one another, the partial positive charge on hydrogen attracts the partial negative charge on oxygen. This weak electrical attraction forms a **hydrogen bond** between the molecules.

In a water-based, or aqueous, solution, hydrogen bonds also form between water molecules and other polar molecules. Similar interactions occur between water and ions. Ions and polar molecules stay in solution because of their interactions with

FIGURE 2.24 Water Is the Most Abundant Molecule in Organisms
Fruits shrink when they are dried because they consist primarily of water.

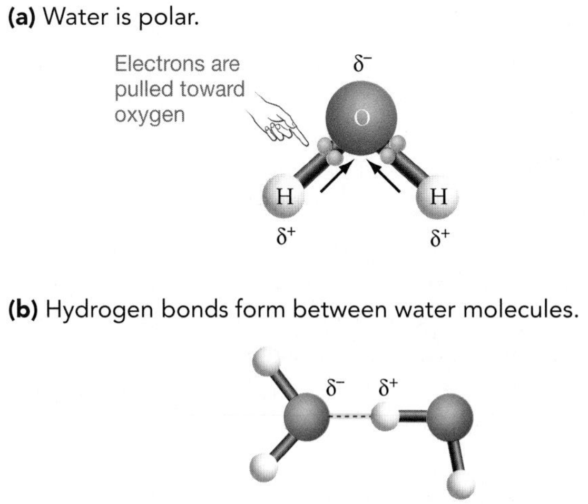

FIGURE 2.25 Water Is Polar and Participates in Hydrogen Bonds
(a) Because of oxygen's high electronegativity, the electrons that are shared when hydrogen and oxygen form a covalent bond are pulled toward the oxygen nucleus. The electrons spend more time close to the oxygen nucleus, so the oxygen atom has a slight negative charge and the hydrogen atom a partial positive charge. **(b)** The electrical attraction that occurs between the partial positive and negative charges on water molecules forms a hydrogen bond.
EXERCISE Label the hydrogen bond in part (b).

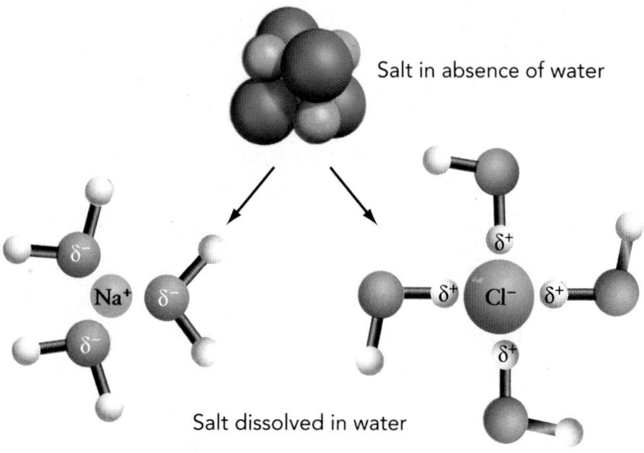
Salt in absence of water
Na⁺
Cl⁻
δ⁻
δ⁺
Salt dissolved in water

FIGURE 2.26 Polar Molecules and Ions Dissolve Readily in Water
Water's polarity makes it a superb solvent for polar molecules and ions.

water's partial charges (**Figure 2.26**). Hydrogen bonding makes it possible for almost any charged or polar molecule to dissolve in water. Although individual hydrogen bonds are not nearly as strong as covalent or ionic bonds, they are much more common. Hydrogen bonding is extremely important in biology because of the sheer number of hydrogen bonds that form in solution and between key molecules.

How Does Water's Structure Correlate with Its Properties?

Water's structure is unusual. Its small size, bent shape, highly polar covalent bonds, and overall polarity are unique among molecules. Because the structure of molecules routinely correlates with their function, it's not surprising that water has some remarkable properties. We've already reviewed the most important of water's chemical attributes—its ability to form hydrogen bonds with ions and polar compounds and its capacity to act as a solvent.

WEB TUTORIAL 2.2 The Properties of Water

Water also has several striking physical properties that are the direct result of hydrogen bonding. In relation to chemical evolution, the most important of these are that water expands as it changes from a liquid to a solid, and that it has an extraordinarily large capacity for absorbing heat. Let's analyze why hydrogen bonding explains these properties, and connect water's physical and chemical behavior with its role in chemical evolution.

Water Is Denser as a Liquid than as a Solid When factory workers pour molten metal or plastic into a mold and allow it to cool to the solid state, the material shrinks. When molten lava pours out of a volcano and cools to solid rock, it shrinks. But when you fill an ice tray with water and put it in the freezer to make ice, the water expands.

Unlike most substances, water is denser as a liquid than it is as a solid. In other words, there are more molecules of water in a given volume of liquid water than there are in the same volume of solid water. **Figure 2.27a** illustrates why this is so. Note that in ice, each water molecule participates in four hydrogen bonds. These hydrogen bonds cause the water molecules to form a regular and repeating structure, or crystal. The crystal

(a) In ice, water molecules form a crystal lattice.

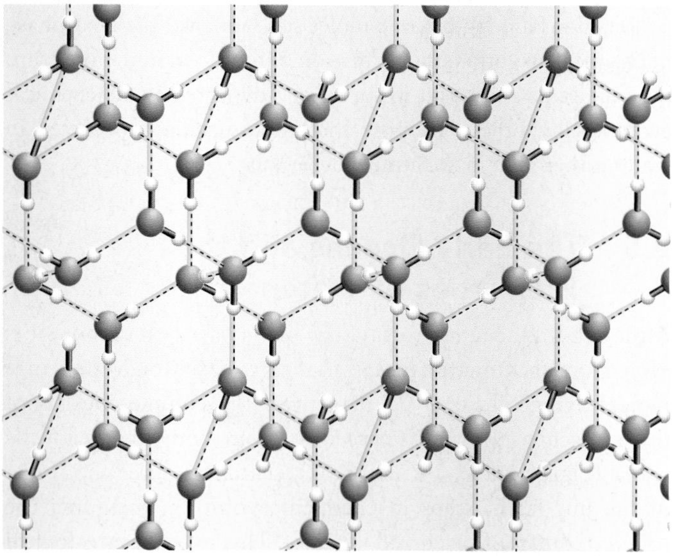

(b) In liquid water, no lattice forms, so liquid water is denser than ice.

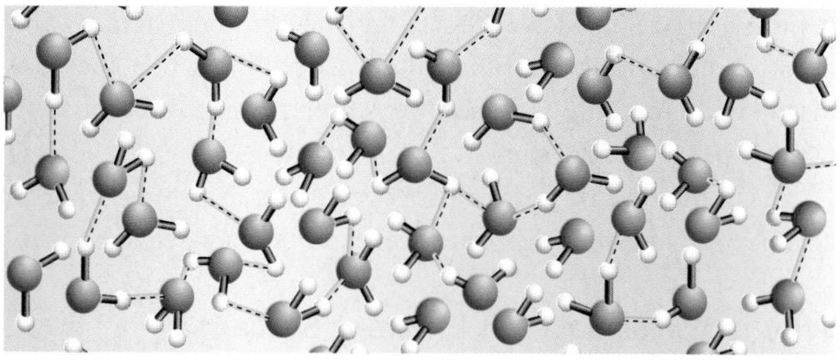

(c) As a result, ice floats.

FIGURE 2.27 Hydrogen Bonding Forms the Crystal Structure of Ice
EXERCISE In liquid water, each molecule can form four hydrogen bonds at one time. (Each oxygen atom can form two; each hydrogen atom can form one.) Choose two molecules in part (b), and draw in all four H bonds.

structure of ice is fairly open, meaning that there is a relatively large amount of space between molecules.

Now compare the extent of hydrogen bonding and the density of ice with that of liquid water, illustrated in **Figure 2.27b.** Although the exact structure of liquid water is not fully understood, it is clearly dynamic. Hydrogen bonds are constantly being made and broken. When hydrogen bonds form, energy is released; when hydrogen bonds break, energy is absorbed.

Overall, the extent of hydrogen bonding in liquid water is much less than that found in ice. As a result, molecules in the liquid phase are packed much more closely together than are molecules of solid water, even though their temperature is higher. Normally, heating a substance causes it to expand as molecules move faster and collide more often and with greater force. But heating ice causes hydrogen bonds to break and the open crystal to collapse. In this way, hydrogen bonding explains why water is denser as a liquid than as a solid.

This property of water has an important result: Ice floats (**Figure 2.27c**). If it did not, ice would sink to the bottom of lakes, ponds, and oceans soon after it formed. The ice would stay frozen in the cold depths. If water wasn't so unusual, it is almost certain that Earth's oceans would have frozen solid before life had a chance to start.

Water Has a High Capacity for Absorbing Energy

Hydrogen bonding is also responsible for another of water's remarkable physical properties: Water has a high capacity for absorbing energy. For example, water has an extraordinarily high specific heat. **Specific heat** is the amount of energy required to raise the temperature of 1 gram of a substance by 1°C. Water's specific heat is high because, when a source of energy such as sunlight or a flame strikes it, hydrogen bonds must be broken before heat can be transferred and the water molecules begin moving faster. As a result, it takes an extraordinarily large amount of energy to change the temperature of water (**Table 2.2**). Similarly, it takes a large amount of energy to break the hydrogen bonds in liquid water and change the molecules from the liquid phase to the gas phase. Water's **heat of vaporization**—the energy required to change 1 gram of it from a liquid to gas—is higher than that of most molecules that are liquid at room temperature. As a result, water has to absorb a great deal of energy to evaporate. Water's high heat of vaporization is the reason that sweating or dousing yourself is an effective way to cool off on a hot day. Because water molecules have to absorb a great deal of energy from your body in order to evaporate, you lose heat.

These properties of water are important to the chemical evolution hypothesis. Sources of energy trigger the formation of reduced carbon compounds such as formaldehyde, but energy inputs can also break them apart. Water's high specific heat insulates dissolved substances from sources of energy including asteroid bombardment, sunlight, and volcanism. Because formaldehyde and hydrogen cyanide dissolve readily in water, they would have rained out of the atmosphere into the ocean—an environment where they were better protected. Similarly, if chemical evolution took place at the edges of ponds or on beaches, the evaporation of water would have kept the surfaces relatively cool. Hydrogen bonding gives water a moderating influence in terms of temperature.

TABLE 2.2 Some Specific Heats of Common Substances

The specific heats reported in this table were measured at 25°C (except for ice, which was measured at −11°C) and are given in units of joules per gram of substance per degree Celsius. (The **joule** is a unit of energy.) Notice that the specific heat of water, in both solid and liquid form, is extremely high. This means that water must absorb a great deal of thermal energy before it begins to change temperature.

Substance	Specific Heat
Air (dry)	1.01
Aluminum	0.90
Copper	0.39
Gold	0.13
Iron	0.45
Mercury	0.14
Table salt	0.86
Water (*s*)	2.03
Water (*l*)	4.18

Source: Table 8.1, p. 312 in John McMurry and Robert C. Fay, *Chemistry*, 4th Edition, ©2004. Reprinted by permission of Pearson Education, Inc., Upper Saddle River, NJ

Acid-Base Reactions and pH

One other aspect of water's chemistry influenced its role in the origin of life. Water is not a completely stable molecule. In reality, water molecules continually undergo a chemical reaction with themselves. This "dissociation reaction" can be written as follows:

$$H_2O \rightleftharpoons H^+ + OH^-$$

The double arrows indicate that the reaction proceeds in both directions.

The molecules on the right-hand side of the expression are the **hydrogen ion** (H^+) and the **hydroxide ion** (OH^-). A hydrogen ion is simply a proton. Substances that give up protons during chemical reactions are called **acids**; molecules or ions that acquire protons during chemical reactions are called **bases.** A chemical reaction that involves a transfer of protons is called an **acid-base reaction.** Acid-base reactions require a proton donor and a proton acceptor, just as redox reactions require an electron donor and an electron acceptor.

In reality, however, protons never exist by themselves. In water, for example, protons associate with water molecules to form the hydronium ion, H_3O^+. Thus, the dissociation of water is more accurately written

$$H_2O + H_2O \rightleftharpoons H_3O^+ + OH^-$$

One of the water molecules on the left-hand side of the expression has given up a proton and acted as an acid, while the other water molecule has accepted a proton and acted as a base. This illustrates another important property of water: Most acids act only as acids, and most bases act only as bases, but water can act both as an acid and as a base.

In a solution, the tendency for acid-base reactions to occur is largely a function of the number of protons present. How many protons are present in water?

Chemists answer this question by measuring the concentration of protons directly. In a sample of pure water at 25°C, the concentration of H^+ is 1.0×10^{-7} M (recall that M represents molarity, or moles per liter). Because this is such a small number (1 ten-millionth), the exponential notation is cumbersome. So chemists and biologists prefer to express the concentration of protons in a solution with a logarithmic notation called the **pH scale**. (The term *pH* is derived from the French *puissance d'hydrogène*, or "power of hydrogen.") By definition, the pH of a solution is the negative of the base-10 logarithm, or log, of the hydrogen ion concentration:

$$pH = -\log[H^+]$$

(The square brackets are a standard notation for indicating "concentration of" a substance in solution.) Taking antilogs gives

$$[H^+] = \text{antilog}(-pH) = 10^{-pH}$$

Figure 2.28 shows the pH scale and reports the pH of some common solutions. The pH of pure water at 25°C is 7. Pure water is used as a standard, or point of reference, on the pH scale. Solutions that contain acidic molecules (molecules that act as acids) have a proton concentration larger than 1×10^{-7} M and thus a pH < 7, because acidic molecules tend to release protons into solution. In contrast, solutions that contain basic molecules (molecules that act as bases) have a proton concentration less than 1×10^{-7} M and thus a pH > 7, because basic molecules tend to accept protons from solution. Solutions with a pH of 7 are considered neutral solutions, neither acidic nor basic. Rainwater is almost pure water, meaning that its pH is close to 7.

What Was Water's Role in Chemical Evolution?

Water's unusual structure and properties gave it a key role in chemical evolution. When the Earth cooled enough for rain to fall and oceans to form, the planet's surface came to be dominated by liquid water. As rain and wind slowly eroded rocks and rivers carried dissolved ions to the ocean, the oceans gradually became saline enough to resemble today's seawater. Organic compounds produced by redox reactions in the atmosphere dissolved in rainwater and accumulated in the ocean.

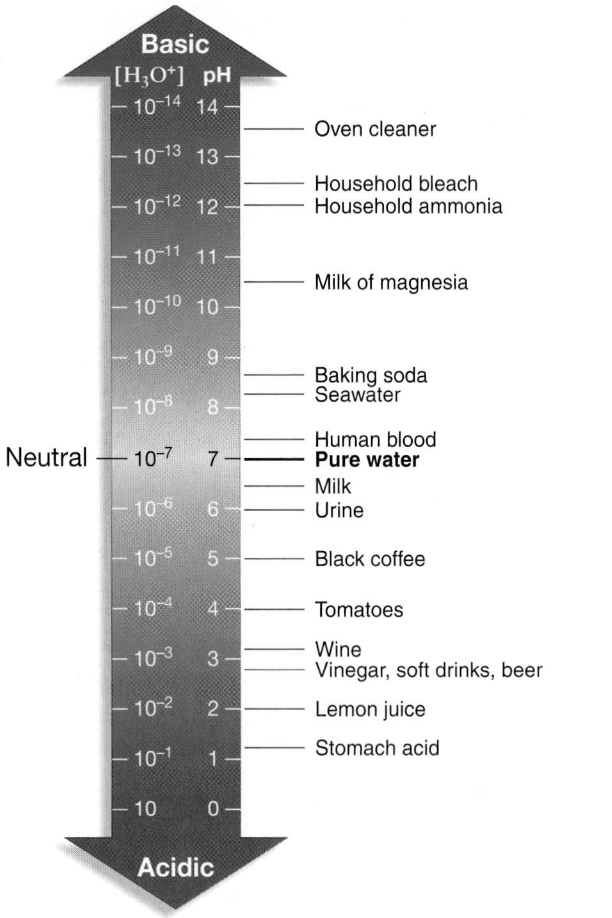

FIGURE 2.28 The pH Scale
Because the pH scale is logarithmic, a change in one unit of pH represents a change in the concentration of hydrogen ions equal to a factor of 10. Coffee has a hundred times more H^+ than pure water has.

Based on the arguments presented in this section, then, the environment from which chemical evolution proceeded was a salty solution of nearly neutral pH. The solution contained at least some compounds with reduced carbon atoms that resulted from redox reactions in the atmosphere, along with simple organic molecules produced when these carbon-containing molecules were heated. Thanks to water's high specific heat, however, these first products of chemical evolution were relatively well protected from further inputs of energy.

What happened next? For chemical evolution to continue, two things had to happen. First, reactions among relatively small and simple organic compounds had to produce the building blocks of the large molecules found in living cells. Second, these building blocks had to link together to form proteins, nucleic acids, and carbohydrates—the large, complex compounds found in organisms. Our task in the next three chapters is to analyze how these events occurred and how proteins, nucleic acids, and carbohydrates function in organisms today. As far as we know, the jump from nonlife to life has occurred only once in the history of the universe. According to the chemical evolution hypothesis, it happened as these molecules began to accumulate in the waters of the ancient Earth.

ESSAY The Search for Extraterrestrial Life

Astronomers who search for life on other planets have recently been invigorated by two findings: (1) evidence that liquid water is present under the ice-crusted surface of Jupiter's moon Europa (**Figure 2.29**); and (2) the discovery of cells living in rock formations hundreds of meters below Earth's surface.

Confirming that liquid water exists elsewhere in the solar system is exciting because life is probably impossible in the absence of liquid water. Finding evidence for liquid water on Europa narrows the search for extraterrestrial life—it gives scientists a promising place to look.

Finding organisms in the deep subsurface of Earth suggested that life could also be found under the surface of Mars, the Moon, or other bodies. Cells have recently been found in Earth rocks located up to 860 meters (about half a mile) below the surface. These discoveries have extended a general realization—that organisms are found in a wide variety of extreme environments on Earth. Species of the single-celled organisms called bacteria and archaea, and sometimes even multicellular (many-celled) animals, can thrive within glaciers, in extremely high pressure water near superheated steam vents deep on the ocean floor, in hot springs, and in ponds that are nearly saturated with salt. The presence of life in extreme environments on Earth has encouraged astronomers to suggest that life may also exist in extreme environments in space. Planets and moons that appear to be lifeless could actually be teeming with organisms just below the surface.

As biologists gain a better understanding of chemical evolution and the diversity of life on ancient Earth, the search for extraterrestrial life has become more focused. Few astronomers expect to find the sophisticated types of life-forms favored by science fiction writers. Instead, astronomers are looking for evidence of water, molecules with reduced carbon atoms, and traces of single-celled life-forms similar to Earth's bacteria and archaea.

Planets and moons that appear to be lifeless could actually be teeming with organisms below the surface.

Finding life elsewhere in the solar system would instantly qualify as a tremendous scientific advance. It would refute the hypothesis that chemical evolution occurred just once and that life is unique to Earth. It would also open up the possibility of finding organisms in many other locations throughout the universe—perhaps in the recently discovered planetary systems that are currently forming around young stars.

(a) Europa (moon of Jupiter)

(b) Closer view of ice crust on Europa

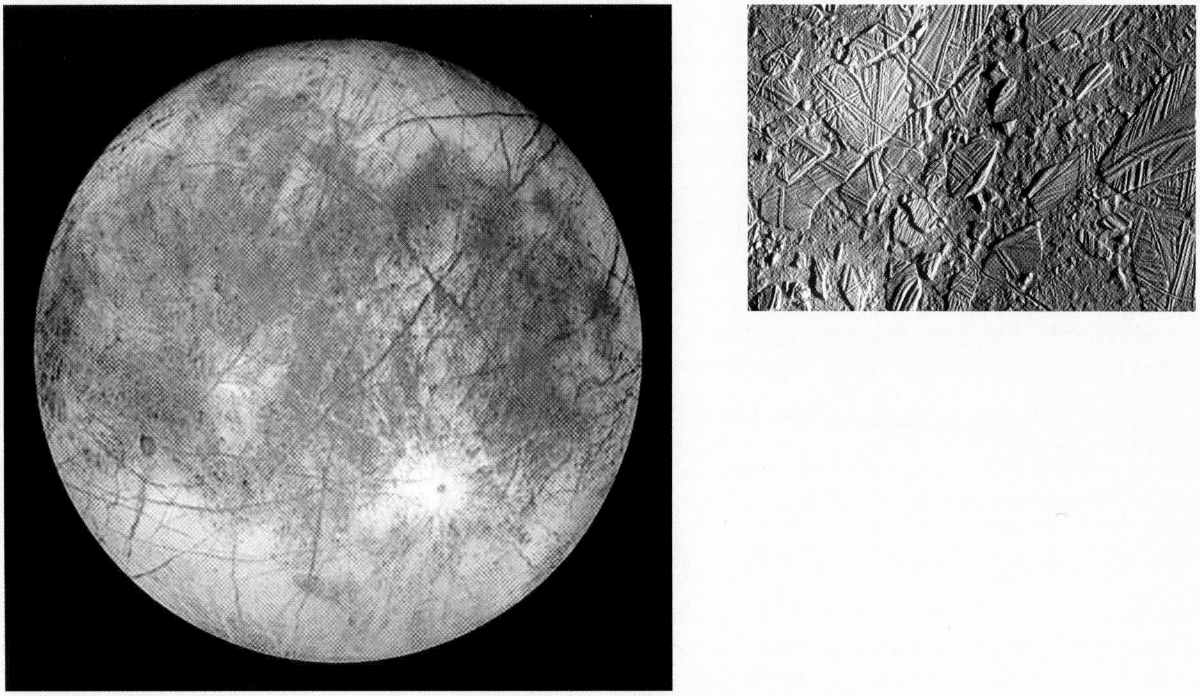

FIGURE 2.29 Europa—a Moon of Jupiter
The surface of Europa is frozen, but data suggest that liquid water exists underneath the ice.

CHAPTER REVIEW

Summary of Key Concepts

Chemical evolution is a hypothesis about the origin of the essential chemical components of life. It maintains that early in Earth's history, large and complex molecules formed from simple precursor compounds in the ancient atmosphere and ocean, as chemical reactions converted the energy in sunlight and other sources into chemical energy. Chemical energy is a form of potential energy and can be stored in bonds between atoms. Based on radiometric dating, advocates of the hypothesis suggest that chemical evolution took place over a span of some 300 million years, beginning about 3.85 billion years ago.

- **Molecules form when atoms bond to each other. Chemical bonds are based on electron sharing. The degree of electron sharing varies from nonpolar covalent bonds, to polar covalent bonds, to ionic bonds.**

When atoms participate in chemical bonds to form molecules such as those produced during chemical evolution, the shared electrons give the atoms full valence shells and thus contribute to the atoms' stability. The electrons may be shared equally or unequally, depending on the electronegativities of the two atoms involved. Nonpolar covalent bonds result from equal sharing; polar covalent bonds are due to unequal sharing; and ionic bonds form when an electron is completely transferred from one atom to another.

- **Chemical reactions tend to be spontaneous if they lead to lower potential energy and higher entropy (more disorder). An input of energy is required for nonspontaneous reactions to occur.**

The reactions involved in chemical evolution resulted in products that had higher potential energy and lower entropy than the reactants. As a result, these reactions were nonspontaneous. They could occur only because there was an input of kinetic energy in the form of heat, sunlight, and lightning.

- **During chemical evolution, (1) energy in sunlight was converted to chemical energy and (2) carbon atoms were reduced.**

Researchers have developed models for how the first step in chemical evolution occurred. This step was the formation of compounds with reduced carbon atoms, such as formaldehyde and hydrogen cyanide, from molecules such as CO_2. The models show that these reactions could occur only if a source of energy, such as sunlight, *and* a source of inorganic compounds with high potential energy, such as ammonia (NH_3), methane (CH_4), and molecular hydrogen (H_2), were present in the ancient atmosphere. Although the composition of the ancient atmosphere is controversial, most investigators now agree that sources of energy were abundant and at least some inorganic molecules with high potential energy existed.

Web Tutorial 2.1 Redox Reactions

- **Water is a small, highly polar molecule. As a result, it is an extremely efficient solvent and has a high capacity for absorbing energy.**

If compounds with reduced carbon atoms formed in the ancient atmosphere, they would have dissolved in water droplets and rained down into the early ocean. Advocates of the chemical evolution hypothesis propose that most subsequent chemical evolution took place in water, primarily because it is such an efficient solvent. Water is the most effective solvent known, because it is polar, meaning that it has partial positive and negative charges. Water is polar because it is bent and has two polar covalent bonds. As a result, polar molecules and charged substances, including ions, interact with water and stay in solution.

The general message of this chapter can be stated as follows: Chemical evolution was possible because the atmosphere and oceans of ancient Earth contained inorganic compounds with high potential energy, abundant and diverse sources of energy, and a large supply of water.

Web Tutorial 2.2 The Properties of Water

Questions

Content Review

1. Which of the following occurs when a covalent bond forms?
   a. The potential energy of electrons drops.
   b. Electrons in valence shells are shared between nuclei.
   c. Ions of opposite charge interact.
   d. Polar molecules interact.

2. If a reaction is exothermic, then which of the following statements is true?
   a. The products have lower potential energy than the reactants.
   b. Energy must be added for the reaction to proceed.
   c. The products have higher entropy (are more disordered) than the reactants.
   d. It occurs extremely quickly.

3. What is thermal energy?
   a. a form of potential energy
   b. the temperature increase that occurs when any form of energy is added to a system
   c. mechanical energy
   d. the kinetic energy of molecular motion, measured as heat

4. What determines whether a chemical reaction is spontaneous?
   a. if it increases the disorder, or entropy, of the substances involved
   b. if it decreases the potential energy of the substances involved
   c. the temperature only—reactions are spontaneous at high temperatures and nonspontaneous at low temperatures
   d. the combined effect of changes in potential energy and entropy

5. Which of the following is *not* an example of an energy transformation?
   a. A shoe drops, converting potential energy to kinetic energy.
   b. A chemical reaction converts the energy in sunlight into the chemical energy in formaldehyde.
   c. The electrical energy flowing through a light bulb's filament is converted into light and heat.
   d. Sunlight strikes a prism and separates into distinct wavelengths.

6. When an atom is reduced, it
   a. gains entropy
   b. loses entropy
   c. gains an electron
   d. loses an electron

Conceptual Review

1. Section 2.2 describes the reaction between carbon dioxide and water, which forms carbonic acid:

$$CO_2(g) + H_2O(l) \rightleftharpoons H_2CO(aq)$$

   In aqueous solution, carbonic acid immediately dissociates to form a proton and the bicarbonate ion, as follows:

$$H_2CO_3(aq) \rightleftharpoons H^+(aq) + HCO_3^-(aq)$$

   Does this reaction raise or lower the pH of the solution? Does the bicarbonate ion act as an acid or a base? If an underwater volcano bubbled additional CO_2 into the ocean, would this sequence of reactions be driven to the left or the right? How would this affect the pH of the ocean?

2. When chemistry texts introduce the concept of electron shells, they emphasize that shells represent distinct potential energy levels. In introducing electron shells, this chapter also emphasized that they represent distinct distances from the positive charges in the nucleus. Are these two points of view in conflict? Why or why not?

3. In using radiometric dating to estimate when a rock formed, geologists must know three quantities. What are they? How are they used to estimate the age of a rock sample? What sources of uncertainty are involved in measuring these three quantities?

4. Why does ice float?

5. Hydrogen bonds form because the opposite, partial electric charges on polar molecules attract. Covalent bonds form as a result of the electrical attraction between electrons and protons. Covalent bonds are much stronger than hydrogen bonds. Explain why, in terms of the electrical attractions involved.

6. Draw a ball-and-stick model of the water molecule, and explain why this molecule is bent. Indicate the location of the partial electric charges on it. Why do these partial charges exist?

Group Discussion Problems

1. Suppose you wanted to use radiometric dating to estimate the age of an archaeological site that was thought to be about 1500 years old. Would you use the uranium-lead system described in Section 2.1, or would you evaluate a radioisotope that had a shorter half-life? Explain your answer.

2. Oxygen is extremely electronegative, meaning that its nucleus pulls in electrons shared in covalent bonds. Because these electrons are close to the oxygen nucleus, they have lower potential energy. Explain the changes in electron position that are illustrated in Figure 2.20 in terms of oxygen's electronegativity.

3. When nuclear reactions take place, some of the mass in the atoms involved is converted to energy. The energy in sunlight is created during nuclear fusion reactions on the Sun. Explain what astronomers mean when they say that the Sun is burning down and that it will eventually burn out.

4. Why do coastal regions tend to have climates with moderate temperatures and lower annual variation in temperature than do inland areas at the same latitude?

Answers to Multiple-Choice Questions 1. b; 2. a; 3. b; 4. d; 5. d; 6. c

www.prenhall.com/freeman is your resource for the following: Web Tutorials; Online Quizzes and other Online Study Guide materials; Answers to Conceptual Review Questions; Solutions to Group Discussion Problems; Answers to Figure Caption Questions and Exercises; and Additional Readings and Research.

3 Protein Structure and Function

KEY CONCEPTS

- Proteins are made of amino acids. Amino acids vary in structure and function because their side chains vary in composition. As a result, proteins vary widely in structure and function.
- In cells, most proteins are enzymes that function as catalysts. Chemical reactions occur much faster when they are catalyzed by enzymes. During enzyme catalysis, the reactants bind to an enzyme's active site in a way that allows the reaction to proceed efficiently.
- In cells, endergonic reactions occur in conjunction with an exergonic reaction involving ATP.

A biologist wearing 3-D goggles studies the structure of a protein, represented by gray ribbons. The researcher is using a computer model to test the hypothesis that a drug can bind to the protein at specific locations..

Chapter 2 introduced the hypothesis that chemical reactions in the atmosphere and ocean of ancient Earth led to the formation of complex carbon-containing compounds. This idea, called chemical evolution, was first proposed in 1923 by Alexander I. Oparin. The hypothesis was published again—independently and six years later—by J. B. S. Haldane. Today, the Oparin-Haldane proposal can best be understood as a formal scientific theory. As noted in Chapter 1, scientific theories typically have two components: a statement about a pattern that exists in the natural world and a proposed mechanism or process that explains the pattern. In the case of chemical evolution, the pattern is that increasingly complex carbon-containing molecules formed in the atmosphere and ocean of ancient Earth. The process responsible for this pattern was the conversion of energy, from sunlight and other sources, into chemical energy in the bonds of large, complex molecules.

Scientific theories are continuously refined as new information comes to light, and many of Oparin and Haldane's original ideas about how chemical evolution occurred have been extensively revised. In its current form, the theory can be broken into four steps, each requiring an input of energy:

1. Chemical evolution began with the production of small compounds with reduced carbon atoms, such as formaldehyde (H_2CO) and hydrogen cyanide (HCN).
2. These simple compounds reacted to form the mid-sized molecules called amino acids, sugars, and nitrogenous bases. These building-block molecules accumulated in the shallow waters of the ancient ocean, forming a complex solution called the **prebiotic soup.**
3. Mid-sized, building-block molecules linked to form the types of large molecules found in cells today, including proteins, nucleic acids, and complex carbohydrates. These large molecules are each made up of different chemical

subunits: Proteins are composed of amino acids; nucleic acids are composed of nucleotides; and complex carbohydrates are composed of sugars.

4. Life became possible when one of these large, complex molecules acquired the ability to make a copy of itself. This self-replicating molecule began to multiply by means of chemical reactions that it controlled. At that point, life had begun. Chemical evolution gave way to biological evolution.

Analyzing how the final three steps in chemical evolution occurred is the subject of this chapter and the next three chapters. Each of these chapters focuses on one of the four primary classes of biological molecules found in cells living today: proteins, nucleic acids, carbohydrates, and lipids. What do the subunits of these large molecules look like, and how could they have been created by chemical evolution? How are the subunits linked to form a protein or nucleic acid or complex carbohydrate or lipid? What do these large molecules look like, and what do they do in living cells? Finally, which type of molecule was responsible for the origin of life? This question is particularly intriguing, because researchers around the world are racing to synthesize a self-replicating molecule—to produce life in a test tube. Which type of molecule are they working with, and why?

Let's begin our analysis of these questions with the workhorse molecules called proteins, and the most famous experiment ever performed on the origin of life.

3.1 Early Origin-of-Life Experiments

In 1953 a graduate student named Stanley Miller performed a breakthrough experiment in the study of chemical evolution. Miller wanted to answer a simple question: Can complex organic compounds be synthesized from the simple molecules present in Earth's early atmosphere and ocean? In other words, is it possible to recreate the first steps in chemical evolution by simulating ancient Earth conditions in the laboratory?

Miller based his experimental design on the assumption that Earth's early atmosphere was dominated by molecules with high free energy when chemical evolution occurred. These molecules are likely to give up electrons, so their presence means that redox reactions can occur.

Miller's experimental setup (**Figure 3.1**) was designed to produce a microcosm of ancient Earth. The large glass flask represented the atmosphere and contained the gases methane (CH_4), ammonia (NH_3), and hydrogen (H_2), all of which have high free energy. This large flask was connected to a smaller flask by glass tubing. The small flask held a tiny ocean—200 milliliters (mL) of liquid water. Miller boiled this water constantly so that water vapor was added to the mix of gases in the large flask. As the vapor cooled and condensed, it flowed back into the smaller flask, where it boiled again. In this way, water vapor circulated continuously through the system. This was important: If the molecules in the simulated atmosphere reacted with one anoth-

Question: Can simple molecules and kinetic energy lead to chemical evolution?

Hypothesis: If kinetic energy is added to a mix of simple molecules with high free energy, redox reactions will occur that produce more complex molecules, perhaps including some with carbon-carbon bonds.

Null hypothesis: If kinetic energy is added to a mix of simple molecules with high free energy, more complex molecules will not be produced.

Experimental setup:

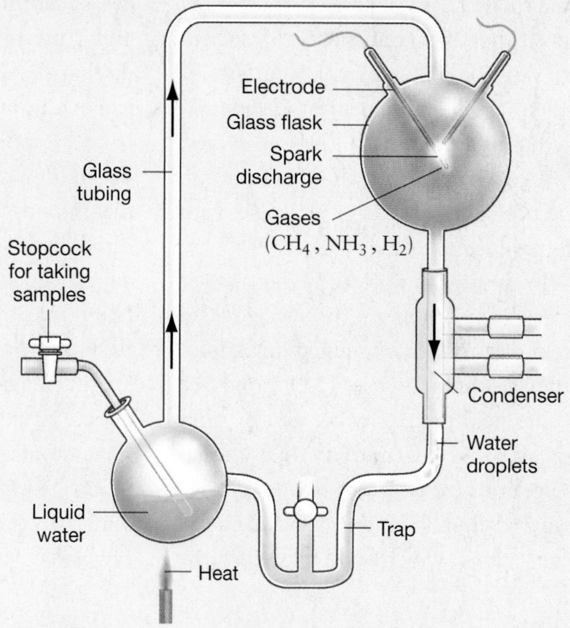

Prediction: Complex organic compounds will be found in the liquid water.

Prediction of null hypothesis: Only the starting molecules will be found in the liquid water.

Results:

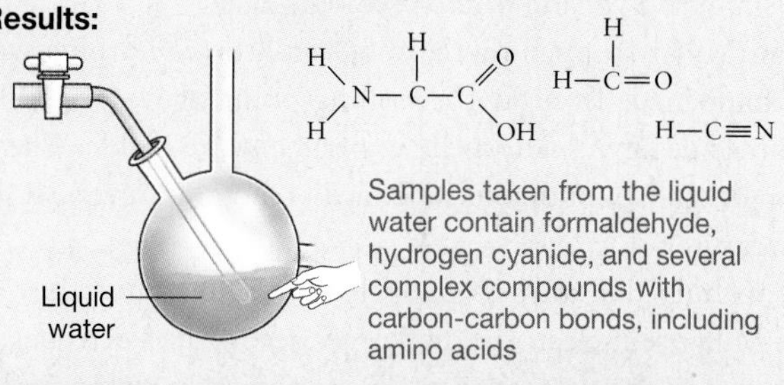

Conclusion: Chemical evolution occurs readily if simple molecules with high free energy are exposed to a source of kinetic energy.

FIGURE 3.1 Miller's Spark-Discharge Experiment
The arrows in the "Experimental setup" diagram indicate the flow of water vapor or liquid, starting with the 200 milliliters in the small, boiling flask. The large glass flask can contain any mixture of gases desired; when a voltage is applied across the electrodes in that flask, a spark jumps across the gap between them. The condenser consists of a jacket with cold water flowing through it. EXERCISE Label the parts of the apparatus that mimic the ocean, the atmosphere, rain, and lightning.

BOX 3.1 Was Miller Correct about Conditions on Ancient Earth?

Miller's work is considered classic, primarily because he showed that hypotheses about chemical evolution can be tested experimentally. But today few researchers consider his experiment an accurate simulation of chemical evolution.

To understand why, recall from Chapter 2 that atmospheric scientists now propose that the atmosphere of ancient Earth was dominated by volcanic gases. If so, then it is likely that most carbon occurred in molecules with low free energy, such as carbon dioxide (CO_2) and carbon monoxide (CO), instead of molecules with high free energy, such as CH_4.

Why is this important? The answer lies in the position of the electrons in these molecules. As explained in Chapter 2, the electrons in the C–O bonds of CO_2 are held tightly by oxygen because of its high electronegativity. In contrast, the electrons in the C–H bonds of CH_4 are shared equally by the carbon and hydrogen atoms, which have relatively low electronegativities. Thus the C–O bonds in CO_2 and CO are much stronger than the C–H bonds of CH_4. As a result, molecules such as CO_2 and CO are much less reactive than forms such as CH_4. This prediction has been confirmed experimentally. When CO_2, N_2, and H_2O are substituted for methane, ammonia, and hydrogen in Miller's spark-discharge experiment, virtually no chemical evolution takes place.

In response, atmospheric scientists began doing different types of experiments with the components of an oxidized atmosphere. For example, when water vapor and carbon monoxide or other volcanic gases are put into a glass flask and exposed to the types of high-energy radiation found in sunlight, a wide variety of compounds with reduced carbon atoms forms, including methane (CH_4), formaldehyde, and acetaldehyde. Similar experiments have altered the ratio of water vapor to carbon monoxide, changed the temperature, extended the length of time the mixture was exposed to light, or included other molecules found in volcanic gases, such as hydrogen (H_2), nitrogen (N_2), or carbon dioxide (CO_2). In each case, a wide variety of organic compounds with high free energy formed.

In addition, some researchers recently asserted that the key events in chemical evolution did not begin in the atmosphere at all. Instead, they suggest that the process also occurred near hydrothermal ("hot-water") vents similar to the volcanic "black smokers" found at the bottom of oceans today (**Figure 3.2**). Black smokers form where molten rock occurs just below the ocean floor. The thermal energy in the liquid rock heats water in the surrounding rocks to temperatures as high as 450°C. Because there is intense pressure at these depths, this "superheated" water does not boil. Instead, it rises up through the crust. As it passes through the crustal rocks, it dissolves enough iron-, sulfur-, nickel-,

er, the "rain" would carry them into the simulated ocean, forming a simulated version of the prebiotic soup.

Had Miller stopped there, however, little or nothing would have happened. Even at the boiling point of water (100°C), the molecules involved in the experiment are stable. They do not undergo spontaneous chemical reactions, even at high temperatures.

Something did start to happen in the apparatus, however, when Miller sent electrical discharges across the electrodes he'd inserted into the atmosphere. These miniature lightning bolts added a crucial element to the reaction mix: pulses of intense electrical energy. After a day of continuous boiling and sparking, the solution in the boiling flask began to turn pink. After a week, it was deep red and cloudy. When Miller analyzed the molecules dissolved in the solution, he found that several complex carbon-containing compounds were present. The experiment, driven by the energy in the electrical discharges, had recreated the start of chemical evolution.

To find out exactly which products resulted from the initial reactions in the simulated atmosphere, Miller drew samples from the apparatus at intervals. In these samples he found large quantities of hydrogen cyanide (HCN) and formaldehyde (H_2CO). The data were exciting because HCN and H_2CO are required for reactions that lead to the synthesis of more complex organic molecules. Indeed, some of these more complex compounds were actually present in the miniature ocean. The sparks and heating had led to the synthesis of compounds that are fundamental to life: amino acids.

3.2 Amino Acids and Polymerization

Based on the presence of amino acids, Miller claimed that his experiment simulated the second stage in chemical evolution—the formation of a prebiotic soup. Although the assumptions behind his experiments, and hence the results, eventually came under fire, follow-up studies have confirmed that the second step in chemical evolution occurred early in Earth's history. On the basis of the data reviewed in **Box 3.1** there is now a strong consensus that amino acids and other components of the prebiotic soup are readily produced under conditions that accurately simulate the atmosphere and oceans of ancient Earth.

and carbon-containing compounds to form a blackened solution that jets out into the surrounding water. At the bottom of the ocean, this surrounding water is frigid—typically 4°C.

Because a carbon- and sulfur-containing molecule called methanethiol (CH_3SH) is often found near black smokers, it appears that chemical evolution is occurring in these environments even today. A research group has shown that, under the temperature and pressure conditions found near black smokers, CO and CH_3SH react to form acetic acid (CH_3COOH), which contains a carbon-carbon bond.

Based on these and other experiments simulating early Earth conditions, most researchers now agree that early in Earth's history, quantities of formaldehyde, hydrogen cyanide, acetic acid, and other carbon-containing molecules were raining down from the skies and bubbling up from hydrothermal vents.

Chemical evolution takes place where superheated water emerges from vents in the otherwise frigid ocean floor

FIGURE 3.2 Black Smokers
The water that erupts from black smokers contains large amounts of iron, carbon, and nickel from the rocks below. This water is superheated, meaning that its temperature is above 100°C. It does not boil because black smokers are found deep in the ocean, where water is under extremely high pressure.

Researchers have also realized that large quantities of amino acids may have rained onto the planet from outer space during the early bombardment phase of Earth's history, described in Chapter 2. This conclusion is based on analyses of space debris that contains carbon. The Murchison meteorite that struck Australia in 1969, for example, contains as many as 18 different amino acids (**Figure 3.3**). These data suggest that about 4 billion years ago, chemical evolution on Earth could have been supplemented by similar reactions occurring in outer space.

Now let's look at the molecules themselves. What are amino acids, and how are they linked to form proteins?

FIGURE 3.3 Some Meteorites Contain Complex Organic Molecules
A 23-gram fragment of the Murchison meteorite. The Murchison meteorite is one of many pieces of space debris that contain abundant and diverse types of complex organic molecules, including amino acids.

The Structure of Amino Acids

The bacterial cells that live on your skin contain several thousand different proteins; as a group, the cells in your body produce tens of thousands of distinct proteins. But most of these proteins are composed of just 20 different building blocks, called **amino acids**. All 20 amino acids have a common structure.

To understand the structure of an amino acid, recall that carbon atoms have four unpaired valence electrons and

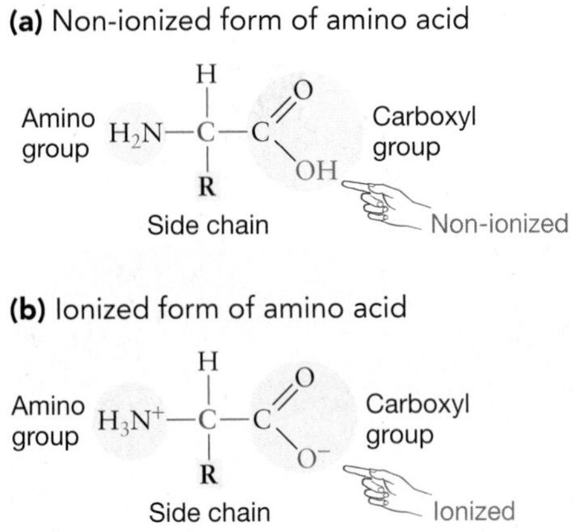
(a) Non-ionized form of amino acid
Amino group
Carboxyl group
Side chain
Non-ionized
(b) Ionized form of amino acid
Amino group
Carboxyl group
Side chain
Ionized

FIGURE 3.4 Amino Acid Structure
All amino acids have the same general structure: a central carbon, shown in red, bonded to an amino functional group, a carboxyl functional group, a hydrogen atom, and a side chain, or R-group.

participate in four bonds. Every amino acid has a carbon that makes the same four bonds. The first bond attaches this carbon to NH_2—the amino functional group (**Figure 3.4a**). The second bond links this carbon to COOH—the carboxyl functional group. The carboxyl group is acidic because its two oxygen atoms are highly electronegative. They pull electrons away from the hydrogen atom, which means that it is relatively easy for this group to lose a proton. The combination of the amino and carboxyl groups inspired the name *amino acid*. A third bond links the highlighted carbon to a hydrogen atom. In all amino acids, then, a carbon atom is bonded to an amino group, a carboxyl group, and hydrogen. As we'll see in a moment, the nature of the fourth attachment is what makes each amino acid unique.

The presence of an amino group and a carboxyl group in amino acids is important. **Figure 3.4b** shows what happens to these functional groups in solution. In water at pH 7, the concentration of protons causes the amino group to act as a base. It attracts a proton to form NH_3^+. The carboxyl group, in contrast, loses a proton to form COO^-. The charges on these functional groups help amino acids stay in solution and add to their chemical reactivity.

The Nature of Side Chains In Figure 3.4, the highlighted carbon atom in the amino acid forms a fourth bond with an atom or a group of atoms abbreviated as "R." Chemists use this symbol to indicate additional atoms called a side chain. In every amino acid, a carbon atom is linked to a hydrogen atom, an amino group, a carboxyl group, and an R-group. The 20 amino acids found in organisms are different because their R-groups are different. These R-groups vary from a single hydrogen atom to large structures containing carbon atoms linked into rings. Several of the side chains found in amino acids contain functional groups (see Chapter 2, Table 2.1).

Figure 3.5 shows the 20 amino acids found in cells and sorts them according to whether their side chain is nonpolar, polar, or electrically charged. This analysis is important because amino acids with nonpolar side chains do not have charged or electronegative atoms capable of forming hydrogen bonds with water. These R-groups are said to be **hydrophobic** ("water-fearing") because water does not interact with them. As a result, hydrophobic side chains tend to coalesce in aqueous solution. In contrast, amino acids with polar or charged side chains interact readily with water and are termed **hydrophilic** ("water-loving"). Hydrophilic amino acids dissolve in water easily. **Table 3.1** ranks the 20 amino acids according to how readily they interact with water.

In addition to affecting the solubility of amino acids, the side chain influences their chemical reactivity. In Figure 3.5, notice that some amino acids contain side chains consisting entirely of carbon and hydrogen atoms. These R-groups rarely participate in chemical reactions. As a result, the behavior of these amino acids depends primarily on their size and shape rather than reactivity. In contrast, amino acids that have hydroxyl, amino, or carboxyl functional groups in their side chains are more reactive. Soon we'll see that amino acids with sulfur atoms (S) in their side chains can help link different parts of large proteins. The point is that variation in the behavior of amino acids results from variation in the structure of their R-groups.

TABLE 3.1 How Amino Acids Interact with Water

The 20 amino acids are ranked according to how likely they are to interact with water. Color codes are based on Figure 3.5.

Isoleucine	Highly hydrophobic
Valine	
Leucine	
Phenylalanine	
Methionine	Moderately hydrophobic
Alanine	
Glycine	
Cysteine	
Tryptophan	
Tyrosine	Mildly hydrophobic
Proline	
Threonine	
Serine	
Histidine	Mildly hydrophilic
Glutamate	
Asparagine	
Glutamine	
Aspartate	
Lysine	Highly hydrophilic
Arginine	

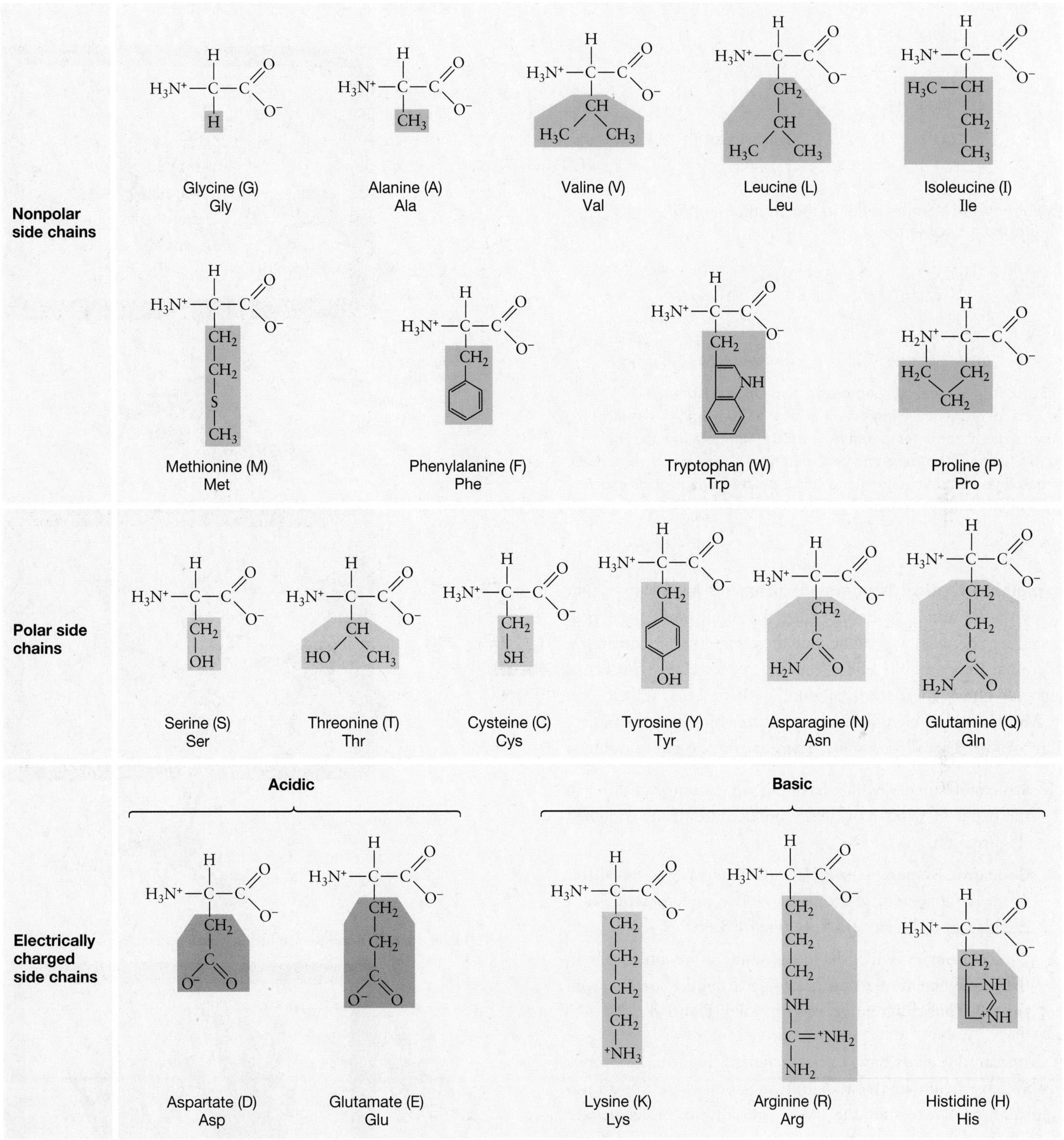
Nonpolar side chains
Glycine (G)
Gly
Alanine (A)
Ala
Valine (V)
Val
Leucine (L)
Leu
Isoleucine (I)
Ile
Methionine (M)
Met
Phenylalanine (F)
Phe
Tryptophan (W)
Trp
Proline (P)
Pro
Polar side chains
Serine (S)
Ser
Threonine (T)
Thr
Cysteine (C)
Cys
Tyrosine (Y)
Tyr
Asparagine (N)
Asn
Glutamine (Q)
Gln
Electrically charged side chains
Acidic
Basic
Aspartate (D)
Asp
Glutamate (E)
Glu
Lysine (K)
Lys
Arginine (R)
Arg
Histidine (H)
His

FIGURE 3.5 The 20 Major Amino Acids Found in Organisms
The structural formulas of the 20 major amino acids found in organisms, at the pH (about 7.0) found in cells. The side chains are highlighted, and standard single-letter and three-letter abbreviations for each amino acid are given. For clarity, the carbon atoms in the ring structures of phenylalanine, tyrosine, tryptophan, and histidine are not shown; each bend in a ring is the site of a carbon atom. The hydrogen atoms in these structures are also not shown. A double line inside a ring indicates a double bond. EXERCISE Label amino acid side chains that are hydrophilic versus those that are hydrophobic. To check your work, examine the data in Table 3.1.

(a) Structural isomers differ in the order which their atoms are attached.

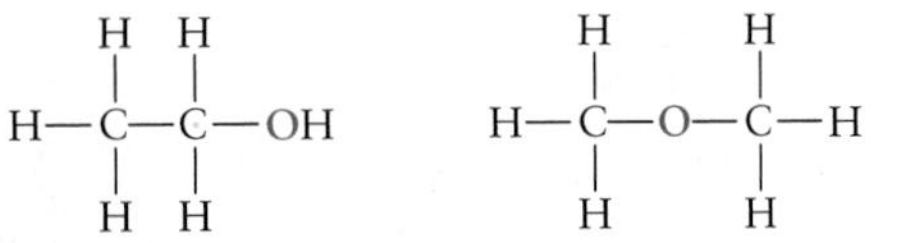

Ethanol (C_2H_6O) Dimethyl ether (C_2H_6O)

(b) Geometric isomers differ in the arrangement of atoms around a double bond.

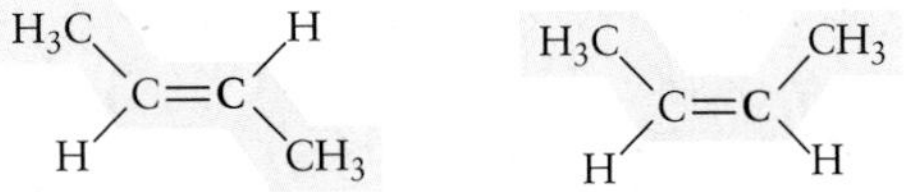

trans-2-butene (C_4H_8) *cis*-2-butene (C_4H_8)

(c) Optical isomers are mirror images of one another—they cannot be exactly superimposed.

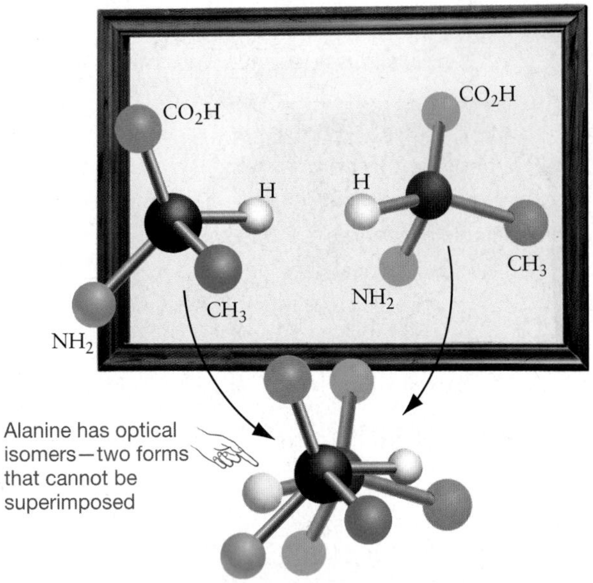

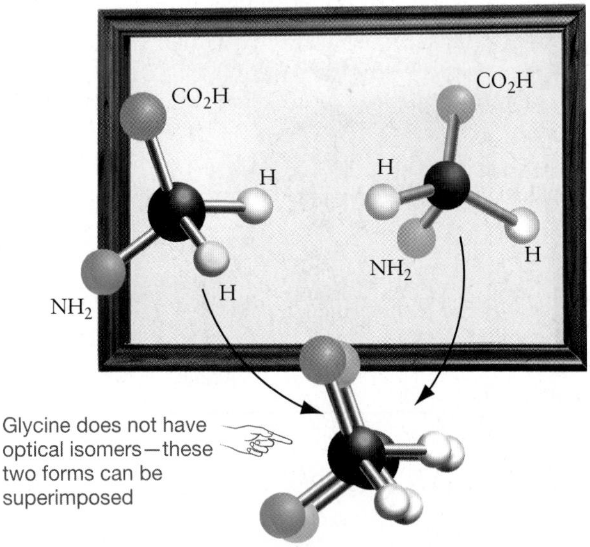

Hands are mirror images, just as optical isomers are.

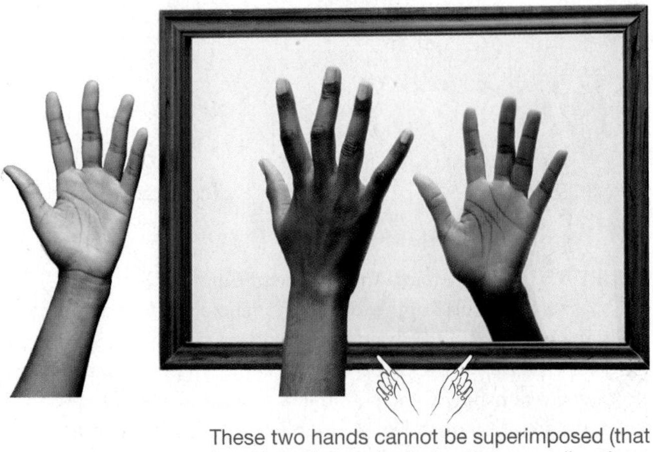

These two hands cannot be superimposed (that is, with both thumbs facing the same direction and both palms facing the same direction)

FIGURE 3.6 Structural, Geometric, and Optical Isomers
(a) Ethanol is the active ingredient in alcoholic beverages; dimethyl ether, a gas at room temperature, is used in refrigeration. **(b)** The molecules *trans*-2-butene and *cis*-2-butene are used in the production of gasoline, synthetic rubber, and solvents. **(c)** All amino acids except glycine have optical isomers.

Explaining Optical Isomers Experiments have shown that many of the amino acids in Figure 3.5 could have been produced by chemical evolution. But these experiments created a dilemma: Every amino acid except glycine exists in two forms, but only one of these forms is found in living cells. Why?

Molecules that have different structures but the same molecular formula are called **isomers.** There are three types of isomers:

1. **Structural isomers,** which have the same atoms but differ in the order in which covalently bonded atoms are attached (**Figure 3.6a**).
2. **Geometric isomers,** which have the same atoms but differ in the arrangement of atoms or groups on either side of a double bond or ring structure (**Figure 3.6b**).
3. **Optical isomers,** which have the same atoms but differ in the arrangement of atoms or groups around a carbon atom that has four different groups attached (**Figure 3.6c**).

Most amino acids have optical isomers. Figure 3.6c shows the arrangements in the two optical isomers of the amino acid alanine. Note that the two forms of the molecule are mirror images of one another, just as your left and right hands are mirror images of each other. Like your left and right hands, the left-handed and right-handed forms of alanine cannot be exactly superimposed. They also do not have a plane of symmetry. Carbon atoms with this feature exist in every amino acid except glycine. In fact, any carbon atom that has four different atoms or groups attached to it has an optical isomer.

The existence of optical isomers is an important issue in biology. Because the structures of each optical isomer of a molecule are different, they have different functions. In cells, only the "left-handed" forms of amino acids exist. If the right-handed form of an amino acid is introduced into cells experimentally, it does not function normally. As in most other molecules, an amino acid's function is determined by its structure.

These observations are a challenge to the chemical evolution hypothesis, because no plausible mechanism has been proposed to explain how the process resulted in the production of only left-handed optical isomers. Was it simply a matter of chance? Or was there something unusual about the chemistry of early Earth that we still don't understand? To date, the issue is unresolved.

How Do Amino Acids Link to Form Proteins?

In the introduction to this chapter, we saw that amino acids link to form proteins. Similarly, the molecular building blocks called nucleotides attach to one another to form nucleic acids, and simple sugars connect to form complex carbohydrates. In general, a molecular subunit such as an amino acid, a nucleotide, or a sugar is called a **monomer** ("one-part"). When monomers bond together, the resulting structure is called a **polymer** ("many-parts"). The process of linking monomers is called **polymerization** (**Figure 3.7a**). Thus, amino acids polymerize to form proteins (**Figure 3.7b**). Biologists also use the word **macromolecule** to denote a very large molecule that is made up of smaller molecules joined together. A **protein** is a linear macromolecule—a polymer—that consists of linked amino acid monomers.

The theory of chemical evolution states that monomers in the prebiotic soup polymerized to form proteins and other types of macromolecules found in organisms. This is a difficult step, because monomers such as amino acids do not spontaneously self-assemble into macromolecules such as proteins. Based on the second law of thermodynamics reviewed in Chapter 2, this is not surprising. Complex and highly organized molecules are not expected to form spontaneously from simpler constituents, because polymerization organizes the molecules involved into a more complex, ordered structure. Stated another way, polymerization decreases the disorder, or entropy, of the molecules involved. In addition, polymers are energetically much less stable than their component monomers. In sum, the ΔH term of the Gibbs free-energy equation is positive and the $T\Delta S$ term is negative, making ΔG positive at all temperatures. Polymerization reactions are nonspontaneous. Monomers must absorb energy in order to link together. How could this have happened during chemical evolution?

(a) Monomers polymerize to form polymers.

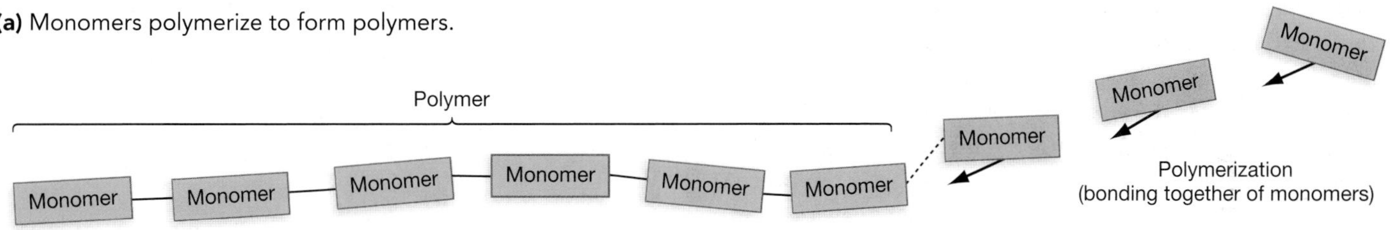

(b) Proteins are polymers that are made up of amino acid monomers.

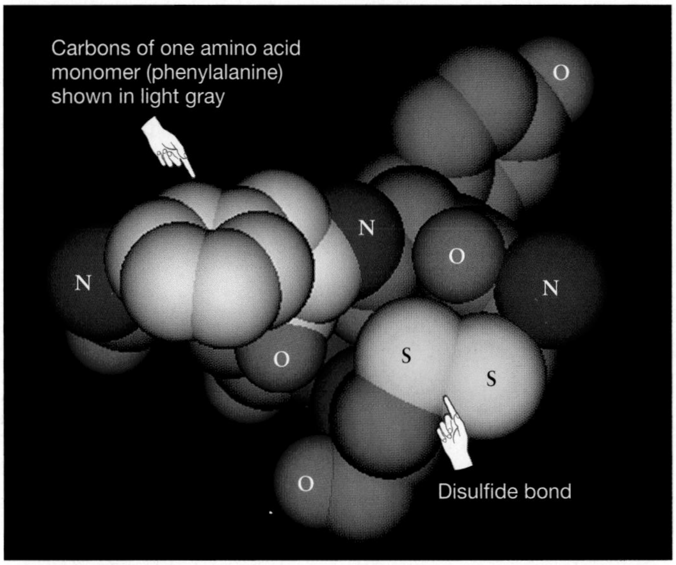

FIGURE 3.7 Monomers Are the Building Blocks of Polymers
(a) Monomers can be linked together to form polymers. **(b)** A polypeptide called ADH, which helps regulate urine formation in your body. One of the nine amino acids in the peptide chain is highlighted.

WEB TUTORIAL 3.1
Condensation and Hydrolysis Reactions

Could Polymerization Occur in the Energy-Rich Environment of Early Earth? Researchers who added heat or electrical discharges to solutions of either amino acids or nucleic acids found that polymerization reactions proceed slowly, if at all. This is because monomers polymerize through **condensation reactions,** also known as **dehydration reactions.** These reactions are aptly named because the newly formed bond results in the loss of a water molecule (**Figure 3.8a**). The reverse reaction, called **hydrolysis,** breaks polymers apart by adding a water molecule (**Figure 3.8b**). The water molecule reacts with the bond linking the monomers, separating one monomer from the polymer chain.

In a solution, such as the prebiotic soup, condensation and hydrolysis represent the forward and reverse reactions of a chemical equilibrium. Hydrolysis dominates because it increases entropy and because it is energetically favorable—it lowers the potential energy of the electrons involved.

According to recent experiments, the key to overcoming hydrolysis during chemical evolution was, quite literally, as common as mud. Researchers have been able to create stable polymers by incubating monomers with tiny mineral particles—the size found in clay or mud. These experiments were based on the hypothesis that growing macromolecules would be protected from hydrolysis if they clung, or adsorbed, to the mineral surfaces.

More specifically, the experiments were designed to simulate events that could have occurred in the prebiotic soup. In one experiment, researchers put amino acids in a solution with tiny mineral particles and allowed them to react. After a day, the researchers separated the mineral particles from the solution. They then put the particles into a fresh solution containing amino acids and a source of energy. After repeating this procedure for several days, they analyzed the mineral particles and found polymers up to 55 amino acids long. These results support the hypothesis that adsorption to mineral particles protects polymers from hydrolysis. Because the experimental procedure was designed to mimic coastal environments where beaches are repeatedly washed with fresh waves or tidal flows, the results make it reasonable to claim that at least some muddy tide pools and beaches became covered with small proteins early in Earth's history.

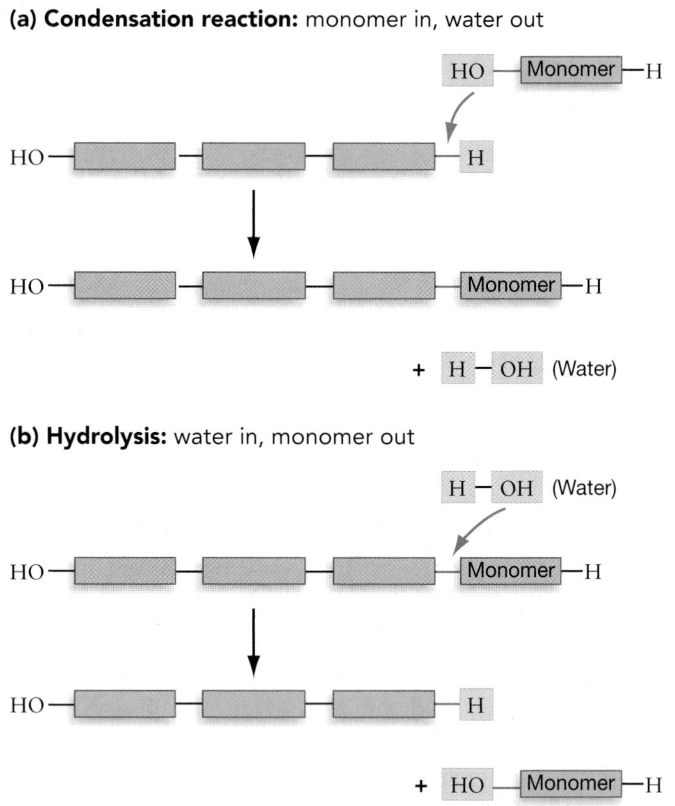

FIGURE 3.8 Polymers Can Be Extended or Broken Apart
(a) In a condensation reaction, a monomer is added to a polymer to make a longer polymer. The new bond that forms results in the formation of a water molecule. **(b)** In hydrolysis, a water molecule reacts with the bond linking the monomers. A monomer is broken off the chain, resulting in a shorter polymer.

The Peptide Bond Exactly how do amino acids polymerize? As **Figure 3.9** shows, amino acids polymerize when a bond forms between the carboxyl group of one amino acid and the amino group of another. The C–N bond that results from this condensation reaction is called a **peptide bond.** This bond is particularly stable because electrons are partially shared between the neighboring carbonyl functional group and the peptide bond. The degree of electron sharing is great enough that peptide bonds ac-

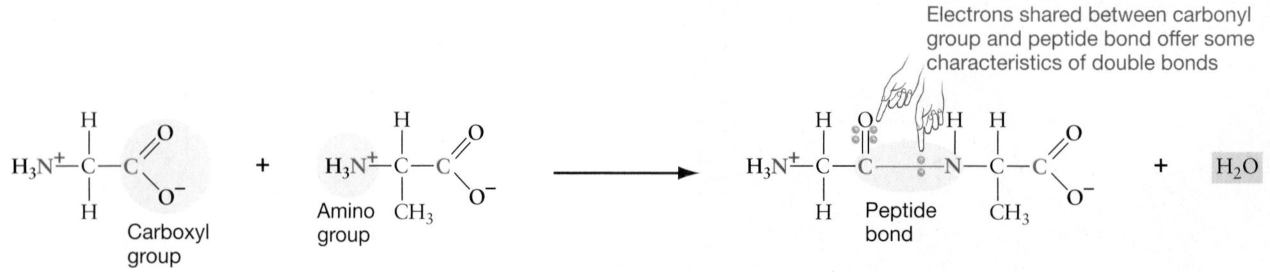

FIGURE 3.9 Peptide Bond Formation
When the carboxyl group of one amino acid reacts with the amino group of a second amino acid, a peptide bond forms. QUESTION Is a peptide bond a hydrogen bond, a nonpolar covalent bond, a polar covalent bond, or an ionic bond?

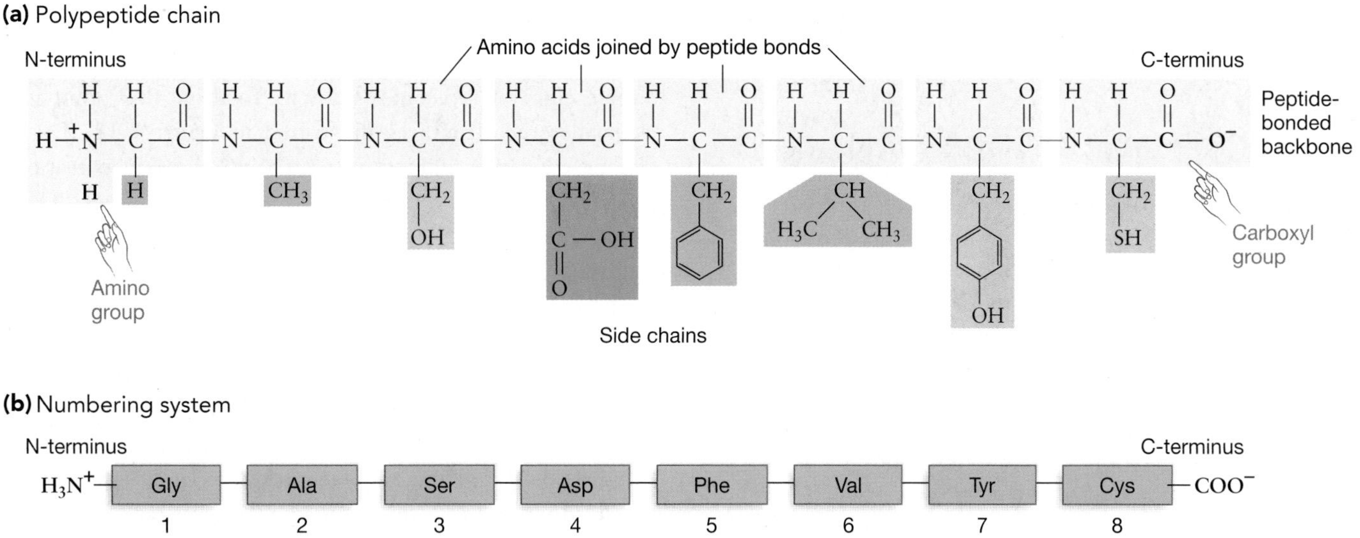
(a) Polypeptide chain
N-terminus
Amino acids joined by peptide bonds
C-terminus
Peptide-bonded backbone
Amino group
Side chains
Carboxyl group
(b) Numbering system
N-terminus
C-terminus
Gly
Ala
Ser
Asp
Phe
Val
Tyr
Cys
1
2
3
4
5
6
7
8

FIGURE 3.10 Amino Acids Polymerize to Form Polypeptides
(a) Amino acids can be linked into long chains, called polypeptides, by peptide bonds. **(b)** By convention, the sequence of amino acids in a polypeptide chain is numbered from the N-terminus to the C-terminus.

tually have some of the characteristics of a double bond. For example, the peptide bond is planar.

When a series of amino acids are linked by peptide bonds into a chain, the amino acids are referred to as *residues* and the resulting molecule is called a **polypeptide**. **Figure 3.10a** shows how the chain of peptide bonds in a polypeptide gives the molecule a structural framework, or a "backbone." Three points are important to note about this backbone: (1) The side chains present in each residue extend out from it, (2) it has directionality, and (3) it is flexible. The backbone is directional because there is an amino group ($-NH_3^+$) on one end of every polypeptide chain and a carboxyl group ($-COO^-$) on the other. By convention, biologists always write amino acid sequences in the same direction. The end of the sequence that has the free amino group is placed on the left and is called the N-terminus, or amino-terminus, and the end with the free carboxyl group appears on the right-hand side of the sequence and is called the C-terminus, or carboxy-terminus. The amino acids in the chain are always numbered starting from the N-terminus (**Figure 3.10b**), because the N-terminus is the start of the chain when proteins are synthesized in cells. Although the peptide bond itself cannot rotate because of its double-bond nature, the single bonds on either side of the peptide bond can rotate. As a result, the structure as a whole is flexible (**Figure 3.11**).

When fewer than 50 amino acids are linked together in this way, the resulting polypeptide is called an **oligopeptide** ("few peptides") or simply a peptide. Polypeptides that contain 50 or more amino acids are formally called proteins. Proteins may consist of single polypeptides or multiple polypeptides that are bonded to each other.

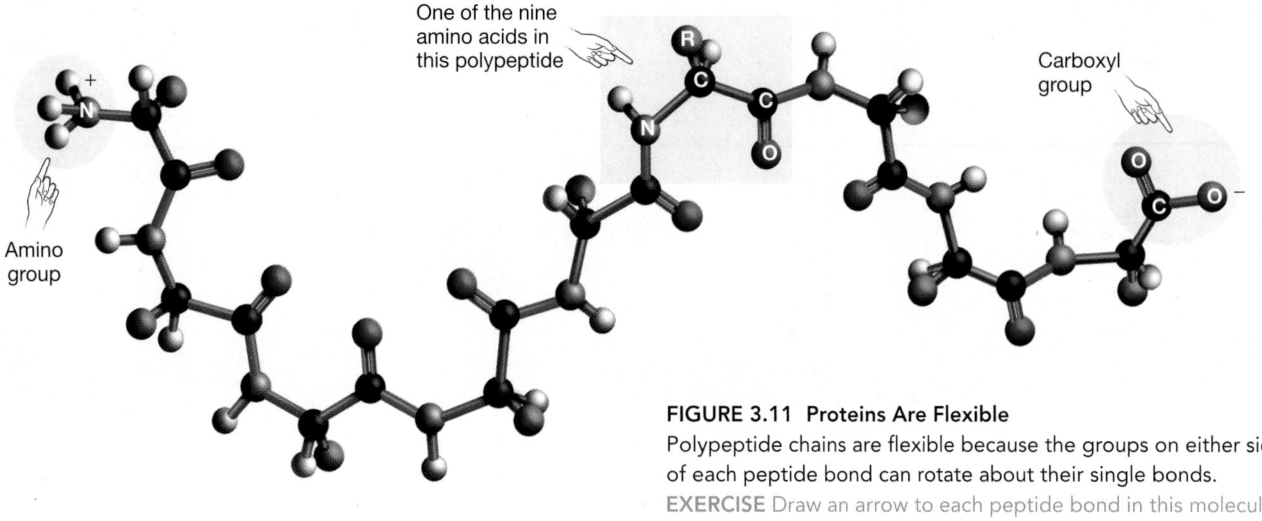
One of the nine amino acids in this polypeptide
Carboxyl group
Amino group

FIGURE 3.11 Proteins Are Flexible
Polypeptide chains are flexible because the groups on either side of each peptide bond can rotate about their single bonds.
EXERCISE Draw an arrow to each peptide bond in this molecule.

Proteins are the stuff of life. Let's take a look at how they are put together, and then at what they do.

✓CHECK YOUR UNDERSTANDING

Amino acids are small molecules with a carbon atom bonded to a carboxyl group, an amino group, a hydrogen atom, and a side chain called an R-group. Each amino acid has distinctive chemical properties because each has a unique R-group. You should be able to draw the general form of an amino acid.

When the carboxyl group of one amino acid reacts with the amino group of another amino acid, a strong covalent bond called a peptide bond forms. Polypeptides are polymers made up of peptide-bonded amino acids. Small polypeptides are called oligopeptides, and large polypeptides are called proteins. You should be able to draw and label two amino acids linked by a peptide bond.

3.3 What Do Proteins Look Like?

With respect to their structure, proteins may be the most diverse class of molecules known. To drive this point home, consider the structures illustrated in **Figure 3.12**. The hormone glucagon, which is regulating the concentration of sugar in your blood right now, is shown in Figure 3.12a. This molecule is just 29 amino acids long, and it consists of a single polypeptide that folds into a simple coil. In contrast, a protein called cytochrome *c* oxidase, which is critical to energy production in your cells, contains over 3600 amino acids. It is made up of 13 distinct polypeptides and is roughly rectangular (Figure 3.12b). Aspartate transcarbamoylase has two types of subunit (shown in light and dark gray in (Figure 3.12c). The complete protein consists of two of the dark subunits and three of the light subunits and is triangular. In many cases, the shape of a protein correlates closely with its function. The TATA-box binding protein (Figure 3.12d) has a groove where DNA molecules fit. Porin is doughnut-

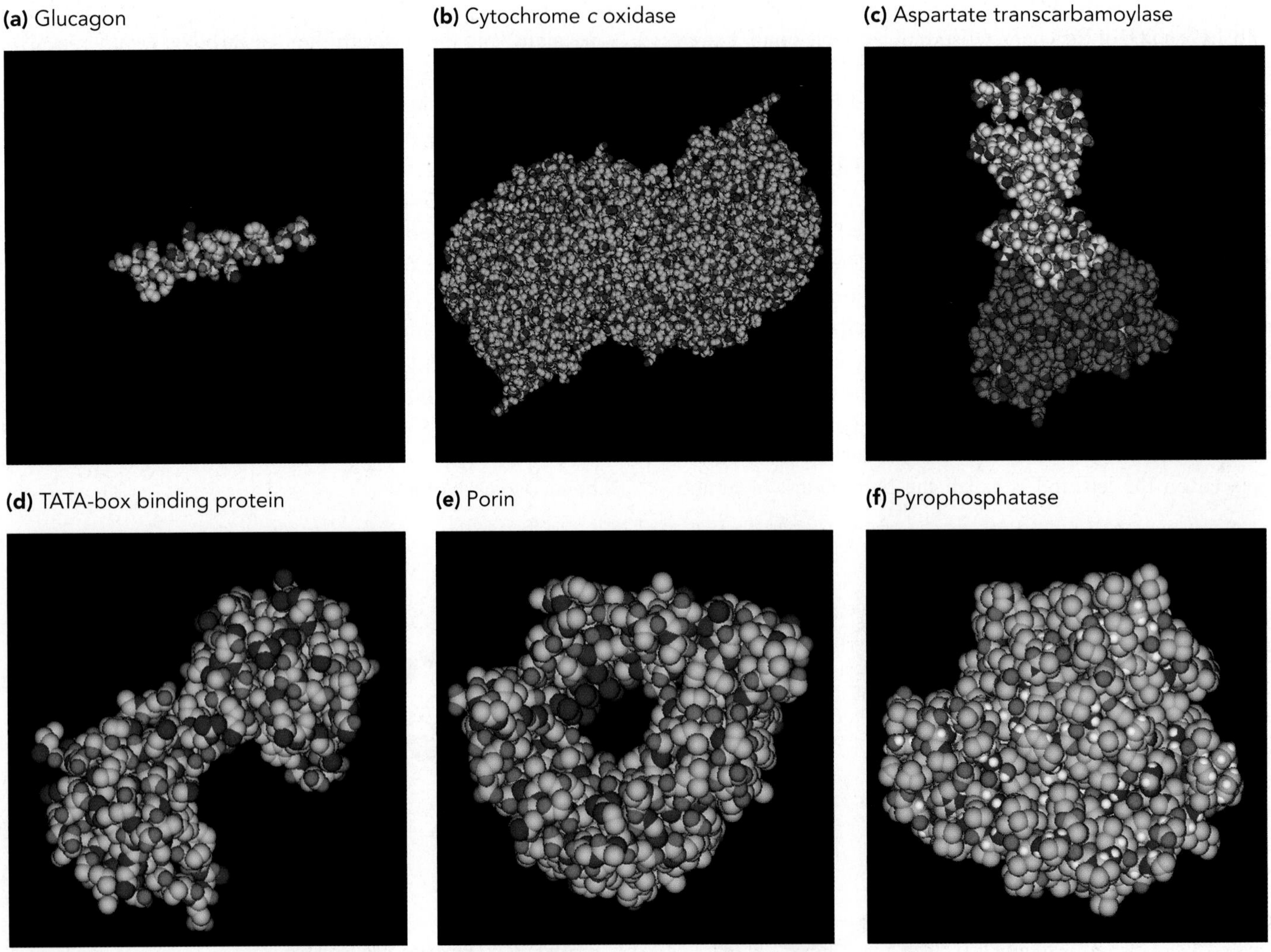
(a) Glucagon
(b) Cytochrome *c* oxidase
(c) Aspartate transcarbamoylase
(d) TATA-box binding protein
(e) Porin
(f) Pyrophosphatase

FIGURE 3.12 In Size and Shape, Proteins Are the Most Diverse Class of Molecules Known
The proteins found in cells may be (a) small or (b) large. In addition, they range from **(a)** coils to **(b)** rectangular, **(c)** triangular, **(d)** butterfly, **(e)** doughnut, and **(f)** globular shapes..

shaped, with a hole that forms a pore (Figure 3.12e). Most of the proteins found in cells function as enzymes and are globular (Figure 3.12f).

How can biologists make sense of this diversity of protein size and shape? Initially, the amount of variation seems overwhelming. Fortunately, it is not. No matter how large or complex a protein may be, its underlying structure can be broken down into just four basic levels of organization.

Primary Structure

Every protein has a unique sequence of amino acids. That simple message was the culmination of 12 years of study by Frederick Sanger and co-workers during the 1940s and 1950s. Sanger's group worked out the first techniques for determining the amino acid sequence of a protein and published the completed sequence of the hormone insulin, a protein that works in conjunction with glucagon to regulate sugar concentrations in the blood of humans and other mammals. When other proteins were analyzed, it rapidly became clear that each protein has a definite and distinct amino acid sequence.

Biochemists call the unique sequence of amino acids in a protein the **primary structure** of that protein. The sequence of amino acids in Figure 3.10, for example, defines that polypeptide's primary structure. With 20 types of amino acids available and a variation in size from two amino acid residues to tens of thousands long, the number of primary structures that are possible is practically limitless. There are, in fact, 20^n different polypeptides of length n. For a polypeptide that is just 10 amino acids long, 20^{10}, or 10,000 billion, primary sequences are possible.

Recall that the R-groups present on each amino acid affect its solubility and chemical reactivity. Based on this observation, it is reasonable to predict that the R-groups present in a polypeptide will affect the polypeptide's properties and function. In some cases, even a single change in the sequence of amino acids can cause radical changes in the way the molecule as a whole behaves. As an example, consider the hemoglobin protein of humans, which is carrying oxygen in your blood right now. In some individuals, hemoglobin has a valine instead of a glutamate at the amino acid numbered 6 in a strand of 146 amino acids (**Figure 3.13a**). Valine's side chain is very different from the R-group in glutamate. The change results in a protein that tends to crystallize instead of staying in solution when oxygen concentrations in the blood are low (**Figure 3.13b**). People whose hemoglobin contains this single amino acid change suffer from the debilitating illness called sickle-cell disease. A protein's primary structure is fundamental to its function.

Secondary Structure

Even though variation in the amino acid sequence of a protein is virtually limitless, it is only the tip of the iceberg in terms of generating structural diversity. The next level of organization in proteins is known as **secondary structure**, which is created by hydrogen bonding. More specifically, secondary structure results from hydrogen bonding that occurs between the carboxyl oxygen of one amino acid residue and the hydrogen on the amino group of another. The oxygen atom in the carboxyl group has a partial negative charge due to its high electronegativity, while the

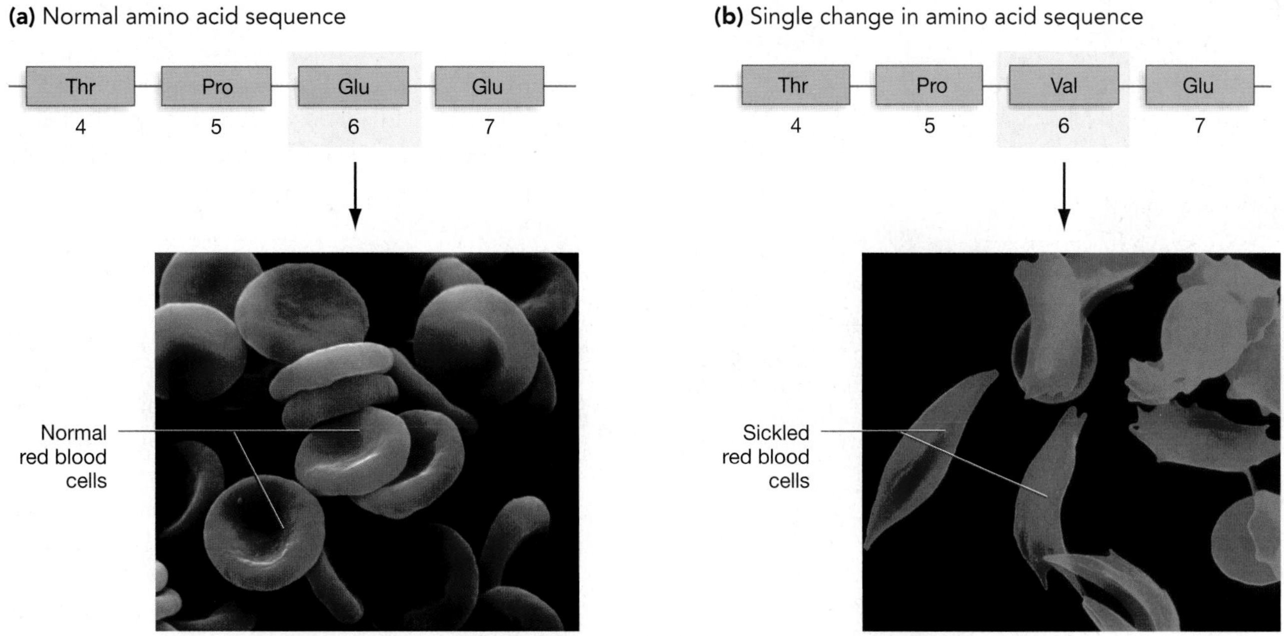
(a) Normal amino acid sequence
Thr
Pro
Glu
Glu
4
5
6
7
Normal red blood cells
(b) Single change in amino acid sequence
Thr
Pro
Val
Glu
4
5
6
7
Sickled red blood cells

FIGURE 3.13 Changes in Primary Structure Affect Protein Function
Compare the primary structure of **(a)** normal hemoglobin with that of **(b)** hemoglobin molecules of people with sickle-cell disease. The single amino acid change causes red blood cells to change from their normal disc shape in (a) to a sickled shape in (b) when oxygen concentrations are low. Each red blood cell contains about 300 million hemoglobin molecules.

hydrogen atom in the amino group has a partial positive charge because it is bonded to nitrogen, which has high electronegativity (**Figure 3.14a**).

Note that secondary structure is not created by interactions among side chains but instead by interactions between atoms that are part of a protein's peptide-bonded backbone. This is a key point, because hydrogen bonding between sections of the backbone is possible only when different parts of the same polypeptide bend in a way that puts carboxyl and amino groups close together. The bending that aligns parts of the backbone and allows these bonds to form occurs in several distinct ways. **Figure 3.14b** shows two of the most important configurations that allow formation of hydrogen bonds: (1) an **α-helix** (alpha helix), in which the polypeptide's backbone is coiled, and (2) a **β-pleated sheet** (beta-pleated sheet), in which segments of a peptide chain bend 180° and then fold in the same plane. In many cases, secondary structure consists of α-helices and β-pleated sheets. Which one forms, if either, is a product of the molecule's primary structure—specifically, the geometry of the amino acids in the sequence. Methionine and glutamic acid, for example, are much more likely to be involved in α-helices than in β-pleated sheets. The opposite is true for valine and isoleucine. Proline, in contrast, is unlikely to be involved in either type of secondary structure.

Although each of the hydrogen bonds in an α-helix or a β-pleated sheet is very weak relative to a covalent bond, the large number of hydrogen bonds in these structures makes them highly stable. As a result, they increase the stability of the molecule as a whole and help define its shape. In terms of overall shape and stability, though, a protein's tertiary structure is even more important.

(a) Hydrogen bonds form between peptide chains.

(b) Secondary structures of proteins result.

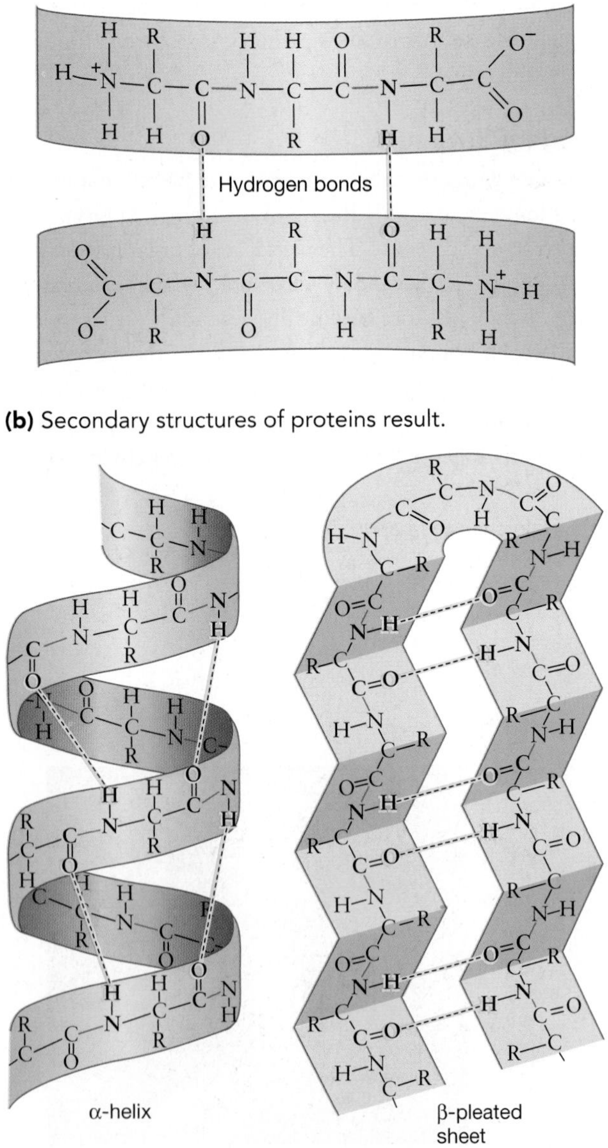

FIGURE 3.14 Secondary Structures of Proteins
(a) The peptide-bonded backbone of a protein can coil or fold in on itself when hydrogen bonds form between amino groups and carboxyl groups. **(b)** The detailed structure of the coils called α-helices and the folds called β-pleated sheets.

Tertiary Structure

Alpha helices and β-pleated sheets form due to interactions between components of a protein's peptide-bonded backbone. In contrast, most of the overall shape, or **tertiary structure**, of a polypeptide results from interactions between R-groups or between R-groups and the peptide backbone. As **Figure 3.15a** shows, side chains can be involved in a wide variety of bonds and interactions. Because each of these contacts causes the peptide-bonded backbone to bend and fold, each contributes to the distinctive three-dimensional shape of a polypeptide.

Four types of interactions that involve side chains are particularly important:

1. Covalent bonds can form between sulfur atoms when a redox reaction occurs between sulfur-containing R-groups, such as those found in cysteine. These **disulfide** ("two-sulfur") **bonds** are frequently referred to as bridges because they create strong links between distinct regions of the same polypeptide.
2. In an aqueous solution, water molecules interact with hydrophilic side chains and force hydrophobic side chains to coalesce. Once the hydrophobic side chains are close to each other, they are stabilized by electrical attractions known as **van der Waals interactions.** These weak attractions occur because the constant motion of electrons gives molecules a tiny asymmetry in charge that changes with time. If molecules get extremely close to each other, the minute partial charge on one molecule induces an opposite partial charge in the nearby molecule and causes an attraction. Although the attraction is very weak relative to covalent bonds or even hydrogen bonds, a large number of van der Waals interactions can occur in a polypeptide when many hydrophobic residues congregate. The result is a significant increase in stability. Such hydrophobic interactions are responsible for folding that results in the globular shape of many proteins.

3. Ionic bonds form between groups that have full and opposing charges, such as the ionized amino and carboxyl functional groups highlighted on the right in Figure 3.15a.
4. Hydrogen bonds form in several ways: between hydrogen atoms and the carboxyl group in the peptide-bonded backbone, and between hydrogen and atoms with partial negative charges in side chains.

With so many interactions possible between side chains and peptide-bonded backbones, it's not surprising that polypeptides vary in shape from rod-like filaments to ball-like masses (**Figure 3.15b**).

Quaternary Structure

The first three levels of protein structure involve individual polypeptides. But many proteins contain several distinct polypeptides that interact to form a single structure. The combination of polypeptides as subunits gives proteins a **quaternary structure.** The individual polypeptides may be held together by bonds or other interactions among R-groups or sections of their peptide backbones.

In the simplest case, a protein with quaternary structure can consist of just two subunits that are identical. The Cro protein found in a virus called bacteriophage λ ("lambda")

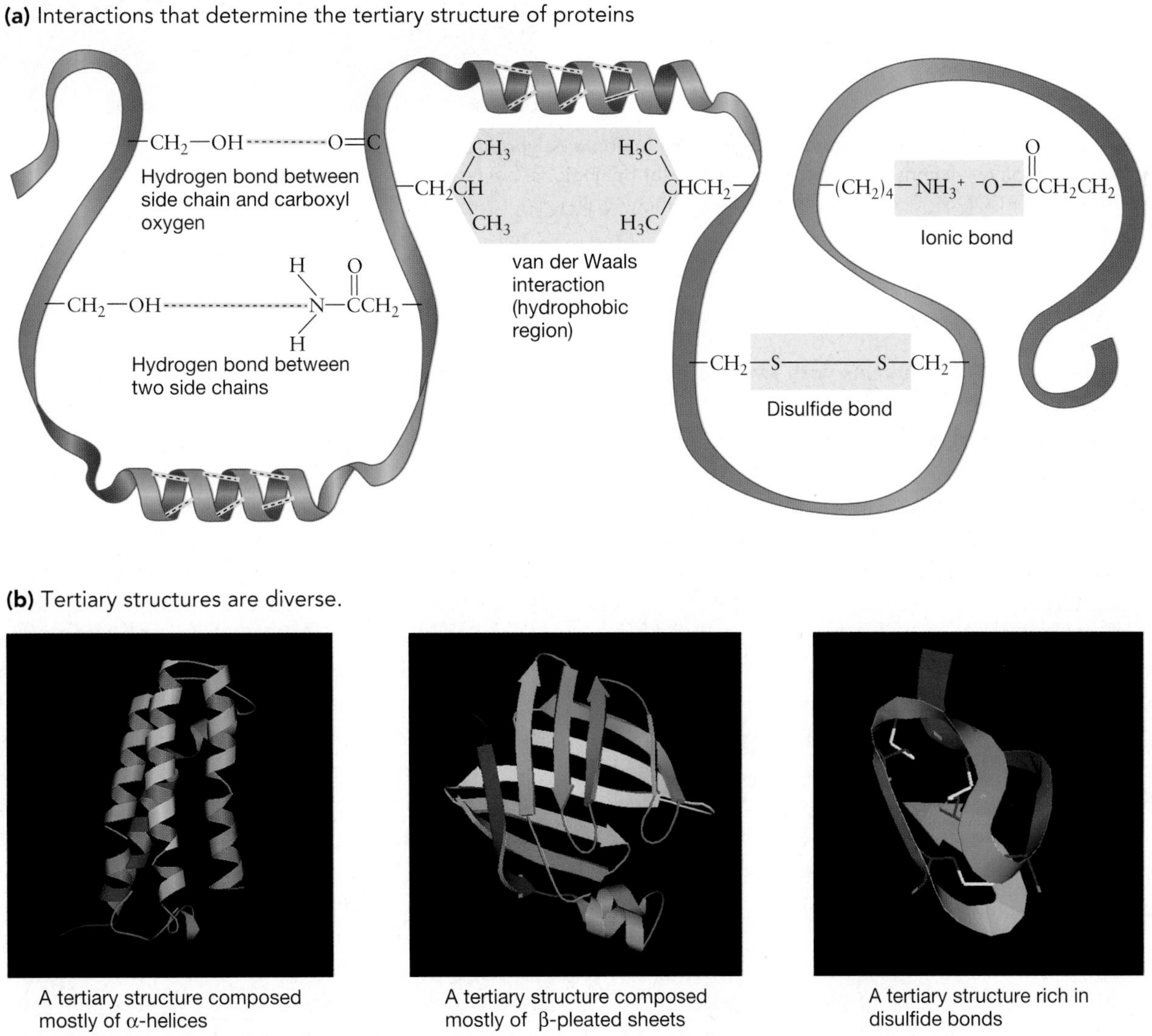

(a) Interactions that determine the tertiary structure of proteins
—CH₂—OH·········O=C
Hydrogen bond between side chain and carboxyl oxygen
—CH₂—OH·········N—CCH₂—
Hydrogen bond between two side chains
CH₃ H₃C —CH₂CH CHCH₂— CH₃ H₃C
van der Waals interaction (hydrophobic region)
—(CH₂)₄—NH₃⁺ ⁻O—CCH₂CH₂—
Ionic bond
—CH₂—S——S—CH₂—
Disulfide bond
(b) Tertiary structures are diverse.
A tertiary structure composed mostly of α-helices
A tertiary structure composed mostly of β-pleated sheets
A tertiary structure rich in disulfide bonds

FIGURE 3.15 Tertiary Structure of Proteins Results from Interactions Involving R-Groups
(a) Each protein has a unique overall shape called its tertiary structure. Tertiary structure is created by bonds and other interactions that cause proteins to fold in a precise way. **(b)** The tertiary structure of these proteins includes interactions between α-helices (indicated by coils) and β-pleated sheets (indicated by flat arrows, with the arrowhead at the carboxyl end). The polypeptide chains are color-coded so that you can follow the chain from one end (red) to the other (dark blue). EXERCISE In part (b): (Left) Label the four α-helices; (center) label the α-helix and one of the β-pleated sheets; (right) label the disulfide bonds, shown in yellow.

(a) Cro protein, a dimer

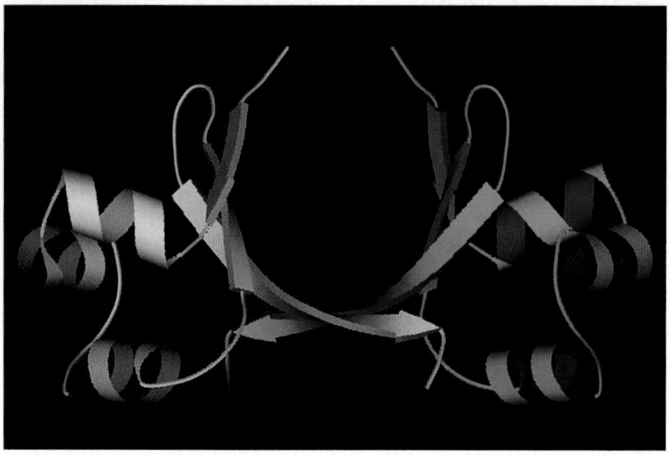

(b) Hemoglobin, a tetramer

FIGURE 3.16 Quaternary Structures of Proteins Is Created by Multiple Polypeptides
(a) The Cro protein is a dimer—it consists of two polypeptides. **(b)** Hemoglobin is a tetramer—it consists of four polypeptides. QUESTION What elements of secondary structure can you identify in each of the two proteins shown here?

is an example (**Figure 3.16a**). Proteins with two polypeptide subunits are called dimers ("two-parts"). More than two polypeptides can be linked into a single protein, however, and the polypeptides involved may be distinct in primary, secondary, and tertiary structure. Hemoglobin, for example, is a tetramer ("four-parts"). As **Figure 3.16b** shows, the completed protein consists of two copies of each of two different polypeptides.

Table 3.2 summarizes the four levels of protein structure, using hemoglobin as an example. The table and preceding discussion have two important messages. First, the combination of primary, secondary, tertiary, and quaternary levels of structure is responsible for the fantastic diversity of sizes and shapes observed in proteins. Second, most elements of protein structure are based on folding of polypeptide chains. Does this folding occur spontaneously? What happens if normal folding is disrupted?

Folding and Function

If you were able to synthesize one of the polypeptides in hemoglobin from individual amino acids, and if you placed the resulting chain in water, it would spontaneously fold into the shape of the tertiary structure shown in Table 3.2. This result seems counterintuitive. Because an unfolded protein has many more ways to move about, it has much higher entropy than the folded version. Folding is spontaneous in some cases, however, because the bonds and van der Waals interactions that occur make the folded molecule more stable energetically than the unfolded molecule. Thus, folding may release free energy.

Folding is also crucial to the function of a completed protein. This point was hammered home in a set of classic experiments by Christian Anfinson and colleagues during the 1950s. Anfinson studied a protein called ribonuclease that is found in many organisms. Ribonuclease breaks ribonucleic acid polymers apart. Anfinson found that the ribonuclease could be unfolded, or **denatured**, by treating it with compounds that break hydrogen bonds and disulfide bonds (**Figure 3.17**). The denatured ribonuclease was unable to function normally—it could no longer break apart nucleic acids. But when Anfinson removed the denaturing agents, the molecule refolded and began to function normally again. These experiments confirmed that ribonuclease folds spontaneously and that folding is essential for normal function.

More recent work has shown that in cells, folding is often facilitated by specific proteins called **molecular chaperones.** Many molecular chaperones belong to a family of molecules called the heat-shock proteins. These compounds are produced in large quantities after cells experience high temperatures or other treatments that make proteins lose their tertiary structure. Heat-shock proteins speed the refolding of other proteins into their normal shape after denaturation has occurred.

To summarize, a protein's function depends on its shape. In most cases, the final shape of a protein depends on folding. To

TABLE 3.2 A Summary of Protein Structure

Level	Description	Stabilized by	Example: Hemoglobin
Primary	The sequence of amino acids in a polypeptide	Peptide bonds	Gly–Ser–Asp–Cys
Secondary	Formation of α-helices and β-pleated sheets in a polypeptide	Hydrogen bonding between groups along the peptide-bonded backbone	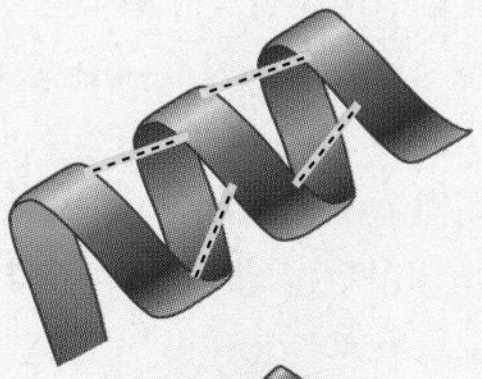
Tertiary	Overall three-dimensional shape of a polypeptide (The model on the right shows one of hemoglobin's subunits. The black and red atoms are in the heme group that carries oxygen; they are not part of the protein itself.)	Bonds and other interactions between R-groups, or between R-groups and the peptide-bonded backbone	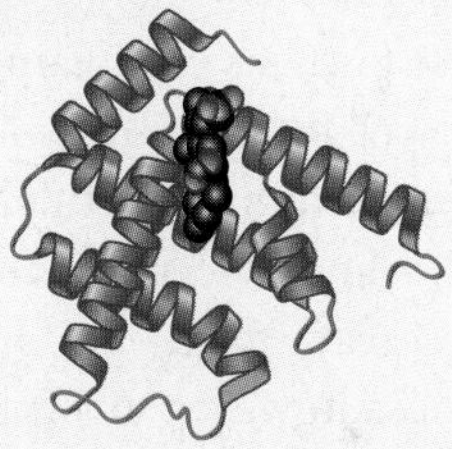
Quaternary	Shape produced by combinations of polypeptides. (The model on the right shows hemoglobin, which consists of four polypeptides.)	Bonds and other interactions between R-groups, and between peptide backbones of different polypeptides	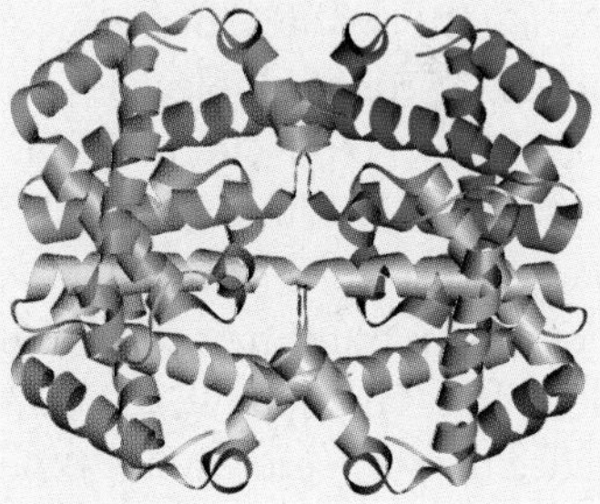

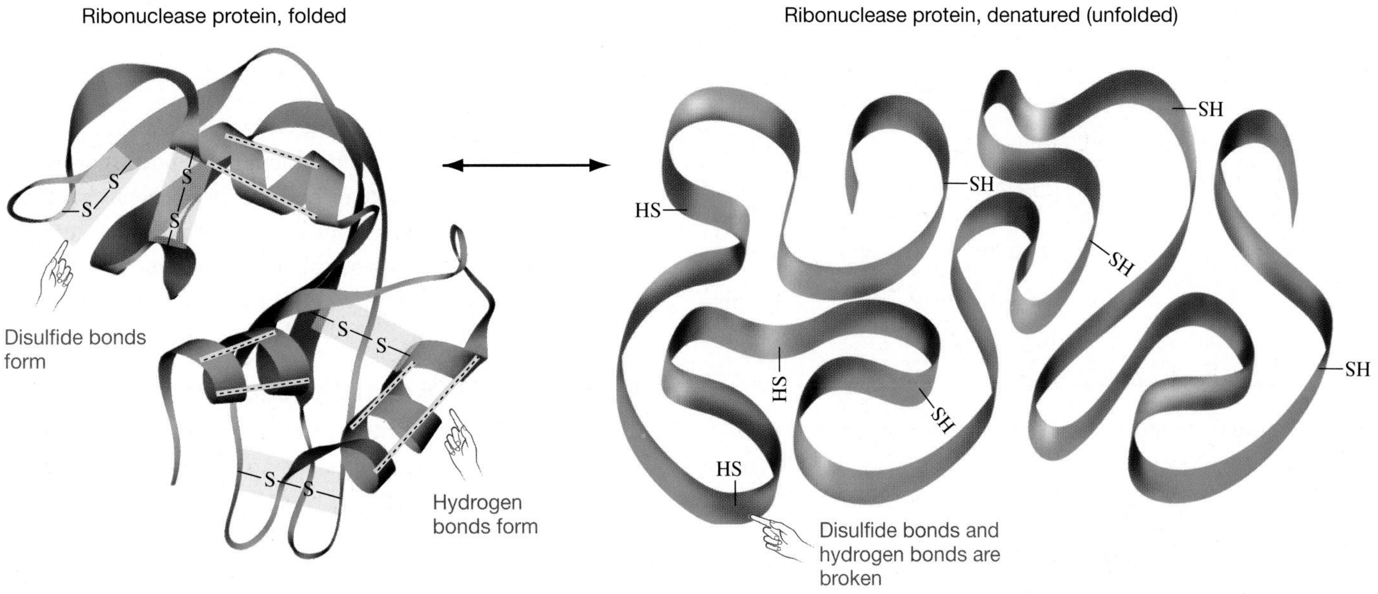

FIGURE 3.17 Proteins Fold into Their Normal, Active Shape
(a) The tertiary structure of ribonuclease is defined primarily by four disulfide bonds, shown in yellow. **(b)** When the disulfide bonds and various non-covalent bonds are broken, the protein denatures (unfolds).

BOX 3.2 Prions

Over the past several decades, evidence has accumulated that certain proteins can act as infectious, disease-causing agents. The proteins involved are called **prions**, or proteinaceous infectious particles. Where do these molecules come from, and how do they work?

According to the prion hypothesis developed by Stanley Prusiner, infectious proteins are improperly folded forms of normal proteins that are present in healthy individuals. The infectious and normal forms do not necessarily differ in amino acid sequence, however. Instead, their shapes are radically different. Further, the infectious form of a protein can induce normal protein molecules to change their shape to the altered form.

Figure 3.18 illustrates the shape differences observed in the normal and infectious forms of the first prion to have been described. The molecule in Figure 3.18a is called the prion protein (PrP) and is a normal component of mammalian cells. Mutant versions of this protein, like the one in Figure 3.18b, are found in a wide variety of species and cause a family of diseases known as the spongiform encephalopathies—literally, "sponge-brain-illnesses." Hamsters, cows, goats, and humans afflicted with these diseases undergo massive degeneration of the brain. Cattle suffer from "mad cow disease"; sheep and goats acquire scrapie (so called because the animals itch so badly that they scratch off their wool or hair); humans develop kuru or Creutzfeldt-Jakob disease. Although some spongiform encephalopathies can be inherited, in many cases the disease is transmitted when individuals eat tissues containing the infectious form of PrP. All of the prion illnesses are fatal.

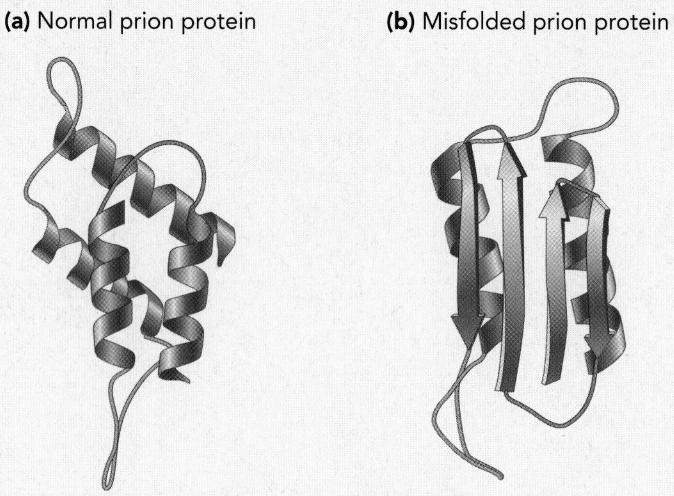

FIGURE 3.18 Prions Are Improperly Folded Proteins
Ribbon model of **(a)** a normal prion protein and **(b)** the misfolded form that causes mad cow disease in cattle. Secondary structure is represented by coils (α-helices) and arrows (β-pleated sheets).

drive this point home, **Box 3.2** investigates the way a change in the shape of the prion ("PREE-on") protein of mammals converts normal molecules into abnormally folded proteins. Misfolded prions cause the brains of humans and other mammals to disintegrate.

Given how proteins are put together and how they are folded, let's explore next what they do. In terms of diversity in function, no other type of molecule even begins to compare with proteins.

✓CHECK YOUR UNDERSTANDING

Proteins have up to four levels of structure. Primary structure consists of the sequence of amino acids. Secondary structure results from interactions between atoms in the peptide-bonded backbone of the same polypeptide, and yields structures such as α-helices and β-pleated sheets. Tertiary structure is a consequence of bonds or other interactions between R-groups or between R-groups and the peptide-bonded backbone of a polypeptide, and produces distinctive folds. Quaternary structure occurs when multiple polypeptides interact to form a single protein. You should be able to draw a hypothetical protein and label the parts that produce each level of structure.

3.4 What Do Proteins Do?

Proteins are the workhorse molecules in organisms. You've already been introduced to hemoglobin, which transports oxygen molecules around the human body. Other proteins protect the body by attacking disease-causing bacteria and viruses; still others make movement possible or form structures that give cells their shape. Let's take a closer look.

Proteins Have Diverse Functions in Cells

Table 3.3 provides a brief overview of protein functions. The table's message is simple: Proteins are crucial to most tasks required for cells to exist. The diverse functions of proteins include the following:

- *Defense.* Proteins called antibodies and complement proteins attack and destroy viruses and bacteria that cause disease.
- *Movement.* Motor proteins and contractile proteins are responsible for moving the cell itself, or for moving large molecules and other types of cargo inside the cell.

- *Catalysis.* Proteins called enzymes catalyze, or speed up, chemical reactions, many of which would not otherwise be able to proceed.
- *Signaling.* Proteins called peptide hormones bind to receptor proteins on particular cells. In response, the activity of the cell with the receptor protein changes. In this way, proteins are involved in carrying and receiving signals from cell to cell inside the body.
- *Structure.* Structural proteins give cells mechanical support. They also create structures such as fingernails and hair.
- *Transport.* Proteins are responsible for allowing particular molecules to enter or exit cells, and for carrying specific compounds throughout the body.

Most of these aspects of protein function will be explored in much greater detail elsewhere in the text. In many if not all cases, the function of the protein involved is closely correlated with its structure. Here let's focus on what may be the most fundamental of all protein functions: **catalysis**, which speeds the chemical reactions that allow cells to stay alive, grow, and reproduce.

An Introduction to Catalysis

In Chapter 2 we reviewed several fundamental ideas about chemical reactions, including how changes in potential energy and entropy determine whether a reaction is spontaneous. Even spontaneous chemical reactions are not necessarily fast, however. Oxidation of iron—the reaction known as rusting—is exothermic and spontaneous, but it occurs slowly. To appreciate why proteins are so important in the life of a cell, it is essential to get a deeper understanding of the factors that limit the speed at which chemical reactions occur.

Reaction rates depend on how the chemical bonds involved are broken and re-formed. Chapter 2 introduced the idea that molecules must collide with a specific orientation and a specific amount of energy for old bonds to break and new bonds to form. This is because the electrons involved in the reactions must interact. But these electrons repel one another as they come into contact. In many cases, for the reaction to proceed, the kinetic energy of the collision must be large enough to overcome this repulsion. If the energy is sufficiently large, then the collision creates a combination of old and new bonds called a **transition state.** The amount of free energy required to reach the transition state is called the **activation energy** of the reaction. The more unstable the transition state, the higher the activation energy and the less likely the reaction is to proceed quickly.

Figure 3.19 pulls these ideas together by showing the changes in free energy that take place during the course of a chemical reaction. In this graph, ΔG indicates the overall change in free energy in the reaction—that is, the energy of the products minus the energy of the reactants. In this case the products have lower potential energy than the reactants, meaning that the reaction is exothermic. But because the activation energy for this reaction, symbolized by E_a, is high, the reaction would proceed slowly.

Reaction rates, then, depend on both the kinetic energy of the reactants and the activation energy of the particular reaction—meaning the free energy of the transition state. If the kinetic energy of the participating molecules is high, then molecular collisions are likely to result in completed reactions. The kinetic energy of molecules, in turn, is a function

WEB TUTORIAL 3.2
Activation Energy and Enzymes

TABLE 3.3 A Summary of Protein Functions

Protein type	Role in cell or organism
Antibodies and complement proteins	Defense—destruction of disease-causing viruses and bacteria
Contractile proteins and motor proteins	Movement
Enzymes	Catalyze chemical reactions
Peptide hormones	Act as signals that help coordinate the activities of many cells
Receptor proteins	Receive chemical signals from outside cell and initiate response
Structural proteins	Provide support for cells and tissues; form structures such as hair, feathers, cocoons, and spider webs
Transport proteins	Move substances across cell membrane; move substances throughout body

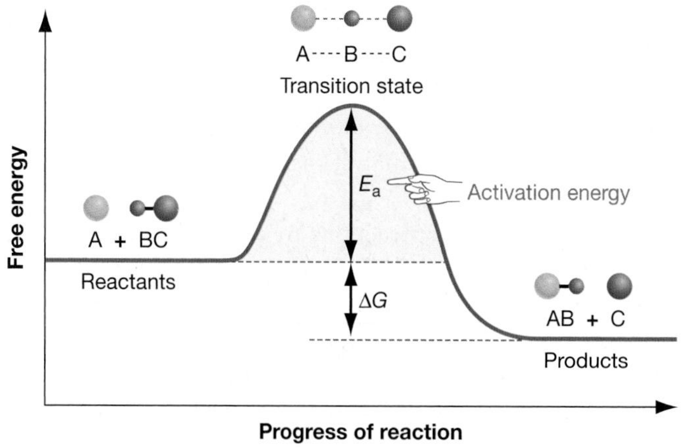

FIGURE 3.19 Changes in Free Energy during a Chemical Reaction The changes in free energy that occur over the course of a hypothetical reaction between an atom A and a molecule containing atoms B and C. The overall reaction would be written as A + BC → AB + C. E_a is the activation energy of the reaction, and ΔG is the overall change in free energy. EXERCISE This graph illustrates an exergonic reaction. Draw the same type of graph for an endergonic reaction.

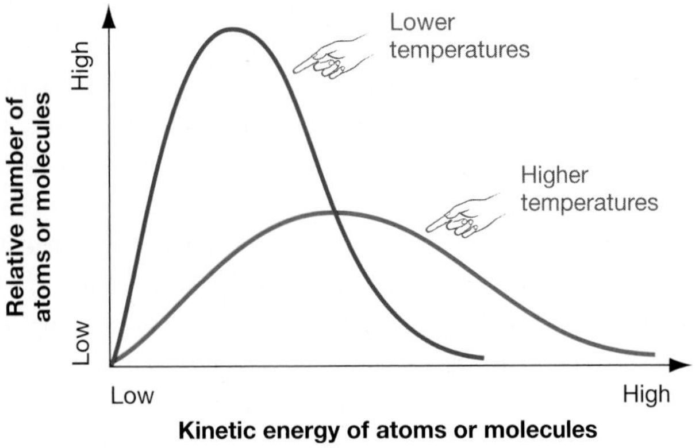

FIGURE 3.20 Atoms amd Molecules Have More Kinetic Energy at High Temperatures than at Low Temperatures
EXERCISE Add a vertical line about two-thirds of the way along the horizontal axis. Label this line "Activation energy for reaction without catalyst." At low and high temperatures, which molecules have enough kinetic energy for this reaction to proceed?

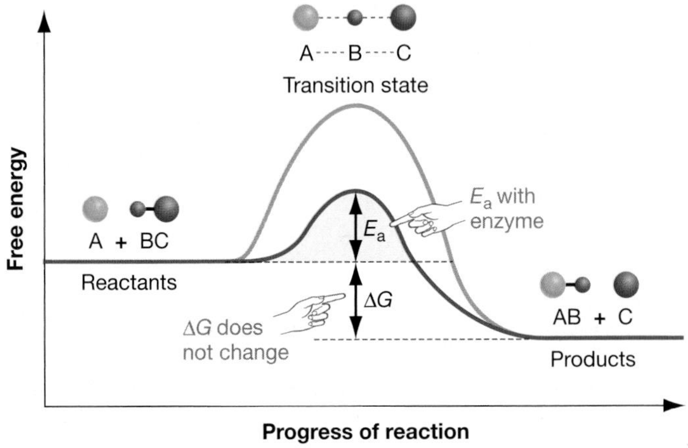

FIGURE 3.21 A Catalyst Changes the Activation Energy of a Reaction
The energy profile for the same reaction diagrammed in Figure 3.19, but with a catalyst present. Even though the energy barrier to the reaction, E_a, is much lower, ΔG does not change. QUESTION Can a catalyst make a nonspontaneous reaction occur spontaneously? EXERCISE Go back to Figure 3.20 and add another vertical line about one-third of the way along the horizontal axis. Label this line "Activation energy for reaction with catalyst." Label the molecules that have enough energy to undergo a reaction with a catalyst but not without a catalyst.

of their temperature (**Figure 3.20**). This is why chemical reactions tend to proceed faster at higher temperatures. But if the activation energy of a particular reaction is also high, then collisions are less likely to result in completed reactions.

In many cases, the electrons in the transition-state molecule can be stabilized when they interact with another ion, atom, or molecule. When this occurs, the activation energy required for the reaction drops and the reaction rate increases. A substance that lowers the activation energy of a reaction and increases the rate of the reaction is called a **catalyst**. A catalyst is not consumed in a chemical reaction, even though it participates in the reaction. The composition of a catalyst is exactly the same after the reaction as it was before.

Figure 3.21 diagrams how catalysts lower the activation energy for a reaction by lowering the free energy of the transition state. Note that the presence of a catalyst does not affect the overall energy change, ΔG, or change the energy of the reactants or the products. A catalyst changes only the free energy of the transition state.

Proteins that catalyze reactions are called **enzymes**. Most enzymes are quite specific in their activity—they catalyze just a single reaction by lowering the activation energy that is required. It's not difficult to appreciate why this is important. Most of the important reactions in biology do not occur at all, or else proceed at imperceptible rates, without a catalyst. In contrast, a single molecule of the enzyme carbonic anhydrase can catalyze over 1,000,000 reactions *per second*. It's not unusual for enzymes to speed up reactions by a factor of a million; some enzymes make reactions go many *trillions* of times faster than they would without a catalyst.

Enzymes are important because they speed up the chemical reactions that are required for life. Now the question is, how do enzymes do what they do?

How Do Enzymes Work?

The initial hypothesis for how enzymes work was proposed by Emil Fischer in 1894. According to Fischer's **lock-and-key model**, enzymes are rigid structures analogous to a lock. The keys are reactant molecules, called **substrates**, that fit into the lock and then react.

Several important ideas in this model have stood the test of time. For example, Fischer was correct in proposing that enzymes bring substrates together in a precise orientation that makes reactions more likely. His model also accurately explained why most enzymes can catalyze only one specific reaction. Enzyme specificity is a product of the geometry and chemical properties of the sites where substrates bind.

As researchers began to test and extend Fischer's model, the locations where substrates bind and react became known as the enzyme's **active site**. When techniques for solving the three-dimensional structure of enzymes became available, it turned out that enzymes tend to be very large relative to substrates and roughly globular, and that the active site is in a cleft or cavity within the globular shape. The enzyme hexokinase, which is at work in most cells of your body now, is a good example. (Many enzymes have names that end with *–ase*.) As the left-hand side of **Figure 3.22** shows, the active site in hexokinase is a small notch in an otherwise large, crescent-shaped enzyme.

As knowledge of enzyme action grew, however, Fischer's model was modified. Perhaps the most important change was based on the realization that enzymes are not rigid and static, but flexible and dynamic. In fact, many enzymes undergo a

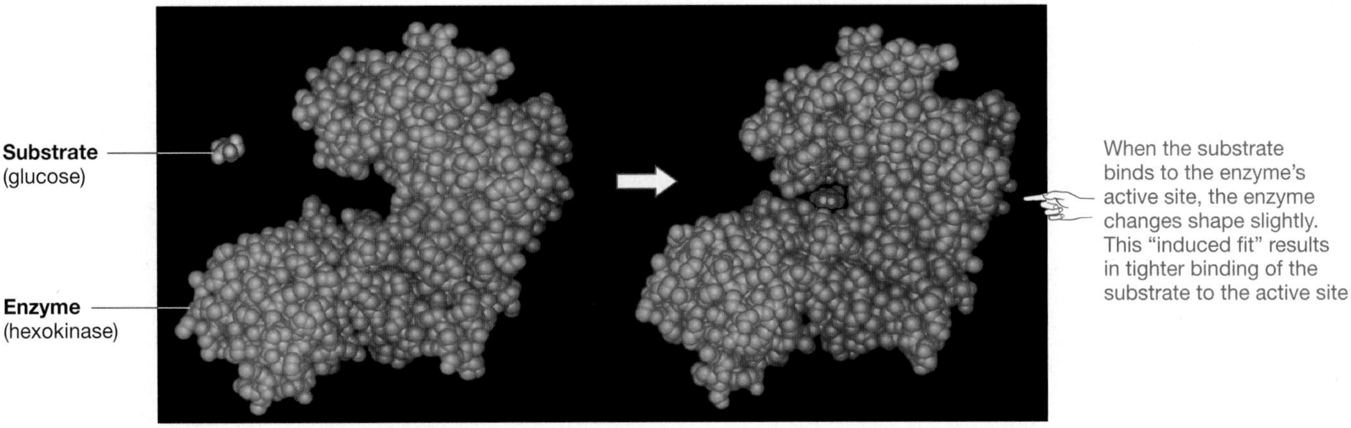
Substrate
(glucose)
Enzyme
(hexokinase)
When the substrate binds to the enzyme's active site, the enzyme changes shape slightly. This "induced fit" results in tighter binding of the substrate to the active site

FIGURE 3.22 Reactant Molecules Bind to Specific Locations in an Enzyme
The reactant molecule, shown in red, fits into a precise location, called the active site, in the green enzyme. In the enzyme shown here and in many others, the binding event causes the protein to change its shape.
EXERCISE Label the active site in the enzyme on the left.

significant change in shape, or conformation, when reactant molecules bind to the active site. This conformational change, called an **induced fit**, is noticeable in the hexokinase molecule on the right of Figure 3.22. As hexokinase binds its substrate—the sugar glucose—the enzyme rocks forward over the active site.

In addition, recent research has clarified the nature of Fischer's key. When one or more substrate molecules enter the active site, they are held in place through hydrogen bonding or other electrical interactions with amino acids in the active site. Once the substrate is bound, one or more R-groups in the active site come into play. The degree of interaction between the substrate and enzyme increases and reaches a maximum when the transition state is formed. Thus, Fischer's key is actually the transition state.

Recall a key point: Interactions with R-groups stabilize the transition state and thus lower the activation energy required for the reaction to proceed. At the atomic level, R-groups that line the active site may form short-lived covalent bonds that assist with the transfer of atoms or groups of atoms from one reactant to another. More commonly, the presence of acidic or basic R-groups allows the reactants to lose or gain a proton more readily.

Figure 3.23 provides a specific example of how substrates interact with an active site. The panels show a substrate fitting into the active site of the enzyme called ribonuclease. Notice that once the substrate has bound, R-groups in the active site interact with the substrate to form the transition state and get the reaction under way.

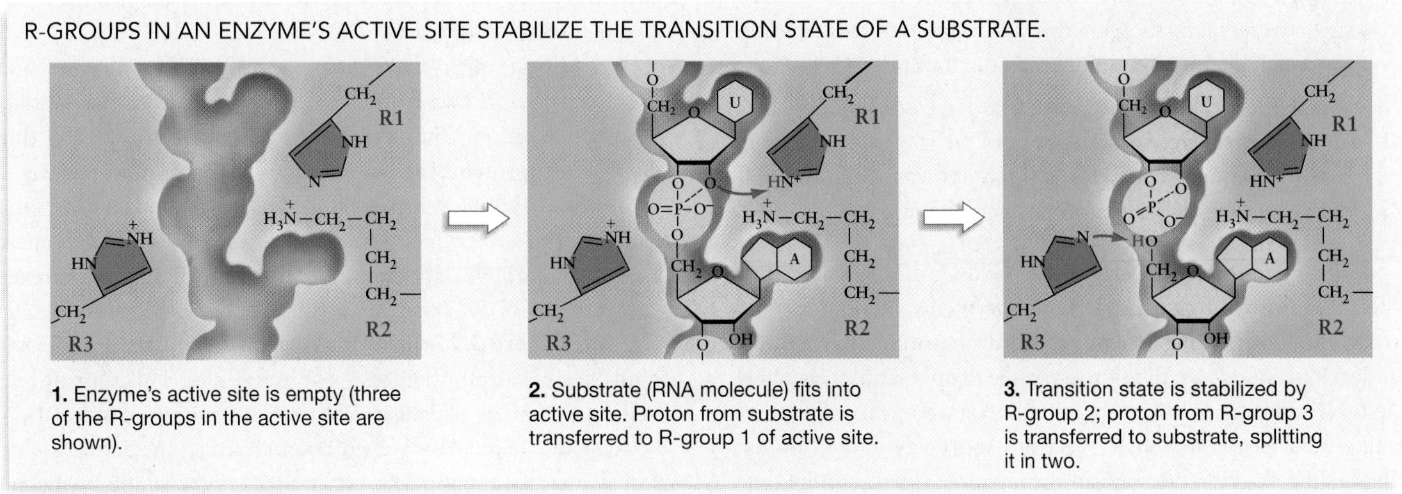
R-GROUPS IN AN ENZYME'S ACTIVE SITE STABILIZE THE TRANSITION STATE OF A SUBSTRATE.
R1
R2
R3
1. Enzyme's active site is empty (three of the R-groups in the active site are shown).
2. Substrate (RNA molecule) fits into active site. Proton from substrate is transferred to R-group 1 of active site.
3. Transition state is stabilized by R-group 2; proton from R-group 3 is transferred to substrate, splitting it in two.

FIGURE 3.23 R-groups in an Enzyme's Active Site Stabilize the Transition State
Part of the reaction sequence that occurs when the enzyme ribonuclease catalyzes a reaction that cuts a substrate in two. Note how precisely the substrate fits into the active site and how specific R-groups in the enzyme are involved in key steps in the reaction.

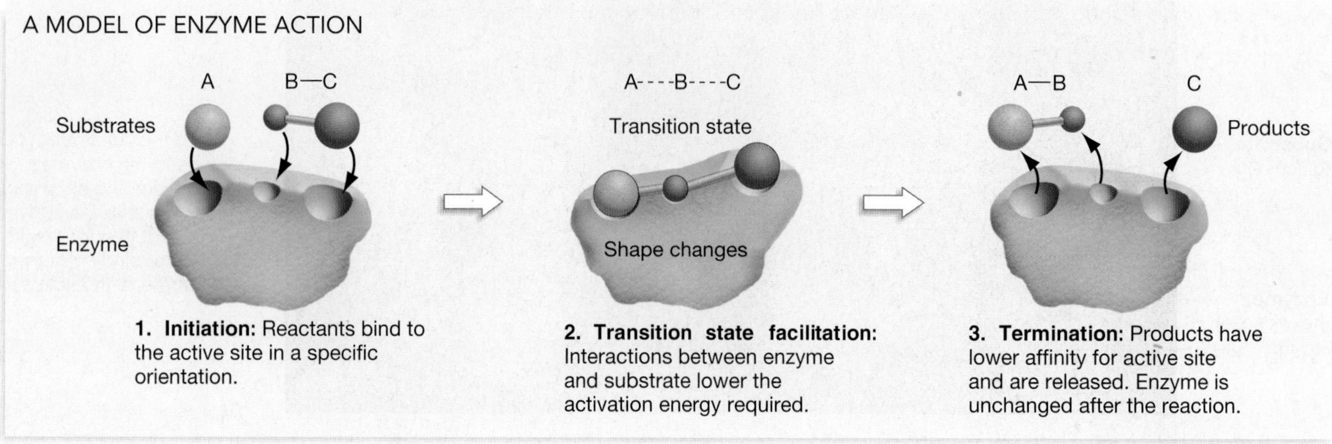

FIGURE 3.24 Enzyme Action Can Be Analyzed as a Three-Step Process

Whatever the mechanism, enzymes induce the formation of the transition state and thereby increase reaction rates. The products of the reaction have a much lower affinity for the active site than do either the reactants or the transition state, however. As a result, they are released from the enzyme once they form.

Figure 3.24 summarizes these principles. Enzyme catalysis can be analyzed as a three-step process:

1. **Initiation.** Instead of reactants occasionally colliding in a random fashion, enzymes orient reactants precisely as they bind at specific locations within the active site.
2. **Transition state facilitation.** The act of binding induces the formation of the transition state. In some cases the transition state is stabilized by a conformational change in the enzyme. The interaction between the substrate and R-groups in the enzyme's active site lowers the activation energy required for the reaction. Inside a catalyst's active site, more reactant molecules have sufficient kinetic energy to reach this lowered activation energy. Thus, the catalyzed reaction proceeds much more rapidly than the uncatalyzed reaction.
3. **Termination.** The reaction products have considerably less affinity for the active site than does the transition state. As a result, binding ends. The enzyme returns to its original conformation, and the products are released.

Fischer's model inspired intensive and fruitful research into the mechanism of enzyme action. Enzymes speed the rate of reactions by orienting reactants and decreasing the reaction's activation energy. Activation energies drop because enzymes destabilize bonds in the reactant, stabilize the transition state, make acid-base and redox reactions more favorable, and/or change the reaction mechanism through a covalent bonding interaction. Enzyme specificity is a function of the active site's shape and the chemical properties of the R-groups present in the active site. In enzymes, as in many molecules, structure dictates function.

Do Enzymes Act Alone? Atoms or molecules that are not part of the enzyme's primary structure are often required for the enzyme to function normally. These **enzyme cofactors** can be either (1) metal ions such as Zn^{2+} (zinc) or Mg^{2+} (magnesium), or (2) small organic molecules called **coenzymes.** In many cases, the cofactor binds to the active site and is thought to play a key role in stabilizing the transition state during the reaction. As a result, the presence of the cofactor is essential for catalysis. To drive this point home, consider that many of the vitamins in your diet are required for the production of enzyme cofactors. Vitamin deficiencies cause enzyme-cofactor deficiencies. Lack of cofactors, in turn, disrupts normal enzyme function and causes disease. For example, thiamine (vitamin B_1) is required for the production of an enzyme cofactor called thiamine pyrophosphate, which is required by three different enzymes. Lack of thiamine in the diet dramatically reduces the activity of these enzymes and causes an array of nervous system and heart disorders collectively known as beriberi.

In addition, most enzymes are regulated by molecules that are not part of the enzyme itself. An enzyme is active or inactive, depending on the presence or absence of molecules that change the protein's structure in some way. In some cases, catalysis is inhibited when a molecule that is similar in size and shape to the substrate(s) binds to the active site. This event is called **competitive inhibition,** because the molecule involved competes with the substrate(s) for access to the enzyme's active site (**Figure 3.25a**). In other cases, a regulatory molecule binds at a location other than the active site. This type of regulation is called **allosteric** ("different-structure") **regulation,** because the molecule involved does not affect the active site directly. Instead, the binding event changes the shape of the enzyme in a way that makes the active site accessible or inaccessible (**Figure 3.25b**). Later chapters will provide detailed examples of how regulatory molecules change the activity of specific enzymes.

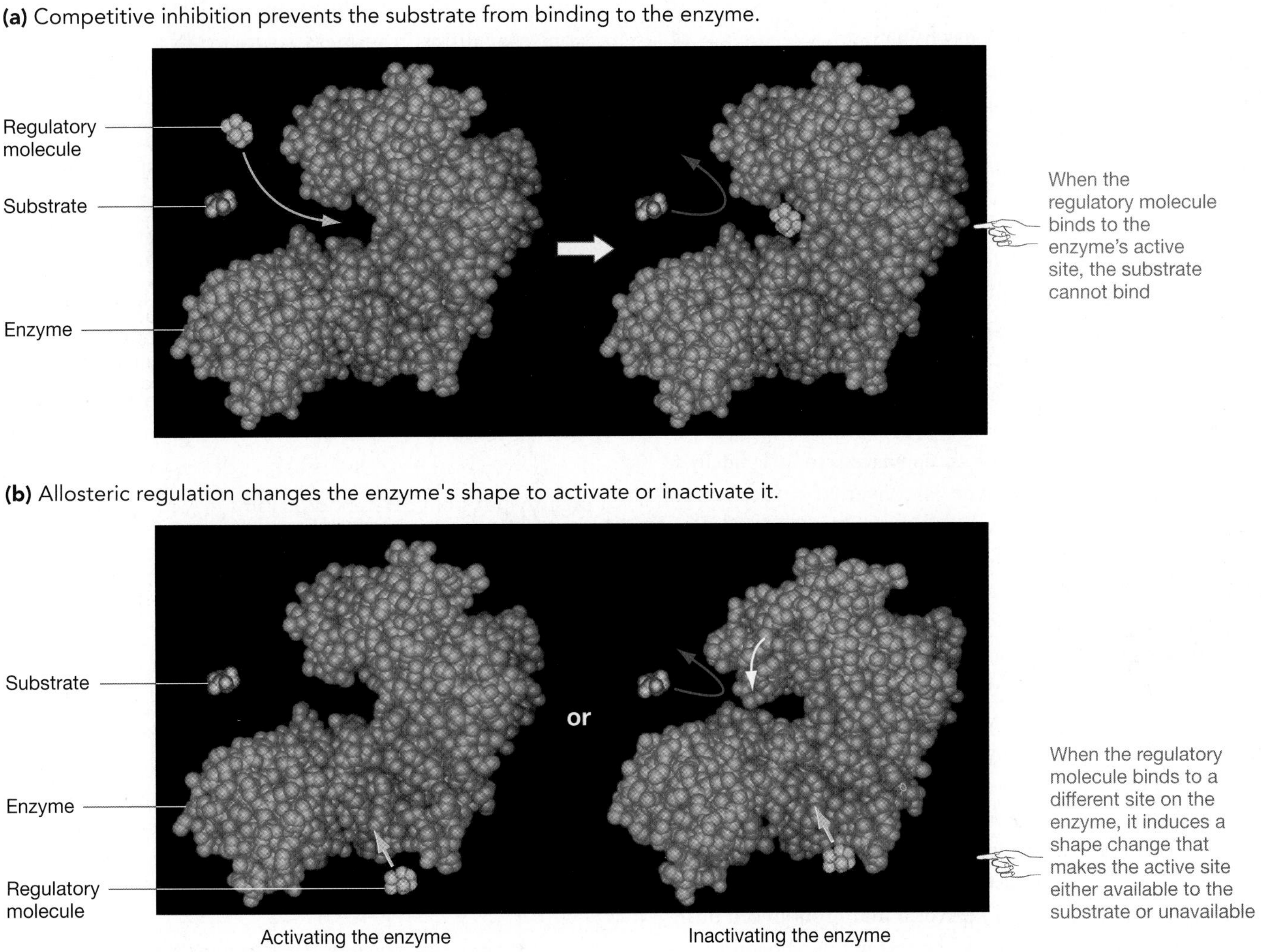
(a) Competitive inhibition prevents the substrate from binding to the enzyme.
Regulatory molecule
Substrate
Enzyme
When the regulatory molecule binds to the enzyme's active site, the substrate cannot bind
(b) Allosteric regulation changes the enzyme's shape to activate or inactivate it.
Substrate
Enzyme
Regulatory molecule
or
When the regulatory molecule binds to a different site on the enzyme, it induces a shape change that makes the active site either available to the substrate or unavailable
Activating the enzyme
Inactivating the enzyme

FIGURE 3.25 An Enzyme's Activity Is Precisely Regulated
Enzymes are turned on or off when specific molecules bind to them. **(a)** In competitive inhibition, binding occurs at the active site—the regulatory molecule competes with substrate(s). **(b)** In allosteric regulation, the regulatory molecule binds elsewhere on the enzyme and causes a shape change that activates or inactives the protein.

What Limits the Rate of Catalysis? For several decades after Fischer's model was published, most research on enzymes focused on rates of enzyme action, or what biologists call enzyme kinetics. Researchers observed that, when the amount of product produced per second—meaning the speed of the reaction—is plotted as a function of substrate concentration, a graph like that shown in **Figure 3.26** results. When the substrate concentrations are low, the reaction speed increases in a linear fashion. As substrate concentrations increase, however, the increase in speed begins to slow. Eventually the reaction rate plateaus at a maximum speed. This pattern is in striking contrast to the situation for uncatalyzed reactions, in which reaction speed tends to show a continuing linear increase with substrate concentration. The "saturation kinetics" of enzyme-catalyzed reactions were taken as strong evidence that the enzyme-substrate complex postulated by Fischer actually exists. The idea was that at some point, active sites cannot accept substrates any faster, no matter how large the concentration

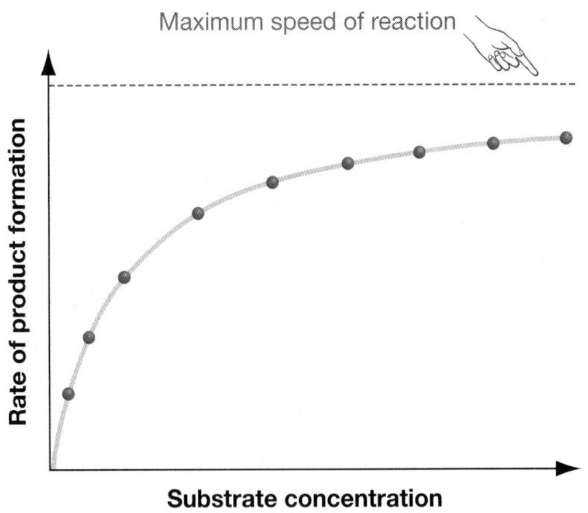
Maximum speed of reaction
Rate of product formation
Substrate concentration

FIGURE 3.26 Kinetics of an Enzyme-Catalyzed Reaction
The general shape of this curve is characteristic of enzyme-catalyzed reactions.

of substrates gets. Stated another way, reaction rates level off because all the available enzyme is being used.

It's important to recognize, however, that even though the general shape of the curve is similar for all enzymes, its position varies. In enzymes that have a high affinity for the substrate, the curve is shifted to the left. This means that substrate concentrations do not have to be very high for these enzymes to function at maximum speed. In enzymes with a lower affinity for the substrate, though, the curve is shifted to the right. This means that substrate concentrations have to be much higher before an enzyme is able to catalyze reactions at its maximal rate.

The processing of ethanol—the active ingredient in alcoholic beverages—in the human liver furnishes an example of these concepts. Ethanol is a poison broken down in two enzyme-catalyzed steps. First ethanol is converted to acetaldehyde (CH_3CHO). Then acetaldehyde is converted to acetic acid (CH_3COOH)—a harmless molecule that gives vinegar its bite. Humans have two versions of the enzyme that catalyzes the second reaction. One version of the enzyme works quickly and the other much more slowly. In some people, the fast-working version of the enzyme has a single amino acid change in its primary sequence that affects the enzyme's active site. In these individuals, the fast-working version of the enzyme is inactivated and only the slow-working version functions normally. Acetaldehyde must be at a high concentration before the slow-working version of the enzyme works at maximum speed. In these individuals, acetaldehyde builds up to high enough concentrations in the bloodstream that uncomfortable symptoms result, including a rapid heartbeat and pronounced flushing of the skin. People who lack the fast-acting version of the enzyme have virtually zero tolerance for alcoholic beverages. Imbibing even a small amount of ethanol makes them feel ill.

How Do Physical Conditions Affect Enzyme Function?

Given that an enzyme's structure is critical to its function, it's not surprising that an enzyme's activity is sensitive to conditions that alter its structure. In particular, the activity of an enzyme often changes drastically as a function of temperature and pH. Temperature affects the movement of the enzyme; pH affects the makeup and charge of amino acid side chains with carboxyl or amino groups, and the active site's ability to participate in proton-transfer or electron-transfer reactions.

Do data support these assertions? **Figure 3.27a** shows how the activity of an enzyme changes as a function of temperature. Data for the enzyme glucose-6-phosphate, which is actively producing energy in your cells right now, are shown for two species of bacteria. Note that in both of the bacterial species illustrated, the enzyme has a distinct optimum—a temperature at which it functions best. One of the bacterial species lives inside your gut, where the temperature is about 40°C, while the other lives in hot springs, where temperatures can be close to 100°C. The temperature optimum for the enzyme reflects these environments. The two types of bacteria have different versions of the enzyme—versions that differ in primary structure. Natural selection has favored different structures that have different functions. The enzymes are adaptations that allow each species to thrive at different temperatures.

Figure 3.27b makes the same point for pH. The enzyme in this graph, called chitinase, protects bacterial cells by digesting a molecule found in the cell walls of fungi that eat bacteria. The data come from a species of bacterium that lives in acidic pools and a bacterial species that lives in the soil under palm trees. The organism that thrives in an acidic environment has a version of the enzyme that performs best at low pH; the organism that lives near palms has a version of the enzyme that functions best near neutral pH. Each enzyme is sensitive to changes in

(a) Enzymes from different organisms may function best at different temperatures.

(b) Enzymes from different organisms may function best at different pHs.

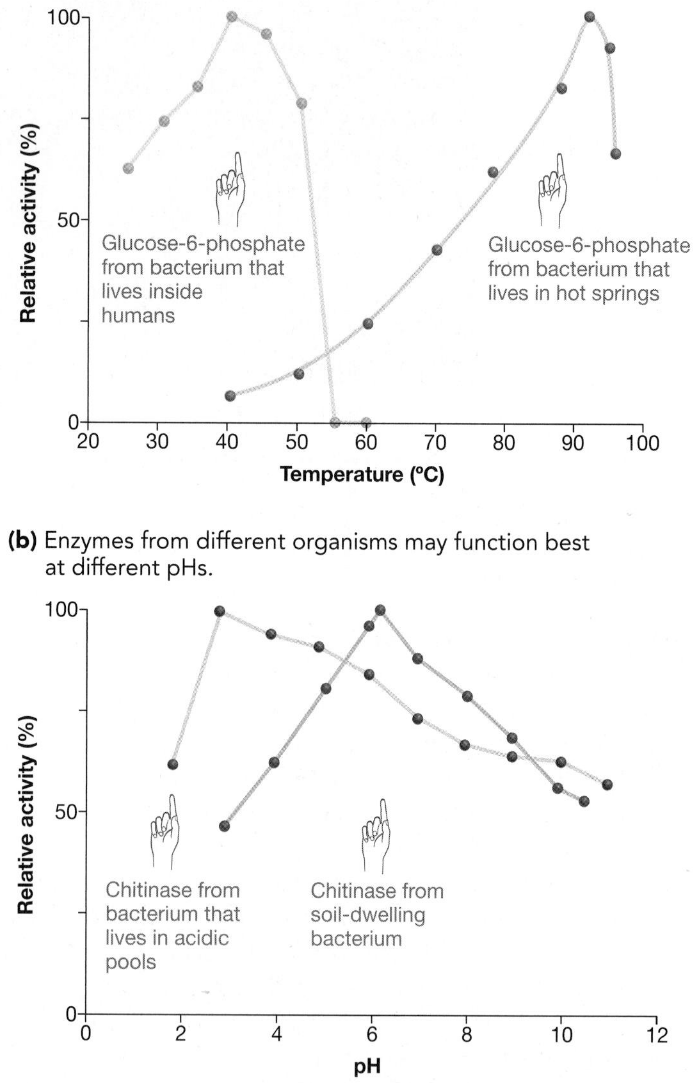

FIGURE 3.27 Enzymes Have an Optimal Temperature and pH
Enzymes are sensitive to changes in temperature and pH. Further, the structures of the enzymes found in a particular organism allow that organism to function well at the **(a)** temperature and **(b)** pH of its environment.

pH, but each species' version of the enzyme has a structure that allows it to function best at the pH of its environment.

How Does ATP Drive Endergonic Reactions? Enzyme catalysis explains why the reactions required for life can proceed quickly. But it is critical to realize that the presence of an enzyme does not change the potential energy or entropy of the reactants or products. Recall from Figure 3.19 that enzymes change only the free energy of the transition state—not the ΔG for the reaction. How, then, do endergonic reactions occur? Answering this question is important for two reasons: (1) Most reactions that take place inside you and in the organisms around you are endergonic; and (2) the vast majority of reactions required for chemical evolution to proceed are endergonic.

The essence of chemical evolution is the production of larger and more complex molecules from simpler ones. The essence of life is the maintenance of highly ordered structures and the ability to grow and reproduce. To understand how these endergonic processes occur, let's return to a reaction that was summarized briefly earlier in the chapter: the production of polypeptides from amino acid monomers, by incubating the reactants on tiny mineral particles.

Polymerization reactions are endergonic because polymers have higher potential energy and lower entropy than do monomers. Recall from Section 3.2 that adsorption onto minerals protected the polypeptides from reactions that break polymers apart. But how was the reaction that first led to a higher free-energy state possible? How were researchers able to produce polypeptides from amino acids?

The answer is that the investigators added chemical energy to the reactants. More specifically, they added **phosphate groups** (PO_4^{3-}) to the amino acids in the reaction mix. The addition of a phosphate group adds two negative charges to an amino acid and raises its free energy enough to make the polymerization reaction proceed. This mechanism is occurring in your cells right now. Many reactions that would otherwise be endergonic, including the polymerization of amino acids into proteins, are being driven by the addition of a phosphate group to a reactant molecule. In your cells, the phosphate group comes from a molecule called **adenosine triphosphate**, or **ATP**. As **Figure 3.28a** shows, ATP contains three phosphate groups. The mechanism works as follows: ATP reacts with the substrate to produce a *phosphorylated* molecule—a molecule that contains a phosphate group—and an **adenosine diphosphate**, or **ADP** molecule. ADP has two phosphate groups rather than three. The reaction that adds a phosphate group to a substrate, called **phosphorylation**, is exergonic (**Figure 3.28b**). Next the "activated" substrate—activated in the sense that it now contains a phosphate group and has high free energy—reacts and forms the product molecule(s). In some cases, the enzyme that

(a) ATP has 3 phosphate groups and high free energy.

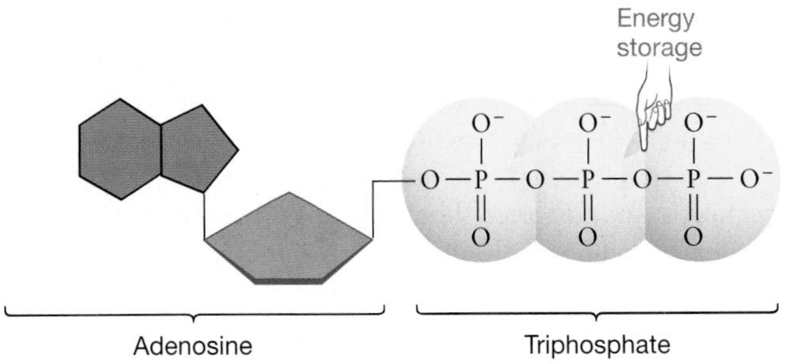

(b) Coupling of exergonic and endergonic reactions: How can monomer B be added to polymer A?

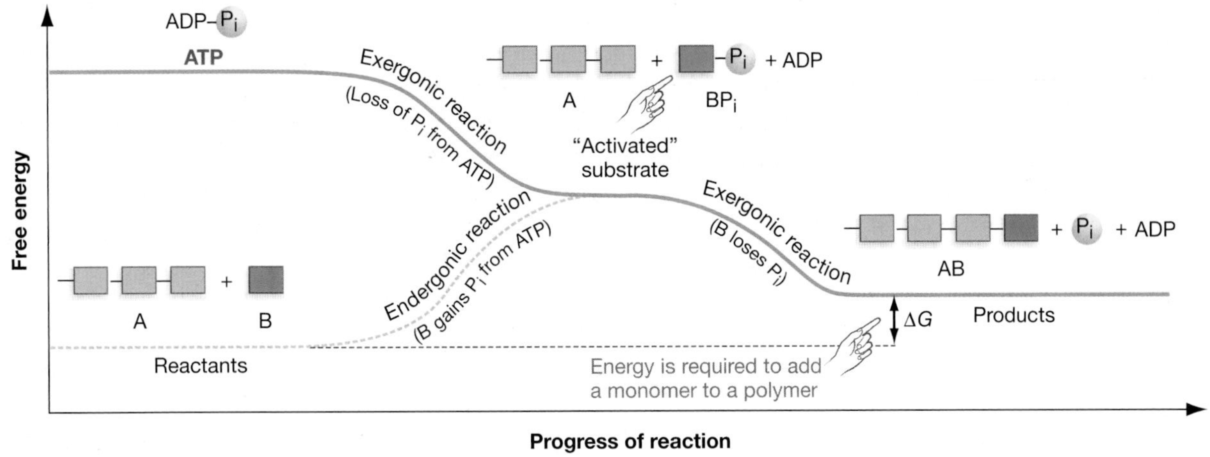

FIGURE 3.28 Chemical Energy from ATP Can Drive Endergonic Reactions
(a) Adenosine triphosphate, or ATP, consists of a molecule called adenosine with three phosphate groups attached.
(b) In cells, endergonic reactions occur if one reactant undergoes phosphorylation—the addition of a phosphate group from ATP. The phosphorylated reactant molecule has high enough free energy that the subsequent reaction is exergonic.

catalyzes the reaction is phosphorylated instead of a reactant. When either a substrate or an enzyme is phosphorylated, the exergonic phosphorylation reaction is said to be "coupled" to an endergonic reaction.

To summarize, two points are absolutely crucial to understanding how biological reactions occur in cells:

1. Phosphorylation makes biological reactions endergonic and thus spontaneous, and
2. Enzyme catalysis makes most biological reactions fast.

During chemical evolution, the phosphorylation of amino acids in the prebiotic soup could have been driven by solar, heat, or electrical energy. The phosphorylated amino acids could then have driven the endergonic polymerization reactions required for continued chemical evolution and the production of macromolecules. If small proteins were produced in this way 4 billion years ago, were they the molecules that triggered the origin of life?

Was the First Living Entity a Protein?

In the introduction to this chapter, we noted that life began with a molecule that could make a copy of itself. This self-replicating entity increased in number and formed a population of individuals. The population then began to evolve by natural selection—the process introduced in Chapter 1 that leads to change in populations of organisms over time. For such a molecule to make a copy of itself efficiently, it must have been a catalyst of some kind. This chapter's examination of protein structure and function raises a key question: Was the first self-replicating molecule on Earth a protein?

Several observations argue that the answer to this question is yes. Experimental studies have shown that amino acids were likely to be abundant in the prebiotic soup and that they could have polymerized to form small proteins. In addition, proteins are the most efficient catalysts known, and a self-replicating molecule had to act as a catalyst during the assembly and polymerization of its copy. These observationss support the hypothesis that the self-replicating molecule was a polypeptide. Indeed, several laboratories that are currently working to create life have focused on synthesizing a self-replicating protein.

To date, however, attempts to simulate the origin of life with proteins have not been successful. Most origin-of-life researchers are increasingly skeptical about the hypothesis that life began with a protein. Their reasoning is that to make a copy of something, a mold or template is required. Proteins cannot furnish this information. Nucleic acids, in contrast, can. How they do so is the subject of Chapter 4.

✓CHECK YOUR UNDERSTANDING

Most proteins are enzymes that make specific chemical reactions occur rapidly. Enzymes catalyze reactions by means of a three-step mechanism: (1) Binding of reactants in a precise orientation inside the active site. (2) Facilitation of the transition state, thus lowering the activation energy required for the reaction. This step often involves a conformational change in the enzyme, resulting in an "induced fit" between active site and substrate. Enzyme cofactors are often required at this step. (3) Release of products, which do not bind tightly to the active site.

The activity of enzymes is controlled through allosteric regulation or competitive inhibition. Different enzymes work at different rates, and enzymes are sensitive to changes in temperature and pH.

Endergonic reactions occur in cells through the mechanism of phosphorylation. When a phosphate group from ATP binds to a substrate or an enzyme, its free energy is raised enough to make the reaction exergonic.

You should be able to diagram the three steps in enzyme catalysis. Your diagram should include (1) coupling with an exergonic reaction involving ATP and (2) regulatory processes.

ESSAY Molecular Handedness and the Thalidomide Tragedy

This chapter introduces the concept of molecular "handedness" in the context of research on chemical evolution. Most amino acids and many other molecules have different forms called optical isomers, which differ in the spatial arrangement of their atoms. The problem facing origin-of-life researchers is to explain why organisms have only one optical isomer of molecules such as the amino acids and the sugar called ribose. But the existence of optical isomers also has urgent practical consequences, especially for the pharmaceutical industry.

Most molecules that chemists synthesize in the laboratory are produced as a mixture of optical isomers in equal amounts. This was true for a sedative developed in the 1950s and marketed under the name thalidomide. People who used this tranquilizer were actually ingesting equal amounts of two mirror-image molecules.

In laboratory testing, thalidomide was considered to be extremely safe, because it had few side effects in experimental animals. As a result, it was widely prescribed in Europe; in Germany it was even available without a prescription. However, the drug was never approved for sale in the United States because

Dr. Frances O. Kelsey (an official with the U.S. Food and Drug Administration) interpreted the laboratory testing data differently. She believed the evidence indicated that the drug might be harmful.

Tragically, Dr. Kelsey was correct. One optical isomer of thalidomide acts as a safe and effective tranquilizer. But the mirror-image molecule can cause severe birth defects. Many women who took the drug during the first three months of a pregnancy bore children without arms or legs or with severely reduced limbs (**Figure 3.29**). Over 12,000 "thalidomide babies" were born before physicians were able to diagnose the cause and have the drug banned.

Over 12,000 "thalidomide babies" were born before physicians were able to diagnose the cause and have the drug banned.

The tragedy made pharmaceutical companies acutely aware of the need to synthesize or isolate single optical isomers of molecules. It also underscores a very general point in molecular biology: Because chemical reactions depend on the orientation of the atoms involved, shape and geometry are fundamentally important characteristics. The function of a molecule is determined by its structure. Later research revealed that one optical isomer of thalidomide damages embryos because it inhibits blood vessel formation. Because cancerous tumors require extensive blood vessel development to sustain their growth, molecules related to thalidomide are now being investigated as possible anticancer drugs.

FIGURE 3.29 An Optical Isomer Caused Abnormal Development
Many mothers who took the drug thalidomide during pregnancy had children with reduced or no limbs.

CHAPTER REVIEW

Summary of Key Concepts

Chemical evolution begins when simple molecules such as methane, ammonia, water, carbon dioxide, and carbon monoxide react to form more complex organic molecules such as formaldehyde and hydrogen cyanide. Experiments using electrical discharges or ultraviolet radiation as energy sources have shown that these nonspontaneous reactions could have occurred on ancient Earth.

Chemical evolution continues when formaldehyde and hydrogen cyanide react to form amino acids, sugars, and nucleotides. These molecules are the building blocks, or monomers, needed to synthesize the complex macromolecules found in living organisms. Experiments have shown that amino acids and other monomers are easily produced under early Earth conditions. Many of these molecules are found in meteorites as well.

- **Proteins are made of amino acids. Amino acids vary in structure and function because their side chains vary in composition. As a result, proteins vary widely in structure and thus in function.**

The polymerization of monomers into macromolecules is the next key event in chemical evolution. Experiments have shown that these condensation reactions occur readily when growing polymers stick to clay minerals. Researchers have observed polypeptide formation from the polymerization of amino acids on clay particles.

Web Tutorial 3.1 Condensation and Hydrolysis Reactions

- **In cells, most proteins are enzymes that function as catalysts. Chemical reactions occur much faster when they are catalyzed by enzymes. During enzyme catalysis, the reactants bind to an enzyme's active site in a way that allows the reaction to proceed efficiently.**

Proteins are far and away the most diverse and efficient class of catalysts known. A catalyst speeds the rate of that reaction by lowering the activation energy required, even though the catalyst itself is unchanged by the reaction. Enzymes are protein catalysts that lower activation energy by stabilizing the transition state of the reaction.

Enzyme-catalyzed reactions take place at a location called the active site of the enzyme. The active site in each type of enzyme has unique chemical properties and a distinctive size and shape. As a result, most enzymes are able to catalyze only one specific reaction. As a group, enzymes are able to catalyze many types of reactions because their chemical and physical structures are so diverse. This diversity is due to the variety of amino acids and the four levels of protein structure. A protein's primary structure, or sequence of amino acids, is responsible for most of its chemical properties. Interactions that take place between carboxyl and amino groups in the same peptide-bonded backbone, between R-groups found in the same polypeptide, and among different polypeptides all help determine the protein's overall shape.

Many enzymes function only with the help of cofactors. In addition, virtually all enzyme activity in cells is regulated by molecules that bind at the active site or at locations on the protein that induce a change in the size or shape of the active site.

Web Tutorial 3.2 Activation Energy and Enzymes

In cells, endergonic reactions occur in conjunction with an exergonic reaction involving ATP.

Almost all of the reactions that occur inside cells are endergonic. When ATP or a phosphate group from ATP is added to a substrate or enzyme that participates in an endergonic reaction, the potential energy of the substrate or enzyme is raised enough to make the reaction exergonic and thus spontaneous. In this way, ATP drives reactions that otherwise would not occur.

The final step in chemical evolution is the formation of a self-replicating molecule. A self-replicating molecule must be able to catalyze polymerization reactions, and it must furnish the template necessary to make a copy of itself. Although enzymes catalyze most of the polymerization reactions observed in cells today, only a handful of researchers currently support the hypothesis that the first living entity was a protein. Proteins are efficient catalysts, but there is no obvious way for them to self-replicate. Instead, most researchers now support the hypothesis that the first molecule to make copies of itself—a fundamental characteristic of life—was a nucleic acid. Chapter 4 explains why.

Questions

Content Review

1. What two functional groups are present on every amino acid?
   a. a carbonyl group and a carboxyl group
   b. an amino group and a carbonyl group
   c. an amino group and a hydroxyl group
   d. an amino group and a carboxyl group
2. Twenty different amino acids are found in the proteins of cells. What distinguishes these molecules?
   a. the location of their carboxyl group
   b. the location of their amino group
   c. the composition of their side chains, or R-groups
   d. their ability to form peptide bonds
3. What determines the primary structure of a polypeptide?
   a. its sequence of amino acids
   b. hydrogen bonds that form between carboxyl and amino groups on different residues
   c. hydrogen bonds and other interactions between side chains
   d. the number, identity, and arrangement of polypeptides that make up a protein
4. In a polypeptide, what is most responsible for the secondary structure called an α-helix?
   a. the sequence of amino acids
   b. hydrogen bonds that form between carboxyl and amino groups on different residues
   c. hydrogen bonds and other interactions between side chains
   d. the number, identity, and arrangement of polypeptides that make up a protein
5. What is a transition state?
   a. the complex formed as covalent bonds are being broken and re-formed during a reaction
   b. the place where regulatory molecules bind to an enzyme
   c. a reactant with high potential energy created by phosphorylation
   d. the shape adopted by an enzyme that has an inhibitory molecule bound at its active site
6. By convention, biologists write the sequence of amino acids in a polypeptide in which direction?
   a. carboxy- to amino-terminus
   b. amino- to carboxy-terminus
   c. polar residues to nonpolar residues
   d. charged residues to uncharged residues

Conceptual Review

1. Explain the lock-and-key model of enzyme activity. Be sure to comment on what the active site of an enzyme does.
2. Isoleucine, valine, leucine, phenylalanine, and methionine are amino acids with highly hydrophobic side chains. Suppose a section of a protein contains a long series of these hydrophobic residues. How would you expect this portion of the protein to behave when the molecule is in aqueous solution?
3. Compare and contrast competitive inhibition and allosteric regulation.
4. Figure 3.8a shows a generalized cartoon of monomers undergoing condensation reactions to form a polymer. Label the type of bond formed when amino acids polymerize to a polypeptide. Does it take energy for polymerization reactions to proceed, or do they occur spontaneously? Why or why not?
5. A major theme in this chapter is that the structure of molecules correlates with their function. Use this theme to explain why proteins can perform so many different functions in organisms and why proteins as a group are such effective catalysts.
6. Explain how endergonic reactions can be driven by phosphorylation.

Group Discussion Problems

1. The essay at the end of Chapter 2 discussed efforts to find life on other components of our solar system. Based on the information in this chapter, is it reasonable to expect that protein-based life forms could exist in extraterrestrial environments that are very hot or highly acidic? Why or why not?
2. Recently, researchers were able to measure movement that occurred in a single amino acid in an enzyme as reactions were taking place in its active site. The residue that moved was located in the active site, and the rate of movement correlated closely with the rate at which the reaction was taking place. Discuss the significance of these findings, using the information in Figures 3.22 and 3.23.
3. Researchers can analyze the atomic structure of enzymes during catalysis. In one recent study, investigators found that the transition state included the formation of a free radical, and that a coenzyme bound to the active site donated an electron to help stabilize the free radical. How would the reaction rate and the stability of the transition state change if the coenzyme were not available?
4. In a recent experiment, researchers found that if they took a mixture containing two optical isomers of a monomer and put a thin layer of the mixture on water, a crystal-like structure formed. Only one of the optical isomers—not both—was included in the crystal. The researchers suggested that a similar phenomenon might occur if both optical isomers of a monomer formed a thin coating on a clay mineral particle. How might this observation relate to the fact that proteins in cells contain only one isomer of each amino acid?

Answers to Multiple-Choice Questions 1. d; 2. c; 3. a; 4. b; 5. a; 6. b

www.prenhall.com/freeman is your resource for the following: Web Tutorials; Online Quizzes and other Online Study Guide materials; Answers to Conceptual Review Questions; Solutions to Group Discussion Problems; Answers to Figure Caption Questions and Exercises; and Additional Readings and Research.

4 Nucleic Acids and the RNA World

KEY CONCEPTS

- Nucleotides are monomers that consist of a sugar, a phosphate group, and a nitrogen-containing base. Ribonucleotides polymerize to form RNA. Deoxyribonucleotides polymerize to form DNA.
- DNA's primary structure consists of a sequence of nitrogen-containing bases, which contain information in the form of a molecular code. DNA's secondary structure consists of two DNA strands running in opposite directions. The strands are held together by complementary base pairing and are twisted into a double helix.
- RNA's primary structure also consists of a sequence of nitrogen-containing bases that contain information in the form of a molecular code. Its secondary structure includes short double helices and structures called hairpins. RNA molecules called ribozymes catalyze important chemical reactions.

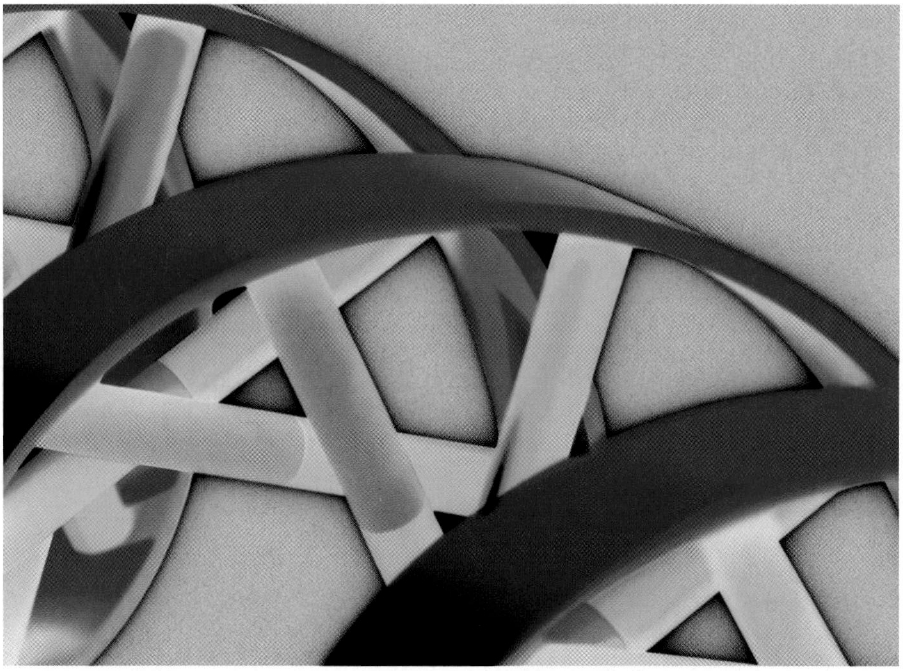

An artist's representation of deoxyribonucleic acid, or DNA. In today's cells, DNA functions as an information repository. But the nucleic acid called ribonucleic acid, or RNA, was probably responsible for the origin of life.

Chapter 3 began with experimental evidence that chemical evolution produced the monomers called amino acids and their polymers called proteins. But the chapter ended by stating that even though proteins are the workhorse molecules of today's cells, few researchers still favor the hypothesis that life began as a protein molecule. Instead, the vast majority of biologists now contend that life began as a polymer called a nucleic acid—specifically, a molecule of ribonucleic acid (RNA). This proposal is called the **RNA world hypothesis.** Just what is a nucleic acid? How could these types of molecules have been produced by chemical evolution? What do the nucleic acids RNA and deoxyribonucleic acid (DNA) do in cells today, and why isn't it plausible to propose that life began with DNA? Before attempting to answer these questions, let's consider an even more basic issue: How would we know when a molecule became alive? Stated another way, What *is* life?

Like many simple questions, the issue of defining life is not easy to answer. In reality, there is no precise definition of what constitutes life. Discussions about the definition of life are still taking place, but most biologists now point to two attributes to distinguish life from nonlife. The first attribute is the ability to reproduce. If something is alive, it can make a copy of itself. (In organisms that reproduce asexually, the copy is exact. In the case of sexually reproducing organisms, the copy is not exact. Instead, traits from male and female individuals are combined to produce an offspring.) In essence, this chapter is about the first molecule that could make a copy of itself.

For something to be alive, however, most biologists insist that it have a second attribute: the ability to acquire particular molecules and use them in controlled chemical reactions that maintain conditions suitable for life and that contribute to growth. In today's organisms, chemical reactions are precisely controlled because enzymes and reactants are bounded by a **plasma membrane.** To many biologists, then, the presence of a plasma membrane enclosing a large suite of coordinated chemical reactions, or **metabolism**, is also required for life.

According to the theory of chemical evolution, these two attributes of life did not emerge simultaneously. Instead, the theory predicts that chemical evolution first led to the existence of a molecule that could make copies of itself. Some researchers argue that this self-replicating molecule was also capable of metabolism, because it had to be able to catalyze the polymerization reactions that allowed copies of itself to form. Later, the theory holds, a descendant of this molecule became enclosed in a membrane. This event created the first cell. The formation of the cell allowed for reactants to be enclosed. As a result, it made precise control over reactions possible.

The key idea is that life started with replication and later became cellular. To distinguish a naked (membrane-less) self-replicator from later, cellular forms of "true" life, this chapter refers to the self-replicator as the first living entity or the first life-form—but *not* as the first organism. The term **organism** is reserved for cellular-based life. This distinction isn't trivial. Naked, self-replicating molecules undoubtedly existed early in Earth's history, and researchers are almost certain to produce one in the laboratory within your lifetime. But it is much less likely that biologists will be able to create cellular life in the foreseeable future. Will humans be able to create life in a test tube? The answer depends, in part, on how you define life.

To understand how life began, it's helpful to delve into some more biochemistry. Let's begin with an analysis of nucleic acids as monomers and end with current research on creating a self-replicating molecule.

4.1 What Is a Nucleic Acid?

Nucleic acids are polymers, just as proteins are polymers. But instead of being made up of monomers called amino acids, **nucleic acids** are made up of monomers called nucleotides. **Figure 4.1a** diagrams the three components of a **nucleotide:** (1) a phosphate group, (2) a sugar, and (3) a nitrogenous (nitrogen-containing) base. The phosphate is bonded to the sugar molecule, which in turn is bonded to the nitrogenous base. A **sugar** is an organic compound with a carbonyl group ($>C=O$) and several hydroxyl (OH) groups. Notice that the "prime" symbols in Figure 4.1 indicate that the carbon being referred to is part of the sugar and not of the attached nitrogenous base.

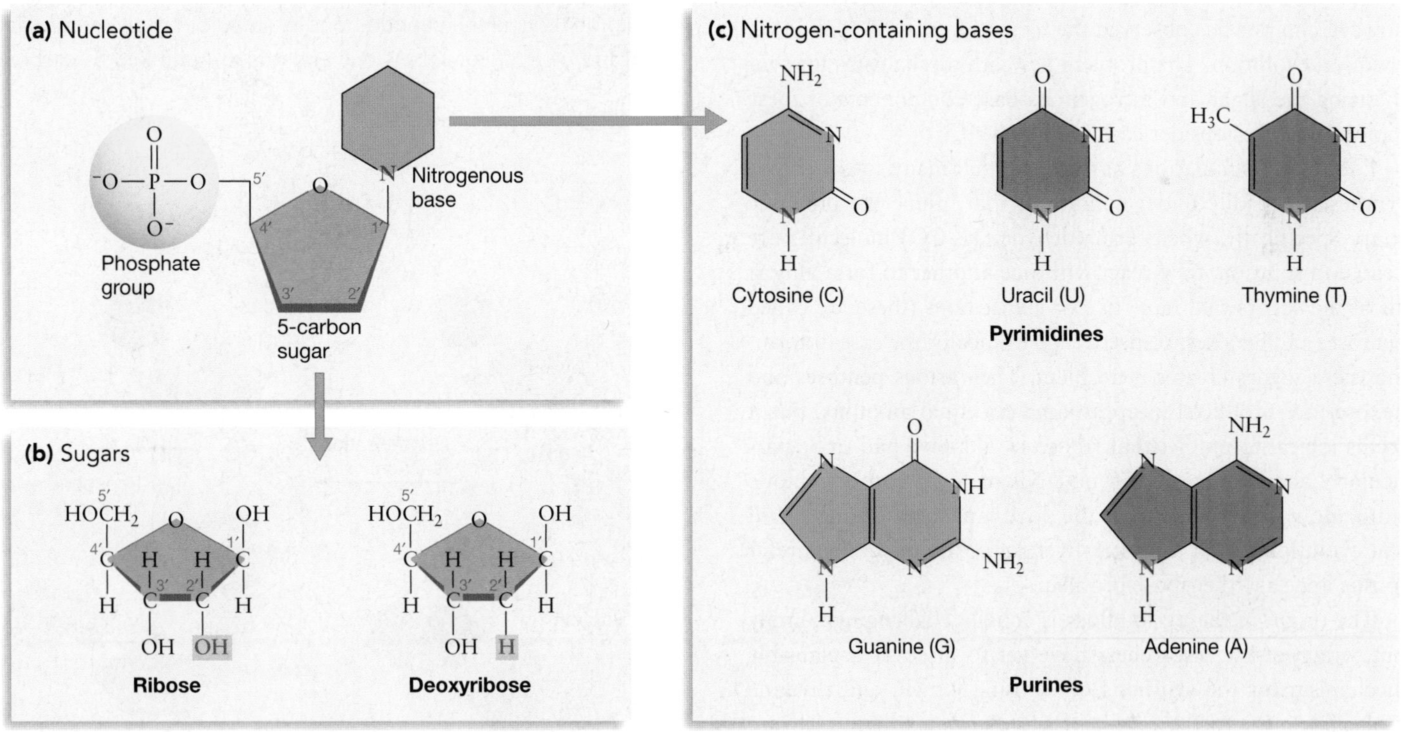

FIGURE 4.1 The General Structure of a Nucleotide
(a) The relationship between the phosphate group, the sugar, and the nitrogenous base of a nucleotide. The numbers indicate the position of each carbon in the sugar's ring. The nitrogenous base is bonded to carbon number 1 in the ring, while the phosphate is bonded to carbon number 5. The bond between the phosphate group and the sugar is called a 5′ linkage. Although hydrogen atoms are bonded to the carbon atoms in the ring (see part b), biologists routinely omit them for clarity. **(b)** Ribose and deoxyribose are similar sugars that are found in nucleotides. **(c)** Purines and pyrimidines are nitrogenous bases. A C–N bond links them to the sugar in a nucleotide. This bond forms at the nitrogen atom that is highlighted in each base. Purines are substantially larger than pyrimidines.

Although a wide variety of nucleotides are found in living cells, origin-of-life researchers concentrate on two types: ribonucleotides and deoxyribonucleotides. In **ribonucleotides**, the sugar is *ribose*; in **deoxyribonucleotides**, it is *deoxyribose*. As **Figure 4.1b** shows, these two sugars differ by a single atom. Ribose has an –OH group bonded to the second carbon in the ring. Deoxyribose has an H instead at the same location. Note that *deoxy-* means "lacking oxygen."

Cells today use four different ribonucleotides, each of which contains a different nitrogenous base. These bases, diagrammed in **Figure 4.1c**, belong to structural groups called **purines** and **pyrimidines.** Ribonucleotides include the purines adenine (A) and guanine (G), and the pyrimidines cytosine (C) and uracil (U).

Similarly, four different deoxyribonucleotides are found in cells today and are distinguished by the structure of their nitrogenous base. Like ribonucleotides, deoxyribonucleotides include adenine, guanine, and cytosine. But instead of uracil, a closely related pyrimidine called thymine (T) occurs in deoxyribonucleotides (**Figure 4.1c**).

Could Chemical Evolution Result in the Production of Nucleotides?

Based on data presented in Chapter 3, most researchers contend that amino acids were abundant early in Earth's history. As yet, however, no one has observed the formation of a nucleotide via chemical evolution. The problem lies with mechanisms for synthesizing the sugar and nitrogenous base components of these molecules. Let's consider each issue in turn.

Laboratory simulations have shown that many sugars can be synthesized readily under conditions that mimic the prebiotic soup. Specifically, when formaldehyde (H_2CO) molecules are heated in solution, they react with one another to form almost all of the sugars that have five or six carbons (these are called pentoses and hexoses, respectively). Ironically, the ease of forming these sugars creates a problem. The various pentoses and hexoses are produced in approximately equal amounts, but it seems logical to predict that ribose would have had to be particularly abundant for RNA or DNA to form in the prebiotic soup. How ribose came to be the dominant sugar during chemical evolution is still a mystery. Origin-of-life researchers refer to this issue as "the ribose problem."

The origin of the pyrimidines is equally challenging. Simply put, origin-of-life researchers have yet to discover a plausible mechanism for the synthesis of cytosine, uracil, and thymine molecules prior to the origin of life. Purines, in contrast, are readily synthesized by reactions among hydrogen cyanide (HCN) molecules. For example, both adenine and guanine have been found in the solutions recovered after spark-discharge experiments.

The ribose problem and the origin of pyrimidine bases are two of the most serious challenges that remain for the theory of chemical evolution. Research on how large quantities of ribose could have formed in the prebiotic soup, along with the pyrimidines, continues. But once formed, how did nucleic acids polymerize to form RNA and DNA? This question has an answer.

How Do Nucleotides Polymerize to Form Nucleic Acids?

Nucleic acids form when nucleotides polymerize. As **Figure 4.2** shows, the polymerization reaction involves the formation of a bond between the phosphate group of one nucleotide and the hydroxyl group of the sugar component of another nucleotide. The result of this condensation reaction is called a **phosphodiester linkage** or a phosphodiester bond. In Figure 4.2, a phosphodiester linkage joins the 5′ carbon on the ribose of one nucleotide to the 3′ carbon on the ribose of the other. When the nucleotides involved contain the sugar ribose, the polymer that is produced is called **ribonucleic acid,** or simply **RNA.** If the nucleotides contain the sugar deoxyribose instead, then the resulting polymer is **deoxyribonucleic acid,** or **DNA.**

Figure 4.3 shows how the chain of phosphodiester linkages in a nucleic acid acts as a backbone, analogous to the peptide-bonded backbone found in proteins. (Compare Figure 4.3 with Figure 3.10a.) The sugar-phosphate spine of a nucleic acid is directional, as is the peptide-bonded spine of a polypeptide. In a strand of RNA or DNA, one end has an unlinked 5′ carbon

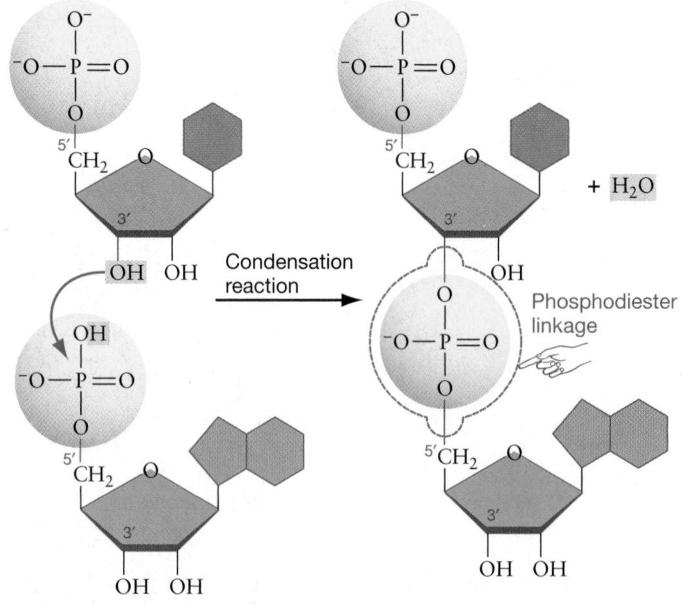

FIGURE 4.2 Nucleotides Polymerize via Phosphodiester Linkages Ribonucleotides can polymerize via condensation reactions. The linkage that results, between the 3′ carbon of one ribonucleotide and the 5′ carbon of another ribonucleotide, is called a phosphodiester linkage.

while the other end has an unlinked 3′ carbon—meaning a carbon that is not linked to another nucleotide. By convention, the sequence of bases found in an RNA or DNA strand is always written in the 5′ → 3′ direction. (In cells, RNA and DNA are always synthesized in this direction. Bases are added at the 3′ end of the growing molecule.) The sequence of nitrogenous bases forms the primary structure of the molecule, analogous to the sequence of amino acids in a polypeptide.

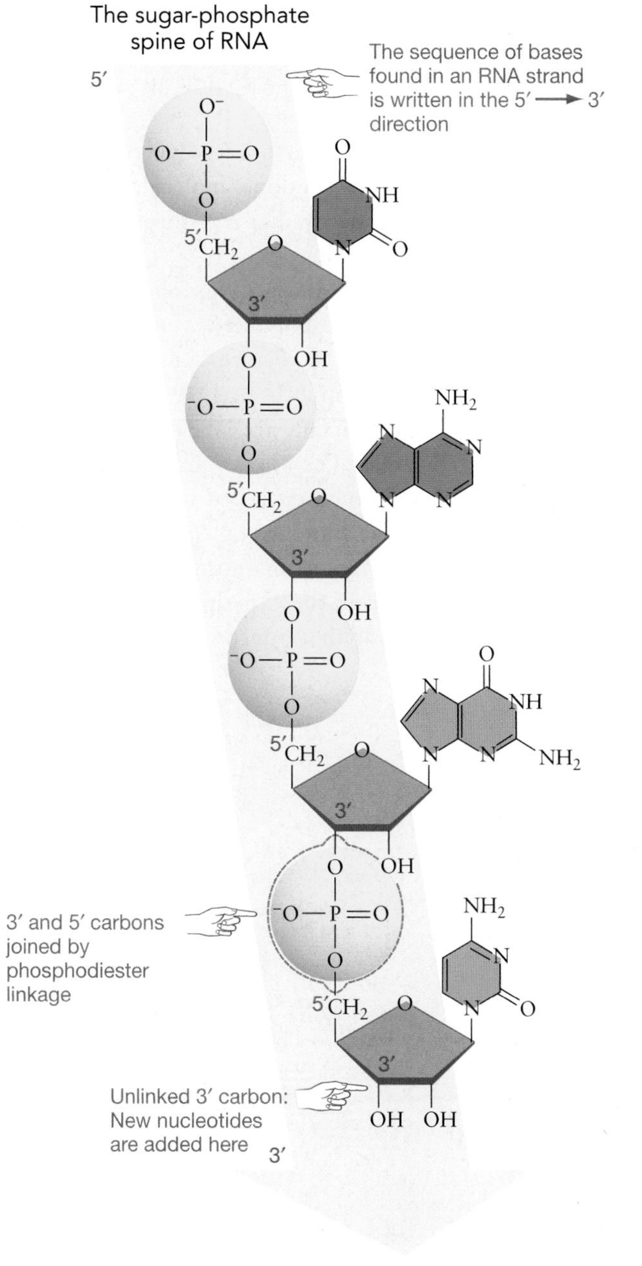

FIGURE 4.3 RNA Has a Sugar-Phosphate Backbone

EXERCISE Identify the four bases in this RNA strand, using Figure 4.1c as a key. Then write down the base sequence, starting at the 5′ end.

In cells, the polymerization reactions that form nucleotides are catalyzed by enzymes. Like other polymerization reactions, the process is endergonic. Polymerization can take place in cells because the free energy of the nucleotide monomers is first raised by reactions that add two phosphate groups to the ribonucleotides or deoxyribonucleotides, creating nucleotide triphosphates (**Figure 4.4**). Recall from Chapter 3 that phosphorylation reactions also make the polymerization of amino acids to polypeptides possible. In both cases, the addition of one or more phosphate groups raises the potential energy of the substrate molecules enough to make the polymerization reaction possible. Researchers refer to the phosphorylated nucleotides or amino acids as "activated."

During chemical evolution, activated nucleotides probably polymerized on the surfaces of clay-sized (very fine-grained) mineral particles. In a suite of experiments analogous to those discussed in Chapter 3 for protein synthesis, researchers have produced RNA molecules by incubating activated ribonucleotides with tiny mineral particles. The hypothesis was that polymerization could occur without an enzyme if the ribonucleotides and growing RNA strands adhered to the clay particles. In one experiment, researchers isolated the clay particles after a day of incubation and then added a fresh batch of activated nucleotides. They repeated this reaction-isolation-reaction sequence for a total of 14 days, or fourteen additions of fresh ribonucleotides. At the end of the two-week experiment, they analyzed the mineral particles, using the techniques introduced in **Box 4.1** (page 78), and found RNA molecules up to 40 nucleotides long. Based on these results, there is a strong consensus that if ribonucleotides and deoxyribonucleotides were able to form during chemical evolution, they would be able to polymerize and form RNA and DNA. Now, what would these nucleic acids look like, and what could they do?

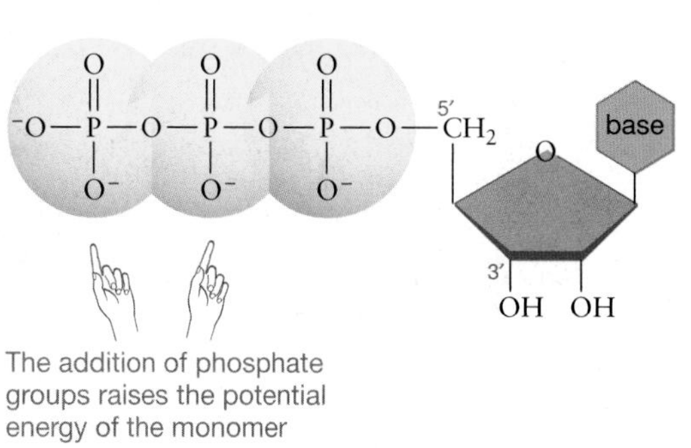

FIGURE 4.4 Activated Monomers Drive Endergonic Polymerization Reactions

Polymerization reactions involving ribonucleotides are endergonic. But polymerization reactions involving ribonucleotide triphosphates are exergonic.

BOX 4.1 Gel Electrophoresis and Autoradiography

In molecular biology, the standard technique for separating and analyzing proteins and nucleic acids is called gel electrophoresis or, simply, electrophoresis ("electricity-moving").This is the technique that was used to determine whether RNA could polymerize when ribonucleotides stuck to clay particles.

The principle behind electrophoresis is fairly simple. Both proteins and nucleic acids carry a charge. As a result, these molecules move when placed in an electric field. Negatively charged molecules move toward the positive electrode (the positive end of the field), and positively charged molecules move toward the negative electrode (the negative end). To separate a mixture of macromolecules so that each can be isolated and analyzed, researchers place the sample in a gelatinous substance. The "gel" consists of long molecules that form a matrix of fibers. This gelatinous matrix has pores through which molecules can pass. When an electrical field is applied across the gel, the molecules in the well move through the gel toward an electrode. Molecules that are smaller or more highly charged for their size move faster than do larger or less highly charged molecules. As they move, the molecules separate by size and by charge.

Figure 4.5 shows the electrophoresis setup used in the polymerization experiment. Step 1 shows how investigators loaded samples of macromolecules, taken from different days during the experiment, into cavities or "wells" at the top of the gel slab. In step 2 the researchers immersed the gel in a solution that conducts electricity and applied a voltage across the gel. After the samples had run down the gel for some time (step 3), they removed the electric field. By then, molecules of different size and charge had separated from one another. In this case, small RNA molecules had reached the bottom of the gel. Above them were larger RNA molecules, which had run more slowly.

Once molecules have been separated in this way, they have to be detected. Often, proteins or nucleic acids can be stained or dyed. In this case, however, the researchers had attached a radioactive atom to the monomers used in the experiment, so the polymers that resulted could be visualized by laying X-ray film over the gel. Because radioactive emissions expose film, a black dot appears wherever a radioactive atom is located in the gel. This technique for visualizing macromolecules is called autoradiography. If the samples are loaded into a rectangular well, as is commonly done, molecules of a particular size and charge form a band in an autoradiograph.

The autoradiograph that resulted from the polymerization experiment is shown in **Figure 4.6**. The samples, taken on days 2, 4, 6, 8, and 14 of the experiment, are labeled along the bottom. The far right lane contains macromolecules of known size; called a size standard or "ladder," this lane is used to estimate the size of the molecules in the experimental samples. The bands that appear in each sample lane

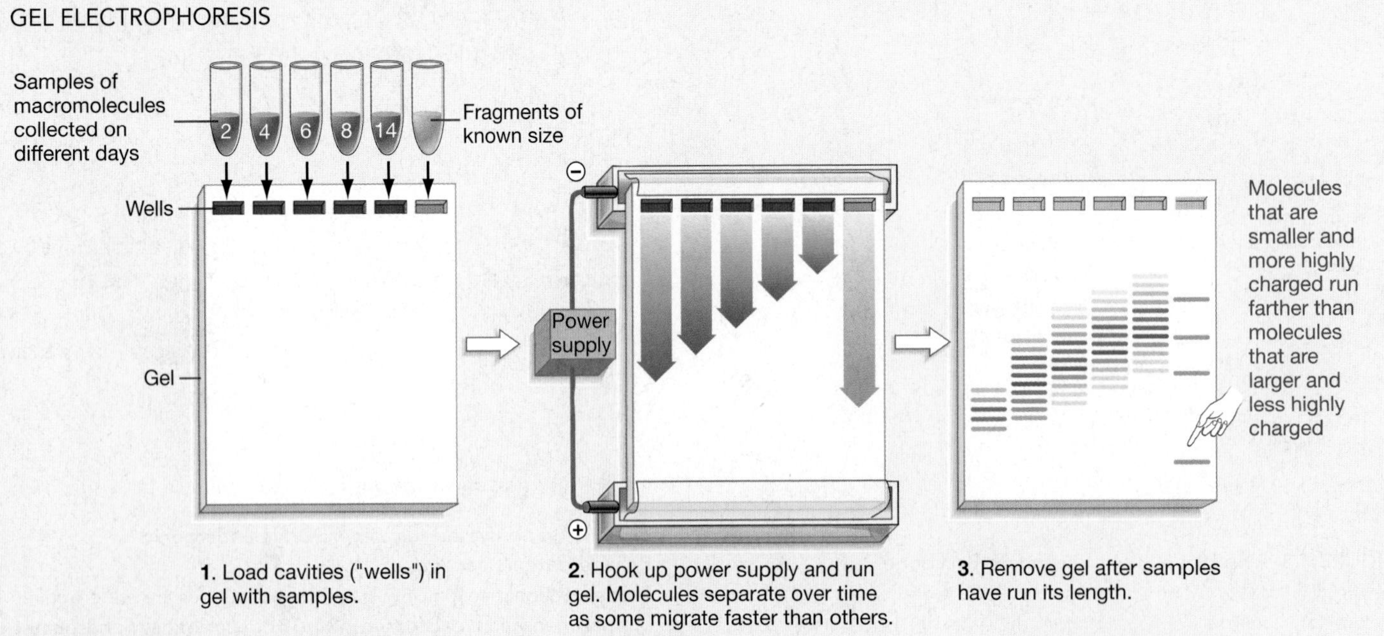
GEL ELECTROPHORESIS
Samples of macromolecules collected on different days
2
4
6
8
14
Fragments of known size
Wells
Gel
Power supply
Molecules that are smaller and more highly charged run farther than molecules that are larger and less highly charged
1. Load cavities ("wells") in gel with samples.
2. Hook up power supply and run gel. Molecules separate over time as some migrate faster than others.
3. Remove gel after samples have run its length.

FIGURE 4.5 Macromolecules Can Be Separated via Gel Electrophoresis

QUESTION DNA and RNA run toward the positive electrode. Why are these molecules negatively charged?

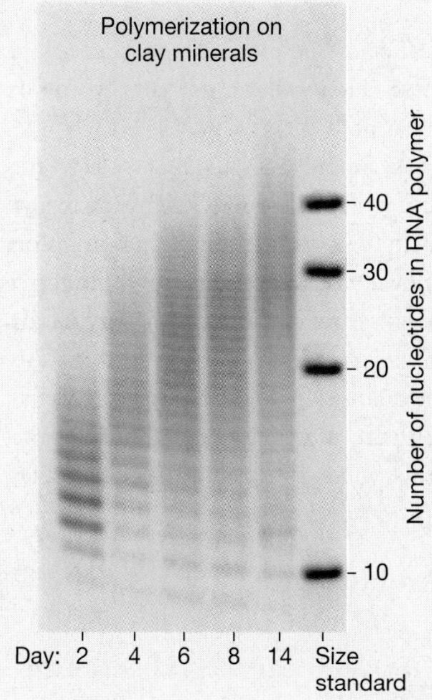
Polymerization on clay minerals
40
30
20
10
Number of nucleotides in RNA polymer
Day: 2 4 6 8 14 Size standard

FIGURE 4.6 Autoradiography Is a Technique for Visualizing Macromolecules
The molecules in a gel can be visualized in a number of ways. In this case, the RNA molecules in the gel exposed an X-ray film because they had radioactive atoms attached. When developed, the film is called an autoradiograph. **EXERCISE** Add labels to the photograph that read: "Top of gel—large molecules" and "Bottom of gel—small molecules."

represent the different polymers that had formed. Darker bands contain more radioactive marker, indicating the presence of many radioactive molecules. Lighter bands contain fewer molecules.

Several conclusions can be drawn from the data. First, a variety of polymers formed at each stage. After the second day, for example, polymers from 12 to 18 monomers long had formed on the clay particles. Second, the overall length of polymers produced increased with time. At the end of the fourteenth day, most of the RNA molecules were between 20 and 40 monomers long. Polymerization had occurred without the aid of an enzyme.

✓CHECK YOUR UNDERSTANDING

Nucleotides are monomers that consist of a sugar, a phosphate group, and a nitrogen-containing base. Nucleotides polymerize to form nucleic acids through formation of phosphodiester linkages between the 3′ carbon on one nucleotide and the 5′ carbon on another. You should be able to draw a simplified cartoon of a nucleotide and a cartoon version of two nucleotides joined by a phosphodiester linkage.

4.2 DNA Structure and Function

Nucleic acids have a primary structure that is somewhat similar to the primary structure of proteins. Proteins have a peptide-bonded backbone with a series of R-groups that extend from it; DNA and RNA molecules have a sugar-phosphate backbone, created by phosphodiester linkages, and a sequence of four nitrogenous bases that extend from it.

Does DNA have secondary structure? The answer to this question, announced in 1953, ranks among the great scientific breakthroughs of the twentieth century. James Watson and Francis Crick presented a model for the secondary structure of DNA in a one-page paper published in *Nature*. At the time, Watson was a 25-year-old postdoctoral fellow and Crick was a 37-year-old graduate student.

Watson and Crick's finding was a hypothesis based on a series of results from other laboratories. They were trying to propose a secondary structure that could explain several important observations about the DNA found in cells:

- Chemists had worked out the structure of nucleotides and knew that DNA polymerized through the formation of phosphodiester linkages. Thus Watson and Crick knew that the molecule had a sugar-phosphate backbone.
- By analyzing the nitrogenous bases in DNA samples from different organisms, Erwin Chargaff had established two empirical rules: (1) The total number of purines and pyrimidines in DNA is the same; and (2) the numbers of T's and A's in DNA are equal, and the numbers of C's and G's in DNA are equal.
- By bombarding DNA with X rays and analyzing how it scattered the radiation, Rosalind Franklin and Maurice Wilkins had calculated the distances between groups of atoms in the molecule (see **Box 4.2**, page 80). The scattering patterns showed that three distances were repeated many times: 0.34 nanometer (nm), 2.0 nm, and 3.4 nm. Because the measurements repeated, the researchers inferred that DNA molecules had a regular and repeating structure. The pattern of X-ray scattering suggested that the molecule was helical or spiral in nature.

Based on this work, understanding DNA's structure boiled down to understanding the nature of the helix involved. What type of helix would have a sugar-phosphate backbone and explain both Chargaff's rules and the Franklin-Wilkins measurements?

Watson and Crick began by analyzing the size and geometry of deoxyribose, phosphate groups, and the nitrogenous bases. The bond angles and measurements suggested that the distance of 2.0 nm probably represented the width of the

WEB TUTORIAL 4.1
Nucleic Acid Structure

BOX 4.2 An Introduction to X-Ray Crystallography

X-ray crystallography is the most widely used technique for reconstructing the three-dimensional structure of molecules. As its name implies, the procedure is based on bombarding crystals of a molecule with X rays. The crystals diffract (that is, scatter) the X rays, producing a diffraction pattern that can be recorded on X ray film or other types of detectors. (The approach is also called *X-ray diffraction analysis*.)

The basic principle is that X rays are scattered in precise ways when they interact with the electrons surrounding the atoms in a crystal (**Figure 4.7**). By varying the orientation of the X-ray beam that strikes a crystal and documenting the diffraction patterns that result, researchers can record a complex set of patterns that reflect the three-dimensional structure of the crystal.

The diffraction pattern can be used to construct a map representing the density of electrons in the crystal. By relating these electron-density maps to information about the primary structure of the nucleic acid or protein, a three-dimensional model of the molecule can be built.

It is important to appreciate the amount of work involved in producing an accurate three-dimensional representation of a nucleic acid or protein. Hundreds or thousands of failed attempts may occur before researchers produce crystals of sufficient quality for X-ray diffraction analysis. But the information contained in an accurate three-dimensional model makes the effort worthwhile.

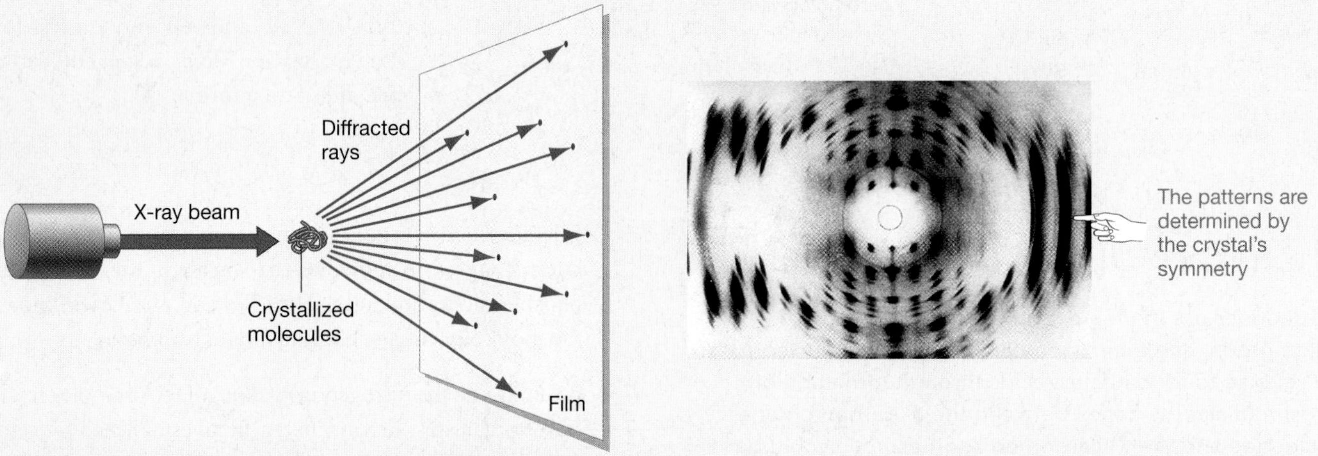

FIGURE 4.7 X-Ray Crystallography Provides Information on the Three-Dimensional Structure of Macromolecules
When crystallized molecules are bombarded with X rays, the radiation is scattered in distinctive patterns. The photograph at the right shows an X-ray film that recorded the pattern of scattered radiation from DNA molecules.

helix and that 0.34 nm was likely to be the distance between bases stacked in a spiral. Now they needed to make sense of Chargaff's rules and the 3.4-nm distance, which appeared to be exactly 10 times the distance between a single pair of bases.

To solve this problem, Watson and Crick constructed a series of physical models such as the one pictured with them in **Figure 4.8**. Building these models allowed them to tinker with different types of helical configurations. After many false starts, they hit on an idea that looked promising. They arranged two strands of DNA side by side and running in opposite directions—meaning that one strand ran in the $5' \rightarrow 3'$ direction while the other was oriented $3' \rightarrow 5'$. Strands with this orientation are said to be antiparallel. Watson and Crick found that if the antiparallel strands were twisted together to form a double helix, the coiled sugar-phosphate backbones ended up on the outside of the spiral and the nitrogenous

FIGURE 4.8 Building a Physical Model of DNA Structure
Watson (left) and Crick (right) did not know if the DNA helix was single, double, or triple stranded, or exactly how the sugar-phosphate backbones in the helix were oriented. To help sort out the possibilities, they built physical models of the four deoxyribonucleotides, using wires with precise lengths and geometries. Then they tested how the nucleotides fit together in different configurations.

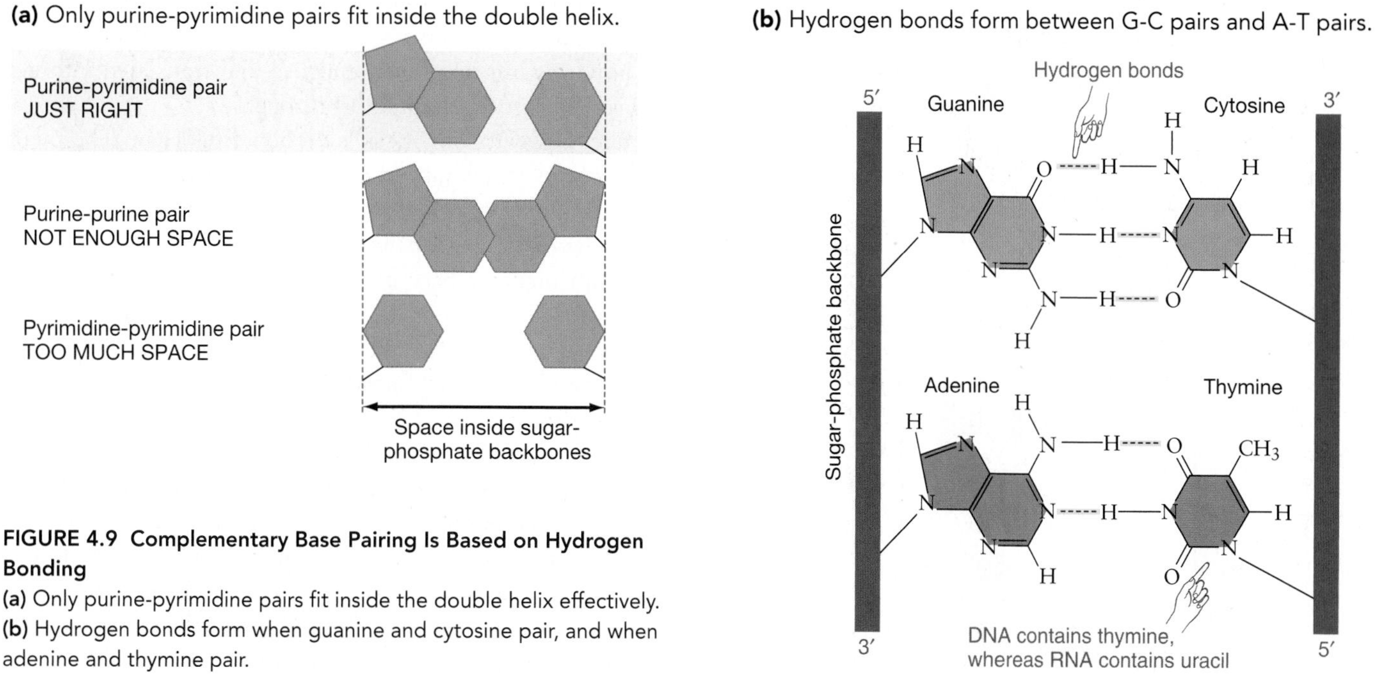
(a) Only purine-pyrimidine pairs fit inside the double helix.
Purine-pyrimidine pair
JUST RIGHT
Purine-purine pair
NOT ENOUGH SPACE
Pyrimidine-pyrimidine pair
TOO MUCH SPACE
Space inside sugar-phosphate backbones
(b) Hydrogen bonds form between G-C pairs and A-T pairs.
Hydrogen bonds
5′
3′
Guanine
Cytosine
Adenine
Thymine
Sugar-phosphate backbone
DNA contains thymine, whereas RNA contains uracil
3′
5′

FIGURE 4.9 Complementary Base Pairing Is Based on Hydrogen Bonding
(a) Only purine-pyrimidine pairs fit inside the double helix effectively. **(b)** Hydrogen bonds form when guanine and cytosine pair, and when adenine and thymine pair.

bases on the inside. But for the bases from each backbone to fit in the interior of the 2.0-nm-wide structure, they had to form purine-pyrimidine pairs (see **Figure 4.9a**). With that came a fundamental insight: Inside the double helix, the bases lined up in a way that allowed hydrogen bonds to form between certain purines and pyrimidines. More specifically, adenine could form hydrogen bonds with thymine, and guanine could form hydrogen bonds with cytosine (**Figure 4.9b**). Because of this specificity, the A-T and G-C bases were said to be complementary. Two hydrogen bonds form when A and T pair, but three hydrogen bonds form when G and C pair. As a result, the G-C interaction is slightly stronger than the A-T bond.

Watson and Crick had discovered the phenomenon known as **complementary base pairing**. In fact, the term *Watson-Crick pairing* is now used interchangeably with the phrase *complementary base pairing*. **Figure 4.10** shows how antiparallel strands of DNA form when complementary bases line up and form hydrogen bonds. As you study the figure, notice that biochemists sometimes say that DNA is put together like a ladder whose ends have been twisted in opposite directions. The sugar-phosphate backbone forms the supports of the ladder; the base pairs represent the rungs of the ladder. The nitrogenous bases in the middle of the DNA helix stack tightly on top of each other. This tight packing forms a hydrophobic interior that is difficult to break apart. The molecule as a whole is water soluble, however, because the sugar-phosphate backbones, which face the exterior of the molecule, are negatively charged and thus are hydrophilic.

It is important to note that the outside of the helical DNA molecule forms two types of grooves and that the types differ in size. The larger of the two is known as the *major groove*, and the smaller one is known as the *minor groove*.

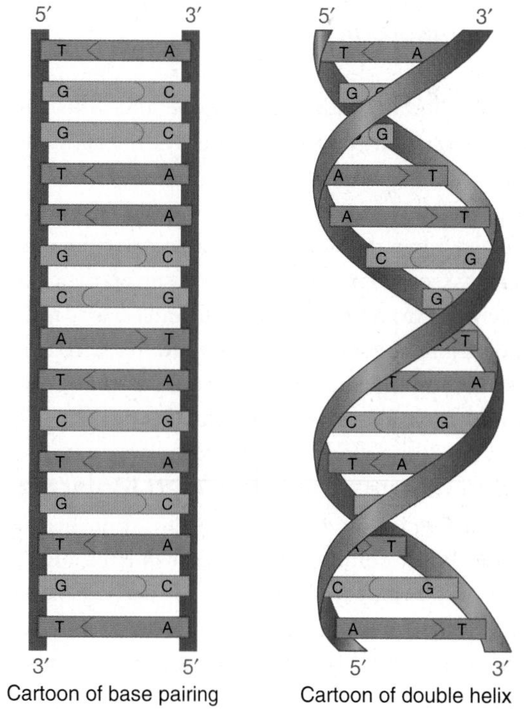
5′
3′
3′
5′
Cartoon of base pairing
Cartoon of double helix

FIGURE 4.10 The Secondary Structure of DNA Is a Double Helix
Complementary base pairing is responsible for twisting DNA into a double helix.

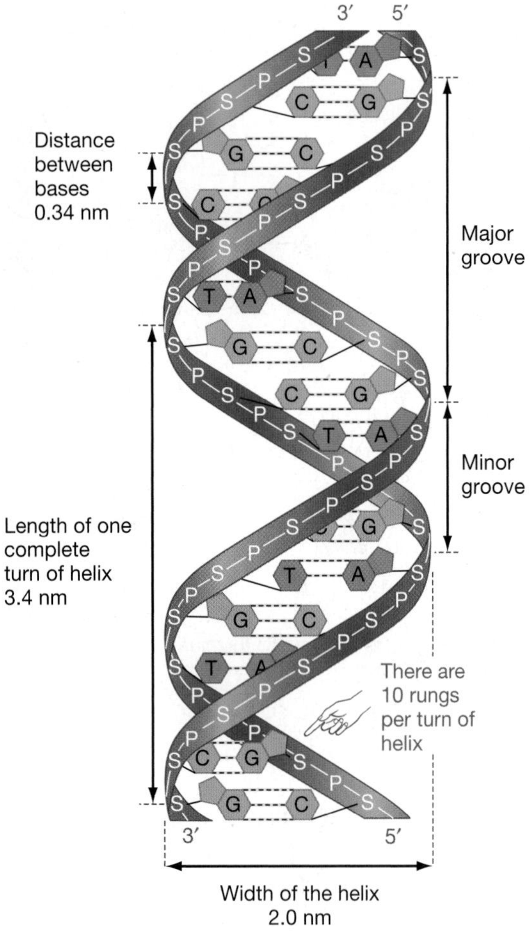

FIGURE 4.11 Dimensions of DNA Secondary Structure
The double-helix hypothesis explains the measurements inferred from X-ray analysis of DNA molecules.

Figure 4.11 highlights these grooves and illustrates how DNA's secondary structure explains the measurements observed by Franklin and Wilkins.

Since the model of the double helix was published, experimental tests have shown that the hypothesis is correct in almost every detail. DNA's secondary structure consists of two antiparallel strands twisted into a double helix. The molecule is stabilized by hydrophobic interactions in its interior and by hydrogen bonding between the complementary base pairs A-T and G-C. How does this secondary structure affect the molecule's function?

DNA Is an Information-Containing Molecule

Watson and Crick's model created a sensation because it revealed how DNA could store and transmit biological information. In literature, information consists of letters on a page. In music, information is composed of the notes on a staff. But inside cells, information consists of a sequence of monomers in a nucleic acid. The four nitrogeneous bases function like letters of the alphabet. A particular sequence of bases is like the sequence of letters in a word—it has meaning. In all cells that have been examined to date, from tiny bacteria to gigantic redwood trees, DNA carries the information required for the organism's growth and reproduction. Exploring how this information is encoded and translated into action is the heart of Chapters 14 through 17.

Here, however, our focus is on how life began. The theory of chemical evolution holds that life began as a self-replicating molecule—a molecule that could make a copy of itself. **Figure 4.12** shows how a copy of DNA could be made by complementary base pairing. Complementary base pairing allows each strand of a DNA double helix to be copied exactly, producing two daughter molecules. In this way, DNA contains the information required for a copy of itself to be made.

Watson and Crick ended their paper on the double helix with one of the classic understatements in scientific literature: "It has not escaped our notice that the specific pairing we have postulated immediately suggests a possible copying mechanism." The central idea is that DNA's primary structure serves as a mold or template for the synthesis of a complementary strand.

In today's cells, however, DNA does not self-replicate spontaneously. Instead, the molecule is copied through a complicated series of reactions catalyzed by a large suite of enzymes. Similarly, it is extremely unlikely that a DNA molecule began copying itself early in Earth's history. Let's explore why.

Is DNA a Catalytic Molecule?

The DNA double helix is a highly structured molecule that is much more stable than RNA and most proteins. DNA is regular and symmetric, with few chemical groups exposed that can participate in chemical reactions. In addition, the lack of a 2′ hydroxl group on deoxyribonucleotides (see Figure 4.1b) makes the polymer less reactive than RNA and highly resistant to chemical degradation. All of these features increase DNA's stability and thus its effectiveness as a reliable information-bearing molecule. Intact stretches of DNA have been recovered from fossils that are tens of thousands of years old. Despite death and exposure to a wide array of pH, temperature, and chemical conditions, the molecules have the same sequence of bases as the organisms had when they were alive.

The orderliness and stability that make DNA such a dependable information repository make it extraordinarily inept at catalysis, however. Recall from Chapter 3 that enzyme function is based on a specific binding event between a substrate and a protein catalyst. Thanks to the enormous diversity of primary through quaternary structures found in proteins, a wide array of binding events can occur. In comparison, DNA's primary and secondary structures are simple. It is not surprising, then, that DNA has never been observed to catalyze any reaction in any organism. Even though researchers have been able to construct single-stranded DNA molecules that can catalyze a few simple reactions in the laboratory, the number and diversity of reactions involved is a minute fraction of the activity catalyzed by proteins.

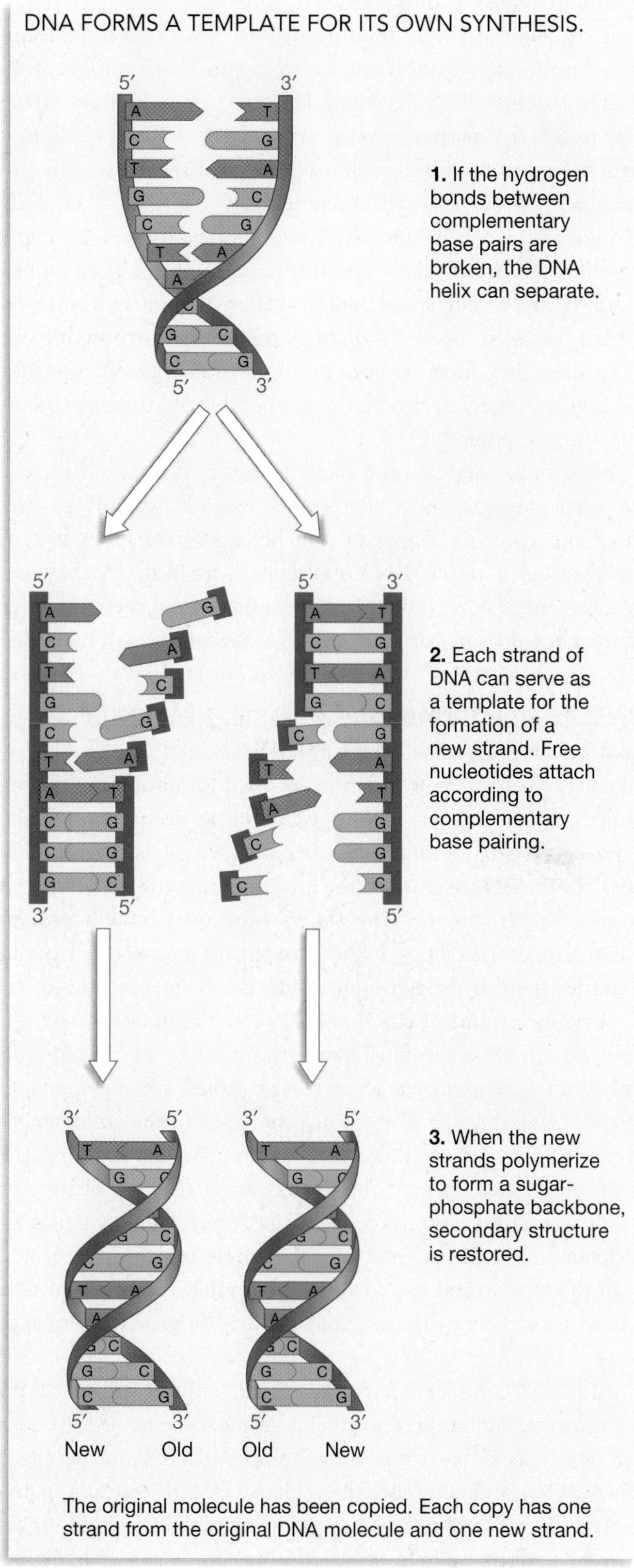

DNA FORMS A TEMPLATE FOR ITS OWN SYNTHESIS.
1. If the hydrogen bonds between complementary base pairs are broken, the DNA helix can separate.
2. Each strand of DNA can serve as a template for the formation of a new strand. Free nucleotides attach according to complementary base pairing.
3. When the new strands polymerize to form a sugar-phosphate backbone, secondary structure is restored.
New
Old
Old
New
The original molecule has been copied. Each copy has one strand from the original DNA molecule and one new strand.

FIGURE 4.12 Making a Copy of DNA
If new bases are added to each of the two strands of DNA via complementary base pairing, a copy of the DNA molecule can be produced.

In short, DNA furnishes an extraordinarily stable template for copying itself. But due to its inability to act as an effective catalyst, virtually no researchers support the hypothesis that the first life-form consisted of DNA. Most biologists who are working on the origin of life support the hypothesis that life began with RNA rather than DNA.

✓CHECK YOUR UNDERSTANDING

DNA's primary structure consists of a sequence of deoxyribonucleotides. Its secondary structure consists of two DNA molecules that run in opposite $5' \rightarrow 3'$ orientations to each other. The two strands are held together through hydrogen bonds between A-T and G-C pairs and hydrophobic interactions among atoms inside the helix. The sequence of deoxyribonucleotides in DNA contains information, much like the sequence of letters and punctuation on this page. Due to complementary base pairing, each DNA strand also contains the information required to form the complementary strand. You should be able to make a sketch of a DNA molecule that (1) uses different symbols to stand for the A, T, G, and C bases; (2) labels the molecule's sugar-phosphate backbone; and (3) indicates the $5' \rightarrow 3'$ orientation of each strand.

4.3 RNA Structure and Function

Like DNA, RNA has a primary structure that consists of a sugar-phosphate backbone formed by phosphodiester linkages and a sequence of four types of nitrogenous base that extend from that backbone. But it's important to recall two significant differences between these nucleic acids:

1. The pyrimidine base thymine does not exist in RNA. Instead, RNA contains the closely related pyrimidine base uracil.
2. The sugar in the sugar-phosphate backbone of RNA is ribose, not deoxyribose as in DNA.

The second point is critical because the hydroxyl (–OH) group on the 2′-carbon of ribose is much more reactive than the hydrogen atom on the 2′-carbon of deoxyribose. This functional group can participate in reactions that tear the polymer apart. The presence of this –OH group makes RNA much more reactive, while its absence makes DNA more stable.

A comparison of secondary structure in RNA and DNA is just as instructive. Like DNA molecules, most RNA molecules have secondary structure that results from complementary base pairing between purine and pyrimidine bases. In RNA, adenine forms hydrogen bonds only with uracil, and guanine again forms hydrogen bonds with cytosine. (Guanine can also bond with uracil, but it does so much less effectively than with cytosine.) Thus, the complementary base pairs in RNA are A-U and G-C. Three hydrogen bonds form between guanine and cytosine, but only two form between adenine and uracil.

How do the secondary structures of RNA and DNA differ? In the vast majority of cases, the purine and pyrimidine bases in RNA undergo hydrogen bonding with complementary bases on the *same* strand, rather than forming hydrogen bonds with complementary bases on a different strand, as in DNA. **Figure 4.13** shows how this works. The key is that when bases on one part of an RNA strand fold over and align with ribonucleotides on another segment of the same strand, the two sugar-phosphate strands are antiparallel. In this orientation, hydrogen bonding between complementary bases results in a stable double helix.

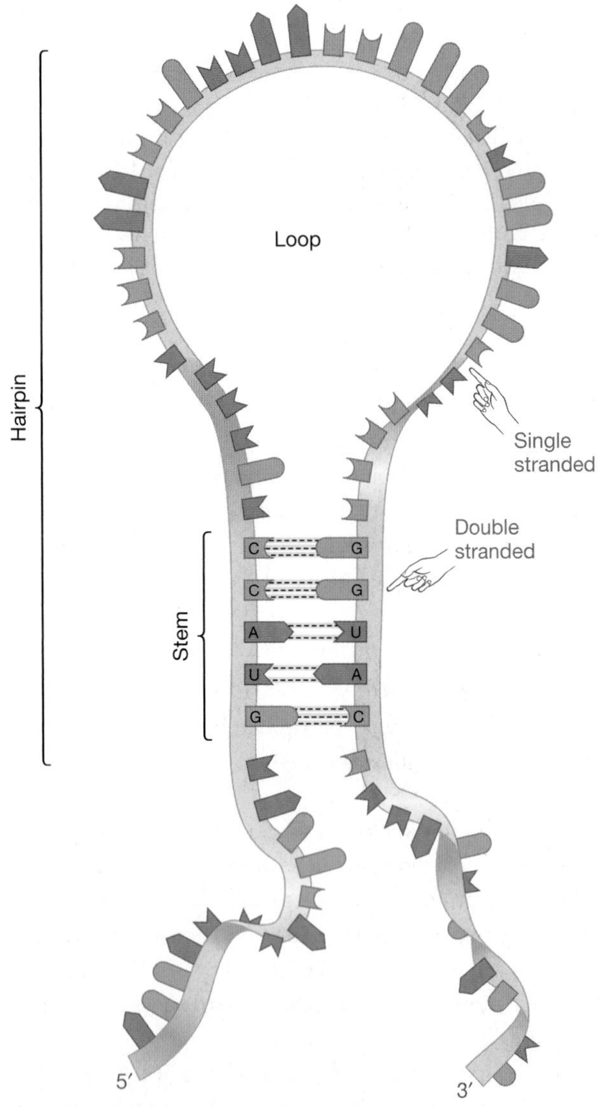

FIGURE 4.13 Complementary Base Pairing and Secondary Structure in RNA: Stem-and-Loop Structures
This RNA molecule has secondary structure. The double-stranded "stem" and single-stranded "loop" form a hairpin. The bonded bases in the stem are antiparallel, meaning that they are oriented in opposite directions.

If the section where the fold occurs includes a large number of unbonded bases, then the stem-and-loop configuration shown in Figure 4.13 results. This type of secondary structure is called a **hairpin**. Several other types of RNA secondary structures are possible, each involving a different length and arrangement of base-paired segments.

Like the α-helices and β-pleated sheets observed in many proteins, RNA secondary structures are stabilized by hydrogen bonding and occur spontaneously. Even though hairpins and other types of secondary structures reduce the entropy of RNA molecules, they form without an input of energy because they are favored energetically. Hydrogen bond formation is exothermic and exergonic.

RNA molecules can also have tertiary structure and quaternary structure, owing to interactions that fold secondary structures into complex shapes or that hold different RNA strands together. As a result, RNA molecules with different base sequences can have very different overall shapes and chemical properties. How does the structure of RNA affect its function in cells?

RNA as an Information-Containing Molecule

Because RNA contains a sequence of bases analogous to the letters in a word, it can function as an information-containing molecule. And because hydrogen bonding occurs specifically between A-U pairs and G-C pairs in RNA, it is theoretically possible for RNA to furnish the information required to make a copy of itself. **Figure 4.14** shows how the copying process might proceed. In steps 1 and 2 of this diagram, free ribonucleotides form hydrogen bonds with complementary bases on the original strand of RNA—also called a **template strand.** As they do, their sugar-phosphate groups form phosphodiester linkages to create a new strand—also called a **complementary strand.** Note that the $5' \rightarrow 3'$ directionality of the complementary strand is opposite that of the template strand. In step 3, the hydrogen bonds between the strands are broken by heating or by a catalyzed reaction. The new RNA molecule now exists independently of the original strand. If these steps were repeated with the new strand used as a template, the resulting molecule would be a copy of the original. An RNA's primary sequence serves as a mold.

Although complementary base pairing allows RNA to carry the information required for the molecule to be copied, RNA is not nearly as stable a repository for particular sequences as is DNA. RNA is much more likely than DNA to degrade or hydrolyze in response to high temperatures, changes in pH, or attack by reactive chemicals—partly because it has 2′ hydroxyl groups and partly because its secondary structure is less extensive. For example, no one has yet found intact RNA in an ancient bone or pollen grain or other type of fossil. The molecule is simply too unstable. In today's cells, RNA molecules that are

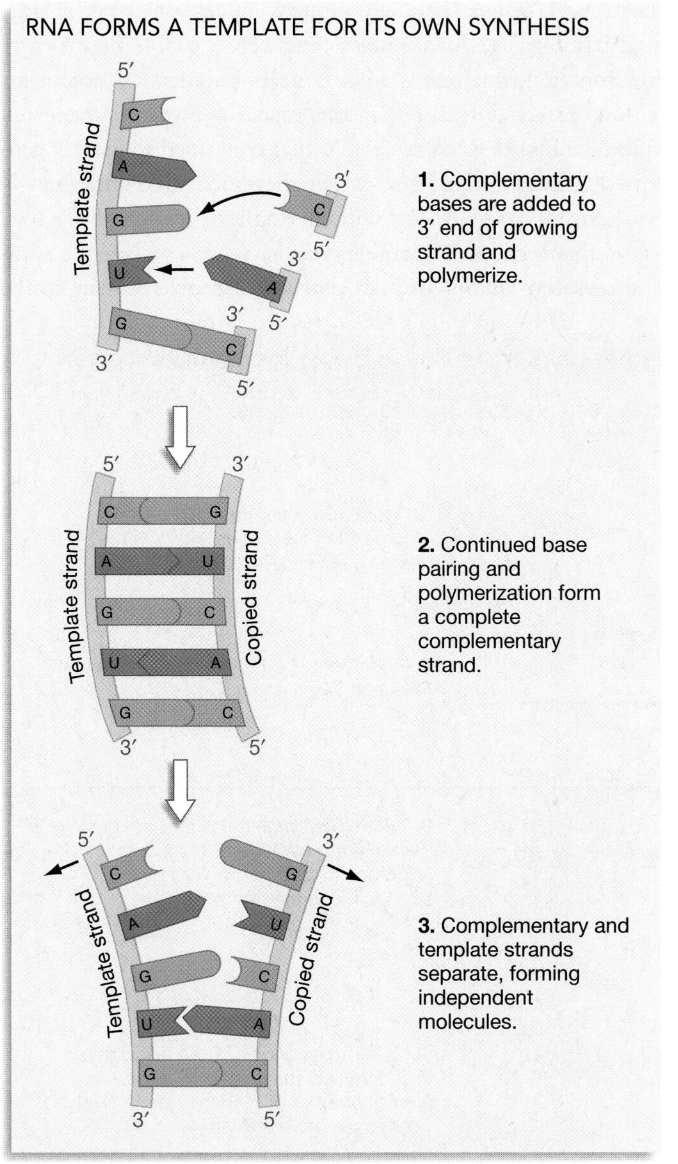
RNA FORMS A TEMPLATE FOR ITS OWN SYNTHESIS
Template strand
1. Complementary bases are added to 3′ end of growing strand and polymerize.
2. Continued base pairing and polymerization form a complete complementary strand.
Copied strand
3. Complementary and template strands separate, forming independent molecules.

FIGURE 4.14 RNA Molecules Can Be Copied Because They Furnish a Template
RNA molecules can be copied when complementary base pairing matches ribonucleotides on a template to form a complementary strand. If the three steps illustrated here were repeated using the copied strand as a template, the newly synthesized molecule would be identical to the original RNA.

shorter lived than DNA ferry the information required to make proteins and other compounds. But in the prebiotic soup, could an RNA molecule have made a copy of itself before it would have degraded?

Is RNA a Catalytic Molecule?

In terms of diversity in chemical reactivity and overall shape, RNA molecules do not begin to match proteins. The primary structure of RNA molecules is much more restricted, because there are only four types of nitrogenous base in RNA versus the 20 types of amino acid found in proteins, and RNA molecules cannot form the variety of bonds that give proteins their extensive tertiary structure. Yet RNA molecules have limited secondary and tertiary structure that allows distinct shapes to form, and their base sequences are variable enough to produce molecules with a certain amount of unique chemical behavior. Because RNA has a degree of structural and chemical complexity, it should be capable of stabilizing a few transition states and catalyzing at least a limited number of chemical reactions. Indeed, biologists have found that RNA *does* act as a catalyst in organisms today. Sidney Altman and Thomas Cech shared the 1989 Nobel Prize in chemistry for showing that RNA enzymes, or **ribozymes**, exist in organisms. The ribozymes they isolated, from a single-celled organism called *Tetrahymena*, could catalyze both the hydrolysis and condensation of phosphodiester linkages. Building on that accomplishment, researchers have discovered ribozymes that catalyze dozens of different reactions in cells. For example, ribozymes catalyze the formation of peptide bonds when amino acids polymerize to form polypeptides. Ribozymes are at work in your cells right now.

The discovery of ribozymes was a watershed event in origin-of-life research. Before Altman and Cech published their discovery, biologists thought that proteins were the only type of molecule capable of catalyzing chemical reactions in organisms. But if a ribozyme in *Tetrahymena* catalyzes the polymerization reactions similar to those diagrammed in steps 1 and 2 of Figure 4.12, it raises the possibility that an RNA molecule could also catalyze the reactions diagrammed in step 3. Such a molecule could copy itself and qualify as the first living entity. Does any experimental evidence support this hypothesis?

4.4 The First Life-Form

The theory of chemical evolution maintains that life began as a naked self-replicator. To make a copy of itself, the first living molecule had to provide a template that could be copied. It also had to catalyze polymerization reactions that would link monomers into a copy of that template. Because RNA is capable of both processes, most origin-of-life researchers propose that the first life-form was made of RNA. As noted in the introduction to this chapter, that proposal is called the *RNA world hypothesis*.

Because no self-replicating RNA molecules exist today, researchers test the hypothesis by trying to simulate the RNA world in the laboratory. The eventual goal is to create an RNA molecule that can catalyze its own replication. When this goal is achieved, a life-form will have been created in the laboratory.

To understand how researchers go about this work, consider a recent set of experiments by Wendy Johnston and others working in David Bartel's laboratory. This research group's goal was ambitious; they wanted to create, from scratch, a ribozyme that was capable of catalyzing the reactions shown in steps 1 and 2 of Figure 4.14. Such a ribozyme would be called RNA replicase. This research program is creating considerable excitement among biologists interested in the origin of life, because adding ribonucleotides to a growing strand is a key attribute of an RNA replicase.

The strategy Bartel's group pursued is outlined in **Figure 4.15**. As you study this figure, note that their procedure essentially mimics the process of natural selection. For example, the researchers began their experiments by establishing a large population of RNA molecules that were variable in structure and function, analogous to a large population of organisms with a variety of features. In this case, the researchers created millions of large RNA molecules that contained randomly generated primary sequences. When they incubated these newly synthesized, large RNAs with a small RNA template, they found that some of the large molecules could catalyze the addition of a few ribonucleotides that were complementary to the

Question: Can a ribozyme follow a template and catalyze the addition of bases to a growing RNA strand?

Hypothesis: Repeated selection among randomly generated RNA sequences can result in an efficient enzyme.

Null hypothesis: Selection cannot produce an efficient enzyme.

Experimental setup:

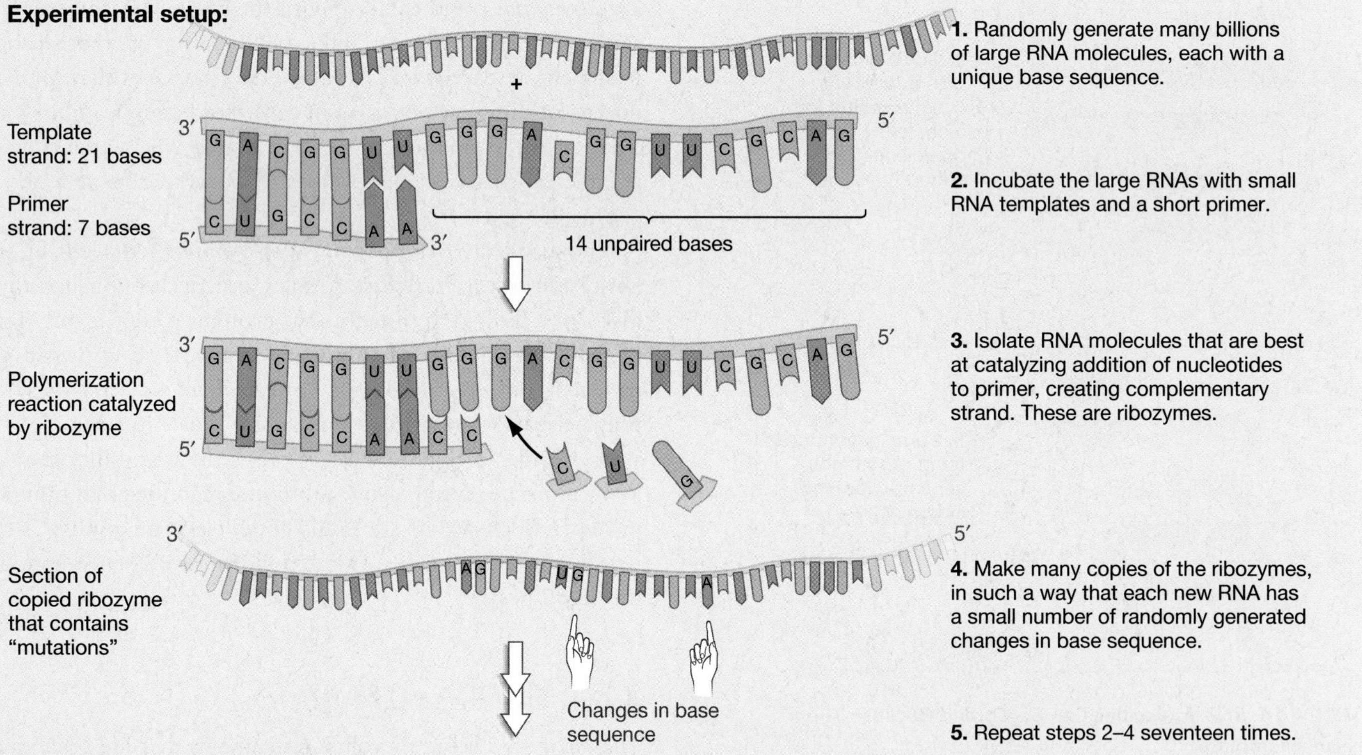

Prediction: After many rounds, the ribozyme will be able to catalyze addition of all fourteen ribonucleotides.

Prediction of null hypothesis: None of the large RNA molecules produced will be able to catalyze addition of ribonucleotides.

Results:

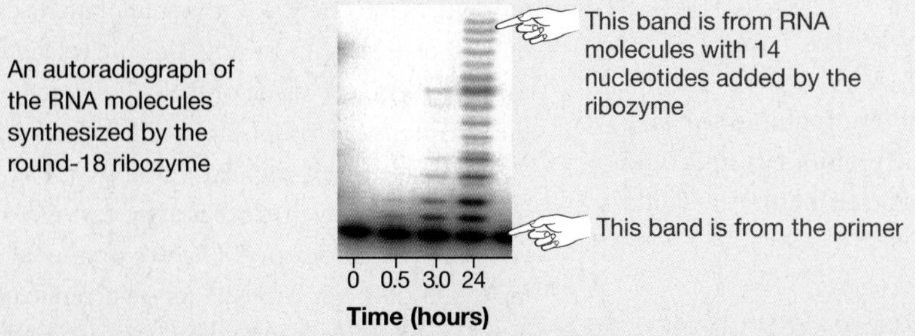

Conclusion: If allowed to react for 24 hours, the round-18 ribozyme can add all fourteen bases to the primer, producing a strand complementary to the template.

FIGURE 4.15 Selection of a Novel Ribozyme
Researchers use variations of this protocol to produce ribozymes that can catalyze a wide variety of reactions.

bases in the template. In this "round-1" experiment, the polymerization didn't happen very quickly or accurately, but it did occur. The researchers had discovered a ribozyme that could catalyze the synthesis of RNA.

To continue the experiment, the researchers isolated the round-1 ribozymes. This step was analogous to natural selection, because only certain variants survived to produce "offspring." To create these offspring, which would represent the next generation of ribozymes, Bartel's team copied the selected molecules in a way that introduced a few random changes to their sequence of bases. These changes were analogous to the types of mutations that occur each generation in natural populations. Mutation is a random process that produces new traits. When the researchers incubated the modified ribozymes with a new template RNA, they found that most of the modified ribozymes worked worse than the original ones. Some worked better, however.

The researchers selected the best of these round-2 ribozymes, copied them in a way that introduced a few additional, random changes in their base sequences, and allowed them to react with yet another template. The best ribozymes in this round 3 were isolated, and then the copying, reaction, and selection process was repeated ... and repeated ... and repeated. By round 18, the group had found a ribozyme that was a much better catalyst than the original molecule. Evolution had occurred. They had created a ribozyme that was reasonably proficient at adding ribonucleotides to a growing strand.

As this book goes to press, the round-18 ribozyme is the closest biologists have come to creating life. Thanks to similar efforts at other laboratories around the world, researchers have produced an increasingly impressive set of ribozymes—an array of molecules capable of catalyzing many of the key reactions responsible for replication and metabolism. Each result provides support for the RNA world hypothesis. Each result also brings research teams closer to the creation of an RNA replicase. If this goal is met, human beings will have created a living entity in a test tube.

ESSAY The Human Side of Research

The discovery of the double helix was a critically important advance in the early 1950s, because DNA had recently been established as the hereditary material. Stated another way, scientists had recently discovered that genes are made of DNA. Before Watson and Crick published, though, it was a complete mystery not only how genes are copied when new cells and offspring are produced but also how the information in genes is used to synthesize the enzymes and other molecules required for life. The discovery of complementary base pairing revealed how genetic information could be copied, and knowledge of DNA's secondary structure stimulated several decades of work aimed at understanding how cells use base sequences as information. It's no exaggeration to claim that the double helix triggered an explosion in research and knowledge. Molecular biology had come of age.

The field's early years were extraordinarily eventful. In the late 1940s and early 1950s, understanding DNA's structure was *the* question in biology. Researchers who were working on the problem knew that the greatest honor in science—the Nobel Prize—would almost certainly be awarded to the person or persons who were first to discover the correct structure. The race was on: Many of the brightest and most famous scientists of the day were involved, working in different laboratories around the world.

Arguably the most important event in the research was Wilkins' and Franklin's ability to produce X-ray pictures of DNA. Recall that the photographs made it possible to make crucial measurements and inferences about the molecule's structure. Although Wilkins and Franklin are justifiably given joint credit for this work, they actually did not cooperate closely. It was Franklin who took the photographs that led to the realization that DNA is helical, but it was Wilkins who showed her photographs to Watson and Crick. Wilkins did so without Franklin's knowledge or permission. Today, such an act would be considered a serious form of professional misconduct. And Linus Pauling—undoubtedly the century's greatest biochemist, who was also working on the problem at the time—was denied permission to travel from the United States to England to see the data. The United States was then in the throes of the Cold War, and the U.S. State Department had revoked Pauling's passport because his political views were considered too liberal.

Scientific research is a very human enterprise.

As expected, Watson and Crick, along with Wilkins, received the Nobel Prize. The award was made in 1962, after experimental work had confirmed that the double-helix model was correct. Tragically, Franklin died of ovarian cancer in 1958, when she was just 37. She never received the Nobel, because the honor is not awarded posthumously.

The moral of the story, if there is one, may be simply that scientific research is a very human enterprise. In addition, times have changed. Although competition can still be intense, biology is an increasingly cooperative and international enterprise; misconduct is usually detected and addressed promptly. Rosalind Franklin's professional life was made difficult at times because she was a woman. Today the majority of graduate students in molecular biology are female.

CHAPTER REVIEW

Summary of Key Concepts

Nucleotides are monomers that consist of a sugar, a phosphate group, and a nitrogen-containing base. Ribonucleotides polymerize to form RNA. Deoxyribonucleotides polymerize to form DNA.

To understand the origin of life and how cells work today, it is critical to understand the structure and function of macromolecules—particularly proteins, nucleic acids, and carbohydrates. Each of these molecules is a polymer that is made up of monomers. In the case of nucleic acids, the monomers involved are nucleotides.

The origin of nucleotides remains an important challenge for the theory of chemical evolution. Experiments have shown that sugars and the nitrogenous bases called purines are easily produced under early Earth conditions. But it is not yet clear how the important sugar called ribose could have been produced in large quantities during chemical evolution or how the nitrogenous bases called pyrimidines could have been synthesized.

The polymerization of monomers into macromolecules is the next key event in chemical evolution. Experiments have shown that these condensation reactions occur readily when growing polymers stick to small mineral particles. For example, researchers have observed RNA formation from the polymerization of ribonucleic acids on clay-sized particles.

DNA's primary structure consists of a sequence of nitrogen-containing bases, which contain information in the form of a molecular code. DNA's secondary structure consists of two DNA strands running in opposite directions. The strands are held together by complementary base pairing and are twisted into a double helix.

DNA is an extremely stable molecule that serves as a superb archive for information in the form of base sequences. DNA is stable because deoxyribonucleotides lack a reactive 2′ hydroxyl group and because antiparallel DNA strands form a secondary structure called a double helix. The DNA double helix is stabilized by hydrogen bonds that form between complementary purine and pyrimidine bases and by hydrophobic interactions between bases stacked on the inside of the spiral. This same structural stability and regularity make DNA ineffective at catalysis, however.

In addition to being stable, DNA is readily copied via complementary base pairing. Complementary base pairing occurs between A-T and G-C pairs in DNA, as well as between A-U and G-C pairs in RNA.

Web Tutorial 4.1 Nucleic Acid Structure

RNA's primary structure also consists of a sequence of nitrogen-containing bases that contain information in the form of a molecular code. Its secondary structure includes short double helices and structures called hairpins. RNA molecules called ribozymes catalyze important chemical reactions.

Compared with proteins and DNA, RNA is exceptionally versatile. The primary function of proteins is to catalyze chemical reactions, and the primary function of DNA is to carry information. But RNA is an "all-purpose" macromolecule that can do both. Most origin-of-life researchers propose that the first self-replicating molecule was RNA, because RNA can catalyze a variety of chemical reactions and because complementary base pairing between nucleotides furnishes a mechanism for making a copy. Recent experiments have succeeded in producing a ribozyme that can catalyze the formation of phosphodiester linkages and the addition of ribonucleotides to a growing strand. Based on results to date, it is reasonable to predict that a self-replicating RNA will be created in the near future.

Questions

Content Review

1. What are the four nitrogenous bases found in RNA?
   a. uracil, guanine, cytosine, thymine (U, G, C, T)
   b. adenine, guanine, cytosine, thymine (A, G, C, T)
   c. adenine, uracil, guanine, cytosine (A, U, G, C)
   d. alanine, threonine, glycine, cysteine (A, T, G, C)

2. What determines the primary structure of an RNA molecule?
   a. the sugar-phosphate backbone
   b. complementary base pairing and the formation of hairpins
   c. the sequence of deoxyribonucleotides
   d. the sequence of ribonucleotides

3. DNA attains a secondary structure when hydrogen bonds form between the nitrogenous bases called purines and pyrimidines. What are the complementary base pairs that form in DNA?
   a. A-T and G-C
   b. A-U and G-C
   c. A-G and T-C
   d. A-C and T-G

4. By convention, biologists write the sequence of bases in RNA and DNA in which direction?
   a. $3' \rightarrow 5'$
   b. $5' \rightarrow 3'$
   c. N-terminal to C-terminal
   d. C-terminal to N-terminal

5. In RNA, when does the secondary structure called a hairpin form?
   a. when hydrophobic residues coalesce
   b. when hydrophilic residues interact with water
   c. when complementary base pairing between ribonucleotides on the same strand creates a stem-and-loop structure
   d. when complementary base pairing forms a double helix

6. The secondary structure of DNA is called a double helix. Why?
   a. Two strands wind around one another in a helical, or spiral, arrangement.
   b. A single strand winds around itself in a helical, or spiral, arrangement.
   c. It is shaped like a ladder.
   d. It stabilizes the molecule.

Conceptual Review

1. Explain how complementary base pairing makes the copying of RNA and DNA molecules possible. Include diagrams that provide examples.
2. Growing strands of nucleic acids are always extended in the $5' \rightarrow 3'$ direction. What do the $5'$ and $3'$ refer to? Draw the bond that forms at these locations to link nucleotides. Why can't nucleic acids polymerize in the $3' \rightarrow 5'$ direction?
3. Make a generalized cartoon of monomers undergoing condensation reactions to form a polymer. Label the type of bond that forms when nucleotides polymerize into RNA or DNA. Does it take energy for polymerization reactions to proceed, or do they occur spontaneously? Why or why not?
4. Why is DNA such a stable molecule?
5. Summarize the types of secondary structure found in RNA. Include a labeled diagram.
6. A major theme in this chapter is that the structure of molecules correlates with their function. Explain why DNA's secondary structure limits its catalytic abilities compared with that of RNA. Why is it logical that RNA molecules can catalyze a modest but significant array of reactions? Why are proteins the most effective catalysts?

Group Discussion Problems

1. Do you agree with the "definition" of life given at the start of this chapter? Why or why not? If you were looking for life on Jupiter's moon Europa, how would you know when you found it?
2. Suppose that experiments like those reviewed in Section 4.4 succeeded in producing a molecule that could make a copy of itself. Outline a one-page opinion piece for your local newspaper that explains the nature of the research and discusses the ethical and philosophical implications of the discovery.
3. Before Watson and Crick published their model of the DNA double helix, Linus Pauling offered a model based on a triple helix. Draw your conception of what a DNA triple helix might look like. What interactions would keep such a secondary structure together? How could such a molecule be copied?
4. Origin-of-life researcher Robert Crabtree maintains that experiments simulating early Earth conditions are a valid way to test the theory of chemical evolution. Crabtree claims that if scientists working in the field agree that an experiment is a plausible reproduction of early Earth conditions, it is valid to infer that its results are probably correct—that the simulation effectively represents events that occurred some 4 billion years ago. Do you agree? Do you find the models and experiments presented in this chapter and previous chapters to be convincing tests of the theory? Explain your answers.

Answers to Multiple-Choice Questions 1. c; 2. d; 3. a; 4. b; 5. c; 6. a

www.prenhall.com/freeman is your resource for the following: Web Tutorials; Online Quizzes and other Online Study Guide materials; Answers to Conceptual Review Questions; Solutions to Group Discussion Problems; Answers to Figure Caption Questions and Exercises; and Additional Readings and Research.

5 An Introduction to Carbohydrates

KEY CONCEPTS

- Sugars and other carbohydrates are highly variable in structure. They perform a wide variety of functions in cells, ranging from energy storage to formation of tough structural fibers.
- Monosaccharides are monomers that polymerize to form polysaccharides, via different types of glycosidic linkages.

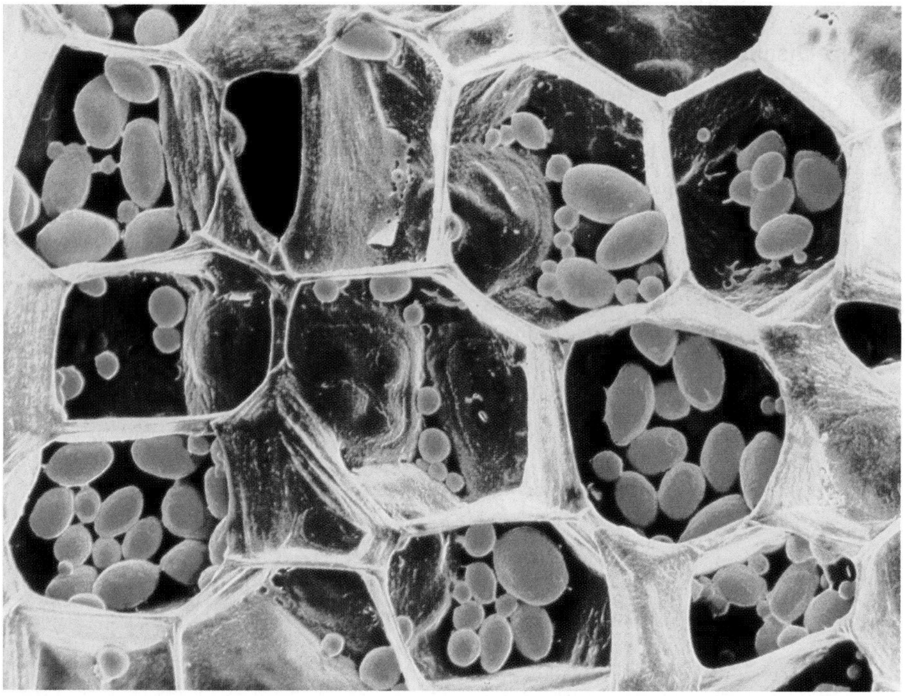

Scanning electron micrograph of plant cell walls (yellow honeycombed structures), which consist primarily of cellulose. The orange granules are stored starch. Both cellulose and starch are carbohydrates. Cellulose provides structural support for these cells, and starch supplies chemical energy.

This unit highlights the four types of macromolecules that are prominent in today's cells: proteins, nucleic acids, carbohydrates, and lipids. Understanding the structure and function of each of these macromolecules is a basic requirement for exploring how life began and how organisms work. In Chapters 3 and 4 we analyzed the way proteins and nucleic acids are put together and what they do. This chapter focuses on carbohydrates; Chapter 6 will introduce lipids.

The term **carbohydrate** encompasses both the monomers called monosaccharides ("one-sugar") and the polymers called polysaccharides ("many-sugars"). The name is logical because the chemical formula of many carbohydrates is $(CH_2O)_n$, where the *n* refers to the number of "carbon-hydrate" groups. The name is also misleading, though, because carbohydrates do not consist of carbon atoms bonded to water molecules. Instead, they are molecules with a carbonyl ($>C=O$) and several hydroxyl (–OH) functional groups, along with several to many carbon-hydrogen (C–H) bonds.

Let's begin by considering the structure of monosaccharides and continue by analyzing how these building blocks polymerize to form polysaccharides. The chapter closes with a look at what carbohydrates do in cells today and an analysis of their role in the origin of life.

5.1 Sugars as Monomers

Sugars are fundamental to life. They provide chemical energy in cells and furnish some of the molecular building blocks required for the synthesis of larger, more complex compounds. Monosaccharides were also important during chemical evolution, early in Earth's history. For example, the sugar called ribose would have been required for the formation of nucleotides. Laboratory simulations have shown that ribose and many other monosaccharides could have been produced in the prebiotic soup. What are these compounds, and how do they differ from one another?

Figure 5.1 illustrates the structure of the monomer called a **monosaccharide**, or simple sugar. Note that the carbonyl groups that serve as one of monosaccharides' distinguishing features can be found either at the end of the molecule, forming an aldehyde sugar (an *aldose*), or within the carbon chain, forming a ketone sugar (a *ketose*). The presence of a carbonyl group along with multiple hydroxyl groups gives sugars an array of functional groups. Based on this observation, it's not surprising that sugars are able to participate in a large number of chemical reactions.

The number of carbon atoms present also varies in monosaccharides. By convention, the carbons in a monosaccharide are numbered consecutively, starting with the end nearest the carbonyl group. Figure 5.1 features three-carbon sugars, or **trioses.** Ribose, which acts as a building block for nucleotides, has five carbons and is called **pentose**; the glucose that is coursing through your bloodstream and being used by your cells right now is a six-carbon sugar, or **hexose.**

In addition to varying in the location of the carbonyl group and the total number of carbon atoms present, monosaccharides can vary in the spatial arrangement of their atoms. There is, for example, a wide array of pentoses and hexoses. Each is distinguished by the configuration of its hydroxyl functional groups. **Figure 5.2** illustrates glucose and galactose, which are six-carbon sugars that are optical isomers—they have the same chemical formula ($C_6H_{12}O_6$) but not the same structure. Although both are aldose sugars with six carbons, they differ in the spatial arrangement of the hydroxyl group at the carbon highlighted in Figure 5.2. Because their structures differ, their functions differ. In cells, glucose is used as a source of chemical energy in the production of ATP. But for galactose to be used as a source of energy, it first has to be converted to glucose via an enzyme-catalyzed reaction. A total of eight different hexoses exist, due to the different ways of configuring the hydroxyl groups in space. In addition, each hexose comes in two forms, each of which is a mirror image of the other. Thus sixteen distinct structures with the molecular formula $C_6H_{12}O_6$ are possible.

Sugars do not usually exist in the form of linear chains as illustrated in Figure 5.1 and Figure 5.2, however. Instead, in aqueous solution they tend to form ring structures. The chain and ring forms exist in equilibrium when simple sugars are in solution, with the ring forms predominating. Glucose serves

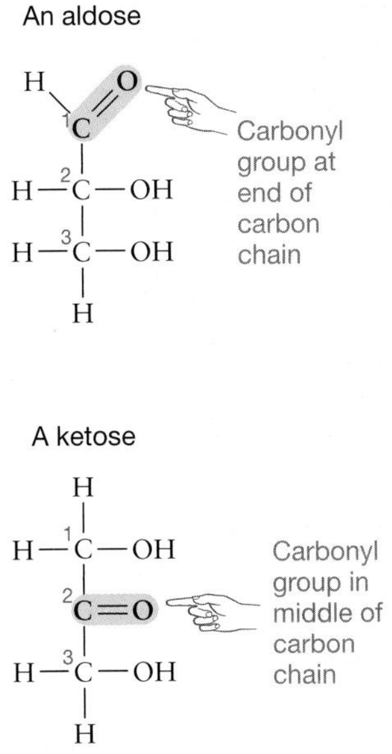

FIGURE 5.1 The Carbonyl Group in a Sugar Occurs in One of Two Configurations
The carbonyl group in a sugar can occur at the end of the molecule, forming an aldehyde sugar (an aldose), or within the carbon chain, forming a ketone sugar (a ketose).

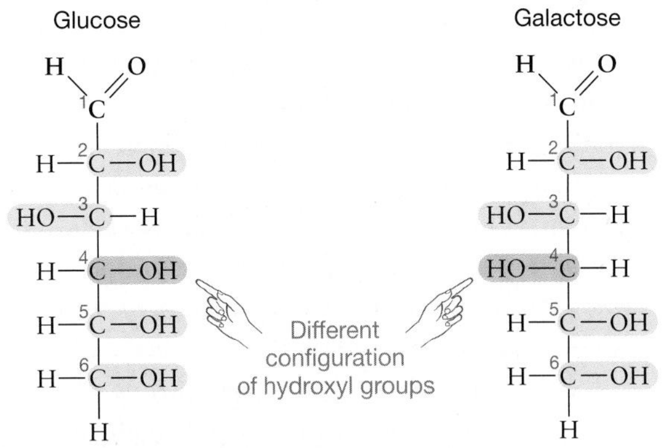

FIGURE 5.2 Sugars May Vary in the Configuration of Their Hydroxyl Groups
The two six-carbon sugars shown here vary only in the spatial orientation of their hydroxyl groups. These molecules are optical isomers, but they are not mirror images of each other. EXERCISE Next to these structures, draw the structural formula of mannose. Mannose is a six-carbon sugar that is identical to glucose, except that the hydroxyl (OH) group on carbon number 2 is switched in orientation.

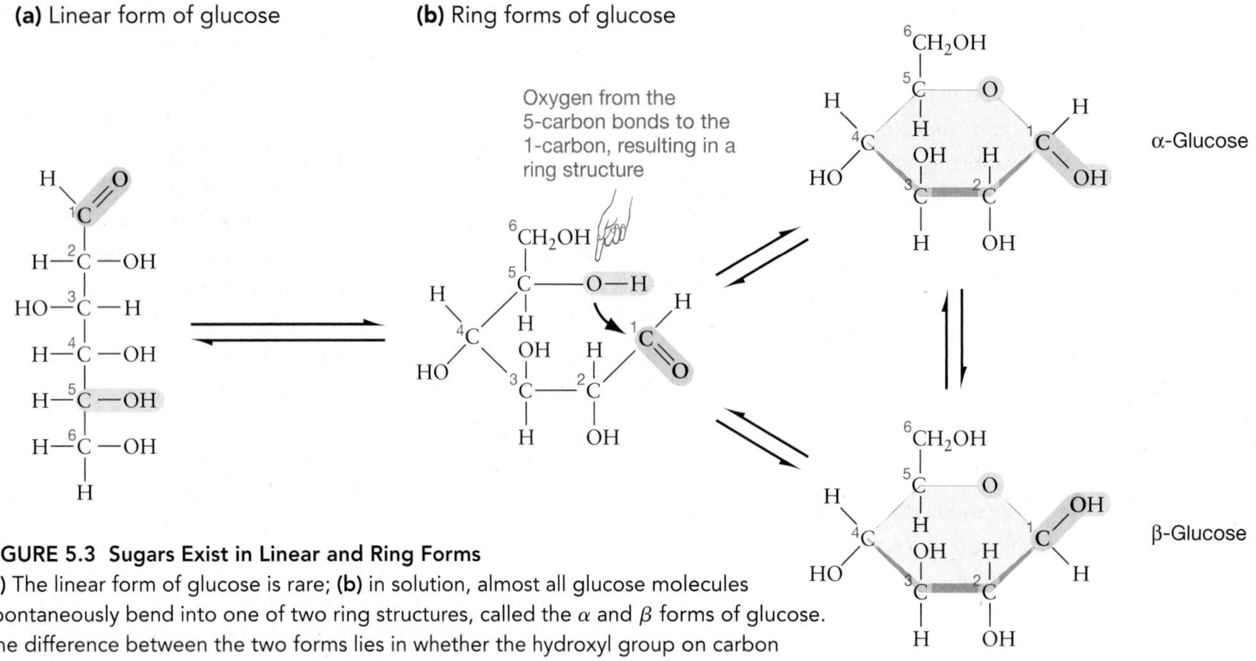

FIGURE 5.3 Sugars Exist in Linear and Ring Forms
(a) The linear form of glucose is rare; **(b)** in solution, almost all glucose molecules spontaneously bend into one of two ring structures, called the α and β forms of glucose. The difference between the two forms lies in whether the hydroxyl group on carbon number 1 is above or below the plane of the ring. The two forms exist in equilibrium, but the β form is more common because it is slightly more stable than the α form.

as the example in **Figure 5.3.** When the cyclic structure forms in glucose, the carbon that is numbered 1 in the linear chain forms a bond with an oxygen atom and with a hydroxyl group that can be oriented in two distinct ways. As the right-hand side of Figure 5.3 shows, the different configurations result in the molecules α-glucose and β-glucose.

To summarize, the existence of alternate ring forms, optical isomers with different arrangements of hydroxyl groups in space and different mirror-image forms, variation in carbon number, and aldose or ketose placement of the carbonyl group makes a very large number of monosaccharides possible. Each has a unique structure and function.

Laboratory simulations have shown that most monosaccharides are readily synthesized under conditions that mimic the prebiotic soup. For example, when formaldehyde (H_2CO) molecules are heated in solution, they react with one another to form almost all of the pentoses and hexoses, as well as some seven-carbon sugars. In addition, researchers recently announced the discovery of the three-carbon ketose illustrated in Figure 5.1, along with a wide array of compounds closely related to sugars, on the Murchison meteorite (introduced in Chapter 3). Based on these observations, investigators suspect that sugars are synthesized on dust particles and other debris in interstellar space, and could have rained down onto Earth as the planet was forming. Most researchers interested in chemical evolution have become increasingly confident that a wide diversity of monosaccharides existed in the prebiotic soup. But it remains a mystery why ribose might have predominated and made the synthesis of nucleotides possible. It also appears highly unlikely that monosaccharides were able to polymerize to form the polysaccharides found in today's cells. Let's explore why.

✓ CHECK YOUR UNDERSTANDING

Simple sugars differ in three respects: (1) the location of their carbonyl group, (2) the number of carbon atoms present, and (3) the spatial arrangement of their atoms—particularly the relative positions of hydroxyl (OH) groups. You should be able to draw the structural formula of a monosaccharide in linear form and then draw other sugars that differ from this one in each of the three aspects listed.

5.2 The Structure of Polysaccharides

Polysaccharides are polymers that form when monosaccharides are linked together. They are also known as complex carbohydrates. The simplest polysaccharides consist of two sugars and are known as **disaccharides.** The two monomers involved may be identical, as in the two α-glucose molecules that link to form maltose. Or they may be different, as in the combination of a glucose molecule and a galactose molecule that forms lactose—the most important sugar in milk. (**Box 5.1** explains how problems with metabolizing lactose or galactose lead to disease in humans.)

Simple sugars polymerize when a condensation reaction occurs between two hydroxyl groups, resulting in a covalent bond called a **glycosidic linkage** (**Figure 5.4**). Glycosidic linkages are analogous to the peptide bonds that hold proteins together and to the phosphodiester bonds that connect the nucleotides in nucleic acids. There is an important difference, however. Peptide and phosphodiester bonds always form at the same location in their monomers. But because glycosidic linkages form between hydroxyl groups, and because every monosaccharide contains

BOX 5.1 Lactose Intolerance and Galactosemia

Most organisms have a large array of enzymes that catalyze reactions involving sugars. A different enzyme processes each type of sugar. This should not be surprising, considering that sugars differ in shape and that the ability of an enzyme's active site to catalyze a reaction depends on the shape of the substrate. As explained in Chapter 3, the specificity of enzyme action is a prime example of how a molecule's structure correlates with its function.

The human enzyme lactase, for example, catalyzes only one reaction, in which lactose is split into glucose and galactose. Another enzyme, called 1-phosphate uridyl transferase, is involved in converting galactose to glucose. When these enzymes do not function normally, lactose or galactose concentrations increase in the body and cause disease.

The condition known as lactose intolerance occurs in most adult humans to some extent. In lactose-intolerant individuals, the enzyme lactase is produced only during infancy—at a time when the primary source of nutrition is mother's milk. Lactase production ends as these individuals mature. This pattern has been observed in most human cultures, where traditional diets did not include cow's milk. Dairy cattle were unknown in Thai culture, for example, and 97 percent of Thai people are lactose intolerant as adults. This pattern is also logical, because it would be a waste of time and energy to produce an enzyme that has no substrate to work on. If these individuals drink milk once lactase production ends, though, the unprocessed lactose causes bloating, abdominal pain, and diarrhea. But in cultures where milk has been an important source of nutrition in adults for many centuries, lactase production continues throughout life. Among Danish people, for example, 97 percent are lactose tolerant throughout life.

To make sense of lactose tolerance and intolerance, recall the concept of adaptation introduced in Chapter 1. An adaptation is a trait that increases the fitness of individuals in a particular environment. Lactose intolerance is an adaptation to environments where milk is not available to adults. Lactose tolerance, in contrast, is an adaptation to environments where milk is readily available to adults.

Unfortunately, problems with galactose metabolism are completely nonadaptive. If a child lacks the enzyme that converts galactose to glucose, galactose concentrations build to high levels in the blood and cause the array of symptoms known as galactosemia ("galactose-sign"). Individuals who suffer from galactosemia are at risk for mental retardation or even death. The only cure is to exclude galactose from the diet. In large enough doses, milk, yogurt, cheese, and other dairy products are life-threatening substances for these individuals.

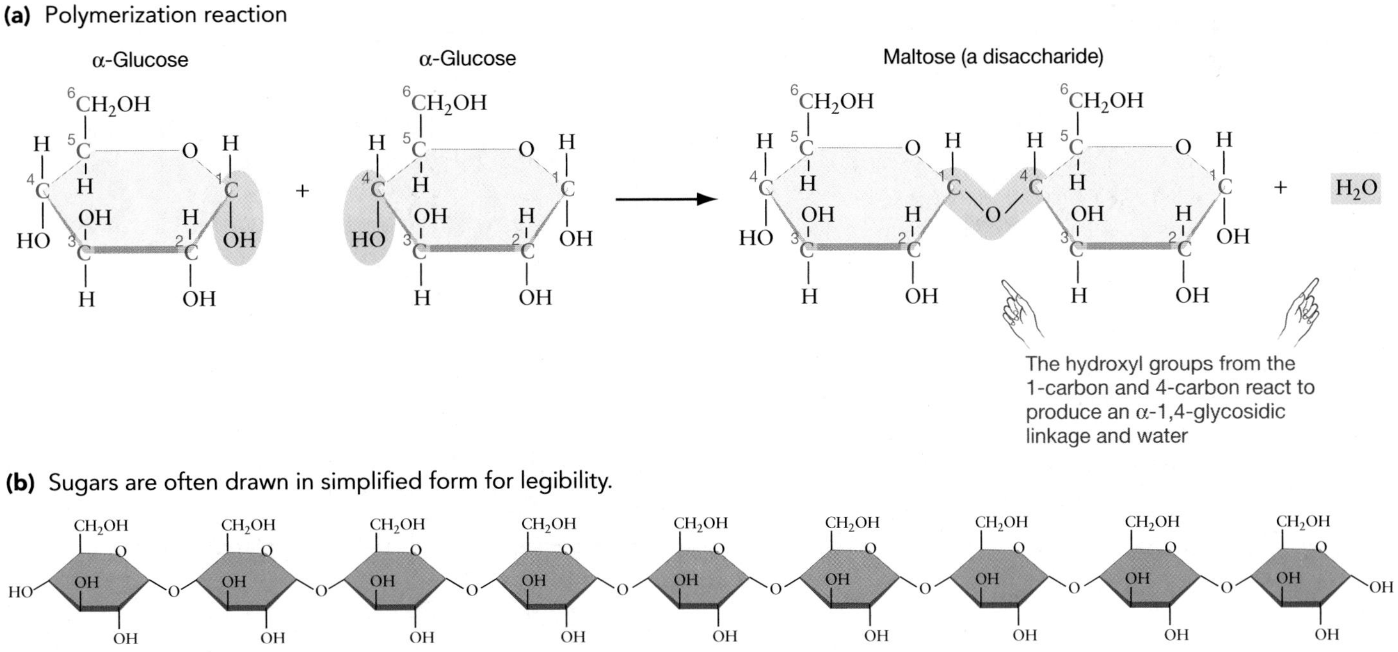

FIGURE 5.4 Monosaccharides Polymerize through Formation of Glycosidic Linkages
(a) A glycosidic linkage occurs when hydroxyl groups on two monosaccharides undergo a condensation reaction to form the bond shown on the right. In this case, the hydroxyl groups involved were on carbons 1 and 4 of the two monomers, which were in the α-ring configuration. The resulting bond is called an α-1,4-glycosidic linkage. **(b)** Polysaccharides consist of a series of monosaccharides linked together by glycosidic bonds. EXERCISE Label the glycosidic linkages in part (b).

at least two hydroxyl groups, the location and geometry of glycosidic linkages can vary widely among polysaccharides. To drive this point home, consider the structures of the most common polysaccharides found in organisms today: starch, glycogen, cellulose, and chitin, along with a modified polysaccharide called peptidoglycan.

Each of these macromolecules can consist of a few hundred to many thousands of monomers, joined by glycosidic linkages at different locations.

Starch: A Storage Polysaccharide in Plants

In plant cells, monosaccharides are stored for later use in the form of starch. **Starch** consists entirely of α-glucose monomers that are joined by glycosidic linkages. As the top panel in **Table 5.1** shows, the angle of the linkages between carbons 1 and 4 causes the chain of glucose subunits to coil into a helix. Starch is actually a mixture of two such polysaccharides, however. One is an unbranched molecule called amylose. The other is a branched molecule called amylopectin. Branching occurs

TABLE 5.1 Polysaccharides Differ in Structure

Polysaccharide	Chemical Structure	Three-dimensional Structure
Starch	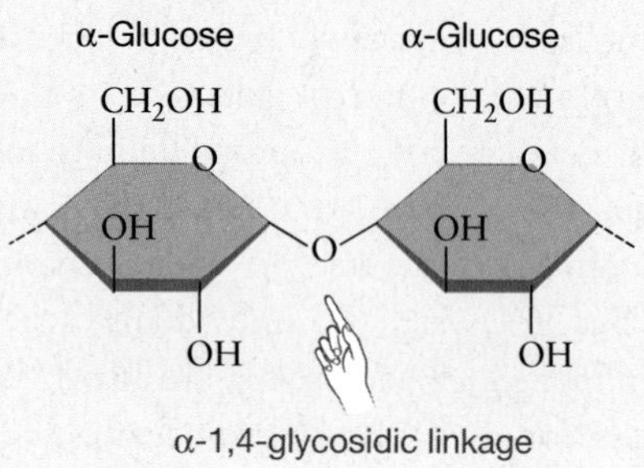	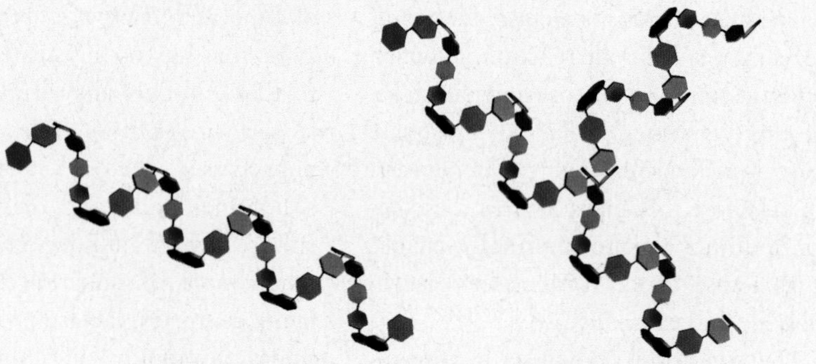 Unbranched helix (amylose); Branched helices (amylopectin)
Glycogen	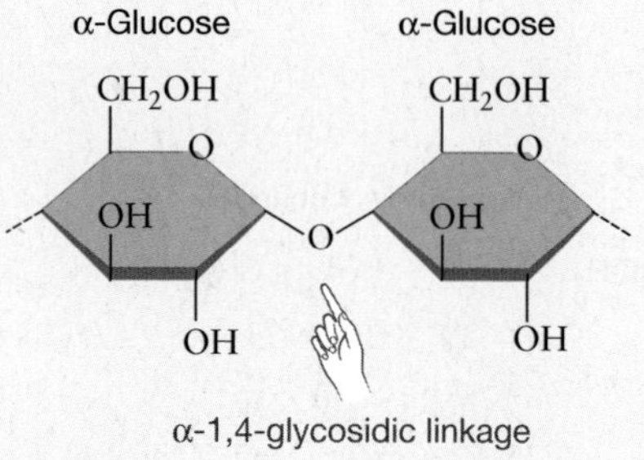	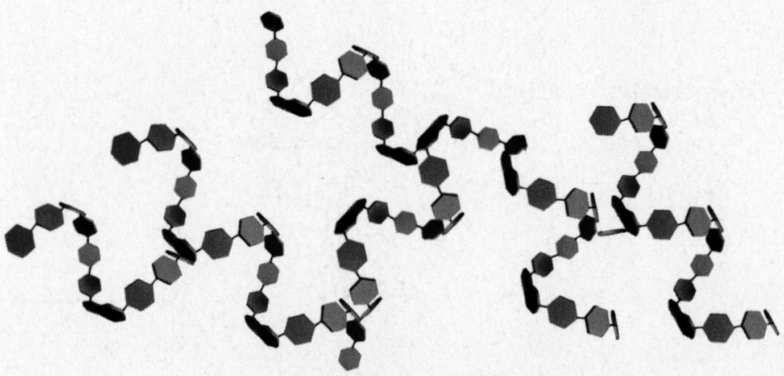 Highly branched helices
Cellulose	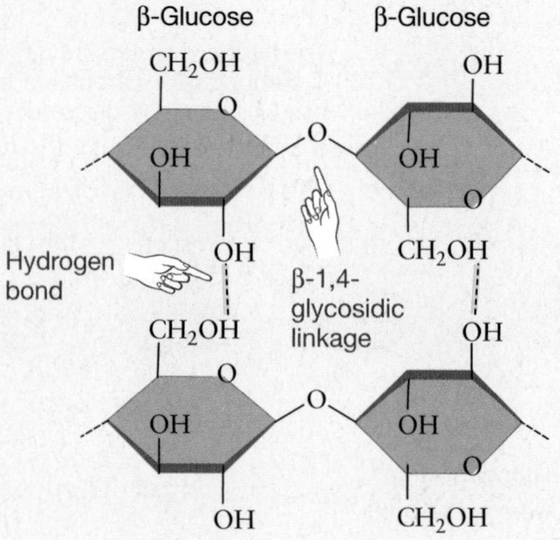	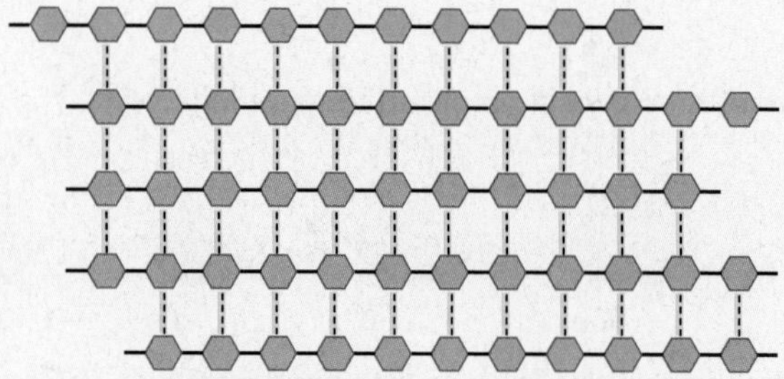 Parallel strands joined by hydrogen bonds

TABLE 5.1 Polysaccharides Differ in Structure (continued)

Polysaccharide	Chemical Structure	Three-dimensional Structure
Chitin	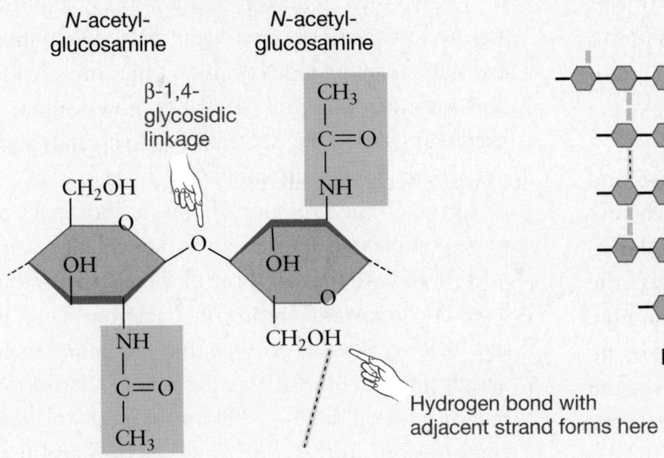	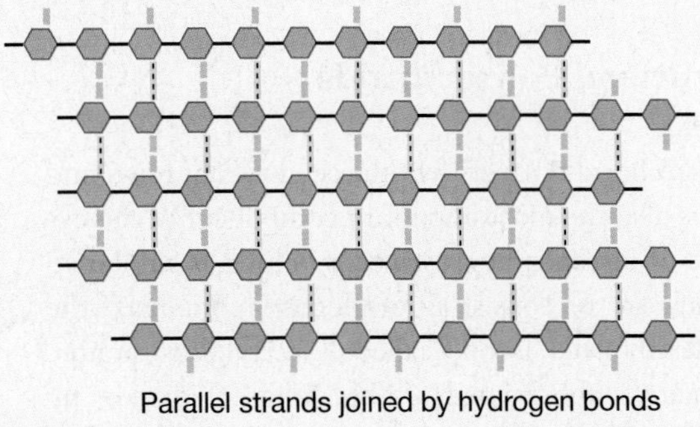
Peptidoglycan	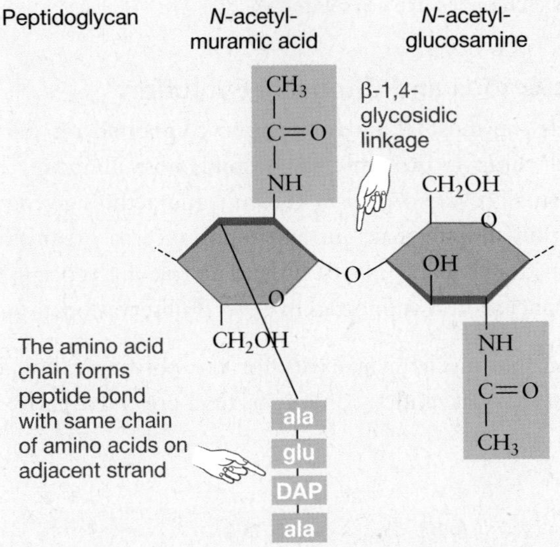	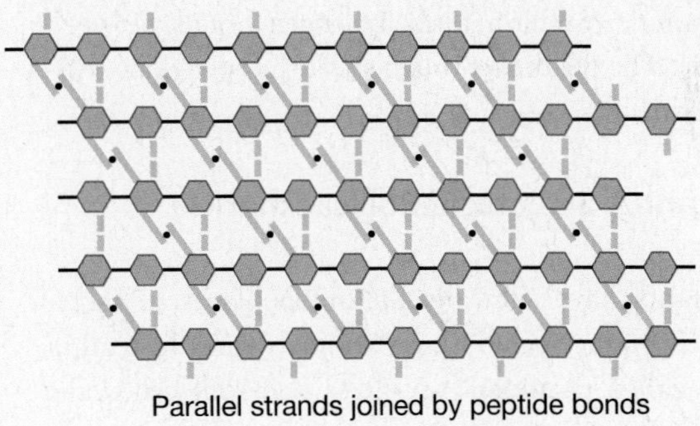

when glycosidic linkages form between carbon 1 of a glucose monomer on one strand and carbon 6 of a glucose monomer on another strand. In amylopectin, branches occur in about one out of every 30 monomers.

Glycogen: A Highly Branched Storage Polysaccharide in Animals

Glycogen performs the same storage role in animals that starch performs in plants. Glycogen is nearly identical to the branched form of starch. But instead of an α-1,6-glycosidic linkage occurring in about one out of every 30 monomers, a branch occurs in about one out of every 10 glucose subunits (see Table 5.1).

Cellulose: A Structural Polysaccharide in Plants

As we saw in Chapter 1, all cells are enclosed by a membrane. In most unicellular and multicellular organisms living today, the cell is also surrounded by a layer of material called a wall. A **cell wall** is a protective sheet that occurs outside the membrane. In algae, plants, bacteria, fungi, and many other groups, the cell wall is composed primarily of one or more polysaccharides.

In plants, cellulose is the major component of the cell wall. **Cellulose** is a polymer of β-glucose monomers, joined by β-1,4-glycosidic linkages. As Table 5.1 shows, the geometry of the bond is such that each glucose monomer in the chain is flipped in relation to the adjacent monomer. This arrangement increases the stability of cellulose strands, because the flipped orientation makes it possible for multiple hydrogen bonds to form between adjacent and parallel strands of cellulose. As Table 5.1 shows, cellulose often occurs in long, parallel strands that are joined by these hydrogen bonds. The linked cellulose fibers are strong and provide the cell with structural support.

Chitin: A Structural Polysaccharide in Animals

Chitin is a polysaccharide that stiffens the cell walls of fungi and many algae. It is also the most important component of the external skeletons of insects and crustaceans. Chitin is similar to cellulose, but instead of consisting of glucose monomers, the monosaccharide involved is one called *N*-acetylglucosamine. These monomers, abbreviated "NAc," are joined by β-1,4-glycosidic linkages (see Table 5.1). As in cellulose, the geometry of these bonds results in every other residue being flipped in orientation.

Like the glucose monomers in cellulose, the subunits of *N*-acetylglucosamine in chitin form hydrogen bonds between adjacent strands. The result is a tough sheet that provides stiffness and protection.

Peptidoglycan: A Structural Polysaccharide in Bacteria

Bacteria, like plants, have cell walls. But unlike plants, in bacteria the ability to produce cellulose is extremely rare. Instead, a polysaccharide called **peptidoglycan** gives bacterial cell walls strength and firmness.

Peptidoglycan is the most complex of the polysaccharides discussed thus far. It has a long backbone formed by two types of monosaccharides that alternate with each other and are linked by β-1,4-glycosidic linkages. In addition, a chain of amino acids is attached to one of the two sugar types. When molecules of peptidoglycan align, peptide bonds link the amino acid chains on adjacent strands. **Box 5.2** describes how certain antibiotics kill bacteria by disrupting the enzymes that catalyze the formation of the peptide-bonded cross-links.

Although the presence of amino acids gives peptidoglycan a more complex structure than either cellulose or chitin, it is important to note that all three of these structural polysaccharides have an important feature in common: They usually exist as sets of long, parallel strands that are linked to each other. This design gives the molecules the ability to withstand forces that pull or push on them—what an engineer would call tension and compression. In this way, the structure and function of structural polysaccharides are correlated.

Polysaccharides and Chemical Evolution

Cellulose is the most abundant organic compound on Earth today, and chitin is probably the second most abundant by weight. Virtually every organism known manufactures glycogen or starch. But despite their current importance to organisms, polysaccharides probably played little to no role in the origin of life. This conclusion is supported by several observations:

- No plausible mechanism exists for the polymerization of monosaccharides under conditions that prevailed early in

BOX 5.2 How Do the Penicillins and Cephalosporins Kill Bacteria?

Antibiotics are molecules that kill bacteria. The first antibiotic to have been discovered and used to cure bacterial infections in humans is produced naturally, by a soil-dwelling fungus called *Penicillium chrysogenum*. Soil is packed with bacteria and fungi that compete for space, water, and nutrients. Some fungi reduce competition by producing and secreting antibiotics, which kill the bacteria surrounding them. The drug we call penicillin is such a molecule, and it was named in honor of the species that produces it.

Penicillin and the drug called cephalosporin, which is produced naturally by species of soil-dwelling fungi in the genus *Cephalosporium*, are closely related in structure and function. Both molecules are effective because they bind very tightly to the enzymes that catalyze the formation of cross-links between individual strands within peptidoglycan. In the absence of these cross-links, the bacterial cell wall begins to weaken. Eventually it fails and tears open, and the cell is destroyed. Penicillin and cephalosporin cause very few side effects in humans, because the binding event between these drugs and the cell-wall enzyme to which they bind is extremely specific. Human cells are not affected by the presence of penicillin or cephalosporin, because they lack the enzymes involved in cross-linking peptidoglycan.

In response to the widespread use of penicillin and cephalosporin, however, many bacterial populations have evolved an enzyme that breaks these drugs apart and renders them ineffective. (Chapter 23 will provide more detail on how drug resistance evolves.) In response, researchers have synthesized molecules that are very closely related to penicillin and cephalosporin structurally and have the same function, but that are not as severely affected by the newly evolved bacterial enzyme. You may, in fact, be familiar with some of these synthetic penicillins and cephalosporins from personal experience. They include widely prescribed molecules such as methicillin, oxacillin, ampicillin, and ceftriaxone.

Earth's history. In cells and in vitro, the glycosidic linkages illustrated in Figure 5.4 and Table 5.1 form only with the aid of specialized enzymes.

- It is highly likely that life began in the form of an RNA molecule. However, researchers have yet to discover a ribozyme that can join simple sugars by catalyzing the formation of glycosidic linkages. Thus, it appears extremely unlikely that polysaccharides were present in significant quantities during the RNA world.
- Even though monosaccharides contain large numbers of hydroxyl and carbonyl groups, they lack the diversity of functional groups found in amino acids. Polysaccharides also have very simple secondary structures, consisting of linkages between adjacent strands. Thus, they lack the structural and chemical complexity that makes proteins, and to a lesser extent RNA, effective catalysts. To date, no reactions have been discovered that are catalyzed by polysaccharides.
- The monomers in polysaccharides are not capable of complementary base pairing. Like proteins but unlike DNA and RNA, polysaccharides are not capable of providing the information required for themselves to be copied. As far as is known, no polysaccharides store information in cells. Thus, no one has proposed that the first living entity might have been a polysaccharide.

Even though polysaccharides probably did not play a significant role in the earliest forms of life, they became enormously important once cellular life evolved. Let's take a detailed look at how they function in today's cells.

✓CHECK YOUR UNDERSTANDING

Polysaccharides form when enzymes catalyze the formation of glycosidic linkages between monosaccharides that are in the α- or β-ring form. Most polysaccharides are long, linear molecules, but some branch extensively. Among linear forms, it is common for adjacent strands to be linked by hydrogen bonding or other types of linkages. You should be able to make a rough sketch that shows the structures of glycogen and cellulose, and label at least two glycosidic linkages in each structure.

5.3 What Do Carbohydrates Do?

Chapter 4 introduced one of the three basic functions that carbohydrates perform in organisms: furnishing the building blocks of molecules needed by cells. Recall that both RNA and DNA contain sugars—the five-carbon sugars ribose and deoxyribose, respectively. In nucleotides, which consist of a sugar, a phosphate group, and a nitrogenous base, the sugar itself acts as a subunit of the larger molecule. But sugars frequently furnish only the raw "carbon skeletons" that are used in the synthesis of important molecules. Amino acids are being synthesized by your cells right now, for example, using sugars as a starting point.

Although the details of how sugars are used in synthesizing amino acids and other complex molecules are beyond the scope of this book, it will be productive to delve into the other three major roles of carbohydrates: indicating cell identity, storing chemical energy, and providing cells with fibrous structural materials.

The Role of Carbohydrates in Cell Identity

As we saw in Section 5.2, polysaccharides do not store information in cells. It's important to recognize, however, that polysaccharides can display important information. **Figure 5.5** shows how this information display happens, on the outer surface of a cell. Certain molecules project outward from this surface, into the environment surrounding the cell. These molecules are called glycoproteins. A **glycoprotein** is a protein that is covalently bonded to a carbohydrate—usually a relatively short chain of sugars.

Glycoproteins are key molecules in what biologists call cell-cell recognition and cell-cell signaling. Each distinct type of cell in a multicellular organism—for example, the nerve cells and immune system cells in your body—displays a different set of glycoproteins on its surface. The same is true for different species of unicellular organisms, such as the approximately 500 species of bacteria that live in your mouth as well as the cells that exist by the millions in a teaspoonful of soil or a drop of pond water. The role of the carbohydrates on the cell surface is to identify the type of cell that displays them. A glycoprotein is a "sugar coating" that acts like the magnetic stripe on the back of a credit card—it immediately identifies the individual that bears it.

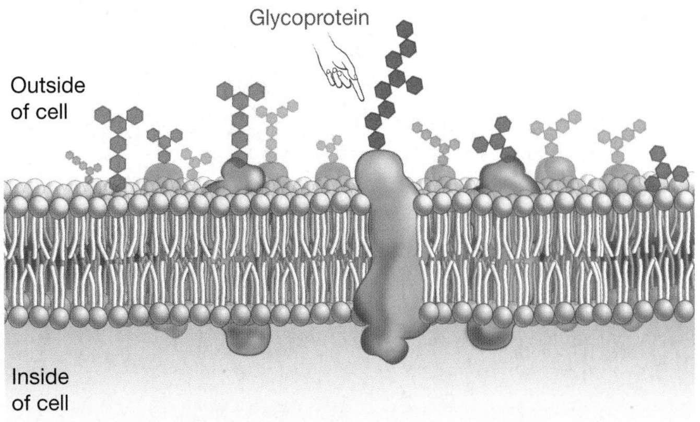

FIGURE 5.5 Carbohydrates Are an Identification Badge for Cells
Glycoproteins contain sugar groups that project from the surface of the plasma membrane. These sugar groups have distinctive structures that identify the type or species of the cell.

In later chapters we'll explore how the identification information displayed by glycoproteins is used by cells and cellular signals. The key point here is to recognize that the enormous number of structurally distinct monosaccharides makes it possible for an enormous number of unique, short polysaccharides to exist. As a result, each cell type and each species can display a unique identity.

The Role of Carbohydrates in Energy Production and Storage

Candy bar wrappers promise a quick energy boost, and ads for sports drinks claim to provide the "carbs" needed for peak activity. If you were to ask friends or family members what carbohydrates do in your body, they would probably say something like "They give you energy." And after pointing out that carbohydrates are also used to establish cell identity, as a structural material, and as a source of carbon skeletons for the synthesis of other complex molecules, you'd have to agree. Carbohydrates store and provide chemical energy in cells. What is it about carbohydrates that makes this possible?

The answer to this query lies in the discussion of redox reactions that began in Chapter 2. Recall that reduction-oxidation reactions involve the transfer of an electron. The atom that receives an electron is said to be reduced, while the atom that loses an electron is said to be oxidized. These points were introduced in Chapter 2 because they helped explain what was happening during the early stages of chemical evolution. You may remember that experimental and computer simulation studies have shown that if molecules with high free energy—such as molecular hydrogen (H_2), methane (CH_4), and ammonia (NH_3)—are available to act as electron donors, and if a source of intense kinetic energy—such as an electrical discharge or high-intensity solar radiation—is present, redox reactions occur. Those reactions result in the production of molecules with reduced carbon atoms, such as formaldehyde (H_2CO) and hydrogen cyanide (HCN). With continued inputs of kinetic energy, these molecules react to form sugars, amino acids, and other complex organic compounds.

Carbohydrates as Electron Donors Today, the key reaction that results in the production of sugar occurs in plants and other photosynthetic organisms. Photosynthesis is a complex process that can be summarized most simply as follows:

$$CO_2 + H_2O + \text{sunlight} \rightarrow (CH_2O)_n + O_2$$

where $(CH_2O)_n$ represents a carbohydrate. **Figure 5.6** shows the structural formulas of the molecules involved and represents the relative positions of their covalently bonded electrons. The figure has two key features:

1. Electrons are closer to the carbon atom in the product molecule $(CH_2O)_n$ than they are to the carbon atom in carbon dioxide (CO_2). Thus, photosynthesis results in the reduction of carbon.
2. The covalently bonded electrons in the C–H bonds of carbohydrates are shared more equally, and thus held less tightly, than they are in the C–O bonds of carbon dioxide. As a result, carbohydrates have much more free energy than carbon dioxide has.

The essence of photosynthesis, then, is that energy in sunlight is transformed into chemical energy that is stored in the C–H bonds of carbohydrates. Because the C–H bonds are unstable and have high free energy, carbohydrates can act as electron donors in redox reactions that lead to the production of chemical energy in the form of ATP—the phosphorylated molecule introduced in Chapter 4. The

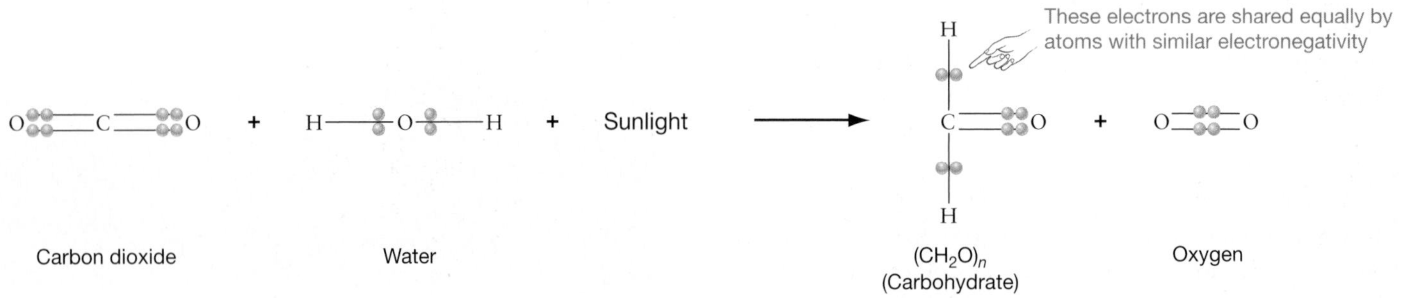

FIGURE 5.6 Carbohydrates Have High Free Energy
In these diagrams, the horizontal lines. indicate covalent bonds. The dots represent the relative positions of electrons in those bonds. QUESTION If carbon becomes reduced in this reaction, which atoms become oxidized? EXERCISE Circle the bonds in this diagram that have the highest free energy.

overall reaction for the oxidation of sugars in cells can be written as follows:

$$CH_2O + O_2 + ADP \rightarrow CO_2 + H_2O + ATP$$

Chemical energy stored in the C–H bonds of carbohydrate is transferred to chemical energy in the form of the third phosphate group in ATP. As indicated in Chapter 3, the free energy in ATP makes it possible for phosphorylation events to drive endergonic reactions, move your muscles, and perform other types of work in cells. Carbohydrates are like the water that piles up behind a dam; ATP is like the electricity, generated at a dam, that lights up your home.

Later chapters will analyze in detail how sugars and other carbohydrates are made in organisms, and how these carbohydrates are then oxidized to provide cells with usable chemical energy in the form of ATP. For now, the important thing is to recognize that carbohydrates store and provide chemical energy to cells because they contain a large number of C–H bonds and because C–H bonds have high free energy.

Figure 5.7 offers another way to think about carbohydrates as molecules that store energy. During chemical evolution and in today's plants, algae, and photosynthetic bacteria, the carbon atom in carbon dioxide is reduced. Stated another way, the carbon atoms in CO_2 molecules gain electrons. This occurs when a proton (H^+) replaces an oxygen in the CO_2 molecule. Compare the structures of carbon dioxide and the carbohydrate shown in Figures 5.7a and 5.7b, and you'll see that hydrogen atoms—an electron plus a proton—have been added to the carbohydrate. As Figure 5.7c shows, the fatty-acid subunits found in fats have even more free energy than carbohydrates have. Fatty acids are made up largely of the long hydrocarbon (C–H) chains introduced in Chapter 2. Compared with carbohydrates, fats contain many more C–H bonds and many fewer C–O bonds. C–H bonds have high free energy because the electrons are shared equally by atoms with low electronegativities. C–O bonds have low free energy because the highly electronegative oxygen atom holds the electrons so tightly. Both carbohydrates and fats are used as fuel in cells. Fats will be discussed in more detail in Chapter 6.

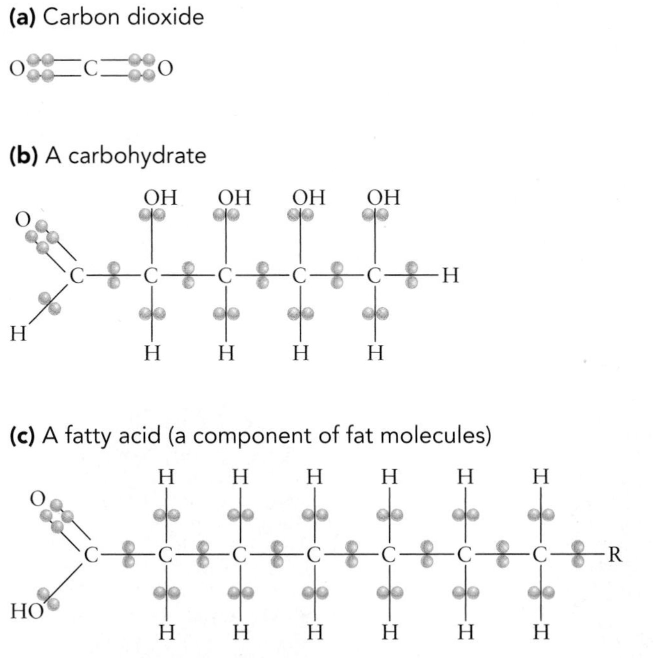
(a) Carbon dioxide
(b) A carbohydrate
(c) A fatty acid (a component of fat molecules)

FIGURE 5.7 Carbon Atoms in Carbohydrates Are Reduced **(a)** The carbon atom in carbon dioxide is highly oxidized. **(b)** In carbohydrates such as the sugar shown here, carbon atoms are highly reduced compared with carbon dioxide. **(c)** The fatty acids found in fat molecules store even more free energy in the form of C–H bonds than carbohydrates do. EXERCISE Label the molecule with the highest amount of chemical energy, and the molecule with the lowest amount. Label which molecules are likely to act as electron donors and which are likely to act as electron acceptors in redox reactions.

How Do Carbohydrates Store Energy? Section 5.2 introduced the structure of the five most important polysaccharides found in organisms: starch, glycogen, cellulose, chitin, and peptidoglycan. Starch and glycogen serve as energy-storage molecules. They can do so because enzymes readily catalyze hydrolysis reactions that break off individual glucose subunits from these polysaccharides. The glucose subunits are then used as electron donors in redox reactions that result in the production of ATP.

The most important enzyme involved in catalyzing the hydrolysis of α-glycosidic linkages in glycogen is called **phosphorylase**. The enzymes involved in breaking these linkages in starch are called **amylases**. Most of your cells contain phosphorylase, so they can break down glycogen to provide glucose on demand. Your salivary glands and pancreas also produce amylases that are secreted into your mouth and small intestine, respectively. These amylases are responsible for digesting the starch that you eat.

Carbohydrates as Structural Molecules

Cellulose and chitin, along with the modified polysaccharide peptidoglycan, are not normally used as a source of chemical energy in cells. Instead, these molecules are structural. They form fibers that give cells and organisms strength and elasticity.

To appreciate why cellulose, chitin, and peptidoglycan are effective as structural molecules, recall that they form long strands and that bonds can form between adjacent strands. Table 5.1 detailed how these bonds form. In the cell walls of plants, a collection of about 80 cellulose molecules, cross-linked by hydrogen bonding, creates a tough fiber. These cellulose fibers, in turn, criss-cross to form a tough sheet (**Figure 5.8a**). Groups of chitin molecules are also organized into fibers held together by hydrogen bonding. These fibers stiffen the cell walls of fungi and algae, and they overlap in the external skeletons of insects to form a dense, waterproof sheet (**Figure 5.8b**). In the cell walls of bacteria, peptidoglycan molecules are linked by peptide bonds and layered in sheets (**Figure 5.8c**).

Almost all organisms have the enzymes required to break the α-1,4- and α-1,6-glycosidic linkages that hold starch and glycogen molecules together. But only a few have enzymes capable of hydrolyzing the β-1,4-glycosidic linkages in cellulose, chitin, and peptidoglycan. The shape and orientation of β-1,4-glycosidic linkages make them difficult to break. Few enzymes have active sites with the correct geometry and reactive groups to do so. As a result, the structural polysaccharides are resistant to degradation and decay. In fact, the cellulose that you ingest when you eat plant cells passes through your gut undigested. It is referred to as dietary fiber, and it helps digestion by adding moisture and bulk to the feces. Cellulose absorbs water, and the addition of mass to fecal material helps it move through the intestinal tract more quickly, avoiding constipation.

In short, then, the remarkable functional difference between the energy-storing polysaccharides and the structural polysaccharides is due to differences in their structures. One key distinction between storage polysaccharides and structural polysaccharides is the presence of α-glycosidic linkages versus β-glycosidic linkages. The α-linkages in storage polysaccharides are readily hydrolyzed to release sugars, while the β-linkages in structural polysaccharides resist enzymatic degradation. Another key distinction between the two types of polysaccharide is that the geometry of the glycosidic linkages in cellulose and chitin make it possible for these macromolecules to form extensive hydrogen bonds. Although each individual hydrogen bond is relatively weak, the combination of many weak bonds allows the structural polysaccharides to form large fibers that are strong yet flexible. Similarly, the presence of amino acid residues in peptidoglycan allows adjacent fibers to cross-link into durable sheets.

Polysaccharides are essential to the formation of all the different types of cell walls observed in organisms today. But if polysaccharides were not produced by chemical evolution, and if all cell walls contain polysaccharides, then it is a virtual certainty that the first form of cellular life lacked a cell wall. It is equally certain, however, that the first cell was surrounded by a membrane. What molecules made up this structure, and how did it function? These are the questions taken up in Chapter 6, "Lipids, Membranes, and the First Cells."

✓ CHECK YOUR UNDERSTANDING

Carbohydrates have several important functions in cells in addition to providing building blocks for the synthesis of more complex compounds. Glycoproteins project from the surface of cells and provide a molecular PIN that identifies the cell's type or species. Starch and glycogen store sugars for later use in reactions that produce ATP. Sugars contain large amounts of chemical energy because they contain reduced carbon atoms, meaning carbon atoms that are bonded to hydrogen. These bonds have high free energy because the electrons are shared equally by atoms with low electronegativity. Polysaccharides such as cellulose, chitin, and peptidoglycan form cell walls, which give cells structural strength. You should be able to describe how the sugars you ate during breakfast today are functioning in your body right now.

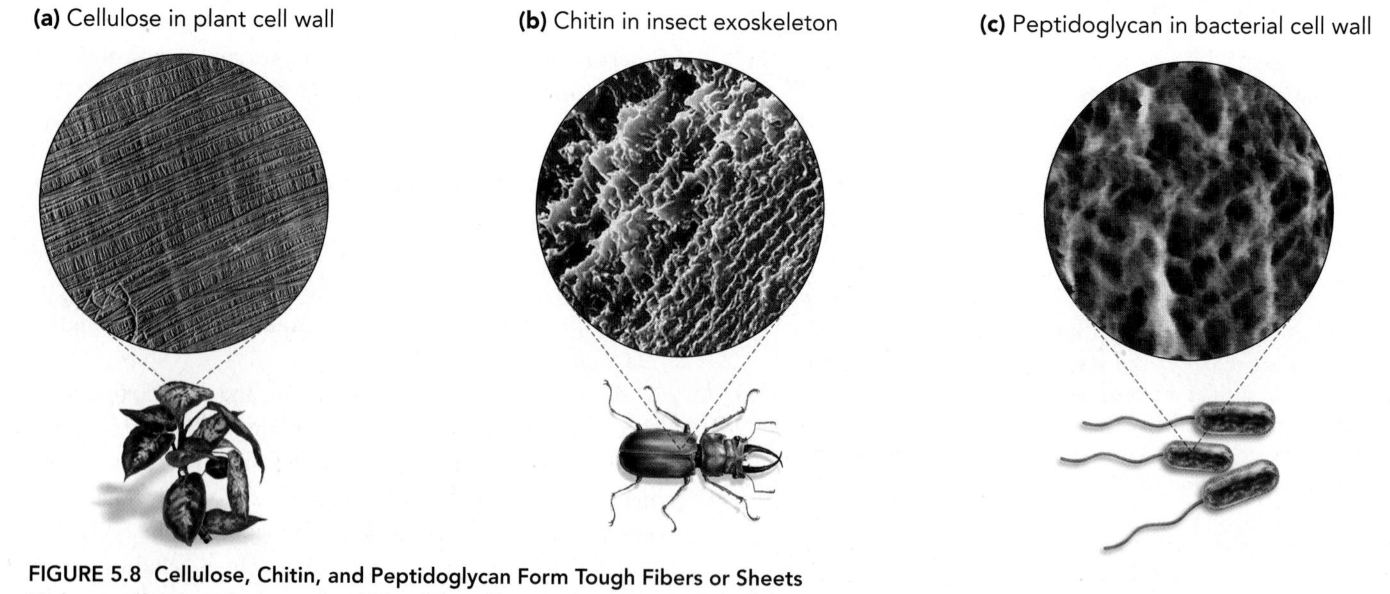
(a) Cellulose in plant cell wall
(b) Chitin in insect exoskeleton
(c) Peptidoglycan in bacterial cell wall

FIGURE 5.8 Cellulose, Chitin, and Peptidoglycan Form Tough Fibers or Sheets
High-magnification photographs of **(a)** cellulose fibers in the cell wall of a plant, **(b)** chitin from an insect's external skeleton, and **(c)** peptidoglycan from the cell wall of a bacterium.

ESSAY Why Do We Have a Sweet Tooth?

If you look at your tongue closely in the mirror, you'll see tiny bumps scattered on the surface. Inside these bumps are small structures called taste buds; inside taste buds are cells called taste receptors. Taste receptors detect the presence of certain molecules and send messages to your brain about them. The sensed food molecules contribute to the tastes we call sweet, sour (acidic), salty, and bitter. Some taste buds respond specifically to amino acids, triggering the taste called umami, or "meaty."

If food in your mouth stimulates your bitter receptors, your brain will probably respond by initiating movements that lead you to spit the food out. But if a mouthful of something stimulates sweet or umami receptors, your brain will probably respond by directing movements that result in your eating more.

Instead of a sweet tooth, then, we actually have sweet receptors. These molecules respond more strongly to some sugars than to others. The six-carbon sugar fructose, for example, is the sweetest of all sugars. If you've read the list of ingredients on food labels, you may have noticed that many foods are sweetened with "high-fructose corn syrup." By adding fructose to foods, manufacturers can make them taste sweeter while lowering the total sugar content. The synthetic sweetener NutraSweet® is even more potent in its ability to stimulate our sweet receptors. It is about 200 times sweeter than sucrose (table sugar).

Why do these receptor molecules and cells exist? Carbohydrates serve so many vital roles in our cells that the ability to sense them and the desire to take in more should benefit individuals. To use the terms introduced in Chapter 1, responding to sweets should be an adaptation that increases the fitness of individuals in most environments. People who sense and respond to sweetness should survive better and reproduce more than do individuals who do not. But having a sweet tooth may have been more adaptive in the past than it is today. Historically, simple sugars were extremely rare in our environment. In many cultures, honey was the only source. But with the recent invention of techniques to refine large amounts of the disaccharide sucrose cheaply, the developed nations are flooded with it. Our taste receptors keep responding strongly, just as they did when our ancestors occupied sugar-poor environments. Now, overdoses of sugar are contributing to obesity and problems with regulating blood sugar levels. Humans love sweets for a reason, but in our current environment we probably love them a little too much.

Historically, simple sugars were extremely rare in our environment.

CHAPTER REVIEW

Summary of Key Concepts

- **Sugars and other carbohydrates are highly variable in structure. They perform a wide variety of functions in cells, ranging from energy storage to formation of tough structural fibers.**

Carbohydrates are organic compounds that have a carbonyl group and several to many hydroxyl groups. They can occur as the simple sugars called monosaccharides or as complex polysaccharides. Many types of monosaccharide exist, each distinguished by one or more of the following features: (1) the location of their carbonyl group—either at the end of the molecule or within it; (2) the number of carbon atoms they contain—from three to many; and (3) the orientation of their hydroxyl groups in the linear chain and/or the ring form. In addition, most monosaccharides have optical isomers.

- **Monosaccharides are monomers that polymerize to form polysaccharides, via different types of glycosidic linkages.**

Monosaccharides can be linked together by covalent bonds, called glycosidic linkages, that join hydroxyl groups on adjacent molecules. Unlike the peptide bonds that form proteins and the phosphodiester bonds of nucleic acids, glycosidic linkages may form at several locations on a monosaccharide. The different disaccharides ("two-sugars") and polysaccharides are distinguished by the type of monomers involved and the location of the glycosidic linkages between them. The most common polysaccharides in organisms today are starch, glycogen, cellulose, and chitin; peptidoglycan is an abundant polysaccharide that has short chains of amino acids attached.

Starch and glycogen are made up of glucose molecules that are in the ring form called α, and that are joined by glycosidic linkages between their first and fourth carbons. In some forms of starch, individual chains are occasionally joined to other chains; in glycogen these links are abundant enough to form a highly branched molecule. Both starch and glycogen function as energy-storage molecules. Because sugars contain many C–H bonds, they contain a significant amount of chemical energy. When cells need energy, enzymes hydrolyze the α-1,4-glycosidic linkages in starch or glycogen. The reaction releases individual glucose molecules, which then act as electron donors in redox reactions. These reactions result in the production of usable chemical energy in the form of ATP.

Cellulose, chitin, and peptidoglycan are made up of monosaccharides joined by β-1,4-glycosidic linkages. When individual molecules of these carbohydrates align side by side, bonds form between them. In cellulose and chitin, the individual molecules are joined by hydrogen bonds; in peptidoglycan, the intermolecular links consist of peptide bonds between amino acid chains that extend from the polysaccharide. Hydrogen bonding produces bundles of cellulose and chitin molecules that form strong, elastic fibers; when peptide bonding occurs between many peptidoglycan strands, the result is a tough fibrous or sheet-like substance. It is no surprise that cellulose, chitin, and peptidoglycan provide cells and organisms with support. In addition, few organisms have enzymes that can degrade β-1,4-glycosidic linkages, which makes cell walls made of cellulose, chitin, and peptidoglycan resistant to attack. In carbohydrates, as in proteins and nucleic acids, structure correlates with function.

Web Tutorial 5.1 Carbohydrate Structure and Function

Questions

Content Review

1. What is the difference between a monosaccharide, a disaccharide, and a polysaccharide?
   a. the number of carbon atoms in the molecule
   b. the type of glycosidic linkage between monomers
   c. the spatial arrangement of the various hydroxyl residues in the molecule
   d. the number of monomers in the molecule
2. What type of bond allows sugars to polymerize?
   a. glycosidic linkage
   b. phosphodiester bond
   c. peptide bond
   d. hydrogen bonds
3. What holds cellulose molecules together in bundles large enough to form fibers?
   a. the cell wall
   b. peptide bonds
   c. hydrogen bonds
   d. hydrophobic interactions between different residues in the cellulose helix
4. What are the primary functions of carbohydrates in cells?
   a. energy storage, cell identity, structure, and building blocks for synthesis
   b. catalysis, structure, and transport
   c. information storage and catalysis
   d. signal reception, signal transport, and signal response
5. Why is it unlikely that carbohydrates played a large role in the origin of life?
   a. They cannot be produced by chemical evolution.
   b. They have optical isomers.
   c. More types of glycosidic linkages are possible than are actually observed in organisms.
   d. They do not polymerize without the aid of enzymes.
6. What is a "quick and dirty" way to assess how much free energy an organic molecule has?
   a. Count the number of oxygen atoms it contains.
   b. Count the number of hydrogen atoms it contains.
   c. Determine whether it contains an amino group.
   d. Determine whether it contains a carbonyl group.

Conceptual Review

1. Draw a six-carbon sugar in linear and ring forms. Draw another six-carbon sugar that differs from the first one. Identify the aspects of these two molecules that are distinct.
2. Draw the ring structure of glucose in the α form. Now add another glucose molecule with an α-1,4-glycosidic linkage, and then a third glucose molecule with an α-1,6-glycosidic linkage.
3. Compare and contrast the structures and functions of starch and glycogen. How are these molecules similar? How are they different?
4. How is it possible for starch and glycogen to function as energy-storage molecules? Explain.
5. How is it possible for cellulose and chitin to function as molecules that provide cells with structural support? Explain.
6. Both glycogen and cellulose consist of glucose monomers that are linked end to end. How do the structures of these polysaccharides differ? How do their functions differ?

Group Discussion Problems

1. A weight-loss program for humans that emphasizes minimal consumption of carbohydrates has recently become popular. What is the logic behind this diet? (Note: This diet plan has caused controversy and is not endorsed by some physicians and researchers.)
2. To treat galactosemia, physicians exclude the monosaccharide galactose from the diet. Why does the disaccharide lactose also have to be excluded from the diet?
3. Amylase, an enzmye found in human saliva, catalyzes the hydrolysis of the α-1,4-glycosidic linkages in starch. If you hold a salty cracker in your mouth long enough, it will begin to taste sweet. Why?
4. Lysozyme, an enzyme found in human saliva, tears, and other secretions, catalyzes the hydrolysis of the β-1,4-glycosidic linkages in peptidoglycan. What effect does contact with this enzyme have on bacteria?

Answers to Multiple-Choice Questions 1. d; 2. a; 3. c; 4. a; 5. d; 6. b

www.prenhall.com/freeman is your resource for the following: Web Tutorials; Online Quizzes and other Online Study Guide materials; Answers to Conceptual Review Questions; Solutions to Group Discussion Problems; Answers to Figure Caption Questions and Exercises; and Additional Readings and Research.

Lipids, Membranes, and the First Cells

6

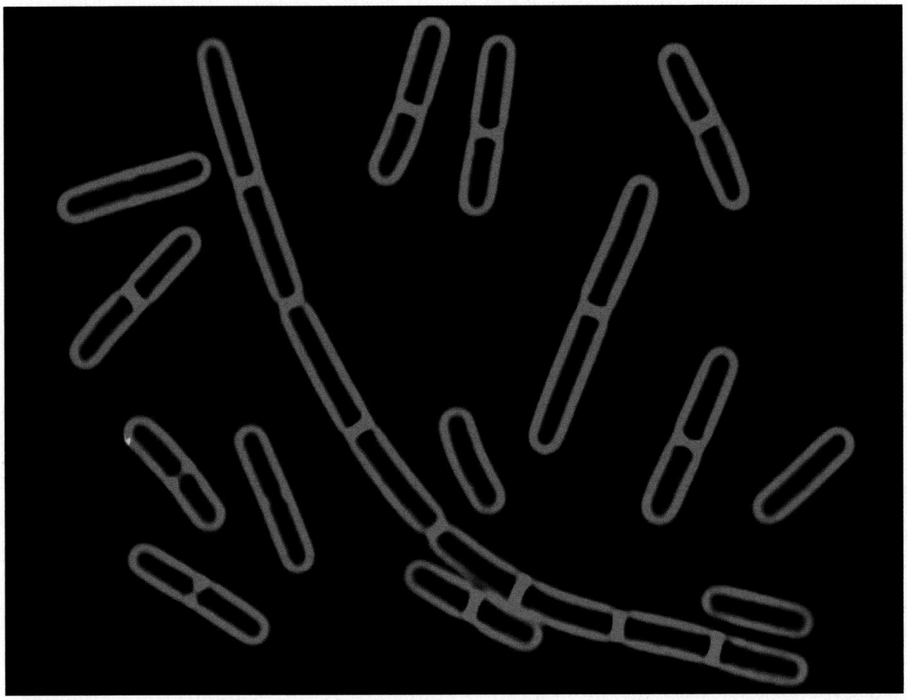

These bacterial cells have been stained with a red compound that inserts itself into the plasma membrane. The plasma membrane defines the basic unit of life. In single-celled organisms like those shown here, the membrane creates a physical separation between life on the inside and nonlife on the outside.

KEY CONCEPTS

- Phospholipids have a hydrophilic region and a hydrophobic region. In solution, they spontaneously form bilayers that are selectively permeable—meaning that only certain substances cross them readily.
- Ions and molecules spontaneously diffuse from regions of high concentration to regions of low concentration. In osmosis, a special case of diffusion, water moves across a selectively permeable membrane from regions of high concentration to regions of low concentration.
- In cells, membrane proteins are responsible for the passage of ions, polar molecules, and large molecules that do not readily cross phospholipid bilayers on their own.

The research discussed in previous chapters suggests that biological evolution began with an RNA molecule that could make a copy of itself. As the offspring of this molecule multiplied in the prebiotic soup, natural selection would have favored versions of the molecule that were particularly stable and efficient at catalysis. Another great milestone in the history of life occurred when a descendant of this replicator became enclosed within a membrane. This event created the first cell and thus the first organism.

The **cell membrane,** or **plasma membrane,** is a layer of molecules that surrounds the cell, separating it from the external environment and selectively regulating the passage of molecules and ions into or out of the cell. The evolution of the plasma membrane was a momentous development because it separated life from nonlife. Before plasma membranes existed, self-replicating molecules probably clung to clay-sized mineral particles, building copies of themselves as they randomly encountered the appropriate nucleotides in the prebiotic soup that washed over them. But the membrane made an internal environment possible—one that could have a chemical composition different from that of the external environment. This was important for two reasons. First, the chemical reactions necessary for life could occur much more efficiently in an enclosed area, because reactants could collide more frequently. Second, the membrane could serve as a selective barrier. That is, it could keep compounds out of the cell that might damage the replicator, but it might allow the entry of compounds required by the replicator. The membrane not only created the cell but also made it into an efficient and dynamic reaction vessel.

The goal of this chapter is to investigate how membranes behave, with an emphasis on how they distinguish an internal environment from the external environment. We begin the chapter by examining the structure and properties of the most abundant molecules in plasma membranes: the "oily" or "fatty" compounds called lipids. We then expand on this introduction by analyzing the way lipids behave when they form membranes. Which ions and molecules can pass through a membrane that consists of lipids? Which cannot, and why? We'll end the chapter by exploring how proteins that become inserted into a lipid membrane can control the flow of materials across the membrane.

6.1 Lipids

Most biochemists are convinced that the building blocks of membranes, called lipids, existed in the prebiotic soup. This conclusion is based on the observation that several types of lipid have been produced in experiments designed to mimic the chemical and energetic conditions that prevailed early in Earth's history. For example, the spark-discharge experiments reviewed in Chapter 3 succeeded in producing at least two types of lipid.

An observation made by A. D. Bangham illustrates why this result is interesting. In the late 1950s, Bangham performed experiments to determine how lipids behave when they are immersed in water. But until the electron microscope was invented, he had no idea what these lipid-water mixtures looked like. As **Box 6.1** explains, **electron microscopes** allow investigators to magnify objects hundreds of thousands of times. When these instruments became available and Bangham was able to take high-magnification photographs of his experimental mixtures, he saw something astonishing: The lipids had spontaneously formed enclosed compartments filled

BOX 6.1 Electron Microscopy

Chapter 1 introduced the light microscope and explained how its invention and use stimulated the development of the cell theory. The electron microscope has had an equally important impact in biology since its invention in the 1950s. Two basic types of electron microscopy are now available: one that allows researchers to examine cross sections of cells at extremely high magnification, and one that offers a view of surfaces at somewhat lower magnification.

Transmission Electron Microscopy

The **transmission electron microscope**, or **TEM**, is an extraordinarily effective tool for viewing cell structure at high magnification. A TEM forms an image from electrons that pass through a specimen, just as a light microscope forms an image from light rays that pass through a specimen.

Biologists who want to view a cell under a TEM begin by "fixing" the cell, meaning that they kill it with a chemical agent that disrupts the cell's structure and contents as little as possible. Then they permeate the cell with an epoxy plastic that stiffens the structure. Once this hardens, the cell can be cut into extremely thin sections with a glass or diamond knife. Finally, the sectioned specimens are impregnated with a metal—often lead. (The reason for this last step is explained shortly.)

Figure 6.1a shows a TEM, outlines how it works, and includes an image produced on a TEM. A beam of electrons is produced by a tungsten filament at the top of a column and directed downward. (All of the air is pumped out of the column so that the electron beam isn't scattered by collisions with air molecules.) The electron beam passes through a series of lenses and the specimen. The lenses are actually electromagnets, which alter the path of the beam much like a glass lens bends light. The lenses magnify and focus the image on a screen at the bottom of the column. There the electrons strike a coating of fluorescent crystals, which emit visible light in response—just like a television screen. When the microscopist moves the screen out of the way and allows the electrons to expose a sheet of black-and-white film, the result is a **micrograph**—a photograph of an image produced by microscopy.

The image itself is created by electrons that pass through the specimen. If no specimen were in place, all the electrons would pass through and the screen (and micrograph) would be uniformly bright. Unfortunately, cell materials by themselves would also appear fairly uniform and bright. This is because the ability of an atom to deflect an electron depends on its density. An atom's density, in turn, is a function of its atomic number. The hydrogen, carbon, oxygen, and nitrogen atoms that dominate biological molecules have low atomic numbers. This is why cell biologists must saturate cell sections with lead solutions. Lead has a high atomic number and scatters electrons effectively. Different macromolecules take up lead atoms in different amounts, so the metal acts as a "stain" that produces contrast. In the TEM, areas of dense metal scatter the electron beam most, producing dark areas in micrographs.

Scanning Electron Microscopy

The **scanning electron microscope**, or **SEM**, is the most useful tool biologists have for looking at the surfaces of cells. Materials are prepared for scanning electron microscopy by coating their surfaces with a layer of metal atoms. This is in contrast to the TEM, which uses sectioned material impregnated with metal atoms. The SEM allows researchers to inspect the surfaces of objects such as cells; the TEM allows researchers to examine the interiors of objects instead.

"Scanning" describes how the SEM works. To create an image of a surface, the instrument scans the surface with a narrow beam of electrons (**Figure 6.1b**). Electrons that are reflected back from the surface or that are emitted by the metal atoms in response to the beam then strike a detector. The signal from the detector controls a second electron beam, which scans a TV-like screen and forms a magnified image of up to 50,000 times. Because the SEM records shadows and highlights, it provides images with a three-dimensional appearance. It cannot clearly magnify objects nearly as much as the TEM can, however.

(a) Transmission electron microscopy: High magnification of cross sections

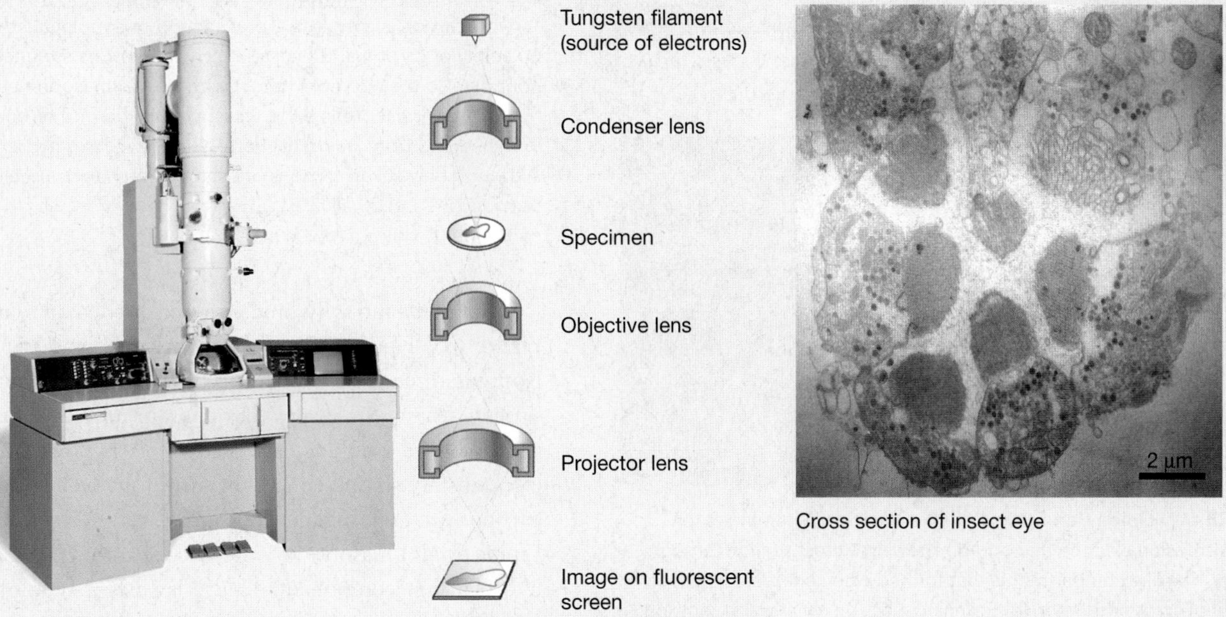
Tungsten filament
(source of electrons)
Condenser lens
Specimen
Objective lens
Projector lens
Image on fluorescent
screen
2 µm
Cross section of insect eye

(b) Scanning electron microscopy: Lower magnification of surfaces

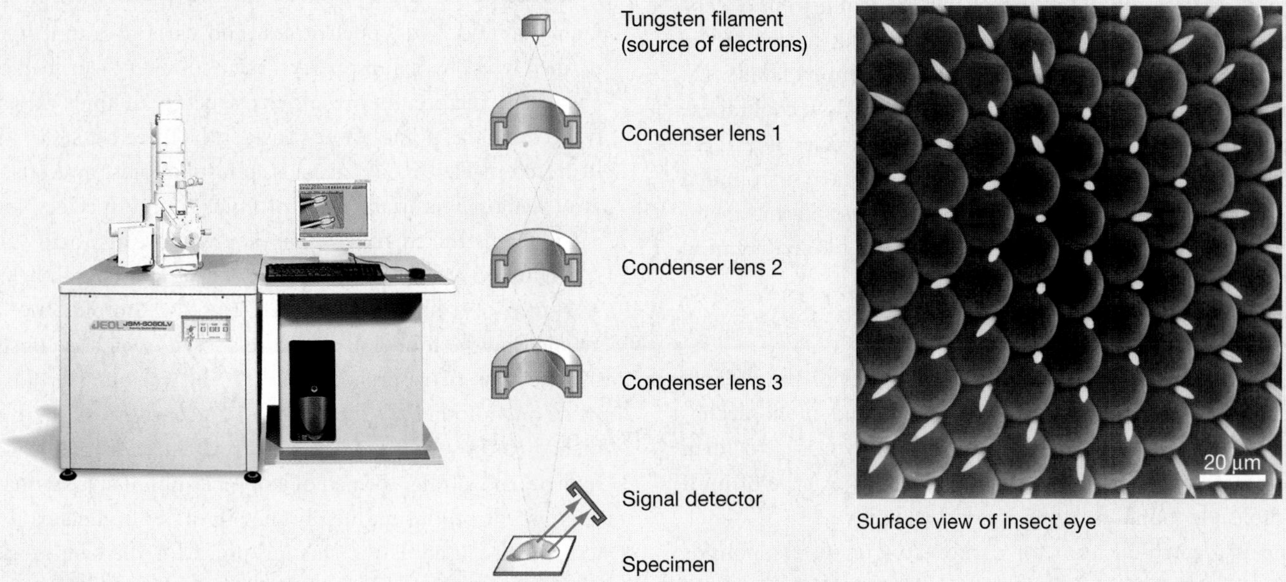
Tungsten filament
(source of electrons)
Condenser lens 1
Condenser lens 2
Condenser lens 3
Signal detector
Specimen
20 µm
Surface view of insect eye

FIGURE 6.1 How Does Electron Microscopy Work?
(a) Transmission electron microscopes provide images of cross sections through cells, often at extremely high magnification. **(b)** Scanning electron microscopes offer images of the surfaces of objects.

(a) In solution, lipids form water-filled vesicles.

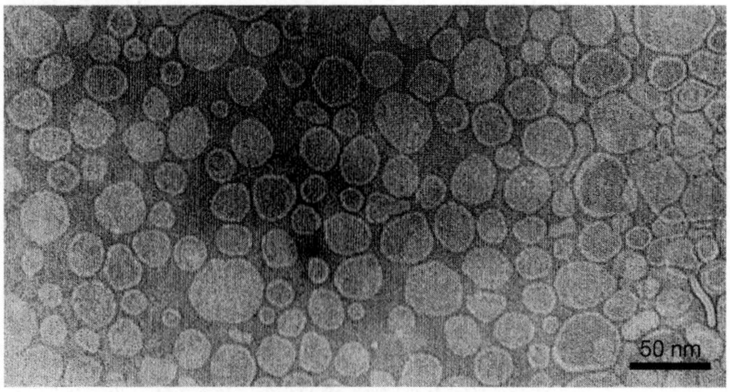

(b) Red blood cells resemble vesicles.

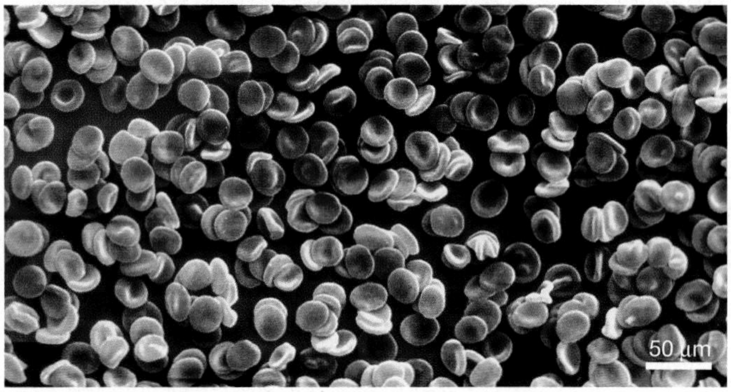

FIGURE 6.2 Lipids Can Form Cell-like Vesicles When in Water
(a) Transmission electron micrograph showing a cross section through the tiny, bag-like compartments that formed when a researcher shook a mixture of lipids and water. **(b)** Scanning electron micrograph showing red blood cells from humans. Note the scale bars in each photo.

with water (**Figure 6.2a**). He called these membrane-bound structures *vesicles* and noted that they were reminiscent of cells (**Figure 6.2b**). Bangham had not done anything special to the lipid-water mixtures; he had merely shaken them by hand.

The experiment raises a series of questions: How could these structures have formed? Is it possible that vesicles like these existed in the prebiotic soup? If so, could they have surrounded a self-replicating molecule and become the first plasma membrane? Let's begin answering these questions by investigating what lipids are and how they behave.

What Is a Lipid?

In earlier chapters we examined the structures of the organic molecules called amino acids, nucleotides, and monosaccharides and explored how these monomers polymerize to form macromolecules. Here we analyze another major type of mid-sized molecule found in living organisms: lipids.

Lipid is a catch-all term for carbon-containing compounds that are found in organisms and are largely nonpolar and hydrophobic—meaning that they do not dissolve readily in water. (Recall from Chapter 2 that water is a polar solvent.) Lipids do dissolve, however, in liquids consisting of nonpolar organic compounds.

(a) Isoprene **(b)** Fatty acid

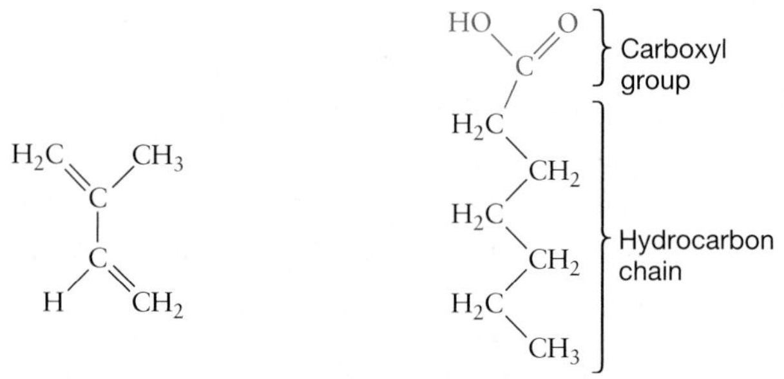

FIGURE 6.3 Hydrocarbon Groups Make Lipids Hydrophobic
A hydrocarbon is a molecule or a section of a molecule that is made up solely of carbon and hydrogen. Hydrocarbons are uncharged and nonpolar, so they do not interact with water. **(a)** Isoprenes are hydrocarbons. Isoprene subunits can be linked end to end to form long hydrocarbon chains. **(b)** Fatty acids have long hydrocarbon tails. Although only seven carbons are shown, fatty acids usually contain between 14 and 20. **EXERCISE** Circle the hydrophobic hydrocarbon components of both molecules.

To understand why lipids do not dissolve in water, examine **Figure 6.3**. Figure 6.3a shows a five-carbon compound called isoprene. It consists of a group of carbon atoms bonded to hydrogen atoms. Molecules that contain only carbon and hydrogen, such as isoprene, are known as **hydrocarbons**. Because electrons are shared equally in carbon-hydrogen bonds, hydrocarbons are nonpolar. This property makes them hydrophobic. Lipids do not dissolve in water, because they have a significant hydrocarbon component. Figure 6.3b is a type of compound called a **fatty acid**, which consists of a hydrocarbon chain bonded to a carboxyl (COOH) functional group. Isoprene and fatty acids are key building blocks of the lipids found in organisms.

A Look at Three Types of Lipids Found in Cells

Unlike amino acids, nucleotides, and carbohydrates, lipids are defined by a physical property—their solubility—instead of their chemical structure. As a result, the structure of lipids varies widely. To drive this point home, consider the three classes of lipids illustrated in **Figure 6.4**: steroids, phospholipids, and fats. These are the most important types of lipids found in cells. Note that each is constructed from either isoprene or fatty-acid subunits.

Figure 6.4a shows the cholesterol molecule, which belongs to a family of compounds called steroids. **Steroids** are distinguished by the four-ring structure shown, which is constructed from chains of isoprene subunits. The various steroids differ from one another by the functional groups or side groups attached to those rings. Cholesterol, which is distinguished by a hydrocarbon "tail" formed of isoprene subunits, is an important component of plasma membranes in many organisms. Cholesterol is also used as the starting point for the synthesis of the mammalian hormones estrogen, progesterone, and testosterone.

Figure 6.4b shows the basic structure of a phospholipid. A **phospholipid** consists of a three-carbon molecule called **glycerol** that is linked to a phosphate group (PO_4^{2-}) and to either two chains of isoprene or two fatty acids. In some cases, the phos-

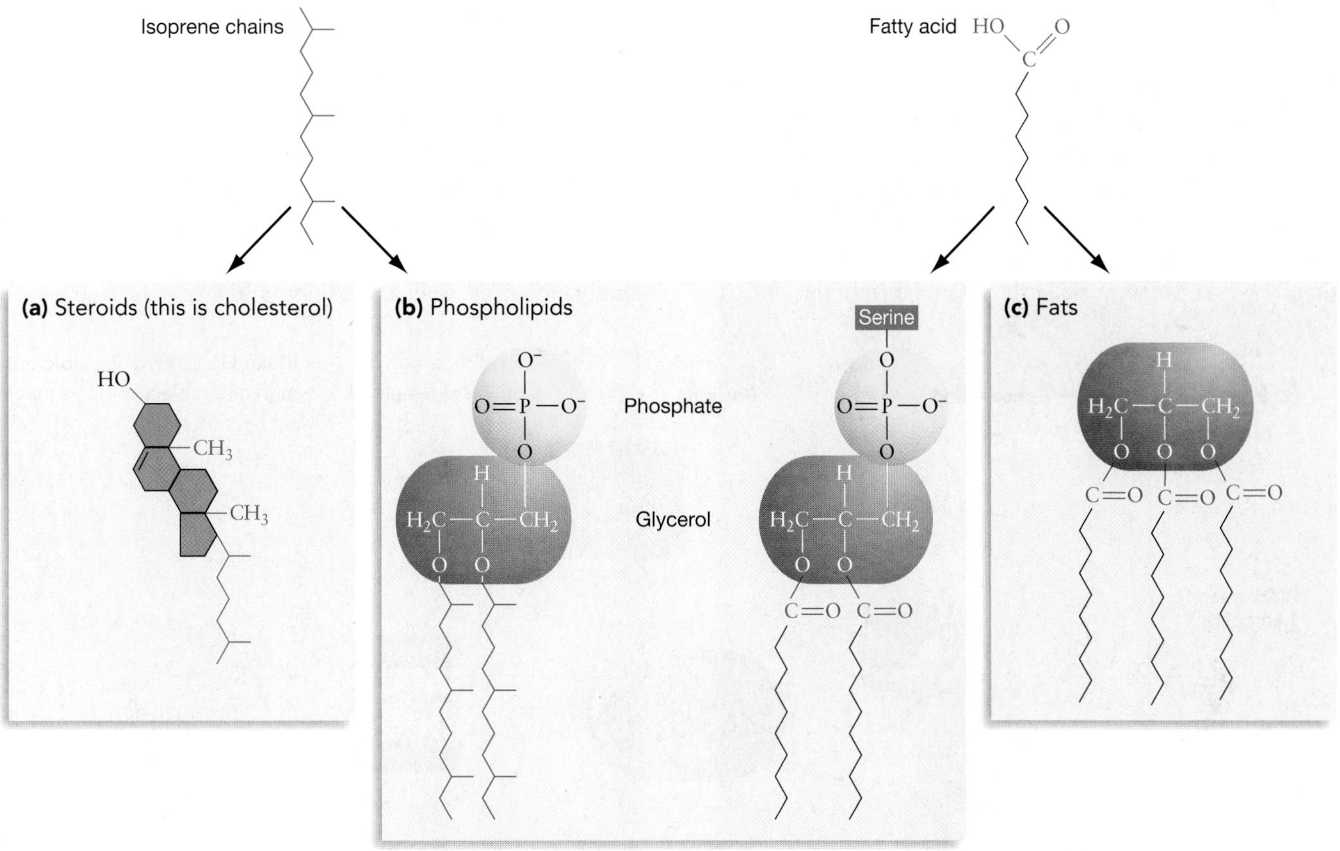

FIGURE 6.4 Major Classes of Lipids
Isoprenes and fatty acids are building blocks for **(a)** steroids, **(b)** phospholipids, and **(c)** fats. Many of the C and H symbols for carbon and hydrogen, respectively, have been omitted for clarity. The red or black lines indicate carbon-carbon bonds.

phate group is bonded to another small organic molecule, such as the amino acid serine in the figure. Phospholipids with isoprene tails are found in the domain Archaea introduced in Chapter 1; phospholipids composed of fatty acids are found in the domains Bacteria and Eukarya.

Fats, such as the molecule in Figure 6.4c, are composed of three fatty acids that are linked to a single glycerol molecule. Because of this structure, fats are also called *triacylglycerols* or *triglycerides*.

Both phospholipids and fats form when a condensation reaction occurs between the hydroxyl group of glycerol and the carboxyl group of a fatty acid (**Figure 6.5a**). The glycerol and fatty-acid molecules become joined by an **ester linkage**, which is analogous to the peptide bonds, phosphodiester bonds, and glycosidic linkages in proteins, nucleic acids, and carbohydrates, respectively. Phospholipids and fats are not polymers, however, and fatty acids are not monomers. As **Figure 6.5b** shows, fatty acids are not linked together to form a macromolecule in

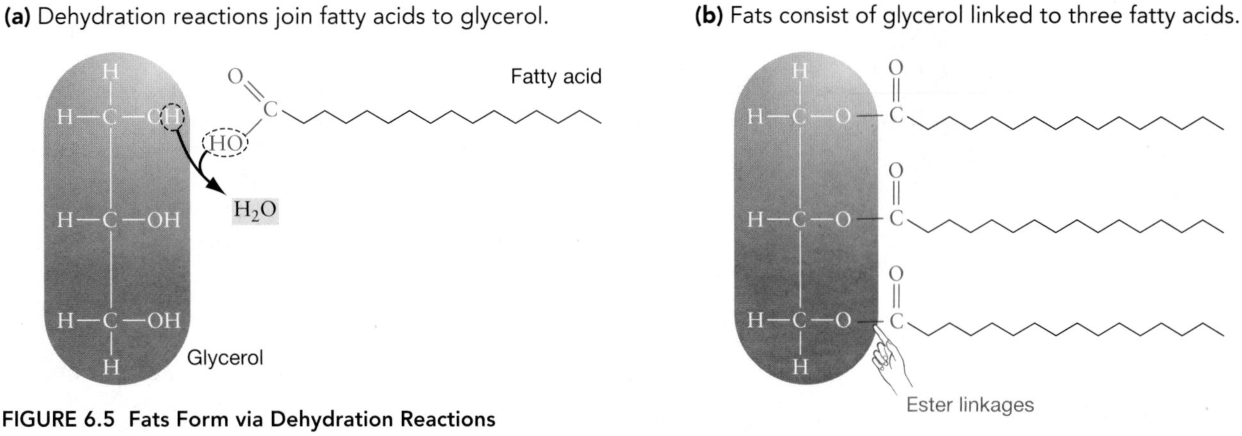

FIGURE 6.5 Fats Form via Dehydration Reactions
(a) When glycerol and a fatty acid react, a water molecule leaves. **(b)** The covalent bonds that result from this reaction are termed an ester linkage.

the way that amino acids, nucleotides, and monosaccharides are.

The classes of molecules in Figure 6.5 will reappear throughout this text, because lipids perform a wide variety of functions in cells. Lipids store chemical energy, act as pigments that capture or respond to sunlight, serve as signals between cells, form waterproof coatings on leaves and skin, and act as vitamins used in an array of cellular processes. The most important lipid function, however, is their role in the plasma membrane.

The Structures of Membrane Lipids

Not all lipids can form the artificial membranes that Bangham and his colleagues observed. In fact, just two types of lipids dominate plasma membranes. Membrane-forming lipids have a polar, hydrophilic region in addition to the nonpolar, hydrophobic region found in all lipids. For example, examine the phospholipid illustrated in **Figure 6.6a**. The molecule has a "head" region, containing covalent bonds that are highly polar, as well as positive and negative charges. The

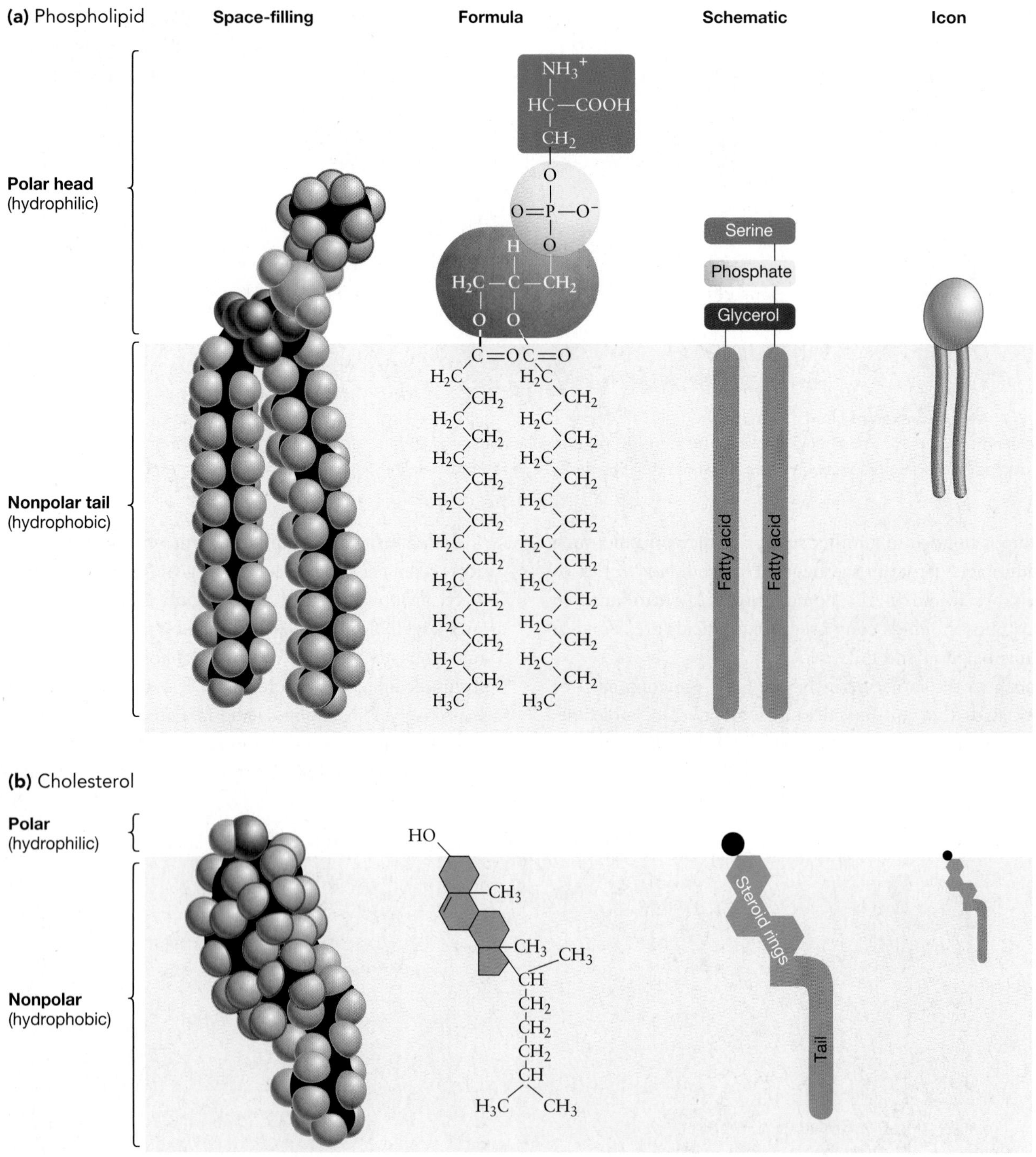

FIGURE 6.6 Amphipathic Lipids Contain Hydrophilic and Hydrophobic Elements
QUESTION If these molecules were in solution, where would water molecules interact with them?

charges and polar bonds in the head region interact with water molecules when a phospholipid is placed in solution. In contrast, the long isoprene or fatty-acid tails of a phospholipid are nonpolar. Water molecules do not interact with this part of the molecule.

Compounds that contain both hydrophilic and hydrophobic elements are **amphipathic** ("dual sympathy"). Phospholipids are amphipathic. As **Figure 6.6b** shows, cholesterol is also amphipathic. It has both hydrophilic and hydrophobic regions.

The amphipathic nature of phospholipids is far and away their most important feature biologically. It is responsible for their ability to form plasma membranes.

6.2 Phospholipid Bilayers

Phospholipids do not dissolve when they are placed in water. Water molecules interact with the hydrophilic heads of the phospholipids, but not with their hydrophobic tails. Instead of dissolving in water, then, phospholipids form one of two types of structures: micelles or lipid bilayers.

Micelles (**Figure 6.7a**) are tiny droplets created when the hydrophilic heads of phospholipids face the water and the hydrophobic tails are forced together, away from the water. Phospholipid bilayers, or simply **lipid bilayers**, are created when two sheets of lipid molecules align. As **Figure 6.7b** shows, the hydrophobic heads in each layer face the solution while the tails face one another inside the bilayer. In this way, the hydrophilic heads interact with water while the hydrophobic tails interact with each other. Micelles tend to form from phospholipids with relatively short tails; bilayers tend to form from phospholipids with longer tails.

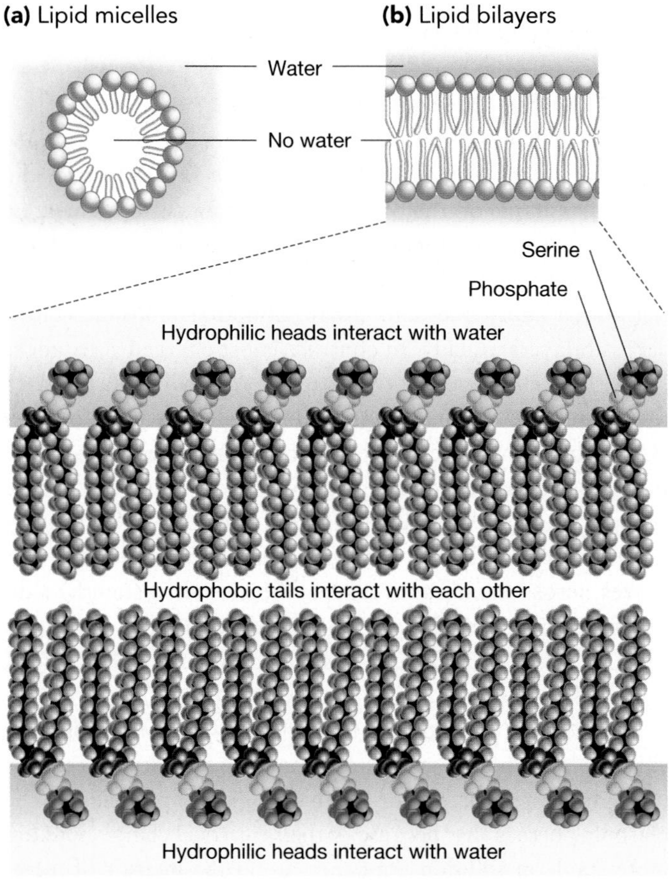

FIGURE 6.7 Phospholipids Form Bilayers in Solution
In **(a)** a micelle or **(b)** a lipid bilayer, the hydrophilic heads of phospholipids face out, toward water; the hydrophobic tails face in, away from water. Plasma membranes consist in part of lipid bilayers.

Energetically, the formation of micelles or lipid bilayers is highly favored. Phospholipids are much more stable in water when they form bilayers than they are when the individual molecules exist independently. As a result, these structures form spontaneously. No input of energy is required.

The key point here is that the amphipathic structure of phospholipids makes it possible for them to function as plasma membranes. The fact that bilayers form spontaneously also explains why Bangham's group was able to make vesicles so easily.

Artificial Membranes as an Experimental System

When lipid bilayers are agitated by shaking, the layers break and re-form as small, spherical structures. This is what happened in Bangham's experiment. The resulting vesicles had water on the inside as well as the outside because the hydrophilic heads of the lipids faced outward on each side of the bilayer.

Researchers have produced these types of vesicles by using dozens of different types of phospholipids. Artificial membrane-bound vesicles like these are called **liposomes.** The ability to create them supports an important conclusion: If phospholipid molecules accumulated during chemical evolution early in Earth's history, they almost certainly formed water-filled vesicles.

The discovery of liposomes sparked intense interest among biologists because lipid bilayers also form the basic structure of plasma membranes. Researchers began an effort to understand the properties of plasma membranes by creating and experimenting with artificial bilayers.

Some of the first questions posed by researchers concerned the permeability of lipid bilayers. The **permeability** of a structure is the structure's tendency to allow a given substance to diffuse across it. Once a membrane forms a water-filled vesicle, can other molecules or ions pass in or out? If so, is this permeability selective in any way? The permeability of membranes is a critical issue, because if certain molecules or ions pass through a lipid bilayer more readily than others, the internal environment of a vesicle can become different from the outside. This difference between exterior and interior environments is a key characteristic of cells.

To explore the permeability of phospholipid bilayers, biologists began performing experiments on artificial membranes.

Figure 6.8 shows the two types of artificial membranes that are used to study the permeability of lipid bilayers. Figure 6.8a shows liposomes, roughly spherical vesicles. Figure 6.8b illustrates **planar bilayers**, which are lipid bilayers constructed across a hole in a glass or plastic wall separating two aqueous (watery) solutions.

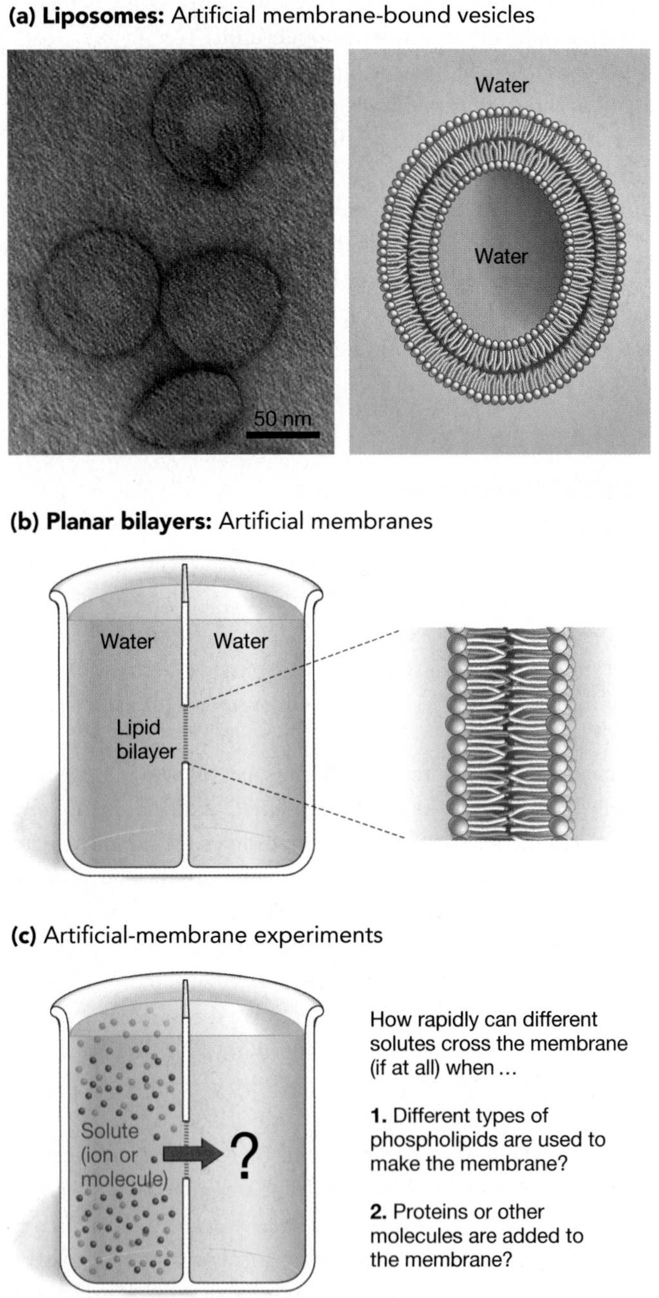

FIGURE 6.8 Liposomes and Planar Bilayers Are Important Experimental Systems
(a) Transmission electron micrograph of liposomes in cross section (left) and a cross-sectional diagram of the lipid bilayer in a liposome. **(b)** The construction of planar bilayers across a hole in a glass wall separating two water-filled compartments (left), and a close-up sketch of the bilayer. **(c)** A wide variety of experiments is possible with liposomes and planar bilayers; a few are suggested here.

Using liposomes and planar bilayers, researchers can study what happens when a known ion or molecule is added to one side of a lipid bilayer (Figure 6.8c). Does the ion or molecule cross the membrane and show up on the other side? If so, how rapidly does the movement take place? What happens when a different type of phospholipid is used to make the artificial membrane? Does the membrane's permeability change when proteins or other types of molecules are added to it?

Biologists describe such an experimental system as elegant and powerful because it gives them precise control over which factor changes from one experimental treatment to the next. Control, in turn, is why experiments are such an effective way to explore scientific questions, as we learned in Chapter 1. A good experimental design allows researchers to alter one factor at a time and determine what effect, if any, each has on the process being studied.

Equally important for experimental purposes, liposomes and planar bilayers provide a clear way to determine whether a given change in conditions has an effect. By sampling the solutions on both sides of the membrane before and after the treatment and then analyzing the concentration of ions and molecules in the samples, researchers have an effective way to determine whether the treatment had any consequences.

Using such systems, what have biologists learned about membrane permeability?

Selective Permeability of Lipid Bilayers

When researchers put molecules or ions on one side of a liposome or planar bilayer and measure the rate at which the molecules arrive on the other side, a strong pattern emerges. Lipid bilayers are *highly* selective. **Selective permeability** means that some substances cross a membrane more easily than other substances can. Small, nonpolar molecules move across bilayers quickly. In contrast, most charged substances cross the membrane slowly, if at all. According to the data in **Figure 6.9**, small, nonpolar molecules such as oxygen (O_2) move across selectively permeable membranes more than a billion times faster than do chloride ions (Cl^-). Larger molecules can also move rapidly if they are nonpolar. Indole, for example, is a large, nonpolar compound that moves across membranes 100 million times faster than do potassium ions (K^+). Very small and uncharged molecules such as water (H_2O) can also cross membranes rapidly, even if they are polar.

The leading hypothesis to explain this pattern is that charged compounds and large, polar molecules can't pass through the nonpolar, hydrophobic tails of a lipid bilayer. The reasoning here is that because of their electrical charge, ions are more stable in solution than they are in the interior of membranes, which is electrically neutral. To test this hypothesis, researchers have manipulated the size and structure of the tails in liposomes or planar bilayers.

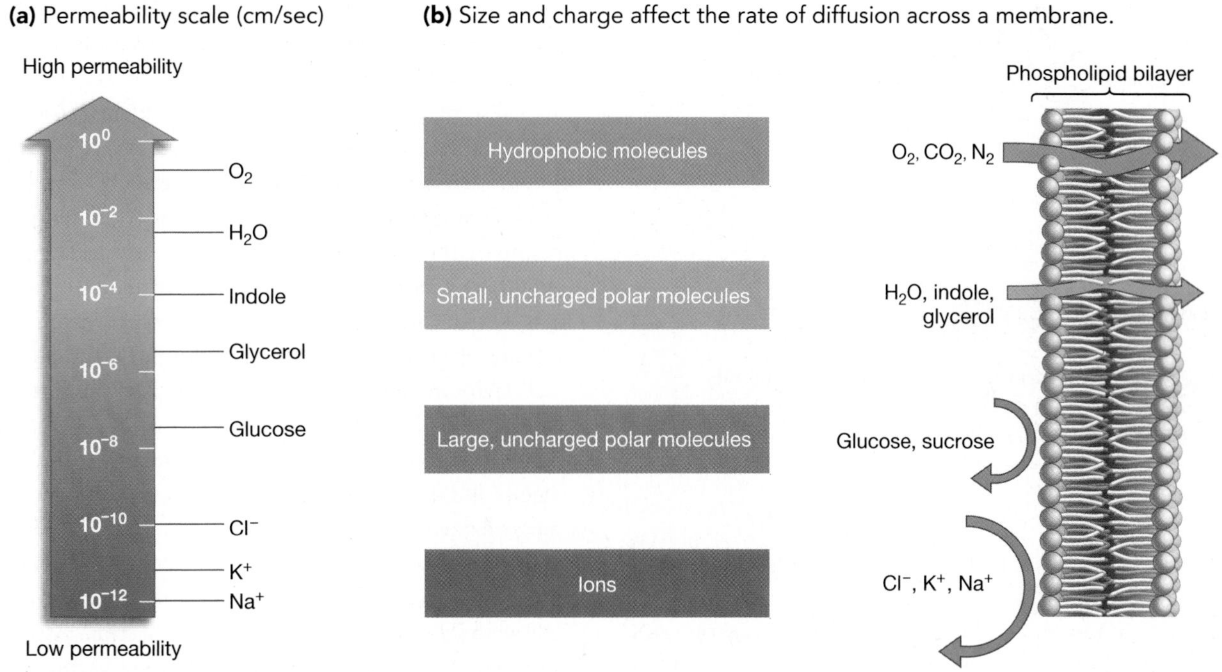

FIGURE 6.9 Selective Permeability of Lipid Bilayers
(a) The numbers represent "permeability coefficients," or the rate (cm/sec) at which an ion or molecule crosses a lipid bilayer. **(b)** The relative permeabilities of various molecules and ions, based on data like those presented in part (a). QUESTION About how fast does water cross the lipid bilayer?

Does the Type of Lipid in a Membrane Affect Its Permeability?

Theoretically, two aspects of a hydrocarbon chain could affect the way the chain behaves in a lipid bilayer: (1) the number of double bonds it contains and (2) its length. Recall from Chapter 2 that when carbon atoms form a double bond, the attached atoms are found in a plane instead of a (three-dimensional) tetrahedron. The carbon atoms involved are also locked into place; they cannot rotate freely, as they do in carbon-carbon single bonds. As a result, a double bond between carbon atoms produces a "kink" in an otherwise straight hydrocarbon chain (**Figure 6.10**).

When a double bond exists between two carbon atoms in a hydrocarbon chain, the chain is said to be **unsaturated.** Conversely, hydrocarbon chains without double bonds are said to be **saturated.** This choice of terms is logical, because if a hydrocarbon chain does not contain a double bond, it is saturated with the maximum number of hydrogen atoms that can attach to the carbon skeleton. If it is unsaturated, then fewer than the maximum number of hydrogen atoms are attached. Because they contain more C–H bonds, which have much more free energy than C=C bonds, saturated fats have much more chemical energy than unsaturated fats do. People who are dieting are often encouraged to eat fewer saturated fats. Foods that contain lipids with many double bonds are said to be *polyunsaturated* and are advertised as healthier than foods with more-saturated fats.

Why do double bonds affect the permeability of membranes? When hydrophobic tails are packed into a lipid bilayer, the kinks created by double bonds produce spaces among the tightly packed tails. These spaces reduce the strength of hydrophobic interactions among the tails. Because the interior of the membrane is "glued together" less

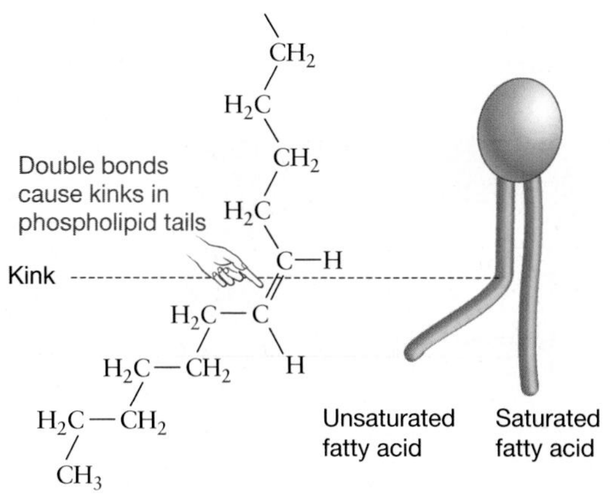

FIGURE 6.10 Unsaturated Hydrocarbons Contain Carbon-Carbon Double Bonds
A double bond in a hydrocarbon chain produces a "kink." The icon on the right indicates that one of the hydrocarbon tails in a phospholipid is kinked and therefore unsaturated. EXERCISE Draw figures analogous to those for an unsaturated fatty acid containing two double bonds.

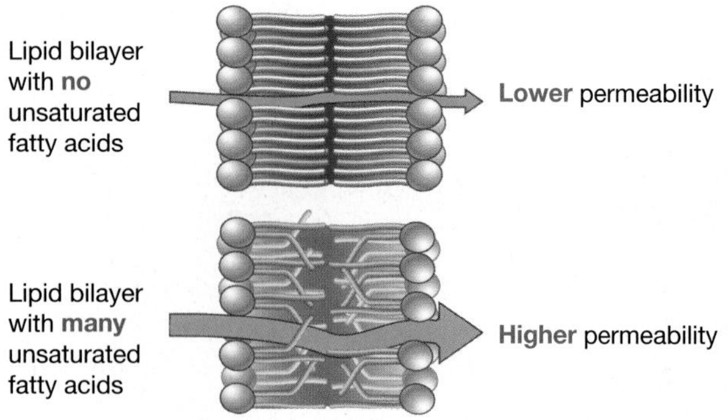

FIGURE 6.11 Fatty-Acid Structure Changes the Permeability of Membranes
Lipid bilayers that contain many unsaturated fatty acids contain more gaps and should be more permeable than are bilayers with few unsaturated fatty acids.

tightly, the structure should become more fluid and more permeable (**Figure 6.11**).

Hydrophobic interactions also become stronger as saturated hydrocarbon tails increase in length. Membranes dominated by phospholipids with long, saturated hydrocarbon tails should be stiffer and less permeable because the interactions among the tails are stronger.

A biologist would predict, then, that bilayers made of lipids with long, straight, saturated fatty-acid tails should be much less permeable than membranes made of lipids with short, kinked, unsaturated fatty-acid tails. Experiments on liposomes have shown exactly this pattern. Phospholipids with long, saturated tails form membranes that are much less permeable than membranes consisting of phospholipids with shorter, unsaturated tails.

The central point here is that the degree of hydrophobic interactions dictates the behavior of these molecules. This is another example in which the structure of a molecule—specifically, the number of double bonds in the hydrocarbon chain and its overall length—correlates with its function.

These data are also consistent with the basic observation that highly saturated fats are solid at room temperature (**Figure 6.12a**). Lipids that have extremely long hydrocarbon tails, as **waxes** do, form stiff solids at room temperature as a result of the extensive hydrophobic interactions that occur (**Figure 6.12b**). Birds, sea otters, and many other organisms synthesize waxes and spread them on their exterior surface as a waterproofing. In contrast, highly unsaturated fats are liquid at room temperature (**Figure 6.12c**). Liquid triacylglycerides are called **oils**.

In addition to exploring the role of hydrocarbon chain length and degree of saturation on membrane permeability, biologists have investigated the effect of adding cholesterol molecules. Because cholesterol is expected to fill spaces between phospholipids, adding cholesterol to a membrane should increase the density of the hydrophobic section. As predicted, researchers found that adding cholesterol molecules to liposomes dramatically reduced the permeability of the liposomes. The data behind this claim are presented in **Figure 6.13**. This graph makes another important point, however: Temperature has a strong influence on the behavior of lipid bilayers.

Why Does Temperature Affect the Fluidity and Permeability of Membranes?

At about 25°C, or "room temperature," the phospholipids found in plasma membranes are liquid and bilayers have the consistency of olive oil. This fluidity, as well as the membrane's permeability, decreases as temperature decreases. As temperatures drop, individual molecules in the bilayer move more slowly. As a result, the hydrophobic tails in the interior of membranes pack together

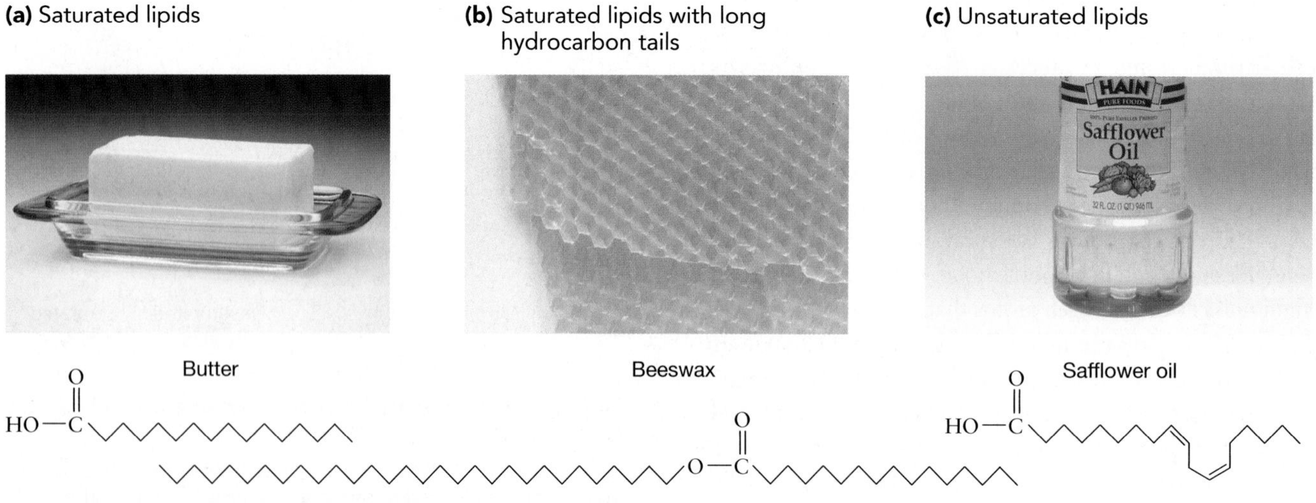

FIGURE 6.12 The Fluidity of Lipids Depends on the Characteristics of Their Hydrocarbon Chains
The fluidity of a lipid depends on the length and saturation of its hydrocarbon chain. **(a)** Butter consists primarily of saturated lipids. **(b)** Waxes are lipids with extremely long hydrocarbon chains. **(c)** Oils are dominated by "polyunsaturates"—lipids with hydrocarbon chains that contain multiple double bonds.

more tightly. At very low temperatures, lipid bilayers begin to solidify. As the graphs in Figure 6.13 indicate, low temperatures can make membranes impervious to molecules that would normally cross them readily.

The fluid nature of membranes also allows individual lipid molecules to move laterally within each layer, a little like a person moving about in a dense crowd. By tagging individual phospholipids and following their movement, researchers have clocked average speeds of 2 micrometers (μm)/second at room temperature (**Figure 6.14**). At these speeds, phospholipids travel the length of a small bacterial cell every second.

Question: Does adding cholesterol to a membrane affect its permeability?

Hypothesis: Cholesterol reduces permeability because it fills spaces in phospholipid bilayers.

Null hypothesis: Cholesterol has no effect on permeability.

Experimental setup:

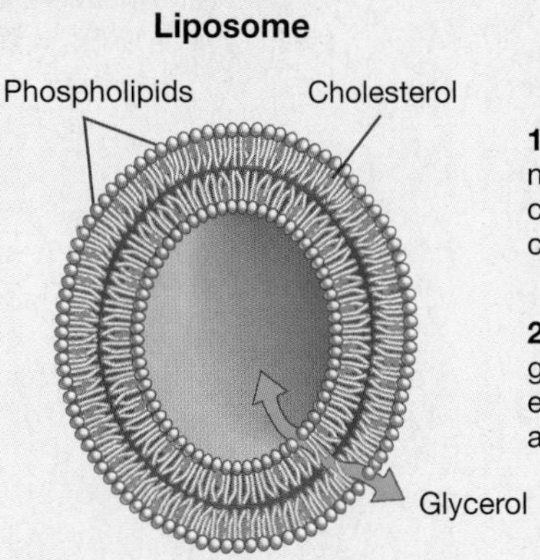

1. Create liposomes with no cholesterol, 20% cholesterol, and 50% cholesterol.

2. Record how quickly glycerol moves across each type of membrane at different temperatures.

Prediction: Liposomes with higher cholesterol levels will have reduced permeability.

Prediction of null hypothesis: All liposomes will have the same permeability.

Results:

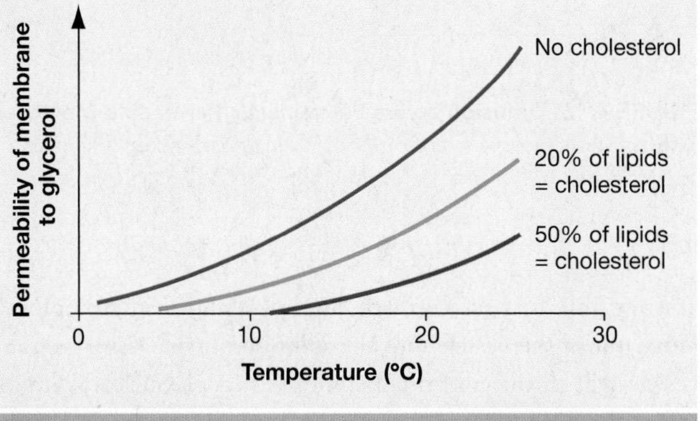

Conclusion: Adding cholesterol to membranes decreases their permeability to glycerol. The permeability of all membranes analyzed in this experiment increases with increasing temperature.

FIGURE 6.13 The Permeability of a Membrane Depends on Its Composition
Experiments like this established that membrane permeability varies with the types of lipids found in the structure.

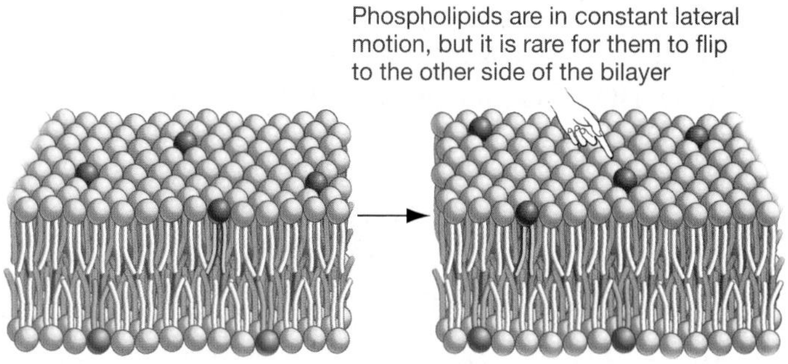

FIGURE 6.14 Phospholipids Move within Membranes
Membranes are dynamic—in part because phospholipid molecules move within each layer in the structure.

These experiments on lipid and ion movement demonstrate that membranes are dynamic. Phospholipid molecules whiz around each layer, while water and small, nonpolar molecules shoot in and out of the membrane. How quickly molecules move within and across membranes is a function of temperature and the structure of the hydrocarbon tails in the bilayer.

Given these insights into the permeability and fluidity of lipid bilayers, an important question remains: *Why* do certain molecules move across membranes spontaneously?

✓CHECK YOUR UNDERSTANDING

In solution, phospholipids form bilayers that are selectively permeable—meaning that some substances cross them much more readily than others can. Permeability is a function of temperature, the amount of cholesterol in the membrane, and the length and degree of saturation of the hydrocarbon tails in membrane phospholipids. You should be able to fill in a chart with (1) rows called "Temperature," "Cholesterol," "Length of hydrocarbon tails," and "Saturation of hydrocarbon tails" and (2) columns named "Factor," "Effect on permeability," and "Reason."

6.3 Why Molecules Move across Lipid Bilayers: Diffusion and Osmosis

A thought experiment can help explain why molecules and ions are able to move across membranes spontaneously. Suppose you rack up a set of blue billiard balls on a pool table containing many white balls and then begin to vibrate the table. Because of the vibration, the balls will move about randomly. They will also bump into one another. After these collisions, some blue balls will move outward—away from their original position. In fact, the overall (or net) movement of blue balls will be outward. This occurs because the random motion of the blue balls disrupts their original, nonrandom position—as they move at random, they are

WEB TUTORIAL 6.1 Diffusion and Osmosis

more likely to move away from each other than to stay together. Eventually, the blue billiard balls will be distributed randomly across the table. The entropy of the blue billiard balls has increased. Recall from Chapter 2 that entropy is a measure of the randomness or disorder in a system. The second law of thermodynamics states that in a closed system, entropy always increases.

This hypothetical example illustrates why molecules or ions located on one side of a lipid bilayer move to the other side spontaneously. The dissolved molecules and ions, or **solutes**, have thermal energy and are in constant, random motion. Because they change position randomly, they tend to move from a region of high concentration to a region of low concentration, along what biologists call a **concentration gradient.** This directed movement of molecules and ions is known as **diffusion.** Diffusion occurs along a concentration gradient because the random motion of molecules and ions causes a net movement from regions of high concentration to regions of low concentration. Diffusion is a spontaneous process because it results in an increase in entropy.

Once the molecules or ions are randomly distributed throughout a solution, equilibrium is established. For example, consider two solutions separated by a lipid bilayer. **Figure 6.15** shows how molecules that pass through the bilayer diffuse to the other side. At equilibrium, molecules continue to move back and forth across the membrane, but at equal rates—simply because each molecule or ion is equally likely to move in any direction. This means that there is no longer a net movement of molecules across the membrane.

What about water itself? As the data in Figure 6.9 showed, water moves across lipid bilayers extremely quickly. Like other substances that diffuse, water moves along its concentration gradient—from higher to lower concentration. The movement of water is a special case of diffusion that is given its own name: **osmosis.** Osmosis occurs only when solutions are separated by a membrane that is permeable to some molecules but not others—that is, a selectively permeable membrane.

The best way to think about water moving in response to a concentration gradient is to focus on the concentration of solutes in the solution. Let's suppose the concentration of a particular solute is higher on one side of a selectively permeable membrane than it is on the other side (**Figure 6.16**, step 1). Further, suppose that this solute cannot diffuse through the membrane to establish equilibrium. What happens? Water will move from the side with a lower concentration of solute to the side with a higher concentration of solute (step 2). It dilutes the higher concentration and equalizes the concentrations on both sides. This movement is spontaneous. It is driven by the increase in entropy achieved when solute concentrations are equal on both sides of the membrane.

This movement of water is important because it can swell or shrink a membrane-bound vesicle. Consider the liposomes illustrated in **Figure 6.17**. If the solution outside the membrane has a higher concentration of solutes than the interior, and the solutes

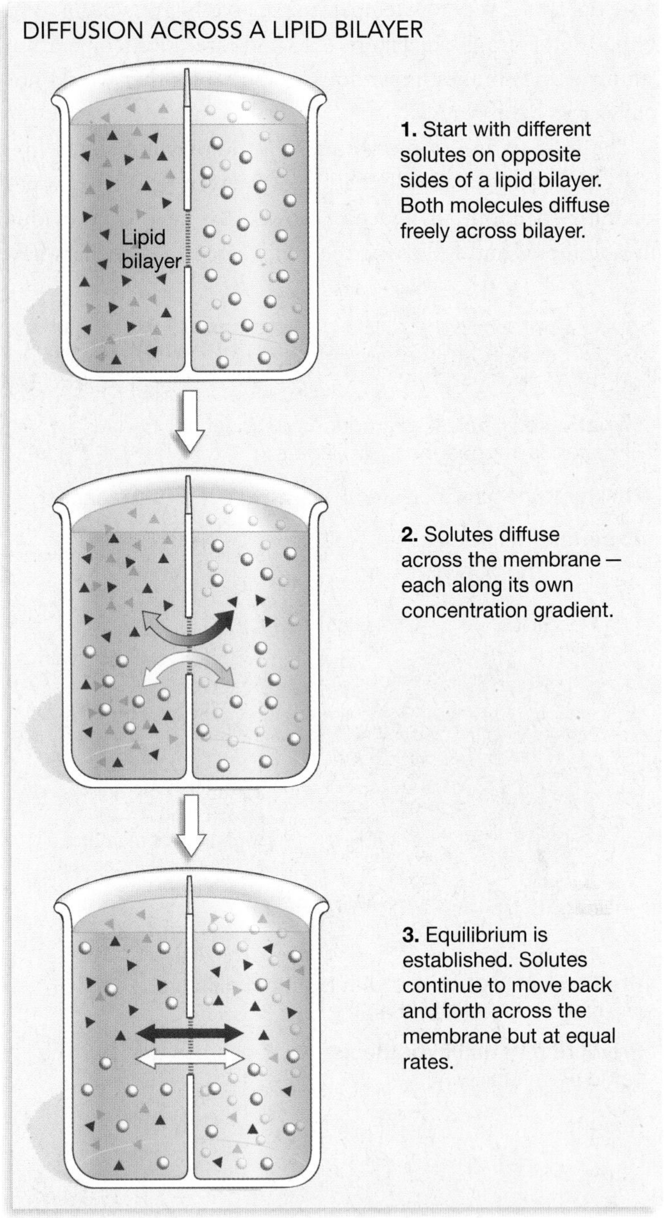
DIFFUSION ACROSS A LIPID BILAYER
Lipid bilayer
1. Start with different solutes on opposite sides of a lipid bilayer. Both molecules diffuse freely across bilayer.
2. Solutes diffuse across the membrane—each along its own concentration gradient.
3. Equilibrium is established. Solutes continue to move back and forth across the membrane but at equal rates.

FIGURE 6.15 Diffusion across a Selectively Permeable Membrane
EXERCISE If a solute's rate of diffusion increases linearly with its concentration difference across the membrane, write an equation for the rate of diffusion across a membrane.

are not able to pass through the lipid bilayer, then water will move out of the vesicle into the solution outside. As a result, the vesicle will shrink and the membrane shrivel. Such a solution is said to be **hypertonic** ("excess-tone") relative to the inside of the vesicle. The word root *hyper* refers to the outside solution containing more solutes than the solution on the other side of the membrane. Conversely, if the solution outside the membrane has a lower concentration of solutes than the interior, water will move into the vesicle via osmosis. The incoming water will cause the vesicle to swell or even burst. Such a solution is termed **hypotonic** ("lower-tone") relative to the inside of the vesicle. Here the word root *hypo* refers to the outside solution containing fewer

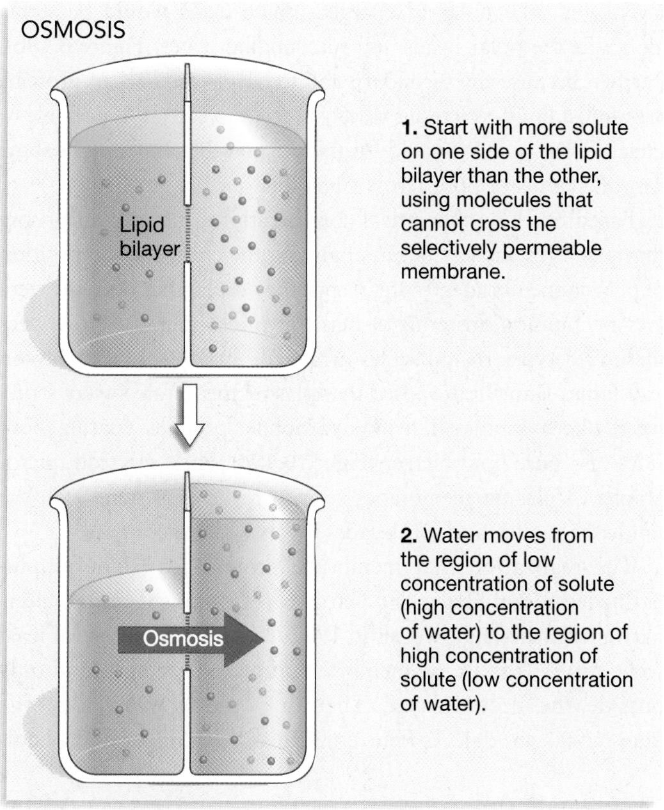
OSMOSIS
Lipid bilayer
1. Start with more solute on one side of the lipid bilayer than the other, using molecules that cannot cross the selectively permeable membrane.
Osmosis
2. Water moves from the region of low concentration of solute (high concentration of water) to the region of high concentration of solute (low concentration of water).

FIGURE 6.16 Osmosis
QUESTION Suppose you doubled the number of molecules on the right side of the membrane (at the start). At equilibrium, would the water level on the right side be higher or lower than what is shown here?

solutes than the solution inside. If solute concentrations are equal on either side of the membrane, the liposome will maintain its size. When the outside solution does not affect the membrane's shape, that solution is termed **isotonic** ("equal-tone").

In sum, diffusion and osmosis move solutes and water across lipid bilayers. What does all this have to do with the first membranes floating in the prebiotic soup? In effect, osmosis and diffusion tend to *reduce* differences in chemical composition between the inside and outside of membrane-bound structures. If liposome-like structures were present in the prebiotic soup, it's unlikely that their interiors offered a radically different environment from the surrounding solution. In all likelihood, the primary importance of the first lipid bilayers was simply to provide a container for self-replicating molecules. Experiments have shown that ribonucleotides can diffuse across lipid bilayers. Further, it is clear that cell-like vesicles grow as additional lipids are added and then divide if sheared by shaking or bubbling or wave action. Based on these observations, it is reasonable to hypothesize that once a self-replicating ribozyme had become surrounded by a lipid bilayer, this simple life-form and its descendants would continue to occupy cell-like structures that grew and divided.

We can now investigate the next great event in the evolution of life: the formation of a true cell. How can lipid bilayers become a barrier capable of creating and maintaining a specialized internal environment that is conducive to life? How could a simple cell become an effective plasma membrane—one that admits ions and molecules needed by the replicator while excluding ions and molecules that might damage it?

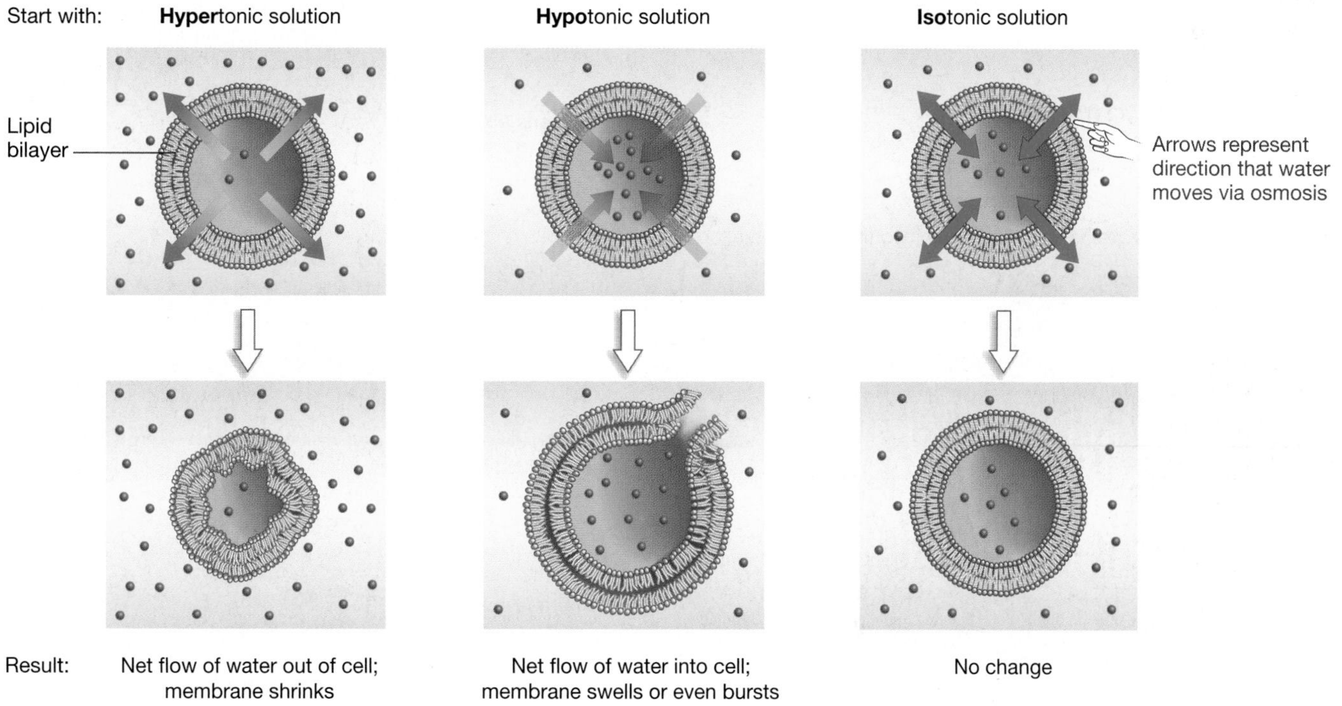
Start with:
Hypertonic solution
Hypotonic solution
Isotonic solution
Lipid bilayer
Arrows represent direction that water moves via osmosis
Result:
Net flow of water out of cell; membrane shrinks
Net flow of water into cell; membrane swells or even bursts
No change

FIGURE 6.17 Osmosis Can Shrink or Burst Membrane-Bound Vesicles
QUESTION Some species of bacteria can live in extremely salty environments, such as saltwater-evaporation ponds. Is this habitat likely to be hypertonic, hypotonic, or isotonic relative to the interior of the cells?

✓CHECK YOUR UNDERSTANDING

Diffusion is the movement of ions or molecules from regions of high concentration to regions of low concentration. Osmosis is the movement of water across a selectively permeable membrane, from a region of low solute concentration to a region of high solute concentration. You should be able to (1) make a sketch analogous to Figure 6.17 and (2) predict what happens when the solution *inside* the cell is hypertonic, hypotonic, and isotonic relative to the outside.

6.4 Membrane Proteins

What sort of molecule could become incorporated into a lipid bilayer and affect the bilayer's permeability? The title of this section gives the answer away. Proteins that are amphipathic can be inserted into lipid bilayers.

Proteins can be amphipathic because they are made up of amino acids and because amino acids have side chains, or R-groups, that range from highly nonpolar to highly polar (or even charged; see Figure 3.5 and Table 3.1). It is conceivable, then, that a protein could have a series of nonpolar amino acids in the middle of its primary structure but polar or charged amino acids on both ends of its primary structure, as illustrated in **Figure 6.18a**. The nonpolar amino acids would be stable in the interior of a lipid bilayer, while the polar or charged amino acids would be stable alongside the polar heads and surrounding water (**Figure 6.18b**). Further, because the secondary and tertiary structures of proteins are almost limitless in their variety and complexity, it is possible to imagine that proteins could form tubes and thus function as some sort of channel or pore across a lipid bilayer.

Based on these theoretical considerations, it is not surprising that when researchers began analyzing the chemical composition of plasma membranes in organisms, they found that proteins were just as common, in terms of mass, as phospholipids. How were these two types of molecules arranged? In 1935 Hugh Davson and James Danielli proposed that plasma membranes were structured like a sandwich, with hydrophilic proteins coating both sides of a pure lipid bilayer (**Figure 6.19a**). Early electron micrographs of plasma membranes seemed to be consistent with the sandwich model, and for decades it was widely accepted.

The realization that membrane proteins could be amphipathic led S. Jon Singer and Garth Nicolson to suggest an alternative hypothesis, however. In 1972, they proposed that at least some proteins span the membrane instead of being found only outside the lipid bilayer. Their hypothesis was called the **fluid-mosaic model.** As **Figure 6.19b** shows, Singer and Nicol-

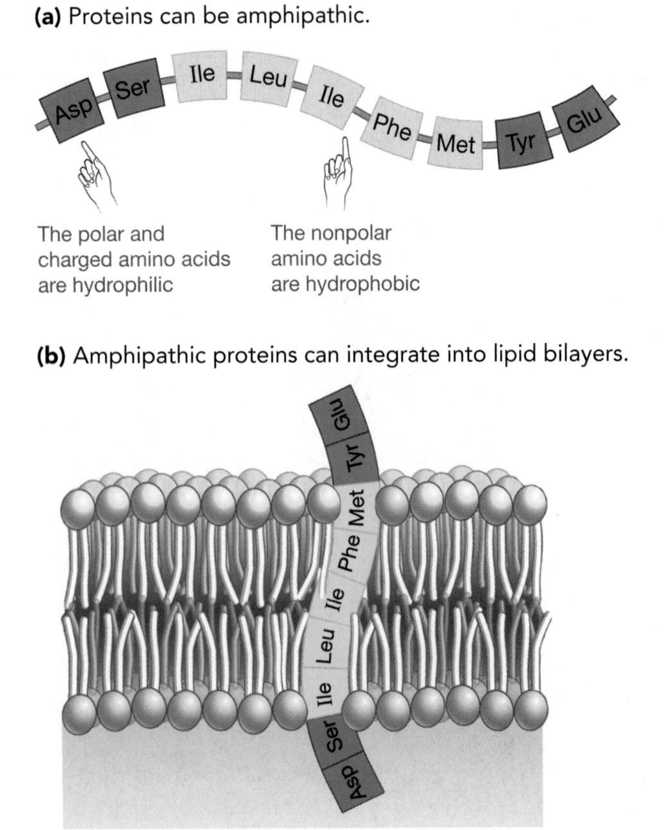
(a) Proteins can be amphipathic.
Asp
Ser
Ile
Leu
Ile
Phe
Met
Tyr
Glu
The polar and charged amino acids are hydrophilic
The nonpolar amino acids are hydrophobic
(b) Amphipathic proteins can integrate into lipid bilayers.
Glu
Tyr
Met
Phe
Ile
Leu
Ile
Ser
Asp

FIGURE 6.18 Proteins Can Be Amphipathic
Note that these drawings are conceptual and not to scale. It actually takes at least 15 amino acids to span a lipid bilayer.

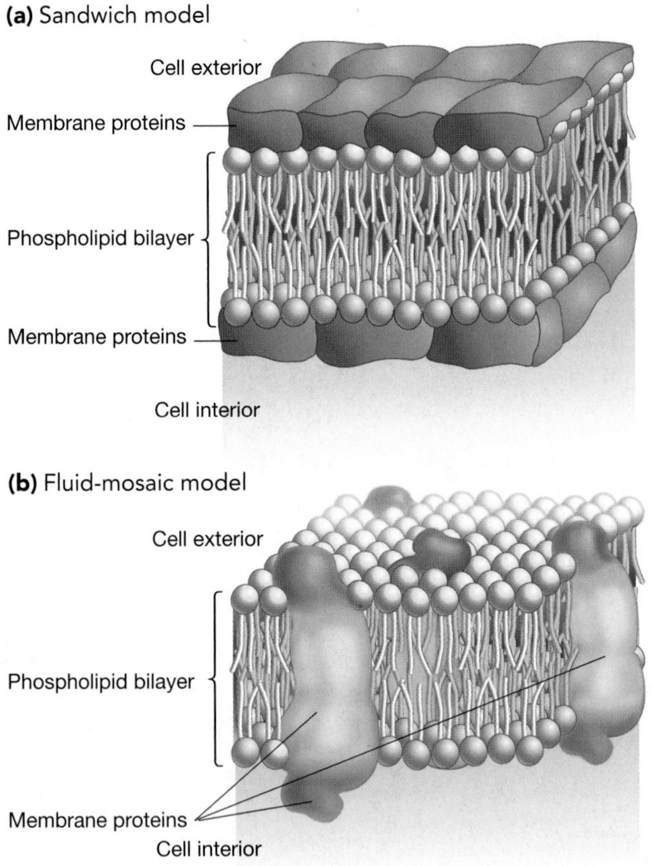
(a) Sandwich model
Cell exterior
Membrane proteins
Phospholipid bilayer
Membrane proteins
Cell interior
(b) Fluid-mosaic model
Cell exterior
Phospholipid bilayer
Membrane proteins
Cell interior

FIGURE 6.19 Past and Current Models of Membrane Structure
(a) The protein-lipid-lipid-protein sandwich model was the first hypothesis for the arrangement of lipids and proteins in plasma membranes. **(b)** The fluid-mosaic model was a radical departure from the sandwich hypothesis.

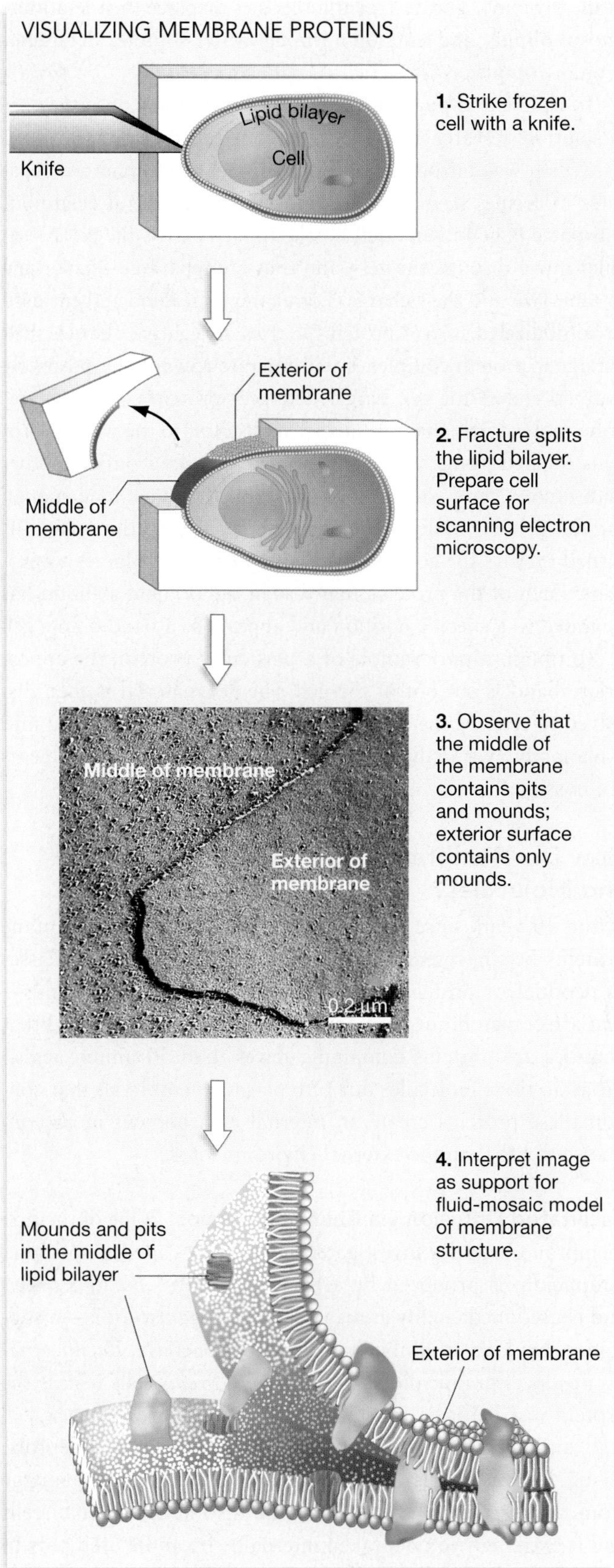
VISUALIZING MEMBRANE PROTEINS
Lipid bilayer
Cell
Knife
1. Strike frozen cell with a knife.
Exterior of membrane
Middle of membrane
2. Fracture splits the lipid bilayer. Prepare cell surface for scanning electron microscopy.
Middle of membrane
Exterior of membrane
0.2 μm
3. Observe that the middle of the membrane contains pits and mounds; exterior surface contains only mounds.
4. Interpret image as support for fluid-mosaic model of membrane structure.
Mounds and pits in the middle of lipid bilayer
Exterior of membrane

FIGURE 6.20 Freeze-Fracture Preparations Allow Biologists to View Membrane Proteins
EXERCISE Draw what the micrograph in step 3 would look like if the sandwich model of membrane structure were correct.

son suggested that membranes are a mosaic of phospholipids and different types of proteins. The overall structure was proposed to be dynamic and fluid.

The controversy over the nature of the plasma membrane was resolved in the early 1970s with the development of an innovative technique for visualizing the surface of plasma membranes. The method is called **freeze-fracture electron microscopy,** because the steps involve freezing and fracturing the membrane before examining it with a scanning electron microscope. As **Figure 6.20** shows, the technique allows researchers to split plasma membranes and view both the middle and the surface of the structure. The scanning electron micrographs that result show pits and mounds studding the inner surfaces of the lipid bilayer. Researchers interpret these structures as the locations of membrane proteins. As step 4 in Figure 6.20 shows, the pits and mounds are hypothesized to represent proteins that span the lipid bilayer.

These observations conflicted with the sandwich model but were consistent with the fluid-mosaic model. Based on these and subsequent observations, the fluid-mosaic model is now widely accepted. **Figure 6.21** summarizes where proteins and lipids are found in a plasma membrane. Note that some proteins span the membrane and have elements facing both the interior and exterior surfaces. Proteins such as these are called **integral membrane proteins,** or **transmembrane proteins.** Other proteins, called **peripheral membrane proteins,** are found only on one side of the membrane. Often these peripheral membrane proteins are attached to an integral membrane protein. The key point is that the arrangement of proteins makes the interior and exterior surfaces of the plasma membrane very different.

What do all these proteins do? In later chapters we'll explore how certain membrane proteins act as enzymes or are involved

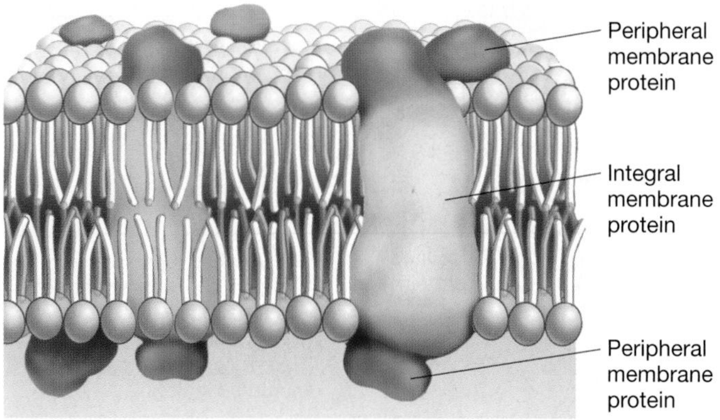
Peripheral membrane protein
Integral membrane protein
Peripheral membrane protein

FIGURE 6.21 Integral and Peripheral Membrane Proteins
Integral membrane proteins are also called transmembrane proteins because they span the membrane. Peripheral membrane proteins are often attached to integral membrane proteins. QUESTION Are the external and internal faces of a plasma membrane the same or different? Explain.

in cell-to-cell signaling or making physical connections between cells. Here, let's focus on how integral membrane proteins are involved in the transport of selected ions and molecules across the plasma membrane.

Systems for Studying Membrane Proteins

The discovery of integral membrane proteins was consistent with the hypothesis that proteins affect membrane permeability. The evidence was considered weak, though, because it was also plausible to claim that integral membrane proteins were structural components that influenced membrane strength or flexibility. To test whether proteins actually do affect membrane permeability, researchers needed some way to isolate and purify membrane proteins.

Figure 6.22 outlines one method that researchers developed to separate proteins from membranes. The key to the technique is the use of detergents. A **detergent** is a small, amphipathic molecule. As step 1 of Figure 6.22 shows, the hydrophobic tails of detergents clump in solution, forming micelles. When detergents are added to the solution surrounding a lipid bilayer, the hydrophobic tails of the detergent molecule interact with the hydrophobic tails of the lipids. In doing so, the detergent tends to disrupt the bilayer and break it apart (step 2). If the membrane contains proteins, the hydrophobic tails of the detergent molecules also interact with the hydrophobic parts of the membrane proteins. The detergent molecules displace the membrane phospholipids and end up forming water-soluble, detergent-protein complexes (step 3).

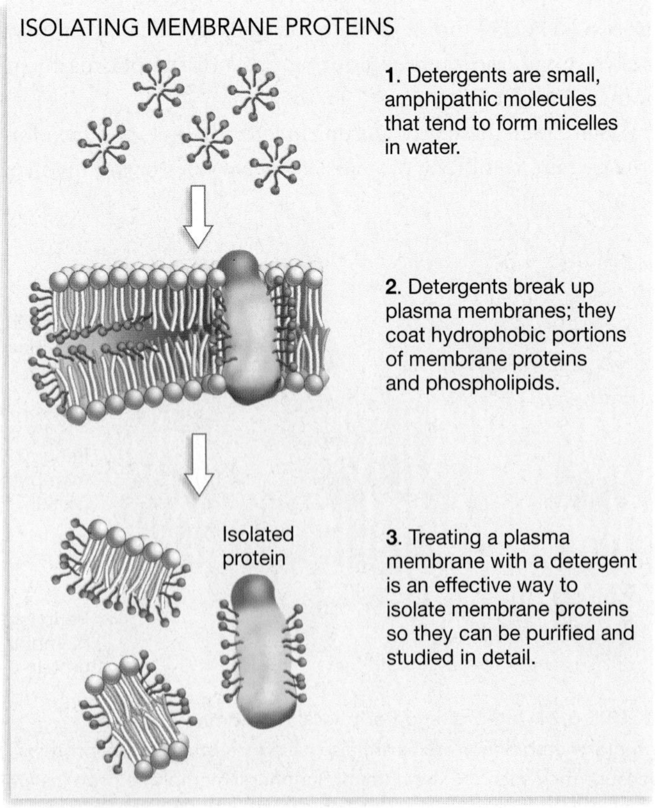
ISOLATING MEMBRANE PROTEINS
1. Detergents are small, amphipathic molecules that tend to form micelles in water.
2. Detergents break up plasma membranes; they coat hydrophobic portions of membrane proteins and phospholipids.
Isolated protein
3. Treating a plasma membrane with a detergent is an effective way to isolate membrane proteins so they can be purified and studied in detail.

FIGURE 6.22 Detergents Can Be Used to Get Membrane Proteins into Solution

To isolate and purify these membrane proteins once they are in solution, researchers use the technique called gel electrophoresis, which was introduced in Box 4.1. **Gel electrophoresis** separates molecules based on their movement through a gelatinous substance (a gel) in an electric field applied across the gel. Molecules move through the gel—and thus are separated—according to their size and their charge. For example, if the detergent used to solubilize (dissolve) proteins carries a negative charge, then detergent-protein complexes will migrate toward the positively charged end of the gel. Larger detergent-protein complexes migrate more slowly than do smaller molecules, so the various proteins isolated from a plasma membrane separate from each other with time. After a voltage has been applied to a gel long enough for the protein molecules to become separated, the voltage is turned off and the gel is stained with a dye that adheres to proteins. Each of the proteins that was in the original solution has migrated to a specific position and appears as a band on the gel.

To obtain a pure sample of a particular protein, the appropriate band is cut out of the gel. The gel material is then dissolved to retrieve the protein. Once this protein is inserted into a planar bilayer or liposome, dozens of different experiments are possible.

How Do Membrane Proteins Affect Ions and Molecules?

In the 40 years since intensive experimentation on membrane proteins began, researchers have identified three broad classes of peptides or proteins—channels, transporters, and pumps—that affect membrane permeability. (Recall from Chapter 3 that peptides are proteins containing fewer than 50 amino acids.) What do these molecules do? Can plasma membranes that contain these proteins create an internal environment more conducive to life than the external environment is?

Facilitated Diffusion via Channel Proteins One of the first membrane peptides investigated in detail is called gramicidin. Gramicidin is produced by a bacterium called *Bacillus brevis* and is used medicinally as an antibiotic ("against-life")—a substance that kills or inhibits the growth of bacteria. *Bacillus brevis* produces gramicidin as a weapon: *B. brevis* cells release the protein just before a resistant coating forms around their cell wall and membrane. The gramicidin wipes out competitors, giving *B. brevis* cells more room to grow when they emerge from the resistant phase. Gramicidin also affects human cells and is extremely toxic if taken internally. It can be used only to fight skin infections.

After observing that experimental cells treated with gramicidin seemed to lose large numbers of ions, researchers became interested in understanding how the molecule works. Could this protein alter the flow of ions across plasma membranes?

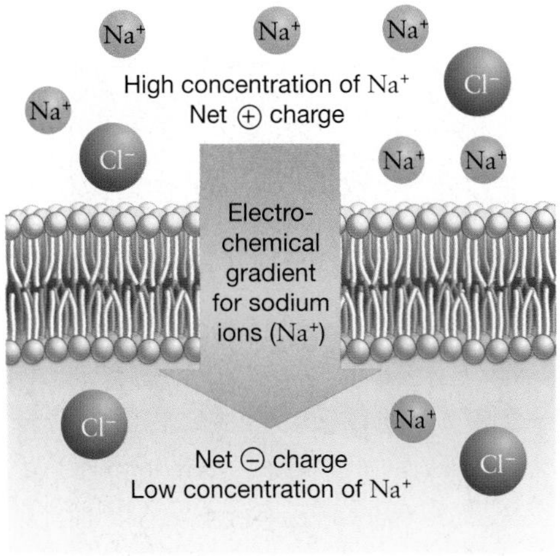

FIGURE 6.23 Electrochemical Gradients
When ions build up on one side of a membrane, they establish a combined concentration and electrical gradient. EXERCISE Add an arrow indicating the electrochemical gradient for chloride ions.

Biologists answered this question by inserting purified gramicidin into planar bilayers. The experiment they performed was based on an important fact about ion movement across membranes: Ions flow not only from regions of high concentration to regions of low concentration but also from areas of like charge to areas of unlike charge (**Figure 6.23**). Stated another way, ions move in response to a combined concentration and electrical gradient, or what biologists call an **electrochemical gradient.**

To determine whether gramicidin affected the membrane's permeability to ions, the researchers measured the flow of electric current across the membrane. Because ions carry a charge, the movement of ions produces an electric current. This property provides an elegant and accurate test for assessing the bilayer's permeability to ions—one that is simpler and more sensitive than taking samples from either side of the membrane and determining the concentrations of solutes present. If gramicidin facilitates ion movement, then an investigator should be able to detect an electric current across planar bilayers that contain gramicidin.

The result? The graph in **Figure 6.24** shows that when gramicidin was absent, no electric current passed through the membrane. But when gramicidin was inserted into the membrane, current began to flow. Based on this observation, biologists proposed that gramicidin is an **ion channel.** An ion channel is a peptide that makes lipid bilayers permeable to ions. Follow-up work corroborated that gramicidin is selective. It allows only positively charged ions, or cations, to pass. Gramicidin does not allow negatively charged ions, or anions, to pass through the membrane. It was also established that gramicidin is most permeable to hydrogen ions (or protons, H^+) and somewhat less permeable to other cations, such as potassium (K^+) and sodium (Na^+).

Researchers gained additional insight into the way gramicidin works when they determined its amino acid sequence

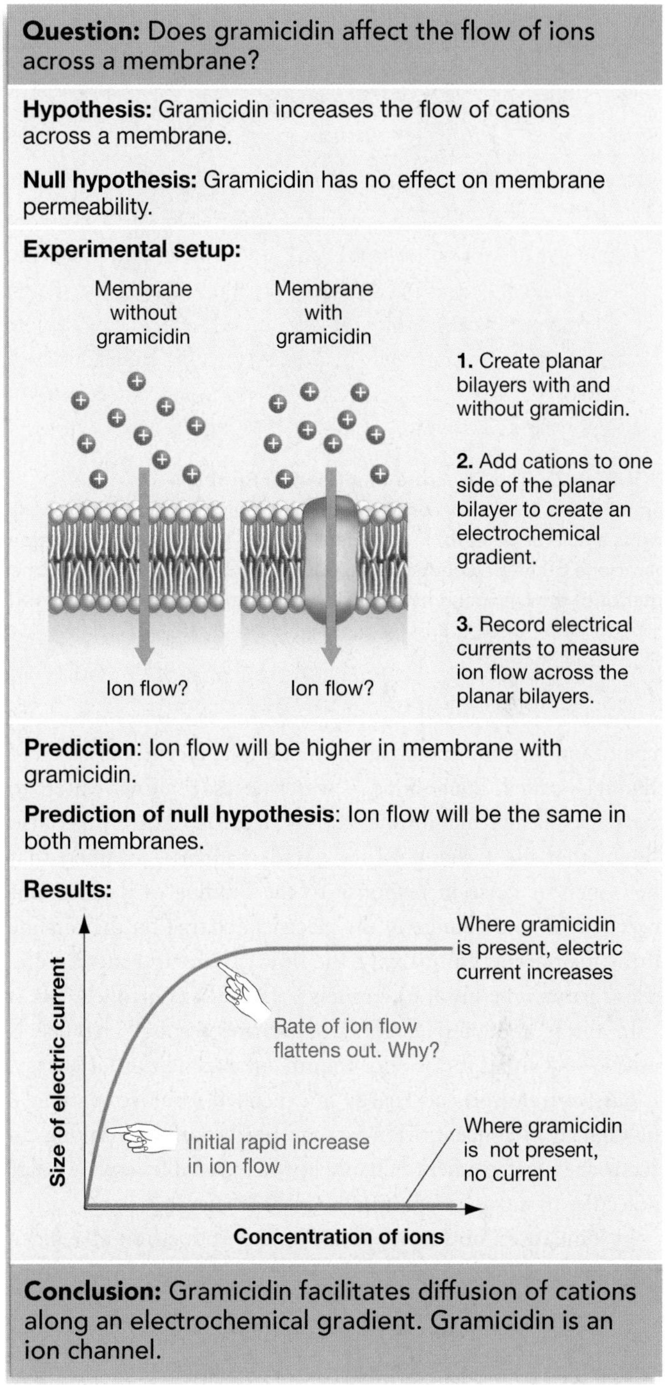

FIGURE 6.24 Measuring Ion Flow through the Channel Gramicidin
Experiment for testing the hypothesis that gramicidin is an ion channel.

(that is, primary structure) and tertiary structure. **Figure 6.25**, page 120, provides a view from the outside of a cell to the inside through gramicidin. The key observation is that the molecule forms a hole. The portions of amino acids that line this hole are hydrophilic, while regions on the exterior (in contact with the membrane phospholipids) are hydrophobic. The molecule's structure correlates with its function.

Cells have many different types of channels in their membranes, each with a structure that allows it to admit a particular

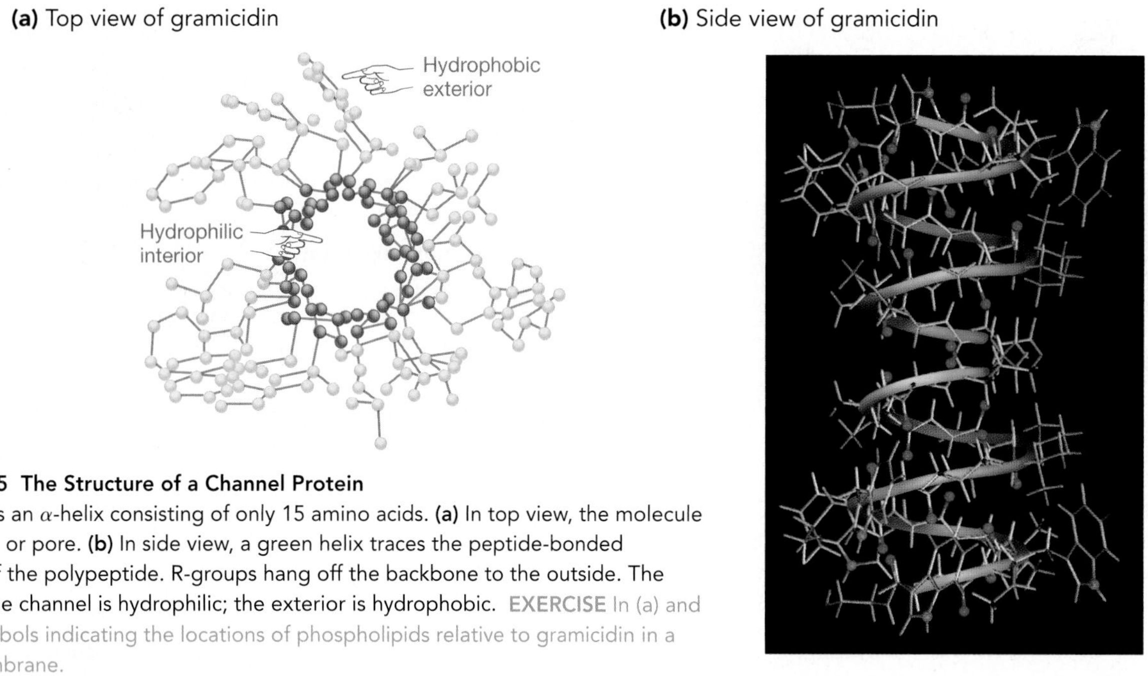
(a) Top view of gramicidin
Hydrophobic exterior
Hydrophilic interior
(b) Side view of gramicidin

FIGURE 6.25 The Structure of a Channel Protein
Gramicidin is an α-helix consisting of only 15 amino acids. **(a)** In top view, the molecule forms a hole or pore. **(b)** In side view, a green helix traces the peptide-bonded backbone of the polypeptide. R-groups hang off the backbone to the outside. The interior of the channel is hydrophilic; the exterior is hydrophobic. EXERCISE In (a) and (b), add symbols indicating the locations of phospholipids relative to gramicidin in a plasma membrane.

type of ion or small molecule. For example, recently discovered channels called **aquaporins** ("water-pores") allow water to cross the plasma membrane rapidly. Recent research has also shown that most channels are **gated channels**—meaning that they open or close in response to the binding of a particular molecule or to a change in the electrical charge on the outside of the membrane. In this way, the flow of ions and small molecules through membrane channels is carefully controlled.

In all cases, the movement of substances through channels is passive—meaning it does not require an expenditure of energy. In **passive transport**, no energy is expended to move materials; the change in concentration is powered by diffusion along an electrochemical gradient. Channel proteins enable ions or polar molecules to move across lipid bilayers efficiently.

In some cases, however, passive transport does not take place through a channel. As **Figure 6.26** shows, the antibiotic valinomycin acts as an **ionophore** ("ion-mover") instead. The protein binds to a potassium (K^+) ion on the membrane's exterior, diffuses across the bilayer, and releases the cation on the interior.

Membrane proteins such as gramicidin and valinomycin circumvent the lipid bilayer's impermeability to small, charged compounds. As a result, their presence *reduces* differences between the interior and exterior. Water molecules and ions are not the only substances that move across membranes through membrane proteins, however. Larger molecules can, too.

Facilitated Diffusion via Transport Proteins Next to ribose, the six-carbon sugar glucose is the most important sugar found in organisms. Glucose was very likely important to the earliest cells simply because virtually all cells alive today use it as a building block for important macromolecules and as a source of chemical energy. But as we saw in Figure 6.9, lipid

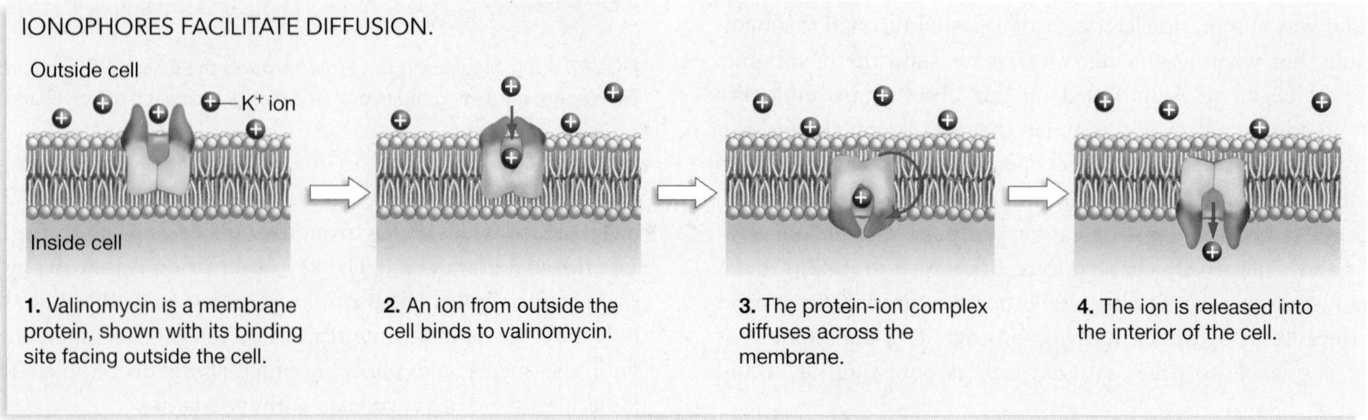
IONOPHORES FACILITATE DIFFUSION.
Outside cell
K⁺ ion
Inside cell
1. Valinomycin is a membrane protein, shown with its binding site facing outside the cell.
2. An ion from outside the cell binds to valinomycin.
3. The protein-ion complex diffuses across the membrane.
4. The ion is released into the interior of the cell.

FIGURE 6.26 The Ionophore Valinomycin
Valinomycin is an ionophore that does not form a channel. It acts instead as a mobile ion carrier.

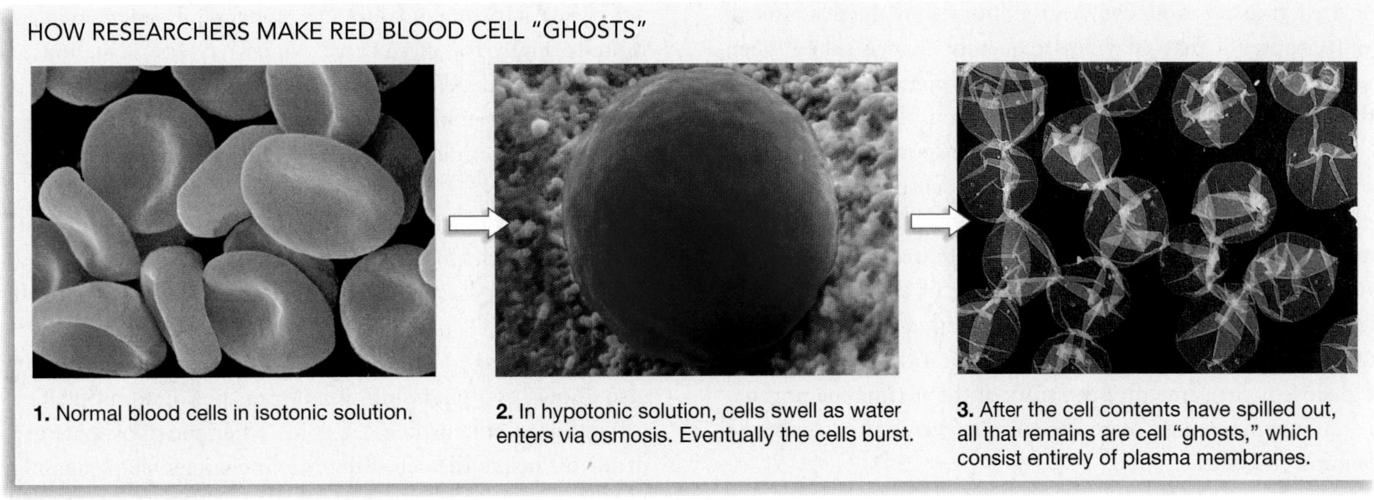
HOW RESEARCHERS MAKE RED BLOOD CELL "GHOSTS"
1. Normal blood cells in isotonic solution.
2. In hypotonic solution, cells swell as water enters via osmosis. Eventually the cells burst.
3. After the cell contents have spilled out, all that remains are cell "ghosts," which consist entirely of plasma membranes.

FIGURE 6.27 Red Blood Cell "Ghosts"
Red blood cell ghosts are simple membranes that can be purified and studied in detail.

bilayers are only moderately permeable to glucose. It is reasonable to expect, then, that plasma membranes have some mechanism for increasing their permeability to this sugar.

This expectation grew into a certainty when researchers compared the permeability of glucose across planar bilayers with its permeability across membranes from cells. The plasma membrane in this study came from human red blood cells, which are among the simplest cells known. Mature red blood cells contain a membrane, about 300 million hemoglobin molecules, and not much else (**Figure 6.27**, step 1). When these cells are placed in a hypotonic solution (step 2), water rushes into them by osmosis. As water flows inward, the cells swell. Eventually they burst, releasing the hemoglobin molecules and other cell contents. This leaves researchers with pure preparations of plasma membranes called red blood cell "ghosts" (step 3). Experiments have shown that these membranes are much more permeable to glucose than are pure lipid bilayers. Why?

After isolating and analyzing many proteins from red blood cell ghosts, researchers found one protein that specifically increases membrane permeability to glucose. When this purified protein was added to liposomes, the artificial membrane transported glucose at the same rate as a membrane from a living cell. This experiment confirmed that a membrane protein was indeed responsible for transporting glucose across plasma membranes. Follow-up work showed that this glucose carrier, a **transport protein** named GLUT-1, transports the "right-handed" optical isomer of glucose but not the left-handed form. This is logical because cells use only the right-handed form of glucose.

Exactly how GLUT-1 works is a focus of ongoing research. Because glucose transport by GLUT-1 is so specific, researchers presume that the mechanism resembles the action of enzymes. One hypothesis is illustrated in **Figure 6.28**. The idea is that glucose binds to GLUT-1 on the exterior of the membrane and that this binding induces a conformational change in the pro-

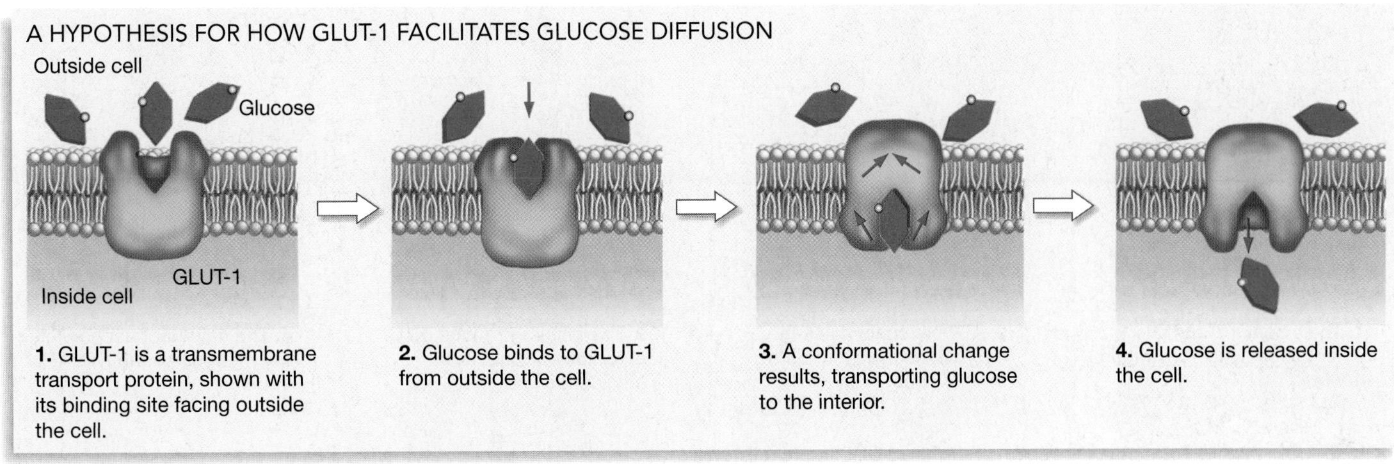
A HYPOTHESIS FOR HOW GLUT-1 FACILITATES GLUCOSE DIFFUSION
Outside cell
Glucose
GLUT-1
Inside cell
1. GLUT-1 is a transmembrane transport protein, shown with its binding site facing outside the cell.
2. Glucose binds to GLUT-1 from outside the cell.
3. A conformational change results, transporting glucose to the interior.
4. Glucose is released inside the cell.

FIGURE 6.28 A Hypothesis to Explain How Membrane Transport Proteins Work
This model suggests that the GLUT-1 transporter acts like an enzyme. It binds a substrate (in this case, a glucose molecule), undergoes a conformation change, and releases the substrate.

tein that transports glucose to the interior of the cell. Recall from Chapter 3 that enzymes frequently change shape when they bind substrates and that such conformational changes are often a critical step in the catalysis of chemical reactions.

Importing molecules into cells via transport proteins is still powered by diffusion, however. When glucose enters a cell via GLUT-1, it does so because it is following its concentration gradient. The movement is called **facilitated diffusion,** because it does not require an expenditure of energy for movement to occur and because it is facilitated—aided—by the presence of a transport protein or **carrier**. If the concentration of glucose is the same on both sides of the plasma membrane, then no net movement of glucose occurs even if the membrane contains GLUT-1.

Active Transport by Pumps Whether diffusion occurs via channel proteins or is facilitated by carriers, it is a passive process that makes the cell interior and exterior more similar. But it is also possible for today's cells to import molecules or ions *against* their electrochemical gradient. Accomplishing this task requires energy, however, because the cell must counteract the entropy loss that occurs when molecules are concentrated. It makes sense, then, that transport against a concentration gradient is called **active transport**. The energy required to move substances against their electrochemical gradient is provided by phosphate groups in adenosine triphosphate, or ATP. In addition to energy, active transport requires a machine—a molecule that is capable of using energy to pump an ion or a molecule against its electrochemical gradient.

Figure 6.29 shows how chemical energy from ATP and a membrane protein called a pump accomplish this task. The figure highlights the first pump that was discovered and characterized: a protein called the **sodium-potassium pump**, or more formally **Na^+/K^+-ATPase.** Na^+/K^+ refers to the ions that are transported; *ATP* indicates that adenosine triphosphate is used; and *–ase* implies that the molecule acts like an enzyme. Unlike the situation with GLUT-1, the mechanism of action in Na^+/K^+-ATPase is now well known. When the protein is in the conformation shown in step 1 of Figure 6.29, three sodium ions from the inside of the cell bind to it. The protein then becomes phosphorylated by the addition of a phosphate group from ATP (step 2) and changes shape in a way that sends these ions to the exterior. This change in conformation also allows two potassium ions from the solution outside the cell to bind to the protein (step 3). When the phosphate group drops off inside the cell, the protein changes shape again and releases the K^+ to the interior (step 4). This movement of ions can occur even if an electrochemical gradient exists that favors the outflow of potassium and the inflow of sodium. By exchanging three sodium ions for every two potassium ions, the outside of the membrane becomes positively charged relative to the inside. In this way, the sodium-potassium pump sets up an electrical gradient as well as a chemical gradient across the membrane.

Similar pumps are specialized for moving protons (H^+), calcium ions (Ca^{2+}), or other ions or molecules. In this way, cells are capable of concentrating certain substances or setting up electrochemical gradients.

Taken together, the lipid bilayer and the proteins involved in passive and active transports enable cells to create an internal environment that is much different from the external one (**Figure 6.30**). When membrane proteins first evolved, then, the early cells acquired the ability to create an internal environment that was conducive to life—meaning that such

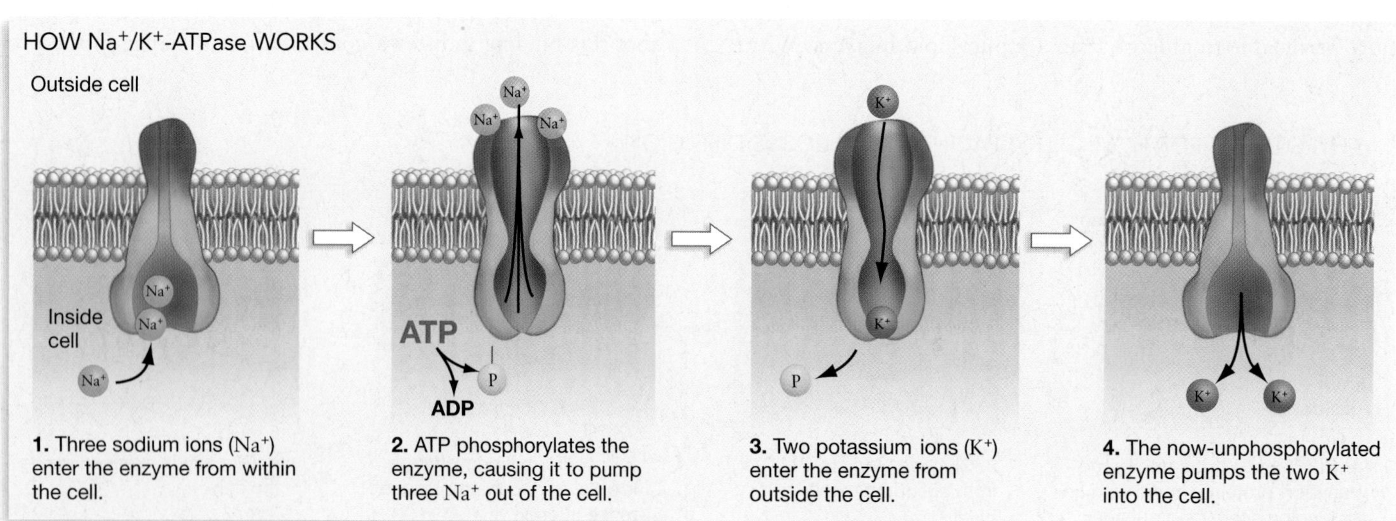
HOW Na^+/K^+-ATPase WORKS
Outside cell
Inside cell
Na^+
ATP
ADP
P
K^+
1. Three sodium ions (Na^+) enter the enzyme from within the cell.
2. ATP phosphorylates the enzyme, causing it to pump three Na^+ out of the cell.
3. Two potassium ions (K^+) enter the enzyme from outside the cell.
4. The now-unphosphorylated enzyme pumps the two K^+ into the cell.

FIGURE 6.29 Active Transport Depends on an Input of Chemical Energy
If chemical energy in the form of a phosphate group is added to the sodium-potassium pump, the pump can transport ions against their electrochemical gradient. The mechanism involves a change in the protein's shape.

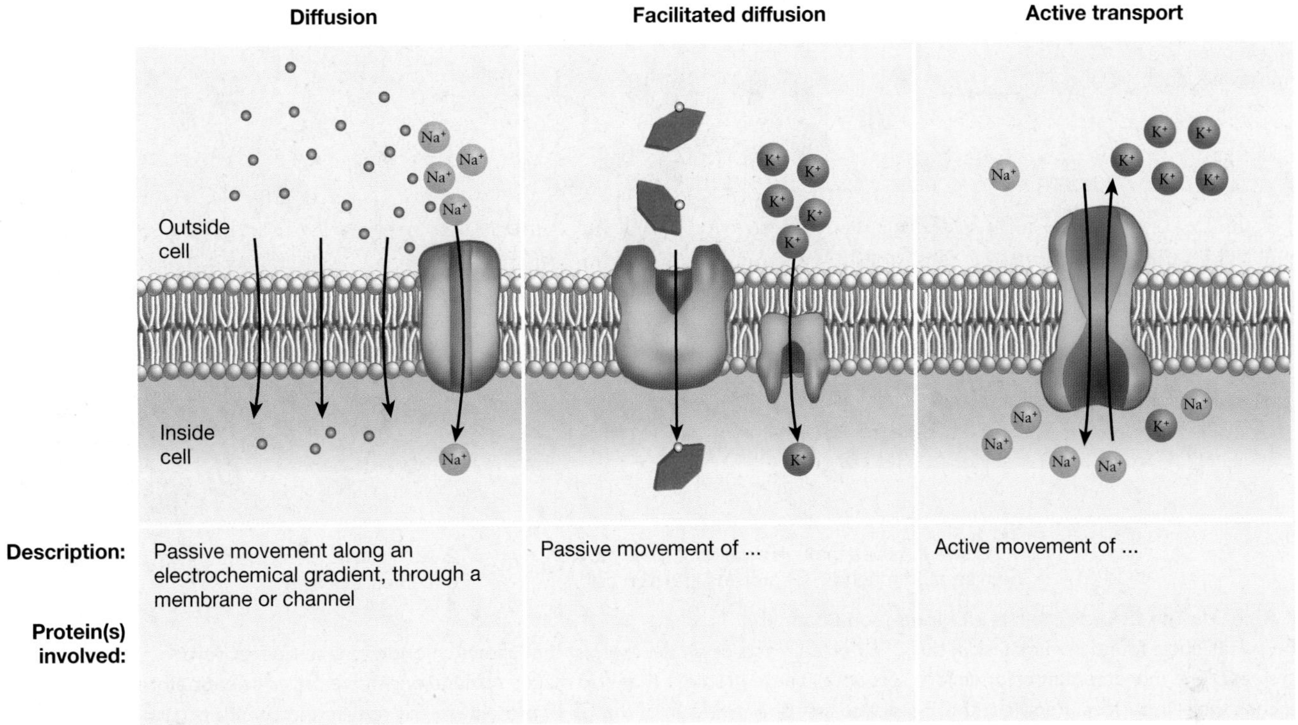

Diffusion
Facilitated diffusion
Active transport
Outside cell
Inside cell
Na+
K+
Description:
Passive movement along an electrochemical gradient, through a membrane or channel
Passive movement of ...
Active movement of ...
Protein(s) involved:

FIGURE 6.30 Mechanisms of Membrane Transport: A Summary
EXERCISE Complete the chart.

an environment contained the substances required for manufacturing ATP and copying ribozymes. Cells with particularly efficient and selective membrane proteins would be favored by natural selection and would come to dominate the population. Cellular life had begun.

Some 3.5 billion years later, cells continue to evolve. What do today's cells look like, and how do they produce and store the chemical energy that makes life possible? How do they use ATP to move pumps and channels and other molecules and machines where they're needed? Answering these and related questions is the focus of Unit 2.

✓CHECK YOUR UNDERSTANDING

Membrane proteins allow ions and molecules that ordinarily do not readily cross lipid bilayers to enter or exit cells. Substances may move along an electrochemical gradient via facilitated diffusion through channel proteins or transport proteins, or they may move against an electrochemical gradient in response to work done by pumps. You should be able to (1) sketch a phospholipid bilayer and (2) indicate how ions and large molecules cross it via each major type of membrane transport protein.

ESSAY The Molecular Basis of Cystic Fibrosis

Cystic fibrosis (CF) is the most common genetic disease in humans of Northern European descent. In these populations, one in every 2500 infants born has CF. Children affected by CF suffer from a progressive deterioration of their lungs and gastrointestinal and reproductive tracts. Although the ability of physicians to manage the disease has improved dramatically in recent years, the median life expectancy for patients with severe forms of CF is still only 32.*

*Half of the numbers in a distribution are found above the median value, and half are found below. A median life expectancy of 32 years means that half of the people with CF live less than 32 years and half live more than 32 years.

CF produces a distinctive set of symptoms. Prominent among them are the production of salty sweat and the chronic development of thickened mucus in the linings of the lungs and associated air passages. The thickened mucus, which is extremely difficult to cough out, leads to two important complications: the direct obstruction of airways and the rapid growth of disease-causing bacteria.

Biologists now understand the molecular basis of cystic fibrosis.

(Continued on next page)

(Essay continued)

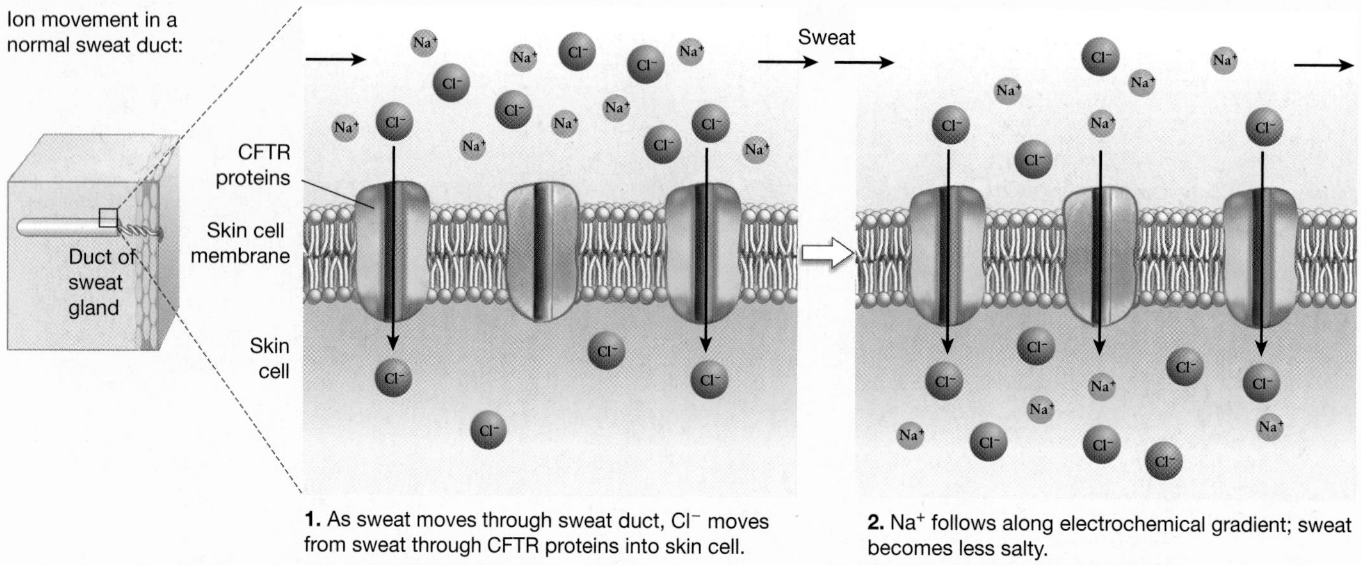
Ion movement in a normal sweat duct:
Sweat
CFTR proteins
Skin cell membrane
Duct of sweat gland
Skin cell
Na+
Cl−
1. As sweat moves through sweat duct, Cl− moves from sweat through CFTR proteins into skin cell.
2. Na+ follows along electrochemical gradient; sweat becomes less salty.

FIGURE 6.31 The CFTR Protein Is a Chloride Ion Channel
In the sweat ducts found in human skin, the CFTR protein acts as an ion channel that allows chloride ions to be reabsorbed from sweat. The movement of chloride ions sets up a charge gradient that also makes sodium ions move out of sweat before it reaches the skin surface. People with CF produce an abnormal form of the CFTR protein, so this movement cannot occur.
EXERCISE Cross out the CFTR protein in the right-hand drawing to represent the situation in an individual with CF. Then redraw the distribution of ions on both sides of the membrane.

What causes these symptoms? Understanding the molecular basis of CF frustrated researchers for decades. Then an important clue came to light in 1983, when P. M. Quinton showed that sweat-duct cells taken from individuals with CF are impermeable to chloride ions (Cl^-). Sweat is produced as a salty solution. But as the solution moves through a sweat duct toward the surface of the skin, chloride ions pass through the membrane and are reabsorbed by the body (**Figure 6.31**). This movement establishes an electrochemical gradient that causes sodium ions (Na^+) to follow. But in people with CF, both ions stay in the sweat duct and are excreted. As a result, these individuals have extraordinarily salty sweat.

Quinton's result suggested that the disease might be caused by a defect in a membrane protein—one that normally allows chloride ions to move across lipid bilayers. Nine years later, Christine Bear and colleagues confirmed this result in spectacular fashion. Bear's group was able to isolate and purify quantities of the normal protein, which had come to be called cystic fibrosis transmembrane conductance regulator (CFTR). The researchers inserted the protein into planar bilayers and showed that it did indeed selectively permit the passage of chloride ions. Subsequent studies showed that people with CF produce abnormal forms of the protein. As a result, their chloride channels do not function.

Biologists now understand the molecular basis of cystic fibrosis. Cells that line the respiratory tract of humans normally release a steady stream of chloride ions, followed by sodium ions. The high concentration of ions on the outside of the cell leads to an outflow of water. But chloride ions are not released in CF individuals, and water remains inside the cells. The result is dramatically thickened mucus in the lungs and air passages, contributing to this fatal syndrome (**Figure 6.32**).

The research that identified the protein's mode of action unleashed a torrent of new studies aimed at finding a cure. As this book goes to press, that work is still in progress. One strategy being tested is to synthesize large quantities of normal protein, package the protein channels into liposomes, and then deliver them to the respiratory lining of CF patients via an inhalator. The hope is that the liposomes will fuse with the plasma membranes, introduce copies of CFTR that can function normally, and alleviate the symptoms of CF.

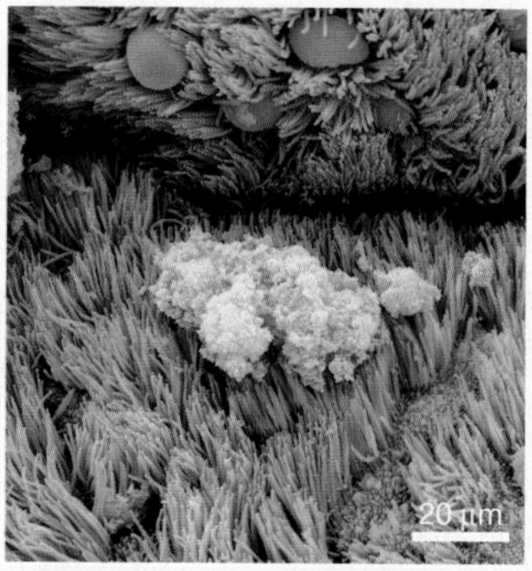
20 µm

FIGURE 6.32 CF Patients Have Thickened Mucus in Their Lungs
Scanning electron micrograph of an airway in the lung of a CF patient. The hairlike structures are cilia; the pink mass is mucus.

CHAPTER REVIEW

Summary of Key Concepts

Phospholipids have a hydrophilic region and a hydrophobic region. In solution, they spontaneously form bilayers that are selectively permeable—meaning that only certain substances cross them readily.

The plasma membrane is a structure that forms a physical barrier between life and nonlife. Molecules called lipids, which do not dissolve in water, are a prominent component of plasma membranes. Lipids called fatty acids and isoprenes, both of which have long hydrocarbon tails, are often found in larger molecules called phospholipids. Phospholipids have a polar head and nonpolar tail. The basic structure of plasma membranes is created by a phospholipid bilayer.

Small, nonpolar molecules tend to move across membranes readily; ions and other charged compounds cross rarely, if at all. The permeability and fluidity of lipid bilayers depend on temperature and on the types of phospholipids present. For example, because phospholipids that contain long, saturated fatty acids form a dense and highly hydrophobic membrane interior, they tend to be less permeable than phospholipids containing shorter, unsaturated fatty acids.

Ions and molecules spontaneously diffuse from regions of high concentration to regions of low concentration. In osmosis, a special case of diffusion, water moves across a selectively permeable membrane from regions of high concentration to regions of low concentration.

Molecules can move across membranes spontaneously because of diffusion. Diffusion is the directed movement of ions or molecules from a region of high concentration to a region of low concentration; it is driven by an increase in entropy. Water also moves across membranes spontaneously if a molecule or an ion that cannot cross the membrane is found in different concentrations on the two sides. In osmosis, water moves from the region with a lower concentration of solutes to the region of higher solute concentration. Osmosis is a passive process driven by an increase in entropy.

Web Tutorial 6.1 Diffusion and Osmosis

In cells, membrane proteins are responsible for the passage of ions, polar molecules, and large molecules that do not readily cross phospholipid bilayers on their own.

The permeability of lipid bilayers can be altered radically by membrane transport proteins, which are scattered throughout the plasma membrane. Channel proteins, for example, are molecules that provide holes in the membrane and facilitate the diffusion of specific ions into or out of the cell. Transport proteins are enzyme-like proteins that allow specific molecules to diffuse into the cell. In addition to these forms of facilitated diffusion, membrane proteins that act as energy-demanding pumps actively move ions or molecules against their electrochemical gradient. In combination, the selective permeability of phospholipid bilayers and the specificity of transport proteins make it possible to create an environment inside a cell that is radically different from the exterior.

Web Tutorial 6.2 Membrane Transport Proteins

Questions

Content Review

1. What does the term *hydrophilic* mean when it is translated literally?
   a. "oil loving"
   b. "water loving"
   c. "oil fearing"
   d. "water fearing"

2. If a solution surrounding a cell is hypotonic relative to the inside of the cell, how will water move?
   a. It will move into the cell via osmosis.
   b. It will move out of the cell via osmosis.
   c. It will not move, because equilibrium exists.
   d. It will evaporate from the cell surface more rapidly.

3. If a solution surrounding a cell is hypertonic relative to the inside of the cell, how will water move?
   a. It will move into the cell via osmosis.
   b. It will move out of the cell via osmosis.
   c. It will not move, because equilibrium exists.
   d. It will evaporate from the cell surface more rapidly.

4. When does a concentration gradient exist?
   a. when membranes rupture
   b. when solute concentrations are high
   c. when solute concentrations are low
   d. when solute concentrations differ on the two sides of a membrane

5. Which of the following must be true for osmosis to occur?
   a. Water must be at room temperature or above.
   b. Solutions with the same concentration of solutes must be separated by a selectively permeable membrane.
   c. Solutions with different concentrations of solutes must be separated by a selectively permeable membrane.
   d. Water must be under pressure.

6. Why are the lipid bilayers in cells called "selectively permeable"?
   a. They are not all that permeable.
   b. Their permeability changes with their molecular composition.
   c. Their permeability is temperature dependent.
   d. They are permeable to some substances but not others.

Conceptual Review

1. Cooking oil is composed of lipids that consist of long hydrocarbon chains. These are not amphipathic molecules. Would you expect these lipids to form membranes spontaneously? Why or why not? Describe, on a molecular level, how you would expect these lipids to interact with water. Your answer should explain the saying, "oil and water don't mix."
2. Ethanol, the active ingredient in alcoholic beverages, is a small, polar, uncharged molecule. Would you predict that this molecule crosses plasma membranes quickly or slowly? Explain your reasoning.
3. Why can osmosis occur only if solutions are separated by a selectively permeable membrane?
4. The text claims that the portion of membrane proteins that spans the hydrophobic tails of phospholipids is itself hydrophobic (see Figure 6.18b). Why does this make sense? Look back at Figure 3.5 and Table 3.1, and make a list of amino acids you would expect to find in these regions of transmembrane proteins.
5. Describe how the structure of phospholipids allows them to function as a plasma membrane. Explain how the amphipathic structure of some proteins allows them to function as integral membrane proteins.
6. Examine the membrane in the accompanying figure. Label the molecules and ions that will pass through the membrane as a result of osmosis, diffusion, and facilitated diffusion. Draw arrows to indicate where each of the molecules and ions will travel.

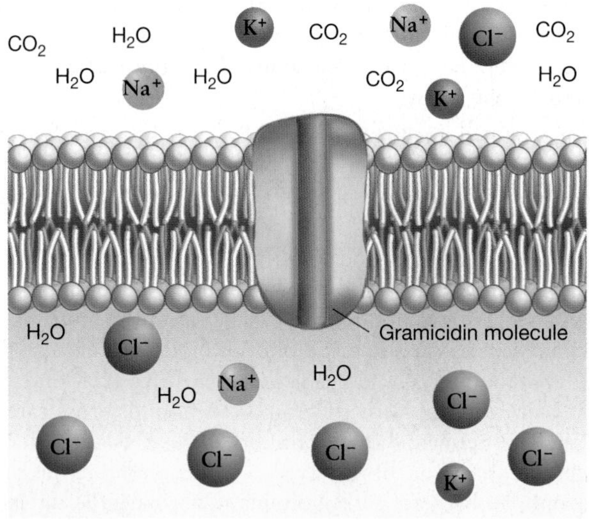

Group Discussion Problems

1. When phospholipids are arranged in a bilayer, it is theoretically possible for individual molecules in the bilayer to flip-flop. That is, a phospholipid could turn 180° and become part of the membrane's other surface. Sketch this process. Based on what you know about the behavior of polar heads and nonpolar tails, predict whether flip-flops are frequent or rare. Then design an experiment, using a planar bilayer made up partly of fatty acids that contain a dye molecule on their hydrophilic head, to test your prediction.
2. Unicellular organisms that live in extremely cold habitats have an unusually high proportion of unsaturated fatty acids in their plasma membranes. Some of these membranes even contain polyunsaturated fatty acids, which have more than one double bond in each hydrocarbon chain. Researchers have proposed that the lack of saturation helps these membranes maintain a semifluid state at low temperatures instead of becoming semisolid or solid. Draw a picture of this type of membrane, analogous to the sketch in Figure 6.11. Using this picture as a resource, comment on the hypothesis that membranes with unsaturated fatty-acid tails function better at cold temperatures than do membranes with saturated fatty-acid tails. How would you test this hypothesis? Make a prediction about the structure of fatty acids found in organisms that live in extremely hot environments.
3. When biomedical researchers design drugs that must enter cells to be effective, they sometimes add methyl (CH_3) groups to make the drug molecules more likely to pass through plasma membranes. Conversely, when researchers design drugs that act on the exterior of plasma membranes, they sometimes add a charged group to decrease the likelihood that the drugs will pass through membranes and enter cells. Explain why these strategies make sense.
4. Advertisements frequently claim that laundry and dishwashing detergents "cut grease." What the ad writers mean is that the detergents surround oil droplets on clothing or dishes, making the droplets water soluble. When this happens, the oil droplets can be washed away. Explain how this happens on a molecular level.

Answers to Multiple-Choice Questions 1. b; 2. a; 3. b; 4. d; 5. c; 6. d

www.prenhall.com/freeman is your resource for the following: Web Tutorials; Online Quizzes and other Online Study Guide materials; Answers to Conceptual Review Questions; Solutions to Group Discussion Problems; Answers to Figure Caption Questions and Exercises; and Additional Readings and Research.

UNIT 2

Cell Structure and Function

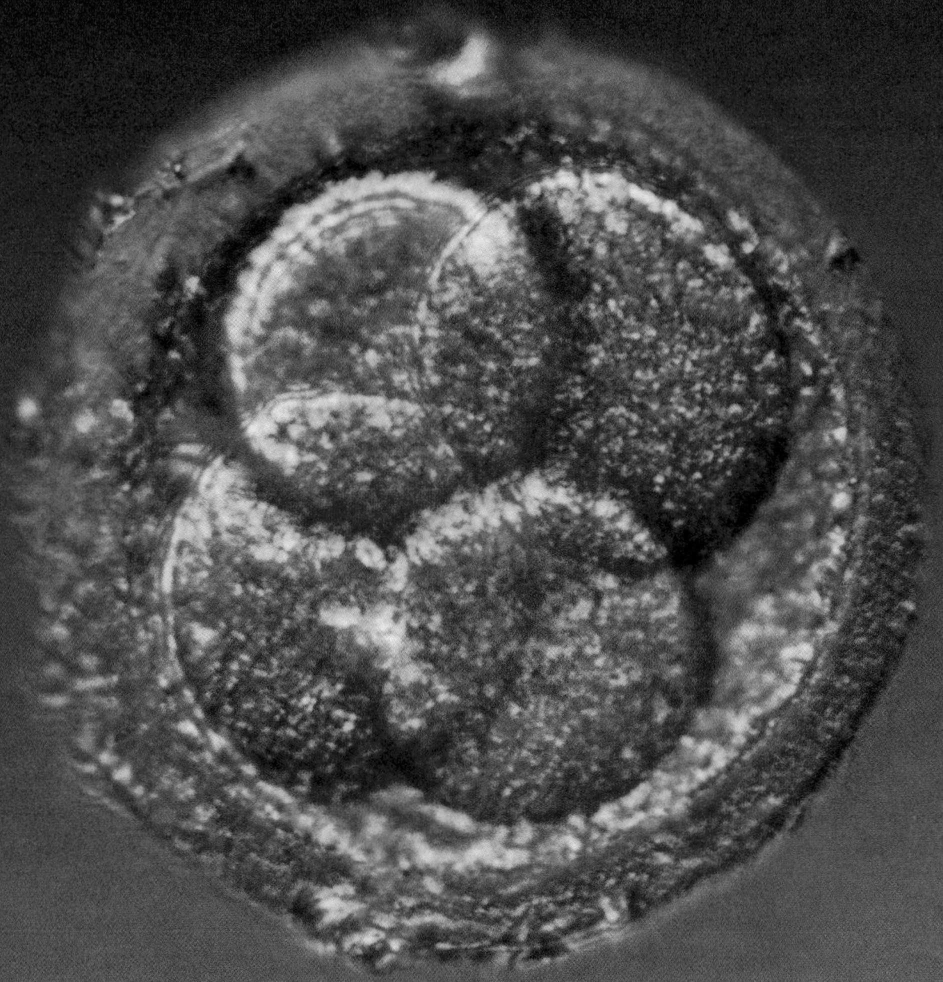

The four cells in this two-day-old human embryo will give rise to the trillions of cells in the adult body.

7 Inside the Cell

KEY CONCEPTS

- The structure of cell components is closely correlated with their function.
- Inside cells, materials are transported to their destinations with the help of molecular "zip codes."
- Cells are dynamic. Thousands of chemical reactions occur each second within cells; molecules constantly enter and exit across the plasma membrane; cell products are shipped along protein fibers; and elements of the cell's internal skeleton grow and shrink.

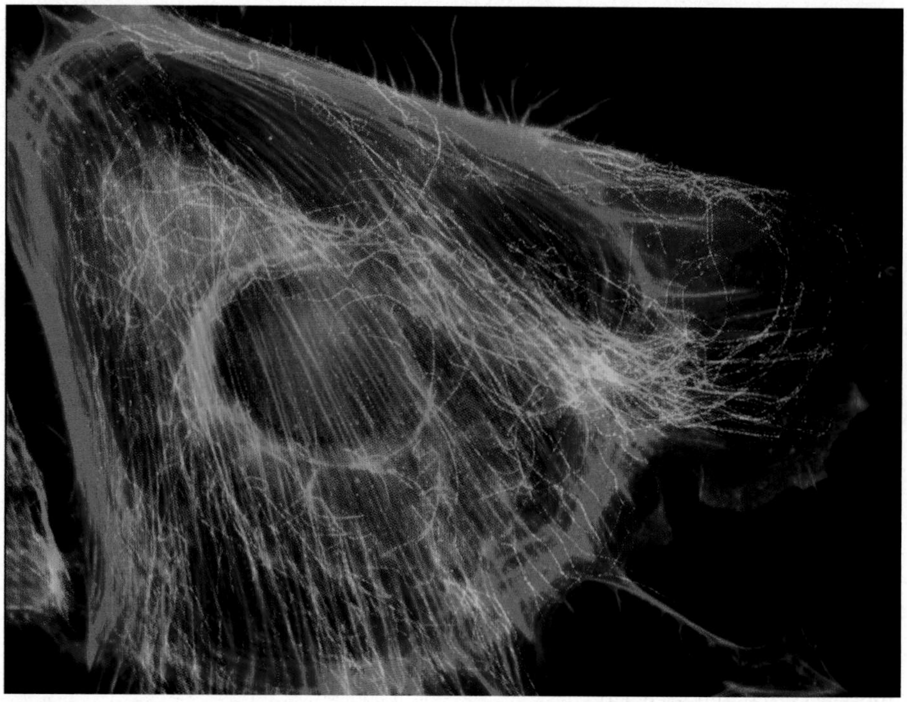

This cell has been treated with fluorescing molecules that bind to its fibrous skeleton. Microtubules (large protein fibers) are green; microfilaments (smaller fibers) are red. The cell's nucleus has been stained blue.

In Chapter 1 you were introduced to the cell theory, which states that all organisms consist of cells and that all cells are derived from preexisting cells. Since this theory was initially developed and tested in the 1850s and 1860s, an enormous body of research has confirmed that the cell is the fundamental structural and functional unit of life. Life on Earth is cellular.

In a very real sense, then, understanding how an organism works is a matter of understanding how cells are structured and how they function. To drive this point home, recall from Chapter 1 that many eukaryotic organisms and virtually all bacteria and archaea are unicellular. In number of individuals present, unicellular organisms dominate life on Earth. For researchers who study these species, understanding the cell is synonymous with understanding the organism as a whole. Even in plants, animals, and other multicellular eukaryotes, complex behavior originates at the level of the cell. For example, your ability to read this page begins with changes in light-sensitive molecules located in cells at the back of your eyes. When these molecules change shape, they trigger changes in the membranes of nerve cells that connect your eyes to your brain. To understand complex processes such as vision, then, researchers often begin by studying the structure and function of the individual cells involved—the parts that make up the whole.

Chapter 6 introduced an essential part of the cell: the **plasma membrane**. Thanks to the selective permeability of phospholipid bilayers and the activity of membrane transport proteins, this structure creates an internal environment that is different from conditions outside the cell. Our task now is to explore the structures that are found inside the membrane and analyze what they do. We'll focus on several particularly dynamic structures and processes and introduce some of the experimental approaches that biologists use to understand them. Let's begin by surveying the basic types of cells, cell structures, and cell processes that biologists have documented to date.

7.1 What's Inside the Cell?

In Chapter 1 you read about the two fundamental types of cells observed in nature. Recall that eukaryotic cells have a membrane-bound compartment called a nucleus, while prokaryotic cells do not. In terms of *morphology* ("form"), then, species fall into the two broad categories: (1) prokaryotes and (2) eukaryotes. But in terms of *phylogeny,* or evolutionary history, organisms fall into the three broad groups called (1) Bacteria, (2) Archaea, and (3) Eukarya. Members of the Bacteria and Archaea are prokaryotic; members of the Eukarya—including algae, fungi, plants, and animals—are eukaryotic.

In the late seventeenth century, biologists began studying the structure of cells with microscopes. Over time, improvements in optics and cell preparation techniques allowed researchers to catalogue the structures reviewed in this section. When electron microscopes became widely available in the 1950s, investigators described the internal anatomy of these structures in more detail. More recent advances in microscopy have allowed investigators to videotape certain types of cell processes in living cells.

What have anatomical studies based on microscopy revealed? Let's look first at the general anatomy of prokaryotic cells and eukaryotic cells, and then consider how the structures that have been identified help cells function.

Prokaryotic Cells

Figure 7.1 shows the general structure of a prokaryotic cell. For most bacterial and archaeal species, the plasma membrane encloses a single compartment—meaning that the cell has few or no substructures delimited by internal membranes. Closer examination reveals a series of intricate structures, however. Let's take a look at a typical prokaryotic cell, starting at the outside and working in.

As Chapter 6 pointed out, the cell membrane, or plasma membrane, consists of a phospholipid bilayer and proteins that either span the bilayer or attach to one side. Inside the membrane, the contents of a cell are collectively termed the **cytoplasm** ("cell-formed"). Because the cytoplasm contains a high concentration of solutes, in most habitats it is hypertonic relative to the surrounding environment. When this is the case, water enters the cell via osmosis and makes the cell's volume expand. In virtually all bacteria and archaea, this pressure is resisted by a stiff **cell wall.** Bacterial and archaeal cell walls are a tough, fibrous layer that surrounds the plasma membrane. In many species the cell wall is made of a carbohydrate-protein complex called *peptidoglycan* or related substances (**Figure 7.2**). The pressure of the plasma membrane against the cell wall is about the same as the pressure in an automobile tire. The cell wall protects the organism and gives it shape and rigidity. In addition, many

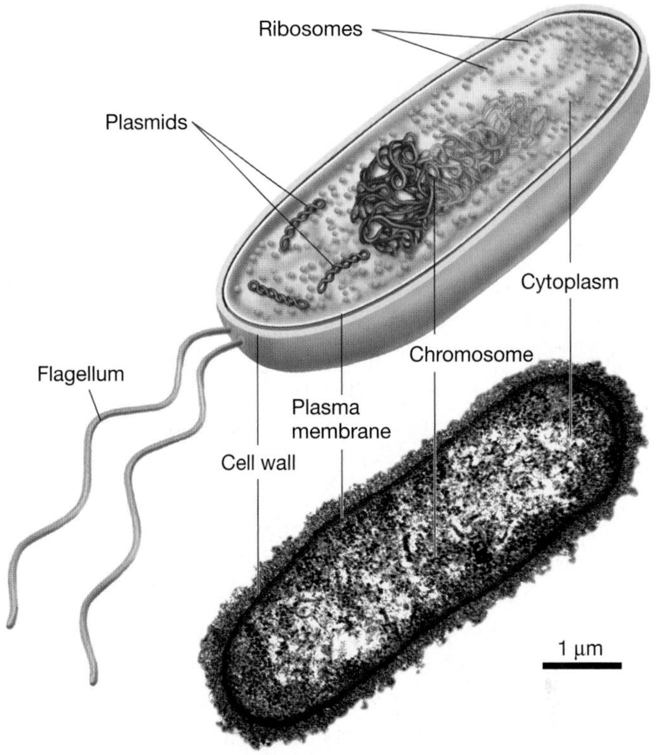

FIGURE 7.1 A Prokaryotic Cell
Prokaryotic cells are identified by a negative trait—the absence of a membrane-bound nucleus. Although there is wide variation in the size and shape of bacterial and archaeal cells, all such cells contain a plasma membrane, chromosomes, and ribosomes; almost all have a stiff cell wall. Some prokaryotes have flagella and/or inner membranes where photosynthesis takes place.

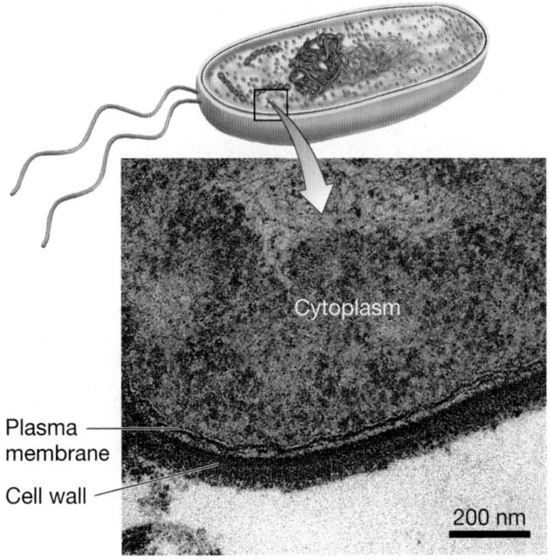

FIGURE 7.2 The Bacterial Cell Wall
In bacteria and archaea, the cell wall consists of peptidoglycan or similar polymers that are cross-linked into tough sheets. The inside of the cell wall contacts the plasma membrane, which pushes up against the wall. The outside of the cell wall makes direct contact with the outside environment, which is almost always filled with competitors and predators.

bacteria have an additional protective layer outside the cell wall that consists of lipids with polysaccharides attached. Lipid molecules that contain carbohydrate groups are termed **glycolipids.**

In the cytoplasm of a prokaryotic cell, the most prominent structure is the chromosome. The prokaryotic **chromosome** consists of a large DNA molecule associated with a small number of proteins. In most species there is a single, circular chromosome, but other species have several circular chromosomes, and a few species—including the bacterium that causes Lyme disease—have one to several linear chromosomes. In prokaryotes and all other organisms, the sequence of nitrogenous bases in DNA acts as a code that contains the genetic, or heredity, information. Stated another way, the primary structure of DNA contains the instructions for making the proteins and other molecules needed by the cell. A **gene** is a segment of DNA that contains the information for building an RNA molecule or a polypeptide. Genes are components of chromosomes.

Prokaryotic chromosomes are found in a localized area of the cell called the **nucleoid.** The genetic material is not separated from the rest of the cytoplasm by a membrane, however. In the well-studied bacterium *Escherichia coli*, the circular chromosome is 500 times longer than the cell itself. This situation is typical in prokaryotes. To fit into the cell, the DNA double helix coils on itself with the aid of enzymes to form the highly compact, "supercoiled" structure shown in **Figure 7.3**. Supercoiled regions of DNA resemble a rubber band that has been held at either end and then twisted.

Depending on the species and population being considered, prokaryotic cells may also contain one to about a hundred small, usually circular, supercoiled DNA molecules called **plasmids.** Plasmids contain genes but are physically independent of the main, cellular chromosome. In most cases the genes carried by plasmids are not required under normal conditions; instead they help cells adapt to unusual circumstances, such as the sudden presence of a poison in the environment. As a result, plasmids can be considered auxiliary genetic elements. They are copied independently of the main chromosome and are passed along to daughter cells when the parent cell divides. Certain plasmid genes also allow a copy of the entire plasmid to be transferred from one cell to another. As a result, plasmids can spread through a population or even be passed between species. Plasmids have been studied intensively because some carry genes that confer resistance to antibiotics. One recently characterized plasmid carries genes that provide resistance to seven distinct antibiotics.

Two other prominent cell structures found in prokaryotes are **ribosomes,** which manufacture proteins, and **flagella** (singular: *flagellum*), which power movement. Ribosomes are observed in all prokaryotic cells and are found throughout the cytoplasm. Bacterial ribosomes are complex structures consisting of a total of three distinct RNA molecules and over 50 different proteins. These molecular components are organized into two major structural elements, called the large subunit and small subunit (**Figure 7.4**). It is not unusual for a single cell to contain 10,000 ribosomes. Both ribosomes and prokaryotic flagella lack a membrane. Not all bacterial species have flagella, however. When present they are usually few in number and are located on the surface of the cell. Over 40 different proteins are involved in building and controlling bacterial flagella. At top speed, flagellar movement can drive a bacterial cell through water at 60 cell lengths per second. In contrast, the cheetah qualifies as the fastest land animal but can sprint at a mere 25 body lengths per second.

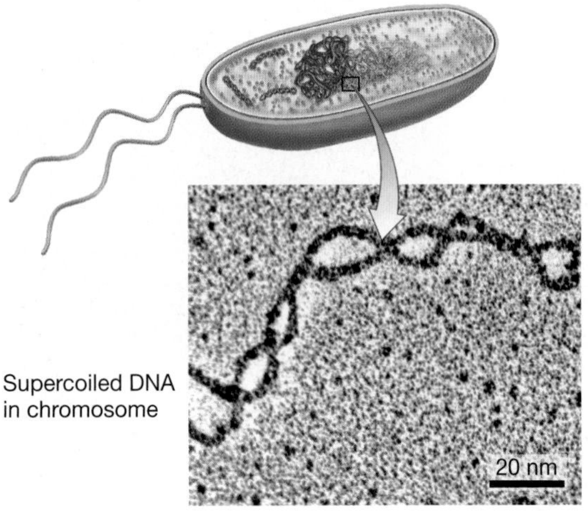
Supercoiled DNA
in chromosome
20 nm

FIGURE 7.3 Bacterial DNA Is Supercoiled
The circular chromosomes of bacteria and archaea must be coiled extensively, into "supercoils," to fit in the cell. For example, the *Escherichia coli* chromosome would be more than 1 mm long if it were linear. This is about 500 times the length of the cell itself.

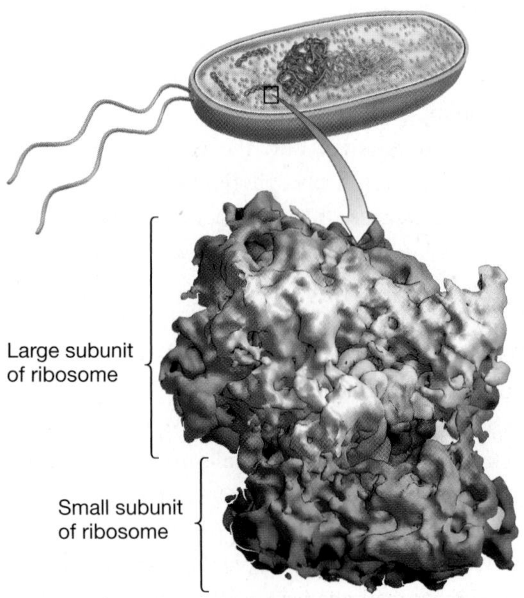
Large subunit
of ribosome
Small subunit
of ribosome

FIGURE 7.4 The Bacterial Ribosome
Bacterial ribosomes are made of RNA and protein molecules that are organized into large and small subunits.

Prokaryotes lack a nucleus, but it is not correct to say that no membrane-bound structures ever occur inside these cells. Many species contain membrane-bound storage containers, and extensive internal membranes occur in bacteria and archaea that perform photosynthesis. The photosynthetic membranes arise as invaginations of the plasma membrane. As the plasma membrane folds in, either vesicles pinch off or the types of flattened stacks shown in **Figure 7.5** form. The internal parts of this membrane contain the enzymes and pigment molecules required to convert the kinetic energy in sunlight into chemical energy in the form of sugar.

In addition, recent research indicates that at least one bacterial species has an internal compartment that qualifies as an organelle ("little organ"). An **organelle** is a membrane-bound compartment in the cytoplasm that contains enzymes specialized for a particular function. The bacterial organelle that was just discovered has proton pumps in its membrane and an acidic environment inside, where calcium ions (Ca^{2+}) are stored.

Recent research has also shown that bacteria and archaea contain long, thin fibers that serve a structural role in the cell. All bacterial species, for example, contain fibers made from the protein FtsZ. These filaments are essential for cell division to take place. Some species also have protein filaments that help maintain cell shape. Protein filaments such as these form the basis of the **cytoskeleton** ("cell skeleton").

Even though internal membranes and some cytoskeletal components are found in prokaryotic cells, their extent pales in comparison with that in eukaryotes. When typical prokaryotic and eukaryotic cells are compared side by side, three outstanding differences jump out: (1) Eukaryotic cells are usually much larger; (2) they contain extensive amounts of internal membrane; and (3) they feature a diverse and dynamic cytoskeleton.

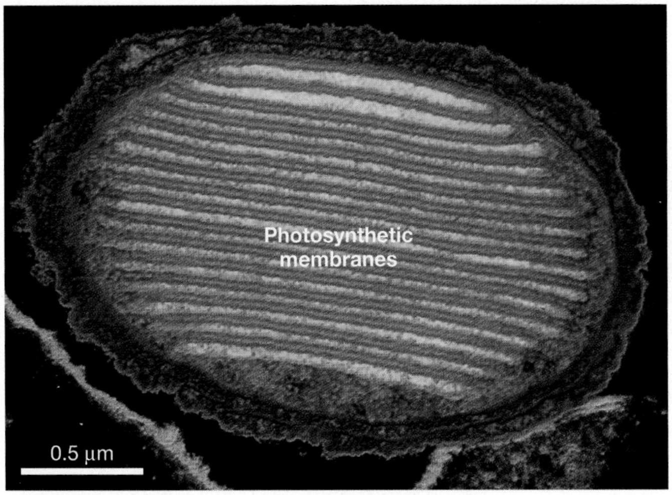

FIGURE 7.5 Photosynthetic Membranes in Bacteria
The green stripes in this photosynthetic bacterium are infoldings of the plasma membrane. They are green because they contain the pigments and enzymes required for photosynthesis.

Eukaryotic Cells

The lineage called Eukarya includes forms ranging from unicellular species to 100-meter-tall redwoods. Brown algae, red algae, fungi, amoebae, and slime molds are all eukaryotic, as are green plants and animals.

The first thing that strikes biologists about eukaryotic cells is how much larger they are on average than bacteria and archaea. Most eukaryotic cells range from about 5 to 100 μm in diameter, while most prokaryotic cells vary between 1 and 10 μm in diameter. A micrograph of an average eukaryotic cell, at the same scale as the bacterial cell in Figure 7.5, would fill this page. This difference in size inspired the hypothesis that when eukaryotes first evolved, they made their living by ingesting bacterial and archaeal cells whole. Stated another way, the evolution of large cell size is thought to have made it possible for eukaryotic cells to act as *predators*—organisms that kill and consume other organisms. Hundreds of eukaryotic species alive today still make their living by surrounding and taking in whole bacterial and archaeal cells.

The evolution of large cells has a downside, however. Ions and small molecules such as adenosine triphosphate (ATP), amino acids, and nucleotides cannot diffuse across a large volume quickly. If ATP is used up on one side of a large cell, ATP from the other side of the cell would take a long time to diffuse to that location. Prokaryotic cells are small enough that ions and small molecules arrive where they are needed via diffusion. In fact, the size of prokaryotic cells is probably limited by the distance that molecules must diffuse or be transported inside the cell.

How do eukaryotic cells solve the diffusion problem? The answer lies in the numerous organelles observed in eukaryotic cells. In effect, the huge volume inside a eukaryotic cell is compartmentalized into a large number of bacterium-sized parts. Because eukaryotic cells are subdivided, the molecules required for specific chemical reactions are often located within a given compartment and do not need to diffuse long distances to be useful. But solving the diffusion problem is not the only advantage conferred by organelles:

- Compartmentalization of the cell allows incompatible chemical reactions to be separated. For example, new fatty acids are synthesized in one organelle while excess or damaged fatty acids are degraded and recycled in a different organelle.
- Compartmentalization increases the efficiency of chemical reactions. First, the substrates required for particular reactions can be localized and maintained at high concentration within organelles. Second, groups of enzymes that work together can be clustered on internal membranes instead of floating free in the cytoplasm. Clustering these molecules increases the speed and efficiency of the reactions, because reactants have shorter distances over which to diffuse or be transported.

Based on their morphological differences, prokaryotic cells can be compared to small machine shops while eukaryotic cells

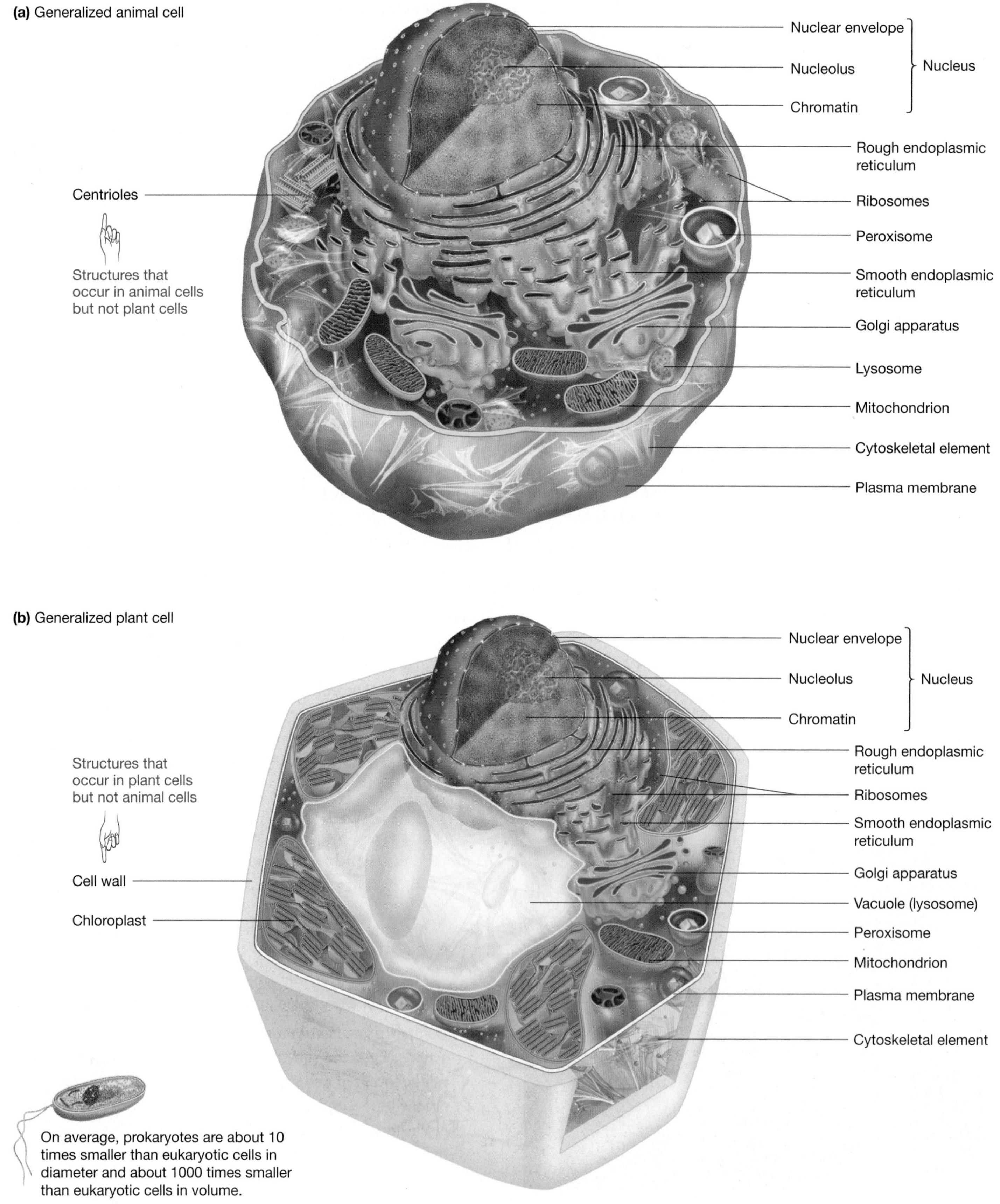
(a) Generalized animal cell
Nuclear envelope
Nucleolus
Chromatin
Nucleus
Rough endoplasmic reticulum
Ribosomes
Peroxisome
Smooth endoplasmic reticulum
Golgi apparatus
Lysosome
Mitochondrion
Cytoskeletal element
Plasma membrane
Centrioles
Structures that occur in animal cells but not plant cells
(b) Generalized plant cell
Nuclear envelope
Nucleolus
Chromatin
Nucleus
Rough endoplasmic reticulum
Ribosomes
Smooth endoplasmic reticulum
Golgi apparatus
Vacuole (lysosome)
Peroxisome
Mitochondrion
Plasma membrane
Cytoskeletal element
Structures that occur in plant cells but not animal cells
Cell wall
Chloroplast
On average, prokaryotes are about 10 times smaller than eukaryotic cells in diameter and about 1000 times smaller than eukaryotic cells in volume.

FIGURE 7.6 Animal and Plant Cells
Generalized or "typical" **(a)** animal and **(b)** plant cells. (Compare with the prokaryotic cell, shown at true relative size at bottom left.)
QUESTION Which organelles are unique to animal cells, and which organelles are unique to plant cells? Which are common to both?

resemble sprawling industrial complexes. The organelles and other structures found in eukaryotes are analogous to highly specialized buildings that act as factories, power stations, warehouses, transportation corridors, and administrative centers. **Figure 7.6** shows how organelles are arranged in a typical animal cell and plant cell. What are these structures, and what do they do?

The Nucleus The **nucleus** is among the largest organelles and is highly organized (**Figure 7.7**). It is enclosed by a unique structure—a complex double membrane called the **nuclear envelope**, which is studded with openings called **nuclear pores**. The inside surface of the nuclear envelope is associated with fibrous proteins that form a lattice-like sheet called the **nuclear lamina**. The nuclear lamina stiffens the envelope and helps organize the chromosomes. Each chromosome occupies a distinct area inside the nucleus and is attached to the nuclear lamina and the inner surface of the nuclear envelope in at least one location. In eukaryotes, chromosomes are linear and consist of DNA that is tightly complexed with a series of ball-shaped *histone* proteins, forming a structure called **chromatin**. Some sections of each chromosome are condensed into a highly compact, supercoiled structure called **heterochromatin**; other sections are unwound into long, filamentous strands called **euchromatin**. The nucleus also includes a distinctive region called the **nucleolus**, where the RNA molecules found in ribosomes are manufactured and the large and small ribosomal subunits are assembled. Section 7.2 discusses the structure and function of the nucleus, and particularly the nuclear envelope, in more detail.

Ribosomes In eukaryotes, the cytoplasm consists of everything inside the plasma membrane excluding the nucleus; the fluid portion of the cytoplasm is called the **cytosol**. Many of the cell's millions of ribosomes are scattered throughout the cytosol. The ribosomes shown in **Figure 7.8** are comprised of two subunits, one small and one large. Each subunit is composed of several different proteins and one large RNA molecule. In eukaryotes the large subunit also contains two small RNA molecules. (In prokaryotes the large subunit has just one small and one large RNA molecule.) Neither ribosomal subunit is enclosed by a membrane. When the two subunits come together, they form a complex molecular machine that synthesizes proteins.

Rough Endoplasmic Reticulum In addition to the ribosomes found free in the cytosol, many ribosomes are associated with membranes. More specifically, hundreds of thousands of ribosomes are attached to a network of membrane-bound sacs and tubules called the **rough endoplasmic reticulum**, or **rough ER**. Translated literally, *endoplasmic reticulum* means "inside-formed network." Notice in Figure 7.6 that the ER is continuous with the outer membrane of the nuclear envelope. From there, the layers of sacs extend into the cytoplasm.

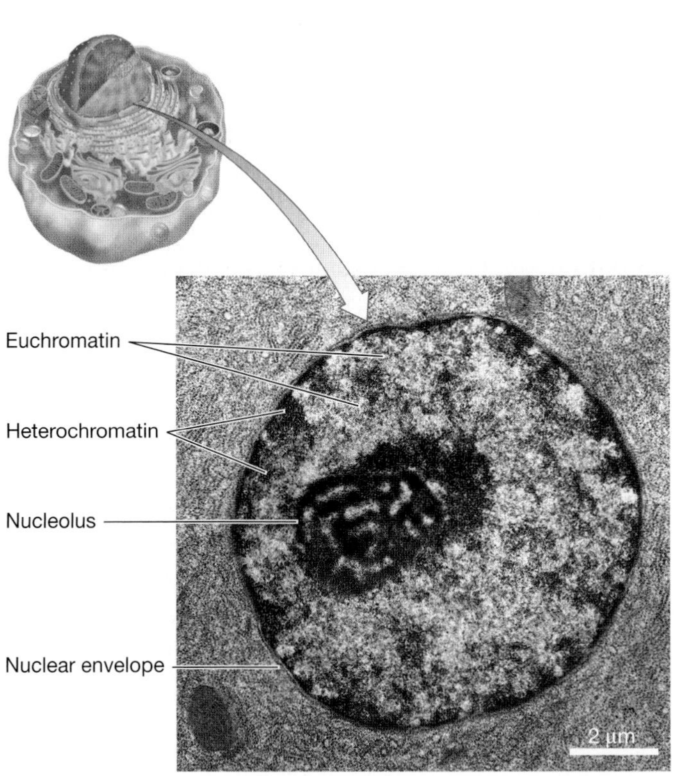
Euchromatin
Heterochromatin
Nucleolus
Nuclear envelope
2 μm

FIGURE 7.7 The Nucleus Is the Eukaryotic Cell's Information Storage and Retrieval Center
The genetic, or hereditary, information is encoded in DNA, which is a component of the chromosomes inside the nucleus.

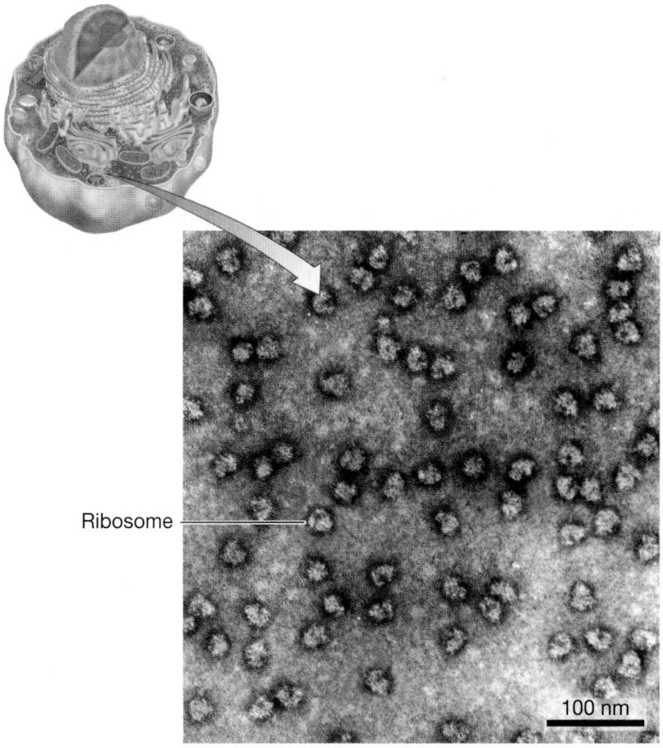
Ribosome
100 nm

FIGURE 7.8 Ribosomes Are the Site of Protein Synthesis
Eukaryotic ribosomes are similar in structure to bacterial and archaeal ribosomes—though not identical. They are comprised of large and small subunits, each of which contains both RNA molecules and proteins.

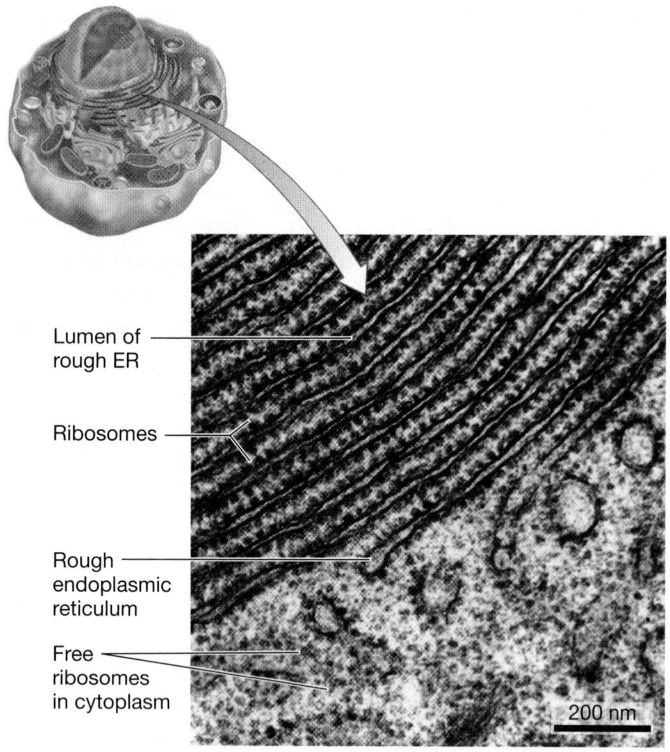
Lumen of rough ER
Ribosomes
Rough endoplasmic reticulum
Free ribosomes in cytoplasm
200 nm

FIGURE 7.9 Rough ER Is a Protein Synthesis and Processing Complex
Rough ER is a system of membrane-bound sacs and tubules with ribosomes attached. It is continuous with the nuclear envelope and with smooth ER.

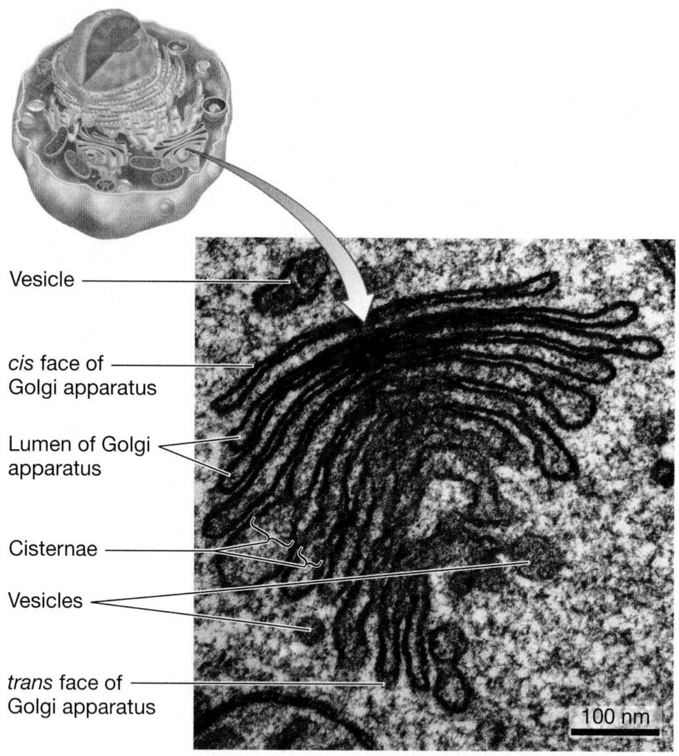
Vesicle
cis face of Golgi apparatus
Lumen of Golgi apparatus
Cisternae
Vesicles
trans face of Golgi apparatus
100 nm

FIGURE 7.10 The Golgi Apparatus Is a Site of Protein Processing, Sorting, and Shipping
The Golgi apparatus is a collection of flattened vesicles called cisternae. The organelle has a *cis* face oriented toward the rough ER and a *trans* face oriented toward the plasma membrane.

The ribosomes associated with the rough ER are responsible for synthesizing proteins that will be inserted into the plasma membrane, secreted to the cell exterior, or shipped to an organelle called the *lysosome*. As they are being manufactured by ribosomes, these proteins move to the interior of the sac-like component of the rough ER (**Figure 7.9**). The interior of any sac-like structure in a cell or body is called the **lumen**. In the lumen of the rough ER, newly manufactured proteins undergo folding and other types of processing.

The proteins produced in the rough ER have a variety of functions. Some carry messages to other cells; some act as membrane transport proteins or pumps; others are enzymes. The common theme is that rough ER products are destined for transport to a distant destination—often to the surface of the cell or beyond.

Golgi Apparatus In many cases, the products of the rough ER pass through the Golgi apparatus before they reach their final destination. The **Golgi apparatus** consists of flattened, membranous sacs called **cisternae** (singular: *cisternum*), which are stacked on top of one another (**Figure 7.10**). The organelle also has a distinct polarity, or sidedness. The *cis* ("this side") surface is closest to the rough ER and nucleus, and the *trans* ("across") surface is oriented toward the plasma membrane. The *cis* side receives products from the rough ER, and the *trans* side ships them out toward the cell surface. In between, within the cisternae, the rough ER's products are processed and packaged for delivery. Micrographs often show "bubbles" on either side of the Golgi stack. These are membrane-bound vesicles that carry proteins or other products to and from the organelle. Section 7.3 analyzes the intracellular movement of products in more detail.

Smooth Endoplasmic Reticulum Not all of the ER is associated with transport of material to Golgi sacs, and not all ER has ribosomes attached. While parts of the ER that contain ribosomes look dotted and rough in electron micrographs, the portions of the organelle that are free of ribosomes appear smooth and even. Appropriately enough, these parts of the ER are called **smooth endoplasmic reticulum** or **smooth ER** (**Figure 7.11**). The smooth ER membrane contains enzymes that are required for reactions involving lipids. Depending on the type of cell, these enzymes may be involved in synthesizing specialized types of lipids needed by the organism or in breaking down hydrophobic molecules that are poisonous to the cell. Smooth ER is the manufacturing site for phospholipids required for the plasma membrane, and smooth ER also functions as a reservoir for calcium ions (Ca^{2+}) that act as a signal inside the cell.

The structure of endoplasmic reticulum correlates closely with its function. Rough ER has ribosomes and functions primarily as a protein-manufacturing center; smooth ER lacks ribosomes and functions primarily as a lipid-processing center.

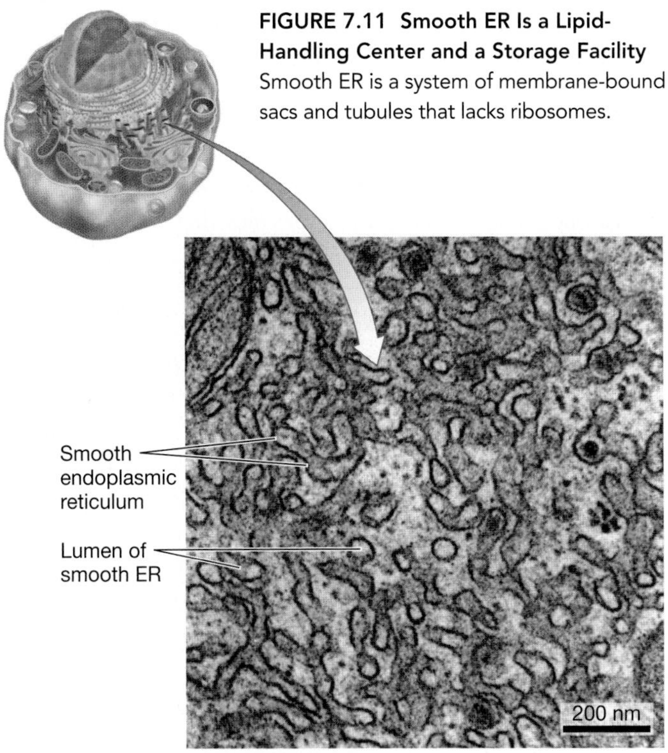
Smooth endoplasmic reticulum
Lumen of smooth ER
200 nm

FIGURE 7.11 Smooth ER Is a Lipid-Handling Center and a Storage Facility Smooth ER is a system of membrane-bound sacs and tubules that lacks ribosomes.

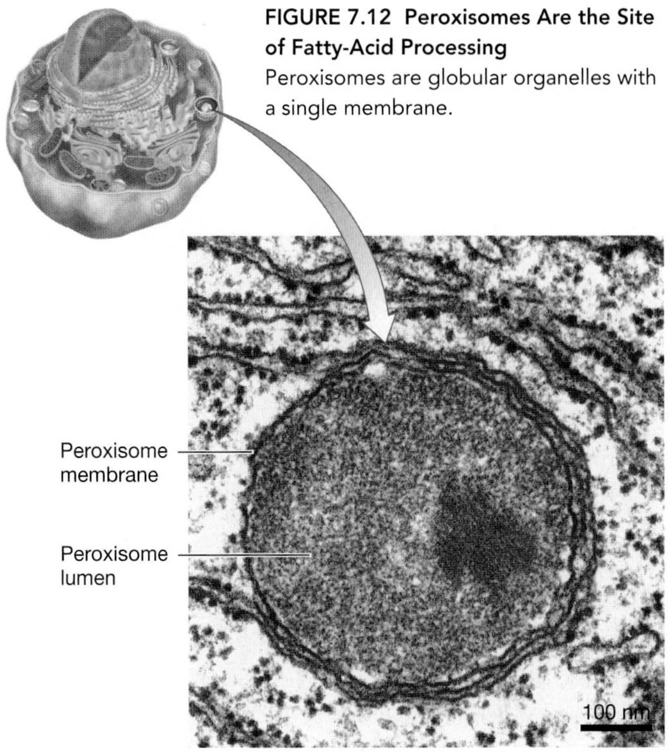
Peroxisome membrane
Peroxisome lumen
100 nm

FIGURE 7.12 Peroxisomes Are the Site of Fatty-Acid Processing Peroxisomes are globular organelles with a single membrane.

Together with the Golgi apparatus and lysosomes, the endoplasmic reticulum forms the **endomembrane system**. The endomembrane ("inner-membrane") system is the primary center for protein and lipid synthesis in eukaryotic cells.

Peroxisomes **Peroxisomes** are globular organelles that are found in virtually all eukaryotic cells (**Figure 7.12**). They have a single membrane and grow and divide independently of other organelles. Although different types of cells from the same individual may have distinct types of peroxisomes, these organelles are united by a common function: Peroxisomes are centers for oxidation reactions. In many cases the products of these reactions include hydrogen peroxide (H_2O_2), which is highly corrosive. If hydrogen peroxide escaped from the peroxisome, the H_2O_2 would quickly damage organelle membranes and the plasma membrane. This is rare, however. Inside the peroxisome, the enzyme catalase quickly converts hydrogen peroxide to water and oxygen.

The various types of peroxisomes that exist contain different suites of enzymes. As a result, each is specialized for oxidizing particular compounds. For example, the peroxisomes in your liver cells contain enzymes that oxidize an array of toxins, including the ethanol in alcoholic beverages. The products of these oxidation reactions are usually harmless and are either excreted from the body or used in other reactions. Other peroxisomes contain enzymes that catalyze the oxidation of fatty acids. These reactions result largely in the production of a molecule called *acetyl CoA*, which is used for the synthesis of important molecules elsewhere in the cell. In plant leaves, specialized peroxisomes called **glyoxisomes** are packed with enzymes that convert one of the products of photosynthesis into a sugar that can be used to produce energy for the cell. Seeds do not perform photosynthesis, so they lack this type of peroxisome. Instead, they have peroxisomes with enzymes that oxidize fatty acids to yield glucose. The glucose is then used by the young plant as it begins to grow. In each case, there is a clear connection between structure and function: The enzymes found inside the peroxisome make a specialized set of oxidation reactions possible.

Lysosomes The major structures involved in solid-waste processing and materials storage in the cell are called **lysosomes.** The size and shape of these organelles vary widely, and in the cells of plants, fungi, and certain other groups they are referred to as **vacuoles**. In animal cells, lysosomes function as digestive centers (**Figure 7.13**). The organelle's interior, or lumen, is acidic

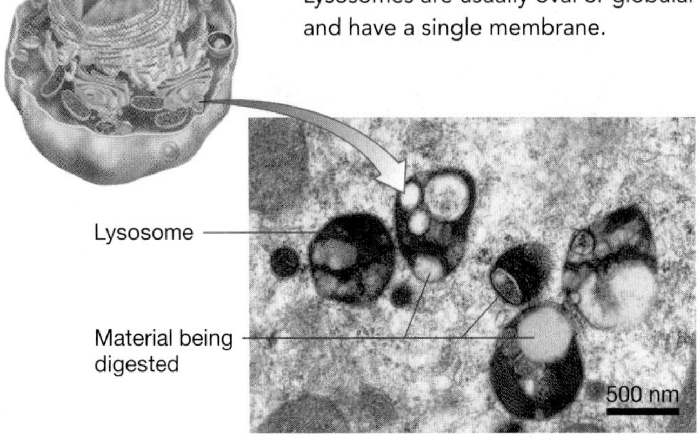
Lysosome
Material being digested
500 nm

FIGURE 7.13 Lysosomes Are Recycling Centers Lysosomes are usually oval or globular and have a single membrane.

because a proton pump in the lysosome membrane imports enough hydrogen ions to maintain a pH of 5.0. This organelle also contains about 40 different enzymes. Each of these proteins is specialized for breaking up a different type of macromolecule—proteins, nucleic acids, lipids, or carbohydrates—into its component monomers. These digestive enzymes are collectively called *acid hydrolases* because they hydrolyze macromolecules most efficiently at pH 5.0. In the cytosol, where the pH is about 7.2, these enzymes are less active.

Figure 7.14 illustrates three ways that materials are delivered to lysosomes in animal cells:

1. When **phagocytosis** ("eat-cell-act") occurs, the plasma membrane of a cell surrounds a smaller cell or a food particle and engulfs it. The resulting structure is delivered to a lysosome, where it is taken in and digested.
2. During **autophagy** ("same-eating"), damaged organelles are surrounded by a membrane and delivered to a lysosome. There the components are digested and recycled.
3. Materials can also find their way into lysosomes as a result of **receptor-mediated endocytosis.** This process begins when macromolecules outside the cell bind to membrane proteins that act as receptors. More than 25 distinct receptors have now been characterized, each specialized for responding to a different macromolecule. Once binding occurs, the plasma membrane folds in and pinches off to form a membrane-bound vesicle called an **early endosome** ("inside-body"). Early endosomes undergo a series of processing steps that include the receipt of digestive enzymes from the Golgi apparatus and the activation of proton pumps that gradually lower their pH. In this way, early endosomes undergo a gradual maturation process that may lead to the formation of a **late endosome** and eventually a fully functioning lysosome.

Regardless of whether materials are delivered to lysosomes via phagocytosis, autophagy, or receptor-mediated endocytosis, the result is similar: Molecules are hydrolyzed. The amino acids, nucleotides, sugars, and other molecules that result from acid hydrolysis leave the lysosome via transport proteins in the organelle's membrane. Once in the cytoplasm, they can be reused.

It is important to note, however, that not all of the materials that are surrounded by membrane and taken into a cell end up in lysosomes. **Endocytosis** ("inside-cell-act") refers to any pinching off of the plasma membrane that results in the uptake of material from outside the cell. Endocytosis can occur in three ways: (1) phagocytosis, (2) receptor-mediated endocytosis, and (3) **pinocytosis** ("drink-cell-act"). Pinocytosis brings fluid into the cytoplasm via tiny vesicles that form from invaginations of the plasma membrane. The fluid inside these vesicles is not transported to lysosomes, but is used elsewhere in the cell. In addition, most of the macromolecules that collect in early endosomes are selectively removed and used long before the structure becomes a lysosome.

Compared with the lysosomes of animal cells, the vacuoles of plant and fungal cells are large—sometimes taking up as much as 80 percent of a plant cell's volume (**Figure 7.15**). Although some vacuoles contain enzymes that are specialized for digestion, most of the vacuoles observed in plant and fungal

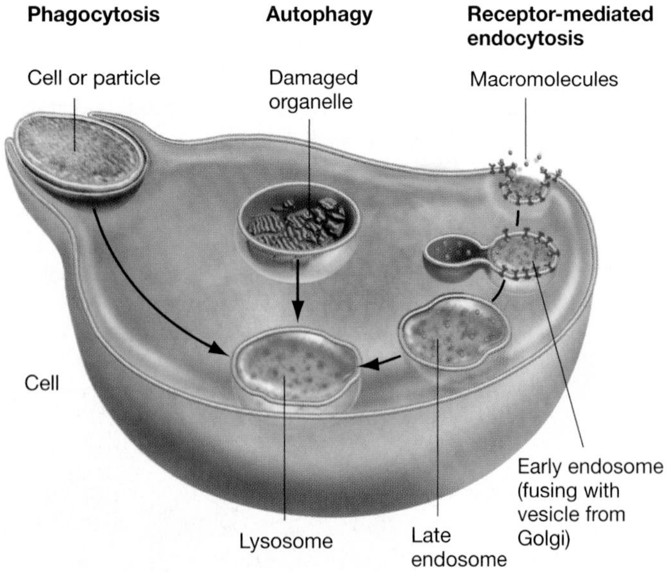
Phagocytosis
Autophagy
Receptor-mediated endocytosis
Cell or particle
Damaged organelle
Macromolecules
Cell
Lysosome
Late endosome
Early endosome (fusing with vesicle from Golgi)

FIGURE 7.14 Three Ways to Deliver Materials to Lysosomes
Materials can be transported to lysosomes after phagocytosis or autophagy. Endosomes created by receptor-mediated endocytosis may mature into lysosomes.

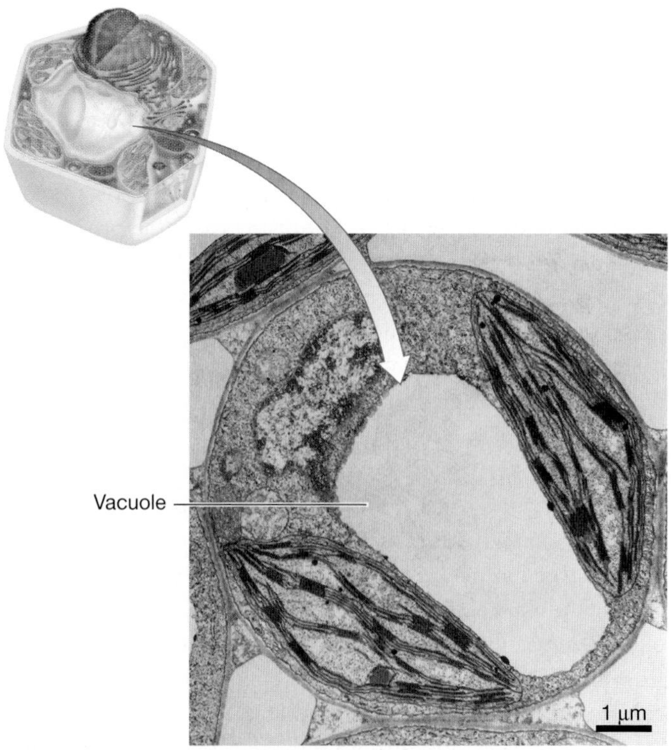
Vacuole
1 μm

FIGURE 7.15 Vacuoles Are Storage Centers
Vacuoles are variable in size and function. Some contain digestive enzymes and serve as recycling centers; most are large storage containers.

cells act as storage depots. In many cases, the stored material is water, which maintains the cell's normal volume, or ions such as potassium (K^+) and chloride (Cl^-). But inside seeds, cells may contain a large vacuole filled with proteins. When germination occurs, enzymes begin digesting these proteins to provide amino acids for the growing individual. In cells that make up flower petals or fruits, vacuoles are filled with colorful pigments. Elsewhere in the plant, vacuoles may be packed with noxious compounds that protect leaves and stems from being eaten by predators. The type of chemical involved varies by species, ranging from bitter-tasting tannins to toxins such as nicotine, morphine, caffeine, or cocaine.

Mitochondria The chemical energy required to build all of these organelles and do other types of work comes from adenosine triphosphate (ATP), most of which is produced in the cell's **mitochondria** (singular: *mitochondrion*). As **Figure 7.16** shows, each mitochondrion has two membranes. The outer membrane defines the organelle's surface, while the inner membrane contacts a series of sac-like **cristae**. The solution inside the inner membrane is called the **mitochondrial matrix**. In eukaryotes, most of the enzymes and molecular machines responsible for providing chemical energy in the form of ATP from food molecules are embedded in the membranes of the cristae or suspended in the matrix. Depending on the type of cell, from 50 to more than a million mitochondria may be present.

Each mitochondrion contains its own small chromosome, independent of the main chromosomes in the nucleus. **Mitochondrial DNA** is a component of a circular and supercoiled chromosome that is similar in structure to bacterial chromosomes. Mitochondria also manufacture their own ribosomes. Like most organelles, mitochondria can grow and divide independently of nuclear division and cell division.

Chloroplasts Most algal and plant cells possess an organelle called the **chloroplast**, in which sunlight is converted to chemical energy during photosynthesis. The chloroplast has a double membrane around its exterior, analogous to the structure of a mitochondrion (**Figure 7.17**). Instead of featuring sac-like cristae that connect to the inner membrane, though, the interior of the chloroplast is dominated by hundreds of membrane-bound, flattened vesicles called **thylakoids**, which are independent of the inner membrane. Thylakoids are stacked like pancakes into piles called **grana** (singular: *granum*). Many of the pigments, enzymes, and molecular machines responsible for converting light energy into carbohydrates are embedded in the thylakoid membranes. Certain critical enzymes and substrates, however, are found outside the thylakoids in the region called the **stroma**.

The number of chloroplasts per cell varies from none to several dozen. Like mitochondria, each chloroplast contains a circular chromosome. **Chloroplast DNA** is independent of the main

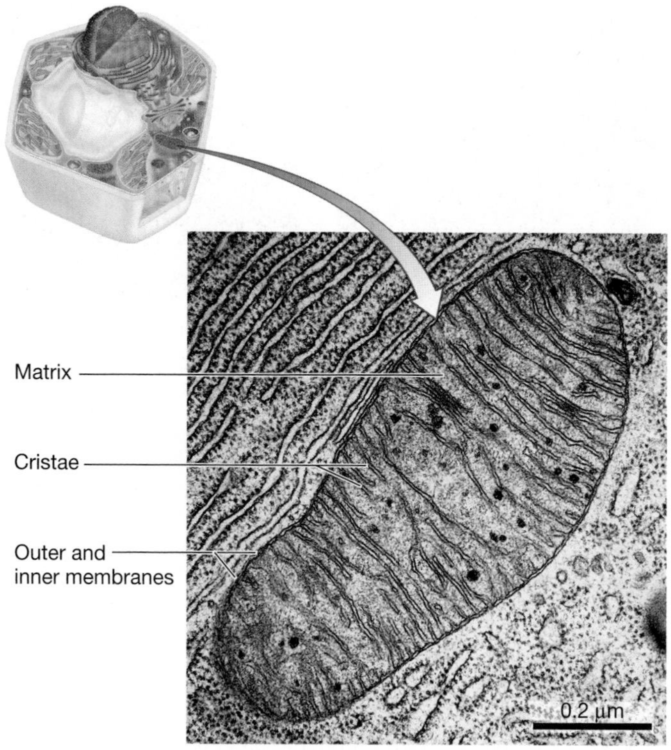
Matrix
Cristae
Outer and inner membranes
0.2 μm

FIGURE 7.16 Mitochondria Are Power-Generating Stations
Mitochondria are variable in size and shape, but all have a double membrane with sac-like cristae inside.

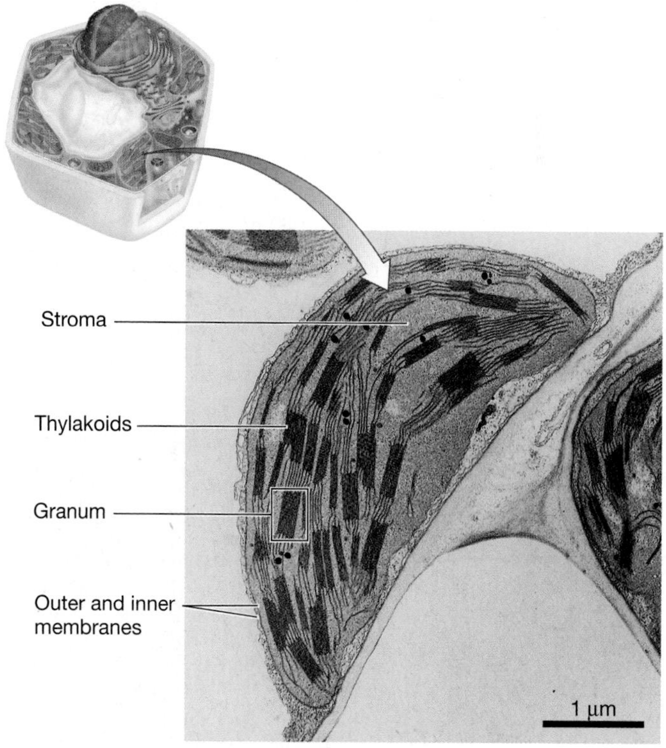
Stroma
Thylakoids
Granum
Outer and inner membranes
1 μm

FIGURE 7.17 Chloroplasts Are Sugar-Manufacturing Centers
In photosynthesis, sunlight is converted to chemical energy in the form of sugar. The enzymes and other molecules required for photosynthesis are located in membranes inside the chloroplast. These membranes are folded into thylakoids and stacked into grana.

genetic material inside the nucleus. Chloroplasts also grow and divide independently of nuclear division and cell division.

Cytoskeleton The final major structural feature that is common to all eukaryotic cells is an extensive system of protein fibers called the **cytoskeleton**. As we'll see in Section 7.4, the cytoskeleton contains several distinct types of proteins and fibers and has an array of functions. In addition to giving the cell its shape, cytoskeletal proteins are involved in moving the cell itself and in moving materials within the cell.

The Cell Wall In fungi, algae, and plants, cells possess an outer **cell wall** in addition to their plasma membrane (**Figure 7.18**). Animals, amoebae, and other groups lack this feature. Although the composition of the cell wall varies among species and even between types of cells in the same individual, the general plan is similar: Rods or fibers composed of a carbohydrate run through a stiff matrix made of other polysaccharides and proteins. In addition, some plant cells produce a secondary cell wall that features a particularly tough molecule called lignin. Lignin forms a branching, cagelike network that is almost impossible for enzymes to attack. The combination of cellulose fibers and lignin in secondary cell walls makes up most of the material we call wood.

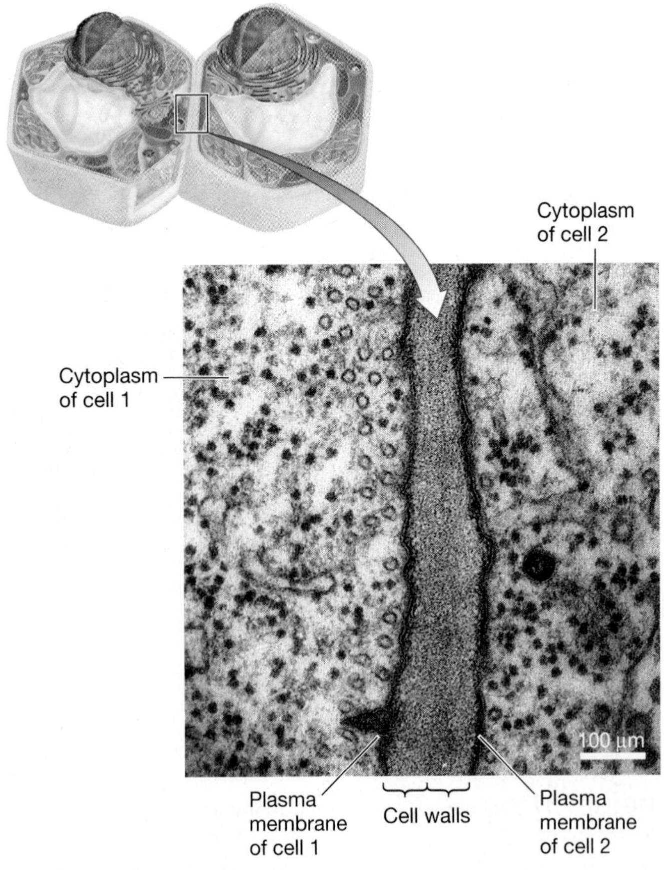
Cytoplasm of cell 2
Cytoplasm of cell 1
100 µm
Plasma membrane of cell 1
Cell walls
Plasma membrane of cell 2

FIGURE 7.18 Cell Walls Protect Plants and Fungi
Plants have cell walls that contain cellulose; in fungi the major structural component of the cell wall is chitin.

How Does Cell Structure Correlate with Function?

The preceding discussion emphasized how the structure of each organelle fits with its role in the cell. As **Table 7.1** indicates, an organelle's membrane and its complement of enzymes correlate closely with its function. The same connection between structure and function occurs at the level of the entire cell. Inside an individual plant or animal, cells are specialized for certain tasks and have a structure that correlates with those tasks. For example, the muscle cells in your upper leg are extremely long, tube-shaped structures. They are filled with protein fibers that slide past one another as the entire muscle flexes or extends. It is this sliding motion that allows your muscles to contract or extend as you run. Muscle cells are also jam packed with mitochondria, which produce the ATP required for the sliding motion to occur. In contrast, nearby fat cells are rounded, globular structures. They consist of little more than a plasma membrane, a nucleus, and a fat droplet. Neither cell bears a close resemblance to the generalized animal cell pictured in Figure 7.6a.

To drive home the correlation between the overall structure and function of a cell, examine the transmission electron micrographs in **Figure 7.19**. The animal cell in Figure 7.19a is from the pancreas and is specialized for the manufacture and export of digestive enzymes. It is packed with rough ER and

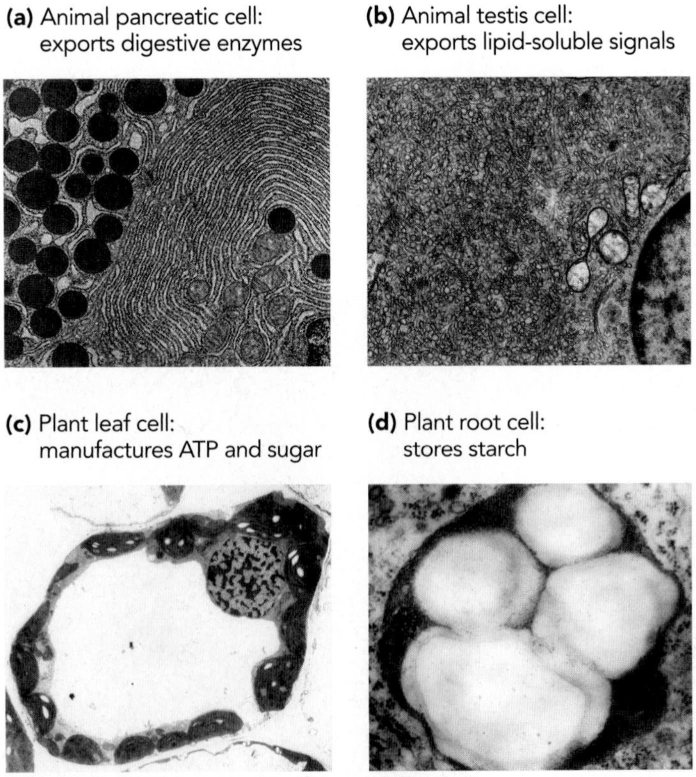
(a) Animal pancreatic cell: exports digestive enzymes
(b) Animal testis cell: exports lipid-soluble signals
(c) Plant leaf cell: manufactures ATP and sugar
(d) Plant root cell: stores starch

FIGURE 7.19 Cell Structure Correlates with Function
EXERCISE In each cell, label ribosomes, rough ER, chloroplasts, the nucleus, smooth ER, mitochondria, vacuole, plasma membrane, and cell wall if they are visible.

TABLE 7.1 A Summary of Cell Components

	Structure		Function
	Membrane	**Components**	
Nucleus	Double ("envelope"); openings called nuclear pores	Chromosomes Nucleolus Nuclear lamina	Genetic information Assembly of ribosome subunits Structural support
Ribosomes	None	Large/small subunits Complex of RNA and proteins	Protein synthesis
Endomembrane system			
Rough ER	Single; contains receptors for entry of selected proteins	Network of branching sacs Ribosomes associated	Protein synthesis and processing
Golgi apparatus	Single; contains receptors for products of rough ER	Stack of flattened cisternae	Processing of proteins
Smooth ER	Single; contains enzymes for synthesizing phospholipids	Network of branching sacs Enzymes for synthesizing lipids	Lipid synthesis
Peroxisomes	Single; contains transporters for selected macromolecules	Enzymes that catalyze oxidation reactions Catalase (processes peroxide)	Processing of fatty acids
Lysosomes	Single; contains proton pumps	Acid hydrolases (catalyze hydrolysis reactions)	Digestion and recycling
Vacuoles	Single; contains transporters for selected molecules	Varies—pigments, oils, carbohydrates, water, or toxins	Varies—coloration; storage of oils, carbohydrates, water, or toxins
Mitochondria	Double; outer contains enzymes for processing pyruvate; inner contains enzymes for ATP production	Enzymes that catalyze oxidation-reduction reactions, ATP synthesis	ATP production
Chloroplasts	Double, plus membrane-bound sacs in interior	Pigments Enzymes that catalyze oxidation-reduction reactions	Production of ATP and sugars via photosynthesis
Cytoskeleton	None	Actin filaments Intermediate filaments Microtubules	Structural support Movement of materials In some species: movement of whole cell
Plasma membrane	Single; contains transport and receptor proteins	Phospholipid bilayer with transport and receptor proteins	Selective permeability—maintains intracellular environment
Cell wall	None	Carbohydrate fibers running through carbohydrate or protein matrix	Protection, structural support

Golgi, which make this function possible. The animal cell in Figure 7.19b is from the testis and synthesizes the lipid-soluble signaling molecule called testosterone. This cell is dominated by smooth ER, where lipid processing takes place. The plant cell in Figure 7.19c is from the leaf of a potato and is specialized for absorbing light and manufacturing sugar; the cell in Figure 7.19d is from a potato tuber (part of an underground stem) and functions as a starch storage container. The leaf cell contains hundreds of chloroplasts, while the tuber cell has a prominent storage vacuole filled with carbohydrate. In each case, the type of organelles in each cell and their size and number correlate with the cell's specialized function.

The Dynamic Cell

Biologists describe the structure and function of organelles and cells by a combination of tools and approaches. Light microscopes and transmission electron microscopes have allowed researchers to see cells at increasingly high magnification. Microscopy allowed biologists to characterize the basic size and shape of organelles and where they occurred in the cell; a technique called **differential centrifugation** made it possible to isolate particular cell components and analyze their chemical composition. As **Box 7.1** (page 140) explains, differential centrifugation is based on breaking cells apart to create a complex mixture and then separating components in a centrifuge. The individual parts of the cell can then be purified and studied in detail.

Although these techniques have led to an increasingly sophisticated understanding of how cells work, they have a limitation. Transmission electron microscopy is based on a fixed "snapshot" of the cell that is to be observed, and differential centrifugation is based on splitting cells into parts that are analyzed independently. Neither technique allows investigators to answer directly questions about how things move from place to place in the cell or how parts interact. The information gleaned from these techniques can make cells seem somewhat static. In reality, cells are dynamic.

BOX 7.1 How Does a Centrifuge Work?

For decades, the centrifuge was among the most common tools used by biologists who study life at the level of molecules and cells. It was vital to early studies of organelles and other cell structures because it can separate cell components efficiently. A centrifuge accomplishes this by spinning cells in a solution that allows molecules and other cell components to separate according to their density or size and shape.

The first step in preparing a cell sample for centrifugation is to release the organelles and cell components by breaking the cells apart. This can be done by putting them in a hypotonic solution, by exposing them to ultrasonic vibration, by treating cells with a detergent, or by grinding them up. Each of these methods breaks apart plasma membranes and releases the contents of the cells.

The pieces of plasma membrane broken up by these techniques quickly reseal to form small vesicles, often trapping cell components inside. The solution that results from the homogenization step is a mixture of these vesicles, free-floating macromolecules released from the cells, and organelles. A solution such as this is called a **cell extract** or **cell homogenate.**

When a cell homogenate is placed in a centrifuge tube and spun at high speed, the components that are in solution tend to move outward, along the dashed line in **Figure 7.20a**. The effect is similar to a merry-go-round, which seems to push you outward in a straight line away from the spinning platform. In response to this outward-directed force, the solution containing the cell homogenate exerts a centripetal ("center-seeking") force that pushes the homogenate away from the bottom of the tube. Larger, denser molecules or particles resist this inward force more readily than do smaller, less dense ones and so reach the bottom of the centrifuge tube faster.

To separate the components of a cell extract, researchers often perform a series of centrifuge runs. Steps 1 and 2 of **Figure 7.20b** illustrate how an initial treatment at low speed causes larger, heavier parts of the homogenate to move below smaller, lighter parts. The material that collects at the bottom of the tube

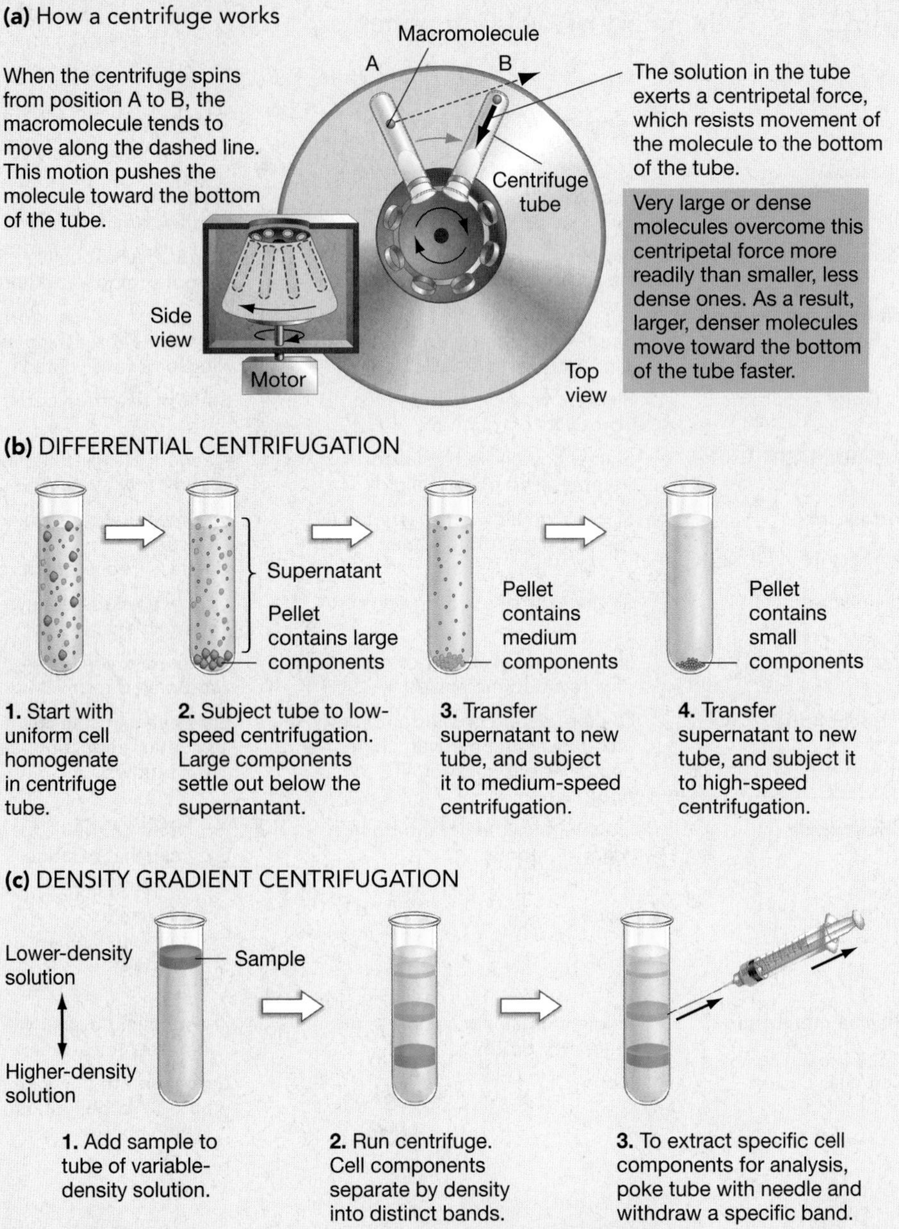

(a) How a centrifuge works
Macromolecule
A
B
When the centrifuge spins from position A to B, the macromolecule tends to move along the dashed line. This motion pushes the molecule toward the bottom of the tube.
The solution in the tube exerts a centripetal force, which resists movement of the molecule to the bottom of the tube.
Centrifuge tube
Very large or dense molecules overcome this centripetal force more readily than smaller, less dense ones. As a result, larger, denser molecules move toward the bottom of the tube faster.
Side view
Motor
Top view
(b) DIFFERENTIAL CENTRIFUGATION
Supernatant
Pellet contains large components
Pellet contains medium components
Pellet contains small components
1. Start with uniform cell homogenate in centrifuge tube.
2. Subject tube to low-speed centrifugation. Large components settle out below the supernantant.
3. Transfer supernatant to new tube, and subject it to medium-speed centrifugation.
4. Transfer supernatant to new tube, and subject it to high-speed centrifugation.
(c) DENSITY GRADIENT CENTRIFUGATION
Lower-density solution
Sample
Higher-density solution
1. Add sample to tube of variable-density solution.
2. Run centrifuge. Cell components separate by density into distinct bands.
3. To extract specific cell components for analysis, poke tube with needle and withdraw a specific band.

FIGURE 7.20 Cell Components Can Be Separated by Centrifugation
(a) Overhead view of a centrifuge, illustrating why cell components separate by being spun. **(b)** Through a series of centrifuge runs made at increasingly higher speeds, an investigator can separate fractions of a cell homogenate according to size by differential centrifugation. **(c)** A high-speed centrifuge run can achieve extremely fine separation among cell components by density gradient centrifugation.

is called the **pellet**, and the solution and solutes left behind comprise the **supernatant** ("above swimming"). The supernatant is placed in a fresh tube and centrifuged at increasingly higher speeds and longer durations. Each centrifuge run continues to separate cell components based on their size and density.

To accomplish even finer separation of macromolecules or organelles, researchers frequently follow up with centrifugation at extremely high speeds. One strategy is based on filling the centrifuge tube with a series of sucrose solutions of increasing density. The density gradient allows cell components to separate on the basis of small differences in size and shape. When the centrifuge run is complete, each cell component comprises a distinct band of material in the tube. A researcher can then collect the material in each band for further study.

The amount of chemical activity and the speed of molecular movement inside cells is nothing short of fantastic. Bacterial ribosomes add up to 20 amino acids per second to a growing polypeptide, and eukaryotic ribosomes typically add two per second. Given that there are about 15,000 ribosomes in each bacterium and possibly a million in an average eukaryotic cell, hundreds or even thousands of new protein molecules can be finished each second in every cell. In the same amount of time, a typical cell in your body uses an average of 10 million ATP molecules and synthesizes just as many. It's not unusual for a cellular enzyme to catalyze 25,000 or more reactions per second; most cells contain hundreds or thousands of enzymes. A minute is more than enough time for each membrane phospholipid in your body to travel the breadth of the organelle or cell where it resides. The hundreds of trillions of mitochondria inside you are completely replaced about every 10 days, for as long as you live. The plasma membrane is fluid, and its composition is constantly changing.

Because humans are such large organisms, it is impossible for us to imagine what life is really like inside a cell. At the scale of a ribosome or an organelle or a cell, gravity is inconsequential. Instead, electrostatic attractions between molecules and the kinetic energy of motion are the dominant forces. At this level, events take nanoseconds, and speeds are measured in micrometers per second. Contemporary methods for studying cells, including those featured in **Box 7.2**, capture this dynamism by tracking how organelles and molecules move and interact over time.

The rest of this chapter focuses on this theme of cellular dynamism and movement. To begin, let's look at how molecules move into and out of the cell's control center—the nucleus. Then we'll consider how proteins move from ribosomes into the lumen of the rough ER and then to the Golgi apparatus and beyond. The chapter closes by analyzing how cytoskeletal elements help transport cargo inside the cell or move the cell itself.

BOX 7.2 Techniques for Studying the Dynamic Cell

Contemporary methods for studying cells allow researchers to see specific molecules moving inside living cells. One of the most popular techniques for tagging molecules of interest relies on a fluorescent molecule called **green fluorescent protein**, or **GFP**. GFP is naturally synthesized in jellyfish that *fluoresce*, or emit light. By affixing GFP molecules to another protein and then inserting it into a cell, investigators can follow its fate over time and even videotape its movement. For example, recent studies have used GFP to tag proteins that are secreted from the cell. Control experiments show that GFP does not affect the behavior of these proteins. Researchers then videotaped the GFP-tagged protein's transport from the rough ER through the Golgi apparatus and out to the plasma membrane. This is cell biology: the movie.

To produce extremely high resolution still images of proteins that are tagged with GFP or other fluorescing tags, researchers often rely on **confocal microscopy**. This technique is based on mounting live cells on a microscope slide and then focusing a beam of ultraviolet light at a specific depth within the specimen. The fluorescing tag emits visible light in response. A detector for this light is then set up at exactly the position where the emitted light comes into focus. The result is a sharp image of a precise plane in the cell being studied (**Figure 7.21**).

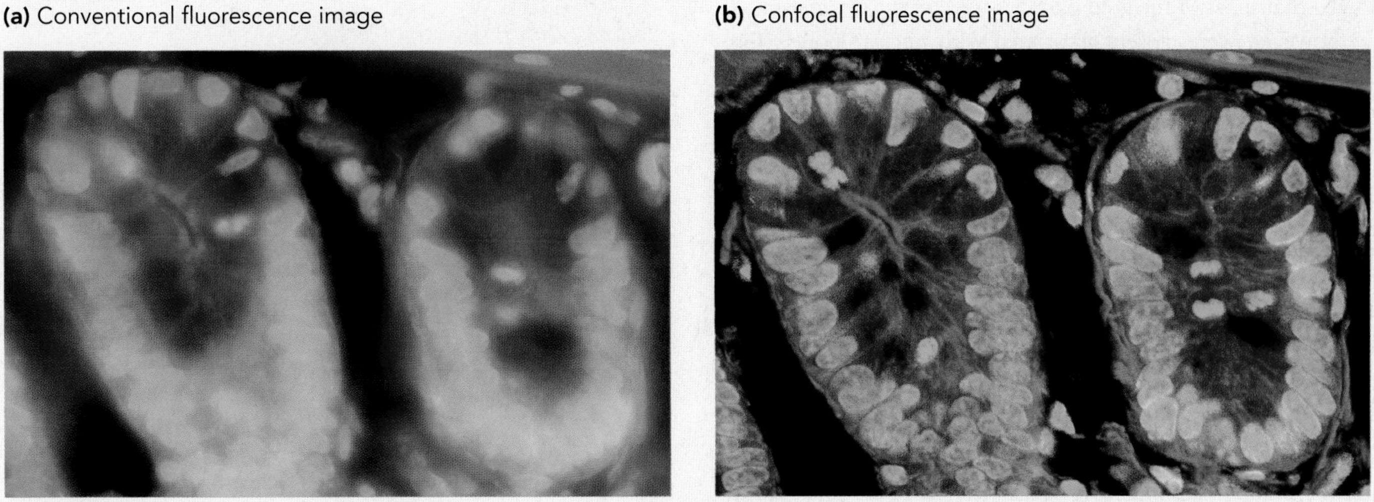
(a) Conventional fluorescence image
(b) Confocal fluorescence image

FIGURE 7.21 Confocal Microscopy Provides Sharp Images of Living Cells
The same living cells from the intestine of a mouse at the same magnification. **(a)** The conventional image is blurred, because it results from light emitted by the entire group of cells. **(b)** The confocal image is sharp, because it results from light emitted at a single plane inside the cells.

7.2 The Nuclear Envelope: Transport Into and Out of the Nucleus

The nucleus is the information center of eukaryotic cells. It is a corporate headquarters, design center, and library all rolled into one. Appropriately enough, its interior is highly organized. The organelle's overall shape and structure are defined by the mesh-like nuclear lamina, which also helps anchor each chromosome. The remainder of each chromosome occupies a well-defined region in the nucleus, and specific centers exist where the genetic information in DNA is decoded and processed. At these locations, large suites of enzymes interact to produce RNA messages from specific genes at specific times. Meanwhile, the nucleolus functions as the site of ribosome synthesis.

Consistent with its role as information repository and processing center, the nucleus is separated from the rest of the cell by the nuclear envelope. Biologists began to understand exactly how the nuclear envelope is structured when electron microscopy became available in the 1950s. As **Figure 7.22a** shows, the nuclear envelope has two membranes, each consisting of a lipid bilayer. The inner membrane and the outer membrane are separated by a space that is continuous with the lumen of the endoplasmic reticulum. Later, electron micrographs showed that the envelope contains thousands of openings called **nuclear pores** (**Figure 7.22b**). Because these pores extend through both inner and outer nuclear membranes, they connect the inside of the nucleus with the cytoplasm. The pore itself consists of over 50 different proteins. These molecules form an elaborate structure called the **nuclear pore complex** (**Figure 7.22c**).

A series of experiments in the early 1960s showed that molecules travel into and out of the nucleus through the nuclear pore complexes. The initial studies were based on injecting tiny gold particles into cells and then preparing them for electron microscopy. In electron micrographs, gold particles show up as black dots. One or two minutes after injection, the micrographs showed that most of the gold particles were in the cytoplasm. A few, however, were closely associated with nuclear pores. Ten minutes after injection, particles were inside the nucleus as well as in the cytoplasm. These data supported the hypothesis that the pores function as the doors to the nucleus. Follow-up work confirmed that the nuclear pore complex is the only gate between the cytoplasm and the nucleus and that only certain molecules go in and out. Passage through the nuclear pore is selective.

WEB TUTORIAL 7.1 Transport into the Nucleus

What substances traverse nuclear pores? DNA clearly does not—it never leaves the nucleus. But information coded in DNA is used to synthesize RNA inside the nucleus. Several distinctive types of RNA molecules are produced, each distinguished by size and function. For example, most ribosomal RNAs are manufactured in the nucleolus, where they bind to proteins to form completed ribosomal subunits. Messenger RNAs, in contrast, carry the information required to manufacture proteins out to the cytoplasm, where protein synthesis takes place. To perform their function, all of the various types of RNA move out of the nucleus. Traffic in the other direction is also impressive. Nucleotide

(a) The nuclear envelope has a double membrane.

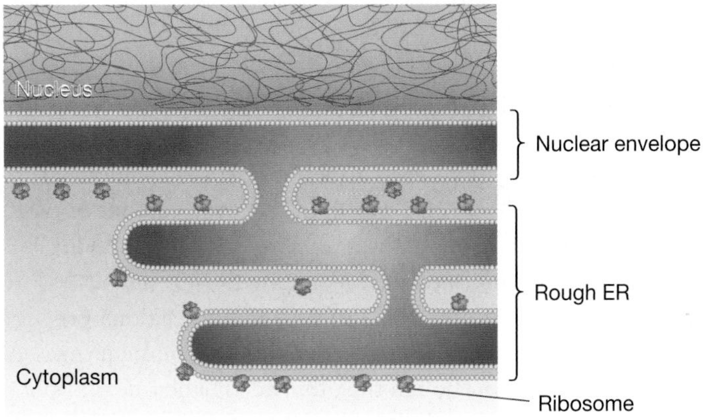
Nucleus
Nuclear envelope
Rough ER
Cytoplasm
Ribosome

(b) Surface view of nuclear envelope

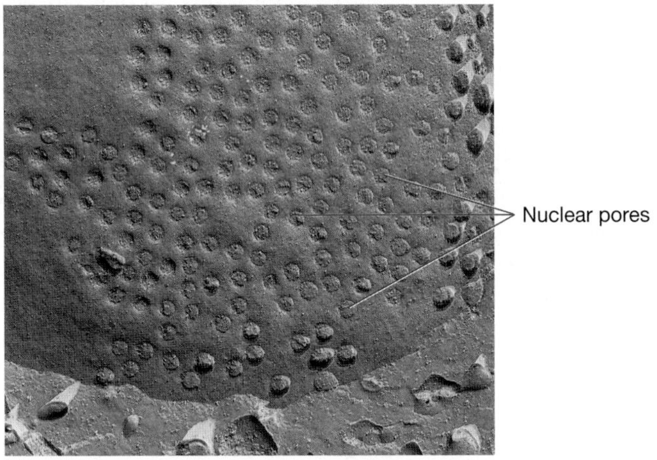
Nuclear pores

(c) Cross-sectional view of nuclear pore

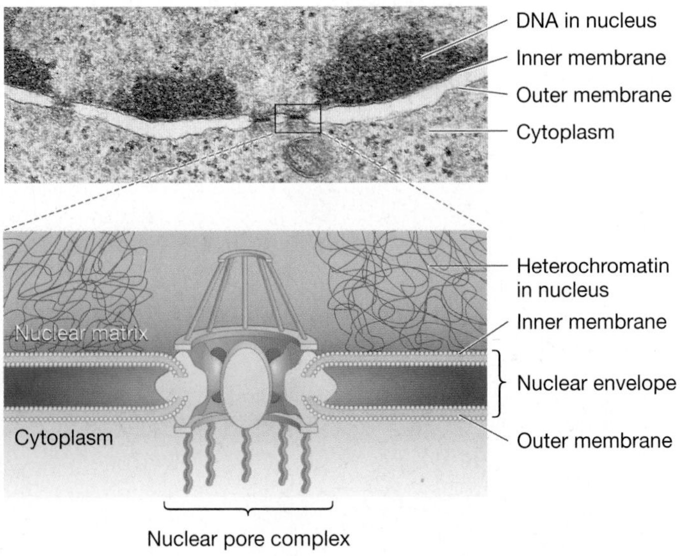
DNA in nucleus
Inner membrane
Outer membrane
Cytoplasm
Heterochromatin in nucleus
Nuclear matrix
Inner membrane
Nuclear envelope
Cytoplasm
Outer membrane
Nuclear pore complex

FIGURE 7.22 The Structure of the Nuclear Envelope
(a) The nuclear envelope is continuous with the endoplasmic reticulum. **(b)** Electron micrograph showing that the surface of the nuclear envelope is studded with nuclear pores. **(c)** The drawing (bottom) is based on electron micrographs of the nuclear pore complex.

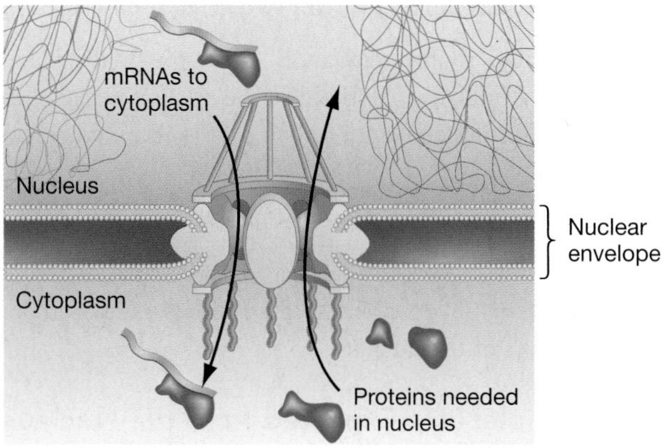
mRNAs to cytoplasm
Nucleus
Nuclear envelope
Cytoplasm
Proteins needed in nucleus

FIGURE 7.23 Molecules Move Into and Out of the Nucleus through the Nuclear Pores
Messenger RNAs are synthesized in the nucleus and must be exported to the cytoplasm. Proteins needed in the nucleus are synthesized in the cytoplasm and have to be imported to the nucleus.
EXERCISE Label the inner and outer membranes of the nuclear envelope. Label the nuclear pore.

triphosphates that act as building blocks for DNA and RNA have to enter the nucleus, as do the proteins responsible for copying DNA, synthesizing RNAs, extending the nuclear lamina, assembling ribosomes, or building chromosomes (**Figure 7.23**).

To summarize, ribosomal subunits and various types of RNAs exit the nucleus; proteins that are needed inside enter it. In a typical cell, over 500 molecules pass through each of the 3000–4000 nuclear pores every second. The traffic is intense. How is it regulated and directed?

How Are Molecules Imported into the Nucleus?

The first experiments on how molecules move through the nuclear pore focused on proteins that are produced by viruses. **Viruses** are parasites that use the cell's machinery to make copies of themselves. When a virus infects a cell, certain of its proteins enter the nucleus. Investigators noticed that if a particular amino acid in one of these proteins happens to be altered, the viral protein is no longer able to pass through the nuclear pore. This simple-sounding observation led to a key hypothesis: Proteins that are synthesized by ribosomes in the cytosol but are headed for the nucleus contain a "zip code"—a molecular address tag that marks them for transport through the nuclear pore complex. The idea was that viral proteins enter the nucleus if they have the same address tag as normal cellular proteins have. This zip code came to be called the **nuclear localization signal (NLS).**

A series of experiments on a protein called nucleoplasmin helped researchers better understand the nature of this signal. *Nucleoplasmin* plays an important role in the assembly of chromosomes and happens to have a distinctive structure: It consists of a globular protein core surrounded by a series of extended protein "tails." When researchers labeled nucleoplasmin with a radioactive atom and injected it into the cytoplasm of living cells, they found that the radioactive signal quickly ended up in the nucleus.

Figure 7.24 outlines how the nuclear localization signal in nucleoplasmin was found. Researchers began by using enzymes called proteases to separate the core sections of nucleoplasmin from the tails. After separating the two components, the researchers labeled each part with radioactive atoms and injected them into the cytoplasm of different cells. When they

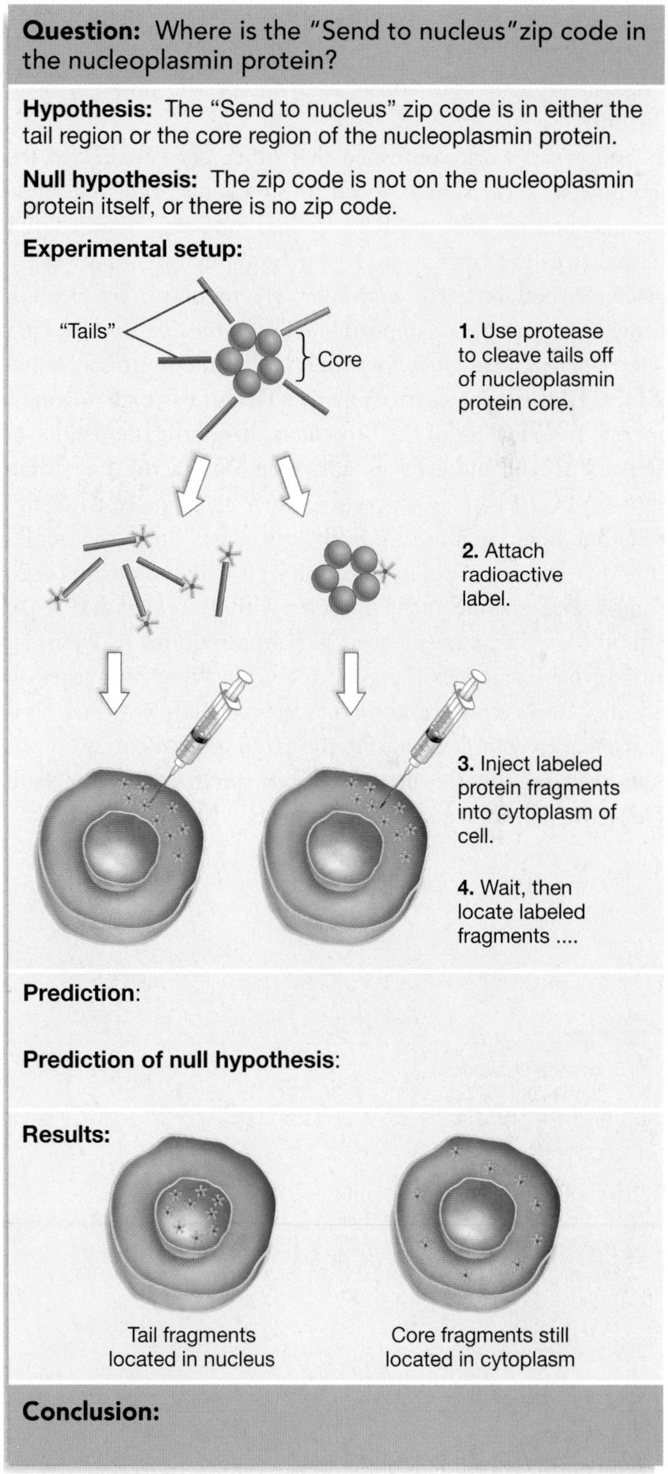
Question: Where is the "Send to nucleus"zip code in the nucleoplasmin protein?
Hypothesis: The "Send to nucleus" zip code is in either the tail region or the core region of the nucleoplasmin protein.
Null hypothesis: The zip code is not on the nucleoplasmin protein itself, or there is no zip code.
Experimental setup:
"Tails"
Core
1. Use protease to cleave tails off of nucleoplasmin protein core.
2. Attach radioactive label.
3. Inject labeled protein fragments into cytoplasm of cell.
4. Wait, then locate labeled fragments ....
Prediction:
Prediction of null hypothesis:
Results:
Tail fragments located in nucleus
Core fragments still located in cytoplasm
Conclusion:

FIGURE 7.24 Where Is the "Send to Nucleus" Zip Code in the Nucleoplasmin Protein?
EXERCISE Without looking at the text, fill in the predictions and conclusion(s) in this experiment.

examined the experimental cells with the electron microscope, they found that tail fragments were transported to the nucleus. Core fragments, in contrast, remained in the cytoplasm. These data suggested that the zip code must be somewhere in the tail part of the protein.

By analyzing different stretches of the tail section, the biologists eventually found a 17-amino-acid-long section that had to be present to direct proteins to the nucleus. The biologists therefore concluded that instead of consisting of five numbers, the NLS zip code consisted of 17 specific amino acids in the tail.

Follow-up work confirmed that other proteins bound for the nucleus have similar localization signals, and that these signals interact with proteins called **importins. Figure 7.25** summarizes the current model for how nuclear import takes place. Several different molecules are involved: the protein being transported, an importin, ATP, *guanosine diphosphate* (GDP) or *guanosine triphosphate* (GTP), and a protein called Ran. GTP is similar in structure to ATP and has high potential energy. If you think of the protein as cargo, the importin as a delivery truck, and ATP as gas, then Ran is the unloading crew and GTP is their supervisor. More specifically, data suggest that when an importin in the cytoplasm binds to a molecule that has a nuclear localization signal, the importin/cargo complex enters the nucleus along with Ran that has GDP bound to it. The movement of the cargo requires ATP. Inside, an enzyme exchanges the GDP for GTP. When this reaction occurs, Ran's conformation changes. It binds to the importin/cargo complex, causing the cargo molecule to drop off. Ran then escorts the importin back out to the cytoplasm. There the import sequence starts anew.

Work on nuclear import carries two general messages:

1. Movement is highly regulated. Although small molecules can diffuse freely into and out of the nuclear pore complex, larger molecules can enter only if they contain a nuclear localization signal.
2. Movement of large molecules is an energy-demanding, active process.

Do the same general principles hold when RNAs and other materials move *out* of the nucleus?

How Are Molecules Exported from the Nucleus?

After several decades' worth of experiments, biologists have come to a satisfying conclusion: Export of ribosome subunits, proteins, and other materials from the nucleus is almost exactly the reverse of import. In almost all cases, Ran and GTP are involved, as are shuttle proteins called **exportins.** Further, proteins that leave the nucleus have a specific zip code—a nuclear export signal. Distinct types of exportins are specialized for binding to the different types of materials and ferrying them to the cytoplasm. The Ran molecules and exportins responsible for nuclear export cycle back and forth, into and out of the nucleus, just as they do during import. Like nuclear import, nuclear export is both highly regulated and energy demanding.

Currently, biologists are focused on understanding how Ran interacts with proteins inside the nuclear pore complex as cargo moves in and out. Investigators are also trying to unravel how traffic is regulated to avoid backups and head-on collisions. The goal is to understand the precise physical mechanisms responsible for moving cargo into and out of the nuclear pore complex.

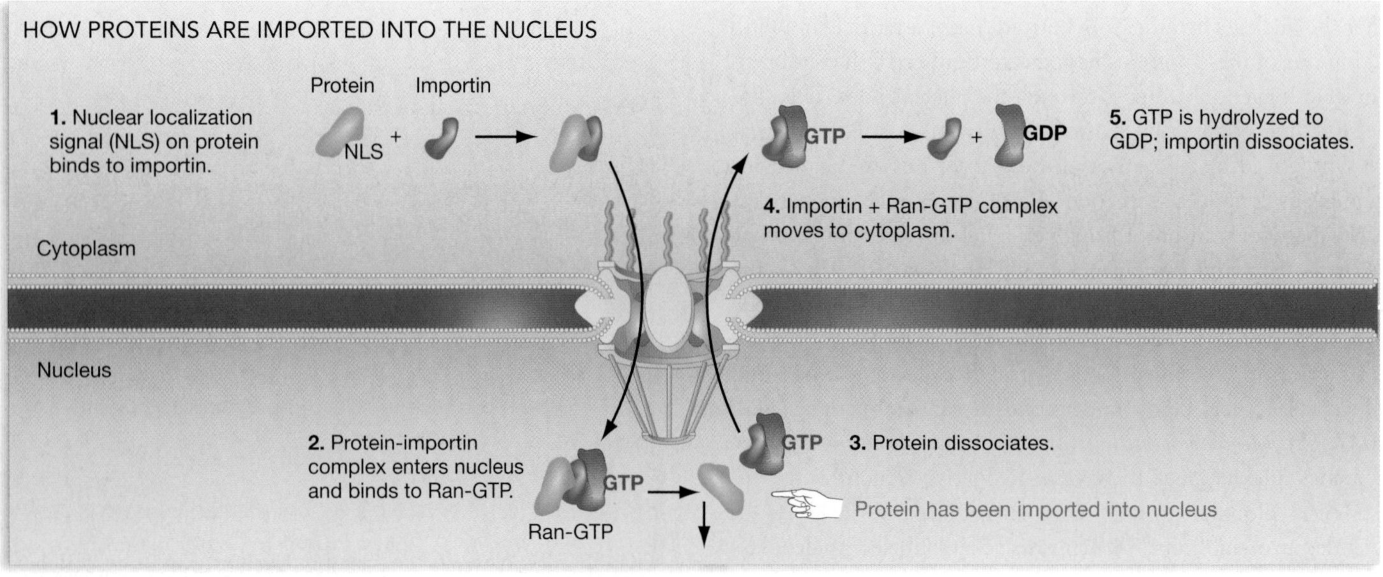

FIGURE 7.25 An Importin, Ran, and GDP Are Required to Import Proteins into the Nucleus
Importin, Ran, and GDP are recycled to the cytoplasm after they deliver cargo to the nucleus.

7.3 The Endomembrane System: Manufacturing and Shipping Proteins

The nuclear membrane is not the only place in cells where cargo moves in a regulated and energy-demanding fashion. For example, Chapter 6 highlighted how specific ions and molecules are pumped into and out of cells or transported across the plasma membrane by specialized membrane proteins. In addition, proteins that are synthesized by ribosomes in the cytosol but are used inside mitochondria or chloroplasts contain special signal sequences, analogous to the nuclear localization signal, that target the proteins for transport to these organelles.

Perhaps the most intricate of all manufacturing and shipping systems, however, involves proteins that are synthesized in the rough ER and move to the Golgi apparatus for processing, and from there travel to the cell surface or other destinations. The idea that materials might move through the endomembrane system in an orderly way was inspired by a simple observation. According to electron micrographs, cells that secrete digestive enzymes, hormones, or other types of products have particularly large amounts of rough ER and Golgi. This correlation led to the idea that these cells have a "secretory pathway" that starts in the rough ER and ends with products leaving the cell (**Figure 7.26**). How does this hypothesized pathway work?

George Palade and colleagues did pioneering research on the secretory pathway with an experimental approach known as a **pulse-chase experiment**. The strategy is based on providing experimental cells with a large concentration of a labeled molecule for a short time. For example, if a cell receives a large amount of labeled amino acid for a short time, virtually all of the proteins synthesized during that interval will be labeled. This "pulse" of labeled molecule is followed by a chase—large amounts of an unlabeled version of the same molecule, provided for a long time. If the chase consists of unlabeled amino acid, then the proteins synthesized during the chase period will *not* be labeled. The general idea is to mark a population of molecules at a particular interval and then follow their fate over time. This approach is analogous to adding a small amount of dye to a stream and then following the movement of the dye molecules.

WEB TUTORIAL 7.2 A Pulse-Chase Experiment

In testing the secretory pathway hypothesis, Palade's team focused on pancreatic cells that were growing in culture, or in vitro.[1] These cells are specialized for secreting digestive enzymes into the small intestine and are packed with rough ER and Golgi. The basic experimental approach was to supply the cells with a 3-minute pulse of the amino acid leucine, labeled with a radioactive atom, followed by a long chase with nonradioactive leucine. Because the radioactive leucine was incorporated into all proteins being produced during the pulse, it labeled them. Then the researchers

[1]The term *in vitro* is Latin for "in glass." Experiments that are performed outside living cells are done in vitro. The term *in vivo*, in contrast, is Latin for "in life." Experiments performed with living organisms are done in vivo.

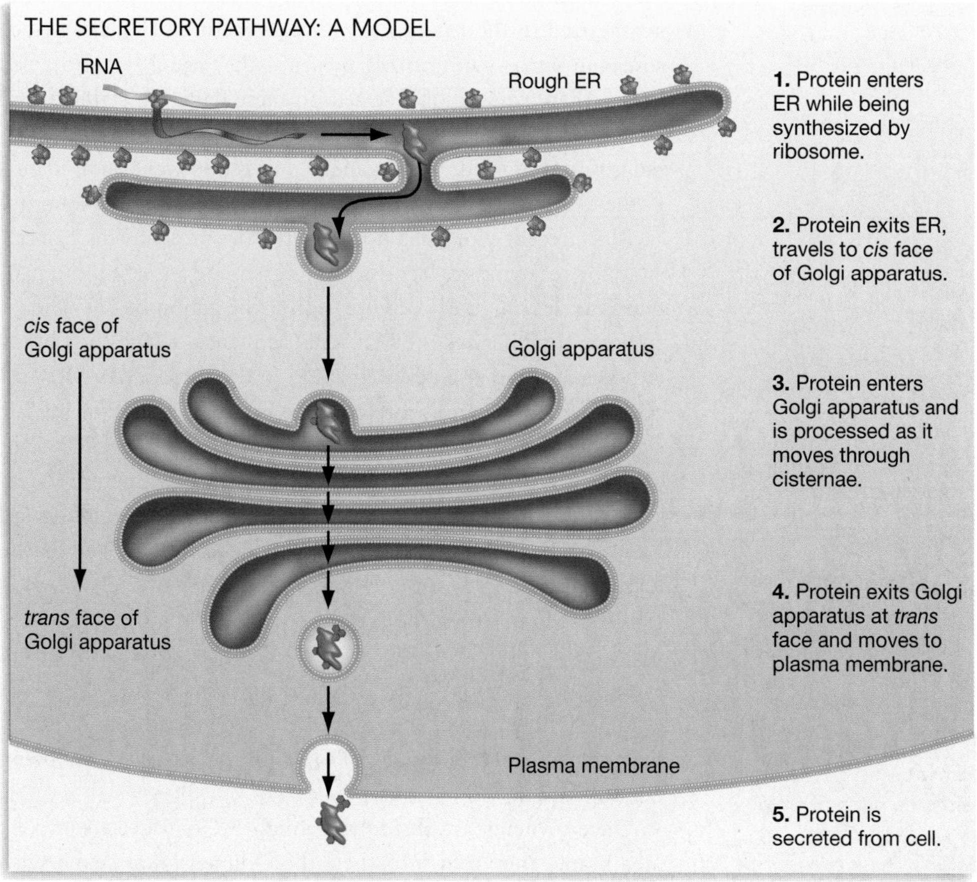
THE SECRETORY PATHWAY: A MODEL
RNA
Rough ER
1. Protein enters ER while being synthesized by ribosome.
2. Protein exits ER, travels to cis face of Golgi apparatus.
cis face of Golgi apparatus
Golgi apparatus
3. Protein enters Golgi apparatus and is processed as it moves through cisternae.
4. Protein exits Golgi apparatus at trans face and moves to plasma membrane.
trans face of Golgi apparatus
Plasma membrane
5. Protein is secreted from cell.

FIGURE 7.26 The Secretory Pathway Hypothesis
The secretory pathway hypothesis proposes that proteins intended for secretion from the cell are synthesized and processed in a highly prescribed set of steps.

prepared a sample of the cells for electron microscopy and autoradiography (see Chapter 4). When they examined cells immediately after the pulse, they found the newly synthesized proteins inside the rough ER (**Figure 7.27a**). Seven minutes later, most of the labeled proteins were in a Golgi apparatus or inside structures called secretory vesicles on the *trans* side of a Golgi apparatus (**Figure 7.27b**). After 80 minutes, most labeled proteins were in secretory vesicles or actually outside the cell (**Figure 7.27c**).

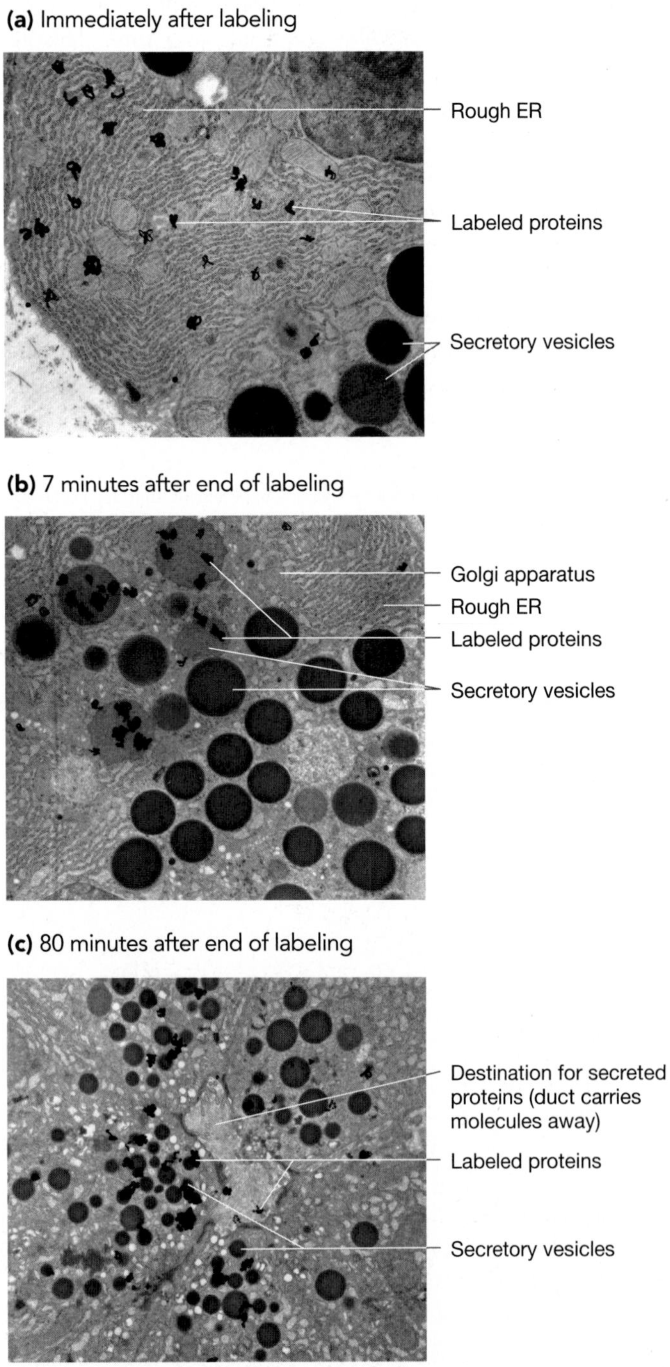
(a) Immediately after labeling
Rough ER
Labeled proteins
Secretory vesicles
(b) 7 minutes after end of labeling
Golgi apparatus
Rough ER
Labeled proteins
Secretory vesicles
(c) 80 minutes after end of labeling
Destination for secreted proteins (duct carries molecules away)
Labeled proteins
Secretory vesicles

FIGURE 7.27 Results of a Pulse-Chase Experiment
The position of labeled proteins immediately after the pulse of amino acids, then 7 and 80 minutes later. QUESTION Do the data support or contradict the secretory pathway hypothesis? Explain your answer.

These results were consistent with the hypotheses that a secretory pathway exists and that the rough ER and Golgi apparatus function as an integrated endomembrane system. Clearly, proteins produced in the rough ER don't float around the cytoplasm aimlessly or drift randomly from organelle to organelle. Instead, traffic through the endomembrane system is highly organized and directed. Now let's break the system down and examine four of the steps in more detail. The ribosomes in rough ER are bound to the outside of the membrane. How do the proteins that they manufacture get into the lumen of the ER? How do they move from the ER to the Golgi apparatus? Once they're inside the Golgi, what happens to them? And finally, how do the finished proteins get to their destination? Let's consider each question in turn.

Entering the Endomembrane System: The Signal Hypothesis

How do proteins enter the endomembrane system? The **signal hypothesis**, proposed by Günter Blobel and colleagues, predicted that proteins bound for the endomembrane system have a zip code analogous to the nuclear localization signal. The idea was that these proteins are synthesized by ribosomes that are attached to the outside of the ER and that the first few amino acids in the growing polypeptide act as a signal that brings the protein into the lumen of the ER.

This hypothesis received important support when researchers made a puzzling observation: When proteins that are normally synthesized in the rough ER are manufactured by naked ribosomes in vitro—with no ER present—they are 20 amino acids longer than usual. Blobel seized on these data. He claimed that the 20 amino acids are the "Send to ER" signal and that the signal is removed inside the organelle. His group went on to identify the exact sequence of amino acids in the **ER signal sequence.**

More recent work has documented the mechanisms responsible for receiving the send-to-ER signal and inserting the protein into the rough ER (**Figure 7.28**). The action begins when a ribosome synthesizes the ER signal sequence, which then binds to a **signal recognition particle (SRP)** in the cytosol. An SRP is a complex of RNA and protein that acts as a receptor for the ER signal sequence. The ribosome + signal sequence + SRP complex then attaches to an SRP receptor in the ER membrane itself. You can think of the SRP as a key that is activated by an ER signal sequence. The receptor in the ER membrane is the lock. Once the lock and key connect, the rest of the protein is synthesized, and then the signal sequence is removed. The finished polypeptide has one of two fates: (1) proteins that will eventually be shipped to an organelle or secreted from the cell enter the lumen of the rough ER; or (2) membrane proteins that remain in the rough ER membrane as they are being manufactured.

Once proteins are inside the rough ER or inserted into its membrane, they fold into their three-dimensional shape with

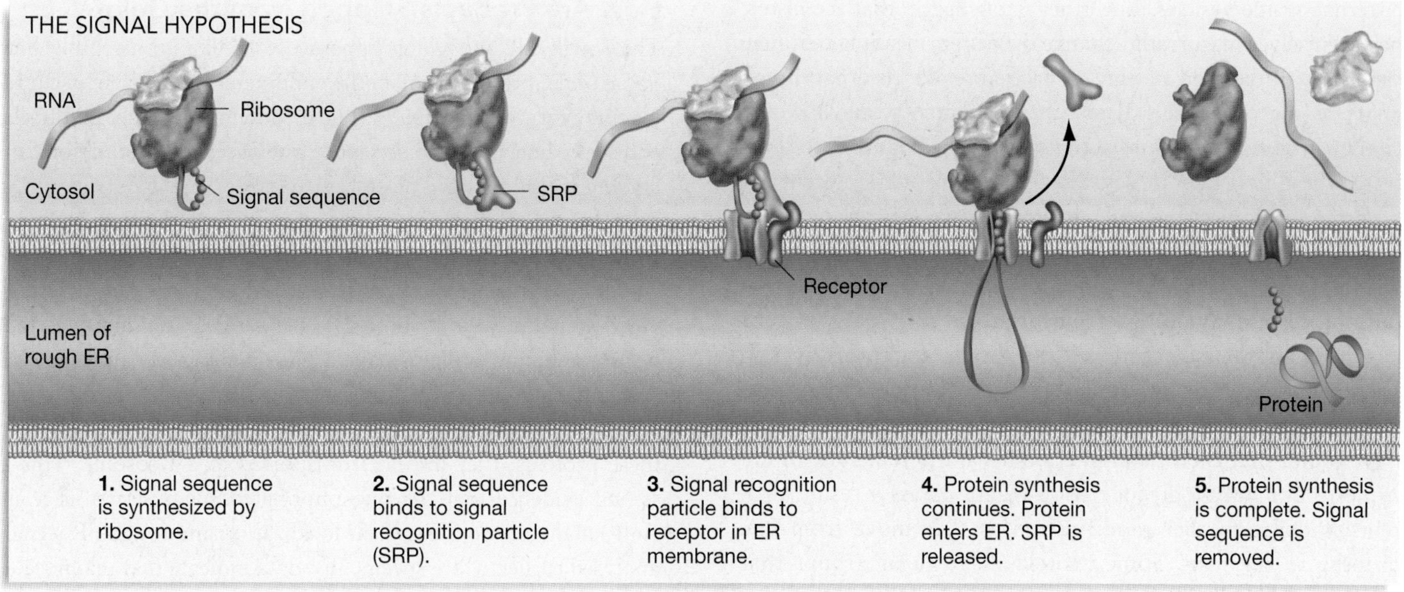
THE SIGNAL HYPOTHESIS
RNA
Ribosome
Cytosol
Signal sequence
SRP
Receptor
Lumen of rough ER
Protein
1. Signal sequence is synthesized by ribosome.
2. Signal sequence binds to signal recognition particle (SRP).
3. Signal recognition particle binds to receptor in ER membrane.
4. Protein synthesis continues. Protein enters ER. SRP is released.
5. Protein synthesis is complete. Signal sequence is removed.

FIGURE 7.28 The Signal Hypothesis Explains How Proteins Destined for Secretion Enter the Endomembrane System According to the signal hypothesis, proteins destined for secretion contain a short stretch of amino acids that interact with a signal recognition particle (SRP) in the cytoplasm. This interaction allows the protein to enter the ER.

the help of chaperone proteins. In addition, proteins that enter the lumen are acted on by enzymes that catalyze the addition of carbohydrate side chains. Because carbohydrates are polymers of sugar monomers, the addition of one or more carbohydrate groups is called **glycosylation** ("sugar-together"). The resulting molecule is called a **glycoprotein** ("sugar-protein"). As **Figure 7.29** shows, proteins that enter the ER often gain a

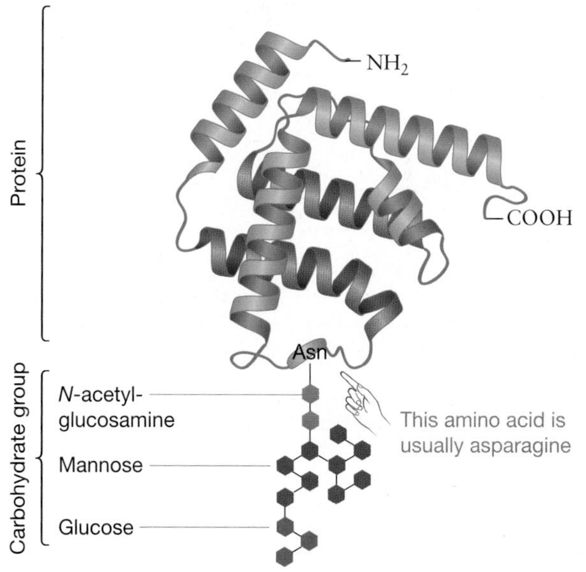
Protein
NH_2
COOH
Asn
Carbohydrate group
N-acetyl-glucosamine
Mannose
Glucose
This amino acid is usually asparagine

FIGURE 7.29 Glycosylation Adds Carbohydrate Groups to Proteins When proteins enter the ER, most acquire the 14 sugar residues shown here. Some of these sugars may be removed or others added as proteins pass through the Golgi apparatus.

specific carbohydrate that consists of 14 sugar residues. Thus, proteins are not only synthesized in the rough ER, they are folded and modified by glycosylation. The completed glycoproteins are ready for shipment to the Golgi apparatus.

Getting from the ER to the Golgi

How do proteins travel from the ER to the Golgi apparatus? Palade's group thought they had the answer, based on data from the pulse-chase experiments that first confirmed the existence of the endomembrane system. When labeled proteins appeared in a region between the rough ER and the Golgi apparatus, they appeared to be inside small membrane-bound structures. Based on these observations, the biologists suggested that proteins are transported between the two organelles in vesicles. The idea was that vesicles bud off of the ER, move away, fuse with the membrane on the *cis* face of the Golgi apparatus, and dump their contents inside.

This hypothesis was supported when other researchers used differential centrifugation to isolate and characterize the vesicles that contained labeled proteins. Using this approach, investigators have established that distinctive types of vesicles carry proteins from the rough ER to the Golgi apparatus and from layer to layer within the Golgi apparatus.

What Happens Inside the Golgi Apparatus?

Recall from Section 7.1 that the Golgi apparatus consists of a stack of flattened vesicles called cisternae, and that cargo enters one side of the structure and exits the other. It is still not clear, however, exactly how material moves through the stack. There is strong evidence that at least some molecules move among the cisternae inside vesicles. But other data suggest that the cisternae themselves mature and change over time, meaning that new cisternae are created at the *cis* face and old cisternae break apart at the *trans* face. If so, then cisternae would have to change in composition and activity over time. **Figure 7.30** illustrates these two hypotheses. Is each cisterna static except for occasional additions or subtractions via vesicle delivery and shipment, or is the entire structure dynamic? The answer is not yet known, and the dichotomy is not necessarily absolute—both processes may occur to some degree.

Although the structure of the Golgi apparatus is still somewhat uncertain, its function is not. By separating individual cisternae and analyzing their contents, researchers have found that each cisterna contains a different suite of enzymes that catalyze glycosylation reactions. As a result, proteins undergo further modification as they move from one cisterna to the next. Some proteins have sugar groups that are phosphorylated in a vesicle near the *cis* face. Later, the carbohydrate group that was added in the rough ER is removed. In other cisternae, various types of carbohydrate chains are attached that may protect the protein or help it attach to surfaces.

How Are Products Shipped from the Golgi?

The rough ER and Golgi apparatus are like an assembly line. Some of the products stay in the endomembrane system itself, replacing worn-out molecules. But if proteins are processed to the end of the line, they will be sent to one of several destinations, including lysosomes, the plasma membrane, or the outside of the cell. How are these finished products put into the right shipping containers, and how are the different containers addressed?

Studies on enzymes that are shipped to lysosomes have provided some answers to both questions. A key finding was that lysosome-bound proteins have a phosphate group attached to a specific sugar subunit on their surface, forming the compound mannose-6-phosphate. If mannose-6-phosphate is removed from these proteins, they are not transported to a lysosome. This is strong evidence that the phosphorylated sugar serves as a zip code, analogous to the nuclear localization and rough ER signals analyzed earlier. More specifically, data indicate that mannose-6-phosphate binds to a protein in membranes of certain vesicles. These vesicles, in turn, have proteins on their surface that interact specifically with proteins in the lysosomal membranes. In this way, the presence of mannose-6-phosphate targets proteins for vesicles that deliver their contents to lysosomes.

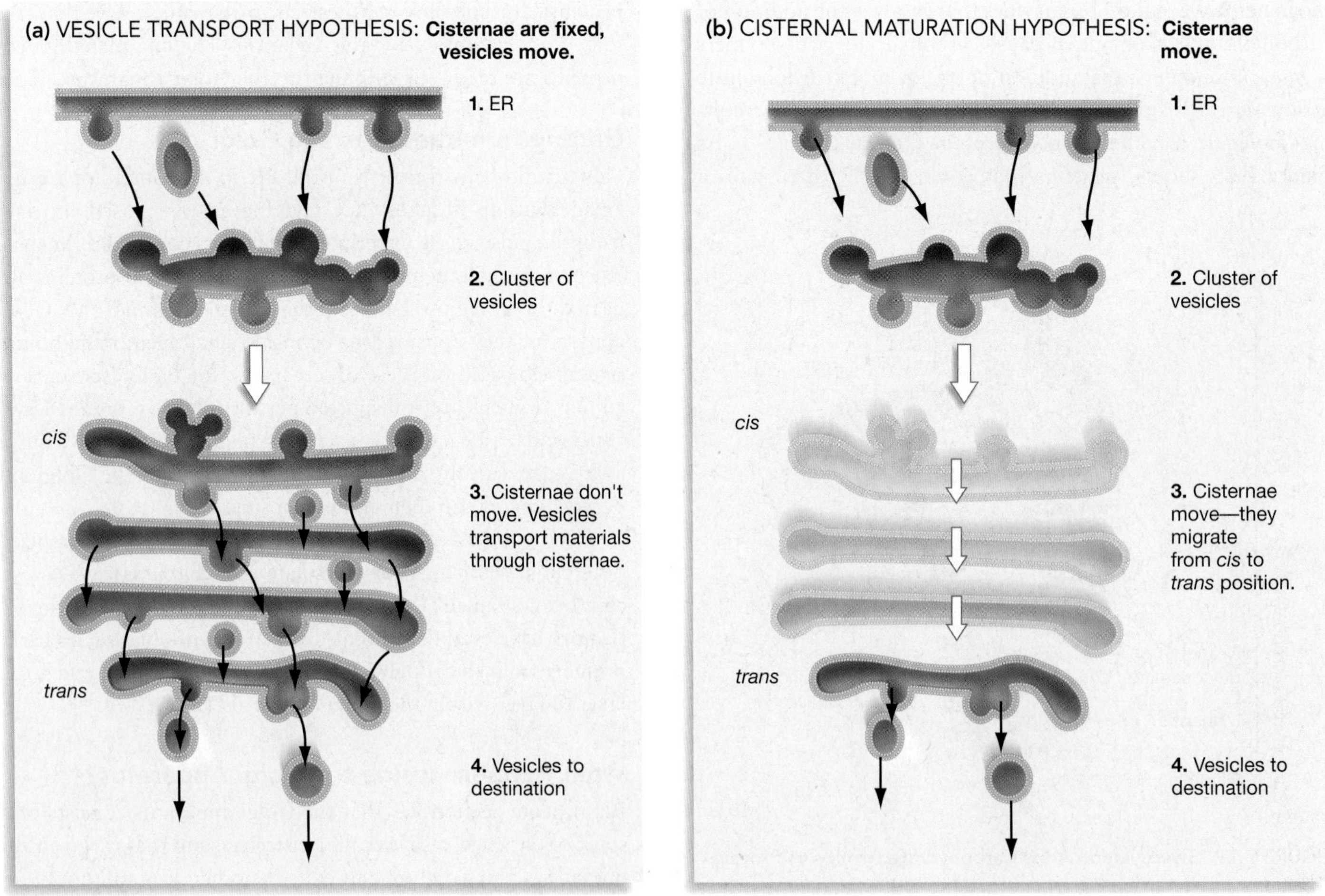
(a) VESICLE TRANSPORT HYPOTHESIS: Cisternae are fixed, vesicles move.
1. ER
2. Cluster of vesicles
cis
3. Cisternae don't move. Vesicles transport materials through cisternae.
trans
4. Vesicles to destination
(b) CISTERNAL MATURATION HYPOTHESIS: Cisternae move.
1. ER
2. Cluster of vesicles
cis
3. Cisternae move—they migrate from cis to trans position.
trans
4. Vesicles to destination

FIGURE 7.30 Two Hypotheses for How Materials Move through the Golgi Apparatus

QUESTION Are these hypotheses mutually exclusive? Explain your answer.

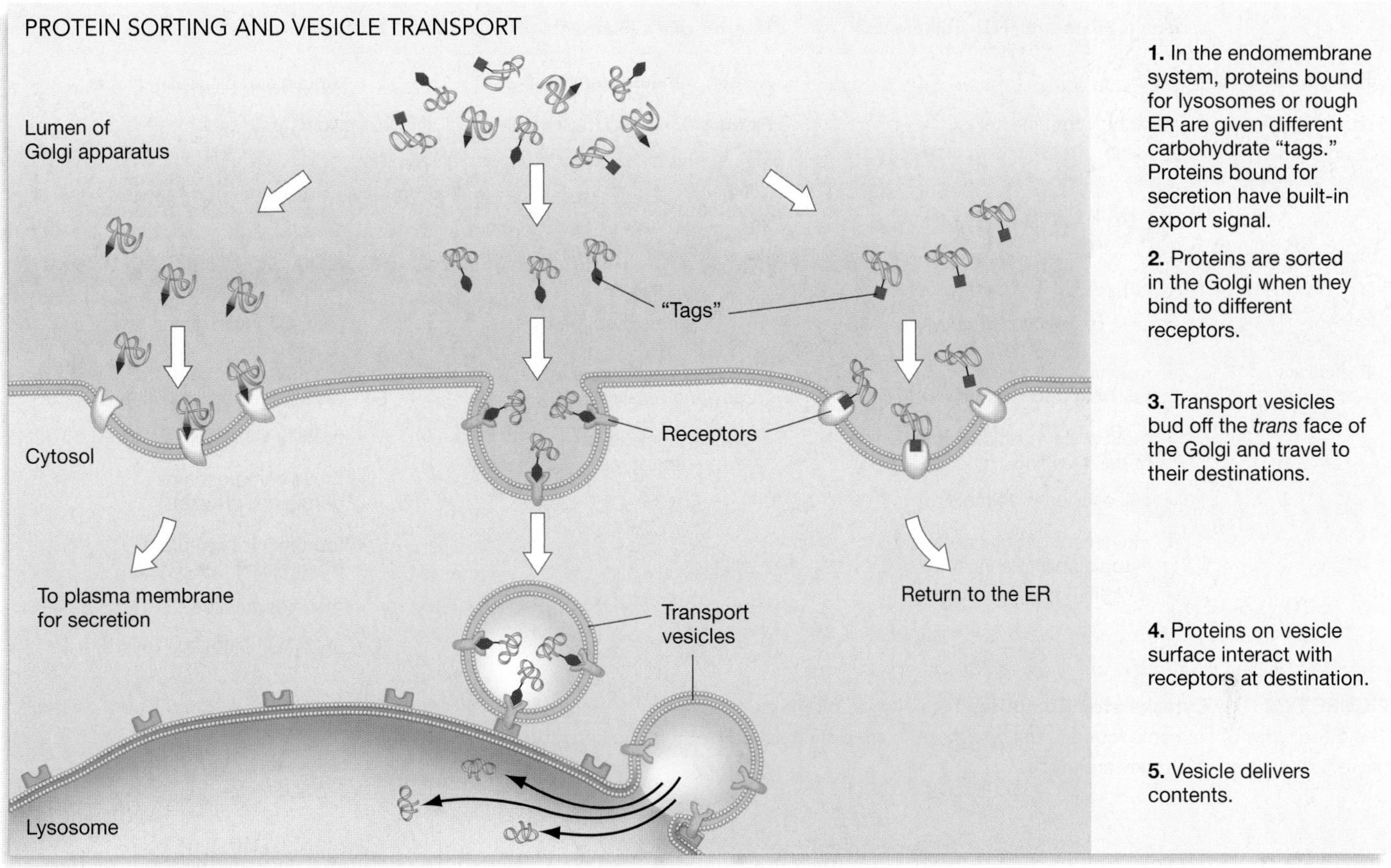

FIGURE 7.31 In the Golgi Apparatus, Proteins Are Sorted into Vesicles That Are Targeted to a Destination
Summary of the current model for how proteins are sorted into distinct vesicles in the Golgi apparatus and how these vesicles are then targeted to their correct destination.

Figure 7.31 pulls these observations together into a comprehensive model explaining how the products of the endomembrane system are loaded into specific vesicles and shipped to their correct destination. Notice that transport vesicles bound for the plasma membrane secrete their contents to the outside. This process is called **exocytosis** ("outside-cell-act"). When exocytosis occurs, the vesicle membrane and plasma membrane make contact and fuse. The two sets of lipid bilayers rearrange in a way that exposes the interior of the vesicle to the outside of the cell. The vesicle's contents then diffuse away from the cell into the space outside the cell. In this way, cells in your pancreas deliver the hormone insulin to your bloodstream.

The general message of this section is that cells have sophisticated cargo production, sorting, and shipping systems. Proteins that are synthesized in the cytoplasm also have zip codes directing them to mitochondria, chloroplasts, or other destinations.

If vesicles function like shipping containers for products that move between organelles, do they travel along some sort of road or track? What molecule or molecules function as the delivery truck, and does ATP or GTP supply the gas? In general, what physical mechanisms are responsible for moving vesicles to their destination?

✓CHECK YOUR UNDERSTANDING

Ions, ATP, amino acids, and other small molecules diffuse randomly throughout the cell, but the transport of proteins and other large molecules is energy demanding and tightly regulated. Proteins must have the appropriate molecular zip code to enter or leave the nucleus, enter the lumen of the rough ER, or become incorporated into vesicles destined for lysosomes or the plasma membrane. In many cases, proteins and other types of cargo are shipped in vesicles that contain molecular zip codes on their surface. You should be able to (1) propose a hypothesis for how proteins are targeted to chloroplasts, and (2) outline an experiment that would test your hypothesis.

7.4 The Dynamic Cytoskeleton

Based on early observations with light microscopes, biologists viewed the cytoplasm of eukaryotic cells as a fluid-filled space devoid of structure. As microscopy improved, however, researchers realized that the cytoplasm contains an extremely dense and complex network of fibers. This cytoskeleton helps maintain cell shape by providing structural support. It's

	Actin Filaments (Microfilaments)	Intermediate Filaments	Microtubules
Protein subunits	Actin	Keratin, vimentin, lamin, others	α-tubulin and β-tubulin dimers
Structure	Strands in double helix 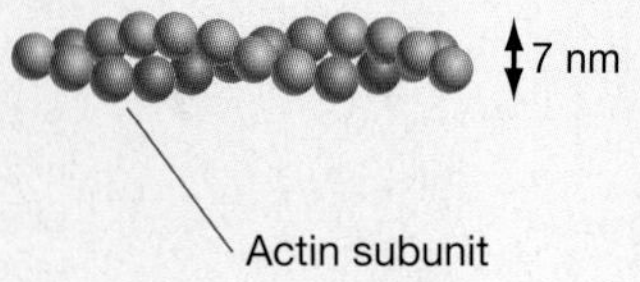	Fibers wound into thicker cables 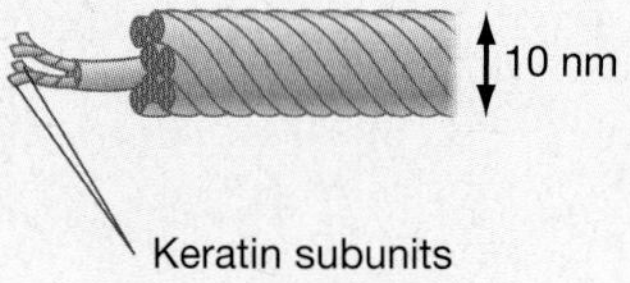	Hollow tube 
Functions	• maintain cell shape by resisting tension (pull) • motility via muscle contraction or cell crawling • cell division in animals • movement of organelles and cytoplasm in plants, fungi, and animals	• maintain cell shape by resisting tension (pull) • anchor nucleus and some other organelles	• maintain cell shape by resisting compression (push) • motility via flagella or cilia • move chromosomes during cell division • formation of cell plate during plant cell division • move organelles • growth of plant cell walls

FIGURE 7.32 The Cytoskeleton Comprises Three Types of Filaments
The three types of filaments found in the cytoskeleton are distinguished by their size and structure, and by the protein subunit of which they are made.

important to recognize, though, that the cytoskeleton is not a static structure like the scaffolding used at construction sites. The fibrous proteins that make up the cytoskeleton move and change to change the cell's shape, to move materials from place to place, and to move the entire structure. Like the rest of the cell, the cytoskeleton is dynamic.

As **Figure 7.32** shows, there are several distinct types of cytoskeletal elements: actin filaments (also known as microfilaments), intermediate filaments, and microtubules. Each of these elements has a distinct size, structure, and function. Let's look at each one in turn.

Actin Filaments

Actin filaments are sometimes referred to as **microfilaments** because they are the cytoskeletal element with the smallest diameter. As Figure 7.32 indicates, **actin filaments** are long, fibrous structures made of a globular protein called actin. In animal cells, actin is often the most abundant of all proteins—typically it represents 5–10 percent of the total protein in the cell. Each of your liver cells contains about half a billion of these molecules.

Actin filaments form when individual actin molecules polymerize. The completed structure resembles two strands that coil around each other. Because each actin monomer in the strand is asymmetrical, the structure as a whole has a distinct polarity. The two ends of an actin filament are different and are referred to as plus and minus ends. Actin filaments tend to grow at the plus end, because polymerization occurs fastest there.

Figure 7.33a shows a fluorescence micrograph of the actin filaments in a mammalian kidney cell. Note that groups of actin filaments are organized into long bundles or dense networks and that actin filaments are particularly abundant just under the plasma membrane. Whether they are arranged in parallel as part of bundles or crisscrossed in networks, individual actin filaments are linked to one another by other proteins. In combination, the bundles and networks of actin filaments help stiffen the cell and define its shape.

Although actin filaments are an important part of the cell's structural support, it would be a mistake to think that they are static. Instead, actin filaments grow and shrink as actin subunits are added or subtracted from each end of the structure. This phenomenon is called *treadmilling*, because the dynamics of the fibers resemble those of a treadmill.

In addition, many cells have actin filaments that interact with the specialized protein **myosin**. When ATP that is bound to myosin is hydrolyzed to ADP, the "head" region of the myosin molecule binds to actin and moves. The movement of this protein causes the actin filament to slide (**Figure 7.34a**). As **Figure 7.34b** shows, the (ATP-powered) interaction between actin and myosin is the basis for an array of cell movements:

- **Cell crawling** occurs in amoebae, slime molds, and certain types of human cells. Cell crawling is based on three processes: a directional extension of actin filaments that pushes the plasma membrane into bulges called **pseudopodia** ("false-feet"), adherence to a solid substrate, and a myosin-driven contraction of actin filaments at the cell's other end. In com-

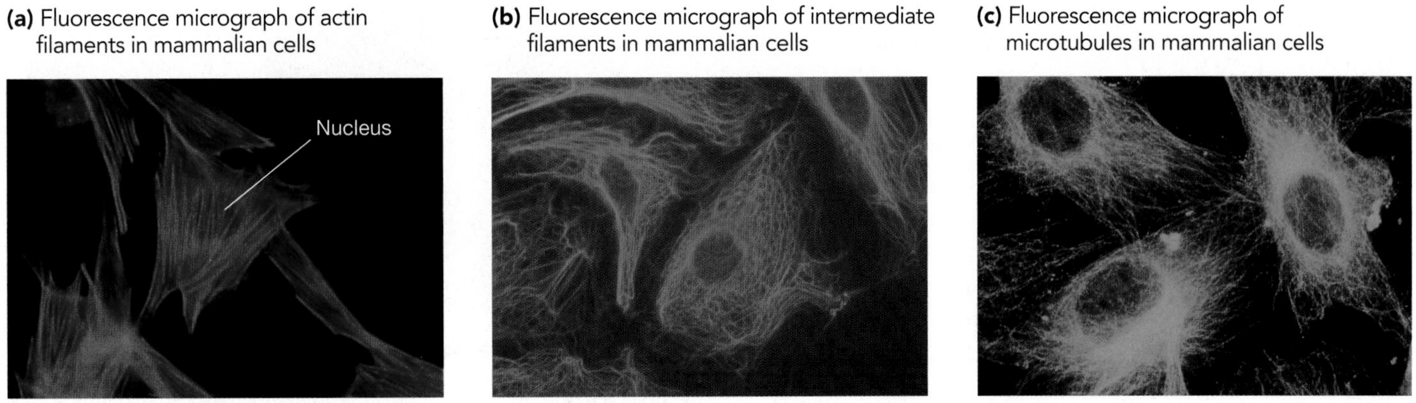
(a) Fluorescence micrograph of actin filaments in mammalian cells
Nucleus
(b) Fluorescence micrograph of intermediate filaments in mammalian cells
(c) Fluorescence micrograph of microtubules in mammalian cells

FIGURE 7.33 How Are Cytoskeletal Elements Distributed in the Cell?
To make these micrographs, researchers attached a fluorescent compound to **(a)** actin, the protein subunit of actin filaments, to **(b)** a protein found in intermediate filaments, and to **(c)** tubulin dimers.

bination, the three events result in directed movement by whole cells.

- **Cytokinesis** ("cell-moving") is the process of cell division in animals. For these cells to divide in two, actin filaments that are arranged in a ring under the plasma membrane must slide past one another. Because they are connected to the plasma membrane, the movement of the actin fibers pinches the cell in two.
- **Cytoplasmic streaming** is the directed flow of cytosol and organelles around plant and fungal cells. The movement occurs along actin filaments and is powered by myosin.

In addition, extension of actin filaments is responsible for the expansion of long, thin fungal cells into soil or rotting wood. The same mechanism causes structures called pollen tubes to grow toward the egg cells of plants, so sperm can be delivered prior to fertilization.

Intermediate Filaments

Unlike actin filaments and microtubules, **intermediate filaments** (**Figure 7.33b**) are defined by size rather than composition. Many types of intermediate filaments exist, each consisting of a different protein. In many cases, different types of cells in the same organism contain different types of intermediate filaments. This is in stark contrast to actin filaments and microtubules, which are made from the same protein subunits in all eukaryotic cells. In addition, intermediate filaments are not polar; instead, each end of these filaments is identical. As a result, intermediate filaments do not treadmill, and they are not

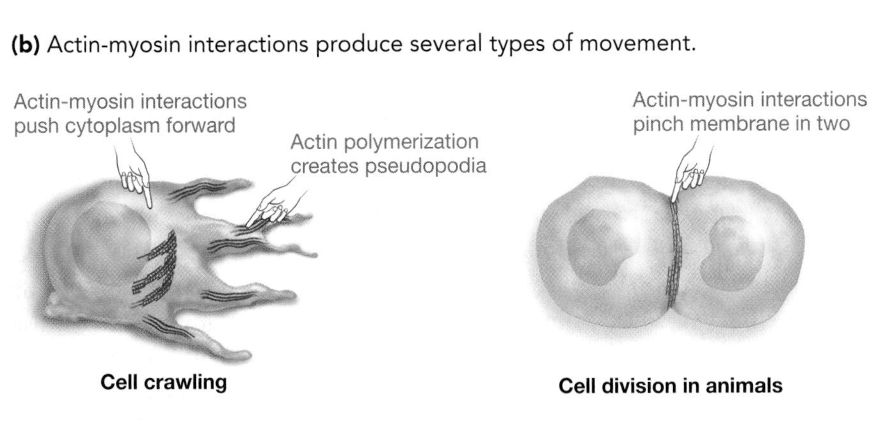
(a) Actin and myosin interact to cause movement.
Myosin
"Head" region
When myosin's "head" attaches to actin and moves, the actin filament slides
Actin
(b) Actin-myosin interactions produce several types of movement.
Actin-myosin interactions push cytoplasm forward
Actin polymerization creates pseudopodia
Cell crawling
Actin-myosin interactions pinch membrane in two
Cell division in animals
Actin-myosin interactions move cytoplasm around cell
Cytoplasmic streaming in plants

FIGURE 7.34 Many Cellular Movements Are Based on Actin-Myosin Interactions
(a) When the "head" region of the myosin protein binds to ATP or ADP, myosin attaches to actin and changes shape. The movement causes the actin filament to slide. **(b)** Actin-myosin interactions can move cells, divide cells, and move organelles and cytoplasm.

involved in directed movement driven by myosin or related proteins. Intermediate filaments serve a purely structural role in eukaryotic cells.

The intermediate filaments that you are most familiar with belong to a family of molecules called the keratins. The cells that make up your skin and that line surfaces inside your body contain about 20 types of keratin. The presence of these intermediate filaments provides the mechanical strength required for these cells to resist pressure and abrasion. Skin cells secrete another 10 distinct forms of keratin. Depending on the location of the skin cell and keratins involved, the secreted filaments form fingernails, toenails, or hair.

Nuclear lamins, which make up the nuclear lamina layer introduced in Section 7.1, also qualify as intermediate filaments. These fibers form a dense mesh under the nuclear envelope. Recall that in addition to giving the nucleus its shape, they anchor the chromosomes. They are also involved in the breakup and reassembly of the nuclear envelope when cells divide. Some intermediate filaments project from the nucleus through the cytoplasm to the plasma membrane, where they are linked to intermediate filaments that run parallel to the cell surface. In this way, intermediate filaments form a flexible skeleton that helps shape the cell surface and hold the nucleus in place.

Microtubules

Microtubules are composed of the proteins α-tubulin and β-tubulin and are the largest cytoskeletal components in terms of diameter (**Figure 7.33c**). Molecules of α-tubulin and β-tubulin bind to form **dimers** ("two-parts"), compounds formed by the joining of two monomers. Tubulin dimers then polymerize to form the large, hollow tube called a **microtubule**. Because each end of a tubulin dimer is different, each end of a microtubule has a distinct polarity. Like actin filaments, microtubules are dynamic and more likely to grow from one end than they are from the other. Microtubules grow and shrink in length as tubulin dimers are added or subtracted.

Microtubules are similar to actin filaments in function as well as structure. Both cytoskeletal elements provide structural support, and both are involved in cell division. Although microtubules are not involved in the physical division of the cell, they are essential for the directed movement of chromosomes to each of the two resulting cells. In animals and fungi, the microtubules involved in chromosome movement emanate from a structure called the **centrosome**. Distinctive structures called **centrioles** are found inside centrosomes (**Figure 7.35**). Centrioles may help organize microtubules; however, they are not essential for cell division to occur. In plants and many other eukaryotes, a region called the **microtubule organizing center** performs the same function as the centrosome.

Microtubules are involved in many other types of cellular movement as well. For the remainder of this chapter, we'll focus on how microtubules function in moving materials inside cells and in moving the entire cell.

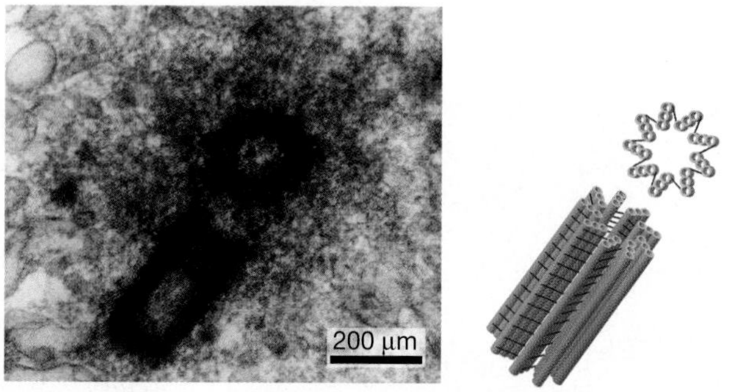

FIGURE 7.35 Centrosomes Are a Type of Microtubule Organizing Center
Microtubules emanate from microtubule organizing centers, which in animals are called centrosomes. The centrioles inside a centrosome are made of microtubules.

Studying Vesicle Transport Materials are transported to a wide array of destinations inside cells. To study how this movement happens, Ronald Vale and colleagues focused on a cell called the giant axon that is found in squid. The giant axon is an extremely large nerve cell that runs the length of a squid's body. If the animal is disturbed, the cell signals muscles to contract so the individual can jet away to safety.

The researchers decided to study this particular cell for three reasons. First, the giant axon is so large that it is relatively easy to see and manipulate. Second, signaling molecules are synthesized in the cell's ER and then transported in vesicles down the length of the cell. As a result, a large amount of cargo moves a long distance. Third, the researchers found that if they gently squeeze the cytoplasm out of the cell, vesicle transport still occurs in the cytoplasmic material.

In short, the squid giant axon provided a cell-free system that could be observed and manipulated efficiently. What did the biologists find out?

Microtubules Act as "Railroad Tracks" To watch vesicle transport in action, researchers mounted a video camera to a microscope. As **Figure 7.36** shows, this technique allowed them to document that vesicle transport occurred along a filamentous track. A simple experiment convinced the group that this movement was an energy-dependent process. If they depleted the amount of ATP in the cytoplasm, vesicle transport stopped.

To identify the type of filament involved, the biologists measured the diameter of the tracks and analyzed their chemical composition. Both types of data indicated that the tracks consisted of microtubules. Microtubules also appear to be required for movement of materials elsewhere in the cell. If experimental cells are treated with a drug that disrupts microtubules, the movement of vesicles from the rough ER to the Golgi apparatus is impaired.

The general message of these experiments is that transport vesicles move through the cell along microtubules. How? Do the tracks themselves move, like a conveyer belt, or are vesicles carried along on some sort of molecular truck?

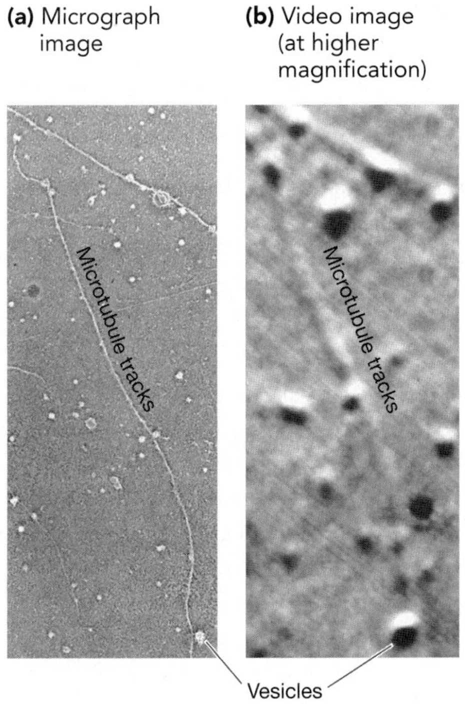
(a) Micrograph image
(b) Video image (at higher magnification)
Microtubule tracks
Microtubule tracks
Vesicles

FIGURE 7.36 Vesicles Move along Microtubule Tracks
Transport vesicles moving along microtubules. The images are of extruded cytoplasm from a squid giant axon. **(a)** An electron micrograph that allowed researchers to measure the diameter of the filaments and confirm that they are microtubules. **(b)** A slightly fuzzy but higher-magnification videomicroscope image, in which researchers actually watched vesicles move.

A Motor Protein Generates Motile Forces To study the way vesicles move along microtubules, Vale's group set out to tear the squid axon's transport system apart and then put it back together. To begin, they assembled microtubule fibers from purified α-tubulin and β-tubulin. Then they used differential centrifugation to isolate transport vesicles. But when they mixed purified microtubules and vesicles with ATP, no transport occurred. Something had been left out—but what?

To find the missing element or elements, the researchers purified one subcellular part after another, using differential centrifugation, and added it to the microtubule + vesicle + ATP system. Through trial and error, they found something that triggered movement. After further purification steps, the researchers finally succeeded in isolating a protein that generated vesicle movement. They named the molecule **kinesin**, from the Greek word *kinein* ("to move"). Like myosin, kinesin is a **motor protein** that converts chemical energy in ATP into mechanical work, just as a car's motor converts chemical energy in gasoline.

Biologists began to understand how kinesin works when X-ray diffraction studies similar to those that revealed the helical nature of DNA revealed the three-dimensional structure of kinesin. As **Figure 7.37a** shows, the protein consists of two intertwined polypeptide chains. It has three major regions: a head section with two globular pieces, a tail, and a stalk that connects the head and tail. Follow-up studies confirmed that the two globular components of the head bind to the microtubule. The tail region binds to the transport vesicle. The kinesin molecule is like a delivery person who carries transport vesicles along microtubule tracks. Cells contain a number of different kinesin proteins, each specialized for carrying a different type of vesicle.

How does kinesin move? More detailed studies of this protein's structure indicated that each of the globular components of the molecule's head has a site for binding ATP as well as a site that binds to the microtubule. To pull these observations together, biologists propose that kinesin transports vesicles by "walking" along a microtubule. The idea is that each part of the head region undergoes a conformational change when it binds ATP. As Chapter 3 showed, these types of shape changes often alter the activity of a protein. As **Figure 7.37b** shows, the ATP-dependent conformational change in kinesin results in a step forward. As each head

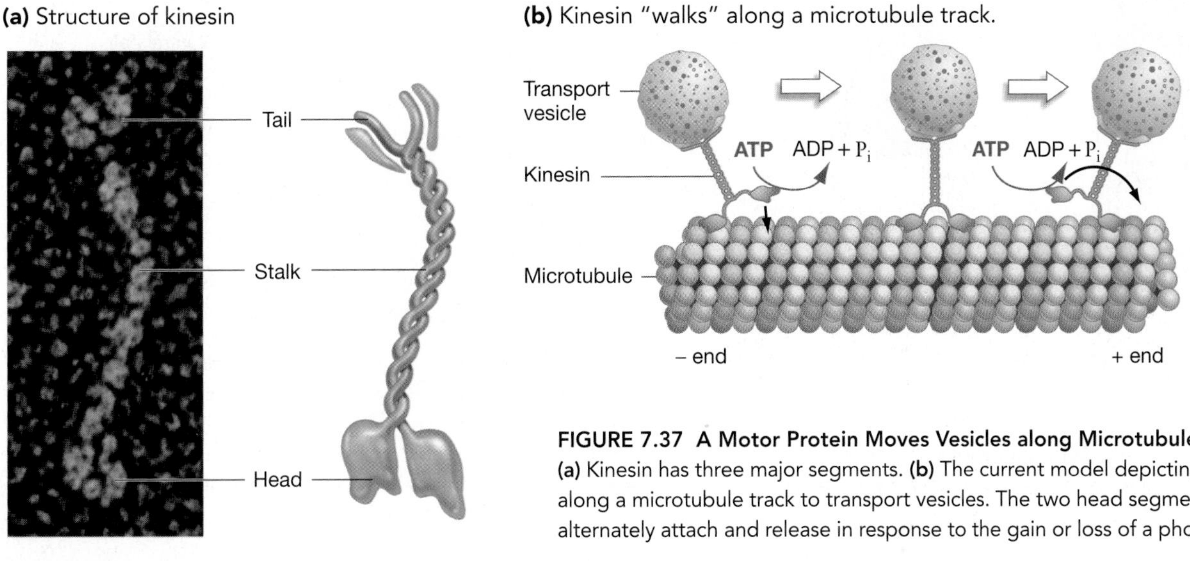
(a) Structure of kinesin
Tail
Stalk
Head
(b) Kinesin "walks" along a microtubule track.
Transport vesicle
Kinesin
Microtubule
ATP
ADP + P_i
ATP
ADP + P_i
– end
+ end

FIGURE 7.37 A Motor Protein Moves Vesicles along Microtubules
(a) Kinesin has three major segments. **(b)** The current model depicting how kinesin "walks" along a microtubule track to transport vesicles. The two head segments act like feet that alternately attach and release in response to the gain or loss of a phosphate group.

alternately binds and hydrolyzes ATP, the protein and its cargo move down the microtubule track.

In short, kinesins move molecular cargo to destinations throughout the cell. They are not the only type of motor protein active inside cells, however. Recall that myosin causes actin filaments to slide, resulting in the movement of cells or cytoplasm. Myosin is also involved in the movement of organelles along tracks made of actin. And a third motor protein, *dynein*, powers the transport of certain organelles as well as swimming movements that move the entire cell. Let's take a closer look at how cells swim.

Cilia and Flagella: Moving the Entire Cell

Flagella are long hairlike projections from the cell surface that function in movement. Flagella are found in many bacteria and eukaryotes. The structure of flagella is completely different in the two groups, however. Bacterial flagella are made of a protein called flagellin; eukaryotic flagella are constructed from microtubules (tubulin). Bacterial flagella move the cell by rotating like a ship's propeller; eukaryotic flagella move the cell by undulating. Eukaryotic flagella are surrounded by plasma membrane; bacterial flagella are not. Based on these observations, biologists conclude that the two structures evolved independently—even though their function is similar.

To understand how cells move, we'll focus on eukaryotic flagella. Eukaryotic flagella are closely related to structures called **cilia** (singular: *cilium*), which are short filamentous projections that are also found in some eukaryotic cells. Unicellular eukaryotes may have either flagella or cilia, while some multicellular organisms have both. In humans, for example, the cells that line the respiratory tract have cilia; sperm cells have flagella.

Flagella are generally longer than cilia, and cells typically have just one or two flagella but many cilia (**Figure 7.38**). But when researchers examined the two structures with the electron microscope, they found that their underlying organization is identical.

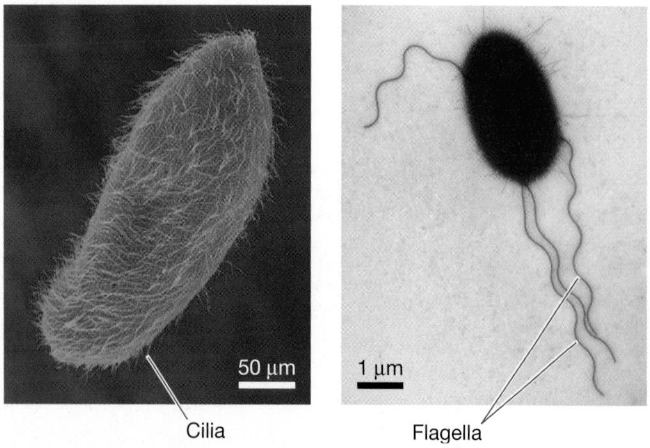
50 µm
1 µm
Cilia
Flagella

FIGURE 7.38 Cilia and Flagella Differ in Length and Number
Cilia are relatively short and large in number; flagella are relatively long and few in number.

How Are Cilia and Flagella Constructed? In the 1950s, anatomical studies established that both cilia and flagella have a characteristic "9 + 2" arrangement of microtubules. As **Figure 7.39a** shows, nine microtubule pairs, or doublets, surround two central microtubules. The doublets, consisting of one complete and one incomplete microtubule, are arranged around the periphery of the structure. The entire 9 + 2 structure is called the **axoneme** ("axle-thread"). The axoneme attaches to the cell at a structure called the **basal body**. The basal body is derived from the centrioles found inside the centrosome. The basal body has a "9 + 0" arrangement of microtubules and plays a central role in the growth of the axoneme.

As electron microscopy improved, biologists gained a more detailed view of the structure. As the sketch in **Figure 7.39b** illustrates, spoke-like structures connect each doublet to the central pair of microtubules. In addition, molecular bridges connect the nine doublets to one another. Finally, each of the doublets has a set of arms that project toward an adjacent doublet. Microtubules are complex. How do their components interact to generate motion?

(a) TEM of axoneme

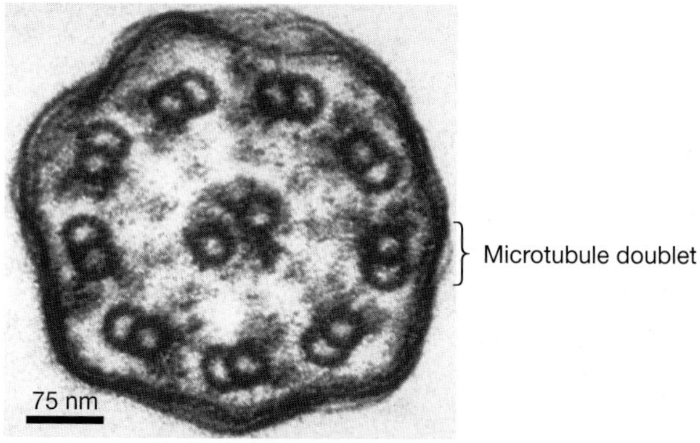
75 nm
Microtubule doublet

(b) Diagram of axoneme

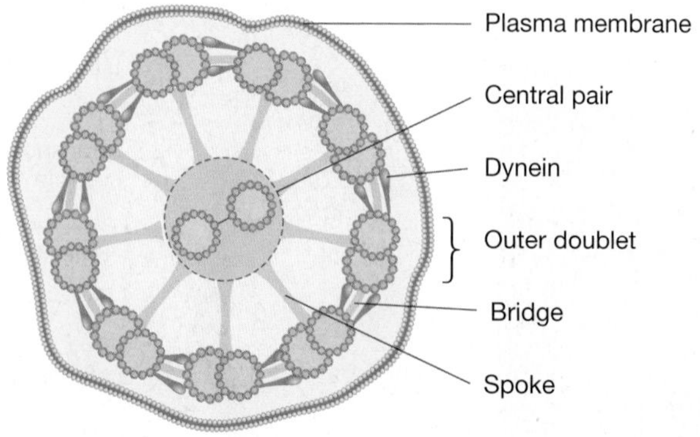
Plasma membrane
Central pair
Dynein
Outer doublet
Bridge
Spoke

FIGURE 7.39 The Structure of Cilia and Flagella
(a) Transmission electron micrograph of a cross section through an axoneme. **(b)** The major structural elements in cilia and flagella. The microtubules are connected by bridges and spokes, and the entire structure is surrounded by the plasma membrane. **EXERCISE** Label the "9 + 2" arrangement of microtubules

A Motor Protein in the Axoneme In the 1960s Ian Gibbons began studying the cilia of a common unicellular eukaryote called *Tetrahymena*, which lives in pond water. Gibbons found that by using a detergent to remove the plasma membrane that surrounds cilia and then subjecting the resulting solution to differential centrifugation, he could isolate axonemes. Further, the isolated structures would beat if Gibbons supplied them with ATP. These results confirmed that the beating of cilia is an energy-demanding process. They also provided Gibbons with a cell-free system for exploring the molecular mechanism of movement.

In an early experiment with isolated axonemes, Gibbons treated the structures with a molecule that affects the ability of proteins to bind to one another. The axonemes that resulted from this treatment could not bend or use ATP. When Gibbons examined them in the electron microscope, he found that the arms had fallen off. This observation led to the hypothesis that the arms are required for movement. Follow-up work showed that the arms are made of a large protein that Gibbons named **dynein** (from the Greek word *dyne*, meaning "force").

Like myosin and kinesin, dynein is a motor protein. Structural and chemical studies have shown that dynein undergoes a conformational change when a phosphate group from ATP attaches to it. More specifically, the end of a dynein molecule changes shape when it is phosphorylated. This shape change moves the molecule along the nearby microtubule. When the protein reattaches, it has succeeded in walking up the microtubule. This walking motion allows the microtubule doublets to slide past one another. But because each of the nine doublets in the axoneme is connected to the central pair of microtubules by a spoke, and because all of the doublets are connected to each other by molecular bridges, the sliding motion is constrained. So if dynein arms on just one side of the axoneme walk while those on the other side are at rest, the result of the constrained, localized movement is bending (**Figure 7.40**). The result of the bending of cilia or flagella is a swimming motion.

Scaled for size, flagellar-powered swimming can be rapid. In terms of the number of body or cell lengths traveled per second, a sperm cell from a bull moves faster than a human world-record-holder does when swimming freestyle. At the level of the cell, life is fast paced.

✓CHECK YOUR UNDERSTANDING

Each component of the cytoskeleton has a unique structure and set of functions. In addition to providing structural support, actin filaments and microtubules work in conjunction with motor proteins to move the cell or materials inside the cell. Intermediate filaments provide structural support. Most elements of the cytoskeleton are dynamic—they grow and shrink over time. You should be able to predict what will happen when experimental cells are treated with drugs that inhibit formation of each type of cytoskeletal filament.

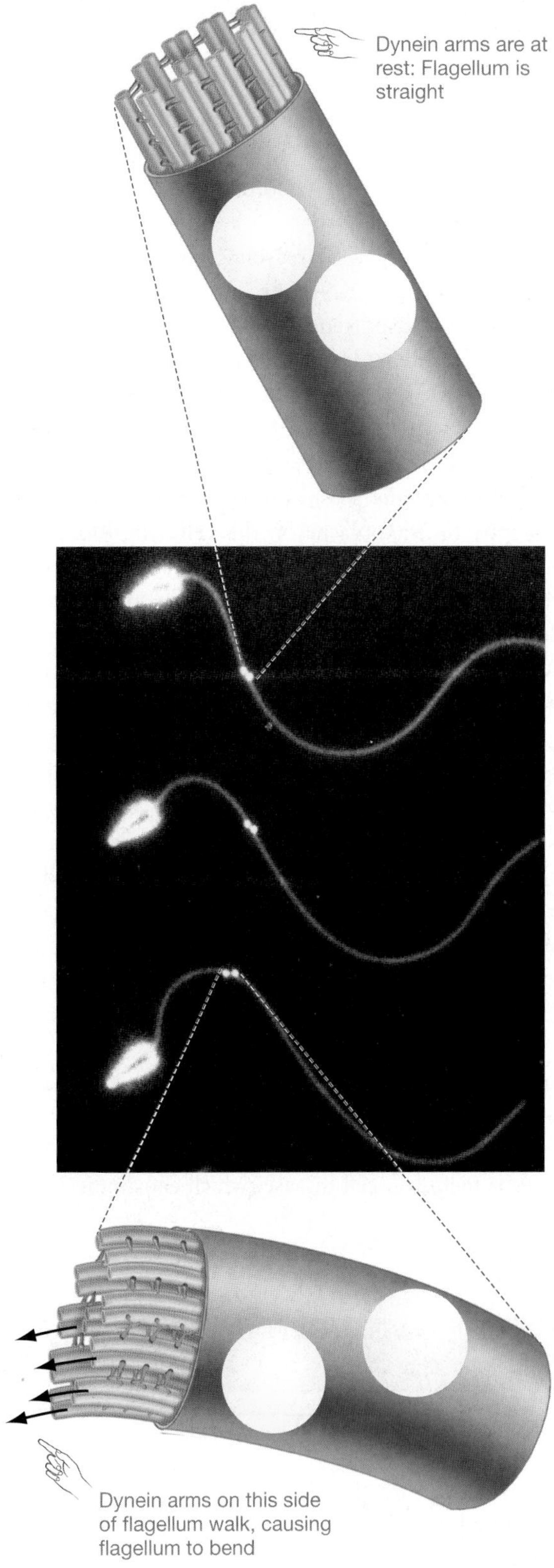
Dynein arms are at rest: Flagellum is straight
Dynein arms on this side of flagellum walk, causing flagellum to bend

FIGURE 7.40 How Do Flagella Bend?
Researchers attached a pair of gold beads to a flagellum and photographed its movement over a short time sequence. As the flagellum bends and beats back and forth, the sperm cell swims forward. When dynein arms walk along the microtubule doublets on one side of a flagellum, the structure bends.

ESSAY Organelles and Human Disease

What happens when organelles malfunction? Given the importance of organelles in the function of eukaryotic cells, it's not surprising that abnormalities in organelles cause disease. In humans, some of the best-understood organelle-based diseases involve lysosomes.

Lysosomes are the cell's recycling center. Defunct organelles and old or damaged RNA, protein, carbohydrate, and lipid molecules are transported to lysosomes. Enzymes inside these organelles then degrade the molecules to simpler components that can be reused. It is essential that these degradative enzymes be confined to the interior of the lysosome. If they escaped into the cytosol, they would threaten to destroy the cell's contents.

Several human diseases result from leakage of lysosomal enzymes into the cytoplasm or the cell exterior. Rheumatoid arthritis is an inflammatory joint disease that is partly caused by the leakage of lysosomal enzymes into the joint fluid. The debilitating lung disease called asbestosis also results from enzyme leakage, even though the population of cells involved is different. Asbestosis is observed in asbestos miners and in construction workers who handled large quantities of asbestos-containing insulation and fireproofing material during their careers. When these individuals breathed asbestos fibers into their lungs, the fibers were localized to the lysosomes of lung cells. But because lysosomal enzymes cannot degrade asbestos fibers, the fibers gradually accumulate inside the organelles. Eventually the buildup of fibers results in damage to the lysosomal membrane and leakage of enzymes into the cell cytoplasm. When many lung cells are harmed in this way, the result is severe coughing and shortness of breath.

In addition to disorders caused by enzyme leakage, about 40 human illnesses are caused by deficiencies in specific lysosomal enzymes. The most severe of these so-called lysosomal storage diseases is inclusion-cell disease, which causes facial and skeletal abnormalities as well as mental retardation. Inclusion-cell disease occurs when most of the degradative enzymes found in lysosomes are missing. Their absence causes the organelles to swell with undigested materials, resulting in structures called inclusions. The swollen lysosomes ultimately cause cell damage, leading to disease symptoms.

What happens when organelles malfunction?

Inclusion-cell disease is caused by a deficiency in a single enzyme—the one that is required for attaching mannose-6-phosphate to proteins. Recall from Section 7.3 that this phosphorylated sugar serves as the zip code that targets proteins to lysosomes from the Golgi apparatus. In the absence of the mannose-6-phosphate zip code, enzymes that are normally shipped to lysosomes are instead secreted from the cell. In fact, researchers originally discovered that mannose-6-phosphate serves as a zip code by comparing lysosomal proteins from healthy individuals with the same proteins secreted by individuals with inclusion-cell disease.

Tay-Sachs disease is another lysosomal storage disease. Unlike inclusion-cell disease, Tay-Sachs results from the absence of one particular enzyme from lysosomes. In normal individuals, this protein degrades a type of glycolipid that is abundant in brain cells. In Tay-Sachs patients, the glycolipid accumulates in these cells and disrupts their function. Symptoms include rapid mental deterioration after about 6 months of age, followed by paralysis and death within 3 years.

Although most organelle diseases cannot yet be cured, drug therapies are beginning to offer some hope. For example, the symptoms associated with certain lysosomal storage diseases show improvement when the normal enzyme is provided in pill form. Research continues on this and other strategies for alleviating the symptoms caused by dysfunctional organelles.

CHAPTER REVIEW

Summary of Key Concepts

- **The structure of cell components is closely correlated with their function.**

Because all organisms consist of cells, many questions in biology can be answered by understanding the structure and function of cells and cell components. There are two basic cellular designs: prokaryotic and eukaryotic. Eukaryotic cells are usually much larger and more structurally complex than prokaryotic cells. Prokaryotic cells consist of a single membrane-bound compartment in which nearly all cellular functions occur. Eukaryotic cells contain numerous membrane-bound compartments called organelles. Organelles allow eukaryotic cells to compartmentalize functions and grow to a large size.

Eukaryotic organelles are specialized for carrying out different functions, and their structure is often correlated closely with their function. Mitochondria and chloroplasts have extensive internal membrane systems, where the enzyme machines responsible for ATP generation and photosynthesis reside. Rough ER is named for the ribosomes that attach to it. Ribosomes are protein-making machines, and rough ER is a site for protein synthesis and processing.

Smooth ER lacks ribosomes because it is a center for lipid synthesis and processing.

Inside cells, materials are transported to their destinations with the help of molecular "zip codes."

The defining organelle of eukaryotic cells is the nucleus, which contains the cell's chromosomes and serves as its control center. For a cell to function properly, the movement of molecules into and out of the nucleus must be carefully controlled. Traffic across the nuclear envelope occurs through nuclear pores, which contain a multiprotein nuclear pore complex that serves as gatekeeper. Both passive and active transports of materials occur through these nuclear pore complexes. Active import and export of proteins and RNAs involves built-in signals that target cargo to the correct compartment.

The endomembrane system is an extensive, interconnected system of membranes and membrane-bound compartments that can extend from the nucleus to the plasma membrane. Two principal organelles in the endomembrane system are the endoplasmic reticulum (ER) and the Golgi apparatus. The ER is the site of synthesis for a wide array of proteins and lipids. Most ER products are shipped to the Golgi apparatus, which serves as a processing and dispatching station. In many proteins the major processing step is glycosylation, or the addition of carbohydrate groups.

The movement of materials through the endomembrane system is highly organized and takes place inside membrane-bound transport organelles called vesicles. Prior to products leaving the endomembrane system, they are sorted with molecular zip codes that direct them to vesicles headed for their final destination. The vesicles contain proteins that interact with receptor proteins on the surface of a target organelle or the plasma membrane, and allow the contents to be delivered.

Web Tutorial 7.1 Transport into the Nucleus

Web Tutorial 7.2 A Pulse-Chase Experiment

Cells are dynamic. Thousands of chemical reactions occur each second within cells; molecules constantly enter and exit across the plasma membrane; cell products are shipped along protein fibers; and elements of the cell's internal skeleton grow and shrink.

The cell is a membrane-bound structure with a highly organized, dynamic interior. Inside the cell, thousands of different chemical reactions take place at incredible speeds. The products of these chemical reactions allow the cell to acquire resources from the environment, synthesize additional molecules, dispose of wastes, and reproduce.

The cytoskeleton is an extensive system of fibers that serves as a structural support for eukaryotic cells. Elements of the cytoskeleton also provide the machinery for moving vesicles inside cells and for moving the cell as a whole through the beating of flagella or cilia, or cell crawling. Both cell motility and the movement of vesicles inside cells depend on motor proteins, which can convert chemical energy stored in ATP into movement. Within the cell, movement of transport vesicles occurs as the motor protein kinesin "walks" along microtubule tracks. Cilia and flagella bend as the motor protein dynein "walks" along microtubule tracks. The bending motion allows these structures to beat back and forth, enabling cells to swim or generate water currents.

The data reviewed in this chapter provide a view of the cell as a dynamic reaction vessel that synthesizes and ships an array of products in a highly regulated manner. How does all this activity inside the cell relate to what is going on outside? This is the issue taken up in Chapter 8.

Questions

Content Review

1. Which of the following best describe the nuclear envelope?
   a. It is continuous with the endomembrane system.
   b. It is continuous with the nucleolus.
   c. It is continuous with the plasma membrane.
   d. It contains a single membrane and nuclear pores.

2. What is a nuclear localization signal?
   a. a stretch of amino acids that directs proteins from the nucleus to the ER
   b. a molecule that is attached to nuclear proteins so that they are retained inside the nucleus
   c. a signal built into a protein that directs it to the nucleus
   d. a component of the nuclear pore complex

3. Which of the following is not true of secreted proteins?
   a. They are synthesized in ribosomes.
   b. They are transported through the endomembrane system in membrane-bound transport organelles.
   c. They are transported from the Golgi apparatus to the ER.
   d. They contain a signal sequence that directs them into the ER.

4. To find the nuclear localization signal in the protein nucleoplasmin, researchers separated the molecule's core and tail segments, labeled both with a radioactive atom, and injected them into the cytoplasm. Why did the researchers conclude that the signal is in the tail region of the protein?
   a. The protein reassembled and folded into its normal shape spontaneously.
   b. Only the tail segments appeared in the nucleus.
   c. With a confocal microscope, tail segments were clearly visible in the nucleus.
   d. The tail and head segments appeared together in the nucleus.

5. Molecular zip codes direct molecules to particular destinations in the cell. How are these signals read?
   a. They bind to receptor proteins.
   b. They enter transport vesicles.
   c. They bind to motor proteins.
   d. They are glycosylated by enzymes in the Golgi apparatus.

6. The number and size of organelles in a cell correlates with that cell's function. Propose a function for cells that contain extensive rough ER.
   a. rapid cell division in growing bones or muscle tissues
   b. production and processing of fatty acids and other lipids
   c. movement via cell crawling
   d. production of proteins that are secreted from the cell

Conceptual Review

1. Compare and contrast the structure of a generalized plant cell, animal cell, and prokaryotic cell. Which features are common to all cells? Which are specific to certain lineages?
2. Draw a diagram that traces the movement of a secreted protein from its site of synthesis to the outside of a eukaryotic cell. Identify all of the organelles that the protein passes through, and indicate the direction of movement.
3. Describe how a motor protein such as kinesin can move a transport vesicle down a microtubule track. Include all necessary steps and components.
4. Describe the logic of a pulse-chase experiment. How was this approach used to document the pattern of protein transport through the endomembrane system?
5. Briefly describe how researchers use centrifugation to isolate particular cell components for further study.
6. Compare and contrast the structure and function of actin filaments, intermediate filaments, and microtubules. Why is it misleading to refer to the cytoskeleton as "scaffolding"?

Group Discussion Problems

1. In addition to delivering cellular products to specific organelles, eukaryotic cells can take up material from the outside and transport it to specific organelles. For example, specialized cells of the human immune system ingest bacteria and viruses and then deliver them to lysosomes for degradation. Suggest a hypothesis for how this material is tagged and directed to lysosomes. How would you test this hypothesis?
2. The leading hypothesis to explain the origin of the nuclear envelope is that a deep infolding of the plasma membrane occurred in an ancient prokaryote. Draw a diagram that illustrates this infolding hypothesis. Does your model explain the existence of the structure's inner and outer membranes? Explain.
3. Propose a function for cells that contain (a) a large number of lysosomes, (b) a particularly extensive cell wall, and (c) many peroxisomes.
4. Suggest a hypothesis or a series of hypotheses to explain why bacteria, archaea, algae, and plants have cell walls. Suppose that mutant individuals from each group lacked a cell wall. How could you use these individuals to test your idea(s)?

Answers to Multiple-Choice Questions: 1. a; **2.** c; **3.** c; **4.** b; **5.** a; **6.** d

www.prenhall.com/freeman is your resource for the following: Web Tutorials; Online Quizzes and other Online Study Guide materials; Answers to Conceptual Review Questions; Solutions to Group Discussion Problems; Answers to Figure Caption Questions and Exercises; and Additional Readings and Research.

Cell-Cell Interactions

8

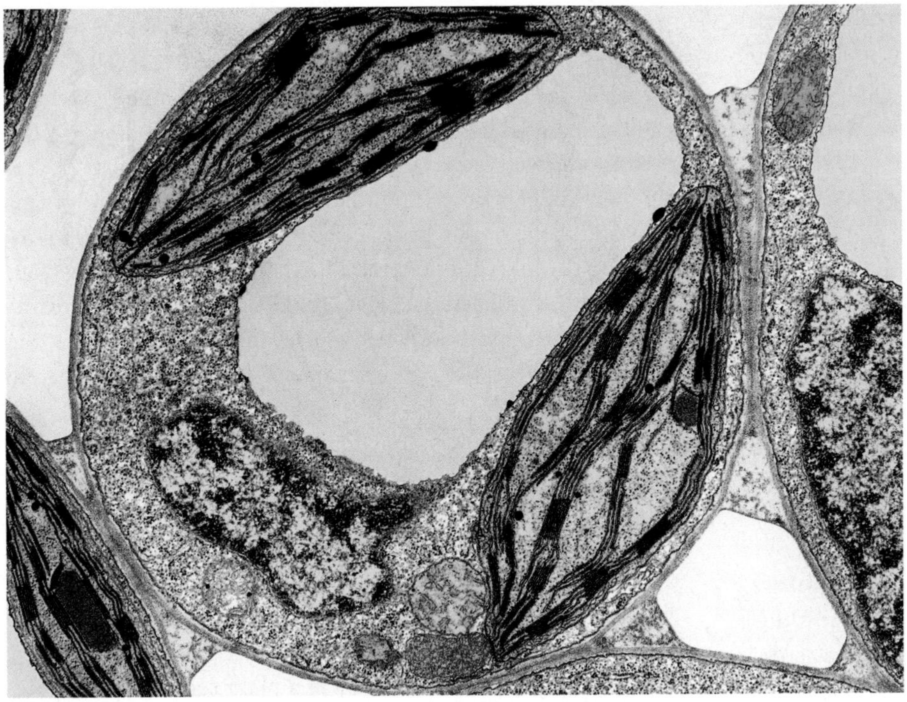

Transmission electron micrograph of the plasma membranes and cell walls (colored yellow) of several plant cells. The cell surface is where adjacent cells join and where signals from distant cells arrive.

KEY CONCEPTS

- Extracellular material strengthens cells and helps bind them together.
- Cell-cell connections help adjacent cells adhere. Cell-cell gaps allow adjacent cells to communicate.
- Intercellular signals are responsible for creating an integrated whole from many thousands of independent parts.

Chapter 6 introduced the structure and function of the plasma membrane, which is the defining feature of the cell. Chapter 7 surveyed the organelles, molecular machines, and cytoskeletal elements that fill the space inside that membrane and explored how cargo moves from sources to destinations within the cell. Both chapters highlighted the breathtaking speed and diversity of events that take place at the cellular level.

The cell is clearly a bustling enterprise. But it would be a mistake to think that cells are self-contained—that they are worlds in and of themselves. Instead, cells constantly interact with other cells. They continuously adjust their activities in response to stimulation from other cells and to changes in environmental conditions.

To understand the life of a cell thoroughly, it is critical to analyze how the cell interacts with the world outside. How do cells obtain information about the outside world and respond to that information? In particular, how do cells interact with nearby cells?

For most unicellular species, the outside environment is made up of soil or water that is teeming with other organisms, either of the same or different species. In your gut, hundreds of billions of bacterial cells are jostling for space and resources. Similar numbers of single-celled organisms are found in every tablespoon of good-quality soil. In addition to interacting with these individuals, each unicellular organism must contend with constantly shifting aspects of the physical environment, such as heat, light, ion concentrations, and food supplies. If unicellular organisms are unable to sense these conditions and respond appropriately, they die.

In multicellular species, the environment outside the cell is made up of other cells, both neighboring and distant. The cells that make up a redwood tree, an *Amanita* mushroom, or your body are intensely social. Although biologists often study cells in isolation, an individual tree, fungus, or person is actually an interdependent community of cells. If those cells do not communicate and cooperate, the whole will break into dysfunctional parts and die.

To introduce the ways that cells interact with each other, this chapter focuses on how cells in a multicellular organism communicate. Let's begin by looking at the plasma membrane and analyzing the molecules involved in cell-cell interactions.

8.1 The Cell Surface

Chapter 6 introduced the currently accepted model for the structure of the plasma membrane. The fluid-mosaic hypothesis contends that the plasma membrane is a phospholipid bilayer with interspersed proteins (**Figure 8.1**). Data from electron micrographs and chemical treatments suggest that these membrane proteins can be either *integral*, meaning that they are embedded in the bilayer, or *peripheral*, meaning that they are attached to one surface. Recall that if an integral protein spans the entire membrane, it is termed a transmembrane protein. Membrane proteins may float free in the lipid bilayer or be anchored by connections to the cytoskeleton or materials outside the cell.

Chapter 6 also analyzed the primary function of the plasma membrane: to create an environment inside the cell that is different from conditions outside. Ions and molecules move across plasma membranes by direct diffusion through the phospholipid bilayer or by means of several types of membrane proteins. Transport of materials across the membrane can be energy demanding or passive, but it is always selective.

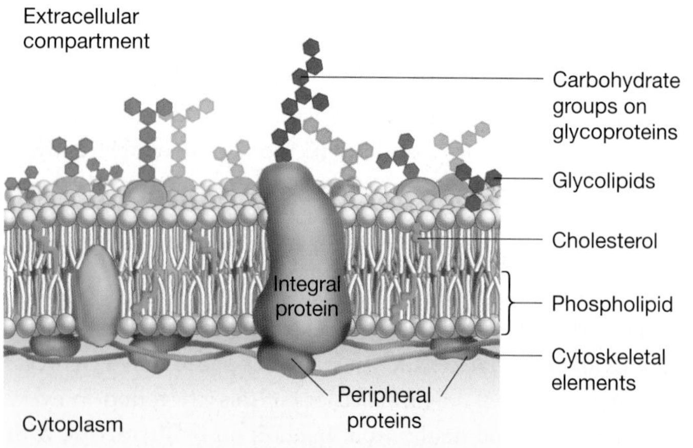

FIGURE 8.1 The Plasma Membrane Is a Mosaic of Phospholipids and Proteins
According to the fluid-mosaic model, the plasma membrane consists of integral and peripheral proteins scattered throughout a phospholipid bilayer.

This picture—of a dynamic, complex plasma membrane that selectively admits or blocks passage of specific substances—is accurate but not complete. The plasma membrane does not exist in isolation. Cytoskeletal elements introduced in Chapter 7 attach to the interior face of the bilayer, and a complex array of extracellular structures exists outside. Let's consider the nature of the material outside the cell first and then analyze how the cell interacts with other cells.

The Structure and Function of an Extracellular Layer

In species from across the tree of life, it is extremely rare for cells to be bounded simply by a plasma membrane. Most cells secrete a layer or wall that forms just beyond the membrane. The extracellular material helps define the cell's shape and either attaches it to another cell or acts as a first line of defense against the outside world. This observation holds for both unicellular organisms and the various types of cells in multicellular species.

Virtually all types of extracellular structures, in turn—from the cell walls of bacteria, algae, fungi, and plants to the extracellular matrix that surrounds most animal cells—follow the same fundamental design principle. Like reinforced concrete and fiberglass, they are "fiber composites": They consist of a cross-linked network of long filaments embedded in a stiff surrounding material, or ground substance (**Figure 8.2a**). The molecules that make up the rods and the encasing material vary from group to group, but the engineering principle is the same.

Fiber composites are a successful design because rods and filaments are extremely effective at withstanding stretching and straining forces, or tension. The steel rods in reinforced concrete and the cellulose fibers in a plant cell wall are unlikely to break as a result of being pulled or pushed lengthwise. Scaled for size, in fact, steel and cellulose are *equally* unlikely to break. Cellulose fibers have the same tensile strength as steel. In addition, the stiff surrounding substance is effective at withstanding the pressing forces called compression. Concrete performs this function in highways, and a gelatinous mixture of polysaccharides achieves the same end in plant cell walls.

Thanks to the combination of tension- and compression-resisting elements, fiber composites are particularly rugged. And in many living cells, the fiber and composite elements are flexible as well as sturdy. When this is the case, the extracellular material is both supple and strong.

What molecules make up the rods and the ground substance found on the surface of plant and animal cells? How are these extracellular layers synthesized, and what do they do?

The Plant Cell Wall

Most plant cells are surrounded by a layer of extracellular material called a cell wall. When new cells first form, they secrete an initial fiber composite designated a **primary cell wall.** As indicated in Chapter 5, the fibrous component of the cell wall consists of long strands of the polysaccharide cellulose, which is cross-linked by polysaccharide filaments and bundled into stout filaments termed microfibrils. Microfibrils form a crisscrossed network that becomes filled with gelatinous polysaccharides (**Figure 8.2b**). Chief among these gel-forming carbohydrates are **pectins**—the molecules that are used to thicken jams and jellies. Pectins are not as strong as cellulose, because the polysaccharides that make up their structure do not form tightly packed filaments strengthened with cross-linkages. Instead, the polysaccharides in pectin are hydrophilic. They attract and hold large amounts of water, forming a gel that helps keep the cell wall moist.

Pectins and other gelatinous polysaccharides that form the ground substance of the cell wall are synthesized in the rough endoplasmic reticulum (ER) and Golgi apparatus and secreted to the extracellular space. The cellulose microfibrils, in contrast, are synthesized by a complex of enzymes in the plasma membrane itself and "spun out" into the exterior.

The primary cell wall defines the shape of a plant cell. Under normal conditions, cytoplasm fills the entire volume of the cell and pushes the plasma membrane up against the wall. Because the concentration of solutes is higher inside the cell than outside, water tends to enter the cell via osmosis. The incoming water inflates the plasma membrane, exerting a force against the wall that is known as **turgor pressure**. Although plant cells experience turgor pressure throughout their lives, it is particularly important in young cells that are actively growing. Young plant cells secrete enzymes named *expansins* into their cell-wall matrix. Expansins catalyze reactions that allow the cellulose microfibrils in the matrix to slide past one another. Turgor pressure then forces the wall to elongate and expand. The result is cell growth.

As certain plant cells mature and stop growing, they secrete a layer of material—designated a **secondary cell wall**—inside the primary cell wall. The structure of the secondary cell wall varies from cell to cell in the plant and correlates with that cell's function. For example, cells on the surface of a leaf have cell walls that are impregnated with waxes that form a waterproof coating. Cells that furnish structural support for the plant's stem have secondary cell walls that contain a great deal of cellulose. In some species, the secondary cell walls of support cells include a complex and extremely tough substance named **lignin**. Lignin forms a rigid network that is exceptionally stiff and

(a) Fiber composites resist tension and compression.

(b) Primary cell walls of plants are fiber composites.

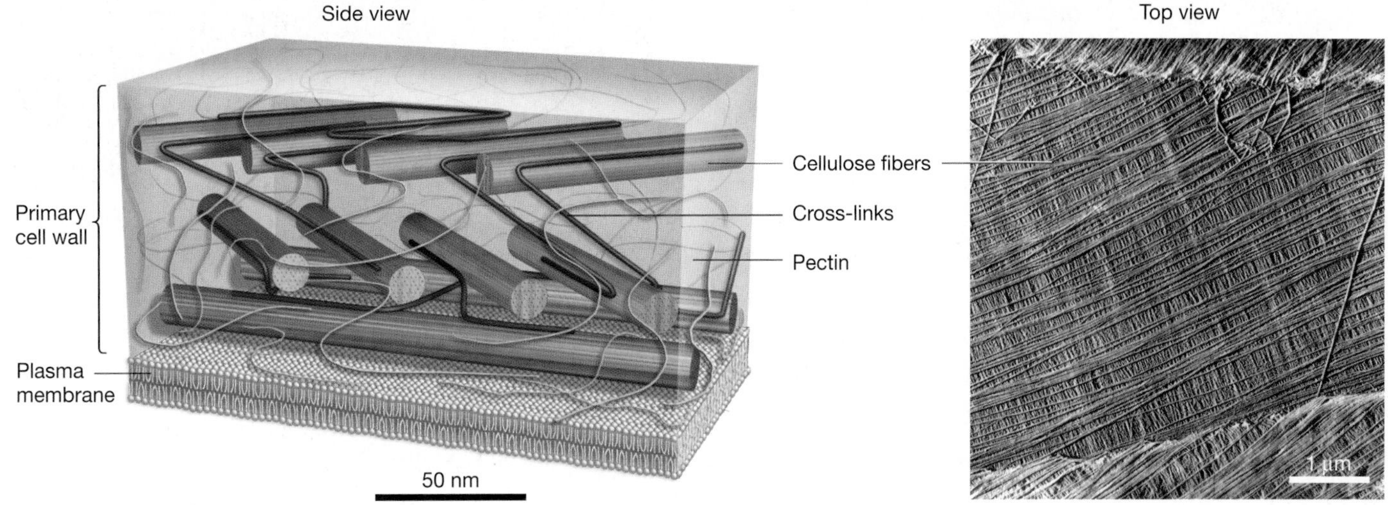
Side view
Top view
Primary cell wall
Plasma membrane
Cellulose fibers
Cross-links
Pectin
50 nm
1 µm

FIGURE 8.2 What Is the Nature of a Fiber Composite?
(a) Fiber composites consist of a massive ground substance that fills spaces between cross-linked rods. In reinforced concrete, the rods or fibers are made of steel, while the ground substance is concrete. The finished material is strong because the fibers resist tension and the ground substance resists compression. **(b)** In the primary plant cell wall, cellulose fibers are cross-linked by polysaccharide chains. The spaces between the fibers are filled with pectin molecules, which form a gelatinous solid.

strong. Cells that form wood secrete a secondary cell wall that contains large amounts of lignin. Cells that have thick cell walls of cellulose and lignin help plants withstand the forces of gravity and wind.

Plant cell walls are also dynamic. If they are damaged by insect attack, they may release signaling molecules that trigger the reinforcement of walls in nearby cells. When a seed germinates (sprouts), the walls of cells that store oils or starch are actively degraded, releasing these nutrients to the growing plant. Cell walls are also degraded in a controlled way as fruits ripen, making the fruits softer and more digestible for the animals that disperse the seeds inside.

Although animal cells do not make a cell wall, they do form a fiber composite outside their plasma membrane. What is this substance, and what does it do?

The Extracellular Matrix in Animals

Virtually all animal cells secrete a fiber composite called the **extracellular matrix (ECM)**. Although the design of this substance follows the same principles observed in the cell walls of bacteria, archaea, algae, fungi, and plants, the fibrous component of the animal ECM consists of protein fibers instead of polysaccharide filaments. **Figure 8.3a** illustrates the most abundant protein found in the ECM—a fibrous, cable-like molecule termed **collagen**. As in cell walls, the matrix that surrounds collagen and the other fibrous components consists of gel-forming polysaccharides. But because collagen and the other common ECM proteins are much more elastic and bendable than cellulose or lignin, the structure as a whole is relatively pliable. Most ECM components are synthesized in the rough ER, processed in the Golgi apparatus, and secreted from the cell via exocytosis. Some of the polysaccharides that form the composite material are synthesized by membrane proteins, however.

The amount of ECM varies among different types of cells in the same organism. In human bone and cartilage, for example, the number of cells is relatively small and the amount of ECM very large (**Figure 8.3b**). Skin cells, in contrast, are packed together with a minimal amount of ECM between them. The composition of the ECM also varies among cell types. For example, the ECM of lung cells contains large amounts of a rubber-like protein called elastin, which allows the ECM to expand and contract during breathing movements. The structure of a cell's ECM correlates with that cell's function.

Wherever it is found in the body, one of the ECM's most important functions is structural support—a common theme in the extracellular materials found in other organisms. ECM provides support because the combination of protein filaments and surrounding polysaccharide gel creates a strong external layer, and because cells "grab" the layer via a chain of proteins. As **Figure 8.4** shows, actin filaments inside the plasma membrane are connected to transmembrane proteins called **integrins**. The integrins bind to nearby proteins in the ECM named **fibronectins**, which in turn bind to collagen fibers. This direct linkage between the cytoskeleton and ECM is critical. In addition to keeping individual cells in place, it helps adjacent cells adhere to each other via their common connection to the ECM.

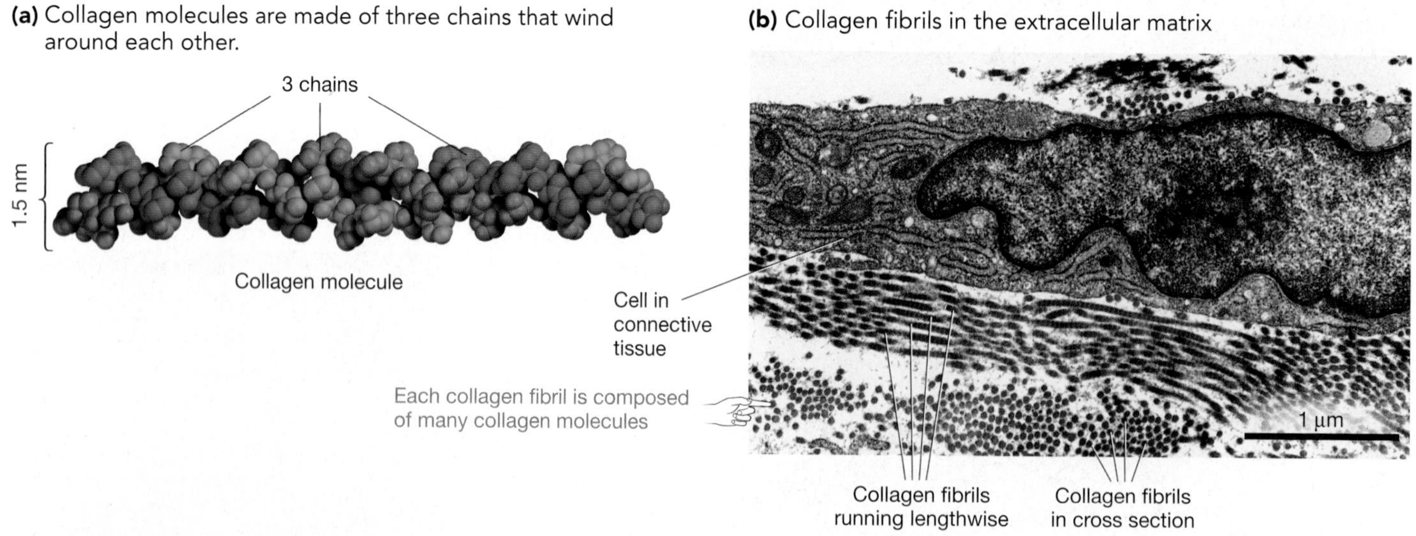
(a) Collagen molecules are made of three chains that wind around each other.
3 chains
1.5 nm
Collagen molecule
(b) Collagen fibrils in the extracellular matrix
Cell in connective tissue
Each collagen fibril is composed of many collagen molecules
1 μm
Collagen fibrils running lengthwise
Collagen fibrils in cross section

FIGURE 8.3 The Extracellular Matrix Is a Fiber Composite
(a) Although several types of protein filaments are found in the ECM of animal cells, the most abundant is collagen. A collagen molecule consists of three strands that wind around each other. Groups of collagen molecules coalesce to form collagen microfibrils, and bundles of microfibrils link up to form collagen fibers.
(b) A cross section of ECM from monkey cartilage tissue, showing a cell surrounded by abundant ECM. The spaces between the collagen fibers are filled with gelatinous polysaccharides.

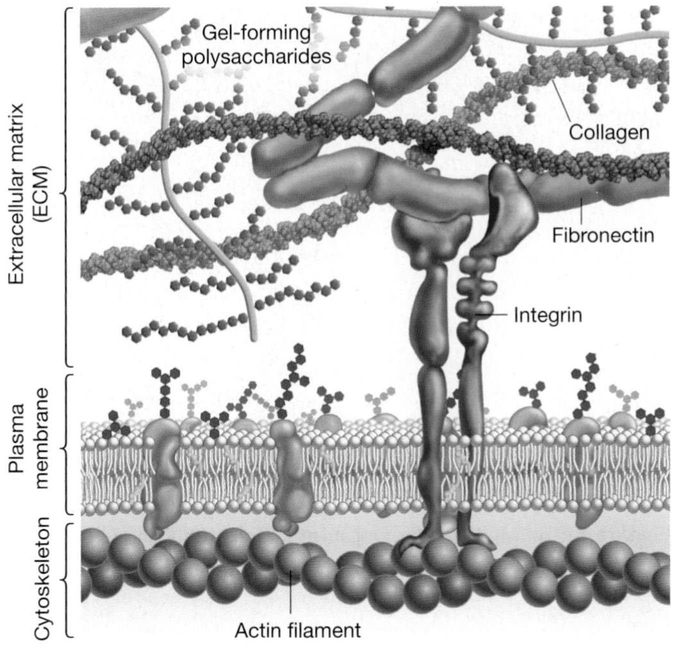

FIGURE 8.4 The Extracellular Matrix Connects to the Cytoskeleton
EXERCISE Circle the direct physical connection among actin filaments, integrins, fibronectin, and collagen fibers.

The cytoskeleton-ECM linkage is also important in the development of cancer. When the cell-ECM connection is lost—through mechanisms that are not well understood—cells that are growing in an uncontrolled fashion can leave the site of a tumor and begin migrating throughout the body. Eventually these cells settle in a new region of the body and seed the formation of new tumors. The result is the development of full-blown cancer. **Box 8.1** details other types of maladies that result from breakdown of the ECM.

✓CHECK YOUR UNDERSTANDING

Most cells secrete a layer of structural material that stiffens the cell and helps define its shape. Although the types of molecules involved vary among species, in almost all cases the extracellular material is a fiber composite—a combination of cross-linked filaments surrounded by a ground substance. You should be able to (1) diagram the general structure of a fiber composite and (2) compare and contrast the molecular composition of a plant cell wall and the ECM of animal cells.

8.2 How Do Adjacent Cells Connect and Communicate?

To understand a cell fully, it is important to appreciate it as a social entity. Cells continually interact with other cells. This is true even for unicellular organisms, which usually live in habitats packed with other unicellular species. For example, the 500 species of bacteria and archaea that inhabit your mouth compete with each other for space and nutrients. Like many other prokaryotes that are adapted for life on a solid surface, certain oral bacteria secrete a hard, polysaccharide-rich substance that accumulates outside their cell walls. This substance creates a **biofilm** that encases the cells and attaches them to the surface.

BOX 8.1 What Happens When the Extracellular Matrix Is Defective?

Given the extent of ECM surrounding animal cells, it is not surprising that defects in this structure have serious consequences. Consider, for example, the effects of collagen breakdown. One of collagen's most common constituents is the amino acid proline. When collagen fibers first assemble in the ECM, many of the proline residues present in the newly synthesized structures are oxidized to hydroxyproline. This reaction depends on the molecule ascorbic acid, also called vitamin C. If vitamin C is lacking in the diet, hydroxyproline formation slows and collagen fibers begin to weaken or disintegrate. The ECM surrounding capillaries is particularly sensitive to this problem. Capillaries are the smallest blood vessels in the body. When collagen fibers in the ECM of capillary cells weaken, the structure as a whole begins to lose its integrity. Blood cells begin leaking out of capillaries, forming bruises. If vitamin C deficiency continues, capillaries weaken to the point where extensive internal bleeding occurs. The potentially fatal disease called *scurvy* results. Scurvy is readily cured by eating citrus fruits or other foods rich in vitamin C.

Skin is the other major organ that is sensitive to collagen damage. Skin cells are layered over a thick, collagen-rich ECM called the **basal lamina**. For reasons that are not completely clear, exposure to intense sunlight signals the cells to produce and secrete the enzyme metalloproteinase-1, which catalyzes a reaction that cleaves collagen fibers. As a result, people who work for long periods in bright sunlight or who cultivate a tanned skin are almost certain to develop what dermatologists called *photoaging*—a premature aging of the skin caused by the loss of collagen.

Defects in ECM components other than collagen can also cause problems. *Marfan syndrome* develops in people who inherit a genetic defect that prevents them from manufacturing the normal form of the protein fibrillin. Along with elastin, fibrillin is a key component of the elastic fibers in the ECM. People with Marfan syndrome tend to have exceptionally long fingers and be extremely tall and thin. They are also susceptible to severe heart problems. Based on his appearance, biomedical researchers have hypothesized that Abraham Lincoln may have had Marfan syndrome.

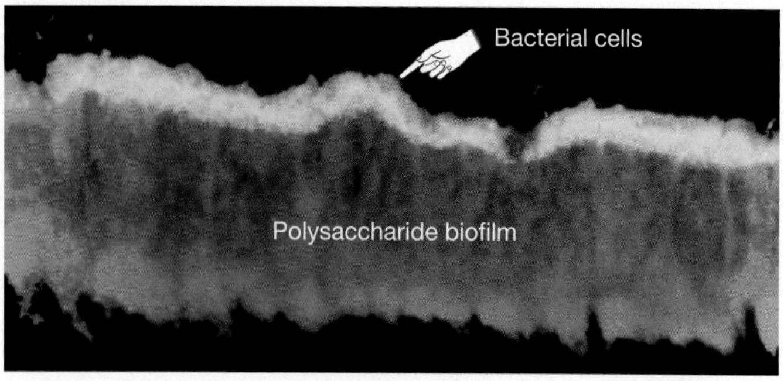
Bacterial cells
Polysaccharide biofilm

FIGURE 8.5 Unicellular Organisms May Form Aggregations
The yellow band consists of *Pseudomonas aeruginosa* cells—a common soil bacterium that also causes infections in cystic fibrosis patients. The red layer is a biofilm secreted by the cells.

Biofilms that form in the mouth are termed **dental plaque**, and they contain an array of species. In some cases, biofilm formation is actually a product of cooperation among members of the same species. When cells of the bacterium *Pseudomonas aeruginosa* settle on a lung cell, for instance, they secrete a signaling molecule that recruits other *P. aeruginosa* cells to the site. The cells then secrete a biofilm that attaches them to the surface of the lung and protects them from attack by your immune system cells or antibiotics (**Figure 8.5**).

Although unicellular species may live in close proximity and even communicate with each other, they do not make physical connections. Physical connections between cells are the basis of **multicellularity** (**Figure 8.6**). This should come as no surprise. The billions of cells that make up a rose bush, a bracket fungus, or a mountain gorilla must be held together somehow. And because cells in multicellular species are specialized for particular tasks, the cells must be able to communicate with each other to function as an integrated whole. It is logical, then, that rose and fungus and gorilla cells are organized into functional units, called tissues. **Tissues** consist of groups of similar cells that perform a similar function. In your body, individual muscle cells are grouped into muscle tissue that contracts and extends to make movement possible. Several tissues, in turn, combine to make up the integrated structures called **organs**. Organs perform specialized functions such as reproduction, digestion, or support. Flowers, roots, and stems are organs; your gonads, small intestine, and bones also are organs.

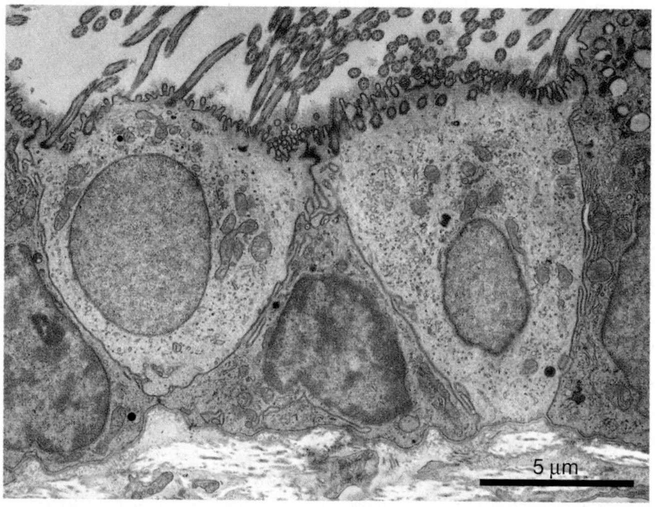
5 μm

FIGURE 8.6 In Multicellular Organisms, Cells Are Connected
A cross section through cells that line the human trachea.
EXERCISE Label two types of cells: ciliated cells that sweep debris from the trachea and cells that secrete mucus, which traps bacteria and viruses.

A look at the phylogeny in **Figure 8.7** should convince you that multicellularity has evolved numerous times, in an array of independent groups. Based on this observation, it is logical to expect that each multicellular group has different types of intercellular ("between-cell") connections and communication mechanisms to organize their tissues and organs.

Many of the structures responsible for cell-cell adhesion and communication in these groups were first described in the 1950s, when transmission electron microscopy first made it possible to examine tissues at high magnification. In the decades that followed, researchers made exciting progress in understanding the molecular nature of cell-cell attachments and openings. Let's look first at the structures that physically

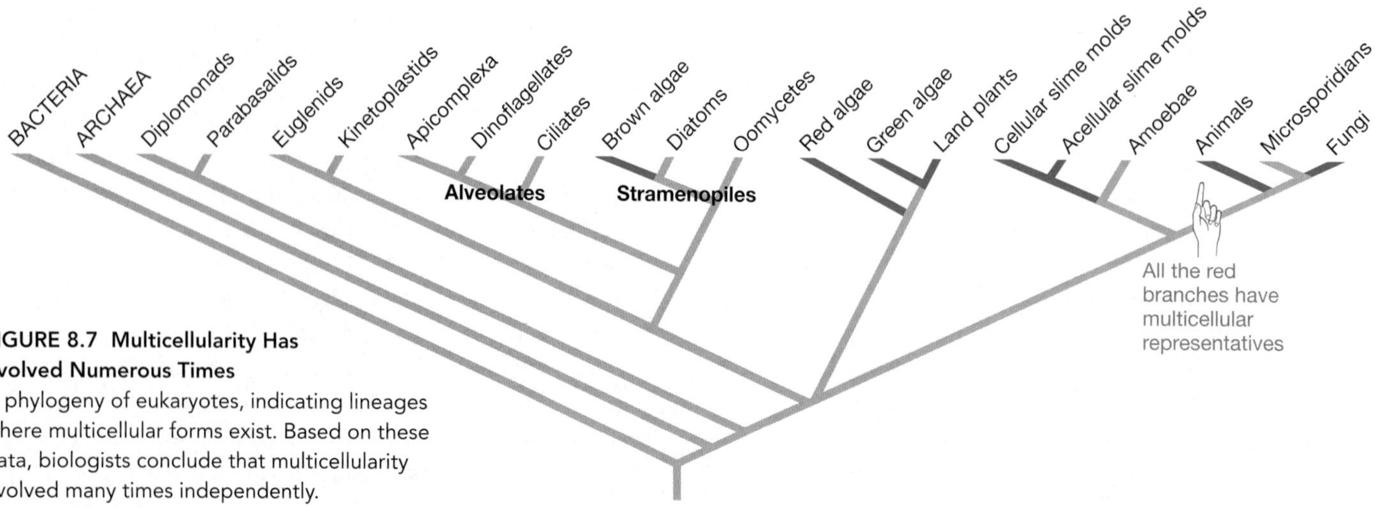
BACTERIA
ARCHAEA
Diplomonads
Parabasalids
Euglenids
Kinetoplastids
Apicomplexa
Dinoflagellates
Ciliates
Brown algae
Diatoms
Oomycetes
Red algae
Green algae
Land plants
Cellular slime molds
Acellular slime molds
Amoebae
Animals
Microsporidians
Fungi
Alveolates
Stramenopiles
All the red branches have multicellular representatives

FIGURE 8.7 Multicellularity Has Evolved Numerous Times
A phylogeny of eukaryotes, indicating lineages where multicellular forms exist. Based on these data, biologists conclude that multicellularity evolved many times independently.

attach adjacent cells to each other, then at the openings that allow nearby cells to exchange materials and information.

Cell-Cell Attachments

The structures that hold cells together vary among multicellular organisms. To illustrate this diversity, consider the intercellular connections observed in the best-studied groups of organisms: plants and animals. Electron microscopy showed that a distinctive layer of material joins plant cells together. The same tool detailed the structure of cell-cell attachment points in animal cells.

As **Figure 8.8** indicates, the extracellular space between adjacent plant cells is composed of three layers. The primary cell walls of adjacent cells sandwich a central layer designated the **middle lamella**, which consists primarily of gelatinous pectins. Because this gel layer is continuous with the primary cell walls of the adjacent cells, it serves to glue them together. If enzymes degrade the middle lamella, as they do when flower petals and leaves detach and fall, the surrounding cells separate.

In many animal tissues, integrins connect the cytoskeleton of each cell to the extracellular matrix (see Section 8.1). A middle-lamella-like layer of gelatinous polysaccharides runs between adjacent cells, so cytoskeleton-ECM connections help hold individual cells together. But in certain animal tissues, the polysaccharide glue is reinforced by protein girders that span the ECM to connect adjacent cells. These structures are particularly important in **epithelia** (singular: **epithelium**)—tissues that form external and internal surfaces. Epithelial cells form a layer that separates parts of the body that contain different solutions—urine versus blood, for example. Epithelial cells must be sealed to prevent mixing of solutions. Although a variety of cell-cell attachment structures exist in epithelia and other tissues, we'll analyze just two: tight junctions and desmosomes.

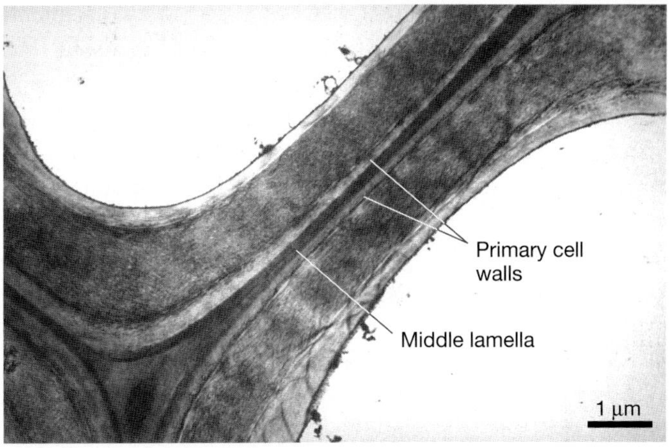

FIGURE 8.8 The Middle Lamella Connects Adjacent Plant Cells
The middle lamella, which holds adjacent plant cells together, is dominated by pectins, a family of gelatinous polysaccharides.

Tight Junctions The electron micrograph in **Figure 8.9a** shows a cross section through cells that are linked by tight junctions. A **tight junction** is a cell-cell attachment composed of specialized proteins in the plasma membranes of adjacent animal cells. As the sketch in **Figure 8.9b** indicates, these proteins line up and bind to one another. The resulting structure resembles quilting, with the proteins acting as stitches.

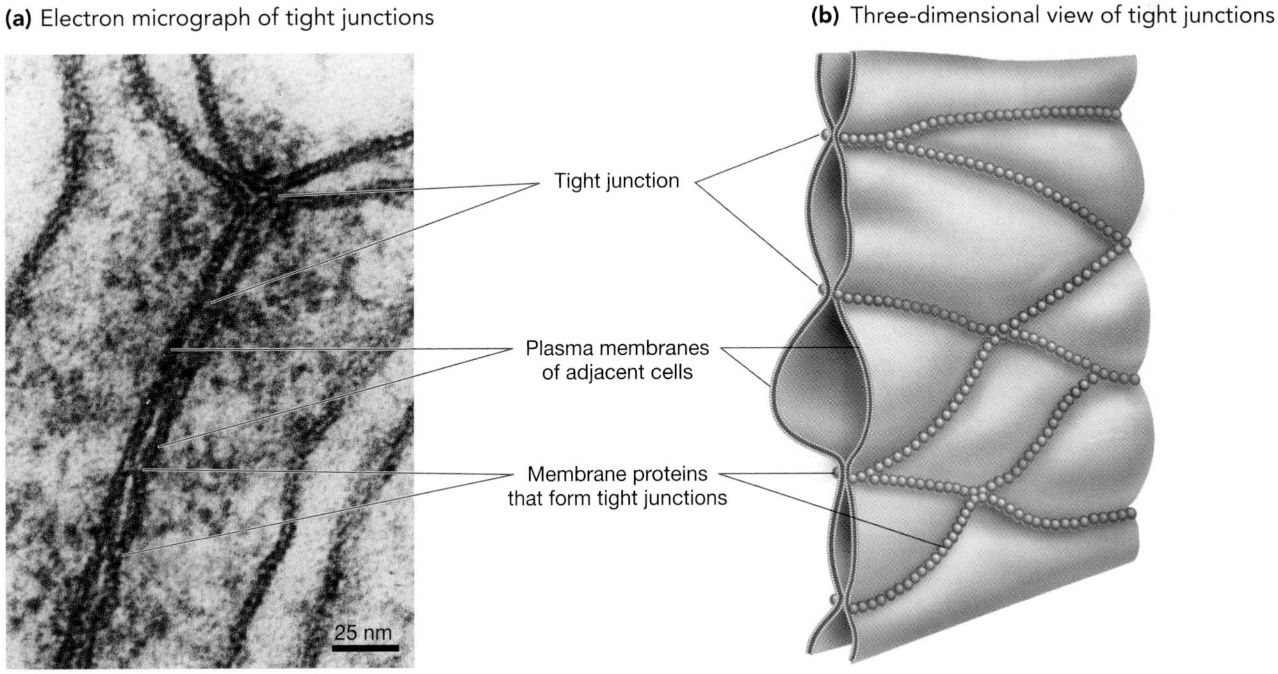

FIGURE 8.9 In Animals, Tight Junctions May Form a Seal between Adjacent Cells
(a) A cross section through adjacent cells that are stitched together by a tight junction. **(b)** Proteins in the membranes of each cell are arranged in a quilted pattern and bind to one another to form the junction.

Tight junctions form a watertight seal between membranes. Based on this observation, it is not surprising that tight junctions are common in cells that form a barrier, such as the cells lining your stomach and intestines. The presence of tight junctions keeps ions and molecules in your gut contents from leaking between stomach cells or intestinal cells and from diffusing into your body. Instead, only selected nutrients enter the cells. These ions and molecules are admitted via specialized transport proteins and channels in the plasma membrane.

Although tight junctions are tight, they are also variable and dynamic. Their structure and function differ in different types of epithelia—in skin versus the lungs, for example. Tight junctions are dynamic because they disassemble and reassemble to allow the passage of immune system cells between epithelial cells.

Desmosomes Figure 8.10a illustrates another type of cell-cell connection found in animals: desmosomes. **Desmosomes** are particularly common cell-cell attachments in epithelial cells and in certain types of muscle cells. The structure and function of a desmosome are analogous to the rivets that hold pieces of sheet metal together. As **Figure 8.10b** indicates, though, desmosomes are extremely sophisticated rivets. At their heart are proteins that form a physical link between the cytoskeletons of the adjacent cells. In addition to binding to each other, these proteins bind to larger proteins that anchor intermediate filaments in the cytoskeletons of the two types of cells. In this way, desmosomes bind together the cytoskeletons of the adjacent cells.

The proteins that form desmosomes may be involved in attaching animal cells long before a desmosome is complete, however. Early in the development of an animal, an embryo's first tissues form when selected cells begin to adhere to each other. Mature desmosomes and adult tissues and organs form much later. Which proteins are involved in these most basic types of cell-cell adhesion, early in development?

Selective Adhesion Long before electron micrographs revealed the presence of desmosomes, biologists realized that some sort of molecule must bind animal cells to each other. This insight grew out of experiments that H. V. Wilson conducted on the cells of sponges in the early 1900s. Sponges are aquatic animals, and the sponge species that Wilson worked with was composed of just two basic types of cells. When Wilson treated adult sponges with chemicals that made the cells dissociate from each other, the result was a jumbled mass of individual and unconnected cells. But when normal chemical conditions were restored, the cells gradually began to move and stick to other cells. With time, cells of each type began to aggregate and adhere to cells of the same tissue type. This phenomenon came to be called **selective adhesion.** In Wilson's experiment, the cells eventually re-formed complete and functional adult sponges.

In an even more dramatic experiment, Wilson dissociated the cells of adult sponges from two differently pigmented

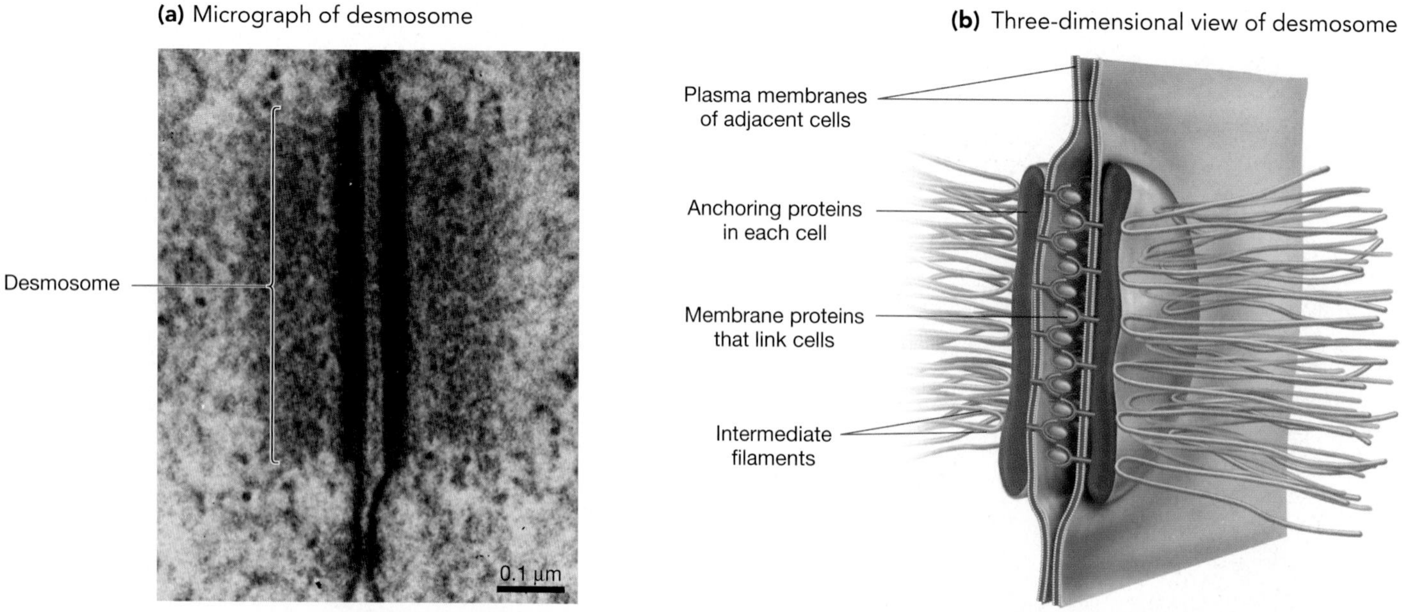
(a) Micrograph of desmosome
Desmosome
0.1 μm
(b) Three-dimensional view of desmosome
Plasma membranes of adjacent cells
Anchoring proteins in each cell
Membrane proteins that link cells
Intermediate filaments

FIGURE 8.10 Adjacent Animal Cells Are Riveted Together by Desmosomes, Binding Together Cytoskeletons

(a) A cross section through a desmosome, showing the connections between adjacent cells and the intermediate filaments that link the desmosome to the cytoskeleton of each cell. **(b)** The major components of a desmosome. EXERCISE Add lines connecting the structures in part (a) to the labels in part (b).

sponge species and randomly mixed them together in a culture dish. As **Figure 8.11** shows, the cells eventually sorted themselves out into distinct aggregates containing cells from only one species and from only one cell type. These results implied that the cells had some way of physically linking to each other and that the linkage was specific to the species and cell type involved. Based on these observations, researchers hypothesized that there must be some sort of molecule on the surface of animal cells that attaches only to cells of the same species and cell or tissue type.

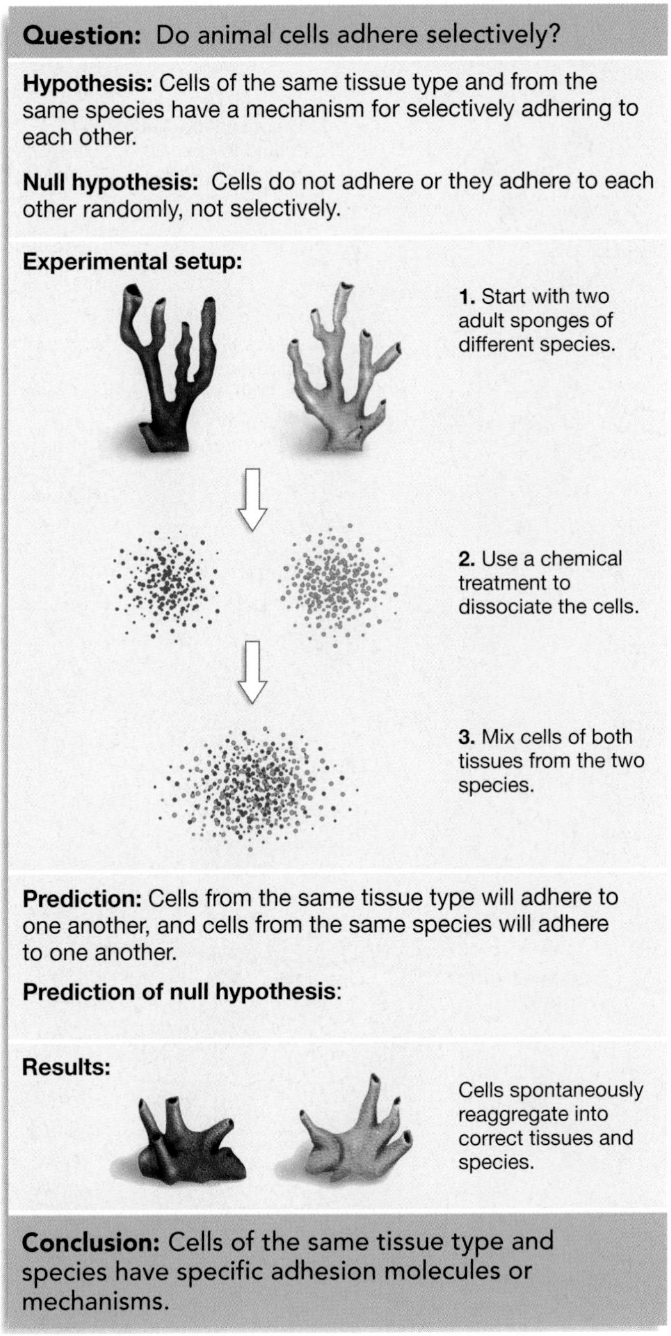
Question: Do animal cells adhere selectively?
Hypothesis: Cells of the same tissue type and from the same species have a mechanism for selectively adhering to each other.
Null hypothesis: Cells do not adhere or they adhere to each other randomly, not selectively.
Experimental setup:
1. Start with two adult sponges of different species.
2. Use a chemical treatment to dissociate the cells.
3. Mix cells of both tissues from the two species.
Prediction: Cells from the same tissue type will adhere to one another, and cells from the same species will adhere to one another.
Prediction of null hypothesis:
Results:
Cells spontaneously reaggregate into correct tissues and species.
Conclusion: Cells of the same tissue type and species have specific adhesion molecules or mechanisms.

FIGURE 8.11 Do Animal Cells Adhere Selectively?
QUESTION What is the prediction made by the null hypothesis?

The Discovery of Cadherins What is the molecular nature of selective adhesion? The initial hypothesis, proposed in the 1970s, was that certain membrane proteins might be involved. The idea was that if different types of cells produce different types of adhesion proteins in their membranes, the molecules might interact in a way that anchors cells of the same type to each other.

This hypothesis was tested through experiments that relied on molecules called antibodies. An **antibody** is a protein that binds specifically to a section of another protein. To test the hypothesis that cell-cell adhesion takes place via interactions between membrane proteins, researchers pursued the following strategy (**Figure 8.12**, page 168):

1. Isolate the membrane proteins from a certain cell type. Create pure preparations of each protein.
2. Inject one of the membrane proteins into a rabbit. The rabbit's immune system cells respond by creating antibodies to the membrane protein. Then purify antibodies that bind tightly to the injected membrane protein. Repeat this procedure for the other membrane proteins that were isolated. In this way, obtain a large collection of antibodies—each of which binds specifically to one (and only one) type of membrane protein.
3. Add one antibody type to the mixture of dissociated cells. Observe whether the cells are able to reaggregate normally. Repeat this experiment with the other antibody types, one type at a time. (Three different antibodies are shown in Figure 8.12, so three experimental treatments are required.)
4. If treatment with a particular antibody prevents the cells from attaching to each other, the antibody is probably bound to an adhesion protein. The hypothesis is that the interaction between antibody and adhesion protein blocks the adhesion protein and keeps it from functioning properly. In effect, the antibody "shakes hands" with the adhesion protein, which prevents the adhesion protein from "shaking hands" with other adhesion proteins to attach cells.

Based on studies such as this, researchers have identified several major classes of cell adhesion proteins. The molecules in desmosomes belong to a group that came to be named **cadherins**. Each major type of cell in the body has a different type of cadherin in its plasma membrane, and each type of cadherin can bind only to cadherins of the same type. In this way, cells of the same tissue type attach specifically to one another.

Animal cells attach to each other in a selective manner because only certain types of cell adhesion proteins can bind and rivet cells together. Cadherins provide the physical basis for most selective adhesion and are a critical component of the desmosomes that join mature cells.

Question: Do animal cells have adhesion proteins on their surfaces?

Hypothesis: Selective adhesion is due to specific membrane proteins.

Null hypothesis: Selective adhesion is not due to specific membrane proteins.

Experimental setup:

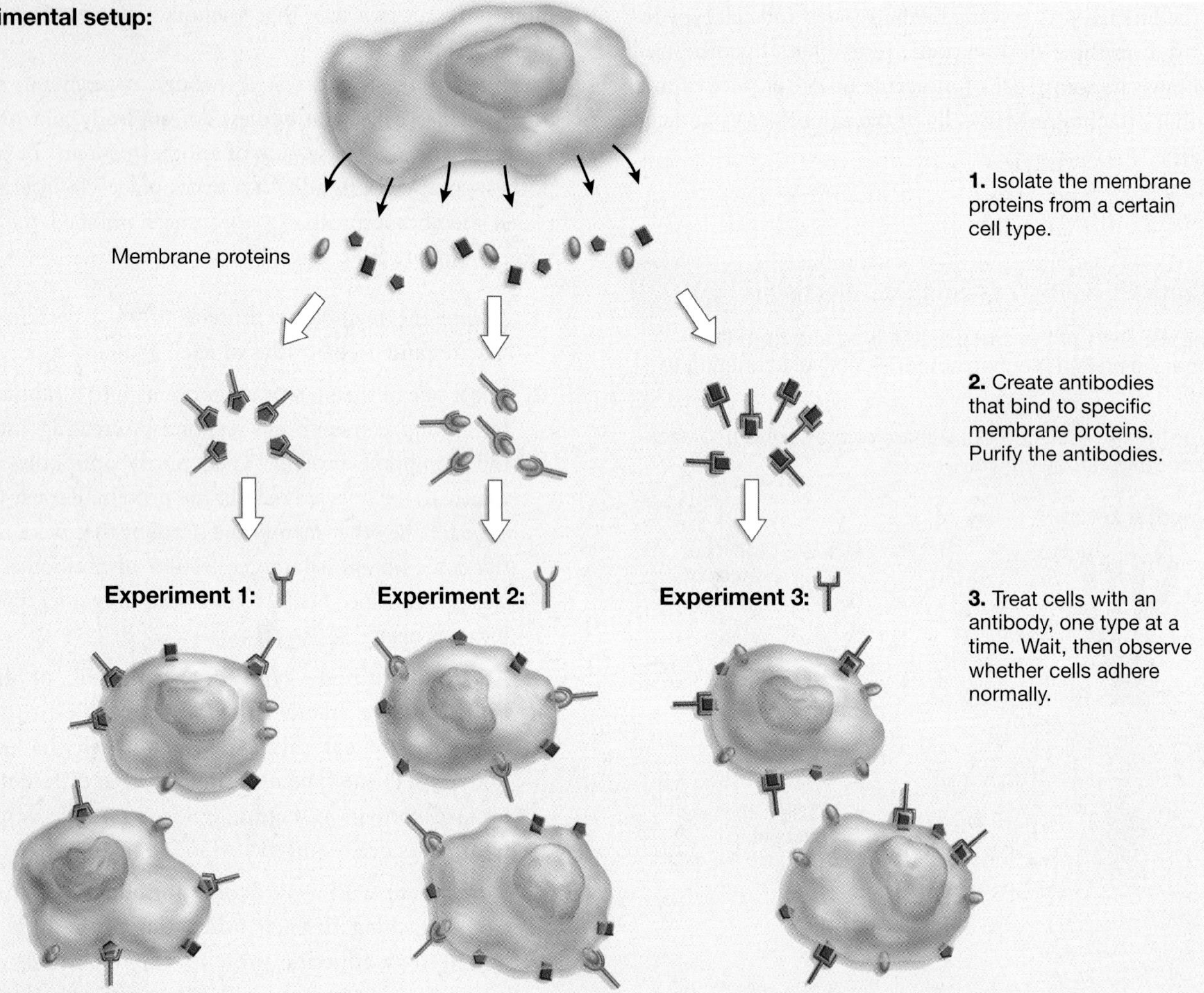
Membrane proteins
1. Isolate the membrane proteins from a certain cell type.
2. Create antibodies that bind to specific membrane proteins. Purify the antibodies.
Experiment 1:
Experiment 2:
Experiment 3:
3. Treat cells with an antibody, one type at a time. Wait, then observe whether cells adhere normally.

Prediction:

Prediction of null hypothesis:

Results:

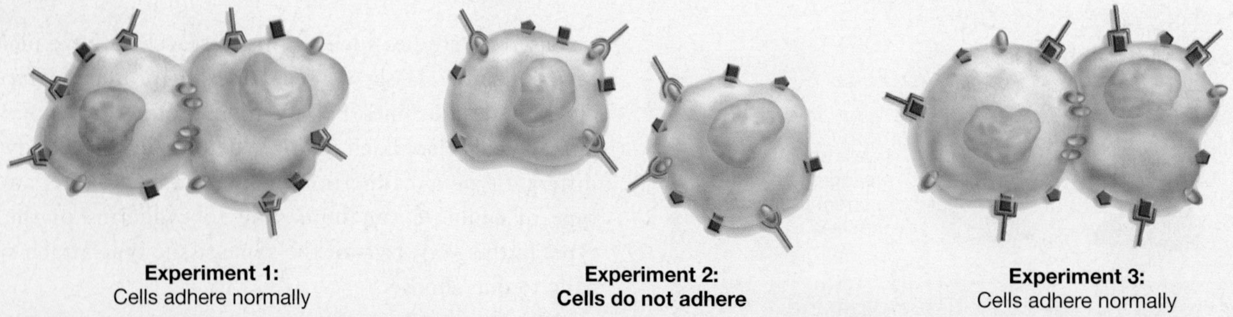
Experiment 1:
Cells adhere normally
Experiment 2:
Cells do not adhere
Experiment 3:
Cells adhere normally

Conclusion: The protein that was blocked in Experiment 2 is involved in cell-cell adhesion.

FIGURE 8.12 Do Animal Cells Have Adhesion Proteins on Their Surfaces?
EXERCISE Fill in the prediction made by each hypothesis.

Cell-Cell Gaps

Once cells of the same tissue type are bound together, how do they communicate? Careful analysis of electron micrographs answered this question. In both plants and animals, direct connections between cells in the same tissue help the cells work in a coordinated fashion.

In plants, gaps in cell walls create direct connections between the cytoplasm of adjacent cells. At these connections, named **plasmodesmata** (singular: *plasmodesma*), the plasma membrane and the cytoplasm of the two cells are continuous. In addition, smooth endoplasmic reticulum (ER) runs through the hole (**Figure 8.13a**). Growing evidence suggests that plasmodesmata also contain proteins that regulate the passage of specific proteins, making the connections similar in function to the nuclear pore complex, which we introduced in Chapter 7. At least some of the proteins that are transported through plasmodesmata are involved in coordinating the activity of adjacent cells.

In most animal tissues, adjacent cells are connected by holes in the ECM and adjacent plasma membranes. These holes, known as **gap junctions**, are created by specialized proteins that line each hole (**Figure 8.13b**). The proteins in gap junctions form channels that admit water, ions, and small molecules such as amino acids, sugars, and nucleotides. The flow of small molecules can help adjacent cells communicate and coordinate their activities by allowing the rapid passage of regulatory ions or molecules. In the muscle cells of your heart, for example, a flow of ions through gap junctions acts as a signal that coordinates contractions. Without this cell-cell communication, a normal heartbeat would be impossible.

(a) Plasmodesmata create gaps that connect plant cells.

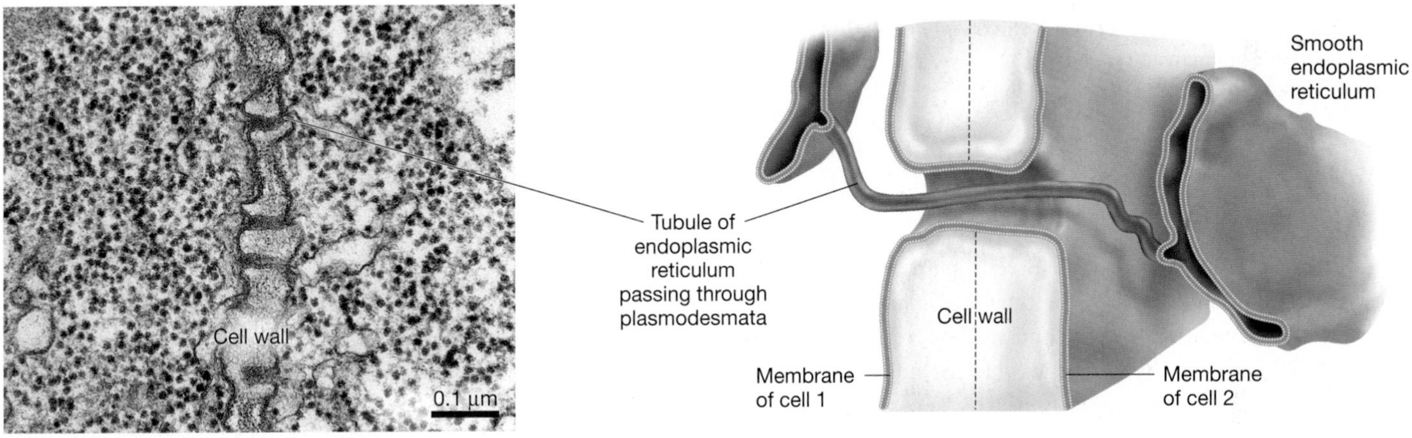

(b) Gap junctions create gaps that connect animal cells.

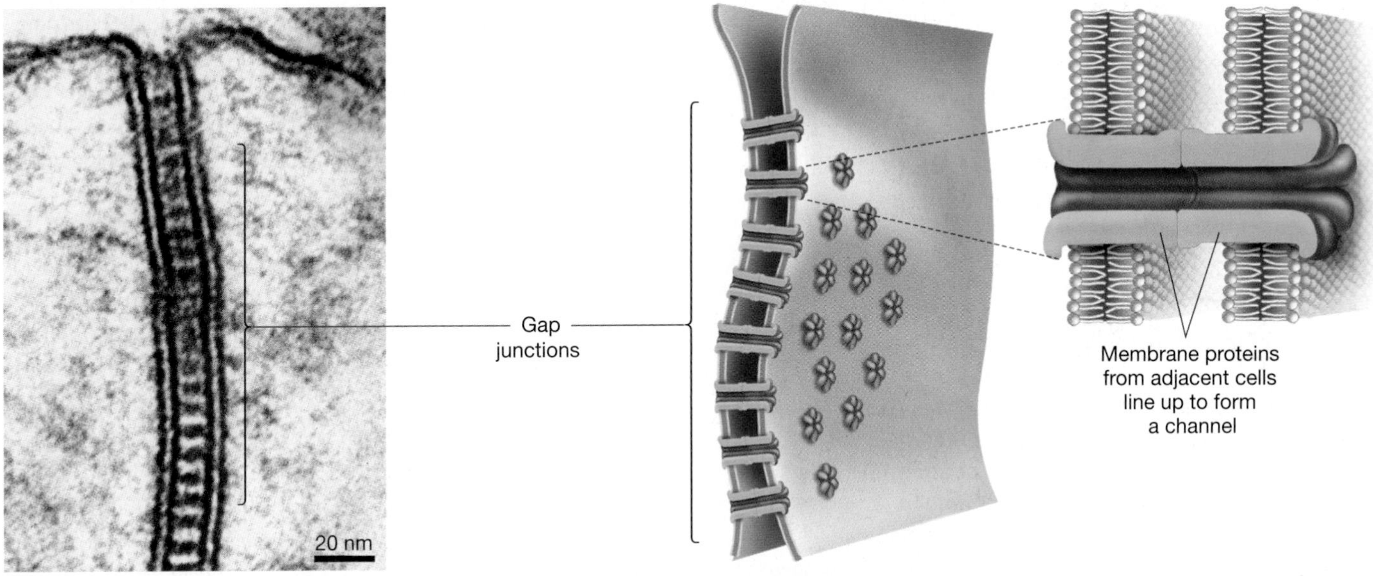

FIGURE 8.13 Adjacent Plant Cells and Adjacent Animal Cells Communicate Directly
(a) Plasmodesmata are holes in plasma membranes and cell walls of adjacent plant cells. A tubule of smooth ER runs through the hole. **(b)** Gap junctions are collections of protein-lined holes in the membranes of adjacent animal cells.

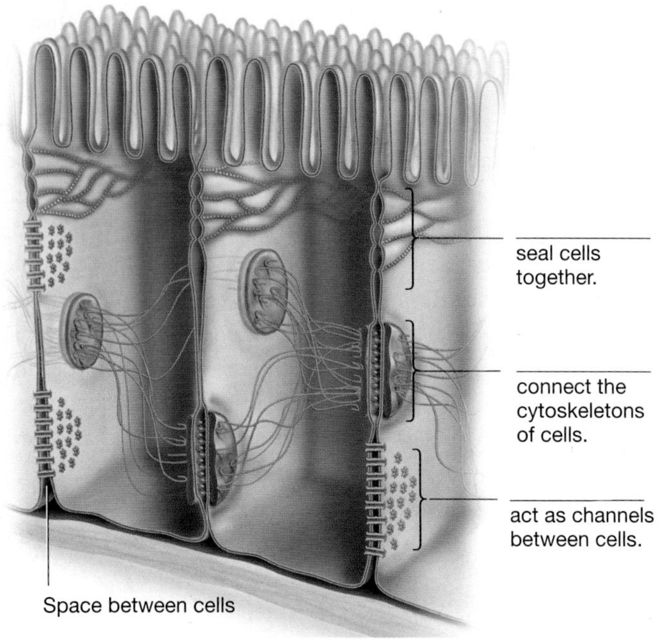
seal cells together.
connect the cytoskeletons of cells.
act as channels between cells.
Space between cells

FIGURE 8.14 An Array of Structures Is Involved in Cell-Cell Adhesion and Communication

EXERCISE In the blanks on the right, write the name of each type of cell-cell attachment.

In effect, plant cells and animal cells share most or all of the ions and small molecules in their cytoplasm but retain their own organelles, proteins, and nucleic acids. The existence of cell-cell gaps makes it possible for adjacent cells to communicate with each other efficiently. A tissue can act as an integrated whole if its component cells are connected by gaps in their extracellular material and in their plasma membranes (**Figure 8.14**).

How do more distant cells in a multicellular organism communicate? For example, suppose that leaf cells in a maple tree are being attacked by caterpillars or that the muscle cells in your arm are exercising so hard that they begin to run low on sugar. How do these cells signal tissues or organs elsewhere in the body to release materials that might be needed to fend off caterpillars or exhaustion?

✓CHECK YOUR UNDERSTANDING

Adjacent plant cells and adjacent animals cells are physically connected. In addition, both plant cells and animal cells communicate with each other through gaps in their plasma membranes. You should be able to (1) compare and contrast the structure and function of the middle lamella of plants and the tight junctions and desmosomes of animals, (2) describe the structure and function of plasmodesmata and gap junctions, and (3) explain the molecular mechanism of selective adhesion.

8.3 How Do Distant Cells Communicate?

As you read this, molecules that carry information about conditions throughout your body are traveling through your bloodstream and arriving at cells from your head to your toes. These molecules might signal that viruses are attacking cells in your nose or that the amount of glucose in your blood is low. The signals are intended for tissues or organs that can respond appropriately. If the signal was triggered by the presence of a virus, then muscles in nearby blood vessels will relax—opening the vessels and increasing blood flow to the affected area. If the molecule indicates that glucose levels are low, cells in your liver that store glucose will release the sugar into your blood. Similar types of molecules course through the plants on your campus. They may carry signals about the direction and rate of growth in stems and roots, the direction of incoming sunlight and the current length of daylight, or the site and severity of attacks by insects or viruses. As in animals, these signals are meant for target cells that can respond in a way that increases the individual's chances of surviving and reproducing.

In a multicellular organism, it is logical that chemical signals convey information from one tissue type or organ to another. If the activities of different cells, tissues, and organs are to be coordinated, then information has to move back and forth among them. How do cells receive and respond to signals from other cells? Researchers began to get a handle on this question when they realized that intercellular signaling is based on a four-step process: (1) receiving a signal, (2) processing it, (3) responding to it, and (4) turning it off.

Signal Reception

Intercellular signals are chemical messengers. They deliver their message by binding to receptor molecules. Any molecule that binds to a specific site on a receptor molecule is a **ligand**. A ligand that acts as an intercellular signal is termed a **hormone**. Hormones are usually small molecules and are often present in minute concentrations, yet they significantly affect the activity of target cells. They are like a fleeting scent or whispered phrase from someone that you are attracted to—a tiny signal, but one that makes your cheeks flush and your heart pound.

As **Table 8.1** indicates, the chemical structure of plant and animal hormones varies widely. Hormones also have a wide array of effects. Despite this diversity in structure and function, all hormones have the same general role in the organism: They coordinate the activities of cells in response to information from outside or inside the body.

The messages carried by hormones are received by **signal receptors**—proteins that bind to a particular signal and change conformation or activity in response. If the signal receptors respond to steroid hormones, then these receptors are located inside the cell. This observation is not surprising, because steroid hormones are lipid soluble and thus readily diffuse through the plasma membrane. But because most hormones are not lipid soluble, most signal receptors are located in the plasma membrane.

TABLE 8.1 Hormones Have Diverse Structures and Functions

Hormone Name	Chemical Structure	Function of Signal
Nitric oxide	NO (a gas)	Produces short-lived messages in animal cells
Ethylene	C_2H_4 (a gas)	Stimulates fruit ripening, regulates aging
Insulin	Protein, 51 amino acids	Stimulates glucose uptake in animal bloodstream
Systemin	Peptide, 18 amino acids	Stimulates plant defenses against herbivores
Estrogen	Steroid	Stimulates development of female characteristics in animals
Brassinosteroids	Steroid	Stimulates plant cell elongation
Prostaglandins	Modified fatty acid	Perform a variety of functions in animal cells
Oligosaccharides	Carbohydrate	Triggers defensive reactions in plants
Thyroxine (T4)	Modified amino acid	Regulates metabolism in animals
FSH	Glycoprotein	Stimulates egg maturation, sperm production in animals
Auxin	Small organic compound	Signals changes in long axis of plant body

No matter where they are located, signal receptors have several important general characteristics:

- Each type of receptor is restricted to particular cell types. Your liver cells, for example, contain receptors for the hormone insulin, which signals that glucose levels in the blood are too high. The presence of insulin receptors on liver cells makes sense because liver cells store glucose. But insulin receptors are not present in bone cells or skin cells or other cell types that cannot help regulate glucose levels.
- Receptors are dynamic. The number of receptors in a particular cell may decline if hormonal stimulation occurs at high levels over a long period of time. The ability of a receptor molecule to bind tightly to a hormone may also decline in response to intensive stimulation.
- Receptors can be blocked. The drugs called beta-blockers, for example, bind to certain receptors for the chemical signal *epinephrine*. Because the binding of epinephrine to these receptors stimulates heart cells to contract, beta-blockers are prescribed to lower heart rate and thus blood pressure.

The most important and most general characteristic of signal receptors, though, is that their physical conformation changes when a hormone binds to them. This is a critical event in cell-cell signaling. The change in receptor structure means that the signal has been received. What happens next?

Signal Processing

Once a cell receives a signal, something has to happen to initiate the cell's response. In some cases the "something" occurs directly. For example, steroid hormones such as testosterone and estrogen diffuse through the plasma membrane and enter the cytoplasm, where they bind to a receptor protein. The hormone-receptor complex is then transported to the nucleus, where it triggers changes in the genes being expressed in the cell (**Figure 8.15**). Early in human development, the arrival of testosterone in cells leads to the expression of genes that trigger the development of male reproductive organs. (In the absence of testosterone, female reproductive organs develop.) Later in life, a surge of testosterone triggers the changes associated with puberty in boys, while the arrival of estrogen molecules spurs the development of adult female characteristics in girls.

The effects of hormones that *cannot* diffuse across the plasma membrane and enter the cytoplasm are not nearly so direct. In this case, the first event triggered by a hormone's arrival at the cell surface consists of **signal transduction**—the conversion of the signal from one form to another. A long and often complex series of events ensues, collectively called a **signal transduction pathway**.

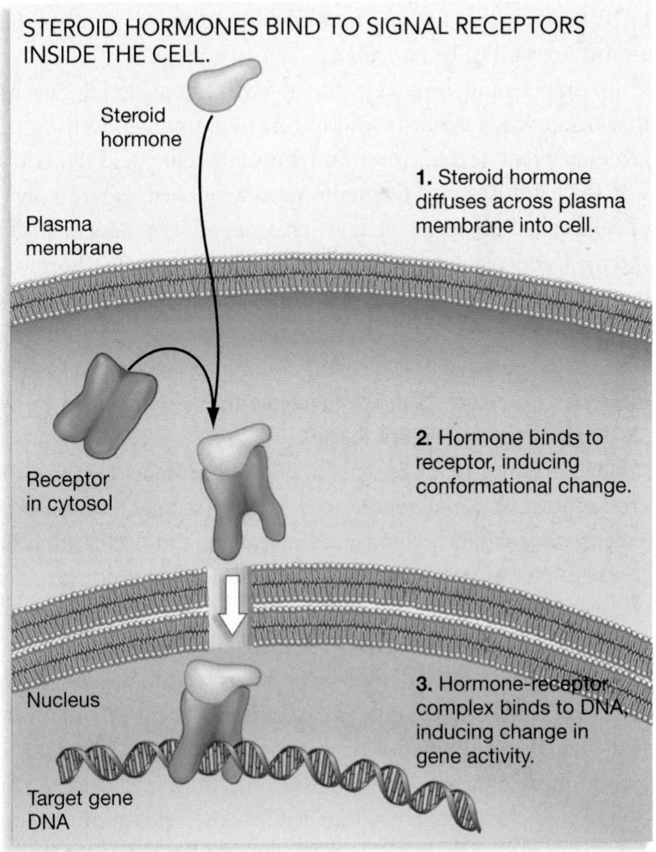

FIGURE 8.15 Some Cell-Cell Signals Enter the Cell and Bind to Receptors in the Cytoplasm
Because they are lipids, steroid hormones can diffuse across the plasma membrane. They then bind to signal receptors inside the cell. The hormone-receptor complex is transported to the nucleus and binds to genes, changing their activity.

WEB TUTORIAL 8.1 Cell-Cell Communication

Signal transduction is a common occurrence in everyday life. For example, the e-mail messages you receive are transmitted electronically from one computer to another over wires or cables. Software in your computer transduces the electronic signals to letters and words on the screen—a form that you can understand and respond to.

Signal transduction pathways work the same way. In a cell, signal transduction converts an extracellular signal to an intracellular signal. As in an e-mail transmission, a signal that is easy to transmit is converted to a signal that is easily understood and that triggers a response. In cells, signal transduction occurs at the plasma membrane and involves one of two major types of membrane proteins: (1) G proteins and (2) receptor protein kinases. Let's look at each in turn.

G Proteins Many signal receptors span the plasma membrane and are closely associated with peripheral membrane proteins inside the cell called **G proteins.** G proteins got their name because they bind guanosine triphosphate (GTP) and guanosine diphosphate (GDP).

G proteins are signal transducers that act like time-delayed light switches—after they are turned on, they later turn themselves off. They are turned on when they bind GTP, and they turn themselves off after they hydrolyze the bound GTP to GDP. Proteins that are on are said to be *activated*; proteins that are off are said to be *inactivated.*

To understand how G proteins work, consider the events that occur when a wheat seed begins to germinate. Cells in the growing plant secrete the hormone gibberellic acid 1 (GA_1). The GA_1 diffuses into the main part of the seed, where a sheet of cells called the *aleurone layer* encloses a large mass of stored starch. Upon reaching cells in the aleurone layer, the hormone binds to receptors in their plasma membranes and activates a G protein by inducing the G protein to release GDP and bind GTP. In response, the G protein splits into two parts. One of these parts activates a nearby enzyme that is embedded in the plasma membrane (**Figure 8.16a**).

The enzymes that are activated by G proteins catalyze the production of small molecules called second messengers. **Second messengers** are non-protein signaling molecules that increase in concentration inside a cell and elicit a response to the received signal. They are intracellular signals that spread the message carried by the extracellular signal—the hormone.

Several types of small molecules act as second messengers in cells. In wheat seed cells, for example, both calcium ions (Ca^{2+}) and cyclic guanosine monophosphate (cGMP) act as second messengers for GA_1. The G protein that is switched on by GA_1 activates enzymes that lead to the release of calcium ions from the smooth ER. In addition, the same G protein activates an enzyme that catalyzes the synthesis of cGMP in the cytoplasm. In response to a different hormone in a different type of cell, diacylglycerol (DAG), inositol triphosphate (IP_3), or cyclic adenosine monophosphate (cAMP) might serve as the second messenger.

Second messengers are effective because they are small and because they diffuse rapidly to spread the signal throughout the cell. In addition, they can be produced in large quantities in a short time. This characteristic is important. Because the arrival of a single hormone molecule can stimulate the production of many second messenger molecules, the signal transduction event amplifies the original signal.

Receptor Tyrosine Kinases Other receptor proteins have a completely different mode of action. Instead of activating a nearby G protein, **receptor tyrosine kinases** form dimers after they bind a signaling molecule. They then phosphorylate each other (**Figure 8.16b**). Recall from Chapter 3 that the addition of a phosphate group often changes a protein's conformation and activity. In many cases, phosphorylation activates a protein. A phosphorylated receptor tyrosine kinase is no exception. Once it is activated by phosphorylation, the receptor tyrosine kinase phosphorylates one or more specific proteins inside the cell, activating them. These phosphorylated proteins then catalyze the phosphorylation of still other proteins, which phosphorylate yet another population of proteins. This sequence of events is termed a **phosphorylation cascade.** Because each enzyme in the cascade catalyzes the phosphorylation of numerous "downstream" enzymes, the original signal is amplified many times over. One activated receptor tyrosine kinase can trigger the activation of thousands of proteins at the end of the cascade.

To summarize, signal transduction occurs via G proteins or receptor tyrosine kinases. The signal transduction event has two results: (1) It converts an easily transmitted extracellular message into an intracellular message, and (2) it amplifies the original message many times over. Although the hormone at the surface of the cell eventually disengages from its receptor and diffuses away or disintegrates, a "louder" intracellular signal—whether it is a large number of second messenger molecules or a large population of phosphorylated proteins—carries the information throughout the cell and triggers a response.

Signal Response

What is the ultimate response to the messages carried by hormones? The answer varies from signal to signal and from cell to cell. Recall that steroid hormones bind to intracellular receptors and directly change which genes are active in the cell. As a result, they change the cell's characteristics. Similar types of changes in gene expression can occur in response to second messengers or a cascade of protein phosphorylation events. In wheat seeds, for example, a rise in cellular cGMP concentrations triggers the production of a protein that activates the gene for a starch-digesting enzyme. As a result, large quantities of the starch-digesting enzyme α-amylase begin to enter the endomembrane system (introduced in Chapter 7). But recall that the G protein activated by GA_1 also triggers a rise in intracellular Ca^{2+}. Recent experiments have shown that the presence of

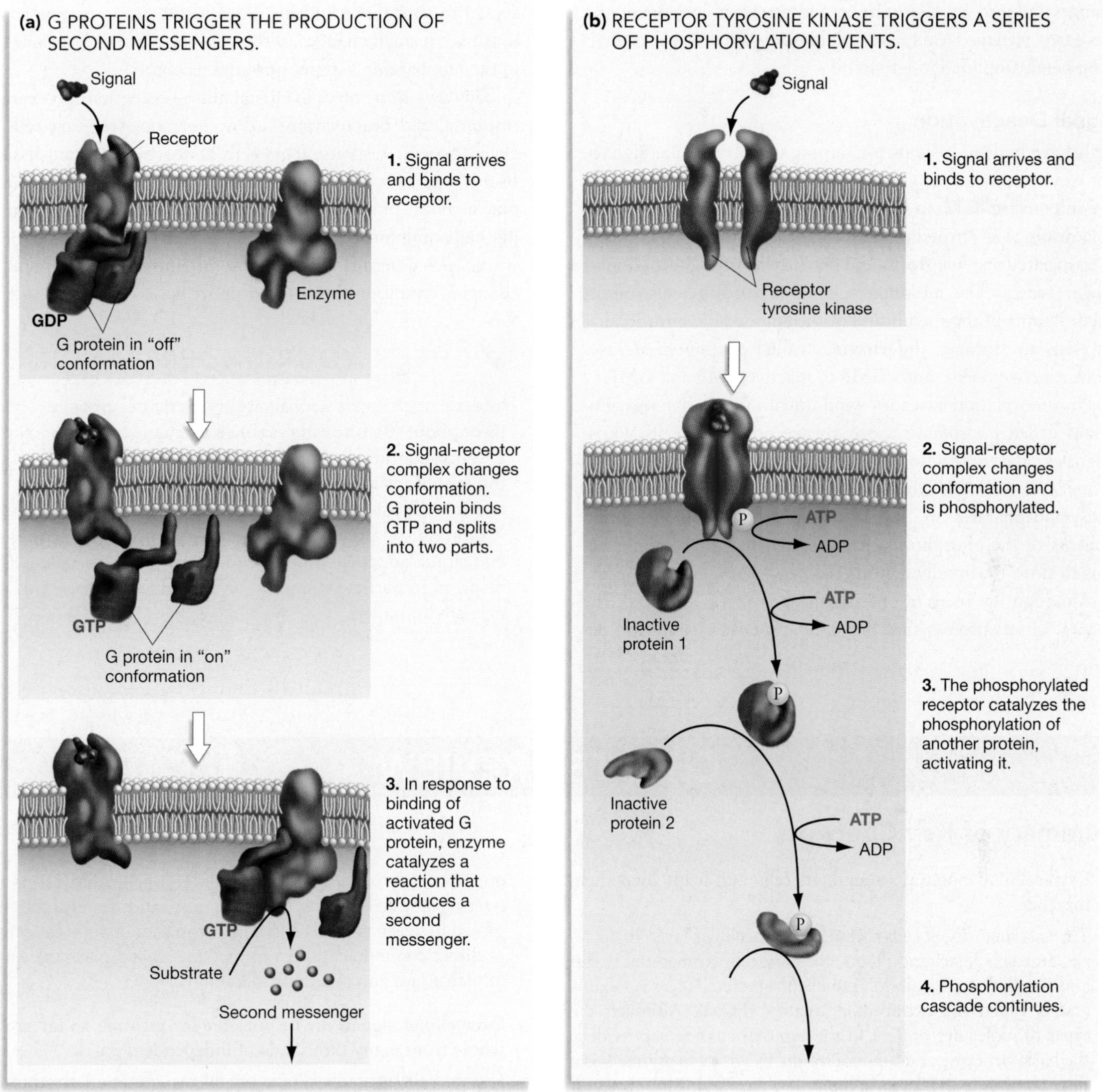
(a) G PROTEINS TRIGGER THE PRODUCTION OF SECOND MESSENGERS.
Signal
Receptor
1. Signal arrives and binds to receptor.
Enzyme
GDP
G protein in "off" conformation
2. Signal-receptor complex changes conformation. G protein binds GTP and splits into two parts.
GTP
G protein in "on" conformation
3. In response to binding of activated G protein, enzyme catalyzes a reaction that produces a second messenger.
GTP
Substrate
Second messenger
(b) RECEPTOR TYROSINE KINASE TRIGGERS A SERIES OF PHOSPHORYLATION EVENTS.
Signal
1. Signal arrives and binds to receptor.
Receptor tyrosine kinase
2. Signal-receptor complex changes conformation and is phosphorylated.
P
ATP
ADP
ATP
ADP
Inactive protein 1
3. The phosphorylated receptor catalyzes the phosphorylation of another protein, activating it.
P
Inactive protein 2
ATP
ADP
P
4. Phosphorylation cascade continues.

FIGURE 8.16 Many Cell-Cell Signals Are Transduced at the Plasma Membrane
(a) If a receptor that is coupled to a G protein is activated by a signal, the G protein binds guanosine triphosphate (GTP) and splits into two parts. The change in the G protein activates a nearby enzyme and results in the production of a second messenger. **(b)** When a receptor tyrosine kinase is activated by a signal, it sets off a cascade of phosphorylation reactions.

Ca^{2+} allows vesicles that are packed with α-amylase to fuse with the plasma membrane and release their contents into the starch storage area. As a result of GA_1's arrival at the plasma membrane, cells in the aleurone layer begin secreting an enzyme that digests stored starch. When the enzyme begins to work, sugars are released. In this way, by sending a GA_1 signal, the germinating embryo has signaled for the release of the nutrients it needs to grow.

Many second messengers or phosphorylation cascades result in the activation or deactivation of a particular target protein that already exists in the cell—typically an enzyme. Either way, the activity of the target cell changes dramatically in response to the arrival of the signal.

But how is the signal turned off? Cells in the aleurone layer of wheat seeds do not need to secrete unlimited quantities of α-amylase, and abnormalities would result if the morphological changes induced by testosterone and estrogen continued past the early teenage years into adulthood. What keeps signals from being amplified indefinitely?

Signal Deactivation

Cells have built-in systems for turning off intracellular signals. For example, activated G proteins can hydrolyze GTP. This reaction converts GTP to GDP and releases a phosphate group. As a result, the G protein's conformation changes, activation of its associated enzyme stops, and production of the second messenger ceases. The messengers in the cytosol are also short lived. Pumps in the membrane of the smooth ER return calcium ions to storage, and enzymes called *phosphodiesterases* convert active cAMP and cGMP to inactive AMP and GMP.

Phosphorylation cascades wind down in a similar way. Enzymes called *phosphatases* are always present in cells. Phosphatases catalyze reactions that remove phosphate groups from proteins. If hormone stimulation of a receptor tyrosine kinase ends, phosphatases are able to dephosphorylate enough components of the phosphorylation cascade that the response begins to slow. Eventually it stops.

Although an array of specific mechanisms is involved, the general observation is that hormone-response systems are designed to shut down quickly. As a result, they are exquisitely sensitive to small changes in the concentration of hormones or in the number and activity of signal receptors.

The four steps of cell-cell signaling—reception, processing, response, and deactivation—allow hormone-secreting cells to elicit a specific response from cells in nearby or distant tissues. In this way, hormones coordinate the activity of cells throughout the body. In wheat plants and in your body, cell-cell signaling helps millions of individual cells function as an integrated whole. As a result, multicellular organisms can respond to changing conditions in an appropriate way.

✓CHECK YOUR UNDERSTANDING

Intercellular signals coordinate the activities of cells throughout the body in response to changes in internal or external conditions. You should be able to (1) describe the four steps involved in signal transduction (signal reception, signal processing via the production of a second messenger or a phosphorylation cascade, signal response, and signal deactivation); (2) explain why only certain cells respond to particular signals; and (3) elucidate how signals are amplified.

CHAPTER REVIEW

Summary of Key Concepts

- **Extracellular material strengthens cells and helps bind them together.**

  The vast majority of cells secrete an extracellular layer. In bacteria, archaea, algae, and plants, the extracellular material is stiff and is called a cell wall. In animals, the secreted layer is flexible and is called the extracellular matrix (ECM). Although the types of molecules present in the external coating vary widely, the basic structure and function of the ECM are the same: It is a fiber composite that defines the cell's shape and helps protect it from mechanical damage. Fiber composites consist of cross-linked filaments that provide tensile strength and a ground substance that fills space and resists compression. In plants the extracellular filaments are cellulose microfibrils; in animals the most abundant filaments are made of the protein collagen. In both plants and animals, the ground substance is composed of gelatinous polysaccharides.

- **Cell-cell connections help adjacent cells adhere. Cell-cell gaps allow adjacent cells to communicate.**

  In both unicellular and multicellular organisms, cell-cell interactions are mediated by molecules in the extracellular layer and plasma membrane. Most interactions among unicellular species are competitive in nature, but the cells of multicellular organisms are intensely social and cooperate with each other. Many cells of multicellular organisms are physically bound to one another via the glue-like middle lamella that forms between plant cells or the tight junctions and desmosomes observed between animal cells. The cytoplasm of adjacent cells is in direct communication, via openings called plasmodesmata in plants and gap junctions in animals.

- **Intercellular signals are responsible for creating an integrated whole from many thousands of independent parts.**

  Distant cells in multicellular organisms communicate through signaling molecules that bind to receptors found on or in specific target cells. Once these signals are received, they are often transduced to a new type of intracellular signal, which is amplified. This internal signal triggers a sequence of events that leads to the activation of certain enzymes, the release or uptake of specific ions or molecules, or a change in the activity of target genes. Because enzymes inside the cell quickly deactivate the signal, the cell's response is tightly regulated. A continued response usually depends on continued stimulation by the signaling molecule. Thus cell-cell communication occurs by a four-step process: (1) signal reception, (2) signal processing, (3) signal response, and (4) signal deactivation. As a result, cells and tissues throughout the body can alter their activity in response to changing conditions, and do so in a coordinated way.

  Web Tutorial 8.1 · Cell-Cell Communication

ESSAY How Do Viagra®, Levitra®, and Cialis® Work?

In the United States alone, 30 million men are estimated to suffer from erectile dysfunction—the inability to produce and sustain an erection when sexually stimulated. In recent years, drugs that treat this disorder effectively have appeared on the market. Among them are molecules that are marketed under the trade names Viagra, Levitra, and Cialis. How do these drugs work?

The development of this drug therapy grew out of a detailed understanding of a signal transduction pathway.

The answer to this question is found in the cell-cell signaling system that leads to erection in males. **Figure 8.17** outlines the sequence of events. When a man is sexually stimulated, nerve impulses from the brain or direct physical contact lead to the release of the chemical messenger nitric oxide (NO) gas from blood-vessel cells in his penis. NO acts as an intercellular signal. Its targets are muscle cells that surround the blood vessels throughout the penis. When it arrives at one of these muscle cells, the NO diffuses through the plasma membrane and binds to the enzyme guanylate cyclase. The activated enzyme catalyzes the production of cyclic guanosine monophosphate (cGMP) from guanosine triphosphate (GTP). cGMP acts as a second messenger. It binds to and activates a protein kinase. This kinase then phosphorylates a protein in the membrane of the smooth ER.

When that protein is activated, it begins pumping calcium ions (Ca^{2+}) out of the cytosol and into the lumen of the smooth ER. When Ca^{2+} levels in the cytoplasm of the muscle cells are low, the muscle tissue that contains these cells relaxes. When muscles that surround blood vessels relax, the vessels widen. As a result, blood flow through the vessels increases. As muscles in the penis relax, blood flows into chambers in the structure and fills them, causing an erection.

How do the drugs Viagra, Levitra, and Cialis fit into this signaling pathway? The muscle cells of the penis contain an enzyme called phosphodiesterase 5 (PDE5), which turns off the NO-induced system. More specifically, PDE5 catalyzes the conversion of cGMP to an inactive form. Because it removes the second messenger, PDE5 diminishes the cells' response to the original signal. Viagra, Levitra, and Cialis inhibit PDE5. When these drugs are in the cytoplasm, the cell's mechanism for deactivating the signal is inhibited and cGMP levels tend to stay high. Because this second messenger is present at a high concentration, each cell's response to the signal tends to be stronger and of longer duration. The development of this drug therapy grew out of a detailed understanding of a signal transduction pathway.

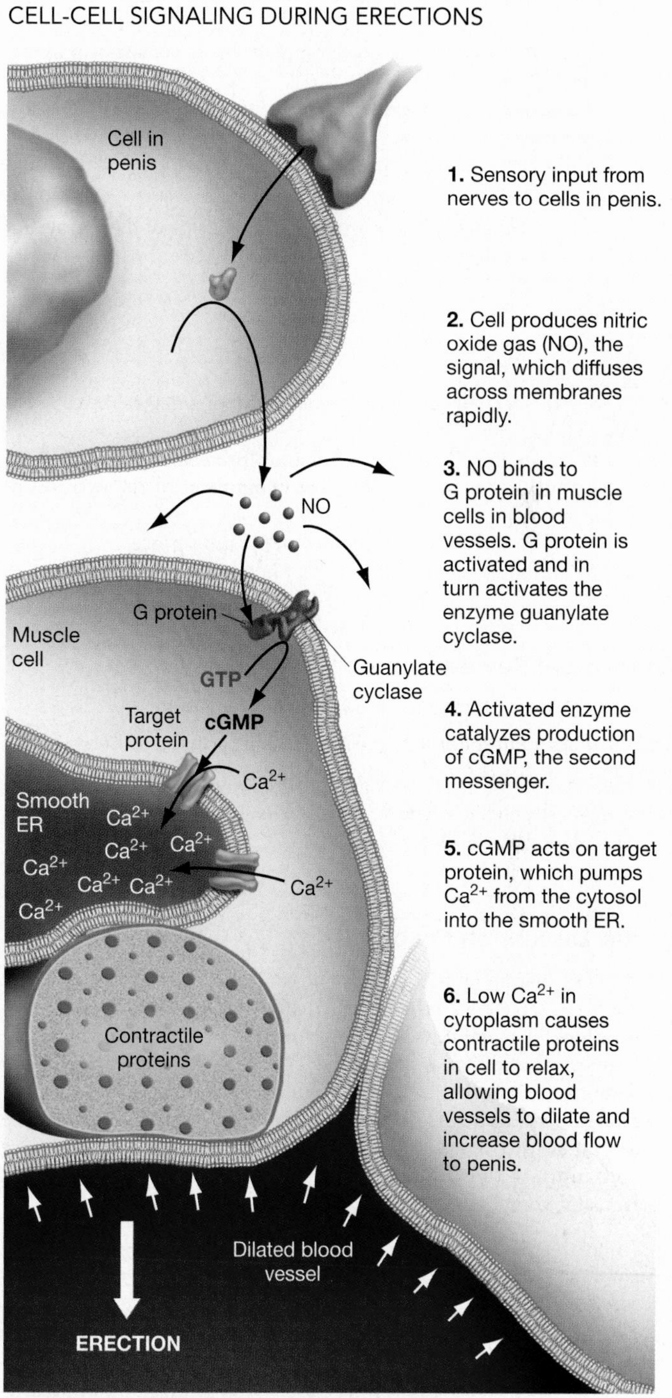
CELL-CELL SIGNALING DURING ERECTIONS
Cell in penis
1. Sensory input from nerves to cells in penis.
2. Cell produces nitric oxide gas (NO), the signal, which diffuses across membranes rapidly.
NO
3. NO binds to G protein in muscle cells in blood vessels. G protein is activated and in turn activates the enzyme guanylate cyclase.
G protein
Muscle cell
Guanylate cyclase
GTP
cGMP
Target protein
4. Activated enzyme catalyzes production of cGMP, the second messenger.
Smooth ER
Ca2+
5. cGMP acts on target protein, which pumps Ca2+ from the cytosol into the smooth ER.
Contractile proteins
6. Low Ca2+ in cytoplasm causes contractile proteins in cell to relax, allowing blood vessels to dilate and increase blood flow to penis.
Dilated blood vessel
ERECTION

FIGURE 8.17 A Signal Transduction Pathway in Male Reproductive Tissue

EXERCISE The enzyme PDE5 converts cGMP to an inactive molecule. Circle the step in the signal transduction pathway that is affected by PDE5. What effect does PDE5 activity have on the response?

Questions

Content Review

1. Which of the following represents a fundamental difference between the fibers found in the extracellular layers of plants and those of animals?
   a. Plant fibers are thicker; they are also stronger because they have more cross-linkages.
   b. Animal fibers consist of proteins; plant fibers consist of polysaccharides instead.
   c. Plant extracellular fibers never move; animal fibers can slide past one another.
   d. Cellulose microfibrils run parallel to each other; collagen filaments crisscross.
2. In plants and animals, where are most components of the extracellular material synthesized?
   a. smooth ER
   b. the rough ER and Golgi apparatus
   c. in the extracellular layer itself
   d. adjacent cells
3. Treating dissociated cells with certain antibodies makes the cells unable to reaggregate. Why?
   a. The antibodies bind to cell adhesion proteins called cadherins.
   b. The antibodies bind to the fiber component of the extracellular matrix.
   c. The antibodies bind to receptors on the cell surface.
   d. The antibodies act as enzymes that break down desmosomes.
4. What does it mean to say that a signal is transduced?
   a. The signal enters the cell directly and binds to a receptor inside.
   b. The physical form of the signal changes between the outside of the cell and the inside.
   c. The signal is amplified, such that even a single molecule evokes a large response.
   d. The signal triggers a sequence of phosphorylation events inside the cell.
5. Why are tight junctions found in only certain types of tissues, while desmosomes are found in a wide array of cells?
   a. Tight junctions are required only in cells where communication between adjacent cells is particularly important.
   b. Tight junctions are not as strong as desmosomes.
   c. Tight junctions have different structures but the same functions.
   d. Tight junctions are found only in epithelial cells that must be watertight.
6. What physical event represents the receipt of an intercellular signal?
   a. the passage of ions through a desmosome
   b. the activation of the first protein in a phosphorylation cascade
   c. the binding of a hormone to a signal receptor, which changes conformation in response
   d. the activation of a G protein associated with a signal receptor

Conceptual Review

1. Why is it difficult to damage a fiber composite?
2. Why does a phosphorylation cascade amplify an intercellular signal?
3. What is the difference between a tight junction and a desmosome?
4. Animal cells adhere to each other selectively. Summarize experimental evidence that supports this statement.
5. Make a sketch showing the reception, processing, and response steps in the signal transduction pathway featured in this chapter's essay.
6. Why do researchers use the term *pathway* in reference to signal transduction?

Group Discussion Problems

1. Suppose that an animal species and a plant species each lacked the ability to secrete an extracellular matrix. What would these organisms look like, and how would they live?
2. Suppose that a signal that is released from one cell triggered a response in another cell in the absence of signal transduction. Diagram the sequence of events that would occur between the arrival of the signal and the arrival of the response. Compare and contrast the response to a signal that does involve signal transduction.
3. According to the types of phylogenetic analyses introduced in Chapter 1, multicellular bodies evolved independently in plants and animals. Both plants and animals have structures that knit adjacent cells together and that form openings between cells. But the molecules in these structures differ markedly between the two groups. Are these observations logical? Explain.
4. Suppose you created an antibody that bound to one of the receptors illustrated in Figure 8.16. How would the signal transduction pathway be affected?

Answers to Multiple-Choice Questions 1. b; 2. b; 3. a; 4. b; 5. d; 6. c

www.prenhall.com/freeman is your resource for the following: Web Tutorials; Online Quizzes and other Online Study Guide materials; Answers to Conceptual Review Questions; Solutions to Group Discussion Problems; Answers to Figure Caption Questions and Exercises; and Additional Readings and Research.

Cellular Respiration and Fermentation

9

When table sugar is heated, it undergoes the uncontrolled oxidation reaction known as burning. Burning gives off heat. In cells, the simple sugar called glucose is oxidized through a long series of carefully regulated reactions. Instead of being given off as heat, some of the energy produced by these reactions is used to synthesize ATP.

KEY CONCEPTS

- Cellular respiration produces ATP from molecules with high potential energy—often glucose.
- Glucose processing has three components: (1) glycolysis, (2) the Krebs cycle, and (3) electron transport coupled with oxidative phosphorylation.
- Fermentation pathways allow glycolysis to continue when the lack of an electron acceptor shuts down electron transport chains.
- Respiration and fermentation are carefully regulated.

Cells are dynamic. Vesicles move cargo from the Golgi apparatus to the plasma membrane and other destinations; enzymes synthesize a complex array of macromolecules; and millions of membrane proteins pump ions and molecules to create an environment conducive to life. How does all this activity occur? The answer lies in the molecule **adenosine triphosphate (ATP)**. ATP has high potential energy and acts as the major energy currency in cells.

In earlier chapters we explored the role that ATP plays in driving endergonic reactions, in moving materials through cells, and in moving cells themselves through the environment. Recall that when a phosphate group is transferred to a molecule, the molecule often becomes more reactive. This action is particularly important in the case of proteins. When ATP or a phosphate group from ATP binds to a specific site on an enzyme, the enzyme's shape and activity can change. When ATP binds to a motor protein such as kinesin or dynein or myosin, the protein responds by moving in a way that transports materials or changes the cell's shape. The transfer of a phosphate group to a protein pump changes its shape in a way that transports ions or molecules across the plasma membrane against an electrochemical gradient (see Chapter 6).

In short, ATP allows cells to do work. Because staying alive takes work, there is no life without ATP.

The goal of this chapter is to investigate how cells make adenosine triphosphate by oxidizing compounds such as sugars that have high potential energy. The energy released during the oxidation is used to transfer a phosphate group to **adenosine diphosphate (ADP)** and generate ATP. Section 9.1 introduces the events involved by reviewing the fundamental principles of reduction and oxidation and summarizing the reactions involved in oxidizing glucose, the most common fuel used by organisms.

Then Sections 9.2 through 9.4 delve into the glucose-oxidizing reactions in detail. Section 9.5 introduces an alternative route for ATP production—called fermentation—that occurs in many bacteria, archaea, and eukaryotes. In closing, the chapter examines how cells shunt certain carbon-containing compounds away from ATP production and into the synthesis of DNA, RNA, amino acids, and other molecules. This chapter is your introduction to **metabolism**—the chemical reactions that occur in cells.

9.1 An Overview of Cellular Respiration

Cells use chemical energy to move molecules, pump ions, complete endergonic reactions, and perform other types of work. Because the electrons in the carbon-hydrogen bonds of carbohydrates and fats contain a great deal of chemical energy, they make it possible for cells to do a great deal of work. But even if a cell has huge amounts of chemical energy stored in the form of carbohydrates and fats, no movement, pumping by membrane proteins, or endergonic reactions occur. Why? Just as the dollar, euro, and yen act as currencies in human economies, ATP acts as the major currency in the economy of a cell. A huge savings account at a bank won't do you any good when you walk up to a vending machine to get a snack—you need cash. Similarly, cells are not able to use the chemical energy stored in carbohydrates or fats to do work. To be useful, the chemical energy must first be transferred to ATP.

It's important to recognize, though, that cells don't hoard large wads of cash in the form of ATP. In general, a cell contains only enough ATP to last from 30 seconds to a few minutes. Like many other cellular processes, the production and use of ATP is dynamic. Because ATP makes every cell function possible, producing a steady supply of it is among the most fundamental of all cell processes.

If carbohydrates and fats act as a savings account for chemical energy, how do cells draw on this account to synthesize ATP? In plants, algae, animals, fungi, and many other organisms, the simple sugar glucose ($C_6H_{12}O_6$) is the key intermediary (**Figure 9.1**). Most glucose is produced by photosynthetic species, which use the energy in sunlight to reduce carbon dioxide (CO_2) to glucose and other carbohydrates. When photosynthetic species decompose or are eaten, they provide glucose to animals, fungi, and many bacteria and archaea. All organisms use glucose as a building block in the synthesis of carbohydrates, fats, and other energy-storage compounds. When a cell needs energy, glucose is used to produce ATP through processes called cellular respiration and fermentation.

Chapter 10 focuses on how plants and other photosynthetic organisms produce glucose from carbon dioxide, water, and sunlight. Here we focus on how glucose is used to produce ATP. Some of the chemistry involved in processing glucose actually predates the evolution of photosynthesis. Indeed, the reactions analyzed early in this chapter were probably occurring as the first cell sprang to life. The ancestry of some of the enzymes introduced in this chapter dates back 3.5 billion years.

Before analyzing what these enzymes are and what they do, though, let's briefly review the general nature of chemical ener-

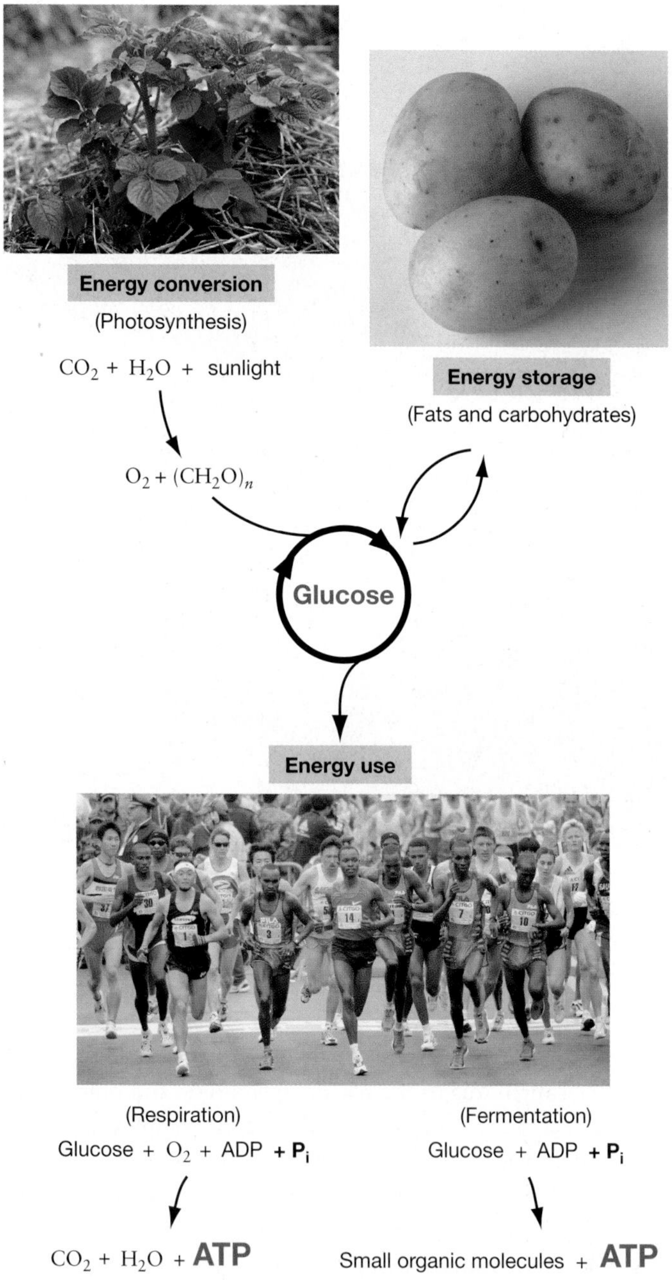
Energy conversion
(Photosynthesis)
$CO_2 + H_2O$ + sunlight
$O_2 + (CH_2O)_n$
Energy storage
(Fats and carbohydrates)
Glucose
Energy use
(Respiration)
Glucose + O_2 + ADP + P_i
$CO_2 + H_2O$ + ATP
(Fermentation)
Glucose + ADP + P_i
Small organic molecules + ATP

FIGURE 9.1 Glucose Is the Hub of Energy Processing in Cells Glucose is the endpoint of energy production via photosynthesis. It is also the intermediary molecule in energy storage. In a cell, when glucose is abundant, it is used to synthesize fats and storage carbohydrates. When glucose is scarce, it is supplied by the breakdown of fats and storage carbohydrates. Glucose is also used to provide chemical energy in the form of ATP, via cellular respiration and fermentation. EXERCISE The CO_2 produced by respiration is used by plants during photosynthesis. The O_2 produced by photosynthesis is used during respiration. Add arrows to this diagram to indicate these relationships.

gy and reduction-oxidation (redox) reactions. Recall from Chapter 2 that redox reactions involve the transfer of an electron from one substance to another. The compound that receives the electron is reduced, while the atom or molecule that gives up the electron is oxidized.

The Nature of Chemical Energy and Redox Reactions

Recall from Chapter 2 that chemical energy is a form of potential energy. Potential energy is energy that is associated with position or configuration. In cells, electrons are the most important source of chemical energy—specifically the electrons in ATP. The amount of potential energy that an electron has is based on its position relative to other electrons and to the protons (positive charges) in the nuclei of nearby atoms. The potential energy of a molecule is a function of the way its electrons are configured.

ATP makes things happen in cells because it has a great deal of potential energy. To understand why this is so, study the molecular formula in **Figure 9.2a**. Note that the three phosphate groups in ATP contain a total of four negative charges and that these charges are confined to a small area. In part because these negative charges repel each other, the potential energy of the electrons in the phosphate groups is extraordinarily high.

When ATP reacts with water during a hydrolysis reaction, the bond between ATP's outermost phosphate group and its neighbor is broken, resulting in the formation of ADP and inorganic phosphate, P_i (**Figure 9.2b**). This reaction is highly exergonic. Under standard conditions of temperature and pressure in the laboratory, a total of 7.3 kilocalories (kcal) of energy, per mole of ATP, is released during the reaction. The hydrolysis of ATP is exergonic because the entropy of the product molecules is much higher than that of the reactants, and because there is a large drop in potential energy when ADP and P_i are formed from ATP. The change in potential energy occurs in part because the electrons from ATP's phosphate groups are now spread between two molecules instead of being clustered on one molecule, meaning that there is less electrical repulsion. In addition, the negative charges on ADP and P_i are stabilized much more efficiently by interactions with the partial positive charges on surrounding water molecules than are the charges on ATP.

For similar reasons, the transfer of a phosphate group from ATP to a protein is usually exergonic. When a phosphate group is transferred from ATP to a protein, two negative charges are added to the protein. In response to this phosphorylation event, the electrons in the protein change configuration and the molecule's shape, or conformation, is usually altered (**Figure 9.2c**).

(a) ATP consists of three phosphate groups, ribose, and adenine.

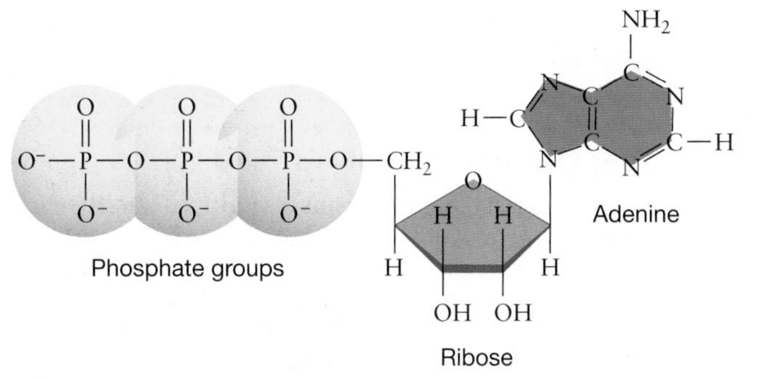

(b) Energy is released when ATP is hydrolyzed.

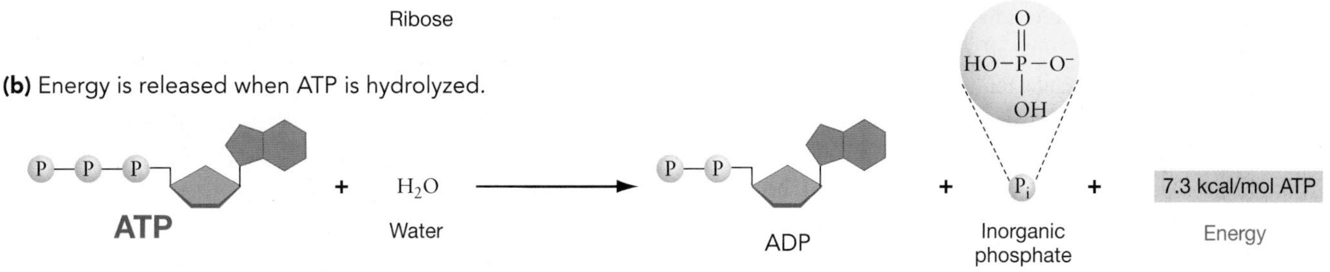

(c) Phosphorylation generates shape change in proteins.

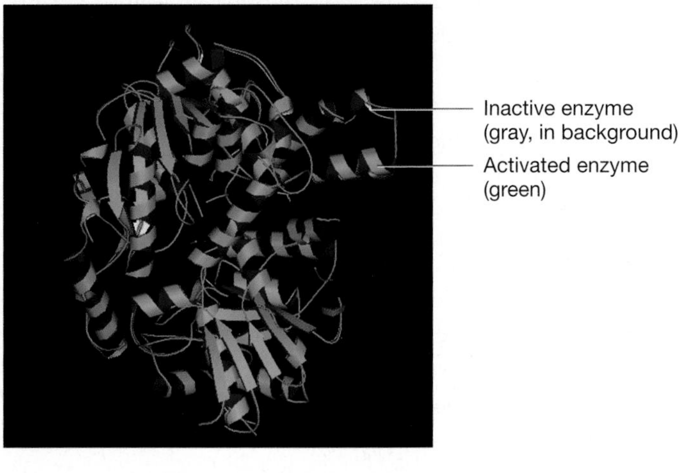

FIGURE 9.2 Adenosine Triphosphate (ATP)
(a) ATP has high potential energy, in part because four negative charges are clustered in its three phosphate groups. The negative charges repel each other, raising the potential energy of the electrons. **(b)** When ATP is hydrolyzed to ADP and inorganic phosphate, a large free-energy change occurs. **(c)** When proteins are phosphorylated or an ATP molecule binds to them, they often change shape in a way that changes their activity. This protein is an enzyme called arsenite-transporting ATPase. EXERCISE Circle portions of the enzyme in part (c) that change conformation between the activated and inactive states.

Part of the protein moves. Movement in response to phosphorylation or to the addition of the entire ATP molecule is what transports materials, powers flagella or cilia, pumps ions, drives endergonic reactions, and completes other kinds of work in the cell.

If a great deal of energy is released when ATP loses a phosphate group, then a great deal of energy must be required to synthesize ATP from ADP by adding P_i. Where does this energy come from? The answer is redox reactions.

Using Redox Reactions to Produce ATP

Recall from Chapter 2 that redox reactions involve a transfer of electrons. When an atom or molecule is reduced, it gains an electron from another atom or molecule. The compound that donates the electron becomes oxidized in the process.

In cells, molecules that gain an electron and become reduced also often gain a proton (H^+) and thus a hydrogen atom. As a result, reduced compounds in cells tend to have many C–H bonds. The C–H bonds have high potential energy, because the electrons in these bonds are not held tightly—recall from Chapter 2 that carbon and hydrogen atoms have low and approximately equal electronegativities. Thus, molecules that have a large number of C–H bonds, such as carbohydrates and fats, store a great deal of potential energy. During photosynthesis, for example, the energy in sunlight is used to add electrons to carbon dioxide (CO_2), resulting in the formation of glucose ($C_6H_{12}O_6$). Glucose contains many C–H bonds and thus a great deal of potential energy.

Conversely, molecules that are oxidized in cells often lose a proton along with an electron. Instead of having many C–H bonds, oxidized molecules in cells tend to have many C–O bonds. Such molecules also have lower potential energy. To understand why, recall from Chapter 2 that oxygen atoms have extremely high electronegativity. Because oxygen atoms hold electrons so tightly, the electrons involved in bonds with oxygen atoms have low potential energy.

To put all of these ideas together, consider what happens when glucose undergoes the uncontrolled oxidation reaction called burning:

$$C_6H_{12}O_6 + 6\ O_2 \rightarrow 6\ CO_2 + 6\ H_2O + \text{energy}$$

In this reaction, electrons are transferred from glucose to oxygen, and protons follow. The carbon atoms in glucose are oxidized, forming carbon dioxide. The oxygen atoms in the oxygen molecule (O_2) are reduced, forming water. Glucose is the molecule that acts as an electron donor and becomes oxidized; oxygen is the molecule that acts as an electron acceptor and becomes reduced. A large drop in potential energy occurs during the oxidation of glucose because the electrons are much more tightly bound in the C–O and H–O bonds of the product molecules than they were in the C–H bonds of the reactant molecules. Entropy also increases as a result of this reaction. When glucose undergoes the uncontrolled oxidation reaction called burning, the change in potential energy is converted to kinetic energy in the form of heat. When one mole of glucose burns, a total of 686 kcal of heat is released.

Glucose does not burn in cells, however. Instead, the glucose in cells is oxidized through a long series of carefully controlled redox reactions. These reactions are occurring, millions of times per minute, in your cells right now. Instead of being given off as heat, much of the energy released by the drop in potential energy is being used to make the ATP you need to think, move, and stay alive. In cells, the change in free energy that occurs during the oxidation of glucose is used to synthesize ATP from ADP and P_i.

How does the oxidation of glucose take place in a way that supports the production of ATP? In many organisms, cellular respiration is a three-step process: (1) Glucose is broken down to pyruvate; (2) pyruvate is oxidized to CO_2; and (3) compounds that were reduced in steps 1 and 2 are oxidized in processes that usually lead to the production of ATP. Let's get an overview of the way these steps work, and then pursue a more detailed analysis in Sections 9.2 through 9.4.

Processing Glucose: Glycolysis

An enormous diversity of organisms use glucose as their primary fuel. In virtually all species examined to date, the first step in the oxidation of glucose is a sequence of 10 chemical reactions that are collectively called **glycolysis** ("sugar breakdown"). These reactions occur in the cytosol of eukaryotic and prokaryotic cells. During glycolysis, one molecule of the six-carbon sugar glucose is broken into two molecules of the three-carbon compound pyruvate. Some of the potential energy released by this sequence of reactions is used to phosphorylate ADP molecules, forming ATP. In addition, one of the reactions in the sequence results in the reduction of a molecule called nicotinamide adenine dinucleotide, symbolized NAD^+. As **Figure 9.3** shows, the addition of two electrons and a proton to NAD^+ produces NADH. We'll soon see that NADH readily donates electrons to other molecules. As a result, it is called an **electron carrier** and is said to have "reducing power."

Figure 9.4a summarizes the results of the reactions of glycolysis. Notice that glucose goes in and that ATP, NADH, and pyruvate come out.

The Krebs Cycle

What happens to the pyruvate produced by glycolysis? If an electron acceptor such as oxygen is present in the cell, pyruvate enters a sequence of reactions called the **Krebs cycle**. In eukaryotic cells, the enzymes involved in the Krebs cycle are located inside mitochondria. Some of these enzymes are found in the

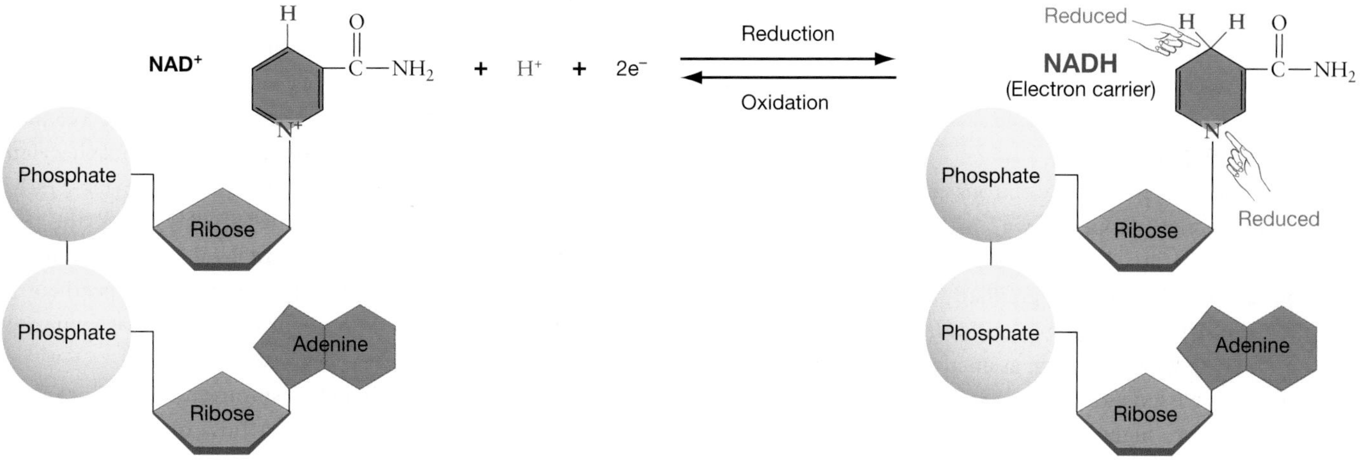

FIGURE 9.3 NAD^+ and NADH
NADH is the reduced form of NAD^+. QUESTION Compare the structure of NAD^+ with the structure of ATP in Figure 9.2. Which portions of the two molecules are identical?

matrix inside the organelle, while others are associated with the surface of the mitochondrion's inner membrane.

In the reactions that lead up to and complete the Krebs cycle, each pyruvate is oxidized to three molecules of carbon dioxide. Some of the potential energy released by these reactions is used to (1) reduce NAD^+ to NADH; (2) reduce another electron carrier, called flavin adenine dinucleotide (FAD), to $FADH_2$; and (3) phosphorylate ADP, yielding ATP (**Figure 9.4b**). Pyruvate enters this sequence of reactions; carbon dioxide, NADH, $FADH_2$, and ATP come out of the sequence.

With the completion of the Krebs cycle, glucose has been completely oxidized to CO_2. This observation raises several questions, though. According to the overall reaction for the oxidation of glucose ($C_6H_{12}O_6 + 6\ O_2 \rightarrow 6\ CO_2 + 6\ H_2O$ + energy), molecular oxygen (O_2) is a reactant. Where does oxygen come

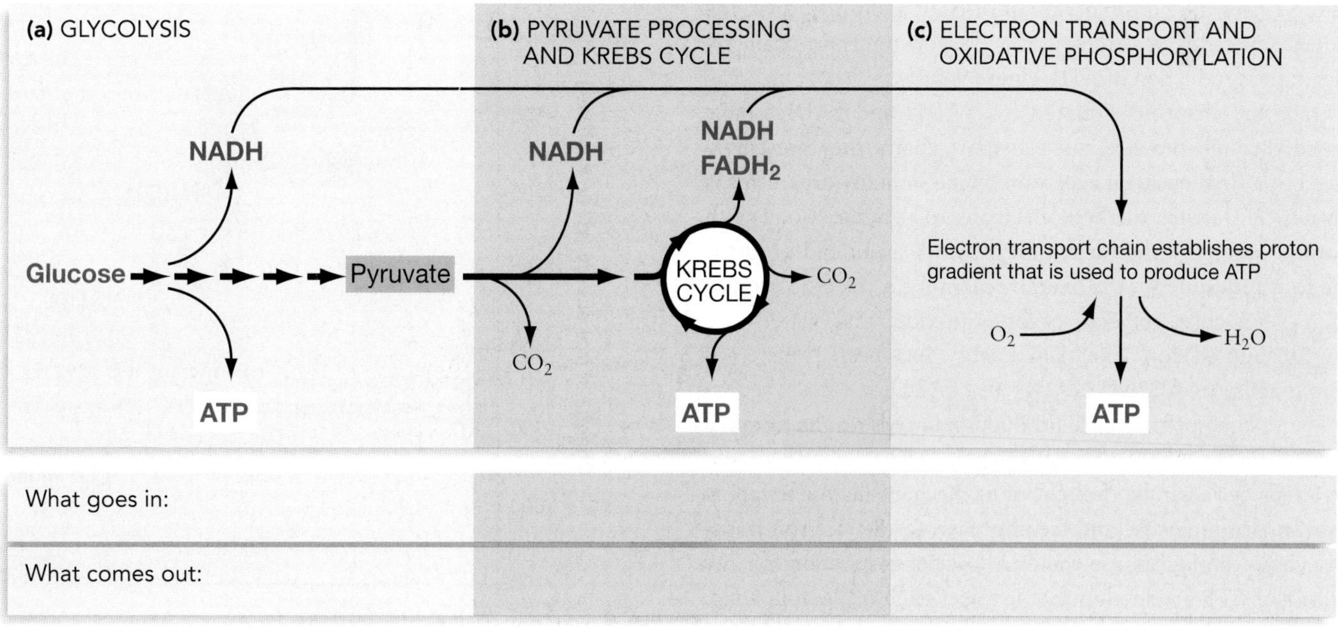

FIGURE 9.4 An Overview of Glucose Oxidation
Cells produce ATP from glucose via a series of processes: **(a)** glycolysis, **(b)** the Krebs cycle, and **(c)** electron transport combined with oxidative phosphorylation. Each component produces at least some ATP. Because the three components are connected, glucose oxidation is an integrated metabolic pathway. Glycolysis and the Krebs cycle are connected by pyruvate, and glycolysis and the Krebs cycle are connected to electron transport chains by NADH and $FADH_2$. EXERCISE Fill in the chart along the bottom.

into play, if not in glycolysis or the Krebs cycle? And what happens to all of the electrons that have been transferred from glucose to NADH and $FADH_2$?

Electron Transport

In cells, the potential energy of electrons carried by NADH and $FADH_2$ is gradually "stepped down" by molecules that participate in a series of redox reactions. Stated another way, the potential energy in the electrons is gradually decreased via a series of step-by-step reactions. The molecules involved in the redox reactions make up what is known as an **electron transport chain.** As electrons are passed along the electron transport chain, they gradually fall from a higher to a lower potential energy.

In eukaryotes, the components of the electron transport chain are located in the inner membrane and cristae of mitochondria; in prokaryotes, they are found in the plasma membrane. The transfer of an electron to each molecule in the electron transport chain changes the molecule's shape and activity. In several cases, the molecule that receives the electron responds by changing conformation in a way that transfers a proton across the mitochondrion's inner membrane. This is crucial. In effect, the movement of electrons through the chain results in the pumping of protons to the outside of the membrane. The subsequent buildup of protons results in a strong electrochemical gradient across the mitochondrial membrane. The gradient drives the protons through a membrane protein called ATP synthase. In eukaryotes, ATP synthase is located in the inner membrane and cristae of mitochondria. The force generated by the flow of protons through ATP synthase makes part of the protein spin. This change in conformation drives the phosphorylation of ADP. Indirectly, then, the sequence of redox reactions in the electron transport chain results in the production of ATP (**Figure 9.4c**).

Once the electrons donated by NADH and $FADH_2$ have passed through the electron transport chain, they are transferred to a final electron acceptor, which in many organisms is oxygen. A transfer of these electrons to oxygen, along with protons, results in the formation of water as an end-product. The four molecules in the overall reaction for the oxidation of glucose—glucose, oxygen, carbon dioxide, and water—are now accounted for. Chemical energy has been transferred from glucose to ATP, via NADH and $FADH_2$.

The vast majority of ATP production in cells results from the reactions in the electron transport chain. Hence, glycolysis and the Krebs cycle can be thought of as mechanisms for stripping electrons from glucose and feeding them to the electron transport chain. Biologists use the term **cellular respiration** for any process of ATP production that involves each of the following: a compound that acts as an electron donor, an electron transport chain, and an electron acceptor.

Methods of Producing ATP

The phosphorylation events that are catalyzed by ATP synthase are powered by a stream of protons. The gradient that drives these protons is set up by an electron transport chain that uses a molecule with low free energy—usually oxygen—as the final electron acceptor. Because this mode of ATP production links the phosphorylation of ADP with the oxidation of NADH and $FADH_2$, it is called **oxidative phosphorylation.** But when ADP is used as a substrate in enzyme-catalyzed phosphorylation reactions during glycolysis or the Krebs cycle, the event is called **substrate-level phosphorylation.** In substrate-level phosphorylation, the energy to produce ATP comes from a phosphorylated substrate and not from a proton gradient.

In many organisms, cellular respiration begins with the oxidation of glucose and ends with a flow of electrons through an electron transport chain to a final electron acceptor. The process starts with glycolysis and the Krebs cycle, which oxidize glucose to CO_2. As **Figure 9.5** shows, these reactions produce dramatic free-energy changes. Note that relative to the other compounds in glycoysis and the Krebs cycle, glucose has a great deal of potential energy and low entropy. During glycolysis and the Krebs cycle, the free energy of each subsequent compound involved drops—at times in small steps and at other times in large steps. The small drops in free energy are coupled with the production of ATP; the large drops are asso-

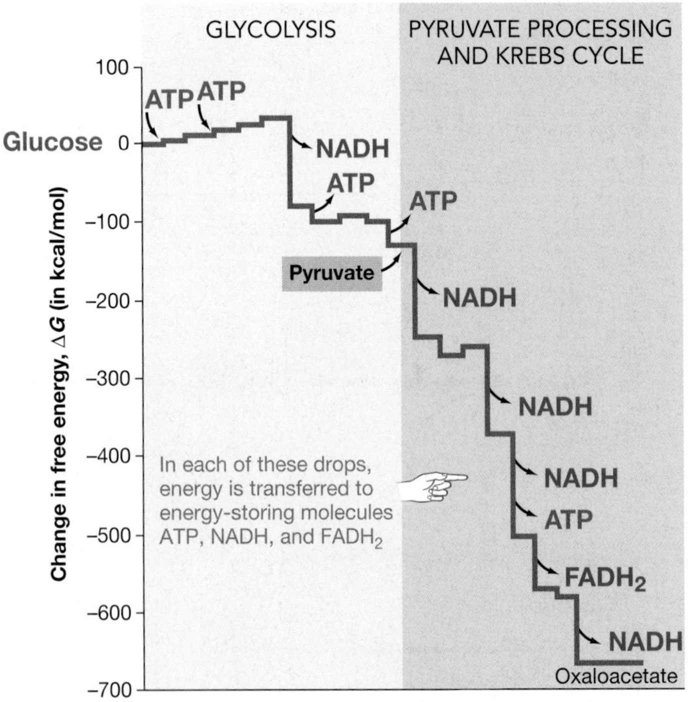

FIGURE 9.5 Free-Energy Changes as Glucose Is Oxidized
Graph of free-energy changes during glycolysis and the Krebs cycle. Drops in free energy are associated with the production of ATP, NADH, and $FADH_2$. Oxaloacetate is the endpoint of the Krebs cycle.
QUESTION Which is associated with larger changes in free energy: production of ATP or production of NADH and $FADH_2$? Add labels indicating where in the cell glycolysis and the Krebs cycle take place.

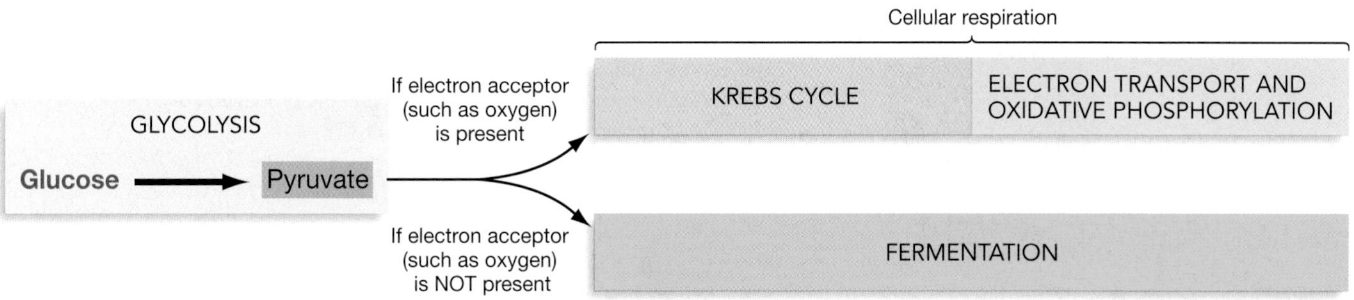

FIGURE 9.6 Cellular Respiration and Fermentation Are Alternative Pathways for Producing Energy
When oxygen or another electron acceptor is present in a cell, the pyruvate produced by glycolysis enters the Krebs cycle and the electron transport system is active. But if no electron acceptor is available, the pyruvate undergoes reactions known as fermentation.

ciated with the synthesis of NADH and $FADH_2$. In effect, then, the free energy present in glucose is transferred to ATP, NADH, and $FADH_2$. In the third (and final) stage of cellular respiration, NADH and $FADH_2$ carry these electrons to the electron transport chain.

If the molecule that functions as the final electron acceptor is unavailable, however, then cellular respiration cannot occur. In this case, alternative pathways called fermentation reactions take over (**Figure 9.6**). In eukaryotes, for example, fermentation occurs when oxygen is lacking. As we'll see in Section 9.5, **fermentation** consists of reactions that allow glycolysis to continue even if pyruvate is not drawn off to the Krebs cycle. In eukaryotes, by-products such as ethanol (the active ingredient in alcoholic beverages) or lactic acid result from various fermentation pathways. In bacteria and archaea, other products are possible.

This overview of cellular respiration is intended simply to introduce the processes responsible for most ATP production in cells. Now we need to explore glycolysis, the Krebs cycle, and electron transport chains in more detail. Let's take a closer look at each process in turn.

9.2 Glycolysis

Glycolysis may be the most fundamental of all metabolic pathways, but it was discovered by accident. In the late 1890s Hans and Edward Buchner were working out techniques for manufacturing extracts of baker's yeast for therapeutic use. (Yeast extracts are still added to some foods as a flavor enhancer or nutritional supplement.) In one set of experiments the Buchners added sucrose, or table sugar, to their extracts. Sucrose is a disaccharide consisting of glucose linked to another six-carbon sugar, called fructose. At the time, sucrose was commonly used as a preservative. But the Buchners found that instead of preserving the yeast extracts, the sucrose was quickly broken down and fermented, with alcohol appearing as a by-product. This was a key finding because it showed that fermentation and other types of cellular metabolism could be studied in vitro—outside the cell. Until then, researchers thought that metabolism could take place only inside cells.

The Buchners and other researchers followed up on this observation by trying to determine how the sugar was being processed. An important early observation was that the reactions could be sustained much longer than normal if inorganic phosphate was added to the mixture. This finding implied that some of the compounds involved were being phosphorylated. Soon after, a molecule called fructose bisphosphate was isolated. (The *bis*–prefix means that two phosphate groups are attached to the molecule at distinct locations.) Subsequent work showed that all but two of the compounds involved in glycolysis—the starting and ending molecules, glucose and pyruvate—are phosphorylated.

A third major finding came to light in 1905, when researchers found that the processing of sugar by yeast extracts stopped if the reaction mix was boiled. Because enzymes were known to be inactivated by heat, this discovery suggested that enzymes were involved in at least some of the processing steps. Years later, investigators realized that each step in glycolysis is catalyzed by a different enzyme.

Over the next 35 years, each of the reactions and enzymes involved in glycolysis was gradually worked out by several different researchers. Because Gustav Embden and Otto Meyerhof identified many of the steps in glycolysis, this process is also known as the Embden-Meyerhof pathway.

A Closer Look at the Glycolytic Reactions

By breaking open cells, separating the cell components via differential centrifugation (introduced in Chapter 7), and testing which components could sustain glycolysis, biologists discovered that all 10 reactions of glycolysis occur in the cytosol.

FIGURE 9.7 Glycolysis Pathway
Glucose is oxidized to pyruvate through this sequence of 10 reactions. Each reaction is catalyzed by a different enzyme. The process begins with an investment of 2 molecules of ATP. Reactions later in the sequence result in the production of 2 molecules of NADH and 4 of ATP, for a net gain of 2 molecules of ATP.

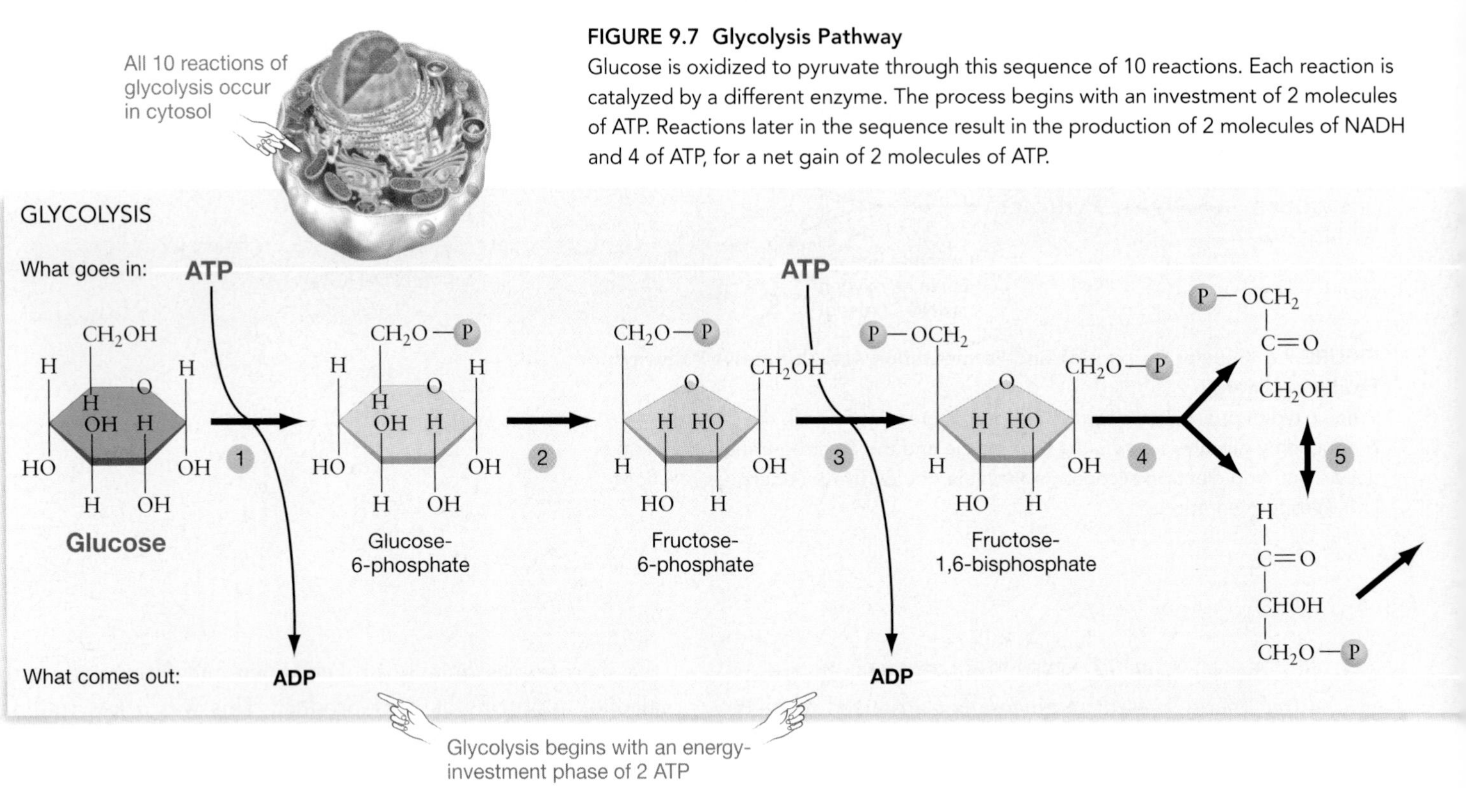

Figure 9.7 details the 10 reactions involved in glycolysis. There are three key points to note about the reaction sequence:

1. Contrary to what most researchers expected, glycolysis starts by *using* ATP, not producing it. In the initial step, glucose is phosphorylated to form glucose-6-phosphate. After an enzyme rearranges this molecule to fructose-6-phosphate in the second step, the third step in the reaction sequence adds a second phosphate group, forming the fructose-1,6-bisphosphate observed by early researchers. Thus, two ATP molecules are used up before any ATP is produced by glycolysis.
2. Once this energy-investment phase of glycolysis is complete, the subsequent reactions represent an energy payoff phase. The sixth reaction in the sequence results in the reduction of two molecules of NAD^+; the seventh produces two molecules of ATP and erases the energy "debt" of two molecules of ATP invested early in glycolysis. The final reaction in the sequence produces another two ATPs. For each molecule of glucose processed, the net yield is two molecules of NADH, two of ATP, and two of pyruvate.
3. The production of ATP during glycolysis occurs by substrate-level phosphorylation. In reactions 7 and 10 of Figure 9.7, an enzyme catalyzes the transfer of a phosphate group from a phosphorylated intermediate in glycolysis directly to ADP.

The discovery and elucidation of the glycolytic pathway ranks as one of the great achievements in the history of biochemistry. Because the enzymes involved have been observed in nearly every bacterium, archaean, and eukaryote examined, biologists infer that the pathway evolved very early in the history of life. It is very likely that the ancestor of all organisms living today made ATP by glycolysis. The reactions outlined in Figure 9.7 are among the most ancient and fundamental of life processes.

How Is Glycolysis Regulated?

Once the glycolytic pathway was worked out, researchers focused on the structures and functions of the enzymes involved and the way the sequence is regulated. It seemed logical to predict that glycolysis does not occur at the maximum rate at all times. Instead, researchers predicted that glycolysis would proceed only when cells needed fresh supplies of ATP. An important advance occurred when biologists observed that high levels of ATP inhibit a key glycolytic enzyme called phosphofructokinase. **Phosphofructokinase** catalyzes step 3 in Figure 9.7—the synthesis of fructose-1,6-bisphosphate from fructose-6-phosphate.

The discovery that ATP inhibits phosphofructokinase was important because this enzyme catalyzes a key step in the reaction sequence. Step 3 is highly exergonic. As a result, its product, fructose-1,6-bisphosphate, cannot easily be converted back into fructose-6-phosphate. In addition, an array of enzymes that aren't involved in glycolysis can convert the products of step 1 and step 2 to products used in other metabolic pathways. Before step 3, then, the sequence can be interrupted and the intermediates used elsewhere. But once fructose-1,6-bisphosphate is synthesized, it is difficult to turn back. Based on these observations, it is logical that the pathway is turned on or off at step 3.

The observation that ATP inhibits phosphofructokinase raises an important question, however: Why would a substrate that is required for the reaction in step 3 also inhibit the reac-

(Figure 9.7 continued)

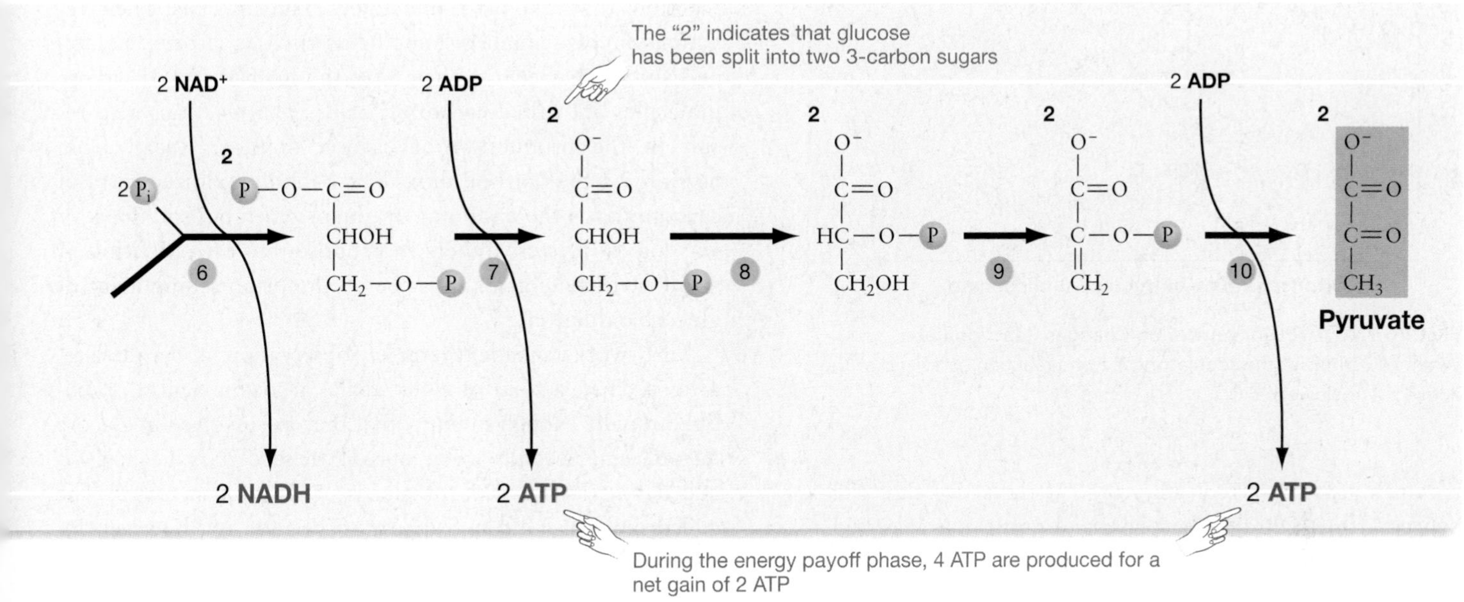

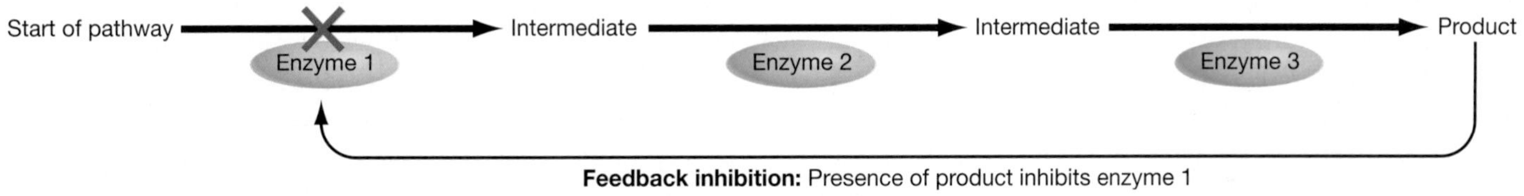

FIGURE 9.8 What Is Feedback Inhibition?
Feedback inhibition occurs when the product of a metabolic pathway inhibits an enzyme that is active early in the pathway. During glycolysis, feedback inhibition occurs when ATP binds to the regulatory site of phosphofructokinase.

tion? In the vast majority of cases, the addition of a substrate *speeds* the rate of a chemical reaction instead of slowing it.

To explain this unusual situation, biologists hypothesized that high levels of ATP are a signal that the cell does not need to produce more ATP. When an enzyme in a pathway is inhibited by the product of the reaction sequence, **feedback inhibition** is said to occur (**Figure 9.8**). The product molecule "feeds back" to stop the reaction sequence when the product is abundant. In glycolysis, cells that are able to stop the glycolytic reactions when ATP is abundant can conserve their stores of glucose for times when ATP is scarce. As a result, natural selection should favor individuals who have phosphofructokinase molecules that are inhibited by high concentrations of ATP.

This hypothesis provided a satisfying explanation for why ATP inhibits the key step of glycolysis, but it did not explain *how* high levels of the substrate inhibit the enzyme. The answer came later, when researchers were able to determine the three-dimensional structure of phosphofructokinase. As **Figure 9.9** shows, phosphofructokinase has two distinct binding sites for ATP. ATP can bind at the enzyme's active site or at a site that changes the

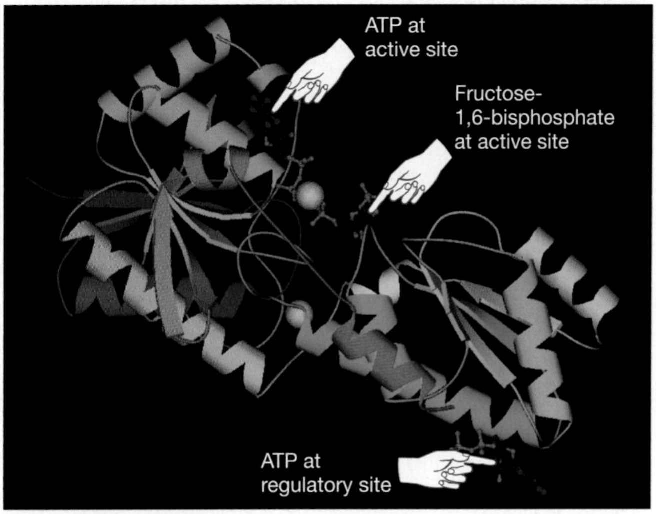

FIGURE 9.9 Phosphofructokinase Has Two Binding Sites for ATP
A model of one of the four identical subunits of phosphofructokinase. Notice the active site, where a phosphate group is transferred from ATP to form fructose-1,6-bisphosphate, and the regulatory site, where ATP binds.

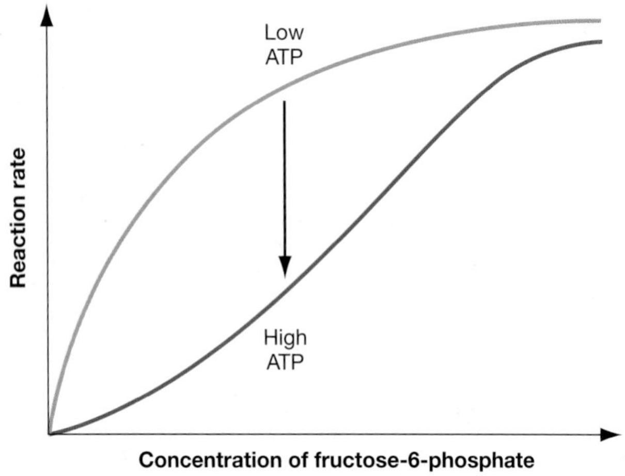

FIGURE 9.10 ATP Concentration Changes Reaction Rate
When ATP binds at the regulatory site, the reaction catalyzed at the active site is slowed.

enzyme's activity—a **regulatory site.** At the active site, ATP is converted to ADP and the phosphate group is transferred to fructose-6-phosphate. This reaction results in the synthesis of fructose-1,6-bisphosphate. But when ATP concentrations are high, the molecule also binds at a regulatory site. When ATP binds at this second location. The enzyme's conformation changes in a way that makes the reaction rate at the active site drop dramatically (**Figure 9.10**). To use the term introduced in Chapter 3, ATP acts as an allosteric regulator.

Thanks to advances such as these, glycolysis is among the best understood of all metabolic pathways. Now the question is, what happens to its product? How is pyruvate processed?

9.3 The Krebs Cycle

While Embden and Meyerhof and others were working out the sequence of reactions in glycolysis, biologists in several other laboratories were focusing on a different set of redox reactions that take place in actively respiring cells. These reactions involve small organic acids such as citrate, malate, and succinate. Because they have the form R-COOH, these molecules are called **carboxylic acids.** The reactions also result in the production of carbon dioxide. Recall from Section 9.1 that carbon dioxide is a highly oxidized form of carbon and is the endpoint of glucose metabolism. Thus, it was logical for researchers to propose that the oxidation of small carboxylic acids must be an important component of glucose oxidation.

Early workers made several key observations about these reactions. First, a total of eight small carboxylic acids are oxidized rapidly enough to imply that they are involved in glucose metabolism. Second, when one of these carboxylic acids is added to cells, the rate of glucose oxidation increases. The added molecules did not appear to be used up, however. Instead, virtually all of the carboxylic acid added seemed to be recovered later. Why this should be so was a puzzle. Even while this observation remained unexplained, however, the biochemists working on the reactions were able to determine the order in which the eight acids were oxidized. The result is the reaction pathway shown in **Figure 9.11a**.

The mystery of why intermediates in the pathway were not used up was resolved when Hans Krebs realized that the reaction sequence might occur in a cyclical fashion instead of a linear pathway. Krebs had another crucial insight when he suggested that the

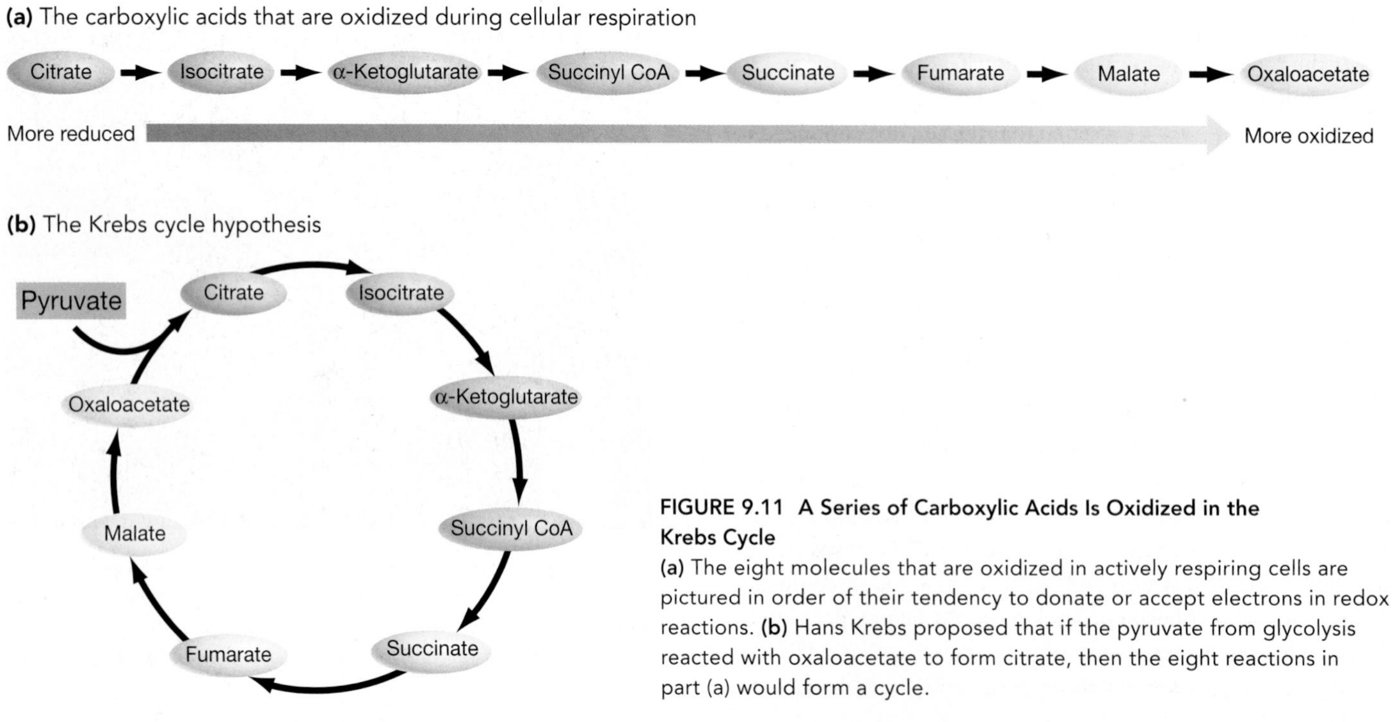

FIGURE 9.11 A Series of Carboxylic Acids Is Oxidized in the Krebs Cycle
(a) The eight molecules that are oxidized in actively respiring cells are pictured in order of their tendency to donate or accept electrons in redox reactions. **(b)** Hans Krebs proposed that if the pyruvate from glycolysis reacted with oxaloacetate to form citrate, then the eight reactions in part (a) would form a cycle.

reaction sequence was directly tied to the processing of pyruvate produced by glycolysis. To test these hypotheses, Krebs and a colleague set out to determine whether pyruvate—the endpoint of the glycolytic pathway—could react with oxaloacetate. Because oxaloacetate was the endpoint of the pathway in Figure 9.11a, the cycle hypothesis predicted that a reaction of oxaloacetate with pyruvate should produce the starting point of the pathway—the six-carbon carboxylic acid called citrate (**Figure 9.11b**).

Their result? As predicted, citrate formed when the biologists added oxaloacetate and pyruvate to cells. Based on this result, Krebs proposed that pyruvate is oxidized to carbon dioxide through a cycle of reactions. In honor of this insight, the pathway became known as the Krebs cycle.*

Converting Pyruvate to Acetyl CoA

When radioactive isotopes of carbon became available in the early 1940s, researchers in several labs used them to confirm the cyclical nature of the Krebs cycle. For example, by adding radioactively labeled citrate or pyruvate to cells and analyzing the radioactive compounds that resulted, it was possible to show that carbon atoms cycle through the sequence of reactions just as Krebs had proposed.

*The Krebs cycle is also known as the *citric acid cycle* because it starts with citrate, which becomes citric acid when protonated. Because citric acid has three carboxyl groups, the reaction sequence is also called the *tricarboxylic acid (TCA) cycle*.

Where do these reactions take place? In bacteria and archaea, the enzymes responsible for glycolysis and the Krebs cycle are located in the cytoplasm. In eukaryotes, though, the pyruvate produced by glycolysis is transported from the cytoplasm to mitochondria. As noted in Chapter 7, mitochondria are organelles found in virtually all eukaryotes. **Figure 9.12** shows a diagram and an electron micrograph of this organelle. Note that it has two membranes and that the interior is filled with layers of sac-like structures called **cristae**, which connect to the inner membrane. The region inside the inner membrane but outside the cristae is called the **mitochondrial matrix.** In eukaryotes, most of the enzymes responsible for the Krebs cycle are located in this matrix.

The pyruvate that links glycolysis and the Krebs cycle moves across the mitochondrion's outer membrane through small pores, then enters the mitochondrial matrix. Entry into the matrix occurs via active transport, through a membrane protein called the pyruvate carrier, located in the inner membrane. Because the transport of pyruvate into the mitochondrion is an active process, it represents an energy-consuming step in glucose oxidation.

What happens to pyruvate once it enters the mitochondrion? After Krebs's hypothesis was confirmed, researchers focused on understanding how pyruvate reacts with oxaloacetate to form citrate and start the cycle. This issue remained unresolved until it was discovered that an organic compound called **coenzyme A (CoA)** serves as a cofactor in a wide variety of reactions catalyzed by cellular enzymes. As we saw in Chapter 3, many enzymes

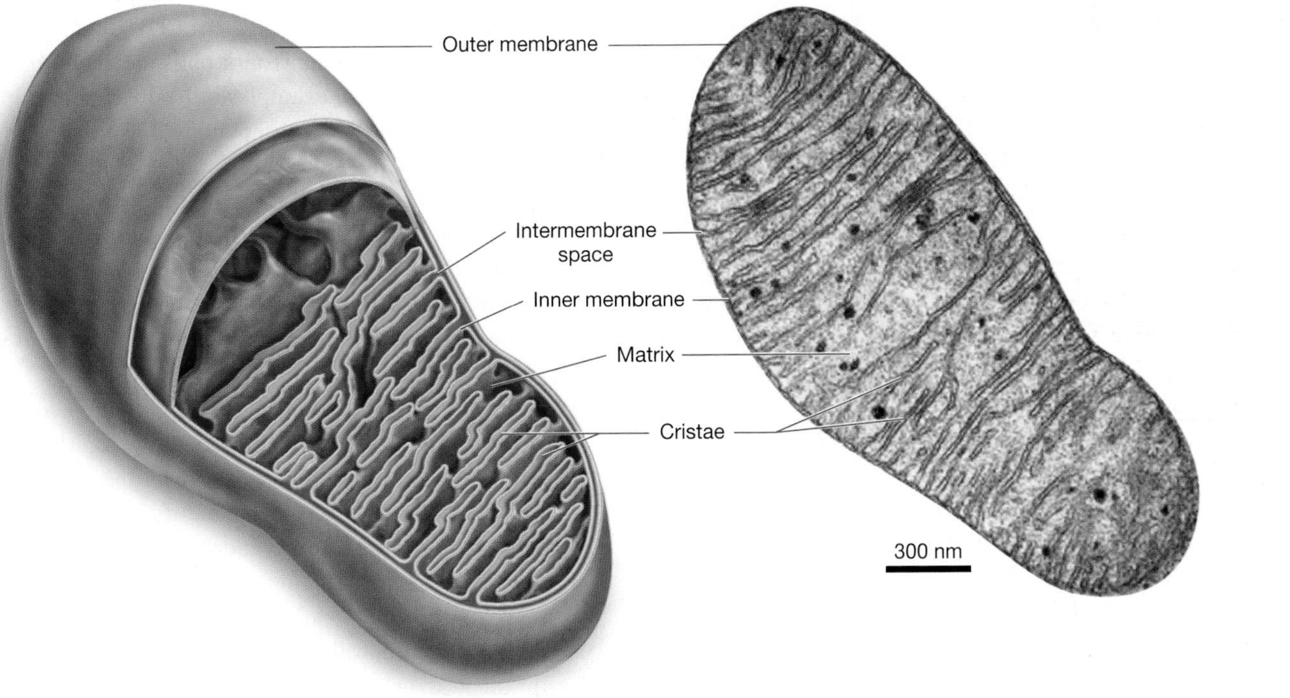

FIGURE 9.12 The Structure of the Mitochondrion
Mitochondria have inner and outer membranes. Saclike structures that contact the inner membrane are called cristae. The reactions of the Krebs cycle occur within the mitochondrial matrix.

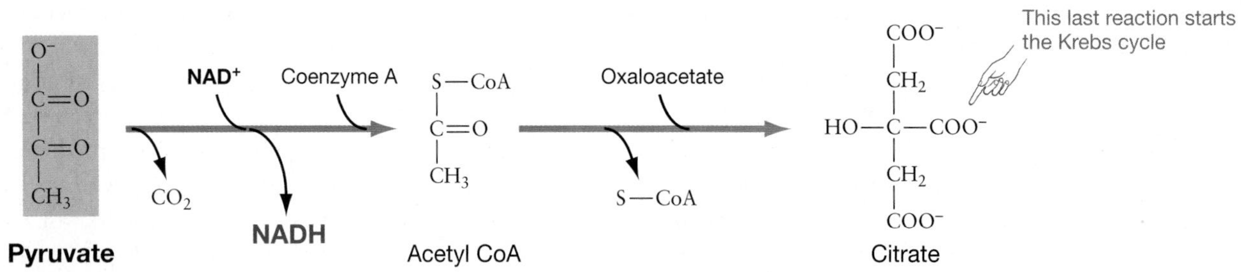

FIGURE 9.13 Pyruvate Is Oxidized to Acetyl CoA, Which Reacts with Oxaloacetate to Start the Krebs Cycle

Once it is inside the mitochondrion, the pyruvate produced by glycolysis is oxidized to acetyl CoA in a reaction catalyzed by pyruvate dehydrogenase. Acetyl CoA then reacts with oxaloacetate to form citrate in the first reaction of the Krebs cycle.

require a metal ion or a relatively small organic molecule called a coenzyme to function. Coenzymes often bind to the enzyme's active site and stabilize the reaction's transition state. Some coenzymes, for example, donate or accept electrons during the reaction; others transfer an acetyl ($-COCH_3$) group or carbon chains. CoA acts as a coenzyme by accepting and then transferring an acetyl group to a substrate. In fact, the "A" in "CoA" stands for *acetylation,* because an acetyl group is often bonded to a sulfur atom on one end of the molecule. When an acetyl group is bound to CoA, the acetyl group becomes activated and can be easily transferred to an acceptor molecule.

Soon after CoA's role as a coenzyme was revealed, other investigators found that pyruvate reacts with CoA to produce a molecule called **acetyl CoA.** Follow-up work showed that the conversion of pyruvate to acetyl CoA occurs in a series of steps inside an enormous and intricate enzyme complex called **pyruvate dehydrogenase.** The pyruvate dehydrogenase complex is located in the inner mitochondrial membrane and consists of three enzymes, which require a total of five cofactors. The coenzymes involved include thiamin, niacin, and pantothenic acid—molecules that you know as the B-complex vitamins.

As pyruvate is being processed, NAD^+ is reduced to NADH and one of the carbons in the pyruvate is oxidized to CO_2. The remaining two-carbon acetyl unit is transferred to CoA. The product, acetyl CoA, goes on to react with oxaloacetate to form citrate—the first compound in the Krebs cycle (**Figure 9.13**).

To summarize, the reactions catalyzed by pyruvate dehydrogenase act as a preparatory step for the Krebs cycle. The acetyl CoA that is produced goes on to react with oxaloacetate, forming citrate. When processing of pyruvate in the mitochondrion is complete, all three carbons in pyruvate are oxidized to carbon dioxide. One of the carbons is oxidized in the pyruvate dehydrogenase complex, and two are oxidized in the Krebs cycle itself. As **Figure 9.14** indicates, the energy released by the oxidation of one molecule of pyruvate is used to produce four molecules of NADH, one of $FADH_2$, and one of **guanosine triphosphate (GTP)** through substrate-level phosphorylation. GTP is then converted to ATP. Pyruvate and CoA enter the sequence of reactions; CoA, carbon dioxide, ATP, NADH, and $FADH_2$ come out.

How Is the Krebs Cycle Regulated?

Like glycolysis, the Krebs cycle is carefully regulated. Reaction rates are high when ATP is scarce; reaction rates are low when ATP is abundant. The major control points are highlighted in **Figure 9.15.** Note that the pyruvate dehydrogenase complex is regulated, as are several enzymes in the cycle itself.

How does regulation occur? When supplies of ATP are abundant, the pyruvate dehydrogenase complex becomes phosphorylated. In this case, phosphorylation changes the shape of pyruvate dehydrogenase in a way that inhibits catalytic activity. High concentrations of other products—specifically, acetyl CoA and NADH—also increase the rate of phosphorylation of the enzyme complex. Similarly, the enzyme that converts acetyl CoA to citrate is shut down when ATP binds to it. Like the control of phosphofructokinase activity in glycolysis, these are examples of feedback inhibition. Reaction products feed back to stop or slow down the pathway.

As Figure 9.15 indicates, feedback inhibition also occurs at two points later in the cycle. At the first of these two points, NADH binds to the enzyme's active site. This is an example of competitive inhibition. At the second point, ATP binds to an allosteric regulatory site. In sum, the reaction pathway can be turned off at multiple points, via several different mechanisms of feedback inhibition.

In addition to these examples of inhibitory control, the presence of NAD^+, CoA, or adenosine monophosphate (AMP)—which indicates low supplies of ATP—speeds the conversion of pyruvate to acetyl CoA. Stated another way, the key steps in the processing of pyruvate are under both positive control and negative control. By speeding up the oxidation of pyruvate when ATP supplies are low and by slowing down the reactions when ATP is plentiful, cells carefully match the rate of cellular respiration to their energy requirements. Natural selection has favored the evolution of enzymes that allow cells to conserve pyruvate or acetyl CoA, just as it has favored glycolytic enzymes whose regulation allows cells to conserve glucose.

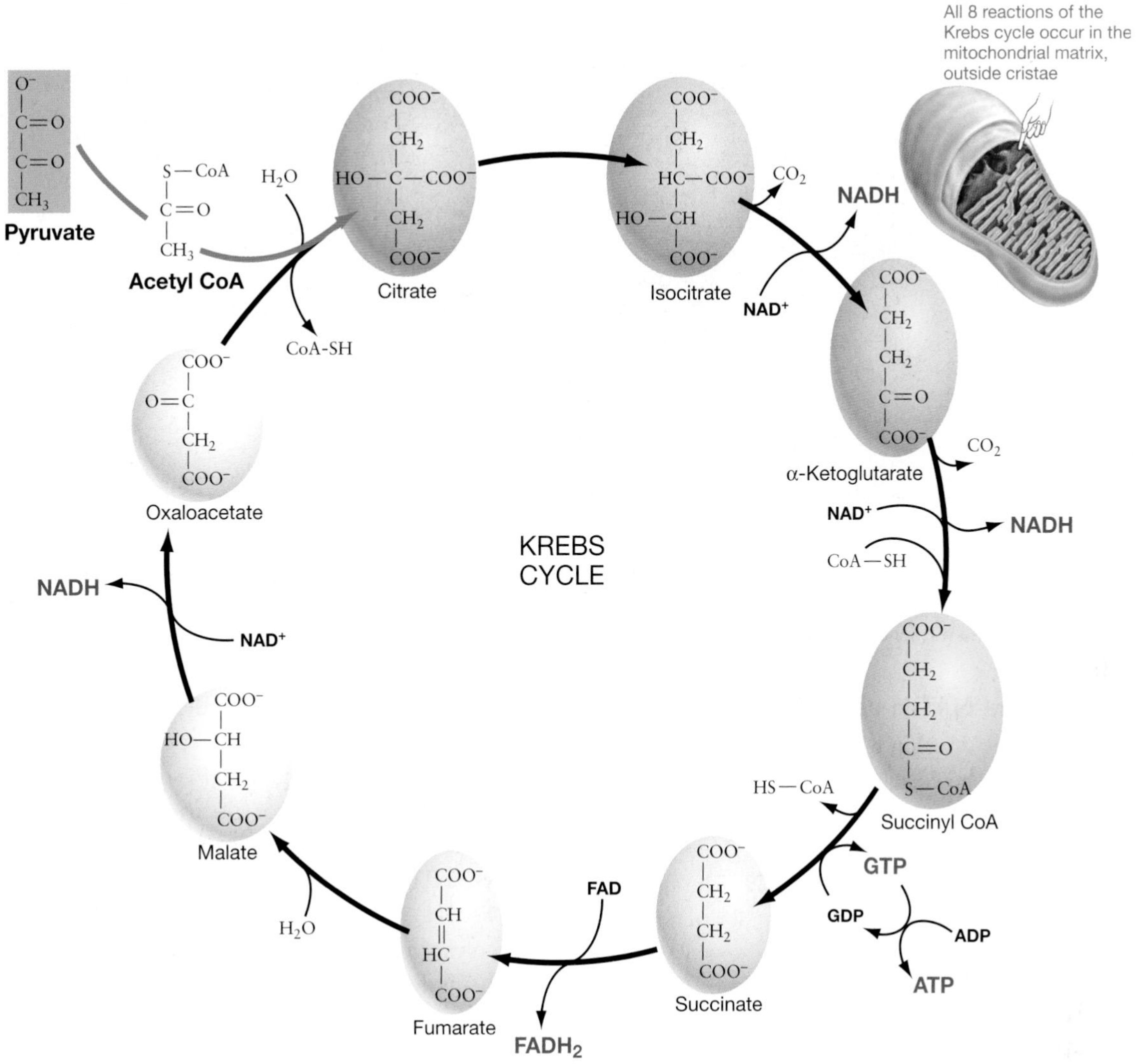

FIGURE 9.14 The Krebs Cycle
Acetyl CoA goes into the Krebs cycle, and carbon dioxide, NADH, $FADH_2$, and GTP come out. The GTP is produced by substrate-level phosphorylation and then converted to ATP. **EXERCISE** Mark one of the carbon atoms in acetyl CoA with a highlighter. Follow this atom from molecule to molecule around the cycle. What happens to it? Is it eventually given off as carbon dioxide?

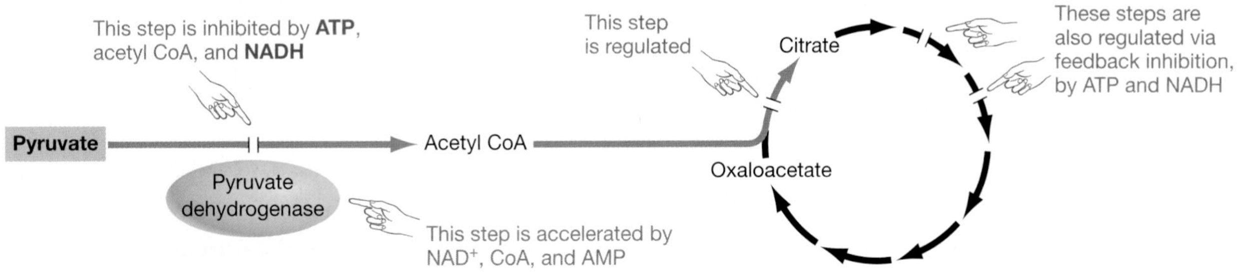

FIGURE 9.15 Pyruvate Dehydrogenase and the Krebs Cycle Are Carefully Regulated
The oxidation of pyruvate is under both negative and positive controls. Several steps in the Krebs cycle itself are also regulated. As a result, the reaction pathway speeds up when ATP supplies are low but slows down when ATP is plentiful.

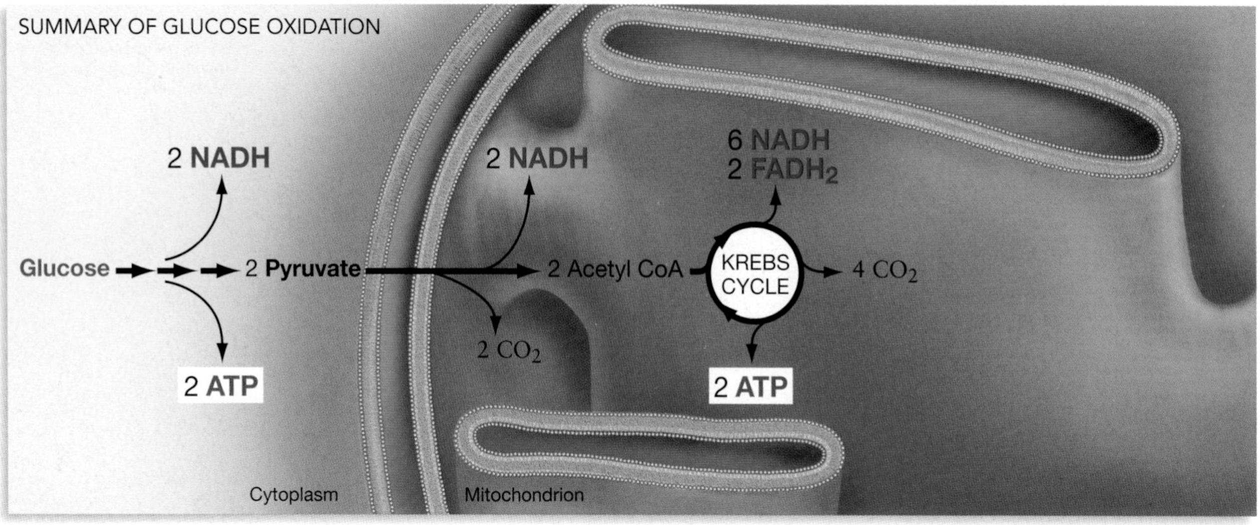

FIGURE 9.16 A Summary of Glucose Oxidation
Glucose is completely oxidized to carbon dioxide via glycolysis, the subsequent oxidation of pyruvate, and then the Krebs cycle. In eukaryotes, glycolysis occurs in the cytoplasm; pyruvate oxidation takes place in the inner mitchondrial membrane; the Krebs cycle occurs in the mitochondrial matrix.

What Happens to the NADH and $FADH_2$?

Figure 9.16 reviews the relationship between glycolysis and the Krebs cycle and identifies where each process takes place in eukaryotic cells. For each molecule of glucose that is fully oxidized to 6 carbon dioxide molecules, the cell produces 10 molecules of NADH, 2 of $FADH_2$, and 4 of ATP. The overall reaction for glycolysis and the Krebs cycle can be written as

$$C_6H_{12}O_6 + 10\ NAD^+ + 2\ FAD + 4\ ADP + 4P_i \rightarrow 6\ CO_2 + 10\ NADH + 2\ FADH_2 + 4\ ATP$$

The ATP molecules are produced by substrate-level phosphorylation and can be used to drive endergonic reactions, power movement, or run membrane pumps. The carbon dioxide molecules are a gas that is disposed of as waste—you exhale it.

What happens to the NADH and $FADH_2$ produced by glycolysis and the Krebs cycle? Recall that the overall reaction for glucose oxidation is

$$C_6H_{12}O_6 + 6\ O_2 \rightarrow 6\ CO_2 + 6\ H_2O + \text{energy}$$

Glycolysis and the Krebs cycle account for the glucose, the CO_2, and—because ATP is produced—some of the chemical energy that results from the overall reaction. But the O_2 and the H_2O that appear in the overall reaction for the oxidation of glucose are still unaccounted for. As it turns out, so is much of the chemical energy. The reaction that has yet to occur is

$$NADH + FADH_2 + O_2 + ADP + P_i \rightarrow NAD^+ + FAD + H_2O + ATP$$

In this reaction, oxygen is reduced to form water. The electrons that drive the redox reaction come from NADH and $FADH_2$. These molecules are oxidized to NAD^+ + FAD.

In effect, glycolysis and the Krebs cycle transfer electrons from glucose to NAD^+ and FAD, creating NADH and $FADH_2$. These molecules then carry the electrons to oxygen, which serves as the final electron acceptor in eukaryotic cells. When oxygen accepts electrons, water is produced. All the components of the overall reaction for glucose oxidation are accounted for.

How does this final part of the process occur? Specifically, how is ATP generated as electrons are transferred from NADH or $FADH_2$ to O_2? In the 1960s—decades after the details of glycolysis and the Krebs cycle had been worked out—a startling answer to these questions emerged.

✓CHECK YOUR UNDERSTANDING

During glycolysis and the Krebs cycle, one molecule of glucose ($C_6H_{12}O_6$) is completely oxidized to six molecules of carbon dioxide (CO_2). You should be able to model these components of cellular respiration. To do this, pretend that a large piece of paper is a cell. Draw a large mitochondrion inside it. Cut out small squares of paper and label them as glucose, glycolytic reactions, Krebs cycle reactions, pyruvate dehydrogenase complex, pyruvate, acetyl CoA, CO_2, ATP, NADH, and $FADH_2$. Put each of the squares in the appropriate location in the cell. Using dimes for electrons, demonstrate how glucose is oxidized to CO_2 and where regulation occurs. Once your model is working, you'll be ready to consider what happens to the NADH and $FADH_2$ you've produced.

9.4 Electron Transport and Chemiosmosis

The answer to one fundamental question about the oxidation of NADH and $FADH_2$ turned out to be relatively straightforward. To determine where the oxidation reactions take place in eukaryotes, researchers isolated mitochondria by using differential centrifugation techniques (introduced in Chapter 7). Then they broke the organelles open and separated the inner and outer membranes from the mitochondrial matrix and the solution in the intermembrane space. The isolated membranes were capable of oxidizing NADH, but the matrix and intermembrane fluid were not. Follow-up work showed that the oxidation process takes place on the inner membrane of the mitochondria and the membranes of cristae. In prokaryotes, the oxidation of NADH occurs in the plasma membrane.

Biologists who analyzed the components of the mitochondrial inner membrane isolated molecules that switch between a reduced and an oxidized state during respiration. The molecules were hypothesized to be the key to processing NADH and $FADH_2$. What are these molecules, and how do they work?

Components of the Electron Transport Chain

Researchers made several fundamental observations about the molecules involved in the oxidation of NADH and $FADH_2$ in mitochondria. Collectively, the molecules involved are designated the **electron transport chain (ETC):**

- Most of these molecules are proteins that contain distinctive chemical groups where the redox events take place. The active groups include ring-containing structures called flavins, iron-sulfur complexes, or iron-containing heme groups. Each subunit is readily reduced or oxidized.
- The inner membrane also contains a molecule called **ubiquinone**, which is not a protein. Ubiquinone got its name because it belongs to a family of compounds called quinones and because it is nearly ubiquitous in organisms. It is also called **coenzyme Q** or simply **Q**. Ubiquinone consists of a carbon-containing ring attached to a long tail made up of isoprene subunits. The structure of Q determines the molecule's function. The long, isoprene-rich tail is hydrophobic. As a result, Q is lipid soluble and can move throughout the mitochondrial membrane efficiently. In contrast, all but one of the proteins in the ETC are anchored in the membrane.
- The molecules involved in processing NADH and $FADH_2$ differ in electronegativity, or their tendency to hold electrons.

Because Q and the proteins can cycle between a reduced state and an oxidized state, and because they differ in electronegativity, investigators realized that it should be possible to arrange these molecules into a logical sequence. The idea was that electrons would pass from a molecule with lower electronegativity to one with higher electronegativity, via a redox reaction. As electrons moved through the chain and formed covalent bonds with each molecule in turn, they would be held more and more tightly. A small amount of energy would be released in each reaction, and the potential energy in each successive bond would lessen.

Using poisons, researchers were able to confirm that electron transport chains actually exist. Investigators found that cyanide, carbon monoxide, and other toxic compounds inhibit particular proteins in the inner membrane. If treatment with a particular toxin caused some proteins to sit in a reduced state while others remained oxidized, by inference the poison had stopped the movement of electrons along the chain. The oxidized molecules had to be "downstream" from the reduced compounds. For example, when an electron transport chain is treated with the drug antimycin A, cytochrome *b* and Q remain reduced; cytochrome *c*, cytochrome *a*, and cytochrome a_3 remain oxidized. This pattern makes sense if electrons flow from NADH and $FADH_2$ to cytochrome *b* and Q, but then stay there because the next element in the chain is poisoned. If they are downstream of the poisoned component in the chain, then cytochrome *c*, cytochrome *a*, and cytochrome a_3 would stay oxidized. This too is logical, because they would pass their electrons on to O_2 but could not pick up any new electrons from the poisoned component. Data from poisoning experiments, along with measurements of the tendency for each protein and Q to gain or lose electrons, allowed researchers to place the molecules in the order shown in **Figure 9.17**.

Experiments with particular poisons also showed that NADH donates an electron to a flavin-containing protein at the top of

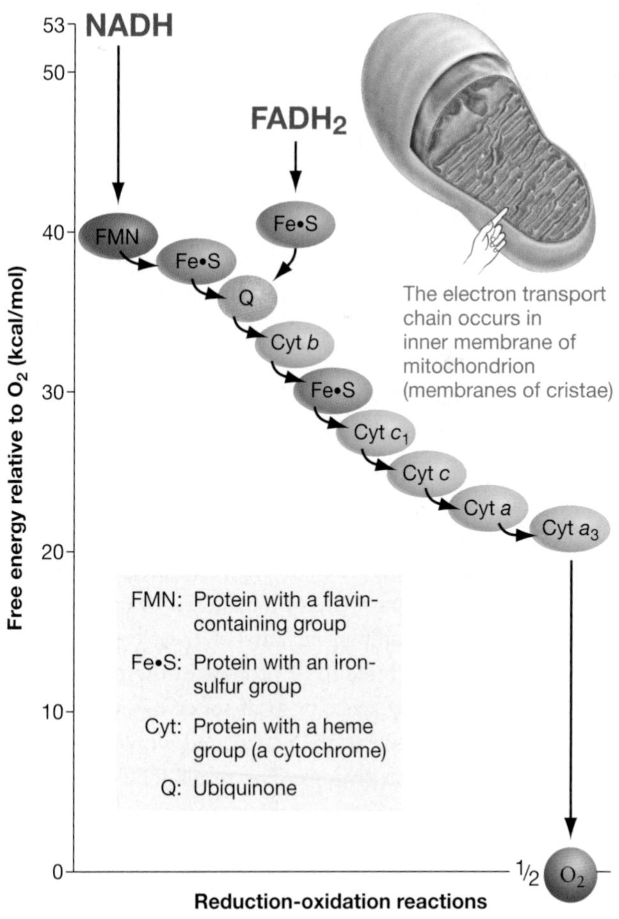

FIGURE 9.17 A Series of Reduction-Oxidation Reactions Occurs in an Electron Transport Chain

Electrons step down in potential energy from the electron carriers NADH and $FADH_2$ through an electron transport chain to a final electron acceptor. When oxygen is the final electron acceptor, water is formed. The overall free-energy change of 53 kcal/mol (from NADH to oxygen) is broken into small steps. EXERCISE Label the most and least electronegative molecules in the figure.

the chain, while $FADH_2$ donates electrons to an iron-sulfur-containing protein that then passes them directly to Q. After passing through each of the remaining components in the chain, the electrons are finally accepted by oxygen. Under standard conditions of temperature and pressure in the laboratory, the total potential energy difference from NADH to oxygen is 53 kilocalories/mole (abbreviated kcal/mol).

To summarize, electrons from NADH and $FADH_2$ pass through an electron transport chain consisting of a series of proteins and Q. Each successive molecule in the chain has slightly higher electronegativity than the previous one. As the electrons move from one link in the chain to the next, they are held tighter and tighter and their potential energy lessens. A molecule with particularly high electronegativity—oxygen, in plants, animals, and fungi—acts as the final electron acceptor. Electron transport takes place in the inner mitochondrial membrane.

Once the nature of the electron transport chain became clear, biologists understood the fate of the electrons carried by NADH and $FADH_2$ and how oxygen acts as the final electron acceptor. All of the electrons that were originally present in glucose were now accounted for. This is satisfying, except for one crucial question: How do these redox reactions generate ATP? Does substrate-level phosphorylation take place in the electron transport chain, just as it does in glycolysis and the Krebs cycle?

The Chemiosmotic Hypothesis

Throughout the 1950s most biologists interested in cellular respiration assumed that electron transport chains include enzymes that catalyze substrate-level phosphorylation. Despite intense efforts, however, no one was able to find a component of the ETC that phosphorylated ADP to produce ATP.

WEB TUTORIAL 9.2 Chemiosmosis

In 1961 Peter Mitchell made a radical break with prevailing ideas by proposing that the connection between electron transport and ATP production is indirect. Mitchell hypothesized that the real job of the electron transport chain is to pump protons from the matrix of the mitochondrion through the inner membrane and out to the intermembrane space or the interior of cristae. According to Mitchell, the pumping activity of the electron transport chain would lead to a buildup of protons in these areas. In this way, the intermembrane space or inside of cristae would become positively charged relative to the matrix and would have a much higher concentration of protons. The result would be a strong electrochemical gradient favoring the movement of protons back into the matrix. This **proton-motive force**, he hypothesized, is used by an enzyme in the inner membrane to synthesize ATP.

Mitchell called the production of ATP via a proton gradient **chemiosmosis**, and his proposal of an indirect link between electron transport and ATP production is the **chemiosmotic hypothesis.** Although proponents of a direct link between electron transport and substrate-level phosphorylation objected vigorously to Mitchell's idea, several key experiments supported it.

One of these experiments demonstrated the existence of a key element in Mitchell's scheme (**Figure 9.18**): A mitochondri-

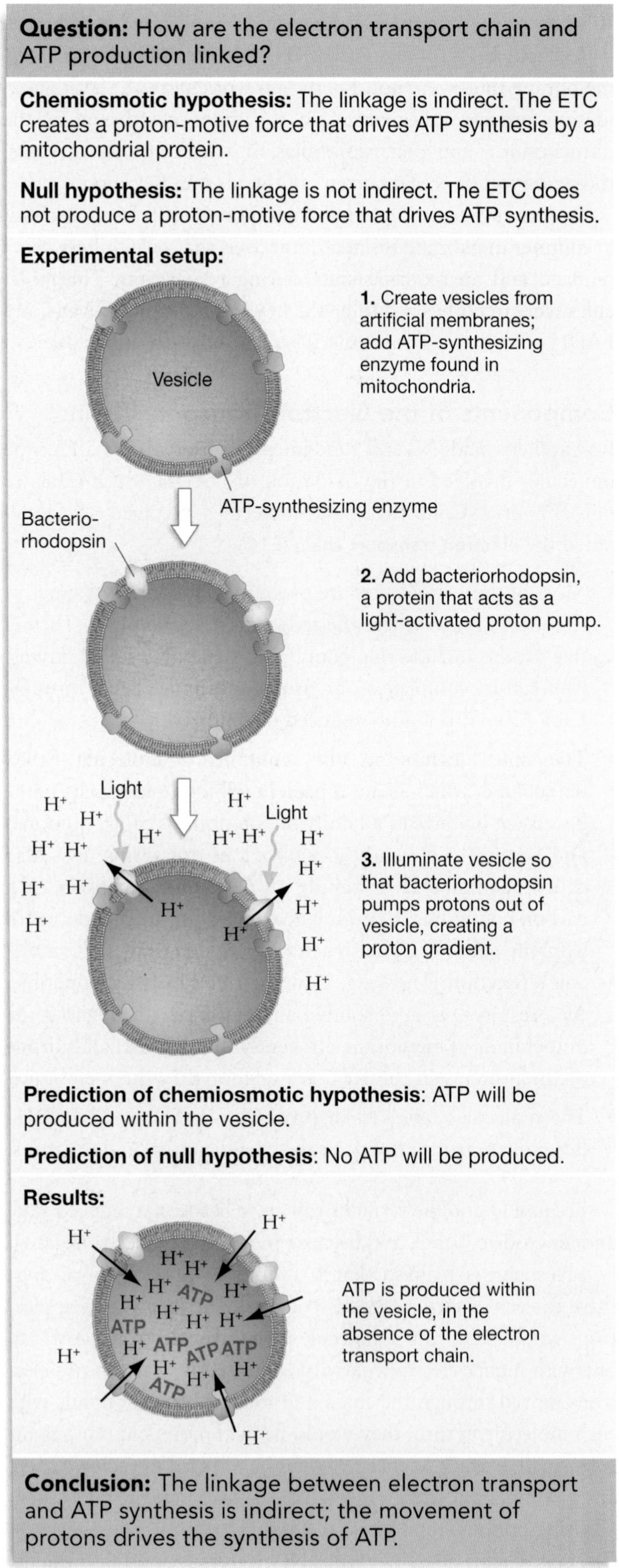

Question: How are the electron transport chain and ATP production linked?
Chemiosmotic hypothesis: The linkage is indirect. The ETC creates a proton-motive force that drives ATP synthesis by a mitochondrial protein.
Null hypothesis: The linkage is not indirect. The ETC does not produce a proton-motive force that drives ATP synthesis.
Experimental setup:
Vesicle
1. Create vesicles from artificial membranes; add ATP-synthesizing enzyme found in mitochondria.
ATP-synthesizing enzyme
Bacterio-rhodopsin
2. Add bacteriorhodopsin, a protein that acts as a light-activated proton pump.
Light
H+
3. Illuminate vesicle so that bacteriorhodopsin pumps protons out of vesicle, creating a proton gradient.
Prediction of chemiosmotic hypothesis: ATP will be produced within the vesicle.
Prediction of null hypothesis: No ATP will be produced.
Results:
ATP
ATP is produced within the vesicle, in the absence of the electron transport chain.
Conclusion: The linkage between electron transport and ATP synthesis is indirect; the movement of protons drives the synthesis of ATP.

FIGURE 9.18 Evidence for the Chemiosmotic Hypothesis

QUESTION Do you regard this as a convincing test of the chemiosmotic hypothesis? Why or why not?

al enzyme can use a proton gradient to synthesize ATP. The biologists who did this experiment began by making vesicles from artificial membranes that contained an ATP-synthesizing enzyme found in mitochondria. Along with this enzyme, they inserted bacteriorhodopsin. Bacteriorhodopsin is a well-studied membrane protein that acts as a light-activated proton pump. When light strikes this protein, the protein absorbs some of the light energy and changes conformation in a way that results in the pumping of protons from the interior of a cell or membrane-bound vesicle to the outside. As a result, the experimental vesicles could establish a strong electrochemical gradient favoring proton movement to the interior. When the vesicles were illuminated to initiate proton pumping, ATP began to be produced from ADP inside the vesicles. This result provided strong support for the chemiosmosis hypothesis. Mitchell's prediction was correct: ATP production depended solely on the existence of a proton-motive force. Consequently, it could occur in the absence of an electron transport chain.

Based on this experiment and others, most biologists accepted the chemiosmotic hypothesis as valid. Instead of being produced by enzymes inside the mitochondrion, ATP is produced by a flow of protons. In combination, the electron transport chain, the mitochondrial membrane, and ATP synthase function much like a hydroelectric dam. The electron transport chain is analogous to a series of gigantic pumps that force water up and behind a dam. The inner mitochondrial membrane functions as the dam, and ATP synthase is like the turbines inside the dam. At a hydroelectric dam, the movement of water makes turbines spin and generate electricity. But in a mitochondrion, protons are pumped instead of water. When protons move through ATP synthase, the protein spins and generates ATP.

Electron transport chains and ATP synthase occur in organisms throughout the tree of life. They are humming away in your cells now. Let's look in more detail at how they function.

How Is the Electron Transport Chain Organized?

Once the predictions of the chemiosmotic hypothesis were verified, researchers focused on understanding the three-dimensional structure of the components of the electron transport chain and on determining how electron transport is coupled to proton pumping. This work has shown that the ETC components are organized into four large complexes of proteins and cofactors (**Figure 9.19**). Protons are pumped by three of the complexes. Q and the protein **cytochrome c** act as shuttles that transfer electrons between complexes.

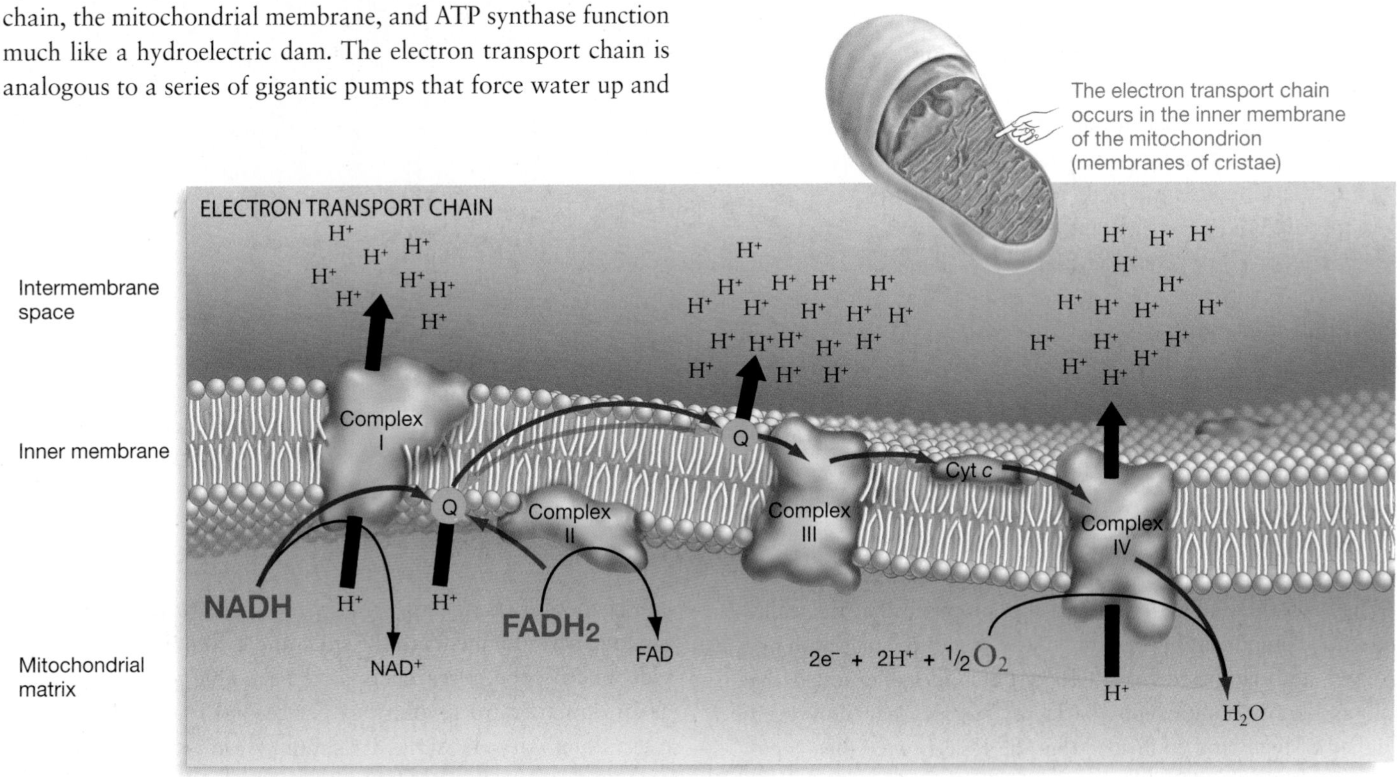

	Complex I	Complex II	Complex III	Complex IV
What goes in:				
What comes out:				

FIGURE 9.19 How Is the Electron Transport Chain Organized?
The individual components of the electron transport chain diagrammed in Figure 9.17 are grouped into large multiprotein complexes. Electrons are carried from one complex to another by Q and by cytochrome *c*. Complexes I, III, and IV use the potential energy released by the redox reactions to pump protons from the mitochondrial matrix to the intermembrane space. EXERCISE Label the solid blue arrows to indicate what they represent. Add an arrow across the membrane and label it "Proton gradient." In the boxes at the bottom, list "What goes in" and "What comes out" for each complex.

The structural studies completed to date confirm that in complexes I and IV, protons actually pass directly through a sequence of electron carriers. The exact route taken by the protons is still being worked out. It is also not clear how the redox reactions taking place inside each complex—as electrons step down in potential energy—make proton movement possible.

The best-understood interaction between electron transport and proton pumping takes place in complex III. Research has shown that when Q accepts electrons from complex I or complex II, it also gains protons. The reduced form of Q then diffuses to the outer side of the inner membrane, where its electrons are used to reduce a component of complex III near the intermembrane space. The protons held by Q are released to the intermembrane space. In this way, Q shuttles electrons and protons from one side of the membrane to the other. The electrons proceed down the transport chain, and the protons released to the intermembrane space contribute to the proton-motive force.

Now, how does this proton gradient make the production of ATP possible?

The Discovery of ATP Synthase

In 1960 Efraim Racker made several key observations about how ATP is synthesized in mitochondrial membranes. When he used these structures to make vesicles, Racker noticed that some happened to form with their membrane inside out. Electron microscopy revealed that the inside-out membranes had numerous large proteins studded along their surfaces. Each of these proteins appeared to have a base in the membrane and a stalk and a knob projecting out from the membrane (**Figure 9.20a**). But when Racker shook the vesicles or treated them with a compound called urea, the stalks and knobs fell off.

Racker seized on this technique to isolate the stalks and knobs and do experiments with them. For example, he found that isolated stalks and knobs could hydrolyze ATP, forming ADP and inorganic phosphate. The vesicles that contained just the base component, without the stalks and knobs, could not process ATP. The base components were, however, capable of transporting protons across the membrane.

Based on these observations, Racker proposed that the stalk-and-knob component of the protein was an enzyme that both hydrolyzes and synthesizes ATP. To test this idea, he added the stalk-and-knob components back to vesicles that had been stripped of them and confirmed that the vesicles were then capable of synthesizing ATP. Follow-up work confirmed his hypothesis that the membrane-bound base component is a proton channel.

As **Figure 9.20b** shows, the structure of this protein complex is now reasonably well understood. The ATPase "knob" component is called the F_1 unit; the membrane-bound, proton-transporting base component is the F_o unit; and the F_1 and F_o units are connected by the stalk. The entire complex is known as **ATP synthase**. According to the current model for the way this enzyme functions, a flow of protons through the F_o unit causes the stalk connecting the two subunits to spin. By attaching long actin filaments to the stalk and examining them with a videomicroscope, researchers have been able to see movement at an estimated 50 revolutions per second. As the F_1 unit rotates along with the stalk, its subunits are thought to change conformation in a way that catalyzes the phosphorylation of ADP to ATP. Understanding exactly how this reaction occurs is currently the focus of intense research. ATP synthase makes most of the ATP that keeps you alive.

(a) A vesicle formed from an "inside-out" mitochondrial membrane

(b) The F_o unit is the base; the F_1 unit is the knob.

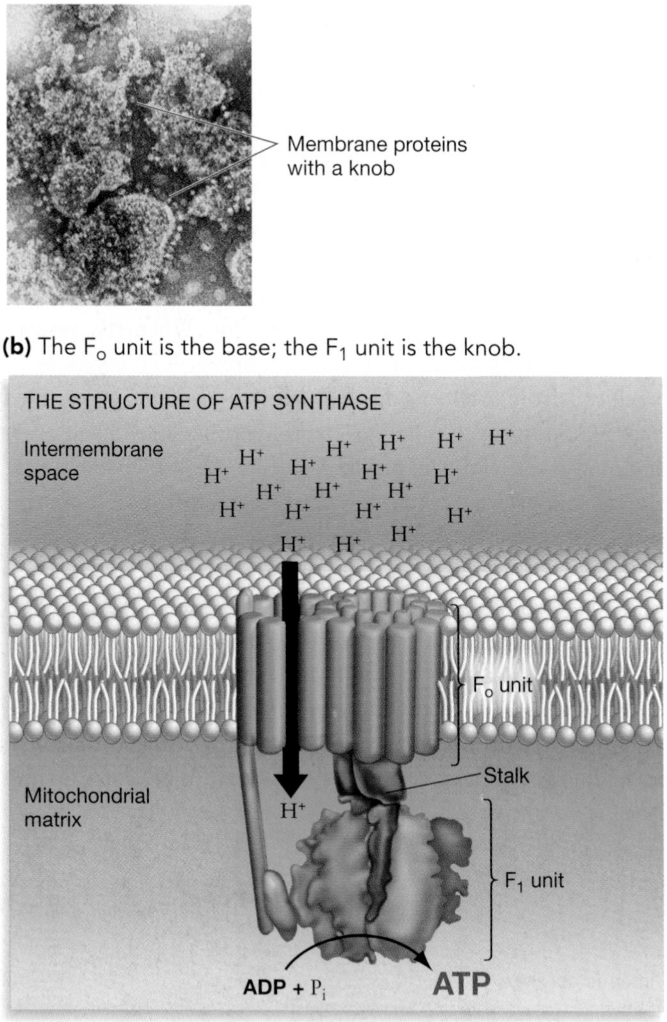

FIGURE 9.20 The Structure of ATP Synthase
(a) When patches of mitochondrial membrane turn inside out and form vesicles, proteins that have a stalk-and-knob structure face outward. Normally, the stalk and knob face inward, toward the mitochondrial matrix. **(b)** ATP synthase has two major components, designated F_o and F_1.

Oxidative Phosphorylation

The formation of ATP through the combination of proton pumping by electron transport chains and the action of ATP synthase is called **oxidative phosphorylation**. This term is appropriate because the phosphorylation of ADP is based on the oxidation of NADH and $FADH_2$.

Figure 9.21 summarizes glucose oxidation and cellular respiration by tracing the fate of the carbon atoms and electrons in glucose. Notice that electrons from glucose are transferred to NADH and $FADH_2$, passed through the electron transport chain, and accepted by oxygen. The pumping of protons during electron transport creates the proton-motive force that drives ATP synthesis. The diagram also indicates the approximate yield of ATP from each component of the process. Recent research has shown that about 30 ATP molecules are produced from each molecule of glucose; of these, 26 ATP molecules are produced by ATP synthase. The fundamental message here is that the vast majority of the "payoff" from the oxidation of glucose occurs as a result of oxidative phosphorylation.

It is important to recognize, however, that cellular respiration can occur without oxygen. Oxygen is the electron acceptor used by all eukaryotes and a wide diversity of bacteria and archaea. Species that depend on oxygen as an electron acceptor are said to use **aerobic respiration** and are called aerobic organisms. (The Latin root *aero* means "air.") Many thousands of bacterial and archaeal species rely on electron acceptors other than oxygen, however, and electron donors other than glucose. As Chapter 27 will show, some bacteria and archaea use H_2, H_2S, CH_4, or other inorganic compounds as electron donors. For species in oxygen-poor environments, nitrate (NO_3^-) and sulphate (SO_4^{2-}) are particularly common electron acceptors. Cells that depend on electron acceptors other than oxygen are said to use **anaerobic** ("no air") **respiration.** Even though the starting and ending points of cellular respiration are different, these cells are still able to use electron transport chains to create a proton-motive force that drives the synthesis of ATP.

Oxygen is the most effective of all electron acceptors, however, because of its high electronegativity. Because oxygen holds electrons so tightly, the potential energy of electrons in a bond between an oxygen atom and a non-oxygen atom is low. As a result, there is a large difference between the potential energy of electrons in NADH and the potential energy of electrons bonded to an oxygen atom (see Figure 9.17). The large differential in potential energy means that the electron transport chain can generate a large proton-motive force. Cells that do not use oxygen as an electron acceptor cannot generate such a large potential energy difference. As a result, they cannot make as much ATP as do cells that use aerobic respiration and thus tend to grow much more slowly. If cells that use anaerobic respiration compete with cells using aerobic respiration, the cells that use oxygen as an electron acceptor almost always grow faster and reproduce more. As a result, species that rely on anaerobic respiration usually live in environments where oxygen is not present, where they do not have to compete with aerobic species.

What happens when oxygen or other electron acceptors are temporarily used up and are unavailable? Without oxygen or another electron acceptor in place, the electrons carried by NADH have no place to go. When this happens, the electron transport chain stops. All of the NAD^+ in the cell quickly becomes NADH. This situation is life threatening. When there is no longer any NAD^+ to supply the reactions of glycolysis, no ATP can be produced. If NAD^+ cannot be regenerated somehow, the cell will die. How do cells cope?

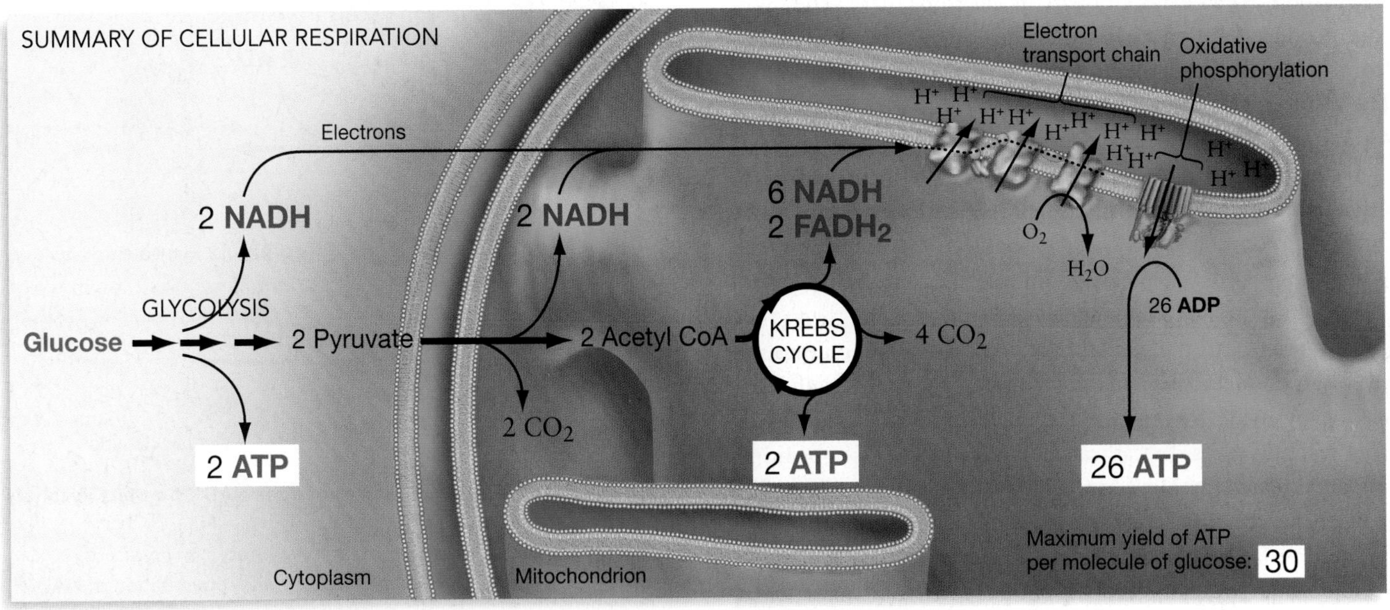

FIGURE 9.21 A Summary of Cellular Respiration

EXERCISE Across the top of the figure, write the overall reaction for the oxidation of glucose. Explain what happens to each reactant. Identify the source of each product.

✓CHECK YOUR UNDERSTANDING

As electrons from NADH and $FADH_2$ move through the electron transport chain, protons are pumped into the intermembrane space of mitochondria. The resulting electrochemical gradient drives protons through ATP synthase, resulting in the production of ATP from ADP. You should be able to add paper cutouts labeled ETC and ATP synthase to the model you made in Section 9.3, and then explain the steps in electron transport and chemiosmosis, using dimes to represent electrons and pennies to represent protons.

9.5 Fermentation

Fermentation is a metabolic pathway that allows continued production of ATP in the absence of an electron acceptor. It occurs when the pyruvate that is produced by glycolysis (or a molecule derived from pyruvate), rather than oxygen, accepts electrons from NADH. When NADH gets rid of electrons in this way, NAD^+ is produced. With NAD^+ present, glycolysis can continue to produce ATP via substrate-level phosphorylation and the cell can stay alive. In many cases, the molecule that is formed by the addition of an electron to pyruvate (or another electron acceptor) cannot be used by the cell. In some cases, this by-product is toxic and is excreted from the cell as waste (**Figure 9.22a**).

In organisms that usually use oxygen as an electron acceptor, fermentation is an alternative mode of energy production when oxygen supplies temporarily run out. For example, if you sprint a long distance, your muscles begin metabolizing glucose so fast that your lungs and circulatory system cannot supply oxygen rapidly enough to keep electron transport chains active. When oxygen is absent, the electron transport chains shut down and NADH cannot donate its electrons there. The pyruvate produced by glycolysis then begins to accepts electrons from NADH, and fermentation takes place (**Figure 9.22b**). The result of this process, called **lactic acid fermentation**, is the formation of lactate and regeneration of NAD^+.

Figure 9.22c illustrates a different type of fermentation pathway. These reactions, called **alcohol fermentation**, occur in the fungus *Saccharomyces cerevisiae*—baker's and brewer's yeast. When fungal cells are placed in an environment such as bread dough or a bottle of champagne and begin growing there, they quickly use up all the available oxygen. They continue to use glycolysis to metabolize sugar, however, by enzymatically converting pyruvate to the two-carbon compound acetaldehyde. This reaction gives off carbon dioxide, which causes bread to rise and creates the bubbles in champagne and beer. Acetaldehyde then accepts electrons from NADH, forming the NAD^+ required to keep glycolysis going. The addition of electrons to acetaldehyde results in the formation of ethanol as a waste product. Ethanol is the active ingredient in alcoholic beverages.

Species of bacteria and archaea exhibit a huge variety of other fermentative pathways. For example, bacteria and ar-

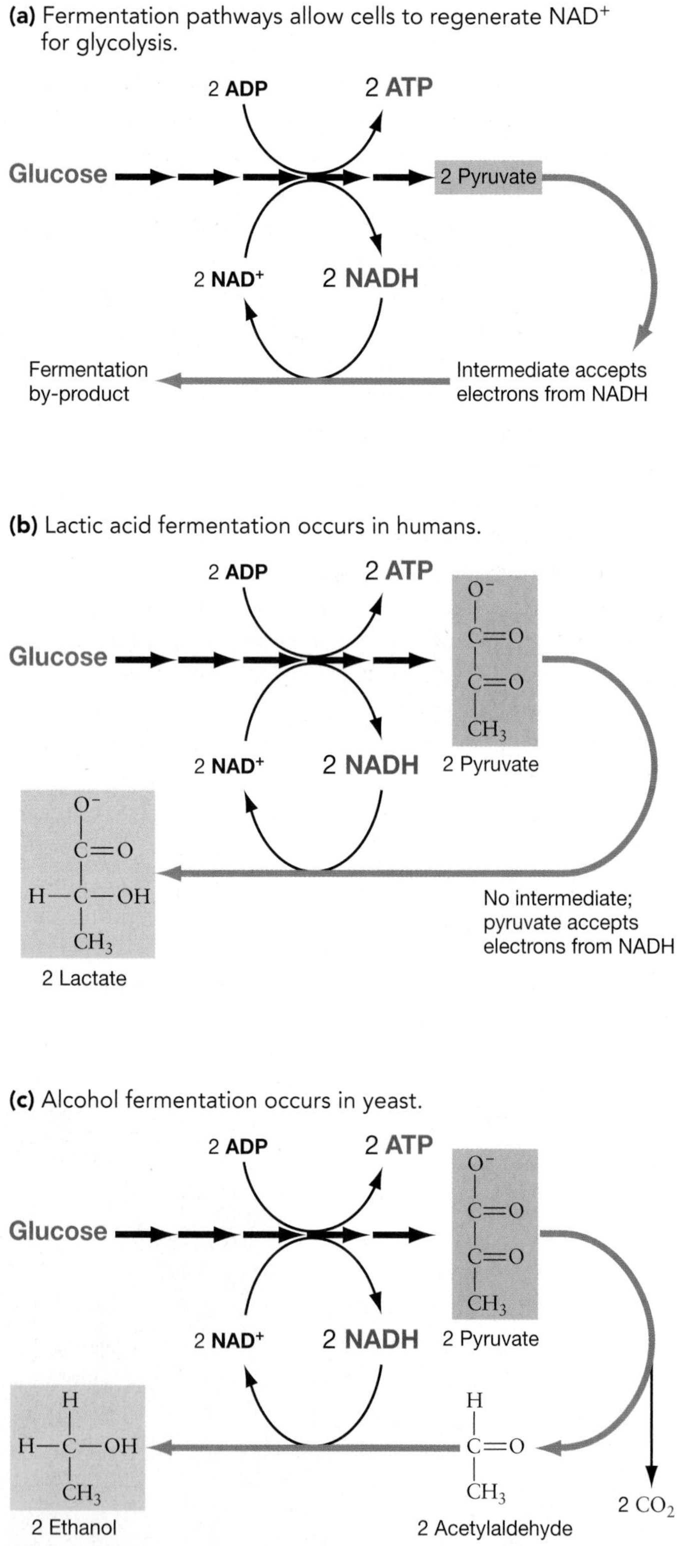

FIGURE 9.22 Fermentation Regenerates NAD^+ So that Glycolysis Can Continue

Among the bacteria, archaea, and eukaryotes, many types of fermentation occur. **(a)** All types result in the regeneration of NAD^+, so glycolysis can continue to produce ATP even when the electron acceptors required by electron transport chains are unavailable. **(b)** Lactic acid fermentation and **(c)** alcohol fermentation are two of the best-studied types.

chaea that exist exclusively through fermentation are present in phenomenal numbers in the oxygen-free environment of your small intestine and in the rumen of cows. The rumen is a specialized digestive organ that contains over 10^{10} (ten billion) bacterial and archaeal cells per *milliliter* of fluid. The fermentations that occur in these cells result in the production of an array of fatty acids. Cows use these fermentation by-products as a source of energy. Cells that employ other types of fermentations are used commercially in the production of soy sauce, tofu, yogurt, cheese, vinegar, and other products.

Even though fermentation is a widespread and commercially important type of metabolism, it is extremely inefficient compared with cellular respiration. Recall that fermentation produces just two molecules of ATP for each molecule of glucose metabolized, while cellular respiration produces about 30—approximately 15 times more energy per glucose molecule than fermentation. The reason for this disparity is that electron acceptors such as pyruvate and acetaldehyde have low electronegativity compared with an electron acceptor such as oxygen. As a result, the potential energy drop between the start and end of fermentation is a tiny fraction of the potential energy change that occurs during cellular respiration.

Based on these observations, it should not be surprising that organisms capable of both processes never use fermentation when an appropriate electron acceptor is available for cellular respiration. Organisms that can switch between fermentation and cellular respiration that uses oxygen as an electron acceptor are called **facultative aerobes**. The term *aerobe* refers to using oxygen, while *facultative* reflects the ability to use this pathway only when oxygen is present.

9.6 How Does Cellular Respiration Interact with Other Metabolic Pathways?

The enzymes, products, and intermediates involved in cellular respiration and fermentation do not exist in isolation. Instead, they are part of a huge and dynamic inventory of chemicals inside the cell. Because metabolism includes thousands of different chemical reactions, the amounts and identities of molecules inside cells are constantly in flux (**Figure 9.23**). Fermentation pathways, electron transport, and other aspects of carbohydrate metabolism may be crucial to the life of a cell, but they also have to be seen as parts of a whole.

To make sense of the chemical inventory inside cells and the full scope of metabolism, it is critical to recognize that cells have two fundamental requirements to stay alive, grow, and reproduce: energy and carbon. More formally, cells need a source of high-energy electrons for generating chemical energy in the form of ATP, and a source of carbon-containing molecules that can be used to synthesize DNA, RNA, proteins, fatty acids, and other molecules. Reactions that result in the breakdown of molecules and the production of ATP are called **catabolic pathways**; reactions that result in the synthesis of larger molecules from smaller components are called **anabolic pathways**.

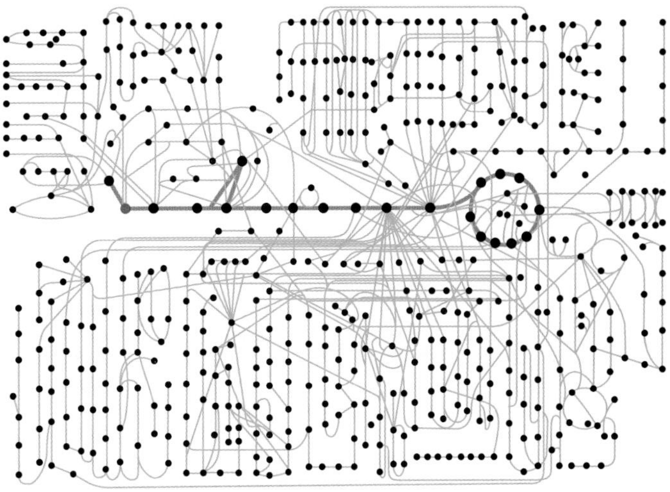

FIGURE 9.23 Metabolic Pathways Interact
A representation of a few of the thousands of chemical reactions that occur in cells. The dots represent molecules, and the lines represent enzyme-catalyzed reactions. EXERCISE Label the red dot, and circle the 10 reactions of glycolysis. Draw a box around the Krebs cycle.

This section introduces the ways in which glycolysis and the Krebs cycle interact with other catabolic pathways and with anabolic pathways. Let's consider how molecules other than carbohydrates are used as fuel in eukaryotes, and then examine how molecules involved in glycolysis and the Krebs cycle are sometimes used as building blocks in the synthesis of cell components.

Processing Proteins and Fats as Fuel

Most organisms ingest, synthesize, or absorb a wide variety of carbohydrates. These molecules range from sucrose, maltose, and other simple sugars to large polymers such as glycogen and starch. As noted in Chapter 5, glycogen is the major form of stored carbohydrate in animals, while starch is the major form of stored carbohydrate in plants. Recall that both glycogen and starch are polymers of glucose, but they differ in the way their long chains of glucose branch. Using enzyme-catalyzed reactions, cells can produce glucose from glycogen, starch, and most simple sugars. Glucose and fructose can then be processed by the enzymes of the glycolytic pathway.

Carbohydrates are not the only important source of carbon compounds used in the catabolic pathways, however. As pointed out in Chapter 6, fats are highly reduced macromolecules consisting of glycerol bonded to chains of fatty acids. In cells, fats are routinely broken down by enzymes to form glycerol and acetyl CoA. Glycerol enters the glycolytic pathway once it has been oxidized and phosphorylated to form glyceraldehyde-3-phosphate—one of the intermediates in the 10-reaction sequence. Acetyl CoA, in contrast, enters the Krebs cycle.

Proteins can also be catabolized. Once they are broken down to their constituent amino acids, the amino ($-NH_2$)

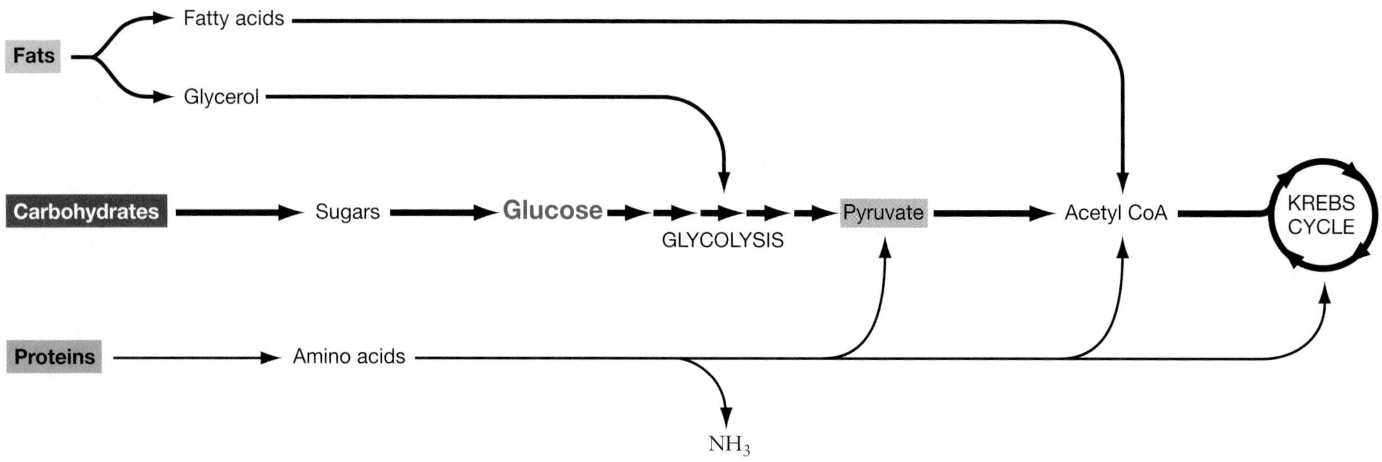

FIGURE 9.24 Proteins, Carbohydrates, and Fats Can All Furnish Substrates for Cellular Respiration
A variety of carbohydrates can be converted to glucose and processed by glycolysis. If carbohydrates are scarce, cells can obtain high-energy compounds from fats or ultimately proteins for ATP production. These are catabolic reactions.

groups are removed in enzyme-catalyzed reactions. These amino groups are excreted in urine as waste. The carbon compounds that remain after this catabolic step are converted to pyruvate, acetyl CoA, and other intermediates in glycolysis and the Krebs cycle.

Figure 9.24 summarizes the catabolic pathways of carbohydrates, fats, and proteins and shows how their breakdown products feed an array of steps in glucose oxidation and cellular respiration. When all three types of molecules are available in the cell to generate ATP, carbohydrates are used up first, then fats, and finally proteins.

Anabolic Pathways Synthesize Key Molecules

Where do cells get the precursor molecules required to synthesize amino acids, RNA, DNA, phospholipids, and other cell components? Not surprisingly, the answer often involves intermediates in carbohydrate metabolism. For example:

- In humans, about half of the 20 required amino acids can be synthesized from molecules siphoned off from the Krebs cycle.
- Acetyl CoA is the starting point for anabolic pathways that result in the synthesis of fatty acids.
- The molecule that is produced by the first reaction in glycolysis can be oxidized to start the synthesis of ribose-5-phosphate—a key intermediate in the production of ribonucleotides and deoxyribonucleotides. These nucleotides, in turn, are required for manufacturing RNA and DNA.
- If ATP is abundant, pyruvate and lactate (from fermentation) can be used as a substrate in the synthesis of glucose. Excess glucose is converted to glycogen and stored.

Figure 9.25 summarizes how intermediates in carbohydrate metabolism are drawn off to synthesize macromolecules. The same molecule can serve many different functions in the cell. As a result, catabolic and anabolic pathways are closely intertwined.

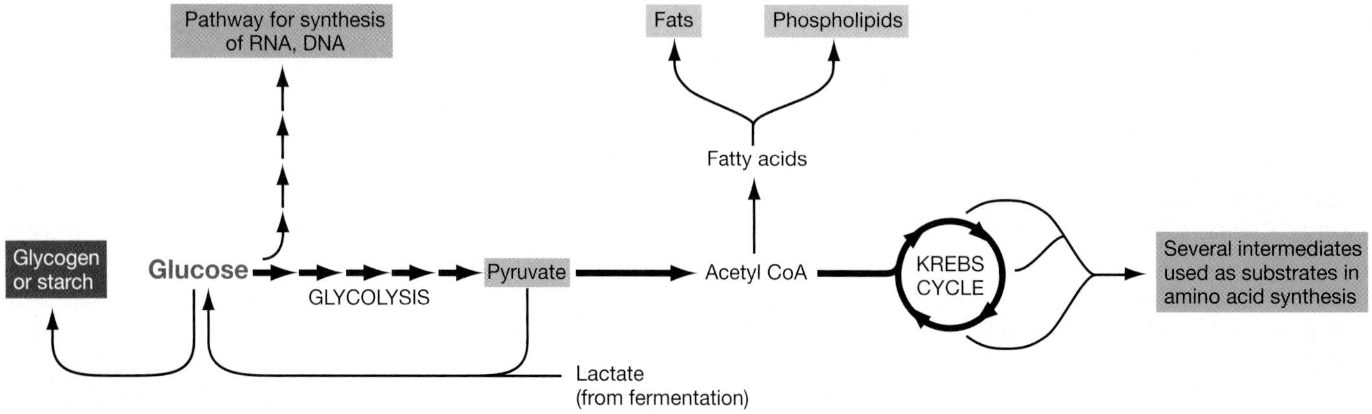

FIGURE 9.25 Intermediates in Carbohydrate Metabolism Can Be Drawn Off to Synthesize Cell Components
Several of the intermediates in carbohydrate metabolism act as precursor molecules in anabolic reactions leading to the synthesis of RNA, DNA, glycogen or starch, amino acids, and fatty acids.

ESSAY ATP Production during Exercise

In an adult animal, the highest demands for ATP are made by exercising muscles. As Chapter 46 will describe, muscles contain a motor protein called myosin. Like the motor proteins kinesin and dynein, introduced in Chapter 7, myosin undergoes conformational changes in response to the addition of ATP and the subsequent release of a phosphate group. This movement by myosin is the physical basis of muscle contraction, and muscle contraction is the physical basis of movement.

On average, a muscle contains only enough ATP to sustain about 15 seconds of intense exercise. For muscle contractions to continue, massive amounts of ATP are required. Depending on the level and duration of activity, the muscles being exercised may produce the ATP they need either by cellular respiration or by fermentation. Sustained periods of moderate activity such as jogging are powered by aerobic respiration. In contrast, short periods of intense activity such as sprinting are powered by a combination of aerobic respiration and fermentation.

To understand the relationship between cellular respiration and fermentation in muscle cells, we'll focus on the constraints experienced by two types of runners. What aspects of cellular respiration affect the performance of a marathoner? What aspects of fermentation limit the duration of sprints?

A marathon course is 26 miles, 385 yards (42 km, 195 m) long. This race was inaugurated with the first of the modern Olympic games, which took place in 1899 in Athens, Greece. The event was designed to commemorate Pheidippides, a Greek soldier who reportedly ran from the plains of Marathon in southeast Greece to Athens immediately after the Greek army had triumphed over the invading Persians in 490 B.C. According to legend, Pheidippides shouted "Victory!" when he reached Athens, then collapsed and died of exhaustion.

Contemporary marathon runners rarely collapse and die, but virtually all experience a phenomenon known as "hitting the wall." What happens when a marathon runner hits the wall? Marathon running is an aerobic activity fueled by cellular respiration. The glucose and glycogen stores in the body are adequate to fuel only about 20 miles of running, however. During the last 6 miles of the race, marathoners rely exclusively on fatty-acid metabolism to keep cellular respiration going. Hitting the wall is a feeling of exhaustion that occurs as the last of the glycogen stores are used up. Because fatty-acid metabolism is markedly slower than glycogen metabolism, ATP delivery is reduced and running becomes much more difficult.

What happens when a marathon runner hits the wall?

Sprinters do not hit the wall. Instead they "feel the burn." During intense activity, metabolism switches from aerobic to anaerobic and a burning sensation is felt in the muscles. This feeling is caused by a buildup of lactic acid due to sustained production of ATP by fermentation. As fermentation proceeds, lactic acid levels can build up, dropping the pH of muscle cells. The acidification may damage enzymes and structural proteins and contribute to pain and fatigue. Formally, muscle fatigue is defined as the inability of a muscle to continue contracting with maximum force. In addition, anaerobic metabolism produces a great deal of heat, which can damage muscle proteins if it is not controlled.

For both sprinters and marathoners, an extended cooldown period after exercise is important for muscle health. Several important events occur during recovery from exercise. Lactic acid is transported from muscles to the liver, where it is converted to pyruvate and then to glucose. Oxygen-rich blood is delivered to muscles, where cellular respiration restores normal levels of ATP. In addition, heat is transferred from muscles to blood. The blood carries the heat to the body surface, where it is radiated to the environment, cooling the body.

Recent research has shown that aerobic training can increase the number of mitochondria in muscle cells and thus increase ATP supplies during exercise. Similarly, sprint training can increase the number of glycolytic enzymes in muscle cells and thus the rate of ATP delivery during bursts of activity. Increasingly, an understanding of cellular respiration and fermentation is being used to design training programs for athletes.

CHAPTER REVIEW

Summary of Key Concepts

Cellular respiration produces ATP from molecules with high potential energy—often glucose.
ATP is the currency that cells use to pump ions, drive endergonic reactions, move cargo, and perform other types of work. When proteins bind ATP or are phosphorylated, they respond by changing conformation, or shape, in a way that allows them to act as enzymes, pumps, or motors.

Cells produce ATP from sugars or other compounds with high free energy by using one of two general pathways: (1) cellular respiration or (2) fermentation. Cellular respiration involves the transfer of electrons from a compound with high free energy, such as glucose, to a molecule with lower free energy, such as oxygen, through an electron transport chain. Fermentation involves the transfer of electrons from one organic compound to another without participation by an electron transport chain.

Glucose processing has three components: (1) glycolysis, (2) the Krebs cycle, and (3) electron transport coupled with oxidative phosphorylation.

In eukaryotes, glycolysis takes place in the cytoplasm. The Krebs cycle occurs in the mitochondrial matrix, and electron transport and oxidative phosphorylation proceed in the inner membranes of mitochondria.

Glycolysis is a 10-step reaction sequence in which glucose is broken down into two molecules of pyruvate. For each molecule of glucose processed during glycolysis, two molecules of ATP are produced by substrate-level phosphorylation and two molecules of NAD^+ are reduced to two molecules of NADH.

Prior to the Krebs cycle, each pyruvate is oxidized to acetyl CoA by a series of reactions that results in the production of one molecule of carbon dioxide and the synthesis of one of NADH. The Krebs cycle begins when acetyl CoA reacts with oxaloacetate to form citrate. The ensuing series of reactions results in the regeneration of oxaloacetate, the production of two molecules of carbon dioxide, the synthesis of one molecule of ATP by substrate-level phosphorylation, the reduction of one molecule of FAD to $FADH_2$, and the reduction of three molecules of NAD^+ to three of NADH.

In essence, glycolysis and the Krebs cycle strip electrons from glucose and use them to reduce NAD^+ and FAD. The resulting electron carriers—NADH and $FADH_2$—then donate electrons to an electron transport chain, which gradually steps the electrons down in potential energy until they are finally accepted by oxygen or by another final electron acceptor, resulting in the formation of water in aerobic organisms. The components of the electron transport chain use the energy released by the oxidation of NADH and $FADH_2$ to pump protons across the inner mitochondrial membrane. The pumping activity creates an electrochemical gradient, or proton-motive force, that ATP synthase uses to produce ATP. Oxidative phosphorylation is the production of ATP by electron transport and the generation of a proton-motive force.

Web Tutorial 9.1 Glucose Metabolism

Fermentation pathways allow glycolysis to continue when the lack of an electron acceptor shuts down electron transport chains.

If no electron acceptor such as oxygen is available, cellular respiration stops, because electron transport chains cannot continue. As a result, all NAD^+ is converted to NADH and glycolysis must stop. Fermentation pathways regenerate NAD^+ when an organic molecule such as pyruvate accepts electrons from NADH. Depending on the molecule that acts as an electron acceptor, fermentation pathways produce lactate, ethanol, or other reduced organic compounds as a by-product.

Web Tutorial 9.2 Chemiosmosis

Cellular respiration and fermentation are carefully regulated.

When supplies of ATP, NADH, and $FADH_2$ are high in the cell, feedback regulation occurs—that is, product molecules bind to and inhibit enzymes involved in ATP production. The glycolytic pathway slows when ATP binds to phosphofructokinase, and the pyruvate dehydrogenase complex is inhibited when it is phosphorylated by ATP. The enzyme that converts acetyl CoA to citrate slows when ATP binds to it, and certain enzymes in the Krebs cycle are inhibited when NADH or ATP bind to them. As a result, ATP is produced only when needed.

Questions

Content Review

1. When does feedback inhibition occur?
   a. when lack of an appropriate electron acceptor makes an electron transport chain stop
   b. when an enzyme that is active early in a metabolic pathway is inhibited by a product of the pathway
   c. when ATP synthase reverses and begins pumping protons out of the mitochondrial matrix
   d. when cellular respiration is inhibited and fermentation begins

2. Where does the Krebs cycle occur in eukaryotes?
   a. in the cytoplasm
   b. in the matrix of mitochondria
   c. in the inner membrane of mitochondria
   d. in the intermembrane space of mitochondria

3. What does the chemiosmotic hypothesis claim?
   a. Substrate-level phosphorylation occurs in the electron transport chain.
   b. Substrate-level phosophorylation occurs in glycolysis and the Krebs cycle.
   c. The electron transport chain is located in the inner membrane of mitochondria.
   d. Electron transport chains generate ATP indirectly, by the creation of a proton-motive force.

4. What is the function of the reactions in a fermentation pathway?
   a. to generate NADH from NAD^+, so electrons can be donated to the electron transport chain
   b. to synthesize pyruvate from lactate
   c. to generate NAD^+ from NADH, so glycolysis can continue
   d. to synthesize electron acceptors, so that cellular respiration can continue

5. When do cells switch from cellular respiration to fermentation?
   a. when electron acceptors are not available
   b. when the proton-motive force runs down
   c. when NADH and $FADH_2$ supplies are low
   d. when pyruvate is not available

6. Why are NADH and $FADH_2$ said to have "reducing power?"
   a. They are the reduced forms of NAD^+ and FAD.
   b. They donate electrons to components of the ETC, reducing those components.
   c. They travel between the cytoplasm and the mitochondrion.
   d. They have the power to reduce carbon dioxide to glucose.

Conceptual Review

1. Explain why NADH and $FADH_2$ are called electron carriers. Where do these molecules get electrons, and where do they deliver them? In eukaryotes, what molecule do these electrons reduce?
2. Compare and contrast substrate-level phosphorylation and oxidative phosphorylation.
3. What is the relationship between cellular respiration and fermentation? Why does cellular respiration produce so much more ATP than fermentation does?
4. Diagram the relationship among the three components of cellular respiration: glycolysis, the Krebs cycle, and electron transport. What molecules connect these three processes? Where does each process occur in a eukaryotic cell?
5. Explain the relationship between electron transport and oxidative phosphorylation. What does ATP synthase look like, and how does it work?
6. Describe the relationship among carbohydrate metabolism, the catabolism of proteins and fats, and anabolic pathways.

Group Discussion Problems

1. Cyanide ($C{\equiv}N^-$) blocks complex IV of the electron transport chain. Suggest a hypothesis for what happens to the ETC when complex IV stops working. Your hypothesis should explain why cyanide poisoning is fatal.
2. The presence of many saclike cristae results in a large amount of membrane inside mitochondria. Suppose that some mitochondria had few cristae. How would their output of ATP compare with that of mitochondria with many cristae? Explain your answer.
3. When yeast cells are placed into low-oxygen environments, the mitochondria in the cells become reduced in size and number. Suggest an explanation for this observation.
4. Most agricultural societies have come up with ways to ferment the sugars in barley, wheat, rice, corn, or grapes to produce alcoholic beverages. Historians argue that this was an effective way for farmers to preserve the chemical energy in grains and fruits in a form that would not be eaten by rats or spoiled by bacteria or fungi. Why does a great deal of chemical energy remain in the products of fermentation pathways?

Answers to Multiple-Choice Questions 1. b; 2. b; 3. d; 4. c; 5. a; 6. b

www.prenhall.com/freeman is your resource for the following: Web Tutorials; Online Quizzes and other Online Study Guide materials; Answers to Conceptual Review Questions; Solutions to Group Discussion Problems; Answers to Figure Caption Questions and Exercises; and Additional Readings and Research.

10 Photosynthesis

KEY CONCEPTS

- Photosynthesis consists of two distinct sets of reactions. In reactions driven by light, ATP and the electron carrier NADPH are produced. In subsequent reactions that do not depend directly on light, the ATP and NADPH are used to reduce carbon dioxide (CO_2) to carbohydrate $(CH_2O)_n$. In eukaryotic cells, both processes take place in chloroplasts.
- The light-dependent reactions transform the energy in sunlight to chemical energy in the form of electrons with high potential energy. Excited electrons either are used to produce NADPH or are donated to an electron transport chain, which results in the production of ATP.
- The light-independent reactions start with the enzyme rubisco, which catalyzes the addition of CO_2 to a five-carbon molecule. The compound that results undergoes a series of reactions that use ATP and NADPH and lead to the production of sugar.

Plants and other photosynthetic organisms convert the energy in sunlight to chemical energy in the bonds of sugar. The sugar produced by photosynthetic organisms fuels cellular respiration and growth. Photosynthetic organisms, in turn, are consumed by animals, fungi, and a host of other organisms. Directly or indirectly, most organisms on Earth get their energy from photosynthesis.

About three billion years ago, a novel combination of light-absorbing molecules and enzymes gave a bacterial cell the capacity to convert light energy into chemical energy in the carbon-hydrogen bonds of sugar. When sunlight is used to manufacture carbohydrate, **photosynthesis** is said to occur. The origin of photosynthesis ranks as one of the great events in the history of life. Since this process evolved, photosynthetic organisms have dominated the Earth in terms of abundance and mass.

The vast majority of organisms alive today rely on photosynthesis, either directly or indirectly, to stay alive. Photosynthetic organisms such as trees and mosses and ferns are termed **autotrophs** ("self-feeders"), because they make all of their own food from ions and simple molecules. Non-photosynthetic organisms such as humans and fungi and the bacterium *Escherichia coli* are called **heterotrophs** ("different-feeders") because they have to obtain the sugars and many of the other macromolecules they need from other organisms. Because there could be no heterotrophs without autotrophs, photosynthesis is fundamental to almost all life. Glycolysis may be the oldest set of energy-related chemical reactions in terms of evolutionary history; but ecologically, photosynthesis is easily the most important.

All organisms perform respiration or fermentation, but only selected groups are also capable of photosynthesis. This chapter presents a step-by-step analysis of how photosynthetic species manufacture sugar from sunlight, carbon dioxide, and water. After studying how plants do this remarkable chemistry, you should have a deeper appreciation for the food you eat and the oxygen you breathe.

10.1 An Overview of Photosynthesis

Research on photosynthesis began very early in the history of biological science. Starting in the 1770s, a series of experiments showed that photosynthesis takes place only in the green parts of plants; that sunlight, carbon dioxide (CO_2), and water (H_2O) are required; and that oxygen (O_2) is produced as a by-product. By the early 1840s enough was known about this process for biologists to propose that photosynthesis allows plants to convert sunlight into chemical energy in the bonds of carbohydrates. Eventually the overall reaction was understood to be

$$CO_2 + 2\,H_2O + \text{light energy} \rightarrow (CH_2O)_n + H_2O + O_2$$

The $(CH_2O)_n$ is a generic carbohydrate. (The "*n*" indicates that different carbohydrates have different multiples of CH_2O.) In essence, energy from light is transformed to chemical energy in the C–H bonds of carbohydrates. When glucose is the carbohydrate produced, the reaction can be written as

$$6\,CO_2 + 12\,H_2O + \text{light energy} \rightarrow C_6H_{12}O_6 + 6\,O_2 + 6\,H_2O$$

How does it happen? Based on the overall reaction, early investigators assumed that CO_2 and H_2O react directly to form CH_2O, and that the oxygen atoms in carbon dioxide are given off as oxygen gas (O_2). Both hypotheses proved to be incorrect, however. Let's see why.

Photosynthesis: Two Distinct Sets of Reactions

During the 1930s two independent lines of research on photosynthesis converged, leading to a major advance. The first research program, led by Cornelius van Niel, focused on how photosynthesis occurs in purple sulfur bacteria. Van Niel and his group found that these cells can grow in the laboratory on a food source that lacks sugars. Based on this observation, he concluded that they must be autotrophs that manufacture their own carbohydrates. But to grow, the cells had to be exposed to sunlight and hydrogen sulfide (H_2S). Van Niel also showed that these cells did not produce oxygen as a by-product of photosynthesis. Instead, elemental sulfur (S) accumulated in their medium. In these organisms, the overall reaction for photosynthesis was

$$CO_2 + 2\,H_2S + \text{light energy} \rightarrow (CH_2O)_n + H_2O + 2\,S$$

Van Niel's work was crucial for two reasons. First, it showed that CO_2 and H_2O do *not* combine directly during photosynthesis. Instead of acting as a reactant in photosynthesis in these species, H_2O is a product of the process. Second, van Niel's data showed that the oxygen atoms in CO_2 are not released as oxygen gas (O_2). This conclusion was logical because no oxygen was produced by the purple sulfur bacteria, even though carbon dioxide participated in the reaction—just as it did in plants.

Based on these findings, biologists hypothesized that the oxygen atoms that are released during plant photosynthesis must come from water. This proposal was supported by experiments with isolated chloroplasts, which produced oxygen in the presence of sunlight even if no CO_2 was present. The hypothesis was confirmed when heavy isotopes of oxygen—^{18}O compared with the normal isotope, ^{16}O—became available to researchers. Biologists then exposed algae or plants to H_2O that contained ^{18}O, collected the oxygen gas that was given off as a by-product of photosynthesis, and confirmed that the released oxygen gas contained the heavy isotope. As predicted, the reaction that produced this oxygen occurred only in the presence of sunlight.

A second major line of research helped support these discoveries. When the radioactive isotope ^{14}C became available in the mid-1940s, Melvin Calvin and others began feeding labeled carbon dioxide ($^{14}CO_2$) to algae and identifying the molecules that subsequently became labeled with the radioisotope. The investigators also noticed that labeled carbon dioxide could become incorporated into complex organic compounds in algae in the dark. This was a remarkable observation. Photosynthesis obviously depended on light, but for some reason the reduction of CO_2 did not. Because Calvin played an important role in detailing the exact sequence of reactions that made this key finding possible, the light-independent component of photosynthesis came to be known as the **Calvin cycle**. The light-independent reactions of photosynthesis reduce carbon dioxide and result in the production of sugar.

To summarize, early research on photosynthesis showed that this process consists of two distinct sets of reactions. One set is dependent on light; the other set can occur in the absence of light. The **light-dependent reactions** result in the production of oxygen from water; the **light-independent reactions** result in the production of sugar from carbon dioxide.

In photosynthesizing cells, the light-dependent and light-independent reactions occur simultaneously. How are these two sets of reactions connected? The short answer is by electrons. More specifically, electrons are released when water is split to form oxygen gas. During the light-dependent reactions, these electrons are transferred to a phosphorylated version of NAD^+, called $NADP^+$, that is abundant in photosynthesizing cells. This reaction forms NADPH. Like NADH, NADPH is an electron

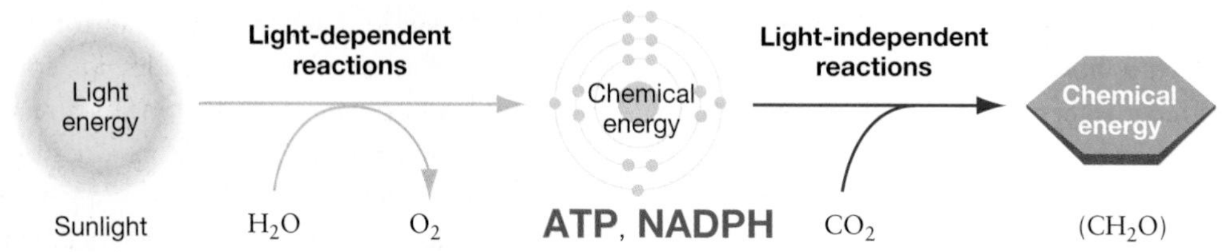

FIGURE 10.1 Photosynthesis Has Two Distinct Components
In the light-dependent reactions of photosynthesis, light energy is transformed to chemical energy. ATP is produced during these reactions. In addition, electrons are removed from water and are used to form NADPH, with oxygen (O_2) being released as a by-product. In the light-independent reactions, the ATP and NADPH produced in the light-dependent reactions are used to reduce carbon dioxide to carbohydrate.

carrier. ATP is also produced in the light-dependent reactions (see the middle portion of **Figure 10.1**). The light-independent reactions of the Calvin cycle use the electrons in NADPH and the potential energy in ATP to reduce CO_2 to carbohydrate (the right-hand side of Figure 10.1). The resulting sugars are used in cellular respiration to produce ATP for the cell. Plants oxidize sugars in their mitochondria and consume O_2 in the process, just as animals and other eukaryotes do.

Where does all this activity take place?

The Structure of the Chloroplast

Experiments with various plant tissues established that photosynthesis takes place only in the green portions of plants. Follow-up work with light microscopes suggested that the reactions occur inside the bright green organelles called **chloroplasts** ("green-formed"). Leaf cells typically contain from 40 to 50 chloroplasts, and a square millimeter of leaf averages about 500,000 chloroplasts (**Figure 10.2a**). When membranes derived from chloroplasts were found to release oxygen after

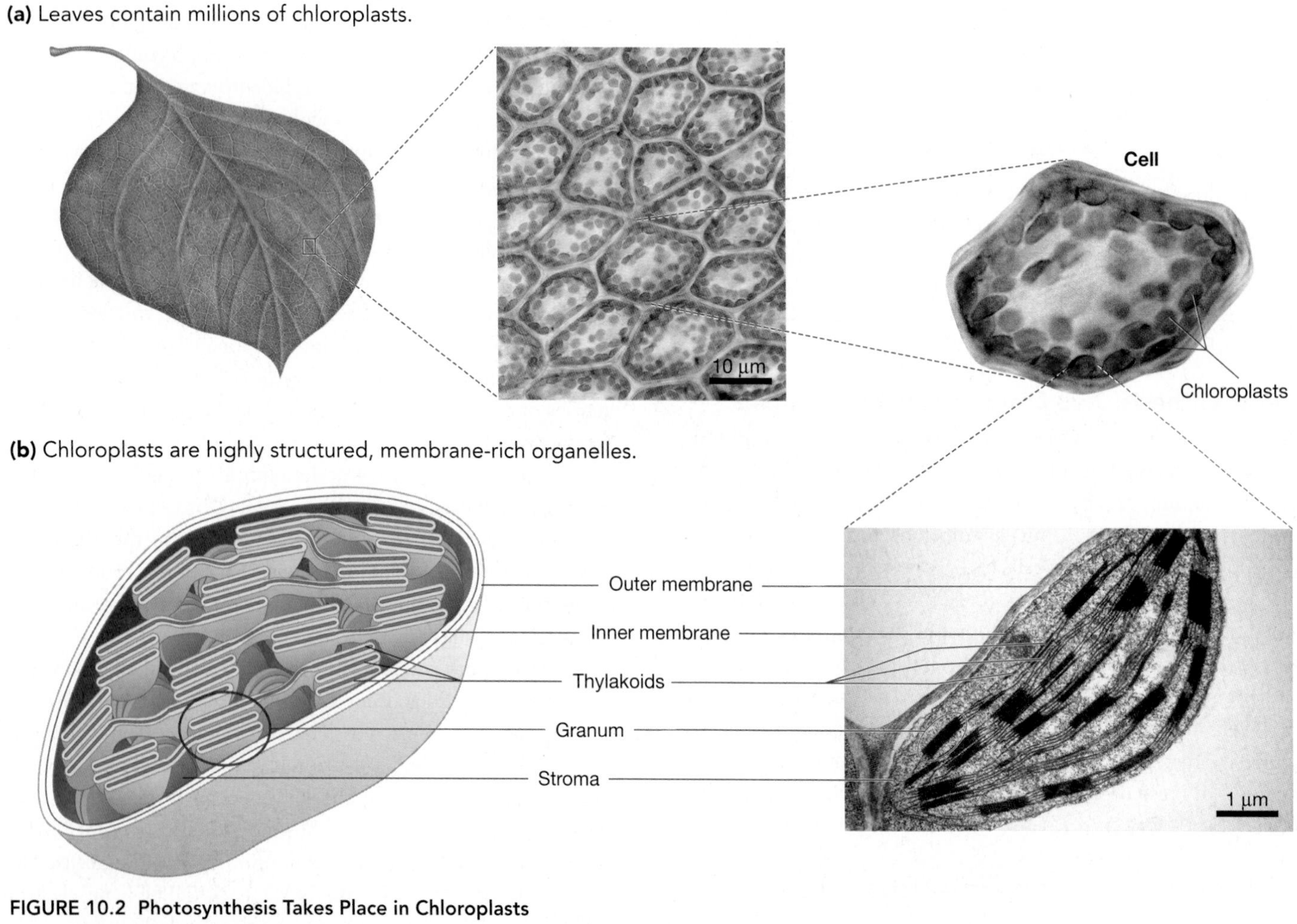

FIGURE 10.2 Photosynthesis Takes Place in Chloroplasts
(a) In plants, photosynthesis takes place in organelles called chloroplasts. **(b)** The internal membranes of chloroplasts form flattened, vesicle-like structures called thylakoids, some of which form stacks called grana.

exposure to sunlight, the hypothesis that chloroplasts are the site of photosynthesis became widely accepted.

When electron microscopy became available in the 1950s, researchers observed that chloroplasts are extremely membrane rich. As **Figure 10.2b** shows, the organelle is enclosed by an outer membrane and an inner membrane. The interior is dominated by vesicle-like structures called **thylakoids**, which often occur in interconnected stacks called **grana** (singular: *granum*). The space inside a thylakoid is its **lumen**. (Recall that *lumen* is a general term for the interior of any sac-like structure. Your stomach and intestines have a lumen.) The fluid-filled space between the thylakoids and the inner membrane is the **stroma**.

When researchers analyzed the chemical composition of thylakoid membranes, they found huge quantities of pigments. **Pigments** are molecules that absorb only certain wavelengths of light—other wavelengths are either transmitted or reflected. Pigments have colors because we see the wavelengths that pass through or bounce off them. The most abundant pigment found in the thylakoid membranes turned out to be chlorophyll. **Chlorophyll** ("green-leaf") absorbs blue light and red light and reflects or transmits green light. As a result, it is responsible for the green color of plants, some algae, and many photosynthetic bacteria.

Developmental studies showed that chloroplasts are derived from colorless organelles called **proplastids** ("before-plastids"). Proplastids are found in the cells of embryonic plants and in the rapidly dividing tissues of mature plants. As cells mature, proplastids develop into chloroplasts or other types of plastids specialized for that cell's particular task (see **Box 10.1**).

BOX 10.1 Types of Plastids

Plastids are a family of double-membrane-bound organelles found in plants. As **Figure 10.3** shows, they develop from small, unspecialized organelles called proplastids. As a cell matures and takes on a specialized function in the plant (that is, *differentiates*), its proplastids also differentiate. If a developing plant cell in a stem or leaf is exposed to light, for example, its proplastids are usually stimulated to develop into chloroplasts.

There are three major types of plastids: (1) chloroplasts, (2) leucoplasts, and (3) chromoplasts. Figure 10.2 detailed the structure of chloroplasts. *Leucoplasts* ("white-formed") often function as energy storehouses. More specifically, leucoplasts may store chemical energy in the bonds of molecules with high potential energy. Some leucoplasts store oils; others synthesize and sequester the carbohydrate called starch; still others store proteins. *Chromoplasts* ("color-formed") are brightly colored because they synthesize and hoard large amounts of orange, yellow, or red pigments in their vacuoles. High concentrations of chromoplasts are responsible for many of the bright colors of fruits and some flowers.

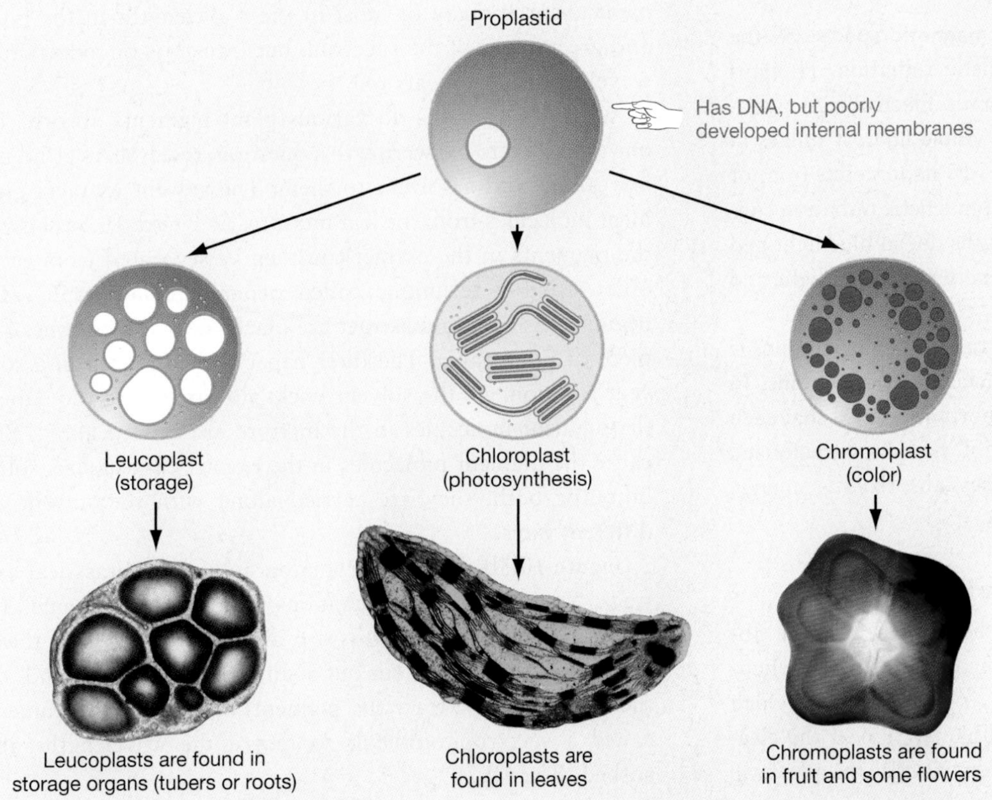

FIGURE 10.3 Chloroplasts Are a Specialized Type of Plastid
Each type of plastid has a distinctive structure and function, but all plastids develop from organelles called proplastids.

Before plunging into the details of how photosynthesis occurs inside a chloroplast, let's consider just how astonishing the process is. Chemists have synthesized an amazing diversity of compounds from relatively simple starting materials, but their achievements pale in comparison to cells that synthesize sugar from just carbon dioxide, water, and sunlight. If it is not *the* most sophisticated chemistry on Earth, photosynthesis is certainly a contender. What's more, photosynthetic cells accomplish this feat in environments ranging from mountaintop snowfields to the open ocean, tropical rain forests, and polar ice caps. Photosynthesis can occur in virtually any habitat where light is available.

10.2 How Does Chlorophyll Capture Light Energy?

Photosynthesis begins with the light-dependent reactions, and these reactions begin with the simple act of sunlight striking chlorophyll. To understand the consequences of this event, it's helpful to review the nature of light. Light is a type of electromagnetic radiation, which is a form of energy. The essence of photosynthesis is converting electromagnetic energy in the form of sunlight into chemical energy in the C–H bonds of sugar.

Physicists describe light's behavior as both wavelike and particle-like. As is true of all waves, including waves of water or air, electromagnetic radiation is characterized by its **wavelength**, the distance between two successive wave crests (or wave troughs). The wavelength determines the type of electromagnetic radiation.

Figure 10.4 illustrates the **electromagnetic spectrum**—the range of wavelengths of electromagnetic radiation. Humans cannot see all these wavelengths, however. Electromagnetic radiation that humans can see is called **visible light**; it ranges in wavelength from about 400 to about 710 nanometers (nm, or 10^{-9} m). Shorter wavelengths of electromagnetic radiation contain more energy than longer wavelengths do, so blue light and ultraviolet light contain much more energy than red light and infrared light do.

To emphasize the particle-like nature of light, physicists point out that it exists in discrete packets called **photons**. In understanding photosynthesis, the important point is that each photon and each wavelength of light has a characteristic amount of energy. Pigment molecules absorb this energy. How?

Photosynthetic Pigments Absorb Light

When a photon strikes an object, the photon may be absorbed, transmitted, or reflected. A pigment absorbs particular wavelengths of light. Sunlight is white light, which consists of all wavelengths in the visible portion of the electromagnetic spectrum at once. If a pigment absorbs all of the

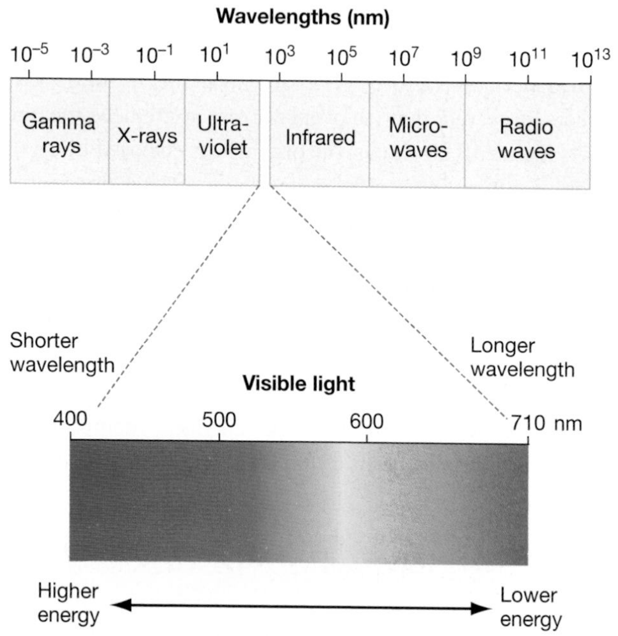

FIGURE 10.4 The Electromagnetic Spectrum
Electromagnetic energy radiates through space in the form of waves. Humans can see radiation at wavelengths between about 400 nm and about 710 nm. The shorter the wavelength of electromagnetic radiation, the higher its energy. QUESTION Does ultraviolet light contain more or less energy than infrared light?

visible wavelengths, no visible wavelength of light is reflected back to your eye, and the pigment appears black. If a pigment absorbs many or most of the wavelengths in the blue and green parts of the spectrum but transmits or reflects red wavelengths, it appears red.

What wavelengths do various plant pigments absorb? In one approach to answering this question, researchers grind up leaves and add a solvent to them. The solvent extracts pigment molecules from the leaf mixture. As **Figure 10.5a** shows, the pigments in the extract can then be separated from each other using a technique called **paper chromatography**. To begin, spots of a raw extract are placed near the bottom of a piece of filter paper. The filter paper is then placed in a solvent solution. As the solvent wicks upward along the paper, the pigment molecules in the mixture are carried along. Because the pigment molecules in the extract vary in size, solubility, or both, they are carried along with the solvent at different rates.

Figure 10.5b shows a chromatograph from a grass-leaf extract. Note that this leaf contains an array of pigments. To find out which wavelengths are absorbed by each of these molecules, researchers cut out a single region (color band) of the filter paper, extract the pigment, and use an instrument called a spectrophotometer to record the wavelengths absorbed (**Box 10.2**).

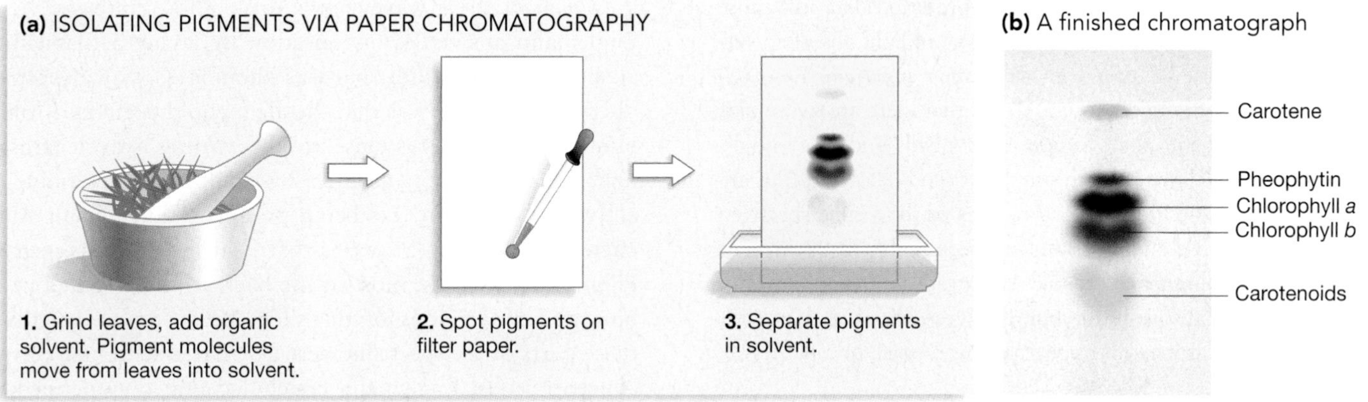
(a) ISOLATING PIGMENTS VIA PAPER CHROMATOGRAPHY
1. Grind leaves, add organic solvent. Pigment molecules move from leaves into solvent.
2. Spot pigments on filter paper.
3. Separate pigments in solvent.
(b) A finished chromatograph
Carotene
Pheophytin
Chlorophyll *a*
Chlorophyll *b*
Carotenoids

FIGURE 10.5 Pigments Can Be Isolated by Using Paper Chromatography
(a) Paper chromatography is an effective way to isolate the various pigments in photosynthetic tissue.
(b) Photosynthetic tissues, including those from the grass leaves used to make this chromatograph, typically contain several pigments. Different species of photosynthetic organisms may contain different types and quantities of pigments.

BOX 10.2 How Do Researchers Measure Absorption Spectra?

Researchers use an instrument called a **spectrophotometer** to measure the wavelengths of light absorbed by a particular pigment. As **Figure 10.6** shows, a solution of purified pigment molecules is exposed to a specific wavelength of light. If the pigment absorbs a great deal of this wavelength, then little light will be transmitted through the sample. But if the pigment absorbs the wavelength poorly, then most of the incoming light will be transmitted. The light that passes through the sample strikes a photoelectric tube, which converts it to an electric current. The amount of the electric current that is generated, as measured by a galvanometer, is proportional to the intensity of the incoming light. High current readings indicate low absorption at a particular wavelength, while low current readings signal high absorption. By testing one wavelength after another, investigators can measure the full absorption spectrum of a pigment.

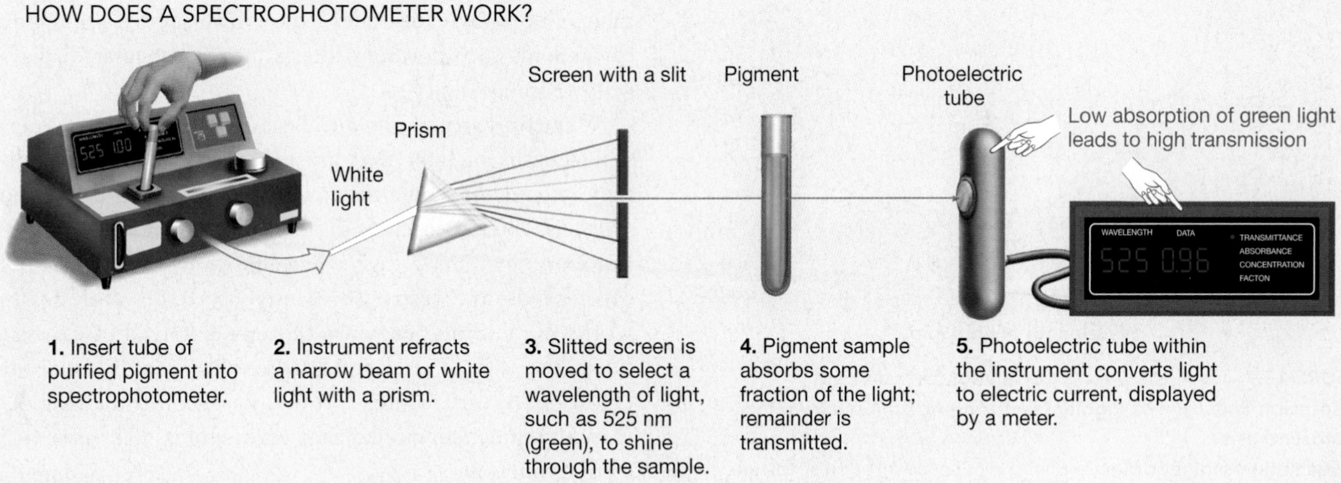
HOW DOES A SPECTROPHOTOMETER WORK?
White light
Prism
Screen with a slit
Pigment
Photoelectric tube
Low absorption of green light leads to high transmission
WAVELENGTH DATA
525 0.96
TRANSMITTANCE
ABSORBANCE
CONCENTRATION
FACTON
1. Insert tube of purified pigment into spectrophotometer.
2. Instrument refracts a narrow beam of white light with a prism.
3. Slitted screen is moved to select a wavelength of light, such as 525 nm (green), to shine through the sample.
4. Pigment sample absorbs some fraction of the light; remainder is transmitted.
5. Photoelectric tube within the instrument converts light to electric current, displayed by a meter.

FIGURE 10.6 Spectrophotometers Can Measure the Light Wavelengths Absorbed by Pigments

QUESTION Suppose that you were analyzing a sample of chlorophyll *b* and that you changed the wavelength selector in the spectrophotometer to 480 nm (blue light). How would the amount of absorption change?

Using this approach, biologists have produced data like those shown in **Figure 10.7a**. This graph is a plot of light absorbed versus wavelength and is called an **absorption spectrum.** Research based on these techniques has confirmed that there are two major classes of pigment in plant leaves: chlorophylls and carotenoids. The **chlorophylls,** designated chlorophyll *a* and chlorophyll *b*, absorb strongly in the blue and red regions of the visible spectrum and therefore reflect and transmit green light. Beta-carotene (β-carotene) and other **carotenoids** constitute a different family of pigments, which absorb in the blue and green parts of the visible spectrum. Thus carotenoids appear yellow, orange, or red.

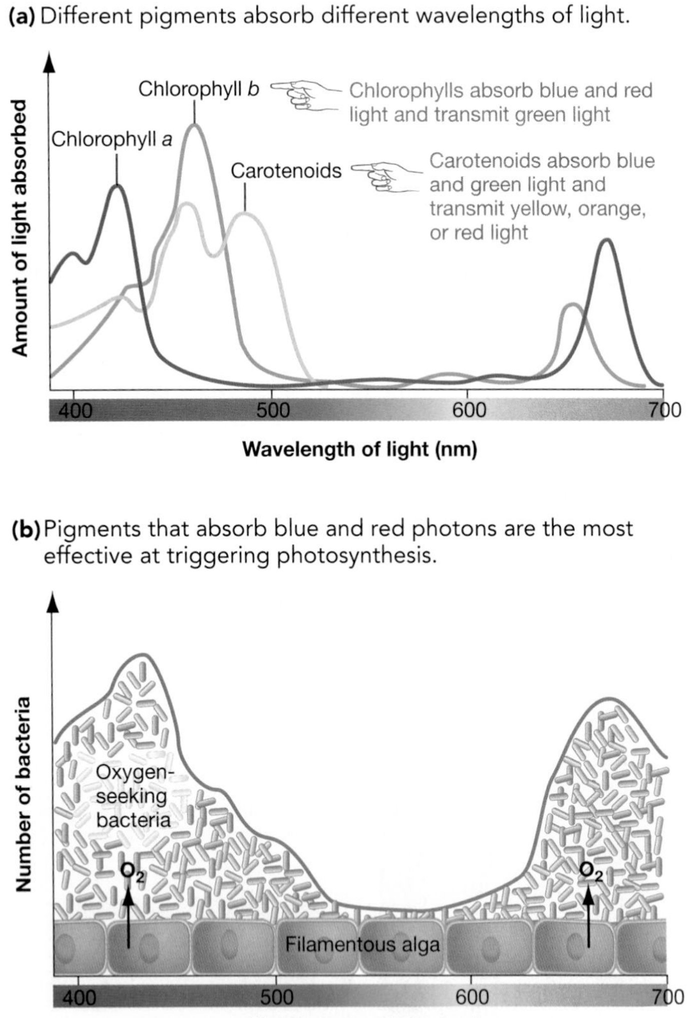

FIGURE 10.7 There Is a Strong Correlation between the Absorption Spectrum of Pigments and the Action Spectrum for Photosynthesis

(a) Each photosynthetic pigment has a distinct absorption spectrum. **(b)** The "action spectrum" for photosynthesis—meaning the wavelengths of light that trigger the process—correlates with the absorption spectra of photosynthetic pigments. EXERCISE Water absorbs strongly in all but the blue and green parts of the spectrum. Add a curve to part (b) predicting the action spectrum for photosynthesis in plants and algae that live at the bottom of lake or ocean habitats.

Which of these wavelengths drive photosynthesis? T. W. Englemann answered this question by laying a filamentous alga across a glass slide that was illuminated with a spectrum of colors. The idea was that the alga would begin performing photosynthesis in response to the various wavelengths of light and produce oxygen as a by-product. To determine exactly where oxygen was being produced, Englemann added bacterial cells from a species that is attracted to oxygen. As **Figure 10.7b** shows, most of the bacteria congregated in the blue and red regions of the slide. Because wavelengths in these parts of the spectrum were associated with high oxygen concentrations, Englemann concluded that they defined the **action spectrum** for photosynthesis. The data suggested that blue and red photons are the most effective at driving photosynthesis. Because the chlorophylls absorb these wavelengths, the data also suggested that chlorophylls are the main photosynthetic pigments.

Before analyzing what happens during the absorption event itself, let's look quickly at the structure and function of the chlorophylls and carotenoids.

What Is the Role of Carotenoids and Other Accessory Pigments? Carotenoids are called **accessory pigments,** because they absorb light and pass the energy on to chlorophyll. The carotenoids found in plants belong to two classes, called **carotenes** and **xanthophylls. Figure 10.8a** shows the structure of β-carotene (beta-carotene), which gives carrots their orange color. A xanthophyll called zeaxanthin, which gives corn kernels their bright yellow color, is nearly identical to β-carotene, except that the ring structures on either end of the molecule contain a hydroxyl (–OH) group.

Both xanthophylls and carotenes are found in chloroplasts. In autumn, when the leaves of deciduous trees begin to die and their chlorophyll degrades, the wavelengths scattered by carotenoids turn northern forests into spectacular displays of yellow, orange, and red.

What do carotenoids do? Because these pigments absorb wavelengths of light that are not absorbed by chlorophyll, they extend the range of wavelengths that can drive photosynthesis. But researchers discovered an even more important function by analyzing what happens to leaves when carotenoids are destroyed. Many herbicides, for example, work by inhibiting enzymes that are involved in carotenoid synthesis. Plants lacking carotenoids rapidly lose their chlorophyll, turn white, and die. Based on these results, researchers have concluded that carotenoids also serve a protective function.

To understand the molecular basis of carotenoid function, recall from Chapter 2 that energy from photons—especially the high-energy, short-wavelength photons in the ultraviolet part of the electromagnetic spectrum—contain enough energy to knock electrons out of atoms and create free radicals. Free radicals, in turn, trigger reactions that degrade molecules. Fortu-

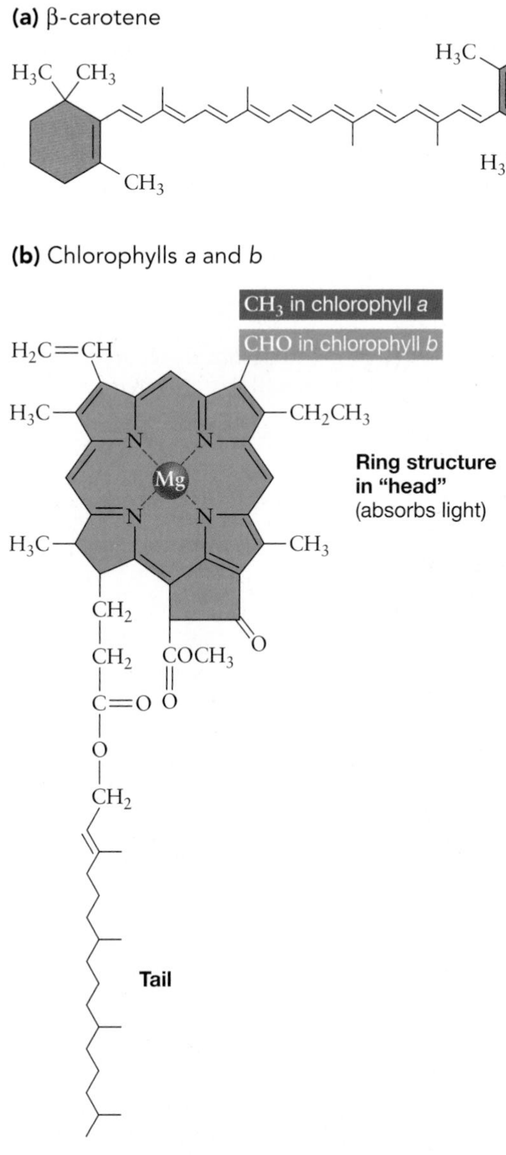

FIGURE 10.8 Photosynthetic Pigments Contain Ring Structures
(a) β-Carotene is an orange pigment found in carrot roots and other plant tissues. **(b)** Although chlorophylls *a* and *b* are very similar structurally, they have the distinct absorption spectra shown in Figure 10.7a.

nately, carotenoids can "quench" or stabilize free radicals and protect chlorophyll molecules from harm. When carotenoids are absent, chlorophyll molecules are destroyed. As a result, photosynthesis stops. Starvation and death follow.

Carotenoids are not the only molecules that protect plants from the damaging effects of sunlight, however. Researchers have recently analyzed individuals of the mustard plant *Arabidopsis thaliana* that are unable to synthesize pigments called flavonoids, which are normally stored in the vacuoles of leaf cells. Because **flavonoids** absorb ultraviolet radiation, individuals that lack flavonoids are subject to damage from UV light. In effect, these pigments function as a sunscreen for leaves and stems. Without them, chlorophyll molecules would be broken apart by high-energy radiation.

The general message here is that the energy in sunlight is a double-edged sword. It makes photosynthesis possible, but it can also lead to the formation of free radicals that damage cells. The role of carotenoids and flavonoids as protective pigments is crucial.

The Structure of Chlorophyll As **Figure 10.8b** shows, chlorophyll *a* and chlorophyll *b* are very similar in structure as well as in absorption spectra. They differ in just three atoms. In land plants, chloroplasts typically have about three molecules of chlorophyll *a* in their internal membranes for every molecule of chlorophyll *b*.

Notice from Figure 10.8b that chlorophyll molecules have two fundamental parts: a long tail made up of isoprene subunits (introduced in Chapter 6) and a "head" that consists of a large ring structure with a magnesium atom in the middle. The tail keeps the molecule embedded in the thylakoid membrane; the head is where light is absorbed.

But just what is "absorption?" Stated another way, what happens when a photon of a particular wavelength—say, red light with a wavelength of 680 nm—strikes a chlorophyll molecule?

When Light Is Absorbed, Electrons Enter an Excited State

When a photon strikes a chlorophyll molecule, the photon's energy can be transferred to an electron in the chlorophyll molecule's head region. In response, the electron is "excited," or raised to a higher electron shell—one with greater potential energy. As **Figure 10.9** shows, the excited electron states that are possible in a particular pigment are discrete—that is, incremental rather than continuous—and can be represented as lines on an energy scale. In the figure, the ground state, or unexcited state, is shown as 0 and the higher energy states are designated 1 and 2. If the difference between the possible energy states is

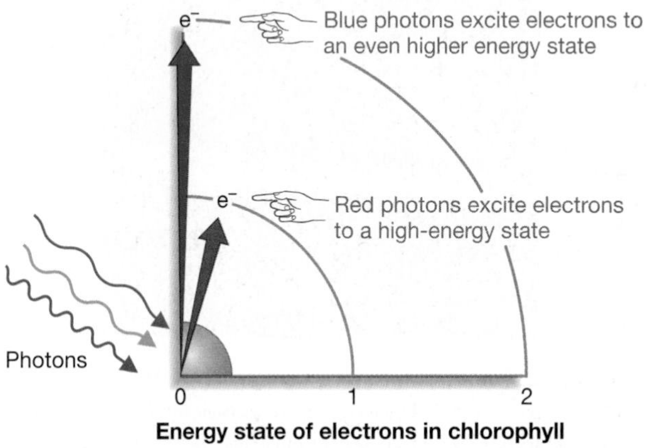

FIGURE 10.9 Electrons Are Promoted to High-Energy States when Photons Strike Chlorophyll
When a photon strikes chlorophyll, an electron can be promoted to a higher energy state, depending on the energy in the photon.

the same as the energy in the photon, then the photon can be absorbed and an electron is excited to that energy state.

In chlorophyll, for example, the energy difference between the ground state and state 1 is equal to the energy in a red photon, while the energy difference between state 0 and state 2 is equal to the energy in a blue photon. Thus chlorophyll can readily absorb red photons and blue photons. Chlorophyll does not absorb green light well because there is no discrete step—no difference in possible energy states for its electrons—that corresponds to the amount of energy in a green photon.

Wavelengths in the ultraviolet part of the spectrum have so much energy that they may actually eject electrons from a pigment molecule. In contrast, wavelengths in the infrared regions have so little energy that in most cases they merely increase the movement of atoms in the pigment, generating heat rather than exciting electrons.

If an electron is excited by a photon, it means that energy in the form of electromagnetic radiation is transferred to an electron, which now has high potential energy. If the excited electron falls back to its ground state, though, some of the absorbed energy is released as heat—meaning molecular movement—while the rest is released as electromagnetic radiation. This is the phenomenon known as **fluorescence**. Because some of the energy in the original photon is transformed to heat, the electromagnetic radiation that is given off during fluorescence has lower energy and a longer wavelength than the original photon does.

Figure 10.10 shows the fluorescence that occurs when a pure solution of chlorophyll is exposed to ultraviolet light. The isolated pigment absorbs the photons but simply fluoresces red in response. The amount of heat given off is equal to the energy difference between the higher-energy ultraviolet photons that are absorbed and the lower-energy red photons that are released.

FIGURE 10.10 Fluorescence
A pure solution of chlorophyll exposed to ultraviolet light. Electrons are excited to a high-energy state but immediately fall back to a low-energy state, emitting red photons and heat. EXERCISE Add to the diagram in Figure 10.9 to show what is happening here. Start with the electron promoted to a high-energy state by blue light; have it fall back to the ground state (0); and indicate the fate of the energy released.

How Do the Chlorophyll Molecules in Leaves Work? When chlorophyll is in a chloroplast, only about 2 percent of the red and blue photons that it absorbs normally produce fluorescence. What happens to the other excited electrons? An answer to this question began to emerge as investigators struggled to interpret the experimental result described in **Figure 10.11**.

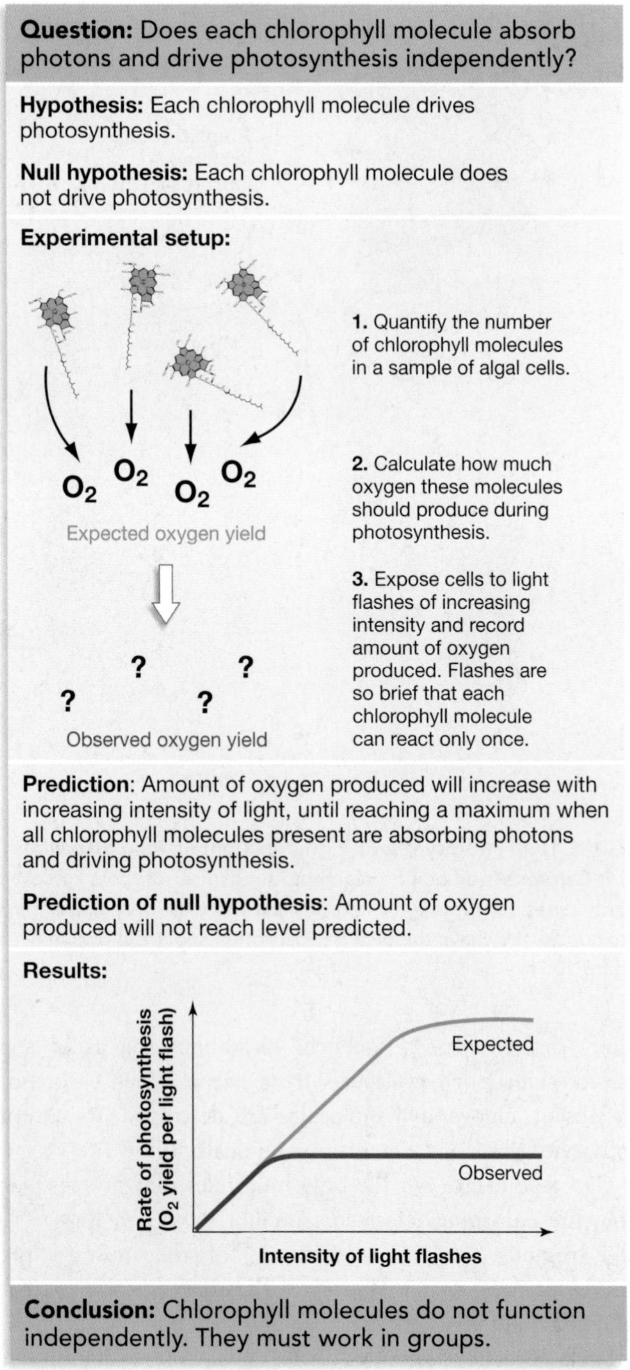
Question: Does each chlorophyll molecule absorb photons and drive photosynthesis independently?
Hypothesis: Each chlorophyll molecule drives photosynthesis.
Null hypothesis: Each chlorophyll molecule does not drive photosynthesis.
Experimental setup:
1. Quantify the number of chlorophyll molecules in a sample of algal cells.
O2 O2 O2 O2
Expected oxygen yield
2. Calculate how much oxygen these molecules should produce during photosynthesis.
3. Expose cells to light flashes of increasing intensity and record amount of oxygen produced. Flashes are so brief that each chlorophyll molecule can react only once.
? ? ? ?
Observed oxygen yield
Prediction: Amount of oxygen produced will increase with increasing intensity of light, until reaching a maximum when all chlorophyll molecules present are absorbing photons and driving photosynthesis.
Prediction of null hypothesis: Amount of oxygen produced will not reach level predicted.
Results:
Rate of photosynthesis (O2 yield per light flash)
Expected
Observed
Intensity of light flashes
Conclusion: Chlorophyll molecules do not function independently. They must work in groups.

FIGURE 10.11 Chlorophyll Molecules Work in Groups, Not Individually
The experimental results outlined here were a complete surprise to the investigators.

Robert Emerson and William Arnold designed this experiment to test the hypothesis that all of the chlorophyll molecules present in a photosynthetic cell absorb photons and drive photosynthesis. The key to the setup was that the researchers could estimate the number of chlorophyll molecules present in an average cell of their study organism—a green alga called *Chlorella*. These data allowed them to predict how much photosynthesis should occur in response to a flash of light. As predicted, the amount of photosynthesis—measured as the amount of oxygen produced—increased steadily as the researchers increased the intensity of the light flashes. But much to Emerson and Arnold's surprise, the total amount of photosynthesis leveled off far below the predicted value. Long before all or even most chlorophyll molecules were active, the total amount of photosynthesis reached a maximum.

What was happening? To make sense of the result, the researchers hypothesized that chlorophyll molecules do not work individually. Instead, chlorophylls must work in groups. Follow-up work has shown that in the thylakoid membrane, 200–300 chlorophyll molecules and accessory pigments such as carotenoids are grouped together in an array of proteins, forming a complex called a **photosystem**. Each photosystem, in turn, has two major elements: an antenna complex and a reaction center.

The Antenna Complex When a red or blue photon strikes a pigment molecule in the **antenna complex**, the energy is absorbed and an electron is excited in response. This energy—but not the electron itself—is passed along to a nearby chlorophyll molecule, where another electron is excited in response. As the energy is transmitted, the original excited electron falls back to its ground state. In this way, energy is transmitted from one chlorophyll molecule to the next inside the antenna complex (**Figure 10.12a**). The system acts like a radio antenna that receives a specific wavelength of electromagnetic radiation and transmits it to a receiver. In a photosystem, the receiver is called the reaction center.

The Reaction Center When energy from the antenna complex reaches the reaction center of a photosystem, an all-important energy transformation event occurs. At the **reaction center**, excited electrons are transferred to a molecule that acts as an electron acceptor (**Figure 10.12b**). When this molecule becomes reduced, the energy transformation event that started with the absorption of light becomes permanent: Electromagnetic energy is transformed to chemical energy. The redox reaction that occurs in the reaction center results in the production of chemical energy from sunlight.

The key to understanding the reaction center is that, in the absence of light, chlorophyll cannot reduce the electron acceptor because the reactions are endergonic. But when light excites electrons in chlorophyll to a high-energy state, the reactions become exergonic.

Now, what happens to these high-energy electrons? Specifically, how are they used to manufacture sugar?

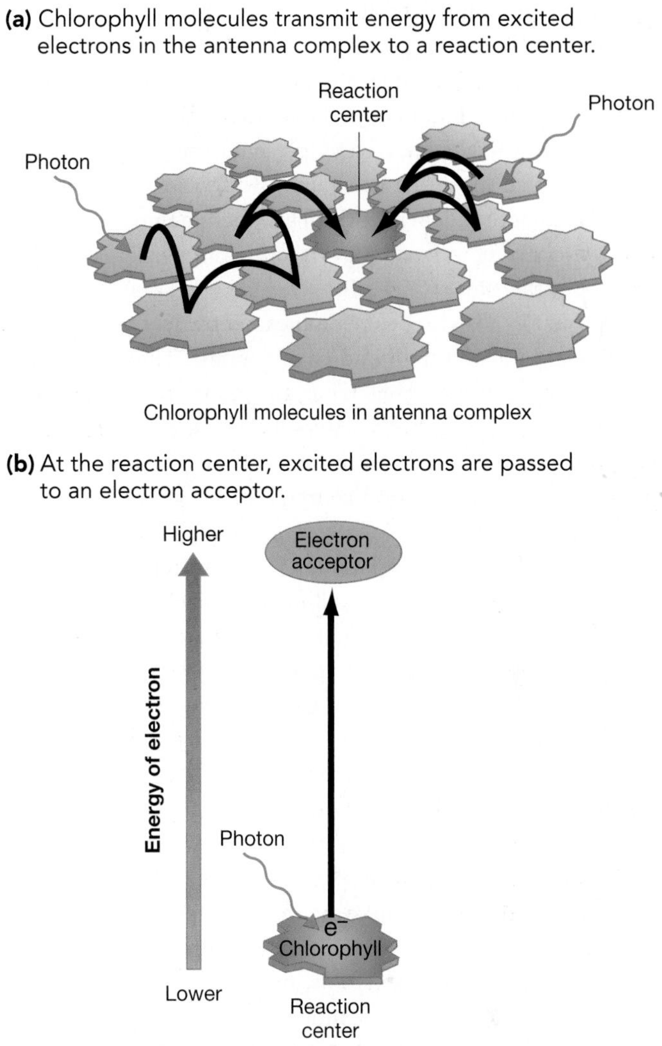

FIGURE 10.12 The Antenna Complex Captures Light Energy and Transmits It to the Reaction Center
(a) The antenna complex is aptly named. It captures specific wavelengths of light and transfers the energy to the reaction center. **(b)** Redox reactions in the reaction center result in the addition of a high-energy electron to an electron acceptor.

10.3 The Discovery of Photosystems I and II

During the 1950s the fate of the high-energy electrons in photosystems was the central issue facing biologists interested in photosynthesis. Ironically, a central insight into this issue came from a simple experiment on how green algae responded to various wavelengths of light.

The experimental setup was based on the observation that photosynthetic cells respond to very specific wavelengths of light. In particular, the algal cells being studied responded strongly to wavelengths of 700 nm and 680 nm, which are in the far-red and red portions of the visible spectrum, respectively. In a key experiment, Robert Emerson found that if

WEB TUTORIAL 10.1 Photosynthesis

cells were illuminated with either far-red light or red light, the photosynthetic response was moderate (**Figure 10.13**). But if cells were exposed to a combination of far-red and red light, the rate of photosynthesis increased dramatically. When both wavelengths were present, the photosynthetic rate was much more than the sum of the rates produced by each wavelength independently. This phenomenon was called the *enhancement effect*. Why it occurred was a complete mystery at the time.

The puzzle posed by the enhancement effect was eventually solved by Robin Hill and Faye Bendall. These biologists synthesized data emerging from an array of labs and proposed that green algae and plants have two distinct types of reaction centers rather than just one. Hill and Bendall proposed that one reaction center, which came to be called **photosystem II**, interacts with a different reaction center, now referred to as **photosystem I**. According to the two-photosystem hypothesis, the enhancement effect occurs because photosynthesis is much more efficient when both photosystems are operating together.

Subsequent work has shown that the two-photosystem hypothesis is correct. In green algae and land plants, thylakoid membranes contain photosystems that differ in structure and function but complement each other.

To figure out how the two photosystems work, investigators chose not to study them together. Instead, they focused on species of photosynthetic bacteria that have photosystems similar to either photosystem I or II, but not both. Once each type of photosystem was understood in isolation, they turned to understanding how the two photosystems work in combination in green algae and land plants. Let's do the same—we'll analyze photosystem II, then photosystem I, and then how the two interact.

How Does Photosystem II Work?

To analyze photosystem II, researchers focused on studying species from the purple nonsulfur bacteria and the purple sulfur bacteria. These cells have a single photosystem that has many of the same components observed in photosystem II of cyanobacteria ("blue-green bacteria"), algae, and plants.

In photosystem II, the action begins when the antenna complex transmits energy to the reaction center and the molecule pheophytin comes into play. Structurally, **pheophytin** is very similar to chlorophyll. The two molecules are identical except that pheophytin lacks a magnesium atom in its head region. Functionally, though, they are different. Instead of acting as a pigment that promotes an electron when it absorbs a photon, pheophytin acts as an electron acceptor. When an electron in the reaction center chlorophyll is excited energetically, the electron binds to pheophytin and the reaction center chlorophyll is oxidized. When pheophytin is reduced in this way, the energy transformation step that started with the absorption of light is completed.

Question: Light at both the red and far-red wavelengths stimulates photosynthesis. How does a combination of these wavelengths affect the rate of photosynthesis?

Hypothesis: When red light and far-red light are combined, the rate of photosynthesis will double.

Null hypothesis: When red light and far-red light are combined, the rate of photosynthesis will not double.

Experimental setup:

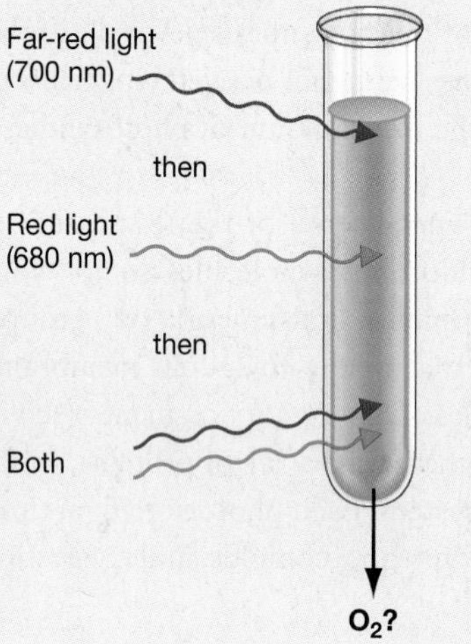

1. Expose algal cells to an intensity of far-red (700 nm) or red (680 nm) light that maximizes rate of photosynthesis at each wavelength. Then expose same algal cells to same intensities of each wavelength at same time.

2. Record the rate of photosynthesis as the amount of oxygen produced.

Prediction: When the wavelengths are combined, the rate of photosynthesis will be double the maximum rate observed for each wavelength independently.

Prediction of null hypothesis: When the wavelengths are combined, the rate of photosynthesis will not be double the maximum rate observed for each wavelength independently.

Results:

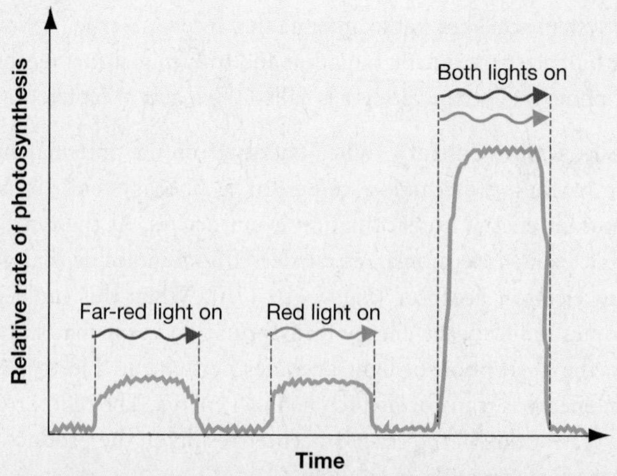

Conclusion: There is an enhancement effect for red and far-red light. The combination of 700 nm and 680 nm wavelengths more than doubles the rate of photosynthesis.

FIGURE 10.13 Discovery of the "Enhancement Effect" of Red and Far-Red Light

Electrons from Pheophytin Enter an Electron Transport Chain Electrons that reach pheophytin are passed to an electron transport chain in the thylakoid membrane. In both structure and function, this group of molecules is similar to the electron transport chain in the inner membrane of mitochondria (Chapter 9). For example, the electron transport chain associated with photosystem II contains several quinones and cytochromes. Electrons in both chains participate in a series of reduction-oxidation reactions and are gradually stepped down in potential energy. In mitochondria as well as chloroplasts, the redox reactions result in protons being pumped from one side of an internal membrane to the other. In both organelles, the resulting proton gradient drives ATP production via ATP synthase.

Figure 10.14a details the sequence of events in the electron transport chain of thylakoids. One of the key molecules involved is a quinone called **plastoquinone**, symbolized PQ. Recall from Chapter 9 that quinones are small hydrophobic molecules. Because plastoquinone is lipid soluble and not anchored to a protein, it is free to move from one side of the thylakoid membrane to the other. When it receives electrons from pheophytin, plastoquinone carries them to the other side of the membrane and delivers them to more electronegative molecules in the chain. These electron acceptors are found in a complex that contains a cytochrome similar to mitochondrial cytochromes.

In this way, plastoquinone shuttles electrons from pheophytin to the cytochrome complex. The electrons are then passed through a series of iron- and copper-containing proteins in the cytochrome complex. The potential energy released by these reactions allows protons to be added to other plastoquinone molecules, which carry them to the lumen side of the thylakoid membrane. As **Figure 10.14b** shows, the protons transported by plastoquinone result in a large concentration of protons in the thylakoid lumen. When photosystem II is active, the pH of the thylakoid interior reaches 5 while the pH of the stroma hovers around 8. Because the pH scale is logarithmic, the difference of 3 units means that the concentration of H^+ is $10 \times 10 \times 10 = 1000$ times higher in the lumen than in the stroma. In addition, the stroma becomes negatively charged relative to the thylakoid lumen. The net effect of electron transport, then, is to set up a large proton gradient that will drive H^+ out of the thylakoid lumen and into the stroma. Based on your reading of Chapter 9, it should come as no surprise that this proton-motive force drives the production of ATP.

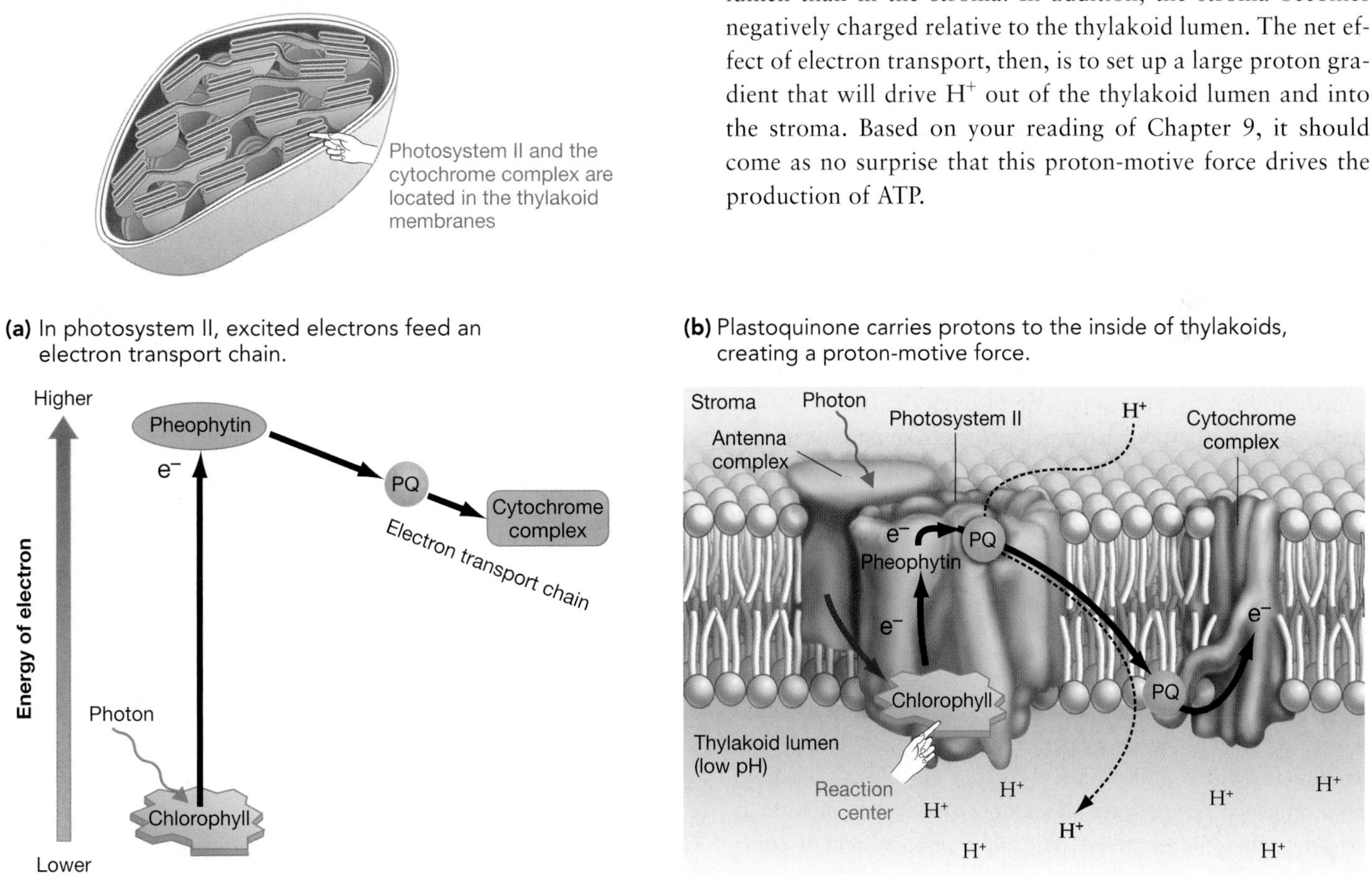

FIGURE 10.14 Photosystem II Feeds an Electron Transport Chain That Pumps Protons
(a) When an excited electron leaves the chlorophyll molecule in the reaction center of photosystem II, the electron is accepted by pheophytin, transferred to plastoquinone (PQ), and then stepped down in energy along an electron transport chain. **(b)** PQ carries electrons from photosystem II along with protons from the stroma. The electrons are passed to the cytochrome complex, and the protons are released in the thylakoid lumen. EXERCISE Add an arrow to part (b), indicating the direction of the proton-motive force.

ATP Synthase Uses the Proton-Motive Force to Phosphorylate ADP In mitochondria, NADH and $FADH_2$ donate electrons to an electron transport chain, and the redox reactions that occur in the chain result in protons being pumped out of the matrix and into the intermembrane space. In photosystem II, pheophytin donates electrons to an electron transport chain, and the redox reactions that occur result in protons being pumped out of the stroma and into the thylakoid lumen.

In both mitochondria and chloroplasts, protons diffuse down the resulting electrochemical gradient. This is an exergonic process that drives the endergonic synthesis of ATP. More specifically, the flow of protons through the enzyme ATP synthase causes conformational changes that drive the phosphorylation of ADP. In photosystem II, then, the light energy captured by chlorophyll is transformed to chemical energy stored in ATP. This process is called **photophosphorylation.** Substrate-level phosphorylation, oxidative phosphorylation, and photophosphorylation all result in the production of ATP.

To summarize, photosystem II starts with an electron being promoted to a high-energy state and ends with the production of ATP. The story is not complete, however, because we haven't accounted for the electrons that flow through the system. The electron that was transferred from chlorophyll to pheophytin in the photosystem II reaction center needs to be replaced. In addition, the electron transport chain of photosystem II needs to donate its electrons to some final electron acceptor.

Where do the electrons required by photosystem II come from, and where do they go? The parallel between the photosystem in purple sulfur bacteria and photosystem II of plants ends here. In purple sulfur bacteria, cytochrome donates an electron back to the reaction center, so the same electron can again be promoted to a high-energy state when a photon is absorbed. In this way, electrons cycle through the system. But in plants, algae, and cyanobacteria, electrons from photosystem II replenish the electrons that leave photosystem I in response to light. How are the electrons that leave photosystem II replaced?

Photosystem II Obtains Electrons by Oxidizing Water To understand where the electrons that feed photosystem II come from, think back to the overall reaction for photosynthesis: $CO_2 + 2\,H_2O + \text{light energy} \rightarrow (CH_2O) + H_2O + O_2$. In the presence of sunlight, carbon dioxide and water are used to produce carbohydrate, water, and oxygen gas. Recall that experiments with radioisotopes of oxygen showed that the oxygen atoms in O_2 come from water, not from carbon dioxide. It turns out that the electrons that enter photosystem II come from water. The oxygen-generating reaction can be written as

$$2\,H_2O \rightarrow 4\,H^+ + 4\,e^- + O_2$$

Because electrons are removed from water, the molecule becomes oxidized. This reaction is referred to as "splitting" water. It supplies a steady stream of electrons for photosystem II and is catalyzed by enzymes that are physically integrated into the photosystem II complex. When excited electrons leave photosystem II and enter the electron transport chain, the photosystem becomes so electronegative that enzymes can strip electrons away from water, leaving protons and oxygen.

Among all life-forms, photosystem II is the only protein complex that can catalyze the splitting of water molecules. Organisms such as cyanobacteria, algae, and plants that have this type of photosystem are said to perform **oxygenic** ("oxygen-producing") **photosynthesis,** because they generate oxygen as a by-product of the process. The purple sulfur and purple nonsulfur bacteria cannot oxidize water. They perform **anoxygenic** ("no oxygen-producing") **photosynthesis.**

It is difficult to overstate the importance of this unique reaction. The oxygen that is keeping you alive right now was produced by it. O_2 was, in fact, almost nonexistent on Earth prior to the evolution of enzymes that could catalyze the oxidation of water. According to the fossil record, oxygen levels in the atmosphere and oceans began to rise only about 2 billion years ago, as organisms that perform oxygenic photosynthesis increased in abundance. This change was a disaster for anaerobic organisms, because oxygen is toxic to them. But as oxygen became even more abundant, certain bacterial cells evolved the ability to use it as an electron acceptor during cellular respiration. This was a momentous development. O_2 is so electronegative that it creates a huge potential energy drop for the electron transport chains involved in cellular respiration. As a result, organisms that use O_2 as an electron acceptor in cellular respiration can produce much more ATP than can organisms that use other electron acceptors. Aerobic organisms grow so efficiently that they have long dominated our planet. Biologists rank the evolution of the oxygen-rich atmosphere as one of the most important events in the history of life.

Despite its fundamental importance, though, the mechanism responsible for the oxygen-generating reaction is not yet understood. Determining exactly how photosystem II splits water may be the greatest challenge currently facing researchers interested in photosynthesis. This issue has important practical applications as well, because if human chemists could replicate the reaction in an industrial setting, it might be possible to produce huge volumes of O_2 and hydrogen gas (H_2) from water. If this could be accomplished inexpensively, the H_2 produced could be used as a clean fuel for cars and trucks.

How Does Photosystem I Work?

Recall that researchers dissected photosystem II by studying similar, but simpler, photosystems in purple nonsulfur and purple sulfur bacteria. To understand the structure and function of photosystem I, they turned to heliobacteria ("sun-bacteria").

Like purple nonsulfur and purple sulfur bacteria, heliobacteria use the energy in sunlight to promote electrons to a high-energy state. But instead of being passed to an electron transport chain that pumps protons across a membrane, the

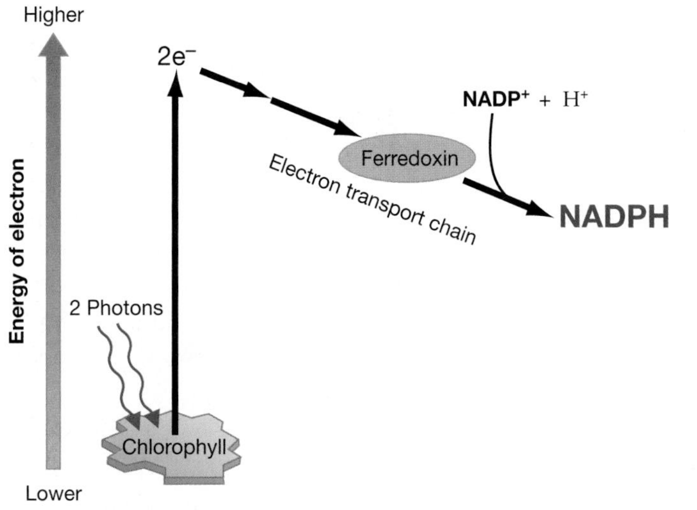

FIGURE 10.15 Photosystem I Produces NADPH
When excited electrons leave the chlorophyll molecule in the reaction center of photosystem I, they pass through a series of iron- and sulfur-containing proteins until they are accepted by ferredoxin. In an enzyme-catalyzed reaction, the reduced form of ferredoxin reacts with $NADP^+$ to produce NADPH.

high-energy electrons in heliobacteria are used to reduce NAD^+. When NAD^+ gains two electrons and a proton, NADH is produced. In the cyanobacteria and algae and land plants, a similar set of reactions reduces a phosphorylated version of NAD^+, symbolized $NADP^+$, yielding NADPH. Both NADH and NADPH function as electron carriers.

Figure 10.15 shows how the system works in photosystem I. When the reaction center in photosystem I absorbs a photon, excited electrons are passed through a series of iron- and sulfur-containing proteins inside the photosystem, and then to a molecule called **ferredoxin.** The electrons then move from ferredoxin to the enzyme ferredoxin/$NADP^+$ oxidoreductase—also called $NADP^+$ reductase—which transfers two electrons and a proton to $NADP^+$. This reaction forms NADPH. The photosystem itself and $NADP^+$ reductase are anchored in the thylakoid membrane; ferredoxin is closely associated with the bilayer.

To summarize, photosystem I results in the production of NADPH, and photosystem II results in the production of a proton gradient that drives the synthesis of ATP. NADPH is similar in function to the NADH and $FADH_2$ produced by the Krebs cycle. It is an electron carrier that can donate electrons to other compounds and thus reduce them.

In combination, then, photosystems I and II produce both chemical energy stored in ATP and reducing power in the form of NADPH. Although several groups of bacteria have just one of the two photosystems, the cyanobacteria, algae, and plants have both. In these organisms, how do the two photosystems interact?

The Z Scheme: Photosystems I and II Work Together

When they realized that photosystems I and II have distinct but complementary functions, Robin Hill and Faye Bendall proposed that these systems interact as shown in **Figure 10.16.** The diagram illustrates a model, known as the **Z scheme**, that furnished a breakthrough in research on photosynthesis. The name was inspired by the shape of the proposed path of electrons through the two photosystems, when that path was plotted on a vertical axis representing the changes that occur in their potential energy.

Following the path of electrons through the Z scheme—by tracing the route of electrons through Figure 10.16 with your finger—will help drive home how photosynthesis works. The process starts when photons excite electrons in the chlorophyll molecules of photosystem II's antenna complex. When

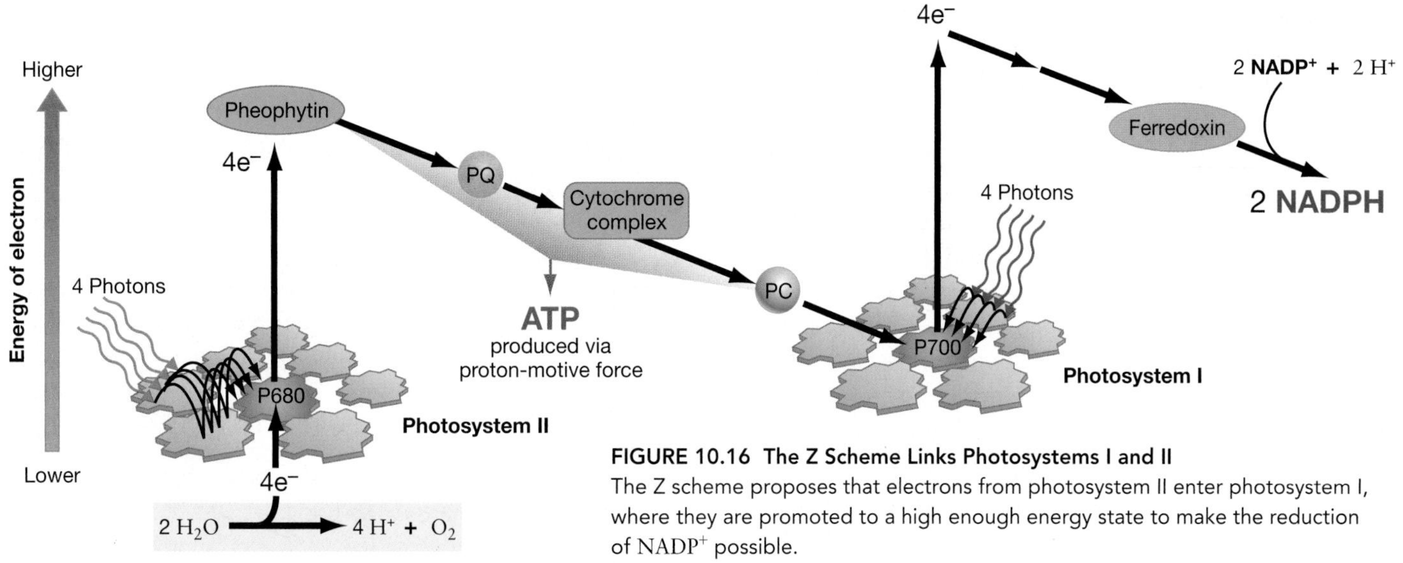

FIGURE 10.16 The Z Scheme Links Photosystems I and II
The Z scheme proposes that electrons from photosystem II enter photosystem I, where they are promoted to a high enough energy state to make the reduction of $NADP^+$ possible.

the energy in the excited electron is transmitted to the reaction center, a special pair of chlorophyll molecules named P680 passes excited electrons to pheophytin. From there the electron is gradually stepped down in potential energy through redox reactions among a series of quinones and cytochromes, which act as an electron transport chain. Using the energy released by the redox reactions, plastoquinone (PQ) carries protons across the thylakoid membrane. ATP synthase uses the resulting proton-motive force to phosphorylate ADP, creating ATP.

When electrons reach the end of photosystem II's electron transport chain, they are passed to a small diffusible protein called **plastocyanin** (symbolized PC in Figure 10.16). Plastocyanin picks up an electron from the cytochrome complex, diffuses through the lumen of the thylakoid, and donates the electron to photosystem I. A single plastocyanin molecule can shuttle over 1000 electrons per second between photosystems. In this way, plastocyanin forms a physical link between photosystem II and photosystem I.

The flow of electrons between photosystems, by means of plastocyanin, replaces electrons that are carried away from a chlorophyll molecule called P700 in the photosystem I reaction center. The electrons that emerge from P700 are eventually transferred to the protein ferredoxin, which then passes electrons to an enzyme that catalyzes the reduction of $NADP^+$ to NADPH. The electrons that initially left photosystem II are replaced by electrons that are stripped away from water, producing oxygen gas as a by-product.

The Z scheme helps explain the enhancement effect in photosynthesis documented in Figure 10.13. When algal cells are illuminated with wavelengths at 680 nm, in the red portion of the spectrum, only photosystem II can run at a maximum rate. The overall rate of electron flow through the Z scheme is moderate because photosystem I's efficiency is reduced. Similarly, when cells receive only wavelengths at 700 nm, in the far red, only photosystem I is capable of peak efficiency; photosystem II is working at a below-maximum rate, so the overall rate of electron flow is reduced. But when both wavelengths are available at the same time, both photosystem II and photosystem I are activated by light and work at a maximum rate, leading to enhanced efficiency.

Recent evidence indicates that a different electron path also occurs in green algae and plants. This pathway, called **cyclic photophosphorylation**, is illustrated in **Figure 10.17**. During cyclic photophosphorylation, photosystem I transfers electrons back to the electron transport chain, to augment ATP generation through photophosphorylation. This "extra" ATP is required for the chemical reactions that reduce carbon dioxide (CO_2) and produce sugars. In this way, cyclic photophosphorylation coexists with the Z scheme and produces additional ATP.

Although the Z-scheme model has held up well under experimental tests, several unresolved questions remain. The precise three-dimensional structure of each photosystem has been determined in bacteria but not yet in eukaryotes. Biologists are also trying to get a better understanding of how the two complexes are situated with respect to one another in the thylakoid membranes. As **Figure 10.18** shows, photosystems I and II are found in different parts of a single granum. Photosystem II is much more abundant in the interior, stacked membranes of grana, while photosystem I is much more common in the exterior, unstacked membranes. This physical separation between the photosystems is perplexing, given that their functions are so tightly integrated according to the Z scheme. Why they are found in different parts of the thylakoid is the focus of intense debate.

In contrast, the fate of the ATP and NADPH produced by photosystems I and II is well documented. Chloroplasts use ATP and NADPH to reduce carbon dioxide to sugar. Your life, and the life of most other organisms, depends on this process. How does it happen?

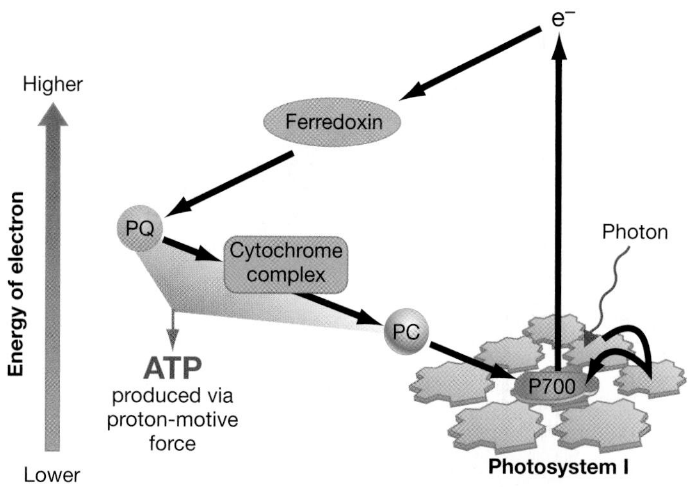

FIGURE 10.17 Cyclic Photophosphorylation Produces ATP
Cyclic electron transport is an alternative to the Z scheme. Instead of being donated to $NADP^+$, electrons cycle through the system and result in the production of additional ATP via photophosphorylation.

✓CHECK YOUR UNDERSTANDING

Photosystem II contributes high-energy electrons to an electron transport chain that pumps protons, creating a proton-motive force that drives ATP synthase. Photosystem I makes NADPH. To check your understanding of the light-dependent reactions, you should be able to make a model of the Z scheme using paper cutouts. On pieces of paper, label the following: the antenna systems of photosystems II and I, pheophytin, plastoquinone and the electron transport chain, plastocyanin, ferredoxin, and the reaction that splits water. Using dimes to represent electrons, explain how they flow through the photosystems.

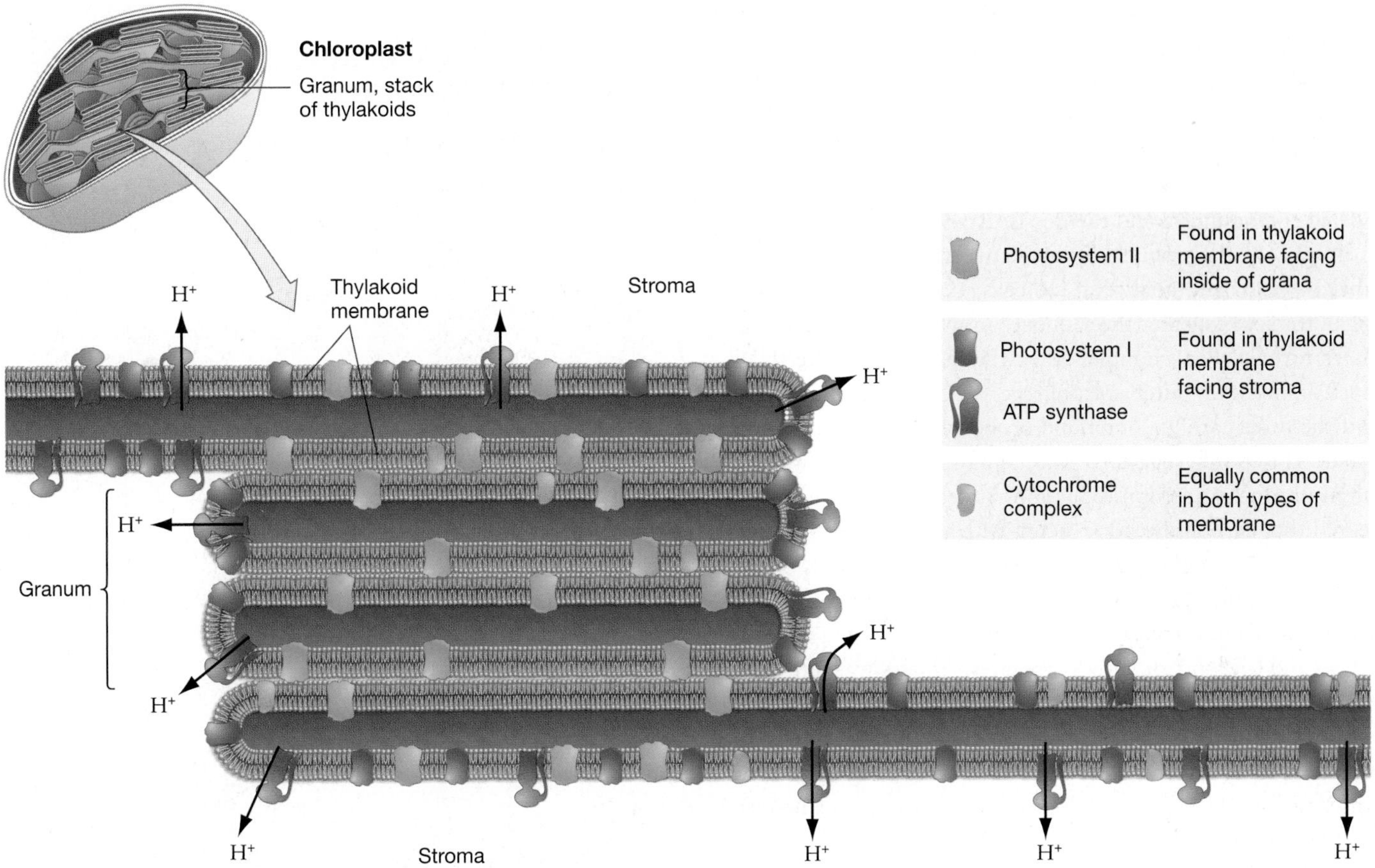

FIGURE 10.18 Photosystems I and II Occur in Separate Regions of Thylakoid Membranes within Grana Virtually all of the active photosytem II is found in membranes facing the inside of the chloroplast's grana. In contrast, virtually all of the photosystem I and ATP synthase are found in membranes that face the stroma. The cytochrome complex, plastoquinone, and plastocyanin are equally common in both types of membranes. EXERCISE On the figure, draw the path of an electron that follows the Z scheme from photosystem II to photosystem I. Then draw the path of an electron that participates in cyclic photophosphorylation.

10.4 How Is Carbon Dioxide Reduced to Produce Glucose?

The reactions analyzed in Section 10.3 occur only in the presence of light. This is logical, because their entire function is focused on energy transformation—the conversion of electromagnetic energy in the form of sunlight to chemical energy in the phosphate bonds of ATP and the electrons of NADPH. The reactions that lead to the production of sugar from carbon dioxide, in contrast, do not depend directly on the presence of light. Although these reactions still require the ATP and NADPH produced by the light-dependent reactions of photosynthesis, it is possible for the reactions that lead to the reduction of CO_2 and production of sugars to occur in darkness.

The realization that the energy transformation and carbon dioxide reduction components of photosynthesis are two separate processes was a fundamental insight. Research on the exact sequence of light-independent reactions gained momentum just after World War II, when radioactive isotopes of carbon became available for research purposes. Between 1945 and 1955, a team led by Melvin Calvin carried out a groundbreaking series of experiments based on exposing green algae to radioactively labeled carbon dioxide ($^{14}CO_2$). By isolating and identifying product molecules that contained ^{14}C, the researchers gradually documented which intermediate compounds are produced as carbon dioxide is reduced to sugar.

The Calvin Cycle

To unravel the reaction sequence that reduces carbon dioxide, Calvin's group used the pulse-chase strategy introduced in Chapter 7. Recall that pulse-chase experiments introduce a pulse of labeled compound followed by a chase of unlabeled compound. The fate of the labeled compound is then followed through time. In this case, the researchers fed green algae a pulse of $^{14}CO_2$ (**Figure 10.19**). After waiting a specified amount of time, they ground the cells up to form a crude extract, separated individual molecules in the extract via paper chromatography, and laid X-ray film over the filter paper. If radioactively labeled molecules were present on the filter paper, the energy they emitted would expose the film and create a dark spot. The labeled compounds could then be isolated and identified.

By varying the amount of time between starting the pulse of labeled $^{14}CO_2$ and analyzing the cells, Calvin and co-workers began to piece together the sequence in which various intermediates formed. For example, when the team analyzed cells almost immediately after starting the $^{14}CO_2$ pulse, they found that the three-carbon compound 3-phosphoglycerate predominated. This result suggested that 3-phosphoglycerate was the initial product of carbon reduction. Stated another way, it appeared that carbon dioxide reacted with some unknown molecule to produce 3-phosphoglycerate.

This was an interesting result, since 3-phosphoglycerate is one of the ten intermediates in glycolysis. The finding that glycolysis and carbon reduction share intermediates was intriguing because of the relationship between the two pathways. The light-independent reactions lead to the manufacture of carbohydrate; glycolysis breaks it down. Because the two processes are related in this way, it was logical that at least some intermediates in glycolysis and CO_2 reduction are the same.

As Calvin's group pieced together the sequence of events in carbon dioxide reduction, an important question remained unanswered: What compound reacts with CO_2 to produce 3-phosphoglycerate? This was the key, initial step. The group searched in vain for a two-carbon compound that might serve as the initial carbon dioxide acceptor and yield 3-phosphoglycerate. Then, while Calvin was running errands one day, it occurred to him that the molecule reacting with carbon dioxide might contain five carbons, not two. The idea was that adding CO_2 to a five-carbon molecule would produce a six-carbon compound, which could then split in half to form two three-carbon molecules.

Experiments to test this hypothesis confirmed that the five-carbon compound **ribulose bisphosphate (RuBP)** is the initial reactant. Eventually the three phases of CO_2 reduction were worked out and became known as the Calvin cycle (**Figure 10.20**):

1. *Fixation phase.* The events begin when CO_2 reacts with RuBP. This phase "fixes" carbon dioxide by attaching it to a more complex molecule. It also leads to the production of two molecules of 3-phosphoglycerate. **Carbon fixation** is the addition of carbon dioxide to an organic compound, putting CO_2 into a biologically useful form.

2. *Reduction phase.* Next, 3-phosphoglycerate is phosphorylated by ATP and then reduced by electrons from NADPH. The product is the phosphorylated sugar **glyceraldehyde-3-phosphate (G3P)**. Some of the resulting G3P is drawn off to manufacture glucose and fructose, which are linked to form the disaccharide sucrose.

Question: What intermediates are produced as carbon dioxide is reduced to sugar?

Hypothesis: (no specific hypothesis)

Experimental setup:

1. Feed algae pulse of $^{14}CO_2$, then CO_2.

2. Wait 5–60 seconds, then homogenize cells by immersing in hot alcohol.

3. Separate molecules by means of paper chromatography.

4. Lay X-ray film on chromatograph to locate radioactive label.

Prediction: (no specific prediction)

Results:

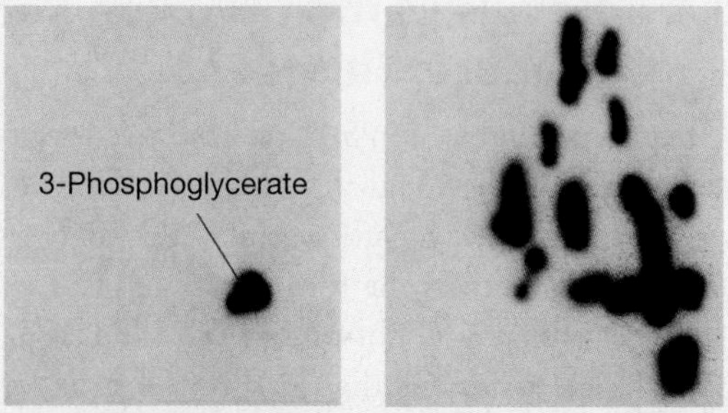

Compounds produced after 5 seconds

Compounds produced after 60 seconds

Conclusion: 3-Phosphoglycerate is the first intermediate product. Other intermediates appear later.

FIGURE 10.19 Experiments Revealed the Reaction Pathway Leading to Reduction of CO_2

QUESTION Why wasn't this experiment based on a specific hypothesis and set of predictions?

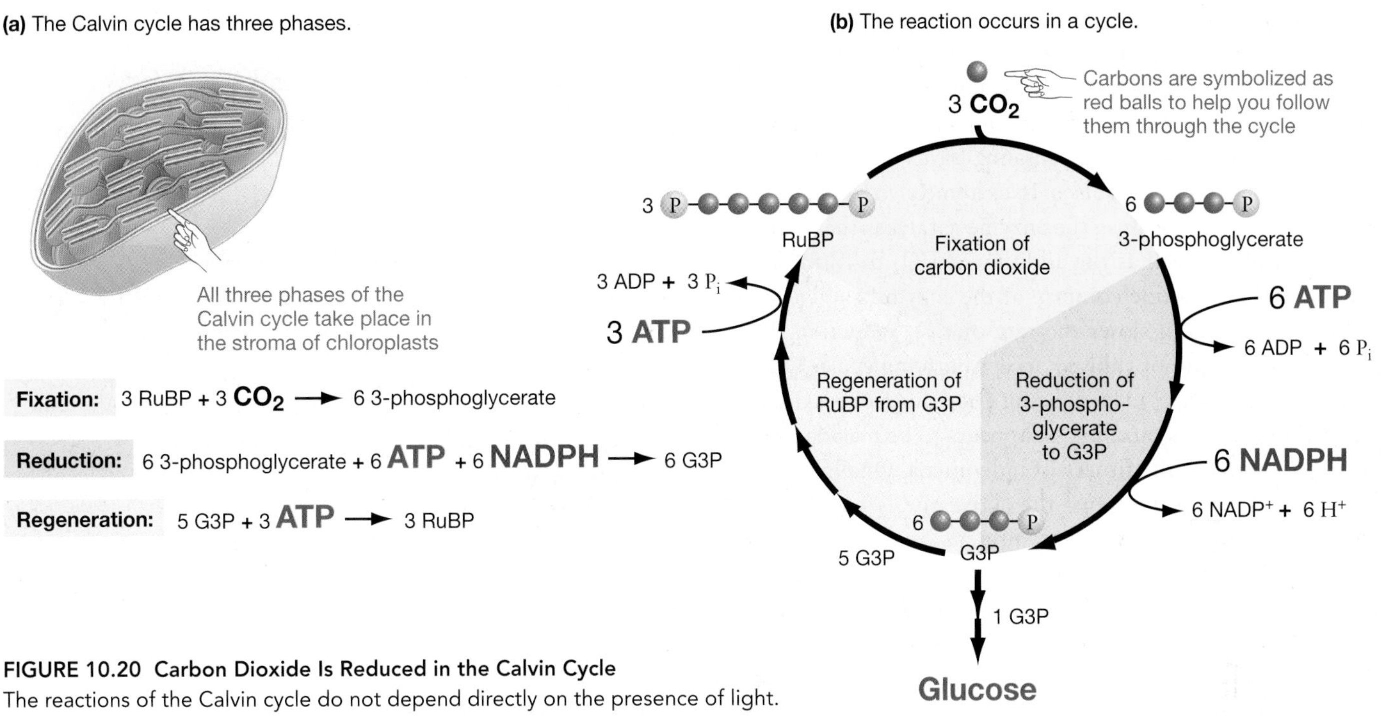
(a) The Calvin cycle has three phases.
All three phases of the Calvin cycle take place in the stroma of chloroplasts
Fixation: 3 RuBP + 3 CO2 → 6 3-phosphoglycerate
Reduction: 6 3-phosphoglycerate + 6 ATP + 6 NADPH → 6 G3P
Regeneration: 5 G3P + 3 ATP → 3 RuBP
(b) The reaction occurs in a cycle.
3 CO2
Carbons are symbolized as red balls to help you follow them through the cycle
3
RuBP
Fixation of carbon dioxide
6
3-phosphoglycerate
3 ADP + 3 Pi
3 ATP
6 ATP
6 ADP + 6 Pi
Regeneration of RuBP from G3P
Reduction of 3-phospho-glycerate to G3P
6 NADPH
6 NADP+ + 6 H+
6
5 G3P
G3P
1 G3P
Glucose

FIGURE 10.20 Carbon Dioxide Is Reduced in the Calvin Cycle
The reactions of the Calvin cycle do not depend directly on the presence of light.

3. *Regeneration phase.* The rest of the G3P keeps the cycle going by serving as the substrate for the third phase in the cycle: reactions that result in the regeneration of RuBP.

All three phases take place in the stroma of chloroplasts.

The discovery of the Calvin cycle clarified how the ATP and NADPH produced by light-dependent reactions allow cells to reduce CO_2 to carbohydrate $(CH_2O)_n$. Because sugars store a great deal of potential energy, producing them takes a great deal of chemical energy—transferred by ATP and NADPH.

Once the reaction sequence in the Calvin cycle was confirmed, attention focused on the initial phase—the reaction between RuBP and CO_2. It is one of only two reactions that are unique to the Calvin cycle. Most reactions involved in reducing CO_2 also occur during glycolysis or other metabolic pathways.

The reaction between CO_2 and RuBP starts the transformation of carbon dioxide gas from the atmosphere to sugars. Plants use sugars to fuel cellular respiration and build leaves, roots, flowers, seeds, tree trunks, and other structures. Millions of non-photosynthesizers organisms—including fish, insects, fungi, and mammals—also depend on this reaction to provide the sugars they need for cellular respiration. Ecologically, the addition of CO_2 to RuBP may be the most important chemical reaction on Earth. The enzyme that catalyzes it is fundamental to all life. What does this molecule look like, and how does it work?

The Discovery of Rubisco

To find the enzyme that fixes CO_2, Arthur Weissbach and colleagues ground up spinach leaves, purified a large series of proteins from the resulting cell extracts, and then tested each protein to see if it could catalyze the incorporation of $^{14}CO_2$ into RuBP to form 3-phosphoglycerate. Eventually they were able to isolate an enzyme that catalyzes the reaction. The enzyme turned out to be extremely abundant in leaf tissue. The researchers' data suggested that the enzyme constituted at least 10 percent of the total protein found in spinach leaves.

The CO_2-fixing enzyme was eventually purified and analyzed. Ribulose-1,5-bisphosphate carboxylase/oxygenase is its full name, but it is commonly referred to as **rubisco**. Rubisco is found in all photosynthetic organisms that use the Calvin cycle to fix carbon. It is thought to be the most abundant enzyme on Earth. Its three-dimensional structure has now been determined (**Figure 10.21**). The molecule is shaped like a cube and has a total of eight active sites where CO_2 is fixed.

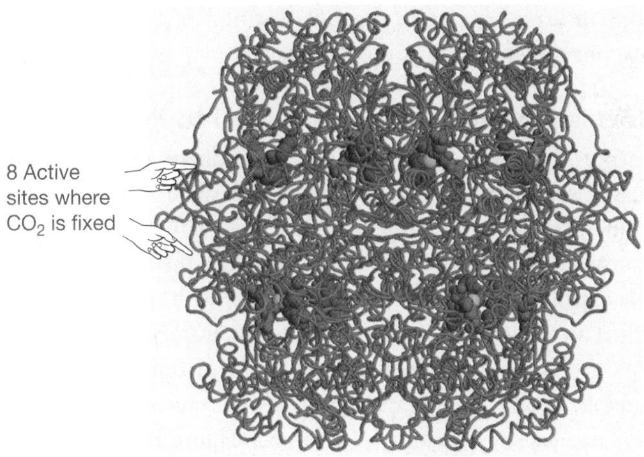
8 Active sites where CO2 is fixed

FIGURE 10.21 Rubisco "Fixes" Carbon Dioxide
A three-dimensional model of rubisco. The red and blue molecules represent substrates at the eight active sites.

Even though it has a large number of active sites, rubisco is a very slow enzyme. Each active site catalyzes just three reactions per second; other enzymes typically catalyze thousands of reactions per second. Plants synthesize huge amounts of rubisco, possibly as an adaptation compensating for its lack of speed.

Besides being slow, rubisco is extremely inefficient. The inefficiency occurs because the enzyme catalyzes the addition of O_2 to RuBP as well as the addition of CO_2 to RuBP. Oxygen and carbon dioxide compete at the enzyme's active sites, and this competition slows the rate of CO_2 reduction. Why would an active site of rubisco accept both molecules? Given rubisco's importance in producing food for photosynthetic species, this detail is puzzling. It appears to be **maladaptive**—a trait that reduces the fitness of individuals. One hypothesis to explain the dual nature of the active site is based on the observation that rubisco was present in photosynthetic organisms long before the evolution of oxygenic photosynthesis. As a result, O_2 was extremely rare in the atmosphere when rubisco evolved. According to this hypothesis, rubisco's inefficiency is a historical artifact. The idea is that rubisco is adapted to an atmosphere that no longer exists—one that was extremely rich in CO_2 and poor in O_2.

Unfortunately, the reaction of O_2 with RuBP does more than simply compete with the reaction of CO_2 at the same active site. One of the molecules that results from the addition of oxygen to RuBP is processed in reactions that consume ATP and release CO_2. Part of this pathway occurs in chloroplasts, and part in peroxisomes and mitochondria. The reaction sequence resembles respiration, because it consumes oxygen and produces carbon dioxide. As a result, it is called **photorespiration**. Because photorespiration consumes energy and undoes carbon fixation, it can be considered a reverse photosynthesis (**Figure 10.22**). When photorespiration occurs, the overall rate of photosynthesis declines.

The oxygenation reaction that triggers photorespiration is favored when oxygen concentrations are high and CO_2 concentrations are low. But as long as carbon dioxide concentrations in leaves are high, the CO_2-fixation reaction is favored and photorespiration is relatively rare.

How Is Carbon Dioxide Delivered to Rubisco?

Carbon dioxide is present in the atmosphere and is continuously used as a reagent in photosynthesizing cells. It would seem straightforward, then, for CO_2 to diffuse directly into plants along a concentration gradient. But the situation is not this simple, because plants are covered with a waxy coating called a cuticle. This lipid layer prevents water from evaporating out of tissues, but it also prevents CO_2 from entering them.

How does CO_2 get into photosynthesizing tissues? A close look at a leaf surface, such as the one in **Figure 10.23a**, provides the answer. The leaf surface is dotted with openings bordered by two distinctively shaped cells. The paired cells are called **guard cells**, the opening is called a **pore**, and the entire structure is

Rubisco catalyzes competing reactions with very different outcomes.

Reaction with carbon dioxide during photosynthesis:

RuBP + CO_2 ⟶ 2 3-phosphoglycerate

↓ used in Calvin cycle

Reaction with oxygen during "photorespiration":

RuBP + O_2 ⟶ 1 3-phosphoglycerate + 1 2-phosphoglycolate

1 3-phosphoglycerate ↓ used in Calvin cycle

1 2-phosphoglycolate ↓ when processed, CO_2 released and ATP used

FIGURE 10.22 Photorespiration Competes with Photosynthesis

QUESTION After studying these reactions, explain why biologists say that photorespiration "undoes" photosynthesis.

(a) Leaf surfaces contain stomata.

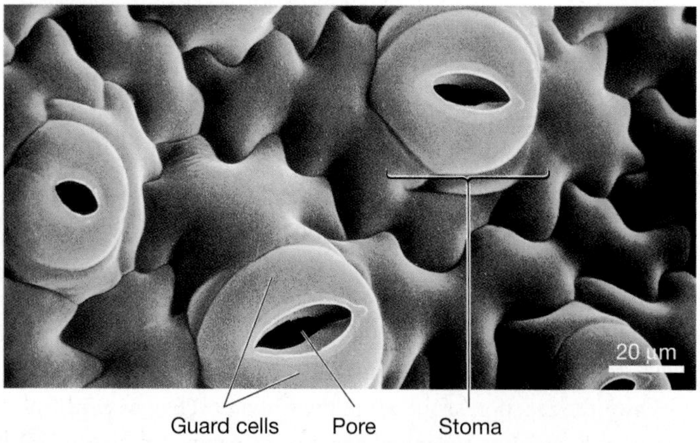
20 µm
Guard cells
Pore
Stoma

(b) Carbon dioxide diffuses into leaves through stomata.

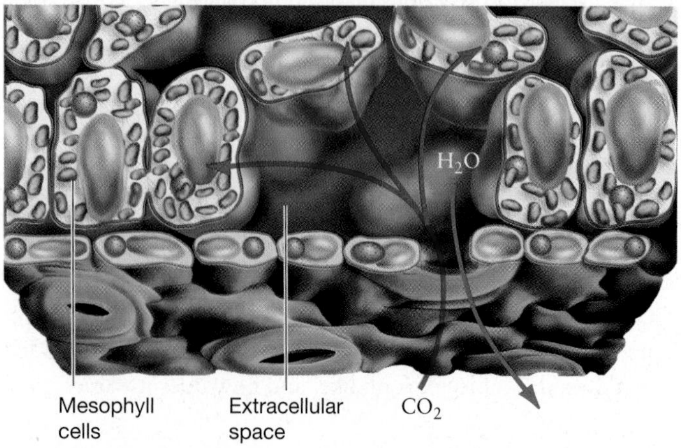
H_2O
Mesophyll cells
Extracellular space
CO_2

FIGURE 10.23 Leaf Cells Obtain Carbon Dioxide through Stomata

(a) Stomata consist of two guard cells and a pore.**(b)** When a stoma is open, CO_2 diffuses into the leaf along a concentration gradient.

called a **stoma** (plural: **stomata**). If CO_2 concentrations inside the leaf are low as photosynthesis gets under way, chemical signals activate proton pumps in the membranes of guard cells. These pumps establish a charge gradient across the membrane. In response, potassium ions (K^+) move into the guard cells. When water follows along the newly created osmotic gradient, the cells swell and create a pore. As **Figure 10.23b** shows, an open stoma allows CO_2 from the atmosphere to diffuse into the air-filled spaces inside the leaf, and from there into the extracellular fluid surrounding photosynthesizing cells. Eventually the CO_2 diffuses along a concentration gradient into the chloroplasts of the cells. A strong concentration gradient favoring entry of CO_2 is maintained by the light-independent reactions, which constantly use up the CO_2 in chloroplasts.

Stomata are normally open during the day, when photosynthesis is occurring, and closed at night. But if the daytime is extremely hot and dry, leaf cells may begin losing a great deal of water to evaporation through their stomata. When this occurs, they must either close the openings and halt photosynthesis or risk death from dehydration. When conditions are hot and dry, then, photosynthesis and growth stop. How do plants that live in hot, dry environments cope? An answer emerged as biologists struggled to understand a surprising experimental result.

C_4 Photosynthesis After the Calvin cycle had been worked out in algae, researchers in a variety of labs used the same pulse-chase approach to investigate how carbon fixation occurs in other species. Just as Calvin had done, Hugo Kortschack and colleagues and Y. S. Karpilov and associates exposed leaves of sugarcane and maize (corn) to radioactive carbon dioxide ($^{14}CO_2$) and sunlight and then characterized the products. Both research teams expected to find the first of the radioactive carbon atoms in 3-phosphoglycerate—the normal product of carbon fixation by rubisco. Instead, they found that in some plant species the radioactive carbon atom ended up in four-carbon compounds such as malate and aspartate—not in three-carbon sugars.

The experiments revealed a twist on the usual pathway for carbon fixation. Instead of creating a three-carbon sugar, it appeared that in some species CO_2 fixation produced four-carbon sugars. The two pathways became known as C_3 and C_4 photosynthesis, respectively (**Figure 10.24**).

Researchers who followed up on the initial reports found that, in some plant species, carbon dioxide can be added to RuBP by rubisco *or* to three-carbon compounds by an enzyme called **PEP carboxylase**. They also showed that the two enzymes are found in distinct cell types within the same leaf. PEP carboxylase is common in **mesophyll cells** near the surface of leaves, while rubisco is found in **bundle-sheath cells** that surround the vascular tissue in the interior of the leaf (**Figure 10.25a**). **Vascular tissue** conducts water and nutrients in plants.

Based on these observations, Hal Hatch and Roger Slack proposed a three-step model to explain how CO_2 that is fixed to a four-carbon sugar feeds the Calvin cycle (**Figure 10.25b**):

1. PEP carboxylase fixes CO_2 in mesophyll cells.
2. The four-carbon organic acids that result travel to bundle-sheath cells.
3. The four-carbon organic acids release a CO_2 molecule that rubisco uses as a substrate to form 3-phosphoglycerate. This step initiates the Calvin cycle.

WEB TUTORIAL 10.2
Strategies for Carbon Fixation

C_3 plants:

RuBP + CO_2 —**Rubisco**→ 2 3-phosphoglycerate (3-carbon sugar)

C_4 plants:

3-carbon compound + CO_2 —**PEP carboxylase**→ 4-carbon organic acids

FIGURE 10.24 Initial Carbon Fixation in C_4 Plants Is Different from That in C_3 Plants

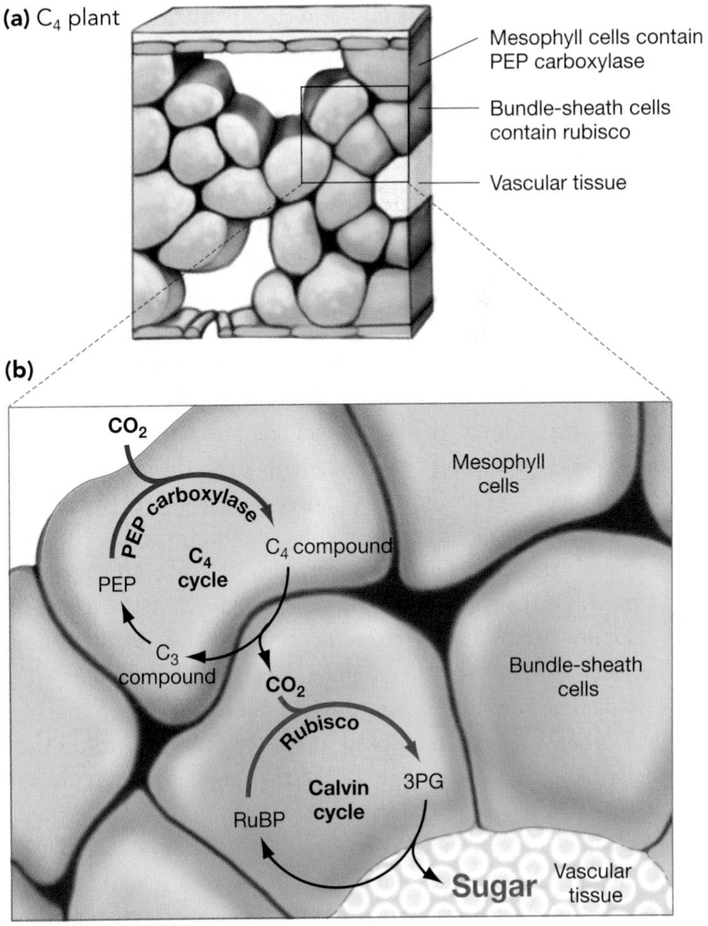
(a) C4 plant
Mesophyll cells contain PEP carboxylase
Bundle-sheath cells contain rubisco
Vascular tissue
(b)
CO2
PEP carboxylase
C4 compound
PEP
C4 cycle
C3 compound
CO2
Rubisco
3PG
Calvin cycle
RuBP
Sugar
Mesophyll cells
Bundle-sheath cells
Vascular tissue

FIGURE 10.25 In C_4 plants, Carbon Fixation Occurs Independently of the Calvin Cycle
(a) The carbon-fixing enzyme PEP carboxylase is located in mesophyll cells, while rubisco is in bundle-sheath cells. **(b)** CO_2 is fixed to the three-carbon compound PEP by PEP carboxylase, forming a four-carbon organic acid.

In effect, then, the C_4 pathway acts as a CO_2 pump. The reactions that take place in mesophyll cells require energy in the form of ATP, but they increase CO_2 concentrations in cells where rubisco is active. Because it increases the ratio of carbon dioxide to oxygen in photosynthesizing cells, less O_2 binds to rubisco's active sites. Stated another way, CO_2 fixation is favored over O_2 fixation when carbon dioxide concentrations in leaves are high. As a result, the C_4 pathway limits the damaging effects of photorespiration. The pathway is an adaptation that keeps CO_2 concentrations in leaves high. Later experiments supported the Hatch and Slack model in almost every detail.

Logically enough, the C_4 pathway is found almost exclusively in plants that thrive in hot, dry habitats. Sugarcane, maize (corn), and crabgrass are some familiar C_4 plants, but the pathway is actually found in several thousand species in 19 distinct lineages of flowering plants. These observations suggest that the C_4 pathway has evolved independently several times. It is not the only mechanism that plants use to continue growth under hot, dry conditions, however.

CAM Plants Some years after the discovery of C_4 photosynthesis, researchers studying a group of flowering plants called the Crassulaceae came across a second mechanism for limiting the effects of photorespiration. This photosynthetic pathway became known as **crassulacean acid metabolism**, or **CAM**. Like the C_4 pathway, CAM is a CO_2 pump that acts as an additional, preparatory step to the Calvin cycle. It also has the same effect: It increases the concentration of CO_2 inside photosynthesizing cells. But unlike the C_4 pathway, CAM occurs at a different time than the Calvin cycle does—not in a different place.

CAM occurs in cacti and other species that occupy environments that are so hot and dry that individuals routinely keep their stomata closed all day. When night falls and conditions become cooler and moister, CAM plants open their stomata and take in huge quantities of CO_2. These molecules are temporarily fixed to organic acids and stored in the central vacuoles of photosynthesizing cells. During the day, the molecules are processed in reactions that release the CO_2 and feed the Calvin cycle.

Figure 10.26 summarizes the similarities and differences between C_4 photosynthesis and CAM. Both function as CO_2 pumps that minimize the amount of photorespiration that occurs when stomata are closed and CO_2 cannot diffuse in directly from the atmosphere. Both are found in flowering plant species that live in hot, dry environments. But while C_4 plants stockpile CO_2 in cells where rubisco is not active, CAM plants store CO_2 at a time when rubisco is inactive. In C_4 plants, the reactions catalyzed by PEP-carboxylase and rubisco are separated in space; in CAM plants, the reactions are separated in time.

Obtaining and reducing CO_2 is fundamental to photosynthesis. In a larger sense, photosynthesis is fundamental to the millions of species that depend on plants, algae, and cyanobacteria for food. What do photosynthetic organisms do with the sugar they synthesize? More specifically, what happens to the G3P that is drawn off from the Calvin cycle?

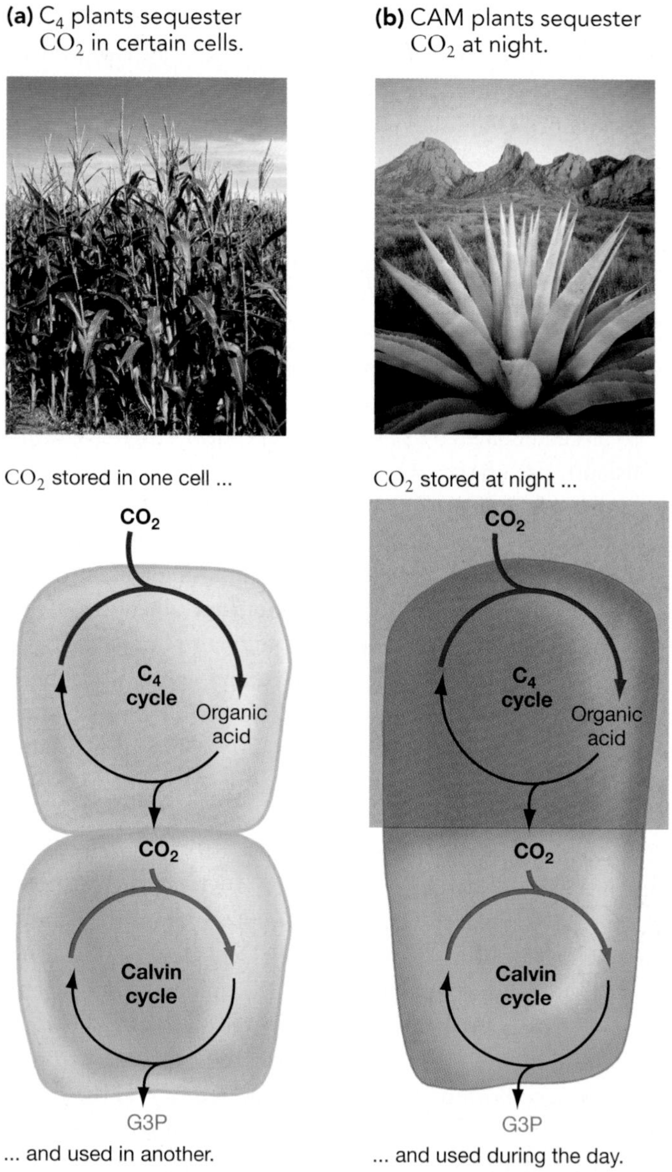

FIGURE 10.26 C_4 Photosynthesis and CAM Accomplish the Same Task in Different Ways
(a) In C_4 plants, CO_2 is fixed to organic acids in some cells and then released to other cells where the Calvin cycle enzymes are located. **(b)** CAM plants open their stomata at night and fix CO_2 to organic acids.

What Happens to the Sugar That Is Produced by Photosynthesis?

The G3P molecules that exit the Calvin cycle enter one of several reaction pathways. The most important of these pathways results in the production of the monosaccharides glucose and fructose, which in turn combine to form the disaccharide su-

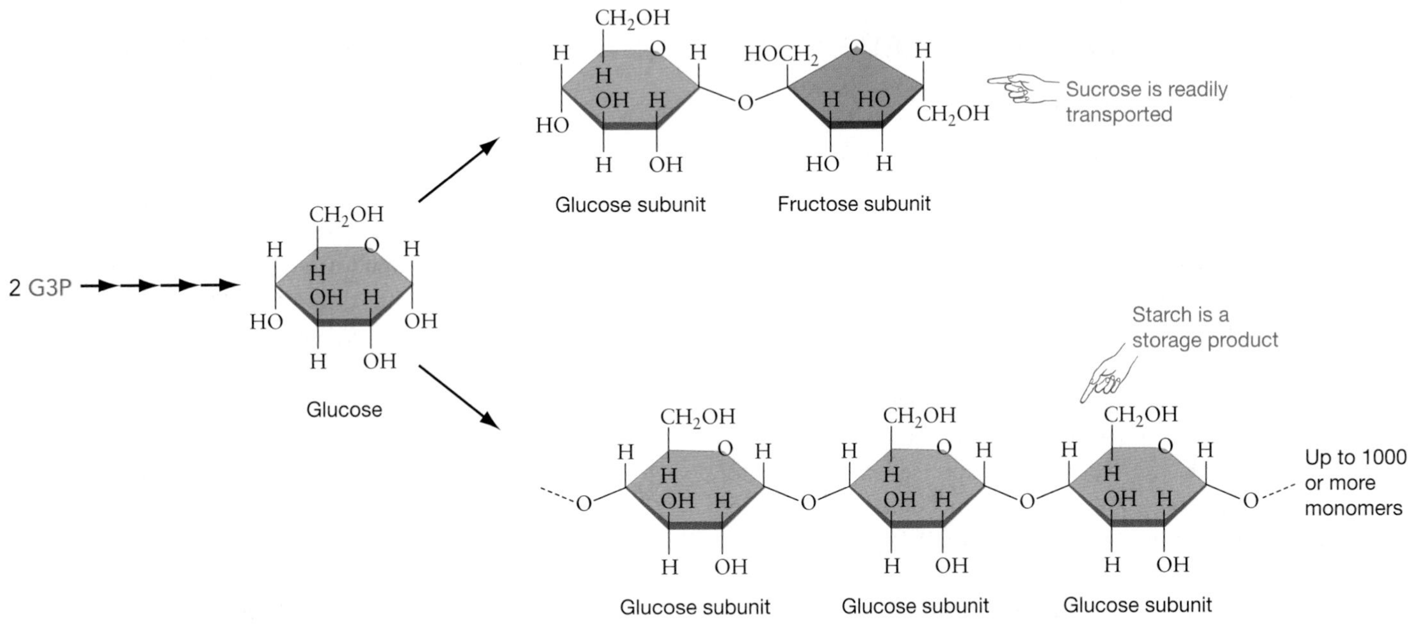

FIGURE 10.27 Sucrose and Starch Are the Main Photosynthetic Products
In plants, sugars are transported in the form of sucrose and stored in the form of starch.
EXERCISE Sucrose can be converted to starch, and starch to sucrose. Add an element to the diagram to indicate this.

crose (**Figure 10.27**). The reaction sequence starts with G3P, involves a series of other phosphorylated three-carbon sugars, includes the synthesis of the familiar six-carbon sugar glucose, and ends with the production of sucrose. An alternative pathway results in the production of glucose molecules that polymerize to form starch. Starch production occurs inside the chloroplast; sucrose synthesis takes place in the cytosol.

All of the intermediates involved in the production of glucose, as well as of G3P, also occur in glycolysis. The general observation here is that many of the enzymes and intermediates involved in glucose processing are common to both respiration and photosynthesis.

When photosynthesis is taking place slowly, almost all the glucose that is produced is used to make sucrose. As noted in Chapter 5, sucrose is a disaccharide ("two-sugar") that consists of a glucose molecule bonded to a fructose molecule. Sucrose is water soluble and is readily transported to other parts of the plant. If the sucrose is delivered to rapidly growing parts of the plant, it is broken down to fuel cellular respiration and growth. If it is transported to storage cells in roots, it is converted to starch and stored for later use.

When photosynthesis is proceeding rapidly and sucrose is abundant, glucose is used to synthesize starch in the chloroplasts of photosynthetic cells. Recall from Chapter 5 that starch is a polymer of glucose. In photosynthesizing cells, starch acts as a temporary sugar-storage product. Starch is not water soluble, so it cannot be transported from photosynthetic cells to other areas of the plant. At night, the starch that is temporarily stored in leaf cells is broken down and used to manufacture sucrose molecules that are then used by the photosynthetic cell in respiration or transported to other parts of the plant. In this way, chloroplasts provide sugars for cells throughout the plant by day and by night.

If a mouse eats the starch that is stored in a chloroplast or root cell, however, the chemical energy in the C–H bonds of the starch is used to fuel the mouse's growth and reproduction. If the mouse is then eaten by an owl, the chemical energy in the mouse's tissues fuels the predator's growth and reproduction. In this way, virtually all cell growth and reproduction can be traced back to the chemical energy that was originally captured by photosynthesis. Photosynthesis is the staff of life.

✓CHECK YOUR UNDERSTANDING

To make sure that you understand the light-independent reactions of photosynthesis, you should be able to summarize the three major phases of the Calvin cycle and explain the relationships among CO_2, G3P, RuBP, rubisco, glucose, and 3-phosphoglycerate. You should also be able to identify three potential sources of CO_2: organic acids in mesophyll cells, organic acids synthesized at night and stored in vacuoles, and direct diffusion through stomata.

ESSAY Are Rising CO_2 Levels in the Atmosphere Affecting the Rate of Photosynthesis?

The concentration of carbon dioxide in the atmosphere has increased dramatically over the past 100 years. In the late 1800s the atmospheric carbon dioxide concentration is thought to have been about 280 μL/L. Today, CO_2 is present in the atmosphere at 360 μL/L. Most of the increase is due to CO_2 that was released when natural gas, gasoline, coal, and other fossil fuels were burned for heat, manufacturing, transportation, and so on or when forests were burned to convert them to agricultural use. If present trends in fossil fuel use and deforestation continue, atmospheric CO_2 levels are expected to increase to 480 μL/L by the year 2050. If this prediction is correct, then CO_2 levels will have increased by 70 percent in just 150 years.

This increase in carbon dioxide concentration is causing dramatic increases in average temperatures around the globe. Global warming is occurring because carbon dioxide in the atmosphere absorbs electromagnetic radiation in the infrared part of the spectrum. These wavelengths radiate from Earth's surface after sunlight strikes it. If they are not absorbed by CO_2, they are lost to space. In this way, carbon dioxide traps heat in the atmosphere.

This process is called the greenhouse effect, because it mimics the effect of glass in a greenhouse. Increases in carbon dioxide concentrations are leading to abnormally large increases in the amount of heat retained in the atmosphere, leading to global warming. How are increases in CO_2 affecting plants?

According to the overall reaction for photosynthesis, increases in a reactant such as CO_2 could lead to increased rates of photosynthesis. Stated another way, plant productivity should rise. Biologists have confirmed this prediction experimentally by increasing CO_2 levels in controlled environmental chambers and in natural habitats. For example, a research team set up a series of experimental plots of land in a 13-year-old pine forest in North Carolina. Some plots were ringed with towers that emitted enough CO_2 to bring average levels inside the plot to 560 μL/L; other plots were left with normal air as a control treatment. As predicted, the growth rate of pine trees in the CO_2-augmented plots increased 25 percent relative to that of the controls.

Experiments like this suggest that rising CO_2 levels may lead to increased rates of photosynthesis in at least some habitats. This is important because increased growth of trees and shrubs removes carbon dioxide from the atmosphere and sequesters the carbon in wood. Similarly, increased growth of algae and photosynthetic bacteria in marine environments sequesters carbon in cellulose or calcium carbonate, which may then drop to the bottom of the ocean when the organisms die. As a result, increased growth by photosynthetic organisms should act as a feedback mechanism that helps offset rising CO_2 levels in the atmosphere. Based on this logic, biologists have suggested that tree planting and forest restoration programs could play a role in a coordinated, worldwide effort to counteract global warming (**Figure 10.28**).

Because many plants respond strongly to augmented CO_2 it is clear that carbon dioxide can be an important limiting nutrient in plant growth. But at some point, plants should stop responding to increased carbon dioxide levels because water availability or some other nutrient—perhaps nitrogen or phosphorus—will become limiting instead. As a result, biologists caution that plant growth rates will eventually stop responding to increased CO_2 availability.

... plant growth rates will eventually stop responding to increased CO_2 availability.

The other major prediction regarding plant responses to increased CO_2 concentrations focuses on desert-dwelling species. To understand this prediction, recall that plants lose water when their stomata are open to admit carbon dioxide for photosynthesis. If carbon dioxide levels rise, then plants will have to open their stomata less to obtain the CO_2 they need, thus losing less water to the atmosphere. As a result, they should be able to grow faster. Consistent with this prediction, a team of biologists found that when they artificially increased CO_2 levels in experimental plots in the Mojave Desert of southwestern North America, plant growth increased. The effect occurred only during a wet year, however—no increased growth was observed in a drought year. What will be the long-term effect of rising CO_2 on desert plants? The answer is not known. Research continues.

FIGURE 10.28 Trees Are a "Carbon Sink"—They Take in Carbon Dioxide and Store Carbon Atoms in Wood

CHAPTER REVIEW

Summary of Key Concepts

Photosynthesis is the conversion of light energy to chemical energy, stored in the bonds of carbohydrates. The sucrose generated by photosynthesis fuels cellular respiration and supplies a substrate for the synthesis of complex carbohydrates, amino acids, fatty acids, and other cell components. As the primary food source for a diverse array of heterotrophs, photosynthetic organisms provide the energy that sustains most life on Earth.

Web Tutorial 10.1 Photosynthesis

- **Photosynthesis consists of two distinct sets of reactions. In reactions driven by light, ATP and the electron carrier NADPH are produced. In subsequent reactions that do not depend directly on light, the ATP and NADPH are used to reduce carbon dioxide (CO_2) to carbohydrate ($(CH_2O)_n$). In eukaryotic cells, both processes take place in chloroplasts.**

The light-dependent reactions occur in internal membranes of the chloroplast that are organized into structures called thylakoids in stacks known as grana. The light-independent reactions, known as the Calvin cycle, take place in a fluid portion of the chloroplast called the stroma.

- **The light-dependent reactions transform the energy in sunlight to chemical energy in the form of electrons with high potential energy. Excited electrons either are used to produce NADPH or are donated to an electron transport chain, which results in the production of ATP.**

The energy transformation step of photosynthesis begins when a pigment molecule in an antenna complex absorbs a photon in the blue or red part of the visible spectrum. When absorption occurs, the energy in the photon is transferred to an electron in the pigment molecule. The electron is raised to an excited state equivalent to the energy in the photon. If the electron falls back to the normal, or ground, state, energy is given off as light (fluorescence) and heat. But in photosynthetic organisms, the energy in the excited electron is eventually transferred to chlorophyll molecules that act as reaction centers. There the high-energy electron is transferred to an electron acceptor, which becomes reduced. In this way, light energy is transformed to chemical energy.

Plants and algae have two types of reaction centers, which are part of larger complexes called photosystem I and photosystem II. Each photosystem consists of an antenna complex with 200–300 chlorophyll and carotenoid molecules, a reaction center, and an electron acceptor that completes energy transformation.

In photosystem II, high-energy electrons are accepted by the electron acceptor pheophytin. Electrons are then passed along an electron transport chain. As electrons move through this chain, they are gradually stepped down in potential energy. The energy released by these reduction-oxidation (redox) reactions is used to pump protons across the thylakoid membrane. The resulting proton gradient drives the synthesis of ATP by ATP synthase. This method of producing ATP is called photophosphorylation. Electrons donated to the electron transport chain by photosystem II are replaced by electrons taken from water, resulting in the production of oxygen as a by-product.

In photosystem I, high-energy electrons are accepted by iron- and sulfur-containing proteins and passed to ferredoxin. In an enzyme-catalyzed reaction, the reduced form of ferredoxin passes electrons to $NADP^+$ to form NADPH. NADPH carries electrons required for the redox reactions that result in the synthesis of sugars and other cell materials. Photosystem I produces the electron carrier NADPH.

The Z scheme describes how photosystems I and II are thought to interact. The scheme begins with the movement of an electron from photosystem II to the electron transport chain. At the end of the chain, the protein plastocyanin carries electrons to photosystem I. There the electrons are promoted to a very high energy state in response to the absorption of a photon, and they are subsequently used to reduce $NADP^+$. Electrons from photosystem I may occasionally be passed to the electron transport chain instead of being used to reduce $NADP^+$, resulting in a cyclic flow of electrons between the two photosystems to produce the additional ATP needed to reduce carbon dioxide.

- **The light-independent reactions start with the enzyme rubisco, which catalyzes the addition of CO_2 to a five-carbon molecule. The compound that results undergoes a series of reactions that use ATP and NADPH and lead to the production of sugar.**

The light-independent reactions of photosynthesis depend on the products of the light-dependent reactions and are called the Calvin cycle. The process of reducing carbon dioxide to sugar begins when CO_2 is attached to a five-carbon compound called ribulose bisphosphate (RuBP). This reaction is catalyzed by the enzyme rubisco. The six-carbon compound that results immediately splits in half to form two molecules of 3-phosphoglycerate. Subsequently, 3-phosphoglycerate is reduced to a sugar called glyceraldehyde-3-phosphate (G3P). Some G3P is used to synthesize glucose and fructose, which combine to form sucrose; the rest participates in reactions that regenerate RuBP so the cycle can continue.

Rubisco catalyzes the addition of oxygen as well as carbon dioxide to RuBP. The reaction with oxygen leads to a loss of fixed CO_2 and ATP and is called photorespiration. Photorespiration is particularly important in hot, dry conditions, when stomata close to prevent excessive water loss. Because the closure of stomata reduces CO_2 levels in photosynthesizing cells, the reaction of O_2 with RuBP is favored. C_4 and CAM plants have distinct but functionally similiar mechanisms for augmenting CO_2 concentrations in photosynthesizing cells and thus for limiting photorespiration.

Web Tutorial 10.2 Strategies for Carbon Fixation

Questions

Content Review

1. What is the stroma of a chloroplast?
   a. the inner membrane
   b. the pieces of membrane that connect grana
   c. the interior of a thylakoid
   d. the fluid inside the chloroplast but outside the thylakoids
2. Why is chlorophyll green?
   a. It absorbs all wavelengths in the visible spectrum, transmitting ultraviolet and infrared light.
   b. It absorbs wavelengths only in the red and far-red portions of the spectrum (680 nm, 700 nm).
   c. It absorbs wavelengths in the blue and red parts of the visible spectrum and transmits wavelengths in the green part.
   d. It absorbs wavelengths only in the blue part of the visible spectrum and transmits all other wavelengths.
3. What does it mean to say that CO_2 becomes fixed?
   a. It becomes bonded to an organic compound.
   b. It is released during cellular respiration.
   c. It acts as an electron acceptor.
   d. It acts as an electron donor.
4. What do the light-dependent reactions of photosynthesis produce?
   a. G3P
   b. RuBP
   c. ATP and NADPH
   d. plastoquinone
5. Why do the absorption spectrum for chlorophyll and the action spectrum for photosynthesis coincide?
   a. Photosystems I and II are activated by different wavelengths of light.
   b. Wavelengths of light that are absorbed by chlorophyll trigger the light-dependent reactions.
   c. Energy from wavelengths absorbed by carotenoids is passed on to chlorophyll.
   d. The rate of photosynthesis depends on the amount of light received.
6. What happens when an excited electron is passed to an electron acceptor in a photosystem?
   a. It drops back down to its ground state, resulting in the phenomenon known as fluorescence.
   b. The chemical energy in the excited electron is released as heat.
   c. The electron acceptor is oxidized.
   d. Energy in sunlight is transformed to chemical energy.

Conceptual Review

1. Explain how the energy transformation step of photosynthesis occurs. How is light energy converted to chemical energy in the form of ATP and NADPH?
2. Explain how the carbon reduction step of photosynthesis occurs. How is carbon dioxide fixed? Why are both ATP and NADPH required to produce sugar?
3. Sketch the Z scheme. Explain how photosystem I and photosystem II interact by tracing the path of an electron through the Z scheme. What molecule connects the two photosystems?
4. In what sense does photorespiration "undo" photosynthesis?
5. Make a sketch showing how C_4 photosynthesis and CAM separate CO_2 acquisition from the Calvin cycle in space and time, respectively.
6. Why do plants need both chloroplasts and mitochondria?

Group Discussion Problems

1. Compare and contrast mitochondria and chloroplasts. In what ways are their structures similar and different? What molecules or systems function in both types of organelles? Which enzymes or processes are unique to each organelle?
2. The Calvin cycle and rubisco are found in lineages of bacteria and archaea that evolved long before the origin of oxygenic photosynthesis. Based on this observation, biologists infer that rubisco evolved in an environment that contained little, if any, oxygen. Some biologists propose that this inference explains why photorespiration occurs today. Do you agree with the hypothesis that photorespiration is an evolutionary "holdover?" Why or why not?
3. In addition to providing their protective function, carotenoids absorb certain wavelengths of light and pass the energy on to the reaction centers of photosystem I and II. Based on their function, predict exactly where carotenoids are located in the chloroplast. Explain your rationale. How would you test your hypothesis?
4. Consider plants that occupy the top, middle, or ground layer of a forest, and algae that live near the surface of the ocean or in deeper water. Would you expect the same photosynthetic pigments to be found in species that live in these different habitats? Why or why not? How would you test your hypothesis?

Answers to Multiple-Choice Questions 1. d; 2. c; 3. a; 4. c; 5. b; 6. d

www.prenhall.com/freeman is your resource for the following: Web Tutorials; Online Quizzes and other Online Study Guide materials; Answers to Conceptual Review Questions; Solutions to Group Discussion Problems; Answers to Figure Caption Questions and Exercises; and Additional Readings and Research.

The Cell Cycle

11

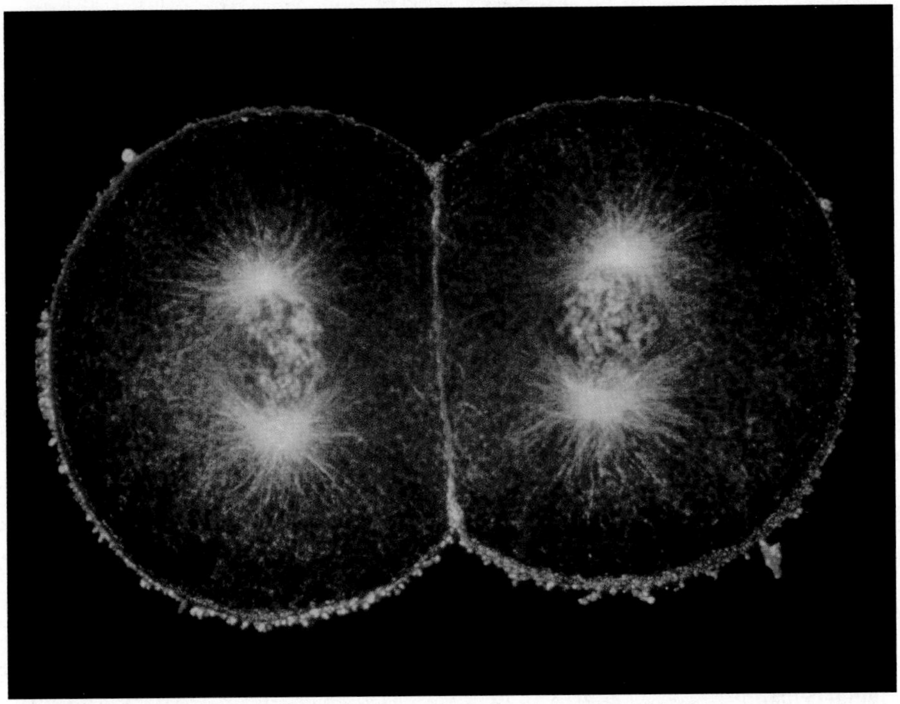

This sea urchin embryo consists of two cells, each of which is beginning to divide. Many thousands more cell divisions will occur as the individual develops into an adult.

KEY CONCEPTS

- After chromosomes are copied, mitosis distributes one chromosome copy to each of two daughter cells. Mitosis and cytokinesis produce two cells that are identical to the parent cell.
- Over their life span, cells go through a life cycle that consists of four carefully controlled phases.
- Uncontrolled cell growth leads to cancer. Different types of cancer result from different types of defects in control over the cell cycle.

In Chapter 1 we considered the cell theory, which maintains that all organisms are made of cells and that all cells arise from preexisting cells. Although the cell theory was widely accepted among biologists by the 1860s, a great deal of confusion remained about how cells reproduced. Most proponents of the cell theory believed that new cells arose within preexisting cells by a process that resembled crystallization. But Rudolf Virchow proposed that new cells arise through the division of preexisting cells—that is, **cell division.**

In the late 1800s, careful microscopic observations of newly developing individuals, or **embryos**, confirmed Virchow's hypothesis. Research documented that multicellular individuals start life as single-celled embryos and grow through a series of cell divisions. As better microscopes became available, biologists were able to describe the division process in more detail. In particular, researchers focused their attention on the nuclei of dividing cells and the fate of the chromosomes. Chromosomes are the carriers of hereditary material—the instructions for building and operating the cell. These early studies revealed two fundamentally different ways that nuclei divide prior to cell division. In animals, one type of nuclear division leads to the production of sperm and eggs, and the other type of nuclear division leads to the production of all other cell types. Sperm and eggs are male and female reproductive cells, termed **gametes**; all other cell types are referred to as **somatic** ("body-belonging") **cells.** In both kinds of cell division, a so-called *parent cell* is said to give rise to *daughter cells.*

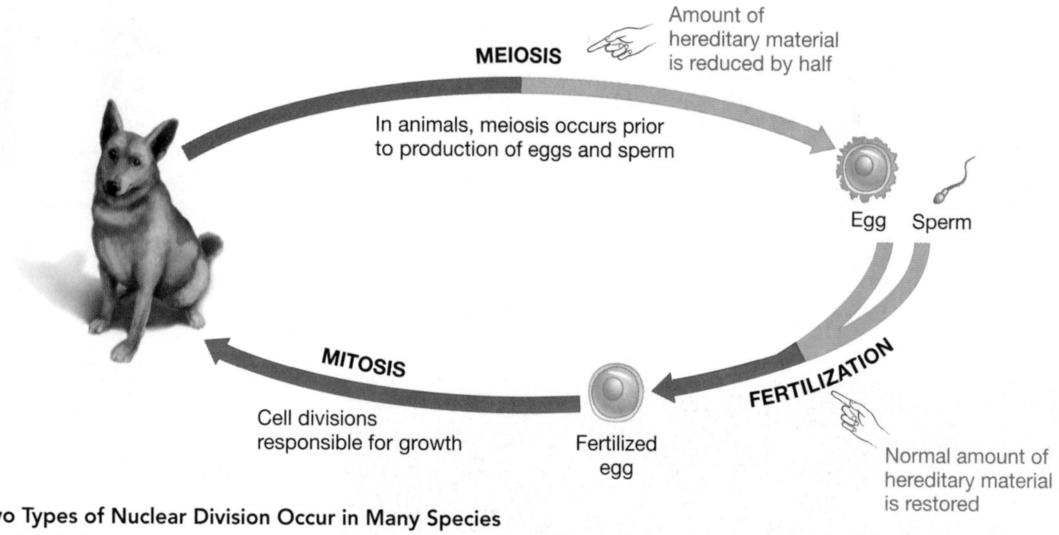

FIGURE 11.1 Two Types of Nuclear Division Occur in Many Species
In animals, meiosis leads to the production of eggs and sperm. Mitosis is responsible for producing somatic cells.

During the type of nuclear division that leads to the production of sperm and eggs, the amount of hereditary material found in the parent cell nucleus is reduced by half. As a result, the daughter cells that become sperm or eggs do not contain the same genetic material as the parent cell. This type of nuclear division is called **meiosis**, and it is involved only in the production of reproductive cells (as in the example shown in **Figure 11.1**). Meiosis is the basis of sexual reproduction and genetic inheritance. Chapter 12 explores the mechanism and consequences of meiosis in detail.

When nuclei divide prior to the formation of new somatic cells, the amount of hereditary material in the original cell and the daughter cells remains constant. **Mitosis** is a division of the genetic material that produces daughter cells that are genetically identical to their parent cell. Mitosis is usually accompanied by **cytokinesis** ("cell movement")—the division of the cytoplasm into the two daughter cells. Mitosis, followed by cytokinesis, supplies the cells required for several key activities in eukaryotes—wound repair, reproduction, and growth:

- New cells are required to repair damaged tissues in multicellular organisms. When you suffer a scrape or cut, the cells that repair your skin and heal the wound are generated via mitosis and cytokinesis.
- When yeast cells greatly increase in number in a pile of bread dough or in a vat of beer, they are reproducing by mitosis and cytokinesis. In both unicellular and multicellular species, mitosis followed by cytokinesis is the basis of asexual reproduction. **Asexual reproduction** results in the production of offspring that are genetically identical to the parent.
- Mitosis and cytokinesis are also responsible for growth in multicellular organisms. For example, the ancestry of the trillions of genetically identical cells that make up your body can be traced back through a series of mitotic divisions to a single fertilized egg—the result of the union of a sperm and an egg from your parents.

For this chapter, our goals are to explore how mitosis occurs and how cell division is regulated. The first section introduces the relationship between mitosis and other major stages in a cell's life cycle. The next two sections provide an in-depth look at each event in mitosis and explore how mitosis and other events in the cell's life cycle are regulated. The chapter concludes by examining why uncontrolled cell division and cancer can result when the systems that regulate cell division break down.

11.1 Mitosis and the Cell Cycle

In the course of studying cell division, nineteenth-century biologists found that certain chemical dyes made threadlike structures visible in the nuclei of dividing cells. In 1879 Walther Flemming followed up on this discovery by documenting how the threadlike structures in salamander embryos change as the cells divide. As **Figure 11.2a** shows, the threads that Flemming observed were paired when they first appeared, just before cell division. Prior to cell division, each pair of threads split to produce single, unpaired threads. Flemming introduced the term *mitosis*, from the Greek *mitos* ("thread"), to describe this division process.

Soon after Flemming made that discovery, similar observations were reported for the roundworm *Ascaris*. In addition to confirming that each pair of threads split in this species, investigators reported that the total number of threads in a cell remained constant during subsequent divisions. Thus, all of the cells in a roundworm's body had the same number and types of threads.

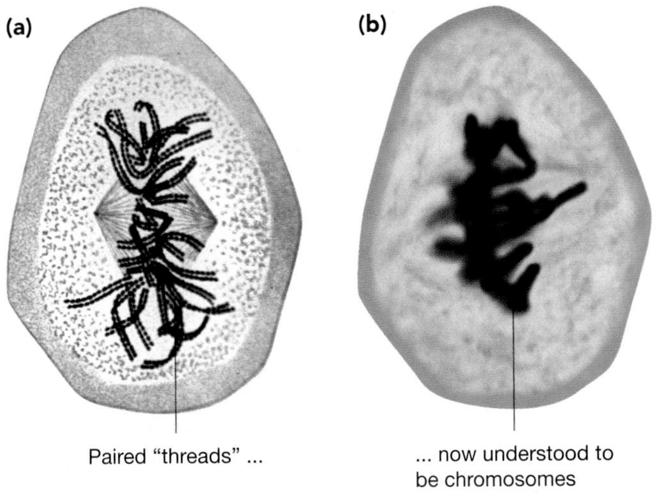
(a)
(b)
Paired "threads" ...
... now understood to be chromosomes

FIGURE 11.2 Chromosomes Move during Mitosis
(a) Walther Flemming's 1879 drawing of mitosis in the salamander larva. The black threads are chromosomes. **(b)** Chromosomes can be stained with dyes and observed in the light microscope.

In 1888 Wilhelm Waldeyer coined the term **chromosome** ("colored-body") to refer to the threadlike structures observed in dividing cells (**Figure 11.2b**). As Chapters 14 and 15 will show, chromosomes are made up in part of deoxyribonucleic acid (DNA). More specifically, a chromosome consists of a single, long DNA double helix that is wrapped around proteins in a highly organized manner. DNA encodes the cell's hereditary information, or genetic material. By observing how chromosomes moved during mitosis, biologists realized that the purpose of mitosis was to distribute the parent cell's genetic material to daughter cells during cell division. Prior to mitosis, each chromosome is copied. During mitosis, one of the copies is distributed to each of two daughter cells.

Before delving into the details of how mitosis occurs, though, let's examine how it fits into the other events in the life of a cell.

The Cell Cycle

As early workers studied the fate of chromosomes during cell division, they realized that even rapidly growing plant and animal cells do not divide continuously. Instead, growing cells cycle between a dividing phase called the **mitotic** (or **M**) **phase** and a nondividing phase called **interphase** ("between-phase"). Chromosomes can be stained and observed with a light microscope only during M phase, when they are condensed into compact structures. Cells actually spend most of their time in interphase, however. No dramatic changes are observed in the nucleus during interphase, when chromosomes are uncoiled into extremely long, thin structures. Even when stains are used, individual interphase chromosomes are not normally visible in the light microscope.

To describe the regular alternation between M phase and interphase, biologists began referring to the cell cycle. The **cell cycle** is the orderly sequence of events that occurs from the formation of a eukaryotic cell, through the duplication of its chromosomes, to the time it undergoes division itself. During the cycle, two key events are (1) the **replication**, or copying, of the hereditary material in chromosomes and (2) the partitioning of copied chromosomes to the two daughter cells. The hereditary material is duplicated, with one copy going to each daughter cell during mitosis. As a result, daughter cells contain genetic information that is identical to that of the parent cell. Let's take a closer look.

When Does Chromosome Replication Occur?

Using the light microscope, nineteenth-century biologists could see chromosomes move to daughter cells during mitosis. Because each daughter cell ended up with the same number of chromosomes as the parent cell, it was logical to infer that the chromosomes were duplicated in the parent cell at some point in the cell cycle. Did the paired threads observed in salamander cells represent the replicated chromosomes? If so, when did this replication step occur? Because no dramatic changes to chromosomes are visible during interphase, some biologists hypothesized that replication must occur early in M phase as chromosomes condense and become visible. But because chromosomes already appear as doublets early in M phase, it was possible that replication occurred during interphase.

Which hypothesis is correct? This question was not answered until the 1950s, when two technical innovations provided the experimental tools needed to answer it. One advance was the ability to grow eukaryotic cells outside the source organism, in culture. Cultured cells are powerful experimental tools because they can be manipulated much more easily than cells in an intact organism can (see **Box 11.1**, page 231). A second advance was the availability of radioactively labeled deoxyribonucleotides—the building blocks of DNA. If radioactive deoxyribonucleotides are present as DNA is being synthesized, they will be incorporated into the new DNA molecule and label it.

To determine when chromosome replication takes place during the cell cycle, Alma Howard and Stephen Pelc exposed cultured cells to radioactive thymidine and nonradioactive adenosine, guanosine, and cytidine. These deoxyribonucleotides are components of DNA and are incorporated into DNA molecules only during replication. In the cultured cells, M phase lasted about 30 minutes, and the entire cell cycle lasted about a day. To test the hypothesis that chromosome replication occurs during M phase, Howard and Pelc let the radioactive thymidine stay in the cultured cells for 30 minutes. If the M-phase hypothesis is correct, then any cells undergoing mitosis during the labeling period would incorporate radioactive thymidine into their replicated chromosomes.

After washing the radioactive thymidine out of the culture, the biologists removed a sample of cells, spread them out, and laid a sheet of X-ray film over them. As the photograph in **Figure 11.3** shows, small black dots appeared where radioactive thymidine molecules had exposed the film.

The key observation in this experiment was that radioactive labels were not found in M-phase nuclei—only in interphase nuclei. Thus, only interphase cells had incorporated radioactive thymidine. Because only replicating DNA would incorporate radioactive thymidine into newly synthesized DNA molecules, the researchers concluded that chromosome replication occurs during interphase. Howard and Pelc had identified a new stage in the cell cycle called **synthesis** (or **S**) **phase**, for DNA synthesis. Their data showed that duplication of the genetic material occurs independently of mitosis—the process that distributes chromosome copies to daughter cells.

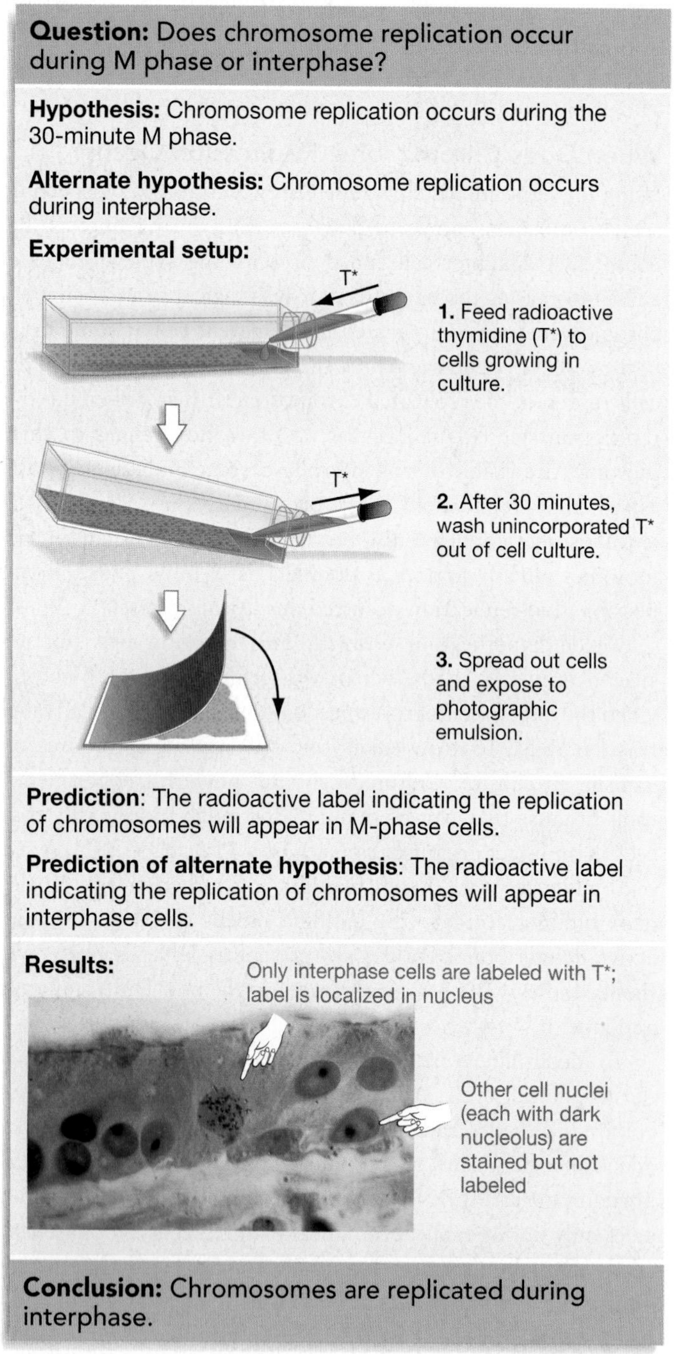

Question: Does chromosome replication occur during M phase or interphase?
Hypothesis: Chromosome replication occurs during the 30-minute M phase.
Alternate hypothesis: Chromosome replication occurs during interphase.
Experimental setup:
T*
1. Feed radioactive thymidine (T*) to cells growing in culture.
T*
2. After 30 minutes, wash unincorporated T* out of cell culture.
3. Spread out cells and expose to photographic emulsion.
Prediction: The radioactive label indicating the replication of chromosomes will appear in M-phase cells.
Prediction of alternate hypothesis: The radioactive label indicating the replication of chromosomes will appear in interphase cells.
Results:
Only interphase cells are labeled with T*; label is localized in nucleus
Other cell nuclei (each with dark nucleolus) are stained but not labeled
Conclusion: Chromosomes are replicated during interphase.

FIGURE 11.3 When Are Chromosomes Replicated?
Because thymidine is incorporated into newly synthesized DNA, the experimental protocol shown established that the chromosomes are replicated during interphase (in particular, the synthesis [S] phase), rather than M phase, of the cell cycle.

Discovery of the Gap Phases

To determine how long it takes a cell to complete S phase, other investigators repeated Howard and Pelc's experiment but waited various lengths of time to examine the cells exposed to a pulse of radioactive thymidine. Recall that these cells take about a day to complete the cell cycle. As **Figure 11.4** indicates, experiments showed that no labeled M-phase cells appeared until 4 to 5 hours after the pulse of radioactive thymidine had ended. Between 5 and 12 hours after the labeling period, though, all of the labeled cells had exhibited the chromosomal changes associated with mitosis.

Cells in these cultures are dividing continuously. Thus, the key to interpreting Figure 11.4 is to realize that the first labeled cells to enter mitosis must represent cells that were just completing chromosome replication when they were labeled. If so, then the 4- to 5-hour time lag between the end of the pulse and the appearance of the first labeled mitotic nuclei corresponds to a time lag that occurs between the end of S phase

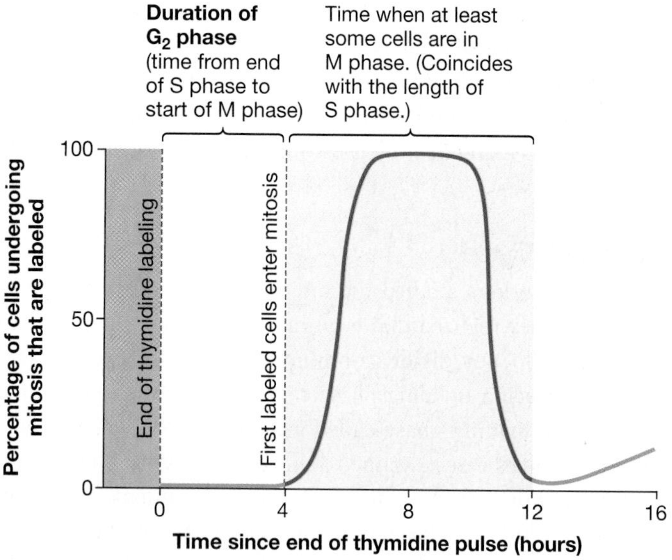

Duration of G_2 phase (time from end of S phase to start of M phase)
Time when at least some cells are in M phase. (Coincides with the length of S phase.)
Percentage of cells undergoing mitosis that are labeled
100
50
0
End of thymidine labeling
First labeled cells enter mitosis
0
4
8
12
16
Time since end of thymidine pulse (hours)

FIGURE 11.4 A Gap Occurs between Chromosome Replication and Mitosis
A plot of the percentage of cells undergoing mitosis that are labeled versus the amount of time that has elapsed since the cells were fed a pulse of radioactive thymidine. If at least some cells are in M phase for 8 hours, it means that S phase lasts 8 hours.

BOX 11.1 Cell-Culture Methods

For researchers, there are important advantages to growing plant and animal cells outside the organism itself. Cell cultures provide homogenous populations of a single type of cell and the opportunity to control experimental conditions precisely.

The first successful attempt to culture animal cells occurred in 1907, when a researcher cultivated amphibian nerve cells in a drop of fluid from the spinal cord. But it was not until the 1950s and 1960s that biologists could routinely culture plant and animal cells in the laboratory. The long lag time was due to the difficulty of recreating conditions that exist in the intact organism.

To grow in culture, animal cells must be provided with a liquid mixture that includes nutrients, vitamins, and hormones that stimulate growth. Initially, this mixture was provided through the use of *serum*, which is the liquid portion of blood; now serum-free media are available for certain cell types. Serum-free media are preferred because they are much more precisely defined chemically than serum. In addition, many types of animal cells will not grow in culture unless they are provided with a solid surface that mimics the types of surfaces to which cells in the intact organisms adhere. As a result, cells are typically cultured in flasks like the one shown in Figure 11.3.

Even under optimal conditions, though, normal cells display a finite life span in culture. In contrast, many cultured cancerous cells grow indefinitely. They also do not adhere tightly to the surface of the culture flask and do not need growth factors in the media.

Because of their immortality and relative ease of growth, cultured cancer cells are commonly used in research. For example, the first human cell type to be grown in culture was isolated in 1951 from a malignant tumor of the uterine cervix. These cells are called HeLa cells in honor of their source, Henrietta Lacks, who died soon thereafter from the cervical cancer. HeLa cells continue to grow in laboratories around the world. They have been used in numerous studies of human cell function, including the experiments that documented chromosome replication occurring independently of mitosis.

and the beginning of M phase. Stated another way, there is a gap in the cycle. The gap represents the period when chromosome replication is complete but mitosis has not yet begun.

This "lag" in the cell cycle came to be called **G_2 phase**, for second gap. It was considered the second gap because data indicate that another gap exists. To follow the logic, note in Figure 11.4 that labeled nuclei undergoing mitosis are observed over a period of about 6 to 8 hours. Because all of these cells had to be in S phase when radioactive thymidine was available, it is logical to conclude that S phase lasts 6 to 8 hours. When the times to get through the S, G_2, and M phases are added up and compared with the 24 hours it takes these cells to complete one cell cycle, though, there is a discrepancy of 7 to 9 hours. This discrepancy represents the gap called **G_1 phase**, for first gap. As **Figure 11.5** shows, G_1 phase occurs after M phase but before S phase. In these cells, the G_1 phase is about twice as long as G_2.

Why do the gap phases exist? In addition to copying their chromosomes during S phase, dividing cells must replicate organelles and manufacture additional cytoplasm. Before mitosis can take place, the parent cell must grow large enough and synthesize enough organelles that its daughter cells will be normal in size and function. The two gap phases provide the time required to accomplish these tasks. They allow the cell to complete all the requirements for cell division other than chromosome replication. In addition, the long G_1 phase is the time when cells perform their normal functions. In your body, for example, most cells have stopped dividing and will spend the rest of your life in G_1.

Given this overview of the major events in the life of a cell, let's turn now to M phase and delve into the process of mitosis. Once the genetic material has been copied, how do cells divide it up between daughter cells? Recall that a **gene** is a segment of DNA that contains the information for synthesizing a particular polypeptide or RNA molecule. How do cells ensure that

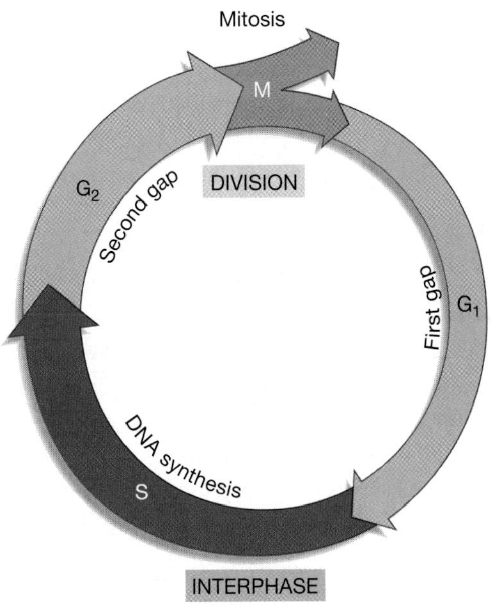

FIGURE 11.5 The Cell Cycle Has Four Phases
A representative cell cycle. The time required for the G_1 and G_2 phases varies dramatically among cells and organisms.

BOX 11.2 How Do Bacteria Divide?

To reproduce, bacterial cells divide into two genetically identical daughters. This process is called **binary fission**. Although the structure of bacterial cells is very different from that of eukaryotic cells, bacteria face challenges similar to those faced by eukaryotes in replicating and partitioning their hereditary material during cell division.

Figure 11.6 sketches the major events in bacterial cell division. Recall that most bacteria contain a single, circular chromosome composed of DNA and that this chromosome is coiled upon itself. Because bacteria lack a nucleus, the bacterial chromosome is located in the cytoplasm. After the chromosome is replicated, the two daughter chromosomes become attached to different sites on the plasma membrane. A contractile ring composed of FtsZ fibers then forms between the two chromosomes. Recall from Chapter 7 that FtsZ fibers are the major component of the bacterial cytoskeleton and are similar in structure to microtubules. As the FtsZ ring closes, the bacterial cytoplasm is divided in two, completing cell division.

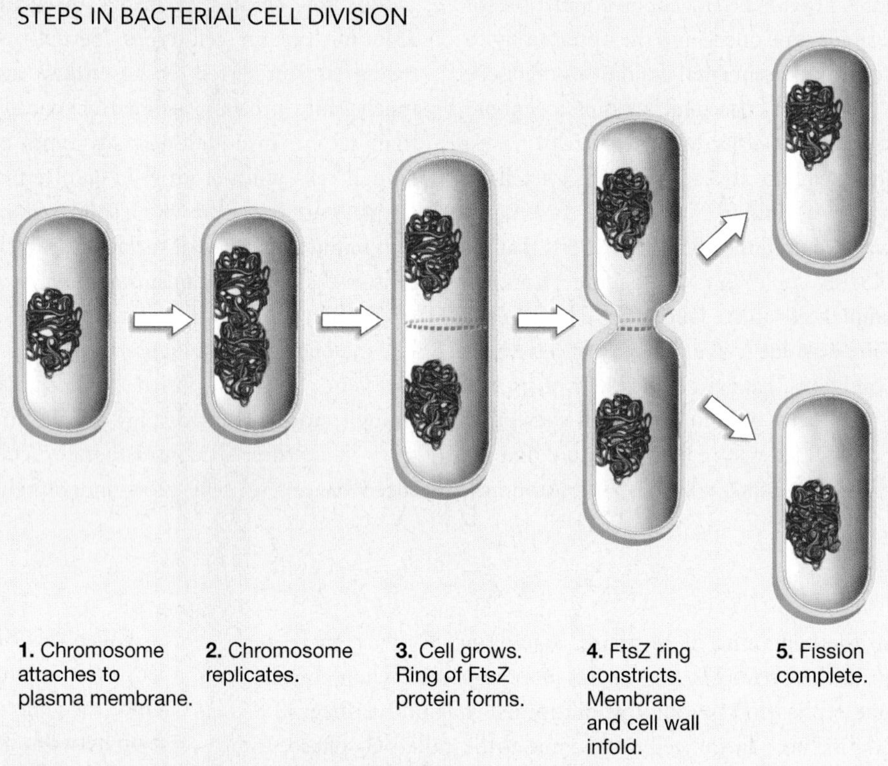
STEPS IN BACTERIAL CELL DIVISION
1. Chromosome attaches to plasma membrane.
2. Chromosome replicates.
3. Cell grows. Ring of FtsZ protein forms.
4. FtsZ ring constricts. Membrane and cell wall infold.
5. Fission complete.

FIGURE 11.6 Steps in Bacterial Cell Division
QUESTION Are the daughter cells of bacterial cell division identical to each other in chromosomal makeup, or are they different?

each daughter cell receives an identical complement of chromosomes and thus an identical complement of genes? (The discussion that follows describes mitosis in plant cells and in animal cells; **Box 11.2** describes cell division in bacteria.)

11.2 How Does Mitosis Take Place?

Early observations of cell division focused on the fate of the parent cell's chromosomes. Because chromosomes that are undergoing mitosis are visible under the light microscope when they are stained, investigators could watch mitosis occur. As a result, the major events in mitosis were well understood long before the cell cycle was fully described.

Recall that mitosis results in the division of chromosomes and the formation of two daughter nuclei. Mitosis is usually accompanied by cytokinesis—cytoplasmic division and the formation of two daughter cells. Let's take a closer look at the events in mitosis, beginning with observations about the nature of a eukaryotic cell's chromosomes.

Events in Mitosis

Figure 11.7a shows the chromosomes found in a hypothetical plant cell or animal cell. The number of chromosomes in each cell varies widely among species. Both humans and potato plants have a total of 46 chromosomes in each cell; a maize (corn) plant has 20, dogs have 66, and fruit flies have 8. In Figure 11.7a there are a total of four chromosomes per cell. (They are shown partially condensed simply to make them visible.) Recall from Section 11.1 that eukaryotic chromosomes normally exist as extremely long, threadlike strands consisting of DNA associated with globular proteins called **histones.** In eukaryotes the DNA-protein complex is called **chromatin.**

Prior to mitosis, the DNA in each chromosome is copied. As mitosis begins, chromatin condenses to form a much more compact structure. **Figure 11.7b** shows what chromosomes look like when they have been replicated and condensed. Replicated, condensed chromosomes correspond to the paired threads that were observed in salamander cells by early biologists (see Figure 11.2a).

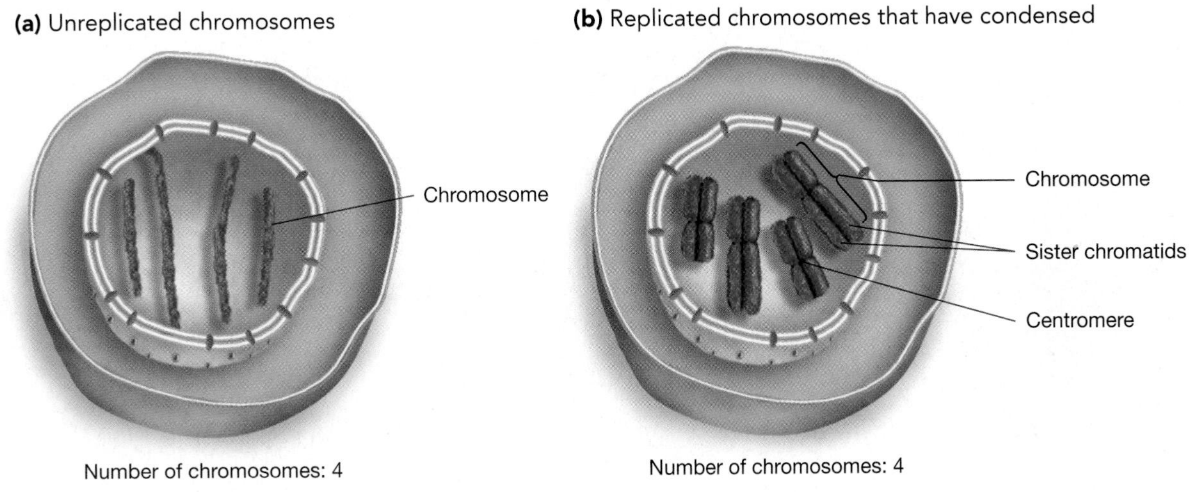

FIGURE 11.7 A Closer Look at Chromosomes
(a) Chromosomes consist of long strands of DNA that are associated with proteins. **(b)** When chromosomes have replicated, the two strands are called sister chromatids and are joined along their entire length as well as at a structure called the centromere. (Recall that chromosomes are visible in the light microscope only when they have condensed and are undergoing mitosis.)

Each of the DNA copies in a replicated chromosome is called a **chromatid**. The two strands are joined together along their entire length as well as at a specialized region of the chromosome called the **centromere**. Chromatids from the same chromosome are referred to as **sister chromatids**. Once replication is complete, each chromosome consists of two sister chromatids. Sister chromatids represent exact copies of the same genetic information. Each chromatid contains one long DNA double helix.

Recall from Section 11.1 that chromosomes are replicated before mitosis begins. At the start of M phase, then, chromosomes consist of two sister chromatids that are attached to one another at the centromere (**Figure 11.8**). During mitosis, the two sister chromatids separate to form independent chromosomes, and one copy of each chromosome goes to each of the two daughter cells. As a result, each daughter cell receives a copy of the genetic information that is contained in each chromosome. Every daughter cell ends up with exactly the same complement of chromosomes as the parent cell had prior to replication, and thus every daughter cell receives the same genetic information.

Although mitosis is a continuous process, biologists routinely identify several subphases within M phase on the basis of distinctive events that occur. These subphases of mitosis are designated prophase, prometaphase, metaphase, anaphase, and telophase. Some students use the mnemonic device IPPMAT to remind themselves that *i*nterphase is followed by the mitotic subphases *p*rophase, *p*rometaphase, *m*etaphase, *a*naphase, and *t*elophase.

To understand how mitosis proceeds, let's look at each subphase in turn.

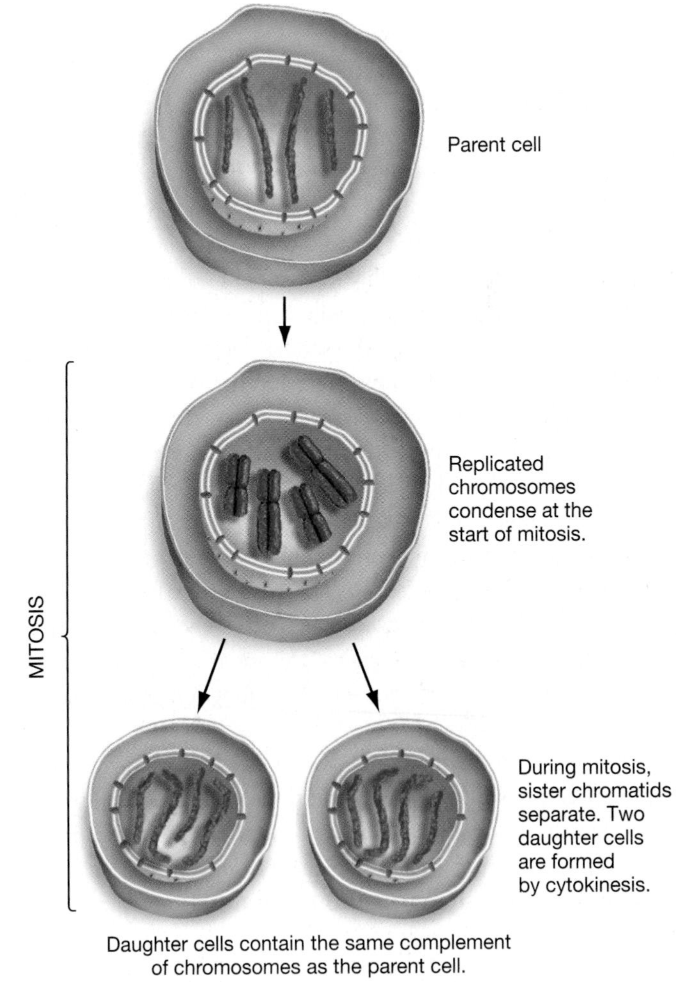

FIGURE 11.8 An Overview of Mitosis
Chromosomes are replicated prior to mitosis. During mitosis, the replicated chromosomes are partitioned to the two daughter nuclei. In most cases, mitosis is followed by cytokinesis.

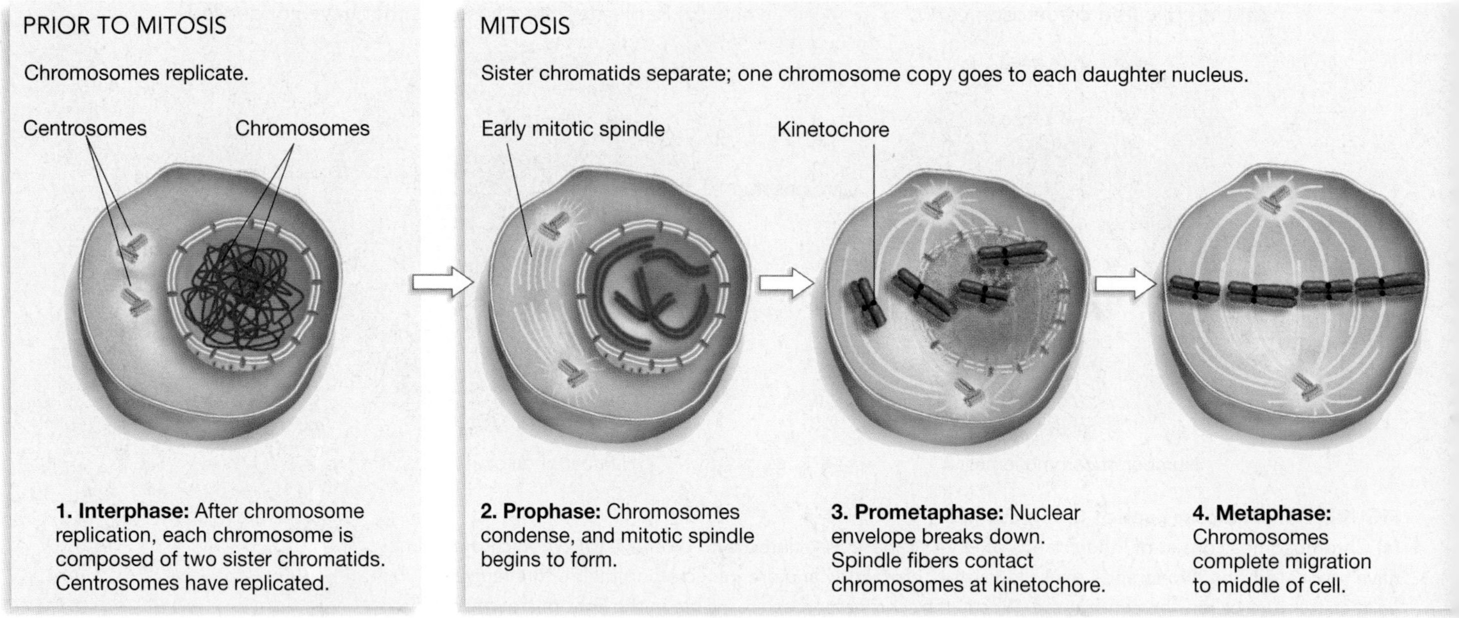
PRIOR TO MITOSIS
Chromosomes replicate.
Centrosomes
Chromosomes
MITOSIS
Sister chromatids separate; one chromosome copy goes to each daughter nucleus.
Early mitotic spindle
Kinetochore
1. Interphase: After chromosome replication, each chromosome is composed of two sister chromatids. Centrosomes have replicated.
2. Prophase: Chromosomes condense, and mitotic spindle begins to form.
3. Prometaphase: Nuclear envelope breaks down. Spindle fibers contact chromosomes at kinetochore.
4. Metaphase: Chromosomes complete migration to middle of cell.

FIGURE 11.9 Mitosis and Cytokinesis

WEB TUTORIAL 11.1 The Phases of Mitosis

Prophase Mitosis begins with the events of **prophase** ("before-phase"), as shown in **Figure 11.9**. The chromosomes have already replicated during interphase (Figure 11.9, step 1); during prophase (step 2), they condense into compact structures. Chromosomes first become visible in the light microscope during prophase.

In the cytoplasm, prophase is marked by the formation of the mitotic spindle. The **mitotic spindle** is a structure that produces mechanical forces that pull chromosomes into the daughter cells during mitosis. The mitotic spindle consists of an array of microtubules—components of the cytoskeleton that were introduced in Chapter 7. Groups of microtubules attach to the chromosomes and are called **spindle fibers.** In fungi, spindle fibers radiate from a microtubule-organizing center called the *spindle pole body*, while in plants the mitotic spindle forms from microtubule-organizing centers associated with the nuclear envelope. In animals, the microtubule-organizing centers responsible for mitotic spindle formation are called **centrosomes.** Recall from Chapter 7 that animal centrosomes contain structures called **centrioles.** Centrioles are not required for mitotic spindle formation, however, and their function is currently a topic of intense research.

The main point is that some sort of microtubule-organizing center is responsible for the formation of the mitotic spindle. Although the nature of this structure varies among groups, the function of the spindle itself is the same. During prophase, the mitotic spindles begin moving to opposite sides of the cell, or they form on opposite sides.

Prometaphase Once chromosomes have condensed, the nucleolus disappears and the nuclear envelope breaks down. After the nuclear envelope has disintegrated, spindle fibers from each mitotic spindle attach to one of the two sister chromatids of each chromosome. These events occur during **prometaphase** ("before middle-phase"); see Figure 11.9, step 3.

The attachment between the spindle fibers and each chromatid is made at a structure called the **kinetochore.** Kinetochores are located at the centromere region of the chromosome, where sister chromatids are attached to each other.

During prometaphase in animals, the centrosomes continue their movement to opposite poles of the cell. In all groups, the microtubules that are attached to the kinetochores begin moving the chromosomes to the middle of the cell.

Metaphase During **metaphase** ("middle-phase"), animal centrosomes complete their migration to the opposite poles of the cell (Figure 11.9, step 4). In all groups, the kinetochore microtubules finish moving the chromosomes to the middle of the cell. When metaphase is over, the chromosomes are lined up along an imaginary plane called the **metaphase plate.** At this point, the formation of the mitotic spindle is complete. Each chromatid is attached to spindle fibers that run from its kinetochore to one of the two poles of the cell. Each chromosome is held by kinetochore spindle fibers reaching to opposite poles and exerting the same amount of tension or pull. A tug of war is occurring, with kinetochore spindle fibers pulling each chromosome in opposite directions.

Anaphase At the start of **anaphase** ("throughout-phase"), the centromeres that are holding sister chromatids together split (Figure 11.9, step 5). Because they are under tension, sister chromatids are pulled apart equally—with the same amount of force—to create independent chromosomes. The kinetochore spindle fibers then begin to shorten. As they do, motor proteins pull the chromosomes to opposite poles of the

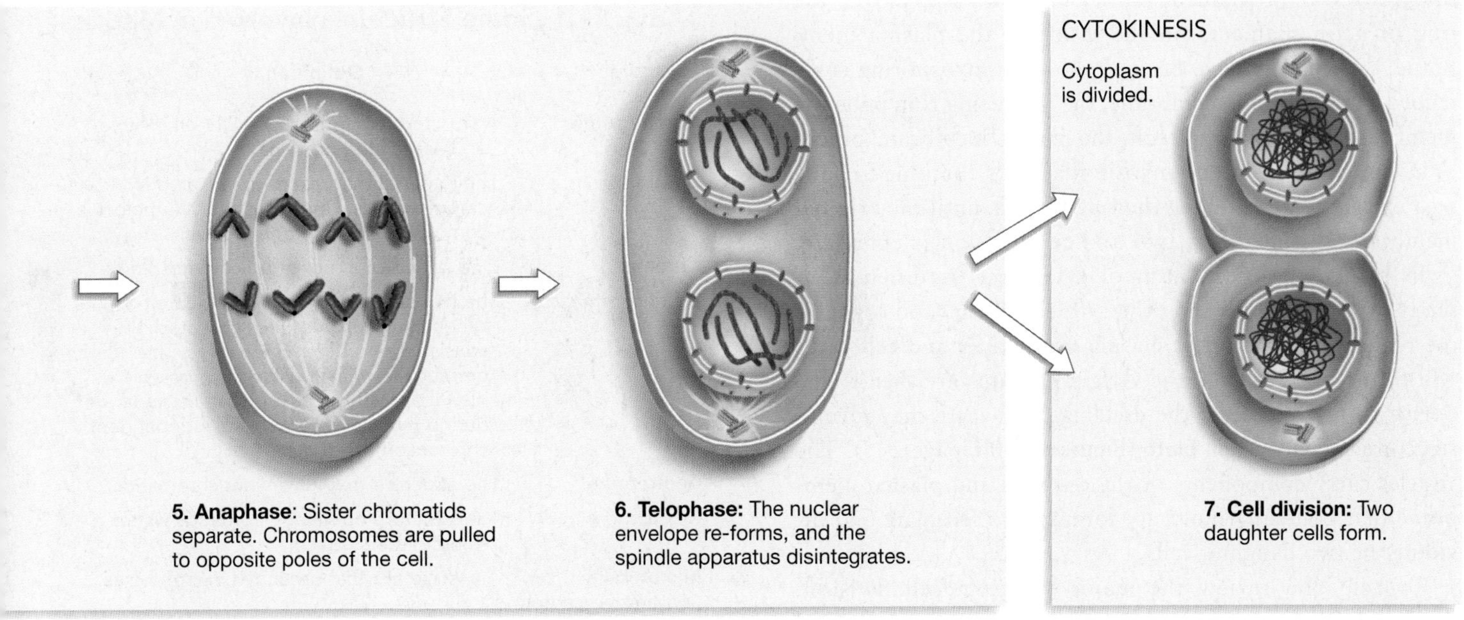
CYTOKINESIS
Cytoplasm is divided.
5. Anaphase: Sister chromatids separate. Chromosomes are pulled to opposite poles of the cell.
6. Telophase: The nuclear envelope re-forms, and the spindle apparatus disintegrates.
7. Cell division: Two daughter cells form.

cell. The two poles of the cell are also pushed away from each other by motor proteins associated with microtubules that are not attached to chromosomes. During anaphase, replicated chromosomes split into two identical sets of unreplicated chromosomes.

The separation of sister chromatids to opposite poles is a critical step in mitosis, because this is the step that ensures that each daughter cell receives the same complement of chromosomes. When anaphase is complete, each pole of the cell has an equivalent and complete collection of chromosomes that are identical to those present in the parent cell prior to chromosome replication.

Telophase During **telophase** ("end-phase"), a nuclear envelope begins to form around each set of chromosomes (Figure 11.9, step 6). The mitotic spindle disintegrates, and the chromosomes begin to de-condense. Once two independent nuclei have formed, mitosis is complete.

Cytokinesis

Prior to the onset of M phase, all of the cell's organelles besides the nucleus have replicated, and the rest of the cell contents have grown. During cytokinesis (Figure 11.9, step 7), the cytoplasm divides to form two daughter cells, each with its own nucleus and complete set of organelles. Although cytokinesis normally occurs immediately following mitosis, in some cases it does not. As the muscle cells in your upper arm grew and developed, for example, repeated mitoses that were not followed by cytokinesis resulted in the formation of cells containing multiple nuclei. Such cells are referred to as *multinucleate*.

In animals, fungi, and slime molds, cytokinesis begins with the formation of a **cleavage furrow** (Figure 11.10a). The furrow appears because a ring of actin and myosin filaments forms just inside the plasma membrane, in a plane that bisects the cell. Recall from Chapter 7 that myosin is a motor protein that binds to actin filaments. When myosin binds to ATP or ADP, part of the protein moves in a way

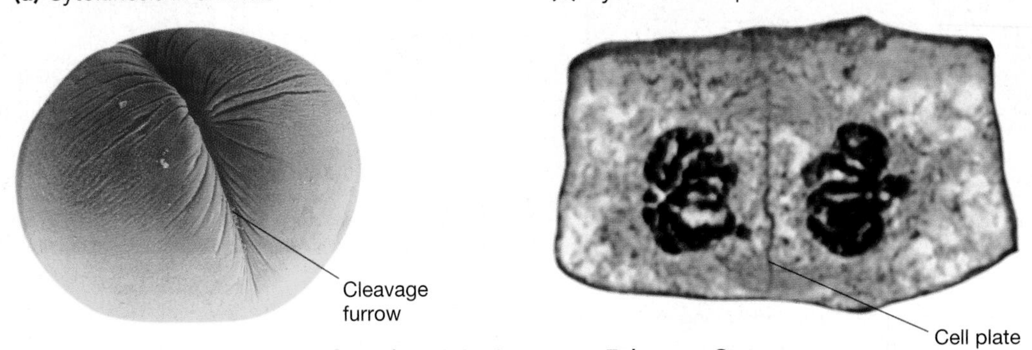
(a) Cytokinesis in animals
(b) Cytokinesis in plants
Cleavage furrow
Cell plate

FIGURE 11.10 The Mechanism of Cytokinesis Varies among Eukaryote Groups
(a) In animals, the cytoplasm is divided by a cleavage furrow that pinches the parent cell in two. **(b)** In plants, the cytoplasm is divided by a cell plate that forms in the middle of the parent cell. [(a) ©Dr. Richard Kessel/Visuals Unlimited]

that causes actin filaments to slide. As myosin moves the ring of actin filaments on the inside of the plasma membrane, the ring tightens. Because the actin-myosin ring is attached to the plasma membrane, the tightening ring pulls the membrane with it. As a result, the plasma membrane pinches inward. The actin and myosin filaments continue to slide past each other, tightening the ring further, until the original membrane is pinched in two and cell division is complete.

In plants, the mechanism of cytokinesis is different. A series of microtubules and other proteins define and organize the region where the new plasma membranes and cell walls will form. Vesicles from the Golgi apparatus are then transported to the middle of the dividing cell, where they form a structure called the **cell plate** (**Figure 11.10b**, page 235). The vesicles carry components of the cell wall and plasma membrane that gradually build up, forming the cell plate and dividing the two daughter cells.

To help you review the major events in cell division, **Table 11.1** summarizes the key structures involved, and **Figure 11.11** shows photographs of cells in interphase and undergoing mitosis and cytokinesis.

TABLE 11.1 Some Structures Involved in Mitosis

Structure	Definition
Chromosome	A structure composed of a DNA molecule and associated proteins.
Chromatin	The material that makes up eukaryotic chromosomes. Consists of a DNA molecule complexed with histone proteins.
Chromatid	One strand of a replicated chromosome.
Sister chromatids	The two strands of a replicated chromosome. When chromosomes are replicated, they consist of two sister chromatids. The genetic material in sister chromatids is identical. When sister chromatids separate during mitosis, they become independent chromosomes.
Centromere	The structure that joins sister chromatids.
Kinetochore	The structure on sister chromatids where spindle fibers attach.
Microtubule organizing center	Any structure that organizes microtubules.
Centrosome	The microtubule organizing center in animals.
Centriole	A paired structure inside animal centrosomes.

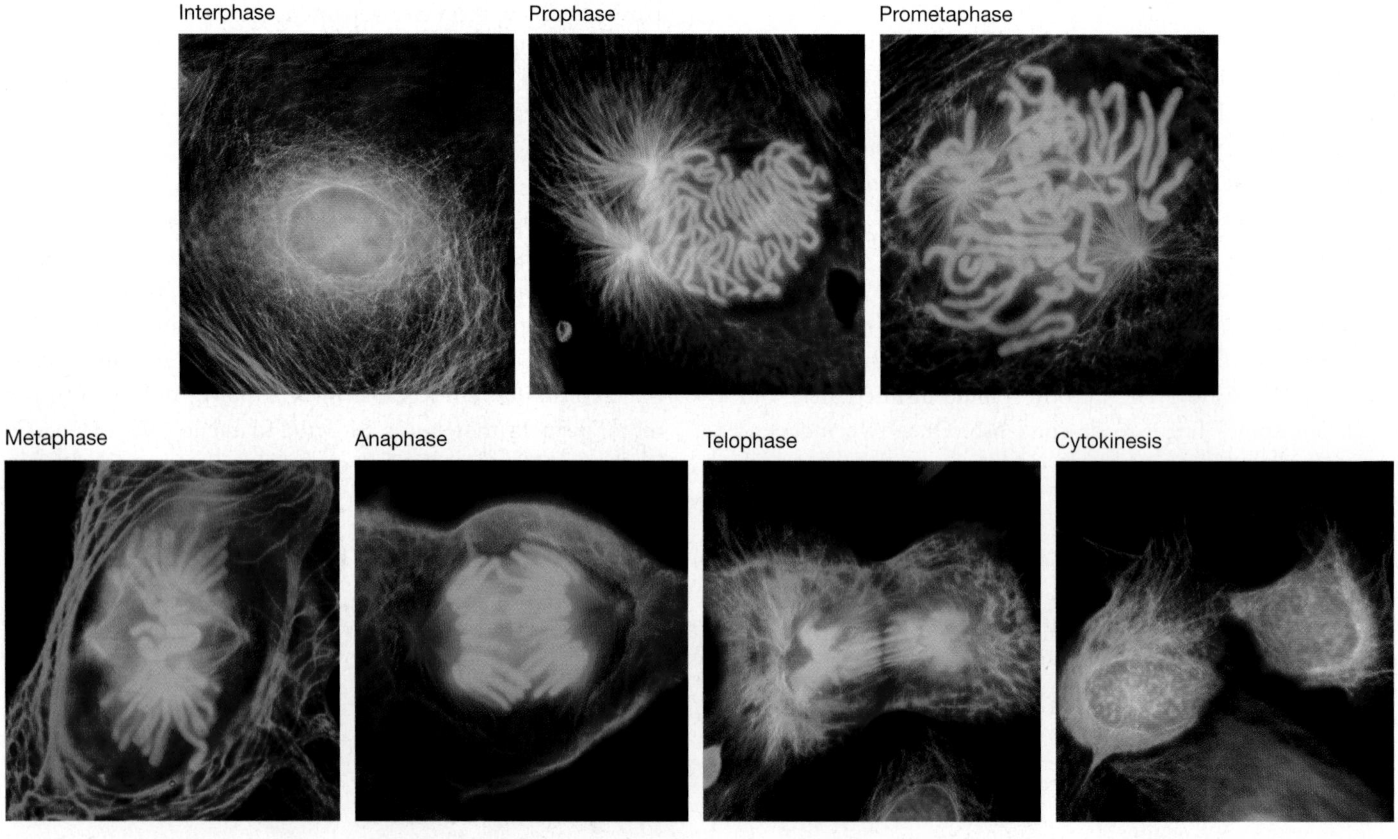

FIGURE 11.11 Mitosis and Cytokinesis in Action
Micrographs showing newt cells in interphase and undergoing mitosis and cytokinesis. Chromosomes are stained blue; microtubules are green, and actin filaments are red. EXERCISE Next to each image, describe what is happening. In at least two of the photographs, label the chromosomes, centrosomes, and mitotic spindle. In the metaphase cell, label the metaphase plate.

Once these structures and processes had been described in detail, biologists turned their attention to understanding the molecular mechanisms that are responsible for them. In particular, biologists wanted to know two things about mitosis: (1) How do sister chromatids separate to become independent chromosomes, and (2) how do those chromosomes move to daughter cells? The exact and equal partitioning of genetic material to the two daughter cells is the most fundamental aspect of cell division. How does this process occur?

How Do Chromosomes Move during Mitosis?

To understand how sister chromatids separate and move to daughter cells, biologists have tried to understand how the mitotic spindle functions. Do spindle microtubules act as railroad tracks, the way they do in vesicle transport? Is some sort of motor protein involved? And what is the nature of the kinetochore, where the chromosome and microtubules are joined?

Mitotic Spindle Forces Spindle fibers are composed of microtubules. Recall from Chapter 7 that microtubules are composed of α-tubulin and β-tubulin dimers, that the length of a microtubule is determined by the number of tubulin dimers it contains, and that microtubules are asymmetric—they have a plus end and a minus end. These observations suggest a mechanism for the movement of chromosomes during anaphase: Is the spindle microtubule shortening due to a loss of tubulin dimers from one end? Alternatively, it is possible that microtubules slide past each other, much as actin and myosin filaments do during cytokinesis in animal cells, and that this sliding action pulls chromosomes toward the poles of the cell.

To test these hypotheses, biologists introduced fluorescently labeled tubulin subunits into prophase or metaphase cells. This treatment made the entire mitotic spindle visible. (See step 1 in the "Experimental setup" section of **Figure 11.12**.) Then, once anaphase had begun, the researchers marked a region of the spindle by irradiating it with a bar-shaped beam of laser light. The laser quenched the fluorescence in the exposed region, making that section of the spindle dark, or "photobleached," although it was still functional (see step 2 of the "Experimental setup"). The "Results" section of Figure 11.12 shows that as anaphase progressed, the photobleached region remained stationary while the chromosomes moved toward this region.

To explain this result, the biologists concluded that the kinetochore microtubules must remain stationary during anaphase. The hypothesis suggested that, instead of sliding past each other as actin fibers do, microtubules shorten at the kinetochore because tubulin subunits are lost from the kineto-

Question: How do microtubules shorten to pull sister chromatids apart at anaphase?

Hypothesis: Microtubules shorten at the kinetochore.

Alternate hypothesis: Microtubules slide past each other like actin filaments.

Experimental setup:

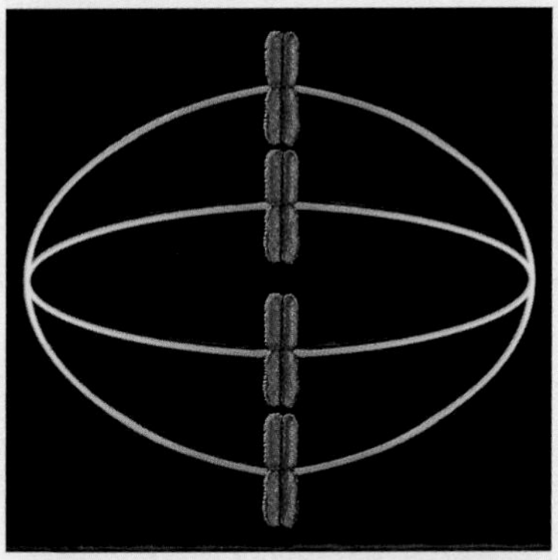

1. Use fluorescent labels to make the metaphase chromosomes fluoresce blue and the microtubules fluoresce yellow.

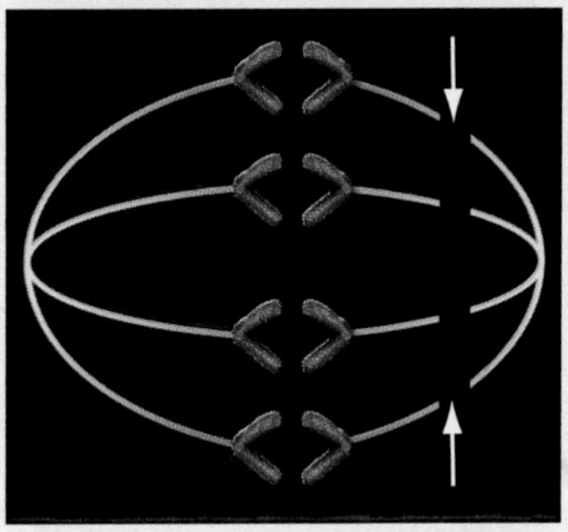

2. At the start of anaphase, photobleach a section of microtubules to mark them without changing their function.

Prediction: The photobleached section will not move. Instead, microtubules on one side of the photobleached section will shorten.

Prediction of alternate hypothesis: The photobleached section will move.

Results:

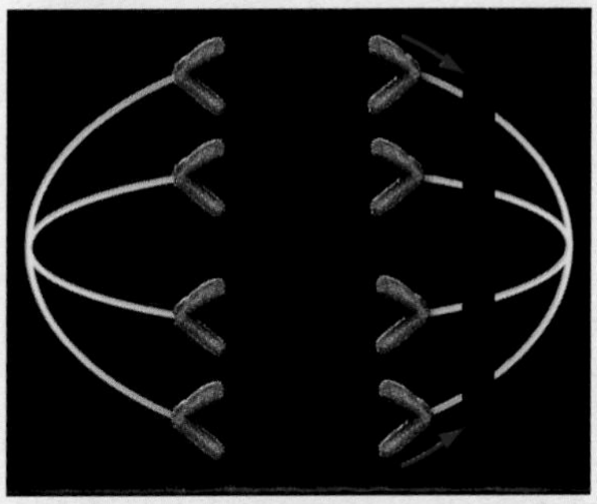

By late anaphase, the distance between chromosomes and photobleached section lessened.

Conclusion: Microtubules shorten near the chromosome.

FIGURE 11.12 During Anaphase, Microtubules Shorten at the Kinetochore

EXERCISE Add drawings to show the outcomes you would expect if microtubules shortened at the end opposite the chromosome or if microtubules slide past each other.

chore ends (**Figure 11.13a**). As kinetochore microtubules shorten at their minus end, the chromosomes are pulled along. How does this happen?

A Kinetochore Motor Recall that kinetochores are structures located at the centromere of replicated chromosomes and that microtubules attach to each sister chromatid at its kinetochore. If microtubules shorten at the kinetochore end, how does the chromosome remain attached to the microtubule? Although the answer is not yet clear, the structure and function of the kinetochore is gradually becoming better understood.

Recent research has shown that the kinetochore contains dyneins and other motor proteins that "walk" chromosomes down microtubules—from their plus ends near the kinetochore toward their minus ends at the spindle (**Figure 11.13b**). This observation suggests that the mechanism of chromosome movement is reminiscent of the way kinesin walks down microtubules during vesicle transport. As explained in Chapter 7, motor proteins such as kinesin and dynein change shape when they are phosphorylated or bind to ATP. These shape changes can produce movement. Biologists hypothesize that, during mitosis, dyneins and other kinetochore motor proteins detach near the chromosome and then reattach to the kinetochore microtubule farther along its length after they move. As this detach-move-reattach cycle repeats, the microtubule shortens and the chromosome is pulled to one end of the mitotic spindle.

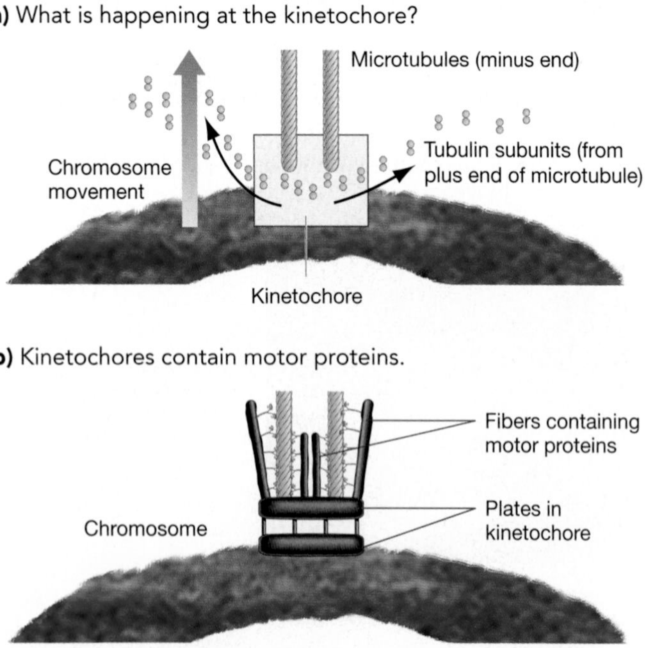

FIGURE 11.13 How Do Microtubules Move Chromosomes during Mitosis?
(a) According to one hypothesis, microtubules shorten as tubulin subunits split off at the kinetochore. **(b)** The kinetochore consists of an inner plate, an outer plate, and associated fibers that contain motor proteins. The fibers are thought to act as "legs" that walk the chromosomes down the length of the kinetochore microtubules.

Efforts to understand spindle structure and movement bring us to the frontier of research on mitosis. Having explored how the process occurs, let's focus on how it is controlled. When does a cell divide, and when does it stop dividing? How is cell division regulated? These questions are fundamental. When cell division occurs in an uncontrolled manner, cancerous tumors can form.

✓CHECK YOUR UNDERSTANDING

After chromosomes replicate, mitosis distributes one copy of each chromosome to each daughter cell. As a result, mitosis and cytokinesis lead to the production of cells with the same genetic material as that of the parent cell. You should be able to (1) draw a replicated chromosome and an unreplicated chromosome and label the sister chromatids and the centromere, and (2) diagram what happens to chromosomes and the nuclear envelope during prophase, prometaphase, metaphase, anaphase, and telophase.

11.3 Control of the Cell Cycle

Although the events of mitosis are virtually identical in all eukaryotes, other aspects of the cell cycle can be extremely variable. For example, the length of the cell cycle can vary enormously among different cell types, even in the same individual. In humans, intestinal cells routinely divide more than twice a day to renew tissue that is lost during digestion; mature human nerve and muscle cells do not divide at all. Most of these differences are due to variation in the length of the G_1 phase. In rapidly dividing cells, G_1 is essentially eliminated. Most nondividing cells, in contrast, are permanently stuck in G_1. Researchers refer to this arrested stage as the G_0 state, or simply "G zero."

A cell's division rate can also vary in response to changes in conditions. For example, human liver cells normally divide about once per year. But if part of the liver is damaged or lost, the remaining cells divide every one or two days until repair is accomplished. Cells of unicellular organisms such as yeasts, bacteria, or archaea divide rapidly only if the environment is rich in nutrients; otherwise, they enter a quiescent (inactive) state.

To explain the existence of so much variability, biologists hypothesized that the cell cycle must be regulated in some way and that regulation varies among cells and organisms. Understanding how the cell cycle is controlled is now the most prominent issue in research on cell division—partly because defects in control can lead to uncontrolled, cancerous growth. What evidence first suggested that regulatory molecules control the cell cycle?

The Discovery of Cell-Cycle Regulatory Molecules

The first solid evidence for cell-cycle control molecules came to light in 1970, when researchers published the results of experiments on fusing pairs of cultured mammalian cells. In the presence of certain chemicals, viruses, or an electric shock, the membranes of two cells can be made to fuse. The hybrid cell that results has two nuclei.

How did cell fusion experiments point to the existence of cell-cycle control molecules? When investigators fused cells that were in different stages of the cell cycle, certain nuclei changed phases. For example, when a cell in M phase was fused with one in interphase, the nucleus of the interphase cell initiated M phase (**Figure 11.14a**). The biologists hypothesized that the cytoplasm of M-phase cells contains a regulatory molecule that induces interphase cells to enter M phase.

This hypothesis was confirmed by experiments on the South African claw-toed frog, *Xenopus laevis*. As the eggs of these frogs mature, they change from a cell called an **oocyte**, which is arrested in a phase similar to G_2, to a mature egg that has entered M phase. The eggs are attractive to study, partly because they are more than 1 mm in diameter. Their large size makes it relatively easy to purify large amounts of cytoplasm and to use microsyringes to inject the eggs with cytoplasm from eggs in different stages of development. When biologists purified cytoplasm from M-phase frog eggs and injected it into the cytoplasm of frog oocytes arrested in the G_2-like phase, the immature oocytes entered M phase (**Figure 11.14b**). But when cytoplasm from interphase cells was injected into G_2 oocytes, the cells remained in the G_2-like phase. The researchers concluded that the cytoplasm of M-phase cells—but not the cytoplasm of interphase cells—contains a factor that drives immature oocytes into M phase to complete their maturation.

This factor was eventually purified and is now called **mitosis-promoting factor**, or **MPF**. Subsequent experiments showed that MPF induces mitosis in all eukaryotes. For example, injecting M-phase cytoplasm from mammalian cells into immature frog eggs results in egg maturation. Human MPF can trigger mitosis in yeast cells. The molecule appears to be a general signal that says "Start mitosis." How does it work?

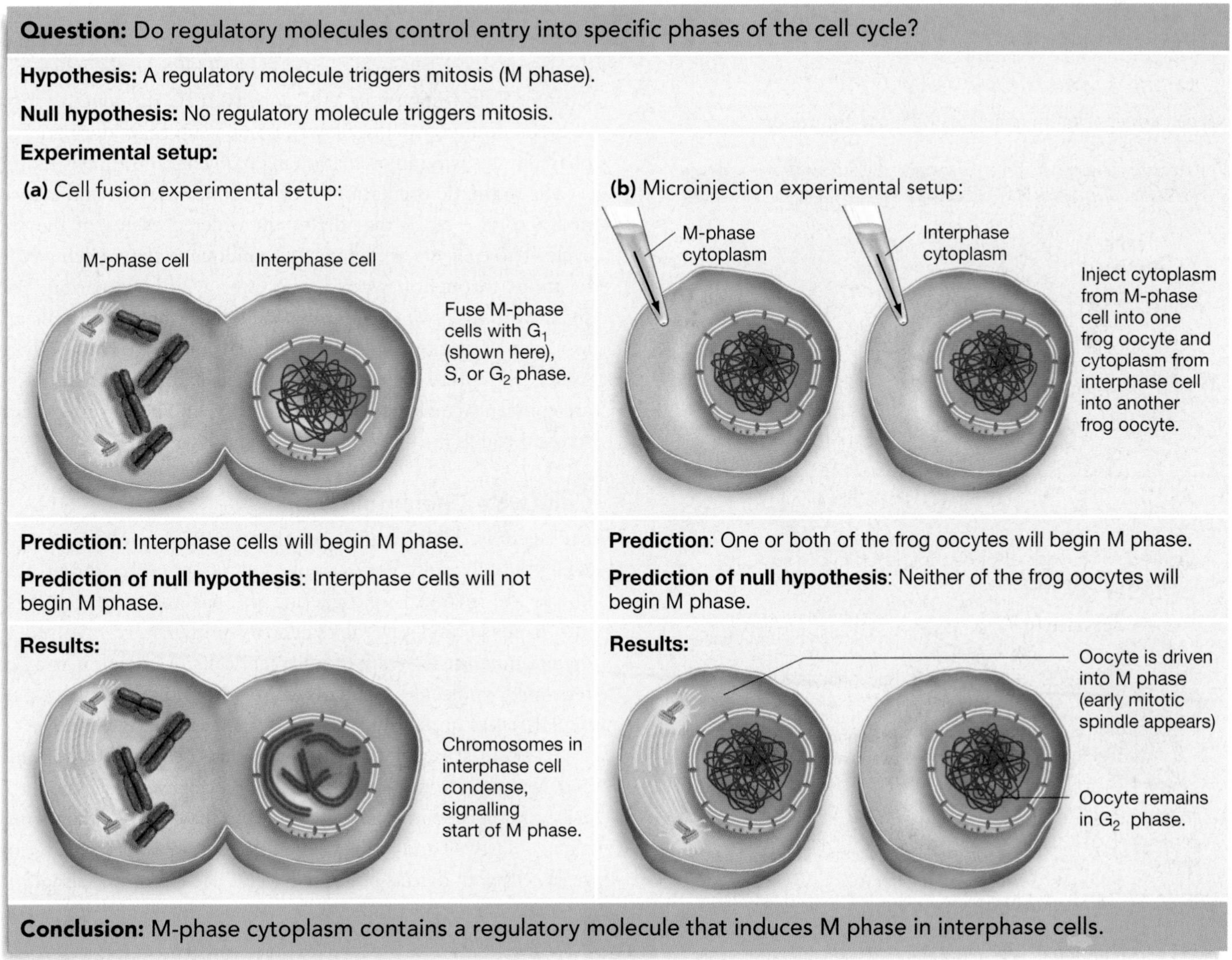
Question: Do regulatory molecules control entry into specific phases of the cell cycle?
Hypothesis: A regulatory molecule triggers mitosis (M phase).
Null hypothesis: No regulatory molecule triggers mitosis.
Experimental setup:
(a) Cell fusion experimental setup:
M-phase cell
Interphase cell
Fuse M-phase cells with G_1 (shown here), S, or G_2 phase.
(b) Microinjection experimental setup:
M-phase cytoplasm
Interphase cytoplasm
Inject cytoplasm from M-phase cell into one frog oocyte and cytoplasm from interphase cell into another frog oocyte.
Prediction: Interphase cells will begin M phase.
Prediction of null hypothesis: Interphase cells will not begin M phase.
Prediction: One or both of the frog oocytes will begin M phase.
Prediction of null hypothesis: Neither of the frog oocytes will begin M phase.
Results:
Chromosomes in interphase cell condense, signalling start of M phase.
Results:
Oocyte is driven into M phase (early mitotic spindle appears)
Oocyte remains in G_2 phase.
Conclusion: M-phase cytoplasm contains a regulatory molecule that induces M phase in interphase cells.

FIGURE 11.14 Experimental Evidence for Cell-Cycle Control Molecules
(a) When M-phase cells are fused with cells in G_1, S, or G_2 phase, the interphase chromosomes condense and begin M phase. **(b)** Microinjection experiments supported the hypothesis that a regulatory molecule induces M phase.

WEB TUTORIAL 11.2 The Cell Cycle

MPF Contains a Protein Kinase and a Cyclin Once MPF had been isolated and purified, researchers found that it is made up of two distinct polypeptide subunits. One of the components is a **protein kinase**—an enzyme that catalyzes the transfer of a phosphate group from ATP to a target protein. Recall from Chapter 3 that proteins can be activated or inactivated by phosphorylation. Because the addition of a phosphate group changes the target protein's shape and activity, protein kinases frequently act as regulatory elements in the cell.

These observations suggested that MPF acts by phosphorylating a protein that triggers the onset of mitosis. But research showed that the concentration of MPF protein kinase is more or less constant throughout the cell cycle. How can MPF trigger mitosis if the protein kinase subunit is always present?

The answer to this question lies in the second MPF subunit, which belongs to a family of proteins called the **cyclins.** Cyclins got their name because their concentrations fluctuate throughout the cell cycle. As **Figure 11.15a** shows, the cyclin associated with MPF builds up in concentration during interphase and peaks during M phase. This increase in concentration is important because the protein kinase subunit in MPF can be active only when it is bound to the cyclin subunit. As a result, the protein kinase subunit of MPF is called a **cyclin-dependent kinase,** or **Cdk**. MPF is a dimer that consists of a cyclin subunit and a cyclin-dependent kinase subunit.

According to Figure 11.15a, the number of complete MPF dimers builds up steadily during interphase. Why doesn't this increasing concentration of MPF trigger the onset of M phase? The answer is that MPF's cyclin unit is phosphorylated when it is initially synthesized. Phosphorylation of the protein changes its conformation in a way that renders the protein inactive. Late in G_2 phase, however, enzymes catalyze the dephosphorylation of cyclin. These reactions activate MPF.

Once MPF is activated, it phosphorylates the proteins listed in **Figure 11.15b**. Chromosomal proteins activated by MPF cause chromosomes to condense into the threads visible during M phase. Microtubule-associated proteins phosphorylated by MPF may be involved in assembling the mitotic spindle apparatus. In this way, MPF triggers the onset of M phase.

MPF also activates an enzyme complex that promotes the degradation of MPF's own cyclin subunit, however. MPF causes its own destruction. This is an example of negative feedback—similar to the mechanisms of feedback inhibition introduced in Chapter 9. In response to MPF activity, then, the concentration of cyclin declines rapidly. Slowly, it builds up again during interphase. In this way, an oscillation in cyclin concentration is set up.

The dramatic oscillation in cyclin concentration and activation acts as a clock that drives the ordered events of the cell cycle. These events are happening millions of times each day in locations throughout your body. Over a 24-hour period, you swallow millions of cheek cells. Millions of additional cells are lost from your intestinal lining each day and leave your body as waste. To replace them, cells in your cheek and intestinal tissue are constantly making and degrading cyclin and pushing themselves through the cell cycle.

(a) Cyclin concentration regulates MPF concentration.

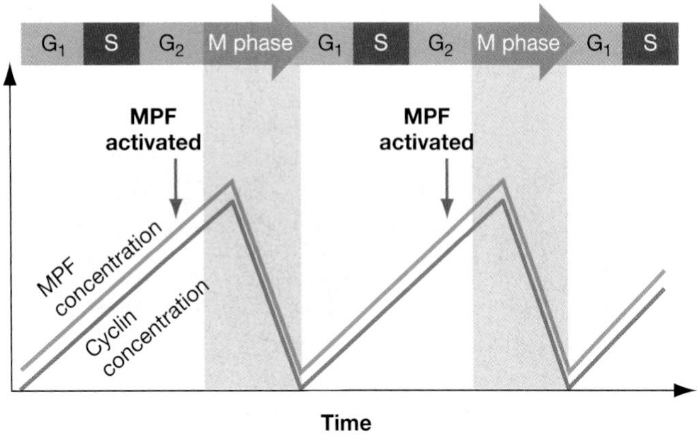

(b) Activated MPF (with unphosphorylated cyclin subunit) has an array of effects.

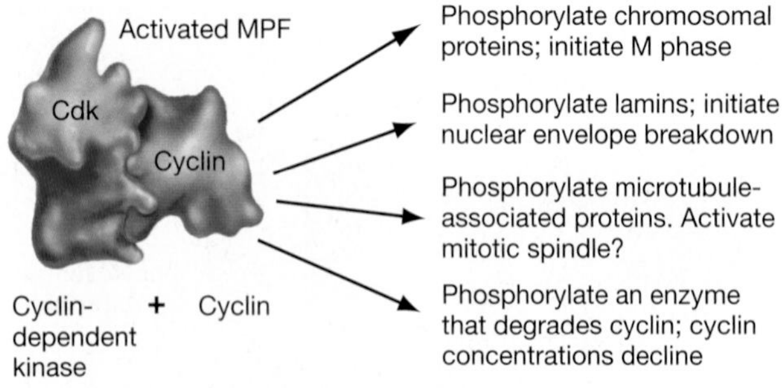

FIGURE 11.15 M-Phase Promoting Factor Is Created When a Cyclin Binds to a Protein Kinase
(a) Cyclin concentrations cycle in dividing cells, reaching a peak in M phase. Cyclin binds to a protein kinase, creating MPF.
(b) MPF activates proteins that initiate M phase, as well as proteins that degrade cyclin.

Cell-Cycle Checkpoints

MPF is only one of many protein complexes involved in regulating the cell cycle. For example, a different cyclin and protein kinase are involved in triggering the passage from G_1 phase into S phase, and several regulatory proteins are involved in maintaining the G_0 state of quiescent cells. A complex array of regulatory molecules is involved in either holding cells in particular stages or in stimulating passage to the next phase.

To make sense of these observations, Leland Hartwell and Ted Weinert introduced the concept of a cell-cycle checkpoint. A **cell-cycle checkpoint** is a critical point in the cell cycle that is regulated. Hartwell and Weinert identified checkpoints by analyzing yeast cells with defects in the cell cycle. The defective cells lacked a specific checkpoint and, as a result, kept dividing under culture conditions when normal cells stopped growing. In the body, cells that keep growing in this way form a mass of cells called a **tumor.**

As **Figure 11.16** indicates, biologists have since obtained evidence of three distinct checkpoints. In effect, interactions

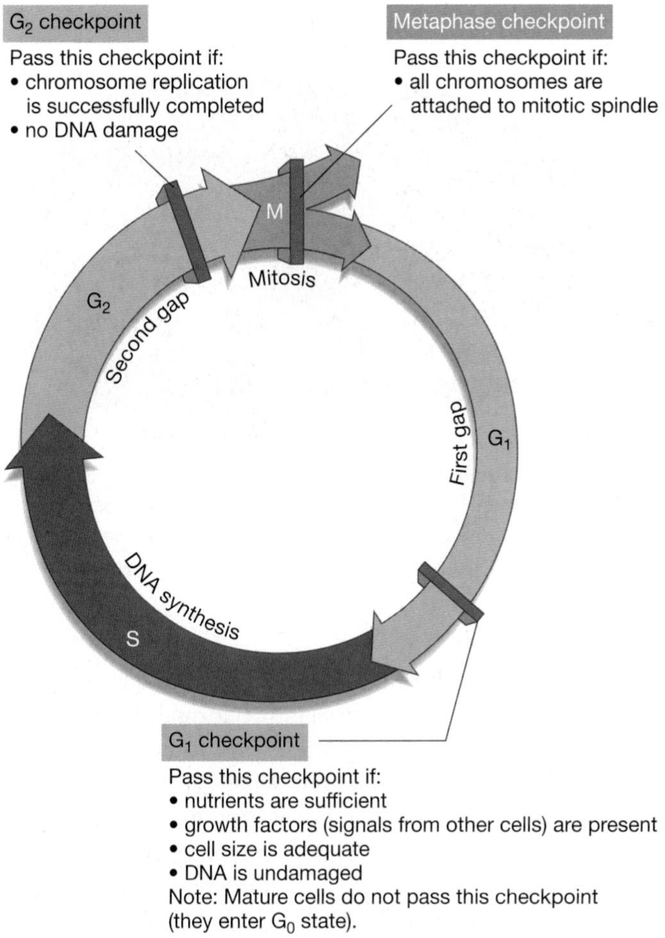
G₂ checkpoint
Pass this checkpoint if:
• chromosome replication is successfully completed
• no DNA damage
Metaphase checkpoint
Pass this checkpoint if:
• all chromosomes are attached to mitotic spindle
M
Mitosis
G₂
Second gap
First gap
G₁
DNA synthesis
S
G₁ checkpoint
Pass this checkpoint if:
• nutrients are sufficient
• growth factors (signals from other cells) are present
• cell size is adequate
• DNA is undamaged
Note: Mature cells do not pass this checkpoint (they enter G₀ state).

FIGURE 11.16 The Three Cell-Cycle Checkpoints

QUESTION Why is it advantageous for a cell to arrest at the M-phase checkpoint if not all chromosomes are attached to the mitotic spindle?

among regulatory molecules at each checkpoint allow a cell to "decide" whether to proceed with division. If these regulatory molecules are defective, the checkpoint may fail. As a result, cells may start growing in an uncontrolled fashion.

The first cell-cycle checkpoint occurs late in G_1. For most cells, this checkpoint is the most important in establishing whether the cell will continue through the cycle and divide. What determines whether a cell passes the G_1 checkpoint?

- Because a cell must reach a certain size before its daughter cells will be large enough to function normally, biologists hypothesize that some mechanism exists to arrest the cell cycle if the cell is too small.
- Unicellular organisms arrest at the G_1 checkpoint if nutrient conditions are poor.
- Cells in multicellular organisms pass through the G_1 checkpoint in response to signaling molecules from other cells, or what are termed *social signals*.
- If DNA is physically damaged, the protein **p53** activates genes that either stop the cell cycle until the damage can be repaired or lead to the cell's programmed, controlled destruction—a phenomenon known as **apoptosis**. In this way, p53 acts as a brake on the cell cycle. Because cancer can develop if "brake" molecules such as p53 are defective, these regulatory proteins are called **tumor suppressors.**

The general message here is that the components of the G_1 checkpoint have the same function: ensuring that the cell is healthy and should proceed to replicating its DNA and dividing.

The second checkpoint occurs after S phase, at the boundary between the G_2 and M phases. Specifically, cells appear to arrest at the G_2 checkpoint if chromosome replication has not been completed properly or if DNA is damaged. Because MPF is the key signal that triggers the onset of M phase, investigators were not surprised to find that it is involved in the G_2 checkpoint. Although much remains to be learned, data suggest that if DNA is damaged or if chromosomes are not replicated correctly, the dephosphorylation and activation of MPF are blocked. When MPF is not activated, cells remain in G_2 phase. Some data indicate that cells at this checkpoint may also respond to signals from other cells and to internal signals relating to their size.

The final checkpoint occurs during mitosis. If not all chromosomes are properly attached to the mitotic spindle, M phase arrests at metaphase. Specifically, anaphase is delayed until all kinetochores are properly attached to mitotic spindle fibers. If this checkpoint did not exist, some chromosomes might not separate properly, and daughter cells would receive an incorrect number of chromosomes during anaphase. Because they would receive too much or too little genetic material, the effect on the daughter cells could be disastrous.

To summarize, the three cell-cycle checkpoints have the same purpose: They prevent the division of cells that are damaged or that have other problems, and they prevent the growth of mature cells that should not grow any more. If one of the checkpoints fails, the affected cells may begin growing in an uncontrolled fashion. For the organism as a whole, the consequences of uncontrolled cell division are dire: cancer.

✓CHECK YOUR UNDERSTANDING

The cell cycle consists of four carefully controlled phases. You should be able to diagram the cell cycle and indicate (1) the locations of cell-cycle checkpoints, (2) the level of Cdk at each stage, (3) the level of cyclin at each stage, and (4) the levels of MPF activity during each stage.

11.4 Cancer: Out-of-Control Cell Division

Few diseases inspire more fear than cancer. The fear springs from the difficulty of treating many forms of cancer, the potentially fatal nature of many cancers, and their frequency. Most of us know someone who has had some form of cancer, and most of us know someone who has died from the disease. According to the American Cancer Society, 50 percent of American men

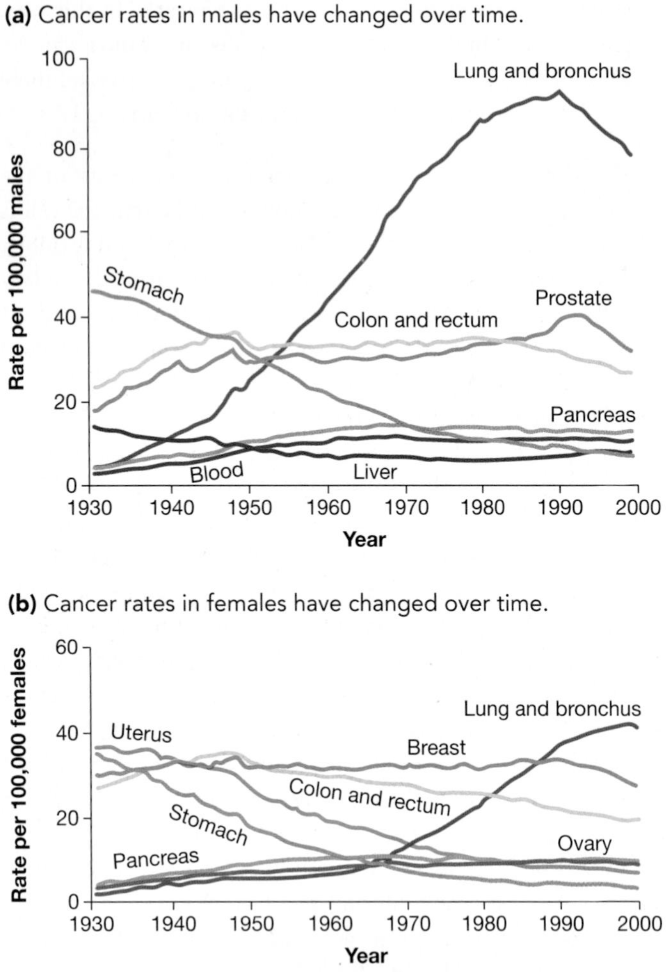

FIGURE 11.17 Changes in Cancer Rates over Time
Changes in the frequencies of various types of cancer in **(a)** men and **(b)** women in the United States.

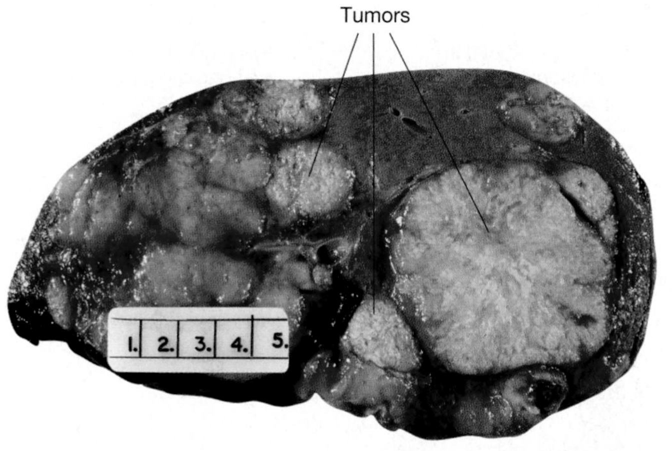

FIGURE 11.18 Cancers May Spread to New Locations in the Body
The liver of a human who died of cancer. A large number of tumors have formed due to the migration of cancerous cells from the original tumor, which occurred in the patient's colon. The numbers provide a scale in centimeters.

and 33 percent of American women will develop cancer during their lifetime (**Figure 11.17**). In the United States, one in four of all deaths are from cancer. It is the second leading cause of death, exceeded only by heart disease.

Humans suffer from at least 200 types of cancer. Stated another way, cancer is not a single illness but a complex family of diseases that affect an array of organs, including the breast, colon, brain, lung, and skin. In addition, several types of cancer can affect the same organ. Skin cancers, for example, come in multiple forms. Some are relatively easy to treat; others are often fatal. Although cancers vary in time of onset, growth rate, seriousness, and cause, all cancers have a unifying feature: They arise from cells in which cell-cycle checkpoints have failed.

To get an understanding of why cancer occurs, let's review general characteristics of the disease and then delve into the details of why regulatory mechanisms become defective.

Properties of Cancer Cells

When even a single cell in a multicellular organism begins to divide in an uncontrolled fashion, a mass of cells called a tumor results. For example, most cells in the adult human brain do not divide. But if a single abnormal brain cell begins unrestrained division, the growing tumor that results may disrupt the brain's function. What can be done? If the tumor can be removed without damaging the affected organ, a cure might be achieved. This is why surgical removal of the tumor is usually the first step in the treatment of a cancer. Often, though, surgery does not cure cancer. Why?

In addition to growing quickly, cancer cells are *invasive*—that is, able to spread throughout the body via the bloodstream or the lymphatic vessels (introduced in Chapter 49). Invasiveness is a defining feature of a **malignant tumor**—one that is cancerous. Masses of noninvasive cells are noncancerous and form **benign tumors**, such as a wart. Noncancerous cells may become malignant, however, if they gain the ability to detach from the original tumor and invade other tissues. By spreading from the primary tumor site where uncontrolled growth originated, cancer cells can establish secondary tumors elsewhere in the body (**Figure 11.18**). This process is called **metastasis.** If metastasis has occurred by the time the original tumor is detected, then secondary tumors have begun to form and surgical removal of the primary tumor will not lead to a cure. As a result, the disease can be very difficult to treat. This is why early detection as the key to treating cancer most effectively.

Cancer Involves Loss of Cell-Cycle Control

If cancer is caused by uncontrolled cell growth, what is the molecular nature of the disease? Recall that when many cells mature, they enter the G_0 phase—meaning their cell cycle is arrested at the G_1 checkpoint. In contrast, cells that do pass through the G_1 checkpoint are irreversibly committed to replicating their DNA and entering G_2. Based on this observation, biologists hypothesized that many or even most types of cancer involve defects in the G_1 checkpoint. To understand the molecular nature of the disease, then, researchers focused on understanding the normal mechanisms that operate at that checkpoint. In this way, cancer research and research on the normal cell cycle have become two sides of the same coin.

Social Control In unicellular organisms, passage through the G_1 checkpoint is thought to depend primarily on cell size and the availability of nutrients. If nutrients are plentiful, cells pass through the checkpoint and grow rapidly. In multicellular organisms, however, cells receive a constant supply of adequate nutrients via the bloodstream. Because nutrients generally are not limiting, most cells in multicellular organisms divide in response to some other type of signal. Because these signals arrive from other cells, biologists refer to *social control* over cell division. The general idea is that individual cells should be allowed to divide only when their growth is in the best interests of the organism as a whole.

The most important signals involved in social control of the cell cycle are called growth factors. **Growth factors** are polypeptides or small proteins discovered in the course of working out techniques for growing cells in culture. When researchers isolated mammalian cells in culture and provided them with adequate nutrients, the cells arrested in G_1 phase. They began to grow only when biologists added **serum**—the liquid that remains after blood clots and the blood cells have been removed. Some component of serum allowed cells to pass through the G_1 checkpoint. What was it?

In 1974 biologists succeeded in identifying one of the serum components responsible for stimulating cell division. The component was a protein called **platelet-derived growth factor (PDGF)**. As its name implies, PDGF is released by blood components called **platelets**, which promote blood clotting at wound sites. PDGF that is secreted by platelets binds to receptor tyrosine kinases in the plasma membranes of cells in the area. When these cells receive the growth signal, they are stimulated to divide. The increased cell numbers facilitate wound healing.

Researchers subsequently found that PDGF is produced by an array of cell types. Investigators also succeeded in isolating and identifying a diverse array of other growth factors. For different types of cells to grow in culture, different combinations of growth factors must be supplied. Based on this result, biologists infer that different types of cells in an intact multicellular organism are controlled by different combinations of growth factors. Cancer cells are another story, however. Cancerous cells can often be cultured successfully without externally supplied growth factors. This observation suggests that the normal social controls on the G_1 checkpoint have broken down in cancer cells.

Social Controls and Cell-Cycle Checkpoints How do growth factors stimulate division in normal cells? Recall from Section 11.3 that for a cell to begin mitosis or to pass through the G_1 checkpoint, a specific cyclin-Cdk complex must be activated. Because activation depends on the presence of a high concentration of the cyclin involved, investigators hypothesized that growth factors trigger cyclin synthesis. This hypothesis turned out to be correct. As **Figure 11.19** shows, the arrival of growth factors initiates cell division by stimulating the production of cyclins.

The link between growth factors and cyclin synthesis also turned out to be important in cancer biology. In some human cancers, the cyclin that is active at the G_1 checkpoint is overproduced. Cyclin overproduction can result from the presence of excessive amounts of growth factors or from cyclin production in the absence of growth signals.

When the G_1 cyclin is overproduced, the Cdk to which it binds phosphorylates target proteins continuously. In many cancers, the key target protein is the retinoblastoma (Rb) protein. This protein was discovered by researchers interested in a childhood cancer called retinoblastoma, which occurs in about 1 in 20,000 children. The disease is characterized by the appearance of malignant tumors in the light-sensing tissue, or retina, of the eye.

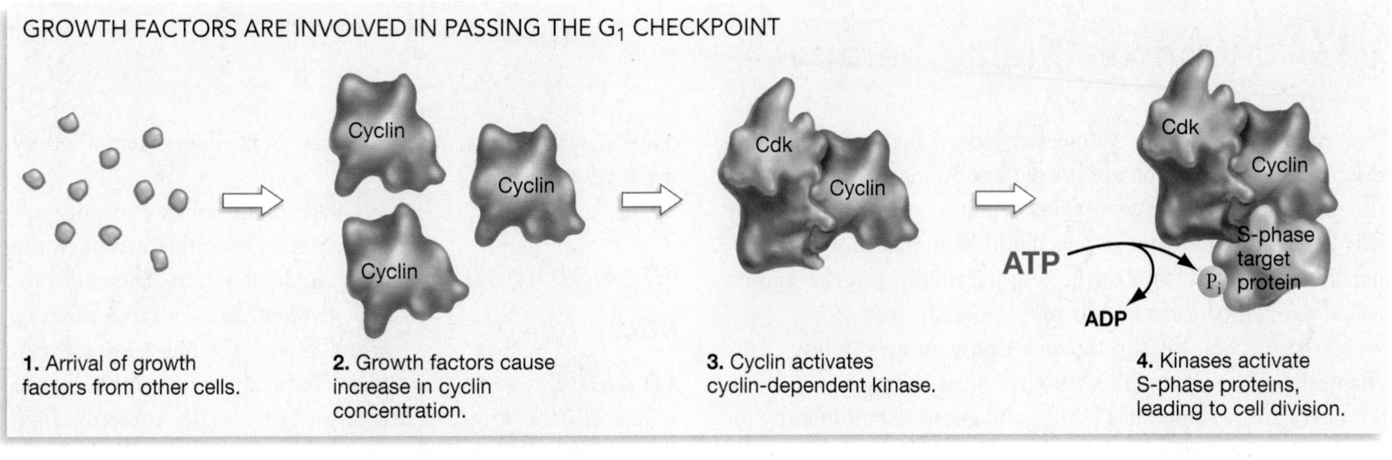
GROWTH FACTORS ARE INVOLVED IN PASSING THE G1 CHECKPOINT
Cyclin
Cyclin
Cyclin
Cdk
Cyclin
Cdk
Cyclin
S-phase
target
protein
ATP
ADP
Pi
1. Arrival of growth factors from other cells.
2. Growth factors cause increase in cyclin concentration.
3. Cyclin activates cyclin-dependent kinase.
4. Kinases activate S-phase proteins, leading to cell division.

FIGURE 11.19 How Is the G_1 Checkpoint Controlled?

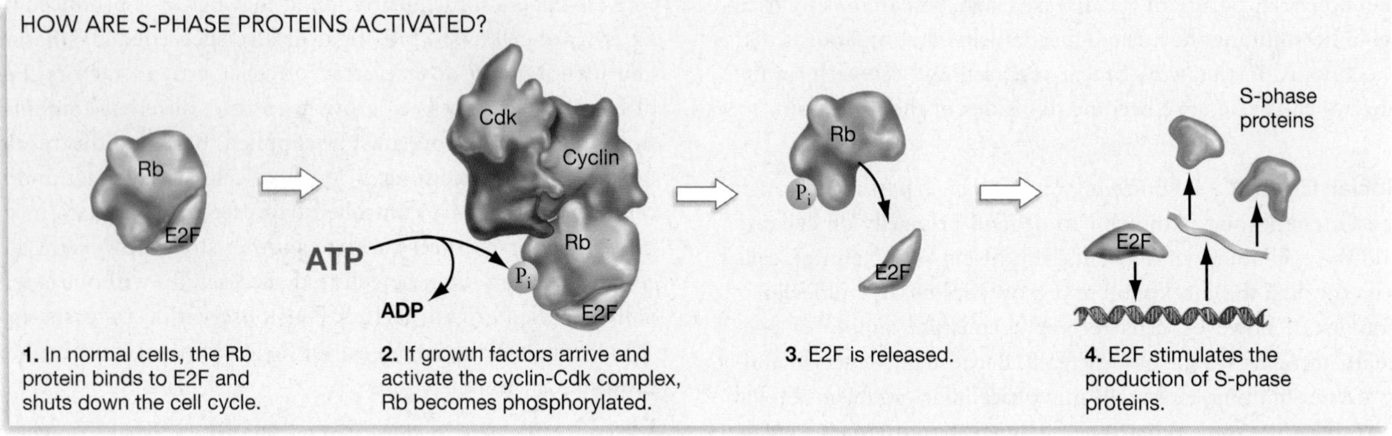

HOW ARE S-PHASE PROTEINS ACTIVATED?
Rb
E2F
Cdk
Cyclin
Rb
ATP
ADP
P_i
E2F
Rb
P_i
E2F
S-phase proteins
E2F
1. In normal cells, the Rb protein binds to E2F and shuts down the cell cycle.
2. If growth factors arrive and activate the cyclin-Cdk complex, Rb becomes phosphorylated.
3. E2F is released.
4. E2F stimulates the production of S-phase proteins.

FIGURE 11.20 The Role of Rb Protein in Cell Division

QUESTION What would happen if the Rb protein were defective or not produced at all? What would happen if cyclin concentrations were high all the time, irrespective of the arrival of growth factors?

The role of Rb protein in controlling normal cell division is diagrammed in **Figure 11.20**. In normal, nondividing cells, Rb binds to a protein called E2F and inhibits it—shutting down the cell cycle. In this way, Rb acts as a stop signal for cell growth. But if a growth factor stimulates the production of cyclin and activates the cyclin-Cdk complex, the Rb protein becomes phosphorylated. In response, it no longer binds to E2F. Once it is free of Rb, E2F begins stimulating the production of molecules needed in S phase.

In effect, a cell's commitment to dividing depends on whether Rb stays bound to the E2F protein. As long as Rb binds to E2F, the cell remains in G_1 phase. Rb acts as a brake on the cell cycle, much as p53 does. Rb is also classified as a tumor suppressor, because its normal function in the cell is to shut down the cell cycle. But if Rb is defective, the E2F protein is always active and cell division is uncontrolled. This is exactly what happens in several types of cancer. In retinoblastoma and some breast, bladder, and lung cancers, the Rb protein is nonfunctional in the cancerous cells or not produced at all.

To summarize, overproduction of cyclins or defects in Rb itself can lead to the inactivation of Rb as a tumor suppressor. The result is failure of the G_1 checkpoint, uncontrolled cell growth, and cancer.

Cancer Is a Family of Diseases One of the broadest messages to come out of research on the Rb protein is that a wide variety of defects can lead to the failure of the G_1 checkpoint and the onset of cancer. In addition, researchers now realize that cancer is seldom due to a single defect. Most cancers develop only after a number of genetic errors combine to break cell-cycle control and induce uncontrolled growth and metastasis. Each type of cancer is due to a unique combination of errors. Stated another way, cancer can be caused by hundreds if not thousands of different defects.

Because cancer is actually a family of diseases with a complex and highly variable molecular basis, there will be no "magic bullet," or single therapy that cures all forms of the illness. Still, recent progress in understanding the cell cycle and the molecular basis of cancer has been dramatic, and cancer prevention and early detection programs are increasingly effective. The prognosis for many cancer patients is remarkably better now than it was even a few years ago.

ESSAY Cancer Chemotherapy

If a malignant tumor is detected before metastasis occurs, doctors may be able to cure the patient by surgically removing the growth. But if the cancer has begun to spread by the time it is detected, stray cancer cells could be hiding anywhere, regardless of the location of the original tumor. For the patient to survive, all cancerous cells must be eradicated.

In these cases, doctors often propose surgery followed by **chemotherapy**—treatment with anticancer chemicals or drugs. You may know someone who has undergone chemotherapy for cancer. If so, you've witnessed the side effects: weakness, acute nausea, weight loss, and hair loss. Why does chemotherapy make people so sick?

Why does chemotherapy make people so sick?

Most of the drugs currently approved for chemotherapy have the same mode of action: They kill rapidly dividing cells or cause them to arrest in the G_1 phase of the cell cycle. To heighten the probability of killing all stray cancer cells, physicians usually prescribe high doses of these drugs. They are applied to the entire body, often through injection into the bloodstream. Besides killing cancer cells, though, anticancer drugs also kill normal cells that divide rapidly. These include blood cells, intestinal cells, and hair follicles. As a result, chemotherapy leads to low blood-cell counts, loss of appetite, and loss of hair, among other symptoms.

Physicians who prescribe chemotherapy must balance the probability of eradicating all cancer cells with the patient's ability to tolerate the side effects of treatment. One approach to achieving this balance is to mitigate the most dangerous side effects, such as low counts of the white blood cells that battle infections. Patients with low blood-cell counts may be given growth factors that stimulate the production of white blood cells. These patients may also be given antibiotics to help them fight infections.

Another approach in chemotherapy is to prevent side effects altogether by targeting the drugs more precisely. If the cancer has not spread far from the original tumor site, anticancer drugs may be applied to just a limited region of the body. This is rarely possible, however, so many researchers are focusing on an alternative strategy—attaching chemotherapy drugs to proteins that bind to specific membrane proteins on the surfaces of cancer cells. One idea is to package drugs inside liposomes that are coated with a protein that binds to cancer cells. The hope is that the liposomes will fuse with the plasma membrane of the cancer cells and deliver the drugs directly to the malignant cells. If this research is successful, chemotherapy may eventually become largely free of side effects and patient suffering.

CHAPTER REVIEW

Summary of Key Concepts

In the late nineteenth century, observations of growing embryos revealed that cells reproduce by division of preexisting cells. Mitosis is the series of events that results in a complete copy of the chromosomes being distributed to daughter nuclei. Cytokinesis is the cell division that usually follows mitosis.

- **After chromosomes are copied, mitosis distributes one chromosome copy to each of two daughter cells. Mitosis and cytokinesis produce two cells that are identical to the parent cell.**

Most unicellular eukaryotes and some multicellular eukaryotes reproduce by mitosis followed by cytokinesis. Mitotic divisions are also responsible for building the bodies of multicellular organisms from a single-celled embryo.

Mitosis can be described as a sequence of five phases:

Prophase Chromosomes condense, and the mitotic spindle apparatus begins to form.

Prometaphase The nuclear envelope disintegrates, and spindle fibers make contact with chromosomes.

Metaphase Spindle fibers move chromosomes to the metaphase plate, an imaginary plane.

Anaphase Spindle fibers pull sister chromatids apart.

Telophase Chromosomes are pulled to opposite poles of the cell, and a nuclear envelope forms around each set.

In most cells, mitosis is followed immediately by division of all cell contents, or cytokinesis.

Web Tutorial 11.1 The Phases of Mitosis

- **Over their life span, cells go through a life cycle that consists of four carefully controlled phases.**

Dividing cells alternate between the dividing phase, called M phase, and a nondividing phase known as interphase. This alternation describes the cell cycle—the ordered sequence of events that occur as a cell duplicates its chromosomes and divides.

Chromosome synthesis and replication, or S phase, occurs during interphase. S phase and M phase are separated by gap phases called G_1 and G_2. Cell growth and replication of nonnuclear cell components occur during the gap phases. The order of cell-cycle phases is $G_1 \rightarrow S \rightarrow G_2 \rightarrow M \rightarrow G_1$ and so forth.

Progression through the cell cycle is regulated at three checkpoints. Passage through the G_1 checkpoint represents an irreversible commitment to divide and is contingent on cell size, nutrient availability, lack of DNA damage, and/or growth signals from other cells. At the G_2 checkpoint, progression through the cycle is delayed until chromosome replication has been successfully completed. At the metaphase checkpoint during M phase, anaphase is delayed until all chromosomes are correctly attached to the mitotic spindle.

An array of different molecules regulates progression through the cell-cycle checkpoints. The most important of these regulatory molecules are cyclin-dependent kinases (Cdks) and cyclins, which are found in all eukaryotic cells. Active Cdks are enzymes that trigger progress through a checkpoint by phosphorylating important target proteins. Cyclins are proteins whose concentrations oscillate during the cell cycle, regulating the activity of Cdks. In multicellular organisms, cyclin concentrations are partially controlled by growth factors from other cells. As a result, the G_1 checkpoint is said to be under social control.

Web Tutorial 11.2 The Cell Cycle

- **Uncontrolled cell growth leads to cancer. Different types of cancer result from different types of defects in control over the cell cycle.**

Cancer is a common disease that is characterized by loss of control at the G_1 checkpoint, resulting in cells that divide in an uncontrolled fashion. Some cancer cells also have the ability to spread throughout the body, which makes treatment difficult. In normal cells, growth-factor signals are required to activate G_1 cyclin-Cdk complexes and trigger division. An important target of activated G_1 cyclin-Cdk complexes is the retinoblastoma (Rb) protein, which regulates the production of enzymes and other proteins required for S phase. Defects in G_1 cyclin and Rb protein are common in human cancers. Cancerous cells usually have defects in several genes involved in cell-cycle control, however.

Questions

Content Review

1. Which statement about the daughter cells of mitosis is correct?
   a. They differ genetically from one another and from the parent cell.
   b. They are genetically identical to one another and to the parent cell.
   c. They are genetically identical to one another but are different from the parent cell.
   d. Only one of the two daughter cells is genetically identical to the parent cell.
2. Progression through the cell cycle is regulated by oscillations in the concentration of which type of molecule?
   a. actin
   b. tubulin
   c. cyclin-dependent kinases
   d. cyclins
3. According to the data in Figure 11.4, the first labeled mitotic cells appear about 4 hours after the labeling period ends. From these data, researchers concluded that G_2 lasted about 4 hours. Why?
   a. The lengths of the total cell cycle and the G_1, S, and M phases were known, so the length of G_2 could be found by subtraction.
   b. It confirmed other data indicating that G_2 lasts about 4 hours in cultured cells of this type.
   c. Cyclins were labeled, so their concentration had to build up for 4 hours.
   d. Labeled cells are in S phase, so 4 hours passed between the end of S phase and the onset of M phase.
4. What major events occur during anaphase of mitosis?
   a. Chromosomes replicate, so each chromosome consists of two identical sister chromatids.
   b. Chromosomes condense and the nuclear envelope disappears.
   c. The chromosomes end up at opposite ends of the cell and two nuclear envelopes form.
   d. Sister chromatids separate, forming independent chromosomes.
5. What evidence suggests that during anaphase, spindle fibers shorten at the kinetochore and not at the base of the mitotic spindle?
   a. Motor proteins are located at the kinetochore.
   b. Motor proteins are located at the kinetochore *and* at the base of the mitotic spindle.
   c. When fluorescing microtubules are bleached in the middle, the bleached segment stays stationary as the fibers shorten.
   d. When fluorescing microtubules are bleached in the middle, the bleached segment moves toward the base of the mitotic spindle as the fibers shorten.
6. The normal function of the Rb protein is to bind to E2F, inactivating it. What are the consequences of defects in Rb?
   a. It binds more tightly to E2F, so E2F cannot stimulate the production of molecules needed in S phase. The cell cannot progress through the cell cycle.
   b. It no longer responds to phosphorylation—it cannot be regulated.
   c. It does not bind to E2F properly, so E2F continuously stimulates the production of molecules needed in S phase.
   d. Rb acts as a tumor suppressor, similar in function to p53.

Conceptual Review

1. Sketch the phases of mitosis, listing the major events of each phase. Identify at least two events that must be completed successfully for daughter cells to share an identical complement of chromosomes.
2. What are the consequences for the cell if the G_1 checkpoint fails? Answer the same question for the G_2 and metaphase checkpoints.
3. Explain how cell fusion and microinjection experiments supported the hypothesis that specific molecules are involved in the transition from interphase to M phase.
4. Why are most protein kinases considered regulatory proteins?
5. Why are cyclins called cyclins? Explain their relationship to cyclin-dependent kinases and to growth factors.
6. Early detection is the key to surviving most cancers. Why?

Group Discussion Problems

1. In multicellular organisms, nondividing cells stay in G_1 phase. For the cell, why is it better to be held in G_1 rather than S, G_2, or M phase?
2. The Rb protein helps regulate the cell cycle in many types of cells. Children with hereditary retinoblastoma have a defective version of the Rb protein but get tumors only in their eyes—not elsewhere. Suggest a hypothesis to explain why tumors start only in the retinas of children. How could you test your hypothesis?
3. Predict the outcome of an experiment involving the fusion of a cell in G_1 phase with a cell in G_2 phase. What would happen to the G_1-phase nucleus? To the G_2-phase nucleus? Why?
4. Cancer is primarily a disease of older people. Further, a group of individuals may share a genetic predisposition to developing certain types of cancer yet vary a great deal in time of onset—or not get the disease at all. Discuss these observations in light of the claim that several defects usually have to occur for cancer to develop.

Answers to Multiple-Choice Questions 1. b; 2. d; 3. d; 4. d; 5. c; 6. c

www.prenhall.com/freeman is your resource for the following: Web Tutorials; Online Quizzes and other Online Study Guide materials; Answers to Conceptual Review Questions; Solutions to Group Discussion Problems; Answers to Figure Caption Questions and Exercises; and Additional Readings and Research.

UNIT 3

Gene Structure and Expression

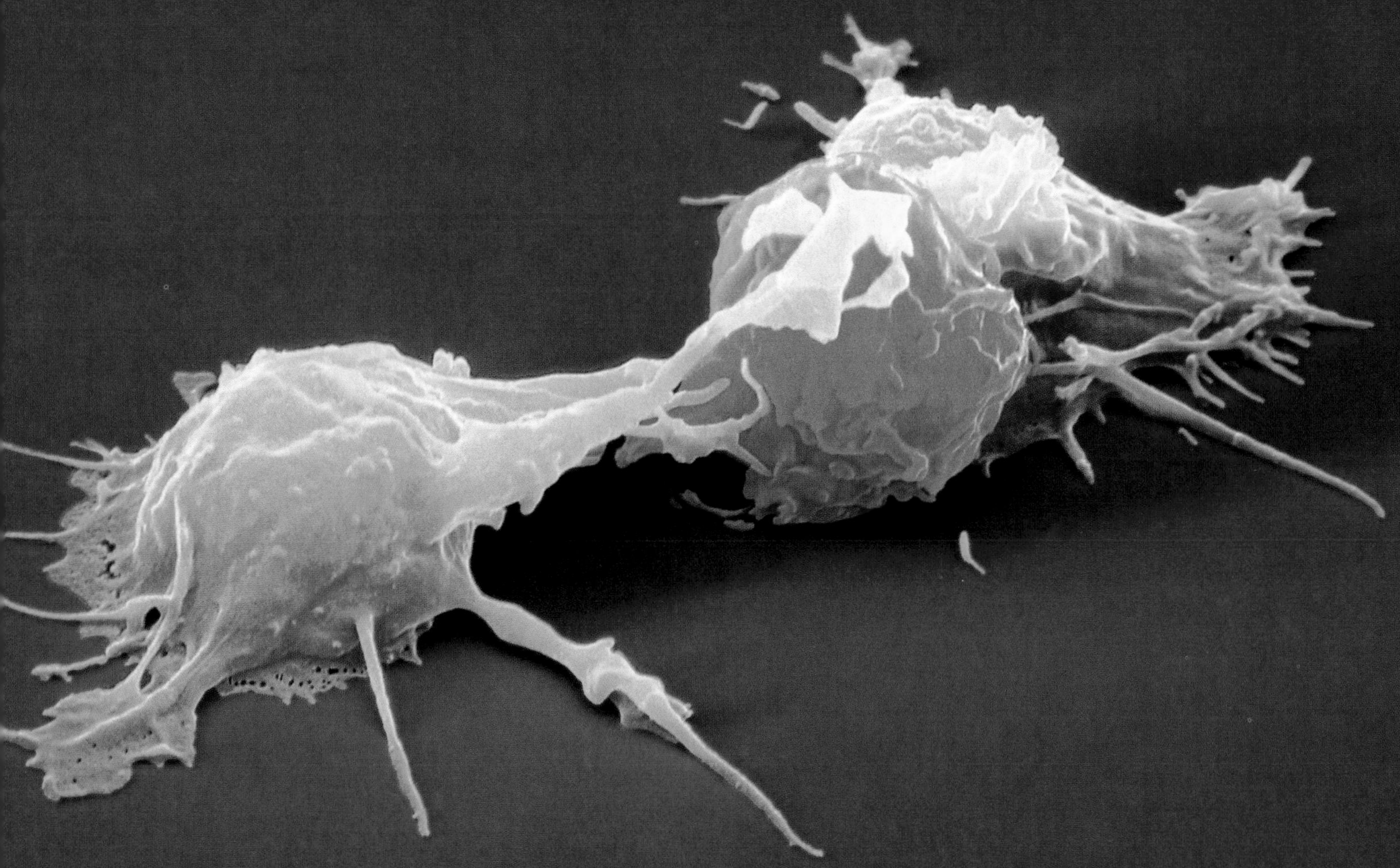

Two immune system cells attack a cancerous cell (colored red) in this micrograph. The three cells contain the same genes. Their shapes and behaviors differ, however, because they are expressing different genes.

12 Meiosis

KEY CONCEPTS

- Meiosis is a type of nuclear division. It results in cells that have half as many chromosomes as the parent cell, and in animals it occurs prior to the formation of eggs and sperm. When an egg and a sperm cell combine to form an offspring, the original number of chromosomes is restored.
- Each cell produced by meiosis receives a different combination of chromosomes. Because genes are located on chromosomes, each cell produced by meiosis receives a different complement of genes. Meiosis leads to offspring that are genetically distinct from each other and from their parents. If offspring are genetically variable, some might thrive if the environment changes.
- If mistakes occur during meiosis, the resulting egg and sperm cells may contain the wrong number of chromosomes. It is rare for offspring with an incorrect number of chromosomes to develop normally.

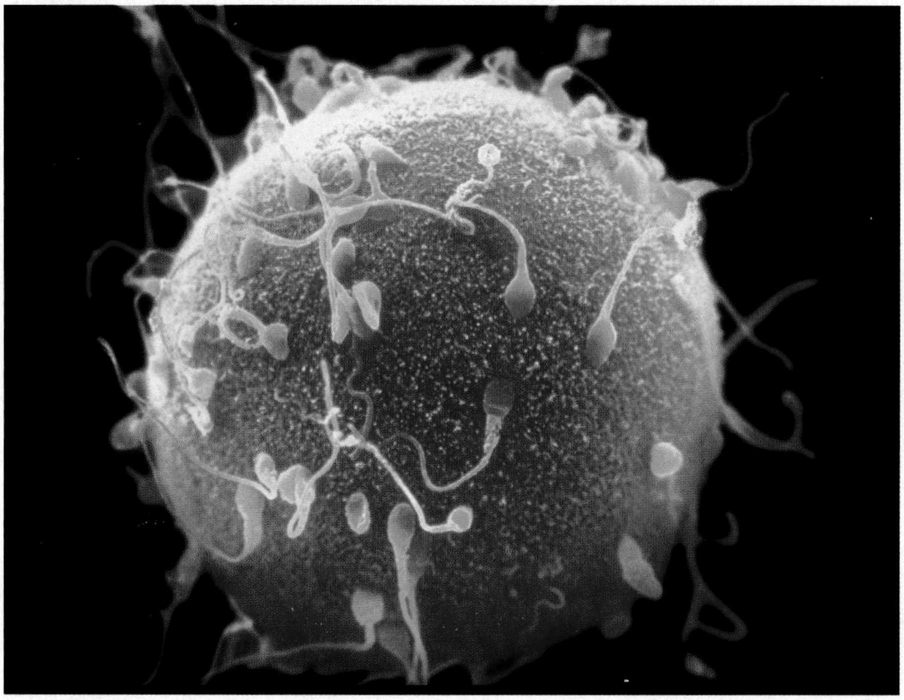

Scanning electron micrograph showing human sperm attempting to enter an egg. This chapter introduces the type of nuclear division called meiosis, which occurs prior to gamete formation in animals.

Why sex?

Simple questions—such as why sexual reproduction exists—are sometimes the best. This chapter asks what sexual reproduction is and why some organisms employ it. Understanding sex has been one of the great mysteries of classical and contemporary biology. After decades of effort, clear answers to the question of why sex exists are finally beginning to emerge.

Before we plunge ahead, it is important to realize that there are two basic levels of explanation in biology: ultimate explanation and proximate explanation. An **ultimate explanation** is evolutionary in nature. It explains *why* something happens. In terms of sex, biologists want to know why it evolved. This issue is important because most organisms do not reproduce sexually or do so only occasionally. Presumably, sexual reproduction increases the ability of certain organisms to produce offspring successfully in particular environments. Why is sex advantageous for some species? This question is the focus of Sections 12.2 and 12.3.

A **proximate explanation** is mechanistic in nature. It explains *how* something happens. In terms of sex, biologists want to know how the process of sexual reproduction occurs. They want to understand the molecular and cellular structures and events that are involved. For example, it has long been understood that during sexual reproduction, a male reproductive cell—a **sperm**—and a female reproductive cell—an **egg**—unite to form a new individual. The process of uniting sperm and egg is called **fertilization**. The first biologists to observe fertilization studied the large, translucent eggs of sea urchins. Due to the semitransparency of the sea urchin egg cell, researchers were able to see the nuclei of a sperm and an egg fuse.

When these results were published in 1876, they raised an important question at the proximate level, concerning the number of chromosomes found in parents, sperm, eggs, and offspring. Cell biologists had already established that the number of chromosomes is constant from cell to cell within a multicellular organism. It was also accepted that chromosome number is the same in the parent and daughter cells of mitosis. Biologists confirmed that all of the cells in a newly growing offspring, or **embryo**, are the products of mitotic divisions, and that all cells in the body are the direct descendants of the nucleus that formed at fertilization. The question is, How can the chromosomes from a sperm cell and an egg cell combine, but form an offspring that has the same chromosome number as its mother and its father?

A hint at the answer came in 1883, when a researcher noted that cells in the body of roundworms of the genus *Ascaris* have four chromosomes, while their sperm and egg nuclei have only two chromosomes apiece. Four years later, August Weismann formally proposed a hypothesis to explain the riddle: During the formation of **gametes**—reproductive cells such as sperm and eggs—there must be a distinctive type of cell division that leads to a reduction in chromosome number. Specifically, if the sperm and egg contribute an equal number of chromosomes to the fertilized egg, Weismann reasoned, they must each contain half of the usual number of chromosomes. Then, when sperm and egg combine, the resulting cell has the same chromosome number as its mother's cells and its father's cells have.

In the decades that followed, biologists confirmed this hypothesis by observing gamete formation in a wide variety of plant and animal species. Eventually this form of cell division came to be called meiosis ("lessening-act"). **Meiosis** is nuclear division that leads to a halving of chromosome number. It precedes the formation of eggs and sperm, and provides a satisfying proximate explanation for how sexual reproduction occurs. Let's begin with a closer look at how meiosis occurs, then shift to the ultimate level of explanation and ask why it exists.

12.1 How Does Meiosis Occur?

When cell biologists began to study the cell divisions that lead to gamete formation, they made an important observation: Each organism has a characteristic number of chromosomes. Consider the drawing in **Figure 12.1**, based on a paper published by Walter Sutton in 1902. It shows the chromosomes of the lubber grasshopper during the cell divisions leading up to the formation of a sperm. There is a total of 24 chromosomes in the cell. Sutton realized, however, that there are just 12 dis-

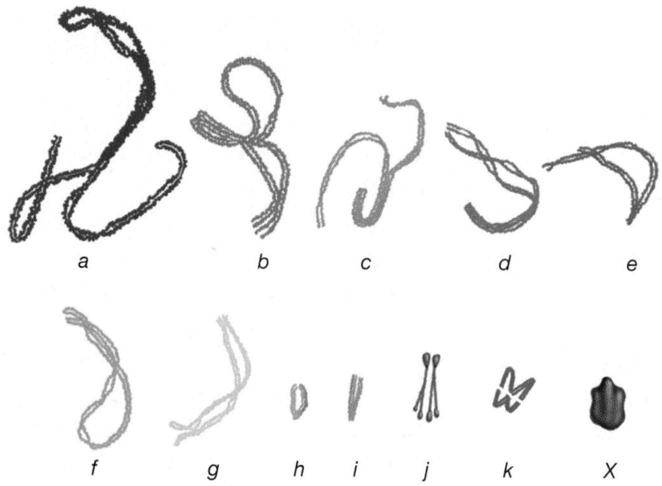

FIGURE 12.1 Cells Contain Different Types of Chromosomes, and Chromosomes Come in Pairs
Letters designate each of the 12 distinct types of chromosomes found in lubber grasshopper cells. There are two of each type of chromosome. The two members of a chromosome pair are called homologs. **EXERCISE** Draw a line over one of the two homologs in each of the 12 pairs shown, to distinguish the two homologs from each other.

tinct types of chromosomes based on size and shape. In this species, there were two chromosomes of each type.

Sutton designated 11 of the chromosomes by the letters *a* through *k* and the twelfth by the letter X. He referred to the pairs of chromosomes as **homologous chromosomes.** (They can also be called **homologs.**) The two chromosomes labeled *c*, for example, have the same size and shape and are homologous.

In Figure 12.1, most of the homologous chromosome pairs happen to be joined at several places. The two chromosomes designated X are joined so tightly that they look like a single chromosome. The two X chromosomes are associated with the sex of the individual, and each is designated a **sex chromosome.** Two types of sex chromosomes, known as X and Y, exist in lubber grasshoppers. Males have one X and one Y chromosome, while females have two X chromosomes. Non-sex chromosomes, such as *a*–*k* in a grasshopper cell, are known as **autosomes.**

Later work showed that homologous chromosomes are similar not only in size and shape but also in content. Homologous chromosomes carry the same genes. A **gene** is a section of DNA that influences one or more hereditary traits in an individual. For example, each copy of chromosome *c* found in lubber grasshoppers might carry genes that influence eye formation, body size, singing behavior, or jumping ability. The versions of a gene found on homologous chromosomes may differ, however. Biologists use the term **allele** to denote different versions of a particular gene. For example, the alleles on each copy of chromosome *c* in a lubber grasshopper might contribute to rounder eyes versus narrower eyes, larger body size versus smaller body size, faster songs

BOX 12.1 Karyotyping Techniques

Although chromosomes maintain their individuality and physical integrity throughout the cell cycle, they are readily visible only during mitosis or meiosis. Early in cell division, chromosomes condense into highly compact structures that can be observed with the light microscope. To describe an individual's karyotype, then, biologists must study cells undergoing cell division.

The first step in generating a karyotype is to obtain a sample of cells from the individual being studied. Cancer researchers might collect cells from a tumor. Physicians who are concerned about the possibility of birth defects might obtain a few cells from the developing embryo inside the mother. The next step is to grow the cells in culture, using techniques introduced in Chapter 11. When the cells are dividing rapidly, they are treated with a compound called colchicine. Colchicine stops mitosis at metaphase by disrupting the formation of the mitotic spindle. At this stage the chromosomes are easy to study, because they are condensed and consist of sister chromatids. The chromosomes of colchicine-treated cells are then stained and examined with the light microscope.

Researchers can distinguish stained chromosomes by size; by the position of the *centromere*, which holds sister chromatids together; and by striping or banding patterns. Subtler differences among chromosomes are apparent when a higher-resolution technique for karyotyping called **spectral karyotyping (SKY)**, or **chromosome painting**, is used. The "painting" is done with fluorescent dyes that are attached to short DNA molecules. The dyed pieces of DNA bind to particular regions of particular chromosomes. By using a combination of dyes, technicians can give each pair of homologous chromosomes a distinctive suite of colors (**Figure 12.2a**). The high-resolution image produced by this technique allows clinicians to diagnose an array of chromosomal abnormalities. Consider some examples:

- **Figure 12.2b** shows chromosomal changes associated with a cancer called *chronic myelogenous leukemia*. The defect involved is a **translocation**—a swapping of chromosome segments. In this case, a small piece of chromosome 9 and a large piece of chromosome 22 have changed places. The translocation causes a genetic change that leads to uncontrolled cell growth.
- **Figure 12.2c** shows the karyotype of an individual with **Klinefelter syndrome**, which develops in people who have two X chromosomes and a Y chromosome instead of a single X and a single Y. People with this syndrome have male sex organs but are sterile. They may also develop some female-like characteristics, such as enlarged breasts.

Karyotyping is an important diagnostic tool. As techniques for generating karyotypes improve, our ability to detect and interpret chromosomal defects also improves.

versus slower songs, and so on. Homologous chromosomes carry the same genes, but each homolog may contain different alleles.

At this point in his study, Sutton had succeeded in determining the lubber grasshopper's **karyotype**—meaning the number and types of chromosomes present (see **Box 12.1**). As karyotyping studies expanded, cell biologists realized that, like lubber grasshoppers, the vast majority of plants and animals have more than one of each type of chromosome. These investigators designated terms to identify the number of chromosome copies they observed. Organisms such as lubber grasshoppers, humans, and cedar trees are called **diploid** (literally, "double-form"), because they have two versions of each type of chromosome. Diploid organisms have two alleles of each gene—one on each of the homologous pairs of chromosomes. Organisms such as bacteria, archaea, and many algae are called **haploid** ("single-form"), because their cells contain just one of each type of chromosome. Haploid organisms do not contain homologous chromosomes. They have just one allele of each gene.

Researchers also invented a compact notation to indicate the number of chromosome sets in a particular organism or type of cell. By convention, the letter n stands for the number of distinct types of chromosomes in a given cell and is called the **haploid number**. If sex chromosomes are present, they are counted as a single type in the haploid number. To indicate the number of complete chromosome sets observed, a number is placed before the n. Thus, a cell can be n, or $2n$, or $3n$, and so on. The combination of a number and n is termed the cell's **ploidy**. Diploid cells or species are designated $2n$, because two chromosomes of each type are present—one from each parent. Haploid cells or species are labeled simply n, because they have just one set of chromosomes—no homologs are observed. In haploid cells, the number 1 in front of n is implied and is not written out.

Later work revealed that it is common for species in some lineages—particularly land plants such as ferns—to contain more than two of each type of chromosome. Instead of having two homologous chromosomes per cell, as many organisms do, **polyploid** ("many-form") species may have three or more of each type of chromosome in each cell. Depending on the number of homologs present, such species are called *triploid* ($3n$), *tetraploid* ($4n$), *hexaploid* ($6n$), *octoploid* ($8n$), and so on.

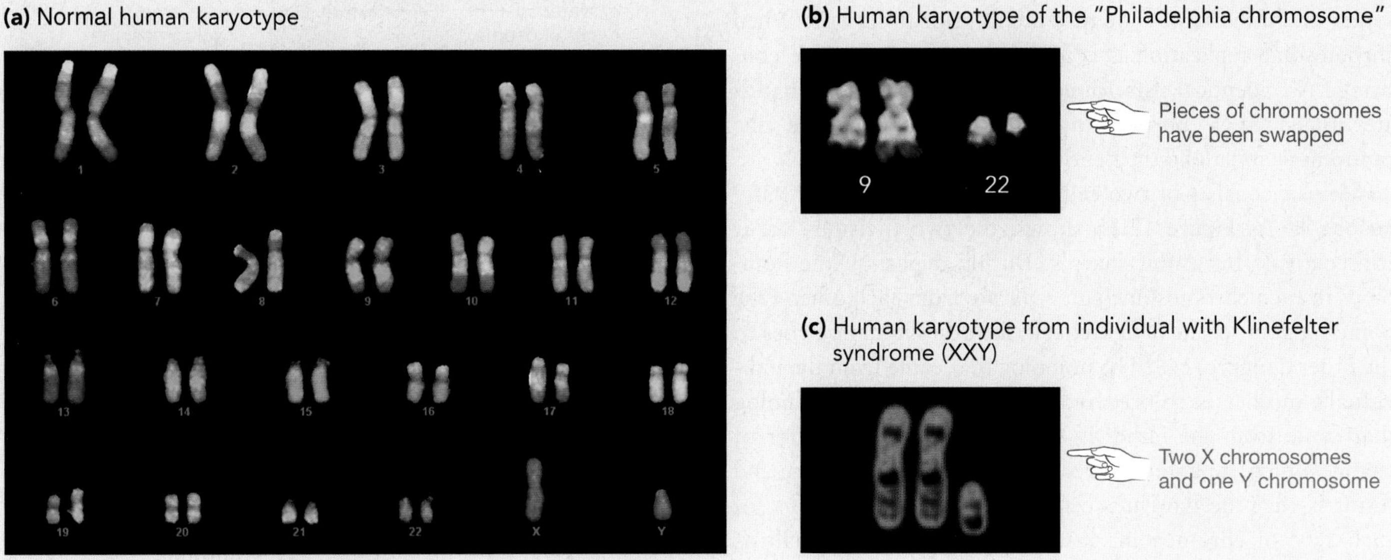
(a) Normal human karyotype
(b) Human karyotype of the "Philadelphia chromosome"
9
22
Pieces of chromosomes have been swapped
(c) Human karyotype from individual with Klinefelter syndrome (XXY)
Two X chromosomes and one Y chromosome

FIGURE 12.2 Human Chromosomes
(a) Condensed chromosomes that are undergoing mitosis are arranged randomly when first observed with the microscope. To determine a karyotype, a technician uses a computer to separate the image of each chromosome. The images of homologous chromosomes are then placed side by side, and the homologous pairs are arranged by number. **(b)** In the individual with this karyotype, the purple sections of chromosome 9 and chromosome 22 have been swapped via translocation. **(c)** Individuals with Klinefelter syndrome have two X chromosomes and a Y chromosome. QUESTION Is the karyotype in part (a) normal, or does it reveal the presence of an extra chromosome or the absence of a chromosome?

Why some species are haploid versus diploid or tetraploid is currently the subject of debate and research.

To summarize, the haploid number n indicates the number of distinct types of chromosomes present. Different species have different haploid numbers (see **Table 12.1**). Human cells contain 23 distinct types of chromosomes, so $n = 23$. In the grasshopper cells that Sutton studied, $n = 12$. In contrast, a cell's ploidy (n, $2n$, $3n$, etc.) indicates the number of each type of chromosome present. Stating a cell's ploidy is the same as stating the number of haploid chromosome sets present. Because most human and grasshopper cells contain two of each type of chromosome, they are diploid. In humans $2n = 46$; in Sutton's grasshoppers $2n = 24$. Ploidy refers to the number of each type of chromosome present; haploid number identifies how many different types of chromosomes occur.

Table 12.2, page 253, summarizes the vocabulary that biologists use to describe the number and types of chromosomes found in a cell.

Sutton and the other early cell biologists did more than describe the karyotypes observed in their study organisms, however. Through careful examination, they were able to track how chromosome numbers change during meiosis. These studies confirmed Weismann's hypothesis that a special type of cell division occurs during gamete formation. This result was a major advance in understanding sex at the proximate level.

TABLE 12.1 The Number of Chromosomes Found in Some Familiar Organisms

Organism	Number of Different Types of Chromosomes (haploid number n)	Diploid Chromosome Number ($2n$)
Humans	23	46
Domestic dog	36	72
Fruit fly	4	8
Chimpanzee	24	48
Bulldog ant	1	2
Garden pea	7	14
Corn (maize)	10	20

WEB TUTORIAL 12.1 Meiosis

An Overview of Meiosis

Recall that cells replicate each of their chromosomes before undergoing meiosis. At the start of the process, then, chromosomes are in the same state they are in prior to mitosis. When chromosome replication is complete, each chromosome consists of two identical **sister chromatids.** These sister chromatids are joined at a portion of the chromosome called the **centromere** and along their entire length (**Figure 12.3a**).

Meiosis consists of two cell divisions, called **meiosis I** and **meiosis II.** As **Figure 12.3b** shows, the two divisions occur consecutively but differ sharply. During meiosis I, the homologs in each chromosome pair separate from each other. One homolog goes to one daughter cell; the other homolog goes to the other daughter cell. The homolog that came from the individual's mother is colored red in Figure 12.3; the homolog that came from the father is colored blue. It is a matter of chance which daughter cell receives which homolog. The end result is that the daughter cells from meiosis I have one of each type of chromosome instead of two, and thus half as many chromosomes as the parent cell has. During meiosis I, the diploid ($2n$) parent cell produces two haploid (n) daughter cells. Each chromosome still consists of two identical sister chromatids, however.

During meiosis II, sister chromatids from each chromosome separate. One sister chromatid goes to one daughter cell; the other sister chromatid goes to the other daughter cell. The cell that starts meiosis II has one of each type of chromosome, but each chromosome has been replicated (meaning it still consists of two sister chromatids). The cells produced by meiosis II also have one of each type of chromosome, but now the chromosomes are unreplicated. Recall from Chapter 11 that a chromosome is considered one chromosome whether it is replicated or unreplicated. This point is crucial. A replicated chromosome consists of two sister chromatids but still represents just one chromosome.

To reiterate, sister chromatids separate during meiosis II, just as they do during mitosis. Meiosis II is actually equivalent to mitosis occurring in a haploid cell. As in mitosis, chromosome movements during meiosis I and II are caused by spindle fibers that attach at the centromere of each chromosome.

Sutton and a host of other early cell biologists worked out this sequence of events through careful observation of cells with the light microscope. Based on these studies, they came to a key realization: The outcome of meiosis is a reduction in chromosome number. For this reason, meiosis is known as a *reduction division*. In most plants and animals, the original cell is diploid and the four daughter cells are haploid. These four haploid daughter cells, each containing one of each homologous chromosome, eventually go on to form egg cells or sperm cells via a process called **gametogenesis** ("gamete-origin"), which is described in Chapter 21. When two gametes fuse during fertilization, a full complement of chromosomes is restored (**Figure 12.3c**). The cell that results from fertilization is diploid

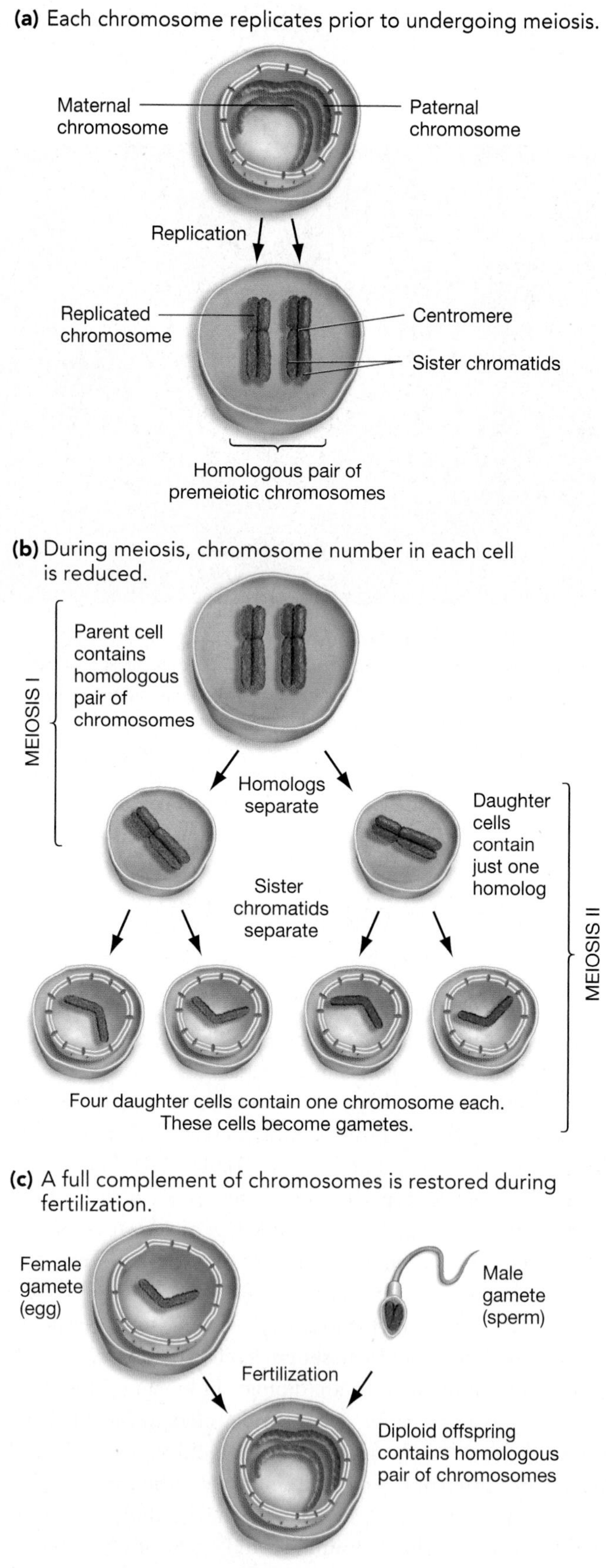
(a) Each chromosome replicates prior to undergoing meiosis.
Maternal chromosome
Paternal chromosome
Replication
Replicated chromosome
Centromere
Sister chromatids
Homologous pair of premeiotic chromosomes
(b) During meiosis, chromosome number in each cell is reduced.
MEIOSIS I
Parent cell contains homologous pair of chromosomes
Homologs separate
Daughter cells contain just one homolog
Sister chromatids separate
MEIOSIS II
Four daughter cells contain one chromosome each. These cells become gametes.
(c) A full complement of chromosomes is restored during fertilization.
Female gamete (egg)
Male gamete (sperm)
Fertilization
Diploid offspring contains homologous pair of chromosomes

FIGURE 12.3 The Major Events in Meiosis
Meiosis reduces chromosome number by half. In diploid organisms, the products of meiosis are haploid. EXERCISE In parts (b) and (c), write *n* or 2*n* next to each cell to indicate its ploidy.

TABLE 12.2 Vocabulary Used to Indicate the Chromosomal Makeup of a Cell

Term	Definition	Example or Comment
Chromosome	Structure made up of DNA and proteins; carries the cell's hereditary information (genes)	Eukaryotes have linear, threadlike chromosomes; bacteria and archaea often have just one circular chromosome
Sex chromosome	Chromosome associated with an individual's sex	X and Y chromosomes of humans: Males are XY, females XX. Also, Z and W chromosomes of birds: Males are ZZ, females ZW.
Autosome	A non-sex chromosome	Chromosomes 1–22 in humans.
Unreplicated chromosome	A chromosome that consists of a single copy (in eukaryotes, a single "thread")	
Replicated chromosome	A chromosome that has been copied; consists of two linear structures joined at the centromere	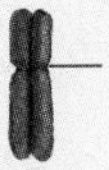
Sister chromatids	The chromosome copies in a replicated chromosome	
Homologous chromosomes (homologs)	In a diploid organism, chromosomes that are similar in size, shape, and gene content	
Non-sister chromatids	The chromosome copies in homologous chromosomes	
Tetrad	Synapsed homologous chromosomes	
Haploid number	The number of different types of chromosomes in a cell; symbolized *n*	Humans have 23 different types of chromosomes ($n = 23$)
Ploidy	The number of each type of chromosome present; equivalent to the number of haploid chromosome sets present	
Haploid	Having one of each type of chromosome (*n*)	Bacteria and archaea are haploid, as are many algae; gametes are haploid
Diploid	Having two of each type of chromosome (2*n*)	Most familiar plants and animals are diploid
Polyploid	Having more than two of each type of chromosome; cells may be triploid (3*n*), tetraploid (4*n*), hexaploid (6*n*), and so on	Seedless bananas are triploid; many ferns are tetraploid; bread wheat is hexaploid

and is called a **zygote.** In this way, each diploid individual receives both a haploid chromosome set from its mother and a haploid set from its father. Homologous chromosomes are therefore referred to as either maternal or paternal in origin. **Maternal chromosomes** come from the mother; **paternal chromosomes** come from the father.

When these studies were published, the mystery of fertilization was finally solved. To appreciate the consequences of meiosis fully, though, we need to analyze the events in more detail and will do so next.

The Phases of Meiosis I

Meiosis begins after chromosomes have been replicated during S phase. Prior to the start of meiosis, chromosomes are extremely long structures, just as they are during interphase of the normal cell cycle. The major steps that occur once meiosis

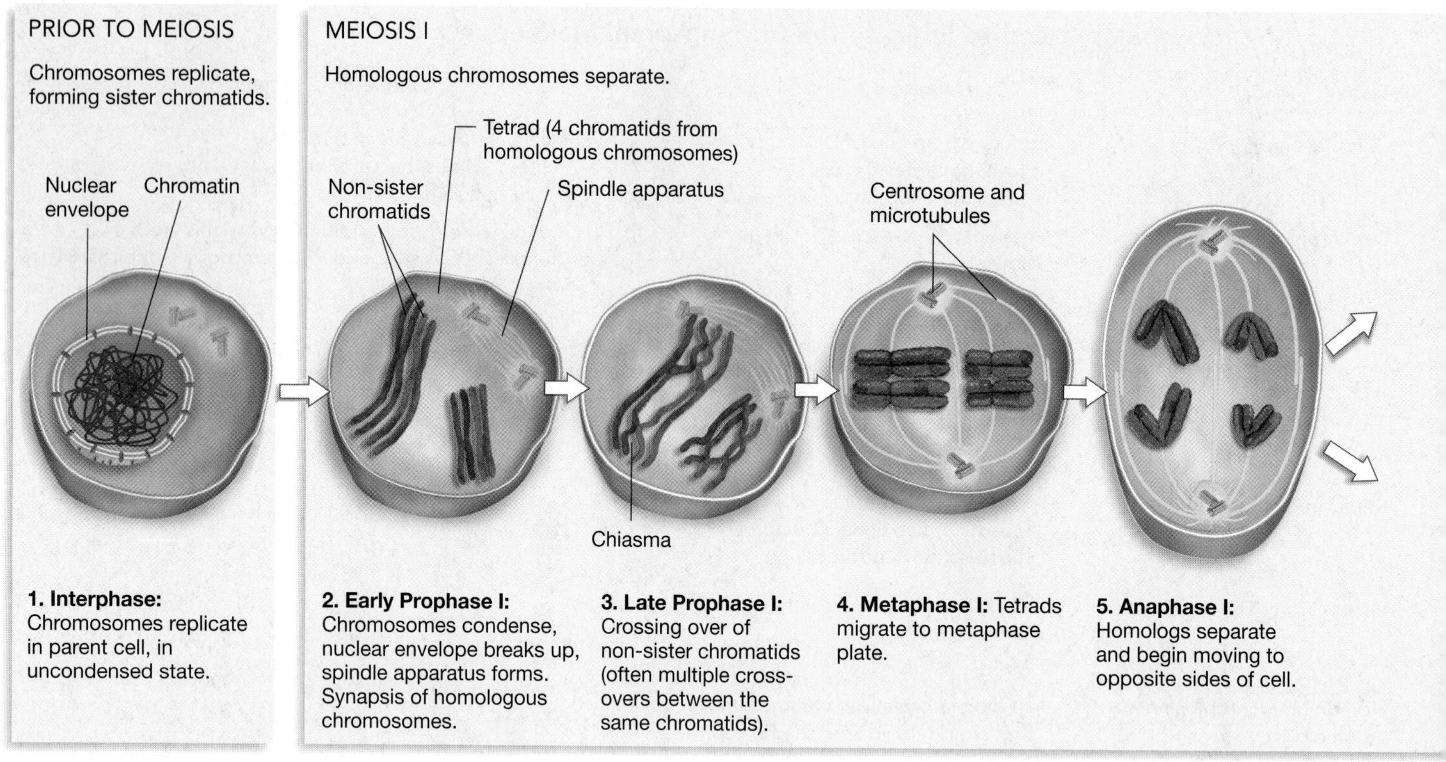

FIGURE 12.4 The Phases of Meiosis

EXERCISE In step 3, circle all of the other chiasmata.

begins are shown in **Figure 12.4**, using a diploid species with a haploid number of 2 as an example ($n = 2$; $2n = 4$). As in Figure 12.3, maternal chromosomes are red and paternal chromosomes are blue.

During **early prophase I** the chromosomes condense, the spindle apparatus forms, and the nuclear envelope begins to disappear. The next event illustrated, still during early prophase of meiosis I, is crucial: Homologous chromosome pairs come together. This pairing process is called **synapsis** and is illustrated in step 2 of Figure 12.4. Synapsis is possible because regions of homologous chromosomes that are similar at the molecular level attract one another, via mechanisms that are currently the subject of intense research. The structure that results from synapsis is called a **tetrad** (*tetra* means "four" in Greek). A tetrad consists of two homologous chromosomes, and each homolog consists of two sister chromatids. The chromatids from the homologs are referred to as **non-sister chromatids.** In the figure, non-sister chromatids are colored red versus blue.

During **late prophase I**, the non-sister chromatids begin to separate at many points along their length. They stay joined at certain locations, however, and look as if they cross over one another. Each crossover forms an X-shaped structure called a **chiasma** (plural: *chiasmata*). (In the Greek alphabet, the letter X is "chi.") Normally, at least one chiasma forms in every pair of homologous chromosomes; usually there are several chiasmata. As step 3 of Figure 12.4 shows, the chromatids involved in chiasma formation are homologous but not sisters. Consistent with this observation, Thomas Hunt Morgan proposed that a physical exchange of paternal and maternal chromosomes occurs at chiasmata. According to this hypothesis, paternal and maternal chromatids break and rejoin at each chiasma, producing chromatids that have both paternal and maternal segments. Morgan called this process of chromosome exchange **crossing over.** In step 4 of Figure 12.4, the result of crossing over is illustrated by chromosomes with a combination of red and blue segments. When crossing over occurs, the chromosomes that result have a mixture of maternal and paternal alleles.

The next major stage in meiosis I occurs during **metaphase I,** when pairs of homologous chromosomes (tetrads) are moved to a region called the **metaphase plate** by spindle fibers from the centrosomes (step 4). Two points are key here: Each tetrad moves to the metaphase plate independently of the other tetrads, and the alignment of maternal and paternal homologs from each chromosome is random. During **anaphase I**, homologs separate and begin moving to opposite sides of the cell (step 5). Meiosis I concludes with **telophase I**, when the homologs finish moving to opposite sides of the cell (step 6). When meiosis I is complete, the cell divides.

The end result of meiosis I is that one chromosome of each homologous pair is distributed to a different daughter cell. As a result of crossing over and the random distribution of maternal and paternal homologues during metaphase, each daughter cell from meiosis I gets a random and unique assortment of maternal and paternal chromosomes.

Study steps 1 through 6 of Figure 12.4 to see that the daughter cells of meiosis I are haploid. In each chromosome, however,

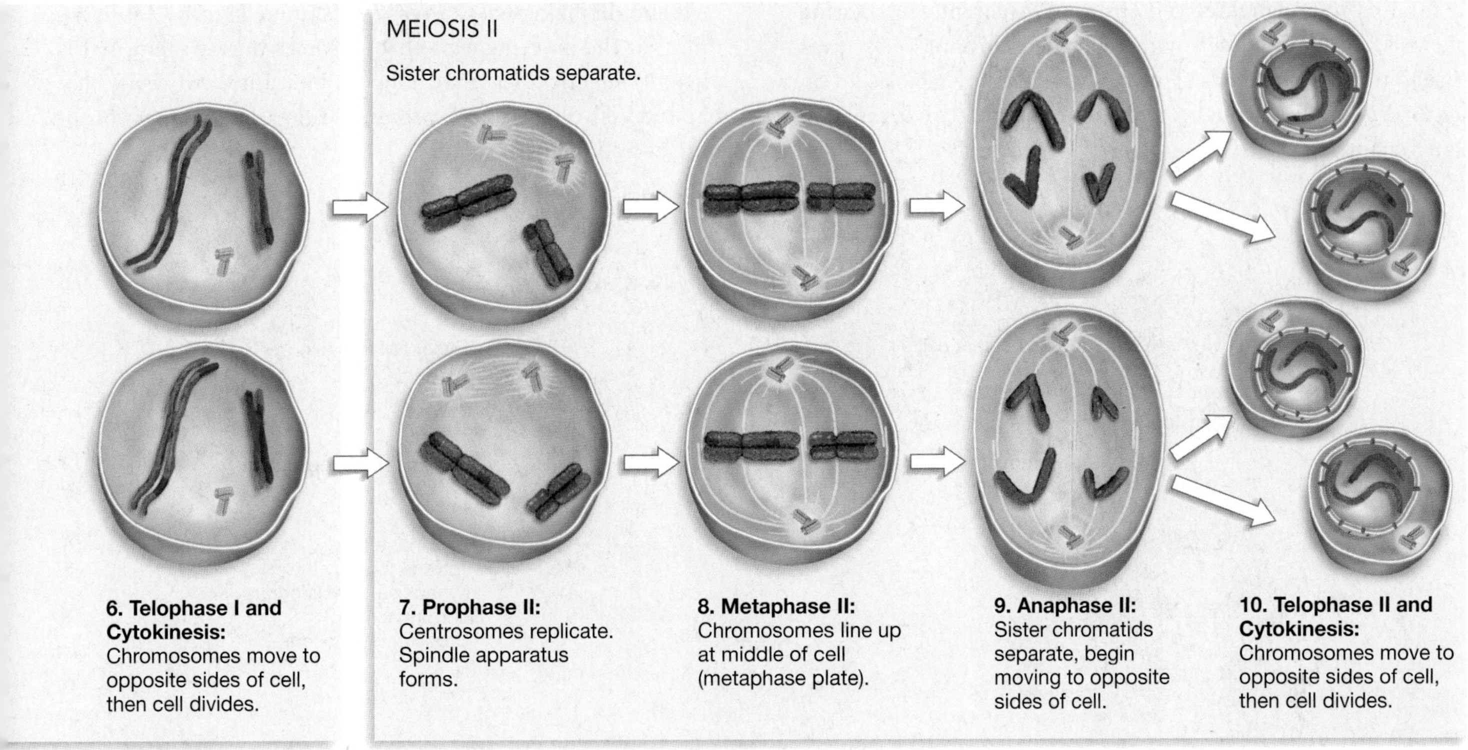
MEIOSIS II
Sister chromatids separate.
6. Telophase I and Cytokinesis:
Chromosomes move to opposite sides of cell, then cell divides.
7. Prophase II:
Centrosomes replicate. Spindle apparatus forms.
8. Metaphase II:
Chromosomes line up at middle of cell (metaphase plate).
9. Anaphase II:
Sister chromatids separate, begin moving to opposite sides of cell.
10. Telophase II and Cytokinesis:
Chromosomes move to opposite sides of cell, then cell divides.

the sister chromatids remain attached. At the end of meiosis I, each daughter cell still contains replicated chromosomes.

Although meiosis I is a continuous process, biologists summarize the events by identifying these distinct phases:

- Early Prophase I—Replicated chromosomes condense, the spindle apparatus forms, and the nuclear envelope disappears. Synapsis of homologs forms pairs of homologous chromosomes (tetrads). Spindle fibers attach to chromosome centromeres.
- Late Prophase I—Crossing over results in a mixing of chromosome segments from maternal and paternal chromosomes.
- Metaphase I—Pairs of homologous chromosomes migrate to the metaphase plate, and homologs line up.
- Anaphase I—Homologs separate and begin moving to opposite ends of the cell.
- Telophase I—Homologs finish moving to opposite sides of the cell. In some species, a nuclear envelope re-forms around each set of chromosomes.

When meiosis I is complete, **cytokinesis** (division of cytoplasm) occurs and two haploid daughter cells form.

The Phases of Meiosis II

Recall that chromosome replication occurred prior to meiosis I. Throughout meiosis I, sister chromatids remain attached. That is, replicated chromosomes remain intact during meiosis I. But no chromosome replication occurs between meiosis I and meiosis II. At the start of meiosis II, each chromosome still consists of two identical sister chromatids. Because only one member of each homologous pair is present, the cell is haploid.

Next, during **prophase II**, a spindle apparatus forms in both daughter cells. Spindle fibers attach to each side of the replicated chromosomes and move toward the middle of each cell (step 7 of Figure 12.4). In **metaphase II**, the replicated chromosomes are lined up at the metaphase plate (step 8). The sister chromatids of each chromosome separate during **anaphase II** (step 9) and move to different daughter cells during **telophase II** (step 10). Once they are separated, each chromatid is considered an independent chromosome. Meiosis II results in four haploid cells, each with one chromosome of each type.

As in meiosis I, biologists routinely designate distinct phases in meiosis II:

- Prophase II—The spindle apparatus forms. If a nuclear envelope formed at the end of meiosis I, it breaks apart.
- Metaphase II—Replicated chromosomes, consisting of two sister chromatids, are lined up at the metaphase plate.
- Anaphase II—Sister chromatids separate. The unreplicated chromosomes that result begin moving to opposite sides of the cell.
- Telophase II—Chromosomes finish moving to opposite sides of the cell. A nuclear envelope forms around each haploid set of chromosomes.

When meiosis II is complete, each cell divides to form two daughter cells. Because meiosis II occurs in exactly the same way in both daughter cells of meiosis I, the process results in a total of four daughter cells from each parent cell. During meiosis, one diploid cell with replicated chromosomes gives rise to four haploid cells. The cells produced by meiosis go on to form gametes, through a series of events that are detailed in later chapters.

It should make sense, after you examine Figure 12.4 in detail, that the movement of chromosomes during meiosis II is virtually identical to what happens in a mitotic division in a haploid cell. **Figure 12.5** provides a detailed comparison of mitosis and meiosis.

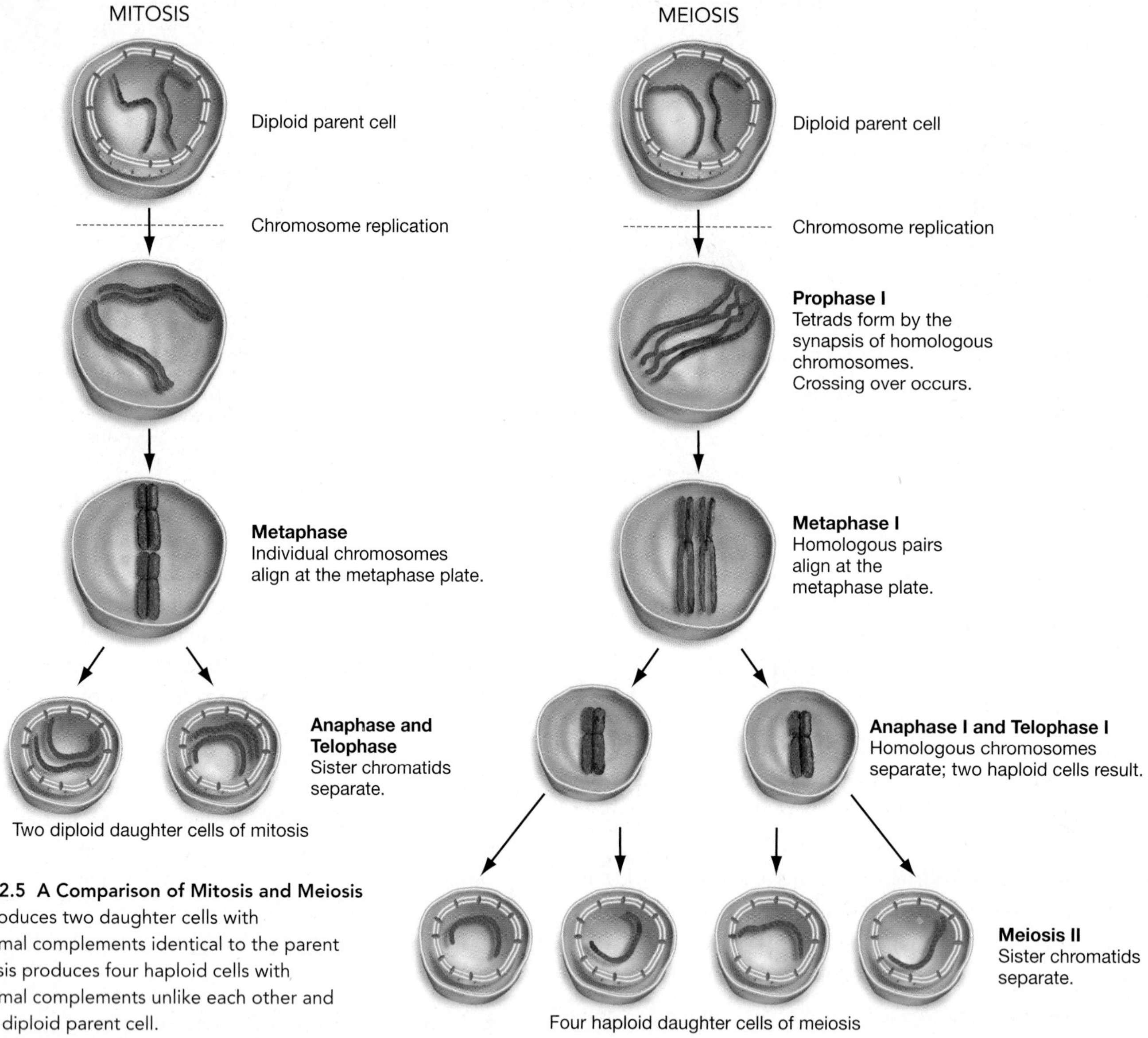

FIGURE 12.5 A Comparison of Mitosis and Meiosis Mitosis produces two daughter cells with chromosomal complements identical to the parent cell. Meiosis produces four haploid cells with chromosomal complements unlike each other and unlike the diploid parent cell.

Feature	Mitosis	Meiosis
Number of cell divisions	One	Two
Number of chromosomes in daughter cells, compared with parent cell	Same	Half
Synapsis of homologs	No	Yes
Number of crossing-over events	None	One or more per pair of homologous chromosomes
Makeup of chromosomes in daughter cells	Identical	Different—only one of each chromosome type present, paternal and maternal segments mixed within chromosomes
Role in life cycle	Asexual reproduction in eukaryotes; cell division for growth of multicellular organisms	Precedes production of gametes in sexually reproducing animals

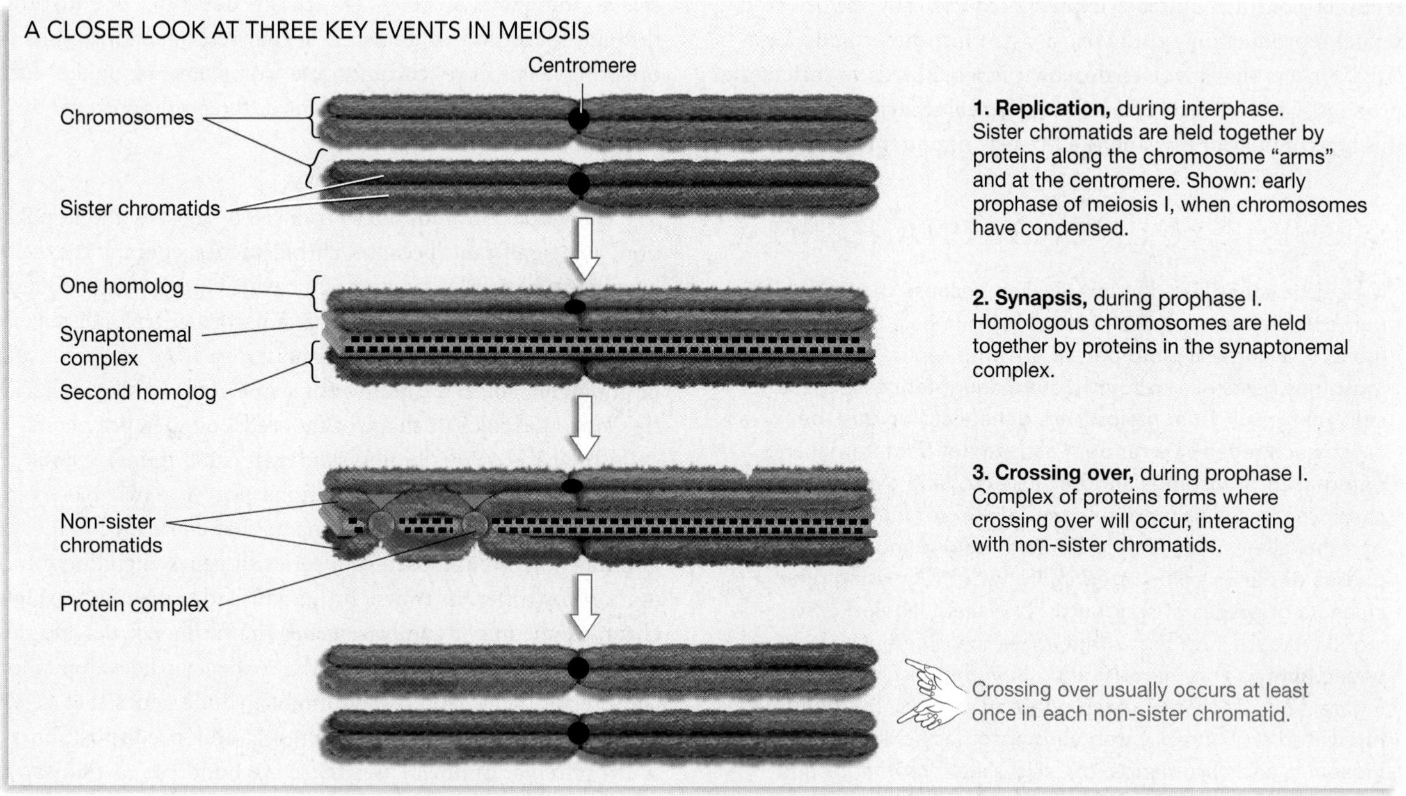
A CLOSER LOOK AT THREE KEY EVENTS IN MEIOSIS
Centromere
Chromosomes
Sister chromatids
1. Replication, during interphase. Sister chromatids are held together by proteins along the chromosome "arms" and at the centromere. Shown: early prophase of meiosis I, when chromosomes have condensed.
One homolog
Synaptonemal complex
Second homolog
2. Synapsis, during prophase I. Homologous chromosomes are held together by proteins in the synaptonemal complex.
Non-sister chromatids
Protein complex
3. Crossing over, during prophase I. Complex of proteins forms where crossing over will occur, interacting with non-sister chromatids.
Crossing over usually occurs at least once in each non-sister chromatid.

FIGURE 12.6 A Closer Look at Chromosome Replication, Synapsis, and Crossing Over

A Closer Look at Key Events in Prophase of Meiosis I

Figure 12.6 provides more detail on how several important events in prophase of meiosis I occur. Step 1 of Figure 12.6 shows that after chromosome replication is complete, sister chromatids stay very tightly joined along their entire length. When homologs synapse, two pairs of non-sister chromatids are brought close together and are held there by a network of proteins called the **synaptonemal complex** (step 2). Crossing over can occur at many locations along the length of this paired structure. Crossing over typically occurs at least once in each pair of homologs and two or three times in some species. It occurs when a complex of proteins cuts the chromosomes and then reattaches segments from homologs (step 3).

This more-detailed look at replication, synapsis, and crossing over should help you interpret the micrographs in **Figure 12.7**, which show grasshopper cells undergoing meiosis. Once you have a solid understanding of how meiosis occurs, you will be

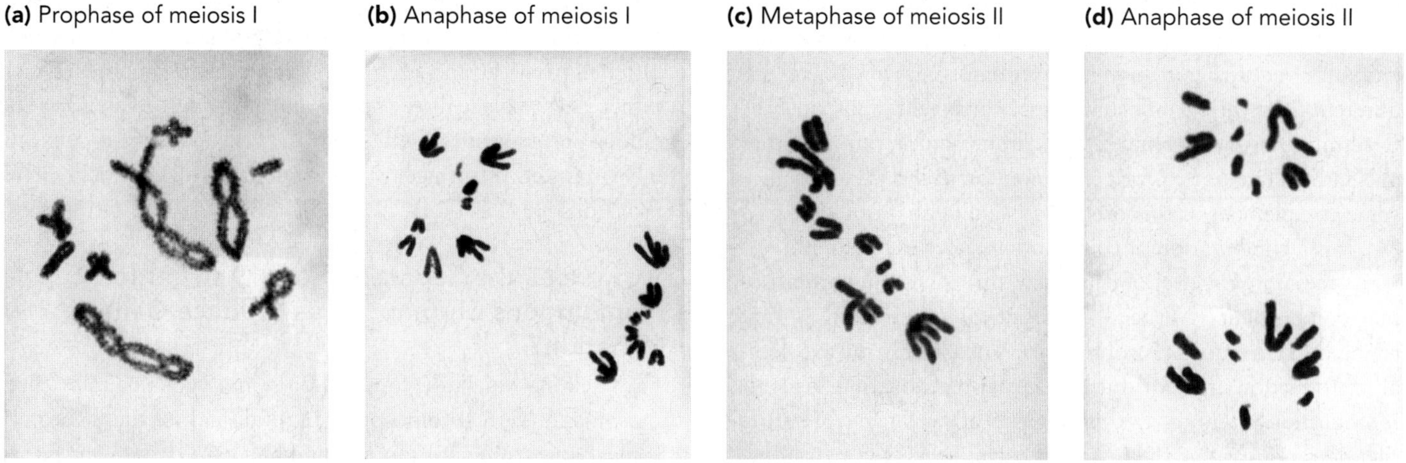
(a) Prophase of meiosis I
(b) Anaphase of meiosis I
(c) Metaphase of meiosis II
(d) Anaphase of meiosis II

FIGURE 12.7 Meiotic Chromosomes in Grasshoppers

EXERCISE (a) Label several chiasmata, sister chromatids, tetrads, and non-sister chromatids; (b) add centrosomes; (c) label sister chromatids and the metaphase plate, and add a spindle apparatus; (d) label the haploid number and ploidy of each nucleus.

ready to consider ultimate explanations for why meiosis and sexual reproduction exist. Meiosis is an intricate, tightly regulated process that involves dozens if not hundreds of different proteins. Given this complexity, it is logical to hypothesize that this type of nuclear division is extremely important. Why?

✓CHECK YOUR UNDERSTANDING

Meiosis is called a reduction division because the total number of chromosomes present is cut in half. During meiosis, a single diploid parent cell with replicated chromosomes gives rise to four haploid daughter cells. The cells that result from meiosis are genetically unique, because each receives a random assortment of maternal and paternal chromosomes and because crossing over leads to chromosomes with random combinations of maternal and paternal alleles. You should be able to (1) demonstrate the phases of meiosis illustrated in Figure 12.4 by using pipe cleaners or pieces of spaghetti; (2) make a labeled drawing like Figure 12.5 that summarizes the differences between mitosis and meiosis; and, (3) in the micrographs of Figure 12.7, (a) identify each phase of meiosis, (b) identify replicated and unreplicated chromosomes, (c) label chromosomes and chromatids, (d) label sister chromatids and non-sister chromatids, (e) label chiasmata, (f) add colors to indicate segments of maternal and paternal chromosomes that exchanged as a result of crossing over, and (g) determine whether each cell is haploid or diploid.

12.2 The Consequences of Meiosis

The cell biologists who worked out the details of meiosis in the late 1800s and early 1900s realized that the process solved the riddle of fertilization. Weissman's hypothesis—that a reduction division precedes gamete formation—was confirmed. This was an important advance in providing a proximate explanation for sexual reproduction. But researchers also understood that meiosis has another important outcome: Thanks to the independent shuffling of maternal and paternal chromosomes and crossing over during meiosis I, the chromosomes in gametes are different from the chromosomes in parental cells. Subsequently, fertilization brings haploid sets of chromosomes from a mother and father together to form a diploid offspring. The chromosome complement of this offspring is unlike that of either parent. It is a combination of contributions from each parent.

At the ultimate level of causation, this change in chromosomal complement is crucial. The critical observation is that changes in chromosome configuration occur only during sexual reproduction and not during asexual reproduction. **Asexual reproduction** refers to any mechanism of producing offspring that does not involve the fusion of gametes. Asexual reproduction in eukaryotes is usually based on mitosis, and the chromosomes in the daughter cells of mitosis are identical to the chromosomes in the parental cell (see Chapter 11). In contrast, **sexual reproduction** refers to the production of offspring through the fusion of gametes. Sexual reproduction results in offspring that have chromosome complements unlike each other and their parents. Why is this difference important?

Chromosomes and Heredity

The changes in chromosomes produced by meiosis and fertilization are significant because chromosomes contain the cell's hereditary material. Stated another way, chromosomes contain the instructions for specifying what a particular trait might be in an individual. These inherited traits range from eye color and height in humans to the number or shape of the bristles on a fruit fly's leg to the color or shape of the seeds found in pea plants.

In the early 1900s biologists began using the term **gene** to refer to the inherited instructions for a particular trait. Chapter 13 explores the experiments that confirmed that each chromosome is composed of a series of genes encoding information for different traits. In humans, for example, a single chromosome might comprise genes that influence height, hair color, the spacing of teeth, and the tendency to develop colon cancer; another chromosome might include genes that affect eye color, susceptibility to allergies, and predisposition to schizophrenia. In most cases there are hundreds or thousands of genes on each chromosome. Recall from Section 12.1 that *allele* refers to a particular version of a gene and that homologous chromosomes may carry different alleles. You might have a chromosome containing alleles that tend to produce medium height, black hair color, and a low predisposition to colon cancer, as well as a homologous chromosome containing alleles that contribute to short height, blond hair color, and a predisposition to colon cancer. Other individuals you know might have alleles associated with extreme height, red hair color, and a moderate predisposition to colon cancer.

Because chromosomes are composed of genes, the offspring produced during asexual reproduction are genetically identical to one another as well as to their parent. The offspring of asexual reproduction are **clones**—or exact copies—of their parent. But the offspring produced by sexual reproduction are genetically different from one another and unlike either their mother or their father. Let's analyze three aspects of meiosis that create variation among chromosomes—and hence the genetic makeup—of sexually produced offspring: (1) separation and distribution of homologous chromosomes, (2) crossing over, and (3) fertilization.

How Does the Separation and Distribution of Homologous Chromosomes Produce Genetic Variation?

Each cell in your body contains 23 homologous pairs of chromosomes and 46 chromosomes in total. Half of these chromosomes came from your mother, and half came from your father. Each chromosome is composed of genes that influence particular traits. For example, one gene that affects your eye color might be located on one chromosome, while one of the genes that affects your hair color might be located on a different chromosome (**Figure 12.8a**).

Suppose that the chromosomes you inherited from your mother contains alleles that tend to produce brown eyes and black hair, but the chromosomes you inherited from your father includes alleles that tend to specify green eyes and red hair. (In reality, many genes with various alleles interact in complex ways to produce human eye color and hair color.) Will any particular gamete you produce contain the genetic instructions inherited from your mother or the instructions inherited from your father?

To answer this question, study the diagram of meiosis in **Figure 12.8b**. It shows that when pairs of homologous chromosomes line up during meiosis I and the homologs separate, a variety of combinations of maternal and paternal chromosomes can result. Each daughter cell gets a random assortment of maternal and paternal chromosomes. In the example given here, meiosis results in gametes with alleles for brown eyes and black hair, like your mother, and green eyes and red hair, like your father. But two additional combinations also occur: brown eyes and red hair, or green eyes and black hair. Four different combinations of paternal and maternal chromosomes are possible when two chromosomes are distributed to daughter cells during meiosis I.

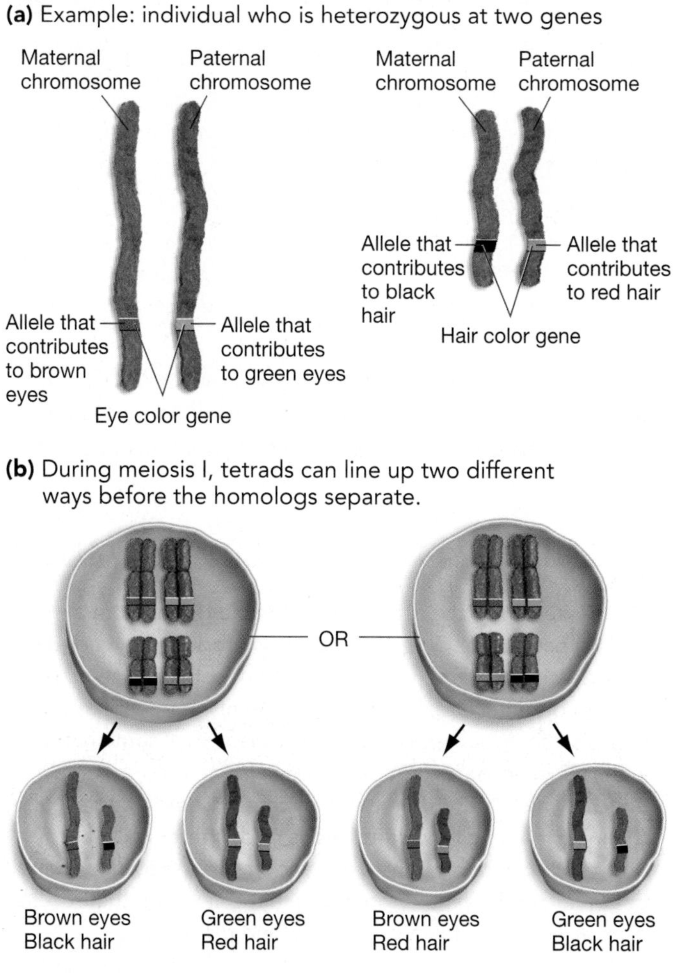

FIGURE 12.8 Separation of Homologous Chromosomes Leads to New Combinations of Genes
(a) A hypothetical example: Genes that influence eye color and hair color in humans are on different chromosomes. **(b)** How gametes with different combinations of genes result from separation of homologous chromosomes during meiosis I.

How many different combinations of maternal and paternal homologs are possible when more chromosomes are involved? In an organism with three chromosomes per haploid set ($n = 3$), eight types of gametes can be generated by randomly grouping maternal and paternal chromosomes. In general, a diploid organism can produce 2^n combinations of maternal and paternal chromosomes, where n is the haploid chromosome number. This means a human ($n = 23$) can produce 2^{23}, or about 8.4 million, gametes that differ in their combination of maternal and paternal chromosome sets. Clearly, the random assortment of whole chromosomes generates an impressive amount of genetic variation among gametes.

The Role of Crossing Over

Recall from Section 12.1 that segments of paternal and maternal chromatids exchange at each chiasma that forms during meiosis I. Thus, crossing over produces new combinations of alleles on the same chromosome—combinations that did not exist in either parent. This phenomenon is recombination. **Genetic recombination** is any change in the combination of alleles on a given chromosome. In species that reproduce sexually, recombination occurs via crossing over during meiosis. But genetic recombination also occurs in haploid organisms such as bacteria, which cannot undergo meiosis. Processes that produce genetic recombination in bacteria are described in later chapters.

Crossing over and recombination are important because they dramatically increase the genetic variability of gametes produced by meiosis. Recall that the separation and distribution of homologous chromosomes during meiosis varies the combination of chromosomes present. But in addition, crossing over varies the combinations of alleles within each chromosome. Recent data on humans, for example, indicate that an average of about 50 chiasmata occur in each cell undergoing meiosis I. As a result, the number of genetically different gametes that an individual can produce is much more than the 8.4 million produced by the separation and distribution of homologs. When crossing over occurs a total of 50 or more times in random locations throughout the entire suite of tetrads, the number of genetically distinct gametes that you can produce is virtually limitless.

Biologists say that meiosis "shuffles" alleles, because each daughter cell gets a random assortment of maternal and paternal chromosomes and because most chromosomes contain both maternal and paternal alleles. As Section 12.3 will show, the genetic variation produced by meiosis has enormous consequences for the ability of offspring to survive and reproduce.

How Does Fertilization Affect Genetic Variation?

Crossing over and the random mixing of maternal and paternal chromosomes ensure that each gamete is genetically unique. Even if two gametes produced by the same individual fuse to

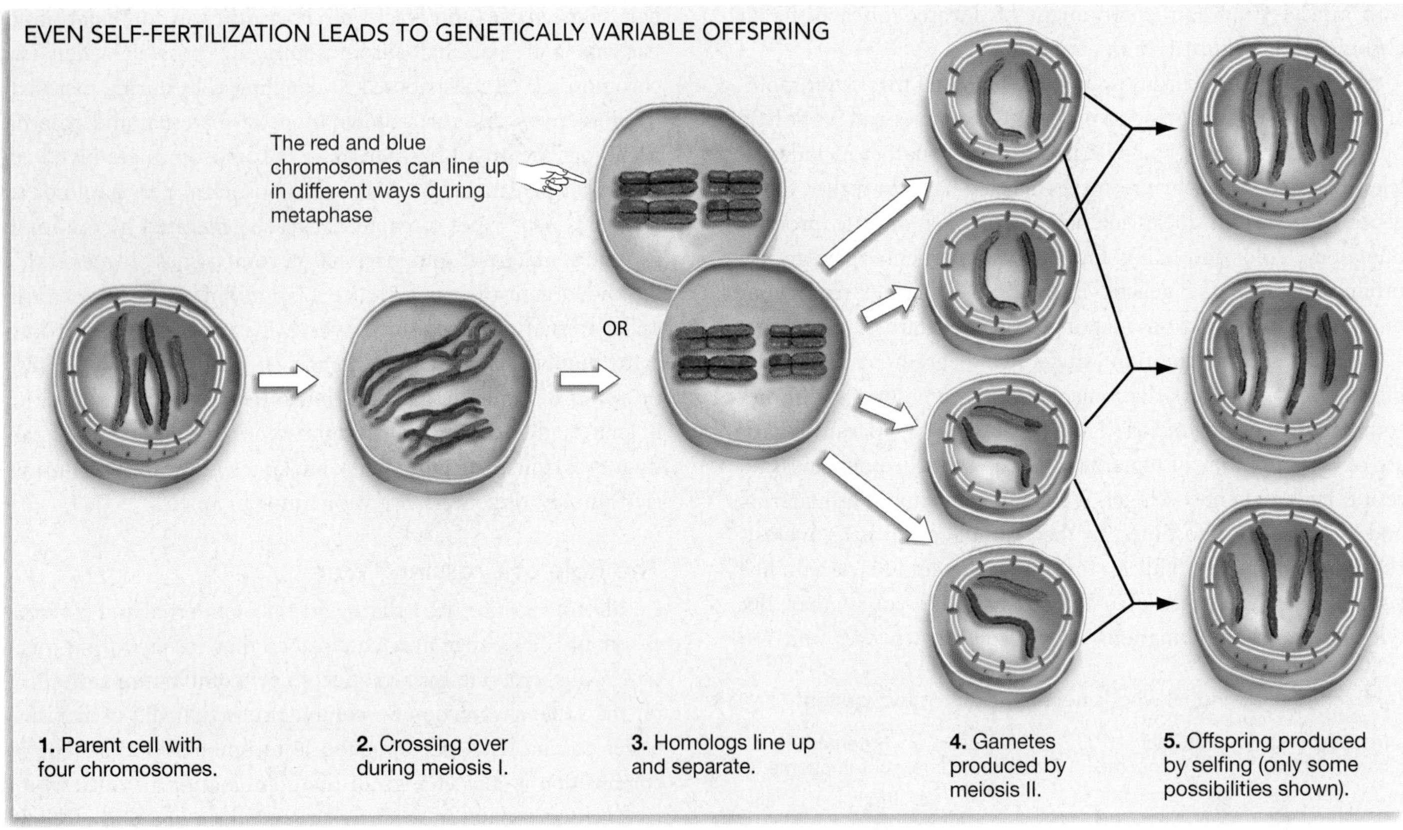
EVEN SELF-FERTILIZATION LEADS TO GENETICALLY VARIABLE OFFSPRING
The red and blue chromosomes can line up in different ways during metaphase
OR
1. Parent cell with four chromosomes.
2. Crossing over during meiosis I.
3. Homologs line up and separate.
4. Gametes produced by meiosis II.
5. Offspring produced by selfing (only some possibilities shown).

FIGURE 12.9 Even If Self-Fertilization Takes Place, Offspring Are Genetically Variable
Some possible results of self-fertilization in an organism with four chromosomes ($2n = 4$). EXERCISE In step 2, label where two chiasmata are causing a double crossover event. Add sketches showing the chromosome complements of offspring produced by other pairings of gametes indicated in step 4.

form a diploid offspring—meaning **self-fertilization**, or "selfing," takes place—the offspring are very likely to be genetically different from the parent (**Figure 12.9**). Selfing is common in some plant species. It also occurs in the many animal species in which individuals contain both male and female sex organs.

Self-fertilization is rare or nonexistent in many sexually reproducing species, however. Instead, gametes from different individuals combine to form offspring. This is called **outcrossing.** Outcrossing increases the genetic diversity of offspring because it combines chromosomes from different individuals, which are likely to contain different alleles.

How many genetically distinct offspring can be produced when outcrossing occurs? Let's answer this question using humans as an example. Recall that a single human can produce about 8.4 million different gametes—even in the absence of crossing over. When a person mates with a member of the opposite sex, the number of different genetic combinations that can result is equal to the product of the number of different gametes produced by each parent. In humans this means that potentially $8.4 \text{ million} \times 8.4 \text{ million} = 70.6 \times 10^{12}$ genetically distinct offspring can result from any one mating. This number is far greater than the total number of people who have ever lived—and the calculation does not even take into account variation generated by crossing over, which occurs at least once along each chromosome. Sexual reproduction results in genetically diverse offspring. Why is this important?

✓CHECK YOUR UNDERSTANDING

Meiosis results in the production of haploid cells from a diploid parent cell. The daughter cells produced by meiosis are genetically different from the parent cell because maternal and paternal homologs align randomly at metaphase of meiosis I and because crossing over leads to recombination within chromosomes. You should be able to (1) draw a parent cell with $n = 3$ (three types of chromosomes) and six of the many genetically distinct types of daughter cells that may result when this parent cell undergoes meiosis; (2) compare and contrast the degree of genetic variation that results from asexual reproduction, selfing, and outcrossing.

12.3 Why Does Meiosis Exist? Why Sex?

Meiosis and sexual reproduction occur in only a small fraction of the lineages on the tree of life. Bacteria and archaea undergo only asexual reproduction; most algae, fungi, and some land

FIGURE 12.10 Asexual Reproduction in Quaking Aspen
Quaking aspen trees send out underground stems, from which new individuals sprout. Both the underground stems and the new shoots are produced by mitosis. As this process continues, a large group of genetically identical individuals forms. The photograph shows such a group in autumn, when quaking aspen leaves turn bright yellow.

plants reproduce asexually as well as sexually. Recall that asexual reproduction in eukaryotes occurs via mitosis. Quaking aspen trees, for example, can produce new individuals by sending up shoots from underground stems (**Figure 12.10**). Asexual reproduction is found even among the vertebrates. For example, several species of guppy in the genus *Poeciliopsis* reproduce exclusively via mitosis.

Sexual reproduction is common among multicellular ("many-celled") organisms, however. It is the major mode of reproduction in lineages such as the insects, with millions of species.

Even though sex plays an important role in the life of many organisms, until recently it was not clear why it occurs. On the basis of theory, biologists had good reason to think that sexual reproduction should not exist. Let's take a closer look.

The Paradox of Sex

In 1978 John Maynard Smith pointed out that the existence of sexual reproduction presents a paradox. Maynard Smith developed a mathematical model showing that because asexually reproducing individuals do not have to produce male offspring, their progeny can produce twice as many grand-offspring as can individuals that reproduce sexually. **Figure 12.11** diagrams this result by showing the number of females (♀) and males (♂) produced over several generations by asexual versus sexual reproduction. In this example, each individual produces four offspring over the course of his or her lifetime. In the asexual population, each individual is a female that produces four offspring. But in the sexual population, it takes two individuals—one male and one female—to produce four offspring.

Based on Maynard Smith's analysis, what will happen when asexual and sexual individuals exist in the same population and compete with one another? If all other things are equal, individuals that reproduce asexually should increase in frequency while individuals that reproduce sexually should decline in frequency. In fact, Maynard Smith's model predicts that sexual reproduction is so inefficient that it should be completely eliminated. At the ultimate level of explanation, the existence of sex is a paradox.

To resolve the paradox of sex, biologists began examining the assumption "If all other things are equal." Stated another way, biologists began looking for ways that meiosis and outcrossing could lead to the production of offspring that reproduce more than asexually produced individuals do.

The Changing-Environment Hypothesis One hypothesis to explain sexual reproduction focuses on the benefits of producing genetically diverse offspring. Here's the key idea: If the

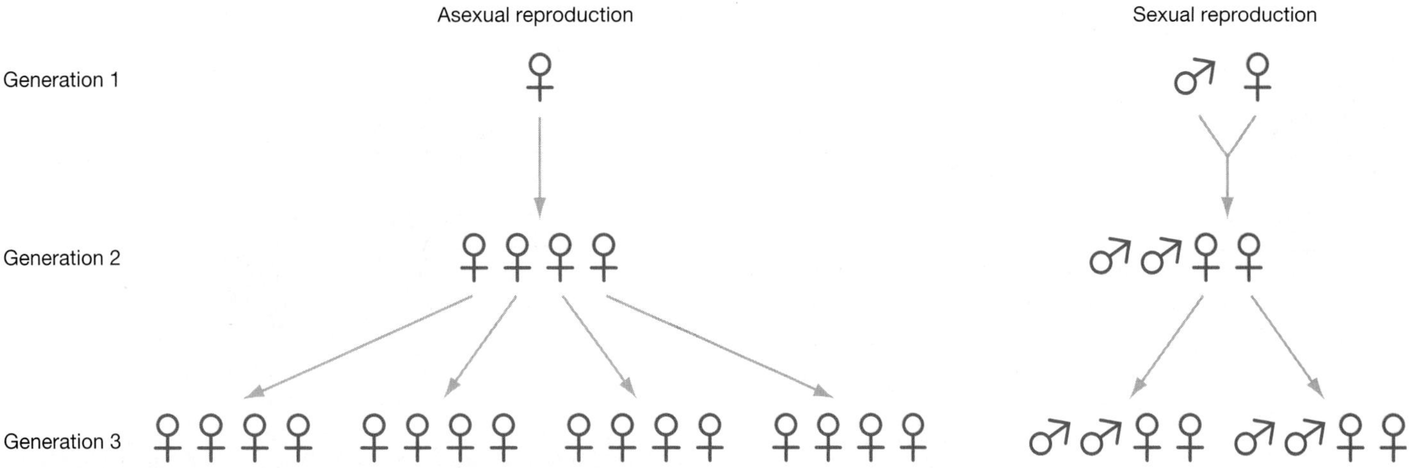

FIGURE 12.11 Asexual Reproduction Confers a Large Numerical Advantage
Each female symbol (♀) and male symbol (♂) represents an individual. In this hypothetical example, every individual produces four offspring over the course of a lifetime, sexually reproducing individuals produce half males and half females, and all offspring survive to breed. **QUESTION** How many asexually produced offspring would be present in generation 4? How many sexually produced offspring?

environment changes from one generation to the next, then offspring that are genetically different from their parents and from each other are more likely to survive and produce offspring of their own. Conversely, offspring that are genetic clones of their parents are less likely to thrive if the environment changes.

What type of environmental change would favor genetically diverse offspring? The possibilities include changes in temperature regimes, moisture, predators, competitors, and food sources. But recently researchers have focused on one particular component of environmental change: the emergence of new strains of disease-causing agents such as bacteria, parasitic worms, the eukaryote that causes malaria, and viruses.

The genetic characteristics of disease-causing organisms and viruses tend to change very quickly over time. In your own lifetime, for example, several new disease-causing agents have emerged that afflict humans; they include the SARS virus and new strains of HIV, the parasite that causes malaria, and the tuberculosis bacterium. In addition to employing drugs to combat these types of agents, humans have hundreds of genes that are involved in defense. At many of these genes, certain alleles help host individuals fight off particular strains of bacteria, parasitic worms, or viruses. As you might predict, the presence of certain disease-fighting alleles is crucial for plants and animals that cannot rely on drug therapies for help.

What happens if all of the offspring produced by an individual are genetically identical? If a new strain of disease-causing agent evolves, then all of the asexually produced offspring are likely to be susceptible to that new strain. But if the offspring are genetically variable, then it is likely that at least some offspring will have combinations of alleles that enable them to fight off the new disease and produce offspring of their own.

The logic of the hypothesis is sound. Do any data support it?

Testing the Changing-Environment Hypothesis Curtis Lively and colleagues tested the changing-environment hypothesis by studying a species of snail that is native to New Zealand. This type of snail lives in ponds and other freshwater habitats and is parasitized by over a dozen species of trematode worms. Snails that become infected cannot reproduce—the worms eat their reproductive organs. In this way, the parasites cause illness. The parasites are rare in some habitats and common in others.

The biologists were interested in working on this snail species because some individuals reproduce only sexually while others reproduce only asexually. If the changing-environment hypothesis for the advantage of sex is correct, then the frequency of sexually reproducing individuals should be much higher in habitats where parasites are common than it is in habitats where parasites are rare (**Figure 12.12**). The logic here is that asexually reproducing individuals should have high fitness in environments where parasites are rare. In contrast, sexually reproducing individuals should have high fitness in habitats where parasites are common.

Question: Why does sexual reproduction occur?

Hypothesis: Sexually produced offspring have higher fitness than do asexually produced individuals in habitats where parasitism is common.

Null hypothesis: There is no relationship between the presence of parasites and method of reproduction.

Study design:

1. Collect snails from a wide array of habitats.

2. Document percentage of males in each population, as an index of frequency of sexual reproduction.

3. Note two types of populations: In one, males are common; in the other, males are almost nonexistent. Infer that sexual reproduction is either common or almost nonexistent.

4. Document percentage of individuals infected with parasites in sexually versus asexually reproducing populations.

Prediction: In populations where sexual reproduction is common, parasitism rates are high. In populations with only asexual reproduction, infection rates are low.

Prediction of null hypothesis: No difference in parasitism rate between populations that reproduce sexually versus asexually.

Results:

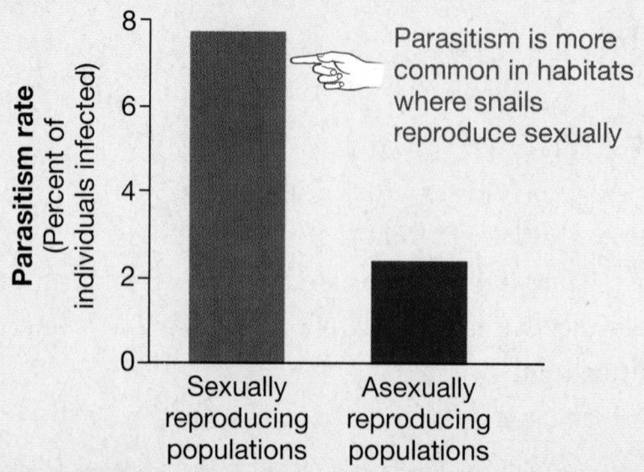

Conclusion: Sexual reproduction is common in habitats where parasitism is common. Asexual reproduction is common in habitats where parasitism is rare.

FIGURE 12.12 Is Sexual Reproduction Favored when Disease or Parasitism Rates Are High?

To test these predictions, the researchers collected a large number of individuals from different habitats. Lively and co-workers examined snails in habitats where parasites were more or less common and calculated the frequency of individuals that reproduce sexually versus those that reproduce asexually. The results are plotted in Figure 12.12. The data show that habitats where parasite infection rates are high have a relatively large number of sexually reproducing individuals compared with habitats that have low parasite incidence.

This result and a variety of other studies support the changing-environment hypothesis. Although the paradox of sex remains an active area of research, more biologists are becoming convinced that sexual reproduction is an adaptation that increases the fitness of individuals in environments where disease-causing organisms are common.

When Do Meiosis and Fertilization Occur during the Life of an Organism?

At the proximate level, all sexually reproducing organisms are thought to undergo meiosis in virtually the same way. At the ultimate level, the changing-environment hypothesis may turn out to be a valid and universal explanation for why sex can be advantageous. But on the question of when meiosis occurs during the course of an organism's lifetime, biologists are faced with a striking amount of variation that remains unexplained.

To bring this question into focus, consider the life cycles observed in various species. A **life cycle** is the sequence of events that occurs during the life span of an individual, from fertilization to the production of offspring. When sexually reproducing species are compared, biologists find that many of the same events occur: meiosis, production of haploid gametes, fertilization and formation of a diploid zygote, and mitosis. What varies among species is the timing of these events and the extent of the haploid and diploid stages. Some of this variation is highlighted in **Figure 12.13**.

In humans and most other animals, individuals are diploid for most of their lives (**Figure 12.13a**). Meiosis occurs in the reproductive organs of an adult, and gametes represent the only haploid phase of the individual's life. In contrast, many species of eukaryotes are haploid for most of their lives. The green alga shown in **Figure 12.13b** illustrates this pattern. In these organisms, the diploid stage consists of a single-celled structure that can resist heat and drying. This cell then undergoes meiosis. The haploid cells that result divide by mitosis to generate a haploid multicellular organism.

Land plants and several groups of algae have yet another type of life cycle, known as **alternation of generations**. This life cycle features a multicellular haploid stage and a multicellular diploid stage (**Figure 12.13c**). In some land plants, such as angiosperms (flowering plants) and ferns, the haploid stage is tiny and relatively short lived. In other species, such as mosses, the haploid stage is larger and longer lived than the diploid plant.

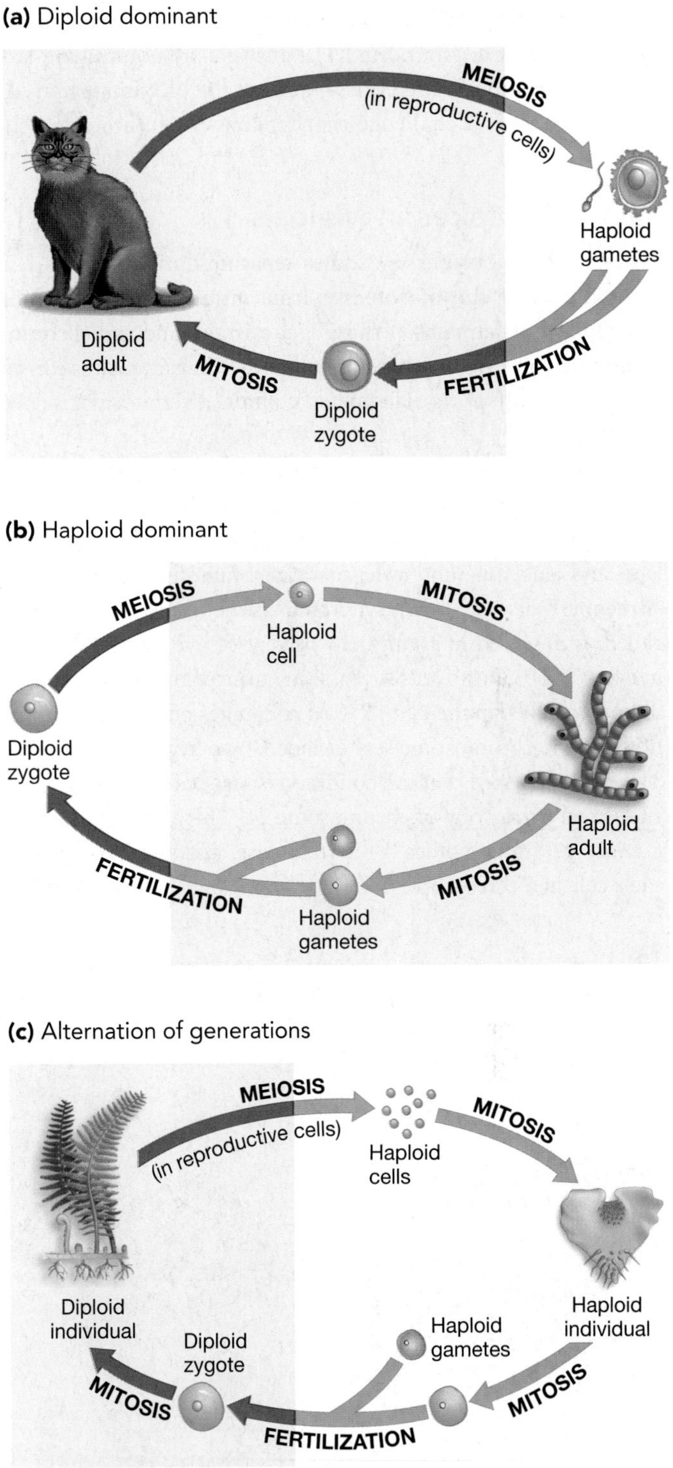

FIGURE 12.13 Life Cycles Differ in the Timing of Meiosis and Fertilization and in the Extent of Haploid versus Diploid Stages
(a) In most animals, gametes are the only haploid cells. Meiosis occurs in special reproductive tissues of adults. **(b)** In many algae, the fertilized egg is the only diploid cell. When this cell undergoes meiosis, the haploid cells that are produced form a multicellular adult. **(c)** In land plants and some algae, there is a multicellular diploid stage and a multicellular haploid stage. Typically, one of these two stages is larger and longer lived than the other. In ferns, the diploid stage is more prominent. QUESTION The green alga illustrated in part (b) is haploid during virtually all of its life. Does it undergo sexual reproduction?

WEB TUTORIAL 12.2 Mistakes in Meiosis

What mechanisms are responsible for all this variation in life cycles, and why does it occur? The answers are not known. Explaining variation in life cycles—at both the proximate and ultimate levels—is a challenge that remains for future research.

12.4 Mistakes in Meiosis

When homologous chromosomes separate during meiosis I, a complete set of chromosomes is transmitted to each daughter cell. But what happens if there is a mistake and the chromosomes are not properly distributed? What are the consequences for offspring if gametes contain an abnormal set of chromosomes?

In 1866 Langdon Down described a distinctive suite of concurrent conditions observed in some humans. The syndrome was characterized by mental retardation, a high risk for heart problems and leukemia, and a degenerative brain disorder similar to Alzheimer's disease. **Down syndrome**, as the disorder came to be called, is observed in about 0.15 percent of live births (3 infants in every 2000). For over 80 years the cause of the syndrome was unknown. Then in the late 1950s a researcher published observations on the chromosome sets of nine Down syndrome children. The data suggested that the condition is associated with the presence of an extra copy of chromosome 21. This situation is called a **trisomy** ("three-bodies")—in this case, trisomy-21—because each cell has three copies of the chromosome. To explain the anomaly, the biologist proposed that the extra chromosome resulted from a mistake during meiosis in one of the parents.

How Do Mistakes Occur?

For a gamete to get one complete set of chromosomes, two steps in meiosis must be perfectly executed. During the first meiotic division in humans, for instance, 23 pairs of homologous chromosomes must separate, or *disjoin*, from each other so that only one homolog ends up in each daughter cell. If both homologs move to the same pole of the parent cell, however, the products of meiosis will be abnormal. This sort of meiotic error, illustrated in **Figure 12.14**, is referred to as **nondisjunction,** because the homologs do not separate or disjoin. Note that two daughter cells have two copies of the same chromosome (blue in Figure 12.14), while the other two lack that chromosome entirely. Gametes that contain an extra chromosome are symbolized as $n + 1$; gametes that lack one chromosome are symbolized as $n - 1$. If an $n + 1$ gamete is fertilized by a normal n gamete, the resulting zygote will be $2n + 1$. This situation is a trisomy. If the $n - 1$ gamete is fertilized by a normal n gamete, the resulting zygote will be $2n - 1$. This situation is called **monosomy.** Cells that have too many or too few chromosomes are said to be **aneuploid** ("without-form").

More rarely, abnormal $n + 1$ and $n - 1$ gametes are produced during nondisjunction in the second meiotic division. If

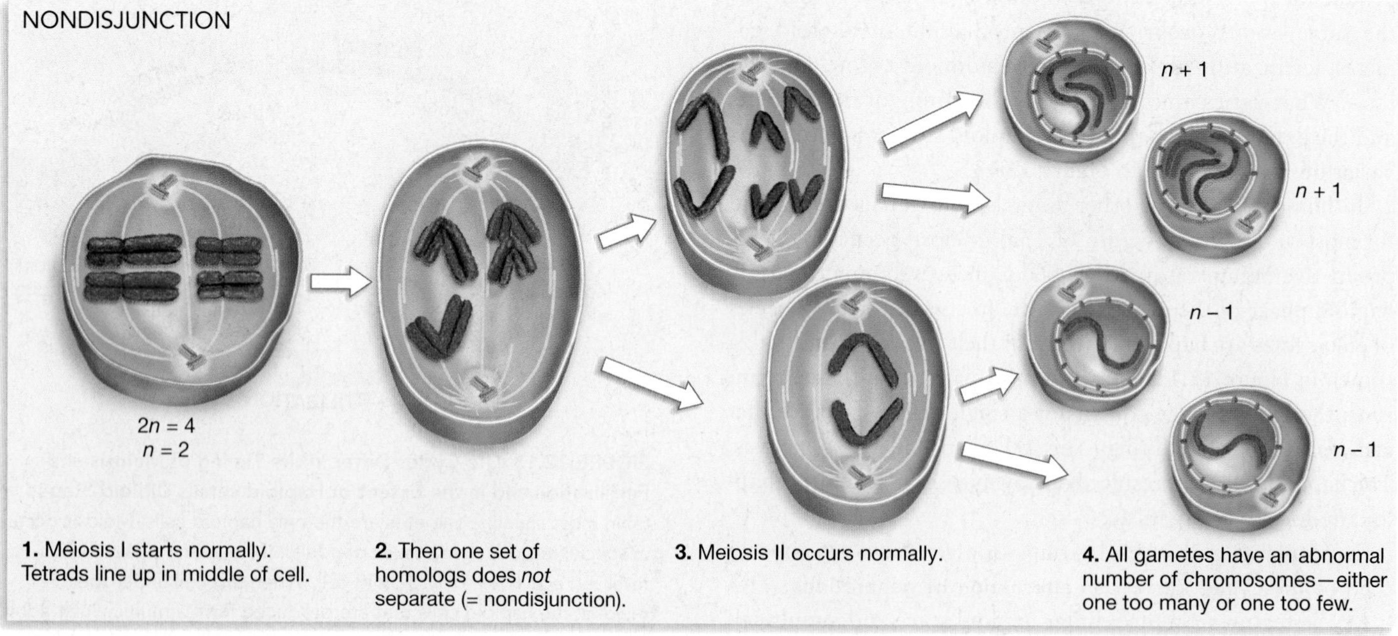
NONDISJUNCTION
2n = 4
n = 2
n + 1
n + 1
n − 1
n − 1
1. Meiosis I starts normally. Tetrads line up in middle of cell.
2. Then one set of homologs does *not* separate (= nondisjunction).
3. Meiosis II occurs normally.
4. All gametes have an abnormal number of chromosomes—either one too many or one too few.

FIGURE 12.14 Nondisjunction Leads to Gametes with Abnormal Chromosome Numbers
If homologous chromosomes fail to separate during meiosis I, the gametes that result will have an extra chromosome or will lack a chromosome. EXERCISE Nondisjunction also results when meiosis I occurs normally but two of the sister chromatids illustrated in step 3 fail to separate. Redraw steps 2–4 to show how this mechanism of nondisjunction occurs.

sister chromatids fail to separate and move to opposite poles of the dividing cell during meiosis II, then the resulting daughter cells will be $n + 1$ and $n - 1$.

Meiotic mistakes occur at a relatively high frequency. In humans, for example, researchers estimate that nondisjunction events occur in as many as 10 percent of meiotic divisions. The types of mistakes vary, but the consequences are almost always severe when defective gametes participate in fertilization. In a recent study of human pregnancies that ended in early embryonic or fetal death, 38 percent of the 119 cases involved atypical chromosome complements that resulted from mistakes in meiosis. Trisomy accounted for 36 percent of the abnormal karyotypes found; the incorrect number of complete chromosome sets, called *triploidy* ($3n$), for 30 percent; aberrantly sized or shaped chromosomes for 4 percent; and monosomy ($2n - 1$) for 2 percent. Mistakes in meiosis are a major cause of spontaneous abortion in humans.

Why Do Mistakes Occur?

The leading hypothesis to explain the incidence of trisomy and other meiotic mistakes is that they are accidents—random errors that occur during meiosis. Consistent with this proposal, there does not seem to be any genetic or inherited predisposition to trisomy or other types of dysfunction. Most cases of Down syndrome, for example, occur in families with no history of the condition.

Even though meiotic errors may be random, there are still strong patterns in their occurrence:

- With the exception of trisomy-21, most of the trisomies and monosomies observed in humans involve the sex chromosomes. Klinefelter syndrome, which develops in XXY individuals—as described in Box 12.1—occurs in about 1 in 2000 live births. Trisomy X (karyotype XXX) occurs in about 1 in 1000 live births and results in females who are healthy and fertile. **Turner syndrome** develops in XO individuals and occurs in about 1 in 5000 live births. Individuals with this syndrome are female but are sterile.
- **Table 12.3** shows data collected on trisomies in the autosomes of human fetuses and infants. Three observations deserve mention: (1) Trisomy is much more common in the smaller chromosomes (numbers 13–22) than it is in the larger chromosomes (numbers 1–12); (2) trisomy-21 is far and away the most common type of trisomy observed; and (3) maternal errors account for most incidences of trisomy. For example, over 90 percent of cases of Down syndrome are due to chromosomal defects in eggs.
- Maternal age is an important factor in the occurrence of trisomy. As **Figure 12.15** shows, the incidence of Down syndrome increases dramatically in mothers over 35 years old.

Why do these patterns occur? Biologists still do not have a good explanation for why most cases of aneuploidy in humans are due to problems with meiosis in females. In addition, it is still a mystery why there is such a strong correlation between maternal age and frequency of trisomy-21. But to explain why most instances of aneuploidy involve chromosome 21 or the sex chromosomes, biologists offer two hypotheses: (1) Individuals with other types of aneuploidy do not develop normally and are spontaneously aborted, long before birth; and (2) the sex chromosomes and chromosome 21 are more susceptible to aneuploidy than are other types of chromosomes. These hypotheses are not mutually exclusive, so both may be correct.

Studies on the mechanism of aneuploidy continue. In the meantime, one overall message is clear: Undergoing meiosis correctly is critical to the health and welfare of offspring.

TABLE 12.3 The Incidence of Trisomy in Humans: Effects of Chromosome Number and Paternal versus Maternal Origin

Trisomy (chromosome number)	Total Number of Cases	Due to Error in Sperm	Due to Error in Egg	Maternal Errors (%)
2–12	16	3	13	81
13	7	2	5	71
14	8	2	6	75
15	11	3	8	73
16	62	0	62	100
18	73	3	70	96
21	436	29	407	93
22	11	0	11	100

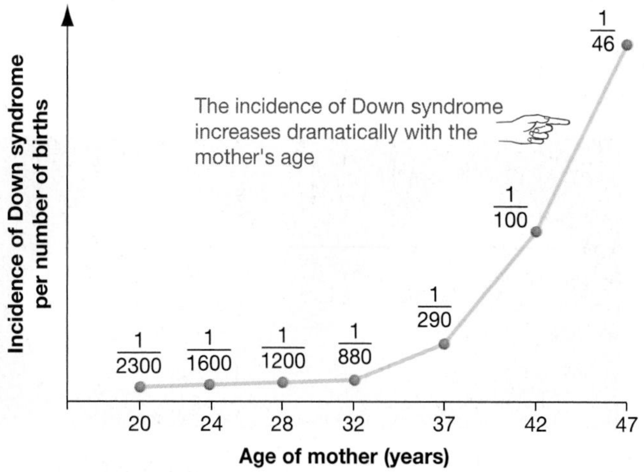

FIGURE 12.15 The Frequency of Down Syndrome Increases as a Function of a Mother's Age

QUESTION Suppose that you are an obstetrician. Based on these data, at what age would you recommend that pregnant mothers undergo procedures to check the karyotype of the embryos they carry?

ESSAY Seedless Fruits

Have you ever sat outside on a hot summer night eating watermelon and having a seed-spitting contest? It is likely that your children will never have the experience, because seedless varieties of watermelon have recently begun to dominate the market (**Figure 12.16**). New varieties of watermelon aren't the only example of a seedless fruit, either. Imagine biting into a banana and crunching down on hard, inedible seeds. The virtually seedless and sterile bananas we eat today are the result of selective breeding and artificial selection. As pointed out in Chapter 1, artificial selection occurs when humans allow only certain offspring of domesticated plants and animals to breed. Bananas have undergone artificial selection for fruit size, fruit sweetness, and seedlessness. The little black specks inside a banana are aborted seeds.

How did these seedless fruits come to be? One set of chromosomes makes the difference between the presence and absence of seeds in bananas and watermelons. In both of these species, diploid ($2n$) or tetraploid ($4n$) varieties make seeds while triploid ($3n$) strains are sterile and seedless.

Many cultivated plant species are polyploid, meaning that they have more than two sets of chromosomes. As long as there is a homolog for each chromosome during synapsis, meiosis I proceeds as it would in a diploid species and results in a halving of the chromosome number. For normal seed production, the key is that there must be an even number of chromosome sets. When the ploidy number is even, homologous chromosomes are able to synapse and later separate correctly. The wheat used in bread making, for example, is hexaploid ($6n$). It generates $3n$ gametes by meiosis. These fuse to form normal $6n$ offspring.

What happens when an individual has an odd number of chromosome sets? Consider triploid bananas and watermelons. Early in meiosis I, each group of three homologous chromosomes tries to pair. Spindle fibers move the three chromosomes to the metaphase plate. When the homologs separate, however, one daughter cell typically receives two copies of a given chromosome while the other daughter cell receives one copy. This asymmetrical distribution of chromosomes occurs randomly for each chromosome in the cell. As a result, virtually all of the gametes that are produced have an unbalanced complement of chromosomes.

What happens when an individual has an odd number of chromosome sets?

In plants, as in animals, additions or deletions of specific chromosomes almost always reduce an individual's ability to produce offspring, and thus its fitness. In bananas and watermelons, gametes with unbalanced chromosome sets cannot produce viable embryos. Because they lack growing embryos, their seeds die and a seedless fruit results. Triploid bananas and watermelons are seedless because they can't complete meiosis correctly.

Where do these triploid, seedless strains come from in the first place? Triploid individuals originate in several ways. For example, massive errors can occur during meiosis I in a diploid parent and lead to all of the chromosomes being pulled to a single daughter cell. If the $2n$ gamete that results fuses with a normal gamete, a triploid offspring results. In other cases, plant breeders cross tetraploid and diploid parents from closely related species to form triploid offspring. Because the triploid is sterile, it has to be propagated through cuttings or some other mode of asexual reproduction.

In tropical regions where bananas are an important export crop, it is routine for thousands of acres to be planted with asexually propagated, genetically identical individuals. Because virtually all of these individuals have the same alleles for genes that fight disease, banana plantations are extremely susceptible to new strains of disease-causing organisms—particularly fungal epidemics. For decades banana farmers have been keeping disease-causing fungi at bay with massive doses of fungicides. But even though these chemicals kill most fungi, some fungal individuals happen to have alleles that make them less susceptible to fungicides. These fungicide-resistant individuals have increased in frequency as fungicide use has continued. In many parts of the world, banana plantations are now being threatened with outbreaks of disease from fungi that have evolved resistance to all of the currently available fungicides. To solve the problem, researchers may have to find diploid and tetraploid species of wild bananas and cross them to make entirely new triploid strains.

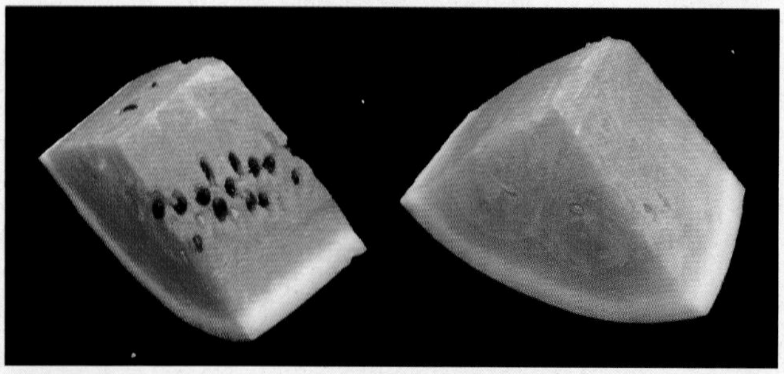

FIGURE 12.16 Seeded Watermelons Are Diploid; Seedless Watermelons Are Triploid
Diploid watermelons undergo meiosis normally, and they produce viable seeds. Triploid watermelons cannot undergo meiosis normally, and they produce aborted seeds.

CHAPTER REVIEW

Summary of Key Concepts

- **Meiosis is a type of nuclear division. It results in cells that have half as many chromosomes as the parent cell, and in animals it occurs prior to the formation of eggs and sperm. When an egg and a sperm cell combine to form an offspring, the original number of chromosomes is restored.**

When biologists confirmed that sperm and egg nuclei fuse during fertilization, it led to the hypothesis that a special type of cell division must precede gamete formation. Specifically, the proposal was that a sperm and egg must each have half of the normal number of chromosomes found in other cells. This hypothesis was confirmed when researchers observed meiosis and established that it results in gametes with half the normal chromosome number.

As the details of meiosis were being worked out, biologists realized that chromosomes exist in sets. In diploid organisms, individuals have two versions of each type of chromosome. One of the versions is inherited from the mother, and one from the father. The similar, paired chromosomes are called homologs. Haploid organisms, in contrast, have just one of each type of chromosome.

Each chromosome is replicated before meiosis begins. At the start of meiosis I, each chromosome consists of a pair of sister chromatids joined along their length and at a centromere. Homologous pairs of chromosomes synapse early in meiosis I, forming a tetrad—a group of two homologous chromosomes. After non-sister chromatids from the homologous chromosomes undergo crossing over, the pair of homologous chromosomes migrates to the metaphase plate. At the end of meiosis I, the homologous chromosomes separate and are distributed to two daughter cells. During meiosis II, sister chromatids separate and are distributed to two daughter cells.

Web Tutorial 12.1 Meiosis

- **Each cell produced by meiosis receives a different combination of chromosomes. Because genes are located on chromosomes, each cell produced by meiosis receives a different complement of genes. Meiosis leads to offspring that are genetically distinct from each other and from their parents. If offspring are genetically variable, some might thrive if the environment changes.**

When meiosis and outcrossing occur, the chromosome complements of offspring differ from one another and from their parents, for three reasons: (1) Maternal and paternal homologs are distributed randomly when chromosomes separate at the end of meiosis I; (2) maternal and paternal homologs exchange segments during crossing over; and (3) outcrossing results in a combination of chromosome sets from different individuals. The consequences of these differences became clear when biologists realized that chromosomes contain the hereditary material. Meiosis leads to genetic differences among offspring and between parents and offspring.

According to the changing-environment hypothesis, meiosis exists primarily because genetically diverse offspring are better able to resist parasites than are genetically uniform offspring. This hypothesis has been tested by studying populations of snails that contain both sexually reproducing and asexually reproducing individuals. The data from this research support the prediction that sexual reproduction is more common in habitats where parasite infection is frequent.

- **If mistakes occur during meiosis, the resulting egg and sperm cells may contain the wrong number of chromosomes. It is rare for offspring with an incorrect number of chromosomes to develop normally.**

Mistakes during meiosis lead to gametes and offspring with an unbalanced set of chromosomes. Children with Down syndrome, for example, have an extra copy of chromosome 21. The leading hypothesis to explain these mistakes is that they are random accidents that result in a failure of homologous chromosomes or sister chromatids to separate properly during meiosis.

Web Tutorial 12.2 Mistakes in Meiosis

Questions

Content Review

1. In the roundworm *Ascaris*, eggs and sperm have two chromosomes, but all other cells have four. Observations such as this inspired which important hypothesis?
   a. Before gamete formation, a special type of cell division leads to a quartering of chromosome number.
   b. Before gamete formation, a special type of cell division leads to a halving of chromosome number.
   c. After gamete formation, half of the chromosomes are destroyed.
   d. After gamete formation, either the maternal or the paternal set of chromosomes disintegrates.

2. What are homologous chromosomes?
   a. chromosomes that are similar in their size, shape, and gene content
   b. similar chromosomes that are found in different individuals of the same species
   c. the two "threads" in a replicated chromosome (they are identical copies)
   d. the products of crossing over, which contain a combination of segments from maternal chromosomes and segments from paternal chromosomes

3. What is a tetrad?
   a. the "X" that forms when chromatids from homologous chromosomes cross over
   b. a group of four chromatids produced when homologs synapse
   c. the four points where homologous chromosomes touch as they synapse
   d. the group of four genetically identical daughter cells produced by mitosis

4. What is genetic recombination?
   a. the new combination of maternal and paternal homologs that results when chromosomes separate at meiosis I
   b. the new combination of maternal and paternal chromosome segments that results when homologs cross over
   c. the new combinations of chromosome segments that result when outcrossing occurs
   d. the combination of a prominent haploid phase *and* a prominent diploid phase in a life cycle

5. What is meant by a paternal chromosome?
   a. the largest chromosome in a set
   b. a chromosome that does not separate correctly during meiosis I
   c. the member of a homologous pair that was inherited from the mother
   d. the member of a homologous pair that was inherited from the father

6. Meiosis II is similar to which process?
   a. mitosis in haploid cells
   b. nondisjunction
   c. outcrossing
   d. meiosis I

Conceptual Review

1. Triploid ($3n$) watermelons are produced by crossing a tetraploid ($4n$) strain with a diploid ($2n$) plant. Briefly explain why this mating produces a triploid individual. Why can mitosis proceed normally in triploid cells, but meiosis cannot?
2. Meiosis is called a reduction division, but all of the reduction occurs during meiosis I—no reduction occurs during meiosis II. Explain why meiosis I is a reduction division but meiosis II is not.
3. Some plant breeders are concerned about the resistance of asexually cultivated plants, such as seedless bananas, to new strains of disease-causing bacteria, viruses, or fungi. Briefly explain their concern by discussing the differences in the genetic "outcomes" of asexual and sexual reproduction.
4. Explain why nondisjunction leads to trisomy and other types of abnormal chromosome complements. In what sense are these chromosome complements "unbalanced?"
5. Examine Figure 12.1, which shows the karyotype of the lubber grasshopper. The researcher drew these chromosomes as they were undergoing meiosis. What event is occurring in the drawing?
6. Lay two pens and two pencils on a tabletop, and imagine that they represent chromosomes in a diploid cell with $n = 2$. Explain the phases of meiosis by moving the pens and pencils around. (If you don't have enough pens and pencils, use strips of paper or fabric.)

Group Discussion Problems

1. The gibbon has 44 chromosomes per diploid set, and the siamang has 50 chromosomes per diploid set. In the 1970s a chance mating between a male gibbon and a female siamang produced an offspring. Predict how many chromosomes were observed in the somatic cells of the offspring. Do you predict that this individual would be able to form viable gametes? Why or why not?
2. Meiosis results in a reassortment of maternal and paternal chromosomes. If $n = 3$ for a given organism, there are eight different combinations of paternal and maternal chromosomes. If no crossing over occurs, what is the probability that a gamete will receive *only* paternal chromosomes?
3. Suppose that while studying coral reefs in Indonesia, you discovered a new species of algae. When you examine the cells of different individuals in the laboratory, you find that some individuals have cells with 20 chromosomes while other, identical-looking individuals have cells with 10 chromosomes. What is going on?
4. The data on snail populations that were used to test the changing-environment hypothesis have been criticized because they are observational and not experimental in nature. As a result, they do not control for factors other than parasites that might affect the frequency of sexually reproducing individuals.
   a. Design an experimental study that would provide stronger evidence that the frequency of parasite infection causes differences in the frequency of sexually versus asexually reproduced individuals in this species of snail.
   b. In defense of the existing data, comment on the value of observing patterns like this in nature, versus under controlled conditions in the laboratory.

Answers to Multiple-Choice Questions: 1. b; 2. a; 3. b; 4. b; 5. d; 6. a

www.prenhall.com/freeman is your resource for the following: Web Tutorials; Online Quizzes and other Online Study Guide materials; Answers to Conceptual Review Questions; Solutions to Group Discussion Problems; Answers to Figure Caption Questions and Exercises; and Additional Readings and Research.

Mendel and the Gene

13

Gregor Mendel performed some of the most brilliant experiments in the history of biology in this garden.

KEY CONCEPTS

- Mendel discovered that in garden peas, individuals have two alleles, or versions, of each gene. Prior to the formation of eggs and sperm, the two alleles of each gene separate so that one allele is transmitted to each egg or sperm cell.
- Genes are located on chromosomes. The separation of homologous chromosomes during meiosis I explains why alleles of the same gene segregate to different gametes.
- If genes are located on different chromosomes, then the alleles of each gene are transmitted to egg cells and sperm cells independently of each other.
- Important exceptions exist to the rules that individuals have two alleles of each gene and that alleles of different genes are transmitted independently. Genes on the same chromosome are not transmitted independently of each other. If a gene is on a sex chromosome, then not every gamete will receive an allele of that gene.

The science of biology is built on a series of great ideas. Two of these—the cell theory and the theory of evolution—were introduced in Chapter 1. The cell theory describes the basic structure of organisms; the theory of evolution by natural selection clarifies why species change through time. These theories explain fundamental features of the natural world and answer some of our most profound questions about the nature of life: What are organisms made of? Where did species come from?

A third great idea in biology addresses an equally important question: Why do offspring resemble their parents? An Austrian monk named Gregor Mendel provided part of the answer in 1865 when he announced that he had worked out the rules of inheritance through a series of experiments on garden peas. The other part of the answer was provided by the biologists who described the details of meiosis during the final decades of the nineteenth century. In 1903 Walter Sutton and Theodor Boveri linked these two parts by formulating the *chromosome theory of inheritance*. This theory contends that meiosis, introduced in Chapter 12, causes the patterns of inheritance that Mendel observed. It also asserts that the hereditary factors called *genes* are located on chromosomes.

This chapter focuses on the evidence for the chromosome theory of inheritance. Let's begin with a basic question: What are the rules of inheritance that Mendel discovered?

13.1 Mendel's Experiments with a Single Trait

Gregor Mendel was a monk who lived and worked in the city of Brünn, located 70 miles north of Vienna. (Brünn was then part of Austria. Today the city is called Brno and is part of the Czech Republic.) Mendel was educated in the natural sciences at the University of Vienna and also studied physics and mathematics under Christian Doppler, who discovered the Doppler effect for sound and light waves.

In Mendel's day, questions about **heredity**—meaning inheritance, or the transmission of traits from parents to offspring—were primarily the concern of animal breeders and horticulturists. A **trait** is any characteristic of an individual, ranging from overall height to the primary structure of a particular membrane protein. In Brünn, for example, there was a particular interest in how selective breeding could result in hardier and more productive varieties of sheep, fruit trees, and vines. To that end, an Agricultural Society had been formed. Its members emphasized the importance of research that would help breeding programs become more efficient. Mendel was an active member of this society; the monastery he belonged to was also devoted to scientific teaching and research.

What Questions Was Mendel Trying to Answer?

Mendel set out to address the most fundamental of all issues concerning heredity: What are the basic patterns in the transmission of traits from parents to offspring? At the time, two hypotheses had been formulated to answer this question. The first, called *blending inheritance*, claimed that the traits observed in a mother and father blend together to form the traits observed in their offspring. For example, blending inheritance contended that black sheep have hereditary determinants for black wool and that white sheep have hereditary determinants for white wool. When these individuals mate, their hereditary determinants blend to form a new hereditary determinant for gray wool—their offspring should be gray. Mendel's scientific mentor, the widely respected Carl Nägeli, was a proponent of the blending-inheritance hypothesis.

The second hypothesis was called the *inheritance of acquired characters*, which claimed that traits present in parents are modified, through use, and passed on to their offspring in the modified form. The classical prediction of this hypothesis is that adult giraffes acquire longer necks by straining to reach leaves high in the tops of trees and that they subsequently produce longer-necked offspring. The idea here is that the genetic determinants present in an individual are modified through use. Jean-Baptiste Lamarck originally formulated this hypothesis in the eighteenth century; in Mendel's day it was championed by Charles Darwin.

These hypotheses were being promoted by the greatest scientists of Mendel's time. But are they correct? What *are* the basic patterns of inheritance?

Garden Peas Serve as the First Model Organism in Genetics

Mendel was certainly not the first scientist interested in studying the basic mechanisms of heredity. Why was he successful where others failed? Several factors came into play. One of the most important was that Mendel chose an appropriate model organism to study. A species that serves as a **model organism** consists of individuals that are usually small, short lived, inexpensive to care for, prolific in producing offspring, and easy to manipulate experimentally. Such species are called models because the conclusions drawn from studying them turn out to apply to many other species as well.

Which model organism did Mendel choose? After investigating and discarding several candidates, he chose the pea plant *Pisum sativum*. His reasons were practical: Peas are inexpensive, are easy to propagate, and have a relatively short reproductive cycle. These features made it possible for Mendel to continue experiments over several generations and to collect data from a large number of individuals. Because of his choice, garden peas became the first model organism in genetics. **Genetics** is the branch of biology that focuses on the inheritance of traits.

Two additional features of the pea made it possible for Mendel to design his experiments: He could control which parents were involved in a mating, and he could arrange matings between individuals that differed in easily recognizable traits such as flower color or seed shape. Why was this important?

How Did Mendel Arrange Matings? **Figure 13.1a** shows a garden pea flower, including its male and female reproductive organs. Sperm cells are produced in **pollen grains**, which are small sacs that mature in the male reproductive structure of the plant. Eggs are produced in the female reproductive structure. Fertilization begins when pollen grains are deposited on a tubelike section of the female reproductive organs. Sperm cells travel down this tube to the egg cells, where fertilization takes place.

Under normal conditions, garden peas pollinate themselves rather than requiring pollen from other pea plants for fertilization to occur. **Self-fertilization** (or *selfing*) takes place when pollen from one flower falls on the female reproductive organ of that same flower. Selfing is common because pollen from other plants rarely reaches the flower—its petals form a compartment that encloses the male and female reproductive organs and tends to exclude bees and other types of pollinating insects.

As **Figure 13.1b** shows, however, Mendel could circumvent this arrangement by removing the male reproductive organs from a flower before any pollen formed. Later he could put pollen from another flower on that flower's female reproductive organ. This type of mating is referred to as a **cross-pollination**, or simply a *cross*. Using this technique, Mendel could control the matings of his model organism.

What Traits Did Mendel Study? Mendel conducted his experiments on varieties of peas that differed in seven traits: seed

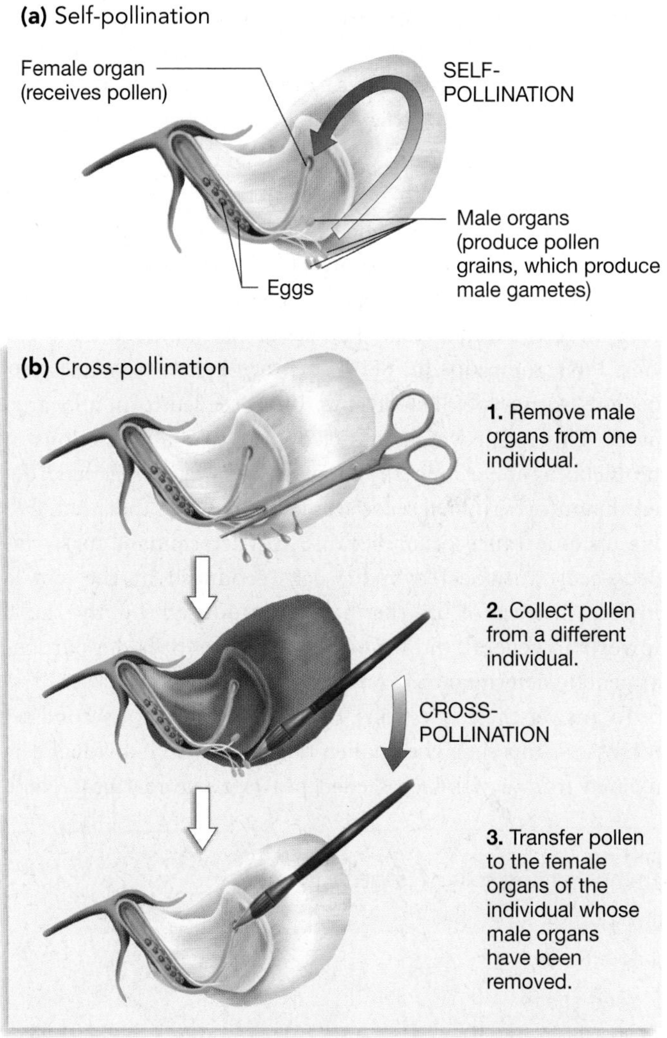
(a) Self-pollination
Female organ (receives pollen)
SELF-POLLINATION
Male organs (produce pollen grains, which produce male gametes)
Eggs
(b) Cross-pollination
1. Remove male organs from one individual.
2. Collect pollen from a different individual.
CROSS-POLLINATION
3. Transfer pollen to the female organs of the individual whose male organs have been removed.

FIGURE 13.1 Peas Can Be Self-Pollinated or Cross-Pollinated
(a) The petals of a pea form an enclosed compartment. As a result, most fertilization takes place when pollen grains from the male reproductive organ of one flower fall on the female reproductive organ of the same flower. **(b)** Mendel arranged matings between individuals by removing the male organs from one flower and then dusting its female organ with pollen collected from a different flower.

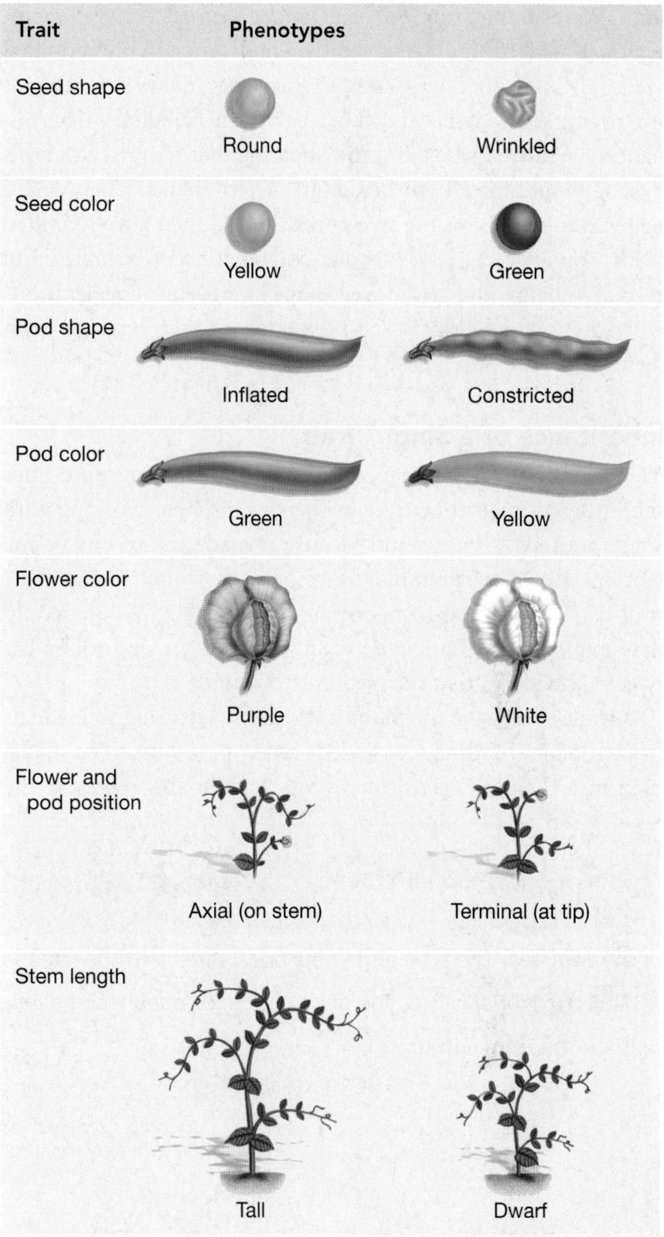
Trait
Phenotypes
Seed shape
Round
Wrinkled
Seed color
Yellow
Green
Pod shape
Inflated
Constricted
Pod color
Green
Yellow
Flower color
Purple
White
Flower and pod position
Axial (on stem)
Terminal (at tip)
Stem length
Tall
Dwarf

FIGURE 13.2 Mendel Studied Seven Traits That Were Variable in Garden Peas
Two distinct phenotypes existed for each of the seven traits that Mendel studied in garden peas.

shape, seed color, pod shape, pod color, flower color, flower and pod position, and stem length. As **Figure 13.2** shows, each trait exhibited one of two forms. Biologists refer to the observable features of an individual, such as the shape of a pea seed or the eye color of a human, as its **phenotype** (literally, "show-type"). In the pea populations that Mendel studied, two distinct phenotypes existed for all seven traits.

Mendel began his work by obtaining individuals from what breeders called pure lines or true-breeding lines. A **pure line** consists of individuals that produce offspring identical to themselves when they are self-pollinated or crossed to another member of the same population. For example, earlier breeders had developed pure lines for wrinkled seeds and round seeds. During two years of trial experiments, Mendel confirmed that individuals that germinated from his wrinkled seeds produced only wrinkled-seeded offspring when they were mated to themselves or to another pure-line individual that germinated from a wrinkled seed; individuals from his round seeds produced only round-seeded offspring when they were mated to themselves or to another pure-line individual from a round seed.

Why is this result important? Remember that Mendel wanted to find out how traits are transmitted from parents to offspring. Once he had confirmed that he was working with pure lines, he could predict how matings within each line would turn out; in other words, he knew what the offspring from these matings would look like. He could then compare these results with the outcomes of crosses between individuals from different pure lines. For example, suppose he arranged matings between an individual with round seeds and an individual with wrinkled seeds. He knew that one parent carried a hereditary determinant for round seeds, while the other carried a hereditary determinant for wrinkled seeds. But the offspring that resulted from this mating would have both hereditary determinants. They would be **hybrids**—a mix of the two types. Would they have wrinkled seeds, round seeds, or a blended combination of wrinkled and round? What would be the seed shape in subsequent generations when hybrid individuals self-pollinated or were crossed with members of the pure lines?

Inheritance of a Single Trait

Mendel's first set of experiments involved crossing pure lines that differed in just one trait, such as seed shape. Working with single traits was important because it made the results of the matings more interpretable. Once he understood how a single trait was transmitted from parents to offspring, Mendel could then explore what happened when crosses were performed between individuals that differed in two traits.

Mendel began his single-trait crosses by crossing individuals from round-seeded and wrinkled-seeded pure lines. The adults used in an initial experimental cross such as this represent the **parental generation.** Their progenies (that is, offspring) are called the **F_1 generation.** (F_1 stands for "first filial"; *filial* refers to the relationship between parent and offspring.) Subsequent generations are symbolized as the F_2 generation, F_3 generation, and so on.

Certain Traits "Recede" In his first set of crosses, Mendel took pollen from round-seeded plants and placed it on the female reproductive organs of plants from the wrinkled-seeded line. As **Figure 13.3** shows, all of the progeny seeds resulting from this cross were round. This was remarkable, for two reasons. First, the traits did not blend together to form an intermediate phenotype. Instead, the round-seeded form appeared intact. This result was in stark contrast to the predictions of the blending-inheritance hypothesis. Second, the genetic determinant for wrinkled seeds seemed to have disappeared. Did this disappearance occur because the determinant for wrinkled seeds was located in the egg (produced by the female flower) instead of in the pollen (produced by the male flower)? In general, did it matter which parent had a particular genetic determinant?

To answer these questions, Mendel performed a second set of crosses—this time with pollen taken from an individual germinated from a wrinkled-seeded pea (see Figure 13.3). These

Question: Is the inheritance of seed shape in peas affected by whether the genetic determinant is in a male or female gamete?

Hypothesis: The type of gamete *does* affect the inheritance of seed shape.

Null hypothesis: The type of gamete *does not* affect the inheritance of seed shape.

Experimental setup:

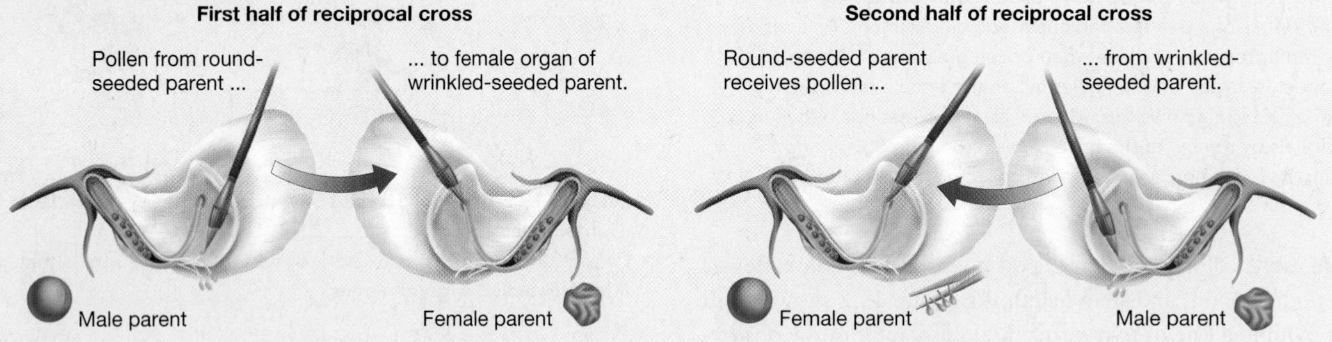
First half of reciprocal cross
Pollen from round-seeded parent ...
... to female organ of wrinkled-seeded parent.
Male parent
Female parent
Second half of reciprocal cross
Round-seeded parent receives pollen ...
... from wrinkled-seeded parent.
Female parent
Male parent

Prediction of "sex matters" hypothesis: Offspring phenotypes will be different in the two crosses.

Prediction of null hypothesis: Offspring phenotypes will be identical in the two crosses.

Results:

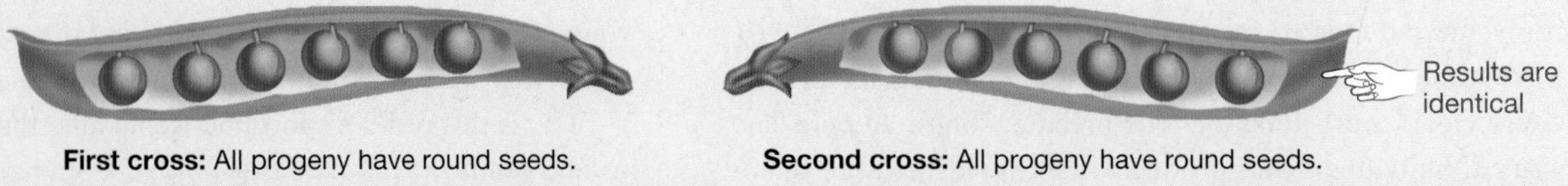
Results are identical
First cross: All progeny have round seeds.
Second cross: All progeny have round seeds.

Conclusion: It makes no difference whether the genetic determinant for seed shape comes from the male gamete or from the female gamete.

FIGURE 13.3 A Reciprocal Cross

QUESTION What is the purpose of a reciprocal cross?

crosses completed a **reciprocal cross**—a set of matings where the mother's phenotype in the first cross is the father's phenotype in the second cross, and the father's phenotype in the first cross is the mother's phenotype in the second cross.

In this case the results of the reciprocal crosses were identical: All of the F_1 progeny in the second cross had round seeds, just as in the first cross. This second cross established that it does not matter whether the genetic determinants for seed shape are located in the male or female parent. But what had happened to the genetic determinant for wrinkled seeds?

Dominant and Recessive Traits Mendel planted the F_1 seeds and allowed the individuals to self-pollinate when they matured. He collected the seeds that were produced by many plants in the F_2 generation and observed that 5474 were round and 1850 were wrinkled. This observation was striking. The wrinkled seed shape reappeared in the F_2 generation after disappearing completely in the F_1 generation! No one had observed this phenomenon before because it had been customary for biologists to stop their breeding experiments with F_1 offspring. Mendel was successful where others failed, in part because he extended his experiments to a second generation.

Mendel invented some important terms to describe this result. He designated the genetic determinant for the wrinkled shape as **recessive**. This was an appropriate term because none of the F_1 individuals had wrinkled seeds—meaning the determinant for wrinkled seeds appeared to recede or temporarily become latent. In contrast, Mendel referred to the genetic determinant for round seeds as **dominant**. This term was apt because the round-seed determinant appeared to dominate over the wrinkled-seed determinant when both were present. It's important to note, though, that in genetics the term *dominant* has nothing to do with the everyday English usage as powerful or superior. Subsequent research has shown that individuals with the dominant phenotype do not necessarily have higher fitness than do individuals with the recessive phenotype. In fact, there are many examples of dominant genetic determinants that lead to the death of the individual carrying them. Nor are dominant genetic determinants necessarily more common than recessive ones. In genetics, the terms *dominance* and *recessiveness* refer *only* to which phenotype is observed in individuals carrying two different genetic determinants.

What is the relationship between these two types of determinants? That is, how do they interact? Mendel made an important start in answering these questions when he noticed that the round and wrinkled seeds of the F_2 generation were present in a ratio of 2.96:1, or essentially 3:1. The 3:1 ratio means that for every four individuals, on average three had the dominant phenotype and one had the recessive phenotype. In other words, about 3/4 of the F_2 seeds were round and 1/4 were wrinkled.

Before trying to interpret this pattern, however, it was important for Mendel to establish that the results were not restricted to inheritance of seed shape. So he repeated the experiments with each of the six other traits of pea plants. In each case, he obtained similar results: The products of reciprocal crosses were the same; one form of the trait was always dominant regardless of the parent it came from; the F_1 progeny showed only the dominant trait and did not exhibit an intermediate phenotype; and in the F_2 generation, the ratio of individuals with dominant and recessive phenotypes was 3 to 1.

How could these patterns be explained? Mendel answered this question with a series of propositions about the nature and behavior of the hereditary determinants. These hypotheses rank as some of the most brilliant insights in the history of biology.

The Nature and Behavior of the Hereditary Determinants

Mendel's results were clearly inconsistent with the hypothesis of blending inheritance. To explain the patterns that he observed, Mendel proposed a competing hypothesis called *particulate inheritance*. He maintained that the hereditary determinants for traits do not blend together or acquire new or modified characteristics through use. In fact, hereditary determinants maintain their integrity from generation to generation. Instead of blending together, they act like discrete entities or particles.

Mendel's hypothesis was the only way to explain the observation that phenotypes disappeared in one generation and reappeared intact in the next. It also represented a fundamental break with ideas that had prevailed for hundreds of years.

What Are Genes, Alleles, and Genotypes? Today geneticists use the word **gene** to indicate the hereditary determinant for a trait. For example, the hereditary factor that determines the difference between round and wrinkled seeds in garden peas is referred to as the gene for seed shape.

Mendel's insights were even more penetrating, however. He also proposed that each individual has two versions of each gene. Today different versions of the same gene are called **alleles**. Different alleles are responsible for the variation in the traits that Mendel studied. In the case of the gene for seed shape, one allele of this gene is responsible for the round form of the seed while another allele is responsible for the wrinkled form. The alleles that are found in a particular individual are called its **genotype**. An individual's genotype has a profound effect on its phenotype—its physical traits.

The hypothesis that alleles exist in pairs was important because it gave Mendel a framework for explaining dominance and recessiveness. He proposed that some alleles are dominant and others are recessive. Dominance and recessiveness identify which phenotype actually appears in an individual when both alleles are present. In garden peas, the allele for round seeds is dominant; the allele for wrinkled seeds is recessive. Therefore, as long as one allele for round seeds is present, seeds are round. When both alleles present are for wrinkled seeds (thus no allele for round seeds is present), seeds are wrinkled.

These hypotheses explain why the phenotype for wrinkled seeds disappeared in the F_1 generation and reappeared in the F_2 generation. But why did round- and wrinkled-seeded plants exist in a 3:1 ratio in the F_2 generation?

The Principle of Segregation To explain the 3:1 ratio of phenotypes in F_2 individuals, Mendel reasoned that the two alleles of each gene must *segregate*—that is, separate—into different gamete cells during the formation of eggs and sperm in the parents. As a result, each gamete contains one allele of each gene. This idea is called the **principle of segregation.**

To show how this principle works, Mendel used a letter to indicate the gene for a particular trait. For example, *R* represents the gene for seed shape. He used uppercase (*R*) to symbolize a dominant allele and lowercase (*r*) to symbolize a recessive allele. (Note that the symbols for genes are always italicized.)

Using this notation, Mendel could describe the genotype of the individuals in the pure line with round seeds (dominant) as *RR*. The genotype of the pure line with wrinkled seeds (recessive) is *rr*. Because *RR* and *rr* individuals have two copies of the same allele, they are said to be **homozygous** for the seed-shape gene (*homo* is the Greek root for "same," while *zygo* means "yoked together"). Pure-line individuals always produce offspring with the same phenotype because they are homozygous—no other allele is present.

Figure 13.4a diagrams what happened to these alleles when Mendel crossed the *RR* and *rr* pure lines. According to his analysis, *RR* parents produce eggs and sperm that carry the *R* allele, while *rr* parents produce gametes with the *r* allele. When two gametes—one from each parent—are fused together, they create offspring with the *Rr* genotype. Such individuals, with two different alleles for the same gene, are said to be **heterozygous** (*hetero* is the Greek root for "different"). Because the *R* allele is dominant, all of these F_1 offspring produced round seeds.

Why do the two phenotypes appear in a 3:1 ratio in the F_2 generation? A mating between parents that are both heterozygous at the gene in question is called a **monohybrid cross.** Mendel proposed that during gamete formation in the F_1 (heterozygous) individuals, the paired *Rr* alleles separate into different gamete cells. As a result, about half of the gametes carry the *R* allele and half carry the *r* allele (**Figure 13.4b**). During self-fertilization, a given sperm has an equal chance of fertilizing either an *R*-bearing egg or an *r*-bearing egg. R. C. Punnett invented a straightforward technique for predicting the genotypes and phenotypes that should appear in the resulting offspring. In a **Punnett square,** each gamete genotype produced by one parent is shown as a row, and each gamete genotype produced by the other parent is shown as a column. Then the boxes of the square or rectangle composed of the rows and columns are filled in with the offspring genotypes that result from fusion of these gametes. Finally, the proportions of each offspring genotype and phenotype can be calculated by tallying the offspring genotypes and phenotypes present in the boxes. (**Box 13.1** explains the logic behind these calculations.) For example, the Punnett square in Figure 13.4b predicts that 1/4 of the F_2 offspring will be *RR*, 1/2 will be *Rr*, and 1/4 will be *rr*. Because the *R* allele is dominant to the *r* allele, 3/4 of the offspring should be round seeded and 1/4 should be wrinkled seeded.

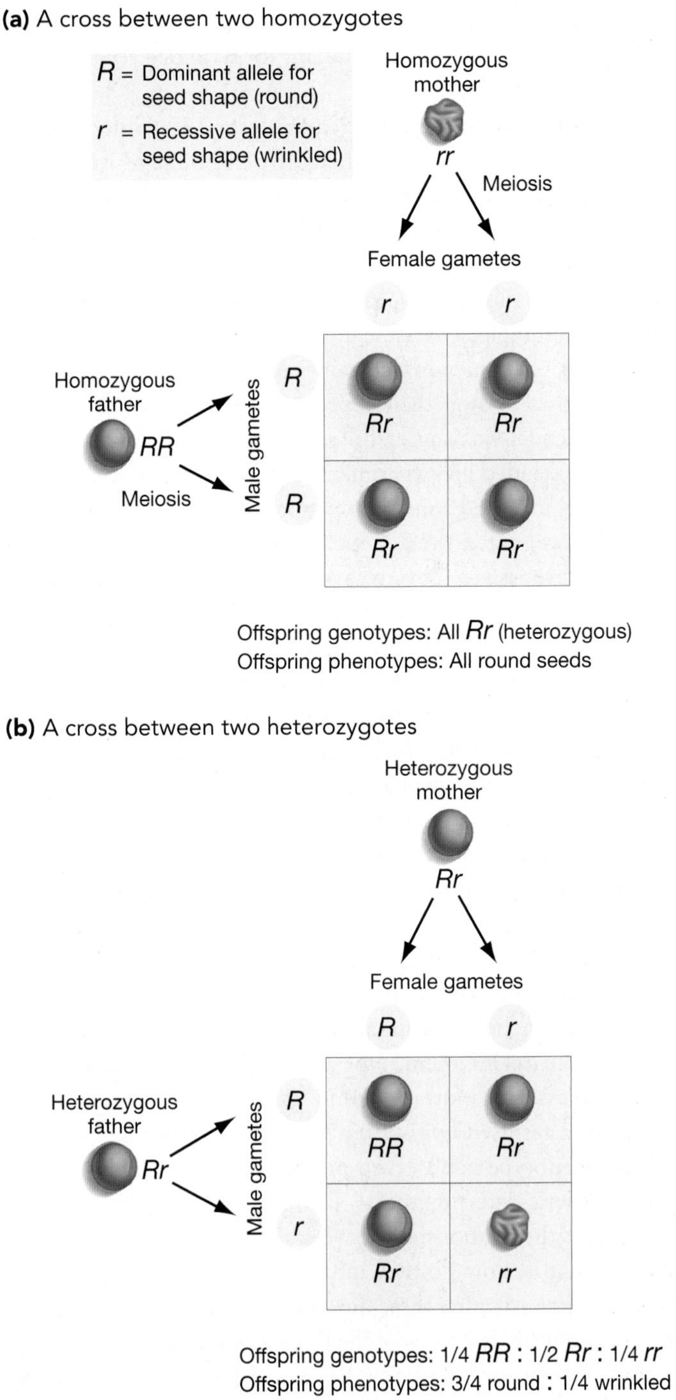

FIGURE 13.4 Mendel Analyzed the F_1 and F_2 Offspring of a Cross Between Pure Lines

QUESTION In constructing a Punnett square, does it matter whether the male or female gametes go on the left or across the top? Why or why not?

BOX 13.1 Why Do Punnett Squares Work?

Punnett squares are powerful tools because they allow biologists to predict the outcome of a particular mating. The calculations involved in setting up and analyzing Punnett squares have a strong theoretical basis. Punnett squares work because they are based on two fundamental rules of probability, called the "both-and rule" and the "either-or rule." Each rule pertains to a distinct situation.

The *both-and rule* applies when you want to know the probability that two or more independent events occur together. Let's use the rolling of two dice as an example. What is the probability of rolling two sixes? These two events are independent, because the probability of rolling a six on one die has no effect on the probably of rolling a six on the other die. (In the same way, the probability of getting a gamete with allele *R* from one parent has no effect on the probability of getting a gamete with allele *R* from the other parent. Gametes fuse randomly.) The probability of rolling a six on the first die is 1/6. The probability of rolling a six on the second die is also 1/6. The probability of rolling a six on both die, then, is $1/6 \times 1/6 = 1/36$. In other words, if you rolled two dice 36 times, on average you would expect to roll two sixes once. It should make sense that the both-and rule is also called the *multiplication rule* or *product rule*. In the case of the cross diagrammed in Figure 13.4b, the probability of getting allele *R* from the father is 1/2 and the probability of getting *R* from the mother is 1/2. Thus, the probability of getting both alleles and creating an offspring with genotype *RR* is $1/2 \times 1/2 = 1/4$.

The *either-or rule*, in contrast, applies when you want to know the probability of an event happening when there are several ways for the event to occur. In this case, the probability that the event will occur is the sum of the probabilities of each way that it can occur. For example, suppose you wanted to know the probability of rolling either a one or a six when you toss a die. The probability of tossing each is 1/6, so the probability of getting one or the other is $1/6 + 1/6 = 1/3$. (The either-or rule is also called the *addition rule* or *sum rule*.) If you rolled a die three times, on average you'd expect to get a one or a six once. In Figure 13.4b, the probability of getting an *R* allele from the father and an *r* allele from the mother is $1/2 \times 1/2 = 1/4$. Similarly, the probability of getting an *r* allele from the father and an *R* allele from the mother is $1/2 \times 1/2 = 1/4$. Thus, the combined probability of getting the *Rr* genotype in either of the two ways is $1/4 + 1/4 = 1/2$. Because a Punnett square shows every possible way to generate a particular genotype, you can calculate the total probability of getting a particular genotype or phenotype simply by calculating how often it occurs in the square.

These results are *exactly* what Mendel found in his experiments with peas. In the simplest and most elegant fashion possible, his interpretation explains the 3:1 ratio of round to wrinkled seeds observed in the F_2 offspring and the mysterious reappearance of the wrinkled seeds.

The term **genetic model** refers to a set of hypotheses that explains how a particular trait is inherited. **Figure 13.5** summarizes Mendel's model for explaining the basic patterns in the transmission of traits from parents to offspring. These hypotheses are sometimes referred to as *Mendel's rules*. His genetic model was a radical break from the hypotheses of blending inheritance and inheritance of acquired characters that previously dominated scientific thinking about heredity.

MENDEL'S MODEL

1. **Peas have two versions, or alleles, of each gene.** This also turns out to be true for many other organisms.
2. **Alleles do not blend together.** The hereditary determinants maintain their integrity from generation to generation.
3. **Each gamete contains one allele of each gene.** The alleles of each gene segregate during the formation of gametes.
4. **Males and females contribute equally to the genotype of their offspring.** When gametes fuse, offspring acquire a total of two alleles for each gene—one from each parent.
5. **Some alleles are dominant to other alleles.** When a dominant allele and a recessive allele for the same gene are found in the same individual, that individual has the dominant phenotype.

FIGURE 13.5 Mendel Created a Model to Explain the Results of a Cross between Pure Lines

QUESTION What is the difference between genes and alleles, a genotype and a phenotype, a homozygous individual and a heterozygous individual, and dominant alleles and recessive alleles?

Testing the Model

Mendel's model explained his results in a logical way. But is it correct? To answer this question, Mendel conducted a series of experiments with the F_2 progeny described in Figure 13.4b. These experiments tested two important predictions:

1. Plants with wrinkled seeds are *rr*. Thus, they should produce only *rr* offspring when they are self-pollinated or crossed with another individual with wrinkled seeds.

WEB TUTORIAL 13.1 Mendel's Experiments

2. Plants with the dominant phenotype may be either *Rr* or *RR*. These two genotypes should be present in the ratio 2:1. (That is, there should be twice as many heterozygotes as homozygotes among individuals with round seeds.) Individuals with the *RR* genotype should produce only *RR* offspring when they are self-pollinated. In contrast, *Rr* individuals should produce offspring with the same 3:1 ratio of round:wrinkled phenotypes observed in the cross diagrammed in Figure 13.4b.

Mendel planted the F_2 seeds and allowed the plants to self-pollinate when they matured. He then examined the phenotypes of the F_3 seeds. He quickly confirmed the first prediction: F_2 plants with wrinkled seeds always produced offspring with wrinkled seeds. This result was consistent with the claim that these F_2 plants have an *rr* genotype.

What about the offspring of parents with the dominant phenotype? Mendel let 565 round-seeded plants self-pollinate. Of these, 193 plants produced only round-seeded offspring. Mendel inferred that these parents had the *RR* genotype. In contrast, 372 of the round-seeded parents produced seeds that were either round or wrinkled. Mendel inferred that in these individuals, the round-seeded parental genotype was *Rr*. In this experiment, the ratio of *Rr* to *RR* in the parents—based on the number of round and wrinkled seeds they produced—was 1.93:1. This is extremely close to the prediction of 2:1. Mendel observed the same patterns when he let F_2 individuals from the other six crosses (involving seed color and the other traits) self-fertilize. These results were a ringing confirmation of his model.

13.2 Mendel's Experiments with Two Traits

Working with one trait at a time allowed Mendel to establish that blending inheritance does not occur. It also allowed him to infer that each gene he was working with had two alleles and to recognize the principle of segregation. His next step was to extend these results. The most important question he addressed was whether the principle of segregation holds true if parental lines differ with respect to two traits.

To explore this issue, Mendel crossed a pure-line parent that produced round, yellow seeds with a pure-line parent that produced wrinkled, green seeds. The F_1 offspring of this cross should be heterozygous at both genes. A mating between parents that are both heterozygous for two traits is called a **dihybrid cross. Figure 13.6** shows how Mendel set up one of his dihybrid crosses.

Mendel's earlier experiments had established that the allele for yellow seeds was dominant to the allele for green seeds; these alleles were designated *Y* for yellow and *y* for green. As Figure 13.6 indicates, two distinct possibilities existed for how the alleles of these two different genes—the gene for seed shape and the gene for seed color—would be transmitted to offspring. The first possibility was that the allele for seed shape and the allele for seed color present in each parent would separate from one another and be transmitted independently. This hypothesis is called **independent assortment,** because the two alleles would separate and sort themselves independently of each other (Figure 13.6a). The second possibility was that the allele for seed shape and the allele for seed color would be transmitted to gametes together. This hypothesis can be called *dependent assortment,* because the transmission of one allele would depend on the transmission of another (Figure 13.6b).

As Figure 13.6 shows, the F_1 offspring of Mendel's mating are expected to have the dominant round and yellow phenotypes whether the alleles were transmitted together or independently. When Mendel did the cross and observed the F_1 individuals, this is exactly what he found. All of the F_1 offspring had round, yellow seeds. All of these individuals were heterozygous at both genes.

In contrast to the situation in the F_1 generation, however, the two hypotheses make radically different predictions about what Mendel should have observed when the F_1 individuals were allowed to self-fertilize and produce an F_2 generation. If the alleles assort independently and combine randomly to form gametes, then each heterozygous parent should produce four different gamete genotypes, as illustrated in Figure 13.6a. A 4-row-by-4-column Punnett square results, and it predicts that there should be 9 different offspring genotypes and 4 phenotypes. Further, the yellow-round, green-round, yellow-wrinkled, and green-wrinkled phenotypes should be present in the frequencies 9/16, 3/16, 3/16, and 1/16, respectively. This is equivalent to a ratio of 9:3:3:1. But if the alleles from each parent stay together, then a 2-row-by-2-column Punnett square would predict only three possible offspring genotypes and just two phenotypes, as Figure 13.6b shows. The hypothesis of dependent assortment predicts that F_2 offspring should be yellow-round or green-wrinkled, present in a ratio of 3:1. Note that the Punnett squares are making explicit predictions about the outcome of an experiment, based on a specific hypothesis about which alleles are present in each parent and how they are transmitted.

When Mendel examined the phenotypes of the F_2 offspring, he found that they conformed to the predictions of the hypothesis of independent assortment. Four phenotypes were present in frequencies that closely approximated the predicted frequencies of 9/16, 3/16, 3/16, and 1/16 and the predicted ratio of 9:3:3:1 (Figure 13.6c). Based on these data, he accepted the hypothesis that alleles of different genes are transmitted independently of one another. This result became known as the **principle of independent assortment.**

As an aside, it's interesting to analyze how these data relate to the 3:1 ratio of phenotypes observed for single traits. According to the data in Figure 13.6c, the ratio of round seeds to

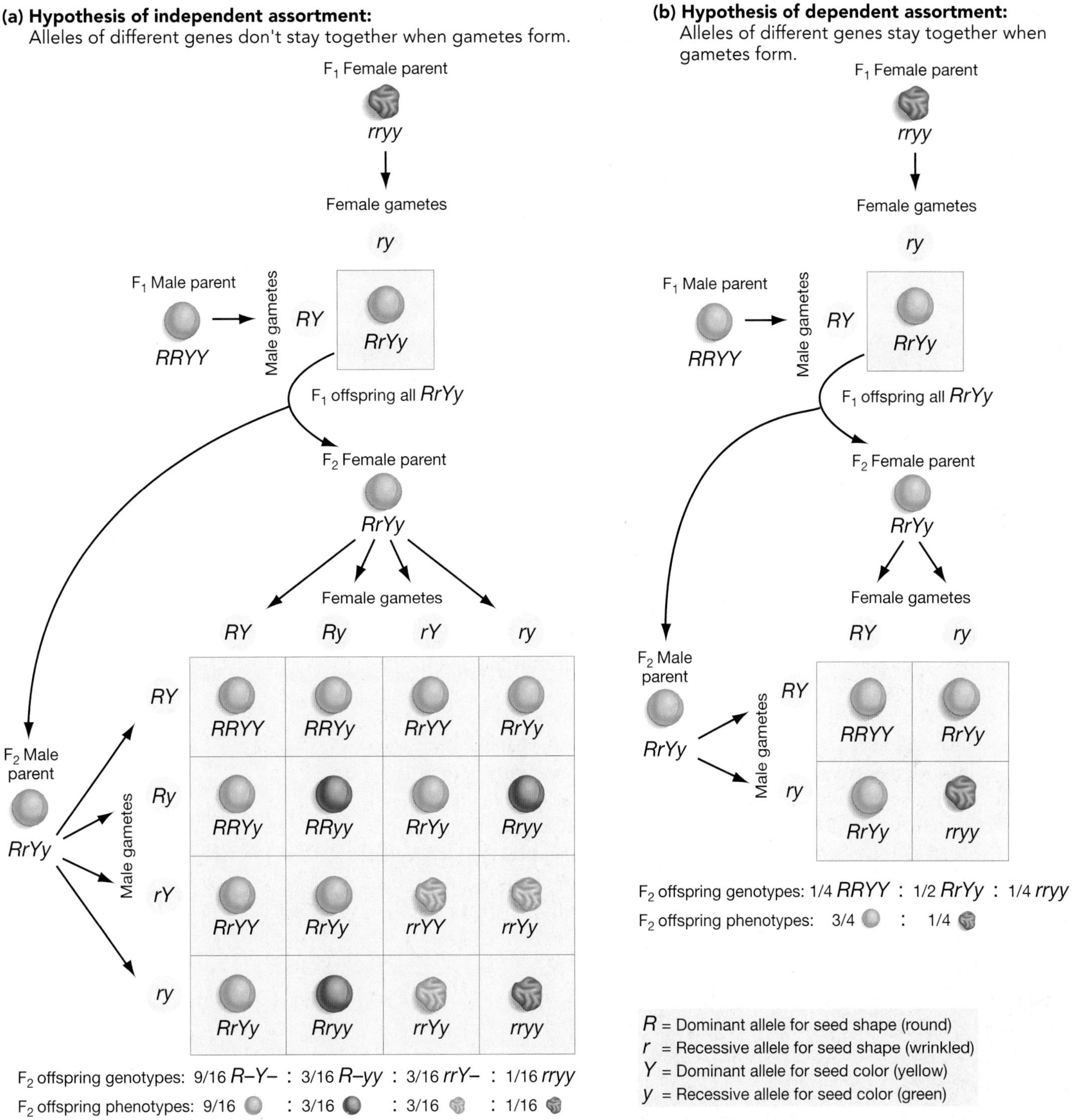

(c) Mendel's results

F_2 generation phenotype					
Number	315	101	108	32	556 total
Fraction of offspring	9/16	3/16	3/16	1/16	These data are consistent with the predictions of independent assortment

FIGURE 13.6 Mendel Analyzed the F_1 and F_2 Offspring of a Cross between Pure Lines for Two Traits
Either of two events could occur when alleles of different genes are transmitted to offspring: The alleles could sort themselves into gametes independently of each other, or alleles from the same parent could be transmitted together, generation after generation.

wrinkled seeds in the F_2 offspring was 423 to 133, or 3.18 to 1. (**Box 13.2** explains why Mendel used such large sample sizes in his experiments.) Likewise, the ratio of yellow to green seeds was 416 to 140—almost exactly 3 to 1. When seed shape and seed color are considered separately, then, the F_2 individuals exhibited the same 3:1 ratio of dominant to recessive phenotypes observed in the first set of experiments. In each case, the dominant phenotype showed up in 3/4 of the F_2 offspring, while the recessive phenotype was observed in 1/4. Because the probability of observing each of the dominant phenotypes is 3/4, the probability of observing both is $3/4 \times 3/4 = 9/16$.

Using a Testcross to Confirm Predictions

Mendel did experiments with combinations of traits other than seed shape and color and obtained results similar to those in Figure 13.6c. Each paired set of traits produced a 9:3:3:1 ratio of progeny phenotypes in the F_2 generation. He even did a limited set of crosses examining three traits at a time. Although all of these data were consistent with the principle of independent assortment, his most powerful support for the hypothesis came from a different type of experiment.

In designing this study, Mendel's goal was to test the prediction that an *RrYy* plant produces four different types of ga-

BOX 13.2 Sample Size and Chance Fluctuations

Each time Mendel designed a cross to answer a question about the inheritance of a trait, he made sure to analyze the results in a large number of offspring. Collecting data from a large sample is a goal of scientific studies. Large sample sizes reduce chance fluctuations in the outcome of an experiment and make it easier to recognize patterns in the data.

Chance fluctuations are inevitable in the outcome of any experiment or observational study. For example, consider the data shown in the accompanying table. They indicate the seed-shape phenotypes that Mendel observed in offspring from heterozygous (*Rr*) parents that self-fertilized.

From parent to parent, the ratios of round and wrinkled seeds fluctuated from almost 1:1 to over 4:1. This is because gametes combine at random and each parent's offspring represent a small sample. For example, even though only half the gametes available when parent plant 4 self-fertilized contained an *r* allele, there happened to be a run of fertilizations between *r*- and *r*-containing gametes, so the ratio of round to wrinkled phenotypes for this parent is lower than usual. Small samples frequently have skewed results like this, just due to chance. But when Mendel pooled the data into a large sample, it became clear that the overall ratio of round to wrinkled phenotypes was about 3:1 (**Figure 13.7**). If Mendel had been able to obtain a larger number of offspring from parent plant 4, it is very likely that the observed ratio of phenotypes from that parent would have been closer to 3:1.

$Rr \times Rr$

Plant Number	Round (*RR* or *Rr*)	Wrinkled (*rr*)	Ratio
1	45	12	3.75:1
2	27	8	3.37:1
3	24	7	3.42:1
4	19	16	1.19:1
5	32	11	2.91:1
6	26	6	4.33:1
7	88	24	3.66:1
8	22	10	2.20:1
9	28	6	4.66:1
10	25	7	3.57:1
Total	**336**	**107**	**3.14:1**

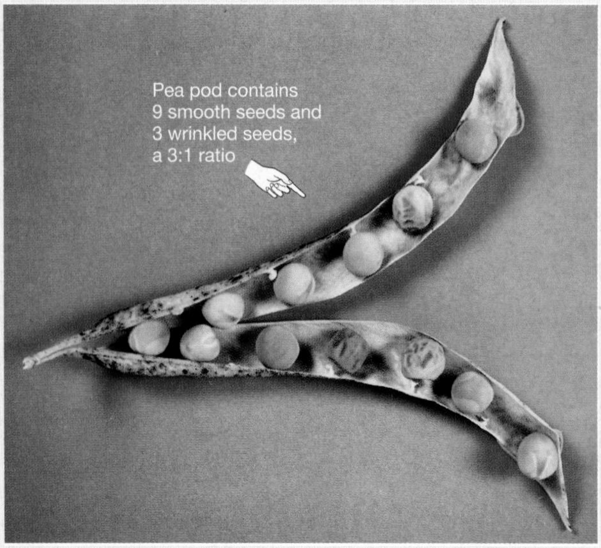
Pea pod contains
9 smooth seeds and
3 wrinkled seeds,
a 3:1 ratio

FIGURE 13.7 When an *Rr* Individual Self-Fertilizes, Offspring Phenotypes Are in a Ratio of 3:1
Garden pea seed pod showing the offspring of an *Rr* individual that was allowed to self-fertilize. In this pod, the offspring phenotypes are in exact 3:1 proportions. QUESTION Do you predict that seed pods with exact 3:1 ratios are observed rarely or frequently in this cross? Explain your answer.

metes in equal proportions. To accomplish this, Mendel invented a technique called a testcross. A **testcross** uses a parent that contributes only recessive alleles to its offspring to help determine the unknown genotype of the second parent. Testcrosses are useful because the genetic contribution of the homozygous recessive parent is easy to predict and analyze. As a result, a testcross allows experimenters to test the genetic contribution of the other parent. If the other parent has the dominant phenotype but an unknown genotype, the results of the testcross allow researchers to infer whether that parent is homozygous or heterozygous for the dominant allele.

In this case, Mendel performed a testcross between parents that were *RrYy* and *rryy*. The types and proportions of offspring that should result can be predicted with the Punnett square shown in **Figure 13.8**. If the principle of independent assortment is valid, there should be four types of offspring in equal proportions.

What were the actual proportions observed? Mendel did this experiment and examined the seeds produced by the progeny. He found that 31 were round and yellow, 26 were round and green, 27 were wrinkled and yellow, and 26 were wrinkled and green. As predicted, these numbers are nearly identical to the 27.5 individuals expected with each genotype, given the total of 110 individuals. The predicted ratio of phenotypes was 1:1:1:1, which matched the observed ratio. The testcross had confirmed the principle of independent assortment.

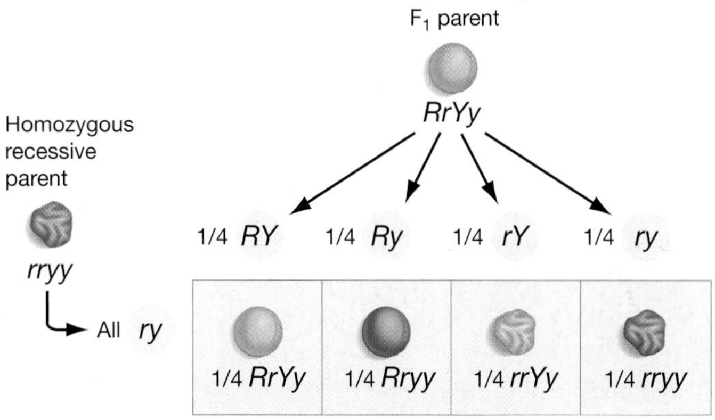

FIGURE 13.8 The Predictions Made by the Principle of Independent Assortment Can Be Evaluated in a Testcross
If the principle of independent assortment is correct and *RrYy* parents produce four types of gametes in equal proportions, then a mating between *RrYy* and *rryy* parents should produce four types of offspring in equal proportions, as this Punnett square shows.

✓CHECK YOUR UNDERSTANDING

Mendel discovered that individuals have two alleles of each gene, that each gamete receives one of the two alleles present in a parent, and that alleles from different genes are transmitted to gametes independently of each other. The alleles he analyzed were either dominant or recessive, meaning heterozygous individuals had the dominant phenotype. You should be able to (1) create and analyze Punnett squares to predict the genotypes and phenotypes that will occur in the F_1 and F_2 offspring of a cross, then calculate the expected frequency of each genotype and phenotype. Given the outcome of a cross, you should also be able to (2) infer the genotypes and phenotypes of the parents. (Use the questions at the end of this chapter to practice these skills.)

Mendel's Contributions to the Study of Heredity

Mendel introduced an approach to studying heredity that is still in use today. This approach includes the following considerations:

- The choice of an appropriate model organism;
- The use of pure lines with discrete differences in traits to explore how traits are transmitted to offspring;
- The examination of large numbers of progeny for each type of cross;
- The application of the rules of probability to predict the numbers and types of progeny produced from crosses;
- The use of reciprocal crosses to test the hypothesis that an allele's transmission depends on the sex of the parent; and
- The use of testcrosses to establish the genotype of a parent.

In short, Mendel's work provided a powerful conceptual framework for thinking about heredity. He was the first individual to describe correctly the basic rules of **transmission genetics**—the patterns that occur as alleles pass from one generation to the next. All of the types and proportions of progeny that Mendel observed in his F_1 and F_2 generations could be explained as the consequence of two processes: (1) the segregation of discrete, paired alleles into separate gametes and (2) the independent assortment of alleles that affect different traits.

Mendel's experiments were brilliant in design, execution, and interpretation. Unfortunately, they were ignored for 34 years. Historians of science frequently debate why Mendel's work was overlooked for so long. It is almost undoubtedly true that his use of probability theory in explaining his results and his quantitative treatment of data were difficult for biologists of that time to understand and absorb. It may also be true that the theory of blending inheritance was so well entrenched that there was a tendency to dismiss his results as peculiar or unbelievable. Whatever the reason, Mendel's work was not appreciated until other biologists, working with a variety of plants and animals, independently reproduced his results in the early 1900s.

13.3 The Chromosome Theory of Inheritance

The rediscovery of Mendel's work, 16 years after his death, ignited the young field of genetics. Mendel's experiments established the basic rules that govern how traits are passed from parents to offspring. They described the pattern of inheritance. But what process is responsible for these patterns? Two biologists, working independently, came up with the answer. Walter Sutton and Theodor Boveri each realized that meiosis could be responsible for

Mendel's rules. When this hypothesis was published in 1903, research in genetics exploded.

Recall from Chapter 12 that meiosis is the type of cell division that precedes gamete formation. The details of the process were worked out in the final decades of the nineteenth century. What Sutton and Boveri grasped is that meiosis not only reduces chromosome number by half, but also explains the principle of segregation and the principle of independent assortment.

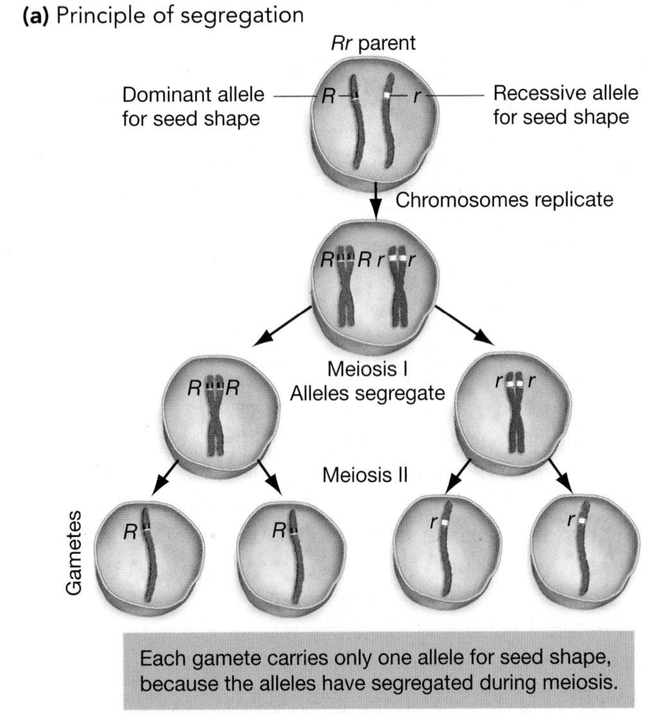

The cell nucleus at the top of **Figure 13.9a** illustrates Sutton and Boveri's central insight—the hypothesis that chromosomes are composed of Mendel's hereditary determinants, or genes. In this example, the gene for seed shape is shown at a particular position along a chromosome. This location is known as a **locus** ("place"; plural: *loci*). A genetic locus is the physical location of a gene. The paternal and maternal chromosomes shown in Figure 13.9a happen to possess different alleles of the gene for seed shape: One allele specifies round seeds (*R*), while the other specifies wrinkled seeds (*r*).

The subsequent steps in Figure 13.9a show how these alleles segregate into different daughter cells during meiosis I, when homologous chromosomes separate. This physical separation of alleles produces Mendel's principle of segregation.

Figure 13.9b follows the fate of the alleles for two different genes—in this case, for seed shape and seed color—as meiosis proceeds. Because these genes are located on different nonhomologous chromosomes, they assort independently of one another at meiosis I. Four types of gametes, produced in equal proportions, result. This is the physical basis of Mendel's principle of independent assortment. Most of the genes that Mendel analyzed assort independently from one another because they are each located on different chromosomes.

Sutton and Boveri formalized these observations in the **chromosome theory of inheritance.** Like other theories in biology, the chromosome theory consists of a pattern—a set of observations about the natural world—and a process that explains the pattern. The chromosome theory states that Mendel's rules can be explained by the independent alignment and separation of homologous chromosomes at meiosis I.

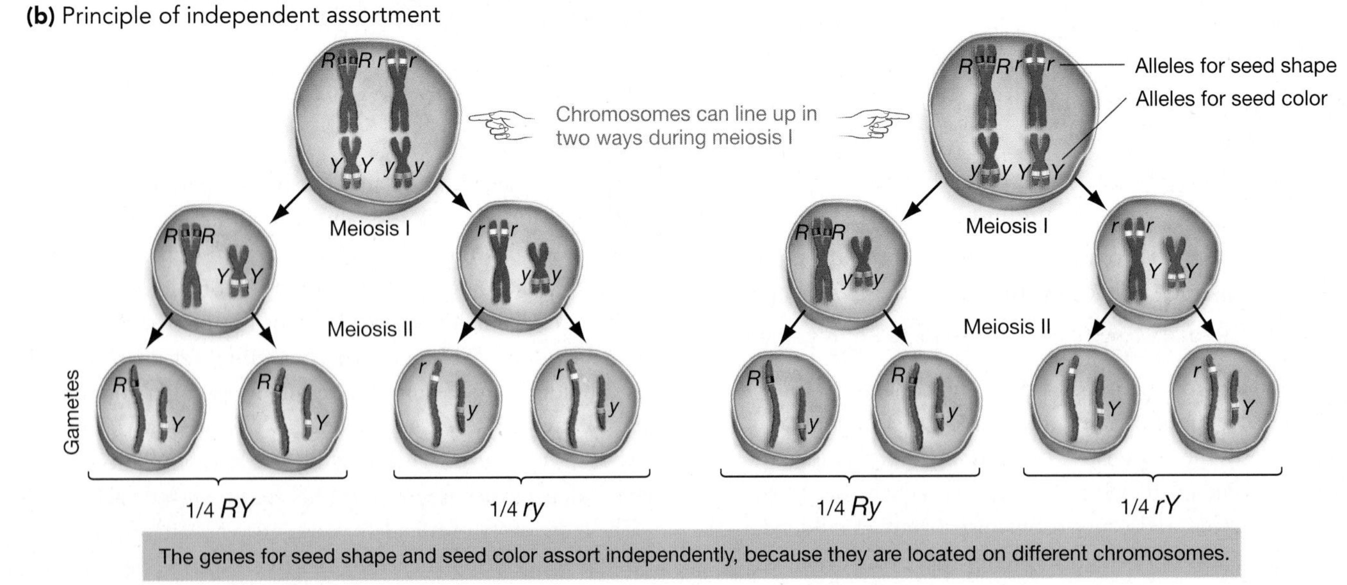

FIGURE 13.9 Meiosis Is Responsible for the Principles of Segregation and Independent Assortment
(a) A parent's two alleles segregate into different gametes, as Mendel hypothesized, because homologous chromosomes separate during meiosis I. **(b)** The alleles for different traits assort independently, again as Mendel hypothesized, because nonhomologous chromosomes assort independently during meiosis I. Maternal and paternal chromosomes are shown in different colors for clarity.

When Sutton and Boveri published their findings, however, the hypothesis that chromosomes consist of genes was untested. What experiments confirmed that chromosomes contain genes?

13.4 Testing and Extending the Chromosome Theory

During the first decade of the twentieth century, an unassuming insect rose to prominence as a model organism for testing the chromosome theory of inheritance. This organism—the fruit fly *Drosophila melanogaster*—has been at the center of genetic studies ever since. *Drosophila melanogaster* has all the attributes of a useful model organism for experimental studies in genetics: small size, ease of rearing in the lab, a short reproductive cycle (about 10 days), and abundant offspring (up to a few hundred per mating). The elaborate external anatomy of this insect also makes it possible to identify interesting phenotypic variation among individuals (**Figure 13.10a**).

The Discovery of Sex-Linked Traits

Drosophila research was pioneered by Thomas Hunt Morgan and his students. But because *Drosophila* is not a domesticated species like the garden pea, Morgan had no readily available phenotypic variants such as Mendel's round and wrinkled seeds.

(a) The fruit fly *Drosophila melanogaster*

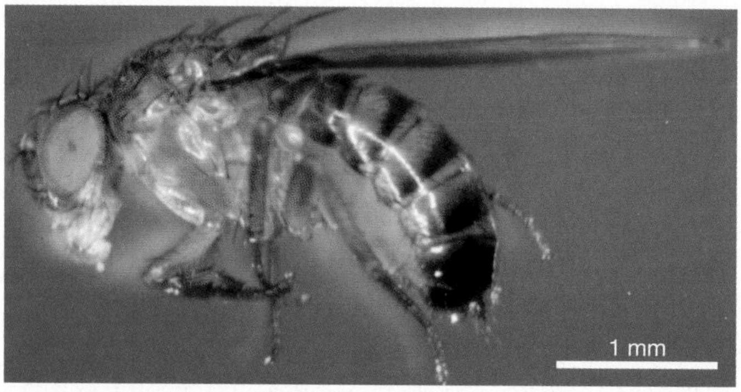

(b) Eye color is a variable trait.

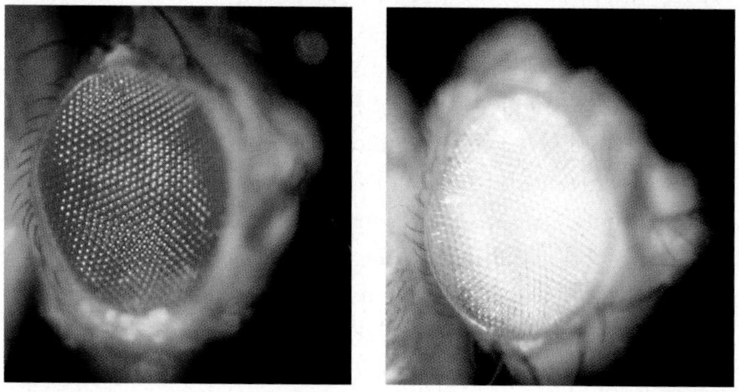

FIGURE 13.10 The Fruit Fly *Drosophila melanogaster* Is an Important Model Organism in Genetics

EXERCISE In part (b), label the phenotype that is considered wild type. Label the phenotype that is a rare mutant.

Consequently, an early goal of Morgan's research was simply to find and characterize individuals with different phenotypes.

Morgan's group referred to individuals with the most common phenotype as **wild type.** But while examining his cultures, Morgan discovered a male fly that had white eyes rather than the wild-type red eyes (**Figure 13.10b**). This individual had a discrete and easy-to-recognize phenotype different from the normal phenotype. Morgan inferred that the white-eyed phenotype resulted from a **mutation**—a change in a gene (in this case, a gene that affects eye color). Individuals with white eyes (or other traits attributable to mutation) are referred to as **mutants.**

To explore how the white-eye trait is inherited in fruit flies, Morgan mated a red-eyed female fly with the mutant white-eyed male fly. Because all of the F_1 progeny had red eyes, Morgan tentatively concluded that the white-eyed phenotype was due to a single allele that was recessive to the wild-type red-eyed allele. To test this hypothesis, Morgan allowed the F_1 males and females to breed with one another. As expected, he observed a 3-to-1 ratio of red-eyed to white-eyed phenotypes in the F_2 offspring. But Morgan noted something peculiar: All of the white-eyed progeny were male. Half of the males in the F_2 generation had white eyes, and half had red. But all of the F_2 females had red eyes. This outcome was remarkable. There appeared to be some sort of association between eye color and sex.

To test the sex-association hypothesis, Morgan crossed F_1 red-eyed females with white-eyed males. Some of the female offspring had white eyes. This told Morgan that females could have white eyes. Obtaining white-eyed females also allowed Morgan to perform the second part of a reciprocal cross. (Recall that Mendel used reciprocal crosses to show that the inheritance of traits in peas is not affected by the transmission of alleles by males versus females.) When Morgan mated white-eyed females to red-eyed males from a pure line, the results were striking: All F_1 females had red eyes, but all F_1 males had white eyes.

Mendel's reciprocal crosses had always given results that were similar to each other. But Morgan's reciprocal crosses did not. The experiment suggested a definite relationship between the sex of the progeny and the inheritance of eye color. Even though the white-eyed phenotype appeared to be recessive in the first set of crosses, it appeared in the F_1 males when the reciprocal cross was performed. How could these observations be reconciled with Mendel's rules of inheritance?

The Discovery of Sex Chromosomes

Nettie Stevens began studying the karyotypes of insects about the time that Morgan began his work with *Drosophila*. One of her outstanding observations was of a striking difference in the chromosome complements of males and females in the beetle *Tenebrio molitor*. In females of this species, diploid cells contain 20 large chromosomes. But diploid cells in males contain 19 large chromosomes and 1 small one. Stevens called the small chromosome the Y chromosome. This Y chromosome paired

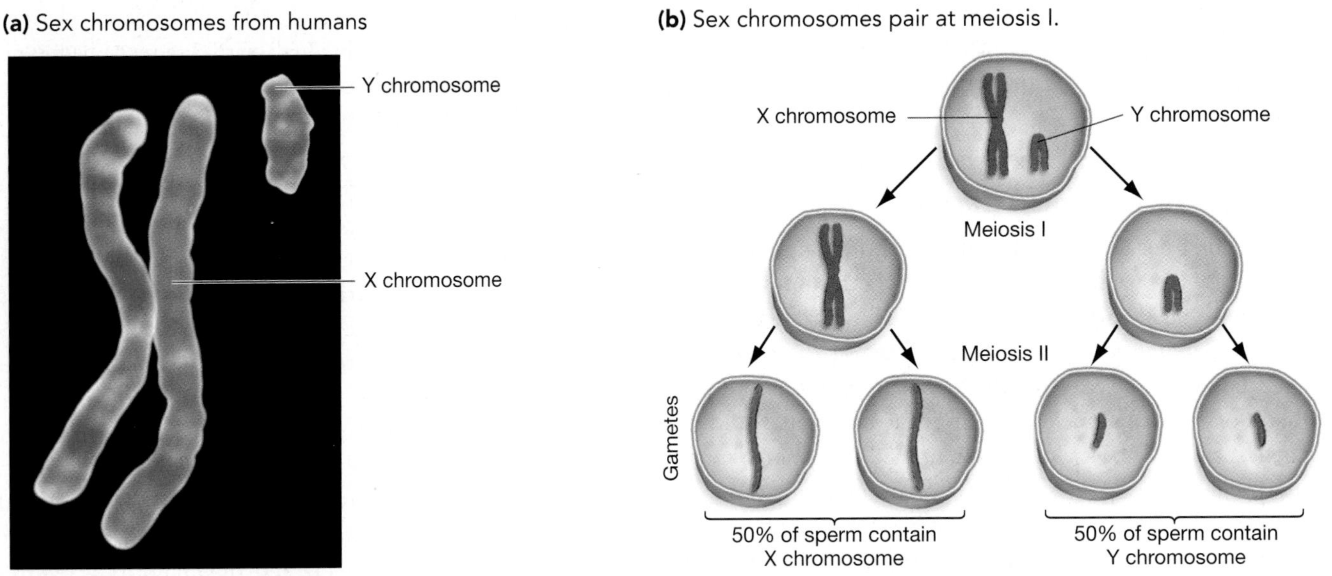

FIGURE 13.11 Sex Chromosomes Pair during Meiosis I, Then Segregate to Form X-Bearing and Y-Bearing Gametes
Sex chromosomes synapse at meiosis I in male fruit flies, even though the X and Y chromosomes differ in size and shape. (No crossing over occurs in male fruit flies.) Thus, half the sperm cells that result from meiosis bear an X chromosome; half have a Y chromosome.

with one of the large chromosomes at meiosis I, which Stevens called the X chromosome. The X and Y were different in size and shape, but they acted like homologs during meiosis.

In addition to discovering the X and Y chromosomes, Stevens observed that all eggs in this species had 10 large chromosomes. Sperm, however, could be divided into two categories. Because the X chromosome and Y chromosome paired up during meiosis and separated into different gamete cells, about 50 percent of the sperm contained 10 large chromosomes, including the X chromosome; the other 50 percent had 9 large chromosomes plus one small (Y) chromosome. Stevens observed similar patterns in other species of insects.

Based on these descriptive studies, Stevens developed a hypothesis to explain how sex determination occurs in these species. She proposed that a male is produced when an egg is fertilized by a sperm carrying a Y chromosome but that a female is produced when an egg is fertilized by a sperm carrying an X chromosome. We now call this pair of chromosomes **sex chromosomes** (**Figure 13.11a**). Sex chromosomes do not carry the same genes. As **Figure 13.11b** shows, the equal ratio of X-bearing and Y-bearing sperm explains why the sexes are produced in nearly equal proportions.

X-Linked Inheritance and the Chromosome Theory

To explain the results of his crosses with white-eyed flies, Morgan put his genetic data together with Stevens's observations on sex chromosomes. *Drosophila* females, like *Tenebrio* females, have two X chromosomes; male fruit flies carry an X and a Y. As a result, half the gametes produced by a male fruit fly should carry an X chromosome; the other half, a Y chromosome.

Morgan realized that the transmission pattern of the X chromosome in males and females could account for the results of his reciprocal crosses. Specifically, he proposed that the gene for white eye color in fruit flies is located on the X chromosome and that the Y chromosome does not carry an allele of this gene. This hypothesis is called **X-linked inheritance** or **X-linkage.** (Proposing that a gene resides on the Y chromosome is called **Y-linked inheritance** or **Y-linkage.** Proposing that a gene is on either sex chromosome is termed **sex-linked inheritance** or **sex-linkage.**) The key observation is that even though the X and Y chromosomes synapse during prophase of meiosis I, they differ in size, shape, and gene content.

According to the hypothesis of X-linkage, a female has two copies of the gene that specifies eye color because she has two X chromosomes. One of these chromosomes came from her female parent, and the other from her male parent. A male, in contrast, has only one copy of the eye-color gene because he has only one X chromosome, which comes from his mother.

The crosses and the Punnett squares in **Figure 13.12** show that Morgan's hypothesis of X-linkage explains his experimental results. In this figure, the allele for red eyes is denoted w^+, while the allele for white eyes is denoted w. (In fruit-fly genetics, the + symbol always indicates the wild-type trait.) When reciprocal crosses give different results, such as those illustrated in Figure 13.12, it is likely that the gene in question is located on a sex chromosome. Recall from Chapter 12 that non-sex chromosomes are called **autosomes.** Genes on non-sex chromosomes are said to show **autosomal inheritance.**

Morgan's discovery of X-linked inheritance carried an even more important message, however. In *Drosophila*, the gene for white eye color is clearly correlated with inheritance of the X chromosome. This correlation was important evidence in support of the hypothesis that chromosomes contain genes. The discovery of X-linked inheritance convinced most biologists that the chromosome theory of inheritance was correct.

Question: Does the sex of the parent affect the inheritance of white eye color in flies?

Hypothesis: The sex of the parent *does* affect the inheritance of white eyes.

Null hypothesis: The sex of the parent *does not* affect the inheritance of white eyes.

Experimental setup:

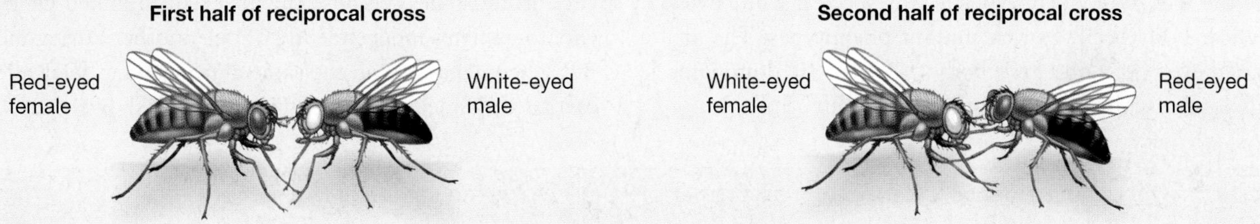

Prediction of "sex matters" hypothesis: Offspring phenotypes will be different in the two crosses.

Prediction of null hypothesis: Offspring phenotypes will be identical in the two crosses.

Results:

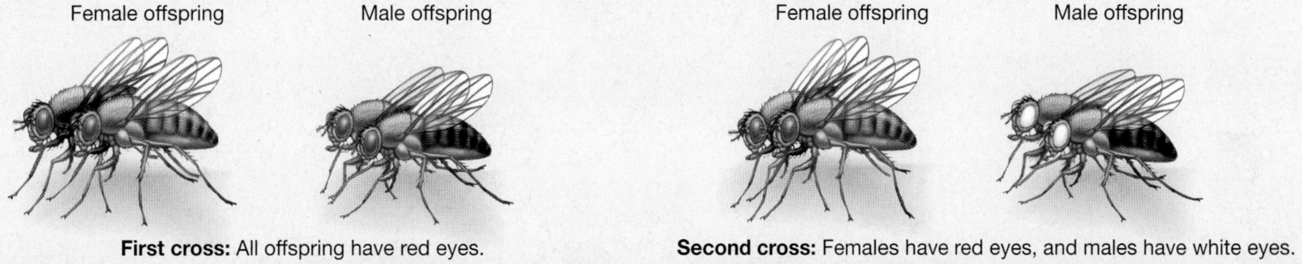

Conclusion: The sex of the parent affects the inheritance of white eye color in flies. The white-eyed gene (w) is on the X chromosome (w^+ = red eyes; w = white eyes). The w^+ allele is dominant to w.

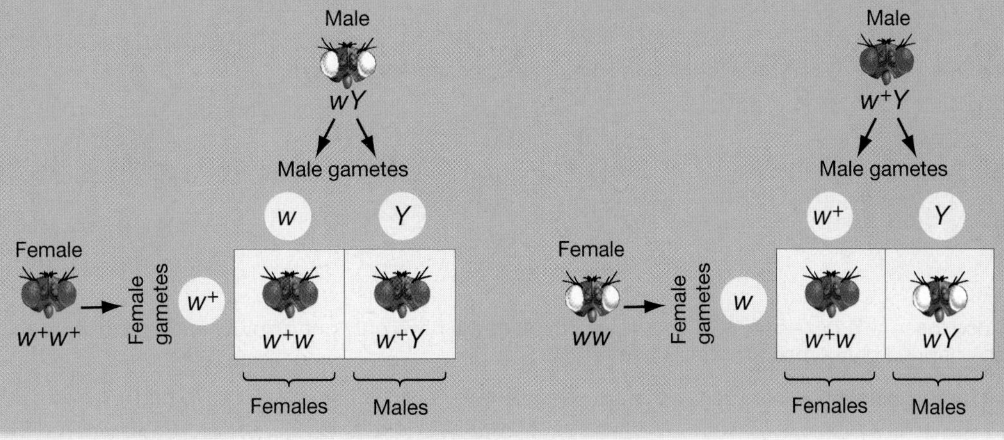

FIGURE 13.12 Reciprocal Crosses Confirm that Eye Color in *Drosophila* Is an X-Linked Trait
When Morgan crossed red-eyed females with white-eyed males and then crossed white-eyed females with red-eyed males, he observed strikingly different results. This was consistent with his hypothesis that eye color is an X-linked trait in fruit flies. **EXERCISE** Morgan also crossed red-eyed w^+w females with red-eyed males, w^+Y. Create a Punnett square to predict the types and proportions of offspring genotypes and phenotypes that would result.

What Happens When Genes Are Located on the Same Chromosome?

When later experiments confirmed that genes are indeed the physical components of chromosomes, the result prompted Morgan and other geneticists to reevaluate Mendel's principle of independent assortment. The key issue was that genes could not undergo independent assortment if they were located on the same chromosome. The physical association of genes that are found on the same chromosome is called **linkage**. (Notice that the terms *linkage* and *sex-linkage* are different in meaning. If two or more genes are linked, it means that they are located on the same chromosome. If a single gene is sex-linked, it means that it makes up part of a sex chromosome.)

The first examples of linked genes involved the X chromosome of fruit flies. After Morgan established that the white-eye gene was located on *Drosophila*'s X chromosome, he and colleagues established that one of the several genes that affects body color is also located on the X. Red eyes and gray body are the wild-type phenotypes in this species; white eyes and a yellow body occur as rare mutant phenotypes. The alleles for red eyes (w^+) and gray body (y^+) also are dominant to the alleles for white eyes (w) and yellow body (y).

Thus, it seemed logical to predict that the linked genes would always be transmitted together during gamete formation. Stated another way, linked genes should violate the principle of independent assortment. Recall from Section 13.2 that independent assortment is observed when genes are on different chromosomes, because the alleles of unlinked genes segregate to gametes independently of one another during meiosis I. But when genes are on the same chromosome, their alleles are carried to gametes together. **Figure 13.13** shows that a female

Question: Will genes undergo independent assortment if they are on the same chromosome?

Linkage hypothesis: Linked genes will violate the principle of independent assortment.

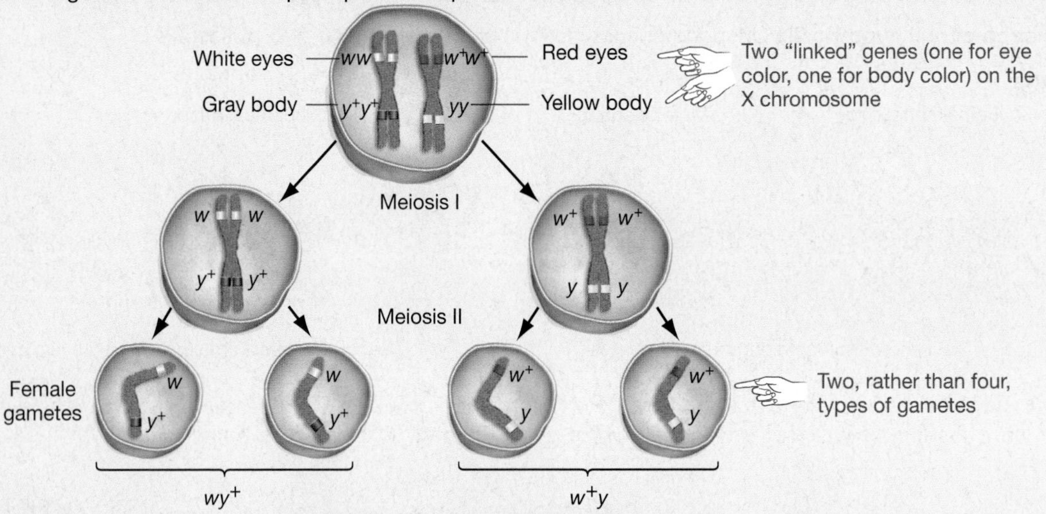
White eyes
ww
w⁺w⁺
Red eyes
Gray body
y⁺y⁺
yy
Yellow body
Two "linked" genes (one for eye color, one for body color) on the X chromosome
Meiosis I
w w
y⁺ y⁺
w⁺ w⁺
y y
Meiosis II
Female gametes
w
y⁺
w⁺
y
Two, rather than four, types of gametes
wy⁺
w⁺y

Null hypothesis: Linked genes will adhere to the principle of independent assortment.

Experimental setup:

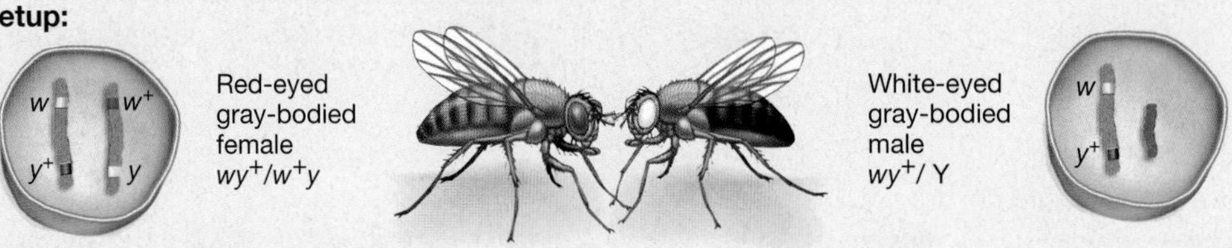
w
w⁺
y⁺
y
Red-eyed gray-bodied female
wy⁺/w⁺y
White-eyed gray-bodied male
wy⁺/ Y
w
y⁺

Prediction: Because these two genes are X-linked, male offspring will have only one copy of each gene, from their mother; the two possible male offspring genotypes are wy^+/Y and w^+y/Y.

Prediction of null hypothesis: Four male genotypes are possible (wy^+/Y : w^+y/Y : wy/Y : w^+y^+/Y) and will occur in equal frequency.

Results:

Male offspring

	Phenotype	Genotype	Number
		wy^+ / Y	4292
		w^+y / Y	4605
Recombinant genotypes		wy / Y	86
		w^+y^+ / Y	44

Four male genotypes were observed (rather than two), but not in equal frequency as predicted by independent assortment

Conclusion: Neither hypothesis is fully supported. Independent assortment does not apply to linked genes; linked genes segregate together except when crossing over and genetic recombination have occurred.

FIGURE 13.13 Linked Genes Are Inherited Together unless Recombination Occurs
Independent assortment does not occur when genes are linked. Morgan crossed pure-line white-eyed, gray-bodied *Drosophila* females with red-eyed, yellow-bodied pure-line males. When he let the F_1 offspring breed, he observed four phenotypes among the male F_2 progeny. Two of the phenotypes were not predicted by complete linkage as in the "Linkage hypothesis" above.

fruit fly with one X chromosome carrying the *w* and y^+ alleles and with a second X chromosome carrying the w^+ and *y* alleles should generate just two classes of gametes in equal numbers during meiosis, instead of the four classes that are predicted under the principle of independent assortment. Is this what actually occurs?

The First Studies of Linked Genes To determine whether linked traits behave as predicted, Morgan performed the cross described in the "Experimental setup" of Figure 13.13. This figure introduces some new notation. When you express a genotype in writing, you are simply listing the relevant alleles that are present. When referring to linked genes, biologists use a slash symbol (/) to separate the alleles that are found on homologous chromosomes. For example, the female fruit flies in Morgan's experiment had wy^+ alleles on one X chromosome and w^+y alleles on the other X chromosome. This genotype is written as wy^+/w^+y.

The interesting result in Figure 13.13 is contained in the "Results" table, which summarizes the phenotypes and genotypes observed in the male offspring of this experimental cross. Most of these males carried an X chromosome with one of the two combinations of alleles found in their mothers: wy^+ or w^+y. Thus, *white* and *yellow* alleles do not segregate independently of each other most of the time. But a small percentage of males had novel phenotypes and genotypes: *wy* and w^+y^+. Morgan referred to these individuals as **recombinant,** because the combination of alleles on their X chromosome was different from the combinations of alleles present in the parental generation.

To explain this result, Morgan proposed that gametes with new, recombinant genotypes were generated when crossing over occurred during prophase of meiosis I in the females. Recall from Chapter 12 that *crossing over* involves a physical exchange of segments from homologous chromosomes. Crossing over occurs at least once in every synapsed pair of homologous chromosomes, and usually multiple times. (Male fruits flies are an exception to this rule. No crossing over occurs in male fruit flies.) As **Figure 13.14** shows, a crossing-over event occurred somewhere between the *w* and *y* genes in the wy^+/w^+y females. The chromosomes that resulted would have the genotypes *wy* and w^+y^+ (see the middle of the bottom row of Figure 13.14). If these chromosomes ended up in a male offspring, they would produce individuals with yellow bodies and white eyes, along with individuals with gray bodies and red eyes. This is exactly what Morgan observed.

The take-home message of Morgan's experiments was that linked genes are inherited together unless crossing over occurs. When crossing over takes place, genetic recombination occurs. These results cemented the connection between the events of meiosis I and Mendel's laws. Linked alleles segregate together unless there is a physical crossover between homologous chromosomes.

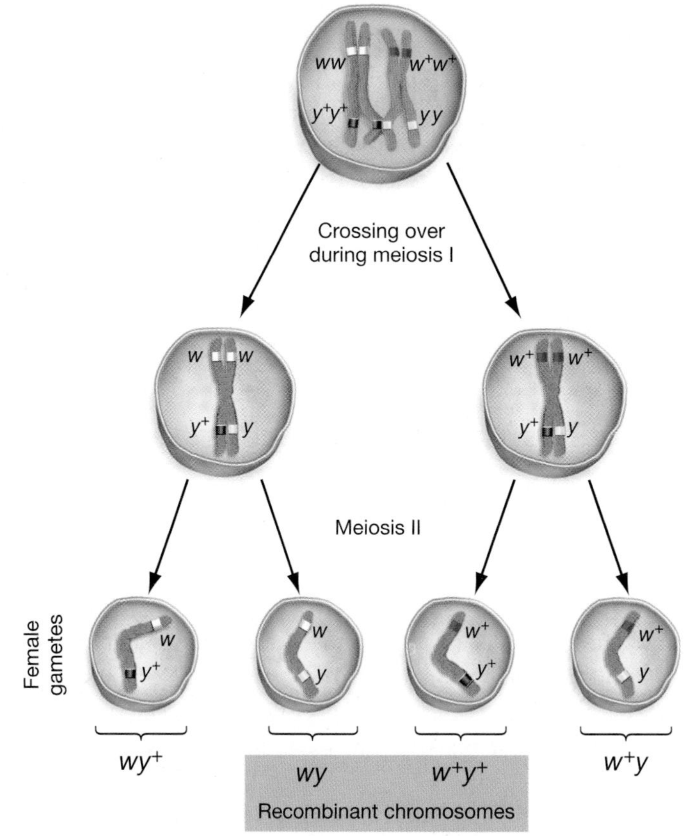

FIGURE 13.14 Genetic Recombination Results from Crossing Over
To explain the results in Figure 13.13, Morgan proposed that crossing over occurred between the *w* genes and the *y* genes in a small percentage of the F_1 females during meiosis I. The recombinant chromosomes that resulted would produce the recombinant phenotypes observed in F_2 males.

Linkage Mapping In the experiment that is summarized in Figure 13.13, about 1.4 percent of the male offspring were recombinant genotypes. But when Morgan and his co-workers performed the same types of crosses with different pairs of X-linked traits, the team found that the fraction of recombinant gametes varied. For example, when crosses involved fruit flies with X-linked genes for a yellow body and a mutant phenotype called singed bristles (*sn*), males with recombinant chromosomes for these two genes were produced about 21 percent of the time.

Morgan explained these observations by making a conceptual breakthrough: He proposed that genes are arranged in a linear array along a chromosome. According to Morgan's hypothesis, the physical distance between genes determines the frequency at which crossing over occurs between them. Recall from Chapter 12 that crossing over occurs at random and can take place anywhere along the length of the chromosome. Based on these observations, it is logical to predict that the shorter the distance between any two genes

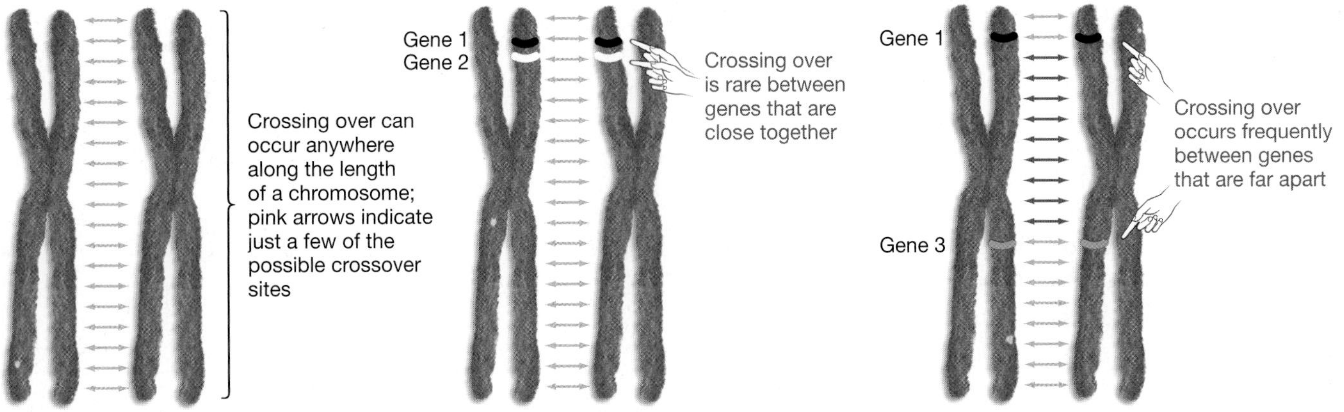

FIGURE 13.15 The Physical Distance between Genes Determines the Frequency of Crossing Over
Crossing over occurs at least once in every meiosis in each homologous pair of chromosomes, and often multiple times.

on a chromosome, the lower the probability that crossing over will take place somewhere in between. The fundamental idea here is that greater physical distance between genes increases the chance that crossing over will take place between them (**Figure 13.15**).

In 1911 A. H. Sturtevant, an undergraduate who was studying with Morgan, realized that variation in recombination frequency had an important implication: If genes are lined up along the chromosome and if the frequency of crossing over is a function of the physical distance between genes, then it should be possible to figure out where genes are in relation to each other based on the frequency of recombinants between various pairs. That is, it should be possible to create a **genetic map**.

Sturtevant proposed that in constructing a genetic map, the unit of distance along a chromosome should simply be the percentage of offspring that have recombinant phenotypes with respect to two genes. Sturtevant called this unit the *centiMorgan* (cM), in honor of his mentor. One map unit, or 1 centiMorgan, represents the physical distance that produces 1 percent recombinant offspring. For example, he proposed that the eye-color and body-color genes of fruit flies are 1.4 map units apart on the X chromosome, because recombination between these loci results in 1.4 percent recombinant offspring (**Figure 13.16a**). The *y* gene for body color and the gene for singed bristles, in contrast, are 21 map units apart. Where is the *white* gene relative to the *singed-bristles* gene? Because recombinants occurred in 19.6 percent of the gametes produced by females that are w^+sn/wsn^+, Sturtevant inferred that the gene for white eyes must be located between the genes for yellow body and singed bristles, as shown in Figure 13.16a.

Figure 13.16b provides a partial genetic map of the X chromosome in *Drosophila melanogaster* and the data on which the map positions are based. Using this logic and these data, Sturtevant assembled the first genetic map.

Do *All* Linked Genes Violate the Principle of Independent Assortment? When crossing over occurs, the new genotypes that are created match the ones that would be produced if the genes involved were located on different chromosomes. Stated another way, crossing over breaks up linkage and makes it appear as if independent assortment is occurring. To convince yourself that this is so, look back at the four gamete genotypes listed at the bottom of Figure 13.14. Recall that these were produced by a female with genotype wy^+/w^+y. The four gamete genotypes listed are wy^+, wy, w^+y^+, and w^+y. It should be clear that these are the same gamete genotypes that meiosis would produce if the female's genotype were w^+wy^+y, meaning the *white* and *yellow* genes were not linked.

If crossing over occurs frequently enough that all four of these gamete genotypes are produced in equal proportions, then the results of a cross involving linked genes would be indistinguishable from the predictions of independent assortment. When genes are 50 or more map units apart on the same chromosome, that is exactly what happens. Genes behave as if they assort independently when 50 percent of the gametes are recombinant with respect to the genes. As it turns out, this situation was occurring in Mendel's experiments. The genes for two of the traits he worked with are located on chromosome number 1 in peas, and at least two of the genes are located on chromosome 4. In each case, however, the genes he analyzed are located far apart. As a result, crossing over was frequent enough to produce the 9:3:3:1 ratios of F_2 phenotypes predicted by the hypothesis of independent assortment.

Before analyzing some extensions to Mendel's rules, let's step back and consider the events we've just surveyed—from the rediscovery of Mendel's work, around 1900, to Sturtevant's chromosome map. In 1899 geneticists did not understand the basic rules of heredity. But by 1911 they could map the locations of genes on chromosomes. A remarkable knowledge explosion had occurred.

(a) Frequency of recombinant offspring can be used to map genetic distance.

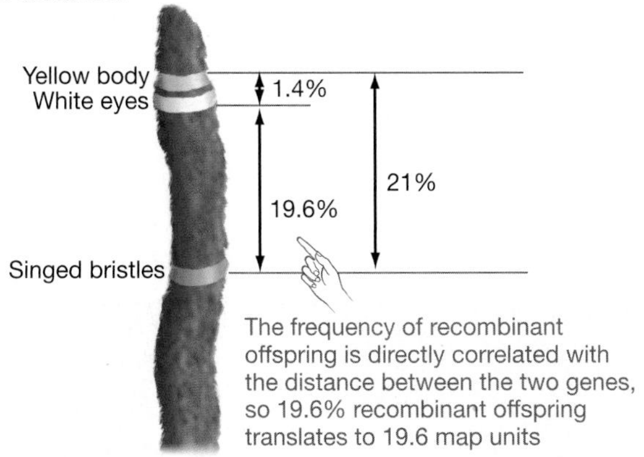

(b) Constructing a genetic map

Frequency of crossing over (%) between some genes on the X chromosome of fruit flies

	Miniature Wings	Crossveinless Wings	Ruby Eyes
Yellow body	36.1	13.7	7.5
White eyes	34.7	12.3	6.1
Singed bristles	15.1	7.3	13.5
Miniature wings	—	22.4	28.6
Crossveinless wings	—	—	6.2

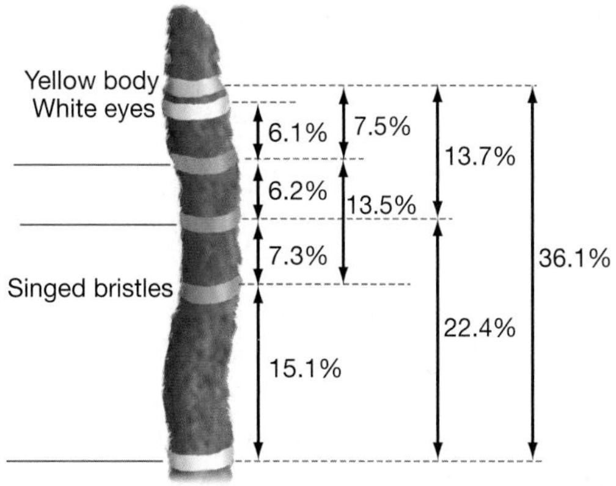

FIGURE 13.16 The Locations of Genes Can Be Mapped by Analyzing the Frequency of Recombination
(a) Recombinants between the yellow-body locus and the white-eye locus occur 1.4 percent of the time. Therefore, these genes are 1.4 map units apart on the chromosome. Recombinants between the yellow-body locus and the singed-bristles locus of another X-linked gene occur 21 percent of the time. But recombinants between the white-eye locus and the singed-bristles locus occur just 19.6 percent of the time. Therefore, the loci must be arranged as shown. **(b)** A partial genetic map of the X chromosome in fruit flies.
EXERCISE On the chromosome in part (b), label the orange, blue, and beige genes.

13.5 Extending Mendel's Rules

Biologists point out that Mendel analyzed the simplest possible genetic system. To understand why it's fair to label the traits and alleles that Mendel studied as simple, consider the following:

- The phenotypes that he observed are influenced by the alleles at a single gene, even though many traits in peas and other organisms are shaped by the effects of multiple genes.
- In the pure lines that he studied, a total of two alleles were present at each gene. Most populations in most species harbor dozens or even hundreds of different alleles at each gene.
- The alleles he studied were completely dominant or recessive to each other. In many cases, alleles don't show complete dominance or recessiveness, as the next section explains.
- Mendel worked with **discrete traits**—characteristics that are qualitatively different. His pea flowers were white or purple; the plants were tall or short; the seeds were wrinkled or round. But it's more common for traits to show what biologists call **quantitative variation**—differences in degree of variation. In humans, for example, characteristics such as height, skin color, and body shape do not show discrete variation. Instead of being either 1.5 m tall or 1.75 m tall or 2.0 m tall, humans may be any height within a wide range of variation. Traits that exhibit quantitative variation are harder to measure and study than discrete traits are.
- Peas do not have sex chromosomes, so sex-linked inheritance did not occur.
- The experimental results were not complicated by linkage, because the linked genes that Mendel analyzed were so far apart that crossing over between them occurred frequently. As a result, alleles from different genes appeared to assort independently.

The key point is that by studying a simple situation, Mendel was able to discover the most fundamental rules of inheritance. This is an extremely important research strategy in biological science. Researchers almost always choose to analyze the simplest situation possible before going on to explore more complex systems. Mendel probably would have failed, as so many others had done before him, had he worked with any of the complications just listed.

Once Mendel's work was rediscovered, though, researchers began to analyze traits and alleles whose inheritance was complex. If experimental crosses produced F_2 progeny that did not conform to the expected 3:1 or 9:3:3:1 ratios in phenotypes, researchers had a strong hint that something interesting was going on. In many cases it turned out that unraveling the cause of surprising results led to new insights into how genes and heredity worked. How can traits that don't appear to follow Mendel's rules contribute to a more complete understanding of heredity?

Incomplete Dominance and Codominance

The terms *dominant* and *recessive* describe which phenotype is observed when two different alleles of a gene occur in the same individual. In all seven traits that Mendel studied, only the phenotype associated with one allele—the "dominant" one—appeared in heterozygous individuals. But consider the flowers called four-o'clocks, pictured in **Figure 13.17a**. In this species, biologists have developed a pure line that has purple flowers and a pure line that has white flowers. When individuals from these strains are mated, all of their offspring are lavender (**Figure 13.17b**). In Mendel's peas, crosses between purple- and white-flowered parents produced all purple-flowered offspring. Why the difference?

Biologists answered this question by examining the phenotypes of F_2 offspring. These are the progeny of self-fertilization in lavender-flowered F_1 individuals. Of the F_2 plants, 1/4 have purple flowers, 1/2 have lavender flowers, and 1/4 have white flowers. This 1:2:1 ratio of *phenotypes* is unlike any we have seen to date, but it exactly matches the 1:2:1 ratio of *genotypes* that is produced when flower color is controlled by one gene with two alleles.

To convince yourself that this explanation is sound, study the genetic model shown in Figure 13.17b. According to the hypothesis shown in the diagram, the inheritance of flower color in four-o'clocks and peas is identical, except that the four-o'-clock alleles show incomplete dominance rather than complete dominance. When **incomplete dominance** occurs, heterozygotes have an intermediate phenotype. In the case of four-o'clocks, neither purple nor white alleles dominate. Instead, the F_1 progeny—all heterozygous—show a phenotype intermediate between the two parental strains.

Codominance Incomplete dominance illustrates an important general point: Dominance is not necessarily an all-or-none phenomenon. In fact, many alleles show a relationship called codominance. When **codominance** occurs, heterozygotes have the phenotype associated with *both* of the alleles present. Whereas incomplete dominance results in a phenotype that is intermediate between the phenotypes associated with both alleles present, codominance results in a phenotype that expresses both alleles.

As an example of codominance, consider the *MN* gene in humans. This gene codes for a membrane protein that is found in the plasma membrane of red blood cells. As a result, the phenotype associated with the *MN* gene is the type of membrane protein present. Different alleles of this gene lead to the appearance of membrane proteins with different amino acid sequences. In most human populations, only two alleles are present. These alleles are designated *M* and *N*. Thus, three genotypes are found: *MM*, *MN*, and *NN*. In *MM* individuals, all of the MN-type membrane proteins present in red blood cells have the primary sequence associated with the *M* allele. In *NN* individuals, all of the MN-type proteins have the *N* allele's primary sequence. But in *MN* individuals, red blood cells have some membrane proteins with the M phenotype and some with the N phenotype. Because both alleles are represented in the phenotype, the alleles are considered codominant.

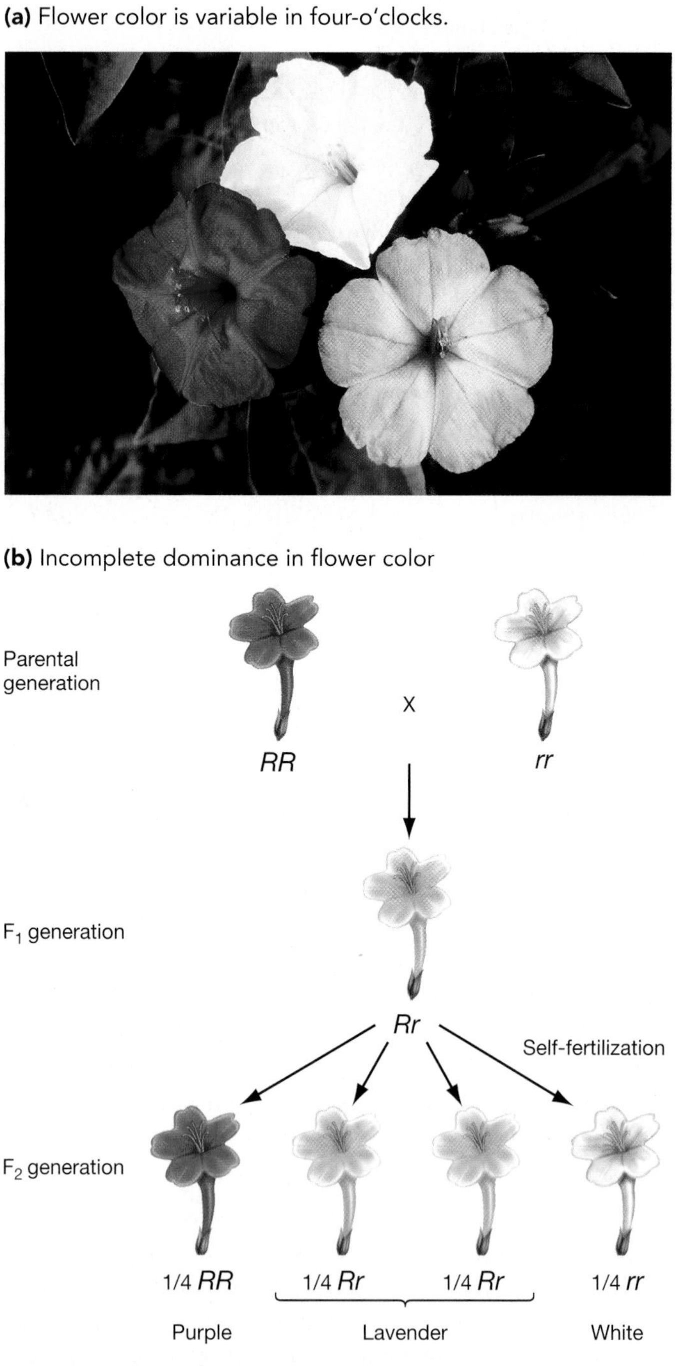
(a) Flower color is variable in four-o'clocks.
(b) Incomplete dominance in flower color
Parental generation
X
RR
rr
F1 generation
Rr
Self-fertilization
F2 generation
1/4 RR
1/4 Rr
1/4 Rr
1/4 rr
Purple
Lavender
White

FIGURE 13.17 When Incomplete Dominance Occurs, Heterozygotes Have Intermediate Phenotypes
(a) Four-o'clocks got their name because their flowers open in the late afternoon. Flower color is highly variable. **(b)** The hypothesis illustrated is that flower color is controlled by a single gene with two alleles, symbolized *R* and *r*. These alleles exhibit incomplete dominance.

Why Do Dominance Relationships Vary among Alleles? Once you understand what genes do on the molecular level, it is straightforward to see why some alleles are dominant, recessive, codominant, or intermediate in dominance. Genes contain the information required to manufacture the molecules that are found in cells. Some of these molecules are proteins that form structures in the cell or body, such as the MN proteins in the plasma membranes of red blood cells. Because the *M* and *N* alleles code for proteins with different amino acid sequences, both are present in the cells of heterozygous individuals. Both gene products are detected in the phenotype, so the alleles are considered codominant. Codominance is fairly common in alleles that are associated with membrane proteins such as the MN protein.

Complete and incomplete dominance tend to occur in alleles that code for enzymes. The plant-height alleles Mendel studied, for example, code for enzymes involved in the synthesis of a key growth hormone. A recessive allele has a defect that keeps the enzyme from functioning. Homozygous recessive individuals lack the growth hormone and are dwarfed. The dominant allele, in contrast, codes for an enzyme that functions normally. In heterozygotes, the one normal allele that is present produces enough enzyme to make a normal amount of growth hormone. Normal height results, and the allele is considered dominant.

The situation in four-o'clock flowers is different. The *R* gene analyzed in Figure 13.17b codes for an enzyme involved in producing a purple pigment in flower petals. The recessive allele codes for a dysfunctional enzyme and leads to a complete lack of pigment. Recessive homozygotes are white as a result. In heterozygotes, though, the one normal copy of the gene can produce enough enzyme to result in a small amount of purple pigment. The lavender phenotype results. Homozygous dominant individuals, in contrast, produce large amounts of normal enzyme and thus enough pigment to produce purple petals.

The message here is that alleles can interact in many different ways, depending on what their products do and how they work. In addition, recall that dominance relationships have nothing to do with fitness—the ability of an individual to produce offspring in a particular environment. Dominant alleles do not necessarily confer high fitness, and they are not necessarily more common than recessive alleles. For instance, the allele that causes Huntington's disease is fatal, dominant, and rare.

Multiple Alleles and Polymorphic Traits

Mendel worked with a total of seven traits and just 14 alleles—two for each trait. In most populations, however, dozens of alleles can be identified at each genetic locus. The existence of more than two alleles of the same gene is known as **multiple allelism.** The gene for the hemoglobin protein in humans is among the most dramatic examples of multiple allelism studied to date. Hemoglobin carries oxygen from the lungs to tissues. As pointed out in Chapter 3, the protein has a quarternary structure made up of four distinct polypeptide chains. The instructions for making one of these polypeptide chains are encoded by the *β-globin* gene. Over the past few decades, biologists have identified and named over 500 different alleles of this gene. Many of these alleles are associated with distinctive phenotypes. Some alleles produce polypeptides with normal oxygen-carrying capacity, while others lead to reduced oxygen-carrying capacity and various types of anemia. Still other *β-globin* alleles are associated with adaptation to living at high altitudes, decreased stability at high temperatures, or resistance to the parasites that infect red blood cells and cause malaria.

TABLE 13.1 The ABO Blood Types in Humans

In humans, the four different ABO blood types are produced by the alleles present at a single locus. Three alleles are common in most populations: i, I^A, and I^B.

Phenotype (blood type)	Genotype
O	ii
A	I^AI^A or I^Ai
B	I^BI^B or I^Bi
AB	I^AI^B

i = recessive
I^A and I^B = codominant

When different combinations of alleles produce more than two distinct phenotypes, the trait is **polymorphic** ("many-formed"). Oxygen-carrying capacity in humans is a highly polymorphic trait. But the first polymorphic trait ever described involves the ABO blood group in humans. Red blood cells contain many membrane proteins in addition to those encoded by the MN gene; the ABO gene is responsible for one of these. Your blood has almost certainly been analyzed and assigned an O, A, B, or AB phenotype. These phenotypes were discovered in 1900; much later it was established that they are caused by carbohydrates located on the membranes of red blood cells.

As **Table 13.1** shows, the ABO phenotypes result from three alleles called i, I^A, and I^B. The i allele is recessive and produces the O phenotype when it is homozygous. The *A* and *B* alleles, in contrast, are dominant with respect to the i allele but codominant with respect to each other.

The ABO blood group involves three alleles, six genotypes, and four phenotypes. This gene is multiallelic, the phenotypes are polymorphic, and the alleles are completely dominant or codominant.

Pleiotropy

As far as is known, the alleles that Mendel analyzed affect just a single trait. The gene for seed color in garden peas, for example, does not appear to affect other aspects of the individual's phenotype. In contrast, many cases have been documented in which a single allele affects a wide variety of traits. A gene that influences many traits, rather than just one trait, is said to be **pleiotropic** ("more-turning"). The gene responsible for

Marfan syndrome in humans is a good example. Although current research suggests that just a single gene is involved, individuals with Marfan syndrome exhibit a wide array of phenotypic affects: increased height, disporportionately long limbs and fingers, an abnormally shaped chest, and potentially severe heart problems. A large percentage of these individuals also suffer from problems with their backbone. The gene associated with Marfan syndrome is pleiotropic.

Genes Are Affected by the Physical Environment and Genetic Environment

When Mendel analyzed height in his experiments, he ensured that each plant received a similar amount of sunlight and grew in similar soil. This was important because individuals with alleles for tallness are stunted if they are deprived of nutrients, sunlight, or water—so much so that they look similar to individuals with alleles for dwarfing. For Mendel to analyze the hereditary determinants of height, he had to control the environmental determinants of height. Let's consider how two aspects of the environment affect phenotypes: (1) the individual's physical surroundings and (2) the alleles present at other genes.

The Physical Environment Has a Profound Effect on Phenotypes The phenotypes produced by most genes and alleles are strongly affected by the individual's physical environment. As a result, it is often the case that an individual's phenotype is as much a product of its physical environment as it is a product of its genotype. To drive this point home, consider two examples in humans:

1. If you have alleles for normal height and normal eyesight but are raised in an environment where the diet is poor in protein and vitamin A, then you are very likely to be short and to have poor eyesight. If one person could be raised in two different environments, it is highly likely that two different phenotypes would result.
2. People with the genetic disease called **phenylketonuria** (PKU) lack an enzyme that helps convert the amino acid phenylalanine to the amino acid tyrosine. As a result, phenylalanine and a related molecule, phenylpyruvic acid accumulate in these people's bodies. The molecules interfere with the development of the nervous system and produce profound mental retardation. But if PKU individuals are identified at birth and placed on a low-phenylalanine diet, then they develop normally. In many countries, all newborns are tested for the defect. PKU is a genetic disease, but it is neither inevitable nor invariant. Because of a simple change in their environment (their diet), individuals with a PKU genotype can have a normal phenotype.

Genetic traits are influenced by more than the physical environment that an individual experiences, however. Phenotypes are also influenced by the action of other genes.

Interactions with Other Genes Have a Profound Effect on Phenotypes In Mendel's pea plants, a single locus influenced seed shape. Further, Mendel's data showed that the seed-shape phenotype does not appear to be affected by the action of genes for seed color, seed-pod color, seed-pod shape, or other traits. The pea seeds he analyzed were round or wrinkled regardless of the types of alleles present at other loci.

In many cases, however, genes are not as independent as the gene for seed shape in peas. As an example, take the inheritance of fruit color in bell peppers. As **Figure 13.18a** shows, bell peppers come in several colors. Are the color variations due to multiple alleles of a single gene, or are several genes involved? To answer this question, consider the data presented in **Figure 13.18b** on the F_1 and F_2 offspring of pure-line parental strains with yellow and brown fruit.

The results are surprising. Like the experimental cross with four-o'clocks, the F_1 phenotype is different from either parental strain. But a fourth phenotype (green), not seen in either parent or in the F_1, appears in the F_2 generation. Further, the progeny ratios in the F_2 generation are in sixteenths, not quarters. This combination of observations is completely different from anything that Mendel encountered.

A moment's thought should convince you that the F_2 data can't be explained by incomplete dominance, because a novel phenotype shows up. The parental generation consisted of pure lines, so multiple allelism can't be involved. The 9:3:3:1 pattern in F_2 phenotypes is interesting, though, because it matches the ratio that Mendel observed when he studied the inheritance of two different traits. These data suggest that two genes interact to produce pepper color. Further, we can infer that the red phenotype, present 9/16 of the time in the F_2, is due to a combination of the dominant alleles from each gene. Green might result from a combination of the recessive alleles for each gene, because this phenotype is present in just 1/16 of the progeny.

Pepper fruits start out with a green color due to the presence of the green pigment called chlorophyll. This observation suggests that one gene affects whether chlorophyll production stops or continues as a pepper develops. If chlorophyll production continues, a green or greenish fruit would result. The second gene might determine the nature of a second pigment in the fruit, such as red or yellow. **Figure 13.18c** shows how several phenotypes could result from the two alleles of each of these genes.

The key point is that the alleles of different genes affect each other in a way that influences the phenotype observed in these individuals. When these types of gene interactions occur, the phenotype produced by an allele depends on the action of alleles of other genes. This phenomenon is known as **epistasis** ("stopping" or "diminishing"). Epistasis occurs when one gene affects the action of another gene. In the case of peppers, the phenotype of an individual with the *R* allele depends on the al-

(a) Fruit color is highly variable in bell peppers.

(b) Crosses between pure lines produce novel colors.

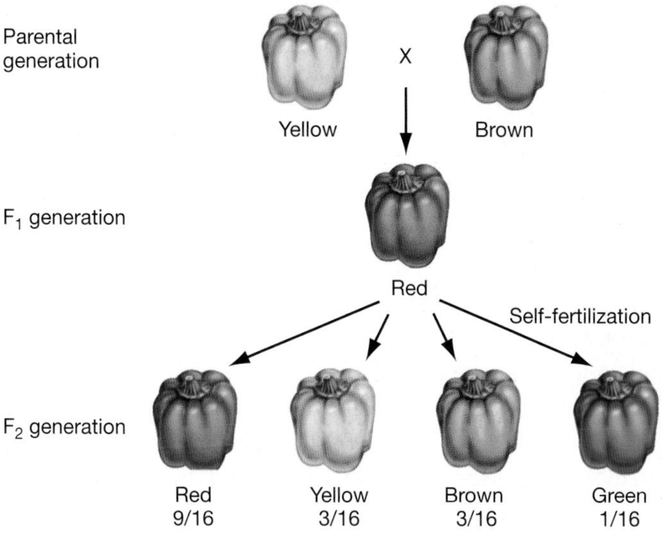

(c) Hypothesis to explain 9 : 3 : 3 : 1 pattern observed above: Two genes interact to produce pepper color.

Gene 1	Gene 2
R = Red	*Y* = Absence of green (no chlorophyll)
r = Yellow	*y* = Presence of green (chlorophyll)
(–) = *R* or *r*	(–) = *Y* or *y*

Genotype	Color	Explanation of color
R–Y–	Red	Red pigment + no chlorophyll
rrY–	Yellow	Yellow pigment + no chlorophyll
R–yy	Brown	Red pigment + chlorophyll
rryy	Green	Yellow pigment + chlorophyll

FIGURE 13.18 Fruit Color in Bell Peppers Results from Interactions among Genes

leles present at the *Y* gene. The term *epistasis* was chosen because, in many cases, the alleles at one gene mask or reduce the effects of alleles at a different gene.

Organelle Genomes

Chapter 7 pointed out that both mitochondria and chloroplasts have small, circular chromosomes that consist in part of DNA. Recall that mitochondria are the ATP-synthesizing centers in eukaryotic cells and that chloroplasts are the sugar-producing centers in photosynthetic eukaryotes. Mitochondria and chloroplasts have their own sets of genes, or genomes. The chromosomes that reside in mitochondria and chloroplasts contain genes that code for molecules used inside each of those organelles.

In most species examined to date, the genes that are located in mitochondria and chloroplasts are transmitted to offspring by only one parent—usually the mother. All of the mitochondrial genes in your cells, for example, came from your mother. In most individual trees and shrubs on your campus, both the mitochondrial genes and the chloroplast genes came from the individual's mother. In these species, none of the organelle genes found in a sperm are passed on to offspring; only the organelle genes found in an egg are imparted. As a result, individuals that are diploid for nuclear genes are haploid for all organelle genes. There is little to no genetic recombination in mitochondrial and chloroplast chromosomes, so all of the genes are inherited as if they were a single allele. And there is no segregation of alleles and no independent assortment. Mitochondrial genes and chloroplast genes are not inherited in a Mendelian fashion.

Quantitative Traits

Recall that Mendel worked with discrete traits. In garden peas, seed color is either yellow or green—no intermediate phenotypes exist. But many traits in peas and other organisms don't fall into discrete categories. In humans, for example, height, weight, and skin color fall anywhere on a continuous scale of measurement. People are not only 160 cm tall or 180 cm tall, with no other heights possible. Height and many other characteristics exhibit quantitative variation—meaning that individuals differ by degree, as described earlier in this section—and are called **quantitative traits.** Like discrete traits, quantitative traits are highly influenced by the physical environment. For example, the effects of nutrition on human height and intelligence have been well documented.

Quantitative traits share a common characteristic: When the frequencies of different values observed in a population are plotted on a histogram, they usually form a bell-shaped curve. This distribution is observed so frequently that it is often called a *normal distribution.* In a normal distribution, high and low values occur at low frequency; intermediate values

FIGURE 13.19 Quantitative Traits Are Normally Distributed
(a) Male undergraduates, sorted by height, at Connecticut Agricultural College in 1914. **(b)** A histogram plotting the heights of the students in part (a). The distribution of height in human populations forms a bell-shaped curve, called a normal distribution. QUESTION The shortest student in the photo is 4′10″ (147 cm), and the tallest is 6′2″ (188 cm). How would a histogram of men in your class compare with the distribution in part (b)?

occur at high frequency. **Figure 13.19** shows a classical example. In this case, the trait is human height and the population is a college class.

In the early 1900s Herman Nilsson-Ehle showed that if many genes each contribute a small amount to the value of a quantitative trait, then a continuous, bell-shaped (normal) distribution results. Nilsson-Ehle established this finding using strains of wheat that differed in kernel color. **Figure 13.20a** shows the results of a cross he performed between pure lines of white wheat and dark-red wheat. Note that the frequency of colors in F_2 progeny forms a bell-shaped curve. To explain these results, Nilsson-Ehle proposed the set of hypotheses illustrated in **Figure 13.20b**:

- The parental strains differ with respect to three genes that control kernel color: *AABBCC* produces dark-red kernels, and *aabbcc* produces white kernels.
- The three genes assort independently. When the *AaBbCc* F_1 individuals self-fertilize, white F_2 individuals would occur at a frequency of 1/4 (*aa*) × 1/4 (*bb*) × 1/4 (*cc*) = 1/64 *aabbcc*.
- The *a*, *b*, and *c* alleles do not contribute to pigment production, but the *A*, *B*, and *C* alleles contribute to pigment production in an equal and additive way. As a result, the degree of red pigmentation is determined by the number of *A*, *B*, or *C* alleles present. Each uppercase allele that is present makes a wheat kernel slightly darker red.

Later work showed that Nilsson-Ehle's model hypotheses were correct in virtually every detail. Quantitative traits are produced by the independent actions of many genes, although it is now clear that some genes have much greater effects on the trait in question than other genes do. As a result, the transmission of quantitative traits is said to result from polygenic ("many-genes") inheritance. In **polygenic inheritance**, each gene adds a small amount to the value of the phenotype.

In the decades immediately after the rediscovery of Mendel's work, analyses of phenomena such as sex-linkage, incomplete dominance, multiple allelism, environmental effects, gene interactions, and polygenic inheritance provided a fairly comprehensive answer to the question of why offspring resemble their parents. **Table 13.2** summarizes some of the key exceptions and extensions to Mendel's rules and gives you a chance to compare and contrast their effects on patterns of inheritance.

✓CHECK YOUR UNDERSTANDING

Meiosis is the process responsible for Mendel's principle of segregation and principle of independent assortment. X-linked genes violate the principle of segregation, because males can have only one allele for a given trait. Genes on the same chromosome violate the principle of independent assortment, because they are not transmitted to gametes independently of each other, unless crossing over occurs between them. If you understand incomplete dominance, codominance, multiple allelism, pleiotropy, environmental effects, epistasis, and polygenic inheritance, you should be able to (1) explain each of these phenomena and give an example of each, and (2) explain why each phenomenon does or does not violate the principle of segregation and the principle of independent assortment.

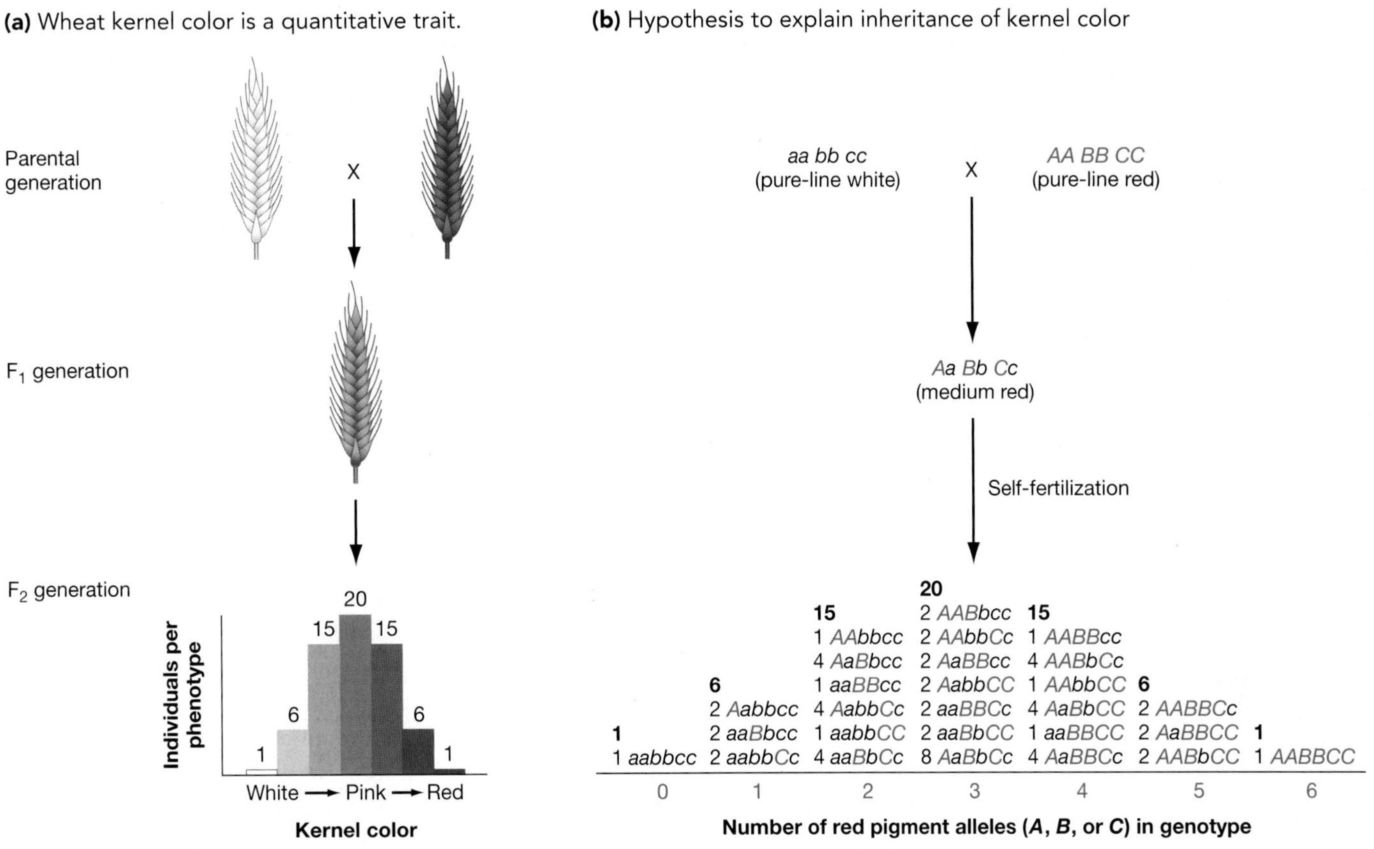

FIGURE 13.20 Quantitative Traits Result from the Action of Many Genes
(a) When wheat plants with white kernels were crossed with wheat plants with red kernels, the F_2 offspring showed a range of kernel colors. The frequency of these phenotypes formed a normal distribution. **(b)** This model attempts to explain the results of part (a). **EXERCISE** Confirm that the distribution of genotypes shown in part (b) is correct by drawing a Punnett square based on the gametes produced by the F_1 parents and then filling in the F_2 offspring genotypes that result. (The square will be 8×8, with 64 boxes!)

TABLE 13.2 Some Extensions to Mendel's Rules: A Summary

Type of Inheritance	Definition	Consequences or Comments
Sex-linkage	Genes located on sex chromosomes.	Patterns of inheritance in males and females differ.
Linkage	Two genes found on same chromosome.	Linked genes violate principle of independent assortment.
Incomplete dominance	Heterozygotes have intermediate phenotype.	Polymorphism—heterozygotes have unique phenotype.
Codominance	Heterozygotes have phenotype of both alleles.	Polymorphism is possible—heterozygotes have unique phenotype.
Multiple allelism	In a population, more than two alleles present at a locus.	Polymorphism is possible.
Polymorphism	In a population, more than two phenotypes present.	Can result from actions of multiple alleles, incomplete dominance, and/or codominance.
Pleiotropy	A single allele affects many traits.	This is common.
Epistasis: Variation in the genetic environment	In discrete traits, alleles of other genes affect phenotype.	Same genotypes at a particular locus can be associated with different phenotypes.
Variation in the physical environment	Phenotype influenced by environment experienced by individual.	Same genotypes can be associated with different phenotypes.
Polygenic inheritance of quantitative traits	Many genes are involved in specifying traits that exhibit continuous variation.	Unlike alleles that determine discrete traits, each allele adds a small amount to phenotype.

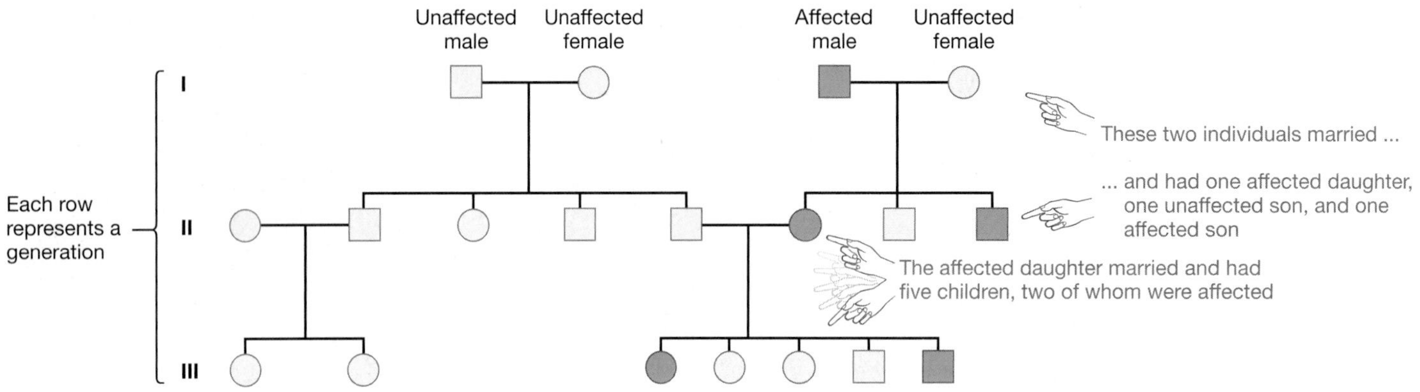

FIGURE 13.21 Reading a Pedigree
Pedigrees use special notation, shown here, as a compact way of communicating a great deal of information about which individuals in a family exhibit a certain phenotype.

13.6 Applying Mendel's Rules to Humans

When researchers set out to study how a particular gene is transmitted in wheat or fruit flies or garden peas, they start out by making a series of controlled experimental crosses. For obvious reasons, this research strategy is not possible with humans. But in many cases it is extremely valuable for biologists and physicians to know the answers to basic questions about the inheritance of certain human genes—particularly those associated with inherited diseases. Do the alleles present in the population have a simple recessive and dominant relationship? Is the gene involved autosomal or sex-linked? Is the trait influenced primarily by alleles at a single gene, or is inheritance polygenic? To answer these questions, investigators have to analyze human crosses that already exist. They do so by constructing a **pedigree**, or family tree, of affected individuals.

A pedigree, such as the one shown in **Figure 13.21**, records the genetic relationships among the individuals in a family along with each person's sex and phenotype with respect to the trait in question. If the trait is due to a single gene, then analyzing the pedigree may reveal whether the trait is due to a dominant or recessive allele and whether the gene responsible is located on a sex chromosome or on an autosome. Let's look at a series of specific case histories to see how this work is done.

Are Alleles Recessive or Dominant?

To analyze the inheritance of a trait that shows discrete variation, biologists begin by assuming that a single autosomal gene is involved and that the alleles present in the population have a simple dominant-recessive relationship. This is the simplest possible situation. If the pattern of inheritance fits this model, then the assumptions—of inheritance via a single gene and simple dominance—are supported. Let's first analyze the pattern of inheritance that is typical of autosomal recessive traits and then examine patterns that emerge in pedigrees for autosomal dominant traits.

Patterns of Inheritance: Autosomal Recessive Traits In analyzing the inheritance of traits, it's helpful to distinguish conditions that *must* be met when a particular pattern of inheritance occurs versus conditions that are *likely* to be met. For example, if a phenotype is due to an autosomal recessive allele, then individuals with the trait must be homozygous. If the recessive allele is rare in the population, then the parents of individuals with the trait are heterozygous for the trait. Heterozygous individuals who carry a recessive allele for an inherited disease are referred to as **carriers** of the disease. These individuals carry the allele and transmit it even though they do not exhibit signs of the disease. When two carriers mate, they should produce offspring with the recessive phenotype about 25 percent of the time.

Figure 13.22a is the pedigree from a family in which an autosomal recessive disease, such as cystic fibrosis, occurs. The key feature to notice in this pedigree is that some boys and girls exhibit the trait even though their parents do not. This is the pattern you would expect to observe under the model of an autosomal recessive trait if the parents of individuals with the trait are heterozygous. In addition, it is logical to observe that when an affected individual has children, those children do not necessarily have the trait. This pattern is predicted if affected people marry individuals who are homozygous for the normal allele, which is likely to be true if the recessive allele is rare in the population.

The recessive phenotype should show up in offspring only when both parents carry that recessive allele and pass it on to their offspring. By definition, a recessive allele produces a given phenotype only when the individual is homozygous for that allele.

Patterns of Inheritance: Autosomal Dominant Traits By definition, individuals who are homozygous or heterozygous for an autosomal dominant trait will have the dominant phenotype. For autosomal dominant traits, individuals with a single copy of the allele must have the dominant phenotype. Even if

(a) Pedigree of a family with an autosomal recessive disease

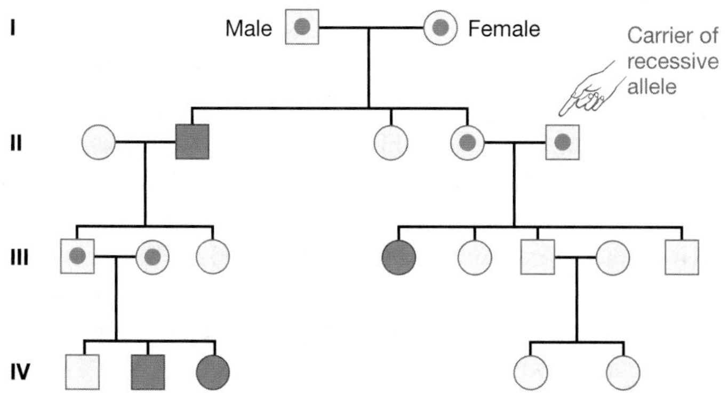

(b) Pedigree of a family with Huntington's disease

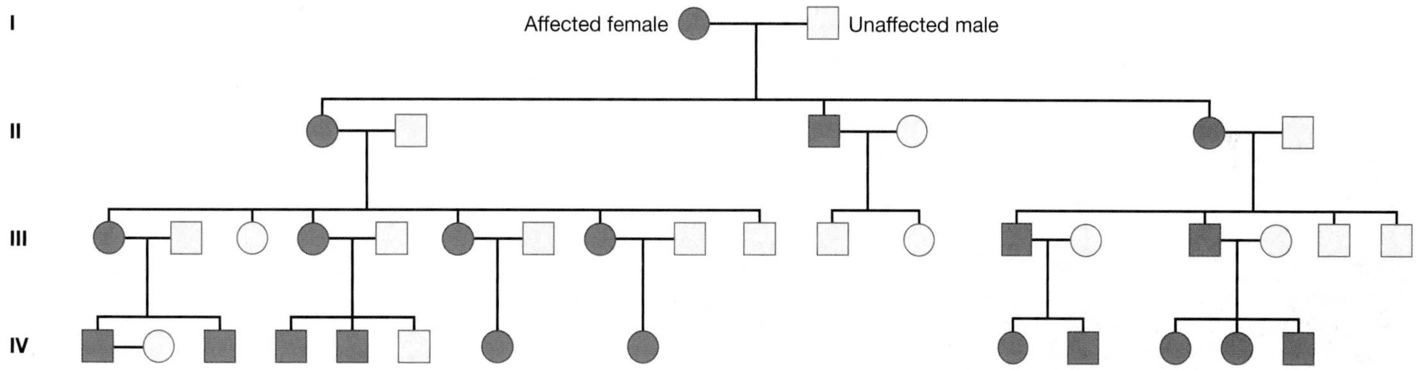

FIGURE 13.22 Pedigrees of Diseases Inherited on Autosomes **(a)** Diseases, such as cystic fibrosis, that are inherited as autosomal recessives appear in both males and females. For an individual to be affected, both parents must carry the allele responsible. If the recessive allele is rare, affected individuals may not appear in every generation. **(b)** Diseases that are inherited as autosomal dominants may appear in both males and females and tend to appear in every generation. QUESTION In a pedigree where an autosomal recessive disease appears, an affected parent frequently has unaffected children. But unaffected parents can have affected offspring. Why?

one parent is heterozygous and the other is homozygous recessive, on average half of their children should show the dominant phenotype. And unless a new mutation has occurred, any child with the trait must have a parent with this trait. The latter observation is in strong contrast to the pattern seen in autosomal recessive traits.

Figure 13.22b shows the consequences of autosomal dominant inheritance, in the pedigree of a family affected by a degenerative brain disorder called **Huntington's disease.** The pedigree has two features that indicate this disease is passed to the next generation as an autosomal dominant allele. First, if a child shows the trait, then one of its parents shows the trait as well. Second, if families have a large number of children, the trait usually shows up in every generation—due to the high probability of heterozygous parents having affected children. Other examples of autosomal dominant traits in humans include a widow's peak, the presence of extra fingers or toes (polydactyly), and achondroplastic dwarfism, which is due to a defect in cartilage structure.

Is the Trait Autosomal or Sex-Linked?

When it is not possible to arrange reciprocal crosses, can data in a pedigree indicate whether a trait is autosomal or sex-linked? The answer is based on a simple premise: If a trait appears about equally often in males and females, then it is likely to be autosomal. For example, the data in the Huntington's disease pedigree indicates that the disease appears in both males and females at about equal rates. But if males are much more likely to have the trait in question than females are, then the gene responsible may be found on the X chromosome. Because so few genes occur on the Y chromosome, Y-linked inheritance is rare.

To understand why a sex bias in phenotypes implicates sex-linked inheritance, recall from Section 13.4 that sex-linked genes are located on one of the sex chromosomes. Because human males have one X chromosome and one Y chromosome, they have just one copy of each X-linked gene. But because human females have two X chromosomes, they have two copies of each X-linked gene. These simple observations are critical. In humans—just as in fruit flies and in every other species that has sex chromosomes—the pattern of inheritance in sex-linked traits is different in men and women, because the complement of sex chromosomes differs in males and females.

What does the pedigree of an X-linked trait look like? To answer this question, let's consider the pedigree of a classic

X-linked trait—the occurrence of hemophilia in the descendants of Queen Victoria and Prince Albert, who were monarchs of England in the mid-to-late 1800s (**Figure 13.23a**). **Hemophilia** is caused by a defect in an important blood-clotting factor. Hemophiliacs are at a high risk of bleeding to death, because even minor injuries result in prolonged bleeding. The key point to notice in the pedigree of Queen Victoria's descendants is that only males developed hemophilia (**Figure 13.23b**). In addition, note the affected male in generation II (the second square from the right in row II). His two sons were unaffected, but the trait reappeared in a grandson. Stated another way, the occurrence of hemophilia skipped a generation.

This pattern, is logical because hemophilia is due to an X-linked *recessive* allele. Because males have only one X chromosome, the phenotype associated with an X-linked recessive allele appears in every male that carries it. Females express the recessive phenotype only if they are homozygous for that allele. Because the defective allele is rare in this and other populations, it is extremely unlikely to appear in the homozygous state.

Further, the appearance of an X-linked recessive trait skips a generation in a pedigree because the affected male passes his only X chromosome on to his daughters (his sons receive his Y chromosome). But because his daughters have two X chromosomes and are most likely to have received a normal allele from their mother, the daughters don't show the trait. They, however, will pass the defective allele on to about half of their sons—the probability is that about half will inherit the X chromosome with the defective allele, and half will inherit the other X chromosome. Thus about half of their sons will develop the trait. Females who can be inferred to be carriers, because one or more of their sons have hemophilia, are highlighted in Figure 13.23b.

In contrast, if an X-linked trait is dominant, it will appear in every individual who has the defective allele. A good indicator

(a) Queen Victoria and Prince Albert

(b) Occurrence of hemophilia in royal families of Europe

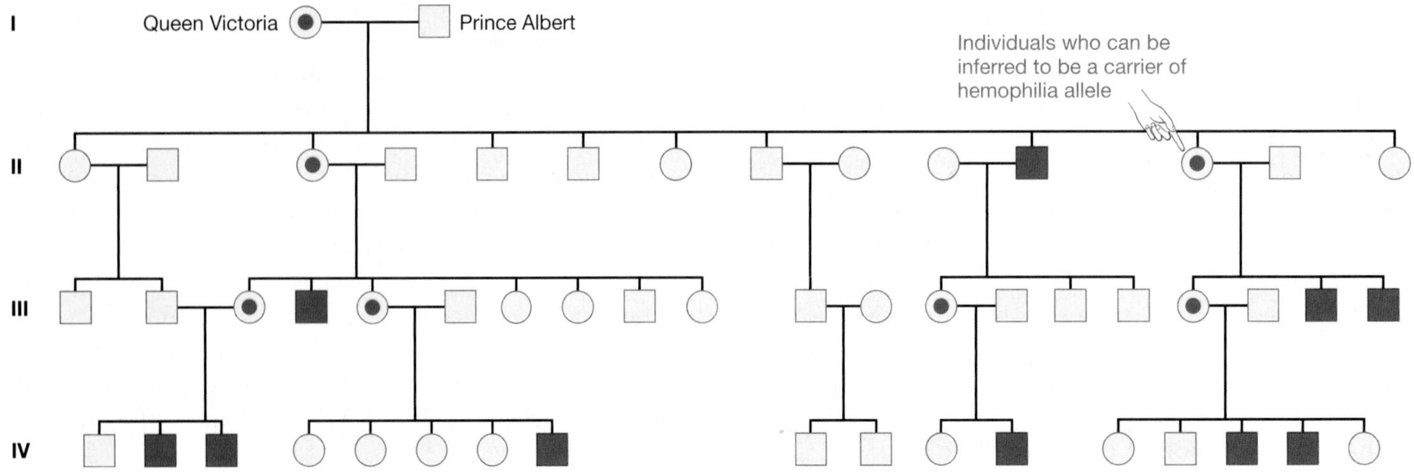

FIGURE 13.23 A Pedigree of an X-Linked Recessive Disease
(a) Queen Victoria, who was a carrier for hemophilia, and Prince Albert. **(b)** A pedigree showing the occurrence of hemophilia in the descendants of Victoria and Albert. In this family, the condition shows up only in males. **EXERCISE** Redraw this pedigree, showing the pattern of inheritance if the allele for hemophilia were X-linked dominant.

of an X-linked dominant trait is a pedigree in which an affected male has all affected daughters but no affected sons. This pattern is logical because every female offspring of an affected father gets an X chromosome from him and will herself be affected, while his sons get their only X chromosome from their unaffected mother. Besides the inherited form of a bone disease called **rickets**, however, very few diseases are known to be due to X-linked dominant alleles.

By analyzing pedigrees in this way, biomedical researchers have been able to characterize how most of the common genetic diseases in humans are inherited. By the 1940s, though, the burning question in genetics was no longer the nature of inheritance, but the nature of the gene itself. What are genes made of, and how are they copied so that parents pass their alleles on to their offspring? These are the questions we turn to in Chapter 14.

ESSAY Does "Genetic Determinism" Exist?

When people say that a trait is genetically determined, they usually mean that an individual with a particular genotype will have a particular phenotype, no matter what environment the individual experiences as it grows and matures. Is seed shape in garden peas genetically determined? Naïvely, the answer would appear to be yes. In Mendel's experiments, individuals with *RR* or *Rr* genotypes had round seeds, while individuals with *rr* genotypes had wrinkled seeds. At the seed-shape gene, then, there is a strong association between genotype and phenotype. There is also a strong resemblance between the phenotypes of parents and offspring. Is the same true for traits in humans, where data confirm a strong resemblance between parents and offspring? Are traits such as your height, weight, intelligence, and personality genetically determined?

The answer is that few traits are "determined" in the sense of being fixed or invariant based on a certain genotype. Among pea seeds, for example, the degree of wrinkling varies widely. There are two reasons for this. First, traits such as seed shape are affected by many aspects of the physical environment. For instance, the availability of water and nutrients and the presence of disease-causing fungi or viruses influence seed development and thus the phenotype. In garden peas, well-watered individuals with the *rr* genotype tend to have more wrinkled seeds than do individuals with the same genotype that experience prolonged water stress. Second, no gene acts alone. Mendel's seven traits also vary depending on the actions of alleles at other genes. Interaction among alleles, or epistasis, is the rule. In garden peas, researchers have identified at least three other genes that may affect the degree of wrinkling in seeds.

Mendel was able to analyze the effect of a single gene on the traits he studied only because he controlled for these effects carefully. He grew individuals in similar environments and analyzed strains with similar genetic backgrounds. (In the process of creating pure lines through breeding, Mendel ensured that individuals in his populations would have similar or identical genotypes for all traits, because they shared common ancestors.)

What does all this have to do with human traits? Each of your characteristics, like seed shape in pea plants, is a combination of your genotype at loci that have a major effect on the trait, environmental influences, and the effect of alleles at dozens or hundreds of other loci. Environmental effects happen to be particularly strong in the case of human height and intelligence.

Are traits such as your height, weight, intelligence, and personality genetically determined?

Over the past 75 years, the aspect of intelligence measured by IQ tests has increased dramatically in the industrialized nations. Yet no biologists support the hypothesis that these changes are a result of changes in the genetic makeup of these populations. Instead, most attribute the increase in IQ scores as a consequence of changes in the environment—specifically, better education.

Where does all this leave us? Few traits are genetically determined in the sense of being fixed or invariant. Similarly, it would be equally careless to refer to a trait as being environmentally determined. All traits have a genetic basis, and all have an environmental component. A genotype defines the range of phenotypes that are possible for an individual. This range is called a **norm of reaction.** Sometimes the range of possibilities is very small, as in the MN proteins of red blood cells or in the development of Huntington's disease. Sometimes the range of possibilities is very large, as in traits such as human height or intelligence. When the range of possibilities is wide, the environment plays a prominent role in determining the eventual phenotype. The relative importance of genetics and environmental effects varies from trait to trait.

In general, then, it is accurate to think of genotypes as contributing to phenotypic tendencies and predispositions or to the potential of an individual. But most genes can "determine" a trait only in the context of a carefully controlled environment and a precisely defined set of alleles at other loci. In the sense that the phrase is used in casual conversation, genetic determinism in complex traits such as intelligence and personality is rare or nonexistent.

CHAPTER REVIEW

Summary of Key Concepts

- **Mendel discovered that in garden peas, individuals have two alleles, or versions, of each gene. Prior to the formation of eggs and sperm, the two alleles of each gene separate so that one allele is transmitted to each egg or sperm cell.**

  Gregor Mendel was the first individual to apply a modern scientific approach to the study of heredity. Mendel proposed two basic principles of transmission genetics. The first was the principle of segregation, which claimed that each trait was specified by paired hereditary determinants (alleles of genes) that separate from each other during gamete formation.

- **Genes are located on chromosomes. The separation of homologous chromosomes during meiosis I explains why alleles of the same gene segregate to different gametes.**

  The chromosome theory of inheritance claimed that genes are located on chromosomes and that the movements of chromosomes during meiosis provide a physical basis for Mendel's observations. The chromosome theory provided a physical basis for the principle of segregation. Alleles of the same gene segregate from each other and end up in different gametes because homologous chromosomes segregate at anaphase of meiosis I.

- **If genes are located on different chromosomes, then the alleles of each gene are transmitted to egg cells and sperm cells independently of each other.**

  Mendel's second basic conclusion was the principle of independent assortment, which stated that the segregation of one pair of genes—controlling a given trait—was not influenced by the segregation of other gene pairs. The chromosome theory provided a physical basis for the principle of independent assortment. Genes located on different chromosomes move to gametes independently of each other during meiosis. The principle of independent assortment holds because chromosomes line up independently of each other at the metaphase plate during meiosis I. Thus, gametes receive random combinations of chromosomes.

  Web Tutorial 13.1 Mendel's Experiments

  Web Tutorial 13.2 The Principle of Independent Assortment

- **Important exceptions exist to the rules that individuals have two alleles of each gene and that alleles of different genes are transmitted independently. Genes on the same chromosome are not transmitted independently of each other. If a gene is on a sex chromosome, then not every gamete will receive an allele of that gene.**

  Thomas Hunt Morgan and colleagues extended Mendel's work by describing X-linked inheritance and by showing that genes located on the same chromosome do not exhibit independent assortment. Studies of X-linked traits helped confirm that genes are found on chromosomes, while studies of linked traits led to the first genetic maps showing the locations of genes on chromosomes. Later studies confirmed that many traits are influenced by the interaction of several genes and that phenotypes are influenced by the environment an individual has experienced as well as by its genotype. Analysis of pedigrees allowed researchers to work out the basic modes of inheritance for most of the common genetic diseases in humans.

Genetics Questions

1. In studies of how traits are inherited, what makes certain species candidates for model organisms?
   a. They are the first organisms to be used in a particular type of experiment, so they are a historical "model" of what researchers expect to find.
   b. They are easy to study because a great deal is already known about them.
   c. They are the best or most fit of their type.
   d. They are easy to maintain, have a short life cycle, produce many offspring, and yield data that are relevant to many other organisms.
2. Why is the allele for wrinkled seed shape in garden peas considered recessive?
   a. It "recedes" in the F_2 generation when homozygous parents are crossed.
   b. The trait associated with the allele is not expressed in heterozygotes.
   c. Individuals with the allele have lower fitness than that of individuals with the dominant allele.
   d. The allele is less common than the dominant allele. (The wrinkled allele is a rare mutant.)
3. The alleles found in haploid organisms cannot be dominant or recessive. Why?
   a. Dominance and recessiveness describe interactions between two alleles of the same gene in the same individual.
   b. Because only one allele is present, alleles in haploid organisms are always dominant.
   c. Alleles in haploid individuals are transmitted like mitochondrial DNA or chloroplast DNA.
   d. Most haploid individuals are bacteria, and bacterial genetics is completely different from eukaryotic genetics.
4. Biologists no longer use the term *pure line* except in a historical context. Instead, they simply refer to "pure-line" individuals as homozygotes. Why?
   a. They are highly inbred and are homozygous at every gene.
   b. Only two alleles are present at each gene in the populations to which these individuals belong.
   c. Calling them "pure" implies some sort of value judgment.
   d. All of the individuals in pure lines are homozygous at the gene in question.

5. The genes for the traits that Mendel worked with either are located on different chromosomes or are so far apart on the same chromosome that crossing over almost always occurs between them. How did this circumstance help Mendel recognize the principle of independent assortment?
   a. Otherwise, his dihybrid crosses would not have produced a 9:3:3:1 ratio of F_2 phenotypes.
   b. The occurrence of individuals with unexpected phenotypes led him to the discovery of recombination.
   c. It led him to the realization that the behavior of chromosomes during meiosis explained his results.
   d. It meant that the alleles involved were either dominant or recessive, which gave 3:1 ratios in the F_1 generation.
6. The text claims that Mendel worked with the simplest possible genetic system. Is this claim legitimate?
   a. Yes—discrete traits, two alleles, simple dominance and recessiveness, no sex chromosomes, and unlinked genes are the simplest situation known.
   b. Yes—the ability to self-fertilize or cross-pollinate made it simple for Mendel to set up controlled crosses.
   c. No—Mendel was unaware of meiosis and the chromosome theory of inheritance, so it was not easy to reach the conclusions he did.
   d. No—Mendel's experimental designs and his rules of inheritance are actually quite complex and sophisticated.
7. Mendel's rules do not correctly predict patterns of inheritance for tightly linked genes or the inheritance of alleles that show incomplete dominance or epistasis. Does this mean that his hypotheses are incorrect?
   a. Yes, because they are relevant to only a small number of organisms and traits.
   b. Yes, because not all data support his hypotheses.
   c. No, because he was not aware of meiosis or the chromosome theory of inheritance.
   d. No, it just means that his hypotheses are limited to certain conditions.
8. The artificial sweetener NutraSweet consists of a phenylalanine molecule linked to aspartic acid. The labels of diet sodas that contain NutraSweet include a warning to people with PKU. Why?
   a. Even though it is an artificial sweetener, NutraSweet stimulates the same taste receptors that natural sugars do.
   b. If an individual cannot metabolize phenylalanine, ingestion of NutraSweet will increase blood levels of phenylalanine.
   c. In people with PKU, phenylalanine reacts with aspartic acid to form a toxic compound.
   d. People with PKU cannot lead normal lives, even if their environment is carefully controlled.
9. When Sutton and Boveri published the chromosome theory of inheritance, research on meiosis had not yet established that paternal and maternal homologs assort independently of each other. Then, in 1913, Elinor Carothers published a paper about a grasshopper species with an unusual karyotype: One chromosome had no homolog (meaning no pairing partner at meiosis I); another chromosome had homologs that could be distinguished under the light microscope. If chromosomes assort independently, how often should Carothers have observed each of the four products of meiosis shown in the following figure? Explain.

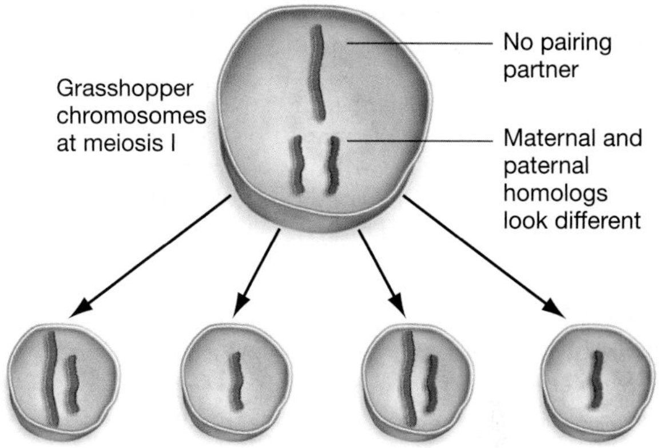
Grasshopper chromosomes at meiosis I
No pairing partner
Maternal and paternal homologs look different
Four types of gametes possible
(each meiotic division can produce only two of the four)

   a. Only the gametes with one of each type of chromosome would occur.
   b. The four types of gametes should be observed to occur at equal frequencies.
   c. The chromosome with no pairing partner would disintegrate, so only gametes with one copy of the other chromosome would be observed.
   d. Gametes with one of each type of chromosome would occur twice as often as gametes with just one chromosome.
10. Consider the alleles involved in quantitative traits such as human height. Is the concept of dominance relevant to these alleles? Why or why not?
   a. Yes—the dominant alleles confer the highest values for the trait and the highest fitness.
   b. Yes—the dominant alleles are the most common.
   c. No—the alleles in quantitative traits are always codominant.
   d. No—dominance can occur only in discrete traits.

Genetics Problems

The best way to test and extend your knowledge of transmission genetics is to work problems. Most genetics problems are set up as follows: You are given some information about the genotypes or phenotypes of one or both parents, along with data on the phenotypes of F_1 or F_2 offspring. Your task is to generate a set of hypotheses—or what biologists call a genetic model—to explain the results. More specifically, you will need to generate a hypothesis to address each of the following questions:

- Is the trait involved discrete or quantitative?
- Is the phenotype a product of one gene or many genes?
- For each gene involved, how many alleles are present—one, two, or many?
- Do the alleles involved show complete dominance, incomplete dominance, or codominance?
- Are the genes involved sex-linked or autosomal?
- If more than one gene is involved, are they linked or unlinked? If they are linked, does crossing over occur frequently?

It's also helpful to ask yourself whether epistasis or pleiotropy might be occurring and whether it is safe to assume that the experimental design carefully controlled for effects of variation in other genes or the environment.

In working the problem, be sure to start with the simplest possible explanation. For example, if you are dealing with a discrete trait, you might hypothesize that the cross involves a single autosomal gene with two alleles that show complete dominance. Your next step is to infer what the parental genotypes are (if they are not already given), and then do a Punnett square to predict the offspring phenotypes and their frequencies based on your hypotheses. Next, check whether these predictions match the observed results given in the problem. If the answer is yes, you have a valid solution. But if the answer is no, you need to go back and change one of your hypotheses, redo the Punnett square, and check to see if the predictions and observations match. Keep redoing these steps until you have a model that fits the data.

11. In the plant species *Plectritis congesta*, known as rosy plectritis, individuals produce fruits that either have or do not have prominent structures called wings. The alleles involved are W^+ = winged fruit; W^- = wingless fruit. Researchers collected an array of individuals from the field and performed a series of crosses. The results are given in the table below. Complete the table by writing down the genotype of the parent or parents involved in each cross. (Write down one genotype if selfing occurs and two genotypes if two different individuals were involved.)

Parental Phenotype(s)	Number of Offspring with Winged Fruits	Number of Offspring with Wingless Fruits	Parental Genotype(s)
Wingless (self-fertilized)	0	80	
Winged (self-fertilized)	90	30	
Winged × wingless	46	0	
Winged × winged	44	0	

12. *Tay-Sachs disease* causes nerve cells to malfunction and results in death by age 4. Two healthy parents know from blood tests that each parent carries a recessive allele responsible for Tay-Sachs. If their first three children have the disease, what is the probability that their fourth child will not? Assuming that they have not yet had a child, what is the probability that, if they have four children, all four will have the disease? If their first three children are male, what is the probability that their fourth child will be male?

13. Two black female mice are crossed with a brown male. In several litters, female I produced 9 blacks and 7 browns; female II produced 57 blacks. What deductions can you make concerning the inheritance of black and brown coat color in mice? What are the genotypes of the parents in this case?

14. Suppose that in garden peas the genes for seed color and seed-pod shape are linked, and that Mendel crossed *YYII* parents (which produce yellow seeds in inflated pods) with *yyii* parents (which produce green seeds in constricted pods).
    - Draw the F_1 Punnett square and predict the expected F_1 phenotype(s).
    - List the genotype(s) of gametes produced by F_1 individuals if no crossing over occurs.
    - Draw the F_2 Punnett square if no crossing over occurs. Based on this Punnett square, predict the expected phenotype(s) in the F_2 generation and the expected frequency of each phenotype.
    - If crossing over occurs during gamete formation in F_1 individuals, give the genotype of the recombinant gamete(s) that result.
    - Add the recombinant gametes to the F_2 Punnett square above. Will any additional phenotypes be observed at low frequency in the F_2 generation? If so, what are they?

15. In Jimson weed, the allele that results in violet flowers (*V*) is dominant to one that results in white flowers (*v*); at another locus, the allele that makes prickly seed capsules (*P*) is dominant to one that results in smooth capsules (*p*). A plant with white flowers and prickly capsules is crossed with one that has violet flowers and smooth capsules. The F_1 consisted of 47 plants with white flowers and prickly capsules, 45 plants with white flowers and smooth capsules, 50 plants with violet flowers and prickly capsules, and 46 plants with violet flowers and smooth capsules. What are the genotypes of the parents? Are these two genes linked?

16. In cats, the *Manx* allele (*M*) causes a short or absent tail, while a recessive allele *m* confers a normal, long tail. Cats of genotype *MM* die as embryos. If two Manx cats mate, what is the probability that each *surviving* kitten has a long tail?

17. A plant with orange-spotted flowers was grown in the greenhouse from a seed collected in the wild. The plant was self-pollinated and gave rise to the following progeny: 88 orange with spots, 34 yellow with spots, 32 orange with no spots, and 8 yellow with no spots. What can you conclude about the dominance relationships of the alleles responsible for the spotted and unspotted phenotypes? Of orange and yellow phenotypes? What can you conclude about the genotype of the original plant that had orange, spotted flowers?

18. As a genetic counselor, you routinely advise couples about the possibility of genetic disease in their offspring based on their family histories. This morning you met with an engaged couple, both of whom are phenotypically normal. The man, however, has a brother who died of Duchenne-type muscular dystrophy, an X-linked condition that results in death before the age of 20. The allele responsible for this disease is recessive. His prospective bride, whose family has no history of the disease, is worried that the couple's sons or daughters might be afflicted.
    - How would you advise this couple?
    - The sister of this man is planning to marry his fiancé's brother. How would you advise this second couple?

19. Study the experimental result illustrated in Figure 13.6a. To further test his hypothesis of independent assortment, Mendel let the F_2 progeny from this experiment self-fertilize. Then he examined the phenotypes of the F_3 generation.
    - Consider the F_2 individuals that have one dominant and one recessive trait (yellow and wrinkled seeds or green and round seeds). When they self-fertilize, what fraction of the progeny will be homozygous for the traits that appeared in the parents?
    - Consider the F_2 individuals that have both dominant traits. When they self-fertilize, what fraction of the progeny will be homozygous for the dominant traits?

20. Suppose that you are heterozygous for two genes that are located on different chromosomes. You carry alleles *A* and *a* for one gene and alleles *B* and *b* for the other. Draw a diagram that illustrates what happens to these genes and alleles when meiosis occurs in your reproductive tissues. Label the stages of meiosis, the homologous chromosomes, sister chromatids, nonhomologous chromosomes, genes, and alleles. Be sure to list all of the genetically different gametes that could form and how frequently each type should be observed. On the diagram, identify the events responsible for the principle of segregation and the principle of independent assortment.

21. In humans, the ABO blood type is a polymorphic trait. Describe the alleles and genotypes responsible for this polymorphism. Suppose a woman with blood type O married a man with blood type AB. What phenotypes and genotypes would you expect to observe in their offspring, and in what proportions?

22. Mr. Spock's father came from the planet Vulcan, and his mother was from planet Earth. A Vulcan has pointed ears and a heart located on the right side of the chest. Mr. Spock has both of these traits, which are known to be determined by two different genes, each with two alleles. Suppose that Mr. Spock married an Earth woman and that they had many children. About half of their children have pointed

ears and a right-sided heart, like Spock, and about half have rounded ears and a left-sided heart, like their mother.

- What would Mendel predict the progeny phenotypes and ratios to be? Explain your answer using formal genetic terminology.
- How do you explain the actual results?

23. In Klingons, one gene determines hair texture; another gene determines whether the individual will have a sagittal crest (a protrusion on the forehead). The two genes are not linked.

K = curly Klingon hair (dominant)

k = silky earthling-like hair (recessive)

S = large sagittal crest (dominant)

s = smooth, flat, earthling-like forehead (recessive)

Kayless is a half-human, half-Klingon with the genotype *KkSs*. He mates with an individual who is also heterozygous for both genes.

- Set up a Punnett square for this dihybrid cross.
- What are the four possible phenotypes that may result from this mating? Include a description of both hair and forehead for each phenotype.
- What is the expected phenotypic ratio from the dihybrid cross?
- What fraction of the progeny are expected to be heterozygous for both genes?
- What fraction are expected to be homozygous for both genes?
- Are Kayless and his mate more likely to see an actual ratio close to the predicted values if they have 16 children or 160? Why? Explain why sample size does or does not affect observed phenotypic ratios.

24. The blending-inheritance hypothesis proposed that the genetic material from parents is irreversibly mixed in the offspring. As a result, offspring should always appear intermediate in phenotype to the parents. Mendel, in contrast, proposed that genes are discrete and that their integrity is maintained in the offspring and in subsequent generations. Suppose the year is 1890. You are a horse breeder and have just read Mendel's paper. You don't believe his results, however, because you often work with cremello (very light-colored) and chestnut (reddish-brown) horses. You know that if you cross a cremello individual from a pure-breeding line with a chestnut individual from a pure-breeding line, the offspring will be palomino—meaning they have an intermediate (golden-yellow) body color. What additional crosses would you do to test whether Mendel's model is valid in the case of genes for horse color? List the crosses and the offspring genotypes and phenotypes you'd expect to obtain. Explain why these experimental crosses would provide a test of Mendel's model.

25. Two mothers give birth to sons at the same time in a busy hospital. The son of couple 1 is afflicted with hemophilia A, which is a recessive X-linked disease. Neither parent has the disease. Couple 2 has a normal son even though the father has hemophilia A. The two couples sue the hospital in court, claiming that a careless staff member swapped their babies at birth. You appear in court as an expert witness. What do you tell the jury? Make a diagram that you can submit to the jury.

26. You have crossed two *Drosophila melanogaster* individuals that have long wings and red eyes—the wild-type phenotype. In the progeny, the mutant phenotypes, called curved wings and lozenge eyes, appear as follows:

Females	**Males**
600 long wings, red eyes	300 long wings, red eyes
200 curved wings, red eyes	300 long wings, lozenge eyes
	100 curved wings, red eyes
	100 curved wings, lozenge eyes

- According to these data, is the curved-wing allele autosomal recessive, autosomal dominant, sex-linked recessive, or sex-linked dominant?
- Is the lozenge-eye allele autosomal recessive, autosomal dominant, sex-linked recessive, or sex-linked dominant?
- What is the genotype of the female parent?
- What is the genotype of the male parent?

27. In parakeets, two autosomal genes that are located on different chromosomes control the production of feather pigment. Gene *B* codes for an enzyme that is required for the synthesis of a blue pigment, and gene *Y* codes for an enzyme required for the synthesis of a yellow pigment. Recessive, loss-of-function mutations are known for both genes. Suppose that a bird breeder has two green parakeets and mates them. The offspring are green, blue, yellow, and albino (unpigmented).

- Based on this observation, what are the genotypes of the green parents? What is the genotype of each type of offspring? What fraction of the total progeny should exhibit each type of color?
- Suppose that the parents were the progeny of a cross between two true-breeding strains. What two types of crosses between true-breeding strains could have produced the green parents? Indicate the genotypes and phenotypes for each cross.

28. The pedigree that follows is for the human trait called osteopetrosis, which is characterized by bone fragility and dental abscesses. Is the gene that affects bone and tooth structure autosomal or sex-linked? Is the allele for osteopetrosis dominant or recessive?

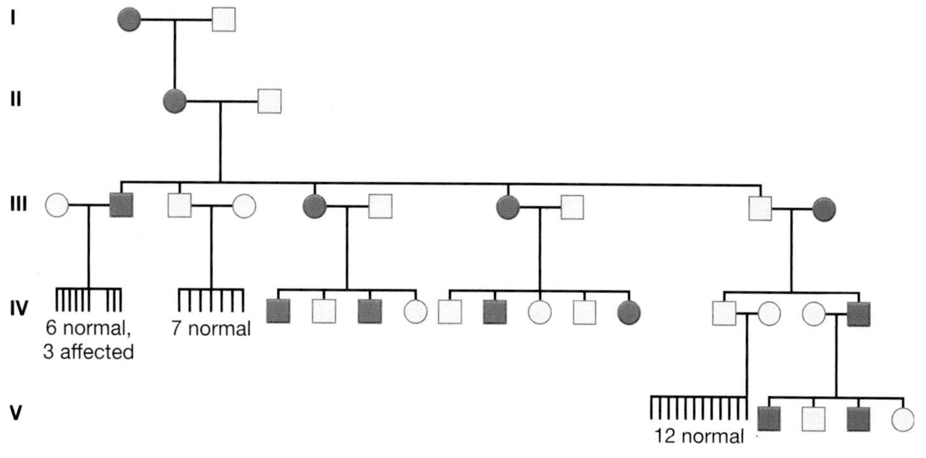
I
II
III
IV
V
6 normal,
3 affected
7 normal
12 normal

Answers to Genetics Questions **1.** d; **2.** b; **3.** a; **4.** d; **5.** a; **6.** a; **7.** d; **8.** b; **9.** b; **10.** d

Answers to Genetics Problems **11.** First cross: W^-W^-; second cross: W^+W^-; third cross: $W^+W^+ \times W^-W^-$; fourth cross: $W^+W^+ \times W^+W^+$ or $W^+W^+ \times W^+W^-$. **12.** 3/4; 1/256; 1/2. **13.** The simplest explanation for the data is that these coat-color phenotypes are caused by a single gene where two alleles are present and where B = black and b = brown. Female I is Bb, female II is BB, and the male is bb. **14.** Your answer should conform to the cross diagrammed in Figure 13.6b, except that different alleles and traits are being analyzed. **15.** Parental genotypes are $vvPp \times Vvpp$. Based on these data, we can't tell if they are linked or not. Both hypotheses make the same prediction. **16.** 1/3. **17.** This is a dihybrid cross that yields progeny phenotypes in a 9:3:3:1 ratio. Let O stand for the allele for orange petals and o the allele for yellow petals; let S stand for the allele for spotted petals and s the allele for unspotted petals. Start with the hypothesis that O is dominant to o, that S is dominant to s, that the two genes are found on different chromosomes so they assort independently, and that the parent individual's genotype is $OoSs$. If you do a Punnett square for the $OoSs \times OoSs$ mating, you'll find that progeny phenotypes should be in the observed 9:3:3:1 proportions. **18.** Let D stand for the normal allele and d stand for the allele responsible for Duchenne-type muscular dystrophy. The woman's family has no history of the disease, so her genotype is almost certainly DD. The man is not afflicted, so he must be DY. (The trait is X-linked, so he has only one allele; the "Y" stands for the Y chromosome.) Their children are not at risk. The man's sister could be a carrier, however—meaning she has the genotype Dd. If so, then half of the second couple's male children are likely to be affected. **19.** 1/2; 1/4. **20.** See website. **21.** Half of their offspring should have the genotype iI^A and the type A blood phenotype. Half of their offspring should have the genotype iI^B and the type B blood phenotype. **22.** Presumably, Vulcans and earthlings are homozygous for both traits. Spock, then, should be heterozygous at both loci. Because he has the Vulcan phenotype, though, we can infer that pointed ears and a right-sided heart are dominant to rounded ears and a left-sided heart. Let P be the allele for pointed ears and p be the allele for rounded ears; let R be the allele for a right-sided heart and r the allele for a left-sided heart. Spock is $PpRr$; his earthling wife is $pprr$. Mendel would predict that their offspring would have the genotypes $PpRr$, $Pprr$, $ppRr$, and $pprr$ in equal proportions. Thus a quarter of their children should have pointed ears and left-sided hearts or rounded ears and right-sided hearts. The only logical explanation for the actual results is that the two genes are closely linked. If so, then Spock's gametes are either PR or pr, and his wife's gametes are all pr. **23.** Four phenotypes should be observed in the ratio 9:3:3:1, as follows: curly and large: curly and smooth: silky and large: silky and smooth. One-quarter of the progeny should be heterozygous for both traits. One-quarter should be homozygous for both traits. They are more likely to see the predicted ratios with a large family, because there would be fewer fluctuations due to chance. **24.** According to Mendel's model, palomino individuals should be heterozygous at the locus for coat color. If you mated palomino individuals, you would expect to see a combination of chestnut, palomino, and cremello offspring. If blending inheritance occurred, however, all of the offspring should be palomino. **25.** See website. **26.** The curved-wing allele is autosomal recessive; the lozenge-eye allele is sex-linked (specifically, X-linked) recessive. Let w^+ be the allele for long wings and w be the allele for curved wings; let r^+ be the allele for red eyes and r the allele for lozenge eyes. The female parent is w^+wr^+r; the male parent is w^+wr^+Y. **27.** Albinism indicates the absence of pigment, so let b stand for an allele that gives the absence of blue and y for an allele that gives the absence of yellow pigment. If blue and yellow pigment blend to give green, then both green parents are $BbYy$. The green phenotype is found in $BBYY$, $BBYy$, $BbYY$, and $BbYy$ offspring. The blue phenotype is found in $BByy$ or $Bbyy$ offspring. The yellow phenotype is observed in $bbYY$ or $bbYy$ offspring. Albino offspring are $bbyy$. The phenotypes of the offspring should be in the ratio 9:3:3:1 as green:blue:yellow:albino. Two types of crosses yield $BbYy$ F_1 offspring: $BByy \times bbYY$ (blue × yellow) and $BBYY \times bbyy$ (green × albino). **28.** Autosomal dominant.

www.prenhall.com/freeman is your resource for the following: Web Tutorials; Online Quizzes and other Online Study Guide materials; Answers to Conceptual Review Questions; Solutions to Group Discussion Problems; Answers to Figure Caption Questions and Exercises; and Additional Readings and Research.

DNA Synthesis

14

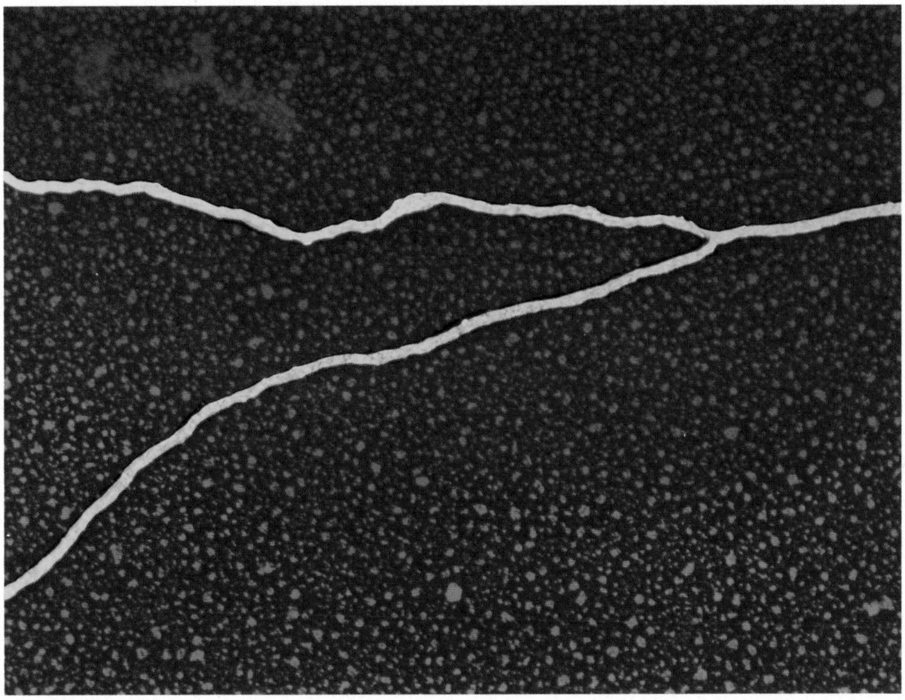

Electron micrograph showing DNA in the process of replication. The "Y" shape results from a structure called the replication fork, where DNA synthesis is taking place. The DNA double helix on the right is being replicated into two DNA double helices on the left.

KEY CONCEPTS

- Genes are made of DNA. When DNA is copied, each strand of a DNA double helix serves as the template for the synthesis of a complementary strand.
- DNA is synthesized only in the $5' \rightarrow 3'$ direction. When a DNA molecule is being copied, a large suite of specialized enzymes is involved in unwinding the double helix, continuously synthesizing the "leading strand" in the $5' \rightarrow 3'$ direction and synthesizing the "lagging strand" as a series of fragments that are then linked together.
- Most mistakes that occur in DNA synthesis are repaired by specialized enzymes. If these repair enzymes are defective, the mutation rate increases. If mutations occur in genes that control the cell cycle, cancer may develop.

What are genes made of, and how are they copied so that they are faithfully passed on to offspring? These questions dominated biology during the middle of the twentieth century. After the chromosome theory of inheritance was published and confirmed in the early 1900s, understanding the molecular nature of the gene became *the* burning question in biological science. But for decades the answer remained a mystery.

During this time, the predominant research strategy in genetics was to conduct a series of experimental crosses, create a genetic model to explain the types and proportions of phenotypes that resulted, and then test the model's predictions through reciprocal crosses, testcrosses, or other techniques. This strategy was extremely productive. It led to virtually all of the discoveries analyzed in Chapter 13, including Mendel's rules, sex-linkage, linkage, genetic mapping, incomplete dominance and codominance, multiple allelism and polymorphic traits, pleiotropy, epistasis, and quantitative inheritance.

Entirely new research strategies came to the fore, however, when the molecular basis of inheritance was finally understood. How was the chemical nature of the gene eventually discovered? How are genes copied, so they can be transmitted to the next generation? Do mistakes ever occur when genes are being replicated prior to mitosis or meiosis?

The goal of this chapter is to answer these questions. We begin with studies that identified deoxyribonucleic acid (DNA) as the genetic material, and we then explore how this molecule is copied during the synthesis phase of the cell cycle.

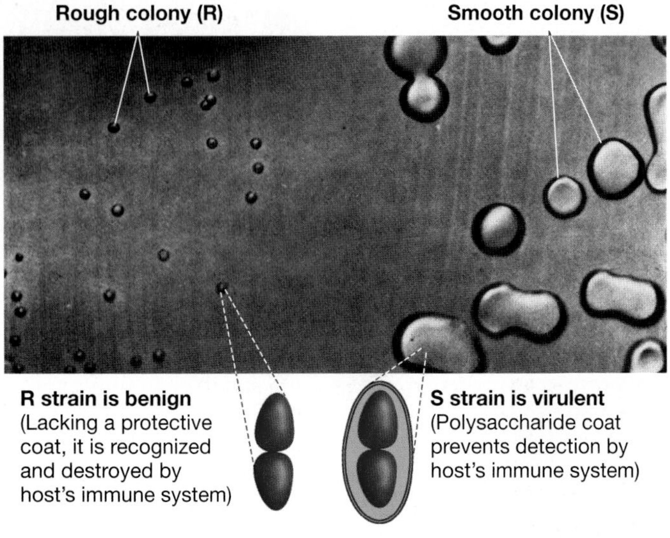
Rough colony (R)
Smooth colony (S)
R strain is benign
(Lacking a protective coat, it is recognized and destroyed by host's immune system)
S strain is virulent
(Polysaccharide coat prevents detection by host's immune system)

FIGURE 14.1 There Are Two Strains of *Streptococcus pneumoniae*
Nonvirulent and virulent strains of *Streptococcus* differ in colony appearance and at the cellular level.

14.1 DNA as the Hereditary Material

The first hint of what genes might be made of was published in 1928, when Frederick Griffith reported the discovery of a mysterious phenomenon involving hereditary traits. Griffith referred to this phenomenon as **transformation.** His transformation experiments appeared to isolate the hereditary material.

In the 1920s Griffith had been doing experiments in an attempt to develop a vaccine against the *Streptococcus pneumoniae* bacterium, which is a leading cause of pneumonia, earaches, sinusitis, and meningitis in humans. For his experiments, however, he worked with strains of *S. pneumoniae* that infect mice. A **strain** is a population of genetically identical individuals. As is the case with the strains that affect humans, the strains that affect mice vary in their **virulence**—their ability to cause disease and death. Virulent strains cause disease; nonvirulent (or benign) strains do not.

As **Figure 14.1** shows, the strains that Griffith happened to work with can be identified by eye when grown on a nutrient medium in a petri dish. Cells from the nonvirulent strain form colonies that look rough; cells from the virulent strain form colonies that look smooth. Logically enough, Griffith called the nonvirulent strain R for rough and the virulent strain S for smooth.

To understand how the strains interact, Griffith carried out the experiment illustrated in **Figure 14.2.** This figure shows four experimental treatments that he designed. In the first treatment, he injected mice with cells of the nonvirulent R strain. As he expected, these mice lived. In the second treatment, he injected mice with cells of the virulent S strain. Not surprisingly, these mice died of pneumonia. So far, so good: These first two treatments were controls showing the effect that each strain of *S. pneumoniae* has on mice. In the third treatment, Griffith killed cells of the virulent S strain by heating them and then in-

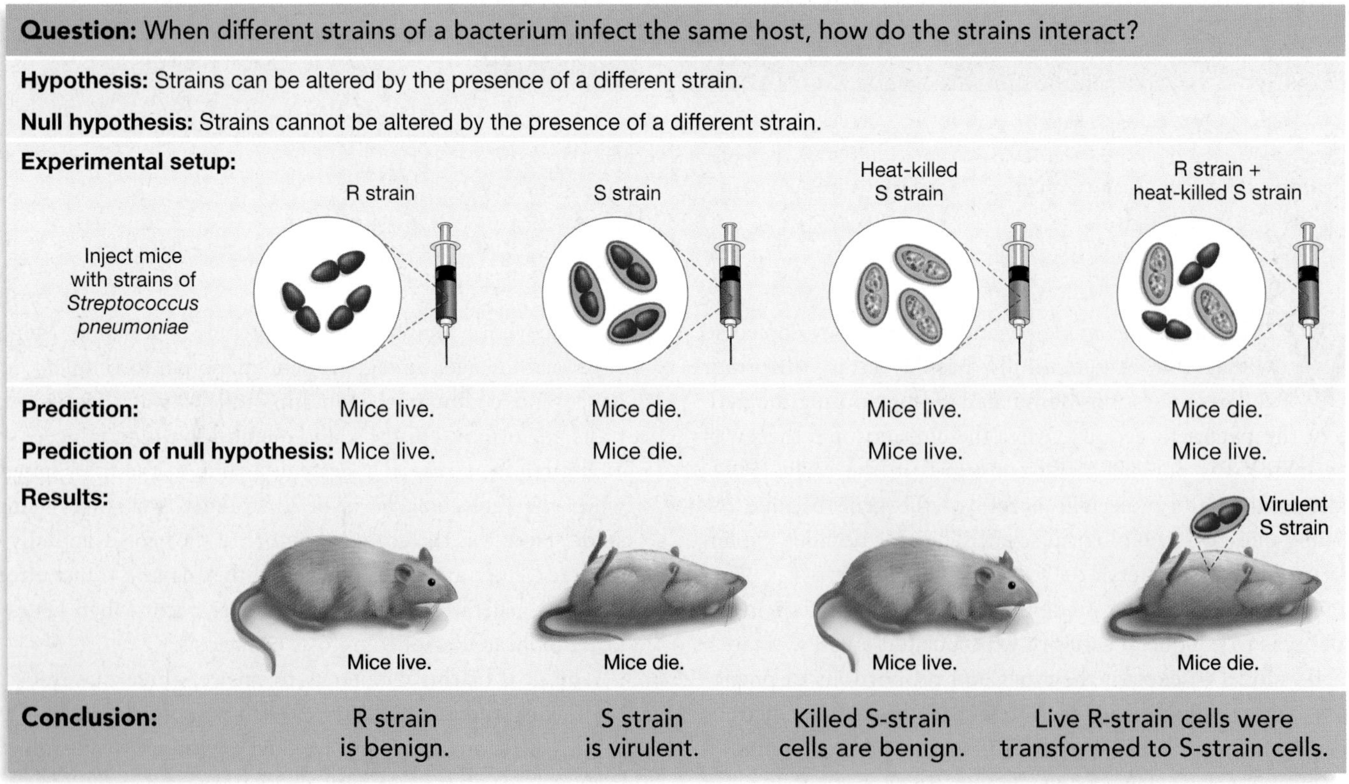
Question: When different strains of a bacterium infect the same host, how do the strains interact?
Hypothesis: Strains can be altered by the presence of a different strain.
Null hypothesis: Strains cannot be altered by the presence of a different strain.
Experimental setup:
R strain
S strain
Heat-killed S strain
R strain + heat-killed S strain
Inject mice with strains of Streptococcus pneumoniae
Prediction: Mice live. Mice die. Mice live. Mice die.
Prediction of null hypothesis: Mice live. Mice die. Mice live. Mice live.
Results:
Virulent S strain
Mice live.
Mice die.
Mice live.
Mice die.
Conclusion:
R strain is benign.
S strain is virulent.
Killed S-strain cells are benign.
Live R-strain cells were transformed to S-strain cells.

FIGURE 14.2 The Discovery of "Transformation"
Mice died after being injected with a combination of live R cells and heat-killed S cells.

jected them into mice. These mice lived. This experimental treatment was interesting because it showed that dead S cells do not cause disease. In the final treatment, Griffith injected mice with heat-killed S cells *and* live nonvirulent R cells. Unexpectedly, these mice died. Autopsies confirmed pneumonia as the cause of death. When Griffith isolated and grew *Streptococcus* from these dead mice, he found virulent S cells, *not* nonvirulent R cells.

What was going on? Griffith proposed that something from the heat-killed S cells had "transformed" the nonvirulent R cells. Stated another way, something had changed the appearance and behavior of the R cells from R-like to S-like. Because this "something" appeared in the growing population of cells that Griffith isolated from the dead mice, it had been passed on to the offspring of the transformed cells. It was clearly some sort of hereditary, or genetic, factor. Next the question became, What is it?

Is DNA the Genetic Material?

The chromosome theory of inheritance, introduced in Chapter 13, proposed that chromosomes are composed of genes. It had been known since the late 1800s that chromosomes are a complex of DNA and proteins. Because the chromosome theory had been confirmed around 1920, it was clear that Griffith's transforming factor had to consist of either protein or DNA.

Initially, most biologists backed the hypothesis that genes are made of proteins. The arguments in favor of this hypothesis were compelling. Hundreds, if not thousands, of complex and highly regulated chemical reactions occur in even the simplest living cells. The amount of information required to specify and coordinate these reactions is almost mind-boggling. With their almost limitless variation in structure and function, proteins are complex enough to contain this much information.

DNA, in contrast, was known to be composed of just four types of deoxyribonucleotides. It was also thought to be a simple molecule with some sort of repetitive and uninteresting structure. So when researchers published the first experimental evidence that DNA was the hereditary material, most biologists had the same reaction: They didn't believe it.

The Avery et al. Experiment In the early 1940s Oswald Avery, Colin M. MacLeod, and Maclyn McCarty set out to understand the molecular basis of Griffith's result. They used an elegant experimental strategy to isolate the transforming factor. To determine whether protein, RNA, or DNA was responsible for transformation, they grew quantities of virulent S cells in culture. A **culture** is a collection of cells that grows under controlled conditions—usually in a liquid suspension or on the surface of a dish on solid growth medium. Avery's group killed the cultured cells with heat, broke them open to create a cell extract, and then used chemical treatments to remove the lipids and carbohydrates from the extracts. These steps left a mixture containing protein, RNA, and DNA from the virulent S cells. The researchers divided the sample into three treatments and used different enzymes to destroy a specific macromolecule in each. One sample was treated with proteases, enzymes that destroy proteins. Another sample was treated with ribonuclease, an enzyme that breaks apart RNA. A third sample was treated with DNAase, an enzyme that cuts up DNA (**Figure 14.3**). When small quantities of the three resulting solutions were added to cultures containing nonvirulent R cells, virulent S cells appeared in all of the cultures that still contained S-cell DNA; no S cells appeared in the sample that lacked DNA. The biologists concluded that DNA—not protein or RNA—must be the transforming factor.

This experiment provided strong evidence that DNA is the hereditary material. But the result was not widely accepted. Researchers who advocated the protein hypothesis claimed that the enzymatic treatments were not sufficient to remove all of the proteins present and that enough protein could remain to transform R cells.

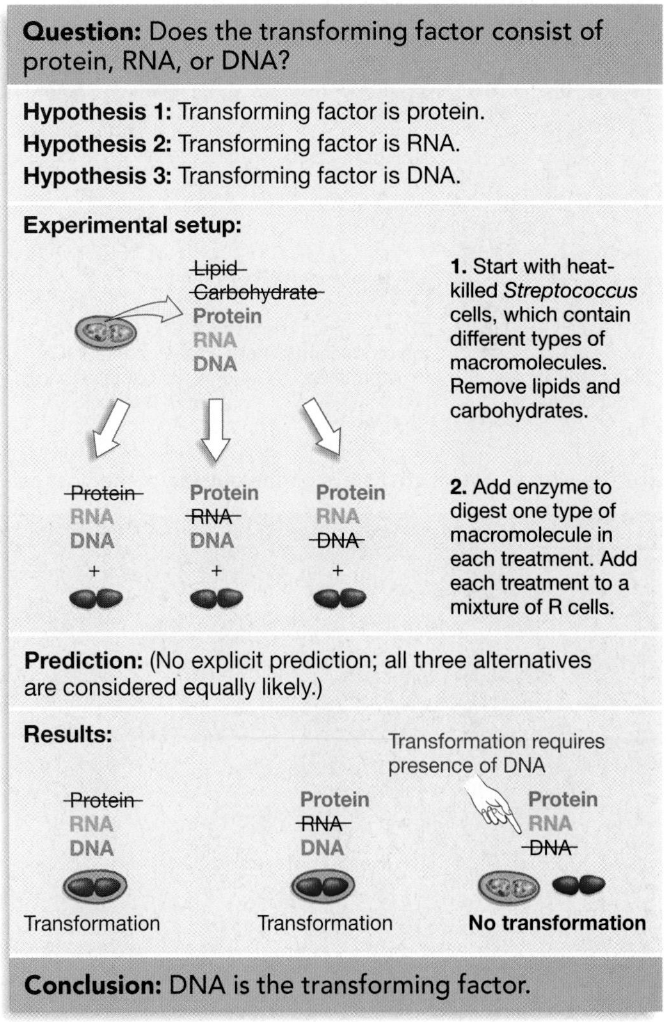

Question: Does the transforming factor consist of protein, RNA, or DNA?
Hypothesis 1: Transforming factor is protein.
Hypothesis 2: Transforming factor is RNA.
Hypothesis 3: Transforming factor is DNA.
Experimental setup:
~~Lipid~~
~~Carbohydrate~~
Protein
RNA
DNA
1. Start with heat-killed *Streptococcus* cells, which contain different types of macromolecules. Remove lipids and carbohydrates.
~~Protein~~ RNA DNA +
Protein ~~RNA~~ DNA +
Protein RNA ~~DNA~~ +
2. Add enzyme to digest one type of macromolecule in each treatment. Add each treatment to a mixture of R cells.
Prediction: (No explicit prediction; all three alternatives are considered equally likely.)
Results:
Transformation requires presence of DNA
~~Protein~~ RNA DNA
Protein ~~RNA~~ DNA
Protein RNA ~~DNA~~
Transformation
Transformation
No transformation
Conclusion: DNA is the transforming factor.

FIGURE 14.3 Experimental Evidence that DNA Is the Transforming Factor QUESTION If genes were made of protein, how would the results of this experiment change?

The Hershey-Chase Experiment Alfred Hershey and Martha Chase took up the question of whether genes were made of protein or DNA by studying how a virus called T2 infects the bacterium *Escherichia coli*. Hershey and Chase knew that T2 infections begin when the virus attaches to the cell wall of *E. coli* and injects its genes into the cell's interior (**Figure 14.4a**). These genes then direct the production of a new generation of virus particles inside the infected cell, which acts as a **host** for the parasitic virus. During the infection, the protein coat of the original, parent virus is left behind, still attached to the exterior of the host cell as a "ghost" (**Figure 14.4b**). Hershey and Chase also knew that T2 is made up almost exclusively of protein and DNA. But was it protein or DNA that entered the host cell and directed the production of new viruses?

Hershey and Chase's strategy for determining which part of the virus enters the cell and acts as the hereditary material was based on two facts: (1) Proteins present in T2 contain sulfur but not phosphorus, and (2) DNA contains phosphorus but not sulfur. As **Figure 14.5** shows, the researchers began their work by growing viruses in the presence of either the radioactive isotope of sulfur (^{35}S) or the radioactive isotope of phosphorus (^{32}P). Because these molecules were incorporated

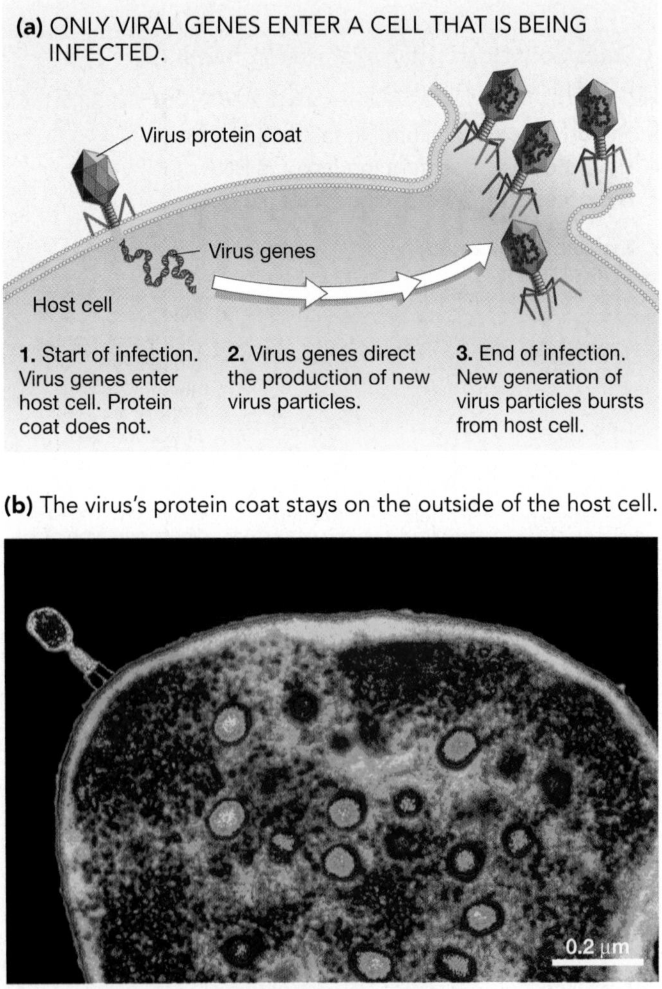
(a) ONLY VIRAL GENES ENTER A CELL THAT IS BEING INFECTED.
Virus protein coat
Virus genes
Host cell
1. Start of infection. Virus genes enter host cell. Protein coat does not.
2. Virus genes direct the production of new virus particles.
3. End of infection. New generation of virus particles bursts from host cell.
(b) The virus's protein coat stays on the outside of the host cell.
0.2 µm

FIGURE 14.4 Viruses Inject Genes into Bacterial Cells and Leave a Protein Coat Behind
(a) Viruses that infect a bacterial cell start by injecting their genes into the cell. **(b)** The protein coat of the original virus is left behind, attached to the bacterial cell wall.

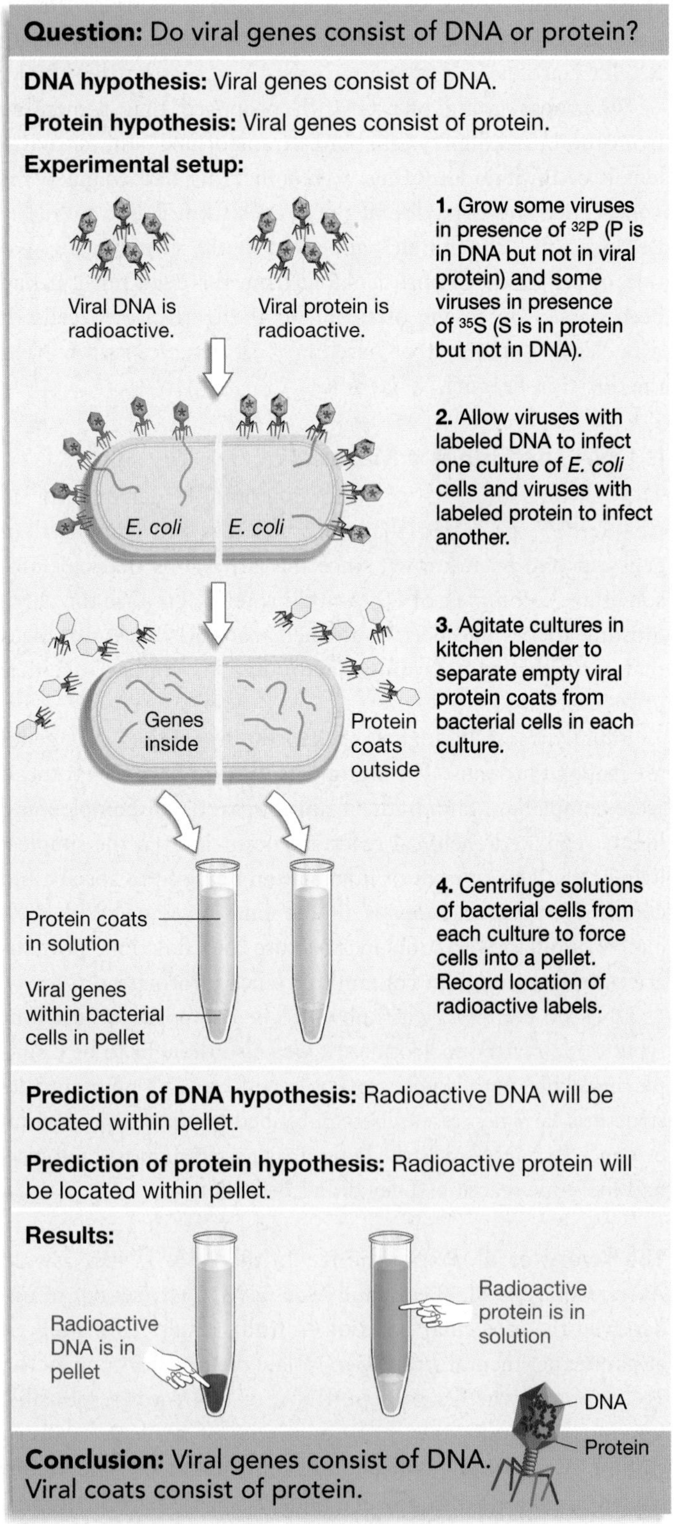
Question: Do viral genes consist of DNA or protein?
DNA hypothesis: Viral genes consist of DNA.
Protein hypothesis: Viral genes consist of protein.
Experimental setup:
Viral DNA is radioactive.
Viral protein is radioactive.
1. Grow some viruses in presence of 32P (P is in DNA but not in viral protein) and some viruses in presence of 35S (S is in protein but not in DNA).
E. coli
E. coli
2. Allow viruses with labeled DNA to infect one culture of E. coli cells and viruses with labeled protein to infect another.
Genes inside
Protein coats outside
3. Agitate cultures in kitchen blender to separate empty viral protein coats from bacterial cells in each culture.
Protein coats in solution
Viral genes within bacterial cells in pellet
4. Centrifuge solutions of bacterial cells from each culture to force cells into a pellet. Record location of radioactive labels.
Prediction of DNA hypothesis: Radioactive DNA will be located within pellet.
Prediction of protein hypothesis: Radioactive protein will be located within pellet.
Results:
Radioactive DNA is in pellet
Radioactive protein is in solution
DNA
Protein
Conclusion: Viral genes consist of DNA. Viral coats consist of protein.

FIGURE 14.5 More Experimental Evidence that DNA Is the Hereditary Material
The experiment shown here convinced most biologists that DNA is indeed the hereditary material.

into newly synthesized proteins and DNA, this step produced a population of viruses with radioactive proteins and a population with radioactive DNA. Then Hershey and Chase allowed each set of radioactive viruses to infect *E. coli* cells. If genes consist of DNA, then the radioactive protein should be found in the empty, or "ghost," protein coats outside the infected host cells, while the radioactive DNA should be located inside the cells. But if genes consist of proteins, the opposite should be true.

To test these predictions, Hershey and Chase sheared the ghosts off the cells by agitating each of the cultures in kitchen blenders. When the researchers spun the samples in a centrifuge, the ghosts stayed in the solution while the cells formed a pellet at the bottom of the centrifuge tube. As predicted by the DNA hypothesis, the biologists found that virtually all of the radioactive protein was in the ghosts, while virtually all of the radioactive DNA was inside the host cells. Because the injected component of the virus directs the production of a new generation of virus particles, it is this component that represents the virus's genes.

After these results were published, proponents of the protein hypothesis had to admit that DNA, not protein, must be the hereditary material. In combination, the evidence from the bacterial transformation experiments and the virus-labeling experiments was convincing. The claim that a seemingly simple molecule contained all the information for life's complexity was finally accepted.

The realization that genes are made of DNA was one of the great advances of twentieth-century biology. The insight raised two crucial questions, however: (1) How did the simple primary structure and secondary structure of DNA hold the information required to make life possible? (2) How is DNA copied, so that genetic information is faithfully passed from one cell to another during growth and from parents to offspring during reproduction? Understanding how DNA contains information is the focus of Chapter 15. The remainder of this chapter concentrates on the question of how genes are replicated prior to mitosis and meiosis.

14.2 Testing Early Hypotheses about DNA Replication

The DNA inside each cell is like an ancient text that has been painstakingly copied and handed down, generation after generation. But while the most ancient of all human texts contain messages that are thousands of years old, the DNA in living cells has been copied and passed down for *billions* of years. And instead of being copied by monks or clerks, DNA is replicated by molecular scribes. What molecules are responsible for copying DNA, and how do they work?

Chapter 4 introduced Watson and Crick's model for the secondary structure of DNA, which was proposed in 1953. Recall that DNA is a polymer made up of monomers called nucleotides, which consist of a deoxyribose molecule, a phosphate group, and a nitrogenous base. The completed DNA molecule is a long, linear polymer that has two major components: (1) a "backbone" made up of the sugar and phosphate groups of nucleotides and (2) a series of nitrogen-containing bases that project from the backbone. Watson and Crick realized that if two of these long strands twist around each other, then certain of the nitrogen-containing bases fit together in pairs inside the spiral. The double-stranded molecule that results is called a **double helix** (**Figure 14.6**). The structure is stabilized by hydrogen bonds that form between the bases adenine (A) and thymine (T) and between the bases guanine (G) and cytosine (C). The specific pairing rules for hydrogen bonding of nitrogen-containing bases is called **complementary base pairing.**

Watson and Crick realized that the A-T and G-C pairing rules suggested a way for DNA to be copied prior to mitosis and meiosis. They suggested that the existing strands of DNA served as a template (pattern) for the production of new strands, with bases being added to the new strands according to complementary base pairing. For example, if the template strand contained a T, then an A would be added to the new strand to pair with that T. Similarly, a G on the template strand would dictate the addition of a C on the new strand.

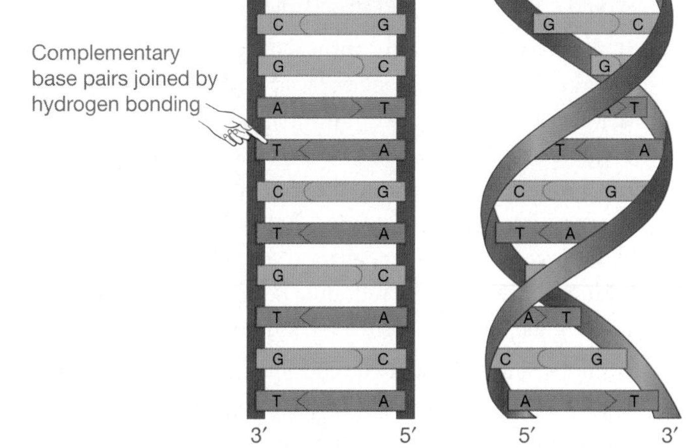

FIGURE 14.6 DNA's Secondary Structure: The Double Helix
(a) Deoxyribonucleic acid normally has a secondary structure consisting of two strands, each with a sugar-phosphate backbone. Nitrogen-containing bases project from each strand and form hydrogen bonds. Only A-T and G-C pairs fit together in a way that allows hydrogen bonding to occur. **(b)** That bonding twists the molecule into a double helix.

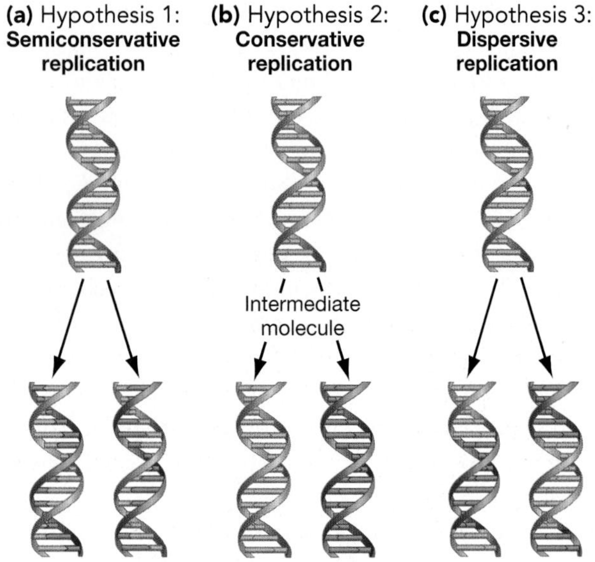
(a) Hypothesis 1: Semiconservative replication
(b) Hypothesis 2: Conservative replication
(c) Hypothesis 3: Dispersive replication
Intermediate molecule

FIGURE 14.7 Alternative Hypotheses for DNA Synthesis
The old strand of DNA is shown in gray, and the newly synthesized strand in red. **(a)** If replication were semiconservative, each strand in a DNA double helix would serve as the template for the synthesis of a daughter strand. **(b)** If replication were conservative, the original molecule would remain intact after synthesis, so the daughter molecule would consist of two newly synthesized strands. **(c)** If dispersive replication occurred, each strand in the daughter DNA would consist of a mixture of old and new DNA.

As **Figure 14.7** shows, however, biologists had three hypotheses about how the old and new strands might interact during replication:

1. If the old strands of DNA separated, each could then be used as a template for the synthesis of a new, daughter strand (Figure 14.7a). This hypothesis is called **semiconservative replication,** because each new daughter DNA molecule would consist of one old strand and one new strand.
2. If the bases temporarily turned outward so that complementary strands no longer faced each other, they could serve as a template for the synthesis of an entirely new double helix all at once. The new strands would bond to each other and be released from the template molecules. This hypothesis is called **conservative replication.** As Figure 14.7b shows, conservative replication results in two parental strands paired to form a double helix and two new daughter strands paired to form a second double helix.
3. If the parent helix was cut and unwound in short sections before being copied and put back together, then new and old strands would intermingle as shown in Figure 14.7c. This possibility is called **dispersive replication.**

Which of these three hypotheses is correct?

The Meselson-Stahl Experiment

Matthew Meselson and Frank Stahl realized that if they could tag parental and daughter strands of DNA in a way that would make them distinguishable from each other, they could determine whether replication was conservative, semiconservative, or dispersive. Soon after their results were published in 1958, the Meselson-Stahl work became recognized as a classic experiment in biological science.

Before they could do any tagging or publishing, however, they needed to choose an organism to study. They decided to work with a common inhabitant of the human gastrointestinal tract, the bacterium *Escherichia coli.* Because *E. coli* is small and grows quickly and readily in the laboratory, it had become a favored model organism in studies of biochemistry and molecular genetics.

Like all organisms, bacterial cells copy their entire complement of DNA, or their **genome**, before every cell division. To distinguish parental strands of DNA from daughter strands when *E. coli* replicates, Meselson and Stahl grew the cells for many generations in the presence of one of two isotopes of nitrogen: either ^{15}N or ^{14}N. Because ^{15}N contains an extra neutron, it is heavier than the normal isotope, ^{14}N.

This difference in mass, which creates a difference in density of ^{14}N-containing and ^{15}N-containing DNA, was the key to the experiment. The biologists reasoned that if different nitrogen isotopes were available in the growth medium when parental and daughter strands of DNA were produced, then the two types of strands should behave differently during centrifugation. More specifically, when intact, double-stranded DNA molecules are added to a solution that forms a gradient from low to high density during centrifugation, DNA strands that contain ^{14}N should form a band in the lower-density part of the centrifuge tube. In contrast, DNA strands that contain ^{15}N should form a band in the higher-density part of the centrifuge tube. Because the highest-density solution is at the bottom of the tube, DNA that contains ^{15}N should be found lower in the tube than DNA containing ^{14}N. In this way, DNA strands containing ^{15}N or ^{14}N should form separate bands. By exposing the centrifuge tube to X-ray film, the biologists could visualize the positions of the bands. How could this tagging system be manipulated to test whether replication is semiconservative, conservative, or dispersive?

Figure 14.8 summarizes Meselson and Stahl's experimental strategy. They began by growing *E. coli* cells in the presence of ^{15}N as the only nitrogen isotope in nitrogen-containing nutrient molecules. They purified DNA from a sample of these cells and transferred the rest of the culture to a growth medium containing only the ^{14}N isotope. After enough time had elapsed for these experimental cells to divide once, Meselson and Stahl removed a sample and isolated the DNA. After the remainder of the culture had divided again, they removed another sample and purified the DNA.

As Figure 14.8 shows, the conservative, semiconservative, and dispersive models make distinct predictions about the makeup of the DNA molecules after replication occurred in the first and second generation. For example, if replication

Question: Is replication semiconservative, conservative, or dispersive?

Hypothesis 1: Replication is semiconservative.

Hypothesis 2: Replication is conservative.

Hypothesis 3: Replication is dispersive.

Experimental setup:

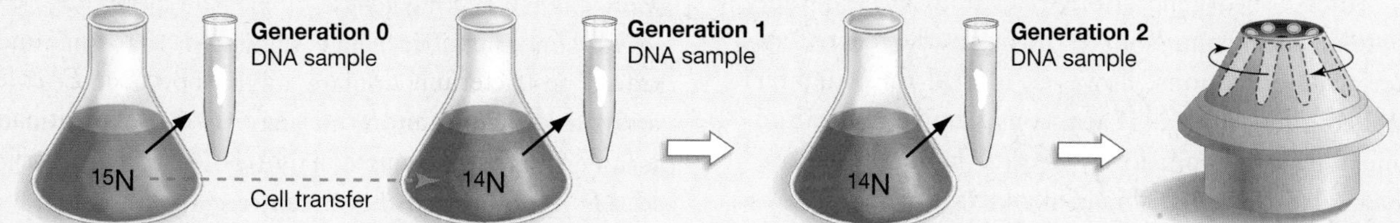
Generation 0
DNA sample
^{15}N
Cell transfer
Generation 1
DNA sample
^{14}N
Generation 2
DNA sample
^{14}N

1. Grow *E. coli* cells in medium with ^{15}N as sole source of nitrogen for many generations. Collect sample and purify DNA.

2. Transfer cells to medium containing ^{14}N. After cells divide once, collect sample and purify DNA.

3. After cells have divided a second time in ^{14}N medium, collect sample and purify DNA.

4. Centrifuge the three samples and compare the locations of the DNA bands.

Predictions:

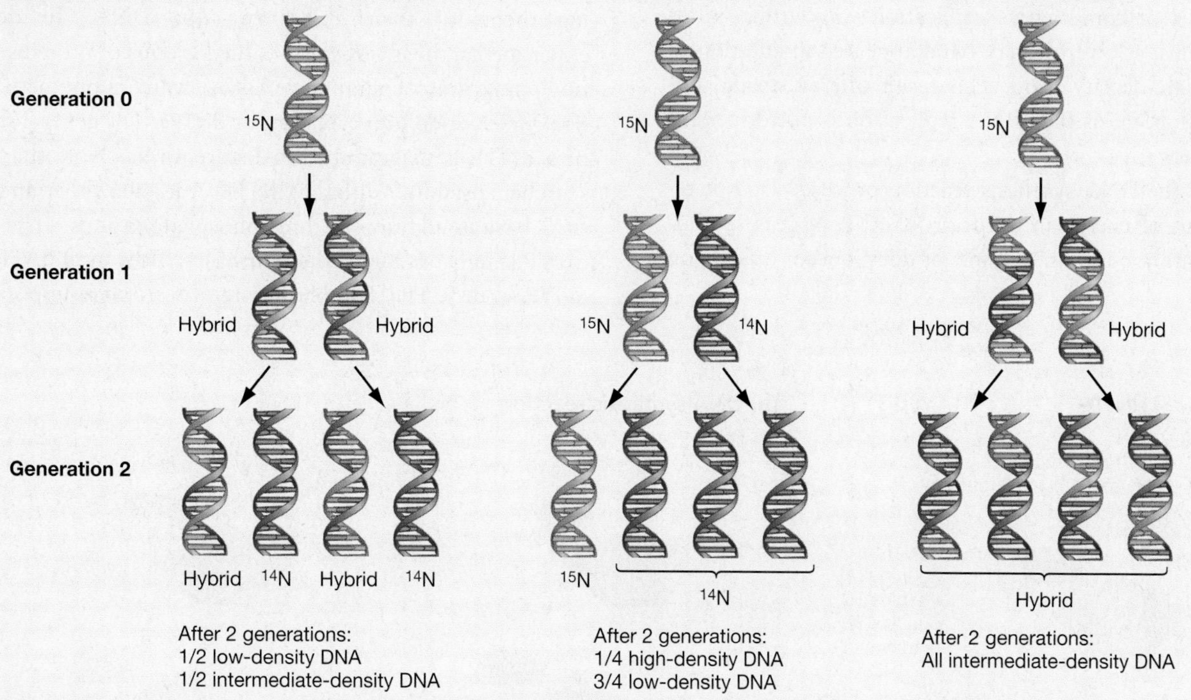
Semiconservative replication
Conservative replication
Dispersive replication
Generation 0
Generation 1
Generation 2
^{15}N
^{15}N
^{15}N
Hybrid
Hybrid
^{15}N
^{14}N
Hybrid
Hybrid
Hybrid ^{14}N Hybrid ^{14}N
^{15}N
^{14}N
Hybrid
After 2 generations:
1/2 low-density DNA
1/2 intermediate-density DNA
After 2 generations:
1/4 high-density DNA
3/4 low-density DNA
After 2 generations:
All intermediate-density DNA

Results:

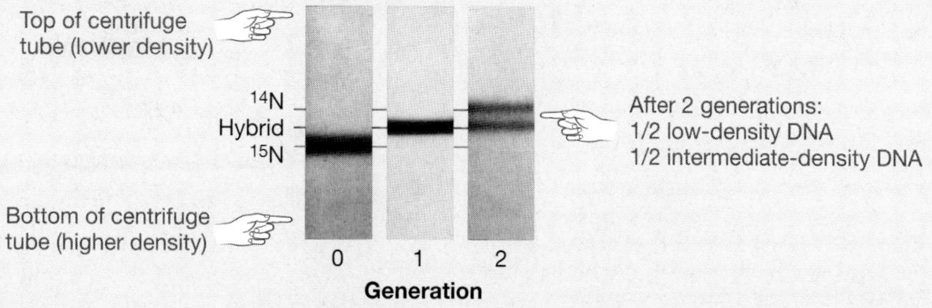
Top of centrifuge tube (lower density)
^{14}N
Hybrid
^{15}N
Bottom of centrifuge tube (higher density)
0 1 2
Generation
After 2 generations:
1/2 low-density DNA
1/2 intermediate-density DNA

Conclusion: Data from generation 1 conflict with conservative-replication hypothesis. Data from generation 2 conflict with dispersive-replication hypothesis. Replication is semiconservative.

FIGURE 14.8 The Meselson-Stahl Experiment

EXERCISE Meselson and Stahl actually let their experiment run for four generations, with cultures continuing to grow in the presence of ^{14}N. Sketch what the data from third- and fourth-generation DNA should look like—that is, where the DNA band(s) should be.

is conservative, then the daughter cells should have double-stranded DNA with either ^{14}N or ^{15}N, but not both. As a result, two distinct DNA bands should form in the centrifuge tube—one high-density band and one low-density band. But if replication is semiconservative or dispersive, then all of the experimental DNA should contain an equal mix of ^{14}N and ^{15}N after one generation. In both cases, one intermediate-density band should form in the centrifuge tube. After two generations, however, half of the daughter cells should contain only ^{14}N if semiconservative replication occurs—meaning a second, lower-density band should appear in the centrifuge tube. This is in contrast to the prediction made by the dispersive model, for which the centrifuge tube should contain only one band at intermediate density.

The photograph at the bottom of Figure 14.8 shows the experiment's results. After one generation, the density of the DNA molecules was intermediate. These data suggested that the hypothesis of conservative replication was wrong. After two generations, a lower-density band appeared in addition to the intermediate-density band. This result offered strong support for the hypothesis that DNA replication is not dispersive but semiconservative.

How does the DNA synthesis reaction proceed? Does it require an input of energy in the form of ATP, or it is spontaneous? Is it catalyzed by an enzyme, or does it occur quickly on its own?

The Discovery of DNA Polymerase

In the mid-1950s Arthur Kornberg and co-workers set out to study how DNA replicates by attempting to generate a cell-free, or in vitro, DNA replication system. Their goal was to resolve a long-standing paradox: If genes contain the information needed for making an organism, then what is responsible for making genes?

Watson and Crick had suggested that the nucleotides paired to a DNA template might be "zippered" together without assistance from an enzyme. Kornberg, in contrast, assumed that some sort of DNA-polymerizing enzyme existed. He hypothesized that a cell-free system capable of replicating DNA would require an enzyme to catalyze the formation of phosphodiester bonds between nucleotides in the newly formed strands.

To begin his search for this molecule, Kornberg and colleagues prepared proteins from *E. coli* cultures and began testing them for their ability to copy DNA. To do these experiments, the investigators added groups of proteins to reaction mixtures containing DNA and monomers called **deoxynucleoside triphosphates,** or **dNTPs.** The general structure of a dNTP is diagrammed in **Figure 14.9a**. Note that the researchers used four different dNTPs, carrying either an A, T, G, or C base in addition to three phosphate groups. (The "N" in dNTP stands for any of the four bases. They used dATP, dTTP, dGTP, and dCTP.) The phosphate groups were important be-

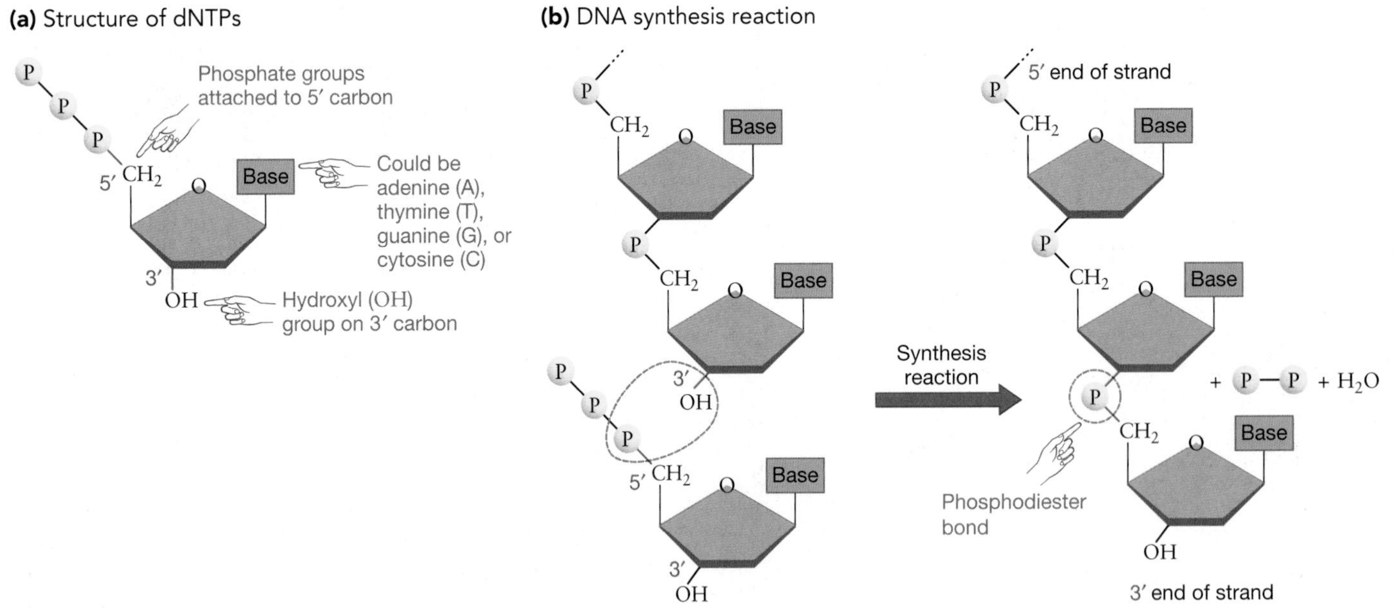

FIGURE 14.9 The DNA Synthesis Reaction
(a) A dNTP (deoxynucleoside triphosphate) monomer **(b)** is added to a DNA polymer when a phosphodiester bond forms between the 3′ carbon on the end of a DNA strand and the 5′ carbon on the dNTP in a condensation reaction. QUESTION According to this reaction, does DNA synthesis proceed from the 5′ end of the new molecule toward the 3′ end, or in the 3′ to 5′ direction? EXERCISE The "P–P" group (pyrophosphate) that is released reacts with water to form two phosphate ions (HPO_4^-). This reaction is highly exergonic. Add this reaction to the figure.

cause the potential energy stored in the bonds between the phosphates could be used to drive the energy-demanding synthesis reaction (**Figure 14.9b**). The biologists also used radioactive versions of these monomers, which provided an assay to detect whether dNTPs had been incorporated into newly synthesized DNA. If Kornberg and co-workers could detect radioactivity in a high-molecular-weight polymer instead of in low-molecular-weight monomers, it meant that DNA synthesis had occurred.

Kornberg and colleagues tested a variety of proteins with this in vitro assay. After many failed attempts, they finally succeeded in identifying an enzyme that allowed DNA synthesis to proceed. They called this molecule **DNA polymerase.** Its name was later changed to DNA polymerase I when researchers realized that there was more than one type of DNA polymerase in *E. coli* and other organisms. Each of these enzymes is specialized to perform a particular function in the cell. The enzyme that is primarily responsible for copying *E. coli*'s chromosome prior to cell division is called DNA polymerase III. When the double helix of bacterial DNA separates, DNA polymerase III adds the complementary bases and catalyzes the formation of phosphodiester bonds in the newly synthesized strand.

Follow-up experiments confirmed that DNA polymerase I and the other DNA polymerases do not catalyze the random addition of monomers to create a DNA molecule. Instead, they synthesize a growing strand of DNA by adding bases that are complementary to a template strand. To capture this point, biologists say that DNA synthesis is template directed.

The discovery of the DNA polymerases ended one chapter in research on DNA synthesis and started another. The pioneering phase of work on DNA replication was over. Researchers had established that DNA replication is semiconservative in nature, that it is enzyme dependent and template directed, and that several DNA polymerases exist. The next challenge was to understand the mechanics of the process in detail.

14.3 A Comprehensive Model for DNA Synthesis

How does DNA polymerase III accomplish DNA synthesis? Three results from the pioneering phase of work on DNA replication helped focus this question and point the way to an answer.

The first result was an observation about how the enzyme works. DNA polymerases catalyze the addition of a dNTP monomer to only the 3′ end of the DNA chain—never to the 5′ end (see Figure 14.9). As a result, DNA synthesis always proceeds in the 5′ → 3′ direction. Nucleotides are added only to the 3′ end of an existing molecule.

The second result, from experiments with in vitro systems, was that DNA polymerase III cannot work if the DNA

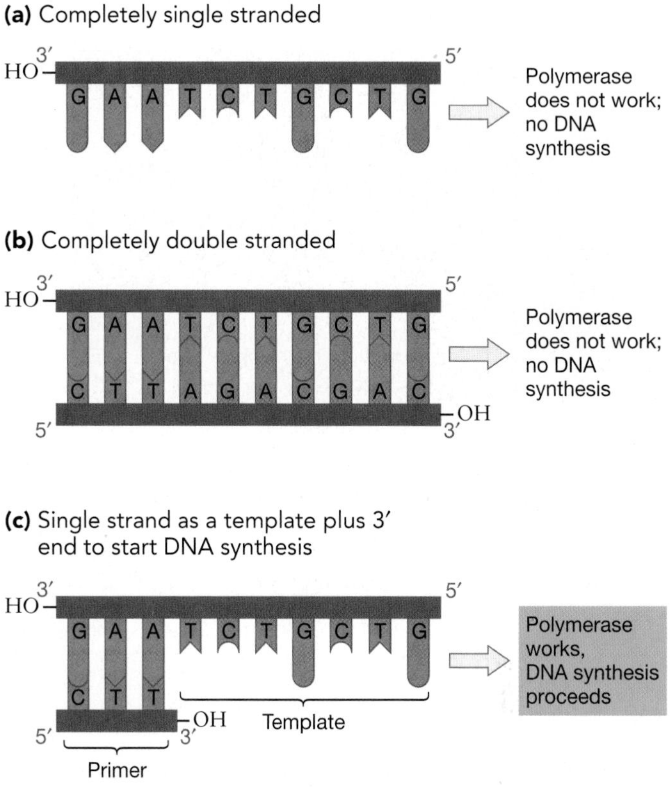

FIGURE 14.10 DNA Polymerase Requires a Primer and a Template
When DNA templates are **(a)** completely single stranded or **(b)** double stranded, no synthesis occurs. **(c)** DNA polymerases work only when both a single-stranded template *and* a 3′ hydroxyl group are available.

that serves as a template is completely single stranded or completely double stranded (**Figure 14.10a, b**). The enzyme begins to work only when a 3′ end *and* a single-stranded template are available (**Figure 14.10c**). In effect, DNA polymerases only add nucleotides that complete a double helix. The existing strand acts as a template that dictates which nucleotide should be added next, while the strand to which nucleotides are being added provides a free 3′ hydroxyl (—OH) group that combines with the incoming dNTP to form a phosphodiester bond.

It's important to note, however, that DNA polymerase III can get started once even a few nucleotides are in place to form a double-stranded section on the parent DNA template. These few nucleotides are called a **primer.** Once a primer is added to a single-stranded template, DNA polymerase III begins working in the 5′ → 3′ direction and adds nucleotides to complete the complementary strand.

The third result that proved to be a major insight into the mechanism of DNA synthesis emerged when electron micrographs caught chromosome replication in action. A "bubble" forms in a chromosome when DNA is actively being synthesized,

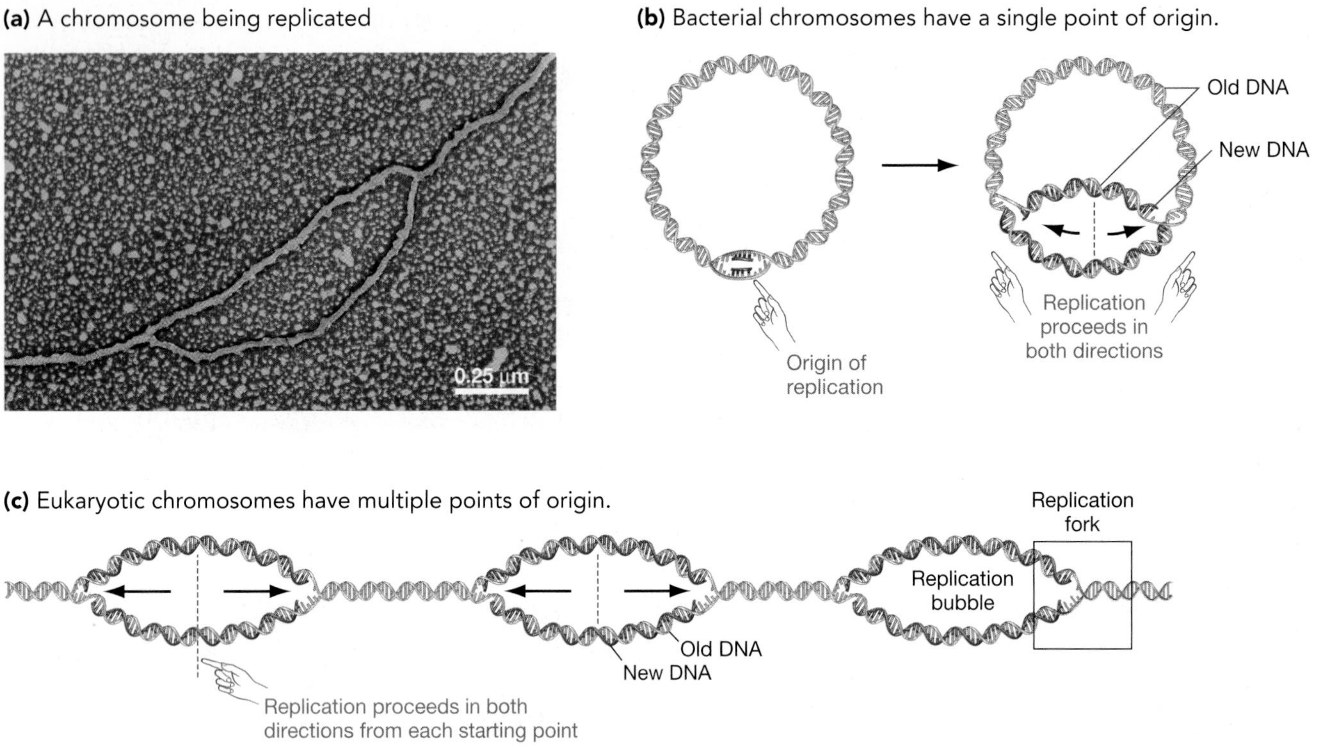
(a) A chromosome being replicated
0.25 μm
(b) Bacterial chromosomes have a single point of origin.
Old DNA
New DNA
Origin of replication
Replication proceeds in both directions
(c) Eukaryotic chromosomes have multiple points of origin.
Replication fork
Replication bubble
Old DNA
New DNA
Replication proceeds in both directions from each starting point

FIGURE 14.11 DNA Synthesis Proceeds in Two Directions from a Point of Origin
(a) A micrograph of a "replication bubble." **(b)** In most bacteria, chromosomes are circular and there is a single point of origin during DNA replication. **(c)** Most eukaryotes have linear chromosomes; each contains several to many points of origin for DNA synthesis. EXERCISE Circle the two replication forks in part (a). Add arrows showing the direction of DNA synthesis in each fork. Note that because replication is bidirectional, each of the two forks should have arrows pointing in opposite directions.

WEB TUTORIAL 14.1 DNA Synthesis

as **Figure 14.11a** shows. Bacterial chromosomes have a single location where the replication process begins, and thus a single bubble forms. Initially, the replication bubble forms a point called the **origin of replication** (**Figure 14.11b**). The bubble grows as DNA replication proceeds, because synthesis is *bidirectional*—that is, it occurs in both directions (though always $5' \rightarrow 3'$, because the strands are antiparallel) at the same time. Eukaryotes also have bidirectional replication; but they have multiple sites along each chromosome where DNA synthesis begins, and thus multiple replication bubbles (**Figure 14.11c**).

To put these observations into a comprehensive picture of how DNA synthesis occurs, let's focus on what happens in the corners of each replication bubble—at the structure called the replication fork. A **replication fork** is a Y-shaped region where the parent DNA double helix is split into two single strands, which are then copied. How does the splitting event occur, and how are each of the resulting strands replicated?

Opening the Helix

Several key events occur at the point where the double helix opens. An enzyme called a **helicase** catalyzes the breaking of hydrogen bonds between nucleotides. This reaction causes the two strands of DNA to separate. Proteins called **single-strand DNA-binding proteins** attach to the separated strands and prevent them from snapping back into a double helix. In combination, then, the helicase and single-strand DNA-binding proteins open up the double helix and make both strands available for copying (**Figure 14.12a**).

The "unzipping" process that occurs at the replication fork creates tension farther down the helix, however. To understand why, imagine what would happen if you started to pull apart the twisted strands of a rope. The untwisting movements at one end would force the intact section to rotate in response. If the intact end of the rope were fixed in place, though, it would eventually begin to coil on itself and kink in response to the twisting forces. This does not happen in DNA, because the twisting stress induced by helicase is relieved by proteins called topoisomerases. A **topoisomerase** (literally, "place-same-part") is an enzyme that cuts and rejoins the DNA downstream of the replication fork. Topoisomerases do this cutting and pasting in a way that undoes twists and knots.

Now, what happens once the DNA helix is open and has stabilized?

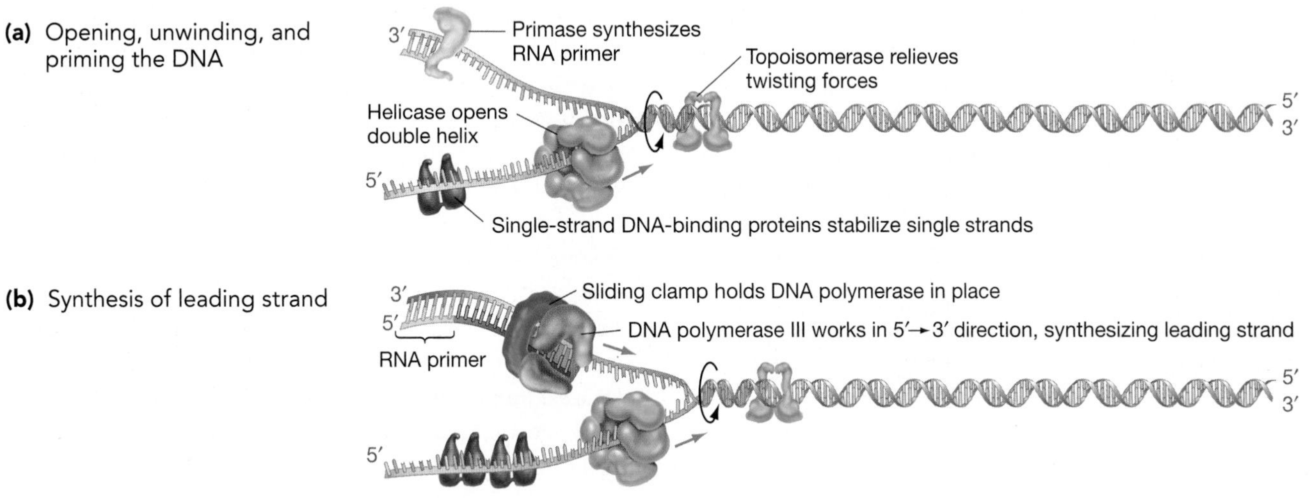

FIGURE 14.12 Synthesizing the Leading Strand during DNA Replication
(a) Once the double helix is opened and stabilized, **(b)** DNA polymerase synthesizes the leading strand in the 5′ to 3′ direction.

Synthesis of the Leading Strand

To understand what happens at the start of DNA synthesis, recall that DNA polymerase III works only in the 5′ → 3′ direction and that it requires a primer to get started. Before DNA synthesis can get under way, an enzyme called **primase** has to synthesize a short stretch of ribonucleic acid (RNA) that acts as a primer for DNA polymerase. Primase is a type of **RNA polymerase**—an enzyme that catalyzes the polymerization of ribonucleotides to RNA. Unlike DNA polymerases, primase and other RNA polymerases do not require a primer. These enzymes can simply match ribonucleotides directly by complementary base pairing on single-stranded DNA. In this way, primase creates a primer for DNA synthesis.

Once the primer is in place, DNA polymerase III begins adding deoxyribonucleotides to the 3′ end of the new strand, in a sequence that is complementary to the template strand. As **Figure 14.12b** shows, the shape of DNA polymerase III grips the DNA strand during synthesis, a little like your hand clasping a rope. Catalysis takes place in the groove inside the enzyme, at an active site between the enzyme's "thumb" and "fingers." As DNA polymerase moves along the DNA molecule, a doughnut-shaped structure behind it, called the **sliding clamp**, holds it in place. Its product is called the **leading strand**, or **continuous strand**, because it leads into the replication fork and is synthesized continuously.

Synthesis of the Lagging Strand

Synthesis of the leading strand is straightforward after an RNA primer is in place—DNA polymerase III chugs along, adding bases to the 3′ end of that strand. The enzyme moves into the replication fork, which "unzips" ahead of it. By comparison, events on the opposite strand are much more involved. To understand why, recall from Chapter 4 that the two strands in the DNA double helix are *antiparallel*—meaning they are parallel to one another but oriented in opposite directions. DNA polymerase works in only one direction, however, so if the DNA polymerase that is synthesizing the leading strand works into the replication fork, then a DNA polymerase must work away from the replication fork to synthesize the other strand in the 5′ → 3′ direction. The strand that is synthesized in the opposite direction of the replication fork is called the **lagging strand**, because it lags behind the fork.

The synthesis of the lagging strand starts when primase synthesizes a short stretch of RNA that acts as a primer. DNA polymerase III then adds bases to the 3′ end of the lagging strand. The key observation is that the enzyme moves *away* from the replication fork. But behind it, helicase continues to open the replication fork and expose new single-stranded DNA on the lagging strand.

These events create a paradox: How can lagging-strand synthesis be completed if new single-stranded template DNA is constantly appearing behind DNA polymerase III—in the direction *opposite* the direction of synthesis? Consider the state of the lagging strand at time 1 and time 2, here:

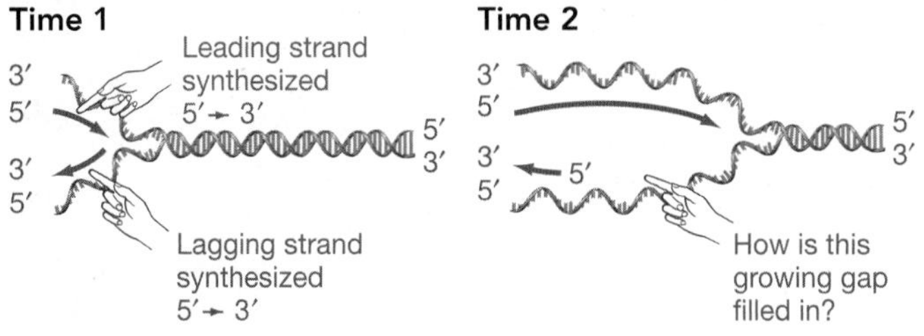

The puzzle posed by lagging-strand synthesis was resolved when Reiji Okazaki and colleagues tested a hypothesis called

discontinuous replication. This hypothesis stated that, once primase synthesizes an RNA primer on the lagging strand (step 1 in **Figure 14.13**), DNA polymerase III might synthesize short fragments of DNA along the lagging strand and that these fragments would later be linked together to form a continuous whole. The idea was that primase would add a primer to the newly exposed single-stranded DNA at intervals and that DNA polymerase III would then synthesize the lagging strand until it reached the fragment produced earlier.

To test this hypothesis, Okazaki's group set out to test a key prediction: Could they document the existence of short DNA fragments produced during replication? Their critical experiment was based on the pulse-chase strategy introduced in Chapter 7. Specifically, they added a short "pulse" of radioactive thymidylate to *E. coli* cells, followed by a large "chase" of nonradioactive thymidylate. Thymidylate is a nucleotide that contains a deoxyribose subunit attached to thymine. It serves as a building block of DNA. According to the discontinuous replication model, some of this radioactive thymidylate should end up in short, single-stranded fragments of DNA.

As predicted, the researchers succeeded in finding these fragments when they purified DNA from the experimental cells and separated the molecules by centrifugation. A small number of labeled pieces of DNA, about 1000 base pairs long, were present. These short sections came to be known as **Okazaki fragments** (steps 2 and 3 of Figure 14.13). Subsequent work showed that Okazaki fragments in eukaryotes are just 100 to 200 base pairs long.

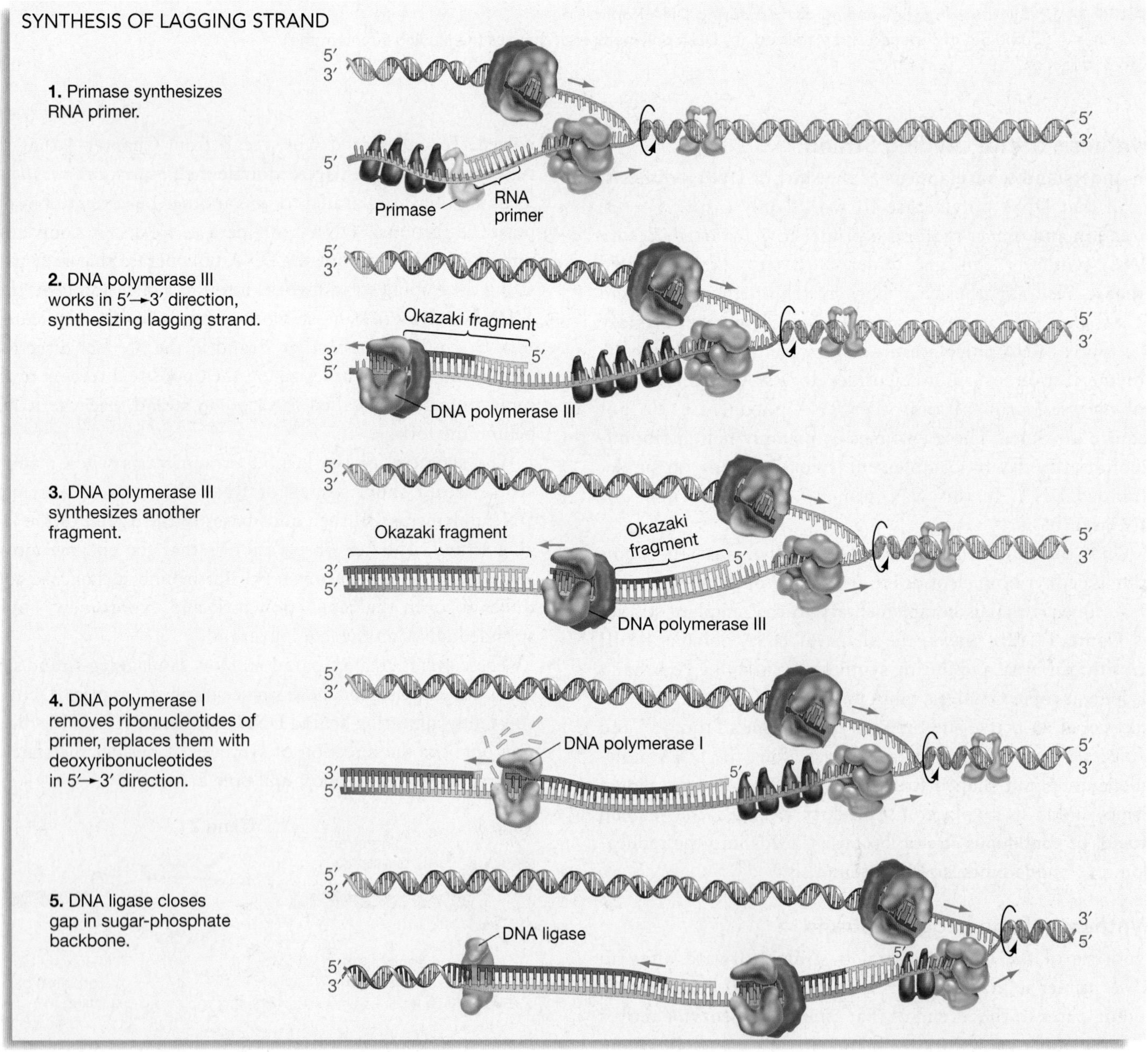
SYNTHESIS OF LAGGING STRAND
1. Primase synthesizes RNA primer.
Primase
RNA primer
2. DNA polymerase III works in 5′→3′ direction, synthesizing lagging strand.
Okazaki fragment
DNA polymerase III
3. DNA polymerase III synthesizes another fragment.
Okazaki fragment
Okazaki fragment
DNA polymerase III
4. DNA polymerase I removes ribonucleotides of primer, replaces them with deoxyribonucleotides in 5′→3′ direction.
DNA polymerase I
5. DNA ligase closes gap in sugar-phosphate backbone.
DNA ligase
5′
3′

FIGURE 14.13 The Completion of DNA Replication
Synthesizing the lagging strand.

How are Okazaki fragments connected into a continuous whole? As step 4 of Figure 14.13 shows, DNA polymerase I removes the RNA primer at the start of each fragment and fills in the appropriate deoxyribonucleotides. Finally, an enzyme called **DNA ligase** catalyzes the formation of a phosphodiester bond between the bases synthesized by DNA polymerase I and DNA polymerase III (Figure 14.13, step 5). Because Okazaki fragments are synthesized independently and joined together later, the lagging strand is also called the **discontinuous strand**.

In combination, then, the enzymes that open the replication fork and manage the synthesis of the leading and lagging strands succeed in producing a faithful copy of the original DNA molecule prior to mitosis or meiosis. Given the number of enzymes and proteins involved in the synthesis machinery, it is not surprising that it took 25 years for biologists to assemble the results summarized in **Table 14.1**.

It is also worth mentioning the following points:

- Although Figures 14.12 and 14.13 show the enzymes involved in DNA synthesis at different locations around the replication fork, in reality most are joined into one large multi-enzyme machine at the replication fork. This complex of proteins opens the double helix and synthesizes both the leading strand and the Okazaki fragments on the lagging strand.
- The basic elements of DNA synthesis were worked out in experiments on *E. coli*. Follow-up research has shown that the fundamental elements of this process are almost identical in eukaryotes. The exceptions are that (1) some of the enzymes at the replication fork in eukaryotes are larger and more complex than similar enzymes in bacteria; and (2) on the lagging strand, two different DNA polymerases do the work done by DNA polymerase III in bacteria. In eukaryotes, enzymes called DNA polymerase δ and DNA polymerase α synthesize the lagging strand. (DNA polymerase δ also synthesizes the leading strand.) In almost every other detail, however, the critical components of DNA synthesis have been highly conserved over the course of evolution.

Most aspects of the model summarized in Figures 14.12 and 14.13 had emerged in the early 1980s, and subsequent decades saw major advances in our understanding of the detailed molecular structures of the major enzymes involved. But in addition to studying the replication process itself, biologists have been exploring other questions about DNA synthesis. This chapter closes with a look at three of these research areas: How are the ends of linear chromosomes replicated? What happens when a DNA polymerase inserts the wrong base into a sequence? And can DNA be repaired if it is damaged? As it turns out, all three questions have practical implications—particularly for the origin of certain types of cancer.

✓CHECK YOUR UNDERSTANDING

DNA synthesis begins at specific points of origin on the chromosome, then proceeds in both directions. Once DNA replication is under way, synthesis can be analyzed as a three-step process that occurs at the replication fork. (1) Helicase, single-stranded DNA-binding proteins, and topoisomerase open the double helix to make single strands available for copying; (2) a DNA polymerase synthesizes the leading strand after primase has added an RNA primer; and (3) a series of enzymes synthesizes the lagging strand. Lagging-strand synthesis cannot be continuous, because it moves away from the replication fork. In bacteria, enzymes called primase, DNA polymerase III, DNA polymerase I, and ligase work in sequence to synthesize Okazaki fragments and link them into a continuous whole. You should be able to draw a replication fork, label the main components, and explain the function of each structure involved.

TABLE 14.1 Proteins Involved in DNA Synthesis

Process	Name	Function
Opening the helix	Helicase	Catalyzes the breaking of hydrogen bonds between base pairs and the opening of the double helix
	Single-strand DNA-binding proteins	Stabilizes single-stranded DNA
	Topoisomerase	Breaks and rejoins the DNA double helix to relieve twisting forces caused by the opening of the helix
Synthesizing the leading strand	Primase	Catalyzes the synthesis of the RNA primer
	DNA polymerase III	Extends the leading strand
	Sliding clamp	Holds DNA polymerase in place during strand extension
Synthesizing the lagging strand	Primase	Catalyzes the synthesis of the RNA primer on an Okazaki fragment
	DNA polymerase III	Extends an Okazaki fragment
	Sliding clamp	Holds DNA polymerase in place during strand extension
	DNA polymerase I	Removes the RNA primer and replaces it with DNA
	DNA ligase	Catalyzes the joining of Okazaki fragments into a continuous strand

14.4 Replicating the Ends of Linear Chromosomes

The circular DNA molecules in bacteria and archaea can be synthesized via the sequence of events diagrammed in Figures 14.12 and 14.13. Similarly, the leading and lagging strands of the linear DNA molecules found in eukaryotes are copied efficiently by the enzymes illustrated in those figures. But how can the *ends* of the linear chromosomes found in garden peas, fruit flies, humans, and other eukaryotes be copied faithfully?

The region at the end of a linear chromosome is called a **telomere** ("end-part"). **Figure 14.14** illustrates the problem that arises during the replication of telomeres. Steps 1 and 2 show that as the replication fork nears the end of a linear chromosome, DNA polymerase synthesizes the leading strand all the way to the end of the parent DNA template. As a result, leading-strand synthesis results in a normal copy of the DNA molecule. But the situation is different on the lagging strand. As step 2 of Figure 14.14 illustrates, primase adds an RNA primer close to the tip of the chromosome. But after this primer is removed and DNA polymerase synthesizes the final Okazaki fragment on the lagging strand (step 3), the enzyme is unable to replace the RNA primer or to add DNA near the tip of the chromosome because DNA polymerase can add bases only to the 3′ end of an existing strand. As step 4 shows, there is no 3′ end available at the end of the lagging strand. As a result, the single stranded DNA that is left once the primer has been removed must stay single stranded.

The single-stranded DNA that remains after telomeres are copied is eventually degraded, meaning the chromosome shortens. If this process were to continue, every chromosome would shorten by 50–100 nucleotides on average each time that DNA replication occurred prior to mitosis or meiosis. Over time, linear chromosomes would be expected to disappear completely.

Bacteria and archaea do not have this end-replication problem, because virtually all of these cells have a single, circular

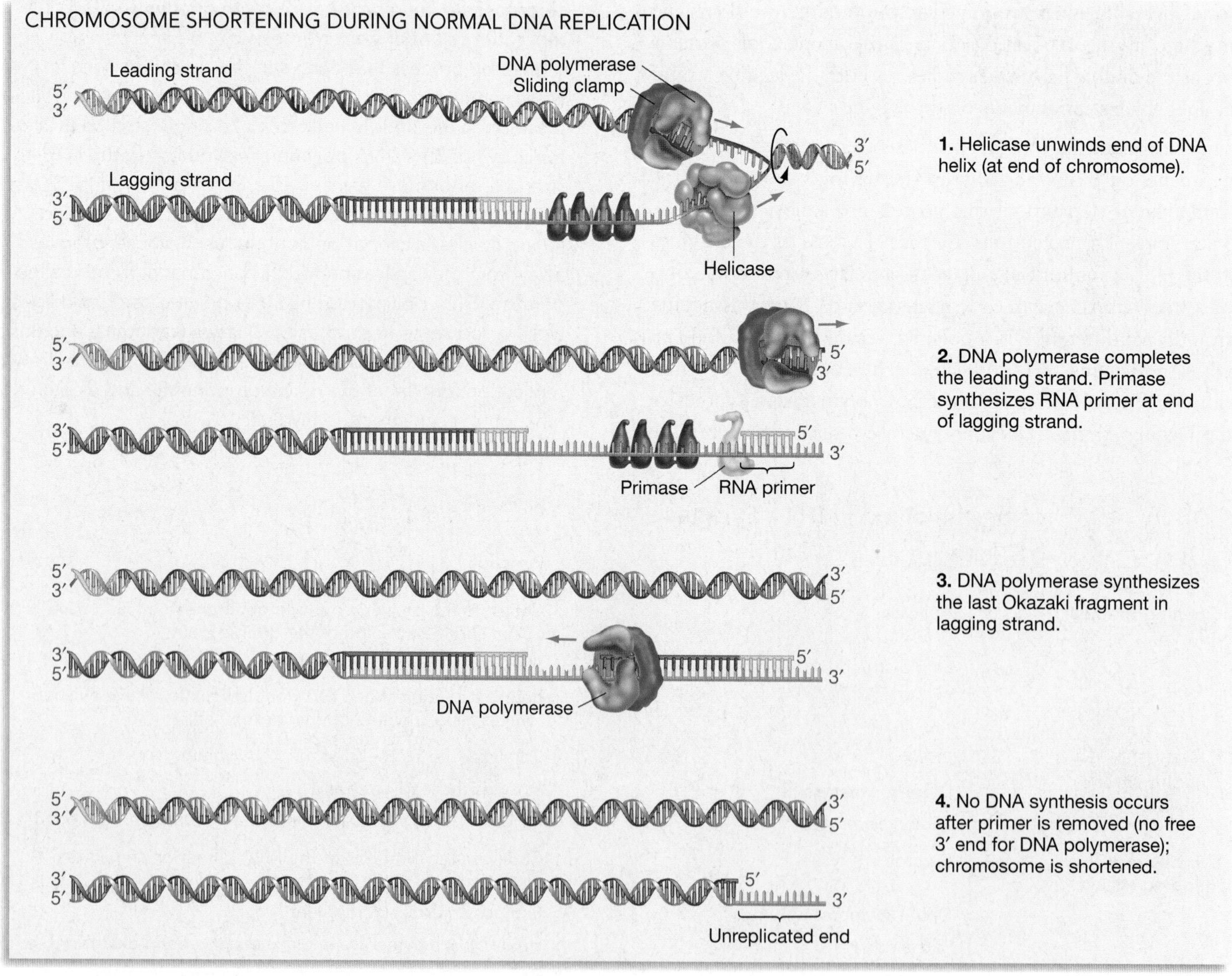

FIGURE 14.14 Telomeres Shorten during Normal DNA Replication
An RNA primer is added to the lagging strand near the end of the chromosome. Once the primer is removed, it cannot be replaced with DNA. As a result, the chromosome shortens.

chromosome. How do eukaryotes maintain the integrity of their linear chromosomes? An answer emerged after two important discoveries were made:

1. Telomeres do not contain genes. Instead, they consist of short stretches of bases that are repeated over and over. In humans, for example, the base sequence TTAGGG is repeated thousands of times. (Because DNA and RNA sequences are always written 5′ → 3′, the base sequence on the complementary strand is CCCTAA rather than AATCCC.) Human telomeres consist of a total of about 10,000 nucleotides with the sequence TTAGGGTTAGGGTTAGGG....
2. An interesting enzyme called telomerase is involved in replicating telomeres. **Telomerase** is remarkable because it catalyzes the synthesis of DNA from an RNA template. The enzyme carries an RNA molecule with it that acts as a built-in template. The sequence of this RNA is complementary to the repeated sequence in telomere DNA. As a result, telomerase can add repeats onto the end of a chromosome.

Figure 14.15 shows how this process works. Step 1 shows the unreplicated segment at the 3′ end of the lagging strand at the telomere. It forms a single-strand "overhang." In step 2, telomerase binds to this overhanging section of single-stranded parent DNA and adds repeats to it. This step lengthens the overhang. The normal machinery of DNA synthesis—primase, DNA polymerase, and ligase—then synthesizes the lagging strand in the 5′ → 3′ direction. The result is that the lagging strand becomes slightly longer than it was originally.

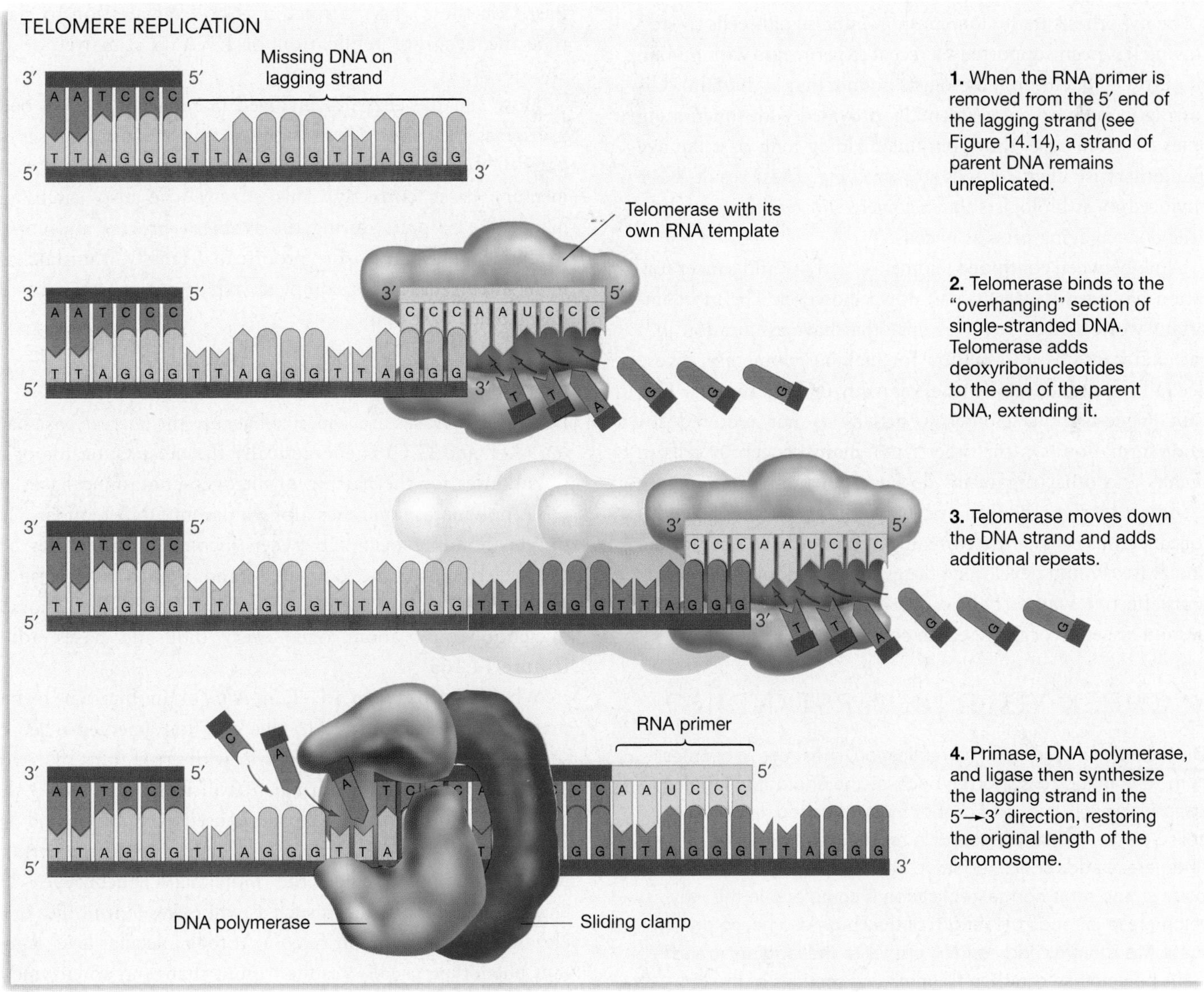

FIGURE 14.15 Telomerase Prevents Shortening of Telomeres during Replication
By extending the number of repeated sequences in the 5′ → 3′ direction, telomerase provides room for enzymes to add an RNA primer to the lagging strand. DNA polymerase can then fill in the missing section of the lagging strand.

It is important to recognize, though, that telomerase is not active in most types of cells. In humans, for example, active telomerase is found primarily in the cells of reproductive organs—specifically, in the cells that eventually undergo meiosis and produce gametes. Cells that are not involved in gamete formation, or what biologists call **somatic cells**, normally lack telomerase. As predicted, the chromosomes of somatic cells gradually shorten with each mitotic division, getting progressively smaller as an individual grows and ages.

These observations inspired a pair of important hypotheses. The first hypothesis was that telomere shortening might eventually cause cells to stop dividing altogether. The idea here was that telomere loss would signal cells to enter the nondividing state typical of adult cells. The second hypothesis was that if telomerase were mistakenly activated in a somatic cell, the maintenance of telomeres with normal length might induce the cell to keep dividing and possibly contribute to uncontrolled growth and cancer.

The hypothesis that telomere shortening signals cells to stop growing has been supported by recent experiments with human cells grown in culture. Biologists found that if human cells growing in vitro are experimentally provided with functioning copies of telomerase, they continue dividing long past the age when otherwise identical cells stop growing. These results have convinced most biologists that telomere shortening is involved in the normal aging process in cells.

A link between continued telomerase activity and cancer formation has been harder to nail down, however. The most important work to date involves mice that have a mutation in a gene that contains information for making telomerase. These mice fail to produce telomerase, yet many develop tumors. This result suggests that telomerase activity is not required for tumor formation. On the other hand, many cancerous cells in humans and other organisms do have functioning telomerase or some other mechanism for maintaining telomere length. Noncancerous cells in these organisms do not have functioning telomerase. Would developing drugs that knock out telomerase be an effective way to fight cancer? To date, the data on this question are equivocal. Research continues.

✓CHECK YOUR UNDERSTANDING

During normal chromosome replication, linear chromosomes in most somatic cells shorten because the end of the lagging strand lacks a primer and cannot be synthesized. In certain cells—particularly those found in reproductive organs—telomerase adds short, repeated DNA sequences to the template strand after normal replication is complete. In this way, telomerase extends the length of the chromosome, so primase has room to add an RNA primer to the lagging strand. DNA polymerase can then fill in missing sections of the lagging strand and restore the chromosome's original length. You should be able to diagram the sequence of events involved in telomere replication, and add labels to indicate the enzymes involved and the polarity of each DNA strand.

14.5 Repairing Mistakes in DNA Synthesis

DNA polymerases work fast. In yeast, for instance, the replication fork is estimated to move at a rate of about 50 bases per second. But the replication process is also astonishingly accurate. In organisms ranging from *E. coli* to animals, the error rate during DNA replication averages less than one mistake per *billion* nucleotides. This level of accuracy is critical. Humans, for example, develop from a fertilized egg that has DNA containing over 6 billion nucleotides. The DNA inside the fertilized egg is replicated over and over to create the trillions of cells that will eventually make up the adult body. If more than one or two mutations occurred during each cell division cycle as a human grew, genes would be riddled with errors by the time the individual reached maturity. Genes that contain errors are often defective. Based on these observations, it is no exaggeration to claim that the accurate replication of DNA is a matter of life and death.

How can the enzymes involved in replicating DNA be as precise as they are? The answer to this question has several parts. DNA polymerase is highly selective in matching complementary bases correctly. This enzyme can also catch mismatched base pairs during the synthesis process and correct them, meaning that it can "proofread." Finally, if mistakes remain after synthesis is complete, repair enzymes can correct mismatched base pairs.

How Is DNA Polymerase Proofread?

DNA polymerases are selective because the correct base pairing (A-T and G-C) is energetically the most favorable of all possibilities for the pairing of nitrogen-containing bases. As DNA polymerase marches along a parent DNA template, hydrogen bonding occurs between incoming nucleotides and the nucleotides on the template strand. Due to the precise nature of these interactions, the enzyme inserts the incorrect nucleotide only about once every thousand bases added (**Figure 14.16a**).

What happens when a G-T, an A-C, or another type of mismatch occurs? An answer to this question emerged when researchers found mutants in *E. coli* with error rates that were 100 times greater than normal. Recall from Chapter 13 that mutants are individuals with traits caused by mutation and that mutation is a change in the gene responsible for those traits. Many mutations change the individual's phenotype. The change may result in a trait such as white eyes in fruit flies or an elevated mutation rate in *E. coli*. At the molecular level, a mutant phenotype usually results from a change in an enzyme or other type of protein.

In the case of *E. coli* cells with high mutation rates, biologists found that the mutation was localized to a particular portion of the DNA polymerase III enzyme, called the ε (epsilon)

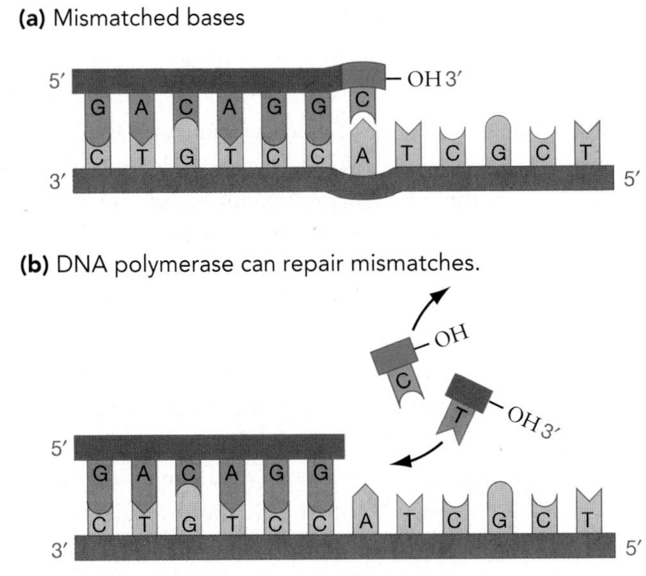

FIGURE 14.16 DNA Polymerase Can Proofread
(a) In bacteria, DNA polymerase adds an incorrect base to a growing strand of DNA about once in every thousand bases added. The result is a mismatch such as the pairing of A with C. **(b)** DNA polymerase can act as a 3′ → 5′ exonuclease, meaning that it can remove bases in that direction. The enzyme then adds the correct base.

subunit. Further analyses showed that this subunit of the enzyme acts as an **exonuclease**—meaning an enzyme that removes nucleotides from DNA (**Figure 14.16b**). The DNA polymerase III exonuclease activity removes nucleotides only from the 3′ end of DNA, and only if they are not hydrogen bonded to a base on the complementary strand. These results led to the conclusion that DNA polymerase III can proofread. If the wrong base is added during DNA synthesis, the enzyme pauses, removes the mismatched base that was just added, and then proceeds with synthesis.

Eukaryotic DNA polymerases have the same type of proofreading ability. Typically, proofreading reduces a DNA polymerase's error rate to about 1×10^{-7} (one mistake per 10 million bases). Is this accurate enough? No. If DNA replication were this sloppy, then at least 600 mistakes would occur every time a human cell replicated its DNA. Given that about a million-billion cell divisions take place in the course of a human's lifetime, an error rate this high would be a disaster. Once replication is complete, other repair mechanisms come into play.

Mismatch Repair

When DNA polymerase leaves a mismatched pair behind in the DNA sequence by mistake, a battery of enzymes springs into action. The proteins responsible for **mismatch repair** were discovered in the same way that the proofreading capability of DNA polymerase III was uncovered—by analyzing *E. coli* mutants. In this case, the mutants had normal DNA polymerase III but abnormally high mutation rates. The first mutant gene that caused a deficiency in mismatch repair was identified in the late 1960s and was called *mutS*. (The *mut* is short for "mutator.") By the late 1980s researchers had identified 10 proteins involved in the identification and repair of base-pair mismatches in *E. coli*.

The observation that repair enzymes fix mismatched bases raised a fundamental question: How do the enzymes know which of the two bases in a mismatched pair is right and which is wrong? A hypothesis to answer this question was inspired by the observation that several minutes after DNA polymerase III completes the synthesis of bacterial DNA, another enzyme adds methyl ($—CH_3$) groups to several of the adenosines on the new strand. The addition of methyl groups is called **methylation.** But until this reaction takes place, there is a marked difference between the parent and daughter strands. The parent DNA template is methylated, but the newly synthesized DNA is unmethylated. When a mismatched pair occurs, the wrong base must be found on the newly synthesized, unmethylated strand. Can the repair enzymes recognize which strand is methylated and use it as a template to fix the mistake?

To test the methyl-directed repair hypothesis, Paul Modrich and co-workers began experimenting with small, circular, double-stranded DNA molecules that they constructed. As **Figure 14.17** indicates, the experimental DNAs had two interesting features: a single base-pair mismatch and an adenine base that had a methyl group attached. The presence of one methylated strand and one unmethylated strand simulated the situation that occurs in *E. coli* immediately after DNA synthesis is complete. When the team added the other enzymes and molecules required for mismatch repair, they observed that repair proceeded only when a methyl group was present. Further, the repair proteins always removed the mismatched base on the unmethylated strand.

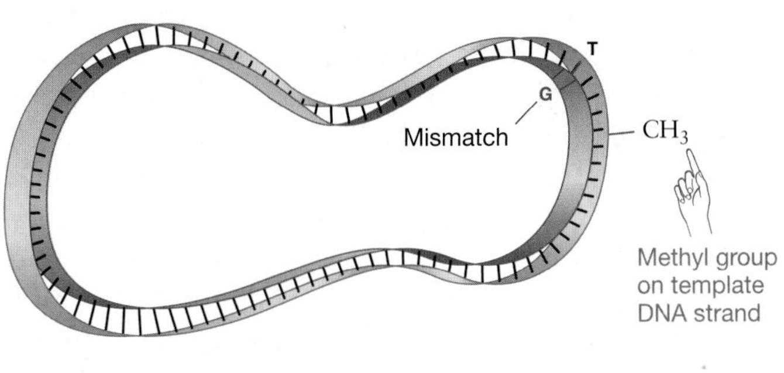

FIGURE 14.17 Methylated DNA Loop
A methyl group ($–CH_3$) on a loop of double-stranded DNA. In bacteria, an enzyme attaches a methyl group to adenosine bases a few minutes after replication is complete.

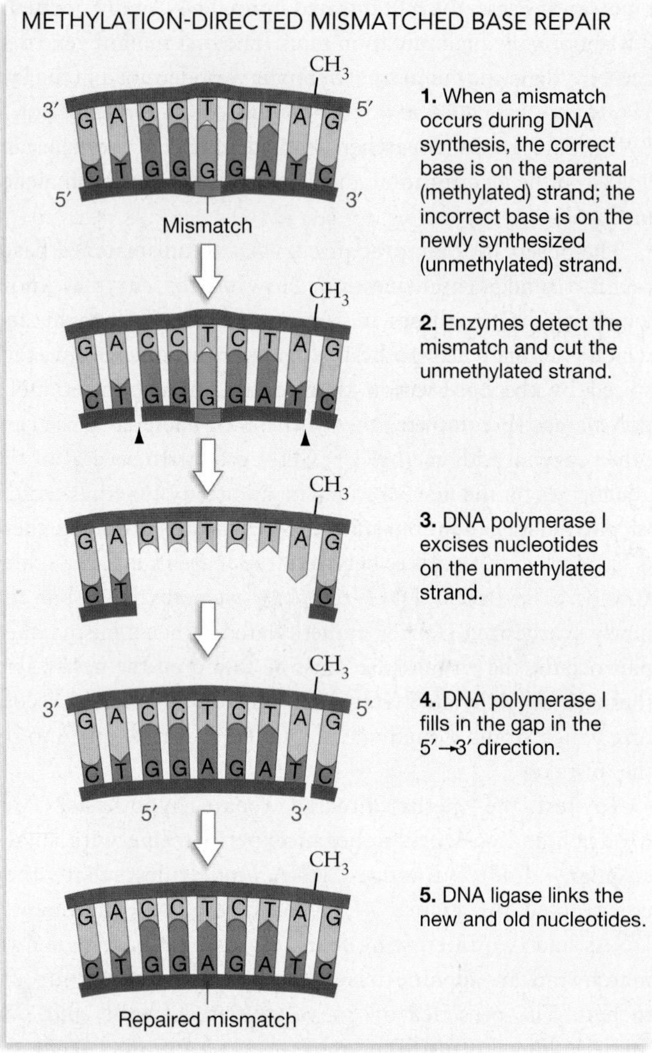
METHYLATION-DIRECTED MISMATCHED BASE REPAIR
CH_3
3′
G A C C T C T A G
5′
C T G G G G A T C
5′
3′
Mismatch
1. When a mismatch occurs during DNA synthesis, the correct base is on the parental (methylated) strand; the incorrect base is on the newly synthesized (unmethylated) strand.
2. Enzymes detect the mismatch and cut the unmethylated strand.
3. DNA polymerase I excises nucleotides on the unmethylated strand.
4. DNA polymerase I fills in the gap in the 5′→3′ direction.
5. DNA ligase links the new and old nucleotides.
Repaired mismatch

FIGURE 14.18 The Role of Methylation in Mismatch Repair

QUESTION Why is it valid to say that the methylated parent strand serves as a template for repair?

Based on results like these, researchers have been able to piece together the picture of mismatch repair that is shown in **Figure 14.18**. This model is called *methylation-directed mismatched base repair*. Although the model works well when applied to bacteria, researchers have realized that mismatch repair in eukaryotes is not based on a methylation signal. Many of the enzymes involved in mismatch repair in bacteria and eukaryotes are similar, but researchers have not been able to determine how the eukaryotic enzymes recognize the newly synthesized strand versus the original strand. Research on the eukaryotic repair pathway continues, and it has recently acquired some added urgency. As the essay at the end of this chapter explains, cancer can develop when the genes involved in the repair pathway are disabled by mutation.

14.6 Repairing Damaged DNA

The DNA that makes up your genome is under constant assault. The chemical bonds that hold DNA together can break spontaneously, sending nucleotides flying or snapping DNA's sugar-phosphate backbone in two. When even small amounts of X-ray, gamma-ray, or ultraviolet radiation bombard cells, one or both strands of DNA can break. In addition, certain molecules found in cells can actively attack and degrade nucleotides. These molecules include the hydroxyl (OH) radicals produced during aerobic metabolism, the aflatoxin B1 found in moldy peanuts and corn, and the benzo[α]pyrene in cigarette smoke.

Under normal conditions, biologists estimate that thousands of nucleotides are altered or lost from the DNA in every human cell every day due to spontaneous, radiation-induced, and chemical-caused damage. If these mistakes are not repaired, they would lead to permanent changes in genes. The vast majority of alterations would be harmful.

To cope with these insults, cells have a sophisticated battery of repair enzymes that act like maintenance crews. These proteins patrol DNA, identify damaged bases or sections of the sequence, and repair them. Collectively, these coordinated groups of molecules are called **excision repair systems.**

DNA Nucleotide Excision Repair: An Overview

Excision repair systems share a common mechanism: They excise a stretch of single-stranded DNA around a damaged site, then resynthesize a new strand based on the information in the intact, complementary strand. The mismatch repair system introduced in Section 14.5, which corrects mistakes made during DNA replication, is an example of an excision repair system.

The symmetry and regularity of DNA's secondary structure makes it possible for the proteins involved in repair to recognize mismatched base pairs or damaged sections after synthesis is complete. As Figure 14.18 showed, kinks in the double helix occur at the site of mismatched pairs, and damaged bases often produce a bulge or other irregularity in the molecule. Once a damaged or error-containing region is recognized, the presence of a DNA strand complementary to the damaged strand provides a template for resynthesis of the defective sequences. In this way, DNA's structure makes accurate repair possible. Because the molecule is repairable, DNA's structure supports its function as the cell's information-processing center.

Many of the proteins involved in the excision repair systems first came to light when researchers isolated strains of *E. coli* or yeast with abnormally high mutation rates. Follow-up studies were able to trace the cause of the defect. Much of this early work focused on how *E. coli* repairs damage induced by exposure to ultraviolet (UV) radiation. Rapid progress in this area inspired studies on DNA repair in a much more unusual model organism. In contrast to *E. coli* and fruit flies and garden peas, this model organism reproduces slowly and is extremely expensive to maintain. The organism? Human beings.

Xeroderma Pigmentosum: A Case Study

Xeroderma pigmentosum (XP) is a rare autosomal recessive disease in humans. Individuals with this condition are extremely sensitive to ultraviolet (UV) light. Their skin develops lesions after even slight exposure to sunlight. In normal individuals, these kinds of lesions develop only after extensive exposure to UV light, X-rays, or other forms of high-energy radiation.

In 1968 James Cleaver proposed a connection between XP and DNA excision repair systems. He knew that in *E. coli*, mutations in certain genes cause DNA excision repair proteins to fail. Cells with these mutations have an increased sensitivity to radiation. Cleaver's hypothesis was that people with XP have similar mutations. He claimed that they are extremely sensitive to sunlight because they are unable to repair the damage that occurs when the nitrogen-containing bases of DNA absorb UV light.

What is the nature of this damage? The most common defect caused by UV radiation is the formation of a covalent bond between adjacent pyrimidine bases. The thymine-thymine pair illustrated in **Figure 14.19a** is an example. This defect is called a *thymine dimer*, and it creates a kink in the secondary structure of DNA. This structural flaw keeps the gene from functioning normally and stalls the movement of the replication fork during DNA replication. If the error is not repaired, the cell may die.

Cleaver's hypothesized connection between DNA damage, faulty error repair, and XP turned out to be correct. Much of the work that he and other investigators did relied on the use of cell cultures. In this case, the researchers collected skin cells from people with XP and from people with a normal phenotype for excision repair. When these cell populations were grown in culture and exposed to increasing amounts of ultraviolet radiation, a striking difference emerged: Cells from XP individuals died off much more rapidly than did normal cells (**Figure 14.19b**).

The connection to excision repair systems was confirmed when Cleaver exposed cells from normal individuals and cells from XP individuals to various amounts of UV light, then fed the cells radioactive thymidine to label DNA synthesized during the

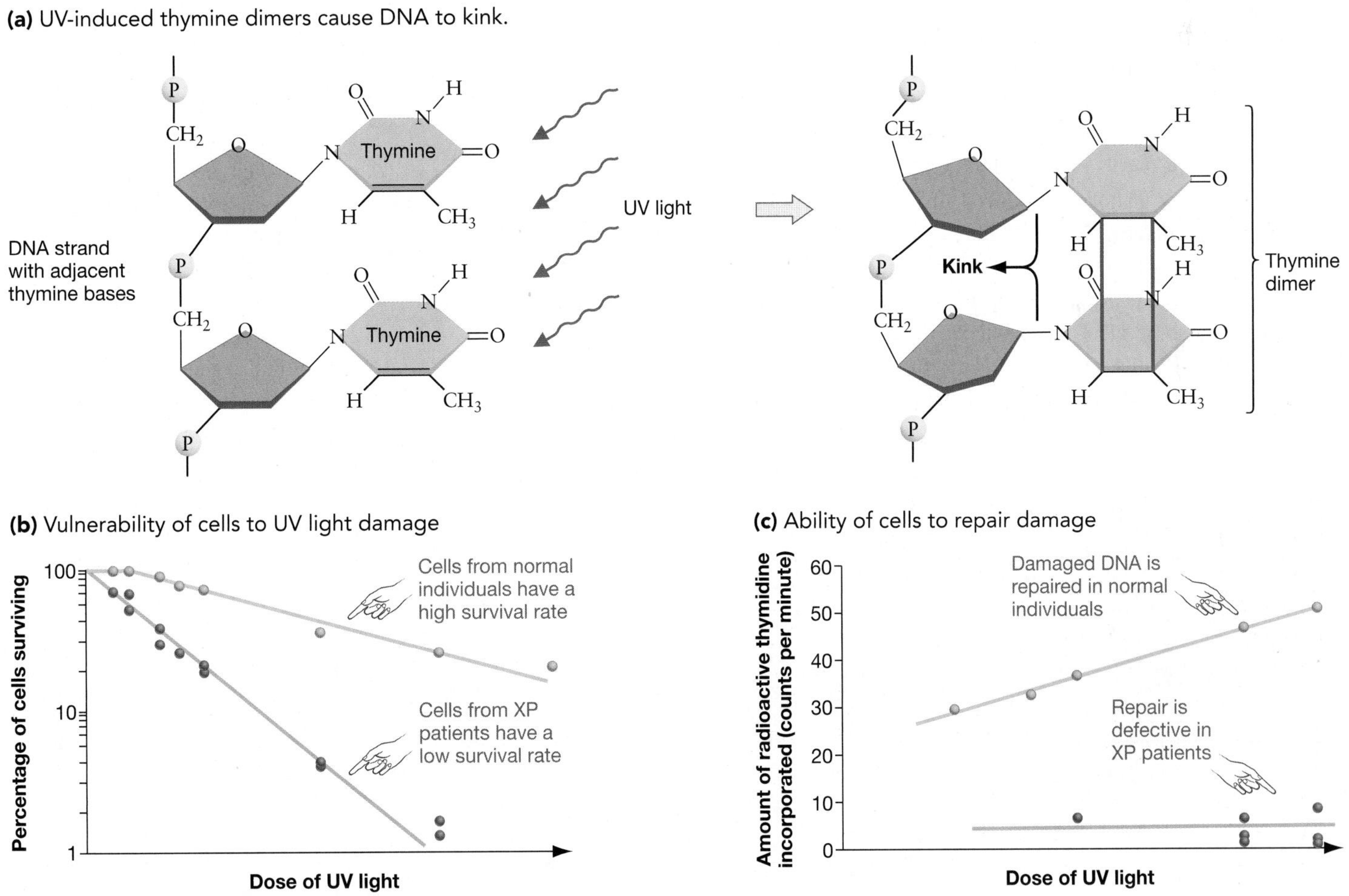

FIGURE 14.19 DNA Damage from UV Light Is Not Repaired Properly in Individuals with XP
(a) When UV light strikes a section of DNA with adjacent thymines, the energy can break bonds within each base and result in the formation of bonds *between* them. The thymine dimer that is produced causes a kink in DNA. **(b)** When cell cultures from normal individuals and from XP patients are irradiated with various doses of UV light, the percentage of cells that survive is strikingly different. **(c)** When cell cultures from normal individuals and from XP patients are irradiated with various doses of UV light and then fed radioactive thymidine, only normal individuals incorporate the labeled base.

repair period. If repair is defective in the XP individuals, then their cells should incorporate virtually no radioactive thymidine into their DNA. Cells from normal individuals, in contrast, should incorporate large amounts of labeled thymidine into their DNA. As **Figure 14.19c** (page 321) shows, this is exactly what happens. These data are consistent with the hypothesis that repair synthesis is virtually nonexistent in XP individuals.

More recently, genetic analyses of XP patients have shown that the condition can result from mutations in any one of seven genes. This result is not surprising in light of the large number of enzymes involved in excision repair even in bacteria.

Finally, as the essay at the end of this chapter points out, defects in the genes required for excision repair are frequently associated with cancer. Individuals with xeroderma pigmentosum, for example, are 1000 to 2000 times more likely to get skin cancer than are individuals with normal excision repair systems. To explain this pattern, biologists suggest that if mutations in the genes involved in the cell cycle (see Chapter 7) go unrepaired, the cell may begin to grow in an uncontrolled manner. Tumor formation could result. Stated another way, if the overall mutation rate in a cell is elevated because of defects in DNA repair genes, then the mutations that trigger cancer become more likely. In this case, research on fundamental aspects of DNA replication and repair, using *Escherichia coli* as a model system, led directly to a major advance in understanding a form of cancer in humans.

ESSAY DNA Mismatch Repair and Cancer

Cancer geneticists have known for decades that families can have predispositions to certain types of cancers, including breast cancer and the skin cancers associated with XP. A common type of colon cancer called **hereditary nonpolyposis colorectal cancer (HNPCC)** also runs in families. Affected individuals frequently develop tumors of the colon, ovary, and other organs before age 50 years (**Figure 14.20**). A major breakthrough in understanding HNPCC occurred in the early 1990s, when techniques introduced in later chapters allowed researchers to determine that a gene associated with susceptibility to HNPCC mapped to a specific region of chromosome 2.

Meanwhile, a different group of investigators was trying to determine whether humans have mismatch repair genes similar to those found in *E. coli*. The pace of this work accelerated when the DNA sequence of the *mutS* gene was determined and when a mismatch repair gene identified in the yeast *Saccharomyces cerevisiae* turned out to be extremely similar to *mutS*. The bacterial and yeast genes were so similar that they were considered **homologous**, meaning that they trace their ancestry to a gene in a common ancestor.

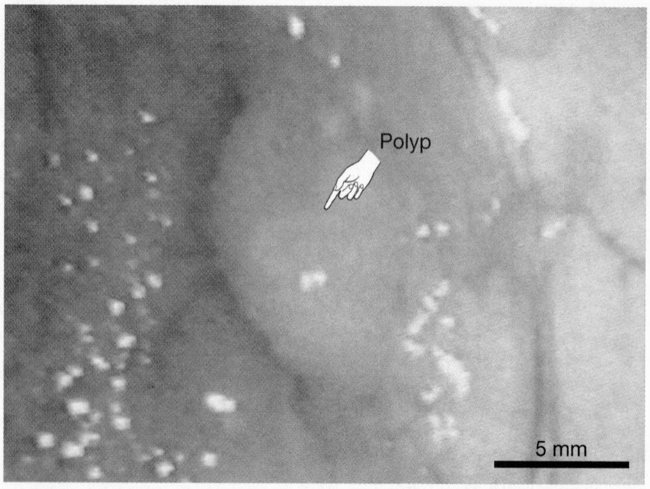

FIGURE 14.20 The Presence of Polyps Indicates HNPCC
This knob of tissue is a polyp that resulted from HNPCC. To diagnose this cancer, physicians try to detect polyps at the earliest stage of their formation.

Using DNA sequence information from these two homologous genes, researchers were able to locate a similar sequence in the human genome. This gene came to be called *hMSH* (for human *mutS* homolog). Sparks really began to fly when biologists discovered that the *hMSH* gene mapped to the same region of chromosome 2 as the HNPCC susceptibility gene.

... because humans, yeast, and bacteria share a common evolutionary history, ... early experiments made a major advance in cancer biology possible.

The link between a defect in mismatch repair and HNPCC was cemented when it was confirmed that individuals with this type of cancer have mutated forms of *hMSH*. As predicted, cells from these patients have mutation rates 100 times that of normal cells when both cell types are grown in culture. Follow-up studies confirmed that this increased mutation rate is indeed due to defective mismatch repair in cells derived from HNPCC patients. People who inherit a nonfunctional copy of the *hMSH* gene have a genetic predisposition for developing HNPCC.

This breakthrough validates the use of model organisms to study basic questions in biology. When experiments on mutated strains of *E. coli* began, they appeared to be pure, or basic, research—without practical application. But because humans, yeast, and bacteria share a common evolutionary history, and because DNA synthesis and repair are so fundamental to the functioning of all cells, the results of those early experiments made a major advance in cancer biology possible. If individuals with mutant forms of *hMSH* can be identified early in life, dietary changes and therapy could significantly reduce their risk of developing cancer.

The HNPCC story also highlights one of the major questions in cancer biology today. Tissues in the colon and certain other organs appear to be extremely sensitive to mutations in *hMSH*. Stated another way, it appears that if cells in those tissues acquire mutations in *hMSH*, they are likely to become cancerous. But if cells in the brain, lung, breast, skin, or many other organs acquire mutations in *hMSH*, they are not likely to become cancerous. Chapter 8 introduced the same phenomenon in the cancer called retinoblastoma, which results from mutations in the *Rb* gene but tends to produce cancer cells only in the retina—not in other tissues. Why are certain types of genetic defects associated with the development of certain types of cancers, but only in certain types of cells? This question brings us to a cutting edge in cancer research. In most cases, the answer is still unknown.

CHAPTER REVIEW

Summary of Key Concepts

- **Genes are made of DNA. When DNA is copied, each strand of a DNA double helix serves as the template for the synthesis of a complementary strand.**

  By labeling DNA with ^{15}N or ^{14}N, Meselson and Stahl were able to validate the hypothesis that DNA replication is semiconservative. Each strand of a parent DNA molecule provides a template for the synthesis of a daughter strand, resulting in two complete DNA double helices.

- **DNA is synthesized only in the $5' \rightarrow 3'$ direction. When a DNA molecule is being copied, a large suite of specialized enzymes is involved in unwinding the double helix, continuously synthesizing the "leading strand" in the $5' \rightarrow 3'$ direction and synthesizing the "lagging strand" as a series of fragments that are then linked together.**

  Kornberg showed that DNA synthesis is an enzyme-catalyzed reaction by purifying DNA polymerase I from *E. coli* and using this enzyme to synthesize DNA in vitro. Follow-up work over several decades showed that DNA synthesis takes place only in the $5' \rightarrow 3'$ direction, requires both a template and a primer sequence, and takes place at the replication fork where the double helix is being opened.

  Synthesis of the leading strand in the $5' \rightarrow 3'$ direction is straightforward, but synthesis of the lagging strand is more complex because DNA polymerase has to work away from the replication fork. By feeding *E. coli* cells a short pulse of radioactive thymidine, Okazaki and co-workers confirmed that short DNA fragments form on the lagging strand. These Okazaki fragments are primed by a short strand of RNA and are linked together after synthesis.

  At the ends of linear chromosomes in eukaryotes, the enzyme telomerase adds short, repeated sections of DNA so that the lagging strand can be synthesized without shortening the chromosome. Telomerase is active in reproductive cells that eventually undergo meiosis. As a result, gametes contain chromosomes of normal length.

  Web Tutorial 14.1 DNA Synthesis

- **Most mistakes that occur in DNA synthesis are repaired by specialized enzymes. If these repair enzymes are defective, the mutation rate increases. If mutations occur in genes that control the cell cycle, cancer may develop.**

  DNA replication is remarkably accurate, because DNA polymerase proofreads and because mismatch repair enzymes excise incorrect bases once synthesis is complete and replace them with the correct sequence. In addition, DNA repair occurs after bases have been damaged by spontaneous breakage of bonds or by chemicals or radiation. Various excision repair systems cut out damaged portions of genes and replace them with correct sequences. Several types of human cancers are associated with defects in the genes responsible for DNA repair.

Questions

Content Review

1. What constitutes an individual's genome?
   a. all of its proteins
   b. all of its mRNAs
   c. all of its DNA
   d. all of its organelles

2. The experiment that removed proteins, RNA, or DNA from *Streptococcus* extracts purported to show that DNA is the hereditary material. Why did it fail to convince many skeptics?
   a. It had no control treatments, so it was poorly designed.
   b. Genes had already been shown to be made of protein.
   c. The trait that was studied—virulence—is not genetic.
   d. Critics argued that the chemical treatments had not removed all of the protein present.

3. How did researchers first isolate DNA polymerase?
   a. They used transmission electron microscopy to see molecules attached to DNA during synthesis.
   b. They isolated mutants that could not synthesize DNA.
   c. They added cell extracts to an in vitro synthesis system and found one that worked.
   d. They used certain recombinant DNA techniques (that is, genetic engineering).

4. Why did the lagging strand get its name?
   a. Replication of the lagging strand is discontinuous.
   b. The lagging strand consists of Okazaki fragments that are ligated late in the replication sequence.
   c. DNA ligase is active on this strand and is the "lagging" enzyme—it is the last enzyme in the replication sequence.
   d. Synthesis of this strand lags behind synthesis of the leading strand.
5. Where and how are Okazaki fragments synthesized?
   a. at the leading strand, oriented in a $5' \rightarrow 3'$ direction
   b. at the leading strand, oriented in a $3' \rightarrow 5'$ direction
   c. at the lagging strand, oriented in a $5' \rightarrow 3'$ direction
   d. at the lagging strand, oriented in a $3' \rightarrow 5'$ direction
6. What does telomerase do?
   a. It adds an RNA primer to the ends of linear chromosomes.
   b. It adds double-stranded DNA to the "blunt end" of a linear chromosome.
   c. It adds double-stranded DNA to the lagging strand at the end of a linear chromosome.
   d. It adds single-stranded DNA to the lagging strand at the end of a linear chromosome.

Conceptual Review

1. Researchers try to design experiments so that an experimental treatment shows the effect that one, and only one, condition or agent has on the phenomenon being studied. Examine Figure 14.2, and decide which treatments in the transformation study acted as experimental treatments and which acted as controls. Write a general statement explaining the role of control treatments in experimental design.
2. Why is the activity of helicase, single-strand DNA-binding proteins, and topoisomerase required during DNA synthesis?
3. What does it mean to say that DNA replication is "bidirectional?"
4. Why is the synthesis of the lagging strand of DNA discontinuous? How is it possible for the synthesis of the leading strand to be continuous?
5. List in chronological order the events that increase the accuracy of DNA replication. Indicate which events involve DNA polymerase III and which involve specialized repair enzymes.
6. Explain how the structure of DNA makes it relatively easy for proteins to recognize base-pair mismatches or damaged bases. How does DNA's secondary structure make it possible for damaged sections or incorrect bases to be removed and repaired?

Group Discussion Problems

1. If DNA polymerase III did not require a primer, which steps in DNA synthesis would differ from what is observed? Would any special enzymes be required to replicate telomeres? Explain your answers.
2. In the late 1950s Herbert Taylor grew bean root-tip cells in a solution of radioactive thymidine and allowed them to undergo one round of DNA replication. He then transferred the cells to a solution without the radioactive nucleotide, allowed them to replicate again, and examined their chromosomes for the presence of radioactivity. His results are shown in the following figure.

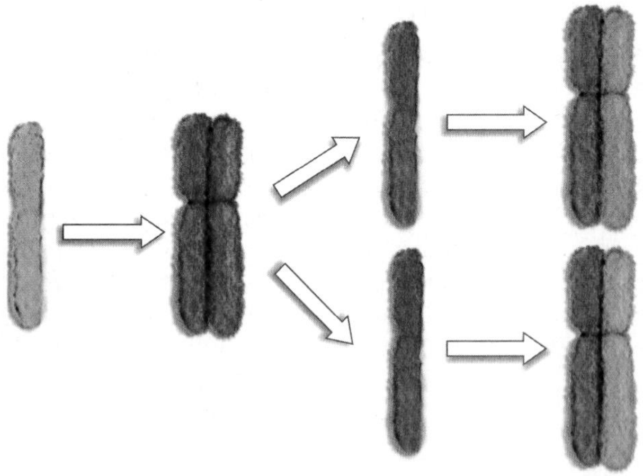

   a. Draw diagrams explaining the pattern of radioactivity observed in the sister chromatids after the first and second rounds of replication.
   b. What would the results of Taylor's experiment be if eukaryotes used a conservative mode of DNA replication?
3. The graph that follows shows the survival of four different *E. coli* strains after exposure to increasing doses of ultraviolet light. The wild-type strain is normal, but the other strains have a mutation in genes called *uvrA*, *recA*, or both *uvrA* and *recA*.

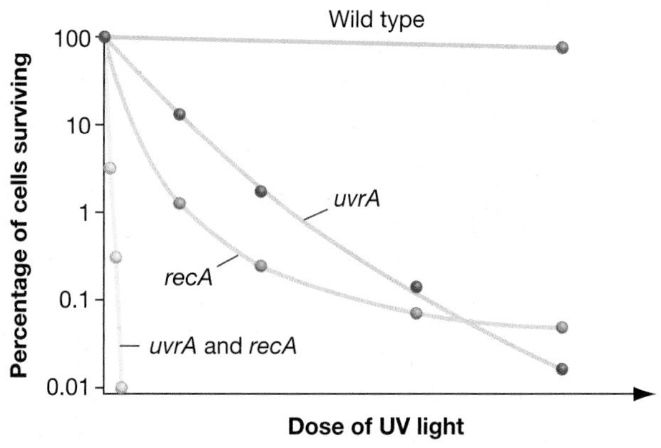

   a. Which strains are most sensitive to UV light? Which strains are least sensitive?
   b. What are the relative contributions of these genes to the repair of UV damage?
4. One widely used test to identify whether certain chemicals, such as pesticides or herbicides, might be carcinogenic (cancer causing) involves exposing bacterial cells to a chemical and recording whether the exposure leads to an increased mutation rate. In effect, this test equates cancer-causing chemicals with mutation-causing chemicals. Why is this an informative test?

Answers to Multiple-Choice Questions 1. c; 2. d; 3. c; 4. d; 5. c; 6. d

www.prenhall.com/freeman is your resource for the following: Web Tutorials; Online Quizzes and other Online Study Guide materials; Answers to Conceptual Review Questions; Solutions to Group Discussion Problems; Answers to Figure Caption Questions and Exercises; and Additional Readings and Research.

How Genes Work

15

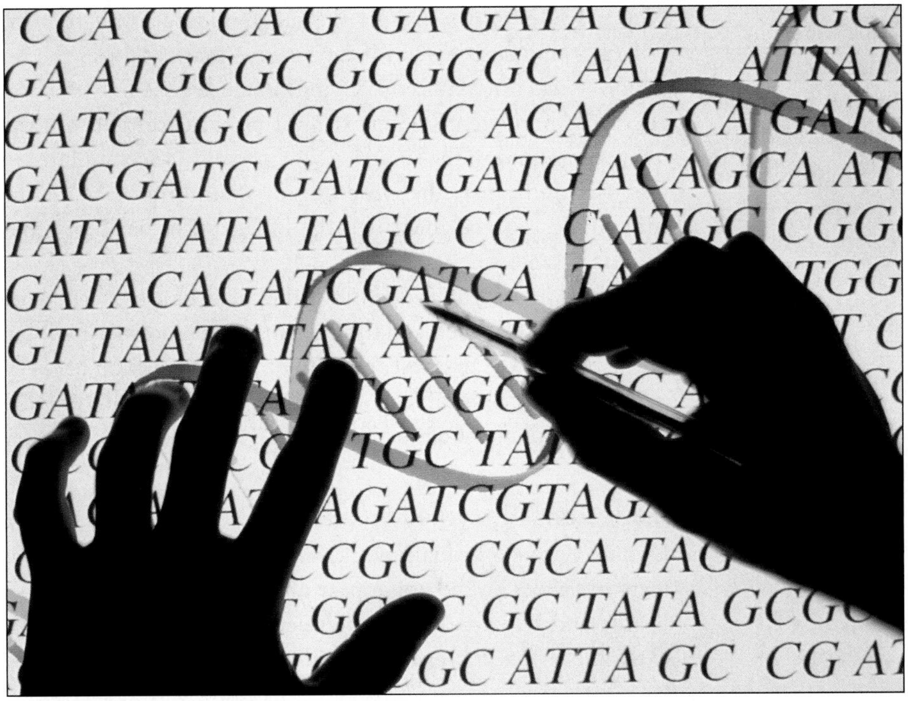

Each A, T, G, and C here represents a nitrogen-containing base in DNA. This chapter explores how biologists determined that genes are made of DNA and how a base sequence in DNA is translated into an amino acid sequence in a protein.

KEY CONCEPTS

- Most genes code for enzymes that catalyze specific chemical reactions in the cell.
- In cells, information flows from DNA to RNA to proteins. DNA is transcribed to messenger RNA by RNA polymerase, and then messenger RNA is translated to proteins in ribosomes. Each amino acid in a protein is specified by a group of three bases in RNA.

DNA has been called the blueprint of life. If an organism's DNA is like a set of blueprints, then its cells are like a construction site. How does the DNA inside each cell specify the types and amounts of lumber, nails, and cement that are needed as the cell grows and as its structure and function change? If the enzymes inside a cell are like construction workers, how does DNA organize them into a team that can build and maintain the cell, and perhaps remodel it when conditions change?

Mendel provided insights that made it possible to ask these questions. He discovered that particular alleles are associated with certain phenotypes and that alleles are transmitted faithfully from parent to offspring. The chromosome theory of inheritance established that genes are found in chromosomes, and the theory detailed how the movement of chromosomes during meiosis explains Mendel's results. Based on these early advances, it was clear that genes carry information—the instructions for making and maintaining an individual. But biologists still did not know how this information is translated into action. How does an organism's genotype specify its phenotype?

This chapter introduces a series of classic results that answered this question and revealed how genes work at the molecular level. Understanding how genes work, in turn, triggered a major transition in the history of biological science. Instead of thinking about genes solely in relation to their effects on eye color in fruit flies or on seed shape in garden peas, biologists could begin analyzing the molecular composition of both genes and their products. The molecular revolution in biology was under way.

15.1 What Do Genes Do?

In 1902 a physician named Archibald Garrod provided the first clue in the mystery of how genes work. A formal hypothesis addressing this mystery was not published until almost 40 years later, however—about the time evidence began to suggest that genes are made of DNA. Let's consider Garrod's results first and then explore a classical experiment that turned his hint into a solid understanding of what genes do.

The Molecular Basis of Hereditary Diseases

Chapter 13 examined how pedigrees can be used to determine patterns of inheritance in human traits. By analyzing the pedigrees of families that were afflicted with particular illnesses, Garrod became convinced that certain human diseases can be inherited. He was particularly interested in a condition known as **alkaptonuria**, which is characterized by arthritis and other symptoms. This condition is easy to diagnose, because an affected person's urine turns black. Garrod observed that people suffering from alkaptonuria excrete huge quantities of a molecule called *homogentisic acid* in their urine. To explain why alkaptonurics accumulate homogentisic acid, Garrod suggested they lack an enzyme that catalyzes a reaction involving this molecule. His idea was that if the enzyme did not function normally, then homogentisic acid would not be broken down or used as a reactant as it normally would. As a result, homogentisic acid would accumulate in the body.

In forming this hypothesis, Garrod was inspired by a fundamentally important concept. Biologists who were studying glycolysis and other processes had established that in cells, chemical reactions frequently take place in a series of steps called a **metabolic pathway**. As **Figure 15.1a** shows, molecules are often formed or broken apart in a distinct sequence of steps, from product to reactant by way of intermediates in a specific order. The glycolytic pathway, Krebs cycle, fermentation pathways, and Calvin cycle introduced in Chapters 9 and 10 are examples of particularly important metabolic pathways.

In most cases, every step in a metabolic pathway is catalyzed by a different enzyme. If the enzyme is not available, the reaction that it catalyzes does not occur, and the substrate for that particular reaction accumulates. For example, **Figure 15.1b** shows part of the metabolic pathway that involves homogentisic acid. If the enzyme that catalyzes the reaction marked with a red "X" in the figure does not function correctly, homogentisic acid cannot be transformed into the next compound in the sequence. As a result, it accumulates in the body and causes the symptoms of alkaptonuria. Using the same logic, Garrod identified the biochemical basis of other genetic diseases.

Garrod had correctly identified the cause of an inherited defect, or what he called an inborn error of metabolism. In doing so, he became the first researcher to draw a connection between genes and the chemical reactions that occur inside cells, or what biologists call **metabolism**.

Genes and Enzymes

Despite Garrod's brilliant work, an explicit hypothesis explaining what genes do did not appear until 1941. That year George Beadle and Edward Tatum published a series of breakthrough experiments on a bread mold called *Neurospora crassa*. The

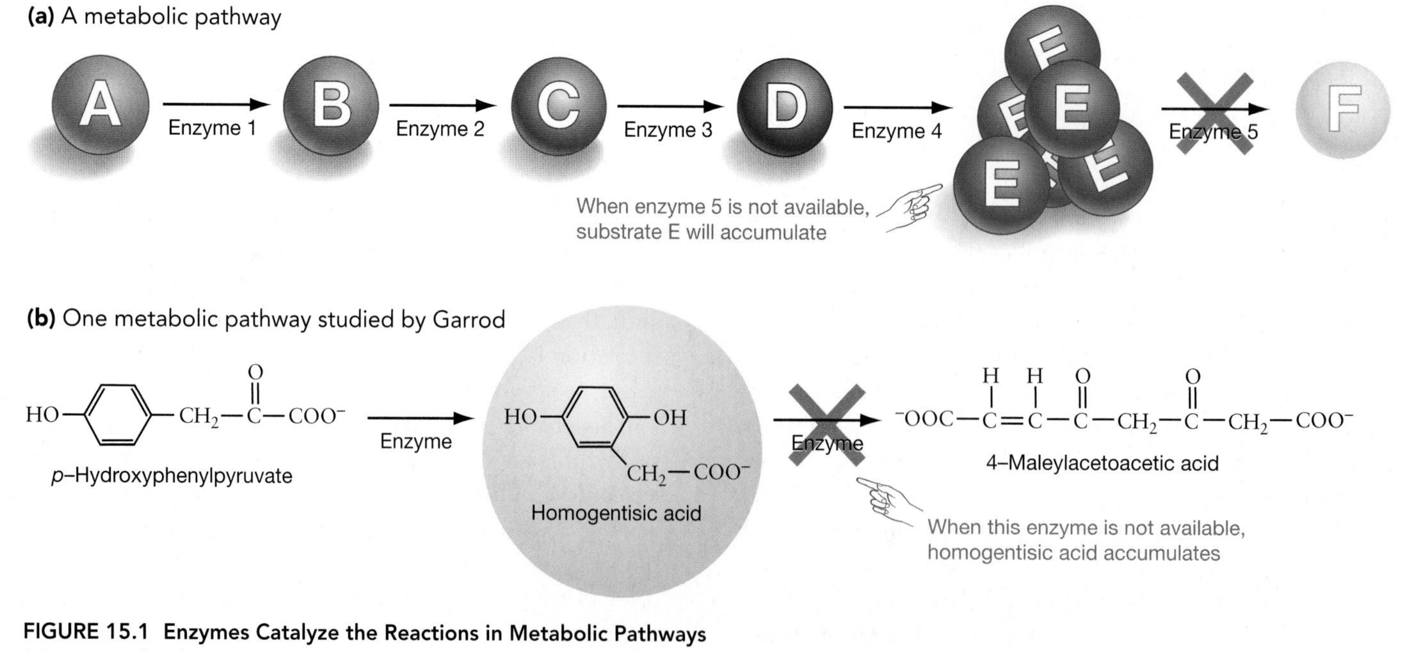

FIGURE 15.1 Enzymes Catalyze the Reactions in Metabolic Pathways
(a) A generalized metabolic pathway. Each reaction is catalyzed by a different enzyme. **(b)** Garrod hypothesized that alkaptonurics lack the enzyme that uses homogentisic acid as a substrate, thus causing the acid to accumulate. Excess homogentisic acid is the cause of alkaptonuria.

work inspired their **one-gene, one-enzyme hypothesis.** This was the idea that genes contain the information needed to make proteins, most of which function as enzymes.

Three years later, Adrian Srb and Norman Horowitz published a rigorous test of the one-gene, one-enzyme hypothesis that focused on the ability of *N. crassa* individuals to synthesize the amino acid arginine. In the lab, normal cells of this bread mold grow well on a laboratory culture medium that lacks arginine. This is possible because *Neurospora* cells are able to synthesize their own arginine.

Earlier, Hans Krebs had established that cells from mammals synthesize arginine via a series of steps. Krebs determined that compounds called ornithine and citrulline act as intermediates in the metabolic pathway that leads to arginine, and that enzymes are required to convert ornithine to citrulline and citrulline to arginine. Srb and Horowitz hypothesized that *N. crassa* cells synthesize arginine via the same pathway, using similar enzymes. More importantly, the researchers hypothesized that specific genes in *N. crassa* cells are responsible for producing each of the enzymes involved. Krebs had established the biochemical aspects of arginine synthesis; Srb and Horowitz set out to link the biochemistry to genetics.

This research was inspired by an idea that was brilliant in its simplicity and expressed concisely by Beadle: "One ought to be able to discover what genes do by making them defective." The logic ran as follows: If *N. crassa* synthesizes arginine via a metabolic pathway, and if the one-gene, one-enzyme hypothesis is correct, then Srb and Horowitz should be able to find individuals with a defective gene for each of the enzymes in the pathway. Their goal was to find out how each gene in the pathway acts by observing what happens when each gene, individually, does *not* act.

In analyzing the effects of defective genes, the early *Neurospora* researchers pioneered a research strategy that is still widely used. Contemporary biologists are often able to infer a great deal about what a gene does by comparing the phenotypes of individuals with and without a functioning copy of that gene. Today, alleles that do not function at all are called **knock-out mutants, null mutants,** or **loss-of-function mutants.** In effect, Srb and Horowitz set out to create knock-out mutants for each of the enzymes involved in arginine synthesis. If this approach were successful, the result would provide strong support for the hypothesis that genes make enzymes.

To pursue that strategy, Srb and Horowitz had to create a large number of mutant individuals. They did this with an approach that Beadle and Tatum had developed: exposing *N. crassa* cultures to high-energy radiation such as X-rays or ultraviolet light. As Chapter 14 indicated, high-energy radiation damages the double-helical structure of DNA—often in a way that makes the affected section nonfunctional. As a result, exposing organisms to radiation is an effective way to create knock-out mutants.

It is important to recognize that most treatments that create mutants are random with respect to the gene involved. High-energy radiation is equally likely to damage DNA in any part of the organism's genome. Most organisms have thousands of genes. Of the tens of thousands of mutants that the biologists created, then, it was likely that only a handful contained knock-out mutations in the pathway for arginine synthesis.

To find cells that contained a knock-out mutation in the pathway for arginine synthesis, the researchers performed what is now known as a genetic screen. A **genetic screen** is a technique for picking certain types of mutants out of many thousands of randomly generated mutants. In this case, Srb and Horowitz began their screen by raising the irradiated cells on a medium that included arginine, the final product in the pathway. Then they grew a sample of each type of cell on a medium that lacked amino acids, and they were *not* able to grow on that medium. This test identified cells in the original cultures that were potential knock-out mutants in the pathway for arginine synthesis. If a new sample of cells was added to a medium containing arginine and the cells were subsequently able to grow, then it was likely that they had a defect somewhere in the metabolic pathway for synthesizing arginine. The biologists followed up by confirming that the offspring of these cells also had this defect. Based on these data, they were confident that they had isolated individuals with mutations in the gene or genes for arginine synthesis.

Did some of these cells have knock-out mutations affecting different enzymes in the metabolic pathway? To answer this question, the biologists grew the mutants on media that contained ornithine, or citrulline, or arginine. The results were dramatic. Some of the mutant cells were able to grow in the presence of some of these compounds but not in the presence of others. More specifically, the mutants fell into three distinct classes, which the researchers called *arg1*, *arg2*, and *arg3*:

	Supplement			
Mutant Type	**no ornithine no citrulline no arginine**	**Ornithine no citrulline no arginine**	**no ornithine Citrulline no arginine**	**no ornithine no citrulline Arginine**
arg1	no growth	GROWTH	GROWTH	GROWTH
arg2	no growth	no growth	GROWTH	GROWTH
arg3	no growth	no growth	no growth	GROWTH

To interpret these data, the biologists drew the following conclusions:

- *arg1* cells lack the enzyme that converts a precursor molecule to ornithine;
- *arg2* cells lack the enzyme that converts ornithine to citrulline; and
- *arg3* cells lack the enzyme that converts citrulline to arginine.

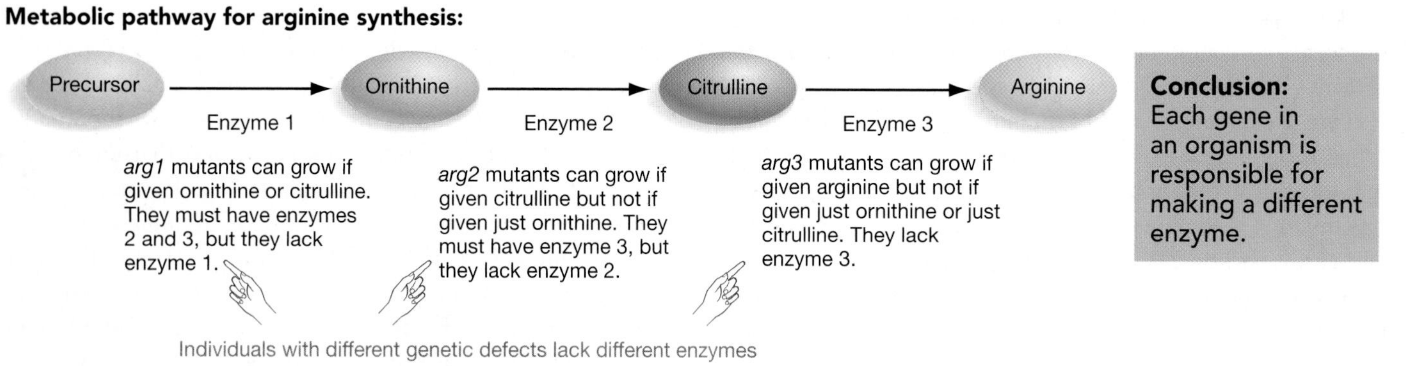
Metabolic pathway for arginine synthesis:
Precursor
Ornithine
Citrulline
Arginine
Enzyme 1
Enzyme 2
Enzyme 3
arg1 mutants can grow if given ornithine or citrulline. They must have enzymes 2 and 3, but they lack enzyme 1.
arg2 mutants can grow if given citrulline but not if given just ornithine. They must have enzyme 3, but they lack enzyme 2.
arg3 mutants can grow if given arginine but not if given just ornithine or just citrulline. They lack enzyme 3.
Conclusion: Each gene in an organism is responsible for making a different enzyme.
Individuals with different genetic defects lack different enzymes

FIGURE 15.2 There Are Several Steps in the Metabolic Pathway for Arginine Synthesis
Different *Neurospora crassa* mutants have defects in different parts of the arginine pathway. This result supports the hypothesis that each gene codes for a different enzyme.

As **Figure 15.2** shows, these are exactly the results predicted by Krebs's finding that ornithine is converted to citrulline and that citrulline is then converted to arginine. The experiments had shown that each type of mutant lacked a different, specific step in a metabolic pathway because of a defect in a particular gene. These results convinced most investigators that the one-gene, one-enzyme hypothesis was correct. Follow-up work showed that genes are responsible for all of the different types of proteins produced by cells—not just enzymes. Biologists finally understood what most genes do: They contain the instructions for making proteins.

15.2 The Genetic Code

How does a gene specify the production of a protein? After Beadle and Tatum's hypothesis had been supported in a variety of organisms, this question became a central one. Part of the answer lay in understanding the molecular nature of the gene. Biochemists knew that the primary components of DNA were four nitrogen-containing bases: the pyrimidines thymine (abbreviated T) and cytosine (C), and the purines adenine (A) and guanine (G). They also knew that these bases were connected in a linear sequence by a sugar-phosphate backbone. Watson and Crick's model for the secondary structure of the DNA molecule, introduced in Chapter 4, revealed that two strands of DNA are wound into a double helix, held together by hydrogen bonds between the complementary base pairs A-T and G-C.

Given DNA's structure, it appeared extremely unlikely that DNA directly catalyzed the reactions that produce proteins. Its shape was too regular to suggest that it could bind a wide variety of substrate molecules and lower the activation energy for chemical reactions. Instead, Crick proposed that the sequence of bases in DNA might act as a code. The idea was that DNA was *only* an information-storage molecule. The instructions it contained would have to be read and then translated into proteins.

Crick offered an analogy with Morse code. Morse code is a message-transmission system using dots and dashes to represent the letters of the alphabet. Crick's idea was that different combinations of bases could specify the 20 amino acids, just as different combinations of dots and dashes specify the 26 letters of the alphabet.

A particular stretch of DNA, then, could contain the information needed to specify the amino acid sequence of a particular enzyme. In code form, the tremendous quantity of information needed to build and run a cell could be stored compactly. This information could also be copied through complementary base pairing and transmitted efficiently from one generation to the next.

DNA's structure correlates closely with this proposed function. As **Figure 15.3** shows, the bases that compose DNA are exposed if the two strands in the double helix are separated. These bases are then able to pair with complementary nucleotides to form two newly synthesized strands of DNA or a new strand of RNA. When this pairing occurs, biologists say that the original strands form a template and are "read."

It soon became apparent, however, that the information encoded in the base sequence of DNA is not translated into the amino acid sequence of proteins directly. Instead, the link between DNA as information repository and proteins as cellular machines is indirect.

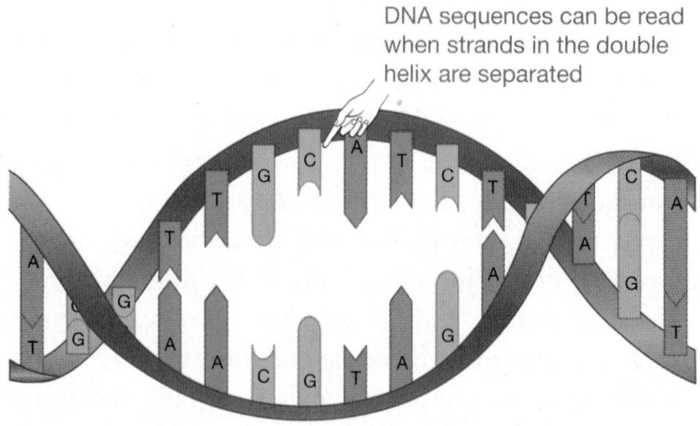
DNA sequences can be read when strands in the double helix are separated

FIGURE 15.3 The Sequence of Bases in DNA Provides Encoded Information
The information encoded in the sequences of bases in DNA can be read when strands are separated.

RNA as the Intermediary between Genes and Proteins

The first clue that biological information does not flow directly from DNA to proteins came from data on the structure of cells. In eukaryotic cells, DNA is enclosed within a membrane-bound organelle called the nucleus (see Chapter 7). But the organelles called ribosomes, where protein synthesis takes place, are outside the nucleus, in the cytoplasm. This observation began to make sense after François Jacob and Jacques Monod suggested that RNA molecules act as a link between genes and the protein-manufacturing centers. Jacob and Monod's hypothesis is illustrated in **Figure 15.4**. They predicted that short-lived molecules of RNA called **messenger RNA**, or **mRNA**, carry information from DNA to the site of protein synthesis. Messenger RNA is one of several distinct types of RNA in cells.

Follow-up research confirmed that the messenger RNA hypothesis is correct. One particularly important piece of evidence was the discovery of an enzyme that catalyzes the synthesis of RNA. This protein is called **RNA polymerase,** because it polymerizes ribonucleotides into strands of RNA. The key observation was that RNA polymerase synthesizes RNA molecules according to the information provided by the sequence of bases in a particular stretch of DNA.

Figure 15.5a shows how the transfer of information from DNA to mRNA works. To determine this connection, researchers created a reaction mix containing three critical elements: (1) the enzyme RNA polymerase; (2) ribonucleotides containing the bases adenine (A), uracil (U), guanine (G), and cytosine (C); and (3) copies of a strand of synthetic DNA that contained deoxyribonucleotides in which the only base was thymine (T). When they allowed the polymerization reaction to proceed, the only base in the RNA molecules that resulted was adenine. This result strongly supported the hypothesis that RNA polymerase synthesizes RNA according to the rules of complementary base pairing introduced in Chapter 4: Thymine pairs with adenine. A sequence of DNA containing thymine as the only base directed the polymerization of a complementary RNA sequence containing adenine as the only base. Synthetic DNAs containing no bases other than cytosine result in the production of RNA molecules containing no bases other than guanine. **Figure 15.5b** summarizes how the sequence of bases in DNA maps to the sequence of bases in RNA. These experiments were possible because RNA polymerase does not require a primer.

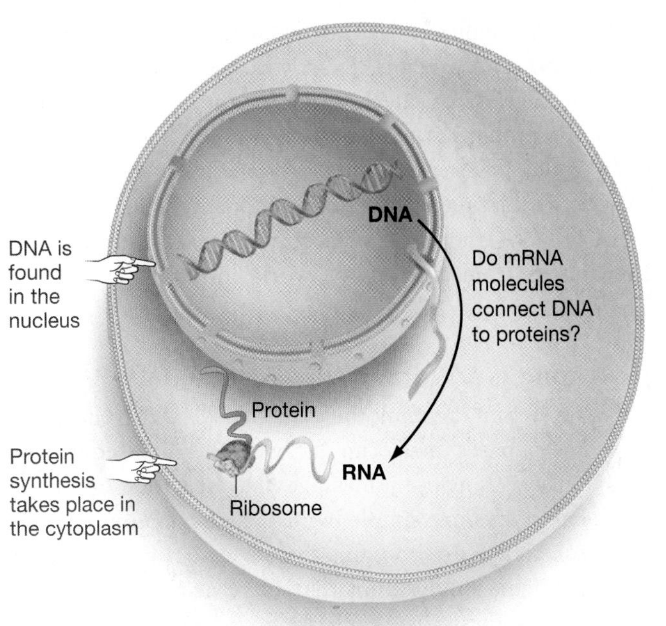

FIGURE 15.4 The Messenger RNA Hypothesis
In the cells of plants, animals, fungi, and other eukaryotes, most DNA is found only in the nucleus, but proteins are manufactured outside the nucleus, at ribosomes. Biologists proposed that the information coded in DNA is carried from inside the nucleus to the ribosomes by messenger RNA (mRNA). QUESTION Why was the choice of the term *messenger* appropriate?

(a) Experimental evidence that RNA is synthesized by complementary base pairing

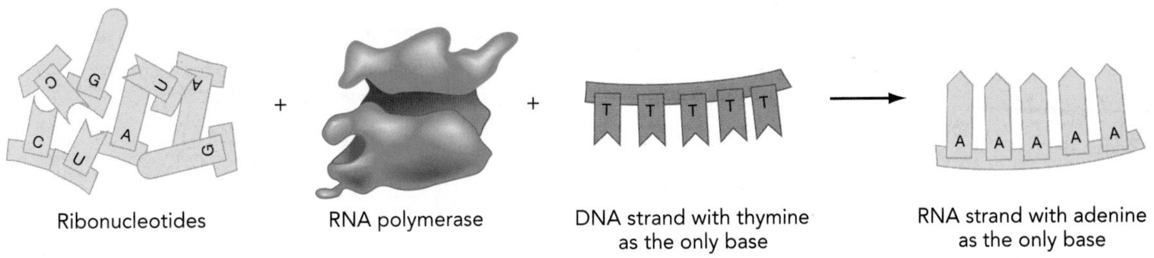

(b) RNA sequences are complementary to DNA sequences.

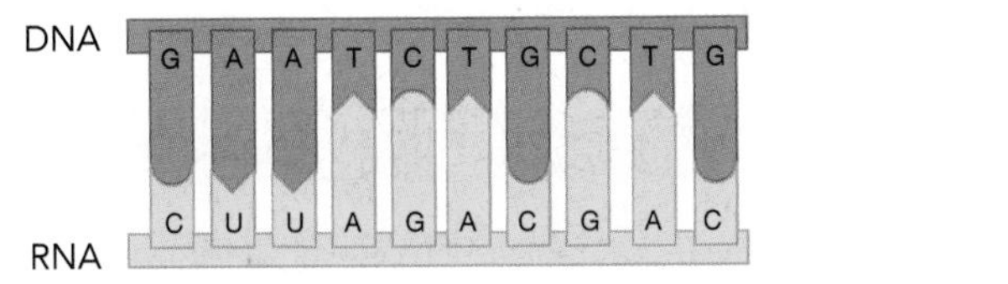

FIGURE 15.5 Complementary Base Pairing Transfers Information from DNA to RNA
(a) This experiment supported the hypothesis that RNA molecules are synthesized with a base sequence that is complementary to the corresponding DNA sequence. **(b)** RNA sequences are complementary to the DNA sequences of genes.

The next challenge was to understand the final link between DNA and proteins: Exactly how does the sequence of bases in a strand of mRNA code for the sequence of amino acids in a protein? If this question could be answered, biologists would have cracked the **genetic code**—the rules that specify the relationship between a sequence of nucleotides in DNA or RNA and the sequence of amino acids in a protein. Researchers from all over the world took up the challenge. The race was on.

How Long Is a Word in the Genetic Code?

The first step in cracking the genetic code was to determine how many bases make up a "word." In a sequence of mRNA, how long is a message that specifies one amino acid? Based on some simple logic, George Gamow suggested that each code word contains three bases. His reasoning derived from the observation that there are 20 amino acids, and the hypothesis that each amino acid must be specified by a particular sequence of mRNA. Because there are only four different bases in ribonucleotides (A, U, G, and C), a one-base code could specify only four amino acids. Similarly, a two-base code could represent just 4 × 4, or 16, amino acids. A three-base code, though, could specify 4 × 4 × 4, or 64, different amino acids. As a result, a three-base code would provide more than enough messages to code for all 20 amino acids (**Figure 15.6**). This three-base code is known as a **triplet code**.

These simple calculations highlight an important point: Gamow's hypothesis predicted that the genetic code is **redundant**. That is, more than one triplet of bases might specify the same amino acid. The idea was that different three-base sequences in an mRNA—say, AAA and AAG—might code for the same amino acid—say, lysine. The group of three bases that specifies a particular amino acid is called a **codon**. According to the triplet code hypothesis, many of the 64 codons that are possible actually specify the same amino acids.

Work by Francis Crick and colleagues confirmed that codons are three bases long. These researchers used chemical dyes called *acridines* to induce a mutation that adds or deletes a single base in the DNA of a virus called T4, which infects *E. coli* cells. As predicted for a triplet code, a 1-base addition or deletion in the base sequence led to a loss of function in the gene being studied. This is because a single addition or deletion mutation throws the sequence of codons, or the **reading frame**, out of register. To understand how this works, consider the sentence "*The fat cat ate the rat.*" The reading frame of this sentence is a three-letter word and a space. If the fourth letter in this sentence—the "f" in "fat"—were deleted but the reading frame stayed intact, the sentence would be transformed into "*The atc ata tet her at.*" This is gibberish.

Figure 15.7a illustrates how the addition or deletion of a single base disrupts the reading frame in a sequence of codons, assuming that the genetic code is read in triplets. The parent DNA sequence (top line) shows the reading frame for a series of AAT codons. The next two lines in Figure 15.7a illustrate how the addition or deletion of a single base causes a different group of triplets to be read. If each codon specifies a distinct amino acid, then the protein produced from the altered sequence will have an entirely different sequence of amino acids than normal. Enzymes with such radically altered primary structure are almost guaranteed to be completely nonfunctional. In this way, the addition or deletion of a single base leads to a knock-out

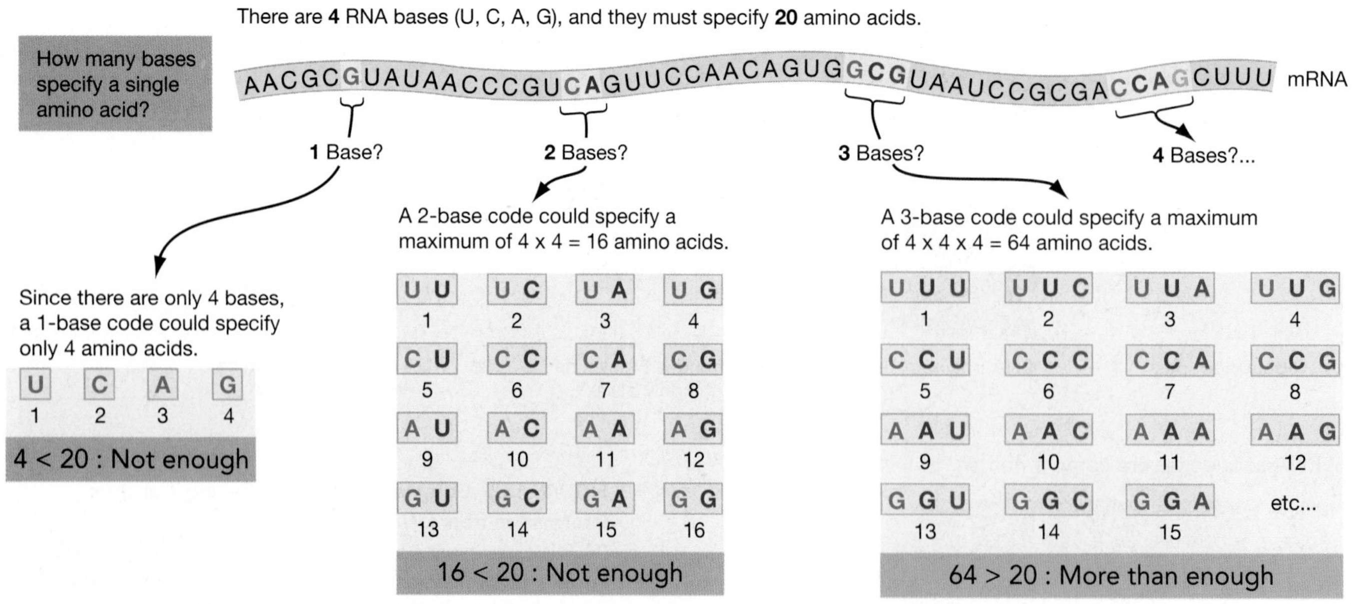

FIGURE 15.6 In the Genetic Code, How Many Bases Form a "Word"?

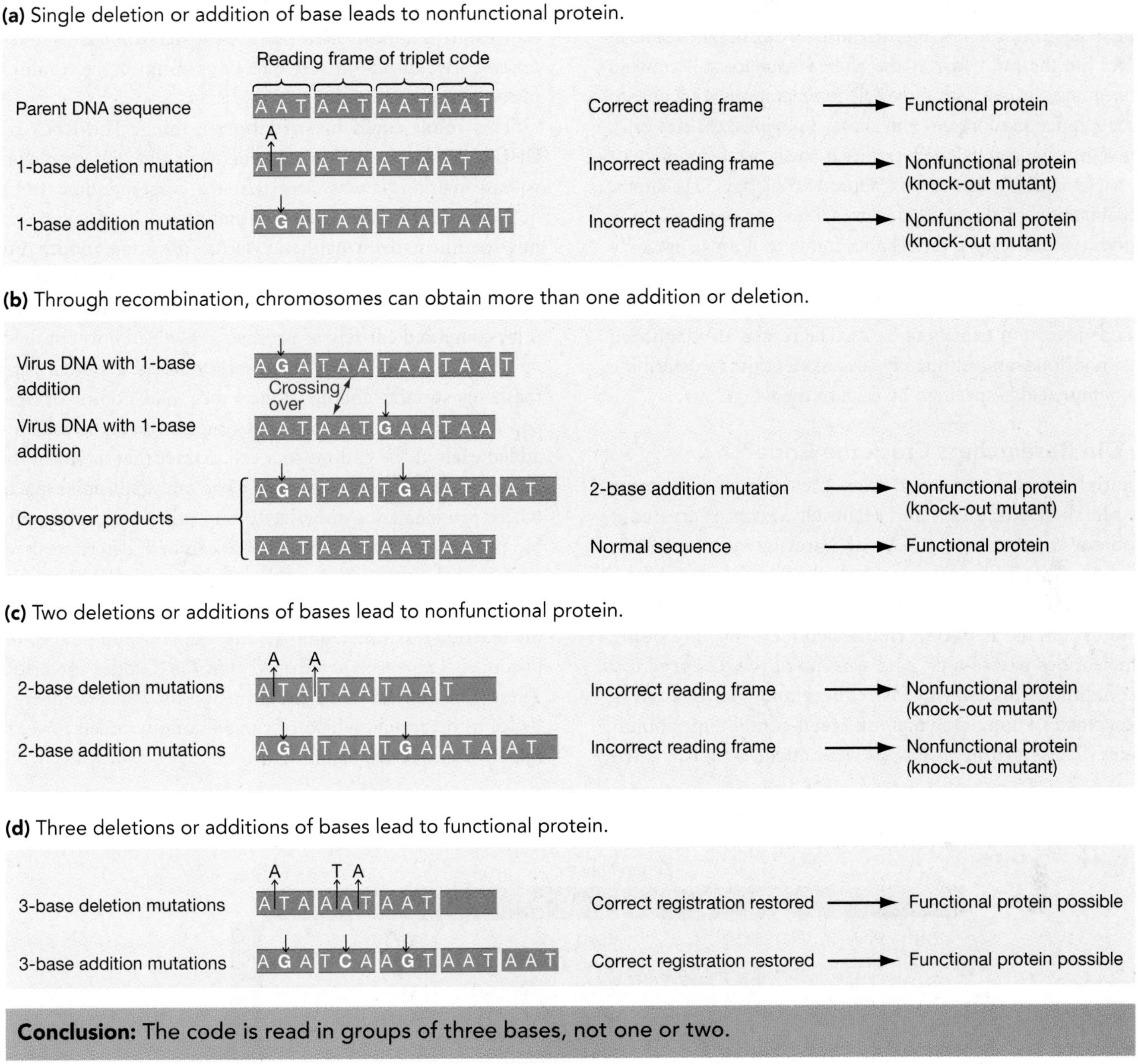

FIGURE 15.7 The Experiments That Confirmed the Triplet Code
(a) If a deletion or addition mutation (indicated by an arrow) occurs, the reading frame of a DNA sequence goes out of register. **(b)** When crossing over occurs between different addition or deletion mutants, the recombinant DNA that results may have multiple addition or deletion mutations or none at all. **(c)** DNA sequences with two addition or deletion mutations are out of register and do not produce a functional protein. **(d)** Three deletion mutations, or three addition mutations, bring the reading frame back into register.

mutation—a complete loss of function. Crick and his colleagues could collect these types of mutants by identifying viruses that were unable to infect *E. coli* cells successfully.

It's important to recognize, though, that the addition or deletion of a single base would also disrupt a code with a reading frame that was two bases long or four bases long. To test the three-base hypothesis rigorously, Crick and co-workers allowed T4 viruses with different single-base-pair additions or deletions to undergo crossing over and recombination. Some of the crossover products contained two 1-base addition mutations (thus, a 2-base addition, **Figure 15.7b**), or two 1-base deletion mutations. These viruses were still unable to infect *E. coli* cells as well as normal viruses could, however (**Figure 15.7c**). This result suggested that the reading frame of the gene involved was still disrupted, leading to a knock-out mutation and nonfunctional protein. From this result, the biologists concluded that the reading frame of the genetic code was not two bases long.

In contrast, viruses carrying two deletion mutations could be allowed to recombine with viruses carrying a single deletion, resulting in a virus chromosome with three deletion mutations. In the sentence "The fat cat ate the rat," the combination of removing one letter from each of the first three words might result in "Tha tca ate the rat." Just as the altered sentence still conveys some meaning, viruses with three deletion mutations were able to produce a functional protein and successfully infect *E. coli* cells. The biologists interpreted these results as strong evidence in favor of the triplet code hypothesis. As **Figure 15.7d** (page 331) shows, the combination of three deletion mutations, or three addition mutations, would restore the reading frame to normal, possibly resulting in a protein with a functional amino acid sequence.

These experiments provided persuasive evidence that the genetic code is read in triplets of bases. The results also launched a long, laborious, and ultimately successful effort to determine which amino acid is specified by each of the 64 codons.

How Did Researchers Crack the Code?

The initial advance in deciphering the genetic code came when Marshall Nirenberg and Heinrich Matthaei created a method for synthesizing RNAs of known sequence. Their method was based on an enzyme called *polynucleotide phosphorylase*, which randomly catalyzes the formation of phosphodiester bonds between ribonucleotides. By providing ribonucleotides whose only base was uracil (U) to a reaction mix containing this enzyme, Nirenberg and Matthaei were able to create a long polymer of uracil-containing ribonucleotides. These synthetic RNAs were added to an in vitro system for synthesizing proteins. The researchers analyzed the resulting amino acid chain and determined that it was *polyphenylalanine*—a polymer consisting of the amino acid phenylalanine.

This result could mean only one thing: The RNA triplet UUU codes for the amino acid phenylalanine. By complementary base pairing, it was clear that the corresponding DNA sequence would be AAA. This initial observation was followed by experiments with synthetic RNAs consisting of only A or C. RNAs consisting of only AAAAA... produced polypeptides consisting of only lysine; poly-C RNAs produced polypeptides composed entirely of proline.

Nirenberg and Philip Leder later devised a system for synthesizing specific codons. Once they had copies of specific codons, they performed a series of experiments in which they added each of the codons to a cell extract that included the 20 different amino acids, ribosomes (the multimolecular machines where proteins are synthesized), and other molecules required for protein synthesis. Then the researchers determined which amino acid was bound to the ribosomes when a particular codon was present. For example, when the codon CAC was in the reaction mix, the amino acid histidine would bind to the ribosomes. This result confirmed that CAC codes for histidine. These ribosome-binding experiments allowed Nirenberg and Leder to determine which of the 64 codons coded for each of the 20 amino acids (**Figure 15.8**).

Second base

First base	U	C	A	G	Third base
U	UUU, UUC: Phenylalanine	UCU, UCC, UCA, UCG: Serine	UAU, UAC: Tyrosine	UGU, UGC: Cysteine	U, C
	UUA, UUG: Leucine		UAA Stop codon; UAG Stop codon	UGA Stop codon; UGG Tryptophan	A, G
C	CUU, CUC, CUA, CUG: Leucine	CCU, CCC, CCA, CCG: Proline	CAU, CAC: Histidine	CGU, CGC, CGA, CGG: Arginine	U, C
			CAA, CAG: Glutamine		A, G
A	AUU, AUC, AUA: Isoleucine	ACU, ACC, ACA, ACG: Threonine	AAU, AAC: Asparagine	AGU, AGC: Serine	U, C
	AUG: Methionine (start codon)		AAA, AAG: Lysine	AGA, AGG: Arginine	A, G
G	GUU, GUC, GUA, GUG: Valine	GCU, GCC, GCA, GCG: Alanine	GAU, GAC: Aspartic acid	GGU, GGC, GGA, GGG: Glycine	U, C
			GAA, GAG: Glutamic acid		A, G

FIGURE 15.8 The Genetic Code

To read the code, match the first base in an mRNA codon, in the red band on the left side, with the second base in a codon, in the blue band along the top, and the third base in a codon, in the green band on the right side. The 64 codons, along with the amino acid or stop (termination) signal that they specify, are given in the boxes. By convention, codons are always written in the $5' \rightarrow 3'$ direction. **EXERCISE** Pick four codons at random. Next to each of them, write the three bases in DNA that specify the codon. (Remember to write the DNA sequence in the $5' \rightarrow 3'$ direction.)

In addition to matching codons to amino acids, researchers discovered that certain codons are punctuation marks that signal "start of message" or "end of message." These codons relay information that the protein chain is complete or that protein synthesis should start at a given codon. There is one **start codon** (AUG), which codes for the amino acid methionine, to signal that protein synthesis should start at that point on the mRNA molecule. There are three **stop codons** (UAA, UAG, and UGA), also known as *termination codons*, which signal that the protein is complete.

The deciphering of the full genetic code, presented in Figure 15.8 is a tremendous achievement. It represents more than five years of work by several teams of researchers. As predicted, the genetic code is redundant. All amino acids except methionine and tryptophan are coded by more than one codon. The code is also unambiguous, however, because one codon never codes for more than one amino acid. Later work showed that the genetic code is also nearly universal: With a few minor exceptions, all codons specify the same amino acids in all organisms. **Box 15.1** explores the evolutionary implications of the universal code.

BOX 15.1 The Evolution of the Genetic Code

Figure 15.9 is a simplified version of the tree of life that was introduced in Chapter 1. Recall that the three largest branches on this tree represent what biologists call the three domains of life: the Bacteria, the Archaea, and the Eukarya. Smaller branches on the tree represent groups of species within each of the three domains, such as the cyanobacteria, α-proteobacteria, land plants, fungi, and animals.

Virtually all of the organisms on this tree use the genetic code shown in Figure 15.8. Exceptions are few. For example, in the single-celled eukaryotes *Tetrahymena* and *Paramecium*, UAA and UAG code for glutamine instead of stop; and in the yeast *Candida cylindracea*, CUG codes for serine instead of leucine. A few species initiate translation at other codons as well as AUG.

To explain why the genetic code is nearly universal, biologists hypothesize that the common ancestor of all species living today used this same code. This common ancestor is indicated at the base of the tree in Figure 15.9. The logic here is that if all species living today use the same genetic code, then it is reasonable to infer that their common ancestor did also.

Given that the genetic code has been in existence for a long time, the question arises, Is the code arbitrary? That is, is the genetic code random with respect to the ability of organisms to survive and produce offspring? A brief examination of Figure 15.8 suggests that the answer is no. Note that when several codons specify the same amino acid, the first two bases in those codons are almost always identical. This observation suggests that the code does not represent a random assemblage of bases, like letters drawn from a hat. In fact, extensive analysis has shown that, compared with randomly generated codes, the existing genetic code is structured in a way that efficiently minimizes the phenotypic effects of small changes in DNA and errors during translation. This result suggests that the genetic code has been honed by natural selection. Stated another way, it is likely that other genetic codes may have existed early in the history of life. But because the code in Figure 15.8 is particularly efficient, it emerged as the most successful version.

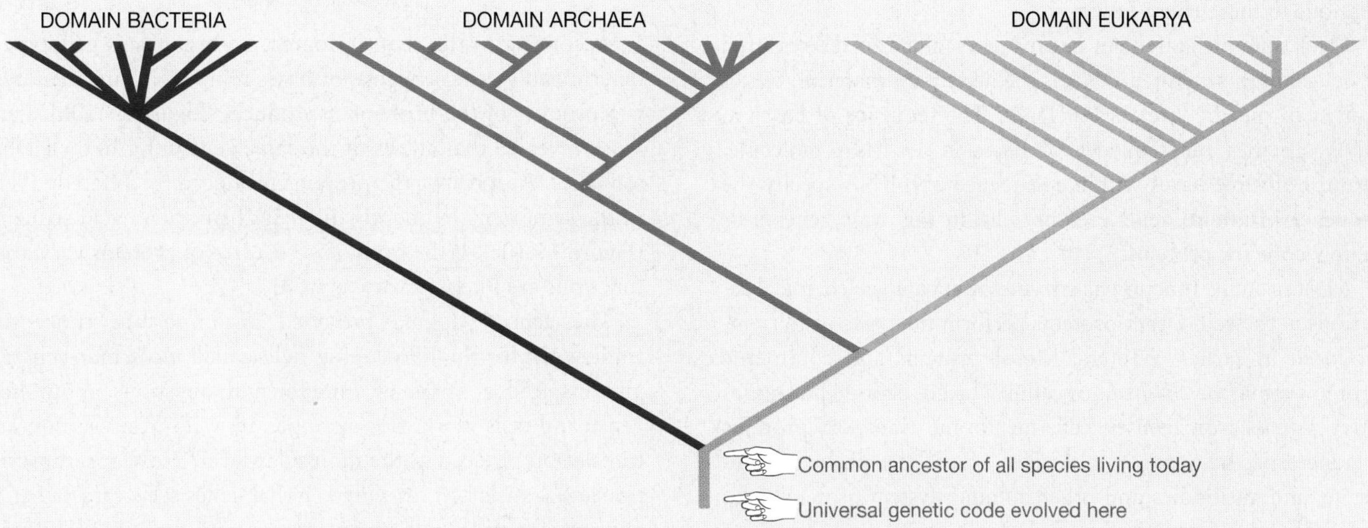

FIGURE 15.9 The Genetic Code Is Universal

✓CHECK YOUR UNDERSTANDING

Genes code for proteins, but they do so indirectly. The sequence of bases in DNA is used to produce messenger RNA (mRNA). The sequence of bases in an mRNA molecule is complementary to the DNA sequence of a gene. The sequence of bases in mRNA forms a code in which particular combinations of three bases lead to the addition of a specific amino acid to the protein encoded by the gene. This code is redundant because 64 combinations of three bases occur, but only 20 amino acids and a stop "punctuation mark"—indicating the end of the coding sequence—need to be specified. You should be able to (1) draw a DNA double helix, (2) choose one of the two strands of the double helixes the coding strand, (3) write the sequence of the mRNA molecule that would be synthesized (in the 5′ → 3′ direction!) from that strand, and (4) write the amino acid sequence of the protein that would be produced by that mRNA.

15.3 The Central Dogma of Molecular Biology

Even before the genetic code was fully revealed, Francis Crick articulated what became known as the central dogma of molecular biology. The **central dogma** summarizes the flow of information in cells. It simply states that DNA codes for RNA, which codes for proteins:

$$\text{DNA} \rightarrow \text{RNA} \rightarrow \text{proteins}$$

Besides clarifying information flow in the cell, the central dogma points out that information flow occurs in only one direction—from DNA to RNA to proteins. RNA does not code for the production of other RNA or DNA, and proteins do not code for the production of RNA or DNA or other proteins. (The essay at the end of this chapter points out important exceptions to these rules, however.)

Crick's simple statement encapsulates much of the research reviewed in this chapter. DNA is the hereditary material. Genes consist of specific stretches of DNA. The sequence of bases in DNA specifies the sequence of bases in an RNA molecule. Groups of three bases within a sequence of mRNA specify the sequence of amino acids in a protein. In this way, genes ultimately code for proteins.

Most proteins function as enzymes that catalyze chemical reactions in the cell. Other proteins perform the types of roles introduced in earlier chapters: Motor proteins and contractile proteins move the cell itself or cellular cargo, peptide hormones carry signals from cell to cell, membrane transport proteins conduct specific ions or molecules across the plasma membrane, and antibodies and other immune system proteins provide defense by recognizing and destroying invading viruses and bacteria.

In addition, some genes code for RNA molecules that do not function as mRNAs. The various types of RNA molecules that exist have an array of functions in the cell. Most of these functions are related to managing the flow of information from DNA to mRNA to proteins.

Biologists use specialized vocabulary to summarize this sequence of events encapsulated in the central dogma. For example, biologists say that DNA is *transcribed* to RNA. In everyday English, the word *transcription* simply means making a copy of information. The scientific use of the term is appropriate because DNA acts as a permanent record—an archive or blueprint containing the information needed to build and run the cell. This permanent record is copied, via **transcription**, to the short-lived form called mRNA. The information is then transferred to a new molecular form—a sequence of amino acids. In everyday English, the word *translation* refers to the transferring of information from one language to another. In biology, the synthesis of protein from mRNA is called **translation**. Translation is the transferring of information from one type of molecule to another—from the "language" of nucleic acids to the "language" of proteins. **Figure 15.10a** summarizes the relationship between transcription and translation, as well as the relationship among DNA, RNA, and proteins. In diagrammatic form,

DNA
(information storage)

↓ transcription

mRNA
(information carrier)

↓ translation

Proteins
(active cell machinery)

According to the central dogma, an organism's genotype is determined by the sequence of bases in its DNA; its phenotype is a product of the proteins it produces (**Figure 15.10b**). Later work revealed that alleles of the same gene differ in their DNA sequence. As a result, the proteins produced by different alleles of the same gene frequently differ in their amino acid sequence (**Figure 15.10c**). If the primary structures of proteins vary, their functions are likely to vary as well.

The central dogma provided an important conceptual framework for the burgeoning field called *molecular genetics*, and inspired a series of fundamental questions about how genes and cells work. For example, how do transcription and translation proceed at the molecular level? How are transcription and translation regulated so that genes are expressed at an appropriate time and in an appropriate amount? The next three chapters are devoted to exploring these questions.

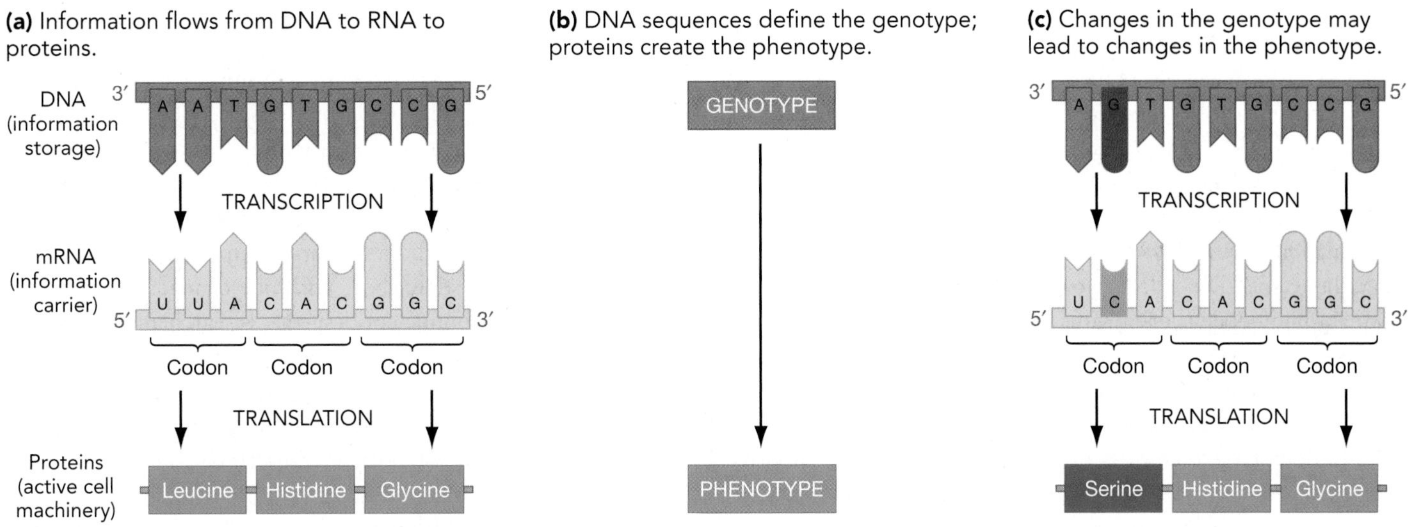
(a) Information flows from DNA to RNA to proteins.
DNA (information storage)
3′ A A T G T G C C G 5′
TRANSCRIPTION
mRNA (information carrier)
5′ U U A C A C G G C 3′
Codon Codon Codon
TRANSLATION
Proteins (active cell machinery)
Leucine Histidine Glycine
(b) DNA sequences define the genotype; proteins create the phenotype.
GENOTYPE
PHENOTYPE
(c) Changes in the genotype may lead to changes in the phenotype.
3′ A G T G T G C C G 5′
TRANSCRIPTION
5′ U C A C A C G G C 3′
Codon Codon Codon
TRANSLATION
Serine Histidine Glycine

FIGURE 15.10 The Flow of Information in the Cell
(a) DNA functions as the information archive in the cell, analogous to a set of blueprints. mRNA is a short-lived copy of the information in DNA, used to manufacture proteins. Most of the machinery and many of the structures in cells consist of proteins. **(b)** The central dogma revealed the connection between genotype and phenotype. **(c)** Alleles of genes differ in their DNA base sequence. As a result, they may produce proteins with different primary structures and thus different functions. In this way, different alleles may be associated with different phenotypes.

ESSAY Exceptions to the Central Dogma

The old adage "Rules are made to be broken" applies to biology. In 1955, just three years after DNA had been confirmed as the hereditary material, Heinz Fraenkel-Conrat discovered that the virus he was studying (the *tobacco mosaic virus*, or TMV; **Figure 15.11**) does not contain DNA. The genes of this virus are made of RNA instead. Fraenkel-Conrat showed that when TMV infects a cell of a tobacco plant, the viral genes are translated directly into viral proteins. Thus, in RNA viruses such as TMV, the flow of information is simply RNA → proteins. Over the years since Fraenkel-Conrat's work was published, researchers have identified hundreds of viruses with RNA genomes. These include the viruses that cause polio, hepatitis, flu, and the common cold in humans.

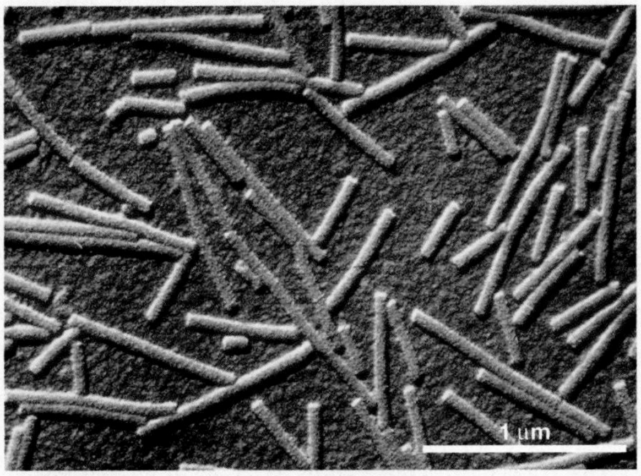
1 µm

FIGURE 15.11 Tobacco Mosaic Virus
The genes inside this virus consist of RNA.

With respect to breaking rules, though, an even bigger bombshell exploded in 1970.

With respect to breaking rules, though, an even bigger bombshell exploded in 1970. That year David Baltimore, Howard Temin, and Satoshi Mizutani announced the discovery of RNA viruses that violate the central dogma's tenet of one-way information flow. The genomes of these viruses include sequences that code for an enzyme called **reverse transcriptase.** This protein catalyzes the formation of DNA from an RNA molecule that acts as a template—with deoxyribonucleotides being matched to the complementary

(Continued on next page)

(Essay continued)

bases in RNA. When one of these viruses infects a host cell, reverse transcriptase "reverse transcribes" the viral RNA genome to DNA. This viral DNA usually incorporates itself into the host's chromosome. Messenger RNAs are then transcribed from the viral DNA and translated into proteins. In these viruses, the direction of information flow is RNA → DNA → RNA → proteins. As a result, they came to be called **retroviruses** ("backward-viruses"). The human immunodeficiency virus (HIV)—the virus that causes AIDS—is a retrovirus.

The experiments leading to this discovery were elegant. To confirm the existence of reverse transcriptase, Temin and Mizutani purified large quantities of the *Rous sarcoma virus* (a virus associated with cancer formation in chickens). Then they disrupted the lipid-containing envelope that surrounds the virus by treating the particles with a detergent. This step released the contents of the virus particles into solution. The researchers added deoxyribonucleotides labeled with a radioactive isotope and then incubated the mixture. Later they observed the formation of radioactive DNA molecules in the mixture. This observation was a smoking gun—strong evidence that the virus contains a molecule capable of synthesizing DNA from an RNA template.

Temin and Mizutani did a follow-up experiment. It was the same experiment just described, with one change: Along with the labeled deoxyribonucleotides, the researchers added ribonuclease, which breaks RNA molecules apart. In this treatment, no DNA formed. This result was consistent with the prediction that the enzyme requires an RNA template.

With the discovery of reverse transcriptase and the retroviruses, biologists could no longer be quite so dogmatic about the central dogma. Although all *cells* conform to the central dogma, some viruses do not.

CHAPTER REVIEW

Summary of Key Concepts

- **Most genes code for enzymes that catalyze specific chemical reactions in the cell.**

The question "What do genes do?" was solved in a series of experiments on the bread mold *Neurospora crassa*. Researchers isolated mutants that cannot grow unless they are supplied with a specific amino acid and then showed that different mutants were unable to make different chemical precursors of the amino acid. The one-gene, one-enzyme hypothesis explains these results by proposing that most genes code for proteins and that most proteins act as enzymes that catalyze specific reactions in metabolic pathways.

Web Tutorial 15.1 The One-Gene, One-Enzyme Hypothesis

- **In cells, information flows from DNA to RNA to proteins. DNA is transcribed to messenger RNA by RNA polymerase, and then messenger RNA is translated to proteins in ribosomes. Each amino acid in a protein is specified by a group of three bases in RNA.**

Experiments confirmed that DNA does not code for proteins directly. Instead, mRNA molecules are transcribed from DNA and then translated into proteins. One-way flow of information, from DNA to RNA to proteins, is called the central dogma of molecular biology. By synthesizing RNAs of known base composition and then observing how they functioned during the process of translation, researchers were able to unravel the genetic code. It is now established that the code is read in triplets and that the code is "redundant"—meaning most of the 20 amino acids are specified by more than one codon.

Web Tutorial 15.2 The Triplet Nature of the Genetic Code

Questions

Content Review

1. What is metabolism?
   a. the chemical reactions that occur in organisms
   b. the synthesis of macromolecules through spark discharges and other reactions during chemical evolution
   c. the in vitro synthesis of macromolecules in a test tube
   d. carefully controlled chemical reactions that occur in an industrial setting

2. What does the one-gene, one-enzyme hypothesis state?
   a. Genes are composed of stretches of DNA.
   b. Genes are made of protein.
   c. Genes code for ribozymes.
   d. A single gene codes for a single protein.

3. DNA's primary structure is made up of just four different bases, and its secondary structure is regular and highly stable. How can a molecule with these characteristics hold all of the information required to build and maintain a cell?
   a. The information is first transcribed, then translated.
   b. The messenger RNA produced from DNA has much more complex primary and secondary structures, and thus holds much information.
   c. A protein produced (indirectly) from DNA has much more complex primary and secondary structures, and thus holds much information.
   d. The information in DNA is in code form.

4. Why did researchers suspect that DNA does not code for proteins directly?
   a. In eukaryotes, DNA is found inside the nucleus, but proteins are produced outside the nucleus.
   b. In prokaryotes, DNA and proteins are never found together.
   c. When DNA was damaged by ultraviolet radiation or other sources of energy, the proteins in the cell did not change accordingly.
   d. There are several distinct types of RNA, of which only one functions as messenger RNA.
5. Which of the following describes an important experimental strategy in deciphering the genetic code?
   a. comparing the amino acid sequences of proteins with the base sequence of their genes
   b. analyzing the sequence of RNAs produced from known genes
   c. analyzing mutants that changed the code
   d. examining the proteins produced when RNAs of known sequence were translated
6. Recessive alleles often contain a knock-out mutation. When a normal allele and a knock-out allele are paired in a heterozygous individual, the normal copy of the gene is frequently able to produce enough functional protein to give a normal phenotype. Why are these facts important?
   a. They explain why some mutants are still able to infect bacterial cells efficiently.
   b. They explain the phenomenon of dominance and recessiveness discovered by Mendel.
   c. They offer a serious challenge to the one-gene, one-enzyme hypothesis.
   d. They illustrate why the central dogma is considered central.

Conceptual Review

1. DNA is referred to as an "information-storage molecule." How can the base sequence of DNA store information?
2. Draw a hypothetical metabolic pathway involving a sequence of five substrates, five enzymes, and a product called BiolSciazine. Number the substrates 1–5, and label the enzymes A–E, in order. (For instance, enzyme A applies to the reaction between substrates 1 and 2.)
   - Suppose a mutation made the gene for enzyme C nonfunctional. What molecule would accumulate in the affected cells?
   - Suppose some individuals with a mutation affecting this metabolic pathway can survive if given substrate 5 in the diet. But they die even if given substrates 1, 2, 3, and 4. State a hypothesis for which enzyme in the pathway is affected by this mutation.
3. Why did experiments with *Neurospora crassa* mutants support the one-gene, one-enzyme hypothesis?
4. When researchers discovered that a combination of three deletion mutations or three addition mutations would restore the function of a gene, most biologists were convinced that the genetic code was read in triplets. Explain the logic behind this conclusion.
5. Why is the genetic code redundant?
6. Explain how a single-base deletion disrupts the reading frame of a gene.

Group Discussion Problems

1. Examine the genetic code as illustrated in Figure 15.8. Note that when several codons specify the same amino acid, the first two bases in those codons are almost always identical. Does this observation support the hypothesis that the genetic code is arbitrary? That is, does the code represent a random assemblage of bases, like letters drawn from a hat, or does it have distinct patterns? (If patterns exist, you could hypothesize that the code is structured in a way that helps transcription or translation occur more efficiently.) How would you test the hypothesis that the code is indeed random?
2. Recall that DNA and RNA are synthesized only in the $5' \rightarrow 3'$ direction and that DNA and RNA sequences are always written in the $5' \rightarrow 3'$ direction. Consider the following DNA sequence:

   5′ T T G A A A T G C C C G T T T G G A G A T C G G G T T A C A G C T A G T C A A A G 3′
   3′ A A C T T T A C G G G C A A A C C A C T A G C C C A A T G T C G A T C A G T T T C 5′

   - Identify the bases in the bottom strand that code for start and stop codons. Write the mRNA sequence that would be transcribed between them if the bottom strand served as the template.
   - Write the amino acid sequence that would be translated from the mRNA sequence you just wrote.
3. Scientists say that a phenomenon is a "black box" if they can describe it and study its effects but don't yet know the underlying mechanism that causes it. In what sense was genetics—meaning the transmission of heritable traits—a black box before the central dogma of molecular biology was understood?
4. One of the possibilities that researchers interested in the genetic code had to consider was that the code was overlapping, meaning that a single base could be part of more than one codon. Make a diagram showing how an overlapping code would work, assuming that each codon is three bases long.

Answers to Multiple-Choice Questions 1. a; 2. d; 3. d; 4. a; 5. d; 6. b

www.prenhall.com/freeman is your resource for the following: Web Tutorials; Online Quizzes and other Online Study Guide materials; Answers to Conceptual Review Questions; Solutions to Group Discussion Problems; Answers to Figure Caption Questions and Exercises; and Additional Readings and Research.

16 Transcription and Translation

KEY CONCEPTS

- After the enzyme RNA polymerase binds to a specific site in DNA with the help of other proteins, it catalyzes the production of an RNA molecule. The base sequence of the RNA produced is complementary to the base sequence of the DNA template strand.
- In eukaryotes, genes consist of coding regions called exons interspersed with noncoding regions called introns. After a eukaryotic mRNA is produced, introns must be spliced out. The ends of the molecule receive a cap and tail.
- Inside ribosomes, mRNAs are translated to proteins via intermediary molecules called transfer RNAs. Transfer RNAs carry an amino acid and have a three-base-pair anticodon, which binds to an mRNA codon. The amino acid carried by the transfer RNA is then added to the growing protein via formation of a peptide bond.
- Many mutations produce changes in the phenotype.

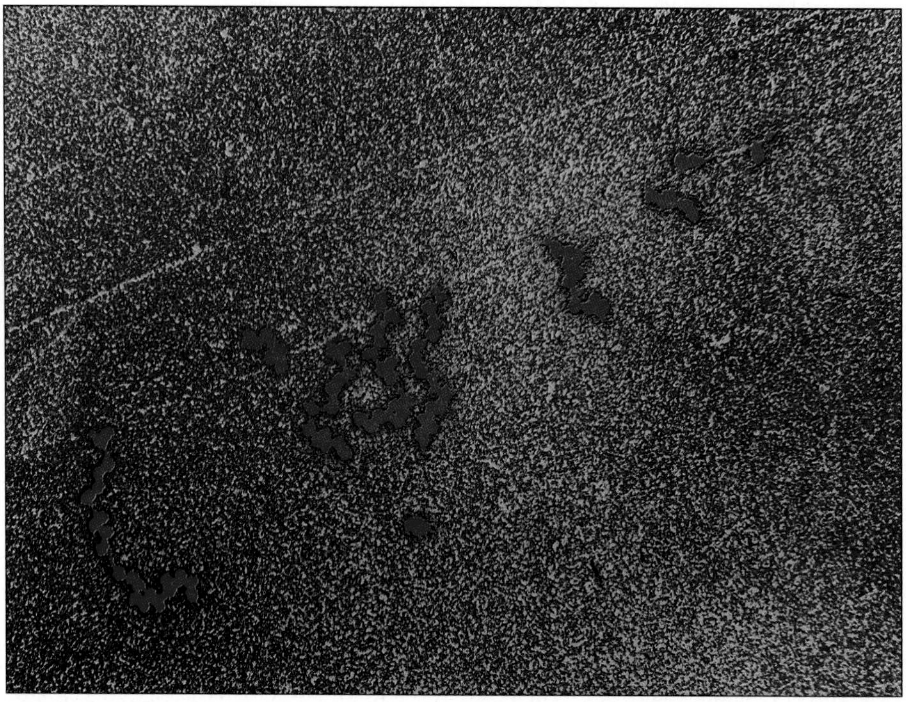

A micrograph showing transcription and translation occurring simultaneously in a bacterial cell. The yellow lines are strands of DNA, and the red structures are ribosomes.

Proteins are the stuff of life. They give cells their shape and control the chemical reactions that go on inside them. No one knows exactly how many different proteins can be made in each of your body's cells, but 100,000 is a reasonable guess. While some of those 100,000 different proteins may not be manufactured at all in certain types of cells, other proteins may be present in quantities ranging from millions of copies to fewer than a dozen. If even one of those types of protein is defective, the cell may sicken or even die.

A cell builds the proteins it needs from instructions encoded in its genome. Chapter 15 explored the basic strategy that cells use to retrieve these instructions and convert them into action in the form of proteins. The first step in the process is the transcription of a gene and the production of a messenger RNA, or mRNA. The RNA message is a short-lived copy of the archived instructions in DNA. The sequence of ribonucleotides in the mRNA is then translated into a sequence of amino acids. This series of events defines the flow of information in the cell:

DNA
(information storage)

↓ transcription

mRNA
(information carrier)

↓ translation

Proteins

The discovery of the relationship between DNA and proteins was one of the great scientific advances of the twentieth century. But once the relationship between transcription and translation had been firmly established, biologists puzzled over how cells actually accomplish the feat of converting genetic information into protein. How do transcription and translation work? For example, how does RNA polymerase know where to start transcribing a gene and where to end? Once an RNA message is produced, how is the linear sequence of ribonucleotides translated into the linear sequence of amino acids in a protein?

This chapter explores the experiments that answered these questions. In addition to providing a deeper understanding of how cells work, the results summarized here have had a significant impact on biomedicine and drug development. Many of the antibiotics prescribed today work by disrupting translation in bacteria that cause disease. In addition, research on transcription and translation set the stage for an explosion of work, introduced in Chapters 17 and 18, on how the expression of specific genes is regulated. Most genes are transcribed and translated only at certain times or in certain types of cells.

Let's begin by examining research on the mechanisms of transcription in bacteria and eukaryotes, then delve into the mechanics of translation. The chapter concludes by examining how changes in DNA can lead to changes in protein structure and function.

16.1 Transcription in Bacteria

The first step in converting genetic information into proteins is to synthesize a messenger RNA version of the instructions archived in DNA. The enzyme RNA polymerase, introduced in Chapters 14 and 15, is responsible for synthesizing mRNA. Like the DNA polymerases introduced in Chapter 14, RNA polymerase performs a template-directed synthesis in the $5' \rightarrow 3'$ direction. (Recall that $5'$ and $3'$ refer to carbons on the sugar subunits of ribonucleotides.) But unlike DNA polymerases, RNA polymerase does not require a primer to begin transcription.

Transcription occurs when RNA polymerase matches the base in a ribonucleotide triphosphate with the complementary base in a gene—a section of DNA. Once a matching ribonucleotide is in place, RNA polymerase catalyzes the formation of a phosphodiester bond between the $3'$ end of the growing mRNA chain and the new ribonucleotide. As this matching-and-catalysis process continues, an RNA that is complementary to the gene is synthesized (**Figure 16.1**). Note that only one of the two DNA strands is used as a template and transcribed, or "read," by RNA polymerase. The strand that is read by the enzyme is called the **template strand**; the other strand is called the **non-template strand.** Because RNA has uracil (U) rather than thymine (T), an adenine (A) in the DNA template strand specifies a U in the complementary RNA strand.

Biologists reached these conclusions by studying RNA synthesis in cell-free, or in vitro, systems. Once the basic chemical

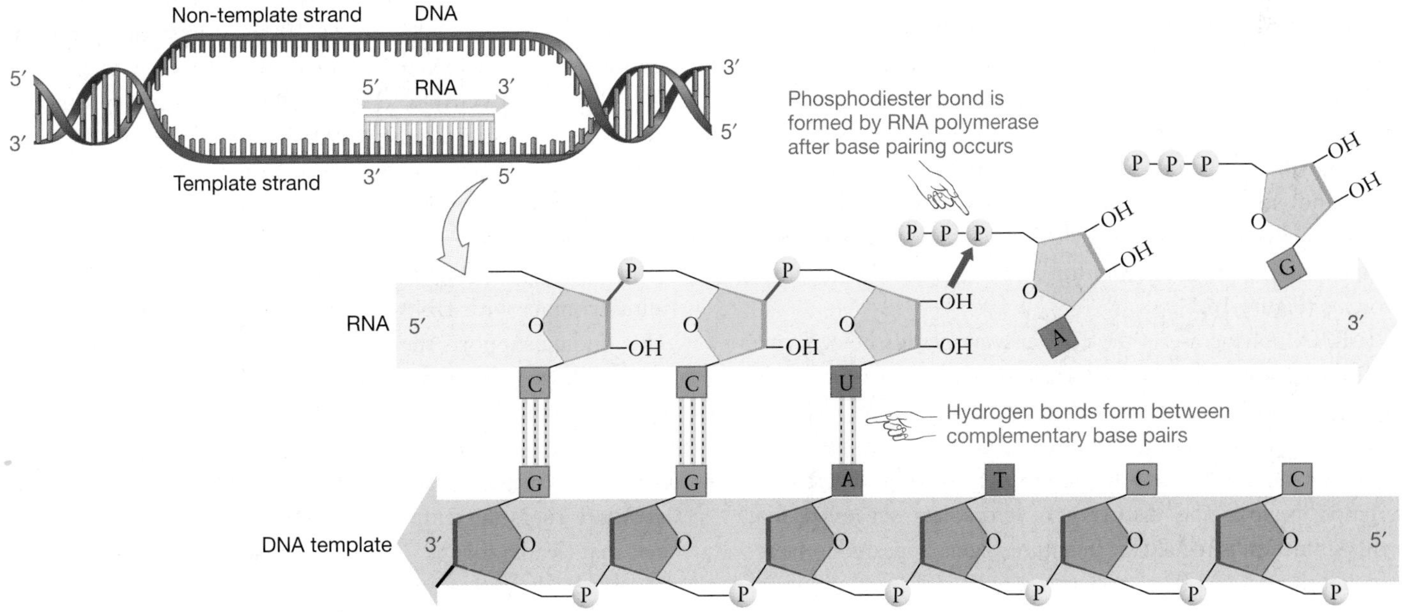

FIGURE 16.1 Transcription Is the Synthesis of RNA from a DNA Template
The reaction catalyzed by RNA polymerase results in the formation of a phosphodiester bond between ribonucleotides. RNA polymerase produces an RNA strand whose sequence is complementary to the bases in the DNA template. QUESTION In which direction is RNA synthesized, $5' \rightarrow 3'$ or $3' \rightarrow 5'$? In which direction is the DNA template "read"?

reaction was understood, an entirely new set of questions arose:

- What does RNA polymerase look like?
- Are any other proteins or factors involved in transcription, or does RNA polymerase act alone?
- How does the enzyme know where to start transcription on the DNA template?

RNA Polymerase Structure and Function

To understand the structure of RNA polymerase, researchers have employed the technique called X-ray crystallography introduced in Chapter 4. Recall that this procedure allows biologists to obtain information about the three-dimensional structure of large, complex molecules.

The most recent models of bacterial RNA polymerase indicate that the enzyme is large and globular and has several prominent channels that run through the interior. The enzyme's active site, where phosphodiester bonds form, is located where several of these channels intersect. The active site contains a strategically placed magnesium atom (Mg^{2+}), which plays a critical role in stabilizing the transition state as the polymerization reaction takes place.

How does this structure correlate with the enzyme's function? It is logical to predict that DNA fits into one of the enzyme's channels and that the two strands in the double helix separate inside the enzyme to expose a single-stranded template at the active site. If so, how do the enzyme and DNA come together?

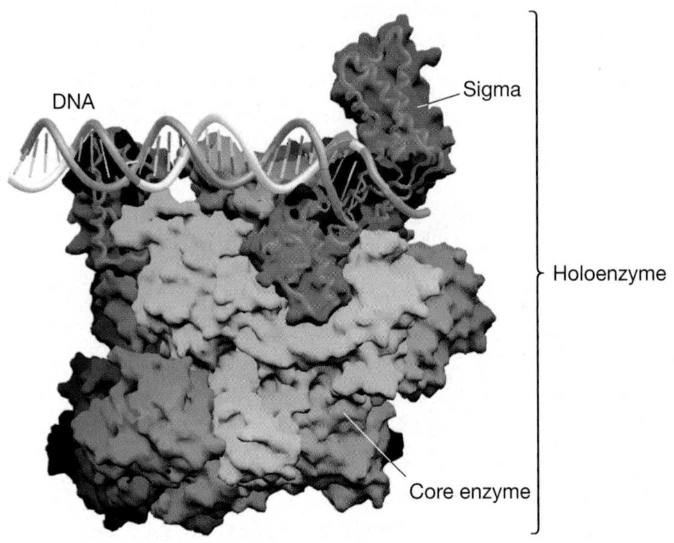

FIGURE 16.2 The Three-Dimensional Structure of RNA Polymerase Bacterial RNA polymerase is a globular enzyme. Sigma is a regulatory component, and the core enzyme contains the active site.

Initiation: How Does Transcription Begin in Bacteria?

WEB TUTORIAL 16.1 Transcription

Soon after the discovery of RNA polymerase, researchers realized that the enzyme cannot initiate transcription on its own. Instead, a detachable protein subunit called **sigma** must bind to RNA polymerase before transcription can begin. RNA polymerase and sigma form what biologists call a **holoenzyme** ("whole enzyme"). A holoenzyme consists of a **core enzyme**, which contains the active site for catalysis, and other required proteins (**Figure 16.2**).

If RNA polymerase is the core enzyme of this holoenzyme, what does sigma do? When researchers mixed RNA polymerase, sigma, and DNA together, they found that the holoenzyme bound tightly to specific sections of DNA. These binding sites were named **promoters**, because they are sections of DNA where transcription begins. The discovery of promoters suggested that sigma's function is regulatory in nature. Sigma appeared to be responsible for guiding RNA polymerase to specific locations where transcription should begin. What is the nature of these specific locations? What do promoters look like, and what do they do?

David Pribnow offered an initial answer to these questions in the mid-1970s. When Pribnow analyzed the base sequence of promoters from a variety of bacteria and from viruses that infect bacteria, he found that the promoters are all 40–50 base pairs long and that a particular section in each promoter looked similar. This similar segment of DNA had a series of bases identical or similar to TATAAT. This six-base-pair sequence is now known as the **−10 box**, because it occurs about 10 bases from the point where RNA polymerase starts transcription. (DNA that is located in the direction that RNA polymerase moves during transcription is said to be **downstream**; DNA in the opposite direction is said to be **upstream**. Thus, the −10 box is 10 bases upstream from the transcription start site.) Soon after, researchers recognized that the sequence TTGACA occurred in these same promoters about 35 bases upstream from the start of the mRNA. This second key sequence is called the **−35 box.**

Follow-up work showed that transcription begins when sigma binds to the −35 and −10 boxes (**Figure 16.3**, step 1). This is a key point: Sigma, and not RNA polymerase, makes the initial contact with DNA that starts transcription in bacteria. This finding supports the hypothesis that sigma is a regulatory protein. Sigma tells RNA polymerase where and when to start synthesizing RNA.

Once sigma binds to a promoter, the DNA helix opens and creates two strands of single-stranded DNA. As shown in step 2 of Figure 16.3, the template strand is threaded through a channel that leads to the active site inside RNA polymerase. Monomers known as *ribonucleoside triphosphates*, or NTPs, enter a channel at the bottom of the enzyme and diffuse to the active site. When an incoming NTP pairs with a complementary base on the template strand of DNA, RNA polymerization begins. The reaction catalyzed by RNA polymerase is exergonic and spontaneous because NTPs have so much potential energy, owing to their three phosphate groups. As step 3 of Figure 16.3

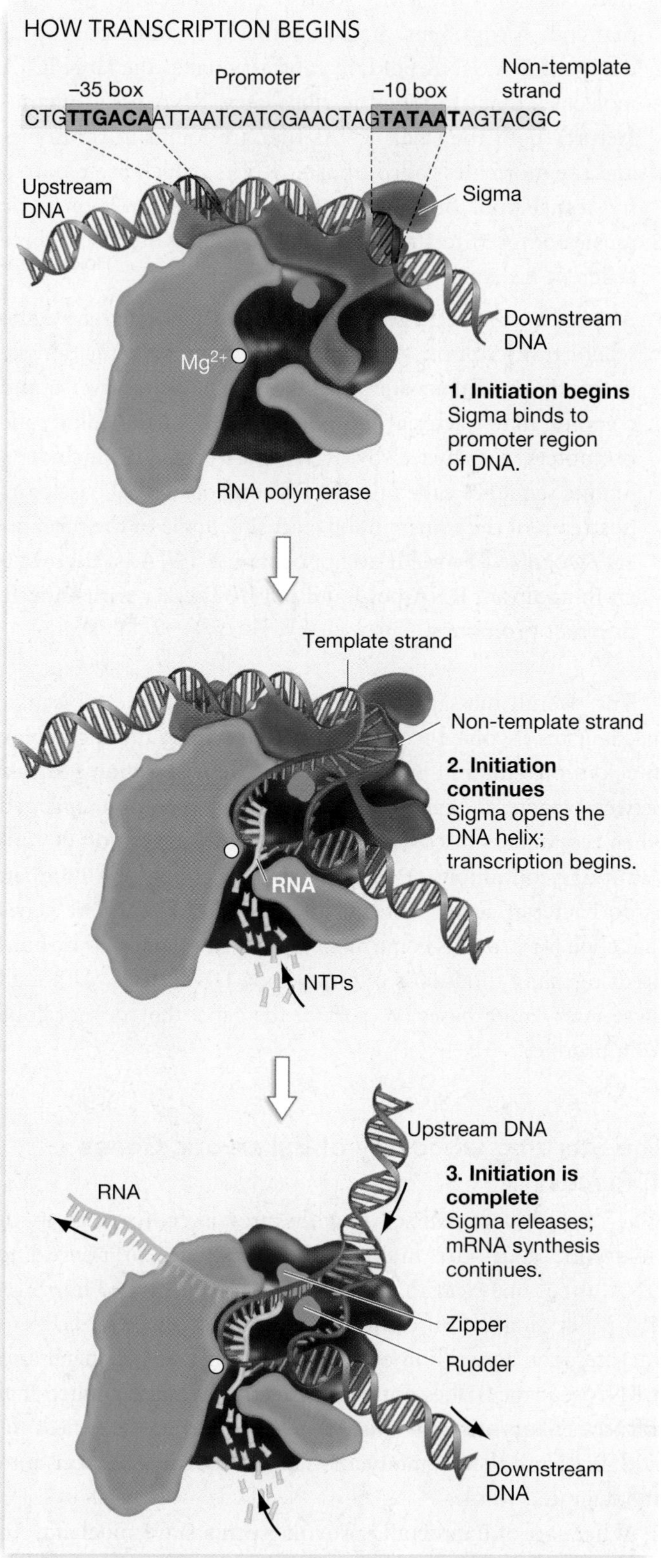

FIGURE 16.3 In Bacteria, Sigma Plays a Key Role in Initiating Transcription
When sigma attaches to RNA polymerase, the holoenzyme is able to bind to the −35 box and the −10 box of a promoter. Once it has bound to a promoter, sigma opens the DNA helix and threads the template strand through the enzyme's active site. When transcription is under way, sigma disengages from RNA polymerase. EXERCISE Circle and label RNA polymerase's active site.

shows, sigma is released once RNA synthesis is under way. The **initiation phase** of transcription is complete.

Recent work has shown that most bacteria have several types of sigma proteins, each with a distinctive structure and function. *Escherichia coli* has seven different sigma proteins, for example, while *Streptomyces coelicolour* has more than 60. Each of these proteins binds to promoters with slightly different DNA base sequences. Although all promoters have a −10 box and a −35 box, the sequences within each box vary somewhat. These variations in DNA sequences affect the ability of different sigma proteins to bind. As a result, each type of sigma protein allows RNA polymerase to bind to a different type of gene. The identity of the sigma protein in the RNA polymerase holoenzyme determines which types of genes will be transcribed. Thus by controlling which sigma proteins are active, bacterial cells control which genes are expressed.

Elongation and Termination

During the **elongation phase** of transcription, RNA polymerase moves along the DNA template in the $3' \rightarrow 5'$ direction of the template strand, synthesizing RNA in the $5' \rightarrow 3'$ direction. In the interior of the enzyme, a group of projecting amino acids called the enzyme's *zipper* helps open the double helix at the upstream end, and a nearby group of amino acids called the *rudder* helps steer the template and non-template strands through channels inside the enzyme. Meanwhile, the enzyme's active site catalyzes the addition of nucleotides to the $3'$ end of the growing RNA molecule at the rate of about 50 nucleotides per second.

Note that during the elongation phase of transcription, three prominent channels or grooves in the enzyme are filled. Double-stranded DNA goes into and out of one groove; ribonucleoside triphosphates enter another; and the growing RNA strand exits to the rear. In this way, the enzyme's structure correlates closely with its function.

Transcription ends with a **termination phase**. In most cases, transcription stops when RNA polymerase reaches a stretch of DNA sequence that functions as a **transcription termination signal**. The bases that make up the termination signal code for a stretch of RNA with an unusual property: As soon as it is synthesized, the RNA sequence folds back on itself and forms a short double helix that is held together by complementary base pairing. The secondary structure that results is called a *hairpin* (see Chapter 4). The formation of the hairpin structure is thought to disrupt the interaction between RNA polymerase and the RNA transcript, resulting in the physical separation of the enzyme and its product.

To summarize, transcription begins when sigma binds to the promoter at the start of a gene. Once binding occurs, RNA polymerase begins to synthesize mRNA. Transcription ends when a termination signal at the end of the gene leads to the formation of a hairpin in the mRNA, disrupting the transcription complex.

16.2 Transcription in Eukaryotes

The data reviewed in Section 16.1 established how transcription occurs in bacteria. How do these results compare with research on the mechanism of transcription in eukaryotes? What similarities and differences have come to light?

One of the first conclusions about transcription in eukaryotes is that it is similar to bacterial transcription in an important respect: RNA polymerase does not bind directly to promoter sequences by itself. Instead, proteins that came to be called **basal transcription factors** initiate eukaryotic transcription by matching the enzyme with the appropriate promoter region in DNA. The function of basal transcription factors observed in eukaryotes is analogous to the function of the sigma proteins in bacteria.

The discovery of basal transcription factors emerged from studies with cell-free, or in vitro, systems. The basic approach was to purify RNA polymerase from human cells, add template DNA, and analyze the RNAs that were produced. But in early experiments, researchers found that RNA polymerase began copying at random locations on the template DNA instead of specifically at promoter regions. In addition, when RNA polymerase was alone, both strands of the DNA were transcribed instead of just one. This observation inspired the hypothesis that eukaryotic cells must contain sigma-like proteins that are required for normal transcription. To test this hypothesis, biologists added proteins isolated from human cells to the cell-free reaction system. By adding one or a few proteins at a time and recording which ones enabled RNA polymerase to bind to promoters and correctly transcribe the template strand, researchers were gradually able to characterize the suite of proteins that came to be called basal transcription factors.

Follow-up work confirmed several important distinctions about how transcription works in bacteria and eukaryotes:

- In bacteria a single sigma protein binds to a promoter and initiates transcription, but in eukaryotes many basal transcription factors are required to initiate transcription.
- Eukaryotes have three distinct types of RNA polymerase (abbreviated RNA pol) instead of just one. As **Table 16.1** shows, RNA pol I, pol II, and pol III each transcribe a discrete class of RNA. Only RNA pol II transcribes the genes that code for proteins. Stated another way, RNA pol II produces mRNA. RNA pol I, in contrast, makes the large RNA molecules that are found in ribosomes. RNA pol III manufactures both the small RNAs that are found in ribosomes and the molecules called transfer RNAs, which are required for translation. **Box 16.1** explains how research on toxic mushrooms led to the discovery that eukaryotes have three different RNA polymerases.
- Although eukaryotic genomes contain promoters that signal where transcription should begin, just as bacteria do, the promoters in eukaryotic DNA are much more diverse and complex than bacterial promoters. Many of the eukaryotic promoters recognized by RNA polymerase II include a unique sequence called the **TATA box**, located 30 base pairs upstream of the transcription start site. Some of the promoters recognized by pol II do not contain a TATA box, however. In addition, RNA pol I and pol III interact with entirely different promoters.

TABLE 16.1 Eukaryotic RNA Polymerases

Name of Enzyme	Type of Gene Transcribed
RNA polymerase I (RNA pol I)	Genes that code for most of the large RNA molecules (rRNAs) found in ribosomes (see Section 16.3)
RNA polymerase II (RNA pol II)	Protein-coding genes (produce mRNAs)
RNA polymerase III (RNA pol III)	Genes that code for transfer RNAs (see Section 16.4), and genes that code for one of the small RNA molecules (rRNAs) found in ribosomes
RNA pol II and RNA pol III	RNA molecules found in snRNPs (see Section 16.2)

The overall message of this research is that the molecular mechanisms involved in transcription are much more complex in eukaryotes than in bacteria. An even more striking contrast between bacterial and eukaryotic gene expression emerged when researchers discovered that eukaryotic genes do not consist of one continuous DNA sequence that codes for a product, as do bacterial genes. Instead, the regions of eukaryotic genes that code for proteins come in pieces that are separated by hundreds or many thousands of intervening DNA bases. Although these intervening bases are part of the gene, they do not code for a product.

The Startling Discovery of Eukaryotic Genes in Pieces

When researchers realized that the protein-coding regions of eukaryotic genes are interrupted by stretches of noncoding DNA, it became clear that the nature of the gene and information processing are different in bacteria and eukaryotes. In eukaryotes, the conversion of information in DNA sequences to mRNA sequences does not occur directly. To make a functional mRNA, eukaryotic cells must dispose of certain sequences inside genes and then combine the separated coding sections into an integrated whole.

What sort of data would provoke such a startling claim? To answer this question, consider work that Phillip Sharp and colleagues carried out in the late 1970s to determine how DNA templates are transcribed. They began one of their experiments by heating DNA molecules enough to break the hydrogen bonds between complementary bases. This treatment separated the two strands. The single-stranded DNA was then incubated with the mRNA encoded by the sequence. The team's idea was to promote base pairing between the mRNA and the single-

BOX 16.1 Toxins and Transcription

Although wild mushrooms can be delicious, dining on certain species may be lethal. For example, in late summer and fall the reproductive structures of the death cap mushroom (*Amanita phalloides*) appear in forests, fields, and backyards in North America and Europe (**Figure 16.4**). The death cap is among the most poisonous of all mushrooms. Worldwide, it is thought to account for 90 percent of mushroom poisoning deaths. Eating a single death cap can kill a healthy adult.

FIGURE 16.4 The Death Cap Mushroom These mushrooms contain the toxin α-amanitin, which at low concentrations poisons the eukaryotic enzyme RNA polymerase II. QUESTION State a hypothesis to explain why the presence of α-amanitin increases the fitness of death cap mushrooms.

The active agent in death caps is a toxin called α-amanitin. By happenstance, researchers who were investigating α-amanitin's mode of action made a crucial discovery about transcription in eukaryotes. The researchers observed that mRNA synthesis stopped when eukaryotic cells were treated with low concentrations of α-amanitin. The production of other types of RNAs—specifically, transfer RNAs (tRNAs) and ribosomal RNAs (rRNAs)—continued, however. At very high concentrations of the toxin, tRNA production also stopped but transcription of rRNA genes continued. These observations eventually led to the discovery that eukaryotes have three RNA polymerases, each with a distinct role in the transcription of mRNA, tRNA, and rRNA. Low concentrations of α-amanitin inhibit RNA polymerase II, which synthesizes mRNAs. As a result, low concentrations of the toxin halt protein synthesis. High concentrations of this poison block RNA polymerase III, which synthesizes tRNAs. RNA polymerase I is not affected by α-amanitin at any concentration.

Toxins such as α-amanitin are useful in research on transcription and translation for the same reason that knock-out mutations, which completely disable genes, are useful in genetics. Toxins allow researchers to understand what happens when a molecule or a particular stage in a process does *not* work. This is similar, in concept, to learning how a car works by removing a part or a system and carefully recording the machine's response; or to learning how the brain works by studying the behavior of people whose brains have been damaged in specific, quantifiable ways by accidents. Toxins and knock-out mutations provide a molecular dissecting kit. They let biologists explore the intricacies of cell function by inactivating key components.

---

stranded DNA. Under these conditions, the research team expected the mRNA to form base pairs with the DNA sequences that act as the template for its synthesis. When the researchers examined the DNA-RNA hybrid molecules with the electron microscope, however, they observed the structure shown in **Figure 16.5a**. Instead of matching up exactly, parts of the DNA formed loops. As **Figure 16.5b** shows, Sharp and his coworkers interpreted these loops as stretches of nucleotides that are present in the DNA template strand but are *not* in the corresponding mRNA.

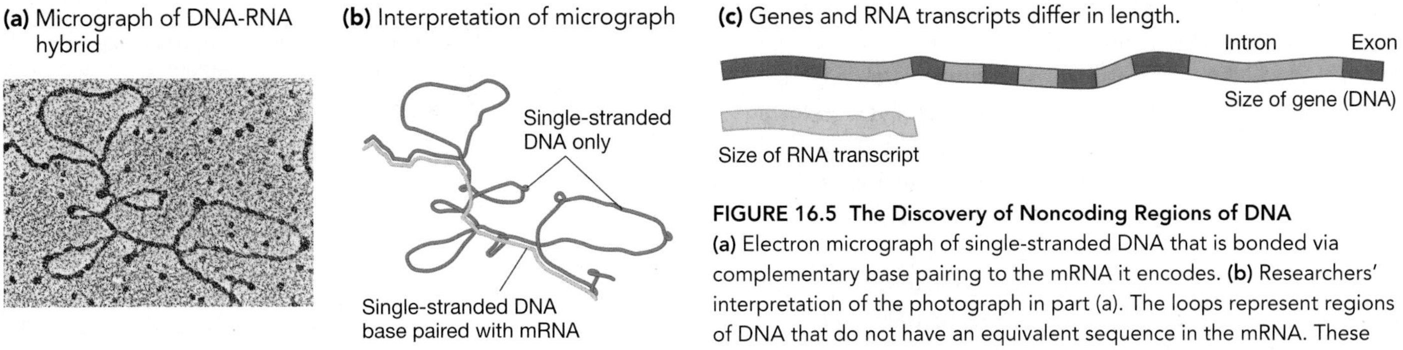
(a) Micrograph of DNA-RNA hybrid
(b) Interpretation of micrograph
Single-stranded DNA only
Single-stranded DNA base paired with mRNA
(c) Genes and RNA transcripts differ in length.
Intron
Exon
Size of gene (DNA)
Size of RNA transcript

FIGURE 16.5 The Discovery of Noncoding Regions of DNA (a) Electron micrograph of single-stranded DNA that is bonded via complementary base pairing to the mRNA it encodes. (b) Researchers' interpretation of the photograph in part (a). The loops represent regions of DNA that do not have an equivalent sequence in the mRNA. These intervening regions are "extra" DNA compared with the sequences in the mRNA. (c) Because eukaryotic genes contain both introns and exons, these genes are much larger than the corresponding mature mRNA.

To make sense of these and other findings, Sharp's group and a team headed by Richard Roberts proposed that there is *not* a one-to-one correspondence between the nucleotide sequence of a eukaryotic gene and its mRNA. By analogy, eukaryotic genes do not carry messages such as "Biology is my favorite course of all time." Instead, eukaryotic genes carry messages that read something like "Biol*τηεπροτεινχοδινγρεγιονσοφγενεσ*ogy is my favor *αρειντερρυπτεδβψνονχοδινγΔ*-ite course of*ανδηαω̄ετοβεσπλιχεδτογετηερ*all time." The sections of noncoding sequence must be removed from the mRNA before it can carry an intelligible message to the translation machinery.

When it became clear that the genes-in-pieces hypothesis was correct, Walter Gilbert suggested that regions of eukaryotic genes that are part of the final mRNA be referred to as **exons** (because they are *ex*pressed) and the untranslated stretches as **introns** (because they are *int*ervening). Introns are sections of genes that are not represented in the final mRNA product. As a result, eukaryotic genes are much larger than their corresponding mature RNA transcripts (**Figure 16.5c**, page 343).

Exons, Introns, and RNA Splicing

The transcription of eukaryotic genes by RNA polymerase generates a **primary RNA transcript** that contains both the exon and intron regions (**Figure 16.6a**). As transcription proceeds, the introns are removed from the growing mRNA strand by a process known as **splicing**. In this phase of information processing, pieces of the primary transcript are removed and the remaining segments are joined together. Splicing occurs while transcription is still under way and results in an mRNA that contains an uninterrupted genetic message.

Figure 16.6b provides more detail about how introns are removed from genes. In the case illustrated here, splicing is catalyzed by a complex of proteins and small RNAs known as small nuclear ribonucleoproteins, or **snRNPs** (pronounced "snurps"). The process begins when snRNPs bind to specific sequences on the primary RNA transcript that define the exon-intron boundaries. Once the initial snRNPs are in place, other snRNPs arrive to form a multipart complex called a **spliceosome**. Recent data have shown that the spliceosomes found in human cells contain about 145 different proteins and RNAs, making them the most complex molecular machines known. Once the spliceosome forms, the intron forms a loop with an adenine ribonucleotide at its base (step 1, Figure 16.6b). The loop breaks when the 2′ hydroxyl group on the adenine attacks a phosphodiester bond at the 5′ end of the intron.

Once the initial cut in the RNA is made, the free 5′ end of the intron becomes attached to the adenine nucleotide and forms a "lariat" of RNA (step 2, Figure 16.6b). Then the free 3′ end of the first exon reacts with the 5′ end of the second exon. This reaction

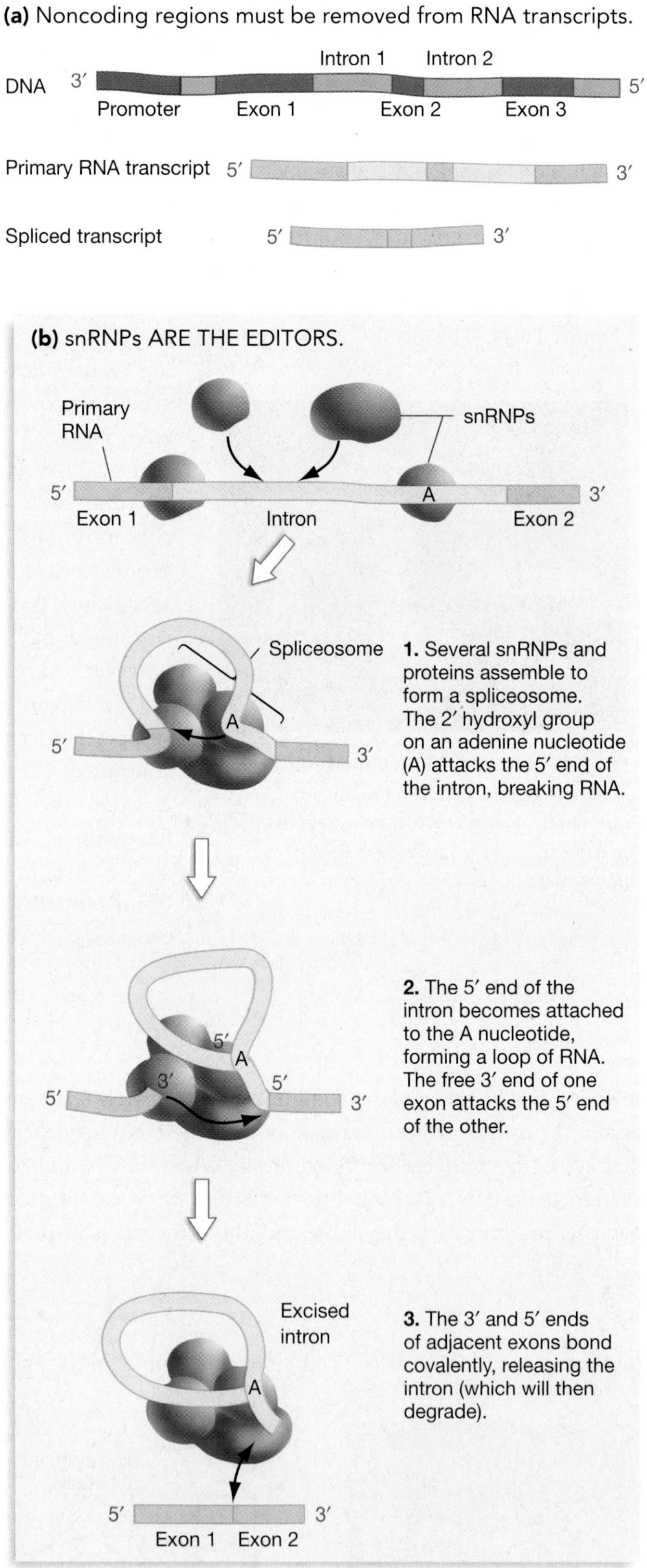
(a) Noncoding regions must be removed from RNA transcripts.
Intron 1
Intron 2
DNA
3′
5′
Promoter
Exon 1
Exon 2
Exon 3
Primary RNA transcript
5′
3′
Spliced transcript
5′
3′
(b) snRNPs ARE THE EDITORS.
Primary RNA
snRNPs
5′
A
3′
Exon 1
Intron
Exon 2
Spliceosome
1. Several snRNPs and proteins assemble to form a spliceosome. The 2′ hydroxyl group on an adenine nucleotide (A) attacks the 5′ end of the intron, breaking RNA.
2. The 5′ end of the intron becomes attached to the A nucleotide, forming a loop of RNA. The free 3′ end of one exon attacks the 5′ end of the other.
Excised intron
3. The 3′ and 5′ ends of adjacent exons bond covalently, releasing the intron (which will then degrade).
Exon 1
Exon 2

FIGURE 16.6 Introns Are Spliced Out of the Original mRNA
(a) RNA polymerase produces a primary RNA transcript that contains exons and introns. Subsequent processing generates a mature transcript. **(b)** snRNPs are responsible for the splicing reactions that take place in the nucleus of eukaryotes.

FIGURE 16.7 In Eukaryotes, mRNAs Are Given a Cap and a Tail Eukaryotic mRNAs have a cap consisting of a molecule called 7-methylguanylate (symbolized as m^7G) bonded to three phosphate groups; the tail is made up of a long series of adenine residues.

breaks the 3′ end of the intron and covalently joins the two exons into a contiguous coding sequence (step 3, Figure 16.6b). Splicing is now complete. In most cases, the excised intron is degraded to ribonucleotide monophosphates.

Current data suggest that both the cutting and rejoining reactions that occur during splicing are catalyzed by RNA molecules in the spliceosome. Recall from Chapter 4 that RNA molecules with catalytic ability are called ribozymes. In eukaryotes, ribozymes play a key role in the production of proteins.

Other Aspects of Transcript Processing: Caps and Tails

Intron splicing is not the only way that primary mRNA transcripts are processed in eukaryotes. As soon as the 5′ end of a eukaryotic mRNA emerges from RNA polymerase, enzymes add a structure called the **5′ cap** (**Figure 16.7**). The cap consists of the molecule 7-methylguanylate and three phosphate groups. When transcription is complete, an enzyme cleaves the 3′ end of most mRNAs, and other enzymes add a long tract of 100–250 adenine nucleotides. This sequence is known as the **poly (A) tail.**

Not long after the caps and tails on eukaryotic mRNAs were described, evidence began to accumulate that they protect mRNAs from degradation by ribonucleases and that they enhance the efficiency of translation. For example, experimental mRNAs that have a cap and a tail last longer when they are introduced into cells than do experimental mRNAs that lack a cap, a tail, or both a cap and a tail. The mRNAs with caps and tails also produce more proteins than do mRNAs without caps and tails.

Recent work has shown why the cap and tail are so critical to mRNAs. The 5′ cap serves as a recognition signal for the translation machinery, and the poly (A) tail extends the life span of an mRNA by protecting the message from degradation by ribonucleases in the cytoplasm.

✓CHECK YOUR UNDERSTANDING

Transcription occurs when RNA polymerase catalyzes the 5′ → 3′ synthesis of an RNA molecule. RNA synthesis is driven by the potential energy stored in nucleotide triphosphates and is based on matching complementary base pairs to the sequence in a template DNA strand. **Table 16.2** summarizes some of the similarities and differences observed in transcription in bacteria versus eukaryotes. You should be able to diagram (1) how transcription initiation, elongation, and termination occur in bacteria and eukaryotes and (2) how mRNAs are processed in eukaryotes.

16.3 An Introduction to Translation

In translation, the sequence of bases in a messenger RNA molecule is converted into a sequence of amino acids in a polypeptide. The genetic code specifies the relationship between the bases of a triplet codon in mRNA and the amino acid it codes for (see Chapter 15). But how are the amino acids assembled into a polypeptide according to the information in messenger RNA?

Studying how translation occurred in cell-free systems proved to be an extremely productive approach to answering this question. Once in vitro translation systems had been developed from human cells, *E. coli*, and a variety of other organisms, it became clear that the basic mechanisms of translation are fundamentally the same throughout the tree of life. The sequence of events that occurs during protein synthesis is similar in bacteria and eukaryotes.

TABLE 16.2 Comparing Transcription in Bacteria and Eukaryotes

Aspect	Bacteria	Eukaryotes
RNA polymerase	One	Three; each produces a different class of RNA
Promoter structure	Typically contains a −35 box and a −10 box	Complex and variable; often includes a TATA box −30 from start of gene
Protein(s) involved in contacting promoter	Sigma; different versions of sigma bind to different promoters	Many basal transcription factors
RNA processing steps	None; translation occurs while transcription is still under way	Extensive; several processing steps occur in nucleus before RNA is exported to cytoplasm for translation: 1. Enzyme-catalyzed addition of 5′ cap 2. Splicing (intron removal) by spliceosome 3. Enzyme-catalyzed addition of 3′ poly (A) tail

(a) In bacteria, transcription and translation are tightly coupled.

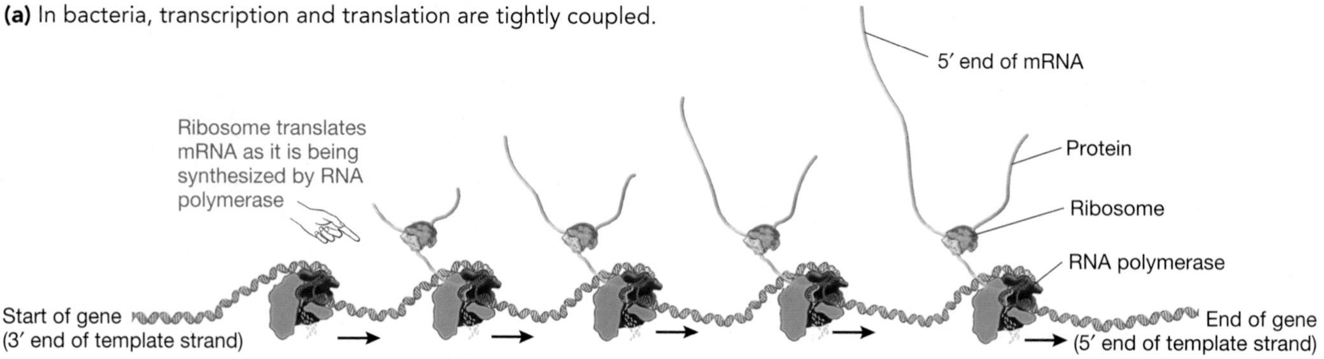
5′ end of mRNA
Ribosome translates mRNA as it is being synthesized by RNA polymerase
Protein
Ribosome
RNA polymerase
Start of gene
(3′ end of template strand)
End of gene
(5′ end of template strand)

(b) In eukaryotes, transcription and translation are separated in space and time.

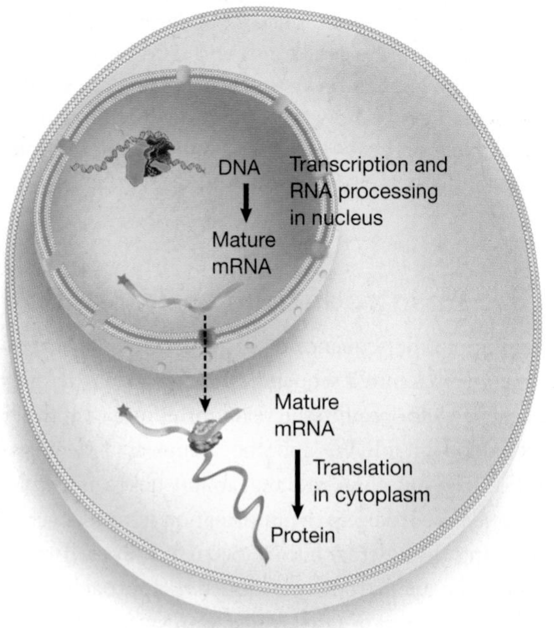
DNA
Transcription and RNA processing in nucleus
Mature mRNA
Mature mRNA
Translation in cytoplasm
Protein

FIGURE 16.8 Transcription and Translation Occur Simultaneously in Bacteria but Not in Eukaryotes
(a) In bacteria, ribosomes attach to mRNA transcripts and begin translation while RNA polymerase is still transcribing the DNA template strand. **(b)** In eukaryotes, mRNA transcripts are processed inside the nucleus before they are exported to the cytoplasm. There, they are translated by ribosomes. EXERCISE In bacteria, many ribosomes attach to an mRNA transcript at the same time—so many copies of the protein product are made simultaneously. Add two ribosomes to the longest mRNA transcript in part (a), along with the protein each is producing.

Ribosomes Are the Site of Protein Synthesis

The first question that biologists answered about translation concerned where it occurs. The answer was inspired by a simple observation: There is a strong, positive correlation between the number of small cellular structures known as **ribosomes** in a given type of cell and the rate at which that cell synthesizes proteins. For example, immature human red blood cells divide rapidly, synthesize millions of copies of the protein hemoglobin, and contain large numbers of ribosomes. In contrast, the same cells at maturity have low rates of protein synthesis and very few ribosomes. Based on this correlation, investigators proposed that ribosomes are the site of protein synthesis in the cell.

To test this hypothesis, Roy Britten and collaborators did a pulse-chase experiment similar in design to experiments introduced in Chapter 7. Recall that the goal of a pulse-chase experiment is to label a population of molecules as they are being produced. The location of the tagged molecules is then followed through time. The tagging is done by supplying a pulse of radioactive atoms that are incorporated into the molecule as it is being synthesized, followed by a chase of unlabeled atoms. Britten's group wanted to label amino acids with radioactive atoms and then track the molecules as translation occurred.

The biologists began by feeding a pulse of radioactive sulfate ($^{35}SO_4^{-2}$) to growing cultures of *E. coli*. They expected the cells to incorporate the radioactive sulfur into the amino acids methionine and cysteine—which contain sulfur (see Chapter 3)—and then into newly synthesized proteins. Fifteen seconds after adding the radioactive sulfate, the researchers "chased" the label by adding a large excess of nonradioactive sulfate to the culture medium. The chase vastly reduced the incorporation of radioactive sulfur into proteins. If the ribosome hypothesis was correct, then the radioactive signal should be associated with ribosomes for a short period of time. Later, all of the radioactivity should be found in proteins independently of the ribosomes.

This is exactly what the researchers found. Soon after the pulse of labeled sulfate ended, the radioactive atoms were found in free amino acids or in ribosomes. Later, all of the radioactive atoms were found on completed proteins. Based on these data, biologists concluded that proteins are synthesized at ribosomes and then released.

About a decade after the ribosome hypothesis was confirmed, electron micrographs like this chapter's opening photograph showed bacterial ribosomes in action. The images confirmed that, in bacteria, ribosomes attach to mRNAs and begin synthesizing proteins even before transcription is complete (**Figure 16.8a**). Transcription and translation can occur concurrently in bacteria because there is no nuclear envelope to separate the two processes. In bacteria, transcription and translation are physically connected. But in eukaryotes, mRNAs are spliced in the nucleus and then exported to the cytoplasm. Once RNA messages are outside the nucleus, ribosomes attach to them and begin translation (**Figure 16.8b**). In eukaryotes, transcription and translation are separated in time and space.

How Does an mRNA Triplet Specify an Amino Acid?

During translation, genetic information is converted from nucleic acids to proteins. In effect, the hereditary instructions are translated from one distinctive chemical language into another. The discovery of the genetic code revealed that each triplet codon in mRNA specifies a particular amino acid in a protein. How does this conversion happen?

One early hypothesis was that mRNA codons and amino acids interact directly. The proposal was that the bases in a particular codon were complementary in shape or charge to the side group of a particular amino acid (**Figure 16.9a**). But Francis Crick pointed out that the chemistry involved didn't make sense. For example, how could the nucleic acid bases interact with a hydrophobic amino acid side group, which does not form hydrogen bonds?

Crick proposed an alternative hypothesis. As **Figure 16.9b** shows, he suggested that some sort of adapter molecule holds amino acids in place while interacting directly and specifically with a codon in mRNA via hydrogen bonding. In essence, Crick predicted the existence of a chemical go-between that produced a physical connection between the two types of molecules. Crick, as it turns out, was right.

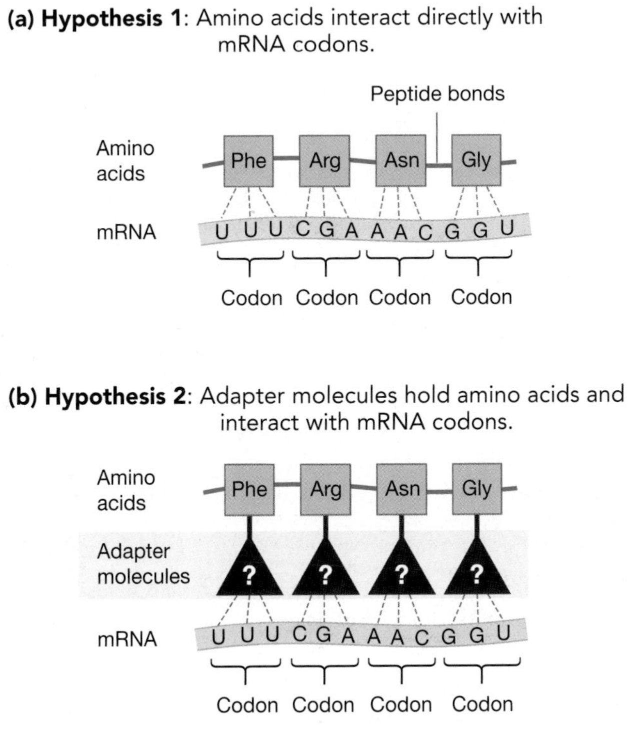

FIGURE 16.9 Two Hypotheses for How mRNA Codons Interact with Amino Acids
(a) An early hypothesis suggested that amino acids and mRNA codons interact directly. **(b)** The adapter-molecule hypothesis predicts that mRNA codons do not interact directly with the amino acids being added to a growing polypeptide. Instead, an intermediary adapter molecule carries an amino acid and aligns it with the appropriate mRNA codon.

16.4 The Role of Transfer RNA

Crick's adapter molecule was discovered by accident. Biologists were trying to work out a cell-free protein synthesis system derived from mammalian liver cells and had discovered that ribosomes, mRNA, amino acids, ATP, and a molecule called guanosine triphosphate, or GTP, had to be present for translation to occur. (GTP is similar to ATP but contains guanosine instead of adenosine.) These results were logical—ribosomes provide the catalytic machinery, mRNAs contribute the message to be translated, amino acids are the building blocks of proteins, and ATP and GTP supply potential energy to drive the endergonic polymerization reactions responsible for forming proteins. But in addition, a cellular fraction that contained a previously unknown type of RNA turned out to be indispensable. If this type of RNA is missing, protein synthesis does not occur. What is this mysterious RNA, and why is it essential to translation?

The novel class of RNAs eventually became known as **transfer RNA (tRNA)**. The role of tRNA in translation was a mystery until some researchers happened to add a radioactive amino acid—leucine—to an in vitro protein synthesis system. The treatment was actually done as a control for an unrelated experiment. To the researchers' amazement, some of the radioactive leucine attached to tRNA molecules. Follow-up experiments showed that the attachment of an amino acid to a tRNA requires an input of energy in the form of ATP. A tRNA molecule that becomes covalently linked to an amino acid, such as any of the molecules observed in the lucky experiment, is called an **aminoacyl tRNA**. More recent research has shown that enzymes called **aminoacyl tRNA synthetases** are responsible for catalyzing the addition of amino acids to tRNAs. For each of the 20 major amino acids, there is a different aminoacyl synthetase and one or more tRNAs. **Figure 16.10**, page 348, summarizes how amino acids are transferred from aminoacyl tRNA synthetases to tRNAs.

What happens to the amino acid that is carried by an aminoacyl tRNA? Biologists who tracked the fate, over time, of radioactive leucine molecules that were attached to tRNAs discovered that the amino acids are transferred from aminoacyl tRNAs to proteins. The data that support this conclusion are

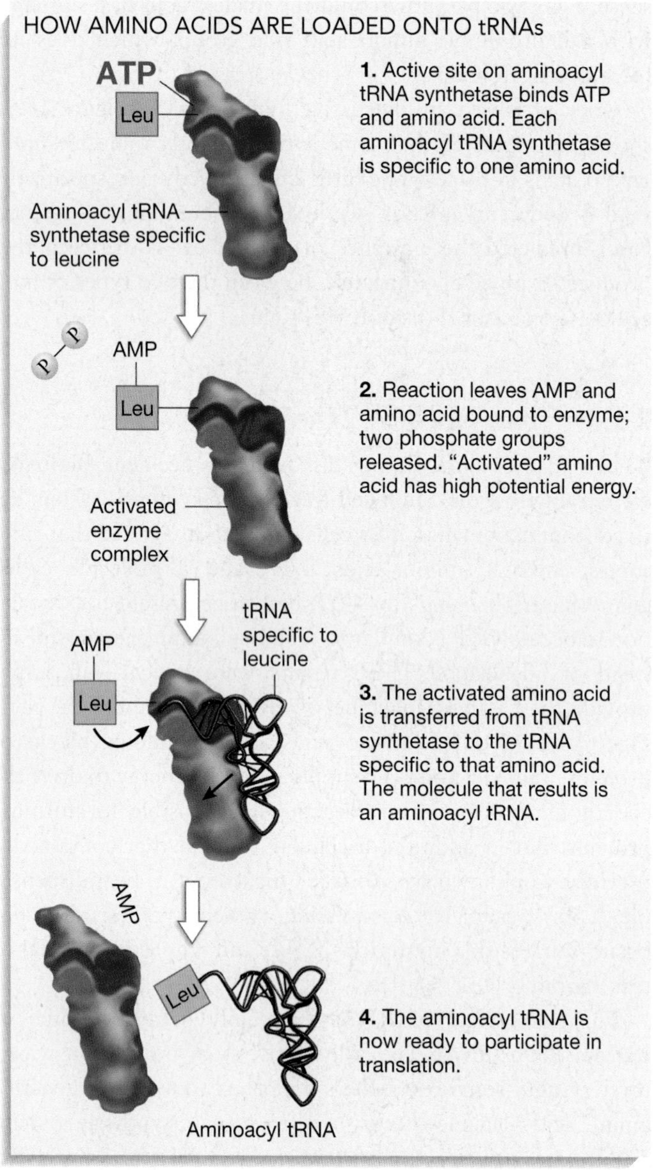
HOW AMINO ACIDS ARE LOADED ONTO tRNAs
ATP
Leu
1. Active site on aminoacyl tRNA synthetase binds ATP and amino acid. Each aminoacyl tRNA synthetase is specific to one amino acid.
Aminoacyl tRNA synthetase specific to leucine
AMP
Leu
2. Reaction leaves AMP and amino acid bound to enzyme; two phosphate groups released. "Activated" amino acid has high potential energy.
Activated enzyme complex
AMP
tRNA specific to leucine
Leu
3. The activated amino acid is transferred from tRNA synthetase to the tRNA specific to that amino acid. The molecule that results is an aminoacyl tRNA.
AMP
Leu
4. The aminoacyl tRNA is now ready to participate in translation.
Aminoacyl tRNA

FIGURE 16.10 Aminoacyl tRNA Synthetases Load Amino Acids onto the Appropriate tRNA
An aminoacyl tRNA synthetase—in this case, for leucine—becomes activated in a reaction that requires ATP. Activated aminoacyl tRNA synthetases transfer an amino acid to a tRNA, producing an aminoacyl tRNA.

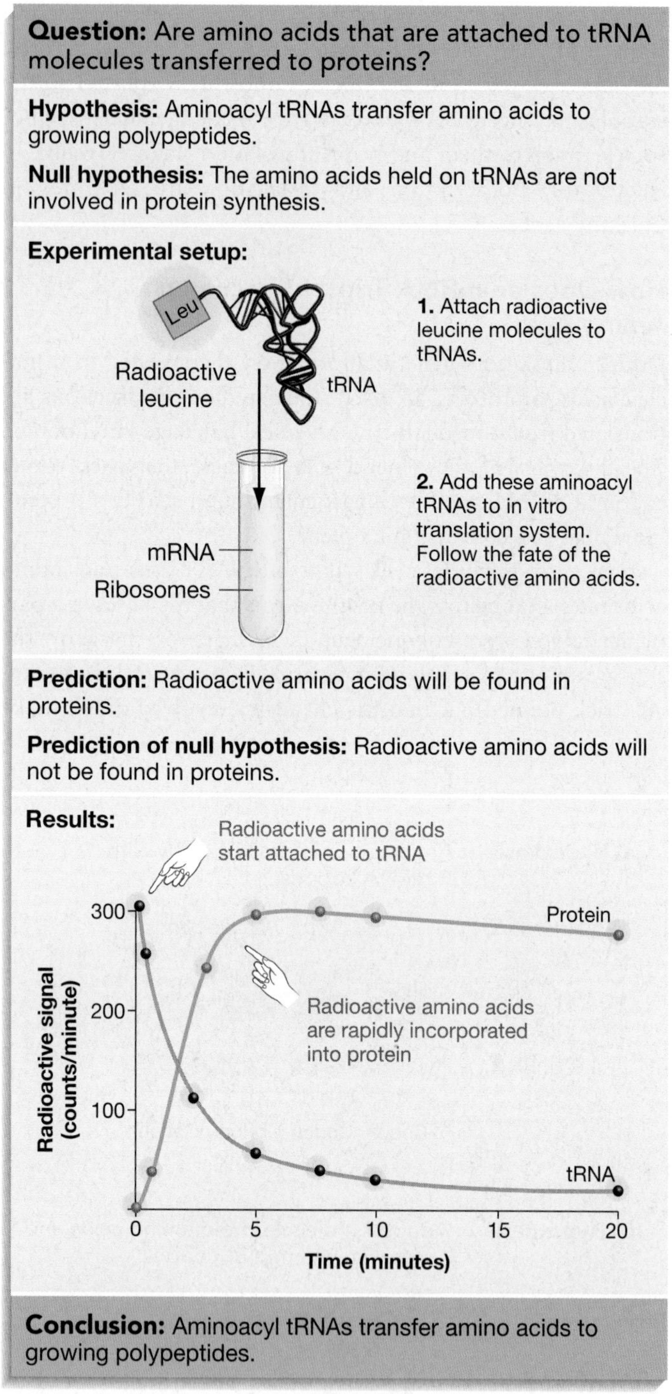
Question: Are amino acids that are attached to tRNA molecules transferred to proteins?
Hypothesis: Aminoacyl tRNAs transfer amino acids to growing polypeptides.
Null hypothesis: The amino acids held on tRNAs are not involved in protein synthesis.
Experimental setup:
Leu
Radioactive leucine
tRNA
1. Attach radioactive leucine molecules to tRNAs.
mRNA
Ribosomes
2. Add these aminoacyl tRNAs to in vitro translation system. Follow the fate of the radioactive amino acids.
Prediction: Radioactive amino acids will be found in proteins.
Prediction of null hypothesis: Radioactive amino acids will not be found in proteins.
Results:
Radioactive amino acids start attached to tRNA
Radioactive amino acids are rapidly incorporated into protein
Protein
tRNA
Radioactive signal (counts/minute)
300
200
100
0
5
10
15
20
Time (minutes)
Conclusion: Aminoacyl tRNAs transfer amino acids to growing polypeptides.

FIGURE 16.11 Experimental Evidence That Amino Acids Are Transferred from tRNAs to Proteins

shown in **Figure 16.11**. The graph's message is that radioactive amino acids are lost from tRNAs and incorporated into polypeptides synthesized in ribosomes. These results inspired the use of *transfer* in tRNA's name. The experiment also confirmed that aminoacyl tRNAs act as the interpreter in the translation process. tRNAs are Crick's adapter molecules.

Connecting Structure with Function

Transfer RNAs serve as the chemical go-betweens that allow amino acids to interact with an mRNA template. But precisely how does this connection occur?

This question was answered by research on tRNA's molecular structure. The initial studies established the sequence of nucleotides in various tRNAs, or what is termed their primary structure. Transfer RNAs are relatively short, ranging from 75 to 85 nucleotides in length. When biologists studied the primary sequence closely, however, they noticed that certain parts of the molecules can form secondary structures. Specifically, some of the bases in the tRNA molecule can form hydrogen bonds with complementary bases in a different region of the

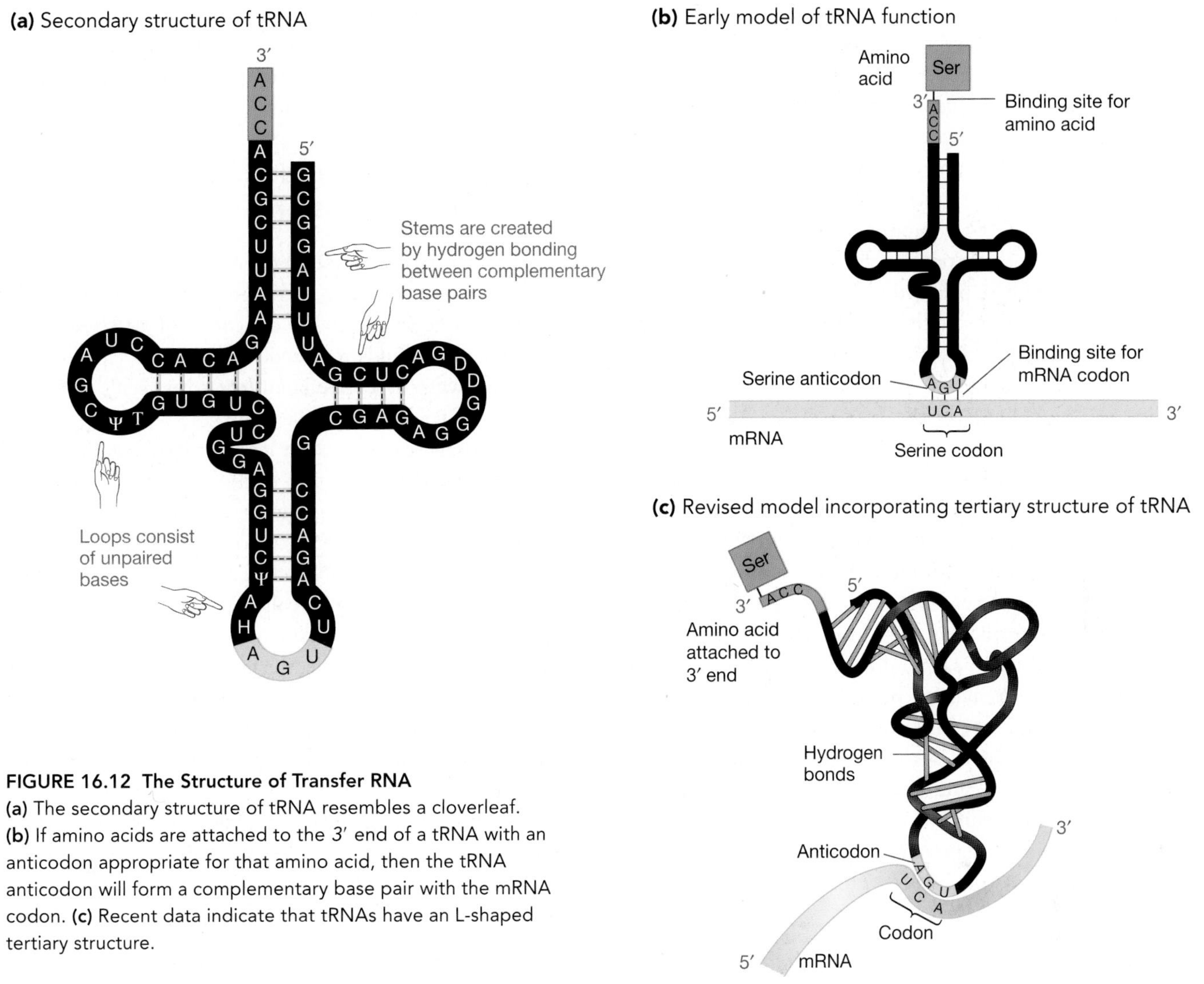
(a) Secondary structure of tRNA
3′
5′
Stems are created by hydrogen bonding between complementary base pairs
Loops consist of unpaired bases
(b) Early model of tRNA function
Amino acid
Ser
Binding site for amino acid
Serine anticodon
Binding site for mRNA codon
mRNA
Serine codon
(c) Revised model incorporating tertiary structure of tRNA
Amino acid attached to 3′ end
Hydrogen bonds
Anticodon
Codon
mRNA

FIGURE 16.12 The Structure of Transfer RNA
(a) The secondary structure of tRNA resembles a cloverleaf. **(b)** If amino acids are attached to the 3′ end of a tRNA with an anticodon appropriate for that amino acid, then the tRNA anticodon will form a complementary base pair with the mRNA codon. **(c)** Recent data indicate that tRNAs have an L-shaped tertiary structure.

same molecule. As a result, the entire tRNA molecule could assume the cloverleaf shape illustrated in **Figure 16.12a**. The stems in the cloverleaf are produced by complementary base pairing between different portions of the molecule, while the loops consist of single-stranded RNA.

Two aspects of this secondary structure proved especially interesting. The CCA sequence at the 3′ end of the molecule offered a binding site for amino acids, while the triplet on a loop on the opposite end of the cloverleaf could serve as an anticodon. An **anticodon** is a set of three nucleotides that forms base pairs with the mRNA codon. **Figure 16.12b** depicts an early model of how the anticodon of a tRNA connects an mRNA codon with the appropriate amino acid.

This model had to be modified, however, when X-ray crystallography studies revealed the tertiary structure of tRNAs. Recall from Chapter 3 that the tertiary structure of a molecule is defined by the three-dimensional arrangement of its atoms and is usually a product of folding. According to the X-ray crystallography data, the cloverleaf structure folds over to produce an L-shaped molecule (**Figure 16.12c**). All of the tRNAs in a cell have the same L-shaped structure, but each has a distinct anticodon and attached amino acid. The tertiary structure of tRNAs is important because it results in a precise separation between the anticodon and the attached amino acid. As we'll see, this separation is key to the positioning of the amino acid and the anticodon in the ribosome.

How Many tRNAs Are There?

When research succeeded in characterizing all of the different types of tRNAs available in cells, a paradox arose. According to the genetic code introduced in Chapter 15, the 20 amino acids found in proteins are specified by 61 different mRNA codons. Instead of containing 61 different tRNAs with 61 different anticodons, though, most cells contain only about 40. How can all 61 mRNA codons in the genome be translated with only two-thirds of the tRNAs required?

To resolve this paradox, Francis Crick proposed what is known as the **wobble hypothesis**. To understand his logic, recall from Chapter 15 that many amino acids are specified by more than one codon. Further, recollect that codons for the same amino acid tend to have the same nucleotides at the first and second positions but a different nucleotide at the third position. For example, both of the codons CAA and CAG code for the amino acid glutamine. (Codons are always written in the $5' \rightarrow 3'$ direction.) Surprisingly, experimental data have shown that a tRNA with an anticodon of GUU can base pair with both CAA and CAG in mRNA. (Anticodons are always written in the $3' \rightarrow 5'$ direction.) The GUU anticodon matches the first two bases (C and A) in both codons, but the U in the third position forms a nonstandard base pair with a G in the CAG codon.

Crick proposed that inside the ribosome, tRNAs with nonstandard base pairing at the third position can still bind successfully to an anticodon. If so, it would allow a limited flexibility, or "wobbling," in the base pairing. According to the wobble hypothesis, a nonstandard base pair in the third position—such as G-U—is acceptable as long as it does not change the amino acid for which the codon codes.

Follow-up research has shown that certain tRNAs are especially prone to pairing with more than one mRNA codon. These tRNAs have a nitrogen-containing base called *inosine* in the third position of their anticodon. (Inosine is derived from adenine by a chemical modification that occurs after the tRNA is produced.) During protein synthesis, an inosine in the anticodon can base pair with a A, U, or C in an mRNA codon. For example, the tRNA with the anticodon GAI carries the amino acid leucine and binds to four of the six codons that specify leucine: CUA, CUC, CUU, and UUA. (The UUA is possible because the G in the anticodon sometimes pairs with U in the mRNA. This interaction between anticodons and codons is one of the only known exceptions to the standard base-pairing rules.) In this way, "wobble" in the third position of a codon allows just 40 or so tRNAs to bind to all 61 mRNA codons.

16.5 Ribosomes and the Mechanism of Translation

During translation, the sequence of bases in an RNA message is converted to a sequence of amino acids in a polypeptide. The conversion begins when the anticodon of a tRNA that carries an amino acid binds to a codon in mRNA. The conversion is complete when a peptide bond forms between that amino acid and the growing polypeptide chain.

Both of these events occur inside a ribosome. Biologists have known since the 1930s that ribosomes contain a considerable amount of protein along with a great deal of **ribosomal RNA (rRNA)**. When high-speed centrifuges became available, researchers found that ribosomes could also be separated into two major substructures, called the *large subunit* and *small subunit*. Each subunit consists of a complex of RNA molecules and proteins. For example, the large subunit in *E. coli* contains two different rRNA molecules and 34 proteins. More recent work has shown that the small subunit holds the mRNA in place during translation and that peptide bond formation takes place in the large subunit.

The three-dimensional model of the ribosome in **Figure 16.13a** shows how all of the molecules and reactions required for translation fit together. Note that during protein synthesis,

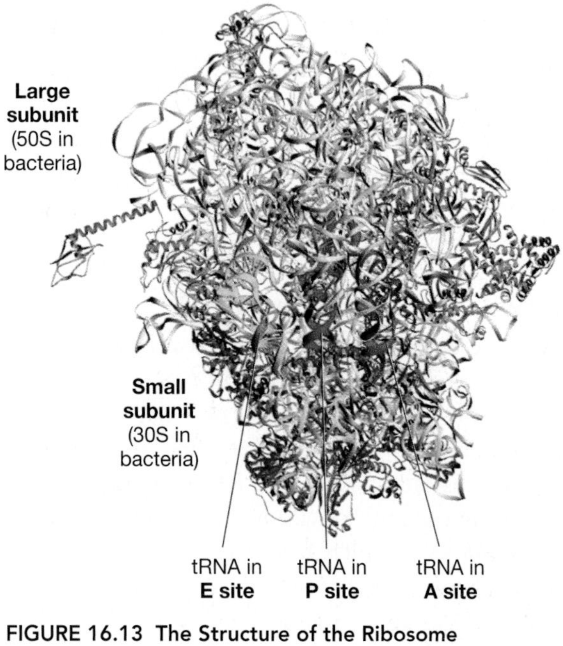
(a) Ribbon model of ribosome during translation
Large subunit (50S in bacteria)
Small subunit (30S in bacteria)
tRNA in E site
tRNA in P site
tRNA in A site

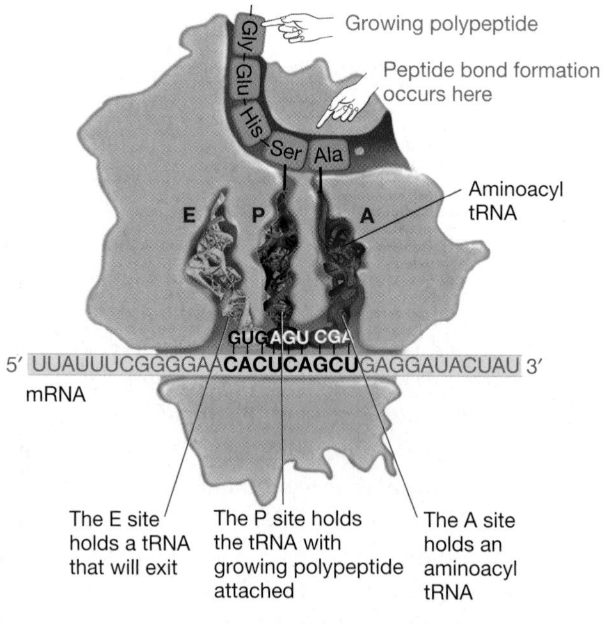
(b) Diagram of ribosome during translation
Growing polypeptide
Peptide bond formation occurs here
Gly
Glu
His
Ser
Ala
Aminoacyl tRNA
E
P
A
GUG AGU CGA
5′ UUAUUUCGGGGAACACUCAGCUGAGGAUACUAU 3′
mRNA
The E site holds a tRNA that will exit
The P site holds the tRNA with growing polypeptide attached
The A site holds an aminoacyl tRNA

FIGURE 16.13 The Structure of the Ribosome

(a) During translation, three tRNAs line up side by side inside the ribosome. **(b)** Different events occur at each of the three sites inside the ribosome where tRNAs are found. EXERCISE Circle and label the active site in the ribosome.

three distinct tRNAs are lined up inside the ribosome. All three are bound to their corresponding mRNA codon at the base of the structure. The tRNA that is on the right (colored red) in the figure carries an amino acid. This tRNA's position in the ribosome is called the **A site**, for acceptor or aminoacyl. The tRNA that is in the middle (green) holds the growing polypeptide chain and occupies the **P site**, for peptidyl, inside the ribosome. (Think of "P" for "peptide-bond formation.") The left-hand (yellow) tRNA no longer has an amino acid attached and is about to leave the ribosome. It occupies the ribosome's **E site**, for exit. **Figure 16.13b** summarizes what happens to each tRNA inside the ribosome. Because all tRNAs have the same secondary and tertiary structure, they all fit equally well in the A, P, and E sites.

The ribosome is a molecular machine that synthesizes proteins in a three-step sequence:

1. An aminoacyl tRNA diffuses into the A site; its anticodon binds to a codon in mRNA.
2. A peptide bond forms between the amino acid held by the aminoacyl tRNA in the A site and the existing polypeptide, which is held by a tRNA in the P site.
3. The ribosome moves ahead, and all three tRNAs move one position down the line. The tRNA in the E site exits; the tRNA in the P site moves to the E site; and the tRNA in the A site switches to the P site.

The protein that is being synthesized grows by one amino acid each time this three-step sequence repeats. The process occurs up to 20 times per second in bacterial ribosomes and about 2 times per second in eukaryotic ribosomes.

This introduction to how tRNAs, mRNAs, and ribosomes interact during protein synthesis leaves a number of key questions unanswered, however. How do mRNAs and ribosomes get together to start the process? Once protein synthesis is under way, how is peptide bond formation catalyzed inside the ribosome? And how does protein synthesis conclude when the ribosome reaches the end of the message? Let's consider each question in turn.

WEB TUTORIAL 16.2 Translation

Initiation

To translate an mRNA, a ribosome must begin at a specific point in the message, translate the mRNA up to the message's termination codon, and then stop. Biologists call these three phases of protein synthesis **initiation**, **elongation**, and **termination**, respectively. One key to understanding initiation is to recall, from Chapter 15, that the codon AUG is found near the 5′ end of all mRNAs and codes for the amino acid methionine. The presence of this start codon is an aspect of initiation that is common to both bacteria and eukaryotes.

Figure 16.14 shows how translation gets under way in bacteria. The process begins when a section of rRNA in the small subunit of a ribosome binds to a complementary sequence on an mRNA. The mRNA region is called the **ribosome binding site** or **Shine-Dalgarno sequence**, after the biologists who discovered it. The site is about six nucleotides upstream from the AUG start codon. It consists of all or part of the bases 5′ AGGAGGU 3′. The complementary sequence in the rRNA of the small subunit reads 3′ UCCUCCA 5′. The interaction between the small subunit and the message is mediated by proteins called **initiation factors.** In eukaryotes, initiation factors bind to the 5′ cap on mRNAs and guide it to the ribosome.

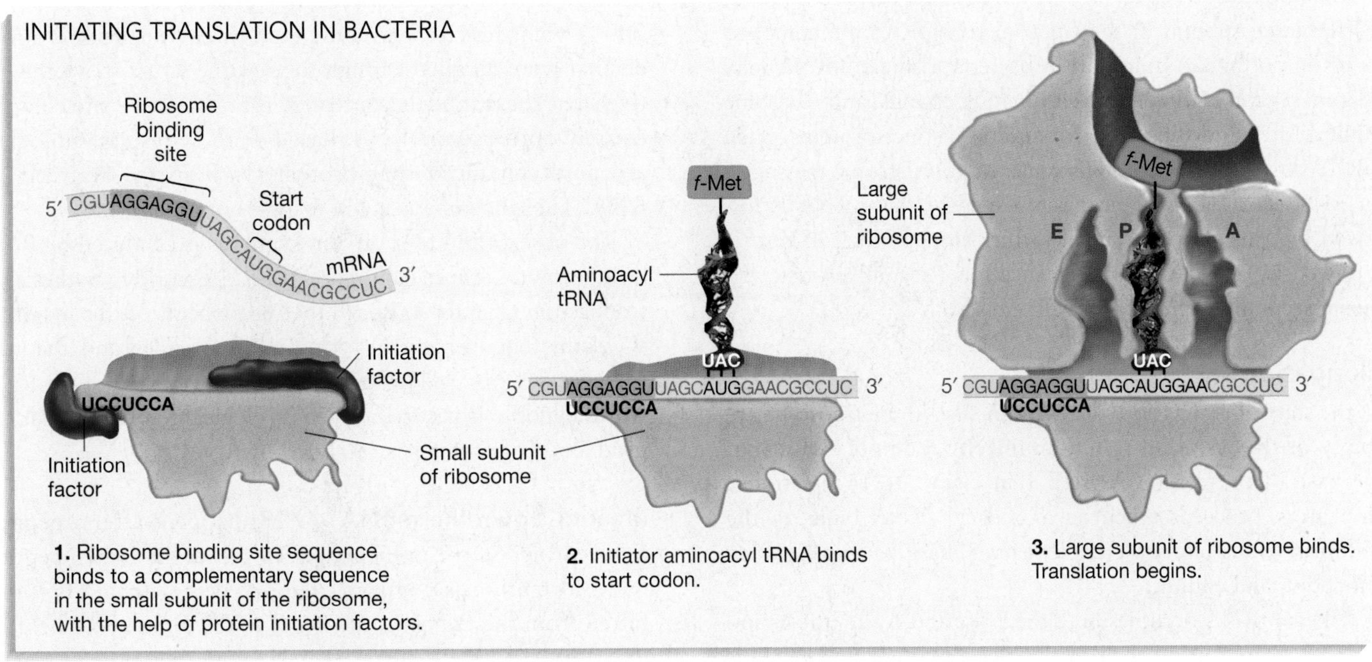

FIGURE 16.14 The Initiation Phase of Translation in Bacteria

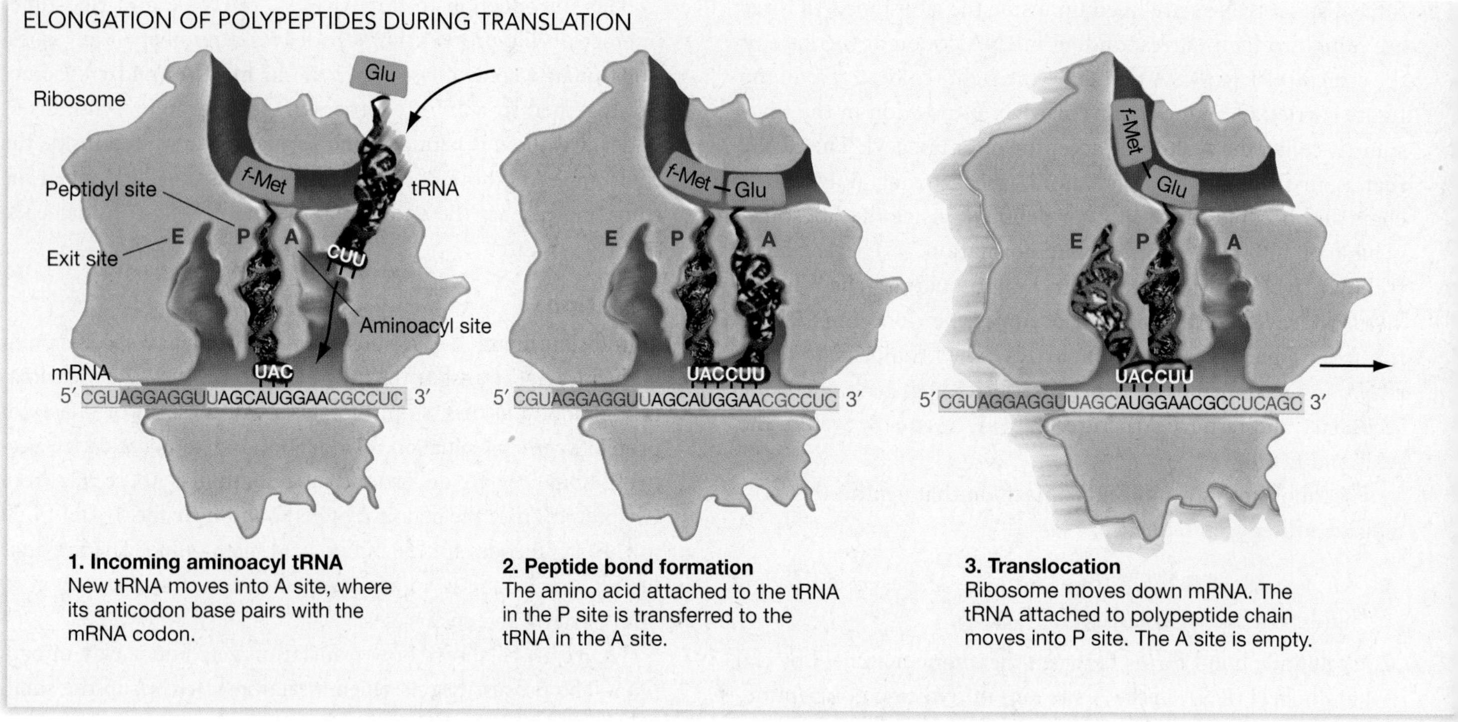

FIGURE 16.15 The Elongation Phase of Translation

Once the Shine-Dalgarno sequence has attached to the small ribosomal subunit, an aminoacyl tRNA bearing a modified form of methionine called *N*-formylmethionine (abbreviated *f*-met) binds to the AUG start codon. Initiation is complete when the large subunit joins the complex. When the ribosome is completely assembled, the tRNA bearing *f*-met occupies the P site.

To summarize, initiation is a three-step process in bacteria: (1) The mRNA binds to the small subunit of the ribosome, (2) the initiator aminoacyl tRNA binds to the start codon, and (3) the large subunit of the ribosome completes the complex.

It's important to note that in bacteria a single mRNA may contain several start codons, termination codons, and ribosome binding sites and thus code for several distinct proteins. Such mRNAs are said to be **polycistronic.** In polycistronic messages, an AUG start sequence near the 5′ end of the mRNA is followed by a termination signal farther along, which in turn is followed by an AUG start sequence for a second protein and then a second termination signal, and so on.

Elongation

At the start of elongation, the E and A sites in the ribosome are empty of tRNAs. As a result, an mRNA codon is exposed at the base of the A site. As step 1 in **Figure 16.15** illustrates, elongation proceeds when an aminoacyl tRNA binds to the codon in the A site via complementary base pairing between anticodon and codon.

When a tRNA occupies both the P site and A site, the amino acids they hold are placed in the ribosome's active site. This is where peptide bond formation—the essence of protein synthesis—occurs. Peptide bond formation is considered one of the most important reactions that takes place in cells, because manufacturing proteins is among the most fundamental of all cell processes. The question is, how does it happen?

Is the Ribosome an Enzyme or a Ribozyme? Because ribosomes contain both proteins and RNA, researchers had argued for decades over whether the active site consisted of protein or RNA. The debate was not resolved until the year 2000, when researchers completed three-dimensional models that were detailed enough to view the structure of the active site. These models confirmed that the active site consists entirely of ribosomal RNA. Based on these results, biologists are now convinced that protein synthesis is catalyzed by RNA. The ribosome is a ribozyme—not an enzyme.

The observation that protein synthesis is catalyzed by RNA is important, because it supports the RNA world hypothesis introduced in Chapter 4. Recall that proponents of this hypothesis claim that life began with RNA molecules and that the presence of DNA and proteins in cells evolved later. If the RNA world hypothesis is correct, then it is logical to observe that the production of proteins is catalyzed by RNA.

Moving Down the mRNA What happens after a peptide bond forms? Step 2 in Figure 16.15 shows that when peptide bond formation is complete, the polypeptide chain is transferred from the tRNA in the P site to the amino acid held by the tRNA in the A site. Step 3 shows the process called **translocation,** which occurs when the ribosome moves down

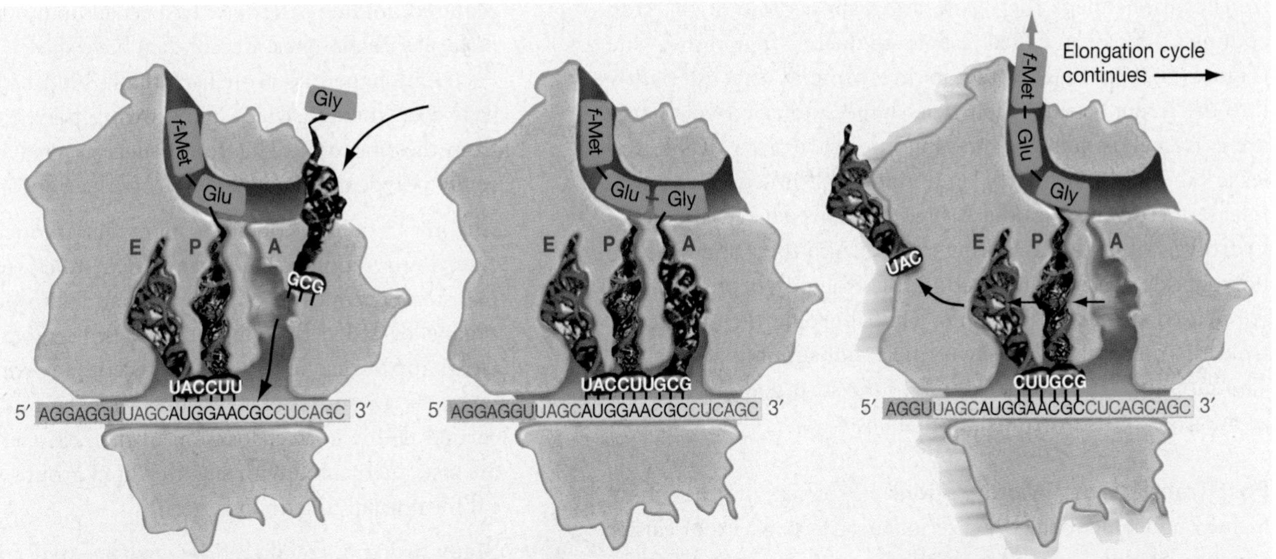

4. Incoming aminoacyl tRNA
New tRNA moves into A site, where its anticodon base pairs with the mRNA codon.

5. Peptide bond formation
The polypeptide chain attached to the tRNA in the P site is transferred to the tRNA in the A site.

6. Translocation
Ribosome moves down mRNA. The tRNA attached to polypeptide chain moves into P site. Empty tRNA from P site moves to E site, where tRNA is ejected. The A site is empty again.

the mRNA in the $5' \rightarrow 3'$ direction. Translocation does several things: It moves the empty tRNA into the E site; it moves the tRNA containing the growing polypeptide into the P site; and it opens the A site and exposes a new mRNA codon. If the E site is occupied when translocation occurs, the tRNA there is ejected into the cytoplasm.

The three steps in elongation—(1) arrival of aminoacyl tRNA, (2) peptide bond formation, and (3) translocation—repeat down the length of the mRNA. By adding specific molecules to in vitro translation systems, researchers have confirmed that each elongation cycle depends on an input of energy from several GTP molecules as well as assistance from proteins called **elongation factors.** Further, recent three-dimensional models of ribosomes in various stages of the translation sequence show that the machine as a whole is highly dynamic during the process. The ribosome constantly changes shape as tRNAs come and go and catalysis and translocation occur. Its structure changes in conjunction with its multipart function: coordinating interactions between tRNAs and an mRNA, and catalyzing peptide bond formation.

Electron microscopy has also confirmed that once elongation is under way, additional ribosomes bind to the start site in the same message and initiate translation. As this process continues, strings of ribosomes called **polyribosomes** assemble along an mRNA. Each ribosome synthesizes a copy of the protein (**Figure 16.16**). Even though it may take a minute or more for a single ribosome to produce a large polypeptide, the presence of polyribosomes can quicken the overall pace of protein production. Polyribosomes are observed routinely in both bacteria and eukaryotes.

Termination

How does protein synthesis end? To answer this question, recall from Chapter 15 that the genetic code includes three stop codons: UAA, UAG, and UGA. In most cells, no aminoacyl tRNA has an anticodon that binds to these sequences. When

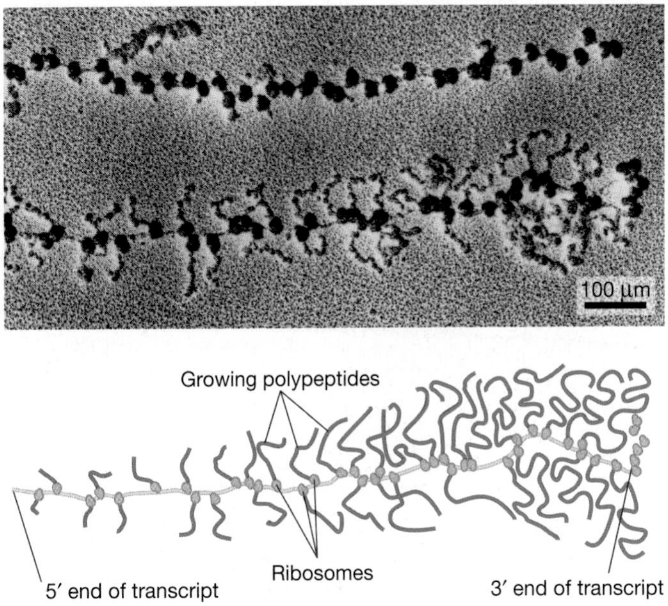

FIGURE 16.16 Polyribosomes
In both bacteria and eukaryotes, many ribosomes synthesize proteins from the same mRNA at the same time. The resulting structures are called polyribosomes.

translocation opens the A site and exposes one of the stop codons, a protein called a **release factor** fills the A site (**Figure 16.17**). Release factors are proteins that fit tightly into the A site because their size, shape, and electrical charge are extremely similar to the same properties in a tRNA. Release factors do not carry an amino acid, however. When a release factor occupies the A site, the active site catalyzes the hydrolysis of the bond linking the tRNA in the P site with the polypeptide chain. This reaction frees the polypeptide, and it is released from the ribosome. The ribosome then separates from the mRNA, and the two ribosomal subunits dissociate. The subunits are ready to attach to the start codon of another message and start translation anew.

Post-Translational Modifications Proteins are not fully formed and functional when termination occurs. From earlier chapters, it should be clear that most proteins go through an extensive series of processing steps, collectively called **post-translational modification**, before they are ready to go to work in a cell:

- Recall from Chapter 3 that a protein's function depends on its shape and that a protein's shape depends on how it folds. Folding actually begins during elongation, long before termination occurs and ribosomes disassemble. Although it occurs spontaneously, in the sense that no energy input is required, folding is frequently speeded up by proteins called **molecular chaperones**. Recent data have shown that, in some bacteria, chaperone proteins actually bind to the ribosome near the "tunnel" where the growing polypeptide emerges from the ribosome. This result suggests that folding occurs as the polypeptide is emerging from the ribosome.
- Chapter 7 pointed out that many eukaryotic proteins are extensively modified after they are synthesized. For example, small chemical groups may be added to proteins in the organelles called the rough endoplasmic reticulum and the Golgi apparatus. Some proteins receive a sorting signal that serves as an address and ensures that the molecule will be carried to the correct location in the cell. Certain proteins are also augmented with sugar or lipid groups that are critical for normal functioning.
- Many proteins are altered by enzymes that add or remove a phosphate group. Phosphorylation (addition of phosphate) and dephosphorylation (removal of phosphate) of proteins were introduced in Chapter 3 and Chapter 8. Recall that because a phosphate group has two negative charges, adding or removing a phosphate group causes major changes in the shape and chemical reactivity of proteins. These changes have a dramatic effect on the protein's activity—often switching it from an inactive state to an active state or vice versa.

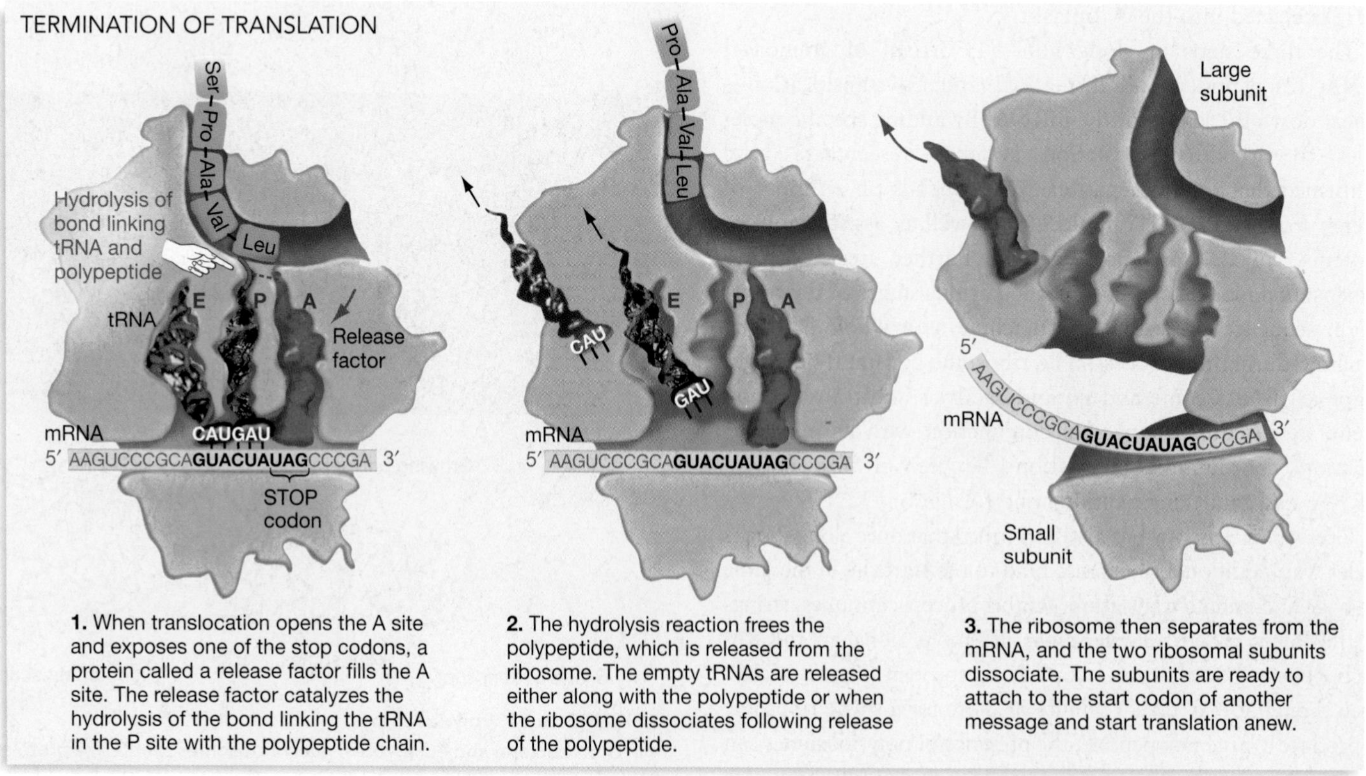

FIGURE 16.17 The Termination Phase of Translation

The overall message is that the manufacture of a completely functional protein depends not only on translation in the ribosome but also on a wide array of other molecules and events that take place throughout the cell.

✓CHECK YOUR UNDERSTANDING

In bacteria, translation begins when the ribosome binding site on an mRNA binds to an rRNA sequence in the small subunit of the ribosome. The initiator aminoacyl tRNA binds to the start codon in the mRNA, and the large subunit of the ribosome attaches to the small subunit to complete the ribosome.

In both bacteria and eukaryotes, polypeptides elongate when the appropriate aminoacyl tRNA enters the A site. A peptide bond forms between the amino acid held by that tRNA in the A site and the polypeptide held by the tRNA in the P site. The ribosome then moves down the mRNA one codon, and the process occurs again. Translation ends when the ribosome reaches a stop codon. You should be able to (1) diagram a ribosome during translation initiation, elongation, and termination; (2) label each major structure in the diagrams; and (3) describe the function of each major structure.

16.6 The Molecular Basis of Mutation

This chapter has focused on how the information in DNA is transcribed to RNA and then translated into a functioning protein.

Although biologists continue to work on issues such as how basal transcription factors in eukaryotes interact with RNA polymerase and how ribosomes change shape during translation, the molecular mechanisms responsible for information flow in the cell, from DNA to RNA to proteins, are now considered fairly well understood.

Before we go on to consider how transcription and translation are regulated in bacteria and eukaryotes in the next two chapters, it will be helpful to tie together some major points covered in the last three chapters by analyzing the molecular basis of mutation. A **mutation** is any change in an organism's DNA. It is a change in a cell's information archive—a change in its genotype. What are the consequences of mutation?

The molecular machinery introduced in this chapter faithfully transcribes and translates mutations in DNA sequences. The result is the production of novel types of proteins. In this way, changes in DNA lead to changes in molecules that affect the organism's phenotype.

Figure 16.18 shows how a common type of mutation occurs. If DNA polymerase mistakenly inserts the wrong base as it synthesizes a new strand of DNA, and if proofreading by DNA polymerase and the mismatch repair system fail to correct the mismatched base, a change in the sequence of bases in DNA results. A single base change such as this is called a **point mutation.**

What happens when point mutations are transcribed and translated? To answer this question, consider the first point mutation ever described. This mutation occurs in the human gene for the protein hemoglobin. Recall that hemoglobin is abundant in red blood cells and that it carries oxygen to tissues.

Figure 16.19a, page 356, shows a small section of DNA sequence from the normal gene for hemoglobin, along with the same sequence from the mutant allele found in individuals who suffer from sickle-cell disease. The mutant form has a thymine in place of an adenine at the second position in the sixth codon specified by this gene. A glance back at the genetic code in Chapter 15 confirms why this point mutation is significant: During protein synthesis, the mutant codon specifies valine instead of glutamic acid in the amino acid chain of hemoglobin. Point mutations that cause changes in the amino acid sequence of proteins are called **missense mutations** or **replacement mutations.** In the case of hemoglobin, the single change in primary structure causes the protein to crystallize when oxygen levels in the blood are low. When hemoglobin crystallizes, it causes red blood cells to become sickle shaped (**Figure 16.19b**). The misshapen cells get stuck in blood vessels, so nearby cells are starved for oxygen. The result is intense pain and anemia.

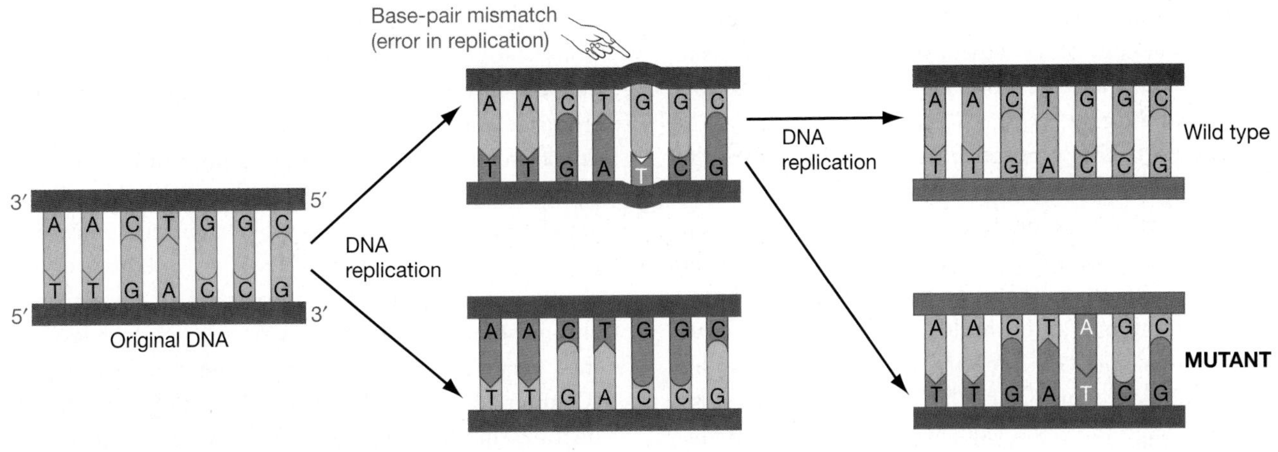

FIGURE 16.18 Unrepaired Mistakes in DNA Synthesis Lead to Point Mutations
QUESTION Why is it logical that the type of mutation illustrated here is termed a point mutation?

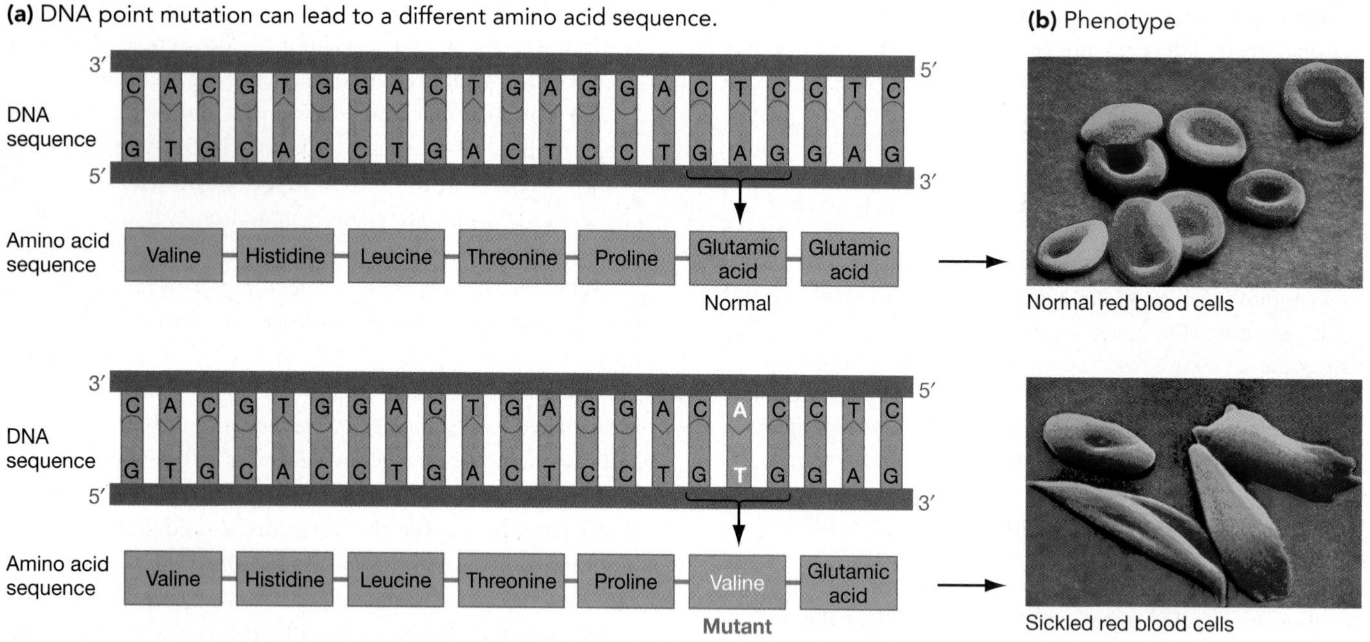

FIGURE 16.19 Sickle-Cell Disease Results from a Point Mutation in the Gene for Hemoglobin **(a)** These normal and mutant genes that code for hemoglobin differ by a single nucleotide. The sequences shown are just a small portion of the gene. **(b)** The mutation in part (a) changes the primary sequence of the gene product. The change in amino acid sequence causes hemoglobin molecules to crystallize when oxygen levels in the blood are low. As a result, red blood cells sickle and get stuck in small blood vessels. **EXERCISE** The mRNA codon GUG specifies the valine at the start (left) of this protein. Label the template strand and the non-template strand of DNA, and indicate their 5′ and 3′ ends.

Missense mutations such as that found in the sickling allele can be **deleterious**, meaning they reduce an individual's fitness—the ability to survive and reproduce. But missense mutations may also be beneficial. For example, the sickling allele confers a fitness advantage in regions of the world where the life-threatening disease malaria is common. The advantageous effect is possible because the parasite that causes malaria infects red blood cells. If an individual has one normal hemoglobin allele and one sickling allele, red blood cells that become infected with the parasite tend to sickle while others function normally. The deformed cells are quickly destroyed by the body, eliminating the parasitic cells inside the blood cells. As a result, individuals who have one copy of the mutant gene have lower parasite loads and tend to be much healthier than people who have two normal copies of the gene.

In contrast to having a negative or positive effect on the organism, many point mutations have virtually no effect at all. To understand why, suppose that at the next site in the hemoglobin gene of Figure 16.19 (the third position in the same sixth codon), a thymine had been substituted for the cytosine. This point mutation would have no consequence. Why? Both GAA and GAG specify glutamic acid. In this case, the amino acid sequence of the gene product does not change even though the DNA sequence is altered due to mutation. This type of alteration in the base sequence is called a **silent mutation.** Silent mutations are said to be **neutral** in their effect on an individual's fitness. It is possible for point mutations to be silent and neutral with respect to fitness because of the redundancy in the genetic code.

The take-home message from this analysis of hemoglobin, sickle-cell disease, and malaria is simple: It is possible for point mutations to be deleterious, beneficial, or neutral. **Figure 16.20a** summarizes the types of point mutations that have been documented and reviews their consequences for the amino acid sequences of proteins.

In addition to documenting various types of point mutations, biologists study mutations that involve larger-scale changes in the composition of chromosomes. Chapter 9 introduced the phenomena called polyploidy and aneuploidy, which involve changes in the number of chromosome complements present and the addition or deletion of a chromosome, respectively. Just as point mutations are caused by random errors made by DNA polymerase, chromosome changes result from chance mistakes in the partitioning of chromosomes during meiosis or mitosis. In addition, errors that occur in crossing over during prophase of meiosis I (Chapter 12) can lead to gene duplications or deletions, as illustrated in **Figure 16.20b**. Finally, when accidental breaks in

(a) Types of point mutations

	Definition	Example	Consequence
		Original DNA sequence — TAT TGG CTA GTA CAT Original polypeptide — Tyr – Trp – Leu – Val – His	
Silent	Change in nucleotide that does not change amino acid specified by codon	TAC TGG CTA GTA CAT Tyr – Trp – Leu – Val – His	Change in genotype but no change in phenotype
Missense (Replacement)	Change in nucleotide that changes amino acid specified by codon	TAT TGT CTA GTA CAT Tyr – **Cys** – Leu – Val – His	Change in primary structure of protein
Nonsense	Change in nucleotide that results in early stop codon	TAT TGA CTA GTA CAT Tyr **STOP**	Premature termination—polypeptide is truncated
Frameshift	Addition or deletion of a nucleotide	TAT TCG GCT AGT ACA T Tyr – **Ser** – **Ala** – **Ser** – **Thr**	Reading frame is shifted—massive missense

(b) Other types of mutations

Gene duplication

Definition: Addition of a small chromosome segment due to an error during crossing over at meiosis I—homologs do not align correctly

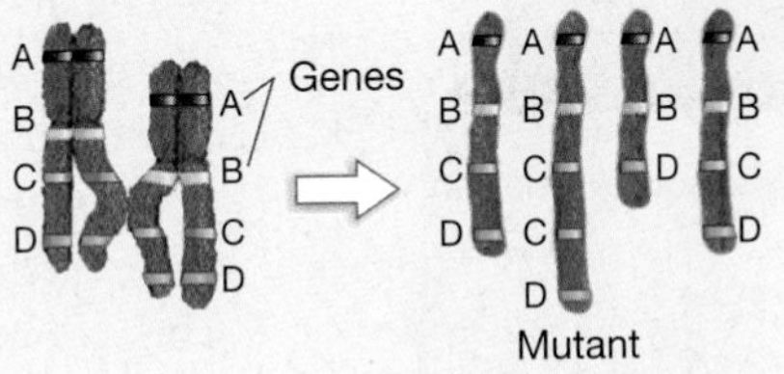

Consequence: Produces an extra copy or deletion of one or more genes. Families of related genes arise by gene duplication.

Chromosome inversion

Definition: Change in a chromosome segment when DNA breaks in two places, flips, and rejoins

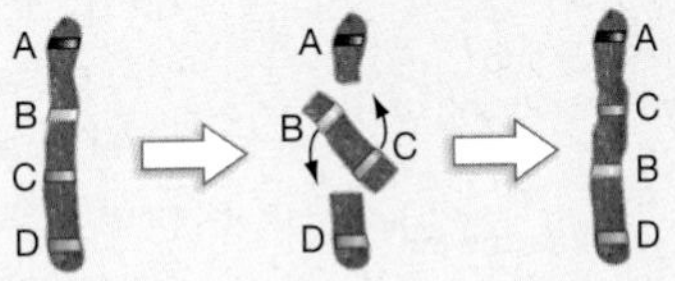

Consequence: Changes gene order along chromosome. Other types of chromosome breaks can lead to deletion or addition of chromosome segments.

FIGURE 16.20 A Mutation Is a Change in an Organism's DNA
(a) Point mutations in DNA produce an array of effects on the amino acid sequences of proteins, ranging from no change to extensive change. **(b)** Many mutations involve large-scale changes in the composition of chromosomes. In addition to gene duplication and chromosome inversions, Chapter 9 introduced polyploidy (multiple chromosome complements), trisomy (additional chromosomes), and aneuploidy (loss or gain of a chromosome).

chromosomes occur, the chromosome segments may become flipped and rejoin—a phenomenon known as a **chromosome inversion** (Figure 16.20b).

When changes in an organism's genotype are transcribed and translated, they may result in changes in its phenotype. In this way, mutations produce new alleles and new traits. At the level of populations, mutations furnish the heritable variation that Mendel and Morgan analyzed and that makes evolution possible. At the level of individuals, mutations can cause disease or death or lead to increases in fitness.

ESSAY Antibiotics That Poison the Ribosome

Antibiotics are molecules that kill bacteria or slow their growth. Since their initial discovery by biomedical researchers in the 1930s, their use has saved tens of millions of human lives that were threatened by bacterial infections. It is no exaggeration to claim that the era of modern medicine began with the widespread availability of inexpensive antibiotics.

(a) Erythromycin binds to ribosomes and plugs the exit tunnel for the polypeptide.

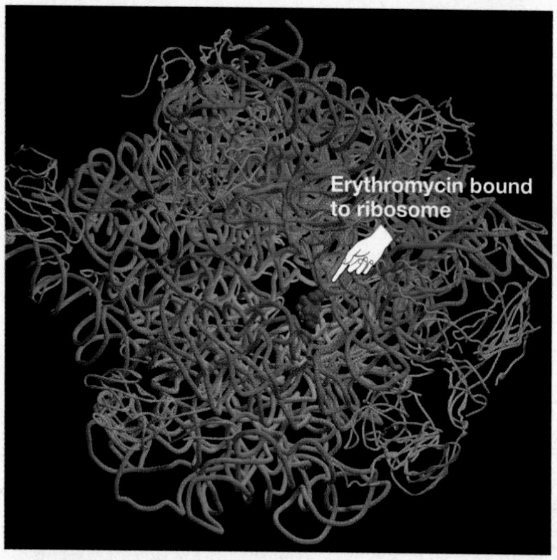
Erythromycin bound
to ribosome

(b) Mode of action of some antibiotics that poison translation

Antibiotic	Mode of Action
Chloramphenicol	Binds at the active site in the large subunit of ribosome
Clindamycin	Binds to A and P sites and prevents correct positioning of tRNAs
Erythromycin, carbomycin A, clarithromycin, roxithromycin, spiramycin, tylosin, azithromycin	Plugs the tunnel where growing polypeptide exits ribosome
Puromycin	Enters A site and bonds to growing polypeptide, but blocks addition of more amino acids
Spectinomycin	Inhibits translocation of ribosome
Streptomycin, parmomycin	Binds to the small subunit of ribosome and prevents accurate reading of the mRNA
Tetracycline	Prevents aminoacyl tRNAs from entering A site in ribosome

FIGURE 16.21 Many Antibiotics Work by Disrupting Translation in Bacteria
[(a) "Structural basis for the interaction of antibiotics with peptidyl transferase centre in eubacteria: by Schluenzen et al." Figure 5 in *Nature* (2001) 413: 814–821 by Schluenzen et al.]

Many of the antibiotics that physicians prescribe today are produced naturally by soil-dwelling fungi. These organisms compete with bacterial cells for space and resources. For a fungus, the ability to synthesize antibiotics is an adaptation that helps individuals destroy competitors and increase their access to nutrients. Antibiotics are weapons in chemical warfare that has gone on among competing soil-dwelling organisms for millions of years. Pharmaceutical companies still search for new antibiotics by studying naturally occurring fungi in the soil.

Recent research on antibiotics has focused on two seemingly distinct topics: the evolution of bacteria that resist them and the molecular mechanism of drug action.

Recent research on antibiotics has focused on two seemingly distinct topics: the evolution of bacteria that resist them and the molecular mechanism of drug action. **Antibiotic resistance** is the ability to grow and reproduce efficiently in an environment containing antibiotics. But as often happens in biological science, efforts to answer questions that seem unrelated produce closely connected answers.

How are evolutionary and molecular studies linked? Early research showed that several medically important antibiotics work by inhibiting translation in bacterial cells. In the presence of certain drugs, bacterial ribosomes stop working, protein synthesis ceases, and the targeted cell dies. Fortunately, bacterial and eukaryotic ribosomes are different enough at the molecular level that ribosome-poisoning antibiotics tend not to disrupt protein synthesis in humans. This is why antibiotics that target bacterial ribosomes have few side effects when administered to humans as a drug.

To see how evolutionary and molecular studies of ribosomes intersect, consider the latest three-dimensional models of bacterial ribosomes from bacteria treated with the antibiotic erythromycin. As **Figure 16.21a** shows, the drug binds to the tunnel where growing polypeptide chains leave the ribosome. When erythromycin binds to this site, it inhibits translation and kills the cell. A number of bacterial strains have evolved resistance to this antibiotic, however. When researchers analyzed the structure of ribosomes from drug-resistant populations of bacteria, they found that the ribosomes contained novel amino acids and rRNA sequences at the site where erythromycin binds. The altered amino acids and bases make erythromycin much less likely to bind to the ribosome. As a result, the drug no longer inhibits translation effectively.

Similar studies have now been done on a wide array of antibiotics. The results have documented the binding site and mode of action of many commonly prescribed drugs (**Figure 16.21b**). In many cases, investigators have also been able to confirm that drug-resistant strains of bacteria have mutations precisely at these binding sites.

To make sense of these findings, biomedical researchers acknowledge that mutations are constantly creating variations in the makeup of ribosomes within bacterial populations. The researchers hypothesize that extensive use of antibiotics has led to intensive natural selection for cells with mutations that make them resistant to one or more drugs. In areas where antibiotic use is widespread, the mutant bacteria have higher fitness and come to dominate their population.

Antibiotic resistance is now so pervasive among disease-causing bacteria that some physicians claim that we have entered a post-antibiotic era. Researchers and physicians have already spent several decades in a race to develop new antibiotics that are effective against drug-resistant strains of bacteria. Research on ribosome structure and the mechanisms of translation has taken on an urgent tone. The more we know about the molecular mechanisms of protein synthesis, the better we can run the race against drug-resistant bacteria.

CHAPTER REVIEW

Summary of Key Concepts

The instructions for making and operating a cell are archived in DNA, transcribed into messenger RNA, and then translated into protein.

- **After the enzyme RNA polymerase binds to a specific site in DNA with the help of other proteins, it catalyzes the production of an RNA molecule. The base sequence of the RNA produced is complementary to the base sequence of the DNA template strand.**

Early experiments established that RNA polymerase begins transcription by binding to promoter sequences in DNA. In bacteria, this binding occurs in conjunction with a regulatory protein called sigma. Sigma recognizes particular sequences within promoters that are located 10 bases and 35 bases upstream from the start of the actual genetic message. These binding sites ensure that the correct DNA strand is transcribed in the correct direction. Eukaryotic promoters are more complex and variable than bacterial promoters, and interact with many more transcription factors.

Web Tutorial 16.1 Transcription

- **In eukaryotes, genes consist of coding regions called exons interspersed with noncoding regions called introns. After a eukaryotic mRNA is produced, introns must be spliced out. The ends of the molecule receive a cap and tail.**

After transcription of a eukaryotic mRNA is complete, several key events occur. Stretches of noncoding RNA called introns are spliced out by complex molecular machines called spliceosomes, a "cap" signal is added to the 5′ end, and a poly (A) tail is added to the 3′ end. Experiments with modified mRNAs established that both the 5′ cap and the poly (A) tail serve as recognition signals for the translation machinery and protect the message from degradation by ribonucleases.

- **Inside ribosomes, mRNAs are translated to proteins via intermediary molecules called transfer RNAs. Transfer RNAs carry an amino acid and have a three-base-pair anticodon, which binds to an mRNA codon. The amino acid carried by the transfer RNA is then added to the growing protein via formation of a peptide bond.**

Experiments with radioactively labeled amino acids confirmed that ribosomes are the site of protein synthesis, and that transfer RNAs (tRNAs) serve as the chemical bridge between the RNA message and the polypeptide product. tRNAs have a predicted secondary structure that resembles a cloverleaf and an L-shaped tertiary structure. One leg of the L contains the anticodon, which forms a base pair with the mRNA codon, while the other leg holds the amino acid appropriate for that codon. Because imprecise pairing—or "wobbling"—is allowed in the third positions of anticodons, only about 40 different tRNAs are required to translate the 61 codons that code for amino acids.

Through intensive study of ribosome structure and function, researchers established that the process of elongating a polypeptide involves three steps: (1) an incoming aminoacyl tRNA occupies a position in the ribosome called the A site; (2) the growing polypeptide chain is transferred from a peptidyl tRNA in the ribosome's P site to the amino acid bound to the tRNA in the A site, and a peptide bond is formed; and (3) the ribosome is translocated to the next codon on the mRNA, accompanied by ejection of the empty tRNA from the E site. Recent data on the three-dimensional structure of the ribosome has confirmed that peptide bond formation is catalyzed by a ribozyme, not an enzyme.

While translation is in progress, proteins fold into their three-dimensional conformation (tertiary structure), sometimes with the aid of chaperone proteins. Some proteins are targeted to specific locations in the cell by the presence of signal sequences, while others remain inactive until modified by phosphorylation or the removal of certain amino acids.

Web Tutorial 16.2 Translation

Many mutations produce changes in the phenotype.

When mutations occur in DNA, the altered sequences are faithfully transcribed and translated into gene products that are part of the organism's phenotype. Depending on the location and type of alteration in DNA and its impact on the resulting RNA or protein product, a mutation can be beneficial, deleterious, or neutral with respect to fitness. Mutations produce novel proteins and RNAs. They are the source of the heritable variation that makes evolution possible.

Questions

Content Review

1. How did the A site of the ribosome get its name?
   a. It is where amino acids are affixed to tRNAs, producing aminoacyl tRNAs.
   b. It is where the amino group on the growing polypeptide chain is available for peptide bond formation.
   c. It is the site occupied by incoming aminoacyl tRNAs.
   d. It is surrounded by α-helices of ribosomal proteins.
2. How did the P site of the ribosome get its name?
   a. It is where the promoter resides.
   b. It is the site where peptide bond formation takes place.
   c. It is where peptidyl tRNAs reside.
   d. It is the site where the growing polypeptide chain is phosphorylated.
3. What is a molecular chaperone?
   a. a protein that recognizes the promoter and guides the binding of RNA polymerase
   b. a protein that activates or deactivates another protein by adding or removing a phosphate group
   c. a protein that is a component of the large ribosomal subunit and that assists with peptide bond formation
   d. a protein that helps newly translated proteins fold into their proper three-dimensional configuration
4. The three types of RNA polymerase found in eukaryotic cells transcribe different types of genes. What does RNA polymerase II produce?
   a. rRNAs
   b. tRNAs
   c. mRNAs
   d. spliceosomes
5. What is an anticodon?
   a. the part of an RNA message that signals the termination of translation
   b. the part of an RNA message that signals the start of translation
   c. the part of a tRNA that binds to a complementary codon in mRNA
   d. the part of a tRNA that accepts an amino acid, via a reaction catalyzed by tRNA synthetase
6. Which of the following questions about transcription was *not* answered by the discovery of the −35 and −10 sequences in bacterial promoters?
   a. Does RNA polymerase act alone to initiate transcription, or are other proteins involved?
   b. How do sigma proteins bind to RNA polymerase?
   c. How does RNA polymerase "know" where to initiate transcription?
   d. How does RNA polymerase "know" which strand of DNA acts as the template?

Conceptual Review

1. Explain the relationship among eukaryotic promoter sequences, basal transcription factors, and RNA polymerase. Explain the relationship among bacterial promoter sequences, sigma, and RNA polymerase.
2. According to the wobble rules, the correct amino acid can be added to a growing polypeptide chain even if the third base in an mRNA codon does not correctly match the corresponding base in a tRNA anticodon. How do the wobble rules relate to the redundancy of the genetic code?
3. Why does splicing occur in eukaryotic mRNAs? Where does it occur, and how are snRNPs involved?
4. Sketch the structure of a tRNA molecule. Label the anticodon and the CCA sequence where amino acids bind. Explain how a tRNA interacts with tRNA synthetase. Sketch the structure of an aminoacyl tRNA. Explain how an aminoacyl tRNA interacts with an mRNA in the ribosome.
5. Explain the sequence of events that occurs during translation as a protein elongates by one amino acid. At each step, specify what is happening in the ribosome's A site, P site, and E site.
6. What evidence supports the hypothesis that peptide bond formation is catalyzed by a ribozyme?

Group Discussion Problems

1. The 5′ cap and poly (A) tail that are attached to eukaryotic mRNAs appear to help the message last longer by protecting it from degradation by ribonucleases. But why is an enzyme such as a ribonuclease in the cell in the first place? What function would an enzyme that destroys messages serve?

2. Look back at Figure 16.1, which shows the formation of a phosphodiester bond during RNA polymerization. Then study the structure of the nucleotide shown below, called cordycepin. If cordycepin triphosphate, which has three phosphate groups bonded to the 5′ hydroxyl group in the figure, is added to a cell-free transcription reaction, the nucleotide is added onto the growing RNA chain. This observation confirms that synthesis occurs by the addition of monomers in the form of triphosphates to the 3′ end of the growing chain. Briefly explain why. Be sure to describe the expected result if synthesis occurred at the 5′ end of the polymer.

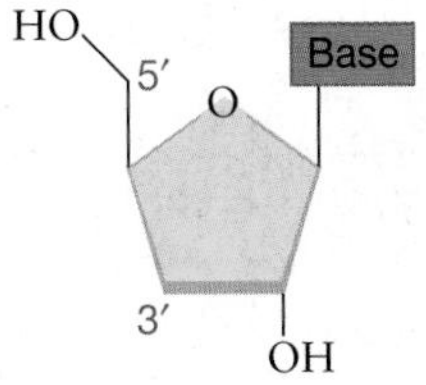

3. Carl Woese has determined the nucleotide sequence of rRNA molecules from a diverse array of organisms and compared them with each other. According to his data, certain portions of the rRNAs in the large subunit are very similar in all organisms. To make sense of this result, Woese suggests that the conserved sequences have an important functional role. His logic is that these conserved sequences are so important to cell function that any changes in the sequence cause death. Woese also claims that the existence of the conserved sequences supports other data indicating that peptide bond formation is catalyzed by the large rRNA, not by ribosomal proteins. Explain the thinking behind his claim.

4. Recent work has revealed why α-amanitin inhibits transcription by RNA pol II. Structural models show that α-amanitin binds to a site inside the enzyme, but not to the active site itself. Based on the model of bacterial RNA polymerase in Figure 16.2, predict where α-amanitin binds and why it inhibits transcription.

Answers to Multiple-Choice Questions 1. c; 2. b; 3. d; 4. c; 5. c; 6. d

www.prenhall.com/freeman is your resource for the following: Web Tutorials; Online Quizzes and other Online Study Guide materials; Answers to Conceptual Review Questions; Solutions to Group Discussion Problems; Answers to Figure Caption Questions and Exercises; and Additional Readings and Research.

17 Control of Gene Expression in Bacteria

KEY CONCEPTS

- Changes in gene expression allow bacterial cells to respond to environmental changes.
- Gene expression can be controlled at three levels: transcription, translation, or post-translation (protein activation).
- Transcriptional control can be positive or negative. In the case of the genes for lactose metabolism in *E. coli*, positive control occurs when a regulatory protein binds to DNA and increases the transcription rate. Negative control occurs because another regulatory protein must be released before transcription can occur.
- Many regulatory proteins bind to specific sites in DNA. Because different regulatory proteins have different amino acid sequences, they bind to different DNA sequences.

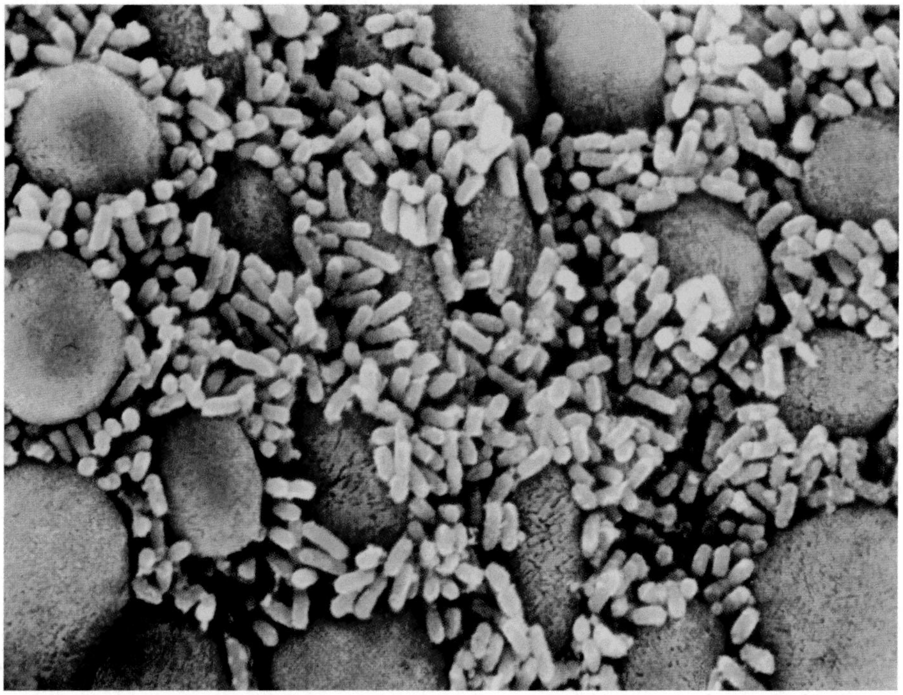

The red mounds in this micrograph are human intestinal cells; the yellow structures are bacteria. In the intestine, the nutrients available to bacteria constantly change. This chapter explores how changes in gene expression help bacteria respond to environmental changes.

Imagine waiting anxiously to hear the opening lines of a wonderfully melodic symphony, played by a renowned orchestra. The crowd applauds as the celebrated conductor comes onstage, then hushes as he takes the podium. He cocks the baton; the musicians raise their instruments. As the baton comes down, every instrument begins blaring a different tune at full volume. A tuba plays "Dixie"; a violinist renders "In-A-Gadda-Da-Vida": and a cellist begins Mexico's national anthem. A snare drum lays down beats for "Baby Got Back" while the bass drum simulates the cannons in the "1812 Overture." Instead of music, there is pandemonium. The conductor staggers off stage, clutching his heart.

A cacophony like this would result if a bacterial cell "played" all its genes at full volume all the time. The *Escherichia coli* cells living in your gut right now have over 4300 genes. If all of those genes were expressed at the fastest possible rate at all times, the *E. coli* cells would stagger off the stage, too. But this does not happen. Cells are extremely selective about which genes are expressed, in what amounts, and when.

This chapter explores how bacterial cells control the activity of their genes. **Gene expression** is said to occur when a protein or other gene product is synthesized and is active in the cell.

Understanding how cells regulate gene expression is a fundamental issue in biological science, for two reasons. First, bacteria are the most abundant organisms on Earth, and they occupy virtually every habitat known. In many cases, bacterial cells are able to grow and reproduce successfully because they respond rapidly to dramatic changes in temperature, pH, light, competitors, and nutrients. Changes in gene expression give bacteria the ability to cope with these types of environmental change. Second, questions about bacterial gene expression have enormous practical significance. For example, biologists are trying to understand which genes are expressed when disease-causing bacteria colonize a person and start an infection in the hope that drugs can be developed that alter the expression of key genes. Bacterial cells are also used commercially to manufacture important products such as human insulin and growth hormone. Efficient use of bacteria depends on a solid understanding of how transcription and translation are regulated.

Understanding gene expression is fundamental to our understanding of life. This chapter's goal is to introduce key concepts in the regulation of gene expression. Let's begin by reviewing some of the environmental challenges that bacterial cells face and then explore how these organisms meet them.

17.1 Gene Regulation and Information Flow

Bacteria are found in virtually every habitat on Earth, from boiling hot springs to alpine snowfields and from the open ocean to crevices miles underground. The millions of bacterial species existing today have evolved a bewildering variety of ways to solve a central problem of living: obtaining the materials and energy required for growth and reproduction. Although some bacteria specialize by using just one type of food, the vast majority of species are able to switch among several different sources of carbon and energy, according to which nutrients are available in the environment at the time. The fundamental question addressed in this chapter is *how* this switching occurs. Each type of nutrient requires a different membrane transport protein to bring the nutrient molecule into the cell and a different suite of enzymes to process it. How does a bacterial cell turn some genes on and others off so that it can take advantage of alternative food sources? More generally, how do bacteria regulate gene expression so that a cell makes only the products that it needs?

As a case study, we explore the strategies of *Escherichia coli*. These cells can use a wide array of carbohydrates to supply the carbon and energy they need. The carbohydrates are in your food. As your diet changes from day to day, the availability of different sugars in your intestines varies. Precise control of gene expression gives *E. coli* the ability to respond to these changes in its environment and use the different sugars.

To appreciate why precise control over gene expression is so important to these cells, it is critical to realize that bacteria can be packed an inch thick along your intestinal walls. The organisms represent many different species that are competing for space and nutrients. For a cell to survive and reproduce in this environment, it must use resources efficiently—particularly resources that provide energy and carbon compounds. For example, recall from Chapter 4 that carbohydrates are polymers made up of simple sugars and from Chapter 9 that carbohydrates provide cells with energy and chemical building blocks. *Escherichia coli* is capable of metabolizing a wide variety of carbohydrates. But it would be energetically wasteful for *E. coli* to produce all of the enzymes required to process all of the various carbohydrates all of the time. Instead, it is logical to predict that the enzymes being produced match the sugars that are available at a given time. Efficient use of resources, via tight control over gene expression, is critical if cells are going to be able to compete successfully for space and nutrients.

On the basis of this reasoning, transcription and translation of individual genes in bacteria should be triggered by specific signals from the environment, such as the presence of specific sugars. Did you drink milk at your last meal, or eat french fries and a candy bar? Each type of food contains different sugars. Each sugar should induce a different response by the *E. coli* cells in your intestine. Just as a conductor needs to regulate the orchestra's musicians, cells need to regulate which proteins they produce at any given time.

Mechanisms of Regulation—an Overview

Gene expression can be controlled at any step between the synthesis of RNA and the activation of the final gene product. Three steps occur as information flows from DNA to proteins, represented by arrows in the following expression:

$$\text{DNA} \rightarrow \text{mRNA} \rightarrow \text{protein} \rightarrow \text{activated protein}$$

The arrow from DNA to RNA represents transcription—the making of messenger RNA (mRNA). The arrow from RNA to protein represents translation, in which ribosomes read the information in mRNA and use that information to synthesize a protein. The arrow from protein to activated protein represents post-translational modifications—including folding, addition of carbohydrate or lipid groups, or perhaps phosphorylation.

How can a bacterial cell conserve ATP and amino acids by avoiding the production of proteins that are not needed at a particular time? A look at the flow of information from DNA to protein suggests three possible mechanisms:

1. The cell could avoid making the mRNAs for particular enzymes. If there is no mRNA, then ribosomes cannot make the gene product. For example, various regulatory proteins affect the ability of RNA polymerase to bind to a promoter and initiate transcription. Genes that are controlled in this way are said to undergo **transcriptional control:**

$$\text{DNA} \not\rightarrow \text{mRNA} \rightarrow \text{protein} \rightarrow \text{activated protein}$$

2. If the mRNA for an enzyme has been transcribed, the cell might have a way to prevent the mRNA from being translated into protein. Mechanisms that alter the length of time an mRNA survives before it is degraded by ribonucleases, that affect translation initiation, or that affect elongation factors and other proteins during the translation process are forms of **translational control:**

$$\text{DNA} \rightarrow \text{mRNA} \not\rightarrow \text{protein} \rightarrow \text{activated protein}$$

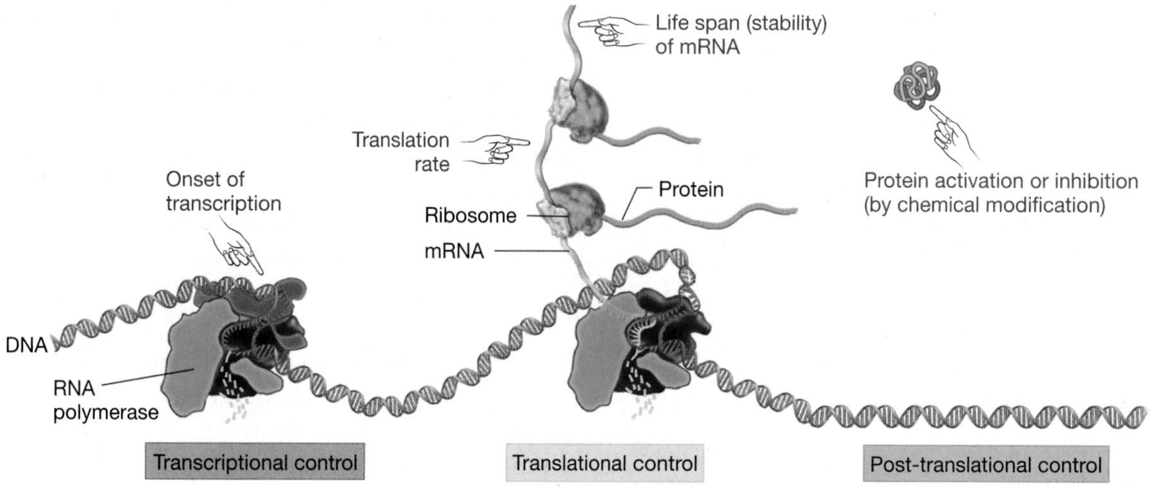

FIGURE 17.1 Gene Expression Can Be Regulated at Three Levels
Although this chapter focuses on how regulatory molecules affect the ability of RNA polymerase to initiate transcription, translational and post-translational controls also occur in bacteria.

3. Chapter 16 pointed out that some proteins are manufactured in an inactive form and have to be activated by chemical modification, such as the addition of a phosphate group. This type of regulation is **post-translational control:**

$$\text{DNA} \rightarrow \text{mRNA} \rightarrow \text{protein} \not\rightarrow \text{activated protein}$$

Which of these three forms of control occur in bacteria? The short answer to this question is "all of the above." As **Figure 17.1** shows, many factors affect how much active protein is produced from a particular gene. Transcriptional control is particularly important due to its efficiency—it saves the most energy for the cell, because it stops the process at the earliest possible point. Translational control is advantageous because it allows a cell to change quickly which proteins are produced. Post-translational control is significant as well, and it provides the most rapid response of all three mechanisms. Among these mechanisms of gene regulation, there is a clear trade-off between the speed of response and the conservation of ATP, amino acids, and other resources. Transcriptional control is slow but efficient in resource use. Post-translational control is fast but energetically expensive.

Although this chapter focuses almost exclusively on mechanisms of transcriptional control in bacteria, it is important to keep in mind that both translational and post-translational controls occur in these organisms. It's also important to recognize that some genes—such as those that code for the enzymes required for glycolysis—are transcribed all the time, or **constitutively.** But the expression of other genes is regulated, meaning that they may be induced or repressed. Finally, it is critical to realize that gene expression is not an all-or-none proposition. Genes are not just "on" or "off"— instead, the level of expression is highly variable. What factors determine how much a particular gene is expressed at a particular time?

Metabolizing Lactose—A Model System

As Chapters 13 through 16 have shown, many of the fundamental advances in genetics have been achieved through the analysis of a model system. Studying the inheritance of seed shape in garden peas revealed the fundamental patterns of gene transmission. Exploring transcription in viruses and *E. coli* led to the discovery of RNA polymerase, transcription factors (such as the sigma proteins), and promoters. In studies of gene regulation, the key model system has been the metabolism of the sugar lactose in *E. coli.*

Jacques Monod, François Jacob, and many colleagues introduced lactose metabolism in *E. coli* as a model system during the 1950s and 1960s. Although they worked with a single species of bacterium, their results had a profound effect on thinking about gene regulation in all organisms. Some details turned out to be specific to the *E. coli* genes responsible for lactose metabolism, but many of Monod and Jacob's results are universal.

Escherichia coli can use a wide variety of sugars for ATP production, via glycolysis and the electron transport chain or via fermentation. These sugars also serve as a raw material in the synthesis of amino acids, vitamins, and other complex compounds. Glucose is *E. coli's* preferred carbon source, however, meaning that it is the source of energy and carbon atoms that the organism uses most efficiently. This observation is logical, because glycolysis begins with glucose and is the main pathway for the production of ATP. Lactose, the sugar found in milk, is also used by *E. coli*, but only when glucose supplies are depleted. Recall from Chapter 5 that lactose is a disaccharide made up of one molecule of glucose and one molecule of galactose.

To use lactose, *E. coli* must first transport the sugar into the cell. Once lactose is inside the cell, *E. coli* can break the sugar down into glucose and galactose by means of the enzyme β-galactosidase. The glucose released by this reaction is used directly via glycolysis; other enzymes convert the galactose to an intermediate in the glycolytic pathway.

In the early 1950s, biologists discovered that *E. coli* produces high levels of β-galactosidase only when lactose is present in the environment. If lactose is absent, little β-galactosidase is synthesized. Because lactose appears to induce the production of the enzyme that breaks itself down, researchers proposed that lactose itself regulates the gene for β-galactosidase. More formally, the hypothesis was that lactose acts as an inducer. An **inducer** is a molecule that stimulates the expression of a specific gene.

In the late 1950s Jacques Monod investigated how the presence of glucose affects the regulation of the β-galactosidase gene. Would *E. coli* produce high levels of β-galactosidase when both glucose and lactose were present in the surrounding environment? **Figure 17.2** summarizes the experimental approach by which Monod found that the answer was no. This meant that β-galactosidase production is induced by the availability of lactose, but only when glucose is unavailable. The key observations can be summarized as follows:

- Cells grown on glucose only: no β-galactosidase production
- Cells grown on glucose + lactose: no β-galactosidase production
- Cells grown on lactose only: β-galactosidase production

Monod teamed up with François Jacob to investigate exactly how lactose and glucose regulate the genes responsible for lactose metabolism. Research on how these genes are regulated opened a window on how genes in all organisms are controlled. Research on this system is still continuing, some 50 years later.

Question: *E. coli* produces β-galactosidase when lactose is present. Does *E. coli* produce β-galactosidase when both glucose and lactose are present?

Hypothesis: *E. coli* does not produce β-galactosidase when glucose is present, even if lactose is present. (Glucose is the preferred food source.)

Alternative hypothesis: *E. coli* produces β-galactosidase whenever lactose is present, regardless of the presence or absence of glucose.

Experimental setup:

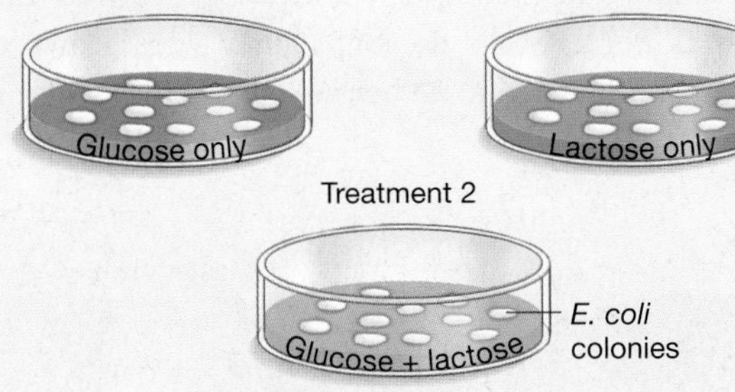

Prediction: β-Galactosidase will be produced only in treatment 3.

Prediction of alternative hypothesis: β-Galactosidase will be produced in treatments 2 and 3.

Results:

Treatment 1	**Treatment 2**	**Treatment 3**
No β-galactosidase	No β-galactosidase	High production of β-galactosidase

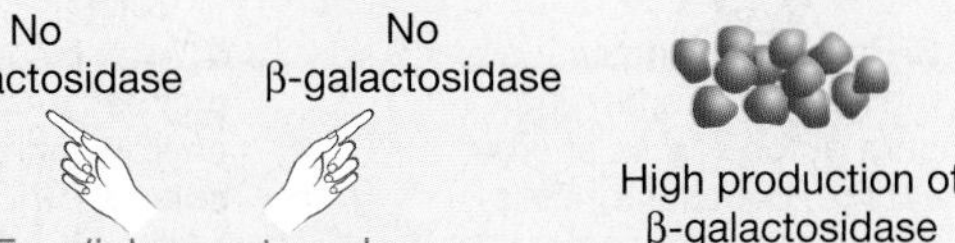

E. coli does not produce β-galactosidase if glucose is present

Conclusion: Glucose prevents the expression of the gene for β-galactosidase. The presence of lactose without glucose stimulates the expression of that gene.

FIGURE 17.2 Glucose Affects the Regulation of the β-Galactosidase Gene

17.2 Identifying the Genes Involved in Lactose Metabolism

To understand how *E. coli* controls production of β-galactosidase and the membrane transport protein that brings lactose into the cell, Monod and Jacob first had to find the genes that code for these proteins. To do this, they employed the same tactic used in the pioneering studies of DNA replication, transcription, and translation reviewed in earlier chapters: They isolated and analyzed mutant individuals. In this case, their goal was to find *E. coli* cells that were *not* capable of metabolizing lactose. Cells that can't use lactose must lack either β-galactosidase or the lactose-transporter protein.

Finding mutants with respect to a particular trait was a two-step process. The researchers' first step was to generate a large number of individuals with mutations at random locations in their genomes. To produce mutant cells, Monod and others exposed *E. coli* populations to X-rays, UV light, or chemicals that damage DNA and increase mutation rates. The second step is to screen the mutants to find individuals with defects in the process or biochemical pathway in question—in this case, defects in lactose metabolism. A technique that allows researchers to identify individuals with a particular type of mutation is called a **screen**.

Monod and colleagues were looking for cells that cannot grow in an environment that contains only lactose as an energy source. Normal cells grow well in this environment. How could the researchers select cells on the basis of *lack* of growth?

Screening Mutants—Replica Plating and Indicator Plates

Replica plating and growth on indicator plates were key techniques in the search for mutants with defects in lactose metabolism. **Replica plating** begins with spreading mutagenized bacteria on a plate that is filled with gelatinous agar containing glucose, as illustrated in **Figure 17.3**. This plate is known as the **master plate**. The bacteria are then allowed to grow, so that each cell produces a single colony. (A bacterial colony consists of a large number of identical cells that are descended from a single cell.) Next, a block covered with a piece of sterilized velvet is pressed onto the master plate. Because of the contact, cells from each colony on the master plate are transferred to the velvet. Then the velvet is pressed onto a plate containing a medium that differs from that of the master plate by a single component. For instance, in Figure 17.3, the second medium has only lactose as a source of carbon and energy. Cells from the velvet stick to the plate's surface, producing an exact copy of the colonies on the master plate. This copy is called the **replica plate**. After these cells grow, an investigator can compare the colonies that thrive on the replica plate's medium with those on the master plate. In this case, colonies that grow on the master plate but that are missing on the replica plate represent mutants deficient in lactose metabolism. By picking these particular colonies from the master plate, researchers build a collection of lactose mutants.

Monod also used an alternative strategy based on **indicator plates**, where mutants with metabolic deficiencies are observed directly. Consider what happens when mutagenized bacteria are grown on agar plates containing lactose and are then sprayed with a solution containing the colorless compound ONPG (o-nitrophenyl-β-D-galactoside). ONPG consists of two subunits—a galactose molecule and a molecule called o-nitrophenol—that are joined by the same β-1,4 glycosidic linkage that joins the galactose and glucose subunits of lactose. Consequently, ONPG acts as an alternative substrate for β-galactosidase. When the enzyme cleaves ONPG, galactose and o-nitrophenol are released. This is significant, because o-nitrophenol is intensely yellow. When colonies exposed to ONPG turn yellow, it means that they have β-galactosidase activity. Colonies that stay white are unable to cleave ONPG, meaning they have a defect in the enzyme or its production.

Different Classes of Lactose Metabolism Mutants

The initial mutant screen yielded the three types of mutants summarized in **Table 17.1**. In one class, the mutant cells were unable to cleave ONPG even if lactose was present inside the cells to induce production of the β-galactosidase protein. The investigators concluded that these mutants must lack a functioning version of the β-galactosidase protein—meaning the gene that encodes β-galactosidase is defective. This gene was designated *lacZ*, and the mutant allele *lacZ*$^-$.

In the second class of mutants, the cells failed to accumulate lactose inside the cell. In contrast, the concentration of lactose in wild-type cells is about 100 times that of lactose in the surrounding environment. To interpret this result, Jacob and Monod hypothesized that the mutant cells had defective copies of the membrane protein responsible for transporting lactose into the cell. This protein was identified and named **galactoside permease**; the gene that encodes it was designated *lacY*.

The third and most surprising class of mutants did not regulate expression of β-galactosidase and galactoside permease normally. For example, when these mutant cells were grown on lactose alone and then sprayed with ONPG, they turned yellow just as normal cells do. But if they were grown on a medium that contained glucose but no lactose, they still turned

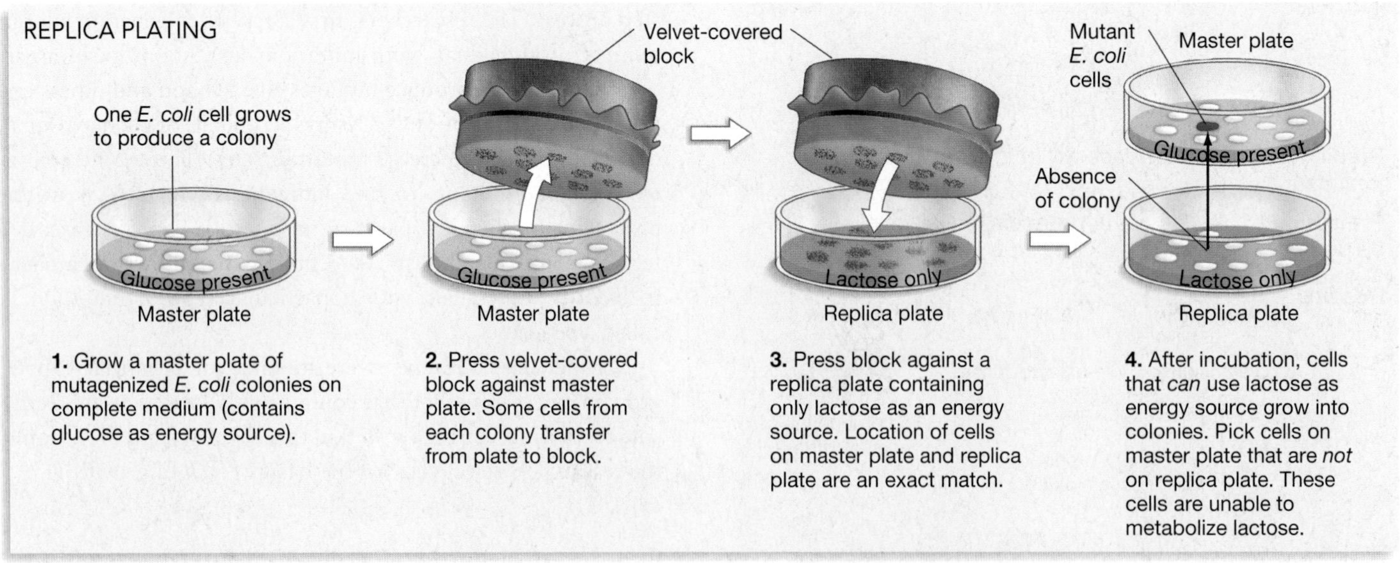
REPLICA PLATING
One E. coli cell grows to produce a colony
Glucose present
Master plate
1. Grow a master plate of mutagenized E. coli colonies on complete medium (contains glucose as energy source).
Velvet-covered block
Glucose present
Master plate
2. Press velvet-covered block against master plate. Some cells from each colony transfer from plate to block.
Lactose only
Replica plate
3. Press block against a replica plate containing only lactose as an energy source. Location of cells on master plate and replica plate are an exact match.
Mutant E. coli cells
Master plate
Glucose present
Absence of colony
Lactose only
Replica plate
4. After incubation, cells that can use lactose as energy source grow into colonies. Pick cells on master plate that are not on replica plate. These cells are unable to metabolize lactose.

FIGURE 17.3 Replica Plating Is a Technique for Identifying Mutant Cells
Replica plating, used here to isolate mutant *E. coli* cells with a deficiency in lactose metabolism. There are two keys to this technique: (1) The location of colonies on the master plate and replica plate must match exactly, and (2) the media in the two plates must differ by just one type of component. QUESTION How would you alter this protocol to isolate mutant cells with a deficiency in the enzymes required to synthesize tryptophan?

TABLE 17.1 Three Distinct Types of Mutants in Lactose Metabolism of *E. coli*

Mutant Phenotype	Interpretation	Inferred Genotype
Cells cannot cleave ONPG even if lactose is present as an inducer.	No β-galactosidase; gene for β-galactosidase is defective. Call this gene *lacZ*.	$lacZ^-$
Cells cannot accumulate lactose inside the cell.	No membrane protein (galactoside permease) required for import of lactose; gene for galactoside permease is defective. Call this gene *lacY*.	$lacY^-$
ONPG is cleaved even if lactose is absent (no inducer).	Constitutive expression of *lacZ* and *lacY*; gene for regulatory protein that shuts down *lacZ* and *lacY* is defective—it does not need to be induced by lactose. Call this gene *lacI*.	$lacI^-$

yellow after being sprayed with ONPG. Normal cells, in contrast, do not produce β-galactosidase under these conditions. Normal cells remain white when grown on glucose and then sprayed with ONPG.

Cells that are abnormal because they produce a product at all times are called **constitutive mutants.** Unlike normal cells, their product is always part of the cell's constitution. The gene that mutated to produce constitutive β-galactosidase expression was named *lacI*. The use of "I" was appropriate because these mutants did not need an inducer to express β-galactosidase or galactoside permease. Recall that, in normal cells, the expression of these genes is induced by the presence of lactose. But in cells with a mutant form of *lacI* ($lacI^-$ mutants), gene expression occurred with or without lactose. This meant that $lacI^-$ mutants have a defect in gene regulation. In these mutants, the gene remains "on" when it should be turned off.

Based on these observations, it is logical to infer that the normal product of the *lacI* gene prevents the transcription of *lacZ* and *lacY* when lactose is absent. Because lactose acts as an inducer for the production of β-galactosidase, it is reasonable to expect that the *lacI* gene or gene product interacts with lactose in some way. (Later work showed that the inducer is a derivative of lactose called *allolactose*. For the sake of historical accuracy and simplicity, however, this discussion refers to lactose itself as the inducer.)

Several Genes Are Involved in Metabolizing Lactose

Jacob and Monod had succeeded in identifying three genes involved in lactose metabolism: *lacZ, lacY*, and *lacI*. They had concluded that *lacZ* and *lacY* code for proteins involved in the metabolism and import of lactose, while *lacI* is responsible for some sort of regulatory function. The *lacI* gene or gene product repressed the expression of *lacZ* and *lacY* in the absence of lactose, but when lactose was present, the opposite occurred—transcription of *lacZ* and *lacY* was induced.

When Jacob and Monod followed up on these experiments by mapping the physical location of the three genes on *E. coli*'s circular chromosome, they found that the genes are close together (**Figure 17.4**). This was a crucial result, because it suggested that both *lacZ* and *lacY* might be controlled by *lacI*. Could one regulatory gene manage more than one protein-encoding gene? If so, how does *lacI* actually work? And why do lactose and glucose have opposite effects on it?

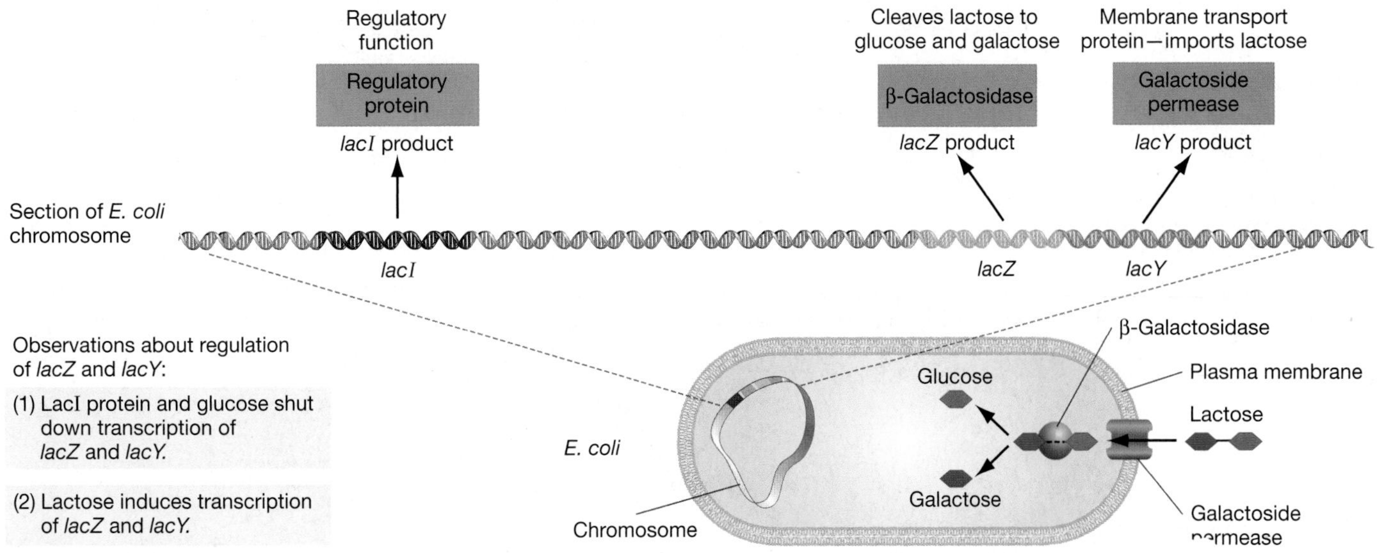

FIGURE 17.4 The *lac* Genes Are in Close Physical Proximity
Early experiments on lactose use in *E. coli* identified the *lacI, lacZ,* and *lacY* genes, documented the function of their protein products, and mapped their locations on the *E. coli* chromosome.

17.3 The Discovery of the Repressor

In principle, there are two ways that transcription can be regulated: via positive control or negative control. **Positive control** is in effect when something must be added for transcription to occur; **negative control** is in effect when something must be taken away for transcription to occur. In starting up a car, positive control occurs when you step on the gas pedal, meaning that you add fuel; negative control occurs when the parking brake is set. It turned out that the *lacZ* and *lacY* genes in *E. coli* are controlled by both a gas pedal and a parking brake—that is, they are under positive control *and* negative control.

The hypothesis that the *lacZ* and *lacY* genes might be under negative control originated with Leo Szilard in the late 1950s. Szilard suggested to Jacques Monod that the *lacI* gene codes for a product that represses transcription of the *lacZ* and *lacY* genes. Recall from Section 17.2 that, in *lacI*$^-$ mutants, the production of β-galactosidase and galactoside permease is constitutive, even if lactose is not available. Szilard proposed that regulation fails because the LacI$^+$ protein—which normally prevents enzyme synthesis—is inactive. Stated another way, the *lacI* gene produces an inhibitor that exerts negative control over the *lacZ* and *lacY* genes. This transcription inhibitor was called a **repressor** and was thought to bind directly to DNA near or on the promoter for the *lacZ* and *lacY* genes (**Figure 17.5a**). To explain how the presence of lactose triggers transcription in normal cells, Szilard and Monod proposed that lactose interacts with the repressor in a way that makes the repressor release from its binding site (**Figure 17.5b**). The idea was that lactose induces transcription by removing negative control. Lactose releases the parking brake. As **Figure 17.5c** shows, it was logical to observe that constitutive transcription occurs in *lacI*$^-$ mutants because a functional repressor is absent—the parking brake is broken.

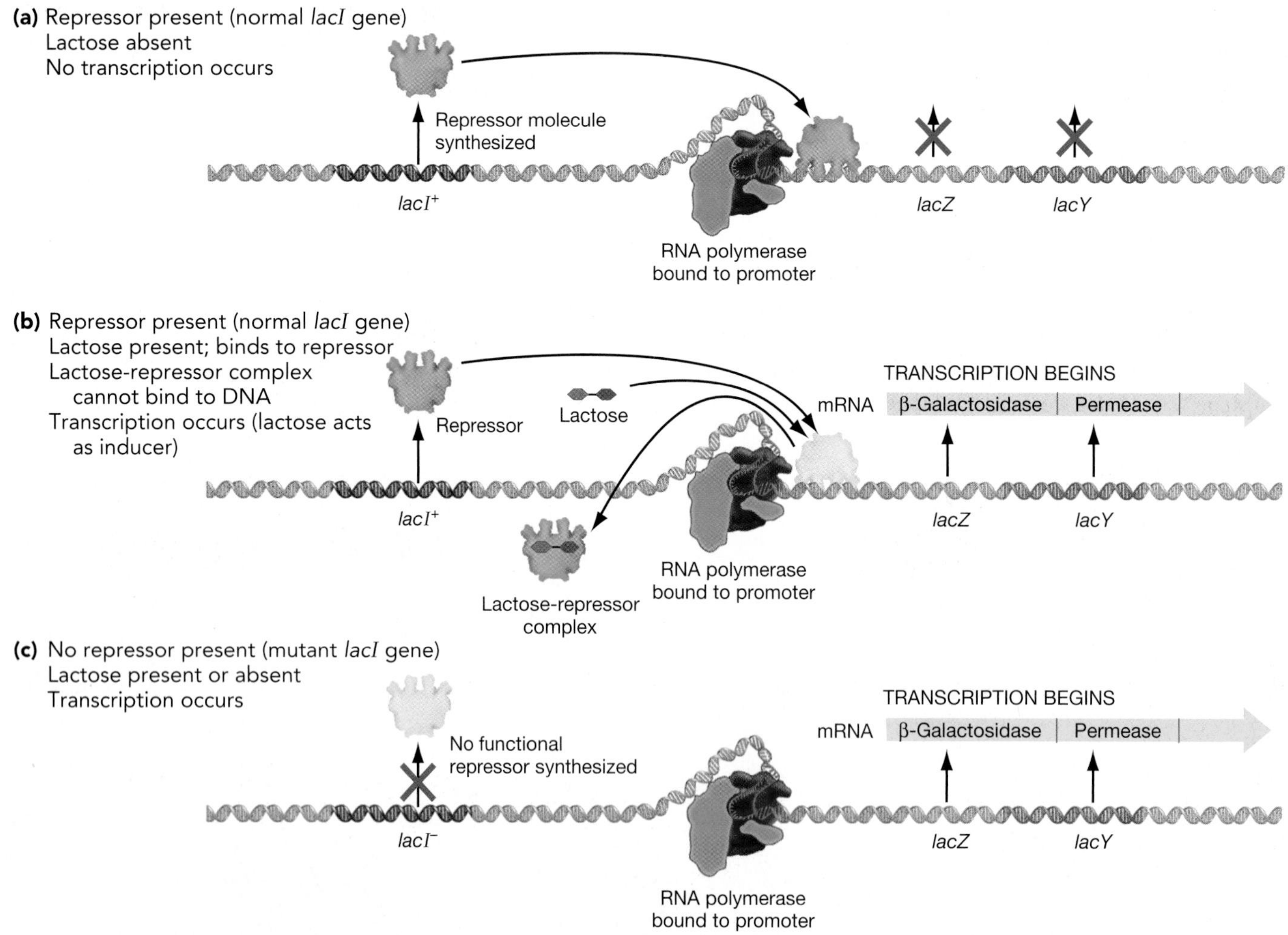
(a) Repressor present (normal lacI gene)
Lactose absent
No transcription occurs
Repressor molecule synthesized
lacI+
lacZ
lacY
RNA polymerase bound to promoter
(b) Repressor present (normal lacI gene)
Lactose present; binds to repressor
Lactose-repressor complex cannot bind to DNA
Transcription occurs (lactose acts as inducer)
Repressor
Lactose
TRANSCRIPTION BEGINS
mRNA
β-Galactosidase
Permease
lacI+
lacZ
lacY
Lactose-repressor complex
RNA polymerase bound to promoter
(c) No repressor present (mutant lacI gene)
Lactose present or absent
Transcription occurs
No functional repressor synthesized
TRANSCRIPTION BEGINS
mRNA
β-Galactosidase
Permease
lacI−
lacZ
lacY
RNA polymerase bound to promoter

FIGURE 17.5 The Hypothesis of Negative Control
The negative-control hypothesis maintains **(a)** that transcription of genes involved in lactose use is normally blocked by a repressor molecule that binds to DNA on or near the promoter for *lacZ* and *lacY*, **(b)** that lactose induces transcription of *lacZ* and *lacY* by interacting with the repressor, and **(c)** that, when a functional repressor is absent, transcription proceeds.

How could researchers assess these hypotheses? This was no trivial task. But in a set of brilliant experiments, each of the predictions of the negative-control hypothesis was tested and confirmed. The key studies were based on creating *E. coli* cells that were diploid for the genes involved in lactose metabolism. For example, consider an experiment that begins with cells that are haploid and have the genotype I^- $lacZ^+lacY^+$. These cells lack the hypothesized repressor but have normal copies of the genes for β-galactosidase or galactoside permease. As predicted by the hypothesis of negative control, they produce β-galactosidase constitutively. But what happens when these cells are made diploid by the addition of DNA with the genotype I^+ $lacZ^-lacY^-$? The second set of genes codes for a functioning copy of the repressor, but not of β-galactosidase or galactoside permease. Once the cells are made diploid, β-galactosidase production declines and then stops. This result supports the hypothesis that the repressor codes for a protein that shuts down transcription. As predicted, the repressor is the "parking brake" on transcription. But if an inducer such as lactose is added to these diploid cells, β-galactosidase activity continues. This result supports the hypothesis that lactose removes the repressor—it releases the parking brake. Stated another way, lactose acts as an inducer by removing negative control.

The *lac* Operon

Jacob and Monod summarized the results of their experiments with a comprehensive model of negative control that was published in 1961. In essence, their experimental data had confirmed the hypotheses illustrated in Figure 17.5. One of their key conclusions was that the genes for β-galactosidase and galactoside permease are controlled together. To capture this point, they coined the term **operon** for a set of coordinately regulated bacterial genes that are transcribed together into one mRNA. Logically enough, the group of genes involved in lactose metabolism was termed the ***lac* operon.**

Three hypotheses are central to the Jacob-Monod model of *lac* operon regulation:

1. The *lacZ* and *lacY* genes are adjacent and are transcribed into one mRNA initiated from the single promoter of the *lac* operon. As a result, the expression of these two genes is coordinated.

2. The repressor is a protein encoded by *lacI* that binds to DNA and prevents transcription of *lacZ* and *lacY*. The binding site for the repressor is a section of DNA in the *lac* operon called the **operator**. Jacob and Monod proposed that *lacI* is expressed constitutively and that its product (the repressor) prevents transcription by blocking RNA polymerase from making contact with the promoter.

3. The inducer (lactose) interacts directly with the repressor by binding to it. As a result, the repressor changes shape in a way that causes it to drop off the DNA strand. Transcription is then able to proceed because RNA polymerase can access the promoter. Recall from Chapter 3 that this form of control over protein function is called *allosteric regulation*. When allosteric regulation occurs, a small molecule binds directly to a protein and causes it to change its shape and activity. When the inducer binds to the repressor, negative control ends.

WEB TUTORIAL 17.1 The *lac* Operon

Testing the *lac* Operon Model

Did Jacob and Monod "get it right"? After their model of negative control was published, Jacob and Monod confirmed the existence of the operator. They did this by finding *E. coli* mutants that expressed *lacZ* constitutively yet had normal forms of the repressor. In these new mutants, the repressor protein was unable to function because the nucleotide sequence of the operator was altered. In 1967 Walter Gilbert and Benno Müller-Hill were able to tag copies of the repressor protein with a radioactive atom. Their experiments showed that the repressor physically binds to the DNA sequences of the operator. This result confirmed that the operator is not a protein or an RNA product, but instead is part of the DNA sequence of the *lac* operon.

The only major modification of the original *lac* operon model is based on the recent recognition that the repressor does not physically block RNA polymerase from contacting the promoter, as Jacob and Monod had proposed. Instead, the repressor prevents the RNA polymerase holoenzyme from opening the DNA helix once that enzyme has bound to the promoter. As a result, the template strand cannot be threaded through the enzyme's active site so that transcription can begin.

In addition, a third gene of the *lac* operon, *lacA*, was discovered after the original model of negative control was published. The *lacA* gene is tightly linked to *lacY* and *lacZ* and is transcribed as part of the same mRNA strand that carries *lacZ* and *lacY*. The gene codes for the enzyme transacetylase, which transfers acetyl (CH_3COO^-) groups to β-galactoside sugars. The protein's function is protective in nature: The acetylation step allows sugars that cannot be metabolized to be exported from the cell instead of accumulating to dangerous levels within the cell.

An mRNA like the one produced by the *lac* operon is known as a polycistronic mRNA. **Polycistronic mRNAs** contain more than one protein-encoding segment. In most cases, bacterial genes that are transcribed as a polycistronic mRNA code for products devoted to a common end, such as the *lacZ* and *lacY* products required for lactose metabolism. Similarly, the polycistronic *trp* (pronounced "trip") operon in *E. coli* consists of five genes, each of which is required for different steps in the synthesis of the amino acid tryptophan. The *trp* operon is also under negative control—but with a twist explained in **Box 17.1**, page 370. Biologists interpret the polycistronic structure of many bacterial genes as an adaptation

BOX 17.1 Almost Parallel Worlds: Similarities and Contrasts in Control of the *trp* and *lac* Operons

The genes in the *lac* operon code for proteins that are responsible for breaking down a compound in the cell, lactose. Logically enough, the expression of these genes depends on the presence of lactose, the compound that is to be broken down. When the inducer, a lactose derivative, binds to the repressor, the repressor comes off the operator and transcription begins.

Now imagine how a cell would regulate an operon that is responsible for synthesizing a molecule instead of breaking it down. In this case, the cell should stimulate gene expression when the molecule is absent but should reduce transcription when the molecule is present. To see how cells accomplish this task, let's explore how *E. coli* regulates the **trp operon**—a set of five cotranscribed genes, each of which is required for the synthesis of the amino acid tryptophan.

If RNA polymerase binds to the promoter for the *trp* operon and begins transcription, all five *trp* operon genes are transcribed and translated together. Once translation is complete, the synthesis of tryptophan commences.

The *trp* operon is under negative control. It has an operator sequence that overlaps the promoter, and a repressor protein that binds to the operator to prevent transcription—as in the *lac* operon.

In the *trp* operon, tryptophan binds to the repressor to regulate transcription, just as allolactose binds to the *lac* operon's repressor to regulate transcription. There is a crucial difference in the two systems, however. If tryptophan binds to the repressor, it means that tryptophan is present in the cell and transcription should not proceed. This is exactly what biologists observe: As **Figure 17.6a** shows, the *trp* repressor binds operator DNA and exerts negative control only when the repressor is complexed with tryptophan. When tryptophan is absent, the repressor does not bind to the operator and the operon is expressed constitutively (**Figure 17.6b**). This is exactly the opposite of how the presence of lactose affects the *lac* repressor. And it makes perfect sense for the *trp* operon: Bacterial cells will have higher fitness if the operon is transcribed when tryptophan levels are low but shut down when tryptophan is abundant.

In addition to offering a new twist on negative control, the *trp* operon has provided other fascinating insights into gene regulation. Experiments have revealed that negative control in the *trp* operon is supplemented by a regulatory mechanism called **attenuation**. Attenuation works like a dimmer switch, because it superimposes fine-tuned control on what would otherwise be a simple on-off system. In the case of the *trp* operon, the dimmer is sensitive to tryptophan levels. When tryptophan levels are high, attenuation causes transcription to terminate early in the operon. When tryptophan levels are low, transcription proceeds through the entire operon. Attenuation that is based on amino acid concentration is seen in many other operons devoted to amino acid synthesis.

After considering how the *lac* and *trp* operons are regulated in *E. coli*, it's difficult not to have new respect for "simple" forms of life. Bacteria may be small and unicellular, but they perform fantastically complex, finely controlled chemistry. Even the simplest cells are marvelous machines.

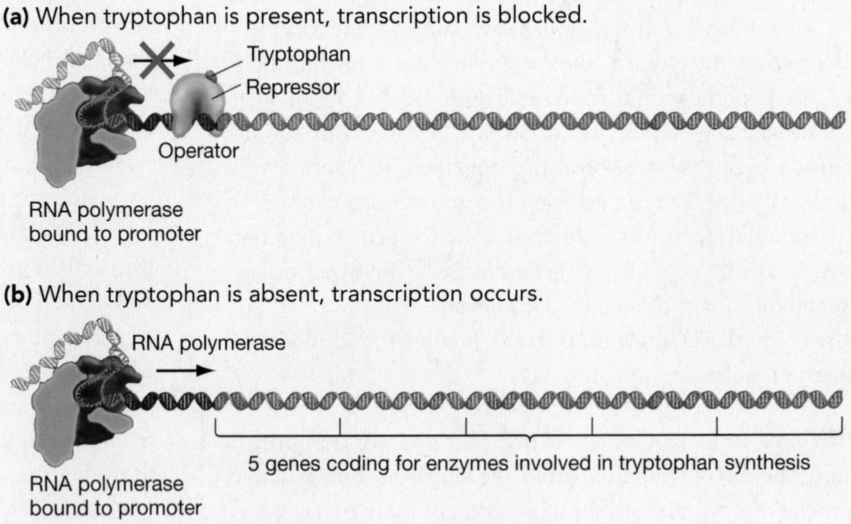

FIGURE 17.6 The *trp* Operon Is under Negative Control
(a) When tryptophan is present, it binds to a repressor protein. The tryptophan-repressor complex binds to the operator and shuts down transcription, which ends the synthesis of tryptophan. **(b)** The *trp* operon contains five coding genes.

that increases the efficiency of gene expression. By binding to one promoter, regulatory proteins can control the expression of several genes that are required for the same process. Polycistronic mRNAs are common in bacteria but rare or nonexistent in most eukaryotes.

The Impact of the *lac* Operon Model

Regulation of the *lac* operon provided an immensely important model system in genetics, partly because aspects of the model turned out to be important in many other species, genes, and operons. For example, follow-up work has shown that numerous bacterial genes are under negative control via repressor proteins. Even more important, the *lac* operon model introduced the idea that gene expression is regulated by physical contact between specific regions in regulatory proteins and specific regulatory sites in DNA. Stated another way, the model introduced the idea that gene expression could be modified by proteins that bind directly to regulatory sequences in DNA. Publication of the *lac* operon model was a watershed event in the history of biological science. Its impact on thinking about gene regulation is hard to overestimate.

As elegant as the model presented in Figure 17.5 is, however, something is missing. Where does glucose fit in? This is the final element of the regulatory system for lactose metabolism. As it turns out, it is also the element that is based on positive control of transcription.

✓CHECK YOUR UNDERSTANDING

Negative control occurs when something must be taken away for transcription to occur. The *lac* operon repressor exerts negative control over three protein-coding genes by binding to the operator site in DNA near the promoter. For transcription to occur, an inducer molecule (a derivative of lactose) must bind to the repressor, causing it to release from the operator. You should be able to (1) explain why it is logical to observe that a molecule derived from lactose induces transcription of the *lac* operon; (2) diagram the *lac* operon, showing the relative positions of the operator, the promoter, and the three protein-coding genes; and (3) show what that operon looks like in the presence and absence of lactose.

17.4 Catabolite Repression and Positive Control

When glucose is present in the environment, transcription of the *lac* operon is reduced. This is true even when lactose is available to induce β-galactosidase expression. Given that glucose is the preferred carbon source, this aspect of control over the *lac* operon is logical. Glucose is produced when β-galactosidase cleaves lactose. When glucose is present, it is not necessary for the cell to cleave lactose and produce still more glucose.

Recall from Chapter 9 that biologists use the term *catabolism* to refer to reactions that break larger molecules into simpler subunits. The hydrolysis of lactose into its glucose and galactose subunits is an example of catabolism (**Figure 17.7a**). In many cases, operons that encode catabolic enzymes are inhibited when the end product of the reaction, the **catabolite**, is abundant. This form of end-product inhibition is **catabolite repression**. In the case of the *lac* operon, glucose is the catabolite. When glucose is abundant in the cell, transcription of the *lac* operon is decreased by catabolite repression (**Figure 17.7b**).

How does glucose prevent expression of the *lac* operon? An answer to this question began to emerge when researchers discovered a second major control element in the *lac* operon. This regulatory switch consists of a DNA sequence known as the **CAP binding site**, which is located just upstream of the *lac* promoter, and a regulatory protein called **catabolite activator protein (CAP)**, which binds to this DNA sequence.

To understand how CAP works, notice that not all promoters are created equal. Strong promoters allow efficient initiation of transcription by RNA polymerase; weak ones support much less efficient initiation of transcription. The *lac* promoter is weak. But when the CAP regulatory protein is bound to the

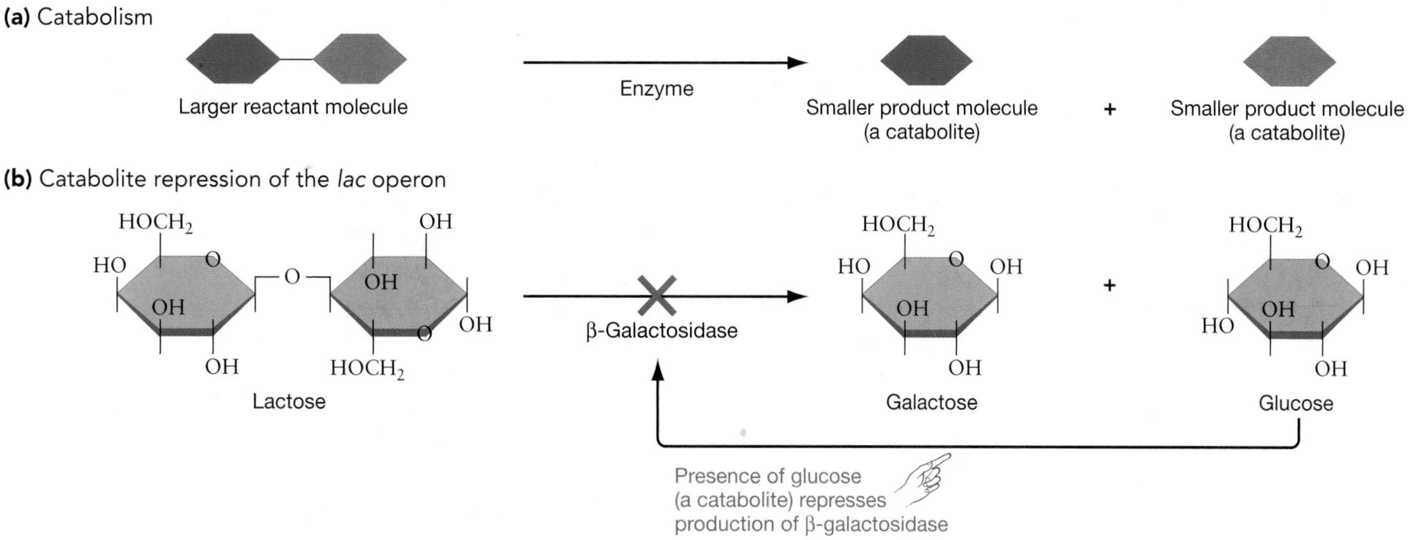

FIGURE 17.7 Catabolite Repression Is a Mechanism of Gene Regulation
(a) A generalized example of catabolism. **(b)** Catabolite repression occurs when one of the small product molecules represses the production of the enzyme responsible for the reaction. In the case of lactose metabolism, the production of β-galactosidase is suppressed when glucose is present.

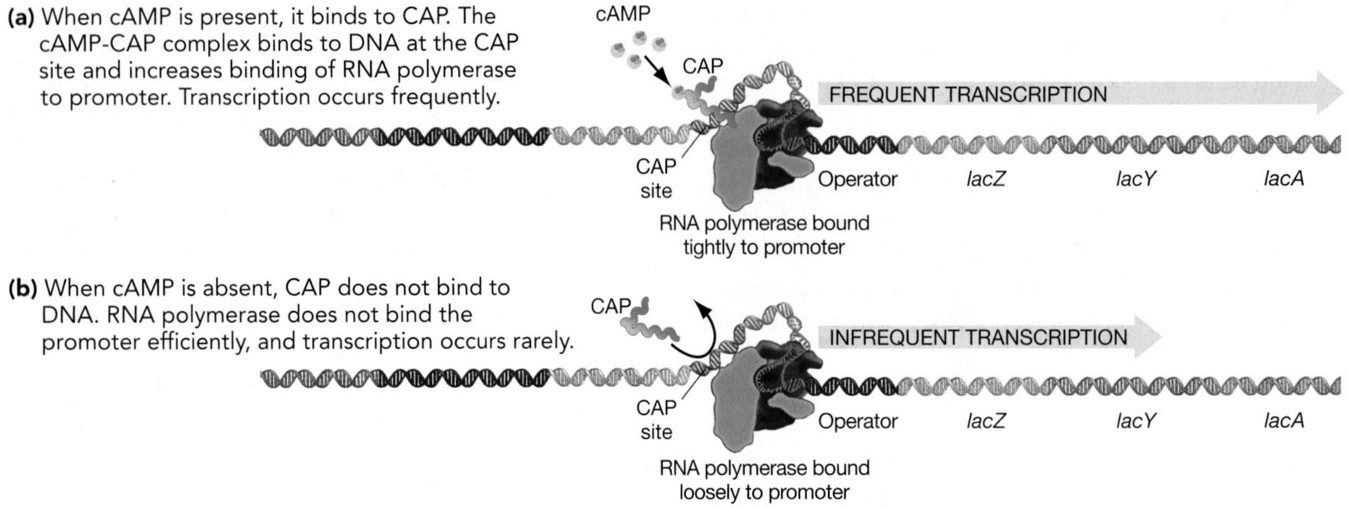
(a) When cAMP is present, it binds to CAP. The cAMP-CAP complex binds to DNA at the CAP site and increases binding of RNA polymerase to promoter. Transcription occurs frequently.
cAMP
CAP
FREQUENT TRANSCRIPTION
CAP site
Operator
lacZ
lacY
lacA
RNA polymerase bound tightly to promoter
(b) When cAMP is absent, CAP does not bind to DNA. RNA polymerase does not bind the promoter efficiently, and transcription occurs rarely.
CAP
INFREQUENT TRANSCRIPTION
CAP site
Operator
lacZ
lacY
lacA
RNA polymerase bound loosely to promoter

FIGURE 17.8 Positive Control of the lac Operon
(a) When glucose levels in an *E. coli* cell are low, cAMP is produced. cAMP then interacts with CAP to increase transcription of the *lac* operon. **(b)** When glucose is abundant, cAMP is rare in the cell and positive control does not occur.

CAP site just upstream of the *lac* promoter, the protein interacts with RNA polymerase in a way that allows transcription to begin much more frequently. CAP binding greatly strengthens the *lac* promoter. In this way, CAP exerts positive control of the *lac* operon. When CAP is active, transcription increases.

Researchers also discovered that CAP, like the repressor protein, is allosterically regulated. CAP changes shape when the regulatory molecule **cyclic AMP (cAMP)** binds to it. Only when cAMP is attached to it can CAP bind to DNA (**Figure 17.8a**). The same type of allosteric regulation goes on with the repressor and allolactose during negative control, except that the repressor binds to DNA only in the *absence* of the inducer. During positive control, the CAP-cAMP complex binds to the CAP binding site. As long as a repressor isn't bound to the operator, the complex increases the efficiency of transcription. If no cAMP is bound to CAP, then CAP has a conformation that does not allow binding to the CAP site (**Figure 17.8b**).

How Does Glucose Influence Formation of the CAP-cAMP Complex?

Where does the elusive molecule glucose fit into this scheme? Glucose's role in positive control of the *lac* operon is indirect. More specifically, it is mediated by the influence that glucose has on the concentration of cAMP. Glucose levels outside the cell and cAMP levels inside the cell are inversely related. When extracellular glucose concentrations are high, intracellular cAMP concentrations are low; when extracellular glucose concentrations are low, intracellular cAMP concentrations are high. This seesaw is driven by the enzyme **adenylyl cyclase**, which produces cAMP from ATP (**Figure 17.9a**). Adenylyl cyclase's activity is inhibited by extracellular glucose. To see the consequences of this fact, imagine a situation in which glucose is abundant outside the cell (**Figure 17.9b**). In this state, adenylyl cyclase activity is low. Therefore, cAMP levels inside the cell are low. CAP is not in a CAP-cAMP complex, so it does not have the conformation that allows it to bind to the CAP site and stimulate *lac* operon transcription. Conversely, when the extracellular concentration of glucose is low, there is an increase in the intracellular level of cAMP. In this case, the CAP-cAMP complex forms, binds to the CAP site, and allows RNA polymerase to initiate transcription efficiently. As long as the repressor is not bound to the operator, transcription of the *lac* operon is initiated frequently and lactose can be employed as an alternative energy source.

It is important to note that the CAP-cAMP system influences many genes in addition to the *lac* operon. CAP sites are found adjacent to the promoters for several operons that are required for the metabolism of sugars other than glucose. When glucose levels fall and cAMP concentrations rise, the effect on gene regulation is similar to ringing an alarm bell. In response to the alarm, genes that encode enzymes required for the use of lactose, maltose, glycerol, and other food sources may be turned on—but only if the metabolite of a particular operon is available to the cell. Conversely, cAMP levels are low when glucose supplies are adequate. In this case, the genes for these enzymes are expressed only at low levels even when the appropriate metabolite is present.

Figure 17.10 summarizes how positive control and negative control combine to regulate the *lac* operon. The general message of this figure is that interactions among regulatory elements produce finely tuned control over gene expression. Because positive and negative control elements are superimposed, *E. coli* fully activates the genes for lactose metabolism only when lactose is available and glucose is scarce or absent. In this way, control over gene expression increases the ability of these cells to compete, grow, and reproduce efficiently.

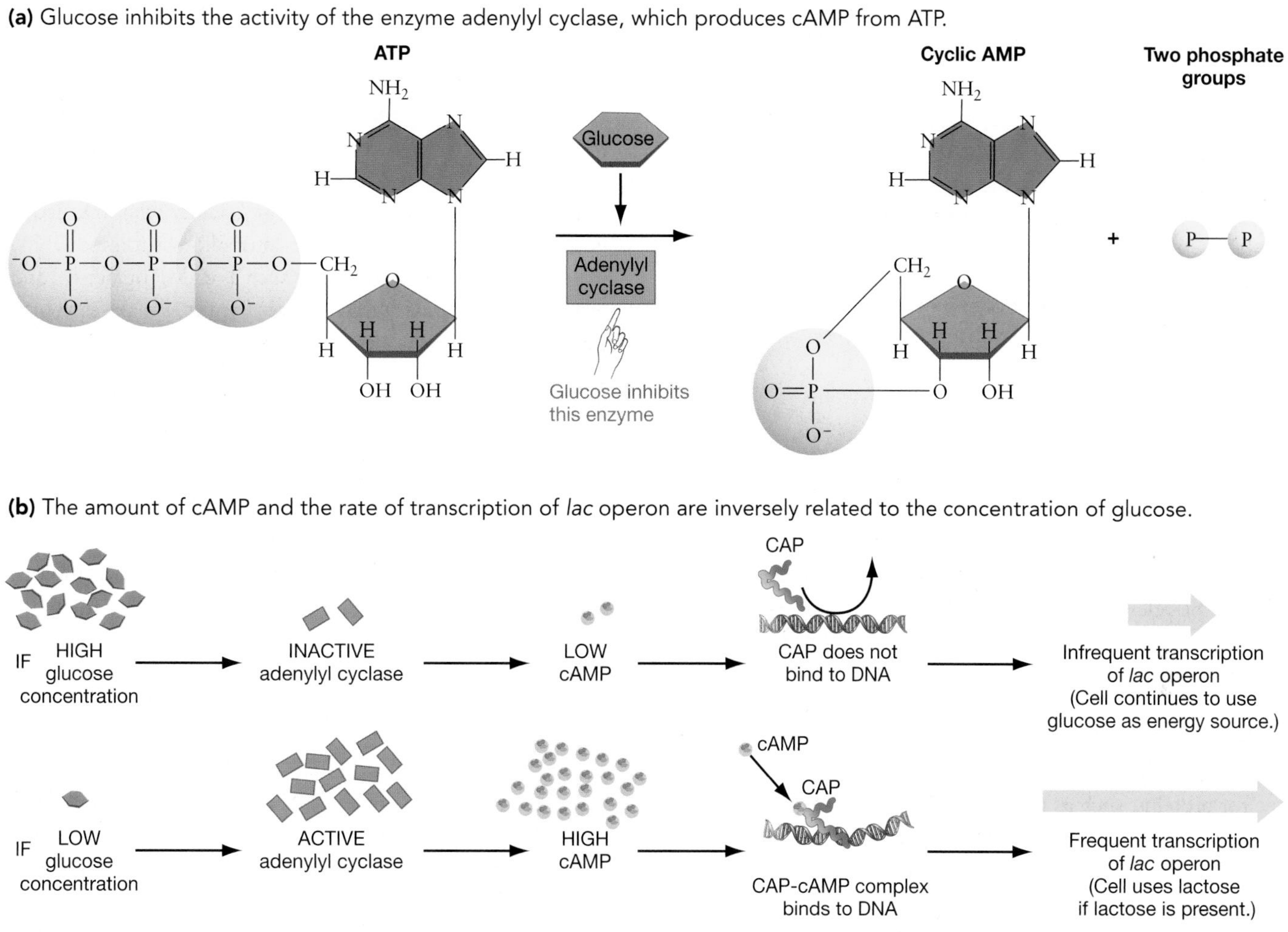

FIGURE 17.9 Cyclic AMP (cAMP) Is Synthesized When Glucose Levels Are Low

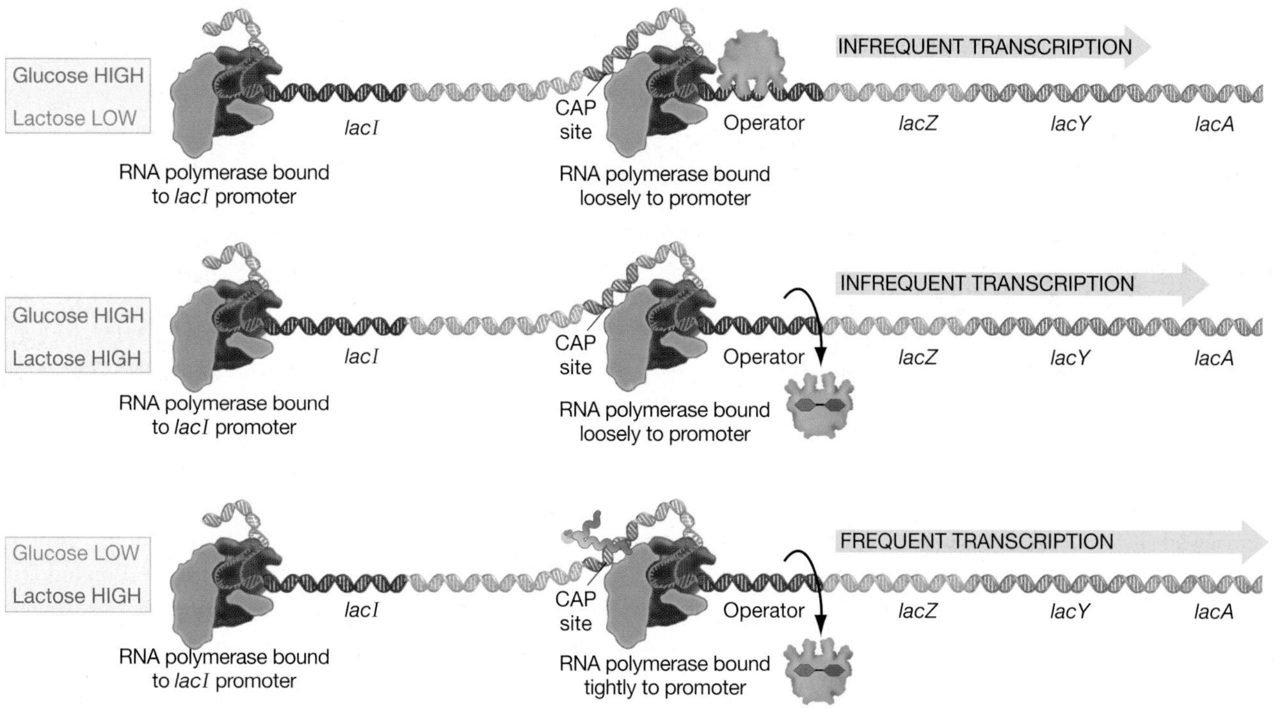

FIGURE 17.10 An Overview of Lactose Use

✓CHECK YOUR UNDERSTANDING

Positive control occurs when something must be added for transcription to occur. CAP exerts positive control over the *lac* operon by binding to the CAP site and increasing the transcription rate. CAP can bind only when it is complexed with cAMP—a signaling molecule indicating that glucose levels are low. You should be able to (1) diagram what the operon looks like when cAMP levels are low and high and (2) explain how positive control and negative control interact, based on the presence or absence of glucose and lactose.

17.5 The Operator and the Repressor—an Introduction to DNA-Binding Proteins

The insights that biologists gained from studying the *lac* operon proved to be relevant to many other genes and species. For example, transcription of bacterial and eukaryotic genes is either constitutive or controlled. When regulation occurs, it is negative, positive, or based on a combination of negative and positive factors. Negative control is based on repressor proteins, such as the *lacI* product, whereas positive control depends on transcription activator proteins, such as CAP. In both bacteria and eukaryotes, as in the *lac* operon, it is particularly common for genes to be regulated by a combination of several control systems. Catabolite repression is also extremely common.

Currently, biologists who are interested in gene regulation are trying to understand how DNA-binding proteins such as CAP and the repressor can control transcription. Precisely how do these proteins interact with DNA? How do small regulatory molecules such as cAMP and lactose induce changes in the conformation of these large DNA-binding proteins? Once again, investigators are using the *lac* operon as a model system that offers insights applicable to all cells.

Before taking a detailed look at the repressor and operator, let's step back and think about what is required for a DNA-binding protein to control transcription. First, the protein must bind to a specific DNA sequence associated with a specific gene or operon. Often an investigator's first task is to identify what that sequence is. Second, the protein's structure must allow it to bind DNA, interact with a regulatory molecule such as cAMP or lactose, and affect RNA polymerase. What structural features make each of these functions possible? As it turns out, the *lac* operon's operator DNA sequence and repressor protein structure have features that are common to many regulatory elements in bacteria and eukaryotes.

Finding the Operator

To understand how the *lac* repressor controls transcription, a logical first step is to identify the DNA sequences it targets: the operator. By mapping mutants with defects in the operator, researchers were able to confirm that the target site is just downstream from the promoter. But what is the actual DNA sequence?

Investigators use a technique called DNA footprinting to find and characterize the sequences targeted by DNA-binding proteins. DNA footprinting was used to characterize the first promoter ever discovered (see Chapter 16); the technique is explained in detail in **Box 17.2**.

Footprinting experiments with the *lac* repressor yielded two new discoveries. First, the *lac* operon actually contains three sites where the repressor protein may bind. As **Figure 17.12a** shows, these sites are called O_1, O_2, and O_3. When the repressor protein is bound to DNA, the protein simultaneously binds O_1 and one of the two other operators. This finding required a modification of the original Jacob-Monod model, which proposed the existence of a single operator. Second, all three

BOX 17.2 DNA Footprinting

DNA footprinting is used to characterize DNA sequences that are bound by regulatory proteins. Researchers begin a footprinting study by obtaining DNA from a region of interest—the *lac* operon of *E. coli*, for example. Then they generate many copies of the sequence, using techniques introduced in Chapter 19. The identical DNA fragments are then radioactively labeled at one end (see step 1 of **Figure 17.11**). Once labeling is complete, half of the fragments are mixed with the DNA-binding protein being studied—in this case, the repressor—and half are left without the DNA-binding protein (step 2). Both sets of fragments are then treated with a nuclease—an enzyme that cleaves DNA in random locations. The concentration of nuclease is so low and the exposure time is so short, however, that most fragments are never cut; those that are cut are cut just once. As step 3 shows, the nuclease treatment produces a population of DNA fragments of different lengths. In the sample containing the DNA-binding protein, the binding site is protected from the nuclease.

The DNA fragments produced by the nuclease treatment are separated by gel electrophoresis and visualized via autoradiography (see Chapter 3) or other techniques. As step 4 shows, the sample with the repressor lacks any cuts and thus any fragments in the segment that was protected by that protein. The gap in the series of fragments is the "footprint" of a DNA-binding protein. By aligning the footprinted region with the gene's DNA sequence, researchers can identify the nucleotide sequence of the binding site.

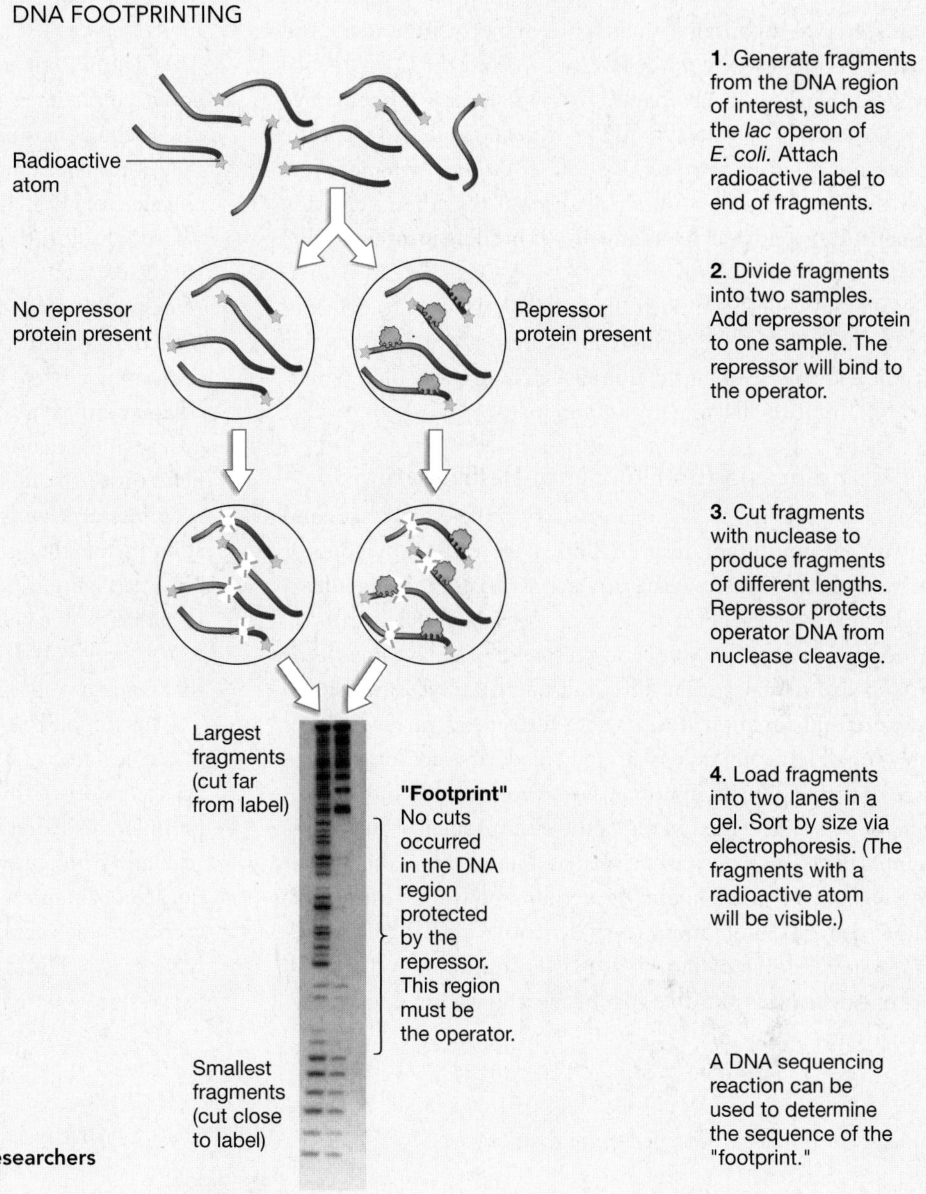
DNA FOOTPRINTING
Radioactive atom
1. Generate fragments from the DNA region of interest, such as the *lac* operon of *E. coli.* Attach radioactive label to end of fragments.
No repressor protein present
Repressor protein present
2. Divide fragments into two samples. Add repressor protein to one sample. The repressor will bind to the operator.
3. Cut fragments with nuclease to produce fragments of different lengths. Repressor protects operator DNA from nuclease cleavage.
Largest fragments (cut far from label)
"Footprint" No cuts occurred in the DNA region protected by the repressor. This region must be the operator.
Smallest fragments (cut close to label)
4. Load fragments into two lanes in a gel. Sort by size via electrophoresis. (The fragments with a radioactive atom will be visible.)
A DNA sequencing reaction can be used to determine the sequence of the "footprint."

FIGURE 17.11 DNA Footprinting Allows Researchers to Identify Sites Where Proteins Bind

(a) Multiple binding sites for repressor

(b) Sequence at O_1 is symmetrical.

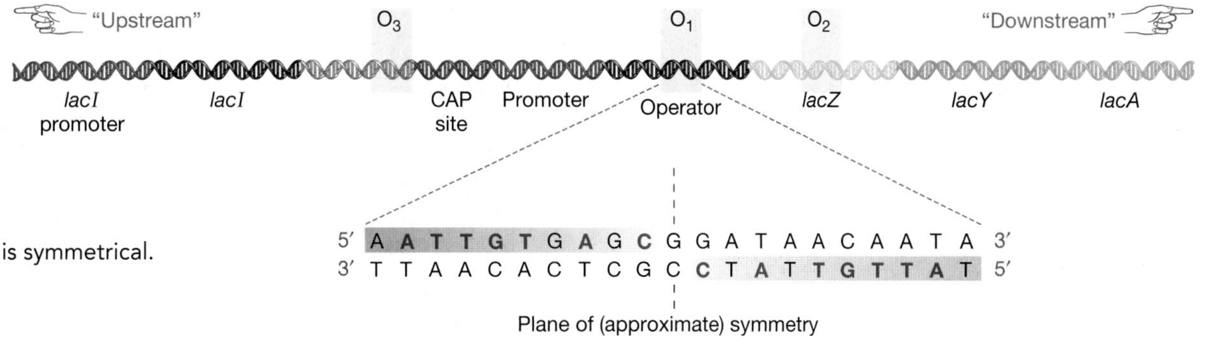
"Upstream"
O_3
O_1
O_2
"Downstream"
lacI promoter
lacI
CAP site
Promoter
Operator
lacZ
lacY
lacA
5′ A A T T G T G A G C G G A T A A C A A T A 3′
3′ T T A A C A C T C G C C T A T T G T T A T 5′
Plane of (approximate) symmetry

FIGURE 17.12 Operator Sequences
(a) The lac operon contains three operator sequences, called O_1, O_2, and O_3, where the repressor can bind. A repressor simultaneously binds O_1 and either O_2 or O_3. **(b)** Many sequences targeted by DNA-binding proteins are symmetrical. In this example, all but three bases are identical on the two sides of the plane of symmetry.

operators have a similar DNA sequence that has an unusual characteristic: It contains an axis of symmetry. To understand what an axis of symmetry in DNA is, look at the O_1 sequence shown in **Figure 17.12b** (page 375). Note that if the sequence were rotated 180°, it would be little changed. This type of symmetry is known as **two-fold rotational symmetry** or **dyad symmetry**. Follow-up work showed that the nucleotide sequences recognized by many DNA-binding proteins have two-fold rotational symmetry.

Now that we've examined the operator, we need to consider the repressor. How does this DNA-binding protein recognize specific sequences along the double helix? In particular, why is two-fold rotational symmetry important?

DNA Binding via the Helix-Turn-Helix Motif

Like other proteins, the *lac* operon's repressor resembles a hand tool in one important respect: Both have distinct regions or parts. Consider a screwdriver, which consists of a handle, a shaft, and a blade. Each of these parts has a distinctive three-dimensional structure and function. Proteins also have regions or parts, called **domains**, that have a distinctive three-dimensional structure and function. Just as screwdrivers of various shapes and sizes all have a handle, shaft, and blade, it is common to observe the same domain in proteins of various shapes and sizes. A domain that is observed in many different proteins is known as a **motif**. Each discrete screwdriver has a handle domain, a shaft domain, and a blade domain. As a group, screwdrivers are characterized by a handle motif, a shaft motif, and a blade motif. The handle and shaft motifs are also found in hammers and socket wrenches, but the blade motif is unique to screwdrivers. Hundreds of different motifs have been characterized in proteins.

When investigators analyzed repressor proteins from different bacteria, they discovered that many of the bacteria contained the domain illustrated in **Figure 17.13a**. This structure is a **helix-turn-helix motif**. As its name indicates, it consists of two α-helices connected by a short stretch of amino acids that forms a turn. Physical models that were constructed during the 1980s suggested that the helix-turn-helix region is the section of the repressor that binds to DNA. This hypothesis was later supported by structural studies based on X-ray crystallography.

To see how a helix-turn-helix domain works, recall from Chapter 4 that the exterior of the DNA double helix has a minor groove and a major groove (**Figure 17.13b**). A protein that has a helix-turn-helix domain can bind to DNA because one of the helices interacts with the sugar-phosphate backbone of a DNA strand while the other helix binds to the base pairs in the major groove (**Figure 17.13c**). DNA does not have to become single stranded for the helix-turn-helix domain of the repressor to bind to the operator. This observation holds for nearly all proteins that recognize specific sequences of double-stranded DNA.

The section of the helix-turn-helix domain that binds inside the major groove is called the **recognition sequence**. The amino acid sequence of this section is crucial, because specific sequences of amino acids bind to specific sequences of DNA. Each type of regulatory protein with a helix-turn-helix motif has a unique sequence of amino acids in its recognition sequence. As a result, each of these regulatory proteins binds to a unique regulatory sequence in DNA. In many cases, precise control over gene expression is based on a precise chemical interactions between

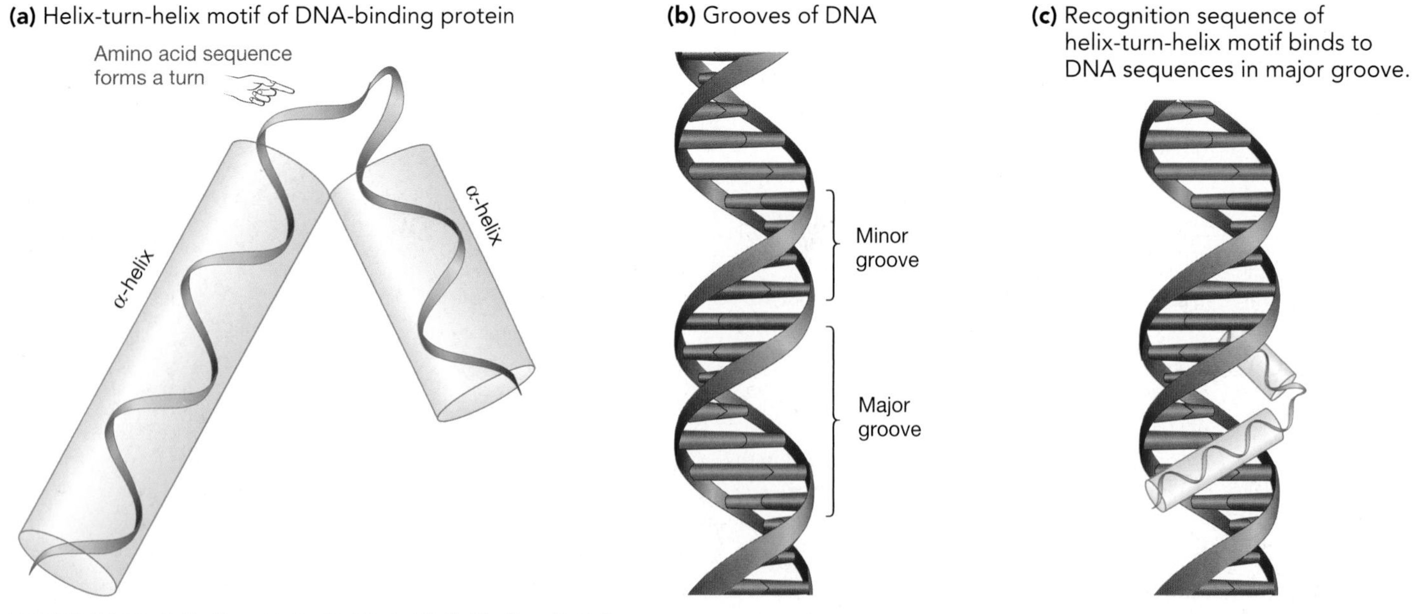
(a) Helix-turn-helix motif of DNA-binding protein
Amino acid sequence forms a turn
α-helix
α-helix
(b) Grooves of DNA
Minor groove
Major groove
(c) Recognition sequence of helix-turn-helix motif binds to DNA sequences in major groove.

FIGURE 17.13 Helix-Turn-Helix Motifs in DNA-Binding Proteins

QUESTION The part of a helix-turn-helix motif that interacts directly with DNA is called the recognition sequence. Why?

the amino acid sequence in a recognition sequence and the DNA sequence in the major groove of a regulatory site.

Biologists used X-ray crystallography to document the precise way that the *lac* repressor binds to the DNA of the operator. Consider the structure of the repressor alone, shown in **Figure 17.14a**. Note that the protein consists of four identical polypeptide chains, and that four helix-turn-helix domains project from the main body of the molecule. When researchers analyzed the structure of the repressor while it was bound to the operator sequence, they confirmed that each of these helix-turn-helix motifs binds operator DNA inside the major groove.

Now recall that the operator essentially is composed of two symmetrical halves. Each half of the operator sequence is bound by one of the repressor's helix-turn-helix motifs. The major grooves in the operator and the recognition helices of the repressor can pair up in this way because the operator is symmetrical and because the repressor consists of identical polypeptides that are flipped 180° with respect to each other. In addition, because the repressor contains four helix-turn-helix domains and two of the four bind per operator, a total of two operators can be bound simultaneously by one repressor. When this occurs, a loop of DNA is created between O_1 and whichever other operator (O_2 or O_3) is bound (**Figure 17.14b**). This type of DNA looping is observed at many other bacterial promoters and at virtually all eukaryotic promoters.

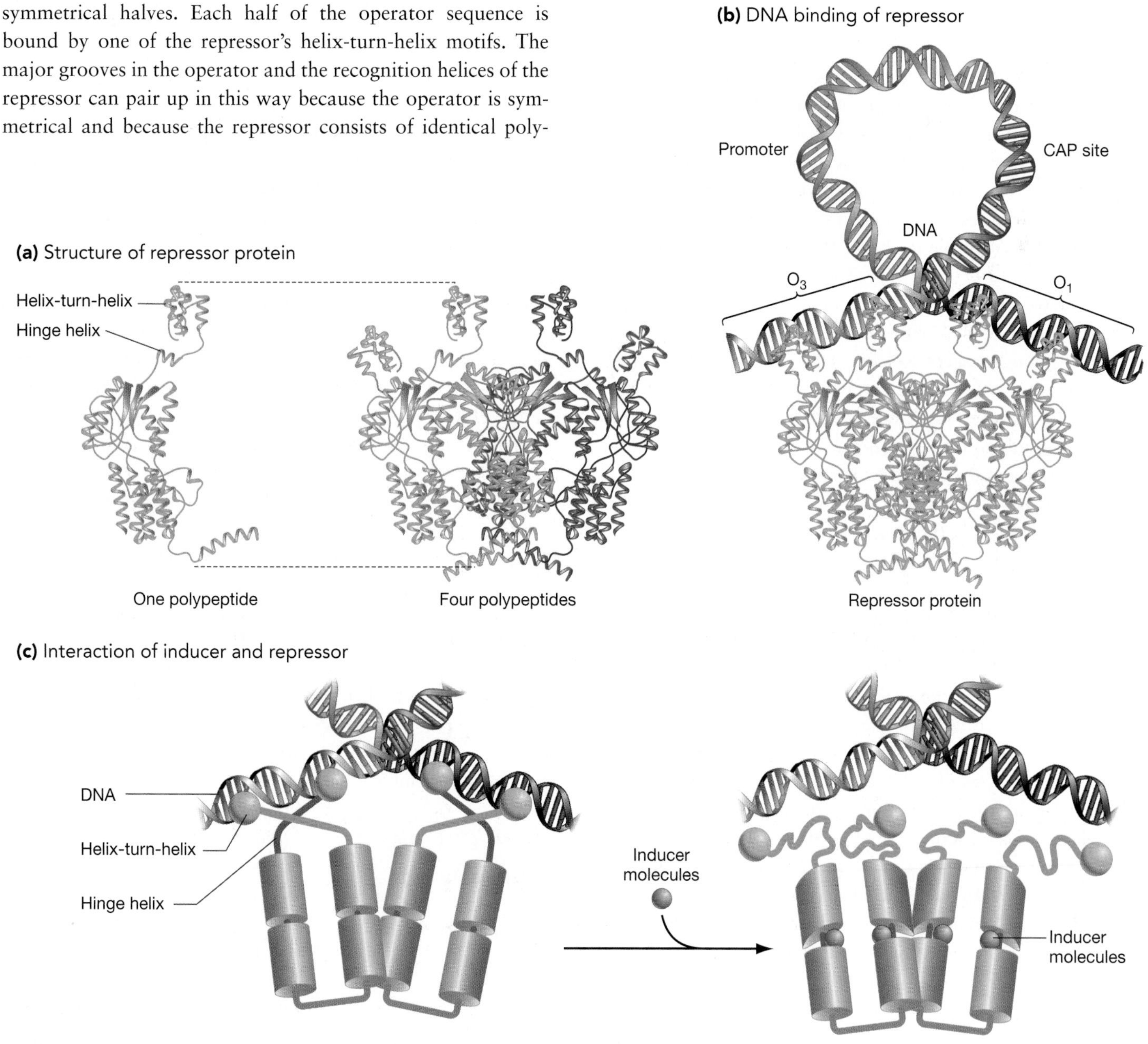
(b) DNA binding of repressor
Promoter
CAP site
DNA
O3
O1
Repressor protein
(a) Structure of repressor protein
Helix-turn-helix
Hinge helix
One polypeptide
Four polypeptides
(c) Interaction of inducer and repressor
DNA
Helix-turn-helix
Hinge helix
Inducer molecules
Inducer molecules

FIGURE 17.14 How Does the Repressor Interact with the Operator and the Inducer?
(a) The repressor is a complex of four polypeptides, each a product of the *lacI* locus. **(b)** The repressor protein has four helix-turn-helix motifs to bind DNA at O_1 and either O_2 or O_3. **(c)** "Hinge helices" in the repressor exist only when the inducer is not present (left). When the inducer binds to the repressor, the hinge helices relax and cause the protein to fall off the operator DNA (right).

How Does the Inducer Change the Repressor's Affinity for DNA?

Negative control of the *lac* operon occurs when the repressor binds to the operator. In this state, the helix-turn-helix motifs of the repressor are attached to DNA. Even though RNA polymerase can bind to the promoter nearby, the enzyme cannot open the double helix and begin transcription. This physical interference is typical of how negative regulators work. Positive regulators, in contrast, work by stabilizing the interaction between RNA polymerase and the promoter.

Negative control continues as long as the repressor is bound to DNA. But when lactose is present, the repressor dissociates from the operator. How does the inducer affect the repressor in a way that disrupts DNA binding?

To answer this question, researchers analyzed the three-dimensional structure of the repressor when it was bound to a molecule called IPTG, which acts as an inducer in the *lac* system. Their data showed that IPTG binds at locations in the middle of the repressor and that it produces an important change in the shape of the protein near the helix-turn-helix domains. To see this change in shape, compare the left and right sides of **Figure 17.14c** (page 377). The diagram on the left summarizes the repressor's shape without the inducer. Pay attention to the structures labeled **hinge helices** near the helix-turn-helix motifs. The hinge helices lock the DNA-binding motifs into place when they contact the operator. As the diagram on the right shows, however, the hinge helices uncoil when IPTG or the inducer binds to the repressor. As a result, the two DNA-binding regions can't bind to the operator at the same time, and the protein releases from the site.

The "unhinging" of the hinge helices that occurs in the *lac* repressor turns out to be a unique mechanism for changing the activity of a regulatory protein. Still, several general lessons emerge from the study of this repressor. First, the shape and activity of a protein may change when a small molecule binds to it. Allosteric shifts like that observed in the *lac* repressor occur in thousands of different proteins. Second, key elements in transcriptional control are regulated at the **post-translational level**. For example, the repressor protein is always present because it is transcribed and translated constitutively at low levels. When a rapid change in *lac* operon activity is required, it does not occur via changes in the transcription or translation of the repressor. Rather, the activity of the already translated protein is altered. In virtually all cases, the activity of key regulatory proteins is controlled by post-translational modifications.

Future Directions

Now that biologists understand the physical basis of negative control and induction, what does the future hold? One promising line of research is focused on how the activity of many genes and operons is coordinated. For example, suppose that an *E. coli* cell in your gut is faced with starvation. Does it adjust the activity of hundreds or even thousands of its genes in order to survive? If so, how are changes in gene expression coordinated throughout the genome? Because biologists have a fairly solid understanding of how individual genes and operons are regulated, the current challenge is to explore how multiple genes and operons are controlled in a way that increases the fitness of bacterial cells. Answering these questions may be important to our understanding of how certain bacteria are able to sustain infections that threaten human health and how some bacterial species are able to detoxify pollutants by using them as food. Chapter 19 introduces some techniques that biologists use to study coordinated changes in gene expression.

ESSAY Control of Gene Expression in Bacterial Pathogens

A staggering number of bacteria are living in and on your body right now. To give just one example, the number of bacterial cells in your mouth at this moment far exceeds the number of people who have ever lived, including the world's current population of more than 6.4 billion. Virtually all of these bacteria are **commensal**—meaning they live in association with their host without causing harm.

As you know, however, a few bacterial species can cause disease. How are certain bacteria able to invade and harm us? How are disease-causing bacteria, or **pathogens**, transmitted from person to person? And how can bacterial infections be controlled or prevented? In many cases, the answers depend on an understanding of how the expression of key genes is controlled. To drive this point home, consider *Vibrio cholerae*, the causative agent of cholera (**Figure 17.15**).

Cholera is a disease characterized by watery diarrhea that, if untreated, can lead to dehydration and death. The disease is currently most prevalent in Africa and Asia, where 5 million cases occur each year. Because *V. cholerae* is transmitted from person to person when diarrhea from an infected individual contaminates drinking water, cholera occurs primarily in regions with poor sanitation.

A staggering number of bacteria are living in and on your body right now.

Vibrio cholerae is extraordinarily flexible regarding the habitats it occupies. For example, near the confluence of the Ganges and Brahmaputra Rivers in the Bengal region of India and Bangladesh, *V. cholerae* can be found as free-swimming cells, inside two different eukaryotes, in dense layers coating rocks and other underwater surfaces, and inside humans. In the Bengal, most incidences of human infection occur during an outbreak that coincides with the end of

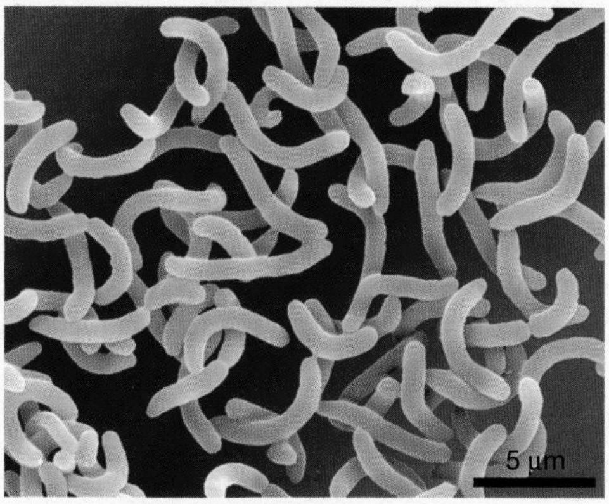

FIGURE 17.15 ***Vibrio cholerae*, the Bacterium That Causes Cholera**
[©Dr. Dennis Kunkel/Visuals Unlimited].

the monsoon rains in October and November each year. This observation suggests that environmental signals trigger *V. cholerae*'s yearly emergence as a human pathogen.

What genes allow *V. cholerae* to act as a human pathogen, and how are these genes regulated? Once it is ingested by a person, *V. cholerae* confronts an environment that is markedly different from its other habitats. In fact, most of the ingested bacteria are killed in the harsh, acidic environment of the stomach. The few that survive enter the small intestine, where they encounter a thick layer of mucus that protects the intestinal cells. Then, in response to a still-unknown signal in the environment, major changes in *V. cholerae* gene expression occur. The new proteins that are produced allow *V. cholerae* cells to move through the mucus layer and make contact with the intestinal cells. Each invading cell then produces a short, hairlike projection called a pilus, which allows the *V. cholerae* cell to attach to an intestinal cell. The proteins required to create this structure result from the coordinately regulated transcription of pilus-specific genes.

Once the bacterial cells have attached to intestinal cells, *V. cholerae* begins to manufacture and secrete a toxin composed of two different polypeptide chains. When intestinal cells take up the toxin, a devastating chain of events occurs. The toxin activates a host protein that stimulates human adenylyl cyclase. Massive amounts of cAMP are produced in response, resulting in a flow of Cl^- ions from the affected intestinal cells into the center (lumen) of the small intestine. Water follows by osmosis. The result is the watery diarrhea symptomatic of cholera.

Although the diarrhea can be fatal to the host, it increases the fitness of *V. cholerae* cells in environments where human sanitation is bad. Because *V. cholerae* cells are shed in diarrhea, they can enter drinking-water supplies if systems for transporting and processing human sewage are in poor repair or nonexistent. In this way, *V. cholerae* cells that induce diarrhea are efficiently transmitted to new hosts. *Vibrio cholerae* cells that have alleles responsible for diarrhea have higher fitness than do *V. cholerae* cells that lack these alleles.

The genes that encode the pilus proteins and cholera toxin are under the control of regulatory genes called the *Tox* (for toxin) loci. As **Figure 17.16** shows, *ToxR* produces a transcription factor that responds to a signal in the intestinal environment. When the ToxR protein is activated, it stimulates transcription of the *ToxT* gene. The *ToxT* product, ToxT, is a DNA-binding protein that activates transcription of several genes required for infection, including the genes that encode the cholera toxin subunits. These genes are called *ctxA* and *ctxB*. As you might suspect, they are part of a polycistronic operon.

Biologists are currently trying to translate research on gene regulation into better tools for fighting cholera. For example, blocking the activity of the ToxR or ToxT proteins should prevent disease. This hypothesis has been supported by the observation that *ToxR*$^-$ mutants are harmless. Researchers are now attempting to develop drugs that prevent ToxR from binding the signal molecule, that stop ToxR from activating *ToxT*, or that prevent ToxT from contacting DNA.

The axiom "knowledge is power" applies here: Although drugs like those proposed earlier are not yet available, research on gene regulation has provided several novel and promising drug targets.

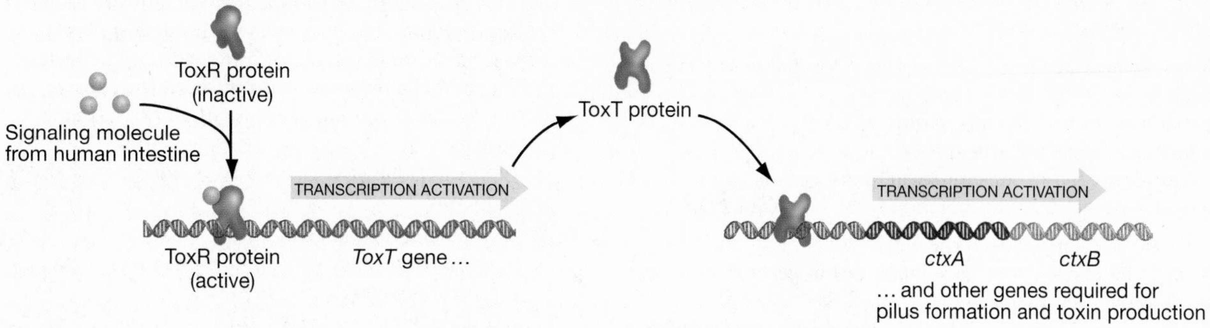

FIGURE 17.16 In the Cholera Bacterium, the Genes Required for Infection Are Regulated
QUESTION Biologists have not been able to identify the signaling molecule that triggers the cascade of events shown here. What properties must this molecule have?

CHAPTER REVIEW

Summary of Key Concepts

Changes in gene expression allow bacterial cells to respond to environmental changes.

Transcription in bacteria is constitutive or inducible. Constitutive expression occurs in genes whose products are required at all times, such as genes that encode glycolytic enzymes. Expression of most genes is induced by environmental signals, however, so gene products are produced on demand.

Precise regulation over gene expression is important to bacterial cells. For example, some species switch between different food sources to take advantage of the nutrients currently available in their environment. Efficient use of nutrients is beneficial because cells are usually competing for access to resources. In the intestinal bacterium *E. coli*, for example, changes in gene expression allow cells to use lactose as a food source when glucose is not available.

Gene expression can be controlled at three levels: transcription, translation, or post-translation (protein activation).

Among the three levels of gene regulation, there is a trade-off between the speed of response to changed conditions and the efficient use of resources. In post-translational control, for example, the change in gene expression is extremely fast but demands a considerable investment of resources to make and regulate the protein in question.

Transcriptional control can be positive or negative. In the case of the genes for lactose metabolism in *E. coli*, positive control occurs when a regulatory protein binds to DNA and increases the transcription rate. Negative control occurs because another regulatory protein must be released before transcription can occur.

Protein-coding genes involved in lactose metabolism are normally transcribed at low levels because a repressor protein binds to a DNA sequence—the operator—near their promoter. But when present, lactose binds to the repressor and induces a conformational change that causes the repressor to fall off the operator. Rapid transcription of the *lac* loci does not begin, however, unless glucose is also lacking. When glucose is scarce, cAMP levels rise in the cell. This is important because the regulatory protein CAP undergoes a conformational change when it is bound to cAMP. The CAP-cAMP complex binds to a control sequence near the *lac* promoter and increases transcription by facilitating binding by RNA polymerase.

Web Tutorial 17.1 The *lac* Operon

Many regulatory proteins bind to specific sites in DNA. Because different regulatory proteins have different amino acid sequences, they bind to different DNA sequences.

The regulatory proteins involved in negative control and positive control are often allosterically regulated, meaning that they have the ability to switch between two different conformations in response to binding a small molecule—often a nutrient or product affected by the gene being regulated. The change in conformation affects the protein's ability to bind to DNA.

Questions

Content Review

1. Genes for enzymes in glycolysis are expressed constitutively. Why?
   a. Their expression is controlled at three levels: transcriptional, translational, and post-translational.
   b. Glycolysis occurs at all times, so they need to synthesize ATP-producing enzymes constantly.
   c. Transcription is activated only in response to signals indicating that ATP supplies are low.
   d. They do not need to be expressed when fermentation instead of cellular respiration is occurring.

2. Why do researchers frequently begin a search for mutants by exposing organisms to UV light or X-rays?
   a. The treatment causes mutants to turn yellow.
   b. The treatment triggers gene expression.
   c. The treatment exposes constitutive mutants.
   d. The treatment increases the mutation rate by damaging DNA.

3. What is the goal of a genetic screen?
   a. to identify individuals that have mutations in genes that are required for a particular process
   b. to produce a large number of mutant individuals, which can later be analyzed in detail
   c. to determine whether the mutations carried by particular individuals are in protein-coding genes or regulatory elements
   d. to determine the fitness or viability of mutant individuals

4. Why are the genes involved in lactose metabolism considered to be an operon?
   a. They occupy adjacent locations on the *E. coli* chromosome.
   b. They have a similar function.
   c. They are all required for normal cell function.
   d. They are under the control of the same promoter.

5. What is catabolite repression?
   a. a mechanism that turns off the synthesis of enzymes responsible for catabolic reactions when the product is present
   b. a mechanism that turns off the synthesis of enzymes responsible for catabolic reactions when the product is absent
   c. repression that occurs because of allosteric changes in a regulatory protein
   d. repression that occurs because of allosteric changes in a DNA sequence

6. What is a helix-turn-helix motif?
   a. a protein domain involved in folding into the active conformation
   b. a protein domain involved in induction
   c. a protein domain involved in DNA binding
   d. a protein domain involved in catalysis

Conceptual Review

1. Explain the difference between positive and negative control over transcription. Why is it advantageous for the *lac* operon in *E. coli* to be under both positive control and negative control? What would happen if only negative control occurred? What would happen if only positive control occurred?
2. In *E. coli*, rising levels of cAMP can be a considered a starvation signal. Explain.
3. Explain the role of post-translational control in the *lac* operon.
4. CAP is also known as the cAMP-receptor protein. Why?
5. Explain how the *lac* repressor binds to DNA. Then explain the relationship between the two-fold rotational symmetry of the operator and the four-part structure of the repressor protein.
6. The galactose released when β-galactosidase cleaves lactose enters the glycolytic pathway in *E. coli* once a series of enzyme-catalyzed reactions has converted the galactose to glucose-6-phosphate. Why, then, is glucose, not lactose, the preferred sugar in *E. coli*?

Group Discussion Problems

1. You are interested in using bacteria to metabolize wastes at an old chemical plant and convert them into harmless compounds. You find bacteria that are able to tolerate high levels of the toxic compounds toluene and benzene, and you suspect that it is due to the ability of the bacteria to break these compounds into less-toxic products. If that is true, these toluene- and benzene-resistant strains will be valuable for cleaning up toxic sites. How could you find out whether these bacteria have enzymes that allow them to metabolize toluene?
2. Assuming that the bacteria you examined in Group Discussion Problem 1 do have an enzymatic pathway to break down toluene, would you predict that the genes involved are constitutively expressed, under positive control, or under negative control? Why? What experiments could you conduct to test your hypothesis?
3. The *lacI* mutants are constitutive because the repressor fails to recognize and bind to the operator region. Other repressor mutants have been isolated that are called *LacI*^{S} mutants. These repressor proteins continue to bind to the operator, even in the presence of the inducer. How would this mutation affect the function of the *lac* operon? Specifically, how well would *LacI*^{S} mutants do in an environment that has lactose as its sole sugar?
4. X-gal is a colorless, lactose-like molecule that can be split into two fragments by β-galactosidase. One of these product molecules is blue. The following photograph is a close-up of *E. coli* colonies growing in a medium that contains lactose.

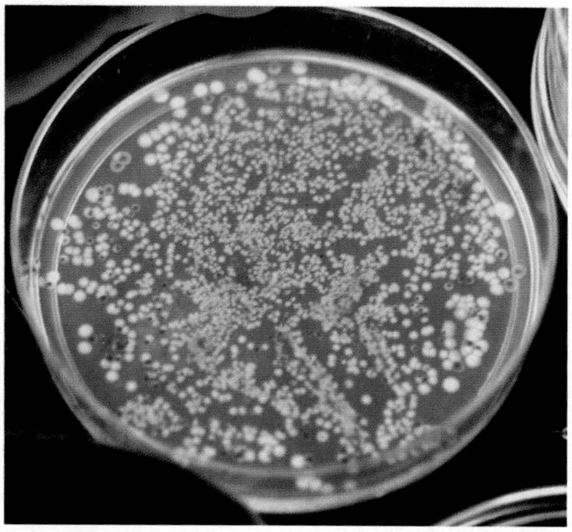

Draw a line to three colonies whose cells have functioning copies of β-galactosidase. Draw a line to three colonies whose cells have mutations in the *lacZ* locus or in one of the genes involved in regulation of the *lacZ*. Suppose you could analyze the sequence of the β-galactosidase gene from each of the mutant colonies. How would these data help you distinguish which cells are structural mutants and which are regulatory mutants?

Answers to Multiple-Choice Questions 1. b; 2. d; 3. a; 4. d; 5. a; 6. c

www.prenhall.com/freeman is your resource for the following: Web Tutorials; Online Quizzes and other Online Study Guide materials; Answers to Conceptual Review Questions; Solutions to Group Discussion Problems; Answers to Figure Caption Questions and Exercises; and Additional Readings and Research.

18 Control of Gene Expression in Eukaryotes

KEY CONCEPTS

- Changes in gene expression allow eukaryotic cells to respond to changes in the environment and cause distinct cell types to develop.
- In eukaryotes, DNA is packaged with proteins into complex structures that must be opened before transcription can occur.
- In eukaryotes, transcription can be initiated only when specific proteins bind to the promoter and to regulatory sequences that may be close to the promoter or far from it.
- Alternative splicing allows a single gene to code for several different products.
- Cancer can develop when mutations disable genes that regulate cell-cycle control genes.

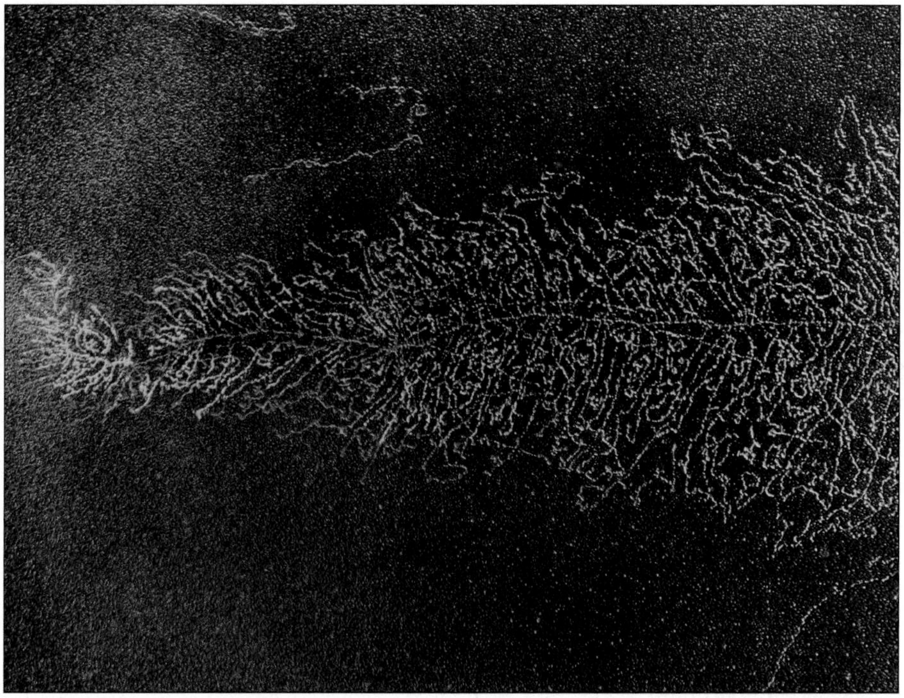

A eukaryotic cell's DNA being transcribed. The horizontal strand in the middle of the micrograph is DNA; the yellow strands above and below it are mRNAs.

In bacteria, precise regulation of genes is essential for cells to respond to changes in their environment, as Chapter 17 indicated. In *Escherichia coli*, the genes that are required to import and cleave lactose are rapidly expressed only when the cells have to rely on lactose as a source of energy—meaning glucose is absent and lactose is present. *Vibrio cholerae* can live in a wide array of environments, but the genes that allow the cells to infect a human are expressed only when the cells find themselves inside someone's small intestine.

Unicellular eukaryotes face similar challenges. Consider the yeast *Saccharomyces cerevisiae*, which is used extensively in the production of beer, wine, and bread. In nature, the cells live on the skins of grapes and other fruits. In the wild, variation in temperature and humidity can be high. In addition, the sugars that are available to the cells change dramatically in type and concentration as the fruit that they are living on ripens, falls, and rots. For yeast cells to grow and reproduce efficiently, gene expression has to be modified in response to these environmental changes. Further, the offspring of a cell that lived on a grape may be blown to a nearby orchard and take up residence on an apple. The genes that are expressed in the cells of grapes versus those expressed in the cells of apples vary markedly.

The cells that make up multicellular eukaryotes face an even greater challenge. Consider that, at one time, your body consisted of a single cell—a fertilized egg. Now you have trillions of cells, each of which has a specialized structure and function. You have heart muscle cells, lung cells, nerve cells, skin cells, and so on. Even though these cells look and behave very differently, each of them contains the same genes. Your bone cells are different from your blood cells not because they contain different genes, but because they express different genes. Your bone cells have blood cell genes—they just don't transcribe them.

How is it possible for the cells inside your body to express unique sets of genes? The answer is that your cells respond to changes in their environment, just as bacteria and unicellular eukaryotes do. But instead of responding to the presence of molecules such as glucose or lactose in the external environment, your cells respond to the presence of signals from other cells—from an internal environment. As a human being or an oak tree develops, cells that are located in different parts of the organism are exposed to different signals. As a result, they express different genes. Differential gene expression is responsible for creating different cell types, arranging them into tissues, and coordinating their activity to form the multicellular society we call an individual.

How does all this happen? Unit 4 explores the nature of the signals that trigger the formation of muscle, bone, leaf, and flower cells in multicellular organisms. Here we need to focus on what happens once a eukaryotic cell receives a signal. Let's begin by surveying the various ways that gene expression can be controlled and then plunge into a detailed look at key points in gene regulation. The chapter closes with a look at how breakdowns in the control of gene expression can help trigger cancer.

18.1 Mechanisms of Gene Regulation—an Overview

Like bacteria, eukaryotes can control gene expression at the levels of transcription, translation, and post-translation (protein activation or inactivation). But as **Figure 18.1** shows, two additional levels of control occur in eukaryotes as genetic information flows from DNA to proteins. The first level involves the DNA-protein complex at the top of the figure. In eukaryotes, DNA is wrapped around proteins to create a protein-DNA complex called **chromatin.** Eukaryotic genes have promoters, just as bacterial genes do. Before transcription can begin in eukaryotes, DNA near the promoter must be released from tight interactions with proteins so that RNA polymerase can make contact with the promoter. To capture this point, biologists say that **chromatin remodeling** must occur prior to transcription.

The second level of regulation that is unique to eukaryotes involves **RNA processing**—the steps required to produce a mature, processed mRNA from a **primary RNA transcript**, the preliminary result of transcription. Recall from Chapter 16 that introns have to be spliced out of primary transcripts. In some cases, carefully orchestrated alterations in the splicing pattern occur. When the splicing events in a particular primary RNA transcript change, a different message emerges. The altered message, in turn, leads to the production of a different product.

How many of the five potential control points shown in Figure 18.1 are actually used? In all cells, every one of them is employed at certain times. The primary focus of this chapter is to explore three of these key transition points: (1) control of packing DNA into chromatin (control over chromatin remodeling), (2) control of transcription, and (3) control of RNA processing.

To appreciate the breadth and complexity of gene regulation in eukaryotes, let's follow the series of events that occurs as an embryonic cell responds to a developmental signal. Suppose a molecule arrives that specifies the production of a muscle-specific protein. What happens next?

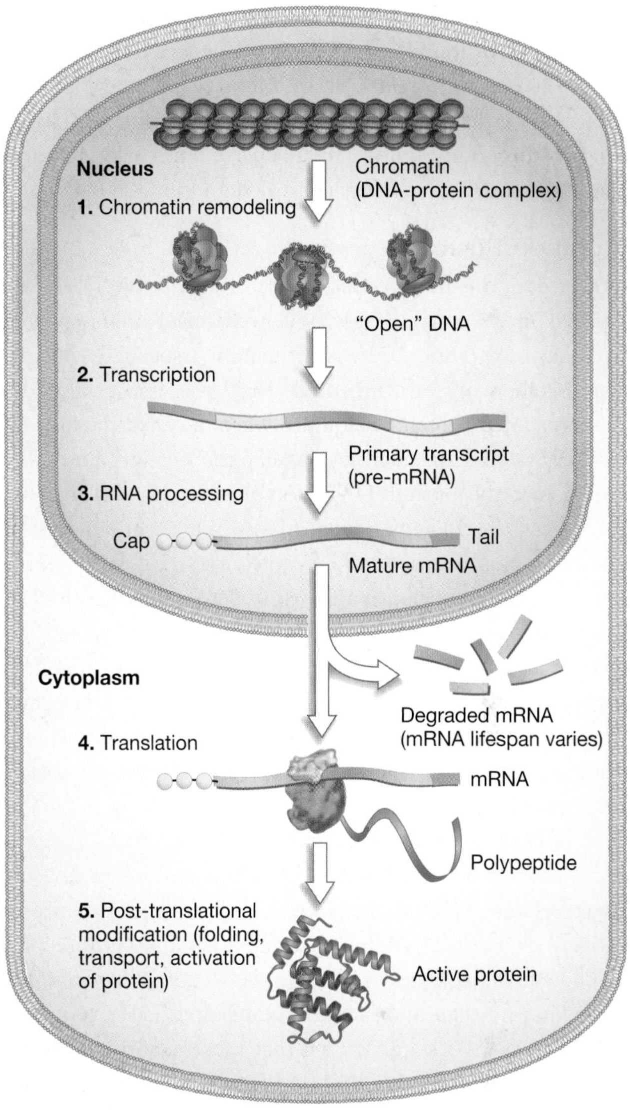

FIGURE 18.1 In Eukaryotes, Gene Expression Can Be Controlled at Many Different Steps
Numerous steps occur during protein synthesis in eukaryotes. Each step provides an opportunity for regulation of gene expression.
EXERCISE Place and label a horizontal line such that mechanisms of transcriptional control fall above the line and mechanisms of post-transcriptional control fall below it.

18.2 Eukaryotic DNA and the Regulation of Gene Expression

If the arrival of a signaling molecule from outside the cell is going to result in the transcription of a particular gene, the DNA around the target gene must be drastically remodeled. To appreciate why, consider that a typical cell in your body contains about six billion base pairs of DNA. Lined up end to end, these nucleotide pairs would form a double helix about 2 m (6.5 feet) long. But the nucleus that holds this DNA is thinner than a sheet of paper. In eukaryotes, DNA is packed inside the nucleus so tightly that RNA polymerase can't access it. Part of this packing is done by supercoiling—meaning the DNA double helix is twisted on itself many times, just as in bacterial chromosomes (see Chapter 7). The supercoiled DNA found in bacteria does not need to be altered extensively before it can be transcribed, however. Eukaryotic DNA is different—it has to undergo a series of significant physical changes before transcription can take place. The reason is that supercoiling is just part of the packing system found in eukaryotes.

Chromatin Structure

The first data on the physical nature of eukaryotic DNA were published in the early 1900s, when chemical analyses established that eukaryotic DNA is intimately associated with proteins. Later work documented that the most abundant DNA-associated proteins belong to a group called the **histones.** In the 1970s electron micrographs like the one in **Figure 18.2a** revealed that the protein-DNA complex, or chromatin, has a regular structure. In some preparations for electron microscopy, chromatin actually looked like beads on a string. The "beads" came to be called **nucleosomes.** More details emerged in 1984 when researchers determined the three-dimensional structure of eukaryotic DNA by using X-ray crystallography (a technique introduced in Chapter 4). The X-ray crystallographic data indicated that each nucleosome consists of DNA wrapped almost twice around a core of eight histone proteins. As **Figure 18.2b** indicates, between each pair of nucleosomes there is a "linker" stretch of DNA associated with a histone called **H1**.

The intimate association between DNA and histones occurs in part because DNA is negatively charged and histones are positively charged. DNA has a negative charge because of its phosphate groups; histones are positively charged because they contain many lysine and/or arginine residues (see Chapter 3).

More recent work has shown that there is another layer of complexity in eukaryotic DNA. As **Figure 18.2c** indicates, H1 histones interact with each other and with histones in other nucleosomes to produce a tightly packed structure. Based on its width, this structure is called the **30-nanometer fiber.** (Recall that a nanometer is one-billionth of a meter and is abbreviated nm.) Often, 30-nm fibers are packed into still larger structures, which organize DNA inside the nucleus.

The elaborate structure of chromatin does more than just package DNA into the nucleus. Chromatin structure has profound implications for the control of gene expression. To drive this point home, consider the 30-nanometer fiber illustrated in Figure 18.2c. If this tightly packed stretch of DNA contains a promoter, how can RNA polymerase bind to it and initiate transcription?

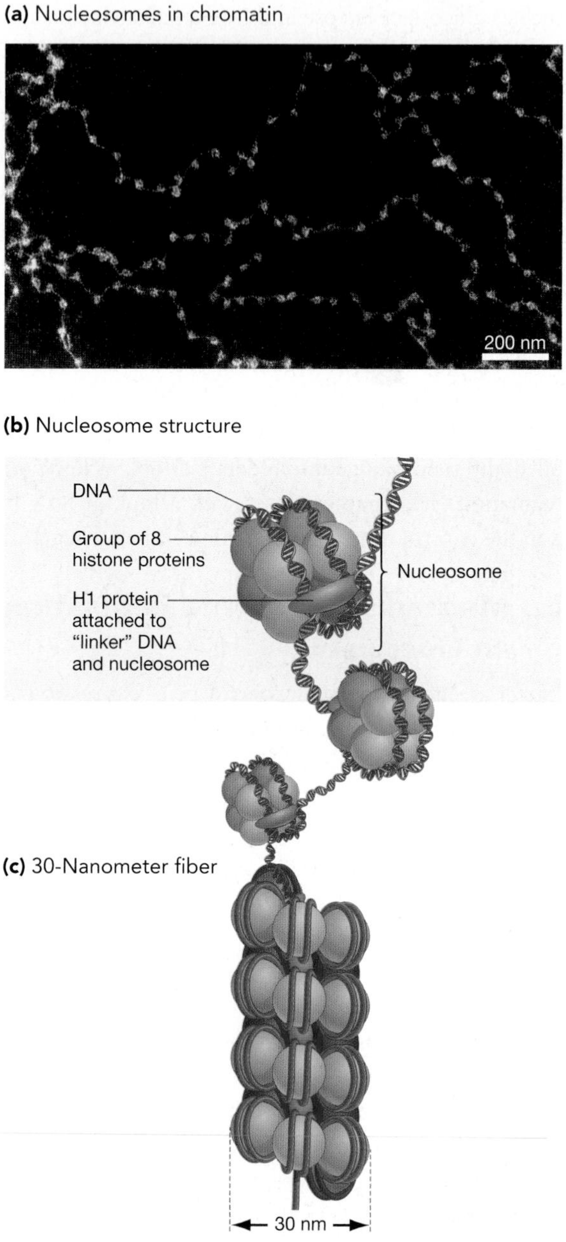

FIGURE 18.2 Chromatin Has Several Levels of Structure
(a) Electron micrograph of chromatin, showing that bead-like structures, called nucleosomes, are linked by short stretches of DNA. **(b)** X-ray crystallographic data revealed that nucleosomes consist of DNA wrapped around a core of eight histone proteins. The linker DNA is associated with a particular histone called H1. **(c)** A model of the structure of the 30-nanometer fiber. It proposes that H1 proteins form a core in the center of the fiber. Each ball here represents a group of eight histones, to make the overall structure of the 30-nanometer fiber clearer.

Evidence that Chromatin Structure Is Altered in Active Genes

Once the nucleosome-based structure of chromatin was established, biologists hypothesized that the close physical interaction between DNA and histones must be altered for RNA polymerase to make contact with DNA. More specifically, biologists hypothesized that a gene could not be transcribed until the chromatin near its promoter was remodeled. The idea was that the chromatin must be relaxed or decondensed for RNA polymerase to bind to the promoter. If so, then chromatin remodeling would represent the first step in the control of eukaryotic gene expression. Two types of studies have provided strong support for this hypothesis.

The first type of evidence comes from studies with the enzyme DNase, which cuts DNA at random locations. DNase cannot cut DNA efficiently if the molecule is tightly complexed with his-

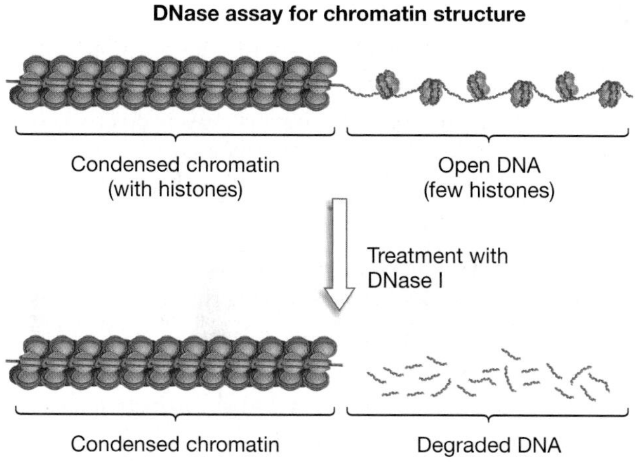

FIGURE 18.3 DNase Cannot Cut Intact Condensed Chromatin
DNase is an enzyme that cuts DNA at random locations.

tones. As **Figure 18.3** shows, the enzyme works effectively only if DNA is in the "open" configuration. Harold Weintraub and Mark Groudine used this observation to test the hypothesis that the DNA of actively transcribed genes is in an open configuration. They performed the test by comparing the structure of chromatin in two genes of the same cell type: the β-globin and ovalbumin genes of chicken blood cells. β-Globin is a protein that is part of the hemoglobin found in red blood cells; ovalbumin is a major protein of egg white. In blood cells, the β-globin gene is transcribed at high levels. The ovalbumin gene, in contrast, is not transcribed at all. After treating blood cells with DNase and then analyzing the state of the β-globin and ovalbumin genes, the researchers found that DNase cut up the β-globin gene much more readily than the ovalbumin gene. They interpreted this finding as evidence that chromatin in blood cells was in an open configuration at the β-globin gene but closed at the ovalbumin gene. Analogous studies using DNase on different genes and in different cell types yielded similar results.

The second type of evidence in support of the chromatin-remodeling hypothesis comes from studies of mutant brewer's yeast cells that do not produce the usual complement of histones. Researchers found that many yeast genes that are normally never transcribed are instead transcribed at high levels at all times in these mutant cells. To interpret this finding, biologists hypothesized that the lack of histone proteins prevented the assembly of normal chromatin. If the absence of normal histone-DNA interactions promotes transcription, then the presence of normal histone-DNA interactions must prevent it.

Taken together, the data suggest that the normal or default state of eukaryotic genes is to be turned off. If so, then gene expression depends on chromatin being opened up in the promoter region.

How Is Chromatin Altered?

Although much remains to be learned about how chromatin is opened up prior to transcription, researchers have made important progress recently. One key finding is that two major types of protein are involved in modifying chromatin structure. One group of proteins creates structures called **chromatin-remodeling complexes**, which reshape chromatin through a series of reactions that are dependent on ATP. It is not known exactly how these ATP-dependent enzymes work, however. The second type of chromatin-modifying protein works by adding small molecules such as acetyl (CH_3COOH) or methyl (CH_3) groups to histones. These processes are known as **acetylation** and **methylation**, respectively.

Figure 18.4 shows how the enzymes called **histone acetyl transferases (HATs)** affect chromatin structure. HATs acetylate the positively charged lysine residues in histones. When a HAT adds an acetyl group to selected histones, the number of positive charges on the histones is reduced. The result is less electrostatic attraction between histones and the negatively charged DNA. The association between nucleosomes and DNA is loosened and the chromatin decondenses.

Figure 18.4 also shows how chromatin is "recondensed" by a group of enzymes called **histone deacetylases (HDACs)**, which remove the acetyl groups added by HATs. Histone deacetylase activity reverses the effects of acetylation, returning chromatin to its default condensed state. If HATs are an on switch for transcription, HDACs are the off switch.

What happens once a section of DNA is opened up and exposed to RNA polymerase? In bacteria, the protein sigma binds to RNA polymerase and allows it to contact the promoter. Then transcription begins if any required positive regulators, such as CAP, are in place and negative regulators, such as the repressor, are absent. Does anything similar happen in eukaryotes?

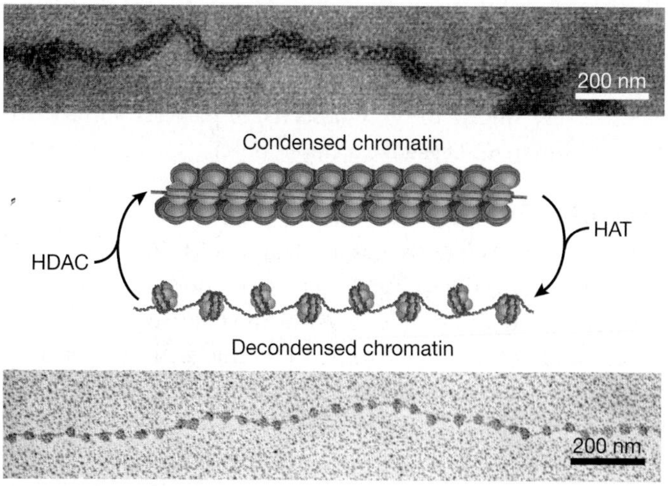

FIGURE 18.4 Chromatin Decondensation and Condensation Are Driven by Histone Acetyl Transferases and Histone Deacetylases
Histone acetyl transferases (HATs) cause chromatin to decondense, and histone deacetylases (HDACs) cause it to condense. EXERCISE In decondensed chromatin, each nucleosome has an acetyl group attached to it. Indicate this by adding "Ac" to several nucleosomes in the bottom part of the drawing.

✓CHECK YOUR UNDERSTANDING

Eukaryotic DNA is wrapped tightly around histones, forming nucleosomes, which are then coiled into dense structures called 30-nm fibers. Before transcription can begin, the chromatin (DNA-protein complex) must be relaxed by chromatin-remodeling complexes and HATs so that RNA polymerase can contact the promoter. You should be able to (1) draw and label the structure of a nucleosome and a 30-nm fiber and explain their function, and (2) compare and contrast the structure of DNA in bacteria versus that in eukaryotes.

18.3 Regulatory Sequences and Regulatory Proteins

Chapter 16 introduced the **promoter**—the site in DNA where RNA polymerase binds to initiate transcription. Eukaryotic promoters are located about 30 nucleotides upstream of the point where RNA polymerase begins transcription. Many contain a specific base sequence, known as a **TATA box**, where a sigma-like protein binds and allows RNA polymerase to contact DNA. Based on their position and function, eukaryotic promoters are much like bacterial promoters. Recall from Chapter 16 that bacterial promoters may vary in sequence and bind different types of sigma proteins. Eukaryotic genes that are transcribed by RNA polymerase II also have promoters that vary in sequence, but all eukaryotic promoters are bound by the same protein: the **TATA-binding protein (TBP).**

If eukaryotic genes have promoters that interact with the same promoter-binding protein, how can transcription be controlled? The answer lies in interactions between regulatory proteins and regulatory sequences. Eukaryotic regulatory proteins are analogous to *E. coli*'s CAP and the repressor protein analyzed in Chapter 17. **Regulatory sequences** are sections of DNA that are involved in controlling the activity of genes, similar to the CAP site and operators introduced in Chapter 17.

As it turns out, some regulatory sequences found in eukaryotes are very similar to those observed in bacteria. Others are radically different. Let's take a closer look.

Promoter-Proximal Elements

The first regulatory sequences in eukaryotic DNA were discovered in the late 1970s, when Yasuji Oshima and co-workers developed an important model system to study control of gene expression in eukaryotes. Much as François Jacob and Jacques Monod had focused on lactose metabolism in *E. coli*, Oshima's team set out to understand how yeast cells control the metabolism of the sugar galactose. When galactose is absent, yeast cells produce tiny quantities of the enzymes required to metabolize it. But when galactose is present, transcription of the genes encoding these enzymes increases by a factor of 1000.

Oshima and co-workers focused their early work on mutants that were unable to use galactose. The team's first major result was the discovery of mutant cells that failed to produce any of the five enzymes required for galactose metabolism, even if galactose was present. To interpret this observation, they hypothesized that the cells had a knock-out (loss-of-function) mutation that completely disabled a regulatory protein. Like CAP, this hypothesized protein was thought to exert positive control over the five genes. The team called the gene *GAL4* and the protein product GAL4. (In *S. cerevisiae*, the names of genes are written in full caps and italicized. As in other species, gene names are italicized but protein names are not.)

This hypothesis was supported when other researchers isolated the regulatory protein and found that it has a DNA-binding domain, analogous to the helix-turn-helix motif introduced in Chapter 17. The domain binds to a 20-base-pair stretch of DNA located just upstream from the promoter for the five genes that GAL4 regulates. The location and structure of this regulatory sequence are comparable to those of the CAP binding site in the *lac* operon of *E. coli*. Similar regulatory sequences have now been found in a wide array of eukaryotic genes and species. Because such sequences are located close to the promoter and bind regulatory proteins, they are termed **promoter-proximal elements.**

As **Figure 18.5** shows, promoter-proximal elements are just upstream from the promoter and the gene's start site; the exons and introns that make up the actual gene are farther downstream. Unlike the promoter itself, promoter-proximal elements have sequences that are unique to specific genes. In this way, they furnish a mechanism for eukaryotic cells to exert precise control over transcription.

The discovery of positive control and promoter-proximal elements provided a satisfying parallel between gene regulation in bacteria and in eukaryotes. This picture changed, however, when researchers discovered a new class of eukaryotic DNA regulatory sequences—sequences unlike anything in bacteria.

Enhancers

Susumu Tonegawa and colleagues made a discovery that may rank as the most startling in the history of research on gene expression. Tonegawa's group was exploring how human immune system cells regulate the genes involved in the production of antibodies. **Antibodies** are proteins that bind to specific sites on other molecules. In your immune system, antibodies bind to viruses and bacteria and mark them for destruction by other

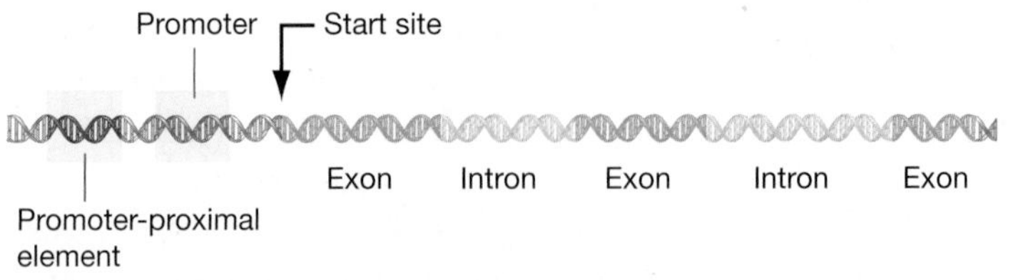
Promoter
Start site
Exon
Intron
Exon
Intron
Exon
Promoter-proximal element

FIGURE 18.5 Promoter-Proximal Elements Regulate the Expression of Some Eukaryotic Genes
All eukaryotic genes have promoters, the sites where RNA polymerase initially is in contact with the gene. Exons and introns are not drawn to scale throughout this chapter. They are typically very large compared with promoters and promoter-proximal (regulatory) elements.

cells. The antibody gene that the researchers were working with is broken into many introns and exons. Introns are DNA sequences spliced out of the primary mRNA transcript; exons are regions of eukaryotic genes that are included in the mature RNA once splicing is complete. The biologists used techniques that will be introduced in Chapter 19 to place copies of an intron in new locations and had found that, when they placed the intron close to a gene, the gene's transcription rate increased. Based on this observation, Tonegawa and co-workers hypothesized that the intron contained some sort of regulatory sequence.

To test this hypothesis, the biologists performed what is now considered a classic experiment (**Figure 18.6**). The protocol was simple in concept: Starting with a human antibody-producing gene that included an intron flanked by two exons, as shown in step 1 of Figure 18.6, the team removed different portions of the intron and observed whether transcription of the gene still took place. The researchers used enzymes to cut out specific pieces of the intron in multiple copies of the gene (step 2) and to ligate the resulting sections back together (step 3). Each of the modified genes was missing a different section of the intron. Then the team

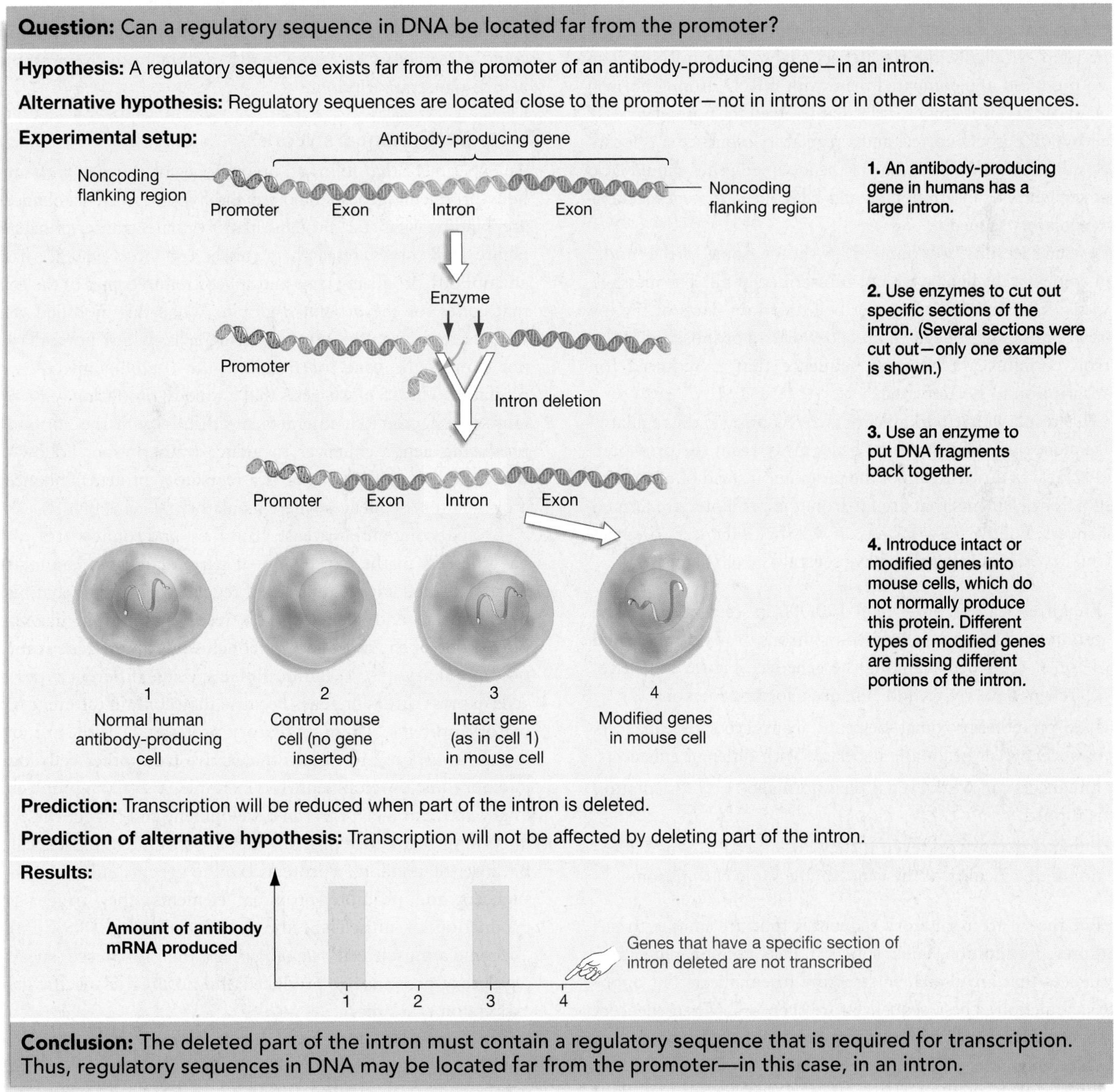
Question: Can a regulatory sequence in DNA be located far from the promoter?
Hypothesis: A regulatory sequence exists far from the promoter of an antibody-producing gene—in an intron.
Alternative hypothesis: Regulatory sequences are located close to the promoter—not in introns or in other distant sequences.
Experimental setup:
Antibody-producing gene
Noncoding flanking region
Promoter
Exon
Intron
Exon
Noncoding flanking region
1. An antibody-producing gene in humans has a large intron.
Enzyme
Promoter
Intron deletion
2. Use enzymes to cut out specific sections of the intron. (Several sections were cut out—only one example is shown.)
3. Use an enzyme to put DNA fragments back together.
Promoter
Exon
Intron
Exon
4. Introduce intact or modified genes into mouse cells, which do not normally produce this protein. Different types of modified genes are missing different portions of the intron.
1
Normal human antibody-producing cell
2
Control mouse cell (no gene inserted)
3
Intact gene (as in cell 1) in mouse cell
4
Modified genes in mouse cell
Prediction: Transcription will be reduced when part of the intron is deleted.
Prediction of alternative hypothesis: Transcription will not be affected by deleting part of the intron.
Results:
Amount of antibody mRNA produced
1
2
3
4
Genes that have a specific section of intron deleted are not transcribed
Conclusion: The deleted part of the intron must contain a regulatory sequence that is required for transcription. Thus, regulatory sequences in DNA may be located far from the promoter—in this case, in an intron.

FIGURE 18.6 Evidence That Enhancers Are Required for Transcription

QUESTION Why did the researchers assess mRNA production in human cells that did *not* receive a modified or intact gene?

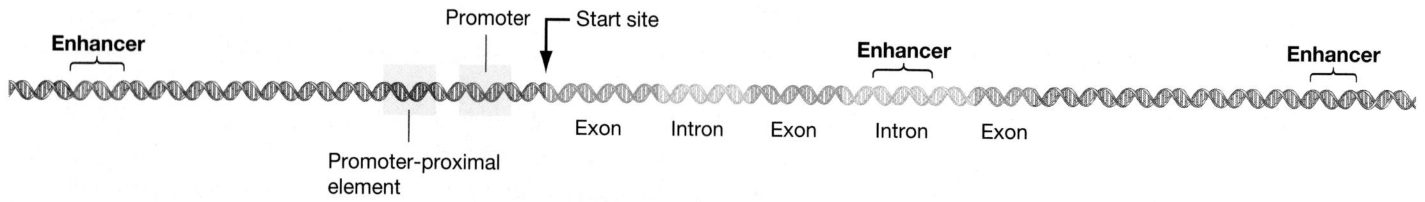

FIGURE 18.7 Enhancers Are Far from the Genes They Regulate
Eukaryotic genes often have more than one enhancer. Enhancers can be located either 5′ or 3′ to the gene they regulate or in introns and can be located tens of thousands of base pairs away from the promoter.

inserted different versions of the gene into mouse cells, which do not normally contain the antibody-producing gene. Some mouse cells received normal copies of the gene; some received copies of the gene missing different portions of the intron; some received no gene (step 4). Finally, the researchers analyzed the mRNAs that were produced and compared them with mRNA from a normal human cell and from mouse cells that received the normal gene. If their hypothesis was correct and a regulatory sequence was located inside the intron, then some of the modified genes should lack that sequence. Transcription should be reduced in genes lacking the regulatory sequence.

As the "Results" of Figure 18.6 shows, some of the modified copies of the gene were not transcribed at all. The message is clear: The gene is not transcribed if certain parts of the intron are missing. Tonegawa and co-workers proposed that the intron contains a regulatory sequence that is required for transcription to be activated.

This result was remarkable for two reasons: (1) the regulatory sequence was thousands of bases away from the promoter, and (2) it was downstream of the promoter instead of upstream. Regulatory elements that are far from the promoter are termed **enhancers.** Follow-up work has shown that enhancers occur in all eukaryotes and that they have several key characteristics:

- Enhancers can be more than 100,000 bases away from the promoter. They can be located in introns or in untranscribed 5′ or 3′ sequences flanking the gene (see **Figure 18.7**). Researchers have yet to find enhancers located in exons.
- Like promoter-proximal elements, many types of enhancers exist. Different genes are associated with different enhancers.
- Enhancers can work even if their normal $5' \rightarrow 3'$ orientation is flipped.
- Enhancers can work even if they are moved to a new location in the vicinity of the gene, on the same chromosome.

Enhancers are regulatory sequences that are unique to eukaryotes. In addition, eukaryotic genomes contain regulatory sequences that are similar in structure to enhancers but opposite in function. These sequences are **silencers.** When silencers are active, they shut down transcription.

Once enhancers and silencers had been characterized, it was clear that they represented a type of regulatory sequence that worked very differently than promoter-proximal elements did. For example, there was good evidence that GAL4 and other types of regulatory proteins bind to promoter-proximal elements and interact directly with TBP or RNA polymerase. These interactions stabilize binding and promote transcription. But how could a regulatory sequence that is distant from the promoter help initiate transcription?

How Do Enhancers Work?

An experiment that followed up on Tonegawa's work with antibody-producing genes supported the hypothesis that enhancers are binding sites for proteins that regulate transcription. As **Figure 18.8** shows, researchers cut the enhancer sequence from an antibody-producing gene and spliced it into copies of the gene that codes for the protein β-globin. When this modified gene was inserted into antibody-producing cells, which normally do not express the gene for β-globin, the β-globin mRNA was produced. This result suggests that antibody-producing cells contain some factor that interacts specifically with the antibody-producing gene's enhancer to induce transcription. Follow-up work showed that the factor is a regulatory protein. Enhancers are regulatory sequences where regulatory proteins bind.

By analyzing mutant yeast, fruit flies, and roundworms that have defects in the expression of particular genes, biologists have identified a large number of regulatory proteins that bind to enhancers and silencers. This research program has supported one of the most general conclusions about gene regulation in eukaryotes: In multicellular species, different types of cells express different genes because they contain different regulatory proteins. These regulatory proteins, in turn, are produced in response to signals that arrive from other cells early in embryonic development. For example, a signaling molecule might arrive at a cell early in development and trigger the production of regulatory proteins that are specific to muscle cells. Because the regulatory proteins bind to specific enhancers and silencers and promoter-proximal elements, they trigger the production of muscle-specific proteins (**Figure 18.9**). If no "become a muscle cell" signal arrives, then no muscle-specific regulatory proteins are produced and no muscle-specific gene expression takes place.

To summarize, differential gene expression is based on the production of specific regulatory proteins. Eukaryotic genes are turned on when specific regulatory proteins bind to enhancers and promoter-proximal elements; the genes are turned off

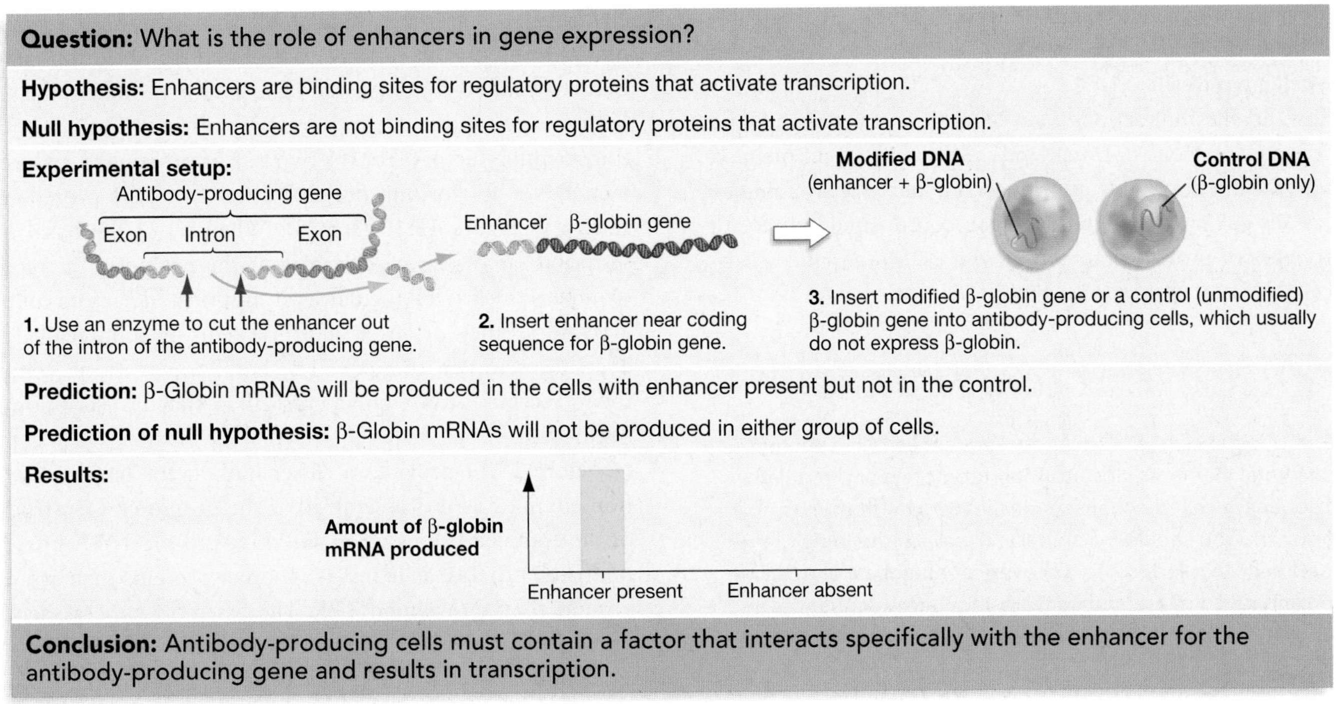
Question: What is the role of enhancers in gene expression?
Hypothesis: Enhancers are binding sites for regulatory proteins that activate transcription.
Null hypothesis: Enhancers are not binding sites for regulatory proteins that activate transcription.
Experimental setup:
Antibody-producing gene
Exon
Intron
Exon
1. Use an enzyme to cut the enhancer out of the intron of the antibody-producing gene.
Enhancer
β-globin gene
2. Insert enhancer near coding sequence for β-globin gene.
Modified DNA
(enhancer + β-globin)
Control DNA
(β-globin only)
3. Insert modified β-globin gene or a control (unmodified) β-globin gene into antibody-producing cells, which usually do not express β-globin.
Prediction: β-Globin mRNAs will be produced in the cells with enhancer present but not in the control.
Prediction of null hypothesis: β-Globin mRNAs will not be produced in either group of cells.
Results:
Amount of β-globin mRNA produced
Enhancer present
Enhancer absent
Conclusion: Antibody-producing cells must contain a factor that interacts specifically with the enhancer for the antibody-producing gene and results in transcription.

FIGURE 18.8 Evidence That Enhancers Are Involved in Tissue-Specific Expression of Genes
Researchers attached the enhancer sequence from an antibody-producing gene to the coding sequence for the β-globin gene. The altered gene was expressed in antibody-producing cells, which normally do not produce β-globin.

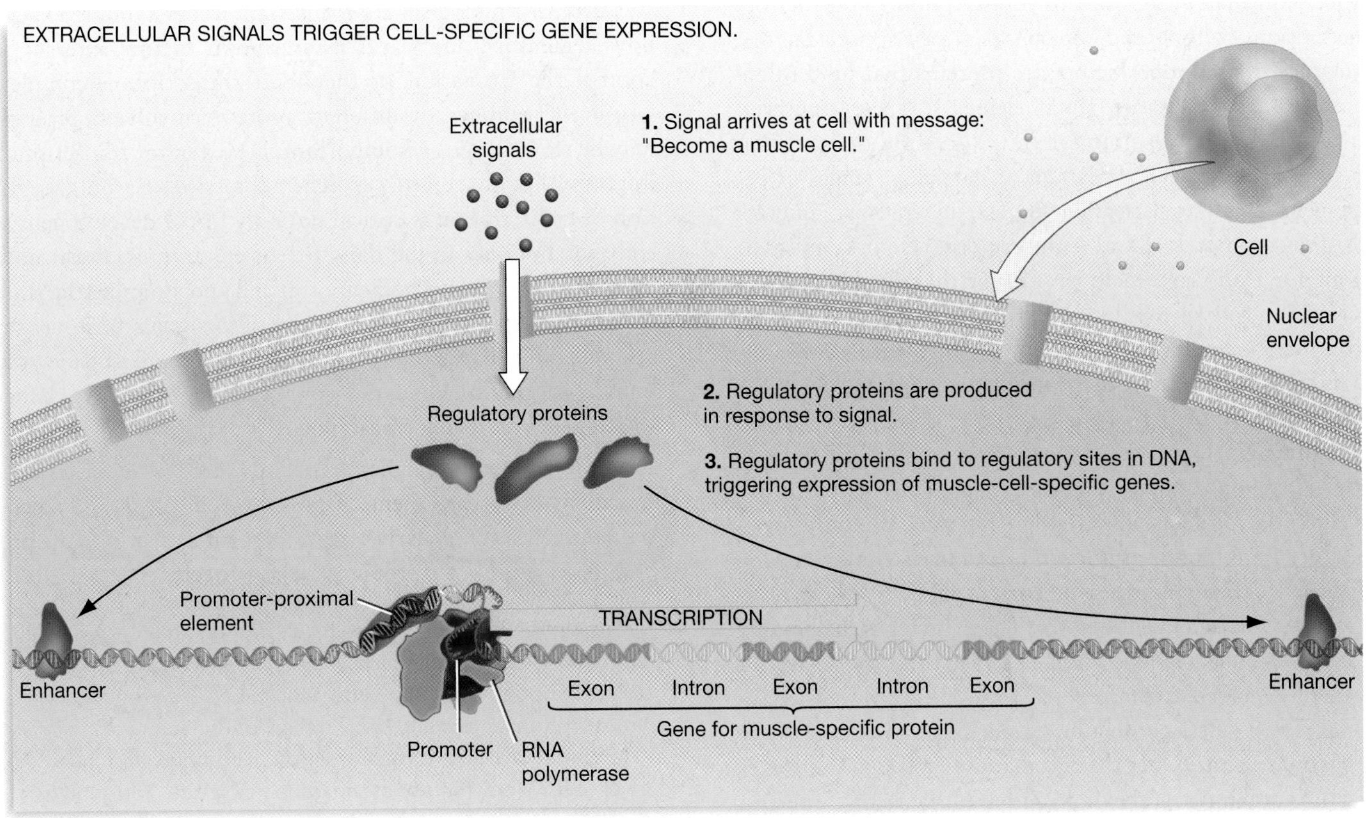
EXTRACELLULAR SIGNALS TRIGGER CELL-SPECIFIC GENE EXPRESSION.
Extracellular signals
1. Signal arrives at cell with message: "Become a muscle cell."
Cell
Nuclear envelope
Regulatory proteins
2. Regulatory proteins are produced in response to signal.
3. Regulatory proteins bind to regulatory sites in DNA, triggering expression of muscle-cell-specific genes.
Promoter-proximal element
TRANSCRIPTION
Enhancer
Exon
Intron
Exon
Intron
Exon
Gene for muscle-specific protein
Enhancer
Promoter
RNA polymerase

FIGURE 18.9 Why Are Certain Proteins Produced Only in Certain Types of Cells?
Signaling molecules from outside the cell trigger the production or activation of cell-specific regulatory proteins, which then influence transcription by binding to enhancers, silencers, or promoter-proximal elements in DNA.

when regulatory proteins bind to silencers. Distinctive regulatory proteins are what make a muscle cell a muscle cell and a bone cell a bone cell.

How do the interactions between regulatory proteins and DNA sequences control transcription? As we'll see in the next section, the initiation of transcription is a marvelously complex process in eukaryotes. It requires the coordinated activity of many types of proteins interacting with each other and a variety of DNA sequences.

✓CHECK YOUR UNDERSTANDING

Eukaryotic genes have regulatory sequences called promoter-proximal elements close to their promoters and regulatory sequences called enhancers or silencers far from their promoters. You should be able to (1) explain the logic of the experiment that led to the discovery of an enhancer in an intron, and (2) compare and contrast regulatory sequences in the DNA of bacteria versus those in eukaryotes.

18.4 Transcription Initiation

After several decades of research, many questions remain about how transcription is initiated in eukaryotes. It is clear, however, that two broad classes of proteins interact with regulatory sequences at the start of transcription: (1) regulatory transcription factors and (2) basal transcription factors. **Regulatory transcription factors** are proteins that bind to enhancers and promoter-proximal elements. These transcription factors are responsible for the expression of particular genes in particular cell types and at particular stages of development. **Basal transcription factors**, in contrast, interact with the promoter. Basal transcription factors must be present for transcription to occur, but they do not provide much in the way of regulation. TBP is a basal transcription factors that is common to all genes. Other basal transcription factors are specific to promoters recognized by RNA polymerase I, II, or III.

WEB TUTORIAL 18.1 Transcription Initiation in Eukaryotes

In addition, proteins called **coactivators** are involved in starting transcription. Coactivators do not bind to DNA. Instead they link the proteins involved in initiating transcription—regulatory transcription factors and basal transcription factors.

Figure 18.10 summarizes the current model for how transcription is initiated in eukaryotes. The first step is the binding of regulatory transcription factors to DNA; these factors recruit chromatin-remodeling complexes and histone acetyl transferases (HATs). The result is chromatin remodeling, a loosening of the chromatin structure. Once the chromatin-remodeling complexes and HATs are in place, they open a broad swath of chromatin that includes the promoter region (step 2).

Once chromatin's grip is relaxed, other regulatory transcription factors can bind enhancers and promoter-proximal elements. When these regulatory transcription factors are bound to DNA, they interact with basal transcription factors. Because enhancers are located far from the promoter, the regulatory transcription factors are normally far from the basal transcription factors. For the two sets of proteins to make contact, DNA has to loop out and away from the promoter, as shown in Figure 18.10 (step 3). When all of the basal transcription factors have assembled at the promoter in response to interactions with regulatory transcription factors and coactivators, they form a multi-protein machine called the **basal transcription complex.** The basal transcription complex then "recruits" RNA polymerase II so that transcription can begin (step 4).

Figure 18.11 provides a closer look at the basal transcription complex as it is assembled at the promoter. Construction of the complex begins when TBP binds to the TATA box. In a multistep process, as many as 60 other proteins then assemble around the DNA-bound TBP. The assembly process depends on interactions with regulatory transcription factors that are bound to enhancers, silencers, and promoter-proximal elements; the result is the formation of a basal transcription complex that can position RNA polymerase II in a way that initiates transcription. Compared with the situation in bacteria, where just 3 to 5 proteins may interact at the promoter to initiate transcription, the state of affairs in eukaryotes is remarkably complicated.

Currently, biologists are focused on understanding exactly how regulatory and basal transcription factors interact to control the formation of the basal transcription complex. Given the number of different proteins involved, progress may be slow. Progress is important, however, for transcription initiation lies at the heart of gene expression. Careful regulation of transcription is critical not only to the development of embryos but also to the daily life of eukaryotes. Right now, cells throughout your body are starting and stopping the transcription of specific genes in response to signals from nearby and distant cells. As the environment inside and outside your body continually changes, your cells continually change which genes are being transcribed.

✓CHECK YOUR UNDERSTANDING

Transcription initiation is a multistep process that begins when regulatory transcription factors bind to DNA and recruit proteins that open chromatin. Interactions between regulatory transcription factors and basal transcription factors result in the formation of the basal transcription complex and the arrival of RNA polymerase at the gene's start site. You should be able to compare and contrast the sigma proteins and regulatory proteins observed in bacteria with the basal and regulatory transcription factors found in eukaryotes.

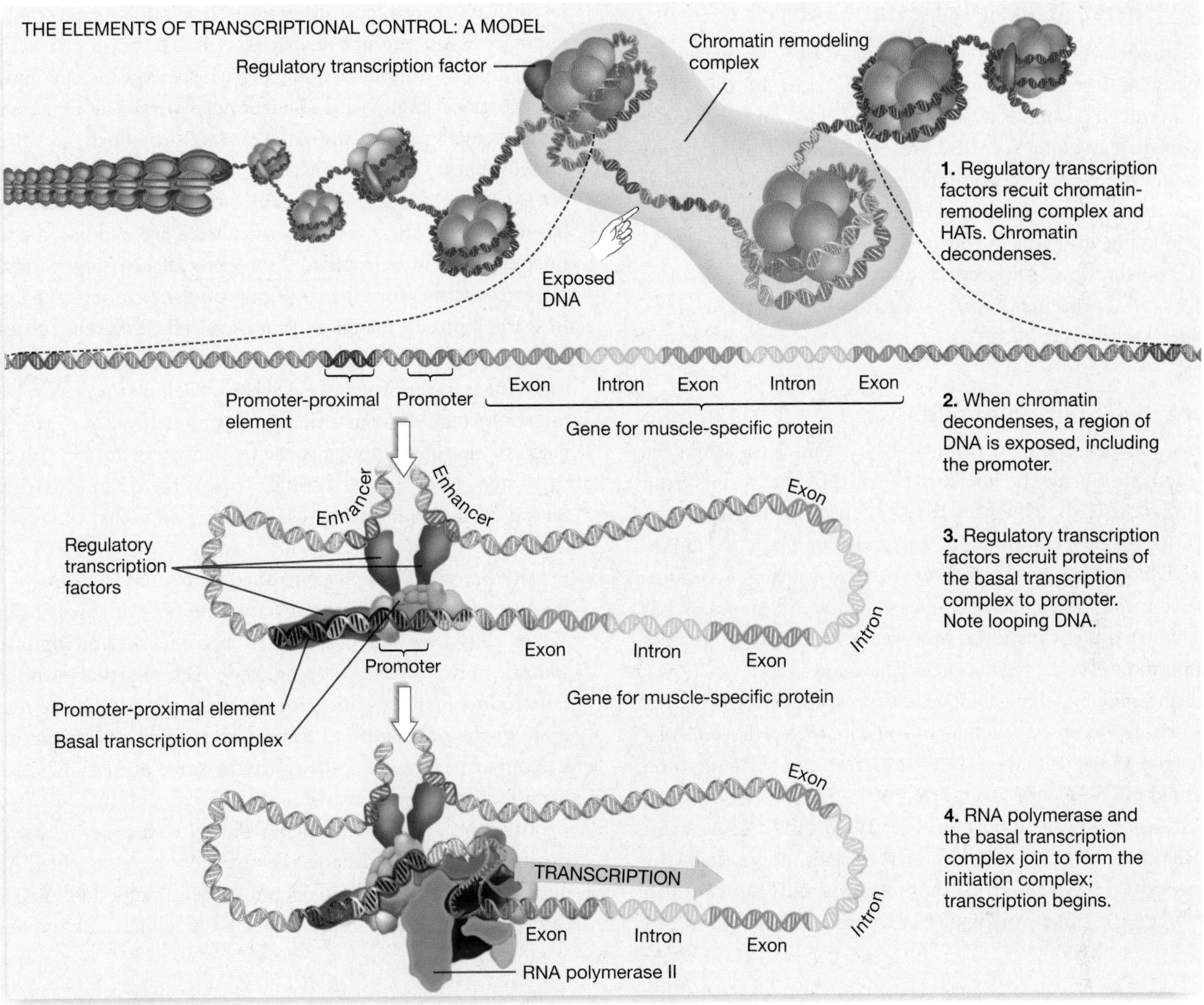

THE ELEMENTS OF TRANSCRIPTIONAL CONTROL: A MODEL
Regulatory transcription factor
Chromatin remodeling complex
Exposed DNA
1. Regulatory transcription factors recuit chromatin-remodeling complex and HATs. Chromatin decondenses.
Promoter-proximal element
Promoter
Exon
Intron
Exon
Intron
Exon
Gene for muscle-specific protein
2. When chromatin decondenses, a region of DNA is exposed, including the promoter.
Enhancer
Enhancer
Regulatory transcription factors
Exon
Intron
Promoter
Exon
Intron
Exon
Gene for muscle-specific protein
Promoter-proximal element
Basal transcription complex
3. Regulatory transcription factors recruit proteins of the basal transcription complex to promoter. Note looping DNA.
Exon
TRANSCRIPTION
Exon
Intron
Exon
Intron
RNA polymerase II
4. RNA polymerase and the basal transcription complex join to form the initiation complex; transcription begins.

FIGURE 18.10 The Elements of Transcriptional Control: A Model
According to the currently accepted model, transcription is initiated through a series of steps.

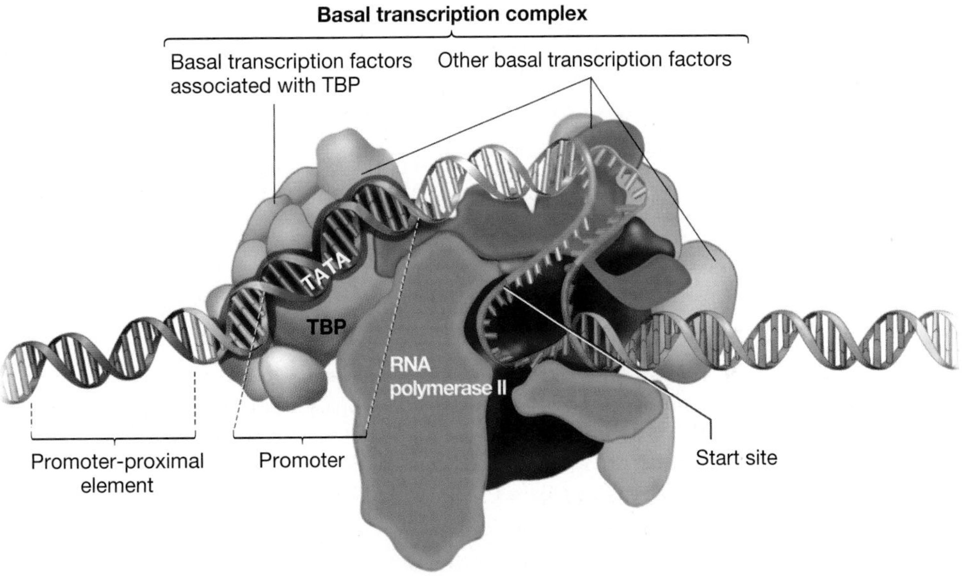

Basal transcription complex
Basal transcription factors associated with TBP
Other basal transcription factors
TATA
TBP
RNA polymerase II
Promoter-proximal element
Promoter
Start site

FIGURE 18.11 Many Proteins Interact at the Start Site for Transcription
Basal transcription factors associate with TBP when it binds to the promoter. The basal transcription complex also includes other basal transcription factors and RNA polymerase II.

18.5 Post-Transcriptional Control

In the process of regulating gene expression, chromatin remodeling and transcription are just the start of the story. Once an mRNA is made, a series of events have to occur if the final product is going to affect the cell. Each of these events offers an opportunity to regulate gene expression, and each is used in some cells at least some of the time. Control points include splicing mRNAs in various ways, altering the rate at which translation is initiated, and modifying the life span of mRNAs and proteins after translation has occurred. Let's consider each in turn.

Alternative Splicing of mRNAs

Introns are spliced out of primary RNA transcripts while the message is still inside the nucleus. The mRNA that results from splicing consists of sequences that are encoded by exons and are protected by a cap on the 5' end and a long poly (A) tail on the 3' end. Recall from Chapter 16 that splicing is accomplished by the molecular machines called *spliceosomes*. What that chapter did not mention, however, is that splicing provides an opportunity for the regulation of gene expression.

During splicing, changes in gene expression are possible because selected exons, as well as introns, may be removed. As a result, the same primary RNA transcript can yield mature, processed mRNAs with several different combinations of transcribed exons. This is important, because, if the sequence of ribonucleotides in the mature mRNAs can vary, then the polypeptides translated from those mature mRNAs can also vary. When the same primary RNA transcript is spliced in different ways to produce different mature mRNAs and thus different proteins, **alternative splicing** is said to occur.

To see how alternative splicing works, consider the muscle-cell protein tropomyosin. The tropomyosin gene is expressed in both skeletal muscle and smooth muscle, which are distinct muscle types. (Skeletal muscle is responsible for moving your bones; smooth muscle lines many parts of your gut and certain blood vessels. The two types of muscle are composed of distinct types of muscle cells.) As **Figure 18.12a** shows, the primary transcript from the tropomyosin gene contains 14 exons. In skeletal muscle and smooth muscle cells, different subsets of the 14 exons are spliced together to produce two different messages for translation (**Figure 18.12b**). Each mature mRNA contains information from a different combination of exons. As a result of alternative splicing, the tropomyosin proteins found in these two cell types are distinct. One of the reasons that skeletal muscle and smooth muscle are different is that they contain different types of tropomyosin.

Alternative splicing is controlled by regulatory proteins that bind to mRNAs in the nucleus and interact with spliceosomes. When cells that are destined to become skeletal muscle or smooth muscle are developing, they receive signals leading to the production of specific proteins that are active in the regulation of splicing. Instead of transcribing different versions of the tropomyosin gene, the cells splice the same primary RNA transcript but in different ways.

Although alternative splicing of RNA was viewed as a curiosity until about a decade ago, attitudes have changed. Alternative splicing is now considered to be a major mechanism in the control of gene expression in multicellular eukaryotes. In

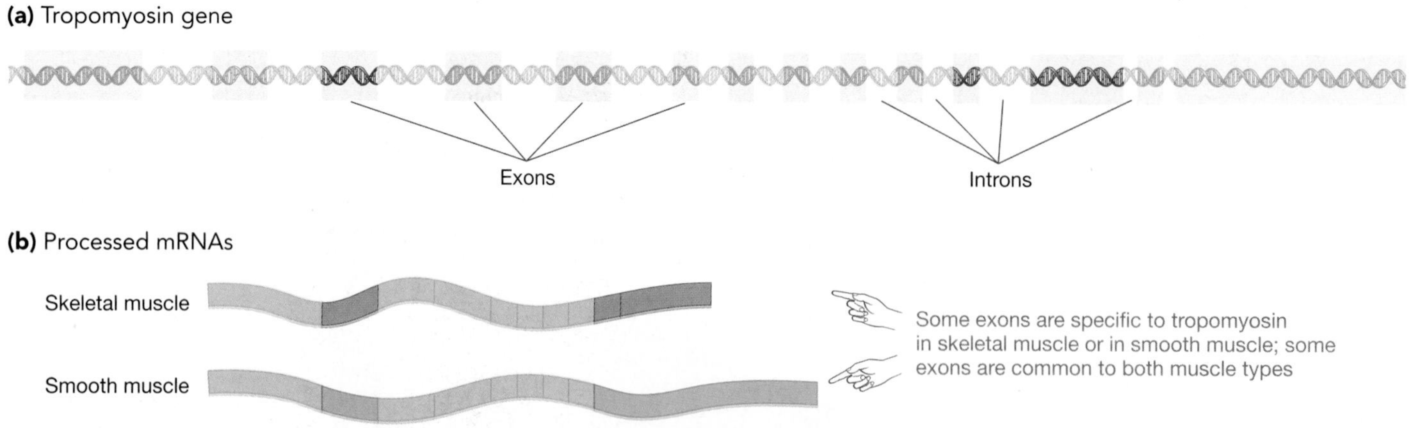
(a) Tropomyosin gene
Exons
Introns
(b) Processed mRNAs
Skeletal muscle
Smooth muscle
Some exons are specific to tropomyosin in skeletal muscle or in smooth muscle; some exons are common to both muscle types

FIGURE 18.12 Alternative Splicing Produces More than One Mature mRNA from the Same Gene
(a) The tropomyosin gene has a large series of exons and introns. The purple exons are found in skeletal muscle tropomyosin, and the blue exons are found in smooth muscle tropomyosin. The orange exons are found in the tropomyosin of both types of muscle cells. **(b)** In skeletal and smooth muscle cells, alternative splicing leads to the production of distinct mature mRNAs for the tropomyosin protein.

fruit flies, for example, the gene that determines gender is transcribed into the same primary RNA transcript in both males and females. A cascade of regulated splicing in females ultimately results in the production of a protein called *doublesex*, which represses the transcription of male-specific genes. In males, an alternative series of splicing events results in a version of the doublesex protein that represses the transcription of female-specific genes.

Alternative splicing makes it possible to produce more than one protein from a single gene. Before the importance of alternative splicing was recognized, a gene was considered to be a nucleotide sequence that encodes a single protein or RNA. Now genes have to be thought of as DNA sequences that are capable of directing the production of one or more related polypeptides or RNAs. As a result, the number of proteins that an organism can produce far exceeds the number of genes it has. For example, at least 35 percent of human genes undergo alternative splicing. Although the human genome contains an estimated 40,000 genes, the number of different proteins that your cells can produce is believed to be between 100,000 and 1 million.

The current record holder for the number of mRNAs derived from one gene is the *Dscam* gene in the fruit fly *Drosophila melanogaster*. The products of this gene are involved in guiding growing nerve cells within the embryo. Because the primary transcript is spliced into about 38,000 distinct forms of mRNA, the *Dscam* gene can produce about 38,000 different products. Researchers are testing the hypothesis that particular combinations of Dscam proteins allow particular nerve cells to connect to each other in the fly's brain.

Current research on alternative splicing is focused on two topics: (1) understanding how regulatory proteins control the process and (2) documenting the extent to which alternative splicing enriches the genetic information stored in a genome. Stay tuned.

Translational Control

Once splicing is complete and processed mRNAs are exported to the cytoplasm, several new regulatory mechanisms come into play prior to translation. The following examples illustrate the array of mechanisms that influence the timing or extent of translation:

- The life span of an mRNA in the cell can vary. For example, the mRNA for casein—the major protein in milk—is produced in the mammary gland tissue of female mammals. Normally this mRNA has a half-life of just over 1 hour, and little casein protein is produced. But when a female mouse is lactating, regulatory molecules increase the mRNA's half-life to 28.5 hours—leading to a huge increase in the production of casein protein.
- In many species of animals, eggs are loaded with mRNAs that are not translated until fertilization occurs. The proteins produced from these messages are involved in directing the early development of the embryo, so they are not needed until fertilization is complete. Translation of the mRNAs in the egg can be prevented when a regulatory protein binds to them, or when the message's cap or tail is modified.
- The overall rate of translation in a cell may slow or stop in response to a sudden increase in temperature or infection by a virus. The slowdown occurs because regulatory proteins add a phosphate group to a protein that is part of the ribosome. As we have pointed out in earlier chapters, phosphorylation frequently leads to changes in the shape and chemical reactivity of proteins. In the case of the phosphorylated ribosomal protein, the shape change slows or prevents translation. For the cell, this dramatic change in gene expression can mean the difference between life and death. If the danger is due to a sudden increase in ambient temperature, shutting down translation prevents the production of improperly folded polypeptides; if the insult is a virus, the cell avoids manufacturing viral proteins.

Post-Translational Control

Control of gene expression may continue even after translation occurs and a protein product is complete. Recall from Chapter 17 that, in bacteria, mechanisms of post-translational regulation are important because they allow the cell to respond to new conditions rapidly. The same is true for eukaryotes. Instead of waiting for transcription and translation to occur, the cell can respond to altered conditions by quickly activating or inactivating existing proteins. Regulatory mechanisms occurring late in the flow of information from DNA to RNA to protein involve a trade-off between speed and resource use, however, because transcription and translation require energy and materials.

A group of regulatory transcription factors that is found in mammals is called **signal transducers and activators of transcription (STATs)** and provides a good example of post-translational control. STATs participate in several signal transduction pathways. Recall from Chapter 8 that signal transduction pathways are interactions that convert a signal at the cell surface into a response by the cell. In white blood cells, STAT activity results in increased cell growth and cell division.

In white blood cells, STATs normally reside in the cytoplasm as single polypeptide chains. In this form, the STATs are inactive. But a chain of events is triggered when a signaling molecule binds to and activates a receptor on the cell surface (**Figure 18.13a**). The activated receptor adds a phosphate to a STAT polypeptide chain, triggering the formation of a **dimer** (a unit made of two parts). The activated STAT dimer then moves into the nucleus, binds to an enhancer, and activates transcription (**Figure 18.13b**). The STAT remains active until the phosphate is removed by another regulatory protein. At this point, the two polypeptide chains come apart and are transported back to the cytoplasm.

Phosphorylation is a very common mechanism of post-translational control over gene expression, particularly in signal transduction pathways. It is only one of several post-translational control mechanisms that have been documented, however. The activity of a protein may also be modified by folding or by enzymes that cleave off a portion of the molecule. But no matter what the mechanism, post-translational control is correlated with particularly rapid changes in gene expression.

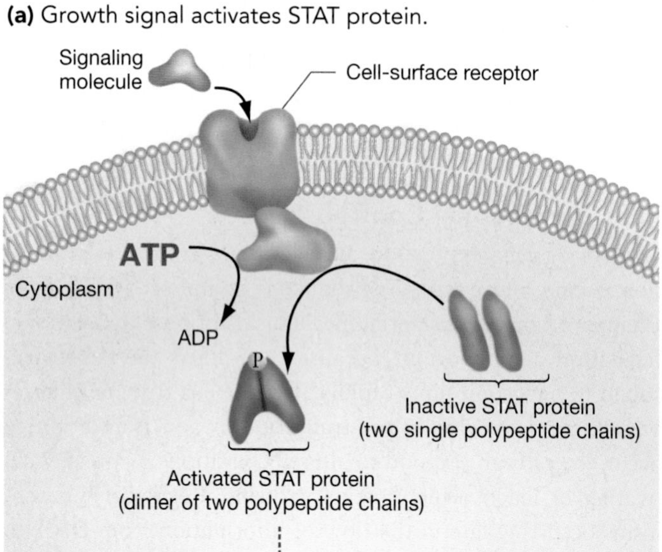

FIGURE 18.13 STATs Activate Transcription after They Are Phosphorylated

(a) A series of events, beginning with the binding of a signaling molecule to a cell-surface receptor, leads to the phosphorylation of STAT proteins. This causes two STAT polypeptide chains to join, forming an activated dimer. **(b)** Activated STAT proteins move to the nucleus, bind to DNA, and activate the transcription of selected target genes.

A 50-Year Retrospective: How Does Gene Expression in Bacteria and in Eukaryotes Compare?

For about five decades, biologists have been studying how gene expression is controlled. Almost as soon as they knew that information in DNA is transcribed into RNA and then translated into proteins, researchers began asking questions about how that flow of information is regulated. Early studies on bacteria established the existence of promoters, the structure and function of regulatory proteins and regulatory sequences, and the principles of positive control and negative control. More recent work has highlighted the importance of chromatin structure and alternative splicing in eukaryotes.

What are the fundamental differences in how gene expression is regulated in bacteria and eukaryotes? Biologists point to four primary differences, two of which involve levels of control that exist in eukaryotes but not in bacteria:

1. *Packaging* The chromatin structure of eukaryotic DNA must be opened in order for TBP, the basal transcription complex, and RNA polymerase to gain access to genes and initiate transcription. A key insight here is that, because eukaryotic DNA is packaged so tightly, the default state of transcription in eukaryotes is off. In contrast, the default state of transcription in bacteria, with their supercoiled DNA and freely accessible promoters, is on.
2. *Alternative splicing* Prior to translation, primary transcripts in eukaryotes must be spliced—an event that does not occur in bacteria. The fundamental consequence is that the one-to-one correspondence between the number of genes and the number of gene products observed in bacteria is not seen in eukaryotes. Instead, each eukaryotic gene may code for one to thousands of distinct products.
3. *Complexity* Transcriptional control is much more complex in eukaryotes than in bacteria. The function of sigma proteins in bacteria is analogous to the role of the basal transcription complex in eukaryotes. Likewise, the function of CAP, the repressor, and other regulatory proteins is analogous to the role of regulatory transcription factors in eukaryotes. But the sheer number of eukaryotic proteins involved in regulating transcription—and the complexity of their interactions—dwarfs those in bacteria.
4. *Coordinated expression* In bacteria, genes that are involved in the same cellular response are organized into operons controlled by a single promoter. Because their polycistronic mRNAs are translated together, several proteins are produced in a coordinated fashion. In contrast, operons are rare in eukaryotes. In these organisms, genes that are physically scattered can be expressed at the same time because a single set of regulatory transcription factors can trigger the transcription of several genes. For example, muscle-specific genes found on several different chromosomes can be transcribed in response to the same muscle-specific regulatory transcrip-

tion factor. In this way, eukaryotes coordinate the expression of functionally related genes.

To explain why these differences exist, biologists point to a fundamental contrast between unicellular and multicellular individuals. In both types of organisms, cells have to respond to environmental changes in an appropriate way. Bacteria and yeast cells change gene expression to respond to changes in the availability of sugars and other nutrients; your muscle cells change gene expression in response to changes in temperature, the availability of oxygen and sugars, and general activity. But in addition, the cells of multicellular organisms have to differentiate as an individual develops. Changes in gene expression are responsible for the differentiation of muscle cells, bone cells, leaf cells, and flower cells in response to signals from other cells. The need for each cell type to have a unique pattern of gene expression is thought to be the reason why control of gene expression is so much more complex in eukaryotes than in bacteria.

One of the two great frontiers in research on gene expression is to understand how developmental signals produce cell-specific gene expression in multicellular organisms. The other major frontier is to understand how certain defects in gene regulation result in uncontrolled cell growth and the suite of diseases called cancer.

18.6 Linking Cancer with Defects in Gene Regulation

Normal regulation of gene expression results in the orderly development of an embryo and appropriate responses to environmental changes. Abnormal regulation of gene expression, in contrast, can lead to developmental abnormalities and diseases such as cancer. Hundreds of distinct cancers exist. These diseases are enormously variable regarding the tissues they affect, their rate of progression, and their outcome. Because the underlying defects, symptoms, and consequences are so diverse, cancer is not a single disease but a family of related diseases. Cancers are related because they all result from uncontrolled cell growth.

From this perspective, the fundamental question in cancer biology is a simple one: What causes uncontrolled cell growth? The short answer is that each type of cancer is caused by a different set of defects that lead to uncontrolled cell growth. The longer and more accurate answer is based on two key observations. First, recall from Chapter 11 that cancer results from defects in the proteins that control the cell cycle. Second, cancer is associated with mutations that knock out key genes. For example, Chapter 14 linked the cancer-causing disorder xeroderma pigmentosum and hereditary nonpolyposis colorectal cancer with defects in DNA repair. This association supports the hypothesis that increased mutation rates are involved in triggering cancer. The hypothesis is also supported by the observation that people who have been exposed to **mutagens**—that is, radiation or chemicals that induce mutation—have an increased risk of developing cancer. For example, cancer rates were extremely high in the tens of thousands of people who were exposed to massive doses of radiation from the atomic bombs dropped on Hiroshima and Nagasaki, Japan, at the end of World War II. The link between mutation and cancer is so strong that suspected **chemical carcinogens** (cancer-causing compounds) are first tested by their ability to cause mutations in bacteria.

The fundamental question of cancer biology comes into sharper focus once the role of cell-cycle regulators and mutation is acknowledged. Now we can ask, Which genes, when mutated, disrupt the cell cycle and trigger uncontrolled cell growth?

Intensive research over the past two decades has shown that many cancers are associated with mutations in regulatory transcription factors or regulatory sequences in DNA. These mutations lead to cancer when they affect one of two classes of genes: (1) genes that stop or slow the cell cycle, and (2) genes that trigger cell growth and division by initiating specific phases in the cell cycle.

Genes that stop or slow the cell cycle are called **tumor suppressor genes.** Their products prevent the cell cycle from progressing unless specific signals indicate that conditions are right for moving forward with mitosis and cell division. If a mutation disrupts normal function of a tumor suppressor gene, then a key "brake" on the cell cycle is eliminated.

Genes that encourage cell growth by triggering specific phases in the cell cycle are called **proto-oncogenes** ("first-cancer-genes"). In normal cells, proto-oncogenes are required to initiate each phase in the cell cycle. They are active only when conditions are appropriate for growth, however. In cancerous cells, defects in the regulation of proto-oncogenes cause these genes to stimulate growth at all times. In cases such as this, a mutation has converted the proto-oncogene into an **oncogene**—an allele that promotes cancer development.

To gain a deeper understanding of how defects in gene expression can lead to cancer, let us consider two examples from recent research. The first, the *p53* gene, illustrates how problems with regulating tumor suppressor genes can lead to uncontrolled cell growth. The second, STAT mutations, demonstrates how defects in regulatory transcription factors can lead to the constant activation of proto-oncogenes, constant cell growth, and tumor formation.

p53—Guardian of the Genome

The gene that is most often defective in human cancers codes for a regulatory transcription factor. This gene is called ***p53***, because the protein it codes for has a molecular weight of approximately 53 kilodaltons (53,000 daltons, or 53,000 amu—see Chapter 2). DNA sequencing studies have revealed that mutant, nonfunctional forms of *p53* are found in over half of all human cancers.

What is the link between a loss of p53 protein activity and cancer? A key observation suggested an answer to this question: When researchers exposed normal, noncancerous human cells to UV radiation, levels of p53 protein increased markedly. Recall from Chapter 14 that UV radiation damages DNA. Follow-up studies confirmed that there is a close correlation between DNA damage and the amount of p53 in a cell. In addition, analyses of the protein's structure showed that it contains a DNA-binding domain.

These observations inspired the hypothesis that p53 is a transcription factor that serves as the master brake on the cell cycle (**Figure 18.14a**). In this model, p53 is activated after DNA damage occurs. The hypothesis contends that the activated protein binds to the enhancers of genes that arrest the cell cycle. Once the cell cycle is arrested, the cell has time to repair its DNA before continuing to grow and divide.

Recent research has shown that this model of p53 function is correct in almost every detail. For example, consider the results of studies based on X-ray crystallography. The three-dimensional models generated by this technique confirmed that p53 binds directly to DNA (**Figure 18.14b**). In addition, researchers who mapped the location of mutations that make p53 defective found that virtually all of the cancer-causing mutations were located in p53's DNA-binding site (see the gray regions of Figure 18.14b). This observation supports the hypothesis that defective forms of the protein can't bind to enhancers. Further, investigators have documented that one of the genes induced by p53 encodes a protein that binds to the cyclin-Cdk complexes introduced in Chapter 11. Binding by this protein prevents the cyclin-Cdk complexes from triggering M (mitosis) phase in the cell cycle.

Recent research has also shown that when a cell's DNA is extensively damaged and cannot be repaired, p53 activates the transcription of genes that cause the cell to take its own life via apoptosis (see Chapter 11). But if mutations in the *p53* gene make the protein product inactive, then damaged cells are not shut down or killed. They continue to move through the cell cycle, except now they are likely to contain many mutations because of the DNA damage they have left unrepaired. If these mutations create oncogenes or inactivate genes for tumor suppressors, the cells have taken a key step on the road to cancer.

To summarize, *p53* functions as a tumor suppressor gene. It prevents cancer onset by stopping the cell cycle when DNA is damaged. When *p53* is functioning normally, mutations that produce oncogenes are repaired or eliminated before uncontrolled cell growth can begin. The role of *p53* in preventing cancer is so fundamental that biologists call this gene "the guardian of the genome." Currently, research on the p53 protein is forging ahead on two fronts: Biologists are striving to identify more of the genes that are regulated by this protein and to find molecules that could act as anticancer drugs by mimicking p53's shape and activity.

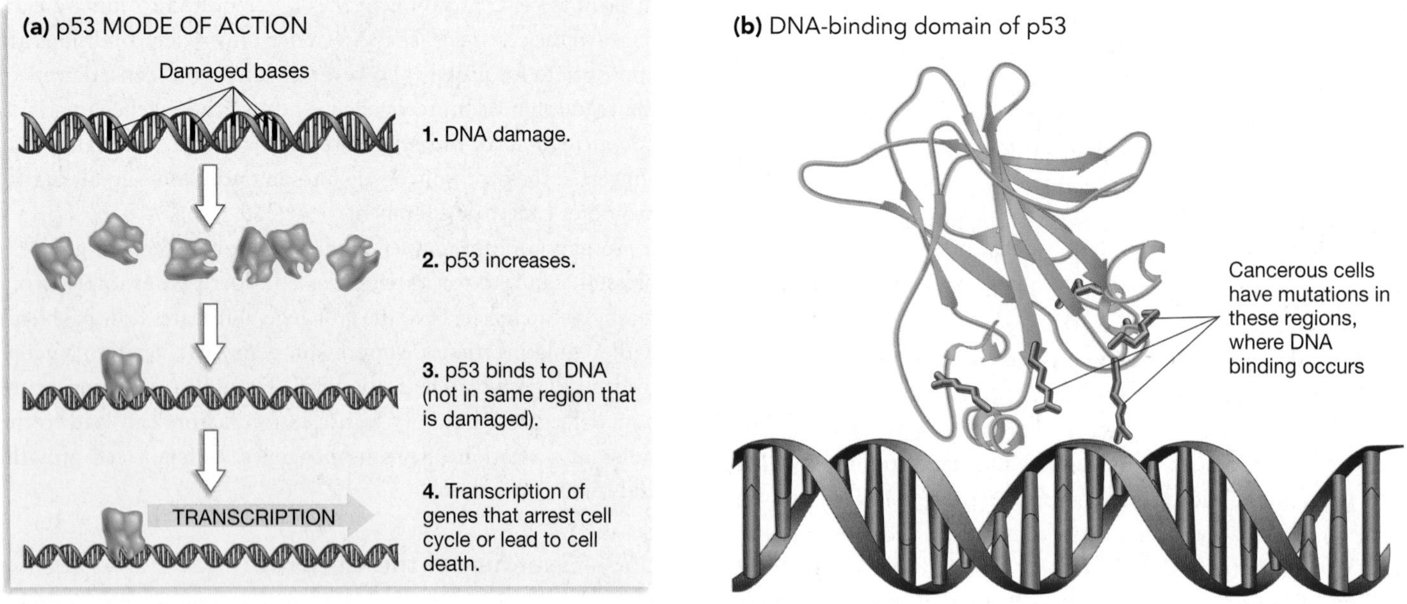
(a) p53 MODE OF ACTION
Damaged bases
1. DNA damage.
2. p53 increases.
3. p53 binds to DNA (not in same region that is damaged).
TRANSCRIPTION
4. Transcription of genes that arrest cell cycle or lead to cell death.
(b) DNA-binding domain of p53
Cancerous cells have mutations in these regions, where DNA binding occurs

FIGURE 18.14 p53 Is a DNA-Binding Protein That Acts as a Tumor Suppressor
(a) Summary of the current model for p53's mode of action. **(b)** Ribbon model showing the DNA-binding domain of p53. In cancer patients, the amino acids in the gray portions of the molecule are often different than the normal versions. QUESTION Why would substituting one amino acid for another at one of the locations highlighted in part (b) affect a protein's ability to bind to DNA? If the model in part (a) is correct, what happens when p53 cannot bind to DNA efficiently?

STAT Mutations in Cancer

The family of transcription activators called STATs is another type of protein implicated in cancer. Recall from Section 18.5 that, in several types of cells, STATs activate genes that promote cell growth in response to signals from outside the cell. In normal cells, STATs must be phosphorylated before they can bind to regulatory sequences in DNA and activate transcription. But in some human cancers, STATs bind to DNA all the time, or constitutively—not just when they are phosphorylated. When this happens, STATs constantly stimulate cell growth and division. These observations suggest that mutations can convert STATs into oncogenes.

What type of mutations could turn STATs into oncogenes? To answer this question, recall that the activated form of a STAT is a dimer—a combination of two polypeptides. For a dimer to form, an amino acid residue at the point of contact between the two STAT proteins must be phosphorylated (**Figure 18.15a**).

These observations inspired a hypothesis that STATs turn into oncogenes if mutations cause them to form dimers without being phosphorylated—that is, without being stimulated by a normal growth signal. More specifically, researchers suggested that if mutations caused cysteine residues to be substituted in the portion of the protein where dimer formation takes place, then disulfide bonds might form between STAT3 monomers, creating constitutively activated dimers (**Figure 18.15b**).

To test this hypothesis, researchers created mutant versions of the STAT3 protein with cysteines in the appropriate location. The researchers expressed the protein in normal (control) cells and analyzed the amount of protein produced by a gene regulated by STAT3. Compared with cells with normal STAT3 proteins, the cysteine-containing mutant cells produced 10 times as much of the target gene product (**Figure 18.15c**). This observation was consistent with the biologists' prediction that cysteine mutants form dimers and bind to DNA constitutively.

To confirm their result, the biologists injected mice either with cells that expressed cysteine-mutant STAT3 or with cells that expressed normal STAT3. After just four weeks, all of the

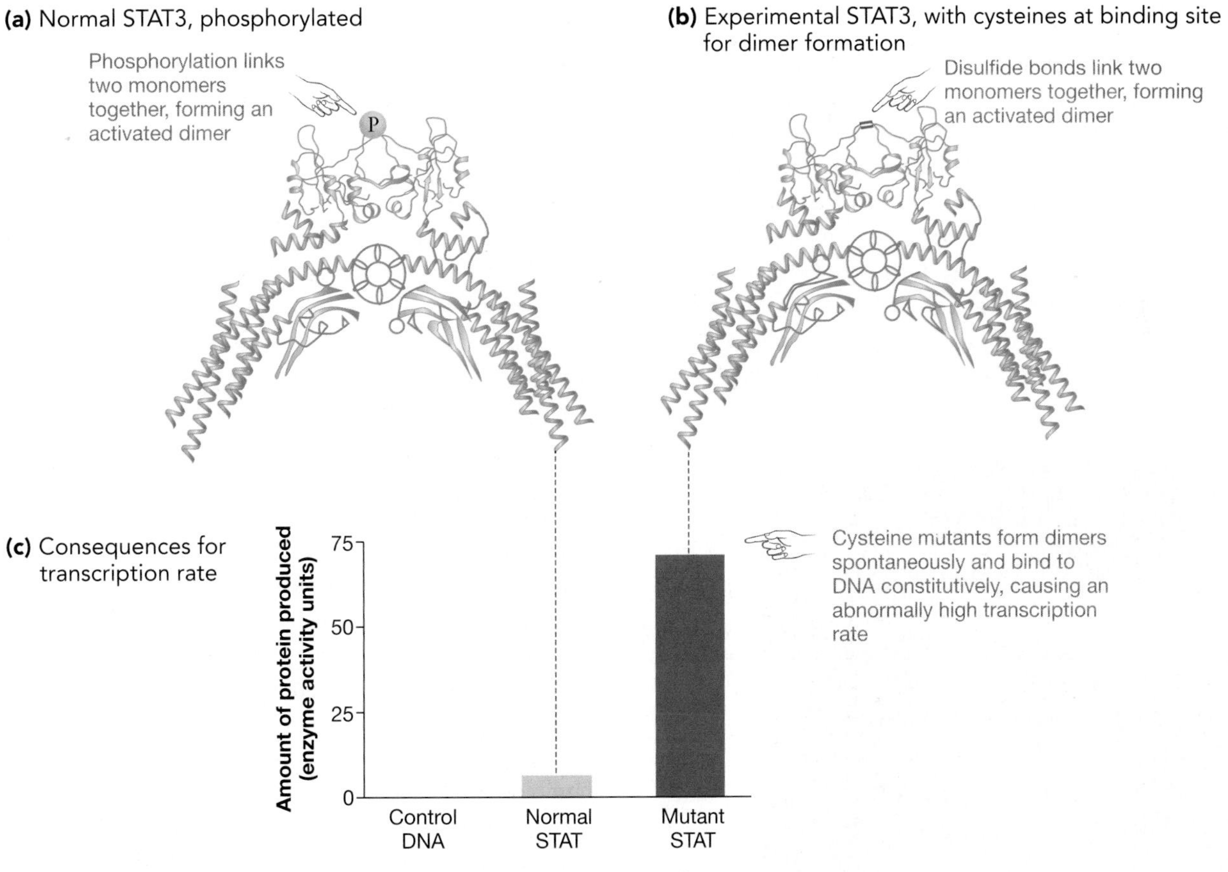

FIGURE 18.15 Generating a Cancer-Causing Mutation Experimentally
(a) Phosphorylated sections of STAT3 proteins bind to one another, forming a dimer. **(b)** Substitution of cysteines at the locations shown in red result in the formation of disulfide bonds between STAT3 proteins. As a result, the STAT proteins are activated at all times. **(c)** Histogram showing the transcriptional activity of a gene that is normally activated by STAT3. The data come from cells that received either a control DNA (a gene that does not encode a transcription activator), a normal *STAT3* gene, or a copy of the cysteine-mutant version of the *STAT3* gene.

mice that had been injected with mutant cells had developed tumors at the site of injection. In contrast, none of the mice that had been injected with normal cells showed evidence of tumors, even after eight weeks.

These experiments have an important message: Any mutation that constitutively forms dimers of STAT proteins causes constitutive activation of STAT target genes. As a result, these mutations set the stage for cancer. An even more global message is that constant activation of a regulatory transcription factor can lead to cancer. Cancers originate with mutations that disrupt the normal regulation of cell-cycle control genes.

✓CHECK YOUR UNDERSTANDING

Cancer is associated with mutations and with loss of control over the cell cycle. These two observations are connected, because mutations can cause cancer when they occur in genes that are involved in regulating cell-cycle control genes. For example, when a protein that activates the cell cycle is activated constitutively because of a mutation in a regulatory gene, then uncontrolled cell growth may occur. You should be able to explain why knock-out (loss-of-function) mutations in *p53* occur in so many cancers.

ESSAY Fly Eyes and Other Strange Tales of Gene Regulation

Imagine a 1950s-vintage science-fiction movie. As thunder cracks and lightning flashes, a mad-eyed scientist working in the bowels of a sinister-looking laboratory inserts a mouse gene into a fruit fly's genome. The next day, the scientist cackles with delight as she discovers eyes sprouting in bizarre locations on the fly. But then the fly escapes. It mates with a normal fly and sires offspring with hundreds of eyes. Thanks to their remarkable 360° vision, no predator or flyswatter can stop the little monsters. Clouds of them soar off, bent on global domination.

Preposterous? A team of not-so-mad-eyed scientists has inserted a mouse gene into fruit flies and observed the formation of eyes on the adult fly's legs, antennae, and torso (**Figure 18.16**). The gene in question is a regulatory transcription factor. When the gene product binds to enhancers, it sets off a cascade of gene expression events that are required for eye development.

Walter Gehring and colleagues were first to isolate and characterize the gene involved. It was called *eyeless*, because it was originally discovered as a mutant allele that causes flies to develop without eyes. When Gehring's team created *Drosophila* embryos that expressed the normal form of the gene in locations that naturally give rise to appendages, eyes developed on the legs and antennae. This result underscores an important conclusion: During development, the fate of a cell depends on the activity of regulatory transcription factors. When *eyeless* is activated in leg cells, the leg cells give rise to eyes.

When eyeless is activated in leg cells, the leg cells give rise to eyes.

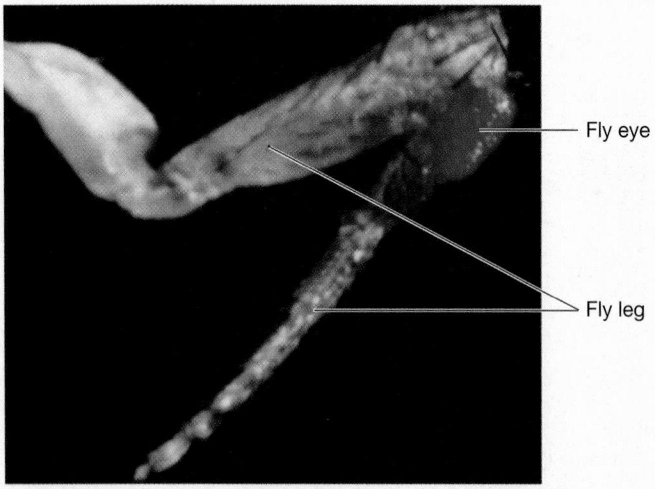

FIGURE 18.16 ***Eyeless*** **Encodes a Transcription Factor That Induces Eye Development**
When *Drosophila* embryos are made to express the *eyeless* gene in cells that gave rise to the legs, adults develop with eyes on their legs.

When the researchers looked for genes in other organisms that were similar to *eyeless* in DNA sequence, the team found them in a wide array of species. Consider a mouse gene called *Small Eye* or *pax6*, which is similar in sequence to *eyeless*. The name *Small Eye* was inspired by the observation that a mutant allele of this gene leads to reduced eye development. Because *eyeless* and *Small Eye* are similar in structure and function, Gehring hypothesized that they are **homologous**—that the genes are descended from a similar gene in a common ancestor of flies and mice, which, according to the fossil record, lived some 550 million years ago. To test this hypothesis, Gehring inserted the mouse gene into *Drosophila* embryos. *Small Eye* induced the development of fly eyes essentially anywhere in the embryo that it was expressed. He got the same result when he inserted a squid gene that was similar to *eyeless*.

What does this tell us about biology? First, it is reasonable to conclude that *eyeless*, *Small Eye*, and like genes in other species are similar in structure and function because they are homologous. Second, these regulatory transcription factors and their recognition sequences in DNA have been so conserved that the proteins are nearly interchangeable among organisms separated by 550 million years of evolution. Third, one key regulatory transcription factor can unleash a cascade of responses that results in a structure as complex as an eye. Fourth, the same starting point—expression of the *eyeless* gene—can lead to very different ends, depending on the context. In flies, the mouse gene *Small Eye* leads to a fly eye, not a mouse eye.

CHAPTER REVIEW

Summary of Key Concepts

- **Changes in gene expression allow eukaryotic cells to respond to changes in the environment and cause distinct cell types to develop.**

In a multicellular eukaryote, different types of cells are radically different in size, shape, and function—even though they have the same DNA. Cells are different not because they have different genes but because they express different genes.

In embryos, cells begin expressing specific genes in response to signals from other cells. Because different cells receive different signals, one cell might begin expressing muscle-specific genes while a nearby cell expresses bone-specific genes. Once cells mature, they continue to receive signals that are released by other cells and that carry information about changes in the environment. In response, changes in gene expression may occur.

In both embryonic cells and mature cells, several types of changes in gene expression may occur. The transcription of specific genes may be initiated or repressed, mRNAs may be spliced in different ways to produce a different product, the life span of specific mRNAs may be extended or shortened, and the life span or activity of particular proteins may be altered.

- **In eukaryotes, DNA is packaged with proteins into complex structures that must be opened before transcription can occur.**

Eukaryotic DNA is wrapped around histone proteins to form a bead-like nucleosome, which is then coiled into 30-nm fibers and higher-order chromatin structures. Transcription cannot be initiated until the interaction between DNA and histones in chromatin is relaxed. These changes depend on the acetylation of histones and the action of chromatin-remodeling complexes.

- **In eukaryotes, transcription can be initiated only when specific proteins bind to the promoter and to regulatory sequences that may be close to the promoter or far from it.**

Regulatory transcription factors are proteins that bind to regulatory sequences called enhancers and silencers, which are often located at a distance from the gene in question, or to promoter-proximal sequences near the start of the coding sequence. The first regulatory transcription factors that bind to DNA recruit proteins that loosen the histones' grip on the gene, making the promoter accessible to basal transcription factors. Interactions between regulatory and basal transcription factors lead to the formation of the basal transcription complex. Once this large, multi-protein complex is intact, RNA polymerase is recruited to the site and transcription begins.

Web Tutorial 18.1 Transcription Initiation in Eukaryotes

- **Alternative splicing allows a single gene to code for several different products.**

Once a message has been transcribed, several other regulatory events come into play. Alternative splicing allows a single locus to produce more than one mRNA and more than one protein. Mechanisms of post-translational control, such as the activation of transcription factors by phosphorylation, trigger rapid responses to signals from outside the cell.

- **Cancer can develop when mutations disable genes that regulate cell-cycle control genes.**

The cell cycle is controlled by specific genes. But if mutations alter the regulatory sequences of cell-cycle control genes or the genes for their regulatory transcription factors, then uncontrolled cell growth and tumor formation may result. For example, the cell-cycle regulator *p53* is responsible for stopping the cell cycle when DNA is damaged. If *p53* is mutated in a way that prevents its protein product from binding to DNA, then the cell cycle is not arrested and the damaged DNA is not repaired—leading to mutations. If the mutations alter genes that control the cell cycle, such as STAT genes, cancer may occur. Mutant STAT proteins that promote cancer bind to DNA at all times and trigger the expression of genes that activate the cell cycle, leading to uncontrolled cell growth.

Questions

Content Review

1. What is chromatin?
   a. the protein core of the nucleosome, which consists of histones
   b. the 30-nm fiber
   c. the DNA-protein complex found in eukaryotes
   d. the histone *and* non-histone proteins in eukaryotic nuclei

2. What is a tumor suppressor?
   a. a gene associated with tumor formation when its product does not function
   b. a gene associated with tumor formation when its product functions normally
   c. a gene that accelerates the cell cycle and leads to uncontrolled cell growth
   d. a gene that codes for a transcription factor involved in tumor formation

3. Which of the following statements about enhancers is correct?
   a. They contain a unique base sequence called a TATA box.
   b. They are located only in 5′-flanking regions.
   c. They are located only in introns.
   d. They are found in a variety of locations and are functional in any orientation.

4. In eukaryotes, why are certain genes expressed only in certain types of cells?
   a. Regulatory transcription factors vary from cell to cell.
   b. The promoter sequence varies from cell to cell.
   c. The location of enhancers varies from cell to cell.
   d. The type of RNA polymerase varies from cell to cell.

5. What is alternative splicing?
   a. the phosphorylation events that lead to different types of post-translational regulation
   b. mRNA processing events that lead to different combinations of exons being spliced together
   c. folding events that lead to proteins with alternative conformations
   d. action by regulatory proteins that leads to changes in the life span of an mRNA

6. What types of proteins bind to promoter-proximal elements?
   a. the basal transcription complex
   b. the basal transcription complex plus RNA polymerase
   c. basal transcription factors
   d. regulatory transcription factors

Conceptual Review

1. Compare and contrast (a) enhancers and the CAP site; (b) promoter-proximal elements and the *lac* operon operator; (c) basal transcription factors and sigma; and (d) eukaryotic promoters and bacterial promoters.
2. Why does chromatin need to be "remodeled" for transcription to occur? Why was it logical to observe that DNA-protein interactions can be loosened by the addition of acetyl groups to histones?
3. Compare and contrast (a) enhancers and silencers; (b) promoter-proximal elements and enhancers; (c) transcription factors and coactivators; and (d) regulatory transcription factors and basal transcription factors.
4. Explain the logic behind the experiment by Tonegawa and colleagues that was responsible for the discovery of enhancers. Which treatments in the experiment served as controls, and what did they control for?
5. When the human genome was first sequenced, researchers estimated that only about 30,000 genes were present. Previously, most investigators expected that the human genome held at least 100,000 genes. Explain how alternative splicing can explain the discrepancy between the observed and predicted numbers of genes.
6. Explain why mutations in regulatory proteins and regulatory sequences can lead to loss of control over the cell cycle and the development of cancer. Include a specific example.

Group Discussion Problems

1. Histone proteins have been extremely highly conserved during evolution. The histones found in fruit flies and humans, for example, are nearly identical in amino acid sequence. Offer an explanation for this observation. (Hint: What are the consequences of a mutation in a histone?)
2. Cancers are most common in tissues where cell division is common, such as blood cells and cells in the lining of the lungs or gut. Why is this observation logical?
3. Levels of p53 protein in the cytoplasm increase after DNA damage. Design an experiment to determine whether this increase is due to increased transcription of the *p53* gene or to activation of preexisting p53 proteins via a post-translational mechanism such as phosphorylation.
4. The frequency of cancer was elevated in individuals who were exposed to the nuclear bombs that were dropped on Japan in World War II. How are these data related to the observation that individuals who experience severe sunburn or deep tanning are more likely to develop skin cancer?

Answers to Multiple-Choice Questions 1. c; 2. a; 3. d; 4. a; 5. b; 6. d

www.prenhall.com/freeman is your resource for the following: Web Tutorials; Online Quizzes and other Online Study Guide materials; Answers to Conceptual Review Questions; Solutions to Group Discussion Problems; Answers to Figure Caption Questions and Exercises; and Additional Readings and Research.

Analyzing and Engineering Genes

19

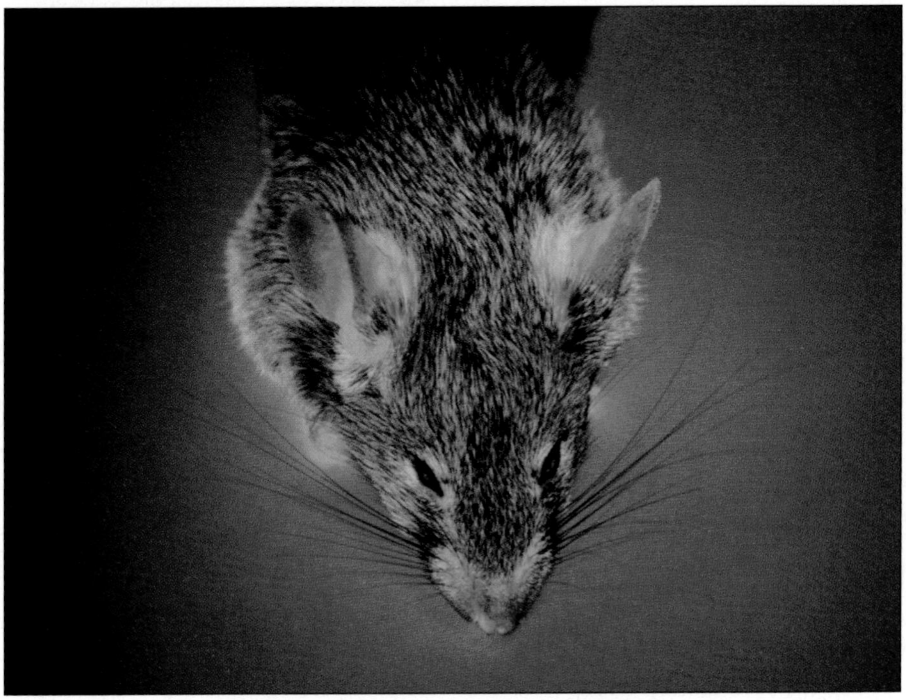

A mouse that has been genetically engineered to express a gene from jellyfish. The jellyfish gene codes for a protein that naturally luminesces, or emits light. The gene's product is called green fluorescent protein, or GFP.

KEY CONCEPTS

- The discovery of enzymes that cut DNA at specific locations, along with enzymes that piece DNA segments back together, gave biologists the ability to move genes from one location to another.
- Biologists can determine a gene's base sequence once they have obtained many copies of the gene—by inserting it into loops of DNA called plasmids in bacterial cells and then allowing the cells to grow or by performing a polymerase chain reaction.
- Researchers use several strategies to find and characterize the genes responsible for specific traits, such as the alleles associated with certain genetic diseases.
- In some cases, it has been possible to insert genes into humans to cure genetic diseases or into plants to provide them with novel traits, such as the ability to resist insect attacks.

Biologists engineer DNA sequences by removing them from a particular organism, manipulating them, and inserting the sequences into another organism of the same species or into one of a different species. These efforts have recently been applied to humans and crop plants in the hope of curing devastating genetic diseases and increasing the quality and quantity of food available to people in poverty-stricken areas of the globe. But manipulating the genetic makeup of organisms raises ethical concerns and has the potential to create harm. Efforts to manipulate DNA sequences in organisms are often referred to as **genetic engineering**. Genetic engineering became possible after the discovery of bacterial enzymes that cut DNA at specific sites and other enzymes that paste DNA sequences together. By mixing and matching sequences from different sources, biologists can create "designer genes." Efforts to manipulate genes usually result in novel combinations of genes along chromosomes, or *genetic recombination* (Chapter 12). Appropriately enough, techniques used to engineer genes are often referred to as *recombinant DNA technology*.

The enzymes used to cut and paste DNA were initially discovered and used during the 1960s and 1970s. Over the subsequent two decades, researchers followed up by developing systems for transferring recombinant genes into various types of organisms and by working out techniques for controlling the expression of the introduced DNA. Since then, research has concentrated on applying these techniques to solve problems in medicine, industry, and agriculture. The goals of genetic engineering are to (1) improve our understanding of how genes work and (2) advance **biotechnology**—the manipulation of organisms to create products or cure disease.

The goal of this chapter is to introduce some of the key techniques and conceptual issues involved in genetic engineering, using a case history approach. Our first example represents one of the initial efforts to use recombinant DNA technology to cure an inherited disease in humans called pituitary dwarfism. The research relied on techniques for identifying genes, copying them, moving them into a new host organism, and then controlling their expression. The remaining sections introduce how biologists determine the DNA sequence of genes, how they located the gene responsible for Huntington's disease in humans, and how novel genes are being placed into humans and crop plants.

In addition to exploring the techniques used to manipulate DNA, it is essential to consider the ethical, economic, ecological, and political issues involved. *Gene therapy* (the introduction of normal alleles to cure genetic defects) and the release of genetically modified plants are under intense scrutiny in the press, and lawmaking bodies around the world are debating legislation to regulate both efforts. What are the potential perils and benefits of introducing recombinant genes into human beings, food plants, and other organisms? This question represents one of the great ethical challenges of the twenty-first century.

19.1 Using Recombinant DNA Techniques to Manufacture Proteins: The Effort to Cure Pituitary Dwarfism

To understand the basic techniques and tools that are used by genetic engineers, let's consider the effort to treat **pituitary dwarfism** in humans. The pituitary gland is a structure at the base of the mammalian brain that produces several important molecules, including a protein that stimulates growth. The molecule, which was found to be just 191 amino acids long, was named **growth hormone**. The gene that codes for this protein is called *GH1*.

Once growth hormone was discovered, researchers immediately suspected that at least some forms of inherited dwarfism might be due to a defect in the GH1 protein. This hypothesis was confirmed when it was established that people with certain types of dwarfism produce little growth hormone or none at all. These people have defective copies of *GH1* and exhibit pituitary dwarfism, type I.

By studying the pedigrees of families in which dwarfism was common, several teams of researchers established that pituitary dwarfism, type I, is an autosomal recessive trait (see Chapter 13). Stated another way, affected individuals have two copies of the defective allele. People who have only one defective allele are carriers of the trait, meaning that they can transmit the defective allele to their offspring but are not themselves affected. Affected individuals grow more slowly than average people, reach puberty from two to ten years later than average, and are short in stature as adults—typically no more than 120 cm (4 feet) tall.

Why Did Early Efforts to Treat the Disease Fail?

Once the molecular basis of pituitary dwarfism was understood, physicians began treating the disease with injections of growth hormone. This approach was inspired by the spectacular success that physicians had achieved in treating type I diabetes mellitus. Diabetes mellitus is due to a deficiency of the peptide hormone insulin, and clinicians had been able to alleviate the disease's symptoms by injecting patients with insulin from pigs. Early trials showed that people with pituitary dwarfism could be treated successfully with growth hormone therapy, but only if the protein came from humans. Growth hormones isolated from pigs, cows, or other animals were ineffective. Until the 1980s, however, the only source of human growth hormone was pituitary glands dissected from human cadavers. Up to 20,000 pituitaries had to be collected from cadavers to supply enough growth hormone to treat the population of affected individuals. As a result, the drug was extremely scarce and expensive.

Meeting demand turned out to be the least of the problems with growth hormone therapy, however. To understand why, recall from Chapter 3 that certain degenerative brain disorders in mammals are caused by infectious proteins called **prions**. Kuru and Creutzfeldt-Jakob disease are prion diseases that affect humans. Certain prion diseases, including some forms of Creutzfeldt-Jakob disease, are hereditary. In most cases, however, people and other animals become infected with prion diseases by directly ingesting prion proteins in food. For example, humans can contract a variant of mad cow disease by ingesting beef products that are contaminated with prion proteins. When physicians found that some of the children treated with human growth hormone developed Creutzfeldt-Jakob disease in their teens and twenties, the physicians realized that the supply of growth hormone was contaminated with a prion protein from the brains of the cadavers supplying the hormone. As a result, the use of growth hormone isolated from cadavers was banned in 1984. What happened next?

Using Recombinant DNA Technology to Produce a Safe Supply of Growth Hormone

Even as the problems with traditional hormone therapy for pituitary dwarfism became apparent, researchers began to work out a genetic engineering approach for producing growth hormone. The idea was to find the gene that encodes the human growth hormone protein and insert it into bacterial cells or yeast cells. The hope was that huge quantities of recombinant *Escherichia coli* or *Saccharomyces cerevisiae* cells could be grown and that the recombinant cells would produce uncontaminated growth hormone in sufficient quantities to meet demand at an affordable price.

To meet this goal, investigators used the steps outlined in **Figure 19.1** to look for the human growth hormone. The first task was to isolate the mRNAs that are produced in the pituitary gland. One of these mRNA transcripts would encode growth hormone. Next, the biologists used the enzyme

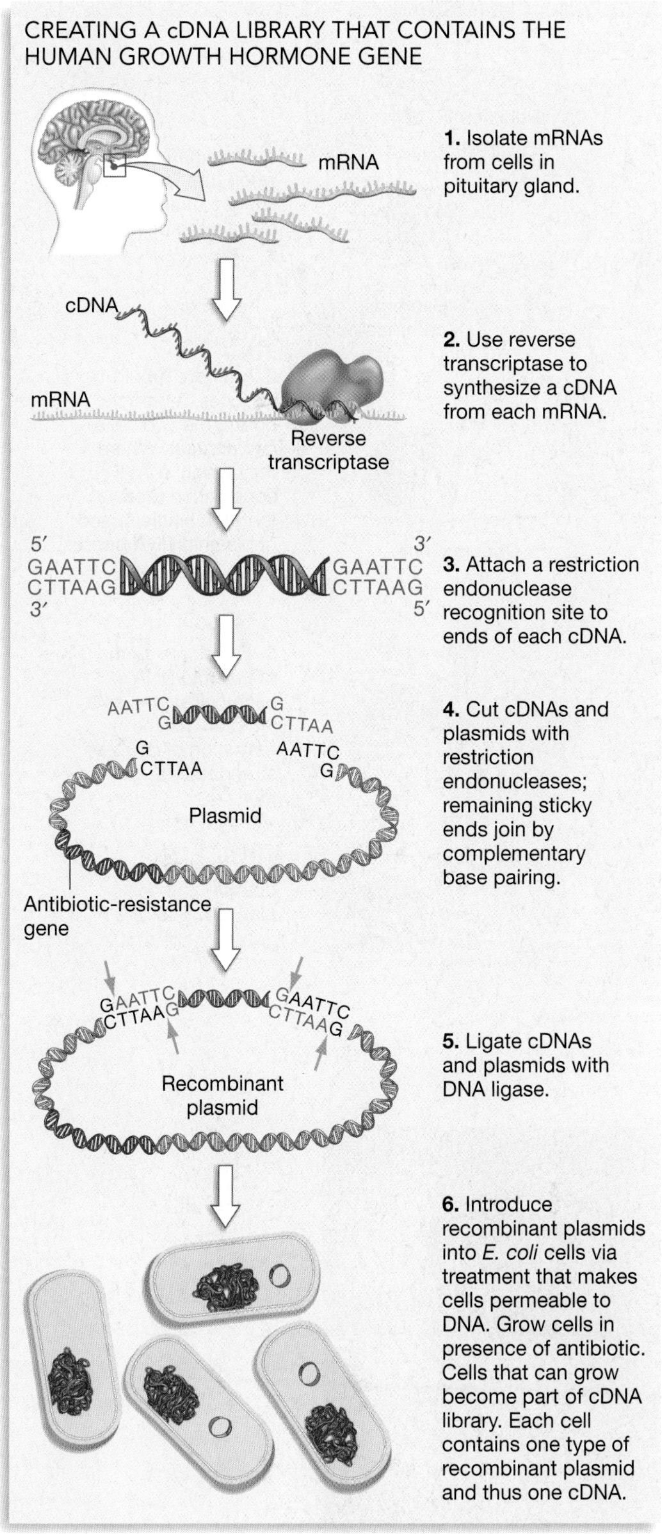
CREATING A cDNA LIBRARY THAT CONTAINS THE HUMAN GROWTH HORMONE GENE
mRNA
1. Isolate mRNAs from cells in pituitary gland.
cDNA
mRNA
Reverse transcriptase
2. Use reverse transcriptase to synthesize a cDNA from each mRNA.
5′
3′
GAATTC
CTTAAG
3′
5′
GAATTC
CTTAAG
3. Attach a restriction endonuclease recognition site to ends of each cDNA.
AATTC
G
G
CTTAA
G
CTTAA
AATTC
G
Plasmid
Antibiotic-resistance gene
4. Cut cDNAs and plasmids with restriction endonucleases; remaining sticky ends join by complementary base pairing.
GAATTC
CTTAAG
GAATTC
CTTAAG
Recombinant plasmid
5. Ligate cDNAs and plasmids with DNA ligase.
6. Introduce recombinant plasmids into *E. coli* cells via treatment that makes cells permeable to DNA. Grow cells in presence of antibiotic. Cells that can grow become part of cDNA library. Each cell contains one type of recombinant plasmid and thus one cDNA.

FIGURE 19.1 Complementary DNA (cDNA) Libraries Represent the mRNAs in a Cell
The hunt for growth hormone gene began with the creation of a cDNA library from pituitary-gland cells. Because growth hormone is produced in this tissue, the cDNA for growth hormone should exist somewhere in the library. QUESTION Would each type of cDNA in the library be represented just once? Why or why not?

reverse transcriptase, which catalyzes the synthesis of a DNA strand from an RNA template. By means of complementary base pairing, reverse transcriptase made a **complementary DNA (cDNA)** of each pituitary mRNA. DNA polymerase was then used to synthesize the complementary strand and yield double-stranded DNA. The product of this step—shown in step 2 of Figure 19.1—was thousands of different cDNAs. But which of these cDNAs encoded the growth hormone protein?

The key to answering this question was to isolate each of the many cDNAs and then obtain enough copies of them so that each of the isolated sequences could be analyzed. The process of producing many identical copies of a gene is called **genetic cloning**. Steps 3 through 6 of Figure 19.1 show how the gene for growth hormone was cloned. These steps also illustrate some of the most basic and widely used techniques in genetic engineering. Let's take a closer look.

WEB TUTORIAL 19.1 Producing Human Growth Hormone

How Are Plasmids Used in Cloning? The small, circular DNA molecules called **plasmids** were initially discovered in bacteria. Recall from Chapter 7 that plasmids are physically distinct from the bacterial chromosome. They are not required by the cell for normal growth and reproduction, and most replicate independently of the chromosome. Some plasmids carry genes for antibiotic resistance or other traits that increase the cell's ability to grow in a particular environment. In natural populations of bacteria, plasmids are sometimes copied and transferred to another bacterial cell via the mechanism explained in **Box 19.1** (page 404). Plasmids have also been discovered in the nuclei of certain unicellular eukaryotes, including baker's yeast.

Researchers quickly realized that if they could splice a loose piece of DNA into a plasmid and insert the modified plasmid into a bacterial cell, then the engineered plasmid would be replicated and passed on to daughter cells as the bacterium grew and divided. If this recombinant bacterium were placed in a liquid nutrient broth and allowed to grow overnight, billions of copies of the plasmid DNA would be produced. Biologists could harvest the recombinant genes by breaking the bacteria open, releasing the DNA, and separating out the plasmids.

Over the past several decades, researchers have modified naturally occurring plasmids to use them as carriers of recombinant genes. For example, the plasmids used in genetic engineering experiments carry at least one gene that readily allows researchers to identify and isolate the bacterial cells that carry these plasmids. In many cases, the gene involved codes for a molecule that confers resistance to a specific antibiotic. Thus, when bacterial cells are grown on plates containing that antibiotic, only cells carrying the plasmid are able to survive. Now, how could researchers cut a plasmid open and insert a specific DNA sequence to create a recombinant plasmid that could be copied and transferred?

BOX 19.1 How Do Plasmids Move between Cells Naturally?

In natural populations of bacteria, plasmids can be transferred from one cell to another through a process called **conjugation**. In *E. coli*, conjugation is possible if a cell has a particular type of plasmid called the **F-plasmid** (the "F" stands for fertility). Cells that have an F-plasmid are referred to as F^+; cells that lack an F^- plasmid are F^-.

Conjugation begins when tubules on the surface of an F^+ cell (the donor) make contact with an F^- cell (the recipient; **Figure 19.2a**, step 1; **Figure 19.2b**). After the tubule attaches, it retracts. As a result, the cell walls of the two cells touch, and a structure called a *conjugation tube* forms between them (step 2 of Figure 19.2a). Then the double-stranded plasma DNA in the F^+ cell is cut, the strands separate, and a single strand of that DNA passes through the conjugation tube into the F^- cell (steps 2 and 3). The single strand of plasmid DNA that is now in each cell serves as a template for the formation of double-stranded DNA, resulting in the formation of two F^+ cells (step 4).

If the F-plasmid involved happens to include a stretch of DNA called an *insertion sequence* (IS), the plasmid can integrate into the bacterial chromosome. Cells in which an F-plasmid has been incorporated into their main chromosome are called **Hfr strains** (for *h*igh *f*requency of *r*ecombination). This name was inspired by the observations that (1) the plasmid DNA retains the ability to mobilize and travel through the conjugation tube; (2) when conjugation occurs between an Hfr cell and an F^- cell, segments of the Hfr cell's chromosomal DNA frequently accompany the plasmid DNA during the transfer; and (3) if the new alleles are inserted into the recipient cell's main chromosome, a recombinant cell—one with a new sequence of genes along the chromosome—results.

(a) CONJUGATION

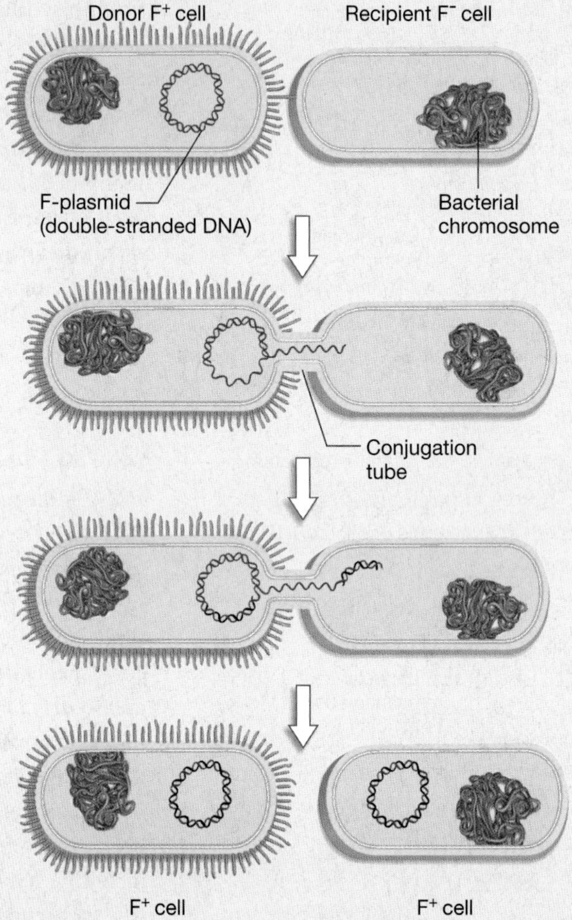
Donor F⁺ cell
Recipient F⁻ cell
F-plasmid (double-stranded DNA)
Bacterial chromosome
Conjugation tube
F⁺ cell
F⁺ cell

1. Conjugation begins when tubules on the surface of an F^+ cell make contact with an F^- cell.

2. After the tubule attaches, it retracts, pulling the two cells into contact. Where they touch, a conjugation tube forms. A single strand of plasmid DNA passes through the tube.

3. The single strand of plasmid DNA in each cell serves as a template for the formation of double-stranded DNA.

4. Both cells have a double-stranded plasmid. Both are F^+.

(b) An F^+ cell contacts an F^- cell.

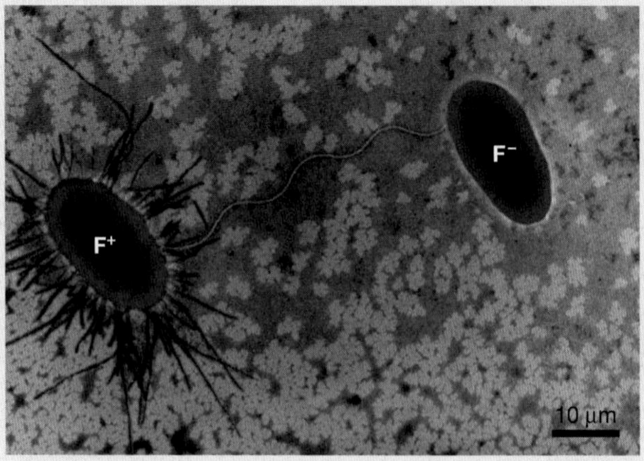
F⁻
F⁺
10 μm

FIGURE 19.2 In *E. coli*, F-Plasmids Are Transferred via Conjugation
(a) After the conjugation tube has formed, one strand of the F-plasmid's DNA is cut and the cut strand travels through the conjugation tube. The transferred strand in the F^- cell and the remaining strand in the F^+ cell are used as templates for DNA replication. **(b)** An F^+ cell and an F^- cell make contact. EXERCISE In part (b), label the tubule connecting the two cells.

Using Restriction Endonucleases to Cut DNA In addition to containing a gene for antibiotic resistance, the plasmids used in genetic engineering contain base sequences that can be cut by enzymes called *restriction enzymes* or (more precisely) restriction endonucleases. A **restriction endonuclease** is a bacterial enzyme that cuts DNA molecules at a specific base sequence. In nature, restriction endonucleases help bacterial cells fend off attacks by viruses. When a virus infects a bacterial cell, the virus injects its genetic material into the cell. In many cases, the bacterial cell's restriction endonucleases cut the viral DNA, inactivating it and rendering the virus harmless. The cell's own DNA is protected from being cut because the specific sequence of bases that is targeted by the enzyme—also called its **recognition site**—has methyl (—CH_3) groups attached (**Figure 19.3**).

Over 400 restriction endonucleases have been identified from a wide variety of bacterial species. The most commonly used enzymes cut DNA only at sites that form palindromes. In English, a word or sentence is a palindrome if it reads the same way backward as it does forward. (*Madam, I'm Adam* is an example.) In biology, a stretch of double-stranded DNA forms a palindrome if the $5' \rightarrow 3'$ sequence of one strand is identical to the $5' \rightarrow 3'$ sequence on the antiparallel, complementary strand. Most of the known restriction endonucleases recognize a unique type of palindromic DNA sequence.

To pull these observations together, consider how researchers inserted pituitary-gland cDNAs into plasmids. To begin, the researchers attached the same palindromic recognition site to the ends of each cDNA produced from mRNAs in the pituitary gland. Step 3 of Figure 19.1 shows the resulting molecules. Note that the $5' \rightarrow 3'$ sequence on one strand of the endonuclease recognition site reads GAATTC and that the $5' \rightarrow 3'$ sequence on the complementary strand also reads GAATTC.

The researchers' next step was to cut the recognition sites at the ends of each cDNA with a restriction endonuclease called EcoRI (for *Escherichia coli* restriction I—this was the first restriction endonuclease discovered in *E. coli*). As the top part of step 4 in Figure 19.1 shows, however, the enzyme involved did not break its recognition sequence neatly in two. Like most restriction endonucleases, it made a staggered cut in the palindrome. Because a staggered cut was made in a palindromic sequence, the DNA fragments that resulted had **sticky ends**. The ends of the fragments are said to be sticky because the single-stranded bases on one fragment are complementary to the single-stranded bases on the other fragment. As a result, the two ends will tend to pair up and hydrogen bond to one another.

When researchers realized that restriction endonucleases created sticky ends in DNA, they had a fundamentally important insight: If restriction sites in different parts of the genome were cut with the same restriction endonuclease, the presence of sticky ends would allow the resulting fragments to be spliced together. This is the essence of **recombinant DNA technology**—the ability to create novel DNA sequences by cutting specific sequences and pasting them into new locations.

Step 5 of Figure 19.1 demonstrates this aspect of genetic engineering. The sticky ends of the plasmid and of the cDNAs bind to each other via complementary base pairing. Researchers then used DNA ligase—the enzyme introduced in Chapter 14 that connects Okazaki fragments during DNA replication—to seal the recombinant pieces of DNA together. The result was a set of recombinant plasmids. Each contained a different cDNA from the human pituitary gland.

Transformation: Introducing Recombinant Plasmids into Bacterial Cells Plasmids serve as a **vector**, meaning a vehicle for transferring recombinant genes to a new host. If a recombinant plasmid can be inserted into a bacterial (or yeast) cell, the foreign DNA will be copied and transmitted to new cells as the

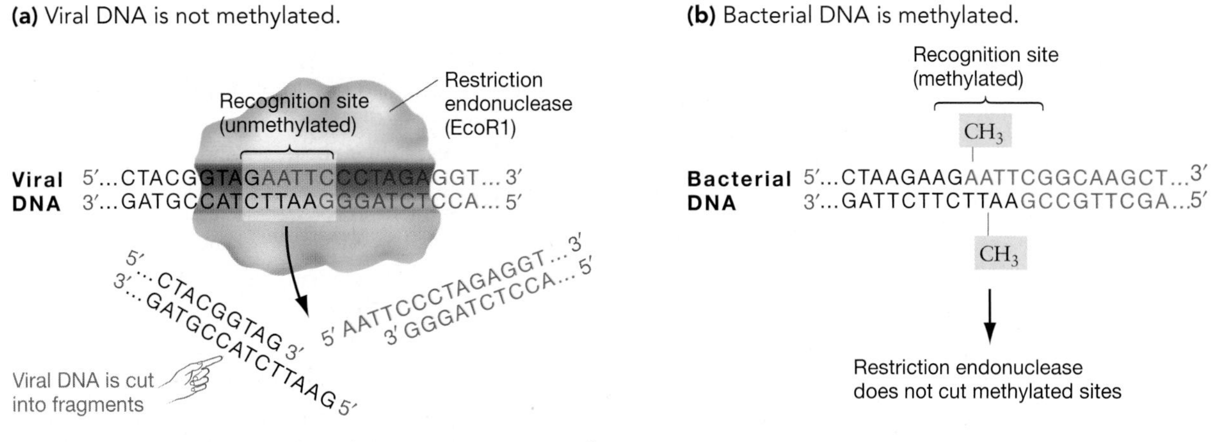

FIGURE 19.3 Restriction Endonucleases Protect Bacteria from Viruses
(a) Viral DNA is not methylated, so it is cut and destroyed by restriction endonucleases. **(b)** Restriction sites in bacteria are not cut by restriction endonucleases, because the sites are methylated.

host cell grows and divides. In this way, researchers can obtain thousands or millions of copies of specific genes.

How can recombinant plasmids be introduced into cells? The natural process of plasmid transfer introduced in Box 19.1 is too inefficient to be useful in a laboratory setting. And even though some species of bacteria have membrane proteins that allow the cells to take up DNA directly from their environment, most do not. Cells that take up DNA from the environment and incorporate it into their genomes are said to undergo **transformation**. In order for most cells to take up foreign DNA efficiently, the permeability of their cell membranes must be increased by means of a specific chemical treatment or an electrical shock. Typically, just a single plasmid enters the cell. Cells that are subjected to this treatment are subsequently spread out on a gelatinous medium (agar) that contains antibiotic and are allowed to grow. Only bacterial cells that have successfully taken up a plasmid carrying the antibiotic-resistance gene are able to grow in the presence of the antibiotic. The resulting collection of transformed bacterial cells represents a cDNA library (step 6 of Figure 19.1). A **cDNA library** is a collection of bacterial cells, each containing a vector with one cDNA from a particular cell type or tissue.

Using Nucleic Acid Hybridization to Find a Target Gene

Which of the cDNAs in the pituitary-gland cDNA library encodes growth hormone? Investigators used a multistep process to answer this question and find the gene. They began by using the genetic code to infer the approximate DNA sequence of the growth hormone gene. This was possible because the sequence of amino acids in the polypeptide was known. Thus, the researchers could infer which mRNA codon and DNA sequences coded for each amino acid. But recall from Chapter 15 that the genetic code is redundant, so it is possible for more than one codon to code for an amino acid. As a result, their inferred sequence for the growth hormone gene was approximate—not exact.

The next step was to synthesize many copies of a short, single-stranded stretch of DNA that was complementary to the inferred sequence. Because these molecules would bind to single-stranded fragments from the actual gene via complementary base pairing, they could act as a probe. A **probe** is a single-stranded fragment of a labeled, known gene that binds to a complementary sequence in the sample being analyzed. Probes are a way to find one specific sequence in a collection of many different sequences. Probes must be labeled in some way so that they can be found after they have bound to the complementary sequence. In this case, researchers attached a radioactive atom to the probe. The labeled DNA was then added to a copy of the cDNA library (**Figure 19.4**). As predicted, the radioactively labeled DNA did bind to its complementary sequence in the cDNA library. The probe had identified the recombinant cell that contained the human growth hormone.

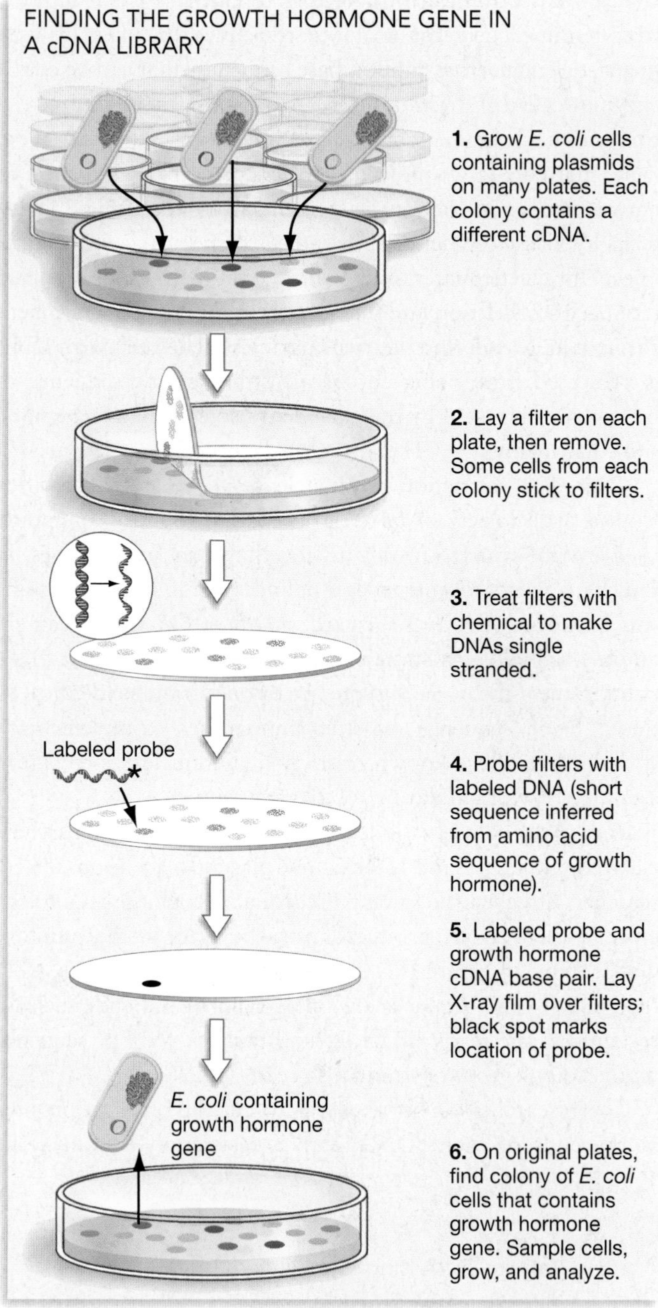
FINDING THE GROWTH HORMONE GENE IN A cDNA LIBRARY
1. Grow *E. coli* cells containing plasmids on many plates. Each colony contains a different cDNA.
2. Lay a filter on each plate, then remove. Some cells from each colony stick to filters.
3. Treat filters with chemical to make DNAs single stranded.
Labeled probe
4. Probe filters with labeled DNA (short sequence inferred from amino acid sequence of growth hormone).
5. Labeled probe and growth hormone cDNA base pair. Lay X-ray film over filters; black spot marks location of probe.
E. coli containing growth hormone gene
6. On original plates, find colony of *E. coli* cells that contains growth hormone gene. Sample cells, grow, and analyze.

FIGURE 19.4 Finding Specific Genes by Probing a cDNA Library Because the amino acid sequence of the growth hormone protein was known, researchers could infer the DNA sequence of the cDNA and use this sequence to create a DNA probe. Once bacteria with the growth hormone cDNA were isolated from the cDNA library, researchers could produce virtually unlimited quantities of the cDNA.

Mass-Producing Growth Hormone To accomplish their goal of producing large quantities of the human growth hormone, the investigators used recombinant DNA techniques to transfer the growth hormone cDNA to a new plasmid. The plasmid in question was one containing a promoter sequence, like those introduced in Chapter 16, that is recognized by *E. coli*'s

RNA polymerase holoenzyme. The recombinant plasmids were then introduced into *E. coli* cells.

The *E. coli* cells that resulted from these steps contained a gene for human growth hormone attached to an *E. coli* promoter. As a result, the transformed *E. coli* cells began to transcribe and translate the human growth hormone gene. (Note that because the cDNAs are synthesized from mature mRNAs, which do not contain introns, cDNAs do not contain introns either. Thus, bacterial cells can translate an mRNA transcribed from a cDNA directly—no splicing is required.)

The human growth hormone accumulated in the cells and could be isolated and purified. Bacterial cells containing the human growth hormone gene are now grown in huge quantities. These cells have proved to be a safe and reliable source of the human growth hormone protein.

Ethical Concerns over Recombinant Growth Hormone

As supplies of growth hormone increased, physicians used it in treating not only people with pituitary dwarfism but also children who were short, although they had no actual growth hormone deficiency. Even though the treatment requires several injections per week until adult stature is reached, growth hormone therapy was popular because it often increased the height of these children by a few centimeters. In essence, the drug was being used as a cosmetic—a way to improve appearance in cultures where height is deemed attractive. But should a parent request growth hormone treatment for genetically normal children to change their appearance? If short people are discriminated against in a culture, is a medical treatment a better solution than education and changes in attitudes? And what if parents wanted a tall child to be even taller to enhance her potential success as, say, a basketball player? Currently, the U.S. Food and Drug Administration has approved the use of human growth hormone for only the shortest 1.2 percent of children, who are projected to reach adult heights of less than 160 cm (5′3″) in males and 150 cm (4′11″) in women.

Growth hormone has also been found to enhance the maintenance of bone density and muscle mass. As a result, it has become a popular performance-enhancing drug for athletes. Part of its popularity stems from the fact that it is currently undetectable in the drug tests administered by governing bodies. Should athletes be able to enhance their physical skills by taking hormones or other types of drugs? And is the drug safe at the dosages being used by athletes? These types of questions are being debated by physicians and researchers, agencies that govern sports, and legislative bodies. Even though these questions are still unanswered, it is clear that, while solving one important problem, recombinant DNA technology created others. Throughout this chapter, an important theme is that genetic engineering has costs that must be carefully weighed against its benefits.

✓CHECK YOUR UNDERSTANDING

The essence of recombinant DNA technology is to cut DNA into fragments with a restriction endonuclease, paste specific sequences together via complementary base pairing of sticky ends and the action of DNA ligase, and insert the resulting recombinant genes into a bacterial (or yeast) cell so that the genes are expressed. You should be able to (1) make a diagram showing how certain restriction endonucleases create DNA fragments with sticky ends, (2) make a diagram showing how DNA ligase can be used to splice sequences with sticky ends into a plasmid or other type of DNA sequence, and (3) explain how a cDNA library is constructed and probed with a known sequence.

19.2 Analyzing DNA: Did Our Ancestors Mate with Neanderthals?

Biologists reached the goal of producing human growth hormone on a commercial scale without knowing the exact sequence of the entire growth hormone gene. This is unusual. Determining the sequence of a gene is usually one of the first things that investigators wish to do. Understanding a gene's sequence is valuable for a variety of reasons:

- Once a gene's sequence is known, it is straightforward to infer the amino acid sequence of its product from the genetic code. Understanding a protein's primary structure, in turn, is usually required to determine its three-dimensional structure and normal function.
- Sequence differences among alleles can be analyzed to understand why some versions of the gene function better or differently than others do.
- A gene sequence can be compared with sequences of genes that have the same function in other species. These comparisons are often interesting. For example, bacteria, yeast, and humans are about as different as organisms can be, yet all three contain DNA polymerase genes with sections that are nearly identical in base sequence. Biologists explain this similarity by hypothesizing that the bacteria inhabiting your gut, the yeast in the bread you eat, and you are descended from a common ancestor that had DNA polymerase with the same sequence. Genes that are similar due to descent from a common ancestor are said to be homologous.
- Gene sequences can be compared to infer how closely related various species are. For example, analyzing DNA sequences allowed researchers to evaluate whether our species, *Homo sapiens*, ever interbred with the species of human called *Homo neanderthalensis*, Neanderthals.

Given that sequence data are important, how can biologists obtain them? This section introduces two common techniques

(a) Primers are required to run PCR.

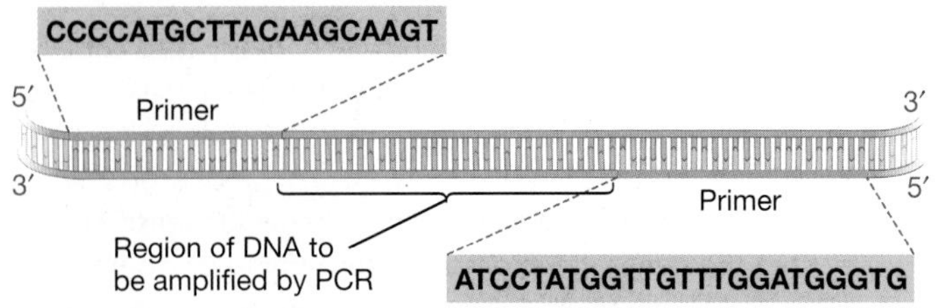

CCCCATGCTTACAAGCAAGT
5′
Primer
3′
3′
5′
Primer
Region of DNA to be amplified by PCR
ATCCTATGGTTGTTTGGATGGGTG

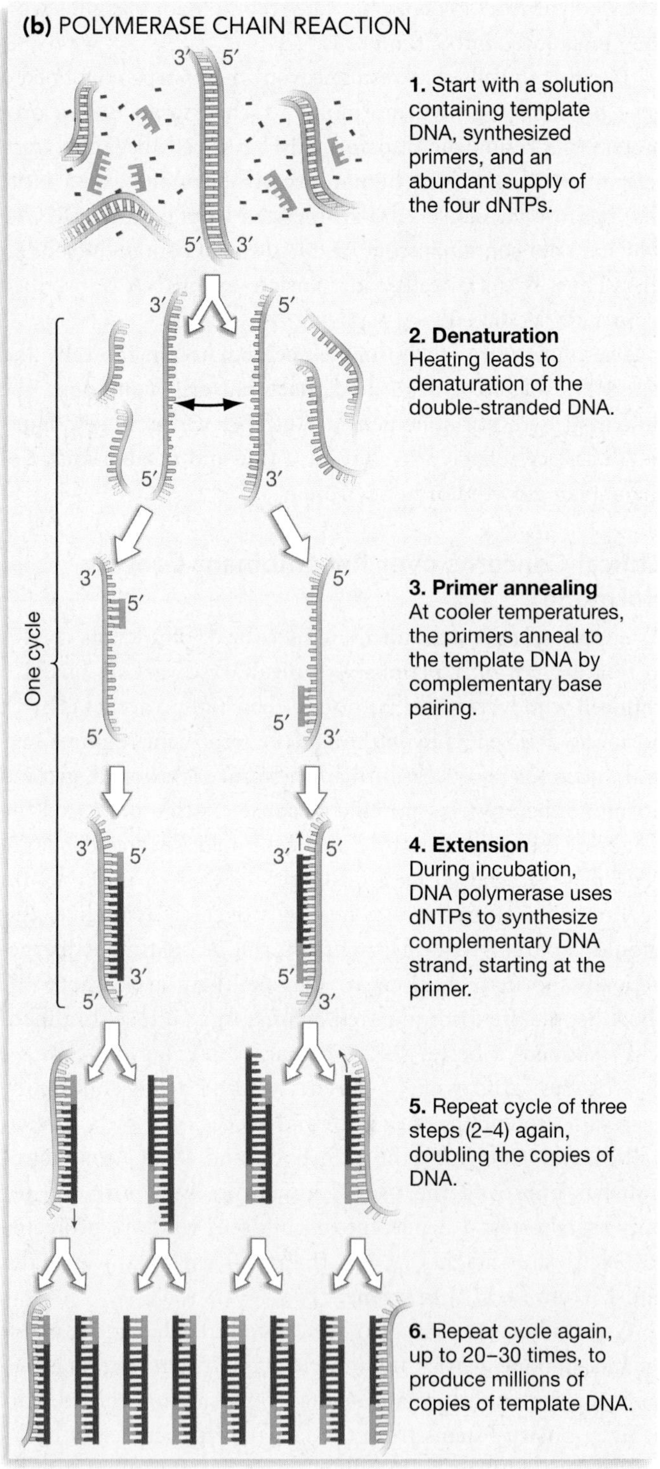

(b) POLYMERASE CHAIN REACTION
1. Start with a solution containing template DNA, synthesized primers, and an abundant supply of the four dNTPs.
2. Denaturation Heating leads to denaturation of the double-stranded DNA.
3. Primer annealing At cooler temperatures, the primers anneal to the template DNA by complementary base pairing.
4. Extension During incubation, DNA polymerase uses dNTPs to synthesize complementary DNA strand, starting at the primer.
5. Repeat cycle of three steps (2–4) again, doubling the copies of DNA.
6. Repeat cycle again, up to 20–30 times, to produce millions of copies of template DNA.
One cycle

FIGURE 19.5 The Polymerase Chain Reaction (PCR) Is a Method for Producing Many Copies of a Specific Sequence
(a) The orange sequences indicate a set of single-stranded primers, which bracket the region of DNA to be amplified. (These are the actual sequences used in a PCR experiment explained later in the chapter.) **(b)** Each PCR cycle (denaturation, primer annealing, and extension) results in a doubling of the number of copies of the sequence between and including the primers. EXERCISE In part (b), draw in the events that produced the sequences in step 5.

used to analyze DNA sequences in the laboratory: the polymerase chain reaction and dideoxy sequencing. Both are variations of in vitro DNA synthesis reactions. The *polymerase chain reaction* (PCR) is a way to create many copies of a particular stretch of DNA. Once many copies of a gene are available, *dideoxy sequencing* allows researchers to determine the exact sequence of bases.

PCR and dideoxy sequencing rank among the greatest of all technological advances in the history of biological science. Their impact is comparable to the development of the light microscope, the electron microscope, and recombinant gene technology.

WEB TUTORIAL 19.2 The Polymerase Chain Reaction

The Polymerase Chain Reaction (PCR)

Researchers have to obtain many copies of a gene in order to analyze its molecular structure. Inserting a gene into a bacterial plasmid is one method for producing many copies of a particular gene. The polymerase chain reaction is another technique for generating these copies. The **polymerase chain reaction (PCR)** is an in vitro DNA synthesis reaction in which a specific section of DNA is replicated over and over to amplify the number of copies of that sequence.

Figure 19.5 illustrates the reaction protocol, which was originally designed by Kary Mullis in 1983. As Figure 19.5a shows, the reaction mix includes two short sequences of single-stranded DNA. These sequences act as the primers that are required for DNA polymerase to work (see Chapter 14). The primers used in a PCR reaction have to bracket the region that will be copied. One primer is complementary to a sequence on one strand of the target DNA; the other primer is complementary to a sequence on the other strand. This is a key feature of PCR. To perform the reaction, a researcher must know the sequences on either side of the region of interest so that the correct primers can be synthesized.

To begin the reaction, Mullis added the primers to a solution containing an abundant supply of the four deoxynucleoside triphosphates (dNTPs; see Chapter 14) and copies of the template DNA—that is, a DNA sample that includes the gene of interest (Figure 19.5b, step 1). Then he heated the mixture to 95°C so that the double-stranded template DNA would *denature*—meaning that the two strands would separate, forming single-stranded templates (step 2). As the mixture cooled to about 60°C, some of the denatured DNA strands re-formed double helices. But some of the primers bonded to complementary portions of the single-stranded template DNA (step 3). This step is called *primer annealing*. The next step (4) was to add DNA polymerase I and incubate the mixture at 37°C, so that the enzyme could synthesize the complementary DNA strand, starting at the primer. This step is called *extension*.

The three steps denaturation, primer annealing, and extension constitute a single PCR cycle. If one copy of the template sequence existed in the sample originally, then there would be two copies at the end of the first cycle. Mullis added more DNA polymerase and repeated the cycle. The two copies present at the start of the second cycle acted as templates, so four copies existed at the end of cycle number 2 (step 5 of Figure 19.5b). As he repeated the cycle again and again, the amount of template sequence doubled each time (step 6). Doubling occurs because each newly synthesized segment of DNA serves as a template in the subsequent cycle. Starting with a single copy, successive cycles resulted in the production of 2, 4, 8, 16, 32, 64, 128, 256 copies, and so on. (A total of n cycles generates 2^n copies.) In just 20 cycles, one sequence can be amplified to over a million copies. By performing up to 30 cycles, Mullis obtained enormous numbers of copies of the template sequence.

Improving the Protocol Mullis's PCR protocol has been modified in several important ways. Current methods were inspired by the discovery of the bacterium *Thermus aquaticus* in a hot spring inside Yellowstone National Park, Wyoming. This organism contains a heat-stable form of DNA polymerase known as ***Taq* polymerase**. Although most DNA polymerases are denatured and destroyed when heated to temperatures high enough to denature DNA, *Taq* polymmerase continues to function normally when heated to 95°C. (This observation is not surprising, because 95°C is the normal temperature at which the species lives.) By switching to *Taq* polymerase, researchers no longer had to add fresh DNA polymerase at each cycle. In current protocols, then, each PCR cycle consists of three steps:

1. Denaturation of DNA (94°C);
2. Annealing of primers to single-stranded DNA (50–60°C);
3. Extension of the complementary strands from the primers (72°C).

A single application of heat-stable *Taq* polymerase at the start of the reaction supports as many cycles as needed. The other major change in Mullis's protocol is that the temperature changes required in each cycle are now automated by PCR machines.

Using PCR to Study Fossil DNA To understand why PCR is so valuable, consider a recent study by Svante Pääbo and colleagues. These biologists wanted to analyze DNA recovered from the 30,000-year-old bones of a fossilized human of the species *Homo neanderthalensis*. The individual involved was actually the first Neanderthal fossil ever discovered, from the Neander valley of Germany. Pääbo's goal was to determine the sequence of bases in the ancient DNA, compare it with DNA from modern humans (*Homo sapiens*), and analyze how similar the two species are. If the base sequences found in Neanderthals were similar enough to those found in modern humans, it would support the hypothesis that some of the DNA in our cells is descended from sequences that existed in Neanderthals. That could happen only if *Homo sapiens* and *Homo neanderthalensis* interbred while they coexisted in Europe about 30,000 years ago. This would mean that some of our ancestors were Neanderthals.

The Neanderthal bone was so old, however, that most of the DNA in it had degraded into tiny fragments too small to sequence. The biologists could recover only a minute amount of DNA that was still in moderate-sized pieces. To have any hope of analyzing the sample, they needed to generate many copies. To accomplish this, they turned to PCR.

Which Neanderthal gene should they try to amplify? Pääbo's group reasoned that, because human cells contain up to a thousand mitochondria, and because each mitochondrion in a cell contains many copies of mitochondrial DNA, there might still be some intact mitochondrial DNA somewhere in the Neanderthal DNA sample. More specifically, the team set out to amplify a locus called the control region of mitochondrial DNA.

To amplify the control region, the team added two single-stranded DNA primers, each about 20 base pairs long, to the sample. These primers had a sequence identical to sequences on either side of the control region in *Homo sapiens*. Using these primers allowed the researchers to amplify just the sequences of the Neanderthal control region, along with the primer sequences. After adding the primers to the reaction mix, the researchers added the four types of dNTPs and *Taq* polymerase, and used a PCR machine to change the temperature of the mix through a large number of cycles automatically. Using PCR, Pääbo's group was able to synthesize many copies of the control region. Obtaining many copies of the gene was essential to meeting their goal of determining the exact DNA sequence of the region.

Dideoxy Sequencing

Once researchers have many copies of a specific stretch of DNA, how can they determine the exact sequence of bases? In 1975 Frederick Sanger developed **dideoxy sequencing** as a clever variation on the basic in vitro DNA synthesis reaction to determine exact base sequences. But saying "clever" is an understatement. Sanger had to link three important insights to make his sequencing strategy work:

1. Sanger chose the name "dideoxy" because he used monomers called dideoxyribonucleoside triphosphates, or ddNTPs, along with deoxyribonucleoside triphosphates (dNTPs). These ddNTPs are identical to the dNTPs found in DNA, except they lack a hydroxyl group at their 3′ carbon (**Figure 19.6a**). Sanger realized that if a ddNTP were added to a growing DNA strand, it would terminate synthesis. Why? A ddNTP has no hydroxyl group available on its 3′ carbon to link to the 5′ carbon on an incoming dNTP monomer. As a result, DNA polymerization stops once a ddNTP is added.
2. Sanger linked this property of ddNTPs to a second fundamental insight. Suppose, he reasoned, that a radioactive primer could be attached to a template DNA, like the control region in Neanderthals that was studied by Pääbo's group. If this labeled template were incubated with DNA polymerase, the four dNTPs, and a small amount of ddGTP, along with a large amount of dGTP, the resulting daughter strands would constitute a limited set of lengths, each ending with a ddGTP. To understand why, consider that the synthesis of each daughter molecule would start at the same point—the primer—but end whenever a ddGTP happened to be incorporated into the growing strand opposite a C in the template. The addition of a ddGTP would stop further elongation. The collection of newly synthesized strands would vary in length as a result, with each specifying the distance from the labeled primer to successive C's in the template strand (**Figure 19.6b**). Analogous reactions done using ddTTP, ddATP, and ddCTP would give the distances between successive A's, T's, and G's, respectively.
3. Finally, Sanger realized that when the fragments produced by the four reactions are lined up by size, they reveal the sequence of bases in the template DNA. For example, when Pääbo and co-workers implemented Sanger's protocol and ran all four reactions, they had a collection of fragments identifying where every A, T, C, and G occurred in the control-region sequence of Neanderthals. To line these fragments up in order of size, they could be separated by gel electrophoresis. As **Figure 19.6c** shows, Pääbo's team could then read the sequence directly from the resulting autoradiograph.

Their result? The control region sequences found in Neanderthal DNA are highly distinct from the control region sequences observed in living humans. Thus, there is no evidence that we inherited some of our DNA from a Neanderthal ancestor. Based on these data, Pääbo and co-workers doubt the hypothesis that Neanderthals and members of our own species interbred. More recent studies, of Neanderthal DNA from Croatia, the Caucasus region of Russia, and a second German individual, support the same conclusion.

(a) ddNTPs terminate DNA synthesis.

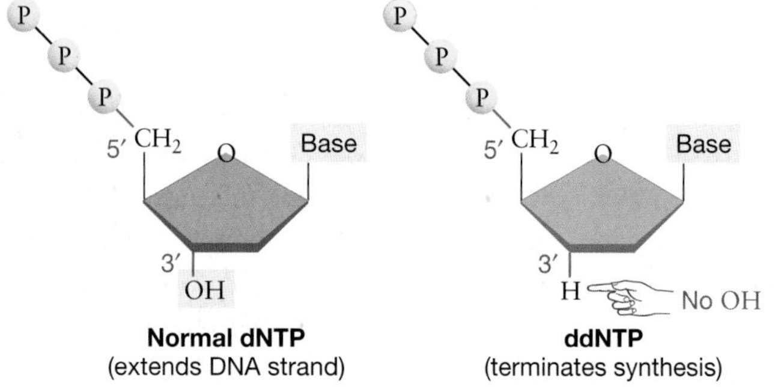
P
P
P
5′ CH2
O
Base
3′
OH
Normal dNTP
(extends DNA strand)
H
No OH
ddNTP
(terminates synthesis)

(b) Daughter strands of different lengths can be produced by using a mix of dNTPs and ddNTPs.

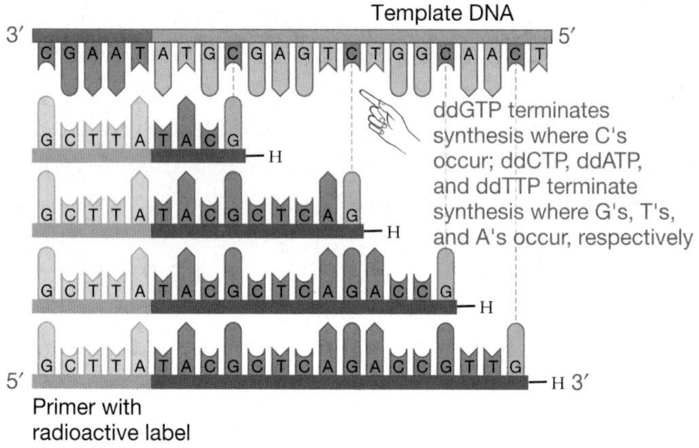
Template DNA
3′
5′
C G A A T A T G C G A G T C T G G C A A C T
G C T T A T A C G
H
G C T T A T A C G C T C A G
H
G C T T A T A C G C T C A G A C C G
H
G C T T A T A C G C T C A G A C C G T T G
H 3′
5′
ddGTP terminates synthesis where C's occur; ddCTP, ddATP, and ddTTP terminate synthesis where G's, T's, and A's occur, respectively
Primer with radioactive label

(c) Different-length strands can be lined up by size to determine DNA sequence.

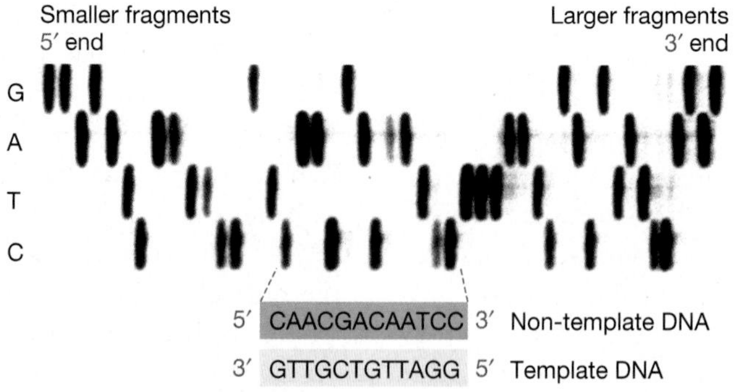
Smaller fragments
5′ end
Larger fragments
3′ end
G
A
T
C
5′ CAACGACAATCC 3′ Non-template DNA
3′ GTTGCTGTTAGG 5′ Template DNA

FIGURE 19.6 Dideoxy Sequencing Is a Method for Determining the Base Sequence of DNA

(a) Unlike dNTPs, ddNTPs have no hydroxyl (OH) group on their 3′ carbon, so DNA synthesis stops if they are inserted into a strand. **(b)** If a small amount of ddGTP is added to an in vitro DNA synthesis reaction along with dNTPs and a short, single-stranded primer that is complementary to the 3′ end of a target sequence, then DNA polymerase will synthesize fragments of varying lengths. **(c)** The fragments that result from separate reactions using ddGTP, ddTTP, ddATP, and ddCTP can be run out on a gel. (The labels along the left-hand side indicate which ddNTP was used for the reaction loaded in each lane.) The DNA sequence of the strand complementary to the template can be read directly off the gel as shown. EXERCISE Starting at the far left end of the gel pictured in part (c), write down the entire sequence shown. Hint: The sequence starts with GG.

Recent innovations in sequencing techniques include the use of fluorescent markers instead of radioactive atoms on ddNTPs. The resulting fluorescently labeled fragments can be read by machine, rather than by eye, to determine the DNA sequence of the template region. Chapter 20 explains how automated sequencing machines are currently being used for large-scale sequencing projects. Even though automation allows today's gene sequencing centers to analyze hundreds of thousands of base pairs of DNA per day, the basic sequencing reaction that is used is still based on Sanger's dideoxy method.

✓CHECK YOUR UNDERSTANDING

PCR is a technique for amplifying a specific region of DNA into millions of copies, which can then be sequenced or used for other types of analyses. You should be able to (1) explain the steps in a PCR cycle; (2) write down the sequence of a DNA strand 50 base pairs long, then design primers that would allow you to amplify the segment via PCR; and (3) explain the logic behind the dideoxy sequencing method and read a base sequence off a sequencing gel.

19.3 Gene Hunting Based on Pedigree Analysis: The Huntington's Disease Story

One of the great promises of DNA sequencing technology is the ability to analyze genes that are associated with inherited diseases of humans and thereby understand the nature of the molecular defect that causes illness. Once the difference between normal and dysfunctional alleles is understood, investigators hope to be able to introduce working copies of alleles into patients with defective alleles and cure these individuals. As the next two sections of this chapter will show, the simple-sounding goal of replacing defective alleles with functioning alleles has been tremendously difficult to achieve.

The effort to cure genetic diseases via recombinant alleles starts with the ability to find the disease genes. As an example of how this type of gene hunt is done, let's consider the first successful effort ever conducted in humans: the search for the gene associated with Huntington's disease.

Huntington's disease is a rare but devastating illness. Typically, affected individuals first show symptoms between the ages of 35 and 45. At onset, an individual appears to be clumsier than normal and tends to develop small tics and abnormal movements. As the disease progresses, uncontrollable movements become more pronounced. Eventually the affected individual twists and writhes involuntarily. Personality and intelligence are also affected—to the extent that the early stage of this disease is sometimes misdiagnosed as the personality disorder schizophrenia. The illness may continue to progress for 10 to 20 years and is eventually fatal.

Because Huntington's disease appeared to run in families, physicians suspected that it was a genetic disease. Recall from Chapter 13 that an analysis of pedigrees from families affected by Huntington's disease suggested that the trait was due to a single, autosomal dominant allele. To test this hypothesis, researchers set out to identify the gene or genes involved and to document that one or more genes are altered in affected individuals. Reaching this goal took over 10 years of intensive effort.

How Was the Huntington's Disease Gene Found?

The search for the Huntington's disease gene was led by Nancy Wexler, whose mother had died of the disease. If the trait was indeed due to an autosomal dominant allele, it meant that there was a 50 percent chance that Wexler had received the allele from her mother and would begin to show symptoms when she reached middle age. Identifying the gene would make it possible for Wexler's research team to reach several important objectives:

- *Genetic Testing* Because the symptoms of Huntington's disease do not appear until middle age, many affected people have married and had children by the time they develop symptoms. If members of families with Huntington's patients could be tested for the defective allele early in life, they might make different decisions about starting their own families and passing on the illness.
- *Therapy* Once the sequence of a gene is known, researchers can better understand the structure and function of its protein product. Information on a protein's mode of action is extremely helpful in developing drugs or other treatments. As we'll see in Section 19.4, finding a gene also makes it possible to consider replacing defective alleles with normal copies.
- *Improved Understanding of the Phenotype* Several fundamental questions about Huntington's disease might be resolved if the causative gene were found. How can one defective protein lead to abnormal movements *and* personality changes? Why does it take so long for symptoms to appear, and why is the illness progressive?

Using Genetic Markers To locate disease-causing genes, researchers start with a **genetic map**, also known as a **linkage map** or **meiotic map** (see Chapter 13). Recall that in fruit flies and other organisms, the relative positions of genes on the same chromosome can be determined by analyzing the frequency of recombination between pairs of genes. A gene that has been mapped can serve as a **genetic marker**, particularly if that gene has two or more alleles present in the population being studied. A genetic marker provides a physical landmark at a known position along a chromosome.

To understand how genetic markers can be used to locate the positions of unknown genes, suppose that you knew the position of a hair-color gene in humans and that various alleles of this gene contributed to the development of black hair, red hair, blond hair, and brown hair in the population that you were studying. Now suppose that you analyzed the pedigree of a large family in which cystic fibrosis was common and you noticed that individuals who had cystic fibrosis almost always had black hair—even though they were just as likely as unaffected individuals to have any other trait. If you observe that two particular alleles are almost always inherited together, it is logical to conclude that the genes involved are close to each other on the same chromosome—meaning that they are closely linked. If they were not closely linked, then crossing over between them would be common and they would not be inherited together. Based on these data, you could infer that the gene for cystic fibrosis is very close to the hair-color gene.

In essence, then, disease-gene hunting boils down to this: Researchers have to find a large number of closely related people who are affected and unaffected, and then attempt to locate a genetic marker with an allele that almost always occurs in the affected individuals but not in the unaffected people. If such a marker is found, the disease gene is almost guaranteed to be nearby.

The types of genetic markers used in gene hunts have changed over time. In the late 1970s and early 1980s, when biologists were searching for the Huntington's disease gene, the best genetic markers available were short stretches of DNA where specific restriction endonucleases cut the double helix. These sequences are known as *restriction endonuclease recognition sites*. (Chapter 20 introduces the types of markers in common use today. Although the nature of the markers employed has changed over time, the logic of how they are used is the same.)

How do restriction endonuclease recognition sites work as genetic markers? The key idea is that the number of recognition sites varies among individuals because the base sequence of their DNA varies. For example, if a mutation alters the DNA sequence of a recognition site, the restriction endonuclease will

BOX 19.2 Southern Blotting

Southern blotting was named after its inventor, Edwin Southern, and it is among the most basic techniques in molecular biology. It is a multistep procedure that allows researchers to identify and characterize specific genes within an organism's genome. If you have a DNA sample from an individual and wish to find a specific gene in that sample, Southern blotting is a way to do it.

The first step in Southern blotting is to obtain DNA from the cells of the organisms being studied and digest it with restriction endonucleases (**Figure 19.7**, step 1). The DNA fragments generated in this way can then be separated by gel electrophoresis (steps 2 and 3). Once the fragments are sorted by size, they are treated with chemicals that break the hydrogen bonds between base pairs, resulting in the formation of single-stranded DNA (step 4). Next the fragments of single-stranded DNA are transferred from the gel to a piece of nitrocellulose or nylon filter, using the blotting technique illustrated in step 5.

The product of these steps is a series of single-stranded DNA fragments, sorted by size and permanently bound to the membrane. In some cases, these fragments represent the entire genome of the organism being studied. To find a certain gene in this collection of fragments, a researcher must have a DNA sequence that is complementary to some region of that gene. Recall that such a sequence is called a

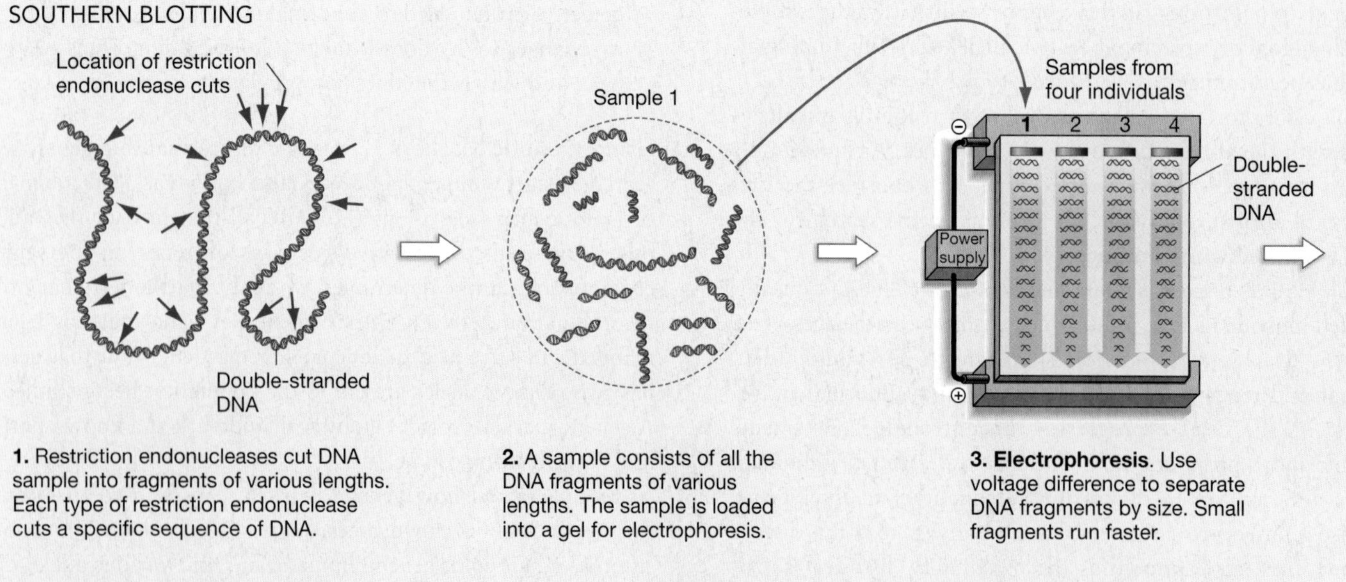
SOUTHERN BLOTTING
Location of restriction endonuclease cuts
Double-stranded DNA
Sample 1
Samples from four individuals
1 2 3 4
Double-stranded DNA
Power supply
1. Restriction endonucleases cut DNA sample into fragments of various lengths. Each type of restriction endonuclease cuts a specific sequence of DNA.
2. A sample consists of all the DNA fragments of various lengths. The sample is loaded into a gel for electrophoresis.
3. Electrophoresis. Use voltage difference to separate DNA fragments by size. Small fragments run faster.

FIGURE 19.7 Southern Blotting Protocol
Southern blotting is a technique for locating a specific gene in a DNA sample containing many genes.

no longer cut at that location. Thus, each recognition site has two alleles: The site is either present or absent.

When an investigator cuts DNA from many individuals with the same restriction endonuclease, the size of the resulting fragments will vary. These size variants are called **restriction fragment length polymorphisms (RFLPs)**. To determine which RFLPs occur in a particular individual, a researcher performs the following four steps, which are explained in more detail in **Box 19.2**:

1. Obtain a DNA sample from individuals in the population being studied (via blood samples in humans). Divide the purified DNA into several subsamples. Allow one of these subsamples to be cut with a particular restriction endonuclease.
2. The fragments from a series of individuals are sorted by size via gel electrophoresis. The separated fragments are then treated chemically to make them single stranded.
3. The fragments are probed with a labeled, single-stranded segment of DNA from a particular genetic marker—meaning a specific location on a chromosome. The probe binds to complementary sequences in the samples being analyzed. The act of binding identifies fragments that include a particular genetic marker or set of markers.
4. The restriction fragments that bind to the probe are visualized via autoradiography or other techniques.

When these steps are complete, a researcher has documented the "phenotype" of a person at a particular genetic marker. Stated another way, the fragment patterns show which allele the person has (recognition site present or absent) at one or more markers.

The bands on the gel are a genetic trait, just like height or hair color or the presence or absence of Huntington's disease. Because

probe. The probe DNA is labeled—typically with a radioactive atom or fluorescent marker—and made single stranded by heating. The probe is then added to a solution bathing the filter (step 6). During incubation, the labeled probe binds to the fragment or fragments on the nylon that have complementary base pairs. This step is called hybridization. In this way, the probe identifies the gene of interest.

To visualize which fragments hybridized with a radioactively labeled probe, a researcher lays X-ray film over the nylon filter. As step 7 of Figure 19.7 shows, radioactive emissions from the probe DNA expose the film. The black band that results identifies the target gene. Fluorescent markers can be visualized and photographed under the appropriate wavelengths of light.

A variation on Southern blotting is based on separating RNAs via gel electrophoresis, transferring them to a filter paper, and probing them with a single-stranded and radioactively labeled DNA probe. This technique is used to identify the RNA fragments produced by a particular gene. It is called **Northern blotting**, in a lighthearted tribute to the protocol from which it was derived. The variation called **Western blotting** involves separating proteins via electrophoresis and then probing the resulting filter with an antibody probe that binds to the protein of interest. The use of antibodies in research is explored in detail in later chapters.

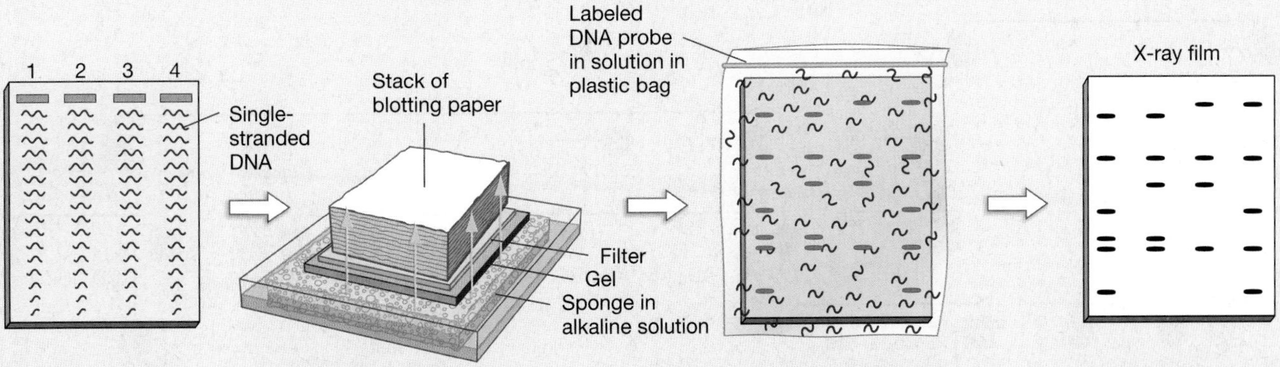

4. The DNA fragments are treated to make them single stranded.

5. Blotting. An alkaline solution wicks up into blotting paper, carrying DNA from gel onto nylon filter, where it is then permanently bound.

6. Hybridization with radioactive probe. Incubate the nylon filter with a solution containing labeled probe DNA. The radioactive probe binds to the fragments containing complementary sequences.

7. Autoradiography. Place filter against X-ray film. Radioactive DNA fragments expose film, forming black bands that indicate location of target DNA.

banding patterns are due to particular sequences in DNA, they are passed from parent to offspring just like any other DNA sequence.

By cutting DNA with different restriction endonucleases and probing the resulting gels with sequences from different locations in the genome, researchers gradually build a large catalog of recognition sites that vary among individuals.

For these RFLPs or other types of molecular markers to be used in a gene hunt, they have to be placed on a map of their species' chromosomes. Genetic markers can be placed on a genetic map by analyzing the frequency of recombination between them. In contrast, a **physical map** of the genome records the absolute position—in numbers of base pairs—along a chromosome. Techniques for creating physical maps are introduced briefly in Chapter 20.

Using a Pedigree Once a genetic map containing many restriction endonuclease recognition sites or other genetic markers has been assembled, biologists need help from families affected by an inherited disease to find the gene in question. Recall that the fundamental goal is to find a genetic marker that is almost always inherited along with the disease-causing allele. Gene hunts are more likely to be successful if large families are involved. Large sample sizes minimize the probability that researchers will observe an association between one or more markers and the disease just by chance—not because they are closely linked. Wexler's Huntington's disease team was fortunate to find a large, extended family affected with the disease living along the shores of Lake Maracaibo, Venezuela.

From historical records, the researchers deduced that the Huntington's disease allele was introduced to this family by a European sailor or trader who visited the area in the early 1800s. When family members agreed to participate in the study, there were over 3000 of his descendants living in the area. One hundred of these people had been diagnosed with Huntington's disease. To help in the search for the gene, family members agreed to donate skin or blood samples for DNA analysis and to furnish information on who was related to whom.

This research resulted in the pedigree shown in **Figure 19.8**. Note that the diagram includes information about the disease phenotype and the particular RFLP pattern observed in each family member. To find the Huntington's disease gene, researchers looked for fragment patterns that were present in affected individuals but absent from unaffected individuals. Several restriction endonuclease recognition sites appeared to be inherited along with Huntington's disease. Were these associations due to chance or to a close physical association between the recognition site and the defective allele? Stated another way, were the recognition site and Huntington's disease gene close to one another on the DNA double helix (**Figure 19.9**)?

To answer this question, the researchers used a statistical analysis to determine the likelihood that an apparent association between a genetic marker and the disease phenotype was due to chance. The analysis supported the hypothesis that the region shown in **Figure 19.10a**—which happened to be on chromosome 4—was close to the Huntington's disease gene. Note that two of the seven restriction endonuclease recognition sites in this region (marked by arrows) are polymorphic, meaning some individuals have one or both sites and some do not. As **Figure 19.10b** shows, four combinations of restriction endonuclease cuts are possible at these two sites. Thus, all four fragment patterns A through D shown in **Figure 19.10c** are possible.

The key finding was that pattern C shown in Figure 19.10c was inherited along with the defective allele often enough to

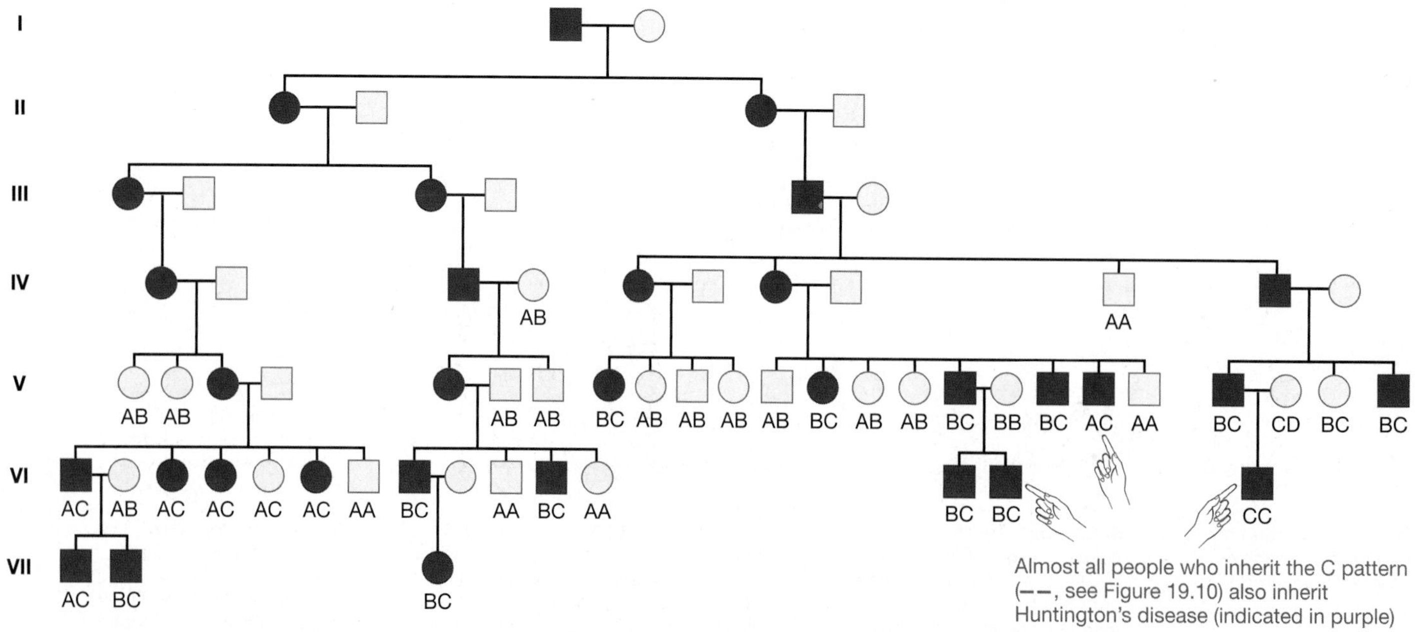

FIGURE 19.8 Certain Genetic Markers Are Inherited with the Huntington's Disease Gene
Part of a pedigree from seven generations (I–VII) of a large extended family. Individuals affected by Huntington's disease inherited the defective allele from a common ancestor. The letters A–D represent the restriction fragment patterns found in each individual.
QUESTION In generation VI, an individual with the genotype AC does not have Huntington's disease. How is this possible?

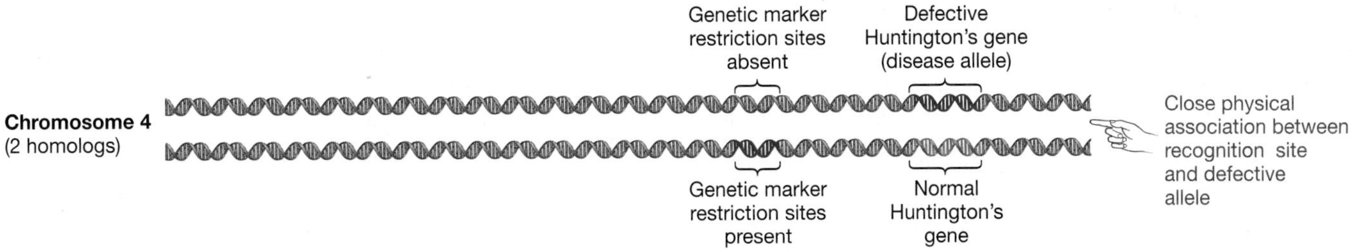

FIGURE 19.9 Genetic Markers and Disease Alleles Are Inherited Together if They Are Closely Linked
To find a gene, researchers look for alleles of a genetic marker that almost always occur in individuals with a specific allele of the gene.

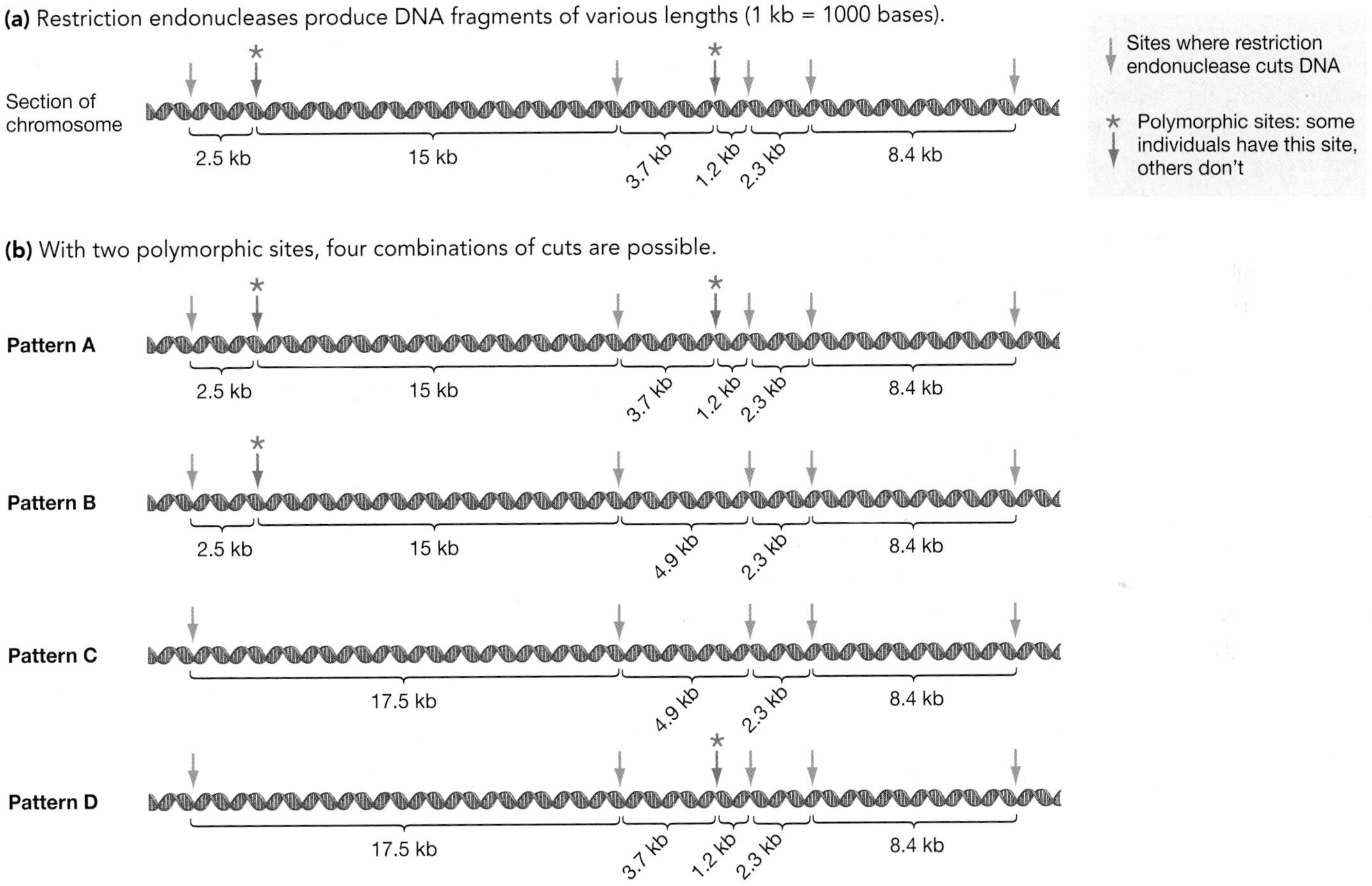

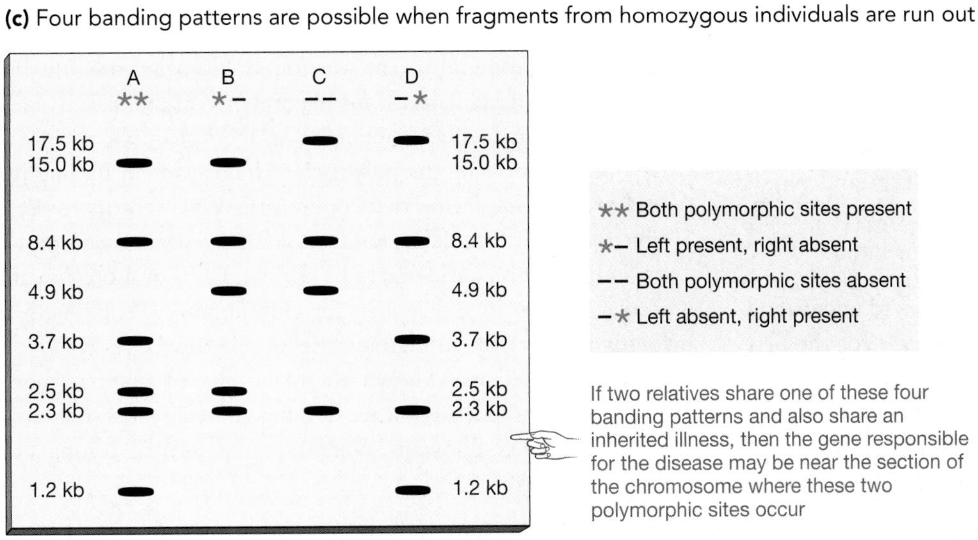

FIGURE 19.10 Restriction Endonuclease Recognition Sites Can Be Used as Genetic Markers
(a) A DNA segment with five restriction sites (green arrows) that occur in all individuals in a given population and two sites (blue starred arrows) that occur in some individuals but not in others. Here "kb" stands for kilobase; 1 kb = 1000 bases. **(b)** Because there are two polymorphic sites, four possible combinations of cuts can occur. **(c)** Four banding patterns predicted for individuals who are homozygous for each pattern.

indicate that the association was not due to chance. Apparently, the English sailor who introduced the Huntington's disease allele also had the "C" allele in the RFLP pattern. The two loci are so close together that recombination between them—which would put an "A" or "B" or "D" allele next to the Huntington's disease allele in his descendants—has been extremely rare. By examining associations between other markers on chromosome 4 with the Huntington's disease phenotype, the team succeeded in narrowing down the location of the Huntington's disease gene to a region about 500,000 base pairs long. Because the haploid human genome contains over 3 billion base pairs, this was a significant accomplishment.

Pinpointing the Defect Once the general location of the Huntington's disease gene was known, biologists looked in that region for exons that encode a functional mRNA. They did this by sequencing exons from diseased and normal individuals, comparing the data, and pinpointing specific bases that were different between the two groups of individuals. In this way, dideoxy sequencing played a key role in the gene hunt.

When this analysis was complete, the research team found that individuals with Huntington's disease have an unusual number of CAG codons near the 5′ end of a particular gene. CAG codes for glutamine. Healthy individuals have 11–25 copies of the CAG codon, while affected individuals have 42 or more copies. Although many genetic diseases are caused by single base changes that alter the amino acid sequence of a protein, diseases caused by the expansion of a particular codon repeat have also been observed. For example, the most common form of inherited mental retardation, **fragile-X syndrome**, is caused by an increase in the number of copies of a CGG codon at a specific site on the X chromosome.

When the Huntington's disease research team confirmed that the increase in the CAG codon was always observed in affected individuals, the team concluded that the long search for the Huntington's disease gene was over. They named the newly discovered locus *IT15* and its protein product huntingtin. In both affected and normal individuals, the huntingtin protein is involved in the early development of nerve cells. It is only later in life that mutant forms of the protein cause disease.

New Approaches to Therapy How has the effort to locate the Huntington's disease gene helped researchers and physicians understand and treat the illness? Biomedical researchers point to three major benefits of a successful gene hunt:

1. The discovery of *IT15* allowed investigators to understand the molecular nature of Huntington's disease for the first time. Autopsies of Huntington's patients showed that their brains actually decrease in size due to the death of neurons and that the brain tissue contains insoluble aggregates of the huntingtin protein. These aggregates are thought to be a direct consequence of the changes in the number of CAG repeats in the *IT15* gene and thus the number of glutamine residues in the protein. Long stretches of polyglutamine (a polymer of glutamine residues) are known to result in the formation of protein aggregates. The leading hypothesis to explain Huntington's disease proposes that a gradual buildup of the huntingtin protein aggregates triggers neurons to undergo apoptosis, or programmed cell death.
2. The discovery of the Huntington's disease gene has allowed scientists to introduce the defective allele into mice via the types of genetic engineering techniques that we will discuss in Section 19.4. These mice are called **transgenic** ("across-genes"), because they have alleles that have been modified by genetic engineering.

   The goal of creating transgenic mice with the defective *IT15* allele was to stimulate research on Huntington's disease. Laboratory animals with disease symptoms that parallel those of a human disease are said to provide an **animal model** of the disease. Transgenic mice that produce defective versions of the huntingtin protein develop tremors and abnormal movements, exhibit higher-than-normal levels of aggression toward litter and cage mates, and experience a loss of neurons in the brain. The transgenic mice get a version of Huntington's disease.

   Animal models are valuable in disease research, because they can be used to test potential treatments before investigators try them on human patients. Now that an animal model for Huntington's disease is available, research groups have begun testing drugs that appear to prevent or reduce the aggregation of the huntingtin protein. Other biologists are attempting to design drugs that prevent the death of cells containing aggregated huntingtin proteins.
3. A third benefit of finding a disease gene is that it makes genetic testing possible. What is genetic testing, and how is it beneficial?

Genetic Testing

When the Huntington's gene was found, biologists used the information to develop a test for the presence of the defective allele. The test involves obtaining a DNA sample from an individual and using the polymerase chain reaction to amplify the chromosome region that contains the CAG repeats responsible for the disease. If the number of CAG repeats is 35 or less, the individual is considered to be normal. Forty or more repeats results in a positive diagnosis for Huntington's. People with a family history of Huntington's disease can now be tested.

As more becomes known about the human genome, more and more tests like this will be developed for diseases that have some genetic component. What types of genetic testing are done currently?

- *Carrier Testing* Before starting their own family, people from families affected by a genetic disease frequently want to know whether they carry the allele responsible. That is especially true for diseases, such as cystic fibrosis (CF), that are due to recessive alleles. If only one of the prospective parents has the allele, then none of the children they have together should develop CF. But if both prospective parents carry the allele, then each child they produce has a 25 percent chance of having CF. Carrier testing can determine whether an individual is a carrier for a genetic disease.
- *Prenatal Testing* Suppose that two parents, both carrying the CF allele, decide to have children but do not want to pass along that allele. Once the mother is pregnant, a physician can obtain fetal cells early in gestation. The cells can be cultured and DNA isolated. The CF allele can then be amplified by PCR and sequenced. Based on the test results, the couple may choose to continue or terminate a pregnancy.
- *Adult Testing* Huntington's disease isn't the only trait that appears in adulthood. For example, about 5 percent of the women who develop breast cancer have inherited a faulty gene. To date, three genes are known to create a predisposition to breast cancer. Adult women with a family history of breast cancer can be tested for the presence of these mutant genes.

It is important to note, though, that the results of genetic tests can be difficult to interpret. If a person has an expanded CAG repeat in the *IT15* gene, it is virtually certain that he or she will develop Huntington's disease. But if a woman has one of the alleles associated with breast cancer, there is just an 80 percent chance that she will develop the illness. In such cases, the benefit of testing is to make the person involved more vigilant about getting regular checkups. Then if breast cancer does develop, it can be caught at an early and potentially curable stage.

Ethical Concerns over Genetic Testing

Genetic testing raises controversial ethical issues. For example, some individuals maintain that it is morally wrong to terminate any pregnancy, even if the fetus involved is guaranteed to be born with a debilitating genetic disease. Also, consider Nancy Wexler's position soon after the discovery of *IT15*. Would you choose to be tested for the defective allele and risk finding out that you were almost certain to develop Huntington's disease? Should physicians agree to test people for genetic diseases that have no cure? Should it be legal for health insurance companies to test clients for genetic diseases that will require expensive treatments so that the companies can refuse coverage?

These questions are still being debated by political and religious leaders, health-care workers, philosophers, and members of the general public. Difficult as they are, the issues raised by genetic testing pale in comparison with some of the questions raised by the effort to create transgenic humans.

✓CHECK YOUR UNDERSTANDING

Genes for particular traits can be located if they are closely linked to a known genetic marker and are thus inherited along with the marker. You should be able to draw a pedigree showing a close association between a particular genetic illness and a specific genetic marker.

19.4 Can Gene Therapy Cure Inherited Diseases in Humans? Research on Severe Immune Disorders

For biomedical researchers interested in curing inherited diseases such as Huntington's, sickle-cell anemia, and cystic fibrosis, the ultimate goal is to replace or augment defective copies of the gene with normal alleles. This approach to treatment is called **gene therapy.**

For gene therapy to succeed, two crucial steps must be completed. First, the wild-type allele for the gene in question and its regulatory sequences must be sequenced and understood. Second, there must be a method for introducing a copy of the normal allele into affected individuals. The DNA has to be introduced in a way that ensures expression of the gene in the correct tissues, in the correct amount, and at the correct time. If the defective allele is dominant, then the introduction step may be even more complicated: In at least some cases, the introduced allele must physically replace or block the expression of the existing dominant allele.

Chapter 18 discussed aspects of gene regulation in humans and other eukaryotes; here, we focus on techniques for introducing DNA into humans. As a case study, let's consider the first successful application of gene therapy in humans.

How Can Novel Alleles Be Introduced into Human Cells?

Section 19.1 reviewed how recombinant DNA sequences are packaged into plasmids and taken up by *E. coli* cells. Humans and other mammals lack plasmids, however, and their cells do not take up foreign DNA in response to chemical or electric treatments as efficiently as bacterial cells do.

How can foreign genes be introduced efficiently into human cells? To date, researchers have focused on packaging foreign DNA into viruses for transport into human cells. A viral infection begins when a virus particle enters or attaches to a host cell and inserts its genome into that host cell. In some cases the viral DNA becomes integrated into a host-cell chromosome. As a result, viruses that infect human cells can be used as vectors to carry engineered alleles into the chromosomes of target cells. Potentially, the alleles delivered by the virus could be expressed and produce a product capable of curing a genetic disease.

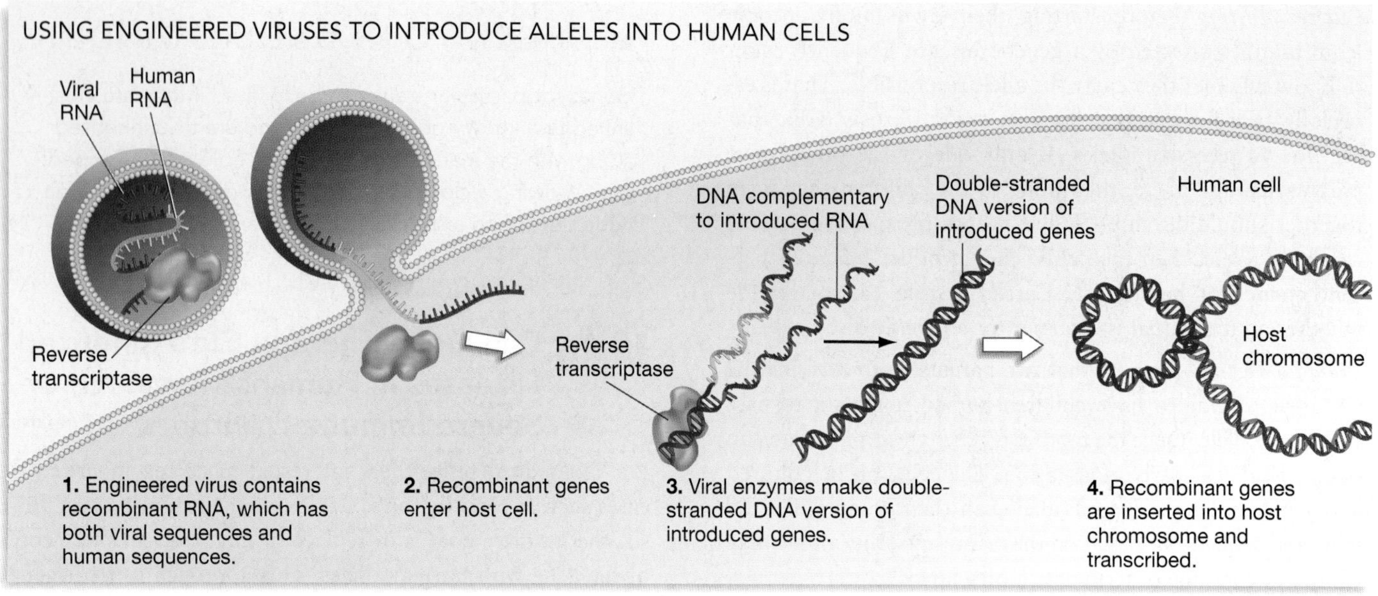
USING ENGINEERED VIRUSES TO INTRODUCE ALLELES INTO HUMAN CELLS
Viral RNA
Human RNA
Reverse transcriptase
Reverse transcriptase
DNA complementary to introduced RNA
Double-stranded DNA version of introduced genes
Human cell
Host chromosome
1. Engineered virus contains recombinant RNA, which has both viral sequences and human sequences.
2. Recombinant genes enter host cell.
3. Viral enzymes make double-stranded DNA version of introduced genes.
4. Recombinant genes are inserted into host chromosome and transcribed.

FIGURE 19.11 Retroviruses Insert Their Genes into Host-Cell Chromosomes
Retroviruses can be used as vectors to carry specific genes into cells and to insert these genes into the chromosomes of target cells. QUESTION What happens if the recombinant DNA is inserted in the middle of a gene that is critical to the cell's normal function?

Two major classes of viruses have proven to be effective vectors for introducing recombinant DNA: (1) retroviruses and (2) adenoviruses. Retroviruses have an RNA genome. The genomes include the gene for the enzyme reverse transcriptase, which catalyzes the production of a double-stranded DNA molecule from a single-stranded RNA template. (Recall from Section 19.1 that researchers use reverse transcriptase to produce cDNAs.) When a retrovirus infects a human cell, reverse transcriptase catalyzes the production of a DNA copy of the viral genome. Other viral enzymes catalyze the insertion of the viral DNA into a host-cell chromosome. If human genes can be packaged into a retrovirus, then the virus is capable of inserting the human alleles into a chromosome in a target cell (**Figure 19.11**).

The adenoviruses have a DNA genome and have also been used to carry recombinant alleles into human cells. Certain adenoviruses infect human lung cells and other tissues in the respiratory tract and cause colds or flu-like symptoms. Because they are specialized for infecting cells in the human respiratory tract, adenoviruses have been tested as vectors for treating individuals who suffer from inherited disorders of the lung, such as cystic fibrosis. Unfortunately, the genes of adenoviruses do not integrate into host-cell chromosomes. As a result, they are expressed in target cells for only a relatively short time. For such reasons, most gene therapy research is focused on retroviruses.

Although viruses would appear to be ideal vectors for the delivery of normal human alleles, there are serious problems associated with their use. Viruses usually cause disease. The retroviruses include the human immunodeficiency virus (HIV), which causes AIDS, as well as an array of viruses that cause cancer in animals. Using these agents for gene therapy requires that sequences responsible for causing disease be inactivated or removed from their genomes. Even if they have been inactivated, the altered particles may be able to recombine with viral DNA that already exists in the individual being treated and lead to the formation of a new infectious strain. In addition, viral proteins trigger a response by the immune system that can cause dangerous side effects during treatment. Finally, if viral genes happen to insert themselves in a position that disrupts the function of an important gene in the target cell, the consequences can be serious. Despite these risks, viruses are still the best vectors currently available for human gene therapy.

Using Gene Therapy to Treat X-Linked Immune Deficiency

Recently, a research team reported the successful treatment of a disease by means of gene therapy. The illness is called **severe combined immunodeficiency (SCID)**. Children who are born with SCID lack a normal immune system and are unable to fight off infections.

The type of SCID treated by the group is designated SCID-X1, because it is caused by mutations in a gene on the X chromosome. The gene is required for the development of immune-system cells called *T cells*. T cells develop in bone marrow, from undifferentiated cells called **stem cells**. Unlike mature T cells, stem cells divide continuously.

The gene responsible for SCID-X1 codes for a polypeptide called γc ("gamma-c"), which is a key component of several

FIGURE 19.12 A "Bubble Child"
Children who are born with SCID cannot fight off bacterial or viral infections. As a result, such children must live in a sterile environment.

plasma membrane proteins. The function of these proteins is to receive signals from growth factors that stimulate the development of T cells from stem cells. Without functioning copies of the receptor proteins, the growth signals are not received and T cells do not mature properly. Without a large supply of mature T cells, infants are helpless to ward off bacterial and viral infections. SCID-X1 is fatal.

Traditionally, physicians have treated SCID-X1 by keeping the patient in a sterile environment, isolated from any direct human contact until the patient could receive bone-marrow tissue transplanted from a close relative (**Figure 19.12**). In most bone-marrow transplants, the T cells that the patient needs then differentiate from the transplanted bone-marrow cells and allow the individual to live normally. In some cases, though, no suitable donor is available. One boy born with SCID had to live in a germ-free "bubble" for over 12 years; when a bone-marrow transplant was finally attempted, it failed. Could gene therapy cure this disease by furnishing functioning copies of the defective gene?

Before any protocol for gene therapy can be implemented in humans, researchers must investigate its safety by showing that the proposed strategy works in animals. To begin their work, then, biologists introduced the γc gene into a retrovirus vector and used the recombinant virus to infect bone-marrow cells from dogs. Follow-up analyses showed that the dog cells that took up the vector were long lived and produced the receptor protein. Further, the receptor recognized the proper growth factor and stimulated the development of T cells. Another set of experiments was carried out on an animal model for SCID—mice that lacked functioning copies of the γc gene. When the team transferred a normal copy of the gene into the marrow cells of these mice, immunodeficiency was prevented.

Based on these results, the team gained approval to treat ten boys, each less than one year old, who had SCID-X1. No suitable bone-marrow donor was available to treat these patients. Even though they'd been isolated in germ-free rooms, several had developed illnesses such as pneumocystic pneumonia and thrush (oral yeast infections), which appear in people with malfunctioning immune systems.

To implement gene therapy, the researchers removed bone marrow from each patient and collected the stem cells that produce mature T cells (**Figure 19.13**). Each day for three days, researchers infected the stem cells with an engineered retrovirus that carried the normal γc gene. About 30 percent of the cells took up the γc gene and manufactured the receptor protein. This recombination step was done in vitro to minimize patient discomfort and to allow researchers to monitor the procedure's success. The stem cells that were producing normal receptor protein were then isolated and transferred back into the patients.

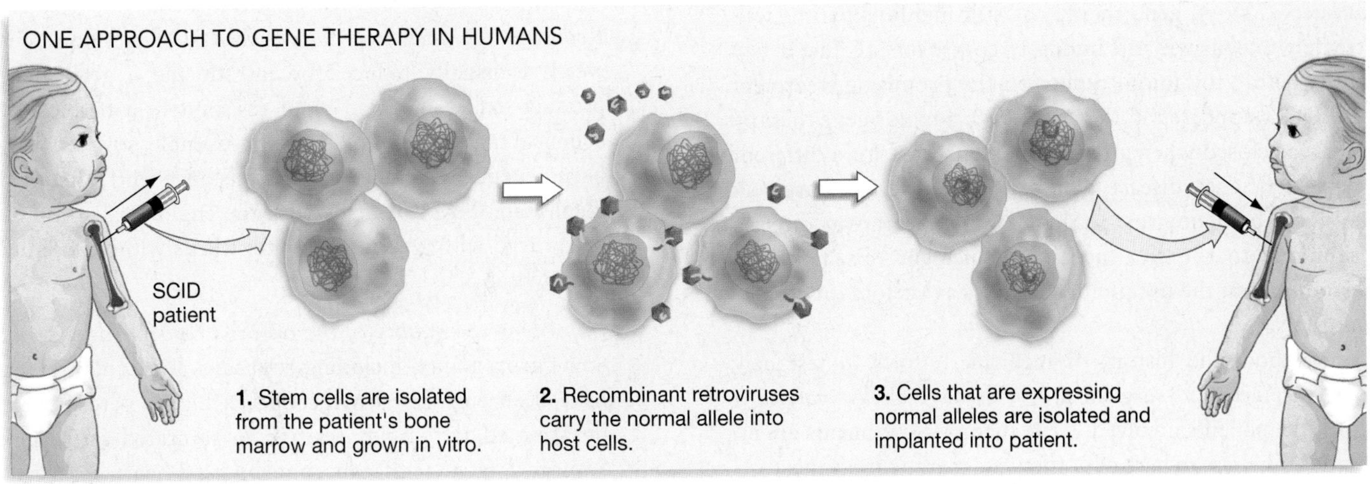
ONE APPROACH TO GENE THERAPY IN HUMANS
SCID patient
1. Stem cells are isolated from the patient's bone marrow and grown in vitro.
2. Recombinant retroviruses carry the normal allele into host cells.
3. Cells that are expressing normal alleles are isolated and implanted into patient.

FIGURE 19.13 A Gene Therapy Protocol
For gene therapy to work, copies of a normal allele have to be introduced into a patient's cells and be expressed.

Within four months after reinsertion of the transformed marrow cells, nine of the boys had normal levels of functioning T cells. T-cell concentrations in the nine children continued to increase for eight months. At that time, the research team had enough confidence in the boys' immune systems to inoculate each boy with polio and tetanus vaccines. To their great relief, they found that the boys' immune systems succeeded in producing antibodies to the polio and tetanus vaccines. Antibodies are the body's normal response proteins. Their production is crucial to fighting off infections. The children now had functioning immune systems. The infections that existed prior to gene therapy also began to clear up.

About three months after the experimental treatment began, the nine patients were removed from germ-free isolation rooms and began residing at home, where they grew and developed normally. During a routine check-up 30 months after gene therapy, however, the research team discovered that two of the boys had developed a type of cancer characterized by unchecked growth of T cells. Follow-up analyses of their bone-marrow cells showed that, in both boys, a viral-borne γc gene had been inserted near the gene for a regulatory transcription factor that triggers the growth of T cells. The viral sequences apparently acted as an enhancer and led to constitutive expression of the transcription factor. As this book goes to press, the two boys have responded to cancer chemotherapy and are healthy.

The tenth boy to receive gene therapy never succeeded in producing T cells at all. For unknown reasons, his recombinant stem cells failed to function normally when they were transplanted back into his bone marrow. Fortunately, physicians were later able to find a bone marrow donor whose cells were a close enough match to the boy's cells to make a successful transplant possible.

Ethical Concerns over Gene Therapy

Although the treatment of SCID-X1 qualifies as an important success story, gene therapy is still highly experimental, extremely expensive, and intensely controversial. Just a few months before the initial report on the promising treatment of SCID-X1 appeared in the year 2000, a major gene therapy center was closed when a patient being treated for a different enzyme deficiency disease died after gene therapy based on the use of an adenovirus. In this case, the death was directly attributable to the treatment. The incident raised serious questions about the use of adenoviruses as vectors for recombinant DNA.

Throughout the history of medicine, efforts to test new drugs, vaccines, and surgical protocols have always carried a risk for the patients involved. Gene therapy experiments are no different. The researchers who run gene therapy trials must explain the risks clearly and make every effort to minimize them. The initial report on the development of cancer in the boys who received gene therapy for SCID-X1 concluded with the following statement: "We have proposed ... a halt to our trial until further evaluation of the causes of this adverse event and a careful reassessment of the risks and benefits of continuing our study of gene therapy." Although recombinant gene therapy holds great promise for the treatment of a wide variety of devastating inherited diseases, fulfilling that promise is almost certain to require many years of additional research and testing, as well as the refinement of legal and ethical guidelines.

19.5 Biotechnology in Agriculture: The Development of Golden Rice

Progress in human gene therapy has been slow, but progress in transforming crop plants with recombinant genes has been rapid. In the United States alone, genetically modified or transgenic strains of 15 major crop species are now approved for commercial use. Farmers growing maize (corn) in the United States currently have over 15 distinct transgenic varieties to choose from. In Mississippi, 97 percent of the land area devoted to cotton is now planted with recombinant strains. Globally, about 67.7 million hectares (167 million acres) of transgenic crops were grown in 2003. The use of genetically modified plants has become a major element of modern agriculture.

Recent efforts to develop transgenic plants have focused on three general objectives:

1. *Reducing losses to herbivore damage* The most popular strategy for reducing losses caused by plant-eating organisms is based on introducing the gene for a naturally occurring insecticide. For example, researchers have transferred a gene from the bacterium *Bacillis thuringiensis* into corn; the presence of the "Bt toxin" encoded by this gene protects the plant from corn borers and other caterpillar pests.
2. *Reducing competition with weeds* Competition with weeds is usually reduced by introducing a gene whose product makes the crop plant resistant to a herbicide (a chemical that destroys plants). For example, soybeans have been genetically engineered for resistance to the herbicide glyphosate. Soybean fields with the engineered strain can be sprayed with glyphosate to kill weeds without harming the soybeans.
3. *Improving the quality of the product consumed by people* Some crop plants, including soybeans and canola plants, have been engineered to produce a higher percentage of unsaturated fatty acids relative to saturated fatty acids. Saturated fatty acids can contribute to heart disease, so crops with less of them are healthier to eat.

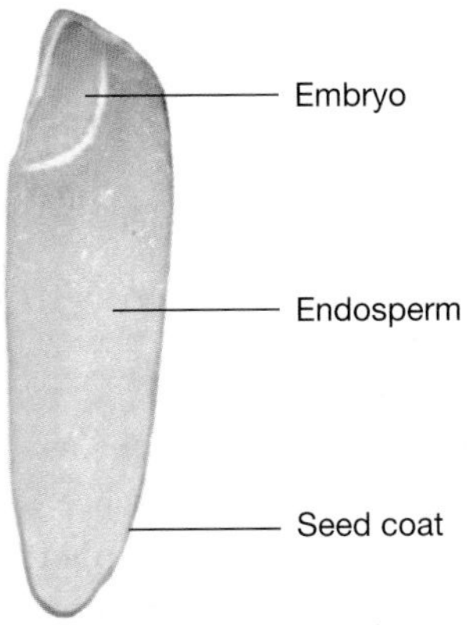

FIGURE 19.14 Endosperm Is a Nutritive Tissue in Seeds
Rice seeds consist of the three regions shown here.

To gain a better understanding of genetic engineering in plants, let's take a detailed look at efforts to produce transgenic rice with improved nutritional qualities.

Rice as a Target Crop

Although almost half the world's population depends on rice as its staple food, this grain is an extremely poor source of certain vitamins and essential nutrients. For example, rice contains no vitamin A. This is a serious issue, because vitamin A deficiency causes blindness in 250,000 Southeast Asian children each year. Vitamin A deficiency also renders children more susceptible to diarrhea, respiratory infections, and childhood diseases such as measles.

Humans and other mammals synthesize vitamin A from a precursor molecule known as β-carotene (beta-carotene). β-Carotene belongs to a family of plant pigments called the carotenoids, which were introduced in Chapter 10. Carotenoids are orange, yellow, and red and are especially abundant in carrots. Rice plants synthesize β-carotene in their chloroplasts but not in the part of the seed that is eaten by humans. Could genetic engineering produce a strain of rice that synthesized β-carotene in the carbohydrate-rich seed tissue called endosperm (**Figure 19.14**)? A research team set out to answer this question. If successful, their research could help to solve an important global health problem.

Synthesizing β-Carotene in Rice

The research team began their work by searching for compounds in rice endosperm that could serve as precursors for the synthesis of β-carotene. They found that maturing rice endosperm contains a molecule called geranyl geranyl diphosphate (GGPP), which is an intermediate in the synthetic pathway that leads to the production of carotenoids. As **Figure 19.15** shows, three enzymes are required to produce β-carotene from geranyl geranyl diphosphate. If genes that

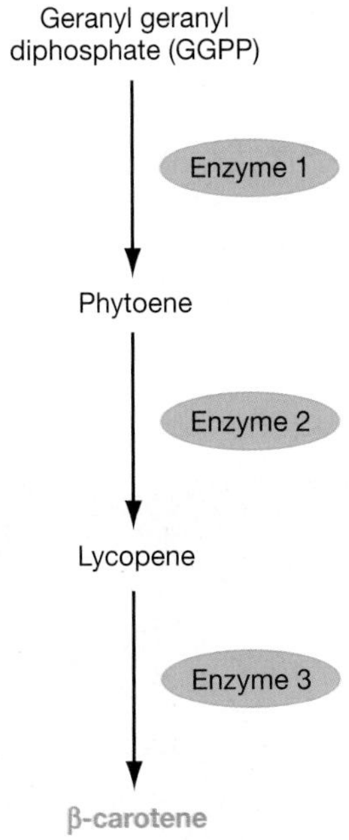

FIGURE 19.15 Synthetic Pathway for β-Carotene
GGPP is a molecule found in rice seeds. Three enzymes are required to produce β-carotene from GGPP.

encode these enzymes could be introduced into rice plants along with regulatory sequences that would trigger their synthesis in endosperm, the researchers could produce a transgenic strain of rice that would contain β-carotene.

Fortunately, genes that encode two of the required enzymes had already been isolated from daffodils, and the gene for the third enzyme had been purified from a bacterium. Because the sequences had been inserted into plasmids and grown in bacteria, many copies were available for manipulation. To each of the coding sequences in the plasmids, biologists added the promoter region from an endosperm-specific protein. This segment included a regulatory site that would promote transcription of the recombinant sequences in endosperm cells.

To develop transgenic rice strains that are capable of producing β-carotene, the three sets of sequences had to be inserted into rice plants. As you know from reading this chapter, introducing recombinant DNA is fairly straightforward into *E. coli* but difficult in humans. How are foreign genes introduced into plants?

The *Agrobacterium* Transformation System

Agrobacterium tumefaciens is a bacterium that infects plants. Plant tissues that become infected with this parasite form a tumorlike growth called a **gall**. Such plants are said to have crown gall disease (**Figure 19.16a**, page 422). When researchers looked into how these infections occur, they found that a plasmid carried by *Agrobacterium* cells, called a

Ti (tumor-inducing) **plasmid**, plays a key role (**Figure 19.16b**, step 1). Ti plasmids contain several functionally distinct sets of genes. One set encodes products that allow the bacterium to bind to the cell walls of a host. Another set, referred to as the virulence genes, encodes the proteins required to transfer part of the Ti DNA, called T-DNA (transferred DNA), into the interior of the plant cell. The T-DNA then travels to the plant cell's nucleus and integrates into its chromosomal DNA (step 2). When transcribed, T-DNA induces the infected cell to grow and divide. The result is the formation of a gall that encloses a growing population of *Agrobacterium* cells (step 3).

Researchers soon realized that the Ti plasmid offered an efficient way to introduce recombinant genes into plant cells. Follow-up experiments confirmed that recombinant genes could be added to the T-DNA that integrates into the host chromosome, that the gall-inducing genes could be removed and that the resulting sequence is efficiently transferred and expressed in its new host plant.

Researchers have also been able to transfer foreign genes into crop plants directly. In this approach, tiny tungsten particles are coated with foreign genes and then shot into plant cells by means of an instrument called a "gene gun."

(a) Plant with crown gall disease

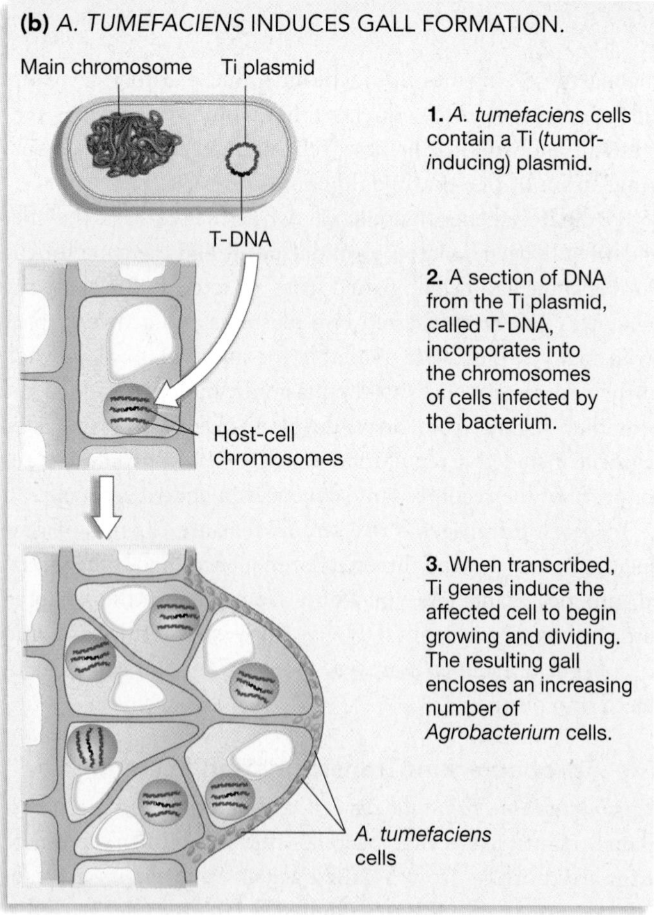
(b) A. TUMEFACIENS INDUCES GALL FORMATION.
Main chromosome
Ti plasmid
1. A. tumefaciens cells contain a Ti (tumor-inducing) plasmid.
T-DNA
2. A section of DNA from the Ti plasmid, called T-DNA, incorporates into the chromosomes of cells infected by the bacterium.
Host-cell chromosomes
3. When transcribed, Ti genes induce the affected cell to begin growing and dividing. The resulting gall encloses an increasing number of Agrobacterium cells.
A. tumefaciens cells

FIGURE 19.16 *Agrobacterium* Infections Introduce Foreign Genes into a Host-Cell Chromosome

(a) The gall, or tumorlike growth at the bottom of this plant's stem, is due to an infection by the bacterium *A. tumefaciens.* **(b)** Ti plasmids of *A. tumefaciens* cells induce gall formation.

Golden Rice

To generate a strain of rice that produces all three enzymes needed to synthesize β-carotene in endosperm, the researchers exposed embryos to *Agrobacterium* cells containing genetically modified Ti plasmids (**Figure 19.17a**). The transformed plants were then grown in the greenhouse. When the transgenic individuals had matured and set seed, the researchers found that some rice grains contained so much β-carotene that they appeared yellow. This "golden rice" is compared with unmodified rice in **Figure 19.17b**. Testing is under way to determine whether the transgenic rice contains enough β-carotene to relieve vitamin A deficiency in children.

While the production of golden rice in the laboratory is promising, there is still much to do before the advance has an impact on public health. The transgenic rice must be grown successfully under normal field conditions. In addition, most or all of the rice strains grown in different parts of the world will need to be crossed with transgenic stock to acquire the appropriate genes. Finally, the transgenic strains must be made available to poor farmers at an affordable price.

In addition to the possibility of alleviating vitamin A deficiency, transgenic strains of rice may also be able to ease other nutritional problems in rice-eating people—specifically, protein deficiency and iron deficiency. Early attempts to develop transgenic rice strains with augmented amounts of stored protein and available iron have been promising.

As the examples reviewed in this chapter show, recent advances in agricultural and medical biotechnology promise to increase the quality and quantity of food and to alleviate at least some inherited diseases that plague humans. Each solution offered by genetic engineering tends to introduce new issues, however. As this chapter's essay points out, researchers and consumer advocates have expressed concerns about the increased numbers and types of genetically modified foods. During this debate, biology students and others who are well informed about the techniques and issues involved will be an important source of opinions and information for the community.

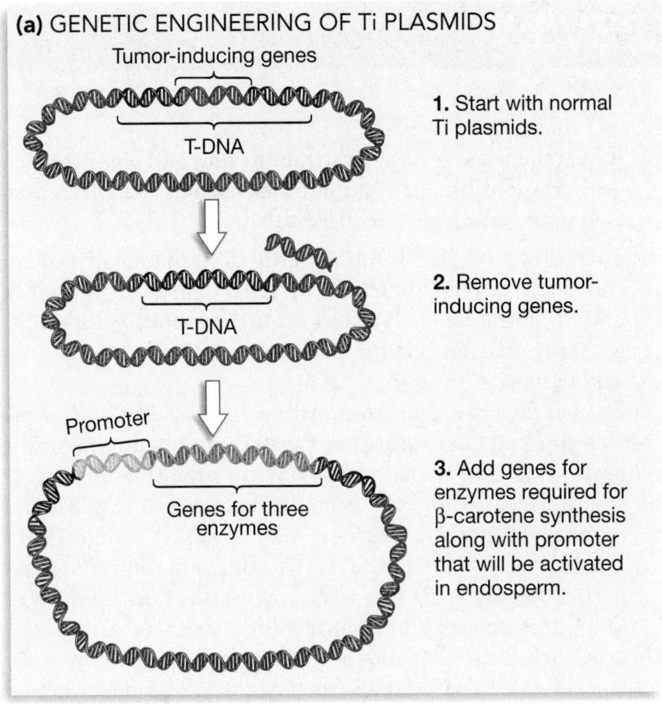
(a) GENETIC ENGINEERING OF Ti PLASMIDS
Tumor-inducing genes
T-DNA
1. Start with normal Ti plasmids.
T-DNA
2. Remove tumor-inducing genes.
Promoter
Genes for three enzymes
3. Add genes for enzymes required for β-carotene synthesis along with promoter that will be activated in endosperm.

(b) Rice plants infected with transformed *Agrobacterium* produce β-carotene in their seeds.

FIGURE 19.17 "Golden Rice" Is a Transgenic Strain
(a) The construction of a Ti plasmid to produce a strain of rice capable of synthesizing β-carotene in the endosperm of their seeds. **(b)** Seeds from individuals that were transformed with the Ti plasmids illustrated in part (a). Chemical analyses confirmed that the golden color in the transformed seeds is due to large quantities of β-carotene.

ESSAY Controversies over Genetically Modified Foods

Enthusiasm about the potential benefits of golden rice and other genetically modified foods is tempered by concern about the risks of releasing large numbers and types of recombinant crop plants. In Europe, public opinion has been extremely skeptical about the wisdom of developing genetically engineered crops. Protesters in Britain, the United States, and elsewhere have even vandalized test plantings of genetically modified crops.

Proponents of genetic engineering in agriculture can point to several important success stories. In 1998 U.S. farmers applied 8.2 million fewer pounds of active pesticide ingredient than they would have in the absence of genetically modified crops, in part because glyphosate-resistant and insecticide-producing crop strains were coming into widespread use. Recall that glyphosate is a herbicide. If crop plants are glyphosate resistant, then the compound can be applied in small amounts to kill weeds after growth starts. If they are not, then other weed killers must be applied in larger amounts earlier in the growing season. In addition, genetic engineering made it possible to develop soybeans and corn strains that produce higher concentrations of the amino acid lysine, improving their nutritional value.

Opponents of genetically modified foods can point to several important concerns, however:

- Research has shown that corn plants genetically engineered to produce the Bt toxin leak the molecule into the soil, where it might kill untargeted organisms.
- Widespread adoption of insecticide-producing crop strains is almost certain to lead to the evolution of pests that are resistant to the compounds. For example, consider an insect species that eats transgenic corn plants producing Bt toxin. If one insect in the population happened to have an allele that reduced the toxicity of the toxin, it would be able to produce many more offspring than could individuals without the allele. As a result, the allele for resistance would spread rapidly.

Are genetically modified crops good or bad?

- Research also shows that crop plants genetically engineered for glyphosate resistance produce pollen carrying the recombinant allele, and that pollen can subsequently be transferred to closely related weeds. Based on these data, there is legitimate concern that the use of these crop strains could lead to the more rapid evolution of herbicide-resistant weeds.

Are genetically modified crops good or bad? There is no easy or pat answer to this question. Like any technology, genetic engineering has costs and benefits. For citizens, the challenge is to become informed about the issues and insist that regulatory agencies strictly enforce measures to maximize the benefits of the technology but minimize the risks involved. For biologists, the challenge is to quantify those benefits and risks as carefully as possible and communicate their results clearly so that good decisions can be made.

CHAPTER REVIEW

Summary of Key Concepts

The discovery of enzymes that cut DNA at specific locations, along with enzymes that piece DNA segments back together, gave biologists the ability to move genes from one location to another.

Genetic engineering is based on inserting a modified version of an allele into an organism. Biologists were able to manipulate genes after the discovery of restriction endonucleases, DNA ligase, and plasmids. Restriction endonucleases allow researchers to cut DNA at specific locations and insert it into plasmids or other vectors with the help of DNA ligase.

Once the gene for the human growth hormone was isolated, it was cloned by introducing it into a plasmid that was taken up by *E. coli* cells. There the gene was expressed. As a result of genetic engineering, pharmaceutical companies have been able to produce large supplies of human growth hormone at an affordable price.

Web Tutorial 19.1 Producing Human Growth Hormone

Biologists can determine a gene's base sequence once they have obtained many copies of the gene—by inserting it into loops of DNA called plasmids in bacterial cells and then allowing the cells to grow or by performing a polymerase chain reaction.

The most common alternative to cloning genes in plasmids is an in vitro DNA synthesis reaction called the polymerase chain reaction (PCR). The PCR technique is based on primers that bracket a target stretch of DNA. With the use of a heat-stable form of DNA polymerase called *Taq* polymerase, a single target DNA sequence can be amplified to millions of copies.

The most common way to analyze the gene copies produced by plasmid-based cloning or PCR is to sequence them by the dideoxy sequencing method. This is an in vitro synthesis reaction that employs dideoxyribonucleotides to stop DNA replication at each base in the sequence. By running the resulting DNA fragments out on a gel, the sequence of nucleotides in a gene can be determined.

Web Tutorial 19.2 The Polymerase Chain Reaction

Researchers use several strategies to find and characterize the genes responsible for specific traits, such as the alleles associated with certain genetic diseases.

To isolate the gene for human growth hormone, investigators created a cDNA library from cells that were transcribing growth hormone mRNA and probed the library with a DNA sequence that encoded a short stretch of the known amino acid sequence.

Locating the gene responsible for Huntington's disease was difficult, because only the disease phenotype and the pattern of inheritance were known at the time the research was started. To find the gene, investigators analyzed a large number of genetic markers and a large pedigree of an affected family. Their goal was to find a marker that was inherited along with the allele responsible for the disease. Once this strategy pinpointed the general area where the gene was located, biologists sequenced exons from normal and diseased individuals to determine exactly where the defect occurred.

In some cases, it has been possible to insert genes into humans to cure genetic diseases or into plants to provide them with novel traits, such as the ability to resist insect attacks.

Once genes are located and characterized, they can be introduced into other individuals or species in an effort to change their traits. Genetic transformation can occur in several ways, depending on the species involved. In humans, recombinant DNA must be introduced by viruses. Because introducing foreign genes into humans is difficult, and because complex ethical and safety issues are involved, progress in human gene therapy has been slow. In bacteria and certain eukaryotes, foreign genes can be inserted into plasmids and the recombinant plasmids can be inserted back into a host cell. An important variation on this approach has made genetic engineering feasible for crop plants. Certain bacteria that infect plants have plasmids that integrate their genes into the host-plant genome. By adding recombinant alleles to these plasmids, researchers have been able to introduce alleles that improve the nutritional quality of crops, make them resistant to herbicides, or allow them to produce insecticides.

Questions

Content Review

1. What do restriction endonucleases do?
   a. They cleave bacterial cell walls and allow viruses to enter the cells.
   b. They join pieces of DNA by catalyzing the formation of phosphodiester bonds between them.
   c. They cut stretches of DNA at specific sites known as recognition sequences.
   d. They act as genetic markers in the chromosome maps used in gene hunts.

2. What is a plasmid?
   a. an organelle found in many bacteria and certain eukaryotes
   b. a circular DNA molecule that in some cases replicates independently of the main chromosome(s)
   c. a type of virus that has a DNA genome and that infects certain types of human cells, including lung and respiratory tract tissue
   d. a type of virus that has an RNA genome, codes for reverse transcriptase, and inserts a cDNA copy of its genome into host cells

3. When present in a DNA synthesis reaction, a ddNTP molecule is added to the growing chain. No further nucleotides can be added afterward. Why?
   a. There are not enough dNTPs available.
   b. A ddNTP can be inserted at various locations in the sequence, so fragments of different length form—each ending with a ddNTP.
   c. The 5′ carbon on the ddNTP lacks a hydroxyl group, so no phosphodiester bond can form.
   d. The 3′ carbon on the ddNTP lacks a hydroxyl group, so no phosphodiester bond can form.
4. Once the gene that causes Huntington's disease was found, researchers introduced the defective allele into mice to create an animal model of Huntington's disease. Why was this model valuable?
   a. It allowed them to test potential drug therapies without endangering human patients.
   b. It allowed them to study how the gene is regulated.
   c. It allowed them to make large quantities of the huntingtin protein.
   d. It allowed them to study how the gene was transmitted from parents to offspring.
5. To begin the hunt for the human growth hormone gene, researchers created a cDNA library from cells in the pituitary gland. What did this library contain?
   a. only the sequence encoding growth hormone
   b. DNA versions of all the mRNAs in the pituitary-gland cells
   c. all of the coding sequences in the human genome, but no introns
   d. all of the coding sequences in the human genome, including introns
6. What does it mean to say that a genetic marker and a disease gene are closely linked?
   a. The marker lies within the coding region for the disease gene.
   b. The sequence of the marker and the sequence of the disease gene are extremely similar.
   c. The marker and the disease gene are on different chromosomes.
   d. The marker and the disease gene are in close physical proximity and tend to be inherited together.

Conceptual Review

1. Explain how restriction endonucleases and DNA ligase are used to insert foreign genes into plasmids and create recombinant DNA. Illustrate some of the key events involved—including an explanation of why sticky ends are important.
2. How do researchers get foreign DNA into cells? Explain how recombinant plasmids are introduced into *E. coli* cells, how viruses are used to transport genes into human cells, and how the Ti plasmid in *Agrobacterium* cells is used to transport genes into plants.
3. What is a cDNA library? How is one created? Give an example of how a cDNA library can be used in research.
4. What are genetic markers, and how are they used to create a genetic map? Explain how researchers combine analyses of pedigrees and genetic markers to narrow down the location of disease genes.
5. Researchers added the promoter sequence from an endosperm-specific gene to the Ti plasmids used in creating golden rice. Why was this step important? Comment on the roles of promoter and enhancer sequences in genetic engineering in eukaryotes.
6. List the molecules that are required for a typical PCR reaction. Then list the function of each molecule required for the replication of DNA in vivo. For each molecule on the in vivo list, explain why it is or is not required for the corresponding in vitro reaction.

Group Discussion Problems

1. The text posed the following questions about Huntington's disease: How can one defective protein lead to abnormal movements *and* personality changes? Why does it take so long for symptoms to appear, and why is the illness progressive? Based on the molecular nature of the disease, offer hypotheses to answer these questions.
2. Discuss some of the ethical issues involved in human gene therapy. Specifically, should therapy be restricted to somatic cells, or should individuals be able to alter their germ-line cells (meaning that they would alter the alleles they pass on to their offspring)? Should gene therapy be approved for disease-causing alleles only, or should parents also be able to pay to transform their children with alleles associated with height, intelligence, hair color, eye color, athletic performance, musical ability, or similar traits?
3. Several organizations are actively trying to stop the development of transgenic crops and implement bans on the marketing of genetically modified foods. Suppose a representative from one of these organizations comes to your door seeking membership support. Explain why you would or would not support the group's efforts.
4. A friend of yours is doing a series of PCR reactions and comes to you for advice. She purchased three sets of primers, hoping that one set would amplify the template sequence shown below. (The dashed lines in the template sequence stand for a long sequence of bases.) None of the three primer pairs produced any product DNA, however.

	Primer a		Primer b
Primer Pair 1:	5′ GTCCAGC 3′	&	5′ CCTGAAC 3′
Primer Pair 2:	5′ GGACTTG 3′	&	5′ GCTGCAC 3′
Primer Pair 3:	5′ GTCCAGG 3′	&	5′ CAAGTCC 3′

Template

5′ ATTCGGACTTG———GTCCAGCTAGAGG 3′
3′ TAAGCCTGAAC———CAGGTCGATCTCC 5′

   a. Explain why each primer pair didn't work. Indicate whether both primers are at fault or just one primer is the problem.
   b. Your friend doesn't want to buy new primers. She asks you whether she can salvage this experiment. What do you tell her to do?

Answers to Multiple-Choice Questions 1. c; 2. b; 3. d; 4. a; 5. b; 6. d

www.prenhall.com/freeman is your resource for the following: Web Tutorials; Online Quizzes and other Online Study Guide materials; Answers to Conceptual Review Questions; Solutions to Group Discussion Problems; Answers to Figure Caption Questions and Exercises; and Additional Readings and Research.

20 Genomics

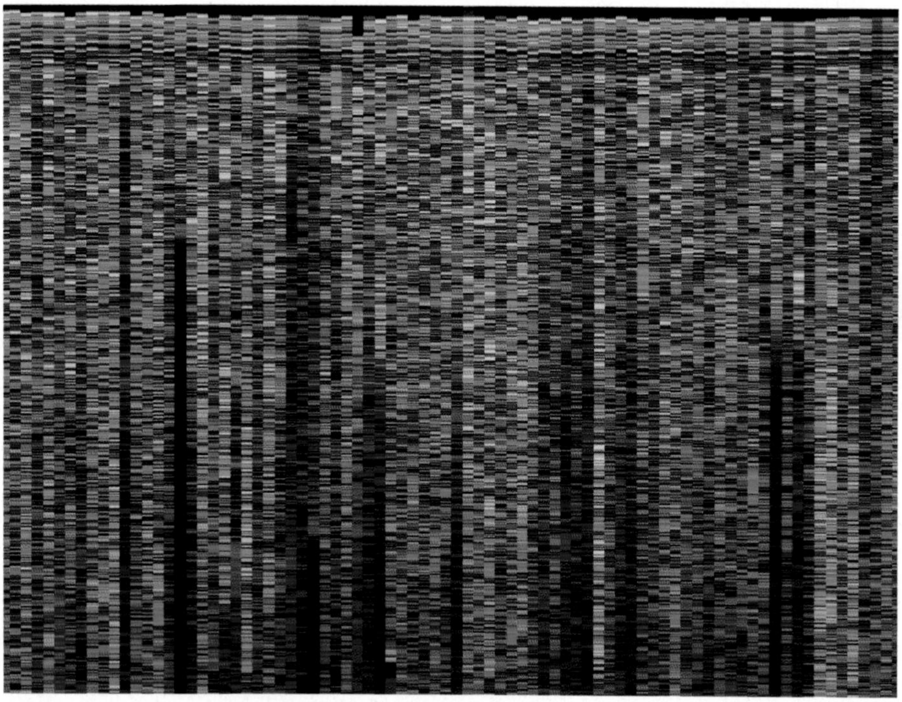

Output from an automated genome sequencing machine, representing about 48,000 bases from the human genome. Each vertical stripe represents the sequence of a stretch of DNA.

KEY CONCEPTS

- Once a genome has been completely sequenced, researchers use a variety of techniques to identify which sequences code for products and which act as regulatory sites.
- In bacteria and archaea, there is a positive correlation between the number of genes in a species and the species' metabolic capabilities.
- In eukaryotes, genomes are dominated by parasitic sequences that do not contribute to the fitness of the organism.
- Data from genome sequencing projects are now being used in the development of new drugs and vaccines.

The first data sets describing the complete DNA sequence, or **genome**, of humans were published in February 2001 as part of the **Human Genome Project.** The achievement was immediately hailed as a landmark in the history of science. It is important to recognize that the Human Genome Project is part of a much larger and ongoing effort to sequence genomes from an array of other eukaryotes, hundreds of bacteria, and dozens of archaea. The effort to sequence, interpret, and compare whole genomes is referred to as **genomics.** The pace of research in this field is nothing short of explosive.

The sudden arrival of genomics at the forefront of biological science prompts a host of questions: Why is whole-genome sequencing attracting so much attention? How do researchers sequence the entire gene complement of an organism? What have biologists learned from the first genomes that have been completed? And what might the future hold in the direction of research? These questions are the foundation of this chapter.

As an introductory biology student, you are part of the first generation trained in the genome era. Genomics is revolutionizing several fields within biological science and will almost certainly be an important part of your personal and professional life. Let's look at what genomics is and why it's being done.

20.1 Whole-Genome Sequencing

Genomics has moved to the cutting edge of research in biology largely because technological advances have increased the speed and driven down the cost of sequencing DNA. Since the Human Genome Project began in 1988, for example, advances in automation and revised experimental protocols have reduced the cost of sequencing by a factor of two every year and a half. As data became less expensive and were obtained faster, the pace of whole-genome sequencing accelerated. The result is that an almost mind-boggling number of sequences are now being generated. As this book goes to press, the primary international repository for DNA sequence data contains over 38 *billion* nucleotides. (With about 3 billion bases, humans have the largest haploid genome sequenced to date.) The size of this database—stored in a publicly funded online service called GenBank—doubles every year.

Which genomes are being sequenced, and why? A little history can shed some light on these questions. The first organismal genome to be sequenced came from a bacterium that lives in the human upper respiratory tract. This bacterium, *Haemophilus influenzae*, has one circular chromosome and a total of 1,830,138 base pairs of DNA. It was an important genome to explore, because this organism causes earaches and respiratory tract infections in children and because one particular strain is capable of infecting the membranes surrounding the brain and spinal cord, causing meningitis. The genome of *H. influenzae* was also small enough to sequence completely with a reasonable amount of time and money, given the technology available in the early 1990s.

The publication of the *H. influenzae* genome in 1995 was quickly followed by the completion of genome sequences from an assortment of bacteria and archaea. The first eukaryotic genome, from the yeast *Saccharomyces cerevisiae*, was finished in 1996. After that breakthrough, complete genome sequences were published from a variety of protists, plants, and animals. To date, genomes have been or are being sequenced from over 110 bacterial species, 16 archaeal species, and 20 eukaryotic species.

Several of the early sequencing projects focused on the need to work out techniques for the Human Genome Project. In other cases, organisms were selected for whole-genome sequencing because they cause disease or have interesting biological properties. For example, genomes of bacteria and archaea that inhabit extremely hot environments have been sequenced in the hopes of discovering enzymes useful for high-temperature industrial applications and understanding how proteins can work under those conditions. Other bacteria and archaea were chosen for sequencing because they produce compounds such as methane (CH_4, natural gas) as a by-product of cellular respiration. The rice genome was sequenced because rice is the main food source for most humans. Finally, species such as the fruit fly *Drosophila melanogaster*, the roundworm *Caenorhabditis elegans*, the house mouse *Mus musculus*, and the mustard plant *Arabidopsis thaliana* were analyzed because they serve as model organisms in biology and because data from well-studied organisms promised to help researchers interpret the human genome.

Before exploring the results of these studies in more detail, it will be helpful to review some of the technological innovations that have made whole-genome sequencing possible.

Recent Technological Advances

Recall from Chapter 19 that researchers use an in vitro DNA synthesis reaction called *dideoxy sequencing* to determine the sequence of bases in a segment of DNA. The gene being sequenced serves as the template, and the sequencing reactions produce a series of DNA fragments that end with an A, a C, a T, or a G. These fragments are then separated via electrophoresis and visualized by autoradiography or by other techniques. This method is still the basis of genome sequencing projects, though important modifications have been adopted. For example, recall that researchers traditionally attached a radioactive isotope to the primer in the reaction mix to label the fragments of DNA produced by a sequencing reaction. Now fluorescent markers are bonded to the dideoxyribonucleoside triphosphates (ddNTPs) used in a sequencing reaction—a different color for each ddNTP. As **Figure 20.1** shows, the switch to fluorescent markers was important for two reasons: (1) Each stretch of DNA could be sequenced with one dideoxy reaction instead of four separate ones, and (2) machines were developed

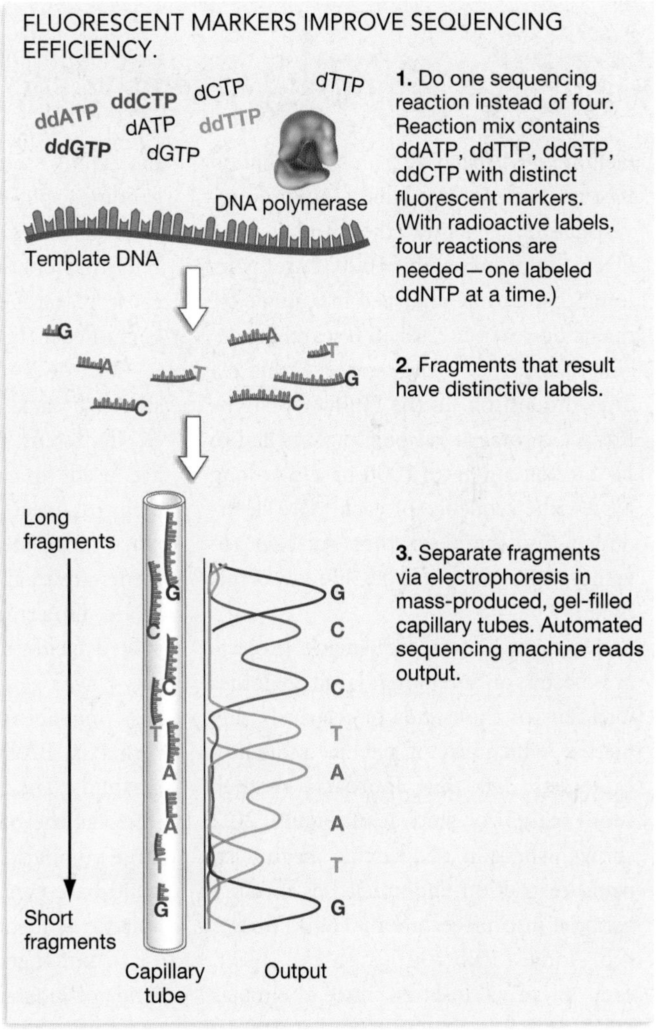

FIGURE 20.1 Use of Fluorescent Markers Simplifies Sequencing
Researchers made DNA sequencing more efficient by attaching distinctive fluorescent markers to the ddATP, ddGTP, ddCTP, and ddTTP used in sequencing reactions. Fluorescent markers are safer than radioactive markers and can be read by a machine.
EXERCISE List two ways that this protocol differs from sequencing methods that rely on radioactive markers.

to detect the fluorescence produced by each fragment and read the output of the sequencing reaction. In addition, instead of separating the fragments generated by sequencing reactions through electrophoresis in hand-prepared gels that are poured between glass plates, researchers now perform the electrophoresis step with mass-produced, gel-filled capillary tubes.

Based on these and other innovations, researchers used automated sequencing machines to establish factory-style DNA sequencing centers such as the one shown at the start of this chapter. Over 20 such centers have now been established in the United States, the United Kingdom, Germany, France, Japan, and China. Some of these laboratories employ dozens of biologists and can conduct 100,000 sequencing reactions daily. **Box 20.1** introduces the way researchers design these reactions so that the resulting data can be assembled into a completed genome sequence for a particular species. That discussion also emphasizes the importance of computer programs in analyzing DNA sequences. The vast quantity of data generated by genome sequencing centers has made **bioinformatics**—the effort to manage, analyze, and interpret information in biology—a key to continued progress in the field.

Which Sequences Are Genes?

Obtaining raw sequence data is just the beginning of the effort to understand a genome. As researchers point out, raw sequence data are analogous to the parts list for a house, except that the list has no punctuation: "windowwabeborogovestaircasedoorjubjub . . . ." Where do the genes for "window," "staircase," and "door" start and end? Are the segments that read

BOX 20.1 Shotgun Sequencing and Bioinformatics

Genomes range in size from a few million base pairs to several billion. How do investigators generate the stretches of DNA, approximately 1000 base pairs long, that can be analyzed in a single sequencing reaction? When researchers decide to sequence a large genome, this is a crucial question. In the Human Genome Project, a total of 3.2 billion bases had to be broken up into 1000-base-pair-long pieces, the sequence of each piece determined, and the pieces then put back together in the correct order. How was this done?

Current genome sequencing projects are based on breaking large genomes into tens of thousands of relatively tiny pieces, which are easier to sequence. Biologists call this approach **shotgun sequencing.** As step 1 of **Figure 20.2** shows, shotgun sequencing begins by using restriction endonucleases to cut a genome into pieces about 160 kilobases (kb) long (1 kb = 1000 bases). Next, each piece is inserted into a circular DNA molecule called a **bacterial artificial chromosome (BAC).** Each BAC is then inserted into a different *Escherichia coli* cell, creating what researchers call a **BAC library** (step 2). A BAC library is an example of a **genomic library:** a set of all the DNA sequences in a particular genome, split into small segments and inserted into a vector. When the *E. coli* cells that contain BACs are allowed to grow, researchers can isolate large numbers of each 160-kb fragment.

Once a BAC library has been constructed, each 160-kb segment is isolated. Different restriction endonucleases are used to cut the segments into an array of pieces, each about 1000 base pairs long (step 3). These small fragments are then inserted into the circular DNA molecules called plasmids and placed inside bacterial cells (step 4). In this way, a genome is broken down into two manageable levels: 160-kb fragments in BACs and 1-kb segments in plasmids. The plasmids are copied many times as the bacterial cells grow into a large population. Large numbers of each 1000-base-pair fragment are then available for sequencing reactions (step 5).

To put together the data from thousands of sequencing reactions, computer programs analyze regions where the ends of each 1000-base-pair segment overlap (step 6). Overlaps occur because different restriction endonucleases were used to fragment the 160-kb segments. The computer mixes and matches segments until an alignment consistent with all available data is obtained. Then the ends of each BAC are analyzed in a similar way (step 7). The computer's goal is to arrange each 160-kb segment in its correct position along the chromosome, based on regions of overlap.

Once complete genome sequences became available, databases that could hold completed sequence information had to be created and managed in a way that made the raw data and a variety of annotations available to the international community of researchers. These sequence databases also had to be searchable, so that investigators could evaluate how similar newly discovered genes were to genes that had been studied previously.

Because the amount of data involved is so large, the computational challenges involved in genomics are formidable. Thus far, sophisticated algorithms and continually improving computer hardware have allowed specialists in bioinformatics to keep pace with the rate of data acquisition. Individuals with a background in biology and experience with software design or development are in high demand in laboratories around the world.

WEB TUTORIAL 20.1 Human Genome Sequencing Strategies

"wabeborogove" and "jubjub" important in gene regulation, or are they simply spacers or other types of sequences that have no function at all?

The most basic task in annotating a genome is to identify which bases constitute genes—segments of DNA that code for an RNA or a protein product. In bacteria and archaea, identifying genes is relatively straightforward. Biologists begin with computer programs that scan the sequence of a genome in both directions. Such programs test each reading frame that is possible on the two strands of the DNA. (Recall from Chapter 15 that a reading frame is the sequence in which codons are read.) With codons consisting of three bases, three reading frames are

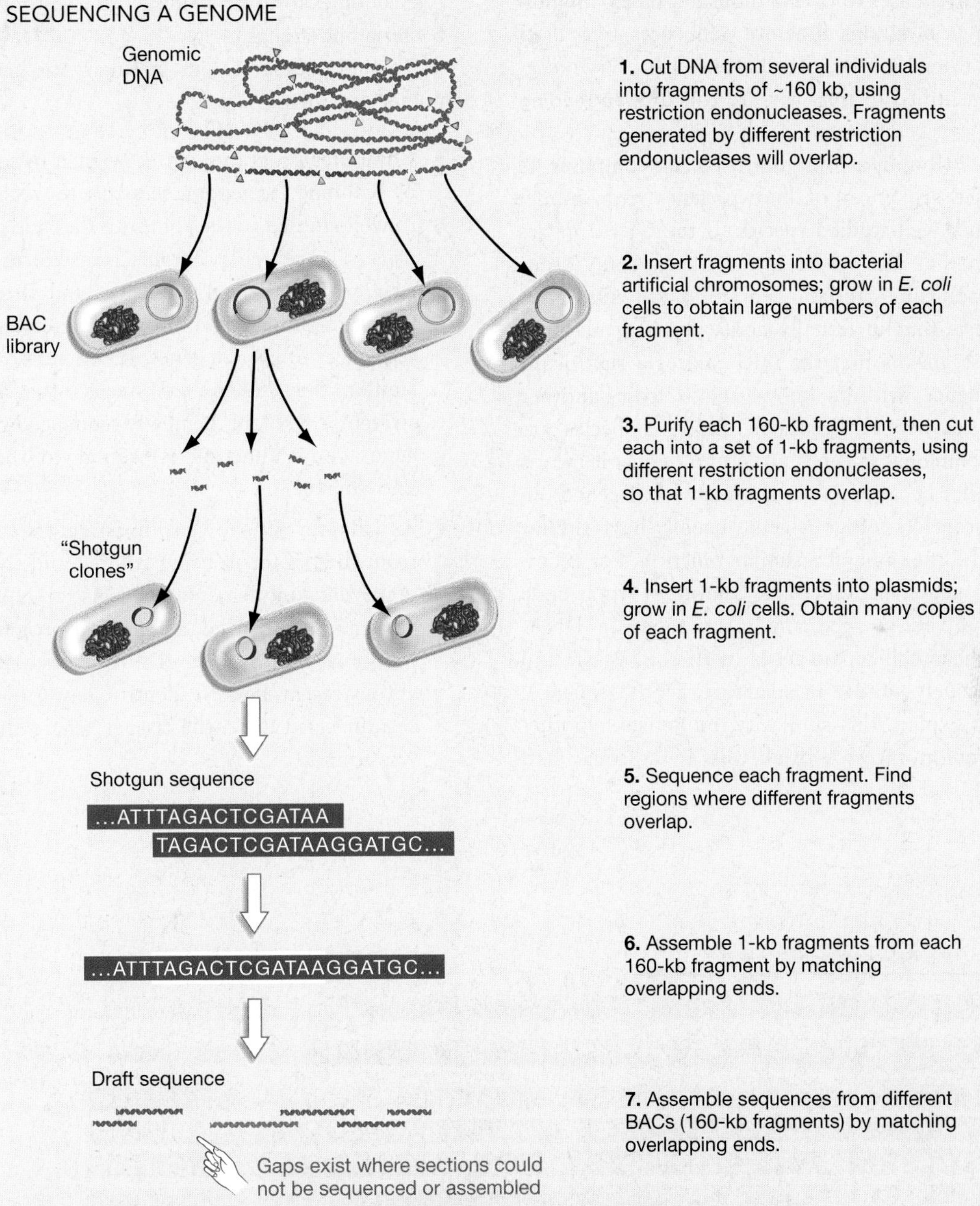
SEQUENCING A GENOME
Genomic DNA
1. Cut DNA from several individuals into fragments of ~160 kb, using restriction endonucleases. Fragments generated by different restriction endonucleases will overlap.
BAC library
2. Insert fragments into bacterial artificial chromosomes; grow in E. coli cells to obtain large numbers of each fragment.
3. Purify each 160-kb fragment, then cut each into a set of 1-kb fragments, using different restriction endonucleases, so that 1-kb fragments overlap.
"Shotgun clones"
4. Insert 1-kb fragments into plasmids; grow in E. coli cells. Obtain many copies of each fragment.
Shotgun sequence
...ATTTAGACTCGATAA
TAGACTCGATAAGGATGC...
5. Sequence each fragment. Find regions where different fragments overlap.
...ATTTAGACTCGATAAGGATGC...
6. Assemble 1-kb fragments from each 160-kb fragment by matching overlapping ends.
Draft sequence
7. Assemble sequences from different BACs (160-kb fragments) by matching overlapping ends.
Gaps exist where sections could not be sequenced or assembled

FIGURE 20.2 Shotgun Sequencing Breaks Large Genomes into Many Short Segments
Shotgun sequencing is a multistep process. A genome is initially fragmented into 160-kb sections with restriction endonucleases and cloned into bacterial artificial chromosomes (BACs). Each 160-kb section is then cut into 1000-base-pair fragments that can be cloned into plasmids, grown to a large number of copies, and sequenced.

possible on each strand, for a total of six possible reading frames (**Figure 20.3**). Because randomly generated sequences contain a stop codon about one in every 20 codons on average, a long stretch of codons that lacks a stop codon is a good indication of a coding sequence. The computer program highlights any "gene-sized" stretches of sequence that lack a stop codon but are flanked by a stop codon and a start codon. Because polypeptides range in size from 50 amino acids to many hundreds of amino acids, gene-sized stretches of sequence range from several hundred bases to over a thousand bases. In addition, the computer programs look for sequences typical of promoters, operators, or other regulatory sites. DNA segments that are identified in this way are called **open reading frames**, or **ORFs.**

Once an ORF is found, a computer program compares its sequence with the sequences of known genes from model organisms or other well-studied species. If the ORF appears to be a gene that has not yet been described in any other species, it is a promising candidate for follow-up studies to test the hypothesis that it actually codes for a product. A "hit," in contrast, means that the ORF shares a significant amount of sequence with a known gene from another species. Similarities between genes in different species are usually due to **homology.** If genes are homologous, it means that they are similar because they are related by descent from a common ancestor. Homologous genes usually have similar base sequences and the same or a similar function. For example, consider the genes introduced in Chapter 14 that code for enzymes involved in repairing mismatches in DNA. Recall that the mismatch repair genes in *E. coli*, yeast, and humans are extremely similar in structure, DNA sequence, and function. To explain this similarity, biologists hypothesize that the common ancestor of all cells living today had mismatch repair genes—thus, the descendants of this ancestral species also have versions of these genes.

Finding and analyzing genes by identifying ORFs does not work well in eukaryotes, however. Most coding regions in eukaryotes are broken up by introns, and most regions in eukaryotic genomes do not actually code for a product. In the human genome, for example, it is estimated that less than 2 percent of the DNA present actually codes for proteins, tRNAs, ribosomal RNAs, or other types of products. As a result, scanning eukaryotic genomes for open reading frames is not as useful as it is in bacteria and archaea.

To find coding regions in eukaryotic genomes, researchers pursue several strategies:

- Computer programs can be written to search for homologs by scanning the genome sequence for regions that are similar to well-studied genes from fruit flies and other model organisms. The search programs translate the sequenced DNA into an amino acid sequence, and then they attempt to match the inferred amino acid sequences to the amino acid sequences of known genes. If two amino acid sequences are similar, then researchers suspect that the corresponding stretch of DNA in the newly sequenced genome codes for a gene whose function is similar to the function of the known gene.
- As Chapter 19 showed, investigators can isolate mRNAs from the organism being studied and then use enzymes to make the complementary DNAs (cDNAs). Once an entire genome is sequenced, a computer program can scan it and pinpoint where each cDNA is located. This approach allows researchers to identify genes that are expressed in certain cell types—the tissues where the original mRNA was found.

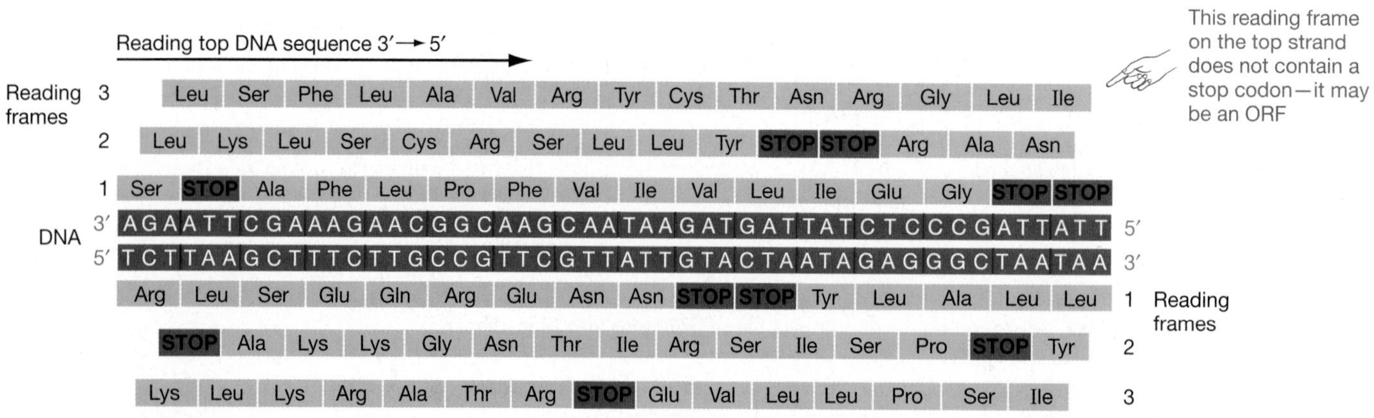

FIGURE 20.3 Open Reading Frames May Be the Locations of Genes
Computer programs scan the three possible reading frames on each strand of DNA and use the genetic code to translate each codon. A long stretch of codons that lacks a stop codon may be an open reading frame (ORF)—a possible gene.

- If genome sequences from closely related species are available, computers are used to compare them and identify sequences that are similar in all of the species involved. Sequences that are shared by closely related species are hypothesized to be located in genes—not in noncoding regions. The logic behind this claim runs as follows: Coding and regulatory sequences are expected to change much more slowly over time than noncoding regions are. This logic is valid because most genes are required for organisms to function and because most proteins work less efficiently when their amino acid sequences change by mutation. Thus, it is logical to expect that natural selection eliminates most mutations in coding sequences and that coding sequences should change slowly over time. Because changes in noncoding sequences do not change gene products, they do not affect the organism's phenotype. Mutations in noncoding regions are much less likely to be eliminated by natural selection, so noncoding regions change relatively rapidly over time.

Although each of these gene-finding strategies has been productive, it will probably be many years before biologists are convinced that they have identified all of the coding regions in even a single eukaryotic genome. As that effort continues, though, researchers are mining the data and making some remarkable observations. Let's first consider what genome sequencing has revealed about the nature of bacterial and archaeal genomes and then move on to eukaryotes. Is the effort to sequence whole genomes paying off?

20.2 Bacterial and Archaeal Genomes

By the time you read this paragraph, the genomes of over 110 bacterial species and 16 archaeal species will have been completely sequenced, and projects focused on sequencing dozens of additional prokaryotic genomes will be under way. In addition to this impressive array of different species, complete genome sequences are now available for several strains of the same bacterial species. For example, researchers have sequenced the genome of a harmless laboratory strain of *Escherichia coli*—derived from the harmless strain that lives in your gut—as well as the genome of a form that causes severe disease in humans. As a result, researchers can now compare the genomes of closely related cells that have different ways of life.

This section focuses on a simple question: Based on data published between 1995 and 2004, what general observations have biologists been able to make about the nature of bacterial and archaeal genomes?

The Natural History of Prokaryotic Genomes

In a sense, biologists who are working in genomics can be compared with the naturalists of the eighteenth and nineteenth centuries. The early naturalists explored the globe, collecting the plants and animals they encountered. Their goal was to describe what existed. Similarly, the first task of a genome sequencer is to catalog what is in a genome—specifically, the number, type, and organization of genes. Several interesting conclusions can be drawn from relatively straightforward observations about the data obtained thus far:

- In bacteria, there is a general correlation between the size of a genome and the metabolic capabilities of the organism. In general, parasites have much smaller genomes than nonparasitic organisms do. **Parasites** live off a host, to the detriment of the host. The smallest genomes are found in parasitic bacteria from the genus *Mycoplasma*. (*Mycoplasma pneumoniae* causes pneumonia in humans.) These bacteria live and multiply inside host cells. Because *Mycoplasma* acquire almost all of their nutrients from their hosts, it is not surprising to observe that they lack the enzymes required to manufacture many essential compounds—these bacteria don't need genes that are being expressed by their hosts. In contrast, the genomes of nonparasitic strains of the bacteria *E. coli* and *Pseudomonas aeruginosa* are 8 to 10 times larger. Their genes code for enzymes that synthesize virtually every molecule needed by the cell. Based on this observation, it is not surprising that *E. coli* is able to grow under a wide variety of environmental conditions. Using similar logic, researchers hypothesize that the large genome of *Pseudomonas* explains why it is able to occupy a wide array of soil types, including marine and marshy habitats, as well as human tissues, where it can cause illness.
- Biologists still do not know the function of many of the genes that have been identified. Although *E. coli* probably qualifies as the most intensively studied of all organisms, the function of over 30 percent of its genes is unknown.
- There is tremendous genetic diversity among prokaryotes. About 15 percent of the genes in each prokaryotic genome appear to be unique to its own species. That is, about one in six genes in a prokaryotic species is found nowhere else.
- Redundancy among genes is common. For instance, the genome of *E. coli* has 86 pairs of genes whose DNA sequences are nearly identical—meaning that the proteins they produce are nearly alike in structure and presumably in function. Although the significance of this redundancy is unknown, biologists hypothesize that slightly different forms of the same protein are produced in response to slight changes in environmental conditions.
- Multiple chromosomes are more common than anticipated. Several species of prokaryotes have two different circular chromosomes instead of one.
- Many species contain the small, extrachromosomal DNA molecules called plasmids. Recall from Chapter 19 that plasmids contain a small number of genes, though not genes that are absolutely essential for growth. In many cases, plasmids can be exchanged between cells of the same species or even of different species.

Perhaps the most surprising observation of all is that in many bacterial and archaeal species, a significant proportion of the genome appears to have been acquired from other, often distantly related, species. In some bacteria and archaea, 15–25 percent of the genetic material appears to be "foreign." This is a remarkable claim. What evidence backs up the assertion that prokaryotes acquire DNA from other species? How could this happen, and what are the consequences?

Evidence for Lateral Gene Transfer

Biologists use two general criteria to support the hypothesis that sequences in bacterial or archaeal genomes originated in another species: (1) when stretches of DNA are much more similar to genes in distantly related species than to those in closely related species and (2) when the proportion of G-C base pairs to A-T base pairs in a particular gene or series of genes is markedly different from the base composition of the rest of the genome. In many cases, the proportion of G-C bases in a genome is characteristic of a genus or species.

How can genes move from one species to another? In at least some cases, plasmids appear to be involved. For example, most of the genes that are responsible for conferring resistance to antibiotics are found on plasmids. Although the physical mechanism involved is not known, researchers have recently documented the transfer of antibiotic-resistance genes between very distantly related species of disease-causing bacteria, by means of plasmids. In some cases, genes from plasmids become integrated into the main chromosome of a bacterium, resulting in genetic recombination (see Chapter 12). The movement of DNA from one species to another species is called **lateral gene transfer** (**Figure 20.4**).

Some biologists hypothesize that lateral gene transfer also occurs when bacteria and archaea take up raw pieces of DNA from the environment—perhaps in the course of acquiring other molecules. This may have occurred in the bacterium *Thermotoga maritima*, which occupies the high-temperature environments near deep-sea hot springs. Almost 25 percent of the genes in this species are extremely closely related to genes found in archaea that live in the same habitats. The archaea-like genes occur in distinctive clusters within the *T. maritima* genome, which supports the hypothesis that the sequences were transferred in large pieces from an archaean to the bacterium.

Similar types of direct gene transfer are hypothesized to have occurred in the bacterium *Chlamydia trachomatis*. This organism is a major cause of blindness in humans from Africa and Asia; it also causes chlamydia, the most common sexually transmitted bacterial disease in the United States. The *C. trachomatis* genome contains 35 genes that resemble eukaryotic genes in structure. Because *C. trachomatis* lives inside the cells that it parasitizes, the most logical explanation for this observation is that the bacterium occasionally takes up DNA directly from its host cell, resulting in a eukaryote-to-bacterium transfer.

In addition to being transferred between species by means of plasmids or DNA fragments, genes can be transported by viruses.

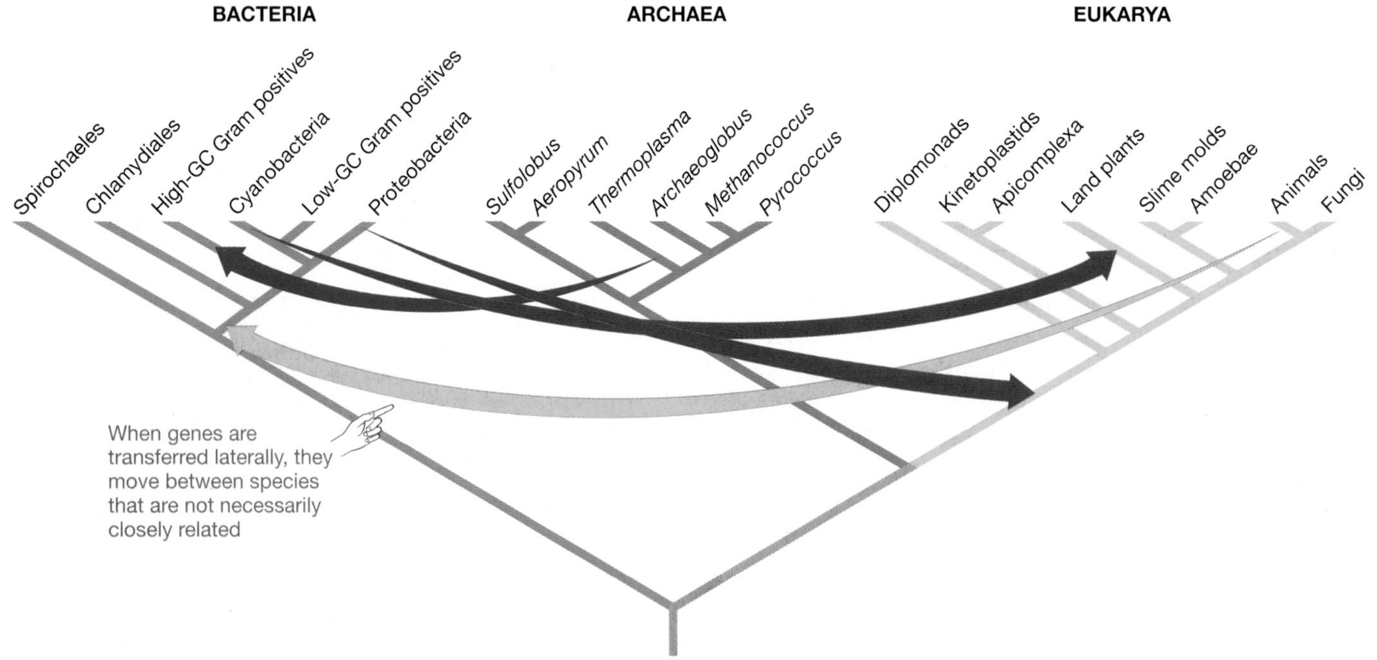

FIGURE 20.4 Lateral Gene Transfer Is Movement of DNA between Species
Lateral gene transfer can occur between very distantly related organisms. EXERCISE List the mechanisms responsible for lateral gene transfer.

For example, investigators who compared the sequences of laboratory and disease-causing strains of *E. coli* found that the pathogenic cells have almost 1400 "extra" genes. Compared with the rest of the genome, most of these genes have a distinctive G-C to A-T ratio. In addition, many are extremely similar to sequences isolated from viruses that infect *E. coli*. Based on these observations, most researchers support the hypothesis that at least some of the disease-causing genes in *E. coli* were brought in by viruses.

To summarize, mutation and gene transfer within species are not the only source of genetic variation in bacteria and archaea. Over the course of evolution, lateral gene transfer has been an important source of new genes and allelic diversity in these domains. This insight would not have been possible without data from whole-genome sequencing. Have efforts to sequence eukaryotic genomes led to similar types of insights?

✓CHECK YOUR UNDERSTANDING

The major surprise that came out of genome sequencing projects involving bacteria and archaea was the extent and importance of lateral gene transfer—that is, the movement of DNA from one species to another. You should be able to (1) summarize evidence that lateral gene transfer is responsible for the presence of a particular DNA sequence and (2) summarize evidence that the size of a prokaryotic genome is correlated with the organism's metabolic capabilities.

20.3 Eukaryotic Genomes

Sequencing eukaryotic genomes presents two daunting challenges. The first is sheer size. Compared with the genomes of bacteria and archaea, which range from 580,070 base pairs in *Mycoplasma genitalium* to over 6.3 million base pairs in *Pseudomonas aeruginosa*, eukaryotic genomes are large. The haploid genome of *Saccharomyces cerevisiae* (baker's yeast), a unicellular eukaryote, contains over 12 million base pairs. The roundworm *Caenorhabditis elegans* has a genome of 97 million base pairs; the fruit-fly genome contains 180 million base pairs; the mustard plant *Arabidopsis thaliana*'s genome has 130 million base pairs; and humans, rats, mice, and cattle contain roughly 3 billion base pairs each.

The second great challenge in sequencing eukaryotic genes is coping with noncoding sequences that are repeated many times. Many eukaryotic genomes are dominated by repeated DNA sequences that occur between genes and that do not code for products used by the organism. These repeated sequences pose serious problems in aligning and interpreting sequence data. What are they? If such sequences don't code for a product, why do they exist?

Natural History: Types of Sequences

In many eukaryotic genomes, the exons, introns, and regulatory sequences associated with genes make up a relatively small percentage of the genome. Recall from Section 20.1 that, in humans, exons constitute less than 2 percent of the total genome while repeated sequences account for well over 50 percent. In contrast, over 90 percent of the DNA sequences in a bacterial or archaeal genome code for a product used by the cell.

When noncoding and repeated sequences were discovered, they were initially considered "junk DNA" that was nonfunctional and probably unimportant and uninteresting. But subsequent work has shown that most of the repeated sequences observed in eukaryotes are actually derived from sequences known as transposable elements. **Transposable elements** are parasitic segments of DNA that are capable of moving from one location to another in a genome. They are similar to viruses, except that they are not transmitted from one host cell or host individual to another by infection. Instead, transposable elements transmit copies of themselves to additional locations within the host genome and are passed on to offspring along with the rest of the genome. Viruses leave a host cell and find a new cell to infect. But transposable elements never leave their host cell—they simply make copies of themselves and move to new locations in the genome. They are examples of what biologists call **selfish genes**—DNA sequences that survive and reproduce but that do not increase the fitness of the host genome.

How Do Transposable Elements Work? Transposable elements come in a wide variety of types and spread through genomes in a variety of ways. Different species typically contain distinct types of transposable elements. The genomes of fruit flies, yeast, and humans each contain distinctive kinds of parasitic sequences.

As an example of how these selfish genes work, let's consider a type called a **long interspersed nuclear element** (LINE), found in humans and other eukaryotes. A LINE is similar to the retroviruses, introduced in Chapter 19. Each LINE consists of a stretch of DNA that contains a gene that codes for the enzyme reverse transcriptase, a gene that codes for the enzyme integrase, and a single promoter that is recognized by RNA polymerase II. Your genome contains tens of thousands of LINEs, each between 1000 and 5000 bases long.

A LINE contains all the sequences required for it to make copies of itself and insert them into a new location in the genome. When a LINE is expressed, reverse transcriptase is translated from the LINE mRNA. This enzyme makes a cDNA version of the LINE mRNA. The second LINE product, integrase, then inserts the newly synthesized LINE DNA into a new location in the genome. In this way, the parasitic sequence reproduces. The current model for how this happens is shown in

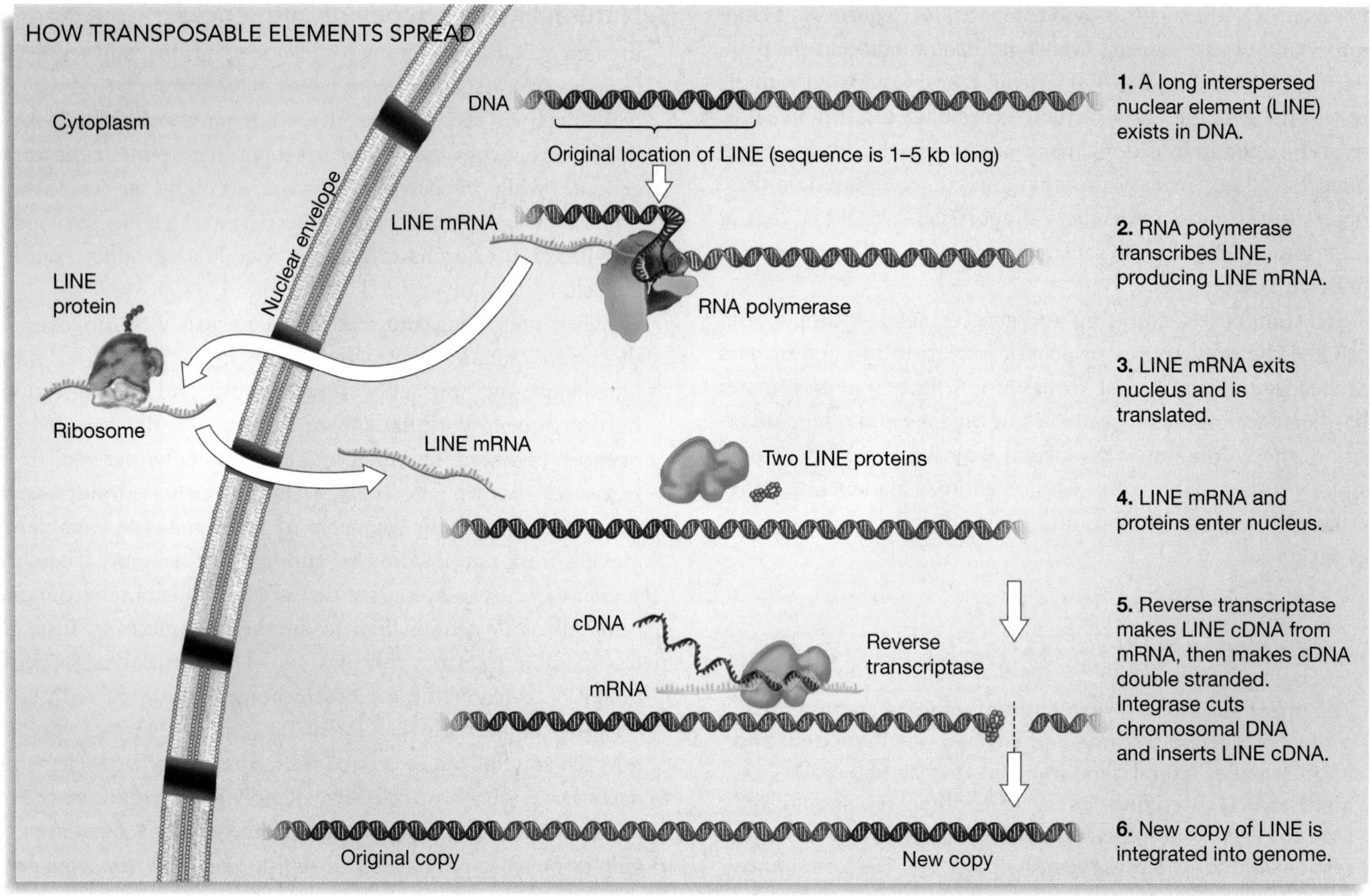
HOW TRANSPOSABLE ELEMENTS SPREAD
Cytoplasm
DNA
Original location of LINE (sequence is 1–5 kb long)
1. A long interspersed nuclear element (LINE) exists in DNA.
Nuclear envelope
LINE mRNA
LINE protein
RNA polymerase
2. RNA polymerase transcribes LINE, producing LINE mRNA.
Ribosome
3. LINE mRNA exits nucleus and is translated.
LINE mRNA
Two LINE proteins
4. LINE mRNA and proteins enter nucleus.
cDNA
mRNA
Reverse transcriptase
5. Reverse transcriptase makes LINE cDNA from mRNA, then makes cDNA double stranded. Integrase cuts chromosomal DNA and inserts LINE cDNA.
Original copy
New copy
6. New copy of LINE is integrated into genome.

FIGURE 20.5 Transposable Elements Spread within a Genome
This sequence of events is the leading hypothesis for how LINEs spread.

Figure 20.5. If the transposition event occurs in reproductive cells that go on to form eggs or sperm, the copied LINE will be passed on to offspring. If the LINE happens to insert itself inside a gene or a regulatory sequence, it causes a mutation that is almost certain to reduce the host's fitness.

Most of the LINEs observed in the human genome are defective, however. Specifically, most of the LINEs in your cells don't contain a promoter or the genes for either reverse transcriptase or integrase. To make sense of this observation, researchers hypothesize that the insertion process illustrated in Figure 20.5 is usually disrupted in some way. Analyses of the human genome have revealed that only a handful of our LINEs appear to be complete and potentially active.

Virtually every prokaryotic and eukaryotic genome examined to date contains at least some transposable elements. They vary widely in type and number, however; bacterial and archaeal genomes have relatively few transposable elements. This observation has inspired the hypothesis that bacteria and archaea either have efficient means of removing parasitic sequences or can somehow thwart insertion events. To date, however, this hypothesis has yet to be tested rigorously.

Research on transposable elements and lateral gene transfer has had a huge impact on how biologists view the genome. Many genomes are riddled with parasitic sequences, and others have undergone radical change in response to lateral gene transfer events. The general picture is that genomes are dynamic. Their size and composition can change dramatically over time.

Repeated Sequences and DNA Fingerprinting In addition to containing repeated sequences from transposable elements, eukaryotic genomes have several hundred loci called **simple tandem repeats.** Repeating units that are just 1 to 5 bases long are known as **microsatellites** or **simple sequence repeats**; repeating units that are 6 to 500 bases long are called **minisatellites** or **variable number terminal repeats** (VNTRs). Both types of repeated sequences make up 3 percent of the human genome. The most common type is a repeated stretch of the dinucleotide AC, giving the sequence ACACACAC .... Both types of satellite sequences are thought to originate when DNA polymerase skips or mistakenly adds extra bases during replication.

Soon after these sequences were first characterized, Alec Jeffreys and co-workers established that microsatellite and minisatellite loci are "hypervariable"—they vary among individuals much more than any other type of sequence does. **Figure 20.6** illustrates one hypothesis for why minisatellites have so many different alleles: These highly repetitive stretches often align incorrectly when

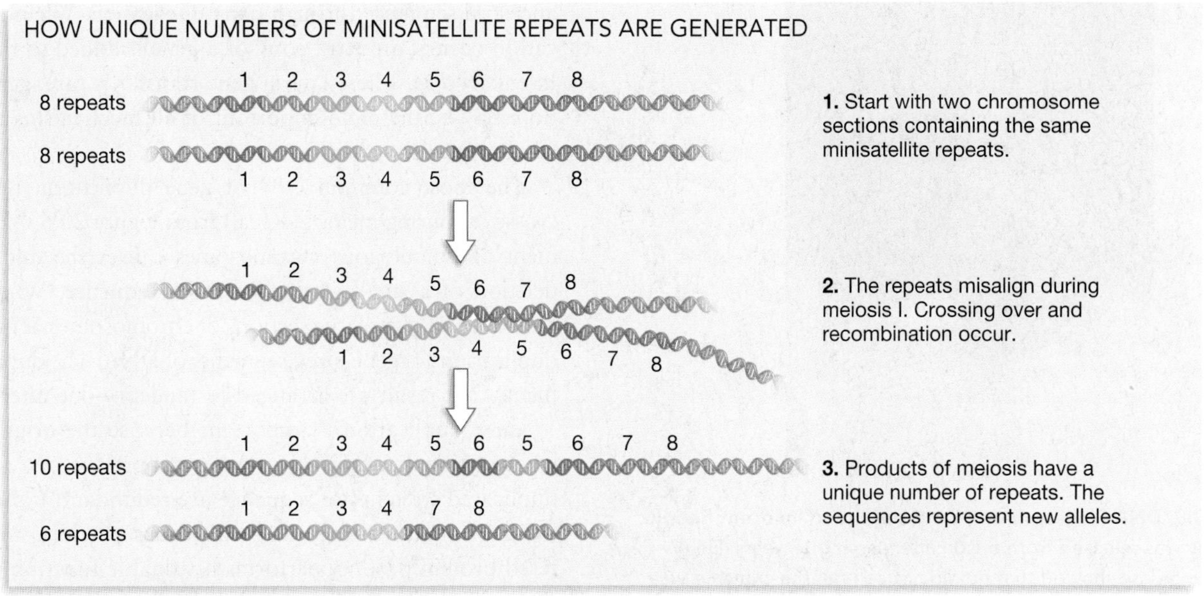

FIGURE 20.6 Unequal Crossover Changes the Numbers of Simple Sequence Repeats
The alignment of homologous chromosomes during prophase of meiosis I is driven by sequence similarity between homologs. Because simple sequence repeats are so similar, they are likely to misalign during synapsis.

homologous chromosomes synapse and cross over during prophase of meiosis I. Instead of lining up in exactly the same location, the two chromosomes pair in a way that matches up bases in different repeated segments. Due to this misalignment, **unequal crossover** occurs. Chromosomes produced by unequal crossover contain different numbers of repeats. The key observation is that, if a particular microsatellite or minisatellite locus has a unique number of repeats, it represents a unique allele. Each of these alleles has a unique length. As with any allele, microsatellite and minisatellite alleles are transmitted from parents to offspring.

Misalignment or errors by DNA polymerase are so common at these loci that, in most populations of eukaryotes, the genome of virtually every individual has at least one new allele. This variation in repeat number among individuals is the basis of DNA fingerprinting. **DNA fingerprinting** refers to any technique for identifying individuals on the basis of unique features of their genomes. Because microsatellite and minisatellite loci vary so much among individuals, they are now the loci of choice for DNA fingerprinting. To fingerprint an individual, researchers obtain a DNA sample and perform the polymerase chain reaction (PCR), using primers that flank a region containing microsatellites or minisatellites. Once many copies of the region are available, they can be analyzed to determine the number of repeats present.

Research on repeated sequences has revealed that the probability of getting a new allele is higher for shorter repeats than for longer repeats. For some two-base-pair repeats, the number of repeats present changes so quickly over time that only very close relatives are likely to share any of the same alleles. This observation has important practical implications. For example, because parents and offspring tend to share alleles at these loci, analyzing microsatellite and minisatellite sequences offers an accurate way to assign paternity in birds, humans, and other species that have these types of sequences (**Figure 20.7**).

DNA fingerprinting has also been enormously useful in **forensic science**—the use of scientific analyses to solve crimes. As an example, consider the case of Lonnie Erby, who was convicted of raping three women in 1985. After Erby had served 17 years of a 115-year prison sentence, a DNA fingerprinting analysis of preserved semen taken from the victims

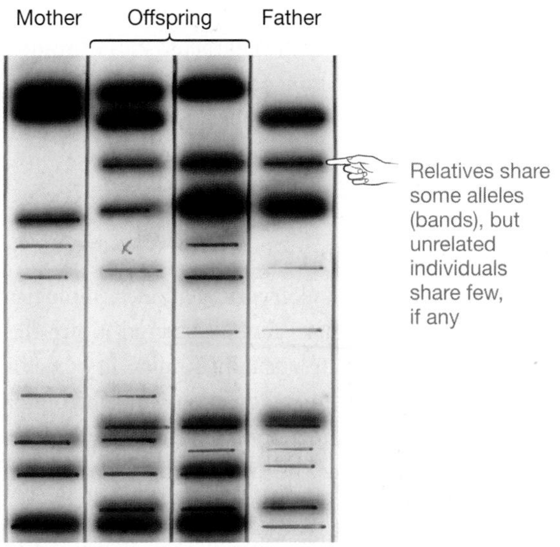

FIGURE 20.7 Close Relatives Have Similar DNA Fingerprints
A gel that contains DNA fragments from parents and offspring and has been probed with sequences from several minisatellite loci. Different individuals have fragments of various lengths if minisatellite loci contain different numbers of repeats. Related individuals tend to share fragment patterns.

FIGURE 20.8 DNA Fingerprinting Data Has Freed Innocent People Lonnie Erby was released from prison after serving 17 years. DNA fingerprinting data showed that he did not commit the crime he was accused of.

was performed. Several microsatellite and minisatellite loci were amplified from the preserved DNA samples, via PCR, and the alleles present were compared with those in Erby's DNA. They did not match. The data showed that there was virtually no chance that Erby had committed the crimes. Once the result was confirmed, he was released (**Figure 20.8**).

A nonprofit group that is coordinating DNA testing of convicted criminals in the United States has documented a total of 136 prisoners exonerated and released to date. Some of the individuals involved had been incarcerated even longer than Erby, for crimes ranging from rape to murder. It is no exaggeration to claim that DNA fingerprinting has given these innocent men and women their lives back.

Now that we've reviewed the characteristics of some particularly prominent types of noncoding sequences in eukaryotes, it's time to consider the nature of the coding sequences in these genomes. Let's start with the most basic question of all: Where do eukaryotic genes come from?

What Is a Gene Family?

In both prokaryotic and eukaryotic genomes, biologists routinely find groups of similar genes clustered along the same chromosome. The genes are usually similar in structural aspects, such as the arrangement of exons and introns, and in their base sequence. The degree of sequence similarity among these clustered genes varies. In the genes that code for ribosomal RNAs (rRNAs) in vertebrates, the sequences are virtually identical—meaning that each individual has many exact copies of the same gene. In other cases, though, the proportion of bases that are identical is 50 percent or less.

Genes that are extremely similar to each other are considered to be part of the same **gene family**. Genes that make up gene families are hypothesized to have arisen from a common ancestral sequence through **gene duplication.** When gene duplication occurs, an extra copy of a gene is added to the genome. In eukaryotes, where lateral gene transfer is rare, gene duplication may be the most important of all mechanisms that generate new genes.

The most common type of gene duplication results from crossover during meiosis. Recall from Figure 20.6 that misalignment of homologous chromosomes causes the duplication or deletion of a small section of DNA sequence. When unequal crossover involves large sections of chromosomes, it can produce duplications of gene-sized segments of DNA. The duplicated segments that result are arranged in tandem—one after the other.

Gene duplication is important because the original gene is still functional and produces a normal product. As a result, the duplicated stretches of sequence are redundant. If mutations in the duplicated sequence alter the protein product, and if the altered protein product performs a valuable function in the cell, then an important new gene has been created. The duplicated gene may also be regulated in a different way, so that it is expressed in novel locations or at novel times. In either case, the duplicated sequences represent new genes and can lead to the evolution of novel traits. Gene duplication produces new genes and creates gene families.

Alternatively, mutations in the duplicated region may make expression of the new locus impossible. For example, a mutation could produce a stop codon in the middle of an exon. When a gene is disabled in some way, the resulting sequence is referred to as a pseudogene. A **pseudogene** is a sequence that closely resembles a working gene but does not produce a working product. Pseudogenes have no function.

As an example of a gene family, consider the human **globin** genes diagrammed in **Figure 20.9**. These genes code for proteins that form part of hemoglobin—the oxygen-carrying molecule in your red blood cells. Analyzing the globin loci illustrates several important points about gene families. In humans, the globin gene family contains several pseudogenes, along with several genes that code for oxygen-transporting proteins. The various coding genes in the family serve slightly different functions. For example, some genes are active only in the fetus or the adult. Follow-up work showed that the proteins encoded by the fetal genes have an exceptionally high affinity for oxygen, compared with the proteins expressed in adults. As a result, oxygen is able to move from the mother's blood to the fetus's blood.

In addition to the gene duplication events that result from unequal crossover, an entire complement of chromosomes may be duplicated as a result of a mistake in either mitosis or meiosis. In this case, the resulting organism contains double the normal complement of chromosomes. Recall from Chapter 12 that species with duplicated chromosome complements are said to be **polyploid.** When polyploidy occurs, every gene in the duplicated genome is free to mutate and possibly to acquire a new function. Polyploidy has been a particularly important source of new genes in plants.

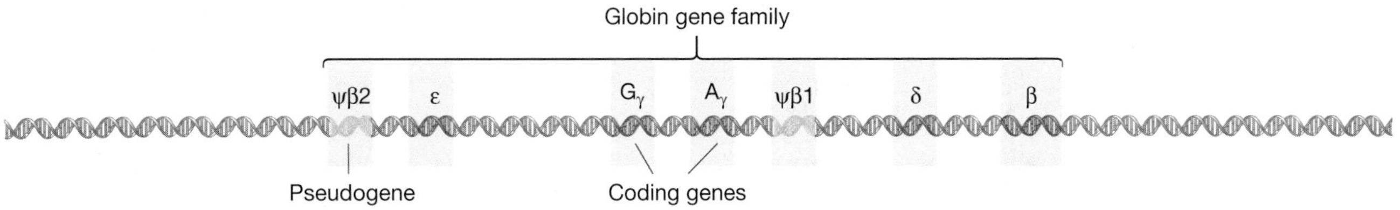

FIGURE 20.9 Gene Families Are Clusters of Closely Related Genes
Genes within the globin family. Red segments represent functioning genes, and yellow segments are pseudogenes. The members of a gene family are arranged one after the other, or in tandem. EXERCISE Add labels to each functioning gene, indicating when it is expressed: ε, 0–6 weeks postconception; G_γ and A_γ, highest 8–36 weeks postconception; δ, very low levels after birth through adulthood; β, low levels 12–36 weeks postconception, increasing to high levels from 6 weeks after birth through adulthood.

✓CHECK YOUR UNDERSTANDING

Eukaryotic genomes are riddled with parasitic sequences that do not contribute to the fitness of the organism. Simple repeated sequences are also common, and many of the coding sequences are organized into families of genes with related functions. You should be able to explain (1) why transposable elements are considered selfish genes, (2) why simple sequence repeats make DNA fingerprinting possible, and (3) how unequal crossover leads to duplicated sequences.

20.4 Comparative Genomics

Analyzing the genomes of individual bacterial, archaeal, and eukaryotic species has provided a wealth of insights, ranging from the size of specific gene families to the nature of transposable elements. Recently, however, biologists have begun to compare the genomes of different species. As the number of completely sequenced genomes expands, and as a wider diversity of species is studied, comparative genomics becomes more and more productive. We've already seen how comparing genome sequences across species led to the realization that there is a positive correlation between genome size and metabolic capabilities in bacteria and archaea, and that lateral gene transfer has played a large role in the evolution of these lineages. Let's delve into comparative analyses, starting with what comparative genomics has to say about the evolution of disease and the importance of changes in gene regulation during human evolution.

Understanding Virulence

Currently, dozens of whole-genome sequencing projects are focused on disease-causing bacteria. As each of these projects is completed, biologists begin comparing the genomes of harmless strains of bacteria with those of pathogenic strains. The goal is to achieve a much more detailed understanding of the genetic basis of **virulence**—the tendency for a parasite to harm its host.

As an example of this work, let's look more closely at what biologists found when they compared the genomes of benign (harmless) and pathogenic strains of *E. coli*. Biomedical researchers are interested in the genome of the disease-causing strain called O157:H7, because it is a common cause of food poisoning. People who eat undercooked meat may develop serious, and occasionally fatal, infections by O157:H7. Cells of this strain have a large number of genes not found in benign strains. The genes found in this strain that are associated with disease originally may have been introduced into the bacterial cells by viruses.

When biologists analyzed the novel genes in O157:H7, they found that several of those genes were extremely similar to sequences from a closely related pathogenic bacterium called *Shigella*. In countries where public water supplies are untreated, *Shigella* is an important cause of dysentery. Some of the *Shigella*-like genes found in O157:H7 code for proteins that poison cells in the human intestine and cause severe diarrhea. This observation suggests that the same genes may be responsible for causing food poisoning by *E. coli* and dysentery by *Shigella*. More specifically, biologists hypothesize that *E. coli* obtained the virulence genes by lateral gene transfer from *Shigella*, via a virus that infects both species. In addition to being intrinsically interesting, this research has an important practical impact: If drugs that neutralize *Shigella* toxins can be developed, they should also be effective against infections by *E. coli* O157:H7.

The data available indicate that the *Shigella–E. coli* story may be typical. In the disease-causing organisms whose genomes have been sequenced thus far, biologists have found that the genes responsible for causing illness or antibiotic resistance tend to occur in discrete clusters. This observation supports the hypothesis that the genes originated in lateral gene transfer events.

The ability to identify genes that are specific to disease-causing strains of bacteria gives investigators a host of new drug targets. If drugs that knock out the protein products of these genes can be developed, they would inhibit disease-causing bacterial strains while leaving helpful forms intact.

Studying Adaptation

Comparing the genomes of different strains or species has been a productive way to identify genes that are responsible for adaptations to particular environments or ways of life. For example, disease-causing organisms such as the cholera bacterium introduced in Chapter 17, *E. coli* O157:H7, and *Shigella* thrive in environments where human populations are dense and sanitation is poor. The presence of virulence genes is thought to increase the ability of these cells to survive and reproduce in those habitats, compared with strains that are not virulent. To use the terms introduced in Chapter 1, virulence genes are thought to increase the fitness of cells in certain types of habitats.

Virulence genes are just one case in which genome sequencing has linked specific genes and specific adaptations. To drive this point home, consider the following examples:

- *Anopheles gambiae* is the mosquito species that transmits the malaria parasite to humans. Female *Anopheles* mosquitoes must obtain a blood meal prior to laying eggs. Thus, it was logical to find that their genome contains 58 genes for fibrinogen-like proteins. In humans, *fibrinogen* forms clots that reduce blood loss from a wound. In contrast, the fibrinogen-like molecules in mosquitoes are thought to prevent blood from clotting when females are sucking blood. This hypothesis is supported by the observation that fruit flies have just 13 genes for fibrinogen-like proteins. Fruit flies are also insects, but they eat rotting fruit instead of blood.
- The single-celled eukaryotes that parasitize vertebrates and cause malaria are in the genus *Plasmodium*. During a *Plasmodium* infection, the host's immune system cells interact with proteins on the surface of the parasitic cells. If host cells recognize *Plasmodium* as foreign, other immune system cells are called in to destroy the parasites. But when the *Plasmodium* genome was sequenced, researchers found that it has 600 to 1000 distinct copies of genes that code for cell-surface proteins—many more than observed in other species. The *Plasmodium* genes are similar in sequence, clustered in the same chromosomal region, and repeated in tandem, suggesting that they are part of the same gene family. To explain this observation, biologists suggest that natural selection has favored the evolution of many cell-surface proteins in *Plasmodium*. The hypothesis is that parasitic cells with many different cell-surface proteins are difficult for the immune system to detect and destroy.
- *Prochlorococcus* is a very abundant photosynthetic bacterium in the oceans. The genomes of three strains have been sequenced. One of the strains is adapted to high light levels and lives near the surface. The other two strains are adapted to moderate or very low light levels and live in deeper water. Researchers who compared the number of genes that code for chlorophyll-binding proteins in the antenna complexes of photosystems I and II (see Chapter 10) found a strong correlation with light availability. The high-light strain contains just one of these genes. The moderate-light strain has two genes for these proteins; the low-light strain has eight. Apparently, natural selection has favored the evolution of particularly large and complex antenna systems in organisms that live in low-light habitats.

Gene Number and Alternative Splicing

Of all observations about the nature of eukaryotic genomes, perhaps the most striking is that particularly complex organisms do not appear to have particularly large numbers of genes. **Table 20.1** indicates the estimated number of genes

TABLE 20.1 Number of Genes in Selected Genomes

Species	Description	Genome Size (Millions of Base Pairs)	Estimated Number of Genes
Mycoplasma genitalium	Parasitic bacterium; causes urogenital-tract infections in humans	0.58	517
Haemophilus influenzae	Bacterium; common resident of human upper respiratory tract; can cause earache and meningitis	1.83	1743
Thermotoga maritima	Bacterium that lives in extremely hot environments	1.89	1877
Vibrio cholerae	Bacterium that lives in saltwater marshes and coastal environments; causes cholera	4.0	3885
Escherichia coli (laboratory strain)	Bacterium that is an important model organism in biochemistry and genetics	4.6	4288
Saccharomyces cerevisiae	Baker's and brewer's yeast; a unicellular fungus; an important model organism in biochemistry and genetics	12	6000
Plasmodium falciparum	Single-celled, parasitic eukaryote; causes malaria in humans	30	6500
Drosophila melanogaster	Fruit fly; an important model organism in genetics and developmental biology	180	13,600
Caenorhabditis elegans	A roundworm; an important model organism in developmental biology	97	19,000
Mus musculus	House mouse; an important model organism in genetics and developmental biology	2500	~30,000
Arabidopsis thaliana	A mustard plant; an important model organism in genetics and developmental biology	119	26,000
Homo sapiens	Humans	3000	~25,000

found in selected genomes. Note that the total number of genes in *Homo sapiens*, which is considered a particularly complex organism, is not that much higher than the total number of genes in fruit flies and roundworms. In fact, gene number in humans is about the same as in mice and in the weedy mustard plant *Arabidopsis*. Before the human genome was sequenced, many biologists expected that humans would have at least 100,000 genes. We now know that we have only 20,000–25,000.

How can this be? Recall that, in prokaryotes, there appears to be a rough correlation between genome size, gene number, a cell's metabolic capabilities, and the cell's ability to live in a variety of habitats. Similarly, it is logical to observe that plants have exceptionally large numbers of genes because they synthesize so many different and complex molecules from just carbon dioxide, nitrate ions, phosphate ions, and other simple nutrients. The idea is that large numbers of genes enable plants to produce large numbers of enzymes. Why isn't there a stronger correlation between gene number and morphological and behavioral complexity in animals?

The leading hypothesis focuses on alternative splicing. Recall from Chapter 18 that the exons of a particular gene can be spliced in ways that produce distinct mature mRNAs. As a result, a single eukaryotic gene can code for multiple transcripts and thus multiple proteins. The alternative-splicing hypothesis claims that at least certain multicellular eukaryotes do not need enormous numbers of distinct genes. Instead, alternative splicing creates different proteins from the same gene. The alternative forms might be produced at different developmental stages or in response to different environmental conditions.

In support of the alternative-splicing hypothesis, researchers have analyzed the mRNAs produced by human genes and have estimated that each gene produces an average of slightly more than three distinct transcripts. If this result is valid for the rest of the genome, the actual number of different proteins that can be produced is more than triple the gene number. Humans have only 20,000–25,000 genes, but these genes may have the ability to produce at least 100,000 different transcripts.

Based on such results, gaining a better understanding of how alternative splicing is regulated has become an urgent research priority. Gene regulation has, in fact, been an increasingly important theme in comparisons of closely related species.

Sequence Similarity and the Importance of Regulatory Sequences

Comparing the numbers of genes found in humans and in mice created a paradox that may be resolved by the alternative-splicing hypothesis. Comparing the base sequences of genes in humans and in other species has created an analogous paradox.

Here is the issue: At the level of base sequences, human beings and chimpanzees are 98.8 percent identical on average. If humans and chimps are so similar genetically, why do they appear to be so different in their morphology and behavior?

The leading hypothesis to resolve this paradox focuses on the importance of regulatory genes and regulatory sequences. Recall from Chapter 18 that a **regulatory sequence** is a section of DNA involved in controlling the activity of other genes; it may be a promoter, a promoter-proximal element, an enhancer, or a silencer. The term **structural gene**, in contrast, refers to a sequence that codes for a tRNA, rRNA, protein, or other type of product.

To resolve the sequence-similarity paradox, biologists propose that even though many structural genes in closely related species, such as humans and chimps, are identical or nearly identical, regulatory sequences and genes that code for regulatory transcription factors might have important differences between the two species. Suppose that the structural gene for human growth hormone and chimp growth hormone are identical in base sequence. But if subtle differences in transcription factors or regulatory sequences change the pattern of expression of that gene—perhaps turning it on later and longer in humans than in chimps—then height and other characteristics will change even though the structural gene is the same. Based on current analyses, biologists suggest that the human genome contains about 3000 different regulatory transcription factors. Subtle mutations in these proteins could have a significant effect on gene expression and thus on the phenotype.

The regulatory hypothesis is logical, but it remains to be tested rigorously. To date, there are no specific examples of changes in regulatory sequences that are responsible for phenotypic differences in humans and chimps or other closely related species.

20.5 Future Prospects

To explain the impact of genomics on the future of biological science, Eric Lander has compared the sequencing of the human genome to the establishment of the periodic table of the elements in chemistry. Once the periodic table was established and validated, chemists focused on understanding how the elements combine to form molecules. Similarly, biologists now want to understand how the elements of the human genome combine to produce an individual.

In essence, a genome sequence is a parts list. Once that list is assembled, researchers delve deeper to understand how genes interact to produce an organism. Let's explore some of the ways in which researchers use whole-genome data to answer fundamental questions about how organisms work.

Functional Genomics and Proteomics

Biologists who are interested in the research field called **functional genomics** look at the lists of protein-coding regions in a genome and ask, How do all these gene products interact? After all, the products of genes do not exist in a vacuum. Instead, groups of proteins act together to respond to environmental challenges such as extreme heat or drought. Similarly, distinct groups of genes are transcribed at different stages as a multicellular eukaryote grows and develops.

Whole-genome sequencing has inspired two important new approaches to studying how genes interact. One method, using DNA microarrays, focuses on analyzing changes in gene expression; the other, proteomics, seeks to understand protein-protein interactions. Let's take a closer look.

DNA Microarrays DNA-based microarrays are beginning to revolutionize how researchers analyze changes in gene activity. A **DNA microarray** consists of a large number of single-stranded DNAs permanently affixed to a glass slide. By probing DNA microarrays with the mRNAs from a particular cell, researchers can determine which genes are being expressed. Using microarrays is exciting, because researchers no longer have to study changes in the expression of genes one at a time. Instead, they can study the expression of thousands of genes at a time and identify which sets of genes are expressed in concert.

As **Figure 20.10** shows, the thousands of different short, single-stranded DNA sequences that make up a microarray are affixed to a glass plate in tiny but separate spots. In many instances, the DNAs on the plate represent short stretches from each exon in a genome. If so, then each spot in the microarray represents a different gene or exon. These types of microarrays give researchers a plate that contains segments from virtually every gene in an organism.

Once such a plate has been manufactured, a variety of experiments are possible. For example, mRNAs can be isolated at different stages in development and converted to cDNAs with reverse transcriptase. The resulting cDNAs are then used as probes. The sequences are labeled with a fluorescent marker, made single stranded, and hybridized with the microarray. Because the labeled cDNAs bind to the complementary sequences on the plate, the results give an accurate picture of which genes are being expressed at a particular stage of development. In this way, researchers can establish which genes are transcribed in response to heat, to the presence of an antibiotic, or to a viral infection.

The sequences in a microarray are attached to the slide via covalent bonds, so they remain affixed after the probe cDNAs are washed off. As a result, the microarray slide can be reused and re-probed—perhaps with cDNAs from a later stage in development or during an attack by a different virus. In this way, researchers collect data on which genes are expressed at different intervals as the organism matures or responds to different types of environmental conditions.

Proteomics **Proteomics** is the large-scale study of protein function. Now that biologists have a catalog of all protein-coding genes in an organism, researchers are working out new techniques for examining protein-protein interactions on a massive scale. One approach is similar to the use of DNA microarrays, except that large numbers of proteins, rather than DNA sequences, are affixed on a glass plate. This microarray of proteins is then treated with an assortment of proteins produced by the same organism. These proteins are labeled with a fluorescent or

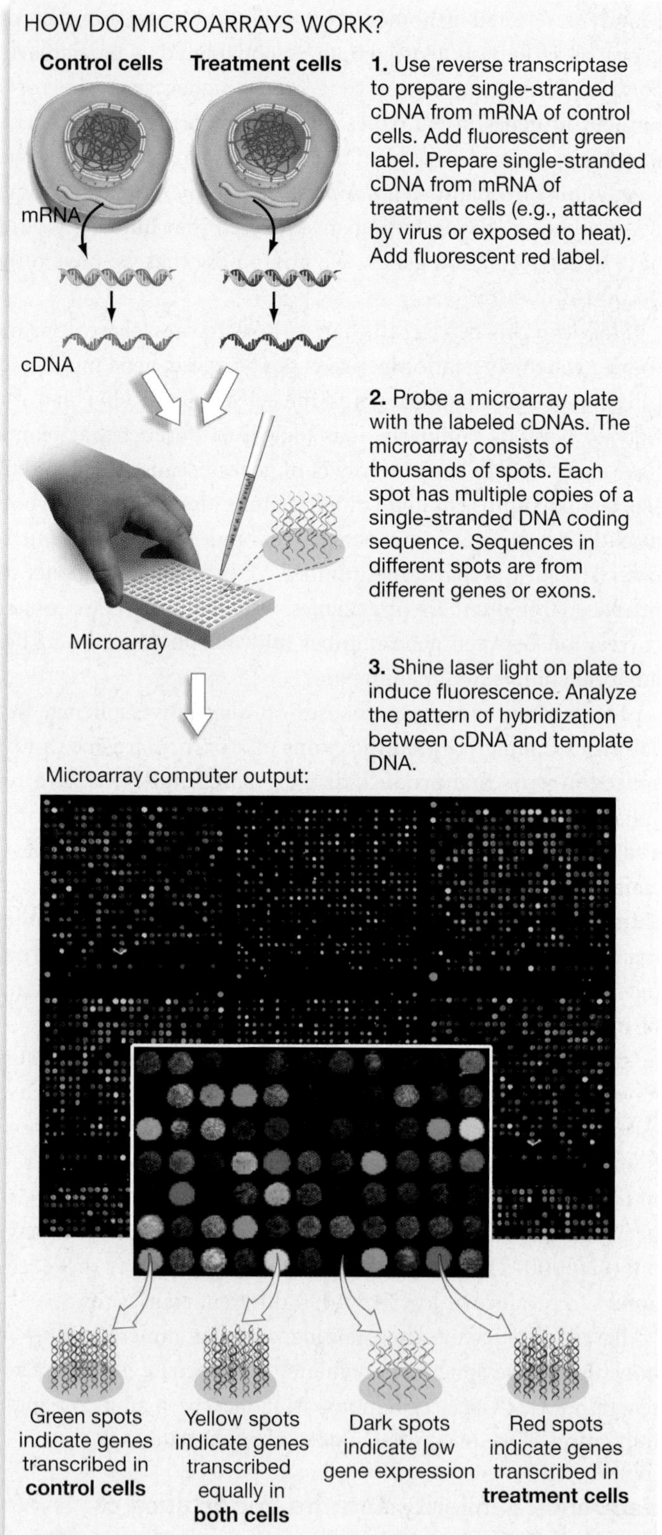
HOW DO MICROARRAYS WORK?
Control cells
Treatment cells
1. Use reverse transcriptase to prepare single-stranded cDNA from mRNA of control cells. Add fluorescent green label. Prepare single-stranded cDNA from mRNA of treatment cells (e.g., attacked by virus or exposed to heat). Add fluorescent red label.
mRNA
cDNA
2. Probe a microarray plate with the labeled cDNAs. The microarray consists of thousands of spots. Each spot has multiple copies of a single-stranded DNA coding sequence. Sequences in different spots are from different genes or exons.
Microarray
3. Shine laser light on plate to induce fluorescence. Analyze the pattern of hybridization between cDNA and template DNA.
Microarray computer output:
Green spots indicate genes transcribed in **control cells**
Yellow spots indicate genes transcribed equally in **both cells**
Dark spots indicate low gene expression
Red spots indicate genes transcribed in **treatment cells**

FIGURE 20.10 DNA Microarrays Represent Every Gene in a Genome
To create a DNA microarray, investigators spot thousands of short, single-stranded DNA sequences from coding sequences onto a glass plate. By probing this microarray with labeled cDNAs synthesized from mRNAs, researchers can identify which coding sequences are being transcribed. Here mRNAs from cells treated with a hormone are red, while mRNAs from untreated (control) cells are green.

radioactive tag. If any labeled proteins bind to the proteins in the microarray, the two molecules may also interact in the cell.

Medical Implications

With the advent of microarray technology, the "periodic tables" provided by genome projects are having an important impact on research into gene expression and protein-protein interactions. But the governments and corporations that fund genome projects have undertaken the expense primarily because of the potential benefits to biomedical research. In this respect, is genomics living up to its promise?

Even though genome data became available only recently, the prospect of improving human health and welfare is clear. To support this statement, consider how whole-genome analyses are being used in drug and vaccine development.

Identifying Potential Drug Targets Many drugs act by disabling a particular enzyme. Because genome sequencing provides a comprehensive catalog of the proteins produced by an organism, it offers a comprehensive list of targets for drug therapy. In addition, comparative genomics can identify which drug targets are found in the pathogen but not in the host. If an enzyme is unique to a disease-causing organism, taking a drug that disables it is less likely to damage nontarget species or the host.

For example, malaria researchers have now isolated and characterized all of the proteins produced by the malaria parasite *Plasmodium* at four stages of its life cycle. The genes that code for these proteins have been identified, and many of the proteins are now slated for drug testing. New drug therapies against malaria are needed urgently, because most *Plasmodium* populations have evolved resistance to the array of drugs currently available.

Designing Vaccines Although efforts to exploit genome data in drug design are still in their infancy, genomics has already inspired important advances in vaccine development. To illustrate how this work is proceeding, consider recent research on the bacterium *Neisseria meningitidis*. This species is a major cause of meningitis and blood infections in children and was one of the first prokaryotes to have its genome sequenced. Although antibiotics can treat *N. meningitidis* infections effectively, the organism grows so quickly that it often injures or even kills the victim before a diagnosis can be made and drugs administered. As a result, biomedical researchers have been interested in developing a vaccine that would prime the immune system and allow children to ward off infections.

Vaccine development has been difficult in this case, however. The immune system usually responds to molecules on the outer surface of bacteria or viruses, and *N. meningitidis* is covered with a polysaccharide that is identical to a compound found on the surface of brain cells. Immune system cells normally do not attack compounds found on the body's own cells, so a vaccine composed of the *N. meningitidis* polysaccharide would elicit no response.

To circumvent this problem, biologists analyzed the genome sequence of *N. meningitidis* and tested 600 open reading frames for the ability to encode vaccine components. The researchers inserted the 600 DNA sequences into *E. coli* cells, following the steps shown in **Figure 20.11**. Later they succeeded in isolating 350 different *N. meningitidis* proteins from the transformed cells. Then the biologists injected these proteins into

Question: Could vaccines be developed from the products of newly discovered genes?

Hypothesis: Some of the genes discovered through genome sequencing code for proteins that can be used in vaccines.

Null hypothesis: None of the genes discovered through genome sequencing code for proteins that can be used in vaccines.

Experimental setup:

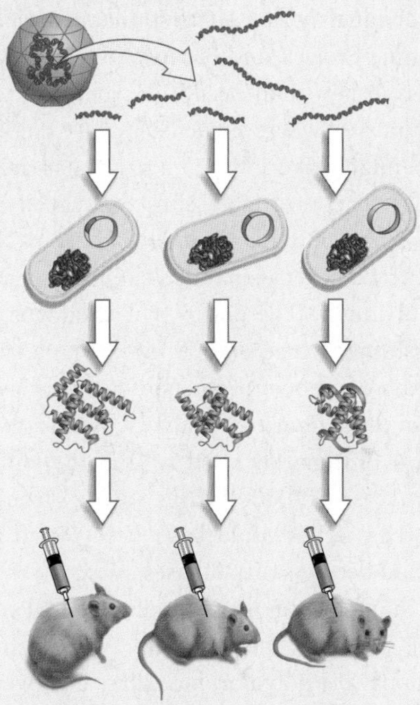
1. Isolate open reading frames (ORFs) from pathogen genome sequence.
2. Introduce ORFs into *E. coli* cells.
3. Isolate proteins that result from transcription and translation.
4. Inject proteins into mice.

Prediction: Some proteins will elicit an immune response similar to that elicited by vaccines.

Prediction of null hypothesis: No proteins will elicit an immune response similar to that elicited by vaccines.

Results:

	Number of mice with immune response:	Number of mice with no immune response:
Strong immune responses	7	343

Conclusion: The seven proteins that elicited an immune response are potential vaccine components. Further research is needed to test their safety and effectiveness.

FIGURE 20.11 Newly Discovered Proteins Can Be Tested for Vaccine Development
Because all potential genes are identified after whole-genome sequencing, virtually all proteins produced by a pathogen can be tested for their ability to provoke an immune response and act as a vaccine.

mice and analyzed whether an immune response occurred. Their results show that seven of the proteins tested evoked a strong immune response and represent promising vaccine components. Follow-up work is now under way to determine whether one or more of these proteins could act as a safe and effective vaccine in humans.

The general message of these case histories is that genome sequencing data are providing biomedical researchers with new drug targets and candidate vaccine components. Although early results are promising, only time will tell if genomics ushers in what some have claimed will be an entirely new era of medical research and practice.

ESSAY Genomics and Genetic Discrimination

Chapter 19 explained how gene hunters analyzed the genetic markers called *restriction fragment length polymorphisms* (RFLPs) to find the allele responsible for Huntington's disease. Data from the Human Genome Project have made this approach to finding disease genes even more powerful. Because DNA from many individuals was used as source material during the Human Genome Project, and because overlapping segments of genes were routinely sequenced, researchers were able to identify 1.42 million sites where single bases vary among individuals. Where you might have a "C" at a particular site, others may have a "T." These variable sites are called *single nucleotide polymorphisms*, or SNPs (pronounced "snips"), and they can serve as genetic markers—mapped sites in the genome that vary among individuals. The group that cataloged these polymorphic sites estimates that 60,000 of them lie within exons and that at least one polymorphic site lies within 5000 bases of every exon in the genome. To realize why this is significant, recall that the RFLP marker used to find the Huntington's disease gene was 500,000 bases away from the Huntington's locus. That gene would have been found much faster if the marker had been just 5000 bases away.

For disease gene hunters, this new catalog of SNPs is an enormously powerful resource. The possibility of analyzing the inheritance of millions of polymorphic sites all over the genome—instead of just a few hundred widely scattered ones such as recognition sequences for restriction endonucleases—makes it much more likely that researchers can locate genes associated with illnesses such as Alzheimer's disease, certain cancers, and inherited forms of mental illness. The database promises to be particularly important in understanding the genetic basis of diseases that involve many different loci. As efforts to annotate the human genome continue, it is very likely that biologists will be able to track down the genes responsible for many or even most inherited diseases.

Suppose that this prediction is correct, and researchers succeed in identifying alleles that make people susceptible to disorders such as heart attack, high blood pressure, diabetes, certain cancers, and mental illness. Now suppose that, in the near future, technology advances to the point at which you will be able to walk into a clinic, request that your DNA be sequenced, and then receive a report detailing the disease-related alleles that you carry. Would you want to know which of your alleles make you susceptible to certain diseases? If you had a family history of the cancer called HNPCC (see Chapter 14), the answer might be yes. The allele associated with this cancer has been identified, so people can learn whether they have a genetic predisposition to the disease. If so, they can change their diets to make tumor development less likely, and they can be screened regularly for the onset of the disorder. But suppose that health insurance companies, life insurance companies, your friends, and even prospective spouses or employers had access to these data. Would you want them to know?

Would you want to know which of your alleles make you susceptible to certain diseases?

Early in the debate about whether government agencies should support the Human Genome Project, questions like these were discussed extensively. As part of the funding proposals that allowed the project to move forward, money was set aside to study the ethical and legal implications of the project. During the rush to sequence the human genome, however, these questions received somewhat less attention than they had earlier. Now that the genome is complete, the discussion over the ethical and legal ramifications of the data is starting anew.

Although actual cases of genetic discrimination have been extremely rare, most U.S. states have passed laws prohibiting the use of genetic information in evaluating applications for health insurance. As this book goes to press, the U.S. Congress is considering legislation that would ban discrimination on the basis of genotype when individuals apply for insurance or employment. But the efforts to understand the social and ethical ramifications of genome sequencing and to regulate the use of the data are still in their infancy. It will be important for you, as a biology student, to help educate the general public and contribute an informed opinion as issues arise.

CHAPTER REVIEW

Summary of Key Concepts

Once a genome has been completely sequenced, researchers use a variety of techniques to identify which sequences code for products and which act as regulatory sites.

Recent technical advances have allowed investigators to sequence DNA much more rapidly and cheaply than before, resulting in a flood of genome data. Researchers annotate genome sequences by finding genes and determining their function. To identify genes in bacteria and archaea, researchers use computers to scan the genome for start and stop codons that are in the same reading frame and that are separated by gene-sized stretches of sequence. Finding such open reading frames (ORFs) is difficult in eukaryotes, because exons are interrupted by introns and because most eukaryotic DNA does not code for a product. One approach to finding eukaryotic genes is to analyze the sequences of complementary DNAs (cDNAs) synthesized from mRNAs and then match these sequences to DNA found in the genome itself. Sequences that are highly conserved among species are also hypothesized to indicate the locations of genes.

Web Tutorial 20.1 Human Genome Sequencing Strategies

In bacteria and archaea, there is a positive correlation between the number of genes in a species and the species' metabolic capabilities.

Species of bacteria and archaea are usually targeted for whole-genome sequencing because they cause disease or have interesting metabolic abilities. In these groups, a general correlation seems to exist between the size of an organism's genome and its morphological complexity or biochemical capabilities. Parasites tend to have small genomes; organisms that live in a broad array of habitats or that use a wide variety of nutrients tend to have larger genomes. Many of the genes identified in bacteria and archaea still have no known function, however, and a significant percentage of them are extremely similar to other genes in the same genome. Another generalization about prokaryotic genomes is that genes are frequently transferred laterally, or between species. Lateral gene transfer appears to be common in genes responsible for causing disease.

In eukaryotes, genomes are dominated by parasitic sequences that do not contribute to the fitness of the organism.

Compared with prokaryotic genomes, eukaryotic genomes are large and contain a high percentage of transposable elements, repeated sequences, and other noncoding sequences. There is no obvious correlation between morphological complexity and gene number in eukaryotes, although the number of distinct transcripts produced may be much larger than the actual gene number in certain species as a result of alternative splicing. By analyzing genes that belong to closely related gene families, researchers have confirmed that gene duplication and polyploidy have been important sources of new genes in eukaryotes.

Data from genome sequencing projects are now being used in the development of new drugs and vaccines.

The availability of whole-genome sequences is inspiring new research programs. Biologists are affixing small amounts of proteins or cDNAs to microarrays in order to study protein-protein interactions or changes in gene expression and in mRNA populations over time. In addition, the availability of whole-genome data has allowed investigators to find novel genes that are closely related to existing drug targets and to find new proteins that may serve as vaccine candidates.

Questions

Content Review

1. What is an open reading frame?
   a. a gene whose function is already known
   b. a DNA section that is thought to code for a protein because it is similar to a complementary DNA (cDNA)
   c. a DNA section that is thought to code for a protein because it has a start codon and a stop codon flanking hundreds of base pairs
   d. any member of a gene family
2. What best describes the logic behind shotgun sequencing?
   a. Break the genome into tiny pieces. Sequence each piece. Use overlapping ends to assemble the pieces in the correct order.
   b. Start with one end of each chromosome. Sequence straight through to the other end of the chromosome.
   c. Use a variety of techniques to identify genes and ORFs. Sequence these segments—not the noncoding and repeated sequences.
   d. Break the genome into pieces. Map the location of each piece. Then sequence each piece.
3. What are minisatellites and microsatellites?
   a. small, extrachromosomal loops of DNA that are similar to plasmids
   b. parts of viruses that have become integrated into the genome of an organism
   c. incomplete or "dead" remains of transposable elements in a host cell
   d. short and simple repeated sequences in DNA
4. What is the leading hypothesis to explain the paradox that large, complex eukaryotes such as humans have relatively small numbers of genes?
   a. lateral transfer of genes from other species
   b. alternative splicing of mRNAs
   c. polyploidy, or the doubling of the genome's entire chromosome complement
   d. expansion of gene families through gene duplication

5. What evidence do biologists use to infer that a gene is part of a gene family?
   a. Its sequence is exactly identical to that of another gene.
   b. Its structure—meaning its pattern of exons and introns—is identical to that of a gene found in another species.
   c. Its composition, in terms of percentage of A-T and G-C pairs, is unique.
   d. Its sequence, structure, and composition are similar to those of another gene in the same genome.

6. What is a pseudogene?
   a. a coding sequence that originated in a lateral gene transfer
   b. a gene whose function has not yet been established
   c. a polymorphic gene—meaning that more than one allele is present in a population
   d. a gene whose sequence is similar to that of functioning genes but does not produce a functioning product

Conceptual Review

1. Explain how open reading frames are identified in the genomes of bacteria and archaea. Why is it more difficult to find open reading frames in eukaryotes?
2. Why is the observation that parasitic organisms tend to have relatively small genomes logical?
3. Review how a LINE sequence transmits a copy of itself to a new location in the genome. Why are LINEs and other repeated sequences referred to as "genomic parasites"?
4. How does DNA fingerprinting work? Stated another way, how does variation in the size of microsatellite and minisatellite loci allow investigators to identify individuals?
5. Researchers can create microarrays of short, single-stranded DNAs that represent many or all of the exons in a genome. Explain how these microarrays are used to document changes in the transcription of genes over time or in response to environmental challenges.
6. Explain the concept of homology and how identifying homologous genes helps researchers identify the function of unknown genes. Are duplicated sequences that form gene families homologous? Explain.

Group Discussion Problems

1. Parasites lack genes for many of the enzymes found in their hosts. Most parasites, however, have evolved from free-living ancestors that had large genomes. Based on these observations, W. Ford Doolittle claims that the loss of genes in parasites represents an evolutionary trend. He summarizes his hypothesis with the quip "use it or lose it." What does he mean?

2. According to eyewitness accounts, communist revolutionaries executed Nicholas II, the last czar of Russia, along with his wife and three children, the family physician, and several servants. Many decades after this event, a grave purporting to hold the remains of the royal family was identified. Biologists were asked to analyze DNA from each adult and juvenile skeleton and determine whether the bodies were indeed those of several young siblings, two parents, and several unrelated adults. If the grave was authentic, describe how similar the DNA fingerprints of each skeleton would be relative to the fingerprints of other individuals in the grave.

3. The human genome contains a gene that encodes a protein called syncytin. This gene is expressed in placental cells during pregnancy. The syncytin gene is nearly identical in DNA sequence to a gene in a virus that infects humans. In this virus, the syncytin-like gene codes for a protein found in the virus's outer envelope. State a hypothesis to explain the similarity between the two genes.

4. A recent study used microarrays to compare the patterns of expression of genes that are active in the brain, liver, and blood of chimpanzees and humans. Although the overall patterns of gene expression were similar in the liver and blood of the two species, expression patterns were strikingly different in the brain. How does this study relate to the hypothesis that most differences between humans and chimps involve changes in gene regulation?

Answers to Multiple-Choice Questions 1. c; 2. a; 3. d; 4. b; 5. d; 6. d

www.prenhall.com/freeman is your resource for the following: Web Tutorials; Online Quizzes and other Online Study Guide materials; Answers to Conceptual Review Questions; Solutions to Group Discussion Problems; Answers to Figure Caption Questions and Exercises; and Additional Readings and Research.

UNIT 4

Developmental Biology

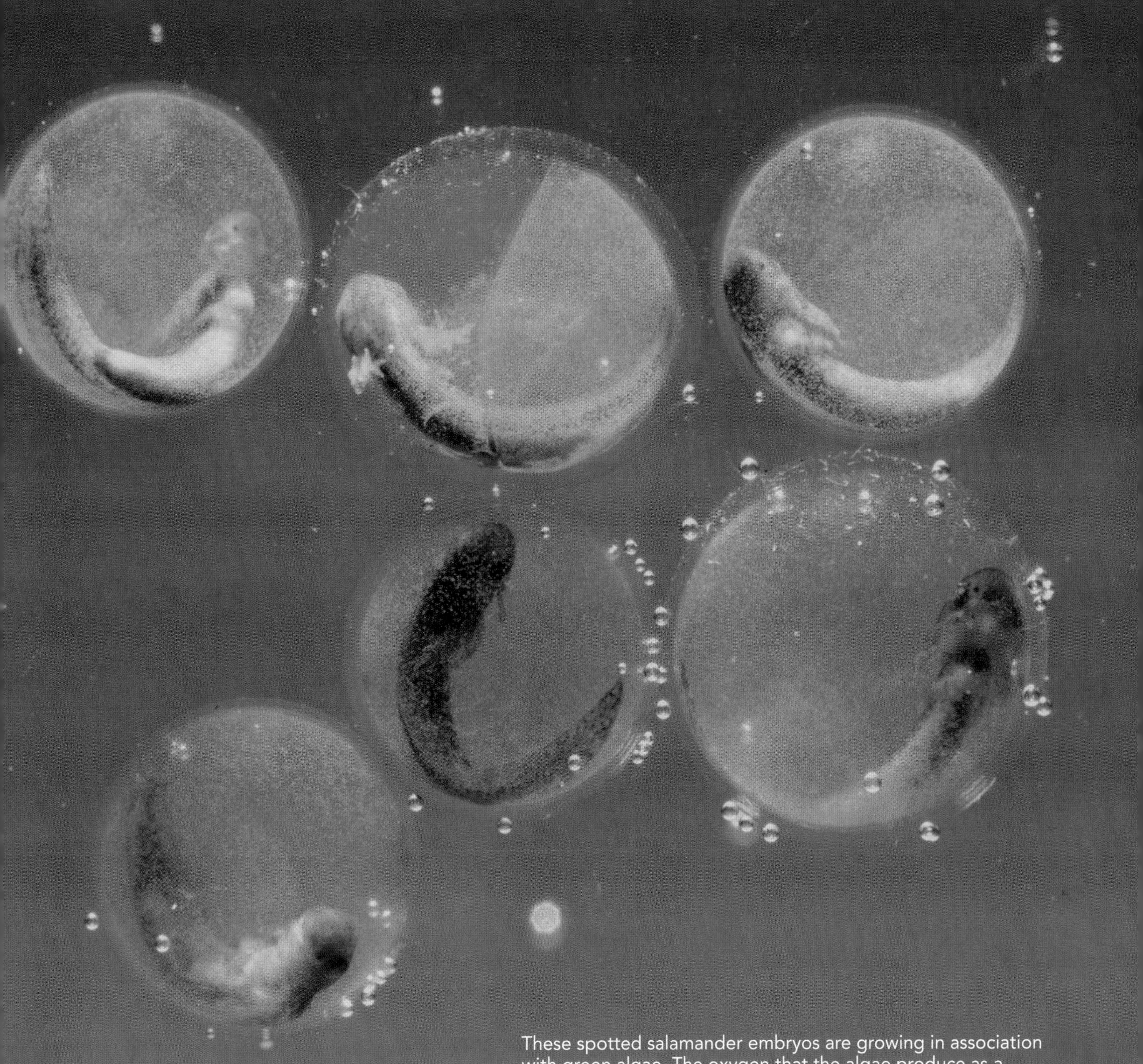

These spotted salamander embryos are growing in association with green algae. The oxygen that the algae produce as a by-product of photosynthesis is used by the embryos in cellular respiration.

21 Early Development

KEY CONCEPTS

- In most animals, fertilization depends on specific interactions among proteins on the plasma membranes of sperm and egg.
- In animals, the earliest cell divisions divide the fertilized egg into a mass of cells. These cells are not all alike; eventually they give rise to different types of cells and tissues as development proceeds.
- In animals as well as plants, the early events in development result in the production of three embryonic tissues and the establishment of the major body axes.

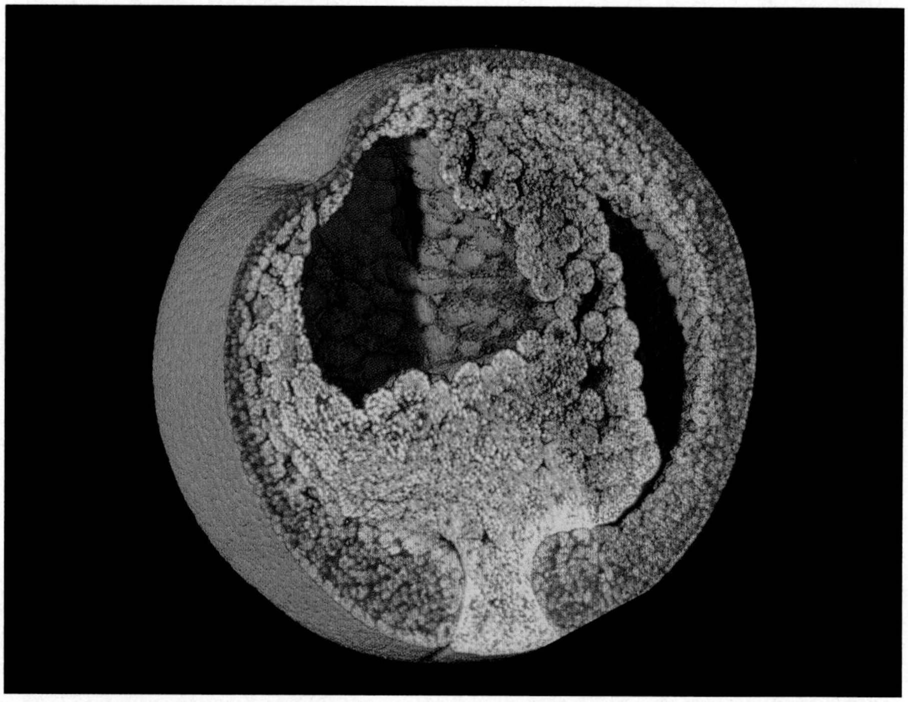

Scanning electron micrograph of a sea urchin embryo, cut in half. As development continues, the cells shown will form the individual's gut, muscle, and skin.

In 1859 Charles Darwin referred to the origin of species as "the mystery of mysteries." At that time the most urgent task confronting biologists was to explain how species come to be. Today, however, Darwin's theory of evolution by natural selection explains many of the most fundamental questions about how species form and how they change through time. What question qualifies as the current mystery of mysteries in biological science? Although there are many candidates, one of the most compelling is the question of how a multicellular individual develops from a single cell—the fertilized egg.

Biologists in the 1800s first began to study how plants and animals develop. Early investigators sought to catalog the events that occur as an individual develops from a fertilized egg to its juvenile and then adult forms. This early work was observational and descriptive in nature and was carried out by means of microscopes.

Based on these studies, development appeared to be highly variable among organisms. In humans, a fertilized egg develops into a newborn child composed of hundreds of cell types and trillions of cells. This process takes nine months. Rapid growth and development continue for another 15 to 18 years, until individuals become adults. In contrast, a fertilized egg from a burr oak tree takes about four months to grow into an **embryo**, or newly developing individual, that is encased in an acorn. After spending the winter in a dormant state, the embryo grows into a seedling. Over the next 250 to 300 years, the tree will continue to grow and develop new leaves, roots, and flowers. At the other extreme, a fertilized egg of the fruit fly *Drosophila melanogaster* transforms into a feeding larva in just a day. About five days later, the larva stops feeding and forms a pupa. In another four days, an adult emerges. The adult fly flies, feeds, courts, mates, and starts the cycle anew.

How did biologists start to make sense of all this variability in the rate and pattern of development? By the late nineteenth century, researchers had started to use experimental approaches to study embryos. Biologists focused on organisms that participate in external fertilization (outside the body) rather than internal fertilization (inside the body), so that embryos could be easily collected and observed. They also chose organisms that developed rapidly and produced large numbers of eggs that were easy to manipulate. Then they removed cells at various stages of development to see which were required for the formation of certain structures, injected dyes into cells to follow their fate through time, and analyzed mutant individuals that developed abnormally. After a century of poking and prodding embryos with these and other techniques, biologists have concluded that several unifying mechanisms and themes underlie the wonderful variety of plant and animal developmental programs.

The goal of this chapter is to explore how common genetic and cellular mechanisms unify the diversity of developmental patterns seen in different groups of plants and animals. We begin by examining *gametogenesis*, the process responsible for producing gametes. Recall from Chapter 12 that **gametes** are haploid reproductive cells; when they differ in size, they are called **sperm** (for males) and **eggs** (for females). As **Figure 21.1** shows, gametogenesis is followed by *fertilization*, the union of a sperm and an egg. In animals, a fertilized egg undergoes a series of rapid cell divisions called *cleavage*, forming a mass of cells known as a *blastula*. In the blastula, massive rearrangements of cells during *gastrulation* result in the formation of a *gastrula*. These processes are described in Sections 21.1 through 21.3. They set the stage for the development of specialized tissues and organs during a process called *organogenesis*. In Section 21.4 we'll survey the diversity of developmental schemes by comparing development in four representative organisms. Let's delve in.

21.1 Gametogenesis

The development of a new individual begins with **gametogenesis**, which is the formation of gametes—sperm and egg—in the reproductive organs of adult organisms. The DNA and cytoplasm in these reproductive cells are the initial components of the new individual. Both sperm and egg contribute an equal number of chromosomes to the offspring—usually a haploid genome containing one copy of each gene. But because egg cells are routinely hundreds or thousands of times larger than sperm cells, an egg contributes much more cytoplasm than does a sperm.

The sequence of mitotic and meiotic cell divisions that lead up to the production of sperm and eggs are discussed in Chapters 40 and 48. Here, let's focus on the structure of the mature reproductive cells. Understanding how these specialized cells are put together will lay the groundwork for addressing questions about fertilization and early development.

Sperm Structure and Function

A mammalian sperm cell has four main compartments: (1) a head, (2) a neck, (3) a midpiece, and (4) a tail. The head region contains the nucleus and an enzyme-filled structure called the **acrosome**; the neck encloses a centriole; the midpiece is packed

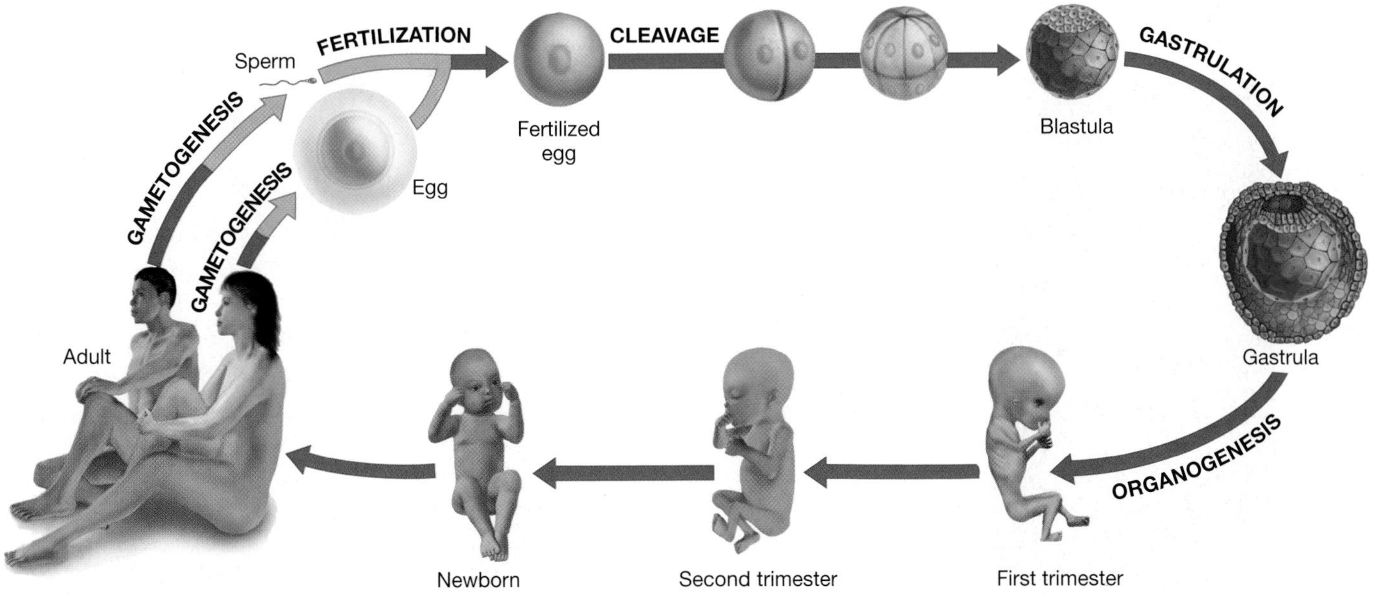

FIGURE 21.1 Development Proceeds in Ordered Phases
In animals, the development of a new individual starts with the formation of gametes and continues with fertilization and early cell divisions (cleavage), resulting in a blastula. Gastrulation then rearranges the blastula into a gastrula. The events through the formation of the gastrula are discussed in this chapter; Chapter 22 explores what happens next—organogenesis.

with mitochondria; and the tail region consists of a flagellum (see Chapter 7). **Figure 21.2a** shows the basic structure of animal sperm. A sperm cell is so highly specialized that it has been called "DNA with a propeller." The propeller is the flagellum, which is powered by ATP manufactured in the mitochondria.

Sperm cells develop after meiosis has resulted in the production of a haploid nucleus. As a sperm cell matures, its chromosomes are condensed into an even more densely packed arrangement than in other cells. The acrosome develops from the Golgi apparatus, and the flagellum arises from the centrioles. Once the sperm reaches an egg, the enzymes in the acrosome are responsible for digesting the outer coverings of the egg cell so that the two cells' plasma membranes can make contact and fuse. In most animal species, the sperm's nucleus, centriole, and mitochondria enter the egg after fusion occurs.

As **Figure 21.2b** shows, the sperm of flowering plants are even simpler in form than the sperm of animals. In most plants, sperm develop from **pollen grains**, which consist of several haploid cells derived from a product of meiosis. An insect or the wind carries a pollen grain to the top of a structure that contains an *ovule*, the female reproductive structure in which a plant's egg cell forms. One of the cells inside the pollen grain then divides by mitosis, producing two **sperm nuclei**. These nuclei move down the **pollen tube**, a structure that grows out of the pollen grain toward the egg cell. Once the pollen tube reaches the egg inside the ovule, the sperm nuclei bud off and move toward the egg. As we'll see in Section 21.2, both sperm cells will participate in fertilization.

Egg Structure and Function

Sperm are small and motile reproductive cells; eggs are large and nonmotile. Eggs are large mainly because they contain the nutrients required for the embryo's early development. Even among mammals, whose embryos start to obtain nutrition through a maternal organ called the *placenta* within a week or two after fertilization, the egg itself must supply the nutrients for early development. In species in which eggs are laid into the environment, stores in the egg are the *only* source of nutrients until organs have formed and a larva or juvenile hatches and begins to feed. In these species, the nutrients required for early development are provided by a **yolk**, the fat- and protein-rich cytoplasm that is loaded into egg cells as they mature. Yolk may be present as one large mass or as many small granules. Plants also provide their embryos with nutritive tissue, which the offspring use until they are large enough to initiate photosynthesis and feed themselves. In contrast to the nutrient-rich yolk of animals, though, these nutrients are not synthesized and stored until after fertilization has occurred. As we'll see in Section 21.2, a flowering plant embryo's nutrient supply is created during seed maturation—not during gametogenesis.

Yolk is not the only specialized material or structure found in animal eggs. Many eggs also contain organelles called **cortical granules**, which are small vesicles filled with enzymes that are involved in fertilization in egg-laying animals. As the egg matures, cortical granules are synthesized, transported to the cell surface, and bound to the inner surface of the plasma membrane. Just outside the plasma membrane, a fibrous, mat-like sheet of

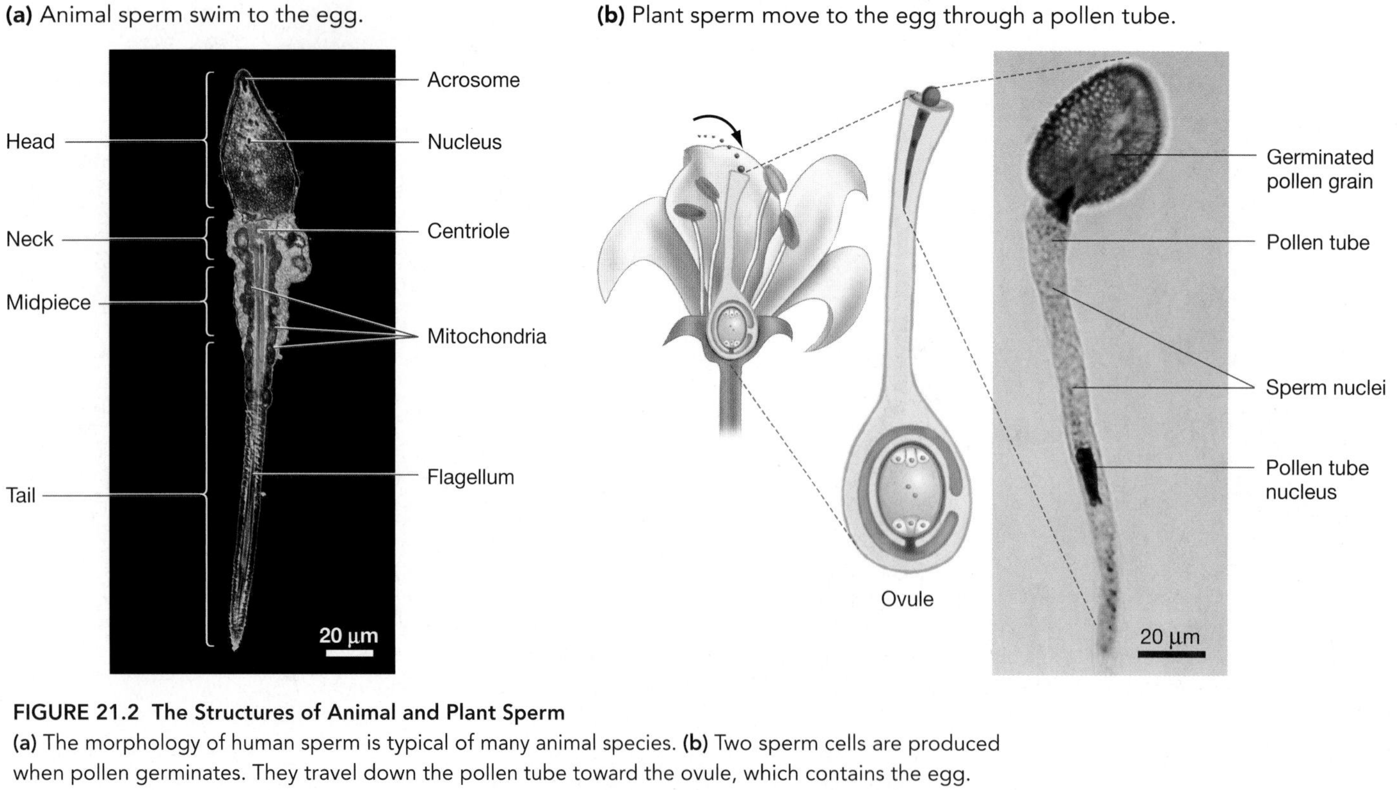

FIGURE 21.2 The Structures of Animal and Plant Sperm
(a) The morphology of human sperm is typical of many animal species. **(b)** Two sperm cells are produced when pollen germinates. They travel down the pollen tube toward the ovule, which contains the egg.

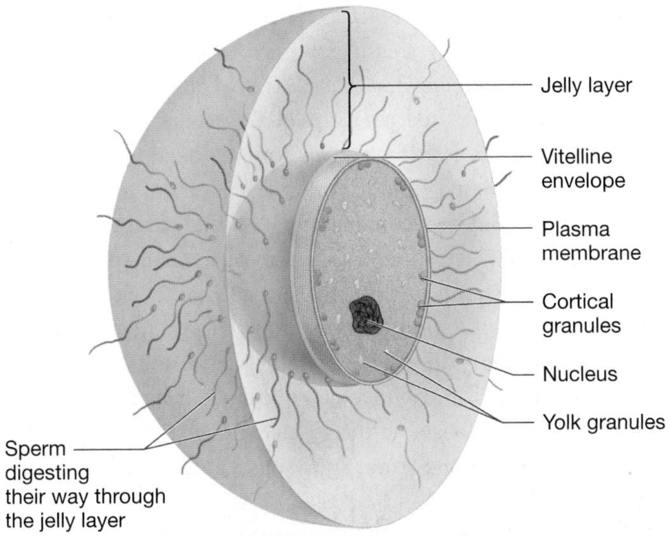
Jelly layer
Vitelline envelope
Plasma membrane
Cortical granules
Nucleus
Yolk granules
Sperm digesting their way through the jelly layer

FIGURE 21.3 Sea Urchin Eggs Are Covered by a Jelly Layer
Animals that lay their eggs in water frequently surround their eggs with a jelly layer, a gelatinous coat.

glycoproteins called the **vitelline envelope** forms and surrounds the egg (**Figure 21.3**). (In humans, the equivalent structure is the unusually thick *zona pellucida*.) In some species, a large gelatinous matrix known as a **jelly layer** also encloses the egg. How does a sperm cell penetrate these coatings to reach the egg's plasma membrane and fertilize the egg?

21.2 Fertilization

Fertilization seems like a simple process: A sperm cell fuses with an egg cell to form a diploid cell known as a **zygote** (a fertilized egg). Upon reflection, though, you should begin to appreciate that the process is extraordinarily complex. For fertilization to take place, sperm and egg cells must be in the same place at the same time and then must recognize and bind to each other. Next they must fuse—even though most of the other cells in the body do not fuse with cells they contact. In most species, fusion must also be limited to a single sperm so that the zygote does not receive extra chromosomes. Finally, the fusion of the two gametes has to trigger the onset of development. This complexity has made fertilization a fascinating research topic. The contact that takes place between sperm and egg at fertilization qualifies as the best studied of all cell-to-cell interactions.

Fertilization in Sea Urchins

Work on fertilization began in earnest early in the twentieth century, when biologists started to study the sperm-egg interaction in sea urchins. Sea urchins are marine invertebrates that perform fertilization externally (**Figure 21.4a**). To maximize the probability that sperm and egg will meet, male and female sea urchins secrete huge quantities of gametes. As a result, this organism provided researchers with large numbers of gametes that could be studied in a test tube or culture dish filled with seawater. Sea urchins continue to be an intensively studied model system in research on fertilization.

In sea urchins, the jelly layer that surrounds the egg contains a molecule that attracts sperm. The attractant is a small peptide that diffuses away from the egg and into the surrounding seawater. Sperm respond to the attractant by swimming toward areas where its concentration is high. The same type of interaction between sperm and eggs has been observed in a wide variety of animals that lay their eggs in seawater.

When a sea urchin sperm and egg meet, the head of the sperm initially encounters the jelly layer of the egg cell (**Figure 21.4b**). To reach the plasma membrane of the egg cell, the sperm head must

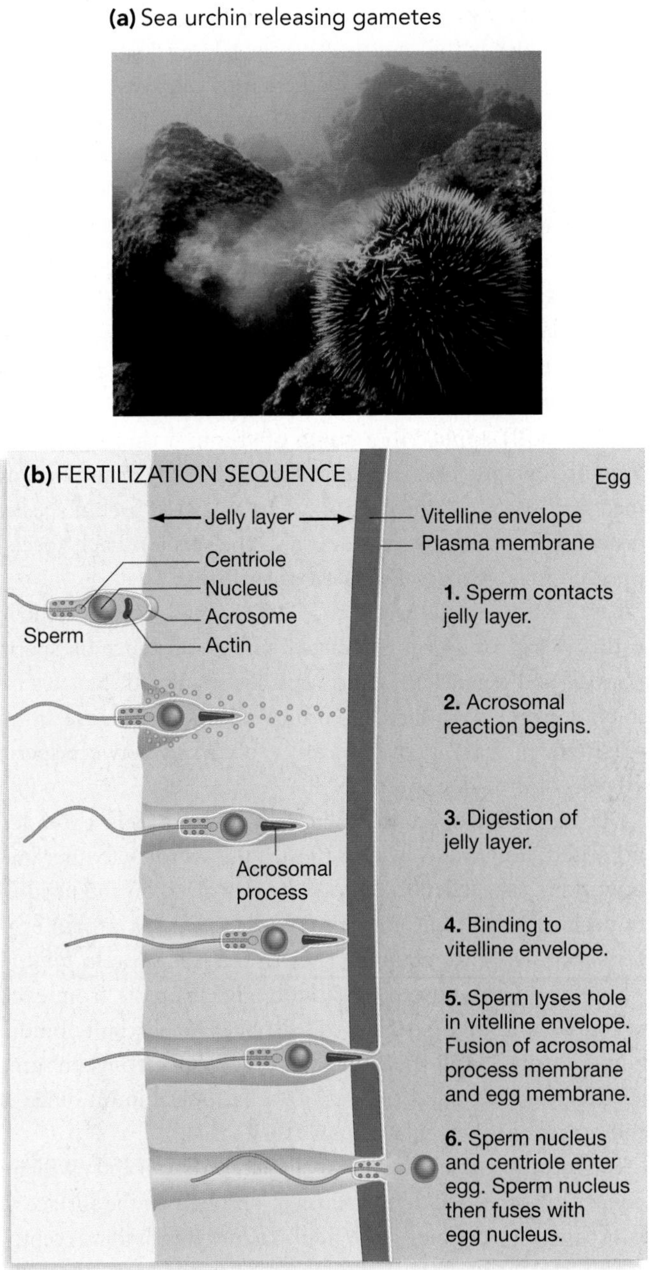
(a) Sea urchin releasing gametes
(b) FERTILIZATION SEQUENCE
Egg
Jelly layer
Vitelline envelope
Plasma membrane
Centriole
Nucleus
Acrosome
Actin
Sperm
1. Sperm contacts jelly layer.
2. Acrosomal reaction begins.
3. Digestion of jelly layer.
Acrosomal process
4. Binding to vitelline envelope.
5. Sperm lyses hole in vitelline envelope. Fusion of acrosomal process membrane and egg membrane.
6. Sperm nucleus and centriole enter egg. Sperm nucleus then fuses with egg nucleus.

FIGURE 21.4 Fertilization in Sea Urchins
(a) Sea urchins release their gametes into the ocean. **(b)** Fertilization involves a complex sequence of events.

digest its way through the egg's jelly layer and vitelline envelope. This process begins with an **acrosomal reaction**, a response that is triggered by contact between the sperm's head and the jelly layer. In the first part of this reaction, the contents of the acrosome—the structure at the tip of the sperm's head—are expelled. The enzymes that are released digest a pathway through the egg's jelly layer and allow the sperm to reach the vitelline envelope of the egg. The second part of the acrosomal reaction involves the polymerization of actin into microfilaments that form the **acrosomal process**, a protrusion that extends until it makes contact with the vitelline envelope. Finally, the plasma membranes of the egg and sperm fuse. The sperm nucleus, mitochondria, and centriole enter the egg, and the sperm and egg nuclei fuse to form the zygote nucleus. Fertilization is complete.

Frank Lillie was the first to recognize that an important question was hidden in this sequence of events: How do gametes from the same species recognize each other? After all, in many habitats sperm and eggs from a particular sea urchin species float in seawater along with eggs and sperm from other sea urchin species and many other organisms. What prevents cross-species fertilization and the production of dysfunctional hybrid offspring?

Species Recognition in Sea Urchins Lillie was the first researcher to identify a substance on the surface of egg cells that appeared to be involved in binding sperm. He called the compound *fertilizin* and showed that if the purified molecule were experimentally applied to a group of sperm, it caused the cells to clump together. He also showed that clumping occurred only when fertilizin from the eggs of a particular sea urchin species was combined with sperm from the same species. Each species appeared to have its own version of fertilizin.

Based on these observations, Lillie proposed that fertilizin on the surface of an egg interacted with a substance on sperm in a lock-and-key fashion. Further, Lillie suggested that this interaction was a product of natural selection, because it increased the probability of fertilization occurring between sperm and eggs of the same species.

Decades passed before Lillie's hypothesis was extended and tested, however. Then, in the 1970s, Victor Vacquier and co-workers succeeded in identifying a protein on the head of sea urchin sperm that binds to the surface of sea urchin eggs in a species-specific manner. They called this protein *bindin*. Follow-up work showed that the bindin proteins from even very closely related species are distinct. As a result, bindin should ensure that a sperm binds only to eggs from the same species. The next question was, To what does bindin bind? If bindin acts as a key, what acts as the lock?

To address these questions, Kathleen Foltz and William Lennarz hypothesized that fertilizin is a protein on the surface of sea urchin eggs that binds to bindin. To find this bindin-receptor protein, the researchers attempted to isolate the part of the receptor that is exposed on the outside of the egg. They predicted that this region of the protein interacts with the bindin on sperm.

Figure 21.5 illustrates Foltz and Lennarz's experimental approach. They began by treating the surface of sea urchin eggs with a **protease**—an enzyme that cleaves peptide bonds. When the investigators isolated the protein fragments that were released from the egg surface, they found one that bound to sperm and to isolated bindin molecules. Further, this binding occurred in a species-specific manner. A protein fragment from the eggs of one species bound to sperm of its own species, but did not bind to sperm of different species. Based on these observations, the biologists claimed that they had found the outward-facing portion of the egg-cell receptor for sperm.

These experiments provided convincing evidence that an egg receptor for sperm exists on the surface of sea urchin eggs. The work also provided important support for Lillie's lock-and-key

Question: If bindin is the protein on the surface of sperm that acts like a key, what is the "lock"?

Hypothesis: The lock is a protein on the surface of sea urchin eggs that binds to bindin.

Null hypothesis: The lock is not a protein on the surface of sea urchin eggs that binds to bindin.

Experimental setup:

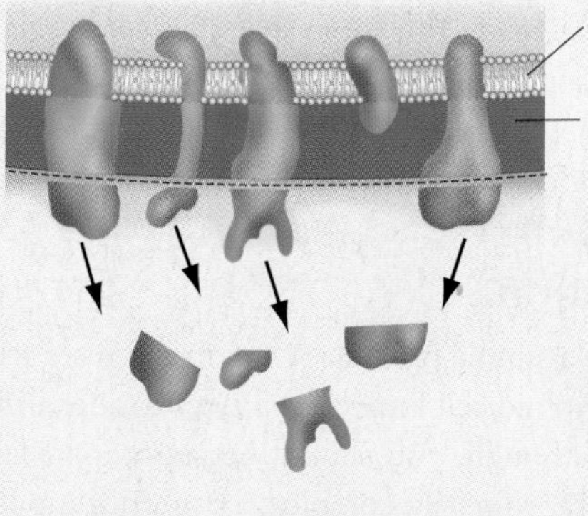
Egg cell's plasma membrane
Vitelline envelope
1. Use protease to release protein fragments from egg surface.
2. Isolate each type of protein fragment and see which ones bind to sperm and to isolated bindin molecules.

Prediction: One of the protein fragments isolated from egg surface will bind to sperm and to isolated bindin molecules.

Prediction of null hypothesis: The protein fragments isolated from egg surface will not bind to sperm or to isolated bindin molecules.

Results:

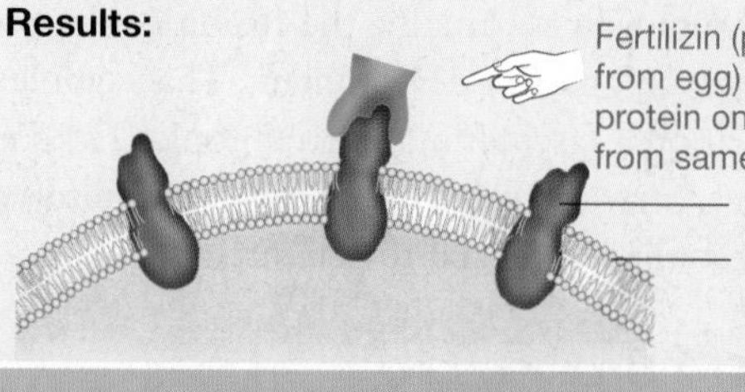
Fertilizin (protein fragment from egg) binds to bindin protein on sperm if sperm is from same species as egg
Bindin
Sperm cell's plasma membrane

Conclusion: Fertilizin is a "lock" protein on the surface of eggs that binds to bindin (the "key") in a species-specific manner.

FIGURE 21.5 The Egg-Cell Receptor for Sperm Was Isolated and Characterized
By treating sea urchin egg cells with a protease, researchers isolated the fragment of the egg-cell receptor that binds to sperm.

hypothesis. During sea urchin fertilization, species-specific bindin molecules on sperm interact with species-specific fertilizin receptors on the surface of the egg. This interaction is required for the plasma membranes of sperm and egg to fuse. As a result, cross-species fertilization is rare.

Blocking Polyspermy Early in the history of studies on sea urchin fertilization, researchers noticed that only one sperm succeeded in fertilizing the egg, even when dozens or even hundreds of sperm were clustered around the vitelline envelope. From the standpoint of producing a viable offspring, this observation was logical. If multiple fertilization, or **polyspermy,** occurred, the resulting zygote would have more than two copies of each chromosome. Sea urchin embryos with more than two copies of each chromosome die. If many sperm are present, how is polyspermy avoided?

Research over the past 70 years has revealed a wide array of mechanisms that block polyspermy in various animal species. In sea urchins, for example, fertilization results in the erection of a physical barrier to sperm entry. As **Figure 21.6a** shows, the entry of a sperm causes calcium ions (Ca^{2+}) to be released from storage areas inside the egg. A wave of ions starts at the point of sperm entry and propagates throughout the egg. In response to this dramatic increase in Ca^{2+} concentration, a series of events occurs in the egg. For example, the cortical granules located just inside the membrane fuse with the egg cell's plasma membrane and release their contents to the exterior. The contents of the cortical granules include proteases that digest the exterior-facing fragment of the egg-cell receptor for sperm. In addition, other compounds from the cortical granules are trapped between the egg cell's plasma membrane and the vitelline envelope and cause water to flow into the space by osmosis. The influx of water then causes the envelope matrix to lift away from the cell and form a **fertilization envelope** (**Figure 21.6b**). The fertilization envelope, in turn, keeps additional sperm from contacting the egg membrane.

To summarize, a century of research on sea urchin fertilization has illuminated one of the most important cell-to-cell interactions in nature. The interaction between sperm and egg triggers a series of remarkable events, including the acrosomal reaction and mechanisms for blocking polyspermy.

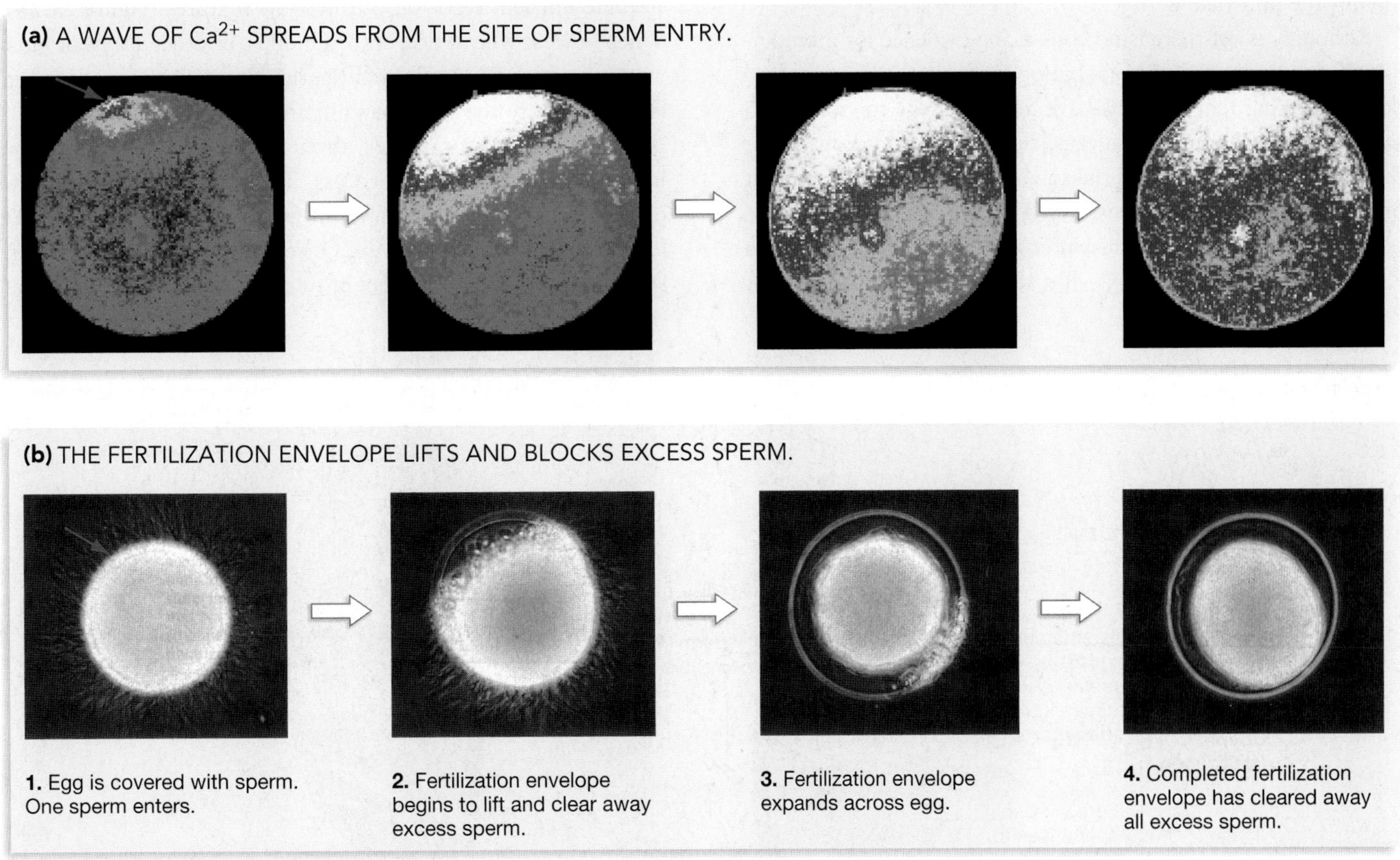
(a) A WAVE OF Ca2+ SPREADS FROM THE SITE OF SPERM ENTRY.
(b) THE FERTILIZATION ENVELOPE LIFTS AND BLOCKS EXCESS SPERM.
1. Egg is covered with sperm. One sperm enters.
2. Fertilization envelope begins to lift and clear away excess sperm.
3. Fertilization envelope expands across egg.
4. Completed fertilization envelope has cleared away all excess sperm.

FIGURE 21.6 A Physical Barrier Erected after Fertilization Prevents Polyspermy
(a) During fertilization, a wave of Ca^{2+} begins at the point of sperm entry (arrow) and spreads throughout the egg. The white dots are from a reagent that reacts with calcium ions. **(b)** In response to increased Ca^{2+} concentrations, cortical granules fuse with the egg cell's plasma membrane and release their contents to the exterior, causing a fertilization envelope to rise and clear away excess sperm.

Fertilization in Mammals

Although sea urchins have acted as a productive model system, an important question remains: How relevant are these findings to other species? In particular, how similar is sea urchin fertilization to events in mammals and plants? Unlike what occurs in sea urchins, fertilization in humans and other mammals occurs internally. This fact makes mammalian fertilization much more difficult to study than fertilization in sea urchins and similar species. With the advent of the *in vitro fertilization* (IVF) methods highlighted in this chapter's essay, however, biologists have finally acquired the ability to study mammalian fertilization under laboratory conditions. What have they found?

In most cases, females actively choose a mate prior to the sperm-egg interaction. As a result, species recognition is not as much of an issue in mammals and other species with internal fertilization as it is in sea urchins and other species with external fertilization. The acrosomal reaction still occurs, however, after the sperm's head reaches the equivalent of the sea urchin vitelline envelope—a gelatinous extracellular matrix known as the **zona pellucida**. The enzymes released from the acrosome digest the zona pellucida. As a result, the sperm's head is able to reach the egg cell's plasma membrane and fuse with it.

Although as yet there is no convincing evidence for a bindin-like protein on the sperm head, biologists have recently presented data suggesting that egg cells have a binding site for sperm. To search for this site in mouse eggs, the researchers analyzed the three glycoproteins found in the zona pellucida. They found that one of the three glycoproteins, called ZP3, binds to the heads of sperm. But as predicted, the binding of mammalian sperm to ZP3 is not species specific. Finally, researchers have found that enzymes released from cortical granules modify ZP3 in a way that prevents binding by additional sperm.

Fertilization in Flowering Plants

It should not be a surprise that marked differences occur in plant and animal fertilization. Considerable variation occurs in developmental processes among groups that have evolved independently for long periods.

Fertilization in flowering plants has been exceptionally difficult to study, because the process takes place inside a female reproductive structure called the **ovule**. Two important observations are worth noting, however. First, an event known as **double fertilization** takes place in flowering plants. As **Figure 21.7a** shows, two sperm enter the ovule via the pollen tube. In the ovule, they encounter the egg cell and a cell that in many species contains two haploid nuclei. One sperm fuses with the egg to form the zygote, while the other sperm fuses with the cell with two haploid nuclei to form a triploid ($3n$) cell. The triploid cell divides repeatedly to form a nutritive tissue called endosperm. **Endosperm** ("inside-seed") provides the proteins, carbohydrates, and fats or oils required for embryonic development, **germination** (resumptions of seed growth), and early seedling growth. In species with large seeds, the endosperm grows into a sizeable nutrient reservoir as the ovule matures (**Figure 21.7b**). When you eat wheat, rice, corn, almonds, or other grains or nuts, you are eating primarily endosperm. Functionally, endosperm is equivalent to the yolk found in animal eggs.

A second major observation about plant fertilization involves the start of the process. The development of a new plant begins when a pollen grain germinates at the top of the female reproductive structure. Once a pollen grain germinates, sperm are produced and begin to move toward the ovule, at

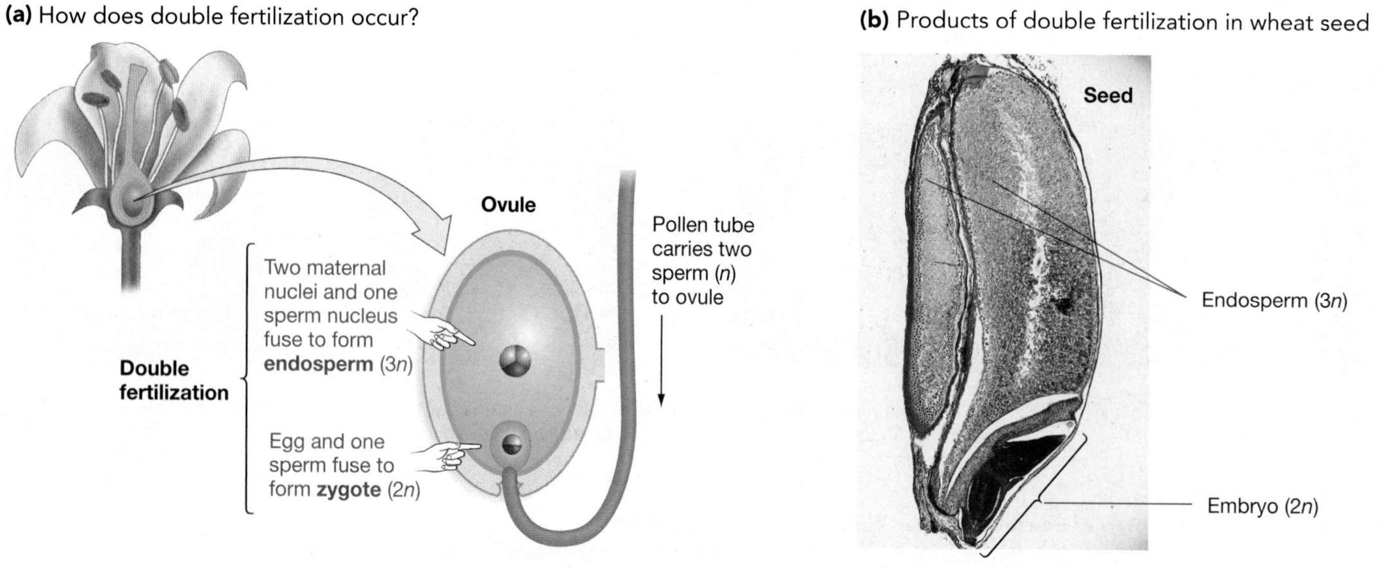

FIGURE 21.7 Double Fertilization Leads to the Formation of Endosperm
(a) In flowering plants, a sperm (n) fuses with two haploid maternal nuclei (n) near the egg to form triploid ($3n$) endosperm tissue. **(b)** Endosperm is a nutritive tissue packed with proteins and carbohydrates and fats or oils.

the base of that structure. Recent research has shown that the germination of a pollen grain involves complex interactions between proteins on the surface of the pollen grain and proteins on the walls and membranes of cells at the top of the female reproductive structure. Some of these interactions are involved in species recognition and ensure that pollen from only the same species is allowed to germinate. But in some species, the interactions also prevent pollen from the same individual from germinating on its own female reproductive parts. In this way, self-fertilization and inbreeding are avoided.

Later chapters will explore the molecular mechanisms involved in self-self recognition in plants, as well as the damaging consequences of inbreeding. For now, let's turn to the cell divisions that take place immediately after fertilization.

✓CHECK YOUR UNDERSTANDING

Fertilization is based on specific interactions between proteins in the plasma membranes of animal sperm and egg cells or on the surfaces of plant pollen grains and female reproductive structures. You should be able to (1) describe the protein-protein interactions in sea urchins, mammals, and flowering plants; and (2) explain why sea urchin eggs have such a sophisticated mechanism to block polyspermy.

21.3 Cleavage and Gastrulation

Cleavage refers to the rapid cell divisions that take place in animals after fertilization. Cleavage partitions the egg cytoplasm without additional cell growth taking place. As **Figure 21.8a** shows, the zygotes of many species simply divide into two cells, then four cells, then eight, and so on, without concurrent growth overall. The cells that are created by cleavage divisions are called **blastomeres.** When cleavage is complete, the embryo consists of a sphere of cells called a **blastula.** The blastula then undergoes *gastrulation*—the coordinated cell movements that result in the formation of a layered body structure. In animals, both cleavage and gastrulation are part of **embryogenesis,** the process by which a single-celled zygote becomes a multicellular embryo.

The exact pattern of cleavage varies widely among species, however. Cells can divide at right angles to each other so that they form tiers, as in Figure 21.8a; this pattern is **radial cleavage.** Instead they may divide at oblique angles so that they pile up as shown in **Figure 21.8b**; this pattern is **spiral cleavage.** In birds, fish, and other species whose eggs have large, membrane-bound structures filled with yolk, cleavage does not split the egg completely but produces a mound of cells around the yolk or on top of it, as shown in **Figure 21.8c**. How are these patterns controlled?

(a) Cleavage divides up the egg cytoplasm.

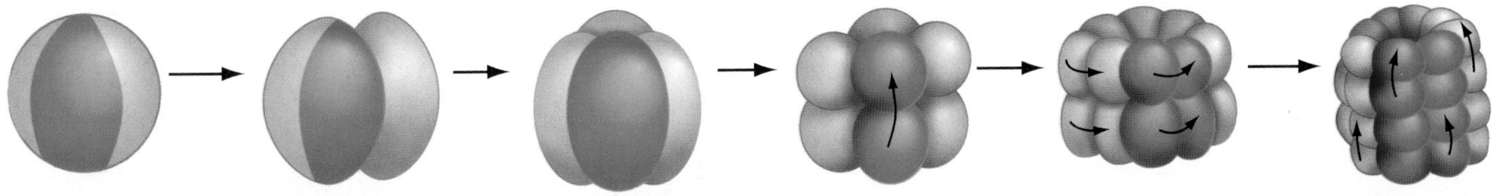

(b) In some species, cleavage creates a spiral group of cells.

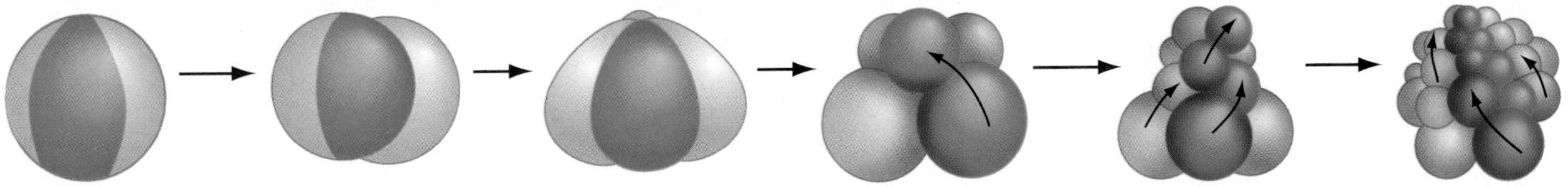

(c) In some species, a disk of small cells forms on top of a large cell containing the yolk.

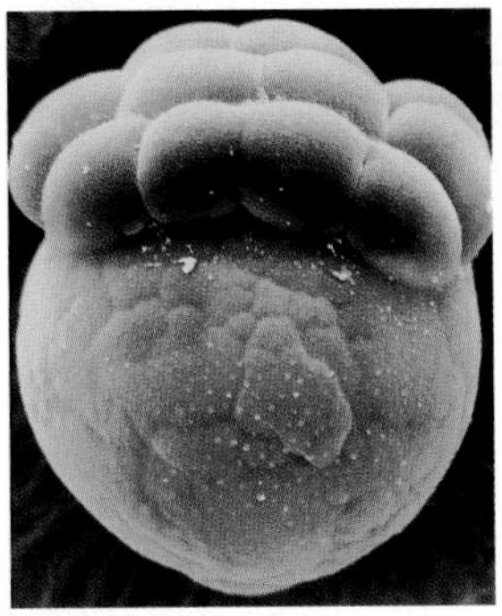

FIGURE 21.8 Cleavage Patterns Vary Widely among Species
(a) Radial cleavage. During cleavage, cells divide but growth does not occur. Some of these cells are colored green simply to help you follow their fate. **(b)** Spiral cleavage. **(c)** Distinctive cleavage patterns occur in fishes, reptiles, birds, and other species with extremely large amounts of yolk.

WEB TUTORIAL 21.1 Early Stages of Animal Development

Controlling Cleavage Patterns

In 1894 H. E. Crampton published a remarkable observation that provided an important clue about how cleavage is controlled. Crampton was studying coiling patterns in the shells of snails. Snail shells coil either to the left or to the right of the opening where the individual's body emerges. Usually, all members of the same species have shells that coil in the same direction. Occasionally, though, Crampton found mutant individuals whose shells coiled in the direction opposite the norm for their species. The most remarkable observation he made is that the direction of cleavage and shell coiling is the same. As **Figure 21.9a** shows, normal members of a given snail species with left-handed coiling undergo cleavage in a way that leads to a left-handed coil of cells in blastulas of that species. But in mutant individuals of the same species, cleavage leads to a right-handed coil of cells as illustrated in **Figure 21.9b**. The mutants also had shells with right-handed coiling.

Another researcher followed up on this observation by arranging matings between right-handed snails and left-handed snails and analyzing the phenotypes of their offspring. He found that the pattern of inheritance could be explained by hypothesizing that a single gene with two alleles controlled the coiling pattern. He called the coiling alleles *D* and *d* and determined that *D* is dominant. Some crosses gave particularly interesting results, however. For example, when *DD* females were crossed with *dd* males, all of the offspring had the *Dd* genotype and had right-handed shells. But when *dd* females were crossed with *DD* males, all of the offspring had the *Dd* genotype but had left-handed shells. To explain these results, the biologist proposed that the direction of cleavage and coiling depends on the mother's genotype and not on the offspring's genotype. That is, mothers with *dd* genotypes produce left-coiling shells in their offspring. Mothers with a *Dd* or *DD* genotype, however, produce offspring with a right-coiled shell.

How is this "maternal effect" possible? The most likely answer is that the cleavage pattern in snails, and thus the coiling pattern of their shells, is established by a molecule that is present in the egg. A molecule that exists in eggs and that helps direct early development is called a **cytoplasmic determinant**. In this case, researchers hypothesized that a molecule produced by the mother and localized in the egg cytoplasm determines the orientation of the mitotic spindle, as shown in Figure 21.9. The mother's genotype determines the coiling pattern because the gene responsible for the pattern is expressed during egg maturation—not during embryo development. This result suggests that the orientation of cleavage is controlled primarily by cytoplasmic determinants from the mother and not by proteins produced in the zygote. What other effects do cytoplasmic determinants have?

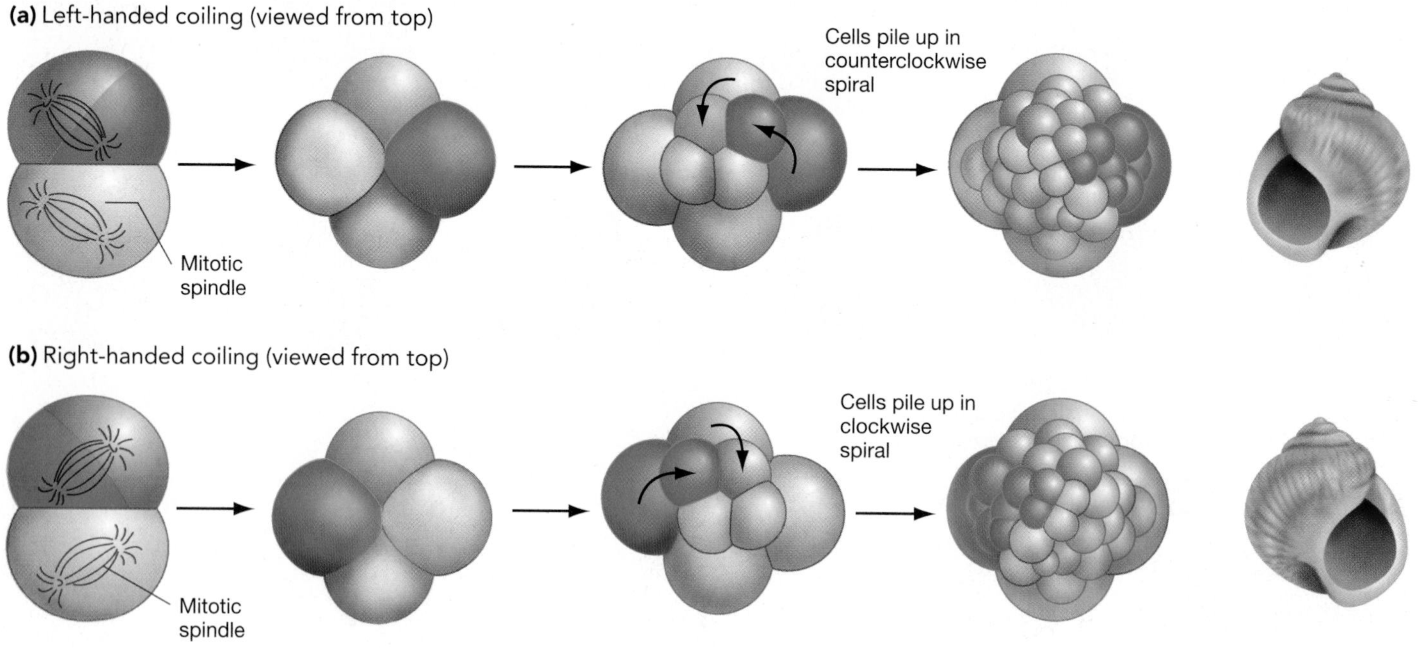

FIGURE 21.9 Cleavage Patterns Correspond to Shell-Coiling Patterns in Snails
(a) In snails, a cytoplasmic determinant sets up the orientation of the mitotic spindles during cleavage. **(b)** Mutant forms of this cytoplasmic determinant switch the orientation of the mitotic spindles, leading to blastomeres and shells that coil in a direction opposite of normal.

The Role of Cytoplasmic Determinants in Development Some of the earliest insights into the nature of cytoplasmic determinants came from studies on a sea squirt (**Figure 21.10a**). Sea squirts are marine organisms that are closely related to the vertebrates. In his experiments, Edwin Conklin focused on sea squirts because their embryos are transparent enough to make individual cells in live specimens visible with just the light microscope. In addition, the cytoplasm inside each sea squirt embryo contains differently pigmented regions. Conklin realized that the presence of different pigments might mark individual cells and allow him to follow their fate as development progressed. When biologists refer to an embryonic cell's **fate**, they mean what that cell is likely to become in the adult individual. Conklin wanted to know what pigmented cells in the embryos of sea squirts were likely to become in the corresponding adult sea squirts.

Conklin began this study by focusing on a band of yellow pigment that is present in fertilized eggs. As **Figure 21.10b** shows, certain embryonic cells inherit the pigment during cleavage. This is a key point: Cleavage divides up the egg cytoplasm in a specific way. As a result, not all blastomeres are identical. Because of their position in the embryo, different blastomeres contain different cytoplasmic determinants or other maternal factors.

By following the yellow-pigmented cells over time, Conklin was able to determine that they were destined to become muscle cells in the **larva**—the sexually immature phase in the life cycle. The yellow pigment, or a molecule associated with it, appeared to be acting as a cytoplasmic determinant. This was a crucial observation, because it showed that cytoplasmic determinants are involved in the process known as differentiation. **Differentiation** is the generation of different cell types from a single cell—the fertilized egg. The muscle cells of sea squirts, for example, have a specialized structure and function and express muscle-specific genes. Conklin's result suggested that a cytoplasmic determinant associated with the yellow pigment was responsible for the differentiation of muscle cells.

By doing the same type of analysis on other pigments in the fertilized egg, Conklin was able to construct a **fate map**—that is, a description of what each cell in the embryo is destined to become. The fate map showed how the positions of cells early in development correlated with their specialized state in larvae.

Follow-up work showed that the role of cytoplasmic determinants is highly variable among species. Cytoplasmic determinants and other types of maternal factors do not appear to play a major role in the early development of mammalian embryos, and it is unclear at present whether they are important in plants. But in millions of animal species, molecules in the egg's cytoplasm direct cleavage and influence the fate of embryonic cells.

The Role of Cell-to-Cell Signals in Development In early studies on cytoplasmic determinants and other maternal factors, researchers turned to the eggs of the African clawed frog *Xenopus laevis*. In eggs of this species, an array of key molecules is asymmetrically distributed. The most obvious of these molecules are components of yolk. About 75 percent of the yolk is localized to just half of the egg. The yolk-rich region is

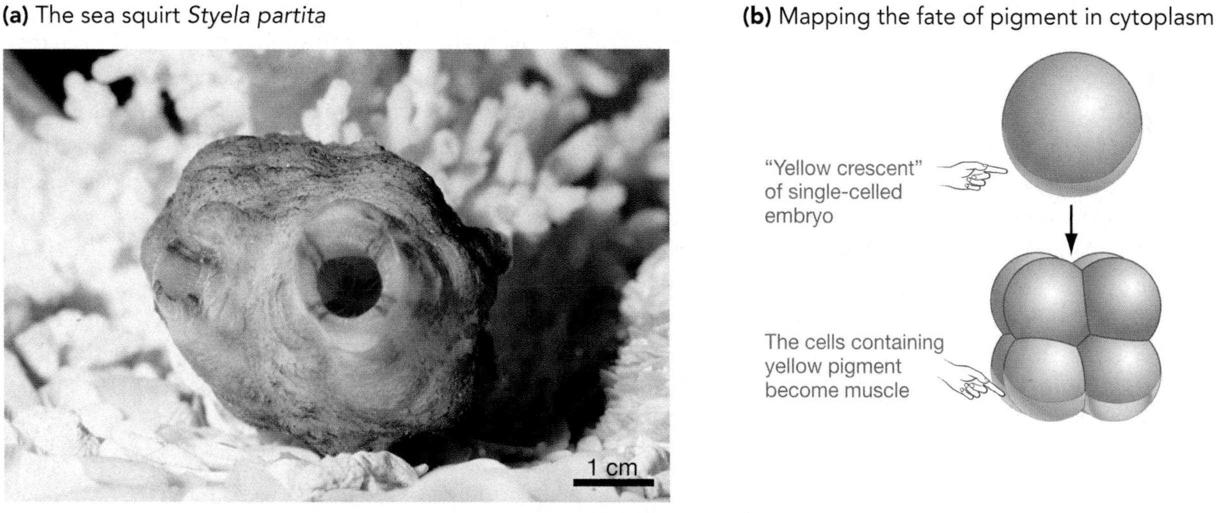
(a) The sea squirt *Styela partita*
1 cm
(b) Mapping the fate of pigment in cytoplasm
"Yellow crescent" of single-celled embryo
The cells containing yellow pigment become muscle

FIGURE 21.10 Molecules in the Egg's Cytoplasm Influence Development
(a) The sea squirt *Styela partita* has eggs that contain distinct pigments. **(b)** The yellow pigment found in sea squirt eggs is associated with the development of muscle cells.

(a) In frogs, yolk is concentrated in the vegetal hemisphere.

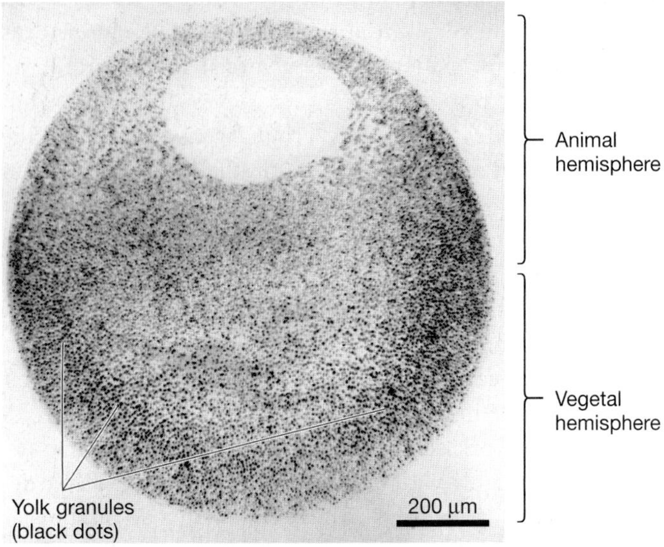
Animal hemisphere
Vegetal hemisphere
Yolk granules (black dots)
200 μm

(b) *Vg1* mRNA is localized to the vegetal pole.

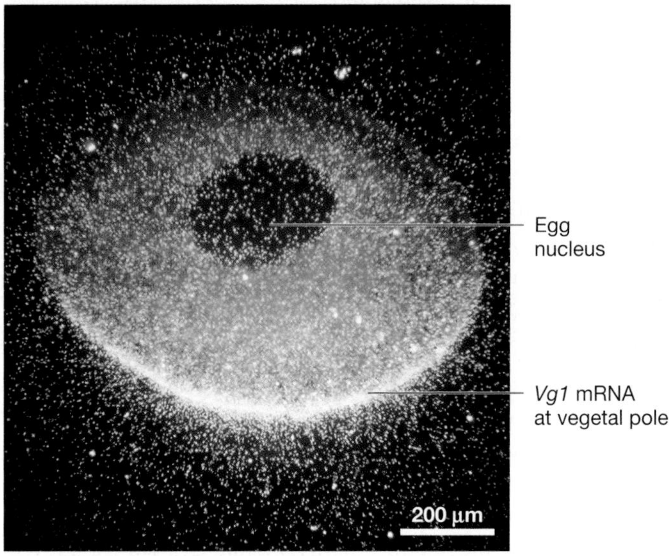
Egg nucleus
Vg1 mRNA at vegetal pole
200 μm

FIGURE 21.11 Eggs Are Highly Asymmetrical
(a) In frogs and in many other species, the yolk is concentrated in half of the egg, called the vegetal hemisphere.
(b) Bright light spots mark the location of *Vg1* mRNA (see Box 21.1).

known as the **vegetal hemisphere;** the yolk-poor half is called the **animal hemisphere** (**Figure 21.11a**).

In the 1980s Douglas Melton discovered a less obvious asymmetrically distributed molecule in frog eggs—an mRNA called *Vg1* (for "vegetal 1"). Using techniques introduced in **Box 21.1**, Melton was able to visualize the mRNA and confirm that it is localized to the vegetal pole of *Xenopus* eggs (**Figure 21.11b**). Researchers in other labs showed that another mRNA, called *VegT*, is also localized in the vegetal pole.

Subsequent work showed that both the *Vg1* and *VegT* mRNAs stay at the vegetal pole during cleavage and that cells inheriting these molecules become a specific type of embryonic tissue. Most animal embryos develop three types of tissues, called **germ layers**, early in development: (1) ectoderm ("outside skin"), (2) mesoderm ("middle skin"), and (3) endoderm ("inner skin"). Embryonic **ectoderm** forms the outer covering and nervous system; **mesoderm** gives rise to muscle, internal organs, and connective tissues such as blood and cartilage; and **endoderm** produces the lining of the digestive tract or gut, along with some of the associated organs. Cells that inherit *Vg1* and *VegT* mRNA become endoderm. Differentiation is triggered by the VegT protein, which is a regulatory transcription factor (Chapter 18). When the VegT protein binds to DNA, it triggers the expression of endoderm-specific genes. VegT is a cytoplasmic determinant.

What does the *Vg1* mRNA do? When the message is translated inside cells that are becoming endoderm, the resulting Vg1 protein is secreted from the cells, via the exocytosis pathway introduced in Chapter 7. In response to Vg1, nearby cells become mesoderm.

Based on these observations, the researchers concluded that the Vg1 protein functions as a **cell-to-cell signal**—a molecule that is secreted from a cell and received by a nearby cell, which changes its activity in response. To capture this point, biologists say that induction has occurred. **Induction** is an interaction between a signaling cell and a nearby recipient cell that changes the activity or fate of the recipient cell. The Vg1 protein produced by endodermal cells induces mesoderm. More specifically, the cells that respond to Vg1 become mesoderm on the **dorsal** (back) side of the embryo.

To summarize, cleavage is organized by molecules that are stored in the egg's cytoplasm. As a result of cleavage, different cytoplasmic determinants and other maternal factors are localized in specific blastomeres. The cytoplasmic determinants found in a cell cause them to differentiate into a particular embryonic tissue. In addition, the cell-to-cell signals produced by certain blastomeres can induce changes in neighboring cells.

Activating the Zygotic Genome

In most animals besides mammals, genes that are present in the zygote are not expressed during cleavage. Instead, early cleavage is directed by cytoplasmic determinants in the egg. The zygotic genome is transcribed for the first time after cleavage is well under way.

These points emerged from a series of studies on the African clawed frog that were published in the 1980s. Researchers injected *Xenopus* embryos with α-amanitin, which is derived from poisonous mushrooms. Recall from Chapter 16 that α-amanitin inhibits transcription. Although the presence of α-amanitin prevented the genomes of the experimental

BOX 21.1 Visualizing mRNAs by in Situ Hybridization

Where are the mRNAs for a particular cytoplasmic determinant localized in an egg? Once the zygote's genome is activated, when and where are certain genes transcribed in the embryo? Answering questions like these became possible when biologists developed a technique called in situ ("in place") hybridization. **In situ hybridization** allows researchers to visualize the location of specific mRNAs in organisms. As a result, biologists are able to infer where particular genes are being transcribed at various times during development.

To perform in situ hybridization for a particular mRNA molecule, a researcher obtains single-stranded DNA or RNA molecules that are complementary in nucleotide sequence to the RNA in question (**Figure 21.12**, step 1). For the DNA or RNA to be used as a probe, some sort of label must be attached to it. The label gives researchers a way of detecting the molecule's location. Radioactive atoms were used as labels when this technique was first developed, but now most researchers prefer to attach enzymes that catalyze color-producing reactions (step 2) for their visibility.

Because in situ hybridization cannot be done in living organisms, the labeled probe is added to a preparation containing preserved tissue. Prior to this hybridization step, the preserved tissue is treated with detergents or other agents that make plasma membranes permeable to small DNA molecules (step 3). As the probe enters the cells, it binds to mRNA molecules whose sequence is complementary (step 4). In this way, the single-stranded, labeled probe identifies a specific mRNA. Excess labeled probe that does not bind to complementary mRNAs is washed away. When the in situ hybridization is complete, the location and intensity of the signal from the label indicate the location and amount of mRNA present in particular regions of the organism (step 5).

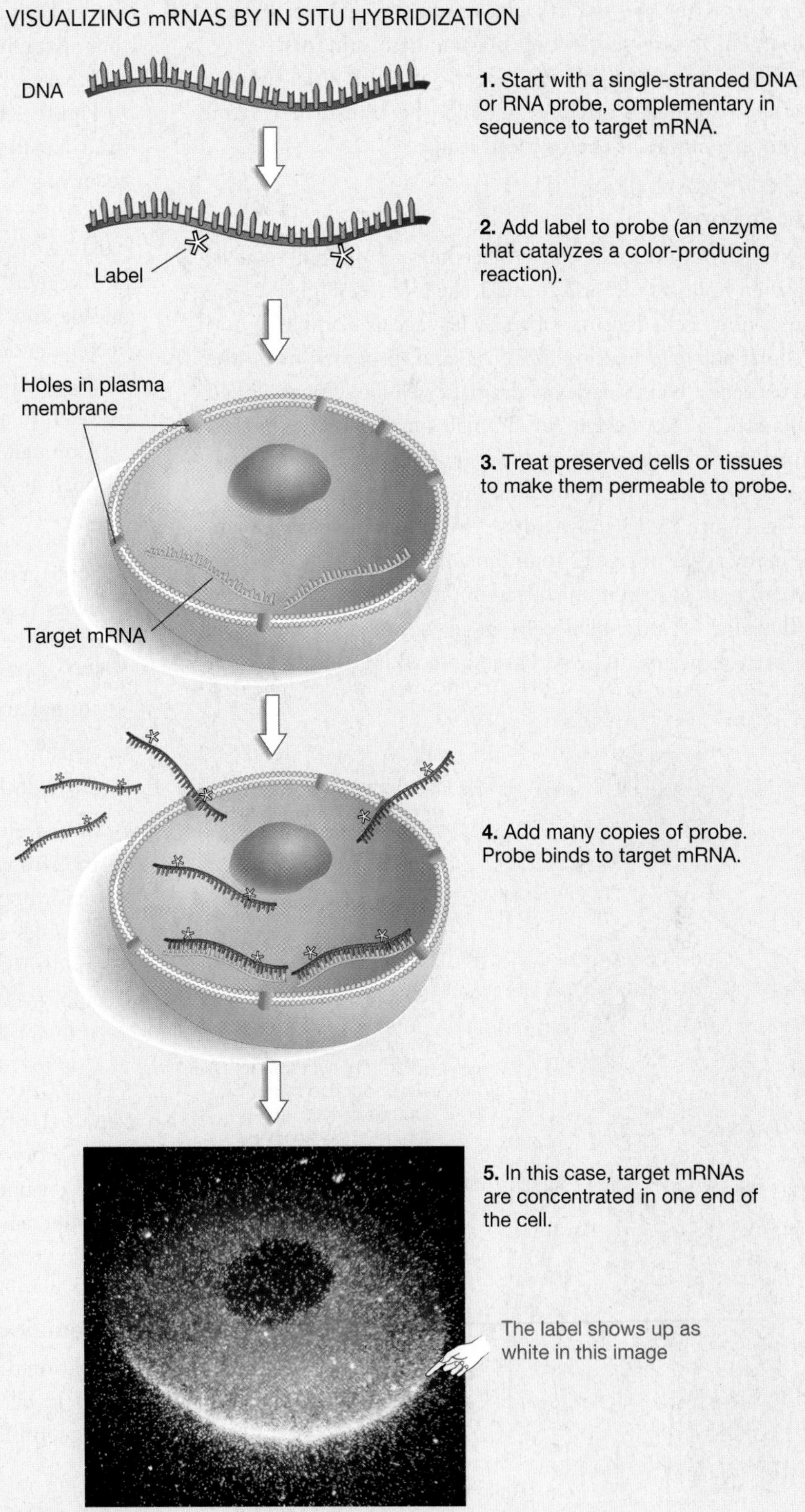
VISUALIZING mRNAS BY IN SITU HYBRIDIZATION
DNA
1. Start with a single-stranded DNA or RNA probe, complementary in sequence to target mRNA.
Label
2. Add label to probe (an enzyme that catalyzes a color-producing reaction).
Holes in plasma membrane
Target mRNA
3. Treat preserved cells or tissues to make them permeable to probe.
4. Add many copies of probe. Probe binds to target mRNA.
5. In this case, target mRNAs are concentrated in one end of the cell.
The label shows up as white in this image

FIGURE 21.12 In Situ Hybridization Allows Researchers to Pinpoint the Location of Specific mRNAs

QUESTION In situ hybridization is usually used to identify cells that are expressing a particular gene. Why is this method valid?

embryos from being expressed, the embryos progressed through early cleavage normally. In follow-up experiments, the researchers showed that new RNAs start to be produced in *Xenopus* after the twelfth cleavage division. Until then, development is directed by cytoplasmic determinants.

Mammals are an important exception to this rule. In mice and other mammals, mRNAs begin to be transcribed from the zygotic genome at the two-cell stage.

Gastrulation

As cleavage continues, the embryo becomes a solid ball of cells, which then hollows out and forms a blastula (see Figure 21.1). As this point, cell division slows. Cleavage is complete, and cells start actively making mRNAs and proteins from the embryo's genes. Next a series of dramatic cell movements called gastrulation, a key event in animal development, begins. **Gastrulation** radically rearranges cells and results in the formation of a **gastrula**, which contains the three embryonic tissue types. As **Figure 21.13** shows, these embryonic tissues give rise to the many types of tissues found in adults.

Research on gastrulation started in the 1920s with efforts to map the fate of individual cells in newt and frog embryos. These early experiments were based on soaking tiny blocks of agar (a gelatinous compound) with a nontoxic dye. The dyed blocks were then pressed against the surface of blastula-stage embryos, so that a small number of blastomeres became marked with dye. By allowing marked embryos to develop and then examining them at intervals during gastrulation, researchers were able to follow the movement of cells.

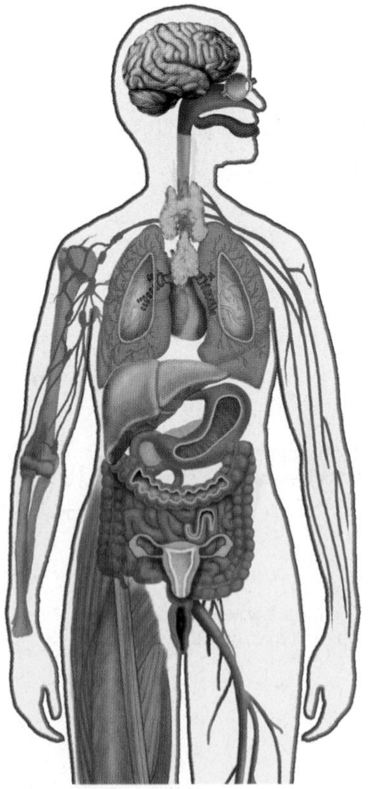
Ectoderm
Nervous system
Cornea and lens of eye
Epidermis of skin
Epithelial lining of mouth and rectum
Mesoderm
Skeletal system
Circulatory system
Lymphatic system
Muscular system
Excretory system
Reproductive system
Dermis of skin
Lining of body cavity
Endoderm
Epithelial lining of:
digestive tract
respiratory tract
reproductive tract
urinary tract
Liver
Pancreas
Thyroid
Parathyroids
Thymus

FIGURE 21.13 The Three Embryonic Tissues Give Rise to Different Adult Tissues and Organs

QUESTION In a gastrula, mesoderm is sandwiched between ectoderm on the outside of the embryo and endoderm on the inside. Relate this observation to the positions of adult tissues and organs derived from mesoderm, ectoderm, and endoderm.

Figure 21.14 summarizes the results of experiments with dyed blastomeres in frogs. The drawings identify blastomeres according to the type of embryonic tissue they will become. The cells at the animal pole are destined to become ectoderm and produce skin and nerve tissues; the cells between the animal and vegetal poles will develop into mesoderm and generate muscle and connective tissue; the cells at the vegetal pole will become endoderm and form the gut and associated organs. As Figure 21.14 shows, each of these three populations of cells moves into a new position during gastrulation, forming layers.

Although the gastrulation pattern varies among species almost as much as do cleavage patterns, certain features are shared:

- In most species, gastrulation begins when an invagination appears in the blastula, similar to the indentation that forms when you push a finger into a balloon. In frogs this invagination becomes round and is called the **blastopore**; it appears about two-thirds of the way between the embryo's animal and vegetal poles and surrounds a plug of yolk cells.
- Cells from the periphery move to the interior of the embryo through the blastopore.
- By the end of gastrulation, (1) the three embryonic tissues are arranged in layers, (2) the gut has formed, and (3) the major body axes have become visible. In frogs the blastopore becomes the anus, and the region just above the blastopore (as drawn in Figure 21.14) becomes the dorsal side of the embryo. In this way, the head-to-tail and back-to-belly axes of the body (discussed next) become apparent.

The discovery of cytoplasmic determinants and cell-to-cell signals clarified how blastomeres become differentiated into the three embryonic tissues. Current research on gastrulation is focused on understanding how cells move in such an organized way. But what about the other major result of gastrulation—the formation of the major body axes?

Organizing the Major Body Axes

Formation of the major body axes is among the most important events in all of development. There are three of these axes, illustrated in **Figure 21.15**:

1. One axis runs **anterior** (toward the head) to **posterior** (toward the tail);
2. One axis runs **ventral** (toward the belly) to **dorsal** (toward the back);
3. One axis runs **proximal** (toward the center of the body) to **distal** (away from the center of the body).

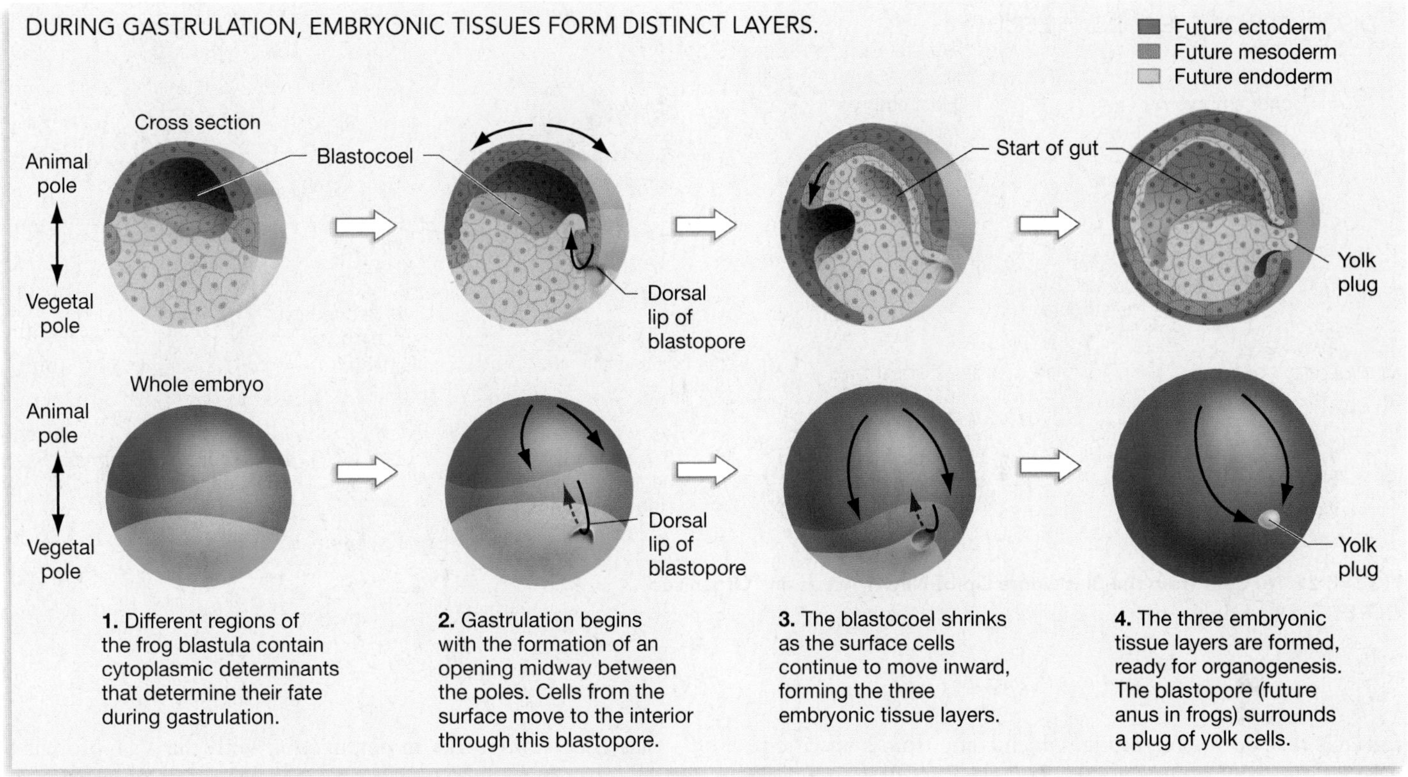
DURING GASTRULATION, EMBRYONIC TISSUES FORM DISTINCT LAYERS.
Future ectoderm
Future mesoderm
Future endoderm
Cross section
Animal pole
Vegetal pole
Blastocoel
Dorsal lip of blastopore
Start of gut
Yolk plug
Whole embryo
Animal pole
Vegetal pole
Dorsal lip of blastopore
Yolk plug
1. Different regions of the frog blastula contain cytoplasmic determinants that determine their fate during gastrulation.
2. Gastrulation begins with the formation of an opening midway between the poles. Cells from the surface move to the interior through this blastopore.
3. The blastocoel shrinks as the surface cells continue to move inward, forming the three embryonic tissue layers.
4. The three embryonic tissue layers are formed, ready for organogenesis. The blastopore (future anus in frogs) surrounds a plug of yolk cells.

FIGURE 21.14 Precise Cell Movements Occur during Gastrulation

EXERCISE In frogs, the initial opening of the blastopore is on what will become the dorsal side of the larva and adult. The blastopore becomes the anus of the larva and adult. On the last drawing, mark where the head, tail, back, and belly will develop.

How are these axes established so that the head, tail, back, and belly of the embryo develop in the correct relationship to each other?

The Spemann-Mangold Experiments

In 1924 Hans Spemann and Hilde Mangold published the results of a dramatic experiment on newt and salamander embryos. These researchers wanted to understand how the body axes of the embryo are established. From fate-mapping studies, they knew that the cells on the animal-pole side of the blastopore became part of the animal's back. Spemann and Mangold predicted that if these cells were designated prior to gastrulation to become back tissue, then the cells should still form back tissue if they were transplanted to a different location in the embryo.

To test this prediction, the investigators transplanted embryonic tissues between embryos of two differently pigmented species of newt. The color differences allowed them to follow the fate of the transplanted cells. For example, **Figure 21.16** (page 460) illustrates an experiment in which they took cells from the dorsal side of a blastopore in a nonpigmented embryo and added those cells to the ventral side of a pigmented embryo. Much to the researchers' amazement, the transplanted cells did not just become back tissue. Instead, a second, "twinned" embryo developed that was joined to the normal host embryo. The second embryo contained both pigmented and nonpigmented cells.

To interpret this result, Spemann and Mangold reasoned that the transplanted cells must be able to alter the fates of host tissues. For example, cells that would otherwise have become part of the belly become part of the back, in response to the transplanted cells. The logic was that, because the second embryo contained both donor and host cells, the transplanted cells must have recruited host cells to form part of the second embryo. Later, Spemann began using the term **organizer** to refer to

WEB TUTORIAL 21.2 The Gray Crescent in Frog Eggs

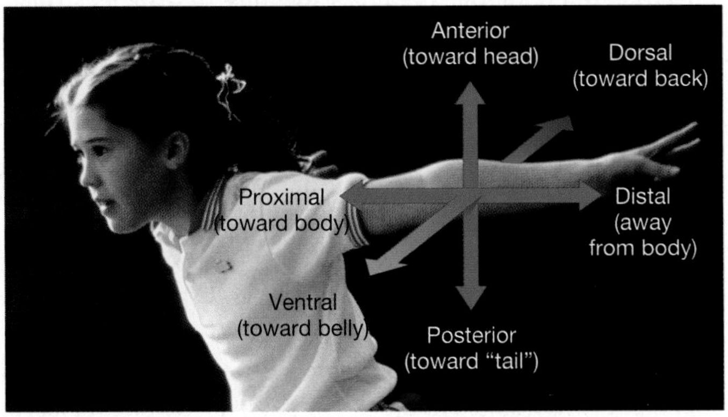
Anterior (toward head)
Dorsal (toward back)
Proximal (toward body)
Distal (away from body)
Ventral (toward belly)
Posterior (toward "tail")

FIGURE 21.15 Most Animals Have Three Major Body Axes

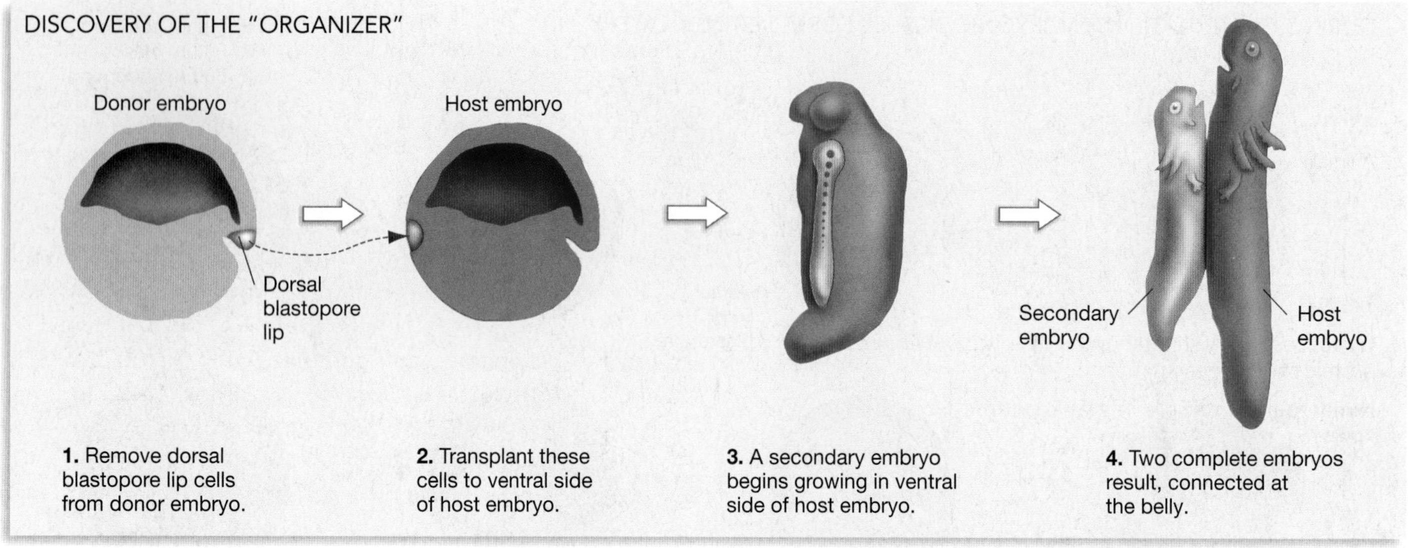
DISCOVERY OF THE "ORGANIZER"
Donor embryo
Host embryo
Dorsal blastopore lip
Secondary embryo
Host embryo
1. Remove dorsal blastopore lip cells from donor embryo.
2. Transplant these cells to ventral side of host embryo.
3. A secondary embryo begins growing in ventral side of host embryo.
4. Two complete embryos result, connected at the belly.

FIGURE 21.16 Cells from the Blastopore Lip of Newts Act as an "Organizer"
QUESTION Why do the twinned embryos end up belly to belly? Why do they have the same head-to-tail orientation?

the cells that do the recruiting, which come from a specific region on the embryo's dorsal side. The name was logical because these cells have the ability to organize host and donor tissues into a second embryo. Spemann also coined the term *induction* to describe the ability of organizer cells to direct the development of other cells. His idea was that the organizer somehow induces the formation of a second embryo.

The organizer appeared to be the key to the movement of cells during gastrulation and the formation of the body axes. Further research has demonstrated that embryos from other vertebrate species have an organizer-like region where gastrulation is initiated. How does this group of cells form, and how does it function?

Forming the Organizer An unusual observation about frog eggs provided an initial clue about the formation of the organizer. Just after fertilization, a region of gray cytoplasm appears in the zygotes of some frog species opposite the point of sperm entry. Because of its shape and color, the region is called the **gray crescent**. Several lines of evidence indicate that the gray crescent corresponds to the region where the organizer forms. For example, fate-mapping studies showed that cells derived from this region become the area where the organizer forms. In addition, Spemann found that when he artificially separated blastomeres at the two-cell stage, blastomeres that lacked the gray crescent developed as a mass of unorganized tissues. In contrast, blastomeres containing the gray crescent developed normally.

How does fertilization lead to the formation of the gray crescent and the organizer? Investigators have shown that, in frogs, fertilization triggers a dramatic rearrangement of the egg cytoplasm. When the cytoplasm moves, a protein called β-catenin in the gray crescent region is activated. Recent work suggests that β-catenin works in conjunction with the Vg1 protein to stimulate the transcription of organizer-specific genes. What are these genes?

Molecules Involved in Organizer Function Cells in the organizer can induce changes in other cells. Based on this observation, biologists hypothesized that cells in the organizer must produce and secrete one or more cell-to-cell signals. Several teams of biologists have been actively searching for the molecules involved in the organizer function and the genes that encode them. This research is driven by the hope that the molecules that are active in frogs will lead to the discovery of similar molecules active in the development of other species with organizer-like regions—including humans.

In frogs, the most common experimental strategy for identifying important organizer molecules is to inject blastulas with particular compounds and observe the response. For example, one research team began a study by isolating the mRNAs found in the organizer regions of frog gastrulas. Then the researchers injected these candidate mRNAs one by one into embryos that had been irradiated with ultraviolet light. This was a clever experimental strategy, because researchers had found that UV light inhibits the egg's cytoplasm from rotating at fertilization. As a result, irradiated embryos do not normally form a gray crescent or organizer. The researchers' goal was to find an mRNA that could restore the ability of these eggs to form an organizer and undergo gastrulation.

Among the many mRNAs that they tested, the biologists found one that resulted in the formation of a complete embryo. This mRNA appeared to code for a protein that is an organizer product (**Figure 21.17**). To support this hypothesis, the researchers used in situ hybridization techniques (see Box 21.1) to locate where the mRNA is produced in intact embryos. As predicted, the mRNA was found only in the organizer region. The investigators also found that if they injected an excess of this mRNA, the treated embryos developed as an enlarged head region with no trunk or tail. As a result, the biologists named the corresponding gene *noggin*. They hypothesized that the noggin protein functions in setting up the major body axes of the embryo. Follow-up experiments have shown that noggin works by binding to signaling molecules called *BMPs* and activating them.

Similar studies have identified other signaling molecules produced by the organizer region. Currently, research is focused on understanding exactly how noggin and these other organizer products act on target cells during and after gastrulation. Noggin and other organizer products are among the most important molecules in all of development. In addition to establishing the major body axes, they set in motion a chain of events that leads to the formation of specialized cells and their organization into tissues and organs.

Question: How do cells in the organizer induce changes in other cells?

Hypothesis: Cells in the organizer produce and secrete a protein that serves as a cell-to-cell signal.

Null hypothesis: Cells in the organizer do not produce and secrete a protein that serves as a cell-to-cell signal.

Experimental setup:

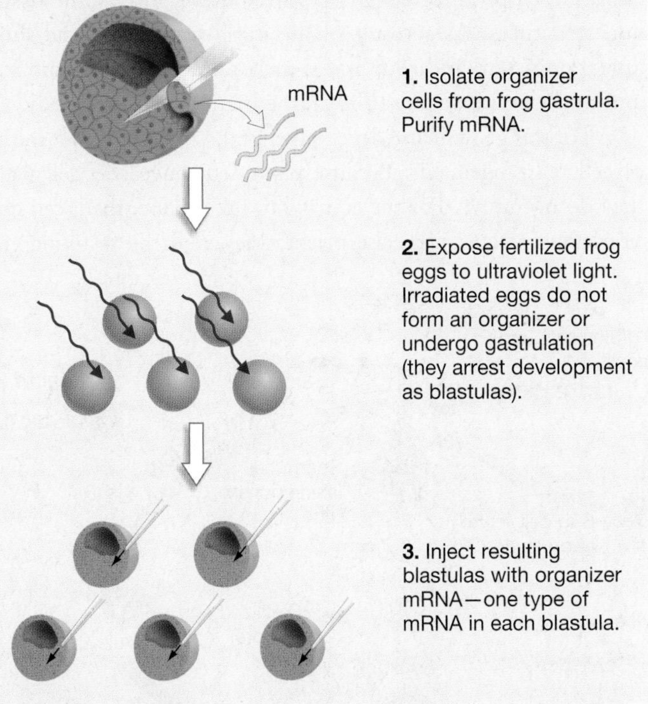
mRNA
1. Isolate organizer cells from frog gastrula. Purify mRNA.
2. Expose fertilized frog eggs to ultraviolet light. Irradiated eggs do not form an organizer or undergo gastrulation (they arrest development as blastulas).
3. Inject resulting blastulas with organizer mRNA—one type of mRNA in each blastula.

Prediction: The embryos containing the mRNA coding for the signal protein will undergo gastrulation.

Prediction of null hypothesis: None of the embryos will undergo gastrulation.

Results:

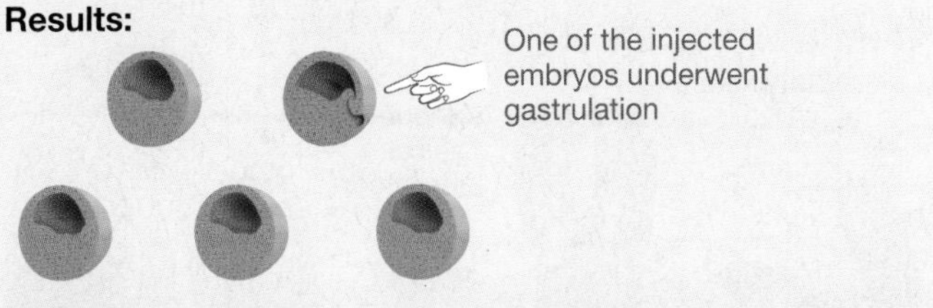
One of the injected embryos underwent gastrulation

Conclusion: Cells in the organizer produce and secrete a protein (called noggin) that induces changes in other cells.

FIGURE 21.17 Finding the Organizer Product

QUESTION Cells that receive cell-to-cell signals secreted by the organizer give rise to a wide array of tissues and structures along the dorsal side of an individual. Do you agree with the hypothesis that the organizer produces a single product? Explain why or why not.

✓CHECK YOUR UNDERSTANDING

Cleavage results in a sphere of cells that then undergo a massive rearrangement during gastrulation. At the end of gastrulation, the three embryonic tissues are arranged in layers and the major body axes are established. You should be able to give an example of how cytoplasmic determinants or cell-to-cell signals direct the pattern of cleavage, the fate of the blastomeres that are created by cleavage, and the location of the initial invagination in gastrulation.

21.4 Patterns of Development

Fertilization, cleavage, and gastrulation produce a multicellular embryo that is organized into several distinct embryonic tissues. Before we go on to explore additional details on how the major body axes are established and how cells become differentiated into specialized types, it will be helpful to review how the overall pattern of development varies among species.

To step back and sample the diversity of developmental patterns observed in plants and animals, let's consider embryonic development in a plant, an invertebrate, and a couple of vertebrates. More specifically, we'll look briefly at the earliest events in the life of the wild mustard plant *Arabidopsis thaliana*, the fruit fly *Drosophila melanogaster*, frogs, and humans. Two vertebrates are featured to illustrate the differences in developmental patterns between species that lay eggs (represented by frogs) and those that give live birth (represented by humans).

Arabidopsis thaliana, *D. melanogaster*, and frogs such as *Xenopus laevis* are among the most important model organisms in developmental biology. These species are relatively easy to rear in the laboratory, and the genetic makeup of individuals can be manipulated through selective breeding or recombinant DNA technology. Experimental results from these species have provided insight into how development works in organisms that are difficult to study experimentally, including humans.

Embryonic Development in *Arabidopsis thaliana*

Arabidopsis thaliana is a flowering plant whose short life span and ease of culture has made it a favorite experimental subject for biologists interested in development and genetics (**Figure 21.18a**). This plant's entire life cycle can take just 6 weeks under laboratory conditions. The cycle begins with gametogenesis (**Figure 21.18b**). Sperm formation occurs in pollen grains, while egg cell formation occurs inside the ovule, within the flower.

After fertilization, the zygote undergoes mitosis and its daughter cells continue dividing mitotically. As a result of these early cell divisions, an embryo begins to develop inside the ovule. As embryogenesis continues, the ovule's covering develops into a **seed coat**, a protective layer that encases both the embryo and the endosperm. The mature seeds are usually dispersed away from the parent plant by wind, an animal, or water. When the seed germinates, the seed coat breaks and the embryo emerges to become a **seedling.** Mature embryos and seedlings have three prominent parts: a **root**, the belowground portion that anchors the plant and absorbs nutrients; initial leaves called **cotyledons**; and a stem-like structure called the **hypocotyl**, which joins the roots to the cotyledons (see Figure 21.18b). Together, the leaves and stem—the aboveground portion of the plant—making up the **shoot.**

How do these embryonic structures form? As **Figure 21.18c** shows, the cells that result from the fertilized egg's first division are asymmetrical in size, orientation, and their eventual function or fate. The bottom, or *basal cell*, is large and gives rise to a column of cells called the *suspensor*, which anchors the embryo as it develops. The small cell above the basal cell, called the *apical cell*, in contrast, is the progenitor of the mature embryo. Initially it gives rise to a simple ball of cells at the tip of the suspensor. As this group of cells grows and develops, the cotyledons begin to take shape. Later, groups of cells called the **shoot apical meristem** and **root apical meristem** form. A **meristem** consists of undifferentiated cells that divide repeatedly, with some of their daughter cells becoming specialized cells. Throughout the individual's life, meristematic tissues continue to produce cells that differentiate into adult tissues and structures. As a result of the initial growth of the shoot apical meristem and root apical meristem, the embryonic root and hypocotyl elongate to form the main axis of the body.

All of this growth and development takes place without the aid of cell rearrangements. Because plant cells have stiff cell walls, they do not move. For the embryo to take shape, then, cell divisions must occur in precise orientations, and the resulting cells

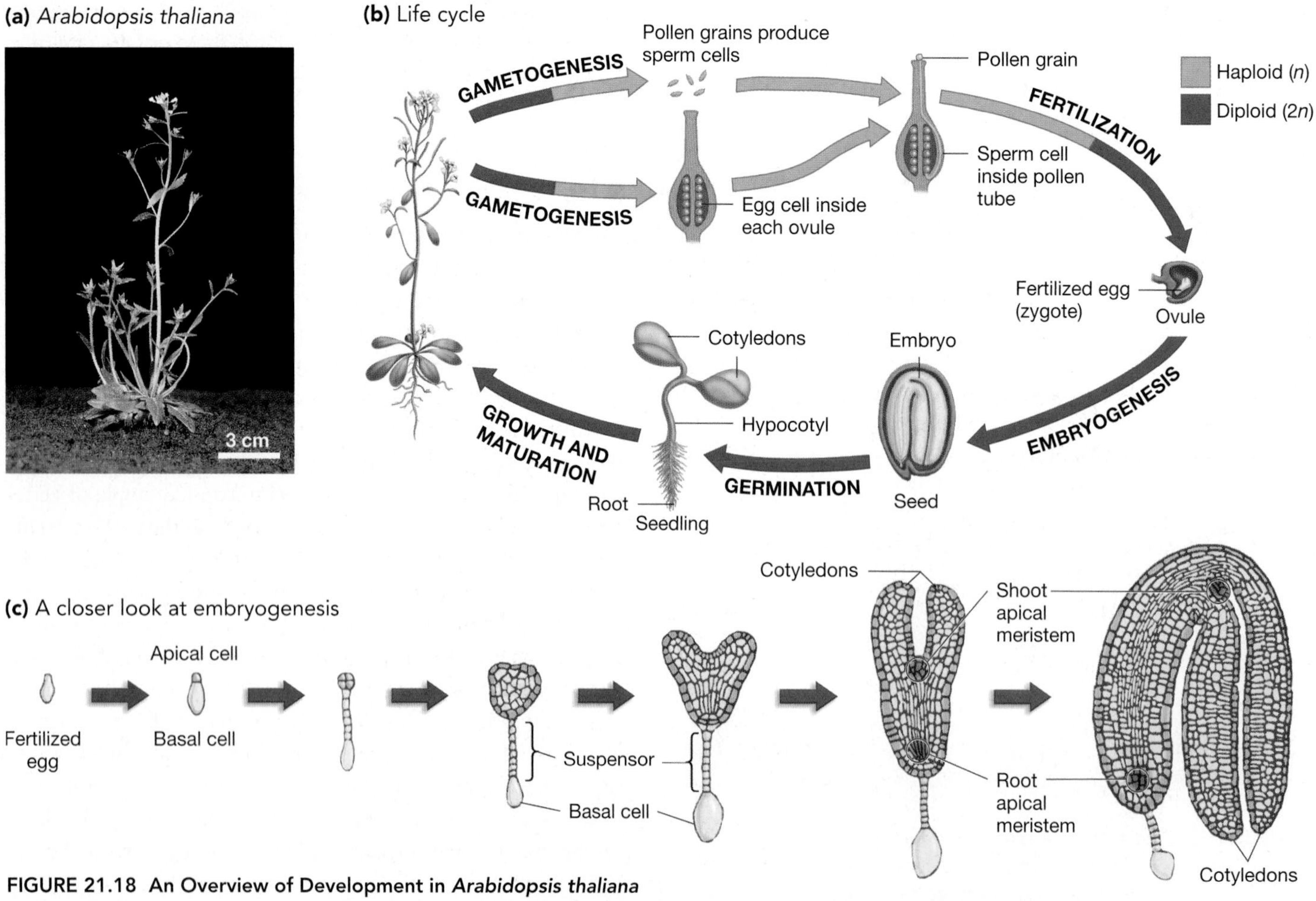

FIGURE 21.18 An Overview of Development in *Arabidopsis thaliana*
(a) *Arabidopsis thaliana* is an important model organism in developmental biology and genetics. **(b)** Its life cycle can take as little as 6 weeks. Fertilization and early embryonic development take place inside the ovule, which matures into a seed. **(c)** During embryogenesis, the root and shoot apical meristems form in addition to the first leaves (cotyledons), the root, and a hypocotyl.

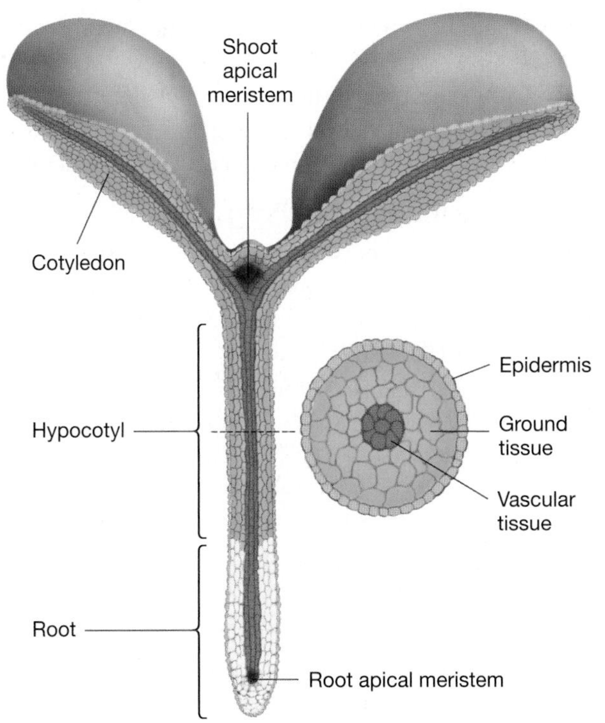

FIGURE 21.19 *Arabidopsis* Has Three Embryonic Tissues
Embryogenesis in plants produces three embryonic tissues: epidermis, ground tissue, and vascular tissue, organized in a radial fashion.

must exhibit differential growth—meaning that some cells grow larger than others. In contrast, animal embryos take shape largely as the result of directed cell movements during gastrulation, with each cell approximately the same size as its neighbors.

The root, hypocotyl, and cotyledons are organs that form along the main axis of the embryo (**Figure 21.19**). The cross-sectional view in the figure shows that early development in *A. thaliana* also produces three embryonic tissues. Note that an embryo contains an **epidermis**—an outer covering of cells, called *epidermal cells*, that protect the individual. Inside this layer of cells is **ground tissue**, a mass of tissue that may later differentiate into cells that are specialized for photosynthesis, food storage, or other functions. The **vascular tissue** in the center of the plant will eventually differentiate into specialized cells that transport food and water between root and shoot. In the embryo, the three tissue systems are arranged in a radial pattern.

Although this pattern of development is typical of many plants, the sequence of events and the structures and tissues that form are very different from those observed during the development of animals. These differences are not surprising, given that both the fossil record and evolutionary trees show that plants and animals evolved from different single-celled ancestors. Stated another way, multicellular bodies evolved independently in plants and animals. As a result, their developmental pathways are distinct. Let's begin an analysis of animal development by looking at what may be the best studied of all multicellular organisms—the fruit fly.

An Invertebrate Model: *Drosophila melanogaster*

In the fruit fly, *Drosophila melanogaster*, gametogenesis leads to the production of sperm and eggs in different individuals—not the same individual, as in *Arabidopsis*. But fertilization occurs inside female reproductive structures, as it does in *Arabidopsis*. A female fly then lays the fertilized eggs in rotting fruit or a similar medium, so they develop outside her body. (In contrast, recall that in *Arabidopsis* the early stages of embryo formation occur while the seed is maturing and still attached to the parent plant.)

As **Figure 21.20** shows, the nucleus of a fertilized fly egg repeatedly undergoes mitosis without cytokinesis occurring. The result is an embryo with many nuclei scattered throughout a

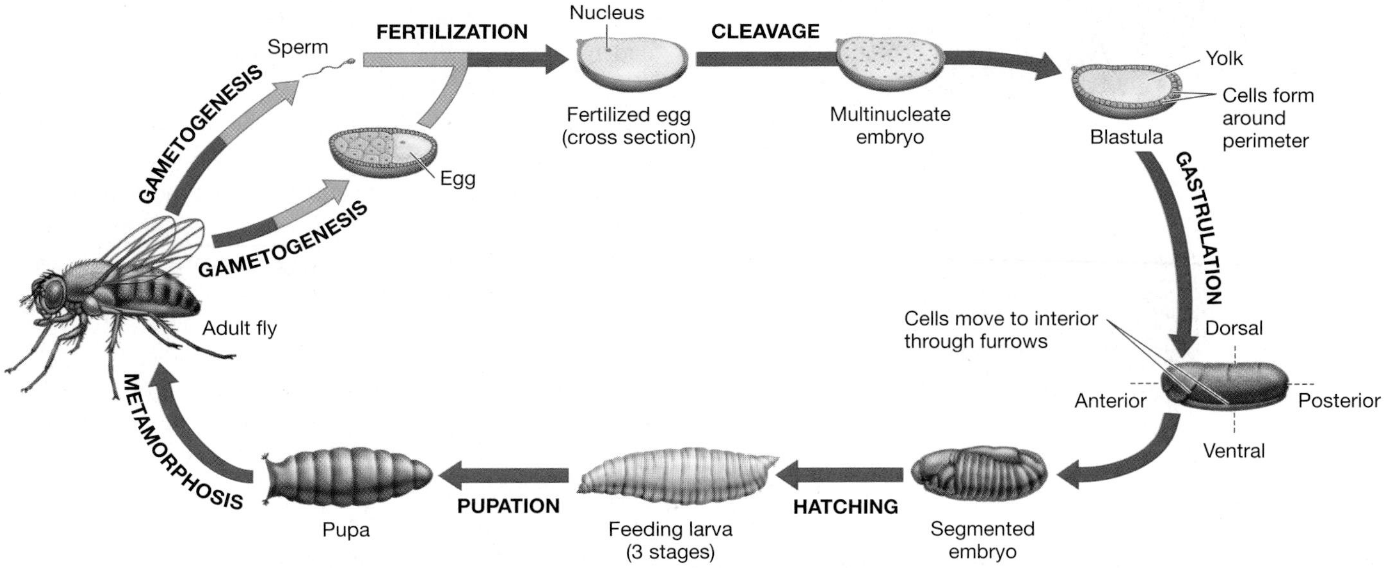

FIGURE 21.20 An Overview of Development in *Drosophila melanogaster*
When fruit-fly embryogenesis is complete, the segmented embryo hatches into a feeding larva. After feeding for several days, the larva stops moving, secretes a case, and becomes a pupa. The entire body is remodeled into the adult form via metamorphosis.

cytoplasm filled with nutrient-rich yolk. At the blastula stage of development, most nuclei migrate to the periphery of the embryo and become surrounded by a plasma membrane.

Gastrulation begins when a long furrow forms on one side of the embryo, and cells on the periphery begin to move through the furrow to the interior. As gastrulation continues, another furrow forms and defines the head region; soon after, furrows form along the length of the body and demarcate the series of body regions called **segments.** A segment is a region of the body that is repeated. Segments contain similar structures and have similar functions.

When gastrulation is complete and the larva matures, the larva has a functioning mouth and digestive tract and begins feeding on rotting fruit or other food sources. Then the fly undergoes **pupation**, the secretion of a hard case around the larva. The result is the formation of a **pupa**, a nonfeeding stage in which the fly's body is remodeled to form an adult. This remodeling process is called **metamorphosis** ("change-form"). It is important to appreciate how remarkably different the larval and adult forms are in many species that undergo metamorphosis. Contrast a maggot and a fly or a caterpillar and a butterfly. In these species, juvenile and adult forms have completely different bodies, live in different habitats, and exploit different food sources.

As **Figure 21.21** indicates, there is a one-to-one correspondence between the segments that appear in the larva and the segments found in the adult. The segments that form in fruit flies are grouped into regions called the *head, thorax*, and *abdomen*. Despite the massive changes in shape and tissue organization that occur during metamorphosis, segmentation is fundamental to the organization of both the larval body and the adult body.

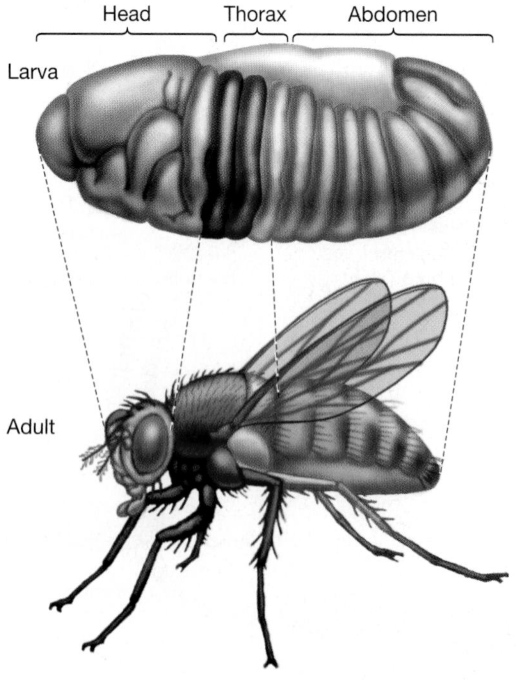

FIGURE 21.21 ***Drosophila* Larvae and Adults Are Segmented**
The segmented body of insects is organized into head, thorax, and abdomen regions.

Finally, it is important to note that following metamorphosis, gastrulation is responsible for segregating cells into the three embryonic tissues of ectoderm, mesoderm, and endoderm.

A Vertebrate Model: The Frog

How does fly development compare with the development of vertebrates such as the frog *Xenopus laevis*? Although the fossil record indicates that fruit flies and frogs last shared a common ancestor more than 600 million years ago, some features of fly and frog development are similar. Features such as asymmetry, segmentation, and three embryonic tissues are also found in frog embryos. These are remarkable observations. Even though many features of development vary among animal species, certain aspects have been highly conserved during evolution. Recent research has shown that many of the underlying genetic mechanisms responsible for development are also shared among some species.

Figure 21.22 shows some prominent stages in the development of a frog. The result of gastrulation in the frog is fundamentally similar to the result of gastrulation in the fly, even though there are many differences in the outward appearance of the embryos produced by these two species. Embryonic tissues separate into distinct layers or compartments, and the major body axes take shape. Notice that as gastrulation ends, a structure called the neural tube begins to form along the embryo's back. The **neural tube** will eventually become the spinal cord and brain of the larva and adult frog. In this way, the dorsal-to-ventral (back-to-belly) axis of the body takes shape. The formation of the neural tube is the key event in preparation for the developmental stage called **organogenesis**, which occurs after gastrulation and results in the formation of specialized cells, tissues, and organs. (An **organ** is a group of tissues that are integrated into a structure with a specific function, such as the heart or lungs.) At this stage, the frog embryo is called a *neurula*. The embryo will hatch into a larva, known as a *tadpole*, which will undergo metamorphosis to become an adult frog.

Frogs, like flies, lay their eggs so that development takes place in the surrounding environment. But in some lizards, sharks, and fish—and in the vast majority of mammals, including humans—early development takes place inside the mother's body. Despite this difference, the early stages of human development are also marked by cleavage, gastrulation, organogenesis, and the formation of body segments.

Early Development in Humans

Like fruit flies and frogs, humans have separate sexes. Gametogenesis in males results in the production of the tiny, motile gametes called sperm; the analogous process in females results in larger, nonmotile eggs. As **Figure 21.23** shows, fertilization and cleavage occur inside the reproductive tract of human mothers, in the fallopian tube. The **fallopian tube** is a structure that connects the ovary and the uterus. The **ovary** is the organ in which the egg matures; release of the egg from the ovary is known as

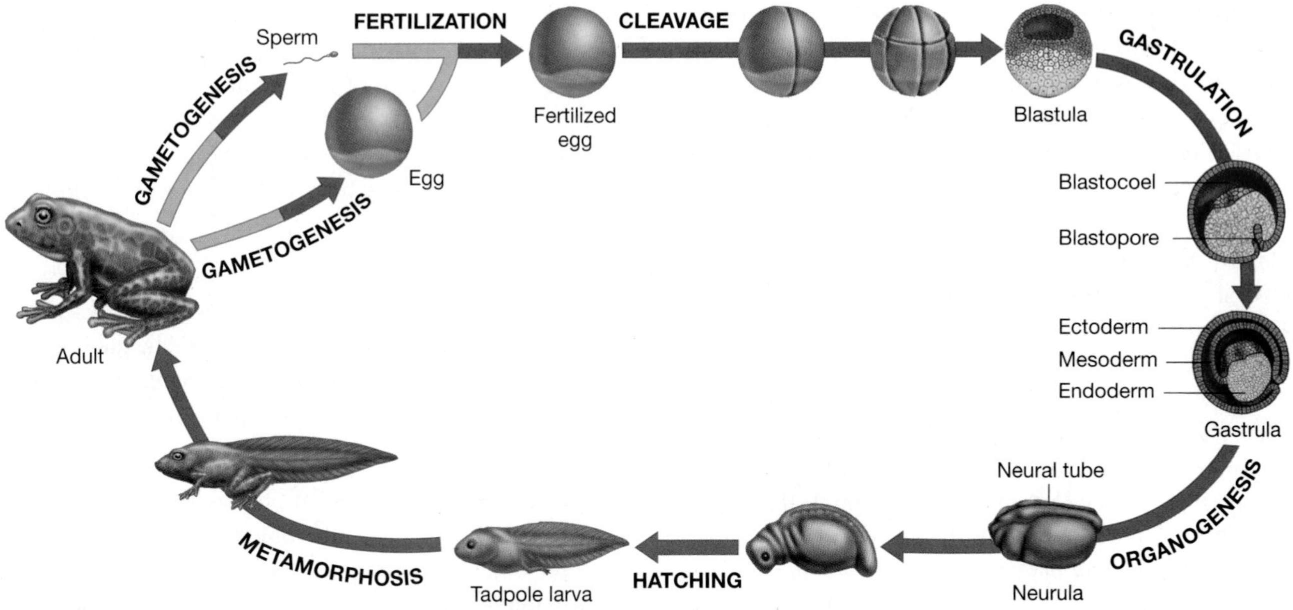

FIGURE 21.22 An Overview of Development in a Frog
In vertebrates, early embryonic development consists of cleavage, gastrulation, and organogenesis. Frogs hatch into a larval stage called a tadpole. Although frogs undergo metamorphosis, they do not form a pupa.

ovulation. The **uterus** is the organ in which the embryo develops. Cleavage occurs as the embryo travels down the length of the fallopian tube.

When cleavage is complete, the embryo, now known as a *blastocyst*, breaks out or hatches from the zona pellucida and implants into the uterine wall. Once this process, called **implantation**, occurs, the placenta begins to form. The **placenta** is an organ derived from a mixture of maternal and embryonic cells. It allows nutrients and wastes to be exchanged between the mother's blood and the embryo's blood. For this reason, human eggs are small compared with the eggs of frogs and largely lack a source of nutrients such as the endosperm of seed plants and the yolk of fly or frog eggs. Gastrulation and the formation of the neural tube occur after the embryo has become implanted into the uterine wall. Later events in pregnancy and birth are detailed in Chapter 48.

Making Sense of Developmental Variation and Similarity

Early descriptive studies of plant and animal development established that the sequence of events varies a great deal among species. At first glance, it seems puzzling that so much variation

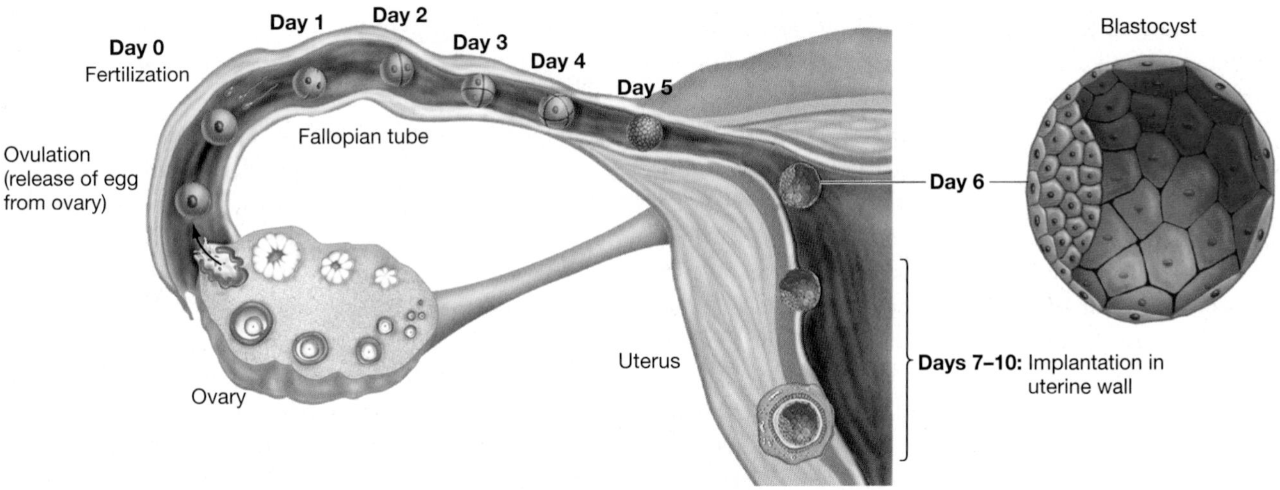

FIGURE 21.23 In Humans, Cleavage Occurs before Implantation into the Uterus
After the blastocyst produced by cleavage is implanted in the uterine wall, gastrulation and organogenesis occur, and the placenta begins to form from a combination of embryonic and maternal cells.

exists in processes as fundamental as gametogenesis, fertilization, and early development. Why do so many differences exist?

Part of the variation observed in developmental patterns is undoubtedly due to natural selection. Recall from Chapter 1 that natural selection occurs when individuals with certain genetically based traits produce more offspring than do individuals without those traits. In the case of development, natural selection can favor certain patterns or events over others. For example, the evolution of larval forms in groups such as insects and frogs was probably favored by natural selection because it allowed juveniles and adults of the same species to exploit different food sources and thus avoid competing with each other. The evolution of the placenta and pregnancy is thought to have been advantageous because of the protection they offer the developing embryo. Similarly, the seed is seen as a trait that protects the embryo during early development and allows it to be dispersed to a location away from the parent plant prior to germination.

Amid this variation, though, several common themes emerge about developmental patterns:

- In both plants and animals, the earliest events in development result in asymmetries in the embryo that establish fundamental aspects of the adult body plan. In plants, an asymmetrical initial division sets up the root-shoot body axis. In most animals, either cleavage or gastrulation makes the head-to-tail and back-to-belly axes of the embryonic and adult bodies visible.
- Although the details of how cleavage occurs vary among animal species, the outcome is the same. Cleavage divides the maternal cytoplasm, produces blastomeres that contain specific cytoplasmic determinants and other maternal factors, and creates a sphere of cells that undergoes gastrulation.
- In both plants and animals, early development results in the formation of three embryonic tissues, which give rise to all the different adult tissues and structures. In animals, gastrulation is responsible for organizing these three populations of cells into distinct layers.

ESSAY Treating Human Infertility

In 1978 the world welcomed the first "test-tube baby" when Louise Brown was born in Britain. Unlike babies born previously, Louise did not develop from a zygote produced by natural conception. Instead, she developed from a zygote created by fertilization in a laboratory dish. Since then, over 100,000 babies have been produced by the technique called **in vitro fertilization (IVF)**.

Many human couples are **infertile**—they are unable to conceive children naturally. For them, IVF can be an important alternative strategy for becoming pregnant. In IVF, egg cells isolated from the woman's ovaries are combined with sperm cells collected from the man's semen.

The gametes are placed in small dishes, under conditions that attempt to duplicate the environment inside the female reproductive tract. Typically, 50,000 to 5 million sperm are used per egg. As in natural conception, fertilization occurs when a healthy sperm penetrates the zona pellucida and fuses with the egg. The resulting embryos are then surgically implanted in the mother's uterus.

IVF is often successful for couples whose infertility results from problems in the woman's reproductive system, such as damaged or absent fallopian tubes. But for up to 40 percent of infertile couples, the problem stems from low sperm counts, poor sperm motility, deformed sperm, or other male problems. Traditional IVF seldom works with such couples, because it requires large numbers of high-quality sperm. In the early 1990s, though, improvements to IVF were introduced that provided new hope for infertile men. In one technique, called **subzonal injection**, or SUZI for short, a concentrated sample of sperm is examined under a microscope. Five to 10 of the healthiest sperm are drawn into a microneedle and then injected directly into the zona pellucida. Depositing the sperm closer to the egg increases the chances of sperm-egg fusion. SUZI may not help men who produce very few sperm, however, and it leads to polyspermy and inviable embryos in about one-third of attempts. A more recent technique called **intracytoplasmic sperm injection** (ICSI, pronounced "*ICK-see*") overcomes these problems. In ICSI, a single sperm is drawn into a microneedle and injected into the cytoplasm of the egg (**Figure 21.24**). In a significant fraction of attempts, ICSI results in the fusion of sperm and egg nuclei and a viable embryo.

Even if IVF, SUZI, or ICSI is successful, only a minority of couples achieve pregnancy when an embryo is transferred to the uterus.

Even if IVF, SUZI, or ICSI is successful, only a minority of couples achieve pregnancy when an embryo is transferred to the uterus. In some cases, failure results from problems with implantation in the uterine wall. To develop in the mother's womb, the embryo must break out or hatch from the zona pellucida and implant into the uterine wall. Eggs from women over age 40 often have problems hatching. To help overcome these problems, a physician may assist hatching by using a micropipette to deliver a small amount of acid or enzyme solution to the zona pellucida. This treatment facilitates hatching by dissolving the treated region of the zona.

Because of IVF and other assisted reproductive technologies, the chances of naturally infertile couples conceiving their own children have greatly increased. An intensive research program undertaken over the past three decades has led to huge advances in our understanding of human fertilization and early development, and has allowed many couples to experience the joy of raising their own children.

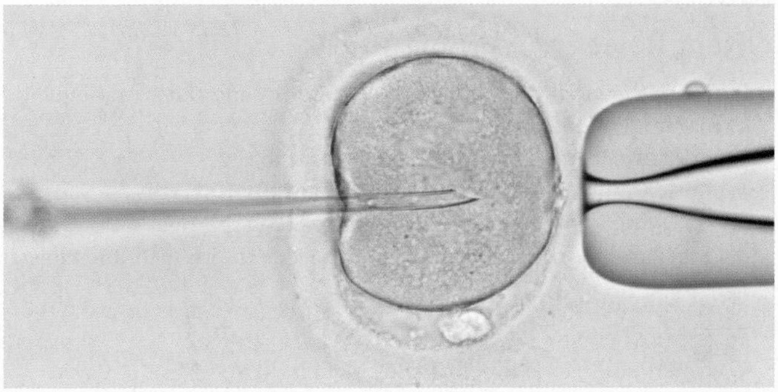

FIGURE 21.24 Human Sperm Can Be Injected Directly into an Egg A human egg undergoing intracytoplasmic sperm injection (ICSI).

CHAPTER REVIEW

Summary of Key Concepts

In most animals, fertilization depends on specific interactions among proteins on the plasma membranes of sperm and egg.

The development of an organism begins with the formation of gametes in its parents. In plants and animals, sperm cells contribute a haploid genome to the embryo; depending on the species involved, the sperm cell may also contribute a few other cell components. Eggs, in contrast, contribute a haploid genome and a large amount of cytoplasm to the embryo.

Fertilization is the best known of all cell-to-cell interactions, primarily because of observational and experimental studies on sea urchin eggs and sperm. When a sea urchin sperm contacts the jelly layer surrounding the egg, the acrosome in the sperm head releases digestive enzymes, and microfilaments polymerize to form a protrusion that extends to the vitelline envelope. When the sperm reaches the egg, a protein called bindin on the sperm-head membrane binds to a receptor on the vitelline envelope surrounding the egg. Because bindin and egg-cell receptors are species specific, cross-species fertilizations are prevented. After the sperm and egg cells' plasma membranes make contact and fuse, a wave of calcium ions is released from stores inside the egg. The increase in Ca^{2+} concentration causes cortical granules to fuse with the egg cell's plasma membrane. The contents of the cortical granules cause a fertilization envelope to lift off the egg cell's plasma membrane and protect against multiple fertilizations.

The fertilization sequence is similar in mammals, except that no bindin-like protein has yet been identified in sperm. Plants, in contrast, undergo a very different series of events. In flowering plants, two sperm nuclei leave the germinating pollen grain and migrate to the ovule. One sperm nucleus fertilizes the egg to form a zygote, while the other sperm nucleus fuses with two maternal nuclei to form the nutritive tissue called endosperm.

In animals, the earliest cell divisions divide the fertilized egg into a mass of cells. These cells are not all alike; eventually they give rise to different types of cells and tissues as development proceeds.

Animal development begins with cleavage, a series of cell divisions that divide the egg cytoplasm into a large number of cells. During cleavage, an array of cytoplasmic determinants and other maternal factors are apportioned to different blastomeres. As a result, different blastomeres have different fates.

Once cleavage is complete, the embryo consists of a sphere of cells. Fate-mapping studies with dyes and other tools have shown that cells undergo massive movements during gastrulation. These movements arrange the three embryonic tissues in layers and make the back-to-belly and head-to-tail axes of the body visible. In vertebrates, a region of cells near the dorsal lip of the blastopore acts as an organizer during gastrulation. Cells in the organizer secrete the protein noggin and other cell-to-cell signals that induce changes in target cells.

Patterns and rates of embryonic development vary widely among species of plants and animals. Cleavage and gastrulation are nearly universal in animals, but gastrulation does not occur at all in plants. Among animals, some species develop into distinct larval and adult stages, while others do not. The time to adulthood may be extremely fast, as in *Arabidopsis thaliana* or *Drosophila melanogaster*, or slow, as in humans.

Web Tutorial 21.1 Early Stages of Animal Development

In animals as well as plants, the early events in development result in the production of three embryonic tissues and the establishment of the major body axes.

Common themes unify the diversity of ways that plants and animals develop. For example, in both plants and animals, the earliest events in development make the major body axes visible and result in the formation of three types of embryonic tissues, which later give rise to distinct larval or adult structures.

Web Tutorial 21.2 The Gray Crescent in Frog Eggs

Questions

Content Review

1. How are the vitelline envelope of sea urchins and the zona pellucida of mammals similar?
   a. Both are gelatinous coats that protect the egg.
   b. Both hold cortical granules, which block polyspermy.
   c. Both hold stores of Ca^{2+}, which block polyspermy.
   d. Both are an extracellular matrix that sperm bind to and digest.
2. What happens during the acrosomal reaction?
   a. Bindin binds to the egg-cell receptor for sperm.
   b. The sperm and egg cells' plasma membranes fuse.
   c. Enzymes that digest the egg's jelly layer are released, and microfilaments in the tip of the sperm head polymerize to form a protrusion.
   d. The centriole released from the sperm orients microtubules in the fertilized egg and causes the cytoplasm to rotate—creating the gray crescent.
3. Many flowering plant species have elaborate mechanisms to prevent an individual's pollen from germinating on its own female reproductive parts. Why?
   a. to prevent self-fertilization and inbreeding
   b. to prevent polyspermy
   c. to prevent cross-species fertilization and the production of dysfunctional hybrid offspring
   d. to prevent double fertilization and the formation of endosperm
4. What happens during cleavage?
   a. The neural tube—precursor of the spinal cord and brain—forms.
   b. Basal and apical cells—precursors of the suspensor and embryo, respectively—form.
   c. The fertilized egg divides without growth occurring, forming a ball of cells.
   d. Massive movements of cells make the primary body axes visible and organize the three embryonic tissues.
5. What happens during gastrulation?
   a. The neural tube—precursor of the spinal cord and brain—forms.
   b. Basal and apical cells—precursors of the suspensor and embryo, respectively—form.
   c. The fertilized egg divides without growth occurring, forming a ball of cells.
   d. Massive movements of cells make the primary body axes visible and organize the three embryonic tissues.
6. In animals, which adult tissues and organs are ectoderm-derived?
   a. lining of the digestive tract and associated organs
   b. blood, heart, kidney, bone, and muscle
   c. nerve cells and skin
   d. blastopore and blastocoel

Conceptual Review

1. List the contributions to offspring that are made by a sperm versus an egg. Given a sample of gametes from a new species, how would you know which gametes were eggs and which were sperm?
2. Why is it logical that sperm-egg interactions are species specific in sea urchins but not in mammals?
3. Compare and contrast yolk and endosperm. How are they similar in structure and function, and how are they different? Be sure to compare when and how they are produced.
4. How did analyses of cleavage and shell-coiling patterns in snails support the hypothesis that cleavage is controlled by cytoplasmic determinants present in the egg? Does the observation that transcription does not start in some animals until cleavage is complete support this hypothesis or challenge it? Explain.
5. Compare and contrast cytoplasmic determinants and cell-to-cell signals. Explain why biologists would say that *Vg1* is a maternal factor that acts as a cell-to-cell signal.
6. What does it mean to say that a cell has a certain fate? Why do some cells in early embryos have different fates than other cells? Support your answer with specific examples.

Group Discussion Problems

1. Many questions about fertilization and early development remain unanswered:
   - Is there a protein on the sperm head of mammals that binds specifically to the glycoprotein ZP3 in the zona pellucida?
   - How are cytoplasmic determinants and factors localized to specific regions of the egg? How is yolk localized?
   - If the zygotic genome is transcribed at the two-cell stage in mammals, do cytoplasmic determinants play a significant role in early development?
   - What forces cause cells to move during gastrulation?

   Choose one of these questions and design an experiment that would contribute to answering it.
2. Suppose you are a physician at a fertility clinic. A couple seeks your assistance with becoming pregnant. The woman is 41 years old, and the man has a low sperm count. Which treatment would you recommend, and why?
3. In a lizard species native to Mexico, some populations lay eggs; others bear live young. How could you use these populations to test the hypothesis that bearing live young is favored by natural selection in certain environments because offspring are better protected?
4. Suppose there is a drug that leads to direct development when administered to frog eggs—meaning that the larval (tadpole) stage is skipped. How could you use this drug as an experimental tool to test the hypothesis that natural selection favors the development of larval forms because of reduced competition with adults for food?

Answers to Multiple-Choice Questions 1. d; 2. c; 3. a; 4. c; 5. d; 6. c

www.prenhall.com/freeman is your resource for the following: Web Tutorials; Online Quizzes and other Online Study Guide materials; Answers to Conceptual Review Questions; Solutions to Group Discussion Problems; Answers to Figure Caption Questions and Exercises; and Additional Readings and Research.

Pattern Formation and Cell Differentiation

22

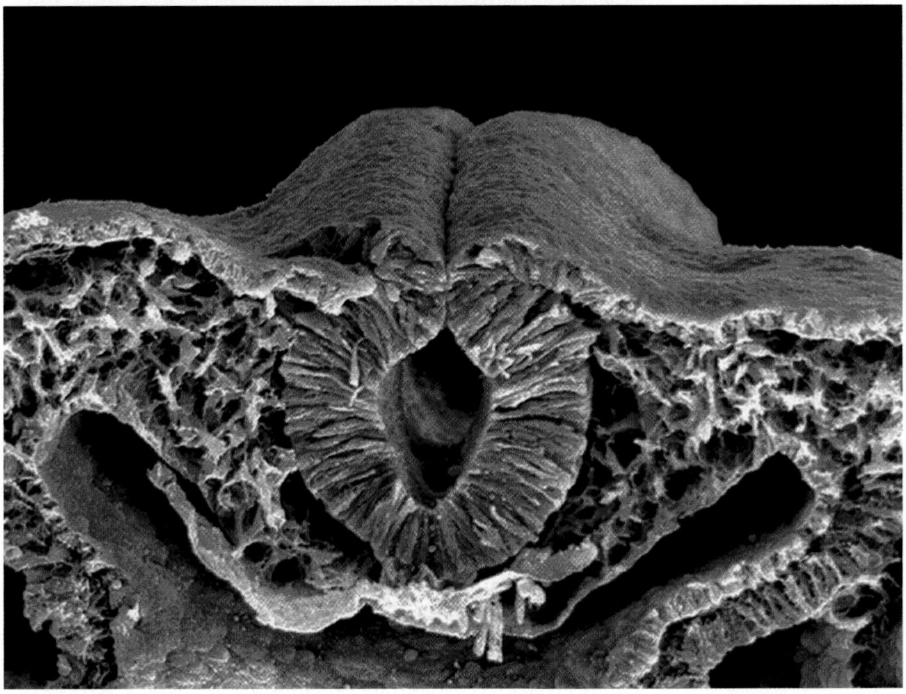

Artificially colored scanning electron micrograph of a chick embryo. Ectodermal cells along the back of the embryo (top of photograph) have folded into a neural tube. The neural tube will develop into the brain and spinal cord of the adult. This chapter explores how cells form specific structures based on their position in the embryo and then differentiate to form specialized cells such as nerve or muscle.

KEY CONCEPTS

- The fate of an embryonic cell—what it will become in the juvenile and adult organism—is established in a series of steps.
- At each step in development, cell-to-cell signals activate regulatory transcription factors that direct the expression of genes in cells throughout the embryo.
- Because virtually all cells in a multicellular organism contain the same genes, development is based on differential gene expression.

When the physician, researcher, and writer Lewis Thomas considered how a human being develops from a fertilized egg, he could only marvel: "You start out as a single cell derived from the coupling of a sperm and an egg, this divides into two, then four, then eight, and so on, and at a certain stage there emerges a single cell which will have as all its progeny the human brain. The mere existence of that cell should be one of the great astonishments of the earth. People ought to be walking around all day, all through their waking hours, calling to each other in endless wonderment, talking of nothing except that cell. It is an unbelievable thing, and yet there it is, popping neatly into its place amid the jumbled cells of every one of the several billion human embryos around the planet, just as if it were the easiest thing in the world to do."[1]

Many biologists share Thomas's feelings. The experience of watching a fertilized egg go through early development inspires awe and wonder. In animals, the zygote divides during cleavage. The resulting cells then begin the massive, coordinated movements of gastrulation, and the embryo is dramatically reorganized. Just a few hours or days later, a recognizable creature with a head, eyes, back, belly, and other structures appears. In plants, the fertilized egg divides to form

[1]Thomas, L. 1979. *The Medusa and the Snail* (New York: Viking Press), p. 156. Note that Thomas was exercising some poetic license here. The brain actually arises from a group of cells in the embryo rather than from a single cell.

two simple-looking cells—one small and one large. The small cell divides repeatedly to form a seemingly disorganized clump of cells. But as this clump grows and lengthens, tiny leaves and a root appear. A mature seedling takes shape a short time later, ready to germinate and begin a life that may last a few weeks or several thousand years.

Understanding how these events happen is among the greatest of all challenges facing biologists today. Chapter 21 introduced this challenge by providing an overview of gametogenesis, cleavage, and gastrulation. It also considered key events in the development of *Arabidopsis*, *Drosophila*, frogs, and humans. As you learned in that discussion, the establishment of the major body axes and the formation of embryonic tissues are among the most important of all early developmental events. Chapter 21 also began exploring how these key events happen, by introducing the role of cytoplasmic determinants and cell-to-cell signals in establishing the fate of embryonic cells. Recall that cytoplasmic determinants are molecules that exist in eggs and that help direct early development.

This chapter begins with a look at how the overall body plan of an animal or plant takes shape. Section 22.1 introduces work on embryonic development in the fruit fly *Drosophila melanogaster*—the best studied of all multicellular organisms. Research on fruit flies has emphasized the importance of a cell's position in determining its fate. Position is important because a cell's fate is specified by localized cytoplasmic determinants and signals from other cells. The same themes carry over to Section 22.2, which examines recent research on the best studied of all plants, the mustard *Arabidopsis thaliana*.

Once the overall form of the body is established and cells are arranged in space, cells start becoming organized into tissues and organs. Eventually cells differentiate and begin to express tissue-specific genes. Section 22.3 explains how tissue-specific gene expression occurs by exploring research on vertebrates—specifically, the transformation of mesoderm into cells that express muscle-specific proteins.

The fundamental messages of this chapter are that patterns of development are established early in embryogenesis, that a single cell's fate during development depends on its position within the embryo, and that its fate is realized in a progression of steps—each mediated by specific regulatory proteins or cell-to-cell signals. For example, a cytoplasmic determinant may mark a blastomere as mesoderm. After gastrulation, other molecules may signal that the cell is now part of the mesoderm on the embryo's dorsal (back) side. Subsequent signals might indicate that this particular cell will become part of the dorsal mesoderm that contributes to muscle or bone. Finally, the cell may receive signals that trigger the production of muscle-specific proteins. A generalized embryonic cell has become a specialized mature cell. How does this process begin? If a cell's fate depends on its position, how does it "know" where it is in the body?

WEB TUTORIAL 22.1
Early Pattern Formation in *Drosophila*

22.1 Pattern Formation in *Drosophila*

Biologists refer to the events that determine the spatial organization of an embryo as **pattern formation**. If a molecule signals that a target cell is in the embryo's head, or tail, or dorsal side, or ventral side, it is involved in pattern formation. Pattern formation is the first step in determining a cell's fate. Before a cell becomes part of a muscle or nerve or gut lining, it receives precise information about where it is positioned in the body.

To understand how cells get information that specifies their position inside the embryo, biologists turned to the fruit fly *Drosophila melanogaster*. Because fruit flies produce large numbers of offspring rapidly, researchers could survey laboratory populations for rare mutant embryos in which the normal spatial relationships among cells are disrupted. In other words, biologists took a genetic approach to studying development. The idea was that identifying individuals with abnormal pattern formation would make it possible to find the genes and gene products responsible for normal development. A *Drosophila* larva that lacks a head, for example, is likely to have a mutation in a gene that helps pattern the head. This approach has turned out to be extraordinarily productive. The genes that direct pattern formation in fruit flies also direct pattern formation in mice, humans, and other animals.

The effort to dissect pattern formation in *Drosophila* began in the 1970s, when Christiane Nüsslein-Volhard and Eric Wieschaus undertook a massive effort to identify pattern-formation mutants. They began by exposing adult flies to treatments that cause mutations by damaging DNA. Later, the researchers examined embryos or larvae descended from these individuals for body plan defects. After intensive effort, Nüsslein-Volhard and Wieschaus were able to identify over 100 genes that play fundamental roles in pattern formation.

The Discovery of *bicoid*

A normal fruit-fly embryo has three types of body segments: head (anterior), thoracic (middle), and abdominal (posterior) segments (**Figure 22.1a**). One of the most dramatic mutations that Nüsslein-Volhard and Wieschaus analyzed is illustrated in **Figure 22.1b**. Embryos with that mutation are missing all of the structures normally found in the anterior end. These mutants have duplicated posterior structures. The gene responsible for this phenotype is called *bicoid*, meaning "two tailed." The mutation is lethal, because larvae cannot develop beyond the stage shown in the figure. Based on its phenotype, Nüsslein-Volhard and Wieschaus suspected that the *bicoid* gene's product plays a role in pattern formation along the anterior-posterior body axis.

In a series of mating experiments with flies that carried the mutant *bicoid* allele, Nüsslein-Volhard and Wieschaus found that the gene displays **maternal effect inheritance**—a pattern of inheritance in which a gene is transcribed in the mother and affects the phenotype observed in offspring. As **Box 22.1** explains, it is the mother's genotype and not the offspring's genotype that determines the expression of the trait. Like the

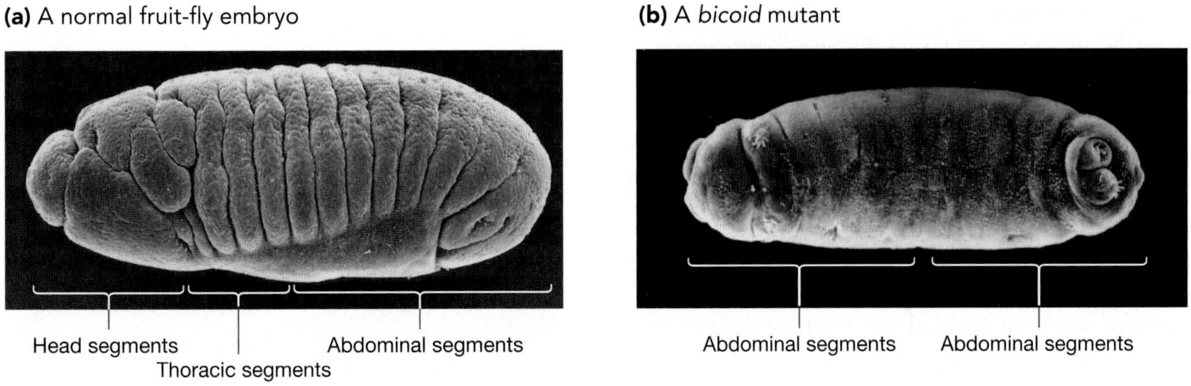
(a) A normal fruit-fly embryo
Head segments
Thoracic segments
Abdominal segments
(b) A *bicoid* mutant
Abdominal segments
Abdominal segments

FIGURE 22.1 Pattern-Formation Mutants Have Misshapen Bodies
(a) Normal embryos have distinct head, thorax, and abdominal regions. **(b)** In *bicoid* mutants, embryos have no head or thoracic segments. Instead they have duplicated sets of the posteriormost regions, making them "two tailed."

BOX 22.1 Maternal Effect Inheritance

When researchers find a gene that appears to be important early in development, one of their first goals is to establish how the gene is inherited. Breeding experiments can answer a fundamental question about the gene: Is it expressed in the mother, meaning it codes for a cytoplasmic determinant present in the egg, or is it expressed in offspring? That is, does the gene have a maternal effect or a zygotic effect?

Crosses are set up to provide an answer. In one example, the mutant gene being considered is recessive. Investigators arrange matings between parents that are heterozygous at this locus. The two sets of crosses show the contrasting patterns of inheritance in a gene that is expressed in offspring (**Figure 22.2a**) versus a maternal effect gene (**Figure 22.2b**). The key point is that the zygotic effect hypothesis and the maternal effect hypothesis make contrasting predictions:

- *Zygotic effect hypothesis*: If the gene is expressed in offspring, then offspring will have the mutant phenotype if they are homozygous for the mutant gene—no matter what their mother's genotype.
- *Maternal effect hypothesis*: If the gene codes for a cytoplasmic determinant, then offspring will have the mutant phenotype only if their mother is homozygous for the mutant gene.

Researchers who performed this experiment with mutant and normal alleles of *bicoid* discovered that the second pattern was correct: *bicoid* displays maternal effect inheritance and has no zygotic effect.

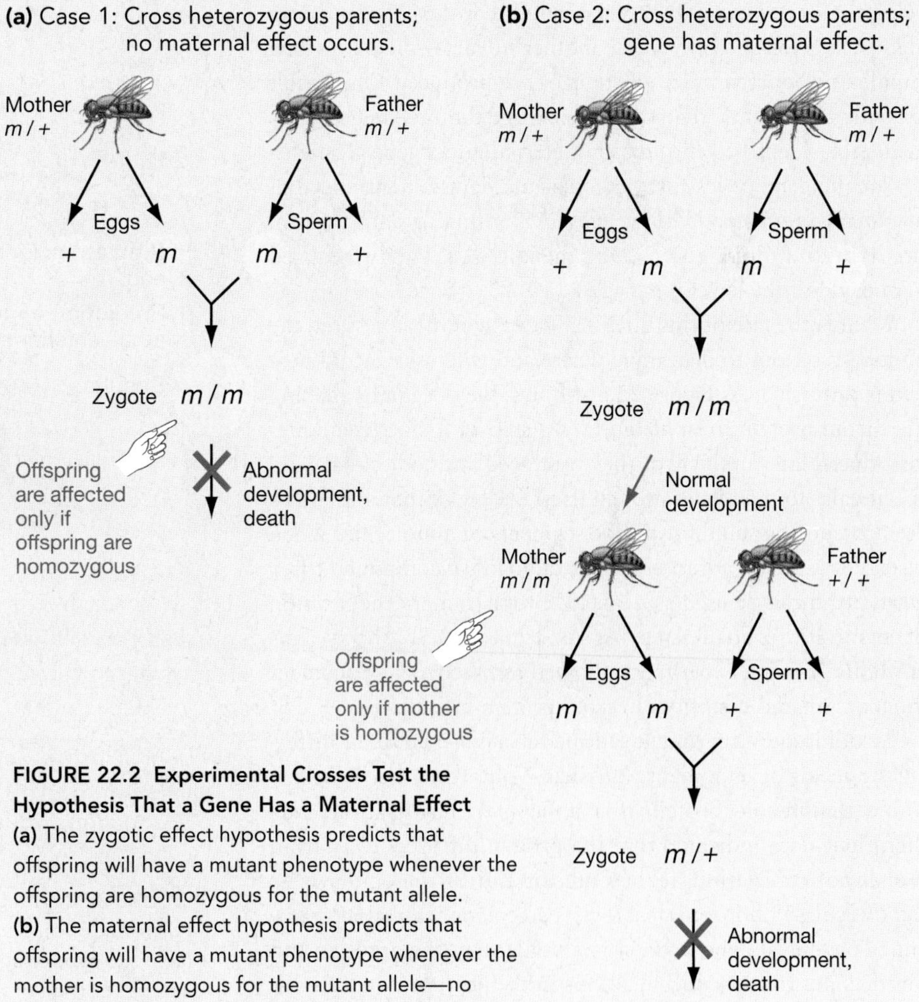
(a) Case 1: Cross heterozygous parents; no maternal effect occurs.
Mother m / +
Father m / +
Eggs
Sperm
+ m m +
Zygote m / m
Offspring are affected only if offspring are homozygous
Abnormal development, death
(b) Case 2: Cross heterozygous parents; gene has maternal effect.
Mother m / +
Father m / +
Eggs
Sperm
+ m m +
Zygote m / m
Normal development
Mother m / m
Father + / +
Offspring are affected only if mother is homozygous
Eggs
Sperm
m m + +
Zygote m / +
Abnormal development, death

FIGURE 22.2 Experimental Crosses Test the Hypothesis That a Gene Has a Maternal Effect
(a) The zygotic effect hypothesis predicts that offspring will have a mutant phenotype whenever the offspring are homozygous for the mutant allele.
(b) The maternal effect hypothesis predicts that offspring will have a mutant phenotype whenever the mother is homozygous for the mutant allele—no matter what the offspring's genotype is.

snail-shell-coiling gene introduced in Chapter 21, *bicoid* affects the development of an embryo because it is expressed in the mother's tissues during egg formation. The Bicoid protein is a cytoplasmic determinant. The *bicoid* mRNA is deposited in the egg along with other maternal determinants and yolk.

What Does *bicoid* Do? To determine how the *bicoid* product works, Nüsslein-Volhard and colleagues crossed individuals carrying the mutant form of *bicoid* with individuals carrying other known genetic markers. This linkage mapping technique, introduced in Chapter 13 and explained in more detail in Chapter 19, allowed the team to map *bicoid*'s location on the *Drosophila* chromosomes. Once the location of the gene was established, the researchers were able to isolate and analyze the DNA sequence that encodes the *bicoid* product. Once they had many copies of the DNA sequence, they used the in situ hybridization technique described in Box 21.1 to locate *bicoid* mRNA in embryos and adults. The investigators added fluorescent or radioactive markers to single-stranded copies of *bicoid* DNA and then treated preserved *Drosophila* adults and embryos with the labeled copies. The probe bound to *bicoid* mRNA in cells where the gene was being transcribed.

When Nüsslein-Volhard's group treated adult flies with labeled copies of *bicoid* DNA, they found that the corresponding mRNA was located only in the mother's ovary—the part of the female reproductive tract where eggs are produced. This result was consistent with the conclusion from the breeding experiments that identified *bicoid* as a maternal effect gene. The idea is that the gene codes for a cytoplasmic determinant and that *bicoid* mRNA is loaded into eggs. According to data from in situ hybridization (**Figure 22.3a**), the mRNA is highly localized in one end of the developing egg.

What is the fate of this mRNA? To answer this question, the biologists set out to determine where and when the Bicoid protein is produced. As **Figure 22.3b** shows, they began by making the Bicoid protein from *bicoid* DNA (step 1). Using techniques introduced later in the text, they produced antibodies that bind specifically to the Bicoid protein (step 2). Recall that antibodies are proteins that bind to specific segments of a molecule. When researchers attach a fluorescent or radioactive compound to an antibody, it can be used as a labeled probe to mark the location of a specific protein (step 3). As Figure 22.3b shows, this antibody-staining experiment allowed researchers to document the location and quantity of Bicoid protein in embryos.

By doing the experiment outlined in Figure 22.3b at different stages of development, Nüsslein-Volhard and co-workers found that Bicoid protein first appears at fertilization. Further, their data indicated that the protein diffused away from the site of translation at the anterior end of the embryo. As the "Results" photograph shows, a steep concentration gradient is produced: The protein is abundant in the anterior end but declines to low concentrations in the posterior end. Recall from Chapter 21 that at this stage in fly development, nuclei

(a) Where is *bicoid* mRNA located?

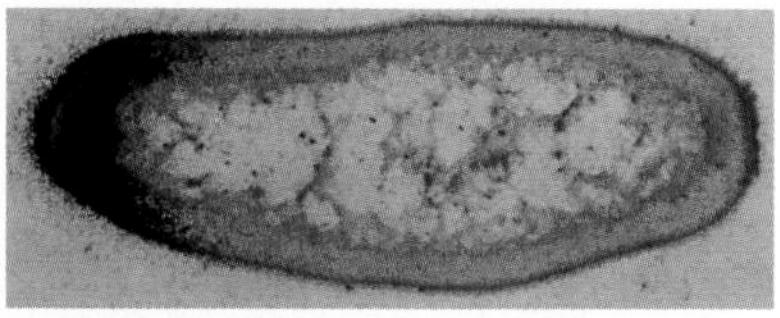

(b) Question: Is Bicoid protein as highly localized as *bicoid* mRNA?

Hypothesis: Bicoid protein is found only in the same location as *bicoid* mRNA.

Null hypothesis: Bicoid protein is not found in the same location as *bicoid* mRNA.

Experimental setup:

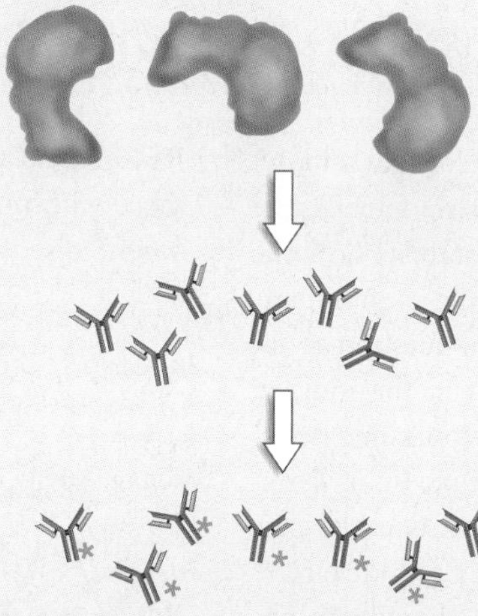

1. Make many copies of Bicoid protein from *bicoid* DNA.

2. Using Bicoid protein, make antibodies that bind specifically to Bicoid protein.

3. Attach a label to antibodies. Add labeled antibodies to embryo, where they will bind to Bicoid protein.

Prediction: Bicoid protein is found only at the anterior end.

Prediction of null hypothesis: Bicoid protein is not found at the anterior end.

Results:

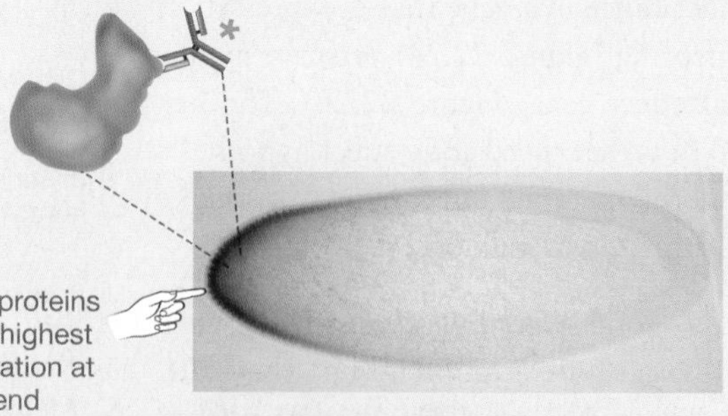

Conclusion: Both hypotheses are partially correct: Bicoid protein forms an anterior-posterior concentration gradient.

FIGURE 22.3 The *bicoid* mRNA Is Localized, and Bicoid Protein Forms a Concentration Gradient
(a) In situ hybridization shows that *bicoid* mRNA is sequestered in the anterior portion of the egg. **(b)** Bicoid protein forms an anterior-posterior concentration gradient.

are scattered throughout the embryo after repeated mitoses occur without cytokinesis. When these nuclei become surrounded by plasma membranes, the Bicoid protein is trapped in the newly formed cells. In other animal and plant species, cytoplasmic determinants like *bicoid* are localized in particular cells during cleavage.

To pull these observations together, Nüsslein-Volhard and co-workers hypothesized that high concentrations of Bicoid protein inside cells lead to the formation of anterior structures such as the head, with progressively lower concentrations giving rise to thoracic segments and the first abdominal segments. Absence of Bicoid, in contrast, results in formation of posterior structures. This hypothesis provided an explanation for the original *bicoid* mutants that the researchers had observed. Because the eggs of mothers homozygous for the mutant form of *bicoid* lack functional mRNAs, no Bicoid protein is produced anywhere in the embryo. The result is a larva in which anterior structures have been replaced by duplicated posterior structures.

Testing the Gradient Hypothesis The hypothesis that a Bicoid gradient sets up the head-to-tail axis of the fly embryo is consistent with the data from antibody-staining experiments. To test the hypothesis more rigorously, Nüsslein-Volhard and colleagues isolated *bicoid* mRNA from wild-type individuals and injected it into eggs from females that lacked a functioning copy of the *bicoid* gene. Because the treated eggs developed normally, the researchers could infer that *bicoid* is responsible for establishing a normal anterior-posterior gradient.

In addition to "rescuing" mutant larvae by injecting *bicoid* mRNA, the researchers tested the gradient hypothesis by artificially altering the distribution of Bicoid protein in developing embryos and determining the effects on the body plan. For example, the biologists produced purified *bicoid* mRNA and injected it into wild-type embryos. The treated individuals developed head structures at the site of injection—even at the posterior pole. Observations such as these provided strong support for the gradient hypothesis.

Bicoid Is a Regulatory Transcription Factor The Bicoid protein appears to provide cells with information about where they are along the anterior-posterior, or head-to-tail, axis of the embryo. In addition, the in situ hybridization and antibody-staining experiments indicated that Bicoid was active when the fly embryo consisted of many nuclei scattered throughout the egg cell's cytoplasm. How does Bicoid act on these target nuclei?

DNA sequencing studies helped answer this question. The *bicoid* gene contains sequences that are typical of regulatory transcription factors—the proteins that bind to enhancers or other regulatory sequences in DNA and control the transcription of specific genes (see Chapter 18). Based on this observation, biologists concluded that the Bicoid protein must enter nuclei, bind to DNA, and either increase or decrease the expression of specific genes. Bicoid turns on genes responsible for forming anterior structures, shuts down the transcription of genes responsible for producing posterior structures, or both.

To summarize, a gradient in Bicoid protein concentration provides cells with information about their position along the anterior-posterior body axis. Because it is sequestered in a precise location in the egg and because it affects the fate of cells later in development, the *bicoid* gene product can be considered a cytoplasmic determinant. The absence of the *bicoid* product leads to a deformed body and death. As other research showed, however, *bicoid* is just one of dozens of pattern-formation genes found in fruit flies.

The Discovery of Segmentation Genes

Among the dozens of mutants isolated by Nüsslein-Volhard and Wieschaus, one called *hunchback* attracted a great deal of attention. Embryos with mutant forms of *hunchback* resemble *bicoid* mutants. Both types of embryos lack a series of consecutive segments from the anterior end. Breeding experiments showed that the *hunchback* mRNA responsible for these mutations is produced in the embryo itself as well as in the mother, however. Unlike *bicoid*, *hunchback* is not a cytoplasmic determinant.

To understand what *hunchback* does, recall from Chapter 21 that the bodies of fly larva and adults are partitioned into a series of segments. The *hunchback* gene turned out to be one of many zygotic genes that affect the identity or arrangement of these segments. Nüsslein-Volhard and Wieschaus were eventually able to identify three general classes of **segmentation genes**—genes that affect segmentation. Except for *hunchback*, all of these genes are transcribed for the first time in the embryo and not in the mother. The three types of segmentation genes are called **gap genes**, **pair-rule genes**, and **segment polarity genes**. Gap gene mutants (such as those with *hunchback*) lack several consecutive segments. Pair-rule mutants lack alternate segments and have just half the normal total number of segments. Segment polarity mutants lack portions of each segment. In many segment polarity mutants, the missing portion is replaced by a mirror-image duplication of the intact part of the same segment.

Figure 22.4 (page 474) shows the patterns of mRNA expression for segmentation genes in normal larvae. According to these in situ hybridizations, gap genes are expressed in broad regions along the head-to-tail axis (Figure 22.4a). Pair-rule genes, in contrast, are expressed in alternate segments (Figure 22.4b), while segment polarity genes are expressed in restricted regions of each segment (Figure 22.4c). If a segmentation gene does not function, the mutant individual lacks segments or parts of segments where the mRNA is found in normal individuals. Based on these data, Nüsslein-Volhard and Wieschaus and co-workers concluded that the segmentation genes are responsible for defining the segmented body plan of the fruit fly.

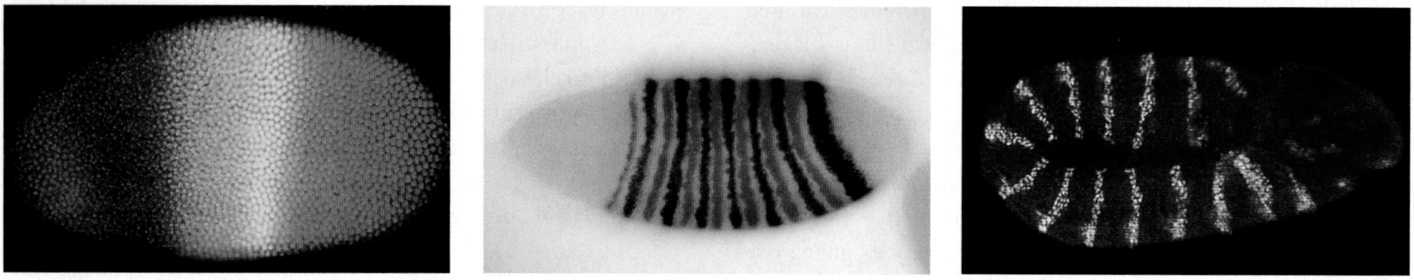
(a) Gap genes: expressed in groups of segments
(b) Pair-rule genes: expressed in alternate segments
(c) Segment polarity genes: expressed in portions of segments

FIGURE 22.4 A Series of Genes Demarcates Body Segments in Fruit Flies
In situ hybridizations show the location of segmentation gene products in fly embryos. The embryos in **(a)** and **(b)** two were stained with antibodies for two different gene products. The embryo in **(c)** was stained for one gene product. QUESTION From (a) to (c), these photographs show embryos at progressively later stages of development. Why is the sequence logical?

Do Segmentation Genes Interact? Researchers who first identified the segmentation genes and described when and where they are expressed noticed that they are expressed in sequence. The gap genes are expressed first, followed by pair-rule genes, followed by segment polarity genes. This observation was logical, because gap, pair-rule, and segment polarity genes are expressed in increasingly restricted regions. The observation also suggested that the segmentation genes might interact in some way. Specifically, investigators hypothesized that the gap genes may activate the transcription of pair-rule genes, which in turn may trigger the expression of other pair-rule genes and segment polarity genes. In this way, the pattern of an embryo's segments would form in a step-by-step fashion.

Although research on interactions among segmentation gene products is continuing, an enormous amount of progress has already been made. One of the key early discoveries was that the Bicoid protein directly activates transcription of *hunchback*. The *hunchback* gene is not expressed until the Bicoid protein binds to its regulatory DNA sequences. As a result, the *hunchback* mRNA protein is found in high concentration in the anterior part of the embryo, where the Bicoid protein is abundant. Subsequent work showed that *hunchback* and other gap genes encode transcription factors that regulate the expression of pair-rule genes. Each gap gene regulates the production of a defined set of pair-rule proteins. In addition, several of the gap genes' products regulate the expression of other gap genes or even their own expression. The pair-rule genes activated by the gap genes also encode regulatory transcription factors, and some of the genes they regulate are segment polarity genes. Based on these data, the step-by-step hypothesis for establishing segment identity appears to be correct.

As **Figure 22.5** shows, this sequence of events can be described as a **regulatory cascade** or hierarchy. The *bicoid* gene, gap genes, pair-rule genes, and segment polarity genes each define a level in the cascade. As development proceeds, genes at levels farther down in the cascade are progressively activated or repressed. The products of these regulatory genes may also interact with genes at the same level.

To appreciate just how complex this regulatory cascade is, consider that the pair-rule gene *even-skipped* has at least five distinct promoters that control its expression. One of these promoters has six binding sites for Krüppel protein, three binding sites for Hunchback protein, three binding sites for Giant protein, and five binding sites for Bicoid protein. For the *even-skipped* gene to be expressed at the maximum level, each of these regulatory transcription factors must be present at a high concentration and bound to the promoter.

The regulatory cascade described in Figure 22.5 results in the progressive subdivision of the body's anterior-posterior axis into smaller and smaller units. A cell first receives information about its position on the body's anterior-posterior axis. Then gap gene products identify the group of segments that will form around the cell. Pair-rule gene products identify the particular segment that will form at that location, while segment polarity gene products provide information on position within that segment.

At each step in the developmental sequence, different cells receive a different combination of signals, and a different set of genes in the cascade is activated or shut down. In early fly embryos, hundreds of different regulatory transcription factors and dozens of different cell-to-cell signals affect the fate of cells. The combination of signals that a particular cell receives depends on its position in the embryo, and each signal causes a change in gene expression. Thus, long before a cell begins expressing tissue-specific proteins that identify it as a muscle cell or a nerve cell, it expresses genes that identify it as a cell in a particular segment or region of the body. This is the essence of pattern formation.

The Nature of Developmental Signals As the in situ hybridization and antibody-staining results in Figures 22.3 and 22.4 show, the gene products that regulate development often

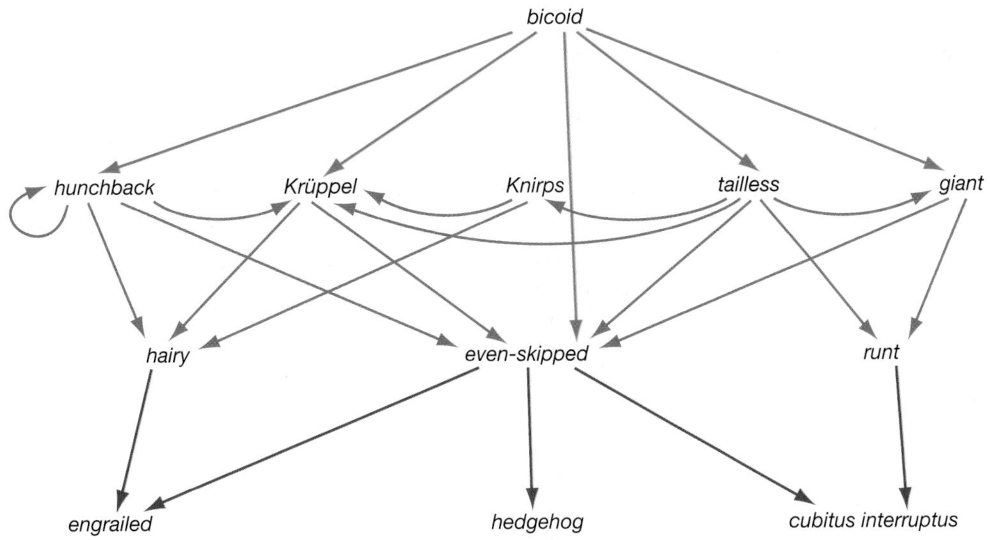
bicoid
hunchback
Krüppel
Knirps
tailless
giant
hairy
even-skipped
runt
engrailed
hedgehog
cubitus interruptus

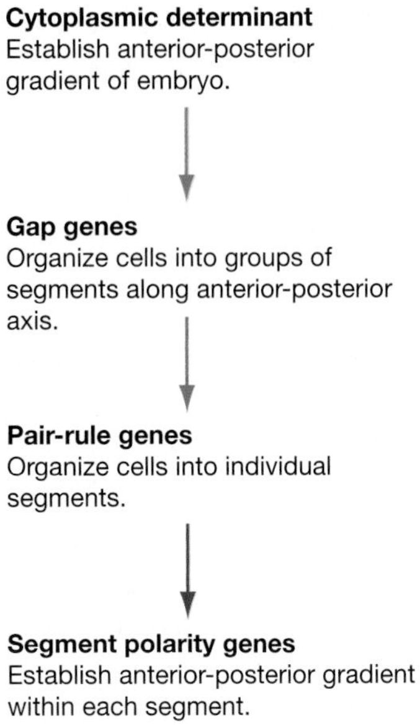
Cytoplasmic determinant
Establish anterior-posterior gradient of embryo.
Gap genes
Organize cells into groups of segments along anterior-posterior axis.
Pair-rule genes
Organize cells into individual segments.
Segment polarity genes
Establish anterior-posterior gradient within each segment.

FIGURE 22.5 Segmentation Genes Regulate Each Other
The arrows indicate a small subset of the known interactions among the segmentation genes and their products. In each case, the outcome depends on the concentration of each protein involved.

form gradients in specific regions of the embryo. This observation suggests that two types of information are being delivered to target nuclei: Both the type and concentration of signals received by cells or nuclei are important. Further, both the type and concentration of signaling molecules vary with respect to position in the embryo.

To grasp why these points are important, think back to the mechanisms of gene regulation introduced in Chapters 17 and 18. Recall that, for a eukaryotic gene to be transcribed efficiently, the regulatory transcription factors that bind to enhancers and activate transcription must be present while transcription factors that bind to silencers must be absent. Differential gene expression occurs because many of the developmental signals received by a cell alter the array of regulatory transcription factors that are present in the nucleus and are active. These transcription factors bind to promoters, enhancers, silencers, or other types of regulatory sequences in DNA and affect the amount and rate of gene expression (**Figure 22.6**). Other developmental signals may lead to alterations in the mRNAs found in target cells or induce changes in proteins that are already present. Which genes are expressed by developing cells at particular places and times? The answer depends on complex interactions among an array of developmental signals and thus transcription factors. The differential gene expression triggered by regulatory proteins shapes embryos.

Given that segmentation genes are responsible for organizing cells and tissues into distinct segments, let's consider the next level of pattern formation. What genes are responsible for producing segment-specific structures?

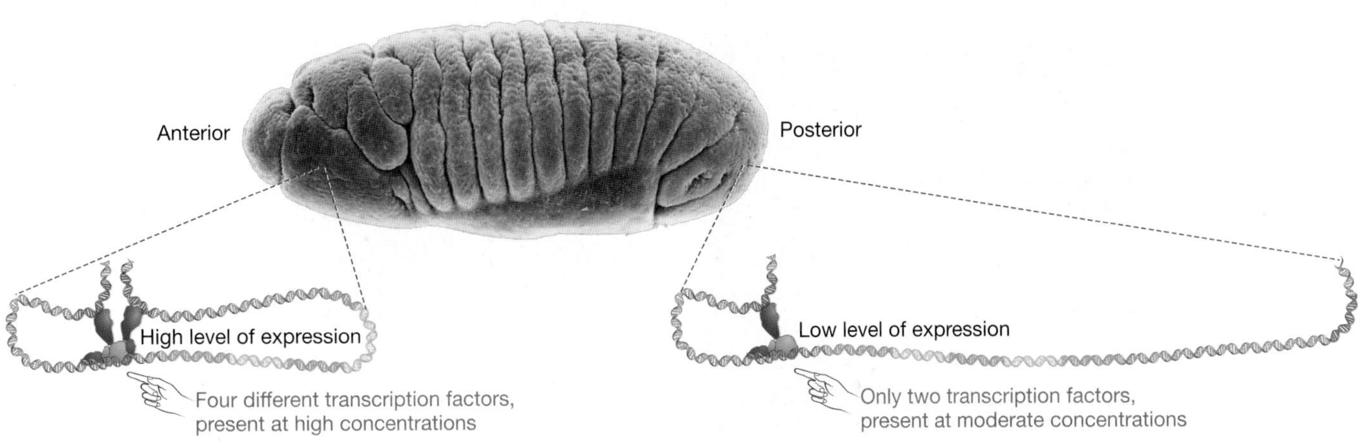
Anterior
Posterior
High level of expression
Low level of expression
Four different transcription factors, present at high concentrations
Only two transcription factors, present at moderate concentrations

FIGURE 22.6 Differential Gene Expression Is Based on the Presence of Different Regulatory Transcription Factors
The genes involved in pattern formation code for cell-to-cell signals or regulatory transcription factors that direct gene expression.

The Discovery of Homeotic Genes

During the 1940s Edward Lewis began a long series of studies that helped researchers pinpoint the genes that identify particular segments. Lewis studied mutations that alter the body pattern of adult flies, much as Nüsslein-Volhard and Wieschaus searched for mutations that affect the body pattern of embryos and larvae. Some of the mutants Lewis and others studied had bizarre phenotypes. As **Figure 22.7** shows, some flies had legs where their antennae should be; others had two sets of wings instead of one.

Lewis realized that these mutants possessed an entire segment or structure that had been transformed into a related segment or structure. This phenomenon had been observed in plants by a series of workers starting in the late 1800s. The replacement of one structure by another is termed **homeosis** ("like-condition"). For example, a mutation or series of mutations could transform thoracic segment number 3 into thoracic segment 2. Instead of bearing a pair of small stabilizer structures called *halteres*, the transformed segment would bear a pair of wings. As a result, the mutant would have four wings instead of two.

The existence of homeotic mutants meant that there must be **homeotic genes**—genes that specify a location and lead to the development of structures appropriate to that location. (Later work showed that homeotic genes interact directly with segmentation genes.) To explain the existence of a four-winged fly, for example, Lewis suggested that the homeotic gene responsible for identifying thoracic segment number 3 was defective. Consequently, the cells in what would normally be thoracic segment 3 developed as if they were in thoracic segment 2. Stated another way, Lewis hypothesized that the products of homeotic genes specify the identity of segments along the body's anterior-posterior axis.

The Homeotic Complex When techniques became available for identifying and sequencing genes, researchers confirmed the homeotic gene hypothesis in spectacular fashion. Studies conducted during the 1970s and 1980s initially identified eight genes in the *Drosophila* genome that lead to homeosis when they are defective. As **Figure 22.8a** shows, the eight genes are found in two clusters on the same chromosome. Investigators who explored where these genes are expressed in *Drosophila* embryos found that five are activated in the anterior part of the embryo while three are expressed in the posterior sections. The five genes expressed in the anterior part of the embryo are known as the *Antennapedia complex*; the three ex-

FIGURE 22.7 Homeotic Mutants in *Drosophila* Have Structures in the Wrong Locations
(a) The head and top view of a normal fruit fly. **(b)** Among the most spectacular homeotic mutants of fruit flies are individuals with legs growing where antennae should be and individuals with wings growing where small, stabilizing structures called halteres should be.

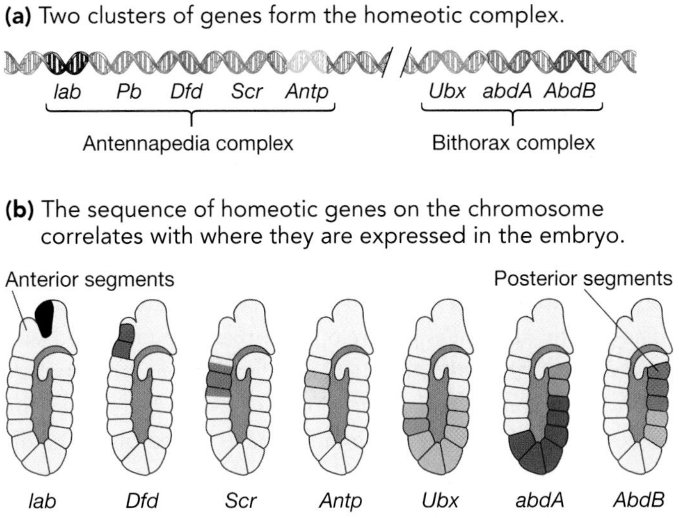
(a) Two clusters of genes form the homeotic complex.
lab Pb Dfd Scr Antp
Ubx abdA AbdB
Antennapedia complex
Bithorax complex
(b) The sequence of homeotic genes on the chromosome correlates with where they are expressed in the embryo.
Anterior segments
Posterior segments
lab
Dfd
Scr
Antp
Ubx
abdA
AbdB

FIGURE 22.8 Organization and Expression of the Homeotic Complex
(a) The arrangement of *HOM-C* genes on the chromosome
(b) Where genes in the homeotic complex are expressed.

pressed in the posterior regions are called the *Bithorax complex*. As a group, the genes became known as the **homeotic complex** (abbreviated ***HOM-C***).

Studies on the timing of gene expression revealed the remarkable pattern illustrated in **Figure 22.8b**. In the embryo, genes in the homeotic complex are expressed in the same sequence as they are found along the chromosome. The gene called *lab*, for example, occurs at one end of the complex and is turned on in the most anterior segment of the embryo. Each subsequent gene in the complex is expressed in slightly more posterior segments, so these genes are expressed in the same "spatial" sequence in the embryo as their order on the chromosome. Why this pattern occurs is currently being debated. The mechanism of gene *HOM-C* action, in contrast, is well understood.

Genes in the Homeotic Complex Encode Proteins That Regulate Gene Expression Sequencing studies revealed that each gene in the homeotic complex has a 180-base-pair DNA sequence called the homeobox. The **homeobox** codes for a DNA-binding domain in the proteins produced by these genes. More specifically, the homeobox codes for a helix-loop-helix motif, which binds to specific sites in the major groove of DNA (see Chapter 17). Based on this observation, biologists hypothesized that *HOM-C* genes code for regulatory transcription factors that bind to DNA and regulate the activity of other genes. More specifically, products from the homeotic complex are thought to activate genes involved in the production of segment-specific structures such as legs, wings, antennae, and halteres. Subsequent work has identified the gene *Distal-less*, which is involved in the formation of legs and other appendages, as an example of a gene regulated by the *HOM-C* loci. Biologists are now working to document other genes targeted by *HOM-C* products.

Where Do Homeotic Genes Fit in the Regulatory Cascade?

The segmentation genes introduced earlier in this chapter demarcate segments along the anterior-posterior axis of the fly embryo. The products of *HOM-C* genes trigger the production of structures in particular segments of adults. Is there a direct relationship between the segmentation genes and *HOM-C* genes? The answer is yes. In flies that carry a mutation in a gap gene or a pair-rule gene, the expression of one or more genes in the homeotic complex is abnormal. Based on this observation, researchers have concluded that the gap and pair-rule genes regulate the transcription of the homeotic loci. The eight *HOM-C* genes represent the next step of the regulatory cascade drawn in Figure 22.5.

Before we leave the topic of homeotic gene action and regulation, it's essential to note that not all homeotic loci are part of the Antennapedia or Bithorax complexes. Many other genes are involved in specifying the structures that develop in particular parts of the fly body; some of these genes cause homeosis when they are defective. The genes discussed in detail here are only part of a large and complicated network of genes that regulate the later stages of pattern formation.

Drosophila as a Model Organism

Based on the experiments reviewed thus far, it is clear that biologists have invested an enormous amount of time and effort in understanding pattern formation in *Drosophila*. Why? In addition to being fascinating and important in its own right, research on *Drosophila* development was intended to be applicable to other species that are difficult to manipulate in the lab. *Drosophila* was used as a model organism—an organism that is studied intensively in the hope that knowledge learned about it can be applied to other organisms.

A variety of model organisms are featured in other chapters. Research on *Escherichia coli* allowed biologists to understand the mechanisms of DNA synthesis and repair in plants and animals. Similarly, work on the yeast *Saccharomyces cerevisiae* led to an understanding of how glycolysis works in virtually every organism on Earth and how the cell cycle is controlled in multicellular species. In the same way, research on pattern formation in *Drosophila* has led to a better understanding of how pattern formation works in many other animal groups. For example, researchers have found that a homeotic complex called the ***Hox* gene** complex occurs in frogs, crustaceans (crabs and their relatives), birds, various types of worms, mice, and humans. Although the number of *Hox* loci varies widely among species, their chromosomal organization is similar to that of *HOM-C* genes of flies, shown in Figure 22.8a.

Recent studies of *Hox* genes have shown that they are expressed along the head-to-tail axis of the mouse embryo in the same sequence as in fruit flies. In addition, experiments have shown that when mouse *Hox* genes are altered by mutation, defects in pattern formation result. Based on these data, biologists conclude that, in flies, mice, and probably humans and most other animals, the *HOM-C* and *Hox* genes appear to play a key role in defining the position of cells along the head-to-tail axis of the body. This conclusion was supported in spectacular fashion when researchers in William McGinnis's lab introduced the *Hoxb6* gene from mice into fruit-fly eggs. The *Hoxb6* gene in mice is similar in structure and sequence to the *Antp* gene of flies. Because it was introduced without its normal regulatory sequences, the *Hoxb6* gene was expressed throughout the treated fly embryos. The resulting larvae had defects identical to those observed in naturally occurring fly mutants in which the *Antp* gene is mistakenly expressed throughout the embryo. This is a stunning result: A mouse allele not only affected the development of a fly but also mimicked the effect of a specific fly allele.

To interpret these observations, biologists hypothesize that the genes in *HOM-C/Hox* complexes are *homologous*—meaning that they are similar because they are descended from genes in a common ancestor. The hypothesis of homology contends that the first *HOM-C/Hox* loci arose before the origin of animals. Since then, the number of *HOM-C/Hox* genes has changed dramatically due to the types of gene duplication events described in Chapter 20, creating a family of related genes. But over this span of about a billion years, the same types of gene products have been involved in directing cells to develop according to their position inside the embryo. In other words, at least some of the molecular mechanisms of pattern formation have been highly conserved during animal evolution. The discovery of these shared mechanisms is one of the most significant results to have emerged from animal development studies.

Even more remarkably, researchers have discovered genes that contain homeoboxes in fungi and plants. No genes similar to those found in the *HOM-C* or *Hox* complexes exist in these groups, however. In plants, the genes involved in pattern formation are different from the segmentation and homeotic genes of animals. Why? Recall from Chapter 21 that multicellular bodies evolved independently in plants and animals. Based on this observation, it is logical to predict that the mechanisms for specifying the positional identity of cells differ between the groups. How does pattern formation occur in plants?

✓CHECK YOUR UNDERSTANDING

Pattern formation occurs as embryonic cells respond to a regulatory cascade of cell-to-cell signals and regulatory transcription factors. In *Drosophila*, this cascade starts with the Bicoid protein and continues with the products of gap genes, pair-rule genes, segment polarity genes, and homeotic genes. At the end of this sequence, each cell in the embryo is expressing genes specific to its exact position in the body. You should be able to describe how an individual cell in a *Drosophila* embryo comes to express genes typical of a specific location—for example, the anterior portion of thoracic segment 2.

22.2 Pattern Formation in *Arabidopsis*

Historically, the field of developmental biology has been dominated by the study of animals. The reason is simple: Plant embryos develop inside the mother plant, but sea urchins, frogs, flies, and many other animals lay eggs outside their bodies. As a result, these animal embryos are accessible and relatively easy to observe and study.

Recently, however, biologists have begun studying how pattern formation occurs in plants by analyzing mutant individuals. The genetic approach to development that was pioneered with research on *Drosophila melanogaster* has proven to be a powerful way of studying plant development. Although the specific genes involved in plant development have turned out to be different from those involved in animal development, the basic logic of pattern-formation mechanisms is often quite similar in plants and animals.

Most work on pattern formation in plants has been done on the weedy mustard plant *Arabidopsis thaliana*. Recall from Chapter 21 that *Arabidopsis* seedlings grow from a seed into a mature plant capable of producing gametes in as little as six weeks. Even though their life span is short, individuals are complex enough to be interesting to study.

An important advantage of studying plants is that pattern formation is not limited to the period of early development. Instead, complex structures such as leaves, roots, branches, and flowers are produced throughout an individual's life. Recall from Chapter 21 that these structures are derived from the apical meristems located in roots and shoots. Stems and roots elongate as cell division takes place in the meristems located at their tips. As these structures grow, certain populations of cells become patterned into leaves, flowers, tubers, or other specialized structures.

To investigate how pattern formation occurs in embryonic and adult plants, let's consider two questions: (1) How is the root-to-shoot body axis established during embryogenesis, and (2) what is the mechanism of pattern formation during flower development? Biologists are addressing both questions by analyzing mutant individuals with defective roots, shoots, or flowers.

The Root-to-Shoot Axis of Embryos

Gerd Jurgens and colleagues set out to identify genes that are transcribed in the zygote or embryo of *Arabidopsis* and that are involved in establishing the root-to-shoot axis of the body. It's no surprise that this effort was similar to the project that Nüsslein-Volhard and Wieschaus had undertaken with *Drosophila*—Jurgens had participated in the work with flies.

The biologists' initial goal was to identify individuals with defects in pattern formation at the seedling stage. More specifically, they were looking for mutants that lacked particular regions along the root-to-shoot axis. The team succeeded in finding several bizarre-looking mutants (**Figure 22.9**). Certain individuals (*apical mutants*) lacked the first leaves, or cotyledons; some individuals (*central mutants*) lacked the embryonic stem, or hypocotyl; and others (*basal mutants*) lacked roots.

To interpret these results, the researchers suggested that each type of *Arabidopsis* mutant had a defect in a different gene and that each gene was involved in specifying the position of cells along the root-to-shoot axis of the body. The idea was that these genes are analogous to gap genes, which specify the identity of cells within well-defined regions along the head-to-tail axis of fruit flies.

What are these *Arabidopsis* genes, and what do they do? To answer these questions, consider the gene responsible for the mutants lacking hypocotyls and roots. This gene has been mapped and sequenced and named *monopterous*. Because its DNA sequence indicates that this gene has a DNA-binding sequence, *monopterous* is hypothesized to encode a regulatory transcription factor that regulates the activity of target genes. The monopterous protein, in turn, is manufactured in response to signals from **auxin**, a cell-to-cell signal molecule that is produced in the apical meristem.

The genes and proteins involved in setting up the root-to-shoot axis of *Arabidopsis* are not yet understood in as much detail as the pattern-formation genes of *Drosophila*. Current evidence suggests that there are strong similarities, however. In both plants and animals, pattern formation is based on cell-to-cell signals. Concentration gradients of regulatory transcription factors are important, and regulatory cascades result in the step-by-step specification of a cell's position and fate.

Many questions remain about pattern formation in plant embryos. How is auxin production turned on as the apical meristem first begins to form in embryos? Once production of the *monopterous* gene product begins, what target genes are affected? What genes other than *monopterous* are found in the regulatory cascade responsible for development along the root-to-shoot axis? Are any of them active in adults as well? Research on pattern formation in plant embryos presents a host of interesting challenges.

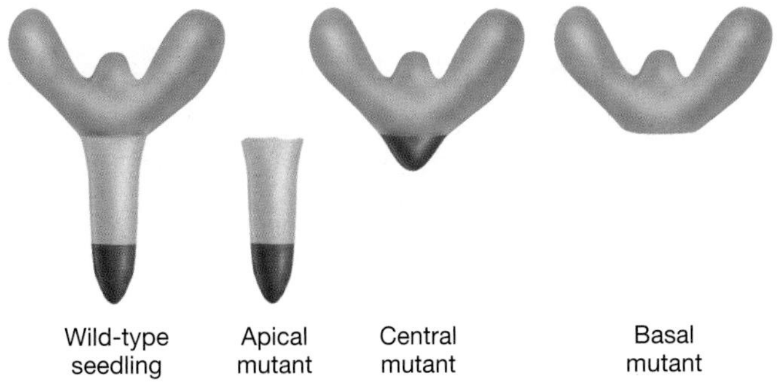
Wild-type seedling
Apical mutant
Central mutant
Basal mutant

FIGURE 22.9 Pattern-Formation Mutants in *Arabidopsis* Embryos Have Misshapen Bodies
Researchers have identified *Arabidopsis* mutant individuals missing certain defined sections of the body along the root-to-shoot axis.

Flower Development in Adults

When receptor proteins in *Arabidopsis* sense that days are getting longer and the temperature is favorable, the shoot apical meristem is stimulated to act as a floral meristem. The **floral meristem** is a modified meristem that produces flowers containing the individual's reproductive organs. During flower development, the floral meristem produces four organs, arranged as shown in **Figure 22.10**: (1) sepals, (2) petals, (3) stamens, and (4) the carpel. **Sepals** are leaflike structures, located around the outside of the flower, that protect the flower. Inside the sepals is a *whorl*, or circular arrangement, of **petals**, modified leaves that enclose the male and female reproductive organs. If insects or other animals pollinate the species in question, the petals may be colored to help advertise the reproductive structures. The pollen-producing organs, or **stamens**, are located in a whorl inside the petals. In the center of the entire structure is the egg-producing reproductive organ, or **carpel**. The question is, How does the floral meristem produce these four organs in the characteristic pattern of whorls within whorls?

The first hint of an answer came in the late 1800s, when researchers realized that several types of mutant flowering plants—including some that were popular garden plants—were homeotic. In the mutant individuals, one kind of floral organ was replaced by another. Instead of having sepals, petals, stamens, and carpels, the flowers had sepals, petals, another ring of petals, and carpels. The mutant phenotype was similar to the transformation of segments later observed in *Drosophila* homeotic mutants.

Over 100 years later, Elliot Meyerowitz and colleagues assembled a large collection of *Arabidopsis* individuals with homeotic mutations in flower structure. The researchers' goal was to identify and characterize the genes responsible for specifying the four floral organs.

Meyerowitz's group found that the mutants could be sorted into three general classes according to the type of homeotic transformation that occurred. Some mutants had only carpels

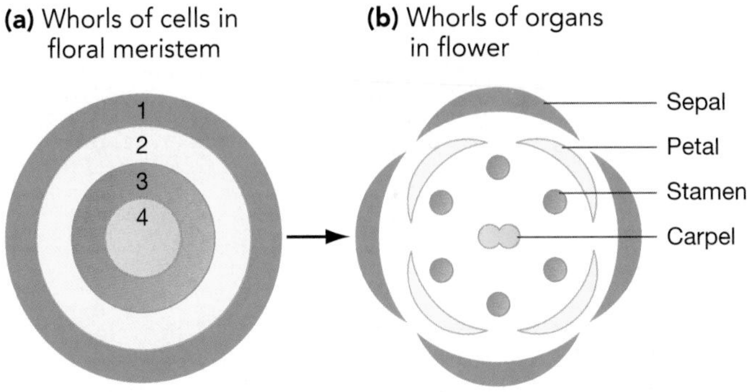
(a) Whorls of cells in floral meristem
(b) Whorls of organs in flower
1
2
3
4
Sepal
Petal
Stamen
Carpel

FIGURE 22.10 Flowers Are Composed of Four Organs
(a) In a floral meristem, whorls of cells produce **(b)** the four whorls of floral organs.

and stamens, as **Figure 22.11a** shows. Others had only sepals and carpels. Still others had only petals and sepals. The key observation was that each type of mutant lacked the elements found in two of the four whorls.

What was going on? Presumably, each of the three classes of homeotic mutation was due to a defect in a single gene. Meyerowitz realized that if three genes are responsible for setting up the pattern of a flower, then the mutants suggested a hypothesis for how the three gene products interact. Because he referred to the three hypothetical genes as *A*, *B*, and *C*, his hypothesis is called the ABC model.

The ABC Model As **Figure 22.11b** shows, there are three basic ideas behind the ABC model of pattern formation in flowers. The first idea is that each of the three genes involved is expressed in two adjacent whorls. The second idea is that

(a) Distinct types of homeotic mutants occur in *Arabidopsis* flowers.

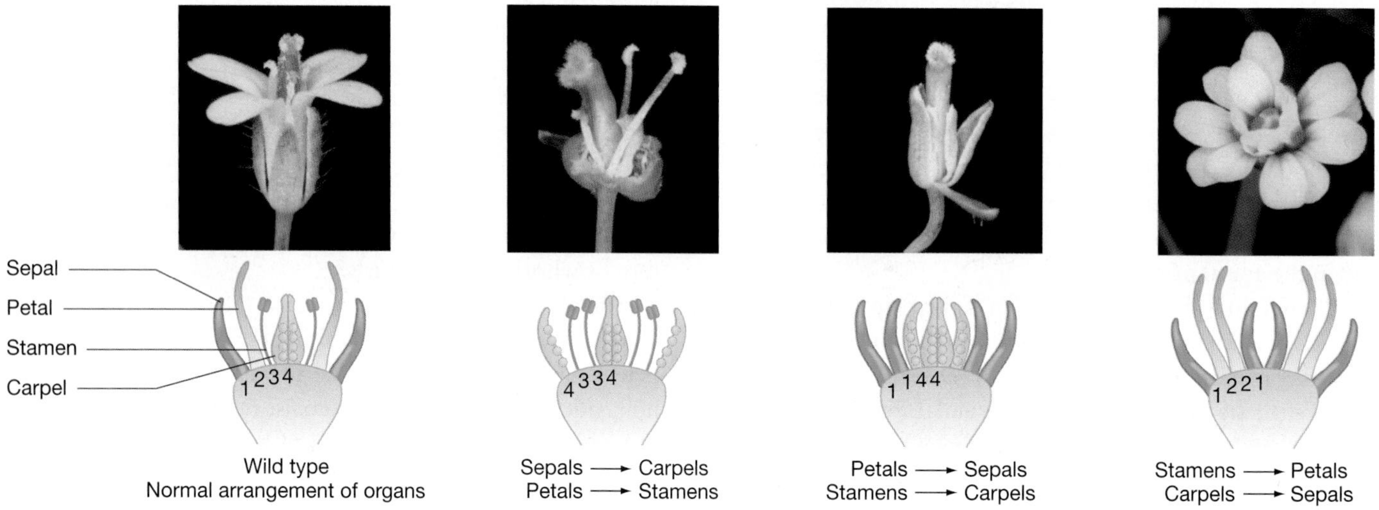

(b) A model to explain why the homeotic mutations occur

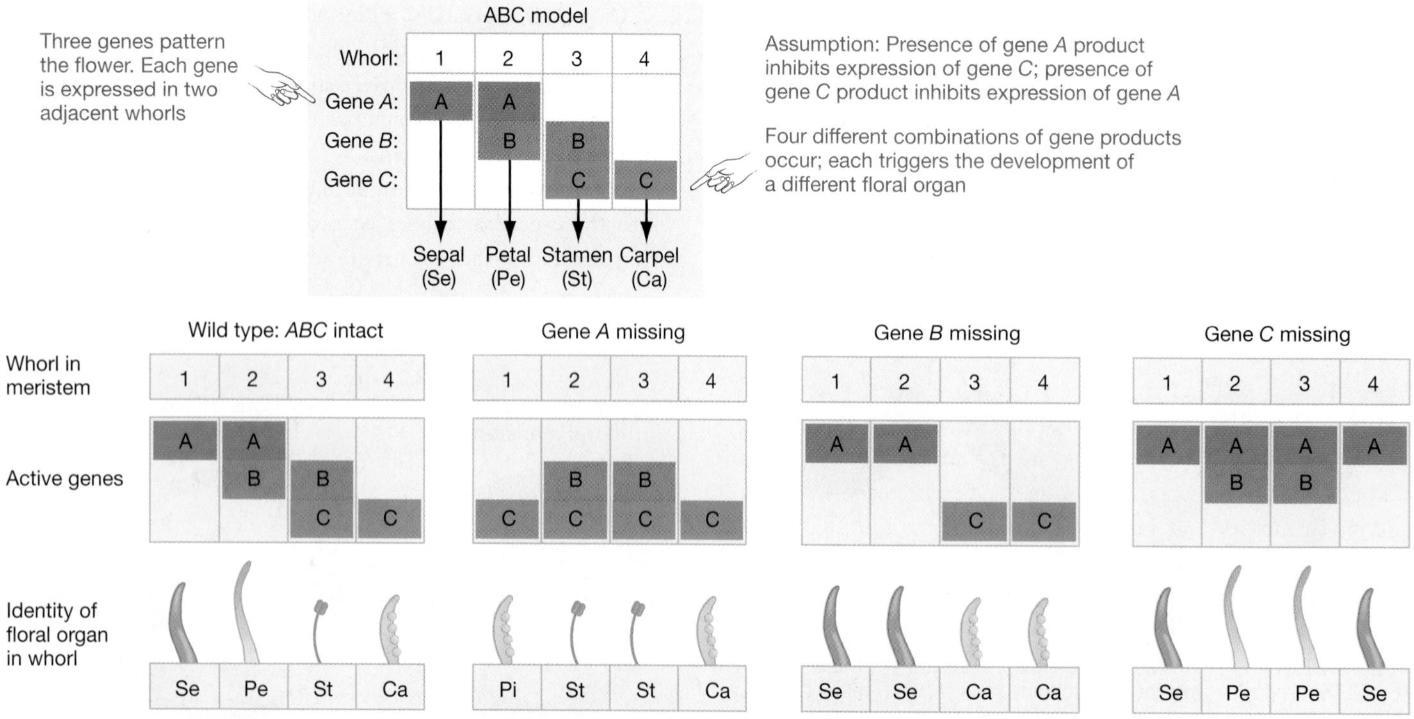

FIGURE 22.11 Homeotic Mutants in *Arabidopsis* Flowers Have Structures in the Wrong Locations
(a) Normal (wild-type) individual and three types of mutants. In each type of homeotic floral mutant, two adjacent whorls of organs are transformed into different organs. Either sepals and petals, petals and stamens, or stamens and the carpel are transformed. **(b)** The ABC model is a hypothesis to explain why three types of homeotic mutants exist.

four different combinations of gene products result from this pattern of expression. The final idea is that each of these four combinations of gene products triggers the development of a different floral organ. Specifically, Meyerowitz proposed that (1) the A protein alone causes cells to form sepals, (2) a combination of A and B proteins sets up the formation of petals, (3) B and C combined specify stamens, and (4) the C protein alone designates cells as the precursors of carpels.

Does this model explain how the three classes of homeotic mutants occur? The answer is yes, if we assume that the presence of the A protein inhibits the production of the C protein and that the presence of the C protein inhibits the production of the A protein. Then the patterns of gene expression and mutant phenotypes in Figure 22.11b correspond. For example, if the *A* gene is disabled by mutation, then it no longer inhibits the expression of the *C* gene and all cells produce the C protein. As a result, cells in the outermost whorl express only C protein and develop into carpels, while cells in the whorl just to the inside produce B and C proteins and develop into stamens.

Although the model is plausible and appeared to explain the data, it needed to be tested directly. To accomplish this, Meyerowitz and co-workers mapped the genes responsible for the mutant phenotypes and identified the DNA sequences involved. Once they had isolated the genes, they were able to obtain and use single-stranded DNAs to perform in situ hybridizations. The goal was to document the pattern of expression of the *A*, *B*, and *C* genes and see if that pattern corresponded to the model's predictions. As anticipated, the mRNAs for each of the three genes showed up in the sets of whorls predicted by the model. The *A* gene is expressed in the outer two whorls, the *B* gene is expressed in the middle two whorls, and the *C* gene is expressed in the inner two whorls.

This result strongly supported the validity of the ABC model. Just as different combinations of *HOM-C* gene products specify the identity of fly segments, different combinations of floral identity genes specify the parts of a flower.

How Do the Floral Identity Genes Work? When Meyerowitz and others analyzed the sequence of the floral organ identity genes, they discovered that all three genes contained a segment that coded for the DNA-binding domain of a protein. Based on this observation, the researchers hypothesized that, like the homeotic genes found in *Drosophila* and other animals, the floral genes appeared to be regulatory transcription factors. But instead of containing a homeobox, the genes that identify floral organs encode a long sequence of 58 amino acids, called the **MADS box**, that binds to DNA. Transcription factors with MADS boxes also occur in fungi and animals. As with research on *Drosophila*, biologists are currently working to identify the genes targeted by the ABC proteins.

To summarize, strong parallels exist between the process of pattern formation in animals and in plants. Although different genes are involved, the logic of how the genes act is the same. In both groups, pattern formation depends on signals that act in a regulatory cascade. In each case, differences in cell fate are mediated by different combinations of regulatory transcription factors. These regulatory proteins are expressed in specific regions of the animal or plant and direct the development of cells located in that region. As **Box 22.2** (page 482) indicates, changes in those regulatory transcription factors can produce morphological change that is important in evolution.

✓CHECK YOUR UNDERSTANDING

In *Arabidopsis*, pattern formation in both embryonic and adult tissues is a step-by-step process that involves a regulatory cascade of cell-to-cell signals and regulatory transcription factors. You should be able to (1) describe two elements of this regulatory cascade in embryos and (2) explain why the *ABC* genes involved in flower development are considered homeotic genes.

22.3 Does the Genetic Makeup of Cells Change as Development Proceeds?

As development in *Arabidopsis*, fruit flies, frogs, and humans proceeds, cells become different from each other. Recall from Chapter 21 that **differentiation** is the process that results in the generation of diverse cell types from a fertilized egg. Cells are differentiated if they produce proteins that are specific to a cell type, such as muscle, nerve, or skin. One of the most basic questions in developmental biology is how differentiation occurs. Does differentiation involve changes in the genetic material of cells or simply in the kinds of genes that are expressed?

In plants, it has long been established that all cells carry similar genetic information. Gardeners and farmers have known for centuries that, in many plant species, new individuals can be produced from a section of root or shoot. Then, during the 1950s and 1960s, a series of researchers succeeded in showing that, in culture, single differentiated cells isolated from mature plants can develop into independent mature plants that function normally. Was this also the case in animals?

Are Differentiated Animal Cells Genetically Equivalent?

In the 1950s Robert Briggs and Thomas King set out to test the hypothesis that different types of differentiated animal cells have the same genetic makeup. To do this, they tested a prediction: If differentiated cells have all the genes that were present in the zygote, then a nucleus from a differentiated cell should be able to direct the development of a complete individual.

Briggs and King developed techniques for isolating nuclei from leopard frog cells and transferring them into unfertilized eggs whose nuclei had been removed. In a series of experiments, the biologists found that nuclei from an embryo undergoing cleavage or organogenesis could direct the development of complete tadpoles. In contrast, donor nuclei taken from

BOX 22.2 Regulatory Evolution

When biologists realized that the homeotic genes found in animals and plants regulate the formation of segments, limbs, flowers, and other key structures, another important realization followed: Changes in the genes responsible for pattern formation could change the size, shape, or number of segments, limbs, flowers, and other structures found in a particular species. As a result, mutations in the genes that regulate development could result in novel types of structures. In turn, the existence of novel structures in a population of organisms could trigger important evolutionary changes.

The link between genes that regulate embryonic development and evolutionary change has been termed **regulatory evolution.** The key idea is that most of the major innovations that have occurred over the course of evolution have been caused by mutations in homeotic genes and other regulatory transcription factors that are active early in development.

As an example of how researchers are exploring regulatory evolution, consider recent work on the leaf shape of tomato plants. As **Figure 22.12a** shows, tomatoes have *compound leaves*—meaning a single leaf blade divided into smaller units called *leaflets*. But leaves of other plant species vary widely in shape. *Simple leaves* consist of a single blade; *palmate leaves* have a series of leaflets that radiate from a single point.

When a new leaf begins to develop, pattern formation is controlled in part by the product of a gene called *PHANTISTICA* (abbreviated *PHAN*). The protein product of *PHAN* has a DNA-binding domain and acts as a regulatory transcription factor. It triggers the expression of genes that cause cells to form the upper surface of leaves. To explore whether changes in the regulation of *PHAN* might have a role in the evolution of various leaf shapes, a team of biologists created transgenic tomato plants in which the *PHAN* gene product was blocked to a moderate or large extent. As **Figure 22.12b** shows, leaf shape in the transgenic individuals changed dramatically. Some of the individuals had cup-shaped (simple) leaves, while others had leaflets radiating from a single point (palmately compound).

These results have inspired the hypothesis that evolutionary changes in leaf size and shape are due to mutations that created new alleles of genes that regulate *PHAN* expression. Alleles that resulted in lowered *PHAN* expression might lead to simple leaves with a single blade; alleles that increased the extent of *PHAN* expression might result in compound leaves like those of normal tomatoes. By altering the genes that regulate development, researchers are beginning to understand the genetic changes that led to novel types of leaves, segments, limbs, and flowers.

(a) Normal leaves—compound
(b) Transgenic leaves (reduced PHAN expression)—simple or palmately compound

FIGURE 22.12 Changes in Pattern-Formation Genes Change the Size and Shape of Key Structures
(a) Normal tomato leaves consist of a series of leaflets. **(b)** When the expression of *PHAN* is reduced experimentally, leaves become cup shaped (left) or have leaflets arising from a single point. [Reprinted by permission from *Nature* (Vol. 424, July 24, 2003, pg. 439, by Kim et al.) ©2004 Macmillan Publishers Ltd. Photographs supplied by Dr. Neelima Sinah.] QUESTION Suppose that *PHAN* were overexpressed. What type of leaf do you predict would form?

tadpoles were unable to do so. These results suggested that cells may lose at least some genetic information late in development, as they become more and more specialized.

The Briggs and King experiment was criticized, however, because the failure of tadpole nuclei to direct development represented a negative result. A **negative result** is the failure to observe a particular pattern or process. Negative results can always be criticized on the grounds that something was wrong with the experimental technique. For example, some researchers contended that the older nuclei used by Briggs and King might have complete genetic information but fail to support early development because they were more easily damaged during transfer.

These criticisms were addressed in a series of experiments that John Gurdon and co-workers carried out during the 1960s using the African clawed frog *Xenopus laevis*. In one dramatic experiment, Gurdon's group showed that nuclei from the intestinal lining of young tadpoles could direct the development of complete tadpoles. In some cases, the tadpoles succeeded in metamorphosing into fertile, adult frogs. Although this experiment seemed to show conclusively that differentiation occurs without changes to the genetic makeup of cells, critics were not satisfied. Some researchers claimed that Gurdon and co-workers had not been careful enough in defining the source of the donor nuclei. This was a legitimate concern, because cells that produce eggs or sperm, which do not undergo differentiation, can be closely associated with the intestines of tadpoles. Other investigators contended that intestinal cells from adults, rather than tadpoles, should have been used. To test more rigorously the hypothesis that differentiation occurs without any loss of genetic information, Gurdon's group cultured skin cells that were actively producing a skin-specific protein called *keratin*—the same protein that makes up the bulk of human hair. When the researchers succeeded in producing tadpoles from eggs that received skin-cell nuclei, the result provided stronger evidence that all mature cells are genetically equivalent.

In 1997 Ian Wilmut and colleagues reinforced this conclusion through nuclear transfer experiments in sheep. As shown in **Figure 22.13a**, these researchers removed mammary-gland cells from a 6-year-old pregnant female and grew them in culture (steps 1 and 2). Later they fused these cells with *enucleated* eggs—whose nuclei had been removed—so that the egg cells contained nuclei of mammary-gland cells (steps 3 and 4). As the drawings show, the eggs came from a black-faced breed of sheep, while the donor nuclei came from a white-faced breed of sheep. After developing in culture, the resulting eggs were

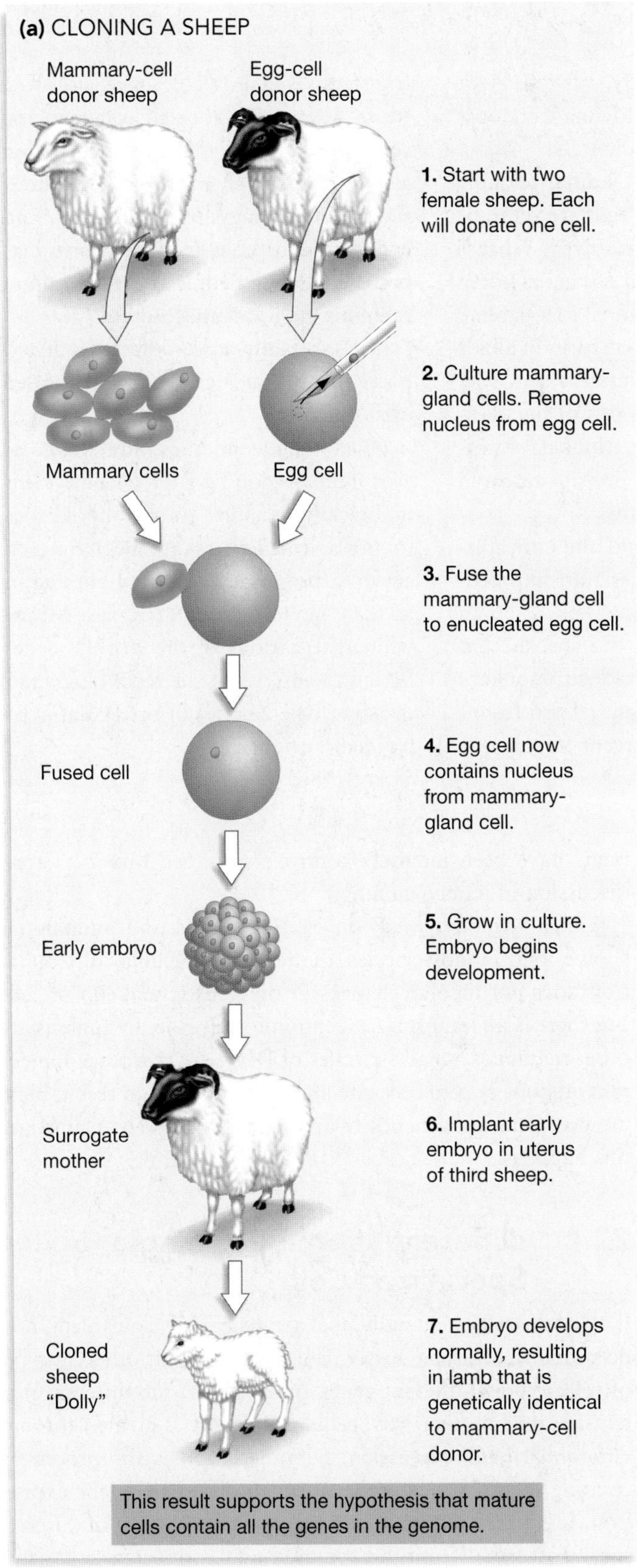
(a) CLONING A SHEEP
Mammary-cell donor sheep
Egg-cell donor sheep
1. Start with two female sheep. Each will donate one cell.
2. Culture mammary-gland cells. Remove nucleus from egg cell.
Mammary cells
Egg cell
3. Fuse the mammary-gland cell to enucleated egg cell.
Fused cell
4. Egg cell now contains nucleus from mammary-gland cell.
Early embryo
5. Grow in culture. Embryo begins development.
Surrogate mother
6. Implant early embryo in uterus of third sheep.
Cloned sheep "Dolly"
7. Embryo develops normally, resulting in lamb that is genetically identical to mammary-cell donor.
This result supports the hypothesis that mature cells contain all the genes in the genome.

(b) Dolly with an offspring of her own

FIGURE 22.13 Mammals Can Be Cloned by Transplanting Nuclei from Mature Cells
(a) When the nucleus from a differentiated mammary-gland cell was transferred to an enucleated egg, normal development occurred. The resulting lamb, Dolly, resembled the breed from which the donor nucleus came, not the breed of the egg donor or the surrogate mother. DNA testing also showed that the lamb was genetically identical to the individual that donated the nucleus. **(b)** Dolly, the sheep produced by cloning, was able to produce her own lamb through normal sexual reproduction.

BOX 22.3 Human Cloning?

Biologists can harvest nuclei from the somatic cells of mice, cows, and sheep and use the nuclei to produce embryos. These nuclear transfer experiments produce offspring that are genetically identical to the individual providing the donor nucleus. A group of genetically identical organisms are clones, so the process of creating individuals by transferring nuclei into enucleated eggs is called cloning. Can humans be cloned?

Before addressing the technical feasibility of cloning humans, let's consider a more practical issue. Why would anyone want to clone a human? Proponents of this technology contend that cloning by nuclear transfer would allow otherwise infertile couples to have offspring that are genetically related—in fact, identical—to one of the parents. Opponents of this technology are concerned that, in an attempt at immortality, dictators or wealthy eccentrics could finance clones of themselves in perpetuity.

Even benign uses of human cloning create ethical dilemmas that are yet to be thoroughly debated or resolved. What is the effect on a family of having children that are genetically identical to a parent? Would cloning be limited to individuals with "desirable" characteristics? Given the number and seriousness of the questions involved, several nations have proposed a moratorium on all research related to cloning humans.

In addition to the legal and ethical issues involved, there are significant technical hurdles to overcome before human cloning is feasible. Of the species that have been successfully cloned, monkeys and mice are most closely related to humans. Less than 3 percent of nuclear transfers result in viable mice, however. Most result in aborted pregnancies; some produce individuals that die soon after birth. Currently, the success rate is similar in cloning sheep and cows, and production of deformed newborns has occurred. Most people regard a failure rate this high as unacceptable for humans, especially given the high likelihood of producing grossly deformed offspring.

As cloning techniques improve, however, human cloning may become technologically feasible. As this book goes to press, researchers are beginning to act on a proposal to attempt cloning in as many as 600 human couples. Meanwhile, the nations of the world are debating the merits of the technology and deciding whether it will be regulated or banned outright.

implanted in the uteri of surrogate mothers (steps 5 and 6). In one of several hundred such transfer attempts, a white-faced lamb named Dolly was born (step 7). Dolly grew into a fertile adult and, by normal mating, produced her own lamb named Bonnie (**Figure 22.13b**). In 1998 other research groups reported similar results in mice and cows; subsequently cats, horses, a monkey, and individuals of several other species have been cloned. Recall from Chapter 12 that a **clone** is a genetically exact copy of a parent.

Although cloning protocols have been successful in animals, they are not free of complications. For example, sheep normally live about 12 years, but Dolly had to be euthanized in 2003 at age 6 because she began to suffer from debilitating diseases typical of old sheep. Researchers now hypothesize that cloned mammals may develop disorders associated with advanced age much sooner than normal. To understand why, recall from Chapter 14 that the enzyme telomerase, which prevents chromosome ends from shortening during DNA synthesis, is not active in somatic cells (cells that are not involved in gamete formation). Recall also that chromosome shortening is thought to be a signal that helps cells stop dividing as they mature. Because Dolly started life with chromosomes from a mature cell, her telomeres (the regions at the ends of linear chromosomes) were much shorter than normal. Telomere shortening is thought to contribute to the aging of cells. These types of complications are one of the many reasons that efforts to clone humans have been intensely controversial. (See **Box 22.3** for a discussion of human cloning.)

Taken together, work on cloning plants and animals has shown that, in most species, the process of cellular differentiation does not involve changes in the genetic makeup of cells. But there is an important exception to this rule. In humans and other mammals, small stretches of DNA are rearranged in certain immune system cells late in development. As a result, many immune cells are genetically unique. Chapter 49 explains how this happens.

22.4 Differentiation: Becoming a Specialized Cell

If all of the cells in an individual are genetically equivalent, how do different cell types arise? The answer is that different types of cells express different genes and manufacture different proteins. Stated another way, cellular diversity is created through **differential gene expression**. Differentiation is the process of creating specialized cell types through differential gene expression. It is based on the highly regulated expression of different subsets of genes from the common and complete set of genes found in all cells of the embryo.

Now, what controls differential gene expression? In a flower petal, the genes that are responsible for synthesizing pigments are activated. In the thorax of a fruit fly, the cells that are destined to become muscles used for flight start producing muscle-specific proteins. How does this final step in development happen?

The key to answering this question is to recognize that a cell's ultimate fate is realized through a four-step series of events:

1. Pattern formation is an early step in differentiation. Recall from Section 22.1 that it serves to organize cells in space. For example, an animal cell might begin to transcribe genes that are typical of a limb in response to signals indicating that the cell is located in a region that will develop into a leg.
2. As development continues, cells become assembled into recognizable tissues and organs in a process called **morphogenesis.** Translated literally, *morphogenesis* means "the generation of form." During pattern formation and morphogenesis, changes in gene expression cause cells to become committed to a particular fate. For example, a limb cell might become grouped with other cells that will become a muscle.
3. **Determination** is the term used to indicate a cell's irreversible commitment to becoming a particular cell type. For instance, when a cell in a muscle becomes determined, that cell is no longer capable of becoming part of a bone or part of blood or cartilage.
4. Finally, cells that are determined begin expressing tissue-specific proteins and begin functioning as muscle or nerve or bone cells. At this point, a cell is differentiated. Our example cell would begin expressing *tropomyosin* and other genes typical of skeletal muscle.

Although this series of events is not strictly linear and sequential, biologists summarize the steps as a progressive process:

$$\text{Pattern formation} \rightarrow \text{Morphogenesis} \rightarrow \text{Determination} \rightarrow \text{Differentiation}$$

To explore how the steps after pattern formation take place, let's consider the determination and differentiation of muscle cells in vertebrates in more detail. What causes undifferentiated mesodermal cells to begin producing the muscle-specific proteins that make breathing, walking, and opening textbooks possible?

Organizing Mesoderm into Somites

To understand how muscle cells become committed to their fate and then differentiated, recall from Chapter 21 that the precursors of muscle cells are located in mesoderm and that, once gastrulation is complete, mesoderm is sandwiched in a layer between endoderm and ectoderm. As **Figure 22.14** shows, the endoderm is internal—it is located toward the interior of the mesoderm and ectoderm. As morphogenesis proceeds, endodermal cells begin to form the digestive tract. At about the same time, the ectoderm on the dorsal (back) side of the embryo folds and forms the **neural tube**, which is the precursor of the brain and spinal cord. What is happening to mesoderm as these events take place? One of the first distinctly mesodermal structures to appear is a rod-like element called the **notochord**, shown in cross section in Figure 22.14. The notochord is unique to the group of animals called the **chordates**, which includes humans and other vertebrates. Cells in or near the notochord produce signaling molecules that induce the neural tube to form; the noto-

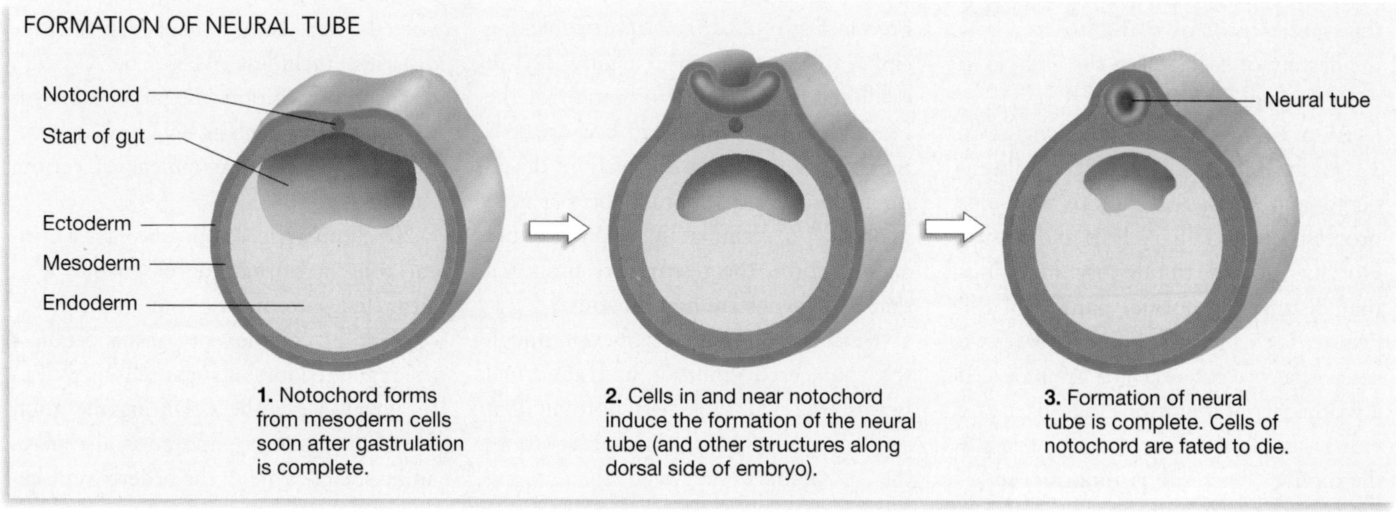

FIGURE 22.14 The Notochord and Neural Tube Form from Dorsal Mesoderm
In vertebrates, the notochord forms from mesoderm cells soon after gastrulation is complete. Molecules produced in or near the notochord induce the formation of the neural tube and other structures along the dorsal (back) side of the embryo.

chord itself also acts as a support as the neural tube folds into its final configuration. The notochord is a transient structure in most species, however, appearing only in embryos. As **Box 22.4** notes, many of the cells in the notochord are fated to die.

Once the neural tube forms, nearby mesodermal cells become organized into blocks of tissue called **somites.** Somites form on both sides of the neural tube, as **Figure 22.16** shows.

Fate-mapping studies have shown that the mesodermal cells in a somite are destined for a variety of structures. Cells from

BOX 22.4 Apoptosis: Programmed Cell Death

As animal tissues and organs take shape, certain cells are fated to die. For example, selected cells in the notochord of vertebrates die and are replaced by cells that form the vertebrae. As the human hand and foot form, cells that are initially present between the fingers and toes die. As a result, separate digits form. In these and many other cases, cell death is a normal part of development—it is carefully timed and regulated, or programmed. Programmed cell death is called **apoptosis.** Translated literally, the term means "falling away."

How does programmed cell death occur? An answer to this question emerged from studies of the roundworm *Caenorhabditis elegans*. This species is a popular research subject, because it has a complex array of organs and tissues but only about a thousand cells. In addition, the cells are transparent. As a result, biologists are able to identify individual cells in both embryos and adults. Thanks to these features, the complete fate map of this species is known. Biologists know the destiny of each cell in the embryo at every stage of development.

As a *C. elegans* individual matures, 131 of its 1090 original somatic cells undergo apoptosis. To explore how this process occurs, Hillary Ellis and Robert Horvitz set out to identify mutations that disrupt the normal pattern of cell deaths. Their initial work uncovered two genes that are essential for apoptosis. If a worm carries mutations that inactivate either gene, cells that would normally die survive. The result is abnormal development and, ironically, the death of the entire organism. Ellis and Horvitz named the genes *ced-3* and *ced-4* (for *c*ell *d*eath abnormal) and proposed that they are part of a genetic program that executes apoptosis. The researchers' hypothesis was that a cell dies if these genes are activated.

Subsequently, Horvitz and colleagues identified a gene that regulates the activity of *ced-3* and *ced-4*. Mutations that inactivate this sequence, called *ced-9*, result in the deaths of many cells that normally survive. Based on this observation, Horvitz's group suggested that the normal function of *ced-9* is to inhibit the suicidal activities of *ced-3* and *ced-4*.

By searching databases of known DNA sequences, biologists found genes in mice called *caspase-9* and *Apaf-1* that are similar to *ced-3* and *ced-4*, respectively. In addition, a mouse protein that inhibits apoptosis, called Bcl 2, is similar to the *ced-9* product. To test the hypothesis that these genes are important in mouse development, Keisuke Kuida and co-workers used genetic engineering techniques to produce mice in which both copies of the *caspase-9* gene are disrupted. In contrast to the normal embryo in **Figure 22.15a**, *caspase-9*-lacking embryos like the one in **Figure 22.15b** exhibited a severe malformation of the brain. The defect occurred because cells that would normally die early in development survived. Because the *Bcl* gene and *ced-9* are similar in both structure and function, the researchers proposed that these genes are homologous.

Kuida's group noted that even though the engineered embryos had abnormal brains, the embryos had normal digit separation. Based on this observation, the researchers suggested that mouse genes other than *caspase-9* are involved in programmed cell death during the formation of the toes.

Are these genes found in humans? The answer is yes—homologs of *Bcl*, the *caspase*s, and all of the other key genes that regulate apoptosis have been identified in the human genome. Normal apoptosis is important in the development of human embryos, and abnormal apoptosis has been implicated in certain diseases of adults. Inappropriate activation of programmed cell death is involved in some neurodegenerative diseases, including ALS (Lou Gehrig's disease). In addition, the failure of negative regulators such as Bcl2 has been implicated in the development of certain cancers.

To summarize, apoptosis plays a critical role in normal development. It is clear that pattern formation in the development of complex organisms requires the regulated loss of some cells as well as the growth of others. During the adult life of an organism, apoptosis also plays an important role in the orderly replacement of cells once their life span has ended. Tissue and organ stability, then, relies as much on programmed cell death as on regulated cell division. Imbalances in either process can lead to disease.

(a) Normal **(b)** Defective

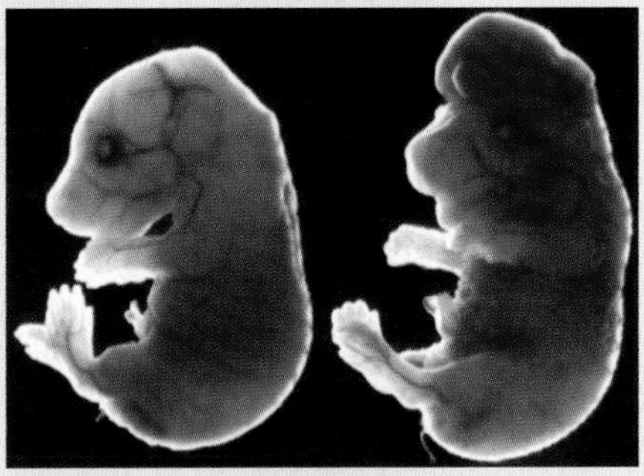

FIGURE 22.15 Defects Occur when Programmed Cell Death Fails
(a) A normal mouse embryo; (b) a mouse embryo that has two defective alleles of the *caspase-9* gene.

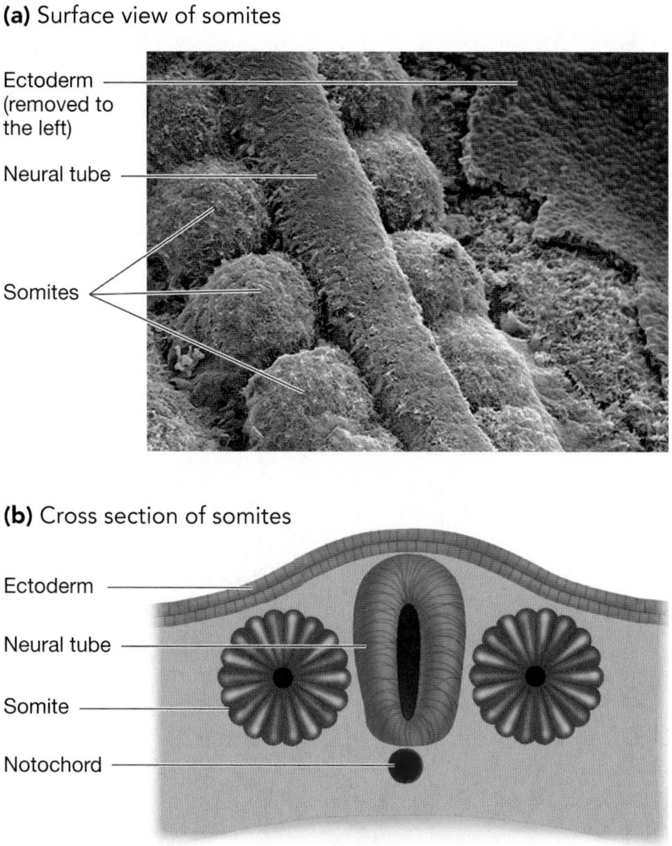
(a) Surface view of somites
Ectoderm (removed to the left)
Neural tube
Somites
(b) Cross section of somites
Ectoderm
Neural tube
Somite
Notochord

FIGURE 22.16 Somites Form on Both Sides of the Neural Tube Somites are made of mesodermal cells.

somites build the vertebrae and ribs, the deeper layers of the skin that covers the back, and the muscles of the back, body wall, and limbs. By transplanting cells from one somite to another, though, researchers have found that initially any cell in a somite can become any of the somite-derived elements of the body. In other words, the cells that form the somite are not determined initially. As a somite matures, though, cells in certain sections of the structure do become committed to a specific fate. For example, biologists transplanted cells from various locations in the somite to the outer part of the somite (farthest from the neural tube) early in that structure's formation. The cells eventually became committed to form the muscles of the limbs (**Figure 22.17a**). This result supports the hypothesis that cells in a somite are initially not determined but later become committed based on their position.

Like the notochord of many species, somites are transient structures that appear relatively briefly during embryonic development. Unlike the notochord cells, however, the cells that make up somites survive and undergo differentiation. Once determination occurs, the cells that make up somites break up into distinct populations and migrate to their final location in the developing embryo. As **Figure 22.17b** shows, each population of cells is committed to becoming a different type of cell.

Recent studies have shown that distinct populations of cells in a somite become determined in response to an array of

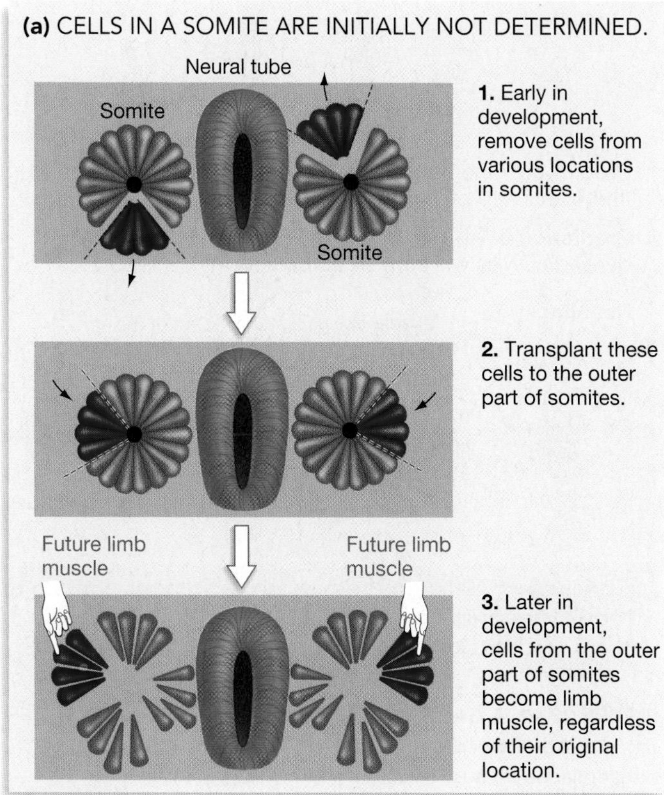
(a) CELLS IN A SOMITE ARE INITIALLY NOT DETERMINED.
Neural tube
Somite
Somite
1. Early in development, remove cells from various locations in somites.
2. Transplant these cells to the outer part of somites.
Future limb muscle
Future limb muscle
3. Later in development, cells from the outer part of somites become limb muscle, regardless of their original location.

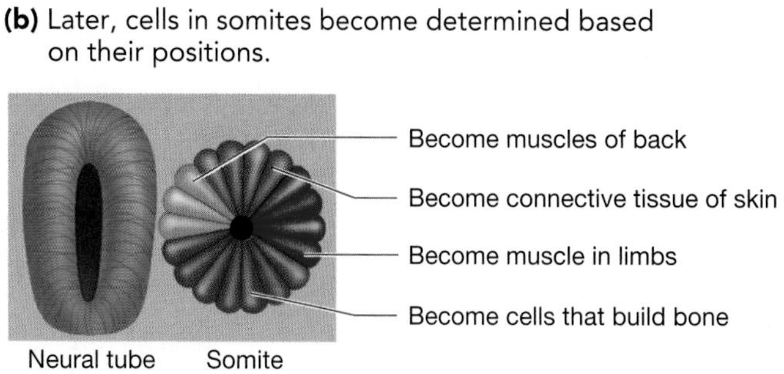
(b) Later, cells in somites become determined based on their positions.
Become muscles of back
Become connective tissue of skin
Become muscle in limbs
Become cells that build bone
Neural tube
Somite

FIGURE 22.17 Cells in the Somite Become Determined
(a) By transplanting cells early in somite formation to new locations in this structure, researchers showed that they are not determined initially. Each somite eventually breaks up into distinct populations of cells, each with a different fate. **(b)** Within a somite, the position of a cell determines its fate.

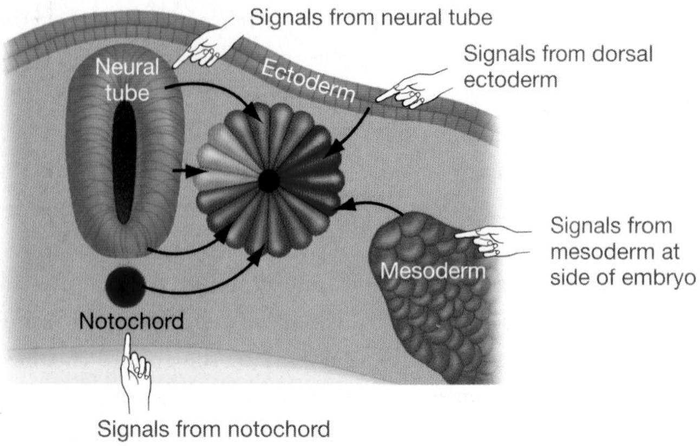

FIGURE 22.18 What Triggers Somite Determination?
Somite cells become determined in response to signals from the neural tube, notochord, and nearby ectoderm and mesoderm.

signals. **Figure 22.18** illustrates how signaling molecules diffuse away from cells in the notochord, the neural tube, and nearby ectoderm and mesoderm and act on target cells in the somite. Each type of signal acts on a distinct population of cells within the structure. In the case of the muscle cells located at the outer part of a somite, how do these signals result in determination?

Determination of Muscle Cells

In a somite, why do cells that are farthest from the neural tube become committed to producing muscle in response to the signals they receive? Harold Weintraub and colleagues answered this question by experimenting with *myoblasts*—cells that are committed to becoming muscle but that have not yet begun producing muscle-specific proteins. Specifically, Weintraub and co-workers hypothesized that myoblasts contain at least one regulatory protein that commits them to their fate. Their idea was that myoblasts begin producing this muscle-determining protein after they receive an appropriate set of signals from nearby tissues.

Figure 22.19 outlines how the biologists went about searching for this hypothetical protein. They began by isolating mRNAs from myoblasts and using reverse transcriptase to convert the mRNAs to cDNAs (steps 1 and 2). Because myoblasts transcribe only the genes required for a muscle cell to function, these cDNAs represented muscle-specific genes. Then the biologists attached a type of promoter to the cDNAs that would ensure the genes' expression in any cell (step 3). Finally, they introduced the recombinant genes into connective tissue cells called *fibroblasts* and monitored the development of the transformed cells (step 4). Just as predicted, one of the myoblast-derived cDNAs converted fibroblasts into muscle cells. Follow-up experiments showed that the same gene could convert pigment cells, nerve cells, fat cells, and liver cells into cells that produced muscle-specific proteins.

Weintraub's group called the protein product of this gene **MyoD**, for *myo*blast *d*etermination. Follow-up work showed

Question: Why do cells on the outside of somites become committed to producing muscle in response to the signals they receive?

Hypothesis: Myoblasts contain at least one regulatory protein that commits them to their fate.

Null hypothesis: Myoblasts do not contain a regulatory protein that commits them to their fate.

Experimental setup:

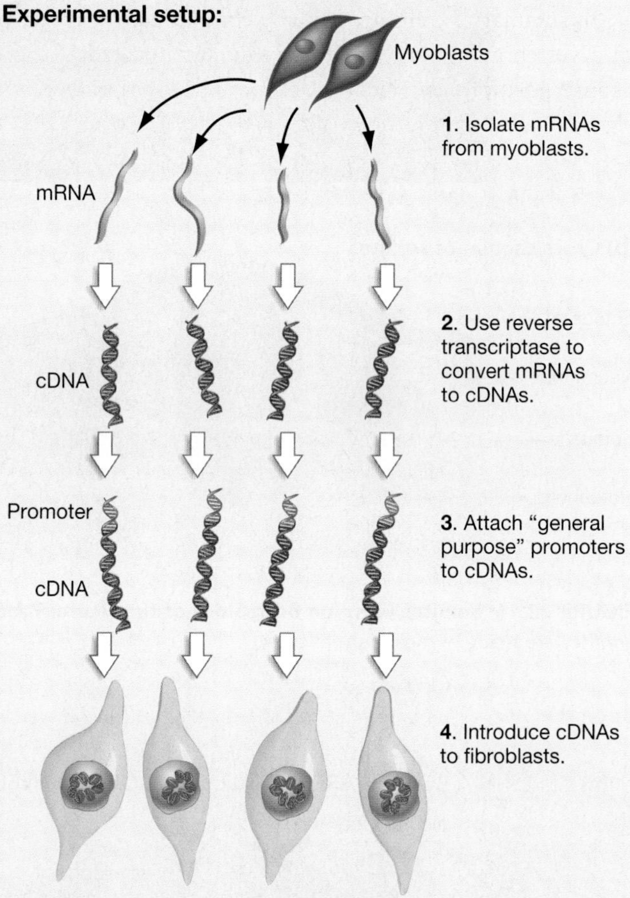

Prediction: One of the myoblast-derived cDNAs will convert fibroblasts into muscle cells.

Prediction of null hypothesis: None of the myoblast-derived cDNAs will convert fibroblasts into muscle cells.

Results:

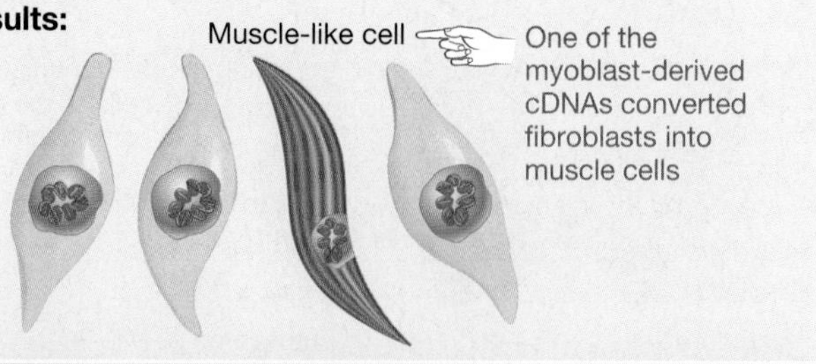

Conclusion: Cells on the outside of somites contain a regulatory protein (later called MyoD) that commits them to differentiate into muscle.

FIGURE 22.19 A Gene That Causes Muscle-Cell Differentiation
Protocol that allowed researchers to find a gene involved in the differentiation of muscle cells. QUESTION Why did the researchers have to attach a "general purpose" promoter to the cDNAs?

that *MyoD* encodes a regulatory transcription factor and that the MyoD protein binds to enhancer elements located upstream of muscle-specific genes. In addition, the MyoD protein activates expression of the *MyoD* gene. This was a key observation because it meant that once *MyoD* is turned on, it triggers its own expression—meaning that it remains on. Other researchers have found that genes closely related to *MyoD* are also required for the differentiation of muscle cells.

To summarize, cells are initially committed to become part of a vertebrate's back when determinants located in the egg change position after fertilization occurs. In response to these signals, cells become committed to become the organizer. (Recall from Section 21.3 that cells in the organizer secrete cell-to-cell signals that induce changes in target cells.) The organizer then induces mesoderm that becomes part of the notochord. Subsequently, signals from the notochord and nearby cells induce the production of MyoD and other muscle-determining proteins in certain populations of cells from somites. In response, these target cells are committed to becoming muscle. The commitment is reinforced by MyoD's ability to enhance its own production.

As a case history, the determination and differentiation of vertebrate muscle cells illustrates several principles common to all cell types. First, it confirms that differential gene expression is at the heart of development. Second, it supports the view that the development of an individual is a stepwise process. Determination occurs when the production of a specific regulatory protein such as MyoD commits a cell to a particular fate. Differentiation occurs when the cell actually begins to produce cell-specific proteins. At that point, the process of development is largely complete. Growth takes over as the most important process in shaping the juvenile organism.

ESSAY Human Stem Cells

In most cases, the fate of a cell is sealed once it has "popped into place." At some point in the step-by-step process, development can become irreversible. When a muscle cell in a human embryo differentiates, for example, it stops growing and begins producing muscle-specific proteins. This cell remains a muscle cell for as long as it lives. It cannot change its fate and become a nerve cell.

There are important exceptions to this rule, however. The cells that make up plant meristems divide continuously and retain the ability to produce an array of cells, tissues, and organs. In many cases, even specialized plant cells retain the ability to "un-differentiate" and give rise to other types of tissues and organs. Gardeners and farmers take advantage of this trait when they take cuttings from stems or roots and use them to propagate new individuals.

Even in adult mammals, certain populations of cells retain the ability to divide and produce an array of cell types. Cells with this ability are called **stem cells**. Stem cells can be found in your muscle, skin, liver, gut lining, and elsewhere. Given that you lose an average of 100 billion cells from the lining of your intestine and from your blood each day, it is not surprising that stem cells exist.

The stem cells in your body continue to divide throughout your life. Some of their daughter cells remain as stem cells; others differentiate to replace cells in nearby tissues that have been lost or damaged. One of the key attributes of your stem cells is that they behave as though they are determined but not differentiated. The stem cells in your skin, for example, can give rise only to skin cells. Although the stem cells in your bone marrow give rise to an impressive array of different cell types, most of their descendant cells become components of blood.

In contrast, stem cells that are isolated from embryos are not as narrowly committed. Cells from the inside of a blastocyst, for example, can give rise to virtually any cell type in the body. Recently, researchers have started to explore whether large numbers of these embryonic stem cells could be grown in culture and then implanted into adults as a treatment for diseases and injuries caused by the loss of differentiated cells. Parkinson's disease and Alzheimer's disease are caused by a loss of brain cells; the most serious form of diabetes is caused by loss of certain cells in the pancreas; muscular dystrophy results from the death of muscle cells. Biomedical researchers hope that implanted stem cells will be able to differentiate and at least partially make up for the loss of the original cells.

Research on human stem cells is intensely controversial. . . .

Research on human stem cells is intensely controversial, however. Because in vitro fertilization techniques (see the Essay in Chapter 21) frequently produce an excess of early embryos, the primary source for embryonic stem cells has been fertility clinics. If these surplus embryos are not discarded, they can be dissected and used as a source of stem cells. Opponents of this practice argue that because the embryos are sacrificed in this process, the procedure represents the taking of a human life.

Currently, nations are debating rules to govern the use of embryonic stem cells in research. Although the practice is banned in Italy and Norway, it is legal or even actively supported in most countries of the European Union. In the United States, public funding for the establishment of new cultures of embryonic stem cells has been banned, but research on cultures of embryonic stem cells that had been established prior to the year 2001 is still being supported. It remains to be seen whether stem cell research will result in important therapies and, if so, whether their use will be approved by regulatory agencies.

CHAPTER REVIEW

Summary of Key Concepts

The fate of an embryonic cell—what it will become in the juvenile and adult organism—is established in a series of steps.

During pattern formation, cells receive information about their location in the embryo. As morphogenesis gets under way, cells change their structure and function in a way that forms an ordered arrangement of cells in space. Determination occurs when a cell becomes committed to a particular fate. A cell is differentiated when it begins producing tissue-specific proteins and functions as a mature, specialized component of the organism.

Researchers have used a genetic approach to study the mechanisms of pattern formation in fruit flies and *Arabidopsis*. By analyzing mutant embryos or adults with defects in pattern formation, biologists have been able to identify a series of genes and gene products that organize cells in space. The pattern-formation genes that have been discovered to date fall into two broad categories, based on the phenotypes of individuals with mutant forms of the genes: (1) those that cause the loss of specific regions in the embryo—for example, the segmentation genes in *Drosophila* or *monopterous* and related genes in *Arabidopsis*—and (2) those that cause homeotic transformations in adults.

The segmentation genes found in fruit flies and the *monopterous*-like genes in mustard plants are thought to define particular segments or regions in the embryo. The *Drosophila* segmentation genes generate a series of segments along the head-to-tail axis of the body; the *monopterous*-like genes of *Arabidopsis* define the cotyledons, hypocotyl, and root along the root-to-shoot axis of the body. The protein products of these genes are thought to carry signals such as "this area is a segmental boundary in the thorax" or "this area is part of the root."

In contrast, when a homeotic transformation takes place, a structure that is usually found at a different place in the body is substituted for the normal structure at a particular location. Flies with legs instead of antennae and flowers with sepals instead of petals are examples of homeotic transformations. To explain why these phenotypes occur, biologists suggest that the products of homeotic genes identify particular locations in the developing organism. For example, a *HOM-C* gene product in *Drosophila* might signal, "This is thoracic segment 2," whereas a MADS-box protein in *Arabidopsis* might signal, "This is the second whorl of floral organs."

Web Tutorial 22.1 Early Pattern Formation in *Drosophila*

At each step in development, cell-to-cell signals activate regulatory transcription factors that direct the expression of genes in cells throughout the embryo.

Research on pattern-formation genes has brought to light several general principles about development. Development is a step-by-step process based on differential gene expression. As pattern formation and other aspects of development proceed, changes in gene expression occur in response to cytoplasmic determinants or signals from other cells. A cell's response depends on the types of signals it receives and their concentration. Often this response occurs through the production of regulatory transcription factors. The type and quantity of signals received by a cell depend on its position in the embryo.

Because virtually all cells in a multicellular organism contain the same genes, development is based on differential gene expression.

The ability to clone entire plants from a cutting or a small group of cells established that mature plant cells contain all of the genes that were present in the fertilized egg. In animals, however, it was much more difficult to establish that adult cells are genetically equivalent. Eventually a series of nuclear transplant experiments culminated in the cloning of sheep, mice, and cows from somatic cell nuclei taken from adults. This result established that virtually all cells in adult plants and animals are genetically identical. In multicellular organisms, cells become specialized through the expression of specific genes—not the loss of specific genes.

One central focus of current research in developmental biology is to understand how differential gene expression leads to determination and differentiation. For example, the mesodermal cells that make up the somites found in vertebrate embryos take on one of several fates, depending on their position in the somite and thus the signals that they receive from nearby structures. Cells on the outside of somites become determined to be muscle because they receive signals that trigger the production of MyoD and other regulatory proteins. In turn, these regulatory proteins cause the cell to differentiate by triggering the transcription of muscle-specific genes.

Questions

Content Review

1. The auxin produced in *Arabidopsis* embryos and Bicoid protein produced in *Drosophila* embryos have what similar function?
   a. Both are regulatory transcription factors that serve as cell-to-cell signals.
   b. Both are cytoplasmic determinants—they are loaded into eggs by cells in the mother or by female reproductive tissues.
   c. Both trigger the transcription of homeotic loci—genes that specify the types of structures produced at a certain position.
   d. Both establish a concentration gradient that demarcates the anterior-posterior axis.

2. In combination, what do the products of gap genes, pair-rule genes, and segmentation polarity genes of fruit flies do?
   a. They trigger the reorganization of the larval body into an adult body.
   b. They define the segmented body plan of the embryo.
   c. They set up the back-to-belly axis of the larval body.
   d. They direct cell movements during gastrulation.

3. What does the ABC model of flower development attempt to explain?
   a. how different combinations of gene products trigger the formation of different floral organs
   b. why petals are found on the inside of the whorl of sepals instead of on the outside
   c. why the four types of floral organs occur in whorls
   d. why apical meristems are converted to floral meristems in response to specific cell-to-cell signals
4. What is a homeotic mutant?
   a. an individual with a structure located in the wrong place
   b. an individual with an abnormal head-to-tail axis
   c. in flies, an individual that is missing segments; in *Arabidopsis*, an individual that is missing a hypocotyl or other embryonic structure
   d. an individual with double the normal number of structures or segments
5. What evidence suggested that the MyoD protein is involved in the differentiation of muscle cells?
   a. The *MyoD* gene is expressed late—not early—in the development of somites.
   b. Expression of *MyoD* DNA can convert nonmuscle cells to muscle-like cells.
   c. *MyoD* mRNA was isolated from muscle cells.
   d. MyoD is part of a closely related family of transcription factors.
6. What is apoptosis?
   a. an experimental technique that biologists use to kill specific cells
   b. programmed cell death that is required for normal development
   c. a pathological condition observed only in damaged or diseased organisms
   d. a developmental mechanism unique to the roundworm *C. elegans*

Conceptual Review

1. What evidence suggests that at least some of the molecular mechanisms responsible for pattern formation have been highly conserved over the course of animal evolution?
2. How did researchers succeed in identifying molecules that have important roles in *Drosophila* and *Arabidopsis* pattern formation?
3. Explain why mRNA injection experiments support the hypothesis that *bicoid* sets up the normal head-to-tail axis of a *Drosophila* embryo. Why did researchers bother to do such experiments? To accept the gradient hypothesis as valid, why wasn't it enough to observe the *bicoid* mutant phenotype, the distribution of *bicoid* mRNA, and the distribution of Bicoid protein?
4. Why is it significant that many of the genes involved in pattern formation and determination encode regulatory transcription factors?
5. What is the difference between determination and differentiation? What is the relationship of these processes to pattern formation?
6. The development of in situ hybridization and antibody-staining technology had a huge impact on biologists' ability to study pattern formation and determination. Why?

Group Discussion Problems

1. Recent research has shown that the products of two different *Drosophila* genes are required to keep *bicoid* mRNA concentrated at the anterior end of the egg. In individuals with mutant forms of these proteins, *bicoid* mRNA diffuses farther toward the posterior pole than it normally does. First, predict what effect these mutations will have on segmentation of the larva. Second, suggest a hypothesis for how these proteins function at a molecular level. How would you test your hypothesis? Finally, predict whether the mutations exhibit maternal effect inheritance. Explain your rationale.
2. In 1992 David Vaux and colleagues used genetic engineering technology to introduce an active human gene for the Bcl 2 protein into embryos of the roundworm *C. elegans*. When the team examined the embryos, they found that cells that normally undergo programmed cell death survive. What is the significance of this observation?
3. Animal cells make massive movements during gastrulation. Similarly, the cells that make up somites break up into distinct populations and migrate to their final position in the embryo after determination has occurred. Plant cells, in contrast, do not move, because they are surrounded by a stiff cell wall. Discuss how this lack of movement might affect pattern formation, differentiation, or determination in plants.
4. According to the data presented in this chapter, segmentation genes set up the segmented body plan of a fruit fly while *HOM-C* genes trigger the production of segment-specific structures such as antennae, legs, and wings. Now consider the differences in the body plans of a spider, a fruit fly, a centipede, and an earthworm. Many biologists hypothesize that differences in the expression of segmentation genes or *HOM-C* genes are responsible for some of the morphological differences observed in these animal groups. Do you find this hypothesis plausible? Why or why not?

Answers to Multiple-Choice Questions 1. d; 2. b; 3. a; 4. a; 5. b; 6. b

www.prenhall.com/freeman is your resource for the following: Web Tutorials; Online Quizzes and other Online Study Guide materials; Answers to Conceptual Review Questions; Solutions to Group Discussion Problems; Answers to Figure Caption Questions and Exercises; and Additional Readings and Research.

UNIT 5

Evolutionary Processes and Patterns

A tiger beetle. Beetles are the most species-rich lineage on the tree of life, with well over 300,000 species named to date.

23 Evolution by Natural Selection

KEY CONCEPTS

- Populations and species evolve, meaning that their characteristics change through time. More precisely, evolution is defined as changes in allele frequencies over time.
- Natural selection occurs when individuals with certain alleles and traits produce the most offspring in a population. An adaptation is a genetically based trait that increases an individual's ability to produce offspring in a particular environment.
- Evolution by natural selection is not progressive, and it does not change the characteristics of the individuals that are selected—it changes only the characteristics of the population. Not all traits are adaptive. All adaptations are constrained by genetic and historical factors.

Alpine skypilots grow at high elevations in the Rocky Mountains of North America. Populations of skypilots that are pollinated by bumblebees have large, sweet-smelling flowers, but populations that are pollinated by flies have small, skunky-smelling flowers. Recent experiments have shown that the differences in these populations are due to evolution by natural selection.

This chapter is about one of the great ideas in science. The theory of evolution by natural selection, formulated independently by Charles Darwin and Alfred Russel Wallace, explains how organisms have come to be adapted to environments ranging from arctic tundra to tropical wet forest. As an example of a revolutionary breakthrough in our understanding of the world, the theory of evolution by natural selection ranks alongside Copernicus's theory of the Sun as the center of our solar system, Newton's laws of motion and theory of gravitation, and Einstein's general theory of relativity.

Like most scientific breakthroughs, this one did not come easily. When Darwin published his theory in 1859 in a book called *On the Origin of Species by Means of Natural Selection*, it unleashed a firestorm of protest throughout Europe. At that time, the leading explanation for the diversity of organisms was a theory called special creation. This theory held that all species were created independently, by God, perhaps as recently as 6000 years ago. The theory of special creation also maintained that species were immutable, or incapable of change, and thus had been unchanged since the moment of their creation. Darwin's ideas were radically different. He proposed that life on Earth was ancient and that species change through time.

To understand the contrast between the theory of special creation and the theory of evolution by natural selection more thoroughly, recall from Chapter 1 that scientific theories usually have two components: a pattern and a process. The first component is either a claim about a pattern that exists in nature or a statement that summarizes a series of observations about the natural world. In short, the pattern component is about facts—about how things *are* in nature. The second component of a scientific theory is a process that produces that pattern or set of observations. For example, the pattern component in the theory of special creation was that species were seen to be created independently of one another and that they do not change through time. The process that explained this pattern was the instantaneous and independent creation of living organisms by a supernatural being.

This chapter begins by examining the pattern component of Darwin's theory. (**Box 23.1** discusses why the theory of evolution became associated primarily with Darwin's name.) Specifically, Section 23.1 considers the evidence behind the claim that species are not independent but are instead related and that they are not immutable but instead have changed through time. Darwin proposed that natural selection explains this pattern; in Section 23.2 we analyze the process. The concluding sections review two recent studies of evolution by natural selection. These case studies illustrate how biologists test Darwin's theory by studying evolution in action. The case studies also help clarify some common misunderstandings about how evolution by natural selection works.

BOX 23.1 Why Darwin Gets Most of the Credit

Although Charles Darwin and Alfred Russel Wallace formulated the same explanation for how species change over time, Darwin's name is much more prominently associated with the theory of evolution because he developed the idea more thoroughly and provided massive evidence for it in *On the Origin of Species*. But historians of science speculate about whether Darwin would have published his theory at all had Wallace not threatened to scoop him (**Figure 23.1**).

Darwin wrote a paper explaining evolution by natural selection in 1842—a full 17 years before the first edition of *On the Origin of Species* came out. He never submitted the work for publication, however. Why? Darwin claimed that he needed time to document all of the arguments for and against the theory and to examine its many implications. There is probably an element of truth in this—Darwin was a remarkably thorough thinker and writer. But many historians of science argue that he held off largely out of fear. Because his theory was inconsistent with the creation story in the Bible's Book of Genesis, Darwin knew that he would be exposed to scathing criticism from religious and scientific leaders. He was also an extremely private person, had a strong religious upbringing, and was frequently in poor health. He responded to stress or personal attacks by suffering long bouts of debilitating illness. The prospect of fighting for his ideas against the most powerful men in Europe was daunting. But Wallace forced his hand.

Wallace was also a native of England, but he had been making a living by collecting butterflies and other natural history specimens in Malaysia and selling them to private collectors. While recuperating from a bout of malaria there in 1858, he wrote a brief article outlining the logic of evolution by natural selection. He sent a copy to Darwin, who immediately recognized that they had formulated the same explanation for how populations change through time. The two had their papers read together before the Linnean Society of London, and Darwin then rushed *On the Origin of Species* into publication a year later. The first edition sold out in a day.

Fortunately for Darwin's health, a fellow biologist and friend named Thomas Huxley publicly defended the theory against criticism, which came from both scientific and religious quarters. Darwin continued to live quietly on his estate in Down, England, and actively continued a brilliant research career.

(a) Charles Darwin **(b)** Alfred Russel Wallace

FIGURE 23.1 The Codiscoverers of Evolution by Natural Selection
(a) Charles Darwin in 1840, four years after he returned from the voyage of the Beagle and two years before he drafted his first paper explaining evolution by natural selection. (b) Alfred Russel Wallace, who in 1858 independently formulated the theory of natural selection.
[(b) Alfred Russel Wallace by unknown artist, after a photograph by Thomas Sims, fl. 1860s. Reg. No.: 1765. National Portrait Gallery, London.]

23.1 The Pattern of Evolution: Have Species Changed through Time?

In *On the Origin of Species*, Darwin repeatedly used the phrase "**descent with modification**" to describe evolution. By this he meant that the species existing today have descended from other, preexisting species and that species are modified or change through time. This view was clearly a radical departure from the pattern of independently created and immutable species proposed by the theory of special creation. But what evidence do we have that Darwin was correct? How can biologists infer that species have changed through time instead of being static entities and that they are indeed related—not independent? Let's consider each question in turn.

Evidence for Change through Time

When Darwin began his work, biologists and geologists had just begun to assemble and interpret the fossil record. A **fossil** is any trace of an organism that lived in the past. These traces range from bones and branches to shells, tracks or impressions, and dung (**Figure 23.2**). The **fossil record** consists of all the fossils that have been found and described in the scientific literature.

Initially, fossils were organized according to their relative ages. This was possible because most fossils are found in **sedimentary rocks**, which form from sand or mud or other materials deposited at locations such as beaches or river mouths, and because sedimentary rocks are known to form in layers. Fossils from rocks underneath other rocks were judged to be older than the fossils found above them. In this way, researchers began putting fossils in an older-to-younger sequence. They also began naming different periods of geologic time, creating the sequence of eons, epochs, and periods called the **geologic time scale**. After the discovery of radioactivity in the early 1900s, researchers used radiometric dating techniques (introduced in Chapter 2) to assign absolute ages to the relative ages in the geologic time scale. According to these data, Earth is about 4.6 billion years old, and the earliest signs of life appear in rocks that formed about 3.85 billion years ago. Instead of being 6000 years old, life on Earth is ancient.

The fossil record continues to expand in size and quality. Three additional observations about this data set—regarding extinction, transitional forms, and environmental changes—support the hypothesis that species have changed through time.

Extinction In the early nineteenth century, researchers began discovering fossil bones, leaves, and shells that were unlike structures from any known animal or plant. At first, many scientists insisted that living examples of these species would be found in unexplored regions of the globe. But as research continued and the number and diversity of fossil collections grew, the argument became less and less plausible. After Baron Georges Cuvier published a detailed analysis of an **extinct** species—that is, a species that no longer exists—called the Irish "elk" in 1812, most scientists accepted extinction as a reality. This gigantic deer was judged to be too large to have escaped discovery and too distinctive to be classified as a large-bodied population of an existing species.

Advocates of the theory of special creation argued that the fossil species were victims of the flood at the time of Noah. Darwin, in contrast, interpreted them as evidence that species are not static, immutable entities, unchanged since the moment of special creation. His reasoning was that if species have gone extinct, then they have changed through time. Recent analyses of the fossil record support the claim that more species have gone extinct than exist today. The data also indicate that species have gone extinct throughout Earth's history—not just in one or even a few catastrophic events.

Transitional Forms Long before Darwin published his theory, researchers reported striking resemblances between the fossils found in the rocks underlying certain regions and the living species found in the same geographic areas. The pattern was so widespread that it became known as the "law of succession." The general observation was that extinct species in the fossil record were succeeded, in the same region, by similar species

(a) 110-Million-year-old ammonite shell
(b) 50-Million-year-old bird tracks
(c) 20,000-Year-old sloth dung

FIGURE 23.2 A Fossil Is *Any* Trace of an Organism That Lived in the Past
In addition to **(a)** body parts such as shells or bones or branches, fossils may consist of **(b)** tracks or impressions, or even **(c)** pieces of dung.

(**Figure 23.3a**). Early in the nineteenth century, the pattern was simply reported and not interpreted. But later, Darwin pointed out that it provided strong evidence in favor of the hypothesis that species had changed through time. His idea was that the extinct forms and living forms were related—that they represented ancestors and descendants.

As the fossil record improved, researchers discovered transitional forms that broadened the scope of the law of succession. A **transitional form** is a fossil species with traits that are intermediate between those of older and younger species. For example, intensive work over the past several decades has yielded fossils that document a gradual change over time from land-dwelling mammals that had limbs to ocean-dwelling mammals that had reduced limbs or no limbs (**Figure 23.3b**). All of the species in this sequence have distinctive types of ear bones that researchers use to identify whales. The oldest whale fossils found to date are from fox-sized animals that had eyes located at the tops of their heads. Based on this observation and the fact that the fossils were found in rocks that form only in ocean deposits, biologists suggest that the earliest whales were semiaquatic animals not unlike hippopotamuses. Over the subsequent 12 million years, the limbs of whale species became more reduced. These observations support the hypothesis that whales gradually became more strictly aquatic and more like today's whales in appearance and lifestyle. Whale species have clearly changed through time.

Similar sequences of transitional forms document changes that led to the evolution of feathers and flight in birds, stomata and vascular tissue in plants, upright posture and large brains in humans, jaws in vertebrates, limbs in amphibians and other vertebrates, and other traits. The discovery of the transitional forms provides strong evidence for change through time.

Environmental Change Several of the whale fossils pictured in Figure 23.3b were found high in the Himalayan Mountains of southern Asia. The rocks surrounding the fossils originated as beach and mud deposits made by oceans, yet are now found thousands of meters above sea level. Similarly, fossilized shells from extinct clams and other ocean-dwelling species are found high in Europe's Alps and in the rock layers that line the Grand Canyon in North America's southwestern desert. In addition, geologists have compiled massive evidence in support of the *theory of plate tectonics*, which holds that Earth's crust is made up of plates and that the positions of the continents change almost continuously (see Chapter 24).

Observations such as these support the hypothesis that Earth's topography and environment have changed dramatically over time. If the Earth itself has changed, it is logical to infer that organisms have changed as well.

To summarize, data from the fossil record indicate that life is ancient and that species have changed through the course of Earth's history. Our planet and its species are dynamic—not static, (unchanging) as claimed by the theory of special

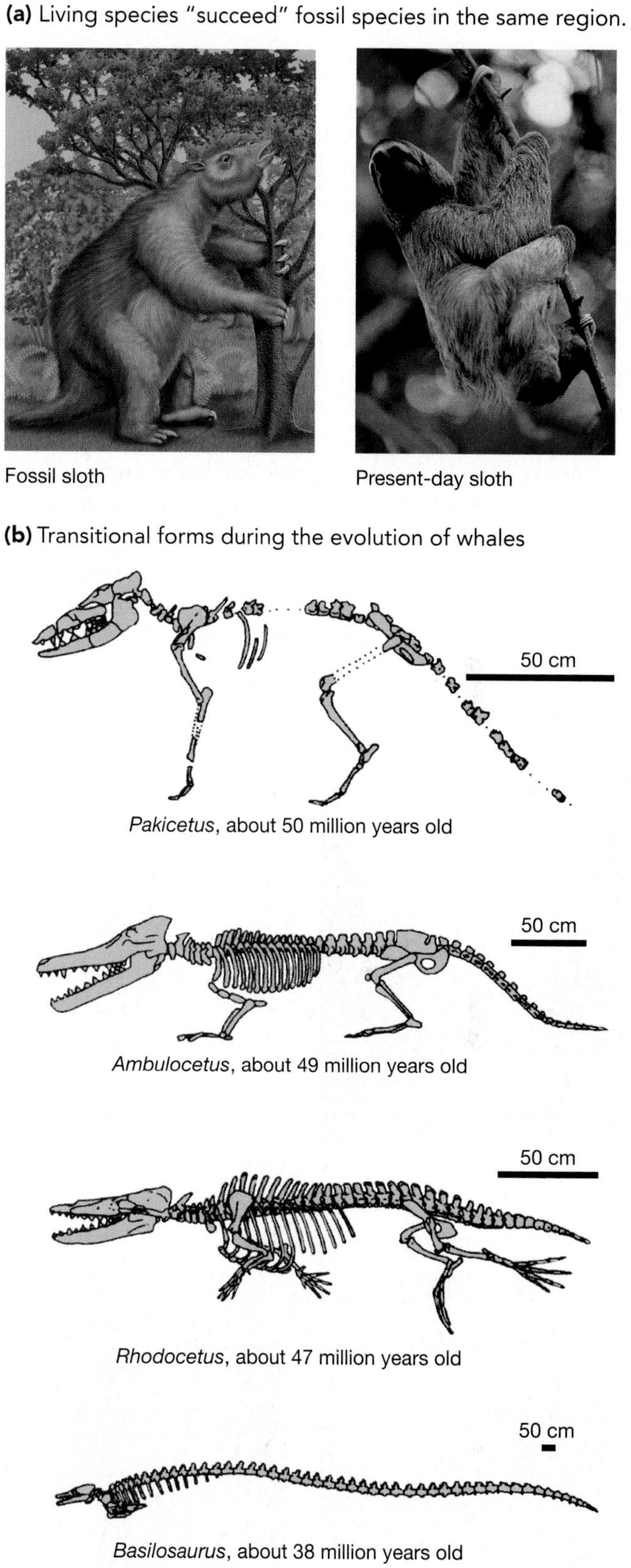
(a) Living species "succeed" fossil species in the same region.
Fossil sloth
Present-day sloth
(b) Transitional forms during the evolution of whales
50 cm
Pakicetus, about 50 million years old
50 cm
Ambulocetus, about 49 million years old
50 cm
Rhodocetus, about 47 million years old
50 cm
Basilosaurus, about 38 million years old

FIGURE 23.3 Evidence That Species Have Changed through Time
(a) Fossil and living sloths are found only in Central America and South America. Darwin interpreted similarities among fossil and living species in the same geographic area as evidence that species had changed through time. **(b)** Transitional forms document the changes that occurred as whales evolved from terrestrial mammals to the aquatic mammals of today.

creation. Two observations about living species support the same conclusion:

1. Darwin was the first to describe and interpret **vestigial traits**—reduced or incompletely developed structures that have no function or reduced function but are clearly similar to functioning organs or structures in other species. Bowhead whales and rubber boas have tiny hip and leg bones that do not help them swim or slither; ostriches have reduced wings and cannot fly. Humans have a variety of vestigial traits. Our appendix is a reduced version of the cecum—an organ found in other vertebrates that functions in digestion. The human coccyx, illustrated in **Figure 23.4a**, is a reduced form of the tailbone found in monkeys and other primates. Even goose bumps, shown in **Figure 23.4b**, are a vestigial trait. Many mammals, including primates, are able to erect their hair when they are cold or excited. But our sparse fur does little to keep us warm, and goose bumps are largely ineffective in signaling our emotional state. In addition, our genome includes dozens of pseudogenes—the functionless DNA sequences introduced in Chapter 20. To make sense of these observations, biologists argue that the structure and function of these traits have changed over time.
2. Researchers have now documented hundreds of instances in which contemporary populations have changed in response to changes in their environment. Bacteria have become resistant to drugs; insects have become resistant to pesticides; and weedy plants have become resistant to herbicides. In several cases, biologists are even documenting the formation of new species, right before our eyes. Change through time continues and can be measured directly.

(a) The human tailbone is a vestigial trait.

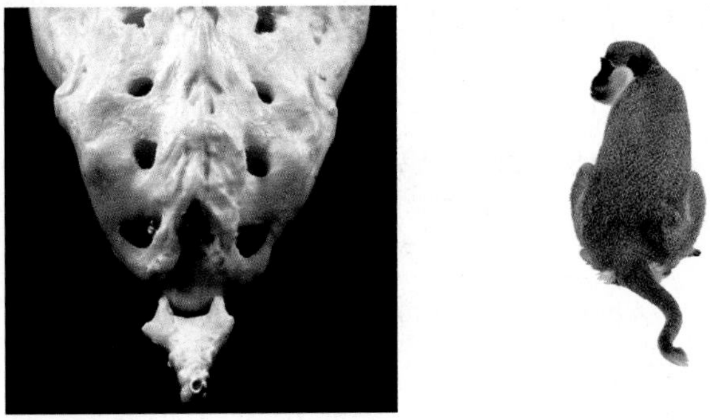

Human coccyx

Vervet monkey tail (used for balance)

(b) Goose bumps are a vestigial trait.

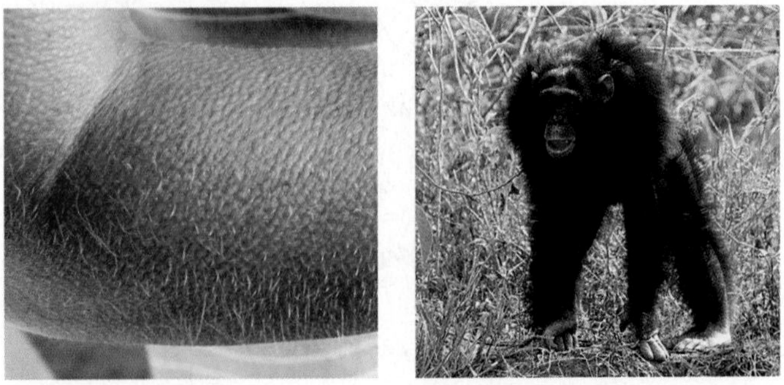

Human goose bumps

Erect hair on chimp (insulation, emotional display)

FIGURE 23.4 Vestigial Traits Are Reduced Versions of Traits in Other Species
(a) The tailbone and **(b)** goose bumps are human traits that have reduced function. They are similar to larger, fully functional structures in other species.

Evidence That Species Are Related

Data from the fossil record and contemporary species refute the hypothesis that species are immutable. What about the claim that species were created independently—meaning that they are unrelated to each other?

Charles Darwin began to formulate an answer during a five-year voyage he took aboard the English naval ship HMS *Beagle*. While fulfilling its mission to explore and map the coast of South America, the *Beagle* spent considerable time in the Galápagos Islands off the coast of present-day Ecuador. Darwin had taken over the role of ship's naturalist and gathered extensive collections of the plants and animals found in these islands. Among the birds he collected were what came to be known as the Galápagos mockingbirds, pictured in **Figure 23.5a**.

Several years after Darwin had returned to England, a naturalist friend in London pointed out that the mockingbirds Darwin had collected on different islands were distinct species, based on differences in coloration and beak size and shape. This struck Darwin as remarkable. Why would species that inhabit neighboring islands be so similar, yet clearly distinct? This turns out to be a very general pattern: In island groups across the globe, it is routine to find very similar, but different, species on neighboring islands.

Darwin realized that this pattern—puzzling when examined as a product of special creation—made perfect sense when interpreted in the context of evolution, or descent with modification. He proposed that the mockingbirds were similar because they had descended from the same common ancestor. If so, then the mockingbird species are part of a **phylogeny**—a family tree of populations or species. Further, the mockingbirds can be placed on a **phylogenetic tree**, a branching diagram that describes the relationships among species, much as a genealogy describes the relationships among individual humans. **Figure 23.5b** illustrates this idea. Darwin's hypothesis was that, instead of being created independently, mockingbird populations that colonized different islands had changed through time and formed new species.

But island forms are not the only species that show strong similarities. Resemblances among species are widespread. Today, these resemblances can be identified and analyzed at three levels, each of which relates to the others. Resemblances among species are structural, developmental, and genetic in nature.

(a) Similar mockingbird species on different Galápagos islands

Nesomimus parvulus
Nesomimus melanotis
Nesomimus trifasciatus
Nesomimus macdonaldi
Galápagos islands
N. parvulus
N. melanotis
N. trifasciatus
N. macdonaldi

(b) Darwin reasoned that they share a common ancestor.

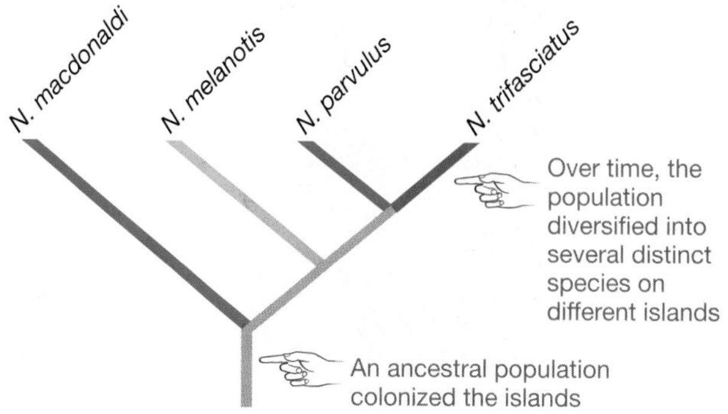
N. macdonaldi
N. melanotis
N. parvulus
N. trifasciatus
Over time, the population diversified into several distinct species on different islands
An ancestral population colonized the islands

FIGURE 23.5 Close Relationships among Island Forms Argue for Shared Ancestry
(a) Darwin collected mockingbirds from several islands of the Galápagos. **(b)** Phylogeny illustrating Darwin's explanation for why mockingbirds from different islands are similar yet distinct.

Structural Homologies Biologists routinely document similarities in the limbs, shells, or flowers of different species. Morphological traits that are similar are termed **structural homologies.** (Translated literally, *homology* means "the study of likeness.") A classic example is the common structural plan observed in the limbs of vertebrates (**Figure 23.6**). In Darwin's own words, "What could be more curious than that the hand of a man, formed for grasping, that of a mole for digging, the leg of the horse, the paddle of the porpoise, and the wing of the bat, should all be constructed on the same pattern, and should include the same bones, in the same relative positions?"

Darwin's point was that an engineer would never use the same underlying pattern to design the structure of a grasping tool, a digging implement, a walking device, a propeller, and a wing. But if all mammals descended from a common ancestor,

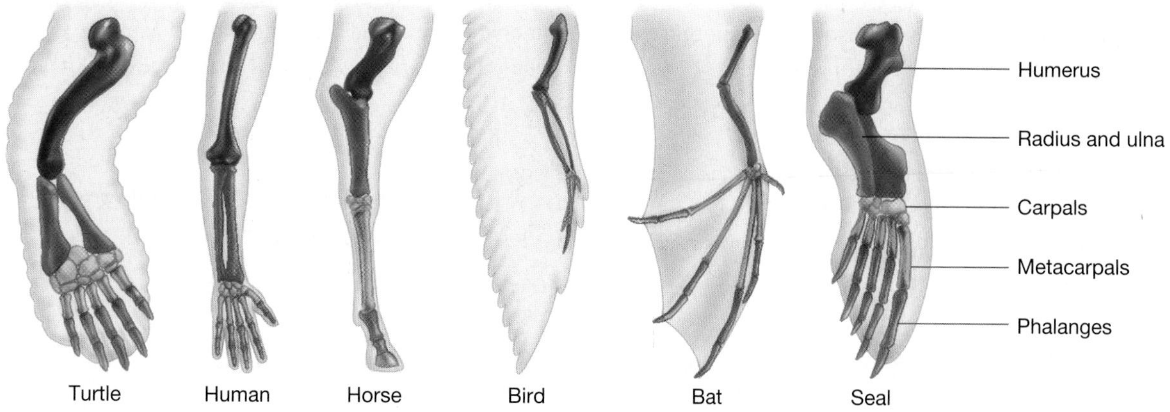
Humerus
Radius and ulna
Carpals
Metacarpals
Phalanges
Turtle
Human
Horse
Bird
Bat
Seal

FIGURE 23.6 Structural Homology: Limbs with Different Functions Have the Same Underlying Structure
Even though their function varies, all vertebrate limbs are modifications of the same number and arrangement of bones. Darwin interpreted structural homologies like these as a product of descent with modification. (These limbs are not drawn to scale.)

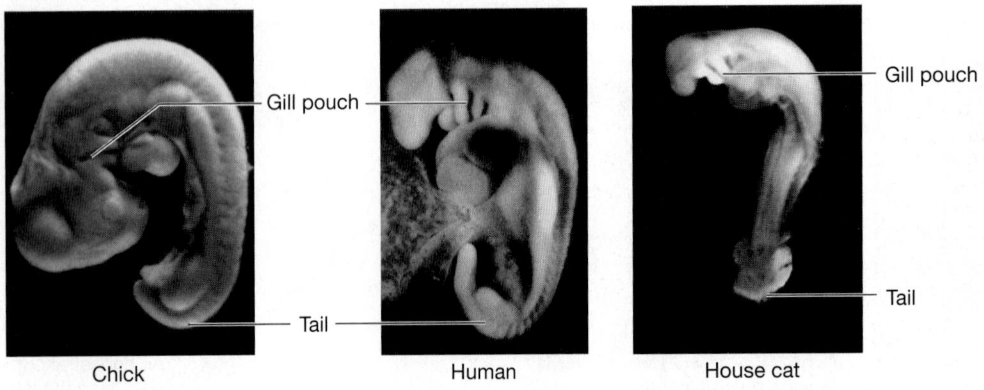

FIGURE 23.7 Developmental Homology: Structures That Appear Early in Development Are Similar
The early embryonic stages of a chick, a human, and a cat, showing a strong resemblance.

and if that ancestor had a limb with the same ground plan shown in Figure 23.6, then it would logical to observe that its descendants had modified forms of the same design.

Developmental Homologies When newly developing embryos of different species have a similar trait, they are said to exhibit a **developmental homology**. Developmental homologies are routinely observed at two levels: in the overall **morphology** (that is, form) of embryos and in the fate of particular embryonic tissues.

Figure 23.7 illustrates the strong general resemblance among the embryos of **vertebrates**—animals with backbones. Early in development, gill pouches and tails form in chicks, humans, and cats. Later in development, gill pouches are lost in all three species and tails are lost in humans. To explain this observation, biologists hypothesize that gill pouches and tails exist in chicks, humans, and cats because they existed in the fish-like species that was the common ancestor of birds and mammals.

Developmental homologies are also observed at the level of specific tissues. Even though the structure of the adult jaw is different in fish and mammals, the same group of embryonic cells develops into the jaw structure in both groups. This observation is logical if fish and mammals descended from the same common ancestor, and if this ancestor also had a jaw that developed from the same population of embryonic cells.

The general point is that, in many cases, traits are similar in different species because the species in question are related to each other by common descent. If species were created independently of one another, these types of similarities would not occur.

Genetic Homologies Today, biologists recognize that structural homologies are observed because developmental homologies exist. The causal connection is logical, because adult structures develop from embryonic structures and processes. In addition, biologists now understand that both structural and developmental homologies occur because genetic homology exists. A **genetic homology** is a similarity in the DNA sequences of genes from different species. Developmental sequences and adult structures cannot be similar in different species unless their genes are similar.

As an example of genetic homology, consider the *eyeless* gene in fruit flies and the *Aniridia* gene in humans. Both genes act in determining where eyes will develop. The genes are so similar in DNA sequence that they code for proteins that are nearly identical in amino acid sequence (**Figure 23.8**). This observation is interesting because eye structure is so different in the two species—fruit flies have a compound eye with many lenses, and humans have a camera eye with a single lens. In most cases, species that have similar alleles also have

Gene:	Amino acid sequence (single-letter abbreviations):
Aniridia (Human)	LQRNRTSFTQEQIEALEKEFERTHYPDVFARERLAAKIDLPEARIQVWFSNRRAKWRREE
eyeless (Fruit fly)	LQRNRTSFTNDQIDSLEKEFERTHYPDVFARERLAGKIGLPEARIQVWFSNRRAKWRREE

Only six of the 60 amino acids in these sequences are different.
The two sequences are 90% identical.

FIGURE 23.8 Genetic Homology: Genes from Different Species May Be Similar in DNA Sequence or Other Attributes
Amino acid sequences from a portion of the *Aniridia* gene product found in humans and the *eyeless* gene product found in *Drosophila*. **QUESTION** What is the relationship among genetic homologies, developmental homologies, and structural homologies?

similar developmental sequences and similar embryonic or adult structures. In this instance, however, homology exists at the genetic level but *not* at the structural level. To explain this observation, biologists propose that fruit flies and humans descended from a common ancestor that had a gene similar to *eyeless* and *Aniridia*, and that this gene was involved in the formation of a simple light-gathering organ. (See Chapter 18 for more detail on homology in genes.)

Perhaps the ultimate similarity among organisms is the genetic code. With a few minor exceptions, the same 64 mRNA codons specify the same amino acids in all organisms that have been studied. To explain the existence of the universal genetic code, biologists hypothesize that today's code also existed in the common ancestor of all organisms alive today. Similarly, all organisms living today have a plasma membrane consisting of a phospholipid bilayer with interspersed proteins, transcribe the information coded in DNA to RNA via RNA polymerase, use ribosomes to synthesize proteins, employ ATP as an energy currency, and make copies of their genome via DNA polymerase. Like the genetic code, these traits undoubtedly existed in the cell that gave rise to all the species alive today.

Although genetic homologies were discovered fairly recently and were unknown to Darwin, structural and developmental homologies were recognized and studied even in the 1700s. Before the theory of evolution by natural selection was developed, no one was sure why striking similarities existed among certain organisms but not others. Darwin, in contrast, saw them as a consequence of descent with modification. In fact, biologists now reserve the term **homology** for similarities that exist *because* of common descent. This distinction is important because traits found in different species can be similar even though they were *not* inherited from a common ancestor. Characteristics that are similar but that did not exist in a common ancestor are said to be **analogous traits**, or **convergent traits.**

How Do Biologists Distinguish Homology from Analogy?

Homology is an exceptionally powerful concept in biology. For example, biomedical researchers study cancer in rats and vision in cats because the genes involved in rat cancers and cat vision are homologous with genes involved in human cancers and human vision. Research findings that improve cancer therapy in rats or that correct vision problems in cats can be successfully transferred to humans; the same underlying mechanisms are involved. The underlying mechanisms are shared because they existed in a common ancestor. Clearly, it is often vital to determine whether a homologous relationship exists among genes and other traits found in different species.

Determining whether homology exists is not always a simple matter, however, because not all similarities among organisms result from shared ancestry. The aquatic reptiles called ichthyosaurs, for example, were strikingly similar to modern dolphins (**Figure 23.9a**, page 502). Both are large marine animals with streamlined bodies and large dorsal fins. Both chase down fish and capture them between elongated jaws filled with dagger-like teeth. But no one would argue that ichthyosaurs and dolphins are similar because the traits they share existed in a common ancestor. As the phylogeny in Figure 23.9a shows, analyses of other traits show that ichthyosaurs are reptiles whereas dolphins are mammals.

Based on these data, it is logical to argue that the similarities between ichthyosaurs and dolphins result from convergent evolution. **Convergent evolution** occurs when natural selection favors similar solutions to the problems posed by a similar way of making a living. But convergent traits do not occur in the common ancestor of the similar species. Streamlined bodies and elongated jaws filled with sharp teeth help *any* species—whether it is a reptile or a mammal—chase down fish in open water. Convergent evolution produces analogous traits. Analogous traits are similar in structure and function, but they are not homologous. They are shared by species even though they were not present in their common ancestor.

In many cases, homology and convergence are much more difficult to distinguish than in the ichthyosaur and dolphin example. How do biologists recognize homology in such cases? As an example, consider the *HOM* genes of insects and the *Hox* genes of vertebrates introduced in Chapter 22. Even though insects and vertebrates last shared a common ancestor some 600–700 million years ago, biologists argue that *HOM* and *Hox* genes are derived from the same ancestral sequences. There are several lines of evidence to support this hypothesis:

- The genes are organized in a similar way. **Figure 23.9b** (page 502) shows that these genes in both insects and vertebrates are found in gene complexes, with similar genes found adjacent to one another on the chromosome. Recall from Chapter 20 that genes with these characteristics are called gene families. The organization of the gene families is nearly identical among insect and vertebrate species.
- All of the *HOM* and *Hox* genes share a 180-base-pair sequence called the *homeobox*, introduced in Chapter 22. The polypeptide encoded by the homeobox is almost identical in insects and vertebrates and has a similar function: It binds to DNA and regulates the expression of other genes.
- The products of the *HOM* and *Hox* genes have similar functions: They specify the locations of cells in embryos. They are also expressed in similar patterns in time and space.

In addition, many other animals, on lineages that branched off between insects and mammals, have similar genes. This is a crucial observation: If genes found in distantly related lineages are indeed related by common ancestry, then similar genes should be found in many intervening lineages on the tree of life. (Examine the phylogeny in Figure 23.9a and ask yourself

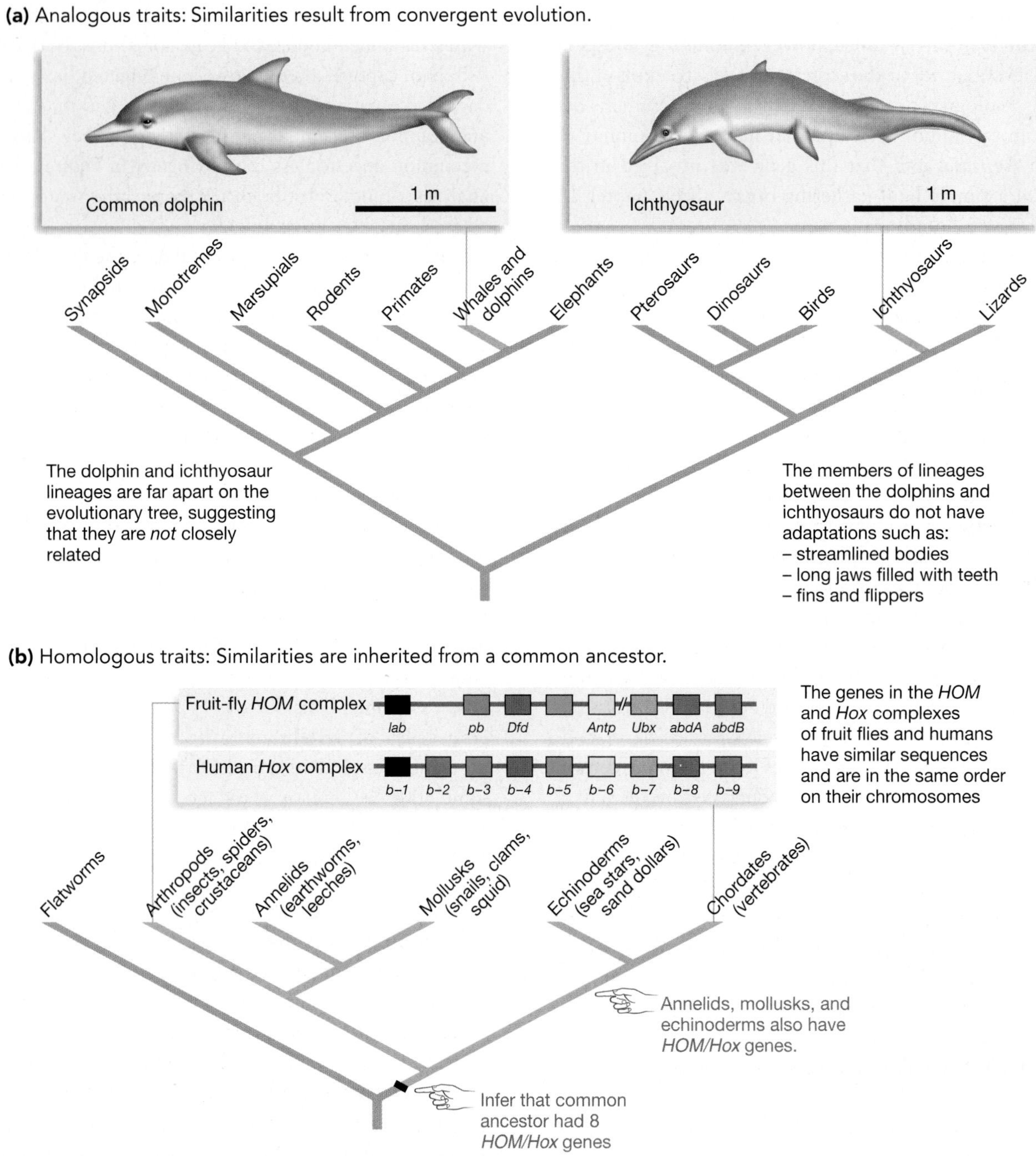
(a) Analogous traits: Similarities result from convergent evolution.
Common dolphin
1 m
Ichthyosaur
1 m
Synapsids
Monotremes
Marsupials
Rodents
Primates
Whales and dolphins
Elephants
Pterosaurs
Dinosaurs
Birds
Ichthyosaurs
Lizards
The dolphin and ichthyosaur lineages are far apart on the evolutionary tree, suggesting that they are *not* closely related
The members of lineages between the dolphins and ichthyosaurs do not have adaptations such as:
– streamlined bodies
– long jaws filled with teeth
– fins and flippers
(b) Homologous traits: Similarities are inherited from a common ancestor.
Fruit-fly *HOM* complex
lab
pb
Dfd
Antp
Ubx
abdA
abdB
Human *Hox* complex
b-1
b-2
b-3
b-4
b-5
b-6
b-7
b-8
b-9
The genes in the *HOM* and *Hox* complexes of fruit flies and humans have similar sequences and are in the same order on their chromosomes
Flatworms
Arthropods (insects, spiders, crustaceans)
Annelids (earthworms, leeches)
Mollusks (snails, clams, squid)
Echinoderms (sea stars, sand dollars)
Chordates (vertebrates)
Annelids, mollusks, and echinoderms also have *HOM/Hox* genes.
Infer that common ancestor had 8 *HOM/Hox* genes

FIGURE 23.9 Analogous Traits and Homologous Traits Are Similar for Different Reasons
(a) Dolphins and ichthyosaurs look similar but are not closely related—dolphins are mammals; ichthyosaurs are reptiles. Because their similar traits did not exist in the common ancestor of mammals and reptiles, biologists infer that traits such as streamlined bodies, sharp teeth, and flippers evolved independently in these two groups. **(b)** All of the animal groups illustrated on this phylogeny have *Hox/HOM* complexes that are similar to those illustrated for fruit flies and humans.

whether the analogous traits found in ichthyosaurs and dolphins fulfill this criterion.)

To summarize, *HOM* and *Hox* genes are considered homologous because they are similar in organization, function, and composition. It is extremely improbable that convergent evolution could produce such a high degree of similarity.

Darwinism and the Pattern Component of Evolution

Biologists draw upon data from several sources to challenge the hypothesis that species are immutable and were created independently. Instead, the data support the idea that species have descended, with modification, from a common ancestor. **Table 23.1** summarizes this evidence.

TABLE 23.1 Evidence for Evolution

Prediction 1: Species Are Not Static, but Change through Time	Prediction 2: Species Are Related, Not Independent
• Many species have gone extinct. • Fossil species frequently resemble living species found in the same area. • Transitional forms document change in traits through time. • Earth is ancient; environments and landscapes have changed through time. • Vestigial traits are common. • Populations and species can be observed changing today.	• Closely related species often live in the same geographic area. • Homologous traits are common and exist at three levels: 1. structural (morphological traits) 2. developmental (embryonic structures and processes) 3. genetic (gene structure and the genetic code)

Even before genetic homologies had been described and many of the most important transitional forms discovered, the vast majority of biologists were convinced that the pattern component of the theory of evolution was valid and that the theory of special creation was incorrect. Among biologists, controversy over the fact of evolution ended more than 120 years ago. As you analyze the evidence, though, recognize that no single observation or experiment instantly "proved" the fact of evolution and swept aside belief in special creation. Rather, Darwin and others argued that the pattern called evolution was much more consistent with the data. Stated another way, descent with modification was a more successful and powerful scientific theory because it explained observations—such as vestigial traits and the close relationships among species on neighboring islands—that special creation could not.

What about the process component of the theory of evolution by natural selection? If the limbs of bats and humans were not created independently and recently, how did they come to be?

23.2 The Process of Evolution: How Does Natural Selection Work?

Darwin's greatest contribution did not lie in recognizing the fact of evolution. Several other researchers had already proposed evolution as a pattern in nature long before Darwin began his work. (**Box 23.2** introduces one of these thinkers.) Instead, Darwin's crucial insight lay in recognizing a process, called natural selection, that could explain the pattern of descent with modification.

BOX 23.2 Evolutionary Theory before Darwin

In the introduction to *On the Origin of Species*, Darwin outlined some of the evidence for the fact of evolution, including many of the points summarized in Section 23.1. Then he wrote that a biologist "might come to the conclusion that each species had not been independently created, but had descended ... from other species. Nevertheless, such a conclusion even if well founded, would be unsatisfactory, until it could be shown *how* the innumerable species inhabiting this world have been modified" (emphasis added). Most scientists would agree with Darwin—recognizing a pattern in nature without understanding the process that caused it means that more work needs to be done.

In 1809, decades before Darwin began his work, Jean-Baptiste de Lamarck proposed that species are not static and independently created entities, but have changed through time and are related by common ancestry. Lamarck was one of several researchers who advocated the reality of evolution before Darwin published *On the Origin of Species*.

More important, Lamarck offered a mechanism to explain how species evolved. He proposed that simple forms of life are continuously being created by spontaneous generation. Lamarck reasoned that these simple cells then evolved to progressively "higher" forms through a process known as the **inheritance of acquired characters**. Specifically, he proposed that evolution was progressive in the sense of always producing larger or more complex or "better" species and that it occurs because individuals change as they develop, in order to meet challenges posed by the environment. Lamarck also proposed that these phenotypic changes are passed on to offspring. A classic scenario is that giraffes that stretched their necks to reach leaves high in treetops would develop longer necks and that they would then produce offspring with elongated necks.

Lamarck's ideas were roundly rejected by the leading scientists of the day, partly because he was not as thorough as Darwin in amassing evidence for the fact of evolution. Today biologists reject Lamarck's ideas because they know that changes in phenotypes that develop as a result of an individual's actions do not produce changes in genotypes and cannot be transmitted to offspring. For example, if you gained increased upper body strength through weight lifting, your offspring would not tend to be stronger as a result. Still, Lamarck deserves credit for bringing evolution to the forefront of scientific debate. In this way, he helped to pave the way for Darwin.

The process of **natural selection** can be broken down into four simple postulates, or steps, in a logical sequence:

1. The individual organisms that make up a population vary in the traits they possess, such as their size and shape.
2. Some of the trait differences are passed on to offspring. For example, tall parents may tend to have tall offspring. This means that the trait has some genetic basis and is heritable.
3. Only a subset of the offspring that are produced in each generation survives long enough to reproduce, and of the individuals that reproduce, not all produce the same number of offspring. Thus, some individuals in the population produce more offspring than others do.
4. The subset of individuals that produce the most offspring is not a random sample of the population. Instead, individuals with certain traits are more likely to produce the greatest number of offspring in a given environment. The individuals with these traits are "naturally selected."

Because the frequency of the selected traits increases from one generation to the next, the outcome of these four steps is **evolution**—a genetically based change in the characteristics of a population over time.

Biologists often combine postulates 1 and 2 by observing that heritable variation exists for most traits in most populations. Postulates 3 and 4 can also be combined by noting that individuals experience differential reproductive success. Evolution, or change in the characteristics of a population over time, results when differential reproductive success is based on heritable variation. The traits that contribute to high reproductive success increase in frequency in the population over time.

To explain the process of natural selection, Darwin referred to successful individuals as "more fit" than other individuals. He gave the word *fitness* a definition quite different from its everyday English usage. **Darwinian fitness** is the ability of an individual to produce offspring, relative to that ability in other individuals in the population. This is a measurable quantity. Researchers study populations in the lab or in the field and estimate the relative fitness of each individual by counting how many offspring it produces relative to other individuals.

The concept of fitness, in turn, provides a compact way of formally defining adaptation. The biological meaning of adaptation, like the biological meaning of fitness, is quite different from its normal English usage. In biology, an **adaptation** is a heritable trait that increases the fitness of an individual in a particular environment relative to individuals lacking the trait. Adaptations increase fitness—the ability to produce offspring.

To help you understand each step in the process of natural selection, Section 23.3 is devoted to data—specifically, to recent studies of how natural selection works in actual populations. Biologists accept the theory of evolution by natural selection not only because of its explanatory power but also because evolution by natural selection has been observed directly.

23.3 Evolution in Action: Recent Research on Natural Selection

Darwin's theory of evolution by natural selection is testable. If the theory is correct, biologists should be able to test the validity of each postulate and actually observe evolution in natural populations.

This section summarizes three examples in which evolution by natural selection has been, or is being, observed in nature. Literally hundreds of other case studies are available, involving a wide variety of traits and organisms. The examples here were chosen because they demonstrate *how* biologists go about studying evolution by natural selection. To begin, let's explore the evolution of drug resistance—one of the great challenges facing today's biomedical researchers and physicians.

How Did *Mycobacterium tuberculosis* Become Resistant to Antibiotics?

Mycobacterium tuberculosis, the bacterium that causes **tuberculosis**, or TB, has long been a great scourge of humankind (**Figure 23.10**). TB was responsible for almost 25 percent of all deaths in New York City in 1804; in nineteenth-century Paris, the figure was closer to 33 percent. To put these num-

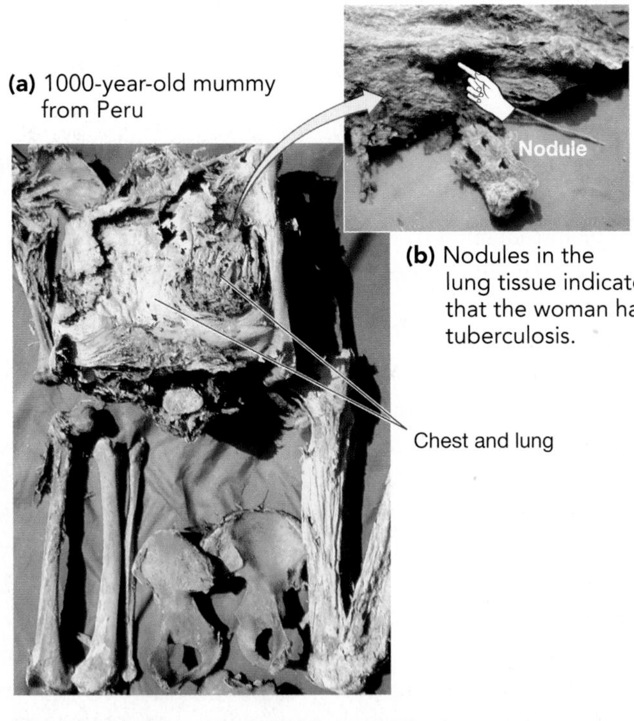
(a) 1000-year-old mummy from Peru
Nodule
(b) Nodules in the lung tissue indicate that the woman had tuberculosis.
Chest and lung

FIGURE 23.10 Tuberculosis Is an Ancient Disease of Humans
(a) A mummy of a middle-aged woman, from a tomb in the coastal deserts of Peru. (The mummy's chest and upper arms are at the top.) **(b)** A close-up of lung tissue from the mummy. The nodules in the tissue are strikingly similar to those found in contemporary TB victims. DNA extracted from nodules in the mummy contained sequences diagnostic of *M. tuberculosis*. This is strong evidence that TB was present in the New World long before the arrival of Europeans.

bers in perspective, consider that all types of cancer, combined, currently account for about 30 percent of the deaths that occur in the United States. TB was once as great a public health issue as cancer is now.

Although tuberculosis still kills more adults than any other viral or bacterial disease in the world, TB attracted relatively little attention in the industrialized nations between about 1950 and 1990. During that time, TB was primarily a disease of developing nations.

The decline of tuberculosis in Western Europe, North America, Japan, Korea, and Australia is one of the great triumphs of modern medicine. In these countries, sanitation, nutrition, and general living conditions began to improve dramatically in the early twentieth century. When people are healthy and well nourished, their immune systems work well enough to stop most *M. tuberculosis* infections quickly—before the infection can harm the individual and before the bacteria can be transmitted to a new host. Equally important, antibiotics such as rifampin started to become available in the industrialized countries in the early 1950s. These drugs allowed physicians to stop even advanced infections, and they saved millions of lives.

In the late 1980s, however, rates of *M. tuberculosis* infection surged in many countries, and in 1993 the World Health Organization (WHO) declared TB a global health emergency (**Figure 23.11**). Physicians were particularly alarmed, because the strains of *M. tuberculosis* responsible for the increase were largely or completely resistant to rifampin and other antibiotics that were once extremely effective. How and why did the evolution of drug resistance occur? The case of a single patient—a young man with AIDS who lived in Baltimore—will illustrate what is happening all over the world.

The story begins when the HIV-positive individual was admitted to the hospital with fever and coughing. Chest X-rays, followed by bacterial cultures of fluid ejected from the lungs, showed that he had an active TB infection. He was given a battery of antibiotics for six weeks, followed by twice-weekly doses of rifampin and isoniazid for an additional 33 weeks. Ten months after therapy started, bacterial cultures from his chest fluid indicated no *M. tuberculosis* cells. His chest X-rays were also normal. The antibiotics seemed to have cleared the infection.

Just two months after the TB tests proved normal, however, the young man was readmitted to the hospital with a fever, severe cough, and labored breathing. Despite being treated with a variety of antibiotics, including rifampin, he died of respiratory failure 10 days later. Samples of material from his lungs showed that *M. tuberculosis* was again growing actively there. But this time the bacterial cells were completely resistant to rifampin.

Drug-resistant bacteria had killed this patient. Where did they come from? Is it possible that a strain that was resistant to antibiotic treatment evolved *within* him? To answer this question, a research team analyzed DNA from the drug-resistant strain and compared it with stored DNA from *M. tuberculosis* cells that had been isolated a year earlier from the same patient. After examining extensive stretches from each genome, the biologists were able to find only one difference: a point mutation in a gene called *rpoB*. This gene codes for a component of the enzyme RNA polymerase. Recall from Chapter 16 that RNA polymerase transcribes DNA to mRNA and that a point mutation is a single base change in DNA. In this case, the mutation changed a cytosine to a thymine, altering the normal codon TCG to a mutant one, TTG. As a result, the RNA polymerase produced by the drug-resistant strain had leucine instead of serine at the 153rd amino acid in the polypeptide chain.

This result is meaningful. The drug that was being used to treat the patient works by binding to the RNA polymerase of *M. tuberculosis*. When the drug enters an *M. tuberculosis* cell

WEB TUTORIAL 23.1 Natural Selection for Antibiotic Resistance

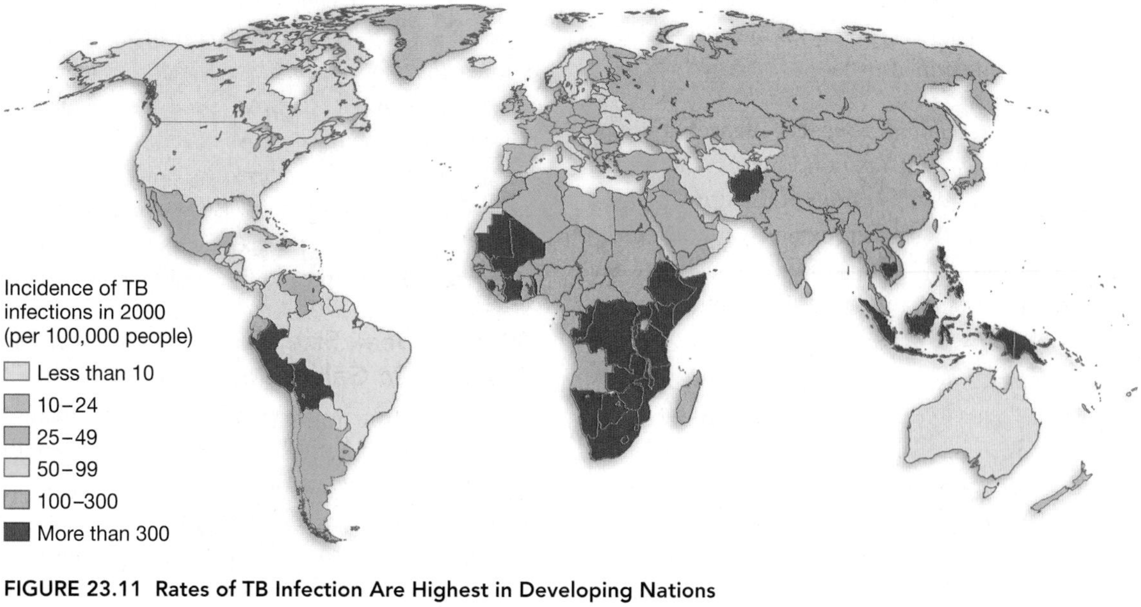
Incidence of TB infections in 2000 (per 100,000 people)
Less than 10
10–24
25–49
50–99
100–300
More than 300

FIGURE 23.11 Rates of TB Infection Are Highest in Developing Nations
In developing nations, TB is responsible for 7 percent of all deaths (children and adults).

and binds to its RNA polymerase, it interferes with transcription. If sufficient quantities of rifampin are present for long enough and if the drug binds tightly, bacterial cells will not be able to make proteins efficiently and will die. But apparently the substitution of a leucine for a serine prevents rifampin from binding efficiently. Consequently, cells with the C → T mutation continue to grow even in the presence of the drug.

These results suggest that a chain of events led to this patient's death. The researchers hypothesized that, by chance, one or a few of the cells present early in the course of the infection happened to have an *rpoB* gene with the C → T mutation. Under normal conditions, mutant forms of RNA polymerase do not work as well as the more common form, so cells with the C → T mutation would not produce many offspring and would stay at low frequency—even while the overall population grew to the point of inducing symptoms that sent the young man to the hospital.

At that point, therapy with rifampin began. In response, cells in the population with normal RNA polymerase began to grow much more slowly or to die outright. As a result, the overall bacterial population declined in size so drastically that the patient appeared to be cured—his symptoms began to disappear. But cells with the C → T mutation had an advantage in the new environment. They began to grow more rapidly than the normal cells and continued to increase in number after therapy ended. Eventually the *M. tuberculosis* population regained its former abundance, and the patient's symptoms reappeared. However, drug-resistant cells now dominated the population. This is why the second round of rifampin therapy was futile. If health-care workers or the patient's family had contracted TB from him, rifampin therapy would have been useless on them, too, and the disease would have continued to spread.

Does this sequence of events mean that evolution by natural selection occurred? One way of answering this question is to review the four postulates listed in Section 23.2 and determine whether each was tested and verified:

1. *Did variation exist in the population?* The answer is yes. Due to mutation, both resistant and nonresistant strains of TB were present prior to administration of the drug. Most *M. tuberculosis* populations, in fact, exhibit variation for the trait; studies on cultured *M. tuberculosis* show that a mutation conferring resistance to rifampin is present in one out of every 10^7 to 10^8 cells.
2. *Was this variation heritable?* The researchers showed that the variation in the phenotypes of the two strains—from drug susceptibility to drug resistance—was due to variation in their genotypes. Because the mutant *rpoB* gene is copied before a *Mycobacterium* replicates, the allele and the phenotype it produces—drug resistance—are passed on to offspring.
3. *Was there variation in reproductive success?* That is, did some *M. tuberculosis* individuals leave more offspring than other *M. tuberculosis* individuals? Clearly, only a tiny fraction of *M. tuberculosis* cells in the patient survived the first round of antibiotics long enough to reproduce—so few that, after the initial therapy, his chest X-ray was normal and his fluid sample contained no *M. tuberculosis* cells. In contrast, drug-resistant bacterial cells were able to survive and keep reproducing after the onset of drug treatment.
4. *Did selection occur?* That is, did a nonrandom subset of the population produce the most offspring? The answer is yes. The *M. tuberculosis* population present early in the infection was different from the *M. tuberculosis* population present at the end. This could have occurred only if cells with the drug-resistant allele had higher reproductive success when rifampin was present than did cells with the normal allele. *Mycobacterium tuberculosis* individuals with the mutant *rpoB* gene had higher fitness in an environment where rifampin was present. The mutant allele produces a protein that is an adaptation when the cell's environment contains the antibiotic.

This study verified all four postulates and confirmed that evolution by natural selection had occurred. The example also neatly illustrates the contemporary definition of **evolution:** a change in allele frequencies in a population and thus its genetic makeup over time. The characteristics of the *M. tuberculosis* population changed over time because the mutant *rpoB* gene increased in frequency. But note that the individual cells themselves did not evolve. When natural selection occurred, the individual cells did not change through time; they simply survived or died, or produced more or fewer offspring. This is a fundamentally important point: Natural selection acts on *individuals*, but only *populations* evolve.

The events just reviewed have occurred many times in other patients. Recent surveys indicate that drug-resistant strains now account for about 10 percent of the *M. tuberculosis*–causing infections throughout the world. And the emergence of drug resistance in TB is far from unusual. Resistance to a wide variety of insecticides, fungicides, antibiotics, antiviral drugs, and herbicides has evolved in hundreds of insects, fungi, bacteria, viruses, and plants. In many cases, the specific mutations that lead to a fitness advantage and the spread of the resistance alleles are known.

Why Are Beak Size, Beak Shape, and Body Size Changing in Galápagos Finches?

The TB example is particularly satisfying, because the molecular basis of both heritable variation and differential reproductive success is understood. But can biologists study natural selection when the alleles responsible for variation are not known and when the molecular mechanism of adaptation has not been identified? The answer is yes. As an example of how this work is done, let's review research led by Peter and Rosemary Grant. These biologists have been investigating changes in beak size, beak shape, and body size that have occurred in finches native to the Galápagos Islands.

The medium ground finch, pictured in **Figure 23.12a**, makes its living by eating seeds. Finches crack seeds with their beaks. For over three decades, the population of medium ground finches on Isle Daphne Major of the Galápagos has been studied intensively by the Grants' team. Because Daphne Major is small—about the size of 80 football fields—the researchers have been able to catch, weigh, and measure all individuals and mark each with a unique combination of colored leg bands.

(a) Medium ground finch

(b) Natural selection during a drought

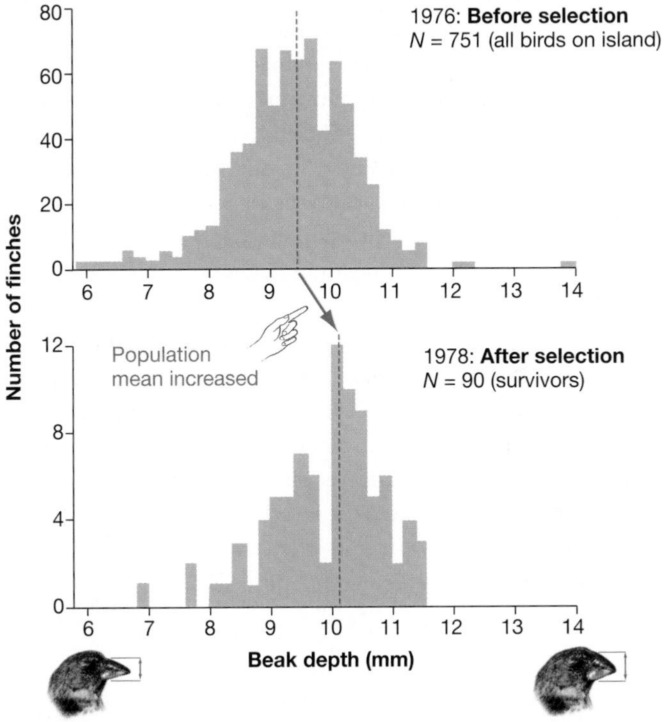

FIGURE 23.12 During a Drought, Finches with Deep Beaks Survived Better
(a) Medium ground finches live only on the Galápagos Islands. **(b)** Histograms showing the distribution of beak depth in medium ground finches before and after natural selection occurred, due to changes in food availability, during the drought of 1977. *N* is the sample size. QUESTION Why was the sample size so much smaller in 1978?

Early studies of the finch population established that beak size and shape and body size vary among individuals, and that beak morphology and body size are heritable. Stated another way, parents with particularly deep beaks tend to have offspring with deep beaks. Large parents also tend to have large offspring.

Not long after the team had established these results, a dramatic selection event occurred. In the annual wet season of 1977, Daphne Major received just 24 mm of rain instead of the 130 mm that normally falls. During the drought, few plants were able to produce seeds, and 84 percent (about 660 individuals) of the medium ground finch population disappeared.

Two observations support the hypothesis that most or all of these individuals died of starvation. The researchers found a total of 38 dead birds, and all were emaciated. Further, none of the missing individuals were spotted on nearby islands, and none reappeared once the drought had ended and food supplies returned to normal.

The research team realized that the die-off presented an opportunity to study natural selection. Were the survivors different from nonsurvivors? When the biologists analyzed the characteristics of each group, they found that survivors tended to have much deeper beaks than did the birds that died. This was an important finding, because the type of seeds that were available to the finches had changed dramatically as the drought continued. At the drought's peak, the tough fruits of a plant called *Tribulus cistoides* served as the finches' primary food source. These fruits are so difficult to crack that they are ignored in years when food supplies are normal. Grant's group hypothesized that individuals with particularly large and deep beaks were more likely to crack these fruits efficiently enough to survive.

As the data in **Figure 23.12b** show, natural selection led to an increase in average beak depth in the population. When breeding resumed in 1978, the offspring that were produced had beaks that were half a millimeter deeper, on average, than the population that existed before the drought. In only one generation, natural selection had led to measurable evolution—a change in the characteristics of the population over time. Alleles that led to the development of deep beaks must have increased in frequency. Large, deep beaks were an adaptation for cracking large fruits and seeds.

In 1983, however, the environment changed again. Over a 7-month period, a total of 1359 mm of rain fell. Plant growth was luxuriant, and finches fed primarily on small, soft seeds that were being produced in abundance. During this interval, small individuals with small, pointed beaks had exceptionally high reproductive success. As a result, the characteristics of the population changed again. In fact, the Grants have documented continued evolution in

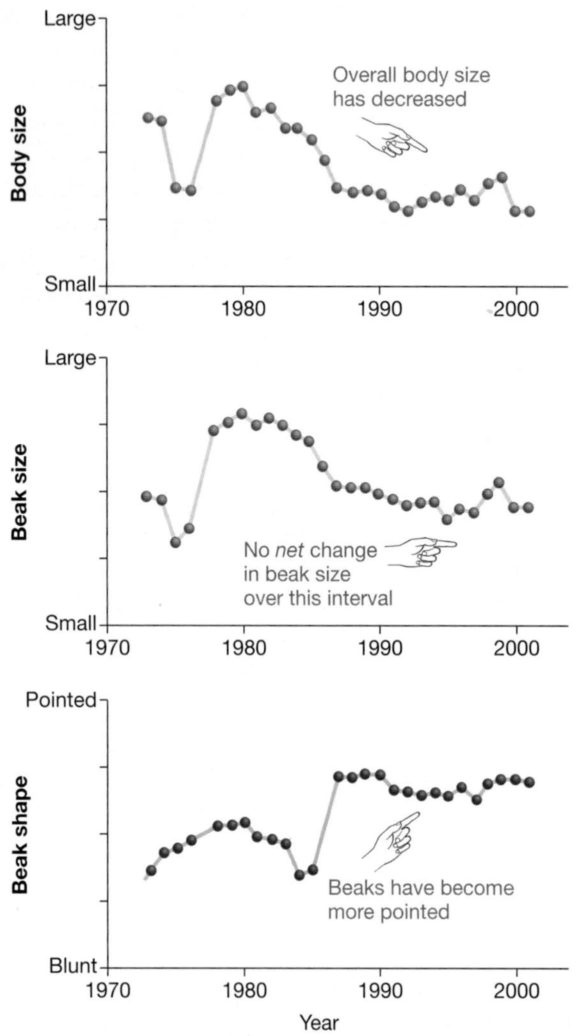

FIGURE 23.13 Body Size, Beak Size, and Beak Shape in Finches Have Changed over 30 Years
Graphs summarizing 30 years of data on changes in the traits of medium ground finches. EXERCISE Label the drought in 1978 and the wet year in 1983. Circle years when (1) average body size increased, (2) average beak size declined, and (3) beaks became pointier.

WEB TUTORIAL 23.2 Natural Selection in Alpine Skypilots

response to continued changes in the environment. **Figure 23.13** documents changes that have occurred in average body size, beak size, and beak shape over the past 30 years. Although beak size has shown no net change when we compare the start and end of this interval, beak shape has changed dramatically. On average, finch beaks have gotten much pointier over the course of the study. In addition, overall body size has gotten smaller.

Long-term studies such as this have been powerful, because they have succeeded in documenting natural selection in response to changes in the environment. But researchers are also using experimental approaches to study natural selection in action.

Can Natural Selection Be Studied Experimentally?

Candace Galen began to do research on variation in the flower size of a Rocky Mountain wildflower after she made a simple but astute observation: Individual alpine skypilot plants found in the treeless, tundra habitats above timberline have flowers that are substantially larger than flowers on individuals growing among the stunted trees at timberline. The tundra-dwellers also have longer stalks, and most have sweet-smelling flowers. Flowers on plants growing at timberline have shorter stalks, and most have an aroma that biologists describe as "skunky."

Galen also noticed that individuals growing above the timberline are pollinated primarily by bumblebees, while individuals below timberline are pollinated primarily by flies. Because bumblebees are much larger than flies and are attracted to sweet smells, Galen thought that the differences in pollinators might be responsible for the difference in skypilot flowers. Specifically, she proposed that natural selection, caused by the bumblebees' preference to land on larger, sweet-smelling flowers and the flies' preference to land on smaller, skunky-smelling flowers, had acted on variation in flower size and produced the differences she observed in the two populations.

To test the hypothesis that large flower size has evolved in the tundra population in response to natural selection by bumblebees, Galen performed the experiment outlined in **Figure 23.14**. Examine the figure and note that she established two treatment groups, called controls and experimentals. The goal of the control treatment was to observe what happens when bumblebees are *not* agents of natural selection; the goal of the experimental treatment was to observe what happens when bumblebees do cause natural selection.

To create the control, Galen pollinated a large group of randomly selected timberline skypilots by hand, using pollen from randomly selected individuals. She collected the resulting seeds, germinated them in the greenhouse, and planted the seedlings out into randomly assigned locations in the timberline habitat. When these individuals matured, she compared their flower sizes with the sizes in offspring of individuals that had been pollinated by bumblebees.

The result? Examine the two histograms in Figure 23.14. They show that the flowers from offspring of bumblebee-pollinated plants were significantly larger than the flowers from offspring of hand-pollinated plants. Because previous experiments had confirmed that flower size is heritable, Galen could conclude that the experimental population, pollinated by bumblebees, was genetically different from the control population, which was randomly pollinated. Natural selection can be documented experimentally.

✓CHECK YOUR UNDERSTANDING

If individuals with certain genetically based traits produce the most offspring in a population, then the alleles responsible for those traits increase in frequency over time, resulting in evolution. You should be able to explain each of Darwin's four postulates by using the examples of selection by drugs on the TB bacterium, changes in seed availability in Galápagos finches, and pollinator preferences in alpine skypilots.

Question: Do pollinators cause natural selection on flower characteristics?

Hypothesis: Bumblebees prefer large flowers. Therefore, natural selection leads to the evolution of large flowers.

Null hypothesis: Bumblebees do not prefer large flowers. Therefore, bumblebees do not cause natural selection on flower size.

Experimental setup:

Experimental group

1. Allow bees to pollinate a large sample of skypilots.

2. Collect seeds; germinate in greenhouse.

3. Plant seedlings in randomly assigned locations in the field.

4. Measure flower sizes and plot data.

Control group

1. Hand pollinate a random sample of skypilots.

2. Collect seeds; germinate in greenhouse.

3. Plant seedlings in randomly assigned locations in the field.

4. Measure flower sizes and plot data.

Prediction: Offspring of bumblebee-pollinated flowers will be larger than offspring of randomly pollinated flowers.

Prediction of null hypothesis: Offspring of bumblebee-pollinated and randomly pollinated flowers will have the same size flowers.

Results:

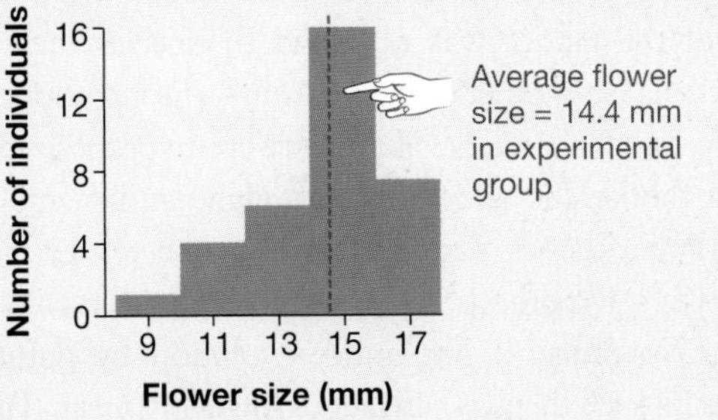

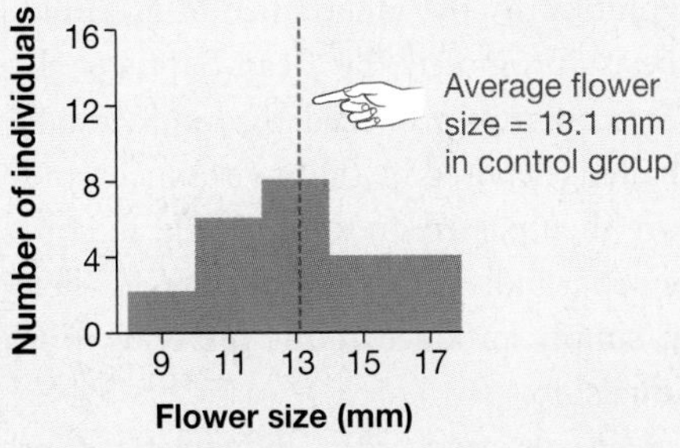

Conclusion: On average, flowers are larger in offspring of bumblebee-pollinated plants than in offspring of hand-pollinated plants. Natural selection by bumblebees favors increased flower size.

FIGURE 23.14 Experimental Evidence for Natural Selection in Action
The histograms show the distribution of flower size in the offspring of alpine skypilots randomly pollinated by hand and in the offspring of those pollinated by bumblebees. The average size of flowers in the experimental treatment is over 1 mm larger than the average size of flowers in the control treatment.
QUESTION Why was it important that hand pollination was carried out randomly with respect to flower size?

23.4 The Nature of Natural Selection and Adaptation

Natural selection appears to be a simple process, but appearances can be deceiving. Research has shown that it is often misunderstood. To help clarify how natural selection works, let's consider some of the more common misconceptions about natural selection in light of data on drug resistance in the TB bacterium, changes in finch populations, and selection on flower size in alpine skypilots.

Selection Acts on Individuals, but Evolutionary Change Occurs in Populations

Perhaps the most important point to clarify about natural selection is that individuals do not change during the process—only the population does. During the drought, the beaks of individual finches did not become deeper. Rather, the average beak depth in the population increased over time, because deep-beaked individuals produced more offspring than shallow-beaked individuals did. Natural selection acted on individuals, but the evolutionary change occurred in the characteristics of the population. In the same way, individual bacterial cells did not change when rifampin was introduced to their environment—only the characteristics of the bacterial population changed. Further, the mutation in the *rpoB* gene that conferred high fitness in the new envirionment was not induced by the presence of rifampin; the gene occurred randomly and just happened to be advantageous. Alpine skypilots did not get bigger and sweeter smelling because bumblebees landed on them. Rather, bumblebees preferred to land on the larger and sweeter-smelling individuals. As a result, individuals with large, sweet-smelling flowers were pollinated frequently and produced the most offspring in habitats where bumblebees are common.

This point highlights a sharp contrast between evolution by natural selection and evolution by the inheritance of acquired characters—the hypothesis promoted by Jean-Baptiste de Lamarck (see Box 23.2). Lamarck proposed that individuals change in response to challenges posed by the environment and that the changed traits are then passed on to offspring. In contrast, Darwin pointed out that individuals do not change when they are selected—they simply produce more surviving offspring than other individuals do.

Lamarck was correct in observing that individuals may change over the course of their lifetime in response to changes in environmental conditions, however. To capture this point, biologists use the terms *adaptation* and *acclimatization* to distinguish between changes that occur in populations and changes that occur in individuals. **Adaptation** occurs when a population changes in response to natural selection. **Acclimatization** occurs when an individual changes in response to a change in environmental conditions. For example, wood frogs native to northern North America are exposed to extremely cold temperatures as they overwinter. When ice begins to form in their skin, their bodies begin producing a sort of natural antifreeze—molecules that protect their tissues from being damaged by the ice crystals. These individuals are acclimatizing to cold temperatures.

In frogs, the ability to produce antifreeze molecules is an adaptation,[1] but actually producing the molecules is acclimatization. You may have observed changes in your own body as you acclimatized to high elevation or to particularly hot or cold environments. You did not adapt to these environments in the evolutionary sense, however. Acclimatization is a short-term change, and this change is not passed on to offspring.

Evolution Is Not Progressive

It is tempting to think that evolution by natural selection is progressive—meaning organisms have gotten "better" over time. (In this context, *better* usually means more complex.) It is true that the groups appearing later in the fossil record are often more morphologically complex or "advanced" than closely related groups that appeared earlier. Flowering plants are considered more complex than mosses, and mammals are more complex than turtles. But there is nothing predetermined or absolute about this tendency. In fact, complex traits are routinely lost or simplified over time as a result of evolution by natural selection. Populations that became parasitic are particularly prone to this trend. Tapeworms, for example, lack any sort of gut. As parasites that live in the intestines of humans and other mammals, they absorb nutrients directly from their environment. But tapeworms evolved from species with a sophisticated digestive tract. Tapeworms lost their digestive tract as a result of evolution by natural selection.

The nonprogressive nature of evolution by natural selection contrasts sharply with the hypothesis of evolution by inheritance of acquired characters (see Box 23.2). According to Lamarck's conception of the evolutionary process, organisms progress over time to higher and higher levels on a "ladder of life." The lowest rung of the ladder was occupied by species that had just appeared via spontaneous generation—the creation of life from nonlife (see Chapter 1). Humans were hypothesized to be at the top of the ladder, and fungi and algae and plants and animals that lack backbones occupied the rungs in between (**Figure 23.15a**). Under this hypothesis, it is sensible to refer to "higher" and "lower" organisms. But under evolution by natural selection, there is no such thing as a higher or lower organism (**Figure 23.15b**). Green algae may be a more ancient group than flowering plants, but neither group is higher or lower than the other. Green algae simply have a different suite of adaptations than do flowering plants, so they thrive in different types of environments.

[1]In some species of frogs, so much extracellular fluid freezes during cold snaps that individuals appear to be frozen solid. Their hearts also stop beating. When temperatures warm in the spring, their hearts start beating again, their tissues thaw, and they resume normal activities.

(a) Lamarckian evolution produces a ladder of life.

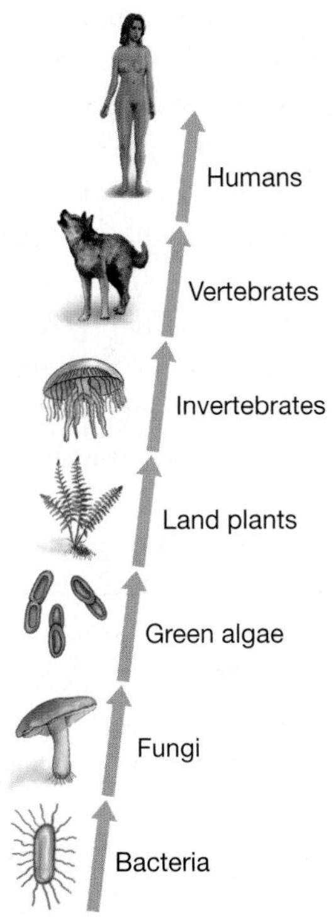
Humans
Vertebrates
Invertebrates
Land plants
Green algae
Fungi
Bacteria
Spontaneous generation

(b) Darwinian evolution produces a tree of life.

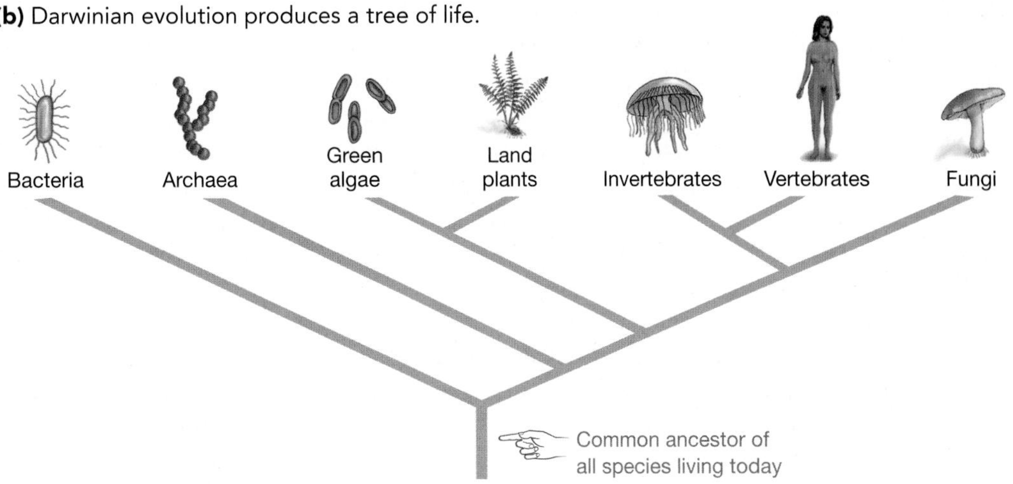
Bacteria
Archaea
Green algae
Land plants
Invertebrates
Vertebrates
Fungi
Common ancestor of all species living today

FIGURE 23.15 Evolution Produces a Tree of Life, Not a Progressive Ladder of Life
(a) Under evolution by the inheritance of acquired characters, species are related according to their position on a ladder of life. As evolution continues, simpler species evolve to become more complex. **(b)** Under evolution by natural selection, species are related according to their position on a tree of life. As evolution continues, species may become simpler or more complex.

The general message here is that all populations have undergone natural selection based on their ability to gather resources and produce offspring. All organisms are adapted to their environment. Evolution by natural selection is not progressive, so no organism is any "higher" than any other organism.

To drive this point home, recall what happened when torrential rains fell on Daphne Major. Instead of continuing to increase in size, the average beak size of the finch population *declined*. Natural selection did not produce a progression of ever-larger beaks. Instead, the population simply responded to whatever change happened to occur in the environment. In many bacterial and viral populations, the frequency of drug-resistant individuals declined dramatically when the drug in question was discontinued. If bumblebees disappeared from alpine habitats and flies began to pollinate tundra alpine skypilots, then tundra individuals with smaller, skunkier-smelling flowers would increase in frequency.

Natural selection is not progressive. It simply favors individuals that happen to be better adapted to their environment.

Not All Traits Are Adaptive

Although organisms are often exquisitely adapted to their environment, adaptation is far from perfect. Vestigial traits such as the human coccyx (tailbone), goose bumps, and appendix do not increase the fitness of individuals with those traits. Many cave-dwelling animals still have structures that correspond to eyes even though the structures have no function and the organisms live in complete darkness. The structures are not adaptive in these populations. They exist simply because they were present in the ancestral population.

Vestigial traits are not the only types of structures with no function. Some adult traits exist as holdovers from structures that appear early in development. For example, human males have rudimentary mammary glands. The structures are not adaptive. They exist only because nipples form in the human embryo before sex hormones begin directing the development of male organs instead of female organs.

Perhaps the best example of nonadaptive traits involves evolutionary changes in DNA sequences. Recall from Chapter 16 that mutation may change a base in the third position of a codon without changing the amino acid sequence of the protein encoded by that gene. Changes such as these are said to be *neutral*, or silent. They occur because of the redundancy of the genetic code (see Chapter 13). Neutral changes in DNA sequences are extremely common, yet not adaptive.

The general point here is that not all traits are adaptive. In addition, structures that do have functions are constrained in a variety of important ways. Let's take a closer look.

Genetic Constraints The Grants' team analyzed data on the characteristics of finches that survived the 1977 drought, and the team made an interesting observation: Although individuals with deep beaks survived better than individuals with shallow beaks, birds with particularly narrow beaks survived better than individuals with wider beaks. This observation made sense because finches crack *Tribulus* fruits by twisting them and because narrow beaks should concentrate the twisting force more efficiently. But narrower beaks did not evolve in the population. To explain why, the biologists noted that parents with deep beaks tend to have offspring with beaks that are both deep and wide. This is a common pattern. Many alleles that affect body size have an effect on all aspects of size—not just one structure or dimension. As a result, selection for increased beak depth overrode selection for narrow beaks, even though a deep and narrow beak would have been more advantageous.

The general point here is that selection was not able to optimize all aspects of a trait. In the case of the finches, wider beaks were not the best possible beak shape for individuals living in an arid habitat. Wider beaks evolved anyway, due to a type of constraint called a **genetic correlation**. In this case, selection on alleles for one trait (increased beak depth) caused a correlated, though suboptimal, increase in another trait (beak width).

Genetic correlations are not the only genetic constraint on adaptation. Lack of genetic variation is also important. Consider that salamanders have the ability to regrow severed limbs and that some eels can sense electric fields. Even though these traits might confer increased reproductive success in humans, they do not exist—because the requisite genes are lacking.

Historical Constraints In addition to being constrained by genetic correlations and lack of genetic variation, adaptations are constrained by history. The reason is simple: All traits have evolved from previously existing traits. For example, the tiny hammer, anvil, and stirrup bones found in your middle ear evolved from bones that were part of the jaw and braincase in the ancestors of mammals. These bones now function in the transmission and amplification of sound from your outer ear to your inner ear. Biologists routinely interpret these bones as adaptations that improve your ability to hear airborne sounds. But are the bones a "perfect" solution to the problem of transmitting sound from one part of the ear to the other? The answer is no. They are the best solution possible, given an important historical constraint. Natural selection was acting on structures that originally had a very different function. Other vertebrates have different structures involved in transmitting sound to the ear.

To summarize, not all traits are adaptive, and even adaptive traits are constrained by genetic and historical factors. In addition, natural selection is not the only process that causes evolutionary change. Chapter 24 introduces three other processes that change allele frequencies over time. Compared with natural selection, these processes have very different consequences.

✓CHECK YOUR UNDERSTANDING

Using the examples of selection by drugs on the TB bacterium, changes in seed availability in Galápagos finches, and pollinator preferences in alpine skypilots, you should be able to explain (1) why individuals do not change when natural selection occurs, (2) why evolution is not progressive, and (3) why genetic and historical constraints prevent adaptations from being "perfect."

ESSAY The Debate over "Scientific Creationism"

The theory of evolution by natural selection is the cornerstone of modern biology. Controversy over the pattern component of the theory (descent with modification) ended in the late nineteenth century, when mounting evidence for the fact of evolution silenced Darwin's critics. Controversy over the process component (natural selection) ended when biologists successfully integrated Mendelian genetics into the theory. The discovery of mutation and genetic recombination explained why heritable variation exists in populations, and Mendel's laws clarified how new alleles and new combinations of alleles are passed on to offspring. This extension of Darwin's theory, called the "Modern Synthesis," occurred in the 1930s. Before the Modern Synthesis, biologists did not know how heritable variation could be maintained in populations so that evolution by natural selection would continue over time.

Although scientific controversy over the theory of evolution by natural selection largely ended over 70 years ago, political and social controversy are alive and well. In the United States, advocates of "scientific creationism" and "intelligent design theory" lobby for a ban on teaching evolution in public schools, for equal time devoted to teaching the theory of special creation, or for disclaimers in textbooks declaring that evolution is "just a theory." Intelligent design theory is based on the claim that, because organisms seem to be well designed or well adapted, they must have been created by an intelligent designer.

Science and religion are compatible because they address different types of questions.

This controversy is puzzling, given that there is no inherent conflict between accepting the validity of the theory of evolution by natural selection and believing in God. Pope John Paul II has stated that evolution by natural selection is compatible with traditional Christian understandings of God, and mainstream Protestant denominations have issued resolutions agreeing with this view.

Science and religion are compatible because they address different types of questions. Biologists ask questions that can be answered by measuring things—by collecting data. Science is about formulating hypotheses and finding evidence that supports or conflicts with the predictions of those hypotheses. Religious faith, in contrast, addresses questions that cannot be answered by data. The questions addressed by the world's great religions focus on why we exist and how we should live.

Because social and political controversy continue in the United States and elsewhere, it will be helpful to review the nature of the arguments promoted by creationists. In objecting to the theory of evolution by natural selection, creationists may stress one or more of the following claims:

1. *Earth is only about 6000 years old.* The assertion here is that life has not existed long enough for natural selection to produce the diversity of organisms seen today. Many creationists do not accept the physical and chemical principles behind radiometric dating, reviewed in Chapter 2, or the principle known as uniformitarianism. **Uniformitarianism** states that the laws of the universe are constant over time and throughout space. In geology, uniformitarianism holds that rock-forming processes occurring today, such as the slow deposition of shells to form limestone, were responsible for rock formations produced in the past, such as the white cliffs of Dover, England, made of massive limestone deposits. Geologists working in the late 1700s realized that under uniformitarianism, Earth must be extremely old.
2. *Organisms are "irreducibly complex."* According to this argument, adaptations such as the vertebrate eye and metabolic pathways are so complex and well integrated that they could not have arisen from a gradual, incremental process such as natural selection. Proponents of this view downplay the importance of the many fossils with characteristics that show transitions between simpler and more complex traits, and ignore analyses that show how complex metabolic pathways have evolved from simpler ancestral pathways.
3. *The theory is unproven.* Creationists argue that the evidence for evolution is inherently "soft" or unsatisfactory because no one was present when life began and because no one actually witnessed major events such as the origin of the flower in plants or the evolution of limbs in vertebrates. In doing so, creationists deny that inferring historical events from contemporary evidence is a valid research program in science. They insist that scientific evidence must be firsthand or direct in nature. To evaluate this viewpoint, consider that for some 150 years after the atomic theory was proposed, no one had actually seen an atom. Yet the theory was widely accepted as correct during this time.

Where does the controversy stand today? The U.S. Supreme Court has repeatedly ruled that legislation banning the teaching of evolution in public schools is unconstitutional on the basis of the First Amendment—the separation of church and state. The Court's opinion is that scientific creationism and intelligent design theory promote a specific religious belief, because they are founded on religious tenets codified in the Bible.

Despite the Court's rulings, however, and even though many religious leaders as well as many scientists see no conflict between evolution and religious faith, the controversy continues.

CHAPTER REVIEW

Summary of Key Concepts

- **Populations and species evolve, meaning that their characteristics change through time. More precisely, evolution is defined as changes in allele frequencies over time.**

Evidence for the fact of evolution—that species are related through shared ancestry and have changed over time—has accumulated since Lamarck, Darwin, Wallace, and others began research on the topic over 200 years ago. An array of observations is inconsistent with the alternative theory that species were formed instantaneously, recently, and independently, and have remained unchanged through time. These observations include the geographic proximity of closely related species such as the Galápagos mockingbirds; the existence of structural, developmental, and genetic homologies; the near universality of the genetic code; resemblances of modern to fossil forms; transitional fossils; the fact of extinction; and the presence of vestigial traits. These data support the hypothesis that populations and species change through time.

- **Natural selection occurs when individuals with certain alleles and traits produce the most offspring in a population. An adaptation is a genetically based trait that increases an individual's ability to produce offspring in a particular environment.**

In addition to articulating evidence for the pattern called evolution, Darwin and Wallace independently conceived of a process responsible for it. Natural selection occurs whenever genetically based differences among individuals lead to differences in their ability to reproduce. Alleles or traits that increase the reproductive success of an individual are said to increase the individual's fitness. A trait that leads to higher fitness, relative to individuals without the trait, is an adaptation. If a particular allele increases fitness and leads to adaptation, the allele will increase in frequency in the population.

Evolution is an outcome of natural selection. It is defined as changes in allele frequencies that occur in populations from one generation to the next. Evolution by natural selection has been confirmed by a wide variety of studies and has long been considered to be the central organizing principle of biology.

Web Tutorial 23.1 Natural Selection for Antibiotic Resistance
Web Tutorial 23.2 Natural Selection in Alpine Skypilots

Evolution by natural selection is not progressive, and it does not change the characteristics of the individuals that are selected—it changes only the characteristics of the population. Not all traits are adaptive. All adaptations are constrained by genetic and historical factors.

Individuals that are naturally selected are not changed by the process—they simply produce more offspring than other individuals do. Traits that increase in frequency under natural selection do so because they improve fitness, not because they are necessarily larger or more complex. Because natural selection acts only on existing traits, it does not lead to perfection.

Questions

Content Review

1. How can Darwinian fitness be estimated?
   a. Document how long different individuals in a population survive.
   b. Count the number of offspring produced by different individuals in a population.
   c. Determine which individuals are strongest.
   d. Determine which phenotype is the most common one in a given population.
2. Why are some traits considered vestigial?
   a. They improve the fitness of an individual who bears them, compared with the fitness of individuals without those traits.
   b. They change in response to environmental influences.
   c. They existed a long time in the past.
   d. They are reduced in size, complexity, and function compared with traits in related species.
3. What is an adaptation?
   a. a trait that improves the fitness of its bearer, compared with individuals without the trait
   b. a trait that changes in response to environmental influences within the individual's lifetime
   c. an ancestral trait—one that was modified to form the trait observed today
   d. a trait that is reduced in size or complexity but increases the fitness of its bearer
4. Why does the presence of extinct and transitional forms in the fossil record support the pattern component of the theory of evolution by natural selection?
   a. It supports the hypothesis that individuals change over time.
   b. It supports the hypothesis that weaker species are eliminated by natural selection.
   c. It supports the hypothesis that species evolve to become more complex and better adapted over time.
   d. It supports the hypothesis that species have changed through time.
5. Why are homologous traits similar?
   a. They are derived from a common ancestor.
   b. They are derived from different ancestors.
   c. They result from convergent evolution.
   d. They result from genetic and developmental similarities.
6. Which of the following statements is correct?
   a. When individuals change in response to challenges from the environment, their altered traits are passed on to offspring.
   b. Species are created independently of each other and do not change over time.
   c. Populations—not individuals—change when natural selection occurs.
   d. The Earth is young, and most of today's landforms were created during the floods at the time of Noah.

Conceptual Review

1. Some biologists encapsulate evolution by natural selection with the phrase "mutation proposes, selection disposes." Explain what they mean, using the formal terms introduced in this chapter.
2. Review the section on the evolution of drug resistance in *Mycobacterium tuberculosis*.
   - In *M. tuberculosis*, how does heritable variation arise for the trait of drug resistance?
   - What evidence do researchers have that a drug-resistant strain evolved in the patient analyzed in their study, instead of having been transmitted from another infected individual?
   - If the antibiotic rifampin were banned, would the mutant *rpoB* gene have lower or higher fitness in the new environment? Would strains carrying the mutation continue to increase in frequency in *M. tuberculosis* populations?
3. Consider the experiment that tested whether the offspring of hand-pollinated plants and bumblebee-pollinated plants differed in flower size. Why did the randomly pollinated population serve as an experimental control?
4. How has the conflict between the everyday and scientific usage of the word *theory* contributed to the controversy over the theory of evolution by natural selection?
5. The evidence supporting the pattern component of the theory of evolution can be criticized on the grounds that it is indirect. For example, no one has directly observed the formation of a vestigial trait over time. Due to the indirect nature of the evidence, it could be argued that structural and genetic homologies are coincidental and do not result from common ancestry. Is indirect evidence for a scientific theory legitimate? Are you persuaded that modification with descent is the best explanation available for the data reviewed in Section 23.1? Why or why not?
6. Why isn't evolution by natural selection progressive? Why don't the strongest individuals in a population always produce the most offspring?

Group Discussion Problems

1. The geneticist James Crow wrote that successful scientific theories have the following characteristics: (1) They explain otherwise puzzling observations; (2) they provide connections between otherwise disparate observations; (3) they make predictions that can be tested; and (4) they are heuristic, meaning that they open up new avenues of theory and experimentation. Crow added two other elements that he considered important on a personal, emotional level: (5) They should be elegant, in the sense of being simple and powerful; and (6) they should have an element of surprise.

   How well does the theory of evolution by natural selection fulfill these six criteria? Think of a theory you've been introduced to in another science course—for example, the atomic theory or the germ theory of disease—and evaluate it by using this list.

2. The average height of humans has increased steadily for the past 100 years in industrialized nations. This trait has clearly changed over time. Most physicians and human geneticists believe that the change is due to better nutrition and a reduced incidence of disease. Has human height evolved?

3. Genome sequencing projects may dramatically affect how biologists analyze evolutionary changes in quantitative traits. For example, suppose that the genomes of many living humans are sequenced and that genomes could be sequenced from many people who lived 100 years ago. (That might be possible with preserved tissue.) If 20 genes have been shown to influence height, how could you use the sequence data from these genes to test the hypothesis that human height has evolved in response to natural selection?

4. Examine Figure 23.11. How would you expect the data on this map to change over the next several decades as drug-resistant strains of *M. tuberculosis* increase in frequency in North America and Western Europe?

Answers to Multiple-Choice Questions **1.** b; **2.** d; **3.** a; **4.** d; **5.** a; **6.** c

www.prenhall.com/freeman is your resource for the following: Web Tutorials; Online Quizzes and other Online Study Guide materials; Answers to Conceptual Review Questions; Solutions to Group Discussion Problems; Answers to Figure Caption Questions and Exercises; and Additional Readings and Research.

24 Evolutionary Processes

KEY CONCEPTS

- Each of the four evolutionary mechanisms has different consequences. Only natural selection produces adaptation. Genetic drift causes random fluctuations in allele frequencies. Gene flow equalizes allele frequencies between populations. Mutation introduces new alleles.
- The Hardy-Weinberg principle acts as a null hypothesis when researchers want to test whether evolution or nonrandom mating is occurring at a particular gene.
- Inbreeding changes genotype frequencies but does not change allele frequencies.

This chapter focuses on how natural selection and other processes are causing cliff swallows and other species to evolve.

Chapter 22 defined evolution as a change in allele frequencies over time. One of the key concepts from that chapter was that even though natural selection acts on individuals, evolutionary change occurs in **populations**. A population is a group of individuals from the same species that live in the same area and regularly interbreed.

Natural selection is not the only process that causes evolutionary change, however. There are actually four mechanisms that shift allele frequencies in populations:

1. *Natural selection* increases the frequency of certain alleles—the ones that contribute to improved reproductive success.
2. *Genetic drift* causes allele frequencies to change randomly. In some cases, drift may even cause alleles that decrease fitness to increase in frequency.
3. *Gene flow* occurs when individuals immigrate into (enter) or emigrate from (leave) a population. Allele frequencies may change when gene flow occurs, because arriving individuals introduce alleles to their new population and emigrating individuals remove alleles from their old population.
4. *Mutation* modifies allele frequencies by continually introducing new alleles. The alleles created by mutation may be beneficial, detrimental, or have no effect on fitness.

This chapter has two fundamental messages: Natural selection is not the only agent responsible for evolution, and each of the four evolutionary processes has different consequences. Natural selection is the only mechanism that consistently results in adaptation. Mutation, gene flow, and genetic drift do not favor certain alleles and rarely produce adaptation. Mutation and drift introduce a nonadaptive component into evolution.

Let's take a closer look at the four evolutionary processes by examining a null hypothesis—what happens to allele frequencies when the evolutionary mechanisms are *not* operating.

24.1 Analyzing Change in Allele Frequencies: The Hardy-Weinberg Principle

To study how the four evolutionary processes affect populations, biologists take a three-pronged approach. First they create mathematical models that track the fate of alleles over time. Then they collect data to test predictions made by the models' equations. Finally, they apply the results to solve problems in conservation biology or human genetics.

This research strategy began with work published independently in 1908 by G. H. Hardy and Wilhelm Weinberg. At the time, it was commonly believed that changes in allele frequency occur simply as a result of sexual reproduction—meiosis followed by the random fusion of gametes (eggs and sperm) to form offspring. Some biologists claimed that dominant alleles would inevitably increase in frequency, while others predicted that two alleles of the same gene would eventually reach a frequency of 0.5.

To test these hypotheses, Hardy and Weinberg analyzed what happens to the frequencies of alleles when many individuals in a population mate and produce offspring. Instead of thinking about the consequences of a mating between two parents with a specific pair of genotypes, as we did with Punnett squares in Chapter 13, Hardy and Weinberg wanted to know what happened in an entire population, when *all* of the individuals—and thus all possible genotypes—bred.

To analyze the consequences of matings among all of the individuals in a population, Hardy and Weinberg invented a novel approach: They imagined that all of the gametes produced in each generation go into a single bin called the **gene pool**. To determine which genotypes would be present in the next generation and in what frequency, they simply had to calculate what happened when two gametes were plucked at random out of the gene pool, many times, and each of these gamete pairs was then combined to form offspring. These calculations would predict the genotypes of the offspring that would be produced, as well as the frequency of each genotype.

The researchers began by analyzing the simplest situation possible—that just two alleles of a particular gene exist in a population. Let's call these alleles A_1 and A_2. We'll use p to symbolize the frequency of A_1 alleles in the gene pool and q to symbolize the frequency of A_2 alleles in the same gene pool. Because there are only two alleles, the two frequencies must add up to 1; that is, $p + q = 1$. Although p and q can have any value between 0 and 1, let's suppose that the initial frequency of A_1 is 0.7 and that of A_2 is 0.3 (**Figure 24.1**, step 1). Thus, 70 percent of the gametes in the gene pool carry A_1, and 30 percent carry A_2 (Figure 24.1, step 2).

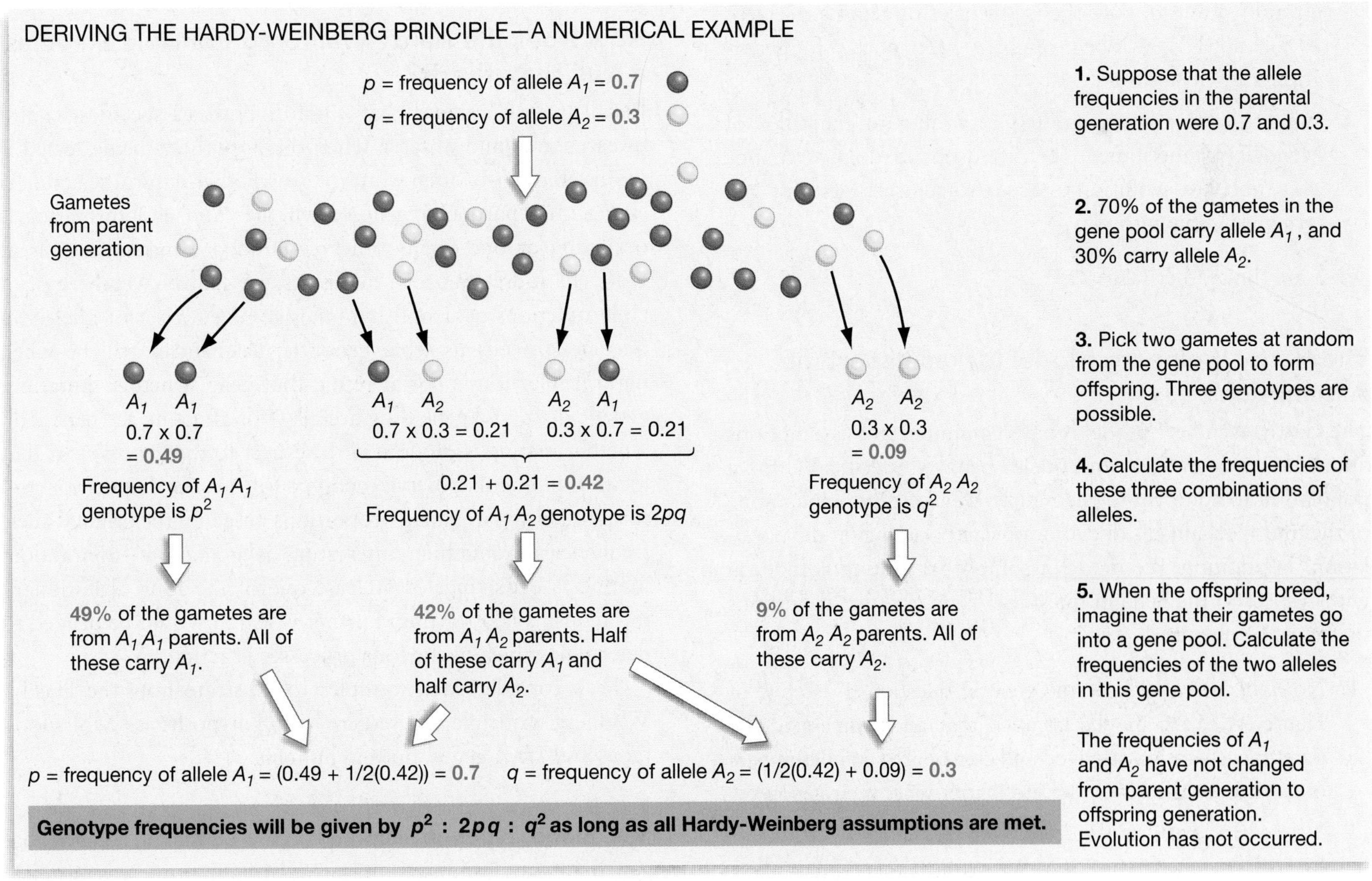

FIGURE 24.1 A Numerical Example of the Hardy-Weinberg Principle

In this population, then, three genotypes are possible: A_1A_1, A_1A_2, and A_2A_2 (Figure 24.1, step 3). What will the frequency of these three genotypes be in the next generation? Figure 24.1, step 4, explains the logic of Hardy's and Weinberg's result:

- The frequency of the A_1A_1 genotype is p^2.
- The frequency of the A_1A_2 genotype is $2pq$.
- The frequency of the A_2A_2 genotype is q^2.

The genotype frequencies in the offspring generation must add up to 1, which means that $p^2 + 2pq + q^2 = 1$. In our numerical example, $0.49 + 0.42 + 0.09 = 1$. Figure 24.1, step 5, shows how the frequencies of alleles A_1 and A_2 are calculated from these genotype frequencies. Hardy and Weinberg showed that, in the next generation, the frequency of allele A_1 is still p and the frequency of allele A_2 is still q. In our example, the frequency of allele A_1 is still 0.7 and the frequency of allele A_2 is still 0.3. Thus, no allele frequency change occurred. Even if A_1 is dominant to A_2, it does not increase in frequency. And there is no trend toward both alleles reaching a frequency of 0.5.

This result is called the **Hardy-Weinberg principle.** It makes two fundamental claims:

1. If the frequencies of alleles A_1 and A_2 in a population are given by p and q, then the frequencies of genotypes A_1A_1, A_1A_2, and A_2A_2 will be given by p^2, $2pq$, and q^2 for generation after generation.
2. When alleles are transmitted according to the rules of Mendelian inheritance, their frequencies do not change over time. For evolution to occur, some other factor or factors must come into play.

What are these other factors?

WEB TUTORIAL 24.1 The Hardy-Weinberg Principle

The Hardy-Weinberg Model Makes Important Assumptions

The Hardy-Weinberg model is based on important assumptions about how populations and alleles behave. Specifically, for a population to conform to the Hardy-Weinberg principle, none of the four mechanisms of evolution can be acting on the population. In addition, the model assumes that mating is random with respect to the gene in question. Thus, the five conditions that must be met are:

1. *No natural selection* at the gene in question. In step 2 of Figure 24.1, the model assumed that all members of the parental generation survived and contributed equal numbers of gametes to the gene pool, no matter what their genotype.
2. *No genetic drift, or random allele frequency changes,* affecting the gene in question. We avoided this type of allele frequency change in step 4 of Figure 24.1 by assuming that we drew alleles in their exact frequencies p and q, and not at some different values caused by chance. For example, allele A_1 did not "get lucky" and get drawn more than 70 percent of the time. Stated another way, there is no random change due to a sampling effect.
3. *No gene flow.* No new alleles were added by immigration or lost through emigration anywhere in Figure 24.1. As a result, all of the alleles in the offspring population came from the original population's gene pool.
4. *No mutation.* We didn't consider that new A_1s or A_2s or other, new alleles might be introduced into the gene pool in step 2 or step 5 of Figure 24.1.
5. *Random mating* with respect to the gene in question. We enforced this condition by picking gametes from the gene pool at random in step 3 of Figure 24.1. We did not allow individuals to choose a mate based on their genotype at this locus.

The Hardy-Weinberg principle tells us what to expect if none of the four evolutionary mechanisms are acting and if mating is random. If there is no selection, no genetic drift, no gene flow, and no mutation affecting a gene, *and* if mating is random with respect to that gene, then the genotypes A_1A_1, A_1A_2, and A_2A_2 should be in the Hardy-Weinberg proportions p^2, $2pq$, and q^2, and no evolution will occur.

How Does the Hardy-Weinberg Principle Serve as a Null Hypothesis?

Recall from Chapter 1 that a null hypothesis specifies what a researcher should observe when the hypothesis being tested is wrong. Biologists often want to test whether natural selection is acting on a particular gene or whether nonrandom mating is occurring or one of the other evolutionary mechanisms is at work. In addressing such questions, the Hardy-Weinberg principle functions as a null hypothesis. Given a set of allele frequencies, it tells us what genotype frequencies will be when natural selection is not affecting that gene; whether mutation, genetic drift, and gene flow are also not affecting the gene; and whether mating is random with respect to the genotype at that locus. If biologists observe genotype frequencies that do *not* conform to Hardy-Weinberg proportions based on the original allele frequencies, something interesting is going on—nonrandom mating is occurring, or at least one of the four evolutionary mechanisms is operating. Further research would be needed to determine which of the four processes is acting.

Let's consider two examples to illustrate how the Hardy-Weinberg principle is used as a null hypothesis: MN blood types and *HLA* genes, both in humans.

MN Blood Types in Humans One of the first genes that geneticists could analyze in natural populations was the MN blood group of humans. Most human populations have two alleles, designated *M* and *N*, at this locus. Because the gene codes for a protein found on the surface of red blood cells, researchers could determine whether individuals are *MM*, *MN*, or *NN* by treating blood samples with antibodies to each protein (this technique was first introduced in Chapter 8). To estimate the frequency of each genotype in a population, geneticists obtain data from a large number of individuals, then divide the number of individuals with each genotype by the total number of individuals in the sample.

Table 24.1 shows the genotype frequencies for populations from throughout the world and illustrates how observed genotype frequencies are compared with the genotype frequencies expected if the Hardy-Weinberg principle holds. The analysis is based on the following steps:

1. Estimate genotype frequencies by observation—in this case, by testing many blood samples for the *M* and *N* alleles. These frequencies are given on the lines labeled "observed" in Table 24.1.
2. Calculate observed allele frequencies from the observed genotype frequencies. In this case, the frequency of the *M* allele is the frequency of *MM* homozygotes plus half the frequency of *MN* heterozygotes; the frequency of the *N* allele is the frequency of *NN* homozygotes plus half the frequency of *MN* heterozygotes. (You can review the logic behind this calculation in step 5 of Figure 24.1.)
3. Use the observed allele frequencies to calculate the genotypes expected according to the Hardy-Weinberg principle. Under the null hypothesis of no evolution and random mating, the expected genotype frequencies are p^2: $2pq$: q^2.
4. Compare the observed and expected values. Although using statistical testing is beyond the scope of this text, researchers must use statistical tests to determine whether the differences between the observed and expected genotype frequencies are small enough to be due to chance or large enough to reject the null hypothesis of no evolution and random mating.

Note that in every case the observed and expected genotype frequencies are almost identical. (A statistical test shows that the small differences observed are probably due to chance.) For every population surveyed, genotypes at the *MN* locus are in Hardy-Weinberg proportions. As a result, biologists conclude that the assumptions of the Hardy-Weinberg model are valid for this locus. The data imply that the *M* and *N* alleles are not affected by the four evolutionary mechanisms and that mating is random with respect to this locus—meaning that humans do not choose mates on the basis of their *MN* genotype.

Before moving on, however, it is important to note that a study such as this does not mean that the *MN* gene has never been under selection or subject to nonrandom mating or genetic drift. Even if selection has been very strong for many generations, one generation of no evolutionary forces and of random mating will result in genotype frequencies that conform to Hardy-Weinberg expectations. The Hardy-Weinberg principle is used to test the hypothesis that currently no evolution is occurring at a particular gene and that, in the previous generation, mating was random with respect to the gene in question.

***HLA* Genes in Humans** A research team recently collected data on the genotypes of 124 individuals from the Havasupai tribe native to Arizona. These biologists were studying two genes that are important in the functioning of the human immune system. More specifically, the genes that they analyzed code for proteins that help immune system cells recognize and destroy invading bacteria and viruses. Previous work had shown that different alleles exist at both the *HLA-A* and *HLA-B* genes,

TABLE 24.1 Do Genotype and Allele Frequencies in the MN Blood Group of Humans Conform to the Hardy-Weinberg Model?

EXERCISE Fill in the values for allele frequencies and expected genotype frequencies for the Ainu people of Japan.

		Genotype Frequencies			Allele Frequencies	
Population and Location		***MM***	***MN***	***NN***	***M***	***N***
Inuit (Greenland)	Observed	0.835	0.156	0.009	0.913	0.087
	Expected	0.834	0.159	0.008		
Native Americans (U.S.)	Observed	0.600	0.351	0.049	0.776	0.244
	Expected	0.602	0.348	0.050		
Caucasians (U.S.)	Observed	0.292	0.494	0.213	0.540	0.460
	Expected	0.290	0.497	0.212		
Aborigines (Australia)	Observed	0.024	0.304	0.672	0.178	0.824
	Expected	0.031	0.290	0.679		
Ainu (Japan)	Observed	0.179	0.502	0.319		
	Expected					

and that the alleles at each gene code for proteins that recognize slightly different disease-causing organisms. Like the *M* and *N* alleles, *HLA* alleles are *codominant*—meaning that both are expressed and create the phenotype (Chapter 13). As a result, the research group hypothesized that individuals who are heterozygous at one or both of these genes may have a strong fitness advantage. The logic is that heterozygous people have a wider variety of HLA proteins, so their immune systems can recognize and destroy more types of bacteria and viruses. They should be healthier and have more offspring than homozygous people do.

To test this hypothesis, the researchers used their data on observed genotype frequencies to determine the frequency of each allele present. When they used these allele frequencies to calculate the expected number of each genotype according to the Hardy-Weinberg principle, they found that the observed and expected values did not match (**Table 24.2**). Specifically, there were too many heterozygotes and not enough homozygotes compared with the predicted values. Statistical tests show that it is extremely unlikely that the difference between the observed and expected numbers could occur purely by chance.

These results supported the team's prediction and indicated that one of the assumptions behind the Hardy-Weinberg principle was being violated. But which assumption? The researchers argued that mutation, migration, and drift are negligible in this case and offered two competing explanations for their data:

1. *Mating may not be random with respect to the* HLA *genotype.* Specifically, people may subconsciously prefer mates with *HLA* genotypes unlike their own and thus produce an excess of heterozygous offspring. This hypothesis is plausible. For example, experiments have shown that college students can distinguish each others' genotypes at loci related to *HLA* on the basis of body odor. Individuals in this study were more attracted to the smell of genotypes unlike their own. If this is true among the Havasupai, then nonrandom mating would lead to an excess of heterozygotes compared with the proportion expected under Hardy-Weinberg.

TABLE 24.2 Do *HLA* Genotype Frequencies of Humans Conform to the Hardy-Weinberg Model?

	Observed Number	Expected Number
HLA-A		
Homozygotes	38	48
Heterozygotes	84	74
HLA-B		
Homozygotes	21	30
Heterozygotes	101	92

Source: T. Markow et al., *HLA polymorphism in the Havasupai: Evidence for balancing selection. American Journal of Human Genetics* 53 (1993): 943–952.

2. *Heterozygous individuals may have higher fitness.* This hypothesis is supported by data collected by a different research team, studying Hutterite people living in South Dakota. In this population, married women who have the same *HLA*-related alleles as their husbands have more trouble getting pregnant and experience higher rates of spontaneous abortion than do women with *HLA*-related alleles different from those of their husbands. The data suggest that homozygous fetuses have lower fitness than do fetuses heterozygous at these loci. If this were true among the Havasupai, selection would lead to an excess of heterozygotes relative to Hardy-Weinberg expectations.

Which explanation is correct? It is possible that both are. But the fact is, no one knows. Using the Hardy-Weinberg principle as a null hypothesis allowed biologists to detect an interesting pattern in a natural population. Research continues on the question of why the pattern exists.

✓CHECK YOUR UNDERSTANDING

The Hardy-Weinberg principle functions as a null hypothesis when researchers test whether nonrandom mating or evolution is occurring at a particular gene. Given a set of genotype frequencies or the numbers of various genotypes observed in a population, you should be able to (1) calculate observed allele frequencies, (2) calculate the genotype frequencies expected according to the Hardy-Weinberg principle, and (3) compare the observed and expected genotype frequencies and suggest whether the data support the null hypothesis of random mating and no selection, genetic drift, gene flow, or mutation.

24.2 Natural Selection and Sexual Selection

Natural selection occurs when individuals with certain phenotypes produce more offspring than individuals with other phenotypes do. If certain alleles are associated with the favored phenotypes, they increase in frequency while other alleles decrease in frequency. The result is evolution.

These concepts were introduced in Chapter 23. It is important to recognize that natural selection occurs in a wide variety of patterns, however, and that each type of natural selection has different causes and consequences. The data in Table 24.2, for example, document a pattern of natural selection called **heterozygote advantage**: When heterozygous individuals have higher fitness than homozygous individuals do, natural selection maintains genetic variation in populations. **Genetic variation** refers to the number and relative frequency of alleles that are present in a particular population.

Lack of genetic variation in a population is usually a bad thing. To understand why, recall from Chapter 23 that selection can occur only if heritable variation exists in a population. If genetic variation is low and the environment changes—perhaps due to the emergence of a new disease-causing virus, a rapid change in climate, or a reduction in the availability of a particular food source—it is unlikely that any alleles will be present that have high fitness under the new conditions. As a result, the average fitness of the population will decline. If the environmental change is severe enough, the population may even be faced with extinction.

Let's examine some of the different types of natural selection with this question in mind: How do they affect the level of genetic variation in the population?

Directional Selection

According to the data introduced in Chapter 23, natural selection has increased the frequency of drug-resistant strains of the tuberculosis bacterium and caused changes in beak shape and body size in medium ground finches. This type of natural selection is called **directional selection**, because allele frequencies changed in one direction.

Figure 24.2a illustrates how directional selection works when the trait in question has a bell-shaped, normal distribution in a population. Recall from Chapter 13 that when many different genes influence a trait, the distribution of phenotypes in the population tends to form a bell-shaped curve. In such cases, directional selection is acting on many different genes at once. In the case of selection on drug resistance in the TB bacterium, however, selection was acting on a single gene.

Most often, directional selection tends to reduce the genetic diversity of populations. If directional selection continues over time, the favored alleles will eventually reach a frequency of 1.0 while disadvantageous alleles will reach a frequency of 0.0. Alleles that reach a frequency of 1.0 are said to be *fixed*; those that reach a frequency of 0.0 are said to be *lost*.

Fixation and loss may not always occur under directional selection, however. To appreciate why, consider recent data on the body size of cliff swallows native to the Great Plains of North America. In 1996 a population of cliff swallows endured a six-day period of exceptionally cold, rainy weather. Cliff swallows are migratory and feed by catching mosquitoes and other insects in flight. Insects disappeared during this cold snap, however, and the biologists recovered the bodies of 1853 swallows that died of starvation. As soon as the weather improved, the researchers caught and measured the body size of 1027 survivors from the same population. As the histograms in **Figure 24.2b** show, survivors were much larger on average than the birds that died. Directional selection, favoring large body size, had occurred. To explain this observation, the investigators suggest that larger birds survived because they had larger fat stores and did not get as cold as the smaller birds. As a result, the larger birds were less likely to die of exposure to cold and more likely to avoid starvation until the weather warmed up and insects were again available.

(a) Directional selection changes the average value of a trait.

(b) For example, directional selection caused average body size to increase in a cliff swallow population.

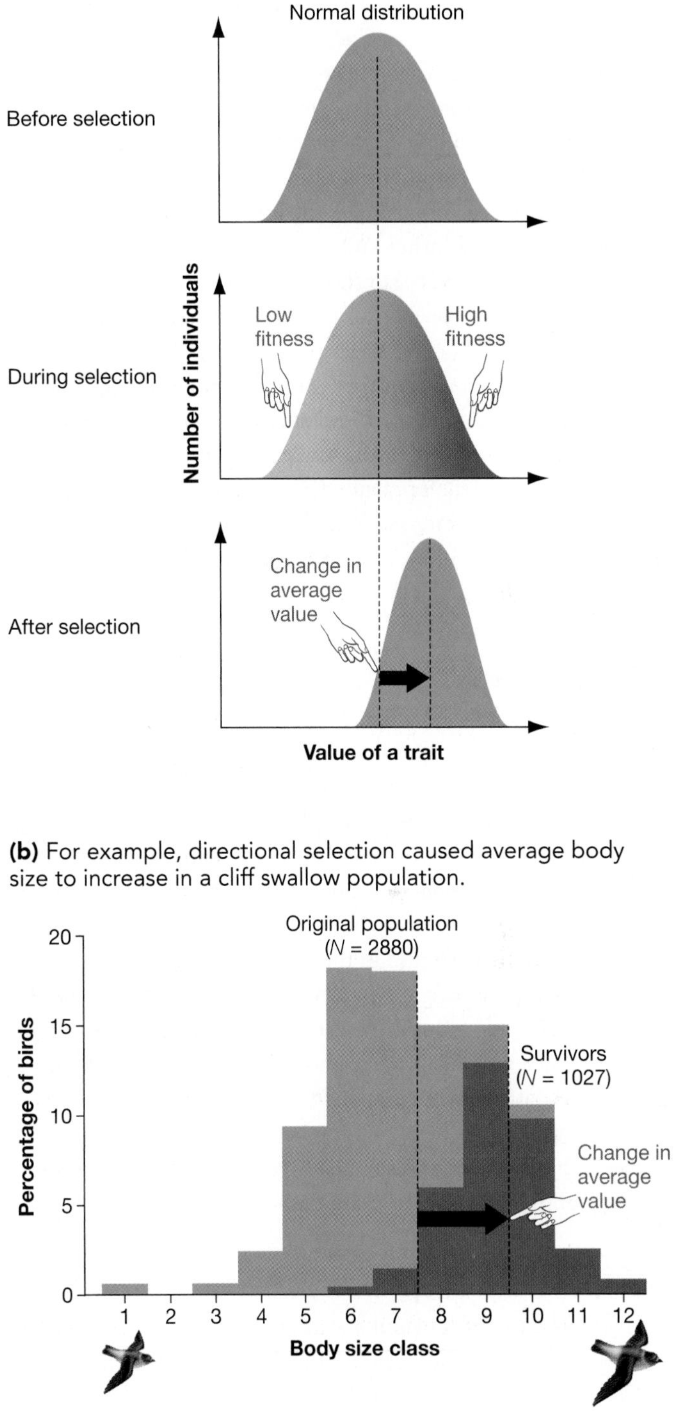

FIGURE 24.2 Directional Selection
(a) When directional selection acts on traits that have a normal distribution, individuals with one set of extreme values experience poor reproductive success. **(b)** The light-green histogram shows the distribution of overall body size in cliff swallows prior to an extended cold snap that killed many individuals. (The various body-size classes were calculated from measurements of wing length, tail length, leg length, and beak size.) The dark green histogram shows the size distribution of individuals from the same population that survived the cold spell. Here *N* indicates the sample size.

WEB TUTORIAL 24.2
Three Modes of Natural Selection

If this type of directional selection continued, alleles that contribute to small body size would quickly be eliminated from the cliff swallow population. It is not clear that this will be the case, however, because directional selection is rarely constant throughout a species' range and through time. By examining weather records, the researchers established that cold spells as severe as the one that occurred in 1996 are rare. Further, research on other swallow species suggests that smaller birds are more maneuverable in flight and thus more efficient when they feed. If so, then selection for feeding efficiency could counteract selection by cold weather. When this is the case, individuals with intermediate body size should be favored. Opposing patterns of directional selection will help maintain genetic variation for this trait.

When studies on a wide array of populations and species are considered, it is common to find that one cause of directional selection on a trait is counterbalanced by a different factor that causes selection in the opposite direction. In such cases, the optimal phenotype is intermediate. The same pattern can result from an entirely different type of natural selection, called stabilizing selection.

Stabilizing Selection

When cliff swallows were exposed to cold weather, selection greatly reduced one extreme in the range of phenotypes and resulted in a directional change in the average characteristics of the population. But selection can also reduce both extremes in a population, as illustrated in **Figure 24.3a**. This pattern of selection is called **stabilizing selection.** It has two important consequences: There is no change in the average value of a trait over time, and genetic variation in the population is reduced.

Figure 24.3b shows a classical data set in humans illustrating stabilizing selection. Biologists who analyzed birth weights and mortality in 13,730 babies born in British hospitals found that babies of average size (slightly over 7 pounds) survived best. Mortality was high for very small babies and very large babies. This is persuasive evidence that birth weight is under strong stabilizing selection in humans.

Disruptive Selection

One of the most important patterns of selection observed in nature has the opposite effect of stabilizing selection. Instead of favoring phenotypes near the average value and eliminating extreme phenotypes, **disruptive selection** eliminates phenotypes near the average value and favors extreme phenotypes (**Figure 24.4a**). As a result, disruptive selection tends to maintain the amount of genetic variation in a population.

Biologists have recently shown that disruptive selection maintains the striking bills of black-bellied seedcrackers, pictured in **Figure 24.4b**. The data plotted in the figure show that

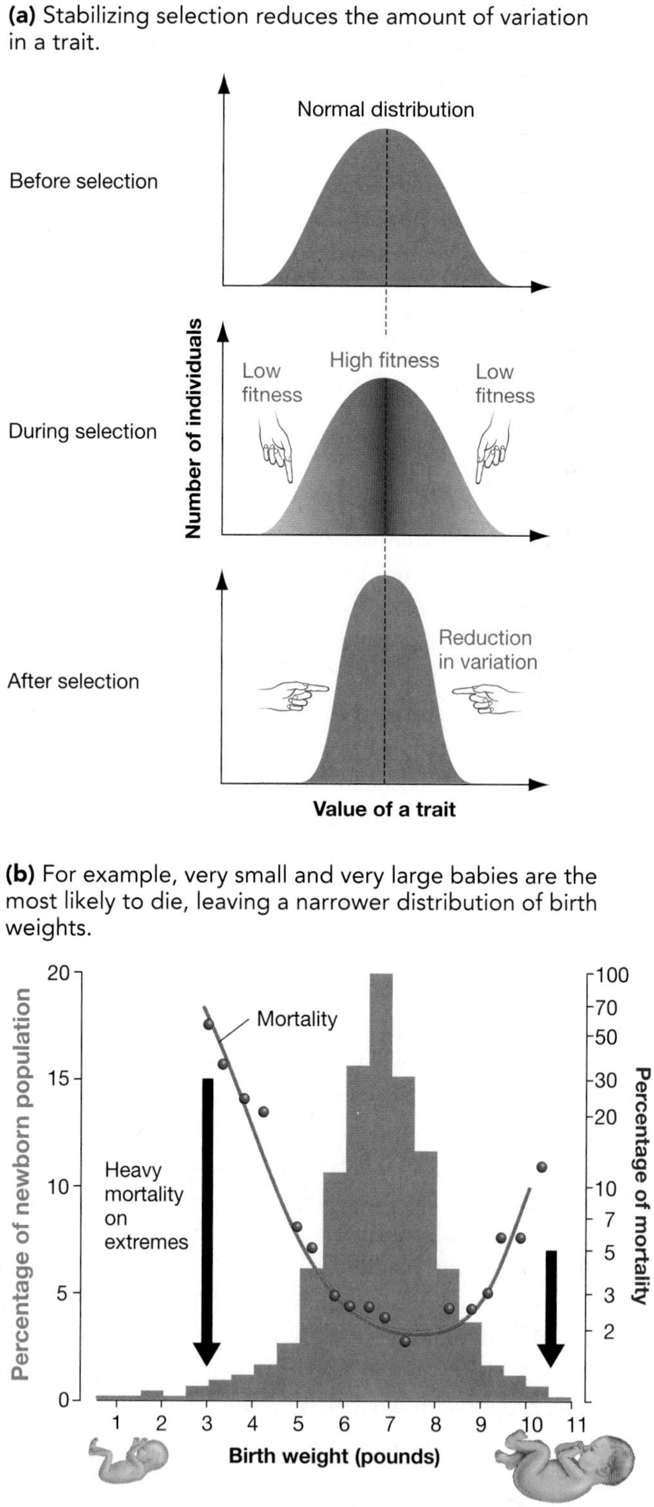

FIGURE 24.3 Stabilizing Selection
(a) When stabilizing selection acts on normally distributed traits, individuals with extreme phenotypes experience poor reproductive success. **(b)** Histogram showing the percentages of newborns with various birth weights on the left-hand axis. The purple dots indicate the percentage of newborns in each weight class that died, plotted on the logarithmic scale shown on the right. The purple line is a function that fits the data points.

(a) Disruptive selection increases the amount of variation in a trait.

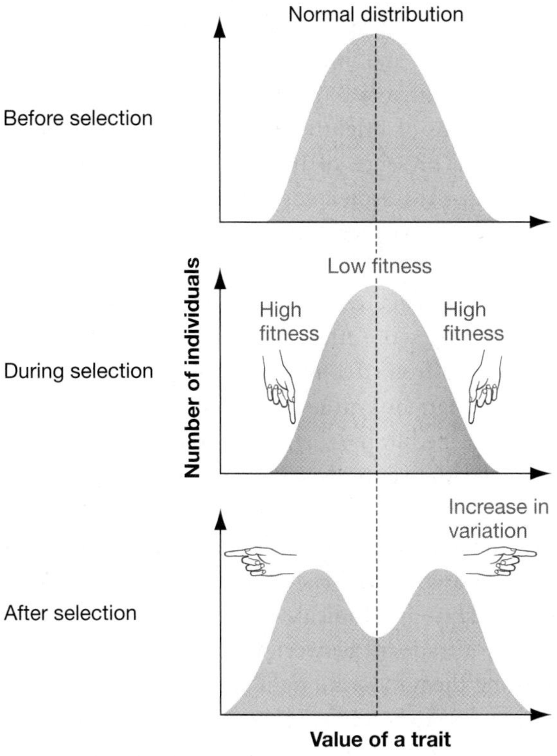

(b) For example, only juvenile black-bellied seedcrackers that had very long or very short beaks survived long enough to breed.

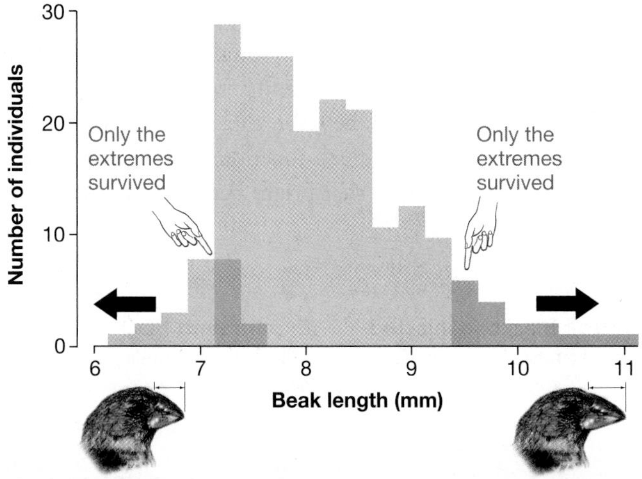

FIGURE 24.4 Disruptive Selection
(a) When disruptive selection occurs on traits with a normal distribution, individuals with extreme phenotypes experience high reproductive success. **(b)** Histogram showing the distribution of beak length in a population of black-bellied seedcrackers from Cameroon, West Africa. The light orange bars represent all juveniles; the dark orange bars, juveniles that survived to adulthood. Only juveniles with the most extreme phenotypes survived long enough to breed.

individuals with either very short or very long beaks survive best and that birds with intermediate phenotypes are at a disadvantage. In this case, the agent that causes natural selection is food. At a study site in south-central Cameroon, West Africa, a researcher found that only two sizes of seed are available to the seedcrackers: large and small. Birds with small beaks crack and eat small seeds efficiently. Birds with large beaks handle large seeds efficiently. But birds with intermediate beaks have trouble with both. Disruptive selection maintains high overall variation in this population.

Disruptive selection is important because it can lead to **speciation**, or the formation of new species. If small-beaked seedcrackers began mating with other small-beaked individuals, all of their offspring would be small beaked and would feed on small seeds. Similarly, if large-beaked individuals chose only other large-beaked individuals as mates, they would produce large-beaked offspring that would feed on large seeds. In this way, selection would result in two distinct populations and eventually two new species. The process of species formation, based on disruptive selection and other mechanisms, is explored in detail in Chapter 25.

In many cases, the process of speciation depends on how individuals choose mates. The process of choosing mates, in turn, causes a distinctive type of natural selection called sexual selection. Let's take a closer look at what sexual selection is and how it works.

Sexual Selection

If female peacocks choose males with the longest and most iridescent tails as mates, then the frequency of alleles that contribute to long, iridescent tails will increase in the population. Charles Darwin was the first biologist to recognize that this form of natural selection, called **sexual selection**, is a mechanism of evolutionary change. Sexual selection occurs when individuals within a population differ in their ability to attract mates. It is a special case of natural selection—selection for enhanced ability to obtain mates.

In 1948 A. J. Bateman contributed a fundamental insight about how sexual selection works. His idea was elaborated by Robert Trivers in 1972. The Bateman-Trivers theory contains two elements: a claim about a pattern in the natural world and a mechanism that causes the pattern.

The pattern component of their theory is that sexual selection usually acts on males much more strongly than on females. As a result, traits that attract members of the opposite sex are much more highly elaborated in males. The mechanism that Bateman and Trivers proposed to explain this pattern can be summarized with a quip: "Eggs are expensive, but sperm are cheap." That is, the energetic cost of creating a large egg is enormous, whereas a sperm contains few energetic resources. Thus, in most species, females invest much more in their offspring than do males. This phenomenon is called the **fundamental asymmetry of sex**. It is characteristic of almost all sexual species and has two important consequences:

1. Because eggs are large and energetically expensive, females produce relatively few young over the course of a lifetime. A female's fitness is limited primarily by her ability to gain the resources needed to produce eggs and rear young—not by the ability to find a mate.
2. Sperm are so simple to produce that a male can father an almost limitless number of offspring. For males, fitness is limited not by the ability to acquire the resources needed to produce sperm, but by the number of females they can mate with.

The theory of sexual selection makes strong predictions. If females invest a great deal in each offspring, then they should be choosy about whom they agree to mate with. Conversely, if males invest little in each offspring, then they should compete with each other for mates. Finally, sexual selection should be much stronger on males than on females. As a result, traits that result from sexual selection—meaning traits that are useful only in courtship or in competition for mates—should be found primarily in males. Do data from experimental or observational studies agree with these predictions? Let's consider each of them in turn.

Sexual Selection Via Female Choice Female birds invest a great deal of time and energy in each offspring. Brown kiwis, which are flightless, ground-dwelling birds that live in New Zealand, offer an extreme example. Female brown kiwis weigh about 2.7 kg (6 lb) and lay an egg that weighs 0.45 kg (1 lb). As **Figure 24.5** shows, each egg represents a considerable investment of resources.

Female kiwis lay a large egg relative to their body size

FIGURE 24.5 In Many Species, Females Make a Large Investment in Each Offspring
X-ray of a female kiwi, ready to lay an egg.

Females that invest this much in each offspring should be choosy about the males they mate with. If so, what criteria do females use to make their choice? Recent experiments have shown that, in several bird species, females prefer to mate with males that are well fed and in good health. These experiments were motivated by three key observations: (1) In many bird species, the existence of brightly colored feathers or a bright beak is due to the presence of the red and yellow pigments called carotenoids. (2) Carotenoids, which were introduced in Chapter 10, function as accessory pigments in plant chloroplasts, where they help to harvest light energy used in photosynthesis. But in animals, carotenoids stimulate the immune system to fight disease more effectively. Their presence also prevents tissue damage from free radicals that form naturally in cells. (3) Animals cannot synthesize carotenoids—these molecules must be obtained in the diet. Taken together, these three observations suggest that birds with the brightest beaks and feathers are those that are well fed. Further, beaks and feathers can be bright only if an individual is healthy enough to allocate carotenoids to these structures instead of using the molecules to fight disease or reduce free radicals.

Is there really a trade-off between using carotenoids to fight disease and using them to look bright and colorful? To answer this question, a research team did an experiment with European blackbirds (**Figure 24.6**). Males of this species have a bright-orange bill due to the presence of carotenoids. When the biologists injected a group of male blackbirds with red blood cells from sheep, it provoked a strong response by the immune system. In addition, beak color in these males dimmed. There was no change in beak color, however, in a control group of males that was injected with a simple salt solution, which did not provoke an immune response. These results suggest that there is indeed a trade-off between the use of carotenoids in beak coloration and in immune function. The results are consistent with the hypothesis that bright beaks are an honest signal of male health.

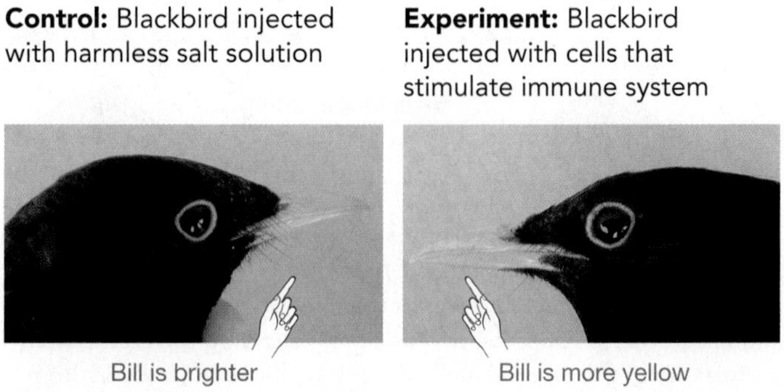
Control: Blackbird injected with harmless salt solution
Experiment: Blackbird injected with cells that stimulate immune system
Bill is brighter
Bill is more yellow

FIGURE 24.6 When the Immune Systems of Male Blackbirds Are Challenged, Their Beaks Dull in Color
These two male blackbirds are brothers. The individual on the right was injected with sheep red blood cells, which activate the immune system in much the same way as a disease-causing bacterium or parasite does. The male on the right is not actually sick, however.

To test the hypothesis that females prefer to mate with bright-beaked birds, a different team of researchers experimented with zebra finches (**Figure 24.7a**). As **Figure 24.7b** shows, they fed one group of male zebra finches a diet that was heavily supplemented with carotenoids. A second group of male zebra finches (the control group), consisting of brothers of males in the first group, was fed a diet that was similar in every way except for the additional carotenoids. As predicted, the males eating the carotenoid-supplemented diet developed brighter beaks than the males fed the carotenoid-poor diet. When females were given a choice of spending time with the two brothers, 9 out of 10 females tested preferred the brighter-beaked male. These results are strong evidence that females of this species are choosy about their mates and that they prefer to mate with healthy, well-fed males.

Enough experiments have been done on other bird species to support a general conclusion: Bright male coloration, along with songs and dances and other types of courtship displays, carry the message "I'm healthy and well fed because I have good alleles. Mate with me." By choosing a colorful male as the father of her offspring, a female is likely to have offspring with alleles that will help the offspring fight disease and feed efficiently.

(a) Zebra finches have bright beaks.

(b) Males fed carotenoids get brighter beaks.

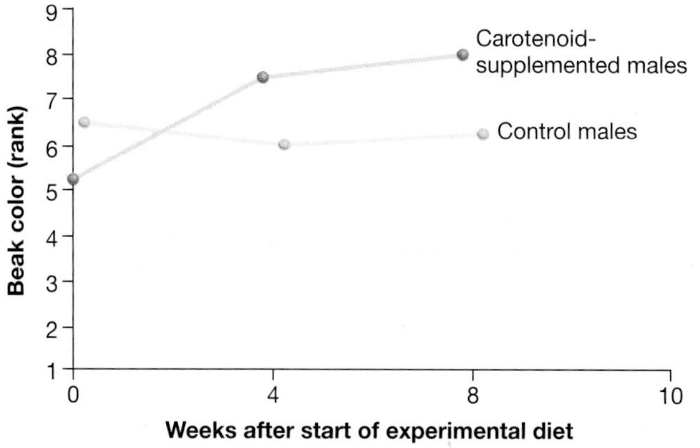

FIGURE 24.7 If Male Zebra Finches Are Fed Carotenoids, Their Beaks Get Brighter
(a) Male and female zebra finches look markedly different—males have brighter feathers. **(b)** In males, beaks brighten in response to an increased presence of carotenoids in the diet.

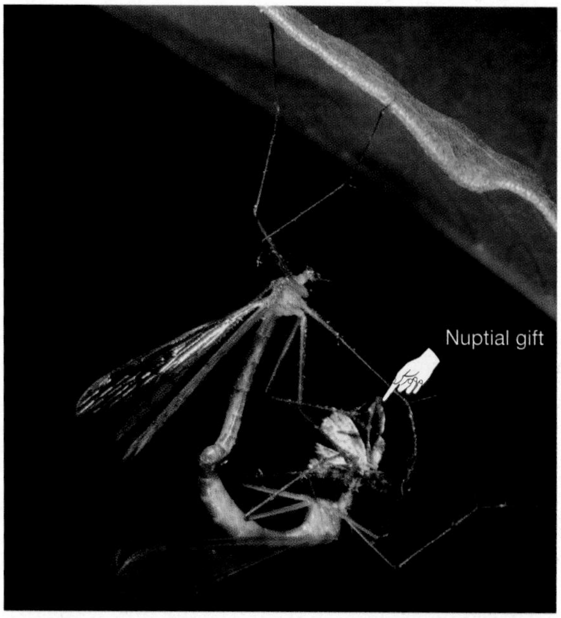

FIGURE 24.8 In Some Insects, Males Present Food to Females as a "Nuptial Gift"
Hanging scorpion flies copulating while the female (below) eats a prey presented to her by the male (above). The larger the prey, the longer she will eat and the longer she will copulate with the male.
QUESTION Most female insects copulate with more than one male before laying eggs. Why would it be advantageous for a male to copulate with a female for a long time?

Extensive experimentation has also shown that in many species, females prefer to mate with males who provide some of the resources required to produce eggs or care for young. In insects, for example, it is common to observe that females will not mate with a male unless he brings a "nuptial gift"—a piece of food that the female devours as they copulate (**Figure 24.8**). In some fish species, females prefer to mate with males that protect a nest site and care for the eggs until they hatch. In humans and many species of birds, males provide food, protection, and other resources required for rearing young.

To summarize, females may choose mates on the basis of physical characteristics that signal male genetic quality, resources provided by males, or both. In some species, however, females do not have the luxury of choosing a male. Instead, competition among males is the primary cause of sexual selection.

Sexual Selection via Male-Male Competition As an example of research on how males compete for mates, consider data from a long-term study of a northern elephant seal population breeding on Año Nuevo Island, off the coast of California. Elephant seals feed on marine fish and spend most of the year in the water. But when females are ready to mate and give birth, they haul themselves out of the water onto land. Females prefer to give birth on islands, where newborn pups are protected from terrestrial and marine predators. Because elephant seals have flippers that are ill suited for walking, females can haul themselves out of the water only on the few beaches that have gentle slopes. As a result, large numbers of females congregate in tiny areas to breed.

Male elephant seals establish territories on breeding beaches by fighting (**Figure 24.9a**). A **territory** is an area that is actively defended and that provides exclusive use by the owner. Males that win battles with other males monopolize matings with the females residing in their territories. Males that lose battles are relegated to territories with few females or are excluded from the beach. Fights are essentially slugging contests and are usually won by the larger male. The males stand face to face, bite each other, and land blows with their heads.

Based on these observations, it is not surprising that male northern elephant seals frequently weigh three tons (2700 kg) and are over four times more massive, on average, than females. The logic here runs as follows: Males that own beaches with large congregations of females will father large numbers of offspring. Males that lose fights will father few or no offspring. As a result, the alleles of territory-owning males will rapidly increase in frequency in the population. If the ability to win fights and produce offspring is determined primarily by body size, then alleles for large body size will have a significant fitness advantage, leading to the evolution of large male size. The fitness advantage is due to sexual selection.

Figure 24.9b provides evidence for intense sexual selection in males. Biologists have marked many of the individuals in the seal population on Año Nuevo to track the lifetime reproductive success of a large number of individuals. As the data show, in this population a few males father a large number of offspring, while most males father few or none. Among females, variation in reproductive success is much lower (**Figure 24.9c**). In this species, most sexual selection is driven by male-male competition rather than female choice.

What Are the Consequences of Sexual Selection? The process component of the theory of sexual selection contends that sexual selection is most intense in the sex that makes the *least* investment in offspring. The biologists who studied the northern elephant seal population have data that is relevant to this prediction. By capturing and weighing females and seal pups throughout the breeding season, the investigators found that adult females weigh about 650 kg on average and give birth to one pup, weighing 50 kg, each year. In their first five weeks of life, the pups routinely gain over 100 kg—entirely from mother's milk. The energy expenditure required to accomplish this growth takes a tremendous toll on the mother. The researchers found that mothers typically lost up to 200 kg, or almost a third of their body mass, during nursing. To breed the following year, females had to regain their lost body weight. The father's parental energy investment, in contrast, was limited to a few minutes spent in copulation and the almost negligible energetic cost of producing sperm.

(a) Males compete for the opportunity to mate with females.

(b) Variation in reproductive success is high in males.

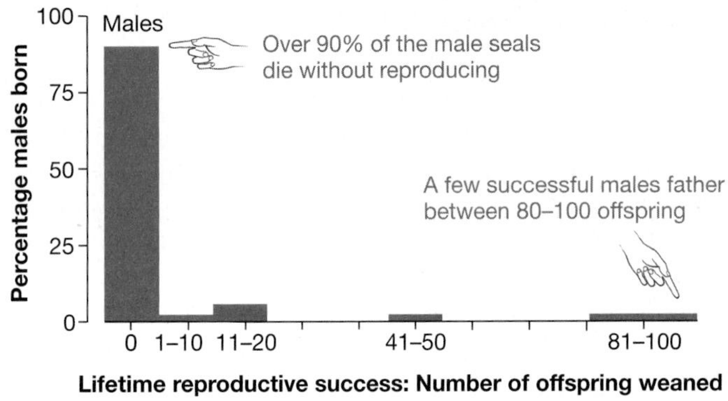

(c) Variation in reproductive success is relatively low in females.

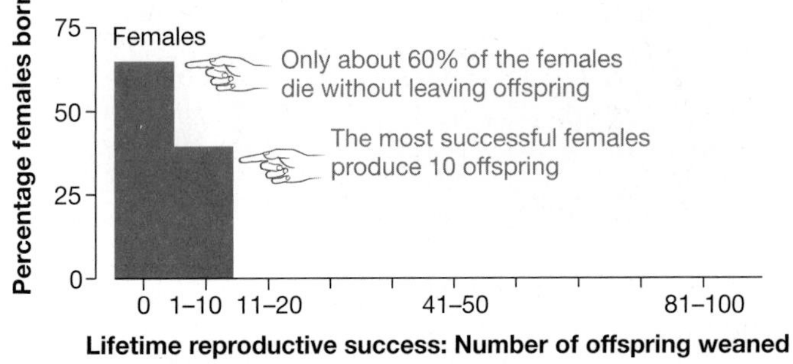

FIGURE 24.9 Sexual Selection on Male Elephant Seals Is Strong
(a) Two bull northern elephant seals compete for the opportunity to mate with females that will breed on this beach. The histograms show that variation in lifetime reproductive success is even higher **(b)** in male northern elephant seals than it is **(c)** in females. Alleles that lead to high reproductive success in males will rapidly increase in frequency.

These data support the hypothesis that the lifetime reproductive ability of elephant seal females is limited by their ability to acquire enough food to support pregnancy and nursing, not by their ability to persuade a male to copulate. In fact, in most animal species, every female that survives to adulthood gets a mate. In contrast, many males do not get any mate at all.

(a) Beetle
(b) Scarlet tanager
(c) Lion
Females
Males

During the breeding season, males of the beetle *Dynastes granti* use their elongated horns to fight over females.

Male scarlet tanagers use their bright coloration in territorial and courtship displays.

Male lions are larger than female lions and have an elaborate ruff of fur called a mane.

FIGURE 24.10 Sexually Selected Traits Are Used to Compete for Mates
Males often have exaggerated traits that they use in fighting or courtship. In many species, females lack these traits. [(b) top ©B. Schorre/VIREO; (b) bottom ©R. & A. Simpson/VIREO] QUESTION How do traits that differ in male and female humans help individuals compete for mates?

The overall message of this analysis is that sexual selection is usually much more intense in males than in females. As a result, it is common to observe that sexually selected traits differ sharply between the sexes. **Sexual dimorphism** ("two-forms") refers to any trait that differs between males and females. **Figure 24.10** illustrates sexually dimorphic traits. They range from weapons that males use to fight over females, such as antlers and horns, to the elaborate ornamentation and behavior used in courtship displays. Humans are sexually dimorphic in size, distribution of body hair, and many other traits. Sexual selection often results in sexual dimorphism. It is a form of natural selection based on the ability to attract mates.

✓CHECK YOUR UNDERSTANDING

Selection can be directional, stabilizing, or disruptive. You should be able to explain how each pattern of natural selection affects genetic variation, and why. Sexual selection occurs on traits that are used in courtship and is based on differential success in obtaining mates. You should be able to define the fundamental asymmetry of sex and explain (1) why males are usually the sex with exaggerated traits used in courtship and (2) how male-male competition for mates differs from female choice of mates.

24.3 Genetic Drift

Both natural selection and sexual selection are directed processes. Natural selection is directed by the environment and results in adaptation. Sexual selection is directed by female choice or male-male competition and results in traits that lead to mating success. **Genetic drift**, in contrast, is undirected. It is any change in allele frequencies that is due to chance in a population. The process is aptly named, because it causes allele frequencies to drift up and down randomly over time. When drift occurs, allele frequencies change due to blind luck—what is formally known as **sampling error.**

To understand why genetic drift occurs, consider the group of people who founded the present population of Pitcairn Island in the South Pacific. The founding event occurred in 1789, when a small band of mutineers led by Fletcher Christian took over the British warship HMS *Bounty* and fled to Pitcairn. The six sailors were joined by two Tahitian men and six Tahitian women. Suppose that six couples formed and raised two children each. How will allele frequencies compare in the founding population that bred versus the next generation?

To answer this question, let's focus on a hypothetical gene A with alleles A_1, A_2, and A_3. Suppose that the six males and six females in the original couples have the genotypes given in

TABLE 24.3 A Thought Experiment on Genetic Drift

The genotypes of the children in this table were generated by flipping a coin to simulate which alleles combined during fertilization. The frequencies of alleles A_1, A_2, and A_3 differ in the two generations, due to genetic drift.

	Father	Mother	Child 1	Child 2
Couple 1	$A_3 A_3$	$A_2 A_3$	$A_3 A_3$	$A_3 A_3$
Couple 2	$A_2 A_3$	$A_2 A_3$	$A_2 A_3$	$A_3 A_2$
Couple 3	$A_1 A_2$	$A_2 A_3$	$A_2 A_3$	$A_1 A_2$
Couple 4	$A_1 A_1$	$A_1 A_2$	$A_1 A_2$	$A_1 A_1$
Couple 5	$A_1 A_2$	$A_3 A_3$	$A_1 A_3$	$A_1 A_3$
Couple 6	$A_1 A_2$	$A_1 A_3$	$A_2 A_3$	$A_2 A_1$

Table 24.3. The gametes formed by each of these parents have an equal chance of carrying either allele. Further, eggs and sperm combine at random when fertilization occurs—meaning that each type of egg and sperm has an equal probability of combining, irrespective of its genotype. To simulate the fertilization process, you can flip a coin to decide which gametes combine. Let heads stand for the first allele listed in the parent's genotype, while tails represents the second allele listed. Doing these coin flips resulted in the offspring genotypes given in the "Child 1" and "Child 2" columns of Table 24.3.

Now let's count up alleles and calculate allele frequencies. Because there are 12 breeding individuals and 12 offspring, the total number of alleles present in each generation (ignoring the two founding individuals who did not breed) is 24. From the data in Table 24.3, the allele frequencies are as follows:

	A_1	A_2	A_3
Allele frequencies in the parents	7/24 = 29.2%	8/24 = 33.3%	9/24 = 37.5%
Allele frequencies in the offspring	7/24 = 29.2%	7/24 = 29.2%	10/24 = 41.6%

Although the frequency of allele A_1 did not change, the frequency of A_2 declined by over 4 percent and the frequency of A_3 increased by over 4 percent, purely by chance. Allele A_3 did not confer higher fitness. Instead, it just got lucky. Random chance caused evolution—a change in allele frequencies in a population. Instead of each allele being sampled in exactly its original frequency when offspring formed, a chance sampling error occurred.

This exercise helps to illustrate several important points:

- *Genetic drift is random with respect to fitness.* The allele frequency changes it produces are not adaptive.
- *Genetic drift is most pronounced in small populations.* If the original population had included only couples 1 and 2, then the frequency of allele A_2 would have declined by 12.5 percent and the frequency of A_3 would have increased by 12.5 percent. The smaller the sample, the larger the sampling error. Conversely, if the population had consisted of 500 couples who produced 1000 offspring (necessitating 2000 coin flips), it is extremely unlikely that drift would produce a change in allele frequency as large as 4 percent. Sampling error is usually insignificant when the sample is large.
- *Over time, genetic drift can lead to the random loss or fixation of alleles.* If allele A_1 continued to be unlucky generation after generation, it would eventually drift to a frequency of 0.0 and be lost. If allele A_3 continued to be lucky for several more generations, it might drift to a frequency of 1.0. When random loss or fixation occurs, genetic variation in the population declines.

Experimental Studies of Genetic Drift

To understand the long-term consequences of genetic drift in small populations, consider an experiment conducted by Warwick Kerr and Sewall Wright in the mid-1950s. These biologists started with a large laboratory population of fruit flies that contained a **genetic marker**—a specific allele that causes a distinctive phenotype. In this case, the marker was the morphology of leg bristles. Fruit flies have bristles on their legs that can be either straight or bent. This difference in leg bristle phenotype depends on a single gene, and Kerr and Wright's lab population contained just two alleles—normal (straight) and "forked" (bent).

To begin the experiment, the researchers set up 96 cages in their lab. Then they placed four adult females and four adult males of the fruit fly *Drosophila melanogaster* in each. They chose individual flies to begin these experimental populations so that the frequency of the normal and forked alleles in each of the 96 starting populations was 0.5. The two alleles do not affect the fitness of flies in the lab environment, so Kerr and Wright could be confident that if changes in the frequency of normal and forked phenotypes occurred, they would not be due to natural selection.

After these first-generation adults bred, Kerr and Wright reared their offspring. Then they randomly chose four males and four females—meaning that they simply grabbed individuals without regard to whether their leg bristles were normal or forked—from each of the 96 offspring populations and allowed them to breed and produce the next generation. The researchers repeated this procedure until all 96 populations had undergone a total of 16 generations. During the entire course of the experiment, no migration from one population to another occurred, and previous studies had shown that mutations from normal to forked are rare. The only evolutionary process operating during the experiment was genetic drift.

Their result? After 16 generations, the 96 populations fell into three groups. Forked leg bristles were found on all of the individuals in 29 of the experimental populations. Due to drift, the forked allele had been fixed in these 29 populations and the normal allele had been lost. In 41 other populations, however, the opposite was true: All individuals had normal bristles. In these populations, the forked allele had been lost due to chance. Both alleles were still present in 26 of the populations. The message of the study is startling: In 73 percent of the experimental populations (70 out of the 96), genetic drift had reduced allelic diversity to zero. Genetic drift decreased genetic variation within populations and increased genetic differences between populations. Is drift important in natural populations as well?

Genetic Drift in Natural Populations

The sampling process that occurs during fertilization and that caused changes in allele frequencies in Pitcairn islanders and fruit flies occurs in every population in every generation. It is particularly important in small populations, where sampling error tends to be high. This is of enormous concern to conservation biologists, because populations all over the planet are being drastically reduced in size by habitat destruction and other human activities. Small populations that occupy nature reserves or zoos are particularly susceptible to genetic drift. If drift leads to a loss of genetic diversity, it could darken the already bleak outlook for some endangered species.

Given enough time, however, drift can be an important factor even in moderately large populations. To drive this point home, consider two types of alleles that were introduced in earlier chapters and that have no effect on fitness. Recall from Chapter 16 that alleles containing silent mutations, usually in the third position of a codon, do not change the gene product. As a result, they have no effect on the phenotype. Yet these alleles routinely drift to high frequency or even fixation over time. Similarly, recall from Chapter 20 that pseudogenes do not code for a product. Although their presence does not affect an individual's phenotype, dozens of pseudogenes in the human genome have reached fixation, due to drift.

It is important to realize that because drift is caused by sampling error, it can occur by *any* process or event that involves sampling—not just the sampling of gametes that occurs during fertilization.

How Do Founder Effects Cause Drift? When a group of individuals emigrates to a new geographic area and founds (establishes) a new population, a **founder event** is said to occur. If the group is small enough, the allele frequencies in the new population are almost guaranteed to be different from those in the source population, due to sampling error. A change in allele frequencies that occurs when a new population is established is called a **founder effect.** As soon as the mutineers and Tahitians set foot on Pitcairn Island and took up residence, evolution had occurred due to drift, via a founder effect. The allele frequencies in the new population were different from those in England or in Tahiti. (Once breeding took place, drift continued to act via sampling errors that occurred during fertilization.)

Fishermen on the island of Anguilla in the Caribbean recently witnessed a founder event involving green iguanas. A few weeks after Hurricane Luis and Hurricane Marilyn swept through the region in September 1995, a large raft composed of downed logs tangled with other debris floated onto a beach on Anguilla. The fishermen noticed green iguanas on the raft and several on shore. Because green iguanas had not previously been found on Anguilla, the fishermen notified biologists. The researchers were able to document that at least 15 individuals had arrived; two years later they were able to confirm that at least some of the individuals were breeding. It is extremely unlikely that allele frequencies in the Anguilla population of green iguanas match those of the source population, thought to be on the islands of Guadeloupe.

Founder events like these have been the major source of populations that occupy islands all over the world, as well as island-like habitats such as mountain meadows, caves, and ponds. Each time a founder event occurs, a founder effect is likely to accompany it, changing allele frequencies via genetic drift.

How Do Population Bottlenecks Cause Drift? If a large population experiences a sudden reduction in size, a **population bottleneck** is said to occur. The term comes from the metaphor of a few individuals passing through the neck of a bottle, by chance. Disease outbreaks, natural catastrophes such as floods or fires or storms, or other events can cause population bottlenecks.

Genetic bottlenecks follow population bottlenecks, just as founder effects follow founder events. A **genetic bottleneck** is a sudden reduction in the number of alleles in a population. Drift occurs during genetic bottlenecks and causes a change in allele frequencies. As an example of a genetic bottleneck, consider the human population of Pingelap Atoll in the South Pacific. On this island, only about 20 people out of a population of several thousand managed to survive the effects of a typhoon and a subsequent famine that occurred around 1775.

The survivors apparently included a person who carried a loss-of-function allele at a gene called *CNGB3*, which codes for a protein involved in color vision. When this allele is homozygous, it causes a serious vision deficit called *achromatopsia*. The condition is extremely rare in most populations, with the homozygous genotype and the affliction occurring in about 0.005 percent of the population. If genotypes at this locus are in Hardy-Weinberg proportions, then $q^2 = 0.00005$ and the frequency of the loss-of-function allele in most populations is about $\sqrt{0.00005}$, or 0.7 percent. In the population that survived the Pingelap Atoll disaster, however, the loss-of-function allele was at a frequency of about 1/40, or 2.5 percent. If the allele was at the typical frequency of 0.7 prior to the population bottleneck, then a huge frequency change occurred during the bottleneck, due to drift.

In today's population on Pingelap Atoll, about 1 in 20 people is afflicted with achromatopsia and the allele is at a frequency of over 20 percent. Because it is extremely unlikely that the loss-of-function allele is favored by directional selection or heterozygote advantage, researchers hypothesize that its frequency in this small population has continued to increase over the past 230 years due to drift.

✓CHECK YOUR UNDERSTANDING

Genetic drift occurs any time allele frequencies change due to chance. Drift occurs during many different types of events, including random fusion of gametes at fertilization, founder events, and population bottlenecks. You should be able to explain why drift leads to a random loss or fixation of alleles and why drift is particularly important as an evolutionary force in small populations.

24.4 Gene Flow

Gene flow is the movement of alleles from one population to another. It occurs when individuals leave one population, join another, and breed. As an evolutionary mechanism, gene flow has one outcome: It equalizes allele frequencies between the source population and the recipient population. When alleles move from one population to another, the populations become more alike.

To see the consequences of gene flow in action, consider data from prairie lupine populations on Mount St. Helens in the state of Washington. Prairie lupine is a perennial, meaning that individuals grow from the same root system year after year. They are fairly short lived, however, with each individual rarely living for more than 5 years. Prairie lupines are most commonly found living at high altitude on volcanoes, in habitats like that shown in **Figure 24.11a.**

The explosion of Mount St. Helens in 1980 created thousands of hectares of mudflows and ash deposits that are ideal habitat for prairie lupines. In 1981 a single individual founded a population on an extensive ash plain in the area most highly affected by the blast. This new population was located over 4 km from the nearest surviving population. In the blast zone near this new population, dozens of other populations eventually appeared.

To study genetic diversity in these populations, biologists collected tissue samples from 532 individuals in 32 populations. The age of each of the study populations was known. The researchers made sure they analyzed tissues from the oldest individuals in populations that had been established just 1 to 3 years prior to the start of the study. These individuals were of special interest, because they were undoubtedly the founders of the new populations. Once tissues from all 532 individuals had

(a) Lupines colonize sites and form populations.

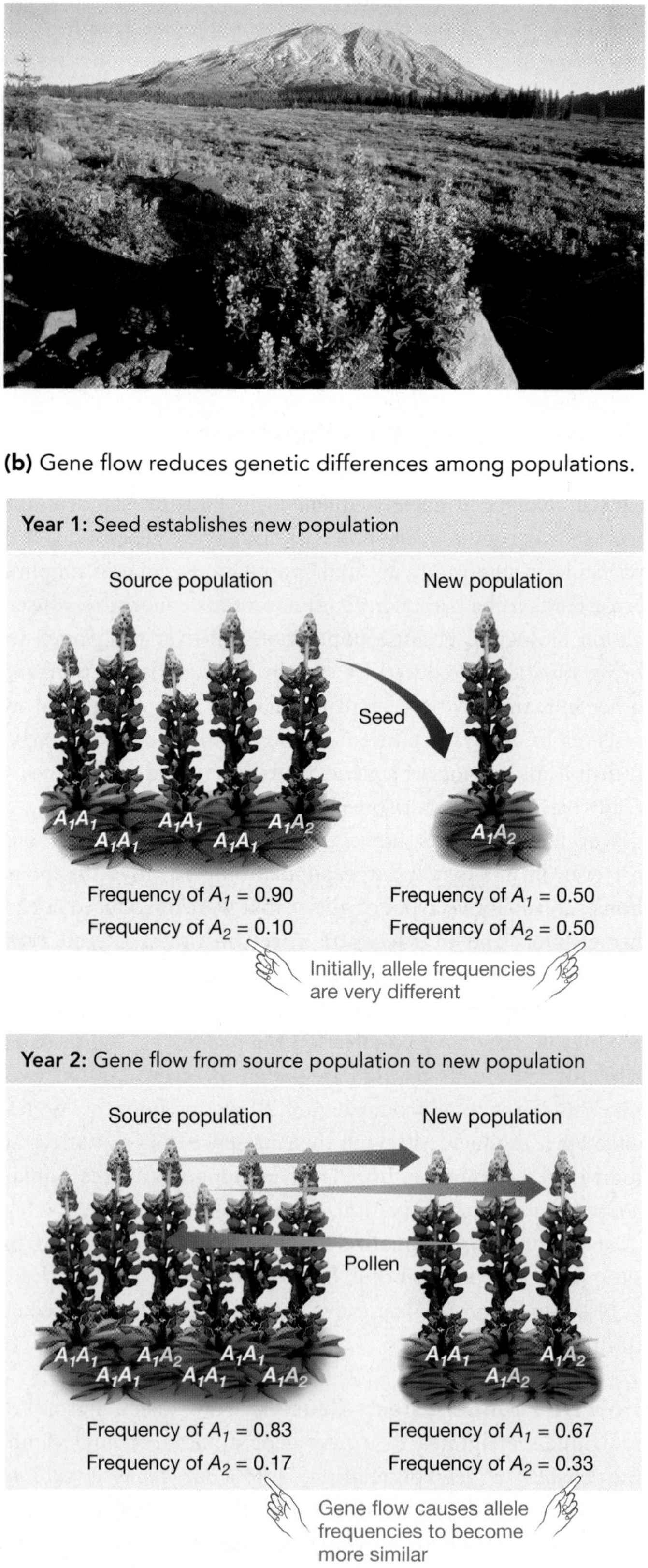

FIGURE 24.11 Gene Flow Equalizes Allele Frequencies between Populations
(a) Several species of lupine grow on the slopes of volcanoes. **(b)** The movement of alleles between populations tends to equalize allele frequencies over time.

been collected, the investigators used the polymerase chain reaction (see Chapter 19) to amplify the alleles of two genes in each individual. These data allowed them to estimate the number of alleles present in each population, as well as the frequency of each allele.

Their results? The data showed that the oldest individuals in newly established populations had allele frequencies that were very different from allele frequencies in the source populations. To interpret this observation, the biologists hypothesize that strong founder effects occurred each time a new lupine population became established. But their data also showed that as populations got older, allele frequencies became progressively more like those in the oldest populations—specifically, like the populations that had survived the eruption. Because selection due to differences in soil and moisture conditions, diseases, and predators present in each population would tend to make populations less similar over time, the researchers doubted that natural selection played a large role in equalizing allele frequencies. Instead, they propose that a continuous flow of alleles between populations, primarily via pollen carried by bees, was responsible for making the populations genetically more alike over time (**Figure 24.11b**).

What impact did gene flow have on the average fitness of individuals in these populations? The answer is not known, but theory suggests that it might vary. If a population had lost alleles due to genetic drift, then the arrival of new alleles via gene flow should increase genetic diversity. In this way, gene flow might increase the average fitness of individuals. But if natural selection had resulted in a population that was highly adapted to a specific habitat, then gene flow from other populations might introduce alleles that have low fitness in that particular environment. In this case, gene flow would have a negative effect on average fitness in the population.

To summarize, the fitness effects of gene flow depend on the situation. But a movement of alleles between populations always tends to reduce genetic differences between them. This latter generalization is particularly important in our own species right now. Large numbers of people from Africa, the Middle East, Mexico, Central America, and Asia are emigrating to the countries of the European Union and the United States. Because individuals from different cultural and ethnic groups are intermarrying frequently and having offspring, allele frequencies in human populations are becoming more similar.

24.5 Mutation

To appreciate the role of mutation as an evolutionary force, let's return to one of the central questions that biologists ask about an evolutionary mechanism: How does it affect genetic variation in a population? Gene flow can increase genetic diversity in a recipient population if new alleles arrive with immigrating individuals. But gene flow can instead decrease genetic variation in the source population if alleles leave with emigrating individuals. The impact of genetic drift on genetic variation is much more clear cut than the impact of gene flow. Genetic drift tends to decrease genetic diversity over time, as alleles are randomly lost or fixed. Similarly, most forms of selection favor certain alleles and lead to a decrease in overall genetic variation.

If most of the evolutionary mechanisms lead to a loss of genetic diversity over time, what restores it? In particular, where do entirely new alleles come from? The answer to both of these questions is **mutation.**

As discussed in Chapter 16, errors by DNA polymerase that occur as a DNA molecule is copied result in changes in the sequence of deoxyribonucleotides. If a mutation occurs in a stretch of DNA that codes for a protein, the changed codon may result in a polypeptide with a novel amino acid sequence. Because errors are inevitable, mutation constantly introduces new alleles into all populations at all genes. Mutation is an evolutionary mechanism that increases genetic diversity in populations.

The second important point to recognize about mutation is that, with respect to the fitness of the affected allele, mutation is a random process. Mutation just happens. Changes in DNA do not occur in a way that tends to increase fitness or decrease fitness. Because most organisms are well adapted to their current habitat, random changes in genes usually result in products that do not work as well as the alleles that currently exist. Stated another way, most mutations result in the production of **deleterious alleles**—alleles that lower fitness. On rare occasions, however, mutation produces a beneficial allele—an allele that allows individuals to produce more offspring. Beneficial alleles should increase in frequency in the population due to natural selection.

Because mutation produces new alleles, it can in principle change the frequencies of alleles through time. That is, the existence of mutation by itself should be able to change allele frequencies and cause evolution. But does mutation occur often enough to make it an important factor in changing allele frequencies of a particular gene? The short answer is no.

Mutation as an Evolutionary Mechanism

To understand why mutation is not a significant mechanism of evolutionary change by itself, consider that the highest mutation rates that have been recorded at individual genes in humans are on the order of 1 mutation in every 10,000 gametes produced by an individual. This rate means that, for every 10,000 alleles produced, on average one will have a mutation. When two gametes combine to form an offspring, then, at most about 1 in every 5000 offspring will carry a mutation at a particular gene. Will mutation affect allele frequencies in a species such as ours? To answer this question, suppose that 195,000 humans live in a population; that 5000 offspring are born one year; and that, at the end of that year, the population numbers 200,000. In a population this size, there is a total of 400,000 copies of each gene. Only one of them is a new allele created by mutation, however.

Over the course of a year, the allele frequency change introduced by mutation is 1/400,000, or 0.0000025 (2.5×10^{-6}). At this rate, it would take 4000 years for mutation to produce a change in allele frequency of 1 percent.

These calculations support the conclusion that mutation does little to change allele frequencies on its own. As an evolutionary mechanism, mutation is extremely slow compared with selection, genetic drift, and gene flow.

Does this conclusion mean that mutation plays *no* role in evolution? The answer is no. For example, because humans have over 30,000 genes, each individual carries 60,000 alleles. Although the rate of mutation per allele may be very low, the total number of alleles is high. Multiplying the estimated number of genes in a human by the average mutation rate per gene suggests that an average person contains 1.6 new alleles created by mutation. In addition, recall from Chapter 12 that these new alleles and existing alleles are constantly being shuffled into new combinations during meiosis and crossing over. Thus, each individual also represents a unique combination of alleles.

The message here is that mutation introduces new alleles into every individual in every population in every generation. And in species that undergo sexual reproduction, meiosis and genetic recombination create variation in terms of the allele combinations present in each individual. Even if selection and drift are eliminating genetic diversity, mutation will act to restore it.

What Role Does Mutation Play in Evolutionary Change?

To get a better appreciation for how mutation affects evolution, consider an experiment conducted by Richard Lenski and colleagues. The experiment was designed to evaluate the role that mutation plays in many genes over many generations. The researchers focused on *Escherichia coli*, a bacterium that, as we have seen, is a common resident of the human intestine. **Figure 24.12** diagrams how Lenski's group set up a large series of populations, each founded with a single cell. Half of the cells contained an allele that allowed them to metabolize the sugar arabinose. The populations founded by these cells were designated Ara^+. The other cells carried an allele that did not allow them to metabolize arabinose. The populations founded by these cells were designated Ara^-. When Ara^+ cells are grown on special plates, they produce colonies that look red. When Ara^- colonies are grown on the same plates, they produce colonies that look white. Under the temperature and nutrient conditions used in the experiment, however, neither Ara^- nor Ara^+ cells had a selective advantage. The presence of the *Ara* allele simply gave the researchers a way to identify particular populations by color. What happened to the descendants of these founding cells over time?

To answer this question, the researchers transferred a small number of cells from each of the populations into a new batch of the same growth medium, under the same light and tempera-

Question: How does average fitness in a population change over time?

Hypothesis: Average fitness increases over time.

Null hypothesis: Average fitness does not increase over time.

Experimental setup:

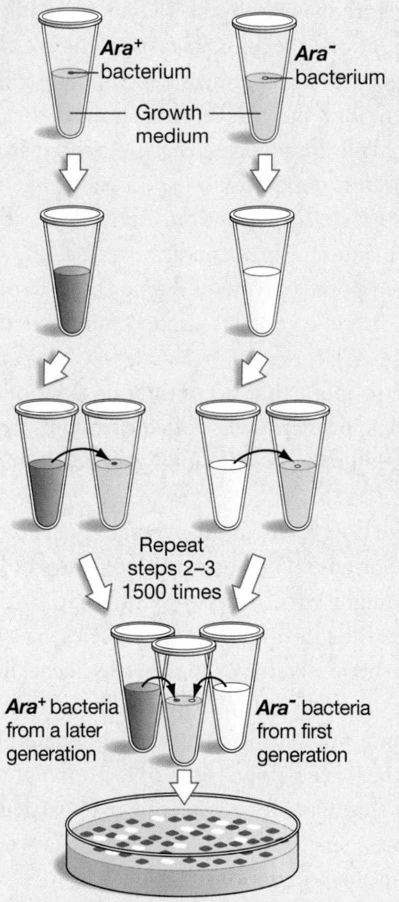

1. Place 10 mL of identical growth medium into many replicate tubes with one bacterium in each, some Ara^+ and some Ara^- so that they can be distinguished by color when grown on special plates.

2. Incubate overnight. Average population in each tube is now 5×10^8 cells. (Cells are red or white when grown on special plates.)

3. Remove 0.1 mL from each tube and move to 10 mL of fresh medium. Freeze remaining cells for later analysis.

4. Put an equal number of cells from generation 1 and a later generation in fresh growth medium (competition experiment).

5. Incubate overnight and count the cells. Which are more numerous?

Prediction: Descendant populations have higher average fitness. (There will be more individuals from descendant than ancestral populations on the plates.)

Prediction of null hypothesis: There will be no difference in fitness between descendant and ancestral populations.

Results:

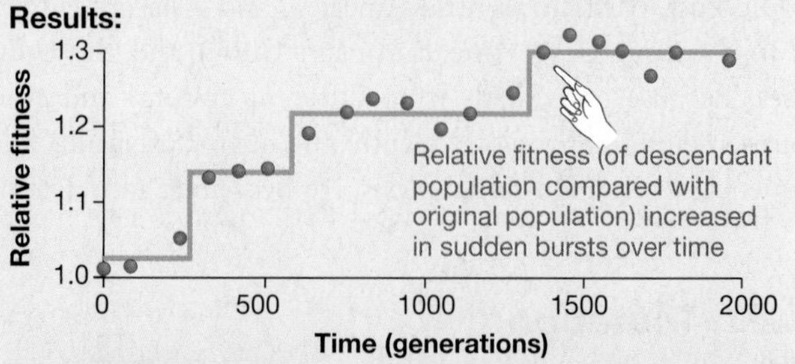

Conclusion: Descendant populations have higher fitness than do ancestral populations.

FIGURE 24.12 An Experiment to Test Changes in Fitness Over Time

TABLE 24.4 A Summary of Evolutionary Mechanisms

Process	Definition and Notes	Effect on Genetic Variation	Effect on Average Fitness
Selection	Certain alleles are favored	Can lead to maintenance, increase, or reduction	Can produce adaptation
Genetic drift	Random changes in allele frequencies: most important in small populations	Tends to reduce, via loss or fixation of alleles	Usually reduces
Gene flow	Movement of alleles between populations; reduces differences between populations	May increase by introducing new alleles; may decrease by removing alleles	May increase by introducing high-fitness alleles or decrease by introducing low-fitness alleles
Mutation	Production of new alleles	Increases by introducing new alleles	Random with respect to fitness; most mutations lower fitness

ture conditions, every day for over four years. In this way, each population grew continuously. Over the course of the experiment, the researchers estimated that each population underwent a total of 10,000 generations. This is the equivalent of over 200,000 years of human evolution. In addition, the biologists saved a sample of cells from each population at regular intervals and stored them in a freezer. The frozen cells served as a fossil record of cells that existed over the 10,000-generation time interval. But because frozen *E. coli* cells resume growth when they are thawed, the "fossil" individuals could be brought back to life.

The strain of *E. coli* used in the experiment is completely asexual and reproduces by cell division. Thus, mutation was the only source of genetic variation in these populations. Although no gene flow occurred, both selection and genetic drift were operating in each population.

Were cells from the older and newer generations of each population different? This question was addressed via competition experiments. Ara^+ cells from one generation and Ara^- cells from a different generation were put together in the same flask. After allowing the cells to grow and compete, the researchers grew them on special plates and counted the number of red versus white colonies. The more numerous population had grown faster, meaning that it was better adapted to the experimental environment. In this way the researchers could measure the fitness of descendant populations relative to ancestral populations. If relative fitness was greater than 1, it meant that recent-generation cells outnumbered older-generation cells when the competition was over.

The data from a series of competition experiments are graphed at the bottom of Figure 24.12. Note that relative fitness increased dramatically—almost 30 percent—over time. But note also that fitness increased in fits and starts. This pattern is emphasized by the solid line on the graph, which represents a mathematical function fitted to the data points.

What caused this stair-step pattern? Lenski's group hypothesizes that genetic drift was relatively unimportant in this experiment because population sizes were so large. Instead, they propose that each jump was caused by a novel mutation that conferred a fitness benefit. Their interpretation is that cells that happened to have the beneficial mutation grew rapidly and came to dominate the population. After a beneficial mutation occurred, the fitness of the population stabilized—sometimes for hundreds of generations—until another random but beneficial mutation occurred and produced another jump in fitness.

The experiment makes an important point: Mutation is the ultimate source of genetic variation. If mutation did not occur, evolution would eventually stop. Without mutation, there is no variation for natural selection to act on.

To summarize, mutation alone is usually inconsequential in changing allele frequencies at a particular gene. When considered across the genome and when combined with natural selection, however, it becomes an important evolutionary mechanism. Mutation is the ultimate source of the genetic variation that makes evolution possible.

Table 24.4 summarizes the four evolutionary forces and their consequences. If one or more of these processes affects a gene, then genotypes will not be in Hardy-Weinberg proportions. But we have yet to consider the effects of another assumption in the Hardy-Weinberg model—that mating takes place at random with respect to the gene in question. What happens when this assumption is violated?

24.6 Inbreeding

In the Hardy-Weinberg model, gametes were picked from the gene pool at random and paired to create offspring genotypes. In nature, however, matings between individuals may not be random with respect to the gene in question. For example, in small populations that are isolated from gene flow, matings between relatives become common. Mating between relatives is called **inbreeding**. Because relatives are likely to share alleles that they inherited from a common ancestor, matings between relatives are nonrandom with respect to all of the genes in the genome. Inbreeding always violates the assumptions of the Hardy-Weinberg principle.

To understand how inbreeding affects populations, let's follow the fate of alleles and genotypes when it occurs. We'll again focus on a single locus with two alleles, A_1 and A_2. Suppose that these

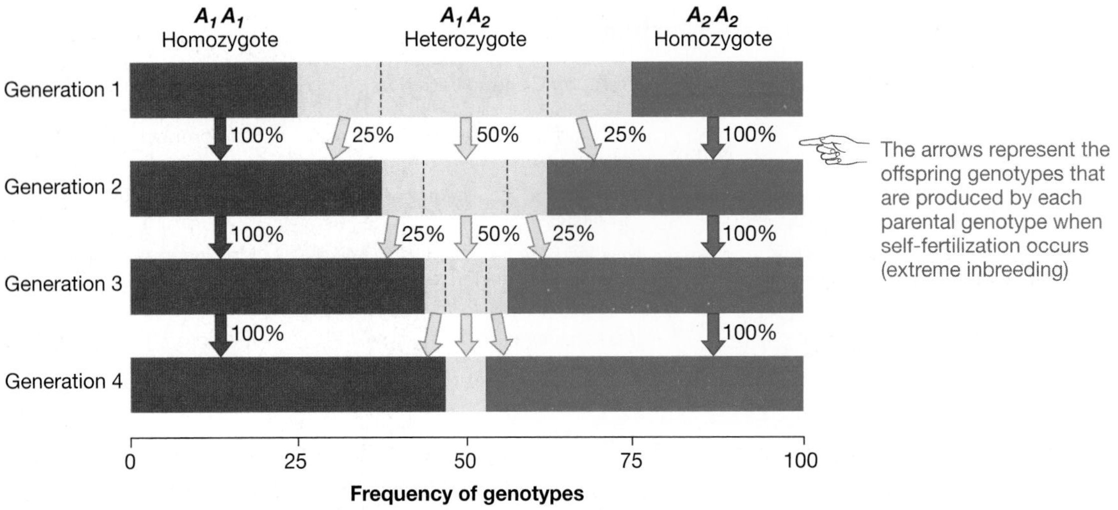

FIGURE 24.13 Inbreeding Increases Homozygosity and Decreases Heterozygosity
The width of the boxes corresponds to the frequency of each genotype. Note that homozygosity has almost doubled in just four generations.

alleles initially have equal frequencies of 0.5. In **Figure 24.13**, the width of the boxes represents the frequency of the three genotypes, which start out at the Hardy-Weinberg ratio of p^2: $2pq$: q^2. But now let's imagine that these individuals don't produce gametes that go into a gene pool. Instead, they self-fertilize. Many flowering plants, for example, contain both male and female organs and routinely self-pollinate. As the arrows in Figure 24.13 show, homozygous parents produce all homozygous offspring. Heterozygous parents, in contrast, produce homozygous and heterozygous offspring in a 1:2:1 ratio. As a result, the homozygous proportion of the population increases each generation, while the heterozygous proportion is halved. At the end of the four generations illustrated in the figure, heterozygotes are rare. No evolution has occurred, however, because allele frequencies have not changed in the population as a whole.

Self-fertilization, or selfing, is the most extreme form of inbreeding; but the same outcome occurs, more slowly, with less extreme forms of inbreeding. Although inbreeding does not change allele frequencies, it does change genotype frequencies. Inbreeding increases the frequency of homozygotes and decreases the proportion of heterozygotes. In essence, inbreeding takes alleles from heterozygotes and puts them into homozygotes.

This outcome is important because of a phenomenon called inbreeding depression. **Inbreeding depression** is a loss of fitness that takes place when homozygosity increases and heterozygosity decreases. Inbreeding depression results from two processes:

1. Many recessive alleles represent loss-of-function mutations. Because these alleles are usually rare, there are normally very few homozygous recessive individuals in a population. Instead, most loss-of-function alleles exist in heterozygous individuals. The alleles have little or no effect when they occur in heterozygotes, because one normal allele usually produces enough functional protein to support a normal phenotype. But inbreeding increases the frequency of homozygous recessive individuals. Loss-of-function mutations are usually deleterious or even lethal when they are homozygous.
2. Many genes—especially those involved in fighting disease—are under intense selection for heterozygote advantage. If an individual is homozygous at these genes, then fitness declines.

As a result, the offspring of inbred matings are expected to have lower fitness than the progeny of outcrossed matings. This prediction has been verified in a wide variety of species, often through laboratory or greenhouse studies that compare the fitnesses of offspring from controlled matings (**Figure 24.14**). As **Table 24.5** shows, inbreeding depression is also pronounced in humans.

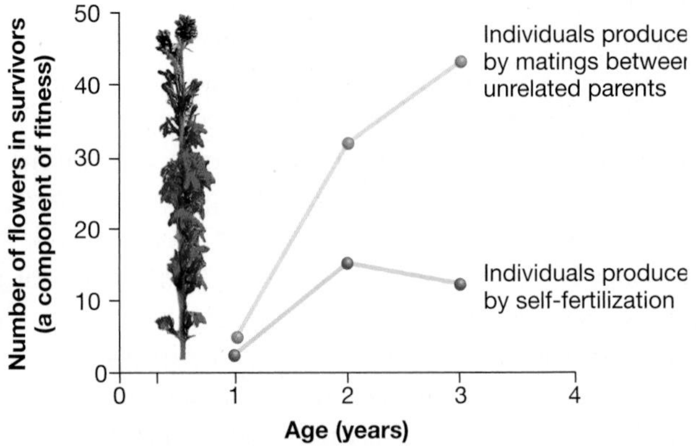

FIGURE 24.14 Inbreeding Depression Occurs in *Lobelia cardinalis*
Inbreeding depression is the fitness difference between non-inbred and inbred individuals. **EXERCISE** Label the parts on this graph that indicate inbreeding depression. Does inbreeding depression increase with age in this species or remain constant throughout life?

Although inbreeding does not change allele frequencies, it is an indirect cause of evolutionary change. Specifically, it increases the rate at which natural selection eliminates deleterious recessive alleles. When natural selection against homozygous individuals and in favor of heterozygous individuals is strong, allele frequencies in an inbred population change rapidly.

Sexual selection also violates the assumption of random mating that underlies the Hardy-Weinberg principle, but its evolutionary consequences are completely different from inbreeding. Inbreeding affects all genes, while sexual selection affects only genes involved in competition for mates. Inbreeding produces changes in genotype frequencies only, but sexual selection produces changes in allele and genotype frequencies. In terms of its evolutionary consequences, then, sexual selection is a form of natural selection rather than a form of nonrandom mating.

Because inbreeding has such deleterious consequences in humans, it is not surprising that many contemporary human societies have laws that forbid marriages between individuals who are more closely related than first cousins. As the end-of-chapter essay shows, analyzing evolutionary processes affects other aspects of human health and welfare as well.

TABLE 24.5 Inbreeding Reduces Fitness in Humans

The percentages reported here give the mortality rate of children produced by first-cousin marriages versus marriages between nonrelatives. In every study, children of first-cousin marriages have a higher mortality rate.

		Deaths	
		Children of First Cousins	Children of Non-relatives
Age	Period	(%)	(%)
Children under 20 (U.S.)	18th–19th century	17.0	12.0
Children under 10 (U.S.)	1920–1956	8.1	2.4
At/before birth (France)	1919–1950	9.3	3.9
Children (France)	1919–1950	14.0	10.0
Children under 1 (Japan)	1948–1954	5.8	3.5
Children 1–8 (Japan)	1948–1954	4.6	1.5

Source: C. Stern, *Principles of Human Genetics* (San Francisco: Freeman, 1973).

ESSAY Evolutionary Theory and Human Health

Even though this chapter is sprinkled with examples of how evolutionary theory can address public health issues, it is still legitimate to ask whether evolutionary theory has practical implications. For example, can efforts to understand the four evolutionary mechanisms help researchers, physicians, and conservationists grapple with real-life problems more successfully?

Later chapters will explore how evolutionary theory is assisting efforts to save endangered species and preserve biodiversity. Here let's examine a case in which an understanding of evolutionary processes helped biomedical researchers better understand an important disease. The illness in question is the genetic disease *cystic fibrosis* (CF), which is caused by a defect in a membrane protein that is important in the functioning of lungs and other organs. The essay at the end of Chapter 6 points out that, in human populations of northern European descent, CF is the most common genetic disease. In these populations, one in 2500 infants is born with CF, and alleles that cause CF are at a frequency of 0.02. Although until recently people with CF rarely lived past their teens and thus were unlikely to reproduce, the incidence of the disease is very high.

Models developed in the mid-1990s suggested that the incidence of 1 in 2500 could not be caused by new mutations alone or by gene flow from other populations. Then why are "defective" alleles at such a relatively high frequency? After all, they are strongly selected against when homozygous. Based on models of allele frequency change, researchers hypothesized that the mutant alleles must have some selective advantage in northern European populations when found in a heterozygous state.

Does evolutionary theory have practical implications?

Shortly after the mathematical analyses of CF's incidence were published, Gerald Pier and coworkers provided dramatic support for the prediction of heterozygote advantage. These biologists showed that the bacterium that causes **typhoid fever**, called *Salmonella typhi*, is not able to enter cells in the gut and cause disease if the cells express mutant forms of the membrane protein involved in CF. Typhoid fever is a potentially fatal illness that is acquired by eating food or drinking water contaminated with feces from an infected person. In the Pier team's experiments, cells in the gut that are homozygous for CF alleles did not allow any *S. typhi* cells to enter, while heterozygous cells admitted only a fraction of

(Continued on next page)

(Essay continued)

the *S. typhi* cells admitted by cells with two copies of the normal allele. The team's experiments suggest that CF alleles are under two distinct types of selection. They are advantageous in the heterozygous state because they protect people against typhoid fever. But they are disadvantageous in the homozygous state, because they cause cystic fibrosis. The observed frequency of 2 percent represents a balance between opposing forms of selection.

The story of CF and typhoid fever is just one of many examples for which an understanding of evolutionary processes has led to a deeper understanding of important diseases. Other chapters in this text highlight evolutionary aspects of the battles against malaria, HIV and AIDS, and antibiotic resistance. What people are calling "Darwinian medicine" is a new but growing topic in biological science.

CHAPTER REVIEW

Summary of Key Concepts

Each of the four evolutionary mechanisms has different consequences. Only natural selection produces adaptation. Genetic drift causes random fluctuations in allele frequencies. Gene flow equalizes allele frequencies between populations. Mutation introduces new alleles.

The four mechanisms of evolution cause allele frequencies to change in populations, but only natural selection results in adaptation. Genetic drift, gene flow, and mutation may introduce a nonadaptive component into evolutionary change.

The Hardy-Weinberg principle acts as a null hypothesis when researchers want to test whether evolution or nonrandom mating is occurring at a particular gene.

Biologists study the consequences of the different evolutionary mechanisms through a combination of mathematical modeling and experimental or observational research. The Hardy-Weinberg principle serves as a null hypothesis in the study of evolution, because it specifies what genotype and allele frequencies are expected to be if mating is random with respect to the gene in question and none of the four evolutionary processes is operating on that gene.

Natural selection occurs in both a wide variety of patterns and a wide variety of intensities. Directional selection may lead to certain alleles becoming fixed—and thus reduces allelic diversity in populations. Stabilizing selection eliminates phenotypes with extreme characteristics and decreases allelic diversity in populations. The rate of evolution under natural selection depends on both the intensity of selection and the amount of genetic variation available.

Sexual selection is a type of selection that occurs when certain traits help males succeed in contests over mates or when certain traits are attractive to females. It is responsible for the evolution of phenotypic differences between males and females.

Genetic drift results from random sampling and is an important evolutionary force in small populations. Drift leads to the random fixation of alleles and tends to reduce overall allelic diversity.

Gene flow is the movement of alleles between populations. Gene flow tends to homogenize allele frequencies and decrease differentiation among populations, but it can also serve as an important source of new variation in populations.

Mutation is too infrequent to be a major cause of allele frequency change. But because mutation continually introduces new alleles at all genes, it is essential to evolution. Without mutation, natural selection and genetic drift would eventually eliminate genetic variation, and evolution would cease.

Web Tutorial 24.1 The Hardy-Weinberg Principle

Web Tutorial 24.2 Three Modes of Natural Selection

Inbreeding changes genotype frequencies but does not change allele frequencies.

Inbreeding, or mating among relatives, is a form of nonrandom mating. Inbreeding does not change allele frequencies, so it is not an evolutionary mechanism. It does, however, change genotype frequencies by leading to an increase in homozygosity and a decrease in heterozygosity. These patterns can accelerate natural selection and can cause inbreeding depression.

Questions

Content Review

1. Why isn't inbreeding considered an evolutionary mechanism?
   a. It does not change genotype frequencies.
   b. It does not change allele frequencies.
   c. It does not occur often enough to be important in evolution.
   d. It does not violate the assumptions of the Hardy-Weinberg principle.

2. Why is random genetic drift aptly named?
   a. It causes allele frequencies to drift up or down randomly.
   b. It is the ultimate source of genetic variability.
   c. It is an especially important mechanism in small populations.
   d. It occurs when populations drift into new habitats.

3. How do sexual selection and inbreeding differ?
   a. Unlike inbreeding, sexual selection changes allele frequencies and affects only genes involved in attracting mates.
   b. Unlike sexual selection, inbreeding changes allele frequencies and involves any mating between relatives—not just self-fertilization.
   c. Unlike inbreeding, sexual selection results from the random fusion of gametes during fertilization. It is particularly important in small populations, where few mates are available.
   d. Inbreeding occurs only in small populations, while sexual selection can occur in any size population.

4. What does it mean when an allele reaches "fixation"?
   a. It is eliminated from the population.
   b. It has a frequency of 1.0.
   c. It is dominant to all other alleles.
   d. It is adaptively advantageous.
5. In what sense is the Hardy-Weinberg principle a null hypothesis, similar to the control treatment in an experiment?
   a. It defines what genotype frequencies should be if nonrandom mating is occurring.
   b. Expected genotype frequencies can be calculated from observed allele frequencies and then compared with the observed genotype frequencies.
   c. It defines what genotype frequencies should be if natural selection, genetic drift, gene flow, or mutation is occurring *and* if mating is random.
   d. It defines what genotype frequencies should be if evolutionary mechanisms are *not* occurring.
6. Mutation is the ultimate source of genetic variability. Why is this statement correct?
   a. DNA polymerase (the enzyme that copies DNA) is remarkably accurate.
   b. "Mutation proposes and selection disposes."
   c. Mutation is the only source of new alleles.
   d. Mutation occurs in response to natural selection. It generates the alleles that are required for a population to adapt to a particular habitat.

Conceptual Review

1. Create a table with three columns and six rows, modeled after Table 24.4. Title the columns "Event," "Effect on Populations," and "Example." In the rows of the "Event" column, write "Mutation," "Gene Flow," "Genetic Drift," "Inbreeding," "Selection," and "None (Hardy-Weinberg Conditions)." Fill in the rest of the table.
2. Directional selection can lead to the fixation of favored alleles. When this occurs, genetic variation is zero and evolution stops. Explain why this rarely occurs.
3. Why does sexual selection often lead to sexual dimorphism? Why are males usually the sex that exhibits exaggerated characteristics?
4. Genotype frequencies occur in Hardy-Weinberg proportions when mutation, selection, drift, and gene flow are not operating on a particular gene, and when mating is random with respect to that gene. Is it possible for one gene in a population *not* to have genotype frequencies that are in Hardy-Weinberg proportions, even though all other genes in the population are in Hardy-Weinberg proportions? Explain why or why not.
5. Explain why small populations become inbred.
6. Explain why genetic drift is much more important in small populations than large populations.

Group Discussion Problems

1. In humans, albinism is caused by loss-of-function mutations in genes involved in the synthesis of melanin, the dark pigment in skin. Only people homozygous for a loss-of-function allele have the relevant phenotype. In Americans of northern European ancestry, albino individuals are present at a frequency of about 1 in 10,000 (or 0.0001). Knowing this genotype frequency, we can calculate the frequency of the loss-of-function alleles. If we let p_2 stand for this frequency, we know that $p_2^2 = 0.0001$; therefore $p_2 = \sqrt{0.0001} = 0.01$. By subtraction, the frequency of normal alleles is 0.99. If these loci conform to the conditions required by the Hardy-Weinberg principle, what is the frequency of "carriers"—or people who are heterozygous for this condition? Your answer indicates the percentage of Caucasians in the United States who carry an allele for albinism.
2. Conservation managers frequently use gene flow, in the form of transporting individuals or releasing captive-bred young, to counteract the effects of drift on small, endangered populations. Explain how gene flow can also mitigate the effects of inbreeding.
3. In 1789 a small band of mutineers led by Fletcher Christian took over the British warship HMS *Bounty*. They fled first to Tahiti and then to tiny Pitcairn Island in the South Pacific, along with a small population of native Tahitians. Some of the sailors married Polynesians there and raised families. All contemporary Pitcairn islanders can trace at least part of their family tree back to the colonization event. What evolutionary forces were at work in this small, isolated population during and after the arrival of the mutineers and Polynesians?
4. You are a conservation biologist charged with creating a recovery plan for an endangered species of turtle. The turtle's habitat has been fragmented into small, isolated but protected areas by suburbanization and highway construction. Some evidence indicates that certain turtle populations are adapted to marshes that are normal, whereas others are adapted to acidic wetlands or salty habitats. Further, some turtle populations number less than 25 breeding adults, making genetic drift and inbreeding a major concern. In creating a recovery plan, the tools at your disposal are captive breeding, the capture and transfer of adults to create gene flow, or the creation of habitat corridors between wetlands to make migration possible. Write a two-paragraph essay outlining the major features of your proposal.

Answers to Multiple-Choice Questions 1. b; 2. a; 3. a; 4. b; 5. d; 6. c

www.prenhall.com/freeman is your resource for the following: Web Tutorials; Online Quizzes and other Online Study Guide materials; Answers to Conceptual Review Questions; Solutions to Group Discussion Problems; Answers to Figure Caption Questions and Exercises; and Additional Readings and Research.

25 Speciation

KEY CONCEPTS

- Speciation occurs when populations of the same species become genetically isolated by lack of gene flow and then diverge from each other due to selection, genetic drift, or mutation.
- Populations can become genetically isolated from each other if they occupy different geographic areas, if they use different habitats within the same area, or if one population is polyploid and cannot breed with the other.
- Populations can be recognized as distinct species if they are reproductively isolated from each other, if they have distinct morphological characteristics, or if they form independent branches on a phylogenetic tree.
- When populations that have diverged come back into contact, several outcomes are possible.

The Bullock's oriole was once considered a member of the same species as Baltimore orioles, because the two populations interbreed in certain regions. Recent analyses, based on the phylogenetic species concept introduced in this chapter, have shown that Bullock's orioles are an independent species.

Although Darwin called his masterwork *On the Origin of Species by Means of Natural Selection*, he actually had little to say about how new species arise. Instead, his data and analyses focused on the process of natural selection and the changes that occur within populations over time. He spent much less time considering changes that occur *between* populations.

Since Darwin, however, biologists have realized that populations of the same species may diverge from each other when they are isolated in terms of gene flow. Recall from Chapter 24 that gene flow equalizes allele frequencies between populations. When gene flow ends, allele frequencies in the isolated populations are free to diverge—meaning that the populations begin to evolve independently of each other. If mutation, selection, and genetic drift cause isolated populations to diverge sufficiently, distinct types, or species, form—that is, the process of **speciation** takes place. Speciation is usually a splitting event that creates two or more distinct species from a single ancestral group (**Figure 25.1**). When speciation is complete, a new branch has been added to the tree of life.

In essence, then, speciation results from genetic isolation and genetic divergence. Isolation results from lack of gene flow, and divergence occurs because selection, genetic drift, and mutation proceed independently in the isolated populations. How does genetic isolation come about? And how do selection, drift, and mutation cause divergence?

This chapter is devoted to exploring these questions. Our first task is to examine how species are defined and identified.

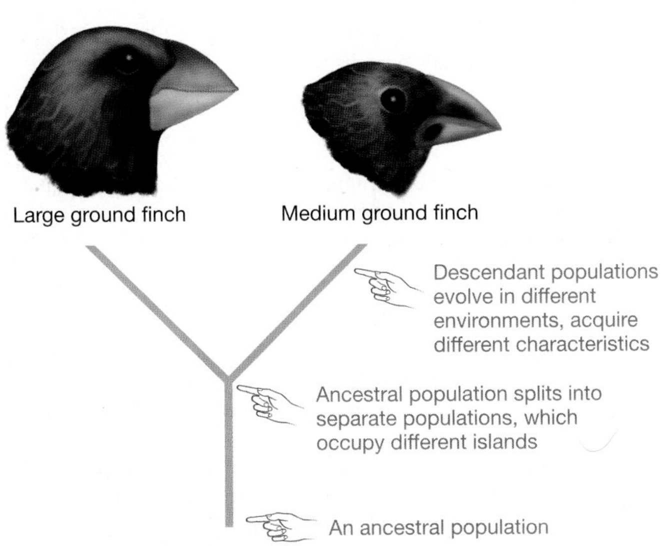
Large ground finch
Medium ground finch
Descendant populations evolve in different environments, acquire different characteristics
Ancestral population splits into separate populations, which occupy different islands
An ancestral population

FIGURE 25.1 Speciation Creates Distinct Populations
The large ground finch and medium ground finch are derived from the same ancestral population. This ancestral population split into two populations that occupied different islands in the Galápagos and were isolated by lack of gene flow. Because the populations began evolving independently, they acquired the distinctive characteristics observed today.

Subsequent sections focus on how speciation occurs in two contrasting situations: when populations occupy the same geographic area and when they are separated into distinct regions. The chapter concludes with a look at a classical question in speciation research: What happens when populations that have been isolated from one another come back into contact? Do they interbreed and merge back into the same species, or do they remain independent and form new species?

25.1 Defining and Identifying Species

Species are distinct types of organisms and represent evolutionarily independent groups. Like the Galápagos finches in Figure 25.1, species are distinct from one another in appearance, behavior, habitat use, or other traits. These characteristics differ among species because their genetic characteristics differ. Genetic distinctions occur because mutation, selection, and drift act on each species independently of what is happening in other populations.

What makes one species "evolutionarily independent" of other species? The answer begins with *lack* of gene flow. As explained in Chapter 24, gene flow eliminates genetic differences among populations. Allele frequencies in populations, and thus the populations' characteristics, become more alike when gene flow occurs between them. If gene flow between populations is extensive and continues over time, it eventually causes even highly distinct populations to coalesce into the unit known as a species. Conversely, if gene flow between populations stops, then mutation, selection, and drift begin to act on the populations independently. If a new mutation creates an allele that changes the phenotype of individuals in one population, there is no longer any way for that allele to appear in the other population. As a result, allele frequencies and other characteristics in the populations diverge. When allele frequencies change sufficiently over time, populations become distinct species.

Formally, then, a **species** is defined as an evolutionarily independent population or group of populations. Even though this definition sounds straightforward, it can be exceedingly difficult to put into practice. How can evolutionarily independent populations be identified in the field and in the fossil record? There is no single, universal answer. Even though biologists agree on the definition of a species, they frequently have to use different sets of criteria to identify them. Three criteria for identifying species are in common use: (1) the biological species concept, (2) the morphospecies concept, and (3) the phylogenetic species concept.

The Biological Species Concept

According to the **biological species concept**, the critical criterion for identifying species is reproductive isolation. This is a logical yardstick because no gene flow occurs between populations that are reproductively isolated from each other. Specifically, if two different populations do not interbreed in nature, or if they fail to produce viable and fertile offspring when matings take place, then they are considered distinct species. Groups that naturally or potentially interbreed, and that are reproductively isolated from other groups, belong to the same species. Biologists can be confident that reproductively isolated populations are evolutionarily independent.

Reproductive isolation can result from a wide variety of events and processes. To organize the various mechanisms that stop gene flow between populations, biologists distinguish (1) **prezygotic** ("before-zygote") **isolation**, which prevents individuals of different species from mating, and (2) **postzygotic** ("after-zygote") **isolation**, in which the offspring of matings between members of different species do not survive or reproduce. In prezygotic isolation, reproductive isolation occurs before mating can occur. In postzygotic isolation, interspecies mating does occur, but any offspring produced have low fitness. **Table 25.1** (page 540) summarizes some of the more important mechanisms of prezygotic and postzygotic isolation.

Although the biological species concept has a strong theoretical foundation, it has disadvantages. The criterion of reproductive isolation cannot be evaluated in fossils or in species that reproduce asexually. In addition, it is difficult to apply when closely related populations do not happen to overlap with each other geographically. In this case, biologists are left to guess whether interbreeding and gene flow would occur if the populations happened to come into contact.

The Morphospecies Concept

How do biologists identify species when the criterion of reproductive isolation cannot be applied? Under the **morphospecies** ("form-species") **concept**, researchers identify evolutionarily

TABLE 25.1 Mechanisms of Reproductive Isolation

Type	Description	Example
Prezygotic Isolation		
Temporal	Populations are isolated because they breed at different times	Bishop pines and Monterey pines release their pollen at different times of the year.
Habitat	Populations are isolated because they breed in different habitats.	Parasites that begin to exploit new host species are isolated from their original population.
Behavioral	Populations do not interbreed because their courtship displays differ.	To attract male fireflies, female fireflies give a species-specific sequence of flashes.
Gametic barrier	Matings fail because eggs and sperm are incompatible.	In sea urchins, a protein called bindin allows sperm to penetrate eggs. Differences in the amino acid sequence of bindin cause matings to fail between closely related populations.
Mechanical	Matings fail because male and female genitalia are incompatible.	In many insects, the male copulatory organ and female reproductive canal fit like a "lock and key." Changes in in either organ initiate reproductive isolation.
Postzygotic Isolation		
Hybrid viability	Hybrid offspring do not develop normally and die as embryos.	When ring-necked doves mate with rock doves, less than 6% of eggs hatch.
Hybrid sterility	Hybrid offspring mature but are sterile as adults.	Eastern meadowlarks and western meadowlarks are almost identical morphologically, but hybrid offspring are largely infertile.

independent lineages by differences in size, shape, or other morphological features. The logic behind the morphospecies concept is that distinguishing features are most likely to arise if populations are independent and isolated from gene flow.

The morphospecies concept is compelling simply because it is so widely applicable. It is a useful criterion when biologists have no data on the extent of gene flow, and it is equally applicable to sexual, asexual, or fossil species. Its disadvantage is that the features used to distinguish species are subjective. In the worst case, different researchers working on the same populations disagree on the characters that distinguish species. For example, some researchers who work on the fossil record of humans argue that the specimens currently named *Homo habilis* and *Homo rudolfensis* (**Figure 25.2**) actually belong to the same species. Disagreements like these often end in a stalemate, because no independent criteria exist for resolving the conflict.

The Phylogenetic Species Concept

The **phylogenetic species concept** is a recent addition to the tools available for identifying evolutionarily independent lineages. It is based on reconstructing the evolutionary history of populations and is increasingly popular among biologists. Proponents of this approach argue that it is widely applicable and precise.

To understand the reasoning behind the concept, recall Darwin's claim, explored in Chapter 23, that all species are related by common ancestry. That chapter also introduced the phylogenetic trees used to represent the genealogical relationships

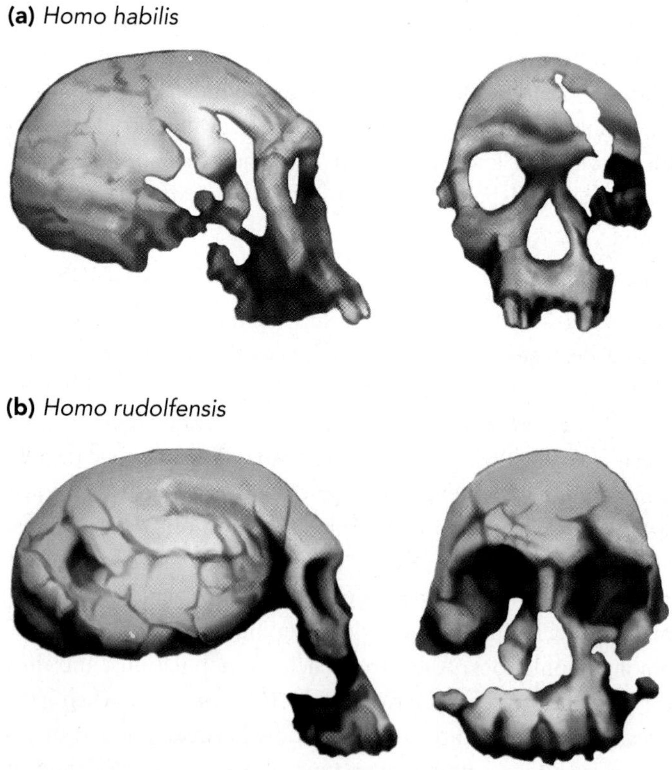

FIGURE 25.2 Morphospecies May Be Difficult to Distinguish from Each Other
Fossils from **(a)** *Homo habilis* and **(b)** *Homo rudolfensis* have been recovered from the same region in Africa, in rocks of the same age. Biologists argue over whether the two populations were distinct enough morphologically to be considered separate species or whether they should be considered the same species.

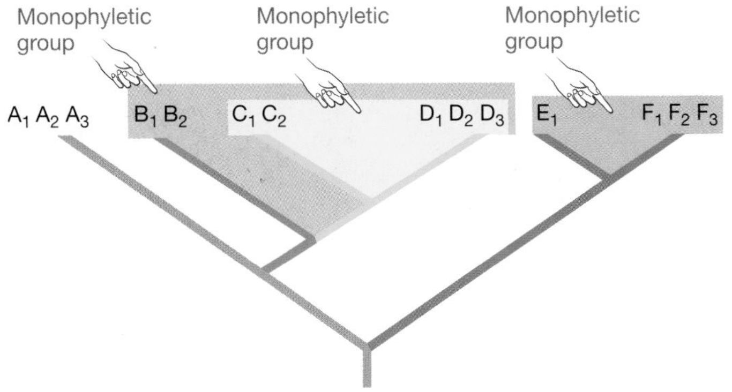

FIGURE 25.3 Monophyletic Groups
The color-coded lineages on this phylogenetic tree are monophyletic because they contain a common ancestor and all its descendants.
EXERCISE This tree has several monophyletic groups other than those indicated. Identify one by picking any branch point (node) on the tree. Then draw a circle around it and all of the branches descending from that point.

among populations. **Figure 25.3** shows such a tree. Here the tips of the branches represent phylogenetic species (A, B, C, D, E, and F), and the clusters at some of the tips (A_1, A_2, A_3, etc.) are populations within those species. On any given evolutionary tree, there are many monophyletic ("one-tribe") groups. A **monophyletic group** consists of an ancestral population, all of its descendants, and *only* those descendants. Three monophyletic groups are color coded on the tree in Figure 25.3, including one (containing species B, C, and D) that overlaps another monophyletic group (the one containing just species C and D). Under the phylogenetic species concept, a species is defined as the smallest monophyletic group in a phylogenetic tree. The populations within each species may be separated geographically, but they are so similar that they do not form independent twigs on the tree.

The phylogenetic species concept has two distinct advantages: (1) It can be applied to any population (fossil, asexual, or sexual), and (2) it is logical because populations are monophyletic only if they are independent of one another and isolated from gene flow. The approach has a distinct disadvantage, however: Carefully estimated phylogenies are available only for a tiny (though growing) subset of populations on the tree of life. Critics of this approach also point out that it would probably lead to recognition of many more species than either the morphospecies or biological species concept. Proponents counter that, far from being a disadvantage, the recognition of increased numbers of species might better reflect the extent of life's diversity.

In actual practice, researchers use all three species concepts summarized here (**Table 25.2**). Conflicts have occurred, however, when different species concepts are applied to the real world. To appreciate this point, let's consider the case of the dusky seaside sparrow.

Species Definitions in Action: The Case of the Dusky Seaside Sparrow

Seaside sparrows live in salt marshes along the Atlantic and Gulf coasts of the United States. Recall from Chapter 1 that scientific names consist of a genus name followed by a species name; the scientific name of this species is *Ammodramus maritimus*. Researchers had traditionally named a variety of seaside sparrow "subspecies" under the morphospecies concept. **Subspecies** are populations that have distinguishing features, such as coloration or calls, but are not considered distinct enough to be called separate species.

Because salt marshes are often destroyed for agriculture or oceanfront housing, by the late 1960s biologists began to be concerned about the future of some seaside sparrow populations. A subspecies called the dusky seaside sparrow (*Ammodramus maritimus nigrescens*) was in particular trouble; by 1980 only six individuals from this population remained. All were males.

At this point government and private conservation agencies sprang into action under the auspices of the Endangered Species Act, a law whose goal is to prevent the extinction of species. The law uses the biological species concept to identify species and calls for the rescue of endangered species through active management. Because current populations of seaside sparrows are physically isolated from one another, and because young seaside sparrows tend to breed near where they hatched, researchers believed that little to no gene flow occurred among populations. Under the biological species concept and morphospecies

TABLE 25.2 A Summary of Species Concepts

Species Concept	Criterion for Recognizing Species	Advantages	Disadvantages
Biological	Reproductive isolation between populations (they don't breed and don't produce viable offspring)	Reproductive isolation = evolutionary independence	Not applicable to asexual or fossil species; difficult to assess if populations do not overlap geographically
Morphospecies	Populations are morphologically distinct	Widely applicable	Subjective (researchers often disagree about how much or what kinds of morphological distinction = speciation)
Phylogenetic	Smallest monophyletic group on phylogenetic tree	Widely applicable; based on testable criteria	Relatively few well-estimated phylogenies are currently available

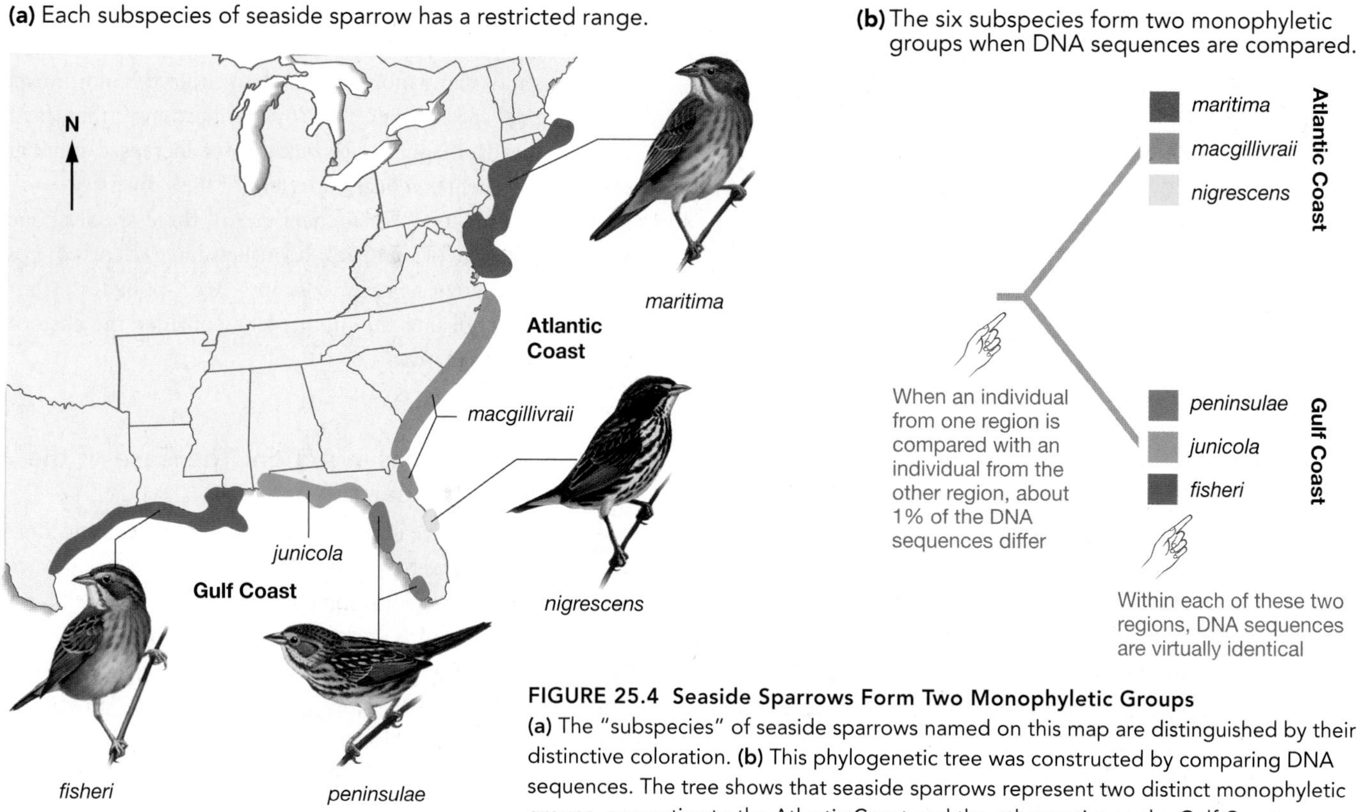

(a) Each subspecies of seaside sparrow has a restricted range.
N
maritima
Atlantic Coast
macgillivraii
nigrescens
junicola
Gulf Coast
fisheri
peninsulae
(b) The six subspecies form two monophyletic groups when DNA sequences are compared.
maritima
macgillivraii
nigrescens
Atlantic Coast
When an individual from one region is compared with an individual from the other region, about 1% of the DNA sequences differ
peninsulae
junicola
fisheri
Gulf Coast
Within each of these two regions, DNA sequences are virtually identical

FIGURE 25.4 Seaside Sparrows Form Two Monophyletic Groups
(a) The "subspecies" of seaside sparrows named on this map are distinguished by their distinctive coloration. **(b)** This phylogenetic tree was constructed by comparing DNA sequences. The tree shows that seaside sparrows represent two distinct monophyletic groups, one native to the Atlantic Coast and the other native to the Gulf Coast.
QUESTION If you were a conservation biologist and could save only two subspecies of seaside sparrows from extinction, would you choose two subspecies from the Atlantic Coast, two from the Gulf Coast, or one from the Atlantic and one from the Gulf? Explain why.

concept, there may be as many as six species of seaside sparrow (**Figure 25.4a**). The dusky seaside sparrow subspecies became a priority for conservation efforts because it was reproductively isolated.

To launch the rescue program, the remaining male dusky seaside sparrows were taken into captivity and bred with females from a nearby subspecies: *A. maritimus peninsulae*. Officials planned to use these hybrid offspring as breeding stock for a reintroduction program. The goal was to preserve as much genetic diversity as possible by reestablishing a healthy population of dusky-like birds. The plan was thrown into turmoil, however, when a different group of biologists estimated the phylogeny of the seaside sparrows by comparing gene sequences. This tree, shown in **Figure 25.4b**, shows that seaside sparrows represent just two distinct monophyletic groups: one native to the Atlantic Coast and the other native to the Gulf Coast. Under the phylogenetic species concept, only two species of seaside sparrow exist. Far from being an important, reproductively isolated population, the phylogeny showed that the dusky sparrow is genetically indistinguishable from other Atlantic Coast sparrows. Further, officials had unwittingly crossed the dusky males with females from the Gulf Coast lineage. Because the goal of the conservation effort was to preserve existing genetic diversity, this was the wrong population to use.

The researchers who did the phylogenetic analysis maintained that the biological and morphospecies concepts had misled a well-intentioned conservation program. Under the phylogenetic species concept, officials should have allowed the dusky sparrow to go extinct and then concentrated their efforts on simply preserving one or more populations from each coast. In this way, the two monophyletic groups of sparrows—and the most genetic diversity—would be preserved. Under the morphospecies concept, however, officials did the right thing by preserving distinct types.

When conservation funding is scarce, life-and-death decisions like these are crucial. Now our task is to consider an even more fundamental question: How do isolation and divergence produce the event called speciation?

✓CHECK YOUR UNDERSTANDING

Species are evolutionarily independent because no gene flow occurs between them and other species. You should be able to describe the advantages and disadvantages of the biological, morphological, and phylogenetic species concepts and explain how each of these approaches to identifying species helps biologists identify populations as evolutionarily independent.

25.2 Isolation and Divergence in Allopatry

Speciation begins when gene flow between populations is reduced or eliminated. Genetic isolation happens routinely when populations become physically separated. Physical isolation, in turn, occurs in one of two ways: dispersal or vicariance. As **Figure 25.5a** illustrates, a population can disperse to a new habitat, colonize it, and found a new population. Alternatively, a new physical barrier can split a widespread population into two or more subgroups that are physically isolated from each other (**Figure 25.5b**). A physical splitting of habitat is called **vicariance.** Speciation that begins with physical isolation via either dispersal or vicariance is known as **allopatric** ("different-homeland") **speciation.** Populations that live in different areas are said to be in **allopatry.**

The case studies that follow address two questions: How do colonization and range-splitting events occur? Answering this question takes us into the field of **biogeography**—the study of how species and populations are distributed geographically. Once populations are physically isolated, how do genetic drift and selection produce divergence?

Dispersal and Colonization Isolate Populations

Peter Grant and Rosemary Grant witnessed a colonization event while working in the Galápagos Islands off the coast of South America. Recall from Chapter 23 that they had been studying medium ground finches on the island of Daphne Major since 1971. In 1982 five members of another species,

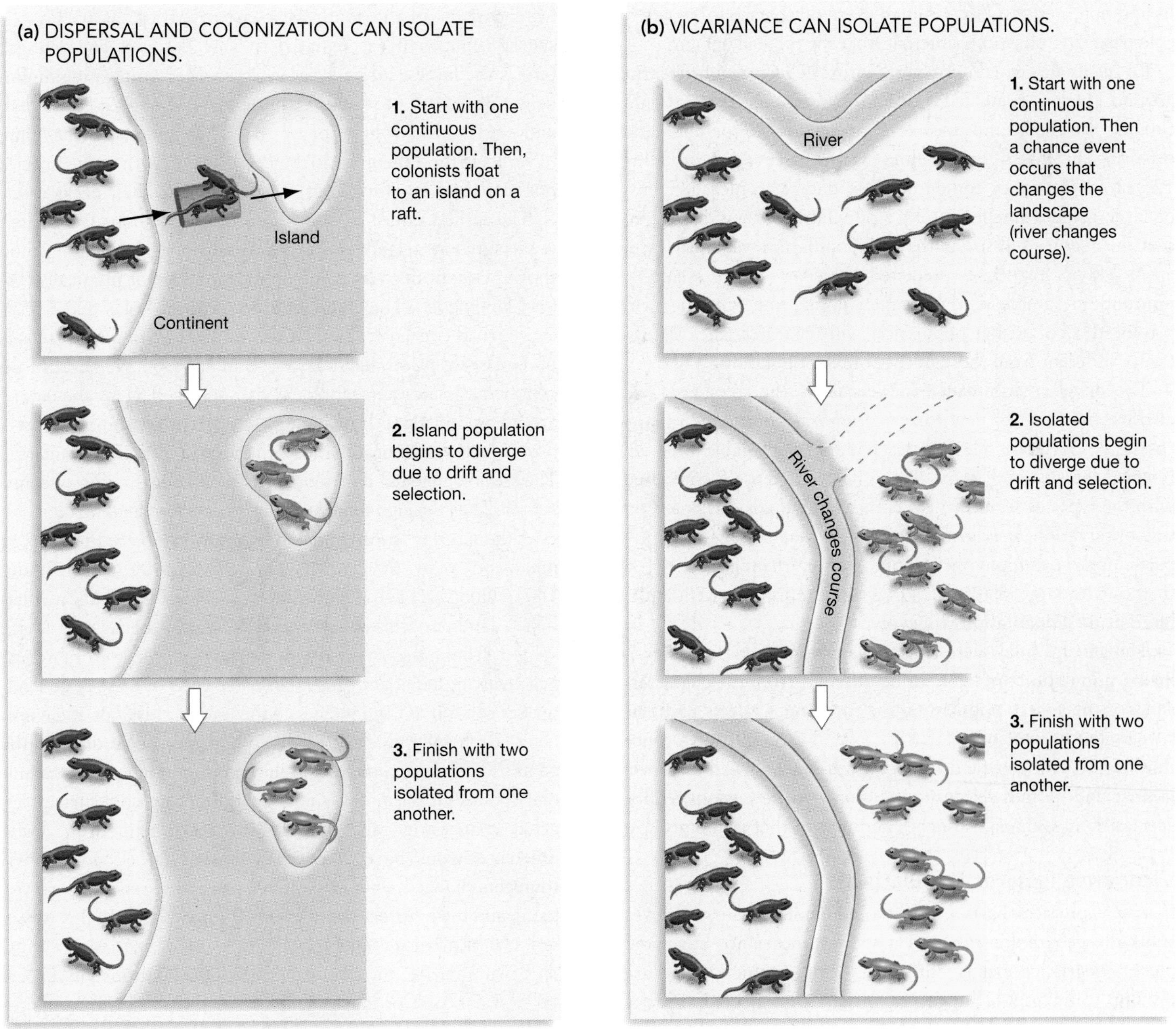

FIGURE 25.5 Allopatric Speciation Begins via Dispersal or Vicariance
(a) When dispersal occurs, colonists establish a new population in a novel location. **(b)** In vicariance, a widespread population becomes fragmented into isolated subgroups.

WEB TUTORIAL 25.1 Allopatric Speciation

called the large ground finch, arrived and began nesting. These colonists had apparently dispersed from a population that lived on a nearby island in the Galápagos. Because finches normally stay on the same island year-round, the colonists represented a new population, allopatric with their source population.

The large ground finches' arrival gave the researchers a chance to test a long-standing hypothesis about dispersal and colonization. Decades ago Ernst Mayr suggested that colonization events are likely to trigger speciation, for two reasons: (1) The physical separation between populations reduces or eliminates gene flow, and (2) genetic drift will cause the old and new populations to diverge rapidly. Drift occurs during the colonization event itself via the founder effect, introduced in Chapter 24. And if the number of individuals in the new population remains small for several generations, genetic drift will continue to alter allele frequencies. In addition, natural selection may cause divergence if the newly colonized environment is different from the original habitat.

To evaluate whether genetic drift occurred when large ground finches colonized Daphne Major, Grant and Grant caught, weighed, and measured most of the parents and offspring produced on Daphne Major over the succeeding 12 years. When they compared these data with measurements of large ground finches in other populations, they discovered that the average bill size in the new population was much larger. As Mayr's hypothesis predicted, the colonists represented a nonrandom sample of the original population. Genetic drift produced a colonizing population with characteristics significantly different from those of the source population.

The novel environment experienced by the colonizers will also expose them to new forms of natural selection. In many cases, the agent of selection is changes in available food. In finches, the size and shape of the beak are closely correlated with the types of seeds or insects that individuals eat. Based on this observation, it is logical to predict that if large seeds are particularly common on Daphne Major, then large ground finches with large beaks will survive and reproduce well, and a large-beaked population will evolve.

The general point here is that the characteristics of a colonizing population are likely to be different from the characteristics of the source population due to founder effects and that subsequent natural selection may extend the rapid divergence that begins with genetic drift. Colonization, followed by genetic drift and natural selection, is thought to be responsible for speciation in Galápagos finches and many other island groups.

Vicariance Isolates Populations

If a new physical barrier such as a mountain range or river splits the geographic range of a species, vicariance has taken place. Vicariance events during the most recent ice age are thought to be responsible for the origin of many of the species observed today. Over the past several million years, glaciers that covered large regions of the northern continents advanced and retreated repeatedly. During most of the glacial advances, the growing ice fields fragmented existing forest and grassland habitats into smaller regions that were isolated from each other by expanses of ice. If populations of the same species had occupied these isolated regions, then an inability of individuals to migrate over the ice fields would have left the populations genetically isolated. The advancing ice would have split turtle, flowering plant, insect, and fish species into geographically separated populations that then might have undergone speciation.

Another example of speciation by vicariance involves the group of large, flightless birds called the ratites. You probably are most familiar with the ostrich; kiwis, emus, rheas, and cassowaries are also ratites. Today, ratites are found in South America, Africa, Australia, and New Zealand. Unfortunately, habitat destruction by humans and hunting recently extinguished ratites called elephant birds that lived on the island of Madagascar, off the southeast coast of Africa, as well as 11 species of moas that were native to New Zealand. The elephant bird is the largest bird species ever recorded, with a maximum height of 3.5 m (11 ft), a maximum mass of 454 kg (1000 lbs), and eggs with a volume of up to 7 L (2 gallons). Moas may have been the tallest birds that ever lived, with some species possibly reaching 4 m (13 ft) when they held their necks erect.

The earliest ratites in the fossil record lived about 150 million years ago, on a landmass called Gondwana. As **Figure 25.6a** shows, Gondwana was made up of a number of physically distinct landmasses. The **theory of plate tectonics** holds that Earth's entire crust—the layer at its surface—is made up of distinct blocks of rock called *plates*. Landmasses, including Gondwana, make up continental plates; crustal blocks that are overlain by the oceans are oceanic plates. All of these plates are in constant motion—probably driven by heat that rises to the crust from Earth's interior. The term **continental drift** is used to describe the motion of continental plates through time, and this process is still going on.

Although the "supercontinent" Gondwana existed for tens of millions of years, the continents in it began to drift apart about 140 million years ago (**Figure 25.6b**). Geologic data indicate that, first, a landmass consisting of today's South America and Africa split off from a slightly smaller landmass composed of Antarctica, Madagascar, India, and Australia. Later, each of these large landmasses split up to form today's configuration of islands and continents. Each splitting event isolated populations of ratites. As the continents moved, climates and the composition of plant and animal species changed. The initial splitting event could have triggered changes in ratite populations via genetic drift. These differences would have been extended by natural selection as environments changed, leading to the evolution of the species observed today and their current distributions (**Figure 25.6c**). In this way, a series of vicariance events triggered a series of speciation events.

To summarize, physical isolation of populations via dispersal or vicariance produces genetic isolation—the first requirement of speciation. When genetic isolation is accompanied by genetic divergence due to mutation, selection, and genetic drift, speciation results.

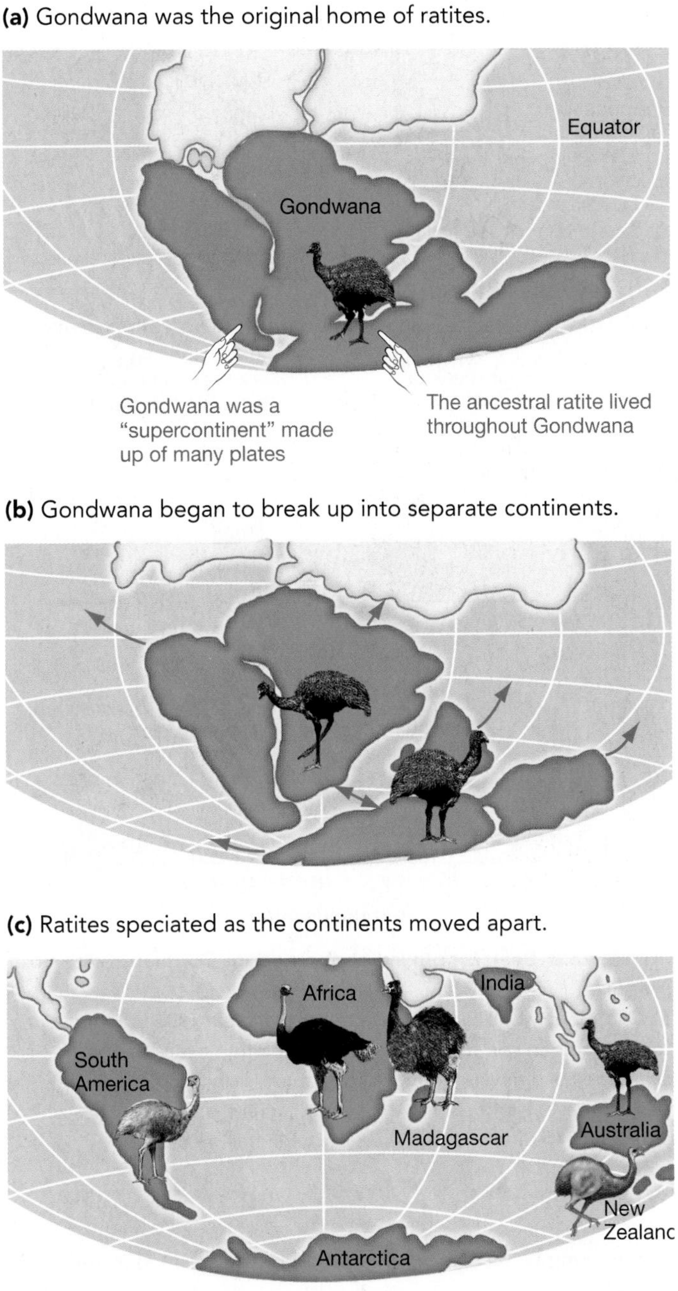
(a) Gondwana was the original home of ratites.
Equator
Gondwana
Gondwana was a "supercontinent" made up of many plates
The ancestral ratite lived throughout Gondwana
(b) Gondwana began to break up into separate continents.
(c) Ratites speciated as the continents moved apart.
Africa
India
South America
Madagascar
Australia
New Zealanc
Antarctica

FIGURE 25.6 Continental Drift Caused Vicariance in Ratite Birds
(a) The ancestors of today's ratites lived on the supercontinent of Gondwana. **(b)** Continental drift led to the breakup of Gondwana. This was a vicariance event that isolated ratite populations. **(c)** Continued continental drift brought the continents and islands into their present positions. Ratites in each area diverged in response to mutation, selection, and drift.

25.3 Isolation and Divergence in Sympatry

When populations or species live in the same geographic area, or at least close enough to one another to make interbreeding possible, biologists say that they live in **sympatry** ("together-homeland"). Traditionally, researchers have predicted that speciation could *not* occur among sympatric populations, because gene flow is possible. The prediction was that gene flow would easily overwhelm any differences among populations created by genetic drift and natural selection.

To illustrate this point, consider research on the water snakes native to the Lake Erie region of North America. Unbanded snakes predominate on islands in the lake (**Figure 25.7a**), but on mainland habitats near the lake, the vast majority of water snakes have a banded coloration (**Figure 25.7b**). Why? In the late 1950s biologists noticed that island snakes bask on limestone rocks. They hypothesized that unbanded snakes are more difficult for predators to distinguish against this surface. A

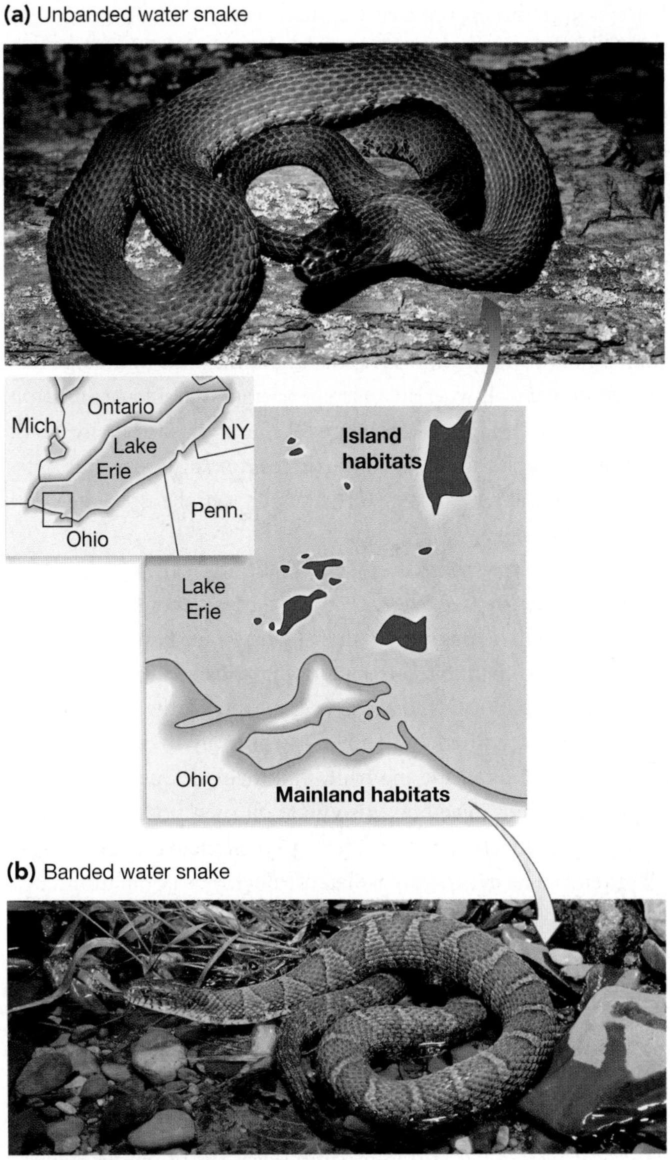
(a) Unbanded water snake
Ontario
Mich.
Lake Erie
NY
Penn.
Ohio
Island habitats
Lake Erie
Ohio
Mainland habitats
(b) Banded water snake

FIGURE 25.7 Gene Flow Prevents Speciation in Island and Mainland Populations of Water Snakes
Water snakes occupy most of the larger islands in Lake Erie, along with mainland habitats around the lake. **(a)** On island habitats, unbanded individuals are common. **(b)** All or most of the water snakes found in mainland habitats are strongly banded. **QUESTION** Banded snakes are most common on the island closest to shore. Why?

researcher recently provided data supporting this hypothesis by showing that, on islands, young unbanded snakes survive much better than do young banded snakes. In mainland habitats, however, limestone rocks are rare. On the shrubby, muddy shores of the mainland, banded snakes may be better camouflaged than unbanded snakes and thus harder for predators to detect.

If natural selection favors different coloration patterns in the two populations, why haven't they diverged to become separate species? The answer is gene flow. Individuals from the mainland swim out and join the island populations on a regular basis. The alleles they introduce when they breed keep the banded coloration pattern at a reasonably high frequency on islands and maintain the two populations as a single species. In this way, gene flow overwhelms the diversifying force of natural selection and prevents speciation. Is this always the case?

Can Natural Selection Cause Speciation Even When Gene Flow Is Possible?

Recently, several well-documented examples have upset the traditional view that **sympatric speciation**—speciation that occurs without physical isolation—is rare or nonexistent. These studies are fueling a growing awareness that, under certain circumstances, natural selection that causes populations to diverge can overcome gene flow and cause speciation. The key realization is that even though populations are not physically isolated, they may be isolated by preferences for different habitats. As an example, let's consider research on the speciation of soapberry bugs.

The soapberry bug is a species of insect, illustrated in **Figure 25.8a**, native to the south-central and southeastern United States. The bugs make their living by feeding on plants in a family called Sapindaceae, including the soapberry tree, serjania vine, and balloon vine. As the figure shows, the bugs feed by piercing fruits with their beaks, reaching in to penetrate the coats of the seeds inside the fruit, and then sucking up the contents of the seeds. The bugs also mate on their host plants. As a result, populations that specialize on a particular type of host plant tend to be genetically isolated from other populations.

The soapberry bug's story began to get interesting when horticulturists brought three new species of sapindaceous plants to North America from Asia in the twentieth century. Two of these *exotic*, or nonnative, species are now cultivated as ornamentals, and one grows as a weed. Soon after these plants were introduced to the New World, soapberry bugs began using them as food. As Figure 25.8a shows, the fruits of the nonnative species are much different from the fruits of native species. Did the arrival of new host plants lead to genetic isolation? If so, have soapberry bugs begun to diverge?

In soapberry bug populations that feed on native host plants, beak length corresponds closely to the size of the host fruit. For example, bugs that feed on species with big fruit tend

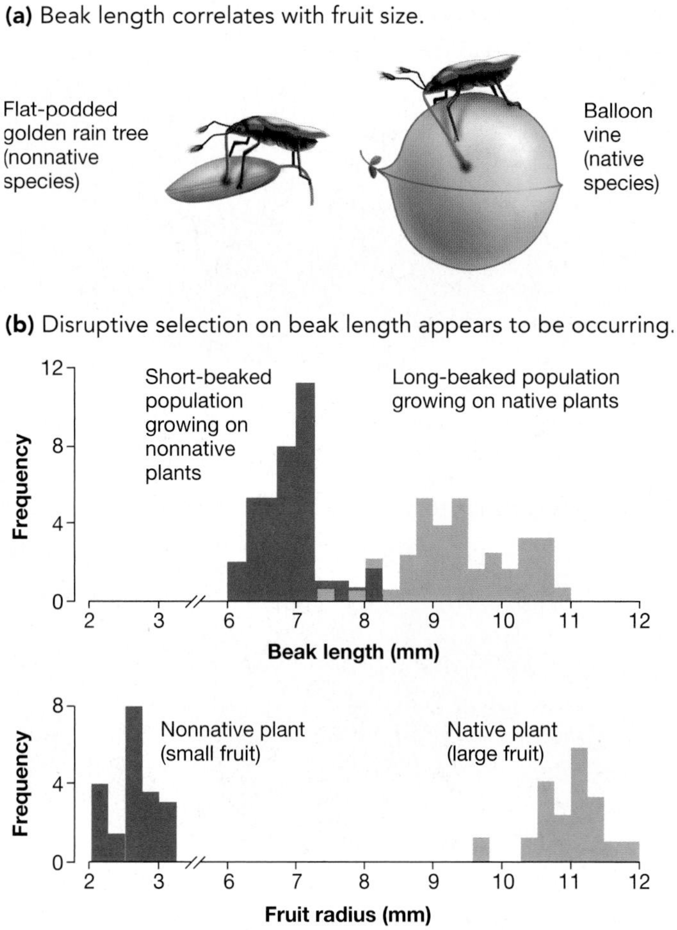

FIGURE 25.8 Beak Length in Soapberry Bugs Correlates with Host Fruit Size
(a) A soapberry bug feeding on a nonnative plant called the flat-podded golden rain tree (left) and one feeding on a native plant called the balloon vine (right). **(b)** Soapberry bug populations that feed on these two host plants have very different beak lengths, which correspond to differences in the size of the host fruit.

to have long beaks. The correlation between fruit size and beak length is logical, because it should allow individuals to reach the seeds inside the fruit efficiently. It also prompted a biologist to ask a simple question: In populations of soapberry bugs that exploit the introduced plant species, have beak lengths evolved to match the sizes of the new fruits? If so, it would imply that populations have become genetically isolated on different host plants and that natural selection is currently causing soapberry bug populations to diverge.

To answer this question, researchers measured large samples of bugs found on both native and nonnative hosts. Some of the resulting data are shown in **Figure 25.8b**. The top histogram shows that bugs collected on native plants growing in south Florida have much longer beaks than those collected on nonnative plants growing in central Florida. The bottom histogram confirms that the fruits of the native species are much larger than the fruits of the introduced species. The data support the argument that soapberry bug populations exploiting exotic species have indeed changed—presumably in response to natural selection for efficient use of host fruits. More specifically, the data support the hypothesis that disruptive selection has occurred on beak length ever since some soapberry bug populations switched to new host plant species.

Will the populations that exploit native and exotic host plants continue to diverge and eventually form new species? Only time, and further research, will tell. But the researchers who are following the story expect that the answer is yes. Because soapberry bugs mate on or near their host plants, switching to a new host species should have reduced gene flow among populations at the same time that it set up disruptive selection. As a result, natural selection may be able to overwhelm gene flow and cause speciation, even when populations are sympatric.

Although the soapberry bug's story might seem localized and specific, the events may be common. Biologists currently estimate that over 3 million insect species exist. Most of these species are associated with specific host plants. Based on these observations, it is reasonable to hypothesize that switching host plants, as soapberry bugs have done, has been a major trigger for speciation throughout the course of insect evolution.

How Can Polyploidy Lead to Speciation?

Based on the theory and data reviewed thus far, it is clear that gene flow and natural selection play important roles in speciation. Can a third evolutionary process—mutation—influence speciation as well? The answer might appear to be no. Chapter 24 emphasized that even though mutation is the ultimate source of genetic variation in populations, it is an inefficient mechanism of evolutionary change. If populations become isolated, it is unlikely that mutation, on its own, could cause them to diverge appreciably.

This view turns out to be naive, however. There is a particular type of mutation, relatively common in plants, that can trigger speciation among sympatric populations. The key is that the mutation reduces gene flow between mutant and normal, or wild-type, individuals. It does so because mutant individuals have more than two sets of chromosomes. This condition is known as **polyploidy**.

To understand why polyploid individuals are genetically isolated from wild-type individuals, consider the mating between a normal individual and a polyploid individual as diagrammed in **Figure 25.9**. In the example given here, the normal individual is diploid, or $2n$, and the polyploid mutant is **tetraploid**, or $4n$. By meiosis, the normal individuals produce haploid gametes, while the mutant individuals produce diploid gametes. These gametes unite to form a **triploid** ($3n$) zygote. Even if this offspring develops normally and reaches sexual maturity, it is rare that it will be able to form functional gametes. The sketch at the bottom of Figure 25.9 illustrates why: When meiosis occurs in a triploid individual, homologous chromosomes cannot synapse and separate correctly. Thus, they are not distributed to daughter cells evenly, and virtually all of the gametes produced by the triploid individual end up with an uneven number of chromosomes. Because its gametes contain a dysfunctional set of chromosomes, the triploid individual is virtually sterile. As a result, the tetraploid and diploid individuals rarely mate and produce fertile offspring. The tetraploid and diploid populations are reproductively isolated.

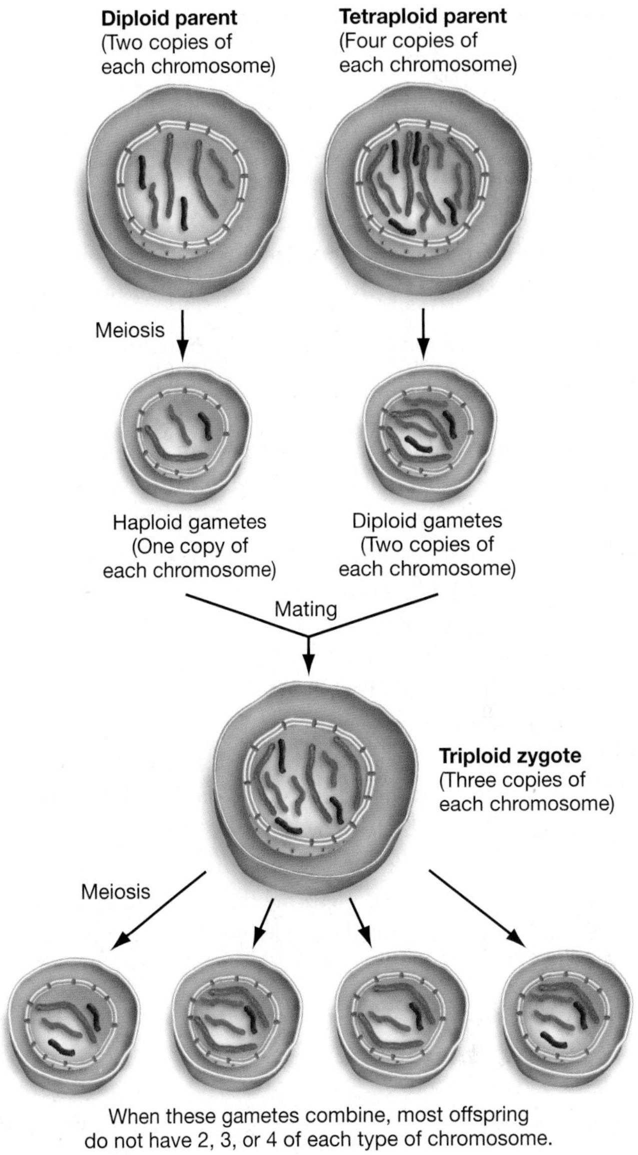
Diploid parent
(Two copies of each chromosome)
Tetraploid parent
(Four copies of each chromosome)
Meiosis
Haploid gametes
(One copy of each chromosome)
Diploid gametes
(Two copies of each chromosome)
Mating
Triploid zygote
(Three copies of each chromosome)
Meiosis
When these gametes combine, most offspring do not have 2, 3, or 4 of each type of chromosome.

FIGURE 25.9 Polyploidy Can Lead to Reproductive Isolation
The mating diagrammed here illustrates why tetraploid individuals are reproductively isolated from diploid individuals. EXERCISE Identify gamete combinations that *would* result in viable offspring. Would the likelihood of triploid individuals producing viable gametes increase or decrease if the number of chromosomes was 10 instead of 4?

How do polyploid individuals form? There are two general mechanisms:

1. **Autopolyploid** ("same-many-form") individuals are produced when a mutation results in a doubling of chromosome number.
2. **Allopolyploid** ("different-many-form") individuals are created when parents that belong to different species mate and produce an offspring with a polyploid chromosome number.

Let's consider specific examples to illustrate how speciation by polyploidy occurs.

Autopolyploidy Although autopolyploidy is thought to be much less common than allopolyploidy, biologists recently documented autopolyploidy in the maidenhair fern. This plant inhabits woodlands across North America. **Figure 25.10** shows that, during the normal life cycle of a fern, individuals alternate between a haploid (n) stage and a diploid ($2n$) stage. Biologists initially set out to do a routine survey of allelic diversity in a population of these ferns. They happened to be examining individuals in the haploid stage and found several individuals that had *two* versions of each gene instead of just one. These individuals were diploid even though they had the "haploid" growth form shown in Figure 25.10. The biologists followed these individuals through their life cycle and confirmed that when the ferns mated, they produced offspring that were tetraploid ($4n$). The researchers had stumbled upon a polyploid mutant within a normal population.

To follow up on the observation, they located the parent of the mutant individuals. This plant turned out to have a defect in meiosis. Instead of producing normal, haploid cells as a result of meiosis, the mutant individual produced diploid cells. These diploid cells eventually led to the production of diploid gametes. Because maidenhair ferns can self-fertilize, they can produce tetraploid offspring. The tetraploid offspring could then self-fertilize or mate with their tetraploid parent. If the process continued, a polyploid population of maidenhair ferns would be established. The polyploid individuals would be genetically isolated from the parental population and thus evolutionarily independent, because tetraploid individuals can breed with other tetraploids but not with diploids. If genetic drift and selection then caused the two populations to diverge, speciation would be under way.

It is no exaggeration to say that this autopolyploidy study documented the critical first step in speciation—the establishment of genetic isolation. In this population of maidenhair ferns, as in soapberry bugs, speciation is under way right before our eyes.

Allopolyploidy New tetraploid species may be created when two diploid species hybridize. **Figure 25.11** illustrates the sequence of events involved. If a diploid offspring that forms from a mating between two different species has chromosomes that do not pair normally during meiosis, the offspring is sterile. But if a mutation occurs that doubles the chromosome number, then homologs synapse, meiosis can proceed, and

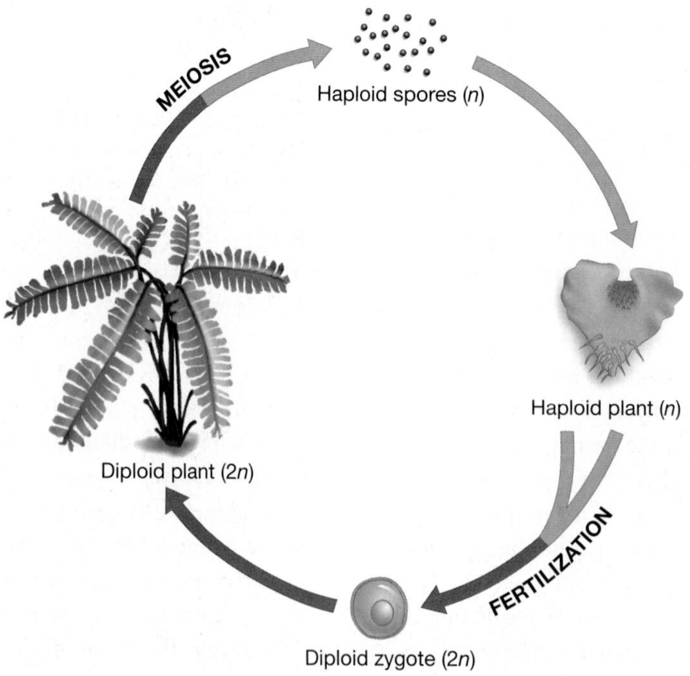

FIGURE 25.10 The Life Cycle of Maidenhair Ferns
Like all ferns, maidenhair ferns have a haploid stage and a diploid stage during their life cycle.

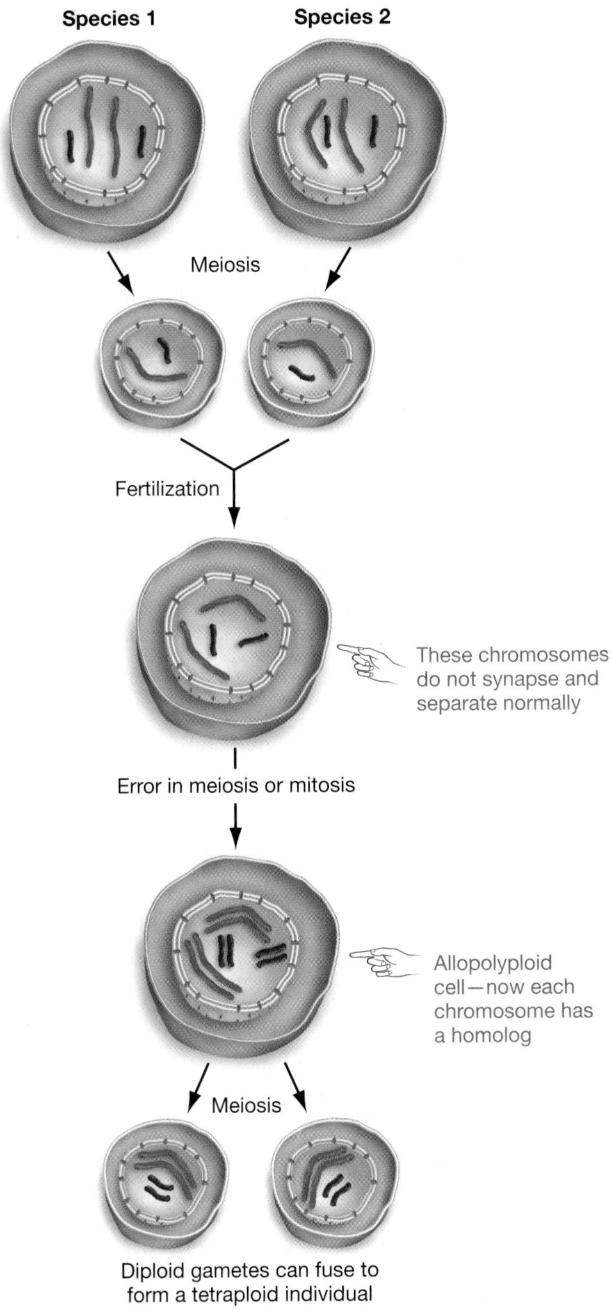
Species 1
Species 2
Meiosis
Fertilization
These chromosomes do not synapse and separate normally
Error in meiosis or mitosis
Allopolyploid cell—now each chromosome has a homolog
Meiosis
Diploid gametes can fuse to form a tetraploid individual

FIGURE 25.11 Allopolyploidy May Occur after Two Species Hybridize Allopolyploid individuals contain chromosomes from two different species.

diploid gametes are produced. When diploid gametes fuse during self-fertilization, a tetraploid individual results.

Exactly this train of events occurred repeatedly after three European species of weedy plants in the genus *Tragopogon* were introduced to western North America in the early 1900s. In 1950 a biologist described the first of two tetraploid species that have been discovered. Both were clearly the descendants of the introduced diploids. Follow-up work has shown that at least one of the new tetraploid species is expanding its geographic range. Further, the nature and amount of genetic variation in the tetraploid species suggests that the allopolyploid sequence occurred repeatedly—meaning each tetraploid species of *Tragopogon* originated independently multiple times.

Because many diploid species of plants have close relatives that are polyploid, speciation by polyploidization is thought to be particularly important in plant evolution. Two properties of plants have been noteworthy in making this mode of speciation possible: (1) In plants, reproductive cells and somatic cells are not separated early in development, as they are in animals. Instead, plant somatic cells that have undergone many rounds of mitosis can undergo meiosis and produce gametes. If sister chromatids fail to separate during anaphase of one of these mitotic divisions, it can result in a tetraploid cell line that later undergoes meiosis to form diploid gametes. (2) The ability of some plant species to self-fertilize makes it possible for diploid gametes to fuse and create genetically isolated tetraploid populations.

To summarize, mutation can create instant species in the form of polyploids. Speciation by polyploidization breaks two long-held "rules" about speciation: It occurs in sympatry, and it is not slow. Speciation by polyploidy is virtually instantaneous.

✓CHECK YOUR UNDERSTANDING

Speciation occurs when populations become isolated genetically and then diverge due to selection, genetic drift, or mutation. You should be able to (1) give an example of at least three events that lead to the genetic isolation of populations and (2) explain why selection and drift cause the populations in each example to diverge.

25.4 What Happens When Isolated Populations Come into Contact?

Suppose two populations that have been isolated come into contact again. If divergence has taken place and if divergence has affected when, where, or how individuals in the populations mate, then it is unlikely that interbreeding will take place. In cases such as this, prezygotic isolation exists. When it does, mating between the populations is rare, gene flow is minimal, and the populations continue to diverge.

But what if prezygotic isolation does not exist, and the populations begin interbreeding? The simplest outcome is that the populations fuse over time, as gene flow erases any distinctions between them. Several other possibilities exist, however. Let's explore three of them: reinforcement, hybrid zones, and speciation by hybridization.

Reinforcement

If two populations have diverged extensively and are distinct genetically, it is reasonable to expect that their hybrid offspring will have lower fitness than their parents. The logic here is that if organisms have evolved distinctive developmental sequences or reproductive systems, then a hybrid offspring will not be able to develop or reproduce normally. Recall from Table 25.1 that hybrid offspring may die early in development or survive to sexual maturity but be infertile. In such cases, postzygotic isolation exists. When it occurs, there should be strong natural selection against interbreeding. The hypothesis is that hybrid offspring represent a wasted effort on the part of parents. Individuals that do not interbreed, due to a different courtship ritual or pollination system or other form of prezygotic isolation, should be favored because they produce more viable offspring.

Natural selection for traits that isolate populations in this way is called **reinforcement**. The name is descriptive because the selected traits reinforce differences that developed while the populations were isolated from one another.

Some of the best data on reinforcement come from laboratory studies of closely related fruit fly species in the genus *Drosophila*. Researchers recently analyzed a large series of experiments that tested whether members of closely related fly species are willing to mate with one another. The biologists found an interesting pattern. If closely related species are sympatric—meaning that they live in the same area—individuals from the two species are seldom willing to mate with one another. But if the species are allopatric—meaning that they live in different areas—then individuals are often willing to mate with one another. This is exactly the pattern that we would expect to observe if reinforcement is occurring. The pattern is logical because natural selection can act to reduce mating between species only if their ranges overlap. Thus, it is reasonable to find that sympatric species exhibit prezygotic isolation but that allopatric species do not. There is a long-standing debate, however, over just how important reinforcement is in groups other than the genus *Drosophila*.

Hybrid Zones

Hybrid offspring are not always dysfunctional. Frequently they are perfectly capable of mating and producing offspring and have features that are intermediate between those of the two parental populations. When this is the case, hybrid zones can form. A **hybrid zone** is a geographic area where interbreeding occurs and hybrid offspring are common. Depending on the fitness of hybrid offspring and the extent of breeding between parental species, hybrid zones can be narrow or wide, and long or short lived. As an example of how researchers analyze the dynamics of hybrid zones, let's consider recent work on two bird species.

Townsend's warblers and hermit warblers live in the coniferous forests of North America's Pacific Northwest. In southern Washington state, where their ranges overlap, the two species hybridize extensively. As **Figure 25.12a** shows, hybrid offspring have characteristics that are intermediate relative to the two parental species. To explore the dynamics of this hybrid zone, a team of biologists examined gene sequences in the mitochondrial DNA (mtDNA) of a large number of Townsend's, hermit, and hybrid warblers collected from forests throughout the region. The team found that each of the parental species has certain species-specific mtDNA sequences. This result allowed the researchers to infer how hybridization was occurring. To grasp the reasoning here, it is critical to realize that mtDNA is maternally inherited in most animals and plants. If a hybrid individual has Townsend's mtDNA, its mother had to be a Townsend's warbler while its father had to be a hermit warbler. In this way, identifying mtDNA types allowed the research team to infer whether Townsend's females were mating with hermit males, or vice versa, or both.

Their data presented a clear pattern: Most hybrids form when Townsend's males mate with hermit warbler females. One of the investigators followed up on this result with experiments showing that Townsend's males are extremely aggressive in establishing territories and that they readily attack hermit warbler males. Hermit males, in contrast, do not challenge Townsend's males. The hypothesis, then, is that Townsend's males invade hermit territories, drive off the hermit males, and mate with hermit females.

The team also found something completely unexpected. When they analyzed the distribution of mtDNA types along the Pacific Coast and in the northern Rocky Mountains, they found that many Townsend's warblers actually had hermit mtDNA. **Figure 25.12b** shows that in some regions—such as the larger islands off the coast of British Columbia—*all* of the warblers had hermit mtDNA, even though they looked like full-blooded Townsend's warblers. To explain this result, the team hypothesized that hermit warblers were once found as far north as Alaska and that Townsend's warblers have gradually taken over their range. Their logic is that repeated mating with Townsend's warblers over time made the hybrid offspring look more and more like Townsend's, even while maternally inherited mtDNA kept the genetic record of the original hybridization event intact.

If this hypothesis is correct, then the hybrid zone should continue moving south. If it does so, hermit warblers may eventually become extinct. In many cases, however, hybridization does not lead to extinction but rather leads to the opposite—the creation of new species.

New Species through Hybridization

A team of researchers recently examined the relationships of three sunflower species native to the American West: *Helianthus annuus*, *H. petiolaris*, and *H. anomalus*. The first two of these species are known to hybridize in regions where their ranges overlap. The third species, *H. anomalus*, resembles these hybrids. In fact, because some gene sequences in *H. anomalus* are remarkably similar to those found in *H. annuus*, while other gene sequences are almost identical to those found in *H. petiolaris*, biologists have suggested that *H. anomalus* originated in hybridization between *H. annuus* and *H. petiolaris* (**Figure 25.13**).

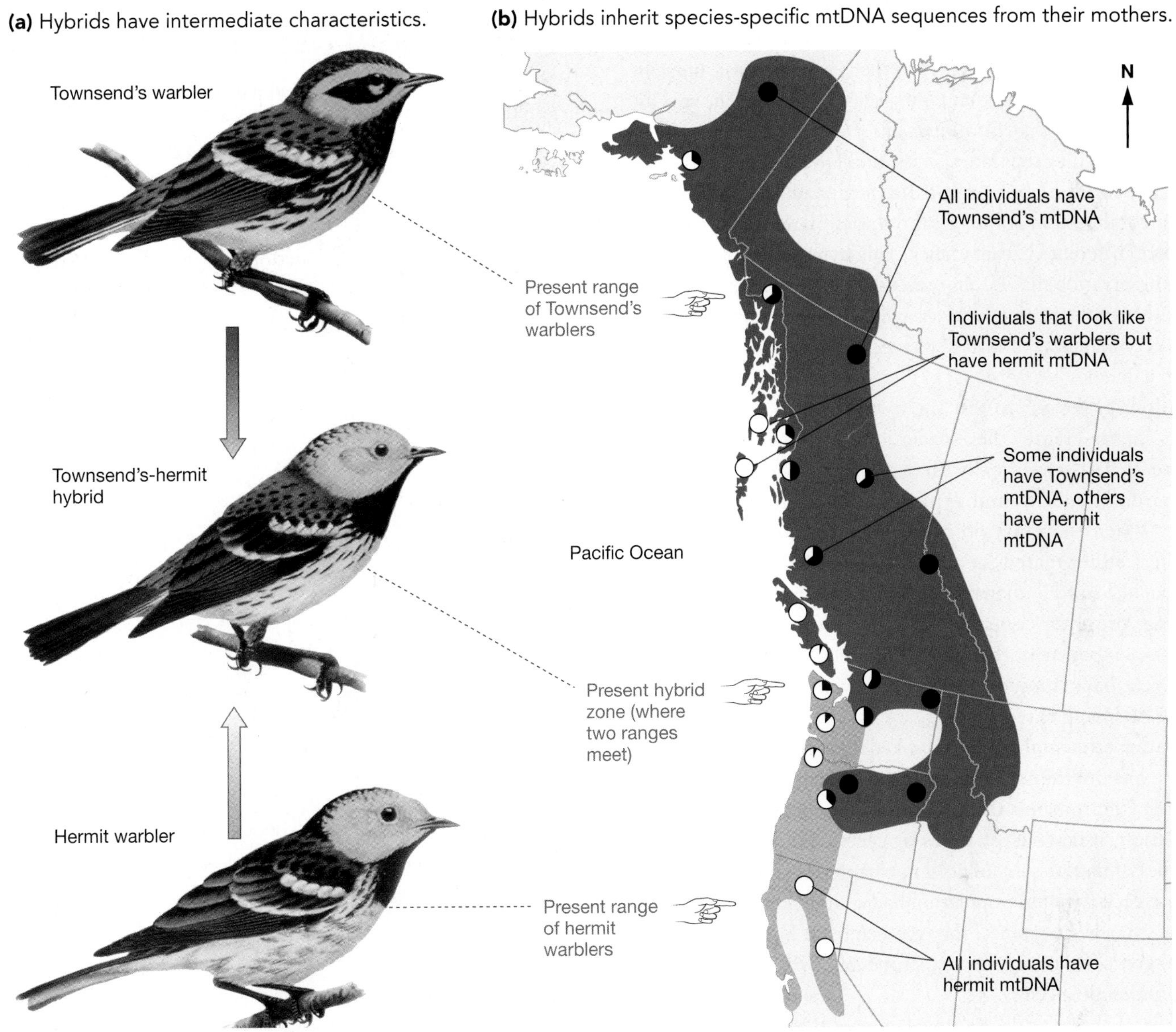
(a) Hybrids have intermediate characteristics.
(b) Hybrids inherit species-specific mtDNA sequences from their mothers.
Townsend's warbler
Townsend's-hermit hybrid
Hermit warbler
N
All individuals have Townsend's mtDNA
Present range of Townsend's warblers
Individuals that look like Townsend's warblers but have hermit mtDNA
Some individuals have Townsend's mtDNA, others have hermit mtDNA
Pacific Ocean
Present hybrid zone (where two ranges meet)
Present range of hermit warblers
All individuals have hermit mtDNA

FIGURE 25.12 Analyzing a Hybrid Zone
(a) When Townsend's warblers and hermit warblers hybridize, the offspring have intermediate characteristics.
(b) Map showing the current range of Townsend's and hermit warblers. The small pie charts show the percentage of individuals with Townsend's warbler mtDNA (in black) and hermit warber mtDNA (in white).

Hybridization hypothesis for the origin of a new species:
x
?
Helianthus annuus
H. petiolaris
H. anomalus

FIGURE 25.13 Hybridization Occurs between Sunflower Species
The sunflower *Helianthus anomalus* may have originated in hybridization events between *H. annuus* and *H. petiolarus.*

All three species have the same number of chromosomes, so neither allopolyploidy nor autopolyploidy was involved.

If this interpretation is correct, then hybridization must be added to the list of ways that new species can form. The specific hypothesis here is that *H. annuus* and *H. petiolaris* were isolated and diverged as separate species, and later began interbreeding. The hybrid offspring created a third, new species that had unique combinations of alleles from each parental species and therefore different characteristics. This hypothesis is supported by the observation that *H. anomalus* grows in much drier habitats than either of the parental species and is distinct in appearance. A unique combination of alleles allowed *H. anomalus* to thrive in certain habitats.

Biologists set out to test the hybridization hypothesis by trying to re-create the speciation event experimentally (**Figure 25.14**). Specifically, they mated individuals from the two parental species and raised the offspring in the greenhouse. When these hybrid individuals were mature, the researchers either mated the plants to other hybrid individuals or "backcrossed" them to *H. annuus* individuals. This breeding program continued for four more generations before the experiment ended. Ultimately, the experimental lines were backcrossed twice, and they were mated to other hybrid offspring three times.

The experimental hybrids looked like the natural hybrid species, but did they resemble them genetically? To answer this question, the research team constructed genetic maps of each population, using a large series of genetic markers, similar to the types of markers introduced in Chapter 19 and Chapter 20. Because each parental population had a large number of unique markers in their genomes, the research team hoped to identify which genes found in the experimental hybrids came from which parental species.

Their results are diagrammed in the "Results" section of Figure 25.14. The bottom bar in the illustration represents a region called S in the genome of the naturally occurring species *Helianthus anomalus*. As the legend indicates, this region contains three sections of sequences that are also found in *H. petiolaris* and two that are also found in *H. annuus*. The top bar shows the composition of this same region in the genome of the experimental hybrid lines. The key observation here is that the genetic composition of the synthesized hybrids matches that of the natural species.

In effect, the researchers had succeeded in re-creating a speciation event. Their results provide strong support for the hybridization hypothesis for the origin of *H. anomalus*. The data also highlight the dynamic range of possible outcomes as a result of secondary contact of two populations: fusion of the populations, reinforcement of divergence, founding of stable hybrid zones, extinction of one population, or the creation of new species.

Question: Can new species arise by hybridization between existing species?

Hypothesis: *Helianthus anomalus* originated by hybridization between *H. annuus* and *H. petiolaris.*

Null hypothesis: *Helianthus anomalus* did not originate by hybridization between *H. annuus* and *H. petiolaris.*

Experimental setup:

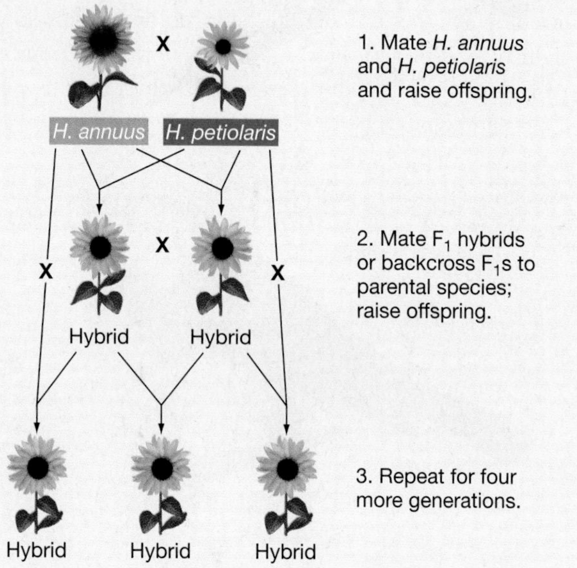
X
H. annuus
H. petiolaris
1. Mate H. annuus and H. petiolaris and raise offspring.
X
X
X
2. Mate F1 hybrids or backcross F1s to parental species; raise offspring.
Hybrid
Hybrid
3. Repeat for four more generations.
Hybrid
Hybrid
Hybrid

Prediction: Experimental hybrids will have the same mix of *H. annuus* and *H. petiolaris* genes as natural *H. anomalus.*

Prediction of null hypothesis: Experimental hybrids will not have the same mix of *H. annuus* and *H. petiolaris* genes as natural *H. anomalus.*

Results:

DNA comparison of a chromosomal region called S:

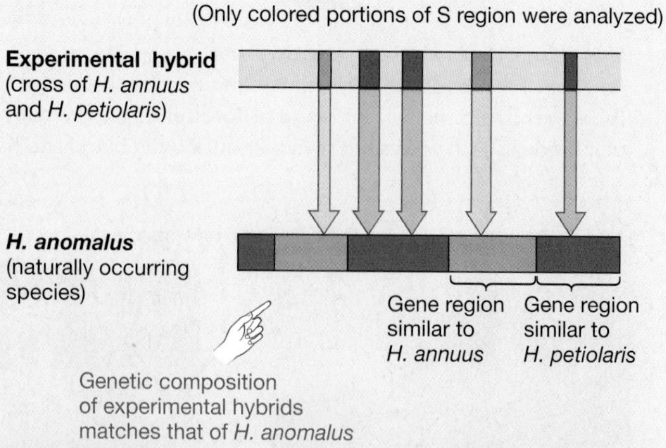
(Only colored portions of S region were analyzed)
Experimental hybrid
(cross of H. annuus and H. petiolaris)
H. anomalus
(naturally occurring species)
Gene region similar to H. annuus
Gene region similar to H. petiolaris
Genetic composition of experimental hybrids matches that of H. anomalus

Conclusion: New species may arise via hybridization between existing species.

FIGURE 25.14 Experimental Evidence That New Species Can Originate in Hybridization Events

ESSAY Human Races

This chapter is built around one fundamental theme: When populations are isolated by lack of gene flow, they can diverge as a result of mutation, genetic drift, and natural selection. Human populations are sometimes called **races** if they are differentiated by physical characteristics such as facial features, the color and texture of hair, skin color, and in some cases height or body build. Presumably, these distinctions arose due to a lack of gene flow in the past. How does the divergence observed among human populations compare with the amount of divergence measured among populations of other organisms? In a biological sense, how profound is the divergence that has occurred among human populations?

To quantify the amount of genetic variation that exists among populations, researchers routinely collect DNA samples from individuals that represent populations from throughout the range of a species. In humans, for example, biologists usually survey genetic diversity by obtaining DNA from aboriginal African, Asian, Australian, European, New Guinean, and Native American people (**Figure 25.15**). Then they analyze the base sequence at one or more loci in the nuclear or mitochondrial genomes of these individuals. To summarize the genetic differences that exist, the researchers calculate the sequence divergence. This number represents the average proportion of bases that are different in two randomly chosen individuals from the populations.

Because of the intense interest in genetic diversity in our own species, numerous studies of humans have been done by using an array of techniques and focusing on an array of loci. And because measuring the amount of genetic divergence among populations is fundamental to understanding a wide variety of questions about speciation and evolution, data sets from dozens, if not hundreds, of plant and animal species are available to compare with data sets of human genes.

Two important conclusions have emerged from these studies:

. . . how profound is the divergence that has occurred among human populations?

1. Genetic divergence among human populations is extremely low compared with that found in other species of animals. For example, when Rebecca Cann and colleagues examined mitochondrial DNA in humans from all over the world, they observed an average sequence divergence of just 0.32 percent. (This means that, on average, only one-third of 1 percent of bases are different in different populations.) When John Avise and co-workers did a similar study of deer mice from North America, they observed an average sequence divergence of 3.1 percent. Deer mouse populations have experienced 10 times as much genetic divergence as human populations.
2. Populations within sub-Saharan Africa exhibit more genetic diversity than do non-African populations. For example, when Sarah Tishkoff and associates examined the distribution of alleles at the *CD4* gene in nuclear DNA, they found that all of the alleles present in humans can occur in African individuals. Populations in other parts of the world have subsets of the array of alleles found in Africa, but no unique alleles. Based on data such as these, Svante Pääbo comments that "in a genetic sense, everyone on this planet looks like an African."

What is the overall message of these studies? Compared with genetic differences observed within other species, genetic differences among human populations are tiny.

(a) African **(b)** Asian **(c)** Australian **(d)** European **(e)** New Guinean **(f)** North American

FIGURE 25.15 At the Genetic Level, How Different Are Aboriginal Populations of Humans?
These individuals represent aboriginal populations from around the world.

CHAPTER REVIEW

Summary of Key Concepts

Speciation occurs when populations of the same species become genetically isolated by lack of gene flow and then diverge from each other due to selection, genetic drift, or mutation.

When gene flow does not occur between populations, natural selection, genetic drift, and mutation act on the populations independently. As a result, the genetic and physical characteristics of the populations change over time. Eventually, the populations become so different that they are recognized as distinct species—evolutionarily independent populations. Speciation is a splitting event in which one lineage gives rise to two or more independent descendant lineages.

Populations can become genetically isolated from each other if they occupy different geographic areas, if they use different habitats within the same area, or if one population is polyploid and cannot breed with the other.

Speciation often begins when small groups of individuals colonize a new habitat or when a large, continuous population becomes fragmented into isolated habitats. Colonization is thought to be a major mode of speciation on islands, while range splitting is thought to be a major mode of speciation in glaciated areas and in species that occupied the supercontinent Gondwana during its breakup.

Contrary to traditional expectations, evidence is mounting that speciation can occur due to disruptive selection even when populations are sympatric. Similarly, mutations that produce polyploidy can trigger rapid speciation in sympatry because they lead to reproductive isolation between diploid and tetraploid populations.

Web Tutorial 25.1 Allopatric Speciation

Web Tutorial 25.2 Speciation by Changes in Ploidy

Populations can be recognized as distinct species if they are reproductively isolated from each other, if they have distinct morphological characteristics, or if they form independent branches on a phylogenetic tree.

Researchers use several criteria to test whether populations represent distinct species. The biological species concept focuses on the degree of hybridization between species to determine whether gene flow is occurring. The morphospecies concept infers that speciation has occurred if populations have distinctive morphological traits. The phylogenetic species concept defines species as the smallest monophyletic groups on evolutionary trees.

When populations that have diverged come back into contact, several outcomes are possible.

If prezygotic isolation exists, populations that come back into contact will probably continue to diverge. Alternatively, interbreeding can cause diverged populations to fuse into the same species. Researchers have also documented that secondary contact can lead to reinforcement and complete reproductive isolation, the formation of hybrid zones, or the creation of a new hybrid species.

Questions

Content Review

1. What distinguishes a morphospecies?
   a. It has distinctive characteristics, such as size, shape, or coloration.
   b. It represents a distinct twig in a phylogeny of populations.
   c. It is reproductively isolated from other species.

2. When does vicariance occur?
   a. when small populations coalesce into a large, continuous population
   b. when a large population is fragmented into isolated subpopulations
   c. when individuals colonize a novel habitat
   d. when individuals disperse and found a new population

3. Why is "reinforcement" an appropriate name for the concept that natural selection should favor divergence and genetic isolation if populations experience postzygotic isolation?
   a. Selection should reinforce high fitness for hybrid offspring.
   b. Selection should reinforce the fact that they are "good species" under the morphological species concept.
   c. Selection acts because hybrid offspring do not develop at all or are sterile when mature.
   d. It reinforces selection for divergence that began when the species were geographically isolated.

4. The biological species concept can be applied only to which of the following groups?
   a. bird species living today
   b. dinosaurs
   c. bacteria
   d. archaea

5. Why are genetic isolation and genetic divergence occurring in soapberry bugs, even though populations occupy the same geographic area?
   a. Members of the different populations feed and mate on different types of fruit.
   b. One population recently became tetraploid, and hybrid offspring cannot undergo meiosis correctly.
   c. A vicariance event occurred when nonnative host plants were introduced.
   d. Beak length has changed due to disruptive selection.

6. When the ranges of different species meet, a stable "hybrid zone" occupied by hybrid individuals may form. How is this possible?
   a. Hybrid individuals may have intermediate characteristics that are advantageous in a given region, relative to traits in the different species.
   b. Hybrid individuals are always allopolyploid and are thus unable to mate with either of the original species. Allopolyploidy forms new species.
   c. Hybrid individuals may have reduced fitness and thus be strongly selected against.
   d. One species has a selective advantage, so as hybridization continues, the other species will go extinct.

Conceptual Review

1. Because speciation is a historical event, it can be difficult to study. Make an outline listing the sections and subsections in this chapter, and then fill it in by describing the experimental and analytical approaches used in the case studies provided for each topic. In your opinion, which approaches to studying speciation described in this chapter are most powerful? Why?
2. In the case of the seaside sparrow, how did the species identified by the biological species concept, the morphospecies concept, and the phylogenetic species concept conflict?
3. Explain how isolation and divergence are occurring in soapberry bugs, even though populations occupy the same geographic area. Of the four evolutionary processes (mutation, gene flow, drift, and selection), which two are most important in causing this event?
4. Why would genetic drift be an especially important evolutionary process causing divergence during colonization events? If colonists occupy a different habitat from that of the source population, why would natural selection also be an important process causing divergence? Explain your answers.
5. Summarize the possible outcomes when populations that have been separated for some time come back into contact and hybridize. In each case, explain why the fitness of hybrid offspring determines the outcome.

Group Discussion Problems

1. A large amount of gene flow is now occurring among human populations due to intermarriage among people from different ethnic groups and regions of the world. Is this phenomenon increasing or decreasing racial differences in our species? Explain.
2. In plants, reproductive cells do not differentiate until late in life. Stated another way, plant cells do not undergo meiosis and produce gametes until they have gone through mitosis hundreds or thousands of times, as the plant grows into an adult. In contrast, animal reproductive cells go through mitosis only a few times before they undergo meiosis and produce gametes. Explain how these observations relate to the following fact: Plants are much more likely to produce diploid gametes and produce polyploid offspring than are animals.
3. Ellen Censky and co-workers recently documented a colonization event. After two hurricanes passed through the Caribbean Sea in the fall of 1995, a large raft made up of fallen trees and other debris, along with green iguanas, drifted to the shore of the island of Anguilla. At least 15 individuals of the green iguana left the raft and went onto the island. Iguanas were not found on Anguilla prior to this event. Based on the wind patterns during the storms, the researchers propose that the 15 iguanas originally lived on the island of Guadalupe.
   - Outline a short-term study designed to test the hypothesis that genetic drift produced allele frequency differences in the two populations (the "old" green iguana population of Guadalupe and the new population on Anguilla).
   - Outline a long-term study designed to test the hypothesis that natural selection will produce changes in the characteristics of the two populations over time.
4. All over the world, natural habitats are being fragmented into tiny islands as suburbs, ranches, and farms expand. In effect, vicariance is occurring at a scale never before seen in the history of life. Based on the data presented in this chapter, do you predict that speciation will take place in the newly isolated populations? Or do you predict that they will be wiped out before speciation can occur?

Answers to Multiple-Choice Questions 1. a; 2. b; 3. d; 4. a; 5. a; 6. a

www.prenhall.com/freeman is your resource for the following: Web Tutorials; Online Quizzes and other Online Study Guide materials; Answers to Conceptual Review Questions; Solutions to Group Discussion Problems; Answers to Figure Caption Questions and Exercises; and Additional Readings and Research.

26 Phylogenies and the History of Life

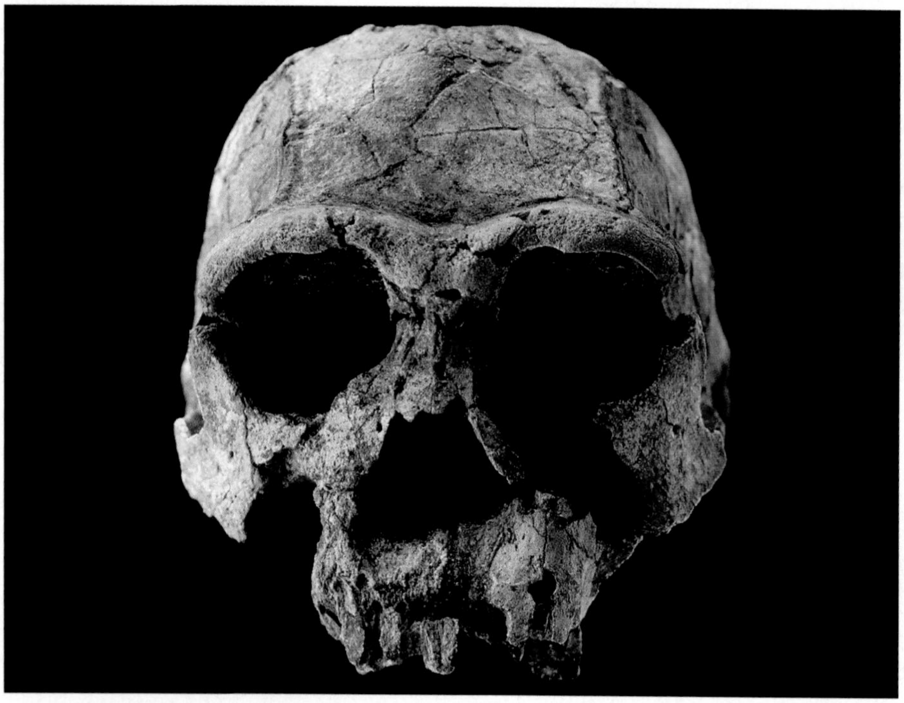

Skull from a species called *Homo erectus* ("upright man"). This individual is from a population that is closely related to the ancestors of modern humans.

KEY CONCEPTS

- Phylogenies and the fossil record are the major tools that biologists use to study the history of life.
- The Cambrian explosion was the rapid morphological and ecological diversification of animals that occurred during the Cambrian period.
- The new field of "evo-devo" is providing insights into how major events in the history of life occurred, by revealing the genetic mechanisms involved.
- Adaptive radiations are a major pattern in the history of life. They are instances of rapid diversification associated with new ecological opportunities and new morphological innovations.
- Mass extinctions have occurred repeatedly throughout the history of life. They rapidly eliminate most of the species alive in a more or less random manner.

This chapter is about the tree of life—the wonderful diversity of organisms that have been produced since life began. More specifically, this chapter is about how the tree of life has grown and changed over time. Life has existed on Earth for some 3.8 billion years. What patterns can be discerned as species appeared and disappeared over the course of history? Why do patterns exist?

Chapters 1 and 2 introduced the size and shape of the tree of life by examining the geologic time scale, radiometric dating, and life's three domains: the Bacteria, Archaea, and Eukarya. Chapter 25 considered speciation, or how individual twigs on the tree of life form. The focus of this chapter falls between the broad scope of the early chapters and the fine level of Chapter 25. The task is to find common themes and patterns in the history of life.

The goal of Sections 26.1 and 26.2 is to introduce the data and analytical tools that researchers use to reconstruct *what* happened in the history of life. Section 26.3 focuses on *how* key changes documented in the fossil record occurred—specifically, how gene duplication events, changes in gene expression, and natural selection cause major evolutionary innovations. A prominent pattern in the tree of life, called adaptive radiation, is the subject of Section 26.4. The chapter closes with a look at what may be the most influential of all events in the history of life: the ecological disasters known as mass extinctions.

26.1 Tools for Studying History: Phylogenies and the Fossil Record

Biologists have two major tools to use when studying the past: phylogenies and the fossil record. A **phylogeny** is the evolutionary history of a group of organisms, and it is usually summarized and depicted in the form of a phylogenetic tree. A **phylogenetic tree** shows the ancestor-descendant relationships among populations or species, just as a genealogy or family tree shows the ancestor-descendant relationships among a group of individuals. A **fossil**, in contrast, is a physical trace of an organism that lived in the past. The **fossil record** is the total collection of fossils that have been found throughout the world. The fossil record is housed in thousands of private and public collections.

Fossils document what organisms looked like in the past and when various groups existed. Phylogenies clarify who is related to whom. Let's first consider how phylogenies are read and how they are estimated and then analyze how the fossil record is collected and interpreted. The remainder of this chapter will explore what these two tools have revealed about the story of life on Earth.

Using Phylogenies

Evolutionary trees have been introduced at various points in earlier chapters—often with just enough information to help you understand a new concept. Now let's focus on the trees themselves. How do biologists read a finished tree, and how are trees put together in the first place? We'll consider each question in turn.

A Field Guide to Reading Phylogenetic Trees Phylogenetic trees are an extremely effective way of summarizing data on the evolutionary history of a group of organisms. They are unusual diagrams, however, and it can take practice to interpret them correctly. To understand how evolutionary trees work, consider **Figure 26.1**. Note that a phylogenetic tree consists of branches, nodes, and tips. **Branches** represent populations through time. **Nodes** (or *forks*) occur where an ancestral group splits into two or more descendant groups (see point B in Figure 26.1). If more than two descendant groups emerge from a node, the node is called a **polytomy** (see node C). **Tips** (or *terminal nodes*) are the tree's endpoints, which represent groups living today or a dead end—a branch ending in extinction. Groups that occupy adjacent branches on the tree are called **sister taxa**. Recall from Chapter 1 that a **taxon** (plural: *taxa*) is any named group of organisms. A taxon could be a single species, such as *Homo sapiens*, or a large group of species, such as Primates.

The phylogenetic trees used in this text are all **rooted**—meaning the bottom, or most basal, node on the tree is the most ancient. To determine where the root on a tree occurs, biologists include an outgroup species when they are collecting data to estimate a particular phylogeny. An **outgroup** is a taxonomic group that is known to have diverged prior to the rest of the taxa in the study. In Figure 26.1, "Taxon 1" is an outgroup of the monophyletic group consisting of taxa 2–6. (Recall from Chapter 25 that a **monophyletic group** consists of an ancestral species and all of its descendants. Monophyletic groups may also be called **clades** or **lineages.**) If you were estimating the phylogeny of human species that lived over the past 5–6 million years, you might use chimpanzees or gorillas as an outgroup in the analysis. Chimps and gorillas would then be the first groups to have split off in the resulting tree, with the various species of humans forming a monophyletic group descended from the common ancestor at the tree's bottom node (**Figure 26.2**, page 558).

To practice how to read a tree, put your finger at the root of the diagram in Figure 26.1 and begin to work your way up. At node A, the ancestral population split into two groups. One of these lineages eventually evolved into Taxon 1. The other lineage gave rise to taxa 2–6. Now continue moving your finger up the tree until you hit node B. It should make sense to you that at this splitting event, one descendant clade eventually gave rise to Taxon 2 and Taxon 3, while the other became the ancestor of taxa 4–6. The polytomy at node C occurs because taxa 4–6 split from one another so quickly that it is not possible to tell which split off earlier or later. You should be able to do the same exercise with the phylogeny of humans shown in Figure 26.2.

Figure 26.3 (page 558) presents a different challenge. Five of the six trees shown in this diagram are identical in terms of the evolutionary relationships they represent. One differs. If you understand how to read a tree, you should be able to determine which one of the six trees is unlike the others.

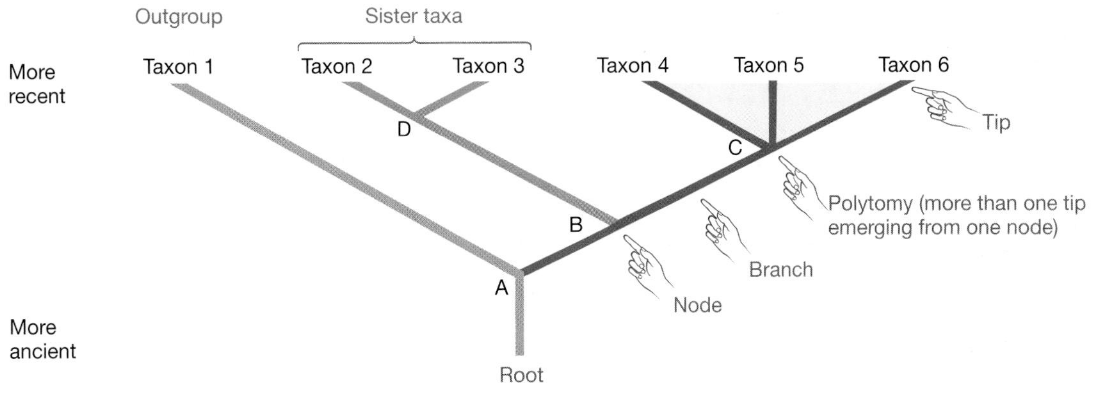

FIGURE 26.1 Reading a Phylogenetic Tree
The parts of a phylogenetic tree, labeled.

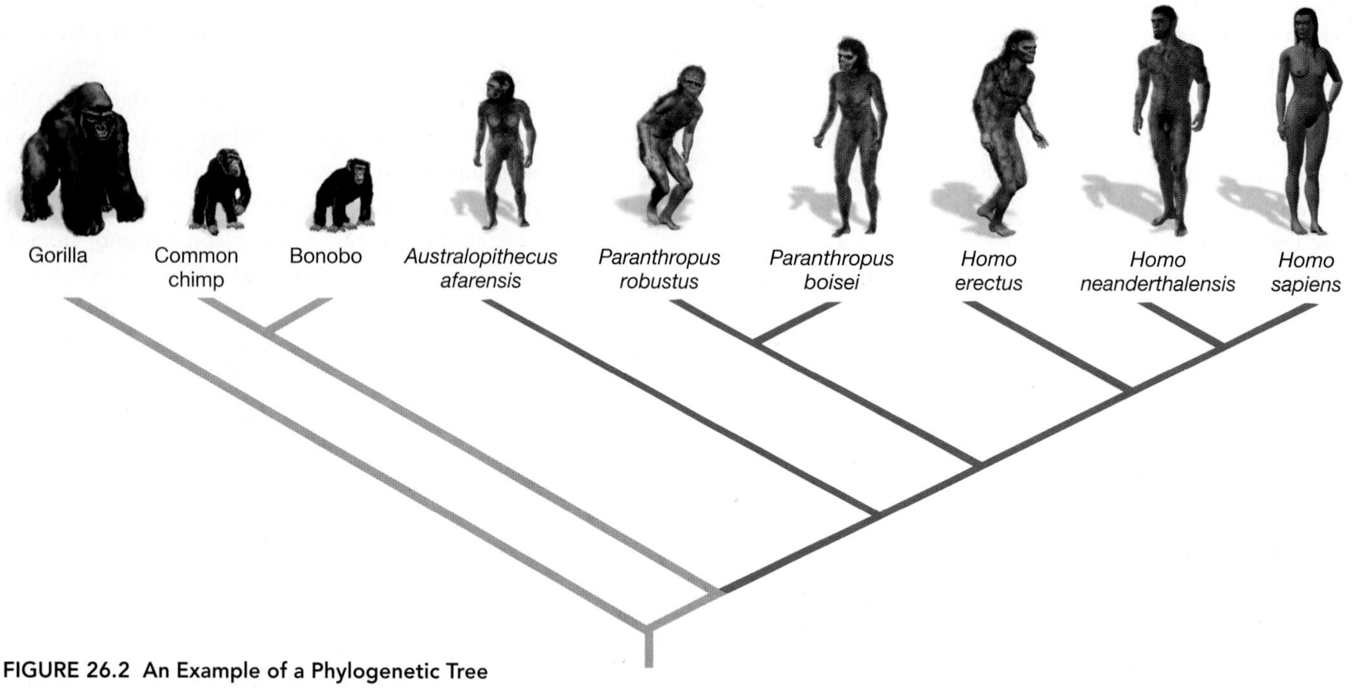

FIGURE 26.2 An Example of a Phylogenetic Tree
A phylogenetic tree showing the relationships among some of the great apes. Chimps and gorillas walk on all four legs, but all other species on this tree walk on two legs and are considered hominins. **EXERCISE** Add a mark on the phylogeny, and label it "origin of walking on two legs." Circle and label a pair of sister species. Circle and label the monophyletic group called hominins. Label one of the outgroups to the hominins.

How Do Researchers Estimate Phylogenies? Like any other pattern or measurement in nature, from the average height of a person in a particular human population to the speed of a passing airplane, the genealogical relationships among species cannot be known with absolute certainty. Instead, the relationships depicted in an evolutionary tree are estimated from data.

To infer the historical relationships among species, researchers analyze the species' morphological or genetic characteristics. For example, to reconstruct relationships among fossil species of humans, scientists analyze aspects of tooth, jaw, and skull structure. To reconstruct relationships among contemporary human populations, investigators usually compare the sequences of bases in a particular gene.

Once the characteristics of different populations have been measured, researchers have several general strategies available for analyzing the data and inferring which species or populations are more closely related or more distantly related. The

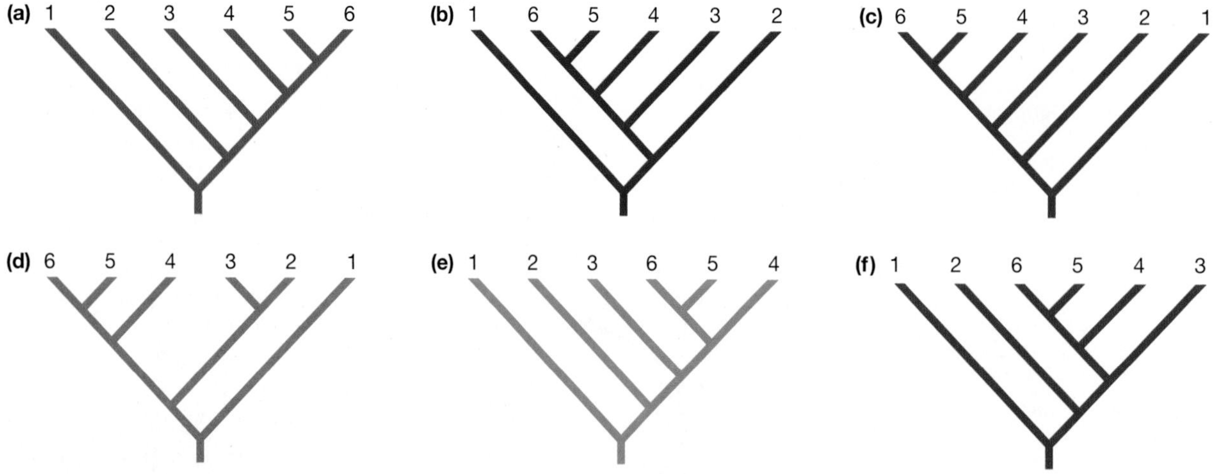

FIGURE 26.3 Alternative Ways of Drawing the Same Tree
QUESTION Five of these six trees describe exactly the same relationships among taxa 1 through 6. Identify the tree that is different from the other five.

fundamental idea is that closely related species should share many of their characteristics, while distantly related species should share fewer characteristics. But there are two general strategies for quantifying the degree of similarity among species: the phenetic approach and the cladistic approach.

The **phenetic approach** to estimating trees is based on computing a statistic that summarizes the overall similarity among populations, based on the data. For example, researchers might use gene sequences to compute an overall "genetic distance" between two populations. A *genetic distance* summarizes the average percentage of bases in a DNA sequence that differs between two populations. A computer program then compares the similarities among populations and builds a tree that clusters the most similar populations and places more divergent populations on more distant branches.

The **cladistic approach** to inferring trees is based on the realization that relationships among species can be reconstructed by identifying shared derived characters, or **synapomorphies** ("union-forms"), in the species being studied. **Figure 26.4** illustrates the logic behind a cladistic analysis. When the ancestral population at the bottom of the figure splits into two descendant lineages at node A, each descendant group begins evolving independently and acquires unique traits. These traits are derived from their common ancestor via mutation, selection, and genetic drift. The traits might be a particular region of DNA that changes as follows:

AAA GCT ACT	ancestral population
AAC GCT ACT	a descendant population
AAA GGT ACT	another descendant population

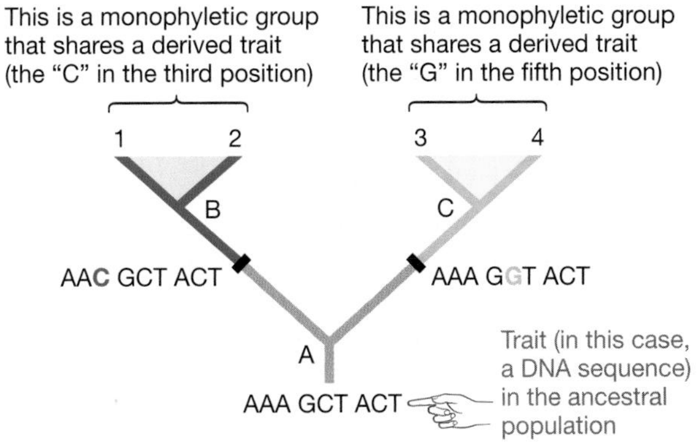

FIGURE 26.4 Synapomorphies Identify Monophyletic Groups Mutation, selection, and drift cause evolutionary changes that are shared in groups derived from the same common ancestor. In this tree, species 1 and 2 can be identified as part of a monophyletic group because base C in the third position of the DNA sequence shown is a synapomorphy. Species 3 and 4 can be identified as part of a different monophyletic group because they have a different synapomorphy—base G in the fifth position of the DNA sequence shown.

When the two lineages themselves split at nodes B and C, the species that result share the derived characteristics. In this way, the ancestor at node B and species 1 and 2 can be recognized as a monophyletic group. Similarly, the ancestor at node C and species 3 and 4 can be recognized as a different monophyletic group. When many such traits have been measured, a computer program can be used to identify which traits are unique to each monophyletic group and then place the groups in a tree in the correct relationship to each other.

Although the logic behind phenetic and cladistic analyses is elegant, problems arise due to convergent evolution. Recall from Chapter 23 that traits in two species may be similar not because they were present in a common ancestor, but because similar traits evolved independently in two distantly related groups. In the example given in Figure 26.4, it is possible that species 2 is not at all closely related to species 1. Its ancestors may have had the sequence AAT GCT AGT, which happened to change to AAC GCT ACT due to mutation, selection, and drift that took place independently of the changes that took place in the ancestors of species 1.

To cope with the existence of convergent evolution, biologists who are using cladistic approaches invoke the logical principle of **parsimony**, which states that the most likely explanation or pattern is the one that implies the least amount of change or the least complexity. For example, a biologist might compare all of the possible branching patterns that could link species 1–4 and count the number of changes in DNA sequences required to produce each pattern. Convergent evolution should be rare compared with similarity due to shared descent, so the tree that implies the fewest overall changes—and thus the least convergence—should be the one that most accurately reflects what really happened during evolution. If the branching pattern in Figure 26.4 is the most parsimonious of all possible trees, then biologists conclude that it is the most accurate.

Whale Evolution: A Case History As an example of how a cladistic approach works, consider the evolutionary relationships of the whales and the group of mammals called the Artiodactyla. Cows, deer, and hippos are artiodactyls. Members of this group have hooves and an even number of toes. They also share another feature: the unusual pulley shape of an ankle bone called the astralagus. The shape of the astralagus is a shared, derived character that identifies the artiodactyls as a monophyletic group. These data support the tree shown in **Figure 26.5a** (page 560).

When researchers began comparing DNA sequences of artiodactyls and other species of mammals, however, the data showed that whales share many similarities with hippos. These results supported the tree shown in **Figure 26.5b**. This tree, supported by the DNA data, conflicts with the morphological data in Figure 26.5a, because it implies that the pulley-shaped astralagus evolved in artiodactyls and then was lost

(a) The astralagus is a synapomorphy that identifies artiodactyls as a monophyletic group.

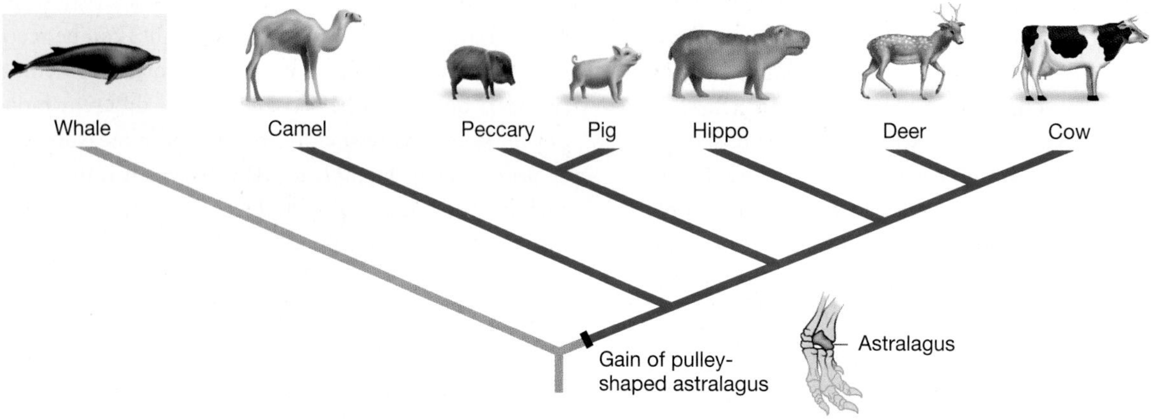

(b) If whales are related to hippos, then two evolutionary changes occurred in the astralagus.

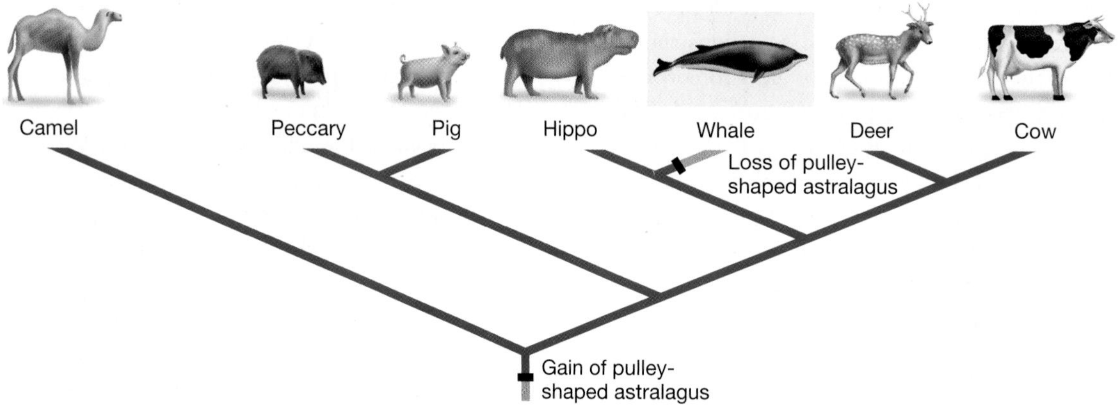

(c) Data on the presence and absence of SINE genes support the close relationship between whales and hippos.

Locus	1	2	3	4	5	6	7	8	9	10	11	12	13	14	15	16	17	18	19	20
Cow	0	0	0	0	0	0	0	**1**	**1**	**1**	**1**	**1**	**1**	**1**	**1**	**1**	**1**	**1**	0	0
Deer	0	0	0	0	0	0	0	**1**	?	**1**	**1**	**1**	**1**	**1**	**1**	?	**1**	**1**	0	0
Whale	**1**	**1**	**1**	**1**	**1**	**1**	**1**	0	?	**1**	0	**1**	**1**	0	0	0	?	**1**	0	0
Hippo	0	?	0	**1**	**1**	**1**	**1**	0	**1**	**1**	0	**1**	**1**	0	0	0	?	**1**	0	0
Pig	0	0	0	?	0	0	0	0	?	0	0	0	?	?	0	0	0	**1**	**1**	**1**
Peccary	?	?	?	?	?	?	?	?	?	?	?	?	?	?	?	?	?	?	**1**	**1**
Camel	0	0	0	0	0	0	0	0	0	0	0	0	0	0	0	0	0	0	0	0

1 = gene present
0 = gene absent
? = still undetermined

Whales and hippos share four unique SINE genes (4, 5, 6, and 7)

FIGURE 26.5 Evidence That Whales and Hippos Form a Monophyletic Group
On the basis of a parsimony analysis of the pulley-shaped astralagus, biologists favored **(a)** a tree excluding whales from the Artiodactyla over **(b)** the hypothesis that whales are artiodactyls and are closely related to hippos. **(c)** Data on the presence and absence of SINE genes supports the "whales + hippos" hypothesis. **EXERCISE** The presence of SINE genes 4, 5, 6, 7 identifies hippos and whales as part of a monophyletic group. The presence of SINE genes 8, 11, 14, 15, 17 identifies deer and cows as part of a monophyletic group. SINE genes 10, 12, 13 identify hippos, whales, deer, and cows as part of a monophyletic group. Map the origin of these SINE genes on the tree in part (b). Where did SINE genes 19 and 20 first insert themselves into the genomes of artiodactyls?

during whale evolution. The tree in Figure 26.5b implies two changes in the astralagus, while the tree in Figure 26.5a implies just one. In terms of the evolution of the astralagus, the "whale + hippo" tree is less parsimonious than the tree in Figure 26.5a.

The conflict between the data sets was resolved when researchers analyzed the distribution of the parasitic gene sequences called SINEs (short interspersed nuclear elements), which occasionally insert themselves into the genomes of mammals (see Table 20.1). As the data in **Figure 26.5c** show, whales and hippos share several types of SINES that are not found in other groups. Specifically, whales and hippos share the SINEs numbered 4, 5, 6, and 7. Other SINE genes are present in some artiodactyls but not in others; camels have no SINE genes at all. To explain these data, biologists hypothesize that no SINEs were present in the population that is ancestral to all of the species in the study. Over the course of evolution, however, different SINEs became inserted into the genomes of descendant populations. As a result, the presence of a particular SINE represents a derived character. Because whales and hippos share four of these derived characters, it is logical to conclude that these animals are closely related.

Based on these data, most biologists accepted the phylogeny that is shown in Figure 26.5b as the most accurate estimate of evolutionary history. According to this phylogeny, whales are artiodactyls and share a relatively recent common ancestor with hippos. This observation inspired the hypothesis that both whales and dolphins are descended from a population of artiodactyls that spent most of their time feeding in shallow water, much as hippos do today. Recently this hypothesis was supported in spectacular fashion. In 2001 two teams of researchers announced the independent discoveries of fossil artiodactyls that were clearly related to whales and yet had a pulley-shaped astralagus. One of these fossils is illustrated in Figure 23.3b. The combination of phylogenetic data and data from the fossil record has clarified how a particularly interesting group of mammals evolved.

Over the past two decades, the effort to infer phylogenetic trees has been revolutionized by the availability of DNA sequence data and powerful computers to analyze large data sets. Methods for collecting and analyzing the data used to estimate phylogenies continue to improve. Biological science is currently witnessing an explosion in the quantity and quality of phylogenetic trees.

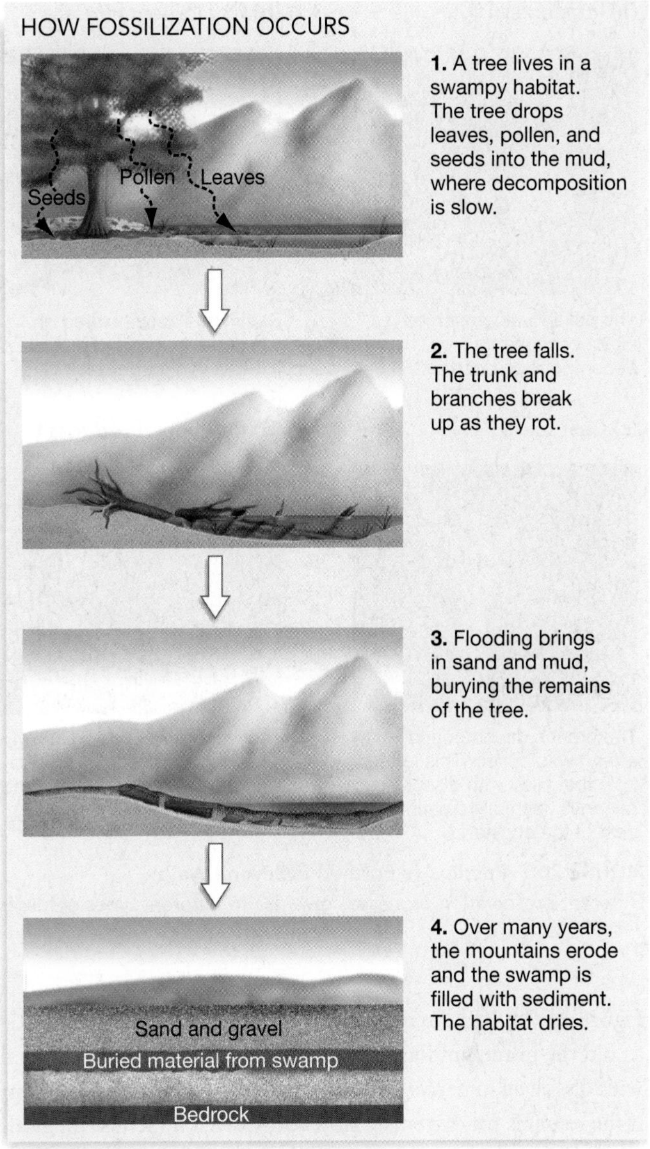
HOW FOSSILIZATION OCCURS
Seeds
Pollen
Leaves
1. A tree lives in a swampy habitat. The tree drops leaves, pollen, and seeds into the mud, where decomposition is slow.
2. The tree falls. The trunk and branches break up as they rot.
3. Flooding brings in sand and mud, burying the remains of the tree.
4. Over many years, the mountains erode and the swamp is filled with sediment. The habitat dries.
Sand and gravel
Buried material from swamp
Bedrock

FIGURE 26.6 Fossilization Preserves Traces of Organisms That Lived in the Past
Fossilization occurs most readily when the remains of an organism are buried in sediments, where decay is slow.

Using the Fossil Record

Phylogenetic analyses are powerful ways to infer the order in which events occurred during evolution and to understand how particular groups of species are related. But only the fossil record provides direct evidence about what organisms that lived in the past looked like, where they lived, and when they existed. Let's review how fossils form, analyze the strengths and weaknesses of the fossil record, and then survey a few of the major events that have taken place in life's 3.8-billion-year history.

How Do Fossils Form? Most of the processes that form fossils begin when part or all of an organism is buried in ash, sand, mud, or some other type of sediment. Consider a series of events that begins when a tree falls into a swamp. **Figure 26.6** illustrates the leaves of a tree falling onto a patch of mud, where they are buried by soil and debris before they decay. Pollen and seeds settle into the muck at the bottom of the swamp, where decomposition is slow. The stagnant water is too acidic and too oxygen poor to support large populations of bacteria and fungi, so much of this material is buried intact before it decomposes. The trunk and branches that sit above the water line rot fairly quickly, but as pieces break off they, too, sink to the bottom and are buried.

Once burial occurs, several things can happen. If decomposition does not occur, the organic remains can be preserved intact—like the fossil pollen in **Figure 26.7a** (page 562). Alternatively, if sediments accumulate on top of the material and become cemented into rocks such as mudstone or shale, the sediments weight can compress the organic material below into a thin, carbonaceous film. This happened to the leaf in

(a) Intact fossil

The pollen was preserved intact because no decomposition occurred.

(b) Compression fossil

Sediments accumulated on top of the leaf and compressed it into a thin carbon-rich film.

(c) Cast fossil

The branch decomposed after it was buried. This left a hole that filled with dissolved minerals, faithfully creating a cast of the original.

(d) Permineralized fossil

The wood decayed very slowly, allowing dissolved minerals to infiltrate the cells gradually and then harden into stone.

FIGURE 26.7 Fossils Are Formed in Several Ways
Different preservation processes give rise to different types of fossils.

Figure 26.7b. If the remains decompose *after* they are buried—as did the branch in **Figure 26.7c**—the hole that remains can fill with dissolved minerals and faithfully create a **cast** of the remains. If the remains rot extremely slowly, dissolved minerals can gradually infiltrate the interior of the cells and harden into stone, forming a *permineralized* fossil, such as petrified wood (**Figure 26.7d**).

After many centuries have passed, fossils can be exposed at the surface by many mechanisms, including erosion, a road cut, or quarrying. If researchers find a fossil, they can prepare it for study by painstakingly clearing away the surrounding rock (**Figure 26.8**). If the species represented is new, researchers describe its morphology in a scientific publication, name the species, estimate the fossil's age based on dates assigned to nearby rock layers, and add the specimen to a collection so that it is available for study by other researchers. It is now part of the fossil record. This is the information database that supports much of the research reviewed in this chapter.

The scenario just presented is based on conditions that are ideal for fossilization: The tree fell into an environment where decomposition was slow and burial was rapid. In most habitats the opposite situation occurs—decomposition is rapid and burial is slow. In reality, then, fossilization is an extremely rare event. To appreciate this point, consider that there are seven specimens of the first bird to appear in the fossil record,

PREPARING A FOSSIL

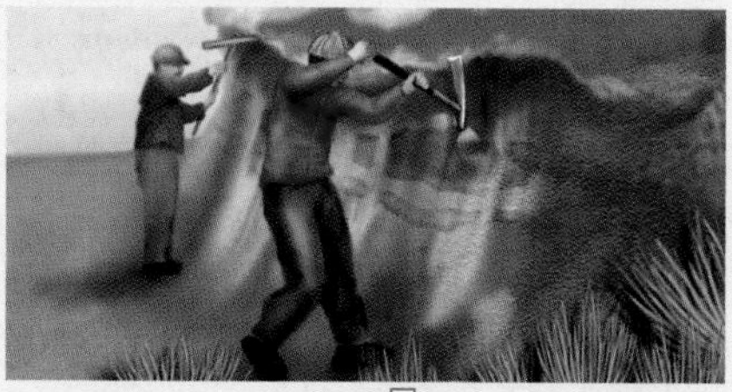

1. Excavate fossil-bearing rock.

2. Split the rock to reveal the fossil. (If fossil is already exposed, remove surrounding rock.)

3. Clean away excess rock to expose as much of the fossil as possible.

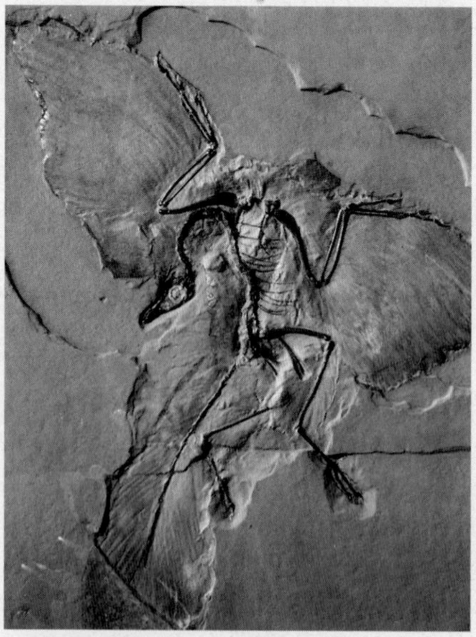

4. The result.

Archaeopteryx—oldest bird in the fossil record

FIGURE 26.8 Preparing a Fossil
EXERCISE Label the head, wings, neck, legs, toes, tail, and ribs on the *Archaeopteryx* specimen. Outline the extent of its feathers.

Archaeopteryx. All were found at the same site in Germany where limestone is quarried for printmaking (the bird's specific name is *lithographica*). If you accept an estimate that crow-sized birds native to wetland habitats in northern Europe would have a population size of around 10,000 and a life span of

10 years, and if you accept the current estimate that the species existed for about 2 million years, then you can calculate that about 2 billion *Archaeopteryx* lived. But as far as researchers currently know, only 1 out of every 286,000,000 individuals fossilized. For this species, the odds of becoming a fossil were almost 40 times worse than your odds are of winning the grand prize in a state lottery.

Limitations of the Fossil Record Before looking at how the fossil record is used to answer questions about the history of life, it is essential to review the nature of this archive and recognize several features:

- Habitat bias: Because burial in sediments is so crucial to fossilization, there is a strong habitat bias in the database. Organisms that live in areas where sediments are actively being deposited, including beaches, mudflats, and swamps, are much more likely to form fossils than organisms that live in other habitats are. Within these habitats, burrowing organisms such as clams are already underground—pre-buried—at death and are therefore much more likely to fossilize. Organisms that live above ground in dry forests, grasslands, and deserts are much less likely to fossilize.
- Taxonomic bias: Slow decay is almost always essential to fossilization, so organisms with hard parts such as bones or shells are most likely to leave fossil evidence. This requirement introduces a strong taxonomic bias into the record. Clams, snails, and other organisms with hard parts have a much higher tendency to be preserved than do worms. A similar bias exists for tissues within organisms. For instance, pollen grains are encased in a tough outer coat that resists decay, so they fossilize much more readily than do flowers. Teeth are the most common mammalian fossil, simply because they are so hard and decay resistant. Shark teeth are abundant in the fossil record, but shark bones, which are made of cartilage, are extremely rare.
- Temporal bias: Recent fossils are much more common than ancient fossils. To understand why, consider that when two of Earth's tectonic plates converge, the edge of one plate usually sinks beneath the other plate. The rocks composing the edge of the descending plate are either melted or radically altered by the increased heat and pressure they encounter as they move downward into Earth's interior. These alterations obliterate any fossils in the rock. In addition, fossil-bearing rocks on continents are constantly being broken apart and destroyed by wind and water erosion. The older a fossil is, the more likely it is to be demolished.
- Abundance bias: Because fossilization is so improbable, the fossil record is weighted toward common species. Organisms that are abundant, widespread, and present on Earth for long periods of time leave evidence much more often than do species that are rare, local, or ephemeral.

To summarize, the fossil database represents a highly nonrandom sample of the past. **Paleontologists**—scientists who study fossils—recognize that they are limited to asking questions about tiny and scattered segments on the tree of life. Yet, as this chapter shows, the record is a scientific treasure trove. Analyzing fossils is the only way scientists have of examining the physical appearance of extinct forms and inferring how they lived. The ancient Egyptians wrote a Book of the Dead to guide individuals into the afterlife. The fossil record is biology's book of the dead—our guide to past life.

Life's Timeline A few of the more significant data points in the fossil record are summarized in **Figure 26.9** (page 564). Most of the data on the timelines are "evolutionary firsts" that document important innovations during the history of life. In almost all cases, the dates given were estimated by using the radiometric dating techniques introduced in Chapter 2. When fossils are not available, key events in the history of life may sometimes be dated by using the molecular approaches introduced in **Box 26.1** (page 566). The eras, eons, and periods in the figure do not represent regular time intervals, however, because they were indentified and named long before radiometric and molecular dating techniques became available. Because the fossil record and efforts to date fossils are constantly improving, the timeline in Figure 26.9 is a work in progress.

As you inspect Figure 26.9, notice that the timeline is broken into four segments and that comments on the types of organisms present, the nature of the climate, major geological events, and the positions of the continents accompany most segments:

- Figure 26.9a outlines the interval between the formation of Earth about 4.6 billion years ago and the appearance of most animal groups about 543 million years ago (abbreviated Ma). The entire interval is called the **Precambrian era,** and it is divided into the **Hadean**, **Archaean**, and **Proterozoic eons.** The important things to note about this time are that (1) life was exclusively unicellular for most of Earth's history and (2) oxygen was virtually absent from the oceans and atmosphere for almost 2 billion years after the origin of life.
- Figure 26.9b summarizes the **Paleozoic** ("ancient life") **era,** which begins with the appearance of many animal lineages and ends with the obliteration of almost all multicellular life-forms at the end of the Permian period. The Paleozoic saw the origin and initial diversification of the animals, land plants, and fungi, as well as the appearance of land animals.
- The **Mesozoic** ("middle life") **era** is outlined in Figure 26.9c. This interval is nicknamed the Age of Reptiles. It begins with the end-Permian extinction events and ends with the extinction of the dinosaurs and other groups at the boundary between the Cretaceous period and Tertiary period.

(a) The **Precambrian era** included the origin of life, photosynthesis, and the oxygen atmosphere.

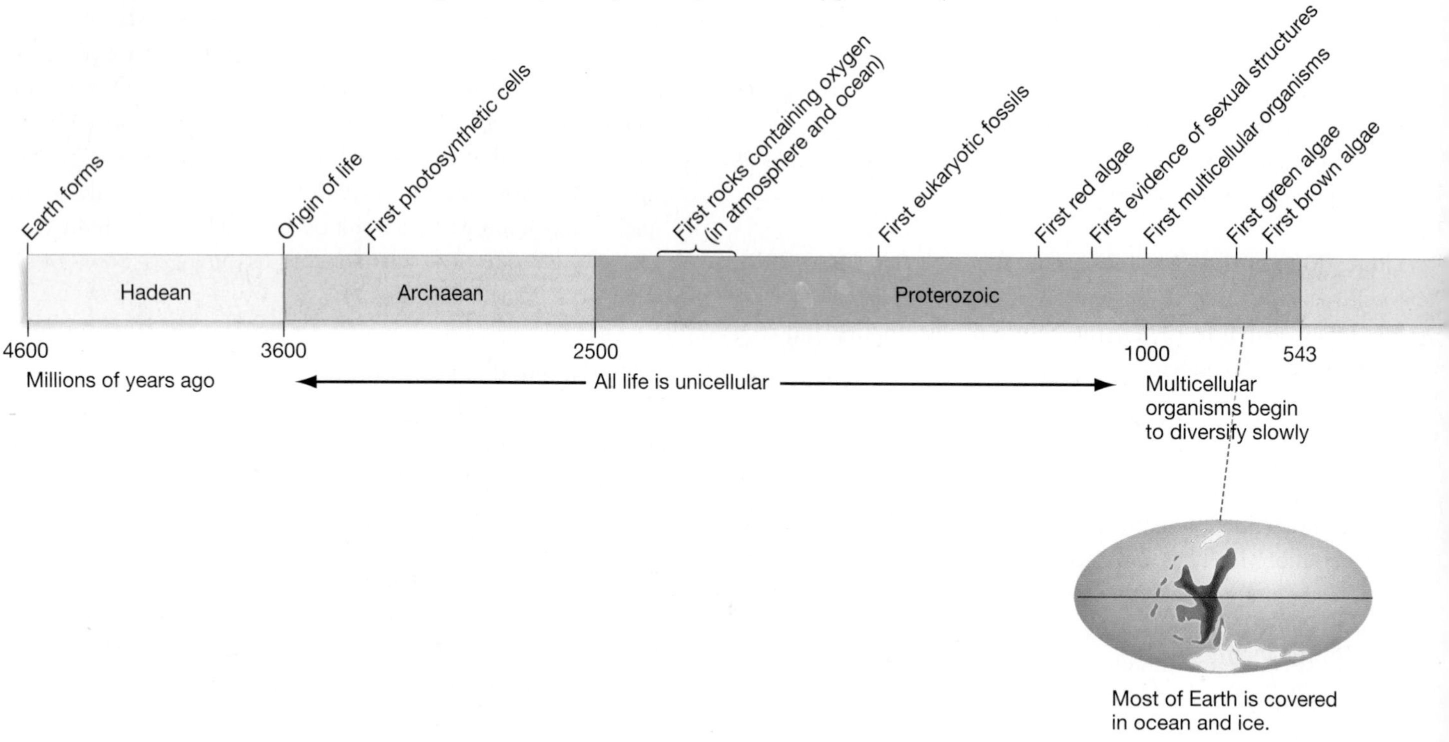

(c) The **Mesozoic era** is sometimes called the Age of Reptiles.

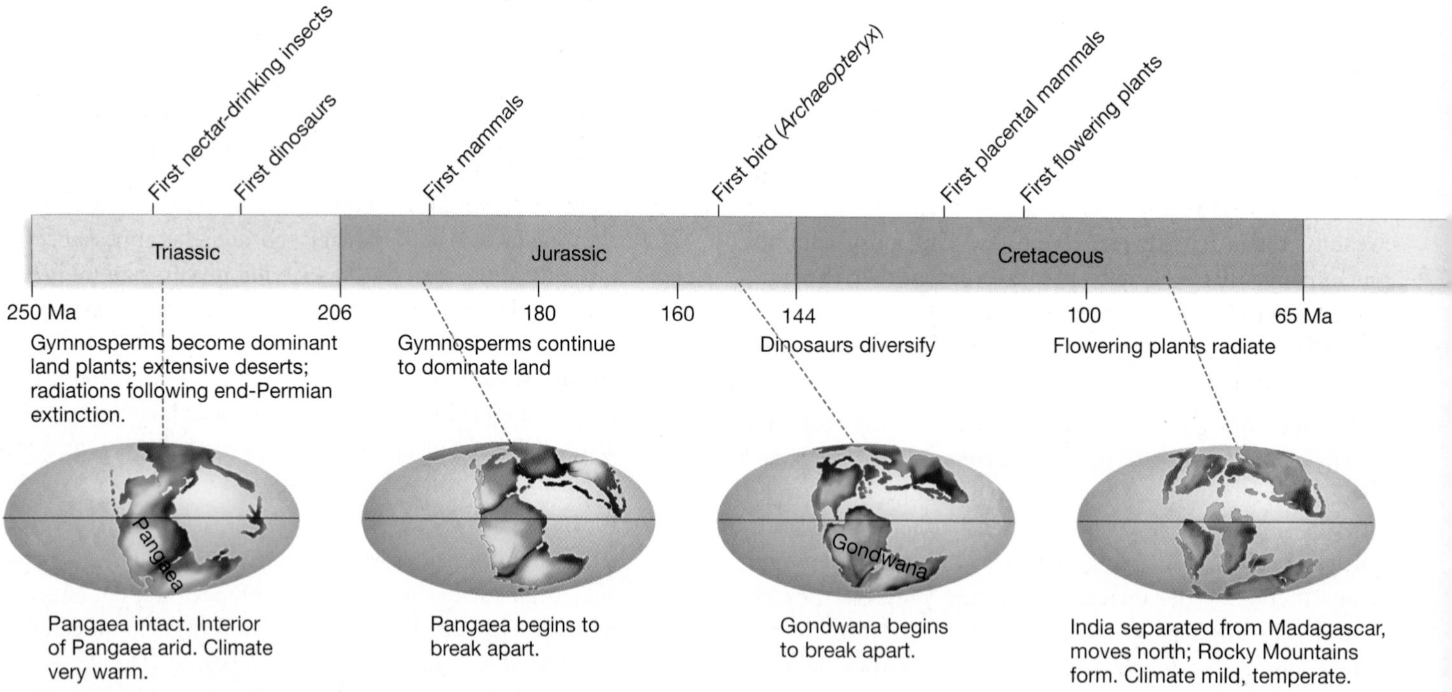

FIGURE 26.9 Life's Timeline

Significant events in Earth's history and the history of life are plotted for the **(a)** Precambrian era, **(b)** Paleozoic era, **(c)** Mesozoic era, and **(d)** Cenozoic era. ("Ma" stands for "million years ago.")

- Figure 26.9d highlights the **Cenozoic** ("recent life") **era**, which is divided into the Tertiary period and the Quaternary period. The Cenozoic is sometimes nicknamed the Age of Mammals, because mammals diversified after the disappearance of the dinosaurs. Events that occur today are considered to be part of the Cenozoic era.

(b) The **Paleozoic era** included the origin and early diversification of animals, land plants, and fungi.

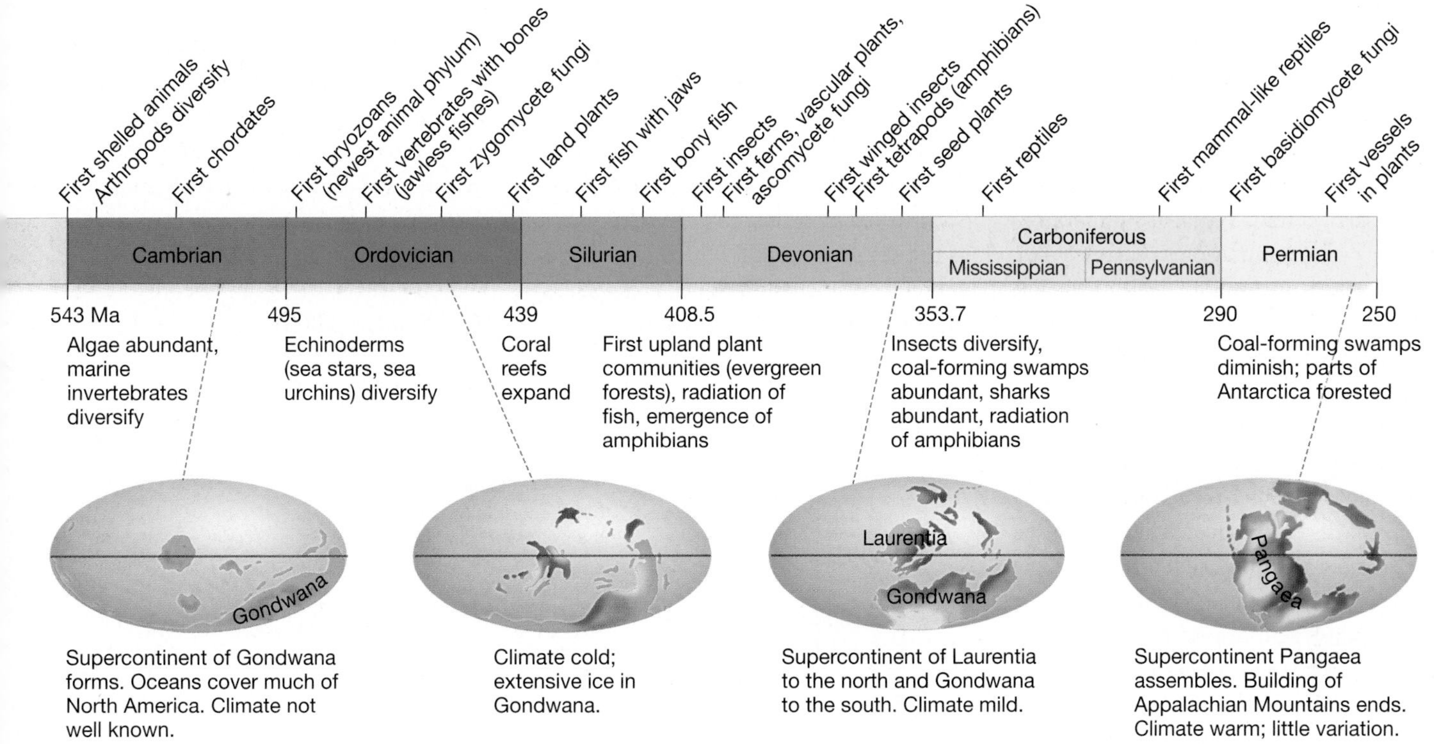

(d) The **Cenozoic era** is nicknamed the Age of Mammals.

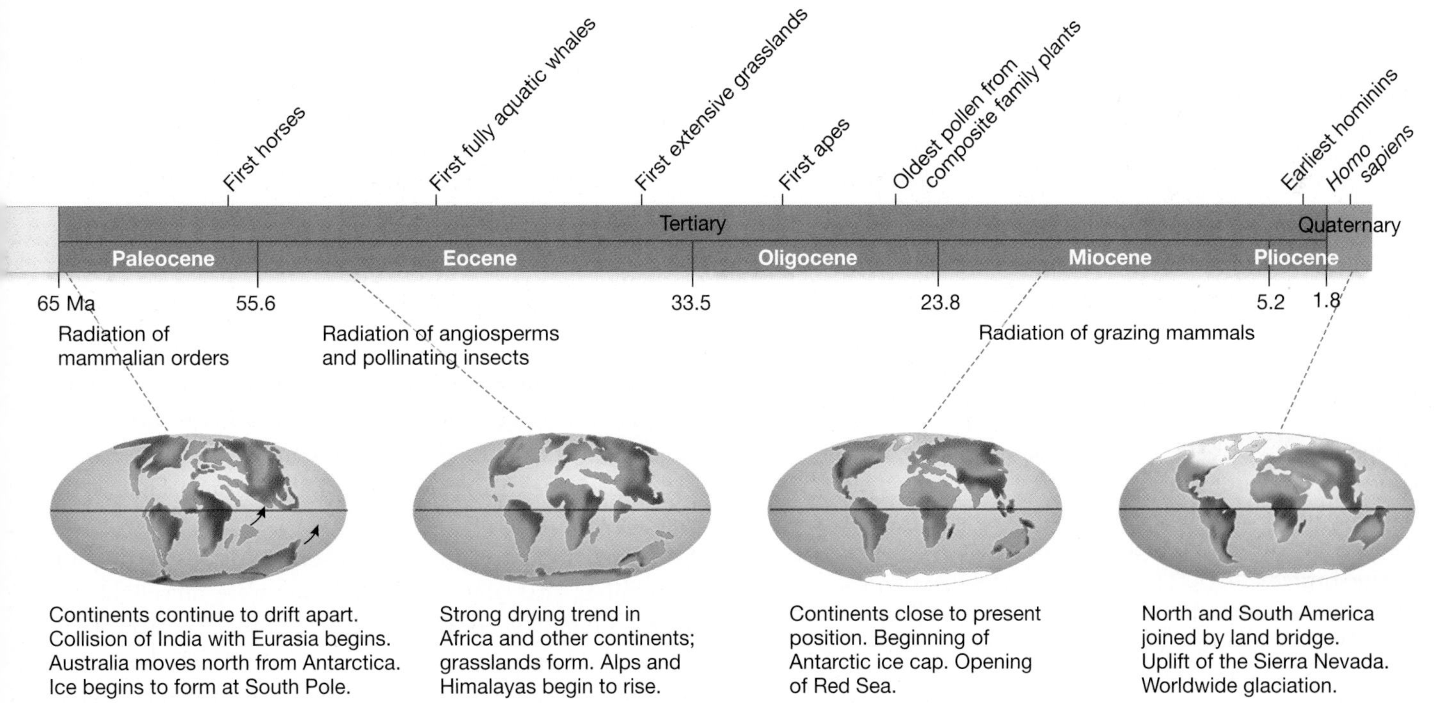

The data in Figure 26.9 show that the nature of life has changed radically since the first cell on Earth appeared. Although the changes that have taken place over the past 3.8 billion years are well documented in the fossil record, the sweep of time involved is difficult for the human mind to comprehend. A semester can seem long to a college student, but biologists who study the fossil record routinely analyze data on events that took place over tens of millions of years. In relation to Earth's history, 100,000 years or even a million years is the blink of an eye.

BOX 26.1 The Molecular Clock

Several researchers have proposed that the fossil record can be supplemented with information on the amino acid and DNA sequences in living species. This hypothesis, called the **molecular clock**, holds that certain types of mutations—such as silent mutations (Chapter 16), or changes in amino acid sequence that do not affect protein function—increase to fixation in populations at a steady rate. For example, consider data on the protein hemoglobin. A biologist compared the amino acid sequence of this molecule in vertebrates ranging from sharks to humans. When he plotted the number of amino acid differences between species against the date that the species diverged, according to the fossil record, the graph shown in **Figure 26.10** resulted. The evolution of hemoglobin seems to proceed at a constant rate over time.

To see how researchers use molecular clocks to supplement the fossil record, let's consider some recent analyses of human evolution. The seven fossil species of *Homo* that have been identified to date, and their estimated interrelationships, are shown in **Figure 26.11.** The lineage leading to both *Homo erectus* and *H. ergaster* and the lineage leading to *H. sapiens* are thought to have split almost 2 million years ago. (The relationship between *H. erectus* and *H. ergaster* is not known.) Two other recent populations of hominins (human-like organisms), called *H. neanderthalensis* and *H. heidelbergensis*, evolved after this split occurred. But it is not yet clear whether these two species represent branches off the *H. erectus* lineage or off the lineage leading to *H. sapiens*. Until more and better-quality fossils can be found and interpreted, researchers are left with a question: When did the most recent common ancestor of our species live? The fossil record indicates that this population existed sometime between the last fossil *H. erectus*, 400,000 years ago, and the first fossil *H. sapiens*, 100,000 years ago.

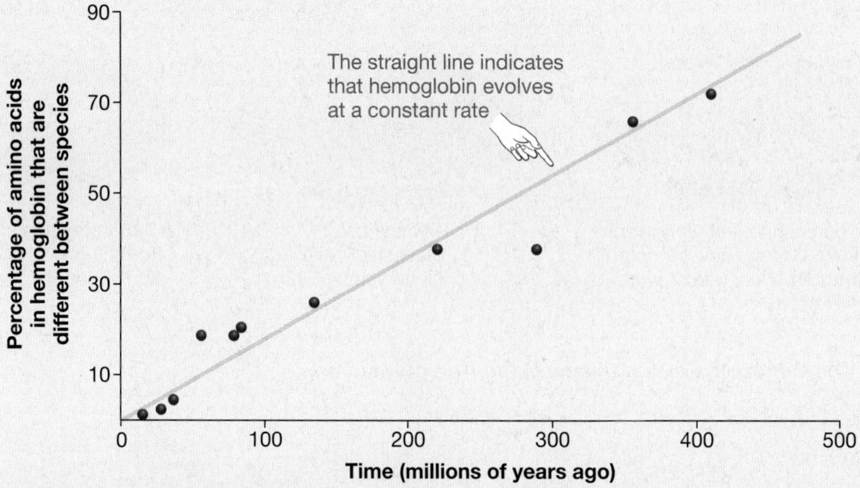

FIGURE 26.10 A Data Set That Supports a Molecular Clock
The rate of change in the hemoglobin protein. To interpret this graph, consider that the point at the upper right represents a comparison of hemoglobin proteins in sharks and mammals. According to the fossil record, sharks branched off from other vertebrates about 450 million years ago. QUESTION How fast does this clock tick? That is, what percentage of amino acids in hemoglobin change per 100 million years, on average?

✓CHECK YOUR UNDERSTANDING

Phylogenies can be estimated by finding synapomorphies that identify monophyletic groups. Using the example of whale evolution, you should be able to explain why the presence and absence of certain SINE genes serve as synapomorphies supporting the hypothesis that hippos and whales recently shared a common ancestor. You should also be able to summarize how whales evolved by synthesizing data from the molecular phylogenies presented in this section with data on the fossil record of whales presented in Chapter 23.

26.2 The Cambrian Explosion

To get a better feel for how biologists use the fossil record to understand the history of life, let's examine recent research on a momentous event—the origin and early diversification of animals. The first animals appear in the fossil record about 570 million years ago. Their early diversification occurred soon after that, at the start of the Cambrian period. This diversification happened so rapidly that it earned a nickname: the **Cambrian explosion.**

It is impossible to appreciate the drama of this event without recalling that almost all life-forms were unicellular for about 2.5 *billion* years after the origin of life. The exceptions were several lineages of multicellular algae, which show up in the fossil record about 1 billion years ago. At that time, and for tens of millions of years afterward, the organisms that were the ancestors of animals were still unicellular, marine creatures with cilia or flagella—probably not too different from some of the aquatic eukaryotes living today. Then, about 565 million years ago, the first animals—sponges, jellyfish, and perhaps simple worms—appear in the fossil record. Just 40 million years later, virtually every major group of ani-

A team of investigators recently used a molecular clock to answer this question. The group compared the complete mitochondrial genomes sequenced from African, European, and Japanese individuals. These data allowed the team to estimate the number of bases in mitochondrial DNA (mtDNA) that had changed as human populations diverged. To estimate how long it took these genetic differences to develop, the researchers also sequenced the mtDNA of our close relatives, the orangutan and African apes. According to the fossil record, orangutans and African apes diverged 13 million years ago. By dividing the number of bases that differ in the mtDNA of orangutans and African apes by 13 million years, the researchers arrived at an estimate that mtDNA changes at the rate of 7×10^{-8} base per base per year.

The total amount of sequence divergence observed between two species should be equal to 2 × (rate of sequence evolution) × (time since divergence occurred). When the researchers used this equation with their estimate of the number of bases in mtDNA that had changed as human populations diverged, they concluded that the last common ancestor of all living humans lived between 125,000 and 161,000 years ago. Large species of mammals typically last in the fossil record for about 1.5 million years prior to extinction. Humans, then, are a relatively young species.

What did hominin populations look like just before this ancestral population split off and the evolution of *Homo sapiens* began? The molecules cannot answer this question. Only the discovery of new fossils can. Even as you read this, it is virtually certain that researchers are at the fossil-rich localities in Africa and Asia that have yielded evidence of our ancestors, looking for traces of the humans who lived about 150,000 years ago.

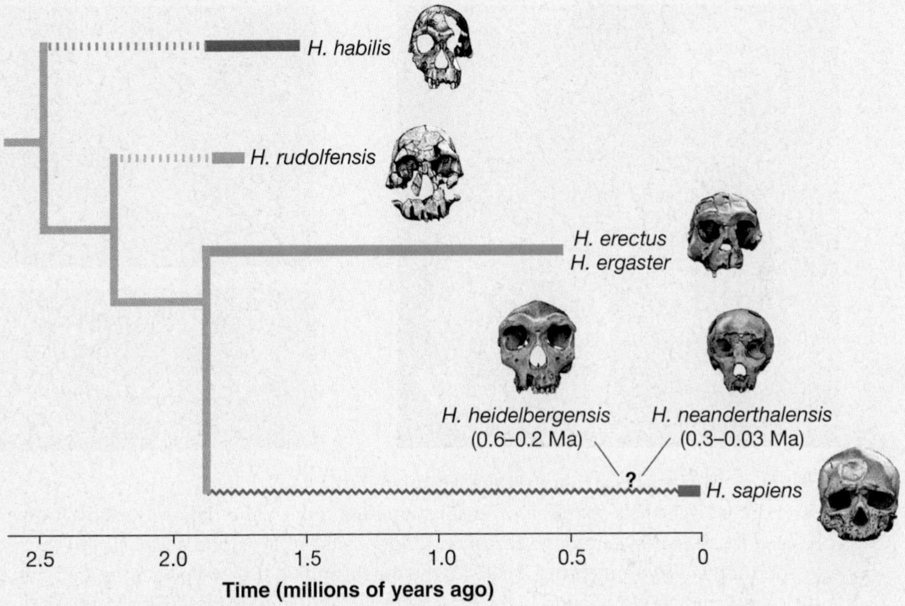

FIGURE 26.11 A Family Tree of Recent Humans
The solid, colored branches on this tree show the range of dates over which fossils of recent hominid species have been found. The dotted lines indicate dates for which researchers expect to find fossils of each species. The line leading to *Homo sapiens* is jagged because it is still not clear where *H. neanderthalensis* and *H. heidelbergensis* fit along this branch.

mals was represented. In a relatively short time, creatures with shells, exoskeletons, legs, heads, tails, and segmented bodies had evolved.

Section 26.3 explores *how* the Cambrian explosion occurred. Here let's focus on *what* happened. By combining evidence from the fossil record and phylogeny reconstruction, researchers have begun to clarify the timing and sequence of key events.

Cambrian Fossils: An Overview

The Cambrian explosion is documented by three major fossil assemblages that record the state of animal life at 570 Ma, at 565 to 544 Ma, and at 525 to 515 Ma. The species collected from each of these intervals are referred to respectively as the Doushantuo fossils (from the Doushantuo formation in China), Ediacaran fossils (from Ediacara Hills, Australia), and Burgess Shale fossils (from British Columbia, Canada), as **Figure 26.12a** (page 568) shows. Fossils from the Ediacaran interval and Burgess Shale interval have now been found at localities throughout the world.

Fortunately for biologists, the glimpses of life provided by the three groups of organisms are extraordinarily clear. Each of the assemblages breaks a cardinal rule of fossil preservation. Soft-bodied animals, which usually do not fossilize efficiently, are well represented in all three. By sheer luck, a few habitats happened to exist at each of these time intervals in which burial occurred so rapidly and decomposition occurred so slowly that organisms without shells were able to fossilize.

To appreciate why this extraordinarily efficient preservation is so important, consider the Burgess Shale. In several localities around the world dated to 525–515 million years ago, fossilization processes were typical—only shelled organisms are found. But soft-bodied organisms also fossilized in the atypical conditions of the Burgess Shale. At this locality, over *five times*

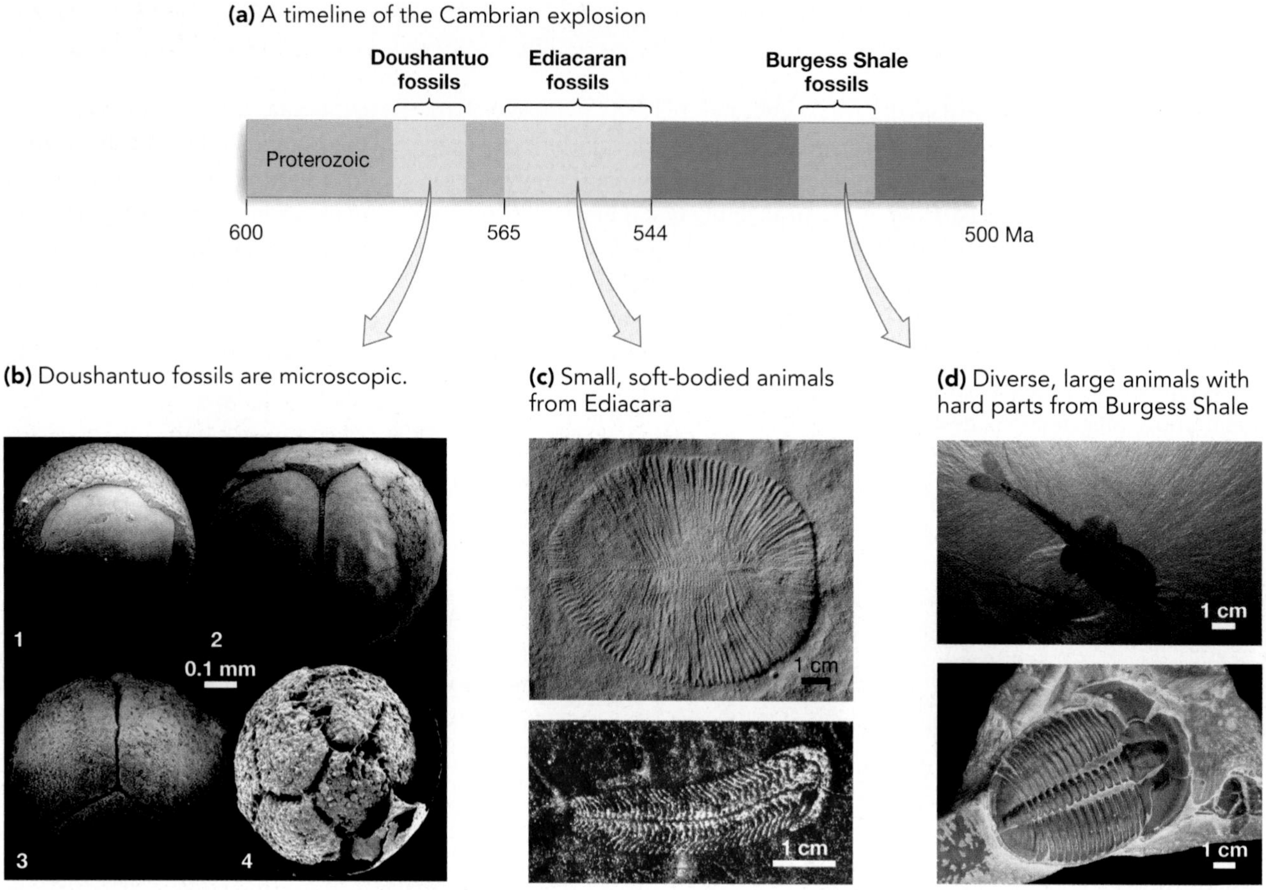

FIGURE 26.12 Fossils Document the Cambrian Explosion
(a) The origin of animals and their diversification during the Cambrian explosion is documented by three major fossil assemblages. **(b)** These fossilized cells from the Doushantuo deposits are arranged according to the hypothesis that they represent (1) a fertilized egg and then embryos at (2) the 2-cell, (3) the 4-cell, and (4) the 16-cell stages. **(c)** These 560 million-year-old fossils of small, soft-bodied animals were found in the Ediacara Hills of Australia. **(d)** Similar to this shrimplike arthropod (top) and trilobite (bottom), many of the animals fossilized in the Burgess Shale had heads, tails, shells (exoskeletons), and appendages.

as many species are represented. The database is 500 percent better than usual.

The presence of these exceptionally rich deposits before, during, and after the Cambrian explosion makes the fossil record for this event extraordinarily complete. Let's take a detailed look at the **faunas**—the animal types—found in the Doushantuo, Ediacaran, and Burgess Shale fossils.

The Doushantuo Microfossils

Two papers, published within days of each other in 1998, introduced the world to the faunas preserved in the Doushantuo formation of southern China. These phosphate rocks, mined extensively for fertilizer, contain abundant submillimeter-sized fossils. Through careful preparation and microscopy, researchers were able to identify several dozen individual sponges, ranging from 150 μm to 750 μm across, in samples dated at approximately 580 million years ago. Several cell types are present in these fossils, and some cell bodies contain spicules (projections) made of silica-containing compounds. Both of these traits are typical of sponges living today.

In slightly younger deposits found in the formation, dated to about 570 million years ago, a different team of biologists found clusters of cells that they interpreted as animal embryos. This conclusion was based on a simple observation: Their samples contained one-celled, two-celled, four-celled, and eight-celled fossils, along with individuals containing larger cell numbers whose overall size was the same (**Figure 26.12b**). Recall from Chapter 21 that this is exactly the pattern that occurs during cleavage in today's animals. In other words, cell number increases but total volume remains constant. What type of animal did such embryos develop into? The answer is still unknown.

The animals isolated from both samples were scattered among abundant cyanobacterial cells, as well as multicellular algae that may represent early members of what you might know as "seaweeds:" the red, brown, and green algal lineages. These bacteria and algae were undoubtedly photosynthetic. The composite picture, then, is of a shallow-water marine habitat dominated by photosynthetic organisms. Scattered among them were tiny animals that may have made their living by filtering organic debris from the water. They were the first animals on Earth.

The Ediacaran Faunas

In the 1940s paleontologists discovered a variety of animal fossils in the Ediacara Hills of southern Australia. The specimens included the compressed bodies of sponges, jellyfish, and comb jellies, and many burrows, tracks, and other traces from unidentified animal species (**Figure 26.12c**). No animals with shells were present, however. In the decades since the initial discovery, similar faunas that are dated between 565 and 544 million years ago have been found at sites around the world.

Taken together, the fossils from this 20-million-year interval indicate that shallow-water marine habitats contained a diversity of animal species. None of the organisms that fossilized during this period have limbs, however, and none have heads or mouths or feeding appendages. As the scale bars in Figure 26.12c show, these organisms were also tiny compared with today's animals. These observations suggest that Ediacaran animals were small individuals that burrowed in sediments, sat immobile on the sea floor, or floated in the water. There is no evidence that they had structures associated with actively hunting and capturing food. Instead, it is likely that Ediacaran animals simply filtered or absorbed organic material from their surroundings.

The Burgess Shale Faunas

The discovery of fossils in the Burgess Shale formation of British Columbia, Canada, early in the twentieth century ranks among the most sensational additions ever made to the fossil record. Combined with the later unearthing of an extraordinary fossil assemblage in the Chengjiang deposits of China, the Burgess Shale gives researchers a compelling picture of life in the oceans 525–515 million years ago.

Few, if any, species in the Ediacaran faunas are also found in the Burgess Shale–type assemblages 20–40 million years later. New species of sponges, jellyfish, and comb jellies are abundant; but entirely new groups are present as well. Principal among these are the arthropods and mollusks. Today, the arthropods include the spiders, insects, and crustaceans (crabs, shrimp, and lobsters); mollusks include the clams, mussels, squid, and octopi. But echinoderms (sea stars and sea urchins), several types of worm and wormlike creatures, and even a chordate—the group ancestral to today's vertebrates—are found in these fossil faunas. In short, virtually every major animal lineage is documented there.

This tremendous increase in the size and morphological complexity of animals, illustrated in **Figure 26.12d**, was accompanied by diversification in how they made a living. The Cambrian seas were filled with animals that had eyes, mouths, limbs, and shells. They swam, burrowed, walked, ran, slithered, clung, or floated; there were predators, scavengers, filter feeders, and grazers. The diversification of the animals created and filled many of the ecological niches found in shallow-water marine habitats today.

The Cambrian explosion still echoes. Animals that fill today's teeming tide pools, beaches, and mudflats trace their ancestry to species preserved in the Burgess Shale.

26.3 The Genetic Mechanisms of Change

The Doushantuo, Ediacaran, and Burgess Shale faunas document relatively rapid and fundamental changes in the size and complexity of multicellular animals. Now the focus shifts to a new question: *How* did the major changes chronicled in the fossil record occur?

A remarkable coalition of scientists is assembling to answer that question. Paleontologists, comparative anatomists, developmental biologists, and molecular geneticists are all contributing data aimed at clarifying the genetic basis for novel structures such as heads, tails, and limbs. This research program is often called **evo-devo**, because it combines evolutionary and developmental studies. To explore this field, let's consider two important types of mutation that can make major innovations possible: gene duplications and changes in gene expression.

Gene Duplications and the Cambrian Explosion

Chapter 22 emphasizes that the proteins coded by *homeotic genes* are responsible for laying out the three-dimensional pattern of multicellular organisms as they develop. Homeotic genes help specify the location and shape of many key structures. As soon as developmental biologists discovered how important these genes are in specifying the tissues and structures found in animals, they began asking about their role in evolution. Did the origin and elaboration of homeotic genes trigger the origin and elaboration of animal body shapes and appendages that occurred during the Cambrian explosion? Stated another way, was the increase in morphological complexity during the Cambrian caused by an increase in the number and complexity of homeotic genes?

To answer this question, biologists are working to determine the number and identity of homeotic genes found in different animal groups and to compare them with the phylogeny of the same groups. The idea is to look for correlations between the evolutionary history of animal groups, their morphology, and their genetic makeup. Many researchers predicted there would be a strong association between the order in which animal groups appeared during evolutionary history, the number of homeotic genes present in each group, and each group's morphological complexity and body size. Some biologists even suggested that each major animal lineage would have unique homeotic genes associated with its unique body plans and appendages.

The logic behind this "new genes, new bodies" hypothesis was that gene duplication events could have occurred before and during the Cambrian explosion and produced new copies of existing homeotic genes. (Look back at Chapter 20 to review how gene duplication events occur.) These new genes would make possible the new body plans and appendages recorded in the Burgess Shale fauna. Again, the idea was that the number of homeotic genes present would correlate directly with morphological complexity.

To see if the predictions made by this hypothesis hold up, use **Figure 26.13** to analyze the homeotic genes called the *Hox* genes. To help interpret the data, recall from Chapter 20 that *Hox* genes are always found in clusters, with genes lined up on the chromosome one after the other. Recall, too, that each gene in the *Hox* cluster has a distinct function as an embryo develops. Specifically, each *Hox* gene is involved in a different aspect of pattern formation—the events that organize cells in the space inside an embryo. Finally, note that each *Hox* gene in Figure 26.13 is color coded to indicate homology with other genes. If genes have the same color, it means their DNA sequences are so similar that researchers are confident the genes are related by common descent.

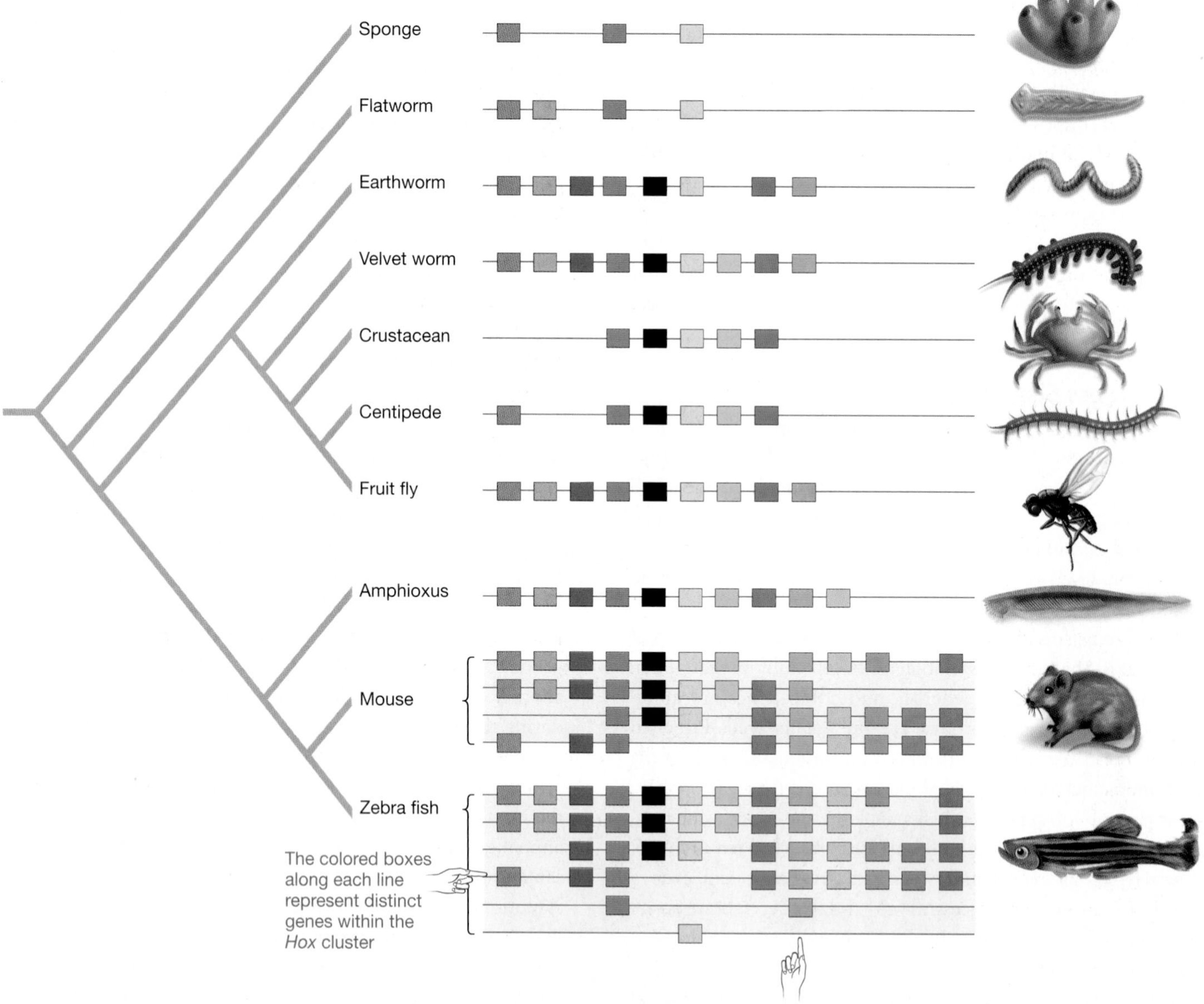
Sponge
Flatworm
Earthworm
Velvet worm
Crustacean
Centipede
Fruit fly
Amphioxus
Mouse
Zebra fish
The colored boxes along each line represent distinct genes within the *Hox* cluster
The same-colored boxes within a vertical column are considered to be homologous—meaning they are related because they are derived from the same ancestral sequence

FIGURE 26.13 ***Hox* Genes in Animals**
Hox clusters, represented by a horizontal line. (Note that mice have four distinct *Hox* clusters, and zebra fish have six.) **EXERCISE** Next to the illustration of each animal on this evolutionary tree, write the number of *Hox* genes it has.

How do the genes found in different lineages map onto the animal tree? Figure 26.13 illustrates the following points:

- As predicted by the new genes, new bodies hypothesis, the number and identity of *Hox* genes varies widely among animals.
- Groups that branched off early and have small, simple bodies, such as sponges, have fewer *Hox* genes than do groups that branched off later—such as vertebrates. The *Hox* cluster appears to have expanded during the course of evolution (from top to bottom in Figure 26.13). This observation also supports the new genes, new bodies hypothesis.
- It is sensible to argue that the new *Hox* genes were created by gene duplication events, because genes within the cluster are similar in their structure and base sequence. For example, the genes that are colored orange in Figure 26.13, which appear first in flatworms, probably originated when a mutation resulted in a duplication of the gold-colored gene, which is present in sponges. The two important points here are that (1) biologists can use the phylogeny to understand the order in which different genes within the cluster appeared, and (2) gene duplication provides a mechanism for the "new genes, new bodies" hypothesis.
- Vertebrates, represented in Figure 26.13 by the mouse and the zebra fish, have several copies of the entire *Hox* cluster. This observation implies that the complete set of 13 genes was duplicated several times. Because all vertebrates examined to date have the duplicated clusters, the mutations probably occurred close to the origin of the lineage. Vertebrates include some of the largest and most complex animals, so this observation also supports the new genes, new bodies hypothesis. But it is important to note from the figure that in some cases, copies of genes within the duplicated clusters have been lost. Mice, for example, have no pink-colored gene in their second set of *Hox* genes.

The data would certainly appear to support a simple-minded version of the new genes, new bodies hypothesis, except for one crucial observation: Within arthropods (represented by the fruit fly, centipede, and crustacean) and within vertebrates, there is no correspondence at all between the number of *Hox* genes and the complexity of the resulting organisms. For example, fish evolved much earlier than mammals did, and the bodies of mammals are considered more complex than are those of fish; but zebra fish have more *Hox* genes than mice do.

Clearly, the situation is more involved than initially predicted. Duplication of *Hox* genes has undoubtedly been important in making the elaboration of animal body plans possible. But new genes are not the whole story: Changes in the expression and function of existing genes have been equally or even more important.

Changes in Gene Expression: The Origin of the Foot

Recent research on the vertebrate foot provides a compelling example of how changes in gene expression can affect evolution. One of the major innovations during vertebrate evolution was the origin of a limb with feet. This led to the evolution of amphibians, reptiles, and mammals—the **tetrapods** ("four-footed"). As data introduced in Chapter 33 will show, the fossil record indicates that the tetrapod limb evolved from the fins of fish. But tetrapods actually have fewer *Hox* genes than fish have. Might changes in the timing or location of homeotic gene expression be responsible for the fin-to-limb transition?

Biologists have explored this question by comparing the expression of the *hoxd-11* gene and a gene called *Sonic hedgehog* (*Shh*) in the zebra fish and the mouse. In tetrapod embryos, *hoxd-11* is expressed as the limbs begin to bud and grow. The gene product marks locations along the long axis of the limb as it grows out. In contrast, the protein produced by *Shh* marks the front-to-back, or anterior-to-posterior, axis of the developing limb. Both genes are involved in pattern formation—the spatial organization of cells inside an embryo.

Are the timing and location of *hoxd-11* and *Shh* expression identical in fish and tetrapods (**Figure 26.14**, page 572)? To answer this question, researchers treated *limb buds*—the tissues that develop into fins and limbs—with a molecule that hybridizes to the mRNA molecule transcribed from *hoxd-11*. (This technique is called in situ hybridization; see Chapter 21.) They also stained limb buds with a molecule that hybridizes to the gene transcript produced by *Shh*. The two treatments allowed them to identify when and where these genes begin producing their protein product.

When the investigators treated the fish and mammal limb buds early in development, they found no differences in the pattern of gene expression. As part (a) in the "Results" of Figure 26.14 shows, *hoxd-11* gene transcripts appear in the hindmost part of the developing limb in both the fish and the mouse. Similarly, *Shh* transcripts are found in similar amounts and locations in fish and mammal limbs at this stage of development.

When the biologists performed the same experiment late in development, however, a striking difference emerged. In mouse limbs, the location of *hoxd-11* transcripts shifts. Study the late limb bud from a mouse embryo, shown in part (b) in the "Results" of Figure 26.14, and note that the *hoxd-11* product localizes to the part of the structure that is away from the body and faces toward the head. There is also a late expression of *Shh*. But neither of these events occurs in fish. Part (b) of the "Results," for example, shows that *hoxd-11* is expressed in small amounts and only in the hindmost cells late in zebra-fish limb development. The data document a dramatic difference in the timing and location of gene expression in zebra fish versus mice.

The researchers who did the study propose that this difference was an innovation that produced the first hand and foot

Question: Does gene expression differ between the fish fin and the tetrapod limb?

Hypothesis: Genes that help organize developing limbs have different patterns of expression in fins and limbs.
Null hypothesis: Patterns of expression in pattern-formation genes are alike in fins and limbs.

Experimental setup: Stain developing limb buds with molecules that attach to *hoxd-11* gene products.

Prediction: The location or the timing of *hoxd-11* gene expression, or both, will differ when fish and limb buds are compared.
Prediction of null hypothesis: The location and the timing of *hoxd-11* gene expression will be alike in fins and limbs.

Results:

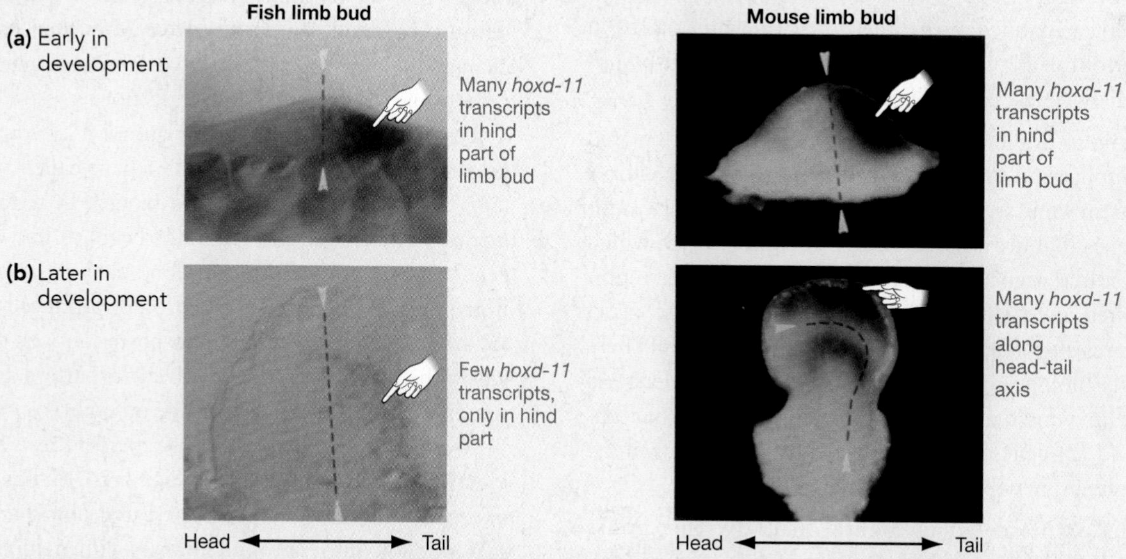

Fish limb bud
Mouse limb bud
(a) Early in development
Many hoxd-11 transcripts in hind part of limb bud
Many hoxd-11 transcripts in hind part of limb bud
(b) Later in development
Few hoxd-11 transcripts, only in hind part
Many hoxd-11 transcripts along head-tail axis
Head
Tail
Head
Tail

Conclusion: There is a late period of *hoxd-11* expression in limbs, oriented in the tail-to-head direction.

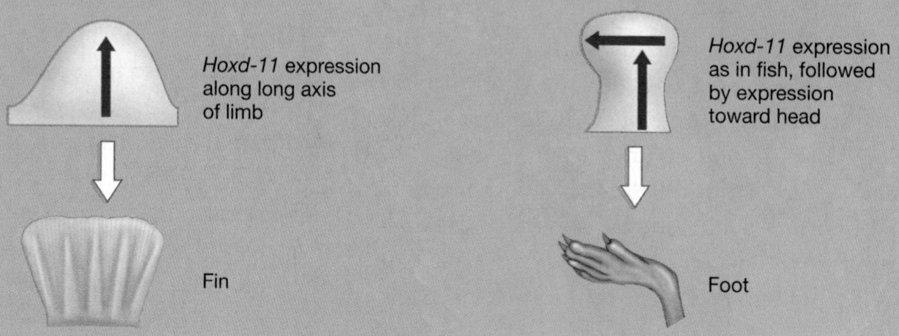

Hoxd-11 expression along long axis of limb
Fin
Hoxd-11 expression as in fish, followed by expression toward head
Foot

FIGURE 26.14 Gene Expression Differs in the Limb Buds of Fish and Mice
(a) In early development, there is no difference in the pattern of gene expression in the fish and mouse limb buds. **(b)** In later development, there is a striking difference in gene expression. This difference may account for the origin of the hand and foot.

tissues in the history of life. The sketches in the "Conclusion" of Figure 26.14 illustrate their hypothesis that the late and "reoriented" expression of *hoxd-11* and *Shh* added an entirely new element to the limb. More specifically, the researchers suggest that the mutations occurred in air-breathing, swamp-dwelling fish that lived about 400 million years ago. They proposed that the mutations were favored by natural selection, because the new structure made movement on land more efficient for individuals in these populations.

The predictions made by this hypothesis are now being tested. For example, the biologists predict that *hoxd-11* and *Shh* expression follows the zebra-fish pattern even in fish groups that branched off very early in their radiation. (Zebra fish belong to a group that emerged millions of years after the tetrapods had evolved from fish). Today the fish lineages most closely related to tetrapods are represented by several lungfish and an unusual species called the coelacanth. Some lungfish use their fins to pull themselves along a substrate. As coelacanths swim, they move their fins in a pattern similar to the walking gait of a tetrapod. How are *hoxd-11* and *Shh* expressed as the limbs of lungfish and coelacanths develop? The experiments needed to answer this question are under way.

✓CHECK YOUR UNDERSTANDING

The emerging field called evo-devo is creating a great deal of excitement among biologists because it offers insight into the genetic changes responsible for major innovations during the history of life. Using the radiation of animals or the origin of the tetrapod foot as an example, you should be able to describe how data from molecular genetics, the fossil record, and phylogenies are being combined to provide a more complete understanding of important evolutionary events.

26.4 Adaptive Radiations

This chapter is about the shape of the tree of life and how and why it has changed through time. Thus far, we have examined the tools biologists use to study the tree of life's past, investigated a major suite of branching events in detail, and probed the genetic mechanisms that underlie major "growth spurts" in the tree of life. Now it's appropriate to ask, If a biologist steps back and looks at the tree as a whole, what broad patterns can be discerned?

When the tree of life is examined from afar, one of the patterns that jumps out is that dense, bushy outgrowths are scattered among the branches. As **Figure 26.15a** shows, this shape results when many large and distinctive groups of organisms branch off from a lineage in a short amount of time. Biologists sometimes call this pattern a *star phylogeny*, because of its starburst shape. Why does this pattern exist? One of the leading causes is a phenomenon known as an adaptive radiation.

An **adaptive radiation** occurs when a single lineage produces many descendant species that live in a wide diversity of habitats and make their living in a variety of ways. When the diversification occurs quickly, it results in a large polytomy or star phylogeny. **Figure 26.15b** shows a few of the Hawaiian honeycreepers—a diverse lineage of songbirds that evolved after a finchlike ancestor happened to colonize the Hawaiian islands. Considering how varied honeycreeper beaks are in size and shape, you should not be surprised that different honeycreepers obtain food by eating insects, sucking nectar, or cracking seeds. **Figure 26.15c** illustrates a few of the Hawaiian silverswords, plants that evolved from a species of tarweed native to California. The silverswords that resulted from the adaptive radiation in Hawaii live in habitats ranging from lush rain forests to austere lava flows, and they range in growth habit from mosslike mat-formers to vines to small trees.

Another classical example of adaptive radiation is the diversification of mammals that took place between 65 and 50 million years ago. During this relatively short interval, the primates (monkeys and apes), bats, carnivores, deer, whales, horses, and rodents originated. The organisms resulting from this rapid divergence represent a remarkable array of adaptive forms. They swim, fly, glide, burrow, swing through trees, walk on four legs, or walk on two legs. They occupy habitats from the open ocean to mountaintops and from rain forests to deserts. They eat fruit, nuts, leaves, twigs, bark, insects, crustaceans, mollusks, fish, and other mammals.

The hallmark of an adaptive radiation is ecological diversification within a single lineage. What makes adaptive radiations occur?

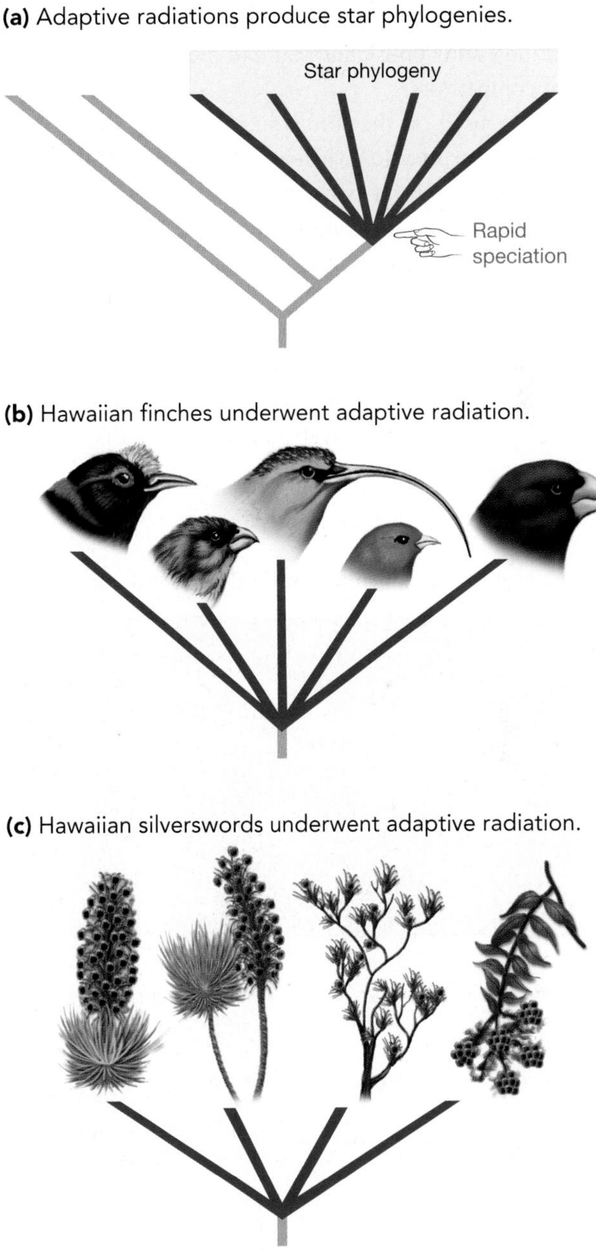
(a) Adaptive radiations produce star phylogenies.
Star phylogeny
Rapid speciation
(b) Hawaiian finches underwent adaptive radiation.
(c) Hawaiian silverswords underwent adaptive radiation.

FIGURE 26.15 Adaptive Radiations Often Produce "Star Phylogenies"
(a) A phylogeny has a starburst shape when rapid speciation has occurred. If speciation is followed by divergence into many different adaptive forms, then an adaptive radiation has taken place. On the Hawaiian islands, adaptive radiations followed the arrival of both a finch from North America and tarweed seeds from California. These radiations produced the Hawaiian **(b)** finches and **(c)** silverswords.

WEB TUTORIAL 26.1 Adaptive Radiation

Colonization Events as a Trigger

One of the most consistent themes in adaptive radiations is opportunity. The radiation of mammals, for example, occurred immediately after the extinction of dinosaurs. As mammals diversified, they took over the ecological roles formerly filled by dinosaurs and swimming reptiles. Adaptive radiations often occur when habitats are unoccupied by competitors.

Recently a group of biologists documented this process in detail. They did not study a radiation that followed an extinction event, however. Instead they analyzed radiations triggered by colonization events on islands that had distinct habitats and were free of competitors.

The study focused on the *Anolis* lizards of the Caribbean. Biologists have interpreted the history of this group of lizards as an adaptive radiation for two reasons: The lineage includes 150 species, and there is a strong correspondence between the size and shape of each species and the habitat it occupies. Most *Anolis* species that are twig-dwellers, for example, have relatively short legs and tails, while those that spend most of their time clinging to broad tree trunks or running along the ground tend to have long legs and tails (**Figure 26.16a**). These data suggest that lizard species have diversified in a way that allows them to occupy many different habitats.

Exactly how did the diversification occur? As the first step in answering this question, the biologists estimated the phylogeny of *Anolis* from DNA sequence data. Then they compared the habitats occupied by each species with their relationships on the phylogenetic tree. The results shown in **Figure 26.16b** for species found on two different islands are typical. Notice that the original colonist on each island belonged to a different ecological type. The initial species on Hispaniola, for example, lived on the trunks and crowns of trees, while the original colonist on Jamaica occupied twigs. From different evolutionary starting points, then, an adaptive radiation occurred on both islands. The key point is that, on both islands, the same four ecological types eventually evolved. New species arose on each island independently, but because both islands had similar habitats, each island ended up with a complement of species that was similar in lifestyle and appearance. The same type of convergent evolution occurred repeatedly. The researchers found the same pattern on many other islands. In other words, a series of "miniature adaptive radiations" occurred—one on each island—within the overall *Anolis* radiation. The small-scale and large-scale radiations were triggered by two conditions: opportunity in the form of both available habitat and lack of competitors.

(a) Species of *Anolis* vary in leg length and tail length. Some species are ground dwelling; others live in distinct regions of shrubs or trees.

(b) The same adaptive radiation of *Anolis* has occurred on different islands, starting from different types of colonists.

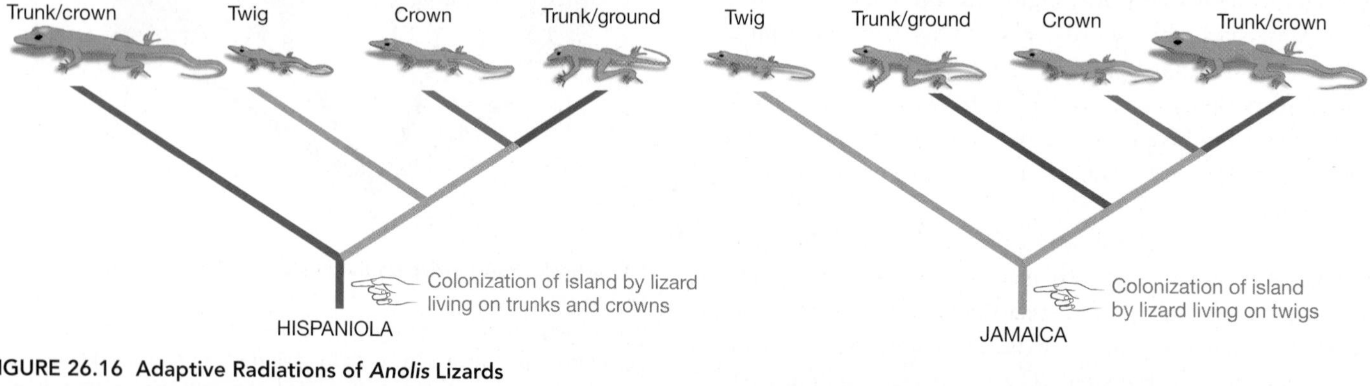

FIGURE 26.16 Adaptive Radiations of *Anolis* Lizards
(a) Short-legged lizard species (left, right) spend most of their time on the twigs of trees and bushes; long-legged species (center) live on tree trunks and the ground. **(b)** The evolutionary relationships among lizard species on the islands of Hispaniola and Jamaica. The initial colonist species was different on these islands; but in terms of how they look and where they live, a similar suite of four species evolved. According to these data, similar adaptive radiations took place independently on the two islands.

(a) Insects have a distinctive body plan.

(b) Flowering plants have a unique reproductive structure.

(c) Cichlids have unique mouths and jaws.

(d) Feathers evolved in the dinosaur ancestors of birds.

FIGURE 26.17 Some Adaptive Radiations Are Associated with Morphological Innovations
(a) The insect body includes several important innovations, including wings, compound eyes, three pairs of legs, and a segmented body organized into three general regions. **(b)** The flower was a morphological innovation that made pollen transfer and reproduction more efficient. **(c)** Cichlid species have distinctive mouthparts, including a pair of specialized jaws located in their throat. [©Don P. Northup www.africancichlidphotos.com] **(d)** The evolution of feathers and flight in dinosaurs triggered the adaptive radiation of birds. Note the long feathers trailing behind the limbs of this theropod dinosaur.

The Role of Morphological Innovation

Other "triggers," besides extinction or colonization events, can ignite adaptive radiations. Foremost among them is morphological innovation—a phenomenon discussed earlier in this chapter. Important new traits such as multicellularity, shells, exoskeletons, and limbs were a driving force behind the adaptive radiation called the Cambrian explosion. The evolution of the limb triggered the diversification of tetrapods. But many of the other important diversification events in the history of life started off with the evolution of a key morphological trait that allowed descendants to live in new areas, exploit new sources of food, or move in new ways:

- The evolution of wings, three pairs of legs, and a protective external skeleton helped make the insects the most diverse lineage on Earth, with perhaps 10 million species today (**Figure 26.17a**).
- Flowers are a unique reproductive structure that triggered the diversification of angiosperms (flowering plants). Today angiosperms are far and away the most species-rich lineage of land plants, with over 250,000 species known (**Figure 26.17b**).
- Cichlids are a lineage of fish that evolved a unique set of jaws in their throat. These second jaws make food processing extremely efficient. Different species have throat jaws specialized for crushing snail shells, shredding tissue from other fish, or mashing bits of algae. Over 300 species of cichlid live in Africa's Lake Victoria alone (**Figure 26.17c**).
- Feathers and wings gave some dinosaurs the ability to fly (**Figure 26.17d**). Today the clade called birds contains about 10,000 species, with representatives that live in virtually every habitat on the planet.

In sum, adaptive radiation, a key pattern in the history of life, is usually associated with a new ecological opportunity or a morphological innovation. After an adaptive radiation, rapid speciation and morphological divergence are tightly coupled.

26.5 Mass Extinctions

Mass extinction events are evolutionary hurricanes. They buffet the tree of life, snapping twigs and breaking branches. They are catastrophic episodes that wipe out huge numbers of species and lineages in a short time, giving the tree of life a drastic pruning. One mass extinction event, about 250 million years ago, nearly uprooted the tree entirely. The end-Permian extinction came close to ending multicellular life on Earth.

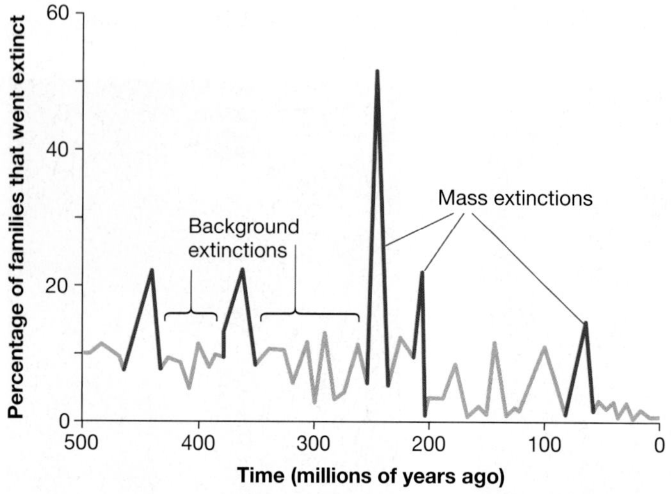

FIGURE 26.18 The Big Five Mass Extinction Events
This graph shows the percentage of families that went extinct over each interval in the fossil record since the Cambrian explosion. The five mass extinction events are drawn in red. **EXERCISE** Using the data in Figure 26.9, label the graph with the names of the geologic periods in which each of The Big Five mass extinctions occurred.

Mass extinction events need to be distinguished from background extinctions. A **mass extinction** refers to the rapid extinction of a large number of lineages scattered throughout the tree of life. More specifically, a mass extinction occurs when at least 60 percent of the species present are wiped out within 1 million years. **Background extinction** refers to the lower, average rate of extinction observed when a mass extinction is not occurring. Although there is no hard-and-fast rule for distinguishing between the two extinction rates, paleontologists traditionally recognize and study five mass extinction events. **Figure 26.18**, for example, plots the percentage of plant and animal families that died out during each stage in the geologic time scale since the Cambrian explosion. Five spikes in the graph—denoting a large number of extinctions within a short time—are drawn in red. These are referred to as "The Big Five."

How Do Background and Mass Extinctions Differ?

Biologists are interested in distinguishing between background and mass extinctions because these events have contrasting causes and effects. Background extinctions are thought to occur when normal environmental change or competition with other species reduces certain populations to the point at which they die out. Mass extinctions, in contrast, are thought to result from extraordinary, sudden, and temporary changes in the environment. During a mass extinction, species do not die out because individuals are poorly adapted to normal conditions. Rather, species die out from exposure to exceptionally harsh, short-term conditions—such as huge volcanic eruptions or rapid sea-level changes. In a general sense, background extinctions are thought to result primarily from natural selection. Mass extinctions, in contrast, function like genetic drift and produce extinctions that are largely random with respect to the fitness of individuals under normal conditions.

To drive these points home, and to see what happens after a mass extinction has occurred, let's examine one of The Big Five in detail. The event we'll analyze, the mass extinction at the end of the Cretaceous period, was not the largest in history. (The mass extinction at the end of the Permian period may have wiped out 90 percent of the multicellular organisms alive at the time.) However, the end-Cretaceous extinction is among the most dramatic: It extinguished the dinosaurs and ushered in the diversification of mammals.

What Killed the Dinosaurs?

The end-Cretaceous extinction of 65 million years ago is as satisfying a murder mystery as you could hope for. Who did it? Decades after biologists recognized the extent of the end-Cretaceous event, the leading hypothesis was that it resulted from climate change. In the early 1970s, however, a researcher offered a dramatic alternative hypothesis. The **impact hypothesis** for the extinction of the dinosaurs claimed that an asteroid struck Earth and caused widespread destruction.

The impact hypothesis for the end-Cretaceous extinction was intensely controversial. As researchers set out to test its predictions, however, support began to grow:

- Worldwide, sedimentary rocks that formed at the Cretaceous-Tertiary (K–T)* boundary were found to contain extremely high quantities of the element iridium. Iridium is extremely rare in Earth rocks, but it is an abundant component of asteroids and meteorites (**Figure 26.19a**).
- Shocked quartz and microtektites are minerals that are found only at documented meteorite impact sites (**Figure 26.19b**). Shocked quartz forms when shock waves from an asteroid impact alter the structure of sand grains. Microtektites form when minerals are melted at an impact site and then cool and resolidify. In Haiti and other locations, both shocked quartz and microtektites have been discovered in abundance in rock layers dated to 65 million years ago.
- A crater the size of Sicily was found just off the northwest coast of Mexico's Yucatán peninsula (**Figure 26.19c**). Microtektites are abundant in sediments from its walls.

Taken together, these data provided conclusive evidence in favor of the impact hypothesis. Researchers now agree that the mystery is solved: An asteroid did it.

Based on currently available data, astronomers and palontologists estimate that the asteroid that struck Earth 65 million years ago was about 10 km across. To get a sense of this

*Geologists use *K* to abbreviate Cretaceous, because C refers to the Cambrian period.

(a) Iridium is present at high concentration in rocks formed 65 million years ago.

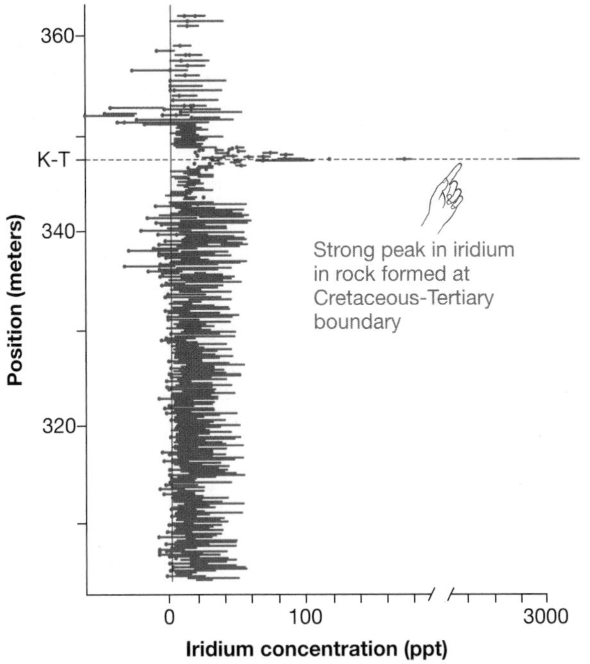

(b) Minerals that form during asteroid impacts

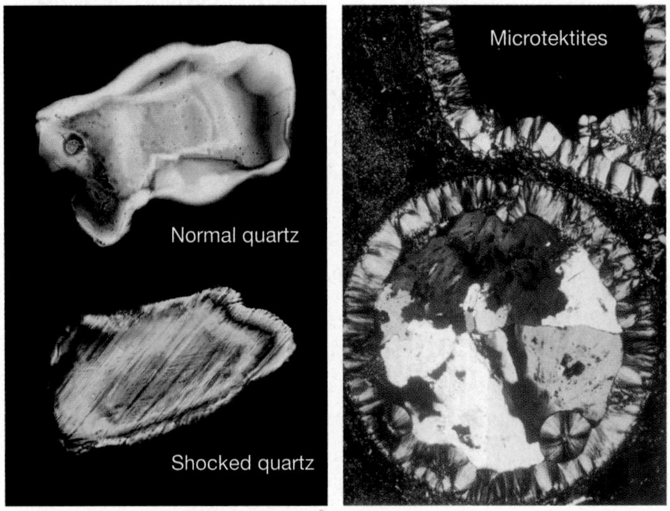

(c) The asteroid left a crater 180 km (112 miles) wide.

FIGURE 26.19 Evidence of an Asteroid Impact 65 Million Years Ago
(a) The concentration of iridium in rocks that formed on either side of the Cretacous-Tertiary (K-T) boundary, in parts per trillion (ppt). **(b)** Normal quartz grains are markedly different from shocked quartz grains. The striations in the shocked quartz are caused by a sudden increase in pressure. Microtektites (right) are tiny glass particles formed when minerals melt at an impact site and then recrystallize. **(c)** Geologists have identified the walls of a crater off the northwest coast of the Yucatán Peninsula, dated to 65 million years ago.
QUESTION Which of these three pieces of evidence for an asteroid impact do you judge to be the strongest, and why?

scale, consider that Mt. Everest is about 10 km above sea level and that planes cruise at an altitude of about 10 km. Imagine Earth being hit by a rock the size of Mt. Everest, or a rock that would fill the space between you and a jet in the sky.

The distribution of shocked quartz and microtektites dated to 65 Ma indicates that the asteroid hit Earth at an angle and splashed material over much of southeastern North America. To understand the impact's consequences, consider the results of the Tunguska event. On June 30, 1908, a piece of a comet about 30 m across, about 2 megatons in mass, exploded at least 5 km above Earth's surface near the Tunguska River in Siberia. The explosion released about 1000 times as much energy as the atomic bomb that destroyed Hiroshima in World War II. The event incinerated vegetation over hundreds of square kilometers, leveled trees over thousands of square kilometers, and significantly increased dust levels across the Northern Hemisphere. A deafening blast was heard 500 km (300 mi) away; people standing 60 km away were thrown to the ground or knocked unconscious. These events pale in comparison to what happened 65 million years ago. The energy the K-T asteroid unleashed was 37 *million* times greater than the Tunguska event.

According to both computer models and geologic data, the consequences of the K-T asteroid strike were nothing short of devastating. A tremendous fireball of hot gas would have spread from the impact site; large soot and ash deposits in sediments dated to 65 million years ago testify to catastrophic wildfires, worldwide. The impact site itself is underlain by a sulfate-containing rock called anhydrite. The SO_4^{2-} released by the impact would have reacted with water in the atmosphere to form sulfuric acid (H_2SO_4), triggering extensive acid rain. Massive quantities of dust, ash, and soot would have blocked the Sun for long periods, leading to rapid global cooling and a crash in plant productivity. The upshot? The fossil record suggests that between 60 and 80 percent of all species went extinct.

Selectivity The asteroid impact did not kill indiscriminately. Perhaps by chance, certain lineages escaped virtually unscathed while others vanished. Among vertebrates, for example, the dinosaurs, pterosaurs (flying reptiles), and all of the large-bodied marine reptiles (mosasaurs, ichthyosaurs, and plesiosaurs) expired; mammals, crocodilians, amphibians, and turtles survived.

Why? Answering this question has sparked intense controversy and debate. For years the leading hypothesis was that the K-T extinction event was size selective. The logic here was that the extended darkness and cold would affect large organisms disproportionately, because they require more food than do small organisms. But extensive data on the survival and extinction of marine clams and snails has shown no hint of size selectivity, and small-bodied and juvenile dinosaurs perished along with large-bodied and adult forms. One hypothesis currently being tested is that organisms that were capable of inactivity for long periods—by hibernating or resting as long-lived seeds or spores—were able to survive the catastrophe. But this aspect of the mystery is still unsolved.

Recovery After the K-T extinction, fern fronds and spore dominate the plant fossil record from North America and Australia. These data suggest that extensive stands of ferns replaced diverse assemblages of cone-bearing and flowering plants after the impact and that terrestrial ecosystems were radically simplified. In marine environments, some invertebrate groups do not exhibit normal levels of species diversity in the fossil record until 4–8 million years past the K-T boundary. Recovery was slow.

The organisms present in the Tertiary were markedly different from those of the preceding period. One prominent change was that mammals took the place of the dinosaurs and marine reptiles. The lineage called Mammalia, which had consisted largely of rat-sized predators and scavengers in the heyday of the dinosaurs, exploded after the impact. Within 10–15 million years, all of the major mammalian orders observed today had appeared—from pigs to primates. Why? A major branch on the tree of life had disappeared. With competitors removed, mammals flourished.

The Cretaceous is sometimes called the Age of Reptiles, and the Tertiary the Age of Mammals. The change was not due to a competitive superiority conferred by adaptations such as fur and lactation. Rather, it was due to a chance event: a massive, low-probability impact event.

✓CHECK YOUR UNDERSTANDING

Mass extinctions have occurred repeatedly throughout the history of life. You should be able to evaluate the strength of the evidence supporting the impact hypothesis for the extinction of the dinosaurs, and explain why mass extinctions wipe out species more or less randomly—much the way genetic drift affects changes in allele frequencies.

ESSAY Is a Mass Extinction Event Under Way Now?

As human populations expand, wildlife habitat shrinks. Peter Ward, an expert on the K-T extinction, has reviewed data on accelerating rates of habitat loss around the globe and warned that a "human meteor" is about to strike the planet. The admonition raises a legitimate scientific question: Are humans currently causing a mass extinction event?

Although this is a difficult question to answer, two types of data are relevant. The first dataset involves extinction events over the past 400 years; the second involves very recent changes in the number of threatened species. Both datasets come from a period when human populations have been expanding rapidly (**Figure 26.20**).

Consider data on recently extinct species, tabulated on the right. Although the numbers and percentages appear small at first glance, recall that the time interval involved is just 400 years. If the rates of extinction recorded here were to continue for even a few thousand years, virtually every species in these groups would be wiped out. Based on these data, it appears that a mass extinction event is well under way.

It is important to note, however, that virtually all the extinct species recorded in the table were island forms. On islands, extinctions usually result from overharvesting and from the introduction of nonnative predators and herbivores, such as rats and goats. Island extinctions have already peaked in intensity. In contrast, most current extinctions are occurring in a new place and have a new cause. The extinction threat has shifted to continental habitats, and the causes have shifted to introduced species and habitat destruction—particularly the conversion of tropical wet forests to pastures

Lineage	Species Extinct since 1600	Total Named Species	Percentage of Extinct Species
Mollusks (snails, clams, mussels)	191	100,000	0.20
Crustaceans (shrimp, crabs, lobsters)	4	4000	0.01
Insects	61	1,000,000	0.005
Vertebrates	229	47,000	0.50
Gymnosperms (conifer trees and other plant groups)	2	758	0.30
Dicotyledons (a group of flowering plants)	120	190,000	0.20
Monocotyledons (a group of flowering plants)	462	52,000	0.20

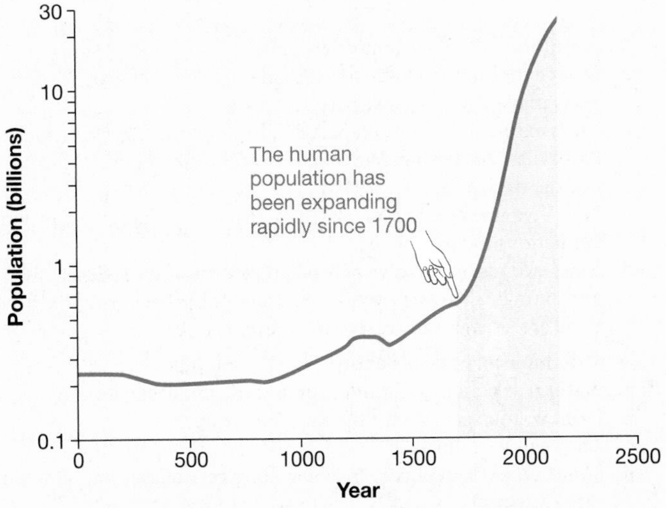

FIGURE 26.20 The Meteoric Rise of Human Populations
The estimated world population size of humans from the year 1 to the year 2150. The line past 2004 is an extrapolation based on the estimate for continued high growth.

and croplands. The destruction of these species-rich habitats is a direct consequence of increasing human populations.

Are humans currently causing a mass extinction event . . . ?

How rapidly are humans currently causing extinctions? The only way to answer this question is to extrapolate from very recent events. From 1986 to 1990, for example, 15 vertebrates were known to go extinct. If this rate of extinction continued, it would take only 7000 years to eliminate half of all known vertebrates. Researchers have also attempted to answer the question by assuming that all species currently considered "threatened"—a formal status conferred by international conservation agencies—will go extinct in the next 100 years. If so, extinction rates in plant and animal lineages would vary from 10 to 100 times their background extinction rates.

The message of these admittedly crude analyses is that the human meteor is probably a real, not an imagined, threat. If present rates of human population growth and habitat destruction continue, a mass extinction event will almost undoubtedly occur.

CHAPTER REVIEW

Summary of Key Concepts

- **Phylogenies and the fossil record are the major tools that biologists use to study the history of life.**

Phylogenetic trees can be estimated by grouping species based on overall similarity in traits or by analyzing shared, derived characters (synapomorphies) that identify monophyletic groups. Phylogenetic trees document the evolutionary relationships among species and identify the order in which events occurred. The fossil record is used in conjunction with phylogenetic analyses because it is the only direct source of data about what extinct organisms looked like and where they lived. Even though the fossil record has strong taxonomic, habitat, and temporal biases, it is an enormously valuable database.

- **The Cambrian explosion was the rapid morphological and ecological diversification of animals that occurred during the Cambrian period.**

The diversification of animals over a 20- to 40-million-year period, starting about 540 million years ago, is perhaps the best-studied event in the history of life. The animals that lived in marine environments just before and after the Cambrian explosion are documented in the Doushantuo, Ediacaran, and Burgess Shale faunas. During this interval the first heads, tails, appendages, shells, exoskeletons, and segmented bodies evolved.

- **The new field of "evo-devo" is providing insights into how major events in the history of life occurred, by revealing the genetic mechanisms involved.**

Data on the number and identity of homeotic genes in different animal lineages suggest that at least some of the animal radiation was possible because gene duplication events created new copies of *Hox* genes. But the correlation between morphological complexity and number of homeotic genes is far from perfect. Changes in the timing and location of existing genes were also responsible for many morphological innovations. For example, changes in gene expression may have been involved in the origin of the hand and foot in tetrapods.

- **Adaptive radiations are a major pattern in the history of life. They are instances of rapid diversification associated with new ecological opportunities and new morphological innovations.**

Adaptive radiations can be triggered by the colonization of a new habitat or the demise of competitors after a mass extinction. Morphological innovations, such as large, complex bodies and limbs with feet and hands, can also initiate adaptive radiations. Speciation events and morphological change occur rapidly during an adaptive radiation, as a single lineage diversifies into a wide variety of ecological roles.

Web Tutorial 26.1 Adaptive Radiation

- **Mass extinctions have occurred repeatedly throughout the history of life. They rapidly eliminate most of the species alive in a more or less random manner.**

Mass extinctions have altered the course of evolutionary history at least five times. They prune the tree of life more or less randomly and have marked the end of several prominent lineages and the rise of new branches. The Cretaceous-Tertiary extinction killed 60 to 80 percent of existing species and was caused by an asteroid impact. After the devastation of a mass extinction, it can take 10–15 million years for ecosystems to recover their former levels of diversity.

Questions

Content Review

1. Choose the best definition of a fossil.
   a. any trace of an organism that has been converted into rock
   b. a bone, tooth, shell, or other hard part of an organism that has been preserved
   c. any trace of an organism that lived in the past
   d. the process that leads to preservation of any body part from an organism that lived in the past

2. Why are the Doushantuo, Ediacaran, and Burgess Shale fossil deposits unusual?
   a. Soft-bodied animals are preserved in them.
   b. They are easily accessible to researchers.
   c. They are the only fossil-bearing rock deposits from their time period.
   d. They include terrestrial, instead of just marine, species.

3. Which of the following best characterizes an adaptive radiation?
   a. Speciation occurs extremely rapidly, and descendant populations occupy a large geographic area.
   b. A single lineage diversifies rapidly, and descendant populations occupy many habitats and ecological roles.
   c. Natural selection is particularly intense, because disruptive selection occurs.
   d. Species recover after a mass extinction.

4. Which of the following is most accurate?
   a. Mass extinctions are due to asteroid impacts; background extinctions may have a wide variety of causes.
   b. Mass extinctions focus on particularly prominent groups, such as dinosaurs; background extinctions affect species from throughout the tree of life.
   c. Only five mass extinctions have occurred, but many background extinctions have occurred.
   d. Mass extinctions involve at least 60 percent of the species present and extinguish groups rapidly and randomly; background extinctions are slower and result from natural selection.

5. Why do molecular clocks exist?
   a. Natural selection is not important at the molecular level.
   b. Homologous genes have the same structure.
   c. They can be calibrated.
   d. Some DNA sequences, in some lineages, change at a steady rate over time.

6. Why is burial a key step in fossilization?
   a. It slows the process of decay by bacteria and fungi.
   b. It allows tissues to be preserved as casts or molds.
   c. It protects tissues from wind, rain, and other corrosive elements.
   d. All of the above.

Conceptual Review

1. The text claims that the fossil record is biased in several ways. What are these biases? If the database is biased, is it still an effective tool to use in studying the diversification of life? Explain.

2. The initial radiation of animals took place over some 40 million years, at the start of the Cambrian period. Why is the radiation called an "explosion"?

3. What is the "new genes, new bodies" hypothesis? Based on the data presented in this chapter, is this hypothesis correct?

4. Give an example of an adaptive radiation that occurred after a colonization event, after a mass extinction, and after a morphological innovation. In each case, provide a hypothesis to explain why the adaptive radiation occurred.

5. Summarize the evidence that supports the impact hypothesis for the K-T extinction.

6. Why are monophyletic groups identified by shared, derived traits?

Group Discussion Problems

1. Suppose that the dying wish of a famous eccentric was that his remains be fossilized. His family has come to you for expert advice. What steps would you recommend to maximize the chances that his wish will be fulfilled?

2. Summarize the nature of the fossils found in the Doushantuo, Ediacaran, and Burgess Shale deposits. (The three fossil assemblages are listed in order, from most ancient to most recent.) According to the fossil record, what trends or general patterns occurred during the early evolution of animals? For example, did animals tend to get larger or smaller? More complex or less complex?

3. Experiments summarized in Section 24.3 suggest that the tetrapod foot resulted from changes in the timing and location of the expression of certain genes. Based on the information presented in Chapter 18 on the regulation of gene expression in eukaryotes, exactly what sort of mutations would lead to changes in when and where genes are transcribed?

4. The synthesis of results from highly disparate fields, such as developmental genetics and fossil studies, is an exciting recent trend in research on the history of life. Comment on whether you find this synthesis valuable. For example, did reading about the "new genes, new bodies" hypothesis and the experiments on gene expression in tetrapod limbs help you understand how important changes have occurred during the diversification of life? Why or why not?

Answers to Multiple-Choice Questions 1. c; 2. a; 3. b; 4. d; 5. d; 6. d

www.prenhall.com/freeman is your resource for the following: Web Tutorials; Online Quizzes and other Online Study Guide materials; Answers to Conceptual Review Questions; Solutions to Group Discussion Problems; Answers to Figure Caption Questions and Exercises; and Additional Readings and Research.

UNIT 6

The Diversification of Life

A red trillium in full bloom. In the lineage that this species belongs to, flower parts tend to occur in multiples of three.

27 Bacteria and Archaea

KEY CONCEPTS

- Bacteria and archaea affect your life. A small percentage of bacteria causes disease. Some species are effective at cleaning up pollution, and photosynthetic bacteria were responsible for the evolution of the oxygen atmosphere. Bacteria and archaea cycle nutrients through both terrestrial and aquatic environments.
- Bacteria and archaea live in a wide array of habitats and use diverse types of molecules in cellular respiration and fermentation. Although they are small and relatively simple in their overall morphologies, these organisms are extremely sophisticated in the chemistry they can do. Many species are restricted in distribution and have a limited diet.

Although this hot spring looks devoid of life, it is actually teeming with billions of bacterial and archaeal cells from a wide variety of species. As this chapter will show, bacteria and archaea occupy virtually every environment on Earth.

Biologists who study organisms in the domains Bacteria and Archaea are exploring the most wide open frontier in biodiversity. So little is known about the extent of these domains that recent collecting expeditions have turned up an entirely new **kingdom** and numerous **phyla** (singular: **phylum**)—names given to major lineages within each domain. To a biologist, this achievement is equivalent to the sudden discovery of a new group of eukaryotes as distinctive as the ferns, sponges, or jellyfish.

Just how many bacterial and archaeal species are alive today? To date, a mere 5000 species have been formally named and described—most by the morphological species concept introduced in Chapter 25. In reality, it is virtually certain that tens of millions exist. Consider that over 400 species of bacteria are living in your gastrointestinal tract right now. Another 500 species live in your mouth, of which just 300 have been described and named. Norman Pace points out that there may be tens of millions of different insect species but notes that "If we squeeze out any one of these insects and examine its contents under the microscope, we find hundreds or thousands of distinct microbial species." Most of these **microbes** (microscopic organisms) are bacteria or archaea, and virtually all are unnamed and undescribed. If you want to discover and name a new species, then study bacteria or archaea.

Although biologists can only guess at the total number of bacterial and archaeal species living today, the abundance of these groups is well documented:

- The approximately 10^{13} (10 trillion) cells in your body are vastly outnumbered by the bacterial cells that live on and in you. An estimated 10^{12} bacterial cells live on your skin, and an additional 10^{14} cells occupy your stomach and intestines. You are a walking, talking habitat—one that is teeming with bacteria and archaea.
- A mere teaspoon of good-quality soil contains *billions* of microbial cells, most of which are bacteria and archaea.
- In sheer numbers, the bacterium *Prochlorococcus*—found in the plankton (surface-dwelling microbes) of the world's oceans—may be the dominant life-form on Earth. Biologists routinely find this organism at concentrations of 70,000 to 200,000 cells per milliliter of seawater. At these concentrations, a drop of seawater contains a population equivalent to that of a large city. Yet *Prochlorococcus* was first described and named fairly recently—in 1988.
- Biologists estimate the total number of individual bacteria and archaea alive today at 5×10^{30}. There is as much carbon in these cells as there is in all of the plants on Earth. In terms of the total volume of living material on our planet, bacteria and archaea are dominant life-forms.

In addition to being diverse and abundant, bacteria and archaea are found almost everywhere. They live in environments as unusual as oxygen-free mud, hot springs, and salt flats. They have been discovered living in bedrock to a depth of 1500 meters below Earth's surface. In the ocean they are found from the surface to depths of 10,000 m and at temperatures ranging from near 0°C in Antarctic sea ice to over 121°C near submarine volcanoes.

The physical world may be mapped and explored, and many of the larger plants and animals are named. But in **microbiology**—the study of organisms that can be seen only with the aid of a microscope—this is an age of exploration and discovery.

27.1 Why Do Biologists Study Bacteria and Archaea?

Bacteria and Archaea form two of the three largest branches on the tree of life (**Figure 27.1**). The third major branch or domain consists of eukaryotes and is called the Eukarya. Virtually all members of the Bacteria and Archaea are unicellular, and all are prokaryotic—meaning they lack a membrane-bound nucleus. These organisms are distinguished by several important features: Species in the domain **Bacteria** have cell walls made primarily of peptidoglycan (see Chapter 7), plasma membranes similar to those found in eukaryotes, and ribosomes and RNA polymerase that are distinct in structure and function from the homologous structures in the Archaea and Eukarya. Species in the domain **Archaea** have cell walls made of polysaccharides not found in bacterial or eukaryotic cell walls, unique plasma membranes (described in detail later), and ribosomes and RNA polymerase similar to those found in eukaryotes.

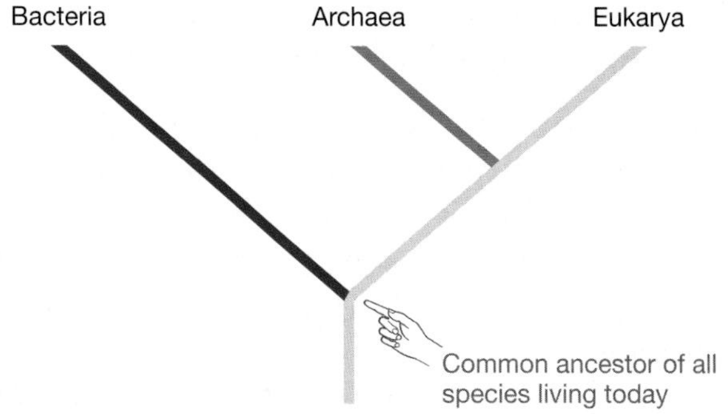

FIGURE 27.1 Bacteria, Archaea, and Eukarya Are the Three Domains of Life
Archaea are more closely related to eukaryotes than they are to bacteria. EXERCISE Circle the prokaryotic lineages on the tree of life. QUESTION Was the common ancestor of all species living today prokaryotic or eukaryotic? Explain your reasoning.

In general, prokaryotic cells are much smaller than eukaryotic cells and are simpler in overall form, or morphology. But in addition to being much more abundant than members of the Eukarya, Bacteria and Archaea are ancestral to eukaryotes and are much more diverse in the habitats they occupy and the molecules they can use as food. These qualities make them exceptionally interesting to study and important agents in medicine, industry, and quality of the environment. Biologists study these organisms not only to understand the diversity of life on Earth but also to improve human health and welfare.

Bacterial Diseases

No archaea are known to cause disease in humans. But of the hundreds or thousands of bacterial species that live in and on your body, a tiny fraction can disrupt normal body functions enough to cause illness. Bacteria that cause disease are said to be **pathogenic** ("disease-producing"). Pathogenic bacteria have been responsible for some of the most devastating epidemics in human history.

Robert Koch was the first biologist to establish a link between a particular species of bacterium and a specific disease. When Koch began his work on the nature of disease in the late 1800s, microscopists had confirmed the existence of the particle-like organisms we now call bacteria, and Louis Pasteur had shown that bacteria and other microorganisms are responsible for spoiling milk, wine, broth, and other foods. Koch hypothesized that bacteria might also be responsible for causing **infectious diseases**, which spread by being passed from an infected individual to an uninfected individual.

Koch set out to test this hypothesis by identifying the organism that causes *anthrax*. Anthrax is a disease of cattle and other grazing mammals that can result in fatal blood poisoning. The disease also occurs infrequently in humans and mice.

To establish a causative link between a specific microbe and a specific disease, Koch proposed that four criteria had to be met:

1. *The microbe must be present in individuals suffering from the disease and absent from healthy individuals.* By careful microscopy, Koch was able to show that the bacterium *Bacillus anthracis* was always present in the blood of cattle suffering from anthrax, but absent from asymptomatic individuals.
2. *The organism must be isolated and grown in a pure culture away from the host organism.* Koch was able to grow pure colonies of *B. anthracis* in glass dishes on a nutrient medium, using gelatin as a substrate.
3. *If organisms from the pure culture are injected into a healthy experimental animal, the disease symptoms should appear.* Koch demonstrated this crucial causative link in mice. The symptoms of anthrax infection appeared, and then the infected mice died.
4. *The organism should be isolated from the diseased experimental animal, again grown in pure culture, and demonstrated by its size, shape, and color to be the same as the original organism.* Koch did this by purifying *B. anthracis* from the blood of diseased experimental mice.

These criteria, now called **Koch's postulates**, are still used to confirm a causative link between new diseases and a suspected infectious agent. Koch's experimental results also became the basis for the **germ theory of disease**. This theory, which laid the foundation for modern medicine, holds that infectious diseases are caused by bacteria and viruses. Some of the bacteria that can cause illness in human beings are listed in **Table 27.1**. The data in the table indicate that pathogenic forms come from many lineages in the domain Bacteria and that pathogenic bacteria tend to affect tissues at the entry points to the body, such as wounds or pores in the skin, the respiratory and gastrointestinal tracts, and the urogenital canal.

The discovery of antibiotics in the late 1920s and their development over subsequent decades gave physicians effective tools to combat most bacterial infections. **Antibiotics** are molecules that kill bacteria. As the essay at the end of this chapter indicates, however, widespread use of antibiotics in the late twentieth century led to the evolution of drug-resistant strains of bacteria. Most of the bacterial species listed in Table 27.1 now include strains that are resistant to one or more of the commonly prescribed antibiotics. Coping with antibiotic resistance in pathogenic bacteria has become a great challenge of modern medicine.

Although biologists began studying bacteria because of the role these prokaryotes play in disease, the scope of research has broadened enormously. Let's consider some topics that are currently inspiring studies of bacteria and archaea.

Bioremediation

Throughout the industrialized world, some of the most serious pollutants in soils, rivers, and ponds consist of organic compounds that were originally used as solvents or fuels but leaked or were spilled into the environment. Most of these compounds

TABLE 27.1 Some Diseases Caused by Bacteria

Bacterium	Lineage	Tissues Affected	Disease
Borrelia burgdorferi	Spirochaetes	Skin and nerves	Lyme disease
Chlamydia trachomatis	Planctomyces	Urogenital canal	Genital tract infection
Clostridium botulinum	Low-GC Gram positives	Gastrointestinal tract, nervous system	Food poisoning (botulism)
Clostridium tetani	Low-GC Gram positives	Wounds, nervous system	Tetanus
Haemophilus influenzae	Proteobacteria (γ group)	Ear canal, nervous system	Ear infections, meningitis
Helicobacter pylori	Proteobacteria (ε group)	Stomach	Ulcer
Mycobacterium leprae	High-GC Gram positives	Skin and nerves	Leprosy
Mycobacterium tuberculosis	High-GC Gram positives	Respiratory tract	Tuberculosis
Neisseria gonorrhoeae	Proteobacteria (β group)	Urogenital canal	Gonorrhea
Propionibacterium acnes	High-GC Gram positives	Skin	Acne
Pseudomonas aeruginosa	Proteobacteria (β group)	Urogenital canal, eyes, ear canal	Infections of eye, ear, urinary tract
Salmonella enteritidis	Proteobacteria (γ group)	Gastrointestinal tract	Food poisoning
Staphylococcus aureus	Low-GC Gram positives	Skin, urogenital canal	Acne, boils, impetigo, toxic shock syndrome
Streptococcus pneumoniae	Low-GC Gram positives	Respiratory tract	Pneumonia
Streptococcus pyogenes	Low-GC Gram positives	Respiratory tract	Strep throat, scarlet fever
Treponema pallidum	Spirochetes	Urogenital canal	Syphilis
Vibrio parahaemolyticus	Proteobacteria (γ group)	Gastrointestinal tract	Food poisoning
Yersinia pestis	Proteobacteria (γ group)	Lymph and blood	Plague

are highly hydrophobic. Because they do not dissolve in water, they tend to accumulate in sediments. If the compounds are subsequently ingested by burrowing worms or clams or other organisms, they can be passed along to fish, insects, humans, birds, and other species. Most of these compounds are toxic to eukaryotes in moderate to high concentrations. Petroleum from oil spills and compounds that contain ring structures and chlorine atoms, such as the family of compounds called dioxins, are particularly notorious because of their toxicity to humans.

It is not unusual for the sediments where these types of compounds accumulate to become **anoxic**—meaning that they lack oxygen. To understand why this is so, recall from Chapter 9 that many organisms use oxygen as an electron acceptor in cellular respiration. Oxygen-using organisms (aerobes) are found among the decomposers that live in soil, river bottoms, and lake sediments. **Decomposers** ingest sugars and other compounds present in dead bodies or other types of organic debris and then use these compounds as electron donors (food) during cellular respiration. If water currents, wave action, or other types of forces are not strong enough to mix sediments frequently, then decomposers that are aerobes use up all of the available oxygen. When sediments become anoxic, the overall rate of decomposition slows dramatically, because organisms must use electron acceptors that are not as effective as oxygen.

Biologists who are responsible for cleaning up sites polluted with organic solvents and fuels face two challenges: (1) The polluted sediments are usually anoxic, so the overall decomposition rate is slow, and (2) at least some of the toxic compounds present are highly resistant to decomposition. Instead of being broken down into harmless compounds, certain toxic molecules tend to just "sit there." These compounds pose a long-term threat to nearby fish, birds, people, and other organisms.

To clean up sites like these, researchers have begun to explore more extensive use of **bioremediation**, the use of bacteria and archaea to degrade pollutants. Bioremediation is often based on complementary strategies:

- *Fertilizing contaminated sites* to encourage the growth of existing bacteria that degrade toxic compounds. After several recent oil spills, researchers fertilized affected sites with nitrogen but left nearby beaches untreated, as controls. In at least some cases, the fertilized sediments cleaned up much faster than the unfertilized sites (**Figure 27.2**). Dramatic increases occurred in the growth of bacteria that use hydrocarbons as an electron donor, probably because the added nitrogen was used to synthesize nucleic acids and amino acids.
- *"Seeding," or adding, specific species of bacteria* to contaminated sites shows promise of alleviating pollution in some situations. For example, researchers have recently discovered bacteria that are able to render certain chlorinated, ring-containing compounds harmless. Instead of being poisoned by the pollutants, these bacteria use ring-containing, chlorinated compounds as electron acceptors during cellular respiration. In at least some cases, the by-product is dechlorinated and nontoxic to humans and other eukaryotes. Because these bacteria do not use oxygen as an electron acceptor, they thrive in anoxic conditions. To follow up on these discoveries, researchers are now growing the bacteria in quantity and testing them in the field, to test the hypothesis that seeding can speed the rate of decomposition in contaminated sediments. Initial reports suggest that seeding may help clean up at least some polluted sites.

FIGURE 27.2 Bacteria and Archaea Can Play a Role in Cleaning Up Pollution
On the left is a rocky coast that was polluted by an oil spill but fertilized to promote the growth of oil-eating bacteria. The portion of the beach on the right was untreated.

Extremophiles

Bacteria or archaea that live in high-salt, high-temperature, low-temperature, or high-pressure habitats are called **extremophiles** ("extreme-lovers"). As an example of these habitats, consider hot springs at the bottom of the ocean, where water as hot as 300°C emerges and mixes with 4°C seawater. At such locations, archaea are abundant forms of life. Researchers recently discovered an archaean that grows so close to these hot springs that its surroundings are at 121°C—a record for life at high temperature. This organism can live and grow in water that is heated past its boiling point (100°C) and at pressures that would instantly destroy a human being.

Extremophiles are attracting a great deal of attention from biologists. The genomes of a wide array of extremophiles have been sequenced, and expeditions regularly seek to characterize new species. Why? Part of the answer is an intrinsic fascination with these organisms. How is it possible for a cell wall and plasma membrane to withstand the water pressure that exists in the deep ocean or the osmotic pressure that exists in a salt pond? What structural changes make it possible for enzymes to function in such extreme cold or heat? In addition, biologists hope that understanding extremophiles will provide insights into how life began. Based on models of conditions that prevailed early in Earth's history, it appears likely that the first forms of life were extremophiles.

There are practical reasons for understanding extremophiles as well. Because enzymes that function at low temperature or high temperature may be useful in industrial processes, extremophiles are of commercial interest. Chapter 19 introduced *Taq* polymerase, which is a DNA polymerase that is stable up to 95°C. Recall that *Taq* polymerase is used to run the polymerase chain reaction (PCR) in research and commercial settings and that its heat stability allowed the PCR procedure to be automated. This enzyme was isolated from a bacterium called *Thermus aquaticus* (literally, "hot water"), which was discovered in hot springs in Yellowstone National Park. In addition, **astrobiologists** ("space-biologists") are using extremophiles as model organisms in the search for extraterrestrial life. The idea is that if bacteria and archaea can thrive in extreme habitats on Earth, it is likely that cells might be found in similar environments on other planets or moons of planets.

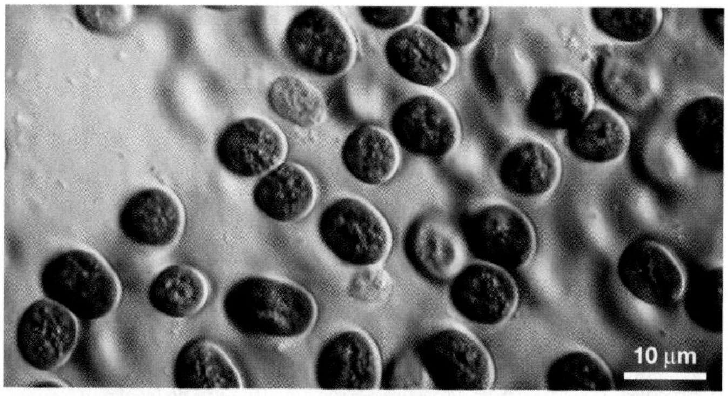

FIGURE 27.3 Cyanobacteria Were the First Organisms to Perform Oxygenic Photosynthesis
Life as we know it today would not have evolved if cyanobacteria had not begun producing oxygen as a by-product of photosynthesis more than 2.7 billion years ago.

Global Change

Bacteria and archaea can live in extreme environments and use a wide variety of compounds as food because they produce extremely sophisticated enzymes. This fact, combined with their numerical abundance, has made them potent forces for global change throughout Earth's history. Bacteria and archaea have altered the chemistry of the oceans, atmosphere, and terrestrial environments for billions of years and continue to do so today.

The Oxygen Revolution Today, oxygen represents almost 21 percent of the molecules in Earth's atmosphere. But climatologists who study the composition of the atmosphere are virtually certain that no free molecular oxygen (O_2) existed for the first 2.3 billion years after the planet formed. This conclusion is based on two observations: (1) There was no plausible source of oxygen at the time the planet cooled to a solid state; and (2) the oldest Earth rocks indicate that, for many years afterward, any oxygen that formed reacted immediately with iron atoms to produce iron oxides, such as hematite (Fe_2O_3) and magnetite (Fe_3O_4). Early in Earth's history, the atmosphere was dominated by nitrogen and carbon dioxide. Where did the oxygen we breathe come from? The answer is cyanobacteria.

Cyanobacteria are a lineage of photosynthetic bacteria (**Figure 27.3**). According to the fossil record, species of cyanobacteria first became numerous in the oceans about 2.7 billion years ago. Their appearance was momentous, because cyanobacteria were the first organisms to perform **oxygenic** ("oxygen-producing") **photosynthesis.** Oxygenic photosynthesis depends on the proteins and pigments in photosystem II. Recall from Chapter 10 that photosystem II includes enzymes capable of stripping electrons from water molecules. The reaction that "splits" water results in the production of oxygen as well as electrons. Although the electrons are required for photosynthesis to continue, the molecules of oxygen are simply released as a waste product.

The fossil record and geological record indicate that oxygen concentrations in the oceans and atmosphere began to increase to significant levels 2.4–2.2 billion years ago. Once oxygen was common in the oceans, cells could begin to use it as an electron acceptor. Aerobic respiration was now a possibility. This was a crucial event in the history of life. Because oxygen is extremely electronegative, it is an efficient electron acceptor. Much more energy is released as electrons fall down electron transport chains with oxygen as the ultimate acceptor than is released with other substances as the electron acceptor. Once oxygen was available, much more ATP could be produced for each electron donated by NADH or $FADH_2$. As a result, the rate of energy production and metabolism could rise dramatically.

Coincidentally, 2.1 billion years ago is about the time the first macroscopic algae appear in the fossil record. Biologists hypothesize that a causal link was involved with the availability of oxygen. The claim is that multicellularity and large body size were made possible by the high metabolic rates and rapid growth fueled by aerobic respiration.

To summarize, data indicate that cyanobacteria were responsible for a fundamental change in Earth's atmosphere—from one dominated by nitrogen gas and carbon dioxide to one dominated by nitrogen gas and oxygen. Never before, or since, have any organisms done so much to alter the nature of our planet.

The Nitrogen Cycle In many environments, fertilizing forests or grasslands with nitrogen results in increased growth. Researchers infer that, in such environments, plant growth and overall productivity are limited by the availability of nitrogen.

Organisms must have nitrogen to synthesize proteins and nucleic acids. Although molecular nitrogen (N_2) is extremely abundant in the atmosphere, most organisms cannot use it. To incorporate nitrogen atoms into amino acids and nucleotides, all eukaryotes and many bacteria and archaea have to obtain N in a form such as ammonia (NH_3) or nitrate (NO_3^-).

The only organisms that are capable of converting molecular nitrogen to ammonia are bacteria. The reactions involved are extremely complex and highly endergonic reduction-oxidation (redox) reactions. The enzymes required to accomplish this feat are found only in selected bacterial lineages.

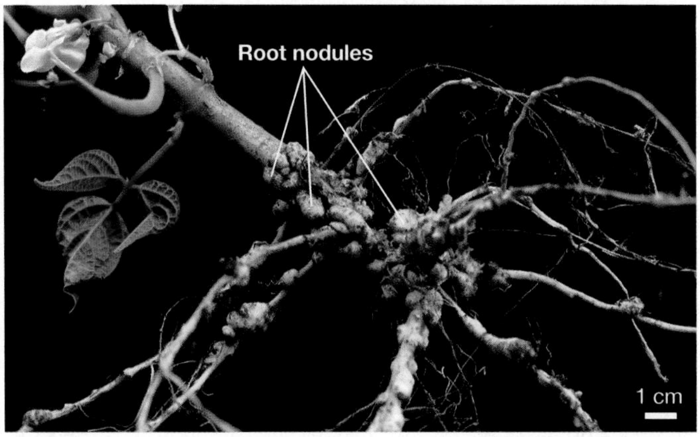
Root nodules
1 cm

FIGURE 27.4 Some Nitrogen-Fixing Bacteria Live in Association with Plants
Root nodules form a protective structure for bacteria that fix nitrogen.

As a whole, the redox reactions that result in the production of ammonia from N_2 are referred to as **nitrogen fixation.** Certain species of cyanobacteria that live in marine plankton or in association with water plants are capable of fixing nitrogen. In terrestrial environments, nitrogen-fixing bacteria live in close association with plants—often taking up residence in special structures in the roots called *nodules* (**Figure 27.4**).

If bacteria could not fix nitrogen, it is virtually certain that only a tiny fraction of life on Earth would exist today. Large or multicellular organisms would probably be rare to nonexistent, because too little nitrogen would be available to make large quantities of proteins and build a large body.

Nitrate Pollution Researchers are fascinated by the nitrogen-fixing bacteria because of their ecological significance, the relationships they form with plants, and their practical importance. If bacterial genes for nitrogen-fixing enzymes could be transferred safely to crop plants such as corn and rice, productivity could increase even if fertilizer use were reduced.

At present, though, most farmers must use synthetic fertilizers to add nitrogen to soils and increase crop yields. In parts of the world, massive additions of nitrogen in the form of ammonia are causing serious pollution problems. **Figure 27.5** shows why. When ammonia is added to a corn field—in the midwestern United States, for example—much of it never reaches the growing corn plants. Instead, a significant fraction of the ammonia molecules is used as food by bacteria in the soil. Bacteria that use ammonia as an electron donor to fuel cellular respiration release *nitrite* (NO_2^-) as a waste product. Other bacteria use nitrite as food and release *nitrate* (NO_3^-). Nitrate molecules are extremely soluble in water and tend to be washed out of soils and carried into groundwater or streams. If nitrate enters groundwater, it can contaminate wells that are used for drinking water. If a person drinks large quantities of nitrate, by-products created inside the body may cause cancer or interfere with hemoglobin's ability to carry oxygen. If nitrate enters streams, it can pollute them and, eventually, the ocean.

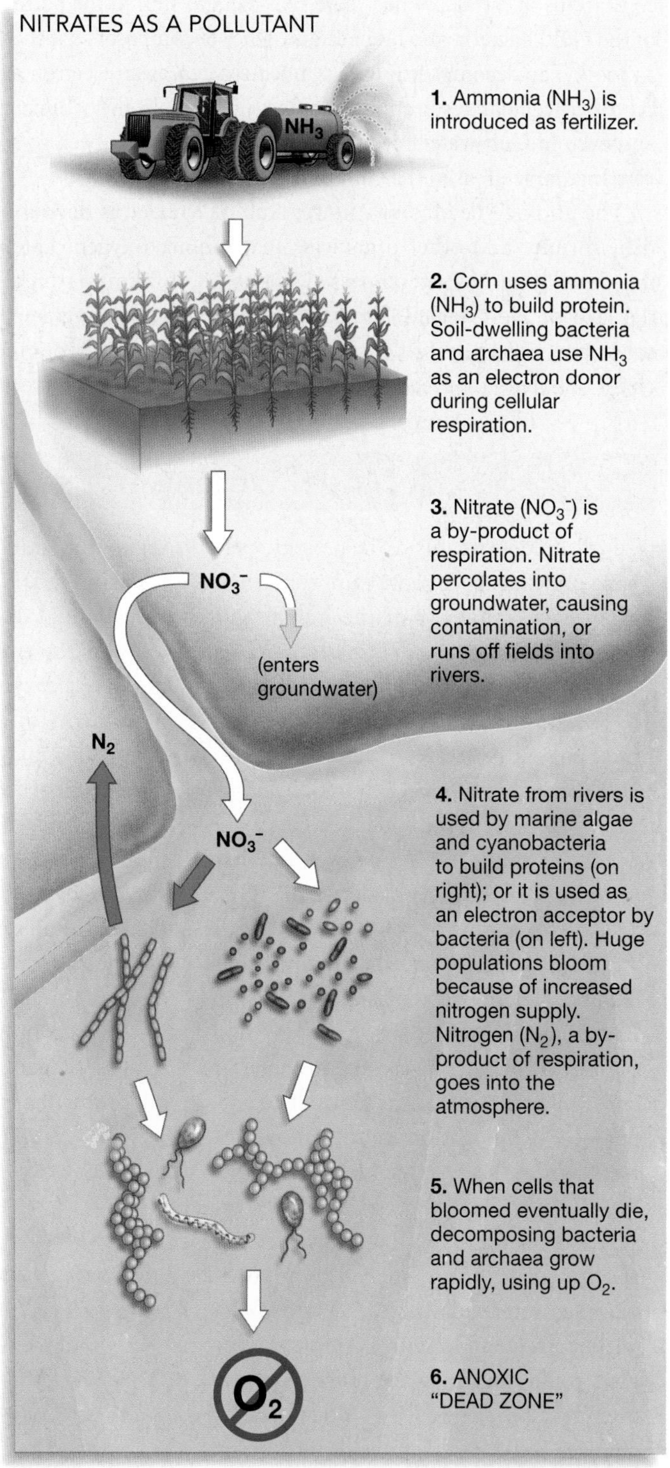
NITRATES AS A POLLUTANT
NH3
1. Ammonia (NH3) is introduced as fertilizer.
2. Corn uses ammonia (NH3) to build protein. Soil-dwelling bacteria and archaea use NH3 as an electron donor during cellular respiration.
NO3-
(enters groundwater)
3. Nitrate (NO3-) is a by-product of respiration. Nitrate percolates into groundwater, causing contamination, or runs off fields into rivers.
N2
NO3-
4. Nitrate from rivers is used by marine algae and cyanobacteria to build proteins (on right); or it is used as an electron acceptor by bacteria (on left). Huge populations bloom because of increased nitrogen supply. Nitrogen (N2), a by-product of respiration, goes into the atmosphere.
5. When cells that bloomed eventually die, decomposing bacteria and archaea grow rapidly, using up O2.
O2
6. ANOXIC "DEAD ZONE"

FIGURE 27.5 Nitrates Act as a Pollutant in Aquatic Ecosystems

To understand why nitrates can pollute the oceans, consider the Gulf of Mexico. Both pollution-related problems and the level of nitrates in the Gulf have increased over the past several decades. Biologists hypothesize that the two events are connected. The hypothesis is that cyanobacteria, along with algae, living in the Gulf of Mexico use as food the nitrates arriving from the Mississippi River. Because the growth of these organisms is limited by nitrogen availability, their numbers explode in response to increased concentrations of nitrate. When these cells die and sink to the bottom of the Gulf, bacteria and archaea and other decomposers use them as food. The decomposers use so much oxygen as an electron acceptor in cellular respiration that oxygen levels in the sediments and even in Gulf waters decline. Nitrate pollution has been so severe that large areas in the Gulf of Mexico are anoxic.

The anoxic "dead zone" in the Gulf of Mexico is devoid of fish, shrimp, and other organisms that require oxygen. Lately the dead zone has encompassed about 18,000 km^2—roughly the size of New Jersey. Similar problems are cropping up in other parts of the world. Virtually every link in the proposed chain of events that leads to nitrate pollution involves bacteria and archaea. But the hypothesized causative link between nitrate pollution and the development of dead zones is controversial and will probably not be resolved until the experimental addition of nitrates is shown to cause anoxic water to form.

27.2 How Do Biologists Study Bacteria and Archaea?

Biologists who study bacteria and archaea claim that they are investigating some of the most interesting and important organisms on Earth. Bacteria have been important model organisms in biological science for many decades (see **Box 27.1**). It

BOX 27.1 A Model Organism: *Escherichia coli*

Of all model organisms in biology, perhaps none has been more important than a common inhabitat of the human gut called *Escherichia coli*. The strain that is most commonly worked on today, called K-12, was originally isolated from a hospital patient in 1922.

Escherichia coli K-12 did not begin to rise to its present prominence as a model organism until 1945, however. In that year, biologists observed that some K-12 cells were capable of conjugation. Recall from Chapter 19 that when conjugation occurs, a copy of a plasmid from one bacterial cell—sometimes joined by one or more genes from the main bacterial chromosome—is transferred to a recipient cell. Conjugation is sometimes referred to as bacterial sex. Its discovery was important, because it gave researchers a mechanism for transferring specific alleles from one cell to another. Conjugation made a wide array of experiments possible. For example, biologists could document the function of the introduced alleles by documenting changes in the recipient cell's phenotype once the transfer process was complete.

During the last half of the twentieth century, key results in molecular biology originated in studies of *E. coli*. These results include the discovery of enzymes such as DNA polymerase, RNA polymerase, DNA repair enzymes, and restriction endonucleases; the elucidation of ribosome structure and function; and the initial characterization of promoters, regulatory transcription factors, regulatory sites in DNA, and operons. In many cases, initial discoveries made in *E. coli* allowed researchers to confirm that homologous enzymes and processes existed in an array of organisms, often ranging from other bacteria to yeast, mice, and humans. Its success as a model for other species inspired Jacques Monod's claim that "Once we understand the biology of *Escherichia coli*, we will understand the biology of an elephant."

The genome of *E. coli* K-12 was sequenced in 1997 (see Chapter 20), and the strain continues to be a workhorse in studies of gene function, biochemistry, and particularly biotechnology.

In the lab, *E. coli* is usually grown in one of two ways: (1) in suspension culture or (2) on plates. In **suspension culture**, cells are introduced to a liquid nutrient medium that is rotated continuously to keep the cells suspended and the nutrient solution mixed (**Figure 27.6a**). When cultured on plates, cells are spread onto **agar**, a gelatinous mix of polysacccharides (**Figure 27.6b**). Under optimal growing conditions—meaning before cells begin to get crowded and compete for space and nutrients—a cell takes just 30 minutes on average to grow and divide. At this rate, a single cell can produce a population of over a million descendants in just 10 hours. With the exception of new mutations, all of the descendant cells are genetically identical.

(a) Growth in liquid medium

(b) Growth on solid medium

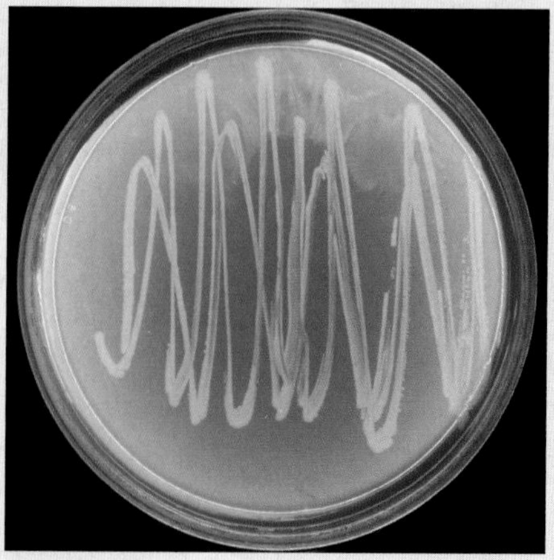

FIGURE 27.6 ***E. coli*** **Is Readily Cultured in the Laboratory**
In the lab, *E. coli* is usually grown **(a)** in suspension culture or **(b)** on plates.
QUESTION *Escherichia coli* is grown at a temperature of 37°C. Why?

is legitimate to state, however, that our understanding of the domains Bacteria and Archaea is advancing more rapidly right now than at any time during the past 100 years—and perhaps faster than our understanding of any other lineages on the tree of life.

As an introduction to the domains Bacteria and Archaea, let's examine a few of the techniques that biologists use to answer questions about them. Some of these research strategies have been used since bacteria were first discovered; some were invented less than 10 years ago.

Using Enrichment Cultures

Which species of bacteria and archaea are present at a particular location, and what do they use as food? To answer such questions, biologists rely heavily on the ability to culture organisms in the lab. Of the 5000 species of Bacteria and Archaea that have been described to date, almost all were discovered when they were isolated from natural habitats and grown under controlled conditions in the laboratory.

One classical technique for isolating new types of bacteria and archaea is called **enrichment culture**. Enrichment cultures are based on establishing a specified set of growing conditions—temperature, lighting, substrate, types of available food, and so on. The idea is to sample cells from the environment and grow them under extremely specific conditions. Cells that thrive under the specified conditions will increase in numbers enough to be isolated and studied in detail.

To appreciate how this strategy works in practice, consider research on bacteria that live deep below Earth's surface. One recent study began with samples of rock and fluid from drilling operations in Virginia and Colorado. The samples came from sedimentary rocks at depths ranging from 860 to 2800 meters below the surface, where temperatures are between 42°C and 85°C. The questions posed in the study were simple: Is anything alive down there? If so, what do the organisms use to fuel cellular respiration?

The research team hypothesized that if organisms were living deep below Earth's surface, such organisms might use hydrogen molecules (H_2) as an electron donor and the ferric ion (Fe^{3+}) as an electron acceptor (**Figure 27.7**). (Recall from Chapter 9 that most eukaryotes use sugars as electron donors and use oxygen as an electron acceptor during cellular respiration.) Fe^{3+} is the oxidized form of iron, and it is abundant in the rocks the biologists collected from great depths. It exists at great depths below the surface in the form of ferric oxyhydroxide. The researchers predicted that, if an organism in the samples reduced the ferric ions during cellular respiration, a black, oxidized, and magnetic mineral called magnetite (Fe_3O_4) would start appearing in the cultures as a by-product of cellular respiration.

What did their enrichment cultures produce? In some culture tubes, a black compound began to appear within a week. Using a variety of tests, the biologists confirmed that the black substance was indeed magnetite. As Figure 27.7 shows, microscopy revealed the organisms themselves—previously undiscovered bacteria. Because they grow only when incubated at

Question: Can bacteria live a mile below Earth's surface?

Hypothesis: Bacteria are capable of cellular respiration deep below Earth's surface by using H_2 as an electron donor and Fe^{3+} as an electron acceptor.

Null hypothesis: Bacteria are not capable of cellular respiration deep below Earth's surface.

Experimental setup:

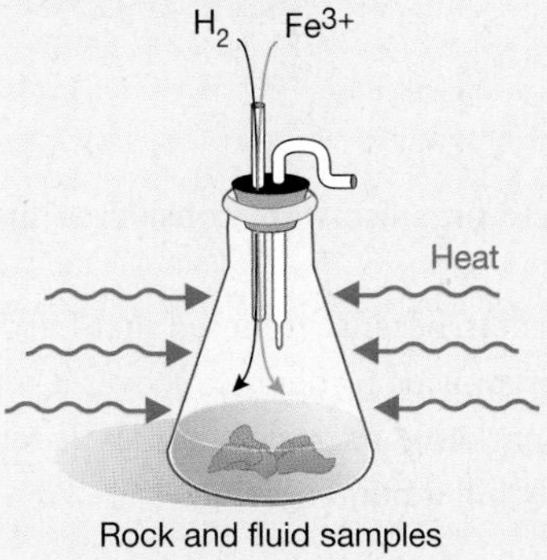

H2
Fe3+
Heat
Rock and fluid samples

Enrichment culture method:

1. Create culture conditions with abundant H_2 and Fe^{3+}; raise temperatures above 45°C.

2. Add rock and fluid samples extracted from drilling operations at depths of about 1000 m below Earth's surface.

Prediction: Black, magnetic grains of magnetite (Fe_3O_4) will accumulate because Fe^{3+} is reduced by growing cells and shed as waste product. Cells will be visible.

Prediction of null hypothesis: No magenetite will appear. No cells will grow.

Results:

Magnetite is detectable, and cells are visible
2 μm

Conclusion: At least one bacterial species that can live deep below Earth's surface grew in this enrichment culture. Different culture conditions might result in the enrichment of different species present in the same sample.

FIGURE 27.7 Enrichment Cultures Isolate Large Populations of Cells That Grow Under Specific Conditions
Bacteria that grow underground can withstand high temperatures. The species enriched in this experiment uses hydrogen (H_2) and ferric ions (Fe^{3+}) as food.

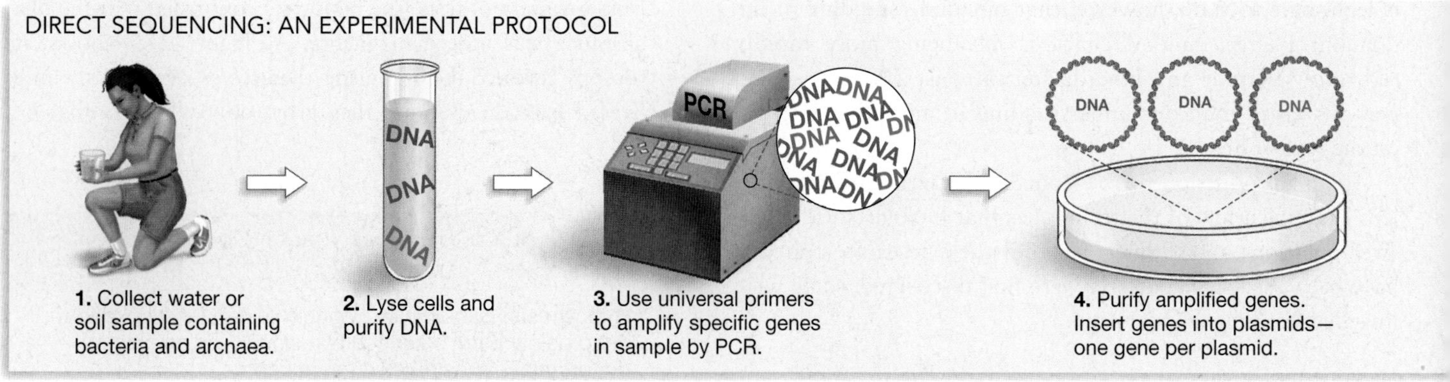
DIRECT SEQUENCING: AN EXPERIMENTAL PROTOCOL
DNA
PCR
1. Collect water or soil sample containing bacteria and archaea.
2. Lyse cells and purify DNA.
3. Use universal primers to amplify specific genes in sample by PCR.
4. Purify amplified genes. Insert genes into plasmids—one gene per plasmid.

FIGURE 27.8 Direct Sequencing Allows Researchers to Identify Species That Have Never Been Seen Direct sequencing allows biologists to isolate specific genes from the organisms present in a sample. The polymerase chain reaction is used to generate enough copies of these genes so that the DNA from different species can be sequenced.

between 45°C and 75°C, these organisms are considered **thermophiles** ("heat-lovers").

This discovery was spectacular, because it hinted that Earth's crust may be teeming with organisms to depths of over a mile below the surface. These remarkable bacteria flourish at temperatures that would instantly kill a human being.

Using Direct Sequencing

Researchers estimate that, of all the bacteria and archaea living today, less than 1 percent have been grown in culture. Although enrichment culture is a valid way to answer questions about which bacteria or archaea are present in a particular sample, it is not an efficient way to get a complete picture of bacterial and archaeal diversity. To augment research based on enrichment cultures, researchers are employing a technique called direct sequencing. **Direct sequencing** is a strategy for documenting the presence of bacteria and archaea that represent phylogenetic species but cannot be grown in culture and studied in the laboratory. Recall from Chapter 25 that a phylogenetic species has enough distinctive characteristics that a phylogenetic analysis places it as an independent twig on an evolutionary tree.

Direct sequencing allows biologists to name and characterize organisms that have never been seen. Direct sequencing has revealed huge new branches on the tree of life and produced revolutionary data on the habitats where archaea are found.

Figure 27.8 outlines the steps performed in a direct sequencing study. To begin, researchers collect a sample from a habitat—a few microliters of water or a few micrograms of soil. Next, the bacterial and archaeal cells in the sample are broken open (lysed), and their DNA is purified. Using the sequence of steps described in Figure 27.8 and methods introduced in Chapter 19, researchers can isolate and analyze the DNA sequences of specific genes from species in the sample. By comparing these data with sequences in existing databases, biologists can determine whether the sample contains previously undiscovered archaea or bacteria.

Direct sequencing studies have produced new and sometimes startling results. For two decades after the discovery of the Archaea, for example, researchers thought that these prokaryotes could be conveniently grouped into just four categories: extreme halophiles, sulfate-reducers, methanogens, and extreme thermophiles. Extreme **halophiles** ("salt-lovers") live in salt lakes, salt ponds, and salty soils. **Sulfate reducers** are species that produce hydrogen sulfide (H_2S) as a by-product of cellular respiration. (H_2S may be familiar because it smells like rotten eggs.) **Methanogens** produce methane (natural gas; CH_4) as a by-product of cellular respiration. Extreme thermophiles grow best at temperatures above above 80°C. Based on these early data, researchers thought that archaea were restricted to extreme environments and that the four phenotypes corresponded to separate lineages within the domain Archaea. As a result of direct sequencing studies, however, these generalizations have been discarded. Beginning in the mid-1990s, direct sequencing revealed archaea in habitats as diverse as rice paddies and the Arctic Ocean. Some of these newly discovered organisms appear to belong to an entirely new lineage, tentatively called the **Korarchaeota**. This group's DNA sequences are so distinctive that it might represent a "kingdom" analogous to plants or animals. Yet the Korarchaeota was identified and named before any of its members had been observed.

Biologists usually know nothing about species that are discovered by direct sequencing, except for their habitat and phylogenetic relationships. Still, the technique has revolutionized our understanding of bacterial and archaeal diversity.

Evaluating Molecular Phylogenies

To put data from enrichment culture and direct sequencing studies into context, biologists depend on the accurate placement of species on phylogenetic trees. Recall from Chapter 1

... AGGCTTACCGTAAC ...
DNA
DNA
DNA
5. Insert plasmids into *E. coli* cells. Each culture contains the gene from a different species in original sample. The cells grow rapidly, producing millions of copies of each gene.
6. Purify genes from plasmids.
7. Sequence the genes; compare with known sequences.

and Chapter 26 that phylogenetic trees illustrate the evolutionary relationships among species and lineages. They are a pictorial summary of which species are more closely or distantly related to others.

Some of the most useful phylogenetic trees for the Bacteria and the Archaea have been based on studies of the RNA molecule found in the small subunit of ribosomes. (See Chapter 16 for more information on the structure and function of ribosomes.) In the late 1960s Carl Woese and colleagues began a massive effort to determine and compare the base sequences of RNA molecules from a wide variety of species. Their goal was to use similarities and differences in the sequence data to infer the evolutionary history of a diversity of organisms. The result of their analysis was shown in Figure 27.1. This diagram came to be known as the *universal tree*, or the *tree of life*.

Woese's tree is now considered a classic result. Prior to its publication, biologists thought that the major division among organisms was between prokaryotes and eukaryotes—between cells that lacked a membrane-bound nucleus and cells that possessed a membrane-bound nucleus (**Figure 27.9**). But based on data from the ribosomal RNA molecule, the major divisions of life-forms are the three groups that Woese named the Bacteria, Archaea, and Eukarya. Follow-up work documented that Bacteria were the first lineage to diverge from the common ancestor of all living organisms. This result means that the Archaea and Eukarya are more closely related to each other than they are to the Bacteria.

Although virtually all biologists accept the three-domains system, it has been very difficult to understand the relationships of major lineages within the Bacteria and Archaea. Analyses of both morphological and molecular characteristics have succeeded in identifying a large series of monophyletic groups within the domains. Recall from Chapter 25 that a **monophyletic group** consists of an ancestral population and all of its descendants. Monophyletic groups can also be called *clades* or *lineages*.

Rapid progress is being made on the question of how the major clades within Bacteria and Archaea are related to each other, and work continues. The phylogenetic tree in **Figure 27.10** (page 592) summarizes recent results but is still considered highly provisional. As more data become available, it is almost certain that at least some of the branches on this tree will change position. In addition, recall that direct sequencing studies have recently led to the discovery of major new groups—meaning that entirely new branches will be added to this tree.

Work on molecular phylogenies continues at a brisk pace. While acknowledging that the tree's resolution will improve as research continues, let's consider some of the innovations that occurred as these groups evolved.

WEB TUTORIAL 27.1
The Tree of Life

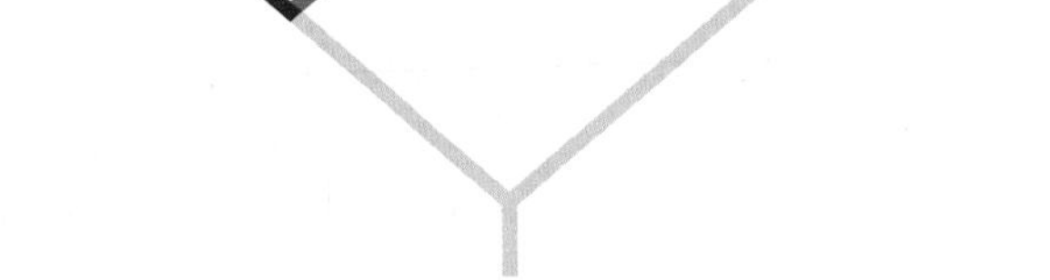
According to morphological similarities, prokaryotes should be closely related
Bacteria
Archaea
Eukarya

FIGURE 27.9 The Tree of Life Based on Morphology Was Incorrect Until recently, the major division among organisms was thought to be between those without a membrane-bound nucleus (the prokaryotes) and those with a membrane-bound nucleus (the eukaryotes). Comparisons of RNA sequences have shown that this tree is not correct. **EXERCISE** Draw the correct tree next to this one. Check your sketch against Figure 27.1. On the correct and incorrect trees, indicate where the nuclear envelope evolved.

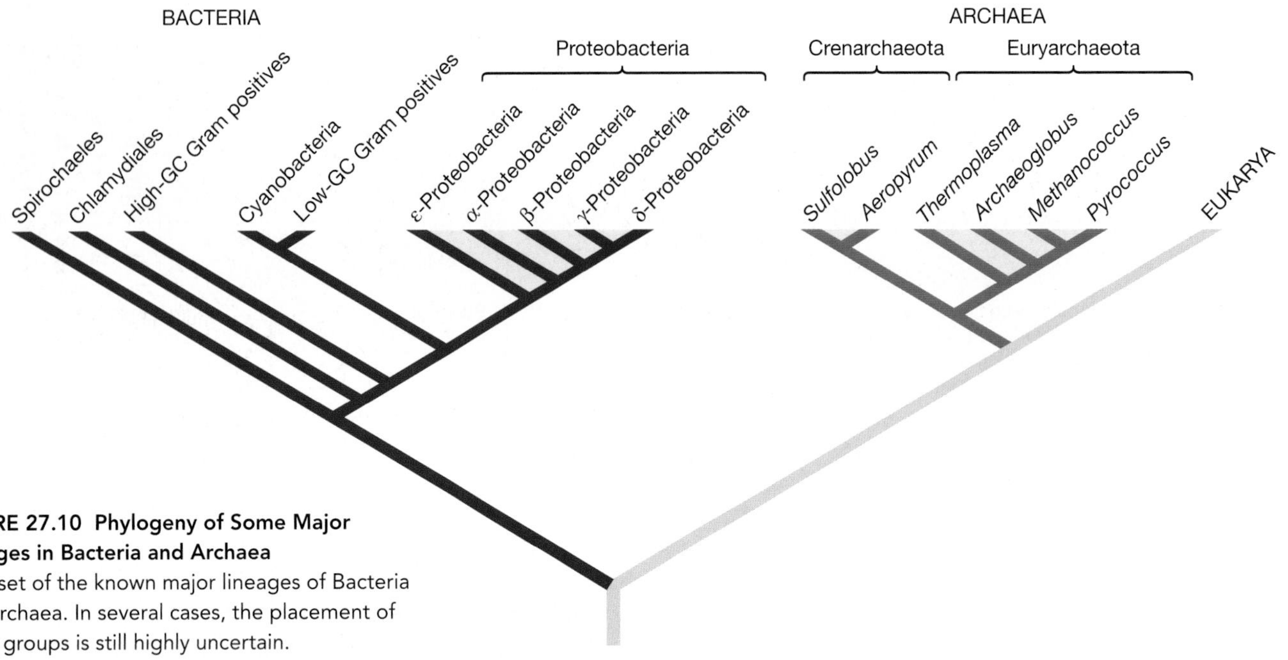

FIGURE 27.10 Phylogeny of Some Major Lineages in Bacteria and Archaea
A subset of the known major lineages of Bacteria and Archaea. In several cases, the placement of major groups is still highly uncertain.

✓CHECK YOUR UNDERSTANDING

Biologists who study bacterial and archaeal diversity use tools called enrichment cultures and direct sequencing. Enrichment cultures are based on setting up specified conditions in the laboratory and isolating the cells that grow rapidly in response. They create an abundant, pure sample of bacteria that thrive under particular conditions and can be studied further. Direct sequencing is based on isolating DNA from samples taken directly from the environment, purifying and sequencing specific genes, and then analyzing where those DNA sequences are found on the phylogenetic tree of Bacteria and Archaea. You should be able to (1) design an enrichment culture that would isolate species that could be used to clean up oil spills, and (2) outline a study designed to identify the bacterial and archaeal species present in a soil sample near the biology building on your campus.

27.3 What Themes Occur in the Diversification of Bacteria and Archaea?

The lineages in the domains Bacteria and Archaea are ancient, abundant, ubiquitous, and diverse. The oldest fossil cells of any type found to date are 3.4-billion-year-old samples that are virtually indistinguishable from contemporary bacteria. Because eukaryotes do not appear in the fossil record until 1.75 billion years ago, and because archaea probably arose at about the same time as eukaryotes, biologists infer that bacteria may have been the only form of life on Earth for almost 1.7 billion years.

Over the past 3.4 billion years, the Bacteria and Archaea have diversified into hundreds of thousands of distinct species that live in every conceivable habitat. What overall patterns and themes help biologists make sense of all the diversity observed among these organisms? Let's answer this question by analyzing morphological and metabolic diversity.

Morphological Diversity

The key morphological and biochemical features of bacteria, archaea, and eukaryotes are summarized in **Table 27.2**. The most important point to note is that even though all of the Archaea and almost all of the Bacteria are unicellular, members of these two domains are strikingly different at the molecular level. Bacteria and Archaea are distinguished by the types of molecules that make up their plasma membranes and cell walls, and by their machinery for transcribing DNA and translating messenger RNA into proteins. More specifically, Bacteria have a unique compound called peptidoglycan in their cell walls, and Archaea have unique phospholipids—containing hydrocarbons called isoprenes in their tails—in their plasma membranes. The structures of the RNA polymerases and ribosomes found in Archaea and Eukarya are distinct from those found in Bacteria and similar to each other. For example, antibiotics that poison bacterial ribosomes do not affect the ribosomes of archaea or eukaryotes.

Within the Bacteria and Archaea, morphological diversity is substantial. These organisms range in size from the smallest of all free-living cells—bacteria called mycoplasmas with volumes as small as $0.03\ \mu m^3$—to the largest bacterium known, *Thiomargarita namibiensis*, with volumes as large as $200 \times 10^6\ \mu m^3$. Over a billion *Mycoplasma* cells could fit inside an individual *Thiomargarita* (**Figure 27.11a**). Bacteria and archaea exhibit a variety of shapes as well, including filaments, spheres, rods, chains, and spirals (**Figure 27.11b**). Many cells

TABLE 27.2 Characteristics of Bacteria, Archaea, and Eukarya

	Bacteria	Archaea	Eukarya
Nuclear envelope?	No	No	Yes
Circular chromosome?	Yes (but linear in some species)	Yes	No
DNA associated with histone proteins?	No	Yes	Yes
Organelles present?	No	No	Yes
Unicellular or multicellular?	Almost all unicellular	All unicellular	Many multicellular
Sexual reproduction?	No*	Not known	Common
Structure of lipids in plasma membrane	Glycerol bonded to straight-chain fatty acids via ester linkage	Glycerol bonded to branched fatty acids (synthesized from isoprene subunits) via ether linkage	Glycerol bonded to straight-chain fatty acids via ester linkage
Cell-wall material	Almost all include peptidoglycan, which contains muramic acid	Varies widely among species, but no peptidoglycan and no muramic acid	When present, usually made of cellulose or chitin
Transcription and translation machinery	One relatively simply RNA polymerase; translation begins with formylmethionine; translation poisoned by several antibiotics that do not affect archaea or eukaryotes	Several relatively complex RNA polymerases; translation begins with methionine	Several relatively complex RNA polymerases; translation begins with methionine

*Sexual reproduction begins with meiosis and often involves the exchange of haploid genomes between individuals of the same species. In bacteria, meiosis does not occur. Small numbers of genes can be transferred from one bacterial cell to another, however, and genetic recombination may occur. For more detail, see Chapter 19.

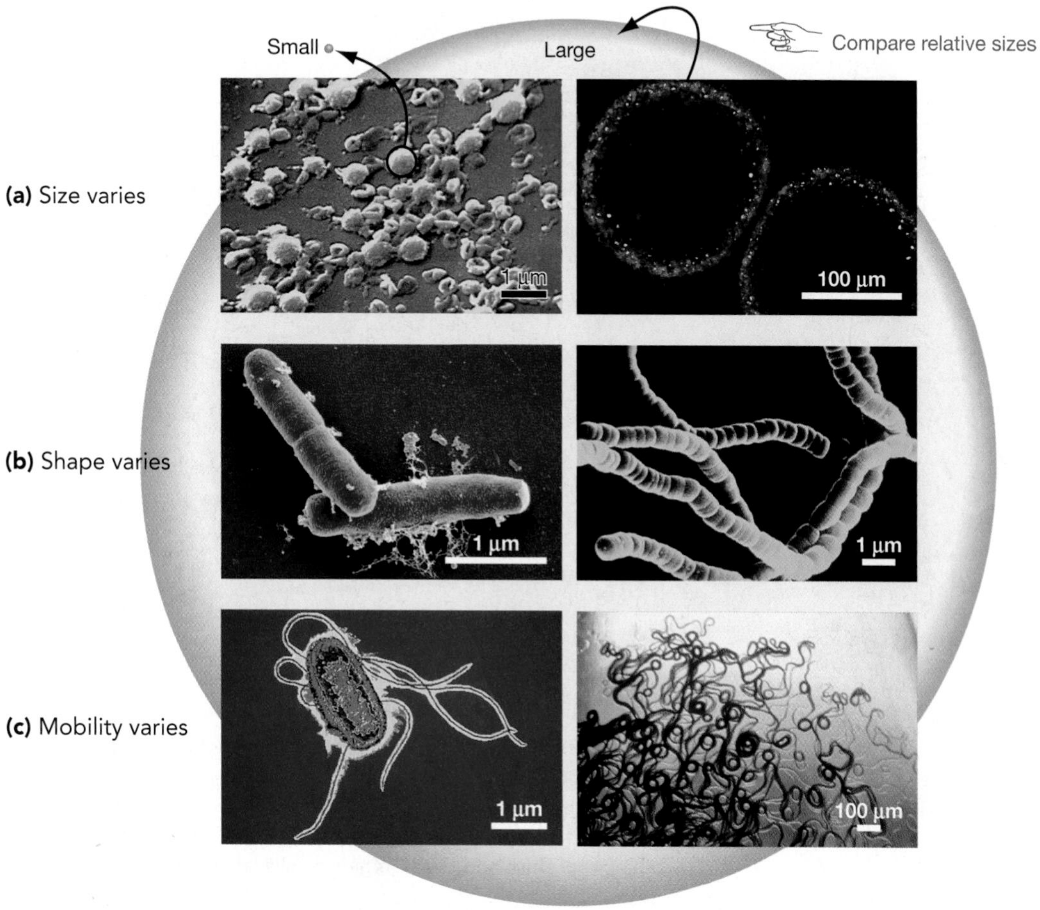

Small
Large
Compare relative sizes
(a) Size varies
(b) Shape varies
(c) Mobility varies
1 µm
100 µm
1 µm
1 µm
1 µm
100 µm

FIGURE 27.11 Morphological Diversity among Bacteria and Archaea Is Extensive
(a) The sizes of bacteria and archaea vary. (Left) *Mycoplasma* cells are about 0.5 μm in diameter, while (right) *Thiomargarita namibiensis* cells are about 0.15 mm in diameter. **(b)** The shapes of bacteria and archaea vary from (left) rods such as *Bacillus anthracis* and spheres to filaments or spirals such as *Rhodospirillum*. In some species, such as (right) *Streptococcus faecalis*, cells attach to one another and form chains. **(c)** (Left) A wide variety of bacteria and archaea use flagella to power swimming movements. (Right) These cyanobacterial cells move by gliding across a substrate.

are motile, with swimming movements powered by flagella or by gliding over a surface (**Figure 27.11c**). Although gliding movement occurs in several groups, the molecular mechanism responsible for this form of motility is still unknown.

Within bacteria, the overall morphology of the cell wall varies. **Gram-positive bacteria** have a plasma membrane surrounded by a cell wall with extensive peptidoglycan (**Figure 27.12a**). **Gram-negative bacteria**, in contrast, have a plasma membrane surrounded by a cell wall that has two components: a thin gelatinous layer containing peptidoglycan, and an outer phospholipid bilayer (**Figure 27.12b**). When treated with a dye called the **Gram stain**, Gram-positive cells look purple but Gram-negative cells look pink (**Figure 27.12c**). The Gram stain is used routinely to aid in the identification of unknown bacterial cells.

All bacteria and archaea are haploid and reproduce by **fission**—the splitting of a cell into two daughter cells. DNA replication precedes fission, so the two daughter cells are genetically identical. In addition, as noted in Chapter 19, bacterial cells are capable of transferring copies of the extracellular loops of DNA called **plasmids**. When **conjugation** occurs, a copy of a plasmid—and potentially one or more genes from the main bacterial chromosome—moves from one cell to a recipient cell (see Box 19.1). The **conjugation tube** that forms between cells that are transferring and receiving a plasmid is a morphological trait that is unique to bacteria and archaea.

To summarize, members of the Bacteria and the Archaea are remarkably diverse in their overall size, shape, and motility, as well as in the composition of their cell walls and plasma membranes. But when asked to name the innovations that were responsible for the diversification of these two domains, biologists do not point to their morphological diversity. Instead, they point to metabolic diversity—variation in the chemical reactions that go on inside these cells.

Metabolic Diversity

Bacteria and archaea are the masters of metabolism. Taken together, they can use almost anything as food, from hydrogen molecules to crude oil. Bacteria and archaea may be small and relatively simple morphologically, but their biochemical capabilities are dazzling.

(a) Cell walls in Gram-positive bacteria have extensive peptidoglycan.

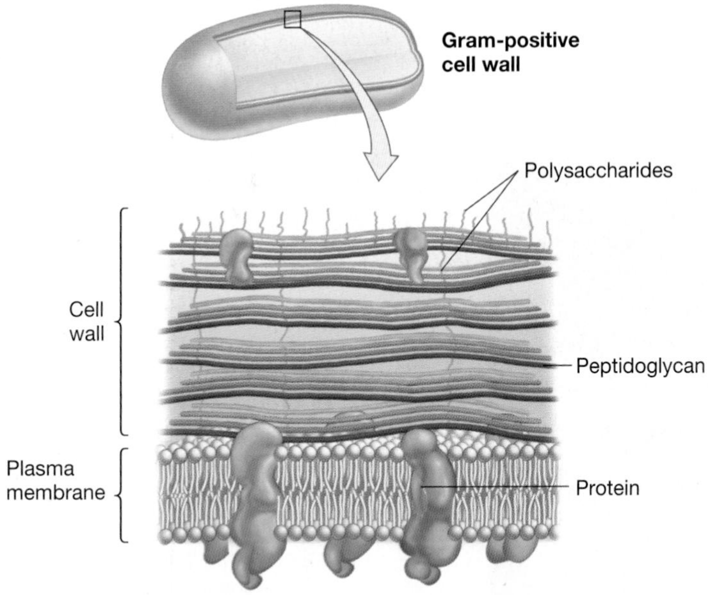

(b) Cell walls in Gram-negative bacteria have some peptidoglycan and an outer membrane.

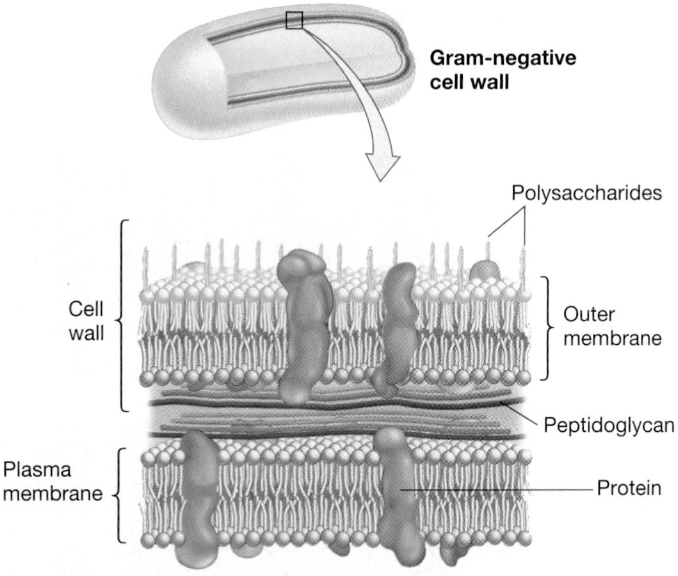

(c) Gram-positive cells retain Gram stain more than Gram-negative cells do.

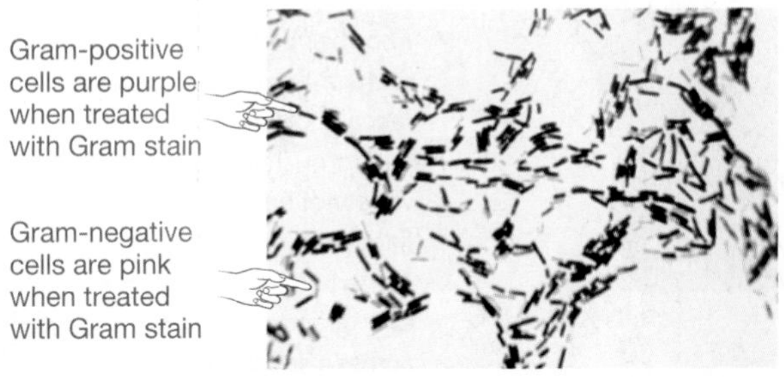

FIGURE 27.12 Gram Staining Distinguishes Two Types of Cell Walls in Bacteria

Both **(a)** the Gram-positive bacteria and **(b)** the Gram-negative bacteria are paraphyletic groups (see Figure 27.10). **(c)** The cell walls of Gram-positive and Gram-negative cells take up the Gram stain differently.

Just how varied are bacteria and archaea when it comes to finding and processing food? To appreciate the answer, recall from Chapters 9 and 10 that organisms have two fundamental nutritional needs: acquiring chemical energy in the form of adenosine triphosphate (ATP) and obtaining carbon in a form that can be used to synthesize fatty acids, proteins, DNA, RNA, and other molecules needed to build cells. Bacteria and archaea produce ATP in three ways:

1. **Phototrophs** ("light-feeders") use light energy to promote electrons to the top of electron transport chains. ATP is produced by photophosphorylation (see Chapter 10).
2. **Organotrophs** oxidize organic molecules with high potential energy, such as sugars. ATP may be produced by cellular respiration—with sugars serving as electron donors—or via fermentation pathways such as those cited in Chapter 9.
3. **Lithotrophs** ("rock-feeders") oxidize inorganic molecules with high potential energy, such as ammonia (NH_3) or methane (CH_4). ATP is produced by cellular respiration, with inorganic compounds serving as the electron donor.

Bacteria and archaea obtain carbon either by processing sources such CO_2 and CH_4 or by absorbing ready-to-use organic compounds from their environment. Organisms that manufacture their own carbon-containing compounds are termed **autotrophs** ("self-feeders"). Organisms that acquire carbon-containing compounds from other organisms are called **heterotrophs** ("other-feeders").

Because there are three distinct ways of producing ATP and two general mechanisms for obtaining carbon, there are a total of six methods for producing ATP and obtaining carbon. **Table 27.3** summarizes these six "feeding strategies"—only two of which are observed in plants, animals, and fungi.

What makes this remarkable diversity possible? Bacteria and archaea have evolved dozens of variations on the most basic themes of metabolism: extracting usable energy from compounds with high potential energy, using light to produce high-energy electrons, and fixing carbon. This metabolic diversity is important because it allows bacteria and archaea to occupy diverse habitats. Let's take a closer look.

Cellular Respiration: Variation in Electron Donors and Electron Acceptors Millions of bacterial, archaeal, and eukaryotic species—including animals and plants—obtain the energy required to make ATP by oxidizing organic compounds such as sugars, starch, or fatty acids. (To review the nature of reduction-oxidation reactions, see Chapter 2.) As shown in Chapter 9, cellular enzymes can strip electrons from organic molecules with high potential energy and transfer them to the electron carriers NADH and $FADH_2$. These compounds, in turn, feed electrons to the molecules that make up electron transport chains. As electrons are stepped down from a high-energy state to a low-energy

TABLE 27.3 Strategies for Obtaining Energy and Carbon: An Overview

		Source of Carbon (For synthesis of organic compounds)	
		Autotrophs	**Heterotrophs**
		Synthesize their own high-potential-energy organic compounds from CO_2, CH_4, or other inorganic sources.	Use organic compounds with high potential energy produced by other organisms.
Source of Energy (For ATP production)	**Light** (phototrophs)	**Photoautotrophs** Cyanobacteria use photosynthesis to produce ATP. They fix CO_2 via the Calvin cycle.	**Photoheterotrophs** Heliobacteria use photosynthesis to produce ATP. They absorb organic molecules with high potential energy from the environment.
	Organic Molecules with High Potential Energy (organotrophs)	**(no term)** *Clostridium aceticum* ferments glucose to produce ATP. It uses reactions called the acetyl-CoA pathway to fix CO_2.	**Chemoorganotrophs** *E. coli* uses fermentation or respiration to produce ATP. It absorbs carbon molecules with high potential energy from the environment.
	Inorganic Molecules with High Potential Energy (lithotrophs)	**Chemolithotrophs** Nitrifying bacteria, including species of *Nitrosomonas*, produce ATP via respiration, using ammonia (NH_3) as an electron donor. They fix CO_2 via the Calvin cycle.	**Chemolithotrophic Heterotrophs** *Beggiatoa* produces ATP via respiration, using hydrogen sulfide (H_2S) as an electron donor. It absorbs carbon molecules with high potential energy from the environment.

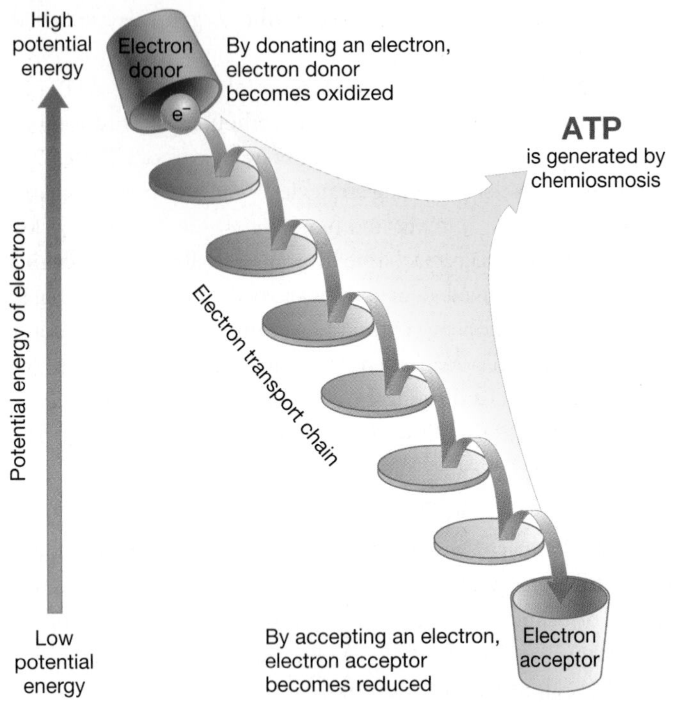

FIGURE 27.13 Cellular Respiration Requires an Electron Donor and an Electron Acceptor
Bacteria and archaea can exploit a wide variety of electron donors and acceptors. Some of these electron donors are more highly reduced than others, and some electron acceptors are more highly oxidized than others. EXERCISE After studying Table 27.4, add the chemical formula for a specific electron donor, electron acceptor, and reduced by-product to this diagram. Do this for a particular species of bacteria or archaea. Then write in the electron donor, electron acceptor, and reduced by-product observed in humans.

state, the components of the electron transport chain generate a proton gradient across a membrane (**Figure 27.13**). As Chapter 9 pointed out, the resulting flow of protons through the enzyme ATP synthase results in the production of ATP, via the process called chemiosmosis.

The essence of this process, called **cellular respiration**, is that a molecule with high potential energy serves as an original electron donor and is oxidized, while a molecule with low potential energy serves as a final electron acceptor and becomes reduced. Much of the potential energy difference between the electron donor and electron acceptor is transformed into chemical energy in the form of ATP.

In introducing respiration, Chapter 9 focused on the role of organic compounds with high potential energy, such as glucose, as the original electron donor and oxygen as the final electron acceptor (see Figure 9.21). Many bacteria and archaea, as well as all eukaryotes, rely on these molecules. When oxygen acts as the final electron acceptor, water is produced as a by-product of respiration.

Many other bacteria and archaea employ an electron donor other than sugars and an electron acceptor other than oxygen, however, and produce by-products other than water. As **Table 27.4** shows, the electron donors used by bacteria and archaea range from hydrogen molecules (H_2) and hydrogen sulfide (H_2S) to ammonia (NH_3) and methane (CH_4). In addition to using oxygen, some organisms use compounds with low potential energy, such as sulfate (SO_4^{2-}), nitrate (NO_3^-), carbon dioxide (CO_2), or ferric ions (Fe^{3+}), as electron acceptors. It is only a slight exaggeration to claim that researchers have found bacterial and archaeal species that can use almost any compound with relatively high potential energy as an electron donor and almost any compound with relatively low potential energy as an electron acceptor.

Because the electron donors and electron acceptors used by bacteria and archaea are so diverse, one of the first questions that biologists ask about a species is how it accomplishes cellular respiration. The best way to answer this question is through the enrichment culture technique introduced in Section 27.2. Recall that, in an enrichment culture, researchers supply specific electron donors and electron acceptors in the medium and try to isolate cells that can use those compounds to support growth.

TABLE 27.4 Some Electron Donors and Acceptors Used by Bacteria and Archaea

Electron Donor	Electron Acceptor	Product	Category*
Sugars	O_2	H_2O	Organotrophs
H_2 or organic compounds	SO_4^{2-}	H_2S	Sulfate reducers
H_2	CO_2	CH_4	Methanogens
CH_4	O_2	CO_2	Methanotrophs
S or H_2S	O_2	SO_4^{2-}	Sulfur bacteria
Organic compounds	Fe^{3+}	Fe^{2+}	Iron reducers
NH_3	O_2	NO_2^-	Nitrifiers
Organic compounds	NO_3^-	N_2O, NO, or N_2	Denitrifiers (or nitrate reducers)
NO_2^-	O_2	NO_3^-	Nitrosifiers

*The name biologists use to identify species that use a particular metabolic strategy.

The remarkable metabolic diversity of bacteria and archaea is important. First, it explains their ecological diversity. Bacteria and archaea are found almost everywhere because they exploit an almost endless variety of molecules as food. Second, it explains why they play such a key role in cleaning up some types of pollution. Species that use organic solvents or petroleum-based fuels as electron donors or electron acceptors may be effective agents in bioremediation efforts if the waste products are less toxic than the solvents or fuels. Finally, metabolic diversity is what makes bacteria and archaea major players in global change. Nitrogen, phosphorus, sulfur, carbon, and other crucial nutrients cycle from one organism to another because bacteria and archaea can use them in almost any molecular form. The nitrite (NO_2^-) that some bacteria produce as a by-product of respiration does not build up in the environment, because it is used as an electron acceptor by other species and converted (via *nitrification*) to molecular nitrate (NO_3^-). This molecule, in turn, is converted (via *denitrification*) to molecular nitrogen (N_2) by yet another suite of bacterial and archaeal species. Bacteria can convert N_2 in the atmosphere to NH_3 in proteins (by bacterial or archaeal fixation and decomposition). Similar types of interactions occur with molecules that contain phosphorus, sulfur, and carbon. In this way, bacteria and archaea play a key role in the cycling of nutrients (**Figure 27.14**). In total, bacteria and archaea contain as much carbon, and 10 times as much nitrogen and phosphorus, as are in all of the plants on Earth.

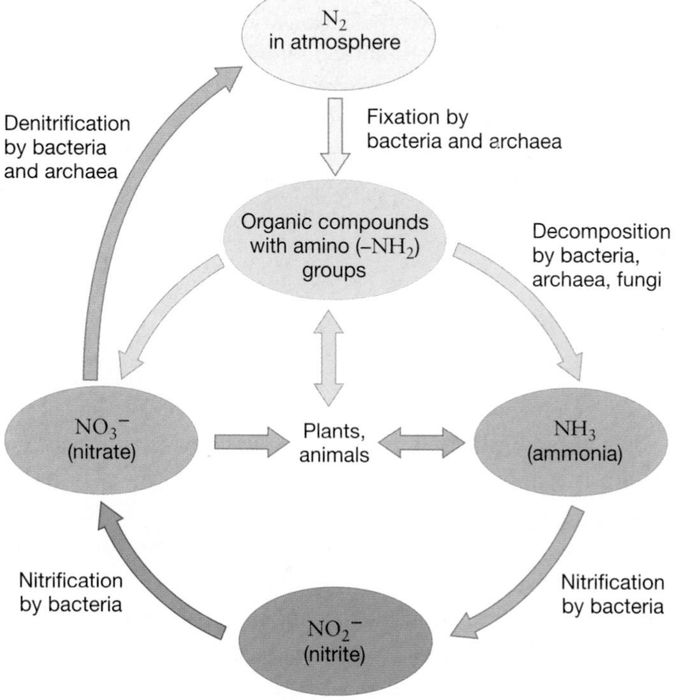

FIGURE 27.14 Nitrogen Atoms Cycle through Environments in Different Molecular Forms

EXERCISE Suppose that bacteria and archaea were no longer capable of fixing nitrogen. Draw an X through the part(s) of the cycle that would be most directly affected.

Fermentation Chapter 9 introduced **fermentation**, a strategy for making ATP that does not involve electron transport chains. In fermentation, no outside electron acceptor is used; the redox reactions are internally balanced. Because fermentation is a less efficient way to make ATP compared with cellular respiration, in many species it occurs as an alternative metabolic strategy when no electron acceptors are available to make cellular respiration possible. In other species, fermentation does not occur at all; in still other species, fermentation is the only way for cells to make ATP.

Although the presentation in Chapter 9 focused on how glucose is fermented to ethanol or lactic acid, some bacteria and archaea are capable of using other organic compounds as fermentable substrates. For example, the bacterium *Clostridium aceticum* can ferment ethanol, acetate, and fatty acids—as well as glucose. Other species of *Clostridium* ferment complex carbohydrates (including cellulose or starch), proteins, amino acids, or even purines. Species that ferment amino acids produce by-products with names such as cadaverine and putrescine. These molecules are responsible for the odor of rotting flesh. Other bacteria can ferment lactose, a prominent component of milk. This fermentation has two end products: propionic acid and CO_2. Propionic acid is responsible for the taste of Swiss cheese; the CO_2 produced during fermentation creates the holes in cheese.

The diversity of enzymatic pathways observed in bacterial and archaeal fermentations extends the metabolic repertoire of these organisms and supports the claim that as a group, bacteria and archaea can use virtually any molecule with relatively high potential energy as a source of energy. Given this diversity, it is no surprise that bacteria and archaea are found in such widely varying habitats. Different environments offer different energy-rich molecules; various species of bacteria and archaea have evolved the biochemical machinery required to exploit most or all of these food sources.

Photosynthesis Instead of using molecules as a source of high-energy electrons, phototrophs pursue a radically different strategy: **photosynthesis**. That is, they use the energy in light to raise electrons to high-energy states. As these electrons are stepped down to lower energy states by electron transport chains, the energy released is used to generate ATP.

Chapter 10 introduced an important feature of photosynthesis: The process requires a source of electrons. Recall that, in cyanobacteria and plants, the required electrons come from water. When these organisms "split" water molecules apart to obtain electrons, they generate oxygen as a by-product. In contrast, many phototrophic bacteria use a molecule other than water as the source of electrons. In many cases, the electron donor is hydrogen sulfide (H_2S); a few species can use the ion

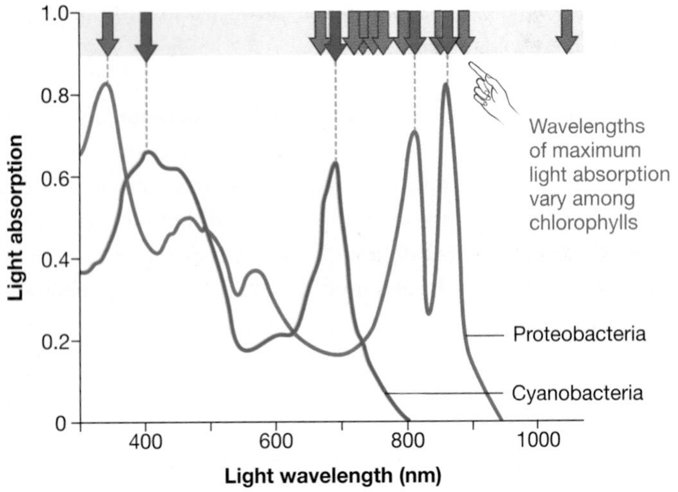

FIGURE 27.15 Highly Diverse Light Wavelengths Are Absorbed by Chlorophylls Found in Photosynthetic Bacteria
The wavelengths of light that are absorbed by the various types of chlorophyll found in different species of photosynthetic bacteria. The complete absorption spectrum is given for the chlorophyll found in cyanobacteria (green curve) and in photosynthetic proteobacteria (blue curve). The red arrows point to absorption peaks observed in chlorophylls isolated from other groups of bacteria.

known as ferrous iron (Fe^{2+}). Instead of producing oxygen as a by-product of photosynthesis, these cells produce elemental sulfur (S) or the ferric ion (Fe^{3+}). They live in habitats where oxygen is completely absent.

Chapter 10 also introduced the photosynthetic pigments found in plants and explored the light-absorbing properties of chlorophylls *a* and *b*. Cyanobacteria have these two pigments. But researchers have isolated seven additional chlorophylls from bacterial phototrophs. As **Figure 27.15** shows, each of these pigments absorbs light best at a different wavelength. Each major group of photosynthetic bacteria has one or more of these distinctive chlorophylls.

Why are bacterial chlorophylls so diverse? The leading hypothesis is that photosynthetic species with different absorption spectra are able to live together without competing for light. If so, then the diversity of photosynthetic pigments observed in bacteria has been an important mechanism for generating species diversity among phototrophs.

Pathways for Fixing Carbon In addition to acquiring energy, organisms must obtain building-block molecules that contain carbon-carbon bonds. Chapters 9 and 10 introduced the two mechanisms that organisms use to procure usable carbon—either making their own or getting it from other organisms. In cyanobacteria and plants, the enzymes of the Calvin cycle transform carbon dioxide (CO_2) to organic molecules that can be used in synthesizing cell material. Animals and fungi, in contrast, obtain carbon from living plants or animals or by absorbing the organic compounds released as dead tissues decay.

Bacteria and archaea pursue these same two strategies, but with some interesting twists. For example, some species can fix carbon from sources other than CO_2. Certain proteobacteria are called **methanotrophs** ("methane-eaters"), because they use methane (CH_4) as their primary electron donor and carbon source. Among the methanotrophs, CH_4 is assimilated via one of two enzymatic pathways, depending on the species. In addition, some bacteria can use sources of carbon other than CO_2 and CH_4, including carbon monoxide (CO) and methanol (CH_3OH). Further, several groups of bacteria can fix CO_2 using pathways other than the Calvin cycle. To date, researchers have identified three distinct biochemical processes other than the Calvin cycle that result in CO_2 fixation.

What do all these observations mean? Although the enzymes of the Calvin cycle are widespread among bacteria and archaea, several species have evolved different solutions to the problem of fixing carbon. This is yet another example of the metabolic diversity observed within the domains Bacteria and Archaea.

✓CHECK YOUR UNDERSTANDING

Compared with the metabolic capabilities of Bacteria and Archaea, those of plants and animals are very limited. You should be able to describe bacteria or archaea that (1) perform cellular respiration by means of an electron donor other than a sugar and an electron acceptor other than oxygen, (2) perform fermentation from a starting material other than pyruvate derived from glucose and yield a by-product other than lactic acid or ethanol, (3) perform photosynthesis by means of an electron source other than water, and (4) fix carbon to sugar via a pathway other than the Calvin cycle.

27.4 Key Lineages of Bacteria and Archaea

Since the phylogenetic tree identifying the three domains of life was first published, dozens of studies have confirmed the result. It is now well established that all organisms alive today belong to one of the three domains, and that archaea and eukaryotes are more closely related to each other than either group is to bacteria.

Although the relationships among the major lineages within Bacteria and Archaea are still uncertain in some cases (see Figure 27.10), most of the clades themselves have been quite well studied. Let's survey the attributes of species from selected major lineages within the Bacteria and Archaea, with an emphasis on their morphological and metabolic diversity, their impacts on humans, and their importance to other species and to the environment.

Bacteria

The name *bacteria* comes from the Greek root *bacter*, for "rod" or "staff." The name was inspired by the first bacteria to be seen under a microscope, which were rod shaped. But as the following descriptions indicate, bacterial cells come in a wide variety of shapes. If the group were to be named now, biologists might use roots from the Latin words *diversus* or *abundantia*.

Biologists who study bacterial diversity currently recognize at least 14 major lineages, or phyla, within the domain. The lineages reviewed here are just a sampling of bacterial diversity.

Spirochaeles (Spirochetes)

The spirochetes are one of the smaller bacterial phyla. They include 13 genera and a total of 62 species. Recent analyses place spirochetes at the base of the bacterial phylogenetic tree, meaning that they are a particularly ancient group (**Figure 27.16**).

Morphological diversity Spirochetes are distinguished by their unique corkscrew shape and unusual flagella. Instead of extending into the water surrounding the cell, spirochete flagella are contained within a structure called the *outer sheath*, which surrounds the cell. When the flagella beat, the cell lashes back and forth and swims forward.

Metabolic diversity Most spirochetes manufacture ATP via fermentation. The substrate used in fermentation varies among species and may consist of sugars, amino acids, starch, or the pectin found in plant cell walls. A spirochete that lives only in the hindgut of termites can fix nitrogen.

Human and ecological impacts The sexually transmitted disease syphilis is caused by a spirochete. So is Lyme disease, which is transmitted to humans by deer ticks. Spirochetes are extremely common in freshwater and marine habitats; many live only under anaerobic conditions.

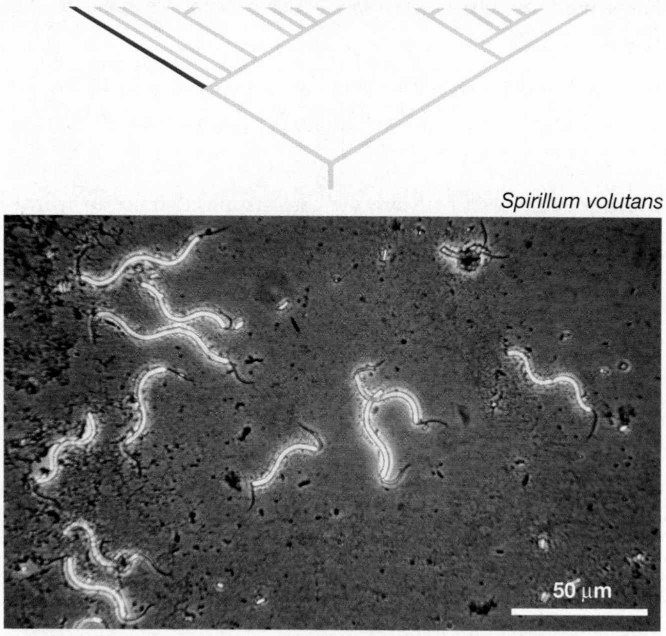

Spirillum volutans
50 μm

FIGURE 27.16 Spirochetes Are Corkscrew-Shaped Cells Inside an Outer Sheath

Chlamydiales

In terms of numbers of species living today, Chlamydiales may be the smallest of all major bacterial lineages. Although chlamydiae are highly distinct phylogenetically, only four species in one genus (*Chlamydia*) are known (**Figure 27.17**).

Morphological diversity Chlamydiae are spherical. They are tiny, even by bacterial standards.

Metabolic diversity All known species live as parasites *inside* host cells and are termed **endosymbionts** ("inside-together-living"). Chlamydiae contain few enzymes of their own and get almost all of their nutrition from their hosts. In Figure 27.17, the chlamydiae are the pink-stained cells, which are living inside blue-stained animal cells.

Human and ecological impacts *Chlamydia trachomatis* infections are the most common cause of blindness in humans. When the same organism is transmitted from person to person via sexual intercourse, it can cause serious urogenital tract infections. One species causes epidemics of a pneumonia-like disease in birds.

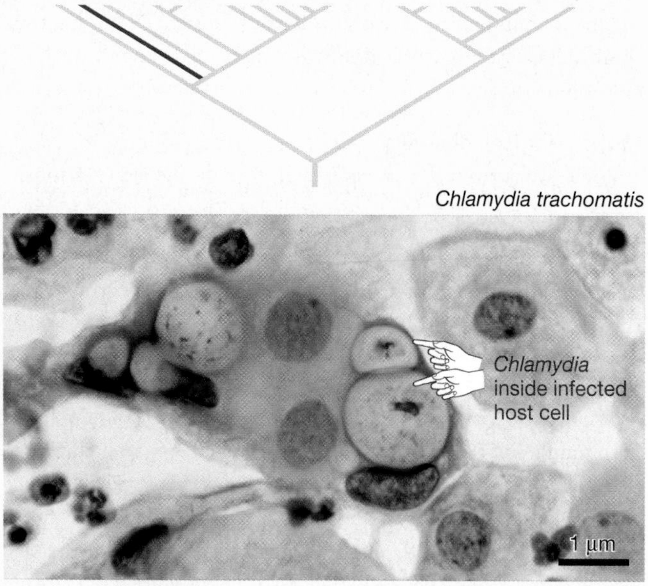

Chlamydia trachomatis
Chlamydia inside infected host cell
1 μm

FIGURE 27.17 Chlamydiae Live Only Inside Animal Cells

High-GC Gram Positives

The name "high-GC Gram positives" is derived from two observations about these bacteria: (1) Their cell-wall material appears purple when treated with the Gram stain, which tests for the absence of a membrane outside the cell wall and is used as an early step in procedures for identifying unknown bacterial cells; and (2) their DNA contains a relatively high percentage of guanine and cytosine. (In some species, G and C represent over 75 percent of the bases present.) Over 1100 species have been described to date (**Figure 27.18**).

Morphological diversity Cell shape varies from rods to filaments. Many of the soil-dwelling species form extensive branched filaments called **mycelia**.

Metabolic diversity Many are heterotrophs that use an array of organic compounds as electron donors and oxygen as an electron acceptor. There are a handful of parasitic species. Like other parasites, they get most of their nutrition from host organisms.

Human and ecological impacts Over 500 distinct antibiotics have been isolated from species in the genus *Streptomyces;* 60 of these—including streptomycin, neomycin, tetracycline, and erythromycin—are now actively prescribed to treat diseases in humans or domestic livestock. Tuberculosis and leprosy are caused by members of this group. One species is critical to the manufacture of Swiss cheese. Species in the genus *Streptomyces* and *Arthrobacter* are abundant in soil and are vital as decomposers of dead plant and animal material. Some species in these genera live in association with plant roots and fix nitrogen; others can break down toxins such as herbicides, nicotine, and caffeine.

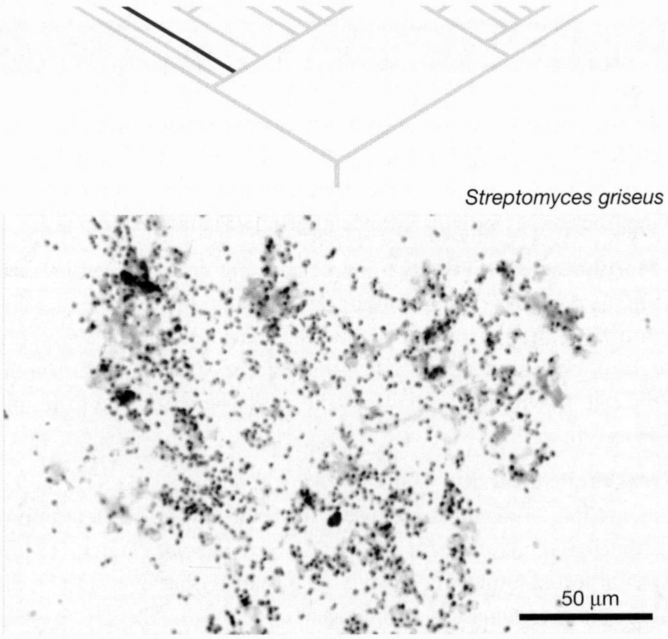
Streptomyces griseus
50 μm

FIGURE 27.18 A *Streptomyces* Species That Produces the Antibiotic Streptomycin

Cyanobacteria

The cyanobacteria were formerly known as the "blue-green algae"—even though algae are eukaryotes. Only about 80 species of cyanobacteria have been described to date, but they are among the most abundant organisms on Earth (**Figure 27.19**). In terms of total mass, cyanobacteria dominate many marine and freshwater environments.

Morphological diversity Cells vary from solitary to colonial. Colonies vary from flat sheets of cells to ball-like aggregations.

Metabolic diversity All perform oxygenic photosynthesis; many can also fix nitrogen. Because cyanobacteria can synthesize virtually every molecule they need, they can be grown in culture media that contain only CO_2, N_2, H_2O, and a few mineral nutrients.

Human and ecological impacts If cyanobacteria are present in high numbers, their waste products can make drinking water smell bad. The compounds are not toxic, however. Cyanobacteria produce much of the oxygen and nitrogen and many of the organic compounds that feed other organisms in freshwater and marine environments. A few species live in association with fungi, forming lichens.

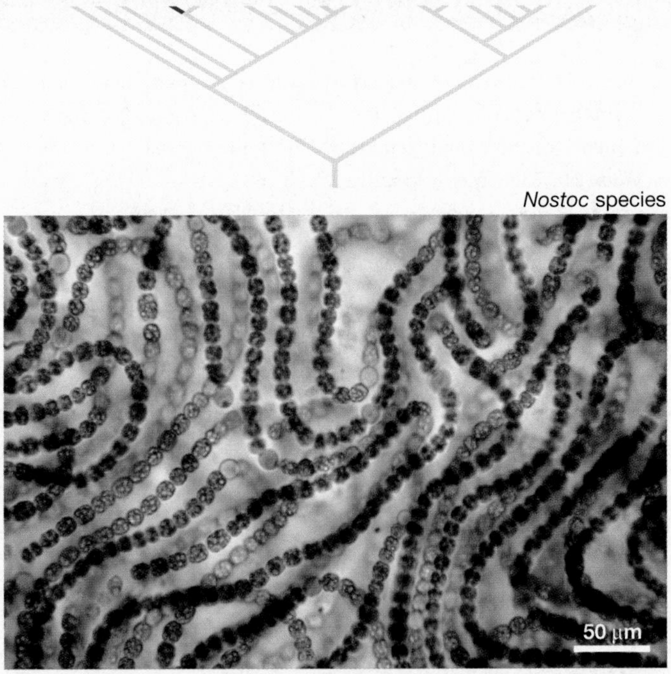
Nostoc species
50 μm

FIGURE 27.19 Cyanobacteria Contain Chlorophyll and Are Green

Low-GC Gram Positives

The low-GC Gram positives react positively to the Gram stain—meaning that they lack a membrane outside their cell wall—but have a relatively low percentage of G and C in their DNA. (In some species, G and C represent less than 25 percent of the bases present.) There are over 1100 species (**Figure 27.20**).

Morphological diversity Most are rod shaped or spherical. Some of the spherical species form chains or *tetrads* (groups of four cells). A few form a durable resting stage called a **spore**. One subgroup lacks cell walls entirely; another synthesizes a cell wall made of cellulose.

Metabolic diversity Some species can fix nitrogen; some perform nonoxygenic photosynthesis. Others make all of their ATP via various fermentation pathways; still others perform cellular respiration, using hydrogen gas (H_2) as an electron donor.

Human and ecological impacts Species in this group cause a variety of diseases, including anthrax, botulism, tetanus, walking pneumonia, boils, gangrene, and strep throat. *Bacillus thuringiensis* produces a toxin that is one of the most important insecticides currently used in farming. Species in the genus *Lactobacillus* are used to ferment milk products into yogurt or cheese. Species in this group are important components of soil.

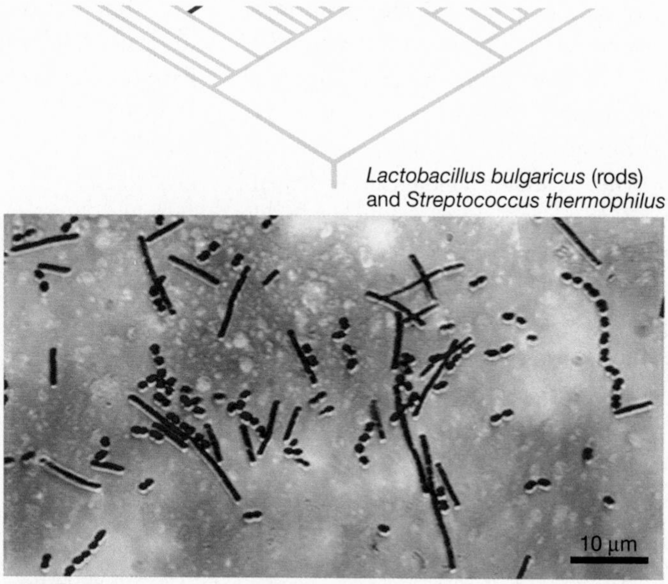
Lactobacillus bulgaricus (rods)
and Streptococcus thermophilus
10 μm
FIGURE 27.20 Low-GC Gram Positives in Yogurt

Proteobacteria

The approximately 1200 species of proteobacteria form five major subgroups, designated by the Greek letters α (alpha), β (beta), γ (gamma), δ (delta), and ε (epsilon).

Morphological diversity Proteobacterial cells can be rods, spheres, or spirals. Some form stalks (**Figure 27.21a**). Some are motile. In one group, cells may move together to form colonies, which then transform into the specialized cell aggregate shown in **Figure 27.21b**. This structure is known as a **fruiting body**. Durable spores are produced at the tips of fruiting bodies.

Metabolic diversity Proteobacteria make a living in virtually every way known to bacteria—except that none perform oxygenic photosynthesis. Various species may perform cellular respiration by using organic compounds, nitrite, methane, hydrogen gas, sulfur, or ammonia as electron donors and oxygen, sulfate, sulfur, or oxygen as an electron acceptor. Some perform nonoxygenic photosynthesis (**Box 27.2**).

Human and ecological impacts Pathogenic proteobacteria cause Legionnaire's disease, cholera, food poisoning, dysentery, gonorrhea, Rocky Mountain spotted fever, ulcers, and diarrhea. Biologists use *Agrobacterium* cells to transfer new genes into crop plants. Certain acid-loving species of proteobacteria are used in the production of vinegars. Species in the genus *Rhizobium* (α-proteobacteria) live in association with plant roots and fix nitrogen. The bdellovibrios are a group in the δ-proteobacteria that are predators—they drill into bacterial cells and digest them. Proteobacteria are critical players in the cycling of nitrogen atoms through terrestrial and aquatic ecosystems.

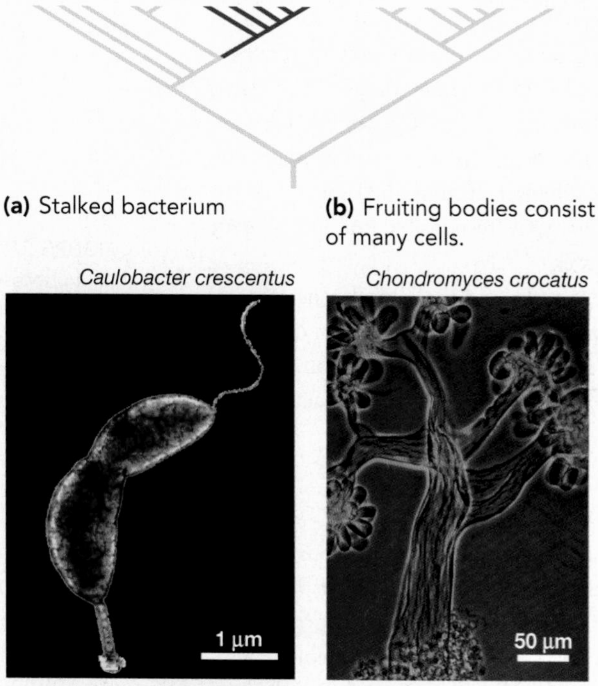
(a) Stalked bacterium
(b) Fruiting bodies consist of many cells.
Caulobacter crescentus
Chondromyces crocatus
1 μm
50 μm
FIGURE 27.21 Some Proteobacteria Grow on Stalks or Form Fruiting Bodies

BOX 27.2 Lateral Gene Transfer and Metabolic Diversity in Bacteria

If you read the notes on metabolic diversity in Section 27.4 carefully, you will realize that species capable of performing various types of photosynthesis are scattered among many bacterial lineages. The same is true for species that can fix nitrogen.

The "scattered among lineages" pattern is interesting because it is reasonable to predict that extremely complex structures and processes such as photosystem I, photosystem II, and nitrogen fixation evolved just once. If so, then it would be logical to predict that photosynthesizers and nitrogen-fixers would each form monophyletic groups, consisting of an ancestral species and all of its descendants. In contrast, a group that consists of a common ancestor and some *but not all* of its descendants is said to be **paraphyletic** ("beside-group"). If species that perform photosynthesis were monophyletic, then they would be arranged in the pattern shown in **Figure 27.22a**. But instead, the data in **Figure 27.22b** shows that photosynthetic species are paraphyletic. Nitrogen fixers are also paraphyletic.

How could the photosystems and nitrogen fixation evolve just once and yet be paraphyletic in their present phylogenetic distribution? **Lateral gene transfer** is a physical transfer of genes from species in one lineage to species in another lineage. Chapter 20 described several of the mechanisms responsible for lateral gene transfer.

According to the lateral gene transfer hypothesis for the distribution of photosynthesis and nitrogen fixation between lineages, it is correct to claim that photosystem I, photosystem II, and the nitrogen-fixing enzymes evolved just once. But over the past 3.4 billion years, the genes responsible for these processes have been picked up by organisms from a wide array of bacterial lineages and incorporated into their genomes. Lateral gene transfer is thought to be an important mechanism for generating metabolic diversity in the Bacteria.

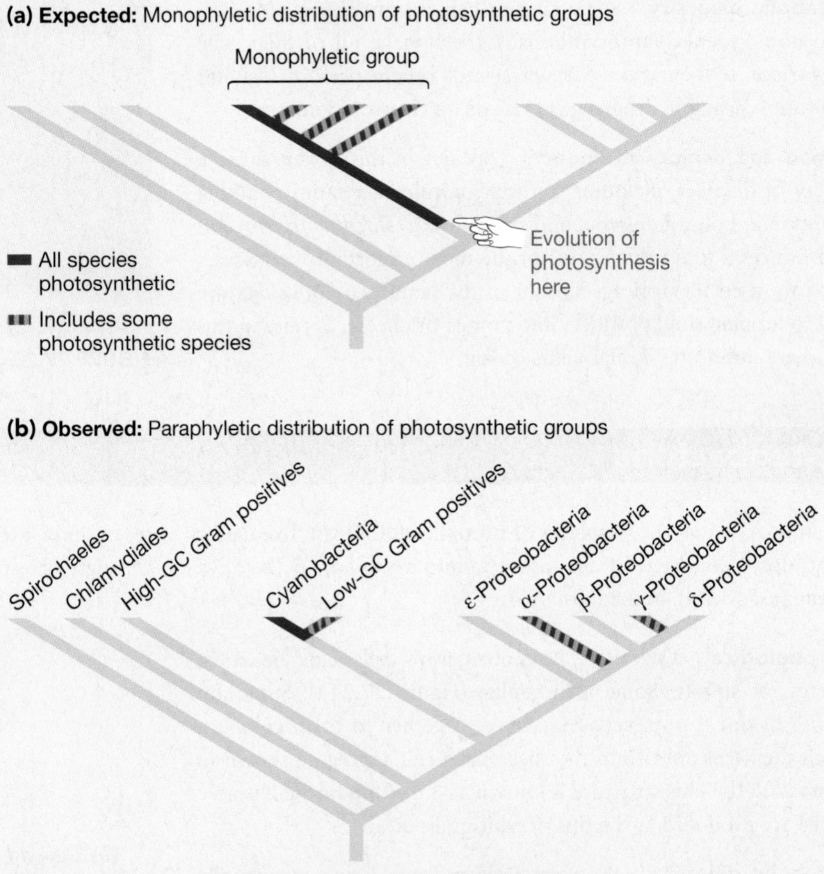
(a) Expected: Monophyletic distribution of photosynthetic groups
Monophyletic group
All species photosynthetic
Includes some photosynthetic species
Evolution of photosynthesis here
(b) Observed: Paraphyletic distribution of photosynthetic groups
Spirochaeles
Chlamydiales
High-GC Gram positives
Cyanobacteria
Low-GC Gram positives
ε-Proteobacteria
α-Proteobacteria
β-Proteobacteria
γ-Proteobacteria
δ-Proteobacteria

FIGURE 27.22 Photosynthetic Bacteria Are Paraphyletic
(a) If photosynthetic bacteria were monophyletic, their distribution on the phylogenetic tree of Bacteria would look something like this. **(b)** Species that can perform photosynthesis are scattered throughout the phylogenetic tree. Two bacterial lineages not shown—green sulfur bacteria and green nonsulfur bacteria—also perform a type of photosynthesis.

Archaea

The name *archaea* comes from the Greek root *archae*, for "ancient." The name was inspired by the hypothesis that this is a particularly ancient group, which turned out to be incorrect. In addition, archaeans were once thought to be restricted to hot springs, salt ponds, and other extreme habitats. But if the group were to be named now, biologists might start with the Latin root *ubiquit*, for "everywhere." Archaea live in virtually every habitat known. As far as biologists currently know, there are no parasitic archaea.

The domain Archaea was discovered so recently that major groups are still being discovered and described. Phylogenies based on DNA sequence data have consistently shown that the domain is composed of at least two major phyla, called the **Crenarchaeota** and **Euryarchaeota**. Some data indicate that a third phylum, tentatively called the Korarchaeota, also exists.

Crenarchaeota

The Crenarchaeota got their name because they are considered similar to the oldest archaeans; the word root *cren* refers to a source or fount. Although only 37 species have been named to date, it is virtually certain that thousands are yet to be discovered (**Figure 27.23**).

Morphological diversity Crenarchaeota cells may be shaped like filaments, rods, discs, or spheres. One species that lives in extremely hot habitats has a cell wall that consists solely of glycoprotein.

Metabolic diversity Depending on the species, cellular respiration may involve organic compounds, sulfur, hydrogen gas, or Fe^{2+} ions as electron donors and oxygen, nitrate, sulfate, sulfur, carbon dioxide, or Fe^{3+} ions as electron acceptors. Some species make ATP exclusively through fermentation pathways.

Human and ecological impacts Although no one understands why, no crenarchaeota are known to cause disease. Crenarchaeota have yet to be used in the manufacture of commercial products. In certain extremely hot, high-pressure, cold, or acidic environments, crenarchaeota may be the only life-form present. Acid-loving species thrive in habitats with pH 1–5; some species are found in ocean sediments at depths ranging from 2500 to 4000 m below the surface.

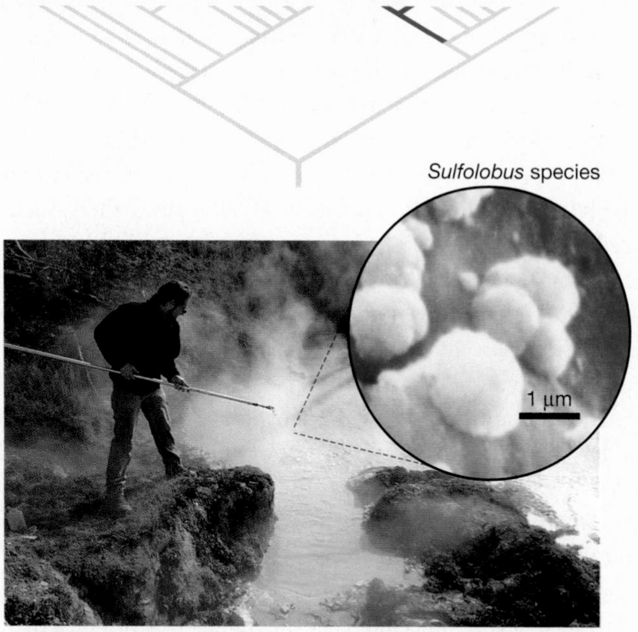
Sulfolobus species
1 µm

FIGURE 27.23 Some Crenarchaetoa Live in Sulfur-Rich Hot Springs

Euryarchaeota

The Euryarchaeota are aptly named, because the word root *eury* means "broad." Members of this phylum live in every conceivable habitat. Some species are adapted to high-salt habitats with pH 11.5—almost as basic as household ammonia. Other species are adapted to acidic conditions with a pH as low as 0. Species in the genus *Methanopyrus* live near hot springs called black smokers that are 2000 m (over 1 mile) below sea level. About 170 species have been identified thus far, and more are being discovered each year (**Figure 27.24**).

Morphological diversity Euryarchaeota cells may be spherical, filamentous, rod shaped, disc shaped, or spiral. Rod-shaped cells may be short or long or arranged in chains. Spherical cells may be found in ball-like aggregations. Some species have several flagella. Some species lack a cell wall; others have a cell wall composed entirely of glycoproteins.

Metabolic diversity The group includes a variety of methane-producing species. These methanogens can use up to 11 different organic compounds as electron acceptors during cellular respiration; all produce CH_4 as a by-product of respiration. In other species of Euryarchaeota, cellular respiration may be based on hydrogen gas or Fe^{2+} ions as electron donors and nitrate or sulfate as electron acceptors. Species that live in high-salt environments can use the molecule retinal—which is responsible for light reception in your eyes—to capture light energy and perform photosynthesis.

Human and ecological impacts Species in the genus *Ferroplasma* (literally, "hot iron") live in piles of waste rock near abandoned mines. As a by-product of metabolism, they produce acids that drain into streams and pollute them. No species in the Euryarchaeota is known to cause disease in humans. Methanogens live in the soils of swamps and the guts of mammals (including yours). They are responsible for adding about 2 billion tons of methane to the atmosphere each year.

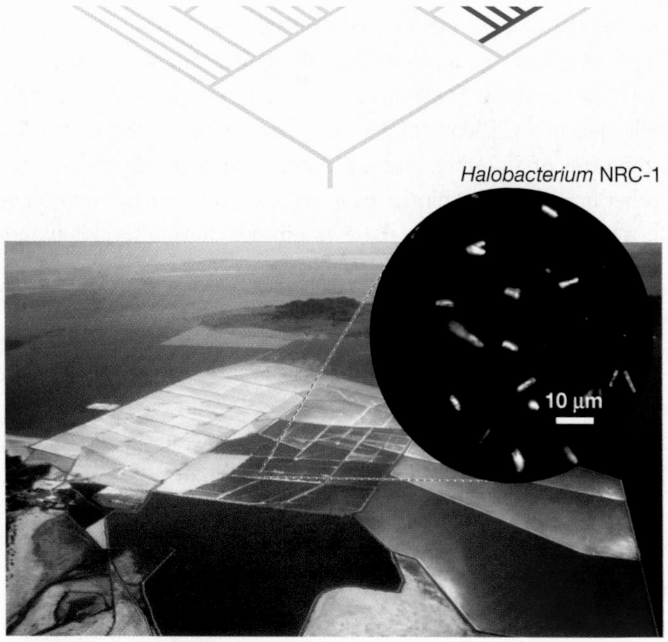
Halobacterium NRC-1
10 µm

FIGURE 27.24 Some Euryarchaeota Live in High-Salt Habitats

ESSAY Antibiotics and the Evolution of Drug Resistance

Antibiotics are molecules that kill bacteria. A wide variety of naturally occurring antibiotics are produced by bacteria or fungi, presumably as a defense against bacterial species that compete with them for space and resources. These molecules work in several ways. Some, such as penicillin, poison enzymes involved in the formation of bacterial cell walls. Others, such as erythromycin, disrupt ribosomes or other parts of the enzymatic machinery involved in protein synthesis. Over 8000 antibiotics are known, and hundreds more are discovered each year. Only a tiny percentage of these is useful in treating human diseases, however.

The first antibiotic that brought important human diseases under chemical control was discovered quite by accident. In 1928 Alexander Fleming was studying the pathogenic bacterium *Staphylococcus* and noticed that cells began dying off in culture plates that had inadvertently been contaminated with a fungus called *Penicillium*. Fleming seized on this observation, confirmed that a molecule from the fungus was responsible for killing the cells, and called it penicillin. In the 1930s a group led by Howard Florey showed that penicillin, administered as a drug, was effective against *Staphylococcus* and *Streptococcus* infections in humans. Florey then led a research consortium that worked out methods for producing the molecule in quantity by the start of World War II.

Penicillin was a "magic bullet"—a drug that killed many different infectious agents without harming the host. The drug saved many millions of lives that would have been cut short by *Staphylococcus* infections or pneumonia. Now, however, the magic is gone. In 1941 physicians could administer 10,000 units of penicillin four times a day for four days and cure pneumonia completely. Now, pneumonia patients who receive 24 *million* units of penicillin a day have a good chance of dying. Similarly, in 1941 all strains of *Staphylococcus aureus* were treated effectively with penicillin. Today, 95 percent of the strains are unaffected by it. Strains of *S. aureus* that are resistant to penicillin and several other important antibiotics have become particularly common in hospitals. In many cases, these strains are causing serious illness in patients who were admitted for other problems. **Figure 27.25** shows how the percentage of patients in intensive care units who are infected with these multiple-drug-resistant strains of *S. aureus* as opposed to nonresistant strains has increased in the United States recently.

The spread of alleles that confer resistance has been favored by natural selection, leading to the evolution of resistant strains. In the case of penicillin and many other antibiotics, however, resistance alleles are found not on the main bacterial chromosome but on plasmids. Because plamids are often transferred from one cell to another, resistance alleles have spread rapidly among strains within pathogenic species and even between species. Antibiotic-resistance genes are a well-documented example of lateral gene transfer—the phenomenon introduced in Chapter 20 and Box 27.2. Penicillin, for example, acts by binding to the enzymes responsible for creating cross-links in bacterial cell walls. This binding prevents cross-linking and weakens the cell walls, which eventually rupture, causing cell death. But a plasmid-borne gene called *β-lactamase* codes for a protein that cleaves the penicillin molecule. This action renders the drug useless, because it can no longer bind to the cross-linking enzymes. Plasmids that carry the *β-lactamase* gene have been transferred to a wide array of pathogenic species. Lateral transfer of *β-lactamase* has been common.

Drug resistance is now a fact of life. . . .

Drug resistance is now a fact of life for drug companies, physicians, and patients. In response, public health agencies are trying to limit the use of antibiotics in animal feed and to reduce the number of antibiotics prescriptions being written by physicians. In Hungary, for example, a steep reduction in penicillin use in treating pneumonia, encouraged by the National Institute of Public Health, led to a sharp drop in the frequency of penicillin-resistant strains of *Pneumococcus*. If antibiotics are to continue being effective, they will have to be prescribed and taken much more carefully than they have been in the past.

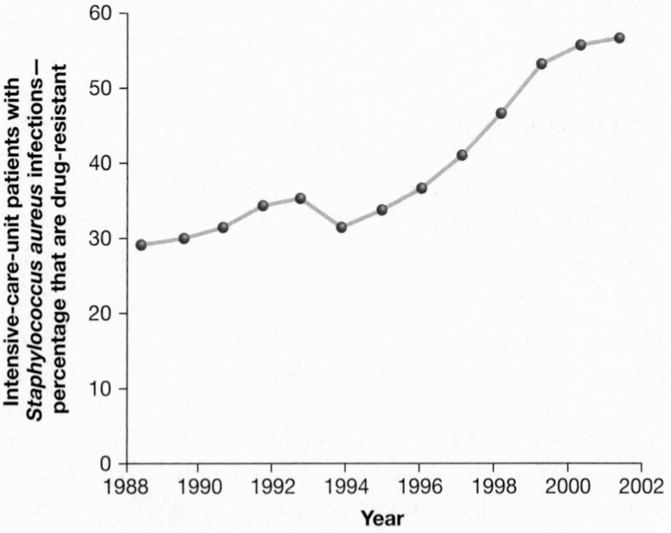

FIGURE 27.25 Drug-Resistant Cells Are Becoming More Common, Particularly in Hospitals
The percentage of *Staphylococcus aureus* infections that are caused in hospital intensive care units by strains resistant to several different antibiotics; from the U.S. Centers for Disease Control and Prevention.

CHAPTER REVIEW

Summary of Key Concepts

Bacteria and archaea affect your life. A small percentage of bacteria causes disease. Some species are effective at cleaning up pollution, and photosynthetic bacteria were responsible for the evolution of the oxygen atmosphere. Bacteria and archaea cycle nutrients through both terrestrial and aquatic environments.

Bacteria and archaea may be tiny, but they have a huge impact on global ecosystems and human health. Cyanobacteria were responsible for producing Earth's first oxygen-containing atmosphere, and nitrogen-fixing bacteria and archaea keep the global nitrogen cycle running. Bacteria also cause some of the most dangerous diseases in humans, including plague, syphilis, botulism, cholera, and tuberculosis. Disease results when bacteria kill host cells or produce toxins that disrupt normal cell functions.

Enrichment cultures are used to grow large numbers of bacterial or archaeal cells that thrive under specified conditions, such as in the presence of certain electron donors and electron acceptors. To study bacteria and archaea that cannot be cultured, biologists frequently take advantage of direct sequencing. In this research strategy, DNA sequences are extracted directly from organisms in the environment—without first culturing the organisms in the lab. By analyzing where these sequences are placed on the tree of life, biologists can determine whether the branches represent new species. If they do, information on where the original sample was collected can expand knowledge about the types of habitats used by bacteria and archaea.

Bacteria and archaea live in a wide array of habitats and use diverse types of molecules in cellular respiration and fermentation. Although they are small and relatively simple in their overall morphologies, these organisms are extremely sophisticated in the chemistry they can do. Many species are restricted in distribution and have a limited diet.

Metabolic diversity and complexity are the hallmarks of the bacteria and archaea, just as morphological diversity and complexity are the hallmarks of the eukaryotes. Like eukaryotes, many bacteria and archaea can extract energy from carbon compounds with high potential energy, such as sugars. These molecules are processed through fermentation pathways or by transferring high-energy electrons to electron transport chains with oxygen as the final electron acceptor. But among the bacteria and archaea, many other inorganic or organic compounds with high potential energy serve as electron donors, and a wide variety of inorganic molecules with low potential energy serve as electron acceptors. Dozens of distinct organic compounds are fermented, including proteins, purines, alcohols, and an assortment of carbohydrates.

Photosynthesis is also widespread among bacteria. In cyanobacteria, water is used as a source of electrons during photosynthesis, and oxygen gas is generated as a by-product. But in other species, the electron excited by photon capture comes from a reduced substance such as ferrous iron (Fe^{2+}) or hydrogen sulfide (H_2S) instead of water (H_2O); the oxidized by-product is the ferric ion (Fe^{3+}) or elemental sulfur (S) instead of oxygen (O_2). These organisms also contain chlorophylls not found in plants or cyanobacteria.

To acquire molecules containing carbon-carbon bonds, some species use the enzymes of the Calvin cycle to reduce CO_2. But biologists have also discovered three additional biochemical pathways in bacteria and archaea that transform carbon dioxide (CO_2), methane (CH_4), or other sources of inorganic carbon into organic compounds such as sugars or carbohydrates.

Web Tutorial 27.1 The Tree of Life

Questions

Content Review

1. How do molecules that function as electron donors and those that function as electron acceptors differ?
   a. Electron donors are organic molecules; electron acceptors are inorganic.
   b. Electron donors are inorganic molecules; electron acceptors are organic.
   c. Electron donors have relatively high potential energy; electron acceptors have relatively low potential energy.
   d. Electron donors have relatively low potential energy; electron acceptors have relatively high potential energy.

2. What do some photosynthetic bacteria use as a source of electrons instead of water?
   a. oxygen (O_2)
   b. hydrogen sulfide (H_2S)
   c. organic compounds (e.g., CH_3COO^-)
   d. nitrate (NO_3^-)

3. What is distinctive about the chlorophylls found in different photosynthetic bacteria?
   a. their membranes
   b. their role in acquiring energy
   c. their role in carbon fixation
   d. their absorption spectra

4. What are organisms called that use inorganic compounds as electron donors in cellular respiration?
   a. phototrophs
   b. heterotrophs
   c. organotrophs
   d. lithotrophs

5. How is direct sequencing used?
   a. to grow unknown organisms in laboratory culture
   b. to determine where organisms that cannot be cultured are located on the tree of life
   c. to determine the metabolic capabilities of newly discovered prokaryotes
   d. to improve the accuracy of phylogenetic trees

6. Koch's postulates outline the requirements for which of the following?
   a. showing that an organism is autotrophic
   b. showing that a bacterium's cell wall lacks an outer membrane and consists primarily of peptidoglycan
   c. showing that an organism causes a particular disease
   d. showing that a disease-causing species also has harmless strains

Conceptual Review

1. Biologists often use the term *energy source* as a synonym for "electron donor." Why?
2. The text claims that the tremendous ecological diversity of bacteria and archaea is possible because of their impressive metabolic diversity. Do you agree with this statement? Why or why not?
3. Suppose that universal PCR primers were available for genes involved in electron transport chains or for some of the different types of chlorophyll found in bacteria. Why would it be interesting to use these genes in a direct sequencing study?
4. The text claims that the evolution of an oxygen atmosphere paved the way for increasingly efficient respiration and higher-energy activities by organisms. Explain.
5. Look back at Table 27.4 and note that the by-products of respiration in some organisms are used as electron donors or acceptors by other organisms. In the table, draw lines between the dual-use molecules listed in the "Electron Donor," "Electron Acceptor," and "Product" columns.
6. Explain the statement, "Prokaryotes are a paraphyletic group."

Group Discussion Problems

1. The researchers who observed that magnetite was produced by bacterial cultures from the deep subsurface carried out a follow-up experiment. These biologists treated some of the cultures with a drug that poisons the enzymes involved in electron transport chains. In cultures where the drug was present, no more magnetite was produced. Does this result support or undermine their hypothesis that the bacteria in the cultures perform anaerobic respiration with Fe^{3+} serving as the electron acceptor? Explain your reasoning.
2. *Streptococcus mutans* obtains energy by oxidizing sucrose. This bacterium is abundant in the mouths of Western European and North American children and is a prominent cause of cavities. The organism (and cavities) is virtually absent in children from East Africa. Propose a hypothesis to explain this observation. Outline the design of a study that would test your hypothesis.
3. Suppose that you've been hired by a firm interested in using bacteria to clean up organic solvents found in toxic waste dumps. Your new employer is particularly interested in finding cells that are capable of breaking a molecule called benzene into less toxic compounds. Where would you go to look for bacteria that can metabolize benzene as an energy or carbon source? How would you design an enrichment culture capable of isolating benzene-metabolizing species?
4. Would you predict that disease-causing bacteria, such those listed in Table 27.1, obtain energy from light, organic molecules, or inorganic molecules? When they perform cellular respiration, which substance would you predict that they use as an electron acceptor? Explain your answer.

Answers to Multiple-Choice Questions 1. c; 2. b; 3. d; 4. d; 5. b; 6. c

www.prenhall.com/freeman is your resource for the following: Web Tutorials; Online Quizzes and other Online Study Guide materials; Answers to Conceptual Review Questions; Solutions to Group Discussion Problems; Answers to Figure Caption Questions and Exercises; and Additional Readings and Research.

Protists

28

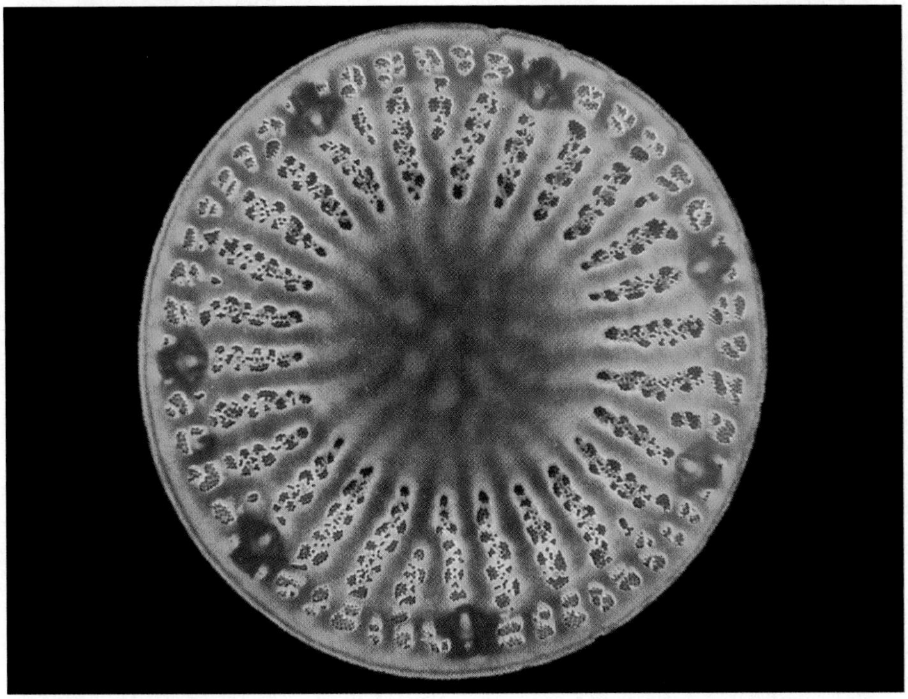

Diatoms are single-celled protists that live inside a glassy case. They may be the most abundant of all eukaryotes found in aquatic environments.

KEY CONCEPTS

- Protists are a paraphyletic grouping that includes all eukaryotes except the green plants, fungi, and animals. Biologists study protists to understand how eukaryotes evolved, because they are important in marine ecosystems and global warming and because some species cause debilitating diseases in plants, humans, and other organisms.
- Protists are diverse morphologically. They vary in the types of organelles they contain; they may be unicellular or multicellular, and they may have a cell wall or other external covering, or no such covering.
- Protists vary widely in terms of how they reproduce and how they find food. Many can reproduce both sexually and asexually. Many species are photosynthetic, while others obtain carbon compounds by ingesting food or parasitizing other organisms.

This chapter introduces the third domain on the tree of life: the **Eukarya**. Eukaryotes range from single-celled organisms that are the size of bacteria to sequoia trees and blue whales. The largest and most morphologically complex organisms on the tree of life—algae, plants, fungi, and animals—are eukaryotes.

Although species in the Eukarya are astonishingly diverse, they share fundamental features. Eukaryotes are defined by the presence of the shared, derived character called the nuclear envelope. All members of the Eukarya possess a nucleus as well as other organelles. All eukaryotes undergo cell division via mitosis, and all have a system of structural and contractile proteins called the cytoskeleton. Like archaea, most eukaryotes have chromosomes where DNA is complexed with proteins called histones.

In introducing the Eukarya, this chapter focuses on an informal grab bag of lineages known as the protists. The term **protist** refers to all eukaryotes that are not green plants, fungi, or animals. Analyzing protist diversity is the key to understanding how eukaryotes came to be and how green plants, fungi, and animals originated.

The first thing to appreciate about protists is their morphological diversity. Many protists are microscopic single cells; others are multicellular organisms up to 60 meters long. Their lifestyles are just as diverse. Some are parasitic, while others are predatory or photosynthetic. They may be stationary all their lives or in virtually constant motion. Their cells can change shape almost continuously or be encased in rigid, glassy shells. The common feature among protists is that they tend to live in

(a) Open ocean:

Surface waters teem with microscopic protists, such as these diatoms.

(b) Shallow coastal waters:

Gigantic protists, such as these kelp, form underwater forests.

(c) Intertidal habitats:

Protists such as these red algae are particularly abundant in tidal habitats.

FIGURE 28.1 Protists Are Particularly Abundant in Aquatic Environments
Protists are abundant in a wide variety of aquatic habitats. In marine environments, they are found in **(a)** the open ocean and **(b)** near-shore and **(c)** intertidal habitats.

moist surroundings such as wet soils, aquatic habitats, or the bodies of other organisms. Protists are crucial in terms of abundance, productivity, and overall impact (**Figure 28.1**).

28.1 Why Do Biologists Study Protists?

Biologists study protists in part because protists are intrinsically interesting, in part because they are so important medically and ecologically, and in part because they are critical to understanding the evolution of plants, fungi, and animals. The remainder of the chapter will focus on why protists are interesting in their own right and how they evolved; here let's consider their impact on human health and the environment.

Impacts on Human Health and Welfare

The most spectacular crop failure in history, the Irish potato famine, was caused by a protist. In 1845 most of the 3 million acres that had been planted to grow potatoes in Ireland became infested with *Phytophthora infestans*—a parasite that belongs to a lineage of protists called Oomycota. Potato tubers that were infected with *P. infestans* rotted in the fields or in storage.

As a result of crop failures in Ireland in 1845 and the following year, an estimated 1 million people out of a population of less than 9 million died of starvation or starvation-related illnesses. Several million others were driven away. Many people of Irish heritage living in North America, New Zealand, and Australia trace their ancestry to relatives who emigrated from Ireland to evade the famine. As devastating as the potato famine was, however, it does not begin to approach the misery caused by the protist *Plasmodium*.

Malaria Physicians and public health officials point to three major infectious diseases that are currently afflicting large numbers of people worldwide: tuberculosis, HIV, and malaria. Tuberculosis was introduced in Chapter 23, and HIV is analyzed in Chapter 34. Here we consider malaria.

Malaria ranks as the world's most chronic public health problem. In India alone, over 30 million people each year suffer from debilitating fevers caused by malaria. At least 300 million people worldwide are sickened by it each year, and over 1 million die from the disease annually. The toll is equivalent to eight 747s, loaded with passengers, crashing every day. Most of the dead are children of preschool age.

Four species of the protist *Plasmodium* are capable of infecting humans and causing malaria (**Figure 28.2**). Because each *Plasmodium* species spends part of its life cycle inside mosqui-

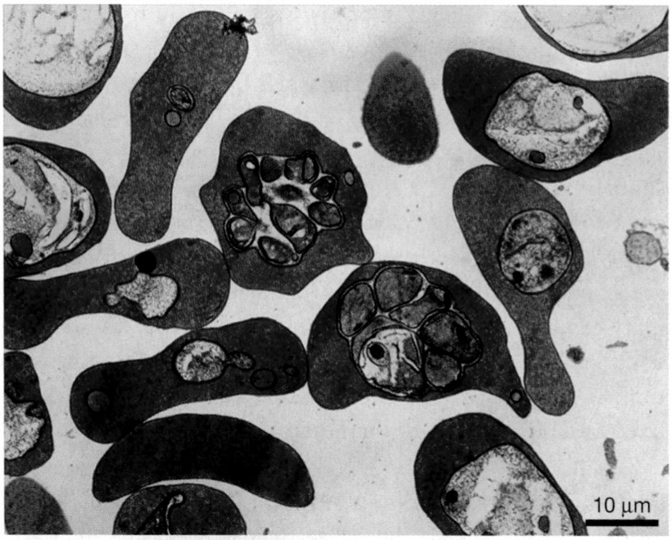

FIGURE 28.2 *Plasmodium* Causes Malaria
Over the course of its life cycle, *Plasmodium falciparum* develops into a series of distinct cell types. Each type is specialized for infecting a different host cell in mosquitoes or humans. The yellow- and orange-stained cells in this micrograph are infecting human red blood cells. The infection kills the host cell, contributing to anemia and high fever.

toes, most antimalaria campaigns have focused on controlling these insects. This strategy has become less and less effective over time, however, because natural selection has favored mosquito strains that are resistant to the insecticides that have been sprayed in their breeding habitats. Further, *Plasmodium* itself has evolved resistance to most of the drugs used to control its growth in infected people. Efforts to develop a vaccine against *Plasmodium* have also been fruitless to date, in part because the parasite evolves so quickly. (Chapter 34 explains why it is difficult or impossible to develop effective vaccines against rapidly evolving viruses and organisms such as *Plasmodium*, cold and influenza viruses, and HIV.) Although *Plasmodium* is arguably the best studied of all protists, researchers have still not been able to devise effective and sustainable measures to control it.

Biologists study the basic biology of *Plasmodium* in the hopes that a better understanding of the parasite will lead to improved drug therapies or even an effective vaccine. In the meantime, the World Health Organization, the World Bank, and several agencies of the United Nations have joined forces to promote the use of insecticide-treated sleeping nets to prevent bites by *Plasmodium*-infected mosquitoes and to urge the use of "combination therapies"—medications that include at least two different antimalarial drugs. Combination therapies have been shown to slow the evolution of drug resistance.

Unfortunately, malaria is not the only important human disease caused by protists. **Table 28.1** lists protists that have been the cause of human misery and economic losses. Parasitic protists affect hundreds of millions of people every year and are a major concern for physicians worldwide. Harmful protists also worry biologists who manage the fisheries that many people depend on for their food or livelihood.

Harmful Algal Blooms When photosynthetic protists experience rapid population growth and reach high densities in marine environments, they are said to "bloom." Unfortunately, a handful of the many species involved in blooms can be

TABLE 28.1 Human Health Problems Caused by Protists

Lineage	Species	Disease
Apicomplexa	Four species of *Plasmodium*, primarily *P. falciparum* and *P. vivax*	Malaria has the potential to affect 40 percent of the world's total population.
	Toxoplasma	Toxoplasmosis may cause eye and brain damage in infants and in AIDS patients.
Dinoflagellates	Many species involved	Toxins released during "red tides" accumulate in clams and mussels and poison people if eaten.
Excavates	*Giardia*	Diarrhea can last for several weeks.
Parabasalids	*Trichomonas*	Trichomoniasis is a reproductive tract infection and one of the most common sexually transmitted diseases. About 2 million young women are infected in the United States each year; some of them become infertile.
Discicristates	*Leishmania*	Leishmaniasis can cause skin sores or affect internal organs—particularly the spleen and liver.
	Trypanosoma gambiense and *T. rhodesiense*	Trypanosomiasis ("sleeping sickness") is a potentially fatal disease transmitted through bites from tsetse flies. Occurs in Africa.
	Trypanosoma cruzi	Chagas disease affects 16–18 million people and causes 50,000 deaths annually, primarily in South and Central America.
Entamoebae	*Entamoeba histolytica*	Amoebic dysentery results from severe infections.
Oomycetes	*Phytophthora infestans*	An outbreak of this protist wiped out potato crops in Ireland in 1845–1847, causing famine.
Microsporidians	*Encephalitozoon cuniculi*	Infections affect the nervous system, respiratory tract, and digestive tract—especially in AIDS patients.

FIGURE 28.3 Harmful Algal Blooms Are Sometimes Called Red Tides Dinoflagellates with red coloration have "bloomed" along this beach.

harmful because they release toxins into the surrounding water or, if they are eaten, into a predator. Although the function of these toxins for the protists is not well understood, they may serve to reduce predation by fish.

When a toxin-producing protist reaches high densities in a particular area, a **harmful algal bloom** is said to occur. Toxin-producing protists named dinoflagellates have high concentrations of red accessory pigments called xanthophylls. Hence these outbreaks are known informally as *red tides* (**Figure 28.3**).

Algal blooms can be harmful to people because clams and other shellfish filter photosynthetic protists out of the water as food. During a bloom, high levels of toxins can build up in the flesh of these shellfish. Typically, the shellfish themselves are not harmed. But if a person eats contaminated shellfish, several types of poisoning can result. For example, paralytic shellfish poisoning occurs when people eat shellfish that have fed heavily on protists that synthesize poisons called saxitoxins. In humans, high dosages of saxitoxins cause unpleasant symptoms such as prickling sensations in the mouth or even life-threatening symptoms such as muscle weakness and paralysis.

No antidote exists to the poisons secreted by the protists during harmful blooms. As a result, biologists prevent poisonings by carefully monitoring protist populations in regions where shellfish are harvested for food. If harmful species begin to bloom, the shellfish beds are immediately closed to harvest until they can recover.

Ecological Importance of Protists

As a whole, the protists represent just 10 percent of the total number of named eukaryote species. Although the species diversity of protists may be relatively low, their abundance is extraordinarily high. The number of individual protists found in some habitats is astonishing. One milliliter of pond water can contain well over 500 single-celled protists that swim with the aid of flagella. Under certain conditions, dinoflagellates can reach concentrations of 60 million cells per liter of seawater.

The great abundance of protists is important to an array of ecological events. For example, photosynthetic organisms take in carbon dioxide from the atmosphere and reduce it to form sugars or other organic compounds with high potential energy (Chapter 10). Photosynthesis transforms some of the energy in sunlight into chemical energy that organisms can use to grow and produce offspring. Species that produce chemical energy in this way are called **primary producers.** Diatoms, for example, are photosynthetic protists that rank among the leading primary producers in the oceans, simply because they are so abundant. Production of organic molecules in the world's oceans, in turn, is responsible for almost half of the total carbon that is fixed on Earth. Why is this important?

Protists Play a Key Role in Aquatic Food Chains Photosynthetic protists live in the surface waters of marine and freshwater habitats, where light is available. Small organisms that live near the surface of oceans or lakes and that drift along or swim only short distances are called **plankton.** The organic compounds that are produced by **phytoplankton**—that is, photosynthetic species of plankton—are the basis of food chains in freshwater and marine environments. A **food chain** describes nutritional relationships among organisms. Thus, it also describes how chemical energy flows within ecosystems.

Figure 28.4 shows a food chain for a marine environment. At the bottom of the chain, primary producers—such as photosynthetic protists and photosynthetic bacteria—are eaten by consumers. Primary consumers are just above primary producers in the chain. Scavengers and decomposers occupy the same consumer level in the food chain, but they feed on primary producers that have died. Organisms at this level of the chain are eaten by secondary consumers, which in turn are eaten by tertiary consumers. The fish and shellfish at the middle levels of the food chain are important sources of protein not only for whales, squid, and very large fish (such as tuna) at the uppermost levels but also for people.

Many of the species at the food chain's base are protists. It is not an exaggeration to say that without protists, most food chains in freshwater and marine habitats would collapse.

How Could Protists Help Reduce Global Warming? As noted in Chapter 10, carbon dioxide levels in the atmosphere are increasing rapidly due to human activities such as the burning of fossil fuels and forests. Because carbon dioxide traps heat that is radiating from Earth back out to space, high CO_2 levels in the atmosphere contribute to global warming—arguably today's most pressing environmental problem.

To understand global warming, it is critical to analyze how carbon atoms move among organisms and molecules in both terrestrial and aquatic habitats. The movement of carbon atoms from carbon dioxide molecules in the atmosphere to organisms, in the soil or the ocean, and then back to the atmosphere is called the **global carbon cycle.** To mitigate global warming, researchers are trying to figure out ways to reduce carbon dioxide concentrations in the atmosphere and increase the amount of carbon stored in terrestrial and marine environments.

Recall from Chapter 10 that in terrestrial environments, trees take CO_2 from the atmosphere and use some of the carbon atoms to make wood. Wood is a storage area, or "sink," for carbon, because carbon atoms tend to remain in wood for decades or centuries. As a result, the growth of trees reduces the amount of carbon in the atmosphere and increases the amount of carbon stored in forests. The essay at the end of Chapter 10 suggested that efforts to plant trees in deforested areas might play a role in efforts to reduce atmospheric CO_2 concentrations and slow the rate of global warming.

In the world's oceans, attempts to fertilize photosynthetic protists and bacteria might have a similar effect. To understand why, consider the carbon cycle diagrammed in **Figure 28.5**. The phytoplankton that dominate marine environments have short life spans—they die or are eaten in a matter of days or weeks. As a result, carbon atoms shuttle rapidly among planktonic species. When those organisms die, they are either consumed or sink to the bottom of the ocean. There they may enter one of two long-lived repositories—sedimentary rocks or petroleum:

1. Several lineages of protists have shells made of calcium carbonate ($CaCO_3$). When these shells rain down from the ocean surface and settle in layers at the bottom, the deposits that result are compacted by the weight of the water and by sediments accumulating above them. Eventually they turn into rock. The limestone used to build the pyramids of Egypt consists of protist shells. Limestones and other carbon-containing rocks lock up carbon atoms for tens of millions of years.
2. Although the process of petroleum (oil) formation is not well understood, it begins with accumulations of dead bacteria, archaea, and protists at the bottom of the ocean. CO_2 that passes from the atmosphere to the body of a photosynthetic protist and then to petroleum is removed from the atmosphere for millions of years—unless humans pump the petroleum out of the ground and burn it.

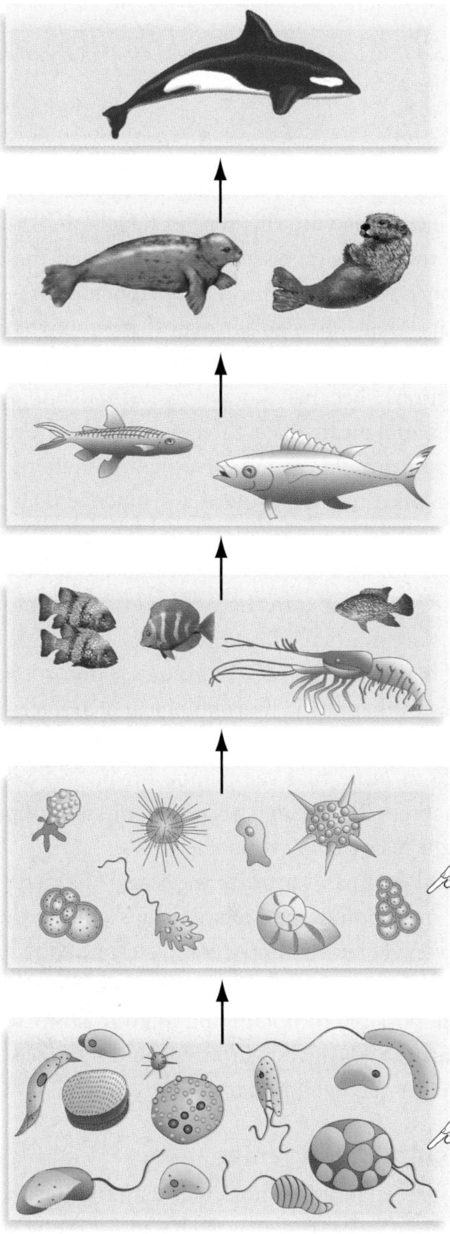

FIGURE 28.4 Protists Are Key Primary Producers in Marine and Freshwater Habitats
In aquatic habitats, photosynthetic protists form the base of the food chain along with photosynthetic bacteria. EXERCISE Indicate the trophic levels at which humans feed.

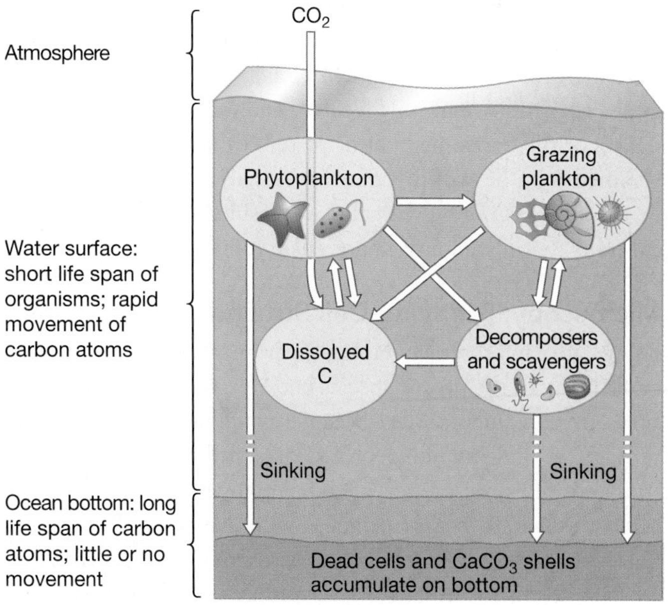

FIGURE 28.5 In the Oceans, Protists Play a Key Role in the Global Carbon Cycle
Carbon atoms tend to shuttle quickly among organisms. But if carbon atoms sink to the bottom of the ocean in the form of shells or dead cells, they may be locked up for long periods in carbon sinks.

Appropriately enough, researchers who study the global carbon cycle refer to long-lived carbon reservoirs as **carbon sinks.**

How do these observations relate to global warming? Recent experiments have shown that when habitats in the middle of the ocean are fertilized with iron, populations of protists and other primary producers increase by a factor of 10. Some researchers hypothesize that, when these blooms occur, the amount of carbon that rains down into carbon sinks in the form of shells and dead cells may increase. If so, then fertilizing the ocean to promote blooms might be an effective way to reduce CO_2 concentrations in the atmosphere. The effectiveness of iron fertilization is hotly debated, however. In addition, some researchers point out that fertilizing the ocean with iron might lead to the formation of anaerobic dead zones (see Chapter 27), toxicity to certain organisms, or other side effects. As a result, many biologists are not convinced that adding iron and inducing massive blooms of protists, bacteria, and archaea would be a good idea. If further research shows that iron fertilization is safe and effective, it could be added to the list of possible approaches for reducing carbon dioxide levels in the atmosphere.

28.2 How Do Biologists Study Protists?

Given that protists are important in medicine and in the world's lakes and oceans, it is not surprising that they have been the focus of intense study. Although great strides have been made in understanding some protist pathogens and the role of protists in the global carbon cycle, it has been extremely difficult for biologists to gain any sort of solid insight into how the group as a whole diversified. The problem is that protists are so diverse that it has been difficult to find any overall patterns in the evolution and diversification of the group.

Recently, researchers have made dramatic progress in understanding protist diversity by combining data on the morphology of key groups and phylogenetic analyses of DNA sequence data. For the first time, a clear picture of how the Eukarya diversified may be within sight. Let's analyze how this work is being done, beginning with classical results on the morphological traits that distinguish major eukaryote groups.

Microscopy: Studying Cell Structure

Using light microscopy, biologists were able to identify and name many of the protist species known today. When transmission electron microscopes became available, a major breakthrough in understanding protist diversity occurred: Detailed studies of cell structure revealed that many protists have a characteristic overall form, organelles with distinctive features, or both. For example, both light microscopy and electron microscopy confirmed that the species that caused the Irish potato famine has reproductive cells with an unusual type of flagellum. Flagella are organelles that project from the cell and whip back and forth to produce swimming movements (Chapter 7). In reproductive cells of *Phytophthora infestans*, one of the two flagella present has tiny, hollow, hairlike projections attached to it. Biologists noted that kelp and other forms of brown algae also have cells with this type of flagellum.

FIGURE 28.6 Species in the Lineage Called Stramenopiles Have a Distinctive Flagellum
The unusual, hollow "hairs" that decorate the flagella of stramenopiles often have three branches at the tip.

To make sense of these results, researchers interpreted these types of distinctive morphological features as **synapomorphies**—shared, derived traits that distinguish major monophyletic groups (see Chapter 26). Species that have a flagellum with hollow, hairlike projections became known as *stramenopiles* ("straw-hairs"); the hairs typically have three branches at the tip (**Figure 28.6**). In recognizing this group, investigators hypothesized that, because an ancestor had evolved a distinctive flagellum, all or most of its descendants also had this trait. The qualifier "most" is important, because it is not unusual for certain subgroups to lose particular traits over the course of evolution, much as humans are gradually losing the tailbone.

Eventually, eight major groups of eukaryotes came to be identified on the basis of diagnostic morphological characteristics. These groups and the synapomorphies that identify them are listed in **Table 28.2**. Note that the plants, fungi, and animals analyzed in Chapters 29 through 33 represent subgroups within two of the eight major eukaryotic lineages.

Although individual groups of protists were well characterized on the basis or morphology, the relationships among the eight eukaryotic lineages remained almost completely unknown. The next major advance in understanding eukaryotic diversity came when it became possible to obtain and analyze DNA sequence data. In the early 1990s, investigators began using molecular traits to reconstruct the evolutionary history of Eukarya.

Evaluating Molecular Phylogenies

When investigators first began using molecular data to infer the phylogeny of eukaryotes, most studies were based on the gene that codes for the RNA molecule in the small subunit of ribosomes (see Chapter 16). Recall from Chapter 27 that analyses of this gene revealed the nature of the three domains of life. When researchers sequenced this gene from an array of eukaryotes and used the data to infer a phylogeny of the domain Eukarya, they

TABLE 28.2 Major Lineages of Eukaryotes

Lineage	Distinguishing Morphological Feature(s)
Excavates	No mitochondrion is present, although genes derived from mitochondria are found in the nucleus. Cells have a pronounced "feeding groove" where prey or organic debris are ingested.
Discicristates	Cells have mitochondria with distinctive disc-shaped cristae.
Alveolates	Cells have sac-like structures called alveoli that form a continuous layer just under the plasma membrane. Alveoli are thought to provide support.
Stramenopiles	If flagella are present, cells usually have two—one of which is covered with hairlike projections.
Cercozoa	Cells lack cell walls, although some produce an elaborate shell-like covering. When pseudopodia project from the cell to move it, they are slender in shape.
Plants	Cells have chloroplasts with a double membrane.
Amoebozoa	Cells lack cell walls. When pseudopodia project from the cell to move it, they form large lobes.
Opisthokonts	Reproductive cells have a single flagellum at their base. The cristae inside mitochondria are flat, not tube shaped as in other eukaryotes. (This lineage includes protists as well as the fungi and the animals. It is discussed in detail in Chapters 30 through 33.)

found that the eight groups identified on the basis of distinctive morphological characteristics were indeed monophyletic groups. This was important support for the hypothesis that the distinctive morphological features were shared, derived characters that existed in a common ancestor of each lineage.

To understand the relationships among the eight lineages and their subgroups, biologists began analyzing sequence data from other genes and combining these results with data from other morphological traits. Although estimating the phylogeny of Eukarya is still a work in progress, the phylogenetic tree in **Figure 28.7** is the current best estimate of the group's evolutionary history. Understanding where the root or base of the tree lies has been problematic. This tree suggests that a group of single-celled organisms called Excavates, which have a distinctive "feeding groove," forms the most basal, or ancient, lineage of the Eukarya. Groups with specialized types of flagella, unusual mitochondria, or chloroplasts evolved later. The tree also confirms that the lineages grouped under the name *protist* (shown in color in Figure 28.7) are **paraphyletic**—meaning they do not constitute all of the descendants of a single common ancestor (see **Box 28.1**, page 614).

As more and more data become available, our understanding of eukaryote evolution continues to improve. Although a great deal remains to be done to clarify the relationships within Eukarya further, several key events in the history of this group are now well understood. As an example, let's consider how biologists have combined morphological data and phylogenetic analyses to understand better the origin of two key organelles found in eukaryotes: the mitochondrion and the chloroplast.

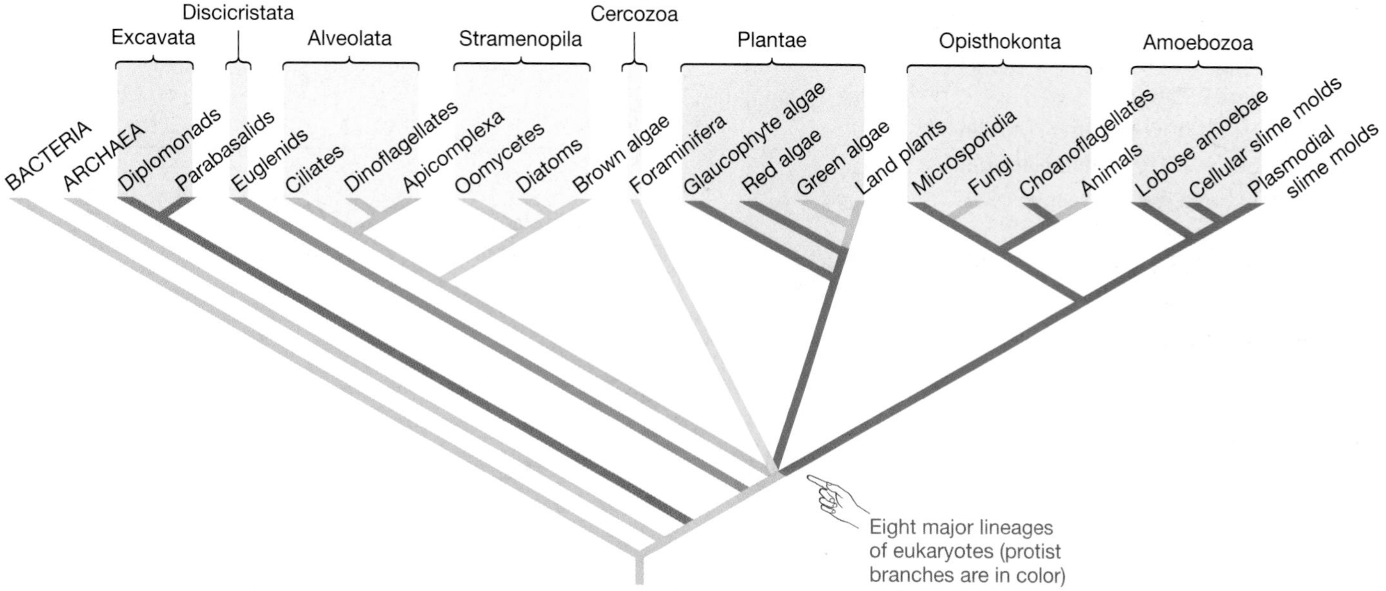

FIGURE 28.7 Phylogenetic Analyses Have Identified Eight Major Lineages of Eukaryotes
Researchers are still evaluating the location of the root on this tree. EXERCISE On the tree, label the node that represents the common ancestor of all eukaryotes. Put a bar and label across the branch where the unusual flagellum of stramenopiles evolved.

BOX 28.1 How Should We Name the Major Branches on the Tree of Life?

Taxonomy is the branch of biology devoted to describing and naming new species and classifying groups of species. Carolus Linnaeus founded this field and published the first work in it in 1735. Linnaeus invented the system of Latin binomials, still being used today, in which each organism is given a unique name consisting of its genus and species. He also invented a hierarchy of more general taxonomic categories that included kingdoms, classes, orders, and families (see Chapter 1). These units were created in an attempt to organize and describe groups of species.

Linnaeus worked long before Darwin had discovered the principle of evolution by natural selection, however. As a result, Linnaeus viewed the groups he described as entities that had been created separately and independently from one another. When biologists came to realize that all organisms are related by common descent, they had to reinterpret the categories that Linnaeus had established. Instead of representing neat "bins" that held increasingly dissimilar species according to a taxonomic (Linnaean) way of thinking (**Figure 28.8a**), biologists recognized that the groupings actually designated twigs, branches, and stems on a tree of life according to a phylogenetic way of thinking (**Figure 28.8b**).

For almost 100 years after Darwin published, however, few reliable techniques existed for reconstructing evolutionary relationships—that is, for estimating phylogenies. As a result, biologists had to continue grouping organisms according to their morphological similarity, much as Linnaeus had done. Biologists were left to hope that these groupings reflected the actual evolutionary relationships on the tree of life.

Over the past 40 years, however, increasingly sophisticated techniques have become available for estimating phylogenies. Chapter 26 introduced these techniques, known as phenetic and cladistic approaches. In general, the old groupings were remarkably accurate in reflecting actual lineages. But there are some important exceptions. Among the protists, for example, old groups such as "protozoans" and "algae" are now understood to be conglomerations of distantly related, though superficially similar, species. Even the name *protist* will probably be abandoned before long. According to rules that are currently being adopted, most biologists assign names only to monophyletic groups—that is, to branches on the tree of life that include all the descendants of a common ancestor. In violation of this rule, the group that is commonly called protists is paraphyletic. The name refers to some, but not all, descendants of the first eukaryote.

According to a Chinese proverb, "The first step in wisdom is to call things by their right name." In taxonomy, a massive effort is under way to do just that. The initial task is to produce accurate estimates of where each branch occurs on the tree of life. Then monophyletic groups can be named, with confidence that each represents a distinct stem, branch, or twig on the tree.

(a) Linnaean thinking

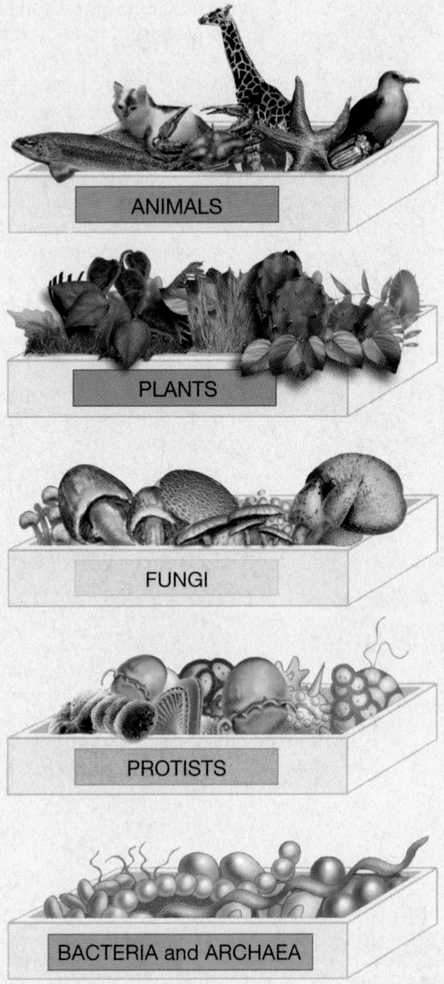
ANIMALS
PLANTS
FUNGI
PROTISTS
BACTERIA and ARCHAEA

(b) Phylogenetic thinking

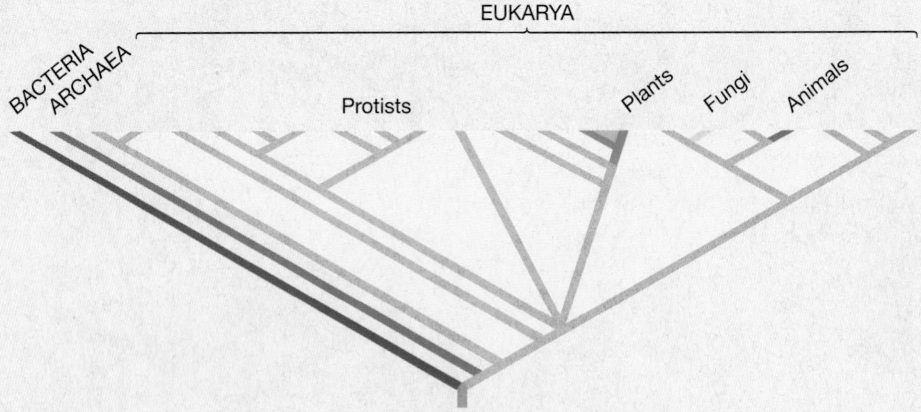
EUKARYA
BACTERIA
ARCHAEA
Protists
Plants
Fungi
Animals

FIGURE 28.8 Contrasting Approaches to Classification
(a) In Linnaean thinking, organisms are grouped according to morphological similarity. **(b)** In phylogenetic thinking, groups are named if they represent distinct lineages on the tree of life.

Combining Data from Microscopy and Phylogenies: Understanding the Origin of Mitochondria and Chloroplasts

What did the earliest eukaryotes look like? Because the most ancient eukaryotic groups are unicellular, and because virtually all bacteria and all archaeans are also unicellular, biologists infer that the first eukaryote was a single-celled organism. Further, all eukaryotes alive today have a nucleus and a cytoskeleton, so researchers conclude that their common ancestor also had these structures. And because the most ancient eukaryotic lineages lack cell walls, biologists suggest that their common ancestor also lacked this feature. In sum, the earliest eukaryotes were probably single-celled organisms with a nucleus and cytoskeleton but no cell wall. What happened next?

The answer is that one ancestral eukaryote acquired the structure that eventually became a mitochondrion. Recall from Chapter 9 that mitochondria are organelles that have a double membrane and that they generate ATP by using pyruvate as an electron donor and oxygen as the ultimate electron acceptor. All living eukaryotes have mitochondria or at least contain genes that are normally found in mitochondria, so biologists contend that the common ancestor of today's eukaryotes also had mitochondria. Where did these organelles come from?

The Endosymbiosis Theory In 1981 Lynn Margulis expanded on a radical hypothesis—first proposed in the nineteenth century—to explain the origin of mitochondria. The **endosymbiosis theory** proposes that mitochondria originated when a bacterial cell took up residence inside a eukaryote about 2 billion years ago. Its name is inspired by the Greek word roots *endo*, *sym*, and *bio* ("inside-together-living"). **Symbiosis** is said to occur when individuals of two different species live in physical contact; **endosymbiosis** occurs when an organism of one species lives inside an organism of another species.

In its current form, the endosymbiosis theory proposes that mitochondria evolved through a series of steps. As **Figure 28.9** shows, the process began when eukaryotic cells started to use their cytoskeletal elements to surround and engulf smaller prey. The theory proposes that instead of being digested, an engulfed bacterium began to live symbiotically within its eukaryotic host. Specifically, the theory maintains that the engulfed cell survived by absorbing carbon molecules with high potential energy from its host and oxidizing them, using oxygen as a final electron acceptor. The host cell, in contrast, is proposed to be a predator capable only of anaerobic fermentation—meaning it could not use oxygen as an electron acceptor in cellular respiration. The relationship between the host and the engulfed cell was presumed to be stable because a mutual advantage existed between them: The host supplied the bacterium with protection and carbon compounds from its other prey, while the bacterium supplied the host with ATP. Recall from Chapter 9 that cells that can use oxgyen during cellular respiration are able to produce much more ATP than are cells that cannot.

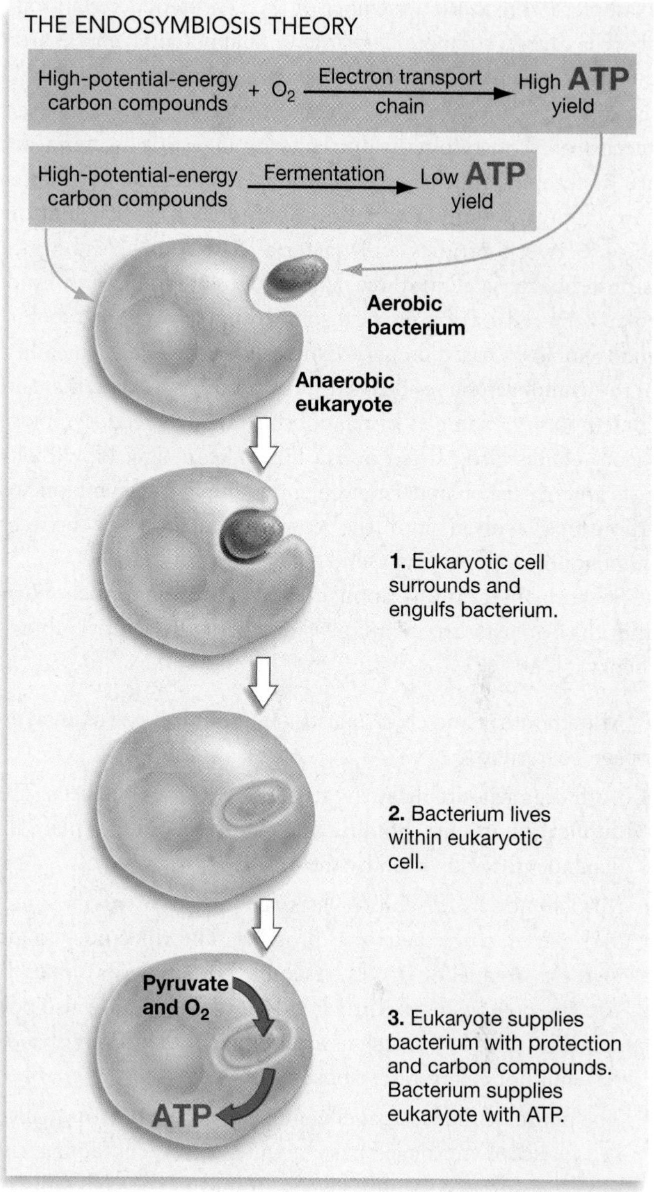
THE ENDOSYMBIOSIS THEORY
High-potential-energy carbon compounds + O_2 Electron transport chain High ATP yield
High-potential-energy carbon compounds Fermentation Low ATP yield
Aerobic bacterium
Anaerobic eukaryote
1. Eukaryotic cell surrounds and engulfs bacterium.
2. Bacterium lives within eukaryotic cell.
Pyruvate and O_2
ATP
3. Eukaryote supplies bacterium with protection and carbon compounds. Bacterium supplies eukaryote with ATP.

FIGURE 28.9 Proposed Initial Steps in the Evolution of the Mitochondrion

QUESTION According to this hypothesis, how many membranes should surround a mitochondrion?

The endosymbiosis theory contends that chloroplasts originated in an analogous way. In this case, however, the theory maintains that a photosynthetic, endosymbiotic cyanobacterium provided its eukaryotic host with oxygen and glucose in exchange for protection. If the endosymbiosis theory is correct, chloroplasts trace their ancestry to cyanobacteria.

Do the Data Support the Endosymbiosis Theory? When biologists assess a theory, they often begin simply by evaluating its plausibility. For example, is it reasonable to suppose that the steps in the endosymbiosis theory actually happened? Given observations that biologists have made about organisms living today, each step in the endosymbiosis theory passes this test. For example, *Paracoccus*, a member of the α-proteobacteria group, absorbs carbon compounds that have high potential energy from the environment and performs aerobic respiration—much like the presumed ancestral mitochondrion. Dozens of contemporary species of protists obtain nutrients by ingesting bacteria and are anaerobic—much like the presumed ancestral host. More convincingly, many examples of endosymbiotic relationships between protists and bacteria exist today. Among the α-proteobacteria alone, three major groups are found *only* inside eukaryotic cells. There are also some examples of contemporary endosymbioses based on photosynthesis. In such cases, members of the cyanobacteria reside inside protist hosts. In short, there are contemporary examples of relationships that the endosymbiosis theory claims also existed over a billion years ago. Now biologists are focused on understanding how simple endosymbiotic relationships evolved into the complex relationship between mitochondria, chloroplasts, and eukaryotic cells.

Several observations about the structure of mitochondria and chloroplasts are also consistent with the endosymbiosis theory:

- Mitochondria and chloroplasts are about the size of an average bacterium.
- Both organelles replicate by fission, as do bacterial cells. The duplication of mitochondria and chloroplasts takes place independently of division by the host cell.
- Mitochondria and chloroplasts have their own ribosomes and manufacture their own proteins. The ribosomes found in these organelles closely resemble bacterial ribosomes in size and composition. Mitochondrial ribosomes are also poisoned by antibiotics such as streptomycin that inhibit bacterial, but not eukaryotic, ribosomes.
- The photosynthetic organelle of one group of protists, called the glaucophyte algae, has an outer layer containing the same constituent (peptidoglycan) found in the cell walls of cyanobacteria.
- Some cyanobacteria even look like chloroplasts. The bacterium *Prochloron*, for example, contains chlorophylls *a* and *b* and has a system of internal membranes similar to the thylakoids found in chloroplasts (see Chapter 10).
- Both mitochondria and chloroplasts have double membranes, consistent with the engulfing mechanism of origin illustrated in Figure 28.9.
- Mitochondria and chloroplasts have their own genomes, which are organized as circular molecules—much like a bacterial chromosome. Mitochondria and chloroplasts have genes that code for the enzymes needed to replicate and transcribe their own genomes.

Although these data are impressive, they are only consistent with the theory. Stated another way, they do not exclude other explanations. Years after Margulis began to champion this theory, however, data emerged that persuaded virtually all biologists that the endosymbiosis theory was correct. These data were important because they tested predictions made by Margulis's idea that contrasted with predictions made by an alternative theory: that both mitochondria and chloroplasts had evolved within eukaryotic cells, separately from bacteria. According to this "within-eukaryotes" theory, the genes found in both mitochondria and chloroplasts had to have been derived from some of the nuclear genes of ancestral eukaryotes. Margulis's theory, in contrast, proposed that the genes found in mitochondria and chloroplasts were bacterial in origin.

These predictions were tested by studies on the phylogenetic relationships of mitochondrial genes and chloroplast genes (**Figure 28.10**). For example, researchers compared gene sequences isolated from the nuclear DNA of eukaryotes, mitochondrial DNA from eukaryotes, and DNA from several species of bacteria. Exactly as the endosymbiosis theory predicted, the mitochondrial sequences turned out to be much more closely related to the sequences from the α-proteobacteria

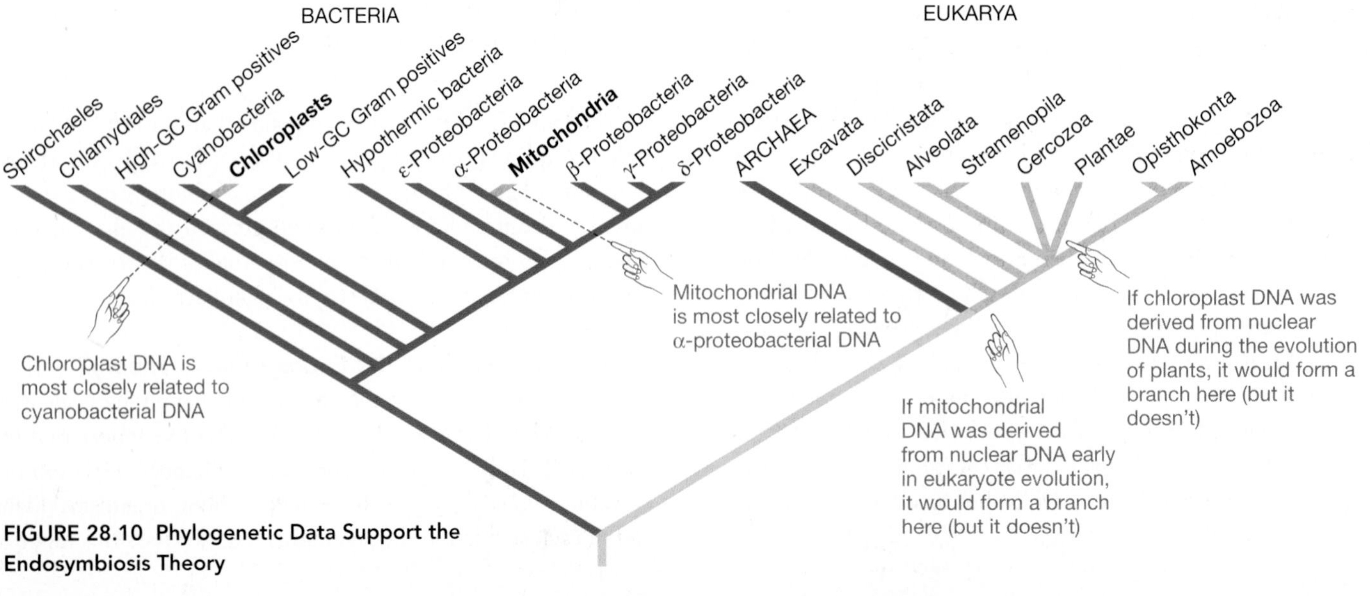

FIGURE 28.10 Phylogenetic Data Support the Endosymbiosis Theory

than to sequences from the nuclear DNA of eukaryotes. The result was considered overwhelming evidence that the mitochondrial genome came from an α-proteobacterium rather than from a eukaryote. A similar analysis showed that chloroplast genes are most closely related to sequences from cyanobacteria, not to those from plants (see Figure 28.10). The endosymbiosis theory was the only reasonable explanation for the data. The results were a stunning vindication of a theory that had once been intensely controversial.

The success of the endosymbiosis theory carries a general message: Protists and other eukaryotes are cut-and-paste jobs. They are chimeras, like the monster from Greek mythology that had the head of a lion, the body of a goat, and the tail of a serpent. The domain diversified after an ancient eukaryote—closely related to archaea—combined with a bacterium. The lineage Eukarya was almost undoubtedly created when two cells began living together. Another major pairing occurred when an ancestor of today's plants engulfed a cyanobacterium.

To put this conclusion into perspective, recall from Chapters 20 and 27 that lateral gene transfer—the transfer of DNA between species that are not closely related—has had a huge impact on the evolution of bacteria and archaea through time and today is playing a large role in the evolution of drug resistance. But even though lateral gene transfer among prokaryotes is impressive, most of the events have involved just one or a handful of genes. These gene movements pale in comparison to the transfer of two entire bacterial genomes to eukaryotes. The evolution of the mitochondrion and the chloroplast resulted in the most massive lateral gene transfers in the history of life.

Discovering New Lineages via Direct Sequencing

What research is currently being done on protist diversity? The answer to this question has several parts. The effort to refine the phylogeny of the Eukarya is continuing, and researchers are developing hypotheses to explain how organelles other than the mitochondria and chloroplasts originated. As **Box 28.2** (page 618) details, protists continue to be used as important model organisms in biology. But of all the research frontiers in eukaryotic diversity, the most exciting may be the one based on the technique called direct sequencing.

As noted in Chapter 27, **direct sequencing** is based on collecting organisms from a habitat and analyzing their DNA without growing larger populations of individuals in laboratory culture. The approach is based on using the polymerase chain reaction (PCR; see Chapter 19) to amplify specific genes in the organisms that have been collected. The resulting DNA sequence data are then used to place the organisms on a phylogenetic tree.

Recall that direct sequencing led to the discovery of previously unknown but major lineages of Archaea. To the amazement of biologists all over the world, the same thing happened when researchers used direct sequencing to survey eukaryotes.

The first direct sequencing studies that focused on eukaryotes were published in 2001 and were motivated by the hypothesis that direct sequencing could detect the presence of previously unknown species. To test this hypothesis, researchers sampled cells at various depths and locations in the oceans, used PCR primers to amplify genes that encode the RNA component of the ribosome's small subunit, sequenced the genes, and compared the data with the sequences of previously studied eukaryotes. One study sampled organisms at depths from 250 to 3000 m below the surface in waters off Antarctica; another focused on cells at depths of 75 m in the Pacific Ocean, near the equator. Both studies found a wide array of distinctive ribosomal RNA sequences—new species under the phylogenetic species concept introduced in Chapter 25.

Investigators who followed up on these results by examining the samples under the microscope were astonished to find that many of the newly discovered organisms were tiny—from 0.2 μm to 5 μm in diameter. Other direct sequencing studies have now confirmed the existence of eukaryotes that are less than 0.2 μm in diameter. These eukaryotic cells overlap in size with bacteria, which range from 0.5 μm to 2 μm in diameter. The take-home message is that eukaryotic cells are much more variable in size than previously imagined. A whole new world of tiny eukaryotes has just been discovered.

The initial analyses suggest that at least some of the new species may represent important new lineages of dinoflagellates, stramenopiles, or alveolates. As additional direct sequencing studies are done, it is virtually certain that our understanding of eukaryotic diversity will change—perhaps radically—and improve.

✓CHECK YOUR UNDERSTANDING

Biologists use data from microscopy and DNA sequencing to estimate phylogenetic trees and study the diversity of protists. Combined use of morphological and molecular data has allowed researchers to conclude that the domain Eukarya comprises eight major lineages, members of which have distinctive aspects of cell structure. Recently, direct sequencing has allowed investigators to recognize large numbers of previously undescribed eukaryotes, some of which are extremely small and do not grow in culture. You should be able to (1) compare and contrast the characteristics of at least four of the eight major lineages of eukaryotes, and (2) state a hypothesis for the origin of the chloroplast, list three predictions that follow from that hypothesis, and summarize data that support or contradict those predictions.

28.3 What Themes Occur in the Diversification of Protists?

The protists range in size from bacteria-sized single cells to giant kelp. They live in habitats from the open oceans to dank forest floors. They are almost bewildering in their morphological and ecological diversity. Because they are a paraphyletic group, they do not share derived characteristics that set them apart from all other lineages on the tree of life.

BOX 28.2 A Model Organism: *Dictyostelium discoideum*

Several species of protists have served as model organisms in biology. Among the most important has been the cellular slime mold *Dictyostelium discoideum*. *Dictyostelium* is not always slimy, and it is not a mold—meaning a type of fungus. Instead, it is an amoeba. **Amoeba** is a general term that biologists use to characterize a unicellular protist that lacks a cell wall and is extremely flexible in shape. *Dictyostelium* has long fascinated biologists because it is a social organism. Independent cells sometimes aggregate to form a multicellular structure.

Figure 28.11 shows the *Dictyostelium discoideum* **life cycle**—the sequence of events that occurs over an individual's life span. Note that under most conditions, *D. discoideum* cells are haploid (n) and move about in decaying vegetation on forest floors or other habitats. They feed on bacteria by engulfing them whole and reproduce asexually, by mitosis. If food begins to run out, however, the cells begin to aggregate. In many cases, tens of thousands of cells cohere to form a 2-mm-long mass called a **slug**. (This is not the slug that is related to snails. After migrating to a sunlit location, the slug stops and individual cells differentiate according to their position in the slug. Some form a stalk; others form a mass of spores at the tip of the stalk. A *spore* is a single cell that is not formed from gamete fusion and that develops into an adult organism. The entire structure, stalk plus mass of spores, is called a **fruiting body**. Cells that form spores secrete a tough coat and represent a durable resting stage. The fruiting body eventually dries out, and the wind disperses the spores to new locations, where more food might be available. On occasion, *Dictyostelium* may also undergo sexual reproduction. When an aggregation is forming, two cells inside it may undergo syngamy—the fusion of gametes—to form a diploid ($2n$) zygote called a *giant cell*. The giant cell grows by feeding on the haploid amoebae in the aggregation. It then secretes a tough, protective coat. Later the zygote undergoes meiosis to form haploid offspring, which undergo mitosis. Eventually the haploid cells break out of the coating and begin to move about in search of food.

Dictyostelium discoideum has been an important model organism for investigating questions about eukaryotes:

- Cells in a slug are initially identical in morphology but then differentiate into distinctive stalk cells and spores. Studying this process helped biologists better understand how cells in plant and animal embryos differentiate into distinct cell types.
- The process of slug formation has helped biologists study how animal cells move and how they aggregate as they form specific types of tissues. In addition, the discovery that aggregating *D. discoideum* cells follow trails of cyclic adenosine monophosphate (cAMP) helped investigators better understand the mechanisms responsible for **chemotaxis**—movement in response to a chemical—in other organisms.
- When *D. discoideum* cells aggregate to form a slug, they stick to each other. The discovery of membrane proteins responsible for cell-cell adhesion helped biologists understand some of the general principles in Chapter 8.
- Cells that aggregate to form a slug are not genetically identical. About 20 percent of the cells help other individuals produce offspring by forming the stalk, but die without producing offspring themselves. Why? Researchers are using *D. discoideum* to answer questions about why some individuals sacrifice themselves to help others.

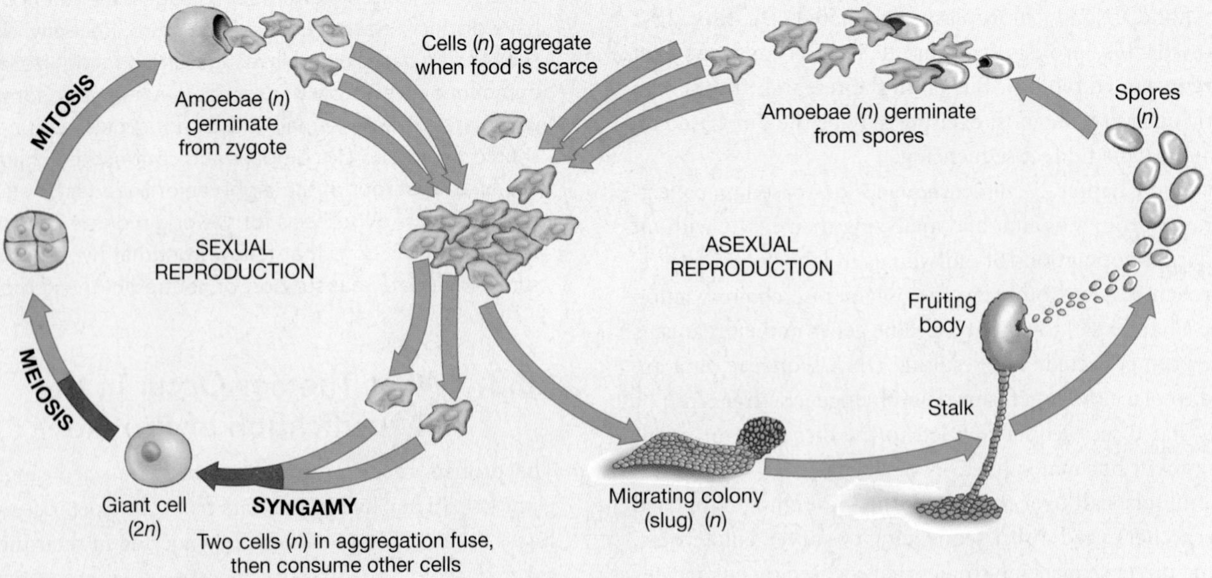

FIGURE 28.11 The Life Cycle of *Dictyostelium discoideum*

Fortunately, several general evolutionary themes tie the diversity of eukaryotes together. The key to understanding the protists is to recognize that a series of important innovations occurred, often repeatedly, as eukaryotes diversified.

Morphological Diversity

Although eukaryotic cells vary widely in size, the domain Eukarya is distinguished by the evolution of the largest cells known. An average-sized eukaryotic cell is 10 times larger in diameter than an average-sized bacterial cell. The evolution of large cells is an important theme in the evolution of the Eukarya.

How is it possible for such large cells to evolve? The answer to this question is not obvious, because large cell size presents an important physical challenge: As cells become larger, their volume increases much more rapidly than does their surface area. This is because a cell's volume increases as the cube of a sphere's diameter (volume $\propto$ diameter^3), while its surface area increases as the square of a sphere's diameter (surface area $\propto$ diameter^2). (The relationship between surface area and volume is explored in more detail in Chapter 41.) As cells get larger, then, the proportion of surface area available gets smaller and smaller relative to the volume present. This creates a problem because food, gases, and waste molecules must diffuse across the cell's surface, while the volume is filled with biosynthetic machinery that requires energy and generates waste. As cell size increases, metabolism in the cell's interior can outstrip the transport and exchange processes that take place along the surface area and that provide the raw materials for metabolism. How do eukaryotes cope?

Organelles Divide a Large Cell into Compartments

Dilemmas regarding surface area and volume are minimized in eukaryotes, because cells are divided into compartments. As an example of how the compartmentalization of the eukaryotic cell works, consider the protist called *Paramecium* (**Figure 28.12**). This organism makes a living by eating bacteria, which it sweeps into an indentation known as the *gullet* and then into the cell mouth. After ingesting a bacterium, a *Paramecium* surrounds it with an internal membrane, forming a compartment called a **food vacuole**. Food vacuoles merge with membrane-bound structures called lysosomes, which hold digestive enzymes, and circulate around the cell. When the food has been digested and nutrients have diffused out of the food vacuole, the vacuole merges with the plasma membrane at a special organelle—the **anal pore**—and expels waste molecules. The food vacuoles provide a large area of internal membrane, allowing nutrients to be delivered efficiently throughout the volume of the cell and waste products to be expelled rapidly.

Although not all eukaryotic cells contain food vacuoles and anal pores, they all have numerous other internal compartments. Each of these internal structures—including the nucleus, lysosomes, peroxisomes, mitochondria, chloroplasts, central vacuole, Golgi apparatus, and rough and smooth endoplasmic reticulum (ER) that were introduced in Chapter 7—has a distinct function. Many are membrane bound and are devoted to the synthesis, transport, and distribution of molecules.

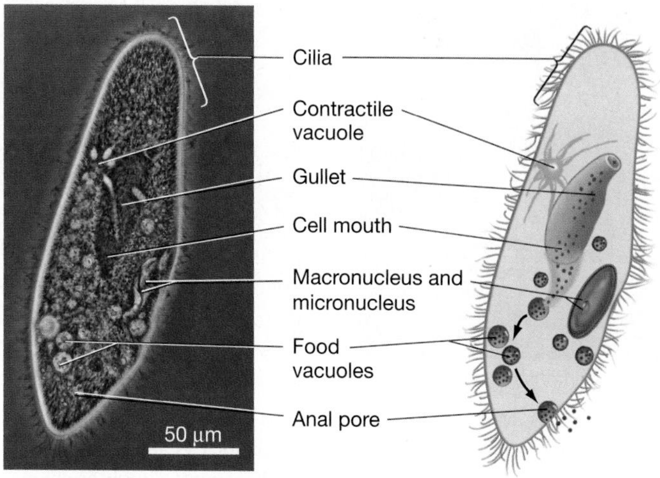

FIGURE 28.12 Protists Have Specialized Intracellular Structures *Paramecium* feeds by sweeping bacteria into a subcellular structure called the gullet. EXERCISE Next to the drawing of *Paramecium*, list other organelles and intracellular structures that are found in protist cells.

In terms of organizing the cell volume, however, the most important feature of eukaryotes may be the cytoskeleton. This system of filaments and tubules, introduced in Chapter 7, provides a scaffolding that supports and organizes the cell and, in many cases, moves it as well. Although some bacteria and archaea have rudimentary cytoskeletal elements, an extensive, highly ordered cytoskeleton is found only in eukaryotes.

In sum, the extensive compartmentalization and differentiation of the eukaryotic cell, along with the ability to generate large amounts of ATP through aerobic respiration, makes large size possible. The evolution of large cell size, specialized organelles, and complex cytoskeletal structures is a major theme in the diversification of protists.

The Evolution of Multicellularity A few species of bacteria are capable of aggregating and forming fruiting bodies similar to those observed in the protist *Dictyostelium discoideum* (see Box 28.2). Because cells in the fruiting bodies of these bacteria differentiate into specialized stalk cells and spore-forming cells, they are considered multicellular. But the vast majority of multicellular species are members of the Eukarya. The evolution of multicellularity is a prominent theme in the evolution of eukaryotes.

Multicellularity is actually a difficult trait to define. To understand why, consider a group of green algae called the Volvocales.

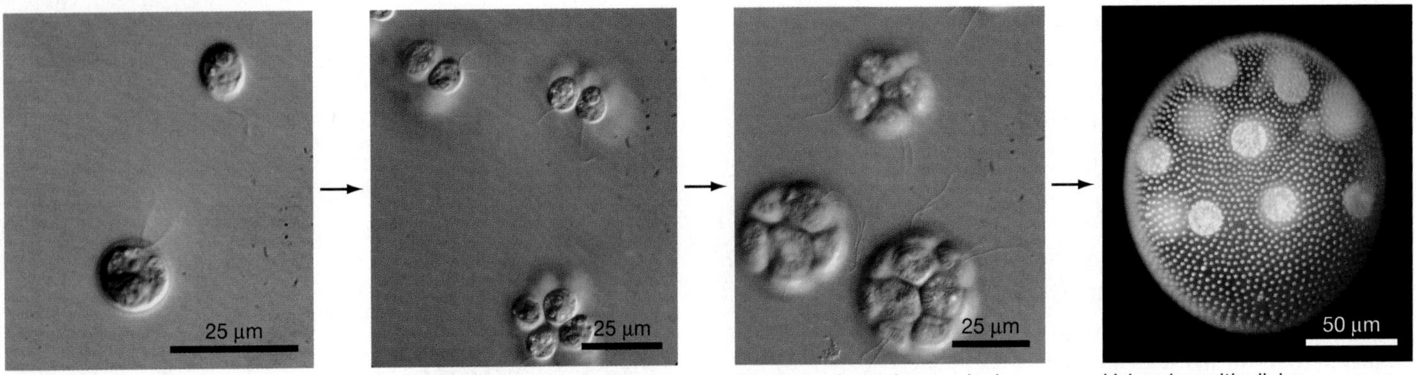

Chlamydomonas is unicellular. *Gonium* forms small colonies. *Pandorina* forms large colonies. *Volvox* is multicellular.

FIGURE 28.13 Multicellularity Evolved in Protists
Members of the green algal family called Volvocales are graded in morphology, from unicellular to colonial to multicellular. The large spheres within the *Volvox* are daughter colonies that are being produced asexually.

As **Figure 28.13** shows, the organisms in this lineage range from single cells such as *Chlamydomonas*, to small clumps of cells such as *Gonium*, to large groupings such as *Pandorina* and the multicellular *Volvox*. In this lineage, there appears to be a smooth continuum between single-celled and multicelled forms. Where do biologists draw the line between the loose aggregations of cells called a **colony** and the more highly structured arrangements of cells that create a body and typify multicellular organisms? For example, why is *Volvox* considered multicellular?

Most biologists would cite two reasons that *Volvox* is multicellular: Individual *Volvox* cells cannot survive on their own, and each cell has a distinct function. Some *Volvox* cells are strictly vegetative, meaning that they perform photosynthesis and supply the organism with ATP and sugar. Sexual reproduction is limited to certain cells that undergo meiosis and form reproductive cells equivalent to sperm and eggs.

The differentiation of cell types is a crucial criterion for multicellulariy. To understand why, recall from Chapter 22 that, when differentiation occurs, only a well-defined subset of each cell's genome is transcribed. Different types of cells express different genes. This feature is used to distinguish **colonial growth**—in which cells aggregate but each performs the same function—from **multicellular growth**. Multicellular organisms have masses of closely interacting cells, distinct cell or tissue types, and large overall size. Multicellularity evolved several times as eukaryotes diversified. This trait arose independently in green plants, red algae, animals, fungi, brown algae, and amoebae.

Structures for Support and Protection Thanks to their cytoskeleton and organelles, all protists have a complex intracellular structure. But in addition, many protists have a rigid internal skeleton or a hard, external structure. Rigid structures inside the plasma membrane, such as the silicon-containing skeletons of protists called heliozoans, support the cell. Many protists have cell walls outside their plasma membrane; others have hard external structures, often called a **test** or a **shell**, that provide support or protection or both.

Shells are found in groups that are scattered about the eukaryotic tree. The structures themselves are equally diverse. As **Figure 28.14** shows, these external skeletons vary from the intricate, chambered calcium carbonate ($CaCO_3$) tests found in foraminifera, to the glass-like, silicon-oxide shells enclosing di-

(a) Foraminiferan **(b)** Diatom **(c)** Dinoflagellate

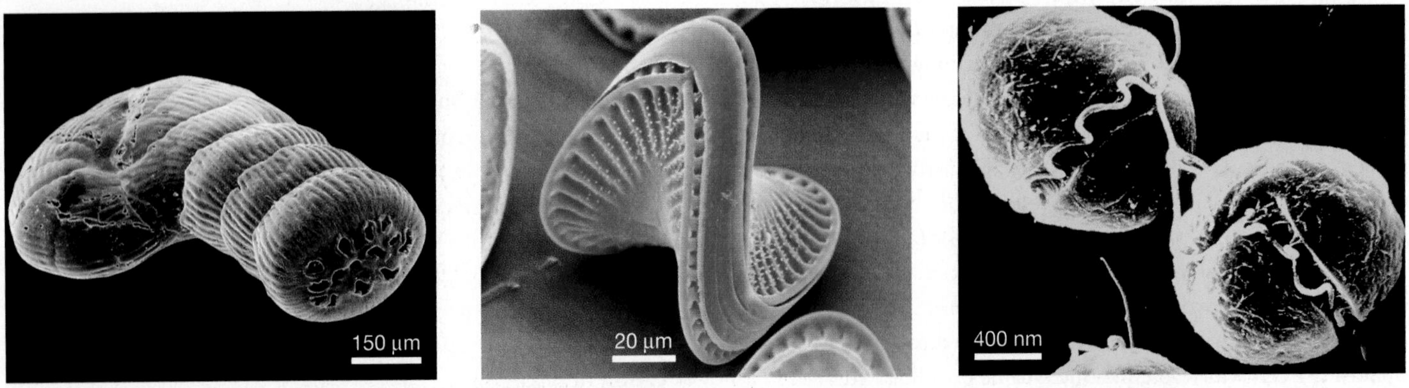

Calcium carbonate test, with chambers — Test made of silicon oxides — Plates made of cellulose

FIGURE 28.14 Hard Outer Coverings in Protists Vary in Composition

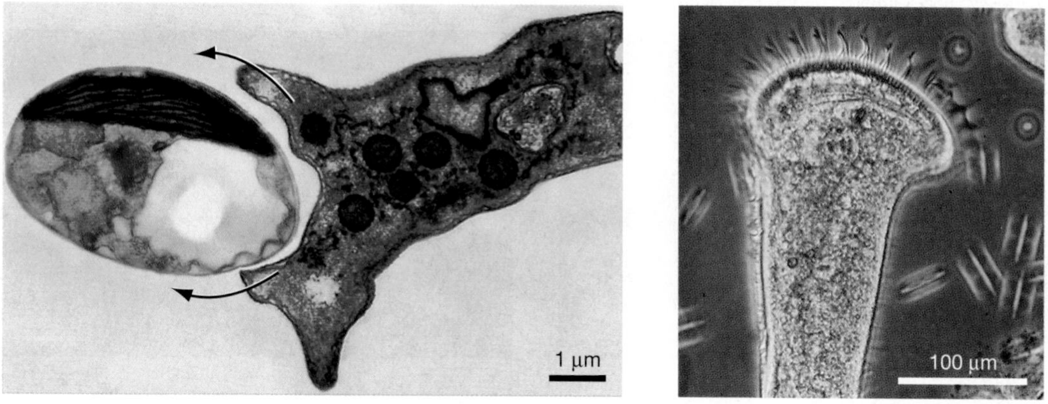
(a) Pseudopodia engulf food
1 μm
(b) Ciliary currents sweep food into gullet
100 μm

FIGURE 28.15 Ingestive Feeding
Methods of prey capture vary widely among ingestive protists. **(a)** Some predators engulf prey with pseudopodia; **(b)** other predators sweep them into their gullets with water currents set up by the beating of cilia.

atoms, to the cellulose plates that surround some dinoflagellates. Some amoebae even cover themselves with tiny pebbles.

To interpret these findings, biologists point out that protists evolved many different ways to support their large cells and protect themselves from predation. Like multicellularity, shells have evolved many times, independently. The evolution of extracellular layers or coatings was an important theme during the evolution of the Eukarya.

How Do Protists Find Food?

Recall from Chapter 27 that bacteria and archaea obtain the chemical energy they need in one of three ways: (1) by absorbing inorganic compounds with high potential energy from the environment; (2) by absorbing high-energy organic compounds, such as sugars, from a host or from decaying organic matter; or (3) through photosynthesis. Among the protists, many species absorb organic compounds for nutrition and many are photosynthetic. Although no protists make a living from inorganic compounds, large cell size and the evolution of the cytoskeleton made a unique feeding strategy possible: Protists can eat bacteria, archaea, or even other protists whole—meaning they can take in packets of food much larger than individual molecules. Thus, eukaryotes feed by (1) ingesting packets of food, (2) absorbing individual molecules, or (3) performing photosynthesis. Let's consider each method.

Ingestive Feeding Ingestive lifestyles are based on eating live or dead organisms or on scavenging loose bits of organic debris. Protists such as the cellular slime mold *Dictyostelium discoideum* are large enough to engulf bacteria and archaea; many protists are large enough to surround and ingest other protists or microscopic animals. The engulfing process is possible because some protists lack a cell wall and because their flexible membrane and cytoskeleton give these species the ability to surround and "swallow" prey with long, fingerlike projections called **pseudopodia** ("false-feet") (**Figure 28.15**).

Losing a cell wall and gaining the ability to move their plasma membrane around objects was an important innovation during the evolution of protists. Recall from Chapter 27 that bacteria and archaea are abundant in wet soils and aquatic habitats, where protists also live. Instead of competing with bacteria and archaea for sunlight or food molecules, protists could eat them.

In addition to hunting down prey and engulfing them, many of the ingestive protists are "sit and wait" predators or scavengers. Species that feed in this way often have cilia that create water currents when they beat. In many cases, these protists attach themselves to a substrate and collect food by sweeping particles into their mouths (**Figure 28.15b**). Organisms that filter food out of water in this way are said to be **filter feeders,** which are also known as **suspension feeders.**

Absorptive Feeding When nutrients are taken up directly from the environment, across the plasma membrane, **absorptive feeding** occurs. Some protists that live by absorptive feeding are decomposers. A **decomposer** is an organism that feeds on dead organic matter, or **detritus.** But many of the protists that absorb their nutrition directly from the environment live inside other organisms. If they damage their host, the absorptive species is called a **parasite.** The protists responsible for the diseases listed in Table 28.1 are parasites (**Figure 28.16**).

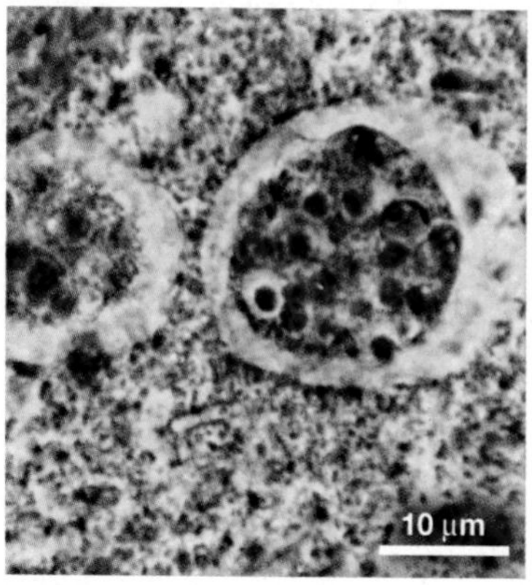
10 μm

FIGURE 28.16 Parasitism
Micrograph of two *Entamoeba histolytica,* which is a common cause of traveler's diarrhea in humans. Parasitic protists get all of their nutrition from their host.

(a) Red: chlorophyll *a* and phycoerythrins **(b)** Brown: chlorophyll *a*, chlorophyll *c*, and xanthins **(c)** Green: chlorophyll *a* and chlorophyll *b* **(d)** Symbiosis

10 μm
100 μm
500 μm
0.5 mm

FIGURE 28.17 Photosynthesis
Many photosynthetic algae are distinguished by their accessory pigments, in addition to chlorophyll *a*. Each accessory pigment intercepts different wavelengths of light. Symbioses are also common in protists. Photosynthetic protists called dinoflagellates live inside the coral on the right and are responsible for its color.

Photosynthesis A wide array of protists are photosynthetic. Plants may be the dominant producers of sugars and other high-energy organic compounds on land, but photosynthetic protists and bacteria are the dominant producers of carbon compounds in oceans and lakes.

When biologists analyze the evolution of photosynthesis in protists, three key themes emerge:

1. The major photosynthetic groups of protists are distinguished by the pigments they contain (**Figure 28.17**). Each of these pigments, in turn, absorbs unique wavelengths of light. As a result, the various groups of photosynthetic protists specialize in harvesting light energy from particular regions of the electromagnetic spectrum. The leading hypothesis to explain this pattern is that photosynthetic species harvest different wavelengths of light to avoid competition. This theme in the diversification of protists echoes the variation in photosynthetic pigments observed in bacteria (see Chapter 27).
2. Many photosynthetic protists live symbiotically with animals or other protists—either inside or outside host cells. Although there are hundreds of examples, perhaps the most important involve the animals called corals. Virtually all reef-building corals harbor single-celled, photosynthetic protists in their skin or gut tissue. The concentrations of endosymbiotic cells can be impressive—up to a million protist cells per square centimeter of coral surface (Figure 28.17).
3. Protists originally obtained chloroplasts by endosymbiosis with a cyanobacterium. But a wide array of photosynthetic protist groups obtained their chloroplasts by ingesting other photosynthetic protists. This process, called **secondary endosymbiosis**, occurs when an organism engulfs a photosynthetic cell and retains its chloroplasts as intracellular symbionts. In groups that obtained chloroplasts via secondary endosymbiosis, the organelle is surrounded by more than two membranes—usually four (**Figure 28.18**). **Figure 28.19**

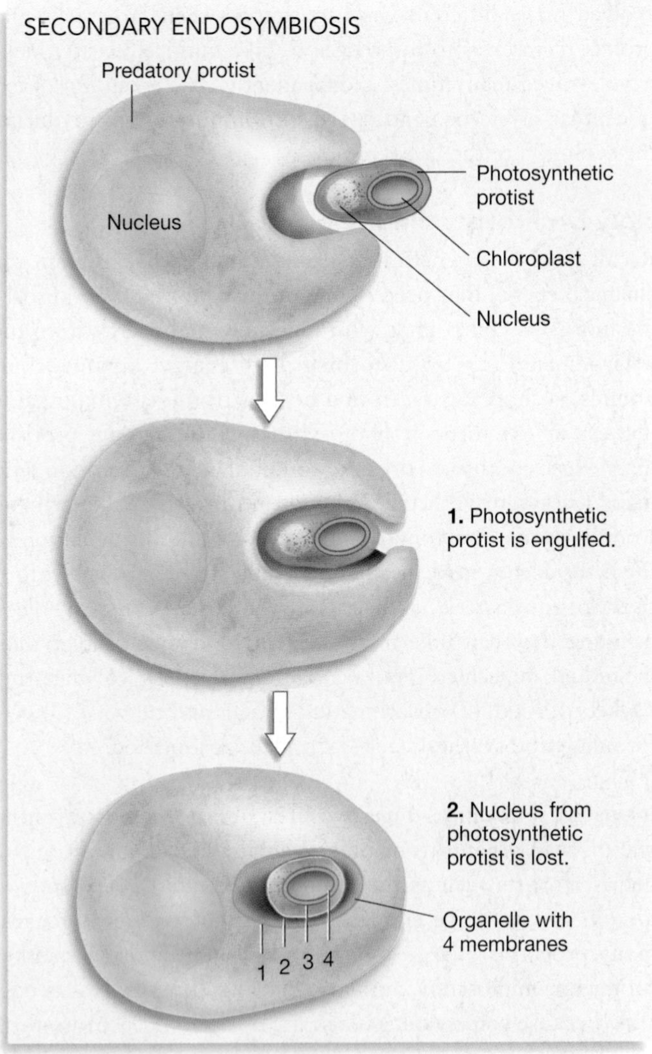
SECONDARY ENDOSYMBIOSIS
Predatory protist
Nucleus
Photosynthetic protist
Chloroplast
Nucleus
1. Photosynthetic protist is engulfed.
2. Nucleus from photosynthetic protist is lost.
Organelle with 4 membranes
1 2 3 4

FIGURE 28.18 Secondary Endosymbiosis Leads to Organelles with Four Membranes
The chloroplasts found in some protists have four membranes and are hypothesized to be derived by secondary endosymbiosis. In species where chloroplasts have three membranes, biologists hypothesize that secondary endosymbiosis was followed by the loss of one membrane.

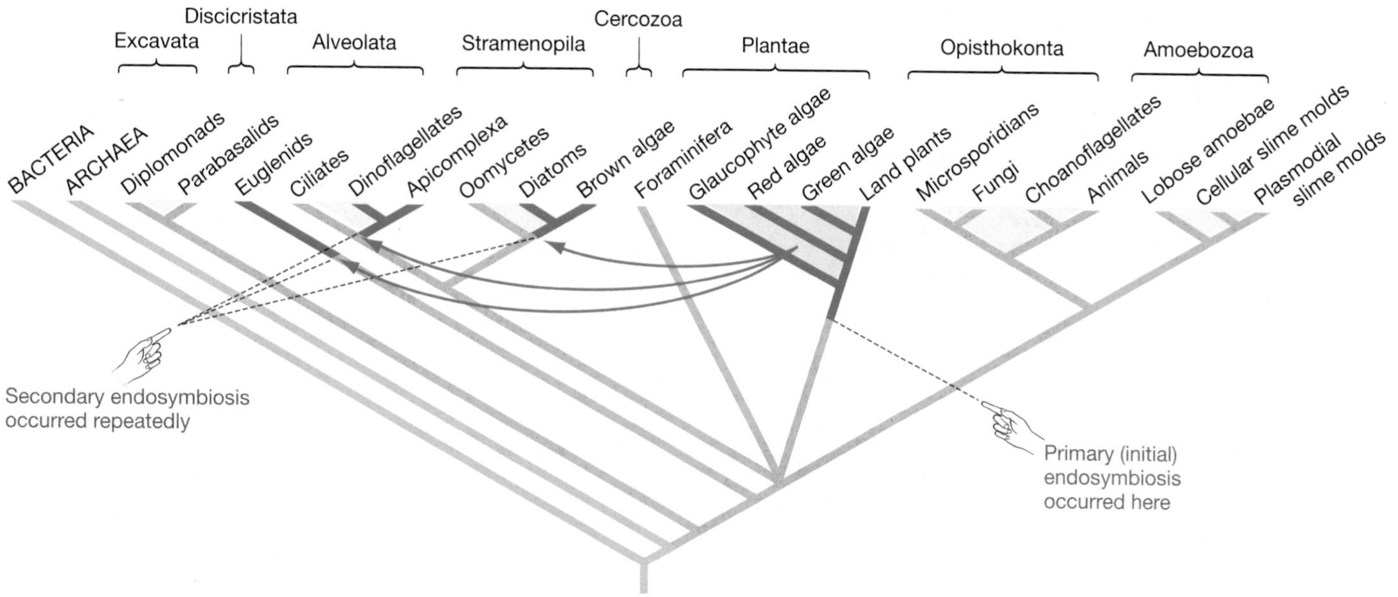

FIGURE 28.19 Photosynthesis Arose in Protists by Primary Endosymbiosis, Then Spread among Lineages via Secondary Endosymbiosis
All of the species above glaucophyte algae on the Eukarya tree have chloroplasts that have two membranes and similar molecular composition, indicating that they arose via primary endosymbiosis. A wide variety of other protist lineages have species with chloroplasts or chloroplast-derived organelles with three or four membranes, indicating that they arose via secondary endosymbiosis.

shows where primary and secondary endosymbiosis occurred on the phylogenetic tree of eukaryotes. The message of this figure is simple: Many lineages of protists acquired the ability to photosynthesize independently of each other.

Diversity in Lifestyles Before leaving the topic of obtaining food in protists, it is important to recognize that all three lifestyles—ingestive, absorptive, and photosynthetic—occur in many different eukaryote lineages. All three types of food getting can occur within a single clade.

To drive this point home, consider the monophyletic group called the alveolates. There are three major subgroups of this lineage, called dinoflagellates, apicomplexa, and ciliates. Within each subgroup, species vary in how they make a living. About half of the dinoflagellates are photosynthetic, while many others are parasitic. The apicomplexa are parasitic. The ciliates include many species that ingest prey, but some ciliates live in the guts of cattle or the gills of fish and absorb nutrients from their hosts. Other ciliate species make a living by holding algae or other types of photosynthetic symbionts inside their cells.

This diversity in modes of nutrition underscores earlier conclusions. Within each of the eight major lineages of eukaryotes, many different morphologies and ways of feeding evolved.

How Do Protists Move?

Many protists actively move to find food. Predators such as *Dictyostelium discoideum* crawl over a substrate in search of prey. Most of the unicellular, photosynthetic species are capable of swimming to sunny locations. How are these crawling and swimming movements possible?

Cell crawling is a sliding motion observed in some protists. In the classic mode of cell crawling illustrated in **Figure 28.20a** (page 624), pseudopodia stream forward over a substrate, with the rest of the cytoplasm, organelles, and plasma membrane following. The motion requires ATP and involves interactions between proteins called actin and myosin inside the cytoplasm. The mechanism is related to muscle movement in animals, which is detailed in Chapter 46. But at the level of the whole cell, the precise sequence of events during cell crawling is still somewhat uncertain.

The other major mode of locomotion in protists involves flagella (**Figure 28.20b**) or cilia (**Figure 28.20c**). Flagella and cilia have identical structures. Recall from Chapter 7 that both consist of nine sets of doublet (paired) microtubules arranged around two central, single microtubules. Flagella, however, are long and are usually found alone or in pairs; cilia are short and usually occur in large numbers on any one cell. Flagella and cilia can also be distinguished by the types of structures associated with the basal bodies where they originate. Both structures are markedly different from the flagella found in bacteria and archaea. The cilia and flagella of protists are made up of microtubules, and dynein is the major motor protein. An undulating motion occurs as dynein molecules walk down microtubules. The flagella of bacteria and archaea, in contrast, are composed primarily of a protein called flagellin (see Chapter 7). Instead of undulating, these flagella rotate to produce movement.

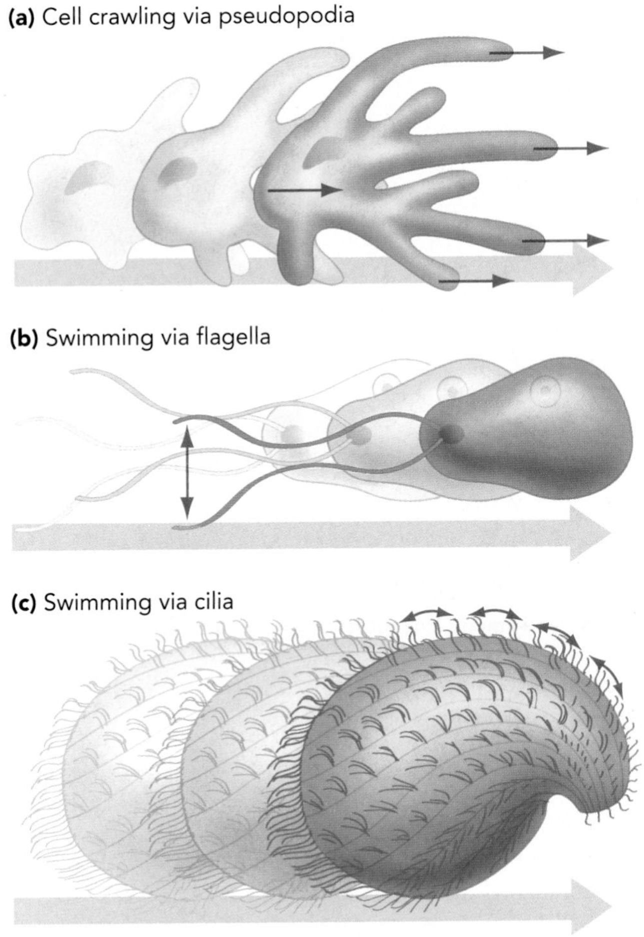
(a) Cell crawling via pseudopodia
(b) Swimming via flagella
(c) Swimming via cilia

FIGURE 28.20 Modes of Locomotion in Protists Vary
(a) In cell crawling, long pseudopodia stream out from the cell. The rest of the cytoplasm, organelles, and external membrane follow. **(b)** Flagella are long and few in number, and they power swimming movements. **(c)** Cilia are short and numerous. In many cases they are used in swimming.

Even closely related protists can use radically different forms of locomotion. For example, consider again the lineage of protists called the alveolates and the subgroups known as ciliates, dinoflagellates, and apicomplexa. The ciliates swim by beating their cilia. Dinoflagellates swim by whipping their flagella. Apicomplexa move by cell crawling, and they have gametes with flagella. Within the other major lineages of protists, it is common to observe variation in how cells move and common to find species that do not exhibit active movement but instead float passively in water currents. Movement is yet another example of diversification that occurred within each of the eight major eukaryote lineages.

How Do Protists Reproduce?

Sexual reproduction evolved in protists. As pointed out in Chapter 12, sexual reproduction can best be understood in contrast to asexual reproduction. Asexual reproduction is based on mitosis in eukaryotic organisms and on fission in bacteria and archaea, and it results in offspring that are genetically identical to the parent. Most protists undergo asexual reproduction routinely. When sexual reproduction occurs, in contrast, offspring are genetically different from their parents and from each other. Sexual reproduction occurs only intermittently in many protists—often at one particular time of year, or when individuals are crowded or food is scarce—and is based on meiosis. The evolution of sexual reproduction ranks among the most significant evolutionary innovations observed in eukaryotes.

Sexual versus Asexual Reproduction Why was the evolution of meiosis important? One hypothesis is that offspring that are genetically variable—meaning their genotypes are different from those of other offspring and from those of their parents—may be able to thrive if the environment changes. For example, offspring with genotypes different from those of their parents may be better able to withstand attacks by parasites that successfully attacked their parents (see Chapter 12). This is a key point, because many types of parasites, including bacteria and viruses, have short generation times and evolve very quickly. Because the genotypes and phenotypes of parasites are constantly changing, natural selection is constantly favoring host individuals with new genotypes. The idea is that new offspring genotypes generated by meiosis may contain combinations of alleles that allow hosts to withstand attack by new strains of parasites. In short, many biologists view sexual reproduction as an adaptation to fight disease.

Variation in Life Cycles A life cycle describes the sequence of events that occurs as individuals grow, mature, and reproduce. The evolution of meiosis not only introduced a new event in the life cycle of protist species, but it also created a distinction between haploid and diploid phases in the life of an individual. Recall from Chapter 12 that diploid individuals have two of each type of chromosome inside each cell, while haploid individuals have just one of each type of chromosome inside each cell. When meiosis occurs in diploid cells, it results in the production of haploid cells.

The life cycle of most bacteria and archaea is extremely simple: A cell divides, feeds, grows, and divides again. Bacteria and archaea do not undergo sexual reproduction, and they are always haploid. In contrast, among protists virtually every aspect of a life cycle is variable—whether meiosis occurs, whether asexual reproduction occurs, and whether the haploid or the diploid phase of the life cycle is the longer and more prominent phase.

Figure 28.21 illustrates some of the variation that occurs in life cycles of protists. Figure 28.21a depicts the haploid-dominated life cycle observed in many unicellular protists. To analyze this life cycle, start with **syngamy**, or fertilization—the fusion of two gametes to form a diploid zygote. Then trace what happens to the zygote. In this case, the diploid zygote undergoes meiosis.

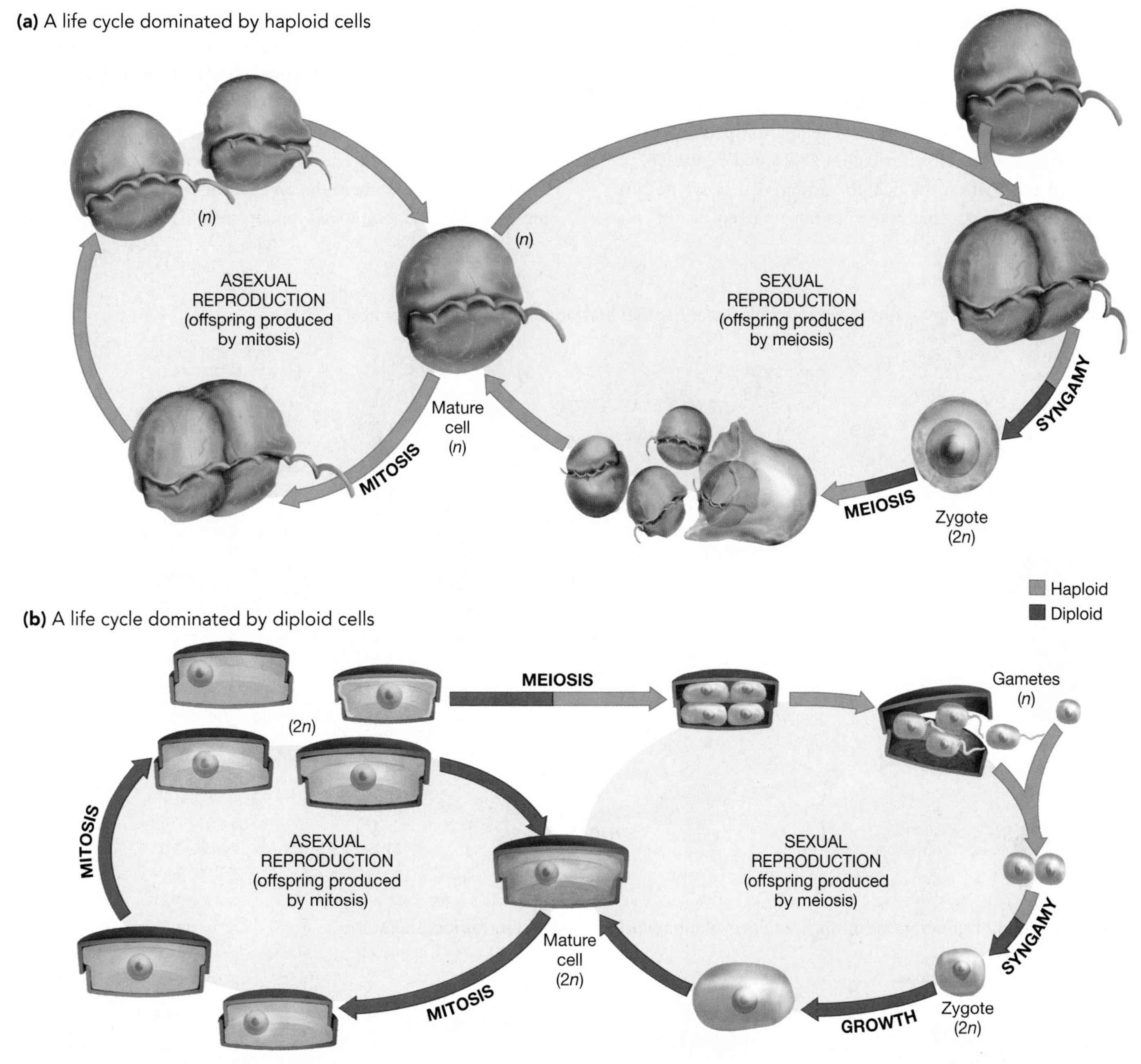
(a) A life cycle dominated by haploid cells
(n)
(n)
ASEXUAL
REPRODUCTION
(offspring produced
by mitosis)
SEXUAL
REPRODUCTION
(offspring produced
by meiosis)
Mature
cell
(n)
MITOSIS
SYNGAMY
MEIOSIS
Zygote
(2n)
Haploid
Diploid
(b) A life cycle dominated by diploid cells
MEIOSIS
Gametes
(n)
(2n)
MITOSIS
ASEXUAL
REPRODUCTION
(offspring produced
by mitosis)
SEXUAL
REPRODUCTION
(offspring produced
by meiosis)
SYNGAMY
Mature
cell
(2n)
MITOSIS
GROWTH
Zygote
(2n)

FIGURE 28.21 Life Cycles Vary Widely among Protists
Many protists can reproduce by both asexual reproduction and sexual reproduction.

The haploid products of meiosis then grow into mature cells that eventually undergo asexual reproduction or produce gametes by mitosis. Contrast that cycle with the diploid-dominated life cycle of Figure 28.21b, in which the diploid zygote develops into a sexually mature adult. In such species, meiosis occurs in the adult rather than in the zygote; gametes are the only haploid cells in the life cycle.

In contrast to the relatively simple life cycle of Figure 28.21, many multicellular protists have one phase in their life cycle that is based on a multicellular haploid form and another phase that is based on a multicellular diploid form. The alternation of multicellular haploid and diploid forms is known as **alternation of generations.** The multicellular haploid form is called a **gametophyte,** because specialized cells in this individual produce gametes by mitosis. The multicellular diploid form is called a **sporophyte,** because it has specialized cells that undergo meiosis to produce haploid cells called spores. Recall that a **spore** is a single cell that is *not* formed by syngamy (the fusion of gametes) and that develops into an adult organism. Gametophytes and sporophytes may be identical in appearance, as in the brown alga illustrated

WEB TUTORIAL 28.1
Alternation of Generations in a Protist

in **Figure 28.22a**. In many cases, however, the gametophyte and sporophyte are different in size or shape or both (**Figure 28.22b**). Among the protists, alternation of generations evolved independently in brown algae and in red algae. To get a better understanding of alternation of generations, compare and contrast the events described in Figure 28.21 and Figure 28.22, starting with fertilization. Notice, for example, that zygotes undergoing alternation of generations do not undergo meiosis immediately. Instead they germinate into diploid individuals: sporophytes. Cells in the sporophyte then undergo meiosis to form haploid spores, which germinate and grow into haploid gametophytes.

Why does so much variation occur in the types of life cycles observed among protists? The answer is not known. Variation in life cycles is a major theme in the diversification of protists. Explaining why that variation exists remains a topic for future research.

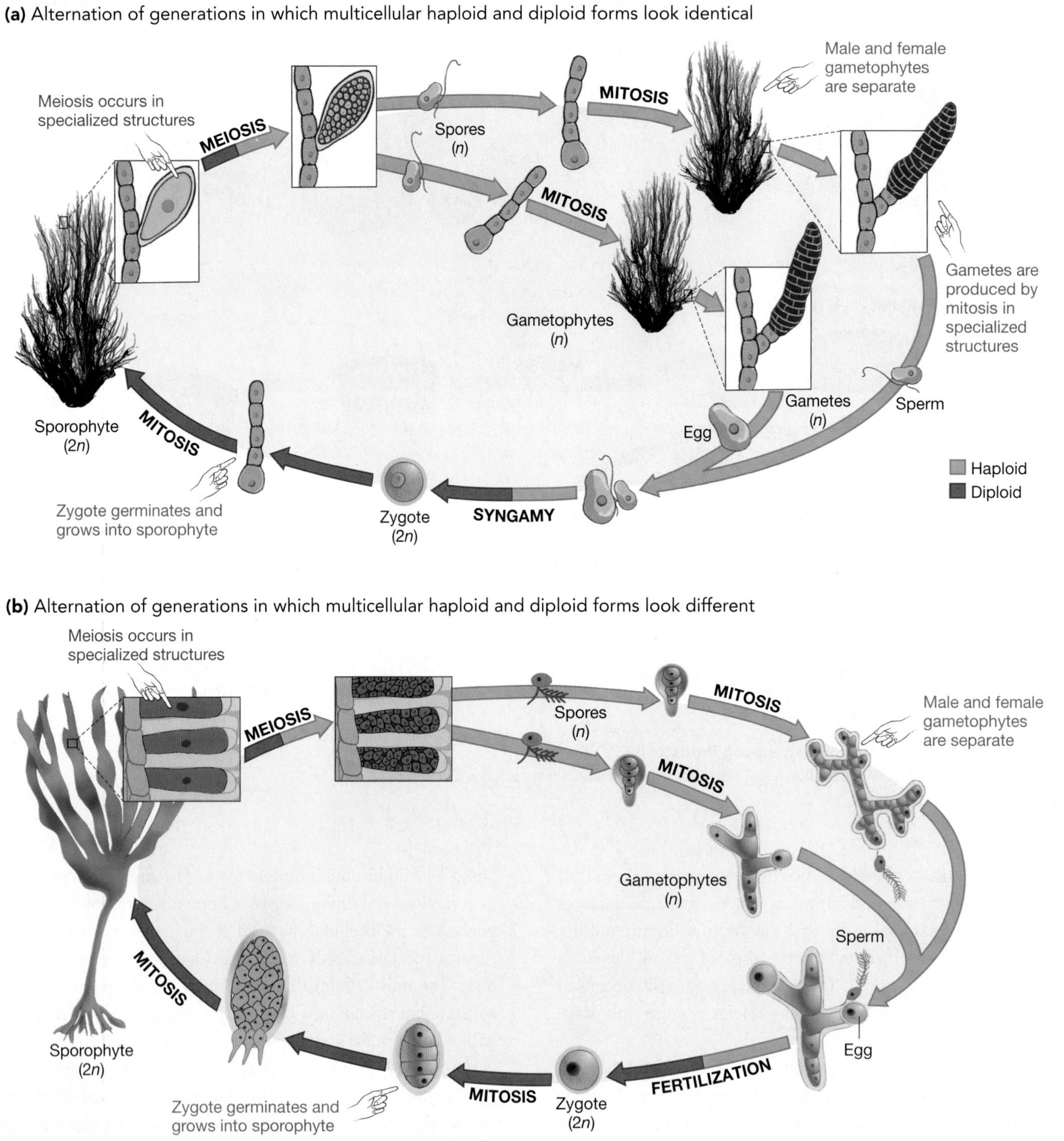

FIGURE 28.22 Alternation of Generations Occurs in an Array of Protist Lineages

✓CHECK YOUR UNDERSTANDING

Protist diversity can be understood by focusing on key evolutionary themes and innovations. The important innovations in morphological evolution were large cell size, made possible by organelles, a cytoskeleton, and aerobic respiration; multicellularity; and exterior layers or coatings. In terms of finding food, the key methods are ingestive, photosynthetic, or absorptive (often parasitic) lifestyles. Regarding movement, protists employ cell crawling, flagella, or cilia. With respect to reproduction, meiosis and sexual reproduction were a crucial evolutionary innovation. You should be able to (1) explain why organelles made the evolution of large cell size possible, (2) give examples of each of the key evolutionary themes and innovations, (3) provide evidence that traits such as multicellularity and photosynthesis arose repeatedly over the course of eukaryote evolution, and (4) diagram the major types of life cycles observed in protists.

28.4 Key Lineages of Protists

An important generalization jumps out of the Eukarya phylogenetic tree and data on variation among protists in morphology, feeding method, locomotion, and mode of reproduction: Each of the eight major Eukarya lineages has at least one distinctive morphological characteristic. But once an ancestor evolved a set of distinctive characteristics, its descendants diversified into a wide array of lifestyles. For example, parasitic species evolved independently in all eight major lineages. Photosynthetic species exist in most of the eight, and multicellularity evolved independently in four. Similar statements could be made about the evolution of life cycles and modes of locomotion.

In effect, each of the eight lineages represents a similar radiation of species into a wide array of lifestyles. In each case, the radiation began with a morphological innovation—often a change in the structure or function of one or more organelles. Let's take a more detailed look at some representative taxa from seven of the eight major groups. The eighth group—the opisthokonts—includes fungi, animals, and several protist lineages and will be discussed in Chapters 30 through 33. The clade called green plants is analyzed in Chapter 29.

Excavata

The unicellular species that form the Excavata are named for the morphological feature that distinguishes them: an "excavated" feeding groove found on one side of the cell. Because all excavates lack mitochondria, they were once thought to trace their ancestry to eukaryotes that existed *prior* to the origin of mitochondria. But researchers have found that excavates either have nuclear genomes containing genes that are normally found in mitochondria or unusual organelles that appear to be vestigial mitochondria. These observations support the hypothesis that the ancestors of excavates had mitochondria, but that these organelles were lost or reduced over time. Several major lineages are included in the group, two of which are detailed here.

Excavata ▶ Diplomonadida

The first species of diplomonad known was described by the early microscopist Anton van Leeuwenhoek, who found *Giardia intestinalis* while examining samples of his own feces. About 100 species have been named to date. Many live in the guts of animal species without causing harm to their host; other species live in stagnant water habitats. In both types of environment, oxygen availability tends to be very low.

Morphological diversity Each cell has two nuclei, which resemble eyes when viewed under the microscope. Some species lack the organelles called peroxisomes and lysosomes in addition to lacking functional mitochondria; all lack a cell wall.

Feeding and locomotion Some are parasitic, but most ingest bacteria whole. They swim using flagella.

Reproduction Only asexual reproduction occurs; meiosis has yet to be observed in members of this lineage.

Human and ecological impacts Both *Giardia intestinalis* and *G. lamblia* (**Figure 28.23**) are common intestinal parasites in humans. *Hexamida* outbreaks cause heavy losses in turkey farms each year.

Giardia lamblia

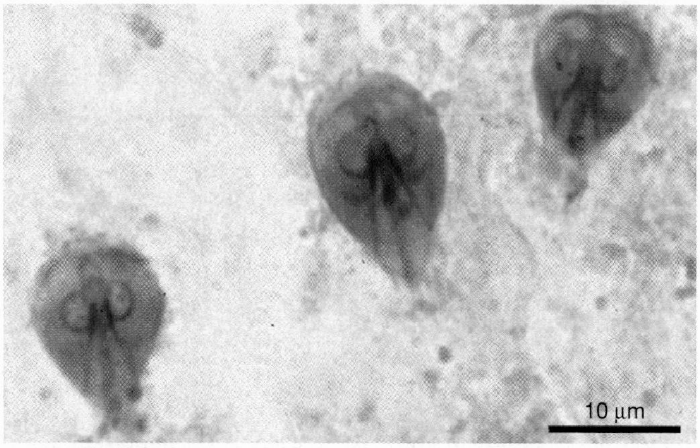

FIGURE 28.23 ***Giardia*** **Causes Intestinal Infections in Humans**

Excavata ▶ Parabasalida

No free-living parabasids are known; all of the species that have been described to date live inside animals. Of the 300 species of parabasalid, several live only in the guts of termites. Termites eat wood but cannot digest it themselves. Instead, the parabasalids produce enzymes that digest the cellulose in wood and release compounds that can be used by the termite host. The relationship between termites and parabasalids is considered mutualistic, because both parties benefit from the symbiosis.

Morphological diversity Cells lack a cell wall and have a single nucleus.

Feeding and locomotion Parabasalids feed by engulfing bacteria, archaea, and organic matter. Although the number of flagella present varies widely, four or five is typical.

Reproduction All reproduce asexually; sexual reproduction has been observed in a few species.

Human and ecological impacts *Trichomonas* infections can sometimes cause reproductive tract problems in humans (**Figure 28.24**), although members of this genus may also live in the gut or mouth of humans without causing harm. *Histomonas meleagridis* is a serious threat to chicken and turkey farms.

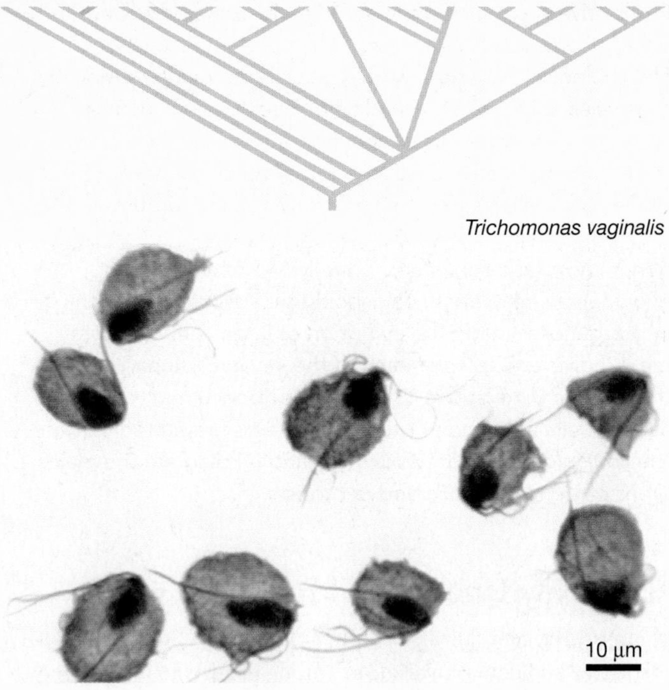
Trichomonas vaginalis
10 µm

FIGURE 28.24 ***Trichomonas*** **Causes the Sexually Transmitted Disease Trichomoniasis**

Discicristata

The discicristates are all unicellular and were named for the distinctive disc shape of the cristae within their mitochondria. Several major subgroups have been identified, such as amoeboid forms and lineages that include the species responsible for the diseases leishmaniasis and trypanosomiasis in humans (see Table 28.1). Because they are extremely common in ponds and lakes, the euglenids are the best-studied discicristates.

Discicristata ▶ Euglenida

There are about 1000 known species of euglenid. Although most live in freshwater, a few are found in marine habitats. Fossil euglenids have been found in rocks that are over 410 million years old.

Morphological diversity Most euglenids lack an external wall but have a system of interlocking protein lying under the plasma membrane.

Feeding and locomotion About one-third of the species have chloroplasts and perform photosynthesis, but most ingest bacteria and other small cells or particles. Some cells have a light-sensitive "eyespot" and use flagella to swim toward light (**Figure 28.25**).

Reproduction Only asexual reproduction is known to occur among euglenids.

Human and ecological impacts Euglenids are important components of freshwater plankton and food chains.

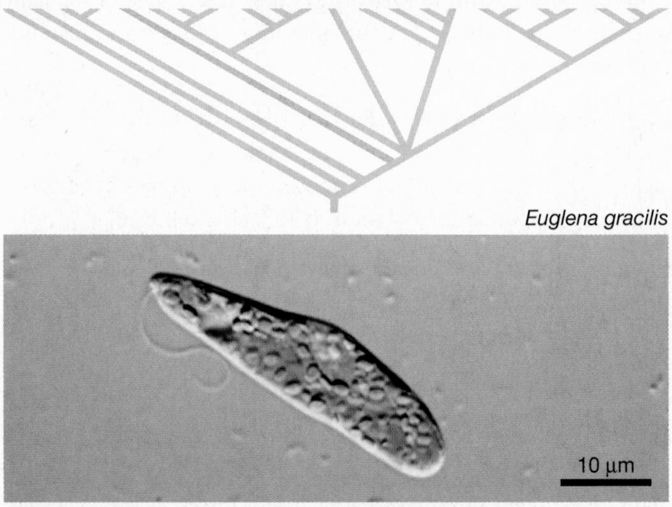
Euglena gracilis
10 µm

FIGURE 28.25 Euglenids Are Common in Ponds and Lakes

Alveolata

Alveolates have small sacs called alveoli that are located just under their plasma membranes. Although all members of this lineage are unicellular, the groups highlighted here—the Ciliata, the Dinoflagellata, and the Apicomplexa—are remarkably diverse in morphology and lifestyle.

Alveolata ▶ Ciliata

Ciliates were named for the cilia that cover them (**Figure 28.26**). Some 12,000 species are known from freshwater habitats, marine environments, and wet soils.

Morphological diversity Cells have two nuclei—a larger *macronucleus* and a smaller *micronucleus*, which is involved only in reproduction. A few ciliates secrete an external skeleton.

Feeding and locomotion Depending on the species, ciliates may be filter feeders, predators, or parasites. They use cilia to swim and have a mouth area where food is ingested.

Reproduction Ciliates divide to produce daughter cells asexually. They also undergo a type of sexual reproduction called **conjugation**. During conjugation, two cells line up side by side and physically connect. Micronuclei exchange between cells and fuse. The resulting nucleus eventually forms a new macronucleus and micronucleus.

Human and ecological impacts Ciliates are abundant in marine plankton and are important consumers. They are common in the digestive tracts of goats, sheep, cattle, and other grazers, where they feed on plant matter and help the host animal digest it.

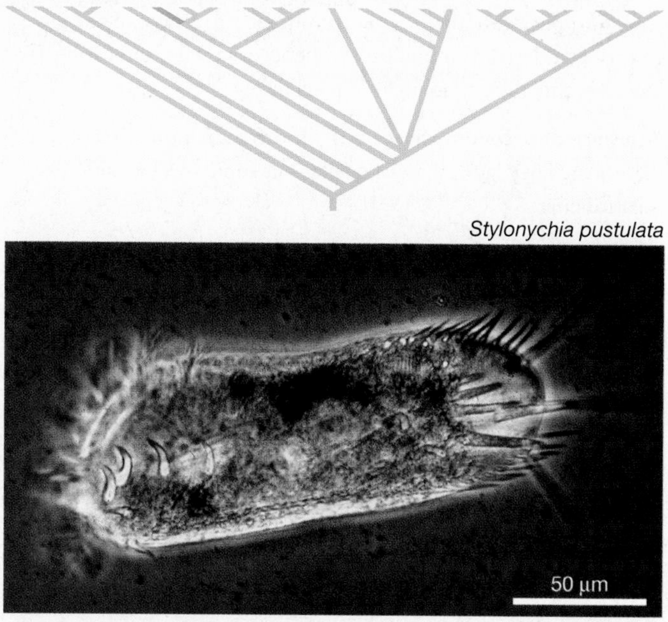
Stylonychia pustulata
50 μm

FIGURE 28.26 Ciliates Are Abundant in Freshwater Plankton

Alveolata ▶ Dinoflagellata

Most of the 4000 species of dinoflagellates are ocean-dwelling plankton, although they are abundant in freshwater as well. Some species are capable of **bioluminescence**, meaning they emit light via an enzyme-catalyzed reaction (**Figure 28.27**).

Morphological diversity Most dinoflagellates are unicellular; some are colonial. Each species has a distinct shape maintained by plates of cellulose inside the cell. Unlike chromosomes of other eukaryotes, chromosomes in dinoflagellates do not contain histones.

Feeding and locomotion About half of the dinoflagellates are photosynthetic. Some of the photosynthetic species live in association with corals or sea anemones, but most are planktonic. The other species are predatory or parasitic. Cells swim in a spinning motion, using two flagella.

Reproduction Both asexual and sexual reproduction occur. Cells that result from sexual reproduction may form tough **cysts**. A cyst is a resistant structure that can remain dormant until environmental conditions improve.

Human and ecological impacts Photosynthetic dinoflagellates are important primary producers in marine ecosystems—second only to diatoms in the amount of carbon they fix per year. A few species are responsible for harmful algal blooms, or red tides.

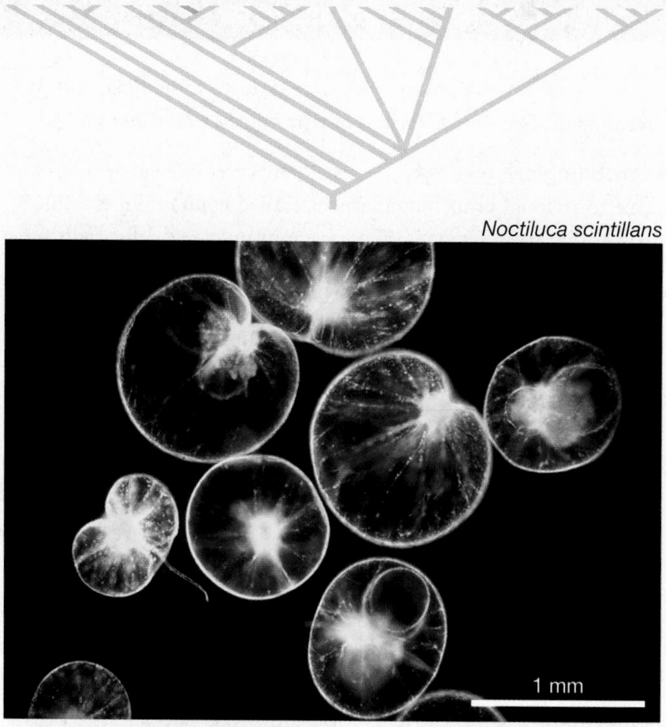
Noctiluca scintillans
1 mm

FIGURE 28.27 Some Dinoflagellates Species Are Bioluminescent

Alveolata ▶ Apicomplexa

All of the 5000 known species of apicomplexa are parasitic.

Morphological diversity Cells have a system of organelles at one end, called the *apical complex*, that is a diagnostic characteristic unique to the group. The apical complex allows the apicomplexan to penetrate the plasma membrane of its host. They have a chloroplast-derived, nonphotosynthetic organelle, indicating that they descended from an algal-like ancestor.

Feeding and locomotion All absorb nutrition directly from their host. They lack cilia or flagella, but some species can move by cell crawling.

Reproduction Reproduction can occur sexually or asexually. In some species, the life cycle involves two distinct hosts, and cells must be transmitted from one host to the next.

Human and ecological impacts Various species in the genus *Plasmodium* infect birds, reptiles, or mammals and cause malaria. *Toxoplasma* is an important pathogen in people infected with HIV (**Figure 28.28**), and chicken farmers estimate that they lose $600 million per year controlling the apicomplexan *Eimeria*. Apicomplexans that parasitize insects have been used to control pests.

Toxoplasma gondii

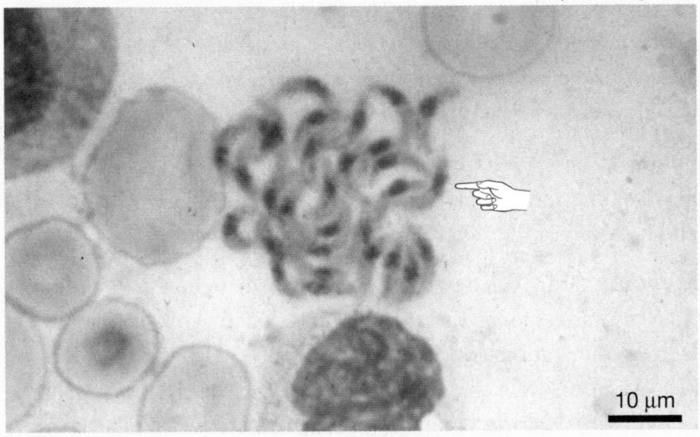

FIGURE 28.28 *Toxoplasma* Causes Infections in AIDS patients

Stramenopila (Heterokonta)

Stramenopiles are sometimes called heterokonts. At some stage of their life cycle, all stramenopiles have flagella that are covered with distinctive hollow "hairs." The structure of these flagella is unique to the stramenopiles. The lineage includes a large number of unicellular forms, although the brown algae are multicellular and include the world's tallest marine organisms, the kelp.

Stramenopila ▶ Oomycota

Based on morphology, oomycetes were thought to be fungi and were given fungus-like names, such as downy mildew and water molds.

Morphological diversity Some oomycetes are unicellular and some form long, branching filaments called **hyphae** (**Figure 28.29**). Species that form hyphae often have multinucleate cells. Cells have walls containing cellulose.

Feeding and locomotion Most species feed on decaying organic material in freshwater environments; a few are parasitic. Mature individuals are sessile.

Reproduction Most species are diploid throughout the majority of their life cycle. In aquatic species, spores that are produced via asexual or sexual reproduction have flagella and swim to find new food sources; in terrestrial species, spores swim in rainwater or are dispersed by the wind in dry conditions.

Human and ecological impacts Oomycetes are extremely important decomposers in aquatic ecosystems. Along with certain bacteria and archaea, they are responsible for breaking down dead bodies and releasing nutrients for use by other species. The organism that caused the Irish potato famine was an oomycete.

Phytophthora infestans

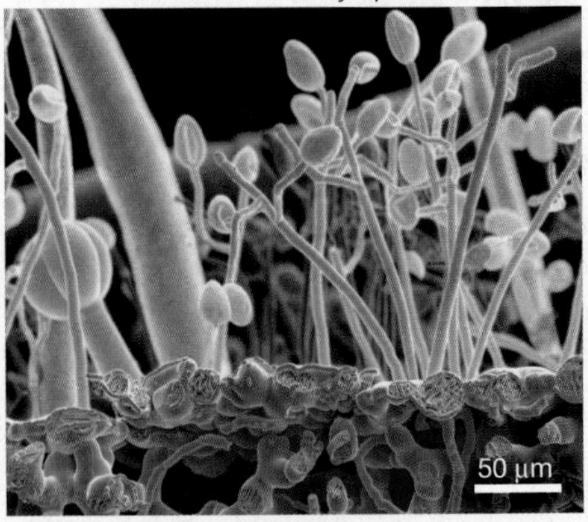

FIGURE 28.29 *Phytophthora infestans* Infects Potatoes

Stramenopila ▶ Diatoms

Diatoms are unicellular or form chains of cells. About 10,000 species have been described to date, but some researchers claim that millions of species exist and have yet to be discovered.

Morphological diversity Cells are supported by external, silicon-rich, glassy shells that form a box-and-lid arrangement (**Figure 28.30**).

Feeding and locomotion Diatoms are photosynthetic. Only reproductive cells have flagella and are capable of powered movement. Photosynthetic cells normally just float in the wate, but species that live on a substrate can glide via microtubules that project from gaps in their shells and that move in response to motor proteins.

Reproduction Diatoms divide by mitosis to reproduce asexually or by meiosis to form gametes that fuse to form a new individual. Many species can produce spores that are dormant during unfavorable growing conditions.

Human and ecological impacts In abundance, diatoms dominate the plankton of cold, nutrient-rich waters. They are found in virtually all aquatic habitats and are considered the most important producer of carbon compounds in fresh- and saltwater. Their shells settle into massive accumulations that are mined and sold commercially as diatomaceous earth, which is used in filtering applications and as an ingredient in polishes, paint, cosmetics, and other products.

Actinoptychus heliopelta

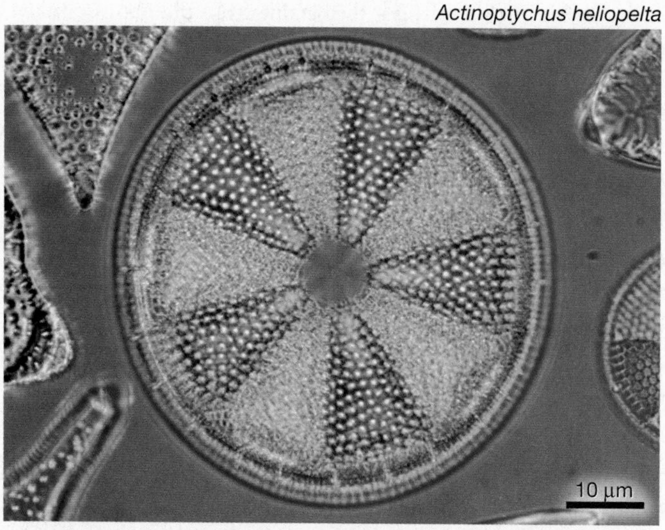

FIGURE 28.30 Diatoms Have Glass-like Shells with a Box-and-Lid Arrangement

Stramenopila ▶ Phaeophyta (Brown Algae)

The color of brown algae is due to their unique suite of photosynthetic pigments. Over 1500 species have been described, most of which live in marine habitats.

Morphological diversity Cells have walls that contain cellulose in addition to other complex polymers. Brown algae are unique among protists, because all species are multicellular. The body typically consists of leaf-like *blades*, a stalk known as a *stipe*, and a root-like *holdfast*, which attaches the individual to a substrate (**Figure 28.31**).

Feeding and locomotion Brown algae are photosynthetic and **sessile**—that is, permanently fixed to a substrate—though reproductive cells may have flagella and be **motile**—capable of locomotion.

Reproduction Sexual reproduction occurs via the production of swimming gametes, which fuse to form a zygote. Most species exhibit alternation of generations. Depending on the species, the gametophyte and sporophyte stages may look similar or different.

Human and ecological impacts In many coastal areas, brown algae form forests or meadows that are important habitats for a wide variety of animals. In the Sargasso Sea, off Bermuda, floating brown algae form extensive rafts that harbor an abundance of animal species. The compound *algin* is purified from kelp and used in the manufacture of cosmetics and paint.

Bull kelp

FIGURE 28.31 Many Brown Algae Have a Holdfast, Stalk, and Leaf-like Blades

Cercozoa

The cercozoans are single-celled amoebae that lack cell walls, though some species produce elaborate shell-like coverings. They move by cell crawling and produce long, slender pseudopodia. A number of important subgroups have been identified and named, including the planktonic organisms called actinopods, which synthesize glassy, silicon-rich skeletons. The most prominent group is the Foraminifera.

Cercozoa ▶ Foraminifera

Foraminifera, or forams, got their name from the Latin *foramen*, meaning hole, because produce tests (shells) that have one or two holes through which pseudopodia emerge. Fossil tests of foraminifera are abundant in sediments. There is a continuous record of fossilized forams that dates back 530 million years. Although they are abundant in marine plankton, their biology is relatively poorly known.

Morphological diversity Cells generally have multiple nuclei. The tests of forams are usually made of organic material stiffened with calcium carbonate, and most species have several chambers (**Figure 28.32**). One species known from fossils was 12 cm long, but most species are much smaller. The size and shape of the test are traits that distinguish foram species.

Feeding and locomotion Like other cercozoans, forams feed by extending their pseudopodia and using them to capture and engulf bacterial and archaeal cells or bits of organic debris, which are digested in food vacuoles. Some species have symbiotic algae that perform photosynthesis and contribute sugars to their host. Forams simply float in the water.

Reproduction Asexual reproduction occurs by mitosis; when meiosis occurs, the resulting gametes are released into the open water, where they fuse to form a new individual.

Human and ecological impacts The tests of dead forams commonly form extensive sediment deposits when they settle out of the water, producing layers that eventually solidify into chalk, limestone, or marble.

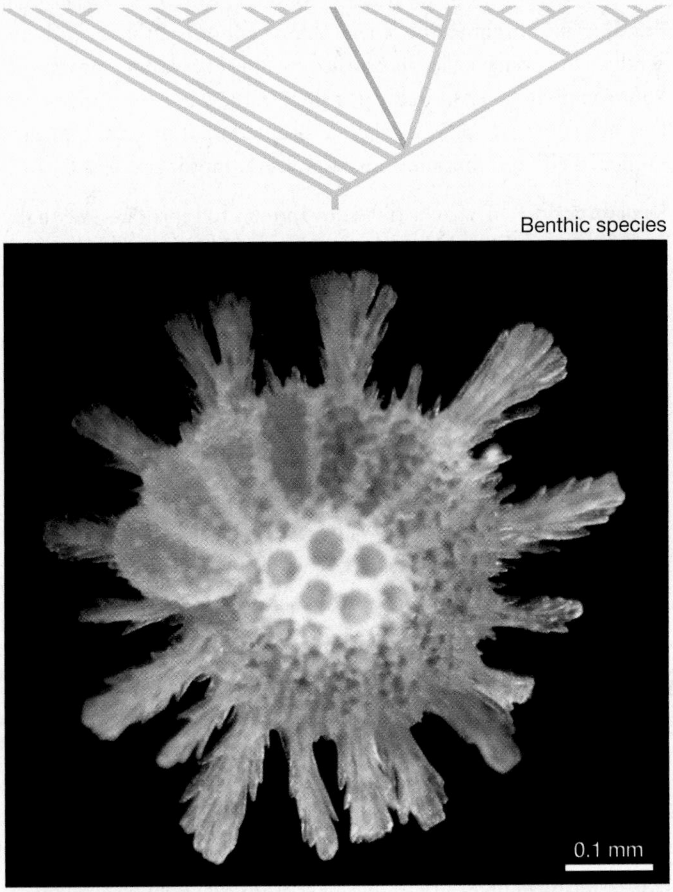
Benthic species
0.1 mm

FIGURE 28.32 Forams Are Shelled Amoebae

Plantae

Biologists are beginning to use the name "Plants" to refer to the monophyletic group that includes red algae, green algae, land plants, and glaucophyte algae. All of these lineages are descended from a common ancestor that engulfed a cyanobacterium, beginning the endosymbiosis that led to the evolution of the chloroplast—their distinguishing morphological feature. This initial endosymbiosis probably occurred in an ancestor of today's glaucophyte algae. To support this hypothesis, biologists point to several important similarities between cyanobacterial cells and the chloroplasts that are found in the glaucophytes. Both have cell walls that contain peptidoglycan and that are susceptible to attack by lysozyme and penicillin. The glaucophyte chloroplast also has a membrane outside its wall that is similar to the membrane found in Gram-negative bacteria (see Chapter 27). Consistent with these observations, phylogenetic analyses of DNA sequence data place the glaucophytes as the most basal, or ancient, group of the Plantae.

The glaucophyte algae are unicellular or colonial and live in plankton or attached to substrates in freshwater environments—particularly in bogs or swamps. Some glaucophyte species have flagella or produce flagellated spores, but sexual reproduction has never been observed in the group. The chloroplasts of glaucophytes have a distinct bright blue-green color. Chapter 29 introduces the green algae and land plants; here we consider just the red algae in detail.

Rhodophyta (Red Algae)

The 6000 species of red algae live primarily in marine habitats. One species lives over 200 m below the surface; another is the only eukaryote capable of living in acidic hot springs. Although their color varies, many species are red because their chloroplasts contain the accessory pigment *phycoerythrin*, which absorbs strongly in the blue and green portions of the visible spectrum.

Morphological diversity Cells have walls that are comprised of cellulose and other polymers. A few species are unicellular, but most are multicellular. Many of the multicellular species are filamentous, but others grow as thin, hard crusts on rocks or coral (**Figure 28.33**). Some species have erect, leaf-like structures called *thalli*. Some species have cells with many nuclei.

Feeding and locomotion The vast majority of red algae are photosynthetic, though a few parasitic species have been identified. Red algae are the only type of algae that lack flagella.

Reproduction Asexual reproduction occurs through production of spores by mitosis. Alternation of generations is common, but the types of life cycles observed in red algae are extremely variable.

Human and ecological impacts On coral reefs, some red algae become encrusted with calcium carbonate. These species contribute to reef building and help stabilize the entire reef structure. Cultivation of *Porphyra* (or nori) for sushi and other foods is a billion-dollar-per-year industry in East Asia.

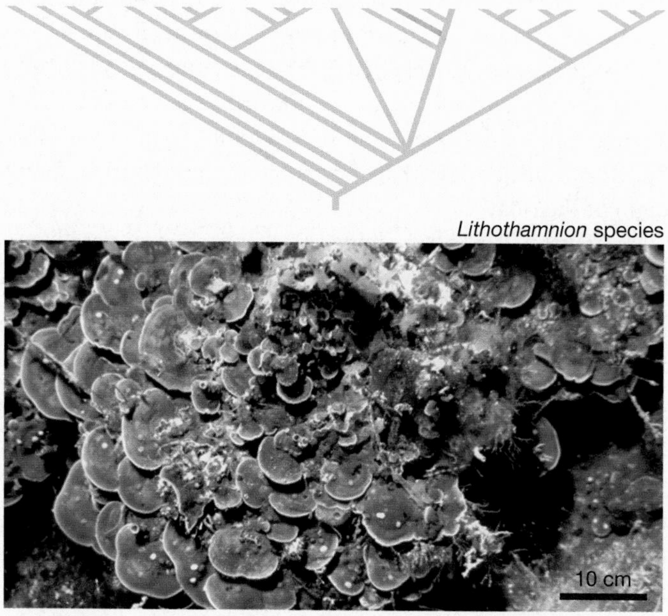
Lithothamnion species
10 cm

FIGURE 28.33 Red Algae Adopt an Array of Growth Forms

Amoebozoa

Species in the Amoebozoa lack cell walls and take in food by engulfing it. They move by means of cell crawling and produce large, lobe-like pseudopodia. The major subgroups in the lineage are lobose amoebae, cellular slime molds, and plasmodial slime molds. The cellular slime mold *Dictyostelium discoideum* was described in detail in Box 28.2. Amoebae are abundant in freshwater habitats and in wet soils; some are parasites of humans and other animals. More details on plasmodial slime molds follow.

Myxogastrida (Plasmodial Slime Molds)

The plasmodial slime molds got their name because individuals form a large, weblike structure that consists of a single cell containing many diploid nuclei (**Figure 28.34**).

Morphological diversity The huge *supercell* form with many nuclei occurs in few protists other than plasmodial slime molds.

Feeding and locomotion Cells feed on decaying vegetation and move by cell crawling.

Reproduction When food becomes scarce, part of the amoeba forms a stalk topped by a ball-like structure in which nuclei undergo meiosis and form spores. The spores are then dispersed to new habitats by the wind or small animals. After spores germinate to form amoebae, two amoebae fuse to form a diploid cell that begins to feed and eventually grows into a supercell.

Human and ecological impacts Like cellular slime molds, plasmodial slime molds are important decomposers in forests. They help break down leaves, branches, and other dead plant material, releasing nutrients that they and other organisms can use.

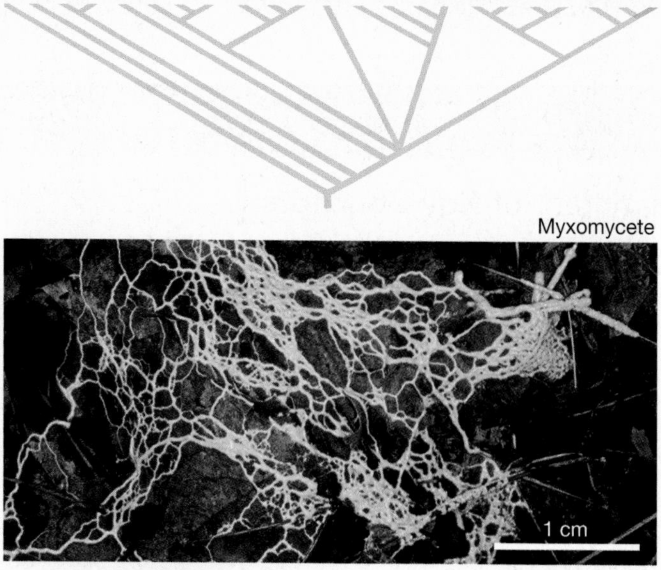
Myxomycete
1 cm

FIGURE 28.34 Plasmodial Slime Molds Are Important Decomposers in Forests

ESSAY Revolutions in Science

In 1962 a philosopher and science historian named Thomas Kuhn published an influential book called *The Structure of Scientific Revolutions*. Most scientific work, Kuhn contends, represents "normal science." In normal science, the questions that scientists ask are inspired by the currently accepted theories and styles of thinking, or what he calls the prevailing paradigm. In Kuhn's view, most scientific results are also "normal" in the sense that they support the prevailing paradigm of their day.

In contrast, Kuhn maintained that entirely new paradigms emerge during "scientific revolutions." Revolutionary science introduces fundamentally new theories and styles of thinking, which eventually displace the older paradigms.

To Kuhn, the Sun-centered theory of the solar system developed by Nicolaus Copernicus represents the classic example of revolutionary science. This theory, published in the early 1500s, maintains that the planets revolve around the Sun rather than the Earth. In addition to being correct, Copernicus's theory defined an entirely new research program in astronomy and physics. It also inspired new ideas in other fields, including religion and philosophy. Copernicus's paradigm inspired new questions, new techniques, and a new style of thinking in science.

This chapter on protists introduced two ideas that could be considered revolutionary. The first is the endosymbiosis theory for the origin of mitochondria and chloroplasts. Prior to this theory's publication, the prevailing paradigm was that the lateral transfer of genes—transfer of genes from one lineage to another—occurred rarely, if ever. The endosymbiosis theory not only inspired a vigorous series of tests, but it also established an entirely new way of thinking about the origin of the eukaryotes and the importance of lateral gene transfer in the history of life.

The second revolutionary idea is the change from taxonomic to phylogenetic thinking, highlighted in Box 28.1. Prior to this revolution, the prevailing paradigm was that prokaryotes and eukaryotes represented the two fundamental types of organisms. Traditional taxonomic schemes aimed to organize and categorize the diversity of life around a handful of kingdoms, often defined as the Monera (prokaryotes), Protista, Plantae, Fungi, and Animalia. Biologists analyzed the diversity of life-forms by placing them into taxonomic categories based on morphological similarity rather than by using morphological and molecular data to infer their history and evolutionary relationships (see Chapter 26).

Revolutionary science introduces fundamentally new theories and styles of thinking....

A change in phylogenetic thinking is currently under way, and it is inspiring the development of a new naming system for the diversity of life. The old five-kingdom system is being replaced by a system of three domains with many kingdoms.

Undoubtedly, 20 years from now you will be able to look back at this book and pinpoint paradigms that have been supplanted by new scientific revolutions. Which of the ideas in this book will be overturned by new insights? What types of data will emerge, showing these ideas to be inadequate? Scientific revolutions are bound to occur. In this sense, they too are "normal."

CHAPTER REVIEW

Summary of Key Concepts

- **Protists are a paraphyletic grouping that includes all eukaryotes except the green plants, fungi, and animals. Biologists study protists to understand how eukaryotes evolved, because they are important in marine ecosystems and global warming and because some species cause debilitating diseases in humans and other organisms.**

  Protists are often tremendously abundant in marine and freshwater plankton and other habitats. As a result, protists provide food for many organisms in aquatic ecosystems and fix so much carbon that they have a large impact on the global carbon budget. Toxin-producing protists that grow to high densities result in a harmful algal bloom. Parasitic protists cause several important diseases in humans, including malaria.

- **Protists are diverse morphologically. They vary in the types of organelles they contain; they may be unicellular or multicellular, and they may have a cell wall or other external covering or no such covering.**

  Several morphological features are common to all eukaryotes: a nucleus, a cytoskeleton, and extensive intracellular structuring that includes organelles. In addition to these common features, many aspects of morphology and lifestyle are extremely variable among the protist lineages. The protists include many unicellular organisms as well as multicellular slime molds, red algae, and brown algae. Multicellularity, swimming via the beating of cilia or flagella, cell crawling, and hard outer coverings have evolved in many different protist groups independently.

Eukaryotes also contain mitochondria or have genes indicating that their ancestors once contained mitochondria. Several types of data support the hypothesis that mitochondria originated as endosymbiotic bacteria, possibly engulfed by an early protist closely related to the archaea. The symbiosis is thought to have been successful because the endosymbiotic bacterium provided its host with ATP while the host provided the bacterium with carbon compounds and protection. Similarly, the chloroplast's size, DNA structure, ribosomes, double membrane, and evolutionary relationships are consistent with the hypothesis that this organelle originated as an endosymbiotic cyanobacterium.

- **Protists vary widely in terms of how they reproduce and how they find food. Many can reproduce both sexually and asexually. Many species are photosynthetic, while others obtain carbon compounds by ingesting food or parasitizing other organisms.**

Protists undergo cell division based on mitosis and reproduce asexually. Many protists also undergo meiosis and sexual reproduction at some phase in their life cycle. Alternation of generations is common in multicellular species—meaning there are separate haploid and diploid forms of the same species. Protists exhibit predatory, parasitic, or photosynthetic lifestyles, which evolved in many groups independently.

Web Tutorial 28.1 Alternation of Generations in a Protist

Questions

Content Review

1. To biologists, the protists comprise all eukaryotes other than which groups?
   a. extinct forms
   b. green plants, fungi, and animals
   c. multicellular forms
   d. those with cells containing a cytoskeleton and nucleus
2. What materials are *not* used by protists to manufacture hard outer coverings?
   a. cellulose
   b. lignin
   c. glass-like compounds that contain silicon
   d. mineral-like compounds such as calcium carbonate ($CaCO_3$)
3. What does cell crawling result from?
   a. interactions among actin, myosin, and ATP
   b. coordinated beats of cilia
   c. the whiplike action of flagella
   d. action by the mitotic spindle
4. According to the endosymbiosis theory, what type of organism is the ancestor of the chloroplast?
   a. a photosynthetic archaean
   b. a cyanobacterium
   c. a primitive photosynthetic eukaryote
   d. a modified mitochondrion
5. Multicellularity is defined in part by the presence of distinctive cell types. At the cellular level, what does this criterion imply?
   a. Individual cells must be extremely large.
   b. The organism must be able to reproduce sexually.
   c. Cells must be able to move.
   d. Different cell types express different genes.
6. Why are protists an important part of the global carbon cycle and marine food chains?
   a. They have high species diversity.
   b. They are numerically abundant.
   c. They have the ability to parasitize humans.
   d. They have the ability to undergo meiosis.

Conceptual Review

1. Why is an advanced cytoskeleton and the lack of a cell wall required for ingestive modes of feeding? How does this mode of feeding relate to increased cell size in protists versus cell size in bacteria and archaea? How does it relate to the acquisition of mitochondria and chloroplasts by endosymbiosis?
2. Compare and contrast sexual reproduction in protists with conjugation in bacteria (see Chapter 19).
3. What is the relationship between meiosis and the alternation of generations?
4. Outline the steps in the endosymbiosis theory for the origin of the mitochondrion. What did each partner provide the other, and what did each receive in return? Answer the same questions for the chloroplast.
5. Provide evidence to support the hypothesis that similar evolutionary innovations occurred within each of the eight major eukaryotic lineages.
6. The text refers to eukaryotes as "chimeras" and draws an analogy with the monster from Greek mythology that was a combination of parts from a lion, a goat, and a snake. Why is this analogy appropriate?

Group Discussion Problems

1. Consider the following:
   - All living eukaryotes have mitochondria or have evidence in their genomes that they once had these organelles. Thus it appears that eukaryotes acquired mitochondria very early in their history.
   - The first eukaryotic cells in the fossil record correlate with the first appearance of rocks formed in an oxygen-rich ocean and atmosphere.

   How are these observations connected? (HINT: Before answering, glance at Chapter 9 and remind yourself what happens in a mitochondrion.)
2. Consider the following:
   - *Plasmodium* has an unusual organelle called an *apicoplast.* Recent research has shown that apicoplasts are derived from chloroplasts via secondary endosymbiosis and have a large number of genes encoded by chloroplast DNA.
   - *Glyphosate* is one of the most widely used herbicides. It works by poisoning an enzyme encoded by a gene in chloroplast DNA.
   - Biologists are testing the hypothesis that glyphosate could be used as an antimalarial drug in humans.

   How are these observations connected?
3. Suppose a friend says that we don't need to worry about global warming. Her claim is that increased temperatures will make planktonic algae grow faster and that carbon dioxide (CO_2) will be removed from the atmosphere faster. According to her, this carbon will be buried at the bottom of the ocean in calcium carbonate tests. As a result, the amount of carbon dioxide in the atmosphere will decrease and global warming will decline. Comment.
4. Biologists are beginning to draw a distinction between "species trees" and "gene trees." A *species tree* is a phylogeny that describes the actual evolutionary history of a lineage. A *gene tree*, in contrast, describes the evolutionary history of one particular gene, such as the gene for chlorophyll *a*. In some cases, species trees and gene trees don't agree with each other. For example, the species tree for green algae indicates that their closest relatives are protists and plants. But the gene tree based on chlorophyll *a* from green algae suggests that this gene's closest relative is a bacterium, not a protist. What's going on? Why do these types of conflicts exist?

Answers to Multiple-Choice Questions **1.** b; **2.** b; **3.** a; **4.** b; **5.** d; **6.** b

www.prenhall.com/freeman is your resource for the following: Web Tutorials; Online Quizzes and other Online Study Guide materials; Answers to Conceptual Review Questions; Solutions to Group Discussion Problems; Answers to Figure Caption Questions and Exercises; and Additional Readings and Research.

Green Plants

29

Mosses are common in moist habitats and share many similarities with the earliest land plants. According to data reviewed in this chapter, the earliest land plants evolved from green algae that inhabited ponds, streams, and other freshwater habitats.

KEY CONCEPTS

- The green plants include both the green algae and the land plants. Biologists study them because humans—along with virtually every other terrestrial organism—rely on plants for food and a healthy environment.
- Land plants were the first multicellular organisms to live on land. A series of key adaptations allowed them to survive with their tissues constantly exposed to dry air. In terms of total mass, plants dominate today's terrestrial environments.
- Once plants were able to grow on land, a sequence of important evolutionary changes made it possible for them to reproduce efficiently—even in extremely dry environments.

The **green plants** consist of the green algae and the land plants. Green algae are important photosynthetic organisms in aquatic habitats—particularly lakes, ponds, and other freshwater settings—while land plants are the key photosynthesizers in terrestrial environments. Before land plants evolved, the only life on the continents consisted of bacteria, archaea, and single-celled protists that thrive in wet soils.

Although green algae have traditionally been considered protists, it is logical to study them along with land plants for two reasons: (1) They are the closest living relatives to land plants, and (2) the transition from aquatic to terrestrial life occurred when land plants evolved from green algae. Land plants were the first organisms that could grow with their tissues exposed to the air instead of being partially or completely submerged. When they evolved, multicellular organisms began to occupy dry terrestrial habitats that had been largely barren of life for over 3 billion years. Land plants made the Earth green. They were pioneers—the first truly terrestrial organisms.

Not long after land plants evolved and began to diversify, fungi and animals also accomplished the feat of moving from aquatic to terrestrial habitats. They could do so, however, only because plants were there first and provided something to eat. Fungi and animals still depend on plants for existence.

By colonizing the continents, plants transformed the nature of life on Earth. In the words of Karl Niklas, the movement of green plants from water to land ranks as "one of the greatest adaptive events in the history of life." How did they do it? Let's delve in and see.

29.1 Why Do Biologists Study the Green Plants?

Biologists study plants because plants are fascinating and because people could not live without them. Along with most other animals and fungi, humans are almost completely dependent on plants for food. People rely on plants for other necessities of life as well—oxygen, fuel, building materials, and the fibers used in making clothing, paper, ropes, and baskets. But we also rely on land plants for important intangible values, such as the aesthetic appeal of landscaping and bouquets. To drive this point home, consider that the sale of cut flowers generates over $1 billion each year in the United States alone.

Based on these observations, it is not surprising that agriculture, forestry, and horticulture are among the most important endeavors supported by biological science. Tens of thousands of biologists are employed in research designed to increase the productivity of plants and to create new ways of using them in ways that benefit people. Research programs also focus on land plants that cause problems for people, either because they are weeds that decrease the productivity of crop plants or newly introduced species that invade and then degrade natural areas. Biologists study plants not only because they are fascinating but also because they keep us alive.

Plants Provide Ecosystem Services

An **ecosystem** consists of all the organisms in a particular area, along with physical components of the environment, such as the atmosphere, precipitation, surface water, sunlight, soil, and nutrients. Plants are said to provide **ecosystem services,** because they add to the quality of the atmosphere, surface water, soil, and other physical components of an ecosystem. Stated another way, plants alter the landscape in ways that benefit other organisms:

- *Plants produce oxygen.* Recall from Chapter 10 that plants perform oxygenic ("oxygen-producing") photosynthesis. In this process, electrons that are removed from water molecules are used to reduce carbon dioxide and produce sugars. In the process of stripping electrons from water, plants release oxygen molecules (O_2) as a by-product. As Chapter 27 noted, oxygenic photosynthesis evolved in cyanobacteria and was responsible for the evolution of the original oxygen atmosphere. The evolutionary success of plants continued this trend by adding huge amounts of oxygen to the atmosphere. Without the green plants, we and other terrestrial animals would be in danger of suffocating for lack of oxygen.
- *Plants build soil.* Leaves and roots and stems that are not eaten when they are alive fall to the ground and provide food for worms, fungi, bacteria, archaea, protists, and other decomposers in the soil. These organisms add organic matter to the soil, which improves soil texture and the ability of soils to hold nutrients and water.
- *Plants hold soil.* The extensive network of fine roots produced by trees, grasses, and other land plants helps hold soil particles in place. And by taking up nutrients in the soil, plants prevent the nutrients from being washed or blown away. When areas are devegetated by grazing, farming, logging, or suburbanization, large quantities of soil and nutrients are lost to erosion by wind and water (**Figure 29.1**).
- *Plants hold water.* Plant tissues themselves take up and retain water. Intact forests, prairies, and wetlands also prevent rain from quickly running off a landscape. There are several reasons why the presence of plants increases the amount of water available in a habitat. Plant leaves soften the physical impact of rainfall on soil; plant roots hold soil particles in place during rainstorms; and plant organic matter builds the soil's water-holding capacity. When areas are devegetated, streams are more prone to flooding and groundwater is not replenished efficiently. It is common to observe streams alternately flooding and then drying up completely when the surrounding area is deforested.

(a) Wind erosion

(b) Water erosion

FIGURE 29.1 Plants Hold Soil in Place
When plants are removed, soils are susceptible to erosion by **(a)** wind and **(b)** water.

- *Plants moderate the local climate.* By providing shade, plants reduce temperatures beneath them and increase relative humidity. They also reduce the impact of winds that dry out landscapes or make them colder. When plants are removed from landscapes to make way for farms or suburbs, habitats become much dryer and are subject to more extreme temperature swings.

Perhaps the most important ecosystem service provided by plants, however, involves food. In addition to their beneficial impact on their physical surroundings, land plants are the dominant primary producers in terrestrial ecosystems. (As Chapter 28 indicated, primary producers convert some of the energy in sunlight into chemical energy.) The sugars and oils that land plants produce by photosynthesis provide the base of the food chain in the vast majority of terrestrial habitats. As **Figure 29.2** shows, plants are eaten by **herbivores** ("plant-eaters"), which range in size from insects to elephants. These consumers are eaten by **carnivores** ("meat-eaters"), ranging in size from spiders to lions. Humans are an example of **omnivores** ("all-eaters"), organisms that eat both plants and animals. Omnivores feed at several different levels in the terrestrial food chain. For example, people consume plants, herbivores such as chicken and cattle, and carnivores such as salmon and tuna.

Tertiary Consumers: Secondary carnivores eat carnivores.
Secondary Consumers: Carnivores eat animals.
Primary Consumers: Herbivores eat plants.
Producers: Plants form the base of the terrestrial food chain.

FIGURE 29.2 Plants Are the Basis of Food Chains in Terrestrial Environments
Virtually every organism that lives on land depends on plants for food, either directly or indirectly. EXERCISE Label the levels in this food chain where humans feed.

Finally, just as photosynthetic protists and bacteria are the key to the carbon cycle in the oceans, green plants are the key to the carbon cycle on the continents. Plants take CO_2 from the atmosphere and reduce it to make sugars. Although both green algae and land plants produce a great deal of CO_2 as a result of cellular respiration, they fix much more CO_2 than they release. The loss of plant-rich prairies and forests, due to fires or logging or suburbanization, has contributed to increased concentrations of CO_2 in the atmosphere. Higher carbon dioxide levels, in turn, are responsible for the rapid warming that is occurring worldwide.

In many cases, plant growth alters the environment in ways that benefit other organisms as well as themselves.

Agricultural Research: Domestication and Selective Breeding

Until the late 1800s, most people made their living by farming. Hence, it is not surprising that many of the early advances in genetics and developmental biology were motivated by a desire to improve domesticated crops and animal breeds. Agricultural research continues to be a crucial branch of biological science.

Agriculture was a key human activity long before the development of biology, however. Most anthropologists agree that the domestication of crops was the single most important event in the evolution of human culture. But this "single most important event" was not actually a single event: There is compelling evidence that plants were domesticated independently by different people living in widely scattered locations around the world.

Data from archaeological sites have shown that the grains that form the basis of our current food supply were derived from wild species between 10,000 and 2000 years ago. In each case, researchers have observed a distinct set of changes over time as wild forms were brought under cultivation. Samples taken from human habitations show that, over time, maize (corn) kernels and other types of *seeds* (plant embryos plus nutritive tissue) became larger and squash rinds became thicker. What was responsible for these changes? The leading hypothesis is that our ancestors were actively selecting seeds from individual plants that exhibited these traits; they used only the selected seeds to plant the next generation of crops. By repeating these steps year after year, humans gradually changed the characteristics of several wild species, resulting in the characters that are now associated with domestication (**Figure 29.3a**. This process is called **artificial selection.** It was introduced in Chapter 1 and is analogous to the process of natural selection analyzed in Chapter 23.

Artificial selection was responsible for the original domestication of crop species, and then for a subsequent diversification of crop varieties. For example, researchers have identified 178 distinct varieties of potato being cultivated in a single farming community in the Andes Mountains of South America. Artificial selection also continues today. Over the past 100 years, artificial selection has been responsible for the dramatic increases in the oil content of corn kernels (**Figure 29.3b**). (Corn oil serves as a cooking oil and is used in other products.) Several tree species are now being artificially selected for disease resistance or rapid growth and production of straight trunks.

To summarize, humans have succeeded in domesticating a wide array of plants for food or ornamental use over the past 10,000 years. We eat the seeds of rice, wheat, beans, and corn; the stalks of celery and bok choi; the buds of brussels sprouts; the leaves of spinach, lettuce, and cabbage; the fruits of tomatoes, squash, bell peppers, strawberries, and oranges; and the storage organs of potatoes, onions, carrots, beets, and cassava. Researchers continue to develop new crop plants and select strains of existing crops that have improved productivity or disease resistance. Many of these current research programs rely on the genetic engineering techniques introduced in Chapter 19.

Plant-Based Fuels and Fibers

In addition to relying on plants for food, humans have historically depended on them for cooking and heating fuels and as a source of fibers for clothing and implements such as ropes and nets. For perhaps 100,000 years, wood burning was the primary source of energy used by humans. As

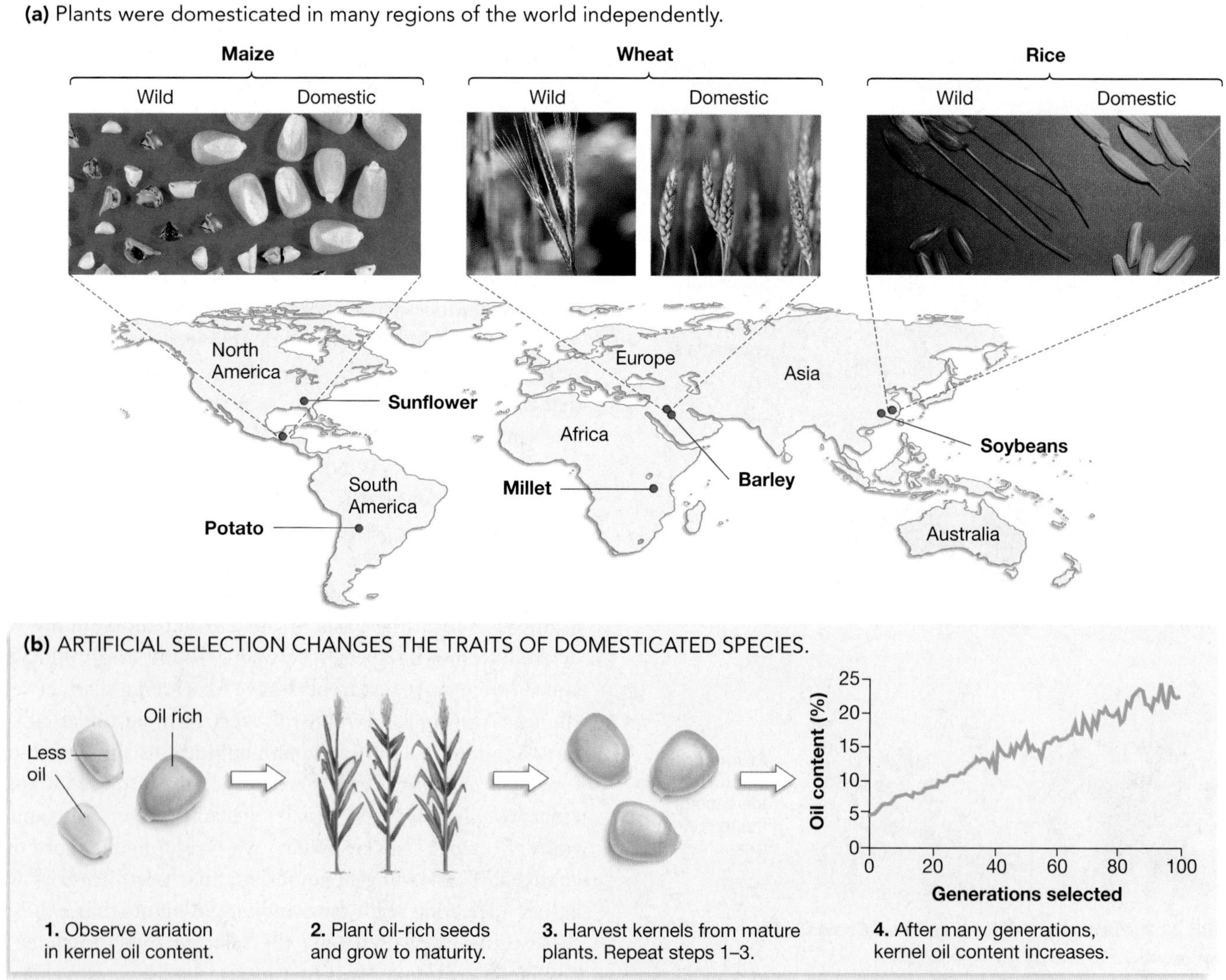

FIGURE 29.3 Crop Plants Are Derived from Wild Species via Artificial Selection
(a) Crop plants were derived from wild relatives in many parts of the world. **(b)** Artificial selection can lead to dramatic changes in plant characteristics.

Figure 29.4a shows, however, wood has been replaced by other sources of energy.

In economies across the globe, the first fuel to replace wood has usually been coal. Coal forms from partially decayed plant material that is compacted over time by overlying sediments and hardened into rock (**Figure 29.4b**). Starting in the mid-1800s, people in England, Germany, and the United States began to mine coal deposits that originally formed during the Carboniferous period some 350–275 million years ago. The coal fueled blast furnaces that smelted vast quantities of iron ore into steel and powered the steam engines that sent trains streaking across Europe and North America. It is no exaggeration to claim that the sugars synthesized by plants about 300 million years ago laid the groundwork for the Industrial Revolution. Coal still supplies about 20 percent of the energy used in Japan and Western Europe and 80 percent of the energy used in China.

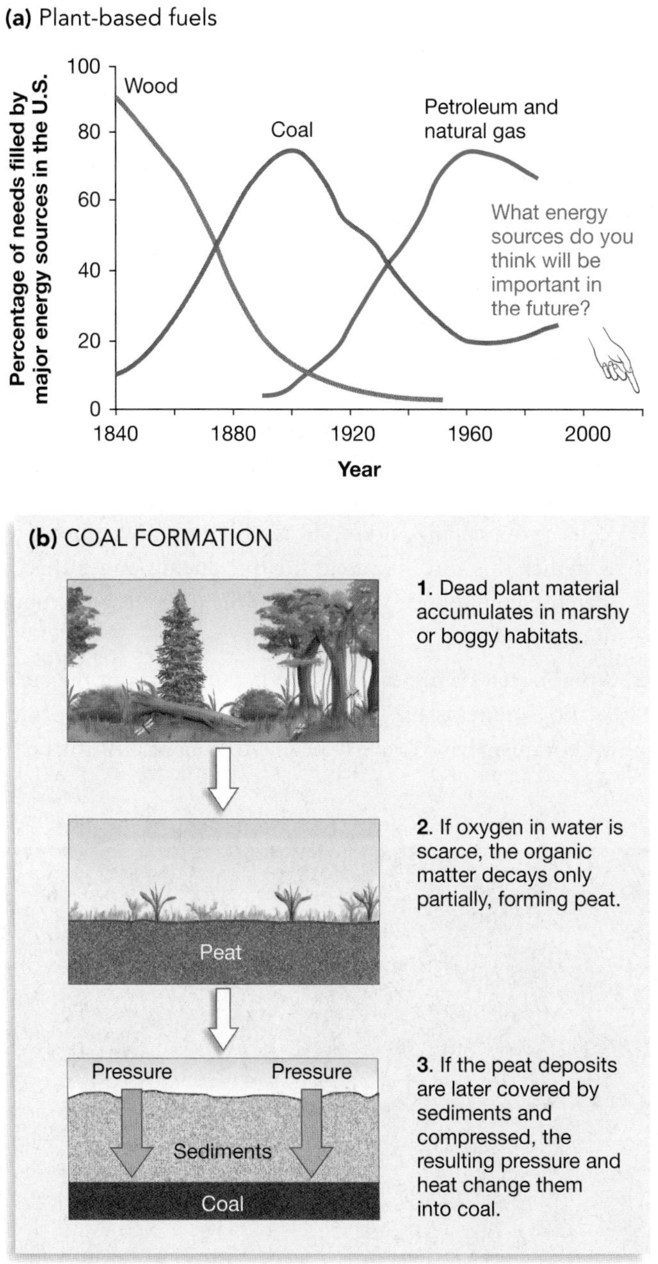
(a) Plant-based fuels
Percentage of needs filled by major energy sources in the U.S.
100
80
60
40
20
0
1840
1880
1920
1960
2000
Year
Wood
Coal
Petroleum and natural gas
What energy sources do you think will be important in the future?
(b) COAL FORMATION
1. Dead plant material accumulates in marshy or boggy habitats.
2. If oxygen in water is scarce, the organic matter decays only partially, forming peat.
Peat
3. If the peat deposits are later covered by sediments and compressed, the resulting pressure and heat change them into coal.
Pressure
Pressure
Sediments
Coal

FIGURE 29.4 Humans Have Relied on Plant-Based Fuels
(a) The percentage of energy needs in the United States filled by various sources since 1840. Wood was replaced by coal, which in turn was replaced by oil and natural gas. Wood is still the primary cooking and heating fuel in many areas of the world. **(b)** Coal is formed by terrestrial deposits of partially decayed organisms.

The transition from wood burning to the use of fossil fuels was important for another reason. When people burn wood, the carbon that is released to the atmosphere is then removed from the atmosphere as forests regrow over the course of a few decades or centuries. As a result, economies that were based on wood burning added little to the concentration of CO_2 in the atmosphere. But carbon that is released to the atmosphere via burning of coal or petroleum cannot be removed from the atmosphere by the formation of coal and petroleum, because the formation processes are so slow. This logic suggests that the current increase in atmospheric CO_2 concentration (and subsequent global warming) traces its origin to the wood-to-coal transition documented in Figure 29.4a.

Will there be yet another generation of plant-based fuels? Although researchers are working on methods to convert corn to liquid fuels such as ethanol or methanol, the answer is probably no. The currently dominant combustible material, petroleum, is derived from the partially decayed remains of marine organisms—primarily photosynthetic bacteria and protists. Oil is also the source material for nylon and polyester fibers, which have supplanted cotton and other types of plant fibers used in traditional clothing and implements. The energy sources that are predicted to be important in the post-petroleum economy are solar power and wind power, combustion of hydrogen, and perhaps nuclear fusion. None of these sources are based on the carbon compounds found in plants. It is doubtful whether land plants could reduce enough carbon dioxide to meet current world energy demands.

Today, the primary commercial interest in woody plants is for building materials and the fibers used in papermaking. Wood excels in both capacities. Relative to its density, wood is a stiffer and stronger building material than concrete, cast iron, aluminum alloys, or steel. The cellulose fibers refined from trees or bamboo and used in paper manufacturing are stronger under tension (pulling) than nylon, silk, chitin, collagen, tendon, or bone—even though cellulose is 25 percent less dense.

Current research in forestry focuses on maintaining the productivity and diversity of forests that are managed for wood and pulp production. The vast majority of old-growth forests in the Northern Hemisphere have now been cut, and mature forests in the Southern Hemisphere are disappearing rapidly due to logging and burning. The current challenge to foresters is twofold: (1) reducing demand for wood products through recycling and more efficient use of existing timber, and (2) sustaining the productivity of forests after they have been logged repeatedly.

Bioprospecting

Because land plants play such a central role in human health and welfare, biologists have started a research program aimed at discovering new uses for plants. The effort to find naturally occurring compounds that can be used as drugs, fragrances, insecticides, herbicides, or fungicides has been called **bioprospecting** or **chemical prospecting.**

Plants have traditionally been a rich source of useful molecules. For example, the prominent anticancer drugs vincristine and vinblastine were first isolated from a plant native to Madagascar (now found throughout the tropics) called the rosy periwinkle. The anticancer drug taxol was discovered in extracts from the Pacific yew tree, which lives only in northwest North America. **Table 29.1** lists some of the more familiar drugs derived from land plants. Overall, it has been estimated that about 25 percent of the prescriptions written in the United States each year include at least one molecule derived from plants.

In most cases, plants synthesize these compounds in order to harm and thus repel leaf-eating beetles, deer, or other types of herbivores. Stated another way, many of the plant molecules that we use for medicine or other purposes are actually compounds that defend plants against their enemies. For example, experiments have confirmed that the presence of morphine, cocaine, nicotine, caffeine, and other compounds that are psychoactive in humans—meaning that they affect the human mind—is an effective deterrent to insect or mammalian consumers. The natural function of tetrahydrocannabinol (THC), the active ingredient in marijuana, is undoubtedly in defense. Later chapters explore the nature and role of plant defense compounds in more detail.

TABLE 29.1 Some Drugs Derived from Land Plants

Compound	Source	Use
Atropine	Belladonna plant	Dilating pupils during eye exams
Codeine	Opium poppy	Pain relief, cough suppressant
Digitalin	Foxglove	Heart medication
Ipecac	Ipecac	Treating amoebic dysentery, poison control
Menthol	Peppermint tree	Cough suppressant, relief of stuffy nose
Morphine	Opium poppy	Pain relief
Papain	Papaya	Reduce inflammation, treat wounds
Quinine	Quinine tree	Malaria prevention
Quinidine	Quinine tree	Heart medication
Salicin	Aspen, willow trees	Pain relief (aspirin)
Steroids	Wild yams	Precursor compounds for manufacture of birth control pills and cortisone (to treat inflammation)
Taxol	Pacific yew	Ovarian cancer
Tubocurarine	Curare vine	Muscle relaxant used in surgery
Vinblastine, vincristine	Rosy periwinkle	Leukemia (cancer of blood)

Researchers are pursuing two paths in the effort to discover new plant compounds that might have medicinal or commercial uses. One path begins in the lab; the other begins in the field:

1. Biologists have been able to stimulate plants that are growing *hydroponically*—that is, in liquid culture instead of in soil—to churn out large quantities of defense chemicals. The researchers expose root cells to a fragment of a bacterial cell wall or a toxin from a fungus, and then collect the defense compounds that are produced in response to the simulated attack. Using this technique, investigators have been able to collect 5000 samples of materials expelled from the roots of 700 plant species. Some of the extracts have been found to kill various types of cancer cells growing in culture. It is still unknown, however, whether this novel strategy for producing and purifying defense compounds will result in the discovery of important new drugs.
2. **Ethnobotanists** are biologists who specialize in the study of how humans use plants (**Figure 29.5**). Teams of ethnobotanists have recently been working at a furious pace

FIGURE 29.5 Ethnobotanists Research New Uses for Plants Traditional healers make most of the medicines they administer from local plants. University-trained ethnobotanists are collaborating with traditional healers to research plant products for possible use in Western medicine.

in rain forests around the world, collecting plants from areas that are threatened with destruction and interviewing traditional healers who make their own medicines from local plants. This work is a race against time, for in many cases both the forests and the indigenous people are being overrun by settlers. If present trends continue, it is likely that within a generation, knowledge about traditional uses of rain forest plants—and many of the plant species themselves—will be lost as ecosystems and cultures change.

The use of plant-derived compounds in medicines underscores a key point: Understanding and preserving plant species diversity can be a life-or-death matter. For humans and other animals, plants are the most important of all organisms. Given the importance of plants to the planet in general and humans in particular, it is not surprising that understanding plant diversity is an important component of biological science. Let's consider how biologists go about analyzing the diversity of green plants; then we'll explore the evolutionary innovations that made green plant diversification possible.

29.2 How Do Biologists Study Green Plants?

To understand the genetics and developmental biology of plants, researchers use thale cress, a weedy mustard, as a model organism (**Box 29.1**). To understand how green plants originated and diversified, biologists use three tools: They (1) compare the fundamental morphological features of various green algae and green plants; (2) analyze the fossil record of the lineage; and (3) assess similarities and differences in molecular traits such as the DNA sequences from selected genes. These data are then used to estimate phylogenies. The three approaches are complementary. As a result, it could be argued that biologists have a better understanding of how green plants diversified than they do for any other group on the tree of life. Let's consider each of these research strategies.

Analyzing Morphological Traits

By comparing and contrasting the morphological traits of green plants, biologists have identified several distinct groups of green algae and 12 major lineages, or **phyla** (singular: **phylum**), of land plants.

BOX 29.1 A Model Organism: *Arabidopsis thaliana*

In the early days of biology, the best-studied plants were agricultural varieties such as maize (corn), rice, and garden peas. When biologists began to unravel the mechanisms responsible for oxygenic photosynthesis in the early to mid-1900s, they relied on green algae—often the unicellular species *Chlamydomonas reinhardii*—as an experimental subject. Although crop plants and green algae continue to be the subject of considerable research, a new model organism emerged in the 1980s and now serves as the preeminent experimental subject in plant biology. That organism is *Arabidopsis thaliana*, commonly known as thale cress or wall cress (**Figure 29.6**).

Arabidopsis is a member of the mustard family, or Brassicaceae, so it is closely related to radishes and broccoli. In nature it is a **weed**—meaning a species that is adapted to thrive in habitats where soils have been disturbed. In Europe, for example, *Arabidopsis* is common along roadsides and the edges of agricultural fields. It is also an **annual** plant, which means that individuals do not live from year to year but overwinter as seeds. (Plants that survive from one year to the next are said to be **perennial**.) Indeed, one of the most attractive aspects of working with *A. thaliana* is that individuals can grow from a seed into a mature, seed-producing plant in just four to six weeks. Several other attributes make it an effective subject for study: It has just five chromosomes, has a relatively small genome with limited numbers of repetitive sequences, can self-fertilize as well as undergo cross-fertilization, can be grown in a relatively small amount of space and with a minimum of care in the greenhouse, and produces up to 10,000 seeds per individual per generation.

Arabidopsis has been instrumental in a variety of studies in plant molecular genetics and development, and is increasingly popular in ecological and evolutionary studies. In addition, the entire genome of the species has now been sequenced, and studies have benefited from the development of an international "*Arabidopsis* community"—a combination of informal and formal associations of investigators who work on *Arabidopsis* and use regular meetings, e-mail, and the Internet to share data, techniques, and seed stocks.

FIGURE 29.6 *Arabidopsis thaliana* Is the Most Important Model Organism in Plant Biology
Like fruit flies and *Escherichia coli*, *A. thaliana* is small and short lived enough to grow easily in the laboratory, yet it is complex enough to be an interesting organism for study.

(a) Green algae are strictly aquatic.

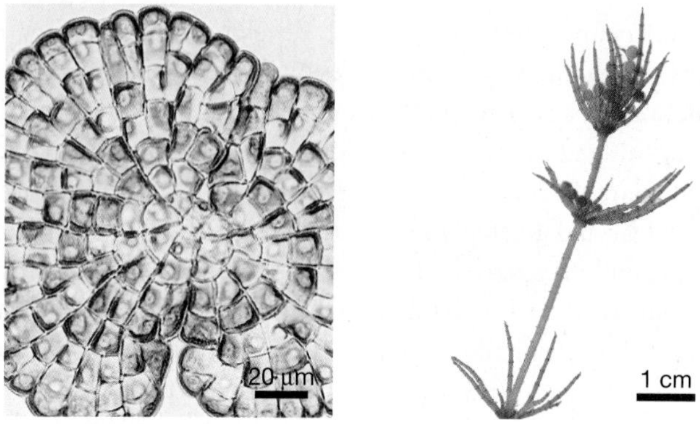

Coleochaetes Stoneworts

(b) Nonvascular plants do not have vascular tissue to conduct water and provide support.

Liverworts

Hornworts Mosses

FIGURE 29.7 Morphological Diversity in Green Plants

The green algae have long been hypothesized to be closely related to land plants on the basis of several key morphological traits. Both green algae and land plants have chloroplasts that contain the photosynthetic pigments chlorophyll *a* and *b* and the accessory pigment β-carotene, as well as similar stacks of the internal, membrane-bound sacs called thylakoids (see Chapter 10). The cell walls of green algae and land plants are almost identical in composition, both groups synthesize starch as a storage product in their chloroplasts, and their sperm and peroxisomes are similar in structure and composition. (Recall from Chapter 7 that peroxisomes are organelles in which specialized oxidation reactions take place.)

The green algae include species that are unicellular, colonial, or multicellular and that live in marine or freshwater habitats. Of all the green algal groups, the two that are most similar to land plants are the Coleochaetales and Charales (**Figure 29.7a**). Because the species that make up these groups are multicellular and live in ponds and other types of freshwater environments, biologists hypothesize that the land plants evolved from multicellular green algae that lived in freshwater habitats.

Although some land plants live in ponds or lakes or rivers, the vast majority of species in this lineage live on land. Based on morphology, the 12 most important phyla are traditionally clustered into three broad categories:

1. **Nonvascular plants**, which include the groups called liverworts, hornworts, and mosses. Liverworts and hornworts lack **vascular tissue**—meaning specialized groups of cells that conduct water or dissolved nutrients from one part of the plant body to another. Some moss species have specialized tissues that conduct water and food, but the cells that make up these tissues do not have the reinforced cell walls that define true vascular tissue. Nonvascular plants are extremely abundant in certain habitats but are usually small and grow close to the ground (**Figure 29.7b**).
2. **Seedless vascular plants**, which have well-developed vascular tissue but do not make seeds. A **seed** consists of an embryo and a store of nutritive tissue, surrounded by a tough protective layer. Horsetails, ferns, lycophytes, and whisk ferns are the major groups of seedless vascular plants (**Figure 29.7c**). Although most of the living representatives of these lineages are relatively small, some fossil and some other living representatives of the seedless vascular plants are tree sized.
3. **Seed plants**, which have vascular tissue and make seeds. Although the members of this group vary a great deal in size and shape, seed plants encompass some of the world's largest organisms. There are five major lineages in the group: cycads, ginkgoes, conifers, gnetophytes, and angiosperms (**Figure 29.7d**). The gnetophytes, cycads, ginkgoes, and conifers are collectively known as **gymnosperms** ("naked-seeds"), because their seeds do not develop in an enclosed structure. In the **flowering plants**, or **angiosperms** ("encased-seeds"), seeds develop inside a protective structure called a **carpel**. Today, angiosperms are far and away the most important lineage of land plants in terms of species diversity. Almost 90 percent of the land plant species alive today, and virtually all of the domesticated forms, are angiosperms.

(c) Seedless vascular plants have vascular tissue but do not make seeds.

(d) Seed plants have vascular tissue and make seeds.

FIGURE 29.7 *(continued)*

This quick overview of morphological diversity in land plants raises a number of questions:

- Does other evidence support the hypothesis that green algae are ancestral to land plants, and that Coleochaetales and Charales are the closet living relatives to land plants?
- Did the green algae evolve first, followed by nonvascular plants, then seedless vascular plants, and finally seed plants?
- Are both green algae and land plants monophyletic—meaning each is a distinct lineage tracing back to a single common ancestor?
- Are the nonvascular plants, seedless vascular plants, and seed plants monophyletic?

Biologists are answering these questions by analyzing the fossil record of the green plants, analyzing the morphology and DNA sequences of living species, and then using these characteristics to infer the group's phylogeny.

Using the Fossil Record

The first green plants that appear in the fossil record are green algae in rocks that formed 700–725 million years ago. The first land plant fossils are much younger—they are found in rocks that are about 475 million years old. Because green algae appear long before land plants, the fossil record supports the hypothesis that land plants are derived from green algae.

The appearance and early diversification of green algae about 700 million years ago is also significant because, at roughly the same time, the oceans and atmosphere were starting to become oxygen-rich, as never before in Earth's history. Based on this time correlation, it is reasonable to hypothesize that the evolution of green algae contributed to the rise of oxygen levels on Earth. The origin of the oxygen atmosphere occurred not long before the appearance of animals in the fossil record, and may have played a role in the origin and early diversification of animals.

The fossil record of the land plants themselves is massive. In an attempt to organize and synthesize the database, **Figure 29.8** breaks it into five time intervals. Each of these time periods encompasses a major event in the diversification of land plants.

The oldest interval begins 475 million years ago, spans 65 million years, and documents the origin of the group. Most of the fossils dating from this period are microscopic. They consist of the reproductive cells called **spores** and sheets of a waxy coating called **cuticle.** Several observations support the hypothesis that these fossils came from green plants that were growing on land. First, cuticle is a watertight barrier that coats today's land plants and helps them resist drying. Second, the fossilized spores are surrounded by a sheetlike coating. Under the electron microscope, the coating material appears almost identical in structure to a watertight material called **sporopollenin,** which encases spores and pollen from modern land plants and helps them resist drying. Third, fossilized spores that are 475 million years old have recently been found in association with spore-producing structures called **sporangia** (singular: **sporangium**). The fossilized sporangia are similar in appearance to the sporangia observed in some of today's liverworts.

The second major interval in the fossil record of land plants is sometimes called the "Silurian-Devonian explosion." In rocks dated between 410 and 360 million years ago, biologists find large fossils from most of the major plant lineages. Virtually all of the adaptations that allow plants to occupy dry, terrestrial habitats are present, including water-conducting cells and tissues and roots.

The third interval in the fossil history of plants is aptly named the Carboniferous period. In sediments dated from about 350 to 290 million years ago, biologists find extensive deposits of coal. Coal is a carbon-rich rock packed with fossil spores, branches, leaves, and tree trunks. These fossils are frequently derived from lycophytes. Although the only living lycophytes are small, forest-dwelling plants, during the Carboniferous this group was species rich and included a wide array of woody, tree-sized forms. Because coal formation is thought to start only in the presence of water, the Carboniferous fossils indicate the presence of extensive forested swamps.

The fourth interval in land plant history is characterized by gymnosperms. Recall that gymnosperms include the gnetophytes, cycads, ginkgoes, and conifers (such as pines and spruces). Because gymnosperms grow readily in dry habitats, biologists infer that both wet and dry environments on the continents became blanketed with green plants for the first time during this interval, which lasted from 275 to 145 million years ago.

The fifth interval in the history of land plants is still under way. This is the age of flowering plants, the angiosperms. The first flowering plants in the fossil record appear about 145 million years ago. The plants that produced the first flowers are the ancestors of today's grasses, orchids, daisies, oaks, maples, and roses.

According to the fossil record, then, the green algae appear first, followed by the nonvascular plants, seedless vascular plants, and seed plants. Organisms that appear late in the fossil record are much less dependent on moist habitats than are groups that appear earlier. For example, the sperm cells of mosses and ferns swim to accomplish fertilization, while gymnosperms and angiosperms produce pollen grains that are transported via wind or insects and that then produce sperm. These observations support the hypotheses that green plants evolved from green algae and that in terms of habitat use, the evolution of green plants occurred in a wet-to-dry trend.

To test the validity of these observations, biologists analyze data sets that are independent of the fossil record. Foremost among these are the DNA sequences and morphological data that are used to infer phylogenetic trees. Does the phylogeny of land plants confirm or contradict the patterns in the fossil record?

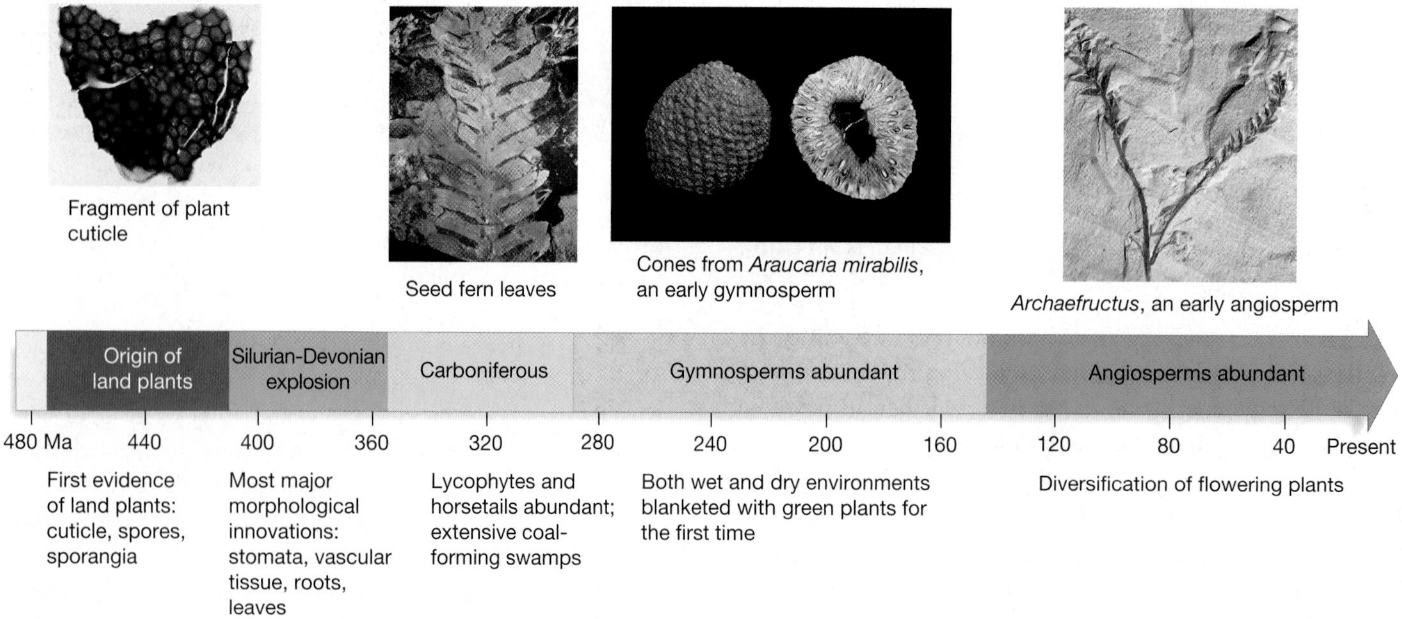
Fragment of plant cuticle
Seed fern leaves
Cones from *Araucaria mirabilis*, an early gymnosperm
Archaefructus, an early angiosperm
Origin of land plants
Silurian-Devonian explosion
Carboniferous
Gymnosperms abundant
Angiosperms abundant
480 Ma
440
400
360
320
280
240
200
160
120
80
40
Present
First evidence of land plants: cuticle, spores, sporangia
Most major morphological innovations: stomata, vascular tissue, roots, leaves
Lycophytes and horsetails abundant; extensive coal-forming swamps
Both wet and dry environments blanketed with green plants for the first time
Diversification of flowering plants

FIGURE 29.8 The Fossil Record of Land Plants Can Be Broken into Five Major Intervals

EXERCISE Add the following to the timeline: (1) first terrestrial vertebrate animals (370 Ma); (2) first mammals (195 Ma).

Evaluating Molecular Phylogenies

Understanding the phylogeny of green plants is a monumental challenge. Fortunately, the challenge is being met by the biologists who are using both morphological traits and DNA sequence data to infer the evolutionary relationships among green plants. In some cases various laboratories are cooperating in analyzing lineages from various parts of the green plant tree, so that in combination a comprehensive picture of green plant evolution can emerge.

The phylogenetic tree in **Figure 29.9** is a recent version of results coming out of laboratories focused on estimating the phylogeny of land plants. There are several important points to note about the relationships implied by this tree:

- The oldest branches on the tree, near the root, lead to various groups of green algae. This result supports the hypothesis that land plants evolved from green algae.
- The green algal group called Charales is the **sister group** to land plants—meaning that Charales are their closest living relative. Because living members of the Charales are multicellular and dwell in freshwater, the data support the hypothesis that land plants evolved from a multicellular ancestor that lived in ponds or lakes or other freshwater habitats.
- The group called "green algae" is paraphyletic. Stated another way, the green algae include some but not all of the descendants of a single common ancestor. The green plants, in contrast, are monophyletic. A single common ancestor gave rise to all of the green algae and land plants.
- The land plants are monophyletic. This result supports the hypothesis that the transition from freshwater environments to land occurred just once. According to the phylogenetic tree, the population that made the transition then diversified into the array of land plants observed today.
- The nonvascular plants—that is, liverworts, hornworts, and mosses—are the earliest-branching, or most basal, groups among land plants. They are not monophyletic, however. Instead, the nonvascular plants represent what biologists call a **grade**—a sequence of lineages that are not monophyletic.
- Mosses have simple water-conducting tissues, and liverworts and hornworts have none at all. This finding suggests that the earliest land plants lacked water-conducting cells and vascular tissue, and that these traits evolved later. The phylogenetic tree supports the hypothesis that water-conducting cells and tissues evolved in a stepwise fashion, with simpler structures preceding the evolution of more complex structures.
- The lycophytes are the sister group to all seedless vascular plants. The whisk ferns, horsetails, and ferns form a monophyletic group, but as a whole the seedless vascular plants form a grade. In contrast, the vascular plants are monophyletic—meaning that vascular tissue evolved once during the diversification of land plants.
- Because they lack roots and leaves, the whisk ferns were traditionally thought to be a basal group in the land plant radiation. Molecular phylogenies challenge this hypothesis and support an alternative hypothesis: The morphological simplicity of whisk ferns is a derived trait—meaning that complex structures have been lost in this lineage.
- The seed plants and the angiosperms are monophyletic, meaning that seeds and flowers evolved once. Although the data available at present suggest that gymnosperms form a monophyletic group, this result has yet to be confirmed.
- The fossil record and the phylogenetic tree agree on the order in which groups appeared. Land plant evolution began with nonvascular plants, proceeded to seedless vascular plants, and continued with the evolution of seed plants.

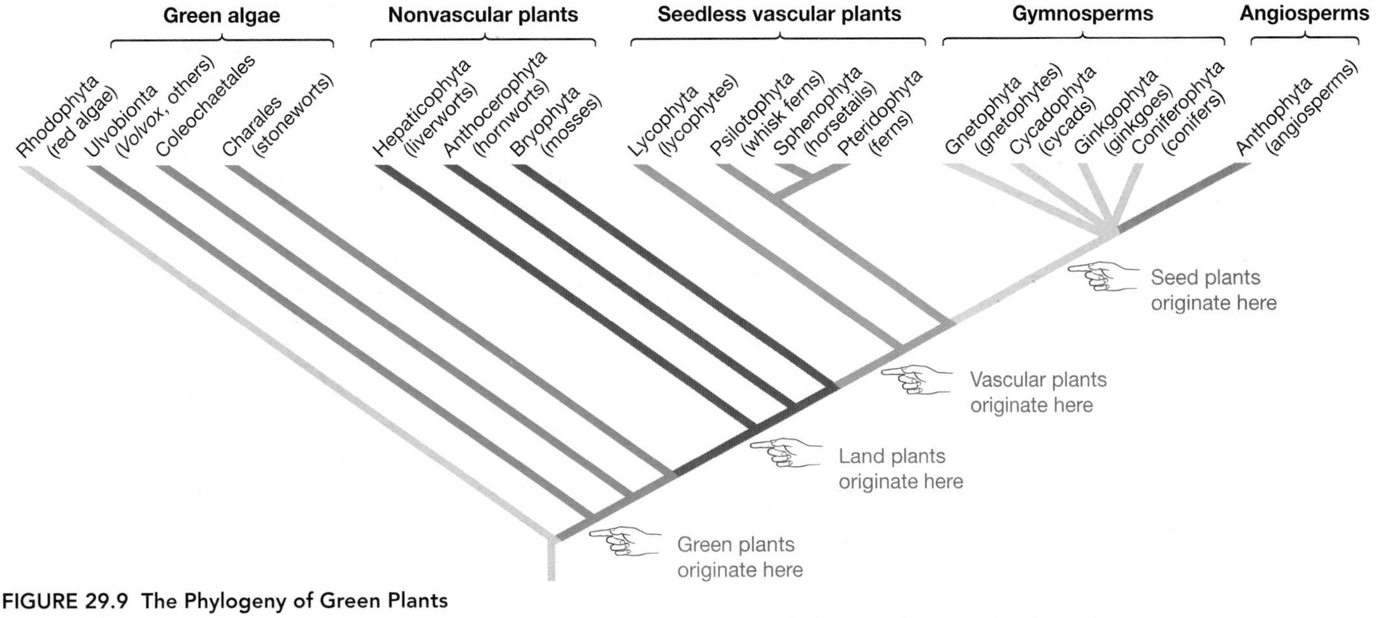

FIGURE 29.9 The Phylogeny of Green Plants

EXERCISE Circle and label the monophyletic groups called (1) land plants, (2) seed plants, and (3) vascular plants. On the figure, identify why the groups called green algae, nonvascular plants, and seedless vascular plants are paraphyletic.

The tree in Figure 29.9 will undoubtedly change and improve as additional data accumulate. Perhaps the most urgent need now is to clarify the relationships among the gymnosperms—the gnetophytes, cycads, ginkgoes, and conifers. Species from these lineages dominated terrestrial environments for over 200 million years, and one of the groups is the sister lineage to the angiosperms—today's most prominent land plant group. If the relationships among gymnosperms can be resolved, the results should improve our understanding of how the flower evolved.

✓CHECK YOUR UNDERSTANDING

Biologists use the fossil record and phylogenetic analyses to study how green plants diversified. Both types of data support the hypotheses that green plants are monophyletic and that land plants evolved from multicellular green algae that inhabited freshwater. The fossil record and molecular analyses agree that the nonvascular plants evolved first, followed by the seedless vascular plants and the seed plants. You should be able to (1) describe three ways that the presence of land plants affects abiotic components of terrestrial habitats; and (2) draw a phylogenetic tree that shows the relationships between the green algae and the three major morphological groups of land plants, and that identifies the monophyletic groups called green plants, land plants, and vascular plants.

29.3 What Themes Occur in the Diversification of Green Plants?

Land plants have evolved from algae that grew on the muddy shores of ponds 475 million years ago to organisms that enrich the soil, produce much of the oxygen you breathe and most of the food you eat, and serve as symbols of health, love, and beauty. How did this happen?

Answering this question begins with recognizing the most striking trend in the phylogeny and fossil record of green plants: The most ancient groups in the lineage are dependent on wet habitats, while more recently evolved groups are tolerant of dry—or even desert—conditions. The story of land plants is the story of adaptations that allowed photosynthetic organisms to move from aquatic to terrestrial environments. Let's first consider adaptations that allowed plants to grow in dry conditions without drying out and dying, and then analyze the evolution of traits that allowed plants to reproduce efficiently when not surrounded by water. This section closes with a brief look at the radiation of flowering plants, which are the most important plants in today's terrestrial environments.

The Transition to Land, I: How Did Plants Adapt to Dry Conditions?

For aquatic life-forms such as the green algae, terrestrial environments are deadly. Compared with a habitat in which the entire organism is bathed in fluid, in terrestrial environments only a portion, if any, of the plant's tissues are bathed in fluid. Tissues that are exposed to air tend to dry out and die.

Once green plants made the transition to survive out of water, though, growth on land offered a bonanza of resources. Take light, for example. The water in ponds, lakes, and oceans absorbs and reflects light. As a result, the amount of light available to drive photosynthesis is drastically reduced even a few feet below the water surface. In addition, the most important nutrient required by photosynthetic organisms, carbon dioxide, is much more readily available in air than it is in water. Not only is it more abundant in the atmosphere, but it diffuses more readily there than it does in water.

Natural selection favored early land plants with adaptations that solved the water problem. These adaptations arose in two steps: (1) prevention of water loss from cells, which kept the cells from drying out and dying; and (2) transportation of water from tissues with direct access to water to tissues without access. Let's examine both of these steps in turn.

Preventing Water Loss: Cuticle and Stomata Section 29.2 pointed out that sheets of the waxy substance called cuticle are present early in the fossil record of land plants, along with encased spores. This observation is significant because the presence of cuticle in fossils is a diagnostic indicator of land plants. Cuticle is a waxy, watertight sealant that covers the aboveground parts of plants and gives them the ability to survive in dry environments (**Figure 29.10a**). If biologists had to point to one innovation that made the transition to land possible, it would be the production of cuticle.

Covering surfaces with wax creates a problem, however, in terms of the exchange of gases across those surfaces. Plants need to take in carbon dioxide (CO_2) from the atmosphere in order to perform photosynthesis. But cuticle is almost as impervious to carbon dioxide as it is to water. Most modern plants solve this problem with a structure called a **stoma** (plural: **stomata**), consisting of an opening surrounded by specialized **guard cells** (**Figure 29.10b**). The opening, called a **pore**, opens or closes as the guard cells change shape. When guard cells become soft, they close the stomata. Pores are closed in this way to limit water loss from the plant. When guard cells become taut, in contrast, they open the pore, allowing CO_2 to diffuse into the interior of leaves and stems where cells are actively photosynthesizing, and allowing excess O_2 to diffuse out. (The mechanism behind guard-cell movement is explored in Chapter 38.)

Stomata are present in living members of two of the earliest-branching groups on the land plant phylogeny: the hornworts and mosses. Some species in the most basal, or earliest-branching, lineage—the liverworts—have pores but no guard cells. These data suggest that the earliest land plants evolved pores that allowed gas exchange to occur at breaks in the cuticle-covered surface. Later, the evolution of guard cells gave land plants the ability to regulate gas exchange—and control water loss—by opening and closing their pores.

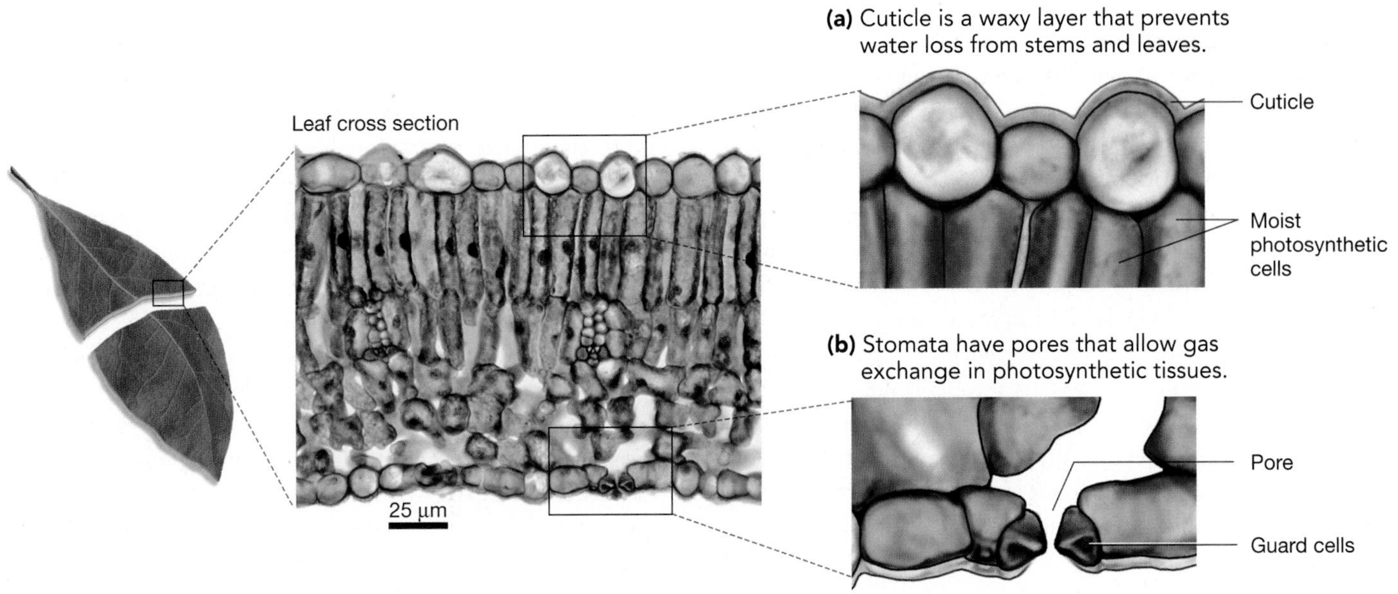
(a) Cuticle is a waxy layer that prevents water loss from stems and leaves.
Cuticle
Moist photosynthetic cells
Leaf cross section
25 μm
(b) Stomata have pores that allow gas exchange in photosynthetic tissues.
Pore
Guard cells

FIGURE 29.10 Cuticle and Stomata Are the Most Fundamental Plant Adaptations to Life on Land
Leaf cells that have been stained blue to make their structure more visible. **(a)** The interior of plant leaves and stems is extremely moist; cuticle prevents water from evaporating away. **(b)** Stomata have pores that open to allow CO_2 to diffuse into the interior of leaves and stems where cells are actively photosynthesizing, and to allow excess O_2 to diffuse out. QUESTION Why was the evolution of guard cells important?

Transporting Water: Vascular Tissue and Upright Growth

Once cuticle and stomata had evolved, plants could keep photosynthesizing while exposed to air. Cuticle and stomata allowed plants to grow in the saturated soils of lake or pond edges. The next challenge? Defying gravity.

Multicellular green algae can grow erect because they float. They float because the density of their cells is similar to water's density. But outside of water, the body of a multicellular green alga collapses. The water that fills its cells is 1000 times denser than air. Although the cell walls of green algae are strengthened by the presence of cellulose, their bodies lack the structural support to withstand the force of gravity and to keep an individual erect in air.

Based on these observations, biologists hypothesize that the first land plants had a low, sprawling growth habit. In addition to lacking rigidity, the early land plants would have had to obtain water through pores or through a few cells that lacked cuticle—meaning they would have had to grow in a way that kept many or most of their tissues in direct contact with moist soil.

The sprawling-growth hypothesis is supported by the observation that the most basal groups of land plants living today (liverworts, hornworts, and mosses) are all low-growing forms. If this hypothesis is correct, then competition for space and light would have become intense soon after the first plants began growing on land. To escape competition, plants would have to grow upright. Plants with adaptations that allowed some tissues to remain in contact with wet soil while other tissues grew erect would have much better access to sunlight compared with individuals that were incapable of growing erect.

For plants to adopt erect growth habits on land, though, two problems had to be overcome. The first is transporting water from tissues that are in contact with wet soil to tissues that are in contact with dry air, against the force of gravity. The second is becoming rigid enough to avoid falling over in response to gravity and wind. As it turns out, vascular tissue solved both problems.

Paul Kenrick and Peter Crane explored the origin of water-conducting cells and erect growth in plants by examining the extraordinary fossils found in a rock formation in Scotland called the Rhynie Chert. These rocks formed about 400 million years ago and contain some of the first large plant specimens in the fossil record (as opposed to the microscopic spores and cuticle found in older rocks). The Rhynie Chert also contains numerous plants that fossilized in an upright position. This indicates that many or most of the Rhynie plants grew erect. How did they stay vertical?

By examining fossils with the electron microscope, Kenrick and Crane established that species from the Rhynie Chert contained elongated cells that were organized into tissues along the length of the plant. Based on these data, the biologists hypothesized that the elongated cells were part of water-conducting tissue and that water could move from the base of the plants upward to erect portions through these specialized water-conducting cells. In addition, some of the water-conducting cells present in these early plants had cell walls with thickened rings containing a molecule called lignin. **Lignin** is a polymer built from six-carbon rings, and it is extraordinarily strong for its weight. It is particularly effective in resisting compressing forces such as gravity. These observations inspired the following hypothesis: The evolution of lignin rings gave stem tissues the strength to remain erect in the face of wind and gravity. The presence of lignin in the cell walls of conducting cells is the defining feature of vascular tissue. In this way, the evolution of vascular tissue allowed early plants both to support erect stems and to transport water to them.

Once simple water-conducting tissues evolved, evolution by natural selection elaborated them. By about 380 million years ago, the fossil record contains the advanced water-conducting cells called tracheids. As **Figure 29.11a** shows, **tracheids** (from the Latin *trachea*, or "windpipe") are elongated cells that are closely packed and organized into a specialized water-conducting tissue. The presence of tracheids marked the evolution of the first vascular plants. Because they die after maturing, tracheids do not contain cytoplasm. These features increased the efficiency of water transport from the base of the plant through vascular tissue to aboveground tissues.

Tracheids that appear later in the fossil record have thickened secondary cell walls with extensive deposits of lignin, in addition to a primary cell wall made of cellulose (see Chapter 8). The evolution of a secondary cell wall gave tracheids the ability to provide better structural support in addition to increased water-conducting capacity.

To summarize, erect growth was possible due to the evolution of water-conducting cells that contained lignin and thus strengthened stems. This innovation was followed by the evolution of the specialized cells called tracheids and the first vascular tissue. Later, tracheids with secondary cell walls, which added strength, appeared.

Finally, in fossils dated to 250–270 million years ago, biologists have documented the most advanced type of water-conducting cells observed in plants, known as vessels. Vessels are shorter and wider than tracheids. The upper and lower ends of vessels have gaps that lack any sort of cell-wall material—making the transport of water extremely efficient (**Figure 29.11b**). Taken together, the data indicate that vascular tissue evolved in a series of gradual steps that provided an increasing level of structural support (**Figure 29.11c**).

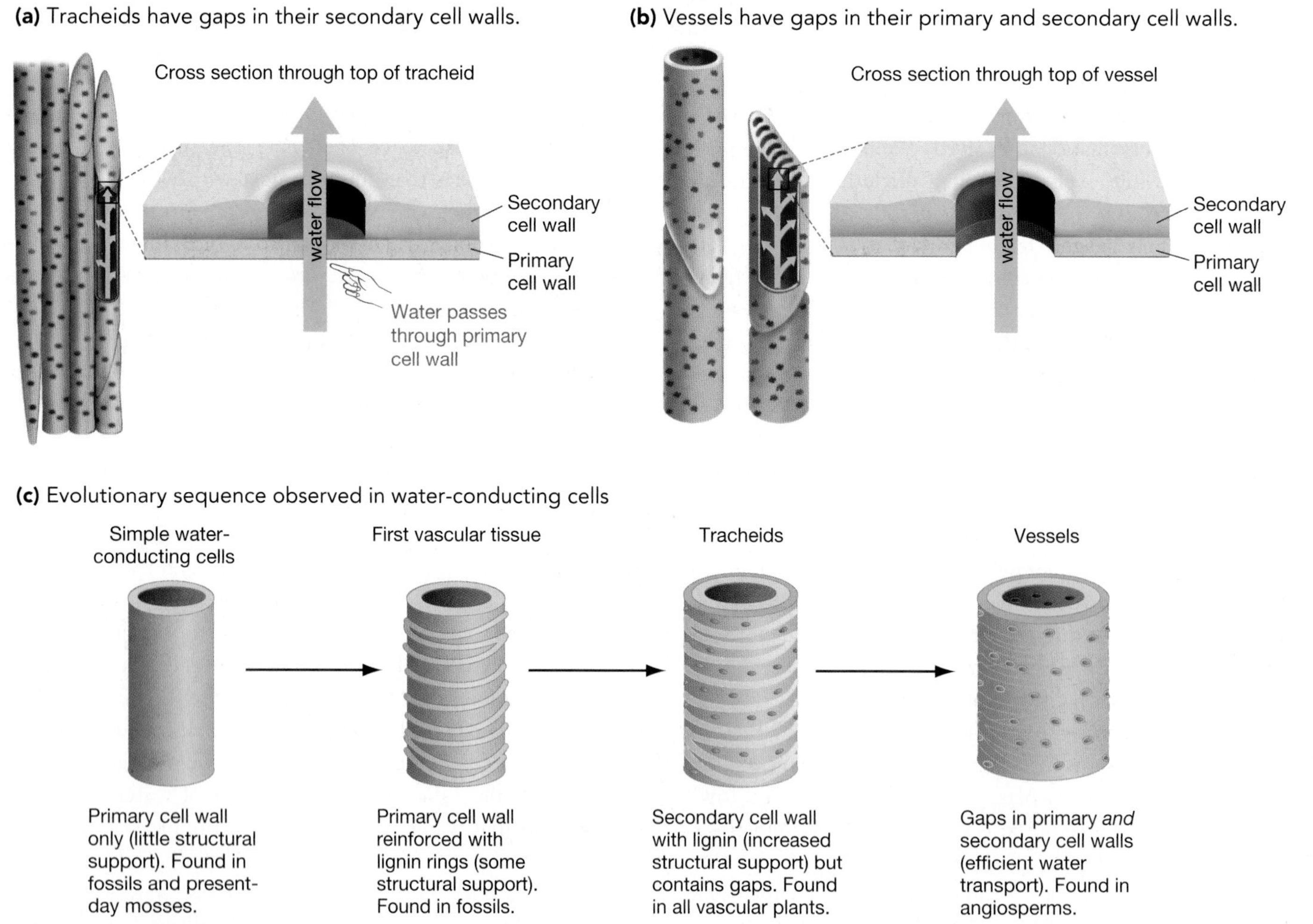

(a) Tracheids have gaps in their secondary cell walls.
Cross section through top of tracheid
water flow
Secondary cell wall
Primary cell wall
Water passes through primary cell wall
(b) Vessels have gaps in their primary and secondary cell walls.
Cross section through top of vessel
water flow
Secondary cell wall
Primary cell wall
(c) Evolutionary sequence observed in water-conducting cells
Simple water-conducting cells
First vascular tissue
Tracheids
Vessels
Primary cell wall only (little structural support). Found in fossils and present-day mosses.
Primary cell wall reinforced with lignin rings (some structural support). Found in fossils.
Secondary cell wall with lignin (increased structural support) but contains gaps. Found in all vascular plants.
Gaps in primary *and* secondary cell walls (efficient water transport). Found in angiosperms.

FIGURE 29.11 Vascular Tissue Is Composed of Tracheids or Vessels
(a) Tracheids are water-conducting cells that are dead and empty at maturity. They have secondary cell walls reinforced with lignin. **(b)** In vessels, there are gaps that lack any cell-wall material. **(c)** According to the fossil record and the phylogeny of green plants, the extent of the secondary cell wall increased as vascular tissue evolved. QUESTION Biologists claim that vessels are more efficient than tracheids at transporting water because vessels are shorter and wider than tracheids. Why does this claim make sense?

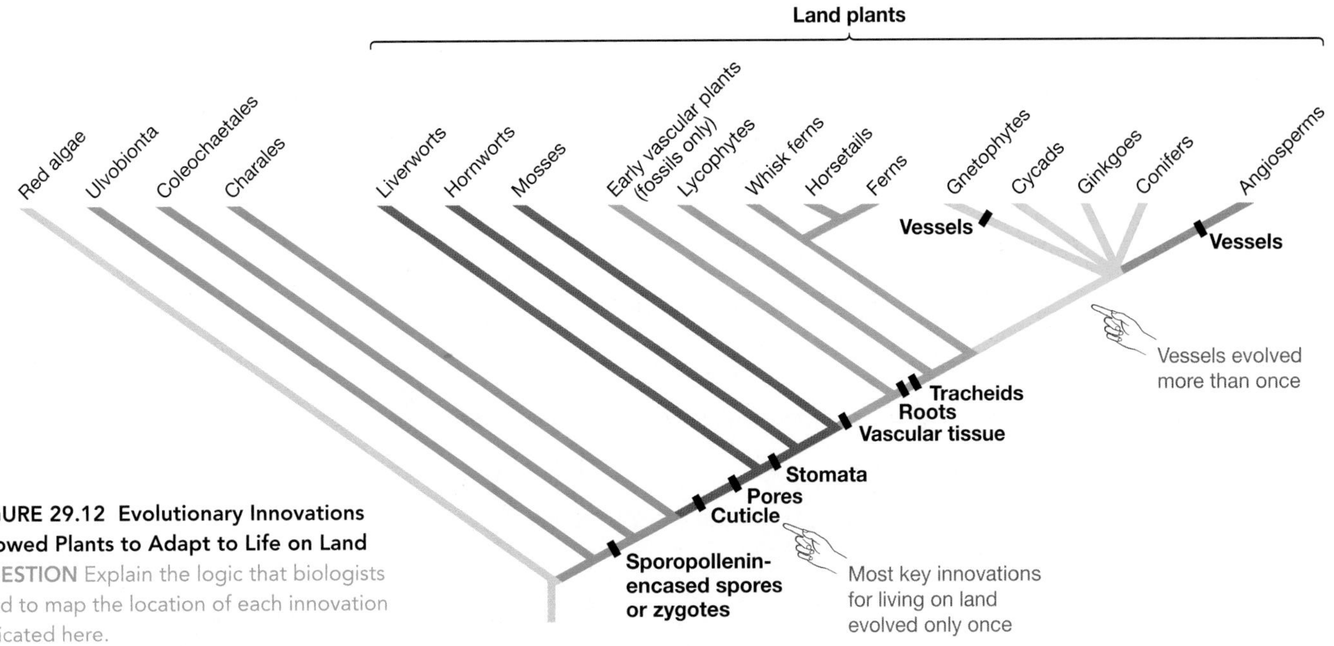

FIGURE 29.12 Evolutionary Innovations Allowed Plants to Adapt to Life on Land
QUESTION Explain the logic that biologists used to map the location of each innovation indicated here.

Figure 29.12 summarizes how land plants adapted to dry conditions by mapping where major innovations occurred as the group diversified. The bars across the branches of the phylogenetic tree show when each of the key innovations occurred, based on the fossil record and their occurrence in species living today. Fundamentally important adaptations to dry conditions, such as the cuticle, pores, stomata, and vascular tissue, evolved just once. In addition, current analyses suggest that vessels may have evolved independently in gnetophytes and angiosperms. The evolutionary history of vessels is controversial, however, and research continues. Leaves also evolved more than once. Lycopods have small leaves called **microphylls**, while ferns, horsetails, and all vascular plants have typically larger leaves known as **megaphylls**, which develop in a different way and have a different structure than microphylls have.

The evolution of the cuticle, stomata, and vascular tissue made it possible for plants to avoid drying out and to grow upright, while moving water from the base of the plant to its apex. But for plants to become free of any dependence on wet habitats, one final challenge remained. The nonvascular plants and the most ancient vascular plant lineages, such as ferns, have male gametes that must swim to the egg to perform fertilization. The land plants made their final break with their aquatic origins when they evolved ways to move their gametes without the aid of water.

The Transition to Land, II: How Do Plants Reproduce in Dry Conditions?

The key to understanding how land plants lost their reliance on swimming gametes is to analyze their life cycles. All land plants undergo the phenomenon known as **alternation of generations,** introduced in Chapter 28. Recall that when alternation of generations occurs, individuals have a multicellular haploid phase and a multicellular diploid phase. The multicellular haploid stage is called the **gametophyte**; the multicellular diploid stage is called the **sporophyte**. The two phases of the life cycle are connected by the distinct types of reproductive cells: gametes and spores.

Although alternation of generations is observed in a wide array of protist lineages and in some groups of green algae, it does not occur in the Coleochaetales or Charales, the algal groups most closely related to land plants. In the Coleochaetales and Charales, the multicellular form is haploid (**Figure 29.13**). Only the zygote is diploid. These data suggest that alternation of generations evolved early in the history of the land plants—soon after they split off from green algae.

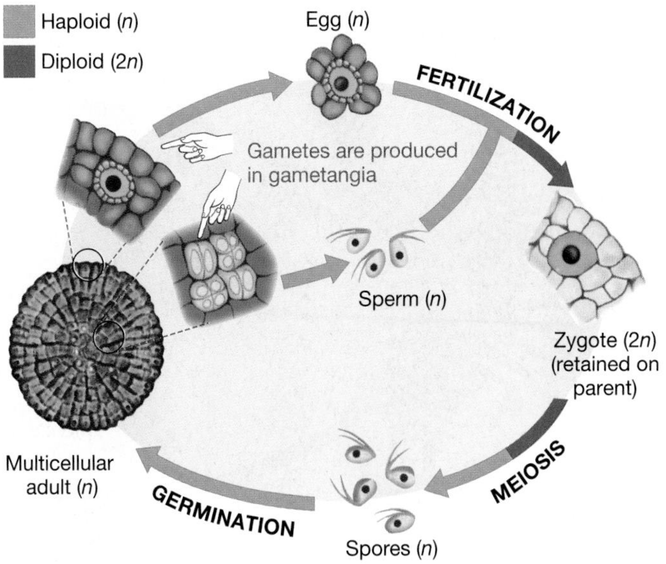

FIGURE 29.13 The Green Algae That Are Closely Related to Land Plants Have Simple Life Cycles
The Coleochaetales and Charales do not have alternation of generations. The multicellular stage is haploid.

(a) Alternation of generations always involves the same sequence of events.

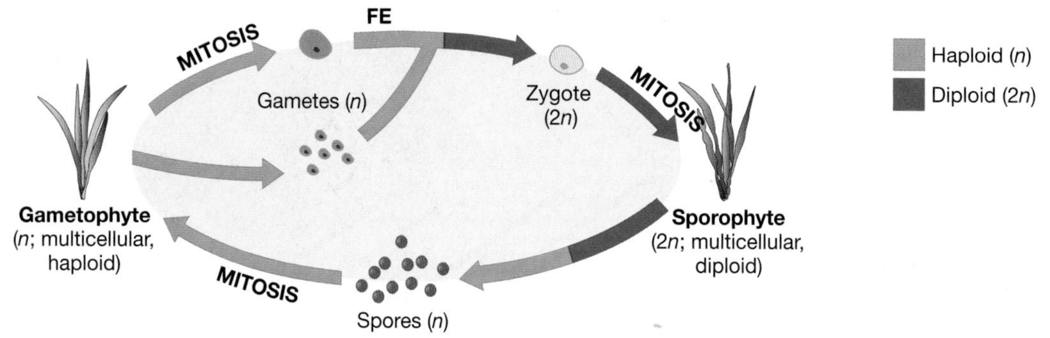

(b) Mosses: Gametophyte is large and long lived; sporophyte depends on gametophyte for nutrition.

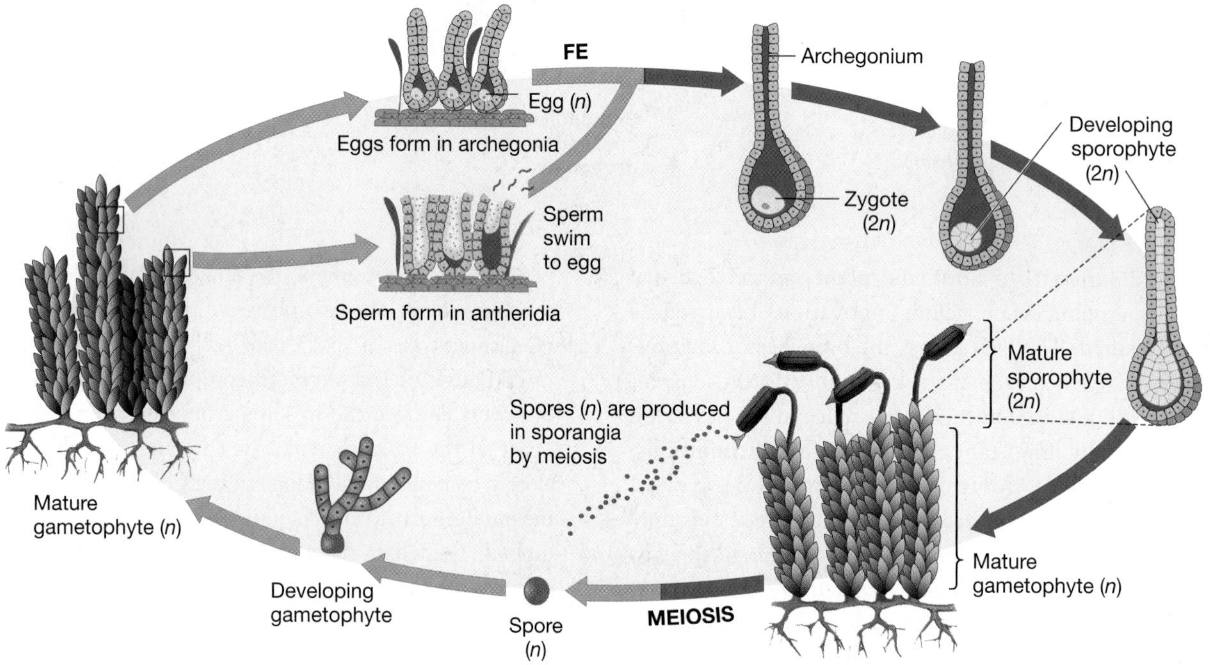

(c) Ferns: Sporophyte is large and long lived but, when young, depends on gametophyte for nutrition.

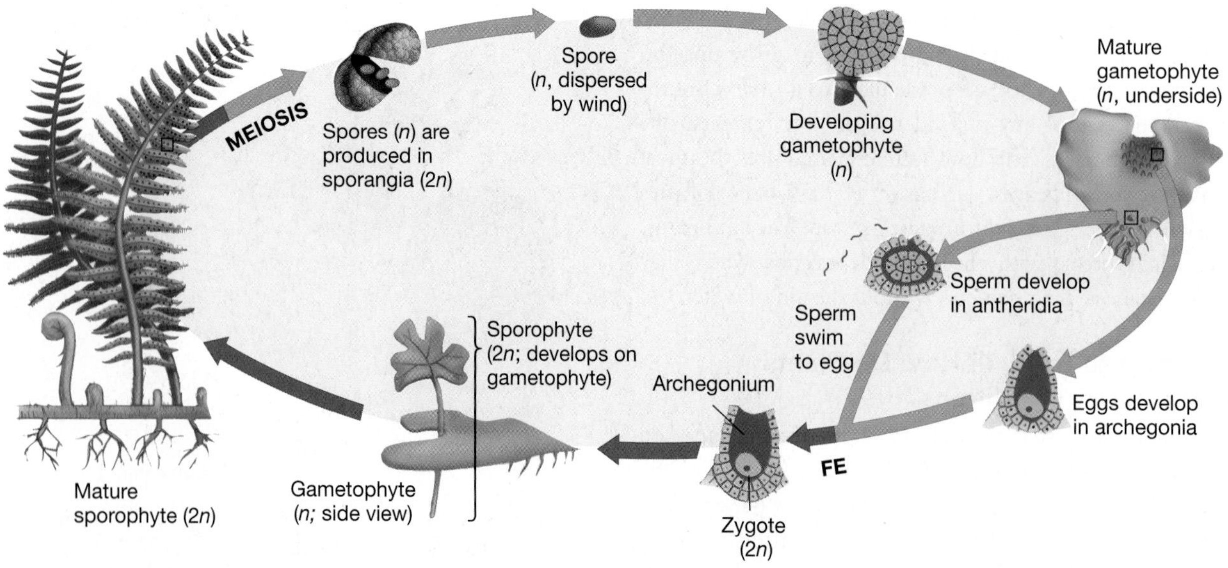

FIGURE 29.14 The Life Cycles of Land Plants

EXERCISE Study part (a). Find each phase of this generalized life cycle in parts (b) through (e).

Alternation of generations always involves the same basic sequence of events (**Figure 29.14a**):

1. Gametophytes produce gametes by mitosis. Both the gametophyte and the gametes are haploid.
2. Two gametes unite to form a diploid zygote.
3. The zygote develops into a multicellular, diploid sporophyte.
4. The sporophyte produces haploid spores by meiosis.
5. Spores develop into a gametophyte.

(d) Conifers: Sporophyte is dominant; gametophyte depends on sporophyte for nutrition.

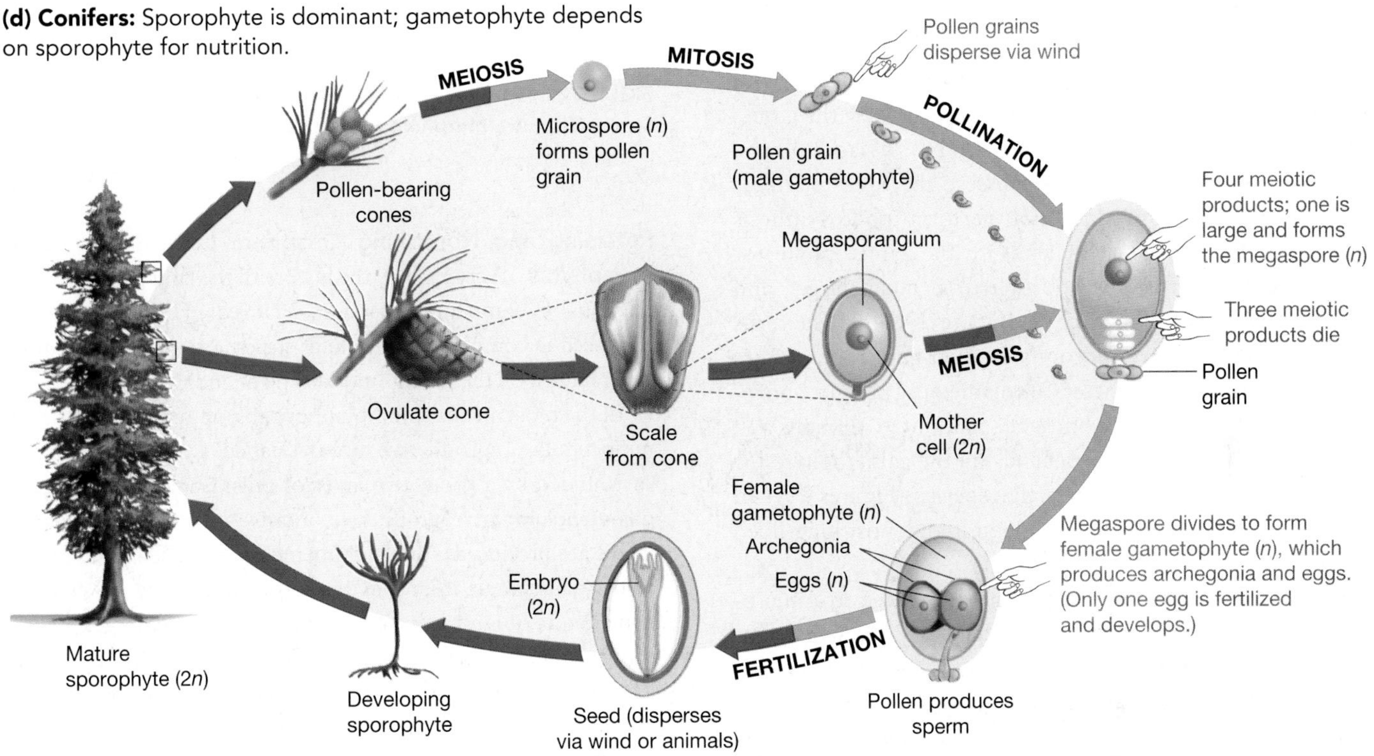

(e) Angiosperms: Similar to gymnosperms, but ovules and seeds form in enclosures called ovaries.

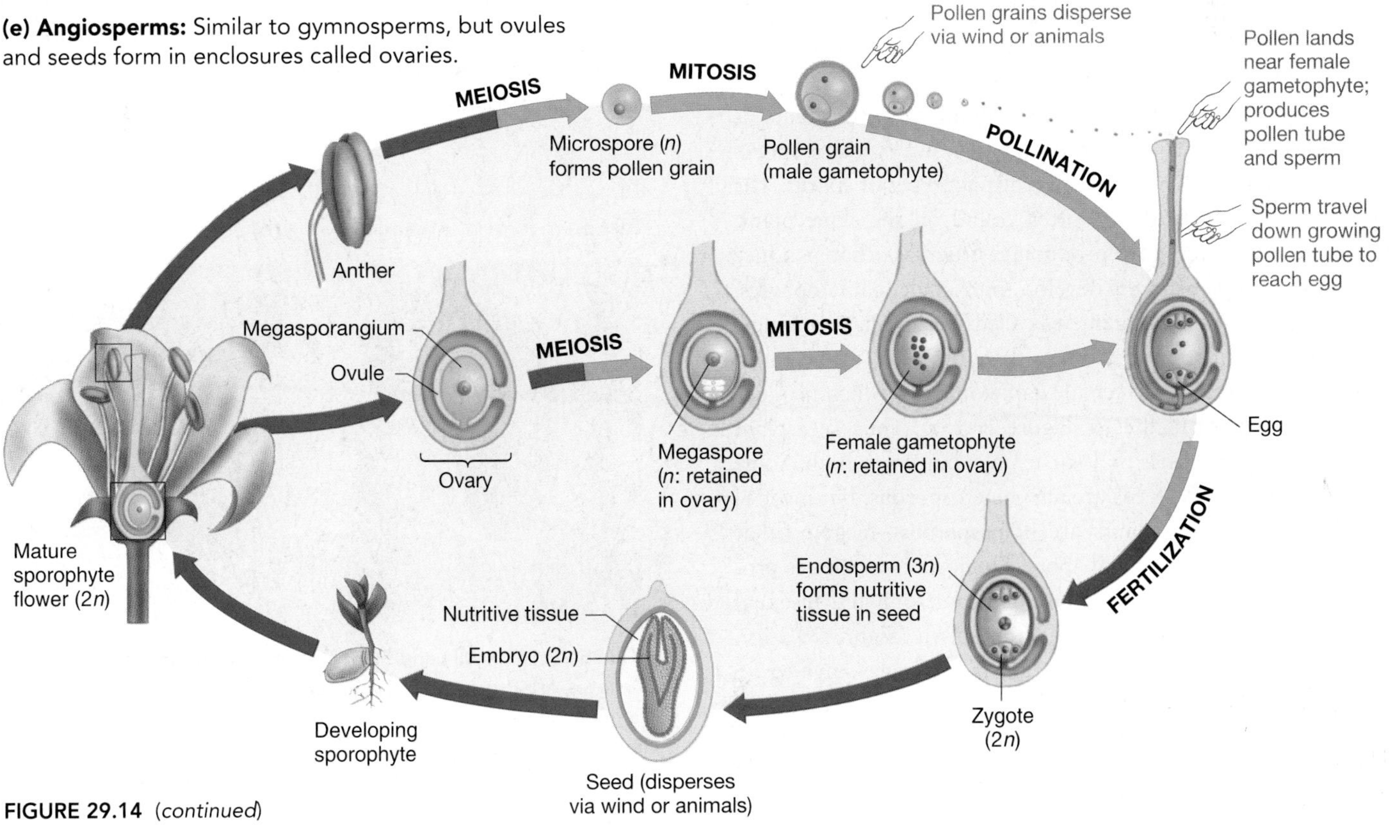

FIGURE 29.14 *(continued)*

Recall from Chapter 28 that a spore is a cell that grows directly into a multicellular individual without fusing with another cell. In nonvascular plants and seedless vascular plants, spores are encased in a tough, watertight coat made of sporopollenin. As a result, these spores resist drying and are an effective way to disperse offspring to new habitats by wind. It's helpful to recognize the distinction between a spore and a zygote: A zygote forms from the fusion of two cells, such as a sperm and an egg, while spores do not.

Although these five steps are common to all species with alternation of generations, the relationship between the gametophyte and sporophyte is highly variable in land plants. In nonvascular plants such as mosses, the sporophyte is small, short lived, and does not perform photosynthesis (**Figure 29.14b**). Instead, it depends on the gametophyte for nutrition. But in ferns, the sporophyte is much larger and longer lived than the gametophyte (**Figure 29.14c**). The ferns you are familiar with are sporophytes; their gametophytes are typically just a few millimeters in diameter. The fern sporophyte still develops directly on the gametophyte, though; and when young, the sporophyte depends on the gametophyte for nutrition. In gymnosperms and angiosperms (**Figures 29.14d** and **29.14e**), the sporophyte is the largest and most long-lived phase in the life cycle.

The male gametophyte of seed plants is reduced to a microscopic structure called a **pollen grain**. Similarly, the female gametophyte is a small structure that is retained in the sporophyte's female reproductive organs and that produces an egg. In gymnosperms and angiosperms, the gametophyte does not perform photosynthesis at all and has to obtain all of its nutrition from the sporophyte. This is the exact opposite of the situation in liverworts and mosses, in which the sporophyte is nourished by the gametophyte.

Another important innovation found in seed plants is **heterospory**—the production of two distinct types of spore-producing structures and thus two distinct types of spores. The two types of structures are often found on the same plant. **Microsporangia** are spore-producing structures that produce microspores. **Microspores** develop into male gametophytes, which produce the small gametes called sperm. **Megasporangia** are spore-producing structures that produce megaspores. **Megaspores** develop into female gametophytes, which produce the large gametes called eggs (**Figure 29.15a**). Thus, seed plant gametophytes are either male or female, but never both. Some lycophytes and a few ferns are also heterosporous, but most of the seedless vascular plants are **homosporous**—meaning that they produce a single type of spore. Homosporous species produce spores that develop into bisexual gametophytes. Bisexual gametophytes produce both eggs and sperm (**Figure 29.15b**).

Now, what does all this variation in life cycles have to do with plants losing their reliance on swimming gametes? The answer involves a series of remarkable evolutionary innovations that occurred as land plants originated and diversified.

(a) Seed plants are heterosporous.

Microsporangia → Microspores → Male gametophyte → Sperm

Megasporangia → Megaspores → Female gametophyte → Eggs

(b) Most other plants are homosporous.

Sporangium → Spores → Bisexual gametophyte → Sperm / Eggs

FIGURE 29.15 Heterosporous Plants Produce Male and Female Spores That Are Morphologically Distinct

Retaining and Nourishing Offspring: Land Plants as Embryophytes Very early in the history of the land plants, two momentous evolutionary changes occurred: (1) Gametes were produced in complex, multicellular structures; and (2) the embryo was retained on the parent (mother) plant and was nourished by it.

In the most ancient land plant groups, gametophytes produce multicellular reproductive organs called gametangia that are surrounded by a protective layer of cells. **Gametangia** (singular: **gametangium**) are reproductive organs—structures in which gametes are produced. Although members of the Charales also develop gametangia that have a protective layer of cells, the gametangia of land plants are larger and more complex. In both groups, the male and female reproductive organs are separate. The sperm-producing structure is called an **antheridium** (plural: **antheridia; Figure 29.16a**), and the egg-producing structure is called an **archegonium** (plural: **archegonia; Figure 29.16b**). The antheridium and archegonium have the same function as the testes and ovaries of animals. The elaboration of the multicellular gametangium was an adaptation that helped protect the gametes of land plants and prevent them from drying out.

(a) Sperm form in antheridia. **(b)** Eggs form in archegonia.

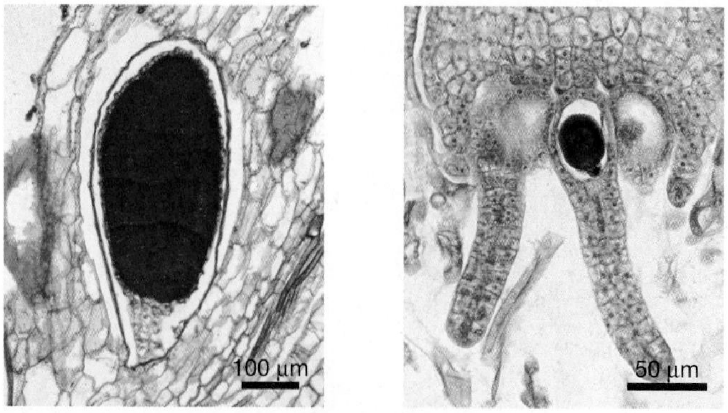

FIGURE 29.16 In All Land Plant Groups but Angiosperms, Gametes Are Produced in Gametangia
Gametangia are complex, multicellular structures that protect developing gametes. **(a)** Sperm-forming gametangia are called antheridia; **(b)** egg-producing gametangia are called archegonia.

(a) Young sporophytes are nourished by the parent plant.

(b) Transfer cells convey nutrients to the embryo.

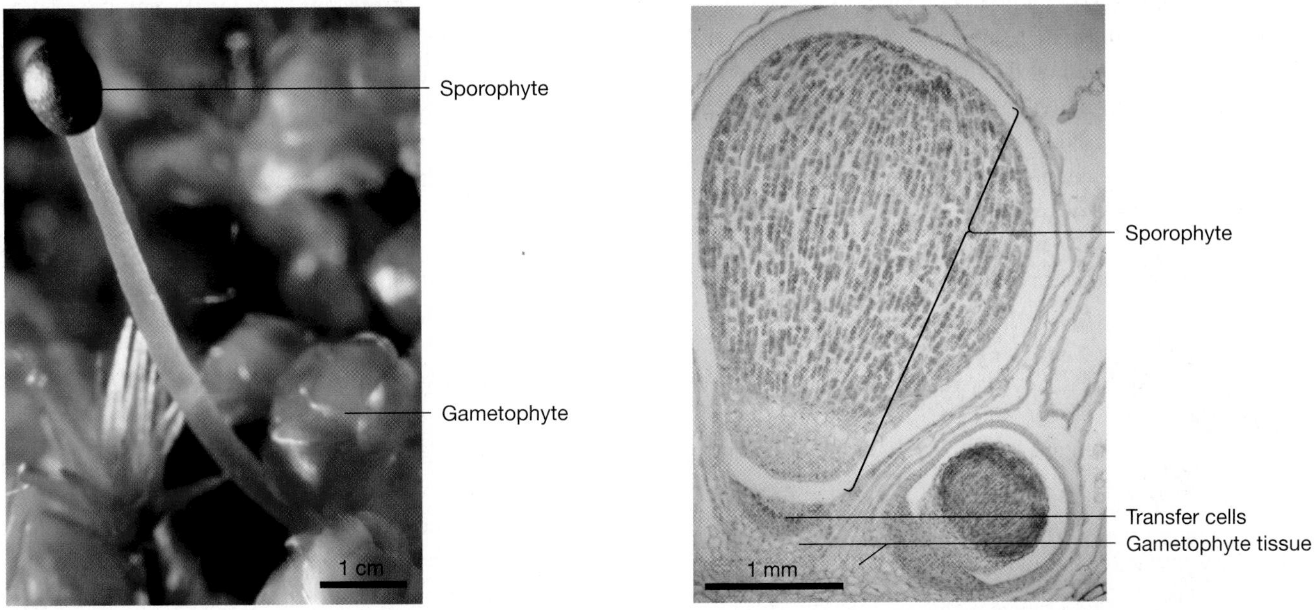
Sporophyte
Gametophyte
1 cm
Sporophyte
Transfer cells
Gametophyte tissue
1 mm

FIGURE 29.17 Land Plants Are Also Known as Embryophytes Because Parents Nourish Their Young
(a) Land plants are known as embryophytes because eggs, zygotes, and embryos are retained on the parent plant. Fertilization and early development take place on the parent plant. **(b)** As land plant embryos develop, transfer cells transfer nutrients from the mother to the offspring.

The second innovation that occurred early in land plant evolution involved the eggs that formed inside archegonia. Instead of shedding their eggs into the water or soil, land plants retain them. Eggs are also retained in the green algal lineages that are most closely related to land plants: In Charales and other closely related groups, sperm swim to the egg, fertilization occurs, and the resulting zygote stays attached to the parent. Either before or after fertilization, the egg or zygote receives nutrients from the mother plant. But because these algae live in northern latitudes, the parent plant dies each autumn as the temperature drops. The zygote remains on the dead tissue from the parent, settles to the bottom of the lake or pond, and overwinters. In spring, meiosis occurs, and the resulting spores develop into haploid adult plants.

In land plants, the zygote is also retained on the gametophyte after fertilization. But in contrast to the zygotes of green algae, the zygotes of land plants begin to develop on the parent plant while it is still alive, forming a multicellular embryo. As a result, the developing embryo can be nourished by the parent (**Figure 29.17a**). This trait is important because it means that land plant embryos do not have to manufacture their own food early in life. Instead, they receive most or all of their nutrients from the parent plant. Based on this innovation, biologists call the land plants **embryophytes**—literally, the "embryo-plants." The retention of the embryo is analogous to pregnancy in mammals, where offspring are retained by the mother and nourished through the initial stages of growth. Land plant embryos even have specialized **transfer cells**, which make physical contact with parental cells and facilitate the transfer of nutrients (**Figure 29.17b**), much like the placenta that develops in a pregnant mammal.

Although the retention of embryos ensures that offspring are well nourished, it has a potential disadvantage: In ferns and horsetails, sporophytes have to live in the same physical location as that of the parent gametophyte. But this limitation is overcome in seed plants, because the embryo is portable and can disperse to a new location.

The Evolution of Pollen During the evolution of land plants, there was clearly a strong trend from gametophyte-dominated life cycles of nonvascular plants to the sporophyte-dominated life cycles of gymnosperms and angiosperms. Why is this trend important? Recall that in the sporophyte-dominated life cycles of gymnosperms and angiosperms, the male sperm-producing structure—the gametophyte—is reduced to a pollen grain. Pollen grains are male gametophytes that are surrounded by a tough, desiccation-resistant coat of sporopollenin. Pollen can be exposed to the air for long periods of time without dying from dehydration. Pollen grains are also tiny enough to be carried to female gametophytes by wind or insects. Upon landing near the egg, the male gametophyte releases the sperm cells that accomplish fertilization.

When pollen evolved, then, the seed plants lost their dependence on water to accomplish fertilization. Instead of swimming to the egg as a naked sperm cell, their tiny gametophytes took to the skies.

The Evolution of the Seed The evolution of large, complex gametangia protected the eggs and sperm of land plants from drying. Embryo retention allowed offspring to be nourished directly by their parent, and pollen enabled fertilization to occur in the absence of water. Seeds were a fourth innovation that allowed plants to reproduce efficiently on land.

A seed is a structure that encloses and protects a developing embryo (**Figure 29.18a**). Seeds allow embryos to be dispersed to a new habitat, away from the parent plant. Like a bird's egg, the seed provides a protective case for the embryo and a store of nutrients provided by the mother. In addition, seeds are often attached to structures that aid in dispersal by wind, water, or animals. As **Figure 29.18b** shows, seed dispersal is aided by structures such as the wings on a milkweed seed.

By the time seeds had evolved about 360 million years ago, land animals were diverse and abundant enough to play a lead role in seed dispersal. The most important connection between seeds and animals occurred during the evolution of angiosperms, however. Angiosperms are the only organisms that make fruits. A **fruit** is a structure that is derived from the **ovary**, a female reproductive organ found in angiosperms. The angiosperms' ovary encloses the **ovules**, structures where female gametophytes are formed and where eggs are produced. When a fruit is mature, it encloses one or more seeds (**Figure 29.19a**). Tissues derived from the ovary are often nutritious and brightly colored. Fruits such as these attract animals that eat the fruits, digest the nutritious tissue around the seeds, and disperse seeds in their feces (**Figure 29.19b**). Animals were also involved in the other major reproductive innovation in angiosperms—the flower.

(a) Seeds package an embryo with a food supply.

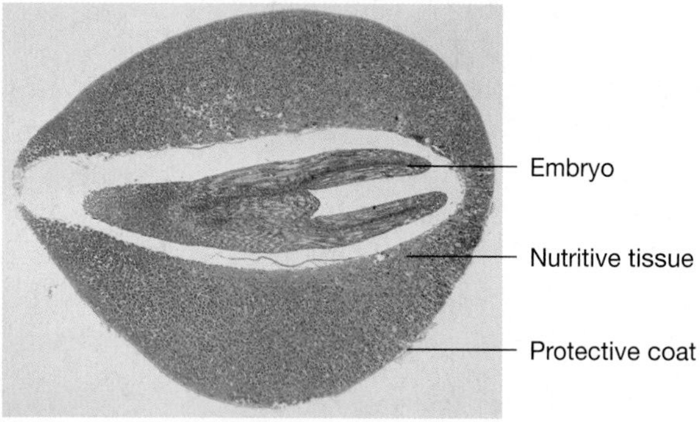
Embryo
Nutritive tissue
Protective coat

(b) Seeds are dispersed by wind, water, and animals.

Wind
Water
Animals

FIGURE 29.18 Seeds Contain an Embryo and a Food Supply and Can Be Dispersed
(a) Seeds consist of an embryo, a store of carbohydrate or oil-rich nutritive tissue, and a protective coat. **(b)** Most seeds have structures that facilitate dispersal by wind, water, or animals.

(a) Fruits are derived from ovaries and contain seeds.

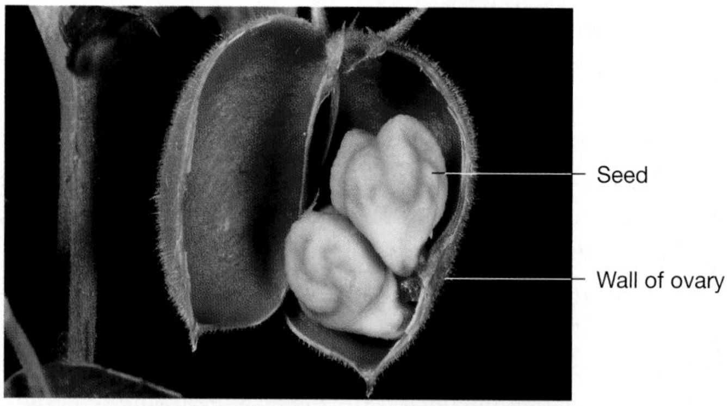
Seed
Wall of ovary

(b) Many fruits are dispersed by animals.

FIGURE 29.19 Fruits Are Derived from Ovaries Found in Angiosperms
(a) A pea pod is one of the simplest types of fruit. **(b)** The ovary wall often becomes thick, fleshy, and nutritious enough to attract animals that disperse the seeds inside.

The Evolution of the Flower Flowering plants, or angiosperms, are the most diverse land plants living today. About 250,000 species have been described, and more are discovered each year. Their success in terms of geographical distribution, number of individuals, and number of species revolves around a reproductive organ: the **flower**. Angiosperms do not produce gametangia. Their gametes are produced by male and female gametophytes, which are derived from microsporangia and megasporangia that develop inside the flower.

As **Figure 29.20** shows, flowers are wonderfully diverse in size, shape, and coloration. They also produce a wide range of scents. Why does all of this variation exist? To answer this question, biologists hypothesize that flowers are adaptations that increase the probability that an animal will perform **pollination**—meaning the transfer of pollen to a plant's female reproductive organ, where eggs are produced and fertilization takes place.

Instead of leaving pollination to an undirected agent such as wind, the hypothesis is that in some circumstances natural selection favored structures that reward an animal—usually an insect—for carrying pollen directly from one flower to another. Under the directed-pollination hypothesis, natural selection has favored flower colors and shapes and scents that are successful in attracting particular types of pollinators. A *pollinator* is an animal that disperses pollen. Pollinators are attracted to flowers because flowers provide the animals with food in the form of protein-rich pollen or a sugar-rich fluid known as **nectar**. In this way, the relationship between flowering plants and their pollinators is mutually beneficial. The pollinator gets food; the plant gets sex.

What evidence supports the hypothesis that flowers vary in size, structure, scent, and color in order to attract different pollinators? The first type of evidence is correlational in nature. In general, the characteristics of a flower correlate closely with the characteristics of its pollinator. A few examples will help drive this point home:

- The carrion flower in Figure 29.20a produces molecules that smell like rotting flesh. The scent attracts carrion flies, which normally lay their eggs in animal carcasses. In effect, the plant tricks the flies. While looking for a place to lay their eggs on a flower, the flies become dusted with pollen. If the flies are already carrying pollen from a visit to a different carrion flower, they are likely to deposit pollen grains near the plant's female gametophyte. In this way, the carrion flies pollinate the plant.
- Flowers that are pollinated by hummingbirds typically have petals that form a long, tubelike structure that corresponds to the size and shape of a hummingbird's beak (Figure 29.20b). Nectar-producing cells are located at the base of the tube. When hummingbirds visit the flower, they insert their beaks and harvest the nectar. In the process, they transfer pollen grains attached to their throats or faces.
- Hummingbird-pollinated flowers also tend to have red petals (Figure 29.20b), while bee-pollinated flowers tend to be purple or yellow (Figure 29.20c). Hummingbirds are attracted to red; bees have excellent vision in the purple and ultraviolet end of the spectrum.

In addition to correlational evidence, there is strong experimental support for the hypothesis that natural selection favors a correspondence between the shapes of flowers and their pollinators. Consider, for example, recent work on two South African populations of the orchid *Disa draconis*. The length of the long tube, or *spur*, located at the back of this flower is very different in the two populations. As predicted by the directed-pollination hypothesis, each population is pollinated by a different insect. Short-spurred plants that grow in mountain habitats are pollinated by horseflies, which have a relatively

(a) Carrion flowers smell like rotting flesh and attract carrion flies.

(b) Hummingbird-pollinated flowers are red and have long tubes with nectar at the base.

(c) Bumble-bee-pollinated flowers are often bright purple.

FIGURE 29.20 Flowers with Different Shapes, Colors, and Fragrances Attract Different Pollinators

(a) Horseflies have a short proboscis and pollinate short-spurred orchids.

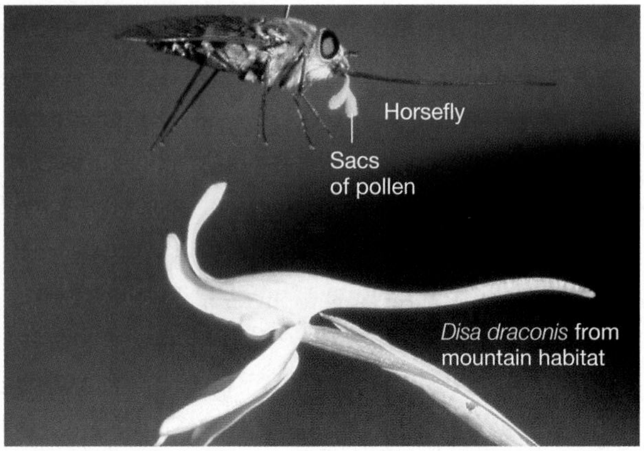
Horsefly
Sacs
of pollen
Disa draconis from
mountain habitat

(b) Tanglewing flies have a long proboscis and pollinate long-spurred orchids.

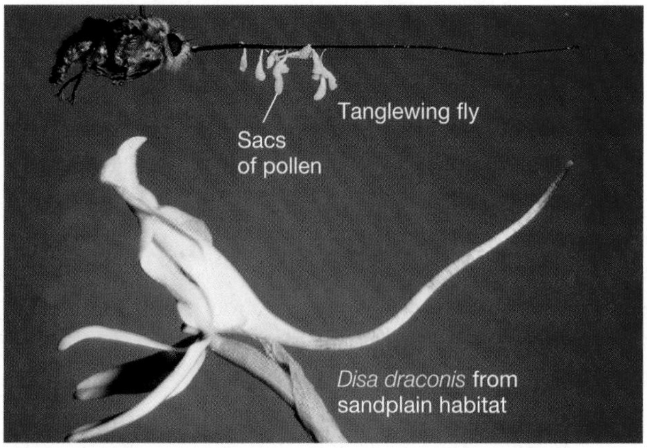
Tanglewing fly
Sacs
of pollen
Disa draconis from
sandplain habitat

FIGURE 29.21 Populations of the Same Orchid Species Have Different Flower Shapes and Different Pollinators
A *Disa draconis* flower **(a)** from a mountainous habitat with its horsefly pollinator and **(b)** from a lower-elevation, sandplain habitat with its tanglewing fly pollinator. Note the correlation between the length of each flower's spur and the length of each pollinator's proboscis.

short proboscis (**Figure 29.21a**). (A *proboscis* is a specialized mouthpart found in some insects. When extended, it functions like a straw in sucking nectar or other fluids.) Long-spurred orchids that grow on low-lying sandplain habitats are pollinated by tanglewing flies, which have a particularly long proboscis (**Figure 29.21b**). Horseflies are common in mountain habitats but rare at lower elevations. Conversely, tanglewing flies are abundant on low-lying sandplains but absent at high elevations.

The directed-pollination hypothesis makes two key predictions about pollination in these populations. First, it predicts that the ability to attract pollinators is an important component of fitness in *D. draconis*. To test this prediction, biologists pollinated a large number of flowers by hand in each population. Hand pollination ensures that large amounts of pollen are transferred. Then they compared the number of seeds that were produced by the hand-pollinated flowers versus normal, insect-pollinated flowers. The data showed that hand-pollinated individuals produced many more seeds than fly-pollinated individuals did. This is a key observation, because it shows that the fitness of these orchids is limited by their ability to attract pollinators. In this species, traits that increase the likelihood of being pollinated should confer a huge fitness advantage.

The second prediction of the directed-pollination hypothesis is that spurs of a particular length are adaptations that increase the probability of being pollinated by horseflies or tanglewing flies. To test this prediction, researchers artificially shortened the spurs of individuals in the long-spurred population (**Figure 29.22**). The researchers did this by tying off the spurs of randomly selected flowers with a piece of yarn, so that flies could not reach the end of the spur. The idea was that flies would not make contact with the flower's reproductive organs when they inserted their proboscis into the shortened tube. The biologists left nearby flowers alone but tied a piece of yarn near the spur. The presence of the yarn near the spur controlled for the possibility that the yarn itself—rather than spur length—affected pollinator behavior. The data collected indicate that flowers with short spurs received much less pollen and set much less seed than did flowers with normal-length spurs. These data strongly support the hypothesis that spur length is an adaptation that increases the frequency of pollination in a particular habitat.

Such results support the hypothesis that plants gain large fitness benefits by attracting pollinators. An even more general implication is that the spectacular diversity of angiosperms resulted, at least in part, from natural selection exerted by the equally spectacular diversity of insect, mammal, and bird pollinators. Recall from Section 29.2 that almost 90 percent of the land plants living today reproduce by making flowers. Once green plants could grow and reproduce in terrestrial environments, the story of their diversification revolved around systems for transferring gametes efficiently and producing the largest number of seeds possible. **Figure 29.23** summarizes the adaptations that enabled plants to reproduce effectively on land.

The Angiosperm Radiation

Angiosperms represent one of the great adaptive radiations in the history of life. As Chapter 26 noted, an **adaptive radiation** occurs when a single lineage produces a large number of descendant species that are adapted to a wide variety of habitats. The diversification of angiosperms is associated with three key adaptations: (1) vessels, (2) flowers, and (3) fruits. In combination, these traits allow angiosperms to transport water, pollen, and seeds efficiently. Based on these observations, it is not surprising that most land plants living today are angiosperms.

On the basis of morphological traits, the 235,000 species of angiosperms identified to date have traditionally been grouped

Question: Does flower shape affect reproductive success?

Hypothesis: A correlation between flower spur length and pollinator proboscis length increases reproductive success in *Disa draconis.*

Null hypothesis: Correlations between flower spur length and pollinator proboscis length have no effect on reproductive success in *Disa draconis.*

Experimental setup:

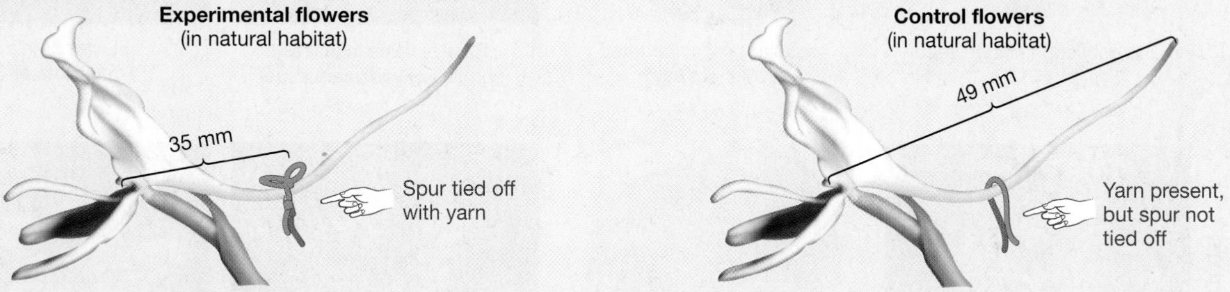

Prediction of hypothesis to be tested: Altering spur length (and thus flower shape) will reduce reproductive success.

Prediction of null hypothesis: Altering spur length (and thus flower shape) will have no impact on reproductive success.

Results:

	Shortened spurs	Normal spurs
What fraction of flowers received pollen? (n = 59)	17%	46%
What fraction of flowers set seed? (n = 56)	18%	41%

Flowers with normal spurs have higher reproductive success

Conclusion: In this species, flower shape—specifically spur length—affects reproductive success.

FIGURE 29.22 The Adaptive Significance of Flower Shape: An Experimental Test
The spurs of experimental *D. draconis* flowers were tied off with yarn 35 mm from the opening; the spurs of control flowers had a piece of yarn loosely tied around them, too, but had normal spurs averaging 49 mm. QUESTION If spur length has no effect on reproductive success, what would the data in this table look like?

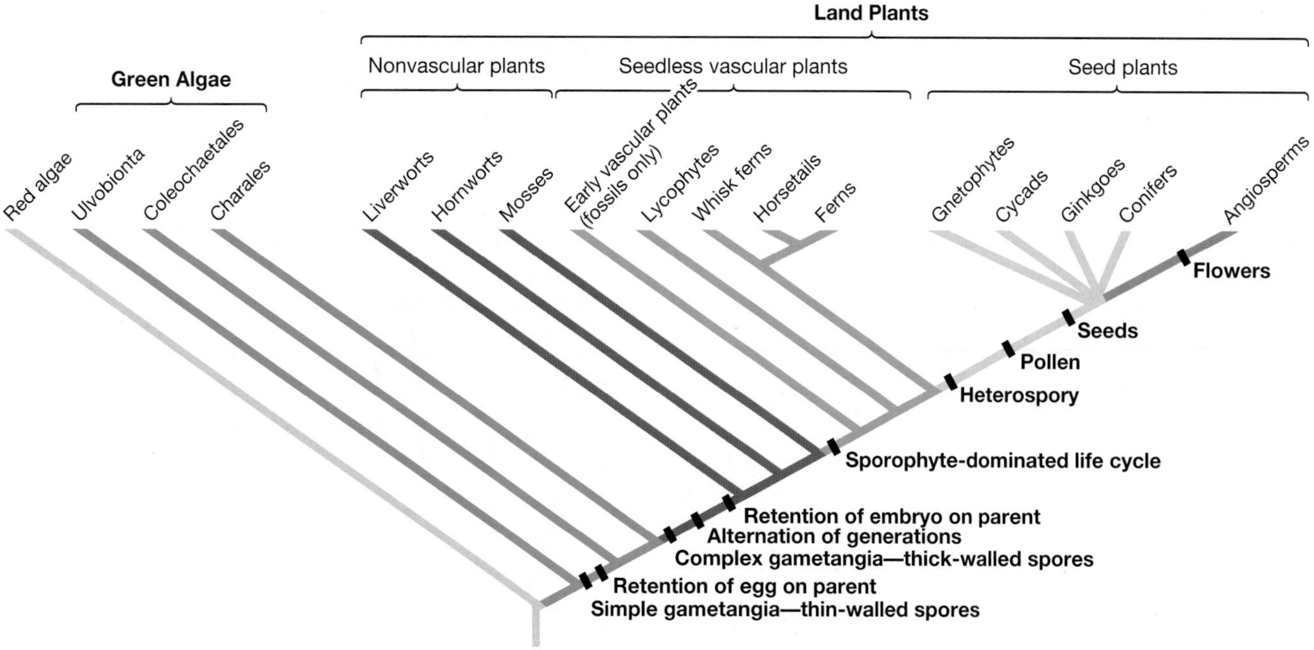

FIGURE 29.23 Evolutionary Innovations Allowed Plants to Reproduce Efficiently on Land
EXERCISE Circle the names of groups (at the tips of the branches) that have flagellated, swimming sperm.

FIGURE 29.24 Four Morphological Differences between Monocots and Dicots

into two major lineages: the *monocotyledons*, or **monocots**, and the *dicotyledons*, or **dicots**. Some familiar monocots include the grasses, orchids, palms, and lilies; familiar dicots include the roses, buttercups, daisies, oaks, and maples. The names of the two groups were inspired by differences in a structure called the cotyledon. A **cotyledon** is the first leaf that is formed in an embryonic plant. As **Figure 29.24** shows, monocots have a single cotyledon (visible inside the seed) while dicots have two cotyledons (visible in newly germinated plants). The figure also highlights other major morphological differences observed in monocots and dicots, concerning the arrangement of vascular tissue and flower characteristics.

It would be misleading, however, to think that all species of flowering plants fall into one of these two groups—either monocots or dicots. Recent work has shown that dicots do not form a natural group consisting of a common ancestor and all of its descendants. To drive this point home, consider the phylogeny illustrated in **Figure 29.25**. These relationships were estimated by comparing the sequences of several genes that are shared by all angiosperms. Notice that species with dicot-like characters are scattered around the angiosperm phylogenetic tree. Based on this analysis, biologists have concluded that dicots are not a monophyletic group. Instead, they are paraphyletic.

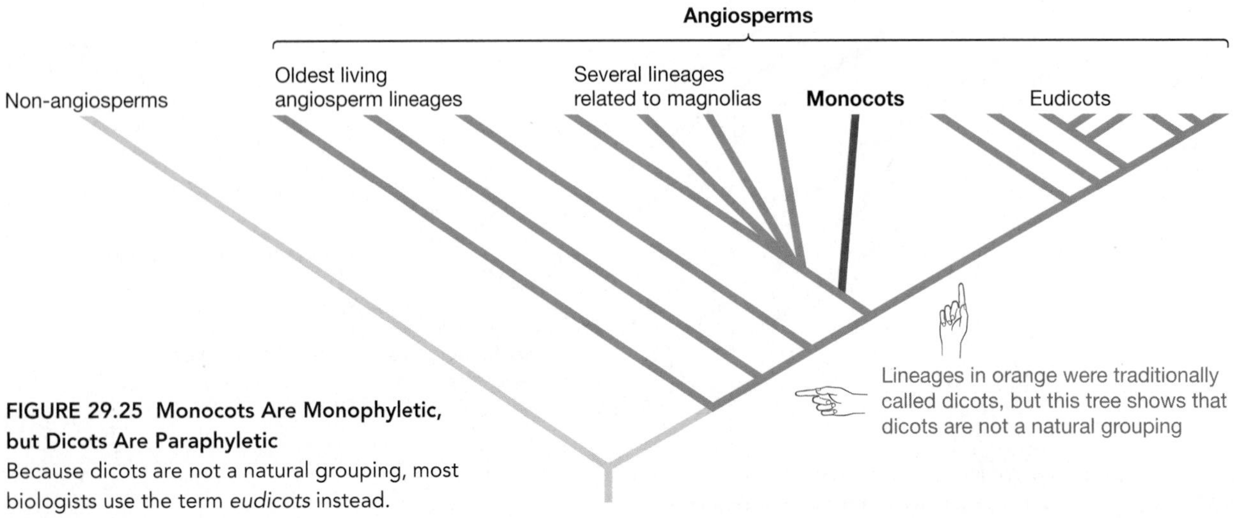

FIGURE 29.25 Monocots Are Monophyletic, but Dicots Are Paraphyletic
Because dicots are not a natural grouping, most biologists use the term *eudicots* instead.

Biologists have adjusted the names assigned to angiosperm lineages to reflect this new knowledge of phylogeny. Now they formally name a series of monophyletic groups in the tree, including the **magnoliid complex** (made up of magnolias and laurels and their relatives), the monocots, and the **eudicots** ("true dicots")—a lineage that includes roses, daisies, and maples. The most basal clades consist of the genus *Amborella*, the water lilies, and the star anise and relatives. Plant systematists continue to work toward understanding relationships at the base of the angiosperm phylogenetic tree, and there will undoubtedly be more name changes as our knowledge grows.

✓CHECK YOUR UNDERSTANDING

Land plants were able to make the transition to growing in terrestrial environments, where sunlight and carbon dioxide are abundant, based on the evolution of cuticle, pores, stomata, water-conducting tissue, and vascular tissue. Reproduction in dry environments became increasingly efficient as land plants diversified, based on a series of evolutionary innovations: gametangia and the retention of embryos, pollen, seeds, and flowers. You should be able to (1) explain why each of these adaptations was important in survival or reproduction, and (2) map where each of these evolutionary innovations occurred on the phylogenetic tree of green plants.

29.4 Key Lineages of Green Plants

The evolution of cuticle, pores, stomata, and water-conducting tissues allowed green plants to grow on land, where resources for photosynthesis are abundant. Once the green plants were on land, the evolution of gametangia, retained embryos, pollen, seeds, and flowers enabled them to reproduce efficiently even in very dry environments. The adaptations reviewed in Section 29.3 allowed the land plants to make the most important water-to-land transition in the history of life.

To explore green plant diversity in more detail, let's take a closer look at some major groups of green algae. Then we'll discuss the 12 major phyla of land plants in the context of their broad morphological groupings: nonvascular plants, seedless vascular plants, and seed plants.

Green Algae

The green algae are a paraphyletic group that totals about 7000 species. Their bright green chloroplasts are similar to those found in land plants, with a double membrane and chlorophylls *a* and *b* but relatively few accessory pigments. And like land plants, green algae synthesize starch in the chloroplast as a storage product of photosynthesis and have a cell wall composed primarily of cellulose. Green algae are important primary producers in nearshore ocean environments and in all types of freshwater habitats. They are also found in several types of more exotic environments, including snowfields at high elevations, pack ice, and ice floes. These habitats are often colored pink or green by large concentrations of unicellular green algae (**Figure 29.26a**). Although these cells live at near-freezing temperatures, they make all their own food via photosynthesis.

In addition, green algae live in close association with an array of other organisms:

- **Lichens** are stable associations between green algae and fungi or between cyanobacteria and fungi. Lichens are found in terrestrial environments—often in locations such as tree bark or bare rock that lack soil (**Figure 29.26b**). The algae or cyanobacteria do not dry out, because they are protected by the fungus. Of the 17,000 species of lichens that have been described to date, about 85 percent involve a unicellular or filamentous green alga. Lichens are explored in more detail in Chapter 30.
- Unicellular green algae are common endosymbionts in planktonic protists that live in lakes and ponds (**Figure 29.26c**). The association is considered mutually beneficial: The algae supply the protists with food; the protists provide protection to the algae.

Green algae are a large and fascinating group of organisms. Let's take a closer look at just three of the many lineages.

(a) Green algae are responsible for pink snow. **(b)** Most lichens are an association between fungi and green algae. **(c)** Many unicellular protists harbor green algae.

FIGURE 29.26 Some Green Algae Live in Unusual Environments

Ulvobionta

The Ulvobionta are a monophyletic group composed of several diverse and important subgroups, with a total of about 4000 species. Members of this lineage range from unicellular to multicellular. Most of the green algae you have seen at the ocean or in lakes or streams belong to this group, as do *Volvox* and related species introduced in Chapter 28.

Many of the large green algae in coastal and intertidal habitats along ocean coastlines are members of the Ulvobionta. *Ulva,* the sea lettuce (**Figure 29.27**), is a representative marine species. But there are also large numbers of unicellular or small multicellular species that inhabit the plankton of freshwater lakes and streams.

Reproduction Most reproduce both asexually and sexually. Asexual reproduction often involves production of spores that swim with the aid of flagella. Sexual reproduction usually results in production of a resting stage—a cell that is dormant in winter. In many species the gametes are not called eggs and sperm, because they are the same size and shape. In most species gametes are shed into the water, so fertilization takes place away from the parent plants.

Life cycle Many unicellular forms are diploid only as zygotes. Alternation of generations occurs in multicellular species. When alternation of generation occurs, gametophytes and sporophytes may look identical or different.

Human and ecological impacts Ulvobionta are important primary producers in freshwater environments and in coastal areas of the oceans.

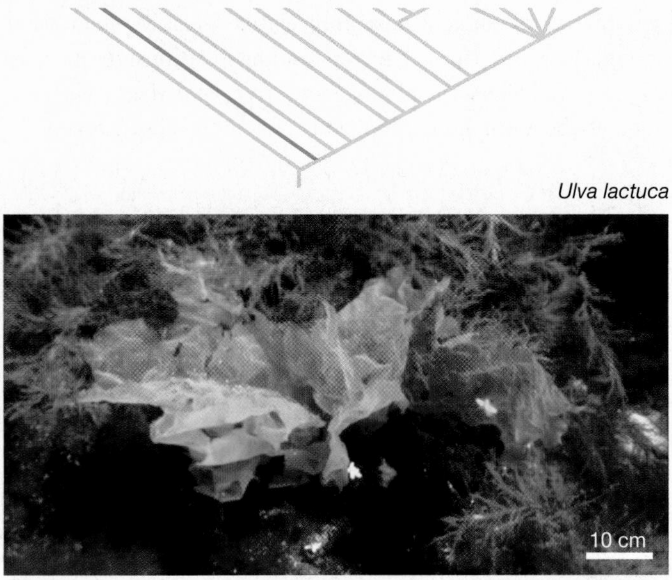
Ulva lactuca
10 cm

FIGURE 29.27 Green Algae Are Important Primary Producers in Aquatic Environments

Coleochaetales

There are 19 species in this group. Most coleochaetes are barely visible to the unaided eye and grow as flat sheets of cells (**Figure 29.28**). They are considered multicellular, because they have specialized photosynthetic and reproductive cells.

The coleochaetes are strictly freshwater algae. They grow attached to aquatic plants such as water lilies or over submerged rocks in lakes and ponds. When they grow near beaches, they are often exposed to air when water levels drop in late summer.

Reproduction Asexual reproduction is common and involves production of flagellated spores. During sexual reproduction, eggs are retained on the parent and are nourished after fertilization with the aid of transfer cells—a situation very similar to that observed in land plants. In some species certain individuals are male and produce only sperm, while other individuals are female and produce only eggs.

Life cycle Alternation of generations does not occur. Multicellular individuals are haploid; the only diploid stage in the life cycle is the zygote.

Human and ecological impacts Because they are very closely related to land plants, coleochaetes are studied intensively by researchers interested in how land plants made the water-to-land transition.

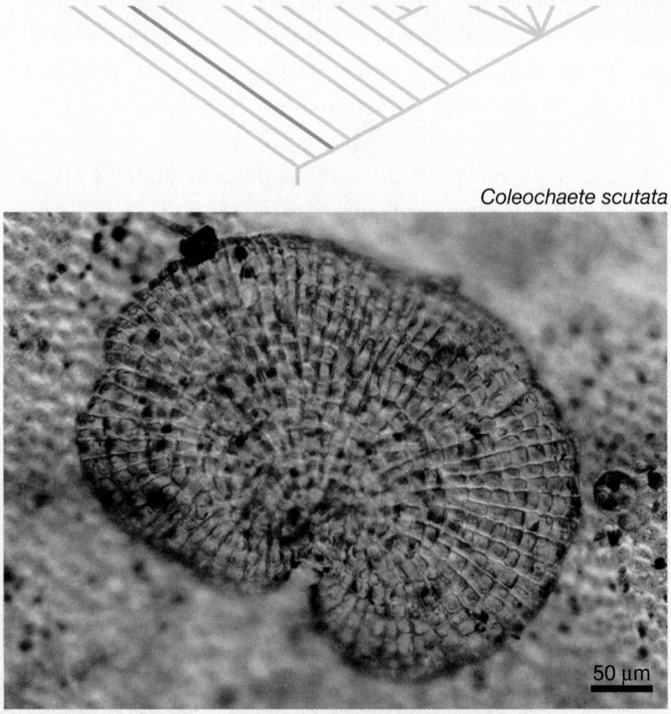
Coleochaete scutata
50 μm

FIGURE 29.28 Coleochaetes Are Thin Sheets of Cells

Charales (Stoneworts)

There are several hundred species in this group. They are collectively known as stoneworts, because they commonly accumulate crusts of calcium carbonate ($CaCO_3$) over their surfaces. All are multicellular. Some species can be a meter or more in length.

The stoneworts are freshwater algae. Certain species are specialized for growing in relatively deep waters, though most live in shallow water near lake beaches or pond edges.

Reproduction Primarily sexual reproduction occurs, with production of prominent, multicellular gametangia similar to those observed in early land plants. Eggs are retained on the parent plant, which supplies eggs with nutrients prior to fertilization.

Life cycle Alternation of generations does not occur. Multicellular individuals are haploid; the only diploid stage in the life cycle is the zygote.

Human and ecological impacts Some species form extensive beds in lake bottoms or ponds and provide food for ducks and geese as well as food and shelter for fish (**Figure 29.29**).

Globularis species (tall) and *C. fibrosa* (short)

FIGURE 29.29 Stoneworts Can Form Beds in Lake Bottoms

Nonvascular Plants ("Bryophytes")

The most basal lineages of land plants are collectively known as nonvascular plants, or **bryophytes**. The three lineages with living representatives (liverworts, hornworts, and mosses) do not form a monophyletic group, but instead represent an evolutionary grade. All of the species present today have a low, sprawling growth habit. In fact, it is unusual to find bryophytes that are more than 5 to 10 centimeters (2 to 4 inches) tall. Individuals are anchored to soil, rocks, or tree bark by specialized structures called **rhizoids.** No bryophyptes have true vascular tissue where cells have reinforced walls. In the lineages present today, simple water-conducting cells and tissues are found only in some mosses. All of the bryophytes have flagellated sperm that swim to eggs through raindrops or small puddles on the plant surface. Spores are dispersed by wind.

Hepaticophyta (Liverworts)

Liverworts got their name because some species native to Europe have liver-shaped leaves. According to the medieval *Doctrine of Signatures*, God indicated how certain plants should be used by giving them a distinctive appearance. Thus, liverwort teas were hypothesized to be beneficial for liver ailments. (They are not.) About 6500 species are known. They are commonly found growing on damp forest floors or riverbanks, often in dense mats (**Figure 29.30**), or on the trunks or branches of tropical trees.

Adaptations to land Liverworts are covered with cuticle. Some species have pores that allow gas exchange; in species that lack pores, the cuticle is very thin.

Reproduction Asexual reproduction occurs when fragments of a plant are broken off and begin growing independently. Some species also produce small structures called **gemmae** asexually, during the gametophyte phase. Mature gemmae are knocked off the parent plant by rain and grow into independent gametophytes. During sexual reproduction, sperm and eggs are produced in gametangia.

Life cycle The gametophyte is the largest and longest-lived phase in the life cycle. Sporophytes are small, grow directly from the gametophyte, and depend on the gametophyte for nutrition. Spores are shed from the sporophyte and are carried away by wind or rain.

Human and ecological impacts When liverworts grow on bare rock or tree bark, their dead and decaying body parts contribute to the initial stages of soil formation.

Calypogeia species

FIGURE 29.30 Liverworts Thrive in Moist Habitats

Anthocerophyta (Hornworts)

Hornworts got their name because their sporophytes have a horn-like appearance (**Figure 29.31**) and because *wort* is the Anglo-Saxon word for *plant*. About 100 species have been described to date.

Adaptations to land Hornwort sporophytes have stomata. Research is under way to determine if they can open their pores to allow gas exchange or close their pores to avoid water loss during dry intervals.

Reproduction Depending on the species, gametophytes may contain only egg-producing archegonia or only sperm-producing antheridia, or both. Stated another way, individuals of some species are either female or male, while in other species each individual has both types of reproductive organs.

Life cycle The gametophyte is the longest-lived phase in the life cycle. Although sporophytes grow directly from the gametophyte, they are green because their cells contain chloroplasts. Sporophytes manufacture some of their own food but also get nutrition from the gametophyte. Spores disperse from the parent plant via wind or rain.

Human and ecological impacts Some species harbor symbiotic cyanobacteria that fix nitrogen.

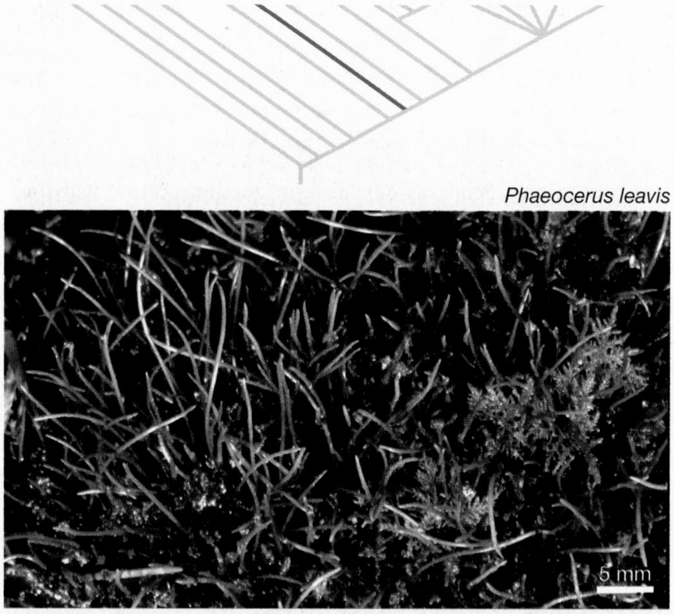
Phaeocerus leavis
5 mm

FIGURE 29.31 Hornworts Have Horn-Shaped Sporophytes

Bryophyta (Mosses)

Over 12,000 species of mosses have been named and described to date, and more are being discovered every year—particularly in the tropics. Mosses are informally grouped with other "bryophytes" (liverworts and hornworts) but are formally classified in their own monophyletic group: the phylum Bryophyta.

Although mosses are common inhabitants of moist forests all over the world, plants in this phylum can also be abundant in more extreme environments, such as deserts and windy, treeless habitats in the Arctic, Antarctic, or mountaintops. In these severe conditions, mosses are able to thrive because their bodies can become extremely dry without dying. When the weather makes photosynthesis difficult, individuals dry out and become dormant or inactive (**Figure 29.32a**). Then when rains arrive or temperatures warm, the plants rehydrate and begin photosynthesis and reproduction (**Figure 29.32b**).

Adaptations to land One subgroup of mosses contains simple conducting tissues consisting of cells that are specialized for the transport of water or food. But because these cells do not have walls that are reinforced by lignin, they are not considered true vascular tissue. As a result, most mosses are not able to grow much taller than a few centimeters.

Reproduction Asexual reproduction often occurs by fragmentation, meaning that pieces of gametophytes that are broken off by wind or a passing animal can begin growing independently. In many species, sexual reproduction cannot involve self-fertilization because the sexes are separate—meaning that an individual plant produces only eggs in archegonia or only sperm in antheridia. A typical sporophyte produces up to 50 million tiny spores. Spores are usually distributed by wind.

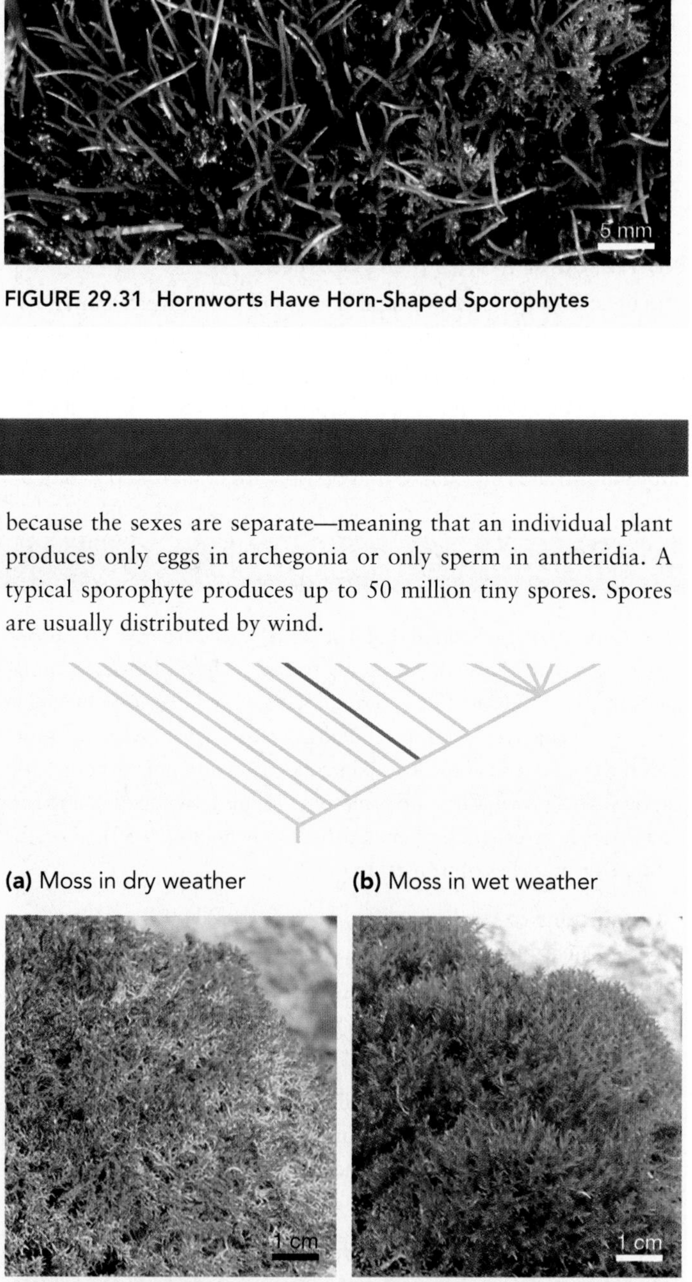
(a) Moss in dry weather
(b) Moss in wet weather
1 cm
1 cm

FIGURE 29.32 Many Moss Species Can Become Dormant When Conditions Are Dry

(Continued on next page)

Bryophyta (Mosses) *(continued)*

Life cycle The moss life cycle is similar to that of liverworts and hornworts: The sporophyte is retained on the much larger and longer-lived gametophyte and gets most of its nutrition from the gametophyte.

Human and ecological impacts Species in the genus *Sphagnum* are the most abundant plant in wet habitats of northern environments. Because *Sphagnum*-rich environments account for 1 percent of Earth's total land area, equivalent to half the area of the United States, *Sphagnum* species are among the most abundant plants in the world. *Sphagnum*-rich habitats have an exceptionally short growing season, however, so the decomposition of dead mosses and other plants is slow. As a result, large deposits of semi-decayed organic matter, known as **peat**, accumulate in these regions. Researchers estimate that the world's peatlands store about 400 billion metric tons of carbon, which makes these peat deposits among the most important carbon sinks in terrestrial environments. If peatlands begin to burn or decay rapidly due to global warming, the CO_2 released will further exacerbate global warming.

Peat is harvested as a traditional heating and cooking fuel in Ireland and other northern countries. It is also widely used as a soil additive in gardening, because *Sphagnum* can absorb up to 20 times its dry weight in water. This high water-holding capacity is due to the presence of large numbers of dead cells in the leaves of these mosses, which readily fill with water via pores in their walls. The same property made dried sphagnum an effective material in wartime for use as bandages.

Seedless Vascular Plants

The seedless vascular plants are a paraphyletic group that forms a grade between the nonvascular plants and the seed plants. All species of seedless vascular plants have conducting tissues with cells that are reinforced with lignin, forming vascular tissue. Tree-sized club mosses and horsetails are abundant in the fossil record, and tree ferns are still common inhabitants of mountain slopes in the tropics. Tree ferns do not make wood but are supported by fibrous roots.

The sporophyte is the larger and longer-lived phase of the life cycle in all three phyla of seedless vascular plants. The gametophyte is physically independent of the sporophyte, however. Eggs are retained on the gametophyte, and sperm swim to the egg with the aid of flagella. Thus, seedless vascular plants depend on the presence of water for reproduction—they need enough water to form a continuous layer that "connects" gametophytes and allows sperm to swim to eggs. Sporophytes develop on the gametophyte and are nourished by the gametophyte when they are small.

Lycophyta (Lycophytes, or Club Mosses)

Although the fossil record documents lycophytes that were 2 m wide and 40 m tall, the 1000 species of lycophytes living today are all small in stature (**Figure 29.33**). Most live on the forest floor or on the branches or trunks of tropical trees. They are often called "ground pines" or club mosses, even though they are neither pines nor mosses.

Adaptations to land Lycophytes are the most ancient land plant lineage with **roots**—a belowground system of tissues and organs that anchors the plant and is responsible for absorbing water and mineral nutrients. Roots differ from the rhizoids observed in bryophytes, because roots contain vascular tissue and thus are capable of conducting water and nutrients from below ground to the upper reaches of the plant. Small leaves called microphylls extend from the stems.

Reproduction Asexual reproduction can occur by fragmentation or gemmae. During sexual reproduction, spores give rise to bisexual gametophytes in most species—meaning that each gametophyte produces both eggs and sperm. Self-fertilization is extremely rare in most of these species, however. Some club mosses have separate male and female gametophytes.

Life cycle The gametophytes of some species live entirely underground and get their nutrition from symbiotic fungi. In certain species, gametophytes live 6–15 years and give rise to a large number of sporophytes over time.

Human and ecological impacts Tree-sized lycophytes were abundant in the coal-forming forests of the Carboniferous period. In coal-fired power plants today, electricity is being generated by burning fossilized lycophyte trunks and leaves.

Lycopodium species
5 cm

FIGURE 29.33 Lycopods Living Today Are Small in Stature

Psilotophyta (Whisk Ferns)

Only two genera of whisk ferns are living today and perhaps six distinct species. Whisk ferns are restricted to tropical regions and have no fossil record. They are extremely simple morphologically, with aboveground parts consisting of branching stems that have tiny, scale-like outgrowths instead of leaves (**Figure 29.34**).

Adaptations to land Whisk ferns lack roots. Some species gain most of their nutrition from fungi (see Chapter 30) that grow in association with the whisk ferns' extensive underground stems called **rhizomes.** Other species grow in rock crevices or are **epiphytes** ("upon-plants"), meaning that they grow on the trunks or branches of other plants—in this case, in the branches of tree ferns.

Reproduction Asexual reproduction is common in sporophytes via the extension of rhizomes and the production of new aboveground stems. When spores mature, they are dispersed by wind and germinate into gametophytes that contain both archegonia and antheridia.

Life cycle Sporophytes may be up to 30 cm tall, but gametophytes are less than 2 mm long and live under the soil surface. Gametophytes absorb nutrients directly from the surrounding soil and from symbiotic fungi. Fertilization takes place inside the archegonium, and the sporophyte develops directly on the gametophyte.

Human and ecological impacts Some whisk fern species are popular landscaping plants, particularly in Japan. The same species can be a serious pest in greenhouses.

Tmesipteris species
5 cm

FIGURE 29.34 Psilotophytes Are Extremely Simple Morphologically

Sphenophyta (or Equisetophyta) (Horsetails)

Although horsetails are prominent in the fossil record of land plants, just 15 species are known today. All 15 are in the genus *Equisetum*. Translated literally, *Equisetum* means "horse-bristle." Both the scientific name and the common name, horsetail, come from the brushy appearance of the stems and branches in some species (**Figure 29.35**). Horsetails may be locally abundant in wet habitats such as stream banks or marsh edges.

Adaptations to land Horsetails have an interesting adaptation that allows them to flourish in waterlogged, oxygen-poor soils. Horsetail stems are hollow, so oxygen readily diffuses down the stem to reach roots that cannot obtain oxygen from the surrounding soil.

Reproduction Asexual reproduction is common in sporophytes and occurs via fragmentation or the extension of rhizomes. From these rhizomes, two types of erect, specialized stems may grow: stems that contain tiny leaves and chloroplast-rich branches and that are specialized for photosynthesis, or stems that bear clusters of sporangia and produce huge numbers of spores by mitosis.

Life cycle Gametophytes perform photosynthesis but are small and short-lived. They normally produce both antheridia and archegonia, but in most cases the sperm-producing structure matures first. This pattern is thought to be an adaptation that minimizes self-fertilization and maximizes cross-fertilization.

Human and ecological impacts Horsetail stems are rich in silica granules. The glass-like deposits not only strengthen the stem but also make these plants useful for scouring pots and pans—hence these plants are often called "scouring rushes."

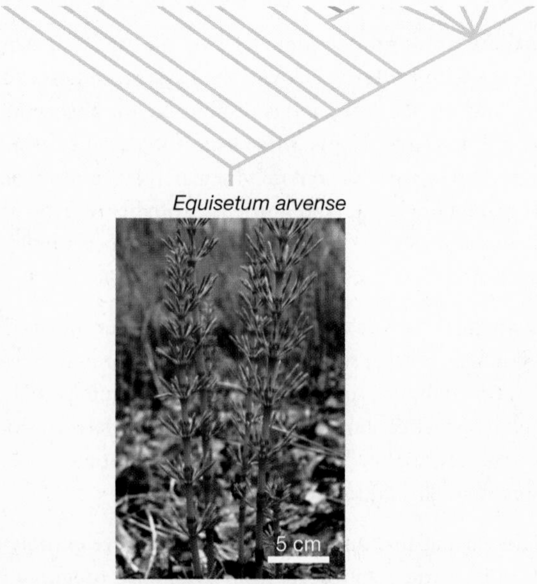
Equisetum arvense
5 cm

FIGURE 29.35 Horsetails Have Separate Reproductive and Vegetative Stalks

Pteridophyta (Ferns)

With 12,000 species, ferns are by far the most species-rich group of seedless vascular plants. They are particularly abundant in the tropics. About a third of the tropical species are epiphytes, usually growing on the trunks or branches of trees. Species that can grow epiphytically live high above the forest floor, where competition for light is reduced, without making wood and growing tall themselves. The growth habits of ferns are highly variable among species, however, and ferns range in size from rosettes the size of your little fingernail to 20-meter-tall trees (**Figure 29.36a**).

Adaptations to land Ferns are the only seedless vascular plants that have large, well-developed leaves—commonly called **fronds**. Leaves give the plant a large surface area with which to capture sunlight for photosynthesis.

Reproduction In a few species, gametophytes reproduce asexually via production of gemmae. Typically, species that can reproduce via gemmae never produce gametes or sporophytes. In most species, however, sexual reproduction is the norm.

Life cycle Although fern gametophytes contain chloroplasts and are photosynthetic, the sporophyte is typically the larger and longer-lived phase of the life cycle. In mature sporophytes, sporangia are usually found in clusters called **sori** on the undersides of leaves (**Figure 29.36b**).

Human and ecological impacts In many parts of the world, people gather the young, unfolding fronds, or "fiddleheads," of ferns in spring as food. Ferns are also widely used as ornamental plants in landscaping.

FIGURE 29.36 Ferns Are the Most Species-Rich Group of Seedless Vascular Plants

(a) Ferns range from sm all to tree sized.

Gonocormus minutus *Dicksonia antarctica*

(b) Fern sporan gia are often located on the underside of fronds.

Polypodium vulgare

Collection of sporangia

Seed Plants

The seed plants are a monophyletic group that consists of the gymnosperms—gnetophytes, cycads, ginkgoes, and conifers—and the angiosperms. Seed plants are defined by (1) the production of seeds and (2) the production of pollen grains. Recall from Section 29.2 that seeds are a specialized structure for dispersing embryonic sporophytes to new locations. Seeds are the mature form of an ovule, the female reproductive structure that encloses the female gametophyte and egg cell. Also recall from Section 29.3 that pollen grains are tiny, sperm-producing gametophytes that are easily dispersed through air as opposed to water. The structure and function of seed plants is the focus of Chapters 35 through 40.

Seed plants are found in virtually every type of habitat, and they adopt every growth habit known in land plants. Their forms range from mosslike mats to shrubs and vines to 100-meter-tall trees. Seed plants can be annual or perennial, with life spans ranging from a few weeks to almost five thousand years.

Of the several major questions remaining about the evolution of seed plants, perhaps the most prominent is whether the gymnosperms are monophyletic or paraphyletic. Although recent analyses of gene sequence data suggest that gymnosperm ancestry may indeed be traced back to a single common ancestor, the issue is still in doubt. As a result, the phylogenetic trees presented in this chapter show the four gymnosperm groups radiating from a single point. If continued research supports the hypothesis that gymnosperms are monophyletic, then the group would be represented as their own, distinct branch on the tree—meaning a major split occurred between gymnosperms and angiosperms. In the meantime, let's consider each of the five phyla of seed plants in more detail.

Gnetophyta (Gnetophytes)

The gnetophytes comprise about 70 species in three genera. One genus consists of vines and trees from the tropics. A second is made up of desert-dwelling shrubs, including what may be the most familiar gnetophyte—the shrub called Mormon tea, which is common in the deserts of southwestern North America (see Figure 29.6d). The third genus contains a single species that probably qualifies as the world's most bizarre plant (**Figure 29.37**): *Welwitschia mirabilis*, which is native to the deserts of southwest Africa. Although it has large belowground structures, the part that is above ground consists of just two strap-like leaves, which grow continuously from the base and die at the tips. They also split lengthwise as they grow and age.

Adaptations to land All of the living species make wood as a support structure.

Reproduction Pollen is transferred by the wind or by insects. Like other gymnosperms, seeds do not form inside an encapsulated structure.

Life cycle The microsporangia and megasporangia are arranged in clusters at the end of stalks, similar to the way flowers are clustered in some angiosperms.

Human and ecological impacts The drug ephedrine was originally isolated from species in the gnetophyte genus *Ephredra*, which is native to northern China and Mongolia. Ephedrine is used in the treatment of hay fever, colds, and asthma.

Welwitschia mirabilis
20 cm

FIGURE 29.37 Welwitschia Is a Bizarre Plant

Cycadophyta (Cycads)

The cycads are so similar in overall appearance to palm trees, which are angiosperms, that cycads are sometimes called "sago palms." Although cycads were extremely abundant when dinosaurs were present on Earth 150–65 million years ago, only about 140 species are living today. Most are found in the tropics (**Figure 29.38**).

Adaptations to land Cycads are woody; many are tree sized.

Reproduction Pollen is carried by insects (usually beetles or weevils) or, in some species, wind. Cycad seeds are large and are often brightly colored. The colors attract birds and mammals, both of which eat and disperse the seeds.

Life cycle Cycads are heterosporous. Each sporophyte individual bears either microsporangia or megasporangia, but not both.

Human and ecological impacts Cycads harbor large numbers of symbiotic cyanobacteria in specialized, above-ground root structures. The cyanobacteria are photosynthetic and fix nitrogen. The nitrogen acts as an important nutrient for nearby plants as well as the cycads themselves. Cycads are popular landscaping plants in some parts of the world.

Cycas revoluta
20 cm

FIGURE 29.38 Cycads Resemble Palms but Are Not Closely Related to Them

Ginkgophyta (Ginkgoes)

Although ginkgoes have an extensive fossil record, just one species is alive today. Leaves from the ginkgo, or maidenhair, tree are virtually identical in size and shape to those observed in fossil ginkgoes that are 150 million years old (**Figure 29.39**).

Adaptations to land Unlike most gymnosperms, the ginkgo is **deciduous**—meaning that it loses its leaves each autumn. This adaptation allows plants to be dormant during the winter or the dry season, when photosynthesis and growth are difficult.

Reproduction Pollen is transported by wind. Sperm have flagella, however. Once pollen grains land near the female gametophyte and mature, the sperm cells leave the pollen grain and swim to the egg cells.

Life cycle Sexes are separate—individuals are either male or female.

Human and ecological impacts Although today's ginkgo trees are native to east Asia, they are planted widely as an ornamental all over the world. The inside of the seed is eaten as a delicacy in some countries.

(a) Fossil ginkgo
Ginkgo huttoni
2 cm
(b) Living ginkgo
Ginkgo biloba
2 cm

FIGURE 29.39 The Ginkgo Tree Is a "Living Fossil"

Coniferophyta (Conifers)

The conifers include the familiar pines, cedars, spruces, and junipers. The group is named for a unique reproductive structure called the **cone**, in which microsporangia and megasporangia are produced (**Figure 29.40**). The 600 species of conifers living today include the world's largest and longest-lived plants. One of the bristlecone pines native to southwestern North America is at least 4900 years old, and redwood trees growing along the Pacific Coast of North America can reach heights of up to 115 m (375 ft) and trunk diameters of over 11 m (36 ft).

Adaptations to land All conifers are trees or large shrubs; many have narrow, needle-like leaves. Conifer leaves have a small amount of surface area, which is not optimal for capturing sunlight and performing photosynthesis. But because the small surface area reduces water loss from leaves, conifers grow in dry habitats or in cold environments where water is often frozen. Each needle lives from two to four years before being replaced.

Reproduction Conifers are wind pollinated. As in all seed plants, the female gametophyte is retained on the parent. Thus, fertilization and seed development take place in the female cone. Depending on the species, conifer seeds are dispersed by wind or by seed-eating birds or mammals.

Life cycle In most species, each individual has both male and female cones. Male cones bear pollen-producing microsporangia. Female cones contain megasporangia, which produce female gametophytes. In pines, pollination and fertilization are separated in time. Once a pollen grain lands on a female cone, it can take over a year for the male and female gametophytes to mature and for fertilization to take place. Seeds mature about 4 months after fertilization occurs. Thus, seeds may be produced almost 18 months after the pollination event.

Human and ecological impacts Conifers dominate forests that grow at high latitudes and high elevations. Their seeds are key food sources for a variety of birds, squirrels, and mice, and their wood is the basis of the building products and paper industries in many parts of the world. The paper in this book was made from the trunk of a conifer.

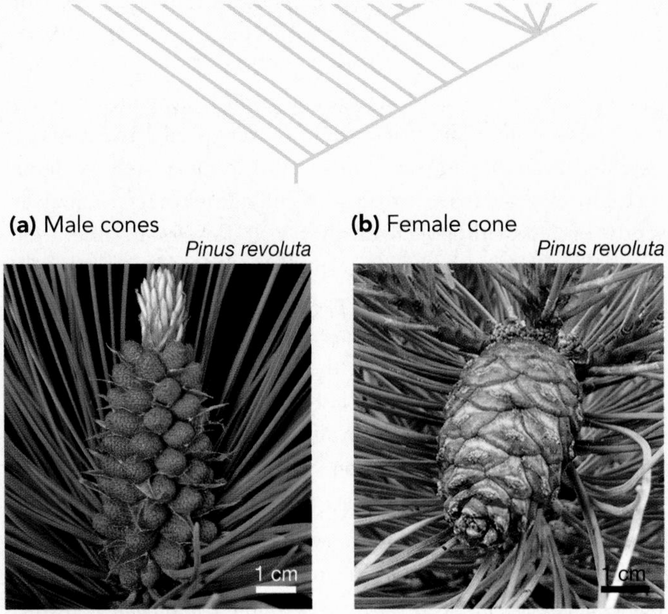
(a) Male cones
Pinus revoluta
1 cm
(b) Female cone
Pinus revoluta
1 cm

FIGURE 29.40 Male Cones Produce Microsporangia; Female Cones Produce Megasporangia

Anthophyta (Angiosperms)

The flowering plants, or angiosperms, are far and away the most species-rich lineage of land plants. Over 235,000 species have already been described. They range in size from *Lemna gibba*—a floating, aquatic species that is less than half a millimeter wide—to massive oak trees. Angiosperms thrive in desert to freshwater to rain forest environments and are found in virtually every habitat except the deep oceans. They are the most common and abundant plants in terrestrial environments other than northern and high-elevation coniferous forests.

The defining adaptation of angiosperms is the flower. **Flowers** are reproductive structures that hold either pollen-producing microsporangia or the megasporangia that produce megaspores and eggs, or both. In most cases, an individual flower contains both male and female sporangia surrounded by brightly colored **petals**, which are modified leaves. When a flower is first developing, the entire structure is enclosed by leaflike structures known as **sepals**, which protect the growing flower. Nectar-producing cells are often present at the base of the flower, and the color of petals helps to attract insects, birds, or bats that carry pollen from one flower to another. Some angiosperms are pollinated by wind, however. Wind-pollinated flowers lack both colorful petals and nectar-producing cells (**Figure 29.41**).

Adaptations to land In addition to flowers, angiosperms evolved vessels, the conducting cells that make water transport particularly efficient. Vessels are hollow tubes with empty ends, not unlike the pipes that conduct water in human habitations. Most angiosperms contain both tracheids and vessels.

Reproduction Unlike gymnosperms, angiosperms have a carpel, a structure within the flower that contains and encloses the ovule, which in turn encloses the female gametophyte. If the egg produced by the female gametophyte is fertilized, the ovule develops into a seed. The presence of enclosed ovules inspired the name angiosperm ("encased-seed"). When the ovary—the bottom portion of a carpel—matures, it forms fruit, which contains the seed or seeds. (The structures of flowers, seeds, and fruits are explored in more detail in Chapter 40.) In most cases, male gametophytes are carried to eggs by pollinators that are inadvertently dusted with pollen as they visit flowers to find food. Depending on the angiosperm species, self-fertilization may be common or absent. When fertilization does occur, it involves two sperm nuclei—one of which fuses with the egg to form the zygote, and a second that fuses with two nuclei in the female gametophyte to form a triploid ($3n$) nutritive tissue called **endosperm**. The involvement of two sperm nuclei is unique to angiosperms and is called **double fertilization.**

Life cycle Male and female gametophytes are small and contain a cell that divides by mitosis to form sperm or eggs. Like other seed plants, angiosperms retain their eggs on the parent plant but then release their embryos with a supply of nutrients in their seeds.

Human and ecological impacts It is almost impossible to overstate the importance of angiosperms to humans and other organisms. In most terrestrial habitats today, angiosperms supply the food that supports virtually every other species. For example, many insects eat flowering plants. Historically, the diversification of angiosperms correlated closely with the diversification of insects, which are by far the most species-rich lineage on the tree of life. It is not unusual for a single tropical tree to support dozens or even hundreds of insect species. Angiosperm seeds and fruits have also supplied the staple foods of virtually every human culture that has ever existed.

(a) Male flower

Acer negundo

(b) Female flower

Acer negundo

FIGURE 29.41 Wind-Pollinated Flowers Lack Colorful Petals and Nectar

ESSAY Genetic Diversity in Crop Plants

The spread of mechanized agriculture throughout the globe has produced some spectacular success stories. Due to increased use of fertilizers and new, high-yielding strains of wheat and rice, India became self-sufficient in grain production during the 1970s after decades of importing food. In India and many other countries, the use of high-yielding crop varieties has been instrumental in allowing food production to keep pace with rapid growth in the human population.

This so-called *green revolution* in agriculture has a downside, however. Because planting programs in numerous countries have relied on just a tiny number of source strains, the genetic diversity of crop systems is extremely low. In the United States, for example, virtually all soybeans that are grown trace their ancestry back to just 12 strains imported from northeast China. Similarly, most winter wheat being planted today is descended from two stocks that originated in Poland and Russia.

Low genetic diversity in crop species can lead to two problems: (1) an inability to improve strains further by selective breeding and (2) susceptibility to catastrophic diseases. Recall from Chapter 23 that for natural or artificial selection to occur, genetic variation must be present. Breeders find it difficult to improve crop populations with low genetic variability, because few alleles are present for selection to act on. Populations with low genetic variability are also more susceptible to disease, because disease resistance in plants depends on the presence of particular alleles. If a disease-causing agent evolves the ability to grow rapidly in a crop strain with low genetic variability, it is unlikely that any individuals of that strain will have alleles that allow them to fend off disease. The Irish potato famine of the mid-1800s (introduced in Chapter 28) was particularly devastating because most Irish farmers were planting the same potato variety, which happened to be particularly susceptible to the parasite that had infested the potato crops. To justify concerns about the lack of genetic diversity in crop plants, agricultural scientists also point to more recent events:

- In 1970 a corn blight wiped out 15 percent of the U.S. corn crop, sending prices skyrocketing and numerous farmers out of business.
- In 1980 over a million tons of sugarcane were destroyed by a single fungal disease in Cuba.
- Banana production in Central America was halved during the 1980s by a fungal disease called black sigatoka.

In response to scientists' concerns, governments ... have established seed banks to serve as repositories for genetic diversity.

In response to scientists' concerns, governments from around the world have established seed banks to serve as repositories for genetic diversity. These are labs where seeds from dozens or hundreds of different crop plant strains are stored. Worldwide, there are now over 700 seed collections with some 2.5 million entries. The United States alone spends $20 million each year to maintain and expand its seed banks.

At present, however, the banks represent a largely untapped resource. Researchers have only begun to use the banks in breeding programs designed to improve the disease resistance or productivity of commercial strains. The wild relatives of species such as tomato and rice have been particularly important in recent breeding programs, however. Using sophisticated maps of the tomato genome to identify the locations of particular alleles, researchers have designed crosses between commercial strains of tomato and wild species that dramatically improved fruit characteristics. With traditional breeding techniques, annual improvement in tomato traits such as fruit yield and percentage of soluble solids (important in the production of tomato paste) has been averaging less than 1 percent. After carefully designed crosses with a wild relative, the same traits improved 29 percent and 22 percent, respectively.

With accelerating human populations and continued loss of agricultural lands to urbanization, this research becomes increasingly urgent. Seed banks are an important way to preserve the biodiversity that exists within species that are particularly important to humans.

CHAPTER REVIEW

Summary of Key Concepts

The green plants include both the green algae and the land plants. Biologists study them because humans—along with virtually every other terrestrial organism—rely on plants for food and a healthy environment.

Plants hold soil and water in place, build soil, moderate extreme temperatures and winds, and provide food. In addition to these ecosystem services, humans depend on plants for food, fiber, and fuel. This dependence became extreme after the domestication of plant species beginning about 10,000 years ago and with the development of agricultural economies beginning about 8000 years ago. Artificial selection techniques have produced huge increases in the yields and dramatic changes in the characteristics of domesticated plants. Lack of genetic diversity in crop populations is a concern that biologists are trying to address.

Land plants were the first multicellular organisms to live on land. A series of key adaptations allowed them to survive with their tissues constantly exposed to dry air. In terms of total mass, plants dominate today's terrestrial environments.

The land plants evolved from green algae and were the first multicellular organisms to grow in terrestrial habitats. The evolution of cuticle allowed their tissues to be exposed to air without dying. The evolution of pores provided breaks in the cuticle and facilitated gas exchange, with CO_2 diffusing into leaves and O_2 diffusing out. Later, the evolution of guard cells allowed plants to control the opening and closing of pores in a way that maximizes gas exchange and minimizes water loss.

Vascular tissue evolved in a series of steps, beginning with simple water-conducting cells and tissues such as those observed in today's mosses. True vascular tissue has cells that are dead at maturity and that have secondary cell walls reinforced with lignin. As a result, vascular tissue conducts water *and* provides structural support that makes erect growth possible. Erect growth is important because it reduces competition for light. Tracheids are water-conducting cells found in all vascular plants; in addition, angiosperms have water-conducting cells called vessels.

Web Tutorial 29.1 Plant Evolution and the Phylogenetic Tree

Once plants were able to grow on land, a sequence of important evolutionary changes made it possible for them to reproduce efficiently—even in extremely dry environments.

All land plants are embryophytes, meaning that eggs and embryos are retained on the parent plant. Consequently, the developing embryo can be nourished by its mother. Seed plant embryos are then dispersed from the parent plant to a new location, encased in a protective housing, and supplied with a store of nutrients.

All land plants have alternation of generations. Over the course of land plant evolution, the gametophyte phase became reduced in size and life span and the sporophyte phase became more prominent. In seed plants, male gametophytes are reduced to pollen grains and female gametophytes are reduced to tiny structures that produce an egg. The evolution of pollen was an important breakthrough in the history of life, because sperm no longer needed to swim to the egg—tiny male gametophytes could be transported through the air via wind or insects.

Questions

Content Review

1. Which of the following groups is monophyletic?
   a. nonvascular plants
   b. green algae
   c. green plants
   d. gymnosperms

2. What is the difference between tracheids and vessels?
   a. Tracheids are dead at maturity; vessels are alive and are filled with cytoplasm.
   b. Tracheids have a thickened secondary cell wall; vessels also have gaps that lack any cell wall.
   c. Only tracheids have a thick secondary cell wall containing lignin.
   d. Only vessels have a thick secondary cell wall containing lignin.

3. Which of the following statement(s) are true?
   a. Phylogenies estimated from both morphological traits and DNA sequences indicate that the ancestor of the land plants was a green alga in the lineage called Charales.
   b. "Bryophytes" is a name given to the land plant lineages that do not have vascular tissue.
   c. The horsetails and the ferns form a distinct clade, or lineage. They have vascular tissue but reproduce via spores, not seeds.
   d. According to the fossil record and phylogenetic analyses, angiosperms evolved before the gymnosperms. Angiosperms have a unique reproductive structure called the flower. They have vessels but not tracheids.

4. The appearance of cuticle and stomata correlated with what event in the evolution of land plants?
   a. the first erect growth forms
   b. the first woody tissues
   c. growth on land
   d. the evolution of the first water-conducting tissues

5. What do seeds contain?
   a. male gametophyte and nutritive tissue
   b. female gametophyte and nutritive tissue
   c. embryo and nutritive tissue
   d. mature sporophyte and nutritive tissue

6. What is a pollen grain?
   a. male gametophyte
   b. female gametophyte
   c. male sporophyte
   d. sperm

Conceptual Review

1. Why are biologists convinced that the spores, cuticle, and sporangia that appear in the fossil record beginning about 475 million years ago represent the remains of the first land plants?
2. The introduction to this chapter claims that the land plants were the first truly terrestrial organisms. But most biologists contend that saturated soils on the continents were undoubtedly teeming with bacteria, archaea, and protists long before land plants evolved. In light of this, what does the phrase "truly terrestrial" mean? After answering this question, list the physiological problems presented by growth on land. Beside each entry in this list, note the adaptation(s) that allowed land plants to overcome the problem.
3. In the experiments with orchid pollination reviewed in Section 29.3, why did the researchers put yarn around the flowers of control individuals?
4. Land plants may have reproductive structures that (1) protect gametes as they develop; (2) allow fertilization to occur in the absence of water, (3) nourish developing embryos, (4) provide a protective coat so that offspring can be dispersed away from the parent plant, and (5) provide nutritious tissue around seeds that facilitates dispersal by animals. Name each of these five structures, and state which land plant group or groups have each structure.
5. Explain the difference between homosporous and heterosporous plants. Some animal species are *hermaphroditic*—meaning that each individual contains both male and female reproductive organs—while other animal species have separate sexes. Which type of animal is most similar to a homosporous plant?
6. Using what you already know, compare and contrast gametangia and the retention of embryos with analogous structures and processes observed in mammalian reproduction.

Group Discussion Problems

1. What is the significance of the observation that some members of the Charales synthesize sporopollenin and lignin?
2. In many vascular plants, the end walls of tracheids form an extreme diagonal (see Figure 29.11a). Regarding the cell's ability to transport water, why does this make sense? (Hint: Consider how the cell's ratio of surface area to volume affects transport functions.)
3. Angiosperms such as grasses, oaks, and maples are wind pollinated. The ancestors of these subgroups were probably pollinated by insects, however. As an adaptive advantage, why might a species "revert" to wind pollination? Why is it logical to observe that wind-pollinated species usually grow in dense stands containing many individuals of the same species? Why is it logical to observe that in wind-pollinated deciduous trees, flowers form very early in spring—before leaves form?
4. You have been hired as a field assistant for a researcher interested in the evolution of flower shape in orchids. Each of the five species she is working on is pollinated by a different insect. Design an experiment to determine which parts of the five flower types are most important in attracting the pollinators. Assume that you can change the flower's color with a dye and that you can remove petals or nectar stores, add particular scents, add nectar by injection, or switch parts among species by cutting and gluing.

Answers to Multiple-Choice Questions 1. c; 2. b; 3. a, b, c; 4. c; 5. c; 6. a

www.prenhall.com/freeman is your resource for the following: Web Tutorials; Online Quizzes and other Online Study Guide materials; Answers to Conceptual Review Questions; Solutions to Group Discussion Problems; Answers to Figure Caption Questions and Exercises; and Additional Readings and Research.

30 Fungi

KEY CONCEPTS

- Fungi are important in part because many species live in close association with land plants. They supply plants with key nutrients and decompose dead wood. They are the master recyclers of nutrients in terrestrial environments.
- All fungi make their living by absorbing nutrition from living or dead organisms. Fungi secrete enzymes so that digestion takes place outside their cells. Their morphology provides a large amount of surface area for efficient absorption.
- Many fungi have unusual life cycles. It is common for species to have a long-lived heterokaryotic stage, in which cells contain haploid nuclei from two different individuals. Although most species reproduce sexually, very few species produce eggs and sperm.

Amanita muscaria lives in association with living tree roots. The mushrooms you see are only reproductive structures—the vast majority of this individual exists underground. *Amanita muscaria* mushrooms contain molecules that can induce hallucinations in mammals. This species has been used in religious ceremonies by some native cultures but is considered highly toxic.

Three major lineages of large, multicellular eukaryotes occupy terrestrial environments: the land plants, the animals, and the fungi. When it comes to making a living, the species in these three groups use radically different strategies. Land plants make their own food through photosynthesis. Animals eat plants, protists, fungi, or each other. Fungi absorb their nutrition from other organisms—dead or alive.

Fungi that absorb nutrients from dead organisms are the world's most important decomposers. For example, although a few types of organisms are capable of digesting the cellulose in plant cell walls, fungi and a few species of bacteria are the only organisms capable of completely digesting both the lignin and cellulose that make up wood. Without fungi, Earth's surface would be piled so high with dead tree trunks and branches that there would be almost no room for animals to move or plants to grow.

Other fungi specialize in absorbing nutrients from living organisms. If fungi absorb these nutrients without providing any benefit in return, then they lower the fitness of their host organism and act as parasites. If you've ever had athlete's foot or a vaginal yeast infection, you've hosted a parasitic fungus.

The vast majority of fungi that live in association with other organisms benefit their hosts, however. In these cases, fungi are not parasites but **mutualists.** Consider that the roots of virtually every land plant in the world are colonized by an array of mutualistic fungi. In exchange for sugars that are synthesized by the host plant, the fungi provide the plant with key nutrients such as nitrogen and phosphorus. Without these nutrients, the host plants grow much more slowly or even starve. It is not possible to overstate the importance of these relationships between living land plants and the fungi that cover their roots. In the soils beneath every prairie, forest, and desert, an underground economy is flourishing. Plants are trading the sugar they manufacture for nitrogen or phosphorus atoms that are available from fungi. These plant-fungal associations are the world's most extensive bartering system. The soil around you is alive with an enormous network of fungi that are feeding the plants you see above ground.

In short, fungi are the master traders and recyclers in terrestrial ecosystems. Some fungi release nutrients from dead plants and animals; others transfer these nutrients to living plants. Because they recycle carbon, nitrogen, phosphorus, and other nutrients, fungi have a profound influence on the productivity and biodiversity of terrestrial ecosystems. In terms of nutrient cycling on the continents, fungi make the world go around.

30.1 Why Do Biologists Study Fungi?

Given their importance to life on land and their intricate relationships to other organisms, it's no surprise that fungi are fascinating to biologists. But there are important practical reasons for studying fungi as well. They nourish the plants that nourish us. They affect global warming, because they are critical to the carbon cycle on land. Unfortunately, a handful of species can cause debilitating diseases in humans and crop plants. Let's take a closer look at some of the ways that fungi affect human health and welfare.

Fungi Feed Land Plants

Fungi that live in close association with plant roots are said to be **mycorrhizal** ("fungal-root"). When biologists first discovered how extensive these fungal-plant associations are, they asked an obvious question: Does plant growth suffer if mycorrhizal fungi are absent? **Figure 30.1** shows a typical result from many experiments and provides a convincing answer. In this case, tree seedlings were grown in the presence and absence of the mycorrhizal fungi normally found on their roots. The photographs document that this tree species grows 3 to 4 times faster in the presence of its normal fungal associates than it does without them. For foresters, farmers, and ranchers, the presence of normal mycorrhizal fungi can mean the difference between profit and loss. Biologists study fungi in part because fungi are critical to the productivity of forests, croplands, and rangelands.

FIGURE 30.1 Plants Grow Better in the Presence of Mycorrhizal Fungi
Typical experimental results when plants are grown with and without their normal mycorrhizal fungi.

Fungi Speed the Carbon Cycle on Land

The introduction to this chapter claims that fungi are master decomposers and recyclers. To back up this claim, consider two particularly dramatic events documented in the fossil record. One episode was based on a lack of fungi, the other on an abundance of fungi:

1. Researchers who examine the fossils present in coal from the Carboniferous period find remarkably few fungi that are capable of degrading dead plant material. Fungi that make their living by digesting dead plant material are called **saprophytes** ("rotten-plants"). Noting the "dip" in the fossil record of saprophytic fungi, researchers hypothesized that their absence was responsible for the enormous buildup of dead plant material that occurred during the Carboniferous period. Recall from Chapter 29 that deposits of compressed, partially decayed plant material are called *peat* and that coal formed when peat was produced during the Carboniferous period was buried under other sediments and subjected to heat and pressure. Peatlands that exist today are highly acidic; so biologists hypothesize that fungi did not grow in the coal-forming swamps of the Carboniferous period, because the pH was too low. The message is that coal exists today because conditions were too acidic for fungi to do their job.
2. At the end of the Permian period, 250 million years ago, the greatest mass extinction in the history of life occurred. It is estimated that in substantially less than a million years, over 90 percent of multicellular species were wiped out. When the end-Permian extinction was first recognized, biologists thought it had affected primarily marine organisms. But then other researchers documented a huge "fungal spike" in rocks that formed in terrestrial

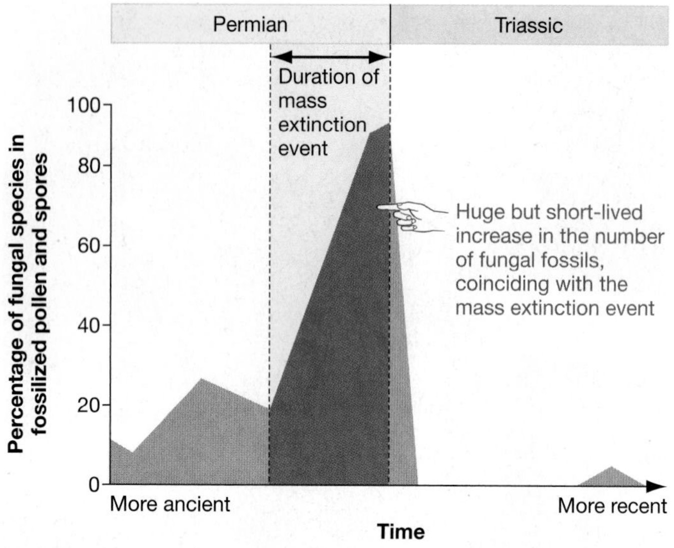

FIGURE 30.2 Fungal Fossils Spike during the End-Permian Extinction
The time interval documents changes in the abundance of fungi before, during, and after the mass extinction event that occurred at the boundary of the Permian and Triassic periods about 250 million years ago.

environments. As **Figure 30.2** shows, the spike is a dramatic but short lived increase in the number of fungal fossils that coincided with the mass extinction event. The explosion of fungal fossils was the first indication that land plants were also devastated during this interval. The hypothesis is that a massive die-off in trees and shrubs produced gigantic quantities of rotting wood and led to the explosion of fungal abundance documented in the spike.

Saprophytic fungi play a key role in today's terrestrial environments as well. To understand why, recall from Chapter 29 that cells in the vascular tissues of land plants have walls containing both lignin and cellulose. Wood forms when stems grow in girth by adding layers of vascular tissue. As mentioned in the introduction to this chapter, the vast majority of organisms that can completely digest wood are fungi. Fungi break down wood into sugars and other small organic compounds that can be used as food by fungi and other organisms.

Figure 30.3 illustrates the consequences of these facts by highlighting the role that fungi play as carbon atoms cycle through today's terrestrial environments. Note that there are two basic components of the **carbon cycle**: (1) the fixation of carbon by land plants—meaning that carbon in atmospheric CO_2 is converted to cellulose, lignin, and other complex organic compounds in the bodies of plants; and (2) the release of CO_2 from plants, animals, and fungi as the result of cellular respiration—meaning the oxidation of glucose and production of the ATP that sustains life. The fundamental point is that, for most carbon atoms, fungi connect the two components. If fungi had not evolved the ability to digest lignin and cellulose soon after land plants evolved the ability to make these compounds, carbon atoms would have been sequestered in wood for millennia instead of being rapidly recycled into glucose molecules and CO_2. Terrestrial environments would be radically different than they are today and probably much less productive. On land, fungi make the carbon cycle turn much more rapidly than it would without fungi. The nutrients that fungi release feed a host of other organisms.

Fungi Have Important Economic Impacts

In humans, parasitic fungi cause athlete's foot, vaginitis, diaper rash, ringworm, pneumonia, and thrush, among other miseries. But even though these maladies can be serious, in reality only about 30 (out of some 80,000) species of fungi regularly cause illness in humans. Compared with the frequency of diseases caused by bacteria, viruses, and protists, the incidence of fungal infections in humans is low. In addition, soil-dwelling fungi have been the source of many of the most important antibiotics currently being prescribed against bacterial infections (see Chapter 27). On balance, fungi have been much more helpful than harmful in human and veterinary medicine.

The major destructive impact that fungi have on people is through our food supply. Fungi known as rusts, smuts, mildews, wilts, and blights cause annual crop losses computed in the billions of dollars. These fungi are particularly troublesome in wheat, corn, barley, and other grain crops (**Figure 30.4a**). Saprophytic fungi are also responsible for enormous losses due to spoilage—particularly for fruit and vegetable growers (**Figure 30.4b**). But fungi have important positive impacts on the human food supply as well. Domesticated by humans for thousands of years, fungi are essential to the manufacture of bread, soy sauce, cheese, beer, wine, whiskey, and other products. Enzymes derived from fungi are used to improve the characteristics of foods ranging from fruit juice and candy to meat.

In nature, recent epidemics caused by fungi have killed 4 billion chestnut trees and tens of millions of American elm trees in North America. The fungal species responsible for these epidemics were accidentally imported on species of chestnut and elm native to other regions of the world. When the fungi arrived in North America, switched hosts, and began growing in chestnuts and elms native to North America, the results were catastrophic. The local chestnut and elm populations had virtually no genetic resistance to the pathogens and quickly succumbed. The epidemics radically altered the composition of upland and floodplain forests in the eastern United States. Before these fungal epidemics occurred, chestnuts and elms dominated these habitats.

Fungi clearly affect a wide range of species, including humans. How do biologists go about studying them? More specifically, what tools are helping researchers understand the diversity of fungi?

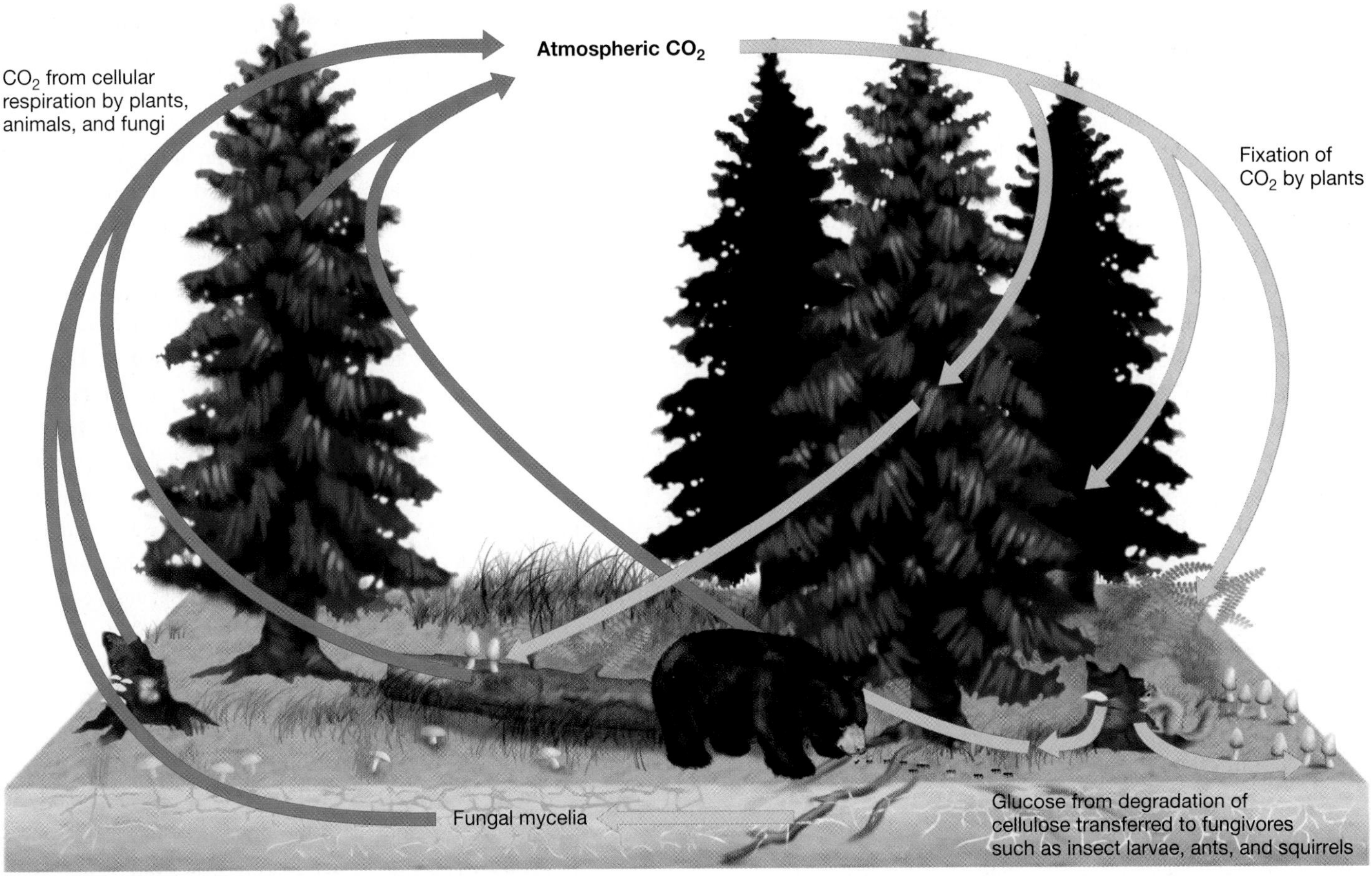
Atmospheric CO_2
CO_2 from cellular respiration by plants, animals, and fungi
Fixation of CO_2 by plants
Fungal mycelia
Glucose from degradation of cellulose transferred to fungivores such as insect larvae, ants, and squirrels

FIGURE 30.3 Fungi Speed Up the Cycling of Carbon Atoms on Land
Carbon atoms cycle through terrestrial ecosystems. If fungi could not degrade lignin to CO_2 and cellulose to glucose, most carbon would eventually be tied up in indigestible woody tissues. As a result, the cycle would slow dramatically.
EXERCISE Draw an X through the arrows that would not exist if fungi could not digest lignin and cellulose.

(a) Parasitic fungi infect corn and other crop plants.

(b) Saprophytic fungi rot fruits and vegetables.

FIGURE 30.4 Fungi Cause Problems with Crop Production and Storage
(a) A wide variety of grain crops are parasitized by fungi. Corn smut is a serious disease in sweet corn, although in Mexico the smut fungus is eaten as a delicacy. **(b)** Fungi decompose fruits and vegetables as well as leaves and tree trunks.

30.2 How Do Biologists Study Fungi?

To date, about 80,000 species of fungi have been described and named; about 1000 more are discovered each year. But the fungi are so poorly studied that the known species are widely regarded as a tiny fraction of the actual total. To predict the actual number of species alive today, David Hawksworth looked at the ratio of vascular plant species to fungal species in the British Isles—the area where the two groups are the most thoroughly studied. According to Hawksworth's analysis, there is an average of six species of fungus for every species of vascular plant on these islands. If this ratio holds worldwide, then the total of 275,000 vascular plant species implies that there are 1.65 million species of fungi.

Although this estimate sounds large, recent data on fungal diversity suggest that it may be an underestimate. Consider what researchers found recently when they analyzed fungi growing on Barro Colorado Island, Panama: Living on the healthy leaves of just two tropical tree species were a total of 418 distinct morphospecies of fungi. (Recall from Chapter 25 that morphospecies are distinguished from each other by some aspect of morphology.) Because over 300 species of trees and shrubs grow on Barro Colorado, the data suggest that tens of thousands of fungi may be native to this island alone. If further work on fungal diversity in the tropics supports these conclusions, there may turn out to be many millions of fungal species.

This viewpoint of fungal diversity was reinforced by a recent analysis of the fungi living in conjunction with the roots of a single species of grass native to Eurasia. In this study, researchers used the direct sequencing approach, introduced in Chapter 27, to analyze the gene that codes for the RNA molecule in the small subunit of fungal ribosomes. The data showed that a total of 49 phylogenetic species were living in conjunction with the grass roots. (The phylogenetic species concept was introduced in Chapter 25.) Most of the species had never before been described, and several represented completely new lineages of fungi. Biologists are only beginning to realize the extent of species diversity in fungi.

Let's consider how biologists are working to make sense of all this diversity, beginning with an overview of fungal morphology.

Analyzing Morphological Traits

Compared with animals and land plants, fungi have very simple bodies. Only two growth forms occur among them:

1. Single-celled forms called **yeasts** (**Figure 30.5a**), and
2. Multicellular, filamentous structures called **mycelia** (singular: **mycelium**; **Figure 30.5b**).

Many species of fungus grow as either a yeast or a mycelium, but some regularly adopt both growth forms.

Because yeasts are small and easy to manipulate in the laboratory, they have been extremely important model organisms in cell biology and genetics (**Box 30.1**). But because most fungi form mycelia and because this body type is so fundamental to the absorptive mode of life, most studies of fungal morphology have focused on them. Let's take a closer look at the structure and function of a fungal mycelium.

The Nature of the Fungal Mycelium If food sources are plentiful, mycelia can be long lived and grow to be extremely large. Researchers recently discovered an individual mycelium growing across 1300 acres (6.5 km^2) in Oregon. This is an area substantially larger than most college campuses. The biologists estimated the individual's weight at hundreds of tons and its age at thousands of years, making it one of the largest and oldest organisms known.

Although most mycelia are much smaller and shorter lived than the individual in Oregon, all mycelia are dynamic. Mycelia constantly grow in the direction of food sources and die back in areas where food is running out. The body shape of a fungus can change almost continuously throughout its life.

The individual filaments that make up a mycelium are called **hyphae** (singular: **hypha**). Most hyphae are haploid or **heterokaryotic** ("different kernel"), meaning each cell contains two haploid nuclei—one from each parent. Hyphae are also similar to one another in structure. As **Figure 30.6a** shows, they are long, narrow filaments that branch frequently. In most fungi, each filament is broken into cell-like compartments by cross-walls called **septa** (singular: **septum**; **Figure 30.6b**). Unlike the cell walls and membranes of plants and animals, septa do not close off segments of hyphae completely. Instead, gaps called **pores** enable a wide variety of materials, even organelles and nuclei, to flow from one compartment to the next. Septa may have single large openings or a series of small gaps that give the septum a sieve-like appearance. Because nutrients, mitochondria, and even genes can flow though the entire mycelium—at least to a degree—the fungal mycelium is intermediate between a multicellular land plant or animal and an enormous single-celled organism. Some fungal species are even **coenocytic**—they lack septa and therefore have cytoplasm and many nuclei that are continuous throughout the mycelium.

Perhaps the most important aspect of mycelia and hyphae, however, is their shape. Because hyphae are tubes and because mycelia are composed of complex, branching networks, the body

(a) Single-celled fungi are called yeasts.

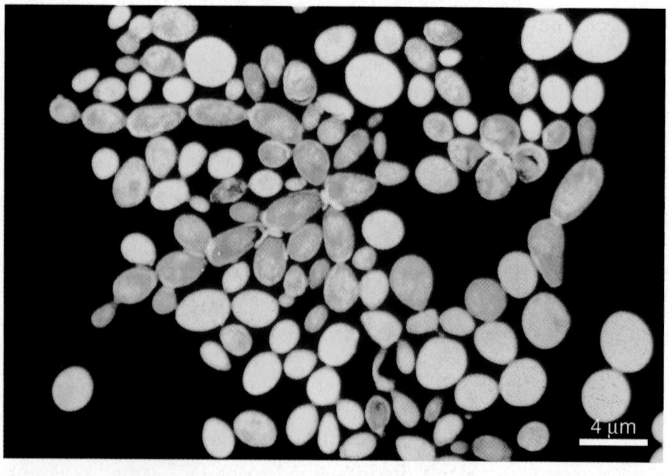

(b) Multicellular fungi have weblike bodies called mycelia.

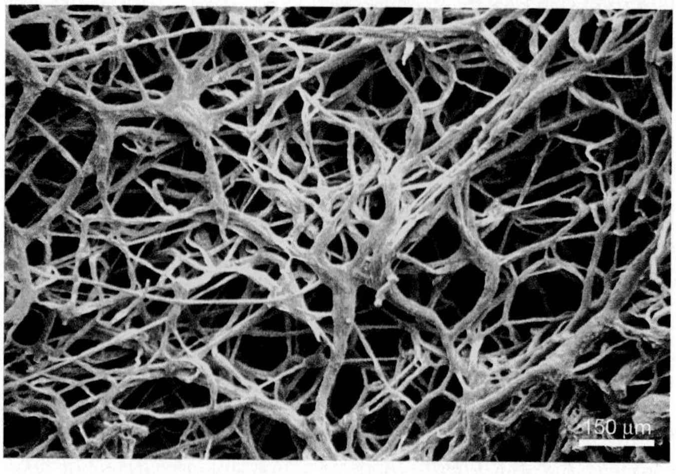

FIGURE 30.5 Fungi Have Just Two Growth Forms
Fungi grow **(a)** as single-celled yeasts and/or **(b)** as multicellular mycelia made up of long, thin, highly branched filaments.

BOX 30.1 A Model Organism: *Saccharomyces cerevisiae*

The yeast *Saccharomyces cerevisiae* (see Figure 30.5a) is a favorite experimental subject for researchers interested in the basic cell biology and molecular genetics of eukaryotes. Because it is unicellular and easy to culture and manipulate in the lab, *S. cerevisiae* has become the organism of choice for research on control of the cell cycle and regulation of gene expression in eukaryotes. The morphology of this organism is so simple that it provides an example of a "pure" eukaryotic cell type—one that is suitable for experiments on how cell division occurs and how particular genes are turned on and off. Just as *Escherichia coli* serves as the model bacterial cell, *Saccharomyces* serves as the model eukaryotic cell. For example, research has confirmed that several genes that control cell division and DNA repair in yeast have homologs in humans that, when mutated, contribute to cancer. Strains of yeast that carry these mutations are now being used to test drugs that might be effective against cancer. *Saccharomyces cerevisiae* was also the first eukaryote whose genome was completely sequenced.

Saccharomyces cerevisiae is a member of the Ascomycota. In nature it lives on the skins of wild grapes and other fruits, which it uses as a source of food. Thousands of years ago, however, humans learned how to use the yeast *Saccharomyces* in baking and brewing. It was the first fungus to be cultivated, and domesticated strains of yeast are still vital to food and beverage production. *Saccharomyces* is useful in these industries because it ferments sugars when it grows in the absence of oxygen. The carbon dioxide that forms as a by-product of fermentation by these cells makes bread rise and gives beer and champagne their fizz. Several *Saccharomyces* species are used in brewing, winemaking, and distilling; the ethanol that they produce as a waste product during fermentation acts as the psychoactive ingredient in beer, wine, and spirits.

of a fungus has the highest surface-area-to-volume ratio possible in a multicellular organism. To drive this point home, consider that the hyphae found in any fist-sized ball of rich soil typically have a surface area equivalent to half a page of this book. This surface area is important because it makes absorption extremely efficient, and fungi make their living via absorption.

The extraordinarily high surface area in a mycelium has a downside, however: Fungi are prone to drying out, because the amount of water that evaporates from an organism is a function of its surface area. Due to the high surface area of mycelia, fungi are most abundant in moist habitats. Fungi often endure dry conditions in the form of tough, watertight spores.

To summarize, mycelia are an adaptation to the absorptive lifestyle of fungi. Thus, it is logical to observe that reproductive organs—not feeding structures—are the only thick, fleshy structures that fungi produce. Mushrooms, puffballs, and other dense, multicellular structures that arise from mycelia do not absorb food. Instead, they function only in reproduction. Typically they are the only part of a fungus that is directly exposed to the air, where drying is a problem. The

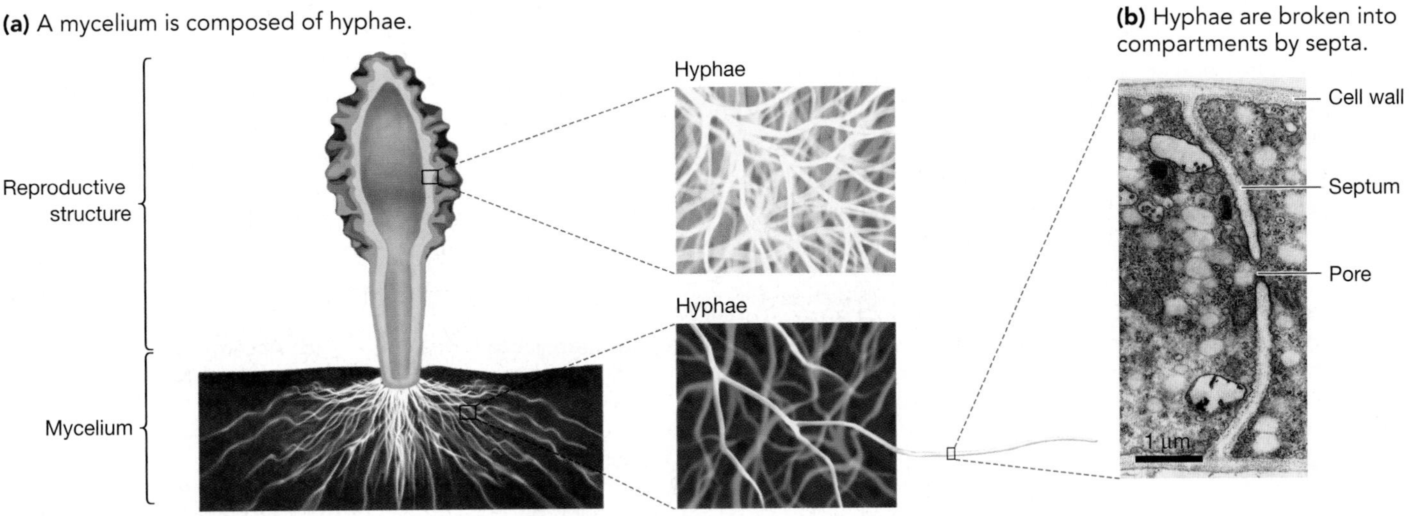
(a) A mycelium is composed of hyphae.
Reproductive structure
Mycelium
Hyphae
Hyphae
(b) Hyphae are broken into compartments by septa.
Cell wall
Septum
Pore
1 μm

FIGURE 30.6 Multicellular Fungi Have Unusual Bodies
(a) The feeding structure of a fungus is a mycelium, which is made up of hyphae. In some species, hyphae come together to form multicellular structures such as mushrooms, brackets, or morels that emerge from the ground. **(b)** Hyphae are often divided into cell-like compartments by partitions called septa, which are broken by pores. As a result, the cytoplasm of different compartments is continuous.

mass of filaments on the inside of mushrooms and other large reproductive structures is protected from drying by the densely packed hyphae forming the surface. However, few species of fungi make the reproductive structures called mushrooms.

Reproductive Structures When biologists study diversity among land plants, protists, or animals, they normally begin by comparing the morphologies of various species. Because fungal mycelia are so simple and are so similar among species, researchers have focused on the distinctive morphological structures that fungi produce during sexual reproduction. On the basis of these structures, fungi fall into four major groups:

1. The **Chytridiomycota (chytrids)** live primarily in water, and they are the only fungi that have motile cells. The spores that chytrids produce during asexual reproduction have flagella, as do the gametes produced by members of this group during sexual reproduction. These swimming cells are reproductive structures that distinguish chytrids from other fungi (**Figure 30.7a**). Structurally, chytrid flagella are similar to the flagella in the sperm cells of animals.
2. The hyphae of **Zygomycota** are haploid and come in several **mating types**. Instead of having morphologically distinct males and females that produce sperm and eggs, hyphae of different mating types look identical but will not combine unless the individuals have different alleles of one or more genes involved in mating. If chemical messengers released by two hyphae indicate that they are of different mating types, the individuals may become yoked together as shown in **Figure 30.7b**. (The Greek root *zygos* means to be yoked together like oxen. Translated literally, *Zygomycota* means "yoked-together fungi.") Cells from the yoked hyphae fuse to form a spore-producing structure called a **zygosporangium**. Yoked hyphae that form a zygosporangium are the reproductive structure unique to this group.
3. Mushrooms, bracket fungi, and puffballs are among the complex reproductive structures produced by members of the **Basidiomycota**, or **club fungi**. Inside these structures, specialized cells called **basidia** ("little-pedestals") form at the ends of hyphae and produce spores (**Figure 30.7c**). Only members of the Basidiomycota produce basidia.
4. Members of the **Ascomycota**, also called **sac fungi**, produce complex reproductive structures—the largest of which are often cup shaped. The tips of hyphae inside these structures produce distinctive sac-like cells called **asci** (singular: **ascus**; **Figure 30.7d**). An ascus is a spore-producing structure found only in Ascomycota.

In sum, morphological studies allowed biologists to describe and interpret the growth habit of mycelia as an adaptation that makes absorption extremely efficient. Careful analyses of morphological features also allowed researchers to identify four major lineages of fungi, based on the nature of their reproductive structures. How are members of these four major groups related? And how are fungi related to other eukaryotes?

Evaluating Molecular Phylogenies

Biologists who observed that fungi had four distinct types of reproductive structures hypothesized that each type of structure identified a monophyletic group. As a result, the Chytridiomycota, Zygomycota, Ascomycota, and Basidiomycota have traditionally been recognized as separate phyla. Based on the assumption that fungi evolved from aquatic protists, researchers hypothesized that chytrids were the most basal group. As **Figure 30.8a** shows, it was also hypothesized that ascomycetes and basidiomycetes—with their more-complex reproductive structures—were the most recent to evolve.

What Are the Relationships among Major Fungal Groups? To test the hypotheses that chytrids were the most basal group and that the four phyla of fungi are monophyletic, biologists have sequenced the gene that codes for the RNA found in the ribosome's small subunit from an array of fungal species. The researchers used these data to estimate the phylogeny of fungi. The results supported the hypothesis of monophyly for the ascomycetes and basidiomycetes, but conflicted with the claim of monophyly for chytrids and zygomycetes. Although many aspects of the tree in **Figure 30.8b** are still uncertain, several important conclusions can be drawn:

- It is likely that a lineage of chytrids forms the most basal group of fungi. This result is consistent with the hypothesis

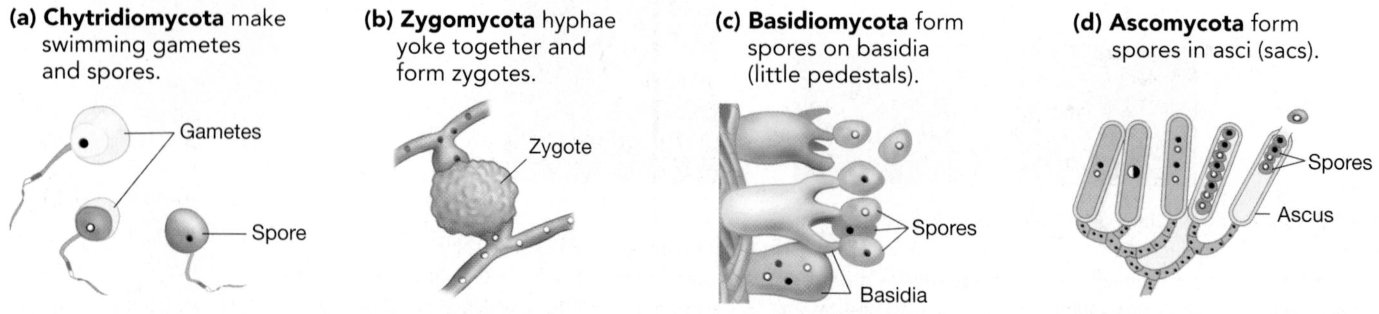
(a) Chytridiomycota make swimming gametes and spores.
Gametes
Spore
(b) Zygomycota hyphae yoke together and form zygotes.
Zygote
(c) Basidiomycota form spores on basidia (little pedestals).
Spores
Basidia
(d) Ascomycota form spores in asci (sacs).
Spores
Ascus

FIGURE 30.7 Four Distinct Reproductive Structures Are Observed in Fungi

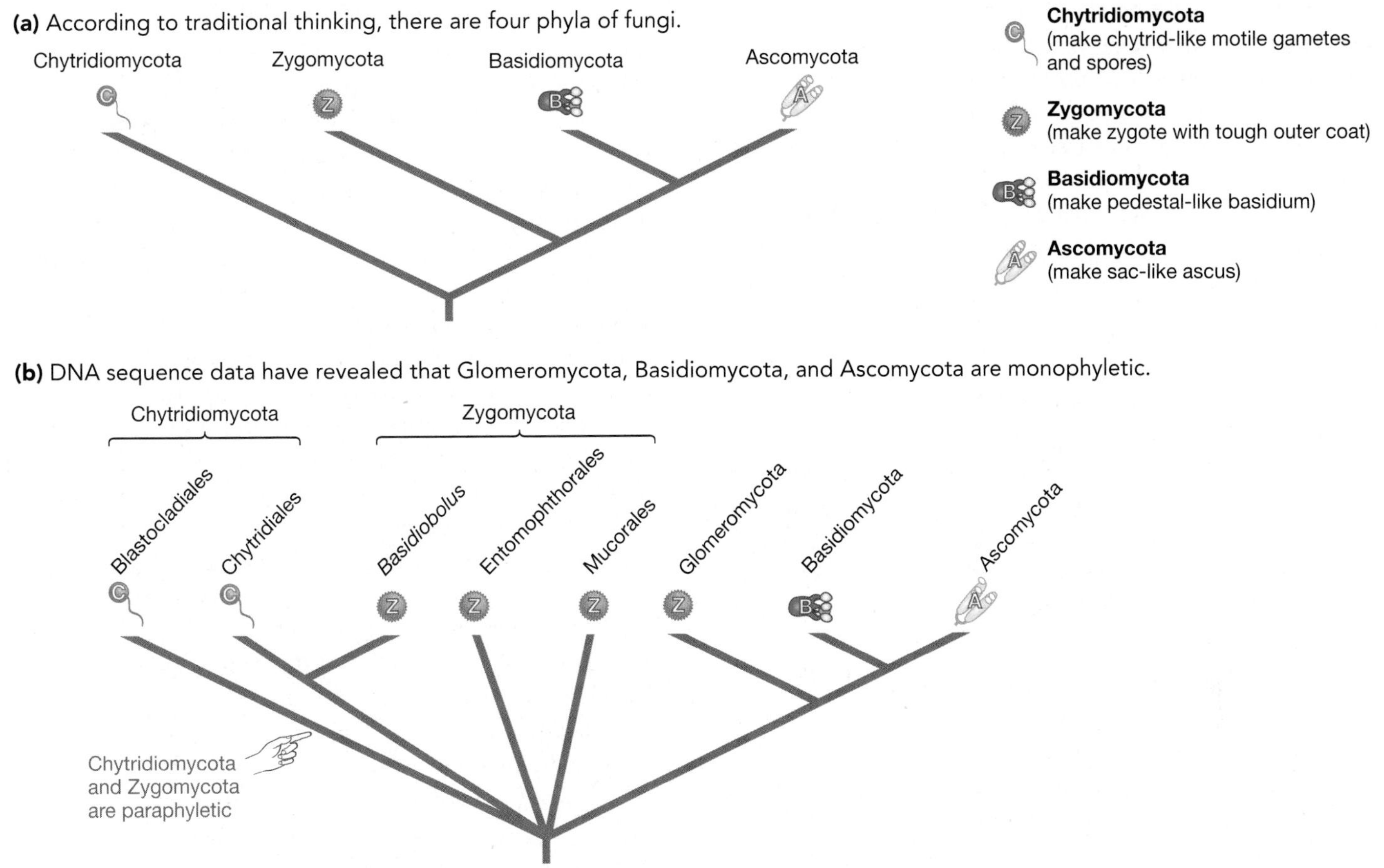
(a) According to traditional thinking, there are four phyla of fungi.
Chytridiomycota
Zygomycota
Basidiomycota
Ascomycota
Chytridiomycota
(make chytrid-like motile gametes and spores)
Zygomycota
(make zygote with tough outer coat)
Basidiomycota
(make pedestal-like basidium)
Ascomycota
(make sac-like ascus)
(b) DNA sequence data have revealed that Glomeromycota, Basidiomycota, and Ascomycota are monophyletic.
Chytridiomycota
Zygomycota
Blastocladiales
Chytridiales
Basidiobolus
Entomophthorales
Mucorales
Glomeromycota
Basidiomycota
Ascomycota
Chytridiomycota and Zygomycota are paraphyletic

FIGURE 30.8 A Phylogeny of the Fungi
(a) A traditional phylogenetic tree based on the four reproductive structures observed in fungi. **(b)** An updated phylogenetic tree based on comparisons of morphological traits and analyses of DNA sequence data. The key indicates the types of sexual reproductive structures in each major lineage.

that fungi evolved from aquatic ancestors. It also suggests that fungi made the transition to land fairly early in their evolution. But according to the data on hand, it is still not clear whether fungi moved from water to land just once or multiple times.

- The Chytridiomycota and Zygomycota appear to be paraphyletic. That is, these groups do not include all descendants of a single common ancestor. Instead, fungal groups with swimming gametes and yoked hyphae are interspersed throughout the basal parts of the tree. If the relationships implied by the molecular data are correct, then either swimming gametes and yolked hyphae evolved more than once, or both were present in a common ancestor but then were lost in certain lineages.
- As predicted by the morphological analyses, the Basidiomycota and Ascomycota are monophyletic and appeared late during the diversification of fungi.

The phylogenetic tree of Fungi is a work in progress. Efforts to understand the relationships of subgroups within the Basidiomycota and Ascomycota are in their infancy, and researchers are still struggling to understand the relationships among the dozens of subgroups of Chytridiomycota and Zygomycota. Fortunately, a coalition of researchers has recently formed to undertake a series of new and more extensive analyses of DNA sequence data, with the goal of working out the details of the fungal phylogeny. The collaboration is important, because a comprehensive phylogeny of the Fungi should clarify exactly how the masters of the absorptive lifestyle diversified and how the diversification of fungi relates to the diversification of land plants.

In the meantime, it is interesting to note that parasitic, saprophytic, and mutualistic species exist within both the Ascomycetes and the Basidiomycetes. This observation supports the hypothesis that the entire range of fungal lifestyles evolved independently in these two lineages. This conclusion is similar to the situation in protists. Recall from Chapter 28 that species that photosynthesize, ingest food, or absorb food are found within each of the eight major protist lineages and that each of these eight monophyletic groups is distinguished by an aspect of cell structure. In both fungi and protists, an array of feeding strategies evolved in lineages that are distinguished by some fundamental morphological feature.

Where Do Fungi Fit on the Tree of Life? Although understanding the relationships among the four major groups of fungi and their subgroups has been difficult, researchers have made rapid progress recently on the question of where fungi fit on the overall tree of life. Based on analyses of similarities and differences in the DNA sequences of a number of genes, researchers are increasingly confident that the phylogenetic tree in **Figure 30.9** is an accurate depiction of how fungi relate to other eukaryotes. The tree supports two conclusions about fungi:

1. Their closest living relatives are unusual, single-celled parasites called **microsporidians.** This was an important result, because microsporidians cause serious disease in bee colonies, silkworm colonies, and people with AIDS. Based on the close relationship between fungi and microsporidians, researchers are testing the hypothesis that **fungicides**—meaning molecules that are lethal to fungi—may prove to be effective in combating microsporidian infections as well.
2. Fungi are much more closely related to animals than they are to land plants. In addition to the DNA sequence data, several morphological traits link animals and fungi. For example, most animals and fungi synthesize the tough structural material called chitin (see Chapter 7). **Chitin** is a prominent component of the cell walls of fungi. Also, the flagella that develop in chytrid spores and in chytrid gametes are very similar to those observed in animals: The flagella are single, are located at the back of the body, and move in a whiplash manner. Further, both animals and fungi store food by synthesizing the polysaccharide glycogen. Green plants, in contrast, synthesize starch as their storage product.

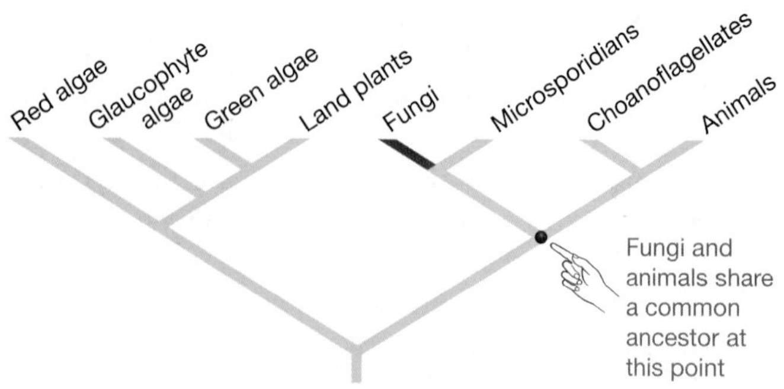

FIGURE 30.9 Fungi Are Closely Related to Animals—More So than to Land Plants
Phylogeny showing the evolutionary relationships among the green plants, animals, fungi, and some groups of protists. (Choanoflagellates are solitary or colonial protists found in freshwater; they are introduced in Chapter 31.) EXERCISE Using the relationships shown here, go back to Figure 30.8 and add microsporidians to the tree in part (b).

Experimental Studies of Mutualism

It is estimated that 90 percent of land plants live in close physical association with certain species of fungi. Stated another way, fungi and land plants often have a **symbiotic** ("together-living") relationship. Although some species of fungi live in association with an array of different land plant species, most documented fungal-plant associations are quite specific. Most fungi live in only a particular type of tissue, in one plant species. Based on this observation, biologists hypothesize that the evolution of symbiotic associations has played a large role in the diversification of fungi. The question is, are these symbiotic relationships *mutualistic*, meaning they benefit both species; *parasitic*, meaning one species benefits at the expense of the other; or *commensal*, meaning one species benefits while the other is unaffected?

Researchers frequently begin studies of plant-fungal associations simply by cataloging the species of fungi present in or on a particular plant tissue. This can be done by analyzing plant tissues directly via light microscopy or electron microscopy. Although growing fungi in the laboratory can be difficult, researchers may also place small samples of plant tissue on a culture plate in the laboratory and then analyze the fungal hyphae that grow out from the tissue into the nutrient medium on the plate.

To understand the nature of the association, biologists turn to experimental approaches. Recall from Section 30.1 that, in early experiments, researchers grew trees or other types of land plants with or without their normal mycorrrhizal fungi. These types of experiments are still done, with experimental treatments lacking fungal symbionts created by sterilizing soils with heat or by treating soils and seeds with fungicides. Presence-absence experiments have generally shown that plants grow much larger with their normal symbiotic fungi than they do without. Similarly, fungi that are typically symbiotic are usually unable to grow and reproduce if their regular host plant is absent.

To explore the nature of fungi-plant symbioses in more detail, researchers have used isotopes as tracers for specific elements (**Figure 30.10**). For example, to test the hypothesis that fungi obtain food in the form of carbon-containing compounds from their plant associates, biologists have introduced radioactively labeled carbon dioxide into the air surrounding plants that do or do not contain symbiotic fungi. The experimental plants are usually grown in pots inside laboratory growth chambers. The labeled CO_2 molecules are incorporated into the sugars produced during photosynthesis, and the location of the radioactive atoms can then be followed over time by means of a device that detects radioactivity. If plants feed their fungal symbionts, then labeled carbon compounds should be transferred from the plant to the fungi.

To test the hypothesis that plants are receiving nutrients from their symbiotic fungi in return for sugars, researchers have added radioactive phosphorus atoms or the heavy isotope of nitrogen (^{15}N) to potted plants that do or do not contain symbiotic fungi. If fungi facilitate the transfer of nutrients from soil to plants, then plants grown in the presence of their symbiotic

Question: Are mycorrhizal fungi mutualistic?

Hypothesis: Host plants provide mycorrhizal fungi with sugars and other photosynthetic products. Mycorrhizal fungi provide host plants with phosphorus or nitrogen from the soil.

Null hypothesis: No exchange of food or nutrients occurs between plants and mycorrhizal fungi. The relationship is not mutualistic.

Experimental setup:

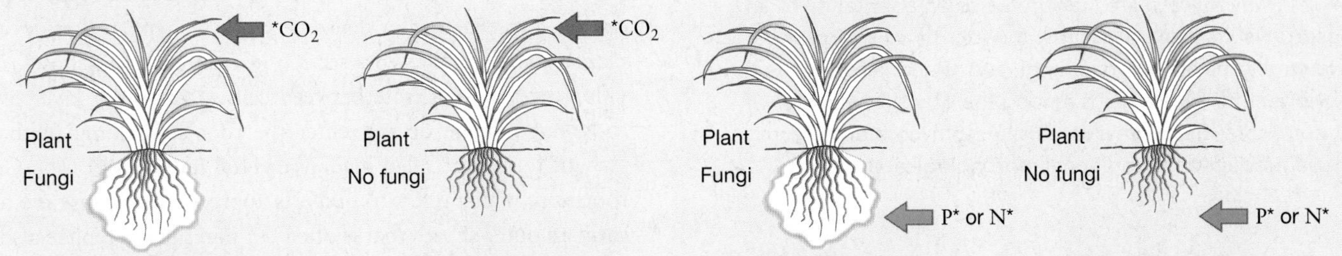

Labeled carbon treatment: Leaves are exposed to radioactive CO_2. Mycorrhizal fungi present.

Labeled carbon control: Leaves are exposed to radioactive CO_2. Mycorrhizal fungi absent.

Labeled P or N treatment: Roots are exposed to radioactive P or heavy isotope of N. Mycorrhizal fungi present.

Labeled P or N control: Roots are exposed to radioactive P or heavy isotope of N. Mycorrhizal fungi absent.

Prediction for labeled carbon: A large percentage of the labeled carbon taken up by the plant will be transferred to mycorrhizal fungi. In the control, little labeled carbon will be present in the soil surrounding the roots.

Prediction of null hypothesis, labeled carbon: There will be no difference between amounts of labeled carbon in mycorrhizal fungi and in soil when fungi are absent.

Prediction for labeled P or N: A large percentage of the labeled P or N taken up by the fungi will be transferred to the plant. In the control, little or no labeled P or N will be taken up by the plant.

Prediction of null hypothesis, labeled P or N: There will be no difference between amounts of labeled P or N found in plant in presence or absence of fungi.

Results:

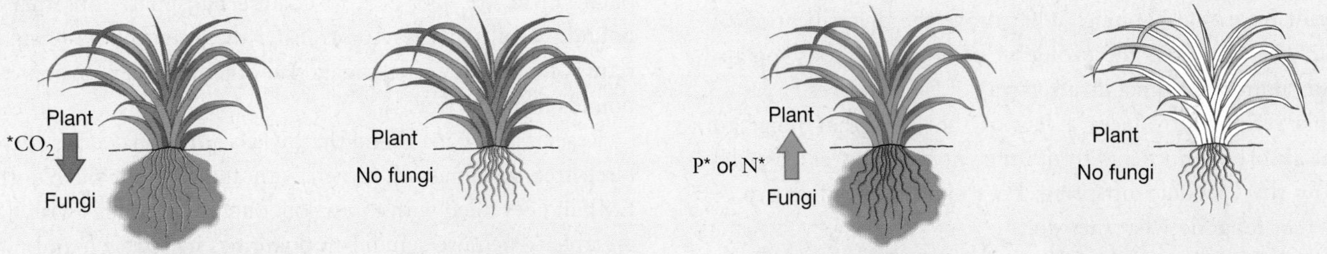

Labeled carbon treatment: Up to 20% of labeled carbon taken up by plant is transferred to mycorrhizal fungus.

Labeled carbon control: Little to no labeled carbon is found in soil surrounding plant roots.

Labeled P or N treatment: Large amount of labeled P or N is found in host plant.

Labeled P or N control: Little labeled P or N is found in host plant.

Conclusion: The relationship between plants and mycorrhizal fungi is mutualistic. Plants provide mycorrhizal fungi with carbohydrates. Mycorrhizal fungi supply host plants with nutrients.

FIGURE 30.10 Experimental Evidence That Mycorrhizal Fungi and Plants Are Mutualistic
Nutrient transfer experiments indicate that sugars flow from plants to mycorrhizal fungi and that key nutrients flow from mycorrhizal fungi to plants.

fungi should receive much more of the radioactive phosphorus or heavy nitrogen than do plants grown in the absence of fungi.

As the "Results" section of Figure 30.10 shows, experiments with isotopes used as tracers have shown that sugars and other carbon-containing compounds produced by plants via photosynthesis are transferred to their fungal symbionts. In some cases, as much as 20 percent of the sugars produced by a plant end up in their symbiotic fungi. In exchange, the symbiotic fungi facilitate the transfer of phosphorus or nitrogen or both from soil to the plant. Because phosphorus and nitrogen are in extremely short supply in most environments, the nutrients supplied by symbiotic fungi are critical to the success of the plant. In this way, studies with isotopes have supported the hypothesis that most relationships between fungi and land plants are mutually beneficial.

✓CHECK YOUR UNDERSTANDING

The bodies of fungi are either single-celled yeasts or multicellular mycelia. During sexual reproduction, different groups of fungi produce distinct reproductive structures. You should be able to (1) describe the nature of a mycelium and explain why mycelia are interpreted as an adaptation to an absorptive lifestyle, (2) identify the four types of reproductive structures observed in fungi, and (3) explain how the phylogenetic relationships among the Chytridiomycota, Zygomycota, Basidiomycota, and Ascomycota differ from those predicted by traditional morphological groupings.

30.3 What Themes Occur in the Diversification of Fungi?

Why are there so many different species of fungi? This question is particularly puzzling given that fungi share a common attribute: They all make their living by absorbing food directly from their surroundings. In contrast to the diversity of food-getting strategies observed in bacteria, archaea, and protists, all fungi make their living in the same basic way. Fungi are similar to plants in this respect. (Recall from Chapter 29 that virtually all land plants make their own food via photosynthesis.)

Chapter 29 showed that the diversification of land plants was driven not by novel ways of obtaining food, but by adaptations that allowed plants to grow and reproduce in a diverse array of terrestrial habitats. What drove the diversification of fungi? The answer is the evolution of novel methods for absorbing nutrients from a diverse array of food sources.

This section introduces a few of the ways that fungi go about absorbing nutrients from different food sources, as well as how they produce offspring. Let's explore the diversity of ways that fungi do what they do.

Fungi Participate in Several Types of Mutualisms

Not long after associations between fungi and the roots of land plants were discovered and shown to be mutualistic, researchers found that there are two distinct types of mycorrhizal interactions. The two types of mycorrhizae involve different phyla, and each type has a diagnostic morphology, geographic distribution, and function. But mycorrhizae are not the only type of symbiotic fungi found in plants. Researchers have also become interested in fungi that live in close association with the aboveground tissues of land plants—their leaves and stems. Fungi that live in the aboveground parts of plants are said to be **endophytic** ("inside-plants"). Recent research has shown that endophytic fungi are much more common and diverse than previously suspected. Further, data indicate that at least some species of endophytes are mutualistic.

These results support the general realization that most plants are covered with fungi—from the tips of their branches to the base of their roots. Many or even most plants are involved in several distinct types of mutualistic relationships with fungi. Let's take a closer look.

Ectomycorrhizal Fungi The type of mycorrhizal fungus illustrated in **Figure 30.11a** is found on virtually all of the tree species that grow in the temperate latitudes of both hemispheres, as well as tree species in northern coniferous forests. In this type of association, mycelia form a dense network of hyphae that covers a plant's roots. As the cross section in Figure 30.11a shows, individual hyphae penetrate between cells in the outer layer of the root, but hyphae do not enter the root cells. Fungi with this growth form are called **ectomycorrhizal fungi (EMF)**. The Greek root *ecto*, which refers to *outer*, is appropriate because the fungi form an outer sheath that is often 0.1 mm thick. Hyphae also extend out from the sheath-like portion of the mycelium into the soil. Most EMF are basidiomycetes; a few are ascomycetes.

How and why do these trees and fungi interact? In the cold, northern habitats where EMF are abundant, the growing season is so short that the decomposition of needles, leaves, twigs, and trunks can be extremely sluggish. As a result, nitrogen atoms tend to remain tied up in amino acids and nucleic acids inside dead tissues instead of being available in the soil. Fortunately, the hyphae of EMF penetrate decaying material and release enzymes (peptidases) that cleave the peptide bonds between amino acids in the dead tissues. The nitrogen released by this reaction is absorbed by the hyphae and then transported to the spaces between the root cells of trees, where it can be absorbed by the plant. EMF are also able to acquire phosphate ions that are bound to soil particles and transfer the ions to host plants. In return, the fungi receive sugars and other complex carbon compounds from the tree.

Researchers have found that, when northern species such as birch tree seedlings are grown with and without their normal EMF in pots filled with forest soil, only the seedlings with EMF are able to acquire significant quantities of nitrogen and phosphorus. Inspired by such data, a biologist has referred to EMF as the "dominant nutrient-gathering organs in most temperate forest ecosystems." The hyphae of EMF are like an army of miners that discover, excavate, and deliver precious nuggets of nitrogen to the trees of northern forests.

Arbuscular Mycorrhizal Fungi (AMF) In contrast to the hyphae of EMF, the hyphae of **arbuscular mycorrhizal fungi (AMF)** grow *into* the cells of root tissue. The name *arbuscular* ("little-tree") was inspired by the bushy, highly branched hyphae, shown in **Figure 30.11b**, that form inside root cells. AMF are also called *endomycorrhizal fungi*, because they penetrate the interior of root cell walls, or *vesicular-arbuscular mycorrhizae* (VAM or VM), because the hyphae of some species form large, balloon-like vesicles inside root cells.

The key point is that the hyphae of AMF penetrate the cell wall and contact the plasma membrane of root cells directly. The highly branched hyphae inside the plant cell wall are thought to be an adaptation increasing the surface area available

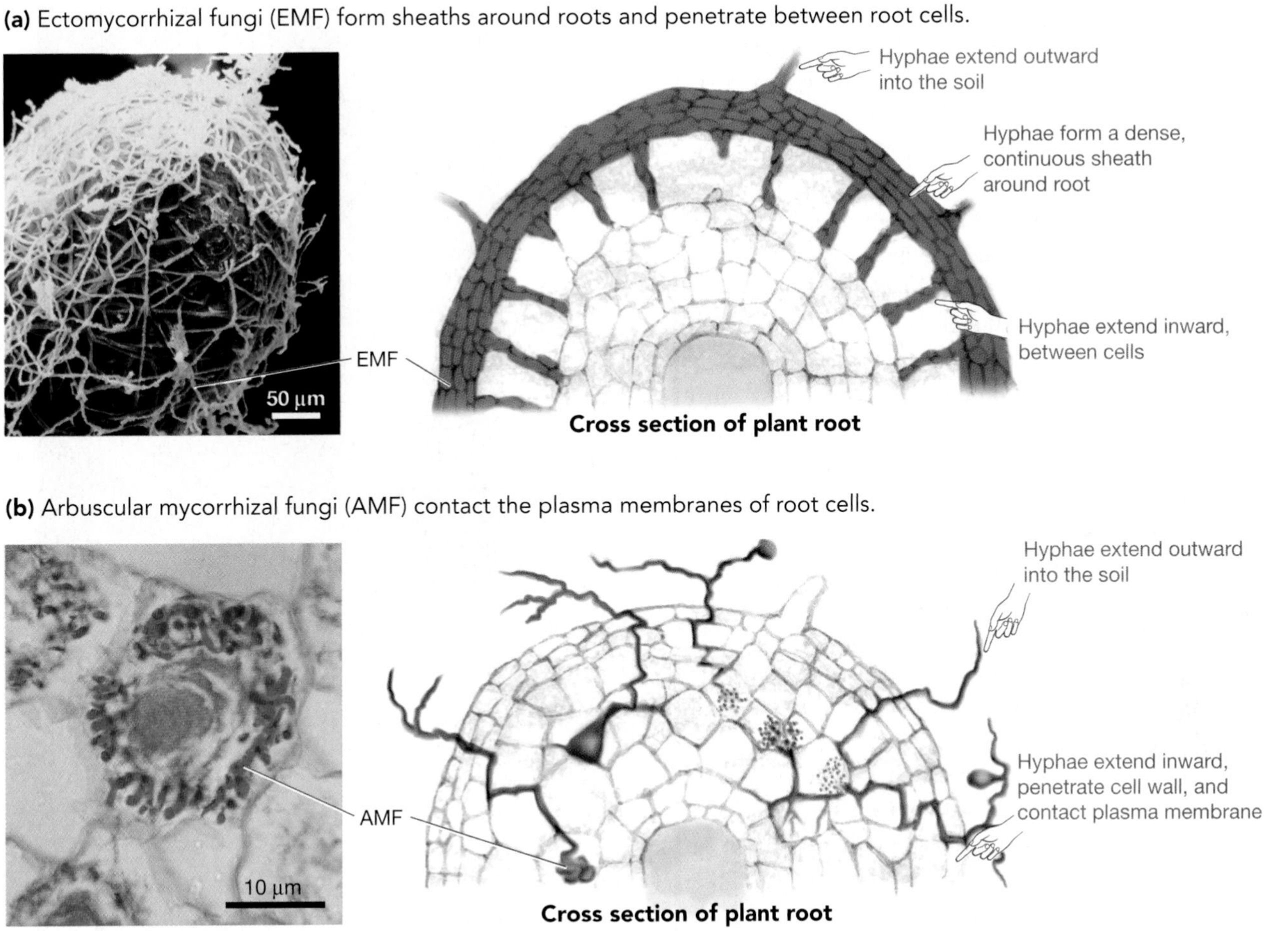
(a) Ectomycorrhizal fungi (EMF) form sheaths around roots and penetrate between root cells.
Hyphae extend outward into the soil
Hyphae form a dense, continuous sheath around root
Hyphae extend inward, between cells
EMF
50 μm
Cross section of plant root
(b) Arbuscular mycorrhizal fungi (AMF) contact the plasma membranes of root cells.
Hyphae extend outward into the soil
Hyphae extend inward, penetrate cell wall, and contact plasma membrane
AMF
10 μm
Cross section of plant root

FIGURE 30.11 Mutualistic Fungi Interact with the Roots of Plants in Two Distinct Ways
(a) Ectomycorrhizal fungi (EMF) form a dense network around the roots of plants. Their hyphae penetrate the intercellular spaces of the root but do not enter the root cells. **(b)** The hyphae of arbuscular mycorrhizal fungi (AMF) penetrate the walls of root cells, where they branch into bushy structures or balloon-like vesicles that contact the root cell's plasma membrane. EXERCISE In the cross sections, add arrows and labels showing the direction of movement of N (nitrogen-containing compounds), P (phosphorus-containing compounds), and C (carbon-containing compounds).

for exchange of molecules via diffusion between the fungus and its host. However, AMF do not form a tight sheath around roots, as do EMF. Instead, they form a pipeline that extends from inside plant cells in the root to the soil well beyond the root.

AMF are zygomycetes, and most belong to the lineage called Glomeromycota (see Figure 30.8b). They are found in a whopping 80 percent of all land plant species and are particularly common in grasslands and in the forests of warm or tropical habitats. Just as EMF are the dominant type of mycorrhizal association in cool, northern environments, AMF are the dominant type in grasslands and in the tropics.

In addition to being extremely common, AMF are ancient. Researchers have found AMF in fossilized root cells that are 400 million years old. This discovery confirms that mycorrhizal associations existed in the most ancient of all land plants. It also supports the hypothesis that plants and fungi colonized terrestrial environments together—meaning that fungi have been nourishing land plants since land plants first evolved.

What do AMF do? In the grasslands and tropical forests where AMF flourish, plant tissues decompose quickly because the growing season is long and warm. As a result, nitrogen is often readily available to plants. In these habitats, phosphorus—not nitrogen—is usually the nutrient that is in greatest demand by plants. Based on these observations, biologists hypothesized that AMF may transfer phosphorus atoms from the soil to the host plant. In most soils, phosphate atoms are found in negatively charged phosphate ions that cling tightly to mineral particles. Experiments with radioactive atoms confirmed the phosphate-transfer hypothesis by showing that, while AMF supply host plants with phosphorus, host plants supply fungi with reduced carbon. EMF mine nitrogen and some phosphorus in the north; AMF mine phosphorus in the south.

Are Endophytes Mutualists? Although endophytic fungi are relatively new to science, they are turning out to be both extremely common and highly diverse. Biologists in Brazil who examine tree leaves for the presence of fungi routinely find several previously undiscovered species of endophytes. Recall from Section 30.2 that a study on fungi in Panama found hundreds of fungal morphospecies living on the leaves of just two tree species. These newly discovered species are endophytes.

Endophytes do not cause disease symptoms in the tissues they invade. Thus, it is clear that they are not parasitic. To test the hypothesis that endophytic fungi are mutualistic, biologists have set up presence-absence experiments with various species of grasses and their endophytes. If endophytic fungi have a mutualistic relationship with their host plants, then host plants should grow more slowly and produce fewer offspring when endophytic fungi are absent. When endophytic fungi are prevented from colonizing grasses, plants that are free of endophytic fungi turn out to be much more susceptible to attack by leaf-eating insects than are plants with a normal array of endophytes. Follow-up work has confirmed that, in grasses, endophytic fungi produce compounds that deter or even kill insect herbivores. The fungus is also hypothesized to benefit by obtaining nutrition from the plant host. Based on these results, biologists have concluded that the symbiotic relationship between endophytes and grasses is mutualistic.

When similar experiments have been done with tree species and their endophytes, however, the results have been much less conclusive. For example, biologists who sprayed spores from endophytic fungi onto the leaves of Emory oak trees confirmed that the treatment dramatically increased the abundance of endophytes. However, the researchers observed no impact on the survival or growth rate of insect larvae that eat the leaves of these trees compared with the growth rate of insects that were eating the leaves of oaks with a normal abundance of endophytes. From these data, the researchers concluded that an increased number of endophytes did not deter insect herbivores in oaks.

Based on such results, the current consensus is that at least some endophytic fungi may be **commensals**—meaning the fungi and the plants simply coexist with no observable effect, either deleterious or beneficial, on the host plant. Research continues, however. Biologists are still working to understand the exact nature of the relationship between endophytic fungi and their tree hosts. In the meantime, it is abundantly clear that most land plants are colonized by a wide array of fungi and that fungi have evolved an array of relationships with their plant symbionts.

Adaptations That Make Fungi Effective Decomposers

The saprophytic fungi are master decomposers and recyclers. Fungi that make their living from dead plant material seek out large, complex molecules such as cellulose, lignin, proteins, and nucleic acids and break them down into hundreds or thousands of smaller compounds. Although bacteria and archaea are also important decomposers in terrestrial environments, fungi and a few bacterial species are the only organisms that can digest wood completely. Given enough time, fungi can turn even the hardest, most massive trees into soft soils that nourish an array of plants (**Figure 30.12**).

FIGURE 30.12 Fungi Recycle Nutrients
The fungi that are decomposing this section of tree trunk are breaking up its proteins, nucleic acids, lignin, and cellulose. In doing so, the fungi release nitrogen, phosphorus, and other nutrients that can be used by other organisms. **EXERCISE** At least two species of basidiomycetes inhabit the area in this photo. Circle their reproductive structures.

As land plants diversified and spread into terrestrial habitats, the ability to digest dead plant materials opened up the same habitats to fungi. As a result, fungi live wherever land plants live. The capacity to degrade plant bodies was an evolutionary breakthrough that helped trigger the diversification of saprophytic fungi. What adaptations made all this possible?

Extracellular Digestion Large molecules such as starch, lignin, cellulose, proteins, and RNA cannot diffuse across the plasma membranes of hyphae. Only sugars, amino acids, nucleic acids, and other small molecules can enter the cytoplasm. As a result, fungi have to digest their food before they can absorb it.

Instead of digesting food inside a stomach or food vacuole, as most animals and some protists do, respectively, fungi synthesize digestive enzymes and then secrete them *outside* their hyphae, into their food. Fungi perform **extracellular digestion**—digestion that takes place outside the organism. The simple compounds that result from enzymatic action are then absorbed by the hyphae.

Extracellular digestion is an important adaptation in fungi. As a case study in how this process occurs, consider the enzymes responsible for digesting lignin and cellulose. These are the two most abundant organic molecules on Earth. The term **lignin** refers to a family of extremely strong polymers built from monomers that are six-carbon rings. Recall from Chapter 29 that lignin is found in the secondary cell walls of plant vascular tissues, where it furnishes structural support. **Cellulose** is a polymer of glucose and is found in the primary and secondary cell walls of plant cells.

Basidiomycetes can degrade lignin completely—to CO_2—as well as digest cellulose. Let's take a closer look at how they do it.

Lignin Degradation How do fungi digest lignin? In addition to its biological significance, this question has enormous practical importance. To make soft, absorbent paper products, manufacturers must find efficient ways to degrade lignin without using caustic, dangerous chemicals. The same problem faces environmental scientists charged with cleaning up waste from old sawmills and paper mills.

To find out how fungi digest lignin, biologists began analyzing the proteins that are secreted into the extracellular space by lignin-digesting fungi. After purifying these molecules, the investigators tested each protein for the ability to degrade lignin. Using this approach, investigators from two labs independently discovered an enzyme called **lignin peroxidase.**

The researchers who followed up on the discovery of lignin peroxidase found that the enzyme catalyzes the removal of a single electron from an atom in the ring structures of lignin. This oxidation step creates a free radical—an atom with an unpaired electron (see Chapter 2). This extremely unstable electron configuration leads to a series of uncontrolled and unpredictable reactions that end up splitting the polymer into smaller units.

Biologists have referred to this mechanism of lignin degradation as *enzymatic combustion*. The phrase is apt because the uncontrolled oxidation reactions triggered by lignin peroxidase are analogous to the uncontrolled oxidation reactions that occur when gasoline burns in a car engine. This mechanism is remarkable, because virtually all of the other reactions catalyzed by enzymes are extremely specific. The lack of specificity in the lignin degradation reaction actually makes sense, however, given the nature of the compound. Unlike proteins, nucleic acids, and most other polymers, lignin is extremely heterogeneous. Over 10 types of linkages are routinely found between the monomers that make up lignin. But once lignin peroxidase has created a free radical in the aromatic ring, any of these linkages can be broken.

The uncontrolled nature of the reactions has an important consequence. Instead of being stepped down an orderly electron transport chain as described in Chapter 9, the electrons involved in these reactions lose their potential energy in large, unpredictable jumps. As a result, the oxidation of lignin cannot be harnessed to drive the production of ATP and fuel the growth of hyphae. This prediction accords with an experimental observation: Fungi cannot grow with lignin as their sole source of food.

If wood-rotting fungi don't use lignin as food, why do they produce enzymes to digest it? The answer is simple. In wood, lignin forms a dense matrix around long strands of cellulose. Degrading the lignin matrix gives hyphae access to huge supplies of energy-rich cellulose. Saprophytic fungi are like miners. But instead of seeking out rare, gem-like nitrogen or phosphorus atoms as do EMF and AMF, the saprophytes use lignin peroxidase to blast away enormous lignin molecules, exposing rich veins of cellulose that can fuel growth and reproduction.

Cellulose Digestion Once lignin peroxidase has softened wood by stripping away its lignin matrix, the long strands of cellulose that remain can be attacked by enzymes called **cellulases.** Like lignin peroxidase, cellulases are secreted into the extracellular environment by fungi. But unlike lignin peroxidase, cellulases are extremely specific in their action. For example, biologists have purified seven different cellulases from the fungus *Trichoderma reesei*. Two of these enzymes catalyze a critical early step in digestion—they cleave long strands of cellulose into a disaccharide called cellobiose. The other cellulases are equally specific and also catalyze hydrolysis reactions. In combination, the suite of seven enzymes in *T. reesei* transforms long strands of cellulose into a simple monomer (glucose) that can be used by fungi as a source of food.

Now that the enzymes responsible for lignin and cellulose degradation have been characterized, researchers would like to pursue a practical question: Can lignin peroxidase and cellulases be used to make paper production more efficient or perhaps speed efforts to clean up waste from old sawmill and paper mill operations? If the answer turns out to be yes, fungi could be used along with bacteria and archaea in the bioremediation efforts introduced in Chapter 28.

Variation in Life Cycles

WEB TUTORIAL 30.1 Life Cycle of a Mushroom

As fungi diversified, they evolved an array of symbiotic relationships with living land plants, as well as the ability to digest dead plant tissue completely. But in addition to being able to absorb nutrition in many ways, fungi reproduce in many ways. Recall from Section 30.2 that fungi may produce swimming gametes and spores, yoked hyphae where nuclei from different individuals fuse to form a zygote inside a protective structure, or specialized spore-producing cells called basidia and asci.

The **spore** is, in fact, the most fundamental reproductive cell in fungi. Spores are the dispersal stage in the fungal life cycle and are produced during both asexual and sexual reproduction. (Recall from Chapter 12 that asexual reproduction is based on mitosis, while sexual reproduction is based on meiosis.) Fungi produce spores in such prodigious quantities that it is not unusual for them to outnumber pollen grains in air samples. If a spore falls on a food source and is able to germinate, a mycelium begins to form. As the fungus expands, hyphae grow in the direction in which food is most abundant. If food begins to run out, mycelia respond by making spores, which are dispersed by wind or animals. Why would mycelia reproduce when food is low? To answer this question, biologists hypothesize that spore production is favored by natural selection when individuals are under nutritional stress. The logic is that spore production allows starving mycelia to disperse offspring to new habitats where more food might be available.

Unique Aspects of Fungal Life Cycles Compared with the life cycles of green plants, protists, and animals, the life cycles of fungi have several unique features. For example, only members of the Chytridiomycota produce gametes, and no species of fungi produce gametes that are so different in size that they are called sperm and egg. In addition, fertilization occurs in two distinct steps in most fungi: (1) fusion of cells and (2) fusion of nuclei from the fused cells. These two steps can be separated by long time spans and even long distances.

In many fungi, the process of sexual reproduction begins when hyphae from two individuals fuse to form a hybrid hypha. When the cytoplasm of two individuals fuses in this way, **plasmogamy** is said to occur (**Figure 30.13a**). But in most cases, the nuclei from the two individuals stay independent and grow into a heterokaryotic mycelium. The two nuclei that are present in heterokaryotic hyphae function independently, even though gene expression must be coordinated in order for growth and development to

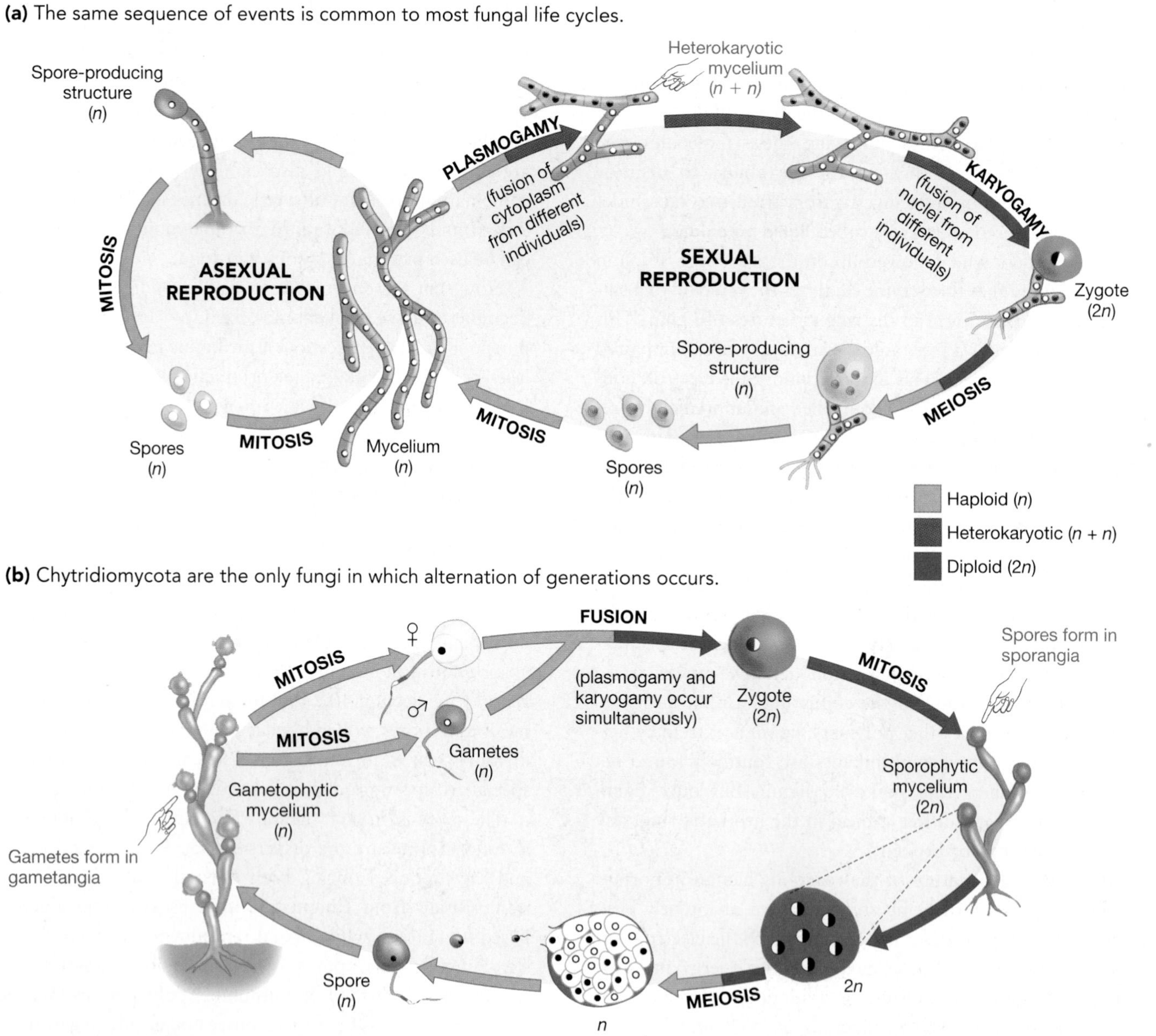

FIGURE 30.13 Fungi Have Unusual Life Cycles

(a) A generalized fungal life cycle, showing both asexual and sexual reproduction. The sexual part of the life cycle in the four major groups of fungi: **(b)** Chytridiomycota, **(c)** Zygomycota, **(d)** Basidiomycota, and **(e)** Ascomycota. Asexual reproduction is particularly common in Zygomycota and Ascomycota. EXERCISE Fungi spend most of their lives feeding. On each drawing, indicate the longest-lived component of the life cycle. In parts (c) through (e), add a loop labeled "Asexual reproduction" similar to that in part (a). This loop should start at the mycelium and return to it, via the growth of asexually produced spores.

(c) Zygomycota form yoked hyphae that produce a zygote.

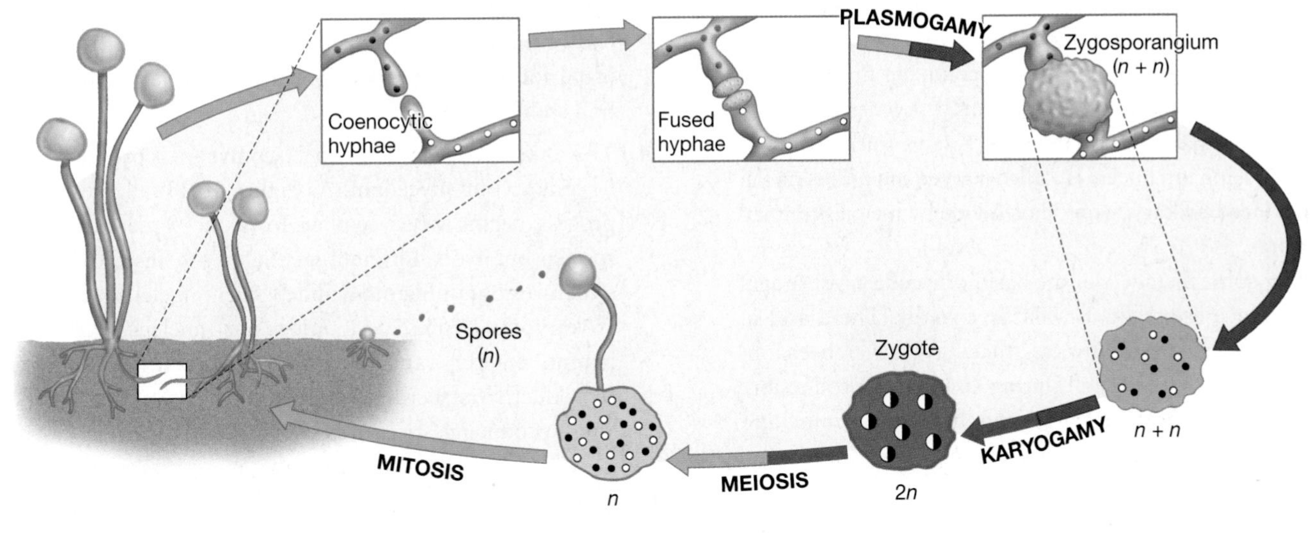

(d) Basidiomycota have reproductive structures with many spore-producing basidia.

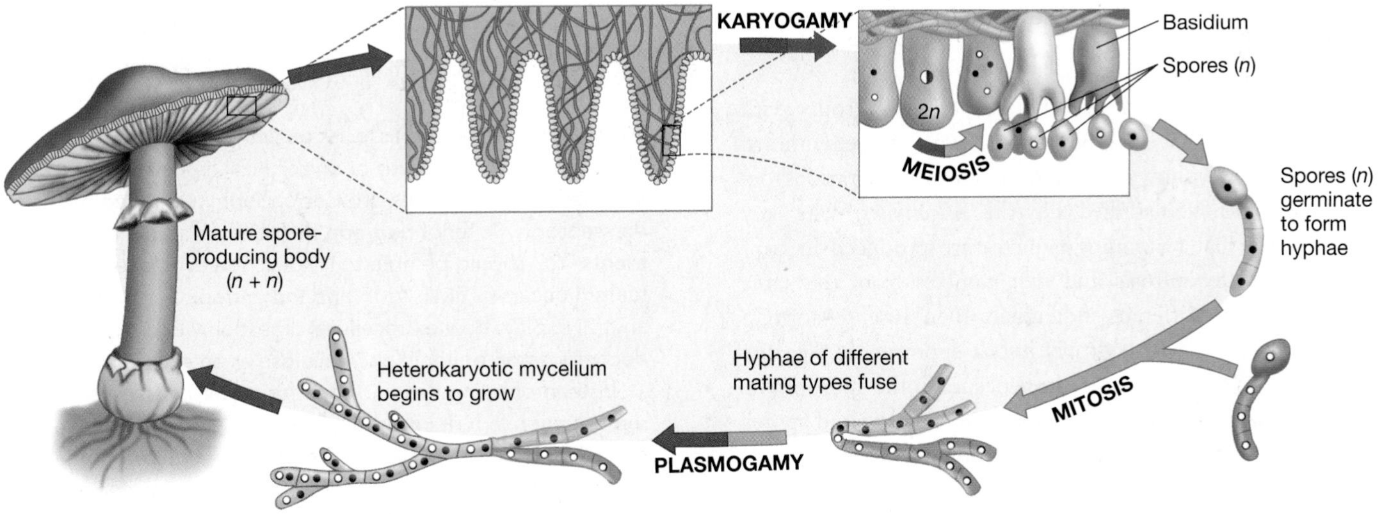

(e) Ascomycota have reproductive structures with many spore-producing asci.

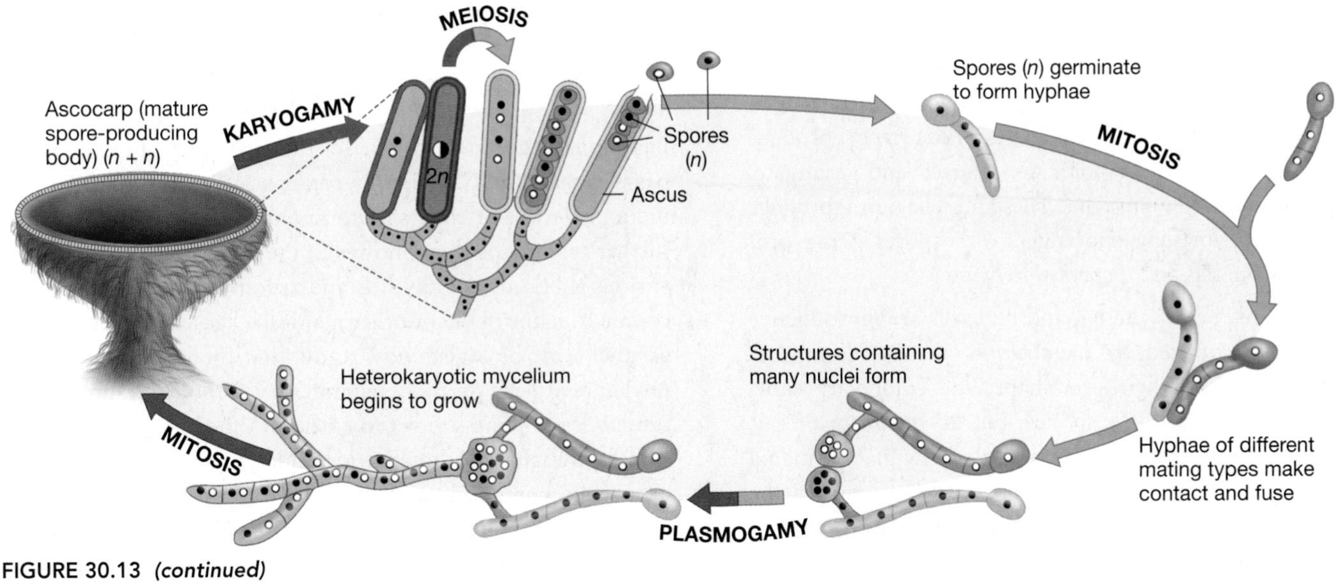

FIGURE 30.13 *(continued)*

occur. For example, the two nuclei divide as the hyphae expand, so each compartment that is divided by a septum contains one of each of the two nuclei. How the activities of the two nuclei are coordinated is almost a complete mystery, and it is an important topic for future research. Eventually, however, one or more pairs of unlike nuclei fuse to form a diploid zygote. The fusion of nuclei is called **karyogamy**. The nuclei that are produced by karyogamy then divide by meiosis to form haploid spores.

Except for some members of the Basidiomycota, most fungal species reproduce asexually as well as sexually. There are, in fact, large numbers of ascomycetes that have never been observed to reproduce sexually. During asexual reproduction, spore-forming structures are produced by a mycelium, and spores are generated by mitosis. As a result, offspring are genetically identical to their parent.

Among the sexually reproducing species of fungi, the nature of the heterokaryotic stage and the morphology of the spore-producing structure vary. Let's take a closer look at each of the four major types of life cycles that have been observed in fungi:

1. The Chytridiomycota are the only fungal group with any species that exhibit alternation of generations. **Figure 30.13b** shows how alternation of generations occurs in the well-studied chytrid *Allomyces*. The key points are that swimming gametes are produced in haploid adults by mitosis and that gametes from the same individual or different individuals then fuse to form a diploid zygote, which grows into a sporophyte. The life cycle is completed when meiosis occurs in the sporophyte, inside a structure called a **sporangium**. The haploid spores produced by meiosis disperse by swimming.
2. In Zygomycota, sexual reproduction starts when hyphae from different individuals fuse, as shown in **Figure 30.13c**. Plasmogamy forms a zygosporangium that develops a tough, resistant coat and is followed by karyogamy. The zygosporangium can persist if conditions become too cold or dry to support growth. When temperature and moisture conditions are again favorable, meiosis occurs. The meiotic products within the sporangium produce haploid spores. When spores are released and germinate, they grow into new mycelia. These mycelia can reproduce asexually by making sporangia, with spores being produced by mitosis and dispersed by wind.
3. Mushrooms, bracket fungi, and puffballs are reproductive structures produced by members of the Basidiomycota (**Figure 30.13d**). Their size, shape, and color vary enormously from species to species, but all members of this group originate from the heterokaryotic hyphae of mated individuals. Inside these reproductive structures, the pedestal-like, spore-producing cells called basidia form at the ends of hyphae. Karyogamy occurs within the basidia. The diploid nucleus that results undergoes meiosis, and haploid spores mature. Spores are then expelled from the end of the basidia and dispersed by the wind. It is not unusual for a single puffball or mushroom to produce a billion spores.
4. The reproductive structure in Ascomycota is produced by a heterokaryotic mycelium. As **Figure 30.13e** illustrates, the process begins when hyphae from the same ascomycete species but from different mating types fuse, forming a structure containing many independent nuclei. A short heterokaryotic hypha, containing one nucleus from each parent, emerges and eventually grows into a complex reproductive structure whose hyphae have the sac-like, spore-producing structures called asci at their tips. After karyogamy occurs inside each sac, meiosis takes place and haploid spores are produced. When the ascus matures, it splits. The spores may be forcibly ejected and are often picked up by the wind and dispersed.

✓CHECK YOUR UNDERSTANDING

Most fungi live in close association with land plants. When plants are alive, two types of mycorrhizal associations occur and many fungal species grow as endophytes. When plants die, saprophytic fungi degrade their tissues and release nutrients. You should be able to (1) describe evidence that mutualism occurs in EMF, AMF, and the endophytes of grasses; and (2) explain how extracellular digestion works, using the decomposition of lignin and cellulose as an example.

Instead of being based on the fusion of gametes, sexual reproduction in fungi is usually based on the fusion of hyphae. You should be able to (3) draw a generalized fungal life cycle; (4) explain the difference between plasmogamy and karyogamy, and relate them to the process of fertilization in plants and animals; and (5) explain what a heterokaryotic mycelium is.

30.4 Key Lineages of Fungi

Based on the current estimate for the phylogeny of Fungi, it appears clear that Chytridiomycota and Zygomycota are not monophyletic and thus should not be named as phyla. Until further research is done, however, the relationships among the various subgroups of chytrids and zygomycetes remain too uncertain to name phyla or other major lineages accurately. Stated another way, biologists now know that the traditional fungal phyla are at least partially incorrect, even though a more accurate understanding has yet to emerge. Although the summaries that follow analyze the chytrids and zygomycetes as coherent groups, keep in mind that they are not. At this time, researchers simply lack the data required to identify the major lineages of fungi more accurately.

Chytridiomycota (Chytrids)

Chytrids are largely aquatic and are particularly common in freshwater environments. Some species live in wet soils, however, and a few have been found in desert soils that are wet only during a short rainy season. Species that are found in dry soils have resting spores that allow them to endure harsh conditions. Resting spores from chytrids have been shown to germinate normally after a resting period of 30 years. Members of this group are the only fungi that produce cells that can move.

Because the animals and protists that are most closely related to fungi are aquatic, it is logical to infer that fungi evolved in water and that chytrids are the most ancient lineage of fungi. The phylogenetic analyses that have been done to date cannot rule out the possibility that a soil-dwelling group of zygomycetes is actually the most basal clade of fungi, however.

Absorptive lifestyle Many species of chytrids have enzymes that allow them to digest cellulose. As a result, these species are important decomposers of plant material in wet soils, ponds, and lakes. Many of the freshwater species are parasitic, and on occasion parasitic chytrids cause disease epidemics in algae or aquatic insects (including mosquitoes). Other species parasitize mosses, ferns, or flowering plants. Mutualistic chytrids are among the most important of the many organisms that live in the guts of deer, cows, elk, and other mammalian herbivores, because the chytrids help these mammals digest their food (**Figure 30.14**). Chytrids produce cellulases that degrade the cell walls of grasses and other plants ingested by the animal, releasing sugars that are used by both the chytrids and their hosts.

Life cycle During asexual reproduction, most chytrid species produce spores that swim to new habitats via a flagellum. A few species reproduce sexually as well as asexually and exhibit alternation of generations. (Recall from Section 30.3 that a few chytrids are the only fungal species to do so.) In species that reproduce sexually, plasmogamy may occur through fusion of gametes, fusion of gamete-forming structures called **gametangia**, or fusion of hyphae.

Human and ecological impacts In addition to their importance as decomposers in aquatic habitats and as mutualists in the guts of large, plant-eating mammals, chytrids have been the focus of intense research. An epidemic of chytrid infections may be responsible for catastrophic declines that recently occurred in frog populations all over the world. For additional details on the amphibian die-off, see the essay at the end of this chapter. Biologists are investigating the possibility of using a chytrid that parasitizes mosquito larvae as a biological control agent, and potato tubers are sometimes invaded and spoiled by a parasitic chytrid that causes black wart disease.

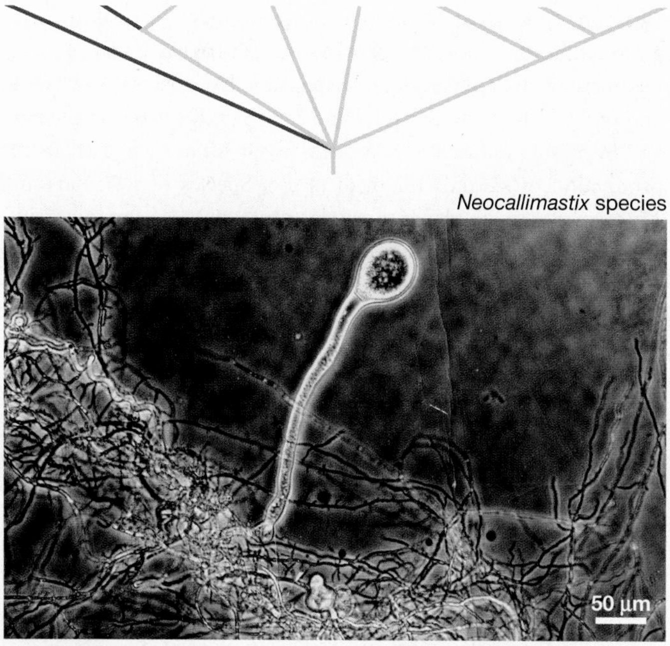
Neocallimastix species
50 µm

FIGURE 30.14 A Chytrid That Lives in the Gut of Cows and Digests the Cellulose in Grass Leaves

Zygomycota ▶ Mucorales and Other Basal Lineages

The Zygomycota ("yoked-fungi") are primarily soil-dwellers. Their hyphae yoke together and fuse during sexual reproduction and then form the durable, thick-walled zygosporangium that is characteristic of the group. Although only a few lineages are highlighted here, over 10 major subgroups have been identified.

In current phylogenetic analyses, most of the major subgroups of Zygomycota cluster together at the base of the fungal tree along with some of the chytrids. These subgroups include Mucorales, Entomophthorales, *Basidiobolus*, and other lineages not discussed here. Because certain chytrids live in wet soils, it is not clear from the phylogeny whether chytrids or zygomycetes were the first fungi to make the transition from aquatic to terrestrial habitats. In addition, it is not yet known whether fungi made the water-to-land transition once or many times. Understanding the relationships among basal lineages with chytrid and zygomycete characteristics is among the most urgent issues facing biologists working on the phylogeny of fungi.

Absorptive lifestyle Many members of zygomycete lineages are saprophytes and live in plant debris. Some parasitize other fungi, however, or are important parasites of insects and spiders. About 65 species in one subgroup are predatory—primarily on amoebae

(Continued on next page)

Zygomycota ▶ Mucorales and Other Basal Lineages *(continued)*

and other unicellular protists, though some are large enough to capture the microscopic animals called roundworms. The predatory zygomycetes trap their prey by means of sticky substances on their cell walls or catch prey in snares consisting of looped hyphae (**Figure 30.15**). Once a prey individual is captured, hyphae invade its body, digest it, and absorb the nutrients that are released.

Life cycle Asexual reproduction is extremely common. Ball-like sporangia are produced at the tips of hyphae that form stalks, and mitosis results in the production of spores, which are dispersed by wind. During sexual reproduction, fusion of hyphae occurs only between individuals of different mating types. Fusion of hyphae from the same individual mycelium does not occur.

Human and ecological impacts The black bread mold *Rhizopus stolonifer* is a common household pest—probably the zygomycete that is most familiar to you. Saprophytic and parasitic members of these lineages are responsible for rotting strawberries and other fruits and vegetables and causing large losses in the fruit and vegetable industries. Some species of *Mucor* are used in the production of steroids for medical use. Species of *Rhizopus* and *Mucor* are used in the commercial production of organic acids, pigments, alcohols, and fermented foods.

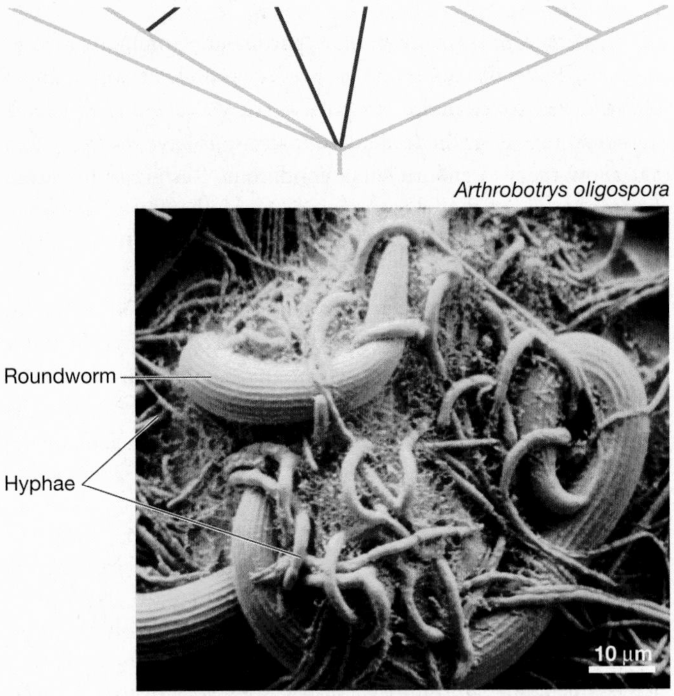

Arthrobotrys oligospora
Roundworm
Hyphae
10 μm

FIGURE 30.15 Some Zygomycetes Are Predatory

Glomeromycota (AMF)

Current phylogenetic analyses place the group of fungi that form arbuscular mycorrhizal associations in their own phylum, indicating that they are a major monophyletic group. Because the Glomeromycota, Basidiomycota, and Ascomycota diverged relatively late in the course of fungal evolution and form the tips of the tree, biologists refer to the three groups as the "crown fungi." Before phylogenies based on molecular characters were available, the lineage now called Glomeromycota was called Glomales and considered part of the Zygomycota.

Absorptive lifestyle All of the arbuscular mycorrhizal fungi (AMF) are members of the Glomeromycota. Recall from Section 30.3 that these fungi absorb phosphorus-containing ions or molecules in the soil and transfer them into the roots of trees, grasses, and shrubs in grassland and tropical habitats (**Figure 30.16**). In exchange, the host plant provides the symbiotic fungi with sugars and other organic compounds.

Life cycle Most species form spores underground. Because these species are difficult to grow in the laboratory, their life cycle is not well known.

Human and ecological impacts Because grasslands and tropical forests are among the most productive habitats on Earth, the AMF are enormously important to both human and natural economies.

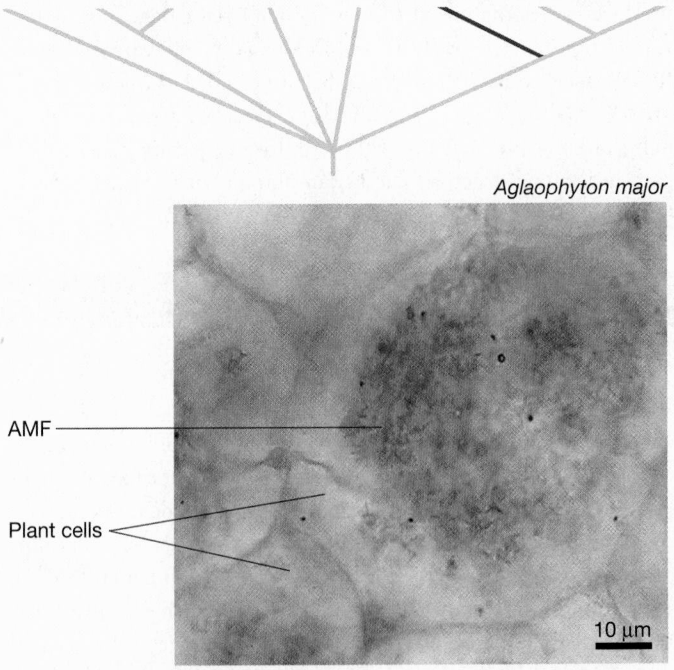

Aglaophyton major
AMF
Plant cells
10 μm

FIGURE 30.16 A Fossilized AMF That Is 398 Million Years Old
[Micrograph courtesy of Hans Kerp and Hagen Hass, Palaeobotanical Research Group, University of Munster, Germany.]

Basidiomycota (Club Fungi)

Although most basidiomycetes form mycelia and produce multicellular reproductive structures, some species have the unicellular growth form. The group is named for basidia, the club-like or pedestal-like cells where meiosis and spore formation occur. About 30,000 species have already been described, and more are being discovered each year.

Absorptive lifestyle Basidiomycetes are important saprophytes. Along with a few soil-dwelling bacteria, they are the only organisms capable of synthesizing lignin peroxidase and completely digesting wood. Other basidiomycetes form ectomycorrhizal associations with trees in temperate and northern forests. One subgroup consists entirely of parasitic forms called *rusts*, including species that cause serious infections in wheat and rye fields. The plant parasites called *smut fungi* are also basidiomycetes. Smuts specialize in infecting grasses; a few infect other fungi.

Life cycle Asexual reproduction through production of spores is common in the Basidiomycota, although not as prevalent as in other groups of fungi. Asexual reproduction also occurs through growth and fragmentation of mycelia in the soil or in rotting wood, resulting in genetically identical individuals that are physically independent. During sexual reproduction, all basidiomycetes—even unicellular ones—produce basidia. In the largest subgroup in this lineage, basidia form in large, aboveground reproductive structures called *mushrooms, brackets*, *earthstars*, or *puffballs* (**Figure 30.17**).

Human and ecological impacts Because temperate and northern forests provide most of the hardwoods and softwoods used in furniture-making, building construction, and papermaking, the EMF are enormously important in forest management. Throughout the world, mushrooms are cultivated or collected from the wild as a source of food. Some of the toxins found in poisonous mushrooms are used in biological research; others have hallucinogenic effects on people and are used and traded illegally.

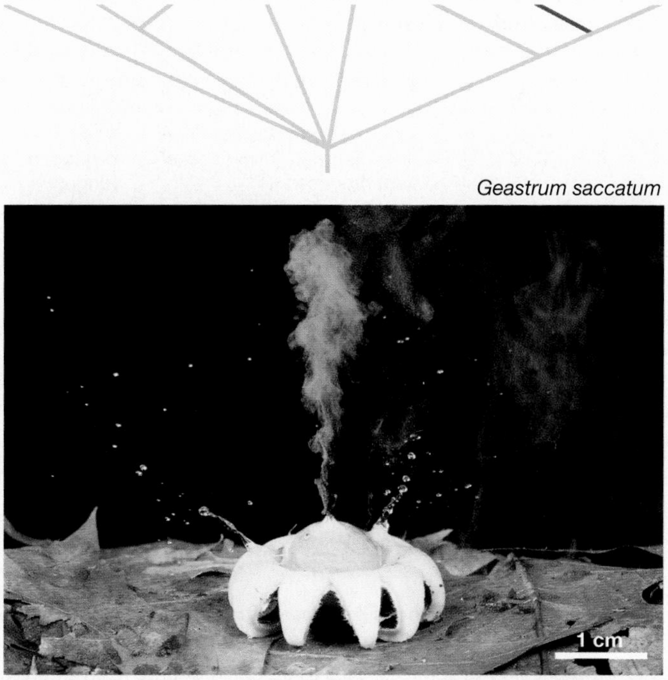
Geastrum saccatum
1 cm

FIGURE 30.17 Earthstars and Puffballs Can Produce Billions of Spores

Ascomycota (Sac Fungi)

About 75 percent of all known fungi belong to the phylum Ascomycota. This group is named for the sac-like cells, or asci, where meiosis and spore formation take place. The group is extremely large and diverse, however, and the phylogenetic relationships among subgroups are largely unknown. As a result, the discussion of "key lineages" that follows is based on two lifestyles observed in ascomycetes and not on distinct monophyletic groups. Work on the phylogenetic relationships within Ascomycota continues.

Ascomycota ▶ Lichen-Formers

About half of the ascomycetes grow in association with cyanobacteria or single-celled members of the green algae, forming the structures called **lichens** (see Chapter 29). Over 15,000 different lichens have been described to date; in most, the fungus involved is an ascomycete (although a few basidiomycetes participate as well). To name a lichen, biologists usually use the genus and species name assigned to the fungus that participates in the association.

Absorptive lifestyle It is not yet clear whether all of the relationships observed in lichens are mutualistic. In habitats where lichens are common, neither partner can exist as a free-living organism. The fungus in lichens appears to protect the photosynthetic cells. The fungal hyphae form a dense protective layer that shields the photosynthetic species and reduces water loss. In return, the cyanobacterium or alga provide carbohydrates that the fungus uses as a source of carbon and energy. However, the hyphae of some lichen-forming fungi have been observed to invade algal cells and kill them. This observation suggests a partially parasitic relationship in at least some lichens. The nature of lichen-forming associations is the subject of ongoing research.

(Continued on next page)

Ascomycota ▶ Lichen-Formers *(continued)*

Life cycle As **Figure 30.18a** shows, lichens reproduce asexually via the production of small structures called **soredia** that contain both symbionts. Soredia disperse to a new location via wind or water and then develop into a new, mature individual via the growth of the algal or bacterial and fungal symbionts. In addition, the fungal partner may form asci, which produce spores that then germinate to form a small mycelium. If the growing hyphae encounter enough algal cells, a new lichen can form.

Human and ecological impacts In terms of their abundance and diversity, lichens dominate the Arctic and Antarctic tundras. They are the major food of caribou in these habitats, as well as the most prevalent colonizers of bare rock surfaces throughout the world (**Figure 30.18b**). Rock-dwelling lichens are significant, because they break off mineral particles from the rock surface as they grow—launching the first step in soil formation. About 10,000 tons of lichens are processed annually and used in perfume production.

(a) Cross section of a lichen, showing three layers

(b) Top view of lichens on a rock

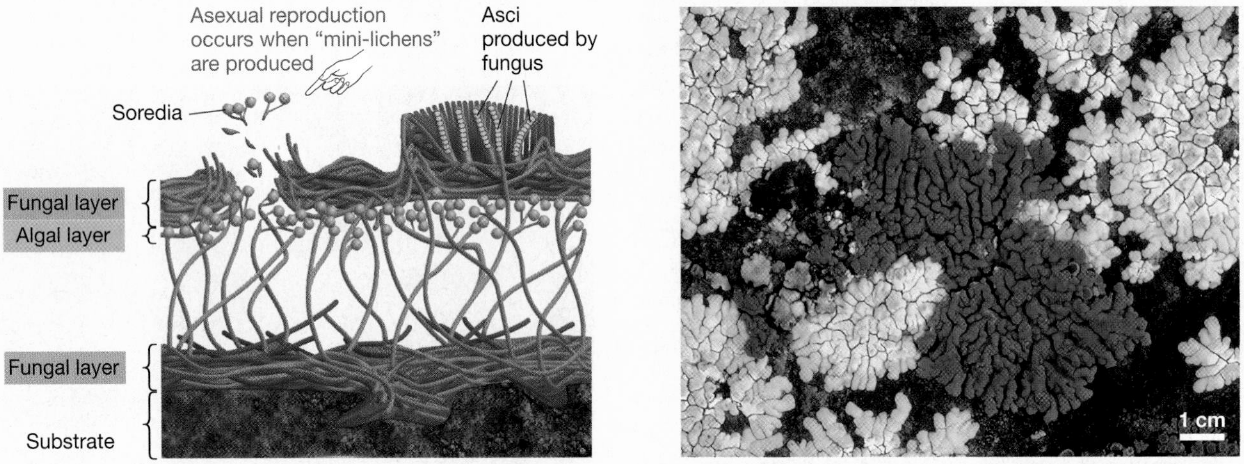

FIGURE 30.18 Lichens Are Associations between a Fungus and a Cyanobacterium or Green Alga
(a) In a lichen, cyanobacteria or green algae are enmeshed in a dense network of fungal hyphae. **(b)** Lichens often colonize surfaces, such as tree bark or bare rock, where other organisms are rare.

Ascomycota ▶ Non-Lichen-Formers

The ascomycetes that do not form lichens are found in virtually every terrestrial habitat, as well as some freshwater and marine environments. Although most ascomycetes form mycelia, the taxonomic group called yeasts is part of the Ascomycota.

Absorptive lifestyle A few members of the Ascomycota form mutualistic ectomycorrhizal associations with tree roots. Ascomycetes are also the most common endophytic fungi on aboveground tissues. Large numbers are saprophytic and are abundant in forest floors and in grassland soils. Parasitic forms are common as well.

Life cycle The aboveground, ascus-bearing reproductive structures of these fungi may be a cup or saucer shape called an **ascocarp** (**Figure 30.19**). For that reason, these ascomycetes are known as **cup fungi**. It is also routine for spores called *conidia* to be produced asexually, at the ends of specialized hyphae called *conidiophores*.

Human and ecological impacts Some saprophytic ascomycetes can grow on jet fuel or paint and are used to help clean up contaminated sites. *Penicillium* is an important source of antibiotics, and *Aspergillus* produces citric acid used to flavor soft drinks and candy. Truffles and morels are so highly prized that they can fetch $400 per pound. A few parasitic ascomycetes cause infections in humans and other animals. In land plants, parasites from this group cause diseases including Dutch elm disease and chestnut blight.

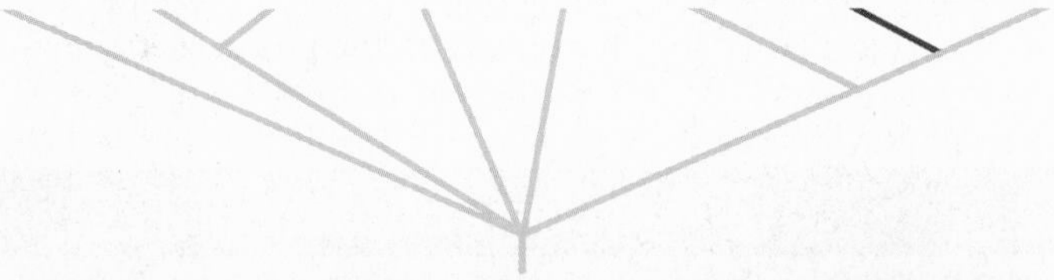

Sarcoscypha coccinea

FIGURE 30.19 Ascomycetes Are Sometimes Called Cup Fungi

ESSAY Why Are Frogs Dying?

Over the past two decades, conservation biologists have become increasingly alarmed about rapid, and in many cases catastrophic, declines in the numbers of frogs and toads around the world. The trend is particularly worrisome because even populations in undisturbed habitats, such as national parks, are being affected. Biologists who have followed the declines closely suspect that several species extinctions have already occurred.

Despite the magnitude of the problem, biologists have had frustratingly few clues as to what has killed the organisms. Researchers would revisit sites where frogs and toads once were abundant and diverse, only to discover that these animals had disappeared. Without dead or dying animals to study, investigators could only speculate about what was happening—until recently.

In the late 1990s an international team of researchers witnessed simultaneous and massive die-offs of amphibians in the rain forests of Australia and Panama. During these events, they collected 120 dead frogs and toads representing 19 species. Although general autopsies showed no obvious tissue abnormalities or lesions, light microscopy and electron microscopy of skin samples were revealing. In every case, the epidermis of the dead frogs was riddled with a parasitic fungus (**Figure 30.20**). An analysis of ribosomal RNA sequences isolated from the fungal cells confirmed what researchers had suspected from the morphology of the spores: The parasite was an undescribed member of the Chytridiomycota. To confirm that this fungus had caused the deaths, the researchers collected spores from the skin of dead frogs, suspended them in water, and added the suspension to aquaria containing six healthy frogs. Within 18 days, all of the experimental frogs had become infected with the fungus and were dead or dying.

This fungus could be the first infectious agent known to cause a species extinction. Biologists who are studying the frog die-off caution, however, that the chytrid is probably not the only agent involved in the worldwide declines. In California, for example, no evidence of fungal infection has been found to date. Instead, pesticides and herbicides are thought to play an important role in the decline of amphibian populations there. It is also entirely possible that well-documented increases in ultraviolet radiation—or other factors—have compromised the immune systems of these organisms, rendering them susceptible to the disease.

This fungus could be the first infectious agent known to cause a species extinction.

Although researchers are convinced that amphibian populations are declining in many parts of the world and that chytrid infections play a role, biologists are far from certain that chytrids are the only causative agent. Research continues.

FIGURE 30.20 Chytridiomycosis Can Be Fatal in Frogs
This frog is heavily infected with a parasitic chytrid. The infected individual is listless and will probably die as a result of the infection.

CHAPTER REVIEW

Summary of Key Concepts

- **Fungi are important in part because many species live in close association with land plants. They supply plants with key nutrients and decompose dead wood. They are the master recyclers of nutrients in terrestrial environments.**

Living plants are covered with fungi. The roots of grasses and trees that grow in warm or tropical habitats are infiltrated by symbiotic fungi that supply the plant with phosphorus. Similar mycorrhizal associations, based on an exchange of nitrogen for sugars and other products of photosynthesis, are found between basidiomycetes and the roots of trees in cold, northern habitats. Many fungi also live in leaves and stems; in grasses, these endophytic fungi secrete toxins that discourage herbivores.

Although mutualistic associations between fungi and living plants are common, so are parasitic relationships. Parasitic fungi are responsible for devastating blights in crops and other plants.

Once plants die, saprophytic fungi degrade the lignin and cellulose in wood and release nutrients from decaying plant material. Because they free up carbon atoms that would otherwise be locked up in wood, fungi speed up the carbon cycle in terrestrial habitats.

All fungi make their living by absorbing nutrition from living or dead organisms. Fungi secrete enzymes so that digestion takes place outside their cells. Their morphology provides a large amount of surface area for efficient absorption.

Several adaptations make fungi exceptionally effective at absorbing nutrients from the environment. The two growth habits found among Fungi—single-celled "yeasts" or mycelia composed of long, filamentous hyphae—give fungal cells extremely high surface-area-to-volume ratios. Extracellular digestion, in which enzymes are secreted into food sources, is another important adaptation in fungi. Perhaps the most important reactions catalyzed by these enzymes are the degradations of lignin and cellulose found in wood. Lignin decomposes through a series of uncontrolled oxidation reactions triggered by the enzyme lignin peroxidase. This "enzymatic combustion" contrasts sharply with the breakdown of cellulose, which occurs in a carefully regulated series of steps, each catalyzed by a specific cellulase.

Web Tutorial 30.1 Life Cycle of a Mushroom

Many fungi have unusual life cycles. It is common for species to have a long-lived heterokaryotic stage, in which cells contain haploid nuclei from two different individuals. Although most species reproduce sexually, very few species produce eggs and sperm.

Many species of fungi have never been observed to reproduce sexually, though most species can produce haploid spores either sexually or asexually. Sexual reproduction usually starts when hyphae from different individuals fuse—an event called plasmogamy. If the fusion of nuclei, or karyogamy, does not occur immediately, a heterokaryotic mycelium forms. Later, heterokaryotic cells produce spore-forming structures where karyogamy and meiosis take place.

The four traditional groupings of fungi are based on distinct reproductive structures: (1) the Chytridiomycota, which are aquatic fungi with motile gametes; (2) the mycorrhizal or saprophytic Zygomycota, which are soil-dwelling fungi with tough sporangia; (3) the mycorrhizal or lignin-decomposing and saprophytic Basidiomycota, which have club-like, spore-forming structures; and (4) the mycorrhizal, saprophytic, or lichen-forming Ascomycota, which have sac-like, spore-forming structures. Recent analyses of DNA sequence data have revealed that only the Basidiomycota and Ascomycota are monophyletic, however. Researchers are still trying to understand how various groups of chytrids and zygomycetes are related.

Questions

Content Review

1. The mycelial growth habit leads to a body with a high surface-area-to-volume ratio. Why is this important?
   a. Mycelia have a large surface area for absorption.
   b. The hyphae that make up mycelia are long, thin tubes.
   c. Hyphae are broken up into compartments by walls called septa.
   d. Hyphae can infiltrate living or dead tissues.
2. What do researchers hypothesize to be the cause of the "fungal spike" at the end-Permian boundary?
   a. unusual soil conditions, which made it more likely that fungal spores and hyphae could fossilize
   b. an adaptive radiation of fungi
   c. a switch to saprophytic lifestyles
   d. a mass extinction event that created a huge quantity of dead plant material
3. The Greek root *ecto* means "outer." Why are ectomycorrhizal fungi, or EMF, aptly named?
   a. Their hyphae form tree-like branching structures inside plant cell walls.
   b. They are mutualistic.
   c. Their hyphae form dense mats that envelop roots but do not penetrate the walls of cells inside the root.
   d. They transfer nitrogen from outside their plant hosts to the interior.
4. The hyphae of AMF form bushy or balloon-like structures after making contact with the plasma membrane of a root cell. Why?
   a. They anchor the fungus inside the root, so the association is more permanent.
   b. They increase the surface area available for the transfer of nutrients.
   c. They produce toxins that protect the plant cells against herbivores.
   d. They appear early in the fossil record of land plants—the first fossil AMF are 390 million years old.
5. Which statement best characterizes a parasitic relationship?
   a. Parasites gain benefits, such as nutrients, and benefit their host in return.
   b. Parasites gain benefits, such as nutrients, and harm their host in the process.
   c. Parasites gain benefits, such as nutrients, without affecting their host.
   d. The participants are not considered symbiotic.
6. Very few organisms besides fungi have a heterokaryotic stage in their life cycle. Which of the following is another usual aspect of the fungal life cycle?
   a. Some fungi exhibit alternation of generations—meaning that there is a multicellular diploid stage and a multicellular haploid stage.
   b. They produce eggs and sperm in approximately equal numbers, instead of many sperm and a few eggs.
   c. Spores have to fuse with each other before they develop into a new mycelium.
   d. Most varieties undergo sexual reproduction without producing eggs or sperm.

Conceptual Review

1. Explain why fungi that degrade dead plant materials are important to the global carbon cycle. Do you accept the text's statement that, without these fungi, "Terrestrial environments would be radically different than they are today and probably much less productive"? Why or why not?
2. Lignin and cellulose provide rigidity to the cell walls of plants. But in most fungi, chitin performs this role. Why is it logical that most fungi don't have lignin or cellulose in their cell walls?
3. Experiments show that AMF contribute phosphate ions to their host plants. Why is this important? Using information from Chapters 3 through 6, include a list of macromolecules found in plants that contain phosphorus.
4. When growing plants in the presence or absence of mycorrhizal or endophytic fungi—to test for the effects of symbiotic fungi on plant growth—researchers sterilize the soil and seeds used in both the presence and absence treatments, and then grow the plants under sterile conditions. Why?
5. Explain why radioactive or heavy isotopes were useful in showing that mycorrhizal fungi have a mutualistic relationship with plants. When researchers introduced radioactive phosphorus or heavy nitrogen atoms into the soil of experimental pots, how could they refute the hypothesis that the nutrients were being taken up by the plant itself—without help from the fungus?
6. Biologists claim that endomycorrhizal fungi are effective mutualists because they have a higher surface area than do plant roots and are more effective at digesting dead organic matter. Provide evidence to back up both claims.

Group Discussion Problems

1. After reading the essay about the amphibian die-off, review the material on Koch's postulates from Chapter 27. These postulates present criteria for implicating an organism as a disease-causing agent. Did the studies on the chytrid found in dead frogs fulfill Koch's postulates? Are you convinced that the fungus is the causative agent? Explain why or why not.
2. Some biologists contend that the ratio of plant species to fungus species worldwide is on the order of 1:6. Explain why you agree or disagree with this claim. In doing so, consider the analyses of endophytic, parasitic, lichen-forming, mycorrhizal, and saprophytic strategies presented in this chapter. Also, consider the diversity of tissues available in plants.
3. Experiments indicate that cellulase genes are transcribed and translated together. If cells are selected to be extremely efficient at digesting cellulose, is this result logical? Would you predict that the gene that codes for lignin peroxidase is transcribed along with the cellulase loci? How would you test your prediction?
4. Many mushrooms are extremely colorful. Why? Fungi do not see, so colorful mushrooms are obviously not communicating with one another. One hypothesis is that the colors serve as a warning to animals that eat mushrooms, much like the bright yellow and black stripes on wasps. Design an experiment capable of testing this hypothesis.

Answers to Multiple-Choice Questions 1. a; 2. d; 3. c; 4. b; 5. b; 6. d

www.prenhall.com/freeman is your resource for the following: Web Tutorials; Online Quizzes and other Online Study Guide materials; Answers to Conceptual Review Questions; Solutions to Group Discussion Problems; Answers to Figure Caption Questions and Exercises; and Additional Readings and Research.

31 An Introduction to Animals

KEY CONCEPTS

- Animals are a particularly species-rich and morphologically diverse lineage of multicellular organisms on the tree of life.
- Major groups of animals are defined by the design and construction of their basic body plan, which differs in the number of tissues observed in embryos, symmetry, the presence or absence of a body cavity, and the way in which early events in embryonic development proceed.
- Recent phylogenetic analyses of animals have shown that there were three fundamental splits during evolutionary history, resulting in two protostome groups (Lophotrochozoa and Ecdysozoa) and the deuterostomes. The most ancient animal group living today is the sponges. The closest living relatives to animals are choanoflagellates, a group of protists.
- Within major groups of animals, evolutionary diversification was based on innovative ways of feeding and moving. Most animals get nutrients by eating other organisms, and most animals move under their own power at some point in their life cycle.

Jellyfish are among the most ancient of all animals—they appear in the fossil record over 560 million years ago. Compared with most animals living today, they have relatively simple bodies. But like most other animals, they make their living by eating other organisms and are able to move.

As a group, animals are distinguished by two traits: They eat and move. Many unicellular protists also ingest other organisms or dead organic material (detritus) but are small, so they are limited to eating microscopic prey. Animals, in contrast, are multicellular. They are the largest and most abundant predators, herbivores, and detritivores in virtually every ecosystem—from the deep ocean to alpine ice fields and from tropical forests to arctic tundras. Animals find food by tunneling, swimming, filtering, crawling, creeping, slithering, walking, running, or flying. They eat nearly every organism on the tree of life.

Over 1.2 million species of animals have been described and given scientific names to date, and biologists predict that tens of millions more have yet to be discovered. To analyze the almost overwhelming number and diversity of animals, this chapter presents a broad overview of how they diversified. It also provides information on the characteristics of the first groups of animals that evolved. In the next two chapters, we'll follow up with a more detailed exploration of two major phylogenetic groups in animals: protostomes and deuterostomes. Chapter 32 explores the protostomes, which include familiar organisms such as the insects, crustaceans (crabs and shrimp), and mollusks (clams and snails). Chapter 33 features the deuterostomes, which range from the sea stars to the vertebrates, including humans.

31.1 Why Do Biologists Study Animals?

If you ask biologists why they study animals, the first answer they'll give is, "Because they're fascinating." It's hard to argue with this statement. Consider ants. Ants live in colonies that routinely number millions of individuals. But colony-mates cooperate so closely in tasks such as food-getting, colony defense, and rearing young that each ant seems like a cell in a multicellular organism instead of an individual. Other species of ant parasitize this cooperative behavior, however. Parasitic ants look and smell like their host species but enslave them, forcing the hosts to rear the young of the parasitic species instead of their own. Ant colonies also vary widely in size and habitat. The smallest ant species forms a colony that would fit inside the brain of the largest ant species. Other species live in trees and protect their host plants by attacking giraffes and other grazing animals a million times their size. There are rancher ants and farmer ants. Rancher ants tend the plant-sucking insects called aphids and eat the sugar-rich honeydew that aphids secrete from their abdomens (**Figure 31.1a**). Farmer ants eat fungi that they carefully plant, fertilize, and cultivate in underground gardens (**Figure 31.1b**). New ant species are discovered every year.

Based on observations like these, most people would agree that ants—and by extension, other animals—are indeed fascinating. But beyond pure intellectual interest, there are other compelling reasons that biologists study animals:

- Animals are **heterotrophs**—meaning they obtain the chemical energy and carbon compounds they need from other organisms. Recall from Chapter 28 that photosynthetic protists and bacteria are **primary producers**, which form the base of the food chain in most marine environments. Land plants play the same role in most terrestrial habitats. Heterotrophs eat producers and other organisms and are called **consumers.** Animals are consumers that occupy the upper levels of food chains in both marine and terrestrial regions. As a result, it is not possible to understand or preserve ecosystems without understanding and preserving animals.
- Animals are a particularly species rich and morphologically diverse lineage of multicellular organisms on the tree of life. Current estimates suggest that there are between 10 million and 50 million species of animals, although only about a million have been formally described and named. Animals range in size and complexity from tiny, sessile (nonmoving) sponges, which contain just a few cell types and no true tissues, to blue whales, which migrate tens of thousands of kilometers each year in search of food and contain trillions of cells, dozens of distinct tissues, an elaborate skeleton, and highly sophisticated sensory and nervous systems. A great deal of evolution has gone on in this lineage. To understand the history of life, it is important to understand how animals came to be so diverse.

(a) "Rancher ants" tend aphids and eat their sugary secretions

(b) "Farmer ants" cultivate fungi in gardens.

FIGURE 31.1 Biologists Study Animals because They Are Fascinating
(a) Some species of ant make their living by protecting aphids from predators and then harvesting the aphids' sugary secretion called honeydew. **(b)** Several species of ant cultivate and eat fungi.

- Humans in every country depend on wild and domesticated animals for food. Horses, donkeys, oxen, and other domesticated animals also provide most of the transportation and power used in preindustrial societies.
- Efforts to understand human biology depend on advances in animal biology. Most drug testing is done on mice, rats, or primates. Current efforts to understand the human genome are based on analyzing the function of genes in model organisms such as mice, zebrafish, and roundworms.

Given that studying animals is interesting and valuable, let's get started. What makes an animal an animal, and how do biologists go about studying them?

31.2 How Do Biologists Study Animals?

The **animals** are a monophyletic group of multicellular eukaryotes. Most animals move under their own power at some point in their life cycle, and all obtain nutrients by eating other organisms or absorbing nutrients from them. The cells of animals lack walls but have an extensive extracellular matrix, which includes proteins specialized for cell-cell adhesion and communication (see Chapter 8). Animals are the only lineage on the tree of life with species that have muscle tissue and nervous tissue. Although many animals reproduce both sexually and asexually, no animals undergo alternation of generations. During an animal's life cycle, adults of most species are diploid; the only haploid cells are gametes produced during sexual reproduction.

Beyond these shared characteristics, animals are almost overwhelmingly diverse—particularly in morphology. Biologists currently recognize about 34 **phyla**, or major lineages, of animals—including those listed in **Table 31.1**. Each animal phylum has distinct morphological features.

Analyzing Comparative Morphology

In essence, animals are moving and eating machines. A quick glance at the diversity of ways that animals find and capture food, like those illustrated in **Figure 31.2**, should convince you that evolution by natural selection has indeed produced a wide array of ways to move and eat. This diversity is possible because of extensive variation in appendages and in mouthparts or other organs used to capture and process food. Limbs and mouths are specialized structures that make particular ways of moving and eating possible.

In contrast to the spectacular diversity observed among animals in their limbs and mouthparts, the basic architecture of the animal body has been highly conserved throughout evolution. Just as there are only a few basic ways to frame a house—with posts and beams, stud walls, or cement blocks, for example—there are just a handful of ways to design and build an animal body. Once a few different ways of developing a body evolved, an extraordinary radiation of species ensued—based on elaborations of limbs and mouthparts or other structures for moving and capturing food.

Based on this overall pattern of animal evolution, biologists have been able to identify the major lineages of animals by analyzing variation in their core body plan. A **body plan** is an animal's architecture—the major features of its structural and functional design. Four features define the basic elements of an animal's body plan: (1) the number of tissue types found in embryos, (2) the type of body symmetry and degree of cephalization (informally, the formation of a head region), (3) the presence or absence of a fluid-filled cavity, and (4) the way in which the earliest events in the development of an embryo proceed. The origin and early evolution of animals was based on the origin and elaboration of these four features. Let's consider each in detail.

The Evolution of Tissues Sponges are the only group of animals that lack tissues. Although sponges have several cell types, these cells are not organized into the tightly integrated structural and functional units called **tissues.** Based on this observation, sponges are sometimes referred to as **parazoans** ("beside-animals"). All other animals have tissues; collectively, they are sometimes referred to as **eumetazoans** ("truly-among-animals").

Among the eumetazoans, the number of tissue layers that exist in an embryo is a key trait. Animals whose embryos have two types of tissues are called **diploblasts** ("two-sprouts"); animals whose embryos have three types are called **triploblasts** ("three-sprouts"). By examining developing embryos with the light microscope, biologists documented that embryonic tissues are organized in layers, called **germ layers.** In diploblasts these germ layers are called **ectoderm** and **endoderm**; the third layer in triploblasts is found between these two and is called **mesoderm.** The Greek roots *ecto*, *meso*, and *endo* refer to *outer*, *middle*, and *inner*, respectively; the root *derm* means "skin." All animal embyros except those of sponges have distinct outer and inner layers, or "skins"; most also have a distinct middle layer.

The embryonic tissues found in animals develop into distinct adult tissues, organs, and organ systems. In triploblasts, for example, ectoderm gives rise to skin and the nervous system. Endoderm gives rise to the lining of the digestive tract. The digestive tract is also called the *gut* or *gastrovascular cavity*. The circulatory system, muscle, and

WEB TUTORIAL 31.1 The Architecture of Animals

(a) Caterpillar mandibles harvest leaves. **(b)** Feather worm tentacles filter debris. **(c)** Shark jaws and teeth capture prey.

FIGURE 31.2 Animals Move and Eat in Diverse Ways
Variation in limbs and mouthparts allows animals to move and harvest food in a wide variety of ways.

TABLE 31.1 An Overview of Major Animal Phyla

Group and Phylum	Common Name or Example Taxa	Estimated Number of Species
Protostomes: Lophotrochozoa		
Porifera	Sponges	5500
Cnidaria	Jellyfish, corals, anemones, hydroids, sea fans	10,000
Ctenophora	Comb jellies	100
Acoelomorpha	Acoelomate worms	10
Rotifera	Rotifers	1800
Platyhelminthes	Flatworms	20,000
Nemertea	Ribbon worms	900
Gastrotricha	Gastrotrichs	450
Acanthocephala	Acanthocephalans	1100
Entoprocta	Entroprocts	150
Gnathostomulida	Gnathostomulids	80
Sipuncula	Peanut worms	320
Echiura	Spoon worms	135
Annelida	Segmented worms	16,500
Mollusca	Mollusks (clams, snails, octopuses)	94,000
Phoronida	Phoronids	20
Ectoprocta	Ectoprocts	4500
Brachiopoda	Brachiopods; lamp shells	335
Protostomes: Ecdysozoa		
Nematoda	Roundworms	25,000
Kinorhyncha	Kinorhynchs	150
Nematomorpha	Hair worms	320
Priapula	Priapulans	16
Onychophora	Velvet worms	110
Tardigrada	Water bears	800
Arthropoda	Arthropods (spiders, insects, crustaceans)	1,100,000
Deuterostomes		
Echinodermata	Echinoderms (sea stars, sea urchins, sea cucumbers)	7000
Chaetognatha	Arrow worms	100
Hemichordata	Acorn worms	85
Chordata	Chordates (tunicates, lancelets, sharks, bony fish, frogs, reptiles, mammals)	50,000

internal structures such as bone and most organs are derived from mesoderm. In general, then, ectoderm produces the covering of the animal and endoderm generates the digestive tract. Mesoderm gives rise to the tissues in between.

Only two groups of diploblastic animals are alive today: the cnidarians and the ctenophorans (**Figure 31.3a**). The Cnidaria (pronounced *ni-DARE-ee-uh*) include jellyfish, corals, hydra, sea pens, and anemones (**Figure 31.3b**). As **Box 31.1** (page 702) indicates, cnidarians have been an important source of model organisms in developmental biology. The Ctenophora

(a) Cnidarians and ctenophores are diploblastic.
Ectoderm
Endoderm
(b) Cnidaria include hydra, jellyfish, corals, and sea pens (shown).
(c) Ctenophora are the comb jellies.

FIGURE 31.3 Diploblastic Animals Have Bodies Built from Ectoderm and Endoderm

(a) Diploblasts have just two tissue types. **(b)** Like most members of the Cnidaria, this sea pen lives in marine environments. **(c)** Comb jellies, belonging to the Ctenophora, are a major component of planktonic communities in the open ocean. This dark blue comb jelly has just swallowed a whitish comb jelly.

BOX 31.1 A Model Organism: *Hydra*

Cnidarians are of particular interest to biologists, because these diploblasts are the most ancient lineage of animals with tissues. The ectoderm and endoderm present in their embryos gives rise to a number of cell types and tissues in the adults. Cnidarian tissues may be composed of sensory cells that initiate electrical signals in response to environmental stimuli, nerve cells that process those electrical signals and conduct them throughout the body, or muscle cells that contract or relax in response to electrical signals. Cnidarians also have a particularly important type of tissue known as epithelium. **Epithelium** consists of a tightly joined layer of cells that is attached to an extensive extracellular matrix. In animals, epithelium covers the outside of the body and lines the surfaces of internal organs.

To understand why the presence of these tissues is interesting, consider the freshwater cnidarian called hydra (**Figure 31.4**). Most species in the genus *Hydra* are about half a centimeter long, live attached to rocks or other firm substrates, and make their living by catching small prey or pieces of organic debris with a cluster of long tentacles. An adult hydra has three major body regions: (1) a *basal disk*, which attaches the individual to a rock; (2) a tubular section that makes up the bulk of the body; and (3) a "head" that contains the mouth and tentacles.

For over 100 years, biologists have been doing experiments based on cutting hydra bodies apart in various ways and studying how missing tissues and body regions **regenerate**—that is, re-form. This work has led to a deeper understanding of how nerve cells, muscles, epithelia, and other specialized cells and tissues arise from unspecialized cells called stem cells (see Chapter 22). In particular, hydra experiments provided fundamental insights into how cell-to-cell signals (1) organize cells into tissues and body segments and (2) trigger the differentiation of cells into specialized cell types. Studying how adult hydra regenerate has helped biologists understand how tissues form and specialized cells arise in animal embryos.

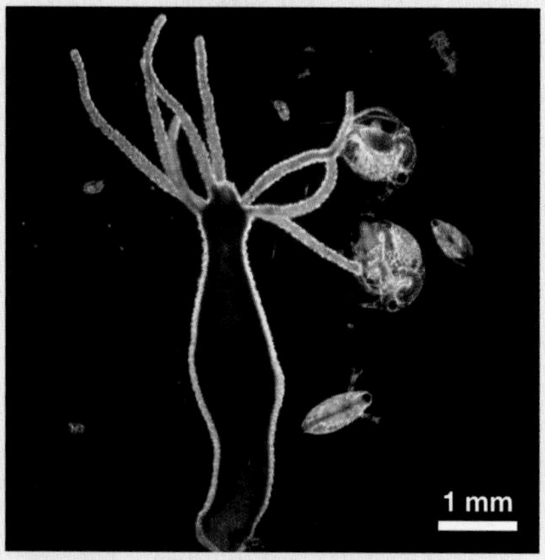

FIGURE 31.4 Hydra Is a Model Organism in Biology
Hydra grow quickly and are relatively easy to maintain in the lab. If an adult is cut into pieces, missing body parts can regenerate in some fragments to form complete adults.

(pronounced *ten-AH-for-ah*) are the comb jellies (**Figure 31.3c**). All other animals, from leeches to humans, are triploblastic.

Symmetry and Cephalization A basic feature of a multicellular body is the presence or absence of a plane of symmetry. An animal's body is symmetrical if it can be divided by a plane such that the resulting pieces are nearly identical. Animal bodies can have 0, 1, 2, or more planes of symmetry. Most sponges, including the one illustrated in **Figure 31.5a**, are **asymmetrical**—that is, having no planes of symmetry. They cannot be sectioned in a way that produces similar sides.

All other animals exhibit radial ("spoke") symmetry or bilateral ("two-sides") symmetry. Organisms with **radial symmetry** have at least two planes of symmetry. Most of the radially symmetric animals living today either float in water or live attached to a substrate. As **Figure 31.5b** shows, their bodies are often cylinder-like. As a result, they can capture prey or react to predators that approach from more than one direction.

Organisms with **bilateral symmetry**, in contrast, face their environment in one direction. Bilaterally symmetric animals have one plane of symmetry and tend to have a long, narrow body with a distinct head end (**Figure 31.5c**). The evolution of bilateral symmetry was a critical step in animal evolution, because it triggered a series of associated changes that are collectively known as **cephalization**: the evolution of a **head**, or *anterior* region, where structures for feeding, sensing the environment, and processing information are concentrated. Bilateral symmetry and cephalization made unidirectional movement possible. Feeding and sensory structures on the head face the environment, while *posterior* regions, at the opposite end of the organism, are specialized for locomotion. With the exception of adult forms of species in the phylum Echinodermata, which have radial symmetry, all triploblastic animals have bilateral symmetry. The echinoderms (pronounced *ee-KINE-oh-derms*) include species such as sea stars, sea urchins, feather stars, and brittle stars. Although their larvae are bilaterally symmetric, adult echinoderms are said to have **pentaradial symmetry**—meaning five planes of symmetry.

To explain the pervasiveness of bilateral symmetry, biologists point out that locating and capturing food is particularly efficient when movement is directed by a distinctive head region and powered by a long posterior region. In combination with the origin of mesoderm, which made the evolution of extensive

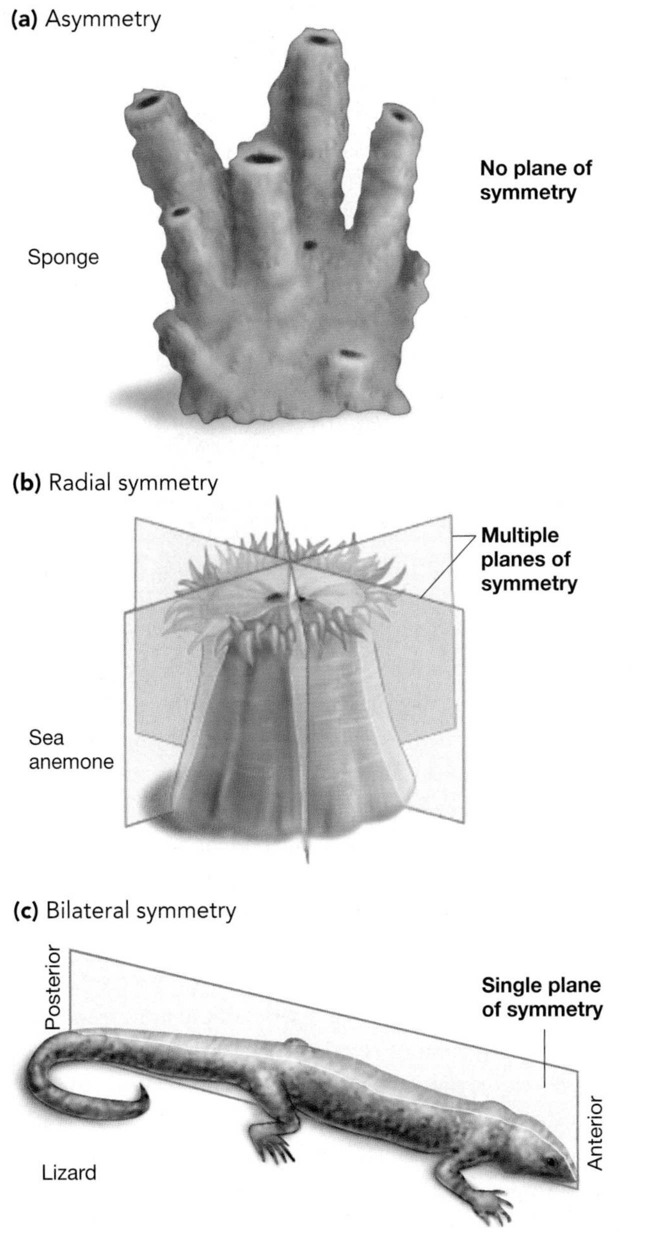
(a) Asymmetry
No plane of symmetry
Sponge
(b) Radial symmetry
Multiple planes of symmetry
Sea anemone
(c) Bilateral symmetry
Posterior
Single plane of symmetry
Anterior
Lizard

FIGURE 31.5 There Are Three Types of Body Symmetry in Animals
QUESTION Are the animals in Figures 31.2a, 31.2c, and 31.3 asymmetric, radially symmetric, or bilaterally symmetric? Answer the same question about humans and sea stars.

musculature possible, a bilaterally symmetric body plan enabled rapid, directed movement and hunting. Lineages with a triploblastic, bilaterally symmetric body had the potential to diversify into an array of formidable eating and moving machines.

Evolution of a Body Cavity A third architectural element that distinguishes animal phyla is the presence of an internal fluid-filled cavity called a **coelom** (pronounced *SEE-loam*). Although diploblasts have a central canal that functions in digestion and circulation, they do not have a coelom. The triploblasts called Platyhelminthes and the Acoelomorpha ("no-cavity-form") also lack a fluid-filled body cavity (**Figure 31.6a**). The remaining triploblasts have a body cavity.

Biologists were able to determine the nature of the body cavity in various phyla through careful observation and dissection of developing embryos and adults. Animals that do not have a coelom are called **acoelomates**; those that possess a coelom are known as **coelomates**. In a few of the coelomate groups, such as the roundworms and rotifers, the enclosed cavity forms between the endoderm and mesoderm layers in the embryo. This design is called a **pseudocoelom**, meaning "false-hollow" (**Figure 31.6b**). The term is unfortunate, because there is nothing false about the fluid-filled cavity—it exists. It simply forms in a different way than a "true" coelom, which forms from within the mesoderm itself and is lined with cells from the mesoderm (**Figure 31.6c**). As a result, muscle and blood vessels can form on either side of the coelomates' body cavity. In this respect, the coelom represents a more effective design than the pseudocoelom and the acoelomate condition in diploblasts.

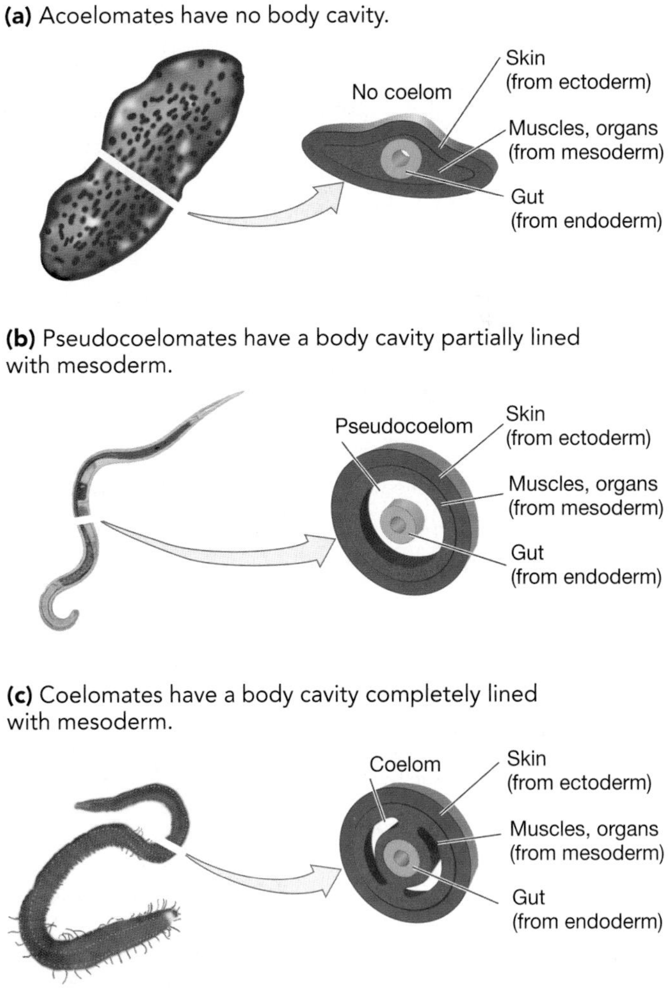
(a) Acoelomates have no body cavity.
No coelom
Skin (from ectoderm)
Muscles, organs (from mesoderm)
Gut (from endoderm)
(b) Pseudocoelomates have a body cavity partially lined with mesoderm.
Pseudocoelom
Skin (from ectoderm)
Muscles, organs (from mesoderm)
Gut (from endoderm)
(c) Coelomates have a body cavity completely lined with mesoderm.
Coelom
Skin (from ectoderm)
Muscles, organs (from mesoderm)
Gut (from endoderm)

FIGURE 31.6 Animals May or May Not Have a Body Cavity

The coelom is important because it creates a container for the circulation of oxygen and nutrients, along with space where internal organs can move independently of each other. In addition, an enclosed, fluid-filled chamber can act as an efficient **hydrostatic skeleton.** Soft-bodied animals with hydrostatic skeletons can move even if they do not have fins or limbs. Movement is possible because the pseudocoelom or coelom of these animals is filled with fluid that is under pressure from the wall of the body cavity—much like a water balloon. The pressurized fluid stiffens the organism, and when muscles in the body wall contract against the pressurized fluid, it moves. When muscles contract, they shorten; when they relax, they lengthen. The shape of the body cavity enclosed by muscles changes in response to muscle movement, because the water inside cannot be compressed. As **Figure 31.7** shows, coordinated muscle contractions and relaxations produce changes in the shape of a hydrostatic skeleton that make writhing or swimming movements possible. By providing a hydrostatic skeleton, the coelom gave bilaterally symmetric organisms the ability to move efficiently in search of food.

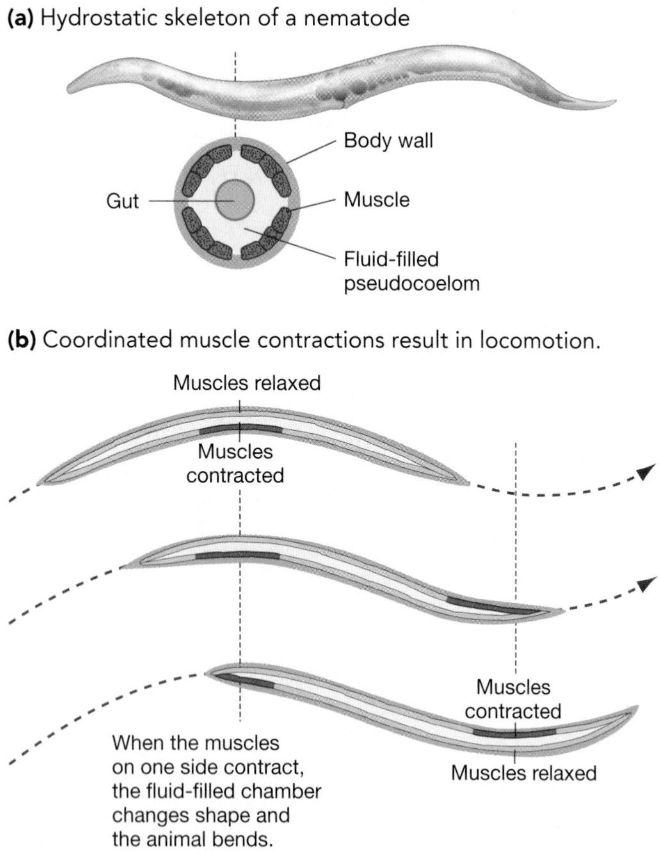
(a) Hydrostatic skeleton of a nematode
Body wall
Gut
Muscle
Fluid-filled pseudocoelom
(b) Coordinated muscle contractions result in locomotion.
Muscles relaxed
Muscles contracted
Muscles contracted
When the muscles on one side contract, the fluid-filled chamber changes shape and the animal bends.
Muscles relaxed

FIGURE 31.7 Hydrostatic Skeletons Allow Limbless Animals to Move

The nematode (roundworm) moves with the aid of its hydrostatic skeleton. In this case, the hydrostatic skeleton is an enclosed, fluid-filled chamber. QUESTION Suppose muscles on both sides of this nematode contracted at the same time. What would happen?

The Protostome and Deuterostome Patterns of Development With the exception of adult echinoderms, all of the coelomates—including juvenile forms of echinoderms—are bilaterally symmetric and have three embryonic tissue layers. This huge group of organisms is formally called the **Bilateria,** because they are bilaterally symmetric at some point in their life cycle. The bilaterians, in turn, can be split into two subgroups based on distinctive events that occur early in the development of the embryo. The two groups are the protostomes and the deuterostomes. The vast majority of animal species, including the arthropods (insects, spiders, crustaceans), mollusks, and annelids (segmented worms), are protostomes. Chordates (ascidians, lancelets, fish, frogs, mammals) and echinoderms are deuterostomes.

To understand the differences in how protostome and deuterostome embryos develop, recall from Chapter 21 that the development of an animal embryo begins with cleavage. **Cleavage** is a rapid series of mitotic divisions that occurs in the absence of growth. Cleavage divides the egg cytoplasm and often results in a hollow ball of cells. In many protostomes, these cell divisions take place in a pattern known as **spiral cleavage.** When spiral cleavage occurs, the mitotic spindles of dividing cells orient at an angle to the main axis of the cells and result in a helical arrangement of cells. In many deuterostomes, the mitotic spindles of dividing cells orient parallel or perpendicular to the main axes of the cells, resulting in cells that stack directly on top of each other; this pattern is called **radial cleavage** (**Figure 31.8a**).

After cleavage has created a ball of cells, the process called gastrulation occurs. **Gastrulation** is a series of cell movements that results in the formation of ectoderm, mesoderm, and endoderm—the three embryonic tissue layers. In both protostomes and deuterostomes, gastrulation begins when cells move into the center of the ball of cells. The invagination of cells creates a pore that opens to the outside (**Figure 31.8b**). In **protostomes,** this pore becomes the mouth. The other end of the gut, the **anus,** forms later. In **deuterostomes,** however, this initial pore becomes the anus; the mouth forms later. Translated literally, protostome means "first-mouth" and deuterostome means "second-mouth."

The final difference between the groups arises as gastrulation proceeds and the coelom begins to form. As **Figure 31.8c** indicates, the coelom of protostomes begins to form via a split within a solid block of mesoderm. In deuterostomes, however, layers of mesodermal cells pinch off from the gut to form the coelom.

To summarize, the protostome and deuterostome patterns of development result from differences in three processes: cleavage, gastrulation, and coelom formation. In essence, the protostome and deuterostome patterns of development represent two distinct ways of achieving the same end—the construction of a bilaterally symmetric body that contains a cavity lined with mesoderm.

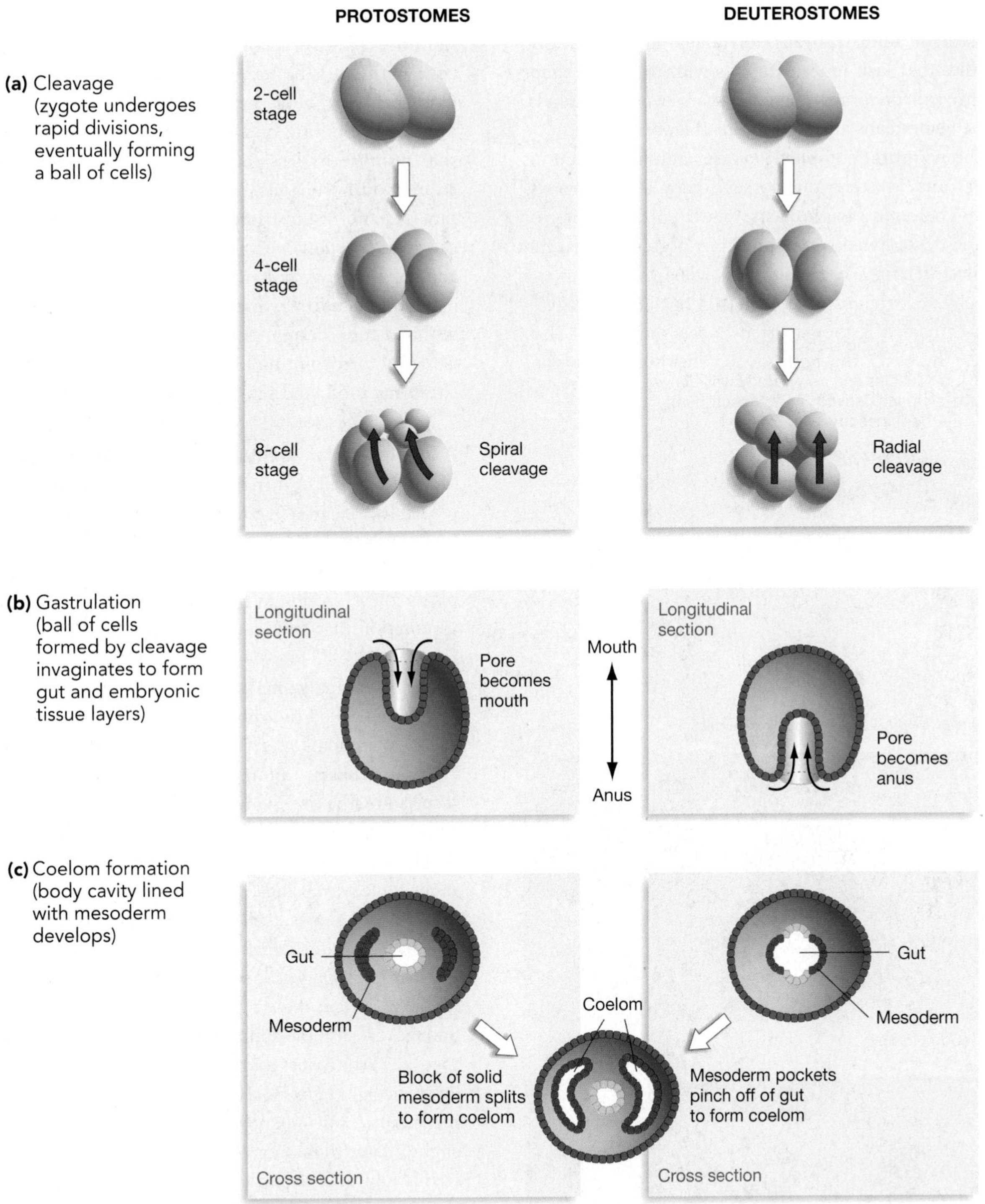
PROTOSTOMES
DEUTEROSTOMES
(a) Cleavage (zygote undergoes rapid divisions, eventually forming a ball of cells)
2-cell stage
4-cell stage
8-cell stage
Spiral cleavage
Radial cleavage
(b) Gastrulation (ball of cells formed by cleavage invaginates to form gut and embryonic tissue layers)
Longitudinal section
Pore becomes mouth
Mouth
Anus
Longitudinal section
Pore becomes anus
(c) Coelom formation (body cavity lined with mesoderm develops)
Gut
Mesoderm
Coelom
Block of solid mesoderm splits to form coelom
Mesoderm pockets pinch off of gut to form coelom
Gut
Mesoderm
Cross section
Cross section

FIGURE 31.8 In Protostomes and Deuterostomes, Three Events in Early Development Differ
The differences between protostomes and deuterostomes show that there is more than one way to build a bilaterally symmetric, coelomate body plan.

The Tube-within-a-Tube Design Over 99 percent of the animal species alive today are bilaterally symmetric triploblasts that have coeloms and follow either the protostome or deuterostome pattern of development. This combination of features has been a spectacularly successful way to design and build a moving and eating machine.

Although it might sound complex to call a certain animal a "bilaterally symmetric, coelomic triploblast with protostome [or deuterostome] development," the bodies of most animals are actually extremely simple in form. The basic animal body is a **tube within a tube**. The inner tube is the individual's gut, and the outer tube forms the body wall, as illustrated in **Figure 31.9a**. The mesoderm in between forms muscles and organs. In several animal phyla, individuals have long, thin, tubelike bodies that lack limbs. Animals with this body shape are commonly called **worms**. There are many wormlike phyla, including the nemerteans and sipunculids (**Figure 31.9b**).

What about more complex-looking animals, such as grasshoppers and lobsters and horses? They are bilaterally symmetric, coelomic triploblasts with protostome or deuterostome development, too, and they aren't worms. But a moment's reflection should convince you that, in essence, the body plan of these animals can also be thought of as a tube within a tube, except that the tube is mounted on legs. Consider that most animals with complex-looking bodies are relatively long and thin. They have an outer body wall that is more or less tubelike and an internal gut that runs from mouth to anus. The body cavity itself is filled with muscles and organs derived from mesoderm. Wings and legs are just efficient ways to move a tube-within-a-tube body around the environment.

Once evolution by natural selection produced the basic tube-within-a-tube design, the diversification of animals was triggered by the evolution of novel types of structures for moving, capturing food, and sensing the environment.

(a) The tube-within-a-tube body plan

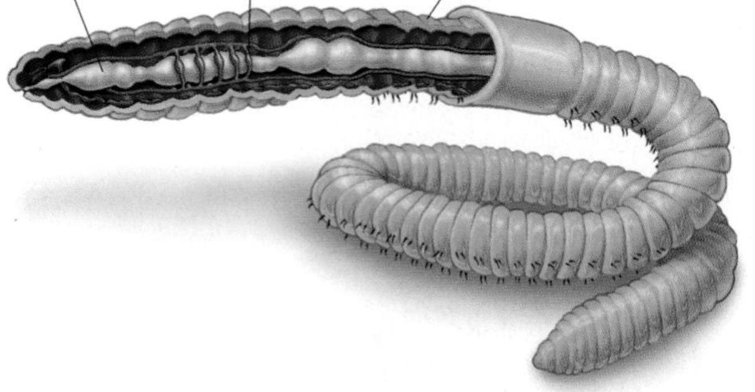
Gut derived from endoderm
Muscles and organs derived from mesoderm
Body wall derived from ectoderm

(b) Many animal phyla have wormlike bodies.

Nemertean (ribbon worm)

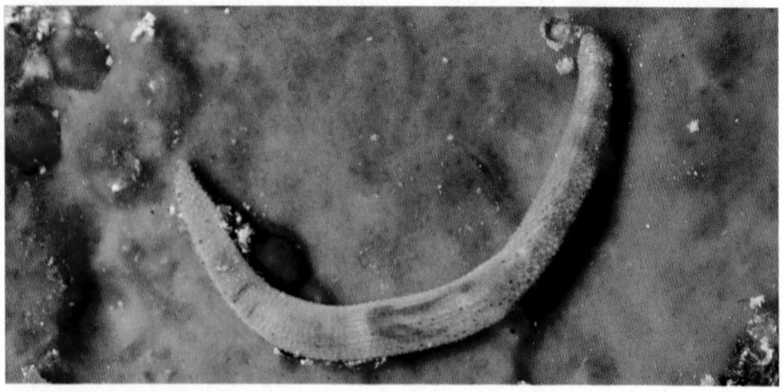

Sipunculid (peanut worm)

FIGURE 31.9 The Tube-within-a-Tube Body Plan Is Common in Animals

(a) Many species in the Bilateria have bodies that are variations on the tube-within-a-tube design. **(b)** The Nemertea and Sipunculida are phyla that have wormlike bodies. QUESTION How is the term *worm* similar to the term *yeast* (see Chapter 30)?

A Phylogeny of Animals Based on Morphology When biologists realized that major groups of animals could be characterized by variation in embryonic tissues, body symmetry, type of body cavity, and early development, they used the data to infer the evolutionary relationships shown in **Figure 31.10**. The tree places these groups in a phylogenetic sequence, based on the assumption that complex body plans are derived from simpler forms.

To begin analyzing this tree, note that it identifies a group of protists called the **choanoflagellates** as the closest living relatives of animals and the **Porifera** (sponges) as the most ancient, or basal, animal phylum. These hypotheses were inspired by the observation that choanoflagellates and sponges share several key characteristics. Both are **sessile**, meaning that adults live permanently attached to a substrate. They also feed in the same way, using cells with nearly identical morphology. As **Figure 31.11** shows, the beating of flagella creates water currents that bring organic debris toward the feeding cells of choanoflagellates and sponges. Sponge feeding cells are called **choanocytes.** In these feeding cells, food particles are trapped and ingested. Sponges are also the animals with the simplest body plans. Recall that they lack tissues and that most are asymmetrical.

Notice in Figure 31.10 that the radially symmetric phyla are placed just up from sponges on the tree, meaning they evolved slightly later than sponges. Their placement at the base of the tree is logical, because cnidarians and ctenophorans have just two embryonic tissues and because radial symmetry is thought to be a simpler design than bilateral symmetry. Among the many bilaterally symmetric phyla, the tree predicts that groups evolved in the following order: acoelomates, then pseudocoelomates, and finally coelomates.

What happened *after* the coelomates split into the protostomes and deuterostomes? A close examination of Figure 31.10 suggests that two major events occurred: (1) Radial symmetry evolved as an adult trait in some echinoderms. (2) A type of body architecture called **segmentation** evolved independently in both protostomes and deuterostomes. When a body is divided into a series of repeated structures, such as an earthworm's segments or

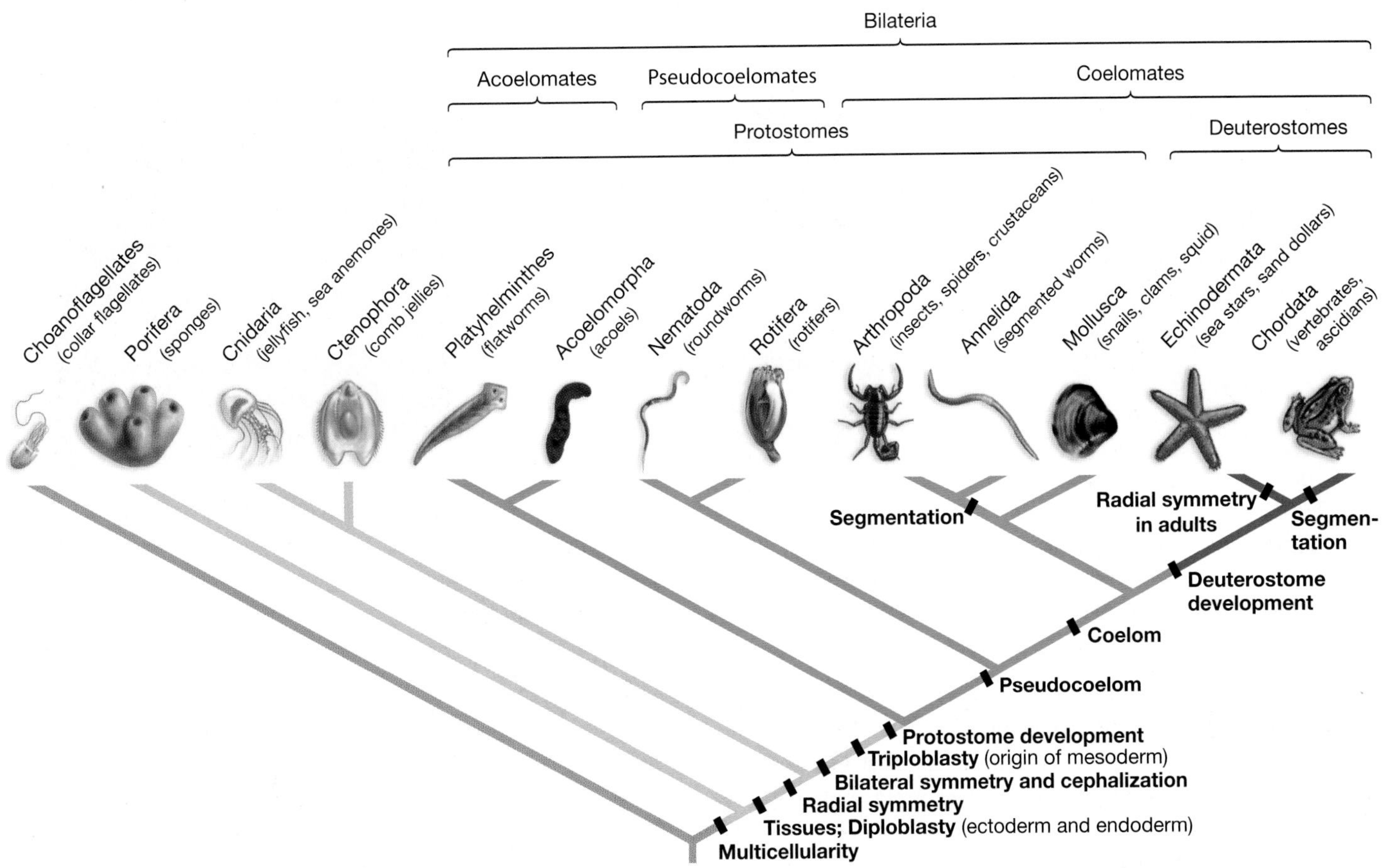

FIGURE 31.10 A Phylogeny of Animal Phyla Based on Morphology
Phylogenetic tree based on similarities and differences in the body plans and developmental sequences of various animal phyla. The bars along the branches indicate when certain traits originated.

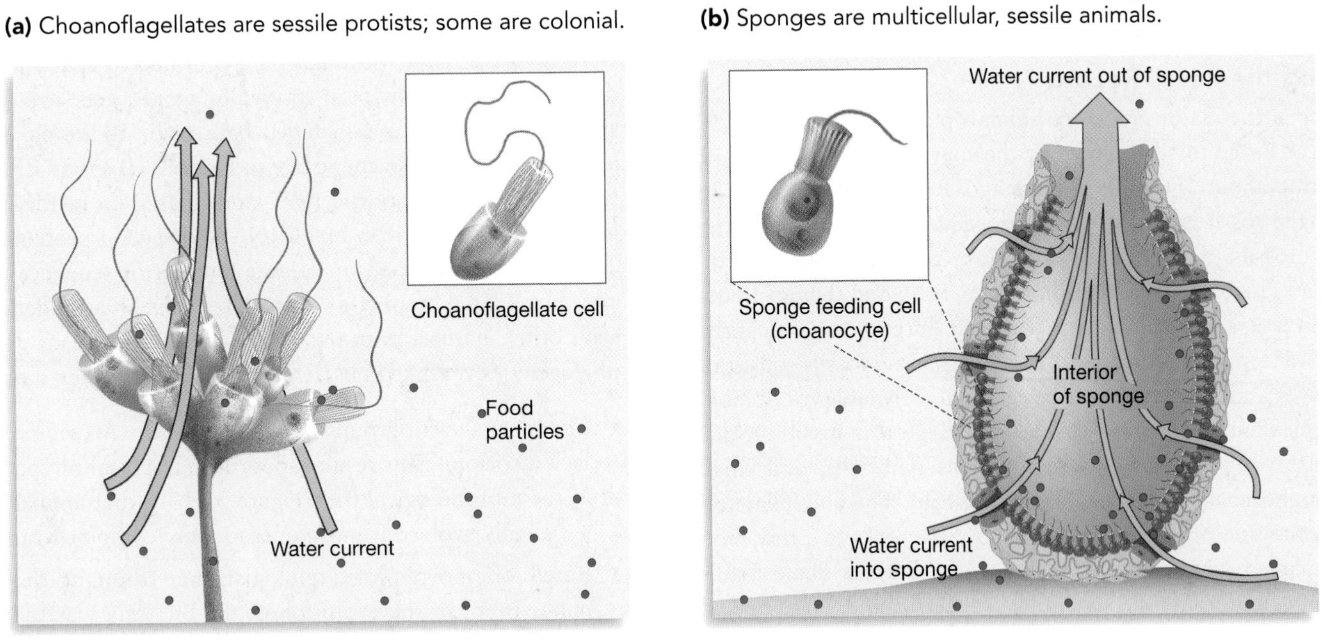

FIGURE 31.11 Choanoflagellates and Sponge Feeding Cells Are Almost Identical in Structure and Function
(a) Choanoflagellates are filter feeders. **(b)** A cross section of a simple sponge. The beating of flagella produces a water current that brings food into the body of the sponge, where it can be ingested by feeding cells.

(a) Sponges are the first animals in the fossil record.
100 µm
Doushantuo fossils: 580–570 million years old
(b) Cnidarians, ctenophores, and other simple forms appear later.
5 cm
Ediacaran fossils: 565–544 million years old
(c) Bilaterians appear still later.
1 cm
Burgess Shale fossils: 525–515 million years old

FIGURE 31.12 The Fossil Record Documents the Origin and Early Evolution of Animals
The fossil record supports the hypothesis that animals evolved in the following order: **(a)** asymmetric species that lacked tissues, **(b)** radially symmetric diploblasts, and **(c)** bilaterally symmetric triploblasts.

a fish's vertebral column and ribs, it is said to be segmented. Segmentation is found in the protostome phylum Annelida and Arthropoda (insects, spiders, crustaceans), as well as in a deuterostome lineage, the vertebrates. **Vertebrates** are a monophyletic lineage defined by the presence of a skull; many vertebrate species also have a backbone. The group called the **invertebrates**, which is defined as all animals that are not vertebrates, is paraphyletic—meaning that they include some, but not all, of the descendants of a common ancestor.

Now, what do other data sets have to say about these relationships? Specifically, do data from the fossil record and from molecular phylogenies agree with the evolutionary tree implied by morphological data, or do they conflict with it?

Using the Fossil Record

Most of the major groups of animals pictured in Figure 31.10 appear in the fossil record over the course of 65 million years, starting about 580 million years ago. Recall from Chapter 26 that the fossil record of animals begins with the Doushantuo microfossils (570 million years ago), continues with the Ediacaran faunas (565–544 million years ago), and then explodes in diversity and complexity with the Burgess Shale deposits (525–515 million years ago). The Doushantuo fossils consist of sponges and what appear to be eggs and early embryos of more complex animals (**Figure 31.12a**). The Ediacaran fossils include an array of sponges, small cnidarians (jellyfish), and small ctenophorans (comb jellies; **Figure 31.12b**). The only bilaterally symmetric organism known from these rocks is a tiny mollusk called *Kimberella*. In contrast, the Burgess Shale fossils include hundreds of bilaterally symmetric and large-bodied species from most major animal phyla. Animals ranging from sponges to chordates are present (**Figure 31.12c**).

In general, then, the fossil record of animal origins is consistent with the overall pattern of evolution described in Figure 31.10. The first animals to appear were the sponges, followed by the diploblasts and then by the bilaterally symmetric triploblasts. The earliest animals in the fossil record were small, and tens of millions of years passed before large-bodied forms evolved. Do data from phylogenetic analyses of DNA sequence data support these conclusions?

Evaluating Molecular Phylogenies

Perhaps the most influential paper ever published on the phylogeny of animals appeared in 1997. Using sequences from the gene that codes for the RNA molecule in the small subunit of the ribosome, Anna Marie Aguinaldo and colleagues estimated the phylogeny of species from 14 animal phyla. The results were revolutionary. Although they continue to spark intense debate, they have now been verified by more recent and extensive analyses that have included data from additional genes and phyla. The phylogenetic tree in **Figure 31.13** is an updated version of the result of the 1997 study, based on further studies of the genes for ribosomal RNA and several proteins. Because this tree is based on a large amount of sequence data, and thus a large number of traits that evolve independently of each other, it represents the best current estimate of animal phylogeny. Several key observations emerge from the data:

- The most ancient groups of triploblasts, the Acoelomorpha, lack a coelom. This result supports an important hypothesis in the morphological tree (Figure 31.10)—that animal bodies usually evolved from simpler to more complex forms.
- Based on morphology, biologists had thought that the major event in the evolution of the Bilateria was the split between the protostomes and the deuterostomes. The molecular data concur but show that an additional, equally fundamental split occurred within protostomes, forming two major subgroups with protostome develop-

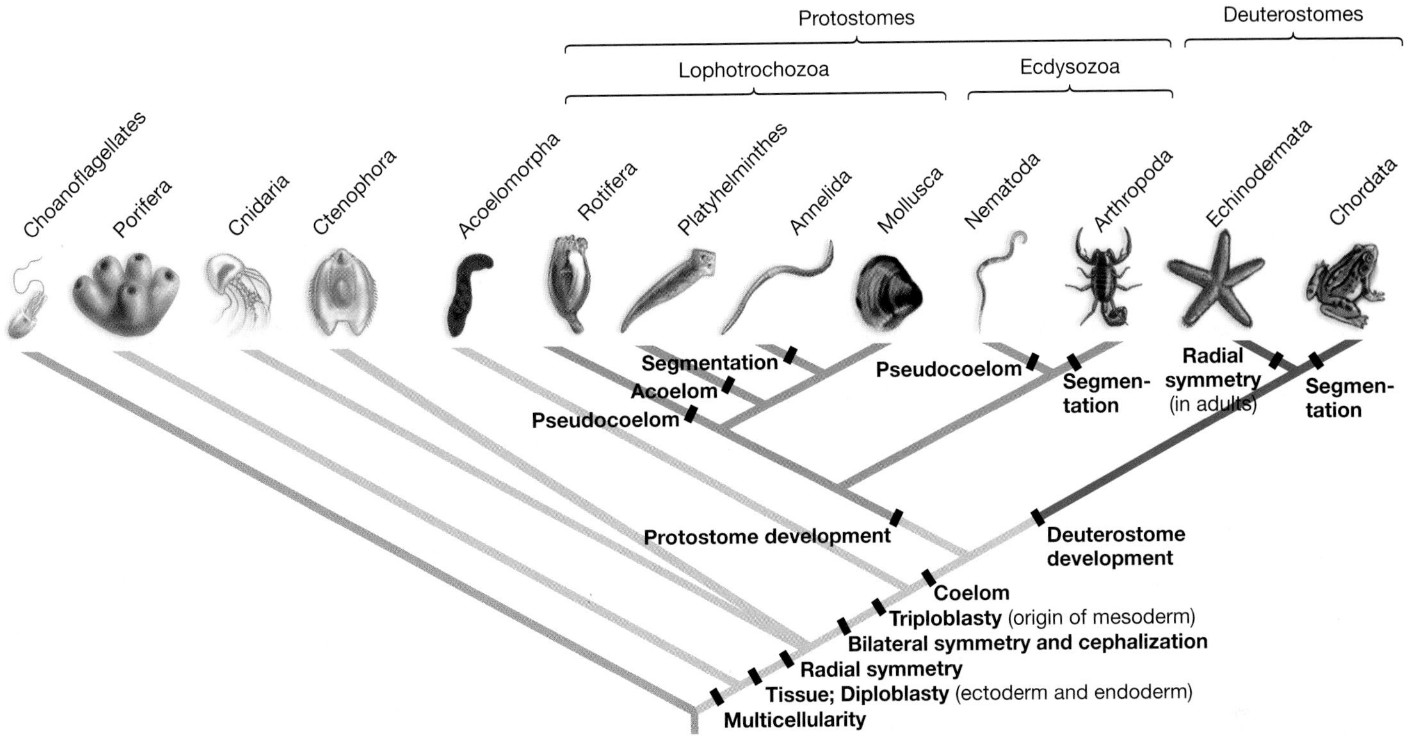

FIGURE 31.13 A Phylogeny of Animal Phyla Based on DNA Sequence Data
Phylogenetic tree based on similarities and differences in the DNA sequences of several genes from various animal phyla. The bars along the branches indicate when certain morphological traits originated. **EXERCISE** Circle and label at least two branches and two bars that differ from those on the tree in Figure 31.10.

ment: (1) The **Ecdysozoa** (pronounced *eck-die-so-ZOH-ah*) includes the arthropods and the nematodes; (2) the **Lophotrochozoa** (pronounced *low-foe-tro-ko-ZOH-ah*) includes the mollusks and the annelids. Ecdysozoans grow by shedding their external skeletons and expanding their bodies, while lophotrochozoans grow by extending the size of their skeletons. In Chapter 32 we explore the differences between these two lineages in more detail.

- Although both annelids (earthworms and other segmented worms) and arthropods (insect, spiders, and crustaceans) have segmented bodies, the molecular phylogeny shows that segmentation evolved independently in the two lineages as well as in vertebrates. Annelids are members of the Lophotrochozoa; arthropods are ecdysozoans.
- Species in the phylum Platyhelminthes (flatworms) do not have a coelom but are lophotrochozoans. To interpret this result, biologists point out that platyhelminths had to have evolved from an ancestor that had a coelom. Stated another way, the acoelomate condition in these species is a derived condition. It represents the *loss* of a complex trait.
- Twice during the course of evolution, bodies with pseudocoeloms arose from ancestors that had "true" coeloms. A change from coelom to pseudocoelom occurred in the ancestors of today's (1) nematodes (roundworms) and (2) rotifers.

Although biologists are increasingly confident that most or all of these conclusions are correct, the phylogeny of animals is still very much a work in progress. As data sets expand, it is likely that new analyses will not only confirm or challenge these results but also contribute other important insights into how the most species-rich lineage on the tree of life originated and diversified. Stay tuned.

✓CHECK YOUR UNDERSTANDING

The origin and early diversification of animals was marked by changes in four fundamental features: body symmetry, the number of embryonic tissues present, the evolution of a body cavity, and protostome versus deuterostome patterns of development. You should be able to (1) explain how the evolution of bilateral symmetry is associated with cephalization, why cephalization was important, and why bilateral symmetry in combination with triploblasty and a coelom is responsible for the "tube-within-a-tube" design observed in most animals living today; (2) make a rough sketch of the phylogeny of animals based on molecular sequences, showing choanoflagellates as an outgroup, sponges and jellyfish as basal groups, Acoelomorpha as the most ancient members of the Bilateria, and the splits that produced the Lophotrochozoa, Ecdysozoa, and deuterostomes; (3) on the tree you sketched, mark the origin of multicellularity, triploblasty, protostome development, deuterostome development, and at least one origin of segmentation; and (4) compare and contrast the tree based on molecular data with the phylogeny implied by morphological traits.

31.3 What Themes Occur in the Diversification of Animals?

Within each animal phylum, the basic features of the body plan do not vary from species to species. For example, mollusks are triploblastic, bilaterally symmetric protostomes with a coelom; their body plan features a muscular *foot*, a cavity called the *visceral mass*, and a structure called a *mantle*. But there are over 100,000 species of mollusk. If animal phyla are defined by a particular body plan, what triggered the diversification of species within each phylum?

In most cases, the answer to this question is the evolution of innovative methods for feeding and moving. Recall that most animals get their food by ingesting other organisms. Animals are diverse because there are thousands of ways to find and eat the millions of different organisms that exist. Let's survey the diverse ways that animals feed, move, and reproduce.

Feeding

The feeding tactics observed in animals can be broken into five general types: (1) suspension feeding, (2) deposit feeding, (3) herbivory, (4) predation, and (5) parasitism. Many animals use more than one of these tactics over the course of their lifetime, because they undergo a **metamorphosis** ("between-forms")—a change in form during development—that allows them to exploit different sources of food as a juvenile than as an adult. Metamorphosis and all five feeding tactics are found among animals fossilized in the Burgess Shale and among species in the most diverse and familiar animal phyla living today—the mollusks, arthropods, and chordates. Let's examine each tactic in turn, then analyze how metamorphosis affects feeding.

Suspension (Filter) Feeding **Suspension feeders,** also known as **filter feeders,** capture food by filtering out particles suspended in water or air. The organisms in **Figure 31.14** illustrate a few of the many variations on this theme.

The clam pictured in Figure 31.14a uses a muscular structure called a **foot** to burrow into sediments. While burrowed, clams extend long tubes called **siphons** to maintain contact with the surface. Inside the clam body, cilia on thin structures called **gills** pump water out one siphon and draw it into the other. The incoming water contains food particles that are trapped on the gills and swept toward the mouth by cilia.

Figure 31.14b shows the small marine animals called krill, which suspension feed as they swim. As individuals move forward, their legs wave in and out. Projections on their legs trap food particles that flow past. The food particles are then moved up the body to the mouth, where they are ingested.

Figure 31.14c illustrates how krill are eaten by a group of suspension feeders called the baleen whales. These whales have a series of long plates, made from a horny material called baleen, hanging from their jaws. Baleen whales feed by gulping water that contains krill, squeezing the water out between their baleen plates, and trapping the krill inside their mouths.

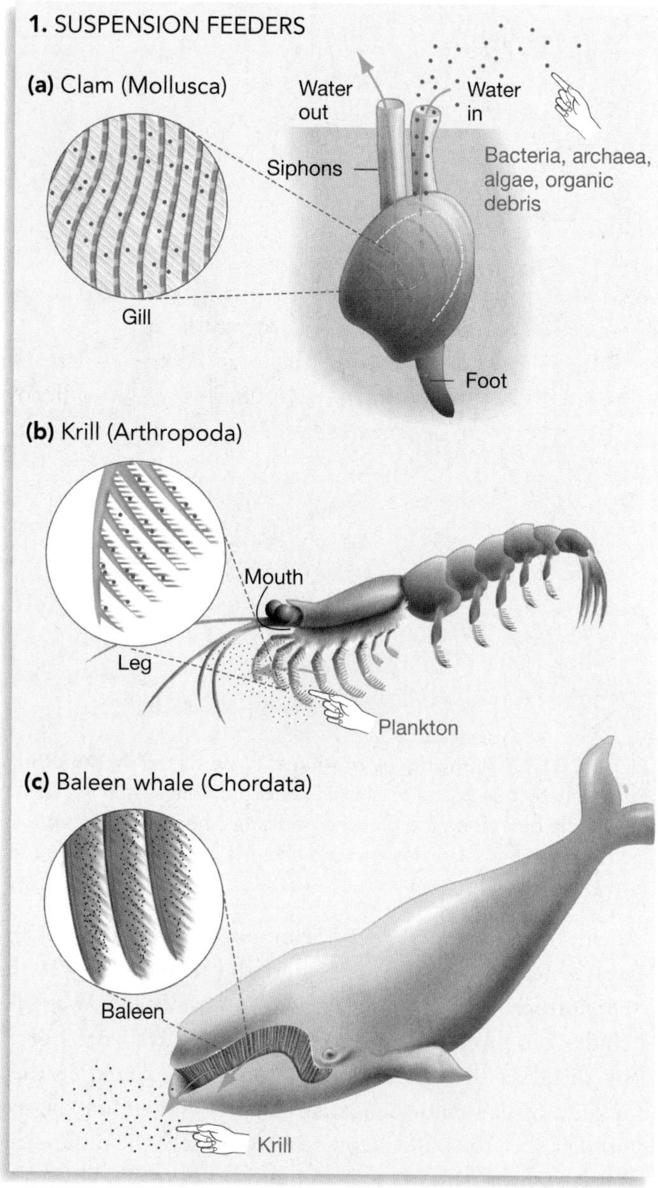

FIGURE 31.14 Suspension Feeders Capture Food by Filtering
(a) Clams, (b) krill, and (c) baleen whales are all suspension feeders. They filter food particles from water, using the trapping structures shown in the close-ups.

It is important to note that suspension feeding is found in a wide variety of animal groups. Clams are mollusks, krill are arthropods, and whales are chordates. A glance at the phylogeny in Figure 31.13 suggests that this strategy has evolved many times, independently. Although the general food-capturing strategy is the same, the mechanism and the type and size of food that is gathered vary from species to species.

Deposit Feeding **Deposit feeders** eat their way through a substrate. Earthworms, for example, are annelids that swallow soil as they tunnel through it. They digest organic matter in the soil and leave behind the mineral material as feces. For these organisms, food consists of soil-dwelling bacteria, protists, fungi, and archaea, along with detritus—the dead and often partially decomposed remains of organisms. Insects that burrow through plant leaves and stems, bore through piles of feces, or mine the carcasses of dead animals or plants can also be considered deposit feeders, because they eat through a substrate. Feeding categories are not rigid, however. Depending on what they eat, deposit feeders can also be considered **herbivores** ("plant-eaters"), parasites, **detritivores** ("detritus-eaters"), or predators. And even though earthworms deposit-feed in soil, they also retrieve and eat dead leaves from the surface.

Unlike suspension feeders, which are diverse in size and shape and use various trapping or filtering systems, deposit feeders are similar in appearance. They usually have simple mouthparts if they eat soft substrates, and a wormlike body shape (**Figure 31.15**). Like suspension feeding, however, deposit feeding occurs in a wide variety of taxonomic groups, including roundworms (Nematoda), segmented worms (Annelida), mollusks (Mollusca), peanut worms (Sipunculida), and chordates such as hagfish.

Herbivory Animals from a diversity of phyla harvest algae or plant tissues. In sharp contrast to suspension feeders and deposit feeders, herbivores have complex mouths with structures that make biting and chewing or sucking possible. The mouthpart structure called a **radula**, for example, which is found in snails and other mollusks, functions like a rasp or a file. The sharp plates on the radula move back and forth to scrape material away from a plant or alga so that it can be ingested (**Figure 31.16a**). The long, hollow proboscis of a moth (**Figure 31.16b**) is used to suck nectar, while the **mandibles** (chewing mouthparts) of

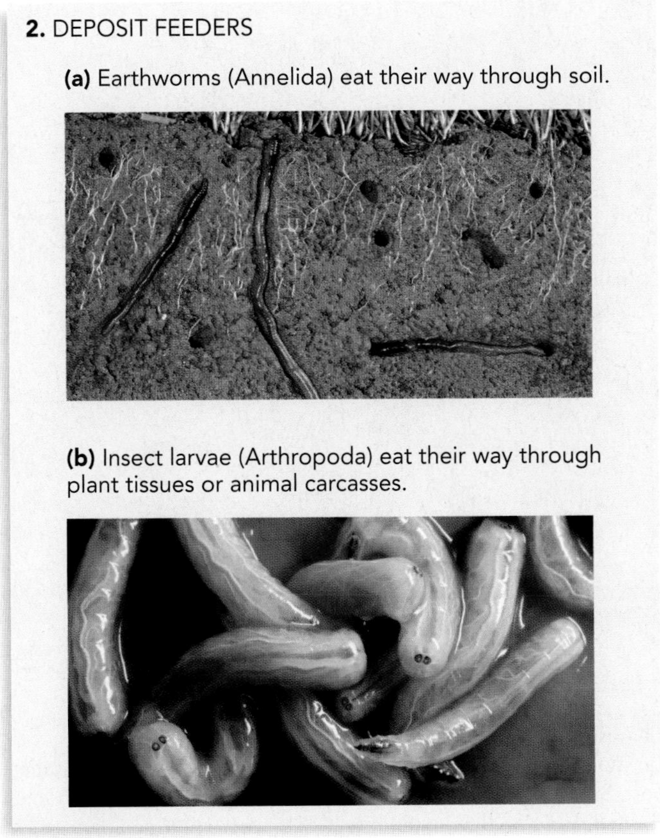

FIGURE 31.15 Deposit Feeders Dig through a Substrate
Deposit feeders, including (a) earthworms and (b) maggots, generally have long, thin bodies. QUESTION Why is it logical that deposit feeders tend to have tubelike bodies?

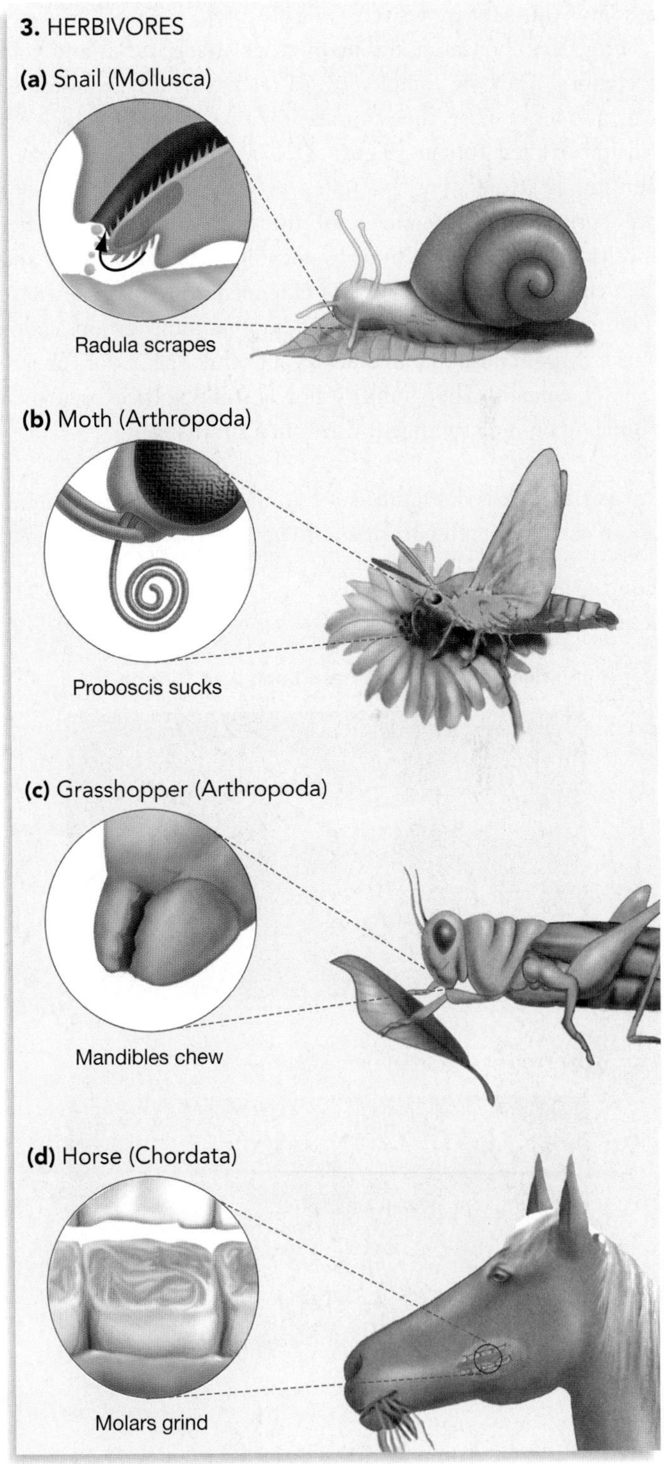

FIGURE 31.16 Herbivores Eat Primary Producers
A tiny subset of methods animals use to harvest algae or plant tissues.

grasshoppers (**Figure 31.16c**, page 711) and grinding molars of horses (**Figure 31.16d**) are used to process leaves or stems. Animal mouthparts are a classic example of how the structures found in organisms correlate with their function in harvesting a particular type of tissue.

Predation Animals use a fascinating variety of strategies and structures to capture and eat other animals. One way to categorize these hunting strategies is to consider whether the **predator** waits for or actively stalks its **prey**.

Frogs are chordates, and many frogs are classic sit-and-wait predators. They sit completely still and wait for an insect or worm to move close, then capture it with a lightening-quick extension of their tongue (**Figure 31.17a**). Stalkers vary in their hunting strategies. For example, wolves and mountain lions prey primarily on members of the deer family. Like other species of dogs, wolves hunt by locating a prey organism and then running it down during an extended, long-distance chase (**Figure 31.17b**). They also live and hunt in family groups called packs. Mountain lions, like nearly all other species of cat, are solitary animals. They hunt by slowly stalking their prey, then pouncing on it or running it down in a short sprint.

Parasitism It is often difficult to draw a sharp distinction between predation and parasitism. In general, parasites are much smaller than their victims and often harvest nutrients without causing death. Predators, in contrast, are typically larger than their prey or about the same size. Predation almost always leads to the death of the victim.

Like the other feeding methods surveyed here, parasitism is practiced by species from a variety of lineages. The strategies employed also vary widely, but they can be grouped into two broad categories: endoparasitism and ectoparasitism.

Endoparasites live inside their hosts. They are often wormlike and can be very simple morphologically. Tapeworms, found in the intestines of humans and other vertebrates, for example, are platyhelminths with no digestive system. Instead of a mouth, they have hooks or other structures on their head, called a *scolex*, that attach to their host's intestinal wall (**Figure 31.18a**). Instead of digesting food in a gut, they absorb nutrients directly from their surroundings. Most endoparasites ingest their food through a mouth and have a digestive tract, however.

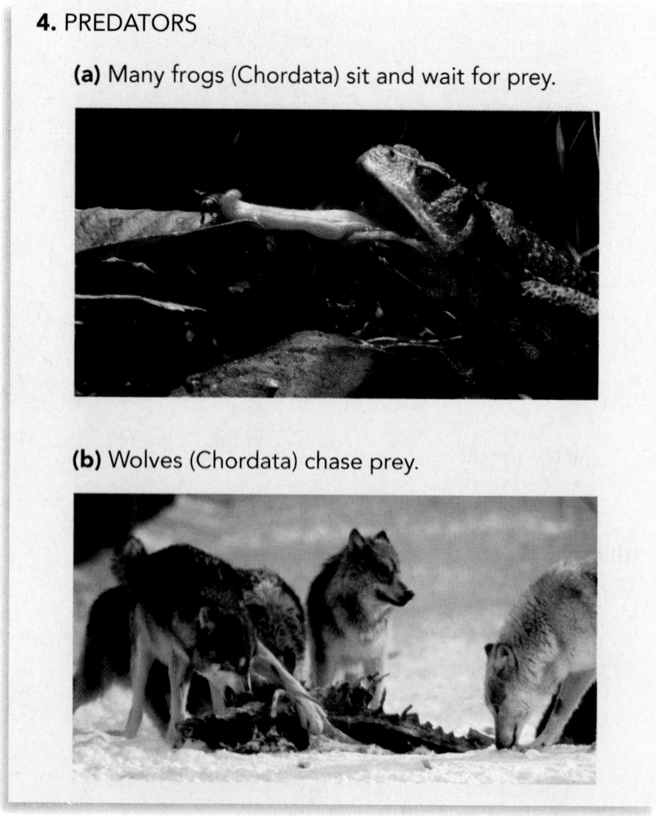
4. PREDATORS
(a) Many frogs (Chordata) sit and wait for prey.
(b) Wolves (Chordata) chase prey.

FIGURE 31.17 Predators Eat Other Animals
(a) Many frogs are sit-and-wait predators. (b) Wolf families hunt together by chasing prey.

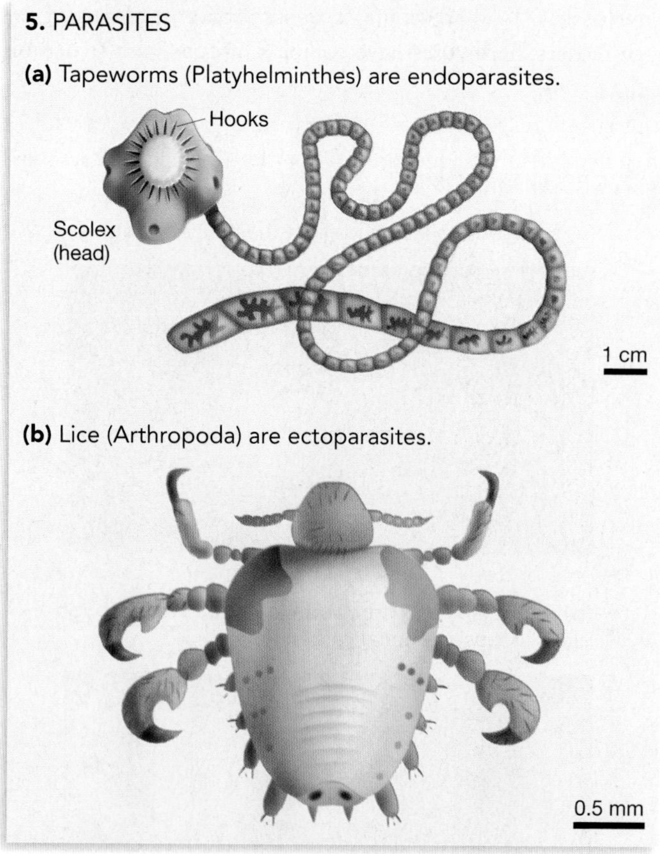
5. PARASITES
(a) Tapeworms (Platyhelminthes) are endoparasites.
Hooks
Scolex
(head)
1 cm
(b) Lice (Arthropoda) are ectoparasites.
0.5 mm

FIGURE 31.18 Parasites Take Nutrients from Living Animals
(a) Tapeworms are common intestinal parasites of humans and other vertebrates. They attach to the wall of the digestive tract, using the barbed hooks on their anterior end, and absorb nutrition directly across their body wall. (b) Lice are insects that parasitize mammals and birds. This louse, *Phthirus pubis*, uses its clawlike legs to attach to the pubic region of humans. The animal pierces the host's skin with its mouthparts and feeds by sucking body fluids.

Ectoparasites live outside their hosts. They usually have limbs or mouthparts that allow them to grasp the host and mouthparts that allow them to pierce their host's skin and suck the nutrient-rich fluids inside. The louse in **Figure 31.18b** is an example of an insect (Arthropoda) ectoparasite that afflicts humans.

Movement

Many animals are sit-and-wait predators, and some are sessile throughout their adult lives. But the vast majority of animals move under their own power either as a juvenile or as an adult. For example, the eggs of sea anemones hatch into larvae that swim with the aid of cilia (**Figure 31.19a**). Adult sea anemones, however, spend most of their lives attached to a rock or another substrate and make their living by capturing and eating fish or other organisms that pass by (**Figure 31.19b**). In species such as these, larvae function as a dispersal stage. They are a little like the seeds of land plants—a life stage that allows individuals to move to new habitats, where they will not compete with their parents for space and other resources.

In animals that move as adults, locomotion has three functions: (1) finding food, (2) finding mates, and (3) escaping from predators. The ways that animals move in search of food and sex are highly variable; as mentioned in the introduction to this chapter, animals burrow, slither, swim, fly, crawl, walk, or run. The structures that power movement are equally variable—they include cilia, flagella, and muscles that attach to a hard skeleton or compress a hydrostatic skeleton, enabling wriggling movements. The hydrostatic skeleton is an evolutionary innovation unique to animals and is responsible for locomotion in the many animal phyla with wormlike bodies. Another major innovation occurred in animals, however, and made highly controlled, rapid movement possible: the **limb**.

(a) Motile larval anemone

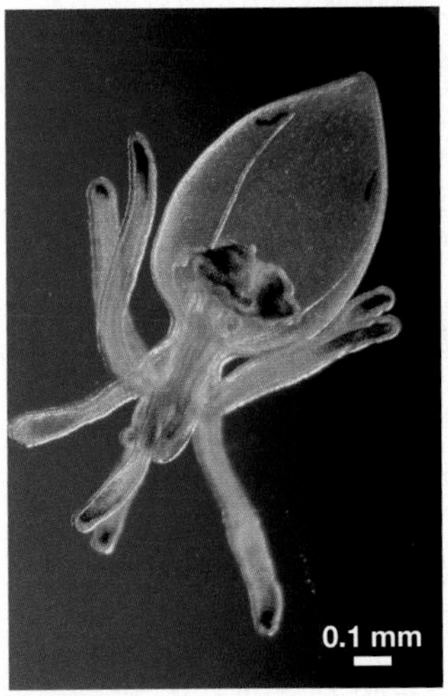

(b) Sessile adult anemone

FIGURE 31.19 If Adults Are Sessile, Then Larvae Disperse
(a) Larval anemones swim under their own power and can disperse to new habitats. **(b)** Anemones are sessile most of their lives. This anemone has captured a blue sea star to eat it.

Types of Limbs: Unjointed and Jointed Limbs are a prominent feature of species in many phyla and are particularly important in two major lineages: the ecdysozoans and the vertebrates. Some members of the ecdysozoa, such as onychophorans (velvet worms), have unjointed, sac-like limbs (**Figure 31.20a**); others, such as crabs and other arthropods, have more complex, jointed limbs (**Figure 31.20b**). Jointed

(a) Onychophorans are ecdysozoans with sac-like limbs.

(b) Crabs (Arthropoda) have jointed limbs.

(c) Polychaetes (Annelida) have parapodia.

(d) Sea urchins (Echinodermata) have tube feet.

FIGURE 31.20 Various Animal Appendages Function in Locomotion
Lineages within the Ecdysozoa have **(a)** sac-like legs or **(b)** jointed limbs. **(c)** Some species in the Lophotrochozoa have small projections called parapodia. **(d)** Echinoderms have unusual structures called tube feet.

limbs make fast, precise movements possible and are a prominent type of limb in vertebrates and arthropods.

In essence, the limbs of arthropods and vertebrates work the same way: Limbs move when muscles that are attached to a skeleton contract or relax. The difference between the two groups is that ecdysozoans have an external skeleton, or **exoskeleton** ("outside-skeleton"), while vertebrates have an internal skeleton, or **endoskeleton** ("inside-skeleton"). But in both cases the skeleton has the same function: It is a stiff structure that resists the forces exerted by muscles. The structure and function of muscles and skeletons are detailed in Chapter 46; here the important point is that muscles, limbs, and skeletons are evolutionary innovations observed only in animals.

Are All Animal Appendages Homologous? Chapter 23 introduced the concept of *homology*, which is defined as similarity in traits due to inheritance from a common ancestor. Traditionally, biologists have hypothesized that appendages used in animal movement evolved independently in a number of groups—meaning that not all animal limbs are homologous. To appreciate the logic behind this hypothesis, it's important to recognize just how diverse animal appendages are. Animals in a wide array of phyla have structures that stick out from the main body wall and function in locomotion. In addition to limbs such as insect and crab legs and the legs and wings of vertebrates, consider the *parapodia* of lophotrochozoans such as polychaetes (**Figure 31.20c**, page 713) and the *tube feet* of echinoderms, such as sea urchins (**Figure 31.20d**). The structure of animal appendages is so diverse that it was logical to maintain that at least some appendages evolved independently of each other. In terms of the concepts introduced in Chapter 23, the low degree of structural homology among the appendages illustrated in Figure 31.20 implies low to nonexistent developmental and genetic homology. Biologists predicted that completely different genes are responsible for each type of appendage.

Recent results have challenged this view, however. The experiments in question involve a gene called *Distal-less*, which was originally discovered in fruit flies. (*Distal* means "away from the body.") *Distal-less*, or *Dll*, is aptly named. In fruit flies that lack this gene's normal protein product, only the most rudimentary limb buds form. The mutant limbs are "distal-less." Based on the morphology of *Dll* mutants, the protein seems to deliver a simple message as a fruit-fly embryo develops: "Grow appendage out this way."

A group of biologists working in Sean Carroll's lab set out to test the hypothesis that *Dll* might be involved in limb or appendage formation in other animals. As **Figure 31.21** shows, they used a fluorescent marker that sticks to the *Dll* gene product to locate tissues where the gene is expressed. When they introduced the fluorescent marker into embryos from annelids, arthropods, echinoderms, chordates, and other phyla, they found that it bound to *Dll* in all of them. More important, the *Dll* gene products were highly localized in cells that form

Question: Is the gene *Dll* involved in limb formation in species other than insects?

Hypothesis: In all animals, *Dll* signals "grow appendage out here."

Null hypothesis: *Dll* is not involved in the development of appendages in species other than insects.

Experimental setup:

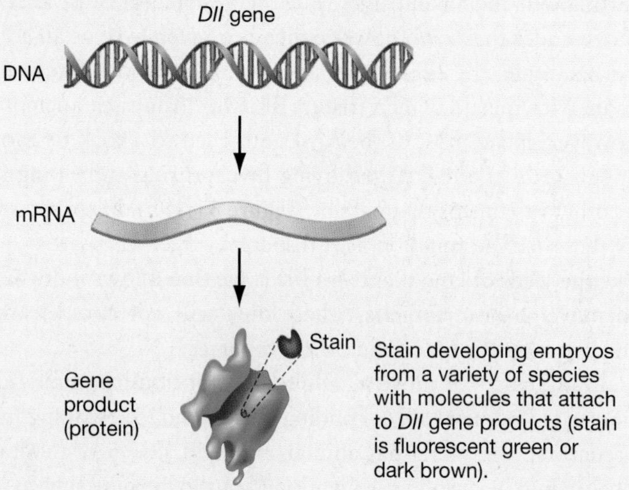

Dll gene
DNA
mRNA
Stain
Gene product (protein)
Stain developing embryos from a variety of species with molecules that attach to Dll gene products (stain is fluorescent green or dark brown).

Prediction: In embryos from a wide array of species, stained *Dll* gene products will be localized to areas where appendages are forming.

Prediction of null hypothesis: Stained *Dll* gene products will be localized to areas where appendages are forming only in insects.

Results:

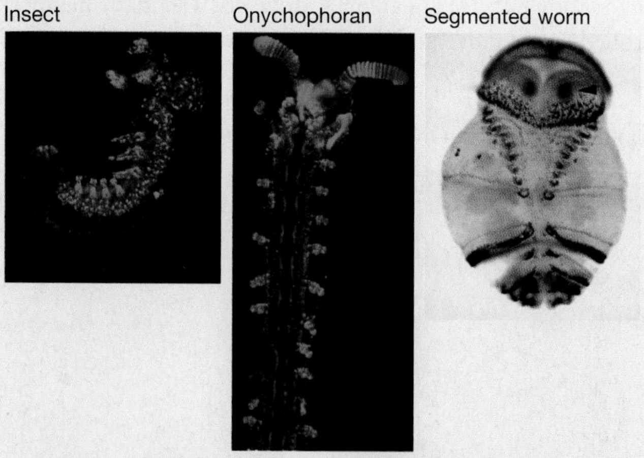

Insect
Onychophoran
Segmented worm

In species representing both Ecdysozoa and Lophotrochozoa, *Dll* is localized in areas of the embryo where appendages are forming.

Conclusion: The gene *Dll* is involved in limb formation in diverse species. The results suggest that all animal appendages may be homologous.

FIGURE 31.21 Experimental Evidence That All Animal Appendages Are Homologous

QUESTION What results would have supported the null hypothesis?

appendages—even in phyla with wormlike bodies that have extremely simple appendages. Other experiments have shown that *Dll* is also involved in limb formation in vertebrates.

Based on these findings, biologists have argued that at least a few of the same genes are involved in the development of *all* appendages observed in animals. To use the vocabulary introduced in Chapter 23, their hypothesis is that all animal appendages have some degree of genetic homology and that they are all derived from appendages that were present in a common ancestor. The idea is that a simple appendage evolved early in the history of the Bilateria and that subsequently, evolution by natural selection produced the diversity of limbs, antennae, and wings observed today. This hypothesis is controversial, however, and research continues at a brisk pace.

Reproduction and Life Cycles

An animal may be efficient at moving and eating, but if it does not reproduce, the alleles responsible for its effective locomotion and feeding will not increase in frequency in the population. As Chapter 23 emphasized, natural selection occurs when individuals with certain alleles produce more offspring than other individuals do. Organisms live to reproduce.

Given the array of habitats and lifestyles pursued by animals, it is not surprising that they exhibit a high degree of variation in how they reproduce. Although animal reproduction will be explored in detail in Chapter 48, a few examples will help drive home just how variable animal reproduction is:

- At least some species in most animal phyla can reproduce asexually, via mitosis, as well as sexually (via meiosis). In the lophotrochozoan phylum Rotifera, a subgroup of species, the bdelloids, reproduces only asexually (**Figure 31.22**). Even certain fish, lizard, and snail species have never been observed to undergo sexual reproduction.

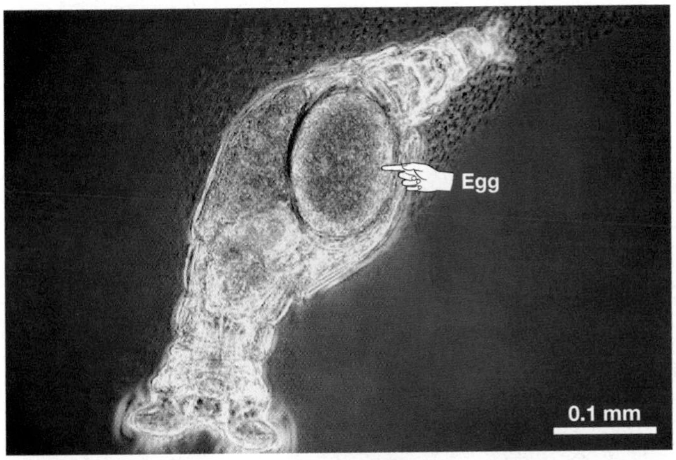

FIGURE 31.22 Bdelloid Rotifers Have Never Been Observed to Reproduce Sexually
Males have never been observed in any of the 370 species of rotifers in the group known as bdelloids. Females produce eggs that hatch into offspring without being fertilized.

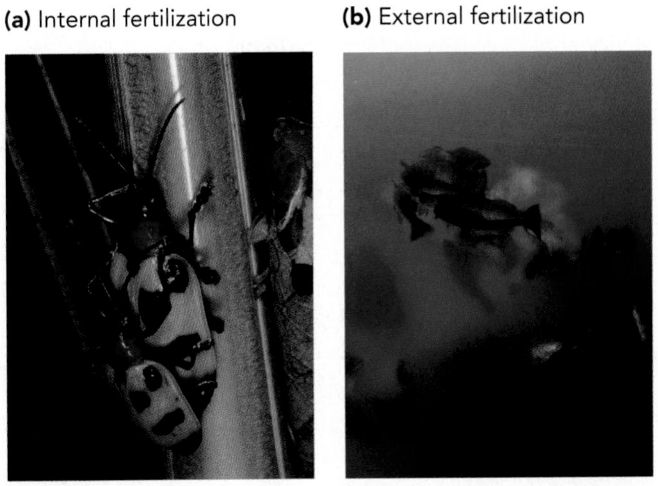

FIGURE 31.23 Fertilization Can Be Internal or External in Animals
(a) Internal fertilization in chrysomelid beetles. **(b)** Male cubera snappers emit sperm that will fertilize a clutch of eggs laid by a female. The water in the photo is cloudy due to a high concentration of snapper sperm.

- When sexual reproduction does occur, fertilization may be internal or external. When internal fertilization takes place, males typically insert a sperm-transfer organ into the body of a female (**Figure 31.23a**). In some cases, males produce sperm in packets, which females then pick up and insert into their own bodies. But in seahorses, females insert eggs into the male's body, where they are fertilized. The male is pregnant for a time and then gives birth to live young. External fertilization is extremely common in aquatic species and involves the fusion of egg and sperm outside the female's body. Females lay eggs onto a substrate or into open water. Males then shed sperm on or near the eggs (**Figure 31.23b**).
- Eggs or embryos may be retained in the female's body during development, or eggs may be laid outside to develop independently of the mother. Species that give birth to live young are said to be **viviparous** ("live-bearing"; **Figure 31.24a**, page 716); species that deposit fertilized eggs are **oviparous** ("egg-bearing"; **Figure 31.24b**); and some species are **ovoviviparous** ("egg-live-bearing"). In ovoviviparous species, the females retain eggs inside their body during early development, but the growing embryos are nourished by yolk inside the egg and not nutrients transferred directly from the mother. In addition to mammals, a few species of sea stars, onychophorans, fish, and lizards are viviparous. The vast majority of animals, however, are oviparous.

In addition to reproducing in a variety of ways, animal life cycles vary widely. Perhaps the most spectacular innovation in animal life cycles involves the phenomenon known as metamorphosis. Recall that metamorphosis is a change from a juvenile to an adult body type.

(a) Viviparity ("live-bearing"): the birth of a shark

(b) Oviparity ("egg-bearing")

FIGURE 31.24 Some Animal Species Give Birth to Live Young, but Most Lay Eggs
(a) A lemon shark is viviparous, meaning its embryos develop for a period of time inside the female's body. **(b)** A corn snake is oviparous, meaning it lays fertilized eggs.

All insects and many other animal species have distinct juvenile and adult stages in their life cycle—often with different body forms and feeding techniques. A juvenile individual is called a **larva** (plural: **larvae**) if it looks substantially different from the adult form or a **nymph** if it looks like a miniature adult. Larvae and nymphs are sexually immature, meaning their reproductive organs are undeveloped and the individual cannot breed. It is important to appreciate just how much larvae can differ from adults in morphology, feeding behavior, and even habitat. For example, the larvae of mosquitoes live in quiet bodies of freshwater, where they suspension-feed on bacteria, algae, and detritus (**Figure 31.25a**). When a larva has grown sufficiently, the individual secretes a protective case. The individual is now known as a **pupa** (plural: **pupae**; **Figure 31.25b**). During **pupation**, the pupa's body is completely remodeled into a new, adult form. In mosquitoes, the adult individual flies and gets its nutrition as a parasite—taking blood meals from mammals and sucking fluids from plants (**Figure 31.25c**).

Fruit flies also undergo **holometabolous** ("whole-change") **metamorphosis**, meaning they undergo a drastic change in form (**Figure 31.26a**). Grasshoppers, in contrast, undergo **hemimetabolous** ("half-change") **metamorphosis.** As a grasshopper grows, it sheds its external skeleton several times and grows. In doing so, it gradually changes from a wingless, sexually immature nymph to a sexually mature adult that is capable of flight (**Figure 31.26b**). Throughout this process, grasshoppers feed on the same food source in the same way: They chew leaves.

In insects, holometabolous metaphorphosis, also known as *complete metamorphosis,* is 10 times more common than hemimetabolous metamorphosis, also called *incomplete metamorphosis*. One hypothesis to explain this observation is based on efficiency in feeding. Because juveniles and adults from holometabolous species feed on different materials in different

(a) Larvae (very different from adult)

(b) Pupae

(c) Adult

FIGURE 31.25 During Metamorphosis, Individuals May Change Form Completely
(a) Juvenile forms of organisms that undergo metamorphosis are called larvae. **(b)** During pupation, the juvenile body is remodeled into the adult form. **(c)** Adult forms are sexually mature.

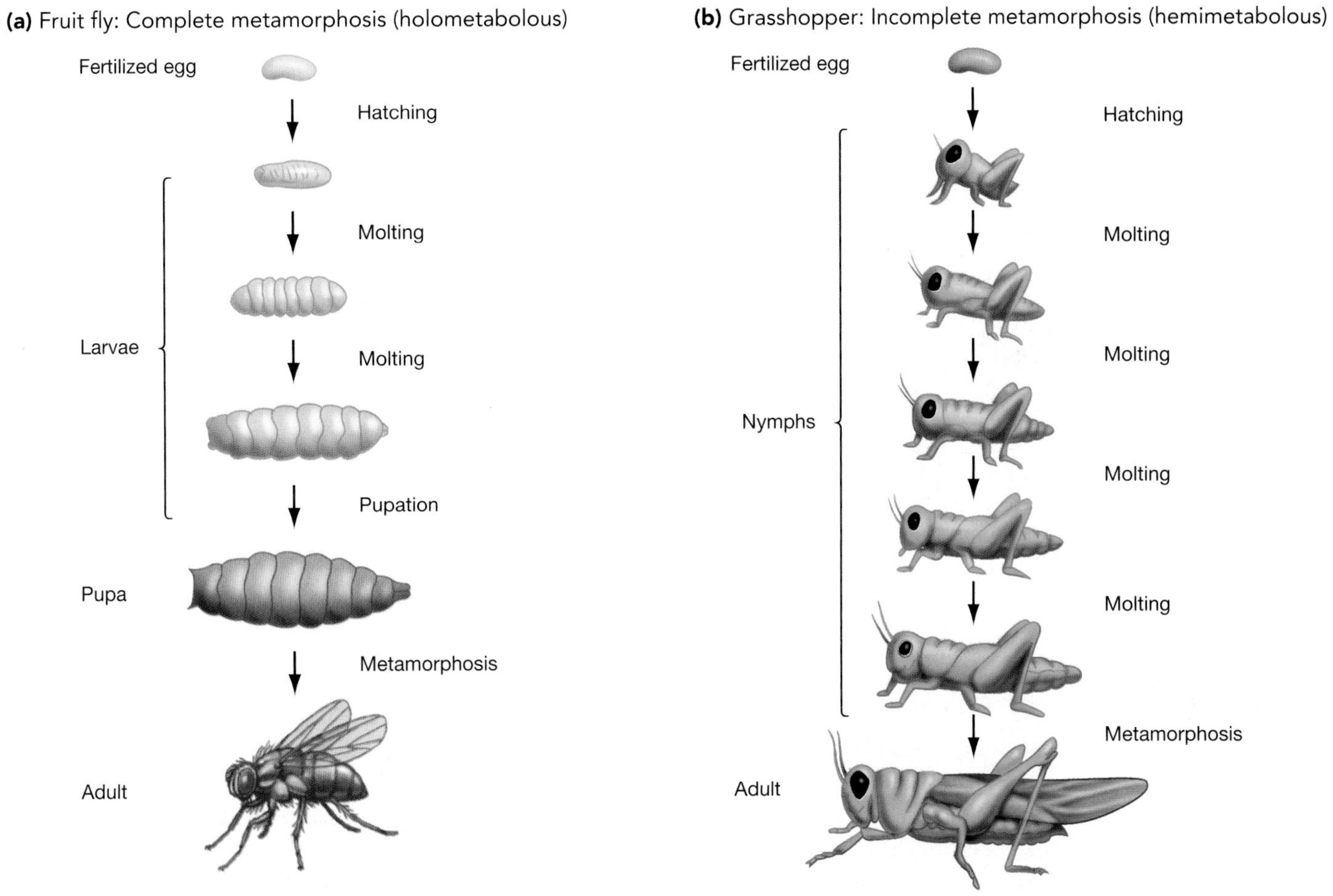

FIGURE 31.26 Incomplete Metamorphosis Does Not Involve a Dramatic Change in Form
(a) In complete metamorphosis, the larval, or sexually immature, form has a different body type than the adult.
(b) In incomplete metamorphosis, nymphs are small versions of sexually mature adults.

ways and sometimes even in different habitats, they do not compete with each other. When complete metamorphosis is part of the life cycle, individuals of the same species but different ages often show dramatic variations in their mode of feeding. An alternative hypothesis to explain the evolutionary success of complete metamorphosis is based on specialization in feeding and mating. In many moths and butterflies, for example, larvae are specialized for feeding, whereas adults are specialized for mating. Larvae are largely sessile, whereas adults are highly mobile. If specialization leads to higher fitness, then complete metamorphosis would be advantageous. Both hypotheses are still being tested, however.

Complete metamorphosis is also extremely common in marine animals. For example, most cnidarians have two distinct body types during their life cycle: (1) A largely sessile form called a **polyp** alternates with (2) a free-floating stage called a **medusa** (plural: **medusae**; **Figure 31.27**, page 718). Polyps live attached to a substrate, suspension feed on detritus, and frequently form large clusters of individuals called colonies. A **colony** is a group of identical cells that are physically attached but do not perform closely coordinated functions. Medusae, in contrast, float freely in the plankton and feed on fish and other, larger prey. Because polyps and medusae live in different habitats, the two stages of the life cycle exploit different food sources.

✓CHECK YOUR UNDERSTANDING

The story of animal evolution is based on two themes: (1) the evolution of a small suite of basic body plans (ways to design and build an eating and moving machine) and (2) a diversification of species with the same basic body plan, based on the evolution of innovative structures and methods for capturing food and moving. You should be able to give at least three examples of fundamental variations in the ways that animals find food, move, and reproduce.

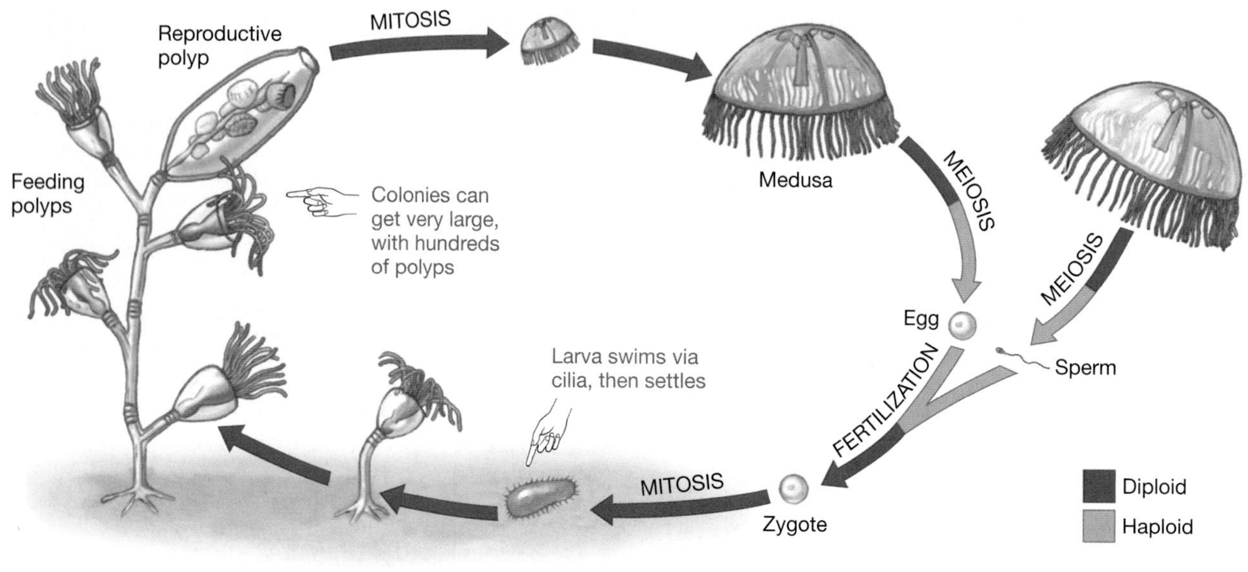

FIGURE 31.27 Cnidarian Life Cycles May Include a Polyp and Medusa Form

31.4 Key Lineages of Animals

The goal of this chapter is to provide a broad overview of how animals diversified into such a morphologically diverse lineage. According to the phylogeny illustrated in Figure 31.13, the closest living relative to the animals are the choanoflagellates, a group of protists. Phylogenetic analyses and the fossil record indicate that the phyla called Porifera (sponges), Cnidaria (jellyfish and others), Ctenophora, and Acoelomorpha are the most ancient of all animal groups. Let's explore the origins of animals by taking a more detailed look at each of these lineages. In Chapters 32 and 33 we'll follow up by examining the Bilateria—the protostomes and deuterostomes, respectively—in more detail.

Choanoflagellates (Collar Flagellates)

The microscopic protists known as "collar flagellates" number about 150 species and are found in both freshwater and marine habitats. They are notable because they are the closest living relative of animals and because their cells are virtually identical to the flagellated feeding cells, or choanocytes, of sponges (see Figure 31.11a). Choanoflagellates are unicellular or occur in colonies. Recall that a colony is an aggregation of individuals that live in close physical proximity.

Feeding As flagella beat, they create water currents that bring bacteria, archaea, and small pieces of organic debris toward the cell. Food particles are then filtered out of the water by slender, hairlike projections of the cell that surround each flagellum (**Figure 31.28**).

Movement Adults attach to a substrate and are sessile. If cells break away from the substrate where they are growing, they can swim via the beating of their flagella.

Reproduction Choanoflagellates reproduce asexually by means of simple fission. Sexual reproduction has never been observed in the group.

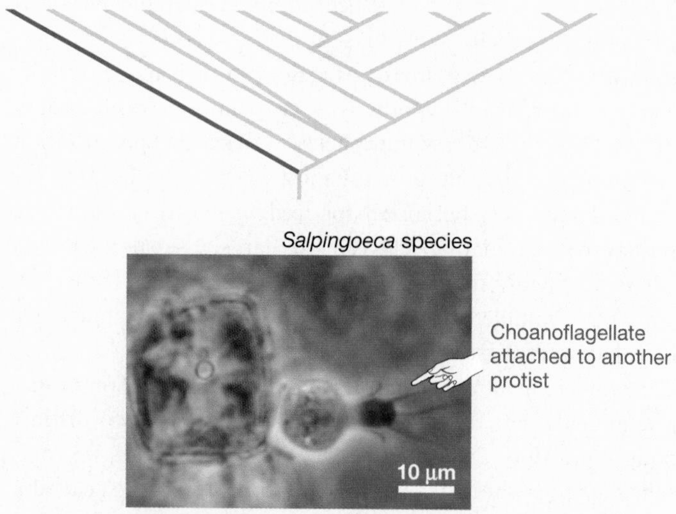

FIGURE 31.28 Choanoflagellates Are Aquatic Suspension Feeders

Porifera (Sponges)

About 5500 species of sponge have been identified to date. Although a few freshwater species are known, most are marine. All sponges are **benthic**, meaning that they live at the bottom of aquatic environments. Sponges are particularly common in rocky, shallow-water habitats of the world's oceans and in coastal areas of Antarctica.

The architecture of sponge bodies is built around a system of tubes and pores that create channels for water currents. Body symmetry varies among sponge species; most are asymmetrical, but some species are radially symmetric (**Figure 31.29**). Sponges are considered multicellular, because they have specialized cell types. They do not, however, have the highly organized, integrated, and specialized groups of cells called tissues. In many species, either flexible collagen fibers or stiff spikes of silica or calcium carbonate ($CaCO_3$) provide structural support for the body. These structures are called **spicules** and serve to stiffen and support the body. One species native to the Caribbean can grow to heights of 2 m.

Sponges have commercial and medical value to humans. The dried bodies of certain sponge species are able to hold large amounts of water and thus are prized for use in bathing and washing. In addition, researchers are increasingly interested in the array of toxins that sponges produce to defend themselves against predators and bacterial parasites. Some of these compounds have been shown to have antibacterial properties or to promote wound healing in humans. The type of bioprospecting being done in plants (see Chapter 29) is also occurring in sponges.

Feeding All sponges are suspension feeders. Their cells beat in a coordinated way to produce a water current that flows through small pores in the outer body wall, into chambers inside the body, and out through a single larger opening. As water passes by feeding cells, organic debris and bacteria, archaea, and small protists are filtered out of the current and then digested.

Movement Adult sponges are sessile but may produce larvae that swim with the aid of flagella.

Reproduction Asexual reproduction occurs in a variety of ways, depending on the species. Sponge cells are *totipotent*, meaning that an isolated adult cell has the capacity to develop into a complete adult organism. Thus, any fragment that breaks off an adult sponge has the potential to grow into a new individual. Although individuals of most species produce both eggs and sperm, self-fertilization is rare because individuals release their male and female gametes at different times. Fertilization usually takes place in the water, but some ovoviviparous species retain their eggs and then release mature, swimming larvae after fertilization and early development have occurred.

Pseudoceratina crassa
10 cm

FIGURE 31.29 Some Sponges Form Radially Symmetric Tubes

Cnidaria (Jellyfish, Corals, Anemones, Hydroids, Sea Fans)

Although a few species of Cnidaria inhabit freshwater, the vast majority of the 11,000 species are marine. They are found in all of the world's oceans, occupying habitats from the surface to the substrate, and are important predators.

Cnidarians are radially symmetric diploblasts consisting of ectoderm and endoderm layers that sandwich gelatinous material known as **mesoglea**, which contains a few scattered ectodermal cells. The gut is *blind*—meaning that there is only one opening to the environment for both ingestion and elimination of wastes.

Most cnidarians have a life cycle that includes both a sessile polyp form (**Figure 31.30a**, page 720) and a free-floating medusa (**Figure 31.30b**). Anemones, hydra, and coral, however, exist only as polyps—never as medusae. Reef-building corals secrete outer skeletons of calcium carbonate that create the physical structure of a coral reef—one of the world's most productive habitats (see Chapter 50).

Feeding The morphological innovation that triggered the diversification of the cnidarians is a specialized cell, called a **cnidocyte**, which is used in prey capture. When cnidocytes brush up against a fish or other type of prey, the cells forcibly eject a barbed, spear-like structure called a *nematocyst* that is coated with toxins. The barbs hold the prey, and the toxins subdue it until it can be brought to the mouth and ingested. Cnidocytes are commonly located near the mouths of cnidarians or on elongated structures called *tentacles*. Cnidarian toxins can be deadly to humans as well as to

(Continued on next page)

Cnidaria (Jellyfish, Corals, Anemones, Hydroids, Sea Fans) *(continued)*

prey organisms; in Australia twice as many people die each year from stings by box jellyfish as from shark attacks. In addition to capturing prey actively, most species of coral and many anemones host photosynthetic dinoflagellates. The relationship is mutually beneficial, because the protists supply the cnidarian host with food in exchange for protection (see Chapter 28).

Movement Both polyps and medusae have simple, muscle-like tissue derived from ectoderm or endoderm. In polyps, the gut cavity acts as a hydrostatic skeleton that works in conjunction with the muscle-like cells to contract or extend the body. Many polyps can also creep along a substrate, using muscle cells at their base. In medusae, the bottom of the bell structure is ringed with muscle-like cells. When these cells contract in a rhythmic fashion, the bell pulses and the medusa moves by **jet propulsion**—meaning a forcible flow of water opposite the direction of movement. Cnidarian larvae swim by means of cilia.

Reproduction Polyps may produce new individuals asexually by (1) budding, in which a new organism grows out from the body wall of an existing individual; (2) fission, in which an existing adult splits lengthwise to form two individuals; or (3) fragmentation, in which parts of an adult regenerate missing pieces to form a complete individual. During sexual reproduction, gametes are usually released from the mouth of a polyp or medusa and fertilization takes place in the open water. Eggs hatch into larvae that become part of the plankton before settling and developing into a polyp.

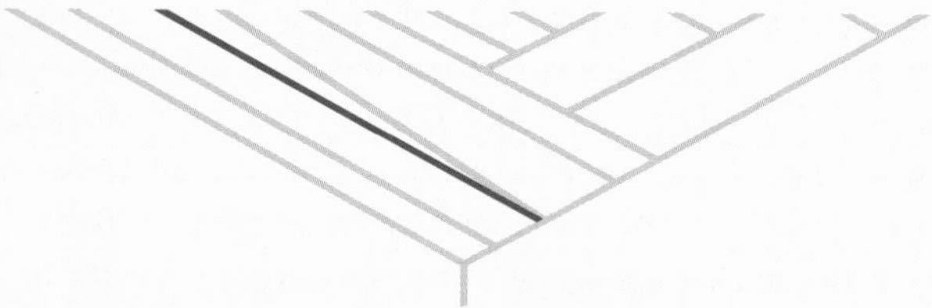

(a) Polyps attach to substrates. *Aurelia aurita*

(b) Medusae float near the water surface. *Aurelia aurita*

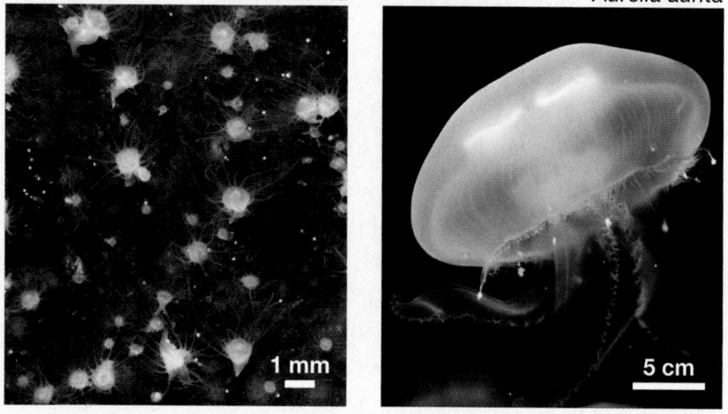

FIGURE 31.30 Most Cnidarians Have a Polyp Stage and a Medusa Stage

Ctenophora (Comb Jellies)

Ctenophores are transparent, ciliated, gelatinous diploblasts that live in marine habitats (**Figure 31.31**). Although a few species live on the ocean floor, most are planktonic—meaning that they live near the surface. Only about 100 species have been described to date, but some are abundant enough to represent a significant fraction of the total planktonic biomass.

Feeding Ctenophores are predators. Feeding occurs in several ways, depending on the species. Some comb jellies have long tentacles covered with cells that release an adhesive when they contact prey. These tentacles are periodically wiped across the mouth so that captured prey can be ingested. In other species, prey can stick to mucus on the body and be swept toward the mouth by cilia. Still other species ingest large prey whole. Figure 31.3c shows a comb jelly that has just swallowed a second comb jelly almost as large as itself.

Movement Adults move via the beating of cilia, which occur in comblike rows that run the length of the body. Ctenophores are the largest animals known to use cilia for locomotion.

Reproduction Most species have both male and female organs and routinely self-fertilize, though fertilization is external. Larvae are free swimming. The few species that live on the ocean floor undergo internal fertilization and brood their embryos until they hatch into larvae.

Pleurobrachia pileus

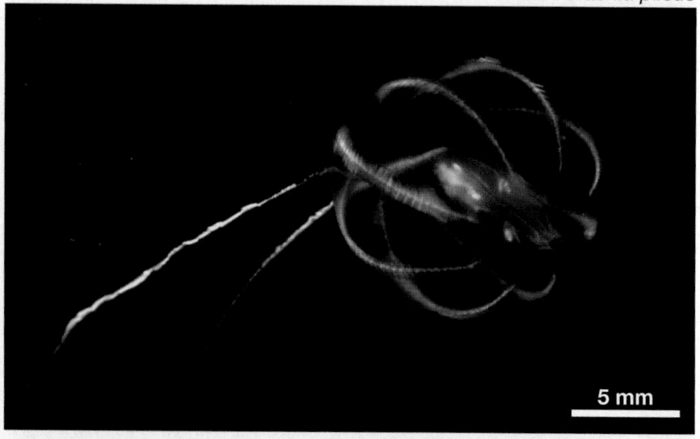

FIGURE 31.31 Ctenophores Are Planktonic Predators

Acoelomorpha

As their name implies, the acoelomorphs lack a coelom. They are bilaterally symmetric worms that have distinct anterior and posterior ends and are triploblastic. They have simple guts or no gut at all. In some species with a gut, the mouth is the only opening for the ingestion of food and excretion of waste products. Most acoelomorphs are only a couple of millimeters long and live in mud or sand in marine environments (**Figure 31.32**).

Feeding Acoelomorphs feed on detritus or prey on small animals or protists that live in mud or sand.

Movement Acoelomorphs swim, glide along the surface, or burrow through substrates with the aid of cilia that cover either the entire body or the ventral surface.

Reproduction Adults can reproduce asexually by fission (splitting in two) or by direct growth (budding) of a new individual from the parent's body. Individuals produce both sperm and eggs. Fertilization is internal, and fertilized eggs are laid outside the body.

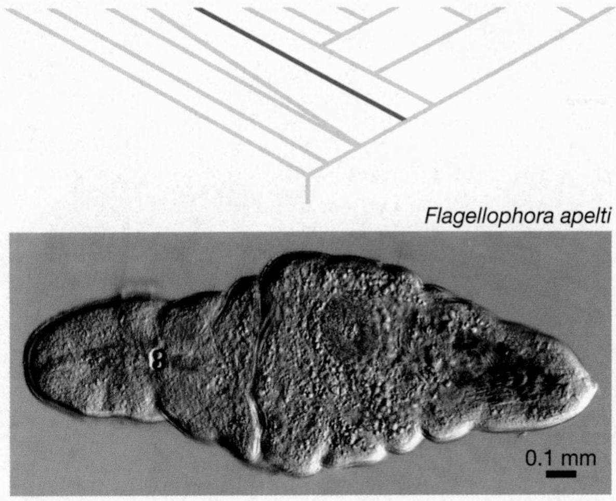
Flagellophora apelti
0.1 mm

FIGURE 31.32 Acoelomorphs Are Small Worms That Live in Mud or Sand

ESSAY Coral Bleaching

Coral reefs are the most productive habitats in the world. They produce more kilograms of new tissue per square meter per year than does any other habitat on Earth, including tropical wet forests. Reefs may also be the most species rich of all habitats. It has been estimated that as many as 9 million species live in the world's coral reefs.

Coral reefs are found in shallow waters around tropical islands and continental areas. Reefs are built by colonies of cnidarians that secrete calcium carbonate skeletons that protect their bodies (**Figure 31.33a**, page 722). When the individuals die, their hard skeletons remain. Coral skeletons then form a substrate for the growth of additional corals, algae, or other organisms.

The cnidarians that build coral reefs make their living by capturing prey, using cnidocytes arranged on tentacles that project from their calcium carbonate skeletons (**Figure 31.33b**). But reef-building corals also host symbiotic dinoflagellates, which perform photosynthesis and contribute sugars and other products to the host animal. The bright colors of coral reefs are due to the photosynthetic pigments in these symbionts. Because cnidarians build the physical structure of coral reefs and house the most important primary producers as symbionts, corals are the essential component of reef ecosystems worldwide.

The future of the world's coral reefs may be in jeopardy, however, due a phenomenon known as coral bleaching. Corals are said to *bleach* when they expel their photosynthetic symbionts—meaning that they eject their symbionts into the surrounding water. As a result, the corals turn white.

The frequency of large-scale bleaching events has increased dramatically since 1980, and data from laboratories and monitoring stations around the world have supported the hypothesis that recent bleaching episodes are correlated with periods of elevated water temperature. The molecular mechanism responsible for bleaching is not known, however, and the link between high temperature and bleaching has yet to be confirmed experimentally. Biologists around the world are focused on the problem because, if bleaching continues for an extended period of time, corals begin to die in large numbers. In addition to monitoring the extent and causes of bleaching, researchers hope that gaining a better understanding of the molecular mechanisms involved may suggest methods to mitigate the damage.

The future of the world's coral reefs may be in jeopardy. . . .

Research on coral reefs has become increasingly urgent. If bleaching is a response to temperature stress, then the extent and frequency of bleaching might continue to increase as global warming continues and ocean temperatures rise. It remains to be seen whether bleaching episodes will increase in duration and frequency enough to threaten the long-term stability of the world's coral reefs.

(Continued on next page)

(Essay continued)

(a) Reefs are made from the calcium carbonate skeletons of corals.

(b) Corals have tentacles lined with stinging cells.

FIGURE 31.33 Coral Reefs Are Highly Productive and Species Rich
(a) Coral reefs are found in shallow, tropical waters. The colors are the result of photosynthetic pigments in symbiotic dinoflagellates. **(b)** Corals feed by stinging passing prey. QUESTION Why are coral reefs restricted to extremely shallow water habitats?

CHAPTER REVIEW

Summary of Key Concepts

- **Animals are a particularly species-rich and morphologically diverse lineage of multicellular organisms on the tree of life.**

The animals consist of about 34 phyla and may number 10 million or more species. Biologists study animals because they are key consumers and because humans depend on them for transportation, power, or food.

- **Major groups of animals are defined by the design and construction of their basic body plan, which differs in the number of tissues observed in embryos, symmetry, the presence or absence of a body cavity, and the way in which early events in embryonic development proceed.**

Sponges are the only animals that are asymmetric and lack tissues. The Cnidarians and Ctenophores have radial symmetry and just two embryonic tissues. A handful of species have bilateral symmetry and three embryonic tissues but lack a body cavity, or coelom. The vast majority of animal species have bilateral symmetry, three embryonic tissues, and a coelom. These design features gave rise to a widespread "tube-within-a-tube" body plan. Depending on the species involved, the tube-within-a-tube design is built in one of two fundamental ways—via the protostome pattern or deuterostome pattern of development.

Web Tutorial 31.1 The Architecture of Animals

- **Recent phylogenetic analyses of animals have shown that there were three fundamental splits during evolutionary history, resulting in two protostome groups (Lophotrochozoa and Ecdysozoa) and the deuterostomes. The most ancient animal group living today is the sponges. The closest living relatives to animals are choanoflagellates, a group of protists.**

Phylogenetic data support the hypothesis that triploblasty, bilateral symmetry, coeloms, and protostome and deuterostome development all evolved just once. However, phylogenetic data also suggest that coeloms were lost in the phylum Platyhelminthes, that pseudocoeloms evolved independently in the phyla Nematoda and Rotifera, and that segmented body plans arose at least three times independently as animals diversified.

- **Within major groups of animals, evolutionary diversification was based on innovative ways of feeding and moving. Most animals get nutrients by eating other organisms, and most animals move under their own power at some point in their life cycle.**

A wide variety of feeding strategies occurs among animals. Suspension feeders filter organic material or small organisms from water; deposit feeders swallow soils or other materials and digest the food particles they contain; herbivores use complex mouthparts to bite, suck, or rasp away plant tissues; predators kill prey by using sit-and-wait or stalking strategies; and parasites can live inside or outside their hosts and take nutrients from the living hosts.

Most animal movement is based on contractions by muscle cells in conjunction with one of three types of skeletons: (1) a hydrostatic skeleton, (2) an exoskeleton, or (3) an endoskeleton. Recent research suggests that, even though the types of appendages used in animal locomotion range from simple saclike limbs to complex lobster legs, all appendages may be homologous.

Questions

Content Review

1. Which of the following is true of *all* animals with bilateral symmetry?
   a. They are triploblastic.
   b. They have a coelom.
   c. They exhibit the protostome pattern of development.
   d. They exhibit the deuterostome pattern of development.
2. Which of the following represents a conflict between phylogenetic trees estimated from morphological data versus DNA sequence data?
   a. Choanoflagellates are the closest living relative of animals.
   b. Protostomes and deuterostomes are a fundamental split within the Bilateria.
   c. Sponges are the most ancient, or basal, lineage of animals.
   d. Body cavities evolved in the following order: acoelomate, pseudocoelomate, coelomate.
3. Which of the following represents a point of agreement between phylogenetic trees estimated from morphological data versus DNA sequence data?
   a. A fundamental split occurred during the evolution of protostomes, creating the Ecdysozoa and the Lophotrochozoa.
   b. Bilateral symmetry evolved once, in conjunction with the evolution of mesoderm.
   c. Segmentation evolved twice, independently, during the evolution of protostomes.
   d. Pseudocoeloms evolved twice, independently, during the evolution of protostomes.
4. Why do some researchers maintain that the limbs of all animals are homologous?
   a. Homologous genes, such as *Dll*, are involved in their development.
   b. Their structure—particularly the number and arrangement of elements inside the limb—is the same.
   c. They all function in the same way—in locomotion.
   d. Animal appendages are too complex to have evolved more than once.
5. In a "tube-within-a-tube" body plan, what is the interior tube?
   a. ectoderm
   b. mesoderm
   c. either the coelom or the pseudocoelom
   d. the gut
6. What is the key difference between choanoflagellates and sponges?
   a. Sponges are multicellular.
   b. Sponges are asymmetrical and do not have tissues.
   c. Choanoflagellates have distinctive flagellated cells that function in suspension feeding.
   d. Choanoflagellates are strictly aquatic.

Conceptual Review

1. Explain the difference between a diploblast and a triploblast. Why was the evolution of a third embryonic tissue layer important?
2. Explain how a hydrostatic skeleton works. How is a hydrostatic skeleton similar to an exoskeleton or endoskeleton? How is it different?
3. Give an example of a suspension feeder that moves and one that is sessile. Describe how these species are able to catch prey.
4. Compare and contrast the types of mouthparts and body shapes you would expect to find in herbivorous insect species that suck plant fluids from stems, bore through stems, bite leaves, and suck nectar from flowers.
5. Explain the differences in the protostome and deuterostome patterns of development. Do you agree with the text's claim that these are just different ways of building a tube-within-a-tube body plan? Explain why or why not.
6. Why was the evolution of cephalization correlated with the evolution of bilateral symmetry? Why were both of these features significant?

Group Discussion Problems

1. Would you expect internal fertilization to be more common in aquatic or terrestrial environments? Explain your answer. What types of differences would you expect to find in the structure of eggs that are laid in aquatic versus terrestrial environments?
2. Ticks are arachnids (along with spiders and mites); mosquitoes are insects. Both ticks and mosquitoes are ectoparasites that make their living by extracting blood meals from mammals. Ticks undergo incomplete metamorphosis, while mosquitoes undergo complete metamorphosis. Based on these observations, would you predict ticks or mosquitoes to be the more successful group in terms of number of species, number of individuals, and geographic distribution? Explain why. How could you test your prediction?
3. Suppose you are walking along an ocean beach at low tide and find an animal that is unlike any you have ever seen before. How would you go about determining how the animal feeds and how it moves? How would you go about determining the major features of its body plan?
4. Suspension feeding is extremely common in aquatic organisms but rare in terrestrial organisms. Generate a hypothesis to explain this observation.

Answers to Multiple-Choice Questions 1. a; 2. d; 3. b; 4. a; 5. d; 6. a

www.prenhall.com/freeman is your resource for the following: Web Tutorials; Online Quizzes and other Online Study Guide materials; Answers to Conceptual Review Questions; Solutions to Group Discussion Problems; Answers to Figure Caption Questions and Exercises; and Additional Readings and Research.

32 Protostome Animals

KEY CONCEPTS

- Molecular phylogenies support the hypothesis that protostomes are a monophyletic group divided into two major subgroups: the Lophotrochozoa and the Ecdysozoa.
- Although the members of many protostome phyla have limbless, wormlike bodies and live in marine sediments, the most diverse and species-rich lineages—Mollusca and Arthropoda—have body plans with a series of distinctive, complex features.
- Key events triggered the diversification of protostomes, including several lineages making the water-to-land transition, a diversification in appendages and mouthparts, and the evolution of metamorphosis in both marine and terrestrial forms.

A small sample of the insects collected from a single tree in the Amazonian wet forest. In numbers of individuals and species richness, protostomes are the most abundant and diverse of all animals.

Protostomes are one of two major monophyletic groups of bilaterally symmetric, coelomate animals. Recall from Chapter 30 that protostomes undergo early development in a dramatically different way than the other major lineage of bilaterians, the deuterostomes, do.

In terms of the total number of species and number of individuals living today, the protostomes are far and away the most successful of all animals. To drive this point home, consider just one lineage in the phylum Arthropoda: the insects. Although biologists can only guess at the total number of insect species, about 925,000 have been formally identified to date. Estimates of the actual number of species that exist range as high as 10 million. Over 33 percent of *all* known animal species on Earth are in the insect lineage called Coleoptera—the beetles. In addition to being a species-rich clade, beetles are abundant. A single acre of pastureland in England is estimated to contain 17,825,000 individual beetles. In tropical rain forests, beetles and other insects make up 40 percent of the total mass of organisms present. The world population of ants is estimated to be 1 million billion individuals.

It is important to recognize, though, that not all protostome lineages have been as spectacularly successful as the Arthropoda. There are just 135 species in the phylum Echiura (spoon worms) and 150 species in the Kinorhyncha (mud dragons). Protostomes range from the most wildly successful phyla on the tree of life to some of the most obscure.

One of the goals of this chapter is to describe this variation in species richness and abundance. Given that each protostome phylum is defined by aspects of its body plan and by innovations that are involved in feeding and locomotion, why have a few lineages been so inordinately successful?

Before delving into this question, let's consider a fundamental issue. Tens of thousands of biologists have devoted their lives to studying protostomes. Why?

32.1 Why Do Biologists Study Protostomes?

If you asked a group of biologists why it's a good idea to study protostomes, they'd probably reply, "Because they're there." Or better yet, because there are so many of them all over the place. Protostomes include some of the most familiar and abundant organisms on Earth (**Figure 32.1**). The phylum Mollusca, for example, is composed of the snails, clams, chitons, and cephalopods (octopuses and squid)—a total of over 93,000 known species. The phylum Arthropoda contains four major subphyla, each of which is extraordinarily diverse and species-rich. The Arthropoda is made up of the insects, chelicerates (spiders and mites), crustaceans (shrimp, lobster, crabs, barnacles), and myriapods (millipedes, centipedes). A fifth major arthropod lineage, the trilobites, went extinct 250 million years ago. Although no one knows for sure, some biologists estimate that there are over 10 million species of arthropods—most of which are unnamed and undescribed.

Protostomes are organisms that you see every day. They live in just about every habitat that you might explore, and they include some of the most important model organisms in all of biological science (**Box 32.1**). If one of biology's most fundamental goals is to understand the diversity of life on Earth, then protostomes demand our attention. They are also important to human health and welfare: Some species transmit diseases or eat crops; others provide food.

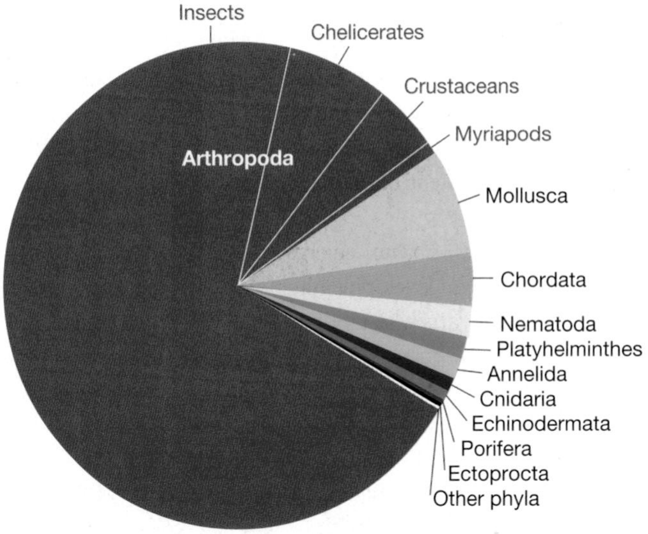

FIGURE 32.1 The Relative Abundances of Animal Lineages
About 70 percent of all known species of animals on Earth are insects, most of them beetles. (Humans and other vertebrates are deuterostomes, in the phylum Chordata.) This chapter focuses on the most species-rich lineages of protostomes.

Because of their diversity and abundance, protostomes play key roles in marine and terrestrial ecosystems and in human economic life. Let's take a brief look at how protostomes affect you and the organisms around you—first in the world's oceans, then on the continents.

BOX 32.1 Model Organisms: *Caenorhabditis* and *Drosophila*

Protostomes include two of the most important model organisms in the life sciences: the fruit fly *Drosophila melanogaster* and the roundworm *Caenorhabditis elegans*. (*Caenorhabditis* is pronounced *see-no-rab-DIE-tiss*.) If you walk into a biology building on any university campus around the world, you are almost certain to find at least one laboratory where fruit flies or roundworms are being studied.

Since the early 1900s, *Drosophila melanogaster* has been a key experimental subject in genetics. It was initially chosen as a focus for study by T. H. Morgan because it can be reared in the laboratory easily and inexpensively (**Figure 32.2a**, page 726), matings can be arranged, the life cycle is completed in less than two weeks, and females lay a large number of eggs. These traits made fruit flies valuable subjects for breeding experiments designed to test hypotheses about how traits are transmitted from parents to offspring (see Chapter 13). More recently, *D. melanogaster* has also become a key model organism in the field of developmental biology. The use of flies in developmental studies was inspired in large part by the work of Christianne Nüsslein-Volhard and Eric Wieschaus, who isolated flies with genetic defects in early embryonic development. By

(Continued on next page)

(Box 32.1 continued)

investigating the nature of these defects, researchers have gained valuable insights into how various gene products influence the development of eukaryotes (see Chapter 22).

Caenorhabditis elegans emerged as a model organism in developmental biology in the 1970s, due largely to work by Sydney Brenner and colleagues. This roundworm was chosen for three reasons: (1) Its cuticle (soft outer layer) is transparent, making individual cells relatively easy to observe (**Figure 32.2b**); (2) adults have exactly 959 nonreproductive cells; and, most important, (3) the fate of each cell in an embryo can be predicted because cell fates are invariant among individuals. When researchers examine a 32-cell *C. elegans* embryo, they know exactly which of the 959 cells in the adult will be derived from each of those 32-embryonic cells. In addition, *C. elegans* are small (less than 1 mm long), are able to self-fertilize or cross-fertilize, and undergo early development in just 16 hours. The entire of genome of *C. elegans*, and that of *D. melanogaster*, has now been sequenced (see Chapter 20).

Model organisms do not need to be large or glamorous. They just need to be effective experimental subjects.

(a) Fruit flies (*Drosophila melanogaster*) can be reared in bottles.

(b) *Caenorhabditis elegans* is transparent and can be reared in petri dishes.

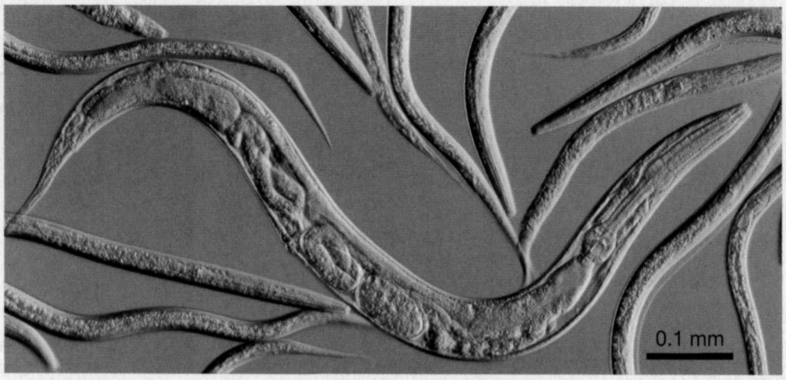

FIGURE 32.2 Model Organisms Can Be Readily Reared and Studied
(a) Fruit flies (*Drosophila melanogaster*) are small enough that they can be raised on a nutrient medium in bottles. **(b)** It is relatively straightforward to observe changes in cells as *Caenorhabditis elegans* develops, because individuals are transparent.

Crustaceans and Mollusks Are Important Animals in Marine Ecosystems

In human cultures all over the world, crabs, lobster, shrimp, clams, oysters, mussels, and squid are among the highest-priced and most sought-after of all seafood. Although a few of these commercially important marine protostomes are cultivated in pens or other types of semidomesticated situations, the vast majority of the harvest represents animals that are trapped or netted in the wild. The crustacean and mollusk harvest is big business. Regulations focused on achieving sustainable harvests have to be written with input from biologists who understand the life cycles, reproductive systems, and habitat requirements of these animals.

In the bigger picture, crustaceans and mollusks are among the protostomes that act as consumers, predators, and scavengers in many marine food chains—particularly those that form in highly productive coastal habitats outside of the tropics. As **Figure 32.3** illustrates, crustaceans, mollusks, polychaete worms, and other protostomes are particularly important in near-shore habitats with either rocky surfaces or soft substrates such as sand or mud. But because the larvae of many crustaceans, mollusks, and other groups are small and able to swim, protostomes are also important consumers and predators in the plankton of near-shore environments. With the exception of coral reefs and kelp forests, protostomes are key players in virtually every shallow-water marine environment on the planet.

Insects, Spiders, and Mites Are Important Animals in Terrestrial Ecosystems

Each year, insects eat about one-third of the crops planted by farmers around the world. That fact alone makes them the most economically important group of protostomes. Biologists who study insects work with farmers, government agencies, and agribusinesses to devise strategies for controlling the impact of insects on food production and storage. **Entomology**, or the study of insects, is among the most important of all applied fields in biological science.

The goal of **Figure 32.4** is to hint at just how important insects and spiders are in terrestrial ecosystems. Like the crustaceans and mollusks of near-shore marine habitats, insects and spiders are the dominant consumers, scavengers, and predators

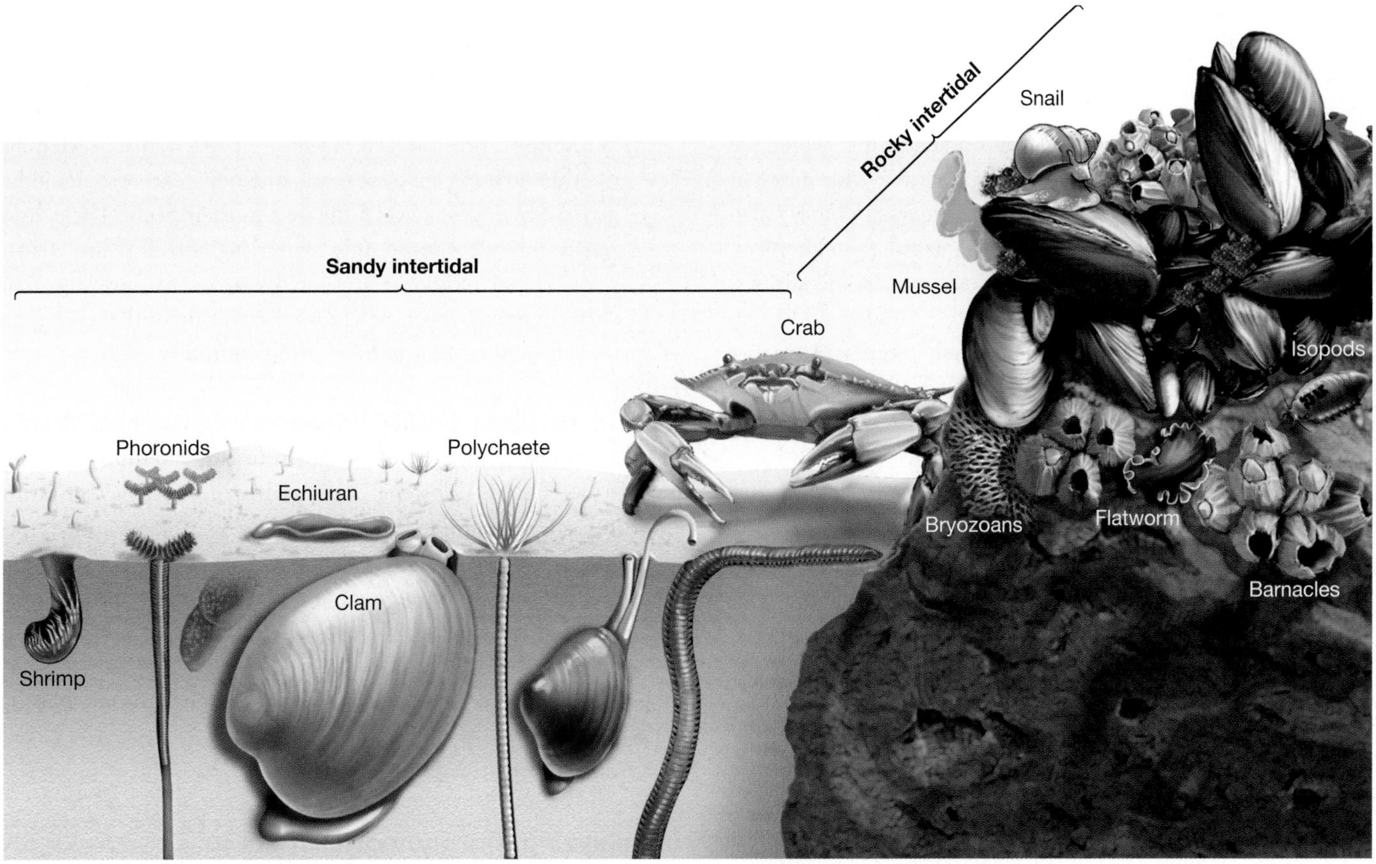

FIGURE 32.3 Protostomes Are Important Consumers and Predators in Shallow-Water Marine Habitats
Along coastlines outside the tropics, protostomes are key consumers of algae and other primary producers. They are also important predators. Protostomes are abundant in habitats with both hard and soft substrates.

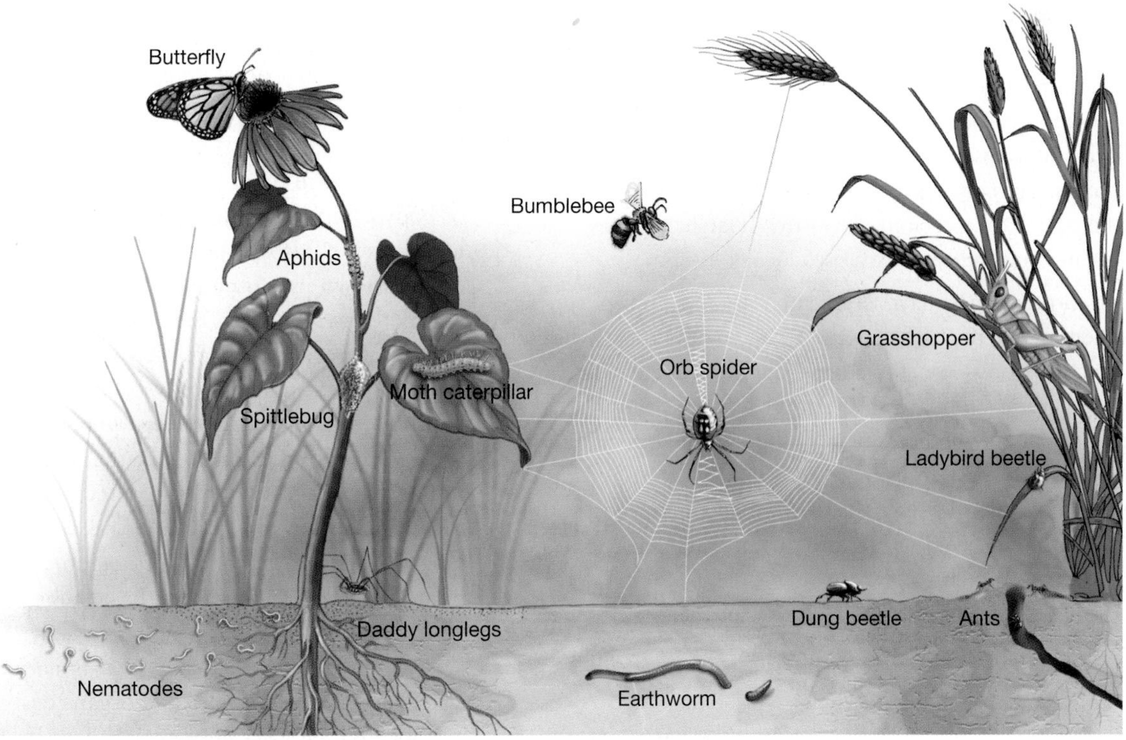

FIGURE 32.4 Protostomes Are Important Consumers and Predators in Terrestrial Habitats
Insects are key consumers of land plants. In terrestrial habitats all over the world, spiders and insects are also the most important predators of insects.

in terrestrial habitats all over the globe. Most flowering plants are pollinated by insects, and many of the leaves, stems, seeds, and roots that they produce are eaten by insects. Insects, in turn, are eaten by other insects and by spiders. Mites, which are closely related to spiders, parasitize or prey on a wide array of insects and other animal species. Insects, spiders, and mites also provide much of the food eaten by birds, lizards, salamanders, sloths, anteaters, and small mammals such as bats and shrews. Although most of the protostomes that live on land are small, their diversity and abundance make their aggregate impact enormous. It is not possible to understand a forest, prairie, or desert fully without understanding the insects, spiders, and mites that live there.

32.2 How Do Biologists Study Protostomes?

Recall from Chapter 31 that analyses of embryonic development, morphology, the fossil record, and phylogenies based on comparisons of DNA sequence data allowed researchers to make several general conclusions about protostomes:

- All protostomes have embryos that develop in a similar way. After an egg is fertilized, the pattern of cell divisions called *spiral cleavage* occurs in many protostome lineages. During gastrulation, the initial invagination that forms in the embryo becomes the mouth. If a **coelom** (a body cavity) forms later in development, it arises when openings form within blocks of mesodermal tissue.
- All protostomes share several common features of their overall body plan. They are bilaterally symmetric organisms that face the environment in one direction. They have a distinctive head region where sensory and information-processing organs are localized and a posterior region that is usually specialized for locomotion. They also have three embryonic tissues. Although most lineages have a coelom, flatworms (Platyhelminthes) are acoelomates and rotifers have a pseudocoelom.
- Certain protostomes have segmented bodies. When **segmentation** occurs, the body has a series of compartments with repeated structures—usually having similar arrangements of muscles, nerves, blood vessels, and gut. Segmentation is a prominent part of the body plan in the Annelida, including earthworms, leeches, and other segmented worms, and the Arthropoda, including insects, myriapods (millipedes and centipedes), crustaceans (crabs and shrimp), and chelicerates (spiders, ticks, and mites).
- In the fossil record, most protostome phyla first appear in the Burgess Shale faunas, dated at about 525–515 million years ago.
- Molecular phylogenies support the hypothesis that protostomes are a monophyletic group. Analyses of DNA sequence data also revealed that a major branching event occurred within the lineage, creating the groups called **Lophotrochozoa** and **Ecdysozoa**. The lophotrochozoans include the mollusks, annelids, and flatworms (Platyhelminthes; pronounced *pla-tee-hell-MIN-theez*). Members of this group have several distinct features, including growth by incremental additions to their body. The ecdysozoans include the roundworms (Nematoda) and arthropods (Arthropoda). Ecdysozoans grow by molting. They shed an exoskeleton or external covering and then expand their body before a new covering solidifies.

WEB TUTORIAL 32.1
Protostome Diversity

To date about 20 phyla of protostomes have been identified and named. Protostomes range from large, familiar animals such as clams, lobsters, and earthworms to lesser-known groups. For example, the unfamiliar gnathostomulids (pronounced *nath-oh-stoh-MEW-lids*) and gastrotrichs (*GAS-troh-triks*) are less than 2 mm long and live in sands and muds along ocean beaches.

The protostomes are easily the most species-rich and morphologically diverse lineage of animals. What do morphological studies, the fossil record, and molecular phylogenies have to say about how this group diversified?

Analyzing Morphological Traits

Decades of careful study have shown that there is little variation in several aspects of the design and construction of the protostome body. All protostomes are triploblastic and bilaterally symmetric. In embryonic development, the initial invagination during gastrulation always forms the mouth. There is relatively little variation among phyla in how cleavage occurs.

In contrast, radical changes occurred in coelom formation as protostomes diversified. Recall from Chapter 31 that the most dramatic change was a reversion to an **acoelomate** body plan—meaning the lack of a body cavity—in Platyhelminthes. Flukes, tapeworms, and other types of flatworms do not have any sort of body cavity, even though their ancestors did. Data reviewed in Chapter 31 also support the hypothesis that the type of body cavity known as a **pseudocoelom**, which forms from an opening that arises between the ectoderm and mesoderm layers of embryos, arose independently in the protostome phylum Rotifera (rotifers) and the ecdysozoans. The Nematoda (roundworms), for example, have a prominent pseudocoelom.

The other important change that occurred in the formation of body cavities was a drastic reduction of the coelom that occurred in certain phyla. To put this change in context, it's important to note that most protostome phyla have wormlike bodies (**Figure 32.5**) with a basic **tube-within-a-tube** design. The outside tube is the skin, which is derived from ectoderm; the inside tube is the gut, which is derived from endoderm (see Chapter 31). Muscles and organs derived from mesoderm are located between the two tubes. In these animals, the coelom is well developed and forms a hydroskeleton that is the basis of movement. But in the most species-rich and mor-

(a) Priapulids are predators that burrow into the substrate and wait for prey.

1 cm

(b) Nematomorpha live in freshwater environments.

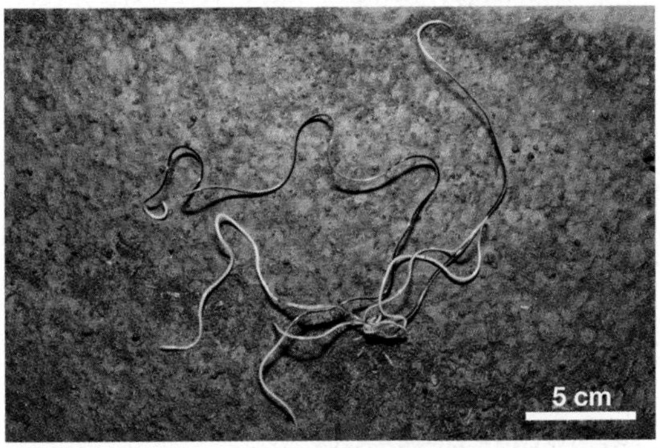
5 cm

FIGURE 32.5 Members of Most Protostome Phyla Have Wormlike Bodies
(a) Priapulids are also called penis worms, because of their superficial resemblance to the male copulatory organ of mammals. (b) Nematomorpha are also called hair worms, because they are extremely long and thin. QUESTION State a hypothesis for how species with soft, wormlike bodies avoid predation. How do species with wormlike bodies move?

phologically complex protostome phyla—the Arthropoda and the Mollusca (snails, clams, squid)—the coelom is drastically reduced. Arthropods and mollusks have distinct and specialized body plans.

Arthropods are segmented and have an exoskeleton made primarily of the polysaccharide chitin (see Chapter 5); in crustaceans, the exoskeleton is strengthened by the addition of calcium carbonate ($CaCO_3$). As **Figure 32.6a** shows, the arthropod body is segmented and has jointed legs and an exoskeleton. Locomotion is based on muscles that apply force against the exoskeleton to move legs or wings. Based on these observations, biologists suggest that the evolution of the exoskeleton and jointed appendages in arthropods made the coelom and hydrostatic skeleton unnecessary. Arthropods do have a spacious body cavity called the **hemocoel** ("blood-hollow"), however, which provides space for internal organs and circulation of fluids. Developmentally, the hemocoel arises as a cavity between an arthropod embryo's ectoderm and mesoderm layers—meaning it corresponds to a pseudocoelom.

Dramatic changes also occurred in mollusks, although the nature of the body plan that evolved is very different. As **Figure 32.6b** shows, the mollusk body plan is based on three major components: (1) the **foot**, a large muscle that is located at the base of the animal and is usually used in movement; (2) the **visceral mass**, the region that contains most of the main internal organs and external gill; and (3) the **mantle**, a tissue layer that covers the visceral mass and that in many species secretes a shell made of calcium carbonate. Mollusks may have one, two, eight, or no shells. When shells are present, they function to protect the animal. Movement is often based on a hydrostatic skeleton, but not one that is created by a coelom. Instead, the muscular foot itself—much like your tongue—may function as a hydrostatic skeleton. In some species, the fluid enclosed in the visceral mass also functions as a hydrostatic skeleton.

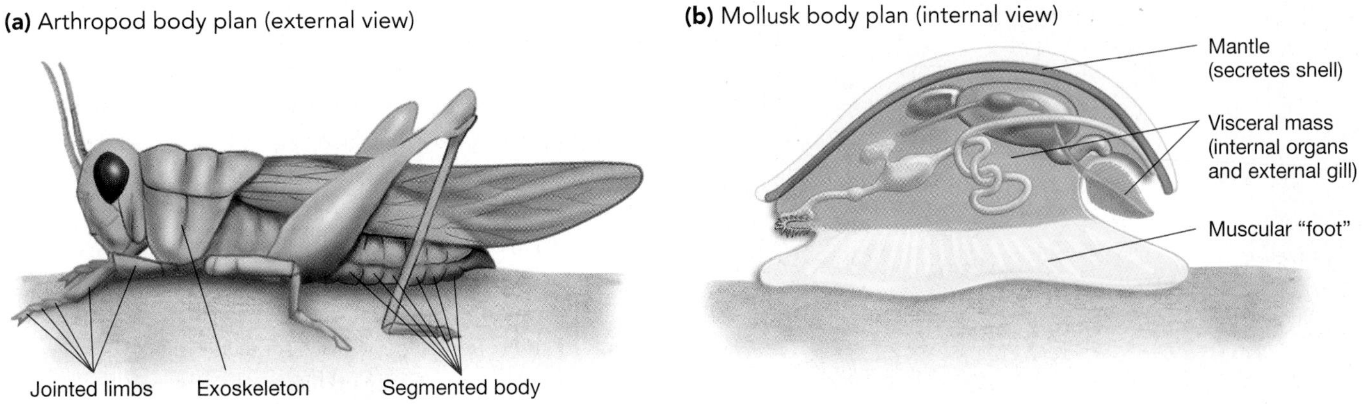
(a) Arthropod body plan (external view)
Jointed limbs
Exoskeleton
Segmented body
(b) Mollusk body plan (internal view)
Mantle (secretes shell)
Visceral mass (internal organs and external gill)
Muscular "foot"

FIGURE 32.6 Arthropods and Mollusks Have Specialized Body Plans
(a) Arthropods have segmented bodies and jointed limbs, which enable these animals to move despite their hard outer covering, the exoskeleton. (b) The mollusk body plan is based on a foot, a visceral mass, and a mantle.

In addition to functioning in movement, the exoskeletons and shells of arthropods and mollusks are important in defense against predation. To support this claim, biologists point to the secretive behavior of ecdysozoans while individuals are molting. When crabs and other crustaceans have shed an old exoskeleton, their new exoskeleton takes several hours to harden. At this time, individuals hide and do not feed or move about. Experiments have shown that it is much easier for predators to attack and subdue individuals that are not protected by an exoskeleton during this intermolt period.

Using the Fossil Record

The Bryozoa (moss animals) pictured in **Figure 32.7** are the only animal phylum that appears after the Cambrian explosion ended about 515 million years ago. Thus, virtually all of the protostome phyla appear very early in the history of animal evolution. Once these phyla were present, two major events are recorded in the fossil record of protostomes: (1) the extinction of a major group of arthropods called the trilobites and (2) the first appearance of insects.

Trilobites are extremely abundant in marine rocks dated from 550–440 million years ago (**Figure 32.8a**). Like other arthropods, trilobites had a hard exoskeleton, segmented bodies, and jointed limbs. Most species lived on the sea floor and made their living by scavenging dead organic matter or deposit feeding; some may have been predators or suspension feeders. Trilobites gradually decline in numbers and species diversity over time, however, and disappear from the fossil record altogether about 250 million years ago, during the great mass extinction that occurred at the end of the Permian period.

The first insects appear in the fossil record about 400 million years ago. The first winged insects, which are the most prominent dominant group today, are observed in rocks that are about 370 million years old (**Figure 32.8b**). Recall from Chapter 29 that these times coincide with the Silurian-Devonian explosion—the initial, rapid diversification of land plants. Given the close association between land plants and insects today, it is logical to observe that insects initially evolved and diversified soon after land plants did.

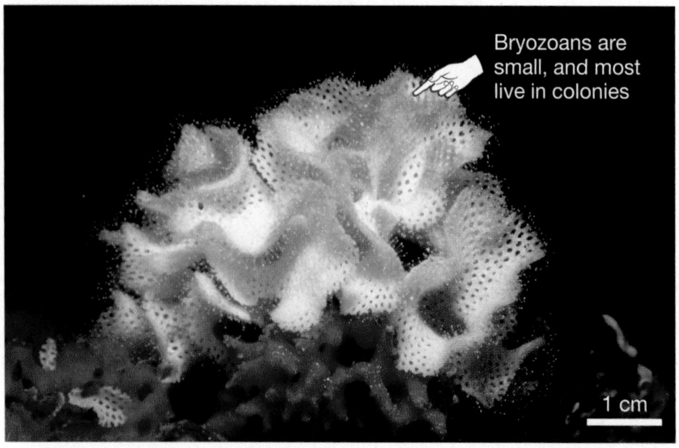
Bryozoans are small, and most live in colonies
1 cm

FIGURE 32.7 Bryozoans Are the Youngest Phylum of Animals
Bryozoans, or moss animals, make their first appearance in the fossil record about 490 million years ago. All other phyla had appeared by about 510 million years ago.

(a) Trilobites are extinct. **(b)** Winged insect, 305 Ma

1 cm
0.5 cm

FIGURE 32.8 The Fossil Record of Protostomes Documents Extinctions as Well as the Emergence of Major New Groups
(a) Trilobites appear during the Cambrian explosion but persisted for only about 260 million years. **(b)** The first winged insects appear in the fossil record just 25 million years after the initial diversification of land plants.

Evaluating Molecular Phylogenies

When molecular phylogenies supported the hypothesis that the protostomes are a monophyletic group, biologists were not surprised. But when analyses of DNA sequence data indicated that two major groups of protostomes existed, it was considered a key insight into animal evolution. Analyses of the gene for the small subunit of ribosomal RNA supported the existence of the monophyletic groups Lophotrochozoa and Ecdysozoa. More recent analyses—based on sequence data from the gene for myosin, a protein involved in muscle contraction—also support this hypothesis. Although work and discussion continue, an emerging consensus holds that the divergence of protostomes into two major subgroups was an important event in animal evolution.

Figure 32.9 summarizes recent results on the phylogeny of the protostomes and indicates where major changes occurred in the evolution of protostome body plans. Only a few of the approximately 20 protostome phyla are included, however, because the relationships of most groups within the Lophotrochozoa and Ecdysozoa are still poorly resolved. Stated another way, it is not yet clear how most of the phyla within these two groups are related. As investigators analyze data from additional genes, a more highly resolved and better-supported tree should gradually emerge.

Until such a tree is available, biologists will be unable to make firm conclusions about the sequence of events that occurred as the Lophotrochozoa and Ecdysozoa diverged. Although much remains to be learned about how protostomes diversified, a great deal is understood about what happened

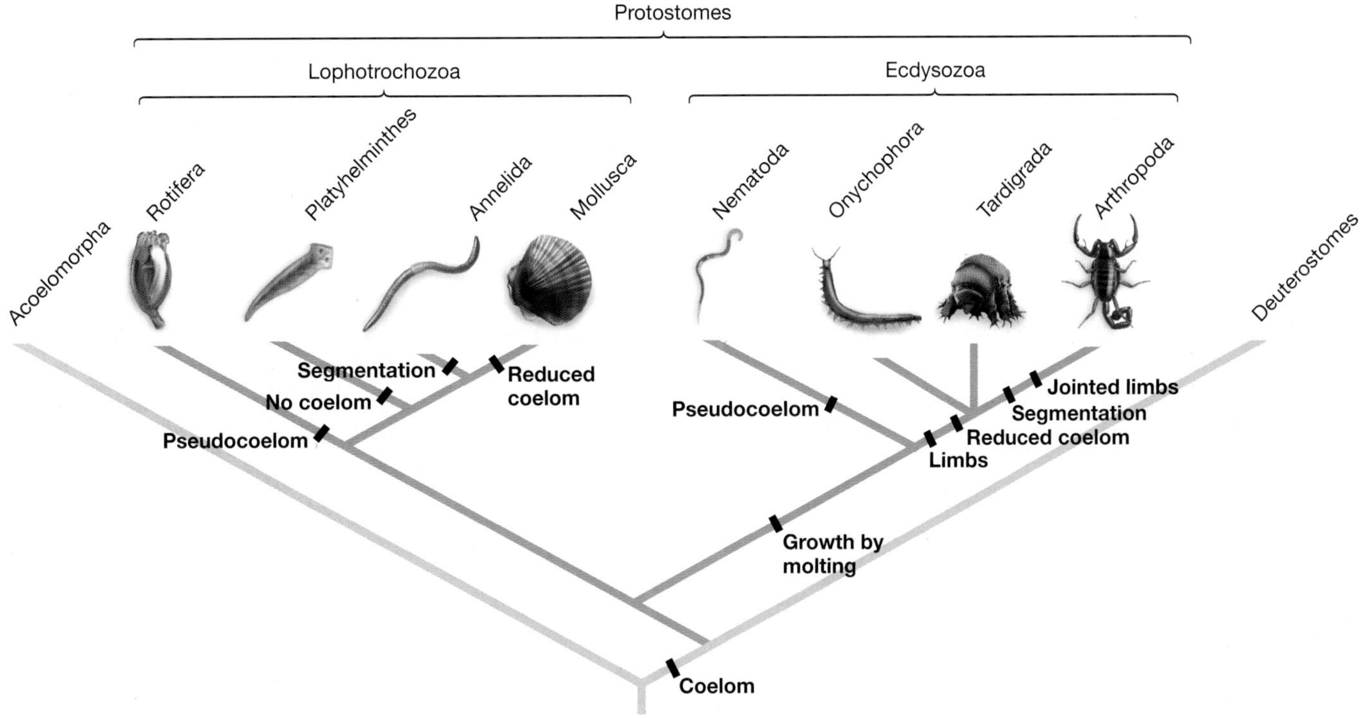

FIGURE 32.9 The Phylogeny of Protostomes
Phylogenetic tree based on analyses of DNA sequences from several genes. The tree does not include protostome phyla that have not yet been included in studies or those with relationships that are still too uncertain.

during their evolution. In addition to evolving the ability to live on land, protostomes underwent dramatic changes in methods for feeding, moving, and reproducing.

32.3 What Themes Occur in the Diversification of Protostomes?

Protostomes are the most abundant animals in the plankton and on the substrates of many marine and freshwater environments. But protostomes are common in virtually every terrestrial setting as well. Like the land plants and fungi, protostomes made the transition from aquatic to terrestrial environments. To help put this achievement in perspective, recall from Chapter 29 that green plants made the move from freshwater to land just once. From Chapter 30, recall that it is not yet clear whether fungi moved from aquatic habitats to land once or several times. Even given present uncertainties in estimating the protostome phylogeny, however, it is clear that a water-to-land transition occurred multiple times as protostomes diversified. The ability to live in terrestrial environments evolved independently in arthropods (at least twice), mollusks, roundworms, and annelids. Repeated water-to-land transitions are a prominent theme in the evolution of protostomes. In each case, the evidence for this claim is based on phylogenetic analyses, which support the hypothesis that the ancestors of the terrestrial lineages in each major subgroup were aquatic.

Why were so many different protostome groups able to move from water to land independently? The short answer is that it was easier for animals to accomplish this than for plants to do so. For example, land plants had to evolve vascular tissue to transport water and support their stems. Similarly, the first terrestrial animals had to be able to support their body weight on land and move about. But the protostome groups that made the water-to-land transition already had hydrostatic skeletons, exoskeletons, appendages, or other adaptations for support and locomotion that happened to work on land as well as water.

To make the transition to land, two of the most important new problems that animals had to solve were (1) exchanging gases and (2) drying out. As detailed in Chapter 44, land animals exchange gases with the atmosphere readily as long as a moist body surface is exposed to the air. The bigger challenge is to prevent the gas-exchange surface and other parts of the body from drying out. Roundworms and earthworms solve this problem by living in wet soils or other moist environments. Arthropods and many mollusks have a watertight exoskeleton or a shell that minimizes water loss, and their respiratory structures are located inside the body. In insects, the openings to respiratory passages can be closed to minimize water loss. And unlike land plants and fungi, all land animals can move to moister habitats if the area they are in gets too dry.

To summarize, the ability to live in terrestrial environments was a key event in the evolution of several protostome phyla. In addition, the evolution of specialized body plans provided a foundation for diversification in the most species-rich lineages of protostomes—arthropods and mollusks. In large part, the remainder of protostome evolution was driven by innovations in feeding, moving, and reproducing.

Feeding

Protostomes include suspension feeders, deposit feeders, herbivores, predators, and parasites. The diversity of feeding strategies observed among protostomes is reminiscent of that among lineages of protists. Recall from Chapter 28 that each of the major protist lineages has at least one distinct morphological characteristic. Once unique aspects of cell or body structure had evolved, each protist lineage diversified into species that pursue a wide array of lifestyles. In a similar way, the protostome pattern of development represents a unique way of building a bilaterally symmetric animal with a coelom and three embryonic tissue types. Once this method of constructing a body had evolved, protostomes diversified into a wide variety of habitats and lifestyles—primarily as a result of variation in feeding strategies.

A wide diversity in feeding strategies is possible because protostomes have a wide variety of specialized mouthparts. Several of the wormlike phyla, for example, have unique jaws or other mouth and throat structures. Echiurans (spoon worms) suspension feed using an extended structure called a **proboscis,** which forms a gutter leading to the mouth (**Figure 32.10a**). Cells in the gutter of echiurans secrete mucus, which is sticky enough to capture pieces of detritus. The combination of mucus and detritus is then swept into the mouth by cilia on cells in the gutter. The nemerteans (ribbon worms), in contrast, have a proboscis that can extend or retract (**Figure 32.10b**). These worms wrap their extended proboscis around small animals to capture them and then pull them into the mouth. In both phyla, the structure of the mouthparts correlates closely with their function in feeding.

Regarding mouthpart diversity within protostome groups, however, arthropods take the prize. The mouthparts observed in insects and crustaceans alone vary in structure from tubes to pincers and allow them to pierce, suck, grind, bite, mop, chew, engulf, cut, and mash (**Figure 32.10c, d, e**). All insects may have the same basic body plan, but their mouthparts and food sources are highly diverse.

In many species, the jointed limbs of arthropods also play a key role in getting food. The crustaceans called krill use their legs to sweep food toward their mouths as they suspension feed; certain insects, crustaceans, spiders, and mollusks use their appendages to capture prey or hold food as it is being chewed or bitten by the mouthparts.

In most cases, juvenile and adult forms of the same species exploit different food sources. Metamorphosis is the predomi-

(a) Echiurans are suspension feeders with a mucus-lined proboscis.
Proboscis
(b) Nemerteans are predators with an extendable proboscis.
Jellyfish being attacked
(c) Leaf-cutter ants cut leaves.
(d) Houseflies lap up fluids.
(e) Ladybird beetles are predators.

FIGURE 32.10 Protostome Feeding Structures Are Diverse
(a) Echiurans and **(b)** nemerteans have similar body plans and burrow in soft marine sediments, but they have distinct mouthparts and feed in completely different ways. **(c)** Leaf-cutter ants, **(d)** houseflies, and **(e)** ladybird beetles are all insects. But each species has a different type of mouthpart structure and feeds on a different source of food.

nant mode of development in protostomes and usually results in larvae and adults that live in different habitats and have different overall morphology and mouthparts.

Movement

In protostomes, variation in movement depends on variation in two features: (1) the presence or absence of limbs and (2) the type of skeleton that is present. Wormlike protostomes that lack limbs move with the aid of a hydrostatic skeleton. This type of movement is also observed in caterpillars, grubs, maggots, and other types of insect larvae—even though adults of the same species have a hard exoskeleton and move with the aid of **jointed limbs** (**Figure 32.11a**). Although insect larvae have a highly reduced coelom, they do have an enclosed, fluid-filled body cavity that functions as a hydrostatic skeleton.

The evolution of jointed limbs made rapid, precise swimming and running movements possible and was a key innovation in the evolution of protostomes. The jointed limb is a major reason that arthropods have been so spectacularly successful in terms of species diversity, abundance of individuals, geographic range, and duration in the fossil record.

But the jointed limb is just one of several evolutionary innovations that allowed protostomes to move in unique ways:

- The insect wing is a marvel of engineering. It is also one of the most important adaptations in the history of life. About two-thirds of the multicellular species living today are winged insects. According to data in the fossil record, insects were the first organisms that had wings and could fly. Like most insects today, the earliest insects had two pairs of wings. In most four-winged insects living today, however, the four wings function as two. Beetles fly with only their hind wings; butterflies, moths, bees, and wasps have hooked structures that make their two pairs of wings move together. Flies appear later in the fossil record and have a single pair of wings (**Figure 32.11b**).

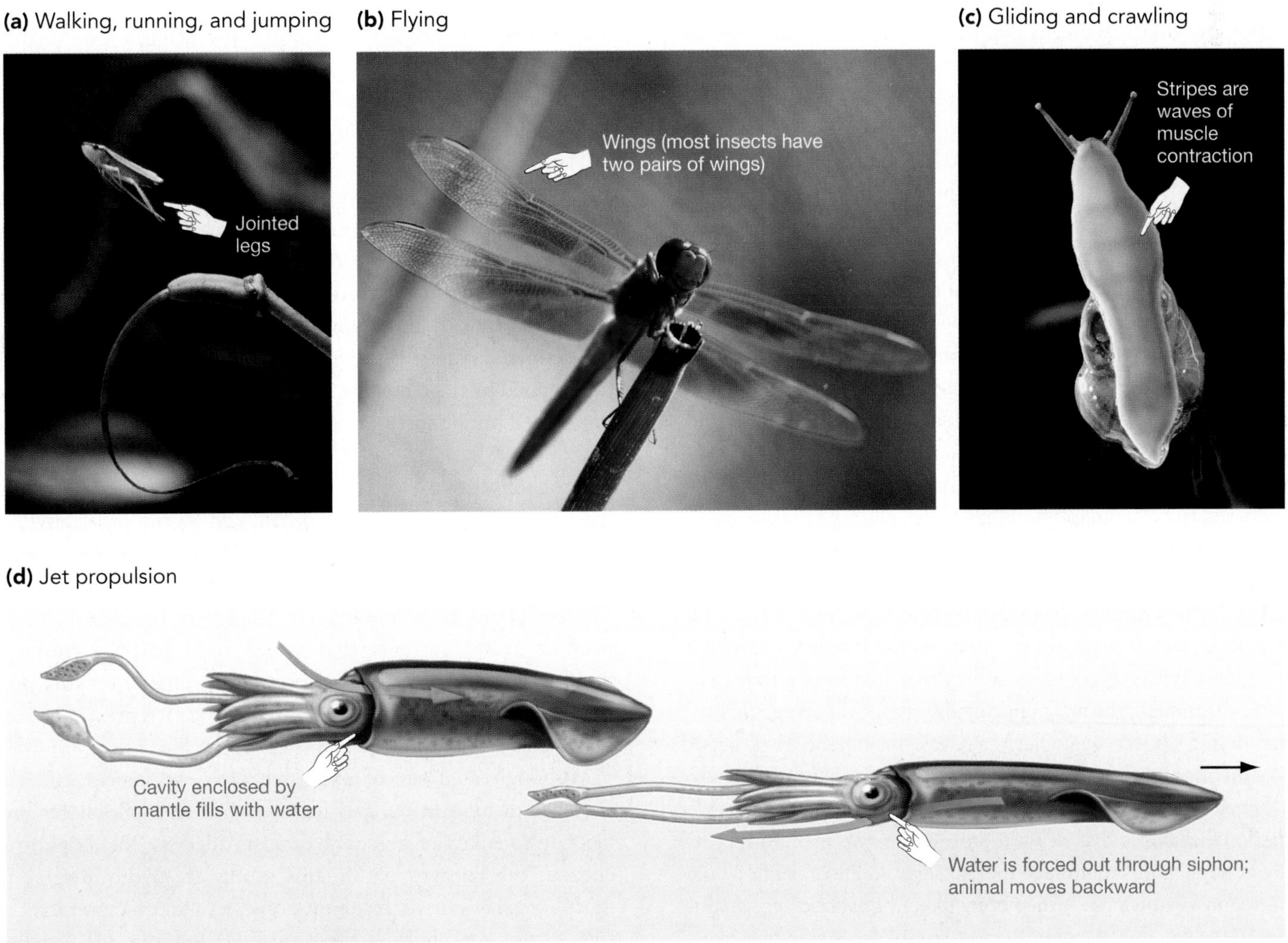
(a) Walking, running, and jumping
Jointed legs
(b) Flying
Wings (most insects have two pairs of wings)
(c) Gliding and crawling
Stripes are waves of muscle contraction
(d) Jet propulsion
Cavity enclosed by mantle fills with water
Water is forced out through siphon; animal moves backward

FIGURE 32.11 Protostome Locomotion Is Diverse
The evolution of **(a)** jointed limbs and **(b)** wings were key innovations in arthropod movement. **(c)** A wave of muscle contractions, down the length of the foot, allows mollusks to glide along a substrate. **(d)** In jet propulsion, muscular contractions force the water out through a moveable siphon. The siphon can aim outgoing water in a specific direction.

- In mollusks, waves of muscle contractions sweep down the length of the large, muscular foot, allowing individuals to glide along a surface (**Figure 32.11c**). Cells in the foot secrete a layer of mucus, which reduces friction with the substrate and increases the efficiency of gliding.
- Squid have a mantle that is lined with muscle. When the cavity surrounded by the mantle fills with water and the mantle muscles contract, a stream of water is forced out of a tube called a **siphon**. The force of the water propels the squid backward (**Figure 32.11d**). This mechanism, **jet propulsion**, evolved in squid long before human engineers thought of using the same principle to power aircraft.

Reproduction and Life Cycles

When it comes to variation in reproduction and life cycles, protostomes do it all. Asexual reproduction by splitting the body lengthwise or by fragmenting the body is common in many of the wormlike phyla. Many crustacean and insect species reproduce asexually via **parthenogenesis** ("virgin-origin")—the production of unfertilized eggs that develop into offspring. Sexual reproduction is the predominant mode of producing offspring in most protostome groups, however. Sexual reproduction usually begins with external fertilization in sessile forms, such as clams, bryozoans, and brachiopods. It begins with internal fertilization in groups that are capable of movement—such as crustaceans, snails, and insects—because males and females can meet. Females of *ovoviviparous* insects and snails bear live young, although they do so by retaining fully formed eggs inside their bodies and then nourishing them via the nutrient-rich yolk that is inside the egg (see Chapter 31).

Two unique reproductive innovations occurred during protostome diversification: (1) the evolution of metamorphosis and (2) an egg that would not dry out on land. Metamorphosis is common in marine protostomes. In these species, metamorphosis is hypothesized to be an adaptation that allows juveniles to disperse to new habitats by floating or swimming in the plankton. Metamorphosis is also common in terrestrial insects (see Figures 31.17 and 31.18). In these species, metamorphosis is hypothesized to be an adaptation that reduces competition for food between juveniles and adults. With respect to the ability to colonize terrestrial environments, though, the most critical adaptation was an egg that resists drying. Insect eggs have a thick membrane that keeps moisture in, and the eggs of slugs and snails have a thin calcium carbonate shell that helps retain water. Thus, protostomes provide additional evidence that desiccation-resistant eggs evolved repeatedly in populations that made the transition to life on land. As we will see in Chapter 33, a membrane-bound *amniotic egg*, analogous to the membrane-bound eggs of insects, evolved in the terrestrial vertebrates. Recall from Chapter 29 that an analogous adaptation arose in land plants, when the evolution of the gametangia helped protect eggs and embryos from drying.

✓CHECK YOUR UNDERSTANDING

Although all protostomes share fundamental aspects of their body plan, they occupy a wide range of habitats and feed and reproduce in a wide variety of ways. A water-to-land transition occurred multiple times in protostomes. That transition was facilitated by the ability to move away from dry areas and toward moister environments in times of water stress, by the existence of watertight shells in mollusks and exoskeletons in arthropods and by the evolution of desiccation-resistant eggs. You should be able to (1) give examples of variation in mouthpart structure and feeding, structures used in locomotion, and metamorphosis in protostomes; and (2) state a hypothesis to explain why arthropods and mollusks are the most species-rich of all animal phyla.

32.4 Key Lineages: Lophotrochozoans

The Lophotrochozoa is named for the group's two distinctive morphological traits: (1) a feeding structure called a lophophore, which is found in three phyla, and (2) a type of larva called a trochophore, which is common to many of the phyla in the lineage.

As **Figure 32.12a** shows, a **lophophore** ("tuft-bearer") is a specialized structure that rings the mouth and functions in suspension feeding. Lophophores consist of tentacles that have ciliated cells along their surface. The beating of the cilia generates a water current that sweeps protists and other alga, detritus, and animal larvae into the region above the mouth, where they are trapped by the tentacles, swept into the mouth opening, and ingested. Lophophores are found in bryozoans, brachiopods, and phoronids.

Trochophores are a type of larvae common to marine mollusks, annelids that live in the ocean, and several other phyla in the Lophotrochozoa. As **Figure 32.12b** shows, a **trochophore** ("wheel-bearer") larva has a ring of cilia around its middle. These cilia aid in swimming. The beating of the cilia may also produce water currents that sweep food particles into the mouth, though many trochophores feed primarily on yolk provided by the mother. Thus, trochophores may represent a feeding stage as well as a dispersal stage in the life cycle.

Although analyses of extensive DNA sequences and the presence of unique, derived features such as lophophores and trochophore larvae helped biologists identify the Lophotrochozoa, the lineages within this group are highly diverse in morphology. To drive this point home, let's take a closer look at four of the key phyla in the group: (1) **Rotifera**, (2) **Platyhelminthes**, (3) **Annelida**, and (4) **Mollusca**.

(a) Lophophores function in suspension feeding in adults.

0.1 mm

(b) Trochophore larvae swim and feed.

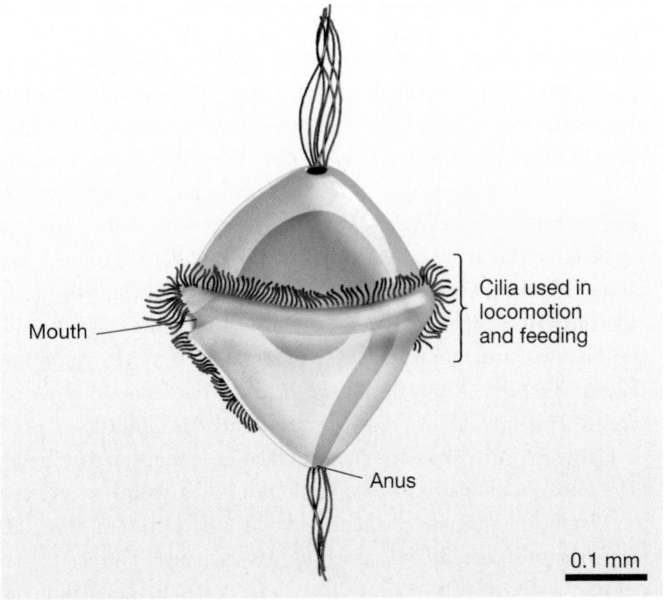
Cilia used in locomotion and feeding
Mouth
Anus
0.1 mm

FIGURE 32.12 Distinctive Traits in Lophotrochozoa
(a) Three phyla of lophotrochozoans have the feeding structure called a lophophore. **(b)** Many phyla of lophotrochozoans have the type of larva called a trochophore.

Rotifera (Rotifers)

The 1800 rotifer species that have been identified thus far live in damp soils as well as marine and freshwater environments. They are important components of the plankton in freshwater and in brackish waters, where rivers flow into the ocean. They have pseudocoeloms, and most species are less than 1 mm long. Although rotifers do not have a lophophore or a trochophore larval stage, extensive similarities in DNA sequence identify them as a member of the lophotrochozoan lineage.

Feeding Rotifers have a cluster of cilia at their anterior end called a **corona** (**Figure 32.13**). In many species, the beating of the cilia in the corona makes suspension feeding possible by creating a current that sweeps microscopic food particles into the mouth. The corona is the signature morphological feature of this group.

Movement Although a few species of rotifers are sessile, most swim via the beating of cilia in the corona.

Reproduction Females produce unfertilized eggs by mitosis, which then hatch into new, asexually produced individuals. Recall from Section 32.3 that the production of offspring via unfertilized eggs is termed *parthenogenesis*. An entire group of rotifers—called the bdelloids—reproduces exclusively asexually; they do so via parthenogenesis. Both sexual reproduction and asexual reproduction are observed in most species, however. Development is direct, meaning that fertilized eggs hatch and grow into adults without going through metamorphosis.

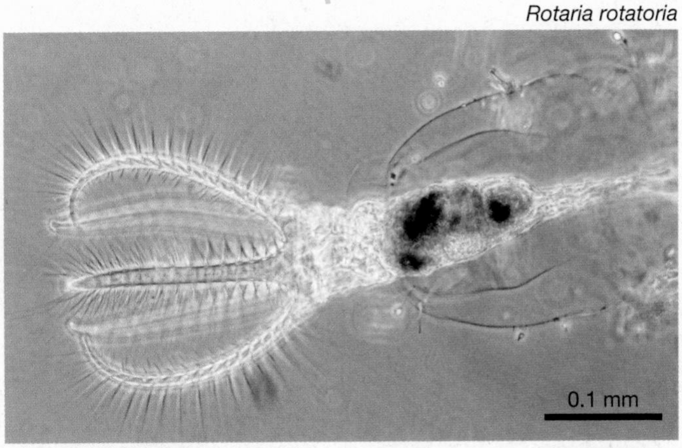
Rotaria rotatoria
0.1 mm

FIGURE 32.13 Rotifers Are Tiny, Freshwater Suspension Feeders

Platyhelminthes (Flatworms)

The flatworms are a large and diverse phylum. More than 20,000 species have been described in three major subgroups. In traditional classification schemes like those described in Chapter 1, each of these lineages was referred to as a *class*. They are (1) the free-living flatworms, (2) the endoparasitic tapeworms, and (3) the endoparasitic or ectoparasitic flukes. The free-living forms are called the Turbellaria (**Figure 32.14a**). Although a few turbellarian species are terrestrial, most live on the substrates of freshwater or marine environments. Tapeworms belong to the subgroup or class called the Cestoda and parasitize fish, mammals, or other vertebrates (**Figure 32.14b**). Most flukes are in a lineage called Trematoda (**Figure 32.14c**). Flukes parasitize vertebrates or mollusks.

Flatworms are named for the broad, flattened shape of their bodies. (The Greek roots *platy* and *helminth* mean "flat-worm.") Species in the Platyhelminthes are unsegmented and lack a coelom. They also lack structures that are specialized for gas exchange. Further, they do not have blood vessels or any other type of system for circulating oxygen and nutrients to their cells. Based on these observations, biologists interpret the flattened bodies of these animals as an adaptation that gives flatworms an extremely high surface-area-to-volume ratio. Because they have so much surface area, a large amount of gas exchange can occur directly across their body wall, with oxygen diffusing into the body from the surrounding water and carbon dioxide diffusing out. Because the volume of the body is so small relative to the surface area available, nutrients and gases can diffuse efficiently to all of the cells inside the animal. This body plan has a downside, however: Because the body surface has to be moist for gas exchange to take place, flatworms are restricted to environments where they are surrounded by fluid.

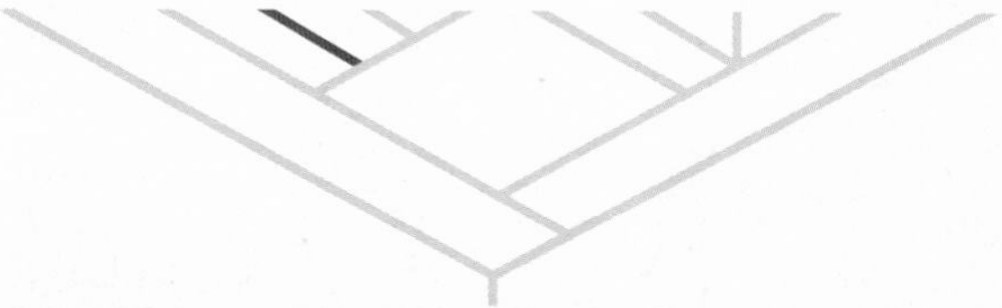

Feeding Platyhelminthes lack a lophophore and have a digestive tract that is "blind"—meaning it has only one opening for ingestion of food and elimination of wastes. Most turbellarians are hunters that prey on protists or small animals; others scavenge dead animals. Cestodes and trematodes, in contrast, are strictly parasitic and feed on nutrients provided by hosts. Trematodes gulp host tissues and fluids through a mouth and have a blind digestive tract. Cestodes lack both structures and obtain nutrients solely by diffusion through their body wall.

Movement Some turbellarians can swim a little by undulating their bodies, and most can creep along substrates with the aid of cilia on their ventral surface. Cestodes and trematodes are largely or completely sessile; adult cestodes have hooked attachment structures at their anterior end that permanently attach them to the interior of their host.

Reproduction Turbellarians can reproduce asexually by splitting themselves in half. If they are fragmented as a result of a predator attack, the body parts can regenerate into new individuals. Most turbellarians contain both male and female organs and reproduce sexually by aligning with another individual and engaging in mutual and simultaneous fertilization. Trematodes and cestodes also reproduce sexually and either cross-fertilize or self-fertilize. In many cases, the life cycle of trematodes and cestodes involves two or even three distinct host species, with sexual reproduction occurring in the **definitive host** and asexual reproduction occurring in one or more **intermediate hosts**. For example, humans are the definitive host of the blood fluke *Schistosoma mansoni*. Fertilized eggs are shed in a human host's feces and enter aquatic habitats if sanitation systems are poor. The eggs develop into larvae that infect snails. Inside the snail, asexual reproduction results in the production of a different type of larva, which emerges from the snail and burrows into the skin of humans who wade in infected water. Once inside a human, the parasite lives in blood and develops into a sexually mature adult.

(a) Turbellarians are free living. *Pseudoceros ferrugineus*

(b) Cestodes are endoparasitic. *Taenia* species

(c) Trematodes are endoparasitic. *Dicrocoelium dendriticum*

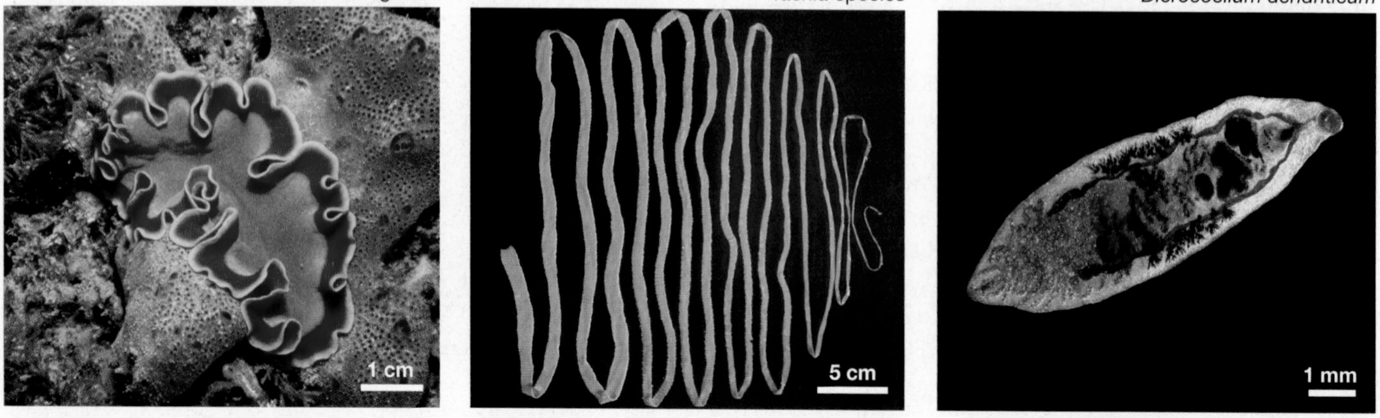

FIGURE 32.14 Flatworms Have Simple, Flattened Bodies

Annelida (Segmented Worms)

All annelids have a segmented body and a coelom that functions as a hydrostatic skeleton. The 16,500 species that have been discovered thus far can be divided into two major lineages, traditionally called classes. Those clades are the Polychaeta and Clitellata:

1. The **Polychaeta** (pol-ee-KEE-ta), or polychaetes ("many-bristles"), are named for their numerous, bristle-like extensions called **chaetae.** The chaetae extend from appendages called **parapodia.** Polychaetes are mostly marine, and they range in size from species that are less than 1 mm long to species that grow to lengths of 3.5 m. They include a large number of burrowers in mud or sand, sedentary forms that secrete chitinous tubes, and active, mobile species (**Figure 32.15a**).
2. The **Clitellata** is composed of the oligochaetes (*oh-LIG-oh-keetes*) and leeches. Oligochaetes ("few-bristles") include the earthworms, which burrow in moist soils (**Figure 32.15b**), an array of freshwater species, and a few marine forms. Oligochaetes lack parapodia and, as their name implies, have many fewer chaetae than do polychaetes. The leeches are members of the **Hirudinea.** Their coelom is much reduced in size compared with that of other annelids and consists of a series of connected chambers. Leeches live in freshwater as well as marine habitats (**Figure 32.15c**).

Feeding Polychaetes have a wide variety of methods for feeding: The burrowing forms deposit feed; the sedentary forms suspension feed with the aid of a dense crown of tentacles; and the active forms hunt small animals, which they capture by everting their throats (that is, turning their throats inside out). Virtually all oligochaetes, in contrast, make their living by deposit feeding in soils. About half of the leeches are ectoparasites that attach themselves to fish or other hosts and suck blood and other body fluids. Hosts are usually unaware of the attack, because leech saliva contains an anaesthetic. The host's blood remains liquid as the leech feeds, because the parasite's saliva contains an anticoagulant. Parasitic leeches are still used by physicians to remove blood from particularly large bruises. The nonparasitic leech species are predators or scavengers.

Movement Polychaetes and oligochaetes crawl or burrow with the aid of their hydrostatic skeletons; the parapodia of polychaetes also act as paddles or tiny feet that aid in movement. Leeches can swim by using their hydrostatic skeletons to make undulating motions of the body.

Reproduction Asexual reproduction occurs in polychaetes and oligochaetes by transverse fission or fragmentation—meaning that body parts can regenerate a complete individual. Sexual reproduction in polychaetes may begin with internal or external fertilization, depending on the species. Polychaetes have separate sexes and usually release their eggs directly into the water. Some species produce eggs that hatch into trochophore larvae; others undergo direct development. In oligochaetes and leeches, individuals produce both sperm and eggs and engage in mutual, internal cross-fertilization. Eggs are enclosed in a mucus-rich, cocoon-like structure where they develop directly into miniature versions of their parents.

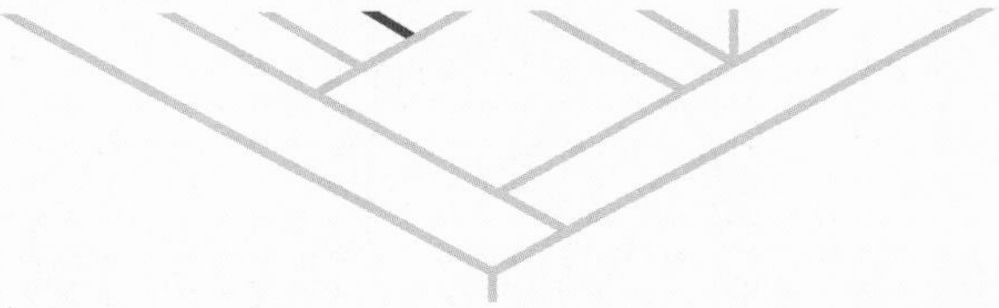

(a) Most polychaetes are marine.

Alvinella pompejana

(b) Most oligochaetes are terrestrial.

Paranais litoralis

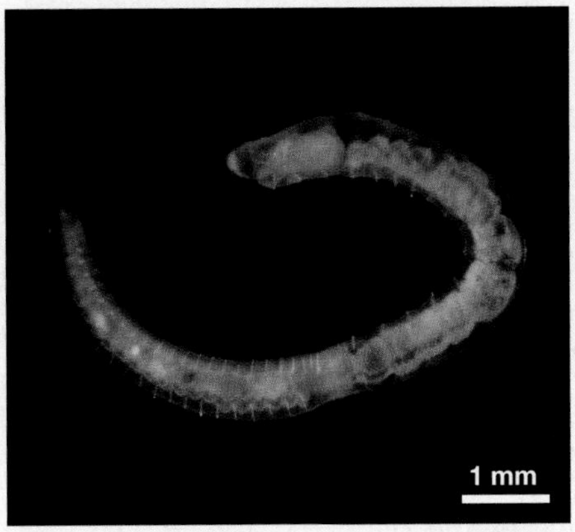

(c) Most leeches live in freshwater.

Helobdella stagnalis

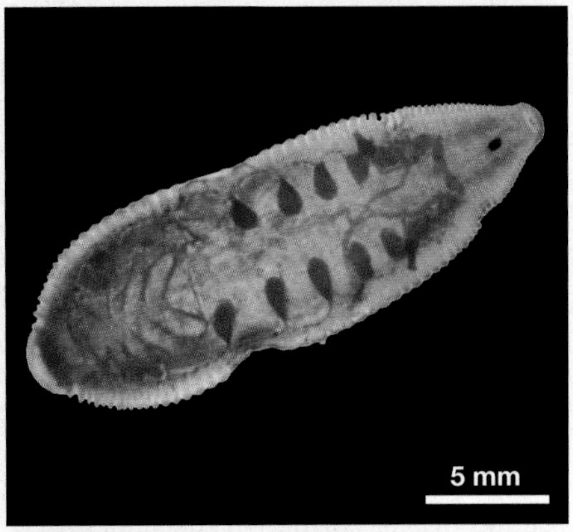

FIGURE 32.15 Annelids Are Segmented Worms

Mollusca (Mollusks)

The mollusks are far and away the most species-rich and morphologically diverse group in the Lophotrochozoa. They have a specialized body plan based on a muscular foot, a visceral mass, and a mantle that may or may not secrete a calcium carbonate shell. The coelom is much reduced or absent. Over 93,000 species have been described thus far. Although most mollusks live in marine environments, there are some terrestrial and freshwater forms.

Because mollusks are so diverse, let's consider each of the major subgroups or classes in the phylum separately and analyze the traits responsible for their diversification. The four most important lineages of mollusks are (1) **bivalves** (clams and mussels), (2) **gastropods** (slugs and snails), (3) **cephalopods** (squid and octopuses), and (4) **chitons.** The bivalves are suspension feeders; the other three groups of mollusks are herbivores or predators.

Mollusca ▶ Bivalvia (Clams, Mussels, Scallops, Oysters)

The bivalves are so named because they have two separate shells made of calcium carbonate secreted by the mantle. The shells are hinged, and they open and close with the aid of muscles attached to them (**Figure 32.16**). When the shell is closed, it protects the mantle, visceral mass, and foot. The bivalve shell is an adaptation that reduces predation.

Most bivalves live in the ocean, though some freshwater forms are known. Clams burrow into mud, sand, or other soft substrates and are sedentary as adults. Oysters and mussels are largely sessile as adults, but most of them live above the substrate, attached to rocks or other solid surfaces. Scallops are mobile and live on the surface of soft substrates. The smallest bivalves are freshwater clams that are less than 2 mm long; the largest bivalves are giant marine clams that may weigh more than 400 kg. All bivalves can sense gravity, touch, and certain chemicals, and scallops even have eyes.

Because most bivalves live on or under the ocean floor and because they are covered by a hard shell, their bodies are often buried in sediment after death. As a result, bivalves fossilize readily, and the Bivalvia lineage has the most extensive fossil record of any animal, plant, or fungal group. This large database has allowed biologists to conduct thorough studies of the evolutionary history of bivalves.

Bivalves are important commercially. Clams, mussels, scallops, and oysters are farmed or harvested from the wild in many parts of the world and used as food by humans. When sold commercially, they are referred to as "shellfish." Pearl oysters that are cultivated or collected from the wild are the most common source of natural pearls used in jewelry.

Feeding Bivalves are suspension feeders that take in any type of small animal or protist or detritus. Suspension feeding is based on a flow of water across gas-exchange structures called **gills**. The gills are part of the visceral mass and consist of a series of thin membranes where particles are trapped. A water current exists because siphons, which are tubes formed by edges of the mantle, extend from the shell and form a plumbing system. The siphons bring water into the gills and then back out of the body, powered by the beating of cilia on the gills (see Figure 31.12a). Bivalves are the only major group of mollusks that lack a radula.

Movement Clams burrow with the aid of their muscular foot, which functions as a hydrostatic skeleton. But they are otherwise sedentary. Scallops are able to swim by clapping their shells together and forcing water to jet out, pushing them backward. Scallop locomotion is similar to the swimming of cnidarian medusae, which move when muscular contractions force water out of their bell (see Chapter 31). Bivalves produce a swimming trochophore larva that is responsible for dispersing individuals to new locations.

Reproduction Only sexual reproduction occurs. Eggs and sperm are shed into the water, and fertilized eggs develop into trochophore larvae. Trochophores then metamorphose into a distinct type of larva called a **veliger**, which continues to feed and swim before settling to the substrate and metamorphosing into an adult form that secretes a shell.

Lima scabra
1 cm

FIGURE 32.16 Bivalves Have Two Shells

Mollusca ▶ Gastropoda (Snails, Slugs, Nudibranchs)

The gastropods ("belly-feet") are named for the large, muscular foot on their ventral side. Terrestrial snails can retract their foot and body into a shell when they are attacked or when their tissues begin to dry out. Land slugs and nudibranchs (pronounced *NEW-da-branks*) lack shells but often contain toxins or foul-tasting chemicals to protect them from being eaten. The bright colors of nudibranchs, or sea slugs, are thought to act as a warning to potential predators (**Figure 32.17**). About 70,000 species of gastropods are known.

Feeding Gastropods and other mollusks have a unique structure in their mouths called a radula. In many species the **radula** functions like a rasp to scrape away algae, plant cells, or other types of food. It is usually covered with teeth made of chitin, which vary in size and shape among species. Although most gastropods are herbivores or detritivores, specialized types of teeth allow some gastropods to act as predators. Species called drills, for example, use their radula to bore a hole in the shells of oysters or other mollusks and expose the visceral mass, which they then eat. Cone snails have highly modified, harpoon-like teeth mounted at the tip of an extensible proboscis and armed with poison. When a fish or worm passes by, the proboscis shoots out. The prey is speared by the tooth, subdued by the poison, and consumed.

Movement Waves of contractions down the length of the foot allow gastropods to move by gliding (see Figure 32.11c). The sea butterflies are gastropods that have a reduced or absent shell but a large, winglike foot that flaps and powers swimming movements.

Reproduction Females of some species can reproduce asexually by producing eggs parthenogenetically, but most reproduction is sexual. Unlike its counterpart in bivalves, sexual reproduction in gastropods begins with internal fertilization. But like bivalves, some marine gastropods produce a trochophore larva that may disperse up to several hundred kilometers from the parent. In most marine species and all terrestrial forms, larvae are not free living. Instead, offspring remain in an egg case while passing through several larval stages, then hatch as miniature versions of the adults.

Purple-spotted nudibranch
1 cm

FIGURE 32.17 Many Gastropods Lack Shells

Mollusca ▶ Polyplacophora (Chitons)

The Greek word roots that inspired the name *Polyplacophora* mean "many-plate-bearing." The name is apt because chitons (pronounced *KITE-uns*) have eight calcium carbonate plates along their dorsal side (**Figure 32.18**). The plates form a protective shell. The approximately 1000 species of chitons are marine. They are usually found on rocky surfaces in the intertidal zone, where rocks are periodically exposed to air at low tides.

Feeding Chitons have a radula and use it to scrape algae and other organic matter off rocks.

Movement Chitons move by gliding on their broad, muscular foot—as gastropods do.

Reproduction Sexes are separate, and fertilization is external. In some species, however, sperm that are shed into the water enter the female's mantle cavity and fertilize eggs inside the body. Depending on the species involved, eggs may be enclosed in a membrane and released or retained until hatching and early development are complete. Most species have trochophore and veliger larvae.

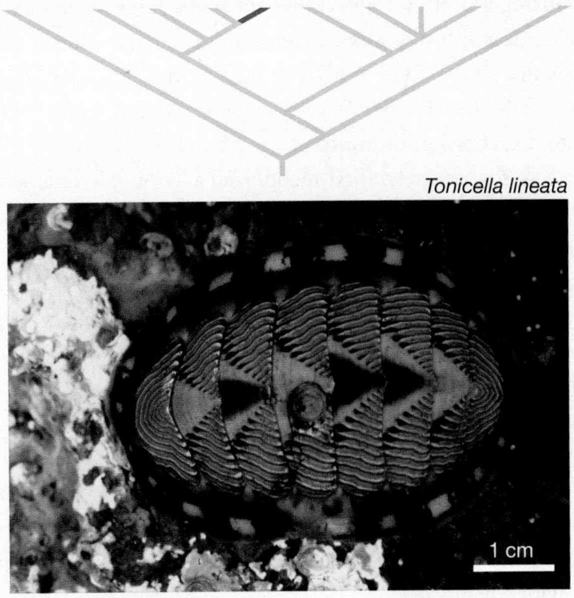
Tonicella lineata
1 cm

FIGURE 32.18 Chitons Have Eight Shell Plates

Mollusca ▶ Cephalopoda (Squid, Nautilus, Octopuses)

The cephalopods ("head-feet") have a well-developed head and a foot that is modified to form tentacles. **Tentacles** are long, thin, muscular extensions that aid in movement and prey capture (**Figure 32.19**). With the exception of the nautilus, they have either highly reduced shells or none at all. They also have large brains and image-forming eyes with sophisticated lenses.

Feeding Cephalopods are highly intelligent predators that hunt by sight and use their tentacles to capture prey—usually fish or crustaceans. They have a radula as well as a structure called a **beak**, which can exert powerful biting forces. Squid and octopuses also inject poisons into their prey to subdue them.

Movement Cephalopods can swim by undulating their bodies and tentacles. They may also move by jet propulsion. They draw water into their mantle cavity and then force it out through a siphon (see Figure 32.11d). Squid are built for speed and hunt small fish by chasing them down. Octopuses, in contrast, crawl along the substrate by using their long, armlike tentacles. They chase down crabs or other crustaceans or pry mussels or clams from the substrate, then use their beaks to crush the exoskeletons of their prey.

Reproduction Cephalopods have separate sexes, and some species have elaborate courtship rituals that involve color changes and interaction of tentacles. When a male is accepted by a female, he deposits sperm that are encased in a structure called a **spermatophore.** The spermatophore is taken up by the female, and fertilization is internal. Females lay eggs. When they hatch, juveniles develop directly into adults.

Octopus dofleini
10 cm

FIGURE 32.19 Cephalopods Have Highly Modified Bodies

32.5 Key Lineages: Ecdysozoans

The lineage Ecdysozoa is named for a distinctive trait shared by members of the group: They grow by **molting**—that is, shedding an exoskeleton or external covering. The Greek root *ecdysis*, which means "to slip out or escape," is appropriate because during a molt, an individual sheds its outer layer—called a **cuticle** if it is soft or an exoskeleton if it is hard—and slips out of it (**Figure 32.20**). The body expands, and a larger exoskeleton or cuticle then forms. As ecdysozoa grow and mature, they undergo a succession of molts.

The ecdysozoans were first recognized as monophyletic when investigators began using DNA sequence data to estimate the phylogeny of protostomes. There are about eight phyla in the lineage (see Table 31.1), including the Onychophora (*on-ee-KOFF-er-uh*) and the Tardigrada (**Figure 32.21**). They are not species-rich phyla, but both are closely related to arthropods. Onychophorans and tardigrades are similar to arthropods in having limbs and a stiff external covering. Unlike arthropods, their bodies are not segmented and their limbs not jointed. The onychophorans, or velvet worms, are small, caterpillar-like organisms that live in moist leaf litter and prey on small invertebrates. Onychophorans have sac-like appendages and segmented bodies with a hemocoel.

The tardigrades, or water bears, are microscopic animals that live in benthic habitats of marine or freshwater environments. Large numbers of water bears can also be found in the film of water that covers moss or other land plants in moist habitats. Tardigrades have a reduced coelom but a prominent hemocoel, and they walk on their clawed, sac-like legs. Most feed by sucking fluids from plants or animals; others are **detritivores** ("detritus-feeders").

Let's take a closer look at the two most diverse and abundant phyla of the Ecdysozoa: (1) **Nematoda** and (2) **Arthropoda.**

FIGURE 32.20 Ecdysozoans Grow by Molting
Once an ecdysozoan has left its old exoskeleton, hours or days pass before the new exoskeleton is hardened. During this time, the individual is highly susceptible to predation or injury.

(a) Onychophorans have lobe-like limbs.

(b) Tardigrades have lobe-like limbs with claws.

50 µm

FIGURE 32.21 Onychophorans and Tardigrades Have Limbs and Are Closely Related to Arthropods
(a) The Onychophora are called velvet worms because they have wormlike bodies and a velvety appearance.
(b) The Tardigrada are known as water bears because they live in moist habitats and, when viewed under the microscope, resemble bears.

Nematoda (Roundworms)

Species in the phylum Nematoda are commonly called nematodes or roundworms (**Figure 32.22**). Roundworms are unsegmented worms with a pseudocoelom, a tube-within-a-tube body plan, and no appendages. They have no muscles that can change the diameter of the body—their body musculature consists solely of longitudinal muscles that shorten or lengthen the body upon contracting or relaxing, respectively. Although some nematodes can grow to lengths of several meters, the vast majority of species are tiny—most are much less than 1 mm long. They lack specialized systems for exchanging gases and for circulating nutrients and wastes. Instead, gas exchange occurs across the body wall, and nutrients and wastes move by simple diffusion.

Nematodes have been studied intensively because several species are common parasites of humans. Pinworms, for example, infect about 40 million people in the United States alone, and *Onchocerca volvulus* causes an eye disease that infects around 20 million people in Africa and Latin America. Advanced infections by the roundworm *Wuchereria bancrofti* result in the blockage of lymphatic vessels, causing fluid accumulation and massive swelling—the condition known as *elephantiasis*.

Parasites actually represent a tiny fraction of the 25,000 nematode species that have been described to date, however. The vast majority of nematodes are free living. They are important ecologically, because they are found in virtually every habitat known.

Nematodes are also fabulously abundant. Biologists have found 90,000 roundworms in a single rotting apple and have estimated that rich farm soils contain up to 9 billion roundworms per acre. The simple nematode body plan has been extraordinarily successful.

Feeding Roundworms feed on a wide variety of materials, including bacteria, fungi, plant roots, small protists or animals, or detritus. In most cases, the structure of their mouthparts is specialized in a way that increases the efficiency of feeding on a particular type of organism or material.

Movement Roundworms move with the aid of their hydrostatic skeleton (see Figure 31.5). Most roundworms live in soil or inside a host, so when contractions of their longitudinal muscles cause them to wriggle, the movements are resisted by a stiff substrate. As a result, the worm pushes off the substrate, and the body moves.

Reproduction Sexes are separate in most species, and asexual reproduction is rare or unknown. Sexual reproduction begins with internal fertilization and culminates in egg laying and the direct development of offspring. Individuals go through a series of four molts over the course of their lifetime.

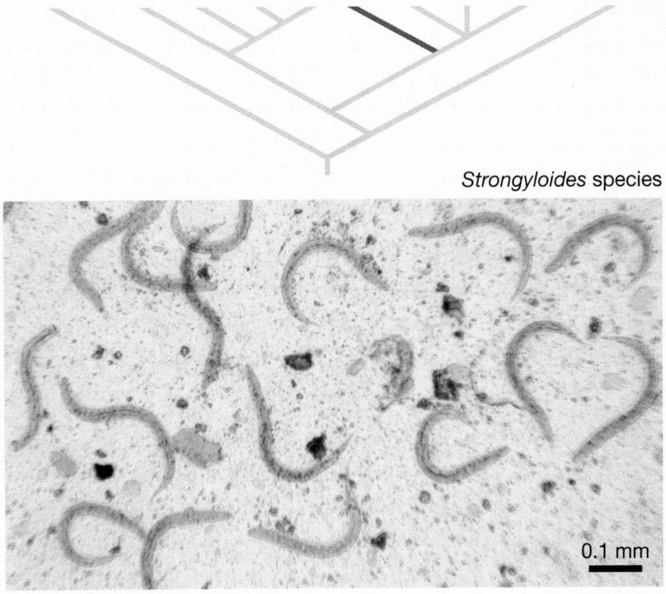
Strongyloides species
0.1 mm

FIGURE 32.22 Most Nematodes Are Free Living, but Some Are Parasites

Arthropoda (Arthropods)

In terms of duration in the fossil record, species diversity, and abundance of individuals, arthropods are easily the most successful lineage of eukaryotes. They appear in the fossil record over 520 million years ago and have long dominated the animals observed in both marine and terrestrial environments. Well over a million living species have been described, and biologists estimate that millions or perhaps even tens of millions of arthropod species have yet to be discovered.

Morphologically, arthropods are distinguished by segmented bodies and a sophisticated, jointed exoskeleton. They have a highly reduced coelom but possess an extensive body cavity called a hemocoel, enclosed by the exoskeleton. The body is organized into distinct **head** and **trunk** regions in all arthropods; in many species there is an additional grouping of segments into trunk regions. The trunk regions are usually called an *abdomen*, and a *thorax* or *tail*.

Metamorphosis is common in arthropods. Their larvae have segmented bodies but may lack a hardened exoskeleton. If metamorphosis occurs, larval and adult forms may grow by molting.

At least some segments along the arthropod body produce paired, jointed appendages. Arthropod appendages have an array of functions: sensing aspects of the environment, exchanging gases, feeding, or locomotion. Arthropod appendages provide the ability to sense environmental stimuli and make sophisticated movements in response. Most species also have sophisticated, image-forming compound eyes. A **compound eye** contains many light-sensing, columnar structures, each with its own lens. (Your eyes are **simple eyes**, meaning they have just one light-sensing structure.) Most arthropods also have a pair of antennae on the head. **Antennae** are long, tentacle-like appendages that are used to touch or smell.

Although the phylogeny of arthropods is still being worked out, most data sets agree that the phylum as a whole is monophyletic; that the myriapods (millipedes and centipedes), chelicerates (spiders), insects, and crustaceans represent four major subphyla within the phylum; and that crustaceans and insects are closely related. The phylum Arthropoda is so large and diverse that a detailed treatment would fill a book this size; space permits just a few notes about the four major lineages.

Arthropoda ▶ Myriapods (Millipedes, Centipedes)

The myriapods have relatively simple bodies, with a head region and a long trunk featuring a series of short segments, each bearing one or two pairs of legs (**Figure 32.23**). If eyes are present, they consist of a few to many simple structures clustered on the sides of the head. The 11,600 species that have been described to date inhabit terrestrial environments all over the world.

Feeding Millipedes and centipedes have mouthparts that can bite and chew. These organisms live in downed, rotting logs and other types of dead plant material that litters the ground in forests and grasslands. Millipedes are detritivores. Centipedes, in contrast, use a pair of poison-containing fangs just behind the mouth to hunt an array of insects. Large centipedes can inject enough poison to debilitate a human.

Movement Myriapods walk or run on their many legs; a few species burrow. Centipedes usually have fewer than 30 segments, with one pair of legs per segment. Some millipedes have over 190 trunk segments, each with two pairs of legs, for a total of over 750 legs.

Reproduction Sexes are separate, and fertilization is internal. Males deposit sperm in packets that are picked up by the female or transferred to her by the male. After females lay eggs in the environment, they hatch into juveniles that develop into adults via a series of molts.

Scolopendra species

FIGURE 32.23 Myriapods Have a Pair of Legs on Each Body Segment

Arthropoda ▶ Chelicerata (Spiders, Ticks, Mites, Horseshoe Crabs, Daddy Longlegs, Scorpions)

Most of the 70,000 species of chelicerates are terrestrial, although the horseshoe crabs and sea spiders are marine. The lineage is considered a class in most of the traditional classification schemes and has several subgroups or subclasses. The most prominent of these chelicerate lineages is the arachnids (spiders, scorpions, mites, and ticks).

The chelicerate body consists of anterior and posterior regions. The anterior region lacks antennae for sensing touch or odor but usually contains eyes. The group is named for appendages called **chelicerae** found near the mouth. Depending on the species, the chelicerae are used in feeding, defense, copulation, movement, or sensory reception.

Feeding Spiders, scorpions, and daddy longlegs capture and sting insects or other prey (**Figure 32.24a**). Although some of these species are active hunters, most spiders are sit-and-wait predators. They create sticky webs to capture prey, which fly or walk into the web and are subsequently trapped. The spider senses the vibrations of the struggling prey, pounces on it, and administers a toxic bite.

Mites and ticks are ectoparasitic and use their piercing mouthparts to feed on host animals (**Figure 32.24b**). Horseshoe crabs eat a variety of protostomes as well as detritus. Most scorpions feed on insects; the largest scorpion species occasionally eat snakes and lizards.

Movement Like other arthropods, chelicerates move with the aid of muscles attached to an exoskeleton. They walk or crawl on their four pairs of jointed walking legs; some species are also capable of jumping. Horseshoe crabs and some other marine forms can swim slowly. Newly hatched spiders spin long, silken threads that serve as balloons and can carry offspring on the wind more than 400 kilometers from the point of hatching.

Reproduction Sexual reproduction is the rule in chelicerates, and fertilization is internal in most groups. Courtship displays are extensive in many arthropod groups and may include both visual displays and the release of chemical odorants. In spiders, males use organs that are located on their legs to transfer sperm to females. These organs fit into the female reproductive tract in a "lock-and-key" fashion, and differences in the size and shape of male genitalia are often the only way to tell closely related species of spider apart.

Development is direct, meaning that metamorphosis does not occur. In spiders, males may present a dead insect as a gift that the female eats as they mate; in some species, the male himself is eaten as sperm is being transferred. In scorpions, females retain fertilized eggs. After the eggs hatch, the young climb on the mother's back. They remain there until they are old enough to hunt for themselves.

(a) Spider *Dolomedes fimbriatus*

(b) Mite *Dermatophagoides* species

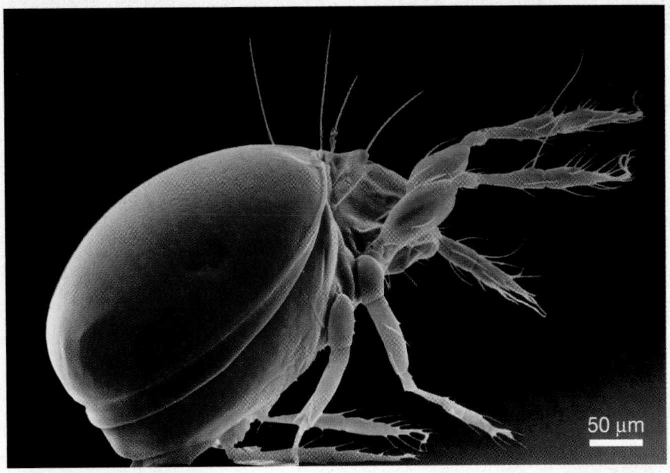

FIGURE 32.24 Chelicerates Have Head and Thorax Regions

Arthropoda ▶ Insecta (Insects)

About 925,000 species of insects have been named thus far, but it is certain that many more exist. In terms of species diversity and numbers of individuals, insects dominate terrestrial environments. In addition, the larvae of some species are common in freshwater streams, ponds, and lakes. Insects are distinguished by having three body regions—(1) the **head**, (2) **thorax**, and (3) **abdomen** (**Figure 32.25a**). Three pairs of walking legs are located on the ventral surface of the thorax. Most species have one or two pairs of wings, mounted on the dorsal side of the thorax. Typically the head contains three sets of mouthparts, a pair of antennae that are used to touch and smell, and a pair of compound eyes.

Feeding Because most insect species have four sets of mouthparts (labrum, mandible, maxilla, and labium) that vary greatly in structure among species (see Figure 32.10), insects are able to feed in every conceivable manner and on almost every type of food source available on land. Larvae have wormlike bodies; most are deposit feeders, though some are leaf eaters. Adults are herbivores, predators, or parasites. The diversification of insects was closely correlated with the diversification of land plants, primarily because so many insects make their living by feeding on plant tissues or fluids. Insect predators usually eat other insects; insect parasites usually victimize other arthropods or mammals.

Movement Insects use their legs to walk, run, or swim, or they use their wings to fly (**Figure 32.25b**). When insects walk or run, the sequence of movements usually results in three of their six legs maintaining contact with the ground at all times.

Reproduction Sexes are separate. Mating usually takes place through direct copulation, with the male inserting a sperm-transfer organ into the female. Most females lay eggs, but in a few species eggs are retained until hatching. Many species are also capable of reproducing asexually, through the production of unfertilized eggs via mitosis. In the vast majority of species, either incomplete or complete metamorphosis occurs, with complete metamophosis being most common.

(a) Swallowtail butterfly *Pterourus* species

(b) Harlequin beetle *Acrocinus longimanus*

Head Thorax Abdomen

FIGURE 32.25 Insects Have Distinctive Body Plans and Move Efficiently

Arthropoda ▶ Crustaceans (Shrimp, Lobster, Crabs, Barnacles, Isopods, Copepods)

The 67,000 species of crustaceans that have been identified to date live primarily in marine and freshwater environments. A few species of crab and some isopods are terrestrial, however. (Terrestrial isopods are known as sowbugs, pillbugs, or roly-polies). Crustaceans are common consumers in planktonic habitats and are also important grazers and predators in shallow-water benthic environments.

The segmented body of a crustacean is divided into two distinct regions: (1) the cephalothorax, which combines the head and thorax, and (2) the abdomen. Many crustaceans have a **carapace**—a platelike section of their exoskeleton that covers and protects the cephalothorax (**Figure 32.26a**). They are the only type of arthropod with two pairs of antennae, and they have sophisticated, compound eyes—usually mounted on stalks. Crustaceans and trilobites were the first arthropods to appear in the fossil record.

Feeding Crustaceans use every type of feeding strategy known. Barnacles (**Figure 32.26b**) and many shrimp are suspension feeders that use feathery structures located on head or body appendages to capture passing prey. Crabs and lobsters are active hunters, herbivores, and scavengers. Typically they have a pair of mouthparts called **mandibles** that bite or chew. Individuals capture and hold their food source with claws or other types of feeding appendages near their mouth, then use their mandibles to shred the food into small bits that can be ingested. As herbivores or detritivores, many species of crustaceans depend on algae for food.

Movement Crustacean limbs are highly diverse, and it is common for species to have many pairs of limbs and more than one type of limb. Limb structures in crustaceans include paddle-shaped

(Continued on next page)

Mollusca ▶ Crustaceans (Shrimp, Lobster, Crabs, Barnacles, Isopods, Copepods) *(continued)*

forms used in swimming, feathery structures used in capturing food that is suspended in water, and jointed appendages that make sophisticated walking or running movements possible. Barnacles are one of the few types of sessile crustaceans. Adult barnacles cement their heads to a rock or other hard substrate, secrete a protective shell of calcium carbonate, and use their legs to create water currents that bring food to the mouth.

Reproduction Each individual is male or female, and sexual reproduction is the norm. Fertilization is usually internal, and eggs are usually retained by the female until they hatch. Most crustaceans pass through several distinct larval stages, and many species include a larval stage called a **nauplius**, which is usually planktonic. A nauplius has a single eye and appendages that develop into the two pairs of antennae and the mouthparts of the adult.

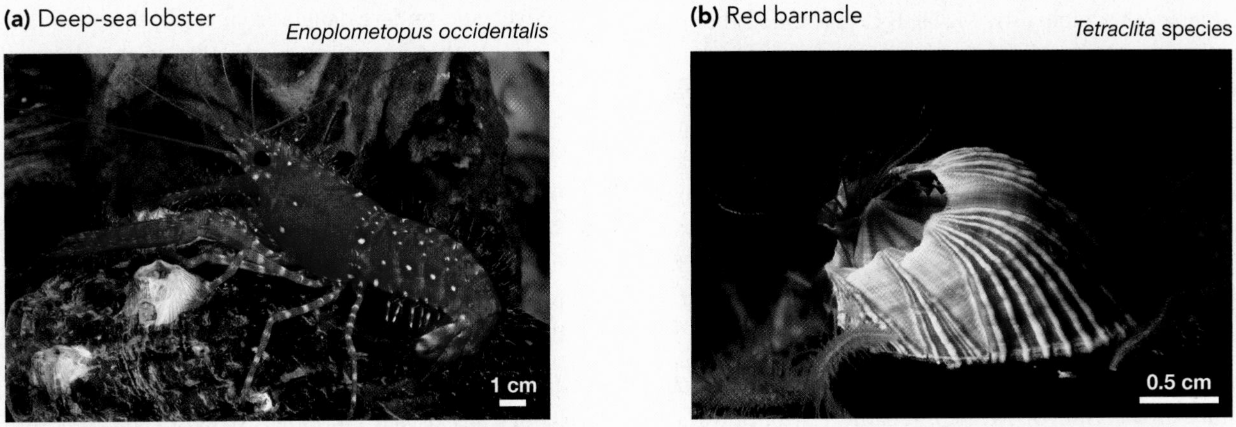
(a) Deep-sea lobster
Enoplometopus occidentalis
1 cm
(b) Red barnacle
Tetraclita species
0.5 cm

FIGURE 32.26 Crustaceans Have a Carapace That Protects Their Head

ESSAY The Role of Natural History in Biological Science

Science begins with questions about the natural world. Physicists ask questions about matter and forces; chemists ask questions about molecules and reactions; geologists ask questions about landforms. Biologists specialize in questions about organisms. Their job is to figure out what organisms are doing, how they do it, and why.

When biologists ask questions that begin with who, what, when, or where, the work that they are doing is descriptive in nature. Biologists spend a lot of time simply describing what exists in nature, carefully and accurately. Work of this type is called **natural history**. Classically, the core task of a naturalist is to describe organismal diversity—to understand who's who in nature. Naturalists are people who know the names and traits of birds, insects, plants, algae, and other organisms. Much of the information you read in this chapter was assembled by people who would describe themselves as naturalists. But when biologists focus on describing the structure and function of genes, organelles, or cell types, they are essentially doing natural history as well.

In the eighteenth and early nineteenth centuries, virtually all biology was descriptive in nature. During that time, the focus of biological science was simply to observe and describe the diversity of cells and organisms that exist. Today, biology is an experimentally driven science that focuses on questions about how genes, cells, and organisms work and why.

Dragonflies can fly forward or backward at speeds of up to 60 mph.

Even though the nature of biological science has changed dramatically over the years, many biologists would argue that natural history still plays a large role in advancing the science. For example, studying natural history remains an effective way to inspire new types of questions about genes, cells, and organisms. People observe things, wonder what's going on, and start doing experiments. In addition, most biologists got started in the field simply because they've loved organisms ever since they were children.

(Continued on next page)

(Essay continued)

To underscore the importance of natural history in contemporary biology, consider an experience that Tarif Awad had as a graduate student. During a break from lab work at the Marine Biological Laboratories in Woods Hole, Massachusetts, Awad wandered out onto nearby mudflats during low tide. He started digging and found interesting-looking polychaete worms. As he was examining them, he was startled by a woman's voice behind him, asking him what he was doing. When he showed the woman the worms, she responded by clapping her hands and exclaiming, "How wonderful! Someone who loves organisms!" The speaker was Barbara McClintock, Nobel laureate and one of the great geneticists of the twentieth century.

Like many other biologists, McClintock has found the study of natural history to be a lifelong source of emotional and intellectual inspiration. The intellectual inspiration springs from a simple premise: Observing organisms not only arouses new questions but also provides a context for understanding why organisms are the way they are. Natural history is the wellspring of biology and an important science in its own right.

The purely descriptive parts of biological science can be fun. Consider the following observations about some protostome animals:

- Every August, spiny lobsters migrate from shallow-water habitats in the Gulf of Mexico and the Caribbean Sea to deeper ocean channels. They do this by forming long single-file lines and marching for several days.
- Dragonflies can fly forward or backward at speeds of up to 60 mph.
- Spider silk is stronger than steel and has twice the elasticity of nylon. Many spiders eat their old or damaged webs and reuse the silk in constructing a new one.
- Rhinoceros beetles are able to carry 100 times their own body weight over short distances and 30 times their body weight for long periods of time.
- In rocks about 425 million years old, biologists find fossils of giant water scorpions that were over 2 meters long.
- Bolas spiders hunt at night, when *Spodoptera* moths are active. The spiders dangle from a silk thread and release a molecule that is similar in structure to the chemical signal that female moths use to attract male moths as mates. If a male moth mistakenly approaches the spider, the spider shoots it with a thread tipped with a sticky liquid, draws it in, and eats it.
- Snapping shrimp are sit-and-wait predators. When small fish or other prey pass their burrow, they snap a clawlike appendage. The movement produces a shock wave that stuns the prey. The shrimp then pounce on the food and devour it.

Organisms are fascinating. More than one biologist has recharged his or her emotional batteries, or come up with a novel question to answer, simply by going outside and looking at the organisms all around.

CHAPTER REVIEW

Summary of Key Concepts

Molecular phylogenies support the hypothesis that protostomes are a monophyletic group divided into two major subgroups: the Lophotrochozoa and the Ecdysozoa.

The protostomes comprise about 20 phyla and were originally identified because their embryos undergo early development in the same way. Only recently did biologists recognize the existence of the lophotrochozoans and ecdysozoans, however. Several phyla in the Lophotrochozoa have characteristic feeding structures called lophophores, and many have a distinctive type of larvae called trochophores. Species in the Ecdysozoa grow by molting—meaning that they shed their old external covering and grow a new, larger one.

Web Tutorial 32.1 Protostome Diversity

Although the members of many protostome phyla have limbless, wormlike bodies and live in marine sediments, the most diverse and species-rich lineages—Mollusca and Arthropoda—have body plans with a series of distinct, complex features.

All protostomes are bilaterally symmetric and triploblastic, but the nature of the body cavity is variable. The wormlike phyla of protostomes have well-developed coeloms that provide a hydrostatic skeleton used in locomotion. Flatworms lack a coelom, however, and both nematodes and rotifers evolved a pseudocoelom independently. The coelom is also much reduced or absent in the Mollusca and the Arthropoda lineages, although a body cavity is present. The mollusk's body is made up of a muscular foot, a visceral mass of organs, and a protective mantle. The arthropod body is segmented, divided into specialized regions such as the head, thorax, and abdomen of insects, and protected by an exoskeleton made of chitin.

- **Key events triggered the diversification of protostomes, including several lineages making the water-to-land transition, a diversification in appendages and mouthparts, and the evolution of metamorphosis in both marine and terrestrial forms.**

The transition to living on land occurred several times independently during the evolution of protostomes. It was facilitated by watertight shells and exoskeletons and the ability to move to moist locations. Diversification in appendages and mouthparts gave protostomes the ability to move and find food in innovative ways. The evolution of metamorphosis was significant because it allowed juveniles in sessile species to disperse to new habitats and because it made it possible for juveniles and adults to find food in different ways.

Questions

Content Review

1. Why is it logical to observe that mollusks have a drastically reduced coelom or no coelom at all?
   a. They evolved from flatworms, which also lack a coelom.
   b. They are the most advanced of all the protostomes.
   c. Their bodies are encased by a mantle, which may or may not secrete a shell.
   d. They move with the aid of a muscular foot, which functions as a hydrostatic skeleton.
2. What major feature or features distinguish most of the wormlike phyla of protostomes from each other?
   a. Their mouthparts and feeding methods vary.
   b. Their modes of locomotion differ—they burrow, swim, crawl, or walk.
   c. Metamorphosis may be lacking, incomplete, or complete.
   d. They have a well-developed coelom and a tube-within-a-tube body plan.
3. What is the function of the arthropod exoskeleton?
   a. Because hard parts fossilize more readily than do soft tissues, the presence of an exoskeleton has given arthropods a good fossil record.
   b. It has no well-established function. (Trilobites had an exoskeleton, and they went extinct.)
   c. It provides protection and functions in locomotion.
   d. It makes growth by molting possible.
4. Why is it logical to observe that Platyhelminthes have flattened bodies?
   a. They have simple bodies and evolved early in the diversification of protostomes.
   b. They lack a coelom, so their body cannot form a rounded tube-within-a-tube design.
   c. A flat body gives the animals high surface area and low volume—necessary because they lack gas-exchange organs and structures that circulate blood.
   d. They are sit-and-wait predators that hide from passing prey by flattening themselves against the substrate.
5. In number of species, number of individuals, and duration in the fossil record, which of the following phyla of wormlike animals has been the most successful? What feature is hypothesized to be responsible for this success?
   a. Annelida—a segmented body plan
   b. Platyhelminthes—bilateral symmetry but acoelomate condition
   c. Phoronida—a lophophore used as a feeding apparatus
   d. Echiura—a grooved proboscis used in suspension feeding
6. Which protostome phylum is distinguished by having body segments organized into distinct regions?
   a. Mollusca
   b. Arthropoda
   c. Annelida
   d. Nematoda

Conceptual Review

1. Describe the traits that distinguish the Lophotrochozoa and Ecdysozoa. Describe the traits that are common to both groups.
2. Define a segmented body plan. Did segmentation evolve once or multiple times during the evolution of protostomes? Provide evidence to justify your answer.
3. The nature of the coelom changed drastically during the evolution of protostomes. Describe the structure and function of the coelom in flatworms, nematodes, rotifers, the wormlike phyla of protostomes other than flatworms, mollusks, and arthropods.
4. Compare and contrast the role of metamorphosis in the life cycle of a sedentary, marine protostome such as a clam and a terrestrial, mobile protostome such as a butterfly.
5. Explain why the evolution of the exoskeleton and the jointed limb was such an important event in the evolution of arthropods.
6. Explain how squids move by jet propulsion.

Group Discussion Problems

1. Recall that the phylum Platyhelminthes includes three major groups: the free-living turbellarians, the parasitic cestodes, and the parasitic trematodes. Because the parasitic forms are so simple morphologically, researchers suggest that they are derived from more complex, free-living forms. Draw a phylogeny of Platyhelminthes that would support the hypotheses that ancestral flatworms were morphologically complex, free-living organisms and that parasitism is a derived condition.
2. Brachiopoda is a phylum within the Lophotrochozoa. Even though they are not closely related to mollusks, brachiopods look and act like bivalve mollusks (clams or mussels). Specifically, brachiopods suspension feed, live inside two calcium carbonate shells that hinge together in some species, and attach to rocks or other hard surfaces on the ocean floor. How is it possible for brachiopods and bivalves to be so similar if they did not share a recent common ancestor?

3. The Mollusca includes a group of about 370 wormlike species called the Aplacophora. Although aplacophorans do not have a shell, their cuticle secretes calcium carbonate scales or spines. They lack a well-developed foot, and some species have a simple radula.
   - Predict where Aplacophora are found in the phylogenetic tree of Mollusca relative to bivalves, gastropods, chitons, and cephalopods. Explain your reasoning.
   - Predict where Aplacophora live and how they move. Explain your reasoning.

4. Consider the following: (a) The first arthropods to appear in the fossil record are marine crustaceans and trilobites, (b) terrestrial insects appeared relatively late in arthropod evolution, (c) the crustacean group called isopods includes both aquatic and terrestrial forms, (d) myriapods are strictly terrestrial, and (e) analyses of DNA sequence data suggest that crustaceans and insects are closely related. Based on these observations, do you agree or disagree with the hypothesis that arthropods made the water-to-land transition more than once? Explain your reasoning.

Answers to Multiple-Choice Questions 1. d; 2. a; 3. c; 4. c; 5. a; 6. b

www.prenhall.com/freeman is your resource for the following: Web Tutorials; Online Quizzes and other Online Study Guide materials; Answers to Conceptual Review Questions; Solutions to Group Discussion Problems; Answers to Figure Caption Questions and Exercises; and Additional Readings and Research.

Deuterostome Animals

33

The flukes (tail lobes) of a humpback whale emerge from the water briefly before the animal dives. The presence of a tail is a diagnostic trait of the vertebrates—one of the most prominent lineages of deuterostomes.

KEY CONCEPTS

- The most prominent deuterostome lineages are the echinoderms and the vertebrate groups called ray-finned fishes and tetrapods.
- Echinoderms and vertebrates have unique body plans. Echinoderms are radially symmetric as adults and have a water vascular system. All vertebrates have a skull, and most have an extensive endoskeleton made of cartilage or bone.
- The diversification of echinoderms was triggered by the evolution of appendages called podia; the diversification of vertebrates was driven by the evolution of the jaw and limbs.
- Humans are a tiny twig on the tree of life. Chimpanzees and humans diverged from a common ancestor that lived in Africa 6–7 million years ago. Since then, at least 14 humanlike species have existed.

The **deuterostomes** include the largest-bodied and most morphologically complex of all animals. They range from the sea stars that cling to dock pilings, to the fish that dart in and out of coral reefs, to the wildebeests that migrate across the Serengeti Plains of East Africa.

Most biologists recognize just three phyla of deuterostomes: the **Echinodermata**, the **Hemichordata**, and the **Chordata**. The echinoderms include the sea stars and sea urchins. The hemichordates, or "acorn worms," are probably unfamiliar—they burrow in marine sands or muds and make their living by deposit feeding or suspension feeding. The chordates include the vertebrates. The vertebrates, in turn, include the sharks, bony fishes, amphibians, reptiles (including birds), and mammals. **Vertebrates** are animals that have a skull. The vast majority of vertebrates also have a backbone or spinal column. Animals that are not vertebrates are collectively known as **invertebrates**.

The deuterostomes were initially recognized because they all undergo early embryonic development in a similar way. When a humpback whale, sea urchin, or human is just beginning to grow, the gut starts developing from posterior to anterior—with the anus forming first and the mouth second. Although the events that take place early in embryonic development are diagnostic features that unite the deuterstomes, their adult body plans and their feeding methods, modes of locomotion, and means for reproduction are highly diverse. Let's explore who the deuterostome animals are and how they diversified, starting with the question of why so many biologists are drawn to studying them.

33.1 Why Do Biologists Study Deuterostome Animals?

If you were to look at data on the number of biologists who study various lineages on the tree of life, you would probably conclude that a disproportionately large number of people study deuterostomes—particularly vertebrates. The charge of disproportionality is valid because, in terms of numbers of species and numbers of individuals, insects and other protostomes are much more successful than deuterostomes. And regarding their ecological impact—meaning their influence on other species and the physical environment—the bacteria, archaea, protists, fungi, and land plants could be considered far more important than the deuterostomes. And yet, thousands of biologists are drawn to the study of vertebrates and other deuterostomes. Why?

The simplest answer to this question is that, because humans are deuterostomes, other species in our lineage interest us. People identify with vertebrates because they are large and mobile. We interact with them extensively as pets or domesticated livestock. We understand their world more intuitively than we do the world of a bacterium or fungus.

In addition to the natural affinity we have for deuterostomes, they are worthy of study because they play important roles in ecological interactions and human economics:

- In marine environments, deuterostomes are key consumers. Echinoderms and ray-finned fishes are important herbivores, and ray-finned fishes, sharks, and mammals important predators. (Ray-finned fishes are distinct from the lobe-finned fishes and coelocanths introduced later in the chapter.) On land, almost all of the large-bodied herbivores and predators are reptiles and mammals. As **Figure 33.1** shows, the upper trophic levels of food chains in most ecosystems are dominated by deuterostomes. As a result, understanding deuterostomes is critical to understanding how energy and nutrients flow through both marine and terrestrial ecosystems.
- Humans depend on vertebrates for food and, in preindustrial economies, for power. Ray-finned fishes and domesticated livestock are key sources of protein in many cultures. For many thousands of years, agriculture was based on the power generated by oxen, horses, water buffalo, or mules. Today, preindustrial societies continue to rely on horsepower or oxpower for transportation and agricultural work. In the industrialized world, millions of people birdwatch, plan vacations around seeing large mammals in national parks, or keep vertebrates as pets. All over the world, large numbers of people make their living by caring for vertebrates and studying them.

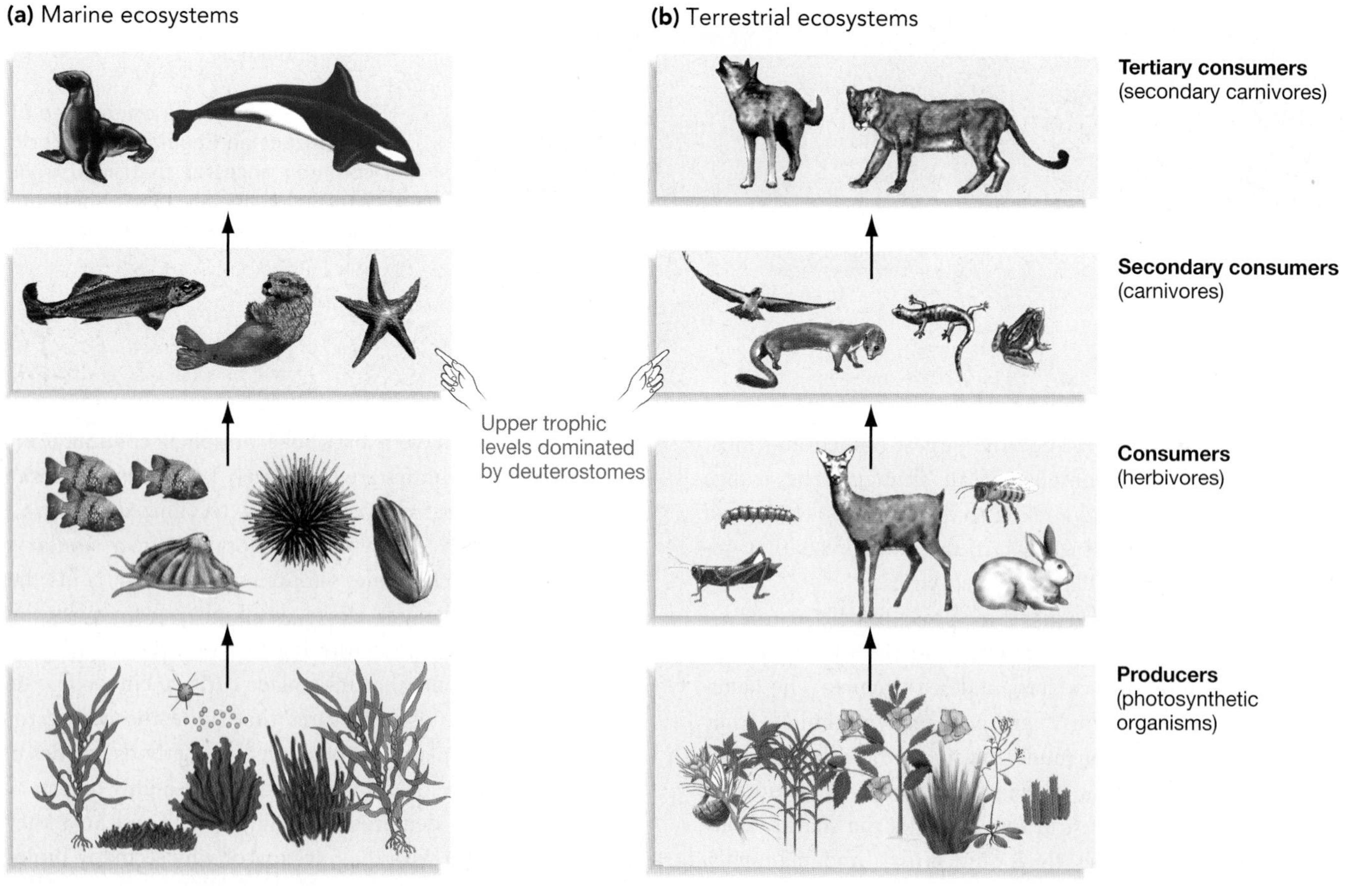
(a) Marine ecosystems
(b) Terrestrial ecosystems
Tertiary consumers
(secondary carnivores)
Secondary consumers
(carnivores)
Consumers
(herbivores)
Producers
(photosynthetic
organisms)
Upper trophic
levels dominated
by deuterostomes

FIGURE 33.1 Deuterostomes Are Important Herbivores and Predators
(a) In marine habitats, echinoderms and ray-finned fishes, both of which are deuterostomes, are the most important large herbivores and predators. **(b)** On land, deuterostomes such as amphibians, reptiles, and mammals play the same role.

(a) Echinoderm larvae are bilaterally symmetric.

(b) Adult echinoderms are radially symmetric.

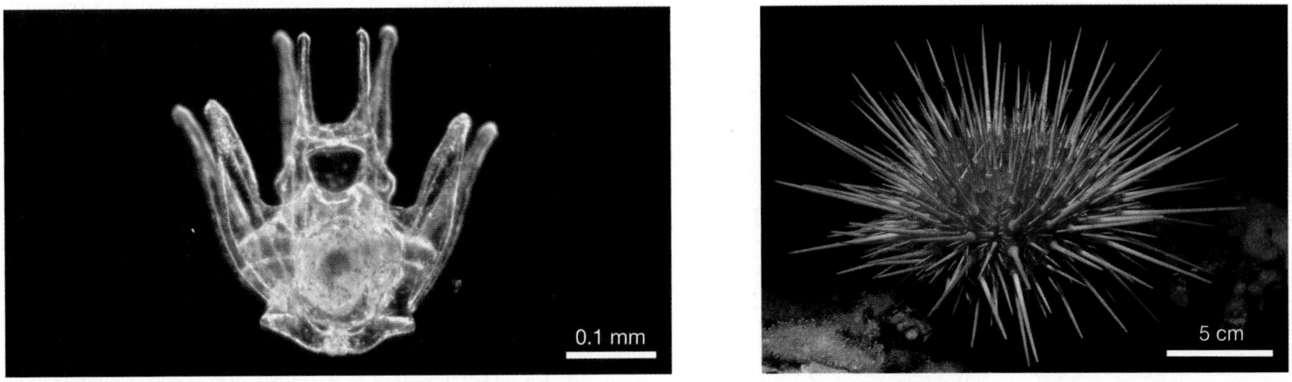
0.1 mm
5 cm

FIGURE 33.2 Body Symmetry Differs in Larval and Adult Echinoderms
(a) The larvae and **(b)** adult forms of the sea urchin *Strongylocentrotus franciscanus*. Bilaterally symmetric larvae undergo metamorphosis and emerge as radially symmetric adults.

33.2 How Do Biologists Study Deuterostomes?

To understand the events that produced the diversity of deuterostomes we see today, biologists study the comparative morphology of species from each of the three deuterostome phyla, analyze the fossil record of the group, and compare DNA sequences and other traits to infer their phylogeny. There are three central issues: (1) understanding the diversity of body plans observed in echinoderms, hemichordates, and the chordates; (2) exploring how vertebrates evolved from invertebrates; and (3) grasping how the vertebrates made the transition from living in aquatic environments to living on land.

Analyzing Morphological Traits

All deuterostomes are triploblastic and have a coelom (see Chapter 31). Patterns of early embryonic development are also similar among all members of the lineage. But in terms of the basic components of body plans, one of the most remarkable events in animal evolution occurred during the evolution of echinoderms: There was a reversion to a type of radial symmetry. Adult echinoderms have bodies with five-sided radial symmetry, even though both their larvae and their ancestors are bilaterally symmetric (**Figure 33.2**).

Recall from Chapter 31 that radially symmetric animals do not have well-developed head and posterior regions. As a result, they tend to interact with the environment in all directions at once instead of facing the environment in one direction. If adult echinoderms are capable of movement, they tend to move equally well in all directions instead of only headfirst.

The evolution of radial symmetry in echinoderms was a remarkable event in animal evolution. If echinoderms do not have a head and tail, how is their body organized?

The Water Vascular System of Echinoderms The unique body plan of echinoderms is based on a unique morphological feature. As **Figure 33.3a**, shows, the echinoderm body contains a series of continuous fluid-filled tubes and chambers called the **water vascular system.** One of the tubes is open to the exterior where it meets the body wall, so seawater can flow into and out

(a) Echinoderms have a water vascular system.

(b) Podia project from the underside of the body.

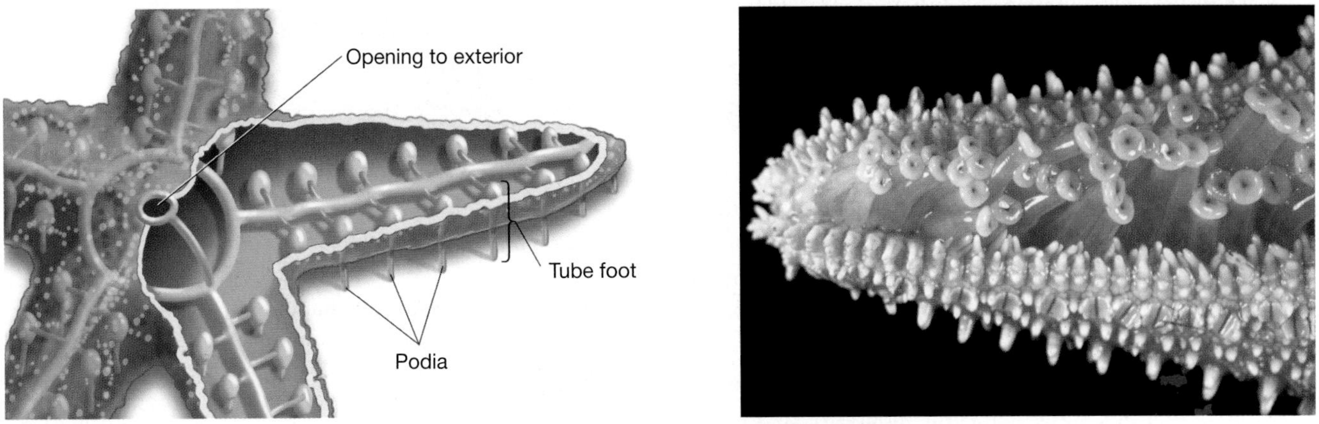
Opening to exterior
Tube foot
Podia

FIGURE 33.3 Echinoderms Have a Water Vascular System
(a) The water vascular system is a series of tubes and reservoirs that extends throughout the body, forming a sophisticated hydrostatic skeleton. **(b)** Podia aid in movement because they extend from the body and can extend or retract.

(a) Live sand dollar (b) Endoskeleton

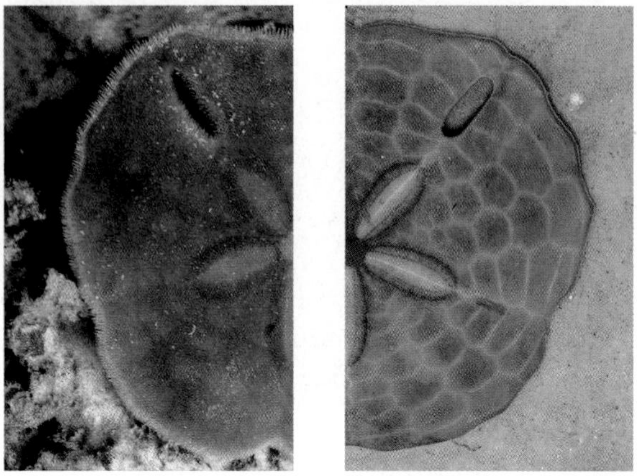

FIGURE 33.4 Echinoderms Have an Endoskeleton
The endoskeleton of an echinoderm is located just under the skin.

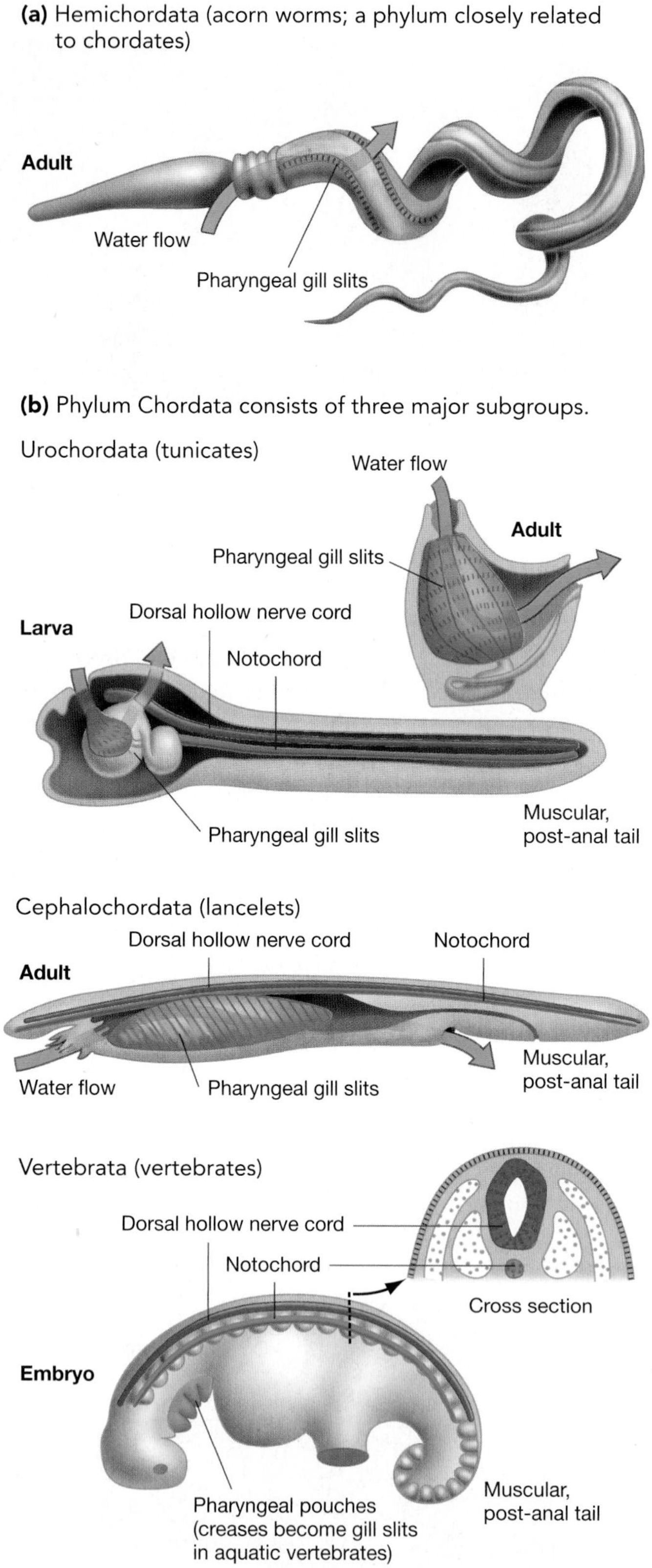
(a) Hemichordata (acorn worms; a phylum closely related to chordates)
Adult
Water flow
Pharyngeal gill slits
(b) Phylum Chordata consists of three major subgroups.
Urochordata (tunicates)
Water flow
Adult
Pharyngeal gill slits
Larva
Dorsal hollow nerve cord
Notochord
Pharyngeal gill slits
Muscular, post-anal tail
Cephalochordata (lancelets)
Dorsal hollow nerve cord
Notochord
Adult
Water flow
Pharyngeal gill slits
Muscular, post-anal tail
Vertebrata (vertebrates)
Dorsal hollow nerve cord
Notochord
Cross section
Embryo
Pharyngeal pouches (creases become gill slits in aquatic vertebrates)
Muscular, post-anal tail

FIGURE 33.5 Four Traits Distinguish the Chordates
(a) Hemichordates, although closely related to echinoderms, have only one of the four traits that distinguish chordates: pharyngeal gill slits. **(b)** In chordates, either larvae or adults or both have notochords and muscular tails in addition to pharyngeal gill slits and dorsal hollow nerve cords. The three major chordate lineages are tunicates, lancelets, and vertebrates.

of the system. Inside, fluids move via the beating of cilia that line the interior of the tubes and chambers. In effect, the water vascular system forms a sophisticated hydrostatic skeleton.

Figure 33.3a highlights a particularly important part of the system called tube feet. **Tube feet** are elongated, fluid-filled structures. **Podia** ("feet") are sections of the tube feet that project outside the body (**Figure 33.3b**) and make contact with the substrate. As podia extend and contract in a coordinated way along the base of an echinoderm, they alternately grab and release from the substrate. As a result, the individual moves.

The other noteworthy feature of the echinoderm body is its **endoskeleton**, which is a hard, supportive structure inside the body (**Figure 33.4**). As an individual is developing, cells secrete plates of calcium carbonate inside the skin. Depending on the species involved, the plates may remain independent and result in a flexible structure or fuse into a rigid case. This type of endoskeleton is a diagnostic feature of echinoderms, along with radial symmetry and the water vascular system.

WEB TUTORIAL 33.1 Deuterostome Diversity

The Origin of Chordates The phylum Chordata is defined by the presence of four morphological features: (1) openings into the throat called **pharyngeal gill slits**; (2) a stiff but flexible rod called a **notochord**, which runs the length of the body; (3) a bundle of nerve cells that runs the length of the body and forms a **dorsal hollow nerve cord**; and (4) a muscular **tail**, which extends past the anus.

The hemichordates are not members of the phylum Chordata, because they have just one of these four features. Hemichordates lack a notochord, a dorsal hollow nerve cord, and a tail, although they do have pharyngeal gill slits that function in feeding and gas exchange (**Figure 33.5a**). Hemichordates are sessile suspension feeders and live in the ocean. As the arrows in Figure 33.5a indicate, water enters the mouths of these animals, flows through structures where oxygen and food particles are extracted, and exits through the pharyngeal gill slits.

In contrast, pharyngeal gill slits, notochords, dorsal hollow nerve cords, and tails are found in each of the three major lineages of chordates (**Figure 33.5b**). In the traditional classification schemes introduced in Chapter 1, these groups are considered subphyla. The three groups are (1) urochordates, (2) cephalochordates, and (3) vertebrates. **Urochordates** are also called tunicates or sea squirts; they are small suspension feeders that as adults live attached to hard substrates in the ocean. **Cephalochordates** are also called lancelets or amphioxus; they are small, mobile suspension feeders that look a little like fish. Vertebrates include the sharks, bony fishes, reptiles, and mammals.

Adult urochordates are sessile, ocean-dwelling suspension feeders. As Figure 33.5b shows, pharyngeal gill slits are present in both larvae and adults and function in both feeding and gas exchange, much as they do in hemichordates. The notochord, dorsal hollow nerve cord, and tail are present only in the larvae, however. Because the notochord stiffens the tail, muscular contractions on either side of a larva's tail wag it back and forth and result in swimming movements. As larvae swim or float in the upper water layers of the ocean, they drift to new habitats where food might be more abundant.

Adult cephalochordates live in ocean-bottom habitats, where they burrow in sand and suspension feed with the aid of their pharyngeal gill slits, much like urochordate larvae. The cephalochordates also have a notochord that stiffens their bodies, so that muscle contractions on either side result in fishlike movement when they swim during dispersal or mating.

In vertebrates, the dorsal hollow nerve cord is elaborated into the familiar spinal cord. Structures called *pharyngeal pouches* are present in all vertebrate embryos. In aquatic species, the creases between pouches open into gill slits and develop into part of the main gas-exchange organ—the **gills**. In terrestrial species, however, gill slits do not develop after the pharyngeal pouches form. A notochord also appears in all vertebrate embryos. Instead of functioning in body support and movement, however, it helps organize the body plan. Recall from Chapter 22 that cells in the notochord secrete proteins that help induce the formation of somites. **Somites** are segmented blocks of tissue that form along the length of the body. Although the notochord itself disappears, cells in the somites later differentiate into the vertebrae, ribs, and skeletal muscles of the back, body wall, and limbs. In this way, the notochord is instrumental in the development of the signature adaptation of vertebrates: an endoskeleton made of bone. To understand how bone evolved, let's delve into the fossil record.

Using the Fossil Record

Echinoderms and vertebrates are present in the Burgess Shale deposits that formed during the Cambrian explosion 544–515 million years ago (**Figure 33.6**; see also Chapter 26). The vertebrate fossils in these rocks show that the earliest members of this lineage lived in the ocean about 530 million years ago, had streamlined bodies like fish, and appear to have had a skull, skeletal elements that reinforced the gills, and a notochord made of cartilage. **Cartilage** is a stiff tissue that consists of scattered cells in a gel-like matrix of polysaccharides and protein fibers. The earliest vertebrates had endoskeletons made of cartilage.

Subsequent to the appearance of vertebrates, the fossil record documents a series of key innovations that occurred as this lineage diversified:

- Fossil vertebrates from the early part of the Ordovician period, about 480 million years ago, are the first fossils to have bone. **Bone** is a tissue that consists of cells and blood vessels encased in a matrix made primarily of calcium phosphate ($CaPO_4$), with a small amount of calcium carbonate ($CaCO_3$)

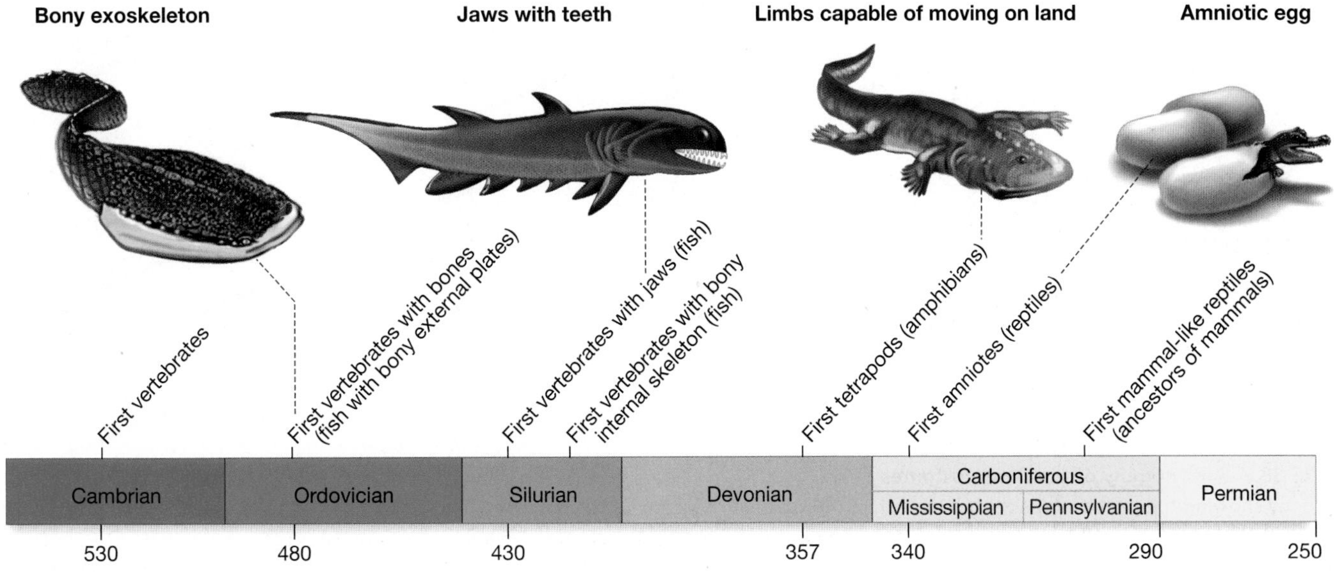

FIGURE 33.6 Timeline of the Vertebrate Fossil Record

and protein fibers. When bone first evolved, it did not occur in the endoskeleton. Instead bone was deposited in scalelike plates that formed an **exoskeleton**—a hard, hollow structure that envelops the body. Based on the fossils' overall morphology, biologists infer that these animals swam with the aid of a notochord, and that they breathed and fed by gulping water and filtering it through their pharyngeal gill slits. Presumably, the bony plates helped provide protection from predators.

- The first vertebrates with **jaws** show up in the fossil record about 430 million years ago. The evolution of jaws was significant because it gave vertebrates the ability to bite, meaning that they were no longer limited to suspension feeding or deposit feeding. Instead, they could make a living as herbivores or predators. Soon after, jawbones with teeth appear in the fossil record (Figure 33.6). With jaws and teeth, vertebrates became armed and dangerous. The fossil record shows that a spectacular radiation of jawed fishes followed, filling marine and freshwater habitats.
- The next great event in the evolution of vertebrates was the transition to living on land. The first animals that had limbs and were capable of moving on land are dated to about 357 million years ago. These were the first of the **tetrapods** ("four-footed")—animals with four limbs (Figure 33.6).
- About 20 million years after the appearance of tetrapods in the fossil record, the first amniotes are present. The **Amniota** is a lineage of vertebrates that includes all tetrapods other than amphibians. They are named for a signature adaptation: the amniotic egg. An **amniotic egg** is an egg that has a watertight shell or case enclosing a membrane-bound food

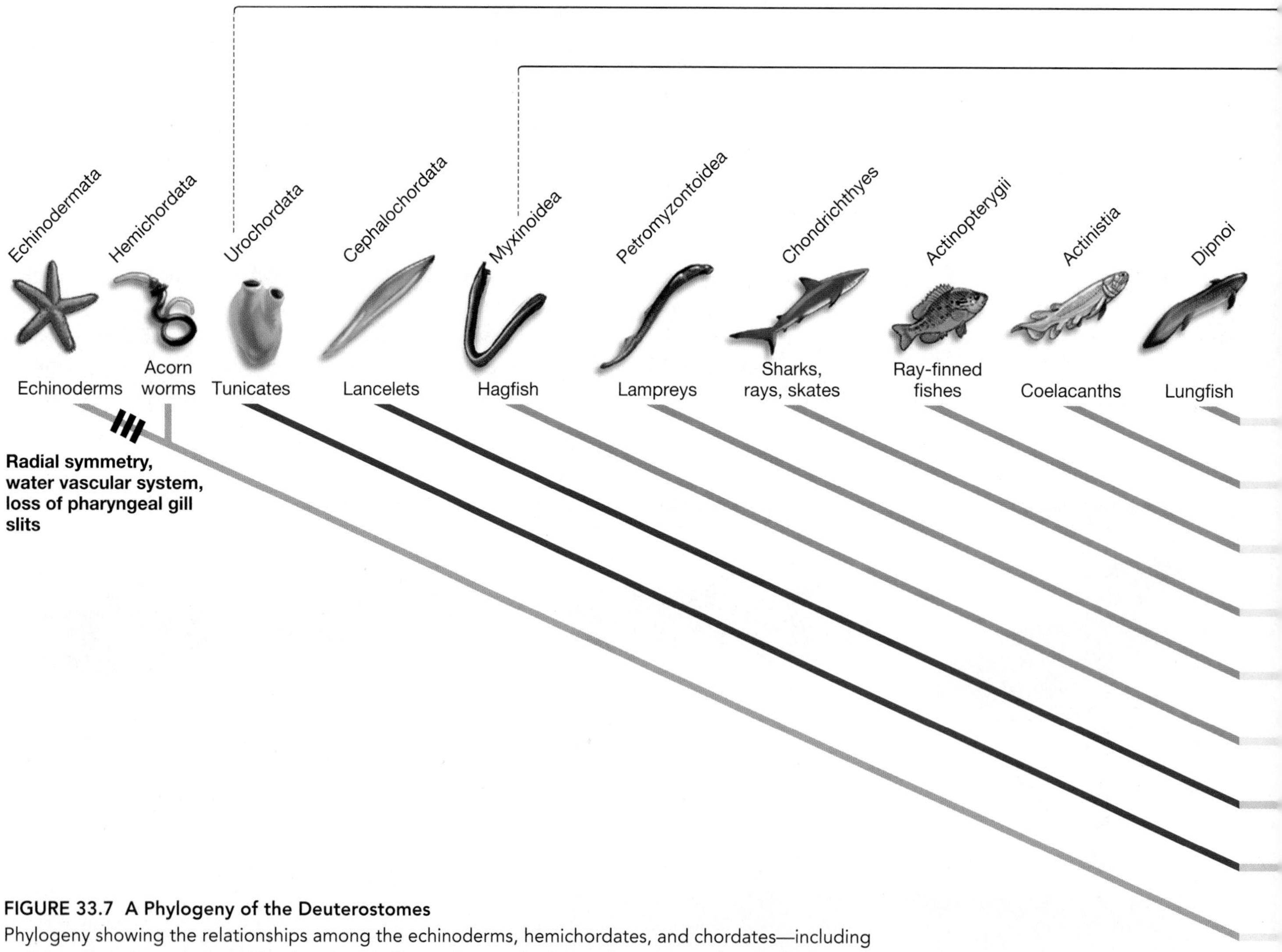

FIGURE 33.7 A Phylogeny of the Deuterostomes
Phylogeny showing the relationships among the echinoderms, hemichordates, and chordates—including major groups of vertebrates. The short bars indicate where major innovations occurred as the lineages diversified. **EXERCISE** Circle the taxa on this tree that comprise the Amniota. Circle and label the groups that are extinct.

supply, water supply, and waste repository. The evolution of the amniotic egg was significant because it gave vertebrates the ability to reproduce far from water. Amniotic eggs resist drying out, so vertebrates that produce amniotic eggs do not have to return to aquatic habitats to lay their eggs.

To summarize, the fossil record indicates that vertebrates evolved through a series of major steps, beginning about 530 million years ago with vertebrates whose endoskeleton consisted of a notochord. The earliest vertebrates gave rise to cartilaginous fishes (sharks and rays) and fish with bony skeletons and jaws. After the tetrapods emerged and amphibians resembling salamanders began to live on land, the evolution of the amniotic egg paved the way for the evolution of the first truly terrestrial vertebrates. The fossil record indicates that a radiation of reptiles followed, along with the animals that gave rise to mammals. Do phylogenetic trees estimated from analyses of DNA sequence data agree or conflict with these conclusions?

Evaluating Molecular Phylogenies

Figure 33.7 provides a phylogenetic tree that summarizes the relationships among deuterostomes, based on morphological traits and DNA sequence data. The labeled bars on the tree indicate where major innovations occurred during the evolution of deuterostomes. Although the phylogeny of deuterostomes continues to be a topic of intense research, researchers are increasingly confident that the relationships described in Figure 33.7 are accurate.

The overall conclusion from this tree is that the branching sequence inferred from morphological and molecular data

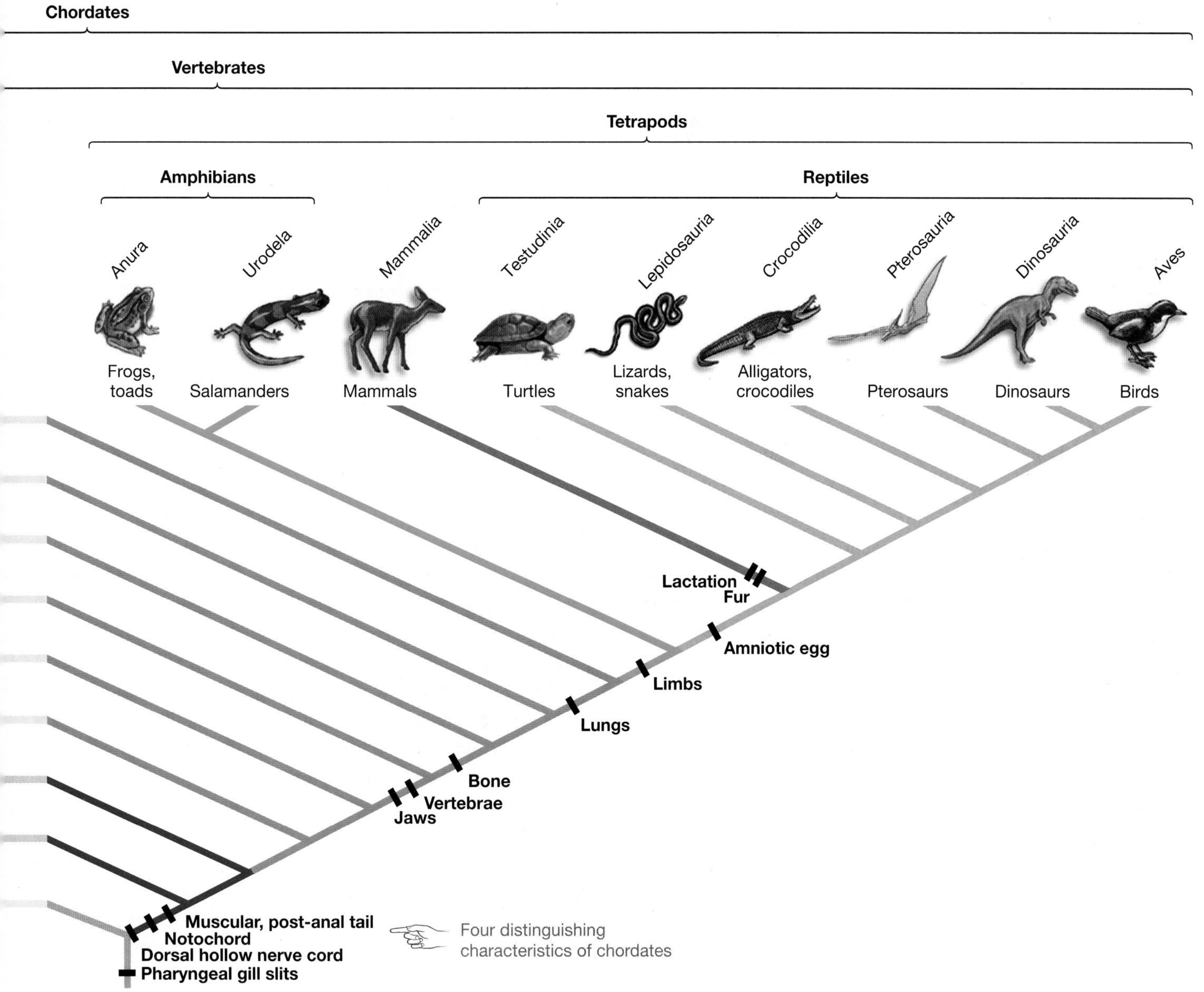

Chordates
Vertebrates
Tetrapods
Amphibians
Reptiles
Anura
Urodela
Mammalia
Testudinia
Lepidosauria
Crocodilia
Pterosauria
Dinosauria
Aves
Frogs, toads
Salamanders
Mammals
Turtles
Lizards, snakes
Alligators, crocodiles
Pterosaurs
Dinosaurs
Birds
Lactation
Fur
Amniotic egg
Limbs
Lungs
Bone
Vertebrae
Jaws
Muscular, post-anal tail
Notochord
Dorsal hollow nerve cord
Pharyngeal gill slits
Four distinguishing characteristics of chordates

correlates with the sequence of groups in the fossil record. Reading up from the base of the tree, it is clear that the three groups of deuterostomes that have traditionally been recognized as phyla are indeed monophyletic, and that hemichordates and echinoderms are more closely related to each other than they are to chordates. The closest living relatives of the vertebrates are the cephalochordates. The most basal groups of chordates lack the skull and vertebral column that define the vertebrates, and the most ancient lineages of vertebrates lack jaws and bony skeletons. To understand what happened during the subsequent diversification of echinoderms and vertebrates, let's explore some of the major innovations involved in feeding, movement, and reproduction in more detail.

✓CHECK YOUR UNDERSTANDING

Echinoderms are distinguished by their fivefold radial symmetry as adults, water vascular system, and calcium carbonate plates that form an endoskeleton. The water vascular system functions in movement, and the endoskeleton provides protection. You should be able to (1) make a rough sketch of a sea star's body; include labels indicating the structure and function of the water vascular system and endoskeleton.

Chordates are distinguished by the presence of a notochord, a dorsal hollow nerve cord, pharyngeal gill slits, and a muscular tail. You should be able to (2) sketch the body of a tunicate larva or cephalochordate, and add labels indicating the structure and function of the four diagnostic features. On a phylogeny of the vertebrates, you should be able to (3) map the origin of key traits such as the cartilaginous skeleton, bony skeleton, jaws, limbs, and amniotic egg.

33.3 What Themes Occur in the Diversification of Deuterostomes?

Deuterostomes evolved in marine environments, but their diversification carried them into all of the world's major habitat types. Echinoderms, hemichordates, urochordates, and cephalochordates are all strictly marine. The most basal groups of chordates are also ocean-dwelling organisms. Lungfish, certain groups of sharks and rays, and about one-third of the ray-finned fishes live in freshwater. All tetrapods, with the exception of marine reptiles and mammals, live in terrestrial habitats, and all must breathe air.

In terms of numbers of species and range of habitats occupied, the most successful lineages of deuterostomes are the echinoderms and the vertebrates. Echinoderms are both widespread and abundant in marine habitats. In some deepwater environments, echinoderms represent 95 percent of the total mass of organisms. Among vertebrates, the most species-rich and ecologically diverse lineages are the ray-finned fishes and the tetrapods (**Figure 33.8**). Ray-finned fishes occupy habitats ranging from deepwater environments, which are perpetually

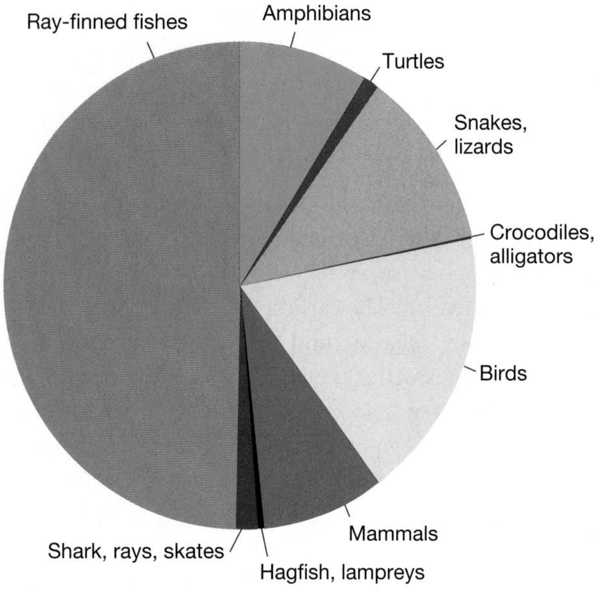

Ray-finned fishes
Amphibians
Turtles
Snakes, lizards
Crocodiles, alligators
Birds
Mammals
Hagfish, lampreys
Shark, rays, skates

FIGURE 33.8 Relative Species Abundance Among Vertebrates
EXERCISE Circle the taxa that are tetrapods.

dark, to shallow ponds that dry up each year. Tetrapods include the large herbivores and predators in terrestrial environments all over the world.

Today there are about 7000 species of echinoderms, 23,700 species of ray-finned fish, and 25,400 species of tetrapods. To understand why these lineages have been so successful, it's important to recognize that they have unique body plans. Recall that echinoderms are radially symmetric and have a water vascular system, and that ray-finned fishes and tetrapods have a bony endoskeleton. In this light, the situation in deuterostomes appears to be similar to that in protostomes. Recall from Chapter 32 that the most evolutionarily successful protostome lineages are the arthropods and mollusks. A major concept in that chapter was that arthropods and mollusks have body plans that are unique among protostomes. Once their distinctive body plans evolved, evolution by natural selection led to extensive diversification based on novel methods for eating, moving, and reproducing.

Box 33.1 highlights model organisms that biologists have used in experimental work on echinoderms, ray-finned fishes, and tetrapods. To understand how these three groups became so species rich and geographically widespread, let's take a closer look at the adaptations that allow them to eat, move, and produce offspring in diverse ways.

Feeding

Animals eat to live, and it is logical to predict that the 7000 species of echinoderms and nearly 50,000 species of ray-finned fishes and tetrapods eat different things in different ways. It is also logical to predict that echinoderms and vertebrates have traits that make diverse ways of feeding possible.

BOX 33.1 Model Organisms: Sea Urchins, Zebrafish, and Mice

In addition to their evolutionary success, the three most prominent deuterostome lineages—echinoderms, ray-finned fishes, and tetrapods—have furnished key model organisms in biological science. More specifically, sea urchins, zebrafish, and mice (**Figure 33.9**) are used as experimental subjects in labs all over the world. Like other model organisms, part of their appeal is practical: They are relatively inexpensive and easy to maintain in the laboratory, they produce large numbers of offspring, and they develop rapidly. In addition, both sea urchins and zebrafish have transparent eggs and embryos, which facilitates observation of cell movements and other developmental events.

Experiments on urchins, zebrafish, and mice have provided fundamental insights into how gene expression is controlled and how early development occurs. Along with fruit flies, these three species provide the most important model systems in animal genetics and developmental biology. Mice are also closely enough related to humans to provide a tractable study organism for biologists interested in the cellular mechanisms responsible for cancer and other diseases.

(a) Purple sea urchin **(b)** Zebrafish **(c)** Mouse

FIGURE 33.9 Model Organisms among the Deuterostomes

Both predictions are correct. Echinoderms have feeding strategies that are unique among marine animals. Many are based on the use of their water vascular system and podia. Ray-finned fishes and tetrapods, in contrast, depend on their jaws.

Feeding Strategies in Echinoderms Depending on the lineage and species in question, echinoderms make their living by suspension feeding, deposit feeding, or harvesting algae or other animals. In most cases, the animal's podia play a key role in obtaining food. Many sea stars, for example, prey on bivalves. Clams and mussels respond to sea star attacks by contracting muscles that close their shells tight. But by clamping onto each shell with their podia and pulling, sea stars are often able to pry the shells apart a few millimeters (**Figure 33.10a**). Once a gap exists, the sea star extrudes its stomach from its body and forces the stomach through the opening between the

(a) Podia adhere to bivalve shells and pull them apart. **(b)** Podia trap particles during suspension feeding.

FIGURE 33.10 Echinoderms Use Their Podia in Feeding
(a) Because they have suckers at their tips, sea stars can use their podia to pry open bivalve shells.
(b) Brittle stars extend their podia when they suspension feed.

bivalve's shells. Upon contacting the visceral mass of the bivalve, the stomach of the sea star secretes digestive enzymes. It then begins to absorb the small molecules released by enzyme action. Eventually, only the unhinged shells of the prey remain.

Podia are also involved in echinoderms that suspension feed (**Figure 33.10b**). In most cases, podia are extended out into the water. When food particles contact them, the podia flick the food down to cilia, which sweep the particles toward the animal's mouth. In deposit feeders, podia secrete mucus that is used to sop up food material on the substrate. The podia then roll the food-laden mucus into a ball and move it to the mouth.

The Vertebrate Jaw The most ancient groups of vertebrates have relatively simple mouthparts. For example, hagfish and lampreys lack jaws and cannot bite algae, plants, or animals. They have to make their living as deposit feeders or as ectoparasites.

Vertebrates were not able to harvest food by biting until jaws evolved. The leading hypothesis for the origin of the jaw proposes that natural selection acted on mutations that affected the morphology of **gill arches**, which are curved regions of tissue between the gills. The jawless vertebrates have bars of cartilage that stiffen these gill arches. The gill-arch hypothesis proposes that mutation and natural selection increased the size of an arch and modified its orientation slightly, producing the first working jaw (**Figure 33.11**). Three lines of evidence, drawn from comparative anatomy and embryology, support the gill-arch hypothesis:

1. Both gill arches and jaws consist of flattened bars of bony or cartilaginous tissue that hinges and bends forward.
2. Unlike most other parts of the vertebrate skeleton, both jaws and gill arches are derived from specialized cells in the embryo called neural crest cells.
3. The muscles that move jaws and that move gill arches are derived from the same population of embryonic cells.

Taken together, these data support the hypothesis that jaws evolved from gill arches.

To explain why ray-finned fishes are so diverse in their feeding methods, biologists point to important modifications of the jaw. In ray-finned fishes, for example, the jaw is protrusible—meaning it can be extended to nip or bite out at food. In addition, several particularly species-rich lineages of ray-finned fishes have a set of pharyngeal jaws. The **pharyngeal jaw** consists of modified gill arches that function as a second set of jaws, located in the back of the mouth. Pharyngeal jaws are important because they make food processing particularly efficient. (For more on the structure and function of pharyngeal jaws, see Chapter 43.)

To summarize, the radiation of ray-finned fishes was triggered in large part by the evolution of the jaw, modifications that made it possible to protrude the jaw, and the origin of the pharyngeal jaw. The story of tetrapods is different, however.

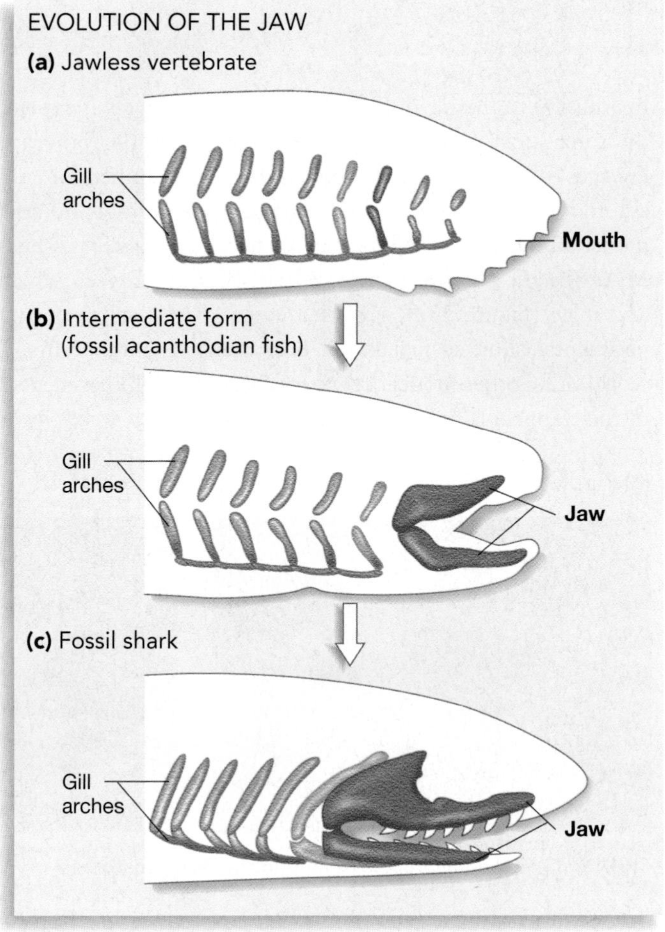

FIGURE 33.11 A Hypothesis for the Evolution of the Jaw
(a) Gill arches support the gills in jawless vertebrates. **(b)** In the fossil record, jawbones appear first in fossil fish called acanthodians. **(c)** Fossil sharks that appeared later had more elaborate jaws.
QUESTION The transition from the gill arches to the jaws of acanthodian fishes is complex, and intermediate forms have yet to be found in the fossil record. Would intermediate stages in the evolution of the jaw have any function?

Although jaw structure varies somewhat among tetrapod groups, the adaptation that triggered their initial diversification involved the ability to get to food, not to bite it and process it.

Movement

The signature adaptations of echinoderms and tetrapods involve locomotion. We've already explored the water vascular system and tube feet of echinoderms; here let's focus on the tetrapod limb.

Most tetrapods live on land and use their limbs to move. But for vertebrates to succeed on land, not only did they have to be able to move out of water, but they also had to breathe air and avoid drying out. To understand how this was accomplished, consider the morphology and behavior of their closest living relatives, the lungfish (**Figure 33.12**). Most living species of lungfish inhabit shallow, oxygen-poor water. To supplement the oxygen taken in by their gills, they have lungs and breathe

FIGURE 33.12 Lungfish Have Limb-like Fins
Some species of lungfish can walk or crawl short distances on their fleshy fins.

air. Some also have fleshy fins supported by bones and are capable of walking short distances along watery mudflats or the bottoms of ponds. In addition, some species can survive extended droughts by burrowing in mud.

Fossils provide strong links between lungfish and the earliest land-dwelling vertebrates. **Figure 33.13** shows three of the species involved. The first, an aquatic animal related to today's lungfishes, is from the Devonian period—about 375 million years ago. The second is one of the oldest tetrapods, or limbed vertebrates, found to date. The third is a more recent tetrapod fossil, dated to about 350 million years ago. The figure highlights the numbers and arrangement of bones in the fossil fish fin and the numbers and arrangement of bones in the limbs of early tetrapods. The color coding emphasizes that each fin or limb has a single bony element that is proximal (closest to the body) and then two bones that are distal (farther from the body) and arranged side by side, followed by a series of distal elements. Because the structures are similar and because no other groups have limb bones in this arrangement, the evidence for homology is strong. Based on the lifestyle of living lungfish, biologists suggest that mutation and natural selection gradually transformed fins into limbs as the first tetrapods became more and more dependent on terrestrial habitats.

The hypothesis that tetrapod limbs evolved from fish fins has also been supported by molecular genetic evidence. Recall from Chapter 26 that several regulatory proteins involved in the development of zebrafish fins and the upper parts of mouse limbs are homologous. Specifically, the proteins produced by *Hox* genes and the homeotic locus *Sonic hedgehog* (*Shh*) are found at the same times and in the same locations in fins and limbs. These data suggest that these appendages are patterned by the same genes. As a result, the data support the hypothesis that tetrapod limbs evolved from fins.

Once the tetrapod limb evolved, natural selection elaborated it into structures that are used for running, gliding, crawling, burrowing, or swimming. In addition, wings and the ability to fly evolved independently in three lineages of tetrapods: the extinct flying reptiles called *pterosaurs* (pronounced *TARE-oh-sors*), birds, and bats. Tetrapods and insects are the only animals that have taken to the skies. **Box 33.2** (page 760) explores how flight evolved in birds.

To summarize, the evolution of the jaw gave tetrapods the potential to capture and process a wide array of foods. With limbs, they could move efficiently on land in search of food. Another challenge remained, however: producing offspring that could survive out of water.

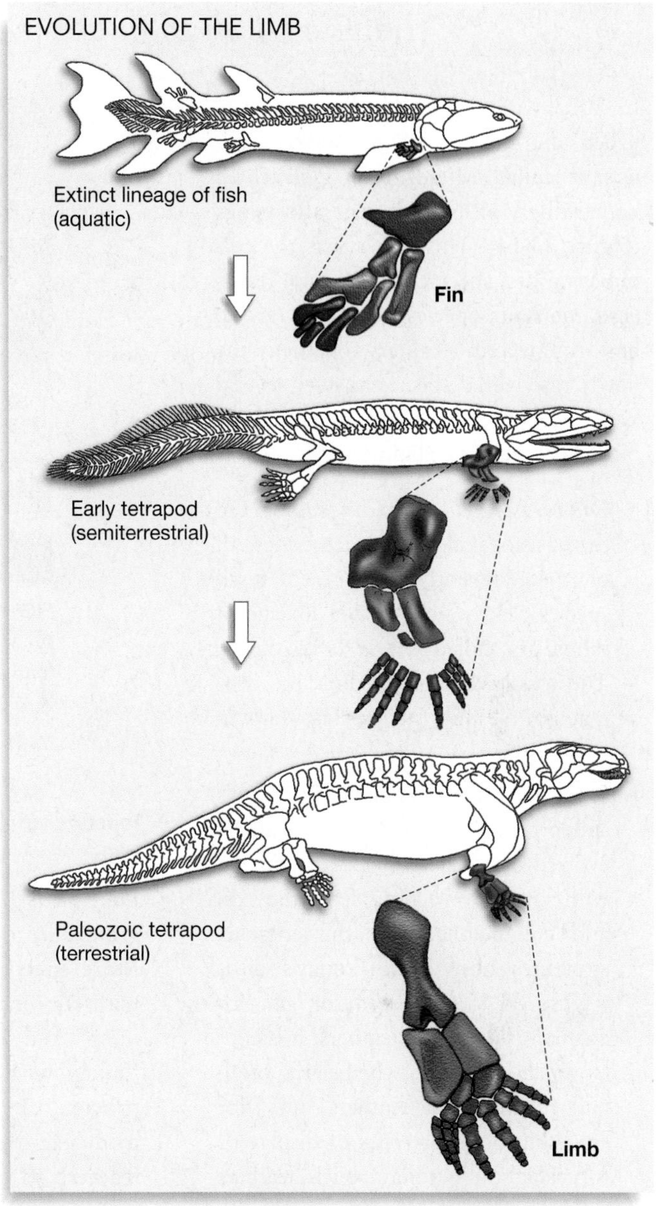
EVOLUTION OF THE LIMB
Extinct lineage of fish
(aquatic)
Fin
Early tetrapod
(semiterrestrial)
Paleozoic tetrapod
(terrestrial)
Limb

FIGURE 33.13 Fossil Evidence for a Fin-to-Limb Transition
Both the fossil fish and the early tetrapod are from the Devonian period, about 375 millon years ago. The number and arrangement of bones in the fins and limbs of these two fossil organisms agree with the general form of Paleozoic tetrapods as well as the modern tetrapod limb. The color coding indicates homologous elements.

BOX 33.2 The Evolution of Flight in Birds

In 2003 Xing Xu and colleagues published an analysis of a fossilized dinosaur called *Microraptor gui*, which had feathers on both its legs and wings (**Figure 33.14**). This paper was the culmination of a spectacular series of feathered dinosaur species that Xu's group has discovered. Taken together, the newly discovered species answer several key questions about the evolution of birds, feathers, and flight:

- *Did birds evolve from dinosaurs?* On the basis of skeletal characteristics, all of these recently discovered fossil species clearly belong to a lineage of dinosaurs called the *dromaeosaurs*. The fossils definitively link the dromaeosaurs and the earliest known fossil birds.
- *How did feathers evolve?* The fossils support Xu's model that feathers evolved in a series of steps, beginning with simple projections from the skin and culminating with the complex structures observed in today's birds (**Figure 33.15**). *Microraptor gui*, for example, has simple feathers. It is controversial, however, whether the original function of feathers was for courtship or other types of display or for insulation. In today's birds, feathers function in display, insulation, and flight.
- *Did birds begin flying from the ground up or from the trees down?* More specifically, did flight evolve with running species that began to jump and glide or make short flights, with the aid of feathers to provide lift? Or did flight evolve from tree-dwelling species that used feathers to glide from tree to tree, much as flying squirrels do today? Because it is unlikely that *Microraptor gui* could run efficiently with feathered legs, Xu and colleagues propose that flight evolved from tree-dwelling gliders.

FIGURE 33.14 Feathers Evolved in Dinosaurs
An artist's depiction of what *Microraptor gui*, a dinosaur that had feathers on its body and all four limbs, might have looked like in life.

Once dinosaurs evolved feathers and took to the air as gliders, the fossil record shows that a series of adaptations made powered, flapping flight

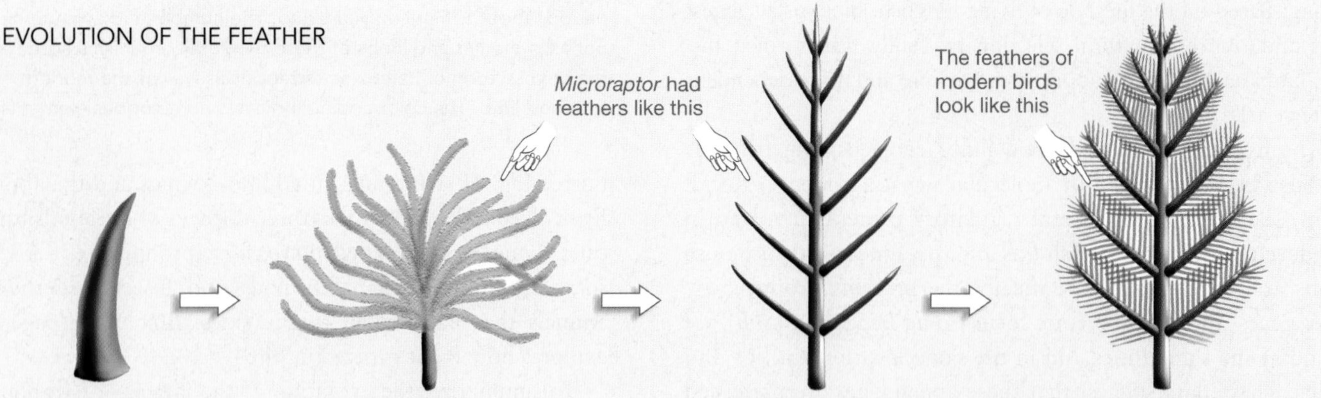
EVOLUTION OF THE FEATHER
Microraptor had feathers like this
The feathers of modern birds look like this
Simple projection
Tufts
Central shaft with pairs of branches
Central shaft with pairs of branches bearing finer branches

FIGURE 33.15 Feathers Evolved through Intermediate Stages
A model for the evolution of feathers that is supported by the fossil record.

increasingly efficient. As **Figure 33.16** shows, the *sternum* (breastbone) of birds has an elongated projection called the *keel*, which provides a large surface area to which flight muscles attach. (On a chicken or turkey, the flight muscles are called "white meat.") Birds are also extraordinarily light for their size, primarily because they have a drastically reduced number of bones and because their larger bones are thin-walled and hollow—though strengthened by bony "struts" (see Figure 33.16). Birds are also capable of long periods of sustained activity year round, because they are **endothermic**—meaning that they maintain a high body temperature by producing heat in their tissues. From dinosaurs that jumped and glided from tree to tree, birds have evolved into extraordinary flying machines.

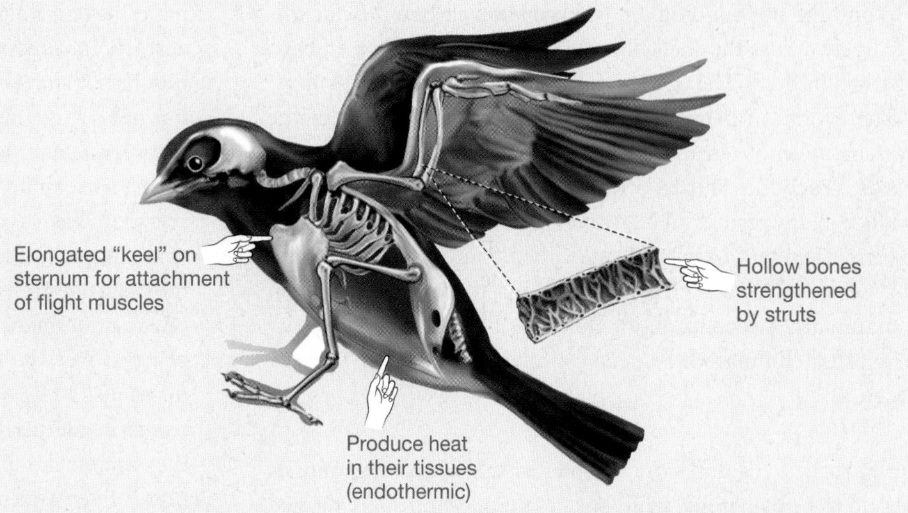
Elongated "keel" on sternum for attachment of flight muscles
Hollow bones strengthened by struts
Produce heat in their tissues (endothermic)

FIGURE 33.16 Birds Have Several Adaptations that Allow for Flight, in Addition to Feathers
The elongated "keel" on the sternum (breastbone), the hollow bones strengthened by crisscrossing struts, and endothermy are all interpreted as adaptations that make bird flight more efficient.

Reproduction

Chapter 31 summarized the diversity of reproductive methods observed in fish. Recall that a few species undergo only asexual reproduction and that when sexual reproduction occurs, it may be based on internal or external fertilization and oviparity, viviparity, or ovoviviparity. In addition, parental care is extensive in some fish species, and often involves guarding eggs from predators and fanning them to keep oxygen levels high. All fish lay their eggs or give birth in water, however. Tetrapods were the first vertebrates that were able to breed on land.

Three major evolutionary innovations gave tetrapods the ability to produce offspring successfully in terrestrial environments: (1) the amniotic egg, (2) the placenta, and (3) elaboration of parental care. Let's explore each of these innovations in turn.

The Amniotic Egg Amniotic eggs have shells that minimize water loss as the embryo develops inside. The first tetrapods, like today's amphibians (frogs, toads, and salamanders), lacked amniotic eggs. Although their eggs were encased by a membrane, the first tetrapods laid eggs that would dry out and die unless they were laid in water. This fact limited the range of habitats that these animals could exploit. Like today's amphibians, the early tetrapods were largely restricted to living in or near marshes, lakes, or ponds.

In contrast, reptiles and the egg-laying mammals produce amniotic eggs. The eggshell is leathery in turtles, crocodiles, and snakes and lizards but hard—due to deposition of the calcium carbonate—in birds and the egg-laying mammals. In addition to having a shell that is largely watertight, an amniotic egg contains a membrane-bound supply of water in a protein-rich solution called **albumen** (**Figure 33.17**). The embryo is enveloped in a protective inner membrane known as the **amnion.** The **yolk sac** is a membranous pouch that contains nutrients for the growing embryo, and the **allantois** is a membranous pouch that holds waste materials. A middle membrane, the **chorion**, separates the

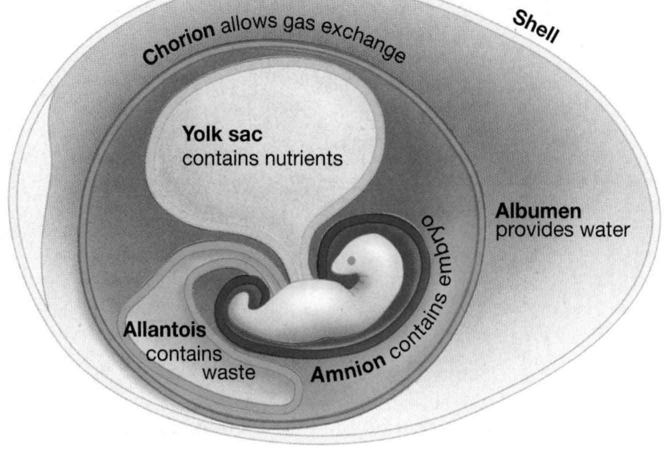
Chorion allows gas exchange
Shell
Yolk sac contains nutrients
Albumen provides water
Allantois contains waste
Amnion contains embryo

FIGURE 33.17 An Amniotic Egg
Amniotic eggs have membrane-bound sacs that hold nutrients, water, and waste and that allow gas exchange.

amnion, yolk sac, and allantois from the albumen and provides a surface where gas exchange can take place between the embryo and the surrounding air. (Oxygen and carbon dioxide diffuse freely across the shell.) Inside an amniotic egg, the embryo is bathed in fluid. The egg itself is highly resistant to drying.

The evolution of the amniotic egg was a key event in the diversification of tetrapods because it allowed turtles, snakes, lizards, crocodiles, birds, and the egg-laying mammals to reproduce in any terrestrial environment—even habitats as dry as deserts. Members of the lineage called Amniota now occupy all types of terrestrial environments. But during the evolution of mammals, a second major innovation in reproduction occurred that eliminated the need for any type of egg laying: the placenta.

The Placenta Recall from Chapter 31 that egg-laying animals are said to be **oviparous**, while species that give birth are termed **viviparous**. In many viviparous animals, females produce an egg that contains a nutrient-rich yolk. Instead of laying the egg, however, the mother retains it inside her body. In these ovoviviparous species, the developing offspring depends on the resources in the egg yolk. In most mammals, however, the eggs that females produce lack yolk. After fertilization occurs and the egg is retained, the mother produces a placenta within her uterus. The **placenta** is an organ that is rich in blood vessels and that facilitates a flow of oxygen and nutrients from the mother to the developing offspring (**Figure 33.18**). After a development period called **gestation**, the embryo emerges from the mother's body.

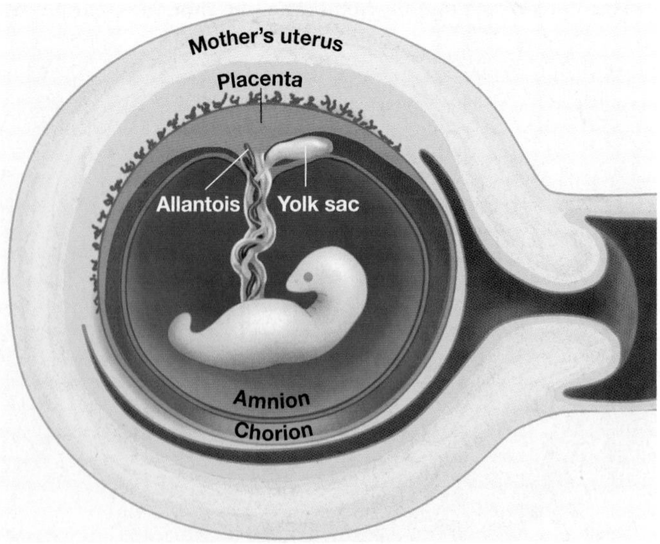

FIGURE 33.18 The Placenta Allows the Mother to Nourish the Fetus Internally

QUESTION Compare the relative positions of the chorion and amnion here with those in the amniotic egg (Figure 33.17). Are they the same or different?

Why did viviparity and the placenta evolve? Biologists have formulated an answer to this question by pointing out that females have a finite amount of time and energy available to invest in reproduction. As a result, a female can produce a large number of small offspring or a small number of large offspring but not a large number of large offspring. Stated more formally, every female faces a trade-off between the quantity of offspring she can produce and their size. In some lineages, natural selection has favored traits that allow females to produce a small number of large, well-developed offspring. Viviparity and the placenta are two such traits. Compared with female insects or echinoderms, which routinely lay thousands or even millions of eggs over the course of a lifetime, a female mammal produces just a few offspring. But because those offspring are protected inside her body and fed until they are well developed, they are much more likely to survive than sea star or insect embryos. Even after birth, many mammals continue to invest time and energy in rearing their young.

Parental Care The term **parental care** encompasses any action by a parent that improves the ability of its offspring to survive, including supplying food, keeping young warm and dry, and protecting them from danger. In some insect and frog species, mothers carry around eggs or newly hatched young; in fishes, parents commonly guard eggs during development and fan them with oxygen-rich water.

The most extensive parental care observed among animals is provided by mammals and birds. In both groups, the mother and often the father continue to feed and care for individuals after birth or hatching—sometimes for many years (**Figure 33.19**). Female mammals also **lactate**—meaning that they produce nutrient-rich fluid called *milk* and use it to feed their offspring after birth. With the combination of the placenta and lactation, placental mammals make the most extensive investment of time and energy in offspring known. The evolution of extensive parental care is hypothesized to be a major reason for the evolutionary success of mammals and birds.

✓CHECK YOUR UNDERSTANDING

The diversification of echinoderms and vertebrates was based on innovations that made it possible for species to feed, move, and reproduce in novel ways. Most echinoderms use their podia to move, but they feed in a wide variety of ways—including using their podia to pry open bivalves, suspension feed, or deposit feed. An array of key innovations occurred during the evolution of vertebrates: Jaws made it possible to bite and process food, limbs allowed tetrapods to move on land, and amniotic eggs could be laid on land. You should be able to (1) explain how the jaw and limb evolved, (2) diagram the structure of an amniotic egg, and (3) explain the role of increased parental care in the evolution of birds and mammals.

(a) Mammal mothers feed and protect newborn young.

(b) Many bird species have extensive parental care.

FIGURE 33.19 Parental Care Is Extensive in Mammals and Birds **(a)** Female mammals feed and protect embryos inside their bodies until the young are well developed. Once the offspring is born, the mother feeds it milk until it is able to eat on its own. In some species, parents continue to feed and protect young for many years. **(b)** In birds, one or both parents may incubate the eggs, protect the nest, and feed the young after hatching occurs.

33.4 Key Lineages: Echinodermata

The echinoderms ("spiny-skins") were named for the spines or spikes observed in many species. They are bilaterally symmetric as larvae but undergo metamorphosis and develop into radially symmetric adults. They all have a water vascular system and produce calcium carbonate plates under their skin to form an endoskeleton.

The echinoderms living today make up five major lineages, traditionally recognized as classes: (1) feather stars and sea lilies, (2) brittle stars and basket stars, (3) sea cucumbers, (4) sea stars, and (5) sea urchins and sand dollars. Most feather stars and sea lilies are sessile suspension feeders (**Figure 33.20a**). Brittle stars and basket stars have five long arms that radiate out from a small central disk (**Figure 33.20b**). They use these arms to suspension feed, deposit feed by sopping up material with mucus, or capture small prey animals. Sea cucumbers are sausage-shaped animals that suspension feed or deposit feed with the aid of tentacles arranged in a whorl around their mouths (**Figure 33.20c**). Sea stars and sea urchins are described in detail in the text that follows.

(a) Feather star

(b) Brittle star (crawling on a sponge)

(c) Sea cucumber

FIGURE 33.20 Three Major Echinoderm Groups: Feather Stars, Brittle Stars, Sea Cucumbers

Echinodermata ▶ Asteroidea (Sea Stars)

The 1700 known species of sea stars have bodies with five or more long arms—in some species up to 300—radiating from a central region that contains the mouth, stomach, and anus (**Figure 33.21**). They range in size from less than 1 cm in diameter when fully grown to 1 m across. They live on hard or soft substrates along the coasts of all the world's oceans. Although the spines that are characteristic of echinoderms are reduced to knobs on the surface of most sea stars, the crown-of-thorns star and a few other species have prominent, upright, movable spines.

Feeding Sea stars are predators or scavengers. Some species pull bivalves apart with their tube feet and evert their stomach into the prey's visceral mass. Sponges, barnacles, and snails are also common prey. The crown-of-thorns sea star specializes in feeding on corals and is native to the Indian Ocean and western Pacific Ocean. Its population has skyrocketed recently—possibly because people are harvesting their major predator, a large snail called the triton, for its pretty shell. Large crown-of-thorns star populations have led to the destruction of large areas of coral reef.

Movement Sea stars crawl with the aid of their tube feet. Any one of the five or more arms may be in "front" as the animal moves.

Reproduction Sexual reproduction predominates, and sexes are separate. At least one sea star arm is filled with reproductive organs that produce massive amounts of gametes—millions of eggs per female, in some species. Species that are native to habitats in the far north, where conditions are particularly harsh, care for their offspring by holding fertilized eggs on their body until the eggs hatch. Most sea stars are capable of regenerating arms that are lost in predator attacks or storms. Some species can reproduce asexually by dividing the body in two, with each of the two individuals then regenerating the missing half.

Pycnopodia helianthoides

FIGURE 33.21 Sea Stars May Have Many Arms

Echinodermata ▶ Echinoidea (Sea Urchins and Sand Dollars)

There are about 800 species of echinoids living today; most are sea urchins or sand dollars. Sea urchins have globe-shaped bodies and long spines and crawl along substrates (**Figure 33.22a**). Sand dollars are flattened and disk-shaped, have short spines, and burrow in soft sediments (**Figure 33.22b**).

Feeding Sand dollars use their mucus-covered podia to collect food particles in sand or in other soft substrates. Most types of sea urchins are herbivores. In some areas of the world, urchins are extremely important grazers on kelp and other types of algae. In fact, when urchin populations are high, their grazing can prevent the formation of kelp forests. Most echinoids have a unique, jaw-like feeding structure in their mouths that is made up of five calcium carbonate teeth attached to muscles. In many species, this apparatus can extend and retract as the animal feeds.

Movement Using their podia, sea urchins crawl and sand dollars burrow. Sea urchins can also move their spines to aid in crawling along a substrate.

Reproduction Sexual reproduction predominates. Fertilization is external, and sexes are separate.

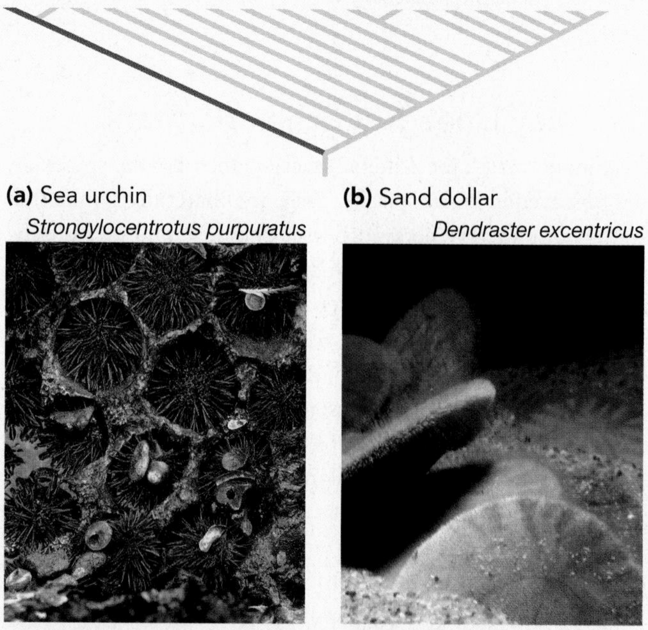
(a) Sea urchin
Strongylocentrotus purpuratus
(b) Sand dollar
Dendraster excentricus

FIGURE 33.22 Sea Urchins and Sand Dollars Are Closely Related

33.5 Key Lineages: Chordata

The chordates comprise three major subgroups or subphyla: (1) the urochordates (also called tunicates or sea squirts), (2) the cephalochordates (lancelets), and (3) the vertebrates. There are about 3000 species of tunicates, 24 species of lancelets, and over 50,000 vertebrates. At some stage in their life cycle, all species in the phylum Chordata have a dorsal hollow nerve cord, a notochord, pharyngeal gill slits, and a muscular tail that extends past the anus.

Tunicates and lancelets were introduced in Section 33.2; here the focus is on the major lineages of Vertebrata. Vertebrates are named for a column of cartilaginous or bony structures, called **vertebrae**, that form along the dorsal sides of most species and that protect the dorsal hollow nerve cord. Not all species of Vertebrata have vertebrae, however. The feature that all species in this lineage have is the **cranium**, or skull—a bony, cartilaginous, or fibrous case that encloses the brain. Biologists once used the term *Craniata* to refer to the vertebrates.

The cranium is thought to be a key innovation in the evolution of vertebrates because it protects the brain along with sensory organs such as eyes. The vertebrate brain develops as an outgrowth of the most anterior end of the dorsal hollow nerve cord and is important to the vertebrate life style. Vertebrates are active predators and herbivores that can make rapid, directed movements with the aid of their endoskeleton. Coordinated movements are possible in part because vertebrates have large brains divided into three distinct parts: (1) a **forebrain**, housing the sense of smell; (2) a **midbrain**, associated with vision; and (3) a **hindbrain**, responsible for balance and hearing. In vertebrate groups that evolved more recently, the forebrain is a large structure called the **cerebrum**, and the hindbrain consists of enlarged regions called the **cerebellum** and **medulla oblongata**. (The structure and functions of the vertebrate brain are analyzed in detail in Chapter 45.) The evolution of a large brain, protected by a hard cranium, was an innovation that defines the vertebrates.

Chordata ▶ Myxinoidea (Hagfish) and Petromyzontoidea (Lampreys)

Although recent phylogenetic analyses indicate that hagfish and lampreys may belong to two independent lineages, some data suggest that they are a single group called the **Agnatha** ("not-jawed"). Because these animals are the only vertebrates that lack jaws, the 110 species in the two groups are still referred to as the jawless fishes. The hagfish and lampreys—several of which are known only from fossils—are the only surviving members of the earliest branches at the base of the vertebrates.

Hagfish and lampreys have long, slender bodies and are aquatic. Most species are less than a meter long when fully grown. Hagfish lack any sort of vertebral column, but lampreys have small pieces of cartilage along the length of their dorsal hollow nerve cord.

Feeding Hagfish are scavengers and predators (**Figure 33.23a**). They deposit feed in the carcasses of dead fish and whales, and some are thought to burrow through ooze at the bottom of the ocean, feeding on polychaetes and other buried prey. Lampreys, in contrast, are ectoparasites. They attach to the sides of fish or other hosts by suction, then use spines in their mouth and tongue to rasp a hole in the side of their victim (**Figure 33.23b**). Once the wound is open, they suck blood and other body fluids.

Movement Hagfish and lampreys have a well-developed notochord and swim by making undulating movements. Lampreys can also move themselves upstream, against the flow of water, by attaching their suckers to rocks and looping the rest of their body forward, like an inchworm. Although lampreys have fins that aid in locomotion, their fins do not occur in pairs as they do in jawed fish.

Reproduction Virtually nothing is known about hagfish mating or embryonic development. Lampreys are **anadromous**, meaning that they spend their adult life in large lakes or the ocean but swim up streams to breed. Fertilization is external, and adults die after breeding once. Lamprey eggs hatch into larvae that look and act like lancelets. The larvae burrow into sediments and suspension feed for several years before metamorphosing into free-swimming adults.

(a) Hagfish
Eptatretus stoutii
(b) Lampreys feeding on fish
Petromyzon marinus
Lampreys

FIGURE 33.23 Hagfish and Lampreys Are Jawless

Chordata ▶ Chondrichthyes (Sharks, Rays, Skates)

The 840 species in this lineage are distinguished by their cartilaginous skeleton (*chondrus* is the Greek word for cartilage), the presence of jaws, and the existence of paired fins. Paired fins were an important evolutionary innovation, because they stabilize the body during rapid swimming—keeping it from pitching up or down, yawing to one side or the other, or rolling.

Most sharks, rays, and skates are marine, though a few species live in freshwater. Sharks have streamlined, torpedo-shaped bodies and an asymmetrical tail—the dorsal portion is longer than the ventral portion (**Figure 33.24a**) In contrast, the dorsal-ventral plane of the body in rays and skates is strongly flattened (**Figure 33.24b**).

Feeding A few species of ray and shark suspension feed on plankton, but most species in this lineage are predators. Skates and rays lie on the ocean floor and ambush passing animals; electric rays capture their prey by stunning them with electric discharges of up to 200 volts. Most sharks, in contrast, are active hunters that chase down prey in open water and bite them. The larger species of shark feed on large fish or marine mammals. Sharks are referred to as the "top predator" in many marine ecosystems, because they are at the top of the food chain—nothing eats them. Yet the largest of all sharks, the whale shark, is a suspension feeder. Whale sharks filter plankton out of water as it passes over their gills.

Movement Rays and skates swim by flapping their greatly enlarged pectoral fins. (*Pectoral fins* are located on the sides of an organism; *dorsal fins* are located on the dorsal surface.) Sharks swim by undulating their bodies and beating their large tail.

Reproduction Sharks have internal fertilization, and fertilized eggs may be shed into the water or retained until the young are hatched and well developed. In some viviparous species, embryos are attached to the mother by specialized tissues in a mammal-like placenta, where the exchange of gases, nutrients, and wastes takes place. Skates are oviparous, but rays are viviparous.

(a) Sharks are torpedo shaped. *Prionace glauca*

(b) Skates and rays are flat. *Taeniura melanospila*

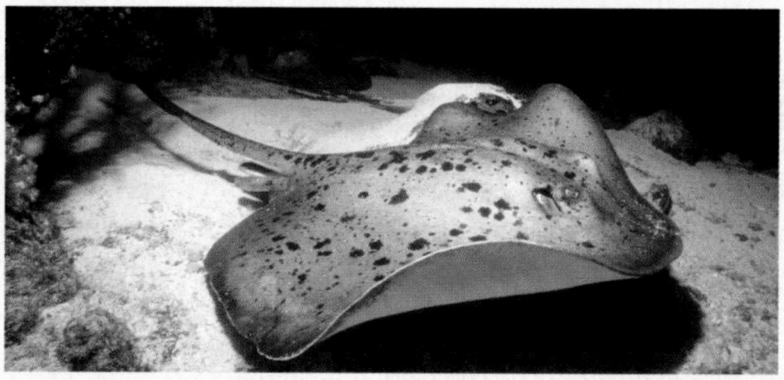

FIGURE 33.24 Sharks and Rays Have Cartilaginous Skeletons

Chordata ▶ Actinopterygii (Ray-Finned Fishes)

Actinopterygii (pronounced *ack-tin-op-teh-RIJ-ee-i*) means "ray-finned." Logically enough, these fish have fins that are supported by long, bony rods arranged in a ray pattern. They are the most ancient living group of vertebrates that have a skeleton made of bone. Their bodies are covered with interlocking scales that provide a stiff but flexible covering, and they avoid sinking in the water with the aid of a gas-filled **swim bladder**. The evolution of the swim bladder was an important innovation, because this structure allowed ray-finned fishes to float. Tissues are heavier than water, so the bodies of aquatic organisms tend to sink. Sharks and rays, for example, have to swim to avoid sinking. But ray-finned fishes have a bladder that changes in volume, depending on the individual's position. Gas is added to the bladder when a ray-finned fish swims down; gas is removed when the fish swims up. In this way, ray-finned fishes maintain neutral buoyancy in water of various depths and thus pressures.

The actinopterygians are the most successful vertebrate lineage based on number of species, duration in the fossil record, and extent of habitats occupied. Almost 24,000 species of ray-finned fishes have been identified. In traditional classifications, Actinopterygii is considered a class.

The most important major lineage of ray-finned fishes is the Teleostei, which includes some 20,000 species. Most of the fish that you are familiar with, such as tuna, trout, cod, and goldfish, are teleosts (**Figure 33.25**).

(Continued on next page)

Chordata ▶ Actinopterygii (Ray-Finned Fishes) *(continued)*

Feeding Teleosts can suck food toward their mouths, grasp it with their protrusible jaws, and then process it with teeth on their jaws and with pharyngeal jaws in their throat. The size and shape of the mouth, the jaw teeth, and the pharyngeal jaw teeth all correlate with the type of food consumed. For example, most predatory teleosts have long, spear-shaped jaws armed with spiky teeth, as well as bladelike teeth on their pharyngeal jaws. In addition to being major predators, ray-finned fishes are the most important large herbivores in both marine and freshwater environments.

Movement Ray-finned fishes swim by alternately contracting muscles on the left and right sides of their bodies from head to tail, resulting in rapid, side-to-side undulations. Their bodies are streamlined to reduce drag in water. Teleosts have a flexible, symmetrical tail, which reduces the need to use their pectoral (side) fins as steering and stabilizing devices during rapid swimming.

Reproduction Most species rely on external fertilization and are oviparous; some species have internal fertilization with external development; still others have internal fertilization and are viviparous. Although it is common for fish eggs to be released in the water and left to develop on their own, parental care occurs in some species. Parents may carry fertilized eggs on their fins, in their mouth, or in specialized pouches to guard them until the eggs hatch. In freshwater teleosts, offspring develop directly, but marine species have larva that are very different from adult forms. As they develop, marine fish larvae undergo a metamorphosis to the juvenile form, which then grows into an adult.

Thunnus thynnus

FIGURE 33.25 Teleosts Are Ray-Finned Fishes that Have a Flexible Tail

Chordata ▶ Actinistia (Coelacanths) and Dipnoi (Lungfish)

Although coelacanths (pronounced *SEEL-uh-kanths*) and lungfish represent independent lineages, they are sometimes grouped together and called **lobe-finned fishes.** Lobe-finned fishes are common and diverse in the fossil record in the Devonian period, about 400 million years ago, but only eight species are living today. They are important, however, because they represent a crucial evolutionary link between the ray-finned fishes and the tetrapods. Instead of having fins supported by rays of bone, their fins are fleshy lobes supported by an array of bones and muscles—similar to those observed in the limbs of tetrapods.

Coelacanths are marine and occupy habitats 150–700 m below the surface (**Figure 33.26**). In contrast, lungfish live in shallow, freshwater ponds (see Figure 33.12). As their name implies, lungfish have lungs and breathe air when oxygen levels in their habitats drop. Some species burrow in mud and enter a quiescent state when their habitat dries up during each year's dry season.

Feeding Coelacanths prey on fish. Lungfish are **omnivorous** ("all-eating"), meaning that they eat algae and plant material as well as animals.

Movement Coelacanths swim by waving their pectoral and pelvic ("hip") fins in the same sequence that tetrapods use in walking with their limbs. Lungfish swim by waving their body, and they can use their fins to walk along pond bottoms.

Reproduction Sexual reproduction is the rule, with fertilization internal in coelacanths and external in lungfish. Coelacanths are ovoviviparous; lungfish lay eggs. Lungfish eggs hatch into larvae that resemble juvenile salamanders.

Latimeria chalumnae

FIGURE 33.26 Coelacanths Are Lobe-Finned Fishes

Chordata ▶ Amphibia (Frogs, Salamanders, Caecilians)

The 4800 species of amphibians living today form three distinct clades traditionally termed orders: (1) frogs and toads, (2) salamanders, and (3) caecilians (pronounced *suh-SILL-ee-uns*). Amphibians are found throughout the world and occupy ponds, lakes, or moist terrestrial environments (**Figure 33.27a**). Translated literally, their name means "both-sides-living." The name is appropriate because adults of most species of amphibian feed on land, but many species lay their eggs in water. When amphibians live on land, gas exchange occurs exclusively or in part across their moist, mucus-covered skin.

Feeding Adult amphibians are carnivores. Most frogs are sit-and-wait predators that use their long, extensible tongues to capture passing prey. Salamanders also have an extensible tongue, which some species use in feeding. Terrestrial caecilians prey on earthworms and other soil-dwelling animals; aquatic forms eat vertebrates and small fish.

Movement Most amphibians have four well-developed limbs. In water, frogs and toads move by kicking their hind legs to swim; on land they kick their hind legs out to jump or hop. Salamanders walk on land; in water they undulate their body to swim like fish. Caecilians lack limbs and eyes; terrestrial forms burrow in moist soils (**Figure 33.27b**).

Reproduction Frogs are oviparous and have external fertilization, but salamanders and caecilians have internal fertilization. Most salamanders are oviparous, but many caecilians are viviparous. In some species of frogs, parents may guard or even carry eggs. In many frogs, young develop in the water and suspension feed on plant or algal material. Salamander larvae are carnivorous. Later the larvae undergo a dramatic metamorphosis into land-dwelling adults. For example, the fishlike tadpoles of frogs and toads develop limbs and replace their gills with lungs.

(a) Frogs and other amphibians lay their eggs in water.
Bufo periglenes

(b) Caecilians are legless amphibians.
Gymnophiona species

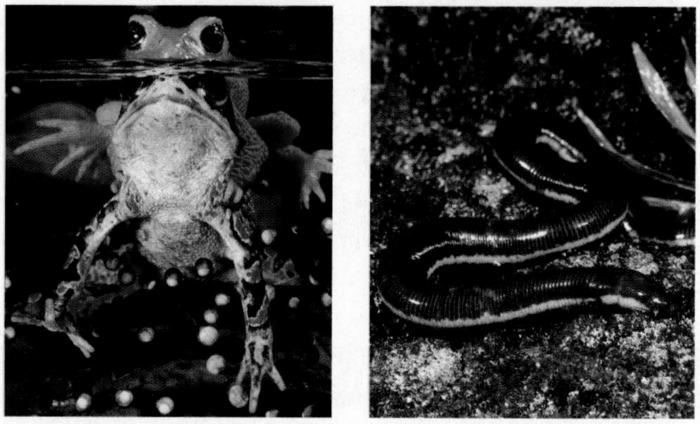

FIGURE 33.27 Amphibians Are the Most Ancient Tetrapods

Chordata: Mammalia (Mammals)

Mammals are easily recognized by the presence of **hair** or **fur**, which serves to insulate the body. Like birds, mammals are endotherms that maintain high body temperatures by oxidizing large amounts of food and generating large amounts of heat. Instead of insulating themselves with feathers, though, mammals retain heat because the body surface is covered with layers of hair or fur. Endothermy evolved independently in birds and mammals. In both groups, endothermy is thought to be an adaptation that enables individuals to maintain high levels of activity—particularly at night or during cold weather.

In addition to being endothermic and having fur, mammals have **mammary glands**—a unique structure that makes lactation possible. The evolution of mammary glands gave mammals the ability to provide their young with particularly extensive parental care. Mammals are also the only vertebrates with facial muscles and lips and the only vertebrates that have jaws formed from a single bone. In traditional classifications, Mammalia is designated as a class.

Mammals evolved when dinosaurs and other reptiles were the dominant large herbivores and predators in terrestrial and aquatic environments. The earliest mammals in the fossil record appear about 195 million years ago; they were small animals that were probably active only at night. Presumably because of their ancestry, most of the 4500 species of mammal living today have good nocturnal vision and a strong sense of smell. The adaptive radiation that gave rise to today's diversity of mammals did not take place until after the dinosaurs went extinct about 65 million years ago. After the dinosaurs were gone, the mammals diversified into lineages of small and large herbivores, small and large predators, or marine hunters—ecological roles that had once been filled by dinosaurs and mosasaurs.

Chordata ▶ Mammalia ▶ Monotremata (Platypuses, Echidnas)

The monotremes are the most ancient lineage of mammals living today, and they are found only in Australia and New Zealand. They lay eggs and have lower metabolic rates than other mammals do. Three species exist: one species of platypus and two species of echidna.

Feeding Monotremes have a leathery beak or bill. The platypus feeds on insect larvae, mollusks, and other small animals in streams (**Figure 33.28a**). Echidnas feed on ants, termites, and earthworms (**Figure 33.28b**).

Movement Platypuses swim with the aid of their webbed feet. Echidnas walk on their four legs.

Reproduction Platypuses lay their eggs in a burrow, while echidnas keep their eggs in a pouch on their belly. The young hatch quickly, and the mother must continue keeping them warm and dry for another four months.

(a) Platypus *Ornithorhynchus anatinus*

(b) Echidna *Tachyglossus aculeatus*

FIGURE 33.28 Platypuses and Echidnas Are Egg-Laying Mammals

Chordata ▶ Mammalia ▶ Marsupiala (Marsupials)

The 275 known species of marsupials live in the Australian region and the Americas (**Figure 33.29**). Although females have a placenta that nourishes embryos during development, the young are born after a short embryonic period and are poorly developed. They crawl from the opening of the female's reproductive tract to the female's nipples, where they suck milk. They stay attached to their mother until they grow large enough to move independently.

Feeding Marsupials are herbivores, omnivores, or carnivores ("meat-eaters"). Many more cases of convergent evolution have occurred than for placental mammals. For example, the recently extinct Tasmanian wolf was a long-legged, social hunter similar to the timber wolves of North America and northern Eurasia. A species of marsupial native to Australia specializes in eating ants and looks and acts much like the South American anteater, which is not a marsupial.

Movement Marsupials move by crawling, gliding, walking, running, or hopping.

Reproduction Marsupial young spend more time developing while attached to their mother's nipple than they do inside her body being fed via the placenta.

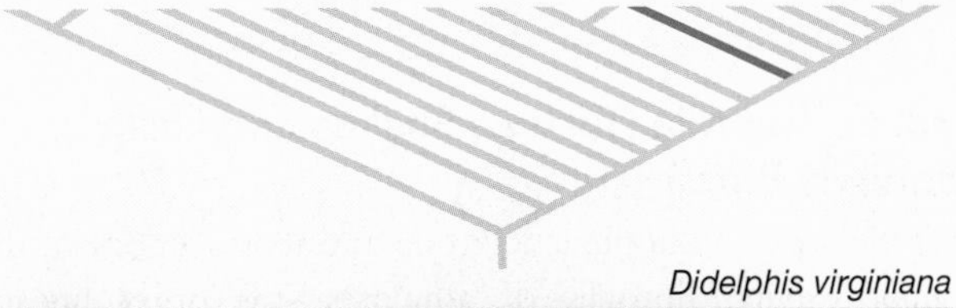

Didelphis virginiana

FIGURE 33.29 Marsupials Give Birth after a Short Embryonic Period

Chordata ▶ Mammalia ▶ Eutheria (Placental Mammals)

The approximately 4300 species of placental mammals are distributed worldwide. They are far and away the most species-rich and morphologically diverse group of mammals.

Biologists group mammals into 18 lineages called orders. The six most species-rich mammalian orders are the rodents (rats, mice, squirrels; 1814 species), bats (986 species), insectivores (hedgehogs, moles, shrews; 390 species), artiodactyls (pigs, hippos, whales, deer, sheep, cattle; 293 species), carnivores (dogs, bears, cats, weasels, seals; 274 species), and primates (lemurs, monkeys, apes, humans; 235 species).

Feeding The size and structure of the teeth correlate closely with the diet of placental mammals. Herbivores have large, flat teeth for crushing leaves and other coarse plant material; predators have sharp teeth that are efficient at biting and tearing flesh. Omnivores, such as humans, usually have several distinct types of teeth. The structure of the digestive tract also correlates with the placental mammals' diet. In some plant-eaters, for example, the stomach hosts unicellular organisms that digest cellulose and other complex polysaccharides.

Movement In placental mammals, the structure of the limb correlates closely with the type of movement performed. Eutherians fly, glide, run, walk, swim, burrow, or swing from trees (**Figure 33.30**). Limbs are reduced or lost in aquatic groups such as whales and dolphins, which swim by undulating their bodies.

Reproduction In mammals, both fertilization and development are internal. Thus, eutherians are viviparous. An extensive placenta develops from a combination of maternal and fetal tissues, and gestation is relatively long. At birth, young are much better developed than in marsupials—some are able to walk or run minutes after birth. All eutherians feed their offspring milk until the young have grown large enough to process solid food. A prolonged period of parental care, extending beyond the nursing stage, is common as offspring learn how to escape predators and find food.

Hylobates lar

FIGURE 33.30 Eutherians Are the Most Species-Rich and Diverse Group of Mammals

Chordata: Reptilia (Turtles, Snakes and Lizards, Crocodiles, Birds)

The reptiles are a monophyletic group and represent one of the two major living lineages of amniotes—the other lineage consisting of the extinct mammal-like reptiles and today's mammals. The major feature distinguishing the reptilian and mammalian lineages is the number and placement of openings in the skull. These openings are used by jaw muscles that make sophisticated biting and chewing movements possible.

Several features adapt reptiles for life on land. Their skin is made watertight by a layer of scales made of the protein keratin, which is the same protein as that found in mammalian hair. They breathe air through well-developed lungs and lay shelled, amniotic eggs that resist drying out. In turtles, the egg has a leathery shell; in other reptiles, the shell is made of stiff calcium carbonate. Because sperm and egg must fuse before the amniotic membrane and shell form, fertilization in reptiles is internal.

The reptiles include the dinosaurs, pterosaurs (flying reptiles), mosasaurs (marine reptiles), and other extinct lineages that flourished from about 250 million years ago until the mass extinction at the end of the Cretaceous, 65 million years ago. Today, the Reptilia are represented by four major lineages, traditionally recognized as subclasses: (1) turtles, (2) snakes and lizards, (3) crocodiles and alligators, and (4) birds. With the exception of birds, all of these groups are **ectothermic** ("outside-heated")—meaning that individuals do not use internally generated heat to regulate their body temperature. It would be a mistake, however, to conclude that reptiles other than birds do not regulate their body temperature closely. Reptiles bask in sunlight, seek shade, and perform other behaviors to keep their body temperature at a preferred level.

Chordata ▶ Reptilia ▶ Testudinia (Turtles, Tortoises)

The 271 known species of turtles and tortoises inhabit freshwater, marine, and terrestrial environments throughout the world. The testudines are distinguished by a shell composed of bony plates that fuse to the vertebrae and ribs (**Figure 33.31**). Their skulls are highly modified versions of the skulls of other reptiles. Turtles and tortoises lack teeth, but their jawbone and lower skull form a bony beak.

Feeding Turtles are either herbivorous or carnivorous—feeding on whatever animals they can capture and swallow. They may also scavenge dead material. Most marine turtles are carnivorous. Leatherback turtles, for example, feed primarily on jellyfish, such as the Portuguese man-of-war, and are only mildly affected by the jellyfish's stinging cnidocytes. Tortoises are plant eaters.

Movement Turtles swim, walk, or burrow. Aquatic species usually have feet that are modified to function as flippers.

Reproduction All turtles are oviparous. Other than digging a nest prior to depositing eggs, parental care is lacking. The sex of a baby turtle is often not determined by sex chromosomes. Instead, in many species the temperature at which the egg develops determines gender. High temperatures produce mostly males, while low temperatures produce mostly females.

Testudo pardalis

FIGURE 33.31 Turtles Have a Shell Consisting of Bony Plates

Chordata ▶ Reptilia ▶ Lepidosauria (Lizards, Snakes)

Most squamates are small reptiles with elongated bodies and scaly skin. Most lizards have well-developed jointed legs, but snakes are limbless (**Figure 33.32**). The hypothesis that snakes evolved from limbed ancestors is supported by the presence of vestigial hip and leg bones in boas and pythons. There are about 6800 species of lizards and snakes alive now.

Feeding Small lizards prey on insects. Although most of the larger lizard species are herbivores, the 3-meter-long monitor lizard from the island of Komodo is a predator that kills and eats deer. Snakes are carnivores; some subdue their prey with the aid of poison that is injected via modified teeth called *fangs*. Snakes prey primarily on small mammals, amphibians, and invertebrates, which they swallow whole (usually headfirst).

Movement Lizards crawl or run on their four limbs. Snakes and limbless lizards burrow through soil, crawl over the ground, or climb trees by undulating their bodies.

Reproduction Although most squamates lay eggs, many are oviviparous. Most squamates are bisexual, but asexual reproduction, via the production of eggs by mitosis, is known to occur in six groups of lizards and one snake lineage.

Morelia viridis

FIGURE 33.32 Snakes Are Limbless Predators

Chordata ▶ Reptilia ▶ Crocodilia (Crocodiles, Alligators)

Only 21 species of crocodile and alligator are known. Most are tropical and live in freshwater or marine environments. They have eyes located on the top of their heads and nostrils located at the top of their long snouts—adaptations that allow them to sit semi-submerged in water for long periods of time (**Figure 33.33**).

Feeding Crocodilians are predators. Their jaws are filled with conical teeth that are continually replaced as they fall out during feeding. The usual method of killing small prey is by biting through the body wall. Large prey are usually subdued by drowning. Crocodilians eat amphibians, turtles, fish, birds, and mammals.

Movement Crocodiles and alligators walk or gallop on land. In water they swim with the aid of their large, muscular tails.

Reproduction Although crocodilians are oviparous, parental care is extensive. Eggs are laid in earth-covered nests that are guarded by the parents. When young inside the eggs begin to vocalize, parents dig them up and carry the newly hatched young inside their mouths to nearby water. Crocodilian young can hunt when newly hatched but stay near their mother for up to three years.

Alligator mississippiensis

FIGURE 33.33 Alligators Are Adapted for Aquatic Life

Chordata ▶ Reptilia ▶ Aves (Birds)

The fossil record provides conclusive evidence that birds descended from a lineage of dinosaurs that had a unique trait: **feathers**. In dinosaurs, feathers are hypothesized to have functioned as insulation and in courtship or aggressive displays. In birds, feathers insulate and are used for display but also furnish the lift, power, and steering required for flight. Birds have other adaptations that make flight possible, including large breast muscles used to flap the wings. Bird bodies are lightweight, because they have a reduced number of bones and organs and because their hollow bones are filled with air sacs, linked to the lungs. Instead of teeth, birds have a horny beak. They are **endotherms** ("within-heating")—they have a high metabolic rate and use the heat produced, along with the insulation provided by feathers, to maintain a constant body temperature. The 9700 bird species today occupy virtually every habitat, including the open ocean (**Figure 33.34**).

Feeding Plant-eating birds usually feed on nectar or seeds. Most birds are omnivores, although many are predators that capture insects, small mammals, fish, other birds, lizards, mollusks, or crustaceans. The size and shape of a bird's beak correlates closely with its diet. For example, predatory species such as falcons have sharp, hook-shaped beaks; finches and other seedeaters have short, stocky bills that can crack seeds and nuts; fish-eating species such as the great blue heron have spear-shaped beaks.

Movement Although flightlessness has evolved repeatedly during the evolution of birds, almost all species can fly. The size and shape of birds' wings correlate closely with the type of flying they do. Birds that glide or hover have long, thin wings; species that specialize in explosive takeoffs and short flights have short, stocky wings. Many seabirds are efficient swimmers, using their webbed feet or wings under water. Ground-dwelling birds such as ostrich and pheasants can run long distances at high speed.

Reproduction Birds are oviparous but provide extensive parental care. In most species, one or both parents build a nest and incubate the eggs. After the eggs hatch, parents feed offspring until they are large enough to fly and find food on their own.

Diomedea melanophris

FIGURE 33.34 Birds Are Feathered Descendants of Dinosaurs

33.6 Key Lineages: The Hominin Radiation

Although humans occupy a tiny twig on the tree of life, there has been a tremendous amount of research on human origins. This section introduces the lineage of mammals called the Primates, the fossil record of human ancestors, and data on the relationships among human populations living today.

The Primates

The lineage called Primates consists of two main groups: prosimians and anthropoids. The **prosimians** ("before-monkeys") consist of the lemurs, found in Madagascar, and the tarsiers, pottos, and lorises of Africa and south Asia. Most prosimians live in trees and are active at night (**Figure 33.35a**). The **anthropoids** ("human-like") include the New World monkeys found in Central and South America, the Old World monkeys that live in Africa and tropical regions of Asia, the gibbons of the Asian tropics, and the great apes—orangutans, gorillas, chimpanzees, and humans (Figure 33.35b). The phylogenetic tree in **Figure 33.35c** shows the evolutionary relationships among these groups.

Primates are distinguished by having eyes located on the front of the face. Eyes that look forward provide better depth perception than do eyes on the sides of the face. Primates also tend to have hands and feet that are efficient at grasping, flattened nails instead of claws on the fingers and toes, brains that are large relative to overall body size, complex social behavior, and extensive parental care of offspring. The anthropoids are distinguished from the prosimians by having a fully *opposable thumb*—meaning a thumb that can touch the tips of all the other fingers—which makes grasping particularly efficient.

The lineage in Figure 33.35c that is composed of the **great apes**, including humans, is known as the **hominids.** From extensive comparisons of DNA sequence data, it is now clear that humans are most closely related to the chimpanzees and that our next nearest living relative is the gorilla.

Compared with most types of primate, the great apes are relatively large bodied and have long arms, short legs, and no tail.

(a) Prosimians are small, tree-dwelling primates.

Loris tardigradus

(b) New World monkeys are anthropoids.

Ateles geoffrogi

(c) Humans are apes.

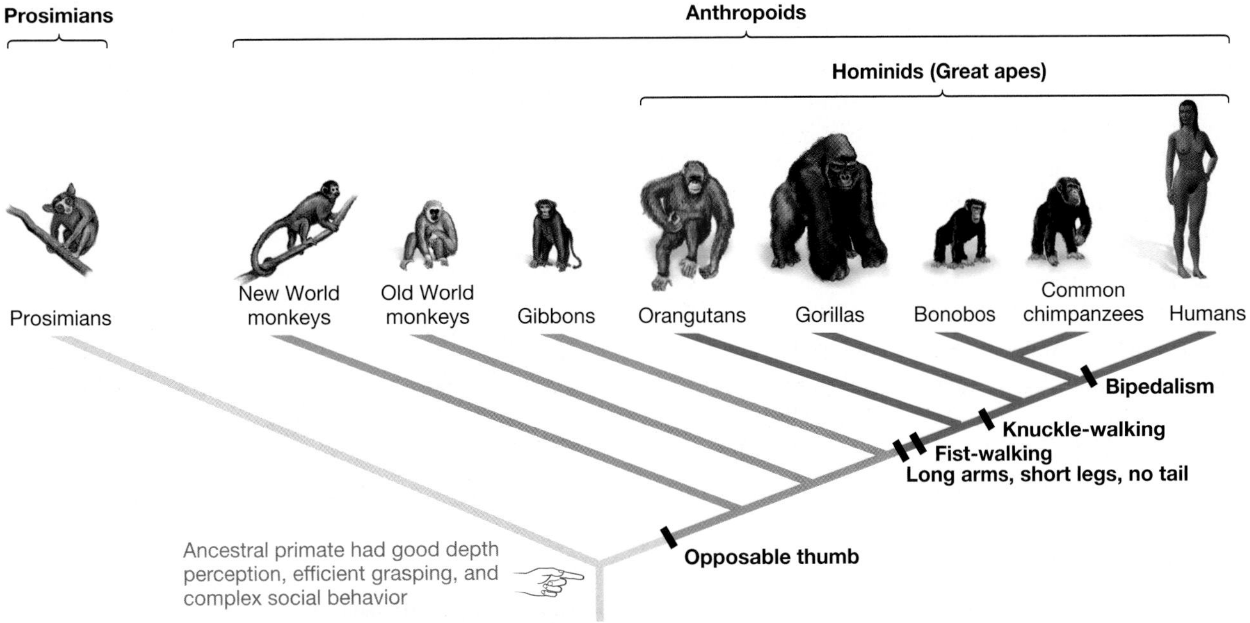

FIGURE 33.35 A Phylogeny of the Monkeys and Great Apes
(a) Prosimians live in Africa, Madagascar, and south Asia. **(b)** Anthropoids include Old World monkeys, New World monkeys, and great apes. **(c)** Phylogenetic tree estimated from extensive DNA sequence data. According to the fossil record, humans and chimps shared a common ancestor 6 to 7 million years ago.

Although all of the great ape species except for the orangutans live primarily on the ground, they have distinct ways of walking. When orangutans do come to the ground, they occasionally walk with their knuckles pressed to the ground. More commonly, though, they "fist walk"—that is, they walk with the backs of their hands pressed to the ground. Gorillas and chimps, in contrast, only "knuckle-walk." They also occasionally rise up on two legs—usually in the context of displaying aggression. Humans are the only great ape that is fully **bipedal** ("two-footed")—meaning they walk upright on two legs. Bipedalism is, in fact, the shared derived character that defines the group called hominins.

Fossil Humans

According to the fossil record, the common ancestor of chimps and humans lived in Africa 6–7 million years ago. As a group, all the species on the branch leading to contemporary humans are called **hominins**. The fossil record of hominins, though not nearly as complete as investigators would like, is rapidly improving. About 14 species have been found, and new fossils that inform the debate over the ancestry of humans are discovered every year. Although naming the hominin species and interpreting their characteristics remain intensely controversial, most researchers agree that they can be organized into four groups:

1. ***Australopithecus*** Four species of small apes called **gracile australopithecines** have been identified to date (**Figure 33.36a**). The adjective *gracile*, or "slender," is appropriate because these organisms were slightly built. The genus name *Australopithecus* ("southern ape") was inspired by the earliest specimens, which came from South Africa. Several lines of evidence support the hypothesis that the gracile australopithecines were bipedal. For example, the hole in the back of their skulls where the spinal cord connects to the brain is oriented downward, just as it is in our species, *Homo sapiens*. In chimps, gorillas, and other vertebrates that walk on four feet, this hole is oriented backward.
2. ***Paranthropus*** Three species are grouped in the genus *Paranthropus* ("beside-human"). Like the gracile australopithecines, these **robust australopithecines** were bipedal. But as **Figure 33.36b** illustrates, their skulls were much broader and more robust than those of the gracile forms. All three species had massive cheek teeth and jaws, very large cheekbones, and a *sagittal crest*—a flange of bone at the top of the skull. Because muscles that work the jaw attach to the sagittal crest and cheekbones, researchers conclude that these organisms made their living by crushing large seeds or coarse plant materials. One species is nicknamed "nutcracker man." The name *Paranthropus* was inspired by the hypothesis that the three known species are a monophyletic group that was a side branch during human evolution—an independent lineage that went extinct.
3. **Early *Homo*** Species in the genus *Homo* are called **humans**. As **Figure 33.36c** shows, species in this genus have flatter

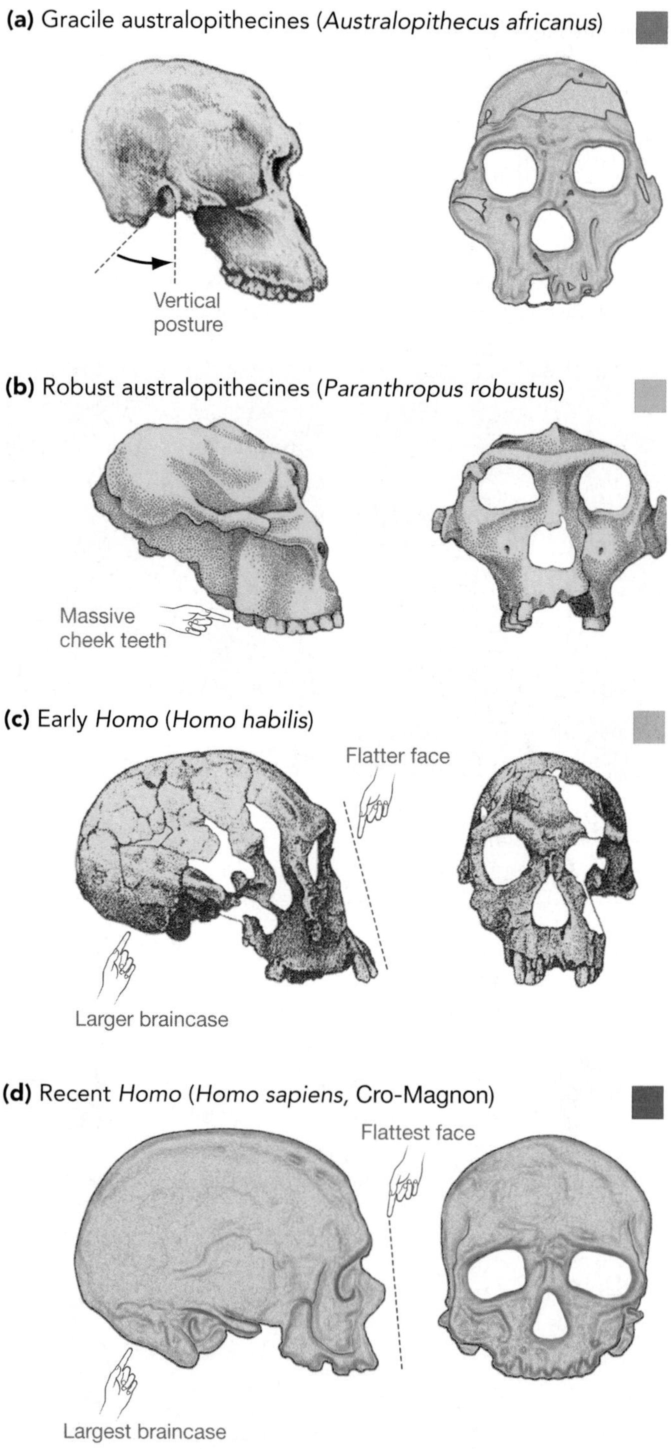
(a) Gracile australopithecines (Australopithecus africanus)
Vertical posture
(b) Robust australopithecines (Paranthropus robustus)
Massive cheek teeth
(c) Early Homo (Homo habilis)
Flatter face
Larger braincase
(d) Recent Homo (Homo sapiens, Cro-Magnon)
Flattest face
Largest braincase

FIGURE 33.36 African Hominins Comprise Four Major Groups
QUESTION The skulls are arranged as they appear in the fossil record, from most ancient to most recent—(a) to (d). How did the forehead and brow ridge of hominins change through time?

TABLE 33.1 Characteristics of Selected Hominins

Species	Location of Fossils	Estimated Average Braincase Volume (cm^3)	Estimated Average Body Size (kg)	Associated with Stone Tools?
Australopithecus afarensis	Africa	450	36	no
A. africanus	Africa	450	36	no
Paranthropus boisei	Africa	510	44	no?
Homo habilis	Africa	550	34	yes
H. ergaster	Africa	850	58	yes
H. erectus	Africa, Asia	1000	57	yes
H. heidelbergensis	Africa, Europe	1200	62	yes
H. neanderthalensis	Middle East, Europe, Asia	1500	76	yes
H. sapiens	Middle East, Europe, Asia	1350	53	yes

and narrower faces, smaller jaws and teeth, and larger braincases than the earlier hominins do. (The **braincase** is the portion of the skull that encloses the brain.) The appearance of early members of the genus *Homo* in the fossil record coincides closely with the appearance of tools made of worked stone—most of which are interpreted as handheld choppers or knives. Although the fossil record does not exclude the possibility that *Paranthropus* made tools, many researchers favor the hypothesis that extensive toolmaking was a diagnostic trait of early *Homo*.

4. **Recent *Homo*** The recent species of *Homo* date from 1.2 million years ago to the present. As **Figure 33.36d** shows, these species have even flatter faces, smaller teeth, and larger braincases than the early *Homo* species do. The 30,000-year-old fossil in the figure, for example, is from a population of *Homo sapiens* called the **Cro-Magnons**. The Cro-Magnons were accomplished painters and sculptors who buried their dead in carefully prepared graves. There is also evidence that **Neanderthal** people (*Homo neanderthalensis*) made art and buried their dead in a ceremonial fashion.

Table 33.1 summarizes data on the geographic range, braincase volume, and body size of selected species within these four groups. **Figure 33.37** provides the time range of each species in the fossil record. Although researchers do not have a solid understanding of the phylogenetic relationships among the hominin species, several points are clear from the available data. First, the shared, derived character that defines the hominins is bipedalism. Second, several species from the lineage were present simultaneously during most of hominin evolution. For example, about 1.8 million years ago there may have been as many as five hominin species living in eastern and southern Africa. Because fossils from more than one species have been found in the same geographic location in rock strata of the same age, it is almost certain that different hominin species lived in physical contact. Finally, compared with the gracile and robust australopithecines and the great apes, species in the genus *Homo* have extremely large brains relative to their overall body size.

Why did humans evolve such gigantic brains? The leading hypothesis on this question is that early *Homo* began using symbolic spoken language along with initiating extensive tool use. The logic here is that increased toolmaking and language use triggered natural selection for the capacity to reason and communicate, which required a larger brain. To support this

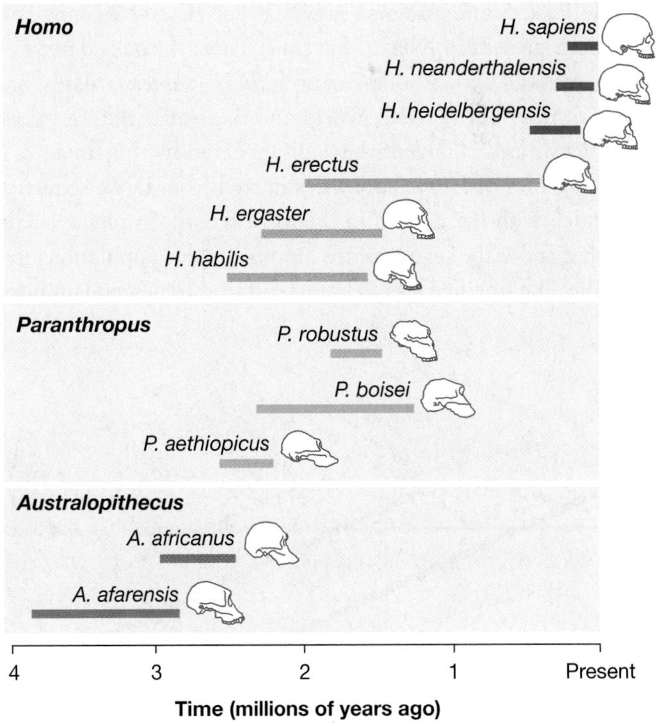

FIGURE 33.37 A Timeline of Human Evolution
Plot of the ages of selected fossil hominins. QUESTION How many species of hominin existed 2.2 million years ago, 1.8 million years ago, and 100,000 years ago?

hypothesis, researchers point out that, relative to the brain areas of other hominins, the brain areas responsible for language were enlarged in the earliest *Homo* species. There is even stronger fossil evidence for extensive use of speech in *Homo neanderthalensis* and early *Homo sapiens*:

- The hyoid bone is a slender bone in the voice box of modern humans that holds muscles used in speech. In Neanderthals and early *Homo sapiens*, the hyoid is vastly different in size and shape from a chimpanzee's hyoid bone. Researchers recently found an intact hyoid bone associated with a 60,000-year-old Neanderthal individual and showed that it is virtually identical to the hyoid of modern humans.
- *Homo sapiens* colonized Australia by boat between 60,000 and 40,000 years ago. An expedition of that type could not be planned and carried out in the absence of symbolic speech.

To summarize, *Homo sapiens* is the sole survivor of an adaptive radiation that took place over the past 3.5 million years. From a common ancestor shared with chimpanzees, hominins evolved the ability to walk upright, make tools, and talk.

The Out-of-Africa Hypothesis

The first fossils of *Homo sapiens* appear in African rocks that date to about 160,000 years ago. For some 100,000 years thereafter, the fossil record indicates that our species occupied Africa while *H. neanderthalensis* resided in Europe and the Middle East. Some evidence suggests that *H. erectus* may still have been present in Asia at that time. Then, in rocks dated between 60,000 and 30,000 years ago, *H. sapiens* fossils are found throughout the Old World and Australia. But *H. neanderthalensis* and *H. erectus* have disappeared by this time.

Phylogenies of *Homo sapiens* estimated from DNA sequence data agree with the pattern in the fossil record. In phylogenetic trees that show the relationships among human populations living today, the first lineages to branch off lead to descendant populations that live in Africa today (**Figure 33.38**). Based on this observation, it is logical to infer that the ancestral population of modern humans also lived in Africa. The tree shows that lineages subsequently branched off to form three monophyletic groups. Because the populations within each of these clades live in a distinct area, the three lineages are thought to represent three major waves of migration that occurred as *Homo sapiens* populations dispersed from Africa to (1) southeast Asia and the South Pacific, (2) central Asia and Europe, and (3) northeast Asia and the Americas (**Figure 33.39**). To summarize, the data suggest that modern humans originated in Africa and then spread throughout the world in a series of three major migrations.

What happened to the Neanderthals and to *Homo erectus* as *H. sapiens* expanded its range? This simple question has provoked years of heated controversy. Currently, the debate boils down to two possibilities: (1) Either *H. sapiens* interbred with the other two hominid groups as it moved into Europe and Asia, or (2) it did not. The first possibility is called the **assimilation hypothesis.** It implies that the genetic composition and morphological features of *H. sapiens* are a mixture of ancient traits from Neanderthals and *H. erectus* and recent traits from *H. sapiens*. The second possibility is called the **out-of-Africa hypothesis.** It contends that *H. sapiens* evolved independently of the European and Asian species of *Homo*—meaning there was no interbreeding between *H. sapiens* and Neanderthals or *H. erectus*. Stated another way, the out-of-Africa hypothesis proposes that *H. sapiens* evolved its distinctive traits in Africa, then dispersed throughout the world. The assimilation hypothesis proposes that *Homo sapiens* acquired its distinctive traits at least in part through extensive interbreeding with *H. neanderthalensis* or *H. erectus*, or both.

If the assimilation hypothesis is correct, then modern humans should contain genes descended from *H. neanderthalensis* or *H. erectus*, or both. Recall from Chapter 19 that researchers have tested this prediction by extracting DNA from the fossilized bones of four Neanderthal individuals from three differ-

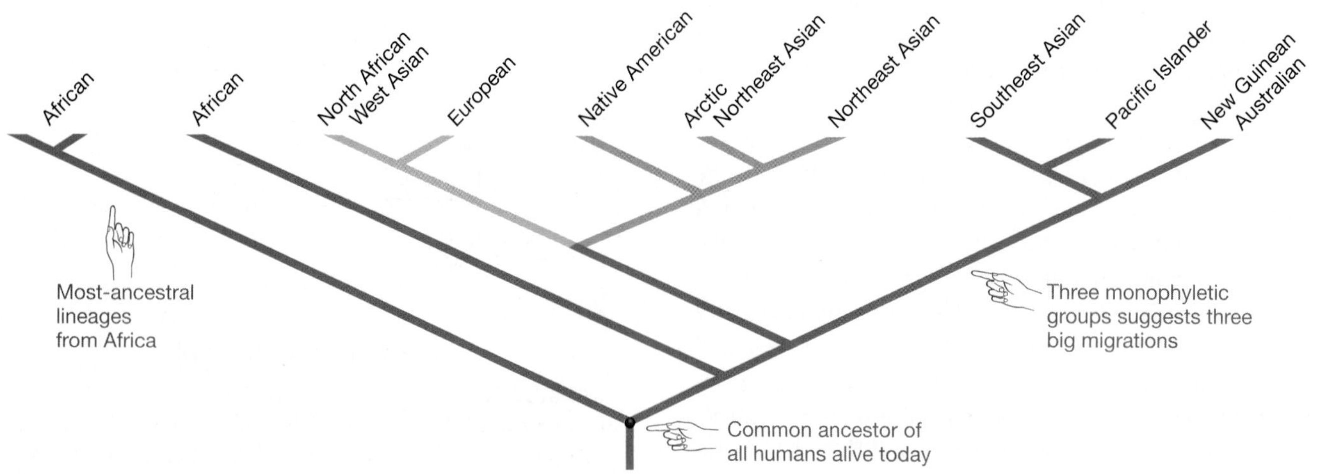

FIGURE 33.38 Phylogeny of Human Populations Living Today
Phylogeny of modern human populations, as estimated from DNA sequence data.

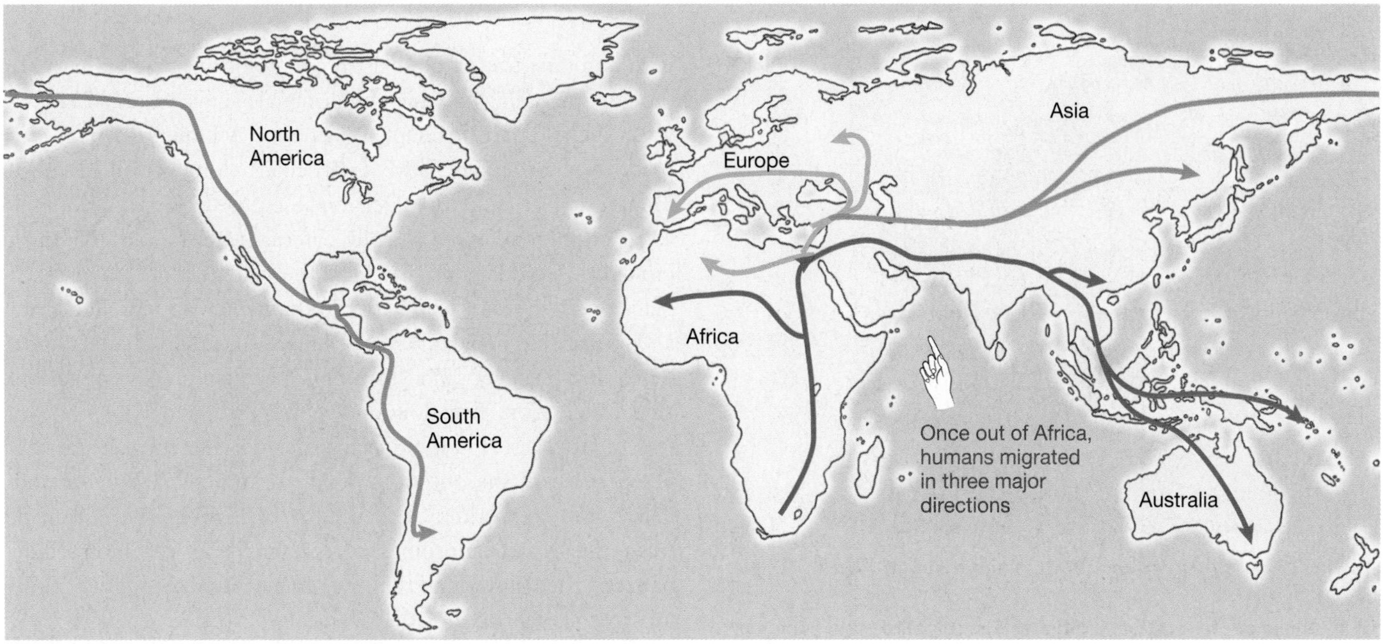

FIGURE 33.39 ***Homo sapiens*** **Originated in Africa and Spread throughout the World**
The phylogeny in Figure 33.38 supports the hypothesis that humans originated in Africa and spread out in three major migrations: to southeast Asia and the Pacific Islands, to Europe, and to northeast Asia and the New World.

ent locations in Eurasia. In each case, the researchers ground samples from the intact bone, extracted DNA, and used the polymerase chain reaction to copy a 379-base-pair section of the mitochondrial genome. When they sequenced the Neanderthal DNA and compared it with sequences from about a thousand humans living today, the results were striking: The Neanderthal DNA was extremely different from the DNA of living humans. For example, when sequences from any two randomly chosen *H. sapiens* are compared with each other, an average of 8 bases are unalike. But when randomly chosen *H. sapiens* sequences are compared to the *H. neanderthalensis* DNA, there is an average of over 25 differences. Further, the sequence differences observed in the Neanderthal DNA are unique—none of the *H. sapiens* sequences contained the nucleotide substitutions found in *H. neanderthalensis*. These results support the hypothesis that *H. sapiens* and *H. neanderthalensis* did not interbreed.

Unfortunately, it has been impossible so far to extract DNA from *H. erectus* fossils and perform the same test. Although the weight of evidence currently tips the scales in favor of the out-of-Africa hypothesis, research continues.

ESSAY So Human an Animal

From a biological point of view, what is a human being? Historically, scientists and philosophers have answered this question by arguing that humans have one or more unique, defining characteristics that set us apart from other organisms. For decades this defining characteristic was thought to be tool use. But then Jane Goodall, who pioneered the study of common chimpanzees in the field, observed that chimps collect and modify twigs to "fish" for termites or ants and eat them (**Figure 33.40**, page 778). Later, Goodall and other biologists reported additional examples of tool use by common chimps, bonobos (also known as pygmy chimps), baboons, sea otters, and various birds.

In response, some observers began to argue that the defining characteristic of humans was not toolmaking but the use of symbolic speech—language based on the abstract symbols we call words and letters. But researchers have since taught chimps rudimentary aspects of the American Sign Language, and the shape of a recently discovered Neanderthal bone located in the larynx suggests that Neanderthals, too, could speak. These results have a key message: The ability to learn and to use abstract symbols and grammatical rules is not unique to humans.

Which 1 to 2 percent of the human genome is responsible for the . . . differences between chimps and humans?

In short, there may be no single characteristic that "defines" humans. Instead, most contemporary biologists would characterize our

(Continued on next page)

(Essay continued)

FIGURE 33.40 Chimpanzees Teach Their Offspring How to Make and Use Tools
Adult female chimp (bottom) using a stick as a tool to "fish" for ants. Her son, sitting on her back, is also using a stick. Tool use is a culturally transmitted behavior in chimpanzees, meaning that it is learned.

species by listing a suite of traits that are not necessarily unique. Humans are an intensely social, bipedal mammal with drastically reduced fur, a brain that is absolutely huge given our body size, and thumbs that are unusually dextrous. If pressed to come up with a unique trait, a biologist might point out that humans are the only mammals in which adult females have enlarged breasts when they are not pregnant or lactating. A biologist would also point out that humans have a particularly long period of sexual immaturity, or childhood, accompanied by extraordinarily high levels of parental and especially paternal care. In addition, most biologists would emphasize that the changes that have taken place over the past 40,000 years of human evolution have probably been driven by **cultural evolution**—meaning changes fashioned by teaching and learning—as much as by mutation, natural selection, gene flow, and genetic drift.

Recently, research in molecular genetics has furnished a new perspective on the question, Who are humans? When Svante Pääbo and co-workers sequenced a 10,156-base-pair segment from the X chromosomes of humans and common chimpanzees, the team found that the sequences were identical at almost 99 percent of the 10,156 sites.

These findings prompt a question: Which 1 to 2 percent of the human genome is responsible for the morphological and behavioral differences between chimps and humans? Will it be possible to identify the alleles responsible for our bipedal posture, large brains, slow rate of sexual maturation, and huge capacity for learning? Researchers have only begun to tackle these questions. But, with a variety of genome analysis projects now under way, answers may soon be forthcoming.

CHAPTER REVIEW

Summary of Key Concepts

- **The most prominent deuterostome lineages are the echinoderms and the vertebrate groups called ray-finned fishes and tetrapods.**

Echinoderms, ray-finned fishes, and tetrapods are the most species-rich groups of deuterostomes and are the most important large-bodied predators and herbivores in marine and terrestrial environments.

Web Tutorial 33.1 Deuterostome Diversity

- **Echinoderms and vertebrates have unique body plans. Echinoderms are radially symmetric as adults and have a water vascular system. All vertebrates have a skull, and most have an extensive endoskeleton made of cartilage or bone.**

Echinoderm larvae are bilaterally symmetric but undergo a metamorphosis into radially symmetric adults. Their water vascular system is composed of fluid-filled tubes and chambers and extends from the body wall in projections called podia. Podia can extend and retract in response to muscle contractions that move fluid inside the water vascular system.

Chordates are distinguished by the presence of a notochord, a dorsal hollow nerve cord, pharyngeal gill slits, and a muscular tail that extends past the anus. Vertebrates are distinguished by the presence of a cranium; most species also have vertebrae. In more recent groups of vertebrates, the body plan features an extensive endoskeleton composed of bone.

- **The diversification of echinoderms was triggered by the evolution of appendages called podia; the diversification of vertebrates was driven by the evolution of the jaw and limbs.**

Most echinoderms move via their podia, and many species suspension feed, deposit feed, or act as predators with the aid of their podia.

Ray-finned fishes and tetrapods use their jaws to bite food and process it with teeth. Species in both groups move when muscles attached to their endoskeletons contract or relax. Ray-finned fishes called teleosts are efficient swimmers because their flexible, symmetrical tail stabilizes their body during rapid movement. Tetrapods can move on land because their limbs enable walking, running, or flying. The evolution of the amniotic egg allowed tetrapods to lay eggs on land. Parental care was an important adaptation in some groups of ray-finned fishes and tetrapods—particularly mammals.

- **Humans are a tiny twig on the tree of life. Chimpanzees and humans diverged from a common ancestor that lived in Africa 6–7 million years ago. Since then, at least 14 humanlike species have existed.**

The fossil record of the past 3.5 million years contains at least 14 distinct species of hominins. Several of these organisms lived in Africa at the same time, and some lineages went extinct without leaving descendant populations. Thus, *Homo sapiens* is the sole surviving representative of an adaptive radiation. The phylogeny of living humans, based on comparisons of DNA sequences, agrees with evidence in the fossil record that *H. sapiens* originated in Africa and later spread throughout Europe, Asia, and the Americas. DNA sequences recovered from the fossilized bones of *H. neanderthalensis* suggest that *H. sapiens* replaced this species in Europe without interbreeding.

Questions

Content Review

1. If you found an organism on a beach, what characteristics would allow you to declare that the organism is an echinoderm?
   a. radially symmetric adults, spines, and presence of tube feet
   b. notochord, dorsal hollow nerve cord, pharyngeal gill slits, and muscular tail
   c. exoskeleton and three pairs of appendages; distinct head and body (trunk) regions
   d. mouth forms second (after the anus) during gastrulation
2. What is the diagnostic trait of vertebrates?
   a. cranium
   b. jaws
   c. endoskeleton constructed of bone
   d. endoskeleton constructed of cartilage
3. Why are the pharyngeal jaws found in many ray-finned fishes important?
   a. They allow the main jaw to be protrusible (extendible).
   b. They make it possible for individuals to suck food toward their mouths.
   c. They give rise to teeth that are found on the main jawbones.
   d. They can bite down on food and help process it.
4. Which of the following lineages make up the living Amniota?
   a. reptiles and mammals
   b. viviparous fishes
   c. frogs, salamanders, and caecilians
   d. hagfish, lampreys, and cartilaginous fishes (sharks and rays)
5. Which of the following does *not* occur in either cartilaginous fishes or ray-finned fishes?
   a. internal fertilization and viviparity or ovoviviparity
   b. external fertilization and oviparity
   c. formation of a placenta
   d. feeding of young
6. Researchers agree that modern *Homo sapiens* originated in Africa and then spread throughout Europe, Asia, and eventually the Americas. What do they disagree about?
   a. whether the original African population of *H. sapiens* left any descendants
   b. whether *H. sapiens* members interbred with *H. erectus* members
   c. whether the Neanderthals represent a distinct species from *H. sapiens*
   d. whether *H. neanderthalensis* and *H. erectus* died out concurrently

Conceptual Review

1. Explain how the water vascular system of echinoderms functions as a type of hydrostatic skeleton.
2. List the four morphological traits that distinguish chordates. How are these traits involved in locomotion and feeding in larvae or adults?
3. Describe evidence that supports the hypothesis that jaws evolved from gill arches in fish.
4. Describe evidence that supports the hypothesis that the tetrapod limb evolved from the fins of lobe-finned fishes.
5. The text claims that "*Homo sapiens* is the sole survivor of an adaptive radiation that took place over the past 3.5 million years." Do you agree with this statement? Why or why not?
6. Explain how the evolution of the placenta and lactation in mammals improved the probability that their offspring would survive. Over the course of a lifetime, why are female mammals expected to produce fewer eggs than do female fish?

Group Discussion Problems

1. A growing number of biologists maintain that only monophyletic groups should be given names. According to these researchers, there is no such thing as a fish. Explain this statement.
2. Compare and contrast adaptations that triggered the diversification of the three most species-rich animal lineages: mollusks, arthropods, and vertebrates.
3. Aquatic habitats occupy 73 percent of Earth's surface area. How does this fact relate to the success of ray-finned fishes? How does it relate to the success of coelacanths and other lobe-finned fishes?
4. Mammals and birds are endothermic. Did they inherit this trait from a common ancestor, or did endothermy evolve independently in these two lineages? Provide evidence to support your answer.

Answers to Multiple-Choice Questions 1. a; 2. a; 3. d; 4. a; 5. d; 6. b

www.prenhall.com/freeman is your resource for the following: Web Tutorials; Online Quizzes and other Online Study Guide materials; Answers to Conceptual Review Questions; Solutions to Group Discussion Problems; Answers to Figure Caption Questions and Exercises; and Additional Readings and Research.

34 Viruses

KEY CONCEPTS

- Viruses are tiny, noncellular parasites that infect virtually every type of cell known. They cannot perform metabolism on their own—meaning outside a parasitized cell—and are not considered to be alive. Different types of viruses are specialized for infecting particular species and types of cells.
- Viruses are highly diverse in overall morphology and in the nature of their genetic material. The genomes of viruses may consist of double-stranded DNA, single-stranded DNA, double-stranded RNA, or one of several types of single-stranded RNA.
- The viral infection cycle can be broken down into five steps: (1) entry into a host cell, (2) replication of the viral genome, (3) the production of viral proteins, (4) the assembly of a new generation of virus particles, and (5) exit from the infected cell.

Photomicrograph created by treating seawater with a fluorescing compound that binds to nucleic acids. The smallest, most abundant dots are viruses. The larger, numerous spots are bacteria and archaea. The largest splotches are protists.

Viruses are parasites that afflict every twig on the tree of life. They are not cells and are not made up of cells, so they are not considered organisms. They cannot manufacture their own ATP or carbon-containing compounds, and they cannot make copies of themselves. Viruses enter a **host cell**, take over its biosynthetic machinery, and use that machinery to manufacture a new generation of viruses. Outside of cells, viruses simply exist—they cannot do anything. A **virus** is an obligate, intracellular parasite. The adjective *obligate* is appropriate because viruses cannot replicate unless they enter the inside of a cell.

Because they are not organisms, viruses are referred to as **particles** or **agents** and are not given scientific (genus + species) names. Most biologists would argue that viruses are not even alive. Yet viruses have a genome, they are superbly adapted to exploit the metabolic capabilities of their host cells, and they evolve. **Table 34.1** summarizes some characteristics of viruses.

The diversity and abundance of viruses almost defy description. Nearly all organisms examined thus far are parasitized by at least one kind of virus. The bacterium *Escherichia coli*, which resides in the human intestine, is afflicted by more than a dozen types of **bacteriophage** (literally, "bacteria-eater"). A bacteriophage is a virus that infects bacteria. The ocean's plankton teems with bacteria and archaea, yet viruses outnumber them in this habitat by a factor of 10 to 1. A wine bottle filled with seawater taken from the ocean's surface contains about 10 *billion* virus particles—almost double the world's population of humans.

TABLE 34.1 Characteristics of Viruses versus Characteristics of Organisms

	Viruses	Organisms
Hereditary material	DNA or RNA; can be single stranded or double stranded	DNA; always double stranded
Plasma membrane present?	No	Yes
Can carry out transcription independently?	No—even if a viral polymerase is present, transcription of viral genome requires use of ATP and nucleotides provided by host cell	Yes
Can carry out translation independently?	No	Yes
Metabolic capabilities	Virtually none	Extensive—synthesis of ATP, reduced carbon compounds, vitamins, lipids, nucleic acids, etc.

34.1 Why Do Biologists Study Viruses?

Any study of life's diversity would be incomplete unless it included a look at the acellular parasites that exploit that diversity. But viruses are also important from a practical standpoint. To health-care workers, agronomists, and foresters, these parasites are a persistent—and sometimes catastrophic—source of misery and economic loss. The nature of viruses has been understood only since the 1940s, but they have been the focus of intense research ever since. Biologists study viruses because they cause illness and death. In the human body, virtually every system, tissue, and cell can be infected by one or more kinds of virus (**Figure 34.1**). Research on viruses is motivated by the desire to minimize the damage they can cause.

In addition, biologists study viruses with the goal of exploiting their ability to enter cells. Recall from Chapter 19 that viruses are being tested as possible therapeutic agents in the treatment of genetic diseases. If viruses can be engineered to carry normal copies of human genes into the cells of patients, it is possible that the gene products could cure symptoms.

Recent Viral Epidemics in Humans

Physicians and researchers use the term **epidemic** ("upon-people") to describe a disease that affects a large number of people at the same time. Viruses have caused the most devastating epidemics in recent human history. During the eighteenth and nineteenth centuries, it was not unusual for Native American tribes to lose 90 percent of their members over the course

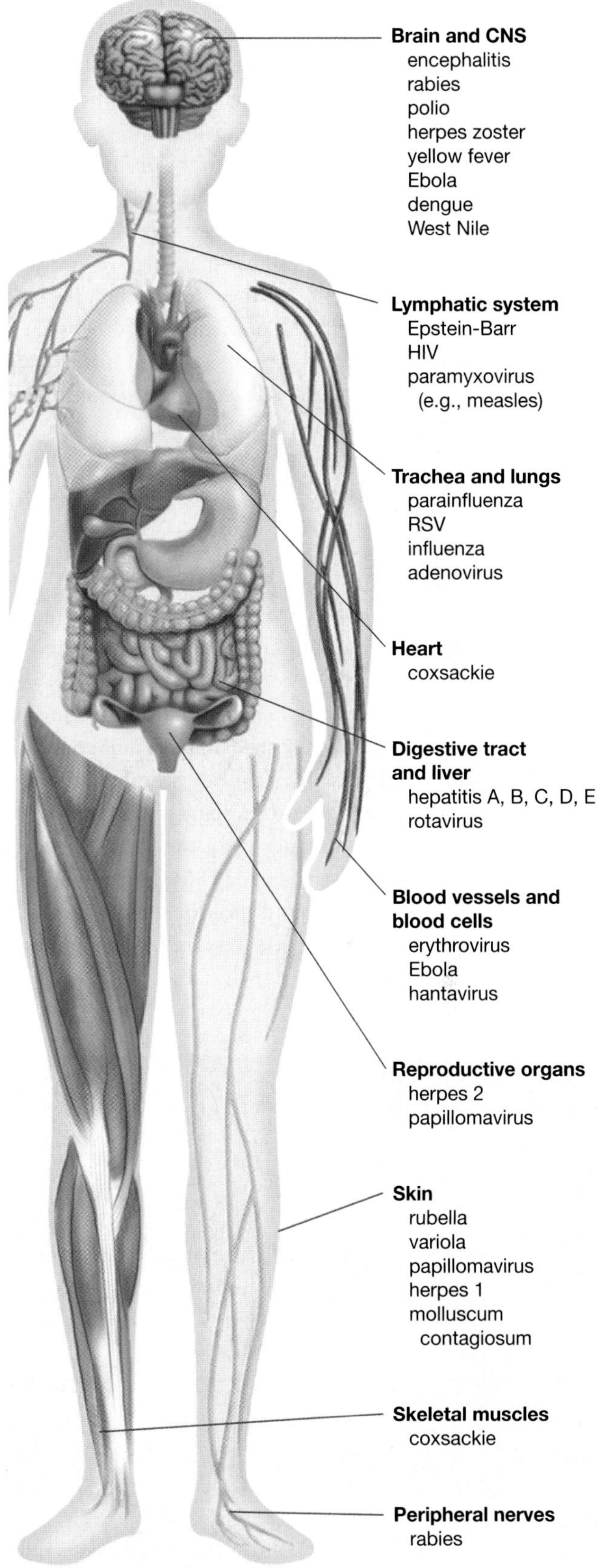

Brain and CNS
encephalitis
rabies
polio
herpes zoster
yellow fever
Ebola
dengue
West Nile
Lymphatic system
Epstein-Barr
HIV
paramyxovirus
(e.g., measles)
Trachea and lungs
parainfluenza
RSV
influenza
adenovirus
Heart
coxsackie
Digestive tract
and liver
hepatitis A, B, C, D, E
rotavirus
Blood vessels and
blood cells
erythrovirus
Ebola
hantavirus
Reproductive organs
herpes 2
papillomavirus
Skin
rubella
variola
papillomavirus
herpes 1
molluscum
contagiosum
Skeletal muscles
coxsackie
Peripheral nerves
rabies

FIGURE 34.1 Human Organs and Systems That Are Parasitized by Viruses

EXERCISE Choose two viruses that you are familiar with. Next to each, write the symptoms caused by an infection with this virus.

of a few years to measles, smallpox, and other viral diseases spread by contact with European settlers.

In terms of global impact, the influenza outbreak of 1918–1919, called "Spanish flu," qualifies as the most devastating epidemic recorded to date. Influenza is a virus that infects the upper respiratory tract. The strain of influenza virus that emerged in 1918 infected people worldwide and was particularly **virulent**—meaning it tended to cause severe disease. Within hours of showing symptoms, the lungs of previously healthy people often became so heavily infected that affected individuals suffocated to death. Most victims were between the ages of 20 and 40. The viral outbreak occurred just as World War I was drawing to a close and killed far more people than did the conflict itself. For example, ten times as many Americans died of influenza than were killed in combat in the war. Worldwide, the Spanish flu may have killed over 50 million people.

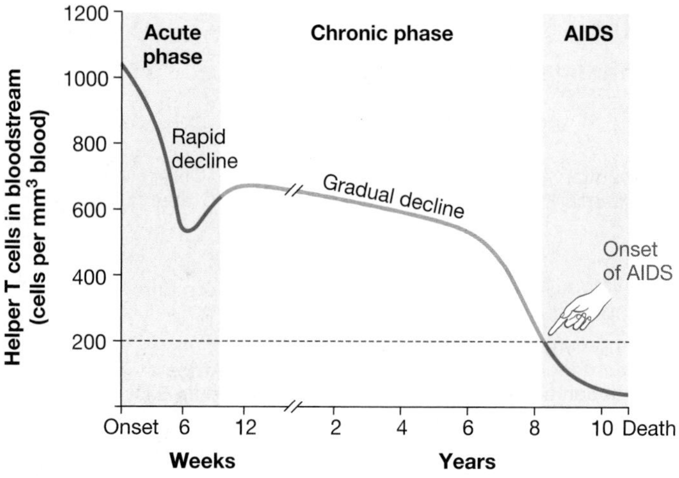

FIGURE 34.3 T-Cell Counts Decline during an HIV Infection
Graph of changes in the number of T cells that are present in the bloodstream over time, based on data from a typical patient infected with HIV. The acute phase of infection occurs immediately after infection and is sometimes associated with symptoms such as fever. Infected people usually show no disease symptoms in the chronic phase, even though their T-cell counts are in slow, steady decline. AIDS typically occurs when T-cell counts dip below 200/mm^3 of blood.

Current Viral Epidemics in Humans: HIV

In terms of the total number of people affected, the measles and smallpox epidemics among native peoples and the 1918 influenza outbreak are almost certain to be surpassed by the incidence of AIDS. **Acquired immune deficiency syndrome** (**AIDS**) is an affliction caused by the **human immunodeficiency virus** (**HIV**; **Figure 34.2**). Relatively little is known about the strain of influenza that caused the Spanish flu epidemic of 1918–1919, but HIV is far and away the most intensively studied of all viruses. Since the early 1980s, governments and private corporations from around the world have spent hundreds of millions of dollars on HIV research. More biologists are working on HIV than on any other type of virus. Given this virus's current and projected impact on human populations around the globe, the investment is justified.

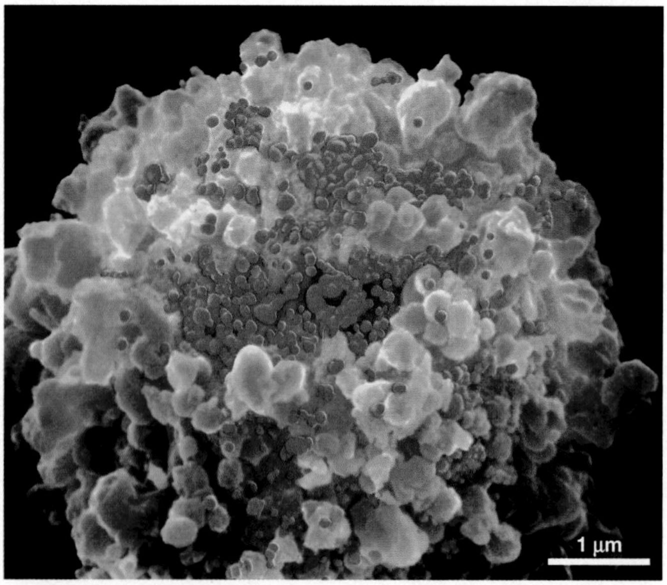

FIGURE 34.2 The Human Immunodeficiency Virus (HIV)
Colorized scanning electron micrograph showing HIV particles (red) emerging from an infected human T cell (green).

How Does HIV Cause Disease? Like other viruses, HIV parasitizes specific types of cells. The cells most affected by HIV are called *helper T cells* and *macrophages*. These cells are components of the **immune system,** which is the body's defense system against disease. Chapter 49 explains just how crucial helper T cells and macrophages are to the immune system's response to invading bacteria and viruses. If an HIV particle succeeds in infecting one of these cells and reproducing inside, however, the cell dies as hundreds of new particles break out and infect more cells. Although the body continually replaces helper T cells and macrophages, the number produced does not keep pace with the number being destroyed by HIV. As a result, the total number of helper T cells in the bloodstream gradually declines as an HIV infection proceeds (**Figure 34.3**). When the T-cell count drops, the immune system's responses to invading bacteria and viruses become less and less effective. Eventually, too few helper T cells are left to fight off pathogens efficiently, and pathogenic bacteria and viruses begin to multiply unchecked. In almost all cases, one or more of these infections prove fatal. HIV kills people indirectly—by making them susceptible to pneumonia, fungal infections, and unusual types of cancer.

What Is the Scope of the AIDS Epidemic? Researchers with the United Nation's AIDS program estimate that AIDS

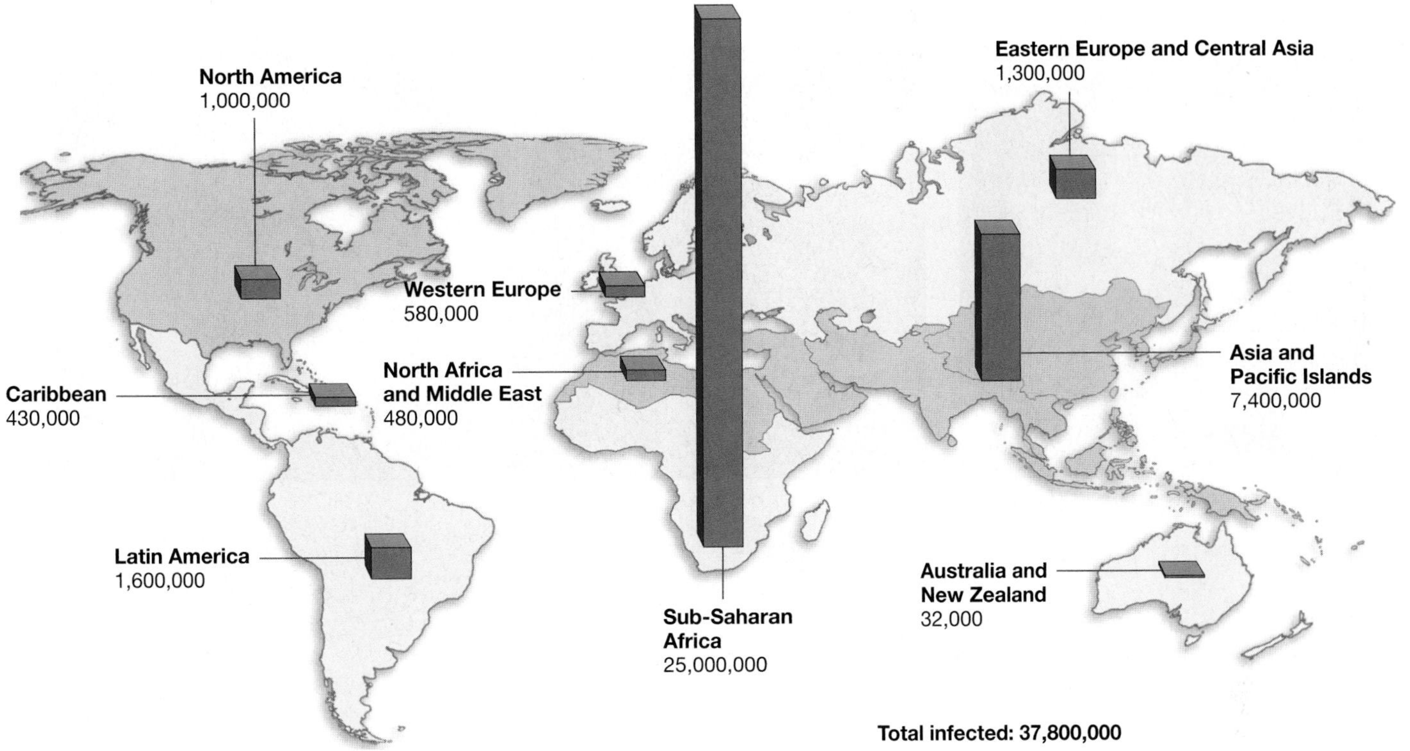

North America
1,000,000
Eastern Europe and Central Asia
1,300,000
Western Europe
580,000
North Africa
and Middle East
480,000
Caribbean
430,000
Asia and
Pacific Islands
7,400,000
Latin America
1,600,000
Australia and
New Zealand
32,000
Sub-Saharan
Africa
25,000,000
Total infected: 37,800,000

FIGURE 34.4 Geographic Distribution of HIV Infections
Data, compiled by the World Health Organization, showing the numbers of people living with HIV in July 2004, by geographic area.

has already killed 25 million people worldwide. HIV infection rates have been highest in east and central Africa, where one of the greatest public health crises in history is now occurring (**Figure 34.4**). In Botswana and elsewhere, blood-testing programs have confirmed that up to a third of all adults carry HIV. Although there may be a lag of as much as 8–12 years between the initial infection and the onset of illness, virtually all people who become infected with the virus will die of AIDS.

Currently, the UN estimates the total number of HIV-infected people worldwide at about 38 million. An additional 5 million people are infected each year, and the epidemic is growing. Researchers are particularly alarmed because the focus of the epidemic is shifting from its historical center of incidence—central and southern Africa—to south and east Asia. Infection rates are growing as much as 17 percent per year in some of the world's most populous countries—particularly India, China, Russia, Ethiopia, and Nigeria. If present trends continue, 80 million more people will be infected by 2010.

Most viral and bacterial diseases afflict the very young and the very old. But because HIV is a sexually transmitted disease, young adults are most likely to contract the virus and die. People who become infected in their late teens or twenties die of AIDS in their twenties or thirties. Tens of millions of people are being lost in the prime of their lives. Physicians, politicians, educators, and aid workers all use the same word to describe the epidemic's impact: staggering.

34.2 How Do Biologists Study Viruses?

Researchers who study viruses focus on two goals: (1) developing vaccines that allow hosts to fight off disease if they become infected and (2) developing antiviral drugs that prevent a virus from replicating efficiently inside the host. Both types of research begin with attempts to isolate the virus in question.

Isolating viruses takes researchers into the realm of "nanobiology," in which structures are measured in billionths of a meter. (One nanometer, abbreviated nm, is 10^{-9} meter.) Viruses are typically 50 to 100 nm in diameter. They are dwarfed by eukaryotic cells; millions of viruses can fit on the head of a pin.

If virus-infected cells can be grown in culture or harvested from a host individual, researchers can usually isolate the virus by passing the cells through a filter. The filters used to study viruses have pores that are large enough for viruses to pass through but are too small to admit cells. To test the hypothesis that the solution that passes through the filter contains viruses that cause a specific disease, researchers expose uninfected host cells to this filtrate. If the virus-causation hypothesis is correct, then exposing host cells to the filtrate will result in infection. In this way, researchers can isolate a virus and confirm that it is the causative agent of infection. Recall from Chapter 27 that these steps are inspired by Koch's postulates, which established the criteria for linking a specific infectious agent with a specific disease.

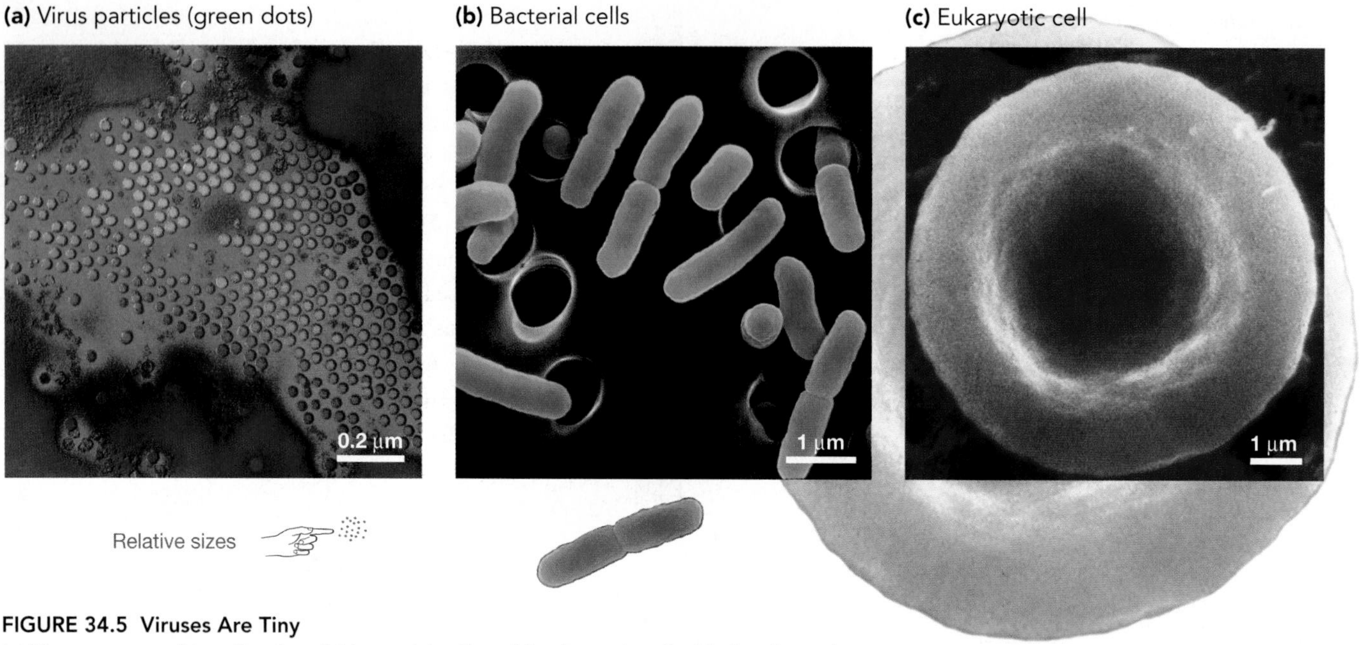
(a) Virus particles (green dots)
0.2 μm
(b) Bacterial cells
1 μm
(c) Eukaryotic cell
1 μm
Relative sizes

FIGURE 34.5 Viruses Are Tiny
(a) Viruses are much smaller than **(b)** bacterial cells or **(c)** eukaryotic cells. Notice the scale bars on each micrograph.

Once biologists have isolated a virus, how do they study and characterize it? Let's begin with morphological traits, then consider how viral life cycles vary.

Analyzing Morphological Traits

To see a virus, researchers usually rely on transmission electron microscopy (**Figure 34.5**). Only the very largest viruses, such as the smallpox virus, are visible with a light microscope. Electron microscopy has revealed that viruses come in a wide variety of shapes, and many viruses can be identified by shape alone (**Figure 34.6**). In overall structure, however, they fall into just two general categories. Viruses can either be (1) enclosed by just a shell of protein called a **capsid** or (2) enclosed by both a capsid and a membrane-like **envelope.** Regarding their morphology, then, the important distinction among viruses is whether they are **nonenveloped** or **enveloped.**

Nonenveloped viruses have an extremely simple structure. They consist of genetic material and possibly one or more enzymes inside a capsid—a protein coat. The nonenveloped virus illustrated in **Figure 34.7a** is an *adenovirus*. You undoubtedly have adenoviruses on your tonsils or in other parts of your upper respiratory passages right now.

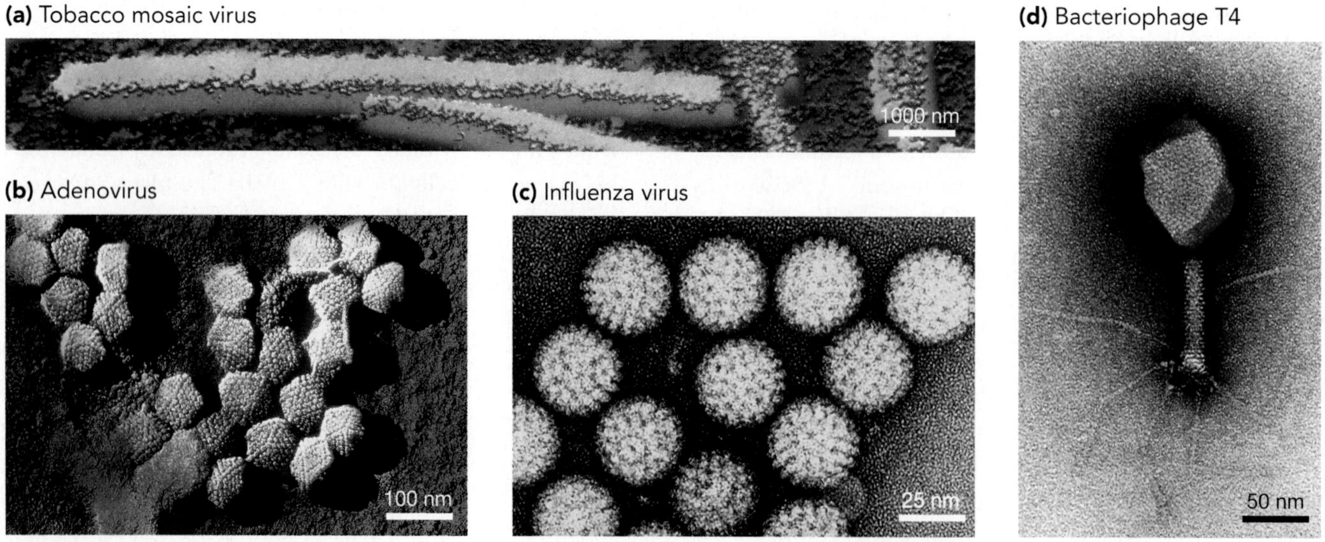
(a) Tobacco mosaic virus
1000 nm
(b) Adenovirus
100 nm
(c) Influenza virus
25 nm
(d) Bacteriophage T4
50 nm

FIGURE 34.6 Viruses Vary in Size and Shape
Virus shapes include **(a)** rods, **(b)** polyhedrons, **(c)** spheres, and **(d)** complex shapes with heads and tails.

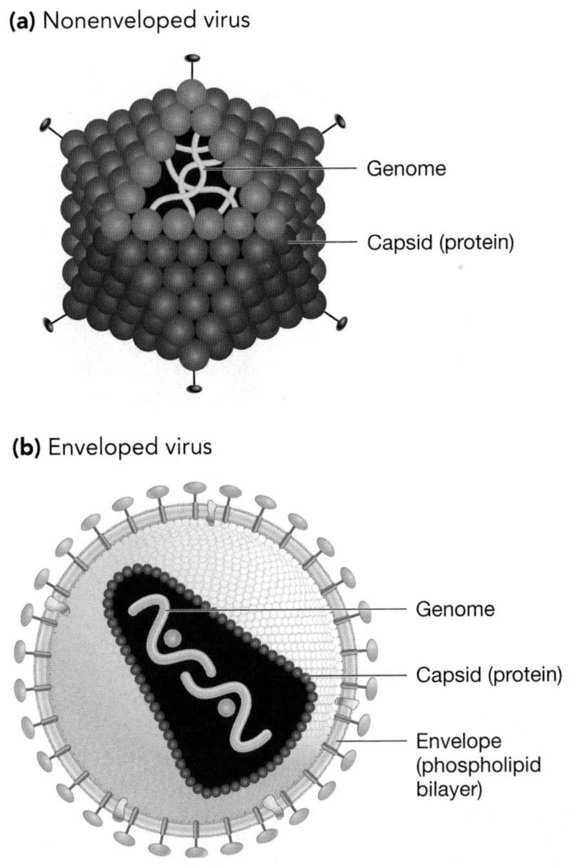
(a) Nonenveloped virus
Genome
Capsid (protein)
(b) Enveloped virus
Genome
Capsid (protein)
Envelope (phospholipid bilayer)

FIGURE 34.7 Viruses Are Nonenveloped or Enveloped
(a) A protein coat forms the exterior of nonenveloped viruses. **(b)** In enveloped viruses, the exterior is composed of a membranous sphere. Inside this envelope, the hereditary material is enclosed by a protein coat.

Enveloped viruses are slightly more complex. They also have genetic material inside a capsid, but the capsid is surrounded by an envelope. The envelope consists of a phospholipid bilayer with a mixture of viral proteins and proteins derived from the plasma membrane of a host cell—specifically, the host cell in which the virus particle was manufactured. The enveloped virus illustrated in **Figure 34.7b** is HIV.

Once a virus has been isolated and its overall morphology characterized, researchers usually focus on understanding the nature of the virus's replication cycle and on attempts to develop a vaccine. A **vaccine** is a preparation that primes a host's immune system to respond to a specific type of virus. **Box 34.1** explains how vaccines work and why it has not been possible to develop a vaccine for certain viruses—particularly the cold virus or HIV. Here let's focus on variation in virus replication cycles.

Analyzing Variation in Replication Cycles: Lytic and Lysogenic Growth

Although it is likely that millions of types of virus exist, they all infect their host cells in one of two general ways: lytic growth or lysogenic growth. All viruses undergo lytic growth; some can grow lysogenically as well. Both types of viral infection begin when part or all of a virus particle enters the interior of a host cell.

Figure 34.9a shows a **lytic replication cycle.** During lytic growth, the viral genome enters the host cell and viral or host enzymes make copies of it, using nucleotides and ATP provided by the host. The host cell also manufactures viral proteins. When synthesis of the viral genome and viral proteins is complete, a new generation

BOX 34.1 How Are Vaccines Developed?

As Chapter 49 will show, one of the primary ways that an individual's immune system responds to parasite attack is through the production of antibodies. An **antibody** ("against-body") is a protein that binds with high specificity to a particular site on another molecule. That is, an antibody will bind only to one particular site on one particular compound. Any molecule that elicits the production of antibodies is considered to be an **antigen** ("against-produce").

Vaccines contain antigens. The antigens are usually components of a virus's exterior—the capsid from a nonenveloped virus or the envelope proteins from an enveloped virus. Exterior proteins are effective antigens because they are the part of the virus that is "seen" by the immune system cells.

To vaccinate a person or a domestic animal, viral antigens are injected or swallowed. Once inside the individual, the antigens stimulate immune system cells called *B cells* (see Chapter 49), which produce antibodies to the antigen (**Figure 34.8**, page 786). Once antigens are coated with antibodies, the antigens are destroyed by other cells and components of the immune system.

Vaccinations are like fire drills or earthquake preparedness drills—they prepare the immune system to respond to a specific type of threat. In effect, they are a faked infection. Immune system cells that are stimulated to produce antibodies by a vaccination remain active in the vaccinated person for a long time—often for life. If a vaccinated person is later exposed to active virus particles, the immune system can respond quickly and effectively enough to stop the infection before it threatens the individual's health.

In many cases, vaccines have been spectacularly successful at curbing or

(Continued on next page)

(Box 34.1 continued)

even eliminating viral diseases in humans and domestic animals. In humans alone, effective vaccines have been developed against smallpox, yellow fever, polio, measles, and some forms of hepatitis. These diseases used to kill or sicken hundreds of thousands of people each year.

There Are Two General Types of Vaccines

Successful vaccines consist of inactivated viruses or attenuated viruses. An **inactivated virus** is not capable of causing an infection, because its genes have been damaged by chemical treatments—often exposure to formaldehyde—or irradiation with ultraviolet light. If you have been vaccinated for hepatitis A or flu, you have received an inactivated virus.

Attenuated viruses are also called **"live" virus vaccines,** because they consist of complete virus particles. The adjective *attenuated* means that they lack virulence. Researchers can make a virus harmless to a host by culturing the virus on cells from species other than that host. In adapting to growth on the atypical cells, viral strains usually lose the ability to grow rapidly in their normal host cells. Although attenuated viruses still provoke a strong immune response, they are not capable of sustaining an infection in a vaccinated individual. The smallpox, polio, and measles vaccines consist of attenuated viruses.

Why Isn't There a Vaccine for Cold Virus, HIV, and the Flu Viruses?

Researchers have not succeeded in developing vaccines against the common cold or HIV and have been only moderately successful in developing a vaccine against flu viruses. The reason is that cold and flu viruses and HIV have exceptionally high mutation rates. HIV actually has the highest mutation rate observed in any organism or virus.

Why does a high mutation rate make vaccine development so difficult? Recall from Chapter 16 that a mutation is defined as a change in DNA. Because the enzymes that copy the genes found in cold and flu viruses and HIV are exceptionally error prone, many of the mature virus particles (virions) that are produced contain mutations in the genes for the virus's envelope proteins. When this DNA is transcribed and translated, the envelope proteins that result are likely to have an altered structure. Due to the high mutation rates of cold and flu viruses and HIV, the envelope proteins in these viruses constantly change through time. Stated another way, new strains of these viruses are constantly evolving. The antibodies produced against the envelope proteins of earlier strains do not work against strains that appear later, because the envelope proteins of the strains are different. A vaccination that protected an individual against certain strains of cold or flu virus or HIV would not help against other strains.

To date, it has not been possible to design antigens from the cold or flu viruses or HIV that protect vaccinated individuals effectively. In the fight against HIV, drugs that inhibit viral enzymes and thus stop or slow viral replication have been much more successful than vaccination efforts.

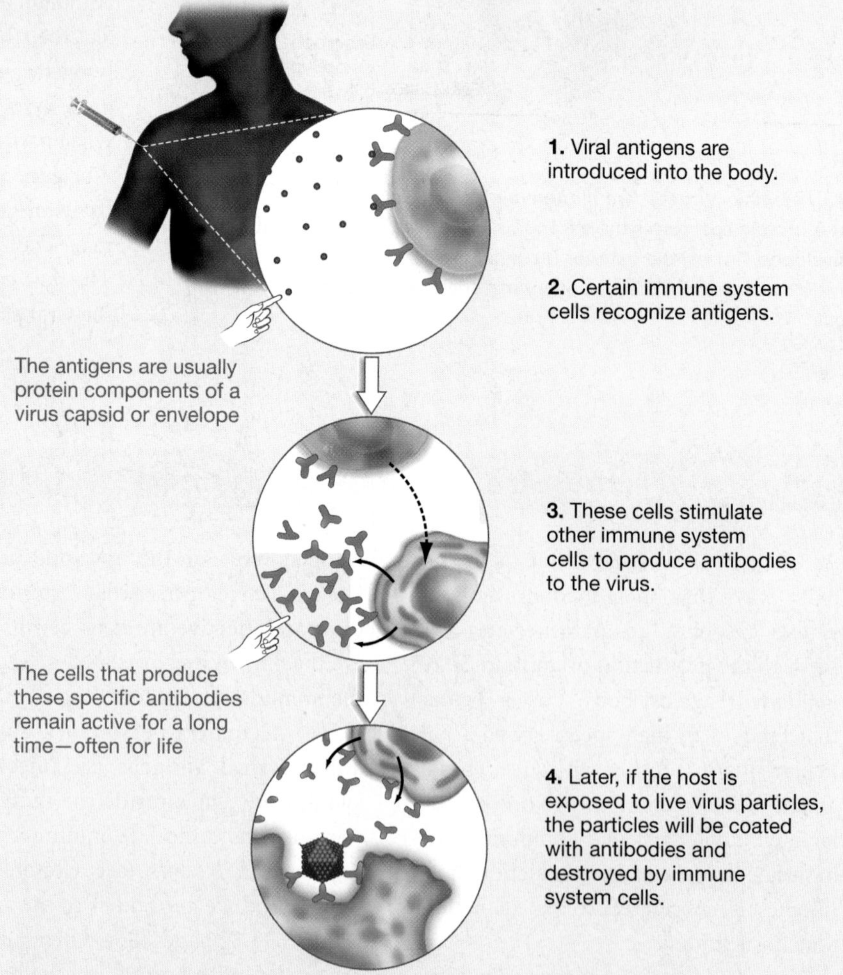
HOW VACCINATION WORKS
1. Viral antigens are introduced into the body.
2. Certain immune system cells recognize antigens.
The antigens are usually protein components of a virus capsid or envelope
3. These cells stimulate other immune system cells to produce antibodies to the virus.
The cells that produce these specific antibodies remain active for a long time—often for life
4. Later, if the host is exposed to live virus particles, the particles will be coated with antibodies and destroyed by immune system cells.

FIGURE 34.8 Vaccination Induces a Response from the Immune System
The immune system cells that respond to a vaccination are "immortalized," meaning that they stay active for a long time. Thus, the body can respond quickly to any future infections by the same virus.

of virus particles assembles inside the host. A mature virus particle is called a **virion**. The infection ends when the viral agents exit the cell—usually killing the host cell in the process.

Figure 34.9b diagrams a **lysogenic replication cycle**, or **lysogeny**. Only certain types of viruses are capable of lysogenic growth. Although HIV and some other viruses that infect humans are capable of lysogeny, most lysogenic viruses infect bacteria. Recall that viruses that infect bacteria are called bacteriophages. In these viruses, lysogeny is an important variation on the lytic replication cycle. During lysogenic growth, viral DNA becomes incorporated into the host's chromosome. Often this integration occurs without serious damage to the host cell. Once the viral genome is in place, it is replicated by the host's DNA polymerase each time the cell divides. Copies of the viral genome are passed on to daughter cells just like one of the host's own genes. In the lysogenic state, a virus is usually latent, or *quiescent*. This means that no new particles are being produced and no unrelated cells are being infected. The virus is transmitted from one generation to the next along with the host's genes.

In bacteriophages, lysogenic growth is typical when an infected bacterium is growing and dividing rapidly; the viral population then grows along with the bacterial population. But if the infected bacterium is damaged by sunlight or a toxin or if the host begins to starve, then the virus switches from lysogenic to lytic growth. To explain this observation, biologists point to the adaptive value of the switch between lytic and lysogenic replication cycles. If a bacterium is growing and dividing rapidly, then a virus's fitness is maximized through lysogenic growth. But if a bacterium is likely to stop growing or die, then a virus's fitness is maximized through lytic growth.

It is not possible to treat a lysogenic infection with drugs, because the viruses are quiescent—they just sit there. But even lytic infections are notoriously difficult to treat, because viruses use so many of the host cell's enzymes during the lytic replication cycle. Drugs that disrupt these enzymes usually harm the host much more than they harm the virus. To understand viral diversity and how antiviral drugs are developed, let's consider each phase of the lytic cycle in more detail.

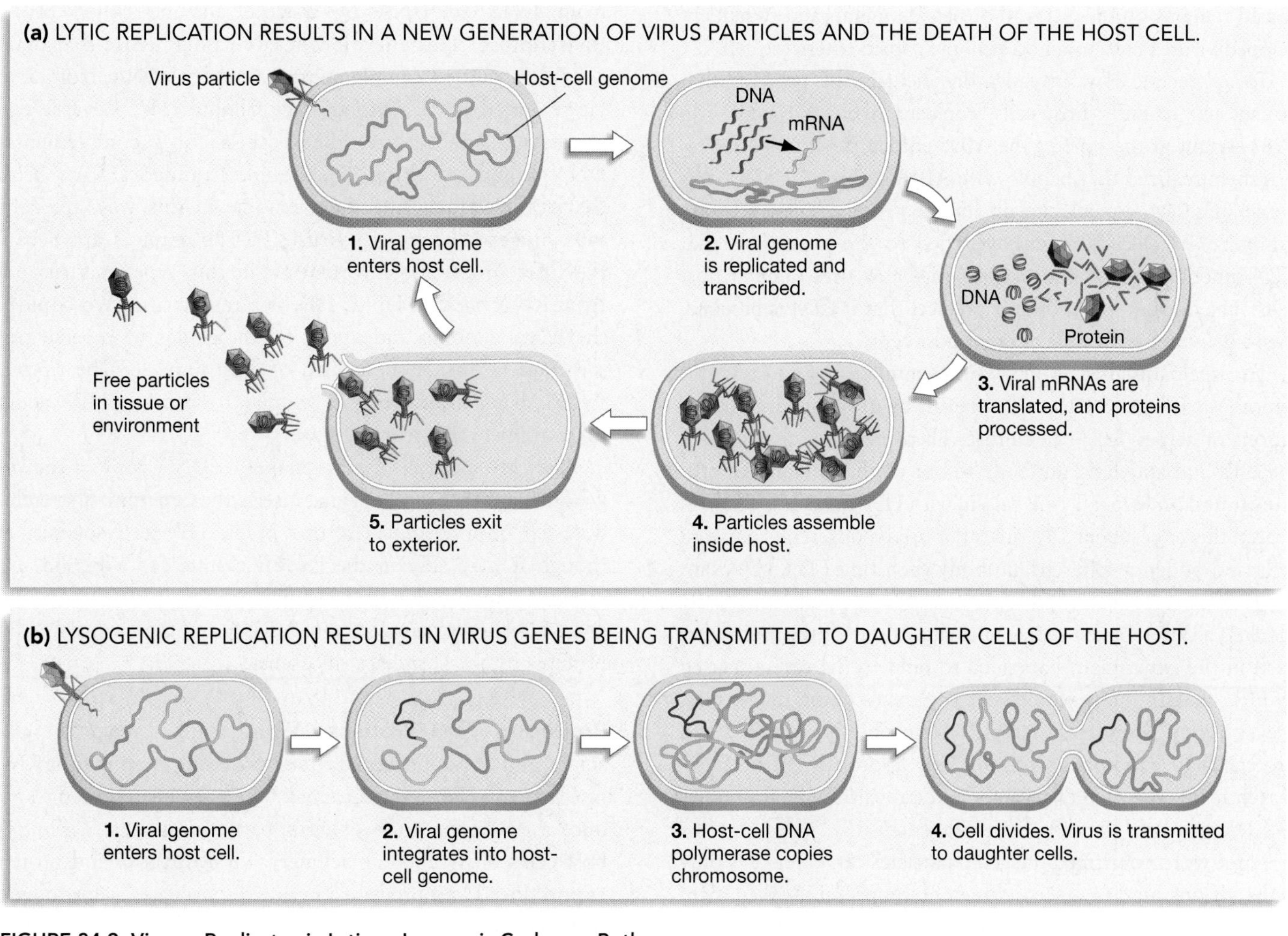

FIGURE 34.9 Viruses Replicate via Lytic or Lysogenic Cycles, or Both
(a) All viruses follow the same general lytic replication cycle. **(b)** Some viruses are also capable of lysogeny, meaning that their genome can become integrated into the host-cell chromosome. EXERCISE Compare and contrast a lysogenic virus to the transposable elements introduced in Chapter 20.

Analyzing the Phases of the Lytic Cycle

Five phases are common to lytic growth in virtually all viruses: (1) entry into a host cell, (2) replication and transcription of the viral genome, (3) production and processing of viral proteins, (4) assembly of a new generation of virions, and (5) exit from the infected cell. Each virus has a particular way of entering a host cell and completing the subsequent phases of the lytic cycle. Let's take a closer look.

How Do Viruses Enter a Cell? The replication cycle of a virus begins when a free viral particle enters a target cell. This is no simple task. All cells are protected by a plasma membrane, and many cells also have a cell wall. How do viruses breach these defenses, insert themselves into the cytoplasm inside, and begin an infection?

Most plant viruses enter host cells after a sucking insect, such as an aphid, has disrupted the cell wall with its mouthparts. In contrast, viruses that parasitize bacterial cells or that attack animal cells gain entry by binding to a specific molecule on the cell wall or plasma membrane. In response to this binding event, the genome of a nonenveloped virus enters the host cell, while the capsid remains on the cell wall or plasma membrane. When enveloped viruses bind to a host cell, the capsid enters the cell.

To appreciate how investigators identify the proteins that viruses use to enter host cells, consider research on HIV. In 1981—right at the start of the AIDS epidemic—biomedical researchers realized that people with AIDS had few or no T cells possessing **CD4**, a particular membrane protein. These cells are symbolized $CD4^+$. This discovery led to the hypothesis that CD4 functions as the "doorknob" that HIV uses to enter host cells. The doorknob hypothesis predicts that if CD4 is blocked, then HIV will not be able to enter host cells.

Two teams used the same experimental strategy to test this hypothesis (**Figure 34.10**). They began by growing large populations of helper T cells in culture. Then they took a sample of the cells and added an antibody to one of the cell-surface proteins found on helper T cells, along with HIV particles. They repeated this experiment 160 times but used a different sample of cells and added a different antibody each time. The key point here is that each of the 160 antibodies bound to and effectively blocked a different cell-surface protein. If one of the antibodies used in the experiment happened to bind to the receptor used by HIV, that antibody would cover up the receptor. In this way, the antibody would prevent HIV from entering the cell and protect that cell from infection. This approach led both research teams to reach exactly the same result: Only antibodies to CD4 protected the cells from viral entry.

Later work confirmed that HIV particles can enter cells only if the virions bind to a second membrane protein, called a **co-receptor**, in addition to CD4. In most individuals, proteins called CXCR4 and CCR5 function as co-receptors. The discovery of these co-receptors inspired a search for compounds that would block them and prevent HIV from entering cells. A drug company recently announced the development of a compound that blocks CCR5. Early reports suggest that the new drug is providing some relief for people infected with HIV.

WEB TUTORIAL 34.1 The HIV Life Cycle

If an HIV particle successfully binds to both CD4 and a co-receptor, however, the lipid bilayers of the particle's envelope and the plasma membrane of the helper T cell fuse. When fusion occurs, HIV has breached the cell boundary. The contents of the virus then enter the cytoplasm, and infection proceeds.

How Do Viruses Copy Their Genomes? Viruses must copy their genes to make a new generation of particles and continue an infection. Many viruses use their own DNA polymerase enzyme to accomplish that crucial step. This protein uses nucleotides provided by the host in synthesizing copies of the viral genome.

In some viruses, however, genes consist of RNA. In most viruses that have an RNA genome, copies of the genome are synthesized by the viral enzyme **RNA replicase**, which is an RNA polymerase. RNA replicase synthesizes RNA from an RNA template, using ribonucleotides provided by the host cell.

In other RNA viruses, however, the genome is transcribed from RNA to DNA by a viral enzyme called **reverse transcriptase**. This enzyme is a DNA polymerase that makes a single-stranded **complementary DNA**, or **cDNA**, from a single-stranded RNA template (see Chapter 19). Reverse transcriptase then catalyzes the synthesis of a complementary DNA strand, resulting in a double-stranded DNA. Viruses that reverse-transcribe their genome in this way are called **retroviruses** ("backward viruses"). The name is apt because the flow of genetic information in this type of virus goes from RNA back to DNA. HIV is a retrovirus. Two copies of the RNA genome and about 50 molecules of reverse transcriptase lie inside the capsid of each particle. The first antiviral drugs that were developed to combat HIV act by inhibiting reverse transcriptase.

After reverse transcriptase makes a cDNA copy of the viral genome, another viral enzyme inserts the copy into a stretch of host-cell chromosome. At this point, HIV is lysogenic. Although it may stay in the lysogenic state for a period, lytic growth is more common. For lytic growth to occur, the viral genes have to be transcribed to mRNA and then translated into proteins by the host cell's ribosomes.

Producing Viral Proteins Viruses cannot make the ribosomes and tRNAs necessary for translating their own mRNAs into proteins. For a virus to make the proteins required to produce a new generation of virus particles, it must exploit the host cell's biosynthetic machinery. Viral mRNAs and proteins are produced and processed in one of two ways, depending on whether the proteins end up in the outer envelope of a particle or in the capsid.

RNAs that code for a virus's envelope proteins follow a route through the cell identical to that of the RNAs of the cell's own

Question: Does the CD4 protein function as the "doorknob" that HIV uses to enter host cells?

Hypothesis: CD4 is the membrane protein that HIV uses to enter cells.

Null hypothesis: CD4 is not the membrane protein that HIV uses to enter cells.

Experimental setup:

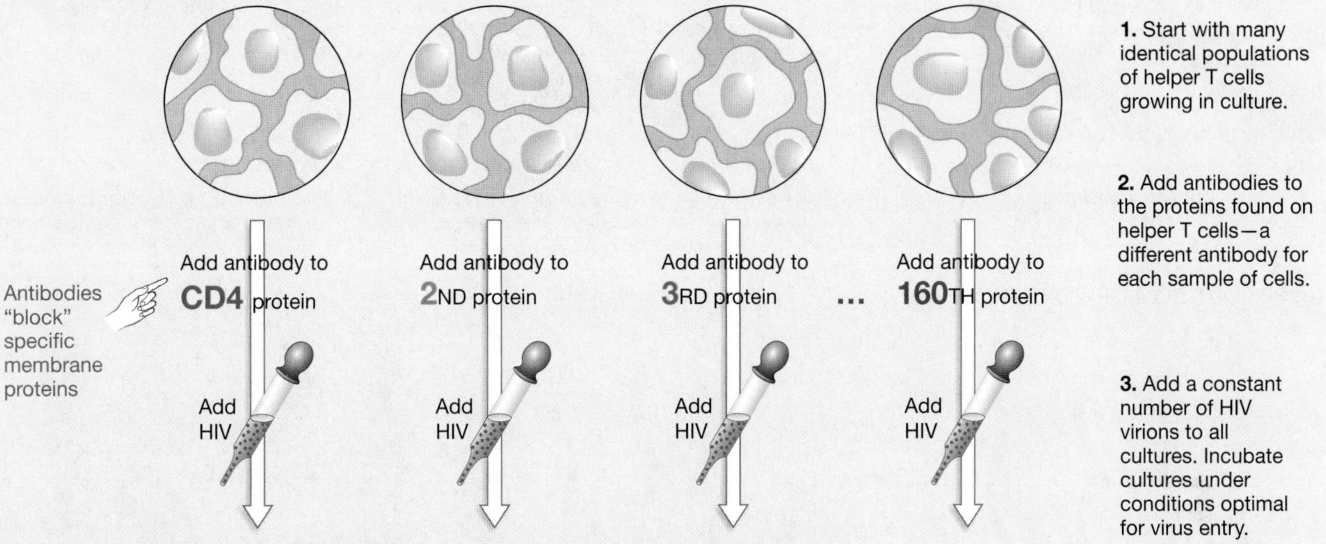
Antibodies "block" specific membrane proteins
Add antibody to CD4 protein
Add antibody to 2ND protein
Add antibody to 3RD protein
...
Add antibody to 160TH protein
Add HIV
Add HIV
Add HIV
Add HIV
1. Start with many identical populations of helper T cells growing in culture.
2. Add antibodies to the proteins found on helper T cells—a different antibody for each sample of cells.
3. Add a constant number of HIV virions to all cultures. Incubate cultures under conditions optimal for virus entry.

Prediction: HIV will not infect cells with antibody to CD4 but will infect other cells.

Prediction of null hypothesis: HIV will infect cells with antibody to CD4.

Results:

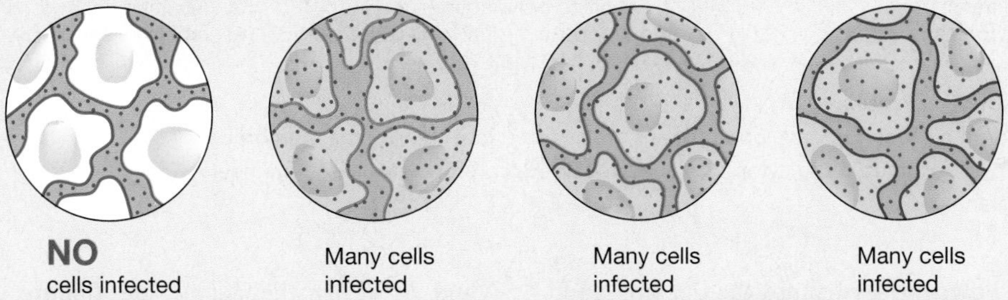
NO cells infected
Many cells infected
Many cells infected
Many cells infected

Conclusion: HIV uses CD4 proteins as the "doorknob" to enter helper T cells. Thus, only cells with CD4 on their surface can be infected by HIV.

FIGURE 34.10 Experiments Confirmed that CD4 Is the Receptor Used by HIV to Enter Host Cells
In this experiment, the antibodies added to each culture bound to a specific protein found on the surface of helper T cells. Antibody binding blocked the membrane protein, so the protein could not be used by HIV to gain entry to the cells.

membrane proteins. How these viral mRNAs are translated by ribosomes attached to the endoplasmic reticulum (ER) is diagrammed in **Figure 34.11a** (page 790). Afterward the resulting proteins are transported to the Golgi apparatus, where carbohydrate groups are attached, producing glycoproteins. The finished glycoproteins are then inserted into the plasma membrane, where they are ready to be assembled into new virions.

In contrast, a different route is taken by RNAs that code for proteins that make up the capsid or inner core of a virus particle, illustrated in **Figure 34.11b**. These RNAs are translated by ribosomes, but in the cytoplasm, just as non-membrane-bound cellular mRNAs are. The long polypeptide sequences that result are later cut into functional proteins by a viral enzyme called **protease**. This enzyme cleaves viral polypeptides at specific locations—a critical step in the production of finished viral proteins. The resulting protein fragments are assembled into a new viral core near the host cell's plasma membrane.

The discovery that HIV produces its own protease triggered a search for drugs that would block the enzyme. This search got a huge boost when researchers succeeded in visualizing the three-dimensional structure of HIV's protease, using the X-ray crystallographic techniques introduced in Chapter 4. The molecule has an opening in its interior that is adjacent to the active site,

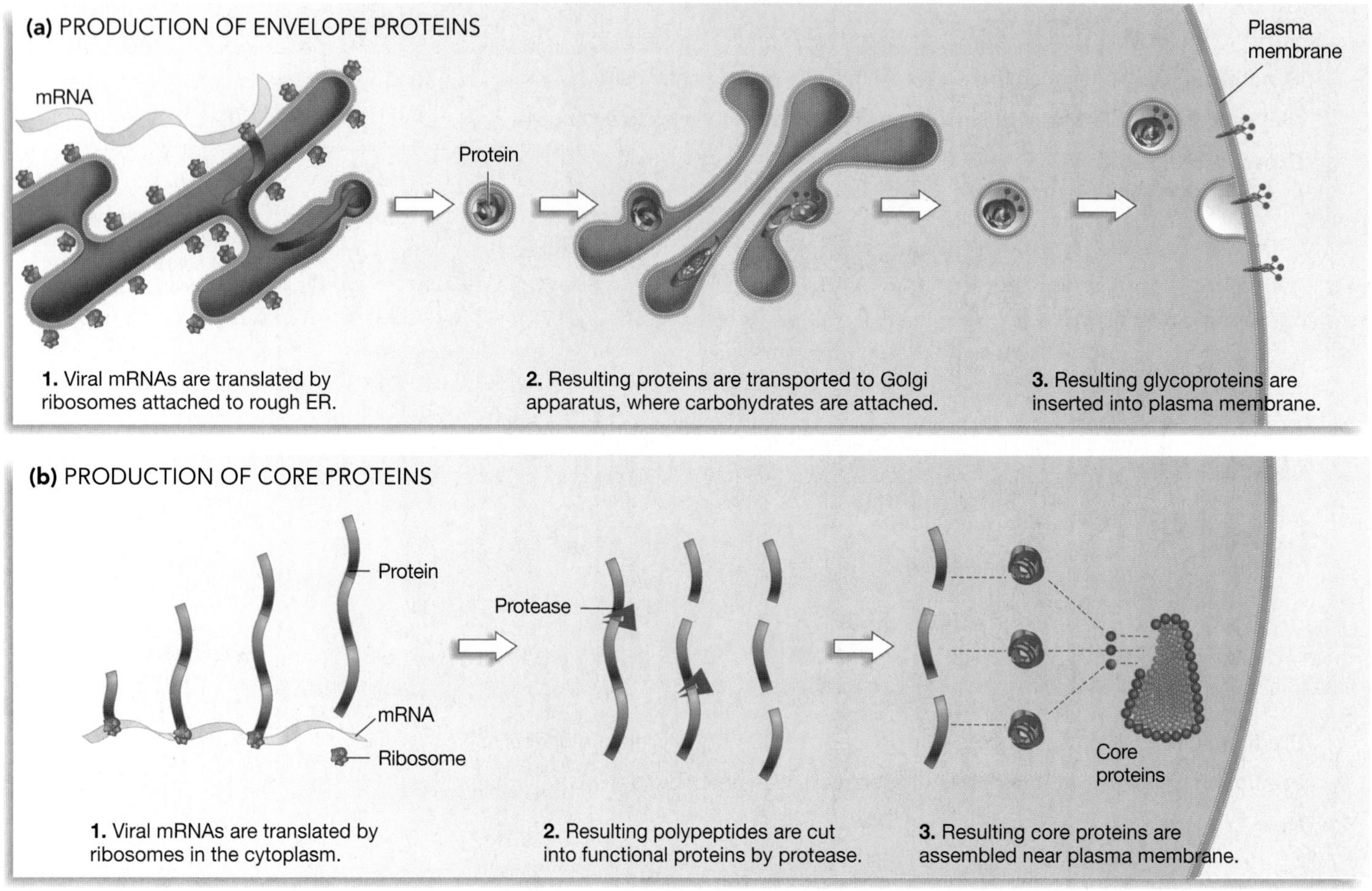
(a) PRODUCTION OF ENVELOPE PROTEINS
Plasma membrane
mRNA
Protein
1. Viral mRNAs are translated by ribosomes attached to rough ER.
2. Resulting proteins are transported to Golgi apparatus, where carbohydrates are attached.
3. Resulting glycoproteins are inserted into plasma membrane.
(b) PRODUCTION OF CORE PROTEINS
Protein
Protease
mRNA
Ribosome
Core proteins
1. Viral mRNAs are translated by ribosomes in the cytoplasm.
2. Resulting polypeptides are cut into functional proteins by protease.
3. Resulting core proteins are assembled near plasma membrane.

FIGURE 34.11 Production of Viral Proteins
(a) After being synthesized on the rough ER, envelope proteins are inserted into the plasma membrane.
(b) After being synthesized by ribosomes in the cytoplasm and processed, core proteins assemble near the host cell's plasma membrane.

as **Figure 34.12a** shows. Polypeptides fit into the opening and are cleaved at the active site. Based on these data, researchers immediately began searching for molecules that could fit into the opening and prevent protease from functioning by binding to and blocking the active site (**Figure 34.12b**). Several HIV *protease inhibitors* are now on the market. **Box 34.2** (page 792) explains why they are often prescribed in combination with other antiviral drugs.

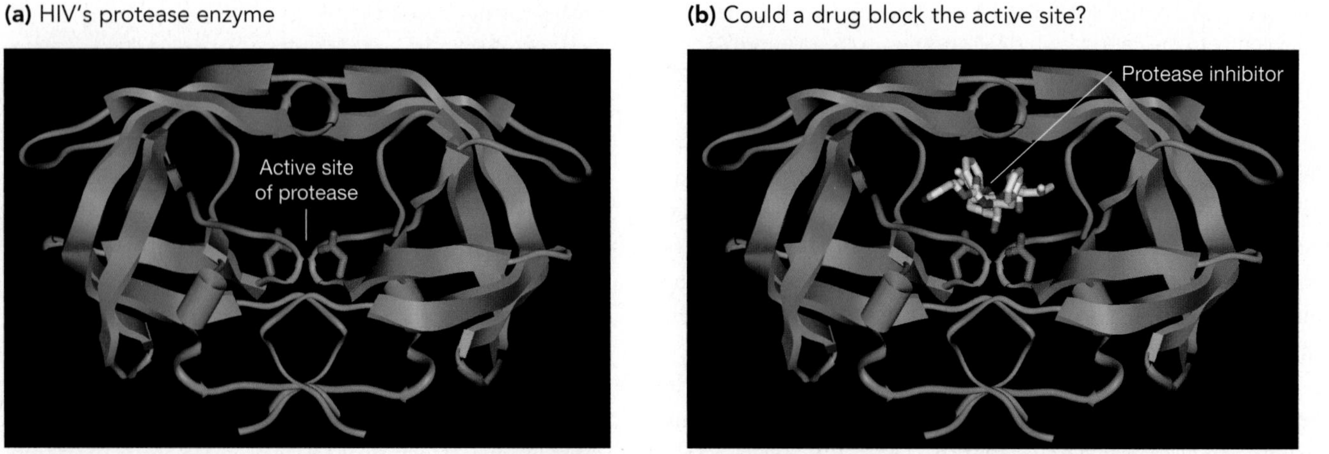
(a) HIV's protease enzyme
Active site of protease
(b) Could a drug block the active site?
Protease inhibitor

FIGURE 34.12 The Three-Dimensional Structure of Protease
(a) Ribbon diagram depicting the three-dimensional shape of HIV's protease enzyme. **(b)** Once protease's structure was solved, researchers began looking for compounds that would fit into the active site and prevent the enzyme from working.

How Are Viruses Transmitted to New Hosts? Viruses leave a host cell in one of two ways: by budding from the plasma membrane or by bursting out of the cell. Viruses that bud from the host cell's plasma membrane take some of that membrane with them. As a result, they incorporate host-cell phospholipids and membrane proteins into their envelope, along with membrane proteins encoded by the viral genome (**Figure 34.13a**). Most budding viruses infect host cells that lack a cell wall. In contrast, viruses that burst erupt from the cell surface, breaking the host cell open in the process, in most cases killing the host cell (**Figure 34.13b**). Most viruses that burst infect host cells that have a cell wall.

Once particles are released, the infection cycle is over. Thousands or millions of newly assembled virions are now in extracellular space. What happens next?

If the host cell is part of a multicellular organism, the new generation of particles begins traveling through the body via the bloodstream or lymph system. There, they may be bound by antibodies produced by the immune system. In vertebrates, this binding marks the particles for destruction. But if a particle contacts an appropriate host cell before it encounters antibodies, then the particle will infect that cell. This starts the replication cycle anew.

What if the virus has infected a unicellular organism, or if the virus leaves a multicellular host entirely? For example, when people cough, sneeze, spit, or wipe a runny nose, they help rid their body of viruses and bacteria. But they also project the pathogens into the environment, sometimes directly onto an uninfected host. From the virus's point of view, this new host represents an unexploited habitat brimming with resources in the

(a) Budding of enveloped viruses

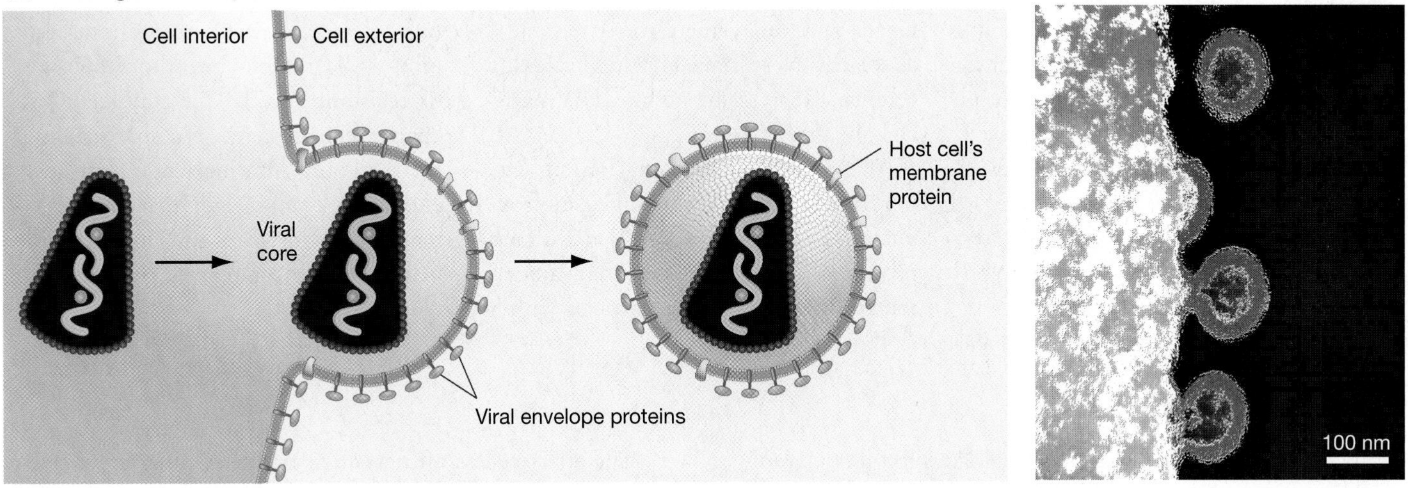

(b) Bursting of nonenveloped viruses

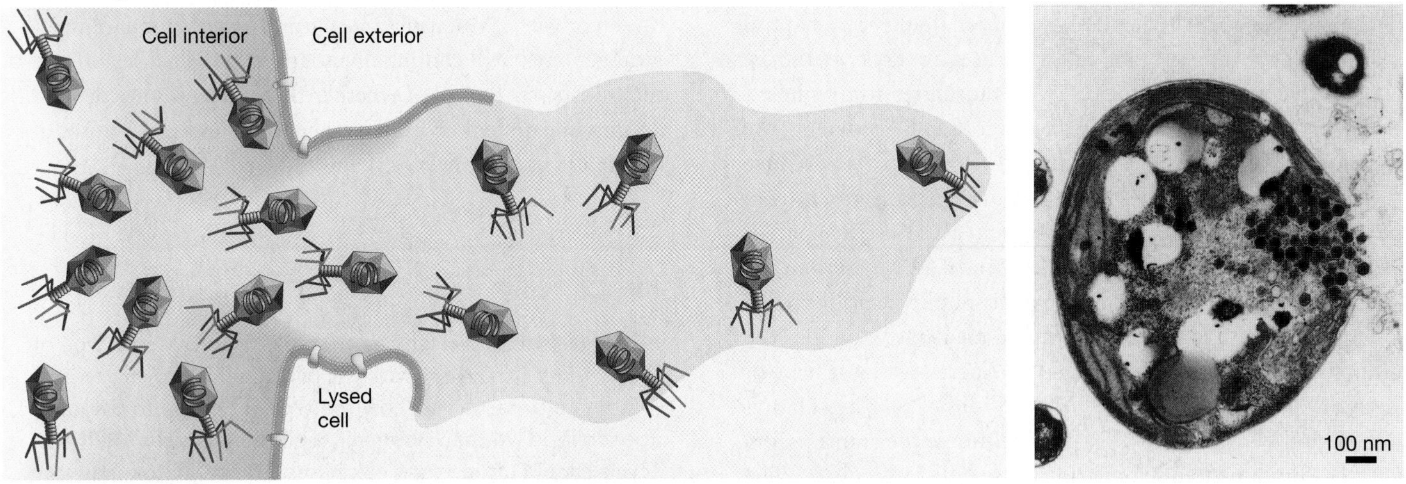

FIGURE 34.13 Viruses Leave Infected Cells by Budding or Bursting
(a) Enveloped viruses bud from a host cell, taking with them a lipid bilayer containing their envelope proteins. **(b)** Nonenveloped viruses burst from a host cell. This event ruptures the plasma membrane (and cell wall, if there is one). QUESTION Both budding and bursting kill the host cell. Propose a hypothesis to explain why infection with a budding virus is fatal.

BOX 34.2 HIV Drug Cocktails and the Evolution of Drug Resistance

HIV protease inhibitors were dispensed widely in North America and Europe beginning in the mid-1990s, with spectacular results. After therapy, many patients no longer had detectable levels of HIV in their blood. The drugs knocked HIV populations down.

Within two years, however, HIV levels in many of the patients taking protease inhibitors began to rebound. To investigate why this was happening, researchers sequenced the HIV protease gene in these patients. The researchers found that a series of mutations had occurred in the protease gene. These mutations led to changed amino acid sequences in the enzyme's active site. Because protease inhibitor molecules did not fit as well into the altered version of the active site, the enzyme could function reasonably well even in the presence of the inhibitor molecules. Almost as soon as a new protease inhibitor went into widespread use, researchers had documented the evolution of resistance to the drug.

The scientific literature abounds with examples of bacteria that have evolved resistance to antibiotics and viruses that have evolved resistance to antiviral drugs. But perhaps no organism or virus has evolved resistance to control agents as quickly as HIV has. The leading hypothesis to explain this observation is that HIV's mutation rate is particularly high. Researchers who assay transcripts produced by HIV's reverse transcriptase find that, on average, the wrong base is inserted once every 8000 nucleotides. (In contrast, *E. coli*'s DNA polymerase incorporates the wrong nucleotide about once in every *1 billion* bases.) This means that, on average, a new mutant is generated every time HIV replicates its genome. Genetically, no two HIV particles are alike.

Why is HIV's high mutation rate important? In infected individuals, approximately 10 billion new viral particles are produced daily. It is likely that, among the 10 billion, there are particles with a mutation in the active sites of reverse transcriptase or protease. As a result, HIV populations are almost certain to contain variants that are at least partially resistant to drugs that cripple most other particles in the population.

The message to researchers and physicians is clear: Due to HIV's high mutation rate, the search for drug therapies promises to be an "arms race"—a constant battle between novel drugs and novel, resistant strains of the virus. In attempting to keep ahead of drug-resistant viruses, physicians prescribe **combination therapy**—drug "cocktails" that include a protease inhibitor and one or more reverse transcriptase inhibitors. If patients begin to show signs of resistant strains, the physician will change the dosage or composition of the cocktail. Although it is extremely expensive, combination therapy has extended the life span and improved the quality of life for tens of thousands of AIDS patients.

form of target cells. The situation is analogous to that of a multicellular animal dispersing to a new habitat and colonizing it. Viruses that successfully colonize a new host replicate and increase in number. The alleles carried by these successful colonists increase in frequency in the total population. In this way, natural selection favors alleles that allow viruses to do two things: (1) replicate within a host and (2) be transmitted to new hosts.

For physicians and public health officials, reducing the likelihood of transmission is often an effective way to reduce the spread of a virus. HIV particles must be transmitted from person to person via body fluids such as blood, semen, or vaginal secretions. Faced with decades of disappointing results in drug and vaccine development, public health officials are aggressively promoting preventive medicine. Condom use reduces sexual transmission of HIV. Aggressive treatment of venereal diseases may also help; the lesions caused by chlamydia, genital warts, and gonorrhea encourage the transmission of HIV-contaminated blood during sexual intercourse. The most effective forms of preventive medicine are sexual abstinence or monogamy. Where blood supplies are routinely screened, HIV is rarely contracted by means other than unprotected sex with an infected person.

The effectiveness of preventive medicine underscores one of this chapter's fundamental messages: Viruses are a fact of life. Every organism is victimized by viruses; every organism has defenses against them. But the tree of life will never be free of these parasites. Mutation and natural selection guarantee that viral genomes will continually adapt to the defenses offered by their hosts, regardless of whether those defenses are devised by an immune system or by biomedical researchers. Viruses are a constant threat for every organism alive.

✓CHECK YOUR UNDERSTANDING

All organisms and cell types are parasitized by some type of virus. Viruses may undergo lytic or lysogenic growth, or both, when they infect a host cell. You should be able to diagram generalized lytic and lysogenic replication cycles. Your lytic cycle should distinguish five phases: (1) entry into a host cell, (2) replication of the viral genome, (3) production of viral proteins, (4) assembly of a new generation of virions, and (5) exit from the infected cell. You should also be able to give an example of how each phase occurs in a particular virus.

34.3 What Themes Occur in the Diversification of Viruses?

If viruses can infect virtually every type of organism and cell known, how can biologists possibly identify themes that help organize viral diversity? The answer is that, in addition to being identified as enveloped or nonenveloped, viruses can be categorized by the nature of their hereditary material—in essence, the type of molecule their genes are made of. The single most important aspect of viral diversity is the variation that exists in their genetic material.

Two other points are critical to recognize about viral diversity: (1) Biologists do not have a solid understanding of how viruses originate, but (2) it is certain that viruses will continue to diversify. After analyzing diversity in viral genes, let's consider hypotheses to explain where viruses come from and recent data on new or "emerging" viruses.

The Nature of the Viral Genetic Material

DNA is the hereditary material in all cells. As cells synthesize the molecules they need to function, information flows from DNA to mRNA to proteins (Chapter 15). Although all cells follow this pattern, which is called the central dogma of molecular biology, some viruses break it. This conclusion traces back to work done in the 1950s, when biologists were able to separate the protein and nucleic acid components of a particle known as the tobacco mosaic virus, or TMV. Surprisingly, the nucleic acid portion of this virus consisted of RNA, not DNA. Later experiments demonstrated that the RNA of TMV, by itself, could infect plant tissues and cause disease. This was a confusing result, because it showed that, in this virus at least, RNA—not DNA—functions as the genetic material.

Subsequent research revealed an amazing diversity of viral genome types. In some groups of viruses, such as the agents that cause measles and flu, the genome consists of RNA. In others, such as the particles that cause herpes and smallpox, the genome is composed of DNA. Further, the RNA and DNA genomes of viruses can be either single stranded or double stranded. The single-stranded genomes can also be classified as "positive sense" or "negative sense" or "ambisense." In a **positive-sense virus**, the genome contains the same sequences as the mRNA required to produce viral proteins. In a **negative-sense virus**, the base sequences in the genome are complementary to those in viral mRNAs. In an **ambisense virus**, some sections of the genome are positive sense while others are negative sense.

Finally, the number of genes found in viruses varies widely. The tymoviruses that infect plants contain as few as three genes, but the genome of smallpox can code for up to 343 proteins. **Table 34.2** summarizes the diversity of viral genome types.

TABLE 34.2 The Diversity of Viral Genomes

Key: ss = single stranded; ds = double stranded; (+) = positive sense (genome sequence is the same as viral mRNA); (−) = negative sense (genome sequence is complementary to viral mRNA).

Genome		Example(s)	Host	Result of Infection	Notes
(+)ssRNA	+	TMV	Tobacco plants	Tobacco mosaic disease (leaf wilting)	TMV was the first RNA virus to be discovered.
(−)ssRNA	−	Influenza	Many mammal and bird species	Influenza	The negative-sense ssRNA viruses transcribe their genomes to mRNA via RNA replicase.
dsRNA	+ / −	Phytovirus	Rice, corn, and other crop species	Dwarfing	Double-stranded RNA viruses are transmitted from plant to plant by insects. Many can also replicate in their insect hosts.
ssRNA or (+)ssDNA that requires reverse transcription for replication	+ or − / or / +	Rous sarcoma virus	Chickens	Sarcoma (cancer of connective tissue)	Rous sarcoma virus was identified as a cancer-causing agent in 1911, decades before any virus was seen.
ssDNA—can be (+), (−), or (+) and (−)	+ or −	$\varphi \times 174$	Bacteria	Death of host cell	The genome for $\varphi \times 174$ is circular and was the first complete genome ever sequenced.
dsDNA	+ / −	Baculovirus Smallpox Bacteriophage	Insects Humans Bacteria	Death Smallpox Death	These are the largest viruses in terms of genome size and overall size.

Where Did Viruses Come From?

No one knows where viruses originated, but many biologists suggest that they are closely related to the plasmids and transposable elements introduced in Chapter 19 and Chapter 20. Viruses, plasmids, and transposable elements are all acellular, mobile genetic elements that replicate with the aid of a host cell. Simple viruses are actually indistinguishable from plasmids except for one feature: The viruses have a protein coat or membrane-like envelope.

Some biologists hypothesize that simple viruses, plasmids, and transposable elements represent "escaped gene sets." This hypothesis states that mobile genetic elements are descended from clusters of genes that physically escaped from bacterial or eukaryotic chromosomes long ago. According to this hypothesis, the escaped gene sets took on a mobile, parasitic existence because they happened to encode the information needed to replicate themselves—at the expense of the genomes that once held them. In the case of viruses, the hypothesis is that the escaped genes included the instructions for making a protein capsid and possibly envelope proteins. According to the escaped-gene hypothesis, it is likely that each of the distinct types of RNA viruses and single-stranded DNA viruses represents a distinct "escape event."

The same researchers contend that DNA viruses with large genomes originated in a very different manner, however. Here the hypothesis is that the large DNA viruses trace their ancestry back to free-living bacteria that once took up residence inside eukaryotic cells. The idea is that these organisms degenerated into viruses by gradually losing the genes required to synthesize ATP, nucleic acids, amino acids, and other compounds. Although this idea sounds speculative, it cannot be dismissed lightly. Chapter 27 introduced species in the genus *Chlamydia*, which are bacteria that live as parasites inside animal cells. And Chapter 28 provided evidence that the organelles called mitochondria and chloroplasts, which reside inside eukaryotic cells, originated as intracellular symbionts. Investigators contend that, instead of evolving into intracellular symbionts that aid their host cell, DNA viruses became parasites capable of destroying the host.

To date, neither of these hypotheses has been tested rigorously. To support the escaped-genes hypothesis, researchers would probably have to discover a brand new virus that originated in this way, or viruses that had so recently derived from intact bacterial or eukaryotic genes that the viral DNA sequence still strongly resembled the DNA sequence of those genes. To support the degeneration hypothesis, researchers would probably have to find strong genetic homologies between viruses and parasitic species of bacteria that live inside cells. But currently, there is no widely accepted view of where viruses came from.

Emerging Viruses, Emerging Diseases

Although it is not known how the various types of virus originated, it is certain that viruses will continue to diversify. With alarming regularity, the front pages of newspapers carry accounts of deadly viruses that are infecting humans for the first time. In 1993 a hantavirus that normally infects mice suddenly afflicted dozens of people in the southwestern United States. Nearly half of the people who developed hantavirus pulmonary syndrome died. Still higher fatality rates were recorded in 1995 when the Ebola virus, a variant of a monkey virus, caused a wave of infections in the Democratic Republic of Congo. By the time the outbreak subsided, over 200 cases had been reported; 80 percent were fatal. During an Ebola outbreak in 1976, 90 percent of cases were fatal.

Hantavirus pulmonary syndrome and Ebola are examples of **emerging diseases**: new illnesses that suddenly affect significant numbers of individuals in a host population. In these cases, the causative agents were considered **emerging viruses** because they had switched from their traditional host species to a new host—humans. Many investigators consider HIV to be an emerging virus because it originated in chimps and was first transmitted to humans in the early to mid-twentieth century (see **Box 34.3**).

Physicians become alarmed when they see a large number of patients with identical and unusual disease symptoms in the same geographic area over a short period of time. The physicians report these cases to public health officials, who take on two urgent tasks: (1) identifying the agent that is causing the new illness and (2) determining how the disease is being transmitted.

Several strategies can be used to identify a pathogen. In the case of the outbreak of hantavirus pulmonary syndrome, officials recognized strong similarities between symptoms in the U.S. cases and symptoms caused by the Hantaan virus native to northeast Asia. The Hantaan virus rarely causes disease in humans; its normal host is rodents. To determine whether a Hantaan-like virus was responsible for the U.S. outbreak, researchers began capturing mice in the homes and workplaces of afflicted people. About a third of the captured rodents tested positive for the presence of a Hantaan-like virus. (The test that was done is explained in Chapter 49.) DNA sequencing studies confirmed that the virus was a previously undescribed type of hantavirus. Further, the sequences found in the mice matched those found in infected patients. Based on these results, officials were confident that a rodent-borne hantavirus was causing the wave of infections.

The next step in the research program, identifying how the agent is being transmitted, is equally critical. If a virus that normally parasitizes a different species suddenly begins infecting humans, if it can be transmitted efficiently from person to

BOX 34.3 Where Did HIV Originate?

The key to discovering where HIV comes from lies in understanding the evolutionary history, or phylogeny, of the virus. Researchers have reconstructed this history by comparing the composition of HIV genes with sequences from closely related viruses that parasitize chimps, monkeys, and other mammals (**Figure 34.14**). The relationships among these gene sequences support several conclusions:

- HIV belongs to a group of viruses called the lentiviruses, which infect a wide range of mammals, including house cats, horses, goats, and primates. (*Lenti* is a Latin root that means "slow"; here it refers to the long period observed between the start of an infection by these viruses and the onset of the diseases they cause.)
- Many of HIV's closest relatives also have *immunodeficiency* in their name. Like HIV, these agents parasitize cells that are part of the immune system. Several of them cause diseases with symptoms reminiscent of AIDS. Curiously, though, HIV's closest relatives don't appear to cause disease in their hosts. These viruses infect monkeys and chimpanzees and are called **simian immunodeficiency viruses (SIVs)**.
- There are two distinct types of human immunodeficiency viruses, called HIV-1 and HIV-2. Although both can cause AIDS, HIV-1 is far more virulent and is the better studied of the two. It serves as the focus of this chapter.
- HIV-1's closest relatives are immunodeficiency viruses isolated from chimpanzees that live in central Africa. In contrast, HIV-2's closest relatives are immunodeficiency viruses that parasitize monkeys called sooty mangabeys. In central Africa, where HIV-1 infection rates first reached epidemic proportions, contact between chimpanzees and humans is extensive. Chimps are hunted for food and kept as pets. Similarly, sooty mangabeys are hunted and kept as pets in west Africa, where HIV-2 infection rates are highest. To make sense of these observations, biologists suggest that HIV-1 is a descendant of viruses that infect chimps, and that HIV-2 is a descendant of viruses that infect sooty mangabeys. The "jumps" between host species probably occurred when a human cut up a monkey or chimpanzee for use as food.
- Several strains of HIV-1 exist. A virus **strain** consists of populations that have distinct characteristics. The most important HIV-1 strains are called O for *outgroup* (meaning, the most basal group relative to other strains), N for *new*, and M for *main*. The existence of these distinct strains suggests that HIV-1 has jumped from chimps to humans several times and may do so again.

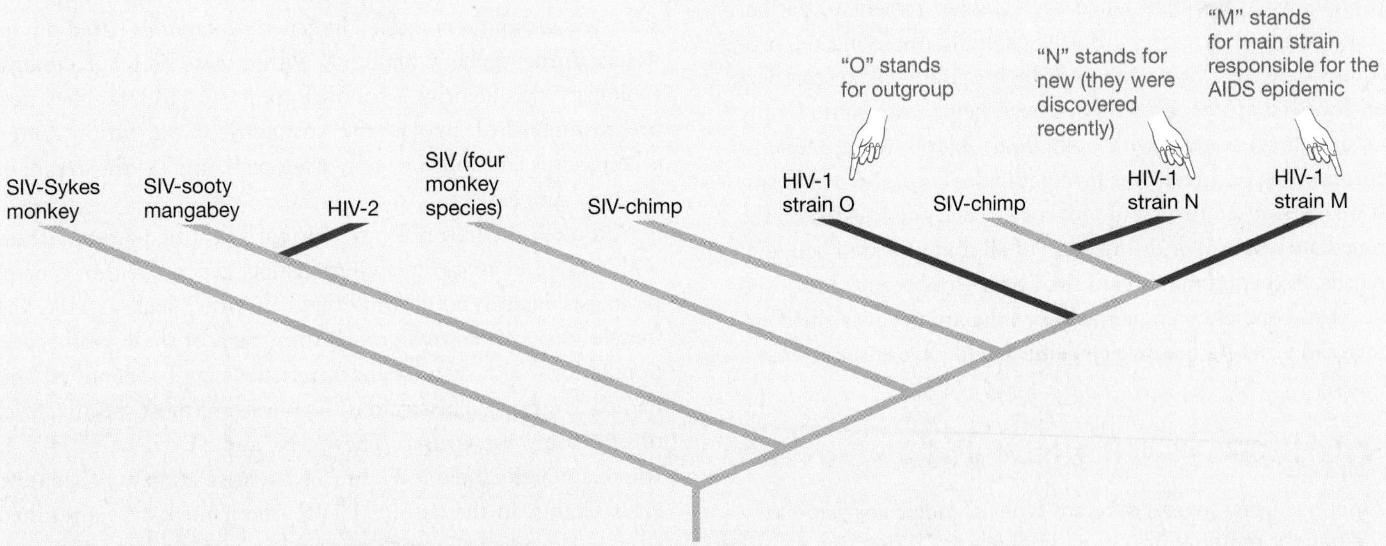

FIGURE 34.14 Phylogeny of HIV Strains and Types
Phylogenetic tree showing the evolutionary relationships among some of the immunodeficiency viruses that infect primates—including chimpanzees, humans, and several species of monkeys. QUESTION Based on this tree, researchers conclude that an SIV has "jumped" from chimps to humans at least twice. Do you agree? Explain why or why not.

person, and if it causes serious illness, then the outbreak has the potential to become an epidemic. But if transmission takes place only between the normal host and humans, as is the case with rabies, then the number of cases will probably remain low.

Determining how a virus is transmitted takes old-fashioned detective work. By interviewing patients about their activities, researchers decide whether each patient could have acquired the virus independently. Were the individuals infected with hantavirus in contact with mice? If so, were they bitten? Did they handle rodent feces or urine during routine cleaning chores or come into contact with contaminated food? Was the illness showing up in health-care workers, implying that it was being transmitted from human to human?

In the hantavirus outbreak, public health officials concluded that no human-to-human transmission was taking place. Health-care workers did not become ill; because patients had not had extensive contact with one another, it was likely that each had acquired the virus independently. The outbreak also coincided with a short-term, weather-related explosion in the local mouse population. The most likely scenario was that people had acquired the pathogen by inhaling dust or handling food that contained remnants of mouse feces or urine. In short, hantavirus did not have the potential to cause an epidemic. The best medicine was preventive: Homeowners were advised to trap mice out of living areas, wear dust masks while cleaning any area where mice might have lived, and store food in covered jars.

The Ebola virus, in contrast, was clearly transmitted from person to person. Many doctors and nurses were stricken after tending to patients with the virus. Infections that originate in hospitals usually spread when carried from patient to patient on the hands of caregivers. But by carefully observing the procedures that were being followed by hospital staff, researchers concluded that the Ebola virus was being transmitted only through direct contact with body fluids (blood, urine, feces, or sputum). The outbreak was brought under control when hospital workers raised their standards of hygiene and insisted on the immediate disposal or disinfection of all contaminated bedding, utensils, and equipment. Had the Ebola virus been transmitted by casual contact, such as touch or inhalation—as is the common cold virus—a massive epidemic could have ensued.

✓CHECK YOUR UNDERSTANDING

Among viruses, several different types of molecules serve as the genetic material. You should be able to (1) describe at least three types of viral genomes other than double-stranded DNA, (2) diagram how the information in viral genes is used to produce a new generation of virions during a lytic replication cycle for one of these types of genomes, and (3) explain why a mutation that allowed Ebola virus to spread via airborne particles coughed out by infected individuals would make the virus more or less dangerous to humans.

34.4 Key Lineages of Viruses

Because scientists are almost certain that viruses originated multiple times throughout the history of life, there is no such thing as the phylogeny of all viruses. Stated another way, there is no single phylogenetic tree that represents the evolutionary history of viruses, as there is for the organisms discussed in previous chapters. Instead, researchers focus on comparing base sequences in the genetic material of small, closely related groups of viruses and using these data to reconstruct the phylogenies of particular lineages. Phylogenetic trees for viruses are usually inferred from comparisons of nucleic acid sequence data, using techniques introduced in Chapter 26.

The phylogenetic tree of the simian immunodeficiency viruses and human immunodeficiency viruses in Figure 34.14 is a good example of how researchers construct and interpret the phylogenies of particular groups of viruses. These types of phylogenies have been important in understanding viral diversity and the sources of emerging viruses. For example, phylogenetic analyses allowed biologists to recognize that there are two distinct types of HIV. The phylogenetic data diagrammed in Figure 34.14 indicate that HIV-1 and HIV-2 are distinct types of virus that originated from different host species (see Box 34.3). These data correlate with the observation that HIV-1 and HIV-2 differ in important ways. For instance, although both types of HIV infect immune system cells and are usually transmitted through sexual contact, HIV-2 is much less virulent and much less easily transmitted from person to person than is HIV-1.

To organize the diversity of viruses on a larger scale, researchers group them into eight general categories based on the nature of their genetic material. Within these broad groupings, biologists also identify a total of about 70 virus families that are distinguished by (1) the structure of the virion (often whether it is enveloped or nonenveloped), and (2) the nature of the host species.

Although they do not have formal scientific names, viruses within families are grouped into distinct genera. Within genera, biologists identify and name types of virus, such as HIV, the measles virus, and smallpox. Within each of these viral types, populations with distinct characteristics may be identified and named as strains. The strain is the lowest, or most specific, level of taxonomy for viruses. The O, N, and M strains of HIV-1, which are highlighted in Figure 34.14, are examples of separate virus strains. In the case of HIV-1, the named strains resulted from independent "jumps" of a simian immunodeficiency virus to a new host: humans. Each of these strains has distinguishing characteristics. The M strain, for example, was named *main*, because it is responsible for most of the HIV infections known to date.

To get a sense of viral diversity, let's survey a few of the major groups that can be identified by the nature of their genetic material.

Double-Stranded DNA (dsDNA) Viruses

The double-stranded DNA viruses are a large group, composed of some 21 families and 65 genera. Smallpox (**Figure 34.15**) is perhaps the most familiar of these viruses. Although smallpox had been responsible for millions of deaths throughout human history, it was eradicated by vaccination programs.

Genetic material As their name implies, the genes of these viruses consist of a single molecule of double-stranded DNA. The molecule may be linear or circular.

Host species These viruses parasitize hosts from throughout the tree of life, with the notable exception of land plants. They include virus families called the T-even and λ bacteriophages, some of which infect *E. coli*. In addition, the pox viruses, herpesviruses, and adenoviruses—some species of which parasitize humans—have genomes that consist of double-stranded DNA.

Replication cycle In most double-stranded DNA viruses that infect eukaryotes, viral genes have to enter the nucleus to be replicated. For that to be true, these viral genes must replicate only during S phase, when the host cell's chromosomes are being replicated. As a result, these types of viruses can sustain an infection only in cells that are actively dividing, such as the cells that line the respiratory tract or urogenital canal.

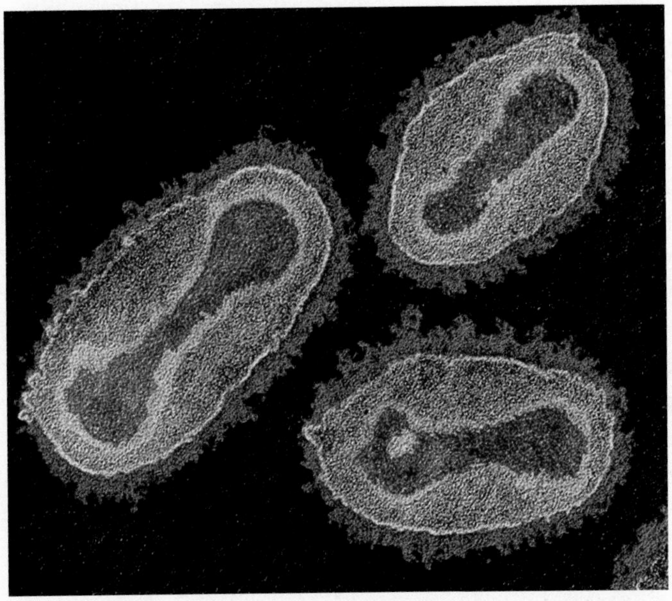

FIGURE 34.15 Smallpox Is a Double-Stranded DNA Virus

RNA Reverse-Transcribing Viruses (Retroviruses)

The genomes of the RNA reverse-transcribing viruses are composed of single-stranded RNA. There is only one family, called the retroviruses.

Genetic material Virus particles have two copies of their single-stranded RNA genome, so they are diploid.

Host species Species in this group are known to parasitize only vertebrates—specifically birds, fish, or mammals. HIV is the most familiar virus in this group. The Rous sarcoma virus, the mouse mammary tumor virus, and the murine (mouse) leukemia virus are other retroviruses that have also been studied intensively. Rous sarcoma virus was the first virus shown to be associated with the development of cancer (in chickens); the mouse viruses were the first viruses known to increase the risk of cancer in mammals (**Figure 34.16**).

Replication cycle Retroviruses contain the enzyme reverse transcriptase inside their capsid. (A typical HIV particle contains about 50 reverse transcriptase molecules.) When the virus's RNA genome and reverse transcriptase enter a host cell's cytoplasm, the enzyme catalyzes the synthesis of a single-stranded cDNA from the original RNA. Reverse transcriptase then makes this cDNA double stranded. The double-stranded DNA version of the genome enters the nucleus with a viral protein called *integrase*. Integrase catalyzes the integration of the viral genes into a host chromosome. The virus may remain lysogenic for a period, but eventually the genes are transcribed to RNA to begin lytic growth and the production of a new generation of virus particles.

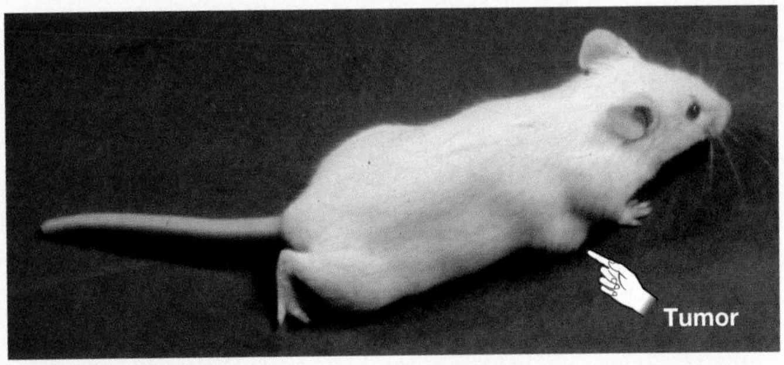

FIGURE 34.16 Some Retroviruses, Such as the Mouse Mammary Tumor Virus, Are Associated with Cancer

Double-Stranded RNA (dsRNA) Viruses

There are 7 families of double-stranded RNA viruses and a total of 22 genera. Most of the viruses in this group are nonenveloped.

Genetic material In some families, virus particles typically have a genome consisting of ten to twelve double-stranded RNA molecules; in other families, the genome is composed of just one to three RNA molecules.

Host species A wide variety of hosts, including fungi, land plants, insects, vertebrates, and bacteria, are victimized by viruses with double-stranded RNA genomes. For humans, the most important viruses in this group are those that cause disease in rice, corn, sugar cane, and other crops. **Figure 34.17** shows rice plants that are being attacked by a double-stranded RNA virus. Infections are also common in *Penicillium*, a filamentous fungus that produces the antibiotic penicillin.

Replication cycle Once inside the cytoplasm of a host cell, the double-stranded genome of these viruses serves as a template for the synthesis of single-stranded RNAs, which are then translated into viral proteins. The proteins form the capsids for a new generation of virus particles. Copies of the genome are created when a viral enzyme makes the original single-stranded RNA double stranded.

Leaves
infected
with virus

FIGURE 34.17 Double-Stranded RNA Viruses, Such as the Ragged Stunt Virus, Parasitize a Wide Array of Organisms—Here, Rice

Negative-Sense Single-Stranded RNA ([−]ssRNA) Viruses

There are 7 families and 30 genera in this group. Most members of this group are enveloped, but some negative-sense single-stranded RNA viruses lack an envelope.

Genetic material The sequence of bases in a negative-sense RNA virus is opposite in polarity to the sequence in a viral mRNA. Stated another way, the single-stranded virus genome is complementary to the viral mRNA. Depending on the family, the genome may consist of a single RNA molecule or up to eight separate RNA molecules.

Host species A wide variety of plants and animals are parasitized by viruses that have negative-sense single-stranded RNA genomes. If you have ever suffered from the flu, the mumps, or the measles, then you are painfully familiar with these viruses (**Figure 34.18**). The Ebola, Hantaan, and rabies viruses also belong to this group.

Replication cycle When the genome of a negative-sense single-stranded RNA virus enters a host cell, a viral RNA polymerase uses that genome as a template to make new viral mRNA. These viral mRNAs are then translated to form viral proteins. The viral mRNAs also serve as a template for the synthesis of new copies of the negative-sense single-stranded RNA genome.

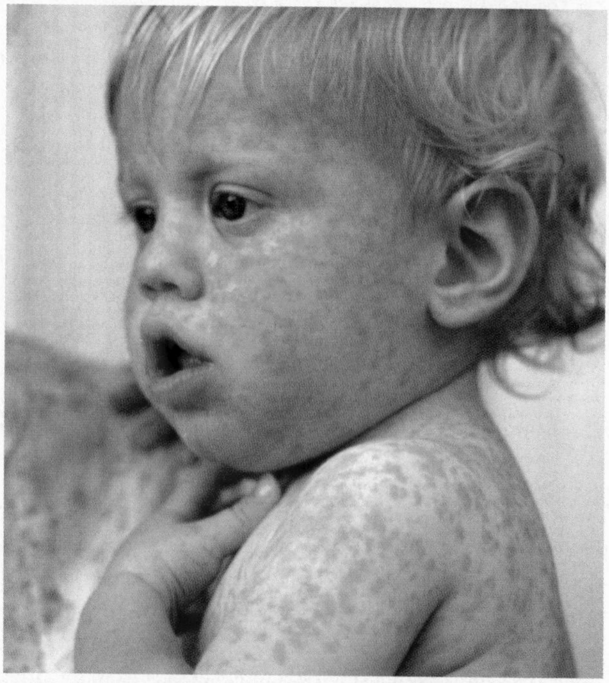

FIGURE 34.18 Negative-Sense Single-Stranded RNA Viruses, Such as the Measles Virus, Cause Some Common Childhood Diseases

Positive-Sense Single-Stranded RNA ([+]ssRNA) Viruses

This is the largest group of viruses known, with 81 genera organized into 21 families.

Genetic material The sequence of bases in a positive-sense RNA virus is the same as that of an mRNA. Stated another way, the genome does not need to be transcribed in order for proteins to be produced. Depending on the species, the genome consists of one to three RNA molecules.

Host species Most of the commercially important plant viruses belong to this group. Because they kill groups of cells in the host plant and turn patches of leaf or stem white, they are often named mottle viruses, spotted viruses, chlorotic (meaning, lacking chlorophyll) viruses, necrotic (meaning, killed cells surrounded by intact tissue) viruses, or mosaic viruses. **Figure 34.19** shows a healthy cowpea leaf and a cowpea leaf that has been attacked by a positive-sense single-stranded RNA virus. Some species in this group of viruses specialize in parasitizing bacteria, fungi, or animals, however. A variety of human maladies, including the common cold, polio, and hepatitis A, C, and E, are caused by positive-sense RNA viruses.

Replication cycle When the genome of these viruses enters a host cell, the single-stranded RNA is immediately translated into viral proteins. These proteins include enzymes that make copies of the genome. The new generation of virus particles is assembled in a complex structure that is associated with the host cell's plasma membrane.

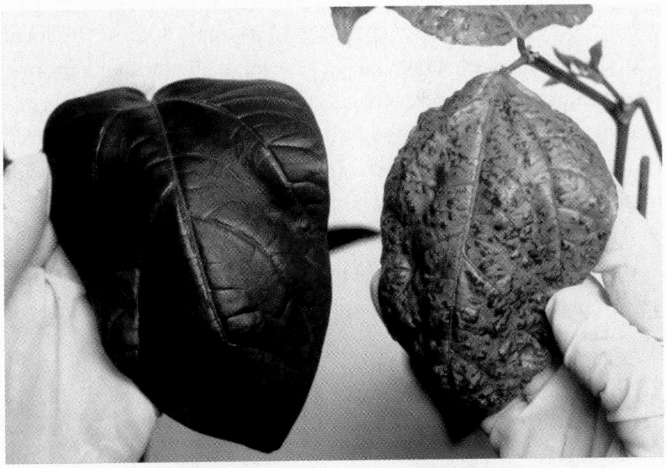
Healthy leaf
Leaf infected with virus

FIGURE 34.19 Positive-Sense Single-Stranded RNA Viruses, Such as the Cowpea Mosaic Virus, Cause Important Plant Diseases

ESSAY The SARS Outbreak

In November 2002, physicians in the southern Chinese province of Guangdong began treating people for an atypical form of **pneumonia**, or inflammation of the lungs. The illness began with a day or two of high fever (over 38°C, or 100.4°F) and progressed to pneumonia and dry coughing, sometimes accompanied by low oxygen levels in the blood. A number of the cases were fatal.

In February 2003 the illness showed up in Hong Kong. At that point, international health officials were alerted. Researchers immediately began to track cases and monitor the spread of the disease. The condition was named **severe acute respiratory syndrome**, or **SARS**. In medicine, a **syndrome** is a group of symptoms that occur together and have the same cause. The syndrome was *severe* because it made people sick enough to require hospitalization and *acute* because it progressed rapidly.

Soon after the disease appeared in Hong Kong, people with the condition began showing up in other countries. By the time SARS was brought under control several months later, a total of 8098 people in 30 countries had fallen ill. Of these, 774—almost 10 percent—had died. This is an extremely high fatality rate. In contrast, the 1918–1919 influenza was fatal in only about 2.5 percent of cases.

To fight the outbreak, public health officials not only quarantined individuals who had been exposed but also restricted access to particularly hard-hit cities or neighborhoods. In addition, many people stopped going to work in affected cities, for fear that they would contract the virus. The Asian Development Bank estimates that this disruption cost the economies of southeast and east Asia $18 billion in gross domestic product, and that an additional $60 billion was lost due to reduced travel and lowered demand for goods and services in the affected regions. These figures work out to an economic loss of about $2 million for each person with SARS.

The SARS outbreak did not become an epidemic, because the virus is not easily transmitted from person to person.

Early in the outbreak, researchers succeeded in isolating the virus, sequencing its genome, and analyzing its place in a

(Continued on next page)

(Essay continued)

phylogeny of closely related viruses. The results showed that SARS was caused by a previously undescribed type of *coronavirus*—a positive-sense single-stranded RNA virus. To determine where the virus originated, researchers followed up on the observation that several of the earliest SARS cases involved workers from restaurants in the city of Shenzhen that specialized in serving exotic meats. Testing showed that the SARS coronavirus was present in several species of wild animals sold in the local exotic meat market, as well as in several of the human vendors. Based on these data, researchers concluded that the SARS coronavirus jumped from a wild animal in the market to an animal vendor or restaurant workers. It is still not known what species in nature is the normal host or "reservoir" of this virus.

The SARS outbreak highlights several key points about virus outbreaks and public health. First, emerging viruses will continue to emerge. Human populations will continually be exposed to novel pathogens, from Ebola virus to Hantaan virus to HIV and SARS coronavirus to viruses that have yet to be identified or even evolve. As this book goes to press, international public health officials continue to monitor pneumonia cases in east Asia for a "reemergence" of SARS. Second, the nature of today's global economy and access to air travel means that emerging diseases can spread worldwide in a matter of weeks or even days. During the SARS outbreak, officials were able to trace the virus's movement to new countries via infected individuals on specific flights of specific airlines. Third, the SARS outbreak did not become an epidemic, because the virus is not easily transmitted from person to person. In the vast majority of cases, people got the virus through extended and direct contact with droplets from an infected person's cough or sneeze. If transmission had occurred through airborne particles instead of water droplets, the virus could have caused a devastating epidemic.

CHAPTER REVIEW

Summary of Key Concepts

- **Viruses are tiny, noncellular parasites that infect virtually every type of cell known. They cannot perform metabolism on their own—meaning outside a parasitized cell—and are not considered to be alive. Different types of viruses are specialized for infecting particular species and types of cells.**

Viruses cause illness and death in plants, fungi, bacteria, and archaea, as well as in humans and other animals. Vaccination is frequently effective in preventing virus epidemics. Unfortunately, it is difficult or impossible to design a vaccine that can prepare the immune system for viruses that mutate very rapidly, such as HIV and the cold and flu viruses.

- **Viruses are highly diverse in overall morphology and in the nature of their genetic material. The genomes of viruses may consist of double-stranded DNA, single-stranded DNA, double-stranded RNA, or one of several types of single-stranded RNA.**

The most important morphological groups in viruses are the enveloped and nonenveloped viruses. Enveloped viruses contain a capsid consisting of protein and are surrounded by a membranous envelope; nonenveloped viruses are surrounded only by a capsid.

Viral genomes are small and can consist of either RNA or DNA, but not both. Viral genomes do not code for ribosomes, and most do not code for the enzymes needed to translate their own proteins or to perform other types of biosynthesis. When a virus infects a host cell, it uses that cell's enzymes, nutrients, and ribosomes to manufacture a new generation of virus particles. This observation explains why viral diseases are difficult to treat with drugs: Molecules that incapacitate enzymes needed by the virus are likely to damage host cells as well.

- **The viral infection cycle can be broken down into five steps: (1) entry into a host cell, (2) replication of the viral genome, (3) the production of viral proteins, (4) the assembly of a new generation of virus particles, and (5) exit from the infected cell.**

A viral infection begins when the contents of a virus particle enter a host cell. For example, when HIV binds to a transmembrane protein called CD4 and a co-receptor, the virus's membrane-like envelope fuses with the host cell's plasma membrane, and the viral genome and proteins enter the cytoplasm. The viral genome is replicated in the second phase of the replication cycle. In HIV, for example, the virus's RNA genome is reverse-transcribed to DNA and integrated into a chromosome of the host cell. In the third phase of the replication cycle, viral proteins are produced. While they are still inside the cell, the viral proteins that result assemble into complete particles (the fourth phase of the infection cycle). In the final phase, the new generation of complete particles buds or bursts from the cell. Viruses that bud from the host cell acquire a lipid bilayer, containing host proteins as well as viral proteins, during the process. Viruses that burst from the host cell do not have this membrane-like envelope. Instead, they are protected only by a protein coat. These coat proteins are coded for by viral genes and assemble inside the host cell's cytoplasm. When a new generation of virus particles buds or bursts from a host cell, the infected cell dies.

Once they are released from the host cell, particles can infect more cells in the same multicellular organism or be transmitted to a new host.

Web Tutorial 34.1 The HIV Life Cycle

Questions

Content Review

1. How do viruses that infect animals enter an animal's cells?
   a. The viruses pass through a wound.
   b. The viruses bind to a membrane protein.
   c. The viruses puncture the cell wall.
   d. The viruses bind to the lipid bilayer.
2. What does reverse transcriptase do?
   a. It synthesizes proteins from mRNA.
   b. It synthesizes tRNAs from DNA.
   c. It synthesizes DNA from RNA.
   d. It synthesizes RNA from DNA.
3. What do host cells provide for viruses?
   a. nucleotides and amino acids
   b. ribosomes
   c. ATP
   d. all of the above
4. When do virus particles acquire a membrane-like envelope, including a lipid bilayer?
   a. during entry into the host cell
   b. during budding from the host cell
   c. as they burst from the host cell
   d. as they integrate into the host cell's chromosome
5. What reaction does protease catalyze?
   a. polymerization of amino acids into peptides
   b. cutting of long peptide chains into functional proteins
   c. folding of long peptide chains into functional proteins
   d. assembly of viral particles
6. Why is it difficult to design a vaccine for viruses with high mutation rates, such as HIV and the cold and flu viruses?
   a. The vaccines tend to be unstable and deteriorate over time.
   b. So many protein fragments are presented by these viruses that the immune system overreacts.
   c. They have no protein fragments that can be recognized by a host cell.
   d. New mutations constantly change viral proteins.

Conceptual Review

1. The outer surface of a virus consists of either a membrane-like envelope or a protein capsid. Which type of outer surface does HIV have? Which type do bacteriophages have? How does the outer surface of a virus correlate with its mode of exiting a host cell, and why?
2. Compare the morphological complexity of HIV with that of bacteriophage T4. Which virus would you predict has the larger genome? Explain the logic behind your hypothesis.
3. Compare and contrast lytic growth and lysogenic growth. When would you expect HIV and bacteriophage T4 to grow lytically versus lysogenically, and why?
4. Explain why viral diseases are more difficult to treat than diseases caused by bacteria.
5. Draw the lytic cycle of a nonenveloped virus with a positive-sense single-stranded RNA genome that infects cells in the roots of rice plants. Describe the modes of action of two drugs that could be developed to treat plants infected with this virus.
6. What type of data convinced researchers that HIV originated when a simian immunodeficiency virus "jumped" to humans? Do you agree with this conclusion? Why or why not?

Group Discussion Problems

1. Suppose you could isolate a virus that parasitizes the pathogenic bacterium *Staphylococcus aureus*. This bacterium causes acne, boils, and a variety of other afflictions in humans. How could you test whether this virus might serve as a safe and effective antibiotic?
2. If you were in charge of the government's budget devoted to stemming the AIDS epidemic, would you devote most of the resources to drug development, vaccine development, or preventive medicine? Defend your answer.
3. Bacteria fight viral infections with restriction endonucleases (bacterial enzymes; introduced in Chapter 19). Restriction endonucleases cut up, or break, viral DNA at specific sequences. The enzymes do not cut a bacterium's own DNA, because the bases in the bacterial genome are protected from the enzyme by methylation (the addition of a CH_3 group). Generate a hypothesis to explain why members of the Eukarya do not have restriction endonucleases.
4. Consider these two contrasting definitions of life:
   a. An entity is alive if it is capable of replicating itself via the directed chemical transformation of its environment.
   b. An entity is alive if it is an integrated system for the storage, maintenance, replication, and use of genetic information.

   According to these definitions, are viruses alive? Explain.

Answers to Multiple-Choice Questions 1. b; 2. d; 3. d; 4. b; 5. b; 6. d

www.prenhall.com/freeman is your resource for the following: Web Tutorials; Online Quizzes and other Online Study Guide materials; Answers to Conceptual Review Questions; Solutions to Group Discussion Problems; Answers to Figure Caption Questions and Exercises; and Additional Readings and Research.

UNIT 7

How Plants Work

This Presidio manzanita leaf is a photosynthetic organ. Leaves sense and respond to light, touch, and insects. [©2004 Susan Middleton & David Liittschwager]

35 Plant Form and Function

KEY CONCEPTS

- The plant body consists of (1) a root system that anchors the individual and absorbs water and key ions, and (2) a shoot system that absorbs carbon dioxide and sunlight.
- The root and shoot systems of plants are composed of several different cell types and tissues, each of which has a specialized structure and function.
- The roots and shoots of all plants increase in length due to cell division in regions called apical meristems. In addition, woody species increase in width due to cell divisions that take place in regions called lateral meristems.

Cross section from the trunk of a redwood tree that lived 1710 years. In this chapter, we explore features of plant anatomy, including how wood forms and why growth rings appear in tree trunks.

Plants do the most remarkable chemistry of any terrestrial organism. Using the energy in sunlight and the simplest of starting materials—carbon dioxide, water, and ions containing nitrogen, phosphorus, or other key atoms—they synthesize thousands of different carbohydrates, proteins, nucleic acids, and lipids and build bodies that may live for thousands of years. The feat is even more impressive when you consider that the starting materials that plants need to grow are not only simple in structure but also diffuse. Carbon dioxide, nitrogen, and other raw materials that plants require are found at low concentrations over a large area. In contrast, animals feed by eating other organisms, which represent highly concentrated and highly localized packets of resources. Instead of moving around and eating concentrated sources of food like animals do, plants stay in one place, gather the nutrients and other resources available on that spot, and make their own food.

In essence, a plant's body is devoted to harvesting diffuse resources and concentrating them in cells and tissues. The structure of the body is dynamic, though, because plants grow throughout their lives. In response to favorable conditions, a plant sends roots and shoots in the most promising directions, seeking out more and more of the diffuse resources it requires.

Even though all plant bodies function in this way, the 300,000 species living today are remarkably diverse in size and shape. Land plants range from redwood trees 90 m (300 ft) tall to sprawling mosses that are just 2 mm (1/16 in.) high. Between these extremes is a bewildering variety of growth forms—from small trees and shrubs to vines, grasses, mat-forming plants, and herbs. Plants go about capturing photons and absorbing water, ions, and carbon dioxide in an extraordinary variety of ways.

This chapter focuses on three fundamental questions: (1) How is the plant body organized? (2) How do plants grow throughout their lives? and (3) Why are plants so diverse in size and shape? Instead of surveying the entire catalog of land plants, though, the focus here is on the angiosperms. Recall from Chapter 29 that angiosperms are the flowering plants. They are the most recent major group of plants to appear in the fossil record and the most abundant and geographically widespread on Earth today. They are also the most species-rich plant lineage and are of huge economic and medical importance to humans. Most of the food we eat and many of the drugs we use are derived from angiosperms.

Exploring questions about the anatomy of flowering plants is vital to understanding the world at large as well as subsequent chapters in this unit. If you look outside or down the produce aisle of a grocery store, you're almost certain to see an impressive diversity of plants or plant products. What are their various parts called, and what does each structure do? Why are the size and shape of leaves, twigs, and roots so variable among species? Learning about plant form and function will not only make you a better biologist, but also it will deepen your appreciation of the richness and complexity of the natural world.

35.1 The Diversity of Plant Form

Chapter 10 detailed how plants—along with algae, cyanobacteria, and a variety of protists—obtain the energy and carbon they need to grow and reproduce. These organisms use photosystem I and photosystem II to harness the radiant energy in sunlight and use it to manufacture ATP and NADPH. The enzymes of the Calvin cycle then use the chemical energy in ATP and NADPH to convert (or "fix") carbon from an oxidized state in the form of the inorganic compound carbon dioxide (CO_2) to a reduced state in the form of an organic compound called glyceraldehyde 3-phosphate (G3P). The G3P produced by the Calvin cycle feeds biochemical pathways that result in the formation of energy-rich carbohydrates such as sucrose and starch.

To perform photosynthesis, plants need large amounts of light and carbon dioxide and a small amount of water as an electron source (**Figure 35.1**). Plants also need large amounts of water to fill their cells and maintain them at normal volume and pressure. To synthesize nucleic acids, enzymes, phospholipids, and the other macromolecules needed to build and run cells, plants must obtain nitrogen (N), phosphorus (P), potassium (K), magnesium (Mg), and a host of other nutrients. Most of these key elements exist in nature as ions that dissolve in water found in soil. Thus, plants must move large quantities of water and ions from the soil surrounding their roots to their leaves.

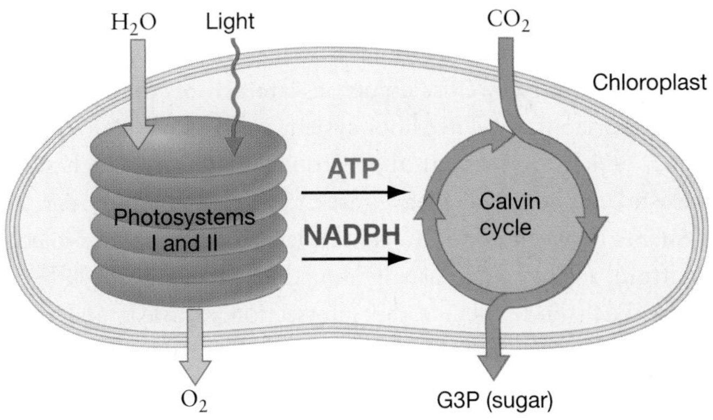

FIGURE 35.1 Plants Need Resources to Perform Photosynthesis
Photosystems I and II require light and water and produce ATP and NADPH. The Calvin cycle requires CO_2 and produces sugar.

The carbon dioxide and light that plants need to perform photosynthesis are available in the atmosphere; the water and ions they need are available in the soil. To acquire these resources, plants use the two basic systems shown in **Figure 35.2**. A belowground portion called the **root system** anchors the plant and takes in water and nutrients from the soil, while an aboveground portion called the **shoot system** harvests light and carbon dioxide. In most plants, the root and shoot systems are connected by **vascular tissue.** Water is transported from roots to shoots through vascular tissue; sugars and other nutrients are transported in both directions.

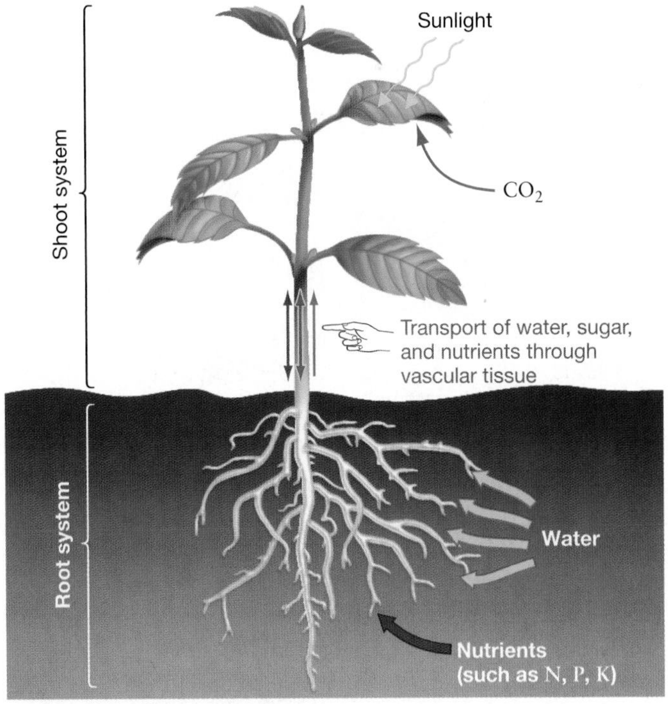

FIGURE 35.2 Root and Shoot Systems Acquire and Transport Resources
Shoot systems are specialized for harvesting light and CO_2; root systems absorb water and key nutrients such as nitrogen (N), phosphorus (P), and potassium (K).

Figure 35.3 provides a more detailed look at the anatomy of roots and shoots. Many root systems have a vertical section called a **taproot**, as well as numerous **lateral roots** that run more or less horizontally. The shoot system consists of one or more **stems**, which are vertical aboveground structures. Each stem consists of **nodes**, where leaves are attached, and **internodes**, or segments between nodes. A **leaf** is a lateral appendage, projecting from a stem, that usually functions as a photosynthetic organ. Each node includes an intersection of a leaf and stem that is the site of a **lateral bud.** If conditions are appropriate, the lateral bud may develop into a **branch**—a lateral extension of the shoot system. At the top of the stem is an **apical bud**, where growth occurs to extend the stem vertically. If conditions are appropriate, the apical buds found on the main stem or branches may develop into flowers or other reproductive structures.

The "generic" plant shown in Figures 35.2 and 35.3 is an **herb**, meaning a seed plant that lacks wood and has a relatively short-lived stem. Most land plants conform to the basic theme illustrated in the figures—meaning they have bodies organized into root and shoot systems joined by vascular tissue. There are many variations on this theme, however. The organization of the vascular tissue is not identical in all plants, and the morphology of root and shoot systems varies widely. Let's take a closer look at diversity in root and shoot systems by focusing on two particularly spectacular examples: North American prairie plants and Hawaiian silverswords.

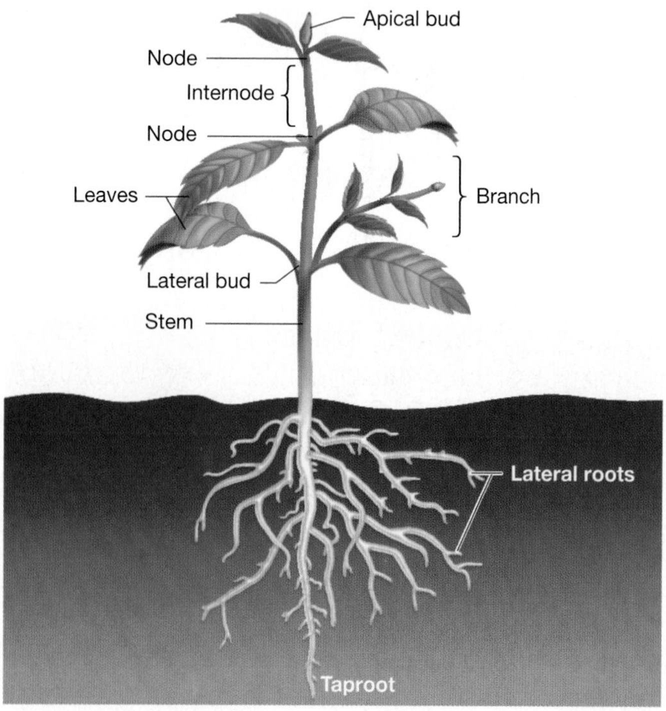

FIGURE 35.3 Most Root and Shoot Systems Have the Same General Structures
Plant bodies include the structures labeled here.

The Diversity of Roots: North American Prairie Plants

Root systems can be impressive in extent. For example, a researcher grew a winter rye plant in a container full of soil for four months, then unearthed the individual and meticulously measured the extent of its roots. He found that the root system of this single individual contained more than 13 million identifiable structures with a combined length of over 11,000 km—almost one-third of Earth's circumference. Other studies have shown that the roots of trees routinely extend wider than their aboveground canopy and that it is not unusual for a plant's root system to represent over 80 percent of its total mass. Many plants devote a great deal of energy and resources to the growth of their root systems.

The morphological diversity observed in root systems is also impressive. As an example, consider the prairie plants of North America. Prairies are grassland ecosystems found in areas of the world such as the Serengeti Plain of East Africa, the Pampas region of Argentina, and the steppes of central Asia. Rain is abundant enough in these areas to support a lush growth of grasses and **herbaceous** (nonwoody) plants but scarce enough to exclude trees and most shrubs (**Figure 35.4a**). The growth of woody species is also discouraged by the fires that regularly sweep through these ecosystems.

Although the aboveground portions of prairie plants die back during the winter or dry season, their root systems are perennial. When spring arrives or the rainy season begins, prairie plants produce a new shoot system from specialized groups of cells located just below the ground surface.

To examine the root systems of prairie plants, researchers performed either of two procedures: They dug deep trenches and examined the root systems that were exposed, or excavated around a particular plant to reveal its roots. Some of the photographs that resulted from this effort are shown in **Figure 35.4b**. The figure shows that the root systems of prairie plants that live side by side can be very different from one another. Compare the dense, **fibrous** root system in the grass called big bluestem with the 6.4-m (21-ft) taproot found in the prairie rose. Other species, such as the compass plant, have thick roots that contain particularly large reservoirs of food stored in the form of starch.

Even though these root systems have similar functions—support of the plant, nutrient and water absorption, and nutrient storage—they are highly diverse in structure. For example, the fibrous portions of roots are where most water absorption and nutrient absorption take place, but the fibrous roots of different species are found at different depths in the soil. To explain this observation, biologists suggest that natural selection has favored a diverse array of root systems in prairie plants to minimize competition for water and nutrients.

The diversity of root systems observed in prairie plants has two important consequences: (1) An array of prairie plants can coexist in the same area, and (2) most individuals can survive intense water stress during drought years. Each type of root

(a) Prairies are dominated by grasses and herbs.

(b) Root systems in prairie plants are diverse.

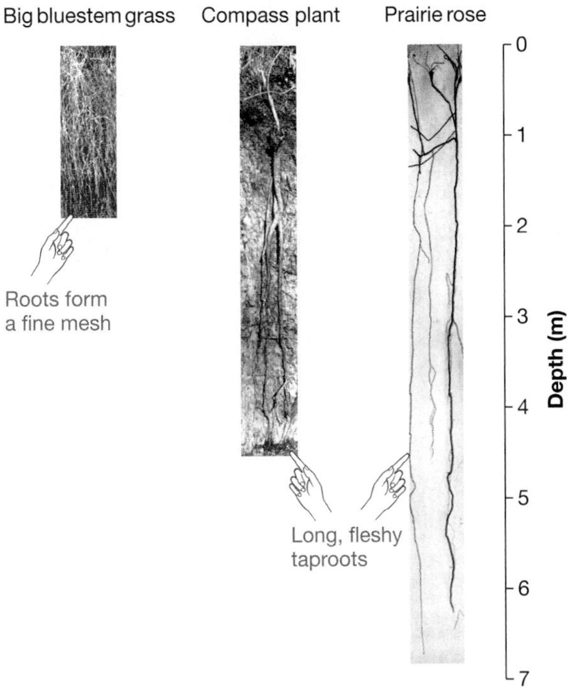

FIGURE 35.4 Prairie Plants Have Diverse Root Systems
(a) Prairies, such as this one in Kansas, are grassland ecosystems. **(b)** The roots of prairie plants that live side by side can be very different. For example, the roots of these three plants reach depths of 6 ft, 14 ft, and 21 ft, respectively. EXERCISE Roots function as storage organs as well as resource-harvesting systems and anchoring devices. In part (b), draw arrows to thickened areas of roots that store starch. How might you confirm that they have high starch concentrations?

(a) Prop roots function in support.

(b) "Knees" function in gas exchange.

FIGURE 35.5 Modified Roots Have Unusual Structures or Functions
(a) The prop roots of corn plants that help stabilize the stem are adventitious—they develop from the shoot system. **(b)** The pneumatophores of mangrove trees allow gas exchange to occur between root tissues and the atmosphere.

system is specialized for acquiring resources from a different region of the soil. Due in part to the diversity of root systems illustrated in Figure 35.4b, a small patch of prairie is able to sustain a relatively large number of species.

The taproots and fibrous roots observed in prairie plants do not begin to exhaust the types of roots found among plants, however. **Figure 35.5a** shows the prop roots of corn, which help brace the plant in windy weather. Prop roots are an example of an **adventitious root:** a root that develops from the shoot system instead of the root system. The roots that grow from the nodes of ivy plants and help them cling to brick walls are another example of an adventitious root. Roots can also be modified for functions other than anchoring the individual and absorbing water and ions from the soil. The "knees" of a mangrove, for example, are actually spongy outgrowths from roots called *pneumatophores* ("breathing-bearers") or air roots, which function in gas exchange (**Figure 35.5b**). Mangroves grow in swamps where water often becomes stagnant and deoxygenated. Their root cells do not suffocate, however, because oxygen from the atmosphere can diffuse into them via the pneumatophores. Unlike most roots, mangrove pneumatophores grow upward—not downward—in response to gravity.

To summarize, the structures and functions of root systems varies widely among plant groups and habitats. In most species, roots function in nutrient and water absorption and in stabilizing the individual. The size and shape of roots that perform these functions are diverse, and in some species roots can arise from shoot systems. In addition, some roots are specialized for oxygen exchange or other tasks.

The Diversity of Shoots: Hawaiian Silverswords

Because most prairie plants are grasses or herbs, prairie ecosystems are not known for the diversity of shoot systems present. To illustrate variation in shoot systems, let's consider the group of closely related species called the silverswords.

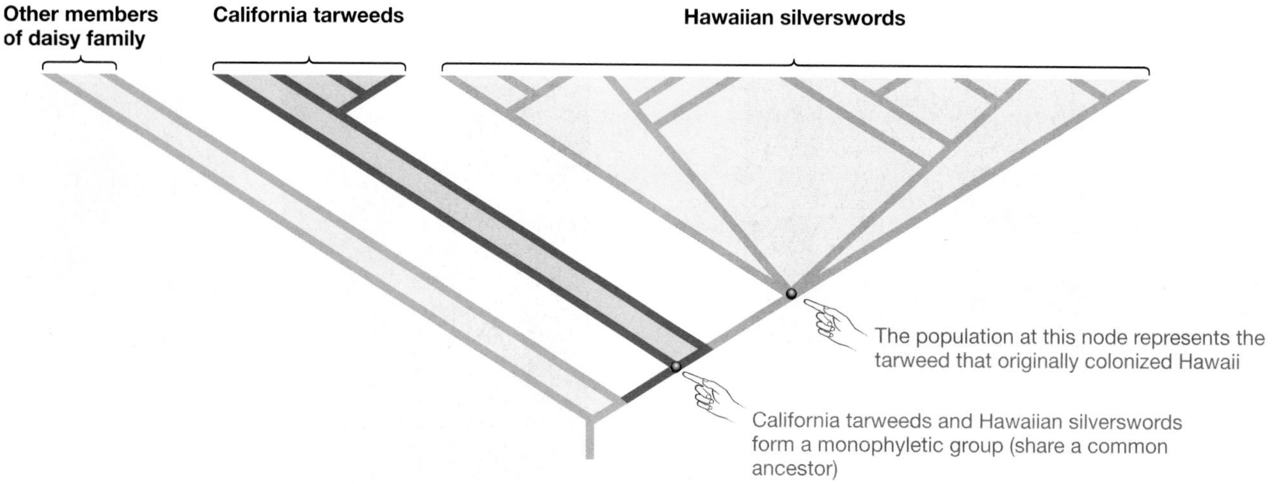
Other members of daisy family
California tarweeds
Hawaiian silverswords
The population at this node represents the tarweed that originally colonized Hawaii
California tarweeds and Hawaiian silverswords form a monophyletic group (share a common ancestor)

FIGURE 35.6 Hawaiian Silverswords Are a Monophyletic Group
Phylogenetic tree estimated by comparing DNA sequence data from silverswords, tarweeds, and closely related members of the daisy family.

Silverswords exist in only one location: the Hawaiian Islands. The phylogenetic tree in **Figure 35.6**, which is based on comparisons of DNA sequence data, supports the hypothesis that the silverswords are a monophyletic group. Stated another way, all silverswords descended from the same common ancestor. (See Chapter 26 for more detail on how biologists infer phylogenies from sequence data.) The diagram also indicates that the closest living relatives of the silverswords are plants called tarweeds that live in California. To interpret these results, biologists infer that a tarweed species colonized Hawaii at some point in the past. Because some of the tarweeds native to California have sticky seeds, researchers suggest that the original tarweed arrived in Hawaii as a seed that was stuck to the legs or feathers of an oceangoing bird. Once this ancestral colonist was established in Hawaii, its descendants underwent adaptive radiation—meaning they evolved into a large number of species that occupy a wide range of ecological niches (see Chapter 26). Today, silverswords can be found from sea level to alpine altitudes and from rain forests to the desertlike conditions of exposed lava flows.

In addition to occupying a wide variety of habitats, silverswords are wonderfully diverse in size, shape, and growth habit. Some have thick, woody stems and grow as shrubs or trees (**Figure 35.7a**). Others form sprawling cushions or mats (**Figure 35.7b**), while several species grow as compact forms called **rosettes** (**Figure 35.7c**). The adaptive radiation of silverswords has been based in part on diversification in shoot systems. Silversword species have evolved a wide variety of ways to capture photons and absorb carbon dioxide.

Modified Shoots Even though they are a single lineage, silverswords illustrate many of the types of shoot systems found among all land plants. But it's important to note that other variations occur on the basic anatomical theme of building a shoot system based on vertical stems. Not all stems grow vertically, and not all stems function in the acquisition of carbon dioxide and photons:

- Many desert cacti have highly modified stems. Instead of functioning primarily to support leaves, cactus stems are often enlarged into water-storage organs (**Figure 35.8a**). Water accounts for up to 98 percent of the weight of a cactus stem. A

(a) Tree-sized silversword
(b) Mat-forming silversword
10 cm
(c) Rosette-forming silversword
50 cm

FIGURE 35.7 Silverswords Have Diverse Shoot Systems
(a) Several of the silverswords, including *Dubautia reticulata*, grow as small trees. **(b)** *Dubautia scabra* is a low-growing silversword that colonizes lava flows and forms extensive mats. **(c)** Rosettes form in species that have extremely reduced internodes between leaves.

(a) Cactus stems store water.

(b) Stolons produce new individuals at nodes above ground.

(c) Rhizomes produce new individuals at nodes below ground.

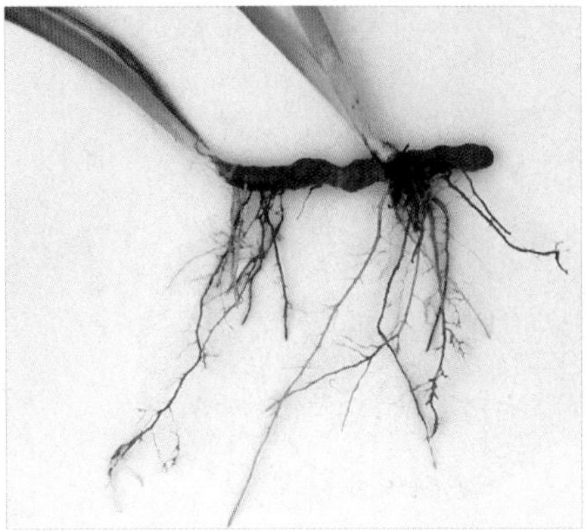

(d) Tubers store carbohydrates.

(e) Thorns provide protection.

FIGURE 35.8 Modified Stems Have Unusual Structures or Functions
In some species, stems are modified to perform functions other than producing leaves that absorb carbon dioxide and sunlight.

cactus stem also contains the plant's photosynthetic tissue, because its leaves are modified into the protective structures called **spines.**

- **Stolons** are modified stems that run over the soil surface, producing roots and leaves at each node (**Figure 35.8b**). Because new plants form at these nodes, stolons function in asexual reproduction (see Chapter 12).
- **Rhizomes** are similar to stolons—they are stems that grow horizontally instead of vertically. But while stolons grow above ground, rhizomes spread *below* ground (**Figure 35.8c**). Rhizomes are also similar to stolons in function: They produce new plants at nodes and thus aid asexual reproduction.
- **Tubers** are rhizomes that are modified as carbohydrate-storage organs (**Figure 35.8d**). The eyes of a potato—a typical tuber—are nodes in the stem where new branches may arise.
- **Thorns** are modified stems that help protect the plant from attacks by large **herbivores**, or plant-eaters, such as deer, giraffe, or cattle (**Figure 35.8e**).

Modified Leaves Leaves are particularly important parts of the shoot system, for a simple reason: In most plant species, the vast majority of photosynthesis occurs in leaves. The total area of leaf produced by a single plant can be enormous—a single tree can have hundreds of thousands of leaves with a total leaf surface area equivalent to that of a football field. All of this area is available for absorbing photons and supporting photosynthesis.

Figure 35.9a (page 810) illustrates the external anatomy of a typical leaf. Leaves are normally composed of just two major structures: an expanded portion called the **blade** and a stalk called the **petiole** that arises from the lateral bud. There are many variations on this theme, however. Grass leaves lack petioles entirely, *compound leaves* are divided into a series of leaflets (**Figure 35.9b**), and *doubly compound* leaves have leaflets that are again divided (**Figure 35.9c**). The arrangement of leaves can vary as well. For example, not all leaves are separated along stems by internodes. The rosette growth form observed in silverswords, dandelions, and cabbages arises when leaves form in the absence of stem growth and internodes.

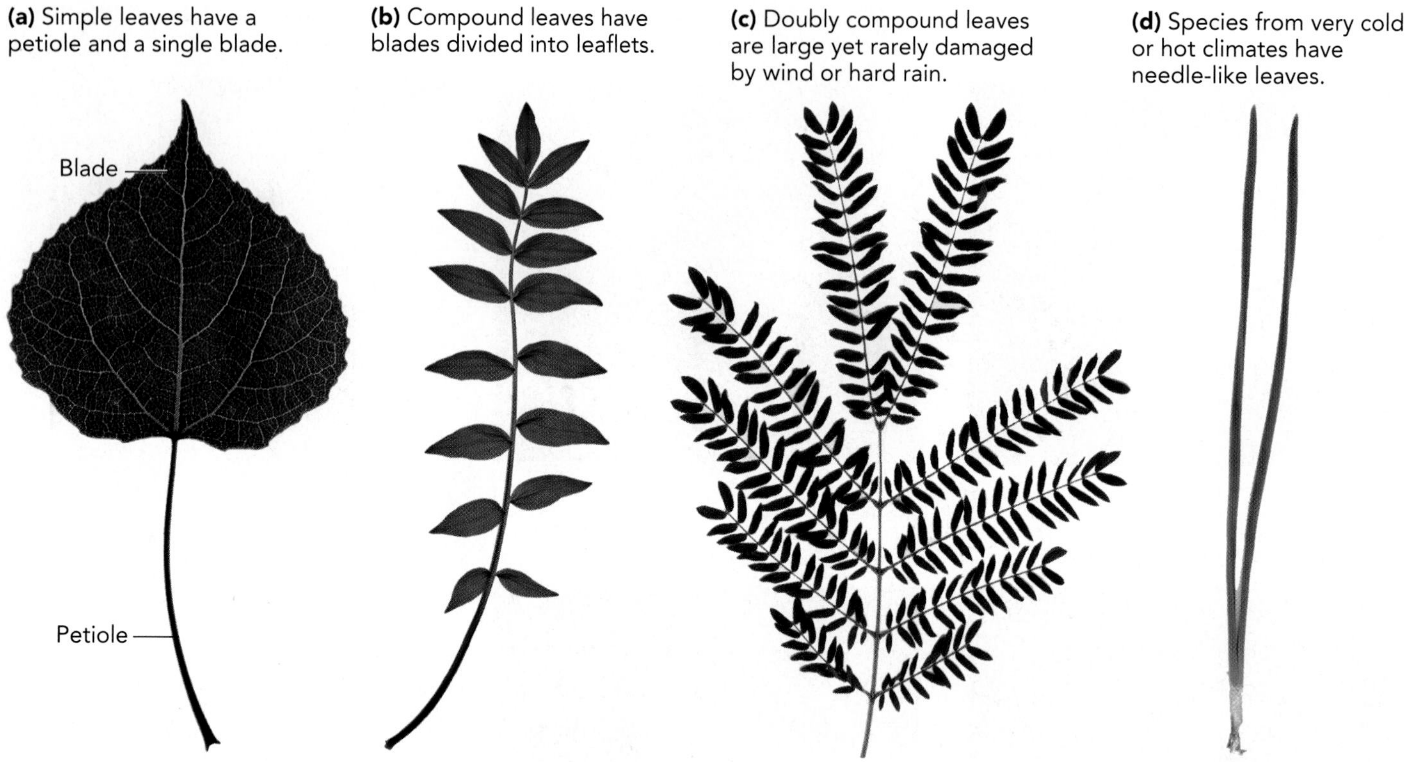

(a) Simple leaves have a petiole and a single blade.
(b) Compound leaves have blades divided into leaflets.
(c) Doubly compound leaves are large yet rarely damaged by wind or hard rain.
(d) Species from very cold or hot climates have needle-like leaves.
Blade
Petiole

FIGURE 35.9 Leaves Vary in Size and Shape
The structures in parts **(a)** through **(c)** represent single leaves; part **(d)** shows two leaves. QUESTION For capturing photons, what is the advantage of having a leaf with a large surface area? In terms of wind damage and water loss, what is the disadvantage of having a leaf with a large surface area?

In addition, not all leaf blades are thin and have a large surface area that is advantageous for absorbing large numbers of photons. For example, plants that thrive in deserts and in northern habitats tend to have leaves that are needle-like (**Figure 35.9d**). The leading hypothesis to explain this pattern is based on two observations: (1) Water is often in short supply in these environments because it is absent in deserts or frozen and thus unavailable in cold habitats, and (2) leaves with large surface areas lose large amounts of water through evaporation from openings called *stomata*—a process called **transpiration**. Thus, needlelike leaves are interpreted as adaptations that minimize transpiration in water-short habitats. Small, narrow leaves are also much less susceptible to wind damage than are large, broad leaves.

It's important to recognize that not all leaves function primarily in photosynthesis. Recall that cactus spines are modified leaves that protect the stem. Onion bulbs consist of thickened leaves, separated by highly condensed internodes, that store nutrients (**Figure 35.10a**). The thick leaves of species called succulents store water (**Figure 35.10b**), the tendrils that garden peas and other vines use to climb are modified leaves (**Figure 35.10c**), and the bright red leaves of poinsettias attract pollinators to the flower nearby (**Figure 35.10d**). The leaves of some carnivorous plants help trap prey as well as performing photosynthesis. The pitcher plant shown in **Figure 35.10e**, for example, has hairy leaves that curl into a tube. When insects enter the tube, they are discouraged from flying up and out by the dark "hood" that covers the opening and discouraged from climbing out by downward-pointing hairs. If they eventually tire and fall into the bottom of the tube, the insects drown and are digested by enzymes secreted by the plant.

The most unusual of all leaf types known, however, may be the balloon-like leaves of the flowerpot plant (**Figure 35.10f**). These leaves are used as homes by ant colonies and develop a network of adventitious roots. These observations inspired Bert Hölldobler and Edward O. Wilson to propose that the adventitious roots help feed the plant by absorbing nutrients from ant feces and other material that accumulates inside the hollow leaf. This "flowerpot leaf hypothesis" was supported when other researchers added radioactive isotopes of carbon and nitrogen to the debris inside leaves, and found that the nutrients were readily taken up by the adventitious roots and incorporated into molecules used in other locations in the plant.

To summarize, the plant body is organized into a root system and a shoot system. This is the most basic anatomical

(a) Onion leaves store food.
(b) Aloe vera leaves store water.
(c) Grape tendrils aid in climbing.
(d) Poinsettia leaves attract pollinators.
(e) Pitcher plant leaves trap insects.
(f) Flowerpot plant leaves collect soil.

FIGURE 35.10 Modified Leaves Have Unusual Structures or Functions
In some species, leaves are modified to perform functions other than photosynthesis.

theme in land plants. It is a theme with many variations, however. Roots and shoots have been modified to absorb water, ions, carbon dioxide, and sunlight efficiently in a wide array of environments. They have also been modified to perform additional functions, ranging from anchoring the plant to storing water and carbohydrates. Given this introduction to how the overall plant body is organized, let's plunge into the interior of plants and explore the structure and function of their cells and tissues.

35.2 Plant Cells and Tissues

Chapter 7 introduced a generalized version of the plant cell, similar to the cell in **Figure 35.11a** (page 812). Recall that plant cells differ from animal cells in several important ways. First, all plant cells are surrounded by a stiff, cellulose-rich **cell wall**. The cell wall supports plant cells and defines their shape. Pressure from fluid-filled cells against the cell wall allows plants that lack wood to stand upright. Second, plant cells often contain several types of organelles that are not found in animals—specifically chloroplasts and a large, membrane-bound organelle called a **vacuole**, which fills most of the cell's volume. Chloroplasts are the site of photosynthesis. Vacuoles, which contain an aqueous solution called **cell sap**, are responsible for digesting waste products. In this sense, they are similar to the lysosomes observed in animal cells. But plant vacuoles also store water and nutrients and may hold pigments that provide color or toxic compounds that deter plant-eating animals (see **Box 35.1**, page 813). Third, the cytoplasm of adjacent plant cells is connected by structures called **plasmodesmata** (singular: **plasmodesma**). Plasmodesmata consist of gaps in the cell wall and plasma membrane. As **Figure 35.11b** shows, segments of smooth endoplasmic reticulum (ER) run through these gaps.

Another important distinction between plant cells and animal cells is that plant cells do not move once they form. Some animal cells move either early in the development of an individual (see Chapter 22) or as mature cells (see Chapters 44 and 49). But plant cells are stationary. They are specialized for capturing diffuse resources, supporting photosynthesis, and concentrating photosynthetic products in vacuoles.

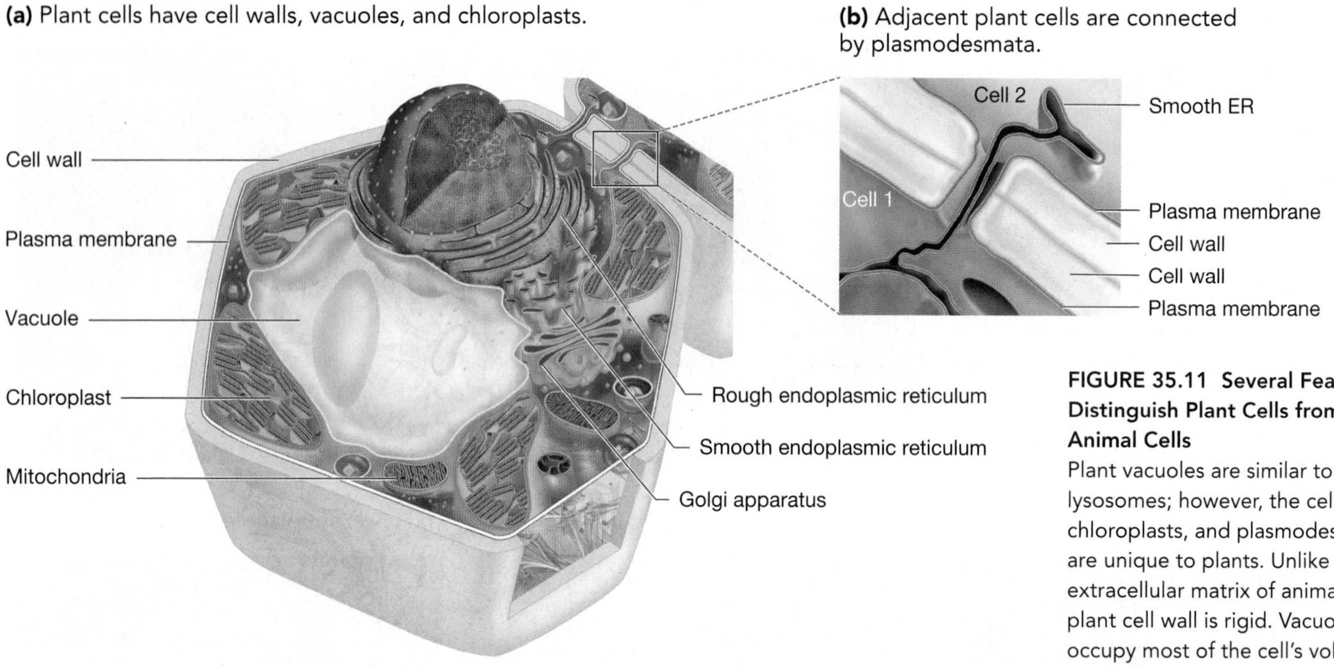

FIGURE 35.11 Several Features Distinguish Plant Cells from Animal Cells
Plant vacuoles are similar to animal lysosomes; however, the cell wall, chloroplasts, and plasmodesmata are unique to plants. Unlike the extracellular matrix of animals, the plant cell wall is rigid. Vacuoles occupy most of the cell's volume.

The Diversity of Plant Cells

Although many plant cells share features of the generalized cell in Figure 35.11, in reality the cells that make up an individual are enormously diverse in size, shape, composition, and function. For example, plant cells are either undifferentiated or mature. Let's consider the structure and function of undifferentiated cells first, then explore the three major types of mature cells.

Meristematic Cells The undifferentiated cells that are responsible for plant growth are called **meristematic cells.** The meristematic cells of plants are similar to the rapidly dividing cells found in early animal embryos or the stem cells found in adult animals. Meristematic cells are said to be undifferentiated because they do not express selected sets of genes that lead to the development of specialized structures and functions.

Meristematic cells divide rapidly when conditions for growth are good. Instead of having specialized organelles or cell walls, they have generalized versions of these structures. For example, they contain undifferentiated organelles called **proplastids.** As the cells produced by the growth and division of meristematic cells mature into cells with specialized functions, proplastids differentiate into photosynthetic chloroplasts, colorful and pigment-filled chromoplasts, or starch-filled amyloplasts (see Chapter 9).

When meristematic cells divide, the daughter cells that form either continue to function as meristematic cells or differentiate to form one of three types of mature, specialized cells: (1) parenchyma (pronounced *pa-REN-ki-ma*), (2) collenchyma (*ko-LEN-ki-ma*), or (3) sclerenchyma (*skle-REN-ki-ma*). Which type of cell develops depends on the position of the new cell in the meristem. Let's consider each of these major types of mature plant cells in turn.

Parenchyma Cells **Parenchyma** cells are the most abundant and versatile plant cells. They have relatively thin cell walls, are alive at maturity, and are **totipotent**, meaning they retain the capacity to divide and develop into a complete, mature plant. The totipotency of parenchyma cells is important in wound healing and in asexual reproduction via stolons or rhizomes. In each case, parenchyma cells may begin to divide, grow, and differentiate to form new roots and shoots. This property of parenchyma cells allows gardeners to clone plants by making cuttings. For example, if you cut a piece of geranium stem and place it in water, parenchyma cells will divide to produce a mass of undifferentiated cells called a **callus.** Roots develop from the callus (**Figure 35.12**), and the new individual can be planted in soil.

FIGURE 35.12 Parenchyma Cells in Cut Stems Can Form Roots
Parenchyma cells in a cut stem divide to form a mass of undifferentiated cells called a callus, which then sprouts roots.

BOX 35.1 Plant Poisons and the Cost of Defense

In addition to being protected by thorns, spines, or other structures, many plants are laced with poisons. Compounds that are toxic to plant-eating insects and mammals are usually concentrated in specialized reservoirs. As a result, the defensive compounds do not poison the plant itself. A few examples will help drive this point home:

- Some species of mint have leaf cells with bulbous *trichomes* that are filled with peppermint oil. Trichomes are specialized outgrowths of cells that make up the surface of the leaf. These appendages form the equivalent of a mine-field for herbivores on the leaf surface but do not threaten the cells themselves.
- The leaves and bark of pines and firs contain *pitch*. Pitch is a complex substance that includes the molecule pinene, which is toxic to bark beetles. Pinene is also partially responsible for the distinctive odor of these trees. Pitch does not harm plant cells, because it is localized in specialized structures called *resin ducts*.
- Poisons are frequently concentrated in cell vacuoles, where they cannot harm enzymes or other molecules in the remainder of the cell. For example, the *tannins* found in a wide array of plant species bind to proteins. Tannins are localized in vacuoles, so they do not harm plant enzymes. But when tannins are ingested by an herbivore, they inactivate the animal's digestive enzymes and make it sick. (In humans, small doses of tannins cause a sharp, astringent sensation in the mouth that is prized in apples, blackberries, teas, and red wine. Because tannins cause structural changes that make proteins less likely to disintegrate over time they are used to tan animal skins and produce leather.)

Secondary Metabolites Most plant defense compounds have structures and compositions that are closely related to molecules that are required for basic cell activities, such as amino acids. To explain this resemblance, biologists hypothesize that, during the evolution of some plant groups, mutations occurred in the genes for enzymes involved in fundamental biosynthetic pathways. This hypothesis claims that the mutations resulted in the production of altered enzymes and the synthesis of new or "secondary" compounds.

The key point is that if these compounds happened to be toxic to herbivores or pathogens, the compounds would have helped mutant individuals survive better and reproduce more. Over time, then, the alleles responsible for encoding the required enzymes would increase in frequency in the population. Based on this logic and the chemical similarity of secondary compounds to required compounds, plant poisons are referred to as **secondary metabolites.**

Secondary metabolites fall into one of three chemical classes: (1) *terpenes*, such as the essential oils found in mint, lemon, basil, and sage, the pinene found in coniferous trees, and the pyrethroids found in the leaves and flowers of chrysanthemums; (2) *phenols*, such as tannins and tetrahydrocannabinol, the active ingredient in marijuana; and (3) *nitrogen-containing compounds*, such as alkaloids. *Alkaloids* are a particularly extensive family of chemicals found in about 20 percent of all plant species. Over 12,000 different alkaloids have been identified in plants to date, including morphine, codeine, cocaine, nicotine, and caffeine. Experiments have shown that these compounds have a wide array of toxic effects at the molecular level and that they are extremely effective at harming a plant's enemies.

The Cost of Defense If secondary compounds are so effective at deterring herbivores, why don't all plants produce a lot of them all the time? To answer this question, researchers rely on a fundamental observation: Every organism has a finite amount of energy to devote to the tasks of surviving and reproducing.

To understand why that simple fact is important, consider a thought experiment about the weedy mustard plant *Arabidopsis thaliana*. Suppose that every individual in a population manufactures an average of 100 million ATP molecules in the course of its 2-month lifetime. Further, suppose that the individuals in a population vary in how they expend the ATP and that this variation is based on genetic variation among individuals. For example, some individuals might spend a large percentage of their energy from ATP making defensive compounds and a relatively small percentage making seeds. Other individuals might make few secondary metabolites and invest most of their energy from ATP in reproduction instead. The question is, Which type of individual will survive and reproduce best?

The short answer to that question is "It depends on conditions in the environment." If herbivores are abundant, then most of the poorly defended plants will die. Well-defended plants will produce more offspring than lightly defended plants do. As a result, alleles that lead to the increased production of secondary metabolites will increase in frequency in the population. But if herbivores are rare, then poorly defended plants will invest the most resources in reproduction and produce the most offspring. Alleles that lead to the high production of defense compounds will then decrease in frequency.

To summarize these ideas, biologists say that trade-offs occur in the way that individuals allocate their resources. The concept of trade-offs suggests an answer to the question of why all individuals don't produce abundant compounds for their own defense all the time: Manufacturing poisons is energetically expensive. Producing these molecules means that less energy is available for growth and reproduction.

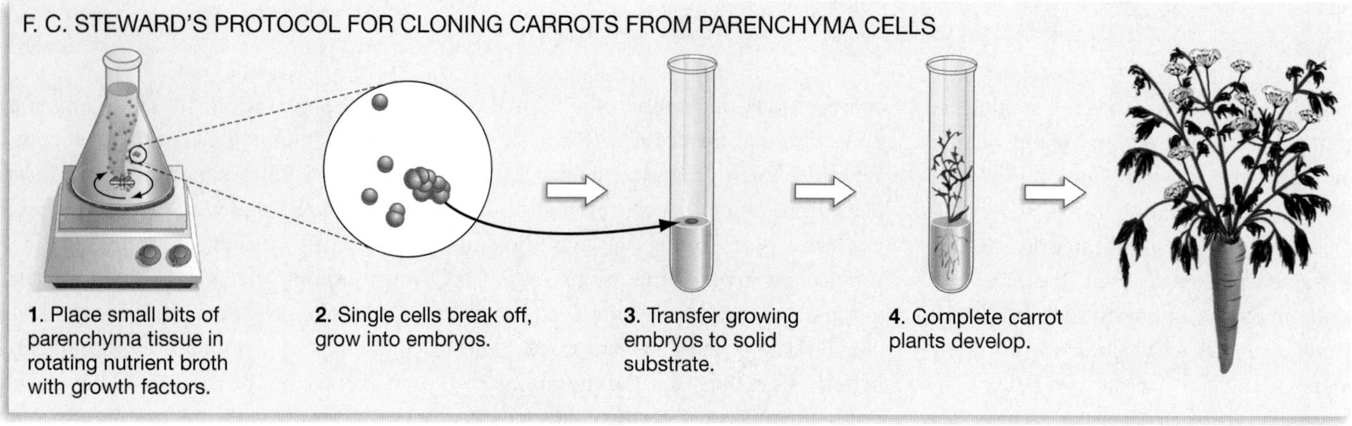
F. C. STEWARD'S PROTOCOL FOR CLONING CARROTS FROM PARENCHYMA CELLS
1. Place small bits of parenchyma tissue in rotating nutrient broth with growth factors.
2. Single cells break off, grow into embryos.
3. Transfer growing embryos to solid substrate.
4. Complete carrot plants develop.

FIGURE 35.13 Plants Can Be Cloned in the Laboratory from Single Parenchyma Cells
Protocol that led to the first successful cloning of a plant.

Bananas and several other commercially important species are propagated entirely by cuttings.

In the 1950s F. C. Steward and colleagues used this property of parenchyma cells to produce cloned copies of carrot plants in the laboratory. As **Figure 35.13** shows, Steward isolated small numbers of parenchyma cells and placed them in a rotating liquid medium containing sucrose, selected inorganic ions, molecules that act as growth factors, and other key nutrients. When single cells broke away from the original group of cells, they divided, producing a mass of cells, and began to form embryos. When Steward placed these growing embryos on a solid substrate, complete root and shoot systems formed. Researchers have since refined this cloning technique and can now routinely produce entire plants from single parenchyma cells.

Parenchyma has many functions besides wound healing and asexual reproduction, however. For example, the parenchyma tissue found in leaves is filled with chloroplasts and is the primary site of photosynthesis (**Figure 35.14a**). The parenchyma cells of roots, in contrast, serve as storage cells for starch deposits (**Figure 35.14b**). When you eat a salad, a potato, or an apple, you are ingesting primarily parenchyma cells. In addition, parenchyma cells make up *phloem*—the type of vascular tissue that transports nutrients throughout the plant body.

Collenchyma Cells **Collenchyma** cells function mainly in support. They have cell walls that are thicker in some areas than others, and their overall shape is longer and thinner than that of parenchyma cells. Even when collenchyma cells are mature, their cell walls retain the ability to stretch and elongate. As a result, collenchyma cells can continue to lengthen as they provide structural support to the growing regions of shoots. Based on these observations, it should not be surprising to learn that collenchyma cells are particularly abundant in elongating stems and in the stalk portions of leaves. The "strings" you may have peeled from a stalk of celery or rhubarb include many strands of collenchyma cells (**Figure 35.15**).

Sclerenchyma Cells All plant cells produce a thin, expandable cell wall called a **primary cell wall** as the plant matures.

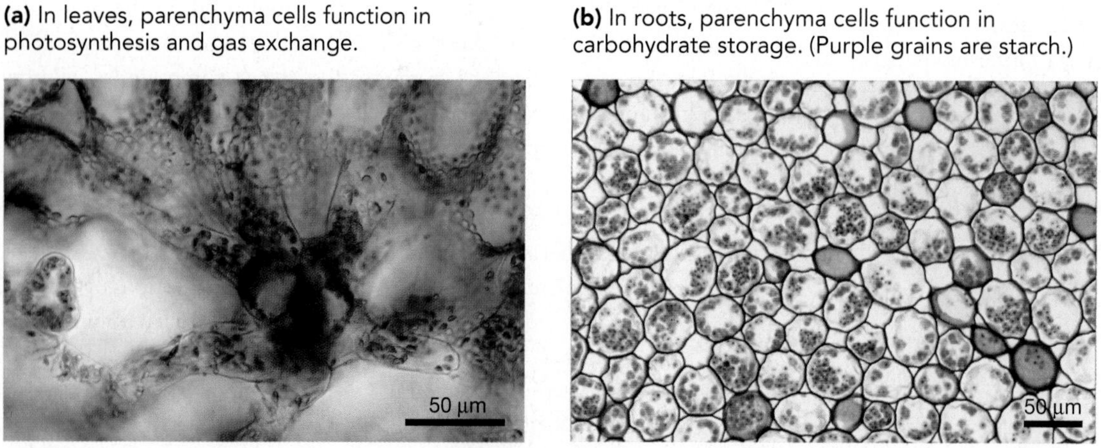
(a) In leaves, parenchyma cells function in photosynthesis and gas exchange.
50 μm
(b) In roots, parenchyma cells function in carbohydrate storage. (Purple grains are starch.)
50 μm

FIGURE 35.14 Parenchyma Cells Perform a Wide Array of Tasks
Parenchyma are "workhorse cells." **EXERCISE** Label chloroplasts in part (a) and starch granules in part (b).

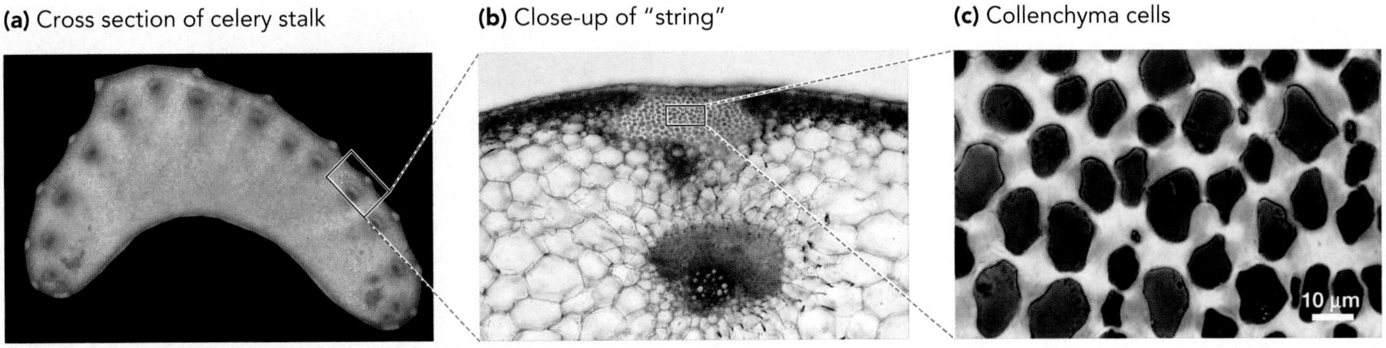
(a) Cross section of celery stalk
(b) Close-up of "string"
(c) Collenchyma cells
10 μm

FIGURE 35.15 Collenchyma Cells Support Growing Tissues
A celery stalk is actually a petiole; the strands you can peel from it are columns of collenchyma cells.

Sclerenchyma cells are specialized support cells that also produce a **secondary cell wall.** In addition to containing cellulose, the secondary cell wall of sclerenchyma cells includes the tough, rigid compound **lignin.** Unlike the primary cell walls of collenchyma, the secondary cell walls of sclerenchyma are thick and non-expandable. Collenchyma supports actively growing parts of the plant; sclerenchyma supports parts of mature stems and other structures after active growth has ceased.

Another key difference between collenchyma and sclerenchyma is that sclerenchyma cells are usually dead at maturity—meaning they have no cytoplasm. There are several types of sclerenchyma. *Tracheids* and *vessel elements* are sclerenchyma cells that function in support and in the transport of water from roots to shoot; we will analyze their structure and function in the next section. The other two types of sclerenchyma cells function only in support:

- **Fibers** are usually associated with vascular tissue and are extremely elongated (**Figure 35.16a**). Depending on the plant species they are isolated from, fibers may be used in the manufacture of paper, hemp or jute ropes, or linen and other fabrics. The fiber cells from ramie plants, also known as China grass, can be over half a meter long.
- **Sclereids** are relatively short, have variable shapes, and often function in protection (**Figure 35.16b**). The tough coats of seeds and the thick shells of nuts are composed of sclereids. The gritty texture of pears is caused by clumps of sclereids.

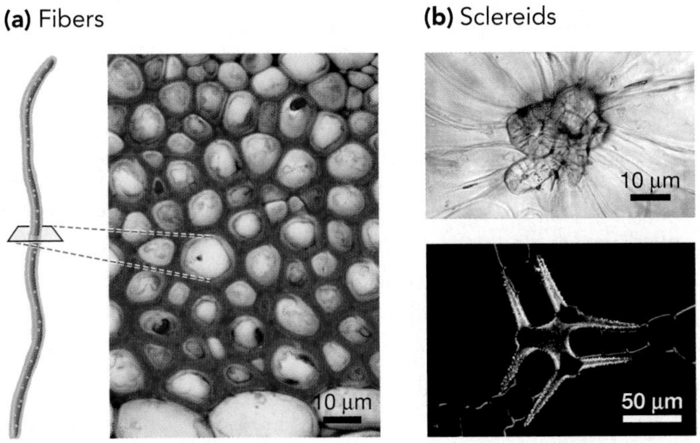
(a) Fibers
(b) Sclereids
10 μm
10 μm
50 μm

FIGURE 35.16 Sclerenchyma Cells Support Mature Tissues
(a) Fibers and (b) sclereids have thickened secondary cell walls. These cells provide support for tissues that are no longer growing.

The Diversity of Plant Tissues

Plant cells do not function in isolation. Instead they are grouped into **tissues:** groups of cells that function as a unit. Plant tissues that consist of a single cell type are called *simple tissues.* For instance, parenchyma, sclerenchyma, and collenchyma cells are found in masses that form the simple tissues called parenchyma, sclerenchyma, and collenchyma tissues, respectively. Tissues that contain several types of cells are termed *complex tissues.* For example, the tissues that transport water and nutrients in plants are made up of several types of sclerenchyma cells, including fibers, tracheids, and vessels. A final level of organization occurs because simple and complex tissues are organized into *tissue systems,* which extend throughout the body.

To understand the internal anatomy of a plant, let's first consider the structure and function of its major tissue systems and then analyze how these tissues are distributed in the plant body.

Meristematic Tissue The rapidly dividing, undifferentiated cells found in meristematic tissues are responsible for growth. Some of the daughter cells produced by mitosis in meristematic tissue remain as meristematic cells, or *initials,* that keep dividing. Other daughter cells, however, continue to divide for a while but then become organized into non-meristematic tissues. These daughter cells, called *derivatives,* begin to differentiate into parenchyma, collenchyma, or sclerenchyma cells, which then function in one of the three types of tissue systems found in mature plants: (1) dermal, (2) ground, or (3) vascular.

Dermal Tissue The **dermal** ("skin") **tissue,** or **epidermis,** usually consists of a single layer of cells that covers the plant body. The epidermis has several tasks, all of which relate to the same overall function: protection.

Perhaps the most important task performed by epidermal tissue is the secretion of the cuticle. Recall from Chapter 29 that **cuticle** is a waxy layer that overlays the epidermal cells of the shoot system. Waxes are lipids and thus highly hydrophobic. As a result, the presence of cuticle on stems and leaves drastically reduces the amount of water that is lost by evaporation from dermal tissue. The water-repellent properties of cuticle also make it a valuable ingredient in polishes and lipsticks. The carnauba wax used in car and floor polishes, for example, is secreted by epidermal cells in the leaves of carnauba palms native to Brazil.

In addition to minimizing water loss, cuticle forms a barrier that protects the plant from virus particles, bacterial cells, and the spores or growing filaments of parasitic fungi. In this way, the plant epidermis forms the first line of defense against disease-causing agents, or **pathogens**. For pathogens to enter the plant body and initiate an infection, they must either secrete enzymes that digest the cuticle or enter via a wound where the cuticle has been torn away.

The waxes found in cuticle reduce gas exchange as well as water loss and pathogen entry, however. This can be a serious problem in parts of the plant where photosynthesis is occurring, because the process depends on the free flow of carbon dioxide to the photosynthetic cells. To allow carbon dioxide to enter photosynthetically active tissues, most land plants have structures called **stomata** (singular: **stoma**). Stomata consist of two **guard cells**, which change shape to open or close an opening known as a **pore** (**Figure 35.17**). If stomata are open, then CO_2, O_2, water vapor, and other gases can move between the atmosphere and the interior of the plant by diffusion. Stomata open when conditions for photosynthesis are favorable; they close at night, as well as in dry conditions, when large amounts of water may be lost to transpiration.

In addition to minimizing water loss and regulating gas exchange, cells in dermal tissue may also be involved in protecting the individual from the damaging effects of intense sunlight and from attacks by herbivores. These functions are accomplished by structures called **trichomes**, which are hairlike appendages from specialized epidermal cells. Trichomes come in a wide variety of shapes, sizes, and abundance. Depending on the species,

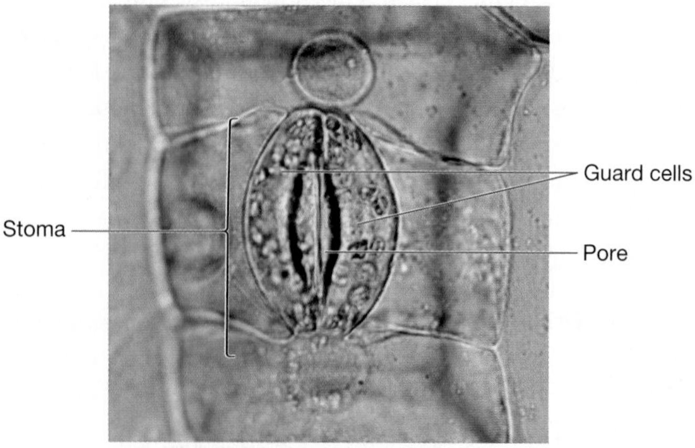

FIGURE 35.17 In Stomata, Guard Cells Regulate the Opening of Pores
Micrograph of guard cells, with the pore closed.

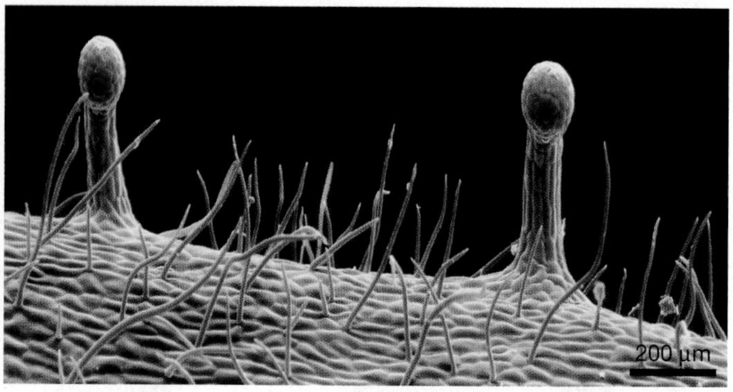

FIGURE 35.18 Epidermal Cells Produce Trichomes That Provide Protection
EXERCISE Label both spearlike trichomes that discourage insects or other small herbivores from eating leaf tissue, and trichomes that are filled with toxic compounds, creating a "minefield" on the leaf surface.

trichomes may (1) help keep the leaf surface cool by reflecting sunlight, (2) help reduce water loss from leaves by forming a dense mat that limits evaporation, or (3) provide barbs or store toxic compounds that thwart herbivores (**Figure 35.18**).

Ground Tissue **Ground tissue** is a complex tissue that typically contains parenchyma and either collenchyma or sclerenchyma. Ground tissue makes up the bulk of the plant body. Most photosynthesis, as well as most carbohydrate storage, takes place in ground tissue. Cells in ground tissue are also responsible for most of the synthesis and storage of specialized products such as the chemical signals called hormones, colorful pigments, and toxins required for defense.

Vascular Tissue Recall that some types of parenchyma and sclerenchyma cells are specialized for support and for the long-distance transport of water and dissolved nutrients. These specialized cells are organized into a complex tissue called **vascular tissue.** There are two types of plant vascular tissue: xylem and phloem. **Xylem** (pronounced *ZYE-lem*) conducts water and dissolved ions from the root system to the shoot system. **Phloem** (*FLO-em*) conducts sugar, amino acids, chemical signals, and other substances throughout the plant body.

Xylem may contain two types of conducting cells: tracheids and vessel elements. All vascular plants contain water-conducting cells called **tracheids** (*TRAY-kee-ids*). In angiosperms and species in the group Gnetophyta, xylem also contains conducting cells called **vessel elements.** Both tracheids and vessel elements are sclerenchyma cells with thick secondary cell walls containing lignin. The secondary cell wall in tracheids and vessel elements is often deposited in ringlike or spiral patterns. Both tracheids and vessel elements are dead at maturity.

As **Figure 35.19a** shows, tracheids are long, slender cells with tapered ends. The sides and ends of tracheids have structures called **pits**, which are interruptions in the secondary cell wall. A pit is a gap where only the primary cell wall is present. When

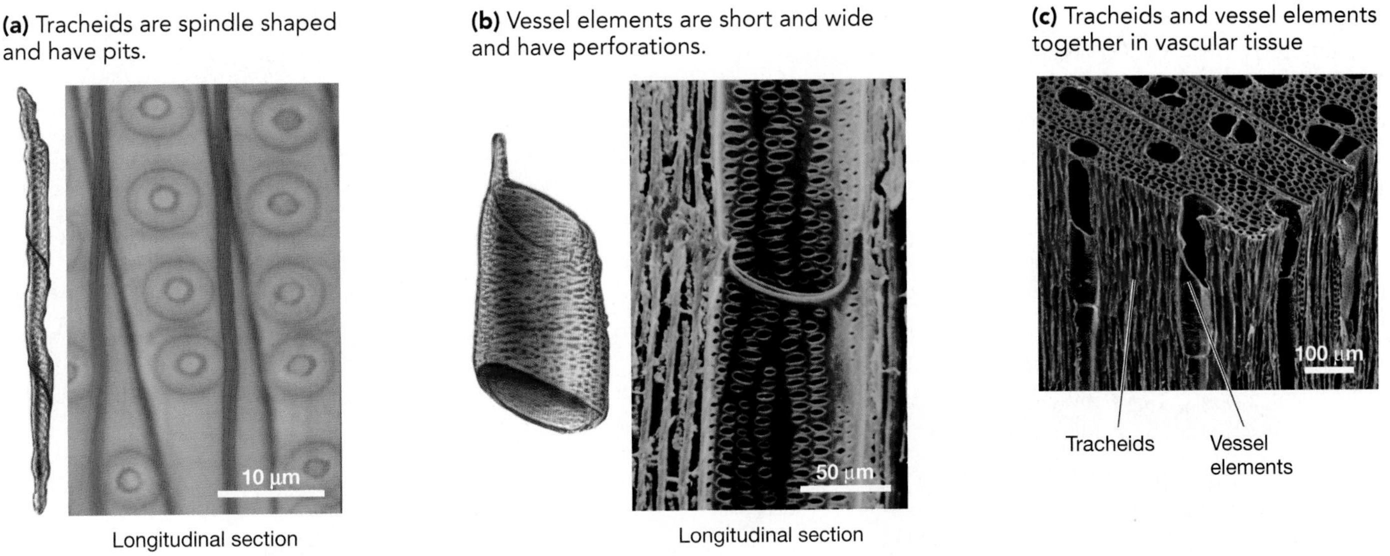

FIGURE 35.19 Xylem May Contain Two Types of Water-Conducting Cells
(a) Tracheids are extremely long and thin compared to **(b)** vessel elements, which are much shorter and wider. **(c)** Both types of water-conducting cells are found in the vascular tissue of angiosperms.

water is moving up a plant through tracheids, it moves from cell to cell both vertically and laterally through pits, because that is where resistance to flow is lowest. Vessel elements, in contrast, are shorter and wider than tracheids (**Figure 35.19b**). Instead of having pits, vessel elements have **perforations**—openings that lack both primary and secondary cells walls. Alternatively, the ends of vessel elements in some species lack any cell wall at all, and stacked cells form open pipes. Because they offer so little resistance to flow, vessel elements conduct water much more efficiently than do tracheids. In angiosperms, tracheids and vessel elements are found adjacent to each other (**Figure 35.19c**).

In contrast to tracheids and vessel elements, the cells that make up phloem are alive at maturity. They are parenchyma cells, and most lack lignified secondary cell walls. Phloem is made up primarily of two types of specialized parenchyma cells: sieve-tube members and companion cells. **Sieve-tube members** (also called *sieve-tube elements*) are long, thin cells that have perforated ends called **sieve plates** (**Figure 35.20**). Sieve-tube members lack nuclei, chloroplasts, and most other major organelles but are directly connected to adjacent **companion cells** by means of numerous plasmodesmata. Companion cells contain all the organelles normally found in a plant cell. Companion cells are not conducting cells but instead provide materials to maintain the cytoplasm and plasma membrane of sieve-tube members. As Chapter 36 will show, companion cells are also involved in the loading and unloading of carbohydrates and other nutrients from the solution inside sieve-tube members.

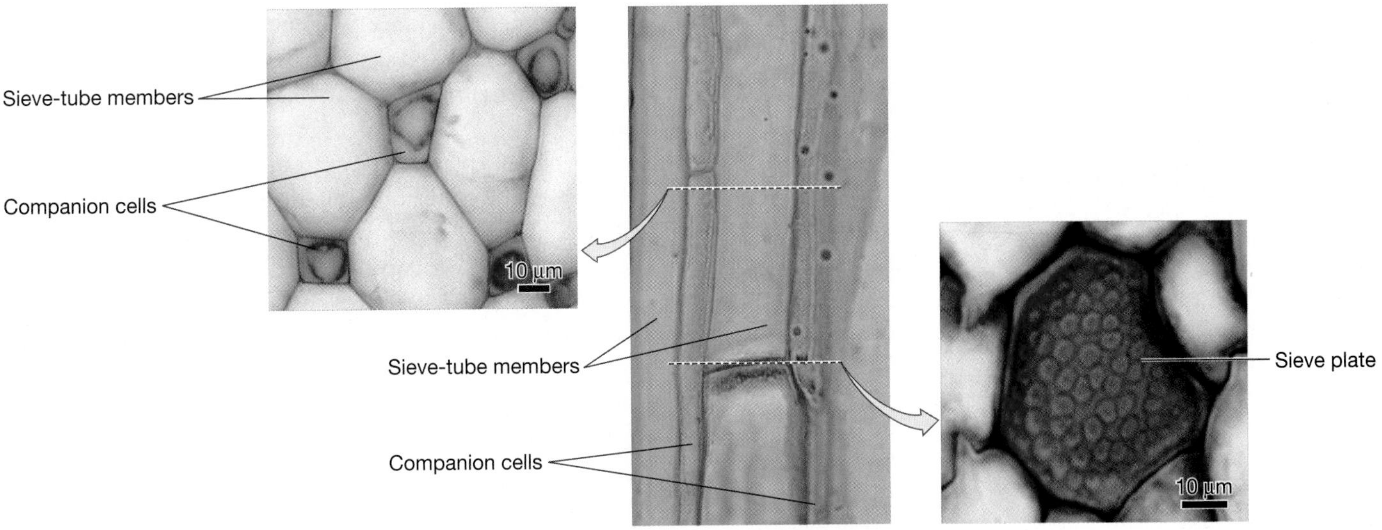

FIGURE 35.20 Phloem Consists of Sieve-Tube Members and Companion Cells
Sieve-tube members conduct sucrose throughout the body; companion cells support sieve-tube members.

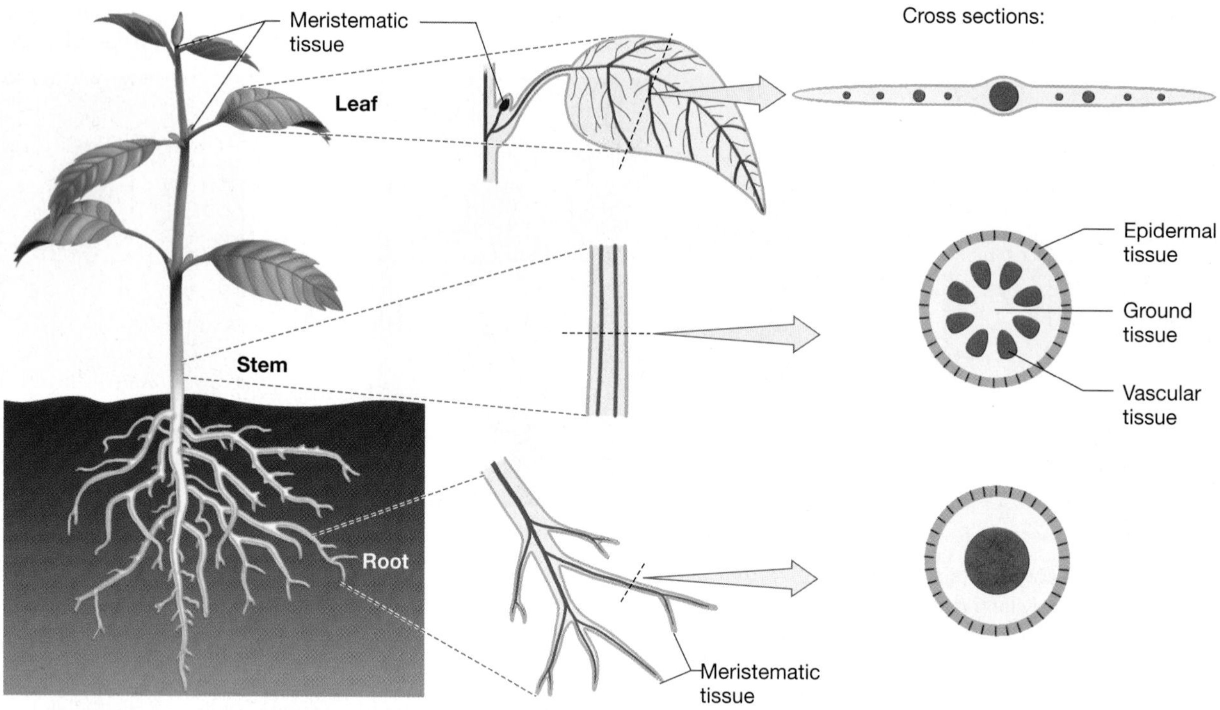

FIGURE 35.21 Meristematic, Dermal, Ground, and Vascular Tissues Have Distinct Distributions in the Plant Body

EXERCISE Next to the label for each tissue type, list at least two cell types found there and describe the tissue's overall function.

Distribution of Tissue Types The meristematic, dermal, ground, and vascular tissues of plants are distributed throughout the root and shoot systems as shown in **Figure 35.21**. Notice that dermal tissue covers the body. As a result, dermal tissue represents the interface between the individual and the external environment. Ground tissue forms the bulk of the body but has vascular tissue running through it. This is an important observation, because it means that the cells that make up ground tissue are adjacent to cells that conduct the water and nutrients they need. Meristematic tissue, in contrast, is highly localized.

Now that you are familiar with the basic structure of a plant body and the cells and tissues inside, let's turn to a different question: How does cell division in localized meristematic tissues lead to the formation of a highly organized body, specialized for harvesting diffuse resources?

✓CHECK YOUR UNDERSTANDING

The plant body is composed of cells with specialized structures and functions. The most important of the cell types are parenchyma, sclerenchyma, and collenchyma cells, which are grouped into tissues. The epidermal, ground, and vascular tissue systems found throughout the plant body consist of an array of distinctive cell and tissue types. You should be able to (1) discuss the structure and function of each major cell type, tissue type, and tissue system, and (2) diagram the distribution of the tissue systems in the plant body.

35.3 The Anatomy of Plant Growth

Plants grow throughout their lives. A bristlecone pine that is over 5000 years old and a giant sequoia that towers 80 m (260 ft) above the ground have roots and shoots that are still growing. To describe this phenomenon, biologists say that plants exhibit **indeterminate growth.**

The growth that occurs in the meristematic tissues of plants is the product of two processes: (1) production of new cells by mitosis and cytokinesis, and (2) expansion of cell volume, or **cell enlargement.** Recall from Chapter 11 that mitosis results in the production of two nuclei and that cell division, via formation of a cell plate, results in two daughter cells. Once these daughter cells form, growth occurs via the addition of cytosol and the growth and division of organelles other than the nucleus. During cell enlargement, water enters cells via osmosis and creates pressure on the cell wall. Enzymes break bonds between cellulose fibers in the cell wall, and additional cellulose is synthesized and added, allowing the cell wall to expand.

In plants, then, the direction of growth and the eventual shape of a plant cell are determined by two factors: the orientation of the cell plate relative to the long axis of the growing root or shoot, and the direction and extent of cell enlargement.

Although all plant growth occurs in the same basic way at the cellular level, a single plant may contain several distinct types of meristematic tissue. **Apical meristems** are found at the tips of shoots and roots and increase their length. Apical meristems consist of small, similar-looking cells that divide rapidly when water and nutrients are plentiful. Growth occurs as cell number and cell size increase. Growth from apical meristems is termed **primary growth.** Primary growth leads to nonwoody, herbaceous stems. Wood forms only in species that have not just apical meristems but also a cylinder of meristematic cells called a **lateral meristem** (or *cambium*) in their roots and shoots. Lateral meristems increase the width of a trunk, root, or stem over time, a phenomenon known as **secondary growth.**

Although the growth of apical and lateral meristems is an important theme in plant anatomy, there are significant variations. For example, many grassland species produce their leaves and stems from **basal meristems,** which are located below ground. Basal meristems allow these species to regenerate their shoot system rapidly if they lose their stems and leaves to fire or grazing. There are also variations on the theme of secondary growth. In palms, bananas, and some other tree-sized species in the group of angiosperms called monocotyledons (see Chapter 29), lateral growth occurs in the absence of a lateral meristem. These species expand in girth as a result of several types of meristems located just below the apical meristem.

Apical and lateral meristems are so fundamental to plant development, however, that it's critical to understand them in detail. Let's first analyze the tissues responsible for primary growth in angiosperms, then dissect the tissues that lead to secondary growth and the formation of wood. How does growth in these meristematic regions relate to the organization of the dermal, ground, and vascular tissues found in mature roots and shoots?

Primary Growth: The Root System

The root system provides the shoot system with water and key elements in the form of ions. It also anchors the plant in the soil and in many species, serves as a storage site for starch and other carbohydrates. To perform these functions—and particularly to furnish cells throughout the body with water—the roots must continually grow into new regions of the soil. How do root tips move through the soil, particularly soil that is compact and stiff?

To answer this question, consider the apical meristems found in the root system. As **Figure 35.22** shows, the growing region of the root is covered by a group of protective cells called the **root cap.** Cells produced by the meristematic region in the tip of the root constantly replenish the cap, which regularly loses cells through abrasion by soil. In addition to protecting the root tip, root cap cells are important in sensing gravity and determining the direction of growth. They also synthesize and secrete a slimy substance called **mucigel,** which helps lubricate the root tip, reducing friction and making movement more efficient.

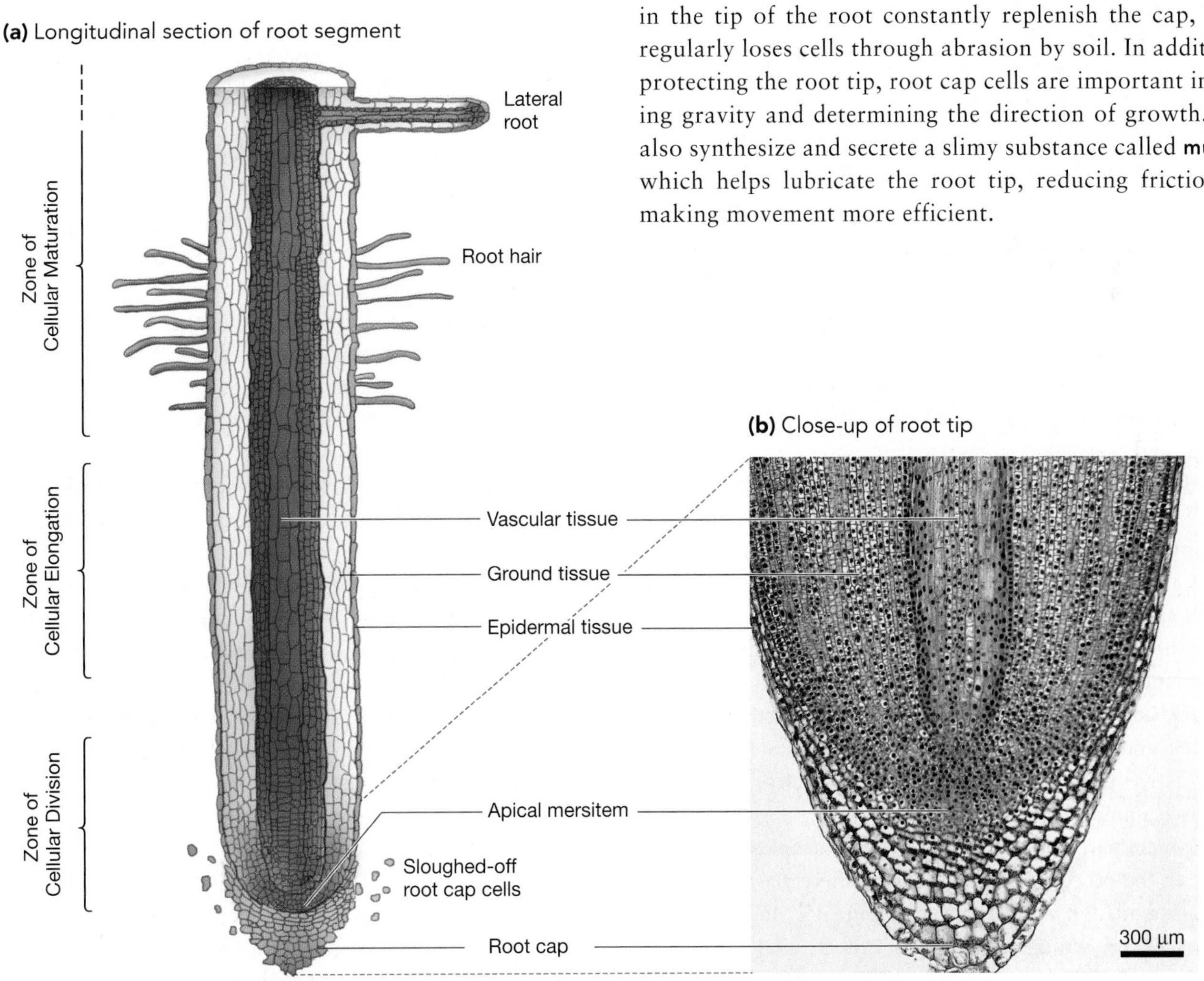

FIGURE 35.22 Roots Extend into the Soil via Growth of Apical Meristems and Cell Elongation
The zone of cellular maturation is actually much larger than can be shown here.

Three distinct populations of cells exist behind the root cap (Figure 35.22a):

1. The **zone of cellular division** (0.5–1.5 mm behind the root tip) contains the apical meristem, where cells are actively dividing;
2. The **zone of cellular elongation** (4–10 mm behind the root tip) is made up of cells recently derived from the apical meristem and which are actively increasing in length; and
3. The **zone of cellular maturation** (1–5 cm from the root tip) is where older cells complete their differentiation into dermal, vascular, and ground tissues.

The zone of cellular elongation is the region responsible for the movement of root tips through the soil. The cells in this region increase in length by taking up water. Their expansion provides the force that pushes the root cap and apical meristem through the soil. When conditions are good, roots can extend by as much as 4 centimeters per day.

Regarding water and nutrient absorption, though, the zone of cellular maturation is the most important root segment. In this region, epidermal cells produce outgrowths called **root hairs**, which greatly increase the surface area of the dermal tissue. Root hairs furnish the actual sites of water and nutrient absorption; the rest of the root system simply provides root hairs with structural support and a transport system. Uptake of water and nutrients in root hairs is so important that large sections of Chapter 36 focus on how these processes occur.

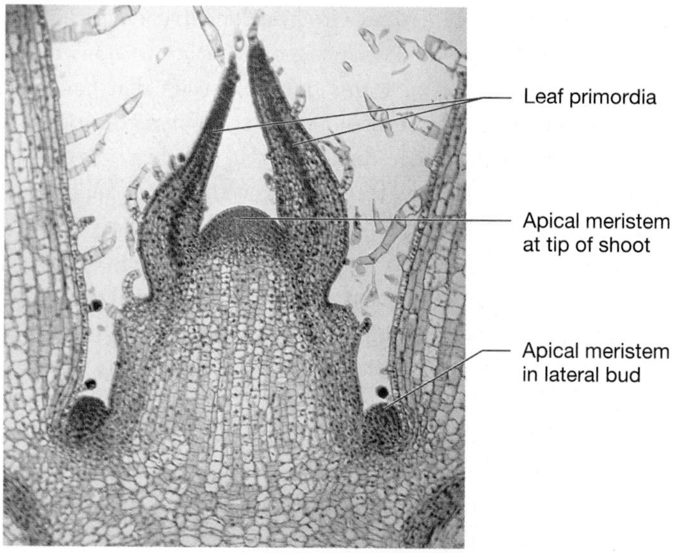

FIGURE 35.23 Shoots Extend via Growth of Apical Meristems Apical meristems consist of small, similar-looking cells that divide rapidly when water and nutrients are plentiful. In the shoot, apical meristems are located at the tip and in lateral buds.

Primary Growth: The Shoot System

If you visit a garden regularly over the course of a growing season, you can only imagine the movement of root tips as they penetrate the soil and expand to form complex networks deep underground. But even a casual observer can watch the growth of shoot systems directly, as the tips of stems extend and branch over time and as new leaves form and expand.

Recall that growth based on extension of the apical meristems in roots and shoots is called primary growth. To appreciate how shoots grow, consider the longitudinal section through a shoot tip shown in **Figure 35.23**. Notice that the top of the stem consists of a rounded dome of small, actively dividing cells—the **shoot apical meristem.** On both sides of this meristematic region, newly developing leaves emerge as **leaf primordia.** Lateral meristems form at the intersections of leaf primordia and the stem. Normally, lateral meristems do not initiate active growth and form branches unless the apical meristem is damaged in some way.

In the main part of the stem, just behind the apical meristem, cells are in the process of enlarging and differentiating to form dermal, ground, and vascular tissues. **Figure 35.24a** shows how these tissues are arranged in the stem of an alfalfa plant—a species widely planted for hay—when they mature. The center of the stem, or **pith**, consists of ground tissue that functions in carbohydrate storage. Vascular tissues are found around the pith in groups called **vascular bundles**, which form strands running the length of the stem. In addition to containing phloem and xylem tissue, vascular bundles usually have collenchyma or sclerenchyma cells that provide support. In alfalfa and most other members of the angiosperm lineage called eudicots (see Chapter 29), vascular bundles are arranged in a ring around the outside of the stem. **Figure 35.24b** details the structure of a vascular bundle in a eudicot. Eudicots also have a layer of ground tissue called the **cortex** between the vascular bundles and the epidermis. In members of the lineage called monocotyledons, however, vascular bundles tend to be scattered throughout the ground tissue (**Figure 35.24c**).

Vascular bundles also form at the tips of developing leaf primordia. Later these bundles connect to vascular tissue in the stem, forming a continuous system for the transport of water and nutrients through both stems and leaves. As **Figure 35.25** shows, vascular bundles run through the ground tissue that occupies the center of a leaf. Two types of parenchyma cells make up this ground tissue. Elongated **palisade mesophyll** ("middle-leaf") cells are packed with chloroplasts and are the site of most photosynthesis. Rounded **spongy mesophyll** cells are loosely packed and are often surrounded by air spaces near stomata. The wet surfaces of spongy mesophyll cells are the site of gas exchange in plants—just as the wet surfaces of cells lining the lungs are the site of most gas exchange in your body. In combination, then, spongy and palisade mesophyll cells make gas exchange and photosynthesis possible in leaves. These cells are protected by a layer of epidermal cells and are in close proximity to both xylem and phloem.

WEB TUTORIAL 35.1 Primary and Secondary Growth

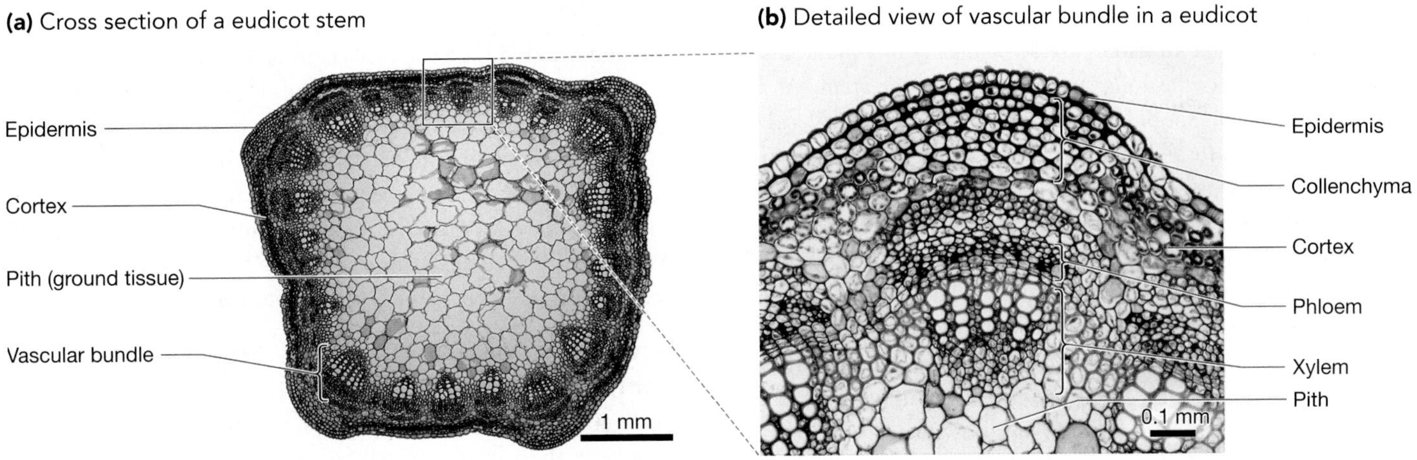
(a) Cross section of a eudicot stem
Epidermis
Cortex
Pith (ground tissue)
Vascular bundle
1 mm
(b) Detailed view of vascular bundle in a eudicot
Epidermis
Collenchyma
Cortex
Phloem
Xylem
Pith
0.1 mm

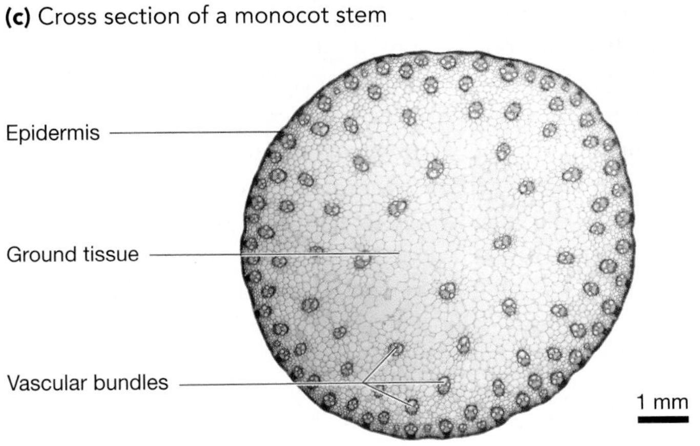
(c) Cross section of a monocot stem
Epidermis
Ground tissue
Vascular bundles
1 mm

FIGURE 35.24 Stems Contain a Variety of Types of Cells and Tissues
Vascular bundles are **(a)** arranged in a ring near the perimeter of eudicot stems and **(b)** contain xylem, phloem, and support cells. **(c)** Vascular bundles are scattered throughout the pith in monocots.

Secondary Growth

Primary growth increases the length of roots and shoots. **Secondary growth** increases their width. The function of primary growth is to extend the reach of the root and shoot system and thus increase a plant's ability to absorb photons and acquire carbon dioxide, water, and ions. The function of secondary growth, in contrast, is to provide the structural support required for extensive primary growth. Secondary growth produces wood. Without the support provided by secondary growth, roots would not be massive enough to anchor extremely large shoot systems, and long stems would fall over or break.

Secondary growth occurs in gymnosperm and angiosperm species that have lateral meristems in addition to apical meristems. Lateral meristems are also called *secondary meristems* or **cambium**. Lateral meristems differ from apical meristems in two ways: (1) Apical meristems are localized at root and shoot tips and are dome shaped. In contrast, lateral meristems form cylinders that run the length of a root or stem. Lateral

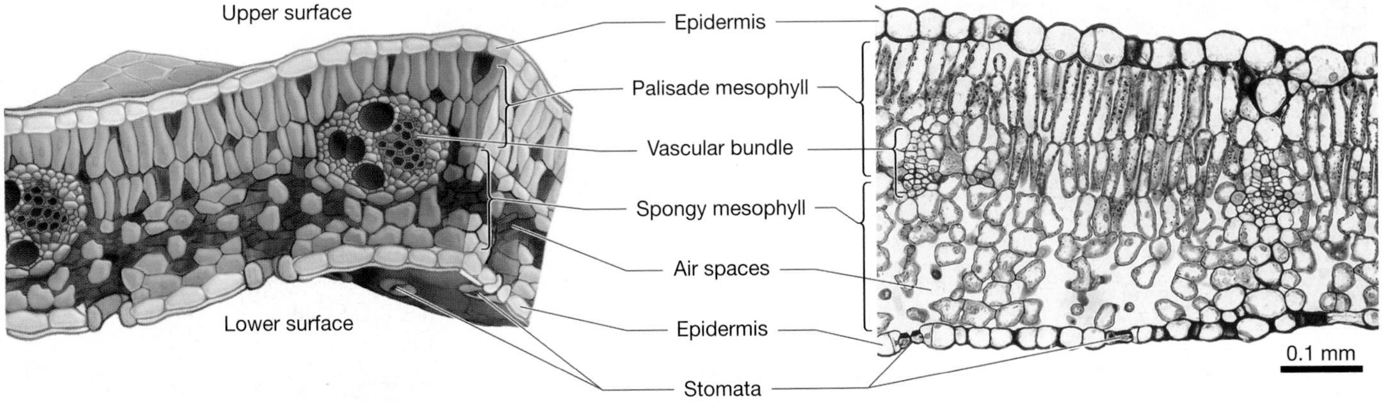
Upper surface
Lower surface
Epidermis
Palisade mesophyll
Vascular bundle
Spongy mesophyll
Air spaces
Epidermis
Stomata
0.1 mm

FIGURE 35.25 Leaves Contain a Variety of Cells and Tissues
Vascular bundles run through the ground tissue of leaves, which consists of palisade mesophyll that is specialized for photosynthesis and spongy mesophyll that is specialized for gas exchange. Notice that stomata are usually located on the underside of leaves and that spongy mesophyll surrounds air spaces inside stomata.

meristems are made up of thin layers of meristematic cells. (2) Cells in an apical meristem divide primarily in a plane that is perpendicular to the long axis of the root or stem. As a result, the root or shoot extends in length when new cells elongate. But in lateral meristems, cells divide in a plane parallel to the long axis of the root or stem. As a result, new cells increase the width of the structure when they elongate.

As **Figure 35.26a** shows, there are two distinct types of lateral meristem in plants that undergo secondary growth. A ring of meristematic cells called the **cork cambium** is located near the perimeter of the stem. A ring of meristematic cells called the **vascular cambium** is found inside the cork cambium. One other observation is critical to understanding how lateral meristems work: The cork cambium primarily produces new cells to the outside. Vascular cambium, in contrast, generates new layers of cells both inside and outside the meristem itself. The result is new rings of cells inside and outside the meristematic tissue. Now let's consider in more detail how each of these meristems works.

How Does Cork Cambium Contribute to Bark? The cork cambium produces **cork cells**. As a woody stem or root matures, the expanding layer of cork cells replaces the original epidermal cells produced by the apical meristem. As they mature, cork cells produce a layer of wax and other molecules inside their cell walls, making them impermeable to water and gases. Cork cells may also produce a tough secondary cell wall containing lignin. The cork cells eventually die, forming a protective outer layer for the mature root or shoot. As the trunk continues to increase in width, the cork layer often cracks and flakes.

Cells produced by the cork cambium are an important component of the protective material we call **bark**. But the term *bark* actually refers to all cells outside the vascular cambium. As a result, it includes some of the cells produced by the vascular cambium as well as the cork cambium and cork cells.

How Does Vascular Cambium Produce Wood and Contribute to Bark? Vascular cambium produces both phloem and xylem. When cells in the cambium divide, new cells that are produced to the outside of the meristem differentiate into phloem, while new cells produced to the inside differentiate into xylem. More specifically, cells produced by vascular cambium develop into **secondary phloem** and **secondary xylem**. Lateral meristems produce secondary phloem and xylem; primary meristems produce primary phloem and xylem. Primary phloem and xylem are found throughout the roots and shoots of all vascular plants; secondary phloem and xylem are found only in woody gymnosperms and angiosperms. Secondary phloem functions in sugar transport and is considered to be part of the bark; secondary xylem functions in water transport and structural support and forms the material we call **wood**.

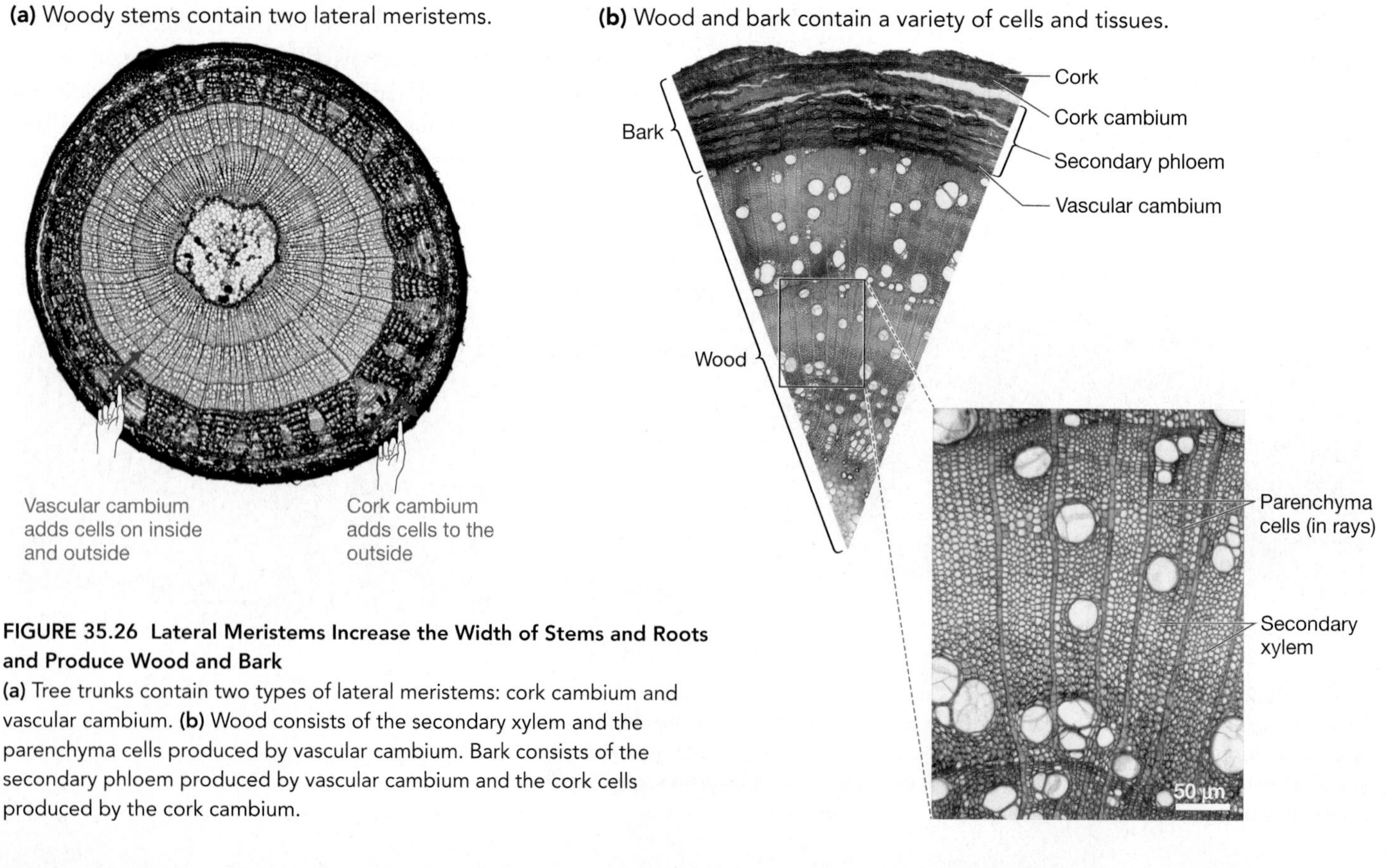
(a) Woody stems contain two lateral meristems.
Vascular cambium adds cells on inside and outside
Cork cambium adds cells to the outside
(b) Wood and bark contain a variety of cells and tissues.
Cork
Cork cambium
Secondary phloem
Vascular cambium
Bark
Wood
Parenchyma cells (in rays)
Secondary xylem
50 μm

FIGURE 35.26 Lateral Meristems Increase the Width of Stems and Roots and Produce Wood and Bark

(a) Tree trunks contain two types of lateral meristems: cork cambium and vascular cambium. **(b)** Wood consists of the secondary xylem and the parenchyma cells produced by vascular cambium. Bark consists of the secondary phloem produced by vascular cambium and the cork cells produced by the cork cambium.

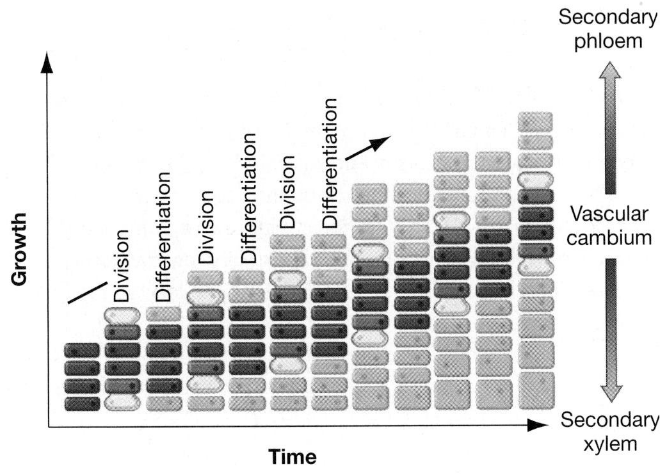

FIGURE 35.27 Secondary Growth in a Tree Trunk
A woody stem grows in width because cells in the vascular cambium undergo a series of mitotic divisions. Each division produces a cell that continues to divide (colored red in the figure) and a derivative cell that later differentiates into a cell of secondary xylem (purple) or a cell of secondary phloem (pink).

Both primary and secondary phloem and xylem are complex tissues. Besides producing conducting cells such as sieve-tube members, tracheids, and vessel elements, the vascular cambium produces fibers for additional strength, along with parenchyma cells. The parenchyma cells produced by the vascular cambium are important because they radiate laterally across the xylem and form structures called **rays** (**Figure 35.26b**). While xylem and phloem transport water and dissolved solutes up a stem, the parenchyma cells in rays transport materials laterally across the stem.

Figure 35.27 details how secondary phloem and xylem are produced. Notice that the stem increases in width as cells produced on the inside of the vascular cambium form secondary xylem and cells produced on the outside form secondary phloem. It's important to realize, though, that the results of cell division in lateral meristems are highly asymmetrical. As the vascular cambium grows, all of the secondary xylem is retained and accumulates, but secondary phloem is eventually degraded and replaced. As a result, mature woody stems are dominated by secondary xylem, or wood.

The Structure of a Tree Trunk Trees are **perennial plants,** meaning they live for many years. (Species that die after a single year or season of growth are called **annual plants.**) In many environments, the vascular cambium of trees ceases growth for a portion of each year. This period of no growth, or **dormancy,** occurs during the winter in cold climates and during the dry season in tropical habitats. When the vascular cambium resumes growth in the spring or at the start of the rainy season, it produces large, relatively thin-walled cells. As the growing season continues and nears its end, however, conditions tend to dry out or become cooler; the secondary xylem cells that are produced tend to be smaller, thicker walled, and darker in appearance. When growth is seasonal, regions of large, thin-walled cells alternate with layers of small, thick-walled cells. As a result, annual growth rings can be observed in cross sections of wood from seasonal habitats (**Figure 35.28a**).

Another important phenomenon occurs as a tree grows in width: The innermost xylem layers stop functioning in water transport. They then begin accumulating protective compounds secreted by other tissues. These compounds form resins, gums, and other complex mixtures. The deposition of these molecules causes the oldest portions of secondary xylem to become darker than the younger portions. The inner xylem region is called **heartwood,** while the outer xylem that still functions in water transport is called **sapwood** (**Figure 35.28b**).

(a) Growth rings result from variation in cell size.

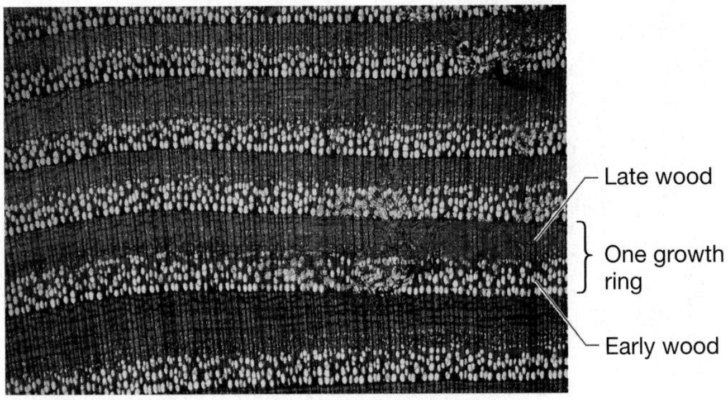

(b) Heartwood and sapwood have different functions.

FIGURE 35.28 Anatomy of a Tree Trunk
(a) Section of wood, stained to show individual cells. Growth rings appear because the size of secondary xylem cells that are produced changes during the course of a growing season. Cells produced at the start of the season are large; those produced near the end of the growing season are small. **(b)** Unstained section of wood, showing the marked color difference between heartwood and sapwood.

Analyzing patterns in tree rings is an important field of study in biology. Because trees grow faster when moisture and nutrients are plentiful, wide tree rings are reliable indicators of wet years. In contrast, narrow rings signal drought years. By studying the tree rings found in fossil trees and extremely old living trees, biologists can gain a better understanding of climate changes that occurred in the past (**Box 35.2**). With continued research, researchers also hope to predict how forests might respond to the global warming that is currently under way.

✓CHECK YOUR UNDERSTANDING

In the shoot and root systems, primary growth occurs at apical meristems and results in a lengthening of both systems. Secondary growth occurs in species with lateral meristems and results in a broadening of the shoot and root systems. You should be able to (1) diagram the structure of a primary meristem and a lateral meristem, and (2) explain how division and differentiation of cells in the vascular cambium lead to the formation of bark and wood.

BOX 35.2 Tree-Ring Studies

In the Sierra Nevada Mountains of California, dead tree trunks decay so slowly that growth rings can still be discerned in plant material that is 3500 years old. To gain a better understanding of how climate and vegetation have changed at this location over this period, researchers have extracted samples of growth rings from a large number of these ancient trunks, using an instrument called an increment borer. By correlating patterns in the fossil growth rings with patterns present in the living trunks of extremely old trees nearby (**Figure 35.29**), the biologists could assign exact or nearly exact ages to each tree in the sample.

The resulting data set represented a nearly continuous record of forest density at these study sites, dating back to the period when the Minoan civilization of Crete dominated the Mediterranean region, Egypt was at the height of its power, and the Shang dynasty ruled China. Although all of the study sites are above the zone where trees are currently able to grow, the tree-ring data showed that the sites were heavily forested during most of the past 3500 years. Will forests once again begin to grow at high altitudes in the Sierra Nevada, now that a period of global warming is under way? According to the fossil growth-ring data, warm periods in the past coincided with heavy forest growth when the climate became wetter as well. Although it is too early to make a strong prediction, tree seedlings have recently become established at these study sites—hinting that in this mountain range, forests may soon begin climbing toward the peaks once again.

(a) Increment borer cuts core from trunk.
(b) Core samples from three nearby trees
These samples are lined up by year; notice similar ring patterns in trees from the same area

FIGURE 35.29 Studying Growth Rings
(a) Using an instrument called an increment borer, researchers can obtain samples of growth rings present in a woody stem. **(b)** Cores from different trees can be correlated by matching wide growth rings, narrow growth rings, and other features of the core.

ESSAY Wood as a Structural Material

Secondary xylem tissue—the substance called wood—helps trees remain upright in the face of hurricane-force winds and keeps them from snapping under heavy loads of snow and ice. Secondary xylem has also provided humans with building material for thousands of years. Why is wood such an effective structural material?

The answer begins at the molecular level. The cellulose fibers that are the primary component of plant cell walls surpass nylon, silk, tendon, and bone in their ability to withstand a pulling force, or tension. In cell walls, sheets of cellulose fibers are laid down at angles of about 67 degrees relative to each other. Research by human engineers has shown that cylindrical materials oriented in this way provide the best combination of resistance to tension (forces that tend to pull a structure apart), compression (forces that push a structure together), and bending. Further, tough lignin molecules are deposited around the strands of cellulose found in the thickened secondary walls of xylem tissue, forming a network. The addition of lignin greatly strengthens the cell walls.

To understand how these tough cell walls help produce wood's most striking structural characteristic, recall that xylem cells are dead at maturity. In effect, then, wood is a honeycombed lattice. The combination of chambers filled with resins or gums and tough walls makes wood lumber exceptionally strong for its weight. Relative to its density, wood is better able to resist bending and twisting than are concrete, cast iron, aluminum alloy, or steel.

Combining high strength and low weight is important, because every component that is used to stiffen or support a structure also adds weight, increasing the total load that the structure must bear. As a structure grows in height, then, it is important that light materials be used to keep the overall load to a minimum. The combination of strength and lightness is the key to wood's success as a structural material.

An additional feature helps tree trunks withstand the bending force exerted by wind. As a tree grows in height, its vascular cambium adds extra width at the base of the trunk. **Figure 35.30** shows how this flare at the bottom gives large trees a shape that is similar to the architectural design used for tall, thin structures such as the Eiffel tower. A broadened base, which has accumulated more annual rings than have regions higher above the ground, helps tree trunks withstand the lateral forces exerted by wind. The resulting structure is so stable that it may stand for decades after the tree's death.

Why is wood such an effective structural material?

From an engineering viewpoint, trees are sophisticated structures. It is not surprising that they are Earth's tallest organisms and are among the longest lived.

FIGURE 35.30 How Can Tall, Thin Structures Be Made Stronger? The flare at the base of large trees and other tall, thin structures helps them withstand the lateral force of wind.

CHAPTER REVIEW

Summary of Key Concepts

- **The plant body consists of (1) a root system that anchors the individual and absorbs water and key ions, and (2) a shoot system that absorbs carbon dioxide and sunlight.**

The root and shoot systems of plants are specialized for harvesting the light, water, and nutrients required for performing photosynthesis and synthesizing sugar. Roots extract water and nutrients such as nitrogen, phosphorus, and potassium from the soil; shoots capture light and carbon dioxide from the atmosphere. Leaves, the major organ for performing photosynthesis, usually consist of a flattened blade that extends from a petiole. Roots, stems, and leaves may be modified to perform a variety of other functions, however, including nutrient storage, water storage, protection, and asexual reproduction.

The overall morphology of root and shoot systems varies widely among plant species. In the prairie plants of North America, for example, root systems range from long, linear taproots to shallow, dense mats. Among the silverswords of Hawaii, shoot systems vary from low mats or rosettes to woody, highly branched tree trunks.

The root and shoot systems of plants consist of several different cell types and tissues, each of which has a specialized structure and function.

Three major types of cells occur in plants. Most parenchyma cells carry out photosynthesis or store food, and most collenchyma and sclerenchyma cells provide the body with support. Parenchyma cells are also involved in the transport of sucrose and other nutrients throughout the plant, however, and sclerenchyma cells make up the tissues responsible for transporting water from roots to shoots.

As a plant grows, its cells become organized into four functional groups: meristematic, dermal, ground, and vascular tissue systems. Meristematic tissues are responsible for growth. The dermal tissue system is usually one cell layer thick and plays a role in protection and water conservation. The ground tissue system performs photosynthesis and stores carbohydrates and other nutrients. The vascular tissue system, which transports materials throughout the plant, consists of xylem and phloem. Xylem transports water and dissolved ions; phloem transports sugars.

The anatomy of roots and shoots is consistent with their role as belowground and aboveground resource-acquisition systems. Root hairs increase the membrane surface area available for water and ion absorption; the region of elongation near the root tip pushes the root into new regions of the soil. In both roots and stems, the organization of vascular tissues into vascular bundles allows efficient transport of water and nutrients. The arrangement of palisade mesophyll, spongy mesophyll, and vascular bundles in a leaf maximizes its ability to harvest sunlight, exchange gases via stomata, and transport materials.

The roots and shoots of all plants increase in length due to cell division in regions called apical meristems. In addition, woody species increase in width due to cell divisions that take place in regions called lateral meristems.

Plants grow from apical meristems that are located at the tips of roots and shoots. Apical meristems are responsible for primary growth, which increases the length of stems and roots. In woody plants, lateral meristems produce secondary xylem, secondary phloem, and cork cells that widen shoots and roots. Wood consists of secondary xylem, while bark consists of all tissue outside of the vascular cambium.

Web Tutorial 35.1 Primary and Secondary Growth

Questions

Content Review

1. Which of the following functions is *not* performed by parenchyma cells?
   a. transporting sucrose
   b. transporting water
   c. performing photosynthesis
   d. storing nutrients
2. How do tracheids differ from vessel elements, in addition to their overall shape?
   a. Tracheids are stacked end to end to form continuous, open columns.
   b. In tracheids, water flows from cell to cell primarily through gaps in the secondary cell wall called pits.
   c. Tracheids are dead at maturity.
   d. Tracheids have secondary cell walls reinforced with lignin.
3. What is a sieve-tube member?
   a. the sugar-conducting cell found in phloem
   b. the widened, perforation-containing, water-conducting cell found only in angiosperms
   c. the nutrient- and water-absorbing cell found in root hairs
   d. the nucleated and organelle-rich support cell found in phloem
4. What four basic tissue types are found in plants?
   a. parenchyma, collenchyma, sclerenchyma, and epidermis
   b. structural, photosynthetic, storage, and protective (defensive)
   c. vascular cambium, cork cambium, apical meristem, and lateral bud meristem
   d. ground, vascular, meristematic, and dermal
5. Which statement best characterizes primary growth?
   a. It does not occur in roots, only in shoots.
   b. It leads to the development of cork.
   c. It produces epidermal, ground, and vascular tissues.
   d. It produces rings of xylem and phloem tissue as well as rings of cork tissue.
6. Which statement best characterizes secondary growth?
   a. It results from divisions of the vascular cambium cells.
   b. It increases the length of the plant stem.
   c. It results from divisions in the apical meristem cells.
   d. It often produces phloem cells to the inside and xylem cells to the outside of the vascular cambium.

Conceptual Review

1. Describe the general function of the shoot system and of the root system. What tissues are continuous throughout these two systems? Suggest a hypothesis to explain why the shoot and root systems of various species are so variable in size and shape.
2. Give examples of (a) adventitious roots; (b) stems that are modified for nutrient storage, water storage, or asexual reproduction; and (c) leaves that are modified for nutrient storage, defense, climbing, water storage, or pollinator attraction.
3. What does cuticle do? What do stomata do? Predict how cuticle and stomata differ in plants from wet habitats versus dry habitats. How would you test your predictions?
4. Describe whether dermal, meristematic, ground, and vascular tissues are considered simple or complex, and why.
5. Consider the following four categories: generalized cell type, specific cell type, generalized tissue type, and specific tissue type. Assign each of the following to one of the four categories: palisade mesophyll, epidermis, tracheid, secondary phloem, companion cell, parenchyma, sclerenchyma, lateral bud meristem, and vascular tissue.
6. Describe how the vascular cambium produces secondary xylem and secondary phloem. Why are growth rings apparent when tree trunks are viewed in cross section? Why do trees that grow in tropical wet forests lack annual growth rings?

Group Discussion Problems

1. Chapter 1 explains how the shoot systems of domesticated broccoli family plants (broccoli, cauliflower, cabbage, kohlrabi, and others) changed in response to artificial selection. Make a sketch of three varieties; then predict which cells, tissues, and systems changed as the ancestral population pictured in Figure 1.3 underwent artificial selection. For example, would you predict that leafy populations such as kale and savoy have more or fewer support cells in their stems and leaves than the wild population has? How would you test your predictions?
2. Look out the nearest window, and pick out a plant to observe carefully. Sketch it and label the prominent features of its external anatomy. Make a cutaway view that predicts the major features of its internal anatomy. If possible, obtain a twig, stem, or section of root. Examine and label its parts.
3. Many tarweed species living today can self-fertilize. Why is this observation relevant to the hypothesis that all of today's Hawaiian silverswords are descended from a tarweed that colonized the islands long ago?
4. Trees are killed by girdling—meaning the removal of bark and vascular cambium in a narrow ring all the way around the tree. Explain why.

Answers to Multiple-Choice Questions 1. b; 2. b; 3. a; 4. d; 5. c; 6. a

www.prenhall.com/freeman is your resource for the following: Web Tutorials; Online Quizzes and other Online Study Guide materials; Answers to Conceptual Review Questions; Solutions to Group Discussion Problems; Answers to Figure Caption Questions and Exercises; and Additional Readings and Research.

36 Water and Sugar Transport in Plants

KEY CONCEPTS

- Water moves along potential energy gradients—from areas of high water potential to areas of low water potential. Water's potential energy in plants is a combination of (1) its tendency to move via osmosis and (2) the pressure exerted on it.
- Plants do not expend energy to replace water that is lost to transpiration when stomata are open and photosynthesis is occurring. Instead, water moves from soil and roots to leaves along a water potential gradient. The gradient exists because water at the air-water surface in leaves is under negative pressure (tension) that is great enough to pull water up from the roots through xylem.
- In phloem, sugars are transported from "sources"—tissues in which sugar is abundant—to "sinks"—tissues in which sugars are being used or stored. Movement occurs because cells expend energy to load sucrose actively into phloem cells near source tissues, creating a pressure gradient that favors the movement of water and sucrose to sinks.

A giant sequoia that is 82.9 m (272 ft) tall and weighs just under 1300 metric tons—about the same as 10 diesel locomotives. The closely related redwoods, which also grow along the Pacific Coast of North America, can be over 100 m in height and are thought to be the tallest organisms that have ever lived. This chapter explores how plants move water from their roots to their leaves and how they transport sugars to all of their tissues, even over such great distances.

On a hot summer day, a single, large deciduous tree can lose enough water to fill three 55-gallon drums. Why does this happen, and how? To answer this question, recall from Chapter 35 that the surfaces of leaves are dotted with structures called stomata. Stomata are open during the day so that gas exchange can occur between the atmosphere and the cells inside the leaf. This exchange is crucial. For photosynthesis and food production to continue, leaf cells must acquire carbon dioxide (CO_2) and release a waste product—oxygen. There's a catch, however. While stomata are open so that gas exchange can take place, the moist interior of the leaf is exposed to the dry atmosphere. As a result, large quantities of water evaporate.

In essence, then, water loss is an inevitable consequence of a plant's need to obtain carbon and release oxygen. Water loss is a side effect of photosynthesis. The evaporation also cools the plant, just as sweating cools your body—which may provide a benefit under extremely hot conditions. Heavy rates of evaporation can actually lower leaf temperatures by as much as 10–15°C, which is one reason that forests often feel cool on a hot day.

The question remains, however: How do plants make up for lost water? If water that is lost to evaporation is not replaced, cells will dry out and die. And in the case of a redwood tree, the leaves that are losing water may be located 100 m from the root hairs that are absorbing water. How do plants transport water a distance equivalent to the length of a football field against the force of gravity? Similarly, how do plants move the sugar they produce from active photosynthetic sites to storage sites in roots? These questions are the heart and soul of this chapter. Answering them is a fundamental part of understanding how plants work.

36.1 Water Potential and Cell-to-Cell Movement

Loss of water via evaporation from the aerial parts of a plant is called **transpiration**. Transpiration occurs whenever two conditions are met: (1) Stomata are open, and (2) the air surrounding leaves is drier than the air inside leaves. The first condition is usually met during the day, when photosynthesis is occurring. The second condition is met whenever atmospheric humidity is less than 100 percent.

Plants replace water that is lost from leaves with water that is absorbed by roots. One of the most astonishing observations about plant biology is that water moves from roots to leaves passively—that is, with no expenditure of ATP. Plants do not need a heart muscle to pump water from roots to shoots. Even in 100-m-tall redwood trees, water flows passively from the root system to the shoot system. The movement occurs because of differences in the potential energy of water. Recall from Chapter 2 that potential energy is stored energy and that changes in potential energy are associated with changes in the position of an object such as a molecule or electron.

Biologists use the term **water potential** to indicate the potential energy that water has in a particular environment compared with the potential energy of pure water at atmospheric pressure and at the same temperature. Water potential is symbolized by the Greek letter ψ (psi, pronounced "sigh"). Differences in water potential determine the direction that water moves, because water always flows from areas of high water potential to areas of lower water potential.

Water potential is key to understanding the movement of water throughout the plant. Because water potential is high in soil and roots but low in leaves and the atmosphere, there is a **water potential gradient** between roots and shoots. Water moves down this gradient to replace the water lost to transpiration.

What Factors Affect Water Potential?

To understand how water moves in a plant, consider the cell in the beaker on the left of **Figure 36.1a**. Notice that it is sitting in a **solution**—a homogenous, liquid mixture containing several substances. In this case, the solution consists of water and dissolved substances, or **solutes**. In the beaker on the left, the solute concentrations in the cell and in the surrounding solution are the same. Recall from Chapter 6 that such a solution is said to be **isotonic** to the cell.

Now consider what happens when the cell is transferred to the beaker on the right of Figure 36.1a. This beaker contains pure water, which has no solutes. As a result, the solution surrounding the cell is strongly **hypotonic** relative to the cell. This is important, because when solutions are separated by a selectively permeable membrane, such as the plasma membrane, water passes through the membrane from regions of low solute concentration to regions of high solute concentration. This movement of water across membranes, in response to differences in solute concentration, is called **osmosis** (see Chapter 6). Water's tendency to move via osmosis is a function of the difference in solute concentrations between two compartments. The size of this tendency is called the **solute potential** (ψ_S) or **osmotic potential.**

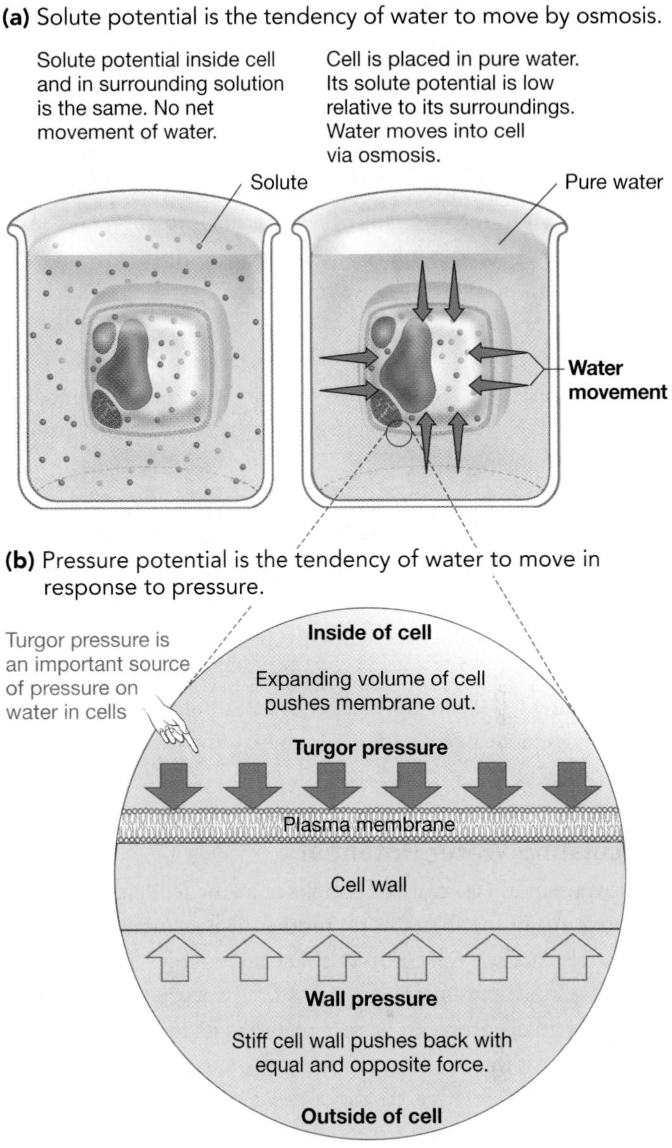
(a) Solute potential is the tendency of water to move by osmosis.
Solute potential inside cell and in surrounding solution is the same. No net movement of water.
Cell is placed in pure water. Its solute potential is low relative to its surroundings. Water moves into cell via osmosis.
Solute
Pure water
Water movement
(b) Pressure potential is the tendency of water to move in response to pressure.
Turgor pressure is an important source of pressure on water in cells
Inside of cell
Expanding volume of cell pushes membrane out.
Turgor pressure
Plasma membrane
Cell wall
Wall pressure
Stiff cell wall pushes back with equal and opposite force.
Outside of cell

FIGURE 36.1 Water Potential Has Two Components: Solute Potential and Pressure Potential
(a) The solute potential of a solution is defined by its solute concentration relative to pure water. If water contains a high concentration of solutes, then it has a low solute potential compared with pure water. **(b)** Turgor pressure defines a cell's pressure potential.

When an animal cell is placed in a hypotonic solution and water enters the cell via osmosis, the volume of the cell increases until the cell bursts. This does not happen to plant cells, however. If a plant cell swells in response to incoming water, its plasma membrane pushes against the relatively rigid cell wall. The cell wall resists expansion of the cell volume by pushing back, much as the walls of a basketball push back when the ball is inflated. The force exerted by the wall is called **wall pressure** (**Figure 36.1b**, page 829). As water moves into the cell, the pressure inside the cell, known as **turgor pressure**, increases until wall pressure is induced. Cells that are firm and that experience wall pressure are said to be **turgid**.

Turgor pressure is important, because it counteracts the movement of water due to osmosis. In the example on the right side in Figure 36.1a, the solute potential favors water moving into the cell. However, the rigid cell walls constrain the amount of water that can enter the cell.

The collective term for any kind of physical pressure on water is termed its **pressure potential** (ψ_P). Inside a cell, the pressure potential consists of turgor pressure.

To summarize, two general conditions affect water potential within plants and thus the tendency of water to move: (1) solute potential and (2) pressure potential. Now, how do these potentials interact? How do biologists determine the direction of water movement when osmosis and pressure affect a cell at the same time?

Calculating Water Potential

Water potential (ψ) can be thought of as the tendency of water to move from one location to another. It is actually a form of potential energy, not unlike chemical energy. Recall from Chapter 2 that potential energy is stored energy. The potential energy of an electron can be thought of as its tendency to move to a new position. The potential energy of an electron is defined by its location relative to the atomic nucleus. Similarly, the water potential of a cell can be thought of as water's tendency to move to a new position. The potential energy of water is defined by the pressure potential and solute potential at the water's location relative to other locations.

When selectively permeable membranes are present, water moves by osmosis from areas of high potential to areas of low potential. When no membranes are present, water moves from areas of high pressure to areas of low pressure, independently of differences in solute potential. When water moves in response to pressure differences, **bulk flow** is said to occur.

If we ignore the effects of gravity, water potential is defined algebraically as

$$\psi = \psi_P + \psi_S$$

In words, the potential energy that water has in a particular location is the sum of the pressure potential and the solute potential that it experiences. Water potential is measured in units called **megapascals** (**MPa**, 10^6 Pa). A **pascal (Pa)** is a unit of measurement commonly applied to pressures—force per unit area. A car tire is inflated to about 0.2 MPa, and the water pressure in home plumbing is usually 0.2 to 0.3 MPa.

By convention, solute potentials (ψ_S) are always negative. They are measured relative to the solute potential of pure water, and because pure water contains no dissolved substances, the solute potential of pure water is defined as 0. As a result, the water inside a cell always has a potential energy lower than that of pure water; water will therefore tend to move *into* the cell. Increasing the concentration of solutes in a cell lowers its water potential even more. In contrast, the pressure potential (ψ_P) from turgor pressure is positive. It increases the potential energy of the water inside by exerting pressure on the water, making it more likely to move *out of* the cell.

To see how solute potential and pressure potential interact, consider the U-shaped tube diagrammed in **Figure 36.2a**. Notice that the two sides of the tube are separated by a selectively permeable membrane and that pure water, which has a solute potential of 0, is on the left side. The system is open to the atmosphere and thus is not under additional pressure, meaning $\psi_P = 0$ MPa. Now suppose that the ψ_S for the solution on the right side of the membrane is -1.0 MPa. Because water potential is higher on the left side of the tube than on the right side, water moves from left to right.

In **Figure 36.2b**, the situation is the same, except that the solution on the right-hand side experiences pressure exerted by a plunger. If the force on the plunger produces a pressure potential of 1.0 MPa on the right side, then the water potential of the right side is -1.0 MPa $+$ 1.0 MPa $=$ 0. In this case, the water potential on both sides of the membrane is equal and there will be no net movement of water. If the force on the plunger were greater than 1.0 MPa, then the solution on the right side would have a *higher* water potential, and water would flow from right to left.

Figure 36.2c introduces an analogous situation in a cell. In this case a cell that is **flaccid**—meaning it has no turgor pressure and thus a pressure potential of 0—has been plunged into a solution of pure water. Because it has low solute potential and the pure water has high water potential, water enters the cell via osmosis. But as **Figure 36.2d** shows, the incoming water creates turgor pressure. When the positive turgor pressure plus the cell's negative solute potential equals 0 MPa—the water potential of pure water—the system reaches equilibrium and there is no additional net movement of water. In this way, turgor pressure acts like the plunger in Figure 36.2b.

Water Potentials in Soils, Plants, and the Atmosphere

Although Figures 36.1 and 36.2 focus on the water potential of individual cells, it's important to realize that ψ can also be measured for tissues, root and shoot systems, and entire plants. The water contained within a leaf, root system, or plant has a pressure potential and a solute potential, just as the water inside a cell does. Likewise, the soil that surrounds the root system has a water potential. So does the air around the shoot system. In the atmosphere, for example, water exists as a vapor with a

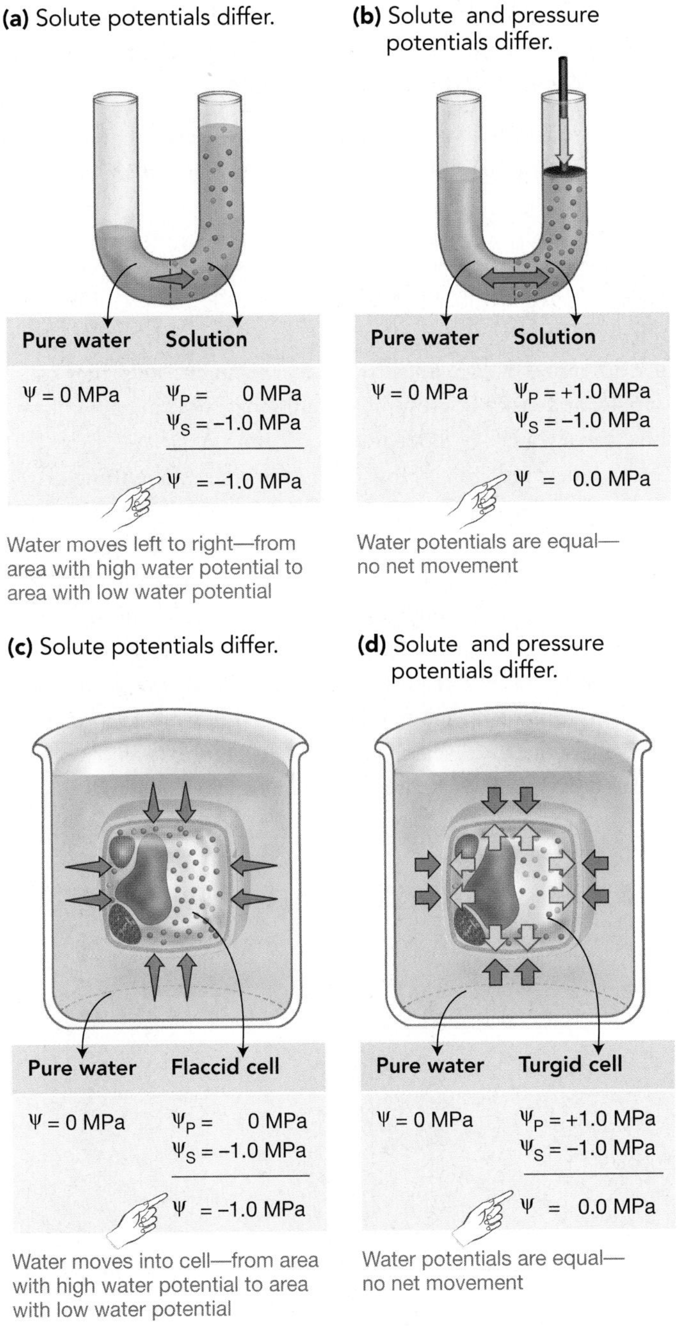
(a) Solute potentials differ.
Pure water
Solution
Ψ = 0 MPa
Ψ_P = 0 MPa
Ψ_S = −1.0 MPa
Ψ = −1.0 MPa
Water moves left to right—from area with high water potential to area with low water potential
(b) Solute and pressure potentials differ.
Pure water
Solution
Ψ = 0 MPa
Ψ_P = +1.0 MPa
Ψ_S = −1.0 MPa
Ψ = 0.0 MPa
Water potentials are equal—no net movement
(c) Solute potentials differ.
Pure water
Flaccid cell
Ψ = 0 MPa
Ψ_P = 0 MPa
Ψ_S = −1.0 MPa
Ψ = −1.0 MPa
Water moves into cell—from area with high water potential to area with low water potential
(d) Solute and pressure potentials differ.
Pure water
Turgid cell
Ψ = 0 MPa
Ψ_P = +1.0 MPa
Ψ_S = −1.0 MPa
Ψ = 0.0 MPa
Water potentials are equal—no net movement

FIGURE 36.2 Solute Potential and Pressure Potential Interact
(a) If solute potentials in solutions on the two sides of a membrane differ, and if there is no pressure potential, then water will move from the side of high solute potential to the side of low solute potential. **(b)** A pressure potential has equalized water potentials on both sides of the membrane and counteracted the movement of water along a solute potential gradient. **(c)** If a flaccid cell is placed in pure water, water enters via osmosis. **(d)** Turgor pressure counteracts the tendency of water to move via osmosis.

solute potential of 0. But the pressure potential of water in the atmosphere can be low or high, depending on the concentration of water molecules.

In soil, the water that fills crevices between soil particles usually contains relatively few solutes and normally is not under much pressure. As a result, its water potential tends to be high relative to the water found in a plant's roots. Similarly, the water potential of air is usually extremely low relative to the moist interior of a leaf. These contrasts set up a series of water potential differences between the soil, a plant, and the atmosphere. Because water moves along this water potential gradient, plants tend to gain water from the soil and lose it to the atmosphere (**Figure 36.3**).

It's important to note, though, that the water potentials of plant tissues, air, and soil are dynamic. They routinely change from day to day and even hour to hour in response to changes in the weather. Water potentials may also vary by location or season. To drive these points home, consider the following examples:

- If a fog rolls in, the pressure potential of water vapor in the atmosphere increases. When the water potential of the atmosphere increases, water is less likely to leave plants.
- When soils dry, droplets of water tend to cling to soil particles. When water adheres to soil in this way, it creates a negative pressure that lowers the pressure potential of soil water. When the water potential in soil is low, water is less

Low water potential
Atmosphere Ψ: −95.2 MPa
(Changes with humidity; usually very low)
Leaf Ψ: −0.8 MPa
(Depends on transpiration rate; low when stomata are open)
Root Ψ: −0.6 MPa
(Medium–high)
Soil Ψ: −0.3 MPa
(High if moist; low if extremely dry)
High water potential

FIGURE 36.3 A Water Potential Gradient Exists between the Soil, Plants, and the Atmosphere
Water moves from regions of high water potential to regions of low water potential. Xylem conducts water along this water potential gradient.

likely to move from soil into plants. **Box 36.1** analyzes how some species of plants can continue gaining water from drying soils by lowering the water potential of their tissues.

- The water in salty soils contains large amounts of solutes. In these types of habitats, soil has a very low solute potential and water potential. **Box 36.2** explores how plants adapt to life in salty soils; the essay at the end of this chapter analyzes why irrigated soils tend to become salty.

Let's return to typical conditions, however. What happens when a water potential gradient exists between the soil, a plant,

BOX 36.1 How Do Plants Adapt to Dry Soils?

Some plant species can thrive in habitats that are dry enough to kill other species. To understand how this is possible, consider recent research on some of the trees and shrubs growing in the Rocky Mountains of western North America. Water availability varies enormously there. Mountainsides that face the south are much drier than northern exposures, and steep slopes are much drier than gentle slopes. To explore how species that occupy dry sites versus wet sites acquire water, biologists measured the water potential and solute potential of leaves from several species over the course of the growing season.

Some of the data that resulted from this research are shown in **Figure 36.4a**. The graph documents how the solute potential of leaf tissue in ninebark shrubs varied during one growing season. This species is found primarily on dry sites. Notice that the solute potential of the leaf tissue is relatively high in June, at the start of the growing season. In the Rockies, June tends to be rainy and cool. July and August are progressively hotter and drier, however. As the summer progressed, ninebark shrubs were able to keep growing because they experienced a huge drop in water potential. Why is this observation significant?

Figure 36.4b plots changes in the water potential of ninebarks and Sitka alder shrubs, which grow on wet north-facing slopes, over the course of a single day during August. Notice that both plants have high water potentials at night, when stomata are closed, and low water potentials during the day. By midday, however, the water potential of ninebark is almost 1.5 MPa lower than that of Sitka alder.

To understand why this difference is important, study **Figure 36.4c** and ask yourself Which direction soil water would tend to move if a Sitka alder and ninebark were growing side by side? Water always moves toward regions of lower water potential, so the ninebark would be able to compete for soil water much more efficiently than the Sitka alder. Its cells are able to tolerate low solute potentials, so the ninebark plant can generate extremely low water potentials without dying. Because ninebarks can extract water from dry soils, they can grow in conditions that might kill a Sitka alder.

How do the enzymes, membrane proteins, and phospholipids of dry-adapted plants allow them to tolerate such low solute potentials? Is there a cost to ninebark and similar species, in terms of how fast they can grow when water is readily available? These questions remain unanswered. Like most good research, the data in Figure 36.4 generate as many questions as they answer.

(a) Changes in solute potential during a growing season

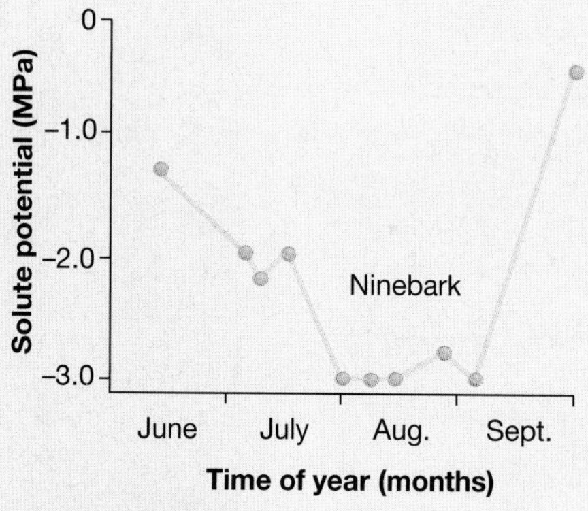

(b) Changes in water potential during a day

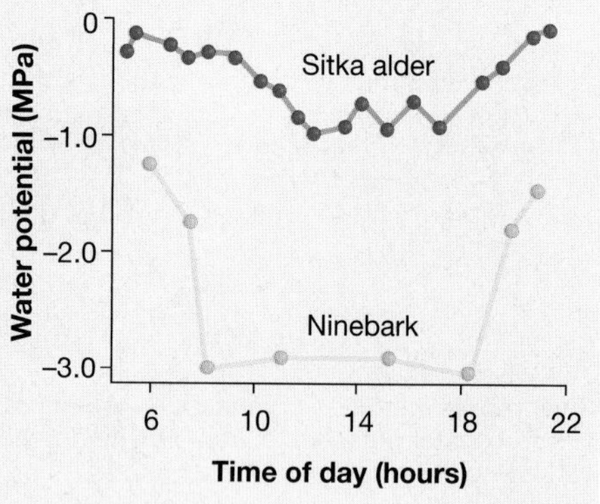

(c) Sitka alder and ninebark compete.

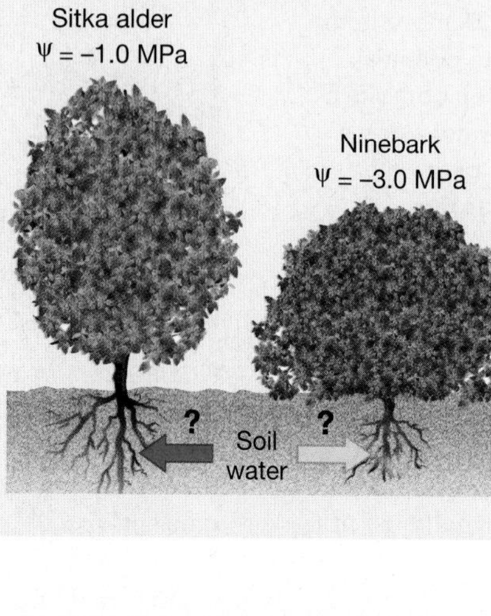

FIGURE 36.4 Plants with Low Solute Potentials Can Grow in Dry Soils
(a) During a growing season, the average solute potential of ninebark leaves drops dramatically. **(b)** On a hot midsummer day, the water potential of ninebark leaves drops much more than does the water potential of sitka alder leaves. **(c)** The water potentials of Sitka alder and of ninebark. QUESTION If Sitka alder and ninebark were competing for water, which plant would gain water more effectively?

BOX 36.2 How Do Plants Adapt to Salty Soils?

Plants that are able to thrive in salty environments are called **halophytes** ("salt-plants"). Salty environments are found throughout the world, primarily along ocean coastlines and in inland deserts. In such habitats, the water potential of the soil is often −4 MPa or less—much lower than the water potential typically found inside plant roots. If water moved along the water potential gradient, it would leave plant roots and enter the soil.

To cope with soils that have extremely low water potentials and to avoid an outflow of water, many halophytes lower the solute potential of their root cells. More specifically, these plants have enzymes and transport proteins that increase the concentration of ions in their vacuoles and increase the concentration of sugars and other organic molecules in the cytoplasm. As a result, they can keep the water potential of their tissues even lower than the water potential of salty soil.

The other major problem that halophytes face is acquiring toxic levels of certain ions present in salty soils—particularly sodium (Na^+) and chloride (Cl^-)—which tend to enter root cells along their electrochemical gradients. Some halophytes have mechanisms for preventing the entry of ions, but others transport them to specialized structures located at the surface of their leaves, where sodium and chloride ions are crystallized as salt (NaCl). Once the ions are out of solution, they are harmless to the plant. Halophiles that crystallize salt often have a grayish appearance due to the salt crystals deposited on their stems and leaves (**Figure 36.5**).

FIGURE 36.5 Some Halophytes Secrete the Salt They Take In
"Big saltbush" with grayish leaves as a result of an accumulation of salt crystals. The crystals form when sodium and chloride ions are secreted from special structures in the leaves.

and the atmosphere? When the cells in a leaf or stem lose water to the atmosphere, their membranes contract and the cells shrivel like grapes drying into raisins. If the cells do not regain turgor soon, they will be at risk of dehydration and death. When an entire tissue loses turgor, it is said to **wilt** (**Figure 36.6**). Unless corrected, wilting leads to the death of the tissue and, eventually, the plant.

Loss of turgor pressure in cells ... leads to wilting.

FIGURE 36.6 Wilting Occurs when Water Loss Leads to Loss of Turgor Pressure
Plants wilt when turgor pressure falls to zero and cells become flaccid. Wilting is a life-threatening condition in plants, analogous to severe dehydration in humans. Turgor pressure is required for growth to occur; without it, cells cannot expand once cell division is complete. In addition, turgor pressure provides structural support—which is why wilted plants droop.

Suppose that you are in charge of caring for the wilted plant shown on the right in Figure 36.6. If you add water to the soil, the water potential of the soil increases and water should begin to move into the plant along a water potential gradient. There are three major hypotheses for how this might happen: (1) root pressure—a pressure potential that develops in roots—could drive water up against the force of gravity; (2) capillary action could draw water up the cells of xylem; and (3) cohesion-tension, which is a force, generated in leaves, that could pull water up from roots. Let's consider each in turn.

✓CHECK YOUR UNDERSTANDING

Water moves from areas where its potential energy is high to areas where its potential energy is low. In areas separated by a selectively permeable membrane, part of water's potential energy is made up of its solute potential—its tendency to move via osmosis. But water also has a pressure potential. In plant cells, for example, the cell wall can exert pressure on water and affect its pressure potential. You should be able to describe the solute potential, pressure potential, and overall water potential of each of the following: wet soils and dry soils; the roots, stems, and leaves of a plant; and the atmosphere.

36.2 Root Pressure and Short-Distance Transport

To understand how water enters a root, consider the cross section through a young buttercup root shown in **Figure 36.7**. Starting at the outside of the root and working inward, notice that several distinct tissues are present:

- The **epidermis** is a single layer of cells. In addition to protecting the root, some epidermal cells produce projections called **root hairs**, which greatly increase the total amount of surface area of the root.
- The **cortex** consists of ground tissue—usually parenchyma cells—and functions in the storage of carbohydrates.
- The **endodermis** is a cylindrical layer of cells that separates the cortex from the vascular tissue.
- The **pericycle** is a layer of cells that can become meristematic. If cells in the pericycle begin dividing, the daughter cells become organized into lateral roots.
- **Vascular tissues** are located in the center of roots in buttercups and other eudicots. Notice that, in these species, phloem is situated between each of four arms formed by xylem, which is arranged in a cross-shaped pattern.

When water enters a root along a water potential gradient, it does so through root hairs. As water is absorbed, it moves through the root cortex toward the vascular tissue—where it can be transported up to the shoot system—through two distinct pathways (**Figure 36.8a**). The first pathway is called the **apoplast** ("away-formed") and lies within cell walls, which are porous. Because the cell walls of adjacent cells are in direct physical contact, they form a continuous structure. In leaves, for example, cell walls may represent 10–15 percent of the structure's total volume. The second pathway is called the **symplast** ("together-formed") and consists of the continuous connection through cells that exists via plasmodesmata (see Chapter 8). In essence, water can flow around cells by the apoplastic route, through cells by the symplastic route, or both.

The situation changes when water reaches the endodermis, however. Endodermal cells are tightly packed and secrete a waxy layer called the **Casparian strip**. This layer is composed primarily of a water-repellent compound called **suberin**. The presence of the Casparian strip blocks the apoplastic pathway. More specifically, it prevents water from creeping through the walls of endodermal cells and proceeding to the vascular tissue (**Figure 36.8b**). This blockage is important, because water and solutes that are moving through the cortex can reach the vascular tissue only by passing through the plasma membrane of an endodermal cell. Endodermal cells, in turn, act as filters. They allow ions such as potassium (K^+) that are needed by the plant to pass through to the vascular tissue but can prevent the passage of ions such as sodium (Na^+) that are not needed or may be harmful. Similarly, the endodermis prevents water from flowing out of the vascular tissue and into the cortex, if conditions are dry and the water potential of the cortex drops.

The endodermis is also responsible for a phenomenon known as **root pressure**. During the night, roots often accumulate ions that are acquired from the soil by epidermal cells. These nutrients move into xylem. The influx of ions lowers the water potential of xylem below the water potential in the surrounding cells. As water flows into xylem in response, a positive pressure is generated that forces fluid up the xylem. Since more water moves up xylem and into leaves than is being transpired from the leaves, water droplets are forced out of the leaves that are closest to the roots. This phenomenon is known as **guttation**. If you are ever up and around early in the morning, you may have observed water drops on leaf tips formed by guttation (**Figure 36.9**).

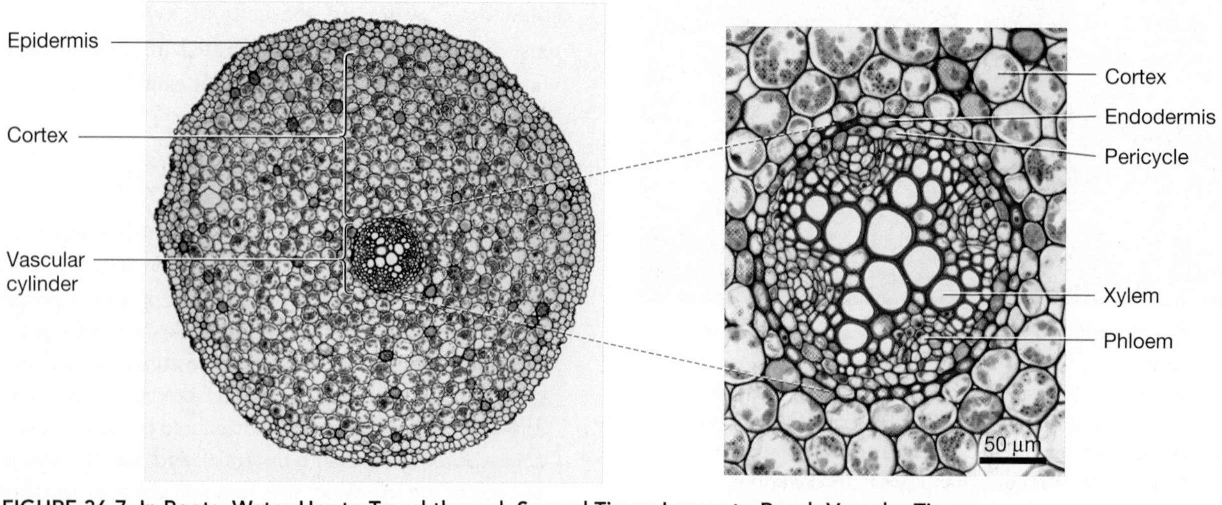

FIGURE 36.7 In Roots, Water Has to Travel through Several Tissue Layers to Reach Vascular Tissue
Cross sections through a buttercup root, showing the anatomy that is typical of roots in eudicots.

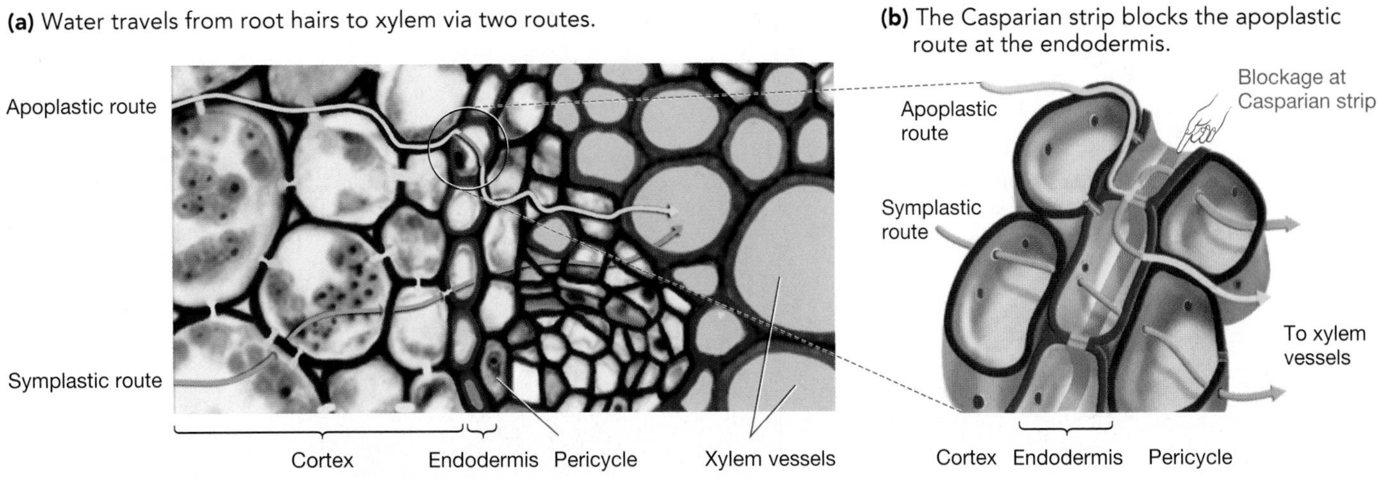
(a) Water travels from root hairs to xylem via two routes.
Apoplastic route
Symplastic route
Cortex
Endodermis
Pericycle
Xylem vessels
(b) The Casparian strip blocks the apoplastic route at the endodermis.
Apoplastic route
Blockage at Casparian strip
Symplastic route
To xylem vessels
Cortex
Endodermis
Pericycle

FIGURE 36.8 Water Moves through Roots via Cell Walls or Cytoplasm
(a) The apoplastic route takes water through cell walls; the symplastic route takes water from cell to cell through cytoplasm via plasmodesmata. **(b)** At the endodermis, the apoplastic route is blocked by suberin in the Casparian strip. As a result, water in the apoplast must pass through the plasma membrane of endodermal cells prior to entering xylem.

At one time, positive root pressure was a leading hypothesis to explain how water moves from roots to leaves in trees. However, careful research showed that root pressure cannot push water all the way up a tall tree. Over long distances, the force of root pressure is not enough to overcome the force of gravity on the water inside xylem. In addition, researchers were able to demonstrate that cut stems, which have no contact with the root system, are still able to transport water to leaves. Biologists concluded that there must be some other mechanism responsible for the long-distance transport of water. What is it?

FIGURE 36.9 Root Pressure Causes Guttation
In low-growing plants, ions may accumulate in the xylem of roots at night. If so, enough water may enter xylem via osmosis to force water up and out of leaves.

36.3 Transpiration and Long-Distance Water Transport

The root-pressure hypothesis was not the only proposed explanation for the long-distance transport of water in plants. Researchers have also evaluated a hypothesis based on the phenomenon of **capillarity**, or movement of water up a narrow tube. When a thin glass tube is placed upright in a pan of water, water creeps up the tube (**Figure 36.10a**, page 836). The movement occurs in response to three forces: (1) surface tension, (2) adhesion, and (3) cohesion. Let's consider each force in turn.

Surface tension is a pull that exists on water molecules at an air-water interface. As **Figure 36.10b** shows, a concave boundary layer called a **meniscus** (plural: **menisci**) forms at most air-water interfaces. Menisci form because of a property of water that was introduced in Chapter 2: hydrogen bonding. In the body of a solution, all the water molecules present are surrounded by other water molecules and form hydrogen bonds in all directions. The water molecules on the surface, however, can form hydrogen bonds in one direction only—with the water molecules below them. As a result, the topmost layer of water molecules is pulled downward and the surface forms a meniscus. Because pulling forces create **tension**, the pulling force on the water molecules at the air-water interface is called surface tension.

Cohesion is a mutual attraction among like molecules, such as the hydrogen bonding that occurs among molecules in water. Because water molecules cohere, the tension at the surface of a thin tube is transmitted downward through the water column, pulling water molecules up. **Adhesion**, in contrast, is the attraction of unlike molecules. In this case, water interacts with a solid substrate, such as the glass walls of a capillary tube or the cell walls of tracheids or vessel elements, through hydrogen bonding. As water molecules cohere to each other and adhere

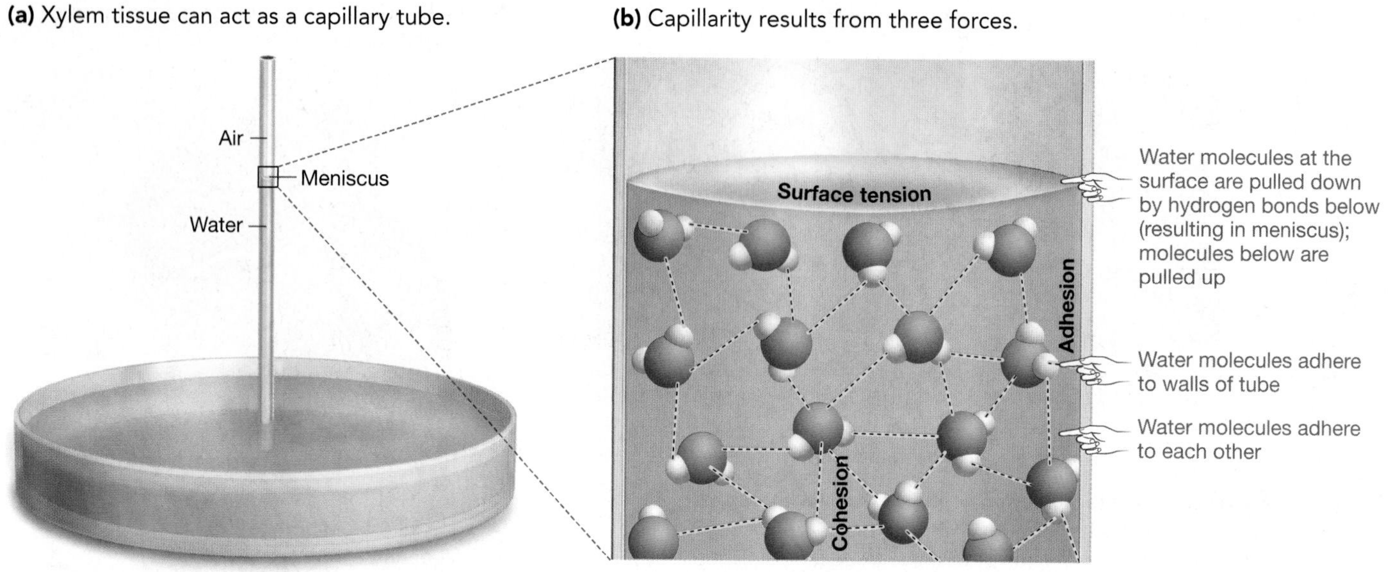

FIGURE 36.10 Water Can Rise in Xylem via Capillarity
(a) Capillarity causes water to rise in thin tubes like xylem. **(b)** Capillarity occurs through a combination of three forces, all of which are generated by hydrogen bonding: the surface tension generated at the meniscus, the adhesion of water molecules to the sides of the tube, and the cohesion of water molecules (which transmits the pulling force of surface tension).

to the side of the tube, they are pulled upward. The result is capillarity.

Like root pressure, however, the amount of water that can be transported by capillarity is limited. Capillary action is responsible for moving water along the surfaces of mosses and other low-growing, nonvascular plants; but capillarity alone is capable of raising the water in the xylem of a vertical stem only about 1 m. Thus, root pressure and capillarity cannot begin to explain how water moves from soil to the top of a redwood tree—an organism that is 5 to 6 stories higher than the Statue of Liberty. No fire company in the world has the ability to pump water that high. The leading hypothesis to explain water movement in vascular plants is not based on pumping or pushing forces. Rather, the cohesion-tension theory claims that water is *pulled* to the tops of trees along a water potential gradient, via forces generated by transpiration.

The Cohesion-Tension Theory

To understand the essence of the **cohesion-tension theory**, consider the leaf cross section shown in **Figure 36.11**. Notice that spaces in the spongy mesophyll region of the leaf are filled with moist air as a result of evaporation from the surfaces of the surrounding cells. When the pore of a stoma opens, this humid air is exposed to the atmosphere. In most cases, the atmosphere is much drier than the space inside the leaf, creating a steep concentration gradient between the leaf interior and its surroundings. Consequently, water vapor exits through the stomatal pore.

As water is lost from the leaf and enters the atmosphere, the humidity of the gas-filled space inside the leaf drops. In response, more water begins to evaporate from the walls of the spongy mesophyll cells, where menisci exist at the air-water interface. If water molecules leave these menisci rapidly due to a steep water potential gradient with the atmosphere, then fewer water molecules are available at the surface than before. The inward pull on the remaining water molecules at the surface becomes stronger, which increases the curve of each meniscus. As menisci deepen, the surface tension increases dramatically—meaning the pull on water molecules below the surface increases. In 1894 Henry Dixon and John Joly hypothesized that the formation of these steep menisci produces a force capable of pulling water up from the roots, dozens of meters into the air.

At first glance, Dixon and Joly's idea seems fantastic. Is it really possible?

The Role of Surface Tension in Water Transport As Figure 36.11 shows, the key idea in the cohesion-tension theory is that the negative force or pull generated at the air-water interface is transmitted through the water present in leaf cells, to the water present in xylem, on to the water in the vascular tissue of roots, and finally on to the water in the soil. This continuous transmission of pulling force is possible because water is present throughout the plant and because all of the water molecules present bond to one another in a continuous fashion. Recall that such bonding between similar molecules is called cohesion.

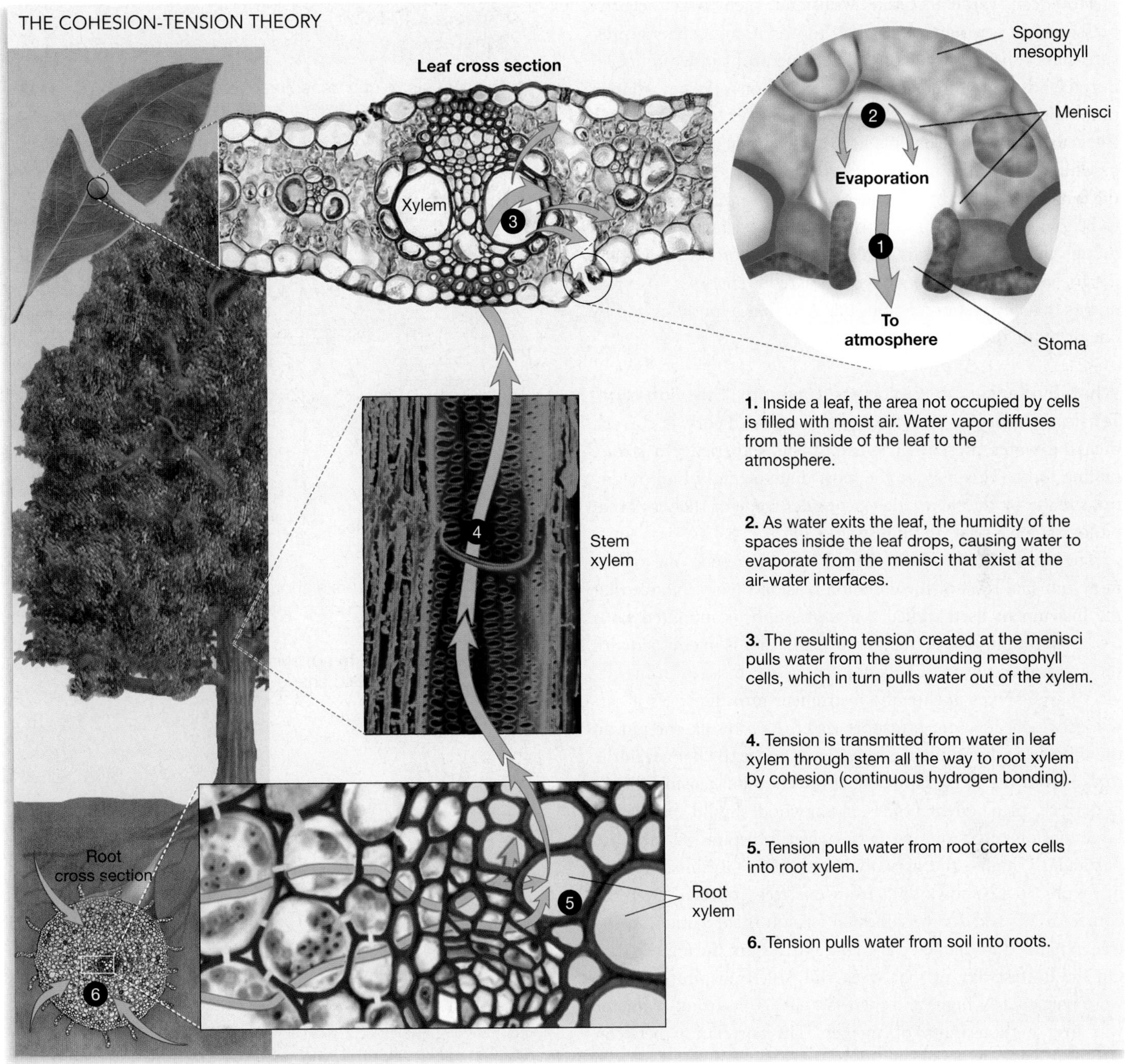
THE COHESION-TENSION THEORY
Leaf cross section
Xylem
3
Spongy mesophyll
2
Menisci
Evaporation
1
To atmosphere
Stoma
1. Inside a leaf, the area not occupied by cells is filled with moist air. Water vapor diffuses from the inside of the leaf to the atmosphere.
4
Stem xylem
2. As water exits the leaf, the humidity of the spaces inside the leaf drops, causing water to evaporate from the menisci that exist at the air-water interfaces.
3. The resulting tension created at the menisci pulls water from the surrounding mesophyll cells, which in turn pulls water out of the xylem.
4. Tension is transmitted from water in leaf xylem through stem all the way to root xylem by cohesion (continuous hydrogen bonding).
Root cross section
5
Root xylem
5. Tension pulls water from root cortex cells into root xylem.
6
6. Tension pulls water from soil into roots.

FIGURE 36.11 Transpiration Creates Tension That Is Transmitted from Leaves to Roots
The essence of the cohesion-tension theory is that a "transpirational pull," created by surface tension in leaves, is transmitted via xylem to roots and the water in soil.

Because it is based on bonding between water molecules and the pulling forces generated by transpiration, the Dixon and Joly proposal is called the *cohesion-tension theory* of water movement. In effect, the theory claims that, because of the hydrogen bonding that occurs between water molecules (cohesion), water is pulled up through xylem in continuous columns. You can also think about the cohesion-tension theory in terms of water potentials. The pulling force generated at menisci lowers the pressure potential of water in leaves enough to create a strong water potential gradient—one that is steep enough to overcome the force of gravity and pull water up long distances.

To appreciate just how great the forces involved are, think of the vessel elements or tracheids in xylem as groups of straws. When you use a straw, the negative pressure or tension that you exert with your mouth causes liquid to rise. Sucking on a straw can create a pressure difference of only about −0.1 MPa, however, which can draw water up a maximum of about 10 m. In contrast, the negative pressure exerted by the menisci in leaves can be as high as −2.0 MPa—enough to draw water up as much as 100 m.

How can vascular tissue withstand such large negative pressures? If you suck on a straw too hard, atmospheric pressure makes the straw collapse. The same problem would occur in tracheids and vessel elements, were it not for a key adaptation: The secondary thickenings characteristic of cell walls in tracheids and vessel elements keeps xylem from collapsing.

Unlike the forces involved in root pressure and capillarity, the forces generated by the cohesion-tension mechanism appear to be great enough to transport water long distances. But even though the theory appears plausible, it is important to evaluate it rigorously. To evaluate a hypothesis, researchers set up experiments that can test predictions made by that hypothesis and by one or more alternative hypotheses.

What Evidence Do Biologists Have for the Cohesion-Tension Theory? If the cohesion-tension theory is correct, then the water present in xylem should experience a strong pulling force. This may be the most fundamental of all predictions made by the theory. Let's consider some of the ways that biologists have gone about testing it.

The apparatus in **Figure 36.12** may represent one of the more unusual tests of the cohesion-tension theory. Notice that the instrument used, called a *dendrograph*, is mounted on a tree. It has a pin that is free to move and that is in contact with the exterior of the tree. This pin is attached to a recording device. Large spikes anchor the instrument into the tree's heartwood, which does not transport water. As a result, the pin on the outside of the trunk can move in response to changes in the thickness of the tree's sapwood. If the cohesion-tension theory is correct, then xylem vessels in sapwood should experience large tension forces on hot days—when transpiration rates are particularly high—and actually shrink. The "Results" section in Figure 36.12 shows data from one tree collected over the course of a week. As the cohesion-tension theory predicts, the tree expanded in diameter at night and shrank during the day. On the hottest day of that week, July 14, transpiration rates were particularly high, and the total change in trunk diameter was over a thousandth of an inch. The correlation between changes in trunk diameter and changes in transpiration suggests that the tension created in xylem is great enough to make tree trunks shrink in diameter.

Simpler experiments have also supported the idea that the fluid in xylem is under tension, as predicted by the cohesion-tension theory. If you find a leaf that is actively transpiring and cut its petiole, the watery fluid in the xylem, or **xylem sap**, withdraws from the edge toward the inside of the leaf. According to the cohesion-tension theory, this observation is due to a transpirational pull at the air-water interface in spongy mesophyll cells of leaves.

Additional evidence supporting the cohesion-tension theory has relied on an instrument called a *pressure bomb*, which measures the water potential of plant tissues. **Figure 36.13** shows how the apparatus works. Researchers place into an air-tight container a leaf or branch that has just been cut from a transpiring plant, apply a steadily increasing external pressure, and record the positive pressure required to push the xylem sap back to the cut surface. This pressure is equal to the negative pressure (tension) exerted on the xylem sap. To understand why, it is critical to realize that the solute potential of the tis-

Question: Do daily changes in the diameter of tree trunks support the cohesion-tension theory?

Hypothesis: When transpiration is occurring, xylem is under enough tension to make tree trunks shrink slightly.

Null hypothesis: When transpiration is occurring, xylem is not under enough tension to make tree trunks shrink slightly.

Experimental setup:

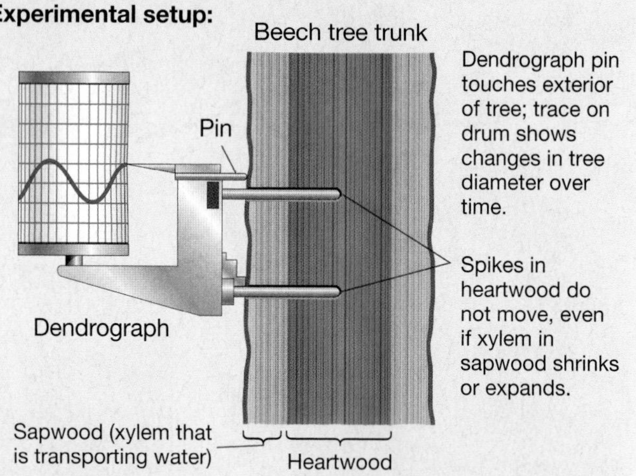

Prediction: Tree trunks should shrink during the day, when transpiration is occurring, and should expand at night, when transpiration stops.

Prediction of null hypothesis: Tree trunks expand only as they grow, not due to changes in transpiration.

Results:

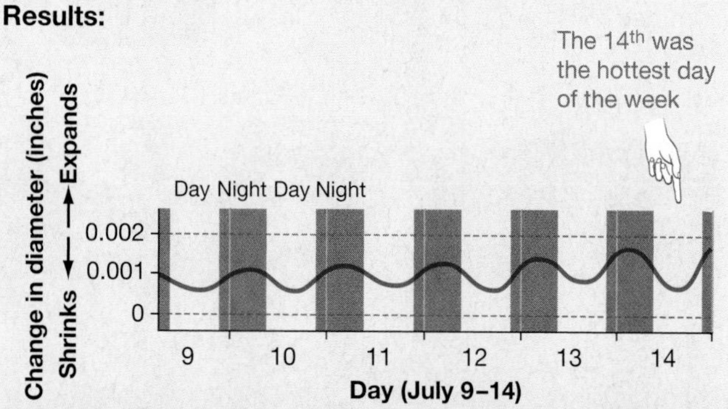

Conclusion: On hot summer days, xylem is under enough tension to shrink tree trunks slightly. This observation supports the cohesion-tension theory.

FIGURE 36.12 Trees Shrink when They Transpire
A dendrograph records changes in the diameter of trees. The data shown represent a trace from a dendrograph monitoring a beech tree in midsummer. QUESTION The pin on the dendrograph touches the exterior of the tree, while the bolts that hold the instrument are anchored to the interior of the trunk. Why don't the exterior and interior of the tree trunk move together?

Question: Do daily changes in the water potential of leaf tissue support the cohesion-tension theory?

Hypothesis: Increased transpiration during daylight hours decreases the water potential of shoot tissues.

Null hypothesis: Transpiration rate does not affect the water potential of shoot tissues.

Experimental setup:

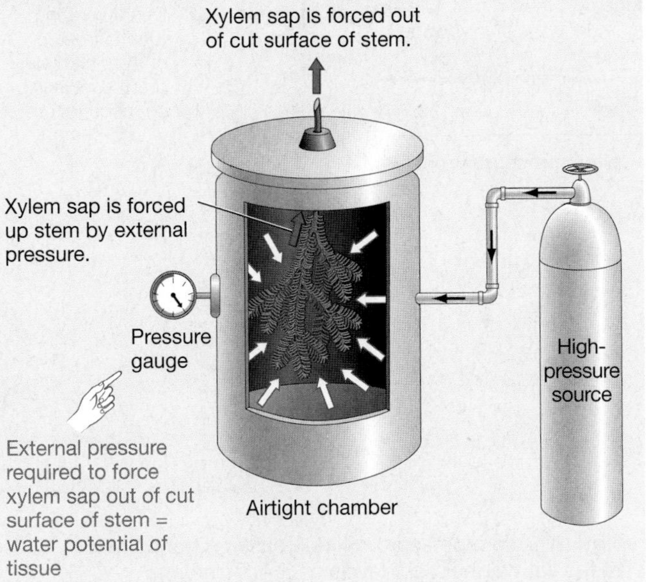

Xylem sap is forced out of cut surface of stem.
Xylem sap is forced up stem by external pressure.
Pressure gauge
High-pressure source
External pressure required to force xylem sap out of cut surface of stem = water potential of tissue
Airtight chamber

Prediction: Water potential of shoots will decrease during the day as transpiration increases.

Prediction of null hypothesis: Water potential will be unaffected by transpiration rate.

Results:

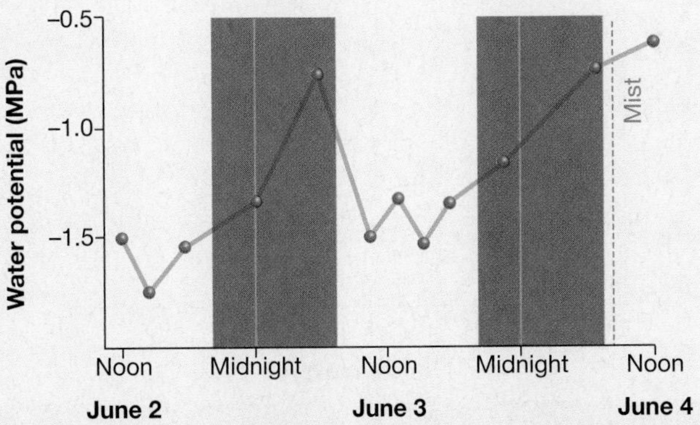

Water potential (MPa)
–0.5
–1.0
–1.5
Mist
Noon
Midnight
Noon
Midnight
Noon
June 2
June 3
June 4

Conclusion: Increased transpiration rates during the day lead to lower water potential in shoot tissues. This observation supports the cohesion-tension theory.

FIGURE 36.13 The Water Potential of Spruce Tree Leaves Varies over the Course of a Day
Data for branches from the tops of spruce trees, over a two-day period in early summer. QUESTION "Mist" indicates that a light rain fell on the morning of June 4. Why didn't the water potential of the tree leaves decrease that morning, as it had the previous two days when it didn't rain?

sues involved does not change during the experiment—only the pressure potential does. If ψ_S is constant, then the positive ψ_P recorded in the experiment is equal to ψ in the leaf or branch. In effect, the pressure that researchers apply to a pressure bomb acts like the plunger in Figure 36.2b.

To see how pressure bombs are used to study water transport, consider data that researchers collected on the water potential of Sitka spruce leaves. In this study, researchers collected branches from the tops of 10-m-tall spruce trees several times a day for several days, used a pressure bomb to measure the water potential of the tissues, and produced the data shown in the "Results" section of Figure 36.13. The fundamental observation is that the water potential of these tissues dropped during the day and rose at night. This result is logical, given that stomata are open during the day, when light is available and photosynthesis is under way. At night, stomata close and transpiration virtually ceases.

Although the observation that xylem sap is under tension was important, the cohesion-tension theory remained controversial. For example, several studies failed to document rapid changes in xylem pressure when temperature and humidity conditions were changed experimentally. Changes in light availability, temperature, or humidity should change either the degree to which stomata open or the transpiration rate or both and thus should lead to rapid changes in xylem pressure. Failure to observe these changes undermined biologists' confidence in the cohesion-tension theory, because the model predicts that the tension on xylem sap should change almost immediately when transpiration slows down or speeds up. But were these negative results due to instruments that could not document small and rapid changes in pressure potential?

Chunfang Wei, Melvin Tyree, and Ernst Steudle addressed these problems by using instruments called a *root bomb* and a *xylem pressure probe*. Their apparatus is diagrammed in the "Experimental setup" section of **Figure 36.14** (page 840). The xylem pressure probe includes an oil-filled glass tube that Wei and co-workers inserted into the xylem of a corn leaf, allowing them to record any changes in xylem pressure instantly and directly.

To test the cohesion-tension theory, the researchers altered xylem pressure in two ways. First, they used the root bomb illustrated in Figure 36.14 to add pressure to the root systems of corn. As predicted by the cohesion-tension theory, pulses of pressure applied through the root bomb produced sharp *increases* in xylem pressure. Second, the investigators released pressure on the root system and began to alter light levels. This experiment was designed to alter transpiration rates in the leaves. By weighing the entire plant (including soil and bomb) and documenting that it became lighter—due to water loss—after experiencing high light intensities, the experimenters confirmed that transpiration rates actually did increase. What did the xylem pressure probe record as the researchers increased light intensities?

Question: Do direct measurements of pressure in xylem tissue support the cohesion-tension theory?

Hypothesis: Pressure applied to plant roots will be quickly transmitted to xylem sap in the leaf.

Null hypothesis: Pressure applied to plant roots will not affect xylem sap in the leaf.

Hypothesis: Increasing transpiration by raising light intensity will lower xylem pressure in leaves.

Null hypothesis: Increasing light intensity will not affect xylem pressure in leaves.

Experimental setup:

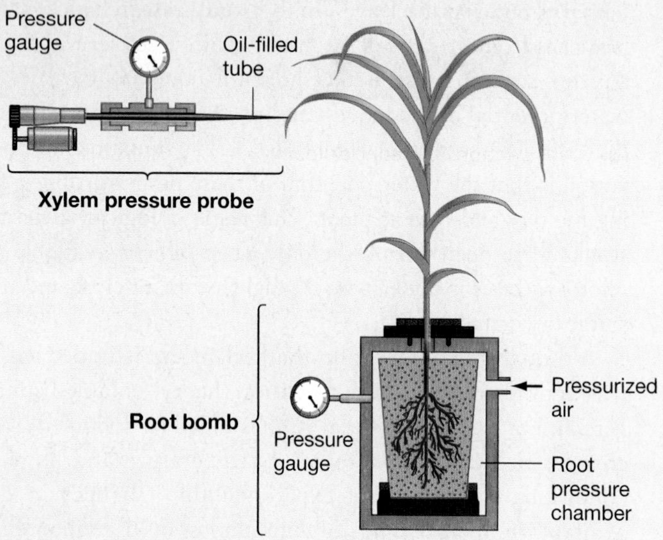

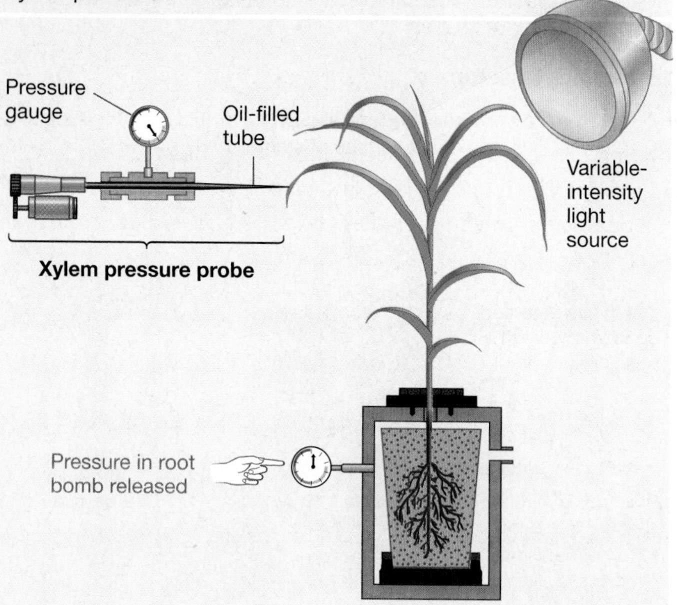

Prediction: As pressure is applied to the roots, the pressure in the xylem will increase.

Prediction of null hypothesis: Pressure applied to the roots will not affect the pressure in the xylem.

Prediction: Xylem pressure will decrease as transpiration increases with higher light levels.

Prediction of null hypothesis: Xylem pressure will not be affected by light intensity.

Results:

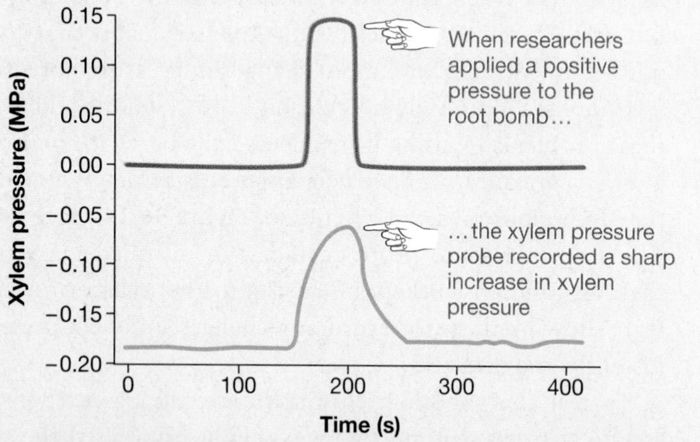

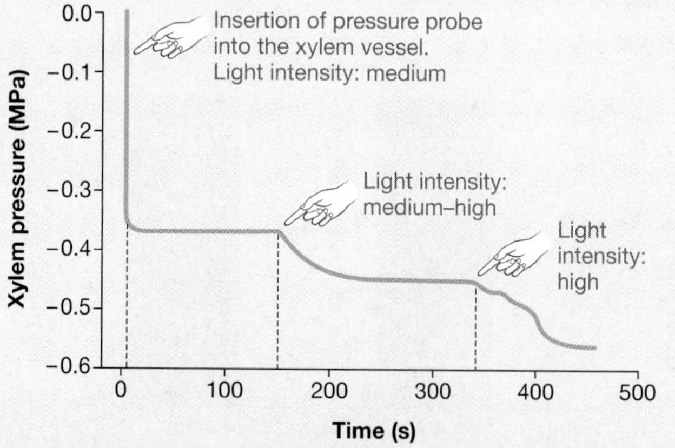

Conclusion: Xylem pressure increases in response to pressure applied to roots. Xylem pressure decreases when light intensity increases. Both observations support the cohesion-tension theory.

FIGURE 36.14 Direct Measurements of Pressure in Xylem Support the Cohesion-Tension Theory

The "Results" section of Figure 36.14 shows the data. While reading this graph, remember that tension is a pull, or a negative pressure. Thus, the xylem pressure probe showed that the tension acting on xylem sap increased each time that light intensity was increased—just as the cohesion-tension theory predicts.

These results were considered convincing evidence that increased transpiration leads to increased tension on xylem sap. Increased tension, in turn, lowers the water potential of leaves enough to pull water up from the roots and soil, where the water potential is high. On the basis of these and other results, most biologists now accept the cohesion-tension theory.

Water Absorption and Water Loss

One of the most important features of the cohesion-tension theory is that it does not require plants to expend energy. Instead, the Sun furnishes the energy required to pull water from roots to shoots. Energy from the Sun heats water molecules at the air-water interface inside leaves enough to break the hydrogen bonds between them and cause transpiration. Rapid transpiration creates deep menisci in the walls of leaf cells, causing tension that lowers the water potential of leaves. Hydrogen bonding between water molecules transmits this tension down to water molecules in the root and soil. Water moves along the resulting water potential gradient by *bulk flow*—a mass movement of molecules along a pressure gradient. Xylem is merely a conduit—a set of pipes that allows water to move from a region of high water potential (the soil) to a region of low water potential (the leaves). Water flows from roots to shoots as long as the water potential of the plant is lower than that of the soil.

If soils begin to dry, however, it may be difficult for plants to replace water being lost via transpiration. If water is not being replaced fast enough, the solute potentials of leaves begin to drop and leaves and branches may begin to wilt. In response, stomata may close down partially or completely to reduce transpiration rates and conserve water. Closing stomata also affects the ability of plants to carry on photosynthesis, however, because CO_2 acquisition slows or stops. The balance between conserving water and maximizing photosynthesis is termed the **photosynthesis-transpiration compromise.** This compromise is particularly delicate for species that grow on dry sites. How do they cope?

Limiting Water Loss Plants that thrive in dry sites have several adaptations that help them slow transpiration and limit water loss. As an example, consider the oleander plant illustrated in **Figure 36.15a**. This shrub is native to the dry shrub-grassland habitats of southern Eurasia. The micrograph in **Figure 36.15b** shows a cross section through an oleander leaf, which has the following special features:

- The upper surface of oleander leaves is covered by a particularly thick cuticle. This waxy layer helps minimize water loss from cells that are directly exposed to sunlight. In general, species that are adapted to dry soils have much thicker cuticles than do species adapted to wet soils.
- The epidermis of most plants is a single cell layer thick, but the epidermis in oleanders is several cell layers deep. The thick epidermis of species found in dry environments is thought to reduce water loss from palisade parenchyma cells, where most photosynthesis takes place.
- The stomata of oleanders are located on the undersides of their leaves, inside deep pits in the epidermis. These pits are shielded from the atmosphere by hairlike extensions of epidermal cells called trichomes (Section 35.2). The leading hypothesis to explain these traits is that they slow the loss of water vapor from stomata to the dry air surrounding the leaf.

(a) Oleander

(b) Cross section of oleander leaf

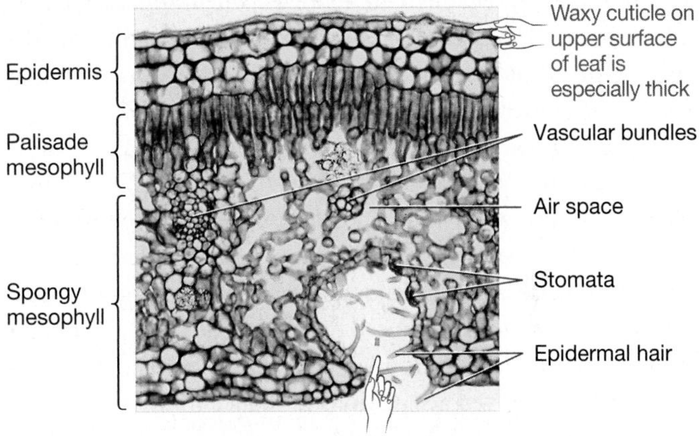

FIGURE 36.15 Species That Are Adapted to Dry Habitats Have Modified Leaf Structures
(a) Oleander grows in desertlike conditions in south-central Asia.
(b) Unlike the stomata of species from wet habitats, oleander stomata are located in pits instead of on the flat leaf surface. Also, the waxy cuticle and layers of epidermal cells on the upper surface of the leaf are particularly thick.

In addition, Chapter 35 noted that many species adapted to water-short habitats—either cold environments where water is often frozen or deserts where rainfall is rare—have needle-like leaves. Such leaf shapes minimize the surface area exposed to sunlight and thus help minimize transpiration. Plants adapted to wetter habitats, in contrast, tend to have broad leaves with a large surface area.

Obtaining Carbon Dioxide under Water Stress Many of the species that thrive in deserts can continue photosynthesizing even when soil moisture content is low. Recall from Chapter 10 that two novel biochemical pathways are found in species native to deserts and other hot, dry habitats: **crassulacean acid metabolism (CAM)** and **C_4 photosynthesis.** Although some of the enzymes involved are different, both CAM and C_4 are mechanisms that allow plants to continue photosynthesizing even if their stomata are closed or partially closed during the day.

CAM plants open their stomata at night and store the CO_2 that diffuses into their tissues, by adding the carbon dioxide molecules to organic acids. When sunlight is available during the day and photosynthesis begins, the CO_2 molecules are released from the organic acids and transferred to **rubisco**—the enzyme that initiates the Calvin cycle. In this way, CAM plants can photosynthesize and grow even with their stomata closed.

C_4 plants, in contrast, can continue photosynthesizing even with their stomata partially closed, because they use CO_2 so efficiently. More specifically, mesophyll cells in C_4 plants take up CO_2 and add it to organic acids. The CO_2 is then transferred to specialized cells called **bundle-sheath cells** (see Chapter 10), where rubisco is abundant. In effect, the C_4 pathway is a mechanism for concentrating carbon dioxide in non-photosynthetic cells, so that CO_2 remains available even when stomata are partially closed. CAM and C_4 photosynthesis are similar to the stomata of oleanders: They are examples of adaptations that help plants conserve water by limiting transpiration.

To summarize, water loss via transpiration is an inevitable side effect of the need to open stomata and obtain CO_2 for photosynthesis. The water that is lost can be replaced by water in the soil without an energy expenditure, however, because transpiration creates a strong enough water potential gradient to pull water up against the force of gravity. Because plants that live in chronically dry soils must limit water losses or risk wilting, they have evolved adaptations to reduce transpiration.

✓CHECK YOUR UNDERSTANDING

Water can move a short distance in xylem via root pressure or capillarity, but long-distance transport of water depends on movement along a steep water potential gradient. This gradient is created primarily by the negative pressure potential of water in leaves, due to surface tension that develops in response to transpiration. You should be able to (1) draw a meniscus like those in the extracellular spaces near spongy mesophyll cells; (2) show how asymmetrical hydrogen bonding and transpiration create a pull on the water in leaves and xylem; and (3) predict what happens to this meniscus when each of the following events occurs: a nearby stoma closes, a nearby stoma opens, a rain shower starts, and weather changes and dry air blows in.

36.4 Translocation

Translocation is the movement of sugars throughout a plant. Sugars move from sources to sinks. In vascular plants, a **source** is a tissue where sugar enters the phloem; a **sink** is a tissue where sugar exits the phloem. Sources contain a high concentration of sugar; sinks have a low concentration of sugar. Thus, translocation is based on the movement of sugar along a concentration gradient.

Where do sources and sinks occur in a plant? The answer often depends on the time of year being considered. During the growing season, mature leaves and stems that are actively photosynthesizing and producing sugar in excess of their own needs act as sources. Sugar moves from these tissues to a variety of sinks, where sugar use is high and production is low: apical meristems, cambium layers (lateral meristems responsible for secondary growth), developing leaves, flowers, developing seeds and fruits, and storage cells in roots (**Figure 36.16**). However, early in the growing season, the situation is different. The storage cells in roots act as sources, and developing leaves act as sinks.

To explore the relationship between sources and sinks in more detail, consider research on sugar-beet plants that were exposed to carbon dioxide molecules containing the radioactive isotope ^{14}C. The goal of these experiments was to track where carbon atoms moved after they were incorporated into sugars via photosynthesis (see Chapter 10). The location of ^{14}C atoms inside a plant can be documented in one of two ways: (1) by measuring the number of radioactive emissions emanating from different tissues or (2) by laying plant parts on X-ray film and allowing the radioactivity to expose and blacken the film.

Do carbon atoms move from sources to sinks, as expected? When a researcher enclosed single leaves of intact plants in a bag and then introduced a fixed amount of radioactive CO_2 for a fixed amount of time, he found that mature leaves retained just over 9 percent of the labeled carbon; growing leaves retained 67 percent. When mature leaves alone were exposed to labeled carbon dioxide, most of the radioactive carbon ended up in young, rapidly growing leaves. These data are consistent with the prediction that fully expanded leaves act as sources of sugar, while actively growing leaves and roots act as sinks. When the researcher exposed all the leaves on a plant to labeled carbon, over 16 percent of the total was translocated to root tissue within 3 hours. This result is consistent with the prediction that, during the growing season, roots act as sinks as well.

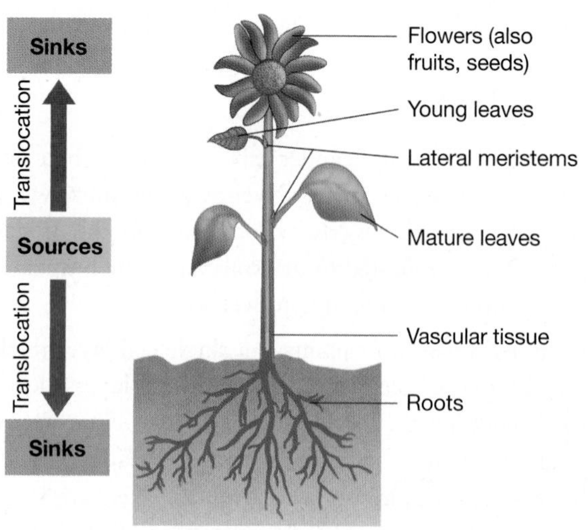

FIGURE 36.16 Sugars Move from Sources to Sinks
Sources are tissues that release sugars to the phloem; sinks are tissues that take up sugars from the phloem. EXERCISE The relationships shown apply during the middle of the growing season. Redraw the figure to indicate the relationship of sources to sinks early in spring, when leaves are first developing and before the plant flowers.

Similar experiments have made two important generalizations possible. First, sugars can be translocated rapidly—typically 50–100 cm/hr. Second, there is a strong correspondence between the physical locations of certain sources and sinks. When mature leaves act as sources, for example, they send sugar to tissues on the same side of the plant (**Figure 36.17a**). In addition, experiments with tall herbaceous plants show that leaves on the upper part of the stem send sugar to apical meristems, but that leaves on the lower part of the plant send sugar to the roots (**Figure 36.17b**). Why would leaves send sugar to tissues on a certain side or part of the body? The answer hinges on understanding the anatomy of phloem.

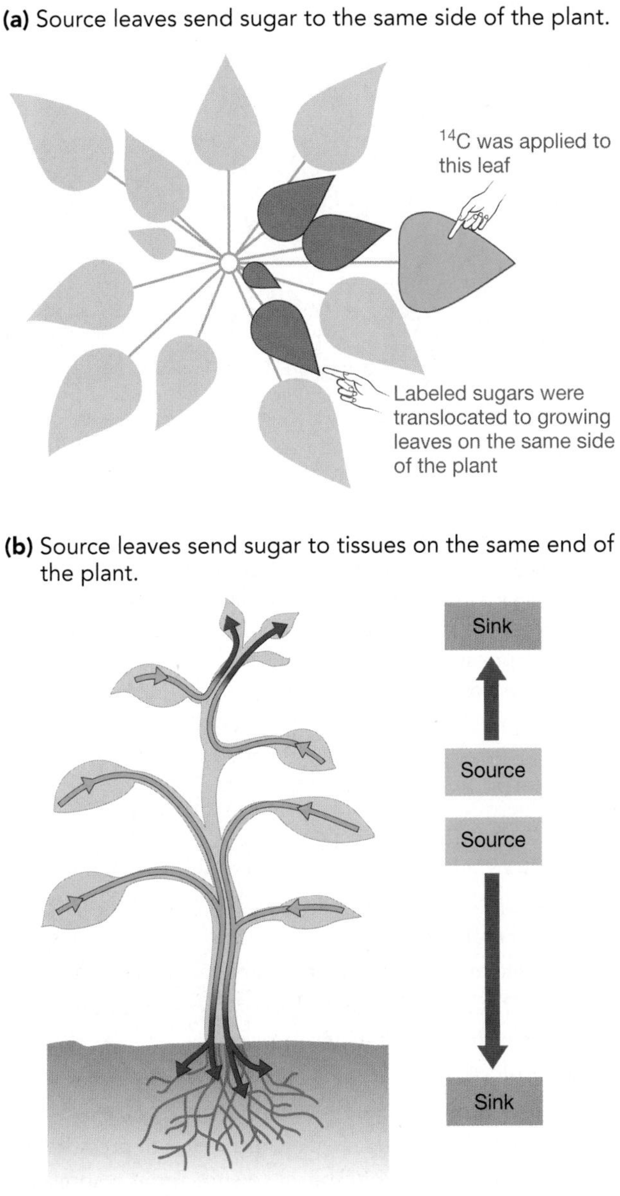
(a) Source leaves send sugar to the same side of the plant.
^{14}C was applied to this leaf
Labeled sugars were translocated to growing leaves on the same side of the plant
(b) Source leaves send sugar to tissues on the same end of the plant.
Sink
Source
Source
Sink

FIGURE 36.17 Sources Supply Sinks on the Same Side and Same End of the Body
Sugar sources supply sinks **(a)** on the same side of the plant and **(b)** on the same end of the plant, because phloem occurs in discrete vascular bundles. Each bundle forms a distinct set of connections inside the plant body.

The Anatomy of Phloem

Chapter 35 introduced the two basic cells types that make up phloem: **sieve-tube members** and **companion cells.** Unlike the tracheids and xylem elements that make up most of the xylem, both sieve-tube members and companion cells are alive at maturity. Recall that, in most plants, sieve-tube members lack nuclei and many major organelles. They are connected to one another, end to end, by perforated structures called **sieve plates** (**Figure 36.18**). The perforations of sieve plates create a direct connection between the cytoplasms of adjacent cells. Companion cells, in contrast, have nuclei and a rich assortment of ribosomes, mitochondria, and other organelles.

Sieve-tube members were discovered by T. Hartig in 1837. Hartig was also the first to establish that a sugary solution called phloem sap moves through phloem. He arrived at this conclusion by removing a ring of bark from a tree trunk during the summer and noting that sugars accumulated above this so-called girdle. It had long been known that cutting away a ring of tree bark would kill a tree; Hartig's result showed why. When phloem tissues are interrupted all the way around a tree, sugars cannot be transported between roots and shoots, and the plant starves.

Subsequent work documented that the phloem present in bark has a high concentration of sugar and that there are direct anatomical connections among the phloem in trunks, stems, branches, and roots. Phloem represents a continuous system for transporting sugar throughout the plant body. Recall from Chapter 35, however, that phloem is restricted to vascular bundles in tissues that do not form wood. Vascular bundles run the length of stems and roots and may branch off into specific branches,

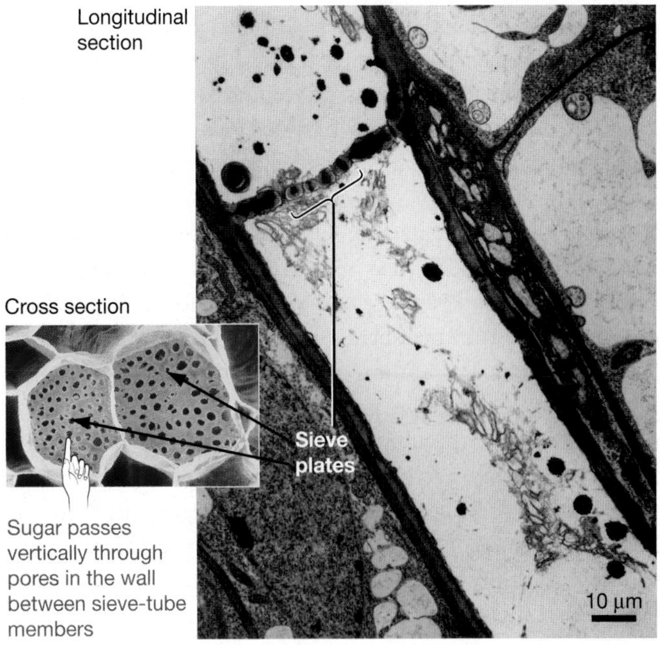
Longitudinal section
Cross section
Sieve plates
Sugar passes vertically through pores in the wall between sieve-tube members
10 µm

FIGURE 36.18 Sieve-Tube Members Are Connected by Pores
Micrographs showing the connections between sieve-tube members in cross section and in longitudinal section.

leaves, and lateral roots. But phloem sap does not move from one vascular bundle to another. Each bundle is independent.

Based on these results, the physical relationships observed between sources and sinks are logical. For example, the phloem in the leaves on one side of a plant connects directly with the phloem of branches, stems, and roots on the same side of the individual, through a specific set of vascular bundles.

By 1930 most major physical components of the sugar transport system in angiosperms had been described in detail, and chemical analyses had shown that phloem sap is often dominated by the disaccharide *sucrose*. Sucrose is found in table sugar and maple syrup, which is actually concentrated phloem sap. Phloem sap can contain small amounts of minerals, amino acids, hormones, and other compounds as well. Now, how do sieve tubes and companion cells interact? The striking morphological differences between the two cell types inspired the hypothesis that sieve-tube members are the conducting element in phloem and that companion cells support them in some way. Is this correct? And what mechanism is responsible for translocating sugars from sources to sinks?

The Pressure-Flow Hypothesis

Ernst Münch proposed a mechanism for translocation in 1926. His idea, called the **pressure-flow hypothesis**, is diagrammed in **Figure 36.19**. The fundamental idea is that events at source tissues and at sink tissues create a steep pressure potential gradient in phloem. The water in phloem sap moves down this gradient, and sugar molecules are carried along by bulk flow. Like the cohesion-tension theory for water transport, the pressure-flow hypothesis is based on movement along a water potential gradient—specifically, a gradient created by changes in pressure potential. Unlike the cohesion-tension theory for water transport, however, the driving force for movement in phloem sap is not provided by transpiration. Instead, the force responsible for movement is generated by large differences in the turgor pressure found in the phloem in source tissues and that found in the phloem in sink tissues.

To understand how Münch's model works, take a closer look at Figure 36.19, starting with the source cell at the upper right. The small red arrows reflect Münch's proposal that sucrose moves from source cells into companion cells and from there into sieve-tube members. As a consequence of this **phloem loading**, the phloem sap near the source has a high concentration of sucrose, resulting in a very low water potential compared with that of the adjacent xylem cells in a vascular bundle. As the blue arrows in the diagram show, water moves from xylem across the selectively permeable plasma membrane of sieve-tube members along a water potential gradient. In response, turgor pressure begins to build in the sieve-tube members nearest the source region.

What is happening at the sink? Münch proposed that cells in the sink (bottom right in Figure 36.19) remove sucrose from the phloem sap. As a result of this **phloem unloading**, the water

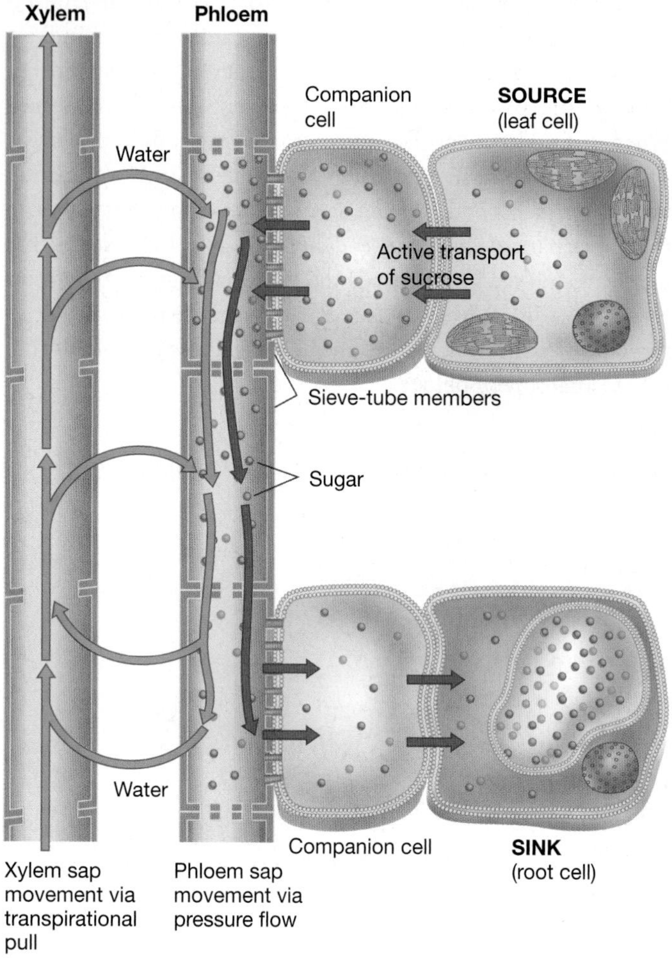

FIGURE 36.19 The Pressure-Flow Hypothesis: High Turgor Pressure Near Sources Causes Phloem Sap to Flow to Sinks
The pressure-flow hypothesis predicts that water cycles between xylem and phloem and that water movement in phloem is a response to a gradient in pressure potential. **EXERCISE** This diagram shows how water moves between xylem and phloem in the middle of the growing season, when leaves are sources and roots are sinks. Add arrowheads to indicate the direction of water and phloem sap flow in spring, when roots act as sources and leaves act as sinks.

potential in sieve-tube members increases until it is higher than the water potential in adjacent xylem cells. As the blue arrow at the bottom of Figure 36.19 shows, water flows across the selectively permeable membranes of sieve-tube members into xylem along a water potential gradient. Turgor pressure in the sieve-tube members near the sink drops in response.

The net result of these events is high turgor pressure near the source and low turgor pressure near the sink, created by the loading and unloading of sugars. This difference in pressure potential drives phloem sap from source to sink and is responsible for a one-way flow of sucrose molecules. All these events produce a continuous loop of water flow, with water supply from the xylem. The cycle illustrated in Figure 36.19 is driven by water potential gradients between xylem and phloem and pressure potential gradients within phloem. In spring, when roots

(a) Just after fluorescent dye is added to phloem sap

(b) Later: fluorescent dye flowing through phloem

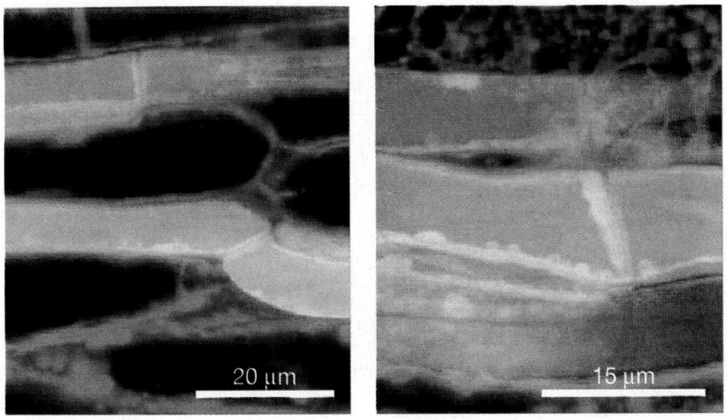

FIGURE 36.20 Direct Observation of Bulk Flow through Sieve-Tube Members
Confocal micrographs showing phloem cells in leaves. **(a)** Immediately after a green fluorescent dye was injected into phloem. **(b)** Researchers could see the green-stained phloem sap flow through sieve-tube members and sieve plates.

act as sources and growing leaves act as sinks, the same cycle occurs in the opposite direction.

The pressure-flow hypothesis is logical, given the anatomy of vascular tissue and the principles that govern water movement. But has any experimental work supported the theory? One recent test was made possible by the development of the confocal microscope. Recall from Chapter 7 that confocal microscopy focuses ultraviolet light at a specific depth in a sample of living tissue that contains fluorescent dyes. Confocal microscopes allow researchers to look at tissues layer by layer. As **Figure 36.20** shows, biologists used a confocal microscope to examine a leaf. They cut a window into a major vein in the leaf, injected the leaf tissue with fluorescent dye, and observed the movement of phloem sap in real time. This experiment provided unequivocal evidence for bulk flow in sieve tubes. An important prediction of the pressure-flow hypothesis was confirmed. But what about the mechanism of phloem loading and unloading? How does sucrose enter and leave phloem in a way that sets up the water potential gradient?

Phloem Loading

In contrast to the cohesion-tension theory for water movement in plants, the pressure-flow hypothesis contends that plants have to expend energy to set up a water potential gradient in phloem. In most crop plants, sugar is loaded into sieve-tube members against a concentration gradient. As a result, loading requires an expenditure of ATP and some sort of membrane transport system. Conversely, sugar must sometimes be unloaded against its concentration gradient at sinks. Unloading requires another expenditure of ATP and a second membrane transport mechanism. Given that phloem loading and unloading require an expenditure of ATP, how does each event occur? What specific membrane proteins are involved?

To answer these questions, let's look first at events at source tissues, where the active transport of sucrose into sieve-tube members results in a high pressure potential and high water potential. The chapter concludes with a look at how the same molecules are offloaded to maintain low water potentials at sinks.

How Are Sugars Concentrated in Sieve-Tube Members at Sources? Molecules and ions can be transported across membranes in a variety of ways. Recall from Chapter 6 that transport is said to be *passive* if ions or molecules move along their concentration gradients and *active* if substances move against their concentration gradients. **Box 36.3** summarizes the mechanisms responsible for passive transport and active transport.

BOX 36.3 How Are Solutes Transported across Membranes?

Passive transport (**Figure 36.21a**, page 846) occurs when ions or molecules move across a plasma membrane by diffusion—that is, with their electrochemical gradient. The adjective *passive* is appropriate because no expenditure of energy is required for the movement to occur.

Recall from Chapter 6 that nonpolar or very small molecules diffuse across phospholipid bilayers rapidly (Figure 36.21a, left). But ions and many large molecules diffuse across phospholipid bilayers very slowly, even when their movement is favored by a strong electrochemical gradient. For charged or large compounds to diffuse across a plasma membrane rapidly, they must avoid the phospholipid bilayer by passing through a membrane protein. Two general types of membrane protein permit the passive diffusion of specific ions or molecules. **Channels** form pores that selectively admit certain ions (Figure 36.21a, middle). **Carriers** are also transmembrane proteins; but instead of forming a pore, they work like enzymes, undergoing a conformational change that transports a bound substrate across the lipid bilayer (Figure 36.21a, right). Carriers are responsible for **facilitated diffusion.**

Active transport (**Figure 36.21b**) occurs when ions or molecules move across a plasma membrane against their electrochemical gradient. The adjective *active* is appropriate because cells must expend energy in the form of ATP to move solutes in an energetically unfavorable direction.

Active transport always involves membrane proteins and can occur di-

(Continued on next page)

(Box 36.3 continued)

rectly or indirectly. **Pumps** are proteins that change conformation when they bind ATP or a phosphate group from ATP and transport ions or molecules against their concentration gradient as they move (Figure 36.21b, left). Chapter 6 introduced the most important pump that is in animal cells: the sodium-potassium pump, which is also known as Na^+/K^+-ATPase. This chapter introduces the most important pumps that are found in the proton pumps, or H^+-ATPases.

In many cases, electrochemical gradients established by pumps are used to transport specific molecules or ions indirectly. Cotransporters, or *symporters*, are membrane proteins that use the energy released when an ion is transported down its electrochemical gradient—a gradient that is established by a pump—to transport a different ion or molecule in the same direction but against its concentration gradient (Figure 36.21b, middle). *Antiporters* work in a similar way, except that the solute that is transported against its concentration gradient moves in the direction *opposite* that of the solute moving down its concentration gradient (Figure 36.21b, right).

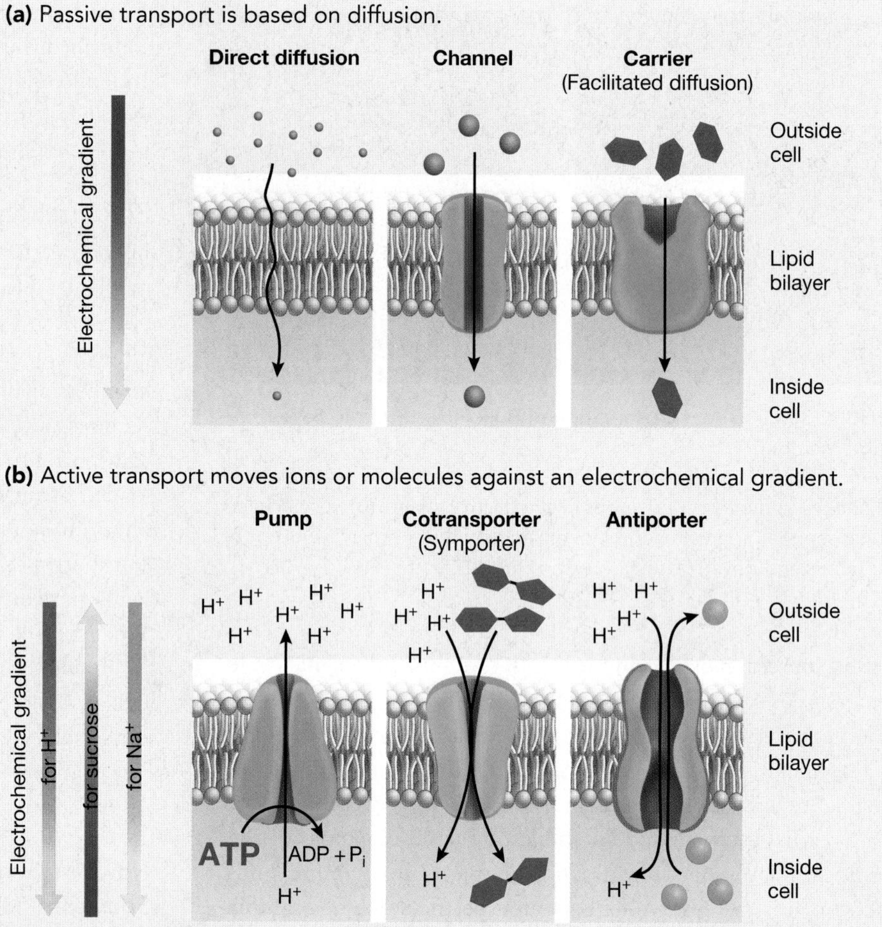
(a) Passive transport is based on diffusion.
Direct diffusion
Channel
Carrier (Facilitated diffusion)
Electrochemical gradient
Outside cell
Lipid bilayer
Inside cell
(b) Active transport moves ions or molecules against an electrochemical gradient.
Pump
Cotransporter (Symporter)
Antiporter
Electrochemical gradient
for H+
for sucrose
for Na+
H+
ATP
ADP + Pi
Outside cell
Lipid bilayer
Inside cell

FIGURE 36.21 Passive and Active Transport Occur via Several Mechanisms
(a) In passive transport, ions or molecules diffuse across membranes—meaning they follow their electrochemical gradient. The movement can occur directly through the phospholipid bilayer or via a channel or carrier protein. **(b)** In active transport, ions or molecules cross a membrane against their electrochemical gradient.

Because measurements showed that sucrose is found in companion cells in higher concentrations than that found in mesophyll cells where it is produced, researchers hypothesized that sucrose transport must be active.

A key observation inspired a hypothesis for how active transport occurs in companion cells: Strong pH differences exist between the interior and exterior of phloem cells. These data suggested that sucrose might enter phloem cells along with protons. **Figure 36.22** explains the logic behind this proposal. The idea was that a membrane protein in phloem cells hydrolyzes ATP and uses the energy released to transport protons (H^+) across the membrane to the exterior of the cell. Such proteins are called **proton pumps**, or more formally **H^+-ATPases.** Proton pumps establish a large difference in charge and in hydrogen ion concentration on the two sides of the membrane. The resulting electrochemical gradient favors the entry of protons into the cell. A second protein, called a **cotransporter**, acts as a conduit for protons to enter the cell. A cotransporter uses the energy released when protons flow *down* their gradient to bring a second molecule into the cell *against* its concentration gradient. In the case of phloem, the second molecule is sucrose.

If phloem loading depends on the activity of a proton pump, then researchers should be able to find and characterize the pump proteins.

Where Are H^+-ATPases Located? Proton pumps are found in the plasma membranes of a wide variety of organisms, including bacteria, fungi, and animals. To analyze the proton pumps that

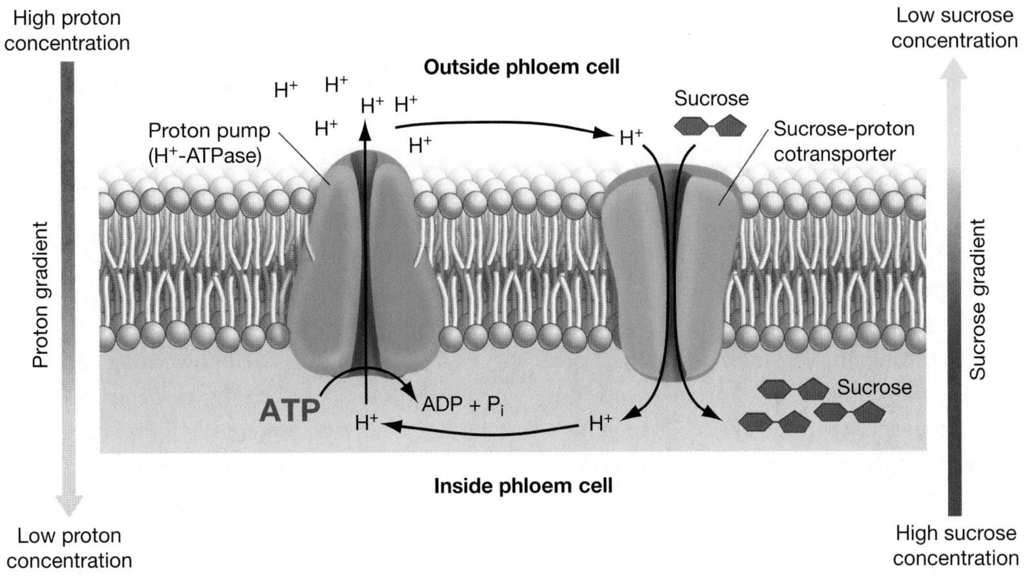

FIGURE 36.22 A Model for Cotransport of Protons and Sucrose
According to the model of cotransport, a proton pump hydrolyzes ATP to move hydrogen ions into the exterior of the cell. The high concentration of H^+ establishes an electrochemical gradient that allows the transport of sucrose into the cell against its concentration gradient.

are found in plants, researchers from several laboratories began focusing on a small member of the mustard family called *Arabidopsis thaliana*. Recall from Chapter 29 that this species serves as a model organism for studies in plant molecular genetics and plant genomics. It is a superb subject for experimental work because it is small, grows and reproduces quickly, and is inexpensive to maintain.

Researchers determined the amino acid sequence of a proton pump purified from the plasma membranes of *Arabidopsis* and used these data to infer the DNA sequence of the corresponding gene. The biologists found that the *Arabidopsis* genome actually codes for 10 different pump proteins. One of these genes, called *AHA3*, appeared to be expressed primarily in vascular tissues.

These observations led Natalie DeWitt and Michael Sussman to hypothesize that *AHA3* encodes the proton pump responsible for phloem loading. To test this hypothesis, they produced antibodies to the *AHA3* protein. Recall from Chapter 18 that an antibody is a protein that binds to a specific location on another molecule—typically, a specific protein. To visualize the antibody, the investigators attached gold particles to it. When viewed with the electron microscope, a gold particle looks like a black dot. DeWitt and Sussman's goal was to treat *Arabidopsis* leaves with the *AHA3* antibody, examine treated leaves under the electron microscope, and determine exactly where the pump proteins are located.

Figure 36.23 shows the resulting data. The key observation is that the proton pumps responsible for phloem loading are found almost exclusively in the membranes of companion

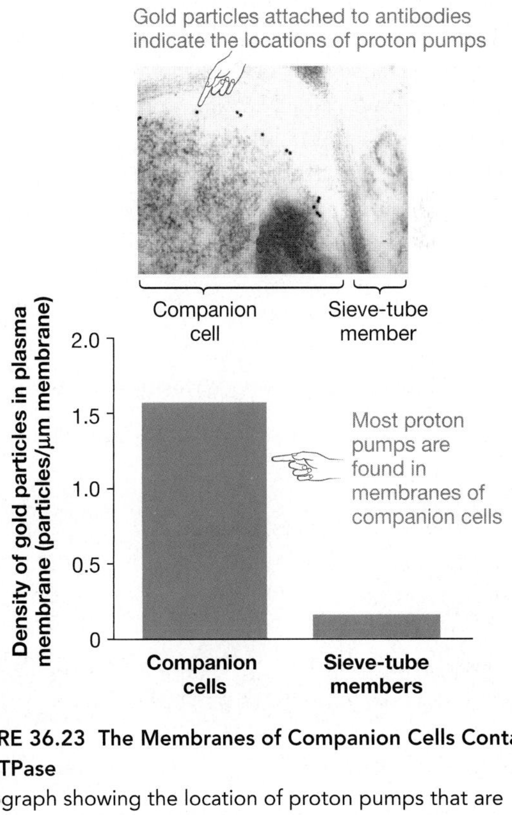

FIGURE 36.23 The Membranes of Companion Cells Contain an H^+-ATPase
Micrograph showing the location of proton pumps that are responsible for phloem loading—in the plasma membranes of companion cells. The histogram compares the average density of proton pumps in the membranes of companion cells versus the membranes of sieve-tube members. These pumps create a proton gradient that allows sucrose to be transported into companion cells against a concentration gradient.

cells. This result supported the following model for phloem loading: (1) proton pumps in the membranes of companion cells create a strong gradient that favors a flow of protons into companion cells; (2) a cotransporter protein in the membranes of companion cells uses the proton gradient to bring sucrose into companion cells from the surrounding cell walls and intercellular spaces around source cells; and (3) once inside companion cells, sucrose travels into sieve-tube members via a direct cytoplasmic connection.

Although work on the mechanism of phloem loading in *Arabidopsis* and other species continues, most researchers are convinced that proton pumps and proton-sucrose cotransporters play a key role. Now, once sucrose has been loaded into sieve-tube members near sources and follows a water potential gradient to sinks, how is it unloaded?

Phloem Unloading

Research on the mechanism of phloem unloading at sinks can be summarized with one general statement: The membrane proteins that are involved in transporting sucrose molecules and the mechanism of movement vary among different types of sinks within the same plant, as well as among different species.

To appreciate this diversity, consider how sucrose is unloaded in the tissues of sugar beets—a crop grown for the storage tissues in its root, which are a major source of the granulated and powdered sugar that are sold in grocery stores. In sugar beets, sucrose crosses the membranes of cells in young, growing leaves along a concentration gradient. This passive transport occurs because sucrose is rapidly used up inside the cells to provide carbon for the synthesis of proteins, nucleic acids, and phospholipids needed by the growing cells (**Figure 36.24a**). In the roots of the same plant, however, an entirely different mechanism is responsible for offloading sucrose. Root cells in this species have a large vacuole that stores sucrose. The membrane surrounding this organelle is called the **tonoplast.** It contains a protein that hydrolyzes ATP and uses the energy released to transport sucrose into the organelle, against its concentration gradient (**Figure 36.24b**). The movement of sucrose into the vacuole keeps the water potential of the nearby phloem sap low, as required by the Münch pressure-flow model.

In short, more than seven decades of research on plants has provided convincing evidence that the pressure-flow hypothesis is fundamentally correct. But understanding how translocation occurs at the molecular level remains an active area of research.

✓CHECK YOUR UNDERSTANDING

Phloem sap moves from areas of high water potential to areas of low water potential. In phloem, high water potential is due to the high turgor pressure observed in the sieve-tube members near source cells. This pressure is created by pumps that actively load sucrose into companion cells against a concentration gradient. At sinks, turgor pressure is much lower than it is at sources because, at sinks, sucrose is removed from phloem sap by growing tissues or by storage cells. You should be able to predict the water potential of phloem sap near leaves and near roots under each of the following conditions: (1) at the start of growth in spring; (2) at midday when growing conditions are ideal in summer; (3) in midwinter; and (4) just after the first frost in autumn.

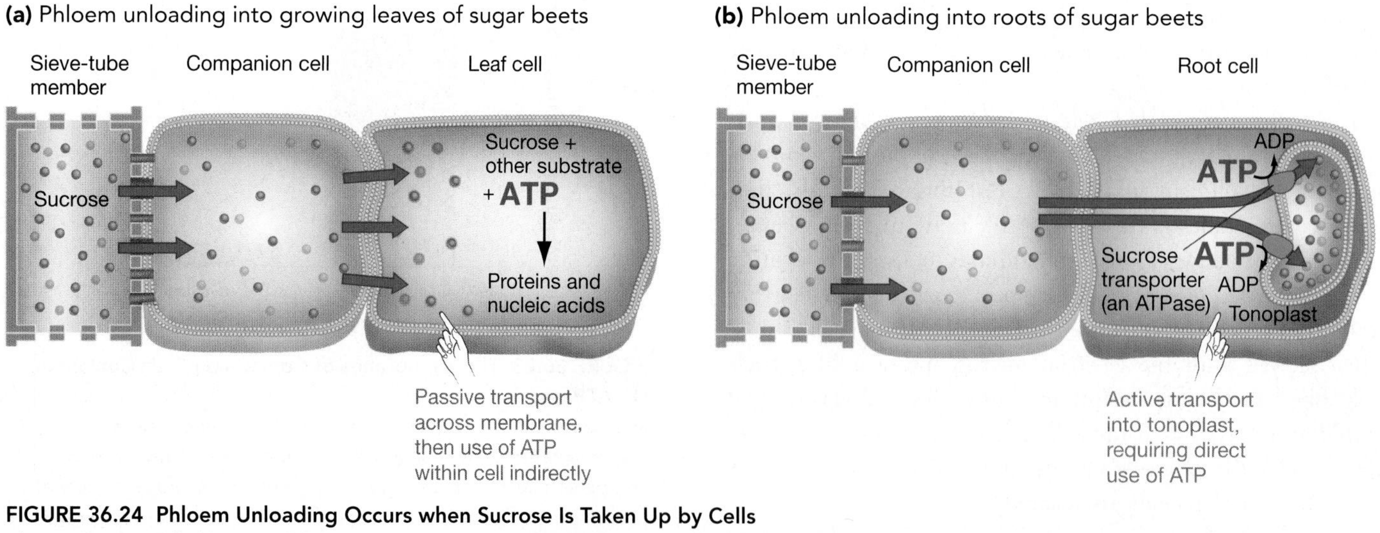
(a) Phloem unloading into growing leaves of sugar beets
Sieve-tube member
Companion cell
Leaf cell
Sucrose
Sucrose + other substrate + ATP
Proteins and nucleic acids
Passive transport across membrane, then use of ATP within cell indirectly
(b) Phloem unloading into roots of sugar beets
Sieve-tube member
Companion cell
Root cell
Sucrose
ADP
ATP
Sucrose transporter (an ATPase)
ATP
ADP
Tonoplast
Active transport into tonoplast, requiring direct use of ATP

FIGURE 36.24 Phloem Unloading Occurs when Sucrose Is Taken Up by Cells
The mechanism of phloem unloading can vary from sink to sink, such as **(a)** young leaves and **(b)** roots within the same plant. EXERCISE In each part, add arrows indicating the concentration gradient that favors the movement of sucrose.

ESSAY Irrigated Agriculture

Most of the crop plant varieties in use today have undergone extensive artificial selection aimed at increasing their yields. The yield or productivity of plants is largely a function of their photosynthetic rate. The rate of photosynthesis, in turn, hinges on a plant's ability to acquire carbon dioxide through its stomata. But open stomata mean increased water loss through transpiration. This is a long-winded way of making a simple point: Crop plants transpire a lot.

How much do crop plants transpire? Through transpiration and evaporation, one hectare of corn growing in midwestern North America uses about 6 million liters of water during a growing season. A hectare is about the size of two football fields; 6 million liters of water is enough to cover this area to a depth of about 2 feet. High-yielding varieties of rice require much more water—up to 18 million liters per hectare.

The correlation between water availability and productivity is strong. Just 16 percent of the world's farmland is irrigated, but irrigated crops account for 33 percent of the world's food. Rice grown in flooded fields produces 50 percent more food than does rice grown under sprinkler irrigation.

It's not surprising, then, that irrigation is frequently promoted as a way to use farmland efficiently, boost productivity, and raise standards of living in food-poor countries. Irrigation is not without costs, however.

The first issue is simply the availability of water. Agriculture already accounts for 87 percent of the water consumed for human use in the United States, and some sources are drying up. The groundwater reservoir underlying south-central North America is tapped extensively to water corn; it is now half the size it was before irrigation began a few decades ago. The Colorado River is used so intensively that it is barely a trickle by the time it reaches the Pacific Ocean. In the Middle East, disputes over access to water are exacerbating political tensions among nations.

Irrigation can also cause significant problems with soil quality, especially in arid regions. When ion-rich water is used to irrigate crops, evaporation leaves large quantities of salt and other inorganic compounds behind. These compounds dissolve in the remaining soil water and lower its solute potential. In extreme cases, the water potential of irrigated soil becomes so low that root cells lose water to the soil instead of gaining it. When this occurs, the land is infertile and has to be abandoned. At the end of the third Punic War with Carthage in 146 B.C., the victorious Romans razed the city and sowed the fields with salt to make them unproductive.

How much do crop plants transpire?

In response to increasing problems with salt buildup, plant breeders are working to develop varieties of plants that require less water or that can tolerate more salinity. In one approach, breeders have inserted a gene into crop plants that codes for the enzyme required to make a molecule called mannitol. Mannitol is sequestered in plant cells and helps them maintain a low water potential. In salty soils, these plants grow substantially better than normal varieties. In another approach, researchers have isolated the gene for a protein that transports salt across the tonoplast into vacuoles. When this gene has been introduced into crop plants, the genetically modified forms have been able to store salt and grow reasonably well in saline soils. In addition, some researchers are trying to reclaim salty soils by growing halophytic plants that sequester salt in their tissue, harvesting the plants, and then disposing of them as a way of lowering salt concentrations in soil.

Only time will tell whether breeders and farmers can successfully outrace the trend toward increasing salt concentrations in irrigated soils worldwide. Meanwhile, the buildup of salts in soil is emerging as a major agricultural and ecological problem.

CHAPTER REVIEW

Summary of Key Concepts

- **Water moves along potential energy gradients—from areas of high water potential to areas of low water potential. Water's potential energy in plants is a combination of (1) its tendency to move via osmosis and (2) the pressure exerted on it.**

  Plants lose water as an inevitable side effect of exchanging gases with the atmosphere. The flow of water from soil to air via plant tissues follows a water potential gradient. Water potential (ψ) is a measure of the tendency of water to move down its potential energy gradient. In plants, water potential has two components: (1) a solute, or osmotic, potential, created by the concentration of solutes in a cell or tissue; and (2) a pressure potential, provided by the cell wall and other factors. The water potential of a cell, tissue, or plant is the sum of its solute potential and pressure potential. When selectively permeable membranes are present, water moves by osmosis from areas of high potential to areas of low potential. When no membranes are present, water moves by bulk flow from areas of high pressure to areas of low pressure, independently of differences in solute potential.

Plants do not expend energy to replace water that is lost to transpiration when stomata are open and photosynthesis is occurring. Instead, water moves from soil and roots to leaves along a water potential gradient. The gradient exists because water at the air-water surface in leaves is under negative pressure (tension) that is great enough to pull water up from the roots through xylem.

According to the cohesion-tension theory, water lost from stomata is replaced without any expenditure of energy by the plant. Water is pulled in one continuous column from the soil to roots to shoots, against the force of gravity, by the surface tension caused by transpiration from leaves. The surface tension occurs at menisci that form as water evaporates from the walls of leaf cells. In this way, the energy in sunlight is responsible for the movement of water from roots to shoots.

Several types of data support predictions made by the cohesion-tension theory. For example, on hot, sunny days when transpiration is rapid, xylem sap is under enough tension to make tree trunks shrink slightly in diameter. The measured water potential of leaf tissue drops during the day, when stomata are open; it rises at night, when stomata are closed. Finally, direct measurements show that the tension on xylem sap changes rapidly in response to changes in light intensity and thus in transpiration rate. All these results support the cohesion-tension theory, because they are consistent with the prediction that xylem sap is under tension.

Plants that occupy dry habitats have several ways of limiting the amount of water they lose to transpiration. In some species, stomata are located in pits on the undersides of their leaves, helping protect the pits from the dry atmosphere. There is also evidence that the tissues of dry-habitat plants can tolerate higher solute potentials than can tissues of wet-habitat plants. As a result, dry-habitat plants have lower water potentials and can compete more successfully for scarce water in the soil. The CAM and C_4 photosynthetic pathways are also interpreted as adaptations that limit water loss in dry habitats.

In phloem, sugars are transported from "sources"—tissues in which sugar is abundant—to "sinks"—tissues in which sugars are being used or stored. Movement occurs because cells expend energy to load sucrose actively into phloem cells near source tissues, creating a pressure gradient that favors the movement of water and sucrose to sinks.

Translocation, the movement of sucrose and other products through the plant, has been studied by labeling sugars with ^{14}C. According to the Münch pressure-flow model, sugars move from sources to sinks via bulk flow along a pressure gradient that develops in the sieve tubes. The gradient is generated by the transport of sugars into sieve-tube members in source tissues, coupled with the transport of sucrose out of sieve-tube members at sink tissues. Water moves from xylem into sieve-tube members near sources and cycles back to xylem near sinks. The membrane proteins responsible for phloem loading in *Arabidopsis* have now been isolated and studied in detail. They include a proton pump and a proton-sucrose cotransporter.

Web Tutorial 36.1 Solute Transport in Plants

Questions

Content Review

1. Under what conditions does the rate of transpiration increase?
   a. in species in which stomata are located in pits on the bottom of leaves
   b. when stomata close at night
   c. during rainstorms, when atmospheric pressure is low
   d. when the weather changes and air becomes drier

2. Which of the following does *not* affect the pressure potential in water?
   a. High sucrose concentrations in companion cells create turgor pressure.
   b. Menisci around spongy mesophyll cells create tension when transpiration rates are high.
   c. Few ions or other solutes are present in soil water.
   d. Water tends to adhere tightly to soil particles when soils dry.

3. The cells of a certain plant species can tolerate extremely high solute potentials. Which one of the following is correct?
   a. Transpiration rates will tend to be extremely low.
   b. It can compete for water effectively and live in dry soils.
   c. It will grow most effectively in soils that are saturated with water year round.
   d. Leaves will wilt easily.

4. What forces are responsible for capillarity?
   a. adhesion of water molecules to the sides of xylem cells, cohesion of water molecules to each other, and surface tension
   b. surface tension created by transpiration and cohesion of water molecules in a continuous flow from leaf to root
   c. high solute potentials created by the entry of ions during the night, when transpiration rates are low, followed by an influx of water
   d. gravity and wall pressure (from the sides of xylem cells)

5. What is a proton pump?
   a. a membrane protein that transports sucrose against a concentration gradient
   b. a membrane protein that transports protons against an electrochemical gradient
   c. a membrane protein that transports protons *with* an electrochemical gradient and sucrose *against* a concentration gradient
   d. any membrane protein that acts as a channel—meaning it does not consume ATP

6. Why is the transport of phloem sap considered an active process?
   a. The manufacture of sucrose via photosynthesis is driven by the energy in sunlight.
   b. Transpiration is driven by the energy in sunlight.
   c. Sucrose is transported into companion cells near sources against a concentration gradient.
   d. In spring, phloem sap moves against the force of gravity.

Conceptual Review

1. Water flows from regions of high water potential to regions of low water potential. Why does this happen?
2. Compare and contrast the forces involved in transporting water in xylem via capillarity, root pressure, and transpiration. Which of these mechanisms is passive? Can they occur at the same time, or are they mutually exclusive?
3. Why are "cohesion-tension" and "pressure-flow" sensible names for the hypotheses analyzed in this chapter?
4. Aphids are insects that prey on plants by inserting an organ called a stylet into phloem and harvesting sap. Suppose Alice and Bernadette are two aphids sitting on the same plant. Alice inserts her stylet into the phloem of a large, mature leaf, while Bernadette prefers to probe the phloem of the young growing tissue near the shoot's apical meristem. Which aphid is attacking a source, and which is preying on a sink? Which aphid is getting a higher concentration of sugar, Alice or Bernadette? Where would you expect aphids to be found on a plant growing in the wild?
5. How does cotransport result in phloem loading? Where does this process occur? What data support your answers to these questions?
6. How does a pressure bomb work, and what does it measure?

Group Discussion Problems

1. The text claims that water loss is an inevitable side effect of gas exchange. What data or observations support or challenge this claim? Would the same statement be true in bacteria? In animals?
2. Suppose that you discovered a mutant pea plant that is unable to make the dominant wax found in cuticle. In terms of water balance, what are the physiological consequences of this mutation?
3. When trees grow on the edge of a forest and a field, one side of their body is partially shaded while the other is exposed to full sunlight. The growth rings in the trunks of such trees are often asymmetric—they tend to be fatter on the sunny side of the trunk and skinnier on the shaded side. Why is this observation logical, given what you've learned about how sugars are translocated from sources to sinks?
4. A seed is a sink when it is developing. When is it a source?

Answers to Multiple-Choice Questions 1. d; 2. c; 3. b; 4. a; 5. b; 6. c

www.prenhall.com/freeman is your resource for the following: Web Tutorials; Online Quizzes and other Online Study Guide materials; Answers to Conceptual Review Questions; Solutions to Group Discussion Problems; Answers to Figure Caption Questions and Exercises; and Additional Readings and Research.

37 Plant Nutrition

KEY CONCEPTS

- In addition to needing carbon dioxide and water, plants require an array of essential nutrients to support growth. These nutrients are available as ions dissolved in soil water and are taken up by roots.
- Nutrient absorption occurs via specialized proteins in root membranes. Most plants also obtain nitrogen or phosphorus from fungi associated with their roots. Toxins that enter roots are excluded from xylem by the Casparian strip or are actively transported into cell vacuoles and stored.
- Some species of plant have specialized methods of obtaining nutrients, including associations with nitrogen-fixing bacteria, parasitism, and carnivory.

Plants obtain most of the nutrients they need from soil, through their root systems.

Obtaining carbon-containing molecules and chemical energy is the most urgent task facing all organisms. Plants fulfill both requirements by producing sugar through the process of photosynthesis. Yet plants cannot live on sugar alone. In addition to making the carbohydrates they need, plants synthesize all of their own nucleic acids, amino acids, enzymes, chlorophylls, enzyme cofactors, and other macromolecules. Plants can do some of the world's most impressive synthetic organic chemistry.

A plant's ability to perform sophisticated reactions depends on its capacity to harvest a wide variety of simple ions and elements as raw materials. In addition to needing carbon dioxide and water, plants have to obtain nitrogen, phosphorus, potassium, sulfur, magnesium, and other elements. Most of these building blocks are available in the soil. The majority exist as ions that are dissolved in soil water at low concentrations—sometimes extremely low concentrations.

The plant body is an efficient machine for harvesting these types of diffuse resources and concentrating them in cells and tissues. Chapter 35 introduced how the organization and growth of the plant root and shoot systems make resource acquisition possible; Chapter 36 focused on how water and nutrients are transported throughout the body. In this chapter, we'll concentrate on how plants take up simple ions and elements from the soil, so that these nutrients can be transported to the cells that need them. Questions about nutrition are fundamental to understanding how plants work and how people who grow plants—from farmers, horticulturists, and foresters to backyard gardeners—manage their land. Let's begin by analyzing the basic nutritional needs of plants—the equivalent of their minimum daily requirements.

37.1 Nutritional Requirements

What do plants need to live? In the early 1600s, Jean-Baptiste van Helmont performed a classic experiment that started a long research tradition focused on answering this question. Van Helmont wanted to know where the mass of a growing plant comes from, and he used a willow tree as a study organism. He began the study by weighing out 200 pounds of soil and placing it in a pot with a 5-pound willow sapling (**Figure 37.1**). He allowed the plant to grow for five years, adding only water. At the end of the experiment, he weighed the willow and the soil. The willow weighed 169 lb, 3 ounces; the soil weighed 199 lb, 14 ounces.

Where had the additional 164 pounds, 3 ounces, of tree come from? Because he was not aware that gases have mass, van Helmont concluded that the new plant material came from water. He also ignored the loss of 2 ounces in the soil, chalking it up to measurement error. As it turned out, his measurements were not the problem—his conclusions were. Follow-up work by other researchers established that most of the mass of the tree came from carbon dioxide in the atmosphere. The 2 ounces removed from the soil contained vital ions and elements—the nutrients that are the focus of this chapter. What are they?

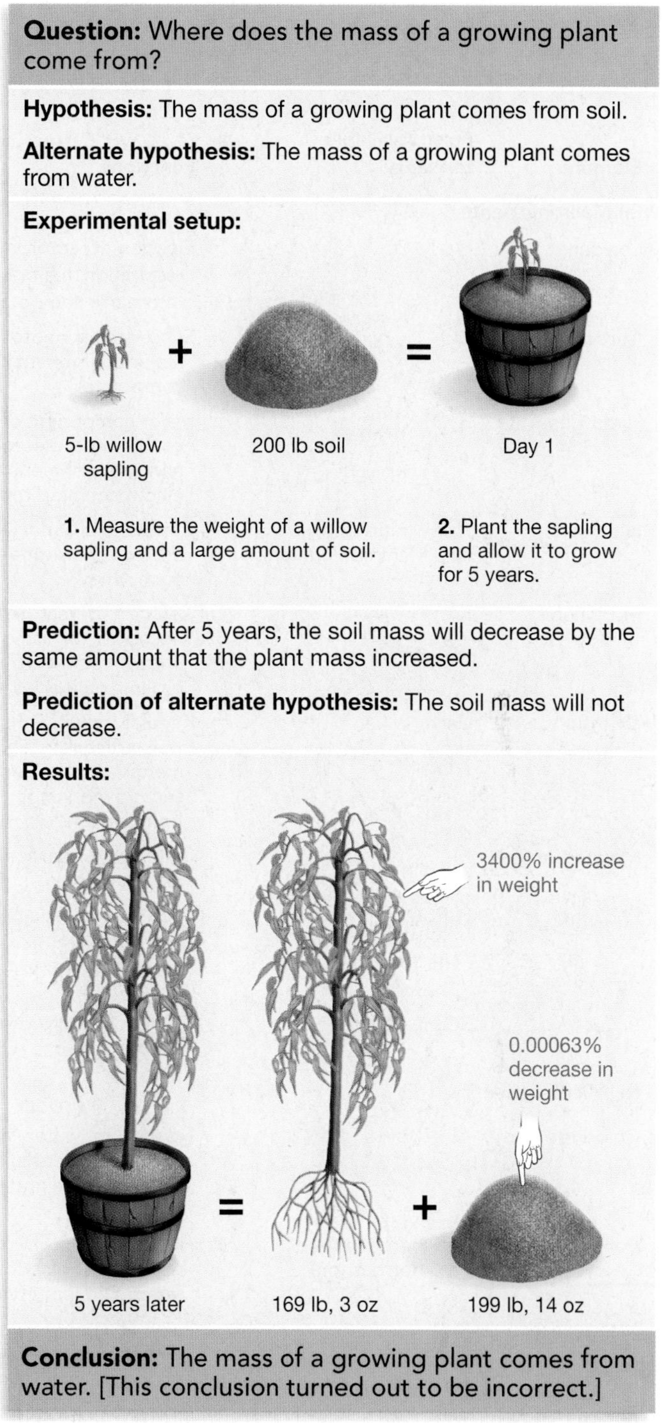

Question: Where does the mass of a growing plant come from?
Hypothesis: The mass of a growing plant comes from soil.
Alternate hypothesis: The mass of a growing plant comes from water.
Experimental setup:
5-lb willow sapling
200 lb soil
Day 1
1. Measure the weight of a willow sapling and a large amount of soil.
2. Plant the sapling and allow it to grow for 5 years.
Prediction: After 5 years, the soil mass will decrease by the same amount that the plant mass increased.
Prediction of alternate hypothesis: The soil mass will not decrease.
Results:
3400% increase in weight
0.00063% decrease in weight
5 years later
169 lb, 3 oz
199 lb, 14 oz
Conclusion: The mass of a growing plant comes from water. [This conclusion turned out to be incorrect.]

FIGURE 37.1 An Early Experiment on the Role of Soil in Plant Nutrition

Essential Nutrients

About half the elements in the periodic table—more than 60 in all—can be found in the tissues of at least one of the plant species analyzed to date. The question that farmers, gardeners, and biologists ask is, which of these elements are essential for growth and reproduction, and in what relative quantities? To answer it, biologists define an **essential nutrient** as an element that fulfills both of the following criteria:

1. It is required for normal growth and reproduction—the plant cannot complete its life cycle without this nutrient; and
2. It is required for a specific structure or metabolic function.

By denying specific nutrients to plants and documenting the effects or lack of effect, biologists have determined that seventeen elements are essential for most vascular plants. Just three of these—carbon, hydrogen, and oxygen—typically make up about 96 percent of the dry weight of a plant.

Although different classification schemes for the essential elements have been proposed, perhaps the most common is to group them as macronutrients and micronutrients. **Macronutrients** are the building blocks of nucleic acids, proteins, carbohydrates, phospholipids, and other key molecules and are required in relatively large quantities. The first three macronutrients listed in **Table 37.1**—oxygen, carbon, and hydrogen—are available in the atmosphere and from water. The other macronutrients have to be extracted from soil.

Nitrogen (N), phosphorus (P), and potassium (K) are macronutrients that commonly act as **limiting nutrients**, meaning that plant growth is limited by their availability. If N, P, and/or K are added to soil as fertilizer, plant growth usually increases.

TABLE 37.1 Essential Nutrients in the Soil

Element	Form Available to Plants	Functions	Average % Dry Weight (Vascular Plants)	Deficiency Symptoms
(a) Macronutrients				
Oxygen	O_2, H_2O	Electron acceptor in cellular respiration; major component of organic compounds	45	Cell death (suffocation)
Carbon	CO_2	Substrate for photosynthesis; major component of organic compounds	45	Cell death (starvation)
Hydrogen	H_2O	Major component of organic compounds; electrical balance and establishment of electrochemical gradients	6	Cell death (desiccation)
Nitrogen	NO_3^- (nitrate) NH_4^+ (ammonium ion)	Component of nucleic acids, proteins, hormones, and coenzymes	1.5	Failure to thrive; chlorosis (yellowing of leaves)
Potassium	K^+	Cofactor for many enzymes; necessary for osmotic adjustment in cells; required for synthesis of organic molecules	1.0	Chlorosis at margins of leaves or in mottled pattern; weak stems; short internodes
Calcium	Ca^{2+}	Regulatory functions; role in cell wall structure; stabilizes membranes, controls movements; second messenger in signal transduction	0.5	Necrosis in meristems; deformation of young leaves; stunted, highly branched root system
Magnesium	Mg^{2+}	Chlorophyll component; activates many enzymes	0.2	Chlorosis between leaf veins; premature leaf drop
Phosphorus	$H_2PO_4^-$ (dihydrogen phosphate ion) HPO_4^{2-} (hydrogen phosphate ion)	Used in energetic bonds (ATP); component of nucleic acids, phospholipids, and several coenzymes	0.2	Stunted growth in young plants; dark green leaves with necrosis (small spots of dead cells)
Sulfur	SO_4^{2-} (sulfate ion)	Component of protein and coenzymes	0.1	Stunted growth; chlorosis
(b) Micronutrients				
Chlorine	Cl^-	Needed for water-splitting step of photosynthesis; functions in water balance and electrical balance	0.01	Wilting at leaf tips; general chlorosis and necrosis of leaves or development of bronze color
Iron	Fe^{3+} (ferric ion) Fe^{2+} (ferrous ion)	Necessary for chlorophyll synthesis; component of cytochromes and ferredoxin; enzyme cofactor	0.01	Chlorosis between veins of young leaves
Manganese	Mn^{2+}	Involved in photosynthetic O_2 evolution; enzyme activator; important in electron transfer	0.005	Chlorosis between leaf veins and small necrotic spots
Zinc	Zn^{2+}	Involved in synthesis of auxin, maintenance of ribosome structure, and enzyme activation	0.002	Small internodes; stunted and distorted ("puckered") leaves
Boron	$H_2BO_3^-$ (borate ion)	Cofactor in chlorophyll synthesis; possible role in sugar transport; aids in regulation of enzyme function	0.002	Black necrosis in young leaves and buds
Copper	Cu^+ (cuprous ion) Cu^{2+} (cupric ion)	Cofactor of some enzymes; present in lignin of xylem	0.0006	Dark-green leaves with necrotic spots; twisted and malformed leaves
Nickel	Ni^{2+}	Cofactor for enzyme functioning in nitrogen metabolism	[no data]	Necrosis at leaf tips
Molybdenum	MoO_4^{2-} (molybdate ion)	Cofactor in nitrogen reduction; essential for nitrogen fixation	0.00001	Chlorosis between veins; necrosis of older leaves

FEEDS 5,000 SQ. FT.
NET WT. 15 LBS.
Glorion
Deluxe
FALL FERTILIZER
10-18-10
GUARANTEED ANALYSIS
GLORION CORPORATION

FIGURE 37.2 Nitrogen, Phosphorus, and Potassium Are the Major Ingredients in Fertilizers
The numbers on this bag indicate the ratio of N to P to K (10-18-10) in a fertilizer.

This observation explains why the leading ingredients in virtually every commercial fertilizer are N, P, and K—usually listed in that order on the bag or box (**Figure 37.2**).

In contrast to macronutrients, **micronutrients** are required in relatively small quantities. Instead of acting as components of macromolecules, they usually function as cofactors for specific enzymes. Even though several are required in amounts so tiny that measurement is difficult, their importance should not be underestimated. For example, a typical plant contains just one molybdenum atom for every 60 million hydrogen atoms in its body, not including water. Yet plants die in the absence of molybdenum, because it functions as a cofactor for several enzymes involved in nitrogen processing. What happens to plants when other essential nutrients are missing?

Nutrient Deficiencies

In at least some cases, biologists can examine a plant that is growing poorly because of a nutrient deficiency and diagnose the problem. In large part, this ability is based on the results of studies that have used hydroponic growth systems to examine the effects of specific nutrient deficiencies. **Hydroponic growth** takes place in liquid cultures, without soil. Because experimental plants are grown in the absence of soil, the availability of nutrients can be precisely controlled.

As an example of this work, consider the experiment on copper deficiency in tomatoes as outlined in **Figure 37.3**. Researchers grew seedlings in two types of treatments. One treatment consisted of flasks containing water and all the essential nutrients listed in Table 37.1 in the relative concentrations that

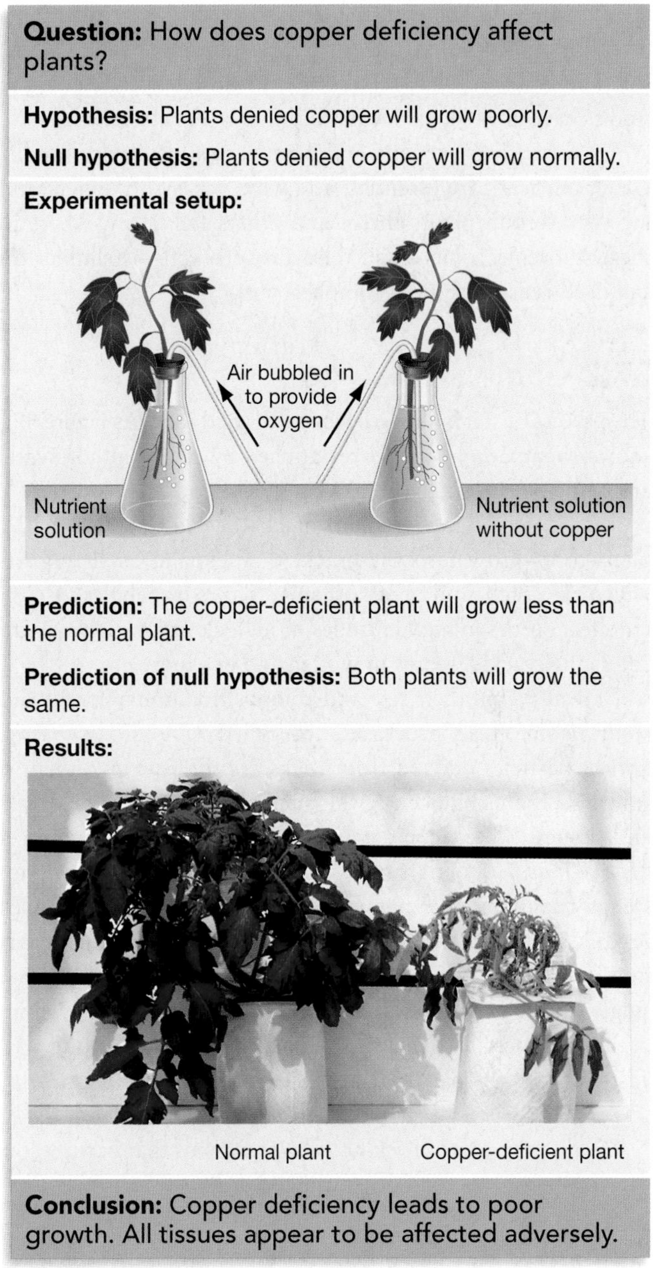
Question: How does copper deficiency affect plants?
Hypothesis: Plants denied copper will grow poorly.
Null hypothesis: Plants denied copper will grow normally.
Experimental setup:
Air bubbled in to provide oxygen
Nutrient solution
Nutrient solution without copper
Prediction: The copper-deficient plant will grow less than the normal plant.
Prediction of null hypothesis: Both plants will grow the same.
Results:
Normal plant
Copper-deficient plant
Conclusion: Copper deficiency leads to poor growth. All tissues appear to be affected adversely.

FIGURE 37.3 Hydroponics Is Used to Study Nutrient Deficiencies
Culturing plants in a solution instead of in soil allows researchers to control precisely which nutrients are available and in what concentrations.

are optimal for tomato growth. The second treatment was identical, except that the nutrient solution lacked copper.

As the "Results" section of Figure 37.3 shows, copper-deprived individuals have stunted shoots, unnaturally light foliage, curled leaves, and no flowers. Given copper's role as a cofactor or component of several enzymes involved in redox reactions required for ATP production, it is understandable that the symptoms were very general in nature, as well as life threatening. All tissues in the plant were affected, and all were affected severely. Likewise, it is reasonable to expect that a relatively small amount of copper would cure the deficiency. In line with this prediction, the researchers found that the symptoms were prevented if the plants were cultured in a solution containing just 0.002 mg/L of this element. Analogous studies have now been done on the other essential nutrients.

For farmers, gardeners, and plant ecologists, understanding which nutrients are essential, and why, is basic to understanding why certain plants thrive and others fail. Now, where do these nutrients come from? The answer—soil—is simple. But soil itself is astonishingly complex.

37.2 Soil

The process of soil building begins with solid rock. As **Figure 37.4** shows, **weathering**—the forces applied by rain, running water, and wind—continually breaks tiny pieces off large rocks. The process is accelerated if small cracks develop in the rock. If plant roots grow into the crack, they expand as they grow, widen the crack, and break off small flakes or pebbles. A similar effect occurs in high latitudes or high elevations when water enters the cracks, freezes in winter, and expands.

Depending on their size and composition, the particles that result from these processes are called *clay, silt, sand*, or *gravel*. Particles derived from rocks are the first ingredient in soil. As organisms occupy the substrate, they add dead cells and tissues. This organic matter is called **humus** (pronounced "*HEW-muss*"). With time, soil eventually becomes an extremely complex and dynamic mixture of inorganic particles, organic particles, and living organisms. It is commonplace to find thousands of species living in the top few centimeters of a single square meter of soil. In addition to plants, soil-dwelling organisms include a variety of fungi and animals, along with vast numbers of bacteria, archaea, and microscopic protists (**Figure 37.5**).

Because the parent rock and the organisms that occupy soils vary from one site to another, the texture of soil varies as well. **Texture** refers to the proportions of different-sized particles present in soil and is important for several reasons:

- Texture affects the ability of roots to penetrate more deeply to obtain water and nutrients, as well as to anchor and support the body. For example, soil that is dominated by clay-sized particles tends to compact and resist root penetration.
- Texture affects a soil's ability to hold water and make it available to plants. Water tends to adhere to clay and silt particles but runs through sands and gravels.
- A soil's texture and its water content dictate the availability of oxygen. Like other eukaryotes, plants have to take in oxygen to use as an electron acceptor during cellular respiration. The oxygen used by plant root cells is found in air pockets among soil particles. This observation explains why overwatering a plant is just as detrimental as underwatering it. Overwatering drowns a plant's roots.

Topsoils that have good texture and large amounts of organic matter can take thousands of years to develop, through the weathering of rocks and the continual addition of humus. Unfortunately, it can take just a few years of abuse by humans to send valuable soils downriver.

Soil Conservation

Soil erosion occurs when soil is carried away from a site by wind or water. Soil erosion occurs naturally when rivers cut away at their banks and carry material downstream, as well as in other situations. In most natural environments, though, soils build up over time, because the rate of soil formation exceeds the rate of soil erosion. Unfortunately, the situation can change dramatically when areas are exploited by humans.

When plant cover is removed for forestry, farming, or suburbanization, soil particles are exposed to wind and rain without plant roots to hold them in place. The results can be devastating.

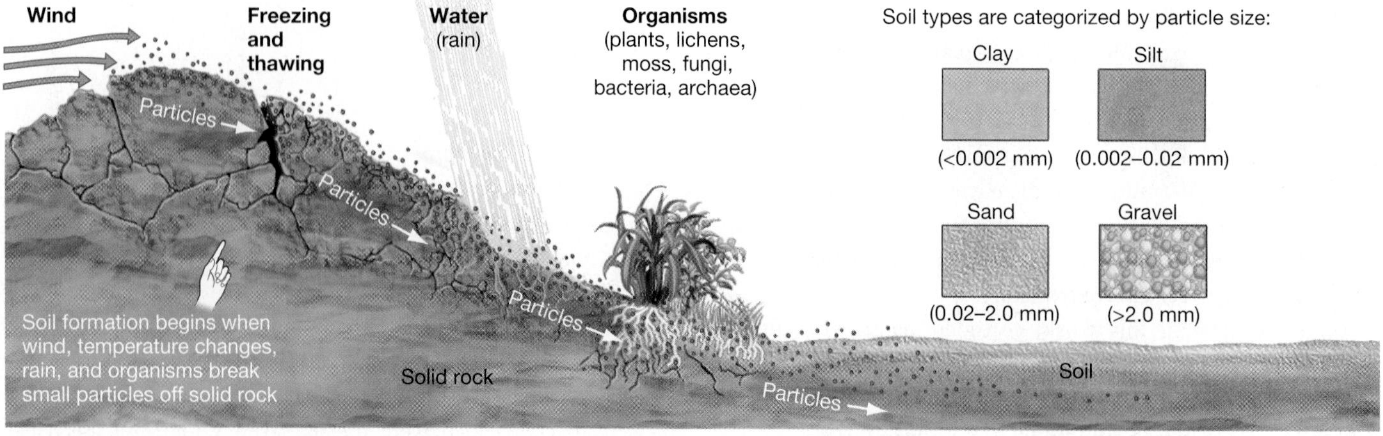

FIGURE 37.4 Soil Formation Begins with Erosion of Rock
Soil formation begins when wind, rain, and organisms break small particles off solid rock.

FIGURE 37.5 Mature Soils Are a Complex Mixture of Organic and Inorganic Components
In addition to mineral particles, soil contains humus—organic material derived from dead organisms—and a wide array of living organisms.

At some locations in the United States, 8–10 cm of topsoil were blown away during the "Dust Bowl" of the 1930s, when drought and poor farming practices left thousands of acres of soil unprotected (**Figure 37.6a**). More recently, deforestation has exposed soils in the tropics and contributed to disasters such as the mudslides and flooding that occurred in the Dominican Republic and Haiti during May and September of 2004, killing close to 5000 people (**Figure 37.6b**). Worldwide, it is estimated that 36 billion tons of soil are being lost to erosion every year. One-sixth of this total is lost in the United States alone.

Erosion rates are increasing as human populations expand and as land that was previously thought to be unsuitable for

FIGURE 37.6 Soil Erosion Can Have Devastating Consequences
(a) Massive soil erosion during the 1930s contributed to the failure of thousands of farms in the United States. **(b)** Deforestation in Haiti and the Dominican Republic contributed to mudslides and flooding that killed thousands of people in May 2004.

agriculture or intensive forestry is brought into production. In addition to outright loss of soils, soil quality can diminish over time in areas subjected to continual farming or tree cutting. The issue is that farmers and foresters remove organic matter that otherwise would have been returned to the soil to form humus.

Fortunately, soil is a renewable resource if managed carefully. Farmers can reduce soil loss dramatically by planting rows of trees as windbreaks and using techniques that minimize the amount of plowing and tilling that needs to be done to control weeds. In addition, they can maintain soil quality by adding organic material in the form of manures and planting cover crops that are plowed in and allowed to decompose. Ranchers and foresters can employ similar management approaches.

Techniques that maintain long-term soil quality and productivity are collectively called **sustainable agriculture**. Productivity, in turn, is influenced by soil texture and organic matter but is primarily a function of nutrient availability. Let's take a closer look.

Nutrient Availability

The elements required for plant growth are not found in the soil as atoms. Instead, they exist as **ions**. The second column of Table 37.1 lists some of the ions that are found in soil and that contain essential nutrients. Notice that some of these nutrients are available as **elemental ions**, such as K^+ or Cl^-, while others exist as **molecular ions**, such as HPO_4^{2-} or NO_3^-.

The ions present in soil tend to behave in one of two ways, depending on their charge (**Figure 37.7**). Anions—ions with negative charges—usually dissolve in soil water, because they interact with water molecules via hydrogen bonding. (Phosphate is an exception to this rule; phosphate ions tend to form insoluble complexes with iron, aluminum, calcium, or other positively charged ions.) Because they exist as solutes, negatively charged ions are readily available to plants for absorption. They are also easily washed out of the soil by rain, however. The loss of nutrients via percolating water is called **leaching**.

In contrast, cations—ions with positive charges—are usually much less readily available than anions. This pattern occurs because cations interact with the negative charges found on two types of soil particles: (1) organic matter that is rich in negatively charged organic acids; and (2) the surfaces of the tiny, sheetlike particles called clay. Organic soils that contain clay tend to retain nutrients, because few positively charged ions leach away and because these soils hold water (and thus anions) better than sandy soils do. The presence of clay also makes cations more difficult for plants to extract and use, however, because they are tightly bound.

Several other factors can influence the availability of essential elements present in soil. Perhaps the most important of these is soil pH. Recall from Chapter 2 that pH indicates the relative concentration of hydrogen ions in a solution and that

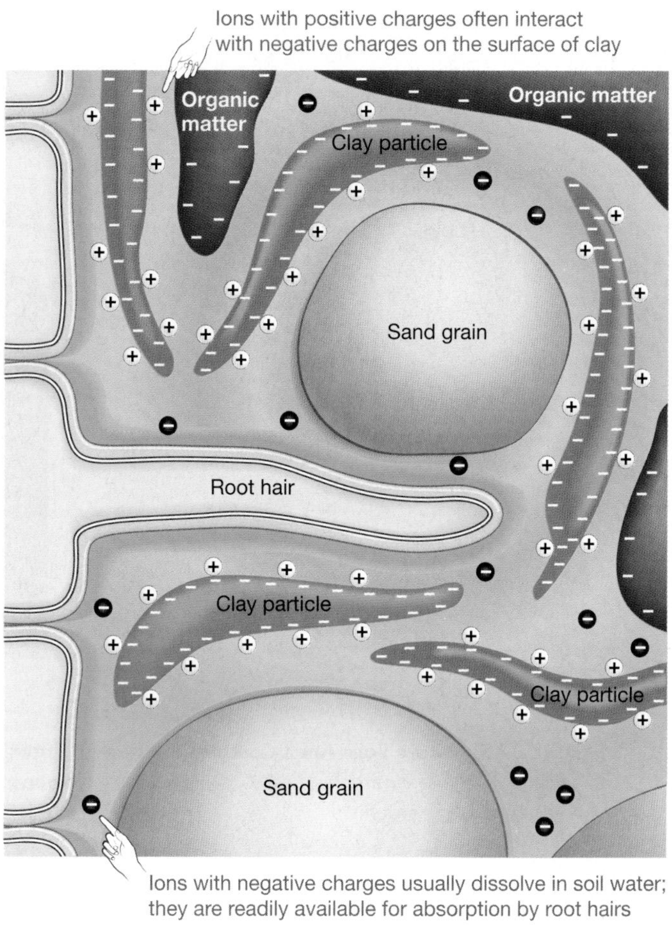

FIGURE 37.7 Cations Tend to Bind to Soil Particles; Anions Stay in Solution
Positively charged ions tend to bind to organic matter in soil as well as to clay particles. Negatively charged ions, in contrast, tend to go into solution. **EXERCISE** Add symbols indicating that positive ions bind to organic matter.

the pH scale ranges from 0 to 14. Soils with low pH have a relatively high concentration of hydrogen ions and are considered acidic. Soils with a high pH contain few hydrogen ions and are termed basic or alkaline. Acidic soils tend to be found in regions where the decomposition of organic matter results in the production of carbonic acid, phosphoric acid, or nitric acid. Alkaline soils are common in regions where limestone ($CaCO_3$) is abundant. When limestone reacts with water, the calcium ions that are released take the place of protons that cling to soil particles. The protons then react with CO_3^{2-} to form bicarbonate ions (HCO_3^-), lowering the hydrogen ion concentration of the soil and raising the pH.

Most plants prefer to grow in soils with a relatively neutral pH between 6 and 7. But species such as tomatoes, rhododendrons, and blueberries grow best in acidic soils (pH around 3 to 5), while some plants require an alkaline soil (pH above 8).

Soil pH affects the availability of plant nutrients in a number of ways. The presence of acids increases the rate at which elemental ions are released from rocks, and many molecular ions are more soluble in solutions with low pH. The presence of protons in soil water can also cause the release of cations that are bound to soil particles, via a process called **cation exchange**. Cation exchange occurs when protons or other cations bind to negative charges on soil particles and cause bound cations such as magnesium or calcium to be released, making those nutrients available to nearby plant roots. Plants influence cation exchange by pumping protons out of root hairs. But if soil is too acidic, then rain can wash cations away before the nutrients can be taken up by roots.

To summarize, negative ions stay in solution in soil water and are readily available to plants. Positive ions tend to bind to soil particles but can be released by cation exchange. Once ions are in solution and available for absorption by plants, how do they actually enter root cells? Nutrients are found at extremely low concentrations in soil but at high concentrations in cells. How are plants able to bring N, P, K, and other key elements into their bodies against a concentration gradient?

37.3 Nutrient Uptake

In most species of plants, the root system is the site of nutrient uptake. The general features of root anatomy were introduced in Chapter 35, and the role of roots in water uptake was analyzed in Chapter 36. Now we need to explore the fine anatomy of roots and analyze events that occur in the plasma membranes of their epidermal cells.

Nutrient uptake occurs just above the growing root tip, in the region called the **zone of maturation**. Recall from Chapter 35 that epidermal cells in this part of the root have extensions called **root hairs** (**Figure 37.8**). Root hairs dramatically increase the surface area available for nutrient and water absorption. They are so efficient at absorbing nutrients from soil that, over time, they create a "zone of nutrient depletion" in their immediate vicinity. The creation of this mined-out region is why continued root growth is vital to a plant's health. Because roots continue to grow throughout a plant's life, the zone of maturation is continually entering new and potentially nutrient-rich areas of soil.

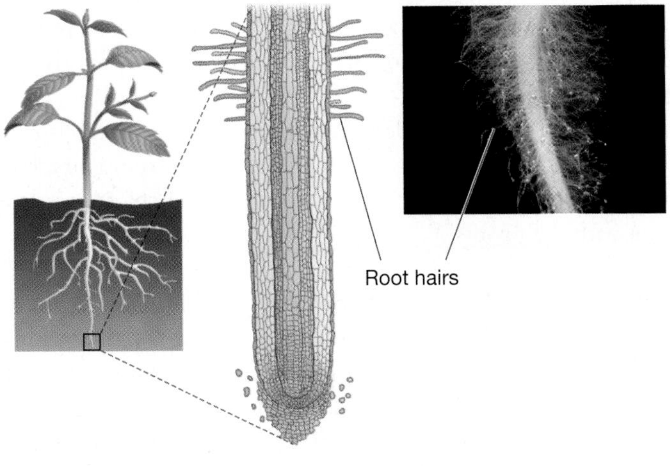

FIGURE 37.8 Root Hairs Increase the Surface Area Available for Nutrient Absorption
EXERCISE On the left, mark where nutrients are absorbed.

If a root hair happens to encounter soil areas where nutrients are available, what happens? How do the ions cross the plasma membrane and enter the cell?

WEB TUTORIAL 37.1
Soil Formation and Nutrient Uptake

Mechanisms of Nutrient Uptake

Ions, small molecules, and even large molecules can pass through the plant cell wall freely. The plasma membrane, in contrast, is highly selective. Recall from Chapter 6 that the plasma membrane is a fluid, sheetlike structure consisting of a phospholipid bilayer studded with proteins. Because the interior of the phospholipid bilayer is nonpolar and hydrophobic, it resists the passage of ions—even ions that are required by the plant. Some membrane proteins span the bilayer, however, and allow specific ions to cross the membrane.

Because root hairs have such a large surface area, they have large numbers of membrane proteins that bring nutrients into the cortex of the root. On their own, though, these membrane proteins cannot import the ions that the plant needs. Why?

Establishing and Using a Proton Gradient Plants harvest diffuse nutrients and concentrate them in their tissues. In a root-hair cell, ions such as potassium, chloride, and nitrate are many times more concentrated than they are in the soil water outside the cell. For these and other ions to enter the cell, they have to cross the plasma membrane against a strong concentration gradient. How is this possible?

As **Figure 37.9a** (page 860) shows, the answer hinges on **proton pumps**, or H^+-ATPases, in the plasma membranes of root hair cells. These H^+-ATPases are similar to the pumps that make it possible for companion cells to load sucrose into phloem against a strong concentration gradient. Recall from Chapter 36 that when a phosphate group from ATP binds to these proteins, they change conformation in a way that allows them to transport protons to the exterior of the cell. The activity of these pumps leads to a strong excess of protons on the exterior of the plasma membrane relative to the interior. As a result, there is a strong concentration gradient favoring the movement of protons into the root hair. In addition, the outside of the root-hair membrane is positively charged relative to the inside. This separation of charges means that there is a **voltage** across the membrane.

Proton pumps are found in all types of plant cells, and all plant cell membranes carry a voltage. Because voltage is a form of potential energy, the charges that are separated by the membrane create a **membrane potential**. By convention, membrane potentials are expressed as inside-cell relative to outside-cell.

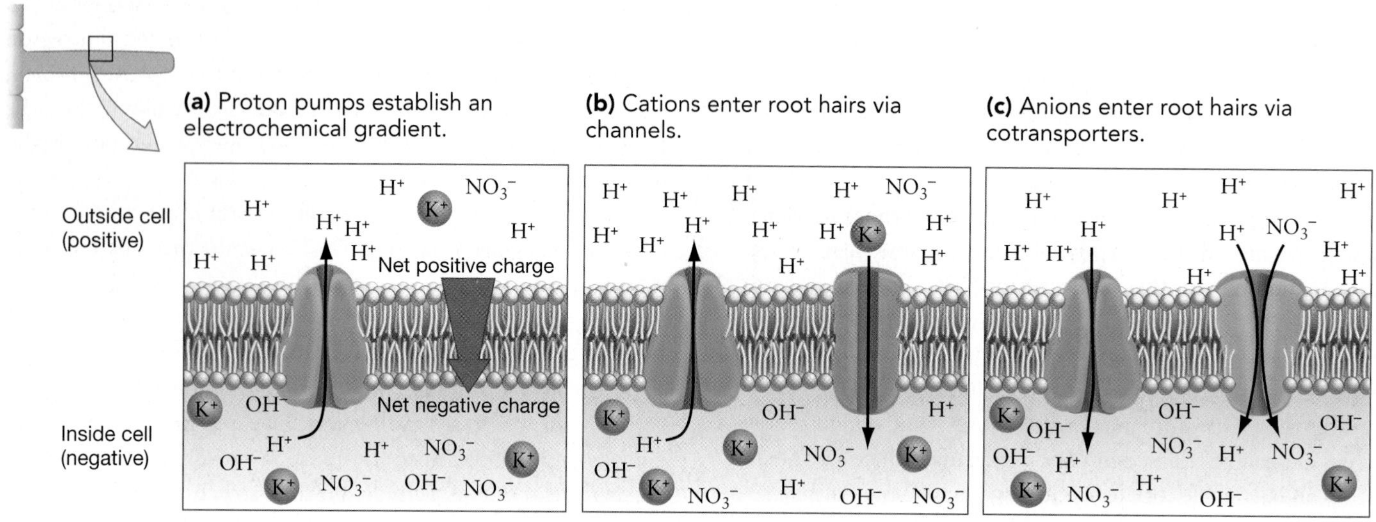

FIGURE 37.9 Ions Enter Roots along Electrochemical Gradients Created by Proton Pumps
(a) Nutrient absorption is possible because proton pumps establish a gradient favoring the entry of protons and other cations. **(b)** Cations enter via channels; **(c)** anions enter via cotransporters. EXERCISE Label the purple arrow in part (a) "Electrical gradient, Proton gradient."

Typically, the membrane potential of an active plant cell is about −200 mV (millivolts).

In addition to concentration gradients that affect nutrients, then, the membranes of root hairs create a strong electrical gradient. The combination of concentration and electrical gradients is called an **electrochemical gradient.** In the membranes of root hairs, the electrical gradient, which favors the entry of positive ions, is strong enough to overcome the concentration gradient, which opposes the entry of these cations. In essence, plant cells are batteries that are charged up to attract nutritionally necessary cations.

To drive this point home, consider recent research on how root-hair membranes import potassium ions. Potassium is an essential nutrient in plants because it is required as a cofactor by over 40 enzymes. In addition, it is found at relatively high concentrations inside plant cells and plays a key role in bringing water into cells via osmosis and maintaining normal turgor pressure. Researchers who added radioactive potassium ions to the solution outside of a root cell and followed their movement found that K^+ does indeed flow into cells along an electrochemical gradient (**Figure 37.9b**). In follow-up experiments, biologists were able to isolate the membrane protein responsible for the uptake of K^+ and to sequence the gene that encodes the protein.

If cations enter root hairs along an electrochemical gradient, how is it possible for anions to enter? The negatively charged interior of the cell should repel these ions. As **Figure 37.9c** shows, though, anions are able to enter root hairs for two reasons: (1) a strong electrochemical gradient favors the entry of protons and (2) membrane proteins called **cotransporters** use the energy released by the entry of a proton *along* its electrochemical gradient to transport nitrate, phosphate ions, or other anions *against* their electrochemical gradients. An array of cations also enters cells via cotransporters. Ions that are missed by root hairs and that move into the root through cell walls can be taken up in a similar fashion by the membranes of cortical cells or of endodermal cells.

To summarize, the electrochemical gradient set up by proton pumps makes it possible for plant roots to absorb a wide array of nutrients via ion channels and cotransporters. Once ions have entered the cytoplasm of a root cell, they are transported across the root cortex via the symplastic or apoplastic pathways introduced in Chapter 36. To enter xylem, however, nutrients have to pass through the plasma membranes of endodermal cells and then diffuse from endodermal cells into the "dead space" of the xylem tubes. Once they become part of xylem sap, the ions are transported passively to tissues throughout the plant.

The vast majority of plants take up nutrients by establishing a proton gradient across the membranes of their root hairs. But most species also get help from fungi that live in close association with their roots.

Nutrient Transfer via Mycorrhizal Fungi To synthesize proteins and nucleic acids, plants need to extract large quantities of nitrogen and phosphorus from the soil. But most plant species can absorb only enough N and P to satisfy their nutritional needs with help from fungi that live in close association with their roots. As a group, these fungal species are called **mycorrhizae** ("fungus-root"). Mycorrhizae and plants are said to be **symbiotic** ("living together")—that is, they live in close association with each other.

Trees and shrubs that live in northern forests receive large quantities of nitrogen from mycorrhizal fungi. The fungal symbionts are particularly efficient at digesting amino acids in decaying plant material and absorbing ammonium ions (NH_4^+) and other forms of usable nitrogen. Because these fungi have hyphae—filaments—that wrap around the epidermal cells of roots and radiate out into the surrounding soil, they are known as **ectomycorrhizal fungi**, or **EMF**; see Figure 30.11a. (The Greek root *ecto* means "outside.") In contrast, plants that live in grasslands and in tropical forests receive much of the phosphorus they need from species of fungi whose hyphae can actually penetrate the walls of plant root cells. Because some of these fungal hyphae form branched clusters that look like little trees when viewed under a microscope, these symbionts are called **arbuscular mycorrhizal fungi**, or **AMF**; see Figure 30.11b. (The Latin root *arbor* means "tree.") Biologists estimate that more than 80 percent of all vascular plant species associate with mycorrhizae.

Chapter 30 presented experimental evidence that EMF and AMF transfer nitrogen and phosphorus, respectively, from soil to plant roots. In exchange, the plants that they associate with transfer sugars and other photosynthetic products to the fungi. Because there is a reciprocal exchange of nutrients, the symbiotic relationship is considered **mutualistic**, or mutually beneficial. Chapter 30 also emphasized two reasons why fungi are particularly efficient at acquiring the nutrients required by plants: First, they form networks of filamentous hyphae that greatly increase the surface area available for absorbing nutrients. Second, they synthesize and secrete a wide array of digestive enzymes. These enzymes break down dead plant material and other sources of nutrients, releasing ions that can be absorbed and used by the fungus itself or by its plant partner. Most plant species would starve if denied their mycorrhizal associates.

Mechanisms of Ion Exclusion

Plants have sophisticated systems for absorbing nutrients via the proton gradients they generate and mutualistic relationships with fungi. Not all ion uptake is beneficial, however. For example, certain types of natural soils—as well as soils that have been contaminated by waste products from mining or smelting operations—contain enough cadmium, zinc, nickel, lead, or other metals to poison most plants. Sodium is also detrimental at high concentrations. An excess of Na^+ inside cells can disrupt enzyme functioning, while a surplus of sodium in extracellular spaces can create a solute potential that pulls enough water out of cells to result in a loss of turgor. How do plants exclude ions that are detrimental?

Passive Exclusion Ions that enter roots via the apoplastic pathway may never make it to xylem. To understand why, recall from Chapter 36 that the Casparian strip forces all solutes that are traveling through the root apoplast to cross the plasma membrane of endodermal cells and enter their cytoplasm. Because the membrane of an endodermal cell contains only certain types of transporters and channels, it acts as a selective filter. In this way, the presence of the Casparian strip prevents some ions from entering the symplast and reaching the xylem. Because excluded ions are not actively pumped out of the root cortex, the Casparian strip is a mechanism for the passive exclusion of ions.

Passive exclusion also occurs in root hairs, the other location where solutes enter the symplast on their way to xylem. As an example, consider recent research on variation in salt tolerance among varieties of corn (*Zea mays*). Some crop species are much more tolerant of salty soils than others are. For example, rice is notoriously sensitive to salt buildup, while barley is relatively tolerant. Even within the same species, the degree of salt tolerance can vary widely. Researchers who grew seedlings from salt-tolerant and salt-intolerant varieties of corn in hydroponic cultures containing a high concentration of NaCl found that salt-intolerant individuals took up almost twice as much Na^+ as did salt-tolerant individuals. To interpret this result, the researchers hypothesized that individuals from salt-tolerant populations have fewer sodium channels in their roots than do salt-intolerant individuals. If this hypothesis is confirmed, it implies that salt tolerance is at least partly based on variation in the abundance of membrane proteins.

Active Exclusion Although passive mechanisms of nutrient exclusion have been confirmed for a variety of ions in addition to sodium, plants also have mechanisms for coping with toxins once they are inside the body. This is important, because all nutrients are toxic if they are present at high concentrations. Consider copper, for example. Plants that grow on soils near copper-mining operations experience large concentration gradients that favor an influx of this nutrient. If copper channels are necessary to admit the small amounts required for normal cell function, it would seem inevitable that a surplus will eventually build up inside the plant body. How do plants neutralize excess nutrients before they begin to poison key enzymes?

One mechanism for coping with toxic concentrations of metals involves small proteins called metallothioneins. **Metallothioneins** bind to metal ions and prevent them from acting as a poison. Genes for these proteins have been found in a wide variety of organisms—bacteria, fungi, and animals as well as plants. Recent research on the mustard-family species *Arabidopsis thaliana* has shown that individuals from populations with a high tolerance for copper produce many more metallothionein proteins than do individuals from populations with a low tolerance for copper. Producing metallothionein proteins requires an expenditure of energy and represents a form of active exclusion.

A second mechanism for actively neutralizing specific toxins involves transport proteins located in the **tonoplast**—the membrane that surrounds the vacuole. Perhaps the best-studied example involves proteins that move sodium ions from the

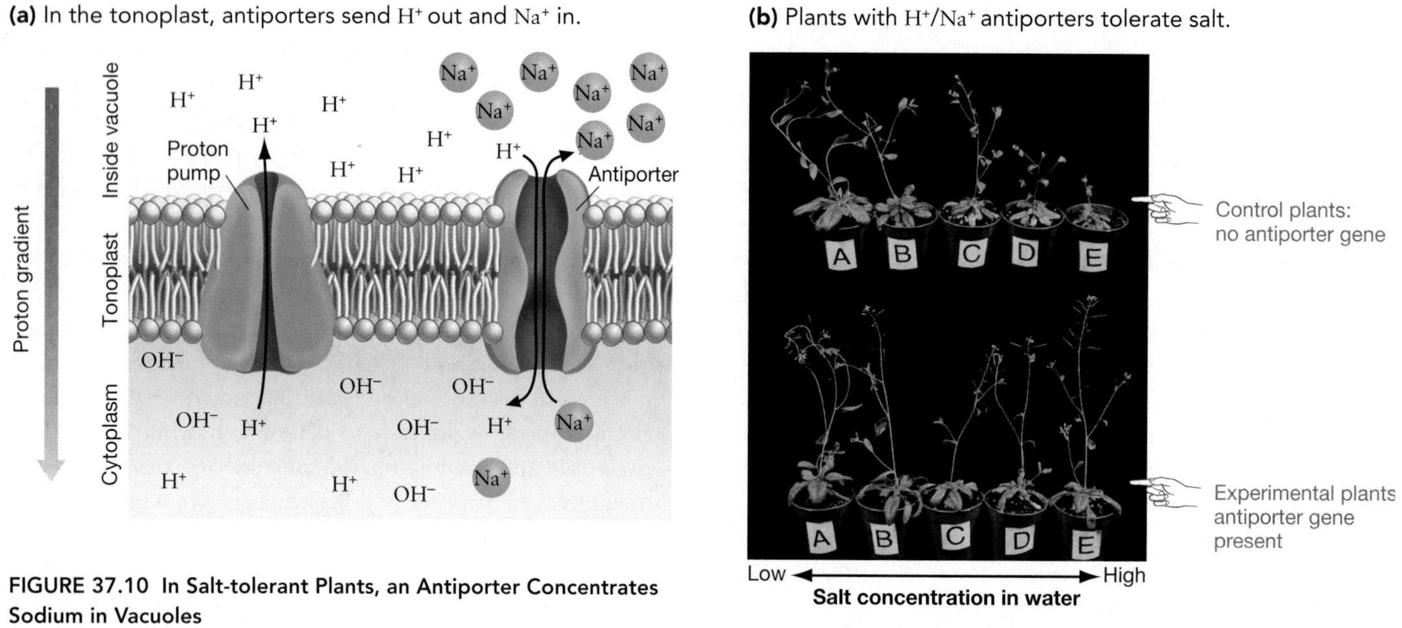
(a) In the tonoplast, antiporters send H+ out and Na+ in.
(b) Plants with H+/Na+ antiporters tolerate salt.
Proton gradient
Inside vacuole
Tonoplast
Cytoplasm
Proton pump
Antiporter
H+
Na+
OH−
Control plants: no antiporter gene
Experimental plants antiporter gene present
A
B
C
D
E
Low
High
Salt concentration in water

FIGURE 37.10 In Salt-tolerant Plants, an Antiporter Concentrates Sodium in Vacuoles

(a) In the tonoplast of salt-tolerant species, a proton gradient is established by proton pumps. This gradient allows antiporters to concentrate sodium ions inside the vacuole, where they do not poison enzymes in the cytoplasm. **(b)** *Arabidopsis thaliana* plants that were transformed with the gene for the antiporter protein grow well even when watered with salty water.

cytosol into vacuoles, where they cannot poison enzymes. **Figure 37.10a** shows how these proteins work. Notice that proton pumps in the tonoplast create an electrochemical gradient that favors the movement of H^+ out of the vacuole. A cotransport protein that functions as an **antiporter** then uses this gradient to conduct protons out of the vacuole and bring sodium ions in. (Recall from Chapter 36 that an antiporter is a protein that transports a solute against its concentration gradient.) This H^+/Na^+ antiporter has created a great deal of excitement among biologists, because it offers a way to genetically engineer crop plants that may grow well in salty soils created by poor irrigation practices. As an example, consider an experiment that exposed normal *Arabidopsis thaliana* plants and *A. thaliana* individuals that were transformed with copies of the gene for the H^+/Na^+ antiporter to different levels of NaCl. As **Figure 37.10b** shows, the genetically transformed individuals were able to grow efficiently even when exposed to high concentrations of salt. **Box 37.1** explains how plants with the ability to sequester toxins in their vacuoles might be used to clean up contaminated soils.

BOX 37.1 Can Phytoremediation Help Clean Up Contaminated Soils?

Some species of plants are able to grow on sites contaminated with toxic cleaning agents, the explosive compound TNT, or metals such as cadmium, zinc, nickel, or lead. Alpine pennycress, for example, can accumulate 300 times as much zinc and 1500 times as much cadmium in its leaves as most plant species can, without showing toxic effects. Pigweed can take up and store large amounts of radioactive cesium in its tissues, and the mustard *Brassica juncea* accumulates lead at high concentrations. If such species can be encouraged to grow on contaminated sites, it might be possible to reduce contamination in the soil gradually by harvesting the aboveground plant parts and disposing of them. Some researchers even suggest that if plants accumulate metal ions and can be harvested like hay, the leaves and stems could be burned and the ash used as a source of metal ore.

Phytoremediation is the use of plants to clean up contaminated soils, just as bioremediation uses bacteria and archaea to reduce concentrations of toxic compounds (see Chapter 27). Researchers are studying plants that accumulate toxins in their tissues in an attempt to understand which genes and proteins are involved. To date, results indicate that these species have specialized transport proteins that bring certain ions or molecules into the xylem and that later concentrate the toxins in the vacuoles of cells found in leaves and stems.

In addition to species that accumulate toxins in aboveground tissues, some plants have enzymes that act on toxic compounds and render them harmless. Poplar trees, for example, are able to break a cancer-causing molecule called trichloroethylene (TCE) into carbon dioxide and chlorine. Poplars are now being tested to see if they can help clean up sites contaminated with TCE.

Based on recent research, it appears that phytoremediation and bioremediation may turn out to be useful tools in the effort to reduce contamination in at least some types of soils.

✓CHECK YOUR UNDERSTANDING

Plants absorb most of the nutrients they need via membrane proteins in root hairs. Some of these proteins pump protons out of the cells, creating an electrochemical gradient that favors the entry of selected ions via carriers or cotransporters. In many species, mycorrhizal fungi are important for bringing nitrogen or phosphorus atoms into the root. You should be able to (1) draw a diagram showing how a membrane potential is established in a root hair and why it sets up an electrochemical gradient that favors the entry of both cations and anions, and (2) describe passive and active mechanisms of ion exclusion.

37.4 Nitrogen Fixation

Nitrogen gas (N_2) makes up 80 percent of the atmosphere. Unfortunately, plants and other eukaryotes cannot use nitrogen in this form. Nitrogen gas is unreactive, because the triple bond between the two nitrogen atoms is extremely strong and hard to break. To synthesize amino acids, nucleic acids, and other nitrogen-containing compounds, plants have to absorb nitrogen in more accessible forms, such as ammonium or nitrate ions. In many soils, these ions are in short supply and plant growth is restricted by the availability of usable nitrogen. Most plants grow much faster when they receive a nitrogen-containing fertilizer.

The use of nitrogen-based fertilizers has drawbacks, however. In many parts of the world, ammonia and other nitrogen-based fertilizers are too expensive for farmers; in more affluent regions, these fertilizers are used so extensively that they are causing serious pollution problems (see Chapter 27). For these and other reasons, there is intense interest in understanding the molecular basis of a phenomenon called nitrogen fixation.

Among all the organisms on the tree of life, only a few species of bacteria and archaea are able to absorb N_2 from the atmosphere and convert it to ammonia, nitrites, or nitrates. This process is called **nitrogen fixation.** Nitrogen fixation requires a series of specialized enzymes and cofactors, including an enzyme complex called *nitrogenase*. The process is extremely energy demanding. An expenditure of up to 24 ATP molecules is required for nitrogenase to reduce one molecule of N_2 to two molecules of NH_3.

In many cases, bacterial cells that are capable of nitrogen fixation take up residence inside plant root cells. The best-studied nitrogen-fixing bacteria are members of the genus *Rhizobium* that associate with plants in the pea family. Members of the genus *Rhizobium* and closely related species are often called **rhizobia**; pea family plants are often called **legumes.**

As **Figure 37.11** shows, nitrogen-fixing bacteria and their host cells form distinctive structures called **nodules** in the roots of legumes. Like mycorrhizal fungi and their host plants, legumes and rhizobia have a mutualistic relationship. The nitrogen-fixing bacteria provide the plant with ammonia, while the legume provides the bacteria with carbohydrates and protection.

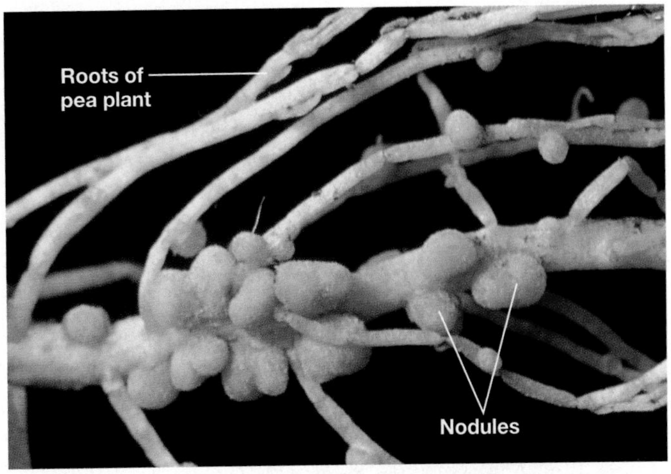

FIGURE 37.11 **In Some Plants, Roots Form Nodules where Nitrogen-Fixing Bacteria Live**

Research groups from all over the world have been trying to understand which genes and proteins are involved as these organisms interact. The research efforts are motivated in part by the hope that the required genes could be introduced into species such as wheat or rice, using techniques described in Chapter 19. The roots of rice and corn could then be occupied by nitrogen-fixing bacteria and the use of nitrogen-based fertilizers reduced.

How Do Nitrogen-Fixing Bacteria Colonize Plant Roots?

When a pea seed germinates, its roots do not contain a population of rhizobia. Instead, the root must make contact with bacterial cells existing in the soil, and the rhizobia must colonize the plant. Colonization is a complex process involving a series of specific interactions between the rhizobia and the legume.

The first event in colonization is a recognition step that occurs between a pea family plant and its symbiotic bacterium. Recognition is possible because the root hairs of pea family plants contain compounds called *flavonoids*, which happen to be closely related to the pigments found in flower petals. When rhizobia contact these compounds, the bacteria respond by producing sugar-containing molecules called **Nod factors** (for *nod*ule-formation). Nod factors, in turn, bind to proteins on the membrane surface of root hairs.

The recognition step in the interaction is species specific. Each legume species produces a different recognition signal, and each rhizobium species responds with one or more unique Nod factors. When investigators have performed experiments based on switching recognition signals or Nod factors between species, the recognition step fails.

How Do Host Plants Respond to Contact from a Symbiotic Bacterium?

When Nod factors bind to the root-hair surface, they set off a chain of events that leads to dramatic morphological changes in the host legume. As **Figure 37.12** shows, rhizobia proliferate around the tip of the root hair, then enter the epidermal cell through an invagination of the root-hair membrane called an **infection thread.** The infection thread invades the root cortex, then bursts and releases bacterial cells into the interior of cortex cells. The infected cortex cells begin to divide rapidly, forming root nodules.

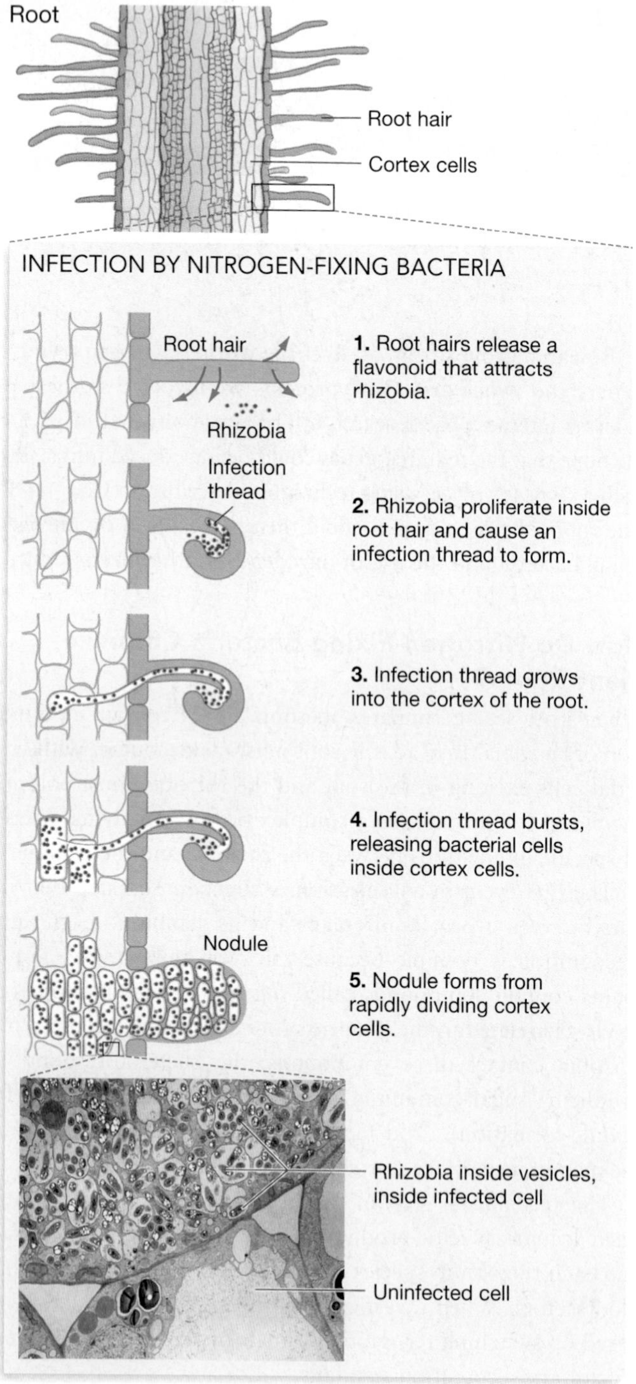

FIGURE 37.12 Infection by Nitrogen-Fixing Bacteria Is a Multistep Process
After rhizobia bind to a root hair, they enter the cytoplasm and travel down an infection thread into the root cortex. If they enter cortex cells, root cells respond by dividing and forming a nodule.

Genetic changes occur in concert with these morphological changes. For example, rhizobia need to obtain oxygen to carry out aerobic respiration and produce the large quantities of ATP required for nitrogen fixation. But the key enzyme complex involved, nitrogenase, is poisoned by the presence of oxygen. The problem is solved by the production of an iron-containing molecule called leghemoglobin (short for legume hemoglobin). **Leghemoglobin** consists of a globin protein complexed with an oxygen-binding heme group (see Chapter 3). The molecule is similar to the hemoglobin in your red blood cells and causes the pinkish color of root nodules (see Figure 37.11). The iron atom in the heme group binds to oxygen molecules and prevents them from poisoning nitrogenase. Instead, oxygen is delivered to enzymes in the electron transport chain of rhizobial cells, where it acts as an electron acceptor during cellular respiration.

To summarize, a complex sequence of steps makes it possible for rhizobia to colonize legume root hairs, form nodules, and fix nitrogen. Each phase of the process involves genes in both the host and the symbiont. Identifying all the critical structural genes and regulatory sequences in DNA and introducing them into wheat, rice, or corn will be no easy task. Although developing a strain of nitrogen-fixing rice or corn would be considered a crowning achievement in biological science, investigators have realized that it will be extremely difficult to replicate the recognition and response steps in nonlegume species.

✓CHECK YOUR UNDERSTANDING

Certain species of bacteria infect plant root cells and fix nitrogen, which is then made available to the plant. You should be able to (1) write a simple equation summarizing the overall reaction of nitrogen fixation and use it as an aid in describing why both bacteria and plants benefit from this interaction; and (2) generate hypotheses addressing two additional questions about nitrogen fixation: Why don't all plants have symbiotic bacteria that fix nitrogen, and Why is the interaction that establishes the infection by nitrogen-fixing bacteria so complex?

37.5 Nutritional Adaptations of Plants

Based on the data available to date, it is reasonable to claim that over 95 percent of vascular plants use proton pumps as a mechanism for importing nutrients from soil. Over 80 percent supplement their "diet" with nutrients acquired from mycorrhizal fungi, and a small but significant fraction associate with nitrogen-fixing bacteria. Perhaps 99 percent of all living plants species make their own sugar through the process of photosynthesis.

What about the small number of plant species that don't follow these rules? Some appear to live on air, some parasitize other plants, and others catch insects and digest them.

Epiphytic Plants

Species from a diverse array of plant lineages do not absorb nutrients from soil. In fact, these species never even make contact with soil. As **Figure 37.13a** shows, they often grow in the leaves or branches of trees. For this reason they are called **epiphytes** ("upon-plants"). The Spanish moss that hangs from oak trees in the southern United States is a familiar example, although this species is actually not a moss but a *bromeliad*—a relative of the pineapple. In the tropics and subtropics, one-third of all ferns grow as epiphytes, and there are thousands of species of epiphytic orchids, bromeliads, and lycophytes. In northern forests, it is common for mosses and ferns to grow epiphytically on tree trunks.

(a) Epiphytes grow on trees.

(b) Water-holding "tanks" formed by leaves of an epiphytic bromeliad

FIGURE 37.13 Epiphytes Are Adapted to Grow in the Absence of Soil
(a) Epiphytic plants are particularly common in tropical wet forests. **(b)** In some epiphytes belonging to the bromeliad family, the leaves curl together at the base to form a "tank" that catches and holds water. These miniature ponds attract insects and algae that provide nutrients.

Epiphytes absorb most of the water and nutrients they need from rainwater, dust, and particles that collect in their tissues or in the crevices of bark. As **Figure 37.13b** shows, some epiphytes have leaves that form "tanks," or rosettes that accumulate and hold water and organic debris. In such cases, nutrients are actually absorbed through the leaves themselves. The flowerpot plant featured in Chapter 36 has modified leaves that fill with soil created by ant feces and that hold a network of adventitious roots.

Parasitic Plants

Organisms that live in close physical contact with individuals from another species and that lower the fitness of those individuals, usually by obtaining nutrients from them, are parasitic. Based on the data currently available, biologists estimate that about 3000 species of angiosperms are parasitic. Most of these species obtain carbohydrates and other nutrients from their host; a few kill the other individual in the process. The number of parasitic plants represents less than 1 percent of the total number of plant species that have been studied and named to date.

Some plant parasites are non-photosynthetic and obtain all of their nutrition by tapping into the phloem of the host individual. But most parasitic plants make their own sugars through photosynthesis and tap the roots systems of other species for water and essential nutrients. The mistletoe shown in **Figure 37.14** is an example. This parasite is green and

Mistletoe haustoria penetrate host xylem
Mistletoe
Host tree

FIGURE 37.14 Some Plant Parasites Tap into the Xylem Tissue of Their Hosts
Mistletoe is a parasite that grows on host trees. It is photosynthetic but also has haustoria that penetrate host xylem and extract water and ions that are used as nutrients. QUESTION Both the bromeliads in Figure 37.13 and mistletoe are epiphytes. How are they different in their relationship to the trees on which they grow?

photosynthetic but has structures called *haustoria* that penetrate a host's xylem and extract water and ions that the mistletoe uses as nutrients.

How detrimental are these root parasites? For example, is parasitism more damaging to the host than competition with an individual of the same species would be? The experiment outlined in **Figure 37.15** illustrates the type of approach researchers use to answer questions about parasitism. In this case, the investigator put two alfalfa seeds into each of a large number of pots and divided the pots into four treatment groups. Some of the pots were left with two alfalfa seeds; some pots received a third alfalfa seed; some received a seed from a root parasite known as red bartsia, while others got a seed from a root parasite called greater yellow rattle. When the plants had grown to maturity, they were dried and weighed. The resulting data showed that the total amount of plant matter produced, or **biomass**, in each pot was much smaller in the pots with parasites than in the pots with only host plants. The results support the conclusion that parasitism is much more damaging than competition is in this host-parasite system. But as researchers also point out, the benefits that parasites experience in nutrient acquisition are offset by a cost in competing with the host plant

Question: How does the presence of a root parasite affect the host plant?

Hypothesis: Parasitism is more damaging to host plants than competition is.

Alternate hypothesis: Competition is more damaging to host plants than parasitism is.

Experimental setup:

Plant two seeds of host plant with other host-plant seeds or with seeds from a parasite. Compare biomass of resulting plants.

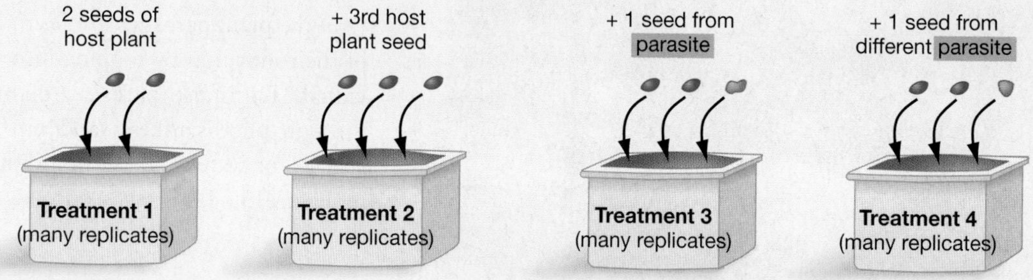
2 seeds of host plant
+ 3rd host plant seed
+ 1 seed from parasite
+ 1 seed from different parasite
Treatment 1 (many replicates)
Treatment 2 (many replicates)
Treatment 3 (many replicates)
Treatment 4 (many replicates)

Prediction: The presence of a root parasite will suppress productivity more than the presence of a competitor.

Prediction of alternate hypothesis: The presence of a competitor will suppress productivity more than the presence of a parasite.

Results:

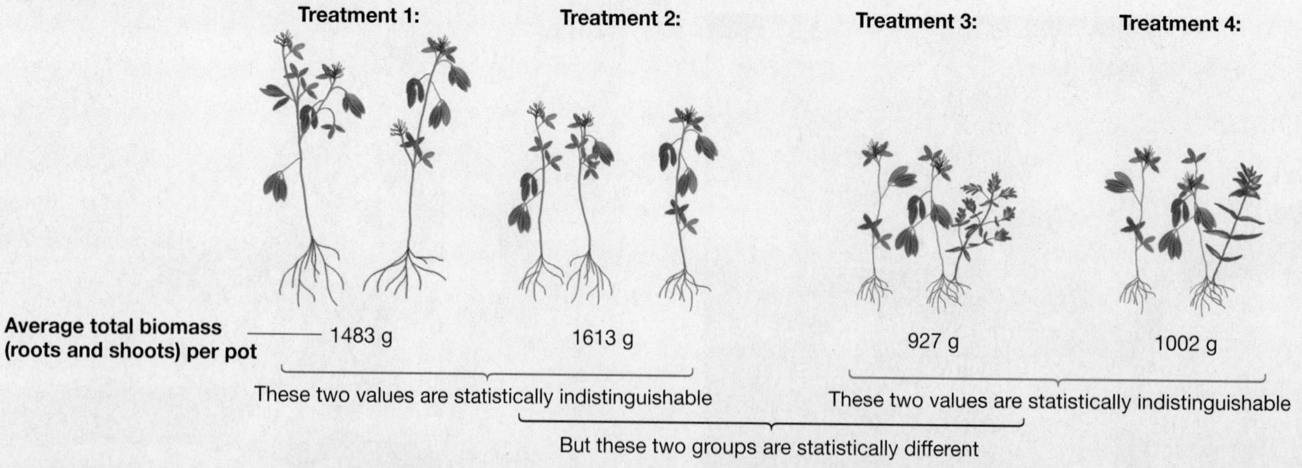
Treatment 1:
Treatment 2:
Treatment 3:
Treatment 4:
Average total biomass (roots and shoots) per pot
1483 g
1613 g
927 g
1002 g
These two values are statistically indistinguishable
These two values are statistically indistinguishable
But these two groups are statistically different

Conclusions: (1) Parasites suppress productivity more than competitors do; (2) the two parasites used here have equal effects.

FIGURE 37.15 The Effects of Root Parasitism Can Be Documented Experimentally
Each of the four treatment groups shown had a different combination of host plants and parasites, and each was replicated 10 times (giving a total of 40 pots). QUESTION Why was it important to have so many replicates of each treatment for the experiment to be meaningful?

for light. If this point of view is correct, then it is not surprising that parasitism is relatively rare in plants.

Carnivorous Plants

Carnivorous plants trap insects and other animals, kill them, and absorb the prey's nutrients. These plant species make their own carbohydrates via photosynthesis but use carnivory as a way to supplement the nitrogen available in the environment. Most are found in habitats where nitrogen is scarce or unavailable, such as the bog environments featured in **Box 37.2**.

Carnivorous plants use modified leaves to trap insects. Chapter 35 highlighted the leaves of pitcher plants, which form tubes that hold water. Prey are enticed to the tube by an attractive odor and either slip on the slick surface of the upper regions of the pitcher or are discouraged from climbing back out by downward-pointing hairs. Eventually they fall into the pool of water below, where they drown. Prey are then digested by resident bacteria and fungi, and the nutrients obtained from their dead bodies are absorbed into the plant.

BOX 37.2 Why Are Bogs So Nutrient Poor?

Bogs are wetlands that have have poor drainage and a pH between 3 and 4. They are most common in northern environments, where winters are long and summers are cool. Because bog water is largely stagnant, the rate of oxygen diffusion into the water may be outpaced by the rate of oxygen consumption by aerobic organisms that live on or in the water. Thus, bog water tends to be oxygen poor. The short growing season and cold, oxygen-poor water discourage the growth of fungi and aerobic bacteria that efficiently decompose dead organic matter. As a result, decomposition rates are slow in bogs, and bogs gradually fill with the partially decayed organic matter called **peat**.

Bogs are nutrient poor for two reasons. First, few nutrients are released by decomposition. Second, the scarcity of oxygen in the water makes it difficult for root cells to make the large amounts of ATP required to power proton pumps and bring nutrients into the body against a concentration gradient. Relatively few plant species thrive in bogs. In addition to carnivorous plants, bogs are home to sphagnum moss, certain shrubs such as cranberries, and trees such as tamarack and black spruce. Although the remains of all these species contribute to peat, the major component is sphagnum moss.

Dried peat has been used as a cooking and heating fuel in northwestern Europe for centuries, and occasionally peat-cutters discover archaeological objects that have been well preserved by the acidic, low-oxygen conditions. Among these finds are the corpses of hundreds of humans, dating from between 8000 BCE and the early medieval period. The body that researchers call Tollund man is a good example of these "bog people" (**Figure 37.16**). Tollund man is so well preserved that when peat-cutters discovered him in 1950, they thought they had a murder case on their hands. He is estimated to be around 2000 years old, however—he lived during the Early Iron Age. He was essentially naked, though his leather cap and belt were intact. Plant material recovered from his intestines indicated that his most recent meal was a soup made of barley and other plants. A leather rope around his neck suggests that he died by strangulation and was thrown in the bog after being murdered or executed. Many of the bog people, in fact, show clear evidence of a violent death.

FIGURE 37.16 Hundreds of Corpses Have Been Found Preserved in Bogs
This corpse, called "Tollund man" was found in a peat bog near Tollund, Denmark.

In sundews, modified leaves develop hairs that exude a sticky substance and function like flypaper (**Figure 37.17**). After insects that become trapped on the surface die, glands near the sundew's trap release enzymes that slowly digest the prey. The leaf then absorbs the nutrients that are released.

In the Venus flytrap, modified leaves trap insects mechanically. Each leaf has several sensory hairs that protrude from the epidermis. When an insect lands on the trap and bumps two or three of these hairs, the leaf responds by snapping shut. If the prey continues to stimulate hairs, the leaf begins to secrete digestive enzymes that slowly digest the prey.

How do the sensory hairs on a flytrap sense the presence of an insect? How is the signal from these hairlike structures transmitted to the rest of the leaf, and how is this information translated into a movement that closes the trap? Chapter 38 focuses on questions like these.

FIGURE 37.17 Sundews Have Modified Leaves with a Sticky Surface That Catches Insects

ESSAY Tropical Soils

The soils that underlie the Amazon basin and much of equatorial Africa are home to the world's most diverse and productive ecosystems: tropical wet forests. Ironically, these soils are also the most nutrient poor in the world.

Why? The same features that encourage luxuriant plant growth—large amounts of rain and high temperatures—also promote rapid leaching of nutrients. Negatively charged ions that dissolve in soil water tend to wash out of soils that are subjected to large amounts of precipitation. Further, a chain of events triggered by tropical climates causes positively charged ions to leach out of tropical soils. This chain begins when hot, wet weather encourages the growth of soil bacteria. As a by-product of cellular respiration, many of these cells produce CO_2, which reacts with soil water to form carboxylic acid. Acidity also accumulates in tropical soils from phosphate-containing molecules and phenolic compounds that are released from decaying vegetation. These chemical groups react with soil water to form phosphoric acid and an array of organic acids. The accumulation of protons is important, because it displaces nutrient-containing cations from their binding sites on clay particles—via cation exchange—and allows them to be leached away.

When intensive leaching continues for hundreds of thousands or even millions of years, what is left behind? Instead of presenting a complex mix of materials, tropical soils typically become dominated by minerals that do not dissolve readily. In fact, soil scientists classify many tropical soils as *oxisols* ("oxygen-soils"), because they contain so many tough iron oxides and aluminum oxides. The iron oxides present in oxisols turn them bright rust red.

The events that lead to the weathering of tropical soils are logical, but an important question remains: If nutrients in tropical soils are in such short supply, how can plant growth be so lush? The answer is that virtually all nutrients in these ecosystems are tied up in living vegetation. When a plant dies in a tropical wet forest, the plant's component molecules are rapidly recycled by the surrounding bacteria, archaea, fungi, and plants.

If nutrients in tropical soils are in such short supply, how can plant growth be so lush?

Unfortunately, one of two things happens when a tropical forest is cleared and burned for agriculture: Nutrients either go up in smoke or sit on the soil surface, in ash, until they are washed away in subsequent rains. Farmers who work tropical soils can get good production from their crops for a year or two after the forest is cleared, but after that the soils are depleted. Trying to raise crops on oxisols is often a losing proposition. Traditionally, native peoples coped with this situation by changing the locations of their gardens every few years and allowing vegetation to regrow on plots before using them again. But it remains to be seen whether modern, intensive farming techniques can be sustained on tropical soils. As an alternative, some land managers are attempting to implement logging practices that protect tropical soils while delivering a sustainable yield of forest products. It is not at all clear, however, that sustainable forestry practices can produce enough revenue to counter the pressure from increasing human populations for more cropland.

CHAPTER REVIEW

Summary of Key Concepts

In addition to needing carbon dioxide and water, plants require an array of essential nutrients to support growth. These nutrients are available as ions dissolved in soil water and are taken up by roots.

Typically about 96 percent of the dry weight of a plant consists of carbon, hydrogen, and oxygen. Plants obtain these elements by absorbing carbon dioxide from the atmosphere and water from the soil. The other 4 percent of the plant body consists of a complex suite of elements. The most important of these are the essential nutrients, which are required for normal growth and reproduction. Essential elements must be absorbed from the soil, generally in the form of ions like nitrate and phosphate.

Soil is a complex and dynamic mixture of inorganic particles such as clay and sand, organic particles, and living organisms. Soil provides plants with oxygen, water, and nutrients as well as a physical substrate for anchoring and supporting the body.

Nutrient absorption occurs via specialized proteins in root membranes. Most plants also obtain nitrogen or phosphorus from fungi associated with their roots. Toxins that enter roots are excluded from xylem by the Casparian strip or are actively transported into cell vacuoles and stored.

Nutrients are present at low concentrations in the soil surrounding roots. Plants import nutrients against a concentration gradient by pumping protons into the extracellular space. A large excess of protons outside the root hair creates a strongly negative membrane potential and thus an electrochemical gradient favoring the entry of positively charged ions. The proton gradient also allows negatively charged ions to cross the plasma membrane via proton cotransporters.

In addition to acquiring nutrients by establishing a membrane potential and a proton gradient, many plants obtain nitrogen or phosphorus from mycorrhizal fungi. The relationship between plants and the fungi is mutually beneficial. The host plant provides the fungi with carbohydrates; the fungi harvest N or P from the surrounding soil and transport it to the root system.

Passive and active systems also exist for excluding certain ions. This is important because all nutrients are toxic if they are present at high concentration. Passive exclusion occurs when ions cannot pass through the Casparian strip. Active exclusion often results from the production of specialized proteins that bind a particular ion, or the active transport of toxic ions into vacuoles.

Web Tutorial 37.1 Soil Formation and Nutrient Uptake

Some species of plant have specialized methods of obtaining nutrients, including associations with nitrogen-fixing bacteria, parasitism, and carnivory.

Certain plants, including those in the pea family, are capable of symbiosis with nitrogen-fixing bacteria. This association begins when flavonoids produced by the host plant are detected by bacteria in the soil. In response, bacterial cells produce sugar-containing molecules called Nod factors, which bind to a receptor protein on a root hair. A suite of host-plant genes is transcribed in response to this binding event, and the proteins that are produced result in the formation of a root nodule. The bacteria eventually take up residence inside the root cells that form the nodule, where they receive protection and sugar in exchange for producing ammonia from nitrogen gas.

Not all plants take up nutrients through membrane proteins or bacterial or mycorrhizal associations. Instead, some species are parasitic, and some are carnivorous. Most parasitic plants produce their own carbohydrates through photosynthesis but steal water and nutrients by infecting the root systems of host plants. Carnivorous plants have evolved mechanisms for trapping and digesting insects. The primary benefit of carnivory is to obtain nitrogen in low-nitrogen environments.

Questions

Content Review

1. Which of the following characteristics defines an element as essential for a particular species?
   a. It has to be added as fertilizer to achieve maximum seed production.
   b. If it is missing, a plant cannot grow or reproduce normally.
   c. If it is present in high concentration, plant growth increases.
   d. If it is absent, other nutrients may be substituted for it.

2. Why is the presence of clay particles important in soil?
   a. They provide macronutrients—particularly nitrogen, phosphorus, and potassium.
   b. They bind metal ions, which would be toxic if absorbed by plants.
   c. They allow water to percolate through the soil, making oxygen-rich air pockets available.
   d. The negative charges on clay bind to positively charged ions and prevent them from leaching.

3. Where does most nutrient uptake occur in roots?
   a. at the root tip, where root tissue first encounters soil away from the zone of nutrient depletion
   b. at the Casparian strip, where ions must enter the symplast prior to entering xylem cells
   c. in the symplastic and apoplastic pathways
   d. in root hairs, in the zone of maturation

4. Why is the activity of proton pumps in root hair membranes important?
   a. They create a strongly positive membrane potential (voltage).
   b. They allow toxins to be concentrated in vacuoles, so the toxins do not poison enzymes in the cytoplasm.
   c. They set up an electrochemical gradient that makes it possible to absorb cations and anions.
   d. They set up the membrane voltage required for action potentials to occur.

5. Phytoremediation is based on which of the following properties?
   a. Fungal hyphae have an extremely large surface area, and they secrete enzymes that digest dead plant material efficiently.
   b. Some plants can take up toxins from the soil and concentrate them in their tissues.
   c. Nitrogen-fixing bacteria live symbiotically *inside* root cortex cells, which divide to form protective nodules.
   d. The cold, oxygen-poor conditions in bogs slow decomposition rates, allowing peat to form.

6. How do epiphytes obtain nutrients?
   a. by trapping and digesting insects
   b. via mycorrhizal fungi
   c. via symbiotic bacteria called rhizobia
   d. from dust or other airborne particles

Conceptual Review

1. A farmer is concerned that her corn crop is suffering from iron deficiency. Design a greenhouse experiment that uses hydroponic cultures to (1) identify the symptoms of iron deficiency in corn and (2) identify the minimum amount of iron required to support normal growth rates.
2. Explain why carnivorous and parasitic plants are most common in nutrient-poor habitats.
3. Sandy soils contain very little clay. They are often nutrient poor because of extensive leaching and tend to be very dry. Why? Would you expect plant productivity to be higher on sandy soils or on soils containing both sand and clay?
4. Sketch the plasma membrane of a root hair. Include proton pumps, ion channels, and cotransporters. Now add ATP to the system, then protons in their correct relative concentration, and ions that move across the membrane via channels or cotransporters. Explain why plant cells have a negative membrane potential.
5. Why is it important for plants to exclude certain ions? Summarize the difference between active and passive exclusion mechanisms.
6. Fill in the table below.

Element	How Used in Cell?	What Form Available to Plants?	How Obtained by Plants?
Nitrogen			
Phosphorus			
Carbon			
Oxygen			
Hydrogen			
Iron			
Magnesium			

Group Discussion Problems

1. There is a conflict between van Helmont's data on willow tree growth and the data on essential nutrients listed in Table 37.1. According to the table, nutrients other than C, H, and O should make up about 4 percent of a willow tree's weight. Most or all of these nutrients should come from soil. But van Helmont claimed that the soil in his experiment lost just 2 ounces, while the tree gained 2627 ounces. If so, then soil contributed just 0.08 percent of the added weight instead of 4 percent. List three hypotheses to explain the conflict. How would you test these hypotheses?
2. Acid rain occurs when sulfur oxides and nitrous oxides released by cars or factories react with water vapor, forming sulfuric and nitric acids, respectively. In areas affected by acid rain, cations quickly leach out of the soil. Explain why.
3. Design an experiment using the drug vanadate, which poisons proton pumps, to test the hypothesis that phosphate ions enter cells via an H^+-PO_4^{2-} cotransporter.
4. Design an experiment to test the hypothesis that spruce trees take up more nitrogen from the soil in the presence of EMF than they do in the absence of EMF.

Answers to Multiple-Choice Questions 1. b; 2. d; 3. d; 4. c; 5. b; 6. d

www.prenhall.com/freeman is your resource for the following: Web Tutorials; Online Quizzes and other Online Study Guide materials; Answers to Conceptual Review Questions; Solutions to Group Discussion Problems; Answers to Figure Caption Questions and Exercises; and Additional Readings and Research.

Sensory Systems in Plants

38

Plants have sophisticated sensory systems. These radish seedlings sense the presence of light and are bending toward it.

KEY CONCEPTS

- Plants can perceive a wide variety of environmental stimuli and respond in ways that increase their ability to grow and reproduce.
- Plants sense several aspects of light. They respond in different ways to wavelengths that are typical of shade versus full sunlight. They also monitor photoperiod (day length).
- Plant shoots and roots sense gravity and respond by growing up and down, respectively.
- Plants sense mechanical stimulation (touch or wind). In some cases, plants respond to touch by generating rapid electrical signals.
- Plants sense disease-causing agents. Certain plant proteins bind to proteins produced by pathogens. When these plant receptors are stimulated, they trigger a defense response.

It is difficult for a human being to understand the sensory world of a plant. But imagine standing in place for several hundred years, like an oak tree. Each spring you produce flowers. All summer you absorb light from the Sun, carbon dioxide from the atmosphere, and water and nutrients from the soil. Water, ions, and sugars flow up and down your vascular tissue. For six months or more your body grows upward, outward, and downward. But as winter approaches, your metabolism slows and you stop growing. Like a hibernating animal, you spend the long, hard months of winter in a state of suspended animation, before awakening the following spring.

If you can imagine living as a tree does even for a moment, then it should come as no surprise that plants have sophisticated systems for sensing the direction, intensity, and quality of the light that strikes their bodies. For a plant, light is life. Plants do not have eyes or ears or noses, but they know when they are being shaded and which leaves are receiving the wavelengths of light that are required to support photosynthesis. They know what season of the year it is and may also be able to tell the time of day. They sense the pull of gravity and the force of wind. Even though they do not have the types of pain receptors that you do, they know when they are being attacked by viruses, bacteria, or insects.

In addition to sensing light, gravity, pressure, and attacks from enemies, plants respond to the information they receive in an appropriate way. They do not jump or swim or run, but their shoot systems grow toward light and become shorter and stockier in response to wind. Plants measure the length of nights and use the information to determine whether they will flower. In response to gravity, shoots grow up and roots grow down. In response to touch, the modified leaves of a Venus flytrap shut fast enough to catch flying insects.

This chapter focuses on how plants sense changes in their environment—specifically, how they gather information about light, gravity, and touch. Chapter 39 follows up by exploring the nature of the chemical signals that plants use to coordinate their responses to this information. The fundamental message of these two chapters is simple: Plants have extremely sophisticated systems for collecting information about their environment and for responding in ways that maximize their chances of surviving, thriving, and producing offspring.

38.1 Sensing Light

Plants can see, in a sense. This conclusion should not be unexpected, given that plants starve in the absence of light. If light is required for life, then plants should be able to gather and process information about it.

What specific qualities of light do plants detect, and what molecular mechanisms are responsible for these abilities? This section of the chapter is devoted to exploring these questions. In doing so, it introduces the fundamental logic of research on the sensory abilities of plants, animals, and other organisms: To show that an organism can perceive a particular aspect of its environment, researchers have to document a response to it. This is exactly what Charles Darwin and his son Francis did in 1881, in the first experiments on plant sensory reception.

What Do Plants See?

The first experiment on the sensory systems of plants involved a paraffin lamp or other light source and the coleoptiles of a plant called reed canary grass. A **coleoptile** is part of an emerging shoot—specifically, a modified leaf that forms a sheath protecting the stems and leaves of young grasses. The Darwins germinated reed canary grass seeds in the dark, placed the young, straight shoots next to a light source, and noted that the shoots bent toward the light (**Figure 38.1a**). You have probably seen the same response in houseplants that are near a window. This type of directed movement in response to light is called **phototropism** ("light-turn"). Plant shoots are positively phototropic, meaning they grow toward light. Because plants respond to light, it was logical to infer that they can sense it.

The Darwins also observed that if they exposed shoots to light filtered through a solution of potassium dichromate, the shoots did not bend toward the light. The researchers did not know why no response occurred, but later it was shown that potassium dichromate solutions filter out wavelengths in the blue part of the visible spectrum. To understand why detecting a source of blue light is important, recall from Chapter 10 that chlorophylls *a* and *b* are the primary photosynthetic pigments and that they absorb strongly in the blue part of the spectrum. Blue light drives photosynthesis. Based on these observations, it is logical to observe that plants sense the presence of blue light and exhibit a phototropic response if blue wavelengths are available—but show no response if blue wavelengths are not present (**Figure 38.1b**). The Darwins' work with potassium dichromate solutions was the earliest indication that plants can sense specific wavelengths of light. When plants see blue light, they respond by bending or growing toward it.

Distinguishing Red and Far-Red Light More recent work has shown that plants are sensitive to wavelengths in the red and far-red portions of the visible spectrum, in addition to blue wavelengths. This sensitivity is interesting, because red wavelengths and far-red wavelengths mean very different things to a plant. Red light (wavelengths from about 660 to 700 nm) drives photosynthesis, just as blue light does. (See Figure 10.7a.) Far-red wavelengths, in contrast, are over 710 nm and are not absorbed strongly by photosynthetic pigments. Instead, they act as an important signal about the local environment. Because far-red wavelengths tend to pass through leaves instead of being absorbed, they are prominent in light that is filtered through tree leaves before it reaches the forest floor. Far-red light indicates shade. A plant that receives a large amount of far-red light is probably being shaded by competitors.

(a) Shoots bend toward full-spectrum light.

(b) Shoots bend specifically toward blue light.

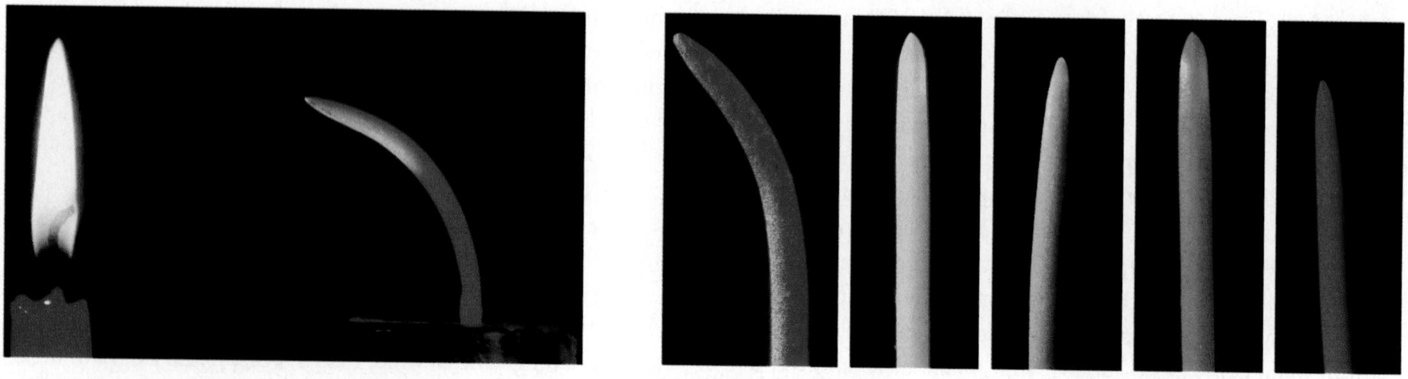

FIGURE 38.1 Experimental Evidence That Plants Sense Specific Wavelengths of Light
(a) In their original experiments, the Darwins used paraffin lamps or other sources of artificial light. **(b)** When shoots are exposed to specific wavelengths of light, phototropism occurs only in response to blue light.

If you have grown bean seeds indoors and watched the shoots get long and spindly, you have seen a response to far-red light firsthand. Beans normally grow in open sunlight. If they are grown indoors, the plants react as though they are being shaded and are attempting to grow high enough to reach open sunlight. To explore this phenomenon more thoroughly, consider the experiment summarized in **Figure 38.2**. Growth chambers with an array of light environments were set up. In each chamber, the availability of the blue and red wavelengths used in photosynthesis was constant, but the amount of far-red light was varied. Each chamber contained two types of plants: species that naturally grow in sunny locations and plant species that grow in shade. According to the "Results" section of Figure 38.2, at least some species from sunny habitats elongate their stems much more strongly in response to shade light than do species from forest-floor habitats. This result prompts several conclusions. First, the result confirms that at least some plants can sense far-red wavelengths. Second, at least some plants that are adapted to growing in high-light conditions respond to far-red wavelengths by elongating their stems.

WEB TUTORIAL 38.1
Sensing Light

Question: How do plants respond to far-red wavelengths of light, which indicate shade?

Hypothesis: Plants adapted to sunny habitats respond to far-red light differently than do shade-adapted plants.

Null hypothesis: Plants adapted to sunny habitats and plants adapted to shady habitats respond similarly to far-red light.

Experimental setup:

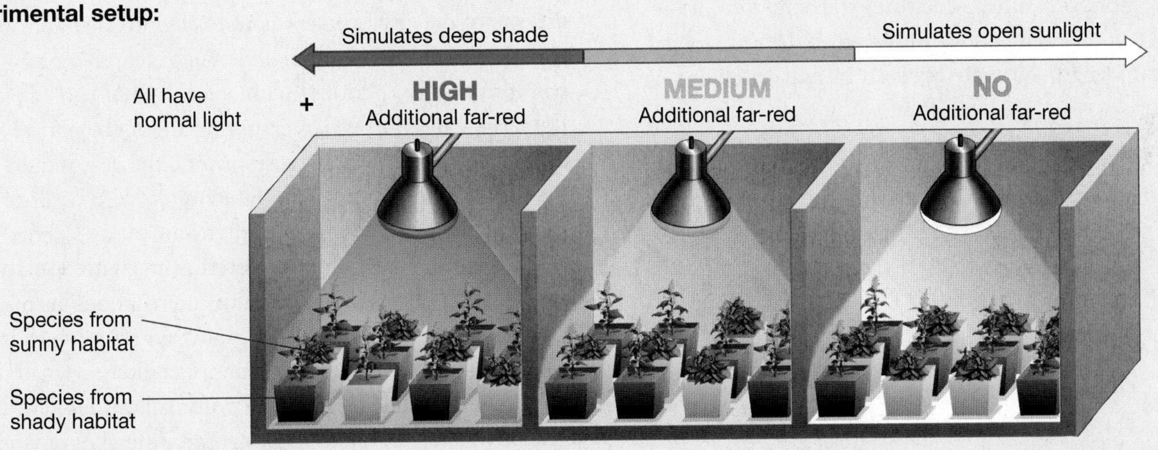
Simulates deep shade
Simulates open sunlight
All have normal light
+
HIGH
Additional far-red
MEDIUM
Additional far-red
NO
Additional far-red
Species from sunny habitat
Species from shady habitat

Prediction: In response to far-red light, species from sunny habitats extend their stems more than do species from shady habitats.

Prediction of null hypothesis: In response to far-red light, species from sunny and shady habitats exhibit similar degrees of stem elongation.

Results:

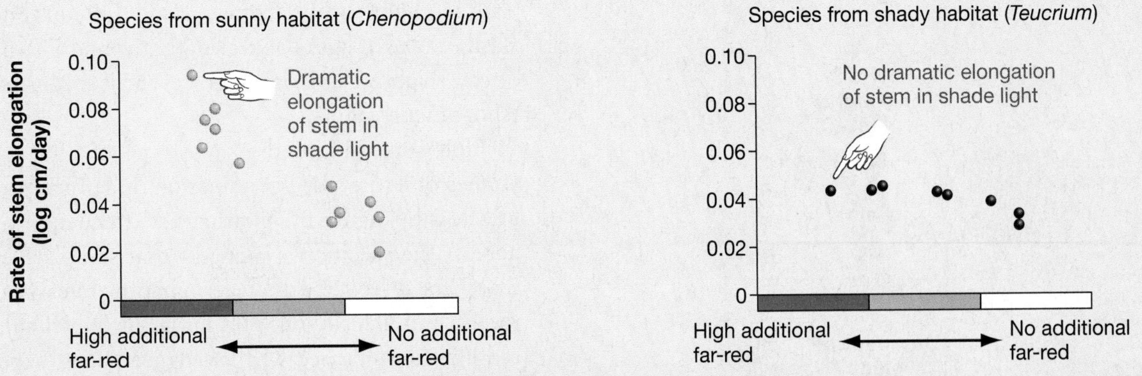
Species from sunny habitat (Chenopodium)
Rate of stem elongation (log cm/day)
0.10
0.08
0.06
0.04
0.02
0
Dramatic elongation of stem in shade light
High additional far-red
No additional far-red
Species from shady habitat (Teucrium)
0.10
0.08
0.06
0.04
0.02
0
No dramatic elongation of stem in shade light
High additional far-red
No additional far-red

Conclusion: At least some species that are adapted to sunny habitats respond more strongly to shade light and show a greater amount of stem elongation than do species adapted to shady habitats.

FIGURE 38.2 Experimental Evidence That Species from Sunny Habitats Elongate Their Stems in Response to Far-Red Light
Each plant received the same quantity of the light wavelengths used in photosynthesis. The only thing that varied among treatments was the relative amount of far-red light. Shady habitats are dominated by far-red light. QUESTION Seedlings were placed in random locations within each growth chamber. Why is this an important part of the experimental design?

To summarize, plants monitor the amount of blue and far-red light they receive and moderate their behavior accordingly. They grow toward blue light. If they are adapted to sunny habitats, they also elongate to get away from far-red light. What other aspects of light can plants sense?

Photoperiodism **Photoperiodism** is any response by an organism that is based on the photoperiod—that is, the relative lengths of day and night. In plants, the ability to measure photoperiod is important because it allows individuals to respond to seasonal changes in climate—for example, to flower when pollinators are available and when resources for producing seed are abundant. Experiments on photoperiodism in plants showed that, in regard to flowering, plants fall into three main categories (**Figure 38.3a**):

1. **Long-day plants** bloom in midsummer, when days are longest and nights shortest. Radishes, lettuce, spinach, corn, irises, and other long-day plants flower only when days are longer than a certain length—usually between 10 and 16 hours.

(a) How do plants respond to: Differences in day length?

(b) How do plants respond to: Nights interrupted by light?

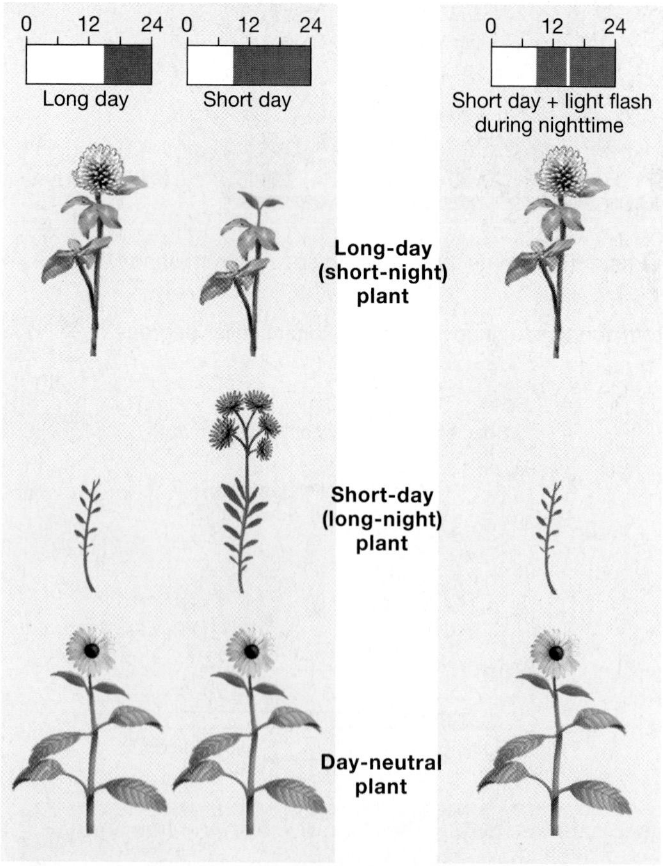

FIGURE 38.3 Different Species Respond to Photoperiod in Different Ways
(a) Flowering responses to photoperiod fall into three categories.
(b) Flashes of light during the night alter the photoperiod response in short-day (long-night) species and long-day (short-night) species.

2. **Short-day plants** bloom in spring, late summer, or fall. Asters, chrysanthemums, poinsettias, and other short-day plants flower only if days are shorter than a certain species-specific length.
3. **Day-neutral plants** flower without regard to photoperiod. Day length has no effect on flowering in roses, snapdragons, dandelions, tomatoes, cucumbers, and many weeds. (A *weed* is a plant that is adapted to grow in disturbed soils.)

Even though researchers still refer to "long-day" and "short-day" species, plants do not actually sense day length. Instead, they monitor the length of the night. This hypothesis was tested in classic experiments on flowering in a short-day plant called cocklebur, which requires less than 16 hours of light to flower. To explore what plants are sensing when they sense photoperiod, researchers grew cocklebur in growth chambers where all aspects of the environment could be controlled and where photoperiod could be manipulated. The researchers found that interrupting day periods with dark made no difference to flowering. In contrast, interrupting the night period with light did matter. To interpret these results, the researchers hypothesized that when nights were interrupted with a short period of light, plants responded to the environment as a short night.

As **Figure 38.3b** shows, experiments with long-day plants showed that they also respond to interruptions in night length. Day-neutral species, in contrast, show no response to interrupted nights. Based on these observations, it could be argued that plants should be classified as short-night instead of long-day, long-night instead of short-day, and night-neutral instead of day-neutral.

The Red/Far-Red Switch When researchers explored the interrupted-night phenomenon in more detail, they made an interesting discovery. In short-day (long-night) plants, flowering is inhibited most effectively by interrupting the dark period with red light—wavelengths around 660 nm. But a subsequent flash of far-red light—wavelengths around 735 nm—erased the effect (**Figure 38.4**). Similar reversals occurred in long-day (short-night) plants.

Biologists noticed the same type of red/far-red switch in studies of lettuce seed germination. In lettuce, seed sprouting is affected by light. By plotting the frequency of germination against the specific wavelengths that seeds are exposed to, researchers discovered that germination rates peak when seeds receive red light (about 660 nm). But the stimulatory effect of red light disappeared when seeds were later exposed to far-red light. Germination was inhibited most effectively by wavelengths near 735 nm.

As the data in **Table 38.1** show, red and far-red light act like an on-off switch for seed germination in lettuce, just as they do for flowering in some plants. In lettuce, red light promotes germination, while far red inhibits it. The last wavelength sensed by the seed determines whether germination occurs at a high rate.

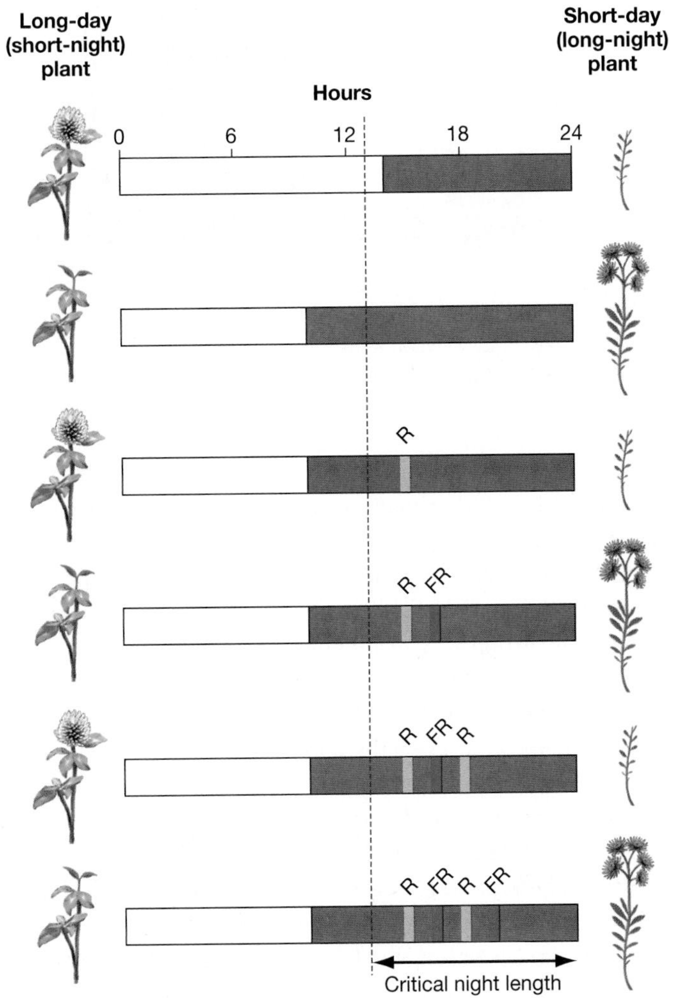

FIGURE 38.4 Flashes of Red Light and Far-Red Light Switch the Photoperiod Response On and Off
The effect of light flashes on photoperiod responses varies with the wavelength of light used. If a series of flashes is given with alternating red light and far-red light during the night, the plant's flowering response correlates with the last light it experienced.

TABLE 38.1 How Do Red Light and Far-Red Light Affect the Germination of Lettuce Seeds?

Biologists exposed lettuce seeds to flashes of light containing one of two wavelengths: red or far red (FR). After exposure to light, the seeds were moistened and held in the dark for several days.
QUESTION According to these data, what is the average germination rate of lettuce seeds that were last exposed to red light? To far-red light? How do these values compare with the germination rate of seeds that are buried underground and receive no light at all?

Light Exposure	Germination (%)
None	9
Red	98
Red + FR	54
Red + FR + Red	100
Red + FR + Red + FR	43
Red + FR + Red + FR + Red	99
Red + FR + Red + FR + Red + FR	54
Red + FR + Red + FR + Red + FR + Red	98

The red/far-red switch is logical, because lettuce requires full sunlight to thrive. Lettuce seed that germinated in shade, in light.

But how would a red/far-red switch work at the molecular level?

How Do Plants See?

To interpret the red/far-red switch in photoperiod and seed germination, biologists hypothesized that the same pigment is involved in triggering both processes and that it exists in two conformations: one that absorbs red light and one that absorbs far-red light. The idea was that switching behavior is observed during experiments on photoperiod and germination because the photoreceptor pigment is induced to switch conformation. Each conformation would be responsible for a different response.

The biologists who suggested these hypotheses called the proposed pigment **phytochrome** ("plant-color"). Phytochrome was thought to be a specialized light receptor, different from any of the pigments involved in absorbing light during photosynthesis. This hypothesis was logical, because photosynthetic pigments are specialized for promoting electrons in response to light absorption and do not show switching behavior, or what the researchers termed **photoreversibility.** In addition, light-dependent responses are observed in seeds that have no chlorophyll and in dark-grown plants, which have so little chlorophyll that they are extremely pale.

Figure 38.5 (page 876) illustrates the photoreversibility hypothesis. The idea was that a conformation of phytochrome called P_r (phytochrome red) absorbs red light. Another conformation of the same molecule, called P_{fr} (phytochrome far-red), absorbs far-red light. According to the photoreversibility hypothesis, each conformation switches to the other when it absorbs its preferred wavelength. Were any of these ideas correct?

Phytochromes as Red and Far-Red Receptors Like the bean seedlings discussed earlier, young corn plants lengthen their stems in response to light deprivation or exposure to an excess of far-red light. This behavior suggested that they have a receptor for far-red light. To follow up on this observation, researchers purified proteins from young corn shoots and isolated one that was photoreversible. Specifically, when the protein was placed in solution and exposed to alternating red and far-red light, the color of the solution switched from blue to blue-green and back. The color switches supported the hypothesis that the protein absorbs red as well as far-red light. Based on these data, the biologists claimed that they had succeeded in isolating phytochrome. The year was 1959.

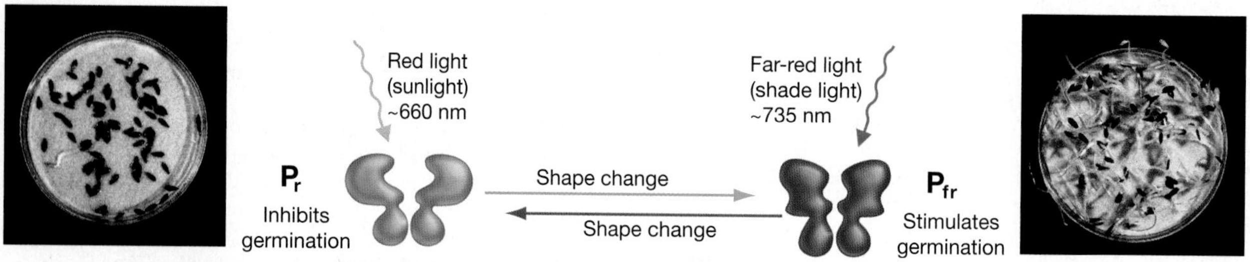

FIGURE 38.5 The Photoreversibility Hypothesis for Phytochrome Behavior
According to the photoreversibility hypothesis, (1) the phytochrome protein has two distinct conformations; (2) the P_r conformation absorbs red light, while the P_{fr} conformation absorbs far-red light; (3) when either form absorbs light, it converts to the other form; and (4) the P_{fr} form stimulates germination, while the P_r form inhibits it.

More recently, the gene that codes for phytochrome has been isolated and sequenced. In fact, the genome of *Arabidopsis thaliana*—the small, weedy mustard plant that is widely used as a model organism in plant biology—has five distinct genes that encode phytochrome proteins. The genes are referred to as phytochromes A through E and are symbolized *PHYA*, *PHYB*, and so on. The proteins encoded by these genes hold a small pigment molecule that absorbs light in the red and far-red parts of the spectrum. In addition, all five proteins appear to be photoreversible. Finally, recent data suggest that different phytochromes within the same individual may be involved in more than one light-dependent process. For example, *Arabidopsis* individuals with mutant forms of *PHYB* have abnormal stem elongation responses, while both *PHYA* and *PHYB* are involved in the regulation of germination. Recent work has also shown that phytochromes are the pigments responsible for monitoring night length and for triggering an appropriate flowering response.

How is the information present in each of the P_r-P_{fr} switches translated into action that affects germination, stem elongation, flowering, and other responses to red and far-red light? This question brings us to the forefront of research on phytochromes. The answer appears to be complex; one study suggests that almost one-third of all *Arabidopsis* genes are involved in the elongated growth characteristic of shade avoidance. Work on linking the phytochrome switch to the elongation and germination responses continues.

Phototropins as Blue-Light Receptors Phytochromes appear to be specialized for monitoring the amount of far-red (shade) light that plants receive. If so, then which pigment is responsible for absorbing blue light and triggering the phototropic response that allows plants to grow toward sunlight?

Charles Darwin and Francis Darwin provided the first hint of an answer in the 1870s, when they established that the receptors responsible for phototropism in reed canary grass are located in the tips of growing shoots. But the key breakthrough came in the early 1990s, when researchers found a membrane protein that is abundant in the tips of emerging shoots and that gains a phosphate group in response to blue light.

To understand why this result was so significant, recall from Chapters 3 and 8 that many proteins switch from inactive to active states, or vice versa, when a phosphate group (PO_4^{2-}) from ATP is added to them or removed. The addition of a phosphate group is called **phosphorylation**; the removal of a phosphate group is termed **dephosphorylation**. Biologists hypothesized that the membrane protein becomes activated when it is phosphorylated in response to blue light and that the activated protein then triggers the phototropic response.

Subsequent work succeeded in isolating the gene that codes for the membrane protein. The gene was named *phot1* and was found by analyzing mutant *A. thaliana* individuals that do *not* show a phototropic response to blue light. **Figure 38.6** summarizes the experiment that convinced most biologists that *phot1* codes for a blue-light receptor. Researchers who inserted copies of the *phot1* gene into insect cells that were growing in culture found that the PHOT1 protein product became phosphorylated in response to blue light. Because no other plant proteins were present in the experimental insect cells, the results suggest that PHOT1 phosphorylates itself in response to blue light. Based on this result and on follow-up experiments, the current consensus is that *phot1* encodes a blue-light detector in plants and that a phototropic response is triggered when this protein is phosphorylated.

The most recent research indicates that there are multiple blue-light receptors related to PHOT1, just as there are multiple phytochromes that act as far-red receptors. Collectively, photoreceptors that detect blue light and initiate phototropic responses are known as **phototropins**. In addition to initiating the bending of stems toward light, phototropins initiate the chloroplast movements within leaf cells that make optimal light absorption possible.

Different types of blue-light receptors trigger plant responses other than phototropism to sunlight. The carotenoid pigment **zeaxanthin**, for example, initiates the opening of stomata in response to blue light, so that carbon dioxide can diffuse into photosynthetically active cells. Photoreceptors called **cryptochromes** ("hidden-colors") are blue-light photoreceptors involved in stem growth under shady conditions and in flower induction. Based on these results, biologists contend that plants have an array of blue-light receptors. Each receptor triggers a response that increases a plant's ability to harvest light, make food, and produce offspring.

From Perception to Response: Signal Transduction

Between the sensing of a signal and an organism's response to that signal, several steps have to occur at the molecular level. In effect, the information that an individual receives about what is going on in the outside world must be translated and communicated in some way. The series of events that occurs between stimulus and response is called **signal transduction**. The name was inspired by the verb *transduce,* which means to convert energy from one form to another, and by the observation that the process of translation and communication often involves an energy transformation. In plants, for example, the energy in blue light is transformed into chemical energy in the form of a phosphorylated PHOT1 protein. In this case, the phosphorylation event is the first step in the signal transduction process. What happens next?

Figure 38.7 illustrates some possible outcomes of signal transduction in various types of cells. It turns out that in many

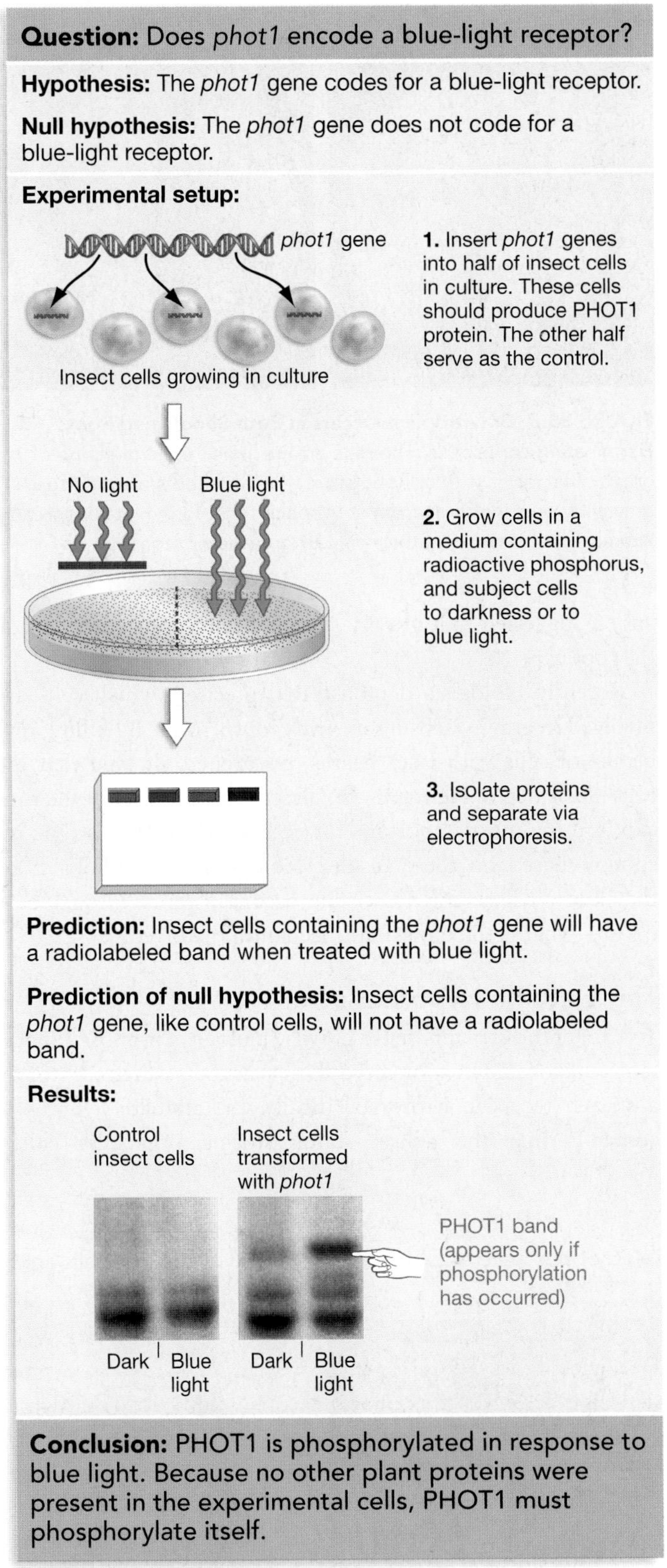

Question: Does phot1 encode a blue-light receptor?
Hypothesis: The phot1 gene codes for a blue-light receptor.
Null hypothesis: The phot1 gene does not code for a blue-light receptor.
Experimental setup:
phot1 gene
1. Insert phot1 genes into half of insect cells in culture. These cells should produce PHOT1 protein. The other half serve as the control.
Insect cells growing in culture
No light
Blue light
2. Grow cells in a medium containing radioactive phosphorus, and subject cells to darkness or to blue light.
3. Isolate proteins and separate via electrophoresis.
Prediction: Insect cells containing the phot1 gene will have a radiolabeled band when treated with blue light.
Prediction of null hypothesis: Insect cells containing the phot1 gene, like control cells, will not have a radiolabeled band.
Results:
Control insect cells
Insect cells transformed with phot1
PHOT1 band (appears only if phosphorylation has occurred)
Dark
Blue light
Dark
Blue light
Conclusion: PHOT1 is phosphorylated in response to blue light. Because no other plant proteins were present in the experimental cells, PHOT1 must phosphorylate itself.

FIGURE 38.6 Experimental Evidence That PHOT1 Is a Blue-Light Receptor That Phosphorylates Itself

The results show that the transformed insect cells produced the PHOT1 protein and that many copies of the PHOT1 protein acquired a radioactive phosphate group if the cells were exposed to blue light.
QUESTION Why did the researchers bother to analyze cells that were not transformed with the *phot1* gene and that were not exposed to blue light?

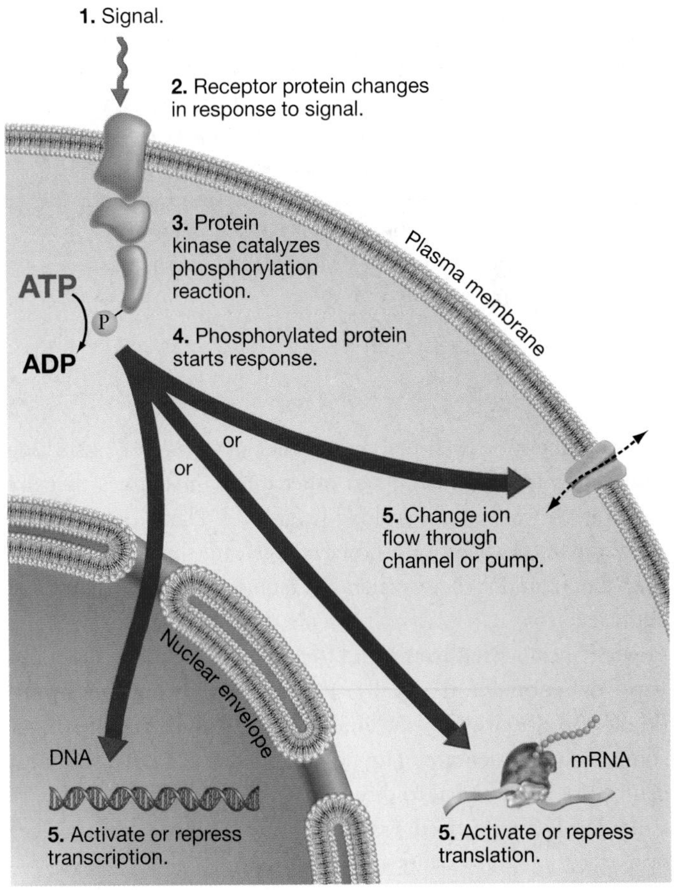

1. Signal.
2. Receptor protein changes in response to signal.
3. Protein kinase catalyzes phosphorylation reaction.
Plasma membrane
ATP
P
ADP
4. Phosphorylated protein starts response.
or
or
5. Change ion flow through channel or pump.
Nuclear envelope
DNA
mRNA
5. Activate or repress transcription.
5. Activate or repress translation.

FIGURE 38.7 Signal Transduction Can Lead to Several General Types of Molecular Responses

Many different responses occur after signal transduction. They range from changes in transcription and translation to alterations in channels, pumps, and other membrane proteins.

cases, the response to a signal involves a phosphorylation event and a subsequent change in the activity of a response protein or a series of proteins (see Chapter 8). The activation of these proteins might lead to an increase or decrease in the transcription of certain genes or a change in how efficiently particular mRNAs are translated into proteins. In other cases, the key response involves the activation or inactivation of a transmembrane channel and a change in ions that flow into or out of the cell. In this way, information from outside the cell can lead to changes in gene expression and a change in the cell's activity. Signal transduction connects an environmental signal with a cellular response.

Is the phototropic response in *Arabidopsis* and other plants triggered by a change in gene expression or ion flows? The answer is still unclear. Understanding the sequence of events that occurs as light signals are transduced to phototropic movements and other responses remains one of the great challenges in research on plant sensory reception. Signal reception is now fairly well understood; the focus of current research is on signal transduction.

✓CHECK YOUR UNDERSTANDING

Plant sensory systems have three major components: (1) a receptor that changes in response to a stimulus; (2) a signal transduction pathway that transforms the stimulus into a chemical or electrical signal inside the plant; and (3) the transcription of target genes, the activation of specific enzymes, or other changes in cell activity. In this way, plants receive information from their environment and respond with appropriate action. You should be able to describe the three components of a system that allow plants to sense and respond to a specific aspect of light.

38.2 Gravity Perception

A plant's body grows into an efficient machine for absorbing sunlight, water, nutrients, and other diffuse resources that are available above ground and below ground. Plants do not grow randomly in all directions, however. Instead, they grow toward light that contains the wavelengths required for photosynthesis, and they grow in a way that orients their body with respect to gravity. Shoots usually respond to gravity by growing in an upward direction; roots usually respond by growing downward (**Figure 38.8**). Gravity is a unique stimulus, because it is constant and unidirectional. How do plants sense gravity, so that it can be used as a signal to direct growth?

In 1881 Charles and Francis Darwin published one of the first experimental results on **gravitropism** ("gravity-bend")—the ability to move in response to gravity—in plants. To understand the result, recall from Chapter 35 that the ends of root tips are covered by a tough, protective collection of cells called the **root cap** (**Figure 38.9a**). The Darwins found that roots stop responding to gravity if the root caps are removed. This observation suggested that gravity sensing occurs somewhere inside the root cap.

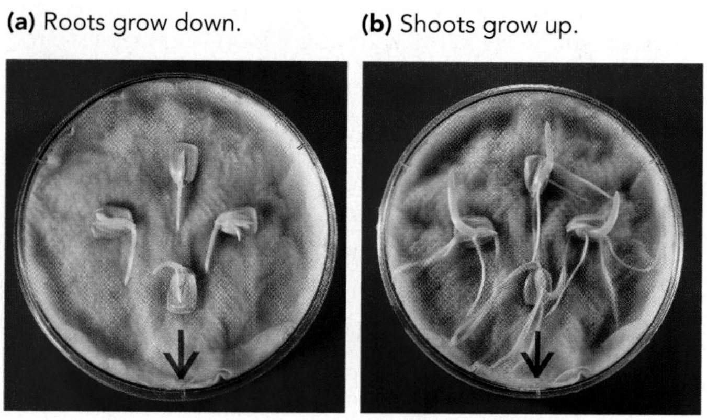
(a) Roots grow down.
(b) Shoots grow up.

FIGURE 38.8 Gravitropism Occurs in Both Shoots and Roots
Germinating corn seeds. The blue arrows indicate the direction of gravity. In species with upright growth habits, shoots respond to gravity by growing away from it. In species that form mats or spread laterally, shoots respond to gravity by growing horizontally.

Recently biologists demonstrated precisely which cells are involved in gravity sensing in *Arabidopsis* roots. By killing tiny blocks of cells with laser beams, researchers showed that the cells illustrated in **Figure 38.9b**, directly at the center of the root cap, are the most important for regulating the gravitropic response. Based on these results, biologists are confident that they know where gravity sensing occurs in roots. How does it occur—what is the molecular mechanism involved?

The Statolith Hypothesis

To explain how plants sense gravity, biologists proposed that a molecule located in root cap cells acts as a pressure receptor or a stretch receptor. More specifically, the **statolith hypothesis** contends that the dense, starch storing organelles called

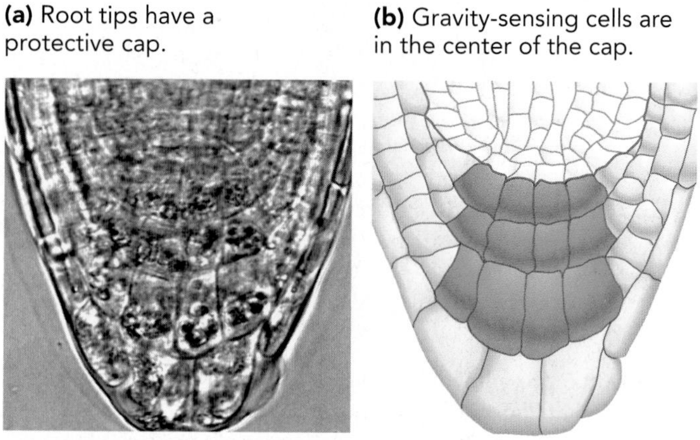
(a) Root tips have a protective cap.
(b) Gravity-sensing cells are in the center of the cap.

FIGURE 38.9 Experimental Evidence that Gravity-Sensing Cells Are Located in the Root Cap
(a) The root cap is a protective structure. **(b)** When the cells marked in red are killed experimentally, the gravitational response in roots is dramatically reduced.

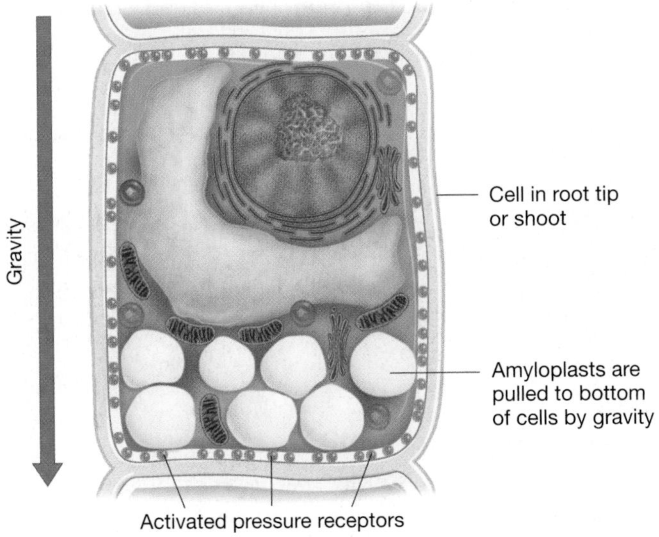

FIGURE 38.10 The Statolith Hypothesis States that Amyloplasts Stimulate Gravity Sensors
Amyloplasts are filled with starch. They are dense, so they sink in response to gravity. The statolith hypothesis predicts that pressure receptors in the plasma membrane become activated as a result.

amyloplasts contain a gravity receptor (**Figure 38.10**). The idea is that gravity pulls the heavy amyloplasts to the bottom of cells, where somehow the weight of the amyloplast activates a receptor that initiates the gravitropic response.

The statolith hypothesis was inspired by the way that many animals sense gravity. Lobsters, for example, take up grains of sand that become positioned in specialized gravity-sensing organs in their antennae. The grains of sand are called **statoliths** ("place-stones"). When the animal tilts or flips over, the statolith moves in response to gravity. Inside the organ, the sand grain ends up pushing against a new receptor cell, which sends an electrical signal to the brain. Because the location of the responding cell tells the brain how the body is now positioned with respect to gravity, the animal can respond by righting itself. Similarly, in plant roots that are reoriented with respect to gravity (turned upside down, for example), the amyloplasts in the root-cap cells settle onto the new "lower" cell walls.

Although recent experiments strongly support the statolith hypothesis, biologists are struggling to interpret several observations. For example, *Arabidopsis* mutants that are incapable of manufacturing starch have amyloplasts that are small and not very dense. Although the starchless mutants have a dramatically reduced sensitivity to gravity, they still show about 25 percent of the normal response. Why they show any response at all is a question. In addition, plant species that lack amyloplasts entirely are still able to sense gravity. To explain this observation, biologists hypothesize that the entire weight of the cell activates a gravity sensor in these species. The most important issue in research on gravity sensing, though, is to find the receptor molecule itself and to explore its mode of action.

Is the Gravity Sensor a Transmembrane Protein?

Some researchers contend that the weight of the statoliths on receptors in the cell's plasma membrane elicits a gravitropic response, while other researchers propose that the key event occurs when amyloplasts pull on cytoskeletal elements, which in turn activate receptors in the plasma membrane. Proponents of both hypotheses are keenly interested in membrane proteins called integrins. Recall from Chapter 8 that in animal cells, **integrins** span the plasma membrane and form a line of communication between the extracellular matrix (ECM) and the interior of the cell. One end of an integrin binds to the fibrous proteins collagen and fibrinogen in the ECM; the other end binds to microfilaments in the cytoskeleton. Biologists have proposed that (1) integrins exist in plants as well as animals, and (2) integrins act as gravity sensors in plants by responding to differences in pressure on the tops and bottoms of root cells. The idea is that integrins at the top of the cell would be under tension (pulling force); integrins at the bottom would experience compression (pushing).

One recent test of the integrin hypothesis used antibodies to animal integrin proteins to locate integrin-like proteins in *Arabidopsis* and the green alga *Chara*. Recall from earlier chapters that antibodies are proteins that bind to specific sites on specific molecules. To use antibodies in localization studies, researchers create an antibody to the protein they want to find, employing techniques introduced in Chapter 49. They attach a radioactive or fluorescent tag to the antibody, treat tissue with the antibody, and then locate the position of the tag.

One recent set of experiments found that antibodies to integrins bound to proteins in the roots and shoots of *Arabidopsis* and in the internodal cells and rhizoids of *Chara*. These polypeptides were associated with plasma membranes; in *Arabidopsis* roots, most of the antibody tags localized to root-cap cells. The researchers who did the study claim that these results provide strong support for the hypothesis that plants have integrins and that plant integrins function as gravity sensors. There is considerable controversy about whether the molecules that became bound by the antibodies are actually integrins, however. Further research is required before biologists can claim that plants and animals share this fundamental aspect of cell structure and communication. Although integrins may qualify as a candidate for a gravity receptor in plants, there is no firm evidence that they actually serve such a function. Research continues.

38.3 How Do Plants Respond to Wind and Touch?

When plants perceive blue light, red light, far-red light, or gravity, the sensor that is involved responds in part by triggering some sort of chemical change in the cell. Often this change is a phosphorylation event that activates a key protein. In this way, plants transduce the kinetic energy in light or the mechanical energy of gravity into changes in chemical energy.

FIGURE 38.11 Plants Get Shorter and Stockier in Response to Wind or Touch
The tomato plants shown here were touched lightly 0, 10, or 20 times per day each day for 10 consecutive days.

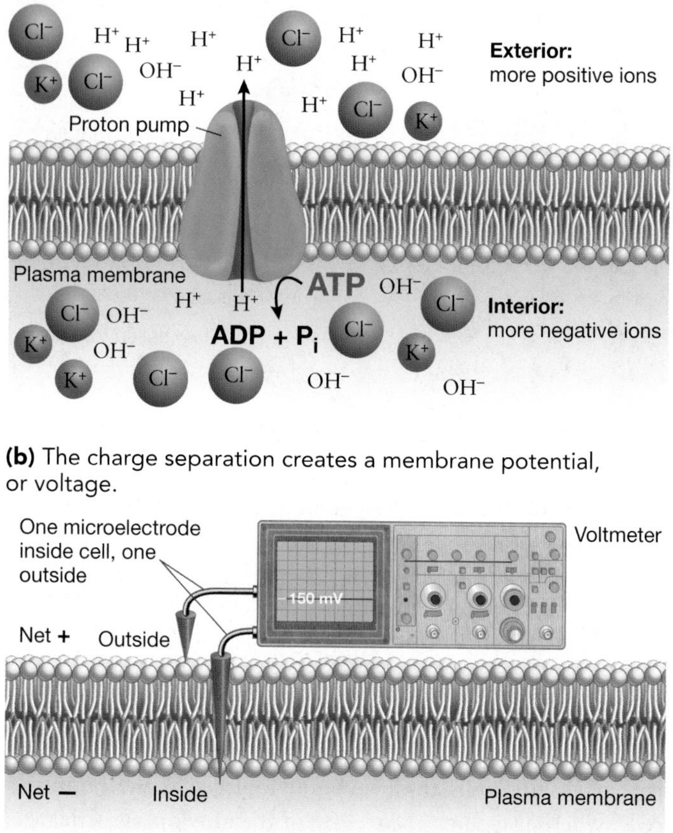

FIGURE 38.12 Proton Pumps Create a Charge Separation across Plasma Membranes in Plants
(a) The insides of plant cells normally have a negative charge relative to the charge on the outside of the cell. **(b)** Any charge separation creates a voltage. The voltage across a plasma membrane can be read by inserting electrodes into the interior and exterior of a cell.

When plants are buffeted by wind, the mechanical force is also transduced by a chemical event. Although the exact receptor and transduction pathway are not known, studies in *Arabidopsis thaliana* have shown that a large suite of genes is transcribed in response to touch or other mechanical stimuli, comparable to that produced by wind. The protein products of some of these genes act to stiffen cell walls, resulting in plants that are shorter and stockier relative to plants that do not experience repeated vibrations or touching (**Figure 38.11**).

In "touch-sensitive" plants, such as mimosas, sundews, and flytrap species, though, plant sensory cells may also transduce the mechanical signal they receive into an electrical signal. Let's consider a deceptively simple question: How does this happen?

An Introduction to Electrical Signaling

The key to understanding electrical signaling in plants is to remember that the interior of most plant cells has a negative charge relative to that in the exterior. This charge difference occurs because the proton pumps introduced in Chapter 37 are active in many cells. Recall several other key points from that chapter and other earlier chapters:

- When H^+-ATPases move protons to the exterior of a plant cell, a charge separation is created across the membrane (**Figure 38.12a**). As a result, most plant cells have a membrane voltage, also called a **membrane potential**.
- The size of a cell's membrane potential is a function of the amount of charge separation. Because membrane potentials are small, they are usually expressed in units called millivolts (mV, or 1/1000 of a volt). By convention, membrane potentials are expressed as the state of a cell's interior relative to the exterior. As a result, the normal membrane potential of an active plant cell is negative—typically −150 mV to −200 mV (**Figure 38.12b**). In contrast, the resting membrane potential of animal nerve cells is typically −70 mV.
- Voltage is a form of potential energy. Potential energy can be thought of as the tendency for something to move. For example, Chapter 2 explained that the chemical potential of an electron describes its tendency to move closer to a nucleus. Chapter 36 pointed out that the water potential of a tissue or substance (such as air or soil) expresses the tendency of water to move toward or away from it. In the same way, an electrical potential is the tendency of charged particles (electrons or ions) to move toward an area of opposite charge.

To pull these observations together, imagine that a sensory receptor in the plasma membrane of a plant cell responds to touch—meaning a mechanical force. Receptors undergo a change in conformation or activity in response to the stimulus. If the change in the receptor causes a change in the cell's membrane potential—perhaps by allowing certain ions to flow across the membrane—then the mechanical signal has been transduced to an electrical signal. If the flow of ions is triggered by a change in a receptor, then changes in membrane potential contain information. This is exactly what happens in a variety of algae and land plants.

As an example of electrical signaling in plants, consider the Venus flytrap, introduced in Chapter 37. The modified leaves of this bog-dwelling species have sensory hairs on their surface. When an insect brushes against at least two of these hairs, the leaves act as a trap and respond by slamming shut. The response to this hair trigger takes just half a second. In 1873 J. Burdon-Sanderson showed that the rapid movement is mediated by an electrical signal that spreads across the trap. Let's take a closer look.

Action Potentials

In animals, nerve impulses travel down the length of specialized cells called neurons. The impulses consist of a flow of charge in the form of ions across the plasma membrane. Because of the charge flow, the voltage across the plasma membrane changes drastically. A voltage change is what Burdon-Sanderson recorded when he inserted an electrode into a Venus flytrap cell and triggered the sensory hairs.

The voltage change that occurs in the Venus flytrap has a characteristic pattern, called an **action potential. Figure 38.13a** shows what an electrode records as an action potential passes through a cell; **Figure 38.13b** diagrams the corresponding events at the plasma membrane. To interpret these figures, begin at the leftmost point of Figure 38.13a. Notice that the potential of the cell where the electrode is placed is −160 mV. At the time marked 0 on the horizontal axis, there is an almost instantaneous change in the voltage across the membrane, with the membrane potential becoming much more positive. This type of change is called a **depolarization,** because the charges on the two sides of the membrane become more alike—that is, the membrane becomes less polar. This rapid change in membrane potential from negative to positive is followed by a gradual change back to negative. As Figure 38.13b shows, the initial, rapid "spike" in membrane potential occurs because positive charges rush into the cell. As the action potential continues, positive ions gradually move back out, causing a dramatic swing in voltage back to negative. The gradual outflow of positive charges restores the polarity on the two sides of the membrane and is called a **repolarization.**

The molecular mechanisms responsible for an action potential in animals are analyzed in much more detail in Chapter 45. In that chapter, we'll explore which ions flow across the membrane during each part of the response in animal neurons, which membrane channels mediate these ion flows, and how the signal propagates down the length of an animal nerve cell. It's important to note several general differences between action potentials in plants and their counterpart in animals, however. For example, action potentials are much slower in plants than they are in animals. Plant action potentials typically last 2 to 5 seconds, while in animals they are over in a millisecond. In addition, plant action potentials involve different ions and ion channels than do action potentials in animals and can pass from cell to cell via the intercellular connections called **plasmodesmata** (see Chapter 8).

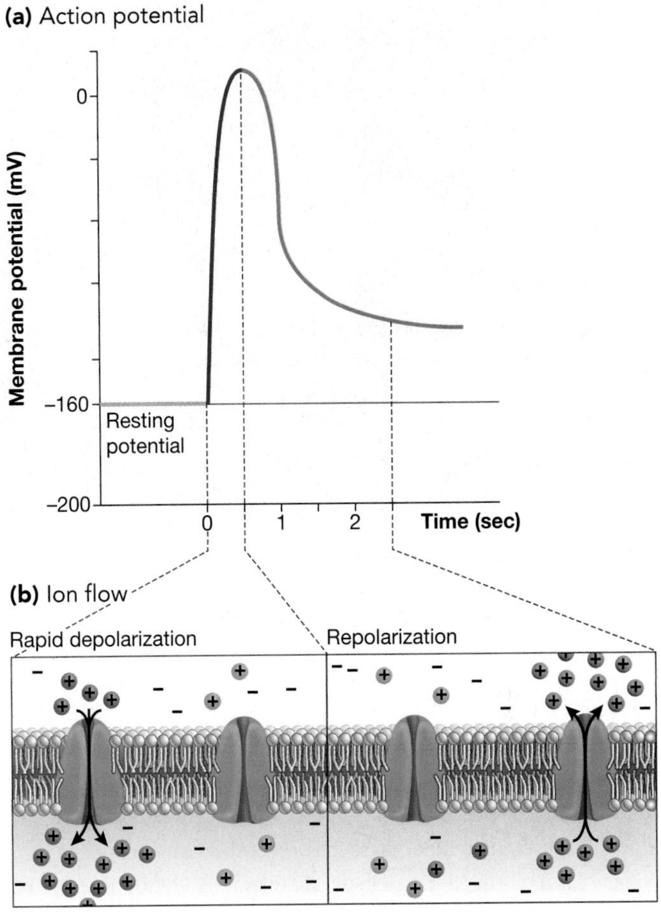

FIGURE 38.13 The Action Potential Is a Change in Membrane Potential That Acts as a Signal

(a) The changes in membrane potential that occur during an action potential. **(b)** Each phase of an action potential is caused by a flow of ions across the plasma membrane through channels.

How Does the Venus Flytrap Close?

For a Venus flytrap to snare an insect, a sequence of three events must occur. First, the mechanical energy that moves the receptor hair on the trap surface needs to be transduced to an electrical signal. Second, this electrical signal must be propagated across the leaf in the form of one or more action potentials. Third, epidermal cells on the outside of the trap must respond

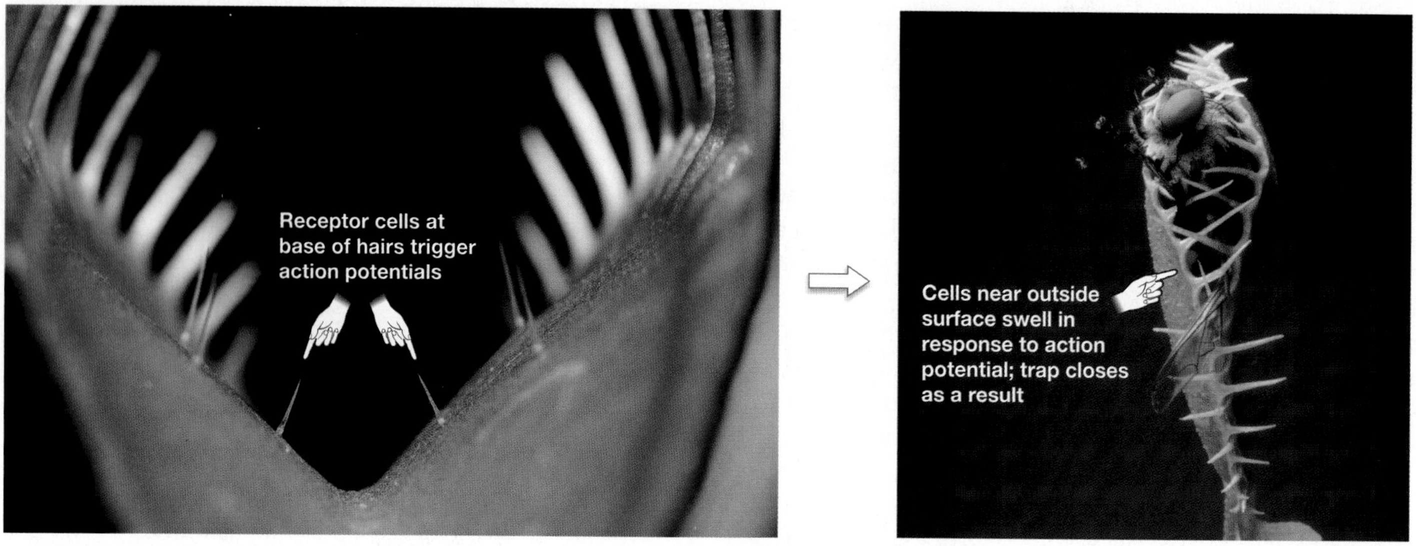

FIGURE 38.14 Venus Flytraps Close in Response to Action Potentials Generated by Sensory Hairs
Action potentials spread across the Venus flytrap from receptor cells at the base of trigger hairs. When these action potentials reach cells on the outside surface of the trap, the cells swell in response and close the trap.

to the action potential by undergoing a rapid change in turgor that moves the leaf and closes the trap (**Figure 38.14**).

Each step in this process is reasonably well understood. For example, biologists have dissected receptor hairs from the trap surface and showed that their membranes depolarize in response to touch. For this depolarization to occur, stretch receptors in these membranes must allow an inflow of positive ions in response to being pulled or pushed. In this way, the mechanical signal generated by touch is transduced to an electrical signal in a sensory cell. If large enough, the depolarization in the receptor cell triggers action potentials in other cells. These signals race across the leaf at a rate of about 10 cm/sec. When the action potentials reach cells on the outer surface of the trap, the cells swell and push the trap shut.

Recent experiments have clarified the mechanism responsible for trap closing. When the electrical signal arrives, H^+ is pumped out of the epidermal cells in response. The subsequent change in extracellular pH causes the cell wall to loosen. It also sets up a strong electrochemical gradient that leads to a rapid influx of calcium ions. When Ca^{2+} enters the cell, water follows by osmosis and causes the cell to swell. The way these cells respond to an action potential is the closest thing to a muscle found in plants.

38.4 How Do Plants Sense Attacks by Parasites?

Like animals and virtually all other organisms, plants are constantly threatened by an array of viruses, disease-causing bacteria, and parasitic fungi. In addition, the roots of plants are susceptible to attacks by nematodes—the soil-dwelling roundworms introduced in Chapter 32. The ability to cause disease is called **virulence**; disease-causing agents are termed **pathogens**. If plants were not able to sense attacks by pathogens and respond to them quickly and effectively, the landscape would be littered with dead and dying vegetation.

Why are most plants healthy? Recall from Chapter 35 that cuticle plays a major role in preventing entry by disease-causing agents. But plant defense systems do not stop at the epidermis. If a virus, bacterium, fungus, or nematode is able to get inside a plant and begins to grow, the cells at site of the infection respond by committing suicide. The rapid and localized death of one or a few infected cells is called the **hypersensitive response (HR)**. If the HR is successful, the pathogen is starved as host cells die around it.

In several respects, the hypersensitive response in plants is similar to the cell-mediated immune response in mammals, which leads to the death of infected cells (see Chapter 49). The HR is also extremely effective. Plants that mount a hypersensitive response rarely succumb to disease. How do plants sense the presence of pathogens, so that the HR can kill them?

The Gene-for-Gene Hypothesis

In the early decades of the twentieth century, crop breeders established that plants have disease-resistance genes that are inherited according to Mendel's rules. These genes came to be known as the **resistance (*R*) loci**. Follow-up research showed that many of the *R* genes are responsible for sensing the presence of pathogens and triggering the HR. The same researchers established that the fungi found to cause disease in wheat, flax, barley, and other crops have alleles that make the fungi either virulent or *avirulent* (not virulent) on certain strains or varieties of these crops. The genes associated with virulence or avirulence in pathogens came to be known as **avirulence (*avr*) loci**.

In 1956 H. H. Flor published data demonstrating a one-to-one correspondence between the resistance alleles found in host plants and the avirulence alleles found in pathogens. In other words, Flor showed that certain *R* alleles and certain *avr* alleles correspond. If an *R* allele in the host corresponds to the *avr* allele in the pathogen, their protein products must also correspond in some way. Stated another way, the *R* gene product appeared to act as a receptor for a particular pathogen. In response to the interaction between *R* gene products and *avr* gene products, a hypersensitive response occurs. If the *R* allele in a host and the *avr* alleles in a pathogen do not correspond, then no HR occurs and the plant succumbs to disease.

The idea that *R* and *avr* gene products interact in a specific way came to be called the **gene-for-gene hypothesis.** What molecular mechanism could be responsible for this pattern? Researchers who followed up on Flor's work suggested that the HR begins when proteins produced by host plants bind to proteins or other molecules produced by the pathogen. The hypothesis was that *R* genes produce receptors and that *avr* genes produce ligands—molecules that bind to receptors. **Figure 38.15** provides a general overview of this model.

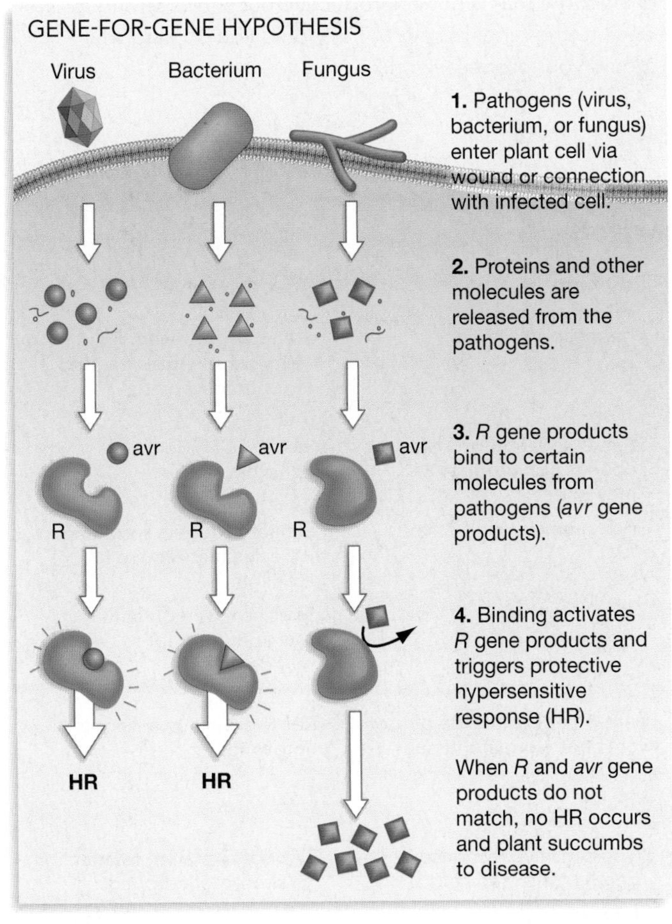
GENE-FOR-GENE HYPOTHESIS
Virus
Bacterium
Fungus
1. Pathogens (virus, bacterium, or fungus) enter plant cell via wound or connection with infected cell.
2. Proteins and other molecules are released from the pathogens.
avr
avr
avr
R
R
R
3. *R* gene products bind to certain molecules from pathogens (*avr* gene products).
4. Binding activates *R* gene products and triggers protective hypersensitive response (HR).
HR
HR
When *R* and *avr* gene products do not match, no HR occurs and plant succumbs to disease.

FIGURE 38.15 The Hypersensitive Response Begins when *R* Gene Products Bind to *avr* Gene Products
The gene-for-gene hypothesis predicts that *R* gene products act as receptors for *avr* gene products and that this interaction initiates the HR, which protects plants from pathogens.

The first breakthrough in testing the gene-for-gene hypothesis occurred when researchers from a variety of laboratories around the world were able to clone and sequence a series of *R* genes from crop plants and *avr* loci from bacterial and fungal pathogens. These results confirmed that *R* genes and *avr* genes exist and that they code for products that could interact at the start of an infection. A second major advance, published in 1996, supported the gene-for-gene hypothesis by showing that *R* products and *avr* products actually do bind to one another.

Why Do So Many Resistance Genes and Alleles Exist?

A large number of *R* genes have now been identified in *Arabidopsis*, tomato, flax, tobacco, and other plants. Two general patterns are emerging as data on these genes accumulate: First, *R* genes that are similar in sequence and structure tend to cluster together on the same chromosome. Second, within a population of plants, there are usually many different alleles at each *R* locus. To use a term introduced earlier in the text, *R* loci are highly polymorphic.

These observations are important because they provide hints about the history and function of these genes. For example, clusters of similar genes, or what biologists call **gene families**, are thought to originate through errors in recombination. Recall from Chapter 20 that **gene duplication** events occur when chromosomes misalign during crossing over. The result of this mutation is a chromosome with an extra copy of a gene. Because the organism with this chromosome already has a functioning copy of the original gene, mutations occurring in the extra copy of the gene do not damage the individual. Instead, the new copy may acquire mutations that make a new function possible. In the case of *R* genes, the hypothesis is that mutations in the new copies gave individuals the ability to recognize and respond to novel *avr* products and thus to new types of pathogens.

Why is it significant that many different alleles exist at each *R* locus? The hypothesis here is that different alleles allow plants to recognize different proteins from the same pathogen. Plants are diploid or polyploid, so they have at least two copies of each *R* gene. If many different alleles exist in a population, then each individual is likely to have at least two different alleles of each gene. The different alleles allow the host to recognize different *avr* products. This is important because new *avr* products constantly arise in pathogen populations via mutation. Plants with different alleles at each of many *R* genes should be able to sense the presence of a wide variety of disease-causing agents and respond by triggering the HR.

Recent work on the HR has established that the response consists of a number of different events. For example, the binding of an *R* gene product to an avr protein triggers the production of nitric oxide (NO) and reactive oxygen intermediates. **Reactive oxygen intermediates (ROIs)** such as hydrogen peroxide (H_2O_2) and superoxide ions (O_2^-) trigger reactions that help reinforce cell walls. In addition, nitric oxide and ROIs disrupt enzymes in cells at the point of infection and kill them. In much the same way, immune system cells in your body frequently use a lethal combination of NO and ROIs to kill cells in your body that have become infected. But plants don't stop there. The R-avr binding event in plant cells also leads to the production of antibacterial and antifungal compounds that are collectively known as **phytoalexins.** As a result, the HR leads to the walling-off of the infected area, the direct killing of pathogens by phytoalexins, and the starvation of pathogens as nearby cells commit suicide. The signal transduction events responsible for these events are not well understood, however, and are currently the focus of intense research.

It is interesting to note that mammals also have genes that are responsible for recognizing pathogens and triggering a gene-for-gene response. As Chapter 49 will show, animals have two types of recognition genes. Each of these recognition genes occurs in gene families, just as *R* genes do. The genes are found in clusters on the same chromosome, and numerous copies are present. Called the immunoglobulin and *MHC* loci, these genes are also highly polymorphic. Within any population, many different alleles of each gene exist. The parallels between plant and mammalian disease-response systems are striking.

Although the sensory and response components of the HR are particularly dramatic, they illustrate a very general point about how plants live: From the research that has been done on plant sensory systems to date, it is abundantly clear that plants don't just sit there. These organisms may be stationary, but they constantly monitor and respond to a wide array of information about their environment (**Table 38.2**). Gaining a better understanding of phototropism, gravitropism, response to disease, and other aspects of plant behavior is an exciting frontier in biology.

✓CHECK YOUR UNDERSTANDING

The presence of *R* gene products allows plants to recognize attacks by parasites such as viruses, bacteria, fungi, and nematodes. When the protein product of an *R* allele binds to a specific *avr* gene product from a pathogen, a signal transduction sequence results in the hypersensitive response. You should be able to explain the fitness advantage of having a wide array of *R* alleles in a single individual. In addition, suppose you are in charge of improving the quality of a popular, high-producing strain of wheat that is susceptible to infection by a particular strain of fungus. Another strain of wheat is available that is not as productive but shows strong resistance to the fungal strain. What would you do, and why would your proposal work?

TABLE 38.2 A Summary of Selected Sensory Systems in Plants

Stimulus	Receptor	Signal Transduction	Response	Adaptive Value
Blue light	PHOT1 in stem and leaves	PHOT1 self-phosphorylates; remainder of signal transduction systems unknown.	Phototropism occurs.	Stems grow toward light with wavelengths needed for photosynthesis.
Blue light	Cryptochromes, zeanthin in stems and leaves.	Details are unknown.	Stem elongation, stomatal opening, chloroplast movement	Increases efficiency of photosynthesis
Red light	Phytochrome in seeds and elsewhere	Phytochrome changes to P_{fr} form and activates responses.	Seed germinates.	Sunlight triggers sprouting in species that require full sunlight.
Far-red light	Phytochrome in stem	Phytochrome changes to P_r form and induces responses; mechanism is unknown.	Stems lengthen.	Species that require full sunlight attempt to escape shade.
Gravity	Integrin proteins located in plasma membrane or in amyloplast membrane?	Details unknown.	Cells on opposite side of root or shoot elongate; tissue curves.	Roots grow down; shoots grow up.
Pathogens	*R* gene products	See Chapter 39.	Hypersensitive response (HR)—death of infected cells.	Pathogens starve, so infection is slowed or stopped.
Touch or wind	Stretch receptors; location unknown	Details unknown, but result is transcription activation in target genes	Stems grow shorter and thicker.	Individual is more resistant to damage.
Touch	Receptor hair cell in Venus flytrap	Depolarization of receptor cell's plasma membrane triggers action potential.	Target cells swell; trap shuts.	Plant can capture prey.

ESSAY Can Plants Tell Time?

The experiments analyzed in this chapter should convince you that plants can see, in the sense that they detect specific wavelengths of light. Plants also know what time of year it is and which direction is up and which is down, as well as perceiving touch or other mechanical forces. But can they tell time?

An experiment conducted by Jean Jacques d'Ortous de Mairan in 1729 inspired the hypothesis that plants have some sort of internal clock. De Mairan observed that mimosa plants fold their leaves at night and open them during the day. When he put individual mimosas into complete darkness, though, he found that the leaf opening and closing movements continued on schedule. This result indicated that plants tell time in some way that is independent of direct information from sunlight.

The same type of rhythmic leaf movement occurs in many pea-family plants. Bean leaves, for example, begin to close before nightfall and start to open again before daybreak. The observation that beans anticipate changes in light reinforces de Mairan's result. It implies that plants have some kind of internal clock or timekeeper.

Leaf movements, the opening and closing of stomatal pores, the release of floral fragrance, and other events that occur on a 24-hour cycle are said to exhibit a **circadian rhythm**. To establish and maintain this rhythm, a biological clock must have two components: There has to be a system for setting and adjusting the clock, and there has to be a mechanism for keeping time. How do these components work?

The leading hypothesis for the clock mechanism maintains that there is a timekeeping protein, called TOC, that oscillates in its expression over the course of 24 hours. Specifically, researchers have proposed that, at dawn, light activates the production of two repressor proteins that, in turn, regulate the expression of the *TOC* gene (**Figure 38.16**). As the day goes on, however, the expression of the repressor proteins decreases. As a result, *TOC* expression increases until it reaches a maximum by the end of the day. To explain what happens next, researchers hypothesize that the *TOC* gene product promotes the expression of the repressor proteins. If so, then their expression would increase during the night until they reached a peak at daybreak, starting the cycle all over again. Systems such as this are described as *negative-regulating transcriptional feedback loops.*

Researchers were able to find these timekeeping genes by analyzing *Arabidopsis thaliana* mutants with abnormal circadian rhythms. One creative new strategy in this gene hunt is turning out to be extraordinarily productive. Researchers insert the gene for the enzyme luciferase into *Arabidopsis* cells in such a way that it is under the direct control of genes expressed with a circadian rhythm. Luciferase is the enzyme responsible for luminescence in fireflies. By videotaping living plant tissues, the investigators can follow changes in the pattern of luminescence over time. If the luminescence pattern is disrupted or unusual, they infer that the individual being studied has a mutant form of one of the timekeeping genes. The researchers then analyze these mutants to identify the altered gene.

... events that occur on a 24-hour cycle are said to exhibit a circadian rhythm.

What about the clock-setting mechanism? Far from the equator, day length varies dramatically over the course of a growing season. To adjust the timekeeping component of the clock, plants must have a molecule that (1) monitors day length and (2) alters the concentration of timekeeping molecules as day length changes. Think about jet lag. If you fly to a location where the night-day period is dramatically different from the one where you live, your internal clock gets out of whack, disrupting your sleep patterns and other rhythms. Within several days, however, your internal clock becomes reset. Evidence suggests that, in *Arabidopsis*, this resetting may be done by a combination of phytochromes and cryptochromes. As evidence for this hypothesis, studies show that if these light-absorbing proteins are disrupted by mutations, circadian behavior is abnormal.

To summarize, plants have timekeeping proteins that undergo 24-hour cycles in abundance and activity, as well as photoreceptors that help set the clock and keep it accurate. With each new experimental result, biologists are more and more confident about adding timekeeping to the list of plants' sensory abilities.

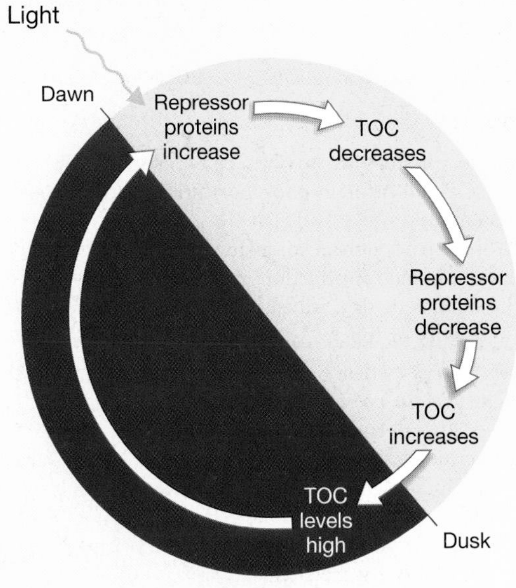

FIGURE 38.16 TOC and Repressor Proteins Oscillate in Concentration Each Day
The current model of timekeeping in plants.

CHAPTER REVIEW

Summary of Key Concepts

- **Plants can perceive a wide variety of environmental stimuli and respond in ways that increase their ability to grow and reproduce.**

  Plant sensory systems have three major components: (1) a receptor that changes in response to a stimulus; (2) a signal transduction pathway that transforms the stimulus into a chemical or electrical signal that triggers a response; and (3) movement, transcription of target genes, or some other action that helps the organism take advantage of the information in the stimulus. In other words, plants receive information from their environment and respond with appropriate action.

- **Plants sense several aspects of light. They respond in different ways to wavelengths that are typical of shade versus full sunlight. They also monitor photoperiod (day length).**

  Far-red wavelengths are enriched in shade light and are perceived by phytochomes, which are photoreceptors that also respond to wavelengths in the red portion of the visible spectrum. Phytochromes are photoreversible, meaning they switch between two distinct conformations when they absorb red or far-red light. These conformational switches trigger different responses by the plant, although the signal transduction mechanisms involved are still under investigation. In response to far-red wavelengths of light, species that are adapted to sunny habitats may elongate their stems or delay germination as seeds.

  Several proteins act as photoreceptors that respond to the blue light required for photosynthesis. Blue-light receptors trigger responses such as phototropism, chloroplast movement, guard cell opening, and stem elongation.

  Plants also monitor night length and may initiate flowering in response to long nights or short nights, depending on the species. Because phytochromes are the photoreceptor responsible for sensing night length, flashes of red or far-red light change the flowering behavior of short-day (long-night) and long-day (short-night) plants.

  Web Tutorial 38.1 Sensing Light

- **Plant shoots and roots sense gravity and respond by growing up and down, respectively.**

  In roots, gravity sensing occurs in root-cap cells. The settling of amyloplasts against the plasma membrane and cell wall is the key stimulus for the gravity sensor—a protein that has yet to be found and characterized.

- **Plants sense mechanical stimulation (touch or wind). In some cases, plants respond to touch by generating rapid electrical signals.**

  If plants are touched repeatedly, sensory cells trigger massive changes in gene expression that lead to shorter, stouter stems. In addition to this slow response to mechanical stimulation, which occurs in a wide array of plant species, the Venus flytrap and selected other species are capable of fast responses to touch via action potentials that originate in sensory cells and spread to cells that cause movement.

- **Plants sense disease-causing agents. Certain plant proteins bind to proteins produced by pathogens. When these plant receptors are stimulated, they trigger a defense response.**

  The *R* gene products of plants bind to specific *avr* proteins produced by viruses, bacteria, fungi, or nematodes. The binding event triggers the hypersensitive response, which leads to the death of infected cells and the starvation of the invading pathogen.

Questions

Content Review

1. What three major steps are involved in a sensory system?
   a. light absorption, protein phosphorylation, and one or more action potentials
   b. gravitropism, mechanical stimulation, and light
   c. perception, signal transduction, and response
   d. length of night or day, wavelength of light received, and forces such as gravity or touch

2. What was the first evidence that plants can sense not only the presence of light, but the presence of particular wavelengths of light?
   a. Growing shoots bend toward light.
   b. Lettuce germinates in the presence of red light but not far-red light.
   c. Short-day (long-night) plants do not flower if night is interrupted by a flash of red light but not far-red light.
   d. Plants bend toward light containing blue wavelengths but not toward light of other wavelengths.

3. Which of the following statements about phytochrome is *not* correct?
   a. It is photoreversible.
   b. Its function was understood long before the protein itself was isolated.
   c. The P_{fr} form activates the responses to light.
   d. It phosphorylates itself in response to specific wavelengths of light.

4. Why was it logical to predict that amyloplasts function as statoliths?
   a. They are dense and sediment to the bottom of gravity-sensing cells.
   b. They are present in all plants, and in both shoot and root tissues where gravity sensing occurs.
   c. They make a direct physical connection with integrins, which have been shown to be the gravity receptor molecule.
   d. Their density changes in response to gravity.

5. If a plant is touched repeatedly over the course of many days or if it experiences long-term exposure to wind, what happens?
   a. Growth of the root system is accelerated, making the plant more stable.
   b. Large-scale changes in gene expression occur, resulting in shoots that are short and stout.
   c. Electrical signals cause leaves to fold up, avoiding damage.
   d. Continued mechanical stimulation indicates that the individual is threatened with destruction, so it initiates flowering in an attempt to reproduce before it dies.

6. What is an action potential?
   a. a lightning-quick response to touch
   b. a rapid change in membrane potential, from negative to positive and back to negative
   c. anything that depolarizes a membrane
   d. a type of potential energy, created by a separation of charge (voltage) across a membrane

Conceptual Review

1. Phytochromes can be considered "shade detectors," while phototropins such as *phot1* can be considered "sunlight detectors." Explain why these characterizations are valid.
2. In the experiment that confirmed the self-phosphorylation hypothesis for the blue-light receptor, researchers inserted the *phot1* gene into insect cells. Why was it important for the gene to be expressed in an organism other than a plant?
3. What does *transduce* mean? Give an example of a signal transduction event in plant sensory systems.
4. Give two reasons why it is important for plants to sense photoperiod. When researchers observed that flashes of red or far-red light alter photoperiod responses in some species, it inspired the hypothesis that phytochromes are involved in sensing photoperiod. Explain why.
5. Summarize the statolith hypothesis for gravity sensing.
6. How do the proton pumps in the plasma membranes of plants create a membrane potential? What is voltage, and why is it considered a form of potential energy?

Group Discussion Problems

1. Suggest a hypothesis for how the force of gravity on a gravity-sensing protein, such as an integrin, could be transduced into an electrical or chemical signal. How would you test this hypothesis?
2. When an insect lands on certain sensitive plants, the leaves collapse in a fraction of a second (see the photos that follow). Design an experiment to test the hypothesis that this response is mediated by an action potential.

3. In general, small seeds that have few food reserves must be exposed to light before they will germinate. (Lettuce is an example.) In contrast, large seeds that have substantial food reserves typically do not depend on light as a stimulus to trigger germination. Why is this contrast logical?
4. Depending on the species, bamboo plants flower every 3, 6, 12, 60, or even 120 years. Generate a hypothesis to explain how bamboo plants keep time. (The actual mechanism is unknown.) How would you test this hypothesis?

Answers to Multiple-Choice Questions 1. c; 2. d; 3. d; 4. a; 5. b; 6. b

www.prenhall.com/freeman is your resource for the following: Web Tutorials; Online Quizzes and other Online Study Guide materials; Answers to Conceptual Review Questions; Solutions to Group Discussion Problems; Answers to Figure Caption Questions and Exercises; and Additional Readings and Research.

39 Communication: Chemical Signals

KEY CONCEPTS

- Hormones are the link between the receipt of information by a sensory receptor and a plant's response to that information.
- Each type of plant hormone tends to have a wide array of effects. These effects usually relate to the general role that a hormone plays in the life of a plant, such as signaling stress or indicating that conditions for growth are good.
- Plant responses to environmental stimuli are usually affected by interactions among several different hormones.

These bananas are identical, except that the bunch on the right was exposed to the plant hormone ethylene, which speeds ripening. Most events in plant growth and development are influenced by hormones.

Plants monitor their environment constantly and respond to light, gravity, touch, and other stimuli in ways that increase the plant's ability to grow and reproduce. Stems sense whether they are in shade or in light and redirect their growth accordingly. Both stems and roots perceive gravity and respond by growing in the appropriate direction.

Although electrical signals do occur in plants, plant hormones provide a much more common link between receptors that receive stimuli from the environment and a response by plants. Plant hormones may be small in size and present in tiny concentrations, but they make shoots bend, flowers bloom, and fruits ripen. They are responsible for regulating virtually every aspect of a plant's life.

39.1 Plant Hormones—an Overview

Chapter 38 analyzed the molecules involved in sensing light and other stimuli, but it left out a key observation: The cells that sense environmental changes are not necessarily the cells that respond to those conditions.

To drive this point home, let's revisit a topic introduced in Chapter 38: the experiments done by Charles Darwin and Francis Darwin on the coleoptiles of reed canary grass seedlings. Recall that a **coleoptile** is a modifed leaf that forms a sheath protecting the stems and leaves of young grasses. The results showed that the coleoptiles are **phototropic**, meaning they bend or grow toward light. Recall that the bending response is triggered specifically by wavelengths in the blue part of the electromagnetic spectrum, detected by photoreceptors known as **phototropins**. The Darwins didn't stop there, however. To find out which part of the coleoptile is responsible for light perception, they removed the coleoptile tips. As **Figure 39.1** shows, the decapitated coleoptiles stopped bending toward the light. To extend this result, the researchers either covered the tips of coleoptiles with opaque covers or put opaque collars below the tips, in the area where bending occurs. As the "Experimental setup and Results" section of Figure 39.1 shows, the coleoptiles with caps did not bend, but the stems with collars did.

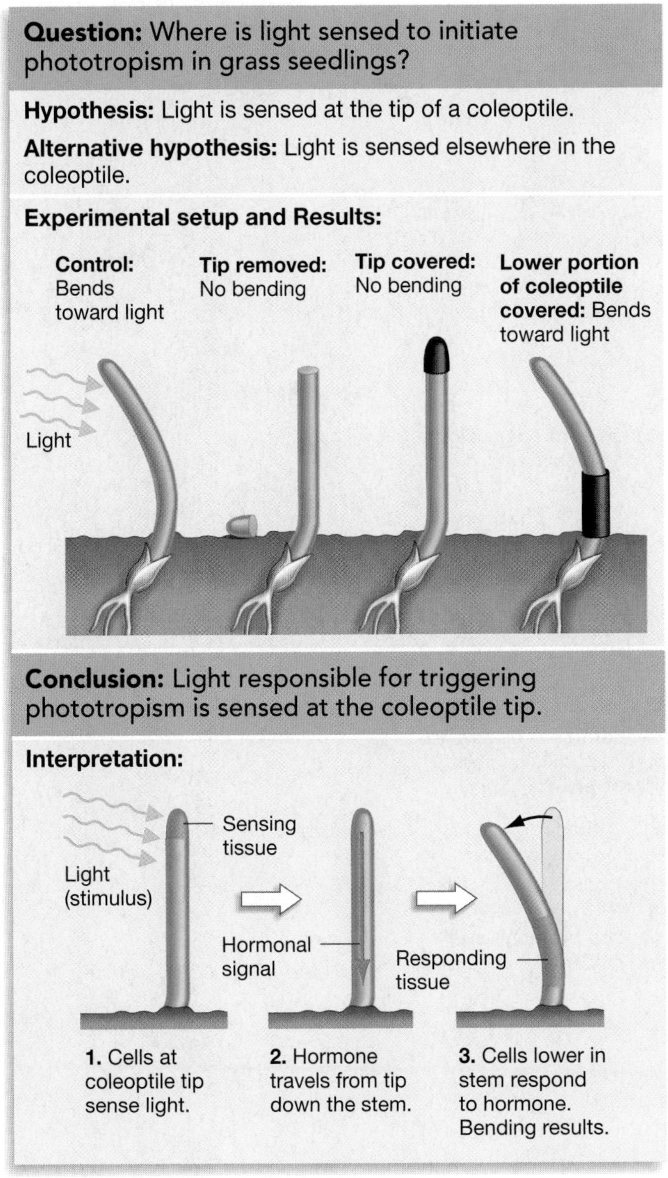
Question: Where is light sensed to initiate phototropism in grass seedlings?
Hypothesis: Light is sensed at the tip of a coleoptile.
Alternative hypothesis: Light is sensed elsewhere in the coleoptile.
Experimental setup and Results:
Control: Bends toward light
Tip removed: No bending
Tip covered: No bending
Lower portion of coleoptile covered: Bends toward light
Light
Conclusion: Light responsible for triggering phototropism is sensed at the coleoptile tip.
Interpretation:
Light (stimulus)
Sensing tissue
Hormonal signal
Responding tissue
1. Cells at coleoptile tip sense light.
2. Hormone travels from tip down the stem.
3. Cells lower in stem respond to hormone. Bending results.

FIGURE 39.1 The Sensory and Response Cells Involved in Phototropism Are Not the Same

QUESTION A critic could argue that this experiment lacked appropriate controls for the treatments labeled "Tip removed" and "Tip covered." Suggest better controls for these treatments than the unmanipulated individual at the far left.

To make sense of these results, the Darwins proposed that phototropism depends on "some matter in the upper part which is acted on by light, and which transmits its effects to the lower part." Their hypothesis was that a substance produced at the tip of the coleoptile acts as a signal and is transported to the area of bending. This "matter" is a **hormone**—an organic compound that is produced in small amounts in one part of a plant and transported to target cells in another region of the individual, where it causes a physiological response.

How do hormones work? To answer a question such as this, biologists start by performing experiments on the simplest possible system or situation. Mendel's inheritance experiments on garden peas are a good example. Unfortunately, there may be no such thing as a simple system with respect to plant hormones. Research on plant hormones is complex for two reasons: (1) A single hormone may affect many different target tissues, meaning there can be an array of responses to the same signal; and (2) it is common to observe that several hormones affect the same response and that these hormones interact with each other instead of working independently. In some cases, the presence of one hormone influences how rapidly another hormone is produced or how it accumulates.

These points are summarized in **Table 39.1** (page 890), which provides notes on the major types of plant hormones. The entry for auxin, for example, states that this molecule can affect numerous aspects of plant growth and development, including cell elongation, root development, and a phenomenon called *apical dominance*, which is the inhibition of lateral branches and stems by cells at the apex or tip of a plant stem. If you study the "Function(s)" column of Table 39.1, you might notice that auxin, the gibberellins, and the brassinosteroids are all involved in cell elongation—a basic physiological response required for growth and movement in stems, roots, and leaves. Plant hormones are challenging to study, because they have more than just a single effect and because they do not act alone.

In addition, hormones are active in tissues at extremely low concentrations and may exist in multiple chemical forms. Because their low concentrations are difficult to measure accurately, it can be hard to collect basic data on how hormone levels change over time. It can also be problematic to determine which form of a hormone actually affects target cells. For example, over 125 structurally distinct gibberellins have been isolated from plants to date, but only a few have been shown to act as hormones. The rest of the molecules are probably either intermediates in the synthesis of active forms or by-products of inactivated hormones.

Despite these challenges, biologists have made remarkable progress in understanding the structure and function of plant

TABLE 39.1 A Summary of Plant Hormones

Hormone	Function(s)	Notes	Chemical Structure
Auxin	Helps define long axis of body (phototropism and gravitropism responses); involved in cell elongation and apical dominance; promotes cell division and leaf expansion; induces ethylene production; development of adventitious roots and secondary growth; differentiation of xylem and phloem.	First plant hormone ever characterized and isolated.	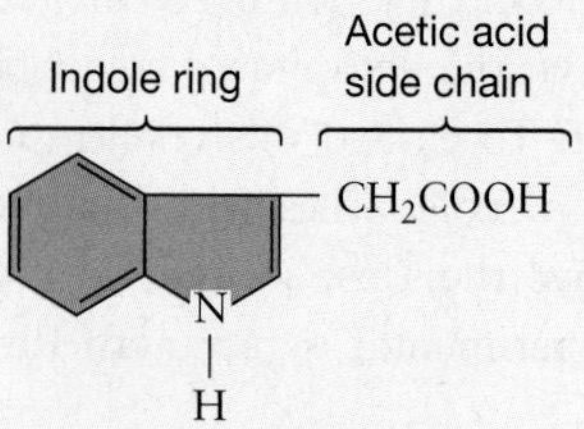Indole ring Acetic acid side chain CH_2COOH N H
Gibberellins (GA)	Promote stem growth via both cell elongation and division; encourage seed germination, involved in flowering.	Fungi that produce gibberellins infect rice plants and induce hyper-elongated stems. Analysis of these fungi led to discovery of gibberellins.	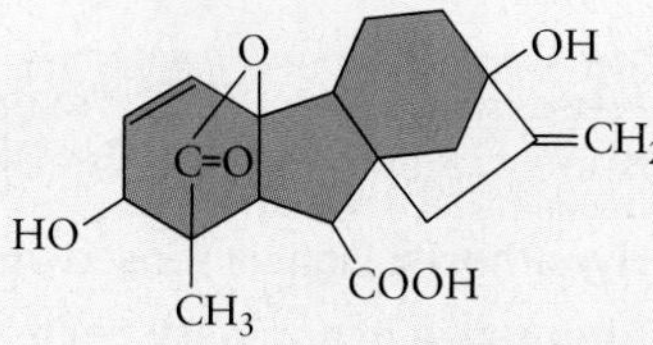O OH C=O $=CH_2$ HO COOH CH_3
Cytokinins	Promote cell division in the presence of auxin. Promote chloroplast development and development of lateral buds; delay senescence (aging).	New data indicate that they may act on cell cycle regulators.	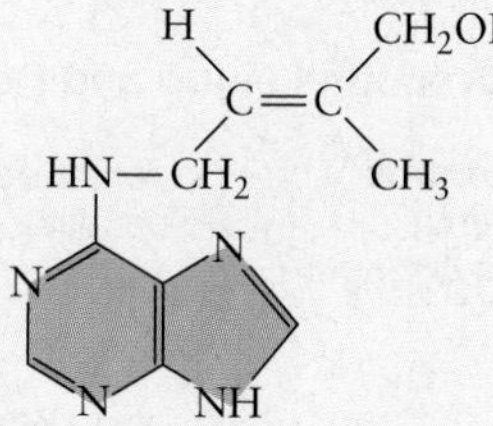H CH_2OH C=C HN—CH_2 CH_3 N N N NH
Abscisic acid (ABA)	Inhibits bud growth and seed germination; induces closure of stomata in response to water stress; involved in leaf abscission (drop).	Acts as a stress hormone analogous to cortisol in humans.	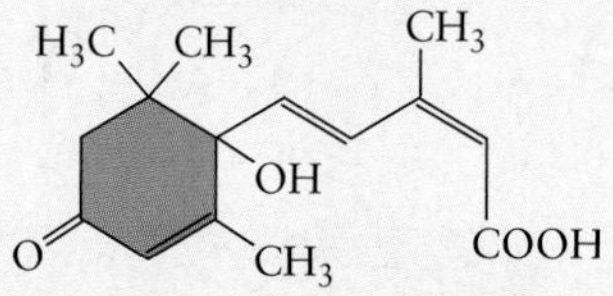H_3C CH_3 CH_3 OH O CH_3 COOH
Ethylene	Involved in fruit ripening; senescence of fruits, flowers, and leaves. Produced when plants are under stress.	A gas; discovered through unusual morphology of plants growing near gas lines for illuminating streets.	$H_2C{=}CH_2$
Brassinosteroids	Promote cell elongation in stems (mutants that lack these hormones or their receptors are dwarfed).	First steroid hormones discovered in plants. Structurally related to steroid hormones in animals.	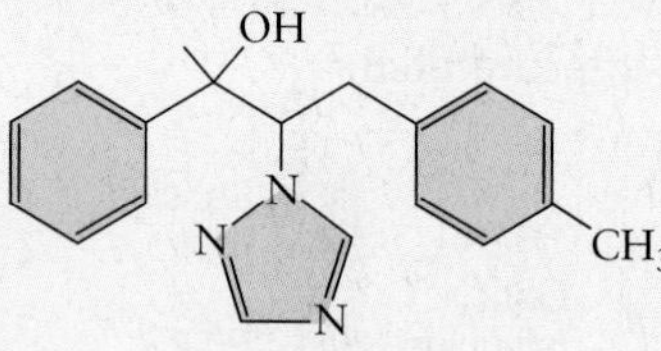OH N N N CH_3
Systemin	Triggers plant defense response to herbivore attack.	The first peptide hormone ever discovered in plants.	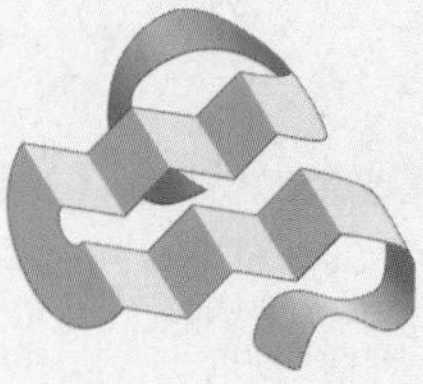

hormones and in determining how hormones interact. This progress is a tribute to the ingenuity and persistence of plant hormone researchers. These biologists recognize that the complexity of hormone interactions in plants also makes them fascinating and that hormone action is so fundamental to how plants function that the payoff for performing successful experiments is high.

Research on plant hormones is also being energized by technical innovations that are opening up promising new frontiers for study. For example, recent advances in applying mass spectrometry to hormone research has resulted in a relatively simple and rapid procedure for accurately quantifying the concentrations of multiple hormones in the same tissue. This advance should make it much more feasible for researchers to analyze how hormones interact and how their concentrations change in response to experimental manipulations.

Currently, researchers are focused on addressing two general questions about plant hormones: (1) How can a single plant hormone produce so many different and unrelated responses? and (2) how is it possible for many different hormones to affect the same response? Let's delve in by analyzing the first plant hormone to be isolated and characterized.

39.2 Auxin and Phototropism

The Darwins may have published their hypothesis about the connection between a chemical signal and phototropism in 1881, but their idea was not tested rigorously until 1913. That year, Peter Boysen-Jensen published the experiments summarized in **Figure 39.2a**. Boysen-Jensen cut the tips off young oat shoots and placed either a porous block of agar or a nonporous substance, such as a flake of the mineral mica, between the tip and the shoot. Because stems treated with the porous agar block showed normal phototropism, Boysen-Jensen concluded that the phototropic signal was indeed a chemical and that it could diffuse. Further, because the agar that he used was a water-based gelatin, it was clear that the molecule was water soluble.

Twelve years later, Frits Went extended these results. He collected the hormone responsible for phototropism by putting gelatinous blocks made of agar under the decapitated tips of oat coleoptiles (**Figure 39.2b**). Then he placed treated or untreated blocks of agar off-center on the decapitated coleoptiles of other individuals. The stems responded by bending—even though the shoots were kept in the dark during the entire experiment. Went had succeeded in producing the phototropic response without the stimulus of light.

Follow-up research showed that the physical basis of the phototropic response is cell elongation (**Figure 39.2c**). Cells on the side of the coleoptile *opposite* to the source of light elongate in response to the phototropic hormone. Because it promotes cell elongation in the shoot, Went named the hormone **auxin** (from the Greek *auxein*, "to increase").

The Cholodny-Went Hypothesis

Went's experiments were a breakthrough in research on plant hormones. He had accomplished the first step in isolating the hormone—it was somewhere in the agar blocks.

Went's experiments also inspired an important hypothesis for how the hormone produces phototropism. Working independently, both N. O. Cholodny and Went proposed that the bending response results from an asymmetric distribution of the hormone. The Cholodny-Went hypothesis contends that auxin produced in the tips of coleoptiles is shunted from one side of the tip to the other in response to light. The auxin is then transported down the shoot. Because auxin concentration is higher on one side than the other, cells on one side of the shoot elongate more than cells on the other side. Bending results.

Other researchers proposed a simpler hypothesis, however. They suggested that auxin is broken down or otherwise inactivated by blue light. If so, then an asymmetric distribution would result not from a transport process but from the destruction of auxin on one side of the tip.

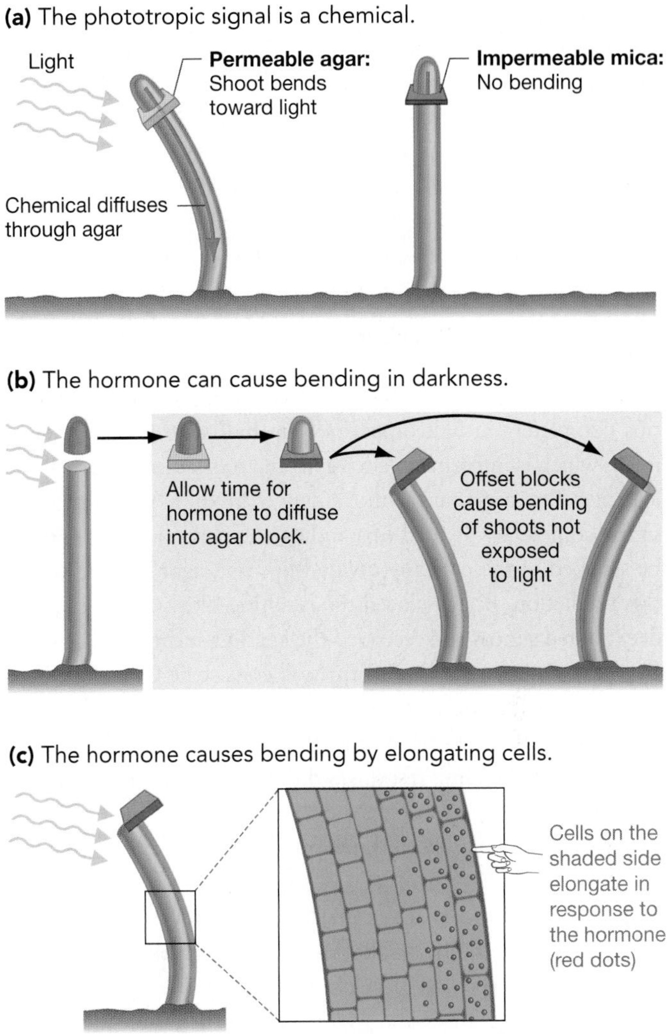

FIGURE 39.2 Experimental Evidence Supports the Hormone Hypothesis for Phototropism
(a) Coleoptiles bend in response to light if substances from the tip are allowed to move downward. **(b)** If bending can take place in darkness, then light is not directly required for the response. Only the hormone is required. **(c)** During the phototropic response, bending occurs because cells on the shaded side of the shoot elongate.

To test these alternatives, Winslow Briggs did a series of experiments. He grew corn seedlings in the dark, cut off their tips, and placed the tips on agar blocks. As **Figure 39.3** shows, he either kept the tips and agar blocks in the dark or exposed them to light from one side. Then he put the agar blocks from each treatment on one side of decapitated coleoptiles. In response, the coleoptiles from each treatment bent the same amount—about 26° from vertical. This result is inconsistent with the auxin destruction hypothesis. If light destroys auxin, then the block exposed to light should be much less effective in inducing bending.

To test the Cholodny-Went model, Briggs divided coleoptile tips completely in half or partially in half with a thin piece of mica, which is impermeable to dissolved molecules (**Figure 39.4**). The idea here was that the movement of auxin would be stopped in the completely divided tips and agar blocks, but it would not be stopped in the partially divided tips and agar blocks. To test this prediction, Briggs placed the resulting blocks on one side of decapitated shoots and recorded the bending response. The result from the completely divided tip was consistent with the earlier result. There was no difference in bending induced by the sunlit or shaded side of the tip. But as the Cholodny-Went hypothesis predicted, the bending responses differed in the partially divided tip. The side away from light induced much more bending, indicating that auxin had been transported from one side of the tip to the other.

Figures 39.3 and 39.4 summarize classic work: Two plausible competing hypothesis made clear but contrasting predictions about auxin's behavior. Briggs's experimental designs elegantly tested the predictions. The asymmetric distribution model has been successful in explaining how auxin produces phototropism, but a basic question remains: Exactly what *is* the hormone?

Isolating and Characterizing Auxin

Went's work gave researchers a place to look for the phototropic hormone—it was somewhere in the agar blocks used in his experiments—along with a way to identify it. If a molecule could be found that produced the bending response in decapitated shoots, it was likely to be the auxin hormone.

After years of effort, researchers in two laboratories independently succeeded in isolating and characterizing auxin. The hormone turned out to be indole acetic acid, or IAA. In a typical plant, about 50 nanograms of IAA are present for every 50 grams of fresh tissue. This is a concentration of about 280 nanomolar (nM). (Recall from Chapter 1 that the prefix *nano-* refers to billionths.)

Why are these observations about auxin's structure and concentration interesting? Like the other plant hormones introduced in this chapter, auxin is a small molecule with a relatively simple structure. It is present in quantities so small that its concentration is difficult to measure. Yet its impact is huge. Auxin can bend stems, in some cases leading to tree trunks that are permanently bowed.

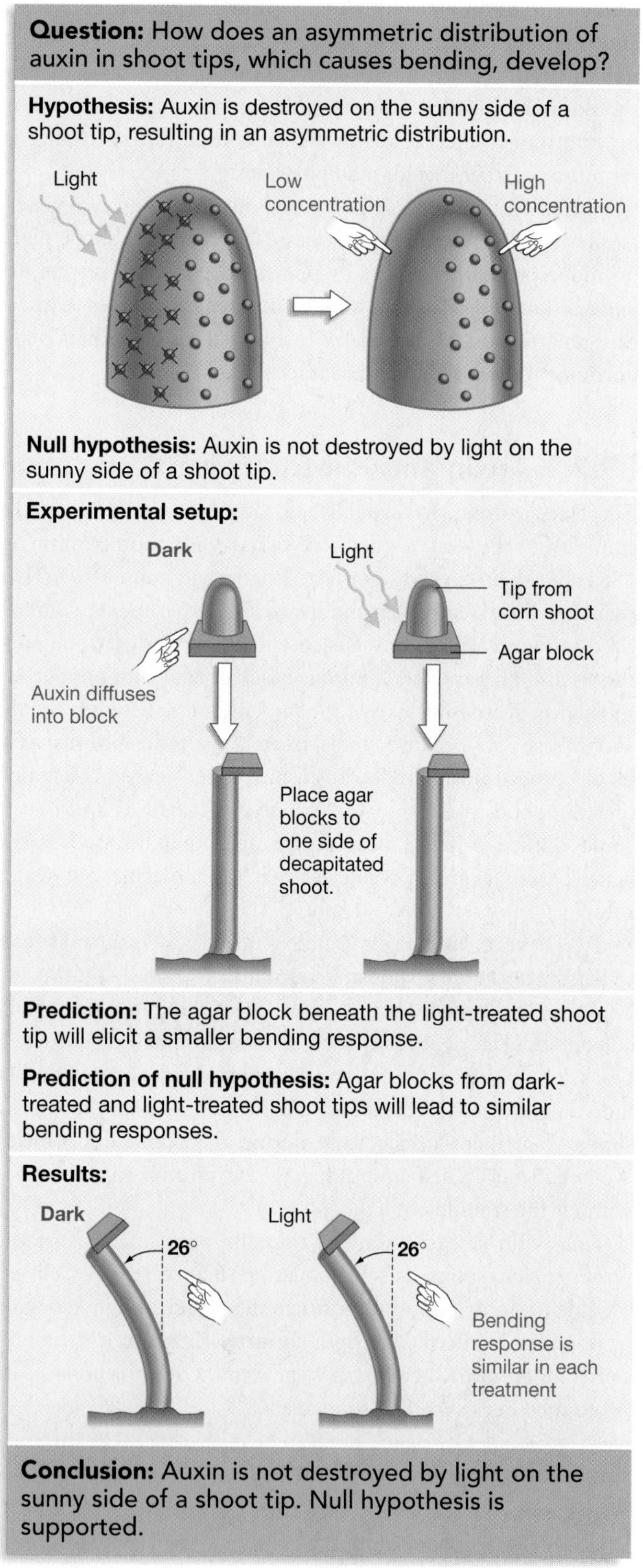
Question: How does an asymmetric distribution of auxin in shoot tips, which causes bending, develop?
Hypothesis: Auxin is destroyed on the sunny side of a shoot tip, resulting in an asymmetric distribution.
Light
Low concentration
High concentration
Null hypothesis: Auxin is not destroyed by light on the sunny side of a shoot tip.
Experimental setup:
Dark
Light
Tip from corn shoot
Agar block
Auxin diffuses into block
Place agar blocks to one side of decapitated shoot.
Prediction: The agar block beneath the light-treated shoot tip will elicit a smaller bending response.
Prediction of null hypothesis: Agar blocks from dark-treated and light-treated shoot tips will lead to similar bending responses.
Results:
Dark
Light
26°
26°
Bending response is similar in each treatment
Conclusion: Auxin is not destroyed by light on the sunny side of a shoot tip. Null hypothesis is supported.

FIGURE 39.3 Experimental Test of the Auxin Destruction Hypothesis

If the auxin destruction hypothesis were correct, much of the auxin in the light-exposed tip would have been destroyed. As a result, the bending response induced by the light-exposed agar block would have been very low.

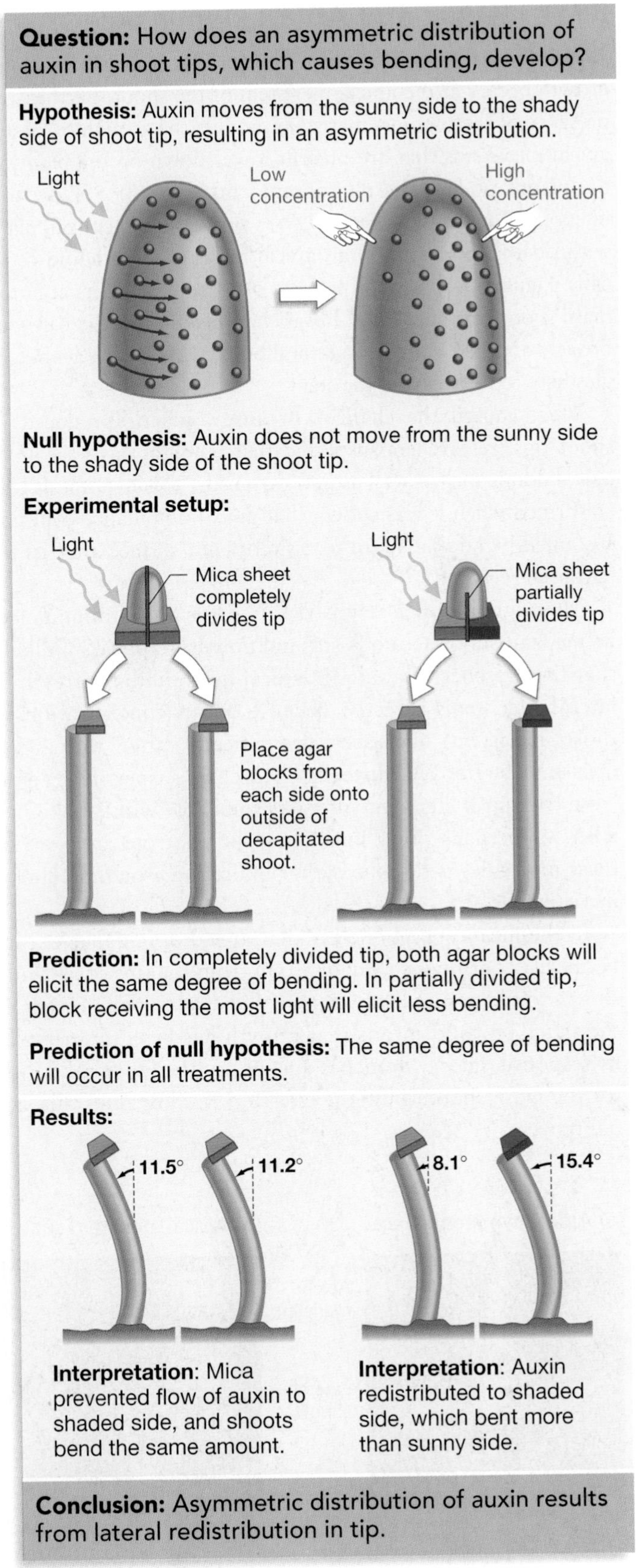
Question: How does an asymmetric distribution of auxin in shoot tips, which causes bending, develop?
Hypothesis: Auxin moves from the sunny side to the shady side of shoot tip, resulting in an asymmetric distribution.
Light
Low concentration
High concentration
Null hypothesis: Auxin does not move from the sunny side to the shady side of the shoot tip.
Experimental setup:
Light
Mica sheet completely divides tip
Light
Mica sheet partially divides tip
Place agar blocks from each side onto outside of decapitated shoot.
Prediction: In completely divided tip, both agar blocks will elicit the same degree of bending. In partially divided tip, block receiving the most light will elicit less bending.
Prediction of null hypothesis: The same degree of bending will occur in all treatments.
Results:
11.5°
11.2°
8.1°
15.4°
Interpretation: Mica prevented flow of auxin to shaded side, and shoots bend the same amount.
Interpretation: Auxin redistributed to shaded side, which bent more than sunny side.
Conclusion: Asymmetric distribution of auxin results from lateral redistribution in tip.

FIGURE 39.4 Testing the Auxin Redistribution Hypothesis
If the auxin redistribution hypothesis is correct, then a mica sheet placed all the way through a shoot tip will prevent a flow of auxin to the shaded side. A mica sheet placed partially through the tip will allow normal flows of auxin to the shaded side.

How Does Auxin Produce the Phototropic Response?

Although biologists understood where auxin is produced, what its molecular structure is, that it causes cells to elongate, and that its asymmetric distribution produces bending, they also wanted to know how the hormone acts at the molecular level. Once the hormone arrives at a cell, how is its message delivered? Once the signal is received, what molecular events lead to elongation and the phototropic response?

The Auxin Receptor Chapter 38 introduced the concept of signal transduction. Signal transduction connects a stimulus arriving at a cell to a response by that cell. When a sensory receptor perceives a change in a stimulus such as light or gravity or touch, the receptor molecule changes its conformation, composition, or activity. The action by the receptor is the first step in a sequence of events that culminates in the cell's response.

In many cases, the sensory cell's response is to produce a hormone—a signal for other tissues to act. When a hormone arrives at a target cell, its message must be received and transduced to produce the appropriate response. That happens when the hormone binds to a receptor in the target cell. Because only certain cells have receptors for specific hormones, hormone action is specific. For this reason, understanding receptors is a basic component of understanding hormone action.

In the case of auxin, the search for a receptor followed a straightforward strategy. Researchers attached a radioactive label to the hormone, treated cells with the labeled molecule, and purified the protein to which it bound. Although purifying the bound molecule turned out to be extraordinarily difficult, in 1985 biologists succeeded in isolating auxin-binding protein 1, or ABP1, from corn plants. Since then, experiments have confirmed that ABP1 is an auxin receptor that is located in the plasma membrane of cells in the stem and leaves. When it binds auxin, it triggers cell elongation. The gene that encodes ABP1 has been sequenced, and researchers are hunting for other proteins that may act as auxin receptors.

How Does Auxin Induce Cell Elongation? To appreciate what happens after auxin binds to ABP1 in the plasma membrane of a target cell, recall from Chapter 38 that the arrival of a signal can trigger several types of responses, from changes in transcription and translation to alterations in channels, pumps, and other membrane proteins. To explain how auxin causes cells to elongate, researchers proposed that once ABP1 has bound to auxin, the receptor triggers a series of events that leads to an increase in the number of membrane H^+-ATPases—the proton pumps that use the energy in ATP to drive protons out of the cell against an electrochemical gradient. Because the pH of the cell wall goes down when H^+-ATPases are active, the idea that these pumps are responsible for cell elongation became known as the **acid-growth hypothesis.**

To understand the rationale behind the acid-growth hypothesis, it's important to realize that two things have to happen for a plant cell to get larger:

1. Water has to enter the cell to increase its volume and generate turgor—meaning pressure on the cell wall (see Chapter 35). Proton pumps could lead to water entry, because potassium (K^+) or other positively charged ions often enter a cell after protons are pumped out and because sugar enters via proton cotransporters. As the concentrations of these ions inside the cell increase, water follows via osmosis.
2. The cell wall has to expand. A connection between acidification and cell-wall expansion is now well established, because researchers have discovered two classes of cell-wall proteins that actively increase cell length when the pH of the wall falls below 4.5. These proteins, called **expansins**, have been found in a wide variety of species and tissues. The genes that code for expansins have been identified and are very similar when compared from species to species. What expansins actually do has yet to be determined, however. One possibility is that they make cell walls extensible by interfering with the ionic bonds between cellulose microfibrils or other wall components. Cell elongation occurs when the existing cell wall is loosened and new cellulose microfibrils are produced to extend the cell's length.

To test the acid-growth hypothesis more rigorously, researchers have used antibodies that bind specifically to H^+-ATPase. Because the antibodies are tagged with a fluorescent molecule, they can be used to quantify the number of proton pumps in a plasma membrane. (This method for visualizing membrane proteins was introduced in Chapters 36 and 38.) Biologists have found that, when corn shoots are treated with additional auxin along with antibodies to H^+-ATPase, the number of proton pumps increased by 80 percent relative to untreated controls. Other researchers have shown that treating cells with auxin can lower the pH of the cell wall by as much as a full unit—from a normal value of 5.5 to 4.5.

WEB TUTORIAL 39.1 Plant Hormones

Based on these results, most biologists accept the acid-growth hypothesis as a general explanation for how plant cells elongate. The chain of events involved in phototropism can be summarized as follows: Auxin is produced by cells in shoot tips when phototropins in those cells absorb blue light. The auxin is redistributed to the shady side of the tip, then transported down the shoot, where it binds to ABP1 receptors in the plasma membrane of target cells. These cells elongate when activated ABP1 leads to the installation of additional proton pumps and the activation of expansin proteins. In this way, auxin acts as the phototropic signal.

Auxin has an array of effects in addition to stimulating bending movements, however. One of the most important is the phenomenon known as apical dominance.

39.3 Auxin and Apical Dominance

Apical dominance is a growth habit in which most of a stem's growth occurs at the apical meristem of the shoot. Further, the presence of the topmost meristem inhibits the growth of lateral apical meristems that are present lower down on the plant, at the plant's nodes. The photographs in **Figure 39.5** show this phenomenon in action. Figure 39.5a shows an intact bean plant with one main shoot and an apical meristem, extending vertically. Figure 39.5b shows the same plant several weeks after the main shoot was cut. Lateral branches have begun to grow vigorously and vertically from **lateral buds**, which form at the angles between leaves and the stem.

What caused the change? Because auxin is produced in shoot tips, researchers suspected that it might have a role in apical dominance as well as phototropism. This hypothesis was confirmed when it was shown that apical dominance could be sustained by adding auxin to a shoot's cut surface after its tip had been removed.

Although auxin clearly is required for apical dominance, its mechanism of action is still unknown. It appears unlikely that the presence of auxin in lateral buds inhibits growth in lateral meristems directly, because auxin concentration in those meristems increases dramatically after the apical meristem is removed. Instead, researchers suspect that auxin from the apical meristem interacts somehow with cytokinins, ABA, or perhaps other hormones and that one or more of these molecules is responsible for inhibiting growth in lateral meristems.

In the meantime, it is clear that apical dominance occurs because a continuous flow of auxin from the tips of growing shoots to the tissues below signals the direction of growth. If the signal stops, it means that growth has been interrupted. In response, lateral branches sprout and begin to take over for the main shoot. Now the question is, How does this signal move?

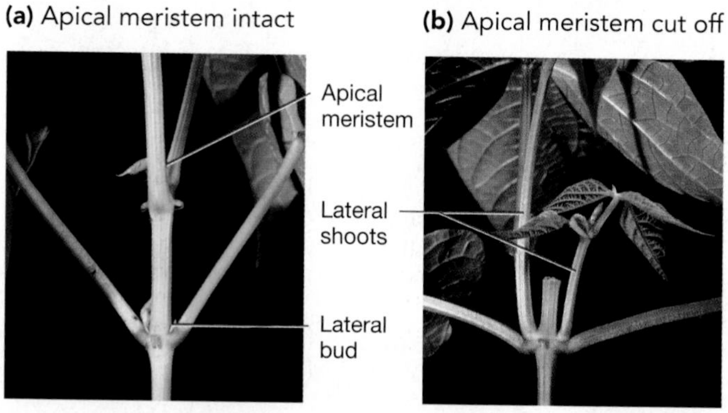

FIGURE 39.5 When Apical Dominance Occurs, Growth of Lateral Buds Is Suppressed
(a) The apical meristem of a bean plant (stem in center). **(b)** When the apical meristem is cut off, lateral buds respond by growing rapidly. The lateral shoots orient themselves vertically.

Polar Transport of Auxin

Auxin transport is **polar**, or unidirectional. If radioactively labeled auxin is added to the top of a cut stem, the hormone is transported toward the base. But if labeled auxin is added to the base of a cut stem, it is not transported toward the apex. Auxin is the only plant hormone known to be transported in one direction only.

Studies with labeled auxin have also shown that the hormone is transported all the way down the stem to the root, via parenchyma cells in the ground tissue and vascular tissue. Labeled auxin moves at about 10 cm/hr—approximately 10 times more slowly than substances traveling in the phloem or xylem. Recent research has shown that auxin enters the apical end of cells via a cotransport protein, which imports the hormone along with two protons that enter the cell along the electrochemical gradient—set up by H^+-ATPase—that favors proton entry. Once inside the cytoplasm, auxin diffuses to the basal portion of the cell, where it is transported out by carrier proteins located in the basal portion of the plasma membrane. In this way, auxin flows down stems in one direction only: into cells via a cotransporter in the apical end and out via a carrier protein in the basal end.

As auxin molecules travel down the long axis of the plant, some are destroyed by enzymes. As a result, polar transport sets up a strong gradient in auxin concentration. Auxin concentrations are much higher in shoots than they are in roots.

After it arrives at the root tip, auxin is transported out toward the epidermal cells and up—in what biologists call the "fountain model" (**Figure 39.6a**). This redistribution in the root is important, because Cholodny and Went predicted that an asymmetrical distribution of auxin in roots causes gravitropism. **Gravitropism** is the movement or growth of a plant in response to gravity. Other researchers suggested that gravity-sensing cells in the root tip alter the direction of auxin transport. Just as an asymmetrical concentration of auxin in shoots—due to auxin production in response to blue light—is responsible for phototropism, might an asymmetrical distribution of auxin in shoots be responsible for gravitropism? As **Figure 39.6b** shows, an asymmetrical distribution of auxin in roots should lead to asymmetrical elongation and the bending response of

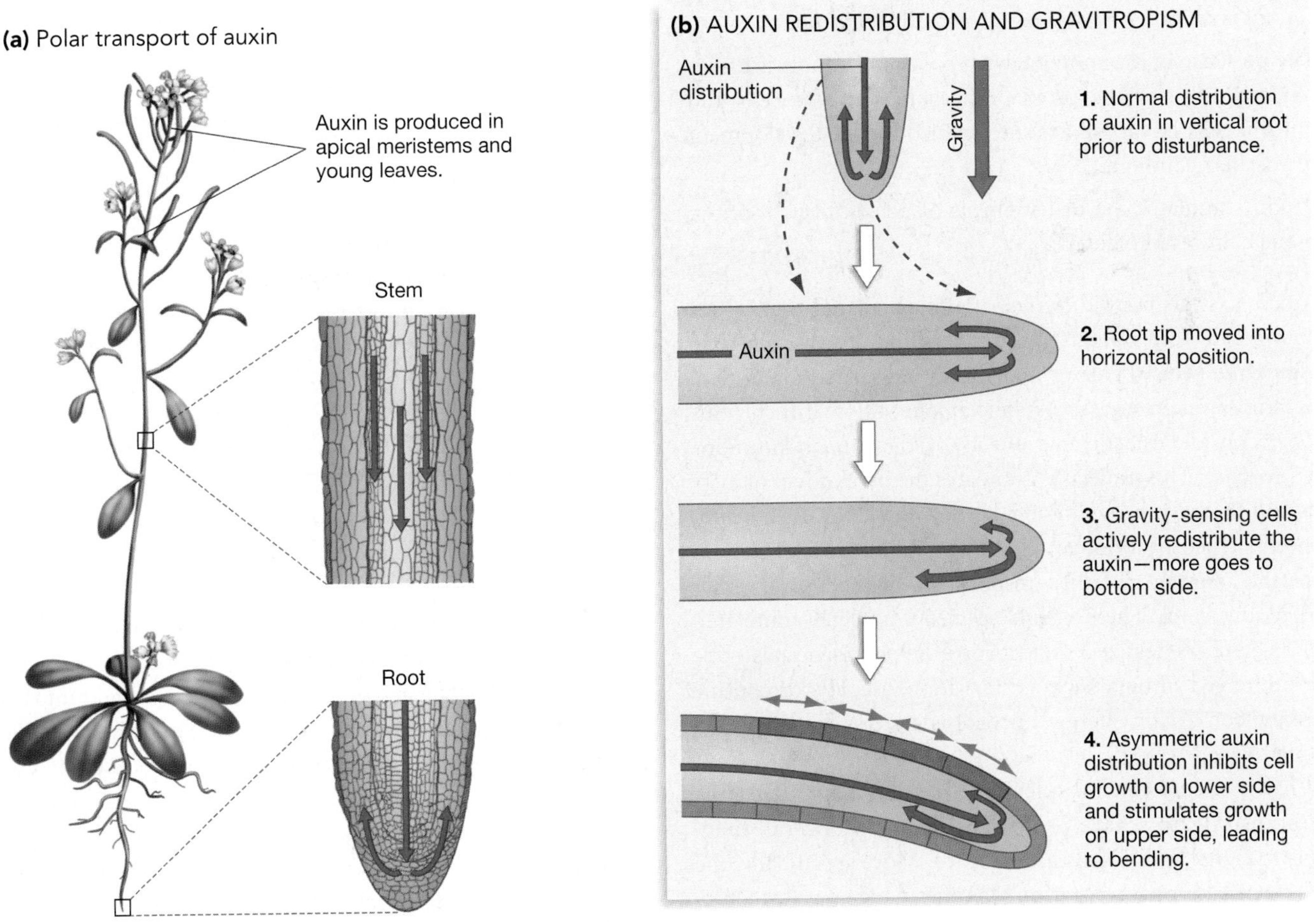
(a) Polar transport of auxin
Auxin is produced in apical meristems and young leaves.
Stem
Root
(b) AUXIN REDISTRIBUTION AND GRAVITROPISM
Auxin distribution
Gravity
1. Normal distribution of auxin in vertical root prior to disturbance.
Auxin
2. Root tip moved into horizontal position.
3. Gravity-sensing cells actively redistribute the auxin—more goes to bottom side.
4. Asymmetric auxin distribution inhibits cell growth on lower side and stimulates growth on upper side, leading to bending.

FIGURE 39.6 Auxin Is Transported from the Apical Meristem to Root Tips
(a) Auxin is transported in one direction down stems. It is redistributed after reaching a root tip, so that it is transported upward and outward. **(b)** The auxin redistribution hypothesis for gravitropism. This sequence of events might begin when a growing root tip hits a rock and is displaced horizontally or when a plant is tipped in a windstorm and is partially uprooted.

gravitropism. In this case, an elevated concentration of auxin on the lower side inhibits growth, while a lowered concentration stimulates growth on the upper side. These effects are possible because high auxin concentrations in root cells stimulate the production of the hormone ethylene (see Section 39.6), which inhibits the growth of root cells.

In short, the polar nature of auxin transport is critical to understanding phototropism, gravitropism, and apical dominance. How auxin is transported through cells and how its message is received are increasingly well understood at the molecular level.

An Overview of Auxin Action

Auxin clearly plays a key role in phototropism, gravitropism, and apical dominance. But it is critical to recognize that this chemical messenger has other important effects as well:

- Fruit development is influenced by auxin produced by seeds within the fruit.
- Falling auxin concentrations are involved in the **abscission**, or the shedding of leaves and fruits, associated with **senescence**, or aging. Auxin's effect on abscission involves interactions with other hormones, principally ethylene.
- The presence of auxin in growing roots and shoots is essential not only for the proper differentiation of xylem and phloem cells in vascular tissue but also for the development of vascular cambium.
- Auxin stimulates the development of adventitious roots in tissue cultures and cuttings.

Auxin has so many different effects on plants that it has been difficult for biologists to understand its overall role. Grasping the general role of a chemical messenger is a critical issue in hormone research. As an example of how difficult this can be, consider the diverse effects of the human hormone called cortisol. This molecule stimulates the breakdown of fatty acids and proteins, providing the body with energy. As a result, the body's glucose stores can be preserved. But cortisol also inhibits the immune system's response to wounds and other short-term insults. These effects seemed completely unrelated until researchers realized that cortisol helps individuals cope with prolonged trauma such as starvation, anxiety, or chronic illness by conserving energy. Cortisol is a stress hormone (see Chapter 47).

What is auxin's general role in the life of a plant? Recently, several investigators have proposed that auxin's overall function is to signal where cells are in space. More specifically, the idea is that auxin concentration identifies where a cell is located relative to the long axis of the plant body. In this way, auxin plays a role in determining the overall shape of a plant. If conditions relating to the long axis change—because of changes in light availability, because a windstorm tips the plant and shifts the position of the roots, or because a deer eats the shoot apex—changes in auxin concentration effectively signal how the individual's tissues should respond. Phototropism, gravitropism, apical dominance, and the production of adventitious roots are all ways of coping with changes in the long axis of the plant.

It is also important to remember that auxin does not act alone. Instead, its effects are modified by the action of other plant hormones. Researchers are particularly interested in how auxin interacts with cytokinins, because there is evidence that plant growth is regulated by an antagonistic relationship between auxin and cytokinins. Auxin travels down from the shoot to the root, suppressing the growth of lateral meristems along the way. In contrast, cytokinins are produced in the roots and travel upward, stimulating growth of the root system. From shoot to root, the two hormones are present in gradients of differing concentrations. Based on this observation, biologists hypothesize that interactions between auxin and cytokinins may help plants maintain an optimal proportion of root growth to shoot growth. Let's take a closer look at what cytokinins are and what they do.

✓CHECK YOUR UNDERSTANDING

Auxin's primary role is to signal the position of cells along the long axis of the plant body. This is possible because auxin is produced in shoot apical meristems and undergoes polar transport to the roots, forming a concentration gradient. You should be able to (1) make labeled diagrams indicating how asymmetries in auxin distribution lead to phototropism and gravitropism; (2) explain the molecular mechanism responsible for cell elongation in response to the stimulation of auxin receptors; and (3) review the bulleted list of "other effects" induced by auxin that appears in the subsection "An Overview of Auxin Action," and propose a connection between one of these effects and the claim that auxin signals cell position along the plant's longitudinal axis.

39.4 Cytokinins and Cell Division

Cytokinins are a group of plant hormones that promote cell division. (*Cyto* is the Greek root for *cell; kinin* refers to *kinesis*, meaning "movement" or "division.") The initial hint that certain molecules can promote cell division in plants was the finding that coconut milk, which stores nutrients used by growing coconut embryos, promotes the growth of cells and embryos in culture. Later, experiments showed that molecules derived from the nitrogenous base adenine also stimulate the growth of cells in culture. Eventually, naturally occurring adenine derivatives that stimulate growth were discovered in corn and apples. These hormones, which biologists found to be similar in structure to the growth-promoting ingredient in coconut milk, were named cytokinins. Cytokinins are synthesized in root tips, young fruits, seeds, growing buds, and other developing organs. **Box 39.1** introduces the use of cytokinins to promote the growth and differentiation of cells in culture.

The cytokinin that has been found in the most species is derived from adenine and is called *zeatin*. The other cytokinins that have been isolated are very similar to zeatin in chemical structure. Most of the zeatin and other cytokinins that are active in plants are synthesized in the apical meristems of roots and transported up into the shoot system via the xylem. A group of closely related proteins that act as cytokinin receptors has now been isolated and characterized. When cytokinin binds to these receptors in the plasma membranes of target cells, the receptors trigger the activation of specific genes.

Recent research on cytokinins has focused on understanding exactly how they function at the molecular level. Specifically, biologists have explored whether cytokinins might affect molecules that regulate the cell cycle. Some of these proteins,

BOX 39.1 The Use of Hormones to Grow Plant Cells in Culture

Recall from Chapter 35 that the parenchyma cells found in plant ground tissue are **totipotent**—meaning they retain the ability to divide and differentiate into a complete, mature plant, including new types of tissue. This property of parenchyma cells is important in wound healing and asexual reproduction. But it also allows researchers to grow complete adult plants from a small number of parenchyma cells.

Biologists who grow plants in **tissue culture** begin by placing parenchyma cells in a liquid or solid medium that contains all the nutrients required for cell maintenance and growth. In the early days of plant tissue culture, investigators found not only that specific hormones were required for successful growth and differentiation but also that the relative abundance of hormones present was critical to success. This was one of the first hints that plant hormones interact with each other to produce a particular outcome.

The earliest experiments on hormone interactions in tissue cultures were done with tobacco cells by Folke Skoog and co-workers. These researchers found that when auxin alone was added to the culture, the cells enlarged but did not divide (**Figure 39.7**). But if the team added roughly equal amounts of auxin and cytokinin to the cells, the cells began to divide and eventually formed a **callus**, or an undiferenitiated mass of parenchyma cells. By varying the proportion of auxin to cytokinins in different parts of the callus and through time, the team could stimulate the growth and differentiation of root and shoot systems and produce whole new plants. A high ratio of auxin to cytokinin led to the differentiation of a root system, while a high ratio of cytokinin to auxin led to the development of a shoot system. Eventually Skoog's team was able to produce a complete plant from just one parenchyma cell.

One way to interpret the response by tissue culture calluses to auxin: cytokinin ratios is to consider a plant's normal response to wounding. If tissue at a wound site has a high concentration of cytokinins relative to auxins, then it is probably located somewhere in the root. In such a case, it is likely that the wound removed shoot material, making it advantageous for cells to develop into shoots. Likewise, a high concentration of auxin relative to cytokinins indicates that cells at the wound are in the shoot and should develop into roots to replace the lost material.

The ability to grow whole new plants in tissue culture from just one cell has been instrumental in the development of genetic engineering (see Chapter 19). Researchers insert recombinant genes into target cells, test the cells to identify those that successfully express the recombinant genes, and then use tissue culture techniques to grow those cells into adult individuals with a novel genotype and phenotype.

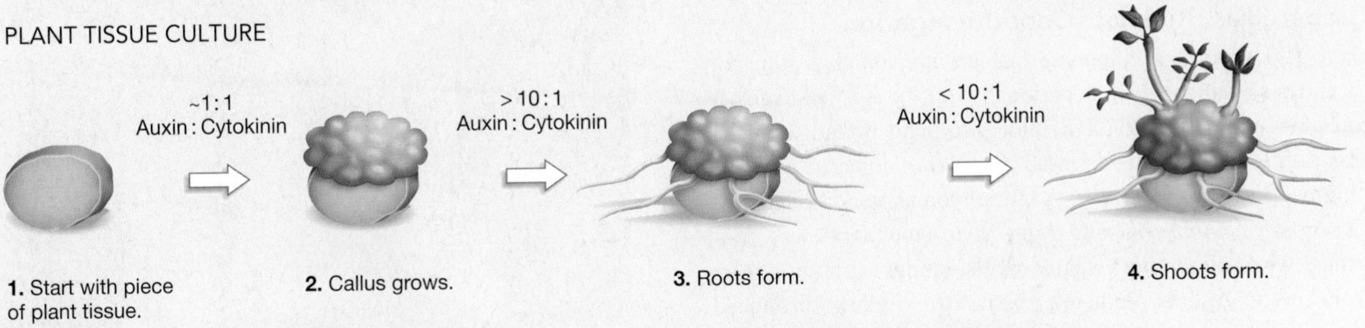
PLANT TISSUE CULTURE
~1:1
Auxin : Cytokinin
>10:1
Auxin : Cytokinin
<10:1
Auxin : Cytokinin
1. Start with piece of plant tissue.
2. Callus grows.
3. Roots form.
4. Shoots form.

FIGURE 39.7 Relative Auxin : Cytokinin Concentrations Affect Root and Shoot Development in Tissue Culture
When auxin and cytokinin concentrations are approximately equal, parenchyma cells grow into an undifferentiated callus. Roots form under high auxin and low cytokinin concentrations; shoots form under relatively low auxin and high cytokinin concentrations.

including the cyclins and the cyclin-dependent kinases (Cdk's), were introduced in Chapter 11. Recall that these proteins must be activated for cells to progress through checkpoints in the cell cycle. In this way, activated cyclins and Cdk's allow cell division to continue.

To assess whether cytokinins affect these genes, researchers grew *Arabidopsis* cells in culture so that nutrients and other molecules available could be carefully controlled. The biologists starved the cells of cytokinins for a day, then added the hormone. When they assessed the level of mRNA from a cyclin gene called *CycD3* in the treated cells, they documented significant increases compared with the level in cells that were not exposed to cytokinins. This is strong evidence that cytokinins act as growth hormones by activating genes that keep the cell cycle going. In the absence of cytokinins, cells arrest at the G_1 checkpoint in the cell cycle and cease growth.

39.5 Gibberellins and ABA: Growth and Dormancy

In northern latitudes and at high elevation, most seeds and mature plants initiate growth in spring. Conditions for growth are good at that time of year, because temperatures are warming and soil moisture levels are usually at their peak. Seedlings and mature plants continue to grow throughout the summer and early fall if moisture and nutrients are still available. But during drought conditions, growth stops. And even though embryos begin to develop as seeds mature, they cease this initial growth in the fall and remain dormant throughout the cold winter months. **Dormancy** is a temporary state of reduced metabolic activity or no metabolic activity.

What signals are responsible for initiating growth in plants, and what signals are responsible for stopping it? Two hormones provide the answer. The first is the family of closely related compounds called the **gibberellins**. The second is **abscisic acid**, commonly abbreviated to **ABA**. In general, gibberellins stimulate growth. ABA inhibits it. In at least some cases, the two hormones interact like stop and start signals.

Gibberellins Stimulate Shoot Elongation

Over 100 years ago, Japanese farmers noticed that some of their rice seedlings grew exceptionally quickly but fell over before they could be harvested. Biologists who investigated the situation found that the diseased plants were infected with the fungus *Gibberella fujikuroi*. A causal connection between infection with *Gibberella* and rapid stem elongation was confirmed when researchers treated rice seedlings with an extract from the fungus. As predicted, the treated seedlings produced abnormally long shoots.

The active component in the extract was eventually isolated and named **gibberellic acid (GA)**, which is a gibberellin. Follow-up research showed that rice plants produce their own gibberellin but respond to applications of additional hormone by elongating their stems. In effect, the infected rice seedlings were suffering from a gibberellin overdose.

Gibberellins make up a large family of structurally related compounds and are found in a wide array of fungi and plants. It is common to observe that a particular plant species produces several different gibberellins that are active as hormones. Even though these molecules have dramatic effects on growth, they are present in vanishingly small concentrations. In stems and leaves, active forms of gibberellin may be present at just one part per billion. In 1 billion grams of tissue, there would be just 1 gram of gibberellin.

Analyzing Stem-Length Mutants The strong association that was initially observed between shoot elongation and gibberellin dosage gave researchers an important tool for dissecting how GAs work. For example, to find the genes that are responsible for producing gibberellins and that encode the proteins involved in the response to GA, biologists focused on analyzing mutant plants with abnormal stem length (**Figure 39.8**). This research strategy was the basis for many of the results analyzed in Unit 3. It begins with a mutant phenotype and attempts to characterize the gene or genes responsible for the defect.

To appreciate how this approach was used to understand gibberellins, consider the dwarfed pea plants introduced in Chapter 13. Recall that Gregor Mendel analyzed a series of hereditary traits with strong effects on the phenotypes of peas and that one of the traits he investigated was stem length. Mendel analyzed the transmission of two alleles at a single

FIGURE 39.8 Dwarfed Individuals May Have Mutations That Affect Gibberellins
Dwarfed plants have much shorter stems than do normal individuals of the same age. By analyzing dwarfed individuals, biologists were able to identify genes involved in gibberellin synthesis.

locus. One allele was associated with the production of tall stems; the other was associated with dwarfed growth. The tall allele was dominant to the dwarf allele.

The gene responsible for the stem-length differences came to be known as *Le* (for *le*ngth). Early work on *Le* mutants showed that they attain normal height if they are treated with the gibberellin called GA_1. This observation suggested that dwarf peas can respond to gibberellins normally but are unable to manufacture normal quantities of their own GA_1. Follow-up experiments focused on treating dwarf peas with a radioactively labeled precursor to GA_1. These plants did not succeed in producing radioactively labeled GA_1, but plants with the normal allele did. Based on these results, researchers became convinced that the *Le* locus encodes an enzyme involved in GA synthesis.

Investigators recently provided support for this hypothesis by finding a gene in peas that encodes an enzyme called 3β-hydroxylase. This enzyme adds a hydroxyl group (–OH) to a gibberellin called GA_{20}, producing the biologically active molecule GA_1. Researchers who compared the DNA sequences of this gene from normal and dwarf pea plants found an important difference: In a part of the enzyme near the active site, the mutant DNA sequence codes for the amino acid threonine instead of alanine. Follow-up tests showed that the mutant enzymes are unable to convert GA_{20} to GA_1. This is strong evidence that the *Le* gene encodes 3β-hydroxylase and that a single amino acid change renders the enzyme largely ineffective and causes dwarfing.

Some 130 years after Mendel's experiments, biologists finally understand the dwarfing trait at the molecular level. Now researchers would like to know how the gibberellins involved in shoot elongation interact with other hormones. In stems, gibberellins appear to promote both cell elongation and rates of cell division. But it is well established that auxin promotes cell elongation and that cytokinins promote cell division. Research continues on how GAs, cytokinins, and auxin interact on the molecular level to control plant growth and development.

Gibberellins and ABA Interact during Seed Dormancy and Germination

Many plants produce seeds that have to undergo a period of drying or a period of cold, wet conditions before they are able to germinate in response to warm, wet conditions. This pattern is logical. A requirement for drying ensures that mature seeds will not sprout on the parent plant; an obligatory cold period protects seeds from germinating just before the onset of winter. Some seeds also have to receive a dose of red light, which indicates that they are in a sunny location (see Chapter 38). In essence, then, seeds have an "off" setting that discourages germination and an "on" setting that initiates growth. The appropriate state for the on/off switch is determined by environmental cues such as temperature, moisture, and light.

By applying hormones to seeds, researchers learned that in many plants, ABA is the signal that inhibits seed germination and gibberellins are the signal that triggers embryonic development. To understand how these messengers work and how they interact, researchers have concentrated on studying a specific event: the production of an enzyme called **α-amylase** (alpha-amylase) in germinating oat or barley seeds.

α-Amylase acts as a digestive enzyme that breaks the bonds between the sugar subunits of starch. (Your saliva contains an amylase that acts on the starch in food. In humans and other mammals, this enzyme initiates carbohydrate digestion in the mouth.) **Figure 39.9a** (page 900) shows that, during the germination of a barley seedling, α-amylase is released from a tissue called the **aleurone layer.** The enzyme diffuses into the carbohydrate-rich storage tissue in the seed and releases sugars that can be transported and used by the growing embryo. Adding GA to the aleurone layer increases the production of α-amylase; adding ABA to that layer decreases α-amylase levels. In this way, α-amylase production is a symptom of dormancy or germination and a model system for understanding ABA and GA action.

How Does GA Activate the Production of α-Amylase?

Once it became clear that GA and ABA interact to turn α-amylase production on or off, biologists turned to figuring out the molecular mechanism involved. Do these hormones activate or repress the transcription of certain genes? Do they alter the translation rate of certain mRNAs? Or do they activate specific membrane channels, carriers, or pumps?

An important hint at the answer emerged when a research group noticed that the promoter sequence near the α-amylase gene resembles the DNA sequences targeted by a class of transcription factors called **Myb proteins.** Mybs (pronounced *mibs*) are DNA-binding proteins that turn gene transcription on or off. To use the vocabulary introduced in Chapter 18, Mybs function as transcription activators or transcription repressors. Based on these observations, the team proposed that a Myb might turn α-amylase on in response to a signal from GA or off in response to a signal from ABA. More specifically, the team proposed that the events diagrammed in **Figure 39.9b** (page 900) occur. It had already been established that, as germination begins, GA produced by the embryo reaches the membranes of cells in the aleurone layer. The researchers hypothesized that a GA receptor on the membrane receives this signal and that the receptor responds by activating the production of a Myb. The Myb, in turn, travels to the nucleus, binds to the α-amylase promoter, and triggers transcription.

To test this idea, the researchers began a search for a Myb that is activated by GA. Their first step was to isolate all the mRNAs produced in the aleurone layer of a germinating barley seed. Then they used the enzyme reverse transcriptase to make a DNA copy of each of these mRNAs. At that point, they had a copy of all the genes being transcribed in activated aleurone

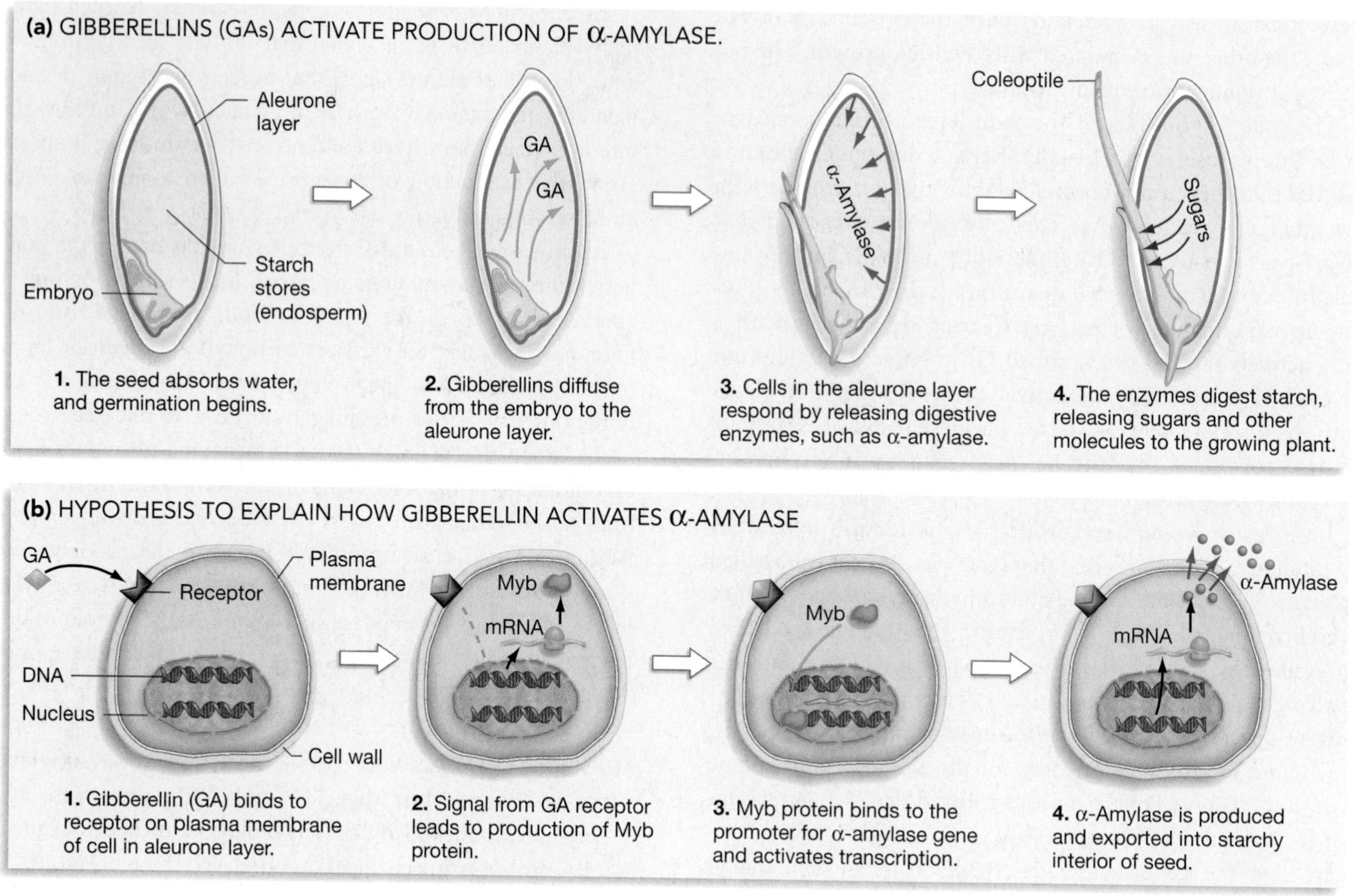
(a) GIBBERELLINS (GAs) ACTIVATE PRODUCTION OF α-AMYLASE.
Aleurone layer
Starch stores (endosperm)
Embryo
GA
GA
α-Amylase
Coleoptile
Sugars
1. The seed absorbs water, and germination begins.
2. Gibberellins diffuse from the embryo to the aleurone layer.
3. Cells in the aleurone layer respond by releasing digestive enzymes, such as α-amylase.
4. The enzymes digest starch, releasing sugars and other molecules to the growing plant.
(b) HYPOTHESIS TO EXPLAIN HOW GIBBERELLIN ACTIVATES α-AMYLASE
GA
Receptor
Plasma membrane
DNA
Nucleus
Cell wall
Myb
mRNA
Myb
mRNA
α-Amylase
1. Gibberellin (GA) binds to receptor on plasma membrane of cell in aleurone layer.
2. Signal from GA receptor leads to production of Myb protein.
3. Myb protein binds to the promoter for α-amylase gene and activates transcription.
4. α-Amylase is produced and exported into starchy interior of seed.

FIGURE 39.9 The Molecular Mechanism of Gibberellin Action
(a) In seeds, the release of gibberellins is an important part of germination. **(b)** To understand how gibberellins lead to the production of α-amylase, researchers hypothesized that they are responsible for activating the transcription factor Myb.

tissue. Were any of these sequences related to the Myb family? To answer this question, the team tested each DNA sequence to see if it would react with the section of Myb DNA that encodes the DNA-binding region. One DNA sequence from barley did. The DNA sequence had to be from a gene that encodes a Myb. The result confirmed that a Myb protein exists in activated aleurone tissue.

Is this transcription factor produced specifically in response to GA, as the researchers' model contends? To answer this question, the team isolated mRNAs from barley aleurone layers that were either exposed or not exposed to GA. They ran these mRNAs out on a gel and then probed the sequences with single-stranded DNA from the *Myb* gene they had discovered in aleurone tissue. (Chapter 19 introduced this experimental technique, which is called **Northern blotting.**) As the gels in **Figure 39.10** show, the mRNA for the Myb protein appeared only in cells that were exposed to GA. Follow-up experiments confirmed that this protein, which they named GAMyb, binds to the α-amylase promoter. Just as their hypothesis predicted, GAMyb acts as a transcription activator that stimulates α-amylase production.

How Do GA and ABA Interact? Research on GA tells only part of the story about dormancy and seed germination. Studies similar to the one we just reviewed indicate that ABA also induces the production of Myb proteins. Preliminary data sug-

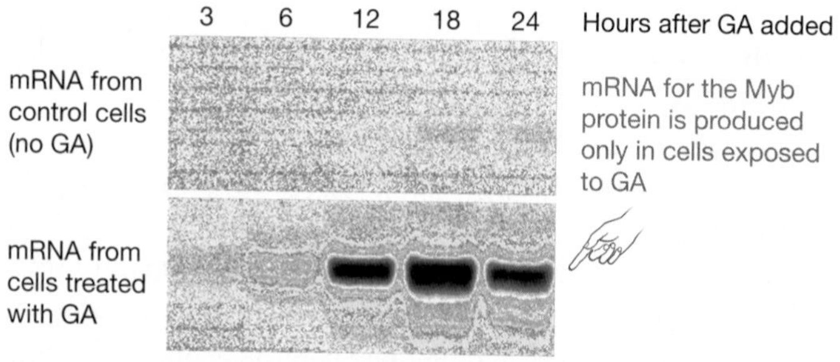
3 6 12 18 24 Hours after GA added
mRNA from control cells (no GA)
mRNA from cells treated with GA
mRNA for the Myb protein is produced only in cells exposed to GA

FIGURE 39.10 Gibberellin-Treated Aleurone Cells Produce mRNAs for a Transcription Factor Called GAMyb
Autoradiographs of gels containing mRNAs isolated from barley aleurone layers. The bands show where radioactively labeled, single-stranded copies of the *GAMyb* gene bind to mRNA. The results show that this gene is transcribed only after cells have been stimulated by gibberellin (GA).

gest that ABA-dependent transcription factors bind to the α-amylase promoter and shut down transcription, just as GA-dependent transcription factors bind to the α-amylase promoter and activate transcription.

If these experiments are confirmed, the results suggest a molecular basis for how hormones with opposite effects interact. The hypothesis is that ABA enforces dormancy by activating transcription repressors that shut down the production of key enzymes, while GA breaks dormancy and supports germination through transcription activators that trigger the production of the same enzymes. If ABA is present in high concentration, repression dominates and dormancy occurs. If GA is present in high concentration, activators prevail and germination proceeds.

The experiments we've reviewed on α-amylase production in seeds carry several important messages about hormone action: (1) A cell's response to a hormone often occurs because specific genes are turned on or off. (2) Hormones rarely act on DNA directly. Instead, a receptor on the surface of a cell usually receives the message and responds by initiating a chain of events that leads to gene activation or repression. (3) Different hormones interact at the molecular level because they induce different transcription activators and repressors. The relative amounts of these regulatory factors determine which proteins are produced by the cell and in what quantities.

ABA Closes Guard Cells in Stomata

Chapter 36 introduced one of the major problems faced by land plants: replacing water that is lost to the atmosphere when structures called **stomata** are open. Because stomata open in response to blue light, they allow gas exchange to occur while the plant is receiving the wavelengths of light used in photosynthesis. Chapter 36 also mentioned that if plant roots are unable to obtain enough water to replace the fluid being lost at the leaves, stomata close. Closing stomata is a logical response when roots cannot find adequate water, because continued transpiration would lead to wilting and potential tissue damage.

How Do Roots Communicate with Leaves about Water Conditions? Early work on stomatal closing suggested that ABA is involved. For example, applying ABA to the exterior of stomata causes them to close. To explore this hypothesis further, researchers performed a series of experiments based on the system diagrammed in **Figure 39.11**. The fundamental idea was to grow plants whose roots had been divided into two containers. Experimental plants were watered in one pot only, while control plants were watered in both pots. During this treatment, investigators documented that the water potential of the leaves remained the same in both control and experimental plants. Yet the stomata of experimental plants began to close. This result suggested that roots from the dry side of the pot were signaling drought stress, even though the leaves were not actually experiencing a water shortage.

Question: Can roots communicate with shoots?

Hypothesis: Roots that are dry can signal shoots to close stomata.

Null hypothesis: Roots cannot communicate with the shoot.

Experimental setup:

1. Divide roots of many plants into two sides.

2. In experimental group, water one side.
3. In control group, water both sides.
4. In both groups, measure water potential of leaves and observe stomata.

Prediction: Stomata in experimental plants will close; stomata in control plants will stay open.

Prediction of null hypothesis: Stomata in both experimental and control plants will stay open.

Results: No difference between experimental and control plants in water potential of leaves.

Stomata began to close.

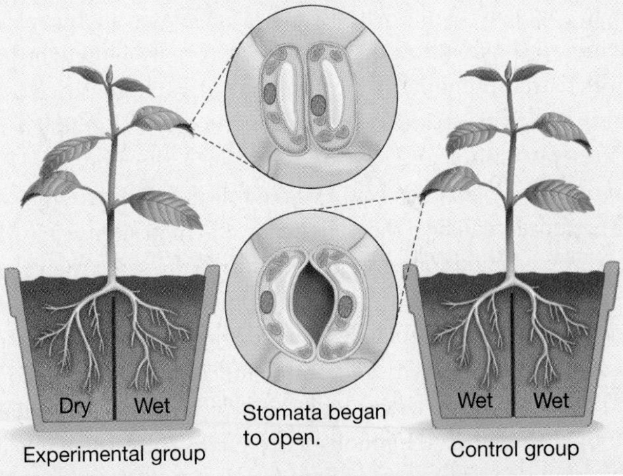

Conclusion: Roots can communicate with shoots. Dry roots signal the shoot and cause stomata to close, even though leaves are receiving sufficient water (from roots on the wet side of the plant).

FIGURE 39.11 Can Roots Communicate with Shoots?
System developed by researchers to assess whether roots communicate information about water availability to shoots. Roots are trained into two sides of the same pot so that they can be watered separately.

Follow-up experiments have supported two important predictions: ABA concentrations in roots on the dry side of the pot are extraordinarily high relative to the watered side, and ABA concentrations in the leaves of experimental plants are much higher than in the leaves of control plants. These results suggested that ABA from roots is transported to leaves and that it serves as an early warning system of drought stress. In doing so, it overrides the signal from the photoreceptor called PHOT, introduced in Chapter 38. Recall that when guard cells absorb blue light, a sequence of events results in the opening of stomata, so that CO_2 can enter cells and support photosynthesis. What is the molecular mechanism involved in the opening and closing of stomata?

How Do Stomata Open and Close? Recall from Chapter 35 that stomata consist of two **guard cells** that surround an opening called a **pore**. When the vacuoles of guard cells are filled with water, the cells are turgid. The shape of turgid guard cells results in an open pore, which allows gas exchange between the atmosphere and the interior of the leaf. But when vacuoles lose water and guard cells become flaccid and lacking in turgor, cell shape changes in a way that closes the pore and stops both gas exchange and loss of water via transpiration.

Based on these observations, the question of how stomata open and close reduces to asking how guard cells become turgid or flaccid, or how water flows into or out of vacuoles. Activation of PHOT by blue light leads to water entry; activation of ABA receptors leads to water exit.

As you might suspect, changes in guard-cell shape are related to changes in the activity of H^+-ATPases in the plasma membrane. Activation of PHOT receptors by blue light leads to increased pumping of protons out of each guard cell. The pump's activity creates a strong electrochemical gradient that brings potassium and chloride ions into the interior of the guard cells (**Figure 39.12a**). Water follows the incoming ions via osmosis, causing the cells to swell. In response to ABA, however, calcium ions are released from storage in the vacuole inside cells. The increase in intracellular calcium triggers two key events: (1) H^+-ATPases and inward-directed potassium channels are inhibited, and (2) channels that allow chloride and other anions to leave along their electrochemical gradients are opened. When the anions leave guard cells, the change in membrane potential causes outward-directed potassium channels to open. Large amounts of K^+ leave the cells, with water following by osmosis. The result is a loss of turgor and closing of the pore (**Figure 39.12b**).

Whether it acts on guard cells or seeds, ABA fulfills a general role in plants as a dormancy or "no-growth" signal. In many cases, its action depends on input from other hormones and photoreceptors. To survive and reproduce successfully, plants have to integrate information from a variety of sources.

✓CHECK YOUR UNDERSTANDING

GAs are general signals to initiate or continue growth; ABA is a general signal to stop growth or remain dormant. You should be able to (1) summarize experimental evidence that supports the claim that mutations in enzymes required for gibberellin synthesis are responsible for dwarfing in garden peas and (2) explain how GAs and ABA interact to control seed germination and dormancy and how signals from blue light and ABA interact to control the opening and closing of stomata.

39.6 Ethylene and Senescence

Senescence is the process of aging, decline, and eventual death of an entire plant or particular plant organs, such as fruits and leaves. Like most other aspects of plant growth and development, senescence is triggered and regulated by complex interactions between several different hormones in response to changes in temperature, light, and other factors.

The hormone most strongly associated with senescence is **ethylene**. Like other plant hormones, ethylene is simple in structure and is active at small concentrations. Unlike other plant hormones, however, ethylene is a gas at normal physiological temperatures. Ethylene is synthesized from the amino acid methionine. Like ABA and unlike the cytokinins and gibberellins, it is the only member of its class. No structurally related compounds in plants have functions similar to those of ethylene.

Ethylene is strongly involved in three aspects of senescence in plants: (1) fruit ripening, which eventually leads to the aging and rotting of fruit; (2) the fading of flowers; and (3) abscission, or the detachment of leaves. In addition, ethylene influences plant growth and is a stress hormone induced by drought and other conditions. The range of physiological responses regulated by ethylene is surprisingly large.

Ethylene was initially "discovered" in ancient China, when fruit growers noticed that burning incense in closed rooms hastened the ripening of pears. Westerners made a similar observation in the late 1800s, when gas street lamps came into wide use in cities and premature leaf drop subsequently occurred in plants that grew near leaky gas lines. Researchers followed up on the latter observation by showing that ethylene in lamp gas was the molecule responsible for the leaf loss. In the 1930s ethylene was found in the gases that are released by ripening apples.

Subsequently, biologists documented sharp spikes in ethylene production during fruit ripening in tomatoes, bananas, apples, and certain other species. Follow-up research on these species showed that ethylene induces the production of some of the enzymes required for the ripening processes, as well as an increase in cellular respiration, which furnishes ATP. During

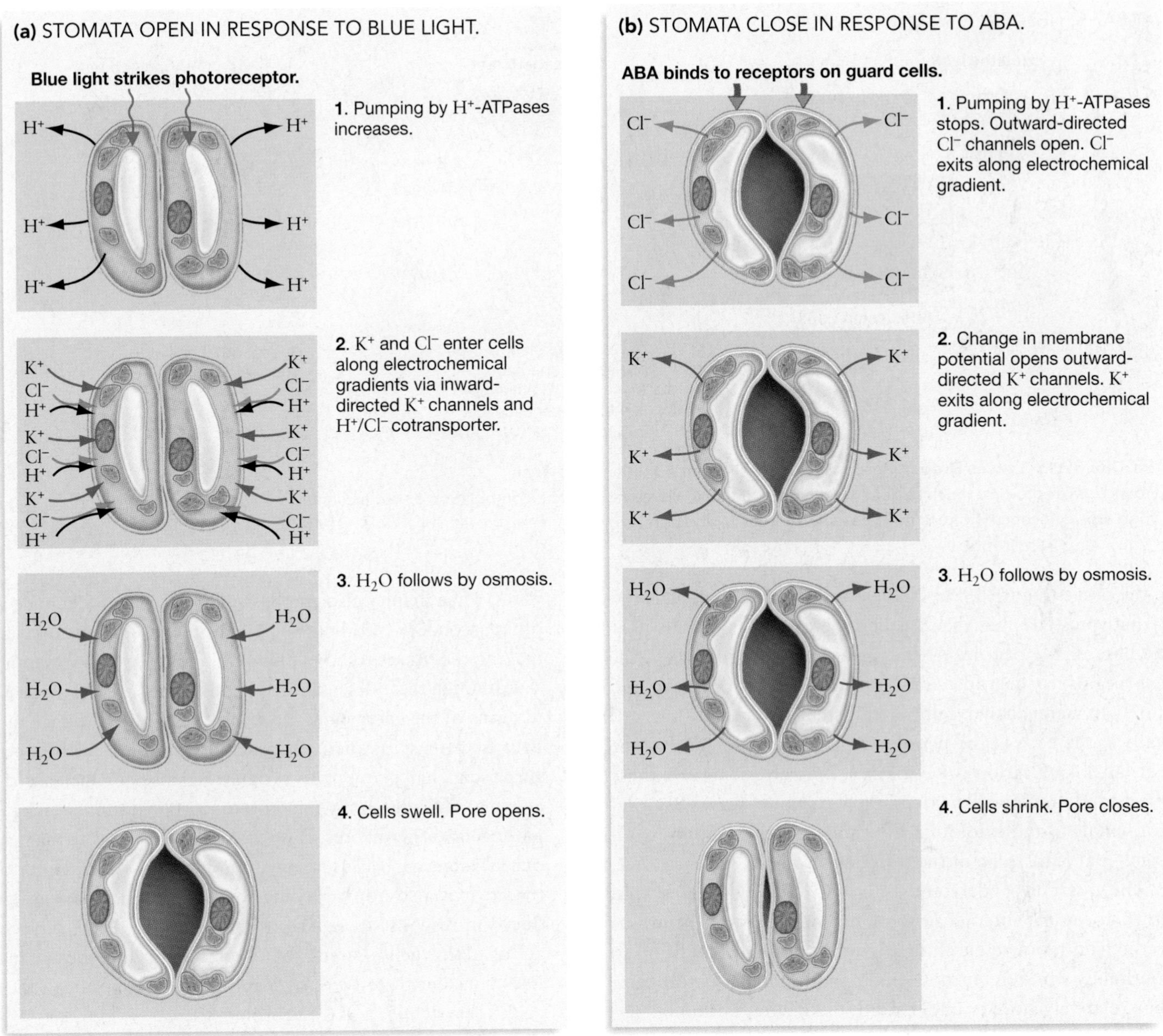
(a) STOMATA OPEN IN RESPONSE TO BLUE LIGHT.
Blue light strikes photoreceptor.
H+
H+
H+
H+
H+
H+
1. Pumping by H+-ATPases increases.
K+
Cl−
H+
2. K+ and Cl− enter cells along electrochemical gradients via inward-directed K+ channels and H+/Cl− cotransporter.
H2O
3. H2O follows by osmosis.
4. Cells swell. Pore opens.
(b) STOMATA CLOSE IN RESPONSE TO ABA.
ABA binds to receptors on guard cells.
Cl−
1. Pumping by H+-ATPases stops. Outward-directed Cl− channels open. Cl− exits along electrochemical gradient.
K+
2. Change in membrane potential opens outward-directed K+ channels. K+ exits along electrochemical gradient.
H2O
3. H2O follows by osmosis.
4. Cells shrink. Pore closes.

FIGURE 39.12 Changes in Ion Flows Are Responsible for the Opening and Closing of Stomata
(a) Activated blue-light receptors trigger ion flows into guard cells. The cells swell when water follows by osmosis. **(b)** Activated ABA receptors trigger ion flows out of guard cells. The cells shrink when water follows by osmosis.

ripening, stored starch is converted to sugar to enhance sweetness, protective toxins are removed or destroyed, cellulose is broken down to degrade cell walls and soften the fruit, chlorophyll is broken down, and pigments and aromas that signal ripeness are produced. As Chapter 40 will show, fruit ripening is interpreted as an adaptation that enhances the attractiveness of fruits to birds, mammals, and other animals that disperse seeds to new locations.

Today, fruit growers treat green bananas with ethylene after the bananas have been shipped to encourage ripening. Conversely, apples are stored in warehouses with high concentrations of CO_2 and low concentrations of O_2 to inhibit ethylene production in the fruit. Apples stored under these conditions can be sold long after their original harvest date, when untreated fruit have rotted.

Both leaf senescence and leaf abscission involve a complex interaction between auxin and ethylene. Healthy leaves produce auxin, which is then transported out through the petiole. In response to age or to changes in ambient temperature or day length, the amount of auxin that is produced by the leaf decreases. As a result, cells in a region of the leaf petiole called the **abscission zone** become more sensitive to ethylene

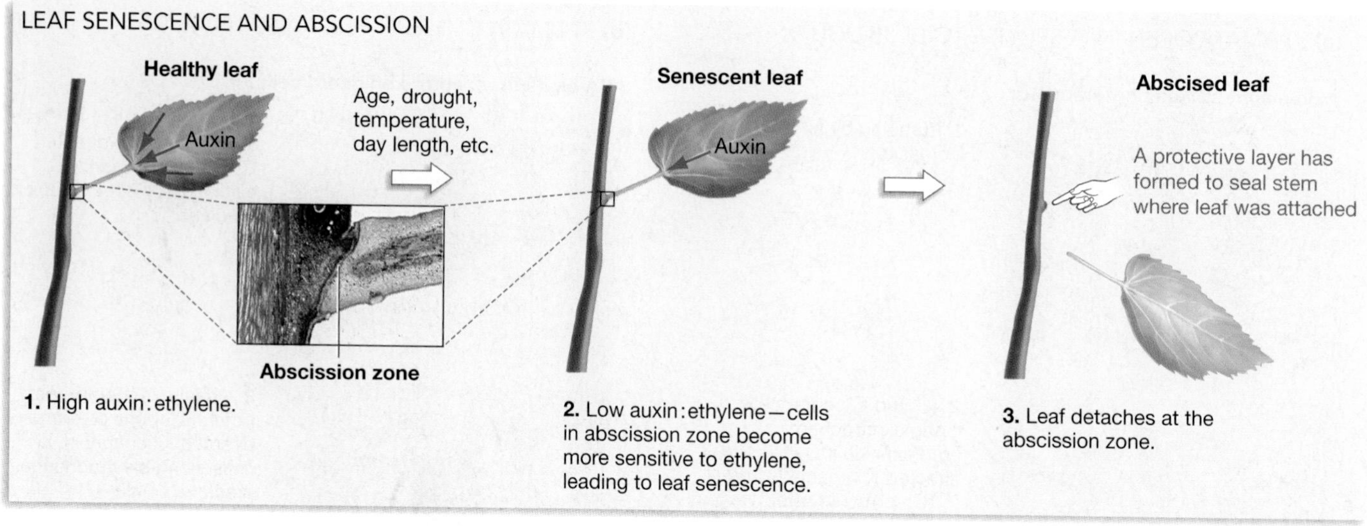

FIGURE 39.13 Leaves Drop in Response to Signals from Auxin and Ethylene
Young leaves produce much larger amounts of auxin than old leaves do. The combination of low auxin and high ethylene concentrations triggers leaf senescence and abscission.

in the tissue (**Figure 39.13**). This increase in ethylene sensitivity activates enzymes that weaken the cell walls of cells near the base of the petiole. At the same time, chlorophyll in the leaf begins to degrade, and nutrients are withdrawn and stored in parenchyma cells in the stem. Eventually the cell walls at the base of the petiole are degraded enough that the leaf falls. Applications of cytokinin, in contrast, reverse these effects and dramatically extend the life span of leaves. As a result, ethylene and cytokinins are thought to have opposite effects on at least some of the processes involved in senescence.

The quest to understand how ethylene, cytokinins, and other hormones interact in so many complex ways brings us to the frontier of research on plant hormones. In addition to continuing work on hormone interactions, biologists recognize that relatively little is understood about how hormonal signals are transduced into changes in gene expression or protein activation. To date, only a handful of genes that are directly regulated by auxin, cytokinins, ABA, GAs, and ethylene have been identified. In short, a great deal remains to be learned about the molecular basis of plant hormone action. The effort to understand chemical signaling in plants is an exceptionally dynamic field of inquiry. Perhaps the most dynamic research, however, focuses on how hormones act in plant defense.

39.7 Salicylic Acid, Systemin, and Pheromones: Defense Signals

Plants cannot run away from danger. Instead, they have to stand and fight. Recall from Chapter 35 that cuticle is an effective barrier to viruses, bacteria, fungi, and other disease-causing agents and that the structures called thorns, spines, and trichomes help protect leaves and stems from damage by herbivores. That chapter also emphasized the role that alkaloids and other secondary metabolites play in protecting plant tissues by making the tissues toxic to plant-eating insects and vertebrates.

Although these defenses are effective, they are also expensive in terms of the energy and materials required to produce them. Because defensive structures and compounds are costly, it is not surprising that plants may produce defenses or increase their existing defenses only in direct response to attacks by pathogens or herbivores. Responses such as these are called **induced defenses**, because they are induced by the presence of a threat. It is also not surprising that induced defenses are produced in response to signals from hormones. These molecules act as alarm bells—sirens that alert cells to imminent danger. Let's consider three types of plant defense signals: (1) salicylic acid, (2) systemin, and (3) pheromones.

Salicylic Acid Extends the Hypersensitive Response to Pathogens

Recall from Chapter 38 that the **hypersensitive response (HR)** is a fast and localized reaction to the presence of viruses, bacteria, or fungi. It leads to the death of infected plant cells and the production of antibacterial and antifungal compounds called *phytoalexins*. In most cases, the HR is supplemented by a slower and more widespread set of events called **systemic acquired resistance (SAR)**. Over the course of several days, SAR primes cells throughout the root or shoot system for resistance to assault by a pathogen—even cells that have not been directly exposed to the disease-causing agent.

Figure 39.14 illustrates how the HR and SAR are thought to work together. The initial event is an interaction between the *R* and *avr* gene products introduced in Chapter 38. In addition to triggering the HR, the activation of an *R* gene product leads to the production of a signal that initiates SAR. This

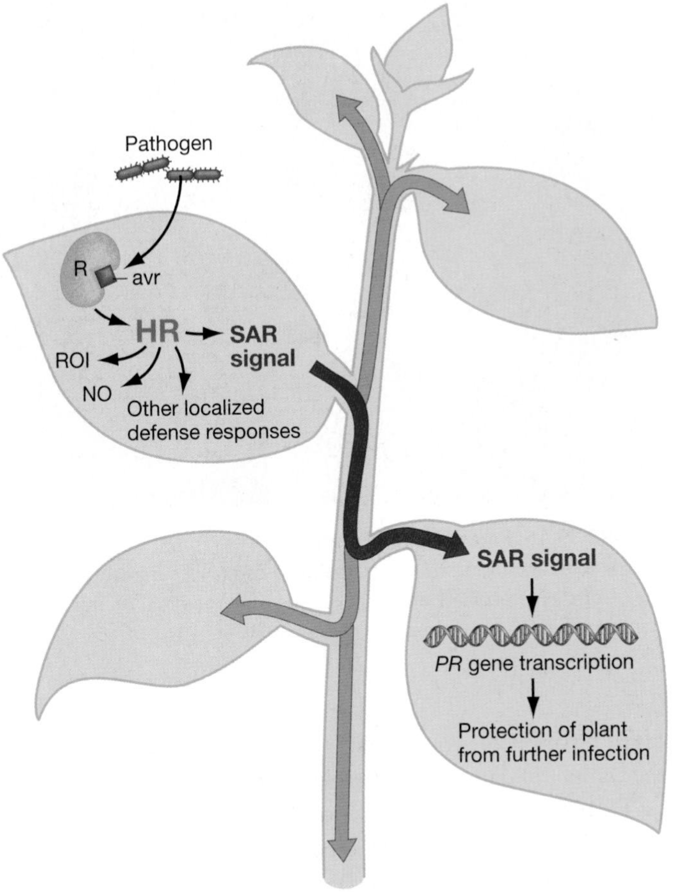

FIGURE 39.14 The Hypersensitive Response Produces a Signal That Induces Systemic Acquired Resistance
Diagram summarizing the current consensus on how the HR and SAR interact.

signal acts globally as well as locally—that is, at the point of infection—and results in the expression of a large suite of genes called the *PR* (pathogenesis-related) genes. The SAR signal qualifies as a hormone, because it carries information from one location to another.

When biologists set out to locate the signaling molecule that triggers SAR, their attention quickly turned to salicylic acid. **Salicylic acid (SA)** is found in a wide variety of plants, and its concentration was found to increase dramatically after tissues are infected with a pathogen. (SA is also the active ingredient in aspirin.) Several lines of evidence suggest that salicylic acid is a defense hormone:

- Direct application of salicylic acid to tissues triggers SAR in a variety of plant species.
- Biologists have introduced into tobacco plants a gene that codes for the enzyme salicylate hydroxylase. This enzyme is not normally found in plants and leads to the breakdown of SA. In tobacco plants that received the gene, both SA accumulation and SAR were abolished. As a result, the transformed plants were susceptible to infection by a wide variety of pathogens.
- Biologists from several research groups have labeled SA with a radioactive atom and confirmed that it is transported from infected to uninfected tissues. This observation is consistent with the hypothesis that SA is a signal that travels throughout the plant.

It is still controversial, however, whether salicylic acid acts as the SAR hormone or is only a local signal that triggers the expression of genes involved in SAR. Research on the mechanisms responsible for systemic acquired resistance continues.

Systemin Leads to the Production of Insecticides

Over a million species of insects have been discovered and named so far. Most of these animals make their living by eating vegetation, seeds, roots, or pollen. How do plants withstand this onslaught of herbivores?

In the course of studies to determine why some plants and plant tissues are more palatable and digestible than others, biochemists discovered that many seeds and some storage organs, such as potato tubers, contain proteins called **proteinase inhibitors.** Proteinase inhibitors have the same effect on herbivores that the alkaloids called tannins do (see Chapter 35). They block the enzymes—found in the mouths and stomachs of animals—that are responsible for digesting proteins. When an insect or a mammalian herbivore ingests a large dose of a proteinase inhibitor, the herbivore gets sick. As a result, herbivores learn to detect proteinase inhibitors by taste and avoid plant tissues containing high concentrations of these molecules.

Although proteinase inhibitors are found in low concentrations in an array of plant tissues, biologists wanted to test the hypothesis that these proteins might also be part of an induced defense. To evaluate this idea, researchers allowed herbivorous beetles to attack one leaf on each of several potato plants. In liquid extracted from the other leaves on the same plants, proteinase inhibitor concentrations averaged 336 μg per mL. In leaves of control plants, where no insect damage had occurred, proteinase inhibitor levels averaged just 103 μg per mL of leaf juice. This result supports the idea that a hormone produced at the site of damage travels to undamaged tissues and induces the production of proteinase inhibitors.

What is this wound-response hormone? After years of effort, biologists succeeded in isolating the molecule by purifying the compounds found in tomato leaves and testing them for the ability to induce proteinase inhibitor production. The hormone turned out to be a polypeptide, just 18 amino acids long, called **systemin.** It was the first peptide hormone ever described in plants. Researchers who labeled copies of systemin with a radioactive carbon atom, injected the hormone into plants, and then monitored its location confirmed that systemin moves from damaged tissues to undamaged tissues.

Currently, work on the production of systemin and proteinase inhibitors is focused on determining each step in the signal transduction pathway that alerts undamaged cells to

WEB TUTORIAL 39.2 Plant Defenses

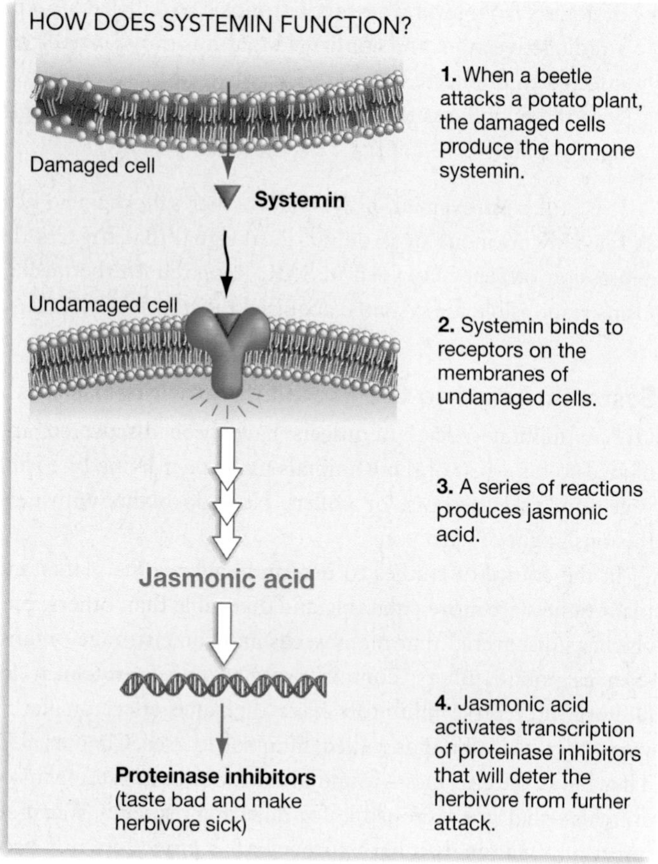
HOW DOES SYSTEMIN FUNCTION?
Damaged cell
Systemin
1. When a beetle attacks a potato plant, the damaged cells produce the hormone systemin.
Undamaged cell
2. Systemin binds to receptors on the membranes of undamaged cells.
3. A series of reactions produces jasmonic acid.
Jasmonic acid
4. Jasmonic acid activates transcription of proteinase inhibitors that will deter the herbivore from further attack.
Proteinase inhibitors
(taste bad and make herbivore sick)

FIGURE 39.15 Signals from Insect-Damaged Cells Prepare Other Cells for Attack
Systemin is a hormone produced by herbivore-damaged cells that initiates a protective response in undamaged cells.

danger. As **Figure 39.15** shows, the data indicate that systemin binds to a receptor on the membrane of an undamaged cell. The activated receptor triggers a long series of chemical reactions that eventually results in the synthesis of a molecule called *jasmonic acid*. Jasmonic acid, in turn, activates the production of at least 15 new gene products, including proteinase inhibitors. In this way, plants build potent concentrations of insecticides in tissues that are in imminent danger of an attack. The HR and SAR help protect plants from attacks by bacteria, viruses, and fungi; proteinase inhibitors help protect plants from herbivores. But in some cases of insect attack, plants also call for help.

Pheromones Released from Wounds Recruit Help from Wasps

In addition to coping with trichomes, spines, secondary metabolites, proteinase inhibitors, and other plant defenses, caterpillars and other herbivorous insects have enemies of their own. In many cases, these enemies are wasps that lay their eggs in the herbivores' bodies. When a wasp egg hatches inside a caterpillar, a wasp larva begins eating its host from inside (**Figure 39.16**). An organism that is free living as an adult but parasitic as a larva is

FIGURE 39.16 Parasitoids Kill Herbivores
Wasp larvae emerging from a parasitized caterpillar. Parasitoids lay their eggs in caterpillars and other types of herbivores. As the parasitoid larvae grow, they devour the host.

called a **parasitoid.** For obvious reasons, parasitoid attacks limit the amount of damage that herbivores do to plants.

The observation that parasitoids seem to be common during herbivore outbreaks in croplands prompted a question: Do wounded plants release compounds that actively recruit parasitoids? If so, the distress signal would furnish a novel and effective plant defense system. More specifically, the hypothesis was that plants might produce **pheromones**—chemical messengers that are synthesized by an individual and released into the environment, and that then elicit a response from a different individual. Pheromones are odorants (odor molecules) that act as chemical messengers. They change the behavior or physiology of another individual.

To explore this idea, researchers began collecting volatile compounds that are released from corn seedlings during attacks by caterpillars. (A *volatile* molecule evaporates rapidly and diffuses in the air.) Biologists who analyzed the compounds chemically found that insect-damaged leaves produced 11 molecules that were not produced by undamaged leaves. These volatile compounds were not produced by leaves that had been cut with scissors or crushed with a tool; only insect damage triggered their production. To follow up on this result, the investigators put female wasps in an arena that contained leaves damaged by insects or leaves that had suffered mechanical damage. In more than two-thirds of the tests that were performed, the wasps preferred to fly toward the insect-damaged leaves. These results support the hypothesis that plants produce wasp attractants in response to attack by caterpillars. Biologists are increasingly convinced that plants can produce pheromones that recruit help in the form of egg-laden wasps. The sophistication and complexity of plant hormones continue to astonish and fascinate biologists.

ESSAY Hormones as Herbicides

The herbicides, or "weed killers," used by farmers and gardeners act in a variety of ways. Some poison specific enzymes. The widely used compound glyphosate, for example, blocks an enzyme in the biosynthetic pathway for a molecule required in the manufacture of some amino acids and a wide variety of other compounds. (Glyphosate is sometimes marketed under the trade name "Roundup.") Paraquat belongs to a family of herbicides that disrupt electron carriers or other components of the photosynthetic apparatus. Trifluralin and oryzalin are examples of herbicides that act by disrupting the polymerization of tubulin during cell division.

Because their mechanism of action involves interfering with such fundamental cellular processes, most of the herbicides just mentioned are referred to as *broad spectrum*—meaning they kill all types of plants. A particularly popular group of herbicides, however, is more selective. These herbicides are synthetic auxins. Their chemical structure is closely related to indole acetic acid (IAA), the auxin that functions naturally in plants. When used at the recommended dosage, synthetic auxins such as 2,4-D selectively kill eudicots but do not harm monocots, such as grasses. As a result, these herbicides have been popular for treating lawns and grass-family crops, such as corn, wheat, and rice, in order to eliminate competition from broad-leaved weeds.

Biologists do not know why the synthetic auxins are selective, however. They appear to be absorbed more readily by dicots than by grasses, are effective at lower doses, and are transported throughout the body more rapidly. Based on these observations, researchers hypothesize that the synthetic auxins lead to hormone overdoses in broad-leaved species. Why excess auxin levels poison cells is also unknown, however. Apparently, high auxin concentrations overstimulate cells in some way.

The compounds in Agent Orange turned out to be harmful to humans as well as to plants.

Synthetic auxins have achieved notoriety, unfortunately, because of their use by the United States military during the Vietnam War. In an effort to clear vegetation used for cover by soldiers of the North Vietnamese army, U.S. forces sprayed extensive areas of forest with a mix of synthetic auxins known as Agent Orange. The compounds in Agent Orange turned out to be harmful to humans as well as to plants. Specifically, a common contaminant of the mixture, dioxin, caused severe skin lesions in people who were exposed to the spray. The same compound is strongly suspected to contribute to cancer formation.

The synthetic auxin that is associated with the contaminant is called 2,4,5-T. Although 2,4-D and other synthetic auxins are still widely and safely used as herbicides, the manufacture and use of 2,4,5-T is now banned in many countries.

CHAPTER REVIEW

Summary of Key Concepts

- **Hormones are the link between the receipt of information by a sensory receptor and a plant's response to that information.**

Certain plant cells can sense changes in water availability, light availability, gravity, or other aspects of the environment. To initiate a response to an environmental change, many receptor cells produce hormones that are transported to target tissues, where they bind to a receptor specific to that hormone.

Plant hormones trigger responses in target cells by changing gene expression, altering translation rates, or activating or shutting down specific membrane pumps, channels, or ion carriers. For example, cells near the tips of shoots sense changes in blue light and respond by altering the distribution of the hormone auxin. Cells on one side of the shoot accumulate high concentrations of auxin, which causes an increase in the number of proton pumps in those cells' plasma membranes. The resulting acidification of their cell walls activates expansin proteins and allows these cells to elongate much more than do cells on the other side of the stem. In this way, plants bend toward sunlight.

Web Tutorial 39.1 Plant Hormones

- **Each type of plant hormone tends to have a wide array of effects. These effects usually relate to the general role that a hormone plays in the life of a plant, such as signaling stress or indicating that conditions for growth are good.**

The general function of auxin is to establish and maintain the long axis of the plant body. In addition to playing a key role in phototropism and gravitropism, auxin is involved in apical dominance. All these responses rely on the polar transport of auxin, which establishes a gradient in auxin from the plant's apex to its roots.

In an analogous way, gibberellins signal that conditions for growth are good and promote the initiation or continuation of growth and development in an array of tissues and organs. Abscisic acid (ABA) signals that environmental conditions are bad by suppressing growth and enforcing dormancy. Cytokinins stimulate cell division. In the absence of cytokinins, cells arrest at the G_1 checkpoint in the cell cycle. Ethylene signals that particular tissues or organs should undergo senescence in response to age or deteriorating environmental conditions.

Plant responses to environmental stimuli are usually affected by interactions among several different hormones.

Regulation of dormancy and growth by abscisic acid (ABA) and by gibberellins (GAs) are examples of how hormones interact. In seeds, ABA activates transcription factors, such as α-amylase, that repress the genes required for germination to proceed. GAs, in contrast, activate regulatory proteins that increase the transcription of enzymes required for growth. ABA also slows growth because it shuts down photosynthesis when soil begins to dry, by stimulating the loss of turgor in guard cells and closing the pores in stomata. In this case, ABA acts in opposition to signals from blue-light receptors, which cause guard cells to open pores in stomata.

Plant defense signals include salicylic acid, a hormone produced as part of systemic acquired resistance (SAR); systemin, a hormone that induces the production of proteinase inhibitors; and pheromones, chemical messengers that recruit help from other individuals, such as wasps.

Web Tutorial 39.2 Plant Defenses

Questions

Content Review

1. Which of the following statements about hormones is *not* correct?
   a. They tend to be small molecules.
   b. They exert their effects on the same cells that produce them.
   c. They can exert strong effects even when they are present in extremely low concentrations.
   d. They trigger a response by binding to receptors in target cells.
2. Which of the following statements about plant hormones is *not* correct?
   a. Gibberellins function as a general signal to start or continue growth.
   b. ABA is a stress hormone. It stops growth or sustains dormancy.
   c. Cytokinins are produced in roots and are transported throughout the plant body via xylem.
   d. Ethylene is a small peptide that stimulates fruit ripening, leaf abscission, and other aspects of senesence.
3. In order for auxin to stimulate cell elongation, what two things have to happen?
   a. Transport proteins on the apical and basal ends of the cell must be activated.
   b. Myb proteins must be activated and α-amylase transcription increased.
   c. Water must flow into the cell, and the cell wall must become more extensible.
   d. Water must flow out of the cell, and the cell wall must become more extensible.
4. According to the auxin redistribution model, how do light and gravity affect auxin?
   a. Receptors for light and for gravity cause auxin to move to one side in the tips of shoots and roots.
   b. Light destroys auxin.
   c. Receptors for light and for gravity increase auxin concentrations throughout the plant.
   d. Light and gravity increase the polar transport of auxin.
5. Why do ABA and GA have opposite effects on the gene for α-amylase?
   a. ABA induces a transcription activator, while GA induces a transcription repressor.
   b. ABA induces a transcription repressor, while GA induces a transcription activator.
   c. Both hormones can bind directly to the gene's promoter.
   d. ABA decreases translation of the α-amylase mRNA; GA increases it.
6. What evidence suggests that ABA from roots can signal guard cells to close?
   a. If roots are given sufficient water, guard cells close anyway.
   b. If roots are dry, guard cells begin to close—even though leaves are not experiencing water stress.
   c. Applying ABA on guard cells directly causes them to close.
   d. If roots are dry, ABA concentrations in leaf cells drop dramatically.

Conceptual Review

1. Summarize the auxin redistribution hypothesis for phototropic and gravitropic responses. Describe the experiments that confirmed and extended the hypothesis. Design an experiment that would use labeled IAA to test the redistribution hypothesis.
2. What evidence suggests that Mendel's dwarf peas had a defect in gibberellin synthesis?
3. A plant's response to a given hormone depends on the cells or tissue involved, the plant's developmental stage or age, the concentration of the hormone, and the concentration of other plant hormones that are present. Provide examples that support each of these claims.
4. Relate the following two observations: (a) Different cytokinin-to-auxin ratios are required to induce the formation of shoot systems and root systems in tissue culture calluses; and (b) auxin is produced in shoot apical meristems and transported to roots, while cytokinins are produced in roots and transported via xylem.
5. Discuss the general role that cytokinins or ethylene serves in plants. Provide evidence that supports your claim.
6. Why is it logical to observe that plant responses usually depend on the action of several hormones?

Group Discussion Problems

1. To explore how hormones function, researchers have begun to transform plants with particular genes. In one experiment, a gene involved in cytokinin synthesis was introduced into tobacco plants. When the recombinant individuals matured, they produced more than the usual number of lateral branches. How does this experiment inform our understanding of how cytokinins function? Plants can also be transformed with antisense genes, which code for products that disable specific mRNAs by binding to them. Suggest an experiment that uses an antisense gene to explore hormone function. Predict the result.
2. In many species native to tropical wet forests, seeds do not undergo a period of dormancy. Instead, they germinate immediately. Make a prediction about the role of ABA in these seeds. How would you test your predictions?
3. Researchers have shown that stomata do not close if ABA is injected into guard cells. Stomata do close, however, if ABA is applied to the surface of guard cells. Based on these results, researchers claim that the ABA receptor must be on the surface of guard cells, not in the interior. Do you agree? Why or why not?
4. Even after years of effort, no one has been able to isolate the gibberellin or ABA receptor. One reason is that individuals with mutant GA or ABA receptors may die prior to germination, making them extremely difficult to study. Why might mutations in these genes be fatal? Suggest an alternative strategy for finding the GA and ABA receptor.

Answers to Multiple-Choice Questions 1. b; 2. d; 3. c; 4. a; 5. b; 6. b

www.prenhall.com/freeman is your resource for the following: Web Tutorials; Online Quizzes and other Online Study Guide materials; Answers to Conceptual Review Questions; Solutions to Group Discussion Problems; Answers to Figure Caption Questions and Exercises; and Additional Readings and Research.

40 Plant Reproduction

KEY CONCEPTS

- Plants undergo alternation of generations, in which a diploid sporophyte phase alternates with a haploid gametophyte phase. Sporophytes produce spores by meiosis. Gametophytes produce gametes by mitosis.
- In angiosperms, male and female gametophytes are microscopic and are produced inside flowers. Male gametophytes are encased in pollen grains and are portable. Female gametophytes are encased in an ovary and are retained in the flower. When pollen grains land on a flower, they produce sperm cells that fertilize the egg produced by the female gametophyte.
- Seeds contain an embyro and a food supply protected by a tough coat. In angiosperms, the walls of the ovary develop into a case that encloses the seed or seeds, forming a fruit.

This chapter focuses on the structure and function of plant reproductive structures, such as the sacred lotus flower.

For humans, flowers symbolize beauty, love, hope, fertility, and abundance. But for plants, flowers do not have symbolic value. **Flowers** are reproductive structures—a collection of organs designed to produce gametes, attract gametes from other individuals, nourish embryos, and develop seeds and fruits. Exploring how these reproductive structures work is a major focus of this chapter.

Plants reproduce in a wide variety of ways. Among the many major groups of plants, only one—the **angiosperms**, or flowering plants—produces flowers. Liverworts, hornworts, mosses, ferns, conifers, and other plant groups reproduce sexually without flowers, and many plants can undergo asexual reproduction as well. Even among the angiosperms, many species are able to reproduce without making flowers or setting seed, through various kinds of asexual reproduction.

It would be difficult to overemphasize the importance of the reproductive organs and processes that are analyzed in this chapter, both to plants and to biologists. For plants, every structure in the body and every physiological process—from the acquisition and transport of water and nutrients, to photosynthesis, to systems for sensing environmental stimuli and responding—exist for just one reason: to maximize the chances that the individual will produce offspring. Reproduction is the unconscious goal of everything that a plant does. For biologists, plant reproduction is not only fundamental to understanding how plants work but also the basis for entire industries. Agriculture, horticulture, forestry, biotechnology, and ecological restoration efforts can be successful only if biologists understand how plants reproduce.

Our exploration of plant reproduction begins with a look at sexuality in plants, a review of the life cycle called alternation of generations, and a discussion of asexual reproduction. Then we'll trace how angiosperms reproduce, beginning with pollination and fertilization and ending with seed dispersal and germination. By the end of this chapter, you'll be able to appreciate the practical aspects of flowers as well as their poetic features.

40.1 An Introduction to Plant Reproduction

The first thing to recognize about plant reproductive structures and processes is that they are highly variable among species. Consider just one aspect of reproductive organs: their size. As **Figure 40.1** shows, flowers vary in size from microscopic to the size of a small child; seeds and fruits range in size from dustlike particles to coconuts.

Fortunately, several basic principles unify this diversity of reproductive systems. Let's begin by defining sex.

Sexual Reproduction

Most plants reproduce sexually. Sexual reproduction is based on meiosis and fertilization. **Meiosis** is analyzed in detail in Chapter 12 and is a type of nuclear division that cuts the chromosome number in half. **Fertilization** is the fusion of haploid cells termed **gametes.** The result of fertilization is the production of offspring called **zygotes.** Male gametes, or **sperm,** are small cells that contribute genetic information in the form of DNA but few or no nutrients to the offspring. Female gametes, or **eggs,** contain not just DNA but also nutrient stores. The important point is that, although both sperm and eggs contribute DNA, they contribute vastly different amounts of other resources to the offspring.

It is also important to recognize that in the majority of plant species, both male and female reproductive organs are produced on the same individual. In angiosperms, flowers containing both male and female parts are referred to as **perfect.** Flowers can also be **imperfect,** meaning they contain either male *or* female structures, but not both. In some cases, separate male and female flowers occur on the same individual. This

(a) Small reproductive structures
Large reproductive structures
Plant
Flower
1 mm
(b) Small seeds
Large seeds
10 cm

FIGURE 40.1 The Size of Plant Reproductive Structures Is Highly Variable
(a) Reproductive structures in plants range in size from (left) the tiny flowers of *Wolffia* to (right) the gigantic flowers of *Rafflesia arnoldii* (top right). **(b)** All seeds contain an embryo and a food supply, but they vary in size from (left) orchid seeds to (right) the two large seeds inside this double coconut fruit.

FIGURE 40.2 Male and Female Flowers Can Occur on the Same Individual or on Different Individuals
(a) The tassels of corn are male flowers; ears are female flowers. **(b)** In willows, male and female flowers are found on different individuals.

condition is termed **monoecy.** The corn plant illustrated in **Figure 40.2a** is **monoecious** ("one-house"). In corn, the tassel is the male flower, and the ear is the female flower. In contrast, some species with imperfect flowers are **dioecious** ("two-houses"), meaning each individual plant produces only male or female flowers. The willows illustrated in **Figure 40.2b** are examples of a dioecious plant: Some individuals produce male flowers only and could be considered "male"; other individuals produce female flowers only and are "female."

The leading hypothesis to explain the evolution of imperfect flowers is that they are an adaptation that eliminates **self-fertilization**, or **selfing**. Selfing occurs when a sperm and an egg from the same individual combine to produce an offspring. Species with perfect flowers can self-fertilize. Species with imperfect flowers, however, must **outcross**—that is, male and female gametes must be exchanged between different individuals.

Plant Life Cycles

In most animals, meiosis leads directly to the formation of gametes. In plants, the situation is much different. Land plants are characterized by life cycles with two distinct multicellular forms—one diploid and one haploid. An individual in the diploid phase of the life cycle is called a **sporophyte**, while an individual in the haploid phase of the life cycle is called a **gametophyte.** Recall from Chapters 28 and 29 that this type of life cycle is called **alternation of generations.** When alternation of generations occurs, meiosis does not lead directly to the formation of gametes. Instead it leads to the production of haploid cells called spores. A **spore** is a cell that grows directly into an adult individual. Alternation of generations occurs in the brown algae, red algae, and green algae, as well as in land plants. It is a life cycle that has evolved several times independently during the history of life.

Several structures and processes are common to most or all plant life cycles:

- Meiosis occurs in sporophytes and results in the production of haploid spores. Unlike zygotes, spores are not produced by the fusion of two cells. Unlike gametes, spores produce an adult without fusing with another cell. Meiosis and spore production occur inside structures called **sporangia.**
- Spores divide by mitosis to form multicellular, haploid gametophytes.
- In many species, mature gametophytes develop reproductive structures known as **gametangia.** Inside gametangia, haploid gametes are produced by mitosis.
- Fertilization occurs when two gametes fuse to form a diploid zygote. The zygote then grows by mitosis to form the sporophyte.

A good way to keep these terms straight is to remember that *sporophyte* means "spore-plant," while *gametophyte* means "gamete-plant." Sporophytes produce spores by meiosis. Gametophytes produce gametes by mitosis.

Recall from Chapter 29 that land plant species vary a great deal in how large and long lived the gametophyte and sporophyte are relative to each other, and in how the two phases relate to each other in terms of nutrition. To drive these points home, consider the life-cycle diagrams in **Figure 40.3.** Figure 40.3a shows a liverwort, which is a nonvascular plant. Notice that the largest stage in the liverwort life cycle is the gametophyte. The sporophyte is dependent on the gametophyte for nutrition. Now consider the angiosperm life cycle in Figure 40.3b. A close examination of this figure should convince you that male and female gametophytes are physically separate. Both are microscopic and completely dependent on the sporophyte for nutrition.

Liverworts are the most ancient or basal lineage on the evolutionary tree of land plants living today. Angiosperms, in contrast, are the most recent lineage of land plants to appear in the fossil record and to branch off on the evolutionary tree. Based on these observations, biologists have concluded that, over the course of land plant evolution, gametophytes became reduced while sporophytes became more conspicuous. Why this trend occurred is still unknown. The adaptive significance of the gametophyte and sporophyte phases remains a challenge for biologists interested in plant reproduction.

Asexual Reproduction

Asexual reproduction does not involve meiosis or fertilization and results in the production of **clones**—genetically identical copies of the parent plant. Some plants extend their life indefi-

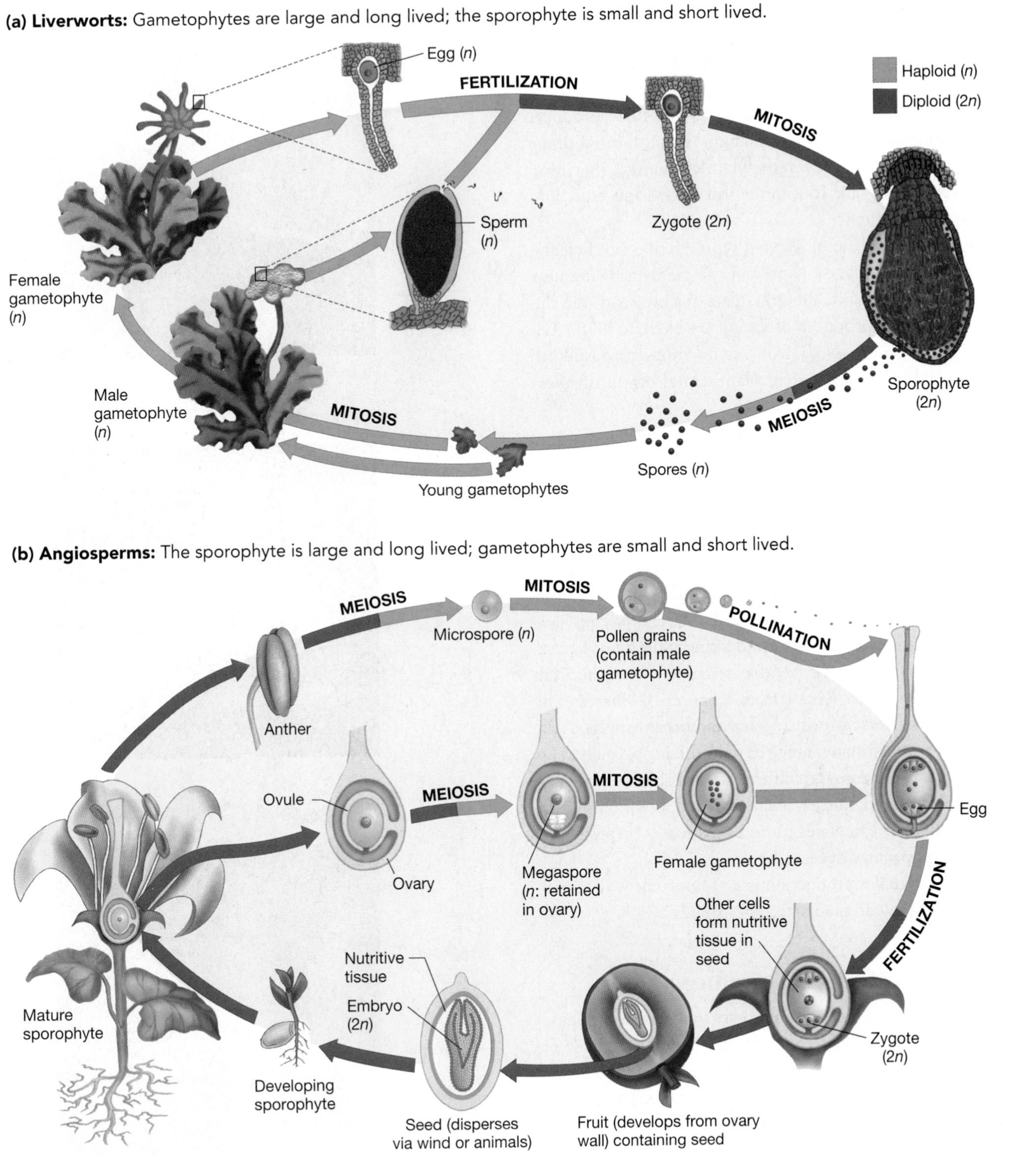

FIGURE 40.3 All Plants Undergo Alternation of Generations
(a) In liverworts and other basal groups of land plants, the sporophyte depends on the gametophyte for nutrition. **(b)** In angiosperms and other more recent groups of land plants, the gametophytes are dependent on the sporophyte for nutrition.

nitely by asexual reproduction. The oldest of all known plants is a ring of creosote bushes in the Mojave Desert of California. This clone originated from a parent plant that is estimated to have germinated 12,000 years ago.

Although all asexual reproduction is based on mitosis, a wide array of mechanisms is involved. Horizontal stems called **rhizomes** connect many of the individuals in a grass sod. Notice that shoots and roots have sprouted from nodes on the rhizomes shown in **Figure 40.4a**. Each of the individuals emerging from these nodes is an asexually produced offspring. Similarly, the gladiolus plant in **Figure 40.4b** has propagated itself via modified stems called **corms**, which grow under the surface of the soil. The kalanchoe pictured in **Figure 40.4c** produces **plantlets**, which form from meristematic tissue located along the margins of its leaves. When the plantlets mature, they drop off the parent plant, take root in the soil below, and grow into independent individuals.

The key characteristic of asexual reproduction is efficiency. If an herbivore or a disease wipes out the plants growing near an individual grass plant, the grass can quickly send out rhizomes. Its asexually produced offspring are likely to fill the unoccupied space before seeds from competitors can establish themselves and grow. The parent plant can also nourish these progeny as they become established.

Although asexual reproduction is extremely common in plants, it does have a downside. A fungus or other disease-causing agent that infects an individual plant will probably succeed in infecting the plant's cloned offspring as well, even if they are not still physically connected. This hypothesis is based on the observation that plants fight disease with a wide variety of molecules (see Chapter 38). Because sexually produced offspring are genetically unlike their parents, these offspring have unique combinations of the disease-fighting molecules. Consequently, the offspring may be able to resist infections that devastate their parents. Recall from Chapter 12 that coping with changes in disease-causing agents and other aspects of the environment is a major advantage of reproducing sexually. This is an important point in agriculture and horticulture, because asexually propagated apples, bananas, and other crops are more susceptible to epidemics caused by diseases or pests than are sexually propagated species.

Given that sexual reproduction is an important way to produce offspring, how do plants accomplish it?

40.2 Reproductive Structures

Each major group of plants, from mosses to angiosperms, has a characteristic life cycle and characteristic male and female reproductive structures. The focus here, though, is on the flower. There are two reasons for this. First, most of the food that humans eat is produced by flowering plants. The staples of our diet—grains such as wheat, corn, and rice—are the seeds of angiosperms. Second, with over 235,000 species named to date, angiosperms are far and away the most species-rich group of plants. Just two families of flowering plants, orchids and sunflowers, contain over 45,000 species, versus about 28,500 species of all nonflowering plants combined.

In both a practical and a biological sense, then, flowers are enormously important structures. Let's begin our analysis of the flower by asking when plants produce them.

(a) Rhizomes

(b) Corms

(c) Plantlets

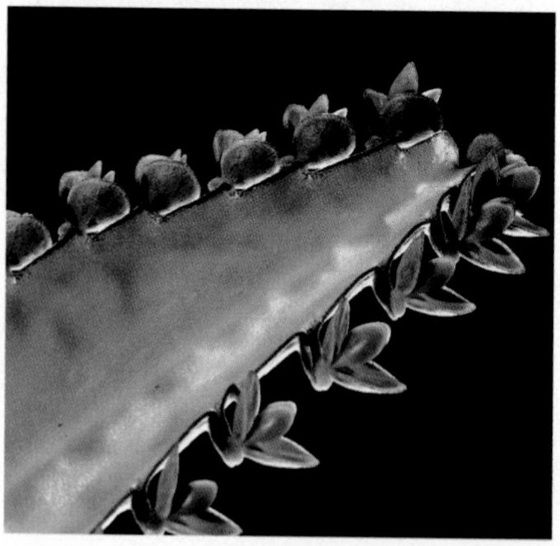

FIGURE 40.4 The Mechanisms of Asexual Reproduction Are Diverse
(a) Many grasses generate horizontal underground stems that produce new shoots and roots. If rhizomes are broken, then the cloned individuals become physically separated. **(b)** Gladioluses spread by producing bulb-like underground stems called corms. **(c)** Kalanchoe produce small plantlets at the margins of their leaves. Plantlets that fall off can grow directly into a new individual if the location is favorable.

When Does Flowering Occur?

Anatomically, a flower is a shoot that develops highly modified leaves. In essence, then, flower formation begins when an apical meristem stops making energy-harvesting stems and leaves and begins to produce the modified leaves that form flowers. Instead of making more food through photosynthesis, a sporophyte commits to investing energy in sexual reproduction—by producing gametophytes that will produce gametes. When does this happen?

Early experiments on the environmental signals that promote flowering focused on the number of hours of light and dark during a day. Recall from Chapter 38 that changes in night length trigger flowering in many species and that phytochrome is the photoreceptor responsible for detecting night length.

For the intensively studied mustard plant, *Arabidopsis thaliana*, which occupies higher latitudes, night length is an external cue containing important information about the time of year. In contrast, for plants growing at the equator, night length provides no information at all, because it does not vary over the course of the year. In desert and tropical habitats, then, it is reasonable to expect that external cues other than night length—such as the onset of seasonal rains—might trigger flowering. Based on observations like these, it is not surprising that different species respond to different external cues to induce flowering.

The situation is complex, though, because some species initiate flowering in response to several independent cues. In *Arabidopsis*, flowering can be triggered by events other than the short night lengths typical of summer. Even if individuals are exposed to long nights, flowering can begin in response to surges of the hormone gibberellin. Recall from Chapter 39 that gibberellins encourage growth. Gibberellins are an internal cue—a signal about the individual's condition.

The advantage of flowering in response to an external cue such as the short nights of summer seems clear. But what sorts of internal conditions might trigger a commitment to flowering? One possibility is the individual's nutritional status. In the case of *Arabidopsis*, flowering could be advantageous when light, water, and nutrient conditions are favorable. An individual that is growing under optimal conditions might be able to flower long before day length signals the arrival of midsummer. If nutritional status or other internal conditions affect the likelihood of flowering, then gibberellins or other hormones could serve as messengers that trigger the production of reproductive organs (see **Box 40.1**). In this way, nutritional status could explain why flowering occurs in the absence of normal external cues.

To summarize, flowering can be stimulated by either internal or external cues, or both. Now the question is, What does the resulting flower look like? How are flowers put together?

WEB TUTORIAL 40.1
Reproduction in Flowering Plants

BOX 40.1 Is There a Flowering Hormone?

Since the 1930s, biologists have known that exposing even one leaf on a plant to the conditions necessary to induce flowering will result in the whole plant flowering (**Figure 40.5a**, page 916). This experimental result suggested that the signal to flower comes from leaves and travels to the apical meristem. Grafting experiments, in which an organ from one individual is physically attached to a different individual, confirmed this result. If you provide one plant with the appropriate flower-triggering night length, then cut off a leaf or stem and graft it onto a second plant that has never been exposed to the correct conditions for flowering, the noninduced plant will flower (**Figure 40.5b**). Like the experiments on phototropism with agar blocks reviewed in Chapter 39, this result supported the hypothesis that some substance or hormone produced in the foreign, transplanted leaf must travel up the plant and trip a developmental switch in the apical meristem, causing it to switch from vegetative growth to flowering mode. Biologists were so convinced that flowering must be induced by a hormone that they gave it the name **florigen**, even though the actual hormone had not been discovered.

Almost 80 years later, however, scientists still have not isolated florigen or any other substance that directly controls flowering. As a result, several alternative hypotheses have emerged. Some biologists are testing the idea that a combination of substances—perhaps gibberellins and cytokinins accompanied by bursts of sucrose and calcium ions—travel up to the bud and trigger flowering. Another possibility is that florigen may consist of one or more small RNA molecules that are transported from cell to cell via plasmodesmata. There is growing evidence that small RNA molecules function in a number of regulative processes in plants and other organisms.

While the search for florigen continues, researchers are increasingly impressed with the complexity of the flowering response. In *Arabidopsis thaliana*, for example, more than 40 genes are now known to be involved in the initiation of flowering. Mutations in some of these genes produce individuals that flower too soon or too late, or that alter the plant's response to night length. Working out the signals, receptors, and response system involved in flowering remains an exciting frontier in biological science.

(Continued on next page)

(Box 40.1 continued)

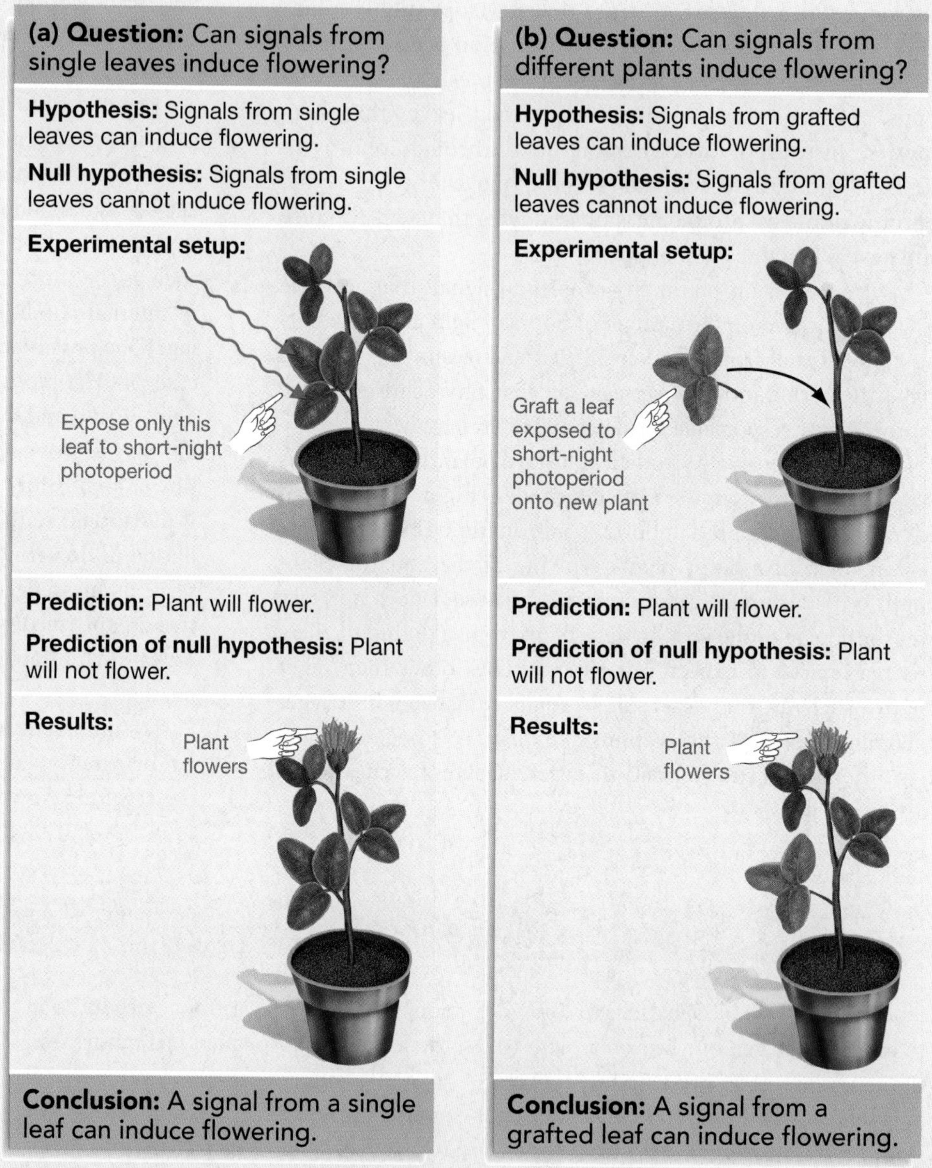
(a) Question: Can signals from single leaves induce flowering?
Hypothesis: Signals from single leaves can induce flowering.
Null hypothesis: Signals from single leaves cannot induce flowering.
Experimental setup:
Expose only this leaf to short-night photoperiod
Prediction: Plant will flower.
Prediction of null hypothesis: Plant will not flower.
Results:
Plant flowers
Conclusion: A signal from a single leaf can induce flowering.
(b) Question: Can signals from different plants induce flowering?
Hypothesis: Signals from grafted leaves can induce flowering.
Null hypothesis: Signals from grafted leaves cannot induce flowering.
Experimental setup:
Graft a leaf exposed to short-night photoperiod onto new plant
Prediction: Plant will flower.
Prediction of null hypothesis: Plant will not flower.
Results:
Plant flowers
Conclusion: A signal from a grafted leaf can induce flowering.

FIGURE 40.5 Experimental Support for the Hypothesis That a Hormone for Flowering Exists

QUESTION For these experiments to support the claim that a flowering hormone exists, control treatments need to be done in each experiment. What are these controls? If the hypothesis is correct, what is the predicted outcome of each control treatment?

The General Structure of the Flower

Structurally, all flowers are variations on a theme. They are made up of four basic parts: (1) sepals, (2) petals, (3) stamens, and (4) one or more carpels (**Figure 40.6a**). Not all four elements are present in all flowers, however. As **Figure 40.6b** shows, the coloration, size, and shape of these four components are fabulously diverse. Let's take a closer look at each part.

Sepals are leaflike structures that make up the outermost parts of a flower. They are arranged in a whorl around the **receptacle**—the site where floral organs attach to the stem. Sepals are usually green and photosynthetic, and they are relatively thick compared with other parts of the structure. Because sepals enclose the flower bud as it develops and grows, biologists hypothesize that they protect young buds from damage by insects or disease-causing agents.

Petals are also arranged around the receptacle in a whorl. They are often brightly colored and scented, and they function to advertise the flower to bees, wasps, hummingbirds, and other animals. In some cases, the color of the petals correlates with the visual abilities of particular animals. Bees, for example, respond strongly to wavelengths in the blue and purple regions of the light spectrum. Flowers that attract bees, in turn, often have blue or purple petals with ultraviolet patches. The ultraviolet sections of these petals, which are invisible to humans but visible to bees, frequently highlight the center of the flower (**Figure 40.7**). Why? In these flowers, the base of the petals contains a gland called a **nectary**. A nectary produces the sugar-rich fluid called **nectar**, which is harvested by many of the animals that visit flowers. Wind-pollinated angiosperms such as oaks, maples, and grasses have flowers that lack nectaries and have small petals or no petals at all.

The entire group of petals in a flower is called the **corolla**. In some species, the petals within the corolla vary in size,

(a) Basic parts of a flower

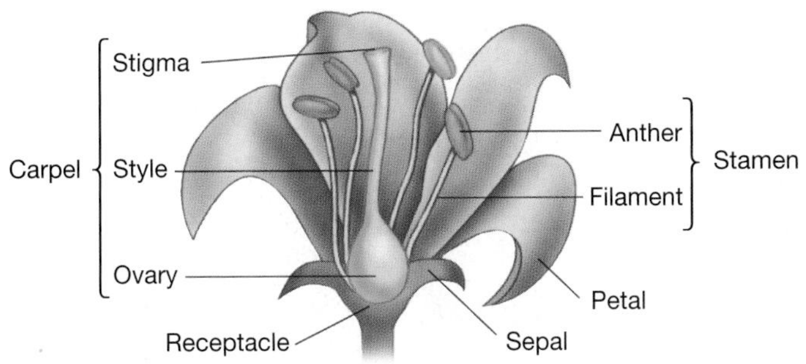

(b) Examples of flower diversity

FIGURE 40.6 The Basic Structures in Flowers Are Highly Variable
(a) The basic parts of a flower are the sepals, petals, stamens, and carpels. Not all four parts are present in all flowers. **(b)** The characteristics of the four parts of a flower vary enormously from species to species.
EXERCISE In the photographs, label as many flower parts as you can.

(a) What you and a bee see **(b)** What a bee sees in addition

FIGURE 40.7 Insects See in the Ultraviolet Range
(a) A normal view of the flower of a black-eyed Susan. **(b)** The same structure, photographed with a camera that records ultraviolet wavelengths that are visible to bees but invisible to humans.

shape, and function. For example, flattened petals may provide a landing pad for flying insects, while elongated, tubelike petals frequently have a nectary at their base. Other petals protect the reproductive organs located inside the corolla. Specialized cells in the petals may also synthesize and release molecules that provide a scent attractive to certain species of pollinating insects.

The reproductive structure that produces male gametophytes is called a **stamen.** The stamen consists of two components: (1) a slender stalk termed the **filament** and (2) the pollen-producing organs called **anthers** (see Figure 40.6a). The reproductive structure that produces female gametophytes is called a **carpel.** The carpel consists of three regions: (1) the **stigma,** which is the moist tip that receives pollen; (2) a slender stalk called the **style**; and (3) the **ovary** at the base of the carpel (see Figure 40.6a). A carpel may contain more than one ovary. Inside the ovary, female gametophytes are produced in structures called **ovules.**

These structures are crucial. Meiosis and gametophyte formation take place in stamens and carpels. Stamens produce male gametophytes; carpels produce female gametophytes. Let's take a closer look.

Producing the Female Gametophyte

What purposes do the three parts of the carpel serve? The function of the stigma and style will become clear in Section 40.3, but for now let's concentrate on what happens inside the ovary. **Figure 40.8** (page 918) provides a cross-sectional view of the inside of a typical angiosperm ovary. Notice that it contains one or more structures called ovules. The ovule contains a diploid cell, called the **megasporocyte,** that will undergo meiosis. The megasporocyte is found inside a structure called the **megasporangium** within the ovule. (The use of "mega" is appropriate, because these structures are much larger than their counterparts in the pollen-producing organs.) The megasporangium is comparable to spore-producing organs found in other plants, such as the sporangia found on the back of fern leaves.

Figure 40.8 illustrates the steps involved in production of the female gametophyte. There are three important things to note:

1. The megasporocyte inside the megasporangium divides by meiosis.
2. Four haploid cells called **megaspores** result from meiosis, but three degenerate. No one is sure how or why this happens.
3. The surviving megaspore divides by mitosis to produce a group of haploid daughter cells. Eight nuclei are produced in the species illustrated here; the number produced varies among species. Notice that the eight nuclei segregate to different positions. Cell walls form around them, resulting in seven cells. (In this species, two of the nuclei stay together within one central cell.) This multicellular, haploid structure is the **female gametophyte**—sometimes referred to as the *embryo sac.*

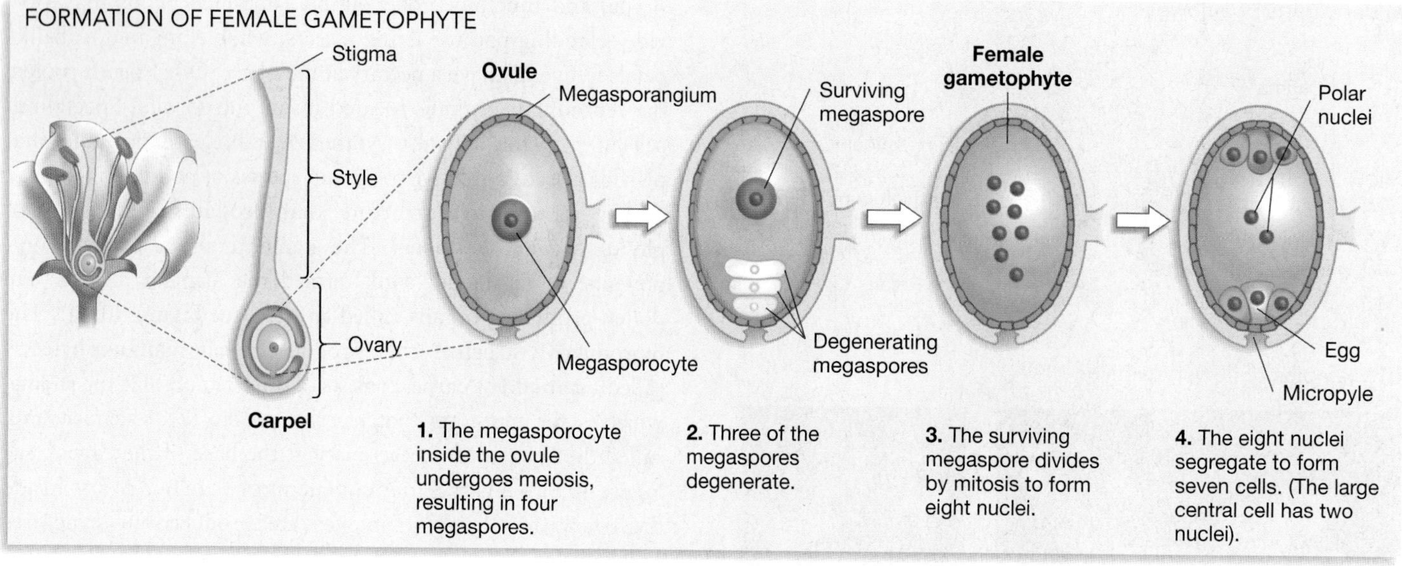

FIGURE 40.8 In Angiosperms, Female Gametophytes Are Produced by Megaspores
In a carpel, the ovary encloses one or more ovules. A megasporocyte inside the ovule divides by meiosis, but only one of the four products, called megaspores, survives. The surviving megaspore divides by mitosis to form the egg and a variety of other cells within the female gametophyte.

To summarize, a diploid cell in the sporophyte divides by meiosis to form a spore, which then divides by mitosis to form the female gametophyte. The most important components of the female gametophyte are the nuclei in the central cell, which are called **polar nuclei**, and the egg. The polar nuclei are found in the largest cell of the ovule. There are two polar nuclei in the species illustrated here and in many other angiosperms, but the number varies among species. The egg is located at one end of the female gametophyte, near an opening in the ovule called the **micropyle** ("little-gate"). When the female gametophyte is mature, a gamete (egg cell) has formed and is ready to be fertilized.

Producing the Male Gametophyte

Figure 40.9 gives a detailed look at the stamen. Notice that it consists of two major parts: an anther and a filament. Inside the anther, structures known as **microsporangia** contain diploid cells called **microsporocytes**, which undergo meiosis. Each haploid cell that results is called a **microspore**. Each of the four resulting microspores divides mitotically to form a haploid **male gametophyte**. The structure composed of the microspore wall and the immature male gametophyte is termed a **pollen grain.**

The wall of a pollen grain develops a tough outer coat that includes the watertight compound called sporopollenin, introduced in Chapter 29. This coat protects the male gametophyte when the pollen is released from the parent plant into the environment.

At this stage, the male gametophyte is considered immature and consists of two cells: a small **generative cell** enclosed within a larger cell called the **tube cell.** The male gametophyte is considered mature when the haploid generative cell produces two sperm cells via mitosis. In some species, maturation occurs while pollen is still in the anther. In other species, sperm cells are not produced until after the pollen grain has been carried to a carpel from the stamen where the pollen developed.

Depending on the species, pollen grains may be dispersed by an animal, the wind, or water currents. Now, what happens when a pollen grain arrives at a mature carpel of the same species, where an egg cell is waiting?

✓CHECK YOUR UNDERSTANDING

In angiosperms, sporophytes produce female and male gametophytes in flowers. The reproductive structures called carpels contain ovaries, which enclose structures called ovules, where female gametophytes are produced. A diploid megasporocyte inside an ovule undergoes meiosis to form a haploid megaspore, which divides by mitosis to form the female gametophyte—including the egg and polar nuclei. Male gametophytes are produced inside reproductive structures called anthers. A diploid microsporocyte undergoes meiosis to form haploid microspores, which divide by mitosis to form the male gametophyte—including a generative cell that will divide by mitosis to form sperm. You should be able to make labeled diagrams showing the major steps in the formation of eggs and sperm in angiosperms, indicating where meiosis occurs, the identity of spores, and which structures are the male and female gametophytes.

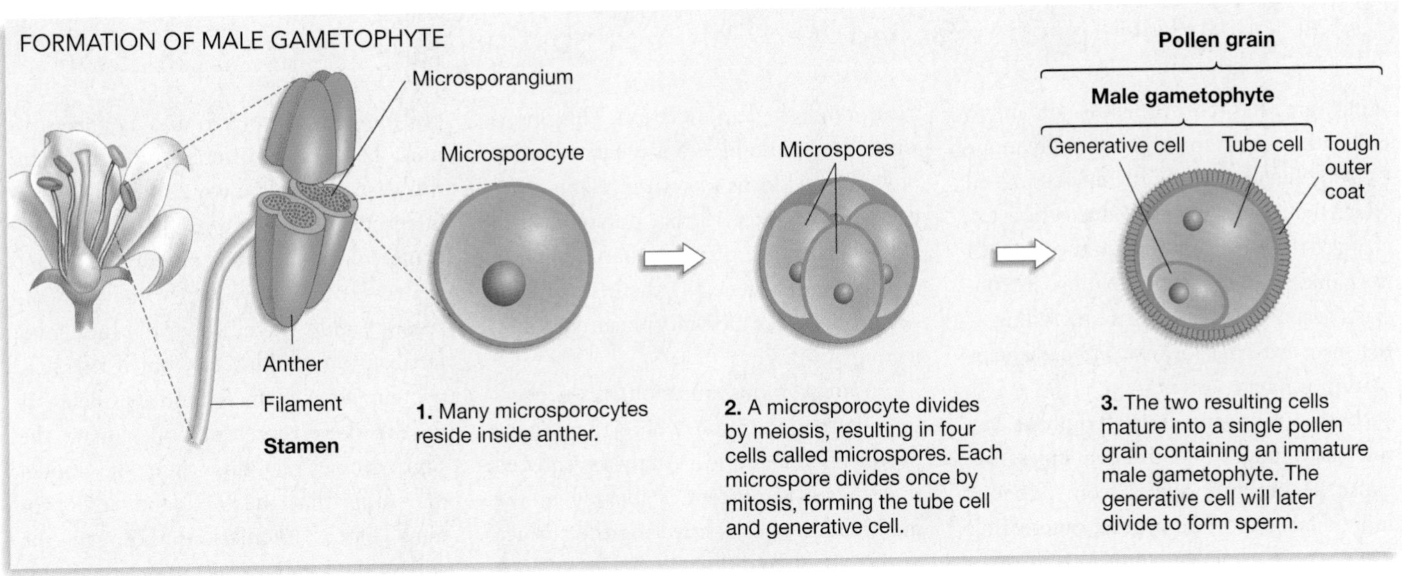
FORMATION OF MALE GAMETOPHYTE
Microsporangium
Microsporocyte
Microspores
Pollen grain
Male gametophyte
Generative cell
Tube cell
Tough outer coat
Anther
Filament
Stamen
1. Many microsporocytes reside inside anther.
2. A microsporocyte divides by meiosis, resulting in four cells called microspores. Each microspore divides once by mitosis, forming the tube cell and generative cell.
3. The two resulting cells mature into a single pollen grain containing an immature male gametophyte. The generative cell will later divide to form sperm.

FIGURE 40.9 In Angiosperms, Male Gametophytes Are Produced by Microspores
Microsporocytes inside the anther divide by meiosis. The resulting cells, called microspores, each divide once by mitosis to form a male gametophyte. A pollen grain consists of the male gametophyte and a tough coat produced by the sporophyte.

40.3 Pollination and Fertilization

Pollination and fertilization are two different events. **Pollination** is the transfer of pollen grains from an anther to a stigma; fertilization occurs when a sperm and an egg actually unite to form a diploid zygote. The two events are separated in space and time.

Pollination is not restricted to angiosperms. The gymnosperms introduced in Chapter 29 also package their male gametophytes into pollen grains. This section will focus on pollination and fertilization in flowering plants, however, because managing pollination and fertilization in angiosperms is a critical challenge for fruit growers and plant breeders. In addition, pollination and fertilization systems are thought to be key to the angiosperm's evolutionary success.

What is so remarkable about pollination and fertilization in flowering plants? Why have these innovations allowed angiosperms to become so successful in terms of their numbers of species?

Pollination

Cross-pollination is said to occur when pollen is carried from the anther of one individual to the stigma of a different individual. **Self-pollination** occurs when pollen falls from the anther of one individual onto the stigma of the same individual. **Box 40.2** (page 920) discusses the advantages and disadvantages of cross- versus self-pollination, along with some of the molecular mechanisms that prevent self-pollination in certain species.

Cross-pollination can be accomplished in a number of ways: Pollen can be carried from flower to flower by abiotic agents such as wind or water or by biotic intermediaries such as insects, birds, or bats. Animals visit flowers to eat pollen grains, harvest nectar, or both. As an animal feeds from a flower, pollen grains adhere to its body incidentally. When the same individual visits another flower of the same species to feed, some of these grains are deposited on a stigma of the second flower. In this way, cross-pollination is accomplished.

In most cases, animal pollination is an example of **mutualism**, a mutually beneficial relationship between two species. Pollinators usually benefit by receiving food; flowering plants gain by having their male gametophytes transferred to a different individual so that outcrossing takes place. As the essay at the end of this chapter shows, however, pollination by animals is not always mutualistic. Certain plant species do not provide a food reward to their pollinators; instead, they deceive them.

What Is the Adaptive Significance of Pollination? In mosses, ferns, and other groups that do not form pollen, sperm have flagella and swim to the egg through droplets of water. In conifers and most other gymnosperms, pollen is transmitted from male to female by wind. But in some of the groups that produce pollen, such as the cycads, gnetophytes, and angiosperms, many species are pollinated by animals—particularly by insects.

When these observations are placed in an evolutionary context, several important patterns emerge. Recall from Chapter 29 that researchers have compared DNA sequences from several genes found in all plants, and then used the observed similarities and differences to reconstruct which plant groups are most closely related and which are most distantly related. The most

BOX 40.2 Selfing, Outcrossing, and Self-Incompatibility

What are the consequences of cross-pollination versus those of self-pollination? Cross-pollination leads to outcrossing, in which male and female gametes are exchanged between different individuals of the same species. Self-pollination, in contrast, leads to self-fertilization. Selfing is the most extreme form of **inbreeding**, or mating among relatives.

Both outcrossing and selfing have distinct advantages and disadvantages. Because it combines alleles from different individuals to form a zygote, outcrossing leads to offspring with much higher genetic variation than offspring that result from selfing have. High genetic variation can lead to offspring that survive better and produce more offspring themselves—meaning they have higher fitness. The disadvantage of outcrossing is that female flowers may never receive pollen and so go unfertilized. Biologists have documented this cost of outcrossing by hand-pollinating plants in natural habitats to ensure pollination. In some species, hand-pollinated plants produce far more seed than do individuals that are pollinated naturally. Selfing makes pollination almost certain but leads to offspring that are genetically very similar to their parents. Selfing can also lead to inbreeding depression—the type of fitness decline analyzed in Chapter 24.

Although selfing is common in angiosperms, most plant species outcross. Outcrossing is obligatory in species that have imperfect flowers. In addition, some species that have perfect flowers are **self-incompatible**—meaning that they have sophisticated mechanisms to enforce outcrossing. The best studied of these self-incompatibility mechanisms involves the products of a group of genes called *S* (for "self") in wild cabbage plants and their domesticated relatives. The *S* locus codes for three proteins that are involved in self-incompatibility, called SRK, SLG, and SCR. By sequencing this gene in many individuals, researchers have established that up to 50 different *S* alleles may be present in a single population. This observation is remarkable—*S* is the most highly polymorphic locus known in plants. (Recall from Chapter 19 that polymorphism is the occurrence of more than one allele at a particular gene.) Is there some connection between polymorphism and self-incompatibility?

To answer this question, researchers set out to determine where the *S* gene products are found. Analyses showed that the SRK protein is located in the membranes of cells covering the surface of the stigma and that the SLG protein is in the walls of stigma cells. *SCR*, in contrast, is expressed by diploid cells in the anther. SCR proteins are found on the surfaces of pollen grains.

These observations inspired a model for how the *S* system works (**Figure 40.10**). The fundamental hypothesis is that, for a cross to be compatible, both alleles of the male parent must be different from both alleles of the female parent. For example, suppose the SRK and SLG proteins on a stigma come from S_1 and S_2 alleles (out of the 50 *S* allele types in the population). If the SCR protein from a pollen grain is from S_3 and S_4 alleles, it must be from a different individual. In this case, the pollen would be allowed to interact with the stigma. Pollen tubes—projections that grow out of the pollen grain—would grow into the stigma, sperm would travel down the tube, and fertilization could occur. But if the SCR protein were from S_1 and S_2 alleles, it would likely represent pollen from the same flower. In this case, the model maintains that the SRK and SLG proteins block interactions between the stigma and pollen grain. In this way, both self-fertilization and inbreeding are avoided. Investigators are now focused on understanding exactly how the S proteins on stigma and pollen interact.

Research on the *S* locus has attracted a great deal of attention, in part because it has important practical implications. If the genes for self-incompatibility could be introduced into crop species that are normally self-compatible, then plant breeders would no longer have to remove the anthers of corn and other self-compatible crop plants by hand to keep them from inbreeding.

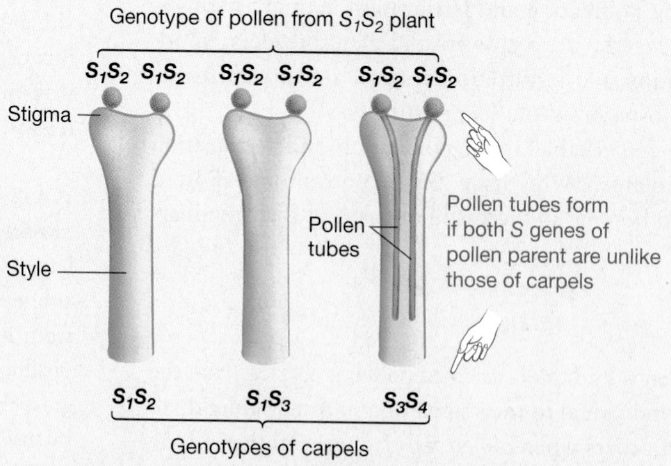
Genotype of pollen from S_1S_2 plant
S_1S_2 S_1S_2 S_1S_2 S_1S_2 S_1S_2 S_1S_2
Stigma
Style
Pollen tubes
Pollen tubes form if both S genes of pollen parent are unlike those of carpels
S_1S_2
S_1S_3
S_3S_4
Genotypes of carpels

FIGURE 40.10 The Products of the *S* Gene Help Cabbage-Family Plants Avoid Inbreeding
If a pollen grain carries S proteins that are encoded by alleles *different* from those for the S proteins found on a stigma, then the pollen is allowed to germinate. If the pollen grain carries S proteins that match the S alleles found in the carpel, then a reaction takes place that prevents the pollen from germinating.

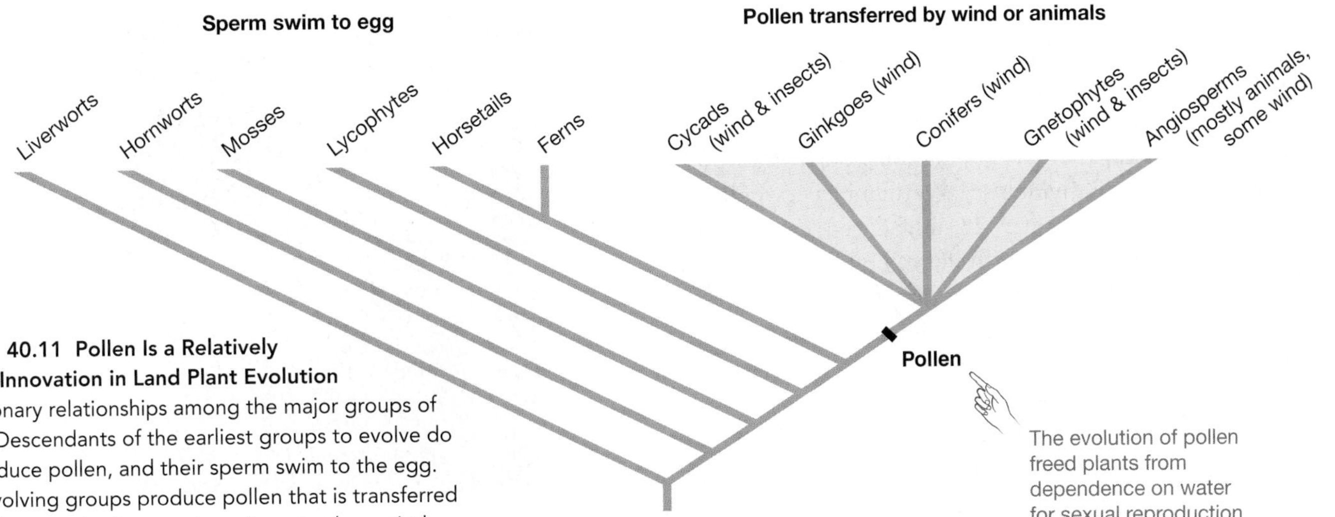
Sperm swim to egg
Pollen transferred by wind or animals
Liverworts
Hornworts
Mosses
Lycophytes
Horsetails
Ferns
Cycads (wind & insects)
Ginkgoes (wind)
Conifers (wind)
Gnetophytes (wind & insects)
Angiosperms (mostly animals, some wind)
Pollen
The evolution of pollen freed plants from dependence on water for sexual reproduction

FIGURE 40.11 Pollen Is a Relatively Recent Innovation in Land Plant Evolution
Evolutionary relationships among the major groups of plants. Descendants of the earliest groups to evolve do not produce pollen, and their sperm swim to the egg. Later-evolving groups produce pollen that is transferred to female reproductive structures by animals or wind.

recent phylogenetic tree estimated from this type of analysis is shown in **Figure 40.11**. Notice that mosses and other groups that do not form pollen were the first lineages of plants to evolve. This conclusion is supported by the order in which species appear in the fossil record. Conifers and other groups that are strictly or primarily wind pollinated evolved later but before angiosperms.

The first important pattern here is that the more recently evolved plant groups do not need water in order for sexual reproduction to occur. As a result, the evolution of pollen has allowed these species to be much less dependent on wet habitats and paved the way for the colonization of drier environments. The second important pattern is that pollination became a much more precise process when animals began to act as pollinators. Wind-borne pollen grains blow about randomly. Consequently, the probability is small that a grain will land successfully on a flower stigma. Insect-borne pollen, in contrast, is much more likely to be successfully transferred to flowers of the same species. Insect pollination is an important adaptation, because it makes sexual reproduction much more efficient.

Does Pollination by Animals Encourage Speciation?

Biologists have recently begun to explore another major consequence of pollination by insects and by other animals: the evolution of new species. In addition to affecting the fitness of individual plants, does pollination by animals make the formation of new species more likely?

As an example of how insect pollination could lead to the evolution of new species, consider the situation shown in **Figure 40.12**. A biologist has documented that two populations of a mountain-dwelling species called the alpine skypilot have flowers with different characteristics. Alpine skypilots that grow in forested habitats at or below timberline have

Timberline habitats
Flowers are small and smell skunky.
Flies are the main pollinators.
Tundra habitats
Flowers are large and smell sweet.
Bumblebees are the main pollinators.

FIGURE 40.12 Interactions between Angiosperms and Animal Pollinators May Lead to Speciation
The characteristics of flowers in alpine skypilot populations from timberline and tundra habitats are very different. Because the flowers attract different pollinators in the two habitats, interbreeding is rare and the two populations may eventually become different enough to be considered different species.

small flowers with short stalks and an aroma described as "skunky." Individuals that grow in the tundra habitats above timberline have large flowers with long stalks and smell sweet. These differences are interesting, because different insects pollinate the two populations. Small flies pollinate the timberline individuals; large bumblebees pollinate the tundra flowers. Chapter 23 introduced experimental data showing that bumblebees prefer to pollinate large flowers. Because flies and bumblebees prefer to visit different types of flowers, the two skypilot populations do not breed together routinely and are evolving distinct characteristics. They are on their way to becoming distinct species.

The message here is that evolutionary changes in the size or food-finding habits of a pollinator affect the angiosperm populations they pollinate. In return, changes in flower size and shape affect the insects pollinating that population. Because mutation continuously introduces variations in these traits, insect and angiosperm populations frequently change, diverge, and form new species. It is no surprise that insects and angiosperms are exceptionally species-rich groups.

It is clear that pollination was a crucial innovation during plant evolution. Now let's get down to mechanics. What happens once a pollen grain is deposited on a stigma?

Fertilization

After landing on the stigma of a mature flower from the same species, a pollen grain absorbs water and germinates. **Germination** is a resumption of growth. The male gametophyte produces a long filament called a **pollen tube**, which grows down the length of the style (**Figure 40.13**). In the species illustrated here, the generative cell travels down the length of the tube and divides to form two sperm, or male gametes. When the pollen tube reaches the micropyle of the ovule, it fuses with the ovule and the sperm move into the female gametophyte. This event sets the stage for fertilization.

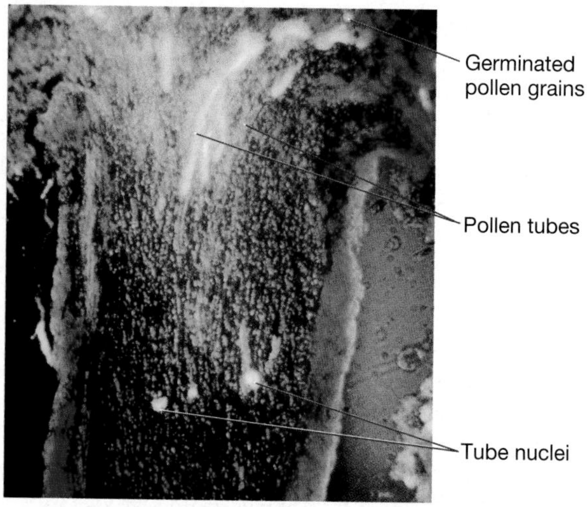

Germinated pollen grains
Pollen tubes
Tube nuclei

FIGURE 40.13 The Pollen Tube Grows toward the Egg
The male gametophyte germinates on the stigma and grows toward the egg. The sperm formed when the generative pollen nucleus divides travel down the tube toward the female gametophyte.

As noted in Section 40.1, fertilization is the fusion of sperm and egg to form a diploid zygote. In most plant groups, fertilization is straightforward—sperm and egg simply combine, and a diploid nucleus is formed. But in angiosperms, an unusual event called **double fertilization** takes place (**Figure 40.14**). One sperm nucleus unites with the egg nucleus to form the zygote. The other sperm nucleus moves

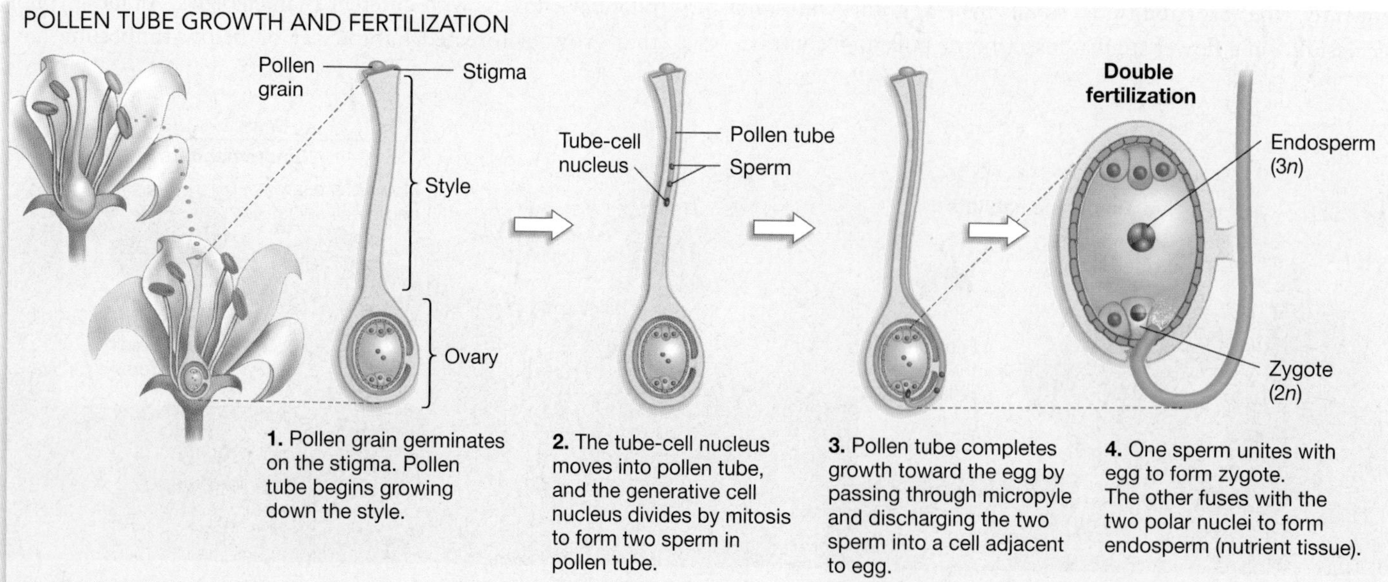

POLLEN TUBE GROWTH AND FERTILIZATION
Pollen grain
Stigma
Style
Ovary
Tube-cell nucleus
Pollen tube
Sperm
Double fertilization
Endosperm (3n)
Zygote (2n)
1. Pollen grain germinates on the stigma. Pollen tube begins growing down the style.
2. The tube-cell nucleus moves into pollen tube, and the generative cell nucleus divides by mitosis to form two sperm in pollen tube.
3. Pollen tube completes growth toward the egg by passing through micropyle and discharging the two sperm into a cell adjacent to egg.
4. One sperm unites with egg to form zygote. The other fuses with the two polar nuclei to form endosperm (nutrient tissue).

FIGURE 40.14 Double Fertilization Produces a Zygote and an Endosperm
When the pollen tube reaches the female gametophyte, one sperm nucleus fertilizes the egg while the second fuses with the polar nuclei to form a polyploid endosperm.

through the female gametophyte and fuses with the polar nuclei. In most cases, two polar nuclei are present and a large triploid ($3n$) cell forms.

The triploid cell resulting from this second fertilization undergoes a series of mitotic divisions that produce a tissue called **endosperm** ("inside-seed"). The function of endosperm is to store nutrients. Endosperm cells are loaded with starch or oils (lipids) plus proteins and other nutrients that will be needed by the plant after it germinates. But before germination can occur, seeds must develop and be dispersed.

✓CHECK YOUR UNDERSTANDING

Pollen grains are carried from one plant to another by wind, water, or animals. When a pollen grain lands on a stigma, the grain germinates. A pollen tube forms and grows until it reaches the ovule. Sperm cells produced by the male gametophyte fertilize the egg and the polar nuclei, forming a diploid zygote and in most cases a triploid endosperm. You should be able to make a labeled diagram showing how pollination and fertilization occur in an angiosperm.

40.4 The Seed

Fertilization triggers the development of a young sporophyte. In angiosperms, the first stage in the sporophyte's life is the maturation of the seed. As a **seed** matures, the embryo and endosperm develop inside the ovule and become surrounded by a tough covering called a **seed coat**. At the same time, the ovary around the ovule develops into a **fruit**, which encloses and helps protect the seed (or seeds, if many ovules are contained within a single ovary). In addition to providing protection, fruits often aid in dispersing seeds away from the parent plant.

The mature seed consists of an embryo, a food supply in the form of endosperm, and a seed coat. Let's begin our analysis of how seeds work with a closer look at how the embryo develops.

Embryogenesis

Recall from Chapter 21 that **embryogenesis** is the process by which a single-celled zygote becomes a multicellular embryo. When an angiosperm zygote divides, it produces the two daughter cells shown in step 1 of **Figure 40.15**. The lower daughter cell, called the **basal cell**, divides to form a row of single cells. This structure provides a route for nutrient transfer from the parent plant to the developing embryo. The upper daughter cell, or **terminal cell**, is the parent of all the cells in the embryo.

As the terminal cell and its progeny divide, the mass of cells in step 2 of Figure 40.15 forms. These cells sort themselves into three groups, each conforming to one of the three adult tissue types introduced in Chapter 35 (step 3). The exterior layer of embryonic cells, called the **protoderm**, is the progenitor of the adult dermal tissue, or epidermis. The **ground meristem**—the cells just within the exterior—gives rise to the ground tissue. And the **procambium** is a group of cells in the core of the embryo that becomes the vascular tissue.

As the embryo continues to develop, the long axis of the plant begins to emerge and several important structures take shape (step 4, Figure 40.15). Particularly important are the **cotyledons**, or *seed leaves*. Recall from Chapter 29 that one prominent group of angiosperms—called the monocotyledons, or monocots—has just one seed leaf, whereas eudicotyledons, or eudicots, have two. In some eudicots, the cotyledons take up the nutrients in the endosperm and store them. In these species, there is no endosperm left by the time the seed matures, and the cotyledons function as the nutrient storage organ. Another important structure within the seed is

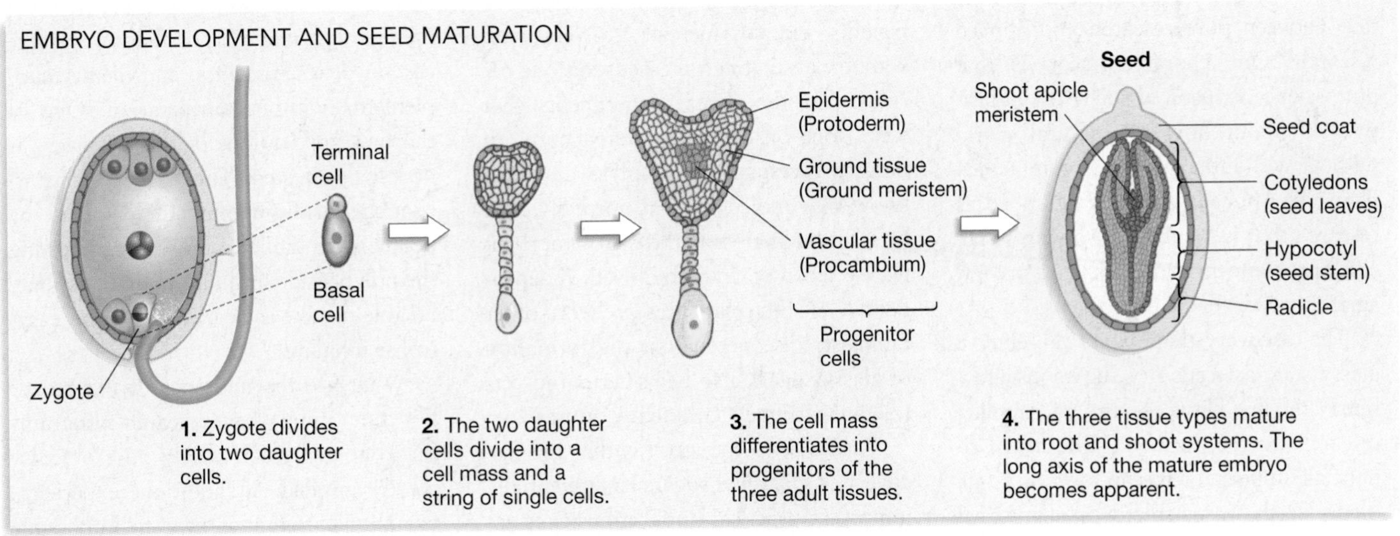
EMBRYO DEVELOPMENT AND SEED MATURATION
Zygote
Terminal cell
Basal cell
1. Zygote divides into two daughter cells.
2. The two daughter cells divide into a cell mass and a string of single cells.
Epidermis (Protoderm)
Ground tissue (Ground meristem)
Vascular tissue (Procambium)
Progenitor cells
3. The cell mass differentiates into progenitors of the three adult tissues.
Seed
Shoot apicle meristem
Seed coat
Cotyledons (seed leaves)
Hypocotyl (seed stem)
Radicle
4. The three tissue types mature into root and shoot systems. The long axis of the mature embryo becomes apparent.

FIGURE 40.15 Embryonic Tissues and Structures Develop Inside Seeds
The embryo inside a seed has the beginnings of root and shoot systems and its first leaves, or cotyledons. The embryonic epidermis and the ground and vascular tissues are organized in distinct layers.

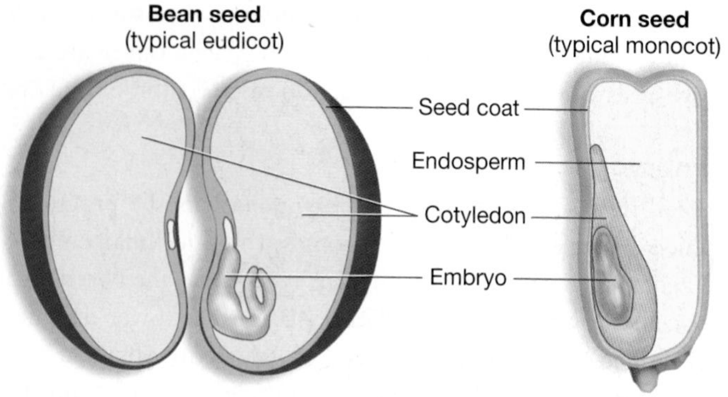
Bean seed
(typical eudicot)
Corn seed
(typical monocot)
Seed coat
Endosperm
Cotyledon
Embryo

FIGURE 40.16 Seeds Contain an Embryo and a Food Supply Surrounded by a Tough Coat
In beans, the nutrients in the endosperm are absorbed by the cotyledons and stored. In corn, the endosperm is intact.

the **hypocotyl** ("under-cotyledon"), or *seed stem*, which is the embyonic stem. The hypocotyl terminates in the **radicle**, or embryonic root. Some embryos also have an **epicotyl** ("above-cotyledon"), which is a portion of the embryonic stem that extends above the cotyledons. **Figure 40.16** compares the seed structure in beans and corn—a representative eudicot and monocot, respectively.

By the time a seed matures, then, the three major tissue types have differentiated in the embryo. The root and shoot systems, along with the first leaves, have formed. Once these events are accomplished, the seed tissues dry and the embryo becomes *quiescent*—meaning it stops growing.

The Role of Drying in Seed Maturation

The seeds of many species dry out as they mature. Water makes up 90 percent of normal plant cells, but dried seeds contain just 5–20 percent water. Loss of water is interpreted as an adaptation that prevents seeds from germinating on the parent plant. In addition, the dry condition of seeds ensures that once they have dispersed from the parent plant, they will not germinate until water is available in the environment. This is logical, because water availability is crucial to the survival of germinated seedlings.

The dry state of seeds raises an important question, however: How do the plasma membranes and proteins in the embryo and endosperm survive? When researchers reduce the amount of water surrounding isolated membranes and proteins to the levels observed in extremely dry seeds, some of the membranes and proteins disintegrate. Clearly, something is happening at the molecular level in seeds to keep these cell components intact.

Researchers have recently established that this "something" involves sugars. As water leaves the seed during drying, sugars act as a replacement for water, maintaining the integrity of plasma membranes and proteins. If drying is extreme, the sugars form an extremely viscous liquid that contains little if any water. Substances such as this are considered vitrified, or glass-like (glass is a liquid solution with the viscosity of a solid). Biologists propose that this glassy, sugary state helps maintain the integrity of plasma membranes and proteins in seeds that experience extremely dry conditions. (**Box 40.3** points out some practical implications of this work.) The researchers have also

BOX 40.3 Pure and Applied Science

People sometimes draw a sharp distinction between pure research and applied research. One characterization is that pure science is inspired solely by inquisitiveness about how the natural world works, without regard to commercial payback. Applied science is described as research that is focused on solving a specific problem in business, technology, agriculture, or medicine.

The discovery that sugars can enter a glassy state as seeds dry shows just how blurry the line between pure and applied research can be, however. The initial financial support for research on the state of dry seeds was inspired by a practical argument: If biologists could discover why certain seeds have such a long life span, the knowledge could have practical benefits for commercial growers who want to preserve certain seeds for long periods. Yet the actual experiments that were done on dry seeds were "pure" in nature. Investigators simply wanted to know how molecules that normally exist in solution maintain their structural integrity when water is removed. Independently of the plant researchers, other biologists discovered that the formation of glassy sugars also helps certain insects tolerate extremely cold temperatures.

The twist to the story is that the discovery of glass-like sugars has potentially important practical applications that are completely different from those originally envisioned. If proteins can be preserved in an extremely dry state by encasing them in sugars, it may be possible to preserve certain drugs, vaccines, and dietary supplements without refrigeration. One of the first applications of this discovery is now being tested. The idea is to sugar-coat the insulin manufactured for use by people with diabetes. Instead of injecting the polypeptide in solution, diabetics may be able to take their medicine as an easy-to-use inhalant.

What are the messages of this story? For one thing, few research programs turn out to be completely pure or completely applied. In addition, experience has shown that it is next to impossible to predict the practical implications of research.

shown that when seeds imbibe water, the glassy sugars dissolve and germination proceeds.

Drying is only one part of the seed maturation process, however. Equally important is the development of tissues surrounding the seed itself. In many cases, these tissues are required for the seed to be dispersed from the parent plant.

Fruit Development and Seed Dispersal

Fertilization initiates not only the development of the seed and embryo in angiosperms but also the development of the ovary wall into the fruit. During maturation of the fruit, the walls of the ovary thicken, forming a structure called the pericarp (**Figure 40.17**). The **pericarp** is the part of the fruit that surrounds and protects the seed or seeds.

Fruits can be dry, as in nuts, or fleshy, as in cherries and peaches. In addition to providing protection from mechanical damage and seed predators, fruits frequently have structures that aid in seed dispersal. Dispersal is important to the fitness of the young sporophyte. This is especially true in long-lived species, in which the offspring may compete with the parent plant for light, water, and nutrients if there is no dispersal.

Dry fruits are dispersed by wind, animals, or mechanical action—or the fruits simply fall to the ground. Seeds that are dispersed by wind are usually tiny; the fruits often have external structures to catch the breeze and extend the distance they travel. The fruits of dandelions and maple trees are familiar examples of wind-dispersed fruits. Other dry fruits have hooks or barbs that adhere to passing animals, while nuts are dispersed by seed predators. Still other types of dry fruits are actually dispersed by the plant via propulsion. The sandbox tree, for example, produces a seed pod that shrinks as it dries (**Figure 40.18**). Eventually the pod splits apart violently, spraying seeds in all directions. The pod bursts with so much force that the plant is sometimes called the dynamite tree. The sound of a seed pod bursting resembles a pistol shot, and seeds can be scattered as much as 40 m away from the parent plant. The dwarf mistletoe fruit, in contrast, fills with sugars as it matures. Enough water follows via osmosis to make the fruit explode and shoot seeds as far as 5 m.

(a) Before bursting **(b)** After bursting

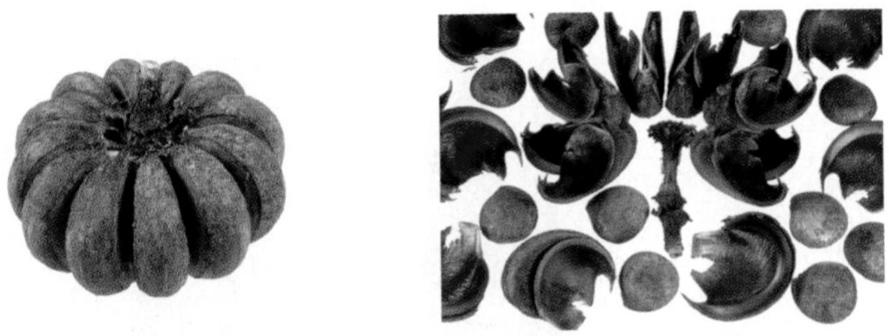

FIGURE 40.18 In Some Species, Seed Dispersal Occurs via Propulsion
A ripe seed pod from a sandbox tree **(a)** before and **(b)** after it has burst open and ejected its seeds.

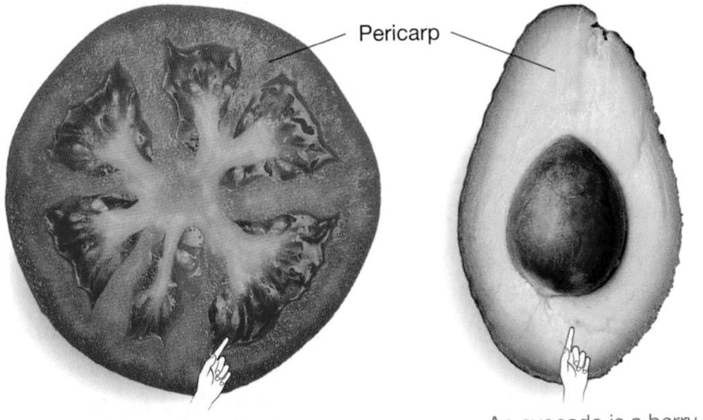
Pericarp
A tomato is a berry with many seeds
An avocado is a berry with one seed

FIGURE 40.17 As Fruits Mature, the Ovary Wall Develops into a Pericarp That Surrounds the Seed or Seeds
A fruit consists of a pericarp and the enclosed seeds.

Animals are the most common dispersal agent for fleshy fruits. In animal-dispersed fruits, the seed coat has to be tough enough to resist the mechanical forces and chemical conditions found in the animal's mouth and digestive tract, so that seeds can emerge unscathed. **Figure 40.19** (page 926) illustrates the development of some common categories of fleshy fruits that are eaten and dispersed by animals. In these species, seed dispersal is an example of mutualism. The plant provides a fruit rich in sugars and other nutrients; in return, the animal carries the fruit to a new location and excretes the seeds along with a supply of fertilizer.

Seed Dormancy

Even after being dispersed from the parent plant, and even though water and oxygen are available, seeds may not germinate for a period of time. This condition is known as **seed dormancy**. Why does dormancy exist? What keeps seeds from germinating, even though conditions appear to be favorable?

In 1995 a team of biologists reported a remarkable finding about the seeds of the sacred lotus plant. The seeds they studied were excavated from an old, dried-up lakebed in China. After germinating several of the seeds successfully, the researchers used radiometric dating techniques like those introduced in Chapter 2 to estimate the age of the seeds. According to their data, the oldest seed that germinated was about 1300 years old. Its parent grew during the height of the Tang dynasty, not long after the founding of Islam and just before Charlemagne ruled France.

Although this example represents an extreme, the important observation is that the seeds of some species can remain dormant for a long time. Dormancy is usually a feature of seeds from species that inhabit seasonal environments, where

WEB TUTORIAL 40.2
Fruit Structure and Development

Flower
Developing fruit
Ripe fruit
APRICOT:
Each flower contains one ovary.
RASPBERRY:
Each flower contains many ovaries.
PINEAPPLE:
Many flowers with many ovaries combine.

FIGURE 40.19 Fruits Vary in Size, Structure, and Color
Fruits come in many types. The structure of a fruit depends on the number of ovules found in each carpel and whether ovaries fuse during fruit maturation. QUESTION When fruits ripen, their color changes in a way that makes them more conspicuous to fruit eaters. State a hypothesis to explain why this color change might increase the fitness of an individual. Design an experiment to test your hypothesis.

conditions may be inhospitable for seedlings for extended periods of time. Based on this observation, dormancy is usually interpreted as an adaptation that allows seeds to remain viable until conditions improve. In contrast, dormancy is rare or nonexistent in seeds that are produced by plants that inhabit tropical wet forests or other areas where conditions are suitable for germination year-round.

Let's consider two questions about dormancy: What molecular mechanisms are responsible for the condition? How does dormancy cease so that germination can begin?

What Role Does ABA Play in Dormancy? Chapter 39 introduced the hormone **abscisic acid (ABA)** and described its role in preventing germination. In some species, seeds that enter dormancy have a high concentration of this hormone. The seeds of desert plants, for example, have high concentrations of ABA in their seed coats. When these seeds are exposed to large amounts of water during rare or seasonal rains, the hormone literally washes out of the seed's outer tissues. Once the ABA concentration is reduced in this way, germination proceeds.

There is not a strict correlation between ABA concentration and degree of seed dormancy, however. In peas and many other species, seeds routinely contain high levels of ABA and yet are not dormant. In *Arabidopsis*, ABA concentrations rise as seeds mature and appear to impose dormancy. ABA levels eventually fall, however, so dormant, mature seeds contain only trace amounts of the hormone.

Faced with observations such as these, researchers have concluded that there is no single, universal mechanism for initiating and maintaining seed dormancy. In some cases, changing ABA levels or the ratio of ABA to gibberellin present controls the initiation or maintenance of the dormant state. In other cases, changes in sensitivity to ABA, rather than the sheer amount of the hormone present, appear to be important. It is also likely that novel mechanisms for initiating and maintaining dormancy are still to be discovered.

How Is Dormancy Broken? The coats of some seeds are thick enough to prevent water and oxygen from physically reaching the embryo. For germination to occur, these seed coats must be disrupted, or **scarified**. Crop seeds that require scarification are placed in large, revolving drums along with pieces of sandpaper. The abrasion from the sandpaper scarifies the seeds. In nature, seed coats can be disrupted by a fire, by the passage of the seed through an animal's digestive tract, or by abrasion against wind- or water-driven soil particles. The basic principle is that the seed coat has to be broken for water to enter the seed.

Other types of seeds must experience particular environmental conditions in addition to exposure to water. As indicated in Chapter 39, species native to northern or alpine habitats often produce seeds that must undergo a time of cool, wet conditions before they will germinate. No one knows how these seeds perceive cold or what molecular mechanisms are involved in breaking dormancy when the cool, wet period ends.

In many species that produce small seeds, successful germination occurs only at or near the soil surface. This observation is logical, because small seeds have few nutrient reserves in their cotyledons or endosperm. As a result, small-seeded species need to germinate near the soil surface, where individuals are exposed to light and can then feed themselves via photosynthesis. As shown in Chapter 38, lettuce seeds and other small seeds must be exposed to red light before they will break dormancy and germinate. Red light is an important environmental cue, because wavelengths in the red portion of the light spectrum support photosynthesis. Red light and blue light indicate that sunlight is abundant.

Finally, many of the seeds produced by species native to habitats where wildfires are frequent, such as the California chaparral and South African fynbos, have an unusual chemical requirement to break dormancy. These seeds must be exposed to fire or smoke before they will germinate. The active chemical agent that triggers the response is unknown, however. (In fact, the commercial food product "liquid smoke" induces germination in these seeds as well as actual smoke does.) Again, it is logical for seeds in this habitat to germinate after fire has cleared away old vegetation and has thinned existing trees and shrubs.

The message here is that dormancy can be broken in response to a wide variety of environmental cues. In general, the cue that triggers germination is a reliable signal that conditions for seedling growth are favorable for a particular species in a particular environment.

Seed Germination

Even if specific environmental signals are required to break dormancy, seeds do not germinate without water. Water uptake is the first event in germination. Once the seed coat allows water penetration, water enters by a steep water-potential gradient, because the seed is so dry.

Figure 40.20 graphs how water uptake occurs in a typical angiosperm seed. The important point is that the curve describing water uptake has three distinct phases. Germination begins with a rapid influx of water. This influx is followed by an extended period in which no further water uptake occurs. The last segment of the graph is marked by water intake starting up again and rising steadily.

During the first phase of water uptake, oxygen consumption and protein synthesis in the seed increase dramatically, but no new messenger RNAs are transcribed. Based on these observations, biologists have concluded that the earliest events in germination are driven by mRNAs that are stored in the seed prior to maturation. During the second phase, when water uptake stops, newly transcribed mRNAs appear and

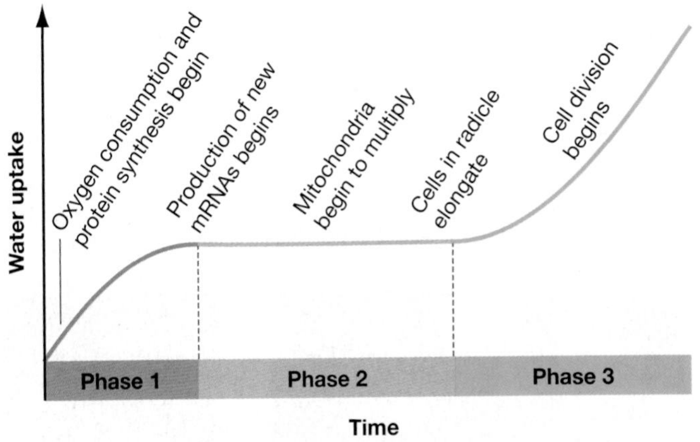

FIGURE 40.20 Germination Begins with Three Distinct Phases
Graph plotting the rate of water uptake as a typical seed germinates. (The graph is conceptual, meaning it represents a general pattern observed in data from many species, so the axes have no units.)

are translated into protein products. Mitochondria also begin to multiply. In effect, seeds take up enough water to hydrate their existing proteins and membranes, and then begin to manufacture the proteins and mitochondria needed to support growth. Water uptake resumes as growth begins. This second phase of water uptake enables cells to develop enough turgor pressure to enlarge. Eventually, the embryo bursts from the seed coat.

Figure 40.21 shows what happens as eudicot and monocot embryos emerge from the seed. The process begins with the emergence of the radicle, which subsequently develops into the mature root system. In eudicots, the shoot system with its cotyledons emerges shortly after the radicle appears. In corn, the radicle and the coleoptile, which covers the young shoot, emerge concurrently. The next major event in the seedling's life occurs when either the cotyledons or the earliest leaves produced by the growing seedling commence photosynthesis. The seedling is said to be established when the young plant no longer relies on food reserves in its endosperm or cotyledons; instead, it receives all of its nourishment from its own photosynthetic products. With this, a new generation is under way.

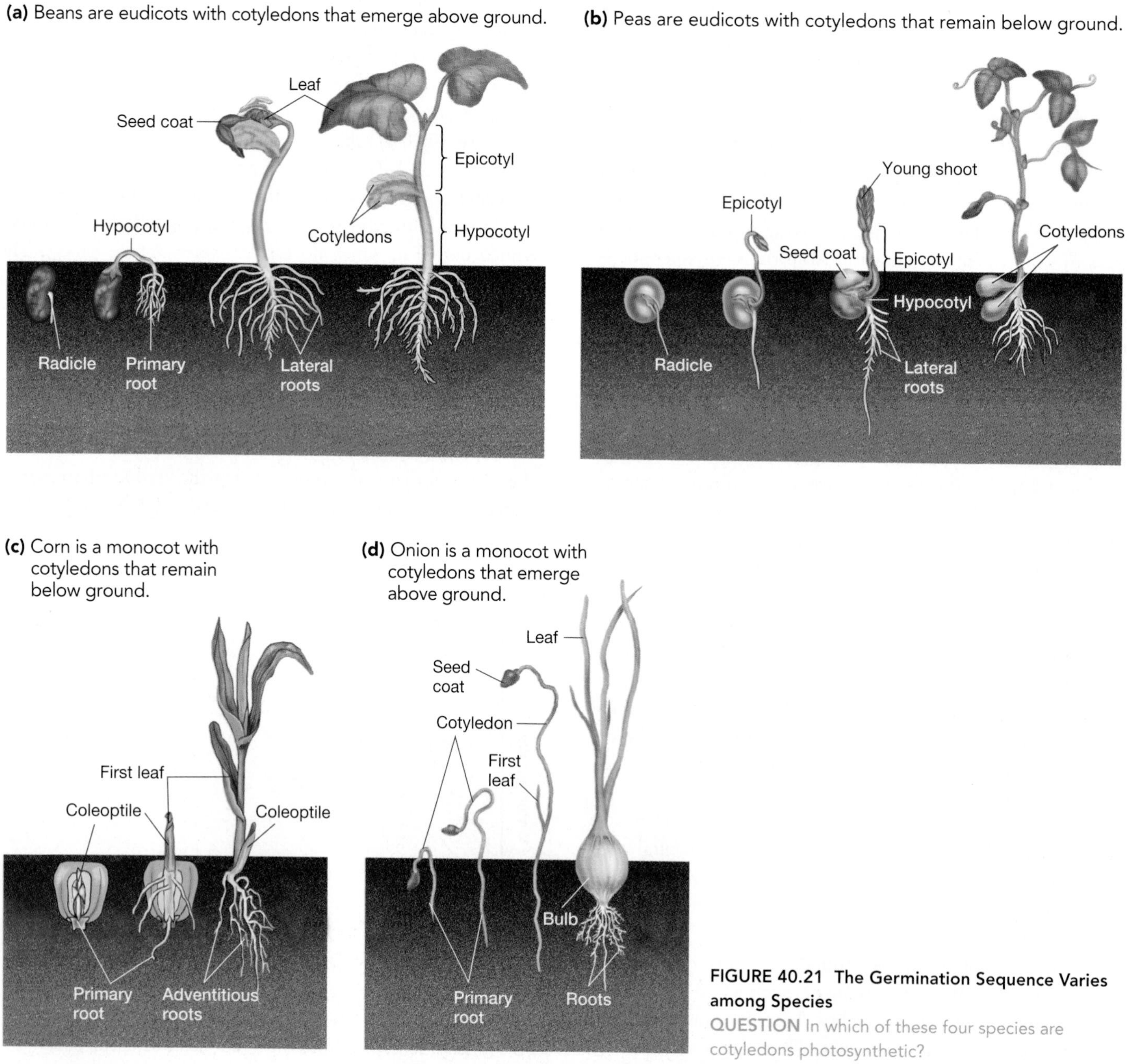

FIGURE 40.21 The Germination Sequence Varies among Species

QUESTION In which of these four species are cotyledons photosynthetic?

ESSAY Why Do Wasps Try to Mate with Hammer Orchids?

Two-thirds of all flowering plant species rely on insects for pollination. In most cases, the relationship between plant and pollinator is mutualistic. But some angiosperm species do not reward visiting animals with nectar or pollen. Instead, these plants rely on trickery to accomplish pollination.

Deceit pollination is particularly common in orchids. Some biologists estimate that as many as 10,000 orchid species accomplish pollination by swindle. In most cases, the deceit involves food. The flowers of food-deceptive orchids usually have a long spur with a bulbous knob at the end. A similar-looking knob, located at the end of the same type of long spur, is found in many other flowering plant species. Typically, the bulb is filled with nectar or oil manufactured by a nearby gland. But in deceit pollinators, there is no nectary. The trick is that a visiting insect picks up the individual's pollen while probing for nectar that isn't there.

An even more spectacular form of deceit pollination involves sex. Consider the hammer orchid (*Drakaea*) illustrated in **Figure 40.22**. This species is native to Australia and is pollinated by wasps in the thynnid family. Female thynnid wasps have no wings and spend the early part of their life feeding on soil-dwelling animals. When the wasps mature, they emerge from the ground, climb up the stem of a plant, and release a chemical signal called a **pheromone**. Male wasps that detect the pheromone follow the scent to its source, land on the female's back, and mate.

Male wasps emerge from below ground before females do, however. At that time of year, hammer orchids are in full bloom. In color, size, and shape, a hammer orchid flower bears a striking resemblance to a female thynnid wasp. The orchid flowers also produce a chemical that is extremely similar to the female wasp's attractant. Male wasps are attracted by the signal and may land on the flower. Just above the landing site, the flower presents a curving column that contains its anthers and stigma. As the male attempts to mate, the flower bends until the male is tipped upside down and his back contacts the reproductive structures. When the wasp flies off, he often does so carrying hammer orchid pollen. If another orchid succeeds in enticing a visit from the same male, it is very likely that the pollen will be deposited on a receptive stigma.

... these plants rely on trickery to accomplish pollination.

Although hammer orchids save large amounts of energy by not producing nectar, deceit pollination has a downside. Most insects are extremely fast learners. If they are fooled once, they are unlikely to be fooled again. But for deceit pollination to work, the same individual insect must be duped twice. In the case of the hammer orchid, the success of sexual deceit relies on the ability of the flower to mimic closely the extremely specific chemical and visual cues from female wasps. In the case of food-deceptive orchids, individuals frequently grow in areas where they are surrounded by high concentrations of similar-looking flowers from species that *do* reward their pollinators with nectar.

FIGURE 40.22 Pollination Occurs by Sexual Deception in Some Orchids
The hammer orchid on the left looks and smells like a female thynnid wasp (right).

CHAPTER REVIEW

Summary of Key Concepts

Most plants reproduce asexually as well as sexually. Asexual reproduction, or cloning, is based on mitosis and results in offspring whose genetic makeup is identical to that of the parent. Sexual reproduction is based on meiosis and results in offspring that are genetically unlike their parent.

- **Plants undergo alternation of generations, in which a diploid sporophyte phase alternates with a haploid gametophyte phase. Sporophytes produce spores by meiosis. Gametophytes produce gametes by mitosis.**

The relative size and life span of the gametophyte and sporophyte phases vary a great deal among plant groups. In the most basal groups of land plants, gametophytes are larger and longer lived than sporophytes, and sporophytes are dependent on gametophytes for nutrition. The opposite is true in angiosperms, or flowering plants, in which sporophytes are the large and long-lived phase where photosynthesis takes place.

- **In angiosperms, male and female gametophytes are microscopic and are produced inside flowers. Male gametophytes are encased in pollen grains and are portable. Female gametophytes are encased in an ovary and are retained in the flower. When pollen grains land on a flower, they produce sperm cells that fertilize the egg produced by the female gametophyte.**

Angiosperms initiate flowering and sexual reproduction in response to external cues from the environment as well as internal cues based on the individual's condition. Frequently, these cues allow individuals to flower when environmental conditions are favorable.

Flowers are made up of sepals, petals, stamens, and one or more carpels. The lower part of the carpel is called the ovary and contains one to many ovules. Within the ovule, a megasporocyte undergoes meiosis, producing a megaspore that develops into the female gametophyte. Similarly, microsporocytes within the anthers of the stamens undergo meiosis. The resulting microspores develop into male gametophytes, which are enclosed in pollen grains. Pollination occurs when pollen grains are transported to the stigma of the carpel.

In many angiosperms, proteins on the surfaces of the pollen grain and stigma interact in a way that reduces the chances of self-fertilization. If allowed to germinate on the stigma, a pollen grain sends a long pollen tube down the style. Two sperm nuclei are produced by mitosis, and double fertilization takes place: One sperm fuses with the egg to form a zygote, while the other fuses with polar nuclei within the female gametophyte. The fusion of sperm and polar nuclei produces endosperm—nutritive tissue that in most species is triploid.

Web Tutorial 40.1 Reproduction in Flowering Plants

- **Seeds contain an embyro and a food supply protected by a tough coat. In angiosperms, the walls of the ovary develop into a case that encloses the seed or seeds, forming a fruit.**

The development of an angiosperm embryo begins with the formation of dermal tissue (epidermis), ground tissue, and vascular tissue layers. This event is followed by the development of the radicle, hypocotyl, and cotyledons. As the embryo develops, endosperm cells divide to form a nutrient-rich tissue. In addition, cells along the outside of the ovules form a tough seed coat, and the ovary develops into a fruit. In many cases, the mature fruit contains structures that aid in dispersal of the mature seed via wind, water, propulsion, or animals.

Many seeds do not germinate immediately but instead experience a period of dormancy. In some species, dormancy is initiated or maintained by the hormone ABA. A wide variety of conditions, ranging from scarification to exposure to red light, may break seed dormancy. In many cases, the event that triggers germination ensures that the seed germinates when environmental conditions are favorable. Germination begins when the seed takes up water, activates existing enzymes, and translates mRNAs already present in the seed. It ends when the radicle breaks the seed coat and begins to penetrate the soil.

Web Tutorial 40.2 Fruit Structure and Development

Questions

Content Review

1. What is the major evolutionary trend in land plant life cycles?
   a. Instead of being approximately the same size and shape, gametophytes and sporophytes began to look different.
   b. Sporophytes became larger and long lived while gametophytes became drastically reduced.
   c. In lineages that evolved more recently, such as angiosperms, spores are no longer produced.
   d. Sporophytes began to rely on gametophytes for all of their nutritional needs.

2. What happens when double fertilization occurs?
   a. Two zygotes are formed, but only one survives.
   b. Two sperm fertilize the egg, forming a triploid zygote.
   c. One sperm fertilizes the egg, while another sperm fuses with the polar nuclei.
   d. One sperm fertilizes the egg, while two other sperm fuse with a polar nucleus.

3. In angiosperms, what is the fundamental difference between a male gamete and a female gamete?
   a. A male gamete is small and contributes few nutrients to the offspring. A female gamete is relatively large and contributes many nutrients to the offspring.
   b. A male gamete is relatively large and contributes many nutrients to the offspring. A female gamete is small and contributes few nutrients to the offspring.
   c. Male gametes swim; female gametes do not.
   d. Male gametes form pollen; female gametes form seeds.
4. Which of the following is a key event during embryogenesis?
   a. The seed coat takes on water so that germination can begin.
   b. Starches are hydrolyzed, providing sugars that fuel the early stages of germination.
   c. The megasporocyte divides by mitosis, forming the cells that will become the embryonic female gametophyte.
   d. Distinct groups of cells form that will become dermal, ground, and vascular tissues.
5. Why is the interaction between angiosperms and pollinators considered mutualistic?
   a. New species can form if mutant flowers attract new types of pollinators.
   b. Flowers may have an array of traits, including corolla shape, color, scent, and the presence of nectar, to attract a specific type of pollinator.
   c. During the interaction, it is not possible for either the plant species or the animal species to gain a fitness advantage through deception.
   d. Angiosperms get their pollen dispersed, while pollinators get food.
6. What happens when outcrossing occurs?
   a. Inbred offspring are produced.
   b. The proteins produced by *S* loci block pollen grains from interacting with the stigma.
   c. Gametes from different individuals fuse to form a zygote.
   d. Gametes from the same individual fuse.

Conceptual Review

1. List three mechanisms of asexual reproduction in plants. In terms of maximizing reproductive success, what is the advantage of asexual reproduction? What is the disadvantage?
2. Draw a general version of an angiosperm life cycle. Label the sporophyte and gametophyte stages and the spores and gametes. Identify the ploidy (n vs. $2n$) of each structure in the drawing, and indicate which cells undergo meiosis.
3. Draw a generalized flower. Indicate the function of each part during reproduction. Provide detailed diagrams of the reproductive structures that produce male gametophytes and female gametophytes. Then diagram flowers that are pollinated by wind, by hummingbirds, and by bumblebees. How is each flower part modified from the generalized flower you first drew, and why?
4. What are the advantages and disadvantages of self-fertilization versus those of outcrossing?
5. Explain the difference between a carpel, an ovary, and an ovule. Be sure to clarify the physical relationship among them (which is enclosed by which), and identify which is responsible for producing the female gametophyte versus the pericarp of a fruit. Are these structures part of the sporophyte, the gametophyte, or a combination of the two?
6. Make a labeled diagram of a bean seed and a corn seed. What is the relationship between the endosperm of corn and the cotyledons of beans? Suggest a hypothesis to explain how the seed coats of these species are broken so that germination can begin.

Group Discussion Problems

1. Suppose you discovered an angiosperm that was new to science. The population grows on an island near the equator. The island is dry 10 months of the year but experiences a two-month period of frequent rains. Predict what cues trigger flowering and seed germination in the newly discovered species. Design experiments to test your predictions.
2. The essay at the end of this chapter points out that some flowering plants "cheat" their pollinators. Likewise, certain pollinators cheat plants by removing nectar from flowers without picking up pollen. (In some cases, they do so by chewing through the petals that hold the store of nectar.) Speculate on the types of mutations that might modify insect behavior and/or plant structure in a way that limits cheating and enforces mutualism.
3. Pollinators frequently deposit pollen from more than one individual on a stigma. When they do, pollen grains from different males compete to fertilize the egg. Design an experiment to test the hypothesis that (a) pollen grains produce a toxin that slows the growth of pollen tubes from different individuals and (b) pollen tubes grow faster when pollen grains germinate in the presence of pollen from a different individual.
4. Consider the following fruits: an acorn, a cherry, a burr, and a milkweed seed. Based on the structure of each of these fruits, predict how the seed is dispersed. How would you test this prediction? Design a study that would estimate the average distance that each type of seed is dispersed from the parent plant.

Answers to Multiple-Choice Questions 1. b; 2. c; 3. a; 4. d; 5. d; 6. c

www.prenhall.com/freeman is your resource for the following: Web Tutorials; Online Quizzes and other Online Study Guide materials; Answers to Conceptual Review Questions; Solutions to Group Discussion Problems; Answers to Figure Caption Questions and Exercises; and Additional Readings and Research.

UNIT 8

How Animals Work

Black-footed ferrets are long, slender animals that run and dig through underground burrows in search of prairie dogs—their major prey. These ferrets are among the rarest mammals in North America. [©1996 Susan Middleton & David Liittschwager]

41 Animal Form and Function

KEY CONCEPTS

- Biologists analyze the structure and function of animals at a variety of levels: molecules, cells, tissues, organs, and systems. These levels form a tightly integrated hierarchy.
- Body size has a strong influence on how animals work. For example, because small animals have a large amount of surface area relative to their volume, they lose heat faster than do large animals. The relationship between surface area and volume also plays a significant role in diffusion-mediated processes, such as gas exchange, nutrient absorption, and waste excretion.
- Animals maintain a relatively constant environment inside their bodies because they have systems that sense changes in internal conditions and trigger responses that return conditions to normal.

Oryx are adapted to desert life. They have an exceptional ability to withstand heat, and they acquire all of their water from the food they eat.

The Sahara and Arabian deserts that cover much of northern Africa and the Arabian peninsula are extreme environments. In some parts of the Sahara, it is not unusual for several years to pass between rainfalls. In a single day, temperatures can fluctuate between –0.5°C and 37.5°C (31°F to 99.5°F). In summer, midday temperatures can exceed 50°C (122°F). Yet few places in the Sahara or Arabian deserts are completely devoid of life. Even large animals, such as the oryx, thrive in parts of both deserts.

How do the organisms native to these areas cope? Plants may endure as seeds for months or years, then germinate, flower, and produce a new generation of seeds immediately after a rain. Small animals avoid the midday heat by retreating to a cool burrow deep underground or to the shade of a small shrub, where they wait until the temperature moderates in the evening. Large mammals have a harder time hiding from the Sun but possess traits that allow them to keep cool and conserve water. The Arabian oryx, for example, never drinks. It gets 86 percent of the water it needs from the vegetation it eats and the other 14 percent from water synthesized as a by-product of cellular respiration—what biologists call **metabolic water.** To conserve what water they have, Arabian oryx produce extremely concentrated urine and exceptionally dry fecal pellets. In addition, oryx do not sweat to cool off in hot weather. Instead they allow their body temperature to rise from a normal 37°C to just over 40°C as temperatures increase during the day. The excess body heat is then given off as temperatures fall during the cool desert nights, and their body temperature returns to normal.

In contrast to oryx, humans die of dehydration when denied water for just a week. And for a human, a body temperature of 40°C is a life-threatening situation—equivalent to running a fever of 104°F. How do oryx do it? More specifically, what aspects of their anatomy and physiology allow them to thrive in such an extreme environment? **Anatomy** is the study of an organism's physical structure. **Physiology** is the study of how the physical structures in an organism function. The anatomy and physiology of an oryx are clearly different from those of a human—or those of a shark, frog, tuna, fruit fly, or crab, for that matter. This chapter is an introduction to how biologists approach the study of anatomy and physiology in animals.

Biologists who study animal anatomy and physiology are studying **adaptations**—traits that allow individuals to survive and reproduce in a certain environment better than do individuals that lack those traits (see Chapter 23). The chapters in this unit explore traits that allow animals to digest food, breathe, move, hear, see, and fight infections. This chapter introduces some basic elements of animal anatomy and identifies fundamental themes that unify the diversity of anatomical and physiological traits observed in animals. Its goal is to help you begin to see animals in a new light—as efficient systems for gathering resources and producing offspring. After millions of years of evolution by natural selection, animals are finely tuned to thrive in a diversity of habitats.

41.1 Form, Function, and Adaptation

Recall from Chapter 23 that adaptation results from evolution by natural selection. Natural selection, in turn, occurs whenever individuals with certain alleles leave more offspring than do individuals with different alleles. Because of this difference in reproductive success, the frequency of the selected alleles increases from one generation to the next. As an example, consider fish that live in the cold ocean waters near Antarctica. In several species, individuals have alleles that code for the production of proteins that act as an antifreeze in blood. Biologists interpret these alleles as an adaptation to the extremely low water temperatures found in this environment. In Antarctic fish populations, individuals with the antifreeze alleles produce more offspring than do individuals without the antifreeze alleles.

But as explained in Chapter 24, natural selection is not the only process that leads to changes in allele frequencies over time. Evolution also occurs through random changes in allele frequencies called *genetic drift,* through *gene flow*—meaning the movement of alleles into and out of a population by migration—and through the constant introduction of new alleles by way of *mutation.* Natural selection is the only mechanism that increases the ability of organisms to survive and reproduce in a particular environment.

Trade-Offs

No adaptation is "perfect." Instead, adaptations are limited by which alleles are present in a population and by the nature of the traits that already exist—because all adaptations are derived from preexisting traits. For example, even though wheels are an extremely efficient structure for movement, animals do not have them. Biologists hypothesize that wheels have not evolved in animals because alleles for wheels do not exist and because limbs, which are the closest trait to a wheel, evolved from appendages in ancestral species that moved back and forth—not in a circle.

Perhaps the most important constraint on adaptation, though, involves the phenomenon known as a trade-off. A **trade-off** is an inescapable compromise between traits. For example, every female animal has a finite amount of time and energy to devote to egg production. Given this constraint, there should be a trade-off between the number of eggs she produces and the quality of those eggs. In species that do not care for their young after laying eggs, egg quality is determined by egg size—specifically by the amount of *yolk*, or nutrient-rich cytoplasm, that the mother puts into the egg. In species that do provide parental care, quality is determined by the amount of food and other resources that parents deliver to each offspring. If the number of offspring is large, then a parent may provide less food to each individual.

The question is, Do any data support the prediction that trade-offs should be an important consideration in animal anatomy and physiology? To answer this question, let's consider experimental work on egg size and egg number in the side-blotched lizard, a reptile that lives in the deserts of western North America (**Figure 41.1**, page 936). To begin, biologists set out to create a large amount of variation in egg size and egg number. For example, they induced the production of small eggs by catching females, surgically removing yolk from their eggs early in development, replacing the eggs, and releasing the mothers back into the wild. To create clutches with small numbers of eggs, biologists caught females and removed all but two or three of their eggs early in development. As an experimental control, the researchers did "sham operations." They caught a large number of females, performed surgery to expose their eggs, but left the eggs alone. These manipulations gave them a study population with a large variation in egg size and egg number.

The graph labeled "Results 1" of Figure 41.1 plots the average mass of eggs that were laid against **clutch size**, meaning the number of eggs laid by each female. The data show that, as egg size increases, clutch size decreases. The pattern confirms that there is a trade-off between egg size and number in this species. It is not possible for a female lizard to produce many large eggs.

What about the prediction that large offspring are higher in quality, meaning that they survive better? To test this idea, the researchers marked 1668 newly hatched lizards in the experimental population, released them, and caught many of

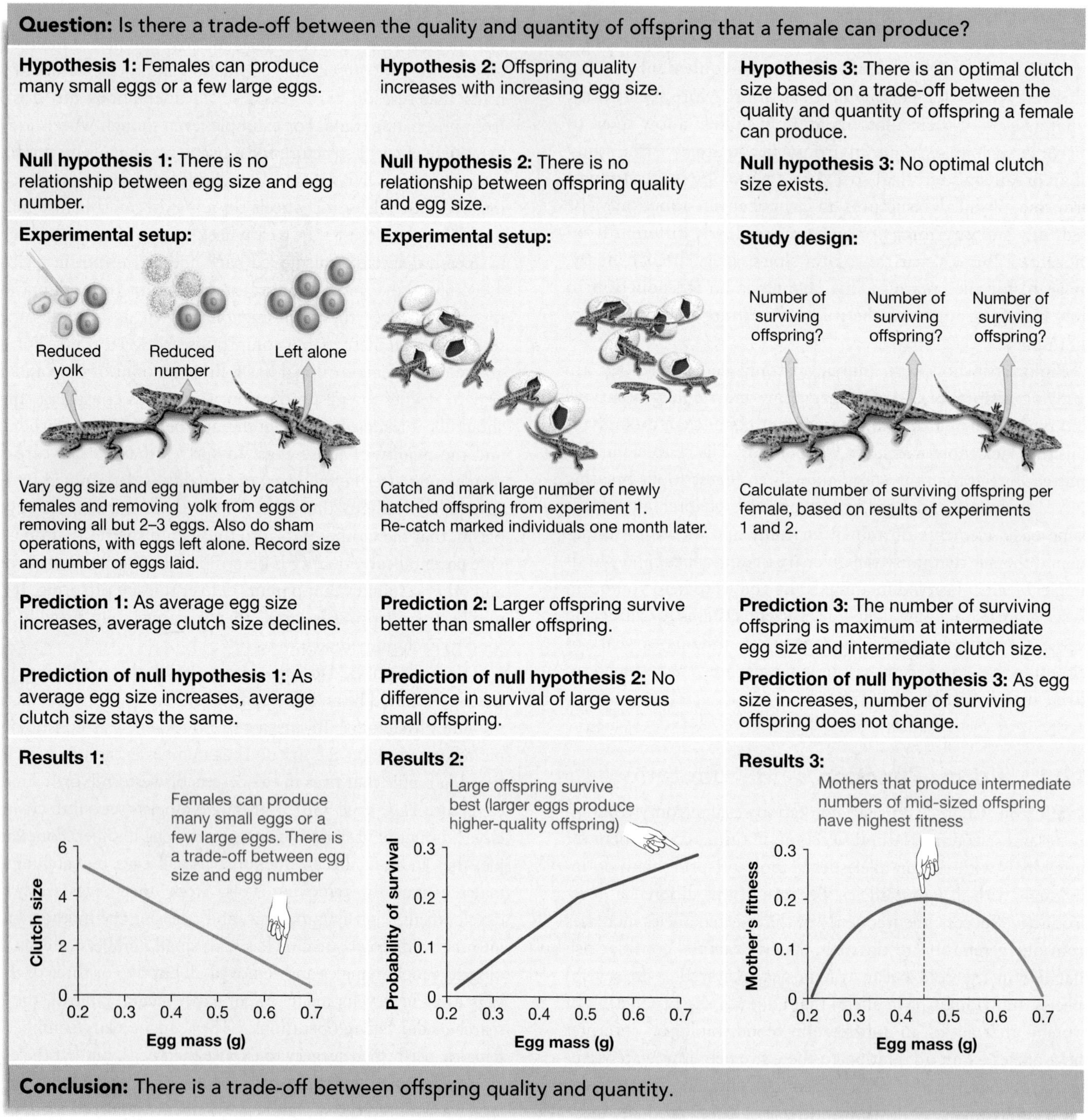
Question: Is there a trade-off between the quality and quantity of offspring that a female can produce?
Hypothesis 1: Females can produce many small eggs or a few large eggs.
Null hypothesis 1: There is no relationship between egg size and egg number.
Experimental setup:
Reduced yolk
Reduced number
Left alone
Vary egg size and egg number by catching females and removing yolk from eggs or removing all but 2–3 eggs. Also do sham operations, with eggs left alone. Record size and number of eggs laid.
Prediction 1: As average egg size increases, average clutch size declines.
Prediction of null hypothesis 1: As average egg size increases, average clutch size stays the same.
Results 1:
Females can produce many small eggs or a few large eggs. There is a trade-off between egg size and egg number
Clutch size
0
2
4
6
0.2 0.3 0.4 0.5 0.6 0.7
Egg mass (g)
Hypothesis 2: Offspring quality increases with increasing egg size.
Null hypothesis 2: There is no relationship between offspring quality and egg size.
Experimental setup:
Catch and mark large number of newly hatched offspring from experiment 1. Re-catch marked individuals one month later.
Prediction 2: Larger offspring survive better than smaller offspring.
Prediction of null hypothesis 2: No difference in survival of large versus small offspring.
Results 2:
Large offspring survive best (larger eggs produce higher-quality offspring)
Probability of survival
0
0.1
0.2
0.3
0.2 0.3 0.4 0.5 0.6 0.7
Egg mass (g)
Hypothesis 3: There is an optimal clutch size based on a trade-off between the quality and quantity of offspring a female can produce.
Null hypothesis 3: No optimal clutch size exists.
Study design:
Number of surviving offspring?
Number of surviving offspring?
Number of surviving offspring?
Calculate number of surviving offspring per female, based on results of experiments 1 and 2.
Prediction 3: The number of surviving offspring is maximum at intermediate egg size and intermediate clutch size.
Prediction of null hypothesis 3: As egg size increases, number of surviving offspring does not change.
Results 3:
Mothers that produce intermediate numbers of mid-sized offspring have highest fitness
Mother's fitness
0
0.1
0.2
0.3
0.2 0.3 0.4 0.5 0.6 0.7
Egg mass (g)
Conclusion: There is a trade-off between offspring quality and quantity.

FIGURE 41.1 Trade-Offs Are Pervasive in Animal Anatomy and Physiology
Summary of an experiment testing whether trade-offs occur in the reproduction of side-blotched lizards.

them again a month later. The graph labeled "Results 2" of Figure 41.1 confirms the prediction that larger offspring survive much better than do smaller offspring.

Given this trade-off between offspring quantity and offspring quality, does natural selection favor mothers that produce large numbers of small offspring or small numbers of large offspring? The graph labeled "Results 3" of Figure 41.1 shows the combined effect of egg number and offspring survival. In this population, mothers that produced an intermediate number of offspring of intermediate size generated the highest total number of surviving offspring.

Trade-offs such as the inescapable compromise between egg size and egg number are pervasive in nature. Desert animals that sweat to cool off are threatened with dehydration. An eagle's beak is superbly adapted for tearing meat but is not useful for weaving nesting materials together. In studying animal anatomy and physiology, biologists are studying compromise and constraint as well as adaptation.

Adaptation and Acclimatization

Although adapation is a key concept in understanding animal form and function, it is important to note that the term is often used incorrectly to describe short-term, reversible responses to environmental fluctuations. Physiological and biochemical changes that take place in response to environmental conditions that change within an individual's lifetime are referred to as **acclimatization**. Adaptation is a genetic change in a population in response to natural selection exerted by the environment; acclimatization is a phenotypic change in an individual in response to short-term changes in the environment.

In practice it is not always easy to tell the difference between adaptation and acclimatization. For example, in response to shortening day lengths and lowered temperatures that signal the onset of winter, some frogs respond by altering gene expression to increase their production of cold-tolerant proteins. This is an example of acclimatization, because it occurs within the lifetime of the individual and is reversible. But suppose natural selection increased the frequency of an allele that coded for a particular protein variant over time, because that protein functioned better at lower temperatures than did the proteins encoded by other alleles of the same gene. In this case, the protein that allowed individuals to acclimatize more efficiently would also represent an adaptation.

41.2 Tissues, Organs, and Systems: How Does Structure Correlate with Function?

If a structure found in an animal is adaptive—meaning that it helps the individual survive and produce offspring—it is common to observe that the structure's size, shape, or composition correlates closely with its function. As an example, consider the size and shape of the bird beaks in **Figure 41.2**. Notice that the structure of the beaks of different species of Galápagos finch correlates with the diet of each species. Birds with large, cone-shaped beaks eat large seeds; birds with small, cone-shaped beaks eat small seeds. Birds with long, tweezer-like beaks pick insects off tree trunks or other surfaces.

Biologists interpret correlations between structure and function as an outcome of evolution by natural selection. If a mutant allele alters the size or shape of a structure in a way that makes it function more efficiently, then individuals who have that allele will produce more offspring than other individuals will. As a result, the allele will increase in frequency in the population over time.

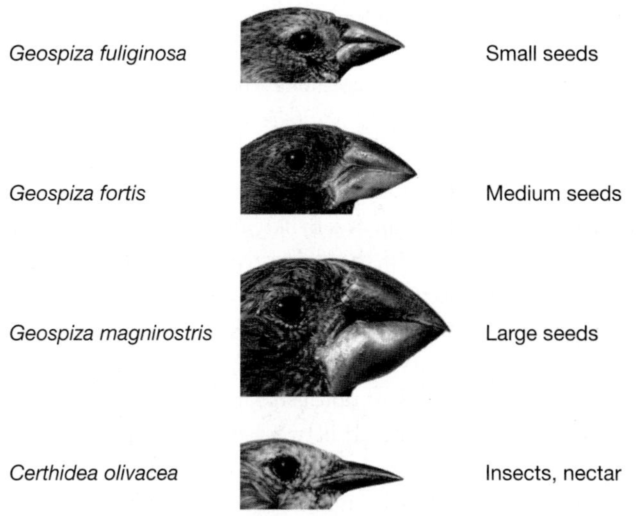

FIGURE 41.2 In Animal Anatomy and Physiology, Form Often Correlates with Function
Among Galápagos finch species, there is a strong correlation between the size and shape of the beak and the food items that the bird prefers to eat.

Correlating structure with function is a pervasive theme in research on animal anatomy and physiology, and it is one of the most basic ways that biologists have to study adaptation. Correlations between form and function occur at many levels in organisms. Chapter 3 emphasized that the shape of proteins correlates with their role as enzymes or structural components of the cell. Similarly, Chapter 7 pointed out that strong correlations exist between the structure and function of the rough ER, Golgi apparatus, mitochondrion, chloroplast, and other organelles. But structure-function correlations are not limited to the molecular and cellular levels. They also exist at the levels of tissues, organs, and systems.

Tissues

Animals are **multicellular**, meaning that their bodies contain distinct types of cells that are specialized for different functions. Sponges are considered to be the simplest of all animals in overall morphology, but even they contain cells that are distinct structurally and that perform specialized functions. In sponges, some types of cells are responsible for covering the body; others are responsible for secreting material to support the body, for feeding, or for reproduction.

In a sponge, cells that are specialized for a particular function operate more or less independently of each other, however. This is not the case in other animals, in which cells that have the same structure are typically organized into highly integrated units called **tissues**. In a tissue, cells work together to perform a specific function.

Chapter 31 introduced the embryonic tissues called ectoderm, mesoderm, and endoderm. As an embryo develops, these tissues give rise to the four adult tissue types illustrated in **Figure 41.3**: (1) connective tissue, (2) nervous tissue, (3) muscle tissue, and (4) epithelial tissue. Let's consider each in turn.

Connective Tissue **Connective tissue** consists of cells that are loosely arranged in a liquid, jellylike, or solid *extracellular matrix*, which is secreted by the connective tissue cells themselves. Each type of connective tissue secretes a distinct type of extracellular matrix. As a result, the structure and function of various connective tissues vary widely. For example, **loose connective tissue** contains an array of fibrous proteins in a soft matrix and serves as a packing material between organs or padding under the skin. Other connective tissues, such as **cartilage** and **bone**, have a firmer extracellular matrix. Cartilage and bone provide structural support for the body or protective enclosures for the brain and other components of the nervous system. **Blood** is a connective tissue that has cells surrounded by a liquid extracellular matrix called *plasma*; blood functions in the transport of materials throughout the body.

Nervous Tissue **Nervous tissue** consists of nerve cells, which are also called **neurons**, and several types of supporting cells. Although they vary widely in shape, all neurons have projections that contact other cells. These projections deliver electrochemical signals, which are produced by changes in the permeability of the neuron's plasma membrane to ions. Although support cells outnumber neurons in the nervous system, they do not conduct electrical signals. Instead, support cells regulate ion concentrations in the space surrounding neurons,

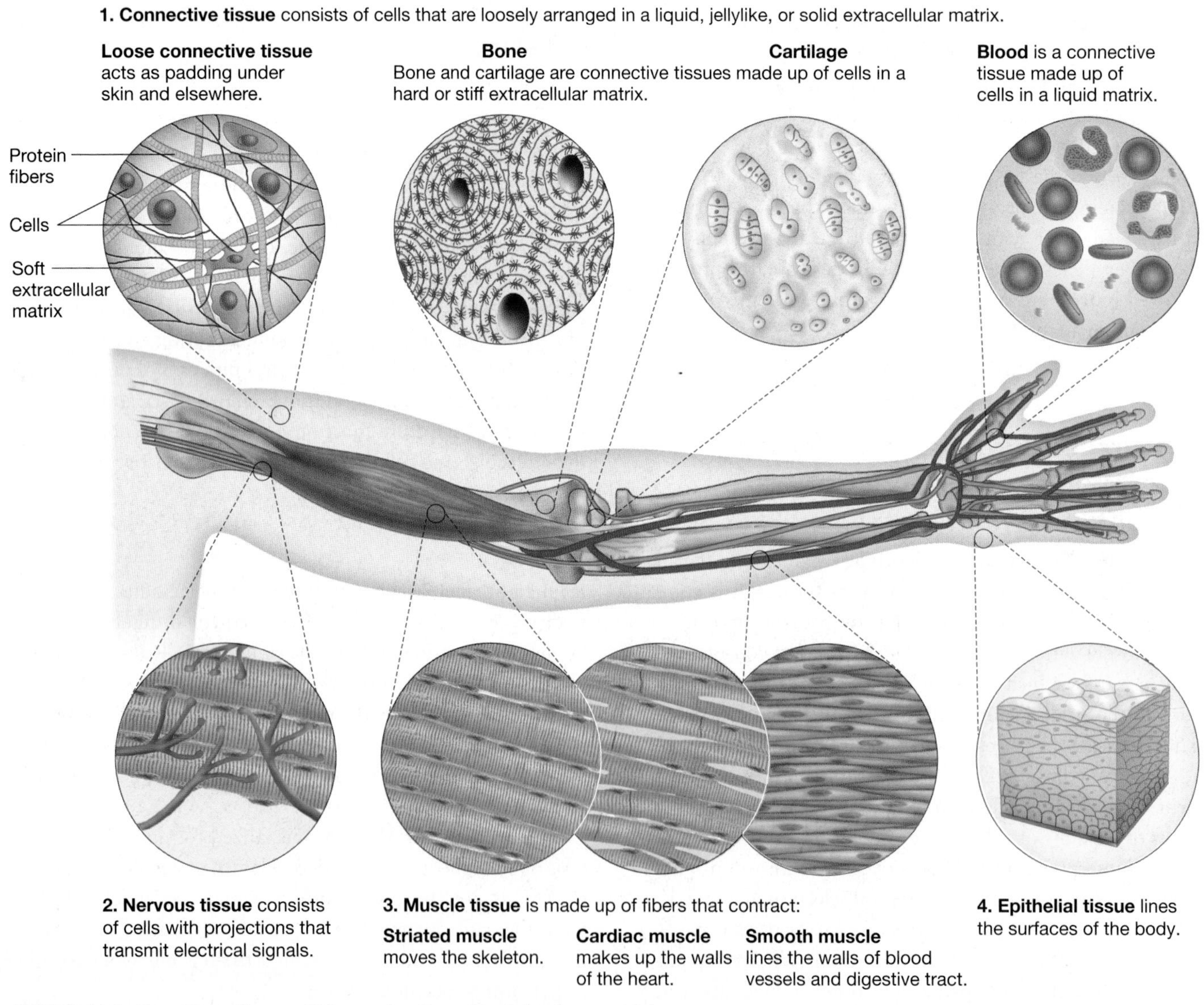
1. Connective tissue consists of cells that are loosely arranged in a liquid, jellylike, or solid extracellular matrix.
Loose connective tissue acts as padding under skin and elsewhere.
Bone
Cartilage
Bone and cartilage are connective tissues made up of cells in a hard or stiff extracellular matrix.
Blood is a connective tissue made up of cells in a liquid matrix.
Protein fibers
Cells
Soft extracellular matrix
2. Nervous tissue consists of cells with projections that transmit electrical signals.
3. Muscle tissue is made up of fibers that contract:
Striated muscle moves the skeleton.
Cardiac muscle makes up the walls of the heart.
Smooth muscle lines the walls of blood vessels and digestive tract.
4. Epithelial tissue lines the surfaces of the body.

FIGURE 41.3 Four Major Types of Tissues Are Found in Animals
QUESTION How does the type of extracellular matrix found in a connective tissue correlate with that tissue's function?

supply neurons with nutrients, or serve as scaffolding or support for neurons.

Muscle Tissue **Muscle tissue** functions primarily in movement. There are three types of muscle tissue:

1. Most muscle tissue is **striated muscle**, which consists of the long cells called **muscle fibers.** Muscle fibers are packed with specialized proteins that move when they hydrolyze ATP. Striated muscle contracts in response to a complex series of events triggered by electrical signals arriving from nerve cells.
2. **Cardiac muscle** makes up the walls of the heart. It is similar to skeletal muscle in structure, except that each cardiac muscle cell branches and makes direct, end-to-end contact with other cardiac muscle cells. These connections help transmit signals from one cardiac muscle cell to another during a heartbeat.
3. **Smooth muscle** cells are tapered at each end and form a muscle tissue that lines the walls of the digestive tract and the walls of the blood vessels. Different types of neurons control the contraction of smooth muscle cells versus striated muscle cells. Smooth muscle is responsible for involuntary movements such as the passage of food down the digestive tract or the dilation (opening) of arteries near the skin in hot weather.

Epithelial Tissues **Epithelial tissues** are also called **epithelia** (singular: **epithelium**). Epithelium covers the outside of the body, lines the surfaces of organs, and forms glands. An **organ** is a structure that serves a specialized function and consists of several tissues; a **gland** is a group of cells that secrete specific molecules or solutions.

Because the primary function of epithelium is to act as a barrier and protective layer, it's not surprising to observe that epithelial cells typically form layers of closely packed cells. In many cases, adjacent epithelial cells are joined by structures that hold them tightly together, such as tight junctions and desmosomes (introduced in Chapter 8).

Figure 41.4 illustrates a key point about epithelium: No matter where it is found in the body, epithelial tissue has a polarity, or sidedness. More specifically, an epithelium has an apical side and a basolateral side. The **apical** side faces away from other tissues and toward the environment, while the **basolateral** side faces the interior of the animal and is connected to other tissues. The apical and basolateral sides of an epithelium have distinct structures and functions. A few examples will help drive this point home:

- The interior space of hollow organs such as your small intestine is called the **lumen.** The epithelial cells that line the lumen of your small intestine have an apical surface that is folded into fingerlike projections, as shown in Figure 41.4. These projections are studded with membrane proteins that take up glucose and other nutrients from the food that you have digested. The epithelial cells on the basolateral side of the small intestine have a flat surface that contains membrane proteins different from those on the apical side. On the basolateral side, glucose and other nutrients are transported across the plasma membrane to nearby blood vessels, which distribute them throughout the body.
- In your trachea (windpipe), the apical side of epithelial cells secretes mucus and is covered with cilia that help sweep away dust, bacteria, and viruses. The basolateral side of tracheal epithelium does not secrete mucus and does not contain cilia.
- Glandular epithelial cells produce secretions that are released onto the apical surface of the epithelium or across the apical surface into the bloodstream. Secretion does not occur across the basolateral surface.

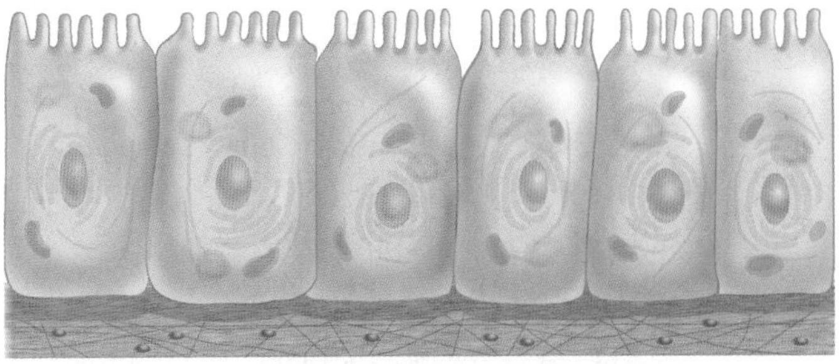
Apical surface of epithelium: faces internal or external environment
Basolateral surface of epithelium: connects to other tissues

FIGURE 41.4 Epithelium Has a Polarity

QUESTION The inner surface of the lungs, stomach, and small intestine are lined with epithelium. What is the nature of the environment faced by the apical surface of these epithelia?

Another major characteristic of epithelium is that epithelial cells tend to have short life spans. The cells that line your esophagus—the tube connecting your mouth and stomach—live for 2 to 3 days, while the cells that line your large intestine live for a maximum of six days. These short life spans occur because epithelial cells are exposed to a harsh environment, where they are likely to be killed or scraped away. The tissue as a whole does not wear away, however, because it includes cells that actively undergo mitosis and cytokinesis—producing new cells to replace those lost on the side that faces the environment.

Organs and Systems

Cells with similar functions are organized into tissues, and tissues are organized into specialized structures called organs. Recall that an organ is a structure that serves a specialized function and consists of several types of tissues. The heart, for example, consists of cardiac muscle tissue, connective tissue, and epithelial tissue. All four of the basic

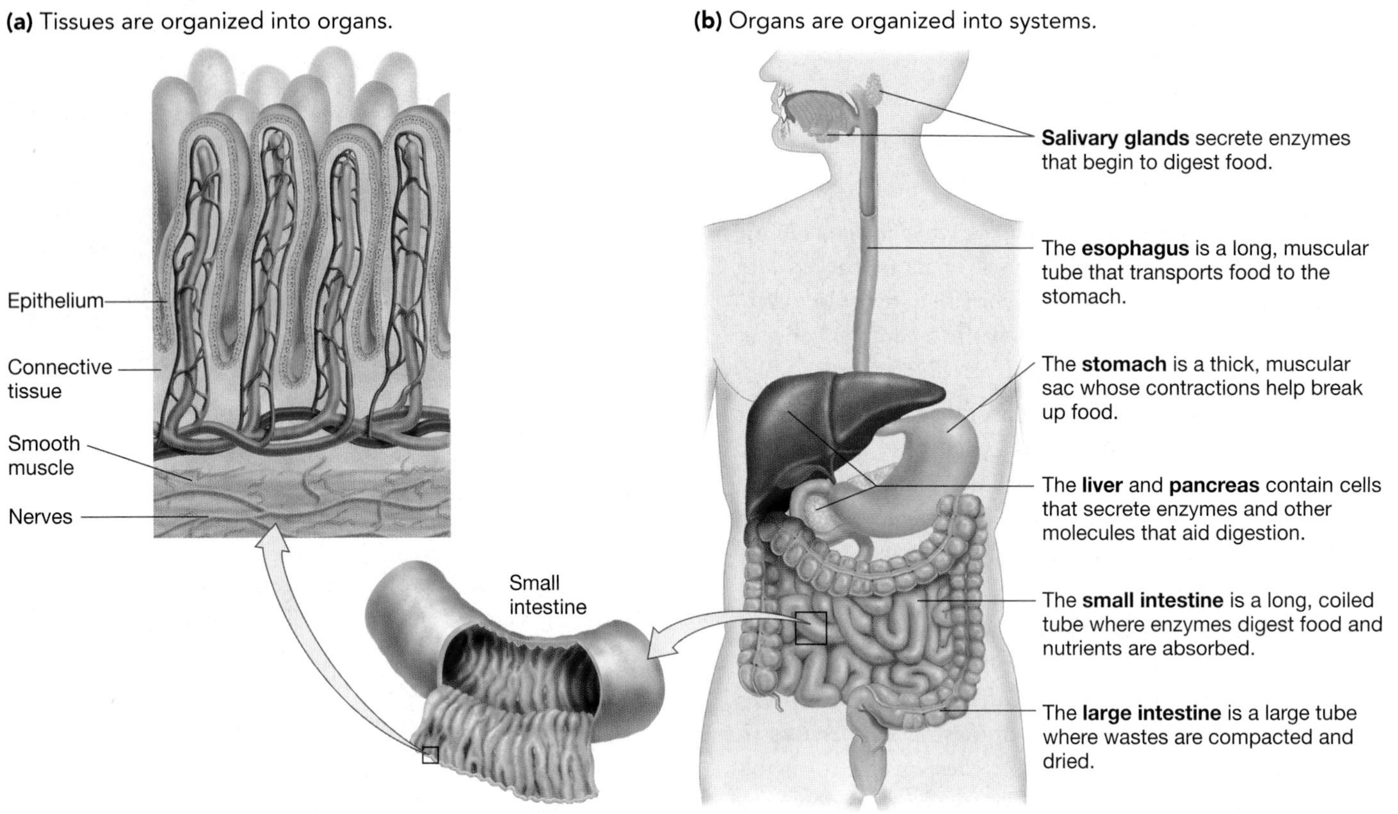
(a) Tissues are organized into organs.
Epithelium
Connective tissue
Smooth muscle
Nerves
Small intestine
(b) Organs are organized into systems.
Salivary glands secrete enzymes that begin to digest food.
The esophagus is a long, muscular tube that transports food to the stomach.
The stomach is a thick, muscular sac whose contractions help break up food.
The liver and pancreas contain cells that secrete enzymes and other molecules that aid digestion.
The small intestine is a long, coiled tube where enzymes digest food and nutrients are absorbed.
The large intestine is a large tube where wastes are compacted and dried.

FIGURE 41.5 Organs Are Composed of Tissues; Systems Are Made Up of Organs
(a) The human small intestine is an organ composed of all four major tissue types: epithelium, connective tissue, smooth muscle, and nerves. **(b)** The human digestive system is essentially one long tube divided into chambers where food is processed and nutrients are absorbed. The salivary glands, liver, and pancreas are organs that secrete specific enzymes or compounds into the tube.

tissue types are found in the small intestine as illustrated in **Figure 41.5a**.

Organs such as the heart or small intestine are also part of larger functional units called systems. A **system** consists of groups of tissues and organs that work together to perform one or more functions. Using the digestive system as an example, **Figure 41.5b** illustrates how the structure of organs correlates with their function and how the components of a system work together in an integrated fashion.

Because an animal's body contains cells, tissues, organs, and systems, biologists who study animal anatomy and physiology must work at various levels of organization to gain a thorough understanding of how that body operates. **Figure 41.6** illustrates these levels of organization, using human neurons and the nervous system as an example. The key point is that, because the structure and function of each component in the body are integrated with other components, and because each level of organization is integrated with other levels of organization, the organism as a whole is greater than the sum of its parts. In other words, an organism is more than just a collection of individual systems, and each system is more than just a collection of individual cells or tissues or even organs.

In effect, each subsequent chapter in this unit focuses on a different system found in animals, beginning with the excretory system and ending with the immune system. The presentations introduce the structures that are found in each system and explore how biologists investigate their functions. Each of these systems can be interpreted as a suite of adaptations. Each system accomplishes a specific task required for survival and reproduction, and each works in conjunction with other systems.

Before we delve into each system, however, it's essential to examine general phenomena that affect all systems in animals. Let's look at how body size affects animal physiology.

✓CHECK YOUR UNDERSTANDING

Biologists study animal anatomy and physiology at the molecular, cellular, tissue, organ, and system levels. Events at these levels interact to form an integrated whole in organisms. You should be able to (1) give an example of a molecule that is part of a cell, which is part of a tissue, which is part of an organ, which is part of a system; (2) compare and contrast the four major types of animal tissues; (3) explain why epithelial tissue is particularly important in animal physiology; and (4) give an example of the polarity of epithelial cells.

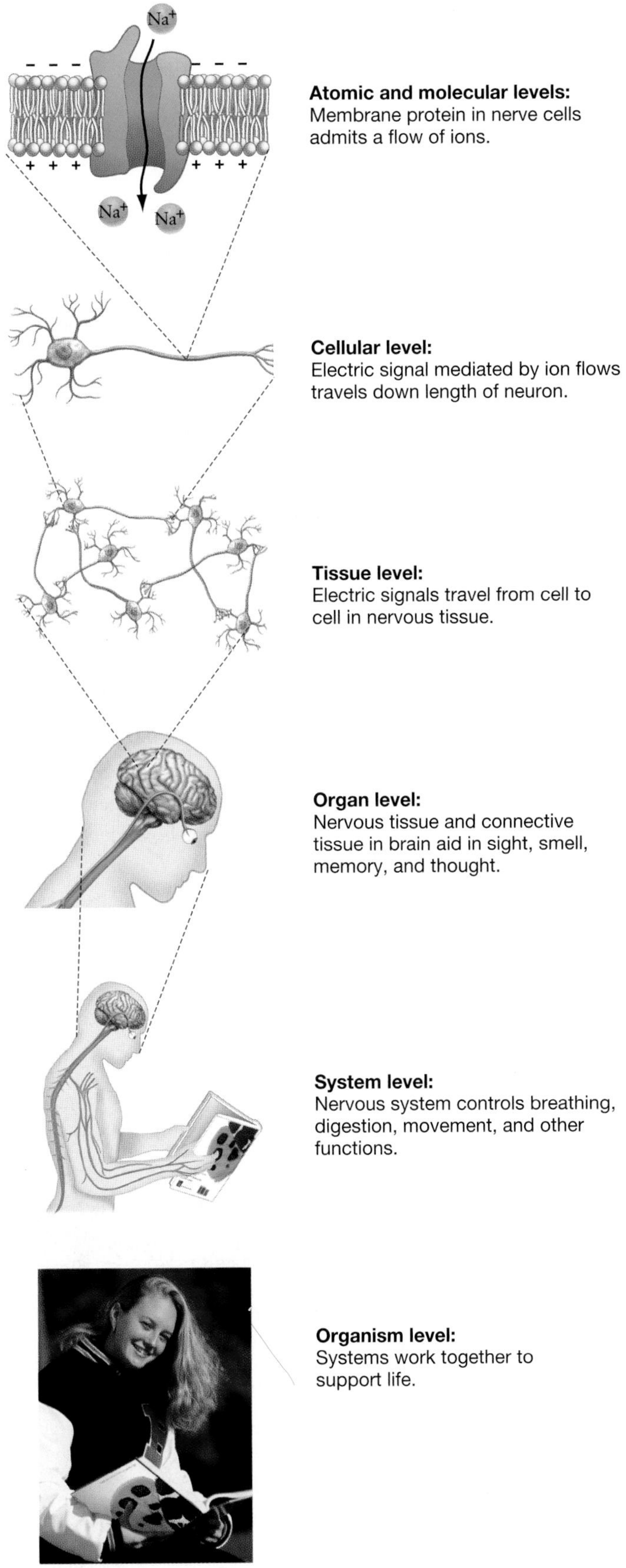
Na+
Na+
Na+
Atomic and molecular levels:
Membrane protein in nerve cells admits a flow of ions.
Cellular level:
Electric signal mediated by ion flows travels down length of neuron.
Tissue level:
Electric signals travel from cell to cell in nervous tissue.
Organ level:
Nervous tissue and connective tissue in brain aid in sight, smell, memory, and thought.
System level:
Nervous system controls breathing, digestion, movement, and other functions.
Organism level:
Systems work together to support life.

FIGURE 41.6 Biologists Study Anatomy and Physiology at Many Levels
The levels of organization within an organism are not independent of each other. Instead, they are tightly integrated.

41.3 How Does Body Size Affect Animal Physiology?

The laws of physics affect the anatomy and physiology of organisms in fundamental ways. The force of gravity, for example, limits how large an animal can be and still move efficiently. Or consider the forces exerted by the medium in which animals live. Because water is much denser than air, it is harder for animals to move through water efficiently. Based on this observation, it is not surprising to find that fish and aquatic mammals have much more streamlined bodies than terrestrial animals do.

Physical laws clearly affect body size and shape; just as clearly, body size and shape have pervasive effects on how animals function. Large animals need more food than small animals do. Large animals also produce more waste, take longer to mature, reproduce more slowly, and live longer. Conversely, small animals are more susceptible to damage from cold and dehydration than large animals are, because they lose heat and water faster. Juveniles and adults of the same species face different challenges simply because their body sizes are different.

Why is body size such an important factor in how animals work? How do biologists study the consequences of size? Let's consider each question in turn.

Surface Area/Volume Relationships

From microscopic roundworms to gigantic blue whales, animals span an incredible range of sizes—a total of twelve orders of magnitude, or a difference in mass of 10^{12} grams. Given this immense size variation, it is logical to predict that small and large animals face vastly different physical and physiological challenges. Many of these challenges are based on the asymmetrical relationship between surface area and volume.

To understand why the relationship between surface area and volume is important, recall from Chapter 6 that diffusion takes place across the surface of the plasma membrane. Oxygen and nutrients such as glucose must diffuse into the cell, and waste products such as urea and carbon dioxide must diffuse out. The rate at which these and other molecules and ions diffuse depends in part on the amount of surface area available for diffusion. In contrast, the rate at which nutrients are used and waste products are produced depends on the volume of the cell.

The contrast between processes that depend on surface area and those that depend on volume is important for a simple reason. As a cell gets larger, its volume increases much faster than its surface area does. Reviewing a little basic geometry will convince you why this is so. As **Figure 41.7a** (page 942) shows, the surface area of a cube increases as a function of its linear dimension *squared*. Because cubes have six sides, the surface area of a cube of length l is $6l^2$ (six times the area of any one side). The volume of the same structure, however, increases as a function of its linear dimension *cubed*. Hence the volume of a cube

WEB TUTORIAL 41.1
Surface Area/Volume Relationships

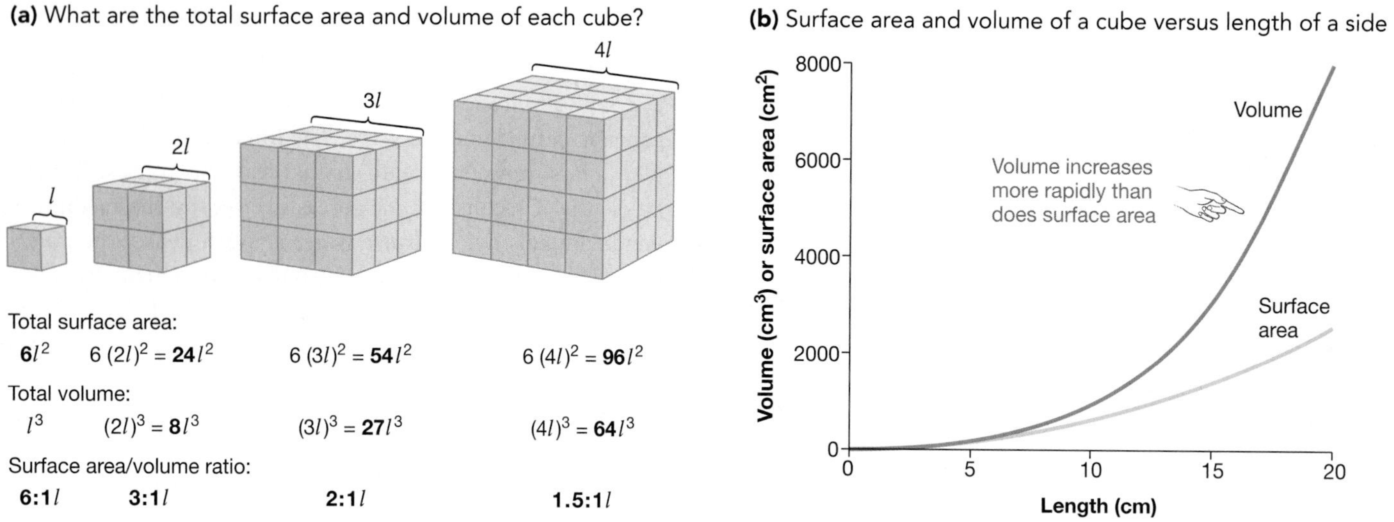

FIGURE 41.7 **Surface Area and Volume Change as a Function of Overall Size**
(a) The total surface area of an object increases as the square of the linear dimension (*l*). The volume increases as the cube of that linear dimension. **(b)** Volume increases much more rapidly than does total surface area as linear dimensions increase. QUESTION Which animals have a higher surface area/volume ratio: large or small ones?

of length *l* is l^3. Area has two dimensions; volume has three. In general:

$$\text{Surface area} \propto (\text{length})^2$$
$$\text{Volume (or mass)} \propto (\text{length})^3$$
$$\text{Surface area} \propto (\text{volume})^{2/3}$$

Figure 41.7b graphs the consequences of these relationships. As a cube gets bigger, its surface area increases more slowly than does its volume (or mass); surface area increases at only the 2/3 power of its volume. The same general relationship holds for cells, tissues, organs, and systems. The general point is that quantities that are based on volume, such as body mass, increase disproportionately fast with increases in linear dimensions.

The relationship between surface area and volume is fundamental. How do changes in this relationship affect the form and function of animals?

Comparing Mice and Elephants The observation that surface area increases as the 2/3 power of volume has important implications for animal design. As an example of how surface area/volume relationships affect an animal's physiology, consider the metabolic rate of mammals. **Metabolic rate** is the overall rate of energy consumption by an individual. Because the consumption and production of energy in mammals depend largely on aerobic respiration, metabolic rate is often measured in terms of oxygen consumption. The metabolic rate is then reported in units of milliliters of O_2 consumed per hour.

Because it is so much larger, an elephant consumes a great deal more oxygen per hour than a mouse does. But what is going on at the levels of cells and tissues? To furnish a more useful way to compare metabolic rates in different species, biologists divide metabolic rate by overall mass and report a mass-specific metabolic rate in units of mL O_2/gram/hour. This mass-specific metabolic rate gives the rate of oxygen consumption per gram of tissue. Because an individual's metabolic rate varies dramatically with its activity, the accepted convention is to report the **basal metabolic rate (BMR)**—the rate at which oxygen is consumed while at rest, with an empty stomach, under normal temperature and moisture conditions.

As **Figure 41.8** shows, small animals have much higher BMRs than do large animals on a per-gram basis. Elephants may be 200,000 times larger than mice, but their mass-specific basal metabolic rate is one-twelfth as large. The leading hypothesis to explain this pattern is based on surface area/volume ratios. Many aspects of metabolism—including oxygen consumption, the di-

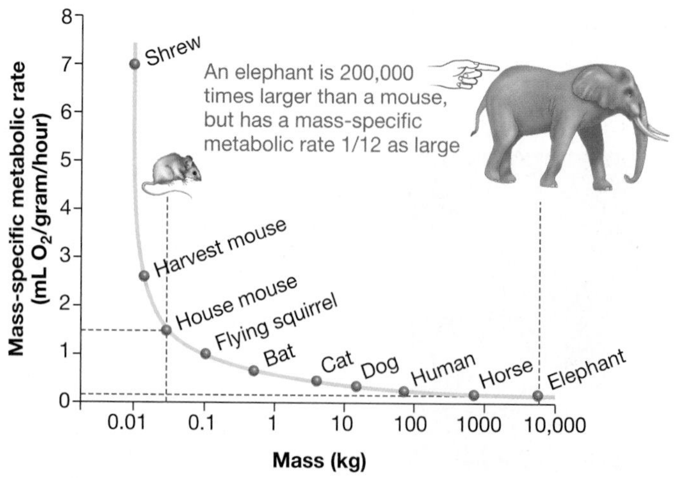

FIGURE 41.8 **Small Animals Have Higher Metabolic Rates than Large Animals Do**
Overall body mass, plotted on a logarithmic scale, versus metabolic rate per gram of tissue.

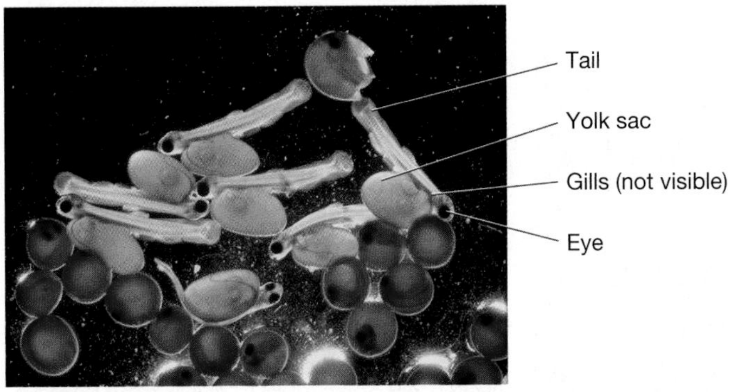
Tail
Yolk sac
Gills (not visible)
Eye

FIGURE 41.9 Newly Hatched Salmon Can Breathe through Their Skin and Their Gills
The gills of newly hatched salmon are so rudimentary that they are not visible here.

gestion of food, the delivery of nutrients to tissues, and the removal of wastes and excess heat—depend on exchange across surfaces. As an organism's size increases, its mass-specific metabolic rate must decrease, or the surface area available for exchange of materials would fail to keep up with the metabolic demands generated by the enzymes in the organism.

The same issues influence the development of an individual over the course of its lifetime. A king salmon, for example, weighs a few milligrams or less at hatching but grows into an adult weighing 50 kg or more. This represents a millionfold increase in body mass. To explore the consequences of this change, biologists have studied how gas exchange—uptake of oxygen and removal of carbon dioxide—occurs in newly hatched Atlantic salmon. Like most fish species, young salmon have rudimentary gills but also exchange gases across their skin and yolk sac (**Figure 41.9**). To document the amount of gas exchange that occurs in the gills versus the general body surface, researchers inserted the heads of individual salmon through a pinhole in a soft rubber membrane (**Figure 41.10**). After waiting for the fish to adjust to the apparatus, the biologists recorded the rate of oxygen uptake on either side of the membrane.

The graph in Figure 41.10 shows the results of such experiments. The data indicate that newly hatched larvae take up most of the oxygen they need by diffusion across the body surface. As an individual grows, however, its skin surface area decreases in relation to its volume. To avoid suffocation, the fish must develop gills to take over the bulk of the gas exchange activity. The individual breathes through its gills for the rest of its life. This outcome raises an interesting question: What makes gills so effective as the site of oxygen uptake?

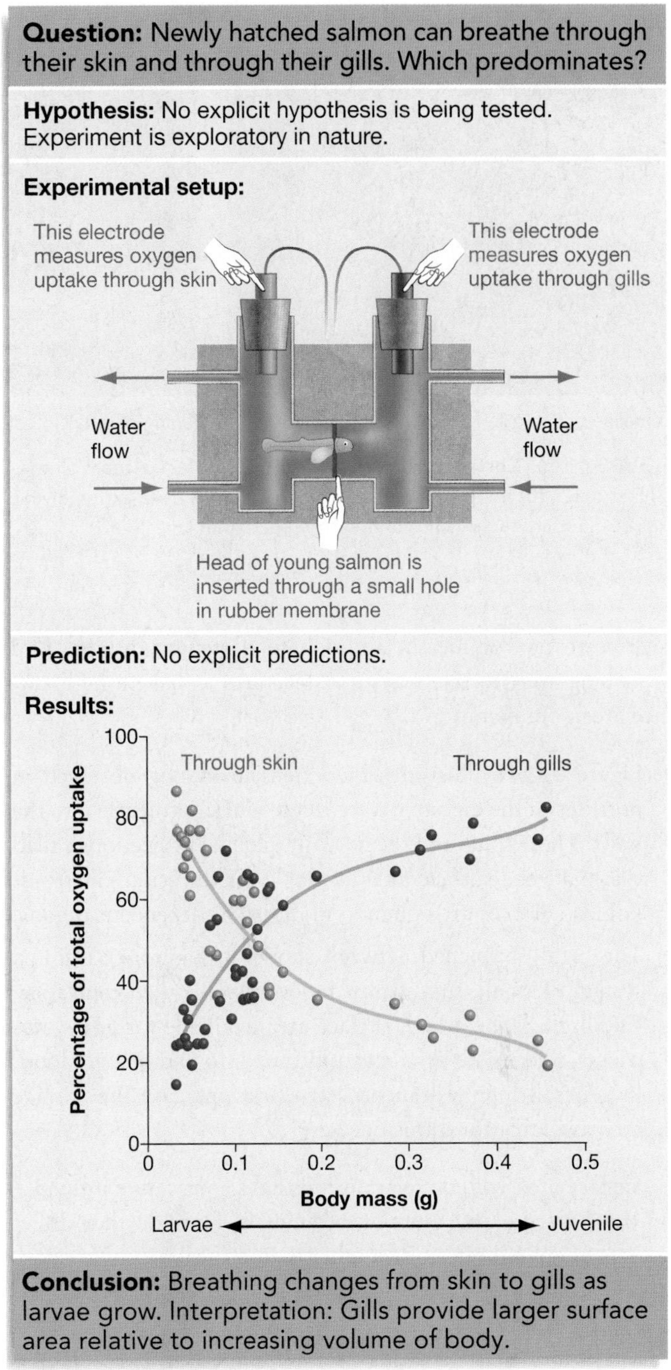
Question: Newly hatched salmon can breathe through their skin and through their gills. Which predominates?
Hypothesis: No explicit hypothesis is being tested. Experiment is exploratory in nature.
Experimental setup:
This electrode measures oxygen uptake through skin
This electrode measures oxygen uptake through gills
Water flow
Water flow
Head of young salmon is inserted through a small hole in rubber membrane
Prediction: No explicit predictions.
Results:
Through skin
Through gills
Percentage of total oxygen uptake
100
80
60
40
20
0
0
0.1
0.2
0.3
0.4
0.5
Body mass (g)
Larvae
Juvenile
Conclusion: Breathing changes from skin to gills as larvae grow. Interpretation: Gills provide larger surface area relative to increasing volume of body.

FIGURE 41.10 Do Young Salmon Breathe Mainly through Their Gills or Their Skin?
Experiment exploring whether the nature of respiration changes as growing salmon experience changes in their surface area/volume ratio.

Adaptations That Increase Surface Area

In aquatic animals, gills are organs that allow the exchange of gases and dissolved substances between the animals' blood and the surrounding water. **Figure 41.11a** (page 944) shows a close-up of a gill. The important points to notice are that a gill consists of sheetlike structures called **lamellae** and that the cells on the surface of the sheets are flattened. These features are significant, because the sheets of flattened cells provide this organ with an extremely high surface area relative to its volume. Because of these traits, gases are able to diffuse across the gills rapidly enough to keep up with the growth in the volume of the developing fish.

In general, if the function of a cell or tissue depends on diffusion, its structure has a shape that increases its surface area

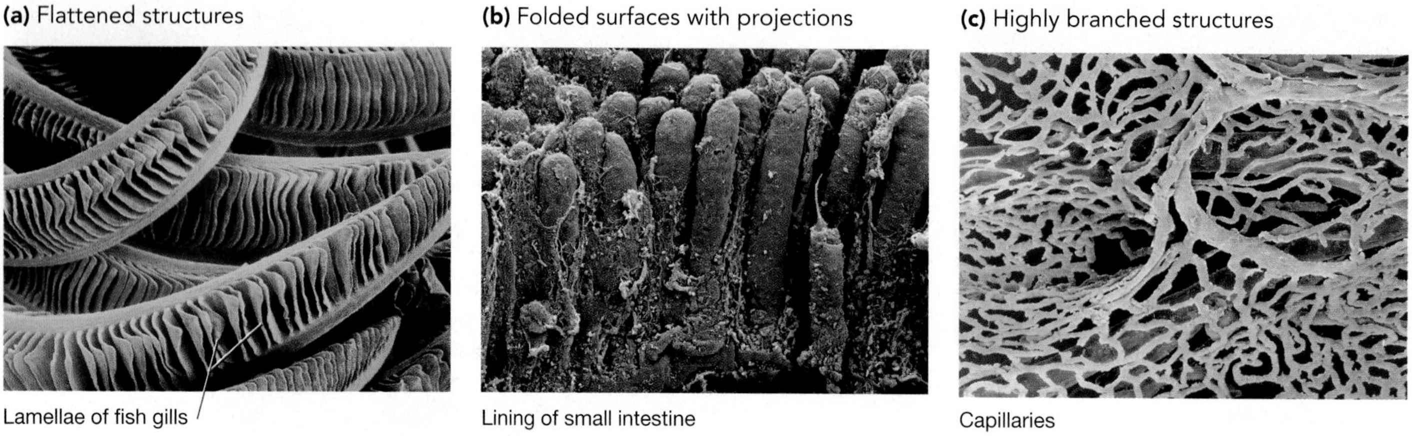

FIGURE 41.11 Certain Structures Have High Surface Area/Volume Ratios
(a) Flattened, **(b)** folded, and **(c)** branched structures have extremely high surface areas relative to their volume.
EXERCISE Identify the diffusion-based process that takes place in each of the structures pictured.

relative to its volume. In addition to flattening, folding and branching are effective ways for structures to have a high surface area/volume ratio:

- **Figure 41.11b** illustrates the extensive folding observed in portions of the digestive tract where nutrients diffuse into the body. These folds and the narrow, tubelike projections called **villi** make the surface area available for diffusion extensive. Folded surfaces are common in diffusion-dependent organs.
- The highly branched network shown in **Figure 41.11c** is a system of small, thin-walled blood vessels called **capillaries.** Capillaries have a high surface area available for gases, nutrients, and waste products to diffuse into and out of blood. In general, highly branched structures increase the surface area available for diffusion.

Surface area/volume relationships have a pervasive influence on the structure and function of animals. How do they affect aspects of animal physiology other than diffusion?

Do All Aspects of an Animal's Body Increase in Size Proportionately?

Biologists have long been fascinated by the fundamental differences between small and large animals. In 1637, for example, Galileo Galilei observed that the bones of large animals are not only thicker than those of small animals, but are also *disproportionately* thicker. Species that are twice the overall mass of other species tend to have bones that are more than twice as thick. In making this observation, Galileo was the first to recognize the existence of allometry. **Allometry** occurs when changes in body size are accompanied by disproportionate changes in anatomical structures or physiological processes.

Figure 41.12 shows the relationship between the overall body mass of terrestrial animals and their skeletal mass. This relationship is considered **allometric** ("different-measure"), because the two quantities do not change at the same rate. If they did change at the same rate, their relationship would be described by the dashed line shown in the figure. Quantities that change at the same rate are considered to be **isometric** ("same-measure").

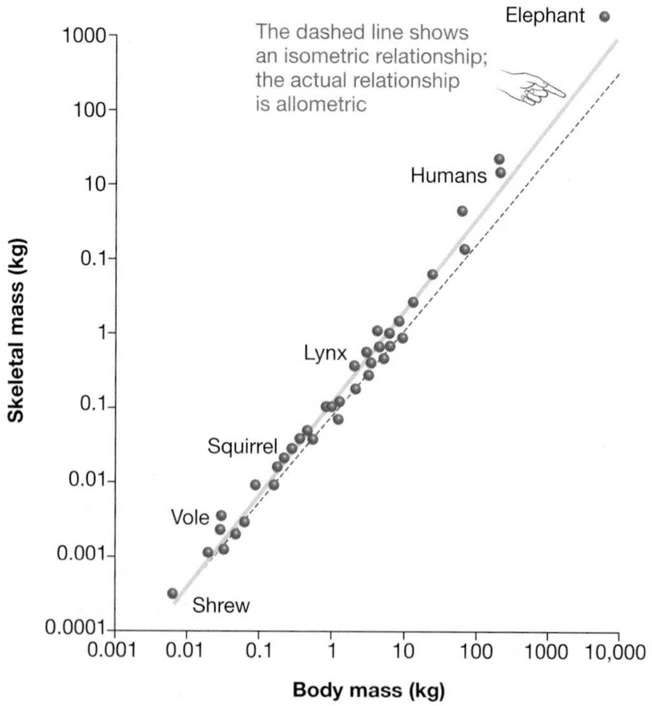

FIGURE 41.12 Among Mammals, Skeletal Mass Increases Much Faster than Overall Body Mass
Overall body mass plotted versus total skeletal mass of terrestrial mammals. Both axes are logarithmic. Large land-dwelling mammals have much heavier skeletons than expected for their size.
QUESTION Would you expect the same relationship in aquatic mammals, whose bodies are supported by water?

Allometry as a Response to Surface Area/Volume Relationships The data in Figure 41.12 show that the relationship between body size and skeletal size in mammals is allometric. Why? Answering this question involves another analysis of area and volume relationships: The volume involved is the mass that must be supported by the skeleton. The area involved is the cross-sectional area of the bones supporting this mass. The strength of these bones, and thus their ability to support weight, is a function of their cross-sectional area. As the animal's mass increases, the amount of bone required for support has to increase disproportionately. Thus, an elephant must have a much higher proportion of its body mass devoted to skeletal support than smaller animals do.

Given these arguments, it is not surprising that elephants can't jump. Even though an elephant's skeleton is disproportionately large compared with smaller animals, its bones are relatively small and fragile given its enormous mass. As a result, an elephant that landed from a height of less than a meter would probably suffer multiple leg fractures.

Allometry as Adaptation Structures that display allometry can sometimes be interpreted as adaptations to a particular lifestyle or environment. As an example, consider the plot of heart size versus overall body size in **Figure 41.13**. Each point on the graph represents the average value for a species in either the dog family or the cat family. Notice that, at any given body size, dogs have much larger hearts than do cats. In these two groups, the relationship between heart size and body size is different.

To explain this pattern, biologists point out that, although both dogs and cats are predators, they have very different hunting styles. Dogs run their prey down during long-distance chases; cats stalk prey and then make a short sprint. Based on this observation, researchers hypothesize that the relatively large hearts of dogs are an adaptation that increases blood flow to muscles, making long-distance chases possible. In this light, the allometric relationship between heart size and body size in these mammals is logical.

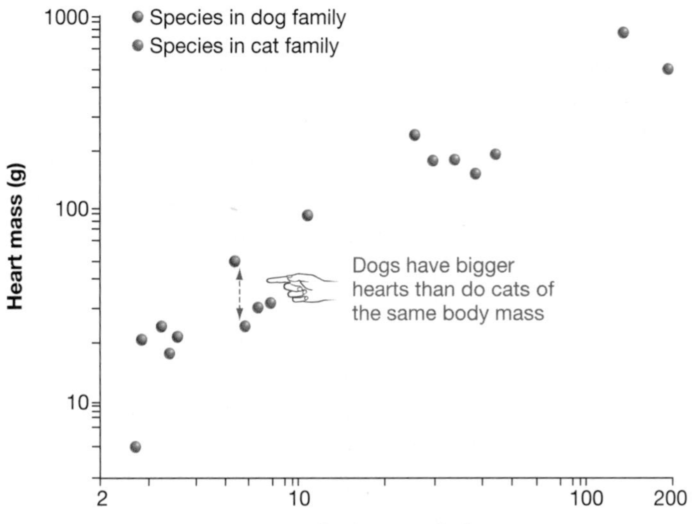

FIGURE 41.13 Scaled for Body Size, Dogs Have Much Larger Hearts than Cats Do
Heart size plotted as a function of overall body size. Each data point represents the average value for a species. Both axes are logarithmic.

✓CHECK YOUR UNDERSTANDING

An animal's overall size is important in part because body mass is acted on by an array of physical forces. In addition, the amount of heat and waste that an animal produces and the amount of food and oxygen that it requires is proportional to its mass. The amount of surface area available relative to that mass is critical, because heat exchange and other important processes take place across surfaces.

You should be able to (1) explain why large animals have a relatively small surface area/volume ratio, (2) explain the relationship between body mass and basal metabolic rate, (3) diagram three structures that increase surface area, and (4) define allometry and give an example.

41.4 Homeostasis

In analyzing the relationship between animal form and function, biologists focus on understanding adaptations as a product of natural selection, exploring the constraints on adaptation, and interpreting the influence of general principles such as surface area/volume ratios. Another general principle that researchers use in analyzing animal physiology is called homeostasis.

Homeostasis ("alike-standing") is the relatively constant chemical and physical conditions in an animal's cells, tissues, and organs. Many of the structures and processes observed in animals can be interpreted as mechanisms for achieving homeostasis with respect to some quantity, from the concentration of nutrients to the pH of blood. It is important, however, to understand that constancy of physiological state can be achieved by two processes: (1) conformation or (2) regulation. The body temperature of Antarctic rock cod is an example of **conformational homeostasis.** The body temperature of these fish closely matches that of the surrounding seawater, which is typically −1.9°C. The rock cod is not regulating its body temperature to match that of seawater. Instead, its body temperature remains constant simply because it conforms to the temperature of its surroundings. **Regulatory homeostasis,** in contrast, requires a physiological mechanism that adjusts the internal state to keep it near a specific value or within limits that can be tolerated, regardless of the external conditions. No matter what environment you are in, you maintain a body temperature of about 37°C. If your body temperature begins to rise, you might shed clothing to maintain homeostasis. If your body temperature begins to drop, you might add clothing to conserve heat and bring your temperature back up to the target value.

Because epithelium is the interface between the internal and external environments, it plays a key role in achieving

WEB TUTORIAL 41.2 Homeostasis

homeostasis. Epithelium is responsible for creating an internal environment that is dramatically different from the external environment, and for maintaining physical and chemical conditions inside an animal that are relatively constant. One of its most basic functions is to control the exchange of materials across its surface in a way that is consistent with homeostasis. As we'll see in subsequent chapters, many epithelial cells are studded with membrane proteins that regulate the transport of ions, water, nutrients, and wastes. No molecule can enter or leave the body without crossing an epithelium. Homeostasis is possible because epithelia control this exchange.

Why is homeostasis important? Much of the answer is based on enzyme function. Recall from Chapter 3 that enzymes are proteins that catalyze chemical reactions within cells and that temperature, pH, and other physical and chemical conditions have a dramatic effect on the structure and function of enzymes. More specifically, most enzymes function best under a fairly narrow range of conditions. Other processes depend on homeostasis, too. Temperature changes affect membrane permeability and how quickly solutes diffuse. To take an extreme case, the expansion of water as it freezes can rip cells apart if tissues are allowed to drop much below 0°C. Conversely, extremely high temperatures can cause proteins to denature—meaning that they lose their tertiary structure and cease to function. When homeostasis occurs, conditions inside the body allow cells, tissues, organs, and systems to function at an optimal level.

Regulation and Feedback

To achieve homeostasis, animals have regulatory systems that constantly monitor internal conditions such as temperature, blood pressure, blood pH, and blood glucose. If one of these variables changes, a homeostatic system acts quickly to modify it. Like the thermostat in a home heating system, each of these systems has a **set point**—a normal or target value for the controlled variable. Animals have a set point for blood pH, tissue oxygen concentration, nutrient availability, and other parameters. In mammals, the set point for body temperature is about 37°C. How does an individual maintain its tissues at the set point in the face of changes in activity and the environment?

The key to answering this question is to recognize that homeostatic systems are based on the three general components shown in **Figure 41.14a**. A **sensor** is a structure that senses some aspect of the external or internal environment. An **integrator** is a component that evaluates the incoming sensory information and "decides" whether a response is necessary to achieve homeostasis. (The word *decides* is in quotation marks because the decision is not a conscious one.) An **effector** is any structure that helps restore the desired internal condition. Without these three elements, homeostatic control systems are unable to maintain a desired set point and homeostasis is impossible.

Homeostatic systems are based on negative feedback. **Negative feedback** is a corrective response in which effectors reduce or oppose the change in internal conditions. For instance, a rise in blood pH triggers effectors that act to reduce that rise. Blood pH returns to the set point in response to this negative feedback.

Subsequent chapters in this unit explore how animals use sensor-integrator-effector systems to achieve homeostasis with respect to the solute concentrations of their cells and tissues, their oxygen supply, and nutrient availability. To explore the concept further, let's focus on how different animals achieve homeostasis with respect to temperature.

41.5 How Do Animals Regulate Body Temperature?

Animals that get too hot or too cold may die. Overheating can cause enzymes and other proteins to denature and cease functioning, and it may lead to excessive water loss and dehydration. A sharp drop in body temperature, in contrast, can slow enzyme function and energy production. In humans, heat stroke and hypothermia ("under-heating") are life-threatening conditions.

The control of body temperature is called **thermoregulation** and is an important aspect of homeostasis. **Figure 41.14b** illustrates how the sensor-integrator-effector components of mammals regulate body temperature. Information about body temperature is constantly monitored by temperature receptors located throughout the body. For example, temperature receptors in the skin respond to cooling or heating by altering the pattern of electrical signals that they send to adjacent neurons. Receptors in the brain region called the *anterior hypothalamus* respond in a similar fashion to changes in blood temperature.

The electrical signals that originate with temperature receptors in the skin, spinal cord, and anterior hypothalamus are transmitted to an integrator located in the brain. Current evidence indicates that separate centers in the hypothalamus of the brain integrate and respond to increases and decreases in body temperature. If a mammal becomes cold, cells in the *posterior hypothalamus* region of the brain send signals to effectors that return body temperature to the set point. For example, signals from the posterior hypothalamus might induce shivering to generate warmth, and fluffing of fur or feathers to improve insulation and retain heat. Signals from the same or nearby cells can also result in the release of blood-borne chemical signals that increase the rate at which cellular respiration is taking place throughout the body—thus generating more body heat. But if the individual becomes too hot, an integrator in the anterior hypothalamus sends signals that initiate sweating or panting—responses that cool the body. Signals from the same or nearby cells can also induce behavioral changes that slow heat production, such as seeking shade or a cool burrow and then resting.

The critical point is to recognize that, in response to either cooling or heating, behavioral and physiological responses move the body temperature back toward the set point via negative feedback. The homeostatic system diagrammed in Figure 41.14b

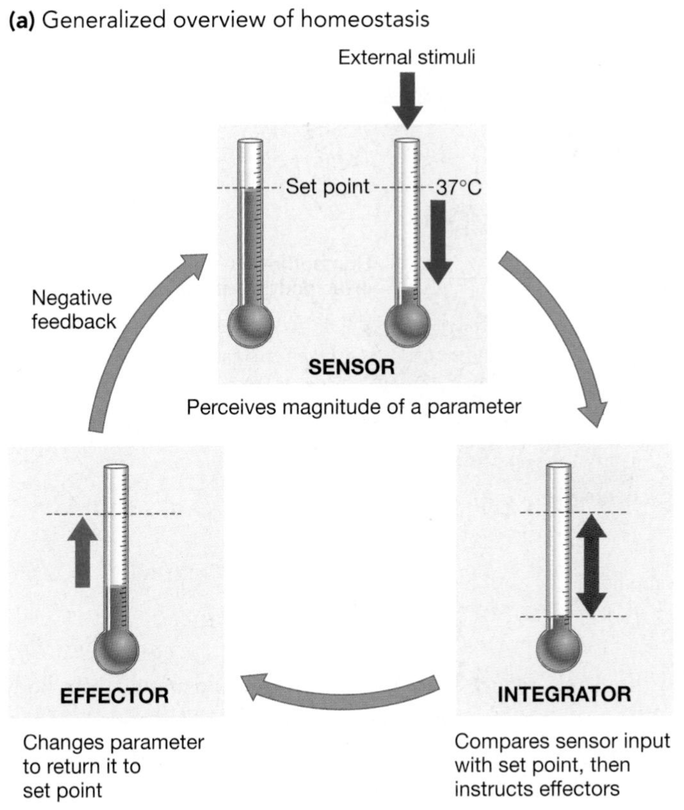
(a) Generalized overview of homeostasis
External stimuli
Set point
37°C
Negative feedback
SENSOR
Perceives magnitude of a parameter
EFFECTOR
Changes parameter to return it to set point
INTEGRATOR
Compares sensor input with set point, then instructs effectors

FIGURE 41.14 Animals Achieve Homeostasis through Negative Feedback
(a) The interactions that occur among the three components of a generalized system for attaining homeostasis. Many animals use homeostatic systems similar to this one to maintain a preferred body temperature. **(b)** In mammals, a set point of 37°C is maintained by a complex negative feedback system that includes integrators in the anterior and posterior hypothalamus and sensors located throughout the body.

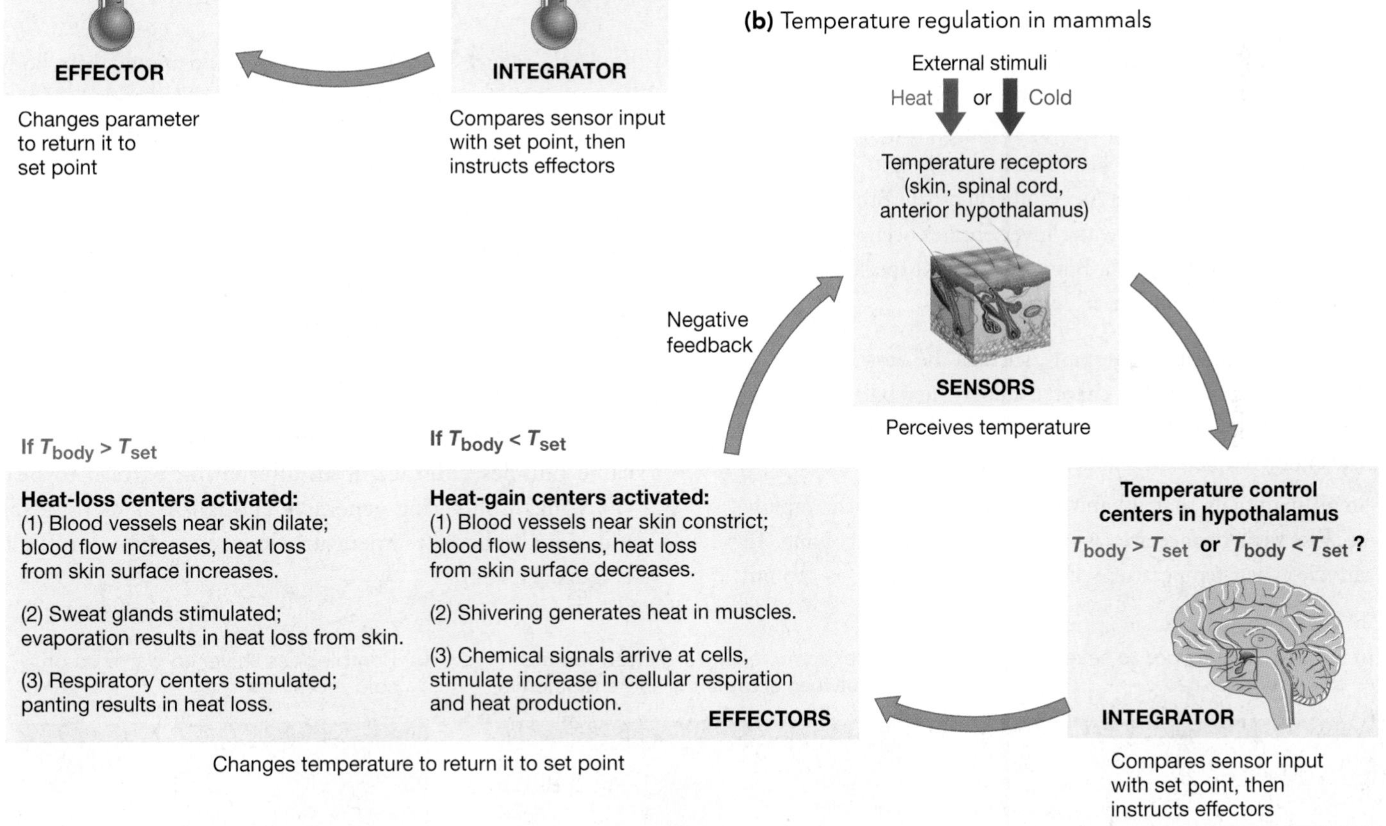
(b) Temperature regulation in mammals
External stimuli
Heat or Cold
Temperature receptors (skin, spinal cord, anterior hypothalamus)
SENSORS
Perceives temperature
Negative feedback
Temperature control centers in hypothalamus
$T_{body} > T_{set}$ or $T_{body} < T_{set}$?
INTEGRATOR
Compares sensor input with set point, then instructs effectors
If $T_{body} > T_{set}$
Heat-loss centers activated:
(1) Blood vessels near skin dilate; blood flow increases, heat loss from skin surface increases.
(2) Sweat glands stimulated; evaporation results in heat loss from skin.
(3) Respiratory centers stimulated; panting results in heat loss.
If $T_{body} < T_{set}$
Heat-gain centers activated:
(1) Blood vessels near skin constrict; blood flow lessens, heat loss from skin surface decreases.
(2) Shivering generates heat in muscles.
(3) Chemical signals arrive at cells, stimulate increase in cellular respiration and heat production.
EFFECTORS
Changes temperature to return it to set point

constantly monitors body temperature relative to the set point and changes the balance between heat production and heat loss appropriately.

Gaining and Losing Heat

To understand temperature regulation in animals, it is critical to recognize that there is wide variation in how these organisms obtain heat: For example, an **endotherm** ("inner-heat") produces adequate heat to warm its own tissues, while an **ectotherm** ("outer-heat") relies principally on heat gained from the environment.

Endotherms and ectotherms represent two opposite extremes along a continuum of heat sources. Many animals are partially endothermic and partially ectothermic. A similar continuum exists in the degree to which animals hold body temperature constant: **Homeotherms** ("alike-heat") keep their body temperature constant, while **heterotherms** ("different-heat") allow their body temperature to rise or fall to some degree depending on environmental conditions.

The sources of heat that an animal uses and the degree to which its body temperature varies can be treated as axes on a

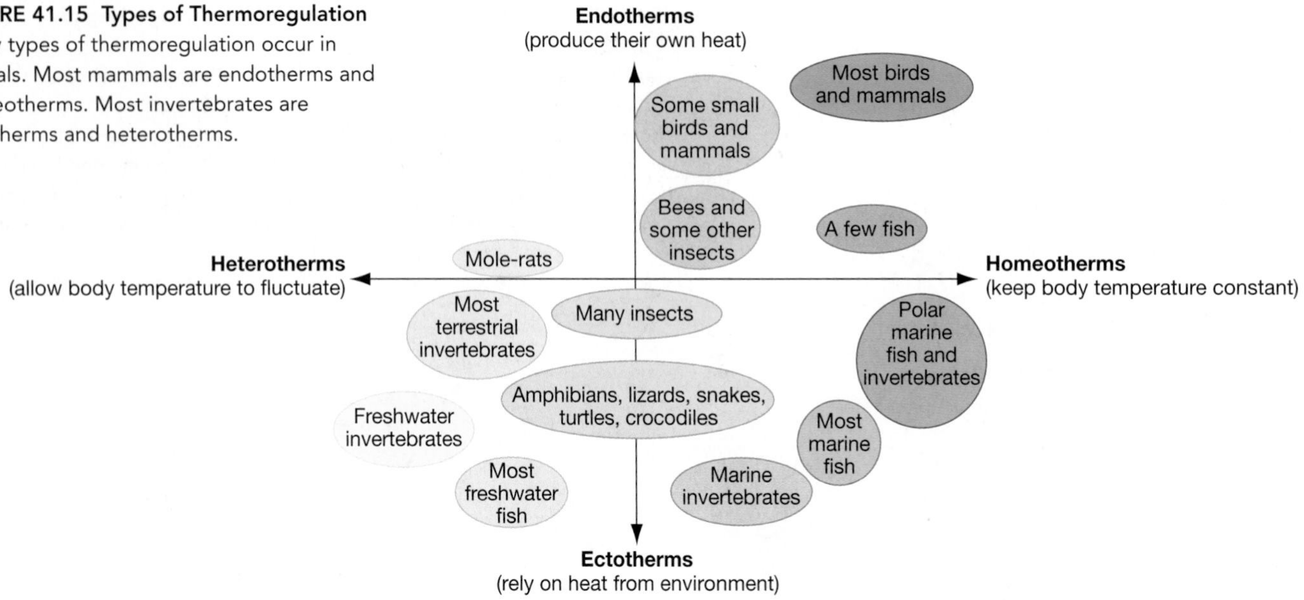
Endotherms
(produce their own heat)
Heterotherms
(allow body temperature to fluctuate)
Homeotherms
(keep body temperature constant)
Ectotherms
(rely on heat from environment)
Most birds and mammals
Some small birds and mammals
Bees and some other insects
A few fish
Mole-rats
Most terrestrial invertebrates
Many insects
Polar marine fish and invertebrates
Amphibians, lizards, snakes, turtles, crocodiles
Freshwater invertebrates
Most marine fish
Most freshwater fish
Marine invertebrates

FIGURE 41.15 Types of Thermoregulation Many types of thermoregulation occur in animals. Most mammals are endotherms and homeotherms. Most invertebrates are ectotherms and heterotherms.

graph, as shown in **Figure 41.15**. As strictly endothermic homeotherms, humans and other mammals, along with most birds, occupy the upper right-hand corner of the graph. These species produce their own heat metabolically and maintain a constant body temperature. As ectothermic heterotherms, mosquito larvae and other freshwater invertebrates occupy the lower left-hand corner of the graph. But many animal species lie somewhere between these extremes:

- Some desert-adapted mammals, such as the oryx featured in the introduction to this chapter, allow their body temperature to rise during the hottest part of the day—meaning they are somewhat heterothermic.
- Small mammals that inhabit cold climates lose heat rapidly because their surface area is large relative to their volume. To survive when temperatures are cold, species such as ground squirrels reduce their metabolic rate and allow their body temperature to drop. This condition is called **torpor**. If torpor persists for weeks or months, it is called **hibernation** (**Figure 41.16a**; see also **Box 41.1**).
- Naked mole-rats are mammals but lack insulation, because they have no fur (**Figure 41.16b**). They live in underground tunnels and allow their body temperature to rise and fall with burrow temperatures. They are heterothermic and intermediate between ectotherms and endotherms.
- On cold mornings, bumblebees "shiver" by contracting their flight muscles in synchrony. (During flight, these same muscles contract in an alternating pattern to beat the wings.) Shivering generates enough heat to raise the individual's body temperature in preparation for flight (**Figure 41.16c**).

(a) Squirrels use torpor to save energy when food is scarce.

(b) Naked mole-rats are intermediate between ectotherms and endotherms.

(c) Bumblebees shiver to warm up on cold mornings.

FIGURE 41.16 Different Strategies for Thermoregulation
(a) Small mammals and birds lose heat quickly, because they have high metabolic rates and a high surface area/volume ratio. Torpor is a way to save energy when food supplies are low. **(b)** The naked mole-rat spends its entire life underground. It lacks insulation and allows its body temperature to match the burrow's temperature. **(c)** Bumblebees are heterothemic. On cold mornings, they shiver until they warm up enough to fly.

BOX 41.1 A Closer Look at Torpor and Hibernation

As temperatures drop and days shorten in the fall, animals that live at high latitudes and high elevations face serious physiological challenges. For an endothermic individual to maintain normal body temperatures when the ambient temperature drops, it has to burn (oxidize) large amounts of food. But for species that eat seasonal foods such as insects, seeds, or nectar, food supplies are declining just as food demands are increasing. The problem is particularly acute for species that lose heat rapidly, because their body size is small. Yet squirrels, voles, and mice are abundant in northern and mountainous habitats. How do these animals cope?

In some cases, small mammals retreat to underground burrows that are much warmer than the ground surface and that have a supply of grasses and seeds stored during the summer. But other species endure by becoming inactive and allowing their body temperature to drop to levels just above ambient temperatures. Hibernating pocket mice and chipmunks expend only enough energy to maintain a body temperature of about 2°C, just above lethal limits, for weeks or months. On cold nights, hummingbirds may allow their body temperature to drop over 25°C, to less than 10°C.

Animals in torpor dramatically reduce their heart rate and oxygen consumption and may have long periods when breathing is temporarily suspended. Torpor results in significant energy savings over a long winter or even a single night.

Arousal from daily torpor or hibernation is energetically costly, however. For an animal in torpor to bring its body temperature back up to normal levels, it must either activate heat production from brown adipose tissue or shiver intensely. The amount of energy consumed during arousal from torpor can deplete energy stores to dangerously low levels. Yet it is clear that torpor confers a fitness advantage, in some mammal and bird species, allowing them to thrive in otherwise inhospitable habitats.

Even in a homeothermic endotherm such as a mammal or bird, body temperature can vary widely in different body regions. When a Canada goose is standing on ice, its feet may be at a temperature of just 9°C, even though its body core is at 35°C. Similar types of variation exist in tuna and mackerel, which are ectotherms but generate heat to warm certain sections of their bodies, such as their eyes or the muscles that are responsible for rapid swimming.

Sources of Body Heat In addition to gaining heat directly by basking in sunlight or lying on warm rocks or soil, ectotherms generate some heat as a by-product of metabolism. The amount of heat they generate is small compared with the amount generated by endotherms, however, because ectotherms have low metabolic rates. Endotherms can warm themselves because their basal metabolic rates are extremely high. They can maintain high body temperatures because they have elaborate insulating structures such as feathers or fur.

The contrast in metabolic rates and heat-producing abilities of endotherms and ectotherms is reflected in the structure of their tissues. For example, researchers find that mitochondrial density and mitochondrial enzyme activity are three to four times higher in endotherms' cells than in cells of ectotherms of similar size. This comparison is meaningful because most reactions that generate ATP and heat occur in mitochondria.

In addition, some endotherms have special heat-generating tissues. The best-studied example is the brown adipose tissue found in some mammals. **Adipose tissue** is a connective tissue whose cells store fats (**Figure 41.17a**). **Brown adipose tissue** is a specialized type of adipose tissue that features a high density of mitochondria as well as stored fats (**Figure 41.17b**).

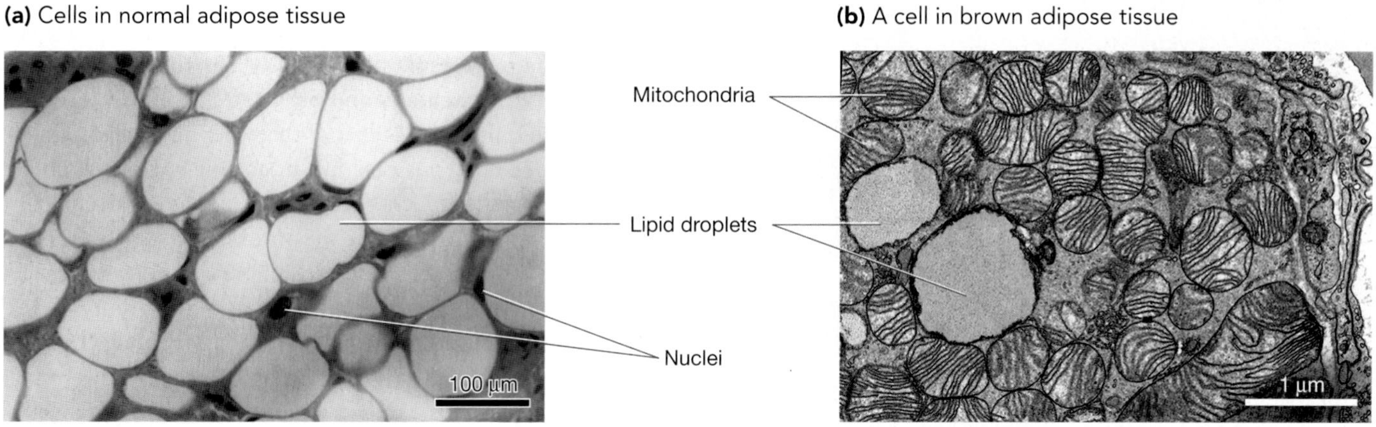
(a) Cells in normal adipose tissue
(b) A cell in brown adipose tissue
Mitochondria
Lipid droplets
Nuclei
100 μm
1 μm

FIGURE 41.17 The Structure of Cells in Brown Adipose Tissue Correlates with Their Function
In contrast to **(a)** cells in normal fat tissue, **(b)** this cell in brown adipose tissue is packed with mitochondria. Instead of producing ATP, these mitochondria produce primarily heat.

When the fat molecules are oxidized by the mitochondria in brown adipose tissue, no ATP is produced. Instead, all of the chemical energy stored in the fats is released as heat. As a result, brown adipose tissue produces almost 10 times as much heat as do other tissues.

Brown adipose tissue is particularly common in small animals and in the infants of large species. In humans, for example, patches of brown adipose tissue are found in the neck and chest regions of newborns. Because infants are small, they have a high surface area/volume ratio and lose heat quickly. Based on such observations, brown adipose tissue is interpreted as an adaptation for small endotherms to achieve homeostasis with respect to temperature.

Exchanging Heat with the Environment All animals exchange heat with their environment constantly. Heat flows "downhill," from regions of higher temperature to regions of lower temperature. If an individual is warmer than its surroundings, it will lose heat; if it is cooler than its environment, it will gain heat.

As **Figure 41.18** shows, animals exchange heat with the environment in four ways: conduction, convection, radiation, and evaporation. **Conduction** is the direct transfer of heat between two physical bodies that are in contact with each other. For instance, when you sit on a cold rock, heat is transferred from your body to the rock. The rate at which conduction occurs depends on the surface area of transfer, the steepness of the temperature difference between the two bodies, and how well each body conducts heat. Water, for example, conducts heat much better than air does. As a result, a person immersed in 15°C water loses heat much faster than does a person exposed to air at the same temperature.

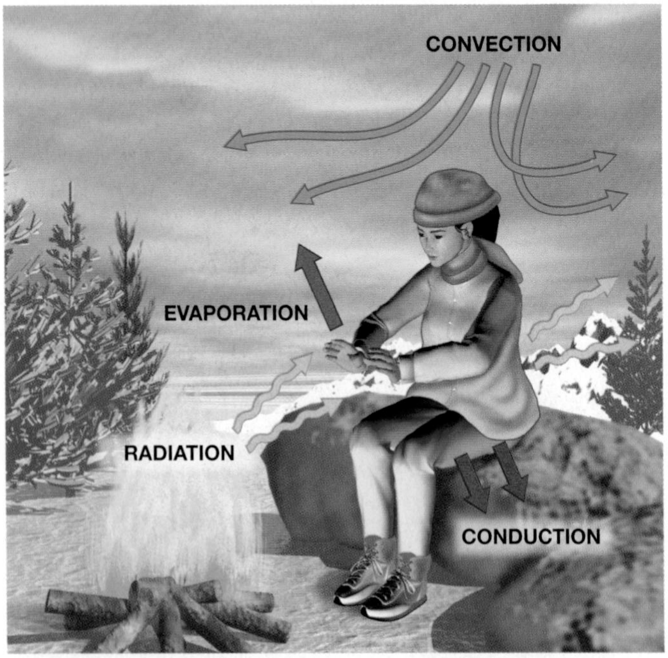

FIGURE 41.18 Four Methods of Heat Exchange
The arrows indicate the direction of heat exchange from the warmer body to the cooler body. **EXERCISE** Sketch a similar diagram that features a butterfly in summer.

Convection occurs when air or water moves over the body surface. The heat loss that occurs when wind blows on your skin is due to convection. In effect, convection increases the rate of heat exchange by conduction. The increase occurs because the air or water in direct contact with the body surface is constantly replaced. As a result, convection maintains a large temperature gradient for conduction. As the speed of the air or water flow increases, so does the rate of heat transfer.

Radiation is the transfer of heat between two bodies that are not in direct physical contact. All objects, including animals, radiate energy as a function of their temperature. A campfire radiates heat; so does your body, although to a much lesser degree. The major source of radiant energy is the Sun.

Evaporation is the phase change that occurs when liquid water becomes a gas. Unlike conduction, convection, and radiation, evaporation leads to heat loss only. You lose heat when water evaporates off your skin. Because of the extensive hydrogen bonding in liquid water, a large amount of energy is needed to heat water and produce evaporation. Consequently, water is an efficient coolant on a hot day. Conversely, getting wet can be deadly on a cold day, because the water on your skin absorbs so much heat.

Conserving Heat In terrestrial environments, heat is readily lost through evaporation, convection, and radiation. Air conducts heat poorly, however, so it is a good insulator. Thus, it is no coincidence that endothermic animals have elaborate external structures that trap air, slow the rate of heat exchange through conduction, and conserve body heat. Birds have down feathers that perform this function; mammals have underfur that forms a fine mesh, traps air, and prevents heat lost by conduction, convection, and radiation.

In aquatic environments, no heat is lost due to evaporation. Little heat is lost or gained due to radiation, because water transmits radiation poorly. But water is such an effective conductor of heat that metabolic heat produced by aquatic organisms is lost rapidly—particularly through structures such as gills, which have a large surface area and a high rate of blood flow. As a result, most aquatic invertebrates and fish are ectotherms. Their body temperature is the same as that of the water they inhabit.

How do endotherms that live in aquatic habitats cope with heat loss due to conduction? If you've ever gone swimming in cold water, you can appreciate the problem faced by seals, otters, and whales. To conserve heat, otters have dense, water-repellent fur that maintains a layer of trapped air next to the skin. Seals and whales are insulated by thick layers of fatty blubber. At least some whales have a feature that minimizes heat loss from their tongue, which is exposed to cold water during feeding. As **Figure 41.19a** shows, the tongue of the gray whale contains a series of arteries and veins that are juxtaposed. Each central artery carries warm, oxygenated blood from the body core to the tongue

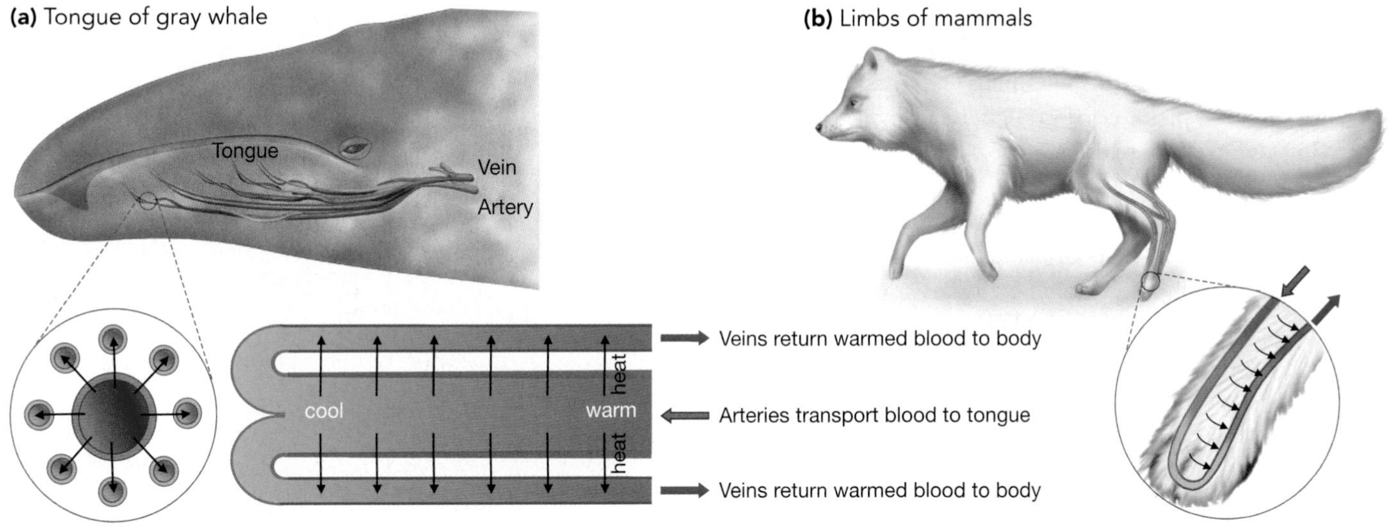
(a) Tongue of gray whale
Tongue
Vein
Artery
cool
warm
heat
heat
Veins return warmed blood to body
Arteries transport blood to tongue
Veins return warmed blood to body
(b) Limbs of mammals

FIGURE 41.19 Countercurrent Exchangers Conserve Heat
(a) In the tongue of a gray whale, blood runs in opposite directions through juxtaposed veins and arteries. Heat flows from arteries to veins. That heat is returned to the body instead of being lost to the surrounding water. **(b)** Countercurrent arrangements are common in the limbs of mammals and birds that live in cold habitats.

and is encircled by several smaller veins, which transport cool blood from the tongue surface back toward the body core. Because the two types of blood vessel are arranged in an anti-parallel fashion, so that they transport blood in opposite directions, heat is transferred efficiently between them. This arrangement is called a **countercurrent heat exchanger.** Similar heat-conserving arrangements of arteries and veins are found in the flippers of whales and dolphins and in the legs of arctic foxes (**Figure 41.19b**).

Countercurrent exchangers are common in animals. The basic design principle is that two adjacent fluids flow through pipes in opposite directions. As the longitudinal section in Figure 41.19a shows, the fluid that enters the countercurrent exchanger is initially warm but steadily transfers heat to adjacent, cooler fluid flowing in the opposite direction. In a countercurrent heat exchanger, the gradient between the two currents is maintained even as one current cools and the other heats. If the two solutions were to run the same direction, the gradient between the two solutions would disappear as the source current cooled and the recipient current heated. Instead, a gradient exists along the entire length of a countercurrent exchanger.

A second key point about countercurrent heat exchangers is that, even though the temperature differential between the solutions at any point along the pipes is relatively small, there is a large temperature differential from one end of each pipe to the other. In effect, small differences in heat along the length of the exchanger sum up to create a large overall temperature gradient. The longer the system, the greater the overall differential.

Ectothermy versus Endothermy

Analyzing animal thermoregulation provides a convenient way to review some of this chapter's major themes. Biologists interpret homeostatic systems as adaptations that have arisen via natural selection. The structure of cells and tissues involved in homeostasis, such as mitochondria-rich brown adipose tissue and countercurrent heat exchangers, correlates with their function. Because heat is gained and lost across surfaces, surface area/volume relationships are important. Small objects lose heat much faster than large objects do, so it is difficult for small endotherms to stay warm without expending prohibitive amounts of energy. Surface area/volume relationships explain why there are no endotherms smaller than a shrew or a hummingbird, even though there are millions of centimeter- or millimeter-sized ectotherms.

Contrasting endothermy with ectothermy also provides a superb example of trade-offs in adaptation. Because endotherms maintain enzymes at optimal temperatures at all times, mammals and birds are able to remain active in winter and at night. Due to their high metabolic rates and insulation, endotherms are also able to sustain very high levels of aerobic activities, such as running or flying. These abilities come at a cost, however: To fuel their high metabolic rates, endotherms have to obtain large quantities of energy-rich food. The energy used to produce heat is then unavailable for other energy-demanding processes, such as reproduction and growth.

Ectotherms, in contrast, are able to thrive with much lower intakes of food. They can also use a greater proportion of their total energy intake to support reproduction. But because chemical reactions are temperature dependent, muscle activity and digestion slow down dramatically as the body temperature of an ectotherm drops. As a result, ectotherms are more vulnerable to predation in cold weather and in general are not as successful as endotherms at inhabiting cold environments or maintaining activity in cool nighttime temperatures.

In short, each suite of adaptations has advantages and disadvantages. Like all adaptations, endothermy and ectothermy involve trade-offs. Analyzing these trade-offs has also inspired a spirited debate on why endothermy evolved in the first place. Birds evolved from ectothermic dinosaurs, and mammals evolved from a lineage of ectothermic reptiles that is now extinct. Why did natural selection favor the evolution of endothermy in each case? Was it to make sustained activity such as flying possible? Or did endothermy allow birds and mammals to colonize cold habitats or remain active at night? These questions are unresolved. Research on animal form and function and the nature of adaptation continues.

✓CHECK YOUR UNDERSTANDING

Two major aspects of temperature regulation vary among animal species: the amount of heat generated by the animal's own tissues and the degree to which body temperature varies over time. You should be able to (1) describe three ways that animals gain heat and four ways that animals lose heat, (2) diagram how a countercurrent heat exchanger works, and (3) discuss the advantages and disadvantages of endothermy and ectothermy.

ESSAY Is Fever Adaptive?

If you've ever experienced a sustained elevation in body temperature called **fever**, you've probably taken aspirin, acetaminophen, or ibuprofen to reduce that fever. But was taking the medicine a good idea? Does fever help or hinder your ability to fight an infection?

High fevers are clearly not adaptive. In adult humans, tissue damage begins to take place if body temperature reaches 40°C (104°F); serious brain injury can occur if temperatures elevate to 41°C (106°F). But it is possible that lower fevers, in the range of 38–39°C, could inhibit the growth of infectious agents or increase the activity of immune system cells that fight infections.

Researchers recently performed an experiment on western fence lizards to test the hypothesis that fever is adaptive. In natural populations of this species, it is not uncommon to observe malaria infections in 15 percent or more of the individuals. Malaria is caused by a eukaryotic parasite in the genus *Plasmodium*. Because *Plasmodium* is transmitted to new hosts by mosquitoes, and because mosquitoes seek hosts in part by sensing heat, it is reasonable to suggest that fever benefits the parasite—*not* the host. The logic here is that a fever would increase the parasite's chances of being taken up by a mosquito and transmitted to a new host.

The experiment that the biologists set up was simple in concept. They brought a large number of lizards into the laboratory and infected them with malaria. Then they divided the infected individuals into two groups. One group was maintained in constant conditions at the preferred, normal body temperature for the species. Individuals in the second group were allowed to change their body temperature by seeking colder or warmer areas in their enclosure. The researchers documented that individuals in the second group actively sought out warm environments and that this behavior induced fever. These results support the hypothesis that lizards with *Plasmodium* infections actively try to raise their body temperatures.

Does fever help or hinder your ability to fight an infection?

The researchers claimed that fever is adaptive in this species because individuals that experienced fever recovered from infection better than did individuals that were maintained at a normal body temperature. Experiments on other ectotherms are consistent with this result and have led to a strong consensus that fever is an adaptive response in ectotherms. Unfortunately, experiments with humans have produced much less conclusive results. The role of fever in the human disease-fighting response is still not clear. The wisdom of using fever-suppressing medications remains controversial.

CHAPTER REVIEW

Summary of Key Concepts

An adaptation is a trait that allows individuals to produce more offspring in a particular environment than can individuals without the trait. Adaptations are genetic changes that occur in populations over time and that result from natural selection exerted by the environment. If the phenotype of an individual changes during its lifetime in response to environmental change, the change in the individual is due to acclimatization—not adaptation. In adaptive traits, there is often a strong correlation between a structure and its function.

- **Biologists analyze the structure and function of animals at a variety of levels: molecules, cells, tissues, organs, and systems. These levels form a tightly integrated hierarchy.**

Cells with a similar structure and common function are grouped together into four general types of tissue: epithelial tissue, connective tissue, muscle tissue, and nervous tissue. Epithelium is a particularly important type of tissue because it defines the interface between the animal's external and internal environments. Epithelial cells and tissues have a distinct polarity.

Organs are structures that are composed of several tissues and that perform specific tasks. Organs are organized into systems that work together in an integrated fashion to perform a specific task.

Body size has a strong influence on how animals work. For example, because small animals have a large amount of surface area relative to their volume, they lose heat faster than do large animals. The relationship between surface area and volume also plays a significant role in diffusion-mediated processes, such as gas exchange, nutrient absorption, and waste excretion.

Many of the effects of body size on the way animals work result from the relationship between the surface area of a structure and its volume. For example, large animals have low metabolic rates, because they have a relatively small surface area for exchanging the oxygen and nutrients required to support metabolism. The relatively high surface area of small animals, in contrast, means that they lose heat extremely rapidly. As a result, there are no tiny endothermic animals.

When structures are compared across species, most structures are found to change allometrically—that is, in a nonlinear fashion—with body mass. Among mammals, for example, skeletal mass increases much faster than overall body mass. Basal metabolic rate, in contrast, increases much more slowly than overall body mass.

Web Tutorial 41.1 Surface Area/Volume Relationships

Animals can maintain a relatively constant environment inside their bodies because they have systems that sense changes in internal conditions and trigger responses that return conditions to normal.

Homeostasis is the maintenance of relatively constant physical and chemical conditions inside the body. Animals have a set point, or target value, for blood pH, tissue oxygen concentration, nutrient availability, and other parameters. For example, mammals have a set point for body temperature of about 37°C. If an individual starts to overheat, it will pant or sweat and seek a cool environment in response. If its body temperature begins to drop, it will respond by shivering, basking in sunlight, or fluffing its fur. Animals have adapted to life in "extreme" environments by evolving mechanisms of temperature regulation to cope with wide fluctuations in environmental temperatures.

Web Tutorial 41.2 Homeostasis

Questions

Content Review

1. How do biologists measure an animal's metabolic rate?
   a. by taking its temperature
   b. by measuring how rapidly it uses oxygen
   c. by measuring how rapidly it uses glucose
   d. by measuring how rapidly it produces wastes
2. How is the structure of a connective tissue most closely correlated with its function?
   a. The density of cells in the tissue correlates with the tissue's function.
   b. The type of cells in the tissue correlates with the tissue's function.
   c. The origin of the tissue (from endoderm, mesoderm, or ectoderm) correlates with the tissue's function.
   d. The nature of the extracellular matrix correlates with the tissue's function.
3. As an animal gets larger, which of the following occurs?
   a. Its surface area grows more rapidly than its volume.
   b. Its volume grows more rapidly than its surface area.
   c. Its volume and surface area increase in perfect proportion to each other.
   d. Its volume increases, but its total surface area decreases.
4. Which of the following best describes the set point in a homeostatic system?
   a. the cells that collect and transmit information about the state of the system
   b. the cells that receive information about the state of the system and that direct changes to the system
   c. the various components that produce appropriate changes in the system
   d. the target or "normal" value of the parameter in question
5. What does it mean to say that an animal is a heterothermic endotherm?
   a. Its body temperature can vary, but it produces heat from its own tissues.
   b. Its body temperature varies because it gains most of its heat from sources outside its body.
   c. Its body temperature does not vary, because it produces heat from its own tissues.
   d. Its body temperature does not vary, even though it gains most of its heat from sources outside its body.
6. Which of the following is an advantage that ectotherms have over endotherms of the same size?
   a. They require much less food.
   b. They can save energy in cold weather by hibernating (entering torpor for long periods).
   c. They can remain active in cold weather or at nighttime—when temperatures cool.
   d. They have higher metabolic rates and grow much more quickly.

Conceptual Review

1. Why is epithelium a particularly important tissue in achieving homeostasis?
2. The metabolic rate of a frog in summer (at 35°C) is about eight times higher than that in winter (at 5°C). Compare and contrast the individual's ability to move, exchange gases, and digest food at the two temperatures. During which season will the frog require more food energy, and why?
3. Consider the following cells, tissues, and organs:

   Nerve cells
   Brown adipose tissue
   Lung tissue in humans
   Absorptive sections of digestive tract
   Capillaries
   Beaks of Galápagos finches
   Fish gills

   How does the structure of these cells relate to their function?
4. Researchers who graph overall body mass versus skeletal mass for mammals ranging from mice to elephants find that skeletal mass increases much more rapidly than body mass does. Explain why this relationship occurs and why it is considered allometric.
5. Consider a day in which daytime temperatures reach 30°C and nighttime temperatures drop to 18°C. Analyze how an ant might gain and lose heat by conduction, convection, radiation, and evaporation to avoid overheating during the day and escape cold-induced lethargy in the early morning and evening.
6. Explain how a countercurrent heat exchanger works.

Group Discussion Problems

1. When people move from sea level to a high elevation, their bodies undergo acclimatization. During this period, heart and lung activity and blood chemistry change to compensate for the low availability of oxygen. Design a study to test the hypothesis that human populations that have lived at high elevations for many centuries are not acclimatized to oxygen scarcity but are adapted. Also, test the idea that they are both acclimatized and adapted.
2. The fossil record documents the existence of tortoises that were the size of small cars, dragonflies that had wingspans of two feet, and predatory sharks that were larger than some of today's whales. Using these examples, discuss the advantages and disadvantages of extremely large size in relation to natural selection. Explain why it is not possible for King Kong, the giant ape of movie fame, to exist.
3. An engineer has to design a system for dissipating heat from a new type of car engine that runs particularly hot. Recall that heat is gained and lost as a function of surface area. Suggest ideas for her to consider that are inspired by biological structures with exceptionally high surface area/volume ratios.
4. Suppose that you are the veterinarian at a zoo and have just been put in charge of caring for six individuals of a new mammal species. Almost nothing is known about the physiology of this animal. Describe how you would go about determining set points for body temperature, blood pH, blood glucose, blood pressure, and heart rate in this species. Is it reasonable to assume that the set points for adult males, adult females, and juveniles are identical? Why or why not?

Answers to Multiple-Choice Questions 1. b; 2. d; 3. b, 4. d; 5. a; 6. a

www.prenhall.com/freeman is your resource for the following: Web Tutorials; Online Quizzes and other Online Study Guide materials; Answers to Conceptual Review Questions; Solutions to Group Discussion Problems; Answers to Figure Caption Questions and Exercises; and Additional Readings and Research.

Water and Electrolyte Balance in Animals

42

Terrestrial animals lose water every time they breathe. For many animals, drinking is an important way to gain water and achieve homeostasis. This chapter explores how terrestrial and aquatic animals maintain water balance.

KEY CONCEPTS

- Freshwater, marine, and terrestrial habitats pose different challenges to animals with regard to maintaining water and electrolyte balance.
- In marine animals, specialized epithelial cells have membrane proteins that remove excess salt (NaCl) from the body so that it can be excreted. The same types of cells are found in the kidneys of mammals.
- Terrestrial insects have key adaptations that limit water loss from the body. The hindgut and Malpighian tubules of insects function in maintaining water and electrolyte balance.
- In terrestrial vertebrates, the kidney is responsible for excreting water-soluble waste products and achieving homeostasis with respect to water and electrolyte concentrations.

The chemical reactions that make life possible occur in an aqueous solution. If the balance of water and solutes in the aqueous solution is disturbed, those chemical reactions—and life itself—may stop. Humans can stay alive for weeks without eating but survive at most 8 days without drinking water. Hurricanes occasionally introduce so much freshwater to the ocean shore that normal salt concentrations are disrupted and marine animals die. Maintaining water balance is a matter of life or death.

An animal achieves water balance when its intake of water equals its loss of water. Maintaining water balance is an important element in homeostasis—the ability to keep cells and tissues in constant and favorable conditions. Water balance is also intimately associated with sustaining a balanced concentration of solutes throughout the body. In many animals, the most abundant solutes are ions: sodium (Na^+), chloride (Cl^-), potassium (K^+), and calcium (Ca^{2+}). As a group, these substances are called electrolytes. An **electrolyte** is a compound that dissociates into ions when dissolved in water. Because cells require precise concentrations of Na^+, Cl^-, K^+, and Ca^{2+} to function normally, maintaining electrolyte balance is crucial to survival. In humans, severe electrolyte imbalances can lead to muscle spasms, confusion, irregular heart rhythms, fatigue, paralysis, or even death.

How do animals maintain water and electrolyte balance? Animals gain water in four ways: (1) by absorbing it via osmosis, (2) by drinking, (3) by eating, and (4) as a by-product of cellular respiration. They lose water via urine and feces, evaporation, or osmotic loss. To maintain water balance, gains and losses must be equal. Ion budgets have to be balanced in a similar fashion.

Our investigation into water and electrolyte balance begins with a look at how water moves into and out of cells. We'll consider why saltwater, freshwater, and terrestrial environments affect animals differently, and then we'll delve into research on the cellular and molecular mechanisms responsible for maintaining water and electrolyte balance in each of these environments. Although aquatic and terrestrial animals face very different challenges, in several cases these challenges are met by common molecular mechanisms.

42.1 Osmoregulation and Osmotic Stress

Chapter 6 introduced diffusion and osmosis. **Diffusion** is the movement of substances from regions of higher concentration to regions of lower concentration. Compounds diffuse along their **concentration gradient.** **Figure 42.1a** illustrates an important example of diffusion. In this case, a dissolved substance, or **solute**, moves down its concentration gradient via diffusion across a selectively permeable membrane. A **selectively permeable membrane**, such as a phospholipid bilayer, is a membrane that some solutes can cross more easily than other solutes can. When the solutes are randomly distributed throughout the solutions on both sides of the membrane, an **equilibrium** is established. At equilibrium, molecules continue to move back and forth across the membrane, but at equal rates.

Water can also move down its concentration gradient. To understand how this happens, notice that the concentration of dissolved substances in a solution, measured in moles per liter, is referred to as the solution's **osmolarity.** When dissolved substances are separated by a selectively permeable membrane and the solutes cannot cross that membrane, water moves from areas where it is more concentrated—that is, solute concentrations are lower—to areas where it is less concentrated—that is, solute concentrations are higher. Stated another way, water moves across a selectively permeable membrane from regions of lower osmolarity to regions of higher osmolarity. The movement of water from areas of higher water concentration to areas of lower water concentration is called **osmosis** (**Figure 42.1b**).

Because diffusion and osmosis affect animals differently in freshwater, marine, and terrestrial habitats, these environments pose different challenges to animals in maintaining water and electrolyte balance. For example, achieving homeostasis with respect to water and electrolyte balance is straightforward in marine invertebrates such as sponges, jellyfish, and flatworms. Seawater is a fairly constant ionic and osmotic environment and nearly matches the electrolyte concentrations found within these animals. Relative to seawater, their tissues are **isotonic** (a term introduced in Chapter 6): Solute concentrations inside and outside these animals

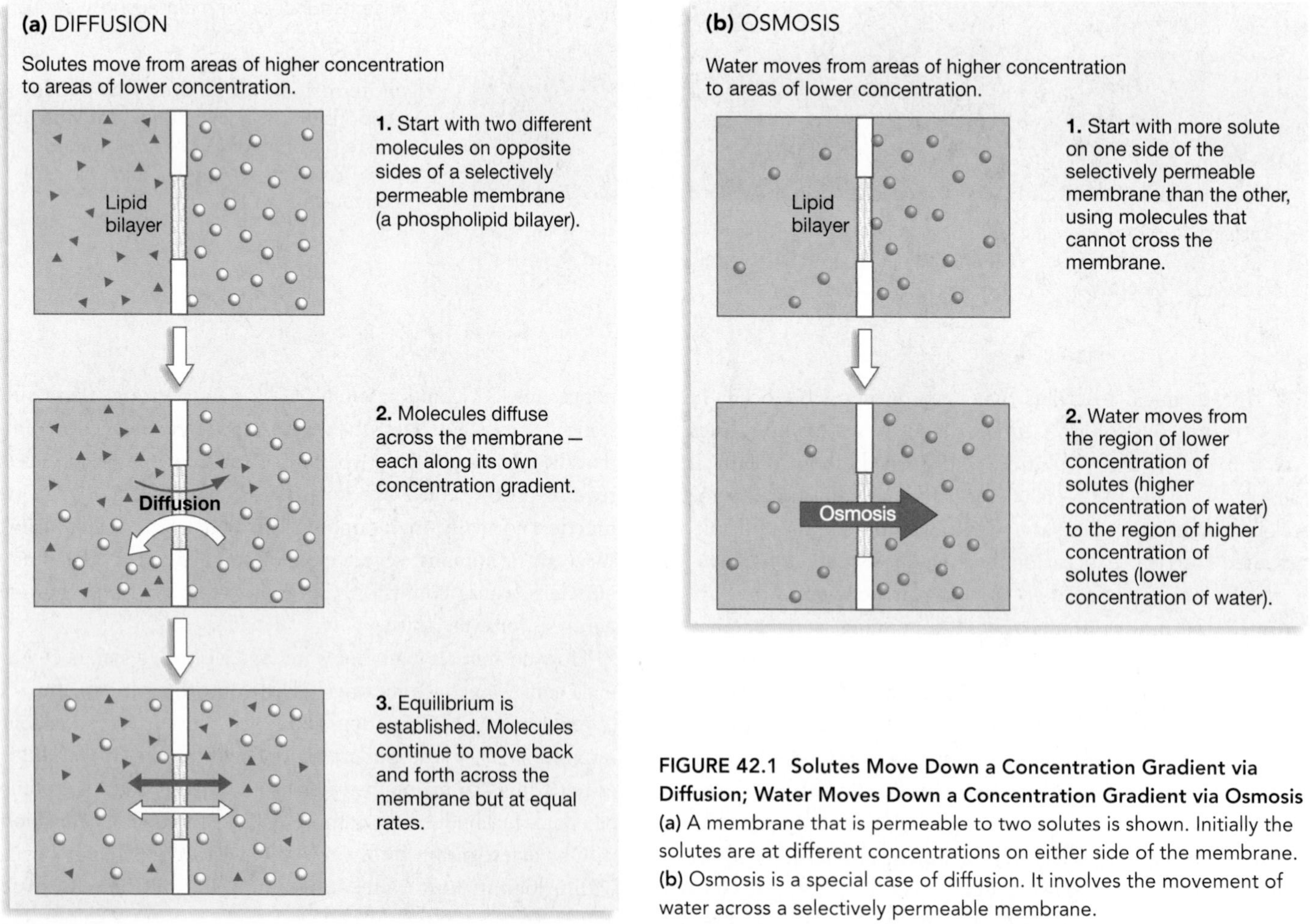
(a) DIFFUSION
Solutes move from areas of higher concentration to areas of lower concentration.
Lipid bilayer
1. Start with two different molecules on opposite sides of a selectively permeable membrane (a phospholipid bilayer).
Diffusion
2. Molecules diffuse across the membrane – each along its own concentration gradient.
3. Equilibrium is established. Molecules continue to move back and forth across the membrane but at equal rates.
(b) OSMOSIS
Water moves from areas of higher concentration to areas of lower concentration.
Lipid bilayer
1. Start with more solute on one side of the selectively permeable membrane than the other, using molecules that cannot cross the membrane.
Osmosis
2. Water moves from the region of lower concentration of solutes (higher concentration of water) to the region of higher concentration of solutes (lower concentration of water).

FIGURE 42.1 Solutes Move Down a Concentration Gradient via Diffusion; Water Moves Down a Concentration Gradient via Osmosis
(a) A membrane that is permeable to two solutes is shown. Initially the solutes are at different concentrations on either side of the membrane. (b) Osmosis is a special case of diffusion. It involves the movement of water across a selectively permeable membrane.

are equal. As a result, diffusion and osmosis are relatively unimportant. Marine invertebrates are **osmoconformers**—they do not actively regulate the osmolarity of their tissues, although they do regulate the composition of solutes inside their cells.

Marine fish, in contrast, are **osmoregulators**—meaning they actively regulate osmolarity inside their bodies to achieve homeostasis. **Osmoregulation** is the process by which living organisms control the concentration of water and salt in their bodies. Osmoregulation is required in marine vertebrates, because their tissues are **hypotonic** relative to salt water—the solution inside the body contains fewer solutes than does the solution outside). The difference in osmolarity is most important in the gills, which are organs involved in gas exchange. The epithelial cells on the surfaces of the gills are in direct contact with seawater. The exposed plasma membranes are selectively permeable, and there is a large difference in the concentration of solutes between the inside of each cell and the seawater outside. As a result, water tends to flow out of the gill epithelium (**Figure 42.2a**). If this water is not replaced, the fish's cells will shrivel and die. Marine fish replace the lost water by drinking large quantities of seawater. Drinking brings in excess electrolytes, however. Ions and other solutes also diffuse into the fish across the gills, following a concentration gradient from seawater to tissues. To rid themselves of excess electrolytes, marine fish excrete urine that has a much higher concentration of electrolytes than seawater does.

Freshwater animals osmoregulate in a dramatically different environment. Marine fish gain salt and lose water, but freshwater animals lose salt and gain water. The gills of freshwater fish, for example, are exposed to a concentration gradient 100 to 1000 times greater than that faced by marine fish. Because the gill epithelium of freshwater fish is **hypertonic** relative to the surrounding water—meaning the solution inside the cells contains more solutes than does the solution outside—the epithelial cells gain water via osmosis (**Figure 42.2b**). This water then moves from the epithelium into adjacent tissues, putting them under osmotic stress. **Osmotic stress** occurs when the concentration of dissolved substances in a cell or tissue is abnormal. If a freshwater fish does not get rid of incoming water, its cells will eventually burst and the individual will die. To achieve homeostasis and survive, freshwater fish excrete large amounts of water in their urine and do not drink.

In addition to gaining water, freshwater fish undergo osmotic stress because they lose electrolytes to the environment. Fresh water is hypotonic to the gill epithelium, so ions and other solutes tend to diffuse out of gill cells. Freshwater animals must replace electrolytes that are lost by obtaining them in food or by actively transporting them from the surrounding water—usually across the gills.

What about land animals? Regarding water balance, terrestrial environments are similar to the ocean. Land animals constantly lose water to the environment (**Figure 42.3**), just as many marine animals do. In this case, however, the process involved is not osmosis but evaporation (see Chapter 41). Evaporation threatens terrestrial animals, because they must exchange

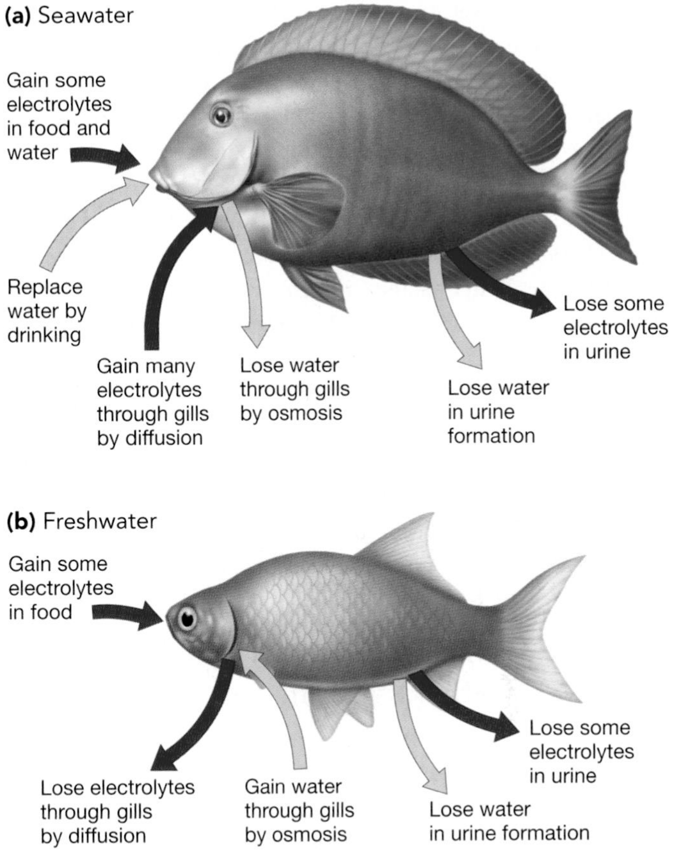

FIGURE 42.2 Saltwater and Freshwater Fish Face Different Challenges in Maintaining Water and Electrolyte Balance
(a) In seawater, fish lose water by osmosis and gain electrolytes by diffusion. **(b)** In freshwater, fish gain water by osmosis and lose electrolytes by diffusion. EXERCISE Add labels indicating whether each environment is hypotonic, hypertonic, or isotonic to the organism illustrated.

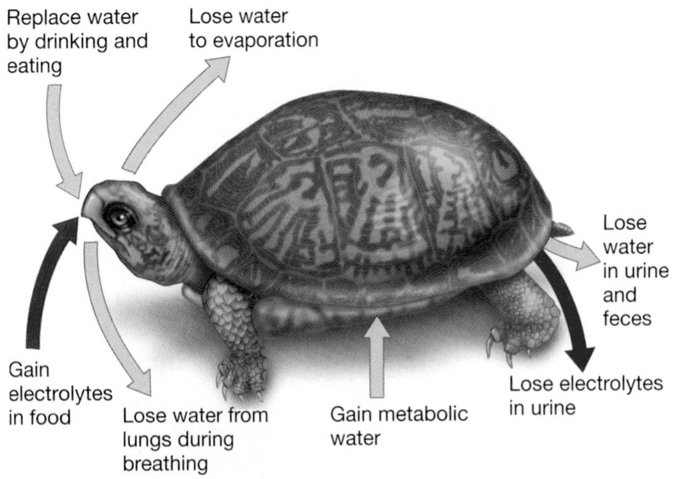

FIGURE 42.3 Maintaining Water and Electrolyte Balance in Terrestrial Environments Is a Challenge
In terrestrial environments, animals lose water by evaporation from the body surface and as water vapor during breathing. Electrolytes are lost primarily in the urine and feces.

gases across a wet surface. The epithelial cells that line a human's lung and a fruit fly's gas exchange surfaces are moistened to facilitate the exchange of oxygen and carbon dioxide through diffusion. (Chapter 44 explores in detail how animals exchange gases.) In this respect, water balance in land animals mirrors the situation in land plants. As Chapter 36 points out, land plants lose huge quantities of water in the course of obtaining CO_2 through their stomata—as an inevitable by-product of gas exchange. Water balance is further complicated in terrestrial animals because some species lose water when they sweat or pant to lower their body temperature and because all species lose water in the form of urine. The lost water has to be replaced by drinking, water contained in food, or metabolic water.

What molecular mechanisms allow animals to cope with these diverse challenges? Let's start answering this question by focusing on how fish from marine or freshwater environments either excrete or take up electrolytes through their gills.

42.2 Water and Electrolyte Balance in Aquatic Environments

Whether they live in seawater or freshwater, virtually all of the 26,000 species of fish living today experience osmotic stress. Early research on how these species maintain water and electrolyte balance focused on sharks as a model organism.

In contrast to most marine fish, sharks and other cartilaginous fishes have tissues that are isotonic to seawater. The situation is complex, though, because the concentrations of sodium ions, potassium ions, and chloride ions in their cells and extracellular fluids are similar to the concentrations in other vertebrates and much less than the concentrations of these ions in seawater. To make up the difference in solute concentrations, shark blood contains large quantities of soluble compounds, such as urea and trimethylamine oxide (TMAO). In combination, the concentration of TMAO, urea, and ions is high enough to match the osmotic concentration of seawater and prevent the loss of water by osmosis. But as **Figure 42.4** shows, sharks still need to secrete salt (NaCl), because sodium and chloride ions diffuse into their gill cells from seawater down their concentration gradients. How do sharks excrete salt?

Research on salt excretion in sharks has been very important. After having worked out the molecular mechanisms of salt transport in sharks, researchers found that the same processes occur in a wide range of marine animals, including oceangoing birds and reptiles. Understanding salt excretion in sharks also had an unforeseen benefit for human biomedical research.

How Do Sharks Osmoregulate?

In studying shark anatomy, biologists discovered an organ called the **rectal gland**, which secretes a concentrated salt solution. To determine how this gland works, researchers studied it in vitro—meaning outside the shark's body. Specifically, they dissected rectal glands and immersed them in a solution with a composition and osmolarity similar to those of normal shark tissue. Then the biologists connected a tube to the blood vessel that carries blood into the rectal gland. They also attached a collecting tube to the end of the gland that empties into the environment. This experimental setup allowed them to introduce solutions of defined composition and evaluate the fluid that the rectal gland produced in response.

To establish that the rectal gland functioned normally in this preparation, the biologists introduced a solution containing Na^+, K^+, and Cl^- in concentrations that might be observed in an intact animal: 280 mM (millimolar), 5 mM, and 270 mM, respectively. The fluid that left the gland contained the same ions in concentrations of 449 mM, 12 mM, and 446 mM instead. Because the product solution was more concentrated than the initial solution, the result gave the researchers confidence that the organ was functioning normally in vitro. And because they got such results only when ATP molecules were added to the preparation, the data supported the hypothesis that salt excretion is an energy-demanding activity. Ions can be concentrated in this way only if they are actively transported against a concentration gradient. (To review how molecules are transported across membranes, see **Box 42.1**, page 960). The question was, How does concentration occur?

FIGURE 42.4 Sharks Face an Unusual Challenge in Maintaining Water and Electrolyte Balance
Water, sodium ions, and chloride ions diffuse into the body of sharks across the gills and the skin along a concentration gradient. EXERCISE Add a label indicating whether seawater is hypotonic, hypertonic, or isotonic to shark tissues.

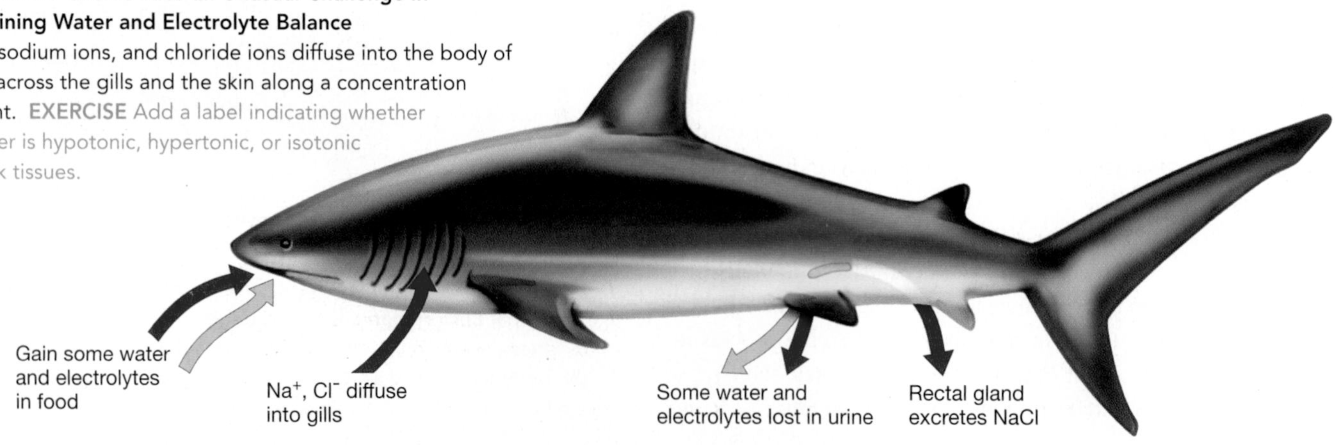
Gain some water and electrolytes in food
Na^+, Cl^- diffuse into gills
Some water and electrolytes lost in urine
Rectal gland excretes NaCl

The Role of Na^+/K^+-ATPase An energy-demanding mechanism for salt excretion implies that a protein in the plasma membrane of epithelial cells is involved in pumping Na^+, Cl^-, or both ions out of the extracellular fluid toward the lumen of the gland, where they are excreted. (Recall that the cavity inside any tube or sac-shaped organ is called the *lumen*.) The best-characterized membrane protein pump involving these ions is the **sodium-potassium pump**, also known as Na^+/K^+-ATPase.

To test the hypothesis that the sodium-potassium pump is involved in salt excretion by sharks, biologists attached a radioactive atom to a plant defense compound called **ouabain** (pronounced *WAA-bane*), which poisons the sodium-potassium pump. Ouabain is toxic to animals because it binds to Na^+/K^+-ATPase and prevents it from functioning. The biologists treated cells from the shark rectal gland with the radioactively labeled poison and then examined the distribution of the radioactive particles with an electron microscope. These studies showed that the ouabain bound to Na^+/K^+-ATPase localized along the basolateral (bottom-side) membrane of the cells, which face the inside of the body and thus the blood supply. No pumps appeared in the apical (top) membrane of the cells, which face the gland's lumen and the secreted solution.

The pumps' location was paradoxical, because Na^+/K^+-ATPase pumps Na^+ out of the cell and K^+ into it. (Three Na^+ go out for every two K^+ that come in.) The paradox arose because the rectal gland cells pump Na^+ on the opposite side of the cell and in the direction opposite to the direction in which those ions are secreted. The researchers also confirmed that Na^+/K^+-ATPase is essential for salt excretion by treating the gland with ouabain in vivo. Rectal glands treated with this drug stopped producing a concentrated salt solution. What was happening?

A Molecular Model for Salt Excretion The paradox of Na^+ being pumped in the direction opposite of excretion was solved by the model shown in **Figure 42.5**. The four-step model hinges on the coordinated action of four different membrane proteins:

1. Na^+/K^+-ATPase pumps sodium ions out of epithelial cells across the basolateral surface and into the surrounding extracellular fluid. The pump creates electrical and concentration (electrochemical) gradients favoring the diffusion of Na^+ into the cell and K^+ out of the cell. These gradients are used to transport other types of ions and molecules.

2. A $Na^+/Cl^-/K^+$ cotransporter, powered by the electrochemical gradient favoring Na^+ diffusion into the cell, brings these three ions from the extracellular fluid into epithelial cells across their basolateral membranes. **Cotransporters** are membrane proteins that transport more than one type of ion or molecule at a time. The cotransporter in the shark rectal gland brings Na^+, $2Cl^-$, and K^+ into the cell together.

3. Although Na^+/K^+-ATPase pumps sodium ions back out, K^+ and Cl^- concentrations build up inside the cell as a result of the cotransport process. A potassium channel located on the basolateral membrane allows K^+ to diffuse back across the basolateral surface after it has been pumped in. A chloride channel located in the apical membrane of the epithelial cells allows Cl^- to diffuse down its concentration gradient into the lumen of the gland.

4. Sodium ions also diffuse into the lumen of the gland, following their charge and concentration gradient. But instead of passing through the epithelial cells as Cl^- does, Na^+ diffuses through spaces between the cells.

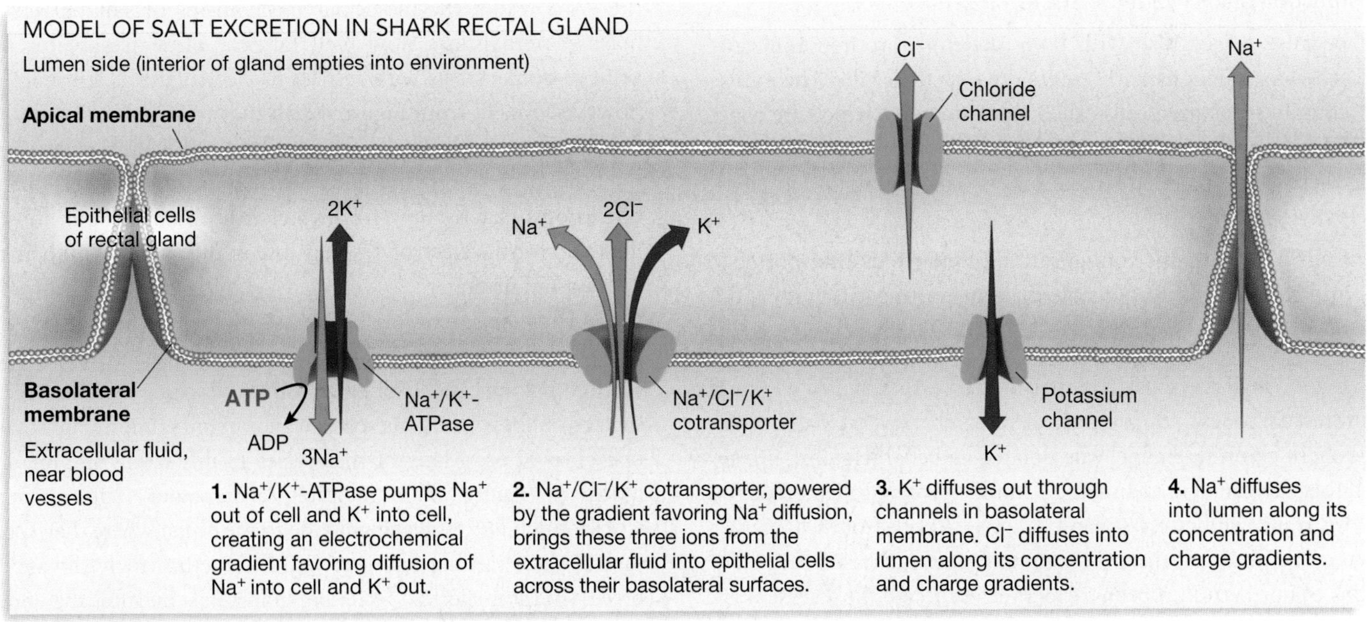
MODEL OF SALT EXCRETION IN SHARK RECTAL GLAND
Lumen side (interior of gland empties into environment)
Apical membrane
Epithelial cells of rectal gland
Basolateral membrane
Extracellular fluid, near blood vessels
2K+
ATP
ADP
3Na+
Na+/K+-ATPase
Na+
2Cl−
K+
Na+/Cl−/K+ cotransporter
Cl−
Chloride channel
Potassium channel
K+
Na+
1. Na+/K+-ATPase pumps Na+ out of cell and K+ into cell, creating an electrochemical gradient favoring diffusion of Na+ into cell and K+ out.
2. Na+/Cl−/K+ cotransporter, powered by the gradient favoring Na+ diffusion, brings these three ions from the extracellular fluid into epithelial cells across their basolateral surfaces.
3. K+ diffuses out through channels in basolateral membrane. Cl− diffuses into lumen along its concentration and charge gradients.
4. Na+ diffuses into lumen along its concentration and charge gradients.

FIGURE 42.5 The Shark Rectal Gland Rids the Body of Excess Salt
The current model to explain how Na^+ and Cl^- move from the shark's body fluids to the lumen of the rectal gland prior to excretion. **EXERCISE** Label membrane proteins that are involved in active transport and that are involved in passive transport.

BOX 42.1 Passive and Active Transport: A Quick Review

Solutes move across membranes by passive or active transport. You might recall from earlier chapters that passive transport is driven by diffusion along an electrochemical gradient and does not require an expenditure of energy in the form of ATP. Active transport, in contrast, occurs when ATP powers the movement of a solute against its electrochemical gradient.

Because ions and large molecules such as glucose do not cross phospholipid bilayers readily, both passive and active transport take place via membrane proteins. In many cases, passive transport occurs through **channels**—proteins that form a pore, or an opening that selectively admits a specific ion or ions. Passive transport also occurs via **carriers**, which are transmembrane proteins that bind a specific ion or molecule and transport it across the membrane by undergoing a conformational change. When solutes move via carriers, **facilitated diffusion** is said to occur.

Active transport is based on membrane proteins called pumps, which change conformation when they bind ATP or are phosphorylated. This energy-demanding change in shape allows them to transport ions or molecules against their concentration gradient. Chapter 6 introduced the sodium-potassium pump, or Na^+/K^+-ATPase, which is the most important type of pump in animals; Chapter 36 introduced the proton pumps, or H^+-ATPases, which are the most important pumps in plants. Once an electrochemical gradient has been established by a pump, a **cotransporter** (also called a symporter) or **antiporter** can use the energy released when an ion is transported *along* that electrochemical gradient to transport a different solute *against* its electrochemical gradient.

A Common Molecular Mechanism Underlies Many Instances of Salt Excretion Once the mechanism of salt excretion illustrated in Figure 42.5 had been worked out, subsequent work showed that the same combination of membrane proteins is found in epithelial cells that transport sodium and chloride ions in many animals:

- Marine birds and reptiles drink salt water and must excrete NaCl. They have salt-excreting glands in their nostrils that function much like the shark rectal gland.
- Because marine fish with bony skeletons are hypotonic to seawater, salt constantly enters through their gills. Their gills contain specialized cells called **chloride cells**, which are configured much like the cells lining the shark rectal gland. These chloride cells excrete excess salt and maintain electrolyte balance.
- Cells with the same configuration of pumps, cotransporters, and channels are responsible for transporting salt in the kidneys of mammals.

Research on the shark rectal gland also turned out to have an unforeseen benefit for biomedical research. Several years after the shark chloride channel was isolated and characterized, investigators succeeded in identifying a human protein called cystic fibrosis transmembrane regulator (CFTR). Recall from Chapter 6 that cystic fibrosis is the most common genetic disease in populations of northern European extraction. Although the disease was known to be associated with defects in the CFTR protein, no one knew what the molecule did. When investigators realized that the amino acid sequence of CFTR is 80 percent identical to that of the shark chloride channel, it was their first hint that CFTR is involved in Cl^- transport. Subsequent studies supported the hypothesis that cystic fibrosis results from a defect in a chloride channel. In this way, studies on water and electrolyte balance in sharks shed light on an important human disease.

How Do Salmon Osmoregulate?

Research on the shark rectal gland and the salt-secreting cells found in the gills of marine fish has been spectacularly successful. As a result, the molecular mechanisms of salt balance in marine animals are now well known. How freshwater fish achieve homeostasis with respect to electrolytes is still unclear, however. Recall from Figure 42.2b that freshwater fish do not drink and that they secrete large volumes of watery urine to rid themselves of water that enters their gills via osmosis. But no one knows exactly how freshwater fish obtain enough ions to make up for the electrolytes they lose to diffusion from their gill cells and in urine.

Recently, an interesting hypothesis to explain how freshwater fish maintain electrolyte balance has emerged from research on salmon. Salmon are a family of fish that include familiar species such as rainbow trout; the species called chum salmon is illustrated in **Figure 42.6a**. Many of the species in this family are anadromous. As **Figure 42.6b** indicates, **anadromous** ("up-running") species hatch in freshwater, migrate to the ocean where they spend several years feeding and growing, then return to freshwater to breed. (In most species of salmon, individuals die after they breed just once.) The question is, How do individuals make the transition from the hypotonic environment where they hatch to the hypertonic environment of the ocean and back again?

(a) Chum salmon

(b) Migration to the ocean and back

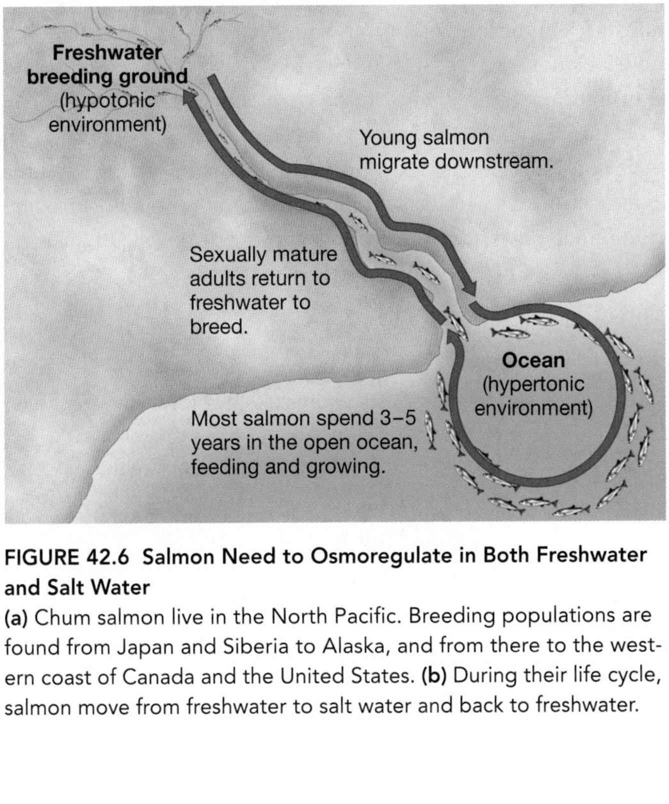

FIGURE 42.6 Salmon Need to Osmoregulate in Both Freshwater and Salt Water
(a) Chum salmon live in the North Pacific. Breeding populations are found from Japan and Siberia to Alaska, and from there to the western coast of Canada and the United States. **(b)** During their life cycle, salmon move from freshwater to salt water and back to freshwater.

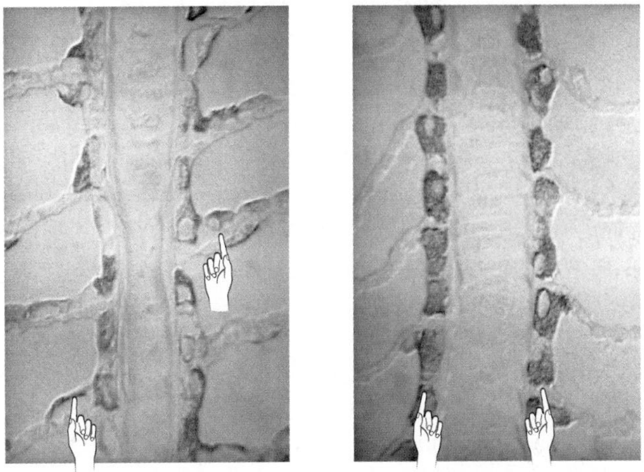

FIGURE 42.7 In Salmon, Different Gill Cells Are Involved in Osmoregulation in Freshwater versus Salt Water
Gills from young salmon that were maintained in **(a)** freshwater versus **(b)** salt water. The gills have been treated with a molecule that binds to Na^+/K^+-ATPase and leaves a dark stain, labeling the chloride cells.

To answer this question, biologists established, that as young salmon prepare to migrate to salt water, there is a significant increase in Na^+/K^+-ATPase activity in their gills, as well as a dramatic increase in the number of chloride cells in the gills. This observation is logical, because these cells are responsible for excreting salt when the fish are living in the ocean. But researchers have also found that the gills of young salmon maintained in freshwater versus salt water have two distinct populations of cells that have Na^+/K^+-ATPase. As **Figure 42.7** shows, individuals that live in freshwater and that have to import electrolytes have cells with Na^+/K^+-ATPase in the sheetlike lamellae that extend from the base of structures called *gill filaments*. In contrast, individuals that live in salt water and excrete electrolytes have a large number of sodium-potassium pumps in cells at the base of their gill filaments. Adult salmon from freshwater versus salt water show the same pattern, and similar changes have been observed in other fish species that switch between freshwater and saltwater habitats.

These observations inspired the hypothesis that salmon have two distinct types of cells involved in maintaining electrolyte balance. Researchers have proposed that chloride cells in the gill lamellae import electrolytes, while chloride cells at the base of gills secrete salt. Both cells contain sodium-potassium pumps, but they operate in opposite directions. This chloride-cell-switch hypothesis implies that salmon sense changes in the osmolarity of their environment and respond by producing or destroying the appropriate populations of chloride cells. The molecular mechanisms involved in the sensory reception and switching response are a mystery, however, and it is still unknown exactly how cells located in the gill lamellae take up electrolytes from freshwater. Identifying the mechanisms of electrolyte uptake in freshwater fish is an important challenge for researchers who want to know how aquatic organisms maintain water and electrolyte balance.

✓CHECK YOUR UNDERSTANDING

Marine fish have to rid themselves of salt, either because they drink salt water to replace the water they lose by osmosis or because sodium and chloride ions diffuse into their cells along a concentration gradient. Freshwater fish, in contrast, have to obtain electrolytes to replace those lost to the surrounding water by diffusion and must rid themselves of excess water. You should be able to (1) explain why Na^+/K^+-ATPase is important in the transport of salt in both saltwater and freshwater fish and (2) predict what happens when cells in the gills of fish or the rectal glands of sharks are treated with ouabain.

42.3 Water and Electrolyte Balance in Terrestrial Insects

To understand how terrestrial invertebrates maintain water and electrolyte balance, researchers have focused on species that inhabit particularly dry environments. The success of this work has validated a general research strategy in biological science: By studying extreme situations or unusual organisms, biologists can often gain insight into how organisms cope with more moderate environments.

In studies on the molecular mechanisms of water and electrolyte balance in terrestrial insects, the most valuable model organisms have been the desert locust (**Figure 42.8**) and the common household pest called the flour beetle. (You may be familiar with the larvae of flour beetles, called mealworms.) Desert locusts and flour beetles rarely, if ever, drink—simply because little or no water is available in the habitats they occupy.

These insects live in environments where osmotic stress is severe. How do they maintain water and electrolyte balance? The answer has two parts: They minimize water loss from their body surface, and they carefully regulate the amount of water and electrolytes that they excrete in their urine and feces. Let's look at each issue in turn.

How Do Insects Minimize Water Loss from the Body Surface?

As Chapter 44 will show, terrestrial animals breathe by exposing an extremely thin layer of epithelium to the atmosphere. Oxygen diffuses into this epithelium, and carbon dioxide diffuses out. But water constantly leaks across the thin respiratory surface and is lost to the atmosphere via evaporation. Evaporation from the body surface itself is another threat—a particular challenge to insects, because they are small. As emphasized in Chapter 41, small organisms have a high surface area/volume ratio. Insects have a relatively large surface area with which to lose water but a small volume in which to retain it.

Desert
locusts rarely,
if ever, drink

FIGURE 42.8 The Desert Locust Has Been a Model Organism for Studying How Insects Limit Water Loss

How do desert locusts, flour beetles, and other insects minimize water loss during gas exchange? In these species, gas exchange occurs across the membranes of epithelial cells that line an extensive system of tubes called **tracheae** (pronounced *TRAY-kee-ee*). The insect tracheal system connects with the atmosphere at openings called **spiracles** (**Figure 42.9a**). Muscles just inside each spiracle open or close the pore, much as guard cells open or close the pores in plant leaves and stems. When investigators manipulated bugs called *Rhodnia* so that their spiracles stayed open and then placed the insects in a dry environment, the bugs died within three days. These data support the hypothesis that the ability to close spiracles is an important adaptation for minimizing water loss. If an insect is under osmotic stress, it may be able to close its spiracles and wait until conditions improve before resuming activity.

Figure 42.9b shows how insects minimize evaporation from their body surfaces. This diagram is a cross-sectional view of the exoskeleton of an insect, which consists of a tough, nitrogen-containing polysaccharide called chitin and layers of protein.

(a) Spiracles can be closed to minimize water loss from tracheae.

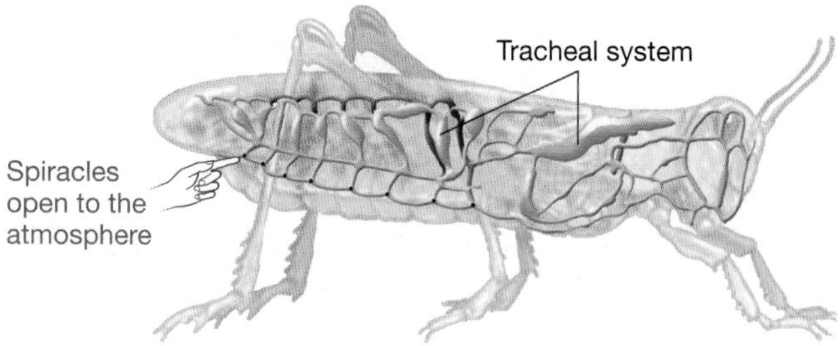
Tracheal system
Spiracles
open to the
atmosphere

(b) The insect body is covered with wax.

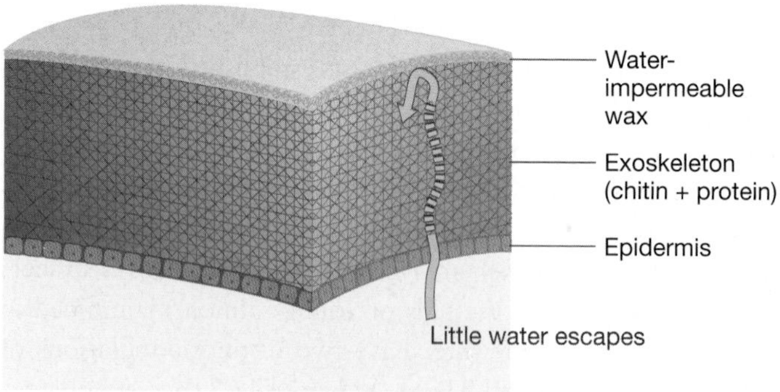
Water-
impermeable
wax
Exoskeleton
(chitin + protein)
Epidermis
Little water escapes

FIGURE 42.9 In Desert Locusts, Adaptations Limit Water Loss during Respiration and from the Body Surface
(a) Transparent view through the insect exoskeleton, showing the extent of the tracheal system. Spiracles are openings in the exoskeleton that lead to the tubelike tracheae, where gas exchange takes place. **(b)** Except at spiracles, the entire insect exoskeleton is covered with wax. **QUESTION** In desert grasshoppers, spiracles are located in a row along the underside of the abdomen. Why would this location help minimize water loss?

This combination of chitin and protein is known as **cuticle.** As the figure shows, the surface of the exoskeleton is covered with a layer of wax. Recall from Chapter 6 that waxes are a type of lipid. They are highly hydrophobic and thus highly impermeable to water. Researchers who removed the wax from insect exoskeletons have confirmed that the rate of water loss from the body surface increases sharply. Based on this observation, the wax layer is interpreted as an adaptation that minimizes evaporative water loss.

Types of Nitrogenous Wastes: Impact on Water Balance

Animal cells contain amino acids and nucleic acids that are used to synthesize proteins, RNA, and DNA. Both amino acids and nucleic acids are nitrogenous (nitrogen-containing) monomers. But if the monomers are present in excess of a cell's needs, they are broken down in catabolic reactions that result in the production of **ammonia** (NH_3). Ammonia is a strong base; it readily gains a proton to form an **ammonium ion** (NH_4^+). Ammonia is toxic to cells, because at high concentrations it raises the pH of intracellular and extracellular fluids enough to poison enzymes.

How do animals get rid of this highly toxic molecule safely and efficiently? Different species solve the problem in different ways. In freshwater fish, ammonia is diluted to low concentration and excreted in a watery urine. In addition, both freshwater and saltwater fish rid themselves of ammonia as it diffuses across their gills into the surrounding water along a concentration gradient. In humans, enzyme-catalyzed reactions convert ammonia to a much less toxic compound called **urea**, which is excreted in urine. In birds, reptiles, and terrestrial arthropods, reactions convert ammonia to **uric acid.** Uric acid is the white, paste-like substance that you have probably seen in bird feces. Compared with urea and ammonia, uric acid is extremely insoluble in water. Its low water solubility explains why washing bird droppings off a car is so difficult, and it provides a mechanism for birds, reptiles, and terrestrial arthropods to rid themselves of ammonia while losing a minimum amount of water. Many birds and some insects, in fact, do not produce any urine at all.

Why do different types of animals have different ways of detoxifying ammonia and excreting nitrogenous wastes? As **Table 42.1** shows, the type of nitrogenous waste produced by an animal correlates with the amount of osmotic stress it endures. For example, most small animals and animals that live in dry habitats minimize the amount of water they must excrete by synthesizing uric acid as their principal waste product. The uric acid produced by locusts, flour beetles, and other insects is interpreted as an important adaptation to minimize water loss.

Maintaining Homeostasis: The Excretory System

For insects, minimizing water loss is only half the battle in avoiding osmotic stress. To maintain homeostasis, insects must also carefully regulate the composition of a blood-like fluid called **hemolymph.** Unlike the situation in vertebrates, hemolymph does not flow through blood vessels in a completely closed circulatory system. Instead, the heart pumps hemolymph through vessels and into the body cavity (see Chapter 44). In this way, hemolymph bathes tissues directly. Nutrients pass from the hemolymph into cells; waste products such as ammonia diffuse out of cells and into the

TABLE 42.1 Attributes of Nitrogenous Wastes Produced by Animals

Attribute	Ammonia	Urea	Uric Acid
Solubility in water	high	medium	very low
Water loss (amount required for excretion of waste)	high	medium	very low
Energy cost (amount of ATP required)	low	high	high
Toxicity	high	low	medium
Groups where it is the primary waste	fish aquatic invertebrates	mammals* sharks	birds† reptiles most terrestrial insects and spiders
Method of synthesis	breakdown of amino acids and nucleic acids	synthesized in liver, starting with amino groups from amino acids	synthesis starts with amino acids and nucleic acids
Method of excretion	in urine and diffuses across gills	in urine (mammals); diffuses across gills (sharks)	in feces (in birds, uric acid is derived from the urine but excreted with the feces)

*Mammals also excrete a small amount of uric acid, synthesized from excess nucleic acids.
†Birds also excrete a small amount of ammonia.

surrounding fluid. Hemolymph also contains a wide variety of electrolytes.

How do insects regulate the composition of the hemolymph? This is an important question, because the nitrogenous wastes present in hemolymph must be removed before they build up to toxic concentrations. Similarly, if excess electrolytes enter the hemolymph after a meal, the ions must be excreted before they lead to osmotic stress. Water balance also must be regulated constantly.

To maintain water and electrolyte balance, insects rely on **Malpighian tubules**, which are an excretory organ, and on their **hindgut**—the posterior portion of their digestive tract (**Figure 42.10a**). Let's consider how each portion of the insect excretory system works in turn.

Pre-Urine Formation in the Malpighian Tubules As the enlarged section of Figure 42.10a shows, Malpighian tubules have a large surface area, are in direct contact with the hemolymph, and empty into the hindgut. Research has shown that the Malpighian tubules are responsible for forming a "pre-urine." This initial urine then passes into the hindgut, where it is processed and modified prior to excretion.

To explore how the Malpighian tubules work, biologists collected fluid from the lumen of Malpighian tubules in mealworms and compared the composition of the pre-urine with that of hemolymph from the same individuals. The analysis determined that the two solutions were roughly isotonic. They were not identical, however. Instead, the concentration of sodium ions inside the tubules was low relative to that of the hemolymph, while K^+ concentrations were extraordinarily high inside the tubules compared with concentrations in the hemolymph. This observation suggested that the epithelial cells in the Malpighian tubules are relatively impermeable to sodium ions but contain a pump that actively transports potassium ions into the organ. To test this hypothesis, researchers dissected the tubules, rinsed them, and bathed their interior and exterior in a solution containing a known concentration of potassium ions. When the biologists measured the concentration of electrolytes on either side of the membrane over time, they found that K^+ accumulated on the inner face of the tubules against its concentration gradient.

This result supported the hypothesis that cells in the membranes of Malpighian tubules contain a pump that transports potassium ions into the lumen of the organ. Subsequent work has shown that a high concentration of potassium ions brings water into the tubules by osmosis. Other electrolytes and nitrogenous wastes then diffuse into the tubules along their concentration gradients.

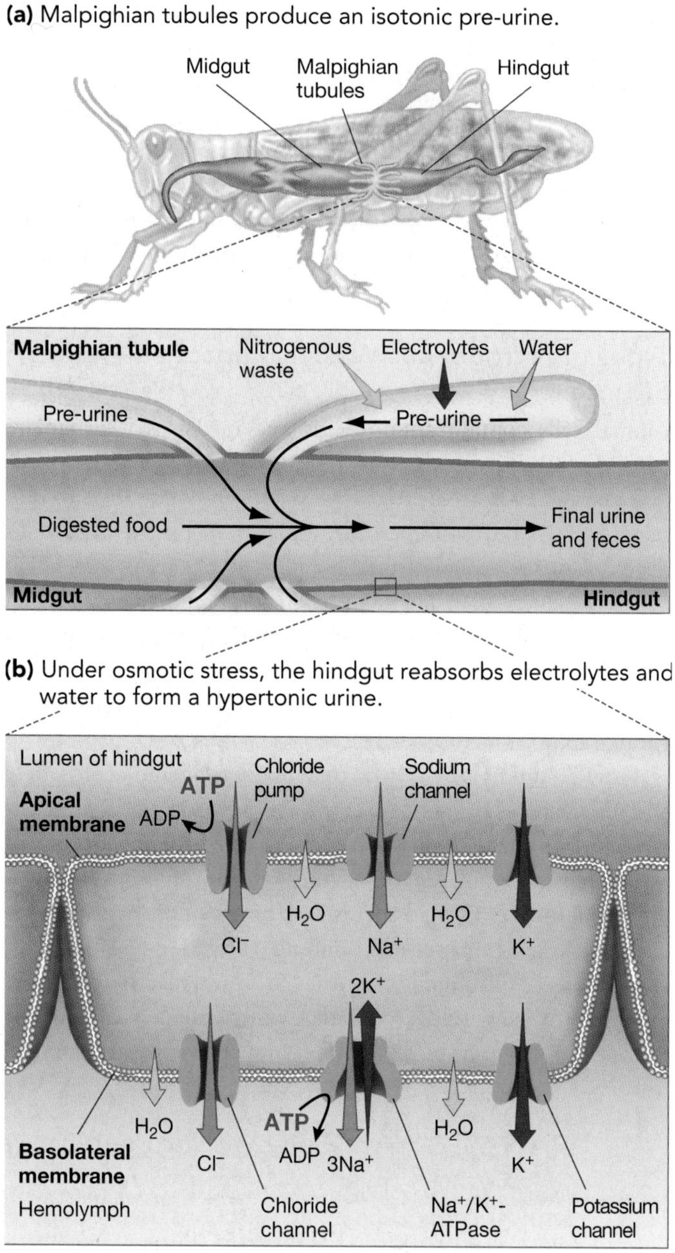

FIGURE 42.10 In Insects, Urine Formation Takes Place in the Malpighian Tubules and Hindgut
(a) The isotonic pre-urine that forms in the Malpighian tubules empties into the hindgut. **(b)** In the hindgut, the primary driving forces for reabsorption of electrolytes and water are chloride pumps in the apical membrane and Na^+/K^+-ATPases in the basolateral membrane. EXERCISE In part (b), add a protein to the apical membrane that cotransports Na^+ and amino acids from the pre-urine. Why would it be advantageous for hindgut cells to recover amino acids from pre-urine?

The Hindgut: Selective Reabsorption of Electrolytes and Water The pre-urine that accumulates inside the Malpighian tubules flows into the hindgut, where it joins material emerging from the digestive tract. If an insect is osmotically stressed due to a shortage of electrolytes and water, electrolytes and water from the pre-urine are reabsorbed in the hindgut and returned to the hemolymph. Reabsorption results in the formation of a hypertonic final urine, the conservation of water, and the efficient elimination of nitrogenous wastes.

In desert locusts, flour beetles, and other species that live in extremely dry environments, 80 to 95 percent of the water in the pre-urine is recovered and kept inside the body. The ability to recover this water allows these insects to live in dry habitats such as deserts and flour bins. How does reabsorption happen?

Experiments on desert locusts established that the mechanism involves a series of specific membrane pumps and channels, not unlike the system found in the chloride cells of fish. The nature of these pumps and channels was revealed by mounting the rectal epithelium from a desert locust as a sheet dividing two solutions. This preparation allowed biologists to manipulate electrolyte concentrations on either side of the rectal wall and then measure changes in the solutions over time. For example, when investigators removed K^+ and Na^+ from the solution on the lumen side of the organ, they observed that water reabsorption stopped. These data established that the hindgut's ability to recover water from urine is dependent on ion movement. More specifically, the epithelial cells in the hindgut must transport ions out of the pre-urine and into the hemolymph. Water would then follow by osmosis, leading to the formation of a concentrated urine.

How do the ions move? By poisoning the experimental membranes with ouabain, biologists confirmed that Na^+/K^+-ATPase is involved in moving ions out of the lumen and into the hemolymph. Ouabain-treated membranes continued to transport Cl^-, however. Later experiments confirmed that the rectal epithelium transports Cl^- against electrical and concentration gradients. These results support the hypothesis that the hindguts of insects have two active pumps: a chloride pump and Na^+/K^+-ATPase.

Years of experiments on the locust hindgut resulted in the model that is diagrammed in **Figure 42.10b**. Notice that Cl^- is pumped into cells from the hindgut lumen, with K^+ following through potassium channels along an electrochemical gradient and water following via osmosis. In the basolateral membrane, Na^+/K^+-ATPase sets up electrochemical and osmotic gradients that favor movement of Cl^-, K^+, and H_2O into the hemolymph. Although the diagram looks complex, it is actually a simplified version of the cells that are involved. There is, for example, strong experimental evidence for the existence of several other channels, pumps, and cotransporters in the epithelium of the insect hindgut. These membrane proteins transport protons, ammonia, amino acids, and other molecules across the tissue and help insects maintain water and electrolyte balance.

Regulating Water and Electrolyte Balance: An Overview

Much remains to be learned about the molecular mechanisms involved in the transport of ions and solutes in the hindgut of insects. Fortunately, the overall pattern in the regulation of water and electrolyte balance in insects is much simpler than the detailed mechanisms involved. Several general principles that have emerged from studies of insect excretion turn out to be relevant to vertebrate systems as well:

- The formation of the pre-urine is not particularly selective. Most of the molecules present in the hemolymph are also present in the Malpighian tubules.
- In contrast to pre-urine formation, reabsorption is highly selective. The protein pumps and channels involved in reabsorption are highly specific for certain ions and molecules. Waste products do not pass through the rectal membrane. Instead, they remain in the urine and feces and are eliminated from the body. Only valuable ions and molecules are reabsorbed.
- Also in contrast to pre-urine formation, reabsorption is tightly regulated. The membrane pumps and channels involved in reabsorption are activated and deactivated in response to osmotic stress. If an insect is dehydrated, then virtually all of the water in the pre-urine is reabsorbed. But if the same individual has plenty to drink, reabsorption does not occur and the urine is watery and hypotonic to the individual's hemolymph. The system is dynamic and allows precise control over water and electrolyte balance.
- Reabsorption of water is based on osmotic gradients created by ion pumps. Insects can form a concentrated urine and conserve water because cells expend energy in the form of ATP to move water out of the pre-urine, making it hypertonic relative to tissues.

Given the success of insects in terms of the numbers of species and individuals that exist and the array of habitats they occupy, it is clear that their systems for maintaining water and electrolyte balance are remarkably effective.

✓CHECK YOUR UNDERSTANDING

Terrestrial insects are prone to dehydration and have a cuticle, a respiratory system, and an excretory system that are designed to conserve water. You should be able to explain how the following traits are involved in water retention: (1) wax, (2) tracheae, (3) the excretion of ammonia in the form of uric acid, and (4) selective reabsorption of electrolytes and water in the hindgut.

42.4 Water and Electrolyte Balance in Terrestrial Vertebrates

On land, vertebrates face the same hazards as terrestrial insects do with respect to water loss. Crocodiles, turtles, lizards, frogs, birds, and mammals lose water from their body surfaces and from the surface of their lungs every time they breathe. To replace water that is lost, most terrestrial vertebrates drink. Vertebrates that live on land also ingest electrolytes in food and excrete them in their wastes. Because sweat formation is based on the production of a salty solution, animals that sweat to cool off lose electrolytes when they do so.

Like other terrestrial organisms, land-dwelling vertebrates must carefully regulate the osmolarity of their tissues. Depending on conditions, terrestrial animals may need either to conserve or to excrete water and electrolytes to achieve homeostasis. How do land-dwelling vertebrates maintain water and electrolyte balance?

Due to intensive research on the mammalian kidney, a detailed answer to this question began to emerge in the mid-1970s. The **kidney** is the organ responsible for regulating water and electrolyte balance in terrestrial vertebrates. In this respect, the kidney's function is similar to the shark rectal gland. In addition, the kidney is where nitrogenous wastes—usually in the form of urea—are removed from the blood and eliminated in urine. As the organ responsible for both water and electrolyte balance and the excretion of water-soluble waste products, the kidney is equivalent to the Malpighian tubules and hindgut of insects.

The Structure of the Kidney

The general anatomy of the mammalian kidney was described in the late nineteenth century. As **Figure 42.11a** shows, kidneys occur in pairs and tend to be bean shaped. A large blood vessel called the *renal artery* brings blood that contains nitrogenous wastes into the organ; a large blood vessel called the *renal vein* carries blood away. The urine that is formed in the kidney is transported via a long tube called the **ureter** to a storage organ called the **bladder**. From the bladder, urine is transported to the body surface via a tube called the **urethra**, then excreted. In most vertebrates, the kidneys are located near the dorsal (back) side of the body.

Most of the kidney's mass is made up of small structures called nephrons. The **nephron** is the basic functional unit of the kidney. Nephrons perform the work involved in maintaining water and electrolyte balance. Most of the approximately one million nephrons in a human kidney are located in the outer region of the organ, or **cortex** (**Figure 42.11b**). As **Figure 42.11c** shows, however, some nephrons extend from the cortex into the kidney's inner region, or **medulla**.

To understand how the kidney maintains water and electrolyte balance, it is essential to understand how the nephron functions. **Figure 42.12a** provides a detailed view of the nephron, with its five major regions labeled, and **Figure 42.12b** shows how each nephron is served by blood vessels. Let's follow the flow of material through each of the five regions of a nephron, starting with the blood that arrives at the structure and ending with the urine that exits it. The changes that occur along the way are like those observed as water and solutes from insect hemolymph pass into the Malpighian tubules and from there to the hindgut. The important point to remember is that urine formation results from three processes: (1) formation of a pre-urine via filtration; (2) reabsorption of water, nutrients, and some solutes; and (3) removal of water from the final urine via osmosis.

Filtration: The Renal Corpuscle

In vertebrates, urine formation begins in a structure called the **renal corpuscle** ("kidney-little-body"). To understand how this

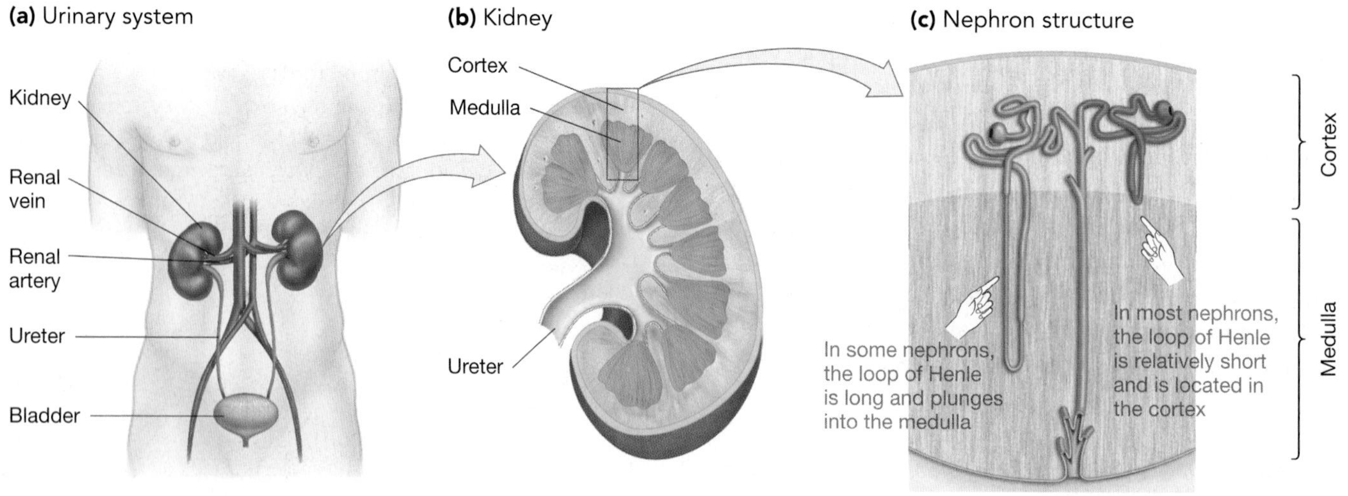

FIGURE 42.11 Anatomy of the Human Urinary System, Kidney, and Nephron
(a) In mammals, kidneys are paired and are located near the spinal column. **(b)** The kidney has an outer region called the cortex and an inner area called the medulla. The cortex and medulla are packed with structures called nephrons. **(c)** Most nephrons are located within the cortex, but some have a long loop that extends into the medulla.

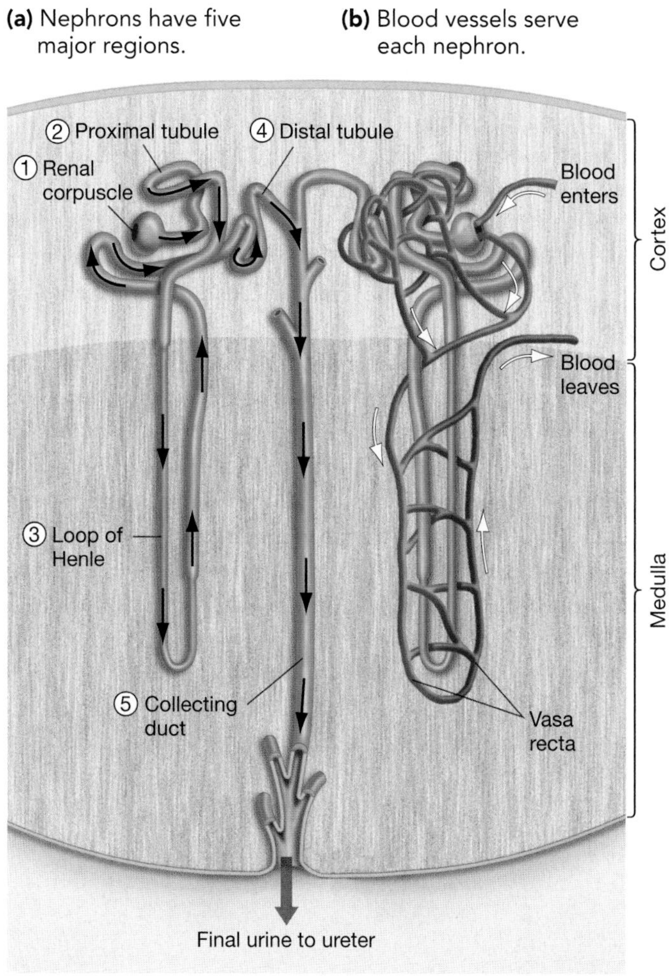

FIGURE 42.12 Nephrons Have Five Major Parts and Are Served by Blood Vessels
(a) Nephrons have the five basic parts labeled here. Urine formation begins in the renal corpuscle and ends in the collecting duct. **(b)** Each nephron is served by blood vessels. EXERCISE Label the parts of the nephron in part (b).

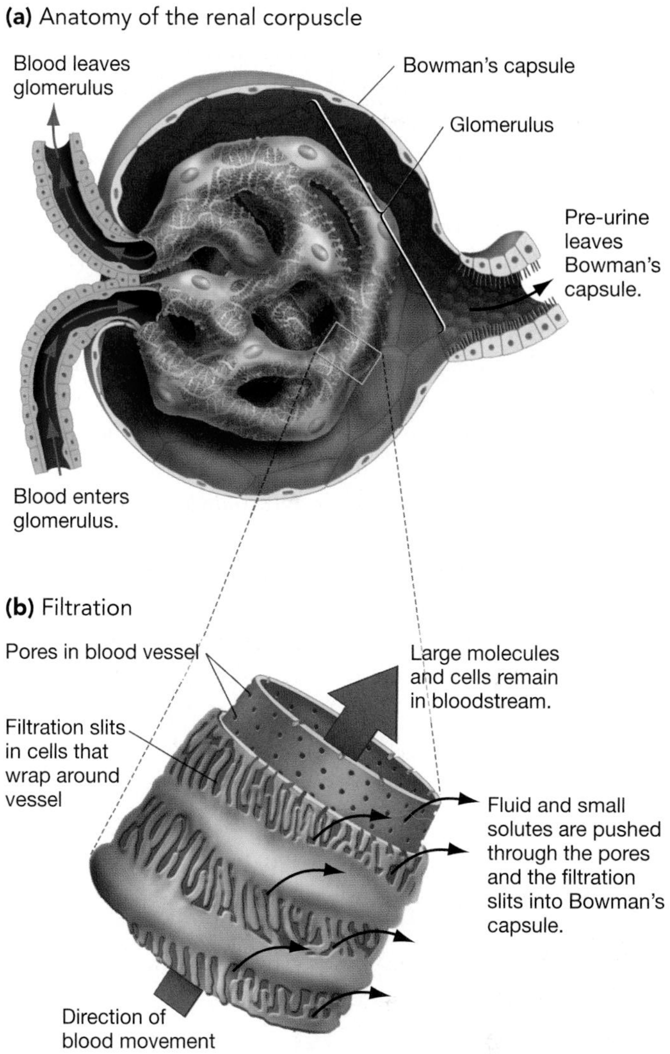

FIGURE 42.13 Urine Formation Begins when Blood Is Filtered in the Renal Corpuscle
(a) The renal corpuscle consists of Bowman's capsule and the glomerulus. **(b)** The capillaries in the glomerulus have pores and are surrounded by cells that have filtration slits. Blood pressure forces water and small molecules out of the capillaries, through the slits, and into Bowman's capsule.

part of the nephron works, examine Figure 42.12a again and notice that the nephron is a tube that is closed at one end and open at the other. The closed end is the beginning of the nephron; the open end is the terminus. As **Figure 42.13a** shows, the closed end forms a capsule that encloses a cluster of tiny blood vessels, or capillaries, that bring blood to the nephron from the renal artery. This cluster of capillaries is called the **glomerulus** ("ball of yarn"), and the region of the nephron that surrounds the glomerulus is named **Bowman's capsule.** The renal corpuscle, then, is composed of the glomerulus and Bowman's capsule.

By carefully sectioning and examining the renal corpuscle, biologists established that the glomerular capillaries have large pores, or openings. These blood vessels are also surrounded by unusual cells whose membranes fold into a series of slits and ridges (**Figure 42.13b**).

Based on these observations, researchers working in the late 1800s proposed that the renal corpuscle serves as a **filtration** device, which allows water and small solutes from the blood to pass into the nephron, and that filtration was based on the size of molecules and ions. The hypothesis was that water and small solutes are pushed out of the capillary's pores, through the slits in the surrounding cells, and into the fluid-filled space inside Bowman's capsule. Because proteins, cells, and other large components of blood would not fit through the pores, they could not enter the nephron. They would stay in the blood instead. Stated another way, urine formation was thought to start with a size-selective filtration step. According to this hypothesis, the force required to perform filtration is supplied by blood pressure. In vertebrates, blood is under pressure because it is pumped by the heart through a closed system of vessels. This pressure is enough to force water and small solutes through the pores in the glomerulus, so the renal corpuscle strains large volumes of fluid without expending energy in the form of ATP. The result is the formation of a **filtrate**—in this case, a **pre-urine**.

In mammals, the filtration hypothesis was confirmed almost 100 years after it was initially proposed. In 1971 biologists were able to make direct measurements of the pressure inside the glomerular capillaries and in the tubule leading away from Bowman's capsule in rats. The investigators found that pressure was much higher inside the capillaries than it was in the surrounding capsule. As the filtration hypothesis predicts, the pressure differential forces water and solutes out of the blood and into the capsule space.

To summarize, the renal glomerulus filters the blood to create a pre-urine. During the formation of this filtrate, up to 25 percent of the water and solutes present in the blood is removed. It is critical to note two facts, though. First, the renal corpuscles of a human kidney are capable of producing about 180 liters of pre-urine per day. This is an impressive volume—especially when you think of 180-liter bottles of soft drink arranged on a supermarket shelf. Second, almost all of the pre-urine is recycled, with only a tiny fraction of the original volume actually excreted. Approximately 99 percent of the volume is reabsorbed and returned to the bloodstream.

Reabsorption: The Proximal Tubule

Where does pre-urine reabsorption occur? Fluid leaves Bowman's capsule and enters a convoluted structure called the **proximal tubule**. Researchers who analyzed samples of the fluid inside this tubule found that it contained the components predicted by the filtration hypothesis: water and small solutes such as urea, glucose, amino acids, vitamins, and electrolytes. It is important to remember that some of these molecules are waste products while others are valuable nutrients.

The first clue to the proximal tubule's function came from anatomical studies. As **Figure 42.14a** shows, the epithelial cells of this tubule have a prominent series of small projections, called **microvilli** ("little shaggy hairs"), facing the lumen. The presence of microvilli greatly expands the

(a) Lumen of proximal tubule has large surface area.

(b) Model of water and solute reabsorption

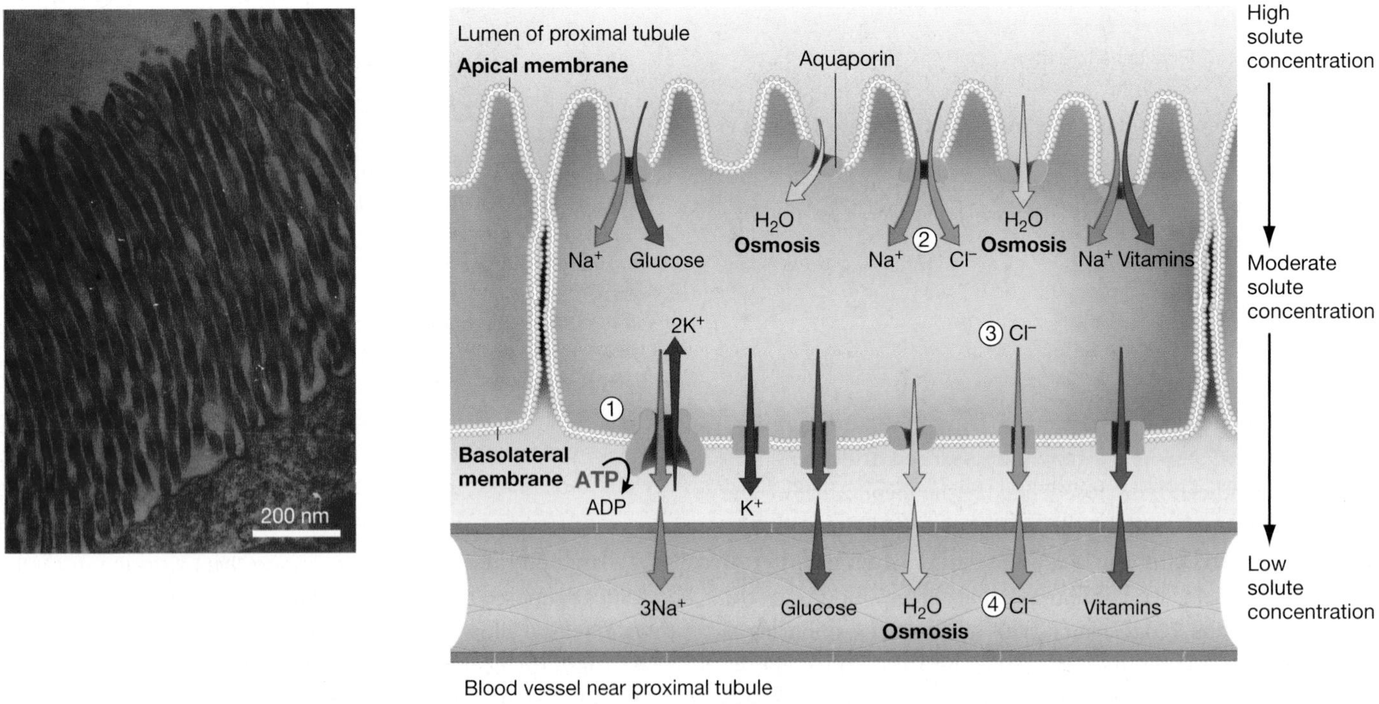

FIGURE 42.14 Water and Electrolytes Are Reabsorbed in the Proximal Tubule
(a) Microvilli from epithelial cells extend into the lumen of the proximal tubule. **(b)** Reabsorption of water and solutes occurs because (1) Na^+/K^+-ATPase in the basolateral membrane of epithelial cells sets up a strong concentration gradient to bring Na^+ into the cell across the apical membrane; (2) Na^+ cotransporters in the apical membrane result in reabsorption of nutrients and electrolytes, with water following via osmosis; (3) solutes also exit the cell through channels, with water following; and (4) solutes and water enter nearby blood vessels. EXERCISE Aquaporins are responsible for water movement in the hindgut of insects as well. Add aquaporins to Figure 42.10b.

surface area of an epithelium. Because greater surface area provides more space for membrane proteins that act as pumps, channels, and cotransporters, it is logical to find that cells with microvilli are often associated with transport processes. The observation that epithelial cells in the proximal tubule are also packed with mitochondria suggests that ATP-demanding **active transport** is occurring. Based on these data, anatomists hypothesized that the proximal tubule functions in the active transport of selected molecules out of the pre-urine.

The selective active-transport hypothesis was supported by experiments on proximal tubules that were dissected from the kidney of rabbits and rats and isolated in vitro. By injecting solutions of known composition into proximal tubules in the presence or absence of ATP, researchers were able to confirm that selected electrolytes and nutrients are actively reabsorbed from the pre-urine that enters the tubules. When these solutes leave the proximal tubule and enter epithelial cells, water follows along the osmotic gradient. In this way, valuable solutes and water are reabsorbed and returned to the body.

Figure 42.14b summarizes the current model of the molecular mechanisms involved in selective reabsorption. Notice that the basolateral membrane of epithelial cells in the proximal tubule contains Na^+/K^+-ATPase and that the apical membrane contains a variety of cotransporters. The key to the model is that the Na^+/K^+-ATPase in the basolateral membranes removes Na^+ from the interior of the cell. The active transport of sodium ions out of the cell creates a gradient favoring the entry of Na^+ from the lumen. In the apical membrane adjacent to the lumen, Na^+-dependent cotransporters use this gradient to remove valuable ions and nutrients selectively from the pre-urine. Specifically, the cotransporters simultaneously bind Na^+ and another solute, such as glucose, an amino acid, or Cl^-. As with other examples of secondary active transport, the movement of Na^+ into the cell, *with* its concentration gradient, provides the means for moving other solutes *against* a concentration gradient.

Another key point is that water follows the movement of these solutes from the pre-urine back to blood, by osmosis. Recent work has shown that water moves from the lumen of the proximal tubule into epithelial cells through specialized membrane proteins called **aquaporins** ("water-pores"). When an aquaporin is open, 3 billion water molecules pass through it per second. Aquaporins are water channels. Aquaporins located in the basolateral membranes of epithelial cells facilitate the movement of water back into the blood.

Careful measurements have confirmed that two-thirds of the NaCl and water that is originally filtered by the renal corpuscle is reabsorbed in the proximal tubule. The osmolarity of the tubular fluid is unchanged despite this huge change in volume, however, because water reabsorption is proportional to solute reabsorption. The electrolytes, nutrients, and water that are reabsorbed eventually enter blood vessels on the basolateral side of the tubule and are returned to the body.

To summarize, the cells that line the proximal tubule act as a recycling center. The pumps and cotransporters in their membranes allow the proximal tubule to recover water, valuable nutrients, and electrolytes from the pre-urine but to leave wastes. Compared with the fluid that enters, the pre-urine that leaves is isotonic but greatly reduced in volume. As it flows into the next part of the nephron, it has a relatively high concentration of waste molecules and a relatively low concentration of nutrients.

WEB TUTORIAL 42.1 The Mammalian Kidney

Creating an Osmotic Gradient: The Loop of Henle

In mammals, the fluid that emerges from the proximal tubule enters a loop. In most nephrons, the loop is short and does not leave the cortex. But in about 20 percent of the nephrons present in a human kidney, the loop is long and plunges from the cortex of the kidney deep into the medulla (look back at Figure 42.11c). Jacob Henle described this part of the mammalian nephron in the early 1860s, and it was named the **loop of Henle.** It was not until 1942, however, that Werner Kuhn offered a hypothesis to explain what the structure does.

Kuhn's proposal was inspired by the use of countercurrent heat exchangers in chemistry and physics. Recall from Chapter 41 that a countercurrent heat exchanger is a system in which two adjacent fluids flow through pipes in opposite directions. Kuhn proposed that the loop of Henle functions as a countercurrent exchanger. But instead of setting up a heat gradient like those diagrammed in Figure 41.19, Kuhn hypothesized that the loop of Henle sets up an osmotic gradient. Although he did not know the molecular mechanisms involved in setting up this gradient, he proposed that the osmolarity of the fluid inside the loop of Henle is low in the cortex and high in the medulla. Further, Kuhn maintained that the osmolarity in tissues surrounding the loop mirrors the gradient inside the loop. This is a key point. Kuhn's hypothesis proposed that an exchange of water and solutes occurs between the cells in each segment of the loop and the cells surrounding the loop of Henle—not between the solutions in the descending and ascending portions of the loop. This situation is in contrast to a conventional countercurrent system, in which exchange occurs between the two segments themselves. According to Kuhn, the exchange results in a steep osmotic gradient that exists both inside and outside the loop of Henle and that runs from the cortex deep into the medulla. Was he correct?

Testing Kuhn's Hypothesis A series of papers published during the 1950s supplied important experimental support for the countercurrent exchange model. Two particularly important data sets, obtained by analyzing the osmolarity of tissue slices cut into sections perpendicular to the loop of Henle, are reproduced in **Figure 42.15**, page 970. The left-hand graph of the figure shows data on the osmolarity of fluid inside the loop of Henle. As predicted by Kuhn's model, a strong gradient in osmolarity exists from the cortex to the medulla. The data in the right-hand graph of Figure 42.15

Question: What is the function of the loop of Henle?

Hypothesis: The loop of Henle sets up an osmotic gradient for countercurrent exchange of water and solutes.

Null hypothesis: The loop of Henle does not set up an osmotic gradient for countercurrent exchange.

Experimental setup:

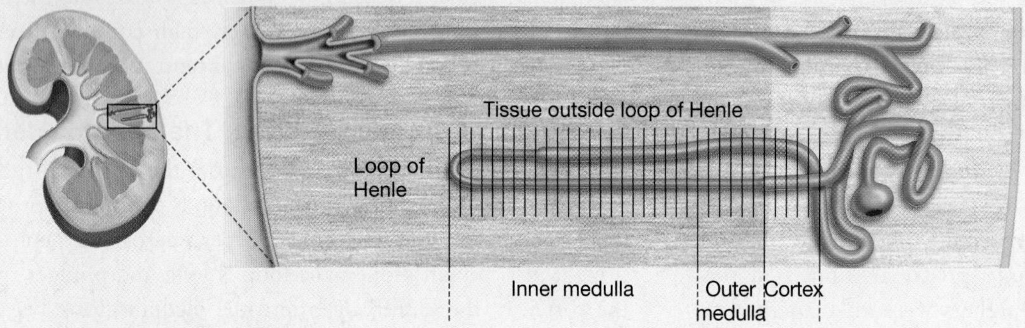

1. Slice kidney tissue in section perpendicular to the loop of Henle.

2. Measure the osmolarity in each section of loop of Henle. (This experiment was performed on a rat nephron.)

3. Measure the concentration of urea, sodium ions, and chloride ions in each section of tissue outside the loop. (This experiment was performed on a dog nephron.)

Prediction: A strong gradient in osmolarity exists from the medulla to the cortex, both within the loop of Henle and in the tissue surrounding the loop of Henle.

Prediction of null hypothesis: A strong gradient in osmolarity does not exist from the medulla to the cortex, both within and outside the loop of Henle.

Results:

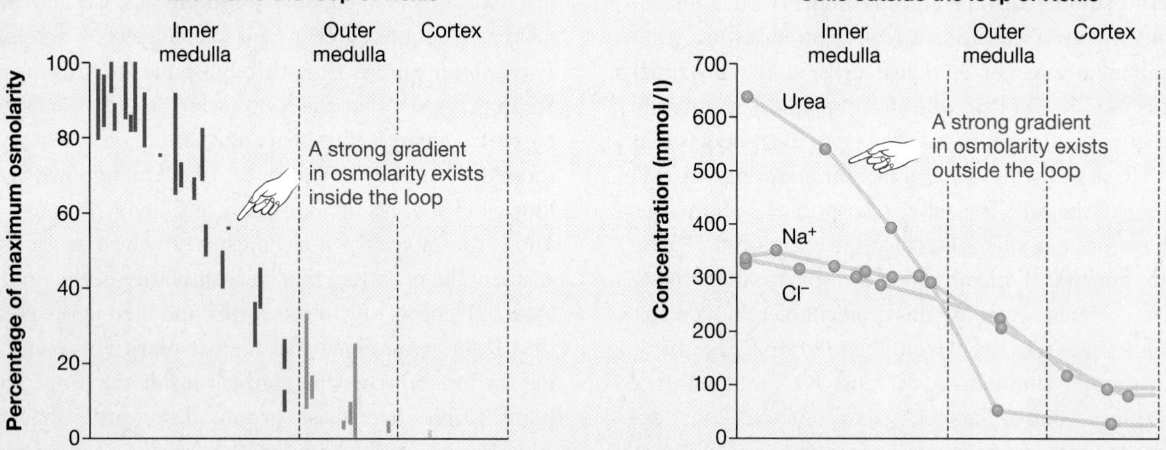

Conclusion: The loop of Henle sets up a countercurrent exchange system for the exchanges of solutes—specifically, urea, sodium, and chloride.

FIGURE 42.15 Data Confirm the Existence of a Strong Osmotic Gradient along the Loop of Henle
The graphs show the relative osmolarity, or solute concentrations, of the fluid (left) inside or (right) outside the nephron as a function of its location in the kidney.

show that outside the loop of Henle, the concentrations of Na^+, Cl^-, and urea also increase sharply from the cortex to the medulla. This was an important observation. It not only confirmed the prediction that an osmotic gradient exists in the tissue surrounding the loop of Henle, but also suggested that the solutes responsible for this gradient are Na^+, Cl^-, and urea. As we'll see, the change in concentration of urea turned out to be particularly important.

Before we consider how the steep osmotic gradient created by the loop of Henle affects urine formation, let's take a look at research on the molecular mechanisms involved in setting up the countercurrent exchange. How do water and solutes move into and out of the loop in a way that establishes the dramatic osmotic gradient that researchers observed?

How Is the Osmotic Gradient Established? Figure 42.16a provides a detailed look at the morphology of the loop of Henle. Notice that there are three distinct regions in the loop: the descending limb, the thin ascending limb, and the thick ascending limb. The thin and thick ascending limbs differ in the thickness of their walls. Do the three regions also differ in their permeability to water and solutes?

It took over 15 years of experiments performed in laboratories around the world to formulate a definitive answer to that

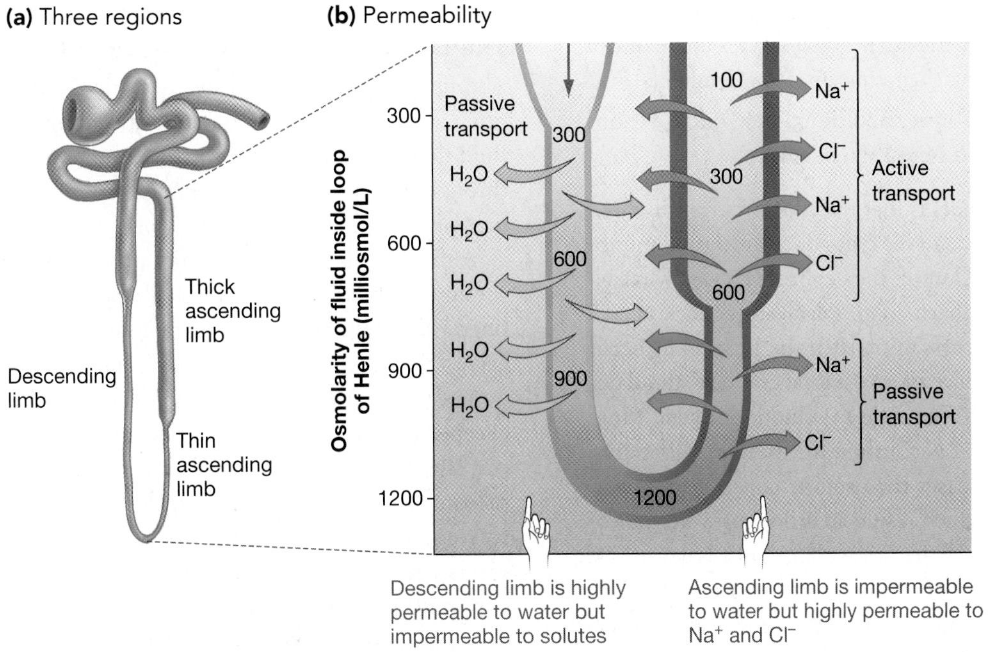
(a) Three regions
(b) Permeability
Thick ascending limb
Descending limb
Thin ascending limb
Osmolarity of fluid inside loop of Henle (milliosmol/L)
300
600
900
1200
Passive transport
H2O
300
600
900
1200
100
300
600
Na+
Cl−
Active transport
Passive transport
Descending limb is highly permeable to water but impermeable to solutes
Ascending limb is impermeable to water but highly permeable to Na+ and Cl−

FIGURE 42.16 The Loop of Henle Establishes an Osmotic Gradient because Water Leaves the Descending Limb and Salt Leaves the Ascending Limb
(a) The loop of Henle has three distinct regions. **(b)** The permeability of the descending limb, thin ascending limb, and thick ascending limb to water and solutes varies dramatically. Active transport occurs only in the thick ascending limb.

question. One important result began to emerge after researchers were able to puncture Henle's loop with a micropipette, analyze the composition of the fluid inside, and compare it with the nephron's final product—urine. Although the two solutions were isotonic at the same cross-sectional level within the nephron, the solutes involved were very different. In the ascending limb of the loop of Henle, Na^+ and Cl^- constituted at least 60 percent of the solutes; urea constituted about 10 percent. But in the final portion of the nephron, both ions were less common. Instead, urea was the major solute. This was an important observation, because it suggested that valuable Na^+ and Cl^- ions were being retained while waste products in the form of urea were being discarded.

What mechanism is responsible for the removal of Na^+ and Cl^-? Because Na^+ and Cl^- were also present at high concentration in the tissue surrounding the thick ascending limb of the loop of Henle, researchers hypothesized that sodium might be actively pumped out of this portion of the nephron. The hypothesis was that the active transport of Na^+ out of the thick ascending limb would create an electrical gradient that would also favor the loss of Cl^- (**Figure 42.16b**).

Follow-up experiments that used ouabain and other poisons supported the hypothesis that sodium ions are actively transported out of the solution inside the thick ascending limb. The epithelial cells responsible for sodium excretion in this portion of the nephron are configured almost exactly like the epithelium of the shark rectal gland (see Figure 42.5). A combination of Na^+/K^+-ATPase, Cl^- channels, and $Na^+/2Cl^-/K^+$ cotransporters in these cells results in the active transport of salt out of the loop and into the surrounding tissue. Once these electrolytes are removed, urea remains as the primary solute in the urine.

What is happening in the descending limb and the thin ascending limb of the loop of Henle? An important series of experiments established how water and solutes move across these two remaining segments. By injecting solutions of known concentration into the nephrons of rabbits, biologists documented that the descending limb is highly permeable to water but almost completely impermeable to solutes. The thin ascending limb of the loop, in contrast, is highly permeable to Na^+ and Cl^-, moderately permeable to urea, and almost completely impermeable to water (Figure 42.16b).

A Comprehensive View of the Loop of Henle All of the observations about the loop of Henle just summarized came together in 1972 when two papers, published simultaneously and independently, proposed the same comprehensive model for how that structure works. To understand the hypothesis, follow the events that occur as fluid leaves the proximal tubule and moves through the loop, using Figure 42.16b:

1. As it flows down the descending limb, urine loses water to the tissue surrounding the nephron. This movement of water is passive, meaning that it does not require an expenditure of ATP. The water moves because it follows an osmotic gradient. This osmotic gradient is created by the ascending limb.
2. At the bottom of the loop, the urine inside the nephron is hypertonic to the surrounding tissues. Once it begins to flow up the ascending limb, it stops losing water, because

the membrane is nearly impermeable to water. Instead, the fluid begins to lose Na^+ and Cl^-. These solutes move out of the loop passively along their concentration gradients.

3. In the thick ascending limb, additional Na^+ and Cl^- ions are actively transported out of the nephron.

The key point in this model is that the countercurrent flow of urine, combined with changes in the types of channels and pumps that are active in the epithelium of the nephron, creates what biologists call a positive feedback loop. **Positive feedback** occurs when the product of a process stimulates the process to occur more. In this case, the presence of an osmotic gradient stimulates water and ion flows that in turn create an osmotic gradient. More specifically, the movement of NaCl from the ascending limb into the surrounding tissue increases the osmotic concentration outside the descending limb, which results in a flow of water out of the water-permeable walls of the descending limb, via osmosis. This loss of water in the descending limb increases the osmolarity in the fluid entering the ascending limb. The presence of a hypertonic fluid at the base of the ascending limb triggers a passive flow of ions out and completes the positive feedback loop.

Although this system is extremely complex, its outcome is extremely simple: the creation and maintenance of a strong osmotic gradient. Due to the countercurrent flow of material through the nephron, the osmotic gradient is maintained with the minimum possible expenditure of energy.

What happens to the water and salt that move out of the loop? They quickly diffuse into the **vasa recta**, a network of blood vessels that runs along the loop. As a result, the water and electrolytes are returned to the body (**Figure 42.17**). The removal of water that leaves the descending limb is particularly important. If it were not drawn off into the bloodstream, it would dilute the concentrated fluid outside the loop of Henle and quickly destroy the osmotic gradient.

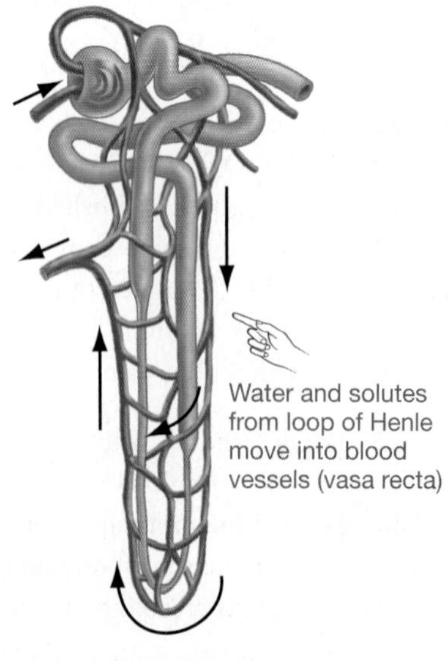
Water and solutes
from loop of Henle
move into blood
vessels (vasa recta)

FIGURE 42.17 Blood Supply to the Loop of Henle
Blood vessels called the vasa recta run along the loop of Henle in a countercurrent fashion.

Regulating Water and Electrolyte Balance: The Distal Tubule and the Collecting Duct

The first three steps in urine formation—filtration, reabsorption, and establishment of an osmotic gradient—result in a fluid that is slightly hypotonic to blood. Once urine has passed through the loop of Henle, the major solutes that it contains are urea and other wastes.

Although the events that take place in the renal corpuscle, proximal tubule, and loop of Henle produce a urine that is distinctly different from the blood that entered the kidney, the events that take place are largely unregulated—meaning they do not vary with changes in the body's condition. There is one exception: The volume of blood filtered at the renal corpuscle can change dramatically, depending on circumstances. But the proximal tubule and loop of Henle function in about the same way at all times. As a result, the composition of the fluid that enters the distal tubule of a nephron is relatively constant in composition over time. This is an important point, because the urine that leaves the collecting duct is highly variable in osmolarity and in Na^+ and Cl^- concentration.

How is it possible for the characteristics of urine to vary over time? The answer is based on two observations about the distal tubule and the collecting duct: Their activity is highly regulated, and their activity changes in response to the osmotic stress that an individual experiences. Although electrolytes and water are always reabsorbed in the distal tubule, the *amount* of Na^+, Cl^-, and water that is reabsorbed in the distal tubule and in the collecting duct varies with the animal's condition.

The activity of the distal tubule and collecting duct is controlled by hormones—signaling molecules that were introduced in Chapter 8 and are explored further in Chapter 47. Specifically:

- If Na^+ levels in the blood are low, the hormone **aldosterone** is released from the adrenal glands and leads to the reabsorption of Na^+ and Cl^- in the distal tubule. Water follows these ions out of the distal tubule via osmosis.
- If an individual is dehydrated, a molecule called **antidiuretic hormone (ADH)** is released from the brain. (The term *diuresis* refers to increased urine production, so *antidiuresis* means inhibition of urine production. ADH is also referred to as vasopressin or arginine vasopressin.) When ADH arrives at epithelial cells lining the collecting duct, the hormone triggers the insertion of aquaporins into the plasma membrane. Because cells become highly permeable to water as a result, large amounts of water are reabsorbed. As **Figure 42.18a** shows, water leaves the collecting duct passively—following the concentration gradient established by the loop of Henle. The result is conservation of water and the formation of urine that is strongly hypertonic in relation to blood. When ADH is absent, however, few aquaporins are found in the epithelium of the collecting duct, and the structure is relatively impermeable to water (**Figure 42.18b**). In this case, a hypotonic urine is produced. People with defective forms of ADH or aquaporins

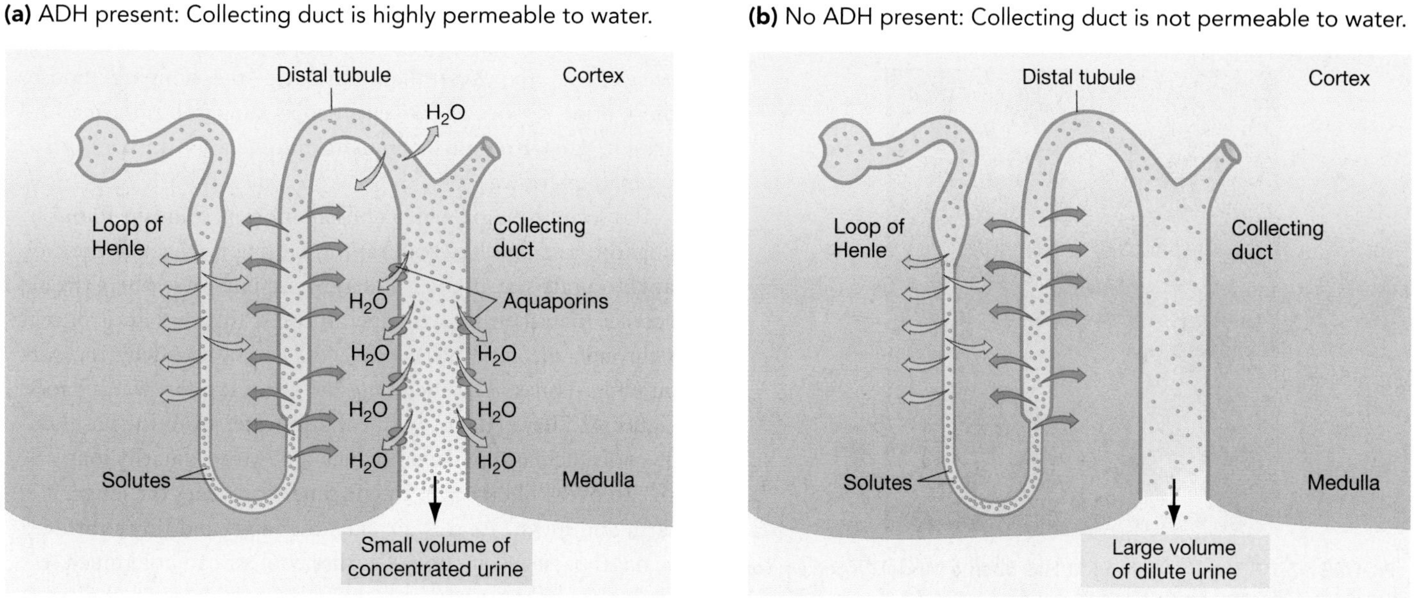

FIGURE 42.18 Water Reabsorption by the Collecting Duct Is Regulated by ADH
(a) Antidiuretic hormone (ADH) makes the collecting ducts permeable to water and leads to the production of small volumes of concentrated urine. The water that leaves the collecting ducts passes into the vasa recta (see Figure 42.17). **(b)** In the absence of ADH, urine increases in volume and contains a low concentration of solutes.

produce copious amounts of urine—up to 30 liters per day. They suffer from the condition **diabetes insipidus.** (*Diabetes* means "to run through"; *insipidus* means "tasteless.")

To summarize, the first three segments of the nephron accomplish two objectives: (1) They effectively concentrate nitrogenous wastes, and (2) they create the possibility for Na^+, Cl^-, and water to be either excreted or reabsorbed by the distal tubule and collecting duct. Of the 180 liters of primary urine that enters the renal corpuscle, all but 1.2 liters is reabsorbed and returned to the body each day. The nephron is a remarkably effective mechanism for regulating water and electrolyte balance and achieving homeostasis.

✓CHECK YOUR UNDERSTANDING

The vertebrate kidney is specialized for the production of hypertonic urine. You should be able to (1) describe the structure and function of the following parts of a nephron: renal corpuscle, proximal tubule, loop of Henle, distal tubule, and collecting duct; (2) explain how the characteristics of urine are regulated by hormones that act in the distal tubule and collecting duct; and (3) predict how the following events would affect urine production: drinking massive amounts of water, eating large amounts of salt, refraining from drinking for 48 hours, and ingesting a drug that inhibits the production of ADH.

ESSAY Life in the Desert

In deserts all over the world, a combination of extreme heat and severe dryness presents extraordinary challenges for animals. To understand how these organisms maintain water and electrolyte balance, researchers have focused on the desert kangaroo rat (**Figure 42.19**, page 974). Like the oryx introduced in Chapter 41, kangaroo rats do not need to drink water. This is fortunate, because water is frequently unavailable where they live. The question is, How do kangaroo rats obtain and conserve water?

Recall from Chapter 41 that oryx get most of the water they need from the leaves and stems that they eat. But kangaroo rats feed on seeds, which are extremely dry. They get significant amounts of water from insects and vegetation consumed, but like oryx they may also depend at least in part on metabolic water. As Chapter 9 showed, the oxidation of sugar during cellular respiration results in the release of water. More specifically, the oxidation of 1 g of carbohydrate yields 0.56 g of metabolic water. Cells synthesize water by burning sugar.

Kangaroo rats do not need to drink water.

Kangaroo rats have two major ways of conserving this precious resource. The first mechanism involves the production of

(Continued on next page)

(Essay continued)

FIGURE 42.19 The Kangaroo Rat Has Been a Model Organism for Studying How Mammals Limit Water Loss
Kangaroo rats are native to the deserts of southwestern North America.

an extremely hypertonic urine. Kangaroo rats have exceptionally long loops of Henle in proportion to their body size. These loops produce steep osmotic gradients in the medulla of their kidneys and make it possible for their kidneys to produce urine that is 15 times more concentrated than their blood. This is the second most hypertonic urine known among animals. (To date, the most concentrated urine known is produced by a mouse species from the Australian Outback—the hopping mouse, whose urine is twice as concentrated as kangaroo rat urine.) As a result, kangaroo rats eliminate nitrogenous wastes with a minimal loss of water.

The second major water conservation mechanism found in kangaroo rats involves respiration. During the heat of the day, kangaroo rats stay in their underground burrows, where the air is cooler than their body temperature. As this cool air is drawn in through the rats' nasal passages, it is warmed by the surrounding tissues. The incoming air also picks up water vapor (**Figure 42.20a**). The loss of heat and water from the nasal tissue cools the nasal passages, which are extraordinarily long relative to overall body size. By the time air reaches the lungs, it is warm and moist. As this air exits the lungs and flows through the nasal passages, it cools. Because cool air can hold much less water vapor than warm air, moisture condenses in the nasal passages before the air is exhaled (**Figure 42.20b**). As a result, most of the water that is lost from the nasal passages during inhalation is recovered during exhalation.

The extraordinarily long nasal passages of a kangaroo rat allow a great deal of cooling and condensation to occur—meaning a minimum of water is lost during respiration. In effect, the nasal passages of the kangaroo rat act as an unusual sort of countercurrent exchanger. Instead of exchanging heat or ions and molecules across a space, as heat exchangers and the loop of Henle do, the nasal passages of kangaroo rats exchange heat and water over time—with each breath.

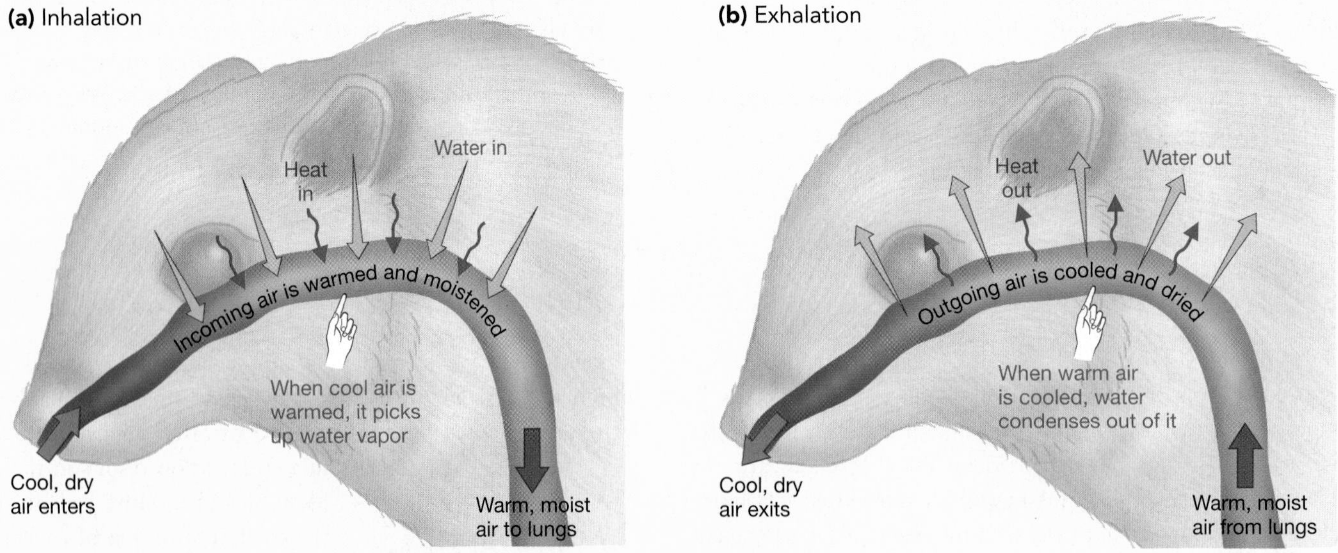

FIGURE 42.20 Countercurrent Exchange Limits Water Loss during Respiration
The nasal passages of the kangaroo rat are particularly long. A strong temperature and moisture gradient from the tip of the nose to the end of the nasal passages establishes suitable conditions for countercurrent exchange each time an individual **(a)** inhales and **(b)** exhales.

CHAPTER REVIEW

Summary of Key Concepts

Homeostasis with respect to water and electrolyte balance is critical to the health of animals.

- **Freshwater, marine, and terrestrial habitats pose different challenges to animals with regard to maintaining water and electrolyte balance.**

The mechanisms involved in regulating water and electrolyte balance vary widely among animal groups, because different habitats present different challenges. Freshwater fish, for example, are strongly hypertonic to their environment and tend to gain water and lose electrolytes. Marine fish, in contrast, are strongly hypotonic in relation to seawater and tend to lose water and gain electrolytes. Terrestrial animals lose water every time they breathe and from their body surfaces by evaporation. In all these organisms, epithelial cells that selectively transport water and electrolytes are responsible for homeostasis.

- **In marine animals, specialized epithelial cells have membrane proteins that remove excess salt (NaCl) from the body so that it can be excreted. The same types of cells are found in the kidneys of mammals.**

Research on the rectal gland of sharks established how marine animals excrete excess salt. Sodium-potassium pumps are located on the basolateral side of epithelial cells in this gland. These pumps establish an electrochemical gradient that carries Na^+ and Cl^- from the blood into the lumen of the gland, where it is excreted. Later research revealed that a similar arrangement of pumps, channels, and cotransporters occurs in a wide variety of animal cells that transport salt. These cells occur in the gills of saltwater fish, the salt glands of marine birds and reptiles, and the kidneys of mammals.

Despite extensive studies, it is still not clear how freshwater fish take up electrolytes from their environment to replace the ions that they lose from their gills. Research on salmon has shown that the gills of freshwater fish contain cells with sodium-potassium pumps. Some researchers suggest that these cells may be involved in the active transport of ions into the gills.

- **Terrestrial insects have key adaptations that limit water loss from the body. The hindgut and Malpighian tubules of insects function in maintaining water and electrolyte balance.**

Terrestrial insects limit water loss in several ways. Their exoskeleton is covered with wax to limit evaporation, and the openings to their respiratory organs close when osmotic stress is severe. Insects are also capable of forming a hypertonic urine that minimizes water loss during the excretion of nitrogenous wastes. A pre-urine that is isotonic with the hemolymph forms in the Malpighian tubules of insects. If K^+ and Cl^- pumps in the epithelium of the hindgut are activated, then electrolytes and water are reabsorbed from the pre-urine and returned to the hemolymph. A hypertonic final urine results.

- **In terrestrial vertebrates, the kidney is responsible for excreting water-soluble waste products and achieving homeostasis with respect to water and electrolyte concentrations.**

Nephrons in the mammalian kidney form a raw filtrate and then reabsorb valuable nutrients, electrolytes, and water. The resulting solution contains urea and electrolytes and flows through the loop of Henle, where changes in the permeability of epithelial cells to water and salt—along with active transport of salt—create a steep osmotic gradient. If ADH triggers changes in the water permeability of the collecting duct of the nephron, water is reabsorbed along the osmotic gradient and a hypertonic urine is produced.

In animals, maintaining water and electrolyte balance is an active, energy-demanding process that is based on the action of membrane proteins. Water and electrolyte excretion and reabsorption are carefully controlled to achieve homeostasis. As a result, animals are able to occupy habitats that present a diversity of osmotic challenges.

Web Tutorial 42.1 The Mammalian Kidney

Questions

Content Review

1. Which one of the following statements is true of fish that live in freshwater?
   a. Their tissues are isotonic with respect to their environment. As a result, they do not require a specialized organ to maintain water and electrolyte balance.
   b. They lose water to their environment primarily through the gills. They replace this water by drinking.
   c. Water enters epithelial cells in their gills via osmosis. Electrolytes leave the same cells via diffusion.
   d. They have specialized cells that actively pump Na^+ and Cl^- from blood into epithelial cells, so the ions can be excreted.

2. Na^+/K^+-ATPase is found in epithelial cells that produce urine or other excretions that are hypertonic relative to tissues. Why?
   a. To form a hypertonic solution, ions or molecules must be actively transported across their concentration gradient.
   b. Cells similar to those found in the shark rectal gland occur in many other species that need to excrete salt (NaCl).
   c. The sodium-potassium pump is directly responsible for moving sodium ions from blood or body tissues into the urine or salt excretion, against a concentration gradient.
   d. They are usually located in the basolateral membrane of those epithelial cells, along with other channels and cotransporters.

3. Regarding urine formation, the function of the insect hindgut is closest to which of the following structures in the vertebrate kidney?
   a. renal corpuscle
   b. proximal tubule
   c. loop of Henle
   d. distal tubules and collecting duct
4. Which of the following gives the correct sequence of the regions of the nephron with respect to fluid flow?
   a. distal tubule → ascending limb of Henle → descending limb of Henle → proximal tubule → collecting duct
   b. ascending limb of Henle → descending limb of Henle → proximal tubule → distal tubule → collecting duct
   c. proximal tubule → ascending limb of Henle → descending limb of Henle → distal tubule → collecting duct
   d. proximal tubule → descending limb of Henle → ascending limb of Henle → distal tubule → collecting duct
5. Which of the following statements about kidney function is *not* correct?
   a. The loop of Henle acts as a countercurrent exchanger, so no energy is required to maintain the osmotic gradient.
   b. The descending limb of the loop of Henle is highly permeable to water.
   c. The thin ascending limb of the loop of Henle is highly permeable to salt.
   d. Reabsorption of water and solutes takes place primarily in the proximal tubule.
6. What effect does antidiuretic hormone (ADH) have on the nephron?
   a. It increases water permeability of the descending limb of the loop of Henle.
   b. It decreases water permeability of the descending limb of the loop of Henle.
   c. It increases water permeability of the collecting duct.
   d. It decreases water permeability of the collecting duct.

Conceptual Review

1. Study the diagram of a salmon's movements in Figure 42.6b. Explain the changes that need to occur in water and electrolyte balance as an individual moves from freshwater to salt water and back. Specifically, state when the animal should drink or not drink, when cells in the gill epithelium should excrete or import electrolytes, and why each change occurs.
2. The chloride cells of fish gills are sometimes called mitochondria-rich cells due to their high density of mitochondria. How does high mitochondrial density relate to the functional role of chloride cells? Would you expect other epithelial cells involved in ion transport to contain large numbers of mitochondria? Explain.
3. Why is it significant that cells involved in transport processes often have microvilli?
4. This chapter introduced a number of features that help terrestrial animals reduce water loss. These traits include the layer of wax found on insect exoskeletons, the ability of insects to close the openings to their respiratory passages, the excretion of nitrogenous wastes as insoluble uric acid, exceptionally long loops of Henle in the mammalian kidney, and the retention of water in the nasal passages of desert mammals. Predict how each of these traits differs in animals that live in very humid versus very dry habitats. How would you test your predictions?
5. In insects, active transport of electrolytes into the Malpighian tubules leads to the formation of a pre-urine that is isotonic with respect to the hemolymph. In mammals, urine formation begins with blood pressure in the renal corpuscle that leads to the formation of a filtered pre-urine that is isotonic with respect to the blood. In insects, the pre-urine is processed in the hindgut; in mammals, the pre-urine is processed in the remainder of the nephron. How are the processing steps in insects and mammals similar? How are they different?
6. Compare and contrast the types of nitrogenous wastes observed in animals. Identify which compound can be excreted with a minimum of water, which is most toxic, and which waste product is found in fish, mammals, and insects.

Group Discussion Problems

1. Examine Figure 42.12b again, and notice that the network of blood vessels that runs along the mammalian nephron is arranged so that blood flows in a countercurrent arrangement relative to the flow of fluid in the loop of Henle. The water and electrolytes that leave the loop of Henle and collecting duct diffuse into this blood vessel and are returned to the body. Why is the countercurrent arrangement for blood flow important?
2. You have isolated a segment of a rat nephron and introduced a solution of known composition. When you compare the fluid collected at the end of the segment with the test solution introduced at the beginning, you find that the volume has decreased by 30 percent and that the Na^+ concentration has decreased by 30 percent, but the urea concentration has increased by 50 percent. What processes in this segment might account for these changes?
3. In some areas of the world, highway maintenance crews make extensive use of salt to keep roads free of ice in winter. Biologists are concerned about the effect of this salt on terrestrial and freshwater organisms. Why?
4. Ethanol is the active ingredient in alcoholic beverages. One of its many toxic effects is to inhibit the release of antidiuretic hormone (ADH) by the brain. How does the ingestion of ethanol affect urine formation? Why do people who ingest large amounts of ethanol suffer from headaches, body aches, and other symptoms of dehydration?

Answers to Multiple-Choice Questions **1.** c; **2.** a; **3.** d; **4.** d; **5.** a; **6.** c

www.prenhall.com/freeman is your resource for the following: Web Tutorials; Online Quizzes and other Online Study Guide materials; Answers to Conceptual Review Questions; Solutions to Group Discussion Problems; Answers to Figure Caption Questions and Exercises; and Additional Readings and Research.

Animal Nutrition

43

A young crocodile has just caught a reed frog. Most animals obtain nutrients by ingesting them.

KEY CONCEPTS

- Animals require an array of nutrients to stay healthy, including specific amino acids, vitamins, and elements, as well as organic compounds that have high potential energy.
- Animals feed on a wide variety of items. In animals that ingest their food, there is usually a close correspondence between the structure of the animal's mouthparts and the function of those mouthparts in capturing and processing food.
- Digestion occurs in the digestive tract, which is compartmentalized into organs that have specialized functions in the digestion of food, absorption of nutrients and water, or excretion of wastes.
- Lack of homeostasis with respect to nutrients such as glucose and cholesterol can cause disease.

If you're like most people, you probably give a fair amount of thought to the types of food that you eat but little thought to what happens to that food in your digestive tract. As a meal moves through the digestive system, its chemical composition and physical characteristics change dramatically. Structurally and chemically, complex raw materials are broken down into simple components that can be used to build and maintain cells. Large packets of food that enter the mouth must be reduced to monomers that can be absorbed by cells and enter the bloodstream. The digestive system is analogous to an oil refinery or a lumber mill, where bulky, complex raw materials are processed into smaller and more usable products.

The types of food that are available to different animals vary widely, however, and food is often in dangerously short supply. Based on these observations, it is logical to predict that animals have a variety of means for obtaining food and that animals are under intense natural selection to make efficient use of the food they have. How do animals get their food, and how do they process it? Which substances in food are used as nutrients, and how do humans and other animals maintain appropriate levels of key nutrients in their bodies?

Research on feeding and digestion is fundamental to understanding many of the most basic aspects of animal anatomy and physiology. But research on animal nutrition has important practical applications as well. For example, this chapter addresses questions about whether certain nutrients can be used to boost athletic performance, and why type II diabetes, heart disease, and other nutrition-related diseases are on the rise in many human populations. Just as plant nutrition is fundamental to the productivity of agricultural and natural ecosystems, nutrition is a basic component of animal health and welfare. Let's begin with a look at what animals must eat in order to live.

43.1 Nutritional Requirements

A **nutrient** is a substance that an organism needs to remain alive. **Food** is any material that contains nutrients. When an animal takes in enough nutrients to perform day-to-day tasks and stay healthy, it is said to be in **nutritional balance.** To maintain this state, an animal must acquire two general types of nutrients: (1) carbon compounds that have high potential energy and can be oxidized to produce ATP, and (2) the elements and molecules that are needed to synthesize body components and sustain cells.

Understanding which nutrients an individual needs, and in what amounts, are basic issues in research on animal nutrition. Determining nutritional requirements is seldom simple, however. To illustrate how biologists go about the task, let's first consider what humans need to maintain good health and then what they need in the special case of enhancing athletic performance.

Meeting Basic Needs

The United States government instituted a program to establish basic nutritional requirements as part of its war effort during World War II. In 1943 the Food and Nutrition Board of the National Academy of Sciences* published the first Recommended Dietary Allowances (RDAs). The goal of the RDAs was to specify the amount of each essential nutrient that an individual must ingest to meet the needs of practically all healthy people.

*The National Academy of Sciences is a group of scientists and engineers that advises the U.S. Congress on scientific and technical matters. Its Food and Nutrition Board is made up of biologists who specialize in animal nutrition.

The recommendations focused on several types of nutrients:

- Proteins provide amino acids and are sometimes oxidized to provide energy. Of the 20 amino acids required to manufacture proteins, humans can synthesize 12. The other eight must be obtained from food and are called **essential amino acids.** The word *essential* refers to nutrients that cannot be synthesized and must be obtained in the diet.
- **Vitamins** are organic compounds that are vital for health but are required in only minute amounts. They have a variety of roles; several function as coenzymes in critical reactions. **Table 43.1** lists a few of the vitamins for which RDAs have been established, notes their functions, and indicates the problems that develop if they are missing in the diet.
- **Essential elements** cannot be synthesized by the body but serve a wide variety of functions. Some, such as calcium and phosphorus, are needed in relatively large quantities. Others, such as iron and magnesium, are required in small or trace amounts. **Table 43.2** provides details on some particularly important elements in the human diet.
- **Electrolytes** are substances that form ions in solution. As ions, they influence osmotic balance and are required for normal membrane function. Sodium (Na^+), potassium (K^+), and chloride (Cl^-) are the major ions in the body.

In the United States, packaged food must display how a serving of the contents meets the RDA values (**Figure 43.1**). In many cases, labels must also furnish information on the amount of chemical energy available in the food. In nutrition research, the **Calorie** is a unit of energy. If a single helping of food contains 150 Calories, then 150 **kilocalories (kcal)** of energy is released when the food is completely oxidized by cells. (It is important to

TABLE 43.1 Essential Vitamins (Required by Humans)

Name	Source in Diet	Function	Symptoms if Deficient
Vitamin B_1 (thiamine)	legumes, whole grains, potatoes, peanuts	formation of coenzyme in Krebs cycle	beriberi (fatigue, nerve disorders, anemia)
Vitamin B_{12}	red meat, eggs, dairy products; also synthesized by bacteria in intestine	coenzyme in synthesis of proteins and nucleic acids; formation of red blood cells	anemia (fatigue and weakness due to low hemoglobin content in blood)
Niacin	meat, whole grains	component of coenzymes NAD^+ and $NADP^+$	pellagra (digestive problems, skin lesions, nerve disorders)
Folate	green vegetables, oranges, nuts, legumes, whole grains; also synthesized by bacteria in intestine	coenzyme in nucleic acid and amino acid metabolism	anemia
Vitamin C (ascorbic acid)	citrus fruits, tomatoes, broccoli, cabbage, green peppers	used in collagen synthesis, prevents oxidation of cell components, improves iron absorption	scurvy (degeneration of teeth and gums)
Vitamin D	fortified milk, egg yolk; also synthesized in skin exposed to sunlight	aids absorption of calcium and phosphorus in small intestine	rickets (bone deformities) in children; bone softening in adults

TABLE 43.2 Essential Elements (Required by Humans)

Name	Source in Diet	Function	Symptoms if Deficient
Calcium (Ca)	dairy products, green vegetables, legumes	bone and tooth formation, nerve signaling, muscle response	loss of bone mass, slow growth
Phosphorus (P)	dairy products, meat, grains	bone and tooth formation, synthesis of nucleotides and ATP	weakness, loss of bone
Sulfur (S)	any source of protein	amino acid synthesis	swollen tissues, degeneration of liver, mental retardation
Magnesium (Mg)	whole grains, green leafy vegetables	enzyme cofactor	nerve disorders
Iron (Fe)	meat, eggs, whole grains, green leafy vegetables, legumes	enzyme cofactor; synthesis of hemoglobin and electron carriers	anemia, weakness
Fluorine (F)	fluoridated water, seafood	maintenance of tooth structure	higher frequency of tooth decay

realize, though, that organisms rarely extract all the calories potentially available, because food is not completely digested.)

Most energy in food is in the form of carbohydrates and fats, which are carbon compounds with high potential energy (see Figure 5.7, page 99). Because fats and other lipids contain more C–H bonds than do carbohydrates such as starch and sugar, lipids provide over twice the energy per gram. Lipids yield 9 kcal per gram when completely oxidized, compared with only 4 kcal/gm for carbohydrates and proteins.

RDAs frequently change as research advances and, in many cases, members of the Food and Nutrition Board are cautious about making recommendations, due to a lack of reliable data. Studying nutrient requirements in humans is difficult. The types of experiments that would be most enlightening would assign different, carefully controlled diets to several groups of subjects. Some of these diets might lead to illness, however, so they are unethical. As a result, researchers rely on observational studies that document the health of people whose intake levels for specific nutrients are known or that monitor the intake of a specific nutrient versus the level maintained in the body over time.

The scientists who publish the RDAs constantly remind the public that a balanced diet fulfills all nutrient requirements. In the vast majority of cases, nutritional supplements are not needed. And if certain types of nutritional supplements are abused, health problems can result. An important exception to this rule exists: Some women may need to take iron supplements as recommended by their physician. These iron supplements are beneficial in some cases because hemoglobin—the protein that carries oxygen in the blood—contains iron. Consequently, significant amounts of iron are lost during menstruation.

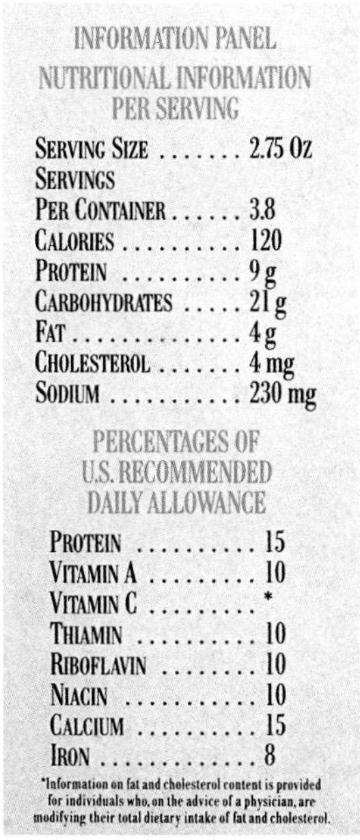
INFORMATION PANEL
NUTRITIONAL INFORMATION PER SERVING

SERVING SIZE	2.75 OZ
SERVINGS PER CONTAINER	3.8
CALORIES	120
PROTEIN	9 g
CARBOHYDRATES	21 g
FAT	4 g
CHOLESTEROL	4 mg
SODIUM	230 mg

PERCENTAGES OF U.S. RECOMMENDED DAILY ALLOWANCE

PROTEIN	15
VITAMIN A	10
VITAMIN C	*
THIAMIN	10
RIBOFLAVIN	10
NIACIN	10
CALCIUM	15
IRON	8

*Information on fat and cholesterol content is provided for individuals who, on the advice of a physician, are modifying their total dietary intake of fat and cholesterol.

FIGURE 43.1 Nutrition Labeling on Food Packaging
A label from a frozen pizza. EXERCISE The label claims that this product contains 120 Calories per serving. Using the Calories-per-gram conversions given in the text, calculate the actual number of Calories per serving.

Nutrition and Athletic Performance

RDAs identify the nutrients that most people need to stay healthy and the amounts that are required to maintain nutrient balance. These recommendations target people with normal needs and activities. How do researchers determine nutrient requirements for extraordinary circumstances? To answer this, let's consider research on athletes. These individuals do extraordinary amounts of physical work. What nutrients do they require to perform at an optimal level?

For a researcher interested in the nutrient requirements of athletes, at least some types of experiments are possible. As long as experimental subjects are provided with a diet that fulfills RDAs, and as long as safe dosages of experimental compounds are used, studies should be able to meet accepted ethical guidelines.

In 1967 Jonas Bergström and colleagues published a classic experiment in nutrition research. They were interested in the performance of endurance athletes, such as cross-country skiers, bicyclists, and distance runners. At the time, most researchers supported the hypothesis that fatty acids provide the fuel required for extended physical labor or athletic activity. This hypothesis was logical, because fatty acids provide more than twice as much energy as do carbohydrates on a per-gram basis.

Glycogen is an important energy storage molecule in animals. Recall from Chapter 5 that glycogen is a polysaccharide made up of glucose molecules and from Chapter 9 that glucose is the preferred starting compound for the production of ATP through cellular respiration. Most of the body's glycogen is stored in skeletal muscles and the **liver**. During exercise, glycogen is catabolized (broken down) into glucose by both types of tissues. Muscle cells use the glucose that is released to manufacture ATP; glucose from the liver enters the bloodstream and is delivered to all cells that are doing work.

Based on these observations, Bergström and associates hypothesized that an individual's ability to run, ski, or bicycle fast and for a long time might depend primarily on the amount of glycogen stored in the body rather than on the amount of fatty acids. The team also predicted that a diet rich in carbohydrates would maximize the amount of stored glycogen.

To test these hypotheses, the biologists set up the experiment outlined in **Figure 43.2**. Nine student volunteers began by eating a mixed diet of carbohydrate, fat, and protein for a day. After the researchers had removed a tiny sample of leg

Question: What types of diet optimize performance in endurance sports?

Hypothesis: Carbohydrate-rich diets should lead to better performance than normal or high-protein–high-fat diets, by maximizing glycogen stores.

Null hypothesis: There is no difference in performance on the basis of diet.

Experimental setup:

1. Start with nine volunteers, split into two groups.
2. Arrange for the volunteers to follow three types of diets:
   - **Mixed normal diet:** Mixed carbohydrate, protein, fat (2800 kcal total)
   - **Protein-fat diet:** 1500 kcal protein, 1300 kcal fat
   - **High-carbohydrate diet:** 2300 kcal carbohydrate, 500 kcal protein
3. Each group eats each type of diet for 1 to 3 days, but in different orders.
4. After eating each diet, volunteers undergo a performance test: Ride bike to exhaustion.

Prediction: Glycogen stores and performance will be best after rider has eaten high-carbohydrate diet before test ride.

Prediction of null hypothesis: No difference in glycogen stores or performance on the basis of diet preceding test.

Results:

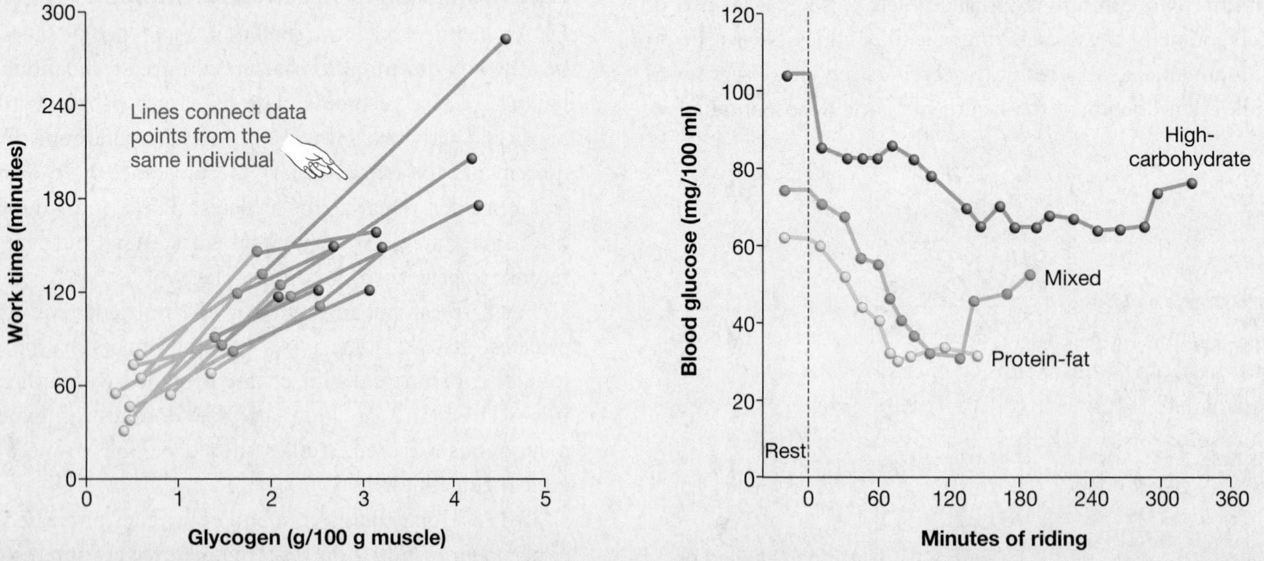

Conclusion: Individuals perform better after eating a diet high in carbohydrates than after eating a normal diet or a diet high in protein and fats.

FIGURE 43.2 Experimental Evidence That Carbohydrate Loading Increases Performance in Endurance Sports
During each test, volunteers rode exercise bikes until they felt exhausted. In the "Results" graphs, the data points correspond with the diets, and each line represents an individual in the study. QUESTION Why were the researchers careful to provide the same amount of energy, in kcal, in each diet?

muscle from each volunteer to measure the amount of stored glycogen, the volunteers rode exercise bikes at the same relative work rate until they were exhausted. During the exercise period, the biologists removed blood samples from each student at regular intervals to check the volunteers' concentration of glucose.

Six students then switched to a diet high in fat and protein for three days. The remaining three students switched to a high-carbohydrate diet. After the researchers checked the volunteers' glycogen stores, all nine volunteers again rode a stationary bike to exhaustion while their blood-glucose levels were recorded. Finally, the volunteers switched diets for a three-day period: The six individuals who had been eating a diet high in fat and protein switched to a high-carbohydrate diet; the three students who had been eating a high-carbohydrate diet switched to a diet high in fat and protein. To conclude the experiment, all nine volunteers had their glycogen reserves documented and then rode the stationary bike to exhaustion as their blood-glucose concentrations were monitored.

The experimental design effectively controlled for two potentially complicating factors: (1) Each volunteer was subjected to all three types of diet, so the outcome of the experiment was not due to a biased sample; and (2) both groups of volunteers experienced the two experimental diets in a different order, to test the hypothesis that the sequence of food types ingested might affect the results.

The left-hand graph of the "Results"section of Figure 43.2 plots the amount of glycogen present in leg muscle versus the "work time"—the amount of time each volunteer was able to ride after consuming each diet. (Each line represents one individual.) Notice that, in every case, both the amount of stored glycogen and the work time were much higher after individuals had eaten a high-carbohydrate diet. Now examine the right-hand graph in the "Results" section of the figure, which plots the changes in blood-glucose levels that occurred in one of the subjects during each of the three rides. These data also support the hypothesis that a diet high in carbohydrate supports optimal performance.

Subsequent experiments, performed in a variety of labs around the world, confirmed that performance in endurance sports is enhanced by a high-carbohydrate diet. Does ingesting glucose or other carbohydrates just before a race also improve performance? A thorough set of studies has failed to support this hypothesis. The current consensus among nutrition researchers is that ingesting sugars right before a contest actually reduces performance.

The experiments that were conducted with student volunteers revolutionized how athletes all over the world train for endurance events. The dietary regime known as carbohydrate loading is now a routine part of race preparations for endurance athletes such as marathoners, triathletes, distance swimmers, and cross-country skiers.

43.2 Obtaining Food: The Structure and Function of Mouthparts

Animals are **heterotrophs.** Instead of making their own food as plants do, animals obtain the energy and nutrients they need from other organisms. The question is, How?

As noted in Chapters 31 through 33, some animals obtain the food they need via suspension feeding or by absorbing it directly from the surrounding environment. Recall that *suspension feeders* (such as sponges and tube worms) filter cells, small multicellular organisms, or bits of organic debris from water by means of cilia or other structures. In contrast, most species that absorb their nutrients are *endoparasites*, meaning that they live inside the body of a host individual. But by far the most common method of heterotrophy in animals is to seize and manipulate food by using mouthparts such as jaws and teeth, beaks, or special toxin-injecting organs.

The types of food that animals harvest range from soupy solutions in decaying carcasses to hard nuts. Given the diversity of food sources that animals exploit, it is not surprising that animals have a wide variety of mouthpart structures involved in capturing and processing food. Recall from Chapter 31 that natural selection has tightly matched the structure of animal mouthparts to their function in obtaining food. The following examples will help drive this point home:

- Some animals, such as the snake in **Figure 43.3a** (page 982), simply open their mouths wide and ingest food items whole. In snakes, the bones that make up the skull are joined in a way that allows the jaws to open extraordinarily wide. In snakes that kill their prey with venom before swallowing, one or more teeth are enlarged into fangs that are hollow or that have deep grooves where poison is conducted into the prey's body. In species that swallow prey that are still struggling, the snake's jaws are lined with sharp teeth that point toward the throat and help prevent the prey from escaping.
- Insects have diverse mouthpart structures (**Figure 43.3b**), ranging from the needlelike proboscis of mosquitoes to the pinching mandibles of leaf-cutting ants. In other insect species, mouthparts are specialized for grasping or filtering small particles or sponging, siphoning, or lapping up liquids.
- **Figure 43.3c** shows the mouthparts of mammals that are adapted to different diets. Mammals are the only vertebrates that evolved the ability to chew their food, and their teeth are modified to serve a variety of functions during chewing. Beavers have long front teeth that are sharp enough to cut wood and that grow constantly. Once the front teeth have bitten off chunks of inner bark, large, flattened molars at the back of the mouth mash the material into particles that can be swallowed. Wolves have bladelike teeth that tear flesh efficiently. Human teeth include sharp canines for tearing meat and flattened molars for crushing seeds, roots, and other plant foods.

(a) Snakes have jaws that are modified for swallowing food whole.

Tiger rattlesnake ingesting a mouse

(b) Arachnid and insect mouthparts are modified for processing a wide array of food sources.

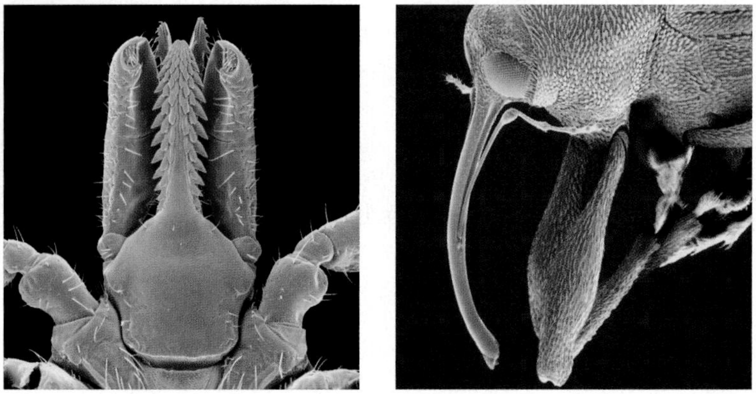

Deer ticks pierce the skin of mammals.

Nut weevils pierce hard fruits.

(c) The jaws and teeth of mammals are modified for eating plant material, other animals, or both.

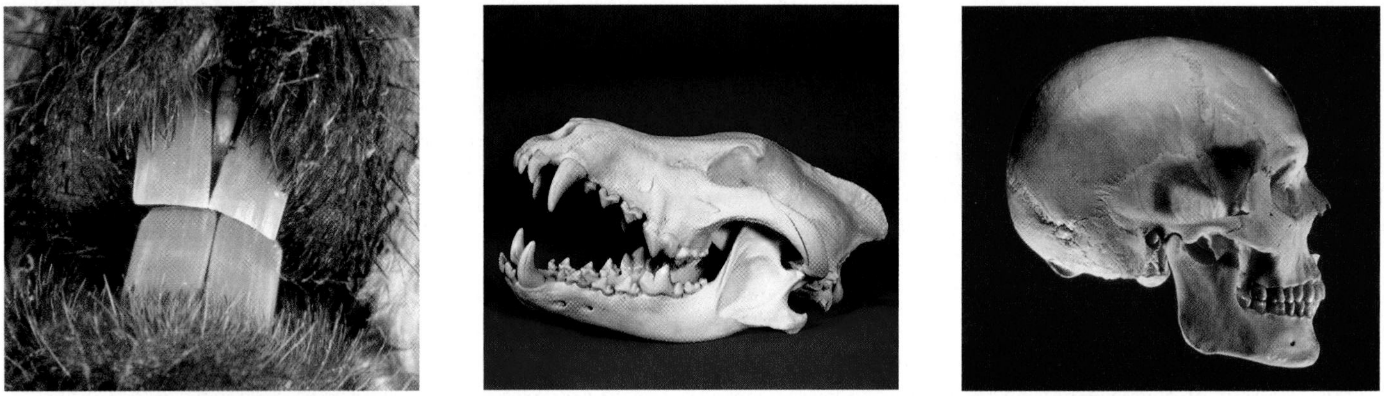

Beavers eat the inner bark of woody plants.

Wolves can subdue large prey by biting.

Humans can process plant and animal tissues efficiently.

FIGURE 43.3 The Structure of Animal Mouthparts Correlates with Their Function in Feeding
(a) Snakes swallow their prey whole instead of biting off chunks of tissue; their skull and jawbones are extraordinarily flexible. **(b)** Insect mouthparts may be modified to function as awls, pincers, knives, extensible sponges, or straws. **(c)** Mammals have a wide array of tooth types.

Let's pursue the correlation between the size and shape of mouthparts and the size and shape of food sources further, by analyzing the structure and function of jaws and teeth in what may be the most diverse lineage among vertebrates: the cichlid fishes of Africa (**Figure 43.4**).

The cichlids that inhabit the Rift Lakes of East Africa are a spectacular example of adaptive radiation. Recall from Chapter 26 that adaptive radiation refers to the diversification of a single ancestral population into many species, each of which lives in a different habitat or employs a distinct feeding method. Lake Victoria, for example, is home to 300 **endemic** cichlids—meaning species that live nowhere else. Most of these species feed on a specific item, but as a group they exploit almost every conceivable food source in the lake: planktonic organisms, algae that grow on surfaces of rocks, algae that have leaflike blades, eggs, fish scales, fish fins, whole fish, plants, insects, and snails.

How can a group of very closely related species exploit so many different food sources? To answer this question, biologists point out that the African cichlids have an extra set of jaws. As **Figure 43.5a** shows, pharyngeal (throat) jaws are located well behind the normal oral jaws. Pharyngeal jaws are actually found in many types of fish besides cichlids, but in non-cichlids the extra jaws are simply used to transport food from the mouth to the back of the throat. In cichlids, however, pharyngeal jaws are anchored against the skull and have muscles attached that allow them to bite. As a result, they are used to process food by crushing, tearing, or compacting it. In effect, then, cichlids are able to feed with particular efficiency because they have two structures instead of just one for ingesting and processing food.

How did the pharyngeal jaws of cichlids originate? In both cichlid and non-cichlid embryos, pharyngeal jaws develop from

Eats snails — Eats small invertebrates — Eats algae

FIGURE 43.4 Rift Lake Cichlids Are a Spectacular Example of Adaptive Radiation
Hundreds of species of cichlids live in the Rift Lakes of East Africa.

(a) Pharyngeal (throat) jaws are located behind the oral jaws. In cichlids, the pharyngeal jaw can bite down on food to help chew it.

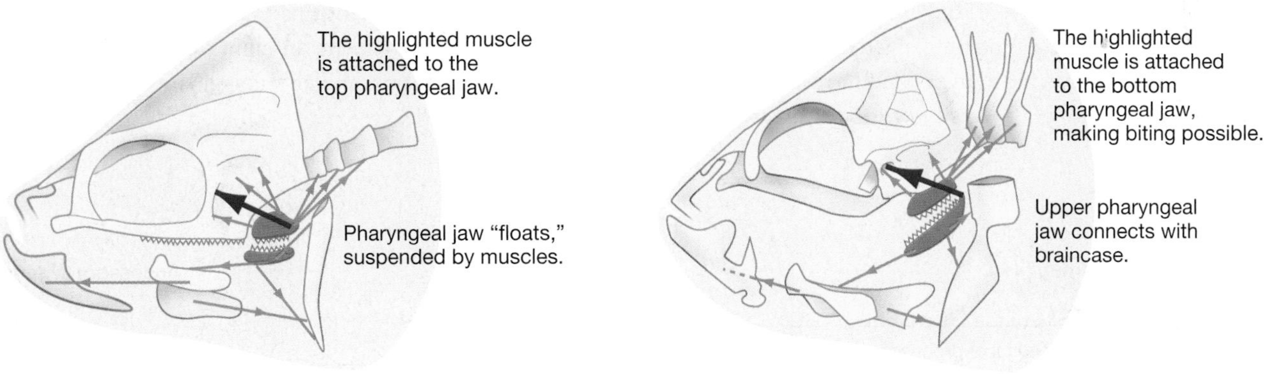
The highlighted muscle is attached to the top pharyngeal jaw.
Pharyngeal jaw "floats," suspended by muscles.
Non-cichlid
The highlighted muscle is attached to the bottom pharyngeal jaw, making biting possible.
Upper pharyngeal jaw connects with braincase.
Cichlid

(b) Teeth on pharyngeal jaws vary among species of cichlids and correlate with the type of food eaten.

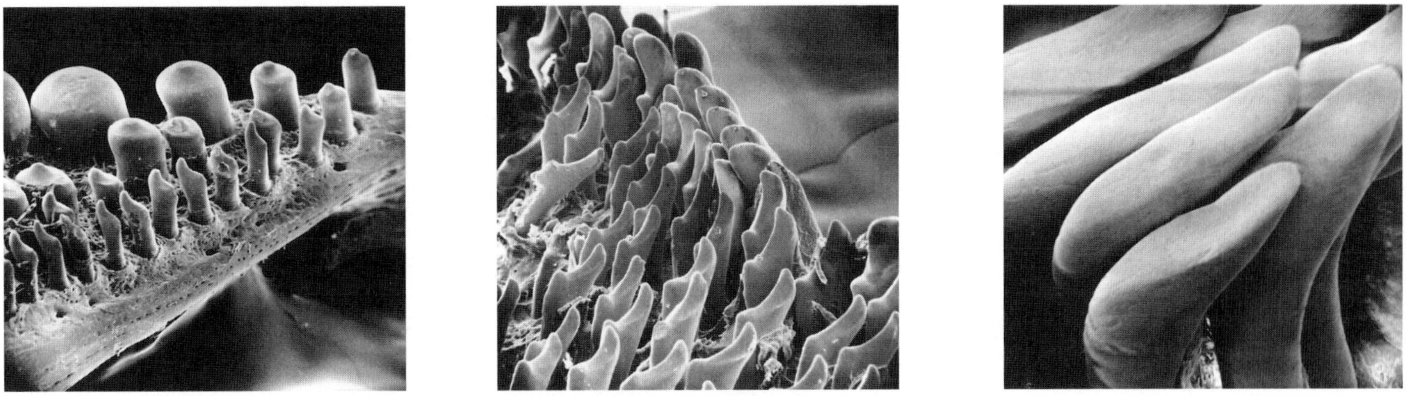

Crushes snails — Eats whole fish — Grazes on algae

FIGURE 43.5 Rift Lake Cichlids Have Pharyngeal Jaws That Make Feeding Particularly Efficient
(a) The arrows indicate where muscles are inserted in and around the pharyngeal jaws. **(b)** Scanning electron micrographs showing the morphology of pharyngeal jaw teeth. **EXERCISE** In the right-hand drawing of part (a), indicate where the teeth illustrated in part (b) are located. Label the oral jaw in the skull on the right of part (a).

stiff structures called gill arches. But at maturity, cichlid pharyngeal jaws have several unique features:

- One end of the cichlid pharyngeal jaw connects with the skull. This is important, because the connection forms a hinge and allows the structure to act as a lever.
- Several muscles that bypass the pharyngeal jaws in other groups attach to the pharyngeal jaws of cichlids. This is a key trait, because the added musculature makes the pharyngeal jaw in cichlids capable of biting.
- The lower pharyngeal jaw is sutured into one solid structure instead of existing as two independent pieces. This feature makes the jaw more effective as a biting device.

In addition, biologists have documented that toothlike protuberances on the pharyngeal jaws of cichlids vary in size and shape. Cichlids have an extra set of toothlike structures, as well as extra jaws. The structure of these "teeth" correlates with their function in crushing snail shells, tearing meat, or compacting algae (**Figure 43.5b**). These observations add to a large body of evidence supporting a general pattern in animal evolution: Mouthparts have diversified in response to natural selection for exploiting a diversity of food sources.

43.3 How Are Nutrients Digested and Absorbed?

Animals ingest just about every type of food conceivable. Ingestion is only the first of four processes needed to obtain energy, however. For an animal to stay alive, ingestion must be followed by digestion, absorption, and excretion.

Digestion is the breakdown of food into small enough pieces to allow for **absorption**—the uptake of specific ions and molecules that act as nutrients. Digestion is a key process in animals that ingest their food, because, unlike plants, unicellular organisms, and certain parasites, animals do not acquire nutrients as individual molecules. Instead, they take in packets of food that must be broken down into small pieces. Nutrients must be extracted from the meal, and waste materials must be eliminated. How and where does this processing occur?

Digestion takes place in a tube called the **digestive tract**—also known as the the *alimentary* ("nourishment)" *canal* or *gastrointestinal* (GI) *tract*. As **Figure 43.6** shows, the digestive tract has two openings: It starts at the **mouth** and ends at the **anus**. The interior of this tube communicates directly with the external environment. In embryos, it derives from the hollow tube that forms as cells invaginate during gastrulation (see Chapter 21).

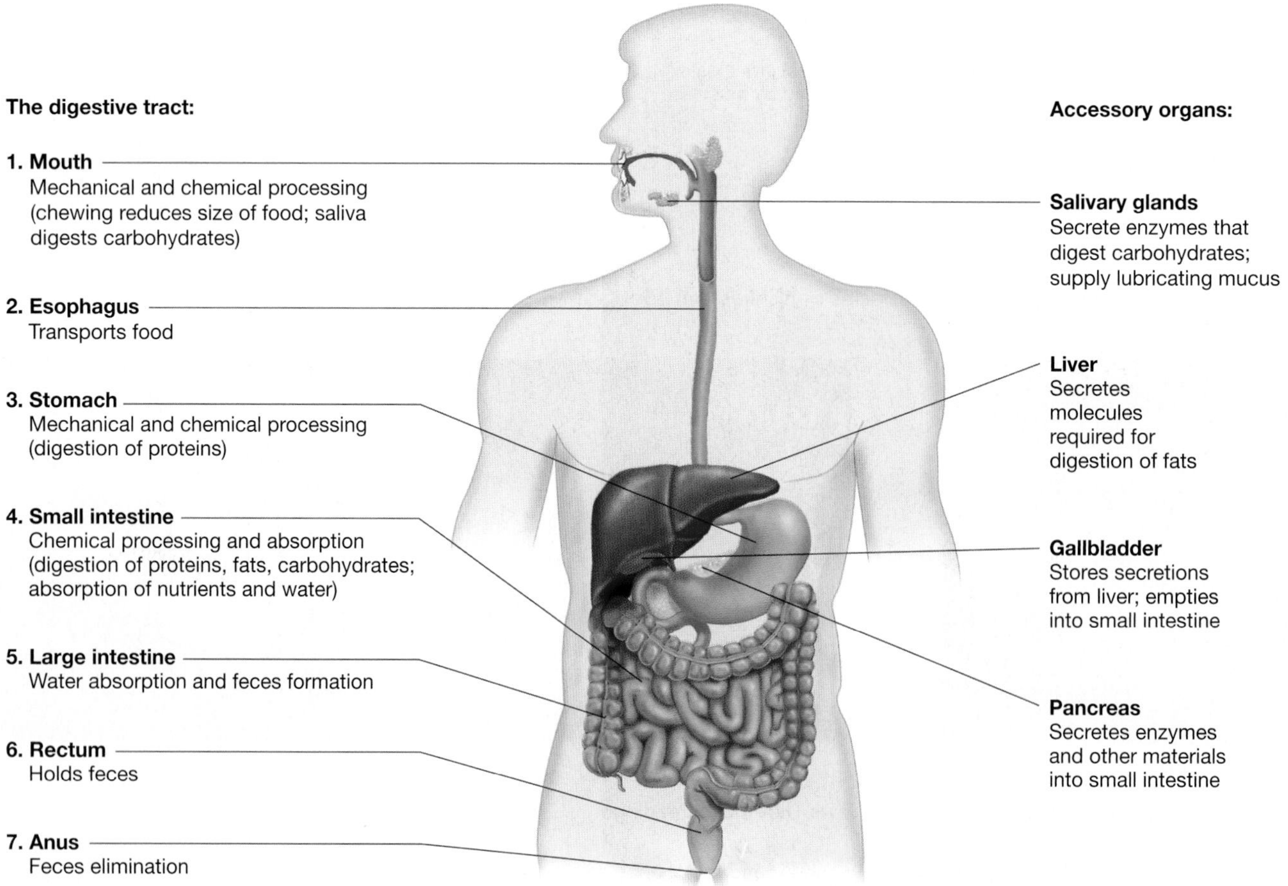
The digestive tract:
1. Mouth
Mechanical and chemical processing (chewing reduces size of food; saliva digests carbohydrates)
2. Esophagus
Transports food
3. Stomach
Mechanical and chemical processing (digestion of proteins)
4. Small intestine
Chemical processing and absorption (digestion of proteins, fats, carbohydrates; absorption of nutrients and water)
5. Large intestine
Water absorption and feces formation
6. Rectum
Holds feces
7. Anus
Feces elimination
Accessory organs:
Salivary glands
Secrete enzymes that digest carbohydrates; supply lubricating mucus
Liver
Secretes molecules required for digestion of fats
Gallbladder
Stores secretions from liver; empties into small intestine
Pancreas
Secretes enzymes and other materials into small intestine

FIGURE 43.6 The Digestive System (Alimentary Canal) and Accessory Organs
The digestive tract is a tube that runs from the mouth to the anus. The salivary glands, liver, gallbladder, and pancreas are not part of the tract itself. Instead, they secrete material into the tract at specific points.

A tubelike digestive tract has three main advantages. First, the digestion of material inside the tract allows animals to feed on large pieces of food. Second, different chemical and physical processes can be separated within the digestive tract, so that they occur independently of each other and in a prescribed sequence. The stomach, for example, provides an acidic environment for digestion. Material is transferred from there to the small intestine, where enzymes are specialized to function in an alkaline environment. Third, a tubelike digestive tract provides a one-way flow of food and allows wastes to exit the body at the anus.

The digestive tract is only part of the entire digestive system, however. Several vital organs and glands are connected to the digestive tract. These structures contribute digestive enzymes and other products to specific portions of the tract. They include the salivary glands, liver, gallbladder, and pancreas (see Figure 43.6).

Before analyzing the function of each component of the digestive system in detail, let's consider the general changes that happen to food on its way through the digestive tract. In this brief overview and in the detailed discussion that follows, humans will serve as a model species—simply because so much is known about human digestion.

Digestion begins with the tearing and crushing activity of teeth. Chewing reduces the size of food particles and softens them. In humans, the mechanical breakdown of food is augmented by the use of knives and cooking. In fact, the invention of cutting tools and cooking, which make food easier to chew, is the leading hypothesis to explain why average tooth size has declined steadily over the past several million years of human evolution.

Figure 43.7 summarizes the chemical changes that occur as food moves through the digestive tract. The chemical breakdown of carbohydrates begins in the mouth, through enzymes in saliva. Chemical processing continues in the stomach and is completed in the small intestine. Each of the three major types of macromolecules—carbohydrates, lipids, and proteins—must be broken down during digestion. The small molecules that result from the digestion of these macromolecules are absorbed in the small intestine,

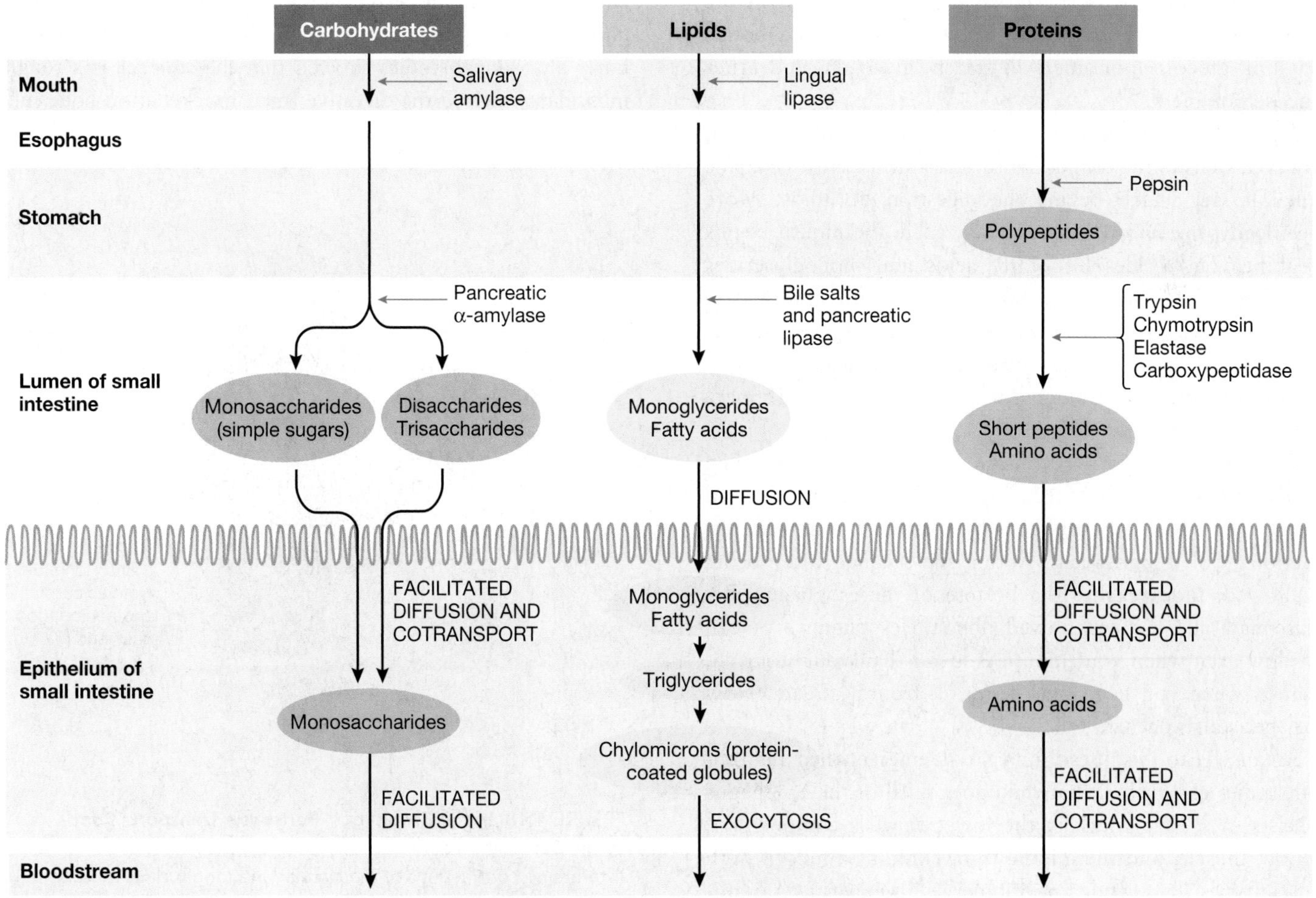

FIGURE 43.7 Carbohydrates, Lipids, and Proteins Are Processed in a Series of Steps
Three basic types of foods enter the digestive system (top of diagram). As they proceed through the digestive tract, they are broken apart by a variety of enzymes. Simple sugars, fatty acids, and amino acids then enter epithelial cells in the small intestine and are transported to the bloodstream.

WEB TUTORIAL 43.1 The Digestion and Absorption of Food

along with water, vitamins, and ions. In the large intestine, or colon, more water is absorbed. The result is **feces**, which are eventually excreted.

For a human to stay healthy, each step in this process must be completed correctly. Problems in the digestive tract can lead to heartburn, ulcers, nausea, constipation, and other maladies. By far the most serious of these is diarrhea, which kills more than 4 million children each year. Because digestion is so important to understanding how animals work, let's analyze each step in more detail.

The Mouth and Esophagus

If you hold a salted cracker in your mouth long enough, it will start to taste sweet. The sensation occurs because starch molecules in the cracker are being hydrolyzed to glucose by an enzyme in saliva. Starch breakdown was actually the first enzyme-catalyzed reaction ever discovered. In the early 1800s several researchers found that a component of certain plant extracts digested starch; in 1831 the same activity was discovered in human saliva. The enzyme responsible is called **amylase**. Amylase ranks as one of the best-studied enzymes; in animals it is the most important catalyst in the breakdown of carbohydrates. Amylase cleaves the bonds that link glucose monomers in starch, glycogen, and other glucose polymers.

Amylase is not the only important ingredient in saliva, however. Cells in the tongue synthesize and secrete an enzyme called **lipase**, which begins the digestion of lipids. More specifically, the *lingual lipase* produced in the mouth begins breaking triglycerides into fatty acids and monoglycerides (see Chapter 6). **Salivary glands** in the mouth also release water and glycoproteins called **mucins**. When mucins contact water, they form the slimy substance called **mucus**. The combination of water and mucus makes food soft and slippery enough to be swallowed.

Once food is swallowed, it enters a muscular tube called the **esophagus**, which connects the mouth and stomach. Food is propelled down the esophagus by a wave of muscular contractions called **peristalsis**. About six seconds after being swallowed, food reaches the bottom of the esophagus. Because material is actively moved along the esophagus, you can swallow even when your mouth is lower than your stomach, such as when you bend over to drink from a stream. How does peristalsis occur?

An answer to this question began to emerge when anatomical studies established that the upper third of the esophagus consists of skeletal muscle, the lower third is composed of smooth muscle, and the middle third contains a mix of both muscle types. By cutting specific nerves in experimental animals, researchers established that peristalsis in the upper third of the esophagus occurs when a series of nerve cells that originate at the base of the brain sends electrical signals in a precise sequence. Each of these nerves terminates at skeletal muscle at a different location along the esophagus. Some of these muscles wrap around the esophagus; others run along its length. In response to nerve signals, the muscle contracts or relaxes in a coordinated fashion (**Figure 43.8**). In this way, a wave of muscle contractions propagates down the tube, propelling the food mass ahead of it. The action of these nerves is not the result of a conscious choice. The system is a **reflex**—an automatic reaction to a stimulus—that is stimulated by the act of swallowing.

Even after decades of experiments, though, the mechanism of peristalsis in the lower third of the esophagus is still not well understood. Researchers can induce normal peristalsis in this section of the esophagus in the absence of stimulation by nerves from the brain or spinal cord. Based on these results, most researchers agree that peristalsis in the lower region of the esophagus is coordinated at least in part by cells in the wall of the tube itself. A clear understanding of the mechanism has yet to emerge, however.

The Stomach

Although little if any digestion occurs in the esophagus, the situation changes dramatically when food reaches the stomach. Early anatomical studies showed that the **stomach** is a tough, muscular pouch in the digestive tract, bracketed on both ends

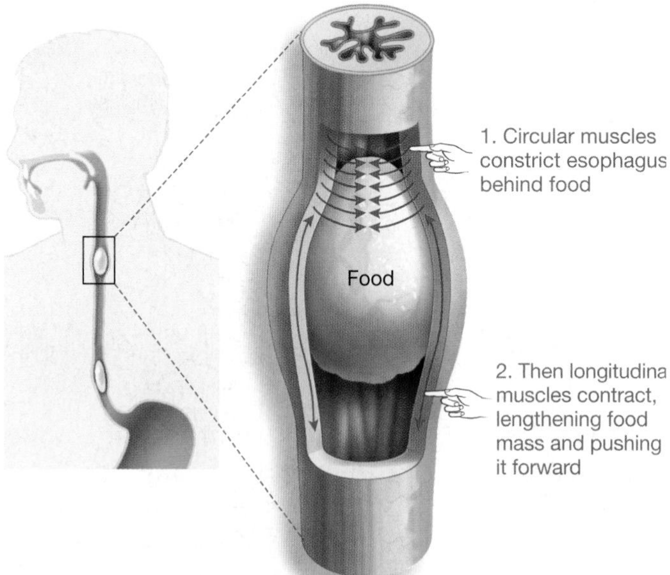
1. Circular muscles constrict esophagus behind food
Food
2. Then longitudina muscles contract, lengthening food mass and pushing it forward

FIGURE 43.8 In the Esophagus, Peristalsis Transports Food to the Stomach
Alternating contractions of the circular and longitudinal muscles in the esophageal wall constrict the lumen of the esophagus behind food and expand the lumen in front of food. The wave of contraction begins at the oral end of the esophagus and propels the food toward the stomach.

by valves called **sphincters** (**Figure 43.9**). When a meal fills the stomach, muscular contractions result in churning that mixes the contents and reduces food to a uniform consistency and osmolarity. A certain amount of mechanical breakdown of food also occurs in response to this churning. The other main function of the stomach is the digestion of proteins.

Compared with the mouth or esophagus (or virtually any other tissue), the stomach is highly acidic. Researchers in the late 1700s and early 1800s documented this fact by analyzing vomit or the contents of sponges that were tied to strings, swallowed, and pulled back up. In the early 1800s chemists were able to document that the predominant acid in the stomach is hydrochloric acid (HCl).

Not long after, a physician named William Beaumont established that digestion takes place in the stomach. He reached this conclusion through an extraordinary series of experiments on a young man named Alexis St. Martin. In 1822, when St. Martin was 19 years old, a shotgun accidentally discharged into his belly area and created a series of wounds. Despite repeated attempts, Beaumont was unable to close a hole in St. Martin's stomach. Eventually Beaumont inserted a small tube through the opening; the tube remained in St. Martin's body for the rest of his life. (Today, biologists insert tubes into various parts of the digestive tract of cows or sheep to study how these animals digest different types of feed.) With the tube in place, Beaumont was able to tie a string onto small pieces of meat or vegetables, insert the food directly into St. Martin's stomach, and draw it out after various intervals. Beaumont was also able to remove liquid from inside the stomach and observe how this *gastric juice** acted on food in vitro. His experiments showed that gastric juice digests food—particularly meat.

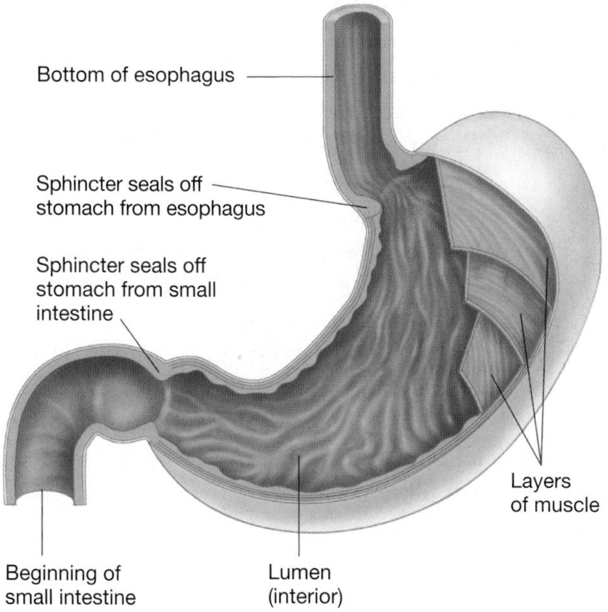

FIGURE 43.9 The Stomach Is an Acidic Environment Where Food Is Mixed and Proteins Are Digested
The stomach is a muscular outpocketing of the digestive tract. Muscle contractions mix food and break it into smaller pieces.

A researcher later purified the enzyme that is responsible for digesting proteins in the stomach and named it **pepsin**. Because it destroys proteins, biologists hypothesized that pepsin must be synthesized and stored in cells while it is in an inactive form—otherwise it would kill the cells where it is synthesized. In 1870 a biologist established through careful microscopy that granules occur in specialized stomach cells called *chief cells*. These granules were hypothesized to be a pepsin precursor. Follow-up work showed that this hypothesis was correct. The precursor compound, which came to be called **pepsinogen**, is converted to active pepsin by contact with the acidic environment of the stomach.

Which Cells Produce Stomach Acid? Researchers who were studying the anatomy of the stomach wall documented the presence of distinctive cells that came to be called parietal cells. As **Figure 43.10a** (page 988) shows, **parietal cells** are located along canals that communicate with the lumen of the stomach. An investigator also noted that the shape and activity of these cells appeared to vary as the digestion of a meal proceeded. On the basis of these observations, he inferred that parietal cells are the source of the HCl in gastric juice, which may have a pH as low as 1.5. Earlier microscopists had shown that another type of cell, called a **goblet cell**, secretes the mucus that is found in gastric juice. Mucus lines the gastric epithelium and protects the stomach from damage by HCl. To summarize, these anatomical studies showed that the epithelium of the stomach contains several types of secretory cells, each of which is specialized for a particular function.

To test whether parietal cells secrete HCl directly or in an inactive form, a biologist injected a dye into the stomach of dogs. The dye that was used in this experiment turns blue in the presence of acid. When the biologist examined the stained stomach tissue with the microscope, she found blue strands running through the canals in the gastric epithelium and connecting to parietal cells. On the basis of these results, she claimed that parietal cells secrete protons and chloride ions directly. But because other researchers found inconclusive results with dye experiments, the hypothesis of direct secretion remained controversial for decades. It was resolved only when researchers began to work out the molecular mechanism of HCl production.

*The Greek root *gastro* refers to the stomach. The root is used in the terms *gastric juice* and *gastrointestinal tract*.

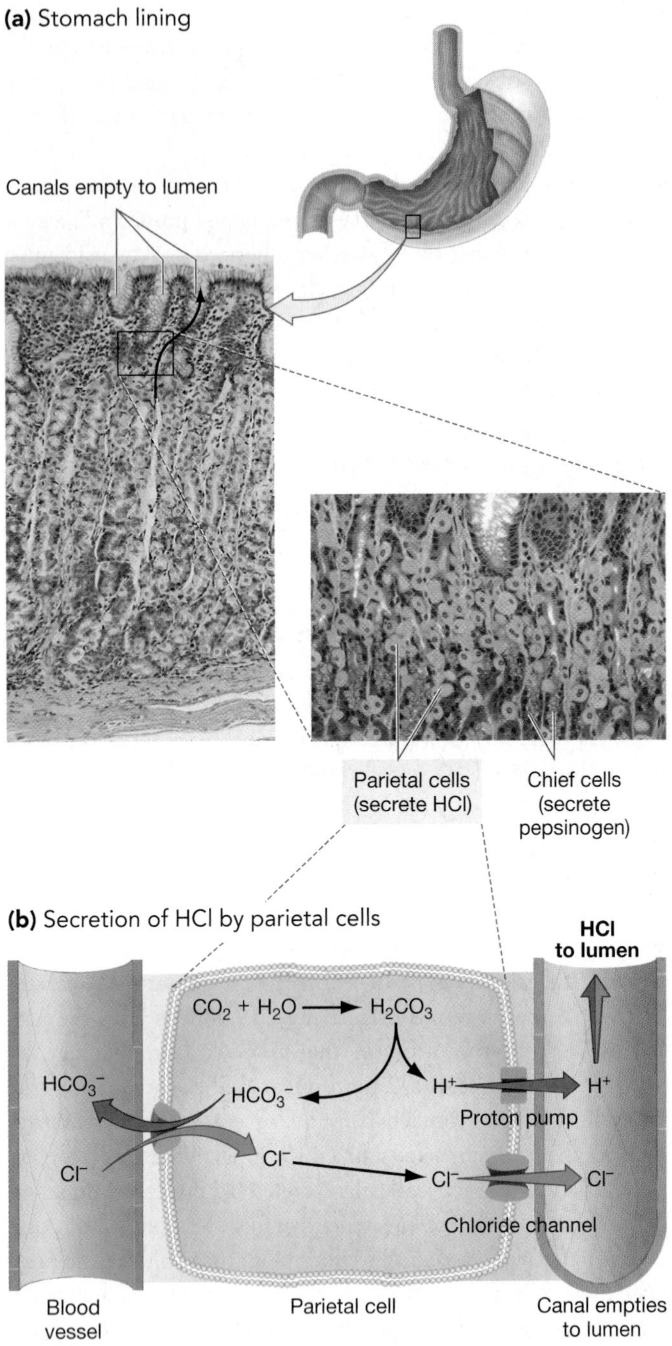
(a) Stomach lining
Canals empty to lumen
Parietal cells (secrete HCl)
Chief cells (secrete pepsinogen)
(b) Secretion of HCl by parietal cells
HCl to lumen
CO2 + H2O → H2CO3
HCO3−
HCO3−
H+
H+
Proton pump
Cl−
Cl−
Cl−
Cl−
Chloride channel
Blood vessel
Parietal cell
Canal empties to lumen

FIGURE 43.10 Parietal Cells in the Stomach Lining Secrete Hydrochloric Acid

(a) Light micrographs showing a section through the stomach wall.
(b) Proposed mechanism for secretion of hydrochloric acid.

EXERCISE In part (a), draw the route of molecules and ions secreted by parietal cells. In part (b), label the reaction catalyzed by carbonic anhydrase.

How Do Parietal Cells Secrete HCl? The first clues about how parietal cells manufacture hydrochloric acid emerged in the late 1930s, when a researcher found a high concentration of an enzyme called carbonic anhydrase in parietal cells. This result was interesting because **carbonic anhydrase** catalyzes the formation of carbonic acid (H_2CO_3) from carbon dioxide and water. In solution, the carbonic acid that is formed immediately dissociates to form a proton and the bicarbonate ion (HCO_3^-):

$$CO_2 + H_2O \longleftrightarrow H_2CO_3 \rightarrow H^+ + HCO_3^-$$

A second clue to the formation of HCl came in the 1950s, when the development of the transmission electron microscope allowed researchers to analyze parietal cells at high magnification. The biologists observed that parietal cells are packed with mitochondria. Based on this observation, researchers suspected that HCl secretion involves active transport. Later work confirmed this hypothesis by showing that the protons formed by the dissociation of carbonic acid are actively pumped into the lumen of the stomach. Subsequent studies showed that chloride ions from the blood enter parietal cells in exchange for bicarbonate ions and then move into the lumen through a chloride channel. The current model for HCl production is diagrammed in **Figure 43.10b**.

Although the acidic environment of the human stomach denatures (unfolds) proteins and makes their digestion highly efficient, it has a downside: When cells of the stomach wall are damaged by the acidity, holes called **ulcers** can develop. Fortunately, the stomach lining is usually well protected by a thick layer of mucus and the rapid replacement of epithelial cells.

The Ruminant Stomach Although the human stomach is a complex and fascinating organ, it is relatively simple compared with the stomachs of cattle, sheep, goats, deer, antelope, giraffe, and pronghorn. These species are called **ruminants.** They are distinct because they have stomachs that are specialized for digesting cellulose—not proteins. Recall from Chapter 5 that cellulose is a polymer of glucose and that it is the main structural component of plant cell walls. Digesting cellulose is not an easy task for ruminants, because mammals lack the enzymes required to digest cellulose. Yet cellulose is the main carbohydrate in the leaves, stems, and twigs that ruminants ingest.

Ruminants are able to harvest glucose from cellulose thanks to a combination of specialized anatomical structures, unique enzymes, and symbiotic relationships with bacteria and unicellular protists. When **symbiosis** occurs, members of two different species live in close physical contact with each other.

As **Figure 43.11** shows, ruminant mammals have four-chambered stomachs that are folded in complex ways. The largest chamber is called the **rumen** and serves as a fermentation vat. The rumen and the adjacent chamber called the reticulum are packed with symbiotic bacteria and protists. These organisms have enzymes capable of breaking apart the chemical bonds in cellulose, yielding glucose. The rumen is an

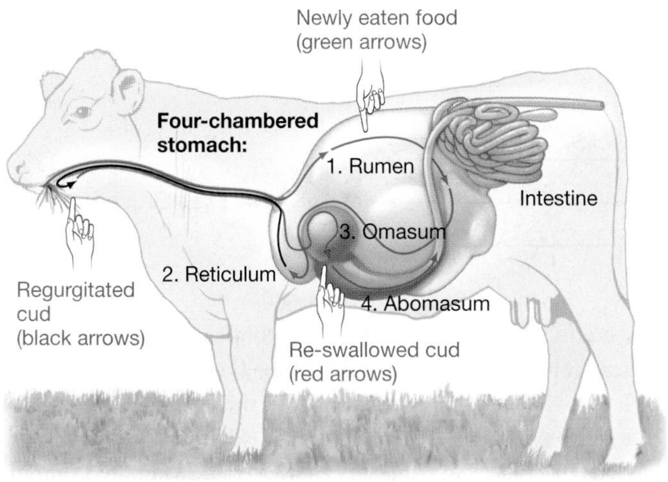
Newly eaten food
(green arrows)
Four-chambered stomach:
1. Rumen
Intestine
3. Omasum
2. Reticulum
4. Abomasum
Regurgitated cud
(black arrows)
Re-swallowed cud
(red arrows)

FIGURE 43.11 Ruminant Stomachs Facilitate the Digestion of Cellulose by Symbiotic Organisms

Ruminants obtain many of their nutrients from symbiotic bacteria and protists that live in the rumen and reticulum chambers of the stomach. QUESTION The ability to regurgitate and re-chew food is thought to be an adaptation that reduces the amount of time that grazing animals have to spend harvesting food (because it can be thoroughly chewed later). Why would this be advantageous?

oxygen-free environment, and the symbiotic organisms produce ATP from this glucose via fermentation, releasing fatty acids as a by-product (see Chapter 9). After the plant material has been partially digested in the rumen and in the reticulum, the ruminant regurgitates portions of that material into its mouth, forming a **cud**. The animal chews that regurgitated material further to enhance mechanical breakdown, then reswallows it. As it passes into the stomach for the second time, the food bypasses the rumen and reticulum and enters the third chamber, or omasum, where water is removed. The final chamber is the abomasum, which contains the individual's own digestive enzymes and corresponds to a true stomach. A key point here is that most of a ruminant's food consists of (1) fatty acids and other compounds that are released as waste products by symbiotic organisms and (2) the symbiotic cells themselves.

Although ruminants are the only animals that house symbionts in a modified stomach, they are not the only animals that use fermentation by symbionts to break down the cellulose in plant material. In rabbits and some species of rodents, for example, fermentation by symbionts occurs in an organ called the cecum. The **cecum** is a blind sac between the small instestine and the colon, and it is greatly enlarged in species that use it as a fermentation chamber for processing cellulose. Specialized digestive systems such as the ruminant stomach and the rabbit cecum are another example of correlations between structure and function in the digestive tracts of animals.

The Small Intestine

In humans, the stomach is responsible for mixing the contents of a meal into a homogenous slurry, mechanically breaking up food material, and providing the acid and enzymes required to partially digest proteins. Peristalsis in the stomach wall then moves small amounts of material through the valve created by a sphincter muscle at the base of the stomach and into the small intestine.

In the **small intestine**, partially digested food mixes with secretions from the pancreas and the liver and begins a journey of about 6 m (20 feet). At the end of the small intestine, digestion is complete and most nutrients—along with large quantities of water—have been absorbed. To understand how these dramatic changes in the composition of food occur, let's explore what happens to proteins, lipids, and carbohydrates as they move through this section of the digestive tract.

Protein Processing by Pancreatic Enzymes The acidic environment of the stomach denatures proteins—meaning their secondary and tertiary structures are destroyed. In addition, pepsin cleaves enough peptide bonds to reduce the macromolecules to relatively small polypeptides. In the small intestine, protein digestion is completed so that individual amino acids can enter the bloodstream and be transported to cells throughout the body.

How does protein digestion occur? By the end of the nineteenth century, it had been established that enzymes in the small intestine digest polypeptides to monomers. In addition, biologists had determined that the key digestive enzymes are synthesized by pancreatic cells in an inactive form. Like the production of inactive pepsinogen by chief cells in the stomach, the production of digestive enzymes in an inactive conformation prevents pancreatic cells from digesting themselves.

It took decades of work to understand how pancreatic enzymes are activated in the small intestine. In 1900 a researcher showed that pancreatic enzymes are activated by contact with juice from the upper part of the small intestine. Because activation did not occur when he heated the intestinal juice, and because heat denatures proteins, he hypothesized that the agent responsible for activating the pancreatic enzymes was also an enzyme. He called the unknown activating the enzyme **enterokinase**. Decades later, a researcher succeeded in purifying a pancreatic enzyme called **trypsinogen** and demonstrated that enterokinase activates it in vitro. Follow-up work showed that enterokinase activates trypsinogen by phosphorylating it. The active enzyme that results from this reaction is called **trypsin**. Trypsin then triggers the activation of other protein-digesting enzymes, such as *chymotrypsin*, *elastase*, and *carboxypeptidase*. These enzymes are also synthesized by the **pancreas** and secreted in an inactive form.

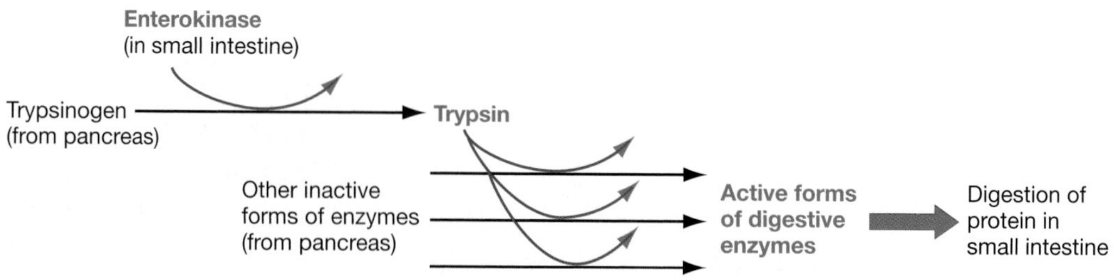

FIGURE 43.12 Enterokinase Triggers an Enzyme-Activation Cascade in the Small Intestine Enterokinase activates trypsin; in turn, trypsin activates other digestive enzymes secreted by the pancreas.

Figure 43.12 summarizes how the sequential activation of digestive enzymes proceeds in the small intestine. Enterokinase triggers the activation of trypsinogen to trypsin, which in turn triggers the activation of the three other protein-digesting enzymes. Once these enzymes are activated in the upper reaches of the small intestine, they begin breaking polypeptides into amino acid monomers.

What regulates the secretion of these proteins from the pancreas? Because digestive enzymes are needed only when food reaches the small intestine, it was logical to predict that their release would be carefully controlled. The leading hypothesis was that pancreatic secretion is controlled by the nervous system, possibly in response to the sight or smell of food. This hypothesis was inspired by an extensive series of experiments based on cutting specific nerves in experimental animals. These experiments had established that the nervous system is involved in regulating an array of digestive processes, including the release of saliva, swallowing, and the secretion of gastric juice in the stomach.

The hypothesis of nervous control in the small intestine was refuted, however, in a classic experiment conducted by William Bayliss and Ernest Starling. To study the mechanism responsible for stimulating secretion from the pancreas, Bayliss and Starling cut the nerves that connect to the pancreas and small intestine of a dog. Nervous control of the two organs was now impossible. But when the researchers introduced a weak HCl solution into the upper reaches of the animal's small intestine to simulate the arrival of material from the stomach, the pancreas secreted in response. This observation was startling. The small intestine had successfully signaled the pancreas that food had arrived, even though the nerves connecting the two organs had been cut.

To explain this observation, Starling hypothesized that a chemical messenger must be involved, and that the chemical messenger must originate in the small intestine and travel to the pancreas—perhaps via the blood. He tested this idea by cutting off a small piece of the small intestine, grinding it up, and injecting the resulting solution into a vein in the animal's neck. Minutes later, the pancreas secreted.

Bayliss and Starling had discovered the first **hormone**—a chemical messenger that influences physiological processes at very low concentrations. The molecule they detected, which they called **secretin**, is produced by the small intestine in response to the arrival of food from the stomach. The discovery of secretin was important because it confirmed that digestion is under both neural control and hormonal control.

Follow-up work showed that secretin's primary function is to induce a flow of bicarbonate ions (HCO_3^-) from the pancreas to the small intestine. The arrival of bicarbonate is important because it neutralizes the acid arriving from the stomach. Researchers also discovered a second hormone produced in the small intestine, called cholecystokinin, that induces secretion from the liver as well as the pancreas. **Cholecystokinin** stimulates the secretion of digestive enzymes from the pancreas and the secretion of molecules from the liver that are involved in the digestion of lipids. Hormones are involved in stomach function as well. For example, after being stimulated by nerves or the arrival of food, certain stomach cells produce the hormone **gastrin.** In response, parietal cells begin secreting HCl.

Digesting Lipids: Bile and Transport The pancreatic secretions include digestive enzymes that act on fats and carbohydrates as well as proteins. Like the lingual lipase added to saliva in the mouth, the enzyme **pancreatic lipase** breaks certain bonds present in complex fats and results in the release of fatty acids and other small lipids.

Before pancreatic lipase can act, the fats that emerge from the stomach must be modified. Recall from Chapter 6 that fats are insoluble in water. As a result, they tend to form large globules as they are churned in the stomach. For lipase to act, these large fat globules must be broken up, or **emulsified.** In the small intestine, emulsification results from the action of small lipids called **bile salts.** As **Figure 43.13** shows, bile salts function like the detergents that researchers use to break up the lipids in plasma membranes (see Chapter 6). Bile salts are synthesized in the liver and secreted in a complex solution called **bile,** which is stored in the **gallbladder.** When bile enters the small intestine, it raises the pH and emulsifies fats. Once fats are broken into small globules with high surface area, they can be attacked by enzymes and digested.

What happens to the monoglycerides and fatty acids released by lipase activity? An answer began to emerge when researchers injected radioactively labeled fatty acids into the small intestines of laboratory rats. The biologists analyzed the epithelial cells that line the small intestine of the experimental animals and found that most of the radioactive label had entered the cells and attached to a protein they named **fatty-acid binding protein.** Later, other researchers established that a

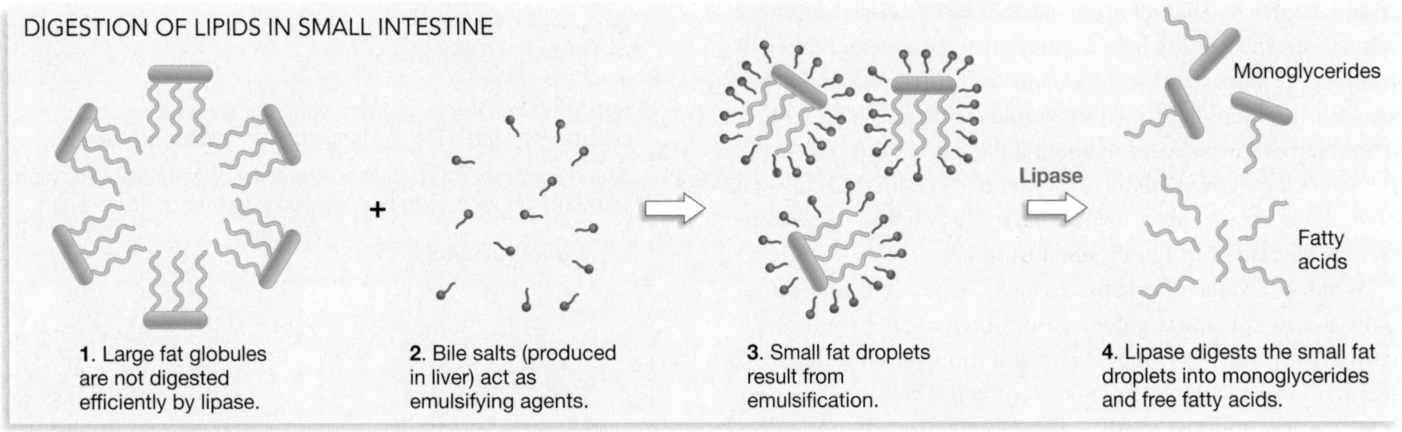
DIGESTION OF LIPIDS IN SMALL INTESTINE
+
Lipase
Monoglycerides
Fatty acids
1. Large fat globules are not digested efficiently by lipase.
2. Bile salts (produced in liver) act as emulsifying agents.
3. Small fat droplets result from emulsification.
4. Lipase digests the small fat droplets into monoglycerides and free fatty acids.

FIGURE 43.13 Emulsifying Agents and Lipases Digest Lipids in the Small Intestine
Bile salts are small lipids that act as emulsifying agents. Once bile salts break up large fat globules, lipase can digest fats efficiently.

fatty-acid binding protein also occurs in the membranes of these cells. Based on these and other results, fatty-acid binding proteins are thought to bring lipids into the cell. There they are processed into protein-coated globules called **chylomicrons** and released for transport to fat-storage cells and other tissues.

Digestion and Absorption of Carbohydrates In addition to manufacturing lipase and protein-digesting enzymes, the pancreas produces nucleases and an amylase that is similar to the salivary enzyme introduced earlier. **Nucleases** digest the RNA and DNA in food, while **pancreatic amylase** continues the digestion of carbohydrates that began in the mouth.

The monosaccharides that are released by amylase and other enzymes are absorbed by epithelial cells that line the small intestine. The surface area available for nutrient absorption in the small intestine is nothing short of remarkable. As **Figure 43.14** shows, the epithelial tissue of this organ is folded and covered with fingerlike projections called **villi** (singular: **villus**). In turn, the cells that line the surface of villi have projections on their apical surfaces called **microvilli** (singular: **microvillus**). (Recall

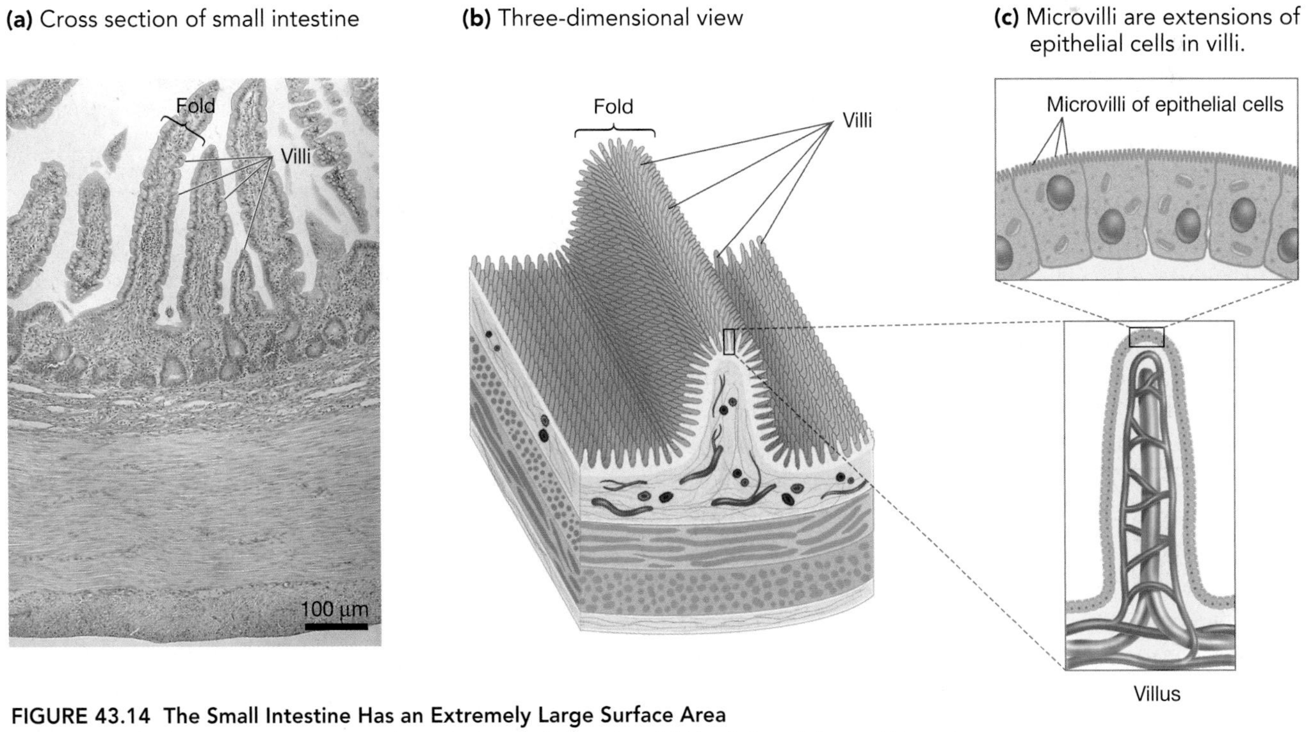
(a) Cross section of small intestine
Fold
Villi
100 μm
(b) Three-dimensional view
Fold
Villi
(c) Microvilli are extensions of epithelial cells in villi.
Microvilli of epithelial cells
Villus

FIGURE 43.14 The Small Intestine Has an Extremely Large Surface Area
(a) Micrograph showing a cross section through the small intestine. **(b)** The lining of the small intestine has extensive folds. **(c)** In a microvillus, epithelial cells are close to blood vessels and to lacteals, vessels that are part of the lymphatic system. These structures transport nutrients and fluids. EXERCISE In part (c), label the lacteal and blood vessels.

from Chapter 41 that an apical surface faces away from other tissues and toward the lumen of an organ or an external environment. A basolateral surface in an epithelium faces other tissues.) If the small intestine lacked folds, villi, and microvilli, it would have a surface area of about 3300 cm^2 (3.6 ft^2). Instead, the epithelium covers about 2 million cm^2 (over 2200 ft^2)—an area about the size of a tennis court. This enormous area increases the efficiency of nutrient absorption.

What molecular mechanisms make it possible for epithelial cells to transport glucose and other nutrients from digested food into the body? The situation is reminiscent of the action of the nephron, where epithelial cells in the proximal tubule and the distal tubule reabsorb valuable electrolytes and water from the urine but leave urea and other waste products so they can be excreted (see Chapter 42). When digestion of food is complete in the small intestine, the resulting slurry is a treasure trove of nutrients and water mixed with undigestible plant fibers and cells from bacteria that live symbiotically in the gut. It is logical to predict two general principles about nutrient absorption: (1) It is highly selective, meaning proteins in the plasma membranes of microvilli are responsible for bringing specific nutrients into the cell; and (2) it is active, meaning it requires an expenditure of ATP to bring nutrients into the epithelium across a concentration gradient.

Work over the past several decades has shown that both predictions are correct. One of the key results grew out of a series of experiments during the 1980s, which established that glucose absorption depends on the presence of an electrochemical gradient favoring an influx of sodium ions into the epithelium. Based on this observation, biologists hypothesized that the apical membranes of these cells must contain a series of cotransporters—proteins that would bring a nutrient molecule into the cell along with sodium ions. Work focused on glucose, because it is such a fundamentally important nutrient. To confirm that a sodium-glucose cotransporter exists, investigators set out to find the gene that codes for the hypothesized membrane protein.

The strategy that the researchers used is outlined in **Figure 43.15**. They began by purifying mRNAs from rabbit intestinal cells, which presumably were transcribing the cotransporter genes. Then the team separated the mRNAs by size via gel electrophoresis and injected one of each type of mRNA into a series of frog eggs—cells that do not normally transport glucose. The frog cells translated the rabbit mRNAs into proteins. When tested, one of the experimental eggs was able to import Na$^+$ and glucose in tandem. Based on these data, the biologists inferred that this egg had received the mRNA for the rabbit Na$^+$-glucose cotransporter. Using techniques introduced in Chapters 14 and 19, the researchers were able to make a DNA copy of the mRNA, determine the sequence of the gene, and infer the amino acid sequence of the membrane protein.

Question: How is glucose transported into epithelial cells of the small intestine?

Hypothesis: Glucose enters epithelial cells along with sodium ions via a Na$^+$-glucose cotransporter protein.

Null hypothesis: Glucose does not enter epithelial cells along with sodium ions via a Na$^+$-glucose cotransporter protein.

Experimental setup:

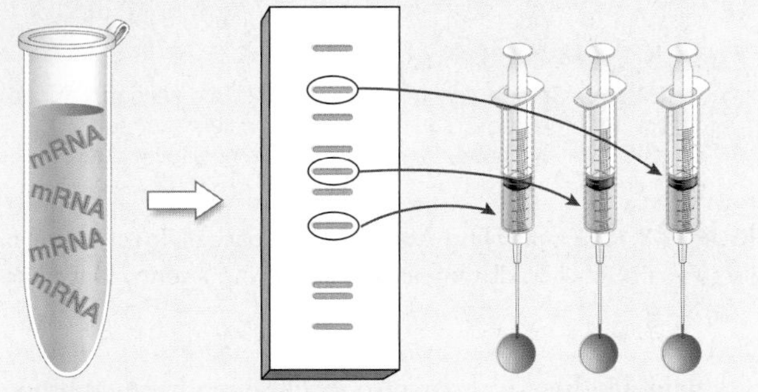
mRNA
mRNA
mRNA
mRNA

1. Purify mRNA from intestinal cells.

2. Separate mRNAs by size via gel electrophoresis.

3. Inject individual mRNAs into frog eggs. Test each egg—can it absorb Na$^+$ and glucose?

Prediction: An egg will be able to absorb Na$^+$ and glucose, because it received the mRNA that codes for the Na$^+$-glucose cotransporter.

Prediction of null hypothesis: None of the eggs will be able to absorb Na$^+$ and glucose.

Results:

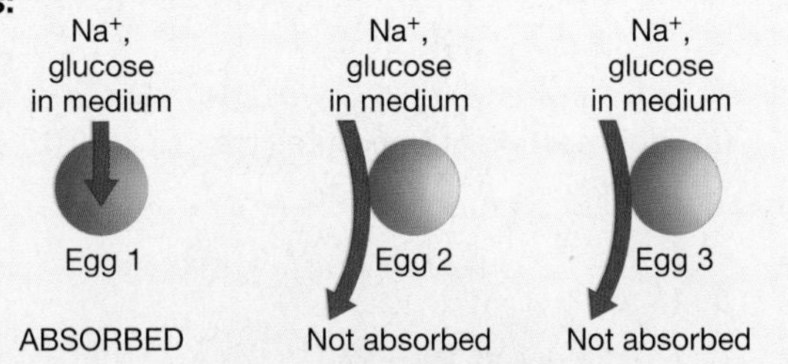
Na+, glucose in medium
Na+, glucose in medium
Na+, glucose in medium
Egg 1
Egg 2
Egg 3
ABSORBED
Not absorbed
Not absorbed

Conclusion: The egg that absorbs Na$^+$ and glucose received the mRNA from the Na$^+$-glucose cotransporter gene.

FIGURE 43.15 The Experimental Protocol for Locating the Na+ /Glucose Cotransporter Gene

The discovery of the Na$^+$-glucose cotransporter inspired the model for glucose absorption that is summarized in **Figure 43.16**. Notice that Na$^+$/K$^+$-ATPase (sodium-potassium pumps) in the basolateral membrane of the epithelial cells creates an electrochemical gradient that favors the entry of Na$^+$. Glucose from digested food enters the cell along with sodium via the cotransporter, then diffuses into nearby blood vessels through a glucose channel in the basolateral membrane. If the configuration of pumps, cotransporters, and channels looks familiar, there is a good reason: The same combination of membrane proteins occurs in the

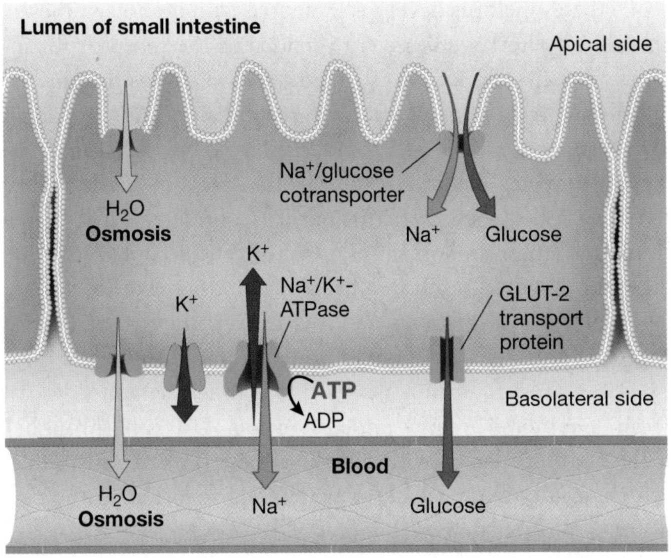

FIGURE 43.16 Glucose Is Actively Transported into Epithelial Cells Diagram summarizing the current understanding of how glucose is absorbed in the small intestine. The GLUT-2 transport protein facilitates the diffusion of glucose across the basolateral membrane, into the bloodstream.

proximal tubule of the kidney, where the proteins are responsible for the reabsorption of sodium and glucose from urine (see Chapter 42).

Follow-up work showed that in the small intestine—just as in the proximal tubule—other cotransporters are responsible for the absorption of other nutrients, with specific channels in the basolateral membrane responsible for their transport to the blood. It is interesting to note, though, that after fatty-acid binding proteins have brought fatty acids into epithelial cells, the complexes that result do not enter the bloodstream. Instead they enter vessels that are part of the lymphatic system, introduced in Chapters 44 and 49. More specifically, fats enter lymphatic vessels called **lacteals** in the core of each villus (see Figure 43.14c).

When solutes are brought into the epithelium via active transport, water follows passively by osmosis. This mechanism of water absorption inspired an important medical strategy called *oral rehydration therapy*. If a patient has diarrhea, clinicians frequently prescribe dilute solutions of glucose to be taken orally. When the glucose in the drink is absorbed in the small intestine through the sodium-glucose cotransporter, enough water follows to prevent life-threatening dehydration. This simple medication saves thousands of lives every year.

The Large Intestine

By the time digested material reaches the large intestine, a large amount of water (approximately 5 liters per day in total) and virtually all of the available nutrients have been absorbed. The primary function of the **large intestine**, or **colon**, is to compact the wastes that remain and absorb enough water to form feces. Feces are held in the **rectum**, which is the final segment of the colon, until they can be excreted. Although the kidneys are responsible for maintaining water balance in the body, water absorption in the large intestine is important for individuals to remain well hydrated.

In an attempt to identify the mechanism of water absorption in the large intestine, researchers have recently focused on the membrane proteins called **aquaporins**, which were introduced in Chapter 42. Recall that aquaporins are water channels in plasma membranes that provide a mechanism for increasing the rate of water movement via osmosis. The best studied of these proteins, called AQP1 for aquaporin 1, is common in the nephrons of the kidney. It is one of the proteins responsible for water reabsorption from urine along the osmotic gradient described in Chapter 42.

To date, four distinct aquaporins have been found in the large intestines of rats, mice, and humans. The aquaporins called AQP3 and AQP4, for example, are located in the basolateral membrane of cells in the epithelium of the large intestine in mice. Researchers are working to unravel exactly how the activity of these aquaporins is regulated, however. To date, the molecular mechanisms responsible for water absorption in the large intestine are not as well understood as are those in the collecting duct of nephrons.

✓CHECK YOUR UNDERSTANDING

In humans and many other animals, digestion begins with the mechanical breakdown of food into small pieces that are then acted on by acids, enzymes, emulsifying agents, and other chemical treatments. Nutrients are actively and selectively absorbed through specific membrane proteins in epithelial cells of the digestive tract. You should be able to (1) diagram the human digestive tract; (2) explain how each compartment in the digestive system aids the digestion of food, absorption of nutrients, and excretion of wastes; and (3) predict the consequences of treating a person with drugs that inhibit the release of bile salts, that inactivate trypsin, or that block the action of Na^+/K^+-ATPase in the epithelial cells of the small intestine.

43.4 Nutritional Homeostasis—Glucose as a Case Study

When digestion is complete, amino acids, fatty acids, ions, and sugars enter the bloodstream and are delivered to the cells that need them. Too much of a nutrient, or too little, can be problematic or even fatal, however. A classic example is the illness called **diabetes mellitus**. People with diabetes experience abnormally high levels of glucose in their blood. Over the course of a lifetime, the chronic glucose imbalance associated with diabetes mellitus can lead to blindness, heart failure, and a reduction of circulation in the legs.

Why does diabetes occur? A key hint came to light in 1879, when researchers removed the pancreas from a dog and observed that diabetes developed. This experiment suggested that the pancreas secretes a compound involved in the removal of glucose from the blood. When other investigators cut up pancreatic tissues and injected extracts into diabetic dogs, however, it did not cure the disease. In fact, the investigators observed no response. Frustrated, researchers realized that digestive enzymes in the pancreas were probably destroying the active agent during the extraction process.

In 1921 Frederick Banting and Charles Best conducted a breakthrough experiment. They began by tying off a dog's *pancreatic duct*—the tube where digestive enzymes collect and flow to the small intestine. The logic was that blocking the secretion of digestive enzymes might kill the cells that synthesize them. The investigators waited several weeks for the cells near the duct to die and then removed the gland. They froze the pancreas tissue, ground it up, and injected an extract into a diabetic dog. To their delight, the dog's blood-sugar levels stabilized and the dog became more active and healthy looking. After Banting and Best repeated the experiment and observed the same result, they grew increasingly confident that they had found the molecule responsible for controlling diabetes. The molecule came to be called insulin.

Insulin's Role in Homeostasis

Within a few years after Banting and Best's discovery, insulin had been isolated in pure form. Drug companies began purifying it from the pancreas of calves and other domestic animals in large enough quantities to administer to people suffering from diabetes mellitus. Later, insulin's mode of action was revealed. **Insulin** is a hormone that is produced in the pancreas when blood-glucose levels are high. It travels through the bloodstream and binds to receptors on cells throughout the body. In response, the cells increase their rate of glucose uptake and processing. For example, cells in the liver and skeletal muscle respond to insulin by synthesizing more glycogen from glucose monomers; cells that store lipids respond by synthesizing more storage forms of fat, using glucose as a precursor. The result is that glucose levels in the blood decline.

WEB TUTORIAL 43.2 Understanding Diabetes Mellitus

If blood-glucose levels fall, as they do after hard exercise or when food is lacking, cells in the pancreas secrete a hormone called **glucagon**. In response to glucagon, cells in the liver catabolize glycogen and produce glucose via **gluconeogenesis** (the synthesis of glucose from other compounds). The result is that glucose levels in the blood rise. **Figure 43.17** summarizes how insulin and glucagon interact to form a negative feedback system capable of achieving homeostasis with respect to glucose concentrations in the blood.

Diabetes mellitus develops in people who do not synthesize insulin or who have defective versions of the receptor for insulin. The first condition is called **type I diabetes mellitus,** or *insulin-dependent diabetes*. The second condition is called **type II diabetes mellitus,** or *non-insulin-dependent diabetes*. Type I diabetes mellitus results from a lack of the hormone itself; type II diabetes mellitus results from detects in the hormone's receptor. In both cases, effector cells do not receive the signal that would result in a drop in blood-glucose levels.

Both types of diabetes mellitus lead to the production of large volumes of urine as the body attempts to clear excess glucose. This condition inspired the illness's name. Recall from Chapter 42 that *diabetes* means "to run through." That chapter also introduced the illness diabetes insipidus, which results from a lack of water reabsorption in the col-

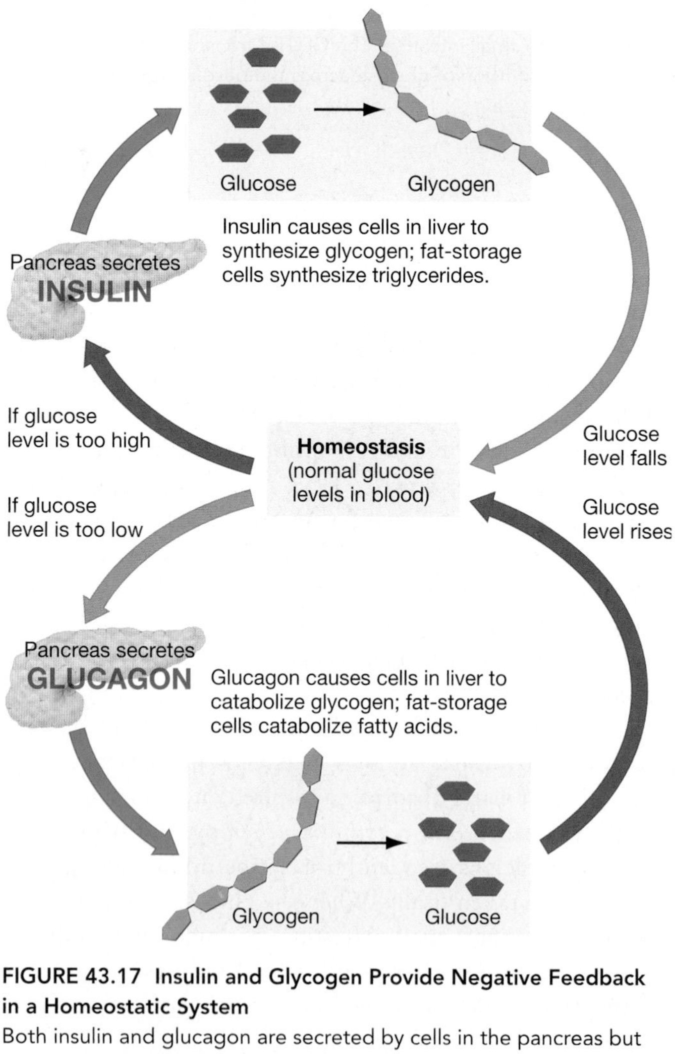

FIGURE 43.17 Insulin and Glycogen Provide Negative Feedback in a Homeostatic System
Both insulin and glucagon are secreted by cells in the pancreas but have opposite effects on blood-glucose concentrations. EXERCISE Draw a large X through the arrow that is disrupted in individuals with diabetes mellitus. Next to the X, jot down notes indicating the nature of the defect in individuals with type I diabetes mellitus versus type II diabetes mellitus.

lecting ducts of the kidney. *Insipidus* means tasteless, while *mellitus* means honeyed (sweet). Before chemical methods of analyzing urine were available, physicians would distinguish between diabetes insipidus and diabetes mellitus by tasting the urine of their patients. This test was effective because a common symptom of diabetes mellitus is the excretion of glucose in urine just after a meal. Normally, signals from insulin keep blood-glucose levels low enough that all of the glucose can be reabsorbed from the pre-urine formed in the kidney.

Currently, type I diabetes mellitus is treated with insulin injections and careful attention to diet; type II diabetes is managed primarily through prescribed diets and the monitoring of blood-glucose levels, as well as by the use of drugs that increase cellular responsiveness to insulin. For physicians and patients, the challenge is to achieve homeostasis with respect to blood glucose in the absence of the body's normal regulatory mechanisms.

The Type II Diabetes Mellitus Epidemic

The discovery of insulin qualifies as one of the great medical advances of the twentieth century, because it led to dramatic improvements in the quality of life for people who suffer from type I diabetes mellitus. Unfortunately, an epidemic of type II diabetes mellitus is currently under way in certain human populations. In the United States, about 6.6 percent of people aged 20 to 74 have type II diabetes. The frequency of this illness is much higher in African American, Latino, and Native American people, however, and the frequency of type II diabetes mellitus in teenagers of all ethnic backgrounds is increasing rapidly. To date, the Pima Indians of the southwestern United States have the highest prevalence ever recorded. Among the Pima, almost 60 percent of adults over 35 years old have type II diabetes mellitus. As **Figure 43.18a** shows, the incidence of this condition in the population has increased rapidly in recent decades.

Because there is a strong association between the prevalence of type II diabetes mellitus in parents and children, researchers have concluded that some individuals have a genetic predisposition for developing the disease. There is also strong evidence that environmental conditions have an important impact. For example, in the Pima and many other populations, type II diabetes mellitus is more prevalent in people who are obese (**Figure 43.18b**). A person who has a body mass index greater than or equal to 25 is considered **obese**. The **body mass index** is calculated as weight (kg) divided by height (m) squared.

Although the incidence of obesity has been rising in many populations, the causes remain controversial. Increases in the amount of fats in the diet, less physical activity, or a combination of fat intake and lack of exercise have been implicated. Because of these environmental changes, nutrition-related diseases such as obesity and diabetes are rapidly becoming a major public health concern—especially in young people.

✓CHECK YOUR UNDERSTANDING

Diabetes mellitus results from a lack of homeostasis in the concentration of the nutrient glucose in the blood. You should be able to (1) explain how insulin and glucagon interact to achieve homeostasis with respect to the blood-glucose concentration and (2) predict what a person with type I diabetes mellitus should do when the blood-glucose level is too high and when the blood-glucose level is too low.

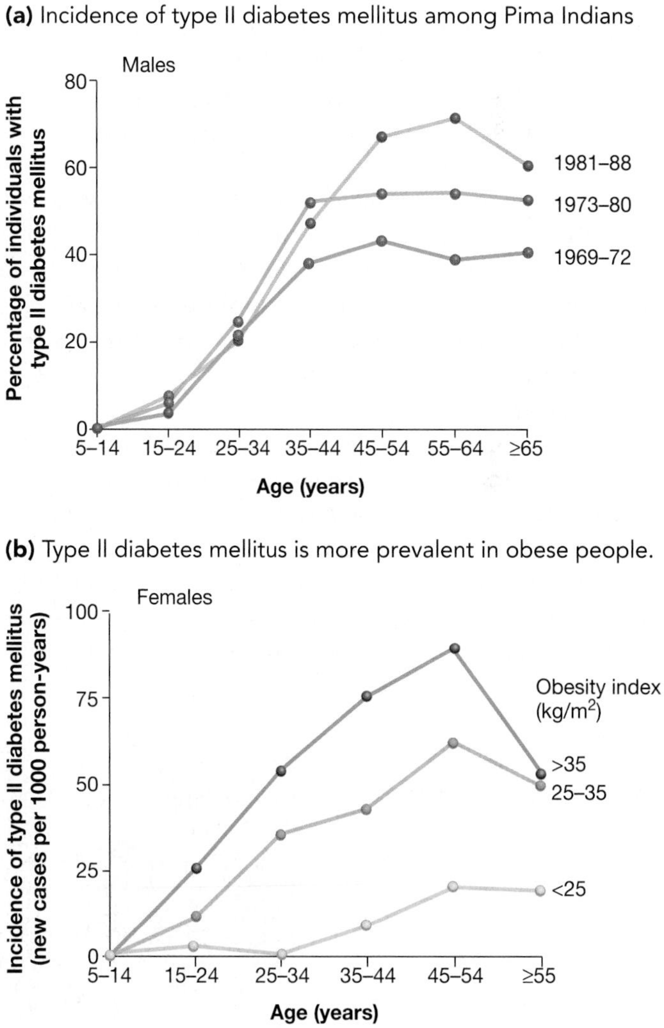

FIGURE 43.18 The Incidence of Type II Diabetes Mellitus Is Increasing in Many Populations
(a) The percentage of Pima Indian males with type II diabetes mellitus during three time intervals. **(b)** Data documenting a strong association between the incidence of diabetes mellitus and obesity (evaluated as weight divided by height squared, in units of kg/m^2).

ESSAY Cholesterol, Diet, and Disease

In Europe and North America, atherosclerosis is responsible for about half of all deaths. **Atherosclerosis** is a disease that develops when lipid deposits accumulate in the body's large arteries—the blood vessels that carry blood away from the heart (**Figure 43.19**). If a portion of a deposit breaks free and begins traveling through the circulatory system, it is likely to get lodged in a blood vessel and block blood flow, potentially killing cells served by that blood vessel. **Heart attacks** occur when such a blockage affects an artery that serves the heart. **Strokes** result from blockages in an artery that serves the brain.

What causes atherosclerosis? By the late 1800s researchers had found two important clues. First, the observation that atherosclerosis tends to run in families implies that some people have a genetic predisposition to the disease. Second, the small lipid called **cholesterol** is a major component of the deposits that characterize atherosclerosis. Cholesterol is an important nutrient and is present in meat, cream, and egg yolks. It is a common component of plasma membranes in animals. It is also required as a substrate for the synthesis not only of the bile salts that emulsify fats in the small intestine but also of steroids, which are molecules that act as hormones. In 1971 William Kannel and co-workers showed that there is a close link between high levels of cholesterol in the blood and a high risk of heart disease. These data suggested that atherosclerosis, like diabetes mellitus, results from a lack of homeostasis with respect to a key nutrient.

Kannel and co-workers made the connection between cholesterol and heart disease by analyzing data that had been collected since 1949 on men and women from Framingham, Massachusetts. Volunteers entered the Framingham study by undergoing a physical examination. If they were found to be free of heart disease, they joined the study and were examined every two years thereafter. Kannel and associates analyzed data on over 5000 people who had participated in the study for at least 14 years. The team found a strong correlation between the concentration of blood cholesterol when individuals entered the study and their risk of subsequently developing **coronary heart disease**—constrictions in the blood vessels supplying the heart—which can lead to heart attacks. The correlation remained significant when the researchers controlled for other factors associated with heart disease, such as smoking, high blood pressure, and diabetes mellitus.

What causes atherosclerosis?

These results inspired a massive effort to monitor and reduce cholesterol levels in the general population and an explosion of research on cholesterol metabolism. One major finding established that cholesterol is transported in the blood in two different forms, called **low-density lipoprotein (LDL)** and **high-density lipoprotein (HDL)**. Cholesterol is transported from the liver to cells throughout the body inside LDL particles. In contrast, HDL particles take up excess cholesterol and return it to the liver. In the liver, cholesterol is used to manufacture bile salts.

One of the more important findings since the Framingham study is that high LDL levels are associated with a high rate of coronary heart disease, while high HDL levels are associated with a low rate of coronary heart disease. It is still not clear, however, exactly how HDL takes up excess cholesterol or how it exerts its protective effects.

A general message of research on cholesterol is that atherosclerosis, like diabetes mellitus, has a strong environmental component as well as a genetic component. Obesity raises LDL, and people who eat a diet high in saturated fat also increase LDL and their risk of developing heart disease. Two of the most prominent diseases in the developed nations—diabetes mellitus and atherosclerosis—are directly related to nutrition.

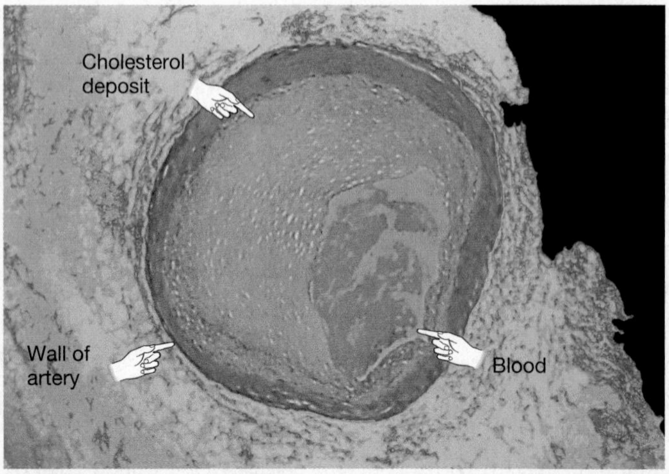

FIGURE 43.19 Cholesterol Forms Deposits That Block Blood Vessels

CHAPTER REVIEW

Summary of Key Concepts

- **Animals require an array of nutrients to stay healthy, including specific amino acids, vitamins, and elements, as well as organic compounds that have high potential energy.**

The diets of animals include fats and carbohydrates that provide energy, proteins that furnish amino acids, vitamins that serve as co-factors for enzymes and perform other functions, ions required for water balance and for nerve and muscle function, and elements that are incorporated into molecules synthesized by cells.

To determine the levels of nutrients that are needed for a particular animal species to sustain normal activities, researchers monitor the relationship between nutrient intake, the levels of nutrients maintained in the body, and health. To determine how nutrition can affect athletic performance in humans, researchers alter the intake of specific nutrients experimentally and assess how athletes respond. For example, investigators have shown that athletes who ingest a high-carbohydrate diet for three days outperform individuals who ingest a normal diet or a high-fat, high-protein diet.

- **Animals feed on a wide variety of items. In animals that ingest their food, there is usually a close correspondence between the structure of the animal's mouthparts and the function of those mouthparts in capturing and processing food.**

All animals are heterotrophs. Some species are suspension feeders, meaning they filter small food items from water using cilia or other structures. Others absorb their nutrition directly across their body wall. But most animals get food by seizing and manipulating it with mouthparts such as teeth, jaws, or beaks or with special toxin-injecting organs. Natural selection has modified mouthparts in different species to act as efficient tools for obtaining particular types of food.

- **Digestion occurs in the digestive tract, which is compartmentalized into organs that have specialized functions in the digestion of food, absorption of nutrients and water, or excretion of wastes.**

Once food is ingested, it is processed in the digestive tract, which begins at the mouth and ends at the anus. Food processing frequently begins with tearing or chewing by a beak, teeth, or other mouthparts. In many species, chemical digestion of food also begins in the mouth. For example, the salivary enzyme α-amylase hydrolyzes the bonds linking glucose monomers in starch, glycogen, and other carbohydrates. Once food is swallowed, it passes down the esophagus via peristalsis.

Digestion continues in the stomach. In humans, the stomach is a highly acidic environment that denatures proteins and in which the enzyme pepsin begins the cleavage of peptide bonds that link amino acids.

Food passes from the stomach into the small intestine, where it is mixed with secretions from the pancreas and liver. Carbohydrate digestion is completed by pancreatic amylase; fats are emulsified by bile salts and digested by lipase; protein digestion is completed by a suite of pancreatic enzymes that are activated by enterokinase. Secretions from the liver and pancreas are triggered by the hormones cholecystokinin and secretin, which are produced in the small intestine.

Cells that line the small intestine absorb the nutrients released by digestion through transporters or cotransporters in their membranes. In many cases, uptake is driven by an electrochemical gradient established by Na^+/K^+-ATPase that favors a flow of Na^+ into the cell. As solutes leave the lumen of the intestine and enter cells, water follows by osmosis. Water reabsorption is completed in the large intestine, where feces form. Although researchers suspect that water channels called aquaporins are involved in water uptake by epithelial cells in the large intestine, the detailed mechanism of water transport in this part of the digestive tract is still unknown.

Web Tutorial 43.1 The Digestion and Absorption of Food

- **Lack of homeostasis with respect to nutrients such as glucose and cholesterol can cause disease.**

Maintaining homeostasis with respect to nutrients is critical to health. Diabetes mellitus develops when concentrations of glucose in the blood are chronically too high. This illness is caused by a defect in the production of insulin—a hormone secreted by the pancreas that promotes the uptake of glucose from the blood—or by a defect in the insulin receptor on the surface of cells. The development of type II diabetes is correlated with obesity and is reaching epidemic proportions in some populations. Atherosclerosis is also associated with diet. Specifically, atherosclerosis results from the lack of homeostasis with respect to the nutrient cholesterol.

Web Tutorial 43.2 Understanding Diabetes Mellitus

Questions

Content Review

1. What does secretin stimulate?
   a. secretion of HCO_3^- from the pancreas
   b. secretion of HCl from the stomach epithelium
   c. secretion of digestive enzymes from the pancreas
   d. uptake of glucose from the bloodstream by cells throughout the body

2. Why does carbohydrate loading increase the performance of distance athletes?
   a. It increases glycogen stores in the muscle and liver.
   b. It increases fat storage.
   c. It increases the blood's capacity to deliver oxygen to tissues.
   d. It increases fatty-acid levels in the blood.

3. Why are the pharyngeal jaws of cichlid fish unique?
   a. No other group of fish has pharyngeal jaws.
   b. They are made of bone instead of cartilage, so they are harder and more durable.
   c. They function only in transporting food from the front of the mouth to the throat.
   d. They articulate with the braincase and have extra muscles attached, so they are capable of biting.
4. How are carbohydrates digested?
   a. by lipases in the small intestine
   b. by pepsin and HCl in the stomach
   c. by aquaporins in the large intestine
   d. by amylases in the mouth and small intestine
5. What role do bile salts play in the digestion of complex fats?
   a. They catalyze the cleavage of bonds leading to the release of fatty acids and other small lipids.
   b. They emulsify lipids, meaning that large masses of fat molecules are broken into smaller masses.
   c. They include fatty-acid binding proteins, which are involved in fat absorption.
   d. They activate the enzymes that are responsible for digesting fats.
6. How is water absorbed in the small intestine?
   a. through aquaporins
   b. The exact mechanism is not known.
   c. by sodium cotransporters
   d. by osmosis (following solutes)

Conceptual Review

1. Describe a human disease that results from a nutritional deficiency. How can researchers establish nutritional requirements for humans, if it is unethical to do experiments that would reduce the levels of specific nutrients?
2. Why is it logical that digestive enzymes are produced in an inactive form and then activated in the lumen of the digestive tract?
3. Do you accept the conclusion that cichlid pharyngeal jaws are adaptations that increase feeding efficiency? Why or why not?
4. How was it established that gastric juice is acidic and that the stomach's primary function is the digestion of proteins?
5. What features are responsible for the large surface area of the small intestine? Why is it important that this organ have such a large surface area?
6. Diabetes mellitus results from the disruption of mechanisms for maintaining homeostasis in blood-glucose concentrations. Explain this statement. Make a diagram explaining how insulin and glucagon provide negative feedback in a homeostatic system.

Group Discussion Problems

1. Predict how the nutritional requirements of female mammals change during pregnancy and breastfeeding. Design studies that would test your predictions, using humans as a study subject and using laboratory mice as a study subject.
2. Predict the physical symptoms that would result from defects in each of the following molecules: pancreatic amylase, pepsin, Na^+-glucose cotransporter, fatty-acid binding protein, and aquaporin.
3. Scientists who backed the hypothesis that secretion from the pancreas is under nervous control strenuously objected to the experiment that led to the discovery of hormonal control. They claimed that the experiment was inconclusive because it was very likely that not all of the nerves innervating the small intestine had been cut. The biologists who did the experiment replied that even if not all the nerves had been cut, the result was still valid. In your opinion, who is correct? Why?
4. Among vertebrates, the large intestine occurs only in lineages that are primarily terrestrial (amphibians, reptiles, and mammals). Propose a hypothesis to explain this observation.

Answers to Multiple-Choice Questions 1. a; 2. a; 3. d; 4. d; 5. b; 6. d

www.prenhall.com/freeman is your resource for the following: Web Tutorials; Online Quizzes and other Online Study Guide materials; Answers to Conceptual Review Questions; Solutions to Group Discussion Problems; Answers to Figure Caption Questions and Exercises; and Additional Readings and Research.

Gas Exchange and Circulation

44

During intense exercise, animal circulatory systems must deliver large amounts of oxygen to tissues and remove large amounts of carbon dioxide. This chapter explores how gas exchange occurs in animals that live in aquatic or terrestrial environments.

KEY CONCEPTS

- Animals have to take in oxygen and expel carbon dioxide to sustain cellular respiration and stay alive. Terrestrial animals and aquatic animals face different challenges in performing gas exchange.
- The organs responsible for gas exchange maximize the rate of oxygen and carbon dioxide diffusion by presenting a large, thin surface area to the environment. These organs also function in a way that maintains a steep partial-pressure gradient that favors entry of oxygen and elimination of carbon dioxide.
- Blood is a specialized tissue that transports gases, along with nutrients and wastes, in some animals.
- Circulatory systems use the pressure generated by one or more hearts to transport blood and other substances throughout the body.

When the mitochondria inside animal cells are producing ATP via cellular respiration, they consume oxygen (O_2) and produce carbon dioxide (CO_2). To support continued ATP production, cells have to obtain oxygen and expel excess carbon dioxide continuously. If carbon dioxide is allowed to build up in tissues, debilitating disease symptoms may appear. And if the flow of oxygen to tissues is interrupted, suffocation or drowning results. Death occurs because ATP production stops.

The goals of this chapter are to examine how gas exchange occurs between an animal's mitochondria and the external environment, to explore how gases and other materials are circulated throughout the body, and to analyze the mechanisms responsible for achieving homeostasis with respect to blood O_2 and CO_2 levels and with respect to blood pressure. Understanding gas exchange and circulation is fundamental to understanding how animals work. If either process fails, the consequences are dire. Tuberculosis, heart disease, lung cancer, malaria, and other diseases of the respiratory tract and circulatory system are leading causes of illness and death in most human populations. Let's delve in by exploring how gases behave in air and water.

44.1 Air and Water as Respiratory Media

How does gas exchange occur between an animal's environment and its mitochondria? In most cases, gas exchange involves the following sequence: (1) **Ventilation** occurs when air or water moves through a specialized gas-exchange organ, such

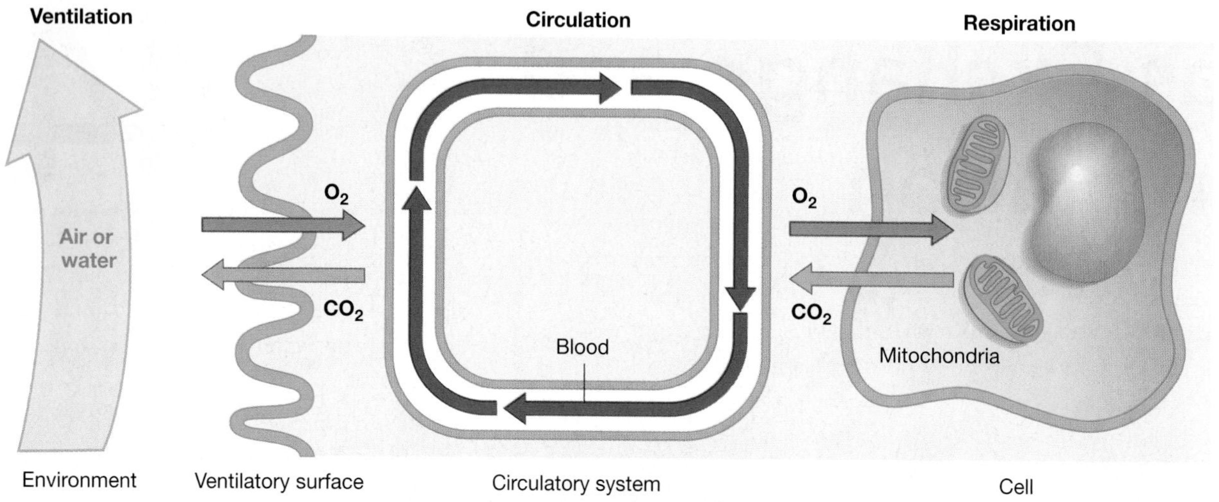

FIGURE 44.1 Gas Exchange Involves Ventilation, Circulation, and Respiration
In most animals, oxygen and carbon dioxide are exchanged across the surface of a lung, a gill, the skin, or some other organ. Once the gases are dissolved in blood (or hemolymph), they are transported to cells—where gas exchange again takes place.

as lungs or gills; (2) O_2 and CO_2 are exchanged between air or water and the blood; (3) **circulation** occurs when the dissolved gases are transported throughout the body via the circulatory system; and (4) cells exchange O_2 and CO_2 with the blood as mitochondria use O_2 and release CO_2 during **respiration.**

These steps, summarized in **Figure 44.1**, are examined in depth in this chapter. But gas exchange between the environment and mitochondria ultimately depends on diffusion. Under normal conditions, oxygen concentrations are relatively high in the environment and low in tissues, while carbon dioxide levels are relatively high in tissues and low in the environment. Thus, oxygen tends to move from the environment into tissues, and carbon dioxide tends to move from tissues into the environment.

To get a more precise understanding of how gas exchange occurs by means of diffusion, it's important to consider some fundamental questions about how diffusion occurs in both terrestrial and aquatic environments. How much oxygen and carbon dioxide are present in the atmosphere versus in the ocean? What factors influence how quickly these gases move by diffusion?

How Do Oxygen and Carbon Dioxide Behave in Air?

As **Figure 44.2a** shows, the atmosphere is composed primarily of nitrogen (N_2) and oxygen, with trace amounts of CO_2. Nitrogen is not important to animals living at sea level and is usually ignored in analyses of gas exchange.

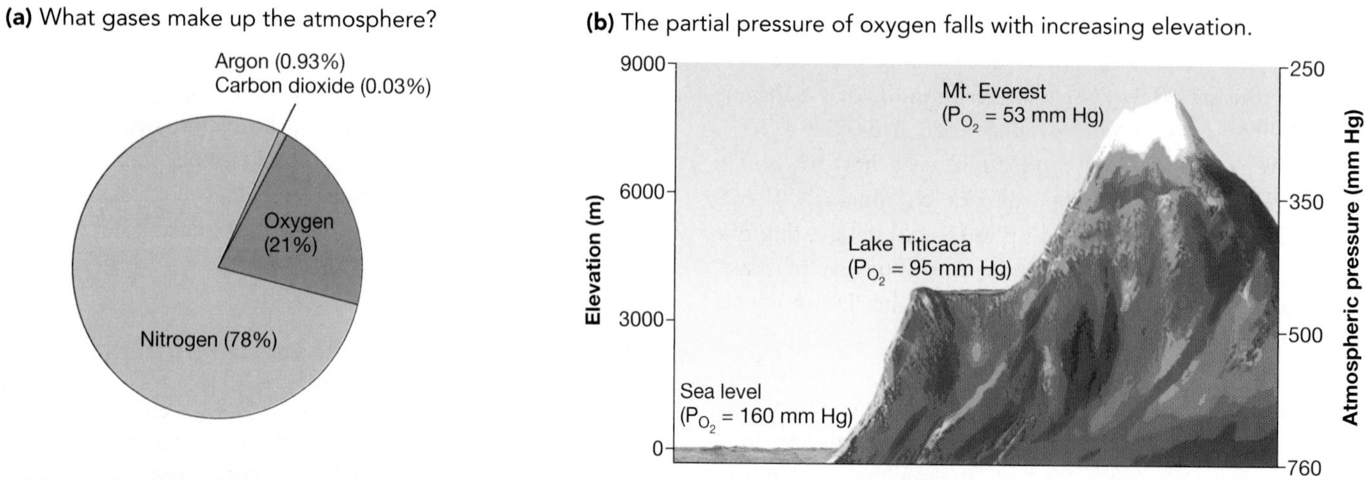

FIGURE 44.2 Oxygen Makes Up 21 Percent of the Atmosphere, but Its Partial Pressure Varies Widely
(a) Earth's atmosphere is dominated by nitrogen and oxygen. Oxygen makes up 21 percent of the atmosphere at all elevations. **(b)** Atmospheric pressure, and thus oxygen partial pressure (P_{O_2}), falls with increasing elevation. Atmospheric pressure is reported in units called millimeters of mercury (mm Hg).

The data in Figure 44.2a are slightly misleading, however. To understand why, consider that the percentage of O_2 in the atmosphere does not vary with elevation. The atmosphere at the top of Mt. Everest is composed of 21 percent O_2, just as it is at sea level. The key difference is that far fewer molecules of oxygen and other atmospheric gases are present at high elevations than at sea level. Air at the top of Mt. Everest is much less dense than air at sea level, so much less oxygen is present.

To understand how gases move by diffusion, it is important to express their presence in terms of partial pressures instead of percentages. Pressure is a type of force. A **partial pressure** is the pressure of a particular gas in a mixture of gases. To calculate the partial pressure of a particular gas, you multiply the fractional composition of that gas by the total pressure exerted by the entire mixture.* For example, **Figure 44.2b** shows that the total atmospheric pressure at sea level is 760 mm Hg (millimeters of mercury). (Hg is the chemical symbol for mercury.) If you multiply this value by 0.21, which is the fraction of air that is O_2, you obtain a partial pressure of oxygen, abbreviated P_{O_2}, at sea level of 160 mm Hg. Because the atmospheric pressure is only about 250 mm Hg at the top of Mt. Everest, the P_{O_2} there is $0.21 \times 250 = 53$ mm Hg.

Oxygen and carbon dioxide diffuse between the environment and cells along their respective partial-pressure gradients, just as solutes diffuse along their concentration gradients. In both air and water, O_2 and CO_2 move from regions of high partial pressure to regions of low partial pressure.

How Do Oxygen and Carbon Dioxide Behave in Water?

To obtain oxygen, water breathers face a much more challenging environment than air breathers do. Aquatic animals live in an environment that contains much less oxygen than the environments inhabited by terrestrial animals. At 15°C, a liter of air can contain up to 209 ml of O_2, while a liter of water may contain up to a maximum of only 7 ml of O_2. To extract a given amount of oxygen, an aquatic animal has to process 30 times more water than the amount of air processed by a terrestrial animal. In addition, water is about a thousand times denser than air and much more viscous. As a result, water breathers have to expend much more energy to ventilate their respiratory surfaces than do air breathers.

The partial pressure of oxygen varies in different types of aquatic habitats, just as it varies with altitude on land. Oxygen and carbon dioxide diffuse into water from the atmosphere, but the amount of gas that dissolves depends on several factors:

- *The solubility of the gas in water* Oxygen has very low solubility in water. Only 0.003 ml of oxygen dissolves in 100 ml of water for each 1-mm-Hg increase in oxygen partial pressure. Because of this low solubility, blood contains a molecule that binds to oxygen and delivers it to tissues. Without this carrier molecule, blood could not deliver enough oxygen to cells to sustain cellular respiration.
- *The temperature of the water* As the temperature of water increases, the amount of gas that dissolves in it decreases. Other things being equal, warm-water habitats have much less oxygen available than cold-water habitats do. For a fish, breathing in warm water is comparable to a land-dwelling animal breathing at high elevation.
- *The presence of other solutes* Because seawater has a much higher concentration of solutes than does freshwater, seawater can hold less dissolved gas. At 10°C, up to 8.02 ml of O_2 can be present per liter of freshwater versus only 6.35 ml of O_2 per liter of seawater. As a result, freshwater habitats tend to be more oxygen rich than marine environments are.
- *The partial pressure of the gas in contact with the water* Increasing the external partial pressure will drive a gas into a liquid, until the partial pressures in both the gas and liquid phases are equal. If the partial pressure in the liquid phase exceeds that in the gas phase, the gas will "boil" out of the liquid. This is what happens when the cap is removed from a bottle of carbonated beverage. The drop in partial pressure in the gas phase allows carbon dioxide to escape the liquid phase.

Other important aspects of oxygen availability in water include the surface area of the water, the production of oxygen by photosynthesis, and the depletion of oxygen by organisms using it in cellular respiration. Surface area is important because it affects the ability of oxygen to diffuse into water. For example, shallow ponds and streams tend to be much better oxygenated than are deep bodies of water, because shallower bodies have a higher ratio of surface area to volume. Rapids, waterfalls, and other types of "whitewater" tend to be the most highly oxygenated of all aquatic environments, because a large surface area is exposed to the atmosphere as water splashes over rocks and logs and because air bubbles are incorporated into the water (**Figure 44.3a**, page 1002). In contrast, oxygen content tends to be extremely low in bogs and other stagnant-water habitats (**Figure 44.3b**).

Photosynthetic algae, plants, and bacteria add oxygen to water when they are alive but deplete oxygen when they die. To understand why this happens, consider how lakes and ponds respond to large influxes of nitrogen- or phosphorus-containing fertilizers. Increased nutrient availability causes dramatic increases, or "blooms," in populations of photosynthetic organisms. These organisms have short life spans and settle to the bottom of the lake or pond when they die. There they provide food for bacteria and other decomposers that use oxygen in cellular respiration. In response to this massive increase in food availability, decomposer populations bloom until the oxygen in the water is used

*This calculation is valid because the total pressure in a mixture of gases is the sum of the partial pressures of all the individual gases. This relationship is called Dalton's law.

(a) Moving, agitated water has high oxygen partial pressure.

(b) Stagnant water has low oxygen partial pressure.

FIGURE 44.3 The Partial Pressure of O_2 in Water Varies Widely
(a) Any process that mixes water and increases its surface area increases the amount of oxygen that diffuses in. **(b)** Stagnant water has low oxygen partial pressure due to poor mixing and O_2 use by decomposers.
QUESTION How are oceans aerated?

up. This chain of events is called **eutrophication.** Eutrophication occurred in hundreds of lakes in the United States and Europe in response to phosphorus pollution during the 1960s and 1970s and is currently occurring in the Gulf of Mexico and other locations due to nitrate pollution (see Chapter 27). It often results in massive die-offs of animals that need oxygen to live.

Now let's consider how ventilatory organs are designed. How do the gills of fish, the tracheae of insects, and the lungs of mammals cope with the differences between air and water as ventilatory media?

✓CHECK YOUR UNDERSTANDING

O_2 and CO_2 move from regions of high partial pressure to regions of low partial pressure. Water breathing is much more difficult than air breathing, in part because the partial pressure of oxygen in water is much lower than its partial pressure in air. You should be able to (1) describe aquatic habitats where oxygen partial pressures are relatively high and relatively low and (2) describe how much air needs to be bubbled into aquaria containing warm water-species, cold-water species, vigorous algal growth, and highly active animals versus sedentary animals.

44.2 Organs of Gas Exchange

Many small animals do not have specialized organs of gas exchange, such as gills or lungs. Instead, they obtain O_2 and eliminate CO_2 by diffusion directly across the body surface. Most of these animals are aquatic or, like earthworms, live in terrestrial habitats that are moist. For these animals, diffusion across the body surface is rapid enough to fulfill their requirements for taking in O_2 and expelling CO_2. Living in wet environments allows them to exchange gases across their outer surface while avoiding catastrophic amounts of water loss.

In contrast, animals that are large or that live in dry habitats need some sort of specialized respiratory organ. To understand why the presence of gills or lungs is associated with large size, recall from Chapter 41 that the volume of an animal's body increases much faster than does its surface area. Respiratory organs provide a vastly increased surface area for gas exchange—enough to meet the demands of a large body filled with cells.

Biologists have long marveled at the efficiency of gills and lungs. To appreciate why, let's first examine the physical factors that control diffusion rates and then look at the structure and function of these respiratory organs.

Design Parameters: The Law of Diffusion

In 1855 Adolf Fick derived an equation regarding diffusion, based on the results of experiments he had performed on the rate at which gases diffuse. As **Figure 44.4** shows, **Fick's law of diffusion** states that the rate of diffusion of a gas depends on five parameters: the solubility of the gas in the aqueous film lining the gas-exchange surface, the temperature, the surface area available for diffusion, the difference in partial pressures of the gas across the gas-exchange surface, and the thickness of the barrier to diffusion.

Fick's law provides a compact way of identifying traits that allow animals to maximize the rate at which oxygen and carbon dioxide diffuse across the surfaces of their gills or lungs. Specifically, the law states that O_2 and CO_2 diffuse in the largest amounts when three conditions are met:

1. A is large, meaning a large surface area is available for gas exchange. Based on Fick's law, it is not surprising that the respiratory surface in the human lungs would cover about 100 m^2—or almost a quarter of a basketball court—if the epithelium were spread flat.

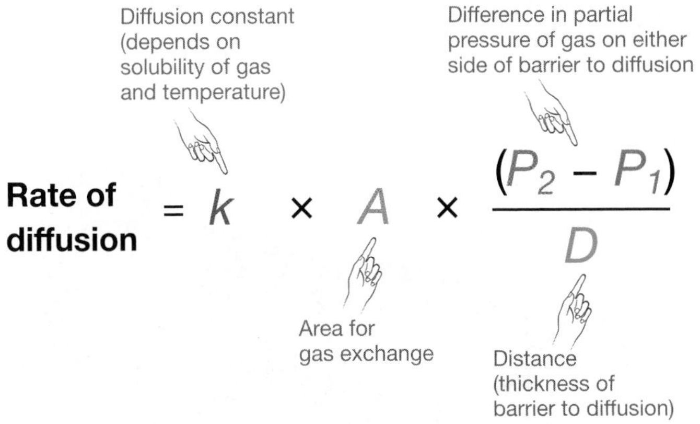
Diffusion constant (depends on solubility of gas and temperature)
Difference in partial pressure of gas on either side of barrier to diffusion
Rate of diffusion $= k \times A \times \frac{(P_2 - P_1)}{D}$
Area for gas exchange
Distance (thickness of barrier to diffusion)

FIGURE 44.4 Fick's Law of Diffusion

QUESTION Suppose you were going to design a system for the rapid diffusion of oxygen across a phospholipid bilayer. According to Fick's law, which aspects of the system should be maximized in order to maximize the diffusion rate? Which should be minimized?

2. D is small, meaning the respiratory surface is extremely thin. To appreciate just how thin the gas-exchange surface is, consider that researchers who asked bicyclists to ride up a steep hill for seven minutes observed a marked increase in blood on the respiratory surface of the athletes' lungs. To explain this observation, the researchers proposed that the cyclists' high heart rates (up to 177 beats/minute during their sprint up the hill) increased blood pressure to the point at which thin-walled vessels in the lungs ruptured and leaked blood into the lung cavity.
3. $P_2 - P_1$ is large, meaning the partial-pressure gradient of the gas across the surface is large. High partial-pressure gradients are maintained in part by having an efficient circulatory system in close contact with the gas-exchange surface. When blood flows close to the respiratory surface, oxygen is rapidly taken away from the area where inward diffusion is occurring, and carbon dioxide is rapidly brought into the area where outward diffusion is occurring. As a result, $P_2 - P_1$ stays high.

What other aspects of gill and lung design conform to the dictates of Fick's law? What other factors affect the structure of these respiratory organs? To answer these questions, let's delve into the anatomy of gills and lungs.

How Do Gills Work?

Gills are outgrowths of the body surface or throat that are used for gas exchange in aquatic animals. To appreciate the structure and function of these organs, it is important to recognize just how challenging water is as a respiratory medium compared with air. Recall from Section 44.1 that water contains much less oxygen than air does and that water is much more difficult to move across respiratory surfaces because of its density and viscosity.

Gills are efficient solutions to the problems posed by water breathing, primarily because they present an extremely large surface area for oxygen to diffuse across an extremely thin epithelium. In some species of invertebrates, such as the nudibranch mollusk pictured in **Figure 44.5a**, gills project from the body surface and contact the surrounding water directly. In other invertebrate species, such as the crayfish in **Figure 44.5b**, gills are located inside the exoskeleton or body wall. If gills are internal, water must be driven over them by cilia, the limbs, or other specialized structures.

In contrast to the diversity of gills found in invertebrates, the gills of bony fishes are all very similar in structure. Fish gills are located on both sides of the head and consist of four

(a) External gills are in direct contact with water.

(b) Internal gills must have water brought to them.

FIGURE 44.5 Gills Can Be External or Internal

(a) Nudibranchs are marine snails with gills that are outside the main body wall. **(b)** Crayfish gills are located inside the main body wall. A portion of the crayfish exoskeleton has been removed to expose the gills.

QUESTION Regarding fitness, what are the advantages and disadvantages of external versus internal gills? Why is it logical to expect that the orange-colored gills in part (a) are protected by toxins?

arches, as **Figure 44.6a** shows. To move water through these structures, most fish open and close their mouths and a stiff flap of tissue, called the **operculum**, that covers the gills. The pumping action of the mouth and operculum creates a pressure gradient that moves water over the gills. In tuna and other fish that are particularly fast swimmers, however, this pumping action does not provide enough water flow to meet the demand for oxygen. Instead, these fish force water through their gills by swimming with their mouths open. This process is called *ram ventilation*.

Regardless of how the gills are ventilated, the flow of water over gills is unidirectional. This is a crucial point. To see why, notice in the detailed sketch in Figure 44.6a that long, thin structures called **gill filaments** extend from each gill arch. Each gill filament is composed of hundreds or thousands of **gill lamellae**. These are sheetlike structures through which a bed of small blood vessels called *capillaries* runs. The flow of blood through the capillary bed is in the opposite direction to the flow of water over the gill surface. As a result, a **countercurrent exchange** system exists. Recall from Chapter 41 that the key point about a countercurrent exchanger is that two adjacent fluids flow in opposite directions (**Figure 44.6b**). In a *co-current* system, in which two adjacent fluids flow through pipes in the same direction, oxygen transfer would stop. That is because the partial-pressure gradient driving diffusion would fall to zero partway along the length of the capillary.

Countercurrent flow makes gills extremely efficient at extracting oxygen from water, because it ensures that the difference in the amount of oxygen and carbon dioxide in water versus blood is large over the entire gas-exchange surface. Stated another way, the effect of countercurrent exchange is to maximize the $P_2 - P_1$ term in Fick's law of diffusion over the entire gill surface. Based on this observation, biologists cite countercurrent exchange as another example of how gills are optimized for efficient gas exchange.

How Do Tracheae Work?

Air and water are dramatically different ventilatory media, because they have different densities, viscosities, and abilities to hold oxygen and carbon dioxide. In addition, the consequences of exposing the gas-exchange surface to air versus water differ. In aquatic habitats, ventilation tends to disrupt water and electrolyte balance, and homeostasis must be maintained by an active osmoregulatory system. As explained in Chapter 42, osmosis causes marine animals to lose water across their gas-exchange surface and freshwater animals to gain water. Diffusion tends to cause marine animals to gain sodium, chloride, and other ions and freshwater animals to lose them. In contrast, breathing leads to a loss of water by evaporation in terrestrial environments.

How do terrestrial animals minimize water loss while maximizing the efficiency of gas exchange? To answer this question, consider the tracheal system of insects. Recall from Chapter 42

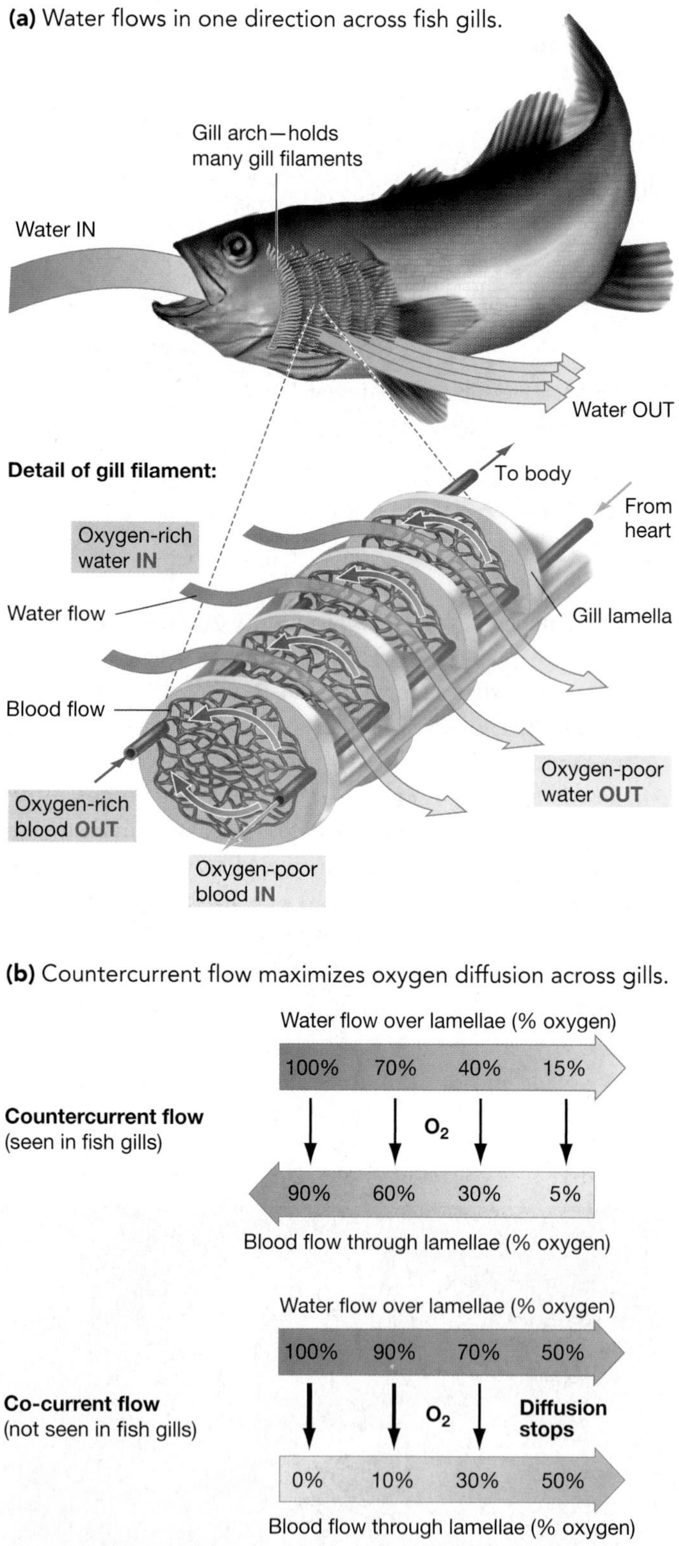
(a) Water flows in one direction across fish gills.
Gill arch—holds many gill filaments
Water IN
Water OUT
Detail of gill filament:
To body
From heart
Oxygen-rich water IN
Water flow
Gill lamella
Blood flow
Oxygen-poor water OUT
Oxygen-rich blood OUT
Oxygen-poor blood IN
(b) Countercurrent flow maximizes oxygen diffusion across gills.
Water flow over lamellae (% oxygen)
100% 70% 40% 15%
Countercurrent flow
(seen in fish gills)
O2
90% 60% 30% 5%
Blood flow through lamellae (% oxygen)
Water flow over lamellae (% oxygen)
100% 90% 70% 50%
Co-current flow
(not seen in fish gills)
O2
Diffusion stops
0% 10% 30% 50%
Blood flow through lamellae (% oxygen)

FIGURE 44.6 Fish Gills Are a Countercurrent Exchange System (a) In fish gills, water and blood flow in opposite directions. (b) In a countercurrent system, oxygen is transferred along the entire length of the capillaries within the lamellae. If co-current flow occurred instead, oxygen transfer would stop, because the partial-pressure gradient driving diffusion would fall to zero partway along the length of the capillary. QUESTION How much more oxygen is extracted from the countercurrent system than from the co-current system?

that **tracheae** are an extensive system of tubes located well within the body. They connect to the exterior through openings called **spiracles**, which can be closed to minimize the loss of water by evaporation.

Another key point about the insect tracheal system is that it consists of a series of tubes that extend throughout the body. The tips of tracheae are tiny, fluid filled, and highly branched. In this way, the tracheal system transports air close enough to cells for gas exchange to take place directly across their plasma membranes. Consequently, insects do not use their circulatory system to transport gases to and from tissues, even though the insect circulatory system transports nutrients and wastes. Insects also lack an oxygen-carrying protein, such as the hemoglobin found in vertebrates.

Recent research on the tracheal system has focused on how air moves into and out of the tubes through spiracles. Is simple diffusion efficient enough to ventilate the system, or is some type of breathing mechanism involved?

The short answer to this question is that breathing movements may play a role in at least some insects. To understand why this claim is valid, consider recent work on the sweet potato hawkmoth. A researcher set out to investigate how the amount of O_2 delivered to muscles changes during flight. To document the partial pressure of oxygen in flight muscles, he inserted a needlelike electrode into the wing muscles of a hawkmoth. The electrode was attached to an instrument that measured the partial pressure of oxygen, and the hawkmoth was tethered to a stand (**Figure 44.7a**).

When the hawkmoth recovered from the procedure and began to beat its wings normally, the biologist began recording P_{O_2} as the insect rested. Then the researcher stimulated the moth to fly by exposing it to wind. The "Results" section of Figure 44.7a shows how P_{O_2} changed during one such experiment. (The procedure was repeated on several individuals, and the same pattern was observed.) Because flight is an energetically demanding activity, it is not surprising that P_{O_2} levels in the flight muscles dropped initially. As flying continued, however, P_{O_2} levels recovered until they were nearly as high as when they were at rest.

These results can be explained by the breathing mechanism proposed in **Figure 44.7b**. The key observation is that the tracheae of flying insects are sandwiched between muscles. When these muscles contract and relax, tracheae should constrict and expand in response. As a result, wing beats could force air into and out of the tracheae and lead to efficient ventilation of the respiratory surface.

How Do Lungs Work?

In most mammals, air enters the body through both the nose and mouth. A tube known as the **trachea** (not to be confused with the tracheae of insects) carries the inhaled air to narrower tubes called **bronchi** (singular: *bronchus*). The bronchi branch off into yet narrower tubes, the **bronchioles**. The organs of

(a) Question: How does air move through the tracheal system?

Hypothesis: Air moves through the tracheal system by some sort of breathing mechanism.

Null hypothesis: Air moves through the tracheal system by diffusion.

Experimental setup:

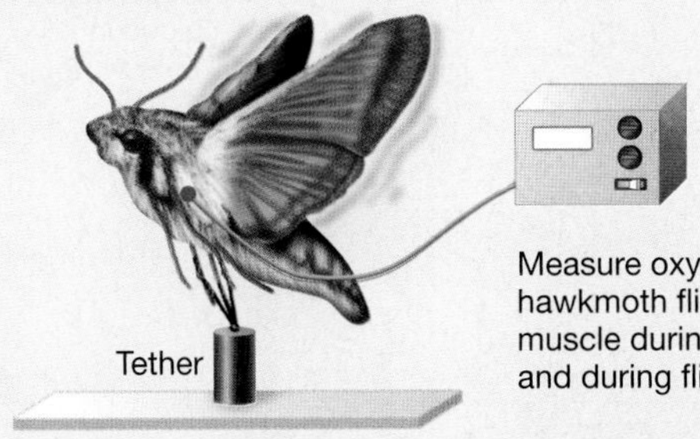

Prediction: Flying will increase ventilation of tracheal system, causing increase in P_{O_2} in wing muscle.

Prediction of null hypothesis: Flying will not increase ventilation of tracheal system; P_{O_2} in flight muscle will decline steadily during flight.

Results:

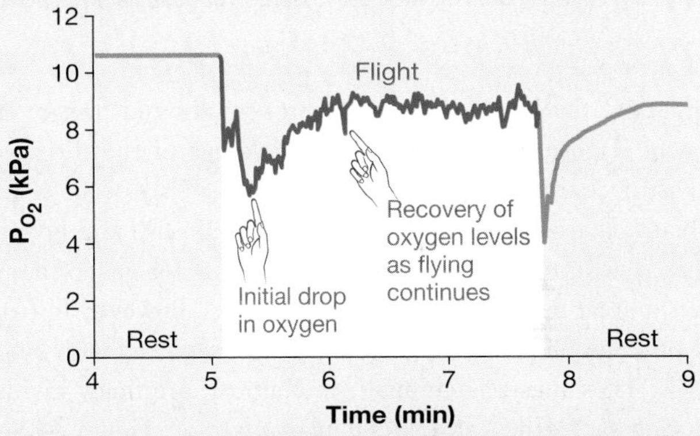

Conclusion: Breathing movements may ventilate the tracheal system in at least some insects.

(b) Model of oxygen delivery during flight

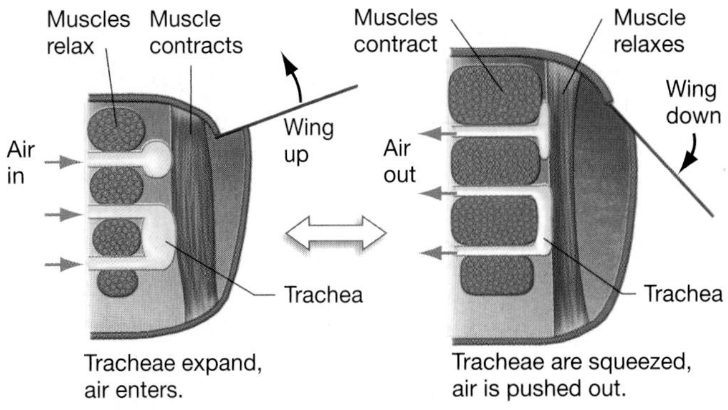

FIGURE 44.7 Muscle Activity May Ventilate Tracheae
(a) Experiment on the partial pressure of oxygen, in kilopascals (kPa), in flight muscles; 1000 kPa = 1 Pa. (A pascal is a unit of pressure—force per unit area.) **(b)** Tracheae change shape as wings beat.

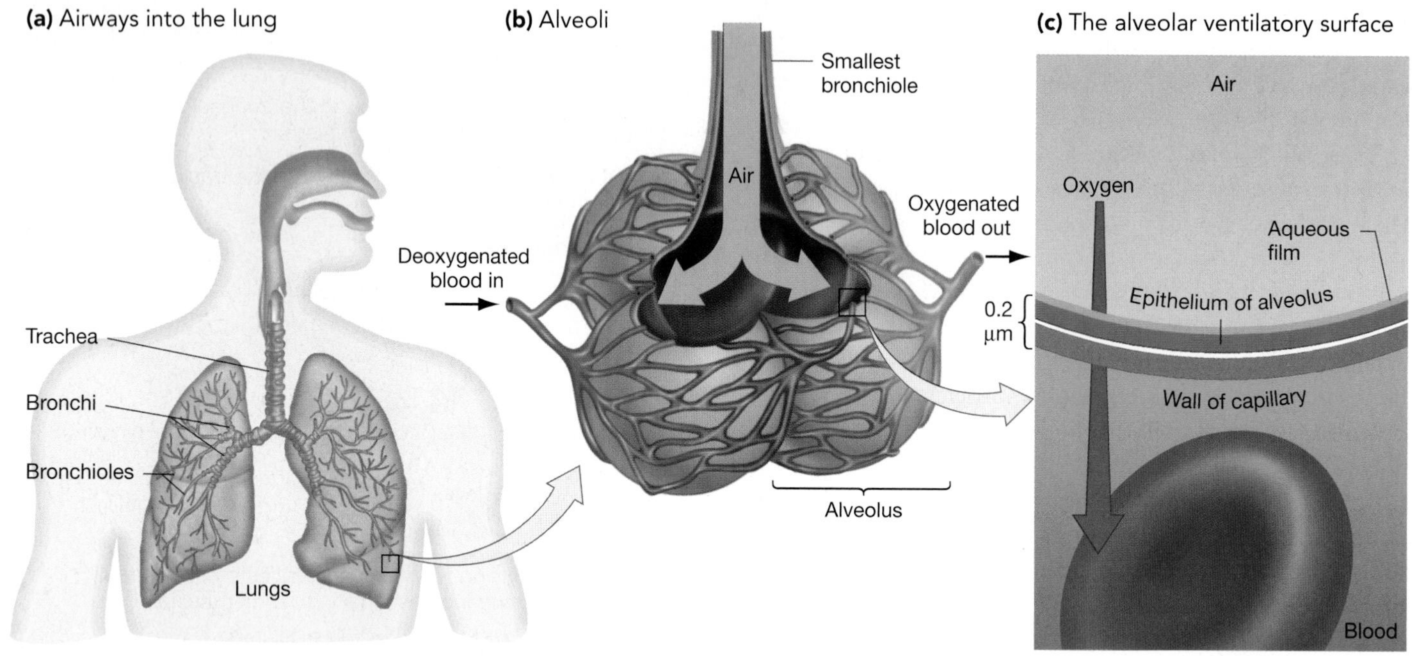

FIGURE 44.8 Lungs Offer a Large, Thin Surface Area for Gas Exchange between Air and Blood
(a) The human respiratory tract branches repeatedly from the largest airway, the trachea, to the smallest bronchioles. Gas exchange does not occur across these airways. **(b)** The system of airways ends in clusters of tiny sacs called alveoli. Alveoli are covered with capillary networks and **(c)** are the site of gas exchange.

ventilation, the lungs, enclose the bronchioles and part of the bronchi (**Figure 44.8a**). **Lungs** are infoldings of the throat or body surface that are used for gas exchange. They occur in amphibians, lizards, birds, mammals, and certain fish and invertebrates. The amount of surface area available for gas exchange varies a great deal among species with lungs, however. In frogs and other amphibians, the lung is a simple sac lined with blood vessels. The lungs of mammals, in contrast, are finely divided into tiny sacs called **alveoli** (singular: *alveolus*; **Figure 44.8b**). Each human lung contains approximately 150 million of these sacs. Due to the presence of alveoli, mammalian lungs have about 40 times more surface area for gas exchange than does an equivalent volume of frog lung tissue.

As **Figure 44.8c** shows, an alveolus provides an interface between air and blood that consists of a thin aqueous film, a layer of epithelial cells, some extracellular material, and the wall of a capillary. In the human lung, this barrier to diffusion is only 0.2 μm thick—about 1/200th of the thickness of this page.

In addition to total surface area, the other major feature of lungs that varies among species is mode of ventilation. In the lungs of snails and spiders, air movement takes place by diffusion only. Vertebrates, in contrast, actively ventilate their lungs by pumping air via muscular contractions. One mechanism for pumping air is **positive pressure ventilation**, used by frogs and related animals. A frog lowers the floor of its throat to draw in air from the atmosphere, through the nasal passages, and into the oral cavity. The animal then closes the nasal passages and contracts its throat muscles. These actions increase the pressure on air in the oral cavity and force it into the lungs.

In effect, frogs push air into their lungs. In contrast, humans and other mammals pull air into their lungs. How does this **negative pressure ventilation** work?

Ventilation of the Human Lung The pressure inside the human chest cavity is about 5 mm Hg less than atmospheric pressure. Even though the lung is **elastic**—it returns to its original, collapsed shape if it is not stretched or compressed—the negative pressure surrounding the lung is just enough to keep it expanded. If a wound opens a hole in the wall of the chest and the pressure differential between the chest cavity and the atmosphere disappears, the lung on the side of the injury will collapse like a deflated balloon.

Humans ventilate their lungs by changing the pressure within the chest cavity between about −5 mm Hg and −8 mm Hg relative to the atmosphere. As **Figure 44.9a** shows, inhalation is based on lowering the negative pressure in the chest cavity. The change is caused by a downward motion of the thin muscular sheet called the **diaphragm** and an outward motion of the ribs. These movements increase the volume of the chest cavity and thus lower its internal pressure. As the pressure surrounding the lungs drops, air flows into the airways along a pressure gradient. Exhalation, in contrast, is a passive process driven by the elastic recoil of the lungs and chest wall as the diaphragm and rib muscles relax. The changes in pressure that occur during negative pressure ventilation are analogous to changing the volume of a jar, as shown in **Figure 44.9b**.

About 450 ml of air moves into and out of the lungs in an average breath. Only about two-thirds of this volume actually

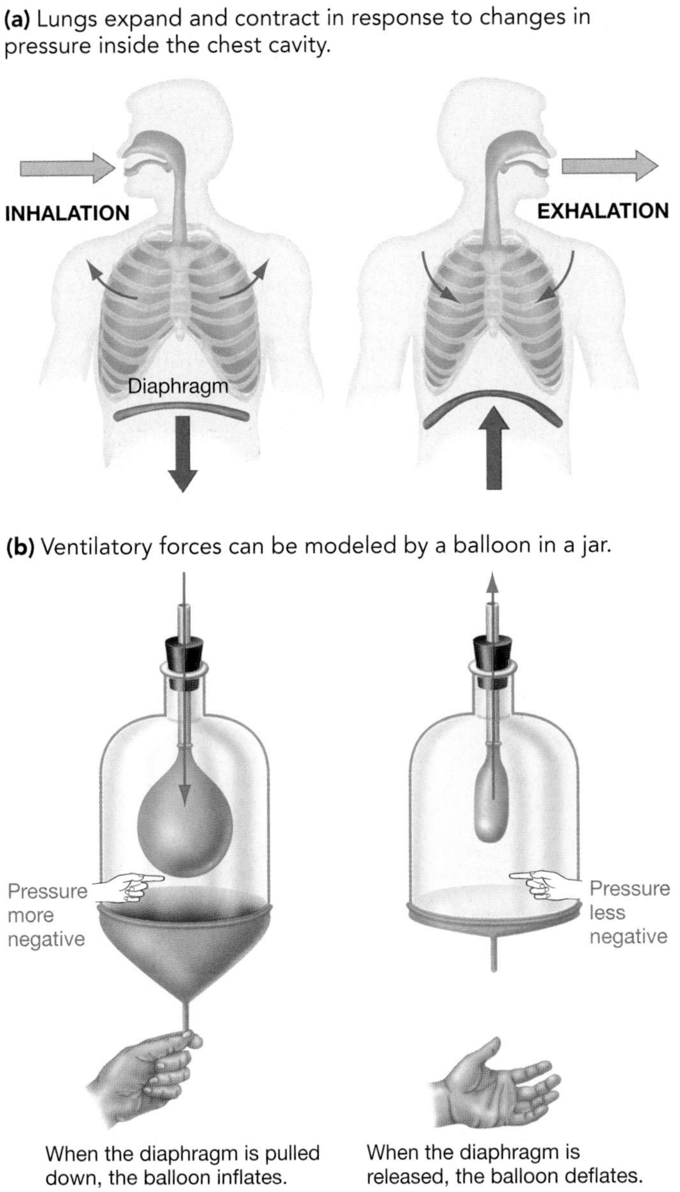
(a) Lungs expand and contract in response to changes in pressure inside the chest cavity.
INHALATION
EXHALATION
Diaphragm
(b) Ventilatory forces can be modeled by a balloon in a jar.
Pressure more negative
Pressure less negative
When the diaphragm is pulled down, the balloon inflates.
When the diaphragm is released, the balloon deflates.

FIGURE 44.9 Changes in the Volume of the Chest Cavity Drive Negative Pressure Ventilation
(a) Inhalation: When the diaphragm and rib muscles contract, the volume of the thoracic cavity increases, causing internal pressure to become more negative. In response, the lungs expand and air flows in. Exhalation: When the diaphragm and rib muscles relax, the volume of the thoracic cavity decreases. Internal pressure becomes less negative, lung volume decreases due to elasticity of the lungs, and air flows out. **(b)** A model of negative pressure ventilation. QUESTION What happens when a person sighs or takes a deep breath?

participates in gas exchange, however, because 150 mL of the air occupies **dead space**—portions of the air passages that do not have a respiratory surface. The trachea and bronchi shown in Figure 44.8a, for example, represent dead space. Breathing is much more efficient during exercise, when the chest cavity undergoes larger changes in volume. When a person is breathing hard, over 2500 mL of air can move with each inhalation-exhalation cycle, but the 150 mL in dead space stays the same.

The volume and pressure changes that occur during a ventilatory cycle are fairly straightforward. But what accounts for the elasticity of the human lung? Stated another way, what force is responsible for returning lung tissue to its former shape? An early hypothesis was inspired by the thin film of fluid that exists on the surface of alveoli. The idea was that hydrogen bonding between water molecules in the film would create a high degree of surface tension and thus create a pull or tension on the alveolar wall (**Figure 44.10**). Once an alveolus was done expanding during inhalation, the contractile force created by surface tension would return the structure to its original size, imparting elasticity to the entire lung.

To test this hypothesis, a biologist realized that he could quantify the elasticity of a lung by measuring the pressure required to force a known volume of air into the structure. If the surface tension hypothesis is correct, then much less pressure should be required to force the same volume of saline solution, rather than air, into the same lung. This prediction is logical, because the presence of the solution would eliminate the air-water interface inside the alveolus and thus eliminate surface tension. The biologist tested this prediction on lungs from a cadaver and showed that it was correct. Much less pressure was required to inflate the lungs with saline solution than with air. When other researchers repeated the experiment and got similar results, biologists accepted the hypothesis that most of the lung's elasticity is due to surface tension. Follow-up work has shown that surface tension within alveoli is so strong that it

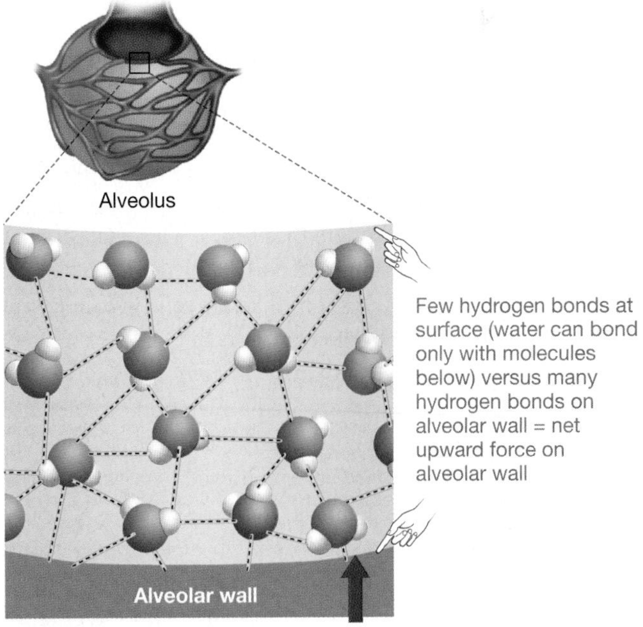
Alveolus
Few hydrogen bonds at surface (water can bond only with molecules below) versus many hydrogen bonds on alveolar wall = net upward force on alveolar wall
Alveolar wall

FIGURE 44.10 Surface Tension Creates a Pull on the Surface of the Lung
Surface tension occurs because water molecules at the surface form hydrogen bonds only with water molecules below them. The bonds create a pulling force, or tension, on the surface. In the lung, this force is transmitted to the walls of alveoli via hydrogen bonds.

BOX 44.1 Respiratory Distress Syndrome in Premature Infants

Because alveoli have a large amount of surface area in contact with air, the watery film that covers them generates a large amount of surface tension. If water were the only molecule present, the pulling force on the alveolar wall would be so great that the structure would not inflate readily during inhalation. Fortunately, cells within each alveolus produce a mixture of soaplike phospholipids and proteins that act as a **surfactant**—that is, they disrupt the aqueous film lining the wall and greatly reduce its surface tension. The presence of the surfactant decreases lung stiffness enough to reduce the amount of work that has to be done by the diaphragm and rib muscles during inhalation.

Human infants acquire the ability to produce adequate amounts of surfactant between 32 and 34 weeks of gestation, and birth normally occurs at 36 weeks. If born prior to 32 weeks of gestation, an infant may have such stiff lungs that it quickly fatigues from the effort of breathing and suffocates. This condition is called **respiratory distress syndrome.** Fortunately, it can be treated by administering a synthetic surfactant mixture directly into the lungs, through a tube placed in the trachea. This relatively simple technology has made it possible for infants to survive even if they are born after as little as 24 weeks of gestation.

must actually be diminished somewhat for breathing to occur at all (see **Box 44.1**).

Ventilation of the Bird Lung Flight is one of the most energy-demanding activities performed by animals. Nevertheless, birds may fly thousands of miles during annual migrations. Even more impressive is the observation that geese regularly fly over the top of Mount Everest. At this elevation, the partial pressure of oxygen is so low that most humans would immediately black out if they were flown there and dropped off. How are birds able to extract enough oxygen from the atmosphere to support long flights and flights at high elevations?

Biologists began working on this question by analyzing the anatomy of the avian ventilatory system. As **Figure 44.11a** shows, a bird's trachea bypasses the lungs almost entirely and enters two large **air sacs** posterior to the lungs. These sacs are connected to the posterior portion of the lung by a series of tubes. The tubes branch into tiny airways, called **parabronchi**, within the lung. The parabronchi pass straight through the lung and terminate in another system of air sacs. These sacs link up with the trachea near the anterior portion of the lung. Gas exchange occurs in millions of spongelike air cells, which extend laterally along the length of the parabronchi.

To explore how this ventilatory system functions, Knut Schmidt-Nielsen and co-workers used a variation on the pulse-chase technique introduced in Chapter 7. In this case, the pulse was a single breath of pure oxygen, followed by a "chase" that consisted of a breath of normal air. To the researchers' surprise, the oxygen-enriched air was not detected at the bird's mouth until the *second* exhalation following the pulse. Follow-up experiments showed that air moves through a bird's ventilatory system as diagrammed in **Figure 44.11b**. The key conclusions were that airflow through the avian lung is (1) *unidirectional*—specifically from posterior to anterior—and (2) *continuous*, meaning air moves through the lung during both inhalation and exhalation.

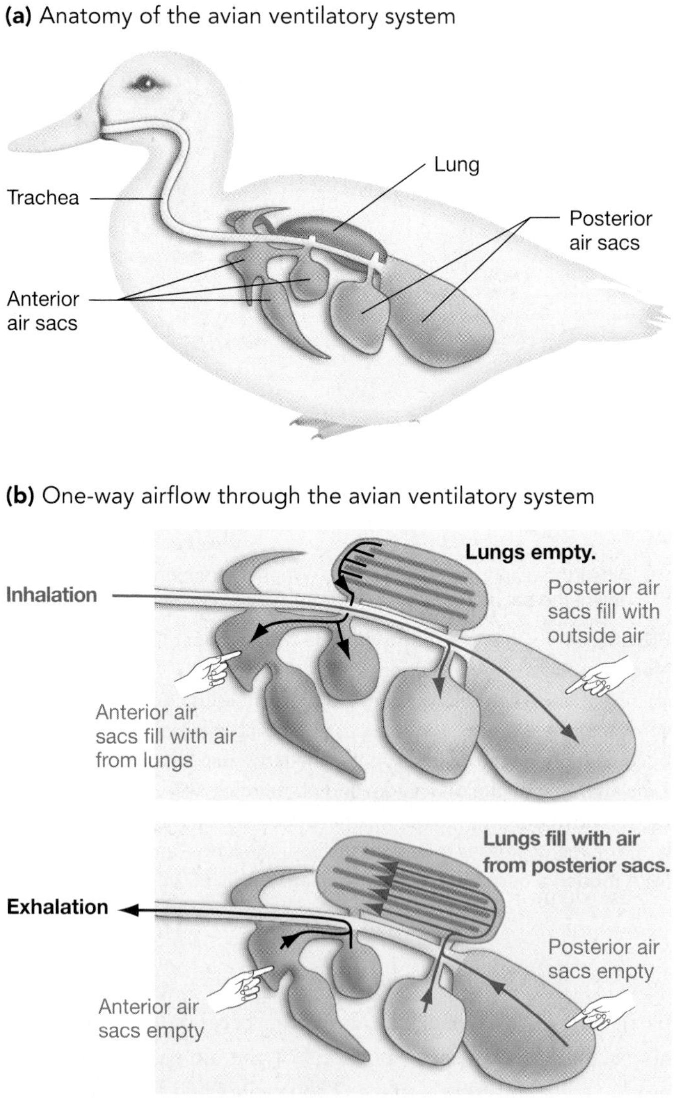

FIGURE 44.11 Air Flows in One Direction through the Bird Ventilatory System
(a) Anatomy of the avian ventilatory system. (b) Diagram of the airflow pattern through the avian ventilatory system, as determined by using pure oxygen as a tracer gas.

To drive home how birds breathe, biologists make an analogy with how Scottish bagpipes are played. Air is introduced intermittently into a large reservoir, such as the bag of a bagpipes or the posterior air sacs of a bird. Then the air is expelled continuously through air passages, such as the drone pipes of the instrument or the parabronchi of the bird lung. In birds, the air expelled from the parabronchi is collected in the anterior air sacs and ultimately exhaled into the trachea through an opening that prevents the mixing of inhaled and exhaled air. Continuous airflow is possible because the chest and air sacs do not compress at the same time but alternately.

Three features make the avian ventilatory system efficient:

1. Dead space in birds is restricted to the short stretch of trachea between the mouth and the opening of the anterior air sacs. As a result, inhaled air is used much more efficiently in birds than in mammals.
2. Birds exchange gases during both inhalation and exhalation. In contrast, no gas exchange occurs during the exhalation half of the ventilatory cycle in mammals. Bird ventilation resembles the continuous ventilation of fish gills in this respect.
3. Blood circulates through the bird lung in capillaries that cross the parabronchi perpendicularly. This crosscurrent pattern is not as efficient as the countercurrent circulation of fish gills, but it is far more efficient than the weblike arrangement of capillaries that surrounds mammalian alveoli.

Homeostatic Control of Ventilation

An animal is in trouble if homeostasis with respect to blood oxygenation or the elimination of carbon dioxide fails. To ensure adequate ATP production, the partial pressures of oxygen and carbon dioxide must be maintained within a narrow range, both at rest and during vigorous exercise. How is ventilation controlled to achieve this critical homeostasis?

In the resting state, the rate of breathing is established by the **medullary respiratory center**, an area at the base of the brain, just above the spinal cord. This center stimulates the rib and diaphragm muscles to contract about 12–14 times per minute in humans. But during exercise, things change. Active muscle tissue takes up more oxygen from the blood. As a result, the partial pressure of oxygen (P_{O_2}) in blood drops. Those same muscles release quantities of carbon dioxide to the blood, tending to raise its partial pressure (P_{CO_2}) in blood. When it reaches the brain, carbon dioxide rapidly diffuses from the blood into the cerebrospinal fluid that bathes the brain. In both blood and cerebrospinal fluid, CO_2 reacts with water to form carbonic acid (H_2CO_3), which then dissociates to release a hydrogen ion (H^+) and a bicarbonate ion (HCO_3^-):

$$CO_2 + H_2O \longleftrightarrow H_2CO_3 \longleftrightarrow H^+ + HCO_3^-$$

The result is a slight drop in the pH of blood and cerebrospinal fluid. The change is sensed by specialized nerves located near the large arteries that travel from the heart into the neck and to the base of the brain itself. When these nerve cells are activated, they signal the medullary respiratory center to increase the breathing rate. The resulting rise in ventilation rate increases the rate of oxygen delivery to the tissues and the rate at which carbon dioxide is eliminated from the body, restoring P_{O_2} and P_{CO_2} to their resting levels.

This control system is so effective that it can maintain stable levels of oxygen and carbon dioxide in the blood even in the face of intense exercise. In addition, it allows the body to regulate blood pH carefully. To see why this is so, it's important to recognize that carbon dioxide is transported in the blood in the form of bicarbonate ions. When CO_2-rich blood reaches the lungs, the bicarbonate ions have to react with hydrogen ions to regenerate carbon dioxide, which then diffuses across the walls of alveoli and is exhaled. When breathing rate increases, then, excess hydrogen ions that are produced during rapid metabolism are eliminated along with the excess CO_2 that is generated, and blood pH is stabilized.

Now let's look in greater depth at blood itself. What is this substance, and how does it transport oxygen and carbon dioxide between the gas-exchange surface and an animal's tissues?

WEB TUTORIAL 44.1
Gas Exchange in the Lungs and Tissues

✓ CHECK YOUR UNDERSTANDING

Most large-bodied animals exchange gases via gills, tracheae, or lungs. You should be able to (1) identify at least three features that are common to all three of these types of organs as well as one trait that is unique to each; (2) explain the contrast between airflow in the mammalian lung and in the bird lung; and (3) make a diagram summarizing the sensor, integrator, and effector(s) involved in homeostatic control of the ventilation rate.

44.3 Blood

Blood is a tissue—a collection of cells that functions as an integrated unit. More specifically, **blood** is a connective tissue that consists of cells in a watery extracellular matrix. In an average human, 50–60 percent of the blood volume is composed of an extracellular matrix called **plasma.** The remainder of the volume comprises a variety of cells and cell fragments that are collectively called **formed elements.**

Given the wide variety of functions that blood serves, it is not surprising that this tissue is complex. In addition to carrying oxygen and carbon dioxide between mitochondria and lungs, blood transports nutrients from the digestive tract to other tissues in the body, moves waste products to the kidney and liver for processing, conveys hormones from glands to target tissues, delivers immune system cells to sites of infection, and distributes heat from deeper organs to the surface.

The formed elements in blood include platelets, several types of white blood cells, and red blood cells. **Platelets** are cell fragments that minimize blood loss from ruptured blood vessels by releasing material that assists in the formation of blockages known as **clots.** **White blood cells** are part of the immune system and fight infections. In number, formed elements are dominated by red blood cells. In humans, red blood cells make up 99.9 percent of the formed elements. **Red blood cells** transport oxygen from the lungs to tissues throughout the body.

In humans, new red blood cells are synthesized at the rate of 2.5 million per second to replace old red blood cells, which are dying at the same rate. These red blood cells live about 120 days. They develop along with white blood cells and platelets from stem cells located in the tissue inside bone (*bone marrow*). As the new red blood cells of mammals mature, the cells lose their nuclei, mitochondria, and most other organelles but become filled with about 280 million molecules of the oxygen-carrying molecule **hemoglobin**. Vertebrates other than mammals transport oxygen in red blood cells that retain their nuclei, and many invertebrates transport oxygen in association with carrier molecules that are not contained within cells at all.

Hemoglobin consists of four polypeptide chains (see Figure 3.16b), each of which binds to a nonprotein group called **heme**. Each heme molecule, in turn, contains an iron ion (Fe^{2+}) that can bind to an oxygen molecule. As a result, each hemoglobin molecule can bind up to four oxygen molecules. The molecule called **myoglobin**, which is found within muscle cells, resembles a single polypeptide chain of the hemoglobin molecule and binds one molecule of oxygen. Myoglobin stores oxygen for use by the mitochondria of muscle cells and imparts the characteristic red color to muscle.

Because of oxygen's low solubility in water, only 1.5 percent of the oxygen carried in blood is dissolved in plasma; 98.5 percent is bound to hemoglobin. What happens when red blood cells carrying oxygen-rich hemoglobin reach cells that contain little oxygen?

Oxygen Delivery: The Bohr Effect

Blood leaving the human lungs has a P_{O_2} of 100 mm Hg, while muscles and other tissues have P_{O_2} levels of about 40 mm Hg at rest. This partial-pressure difference drives the unloading of O_2 from the hemoglobin to the tissues through diffusion.

Researchers who studied the dynamics of O_2 unloading in tissues found the pattern shown in **Figure 44.12a**. This graph, which plots the percentage saturation of hemoglobin in red blood cells versus the P_{O_2} levels in the blood within tissues, is called the oxygen-hemoglobin **dissociation curve**. Each 25 percent change in saturation corresponds to an average of one additional oxygen molecule bound per hemoglobin molecule or one oxygen molecule delivered to tissues. The remarkable feature of the relationship is that it is *sigmoidal*, or S-shaped. This pattern occurs because the binding of each successive oxygen molecule to a subunit of the hemoglobin molecule causes a conformational change in the protein that makes the remaining subunits much more likely to bind oxygen. This phenome-

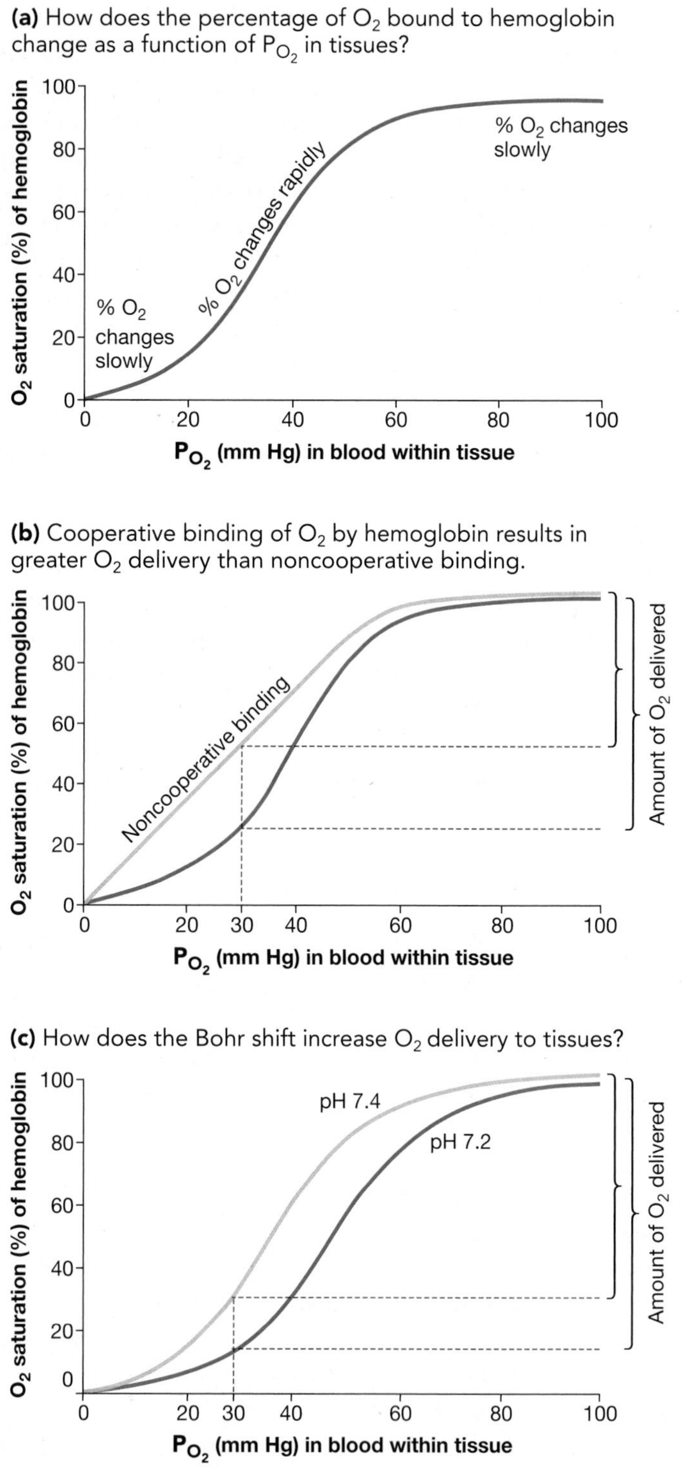

FIGURE 44.12 Oxygen-Hemoglobin Saturation Curves
(a) Partial pressure of oxygen in blood and tissues versus the percentage of binding sites on hemoglobin that hold oxygen molecules. The sigmoidal curve has three distinct regions. **(b)** Effect of cooperative versus noncooperative oxgyen binding. As blood moves from a P_{O_2} of 100 mm Hg in the lungs to 30 mm Hg in exercising tissue, much more oxygen is dropped off in tissues with cooperative binding to hemoglobin. **(c)** As pH drops, oxygen becomes less likely to stay bound to hemoglobin at all values of tissue P_{O_2}. Exercising tissues have lower pH than resting tissues have and thus receive more oxygen from hemoglobin.

non is **cooperative binding**. If cooperative binding did not occur, all four subunits of hemoglobin would load or unload oxygen independently of each other. They would lose or gain oxygen in direct proportion to the partial pressure of oxygen in the blood, until the molecule was either completely saturated or unsaturated with oxygen. The relationship shown by the yellow line in **Figure 44.12b** would result instead.

Why is cooperative binding important? The answer lies in the difference between the curves in Figure 44.12b and in the observation that tissue P_{O_2} ranges between about 25 and 50 mm Hg. In this range of partial pressures, a relatively small change in tissue P_{O_2} results in a large change in hemoglobin saturation. Consider what happens to hemoglobin saturation when tissue P_{O_2} changes from its resting level of about 40 mm Hg to 30 mm Hg, as might be seen during exercise. In response to this small change in tissue P_{O_2}, there is an enormous change in the percentage saturation of hemoglobin—from about 70 percent to about 30 percent. This means that when blood encounters tissues at rest, hemoglobin gives up about 30 percent of the oxygen it is carrying (from almost 100 percent saturation to about 70 percent). But when blood encounters tissues during exercise, it gives up about 70 percent of its oxygen. If cooperative binding did not occur, much less oxygen would be delivered (see Figure 44.12b). Because of the sigmoidal relationship characteristic of cooperative binding, hemoglobin responds quickly and effectively to small changes in oxygen demand. When one oxygen molecule leaves hemoglobin, others are more likely to follow.

Cooperative binding is only part of the story behind oxygen delivery, however. Hemoglobin—like other proteins—is sensitive to changes in pH and temperature. During exercise, the temperature and partial pressure of CO_2 in active muscle tissue rise. The CO_2 produced by exercising muscle reacts with the water in blood to form carbonic acid, which dissociates and releases a hydrogen ion. As a result, the pH of the blood in exercising muscle drops.

Decreases in pH and increases in temperature alter hemoglobin's conformation. As the graph in **Figure 44.12c** shows, these shape changes make hemoglobin more likely to release O_2 at any given value of tissue P_{O_2}. This phenomenon is known as the **Bohr shift**. The name honors Christian Bohr, who published on the phenomenon in 1904, and highlights the rightward shift of the oxygen-hemoglobin dissociation curve with decreasing pH. The Bohr shift is important because it makes hemoglobin more likely to release oxygen during exercise or other conditions in which P_{CO_2} is high and tissues are under oxygen stress. To convince yourself that this is true, study Figure 44.12c and ask, When tissues have a P_{O_2} of 30 mm Hg, how much more oxygen is released from hemoglobin in the blood at pH 7.2 than at pH 7.4?

To see the effect of cooperative binding and the Bohr shift in action, consider a recent experiment on how the oxygen transport system in rainbow trout responds to sustained exercise. To explore changes in blood O_2 content during exercise, biologists had fish swim continuously against a current in a water tunnel. As the researchers increased the speed of the current and thus the swimming speed of the fish, they periodically sampled the O_2 content of both the arterial and venous blood of experimental individuals. *Arterial blood* is freshly oxygenated and moving away from the lungs; *venous blood* is returning to the lungs from the rest of the body. Not surprisingly, the biologists found that arterial O_2 levels remained fairly constant as swimming speed increased—meaning the gills worked efficiently enough to saturate hemoglobin with oxygen. In contrast, the O_2 content of venous blood, which had undergone gas exchange with the tissues, dropped steadily as swimming speed increased. When the fish had reached their maximum sustainable speed, virtually all the oxygen available in hemoglobin had been extracted. The data show that in hard-working tissues, the combination of increased temperature, lower pH, and lower P_{O_2} caused hemoglobin to become almost completely deoxygenated. Oxygen-delivery systems based on hemoglobin are extremely efficient.

We have yet to consider the other side of gas exchange, however: What happens as CO_2 is released from tissues to the blood? How does the body cope with the decline in pH that results? After all, if hemoglobin's conformation and activity are affected so dramatically by pH changes, it is likely that many other proteins will be altered as well. If these proteins function less well at low pH, then carbon dioxide buildup should result in illness.

CO_2 Transport and the Buffering of Blood pH

The carbon dioxide that is produced by cellular respiration in the tissues enters the blood, where it reacts with water to form carbonic acid. Recall that the resulting drop in blood pH stimulates an increase in breathing rate. Rapid exhalation of CO_2 then counteracts the drop in blood pH.

Changes in blood pH are also diminished by an elegant series of reactions that takes place inside red blood cells. Biologists were able to work out what was happending when they discovered large amounts of the enzyme **carbonic anhydrase** in red blood cells. Recall from Chapter 43 that carbonic anhydrase catalyzes the formation of carbonic acid from carbon dioxide in water. Consequently, CO_2 that diffuses into red blood cells is quickly converted to bicarbonate ions and protons. This is the mechanism responsible for the observation, made in Section 44.2, that most CO_2 is transported in blood (specifically in plasma) in the form of HCO_3^-—the bicarbonate ion. The same reaction occurs in the plasma surrounding red blood cells, but much more slowly in the absence of the enzyme.

Why is the carbonic anhydrase activity in red blood cells so important? The answer has two parts. First, the protons produced by the enzyme-catalyzed reaction induce the Bohr shift, which makes hemoglobin more likely to release oxygen. Second, the partial pressure of CO_2 in blood drops when carbon dioxide is converted to soluble bicarbonate ion. Thus, carbonic anhydrase activity maintains the strong partial-pressure gradient favoring the entry of CO_2 into red blood cells, making CO_2 uptake from tissues more efficient (**Figure 44.13**).

Once bicarbonate ions form in the red blood cell, they diffuse into the blood plasma along a concentration gradient. Bicarbonate ions move via an antiporter that exchanges bicarbonate with chloride ions. In contrast, the protons produced by the reaction stay inside red blood cells. Relatively small amounts of H^+ build up in the plasma.

What happens to the protons inside red blood cells? The answer to this question emerged when researchers studied the composition of deoxygenated hemoglobin. These analyses showed that when hemoglobin is not carrying oxygen molecules, it has a high affinity for protons. As a result, much of the H^+ that is produced by the dissociation of carbonic acid is taken up by hemoglobin. In this way, hemoglobin acts as a **buffer**—a compound that minimizes changes in pH. In addition, analyses of hemoglobin in deoxygenated blood revealed that the CO_2 that is not transported as HCO_3^- binds to amino groups on the protein portion of the molecule. The overall effect of these events is efficient transport of CO_2, with a minimal effect on blood pH.

When deoxygenated blood reaches the alveoli, however, the environment changes dramatically. In the lungs, a strong partial-pressure gradient favors the diffusion of CO_2 from plasma and red blood cells to the atmosphere. As CO_2 diffuses from the blood into the alveoli, P_{CO_2} in blood declines. The drop promotes the reversal of the chemical reactions that occurred in tissues: Hydrogen ions leave their binding sites on hemoglobin and combine with bicarbonate to form CO_2, which then diffuses into the alveoli and is exhaled from the lungs. In the meantime, hemoglobin has picked up O_2. The cycle begins anew.

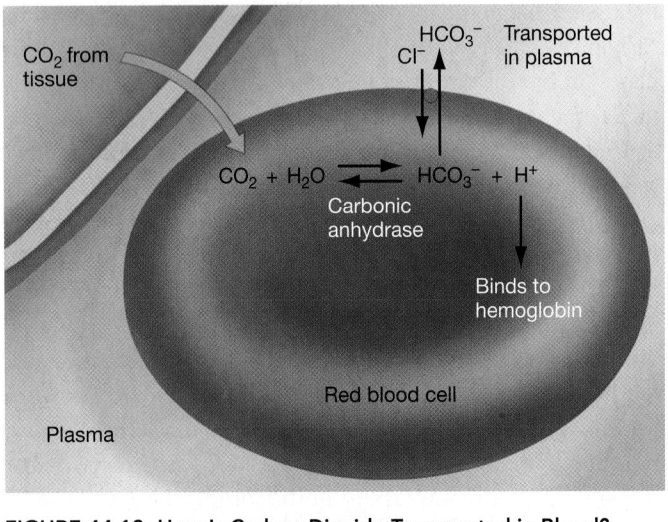

FIGURE 44.13 How Is Carbon Dioxide Transported in Blood? When CO_2 diffuses into red blood cells, carbonic anhydrase quickly converts it to carbonic acid, which dissociates into bicarbonate ion (HCO_3^-) and a proton (H^+). This reaction maintains the partial-pressure gradient favoring the entry of CO_2 into red blood cells. The protons produced by the reaction bind to deoxygenated hemoglobin. Most CO_2 in blood is transported to the lungs in the form of HCO_3^-. **QUESTION** This diagram shows the sequence of events in tissues. Explain what happens when the red blood cell in the diagram reaches the lungs.

Homeostatic Control of Oxygen-Carrying Capacity

Homeostasis with respect to carbon dioxide levels in the blood is tightly regulated through the monitoring of blood pH and the control of ventilation rate. But what about the oxygen-carrying capacity of blood? When environmental conditions change, how does the body keep the oxygen-delivery capacity of blood at levels that maintain health?

The answer is that the total mass of red blood cells is regulated. Total red blood cell mass increases in response to situations in which oxygen delivery to tissues is impaired—in people living at high elevations and in cigarette smokers, for example. If an individual moves to sea level or stops smoking, red blood cell mass drops.

An unusual observation helped biologists discover the mechanism responsible for homeostasis with respect to oxygen-carrying capacity: Red blood cell mass falls to very low levels in people with kidney failure. If a functional kidney is transplanted back into these individuals, however, red blood cell mass rapidly returns to normal. Follow-up analyses revealed that cells in the kidney sense blood oxygen levels and that these cells release a chemical messenger called **erythropoietin** (EPO) when oxygen levels are low. Erythropoietin travels via blood to the bone marrow, where it stimulates additional red blood cell production. The end result is an increase in tissue oxygenation.

Following the discovery of erythropoietin, biomedical researchers used the types of recombinant DNA techniques introduced in Chapter 19 to produce it commercially. As a drug, erythropoietin has been responsible for dramatic improvements in the quality of life of people with kidney failure. Erythropoietin increases red blood cell mass considerably in these patients and results in a heightened exercise capacity and sense of well-being. Unfortunately, the commercial production of erythropoietin also paved the way for its use as an illegal performance-enhancing drug by athletes (see **Box 44.2**).

To summarize, blood is a sophisticated system for delivering oxygen to tissues and removing carbon dioxide. Oxygen delivery is efficient because hemoglobin's subunits bind oxygen cooperatively and because lowered pH and increased temperature make the molecule more likely to give up bound oxygen. Carbon dioxide is transported efficiently as highly soluble bicarbonate ions, while binding of protons by oxygen-deficient hemoglobin minimizes changes in blood pH. Regulation of the breathing rate and the red blood cell count also play key roles in achieving homeostasis with respect to blood and tissue oxygen levels and blood pH. Now, how is blood transported from lungs to tissues and back?

BOX 44.2 Blood Doping and Erythropoeitin Abuse by Athletes

To win a marathon or a long-distance bicycle or cross-country ski race, an athlete's blood has to deliver large quantities of oxygen to a large number of muscles for a long period of time. As a result, training for these events often centers around techniques for increasing the oxygen-carrying ability of blood. Only some of these techniques are legal, however.

One widespread illegal strategy is referred to as **blood doping.** In this practice, athletes have blood withdrawn at intervals as they begin training for a competition. The red blood cells are isolated and preserved by refrigeration, then transfused back into the athlete right before the race. The idea is to provide a short-term boost in oxygen-carrying capacity by increasing red blood cell count.

The same end can be achieved, over a longer period of time, by increasing erythropoeitin levels in the blood. This can be done legally by inducing low oxygen levels in the blood—for example, by training at high altitudes or by sleeping in low-oxygen enclosures. It can also be done illegally by injecting human erythropoeitin, or EPO, produced by recombinant DNA techniques.

For some athletes, EPO has become a favorite performance-enhancing drug, because the commercially available molecule is identical to the molecule produced naturally in the kidney. Thus, it is difficult for officials from sports-regulatory agencies to detect illegal use of EPO in drug screening. Although data are extremely difficult to obtain, it is widely believed that EPO abuse is rampant in some sports.

The use of synthetic EPO has been banned for two reasons: (1) It gives some competitors an unfair advantage, and (2) it entails serious health risks. Risk is present because the viscosity of blood increases when red blood cell mass increases. This increased viscosity is accentuated during a race, when blood plasma volume drops due to dehydration. The combination of high blood viscosity and low plasma volume can greatly impair blood flow through capillaries. Impaired blood flow can lead to cramping and muscle damage, due to low oxygen delivery. In addition, blood clotting is much more likely to occur when blood flow slows. If clots form in blood vessels that lead to the brain or heart, stroke or heart attack can occur. EPO abuse is thought to be responsible for the collapse and death of several cyclists during races in the mid-1990s. There is little question that EPO use can give the sports phrase "sudden death" new meaning.

✓CHECK YOUR UNDERSTANDING

In blood, oxygen is bound to hemoglobin and transported inside red blood cells, while carbon dioxide is converted to bicarbonate ions and transported in plasma. Hemoglobin has several properties that make it an effective transport protein, including cooperative binding of oxygen, the Bohr shift response to low pH, and the ability to bind protons generated when carbon dioxide is converted to bicarbonate ions. You should be able to draw an oxygen-hemoglobin saturation curve and explain how it increases the efficiency of oxygen delivery to tissues, relative to a saturation curve in which cooperative binding does not occur.

44.4 The Circulatory System

Recall from Chapter 41 that large animals have a much smaller amount of surface area relative to their volume than do small animals. In most animals that are more than a few millimeters in size, the available surface area is too small for gases, nutrients, and wastes to diffuse efficiently between the animal's surface and its interior. As a result, diffusion across the individual's exterior surface is not rapid enough to support metabolism. In large animals, this problem is solved by a circulatory system.

The function of a circulatory system is to carry a transport fluid, such as blood or the blood-like fluid called *hemolymph*, into close contact with every cell in the body. In this case, "close contact" is a distance of about 0.1 mm between the smallest blood vessels and cells within tissues. Diffusion is efficient at this scale. But an enormous amount of tubing is required to distribute blood this close to cells throughout the body. If all the blood vessels in a human body were laid end to end, they would stretch about 100,000 km (over 60,000 miles).

The most sophisticated circulatory systems consist of one or more pumps called **hearts**; tough, thick-walled **arteries**, which are vessels that take blood away from the heart under high pressure; vessels called **capillaries**, whose walls are just one cell thick, allowing gases and other molecules to exchange with tissues in networks called **capillary beds**; and **veins**, which are vessels that return blood to the heart under low pressure. It's important to recognize that animals may lack some or all of these components or may have no circulatory system at all. In jellyfish and corals, for example, a large, internal gastrovascular cavity offers enough surface area that adequate exchange of molecules with the environment occurs via diffusion. Circulatory systems are also lacking in flatworms and tapeworms, whose bodies are flattened enough that their surface area/volume ratio is high. In these animals, molecules are exchanged with the environment directly across the outer body surface. Roundworms lack hearts but facilitate exchange between their fluid-filled body cavity and the environment through muscular contractions in their body wall. In effect, roundworms circulate fluids by sloshing them back and forth.

To explore how more sophisticated circulatory systems work, let's start by distinguishing the two most general types of systems—open and closed.

Open Circulatory Systems

In an **open circulatory system**, hemolymph is actively pumped throughout the body but is not confined exclusively to blood vessels. **Hemolymph** is a blood-like tissue that transports wastes and nutrients in most invertebrates. Depending on the species, it may also contain oxygen-carrying pigments, some cells, and clotting agents. The key observation about hemolymph is that it comes into direct contact with the tissues—meaning the molecules being exchanged between hemolymph and tissues do not have to diffuse across the wall of a blood vessel.

Figure 44.14 illustrates the general features of open circulatory systems in grasshoppers and clams. Notice that in both cases, one or more hearts pump hemolymph into an artery that empties into an open, fluid-filled space. Hemolymph is returned to the heart when it relaxes and its internal pressure drops below that in the body cavity. General body movements also move hemolymph to and from the heart(s).

One salient characteristic of an open circulatory system is that hemolymph is under relatively low pressure. As a result, hemolymph flow rates are also low. This feature makes open systems most suitable for relatively sedentary organisms, which do not have high oxygen demands. Insects, with their rapid movements and more active lifestyles, are an obvious exception to this rule. Insects overcome the limitations imposed by low hemolymph pressure via their tracheal ventilatory system, which delivers oxygen directly to the tissues.

Another characteristic of open circulatory systems is that without discrete, continuous vessels, the flow of hemolymph cannot be directed toward tissues that have a high oxygen demand and CO_2 buildup. An open circulatory system moves hemolymph throughout an animal's body in much the same way that a ceiling fan moves air throughout a room in a house.

Closed Circulatory Systems

Open circulatory systems are the norm in phyla that appear prior to vertebrates on the evolutionary tree. There are exceptions, however. Earthworms and other annelids have closed circulatory systems. Annelids exchange gases with the environment across their thin, moist skin, which is perfused with capillaries. With this ventilatory surface and closed circulatory system, annelids are able to obtain and circulate enough oxygen to support intense muscular activity. Based on these observations, it is logical to find that most annelids live as active burrowers and hunters. A similar situation occurs in squid, octopuses, and other cephalopods that hunt down prey. The closed circulatory system of these animals generates high hemolymph flow rates, which oxygenates their muscles well enough to support rapid movements and a predatory lifestyle. And even though the circulatory system of crustaceans is classified as open, crustaceans have a fine network of vessels that can preferentially send blood to tissues with the highest oxygen demands.

The defining feature of a **closed circulatory system** is that blood is completely contained within blood vessels. The tissue flows in a continuous circuit through the body under pressure generated by the heart. Because blood is confined to vessels, the system can generate enough pressure to maintain a high flow rate. Blood flow can also be directed in a precise way. Regulatory systems can shunt blood to specific vessels and thus to specific locations, such as muscles during exercise, the intestines after a meal, or regions of the brain engaged in particular mental tasks.

Closed circulatory systems contain an array of blood vessel types, each of which has a distinct structure and function. Let's review some of these major types and then consider how the vessels of a closed circulatory system interact with the lymphatic system.

Types of Blood Vessels In a closed circulatory system, the heart ejects blood into a large artery, usually called the **aorta**. Although all arteries have both muscle fibers and elastic fibers in their walls, the walls of the aorta are dominated by elastic fibers.

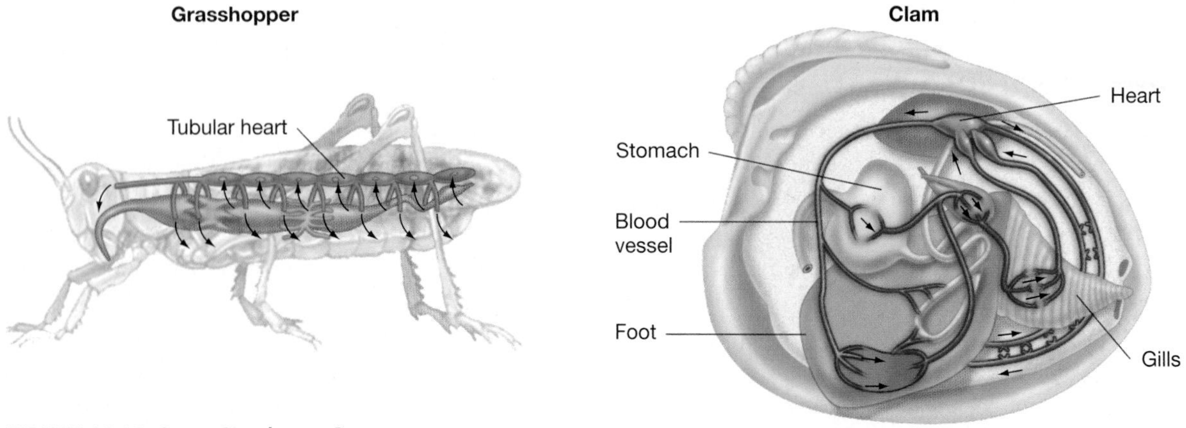
Grasshopper
Tubular heart
Clam
Heart
Stomach
Blood vessel
Foot
Gills

FIGURE 44.14 Open Circulatory Systems
The arrows in these diagrams indicate the direction of blood flow. QUESTION What is "open" about an open circulatory system?

As a result, the aorta can expand when blood enters it under high pressure from the heart. When contraction of the heart ends, the diameter of the aorta returns to its resting state. This elastic response propels blood away from the heart and augments the force generated by heart contraction. Similar types of secondary pumping action occur to some extent in other large and medium-sized arteries as well and help maintain forward blood flow in the period between cardiac contractions.

In the walls of the smallest arteries, called **arterioles**, muscle fibers wrap around the circumference of the vessel. These muscle fibers are called **sphincters** and allow the diameter of the vessel to be carefully regulated in response to signals from the nervous system. When the muscle fibers are relaxed, the arteriole diameter increases, resistance to flow in the arteriole is reduced, and blood flow increases in the tissues served by the vessel. But when these arteriolar sphincters constrict, they increase resistance to flow, slow blood passage, and thus divert blood flow for use by other tissues.

Capillaries are the smallest blood vessels. They have walls that are only one cell layer thick, and they are just wide enough to let red blood cells through one at a time (**Figure 44.15a**). Because capillaries are extremely thin and because they form an extremely dense network throughout the body, they are the site where gases, nutrients, and wastes are exchanged between the blood and other tissues. In some organs, such as the liver, the walls of capillaries contain numerous small openings, which further diminish the barrier to diffusion between blood and the tissues. Despite their thinness, it is rare for capillaries to rupture, because blood pressure drops dramatically as blood passes through arterioles on its way to capillary beds.

Veins carry blood back to the heart after the blood passes through the capillaries. Because blood is under relatively low pressure by the time it exits the tissues, veins have much thinner walls and much larger interior diameters than arteries do (**Figure 44.15b**). Large veins are compressed by muscle activity in the extremities (say, the hands and feet), which helps speed the return of blood to the heart. Larger veins also contain **valves**, which are thin flaps of tissue that prevent any backflow of blood, and all veins contain some muscle fibers. These fibers contract in response to signals from the nervous system, decreasing the diameter and overall volume of the vessels. Blood pressure in a closed circulatory system is regulated, in part, by active adjustment of the volume of blood contained within the veins.

Blood, Interstitial Fluid, and the Lymphatic System The relatively high operating pressure of closed circulatory systems, combined with the thinness of capillaries, produces a small but steady leakage of fluid from the blood vessels into the surrounding space. The area between cells is called **interstitial space**, and the fluid that fills it is called **interstitial fluid.** Because blood cells are retained within capillaries, the interstitial fluid resembles plasma in its electrolyte composition.

(a) Capillaries are small and extremely thin walled.

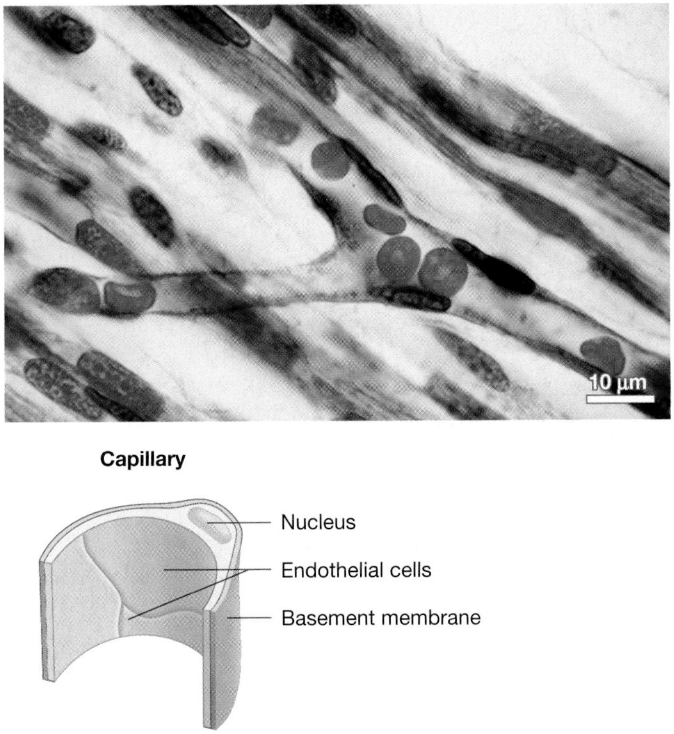

(b) Veins and arteries differ in structure.

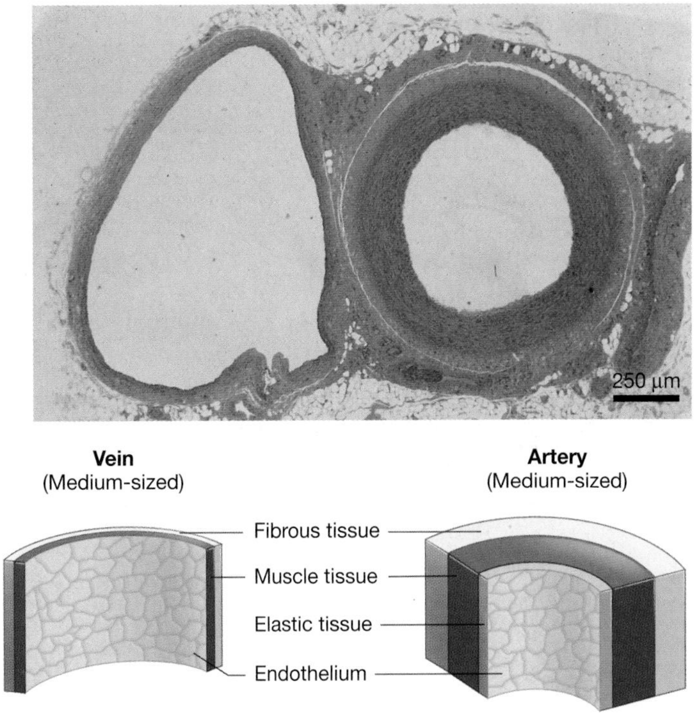

FIGURE 44.15 Capillaries, Veins, and Arteries
Cross sections through **(a)** a capillary and **(b)** a vein and an artery. Notice the difference in relative wall thickness and overall size of each type of vessel.

Ernest Starling proposed that two forces affect fluid flow across capillaries (**Figure 44.16**). The first is the outward-directed hydrostatic force created by the pressure on blood generated by the heart. This force is analogous to the pressure that drives water through the wall of a leaky garden hose. The second is an inward-directed osmotic force, created by the higher concentration of solutes in the blood plasma than in the interstitial space.

Starling reasoned that at the end of the capillary nearest to an arteriole, the hydrostatic force would exceed the osmotic force and fluid would move out of the capillary into the interstitial space. But because hydrostatic pressure drops as fluid passes through a long, thin tube, he proposed that the osmotic force would exceed the hydrostatic force at the venous end of the capillary. Thus, the fluid lost at the arteriolar end of the capillary would be largely reclaimed at the venous end.

Starling also proposed that any interstitial fluid that was not reclaimed in the capillary would be collected in the system of thin-walled, branching tubules called **lymphatic ducts.** It had been known for many years that these tiny ducts permeate all tissues and that they eventually join with one another like the tributaries of a river to form larger vessels. The largest lymphatic vessels return excess interstitial fluid, or **lymph**, to the major veins entering the heart. Starling's model was a comprehensive proposal for how blood, lymph, and the lymphatic ducts interact.

This model rested on a key assumption, however: For Starling's hypothesis to work, the total solute concentrations in plasma and interstitial fluid must be different enough to bring fluid into capillaries via osmosis. For this to be true, capillaries would have to act as filters. Specifically, large protein molecules or other solutes would have to be retained inside capillaries as fluid leaked out. To test this assumption, a group of biologists placed small tubes in the lymphatic ducts of animals so that they could collect lymph. Then the team introduced radioactively labeled proteins of various sizes into the blood inside veins. They found that even though small proteins crossed readily from the blood to the lymph, larger proteins did not. Enough large proteins were retained in the capillaries to set up the osmotic gradient required by Starling's model.

Follow-up work has shown that the most important of these large proteins is **albumin.** Albumin's large size and net negative charge effectively excludes it from interstitial fluid. The presence of albumin in blood keeps solute concentrations high and restricts the total daily formation of lymph to just 2–5 percent of plasma volume. This amount is easily taken up by the **lymphatic system** and returned to the blood.

Early work on the structure and function of the lymphatic system also highlighted a fundamental distinction between open and closed circulatory systems: In an open system, there is no distinction between blood and the fluid surrounding tissues. There is only one type of circulating fluid—hemolymph. But in a closed system, blood and tissue fluid are distinct, even though they interact via the formation of lymph.

The Heart

In animals with closed circulatory systems, the heart contains at least two chambers. There is at least one thin-walled **atrium** (plural: *atria*), which is a chamber that receives blood, and at least one thick-walled chamber called a **ventricle**, which generates the force required to propel blood out of the heart and through the system. Atria are separated from ventricles by atrioventricular valves.

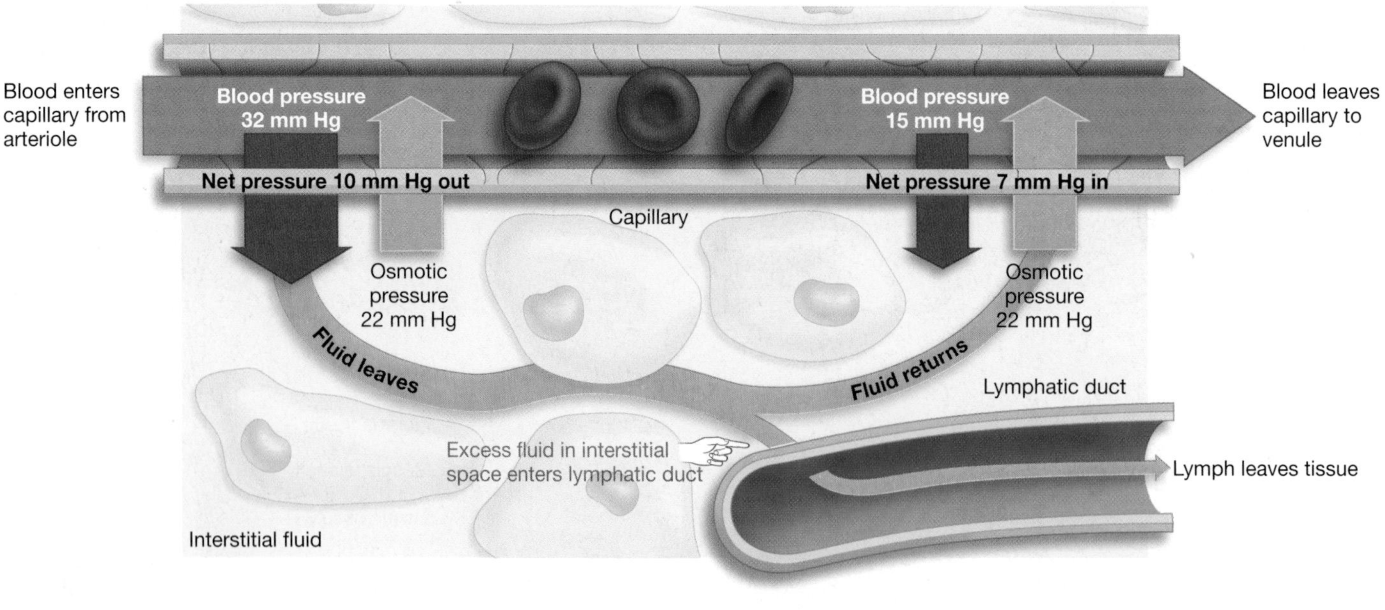

FIGURE 44.16 Pressure Differences in Capillaries Create Interstitial Fluid and Lymph
The balance of blood pressure and osmotic forces favors fluid loss from the beginning (inflow end) of capillaries and fluid recovery at the other (outflow) end. Fluid that is not recovered by the capillaries is transported out of the tissue as lymph.

When studies of vertebrate hearts and circulatory systems had advanced enough to make comparisons among groups possible, biologists noticed several important patterns. To understand these, study **Figure 44.17**. This phylogenetic tree shows the evolutionary relationships among some major vertebrate groups. The sketches for each group offer a simplified version of the heart and circulatory system found in each lineage. Two points are particularly important to note:

1. The number of distinct chambers in the heart increased as vertebrates diversified. Fish hearts have two chambers; amphibians have three; turtles and lizards have five (if the partially divided ventricle is counted as three chambers); crocodiles, birds, and mammals have four.
2. In fish, the circulatory system forms a single circuit—one loop. In other groups, there are separate circuits to the lungs and to the body.

To understand these evolutionary patterns, let's first consider the relatively simple heart and circulatory system found in fish. A two-chambered heart and single circuit are adequate in fish, because they live in the neutrally buoyant environment of water. High pressures are not required to circulate blood throughout the body. The drop in blood pressure that occurs as blood passes through the gills is not enough to prevent blood from being circulated to the rest of the body. In contrast, the blood pressure drop that occurs as blood moves through the lungs of land-dwelling vertebrates could prevent efficient circulation through the rest of the body—particularly the elevated portions. This problem cannot be solved by increasing the force generated at the heart, because the capillaries and alveoli of the lungs are too thin to withstand such high pressures.

The successful solution used by land-dwelling vertebrates was the evolution of two separate pumping circuits. The **pulmonary circulation** is a lower-pressure circuit to the lung, while the **systemic circulation** is a higher-pressure circuit to the rest of the body. The pulmonary and systemic circulations are completely separated in the four-chambered hearts of birds and mammals, which are essentially two fish hearts side by side. The two circulations are only partially separated

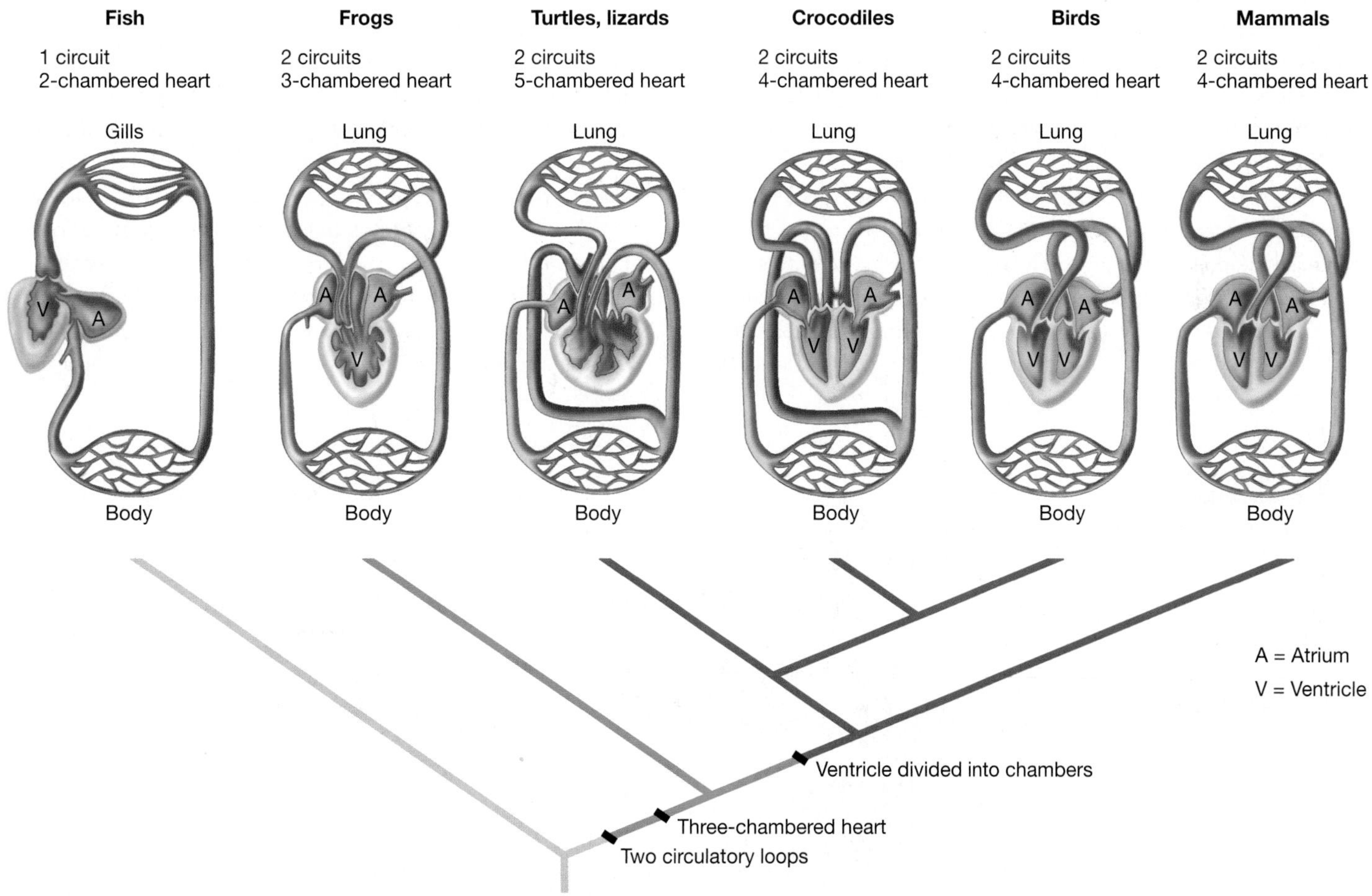

FIGURE 44.17 The Evolution of the Vertebrate Circulatory System
"A" denotes the atria—chambers that receive blood coming into the heart from the body and the lungs; "V" denotes the ventricles—chambers that pump blood out to the lungs and the body. EXERCISE Turtles (and relatives) and crocodiles have blood vessels that shunt blood between their two circuits, as needed. Label these shunts on the diagram.

in the three-chambered hearts of amphibians and the "five-chambered" hearts of turtles and lizards, however. Some of these animals also have a bypass vessel running from the right side of the ventricle directly into the systemic circulation. This trait is also observed in the unusual four-chambered hearts of crocodiles. The bypass vessel allows blood to be shunted from the pulmonary to the systemic circulation when the animal is under water.

The Human Heart Your heart is located in the chest cavity between your lungs and is roughly the size of your fist. As **Figure 44.18** shows, the circulatory system returns blood from the body to the right atrium of the human heart through two large veins called the inferior (lower) and superior (upper) **venae cavae** (singular: **vena cava**). This blood is low in oxygen. When the muscles that line the right atrium contract, this deoxygenated blood is sent to the right ventricle. The right ventricle, in turn, contracts and sends blood out to the lungs, via the **pulmonary artery**. In this way, the right ventricle powers the movement of blood through the pulmonary circulation.

Blood flows from atrium to ventricle to artery only one way, because valves separate the heart's chambers from each other and from the adjacent blood vessels. As Figure 44.18 indicates, the valves are flaps, oriented in a way that ensures a one-way flow of blood with little or no backflow. If heart valves are damaged or defective, the resulting backflow can be heard through a stethoscope. The result is called a **heart murmur**.

After blood has circulated through the capillary beds in the lung's alveoli and becomes oxygenated, it returns to the heart through the **pulmonary veins**. The oxygenated blood enters the left atrium. When this chamber contracts, blood is pushed into the left ventricle. The walls of the left ventricle are so thick with muscle cells that their contraction sends oxygenated blood at high pressure out the aorta and into the arteries and capillaries that form the systemic circulation.

Figure 44.19 summarizes the flow pattern through the human circulatory system and the blood gas concentrations at various points in the pulmonary and systemic circulations. Notice that blood vessels are called arteries or veins according to the direction of blood flow relative to the heart, not because of the oxygen content of the blood in them. Thus, the pulmonary artery is called an artery because it takes blood away from the heart, even though this blood is low in oxygen content. The data on gas partial pressures in the figure make another important point: In healthy individuals, gas exchange in the lungs and tissues is rapid relative to the rate at which blood flows through capillaries. This situation ensures that the partial pressures of oxygen and carbon dioxide in the alveoli and the pulmonary veins are equal. Similarly, the partial pressures of oxygen and carbon dioxide in the tissues and the systemic veins are equal.

The Cardiac Cycle Although the preceding discussion suggests that each chamber in the heart acts independently, this is not the case. The contraction phase of the atria and the ventricles, called **systole**, is closely coordinated with their relaxation phases, or **diastole**. For example, both atria contract simultaneously about 0.1 second ahead of ventricular contraction. Because both ventricles are still in diastole and relaxed when the atria contract, the ventricles fill with blood from the atria just before they enter systole and contract. This sequence of contraction and relaxation is called the **cardiac cycle**. It consists of one diastole and one systole from both atria and ventricles.

Ventricular contraction (ventricular systole) leads to a rapid increase in pressure within both ventricles (**Figure 44.20**). Blood is ejected into the pulmonary artery and the aorta when

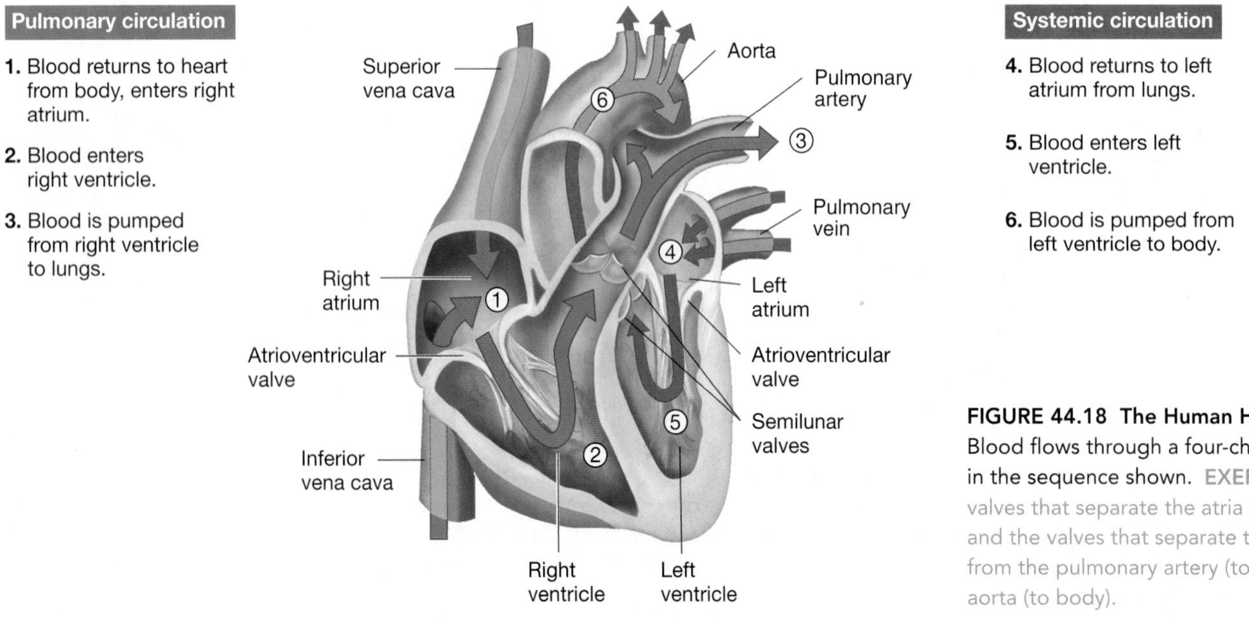
Pulmonary circulation
1. Blood returns to heart from body, enters right atrium.
2. Blood enters right ventricle.
3. Blood is pumped from right ventricle to lungs.
Superior vena cava
Aorta
Pulmonary artery
Pulmonary vein
Right atrium
Left atrium
Atrioventricular valve
Atrioventricular valve
Semilunar valves
Inferior vena cava
Right ventricle
Left ventricle
Systemic circulation
4. Blood returns to left atrium from lungs.
5. Blood enters left ventricle.
6. Blood is pumped from left ventricle to body.

FIGURE 44.18 The Human Heart
Blood flows through a four-chambered heart in the sequence shown. EXERCISE Circle the valves that separate the atria and ventricles, and the valves that separate the ventricles from the pulmonary artery (to lungs) and aorta (to body).

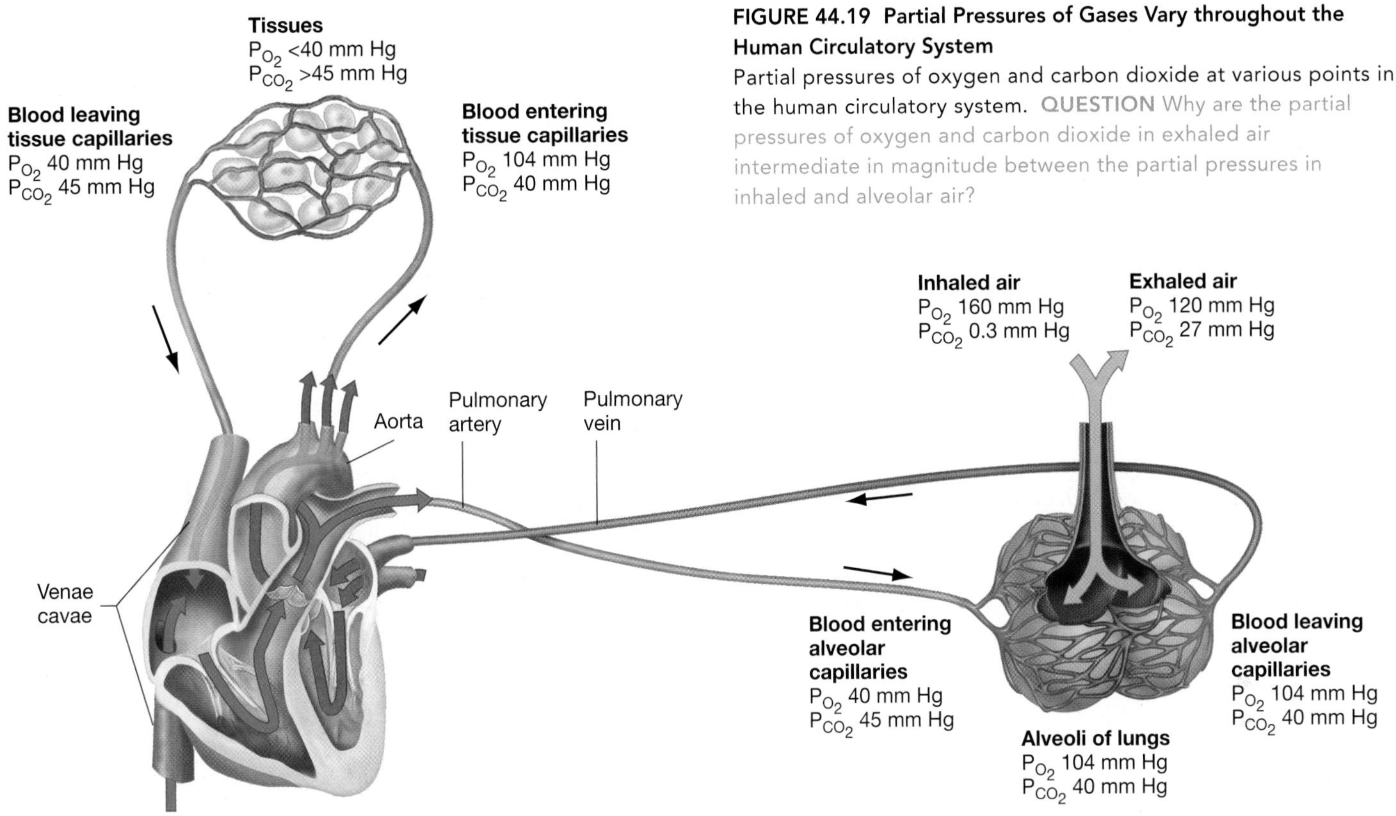

FIGURE 44.19 Partial Pressures of Gases Vary throughout the Human Circulatory System

Partial pressures of oxygen and carbon dioxide at various points in the human circulatory system. QUESTION Why are the partial pressures of oxygen and carbon dioxide in exhaled air intermediate in magnitude between the partial pressures in inhaled and alveolar air?

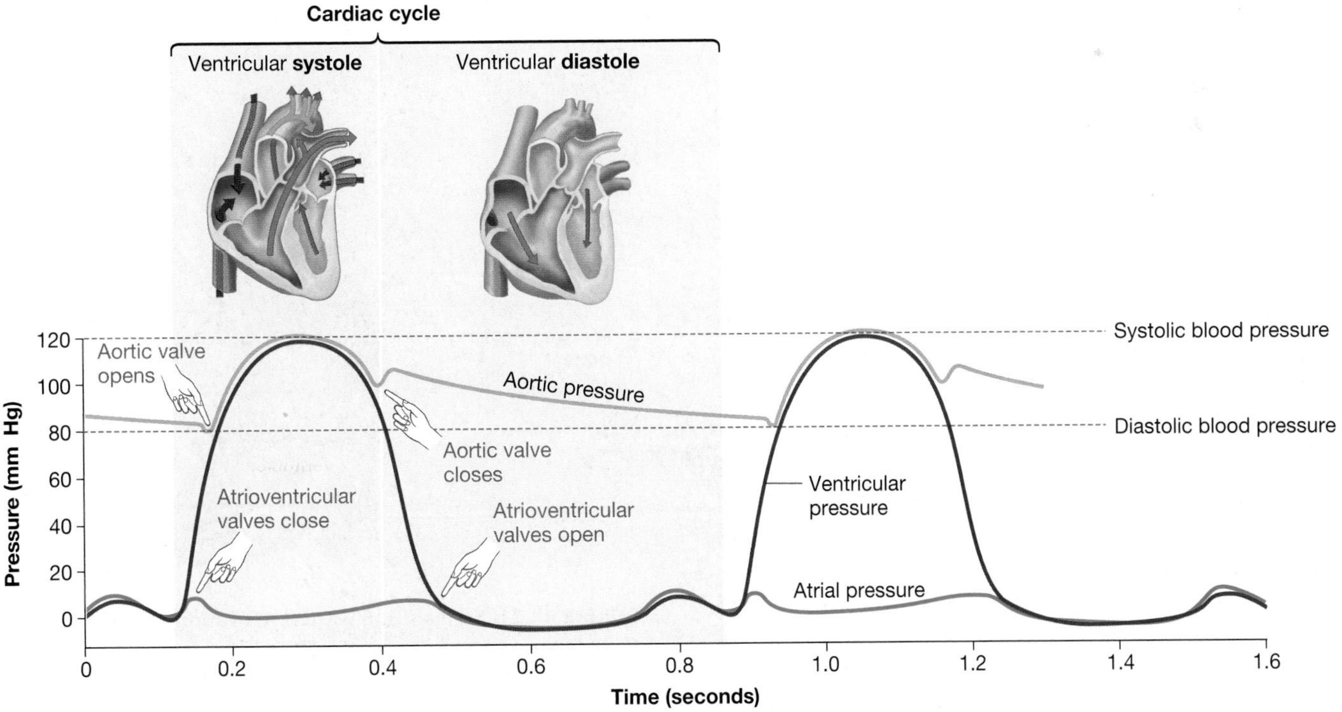

FIGURE 44.20 Blood Pressure Changes during the Cardiac Cycle

Pressures created by the left ventricle and aorta in the cardiac cycle. In this case, the blood pressure measured in the upper arm would be 120/80 mm Hg. The right ventricle and pulmonary artery pressures follow a similar pattern, but the blood pressure measured in the pulmonary artery would be closer to 25/8 mm Hg. QUESTION Why does the left aortic valve open shortly after the beginning of ventricular systole and close slightly before the end of systole?

BOX 44.3 Hypertension and the Measurement of Blood Pressure

To measure a person's blood pressure, a clinician wraps an inflatable cuff attached to a pressure gauge around the upper arm. When the cuff is fully inflated, the pressure it exerts is transmitted evenly throughout the arm and closes down the main artery in this part of the body. The clinician then uses a stethoscope to listen for the sounds of blood flow downstream from the cuff.

At first, the clinician hears no pulse sounds through the stethoscope, because no blood is flowing downstream of the cuff. As the clinician slowly releases air from the cuff, however, the pressure in the cuff begins to fall until it is just below the systolic pressure in the artery. At this point, blood begins to spurt through the artery with each systolic peak. The flow is turbulent, so the clinician can detect a thumping pulse sound in the stethoscope. The cuff pressure when sound is first detected is equal to the systolic pressure. As more air is released from the cuff, its pressure falls below the diastolic pressure in the artery. At this point, blood flow in the artery is smooth and the sound of turbulent flow fades away. The cuff pressure when the sound becomes muffled or disappears represents the diastolic pressure.

A blood pressure reading—especially when combined with a measurement of resting pulse rate and the cholesterol counts introduced in Chapter 43—gives clinicians relatively quick, accurate, and inexpensive data on the health risk to a patient's heart and circulatory system. Blood pressure measurements are particularly helpful in diagnosing hypertension, which is a common condition. Hypertension is a serious concern to physicians, because it can lead to a variety of defects in the heart and circulatory system. Abnormally high blood pressure puts mechanical stress on arteries and thus increases the risk of heart attack, stroke, kidney failure, and burst or dilated blood vessels.

ventricular pressure equals the pressure within each respective artery. Blood pressure measured in the systemic arterial circulation at the peak of ventricular ejection into the aorta is called the **systolic blood pressure.** Blood pressure measured just prior to ventricular ejection is called the **diastolic blood pressure.** Clinicians report blood pressure measurements in fractional notation, with systolic pressure as the numerator and diastolic pressure as the denominator. For the example shown in Figure 44.20, blood pressure is 120/80 mm Hg. People with blood pressures in excess of 150/90 mm Hg are considered to have high blood pressure, or **hypertension.** The measurement of blood pressure and some of the health consequences of hypertension are discussed further in Box 44.3.

Electrical Activation of the Heart How is the activity of each chamber of the heart coordinated? Cardiac muscle cells, like other muscle cells, contract in response to electrical signals. In invertebrates, the electrical impulses that trigger contraction come directly from the nervous system. In vertebrates, a group of cells in the heart itself is responsible for generating the initial signal. The cells that initiate contraction, known as **pacemaker cells,** are located in a region of the right atrium called the **SA (sinoatrial) node** (Figure 44.21). The SA node and the muscle cells of the heart receive input from the nervous system and from chemical messengers carried in the blood. These inputs are important for regulating both the heart rate and the strength of ventricular contraction. However, a vertebrate heart will beat even if all nerves supplying it are severed.

The electrical impulse generated by the SA node is rapidly conducted throughout the right and left atria. The signal spreads quickly from cell to cell because of a striking property of cardiac muscle cells: They branch to contact several other cells, are joined end to end with these neighboring cells, and are connected by specialized structures called **intercalated discs.** Because these discs contain numerous gap junctions—cell-to-cell connections described in Chapter 8—electrical signals pass directly from one

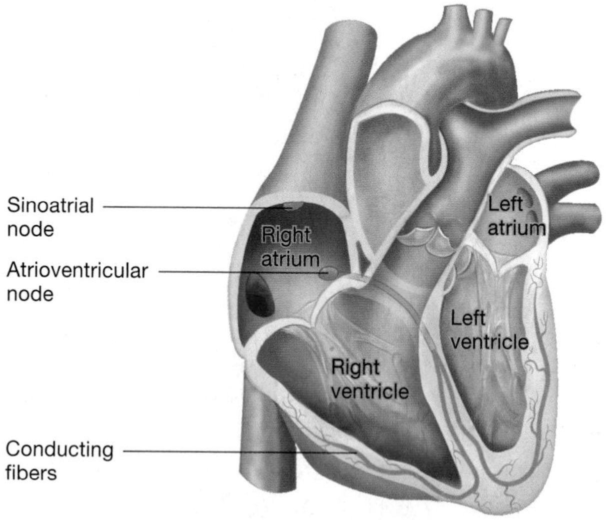

FIGURE 44.21 Sequential Electrical Activation Leads to Coordinated Contraction of the Heart
Electrical activation of the heart begins with the pacemaker cells in the sinoatrial node, spreads through the electrically connected muscle cells of the atria, and converges on the atrioventricular node, where it is delayed long enough to allow the atria to eject blood into the ventricles. The ventricles are then activated rapidly by the spread of electricity through specialized conducting fibers in the ventricular walls. EXERCISE In this diagram, the heart is sectioned through the pulmonary artery. Circle the conducting fibers just dorsal to this artery.

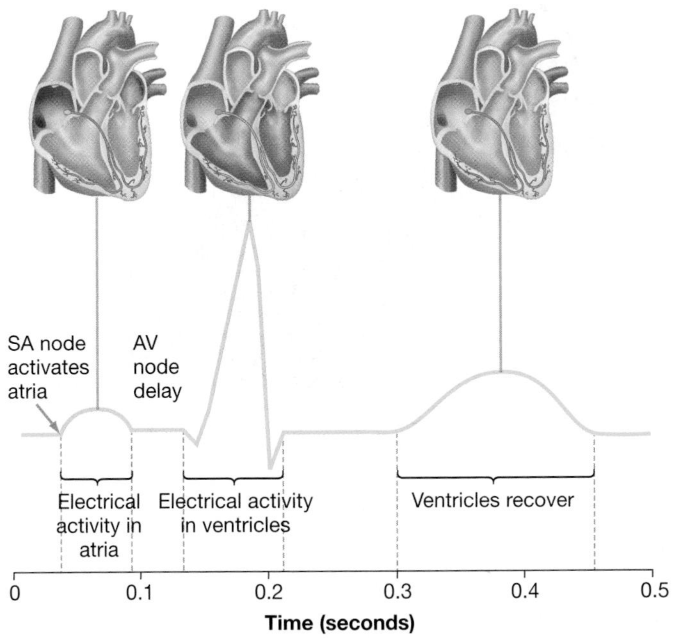

FIGURE 44.22 EKGs Record Electrical Events Associated with Cardiac Muscle Contraction

QUESTION The electrical activity generated by ventricular contraction is much larger and more compressed than the electrical activity generated by atrial contraction. Why?

cell to the next. (In Chapter 45, we'll explore how electrical signals in animals are initiated and propagated.) As a result, the atria contract simultaneously and fill the ventricles.

Once the electrical signal has swept over the atria, however, it is conducted to an area of the heart called the **atrioventricular (AV) node**, which delays the signal slightly before passing it to the ventricles. This delay allows the ventricles to fill completely before they contract. After the delay, the electrical impulse is rapidly transmitted through specialized conducting fibers that lie within the muscular wall that separates the ventricles. The impulse then spreads evenly through both ventricles, causing them to contract as the atria relax.

The electrical events responsible for the synchronized contraction of the atria and ventricles are summarized in **Figure 44.22**. This figure also shows the recording known as an **electrocardiogram**, or **EKG**, which corresponds to the electrical activity associated with cardiac muscle contraction. An EKG trace is generated by amplifying the overall electrical signal conducted from the heart to the chest wall through the tissues of the body. The EKG allows physicians to diagnose disturbances of heart rhythm and detect damage to the heart muscle.

Regulation of Blood Pressure and Blood Flow

As blood moves through capillaries, blood pressure drops dramatically. **Figure 44.23** shows why. The key point is that as arteries branch, rebranch, and eventually form networks of capillaries, the total cross-sectional area of blood vessels in the circulatory system increases—even though the size of individual blood vessels decreases as blood moves away from the heart. Because the same amount of fluid passes through a larger and larger area, the pressure exerted by the fluid drops. In addition, the pulsing nature of the pressure noted in the larger arteries diminishes. Finally, the velocity of blood flow decreases significantly as the circulatory cross-sectional area increases, allowing sufficient time for gases, nutrients, and wastes to diffuse between tissues and blood in the capillaries.

Another factor contributing to the drop in magnitude and the pulsing nature of blood pressure is the resistance to blood flow imposed by the progressively narrowing arteries. The greatest resistance to flow, and consequent greatest drop in blood pressure, occurs in the arterioles that deliver blood directly into the capillary beds of tissues. As noted earlier, the resistance imposed by the arterioles is variable and depends on the activity of the muscular sphincters in their walls. This means that the nervous system, along with certain chemical messengers in the circulation, can accurately control blood flow to various tissues by contracting or relaxing the arteriolar sphincters. For example,

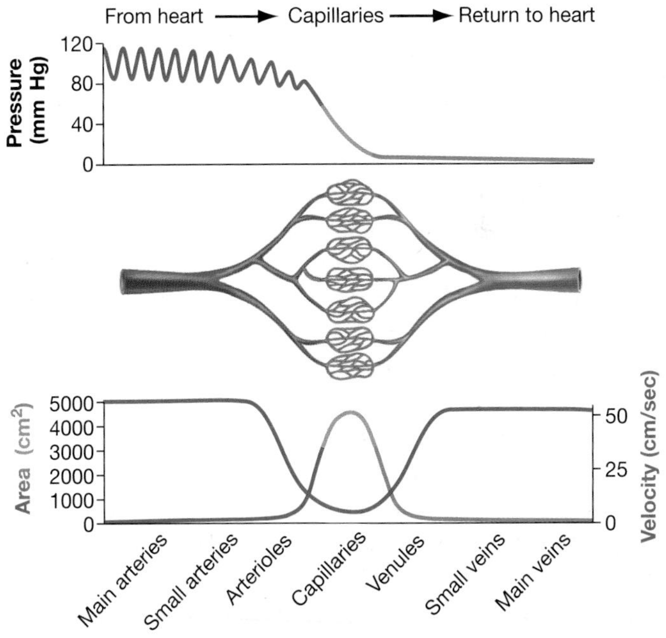

FIGURE 44.23 Blood Pressure Drops Dramatically in the Circulatory System

The top graph shows how blood pressure changes as blood leaves the heart, travels through a succession of branching arteries to capillaries, and moves from capillaries into a succession of veins that eventually return blood to the heart, as in the branching pattern of the middle diagram. In arteries near the heart, each heartbeat causes fluctuations in blood pressure. These pressure pulses disappear in the capillaries, so blood flows there at a steady speed. The bottom graph plots the total area of blood vessels shown in the diagram, as well as the velocity of blood flow through the vessels. EXERCISE Label the systolic and diastolic blood pressures in the top graph.

arterioles in the skin may dilate during exercise. When they do, they to divert blood flow to the skin for the purpose of eliminating excess heat. This accounts for the flushed facial appearance often induced by vigorous exercise.

A particularly striking example of blood flow regulation is provided by the Baikal seal. Lake Baikal, in Siberia, is the world's deepest lake, and the resident seals are capable of extended dives in search of food. Seals are mammals; in some species of seals, dives can last 20–60 minutes and take individuals to depths of 500 m. How do seals deliver enough oxygen to the brain to maintain consciousness during extremely long dives?

An answer to this question began to emerge when biologists found that the Baikal seals are able to store huge amounts of oxygen bound to myoglobin in the muscles, as well as in hemoglobin contained within their unusually large blood volume. This observation led to the hypothesis that these stores are carefully rationed during long dives, with priority for oxygen delivery given to the brain and the heart. (It was also known that the heart beats much more slowly during a dive, diminishing its oxygen requirement significantly.) The logic here is that if oxygen levels in the blood become low during particularly long dives, only the most vital organs should receive it. Restricting blood flow to the digestive system might lead to indigestion, but restricting blood flow to the brain or heart would result in unconsciousness and death.

To test this hypothesis in Baikal seals, researchers injected a radioactive atom into several individuals. Because the atom chosen is physiologically inactive, it was simply carried around in the blood as a marker of flow and did not harm the animals. The researchers then extracted blood samples from different tissues at intervals and measured the amount of radioactivity present.

As **Figure 44.24** shows, the researchers' data showed a strong contrast in circulation during resting and diving. When the seals were inactive, the amount of radioactive marker did not vary among tissue groups. This observation indicated that blood was flowing evenly throughout the body. When the biologists sampled the same tissues during dives, however, they found that blood flow to the head was normal but declined dramatically—in several locations to zero—everywhere else they sampled. This result suggests that virtually all of the blood vessels in the animal's trunk and limbs constricted during the dive and that blood flow was highest in the circuit between the heart and brain.

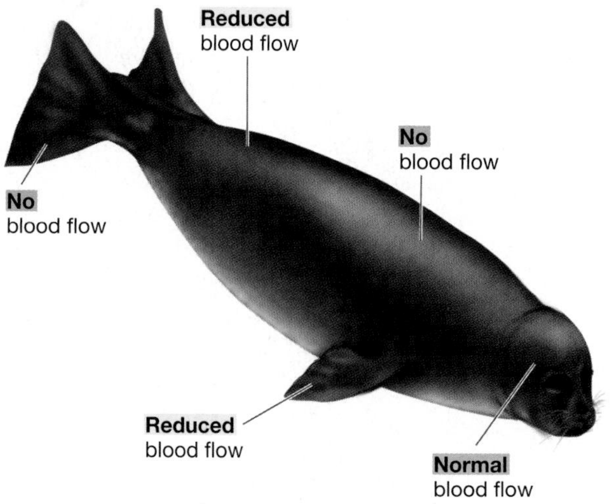

FIGURE 44.24 In a Closed Circulatory System, Blood Can Be Shunted to Certain Locations
The labels indicate how blood flows during a dive compared with blood flow in the same locations at rest. QUESTION In humans, blood is shunted between three major compartments: the brain, digestive system, and muscles. How would you predict that blood flow changes after a large meal?

Circulatory Homeostasis under Stressful Circumstances

Other situations besides extended dives challenge the circulatory system's ability to maintain blood pressure and the flow of blood to vital organs. Suppose you sit long enough for blood to pool in your lower extremities, under the influence of gravity. If you stand up rapidly, your blood pressure can drop enough to limit blood flow to the brain and cause dizziness, lightheadedness, or even a blackout. More serious drops in blood pressure, due to severe dehydration or blood loss following an injury, can lead to death. Fortunately, decreases in blood pressure elicit a powerful homeostatic response.

Recall from Chapter 41 that all homeostatic responses involve (1) sensors that detect the change in condition, (2) an integrator that processes information about the change, and (3) effectors that diminish the impact of the change. In the case of the circulatory system, falling blood pressure is detected by specialized pressure-sensing nerve cells called **baroreceptors**. Baroreceptors are found in the walls of the heart and the major arteries, both as they leave the heart and as they enter the neck. This distribution is logical, because the head tends to be the most elevated part of land-dwelling animals and because the preservation of blood flow to the brain is the highest priority.

When baroreceptors transmit signals to the brain indicating a serious fall in blood pressure, a rapid, three-component effector response ensues:

1. An increase in cardiac output, meaning the volume of blood leaving the left ventricle, is brought about by both an increase in heart rate and an increase in stroke volume, or the amount of blood ejected from the ventricles during each cardiac cycle. (Cardiac output = heart rate × stroke volume.)

2. The constriction of arterioles serving the capillaries of tissues such as the skin and intestines, which are not essential for short-term survival. As in the diving Baikal seal, this arteriolar constriction diverts blood to more-critical organs.

3. A generalized constriction of the veins, resulting in a decrease in their overall volume. Since more than half of the blood in the circulatory system is contained within the veins, constriction of these vessels shifts blood volume toward the heart and arteries to maintain blood pressure and flow to vital organs.

This coordinated response is mediated both by a portion of the nervous system called the *sympathetic nervous system* (Chapter 45) and by *hormones*, or chemical messengers (Chapter 47), produced by the adrenal glands. Sympathetic nerves and the hormones involved in blood pressure regulation deliver their messages directly to the SA node to increase heart rate, the ventricles of the heart to increase stroke volume, the sphincters of the arterioles to modify their resistance, and the muscular walls of the veins to modify their total volume.

The homeostatic response to falling blood pressure resembles the *fight-or-flight response*, which is triggered by a threatening stimulus (see Chapter 47). During the fight-or-flight response, an organism prepares to defend itself or make a rapid escape from danger. Part of this response involves directing blood to the brain and muscles in preparation for quick action. The side effects include the "cold sweat" and nausea often induced by a fearful situation. These feelings occur because blood flow is shunted away from the skin and intestines.

✓CHECK YOUR UNDERSTANDING

Animal circulatory systems may be open or closed, but both types of systems circulate blood or hemolymph via pressure generated by one or more hearts. In closed systems, regulated changes in the diameter of blood vessels can direct blood to specific regions, and overall blood pressure is carefully regulated through changes in cardiac output. You should be able to (1) explain how the mammalian lymphatic system and circulatory system interact, (2) make a labeled diagram showing how blood circulates through the mammalian heart, and (3) explain why blood pressure drops dramatically as blood moves from arterioles through capillary beds.

ESSAY Smoking and Lung Function

People use tobacco primarily because they are addicted to a molecule called *nicotine*. **Addiction** is characterized by compulsive drug seeking, even when the user recognizes adverse health consequences. Recent research has shown that, within 10 seconds of inhaling cigarette smoke, blood transports nicotine from the lungs to the brain. Once in the brain, nicotine activates cells associated with feelings of reward or pleasure. Because these effects are short lived, cigarette smokers dose themselves repeatedly throughout the day. A typical addict takes 10 puffs from 30 cigarettes a day, meaning he or she acquires an average of 300 "hits" of nicotine.

How do the other 4000 ingredients in cigarette smoke affect the body? Carbon monoxide, ammonia, nitrogen oxides, and formaldehyde are among the many toxic ingredients of tobacco smoke. Using test systems that quantify the tendency of particular compounds to cause mutations, researchers have determined that at least 43 ingredients in cigarette smoke are carcinogenic.

Tobacco smoke also has a dramatic effect on the ventilatory surfaces. Soon after a person begins to smoke regularly, the lung tissue starts to swell. As smoking continues and the swelling worsens, airflow into the alveoli is reduced and gas-exchange rates decrease. Mucus-producing cells that line the airways become hyperactive, producing excessive amounts of mucus. Toxins in smoke also kill or paralyze cilia that normally sweep debris and mucus out of the respiratory passages. As a result, mucus and debris build up in the lungs, leading to smoker's cough—a chronic and futile effort to clear the airways. Because accumulations of mucus and debris promote the growth of bacteria, smokers suffer from an increased incidence of respiratory infections.

... at least 43 ingredients in cigarette smoke are carcinogenic.

Lifelong smokers are at high risk for developing **emphysema**, a disease that occurs when alveoli break down and lungs lose their elasticity, making exhalation increasingly inefficient. As emphysema advances, breathing is so impaired that physical activity becomes difficult. Eventually, the patient slowly suffocates to death. Male smokers are also 22 times more likely to develop lung cancer than nonsmokers are, while females are 12 times more likely. In the United States, lung cancer is the most deadly of all cancers.

When all the effects of nicotine addiction are taken into account, public health officials in the United States attribute over 430,000 deaths each year to tobacco use. This figure is larger than deaths attributed to alcohol, cocaine, heroin, homicide, suicide, car accidents, fire, and AIDS combined. Nicotine addiction also continues to grow as a public health problem. Even though it is not legal for minors in the United States to obtain tobacco products, surveys indicate that 28 percent of U.S. high school students smoke. Currently, it is estimated that nearly 3000 minors start smoking each day in the United States.

CHAPTER REVIEW

Summary of Key Concepts

Animals have to take in oxygen and expel carbon dioxide to sustain cellular respiration and stay alive. Terrestrial animals and aquatic animals face different challenges in performing gas exchange.

To support the production of ATP in mitochondria, animals have to obtain oxygen and rid themselves of carbon dioxide. As media for exchanging these gases, air and water are dramatically different. Compared with water, air contains much more oxygen and is much less dense and viscous. As a result, terrestrial animals have to process a much smaller volume to extract the same amount of O_2 and don't have to work as hard to do so.

Both aquatic and terrestrial animals pay a price for exchanging gases. Land-dwellers lose water to evaporation during ventilation. Freshwater animals lose ions through diffusion and gain excess water by osmosis. Marine animals gain sodium and chloride through diffusion and lose water by osmosis.

The organs responsible for gas exchange maximize the rate of oxygen and carbon dioxide diffusion by presenting a large, thin surface area to the environment. These organs also function in a way that maintains a steep partial-pressure gradient that favors entry of oxygen and elimination of carbon dioxide.

Natural selection has resulted in the evolution of gills, tracheae, lungs, and other gas-exchange organs that minimize the cost of ventilation while maximizing the rate at which O_2 and CO_2 diffuse. Consistent with predictions made by Fick's law of diffusion, respiratory epithelia tend to be extremely thin and are highly folded to increase surface area. In fish gills, countercurrent exchange ensures that the partial-pressure differences between O_2 and CO_2 in water and blood are high over the entire length of the ventilatory surface. In bird lungs, structural adaptations lead to a much greater ratio of useful ventilatory space to dead space than is observed in the lungs of mammals. The rate of breathing is regulated by an area in the brain to keep the carbon dioxide content of the blood stable during both rest and exercise.

Blood is a specialized tissue that transports gases, along with nutrients and wastes, in some animals.

About half the blood volume in a typical human consists of red blood cells that are specialized for gas exchange. Red blood cells lack most organelles but contain hundreds of millions of hemoglobin molecules. Each hemoglobin molecule can carry up to four oxygen molecules. The tendency of hemoglobin to give up these oxygen molecules varies as a function of the P_{O_2} in surrounding tissue in a sigmoidal fashion. As a result, a relatively small change in tissue P_{O_2} causes a large change in the amount of oxygen released from hemoglobin. In addition, oxygen binds less tightly to hemoglobin when pH is low. Because CO_2 tends to react with water to form carbonic acid, the existence of high CO_2 partial pressures in exercising muscle tissues lowers their pH and makes oxygen less likely to stay bound to hemoglobin.

The CO_2 that diffuses into red blood cells from tissues is rapidly converted to carbonic acid by the enzyme carbonic anhydrase. The protons that are released as carbonic acid dissociates bind to deoxygenated hemoglobin. In this way, hemoglobin acts as a buffer that takes protons out of solution and prevents large changes in blood pH.

Web Tutorial 44.1 Gas Exchange in the Lungs and Tissues

Circulatory systems use the pressure generated by one or more hearts to transport blood and other substances throughout the body.

In many animals, blood or hemolymph moves through the body via a circulatory system consisting of a pump (heart) and vessels. In open circulatory systems, overall pressure is low and tissues are bathed directly in hemolymph. In closed circulatory systems, blood is contained in vessels that form a continuous circuit. Containment of blood allows higher pressures and flow rates, as well as the ability to direct blood flow accurately to tissues that need it the most. In organisms with a closed circulatory system, a lymphatic system returns excess fluid that leaks from the capillaries back to the circulation.

In mammals and birds, a four-chambered heart pumps blood into two circuits, which separately serve the lungs and the rest of the body. Although the heart rate is controlled by electrical signals that originate in the heart itself, heart rate, cardiac output, and contraction of both arterioles and veins are constantly modified by signals from the sympathetic nervous system and the adrenal glands. These signals act to preserve blood flow to vital organs in response to falling blood pressure or situations that are perceived as threatening.

Web Tutorial 44.2 The Human Heart

Questions

Content Review

1. O_2 will diffuse from blood to tissue faster in response to which of the following conditions?
   a. an increase in the P_{O_2} of the tissue
   b. a decrease in the P_{O_2} of the tissue
   c. an increase in the thickness of the capillary wall
   d. a decrease in the surface area of the capillary

2. Which of the following does blood *not* do?
   a. transport O_2 and CO_2
   b. distribute body heat
   c. produce red blood cells and other formed elements
   d. buffer against pH changes

3. How is most carbon dioxide transported in the blood?
   a. as a gas in the plasma
   b. bound to amino groups on hemoglobin
   c. bound to the heme group of hemoglobin
   d. as bicarbonate ions (HCO_3^-) in plasma
4. Which of the following are advantages of breathing air over breathing water?
   a. Air is less dense than water, so it takes less energy to move during ventilation.
   b. Oxygen diffuses faster through air than it does through water.
   c. The oxygen content of air is greater than that of an equal volume of water.
   d. All of the above are advantages.
5. Which of the following promotes oxygen release from hemoglobin?
   a. a decrease in temperature
   b. a decrease in CO_2 levels
   c. a decrease in pH
   d. a decrease in carbonic anhydrase
6. An open circulatory system is less efficient than a closed circulatory system in what respect?
   a. It is harder to deliver O_2 to specific tissues based on need.
   b. Hemolymph does not contain respiratory pigments, such as hemoglobin.
   c. There is no heart to pump the blood.
   d. In closed systems, body movements cannot help circulate the blood.

Conceptual Review

1. Compare and contrast the ventilatory and circulatory systems of an insect and a human. What are some advantages and disadvantages of each system?
2. Why is the sigmoidal shape of the oxygen-hemoglobin dissociation curve significant? Describe the changes in oxygen delivery that occur as a person changes from a resting state to intense exercise. What is the Bohr shift?
3. Explain how carbon dioxide is transported in the blood. In humans, why doesn't intense exercise and rapid production of CO_2 lead to a rapid reduction in blood pH?
4. Draw a diagram summarizing how bird lungs are ventilated. Why is this pattern of ventilation considered much more efficient than the ventilatory system of humans and other mammals?
5. Review Fick's law of diffusion. Explain how each parameter in the equation is reflected in the structure of ventilatory organs.
6. Explain why surface tension helps make the lungs of mammals elastic. Why does the large surface area in alveoli increase elasticity?

Group Discussion Problems

1. Researchers who compared the total amount of ventilatory surface area in the gills of various species of fish found that species that do a lot of high-speed swimming have a larger total surface area than do slow-moving species of the same size. Interpret this pattern in light of the theory of evolution by natural selection.
2. At high elevations, certain cells in humans increase the production of 2,3-diphosphoglycerate (DPG). In blood, DPG shifts the oxygen-hemoglobin dissociation curve to the right. How does this shift affect O_2 unloading and loading?
3. Carp are fish that thrive in stagnant-water habitats with low oxygen partial pressures. Compared with the hemoglobin of many other fish species, carp hemoglobin has an extremely high affinity for O_2. Is this trait adaptive? If so, why?
4. The carbon monoxide (CO) found in furnace and engine exhaust and in cigarette smoke binds to the heme groups in hemoglobin 210 times more tightly than does O_2. Explain why exposure to large doses of CO can lead to suffocation.

Answers to Multiple-Choice Questions 1. b; 2. c; 3. d; 4. d; 5. c; 6. a

www.prenhall.com/freeman is your resource for the following: Web Tutorials; Online Quizzes and other Online Study Guide materials; Answers to Conceptual Review Questions; Solutions to Group Discussion Problems; Answers to Figure Caption Questions and Exercises; and Additional Readings and Research.

45 Electrical Signals in Animals

KEY CONCEPTS

- Neurons are cells that transmit electrical signals. They have a membrane potential, which is an electric potential, or voltage, due to differences in the concentrations of ions on the two sides of the plasma membrane.
- Electrical signals are transmitted as all-or-none changes in membrane voltage called action potentials. During an action potential, an inflow of sodium ions is followed by an outflow of potassium ions.
- At synapses, an electrical signal from one neuron is converted to a chemical signal in the form of a neurotransmitter. When it arrives at an adjacent neuron, the neurotransmitter causes a change in that cell's membrane potential.
- Vertebrates have a central nervous system (CNS) and a peripheral nervous system (PNS). PNS neurons receive sensory information and transmit it to the CNS for processing. The CNS then sends signals to muscles, glands, or other target tissues via other PNS neurons.

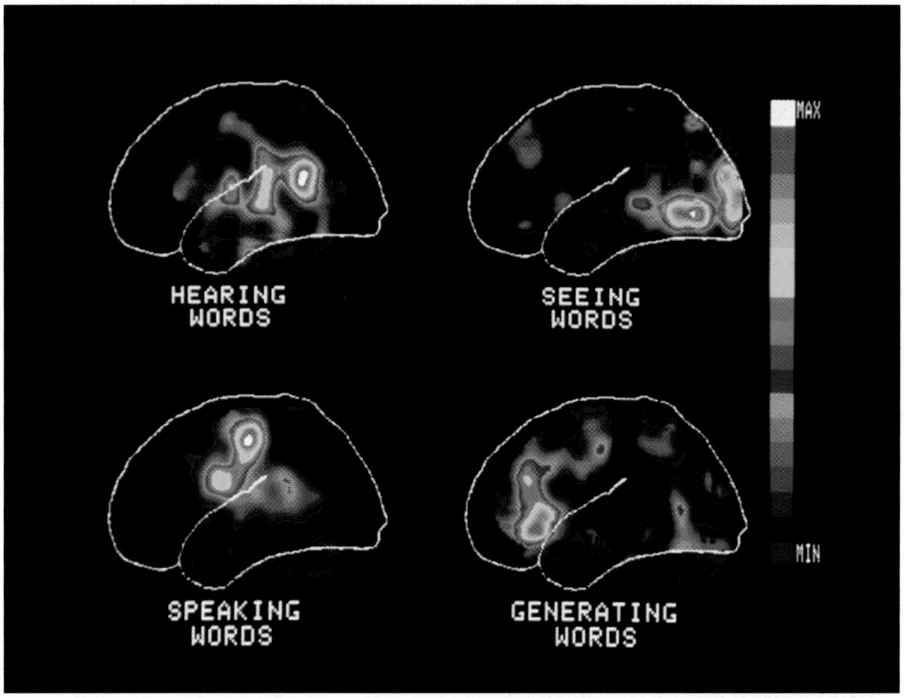

Positron-emission tomography (PET) scans showing changes in the activity of brain cells during different tasks, such as remembering words and speaking them aloud. Different brain areas are specialized for performing certain tasks. This chapter explores electrical signaling at a variety of levels, from individual ion channels to whole cells to regions of the brain.

If you are unfortunate enough to touch a hot stove burner, your hand will jerk away before your brain perceives pain. When a trout realizes that it is about to be attacked by an osprey, the trout's tail thrashes so rapidly that the movement is a blur. These and other animal movements are triggered by electrical impulses. Sensory cells in your finger and in the trout's eye send information to the brain and spinal cord, which send electrical signals to muscles in your arm and the trout's tail. Electrical signals are fast and precise. They are conducted from point to point by nerve cells, called **neurons**, at speeds of up to 100 m/sec (225 mph). The result is movement that saves your finger from extensive tissue damage—or saves the trout's life.

The goal of this chapter is to explore how electrical signaling occurs. Section 45.1 introduces the general anatomy of a nerve cell, discusses why a voltage exists across the cell's plasma membrane, and investigates how electrical currents are produced in cells. The next three sections focus on how neurons generate and propagate electrical signals, how information is transferred from cell to cell, and how neurons work in groups to integrate information from different sources. One of this chapter's most fundamental messages is that complex processes such as seeing, moving, and thinking are based on seemingly simple events: flows of ions across plasma membranes.

45.1 Principles of Electrical Signaling

Nerve cells, or neurons, transmit information. In many cases, the information they carry originates in sensory cells that are located on the periphery of the body. Sensory receptors inside the skin, eyes, ears, and nose transmit streams of data about the external environment in the form of electrical signals. Similarly, sensory cells inside the body monitor conditions that are important in homeostasis, such as blood pH and oxygen levels. In this way, sensory cells monitor conditions both outside and inside the body.

As **Figure 45.1a** shows, a sensory receptor cell transmits the information it receives by means of a nerve cell known as a **sensory neuron**. (In many cases, the receptor cell itself also acts as a sensory neuron.) In vertebrates, the sensory neuron sends the information to neurons in the brain or spinal cord via nerves. **Nerves** are long, tough strands of nervous tissue, containing thousands of neurons, that carry information to and from the brain and spinal cord. Together, the brain and spinal cord form the **central nervous system**, or **CNS**. One function of the CNS is to integrate information from many sensory neurons. Cells in the CNS called **interneurons** ("between-neurons") perform this integration. Interneurons make connections between sensory neurons and **motor neurons**, which are nerve cells that send signals to effector cells in glands or muscles. Recall from Chapter 41 that effectors are structures that restore a desired internal condition. All the components of the nervous system that are outside the central nervous system are considered part of the **peripheral nervous system**, or **PNS**. (Section 45.4 describes the structure and function of the PNS in more detail.)

Typically, sensory information from receptors in the PNS is sent to the CNS, where it is processed. Then a response is transmitted back to appropriate parts of the body. When you stub your toe, pain receptors in your toe relay sensory information rapidly to the brain, which then modifies your movements to avoid further injury. You might limp to avoid further contact with that toe. In some cases, however, information from a sensory neuron initially bypasses the brain. This is what happens when you touch a hot burner and jerk your hand back (**Figure 45.1b**). When pain receptors are stimulated, sensory neurons activate interneurons in the spinal cord that directly activate motor neurons serving muscles in the arm. The signals from these motor neurons stimulate muscles that pull the lower part of your arm toward the body and inhibit muscles that extend the lower arm. The movement is a **reflex**—a direct response to a signal that bypasses the brain. It is fast and automatic. Later, when your brain integrates information from pain receptors in your hand and from light receptors as you look at the burn, other motor neurons might evoke a response such as plunging your hand into cold water. What do the cells that carry and process this information look like?

The Anatomy of a Neuron

Compared with studying cells in the kidney, digestive system, or lungs, documenting the anatomy of neurons was difficult. Neurons are small, transparent, and morphologically complex.

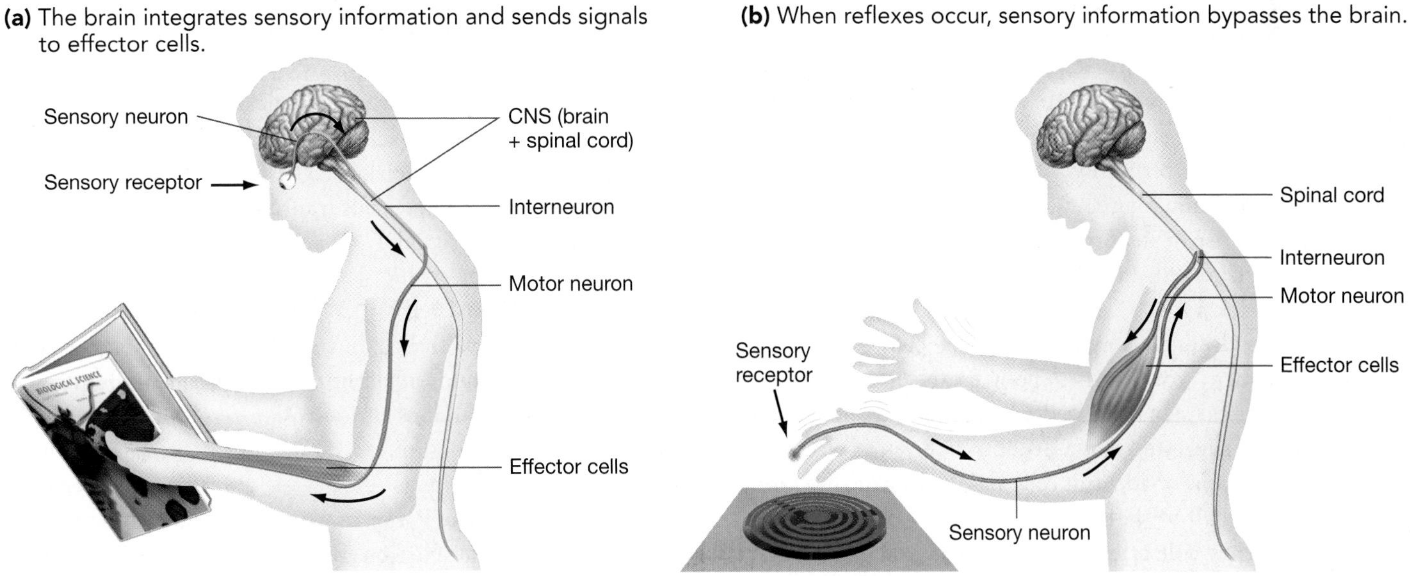

FIGURE 45.1 How Does Information Flow through the Nervous System?
(a) In most cases, sensory neurons send information to the brain. There, the electrical signals are integrated with information from other sources. Once integration is complete, a response is sent to effector cells through motor neurons. **(b)** In a spinal reflex, sensory information triggers a direct response by effector cells—without integration by the brain.

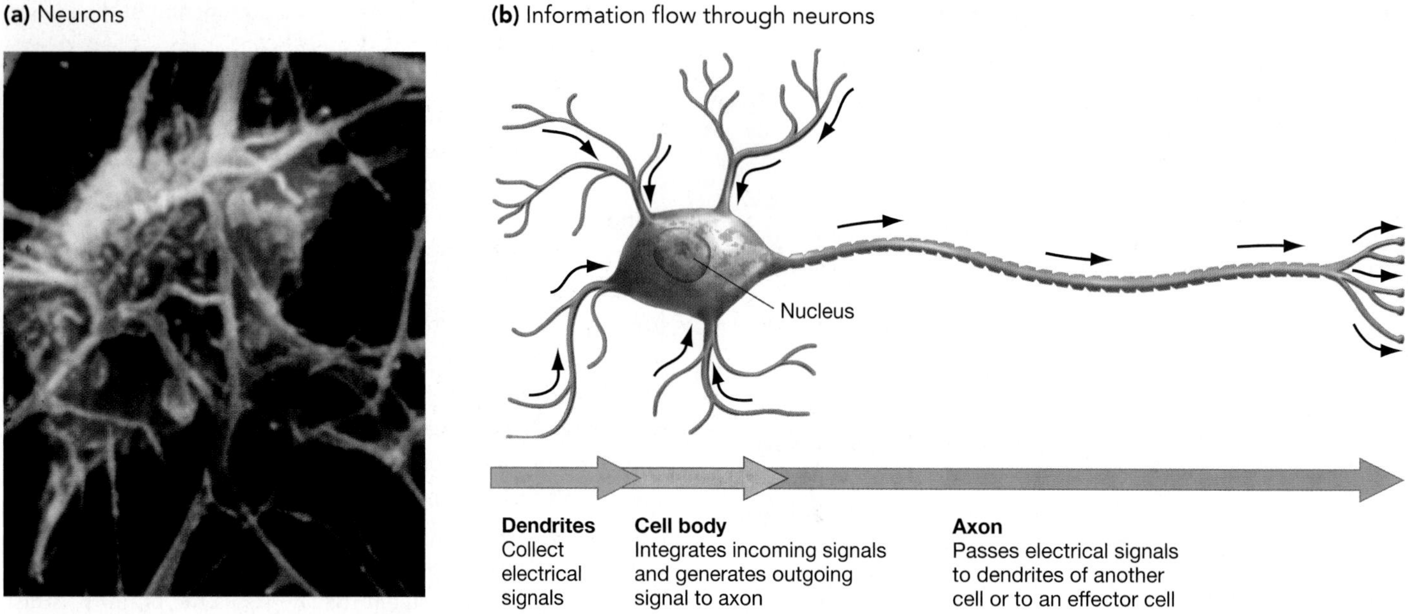
(a) Neurons
(b) Information flow through neurons
Nucleus
Dendrites
Collect electrical signals
Cell body
Integrates incoming signals and generates outgoing signal to axon
Axon
Passes electrical signals to dendrites of another cell or to an effector cell

FIGURE 45.2 How Does Information Flow in a Neuron?
(a) Neurons growing in cell culture. **(b)** The structure of a generalized neuron. EXERCISE Choose a neuron in part (a). Label the cell body and axon. Circle points where the cell makes contact with other neurons.

So when Camillo Golgi discovered that some neurons became visible when samples of preserved tissue were treated with a solution containing silver nitrate, his finding was a major advance (**Figure 45.2a**).

Golgi published his initial observations in 1898. Through the early decades of the twentieth century, his work and that of Santiago Ramón y Cajal revealed several important points about the anatomy of neurons. Most neurons have the same three parts, shown in **Figure 45.2b**: (1) a cell body, (2) a highly branched group of short projections called dendrites, and (3) one or more long projections called axons. Dendrites are rarely more than 2 mm long, but axons can be over a meter in length. The number of dendrites and their arrangement vary greatly from cell to cell. Further, many brain cells have only dendrites and lack axons.

What do dendrites and axons do? Long before the anatomists began their work, it had been established that neurons carry electrical signals. Based on observations of stained cells, researchers suggested that a **dendrite** receives an electrical signal from the axons of adjacent cells and that a neuron's **axon** then sends the signal to the dendrites and cell bodies of other neurons. Dendrites collect electrical signals; axons pass them on. The **cell body**, or **soma**, which includes the nucleus, integrates the incoming signals and generates an outgoing signal in the axon.

Golgi and Cajal disagreed about how these signals pass from neuron to neuron, however. Golgi thought that the cytoplasm of all neurons was directly connected. His hypothesis was that neurons form a continuous network. In contrast, Cajal maintained that the plasma membrane of each neuron is distinct and that the membranes of axons and dendrites meet at junctions called *synapses*. The dispute wasn't completely settled until the 1950s, when images from electron microscopes confirmed that only a small subset of neurons are directly connected; most interactions occur where the plasma membranes of two neurons meet at synapses.

The basic message of early anatomical studies was that neurons are highly branched and that each neuron makes many connections with other neurons. The focus of research shifted to a new question: If these cells transmit electrical signals, how do they create these signals?

An Introduction to Membrane Potentials

Ions carry an electric charge. In virtually all cells, the cytoplasm and extracellular fluids adjacent to the plasma membrane contain unequal distributions of ions. As a result, cells are inherently electrical in nature. A difference of electrical charge between any two points creates a difference in **electrical potential**, or a **voltage**. For example, if the positive and negative charges on ions that exist on the two sides of a plasma membrane do not balance each other, the membrane will have an electrical potential. When an electrical potential exists across a plasma membrane, the separation of charges is called a **membrane potential**. If there is a large separation of charges across the membrane, then the membrane potential is large. It is important to remember that membrane potentials refer only to a separation of charge immediately adjacent to the plasma membrane, on both sides of the membrane. Even if there is a large membrane potential, there may be no charge separation slightly farther from the membrane.

Membrane potentials are measured in units called millivolts. A **millivolt (mV)** is 1/1000 of a volt; a **volt** is the standard unit of electrical potential. To help you get a better understanding of how large a millivolt is, a typical flashlight (AA) battery has an electrical potential of 1500 mV between its + and − terminals. In neurons, membrane potentials are typically about 70–80 mV.

Membrane potentials are a form of potential energy. When a membrane potential exists, the ions on both sides of the membrane have potential energy. To convince yourself that this is true, consider that if the membrane were removed, ions would spontaneously move from the area of like charge to the area of unlike charge. A flow of charge is called an **electric current.** This movement would occur because like charges repel and unlike charges attract.

As emphasized in Chapter 6, however, ions move across membranes in response to concentration gradients as well as charge gradients. Therefore, a membrane potential includes energy stored as the concentration gradient of charged ions on the two sides of the membrane. Recall that the combination of an electric gradient and a concentration gradient is called an *electrochemical gradient*. What do all of these facts have to do with neurons?

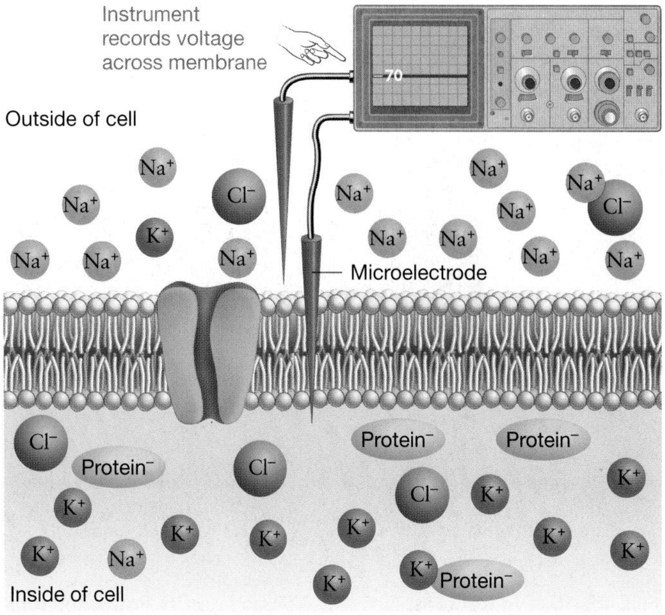

FIGURE 45.3 Neurons Have a Resting Potential
In resting neurons, the membrane is selectively permeable to K^+. As K^+ leaves the cell along its concentration gradient, the inside becomes negatively charged relative to the outside. To measure a neuron's membrane potential, researchers insert a microelectrode into the cell and compare that reading with the reading outside the cell.

The Resting Potential

When a neuron is not transmitting an electrical signal but is merely sitting in extracellular fluid at rest, the membrane potential across its membrane is called the **resting potential.** To understand why the resting potential exists, consider the distribution of the various ions and other charged molecules on the two sides of the neuron's plasma membrane. Most of the large molecules inside cells are acids, which tend to release a proton. Proteins, for example, are made up of amino acids. When acids release a proton, they acquire a negative charge. Intracellular proteins that tend to be negatively charged are referred to as *organic anions*. These molecules rarely cross the plasma membrane. The interior of a neuron also contains low concentrations of Na^+ and Cl^-. Inside a neuron, the dominant positively charged ion is the potassium ion (K^+). In the extracellular fluid, however, the sodium ion (Na^+) and chloride ion (Cl^-) predominate.

Figure 45.3 shows the relative concentrations of organic anions, K^+, Na^+, and Cl^- inside and outside a resting neuron. If each type of ion diffused across the membrane in accordance with its concentration gradient, organic anions and K^+ would leave the cell while Na^+ and Cl^- would enter. But ions cross membranes readily in only three ways: (1) flowing along their electrochemical gradient through an ion channel; (2) carried, via a membrane cotransporter, with an ion that experiences a strong electrochemical gradient; or (3) pumped against an electrochemical gradient by a membrane protein that hydrolyzes ATP.

In each case of ion movement across the plasma membrane, a membrane channel or pump is involved. Recall from Chapter 6 that **ion channels** are pores in the membrane that allow only specific ions to pass through. When a neuron is not actively involved in transmitting an electrical signal, the most common type of channel that is open is one that admits potassium ions. Therefore, resting neurons are most permeable to K^+, which cross the membrane easily along their concentration gradient. To capture this point, biologists say that the membrane is **selectively permeable**—it permits a much greater flow of K^+ ions compared with Cl^- or Na^+ ions. The potassium channels involved are sometimes called **leak channels**, because they allow K^+ to leak out of the cell.

As K^+ moves from the interior of the cell to the exterior through potassium channels, the inside of the cell becomes more and more negatively charged relative to the outside. Eventually, the buildup of negative charge inside the cell begins to attract K^+ and counteract the concentration gradient that had favored the movement of K^+ out. As a result, the membrane reaches a voltage at which there is an equilibrium between the concentration gradient that moves K^+ out and the electrochemical

WEB TUTORIAL 45.1
Membrane Potentials

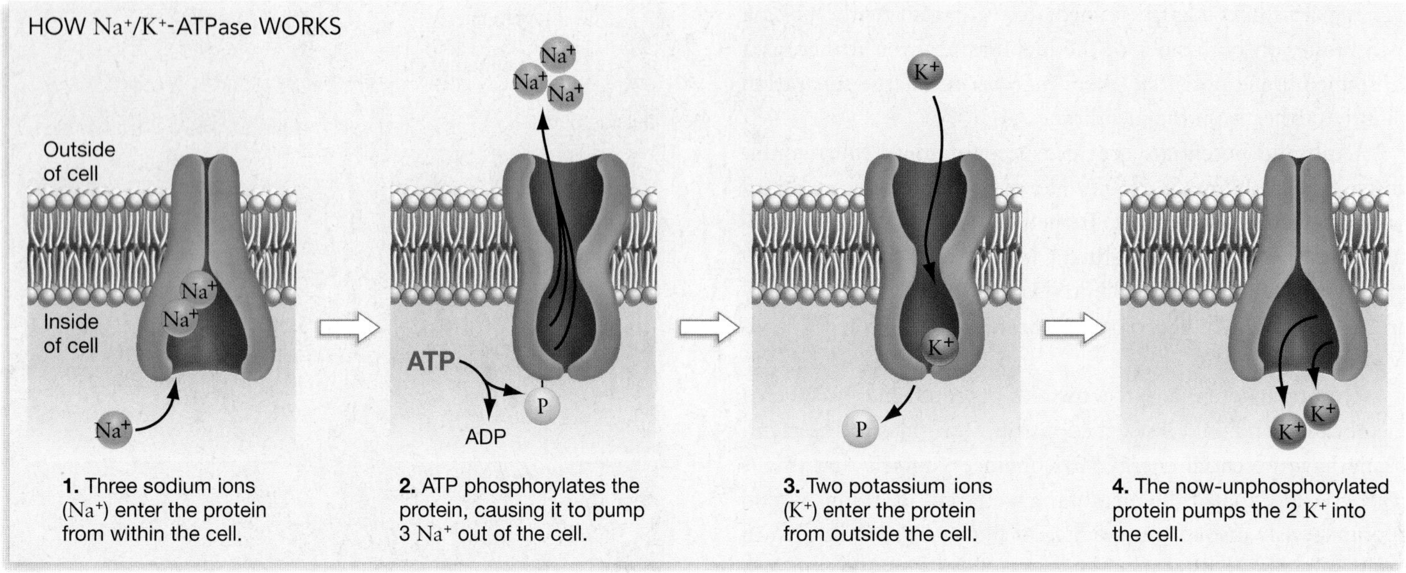
HOW Na+/K+-ATPase WORKS
Outside of cell
Inside of cell
Na+
ATP
ADP
P
K+
1. Three sodium ions (Na+) enter the protein from within the cell.
2. ATP phosphorylates the protein, causing it to pump 3 Na+ out of the cell.
3. Two potassium ions (K+) enter the protein from outside the cell.
4. The now-unphosphorylated protein pumps the 2 K+ into the cell.

FIGURE 45.4 In Neurons, Na^+/K^+-ATPase Imports Potassium Ions and Exports Sodium Ions
A model of how the pump works. By following radioactive Na^+ and K^+, biologists found that the pump transports 3 Na^+ out of the cell for every 2 K^+ brought in. Notice that this pump operates via a conformational change.

gradient that moves K^+ in. At this voltage, there is no longer a net movement of K^+. This voltage is called the **equilibrium potential** for K^+. Although Cl^- and Na^+ cross the plasma membrane much less readily than does K^+, some movement of these ions also occurs through a small number of open ion channels that are selective for each ion.

In addition to the passive movement of ions that takes place when neurons are at rest, Na^+/K^+-ATPase actively pumps Na^+ out of the cell and K^+ into the cell. More specifically, the sodium-potassium pump uses the energy released from the breakdown of ATP to remove three Na^+ ions that had leaked into the cell and simultaneously recover two K^+ ions that had leaked out of the cell (**Figure 45.4**). In this way, active transport via Na^+/K^+-ATPase ensures that eventually the concentration of K^+ is much higher inside the cell than outside, while the concentration of Na^+ is lower inside than outside.

As a result of these concentration differences on the two sides of the membrane, Na^+ ions would readily enter the cell and K^+ ions would rapidly exit the cell if they could cross the plasma membrane easily. In addition, the inside of the neuron is negatively charged with respect to the extracellular environment. Thus, the neuron has a negative resting membrane potential. The resting potential represents energy stored as concentration gradients in a series of ions. How do researchers calculate the size of this voltage?

Using the Nernst and Goldman Equations

To calculate a membrane voltage created by several ions, biologists begin by analyzing the membrane potentials produced by individual ions. Specifically, they begin by calculating the equilibrium potential for each ion involved. The equilibrium potential for an ion is the voltage that is required to counteract the tendency for that ion to cross a membrane in response to its concentration gradient. This is a crucial point. In the plasma membrane of a neuron, an electrical potential for a given ion exists because there is a concentration gradient for that ion. Ion concentration gradients, in turn, are derived from the chemical energy stored in ATP, via the action of Na^+/K^+-ATPase. In this way, chemical energy in the form of ATP is converted to electrical energy in the form of a membrane potential. When energy stored in the concentration gradient is converted to energy stored in an electrical potential, no energy is lost—in accordance with the first law of thermodynamics (see Chapter 2).

To calculate the amount of energy involved, biologists use the **Nernst equation**—a formula that converts the energy stored in a concentration gradient to the energy stored as an electrical potential. The concentration gradient is symbolized $\text{ion}_o/[\text{ion}]_i$, where $[\text{ion}_o]$ and $[\text{ion}]_i$ are the concentrations of the ion outside and inside the cell, respectively. The electrical potential for that ion is symbolized (E_{ion}). The Nernst equation specifies the equilibrium potential for a type of ion as

$$E_{\text{ion}} = 2.3\frac{RT}{zF}\log\frac{[\text{ion}]_o}{[\text{ion}]_i}$$

In this expression, z is the valence of the ion (for instance, +1 for potassium) and the expression RT/F is known as the *thermodynamic potential.* The three terms in the thermodynamic potential are the gas constant (R), which acts as a con-

TABLE 45.1 Concentration of Important Ions across a Neuron's Plasma Membrane at Rest

Ion	Cytoplasm Concentration	Extracellular Concentration	Equilibrium Potential
(a) Squid neuron			
Na^+	50 mM	440 mM	+55 mV
K^+	400 mM	20 mM	−75 mV
Cl^-	51 mM	560 mM	−60 mV
Organic ions$^-$	385 mM	—	—
(b) Mammalian neuron			
Na^+	10 mM	145 mM	+68 mV
K^+	140 mM	5 mM	−85 mV
Cl^-	10 mM	110 mM	−60 MV

stant of proportionality; the absolute temperature (T), measured in kelvins; and the Faraday constant (F), which specifies the amount of charge carried by one mole of an ion with a valence of +1 or −1. The thermodynamic potential specifies the voltage required to balance a tenfold concentration ratio across the membrane.

It is important to remember that the Nernst equation applies to only a single ion. For example, suppose that the potassium concentration inside the axon of a squid neuron has been measured as 400 mM, yet the outside concentration for this ion is only 20 mM (**Table 45.1**). Potassium has a charge of +1, so at 20°C the RT/zF part of the equation yields +25 mV. In this case, the equilibrium potential for potassium becomes

$$E_K = 2.3 \times 25 \text{ mV} \times \log \frac{20 \text{ mM}}{400 \text{ mM}}$$

$$E_K = 58 \text{ mV} \times \log(0.05) = -75 \text{ mV}$$

The minus sign indicates that the interior of the axon is negatively charged with respect to the exterior. Repeating this process for Na^+ ions and Cl^- ions yields the following results for the squid axon:

$$E_{Na} = 58 \text{ mV} \times \log\left(\frac{440 \text{ mM}}{50 \text{ mM}}\right) = +54 \text{ mV}$$

$$E_{Cl} = -58 \text{ mV} \times \log\left(\frac{560 \text{ mM}}{51 \text{ mM}}\right) = -60 \text{ mV}$$

Again, notice that the equilibrium potential given by the Nernst equation is calculated independently for each ion. The equilibrium potential calculated with the Nernst equation is also unrelated to the resting membrane potential. For example, in the case of the squid axon just considered, the resting membrane potential is approximately −70 mV. This is a key point, because, when the equilibrium potential for an individual ion does not match the resting membrane potential of the cell, that ion is not at equilibrium.

How do individual equilibrium potentials relate to the overall resting membrane potential? The Nernst equation cannot be used to calculate resting membrane potentials, because no cell is permeable to only a single ion. David Goldman realized that, to calculate a membrane potential when K^+, Na^+, and Cl^- are each at their equilibrium potential, the permeability of the membrane to the three ions had to be considered. If these permeabilities are symbolized as P_K, P_{Na}, and P_{Cl}, then the membrane potential is given by the following expression:

$$V_m = 58 \text{ mV} \times \log\left(\frac{[K^+]_o + (P_{Na}/P_K)[Na^+]_o + (P_{Cl}/P_K)[Cl^-]_i}{[K^+]_i + (P_{Na}/P_K)[Na^+]_i + (P_{Cl}/P_K)[Cl^-]_o}\right)$$

In a squid's neuron, the relative permeabilities for K^+, Na^+, and Cl^- are described by the ratio 1.0 : 0.03 : 0.1. Given these ratios and the observed concentration of each ion inside and outside the axon, the resting potential of the neuron can be calculated as follows:

$$V_m = 58 \text{ mV} \times \log\left(\frac{10 + (0.03)460 + (0.1)40}{400 + (0.03)50 + (0.1)540}\right) = -70 \text{ mV}$$

This value is very close to the observed value.

Neurons, however, are rarely at rest—their membrane potentials are constantly fluctuating. These fluctuations in membrane potential are the essence of electrical signaling. How do biologists measure changes in membrane potentials through time?

Using Microelectrodes to Measure Membrane Potentials

During the 1930s and 1940s, A. L. Hodgkin and Andrew Huxley helped pioneer the study of electrical signaling in animals. Their work focused on what has become a classic model system in the study of electrical signaling: the axons of squid. Squid live in the ocean and are preyed on by fish and whales. When a squid is threatened, electrical signals travel down the axon to muscle cells. When these muscles contract, water is expelled from a cavity in the squid's body. As a result, the squid lurches away from danger by jet propulsion (see Chapter 31).

Hodgkin and Huxley decided to study the squid's axon simply because it is so large. Many of the axons found in humans are a mere 2 μm in diameter, but the squid axon is about 500 μm in diameter. The squid axon was large enough that the researchers could record membrane potentials by inserting a wire down its length. That was the best approach available at the time for measuring voltages in cells and required an axon with an exceptionally large diameter.

By measuring the voltage difference between the wire and an electrode outside the cell, Hodgkin and Huxley could estimate the voltage that existed across the neuron's plasma membrane and observe how it changed through time. Later, other researchers developed glass microelectrodes that were small enough to be inserted into smaller nerve cells to record membrane voltage.

With their relatively simple early equipment, Hodgkin and Huxley were able to record the axon's resting potential and document that it can be disrupted by an event called the action potential.

What Is an Action Potential?

An **action potential** is a rapid, temporary change in a membrane potential. **Figure 45.5** shows the form of the action potential that Hodgkin and Huxley recorded from the squid's giant axon. The action potential has three distinct phases:

1. The initial event is a rapid **depolarization** of the membrane. (A membrane is said to be polarized if the charges on the two sides are different. Depolarization means that the membrane becomes less polarized than before.) For an action potential to begin, the membrane potential must shift from its resting potential of −70 mV to about −55 mV. If the membrane depolarizes less than that, an action potential does not occur. But if this **threshold** potential is reached, certain channels in the axon membrane open and ions rush into the axon, following their electrochemical gradients. The current flow causes further depolarization. More specifically, the inside of the membrane becomes less negative and then positive with respect to the outside.
2. When the membrane potential reaches about +40 mV, an abrupt change occurs. This change is triggered by the closing of certain ion channels and the opening of other ion channels in the membrane. During this second phase of the action potential, the membrane experiences a rapid **repolarization** as ions flow out of the axon.
3. The repolarization event actually results in the membrane becoming more negative than the resting potential. For this reason, biologists call the final part of the action potential the **undershoot**.

Taken together, all three phases of an action potential occur in about a millisecond.

Hodgkin and Huxley made other important observations about the action potential. In addition to being fast, it is an all-or-none event. There is no such thing as a partial action potential, and all action potentials for a given neuron are identical in magnitude and duration. The biologists also found that action potentials could be triggered by artificially depolarizing the membrane with an injection of electrical current through a microelectrode. Finally, action potentials are propagated down the length of the axon. When an impulse was recorded at a particular point on a squid axon, an action potential that was identical in shape and size would be observed farther down the same axon soon afterward. Neurons are said to have **excitable membranes**, because they are capable of generating action potentials that propagate rapidly along the length of their axons.

Taken together, these observations suggested a mechanism for electrical signaling. In the nervous system, information is coded in the form of action potentials that travel along axons. The frequency of action potentials—not their size—is the meaningful signal. In the squid's giant axon, action potentials signal muscles to contract. As a result, the animal escapes from danger.

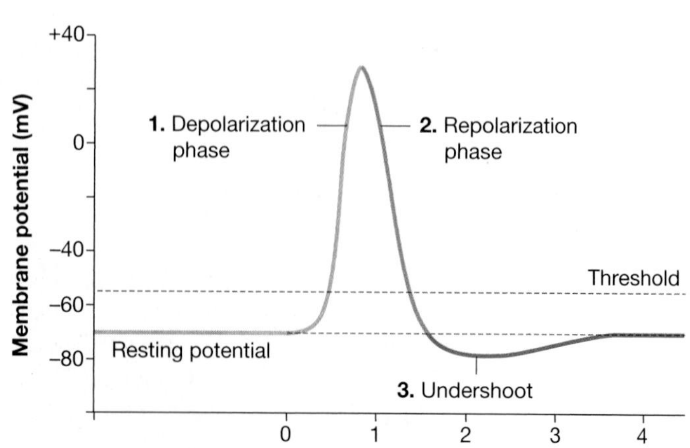

FIGURE 45.5 The Action Potential Is a Stereotyped Change in Membrane Potential

All action potentials have the same general shape shown.

✓ CHECK YOUR UNDERSTANDING

The plasma membranes of neurons carry a resting potential because Na^+/K^+-ATPase pumps Na^+ out of the cell and K^+ into the cell and because the membrane is selectively permeable to K^+ ions. You should be able to (1) explain the Nernst equation and Goldman equation in words, (2) use the formulas to calculate the equilibrium concentration for each ion listed in Table 45.1b, as well as the resting membrane potential of a mammalian neuron, and (3) predict what would happen to the resting potential of a squid axon if potassium channels were blocked.

45.2 Dissecting the Action Potential

Once it became possible to record rapid changes in membrane potential accurately, an entirely new field of research opened up. The most urgent question was to understand which ions are involved in the currents that form the action potential. Is Na^+, Cl^-, or K^+ responsible for the depolarization and repolarization phases of the event? One early hypothesis was that the action potential resulted from a temporary breakdown in the membrane's selective permeability. The idea was that all channels opened and allowed a free flow of Na^+ and Cl^- as well as K^+. But the all-channels-open hypothesis ran into trouble when researchers were able to measure the concentrations of each ion on the two sides of the squid's axon membrane. Given the observed concentrations of ions when the neuron was at rest, the all-channels-open hypothesis predicted that the peak of the action potential should occur at about 0 mV, not the +40 mV observed.

In contrast, Hodgkin noted that +40 mV corresponded to the equilibrium potential for Na^+. This insight led to the hypothesis that the depolarization phase of the action potential resulted from an influx of sodium ions. How could this hypothesis be tested?

Distinct Ion Currents Are Responsible for Depolarization and Repolarization

To understand the currents responsible for the action potential, Hodgkin and Huxley recorded electrical activity in a squid axon bathed in seawater. Then they replaced the seawater with an isotonic solution containing an uncharged and metabolically inert sugar called dextrose. **Figure 45.6** shows the result of this experiment. Washing Na^+ out of the solution surrounding the axon abolished action potentials. But when the biologists replaced the dextrose solution with seawater, action potentials again began traveling down the neuron. Further, when they did the same experiment using solutions with various concentrations of Na^+ instead of seawater, Hodgkin and Huxley found that the peak of the action potential tracked the equilibrium concentration of Na^+ based on its concentration outside the cell. Taken together, these experiments furnished strong support for the hypothesis that the action potential begins when Na^+ flows into the neuron. Sodium ions are responsible for the depolarization phase.

What flow of ions is responsible for the repolarization phase of an action potential? Using radioactive K^+, Hodgkin and Huxley showed that there was a strong flow of potassium ions out of the cell during the repolarization phase. Thus, the action potential consists of a strong inward flow of sodium ions followed by a strong outward flow of potassium ions. Which ion channels are responsible for these currents, and how do they work?

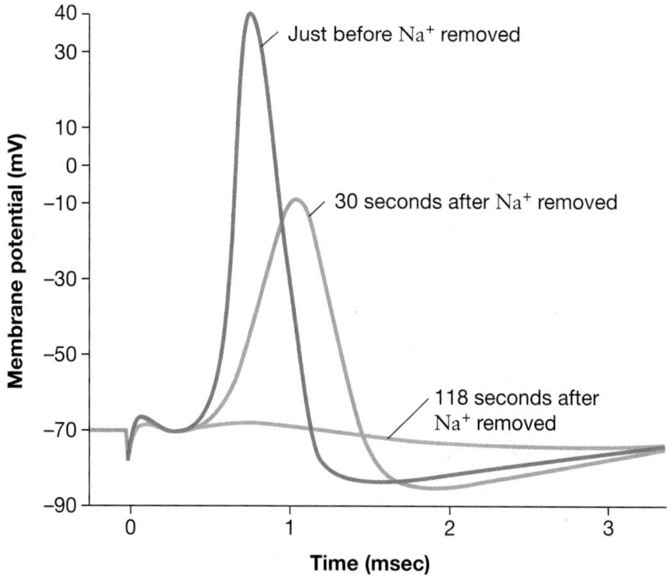

FIGURE 45.6 Action Potentials Diminish when Na^+ Is Removed from the Solution Surrounding a Neuron

Action potentials recorded at three different intervals after a solution containing Na^+ was flushed away from a squid axon with a solution containing isotonic dextrose. QUESTION Why was it important that the researchers used an isotonic solution of dextrose?

Voltage-Gated Channels

When ion channels were introduced in Chapter 6, they were presented as simple, static proteins that facilitate a constant stream of ions. But research on the action potential suggests that certain ion channels open and close in response to changes in membrane voltage. These types of proteins are called **voltage-gated channels.** The conformation of voltage-gated channels changes in response to the charges present at the membrane surface. These conformational changes lead to the opening or closing of the channels.

To study the relationship between membrane voltage and channel activity, Kenneth Cole and his co-workers came up with a way to hold the membrane potential of a neuron constant, even while ions were flowing through channels in the neuron's plasma membrane. Their experimental apparatus employed an electrical feedback system in which they injected an artificial current that was equal in size and opposite in direction to the current that flowed naturally across the membrane. Using this technique, called **voltage clamping**, Cole and associates could hold an axon at any voltage they desired and record the electrical currents generated by flows of Na^+ and K^+ across the membrane.

Hodgkin and Huxley quickly adopted the new equipment devised by Cole's team and produced data such as those shown in **Figure 45.7** (page 1034). The top record is from an experiment in which they voltage-clamped a squid axon at −9 mV. The next record down from the top shows the current that resulted from the depolarization. Notice that this trace indicates a rapid inward-directed current followed by a slower outward-directed current. The bottom two records break this overall current down into the components due to an inward flow of Na^+ and outward flow of K^+, respectively.

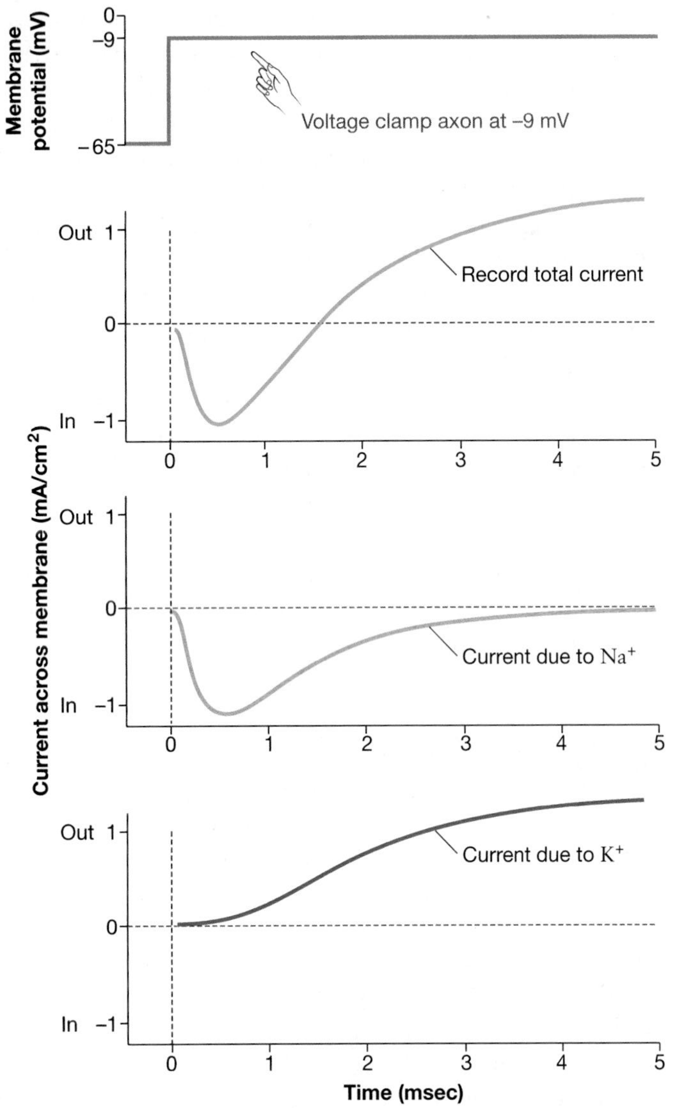

FIGURE 45.7 Voltage Clamping Allows Researchers to Record Current Flows across Membranes

When Hodgkin and Huxley voltage-clamped a squid axon at −9 mV, they recorded the flow of current shown in the second trace. Current initially flowed into the axon and then back out. As the bottom two traces show, this overall current is due to a rapid flow of Na^+ into the cell and a slower flow of K^+ out of the cell.

This experiment, and others like it, confirmed that Na^+ and K^+ channels open in response to changes in voltage. When the plasma membrane in the region of the axon is depolarized, voltage-gated Na^+ channels open for less than a millisecond. Sodium ions rush into the cell through these open channels, because the Na^+ concentration is much higher outside the cell and because Na^+ ions are attracted to the negatively charged interior. The result is that the inside of the membrane becomes more positively charged. More specifically, it goes from a resting potential of −70 mV to −50 mV. The change in electrical potential reaches a threshold level that triggers many additional Na^+ channels to open. As a result, Na^+ ions flood the interior, generating an action potential. Shortly after the action potential is under way, the voltage-gated Na^+ channels close and voltage-gated K^+ channels open.

Hodgkin and Huxley's data also answered an important question: If Na^+ flows in and K^+ flows out, why don't the two currents cancel each other out? The key observation is that the potassium current is delayed. The Na^+ channels close as the membrane potential approaches +54 mV and as the voltage-gated K^+ channels open. At this point, the interior of the membrane contains excess positive charge and favors the outward flow of K^+ through the open K^+ channels. As K^+ exits the cell, the membrane potential drops back toward resting levels.

To summarize, early work on electrical signaling showed that action potentials result from the staggered activity of voltage-gated Na^+ and K^+ channels. Once these results had been established, attention turned to the channels themselves. How do these voltage-gated proteins work?

Patch Clamping and Studies of Single Channels

Studying individual ion channels became possible when Erwin Neher and Bert Sakmann perfected a technique known as **patch clamping.** As **Figure 45.8a** shows, the researchers touched a membrane with a fine-tipped microelectrode and then applied suction to capture a single ion channel within the tip of the microelectrode. The suction seals the membrane against the glass so that no current leaks out. Using this technique, the researchers documented the currents that flowed through individual channels. The recordings shown in **Figure 45.8b** are typical, and they make several important points about the nature of voltage-gated channels:

- The shape of the current traces supports the hypothesis that voltage-gated channels are either open or closed. There is no gradation in channel behavior.
- Sodium channels open quickly after depolarization. They stay open for about a millisecond and then close. As Na^+ rushes into the axon, the inside becomes electrically positive. Shortly after the initial depolarization, the sodium channels close and remain inactive for 1 to 2 milliseconds. That explains why the cell can repolarize: Once the sodium channels close, they have a lag before they can open again. **Figure 45.8c** shows a simplified model of how voltage-gated sodium channels open.
- Potassium channels open with a delay after depolarization. They continue to flip open and closed until the membrane repolarizes. Once the membrane returns to the resting potential, these channels remain closed.

Observations of the behavior of Na^+ channels also helped explain why the action potential is an all-or-none event. The key observation is that Na^+ channels are more likely to open as a membrane depolarizes. An initial depolarization thus leads to

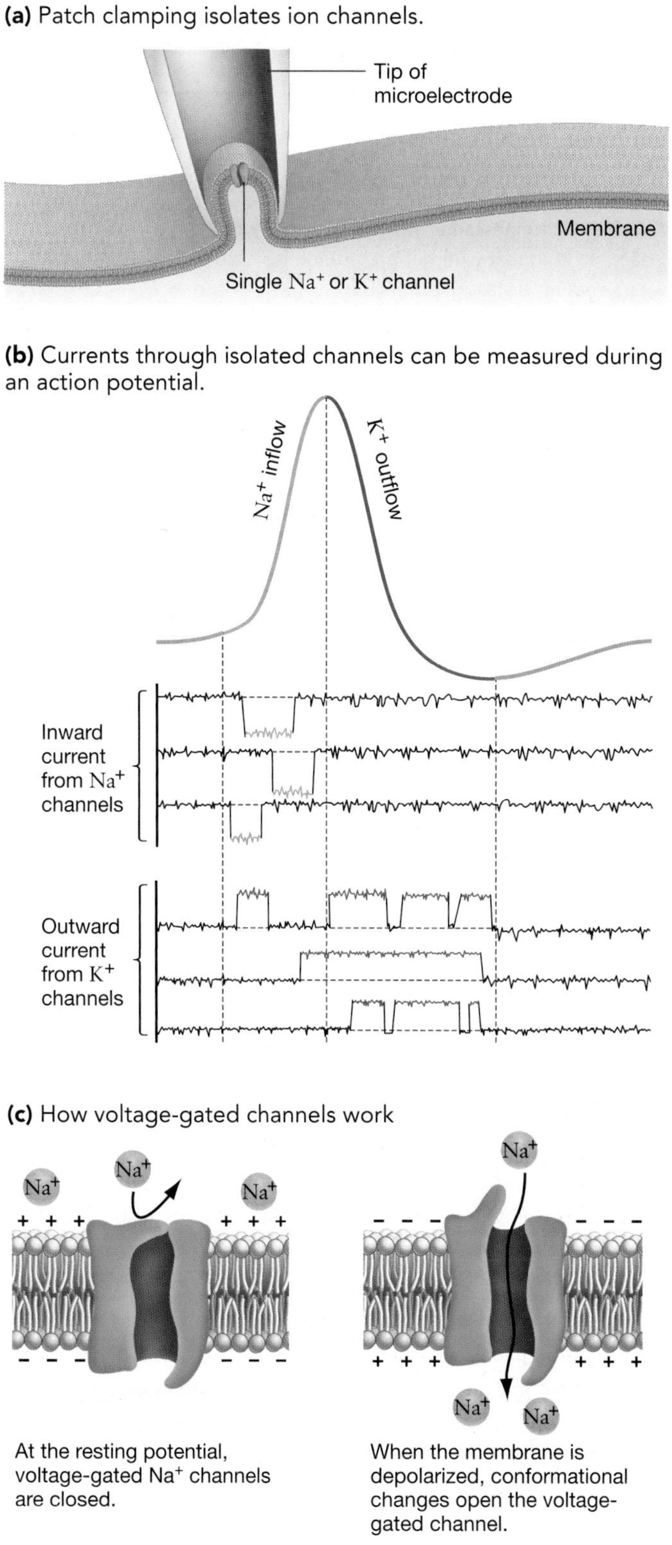

FIGURE 45.8 Patch Clamping Allows Researchers to Record from Individual Channels

(a) Patch clamping depends on the use of extremely fine-tipped microelectrodes. The goal is to isolate one channel and record from it. **(b)** Current traces from three voltage-gated sodium channels and three voltage-gated potassium channels. No current flows through either type of channel at the resting potential. **(c)** Changes in the conformation of voltage-gated channels are responsible for changes in a neuron's permeability to Na^+ and K^+.

the opening of more Na^+ channels, which depolarizes the membrane further and leads to the opening of additional Na^+ channels. In this way, the opening of Na^+ channels exhibits positive feedback. Recall from Chapter 42 that **positive feedback** takes place when the occurrence of an event makes the same event more likely to recur. When a fuse is lit, for example, the heat generated by the oxidation reaction accelerates the reaction itself, which generates still more heat and leads to additional oxidation reactions. Positive feedback is rare in organisms—the opening of Na^+ channels during an action potential is one of the few examples known.

Voltage clamping and patch clamping were not the only technical advances that helped researchers explore the dynamics of voltage-gated channels. The discovery of poisons that target neurons, from sources as diverse as venomous snakes and foxglove plants, also provided important experimental tools.

Using Neurotoxins to Identify Channels and Dissect Currents Many of the toxins produced by poisonous animals and plants cause convulsions, paralysis, or unconsciousness when they are ingested. Based on these symptoms, physicians and researchers suspected that at least some poisons affect neuron function; such poisons are known as **neurotoxins.** This hypothesis was supported when biologists treated giant axons from lobsters with the tetrodotoxin found in puffer fish.* Although these experiments showed that the resting potential in treated neurons was normal, action potentials were abolished. More specifically, the researchers documented that the outward-directed K^+ current was normal but that the inward-directed Na^+ flow was wiped out. Based on this result, they concluded that puffer-fish toxin specifically blocks the voltage-gated Na^+ channel. The hypothesis is that the toxin binds to a specific site on the channel protein, preventing the channel from functioning. Thus, researchers use neurotoxins to block only certain channels.

In contrast, when researchers treated neurons with the venom of black mamba snakes, preliminary data hinted that K^+ channels were blocked. To test this finding more rigorously, researchers voltage-clamped neurons that had been dissected from frogs, bathed the neurons in a solution containing black-mamba poison, varied the membrane potential, and studied the K^+ currents that resulted. As predicted, the poison had a dramatic effect on membrane permeability to K^+. However, the venom did not stop the current completely. Instead, the data suggest that a specific K^+ channel was poisoned but that other K^+ channels still allowed an outflow of

*Puffer fish are considered a delicacy in Japan. The toxin that these fish synthesize and sequester in their tissues is so dangerous that professional chefs must be trained and certified in safe preparation techniques. In one recent 10-year study period, however, 646 cases of puffer-fish poisoning were reported in Japan, with 179 fatalities. Most of the poisoning events involved nonprofessionals who were attempting to prepare puffer fish at home.

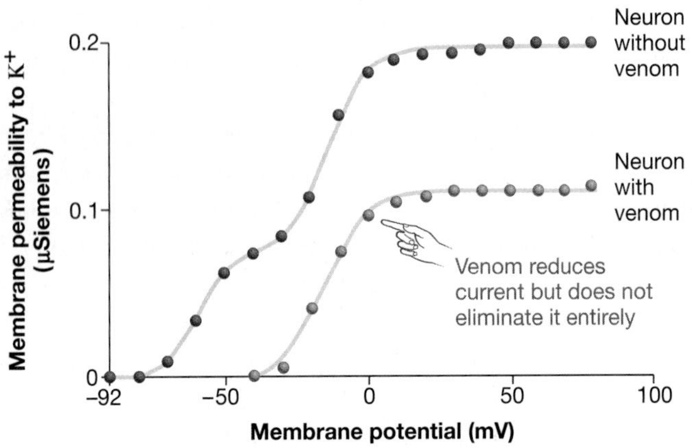

FIGURE 45.9 Experiments with Poisons Showed that There Is More than One Potassium Channel
Neuron permeability to K^+ as a function of membrane voltage. Each data point was generated by measuring membrane permeability while the membrane was voltage-clamped at a particular potential.

current (**Figure 45.9**). The existence of multiple K^+ channels in axons was confirmed by patch-clamping studies that documented K^+ flowing through several types of K^+ channels. In this way, experiments with neurotoxins helped establish that most neurons contain just one type of sodium channel but several types of potassium channels.

The Role of the Sodium-Potassium Pump

Of all the toxins that poison ion channels, perhaps the most important is a molecule from foxglove plants. This compound, introduced in Chapter 42, is called **ouabain**. It wipes out the resting potential of neurons. To understand why this observation was important, recall that Na^+ enters the neuron during an action potential and that K^+ leaves. Although the number of ions that flow across the membrane during an action potential is tiny, the concentration gradients that favor an inflow of Na^+ and an outflow of K^+ would eventually decline if enough action potentials occurred—that is, if enough sodium ions entered the cell and enough potassium ions left. In addition, recall that K^+ slowly leaks out of neurons when they are at rest.

To maintain the concentration gradients in sodium and potassium ions, Na^+/K^+-ATPase must pump Na^+ out and K^+ in. This is an energy-demanding process, because the ions are moved against their electrochemical gradients. The sodium-potassium pump is responsible for maintaining the resting potential.

Does ouabain poison the sodium-potassium pump? A researcher was able to answer this question when he succeeded in isolating a protein from crab neurons that began hydrolyzing ATP if both Na^+ and K^+ were present (**Figure 45.10**). When he added ouabain to this reaction mix, the enzyme's activity stopped. Soon after, other researchers used radioactive Na^+ and K^+ to establish that the enzyme is an **Na^+/K^+-ATPase**, which transports three sodium ions out of cells for every two potassium ions transported inward (see Figure 45.4).

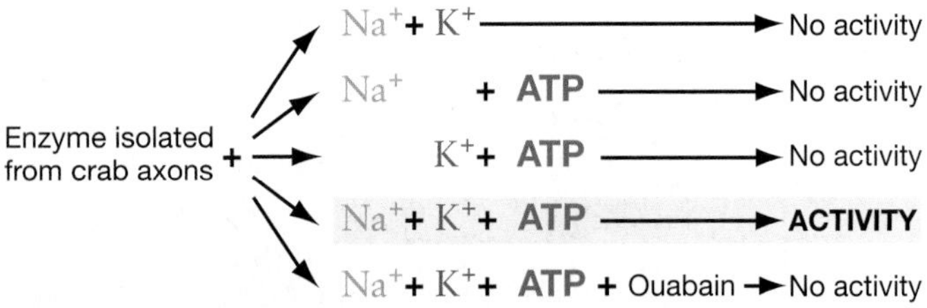

FIGURE 45.10 Tests Led to the Discovery of Na^+/K^+-ATPase
Reactions that test whether an enzyme isolated from crab neurons was a sodium-potassium pump.

With the discovery and characterization of Na^+/K^+-ATPase, researchers added an important piece to the puzzle of how action potentials are generated. The sodium-potassium pump restores normal concentrations of Na^+ and K^+ after action potentials have brought sodium ions into the cell and sent potassium ions out. Therefore, the pump is responsible for maintaining the concentration gradients that create the resting potential.

The experiments reviewed thus far haven't considered a fundamental question about electrical signaling, however: How do action potentials move down the axon?

How Is the Action Potential Propagated?

To explain how action potentials propagate down an axon, Hodgkin and Huxley hypothesized that the influx of Na^+ causes charge to spread away from sodium channels. As **Figure 45.11a** shows, positive charges inside the cell are repulsed by the influx of Na^+, and negative charges are attracted to the Na^+. As these positive charges are pushed farther from the initial sodium channel, they create a local current that causes new sections of the membrane, close to the initial site of an action potential, to become depolarized. Voltage-gated Na^+ channels that are nearby pop open in response, positive feedback occurs, and a full-fledged action potential results. In this way, an action potential is continuously regenerated as it moves down the axon (**Figure 45.11b**). The signal does not diminish as it moves, because the response is all or none.

Why don't action potentials propagate back up the axon? The answer is that Na^+ channels are **refractory**—that is, once they have opened and closed, they are less likely to open again for a short period of time. Action potentials are propagated in one direction only, because sodium channels "downstream" of the site are not in the refractory state. The undershoot phase, in which the membrane is briefly hyperpolarized relative to the resting potential, also keeps the charge that spreads "upstream" from triggering an action potential in that direction.

Understanding how the action potential propagates helped researchers explain why the squid's axons are so large. Because charge spreads farther in a cell with a large diameter, the squid's giant axon and other large neurons transmit action potentials much faster than small axons can.

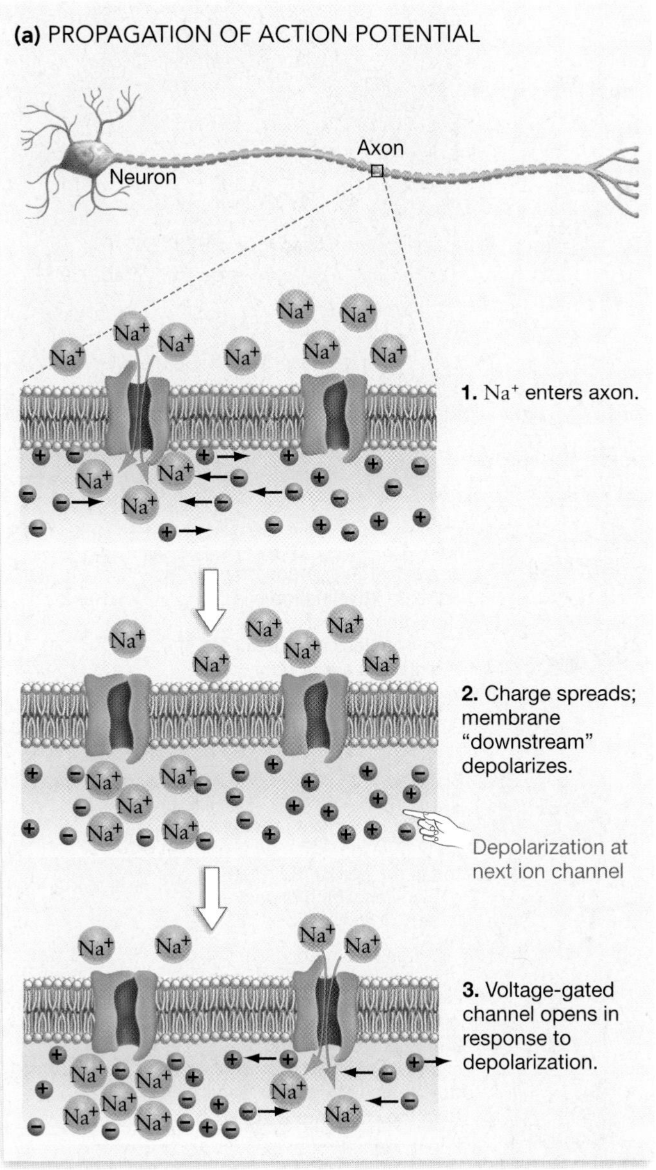
(a) PROPAGATION OF ACTION POTENTIAL
Axon
Neuron
1. Na⁺ enters axon.
2. Charge spreads; membrane "downstream" depolarizes.
Depolarization at next ion channel
3. Voltage-gated channel opens in response to depolarization.

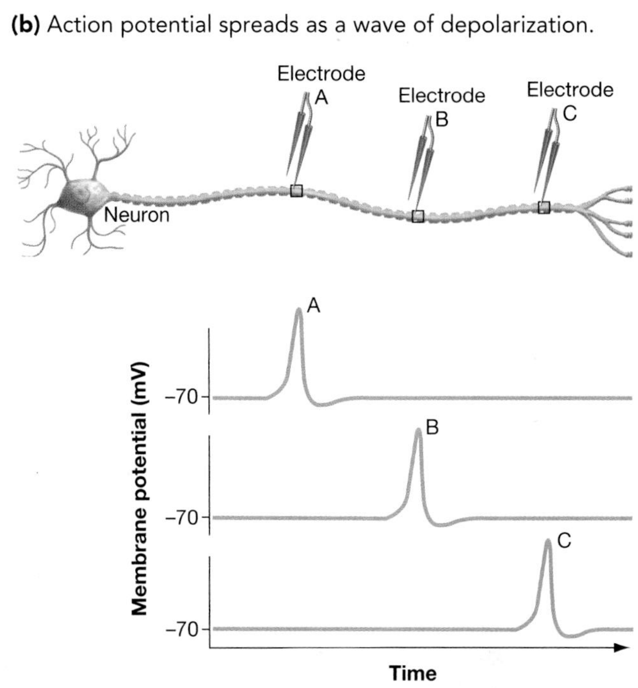
(b) Action potential spreads as a wave of depolarization.
Electrode A
Electrode B
Electrode C
Neuron
Membrane potential (mV)
-70
A
B
C
Time

FIGURE 45.11 Action Potentials Propagate because Charge Spreads down the Membrane
(a) An action potential starts with an inflow of Na^+. The influx of positive charge attracts negative charges inside the cell and repels positive charges. As a result, positive charge spreads away from the channel where the Na^+ enters, and depolarizes nearby regions of the neuron. Voltage-gated Na^+ channels open in response. **(b)** The action potential spreads down the axon as a wave of depolarization, but there is no loss of signal because the all-or-none action potential regenerates itself as it travels.

Analyzing charge spread also helped biologists explain the phenomenon called myelination. Vertebrates do not have giant axons. Instead, the membranes of specialized accessory cells, called **Schwann cells** (Figure 45.12a, page 1038) wrap around the axons of neurons outside the brain and spinal cord—neurons that are part of the peripheral nervous system (PNS), described later in this chapter. Schwann cells are not neurons but instead are one of several types of accessory cells in the nervous system that support neurons. Collectively, these accessory cells are called **glia.** When Schwann cells wrap around an axon, they form a **myelin sheath,** which acts as electrical insulation. As charge spreads down an axon, the presence of the myelin sheath prevents charge in the form of ions from leaking out across the plasma membrane of the neuron. Consequently, the influx of charge that results from an action potential is able to spread unimpeded until it hits an unmyelinated section of the axon, called a **node of Ranvier** (Figure 45.12b). The node has a dense concentration of voltage-gated Na^+ channels and supports action potentials. In this way, electrical signals jump down the axon much faster than they can move down an unmyelinated cell.

To drive this point home, consider what happens when myelination is decreased. If myelin degenerates, the transmission of electrical signals slows considerably. The disease **multiple sclerosis (MS)** develops as damage to myelin increases and electrical signaling is impaired, causing the muscles to weaken and coordination to lessen. Although the symptoms of MS are highly variable, in severe cases the disease progresses and can be crippling.

What happens once an action potential has traveled the length of the axon? Cajal proposed, and electron microscopy confirmed, that in most neurons the membrane at the end of the axon approaches the membrane of another neuron's dendrite. The surfaces of the two membranes are separated by a tiny gap. What happens when an action potential arrives at this interface between cells?

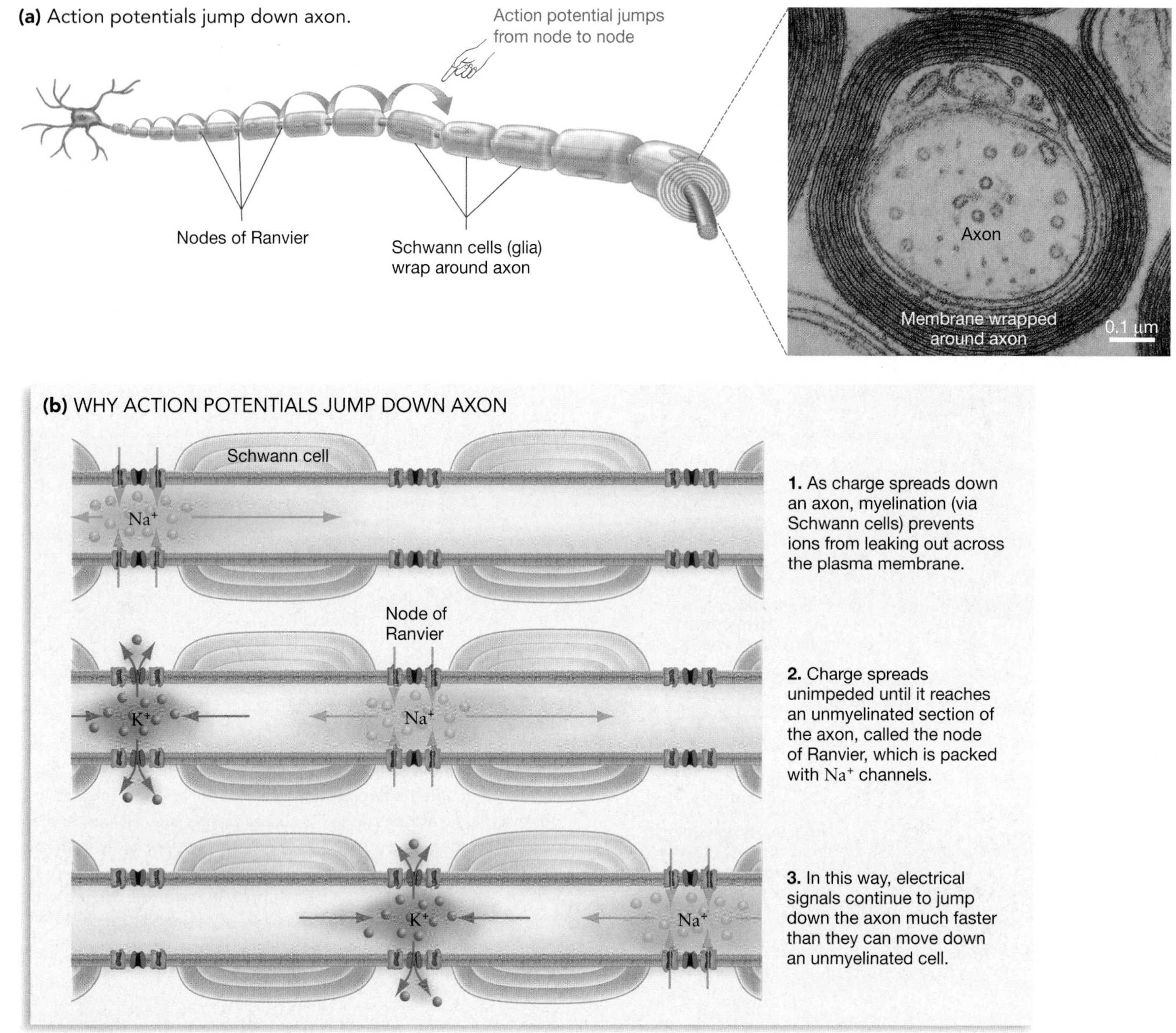
(a) Action potentials jump down axon.
Action potential jumps from node to node
Nodes of Ranvier
Schwann cells (glia) wrap around axon
Axon
Membrane wrapped around axon
0.1 μm
(b) WHY ACTION POTENTIALS JUMP DOWN AXON
Schwann cell
Na+
1. As charge spreads down an axon, myelination (via Schwann cells) prevents ions from leaking out across the plasma membrane.
Node of Ranvier
K+
Na+
2. Charge spreads unimpeded until it reaches an unmyelinated section of the axon, called the node of Ranvier, which is packed with Na+ channels.
K+
Na+
3. In this way, electrical signals continue to jump down the axon much faster than they can move down an unmyelinated cell.

FIGURE 45.12 Action Potentials Propagate Quickly in Myelinated Axons
(a) Schwann cells wrap their membranes around PNS axons. **(b)** In myelinated axons, no charge leaks across the membrane as the action potential spreads down the axon. Consequently, action potentials jump down the axon rapidly.

✓ CHECK YOUR UNDERSTANDING

During an action potential, membrane voltage undergoes rapid changes due to an influx of sodium ions, followed by an outflow of potassium ions. You should be able to (1) make labeled diagrams that explain the changes in membrane voltage that occur during the action potential and why an action potential spreads down an axon; (2) explain how voltage gating works, why the action potential is an all-or-none phenomenon, why the refractory nature of sodium channels is important, and why myelination increases the speed of electrical signals; and (3) predict what would happen if sodium channels continued to open once membrane depolarization was complete.

45.3 The Synapse

The cytoplasm of most neurons is not directly connected to the cytoplasm of other neurons. Based on this observation, there must be some indirect mechanism that transmits electrical signals from cell to cell, across their plasma membranes. In the 1920s Otto Loewi showed that the indirect mechanism involves molecules called neurotransmitters. **Neurotransmitters** are molecules that transmit information from one neuron to another neuron or from a neuron to a target cell in a muscle or gland.

Loewi was interested in how the nervous system affects the heart rate, and he knew that signals from the vagus nerve slow the heart. He performed the experiment diagrammed in **Figure 45.13** to test his hypothesis that the signal from nerve

Question: How is information transferred from one neuron to another?

Hypothesis: Molecules called neurotransmitters carry information from one neuron to the next.

Null hypothesis: Molecules do not carry information from one neuron to the next.

Experimental setup:

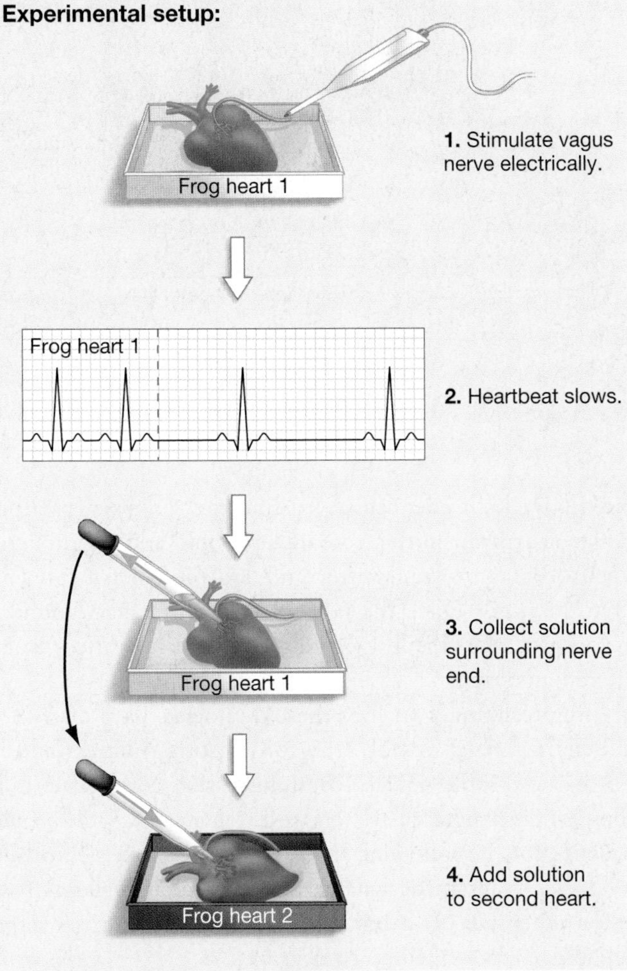
Frog heart 1
1. Stimulate vagus nerve electrically.
Frog heart 1
2. Heartbeat slows.
Frog heart 1
3. Collect solution surrounding nerve end.
Frog heart 2
4. Add solution to second heart.

Prediction: The heartbeat will slow.

Prediction of null hypothesis: There will be no change in heartbeat.

Results:

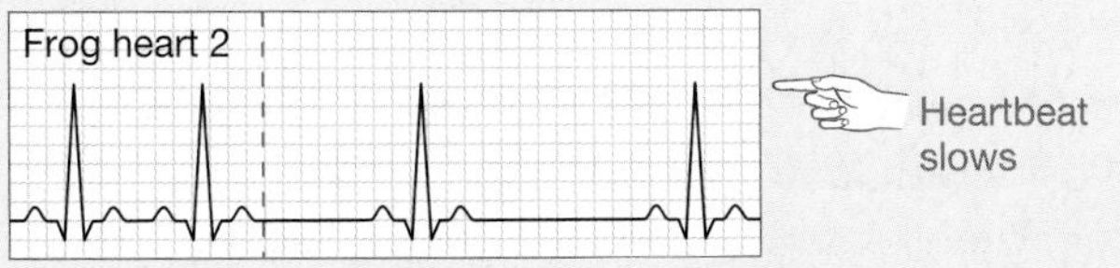
Frog heart 2
Heartbeat slows

Conclusion: The vagus nerve releases molecules that slow heartbeat. Neurotransmitters carry information.

FIGURE 45.13 Experimental Evidence for the Existence of Neurotransmitters

QUESTION Prior to this experiment, some researchers contended that most neurons communicated with each other electrically, not chemically. Why does this experiment fail to support the hypothesis of electrical transmission? (Note: It turns out that a few neurons do make electrical connections with one another.)

to muscle is delivered by a chemical. First, he isolated the vagus nerve and heart of a frog. As predicted, the heart rate slowed when he stimulated the nerve electrically. But when he took the solution that bathed the first heart and applied it to another, isolated heart, the second heart rate slowed as well. This result provided strong evidence for the chemical transmission of electrical signals. The vagus nerve had released a neurotransmitter into the bath.

How are neurotransmitters delivered? When transmission electron microscopy became available in the 1950s, biologists finally understood the physical nature of the interface between neurons. This interface is called a **synapse**. As **Figure 45.14** shows, the membranes of axons and dendrites juxtapose closely. Just inside the synapse, the axon contains numerous sac-like structures called **synaptic vesicles**. These structures were hypothesized to be storage sites for neurotransmitters.

Anatomical observations such as these, combined with chemical studies of the synapse, led to the model of synaptic transmission illustrated in **Figure 45.15** (page 1040). Notice that the "sending" cell is called the **presynaptic neuron** and the "receiving" cell is called the **postsynaptic neuron**. The sequence of events begins when the depolarization created by an action potential opens voltage-gated calcium channels located near the synapse, in the presynaptic membrane. The electrochemical gradient for Ca^{2+} results in the inflow of calcium ions through the open channels. In response to the increase in calcium concentration inside the axon, synaptic vesicles fuse with the membrane and release neurotransmitters into the gap between the cells. This gap is called the **synaptic cleft**. The delivery of neurotransmitters into the cleft is an example of *exocytosis*, a process introduced in Chapter 8.

WEB TUTORIAL 45.2
Action Potentials

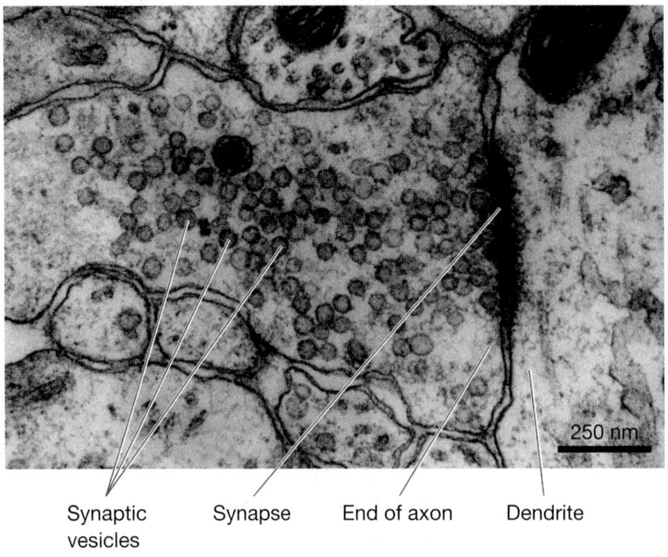
250 nm
Synaptic vesicles
Synapse
End of axon
Dendrite

FIGURE 45.14 Cross Section of Synapse

Transmission electron micrograph of the site where an axon meets a dendrite.

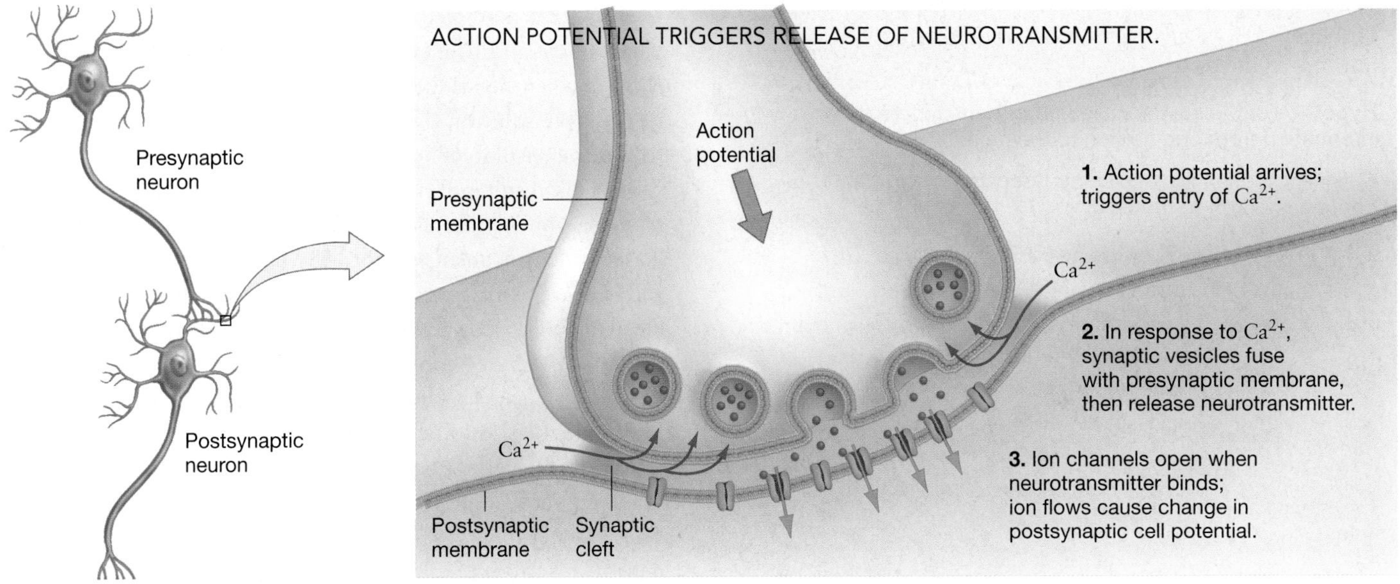

FIGURE 45.15 Neurons Meet and Transfer Information at Synapses
The sequence of events that occurs when an action potential arrives at a synapse.

What happens next? Biologists hypothesized that neurotransmitters bind to receptors in the membrane of the next cell in the circuit. As step 3 in Figure 45.15 shows, the idea was that interactions between neurotransmitters and receptors on the postsynaptic cell would cause a change in the membrane potential of the postsynaptic cell and possibly trigger the start of an action potential there. Is this part of the model correct?

What Do Neurotransmitters Do?

Once Loewi had established that neurotransmitters exist, investigators were able to isolate and purify them from a variety of neurons. They accomplished this by stimulating a neuron, collecting the neurotransmitter that was released, and analyzing it chemically. To find the receptor for a particular neurotransmitter, researchers could add a radioactive label to the neurotransmitter and add the labeled transmitter to neurons. Once the labeled transmitter had bound to its receptor, the receptor protein could be identified, isolated, and analyzed.

Using techniques such as these, biologists have discovered and characterized a wide array of neurotransmitters and receptors (see **Table 45.2**). To establish that a molecule functions as a neurotransmitter, researchers have to provide evidence for the following three criteria: (1) The neurotransmitter is present at the synapse and released in response to an action potential, (2) it binds to a receptor on a postsynaptic cell, and (3) it is taken up or degraded.

TABLE 45.2 Categories of Neurotransmitters

Neurotransmitter	Site of Action	Action
Acetylcholine	Neuromuscular junction, some CNS pathways	Excitatory (inhibitory in some parasympathetic neurons)
Monoamines		
Norepinephrine	Sympathetic neurons, some CNS pathways	Excitatory or inhibitory
Dopamine	Many CNS pathways	Primarily excitatory
Serotonin	Many CNS pathways	Inhibitory or modulatory
Amino Acids		
Glutamate	Many CNS pathways	Excitatory
Gamma-aminobutyric acid (GABA)	Some CNS pathways	Inhibitory
Peptides		
Endorphins, enkephalins, substance P	Used in sensory pathways (pain)	Inhibitory

By patch-clamping receptors, biologists confirmed that many neurotransmitters function as ligands. A **ligand** is a molecule that binds to a specific site on a receptor molecule. Many neurotransmitters are ligands that bind to receptors called **ligand-gated ion channels.** These are channel proteins that open in response to binding by a specific ligand. When a neurotransmitter binds to a ligand-gated ion channel in the postsynaptic membrane, the channel opens and admits a flow of ions. In this way, the neurotransmitter's chemical signal changes form. It is transduced to an electrical signal—a change in the membrane potential of the postsynaptic cell.

Not all the receptors to which neurotransmitters bind are ion channels, however. Instead, some receptors activate enzymes that lead to the production of a second messenger molecule in the postsynaptic cell. Recall from Chapter 8 that **second messengers** are chemical signals produced inside a cell in response to a chemical signal that arrives at the cell surface. In the case of some neurotransmitters, the arrival of a chemical signal triggers the production of another chemical signal. The second messengers induced by neurotransmitters may trigger changes in enzyme activity, gene transcription, or membrane potential. The role of second messengers in cells is explored in detail in Chapter 47.

Postsynaptic Potentials and Summation

What happens when a neurotransmitter binds to a receptor and opens an ion channel? As the scanning electron micrograph in **Figure 45.16a** indicates, numerous synapses can be found in the same region of a dendrite or cell body. And as **Figure 45.16b** illustrates, these synapses can be one of two general types. If the receptors at the synapse admit sodium ions in response to the arrival of the neurotransmitter, then the postsynaptic membrane depolarizes in response to an action potential in the presynaptic cell. In most cases, depolarization makes an action potential in the postsynaptic cell more likely. Changes in the postsynaptic cell that make action potentials more likely are called **excitatory postsynaptic potentials (EPSPs).** Other receptors, in contrast, lead to an outflow of potassium ions or an inflow of chlorine ions in the postsynaptic cell. These events hyperpolarize the membrane and make action potentials less likely to occur in the postsynaptic cell. Changes in the postsynaptic cell that make action potentials less likely are called **inhibitory postsynaptic potentials (IPSPs).**

It is critical to realize that, unlike action potentials, EPSPs and IPSPs are not all-or-none events. Instead, they are graded in size and short lived. The size of an EPSP or IPSP depends on the amount of neurotransmitter that is released at the synapse. Both types of signal are short lived because neurotransmitters do not bind irreversibly to channels in the postsynaptic cell. Instead, they are quickly inactivated or taken up by the presynaptic cell and recycled. If either the amount or life span of neurotransmitters is altered, the normal functioning of neurons is altered. The street drugs cocaine and amphetamine, for example, exert their effects by inhibiting the uptake and recycling of particular neurotransmitters.

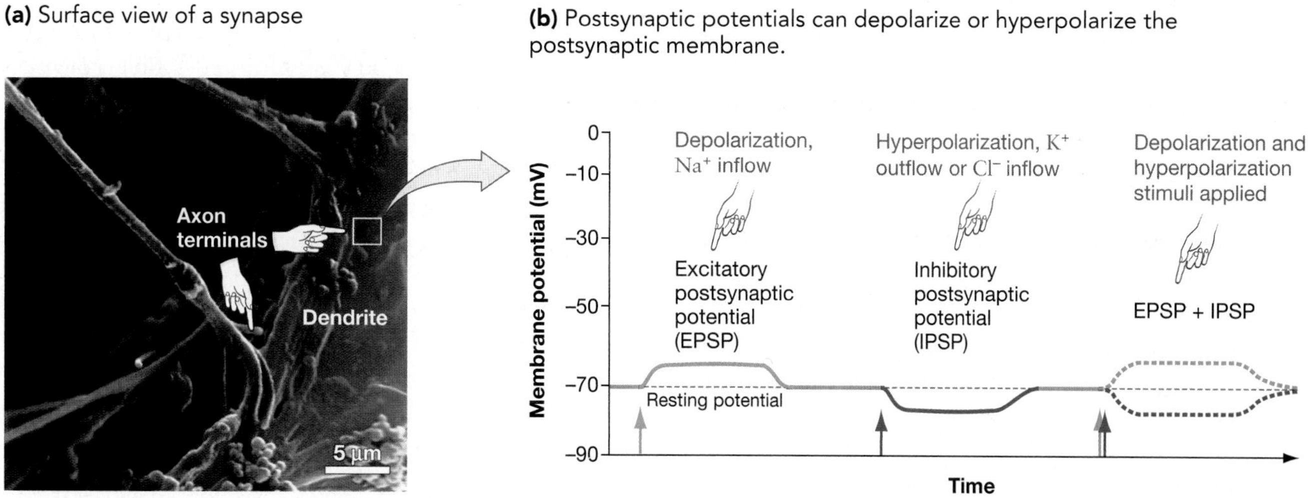

FIGURE 45.16 There Are Two Types of Synapses, Distinguished by whether They Lead to Depolarization or Hyperpolarization of the Membrane

(a) Scanning electron micrograph showing axons from one neuron synapsing with the cell body of a second neuron. **(b)** Records showing that the arrival of an action potential causes a short-lived depolarization or hyperpolarization of the postsynaptic plasma membrane. The change in postsynaptic potential can be excitatory or inhibitory, depending on whether it makes action potentials more or less likely. If both excitatory and inhibitory signals arrive at the same time, they cancel one another and the cell remains at resting level.

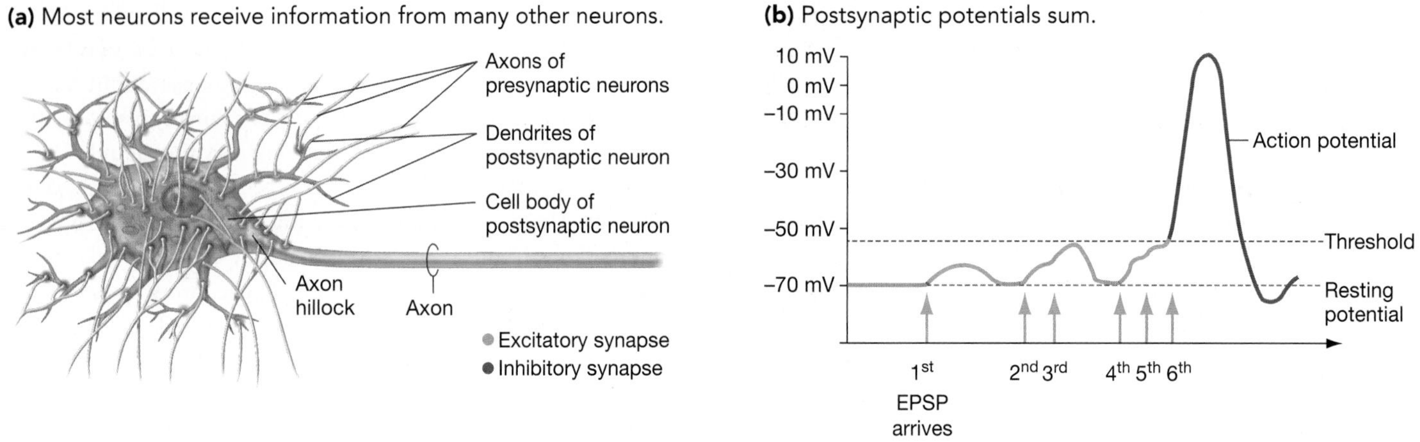

(a) Most neurons receive information from many other neurons.
Axons of presynaptic neurons
Dendrites of postsynaptic neuron
Cell body of postsynaptic neuron
Axon hillock
Axon
Excitatory synapse
Inhibitory synapse
(b) Postsynaptic potentials sum.
10 mV
0 mV
−10 mV
−30 mV
−50 mV
−70 mV
Action potential
Threshold
Resting potential
1st EPSP arrives
2nd 3rd
4th 5th 6th

FIGURE 45.17 Neurons Integrate Information from Many Synapses
(a) The dendrites and cell body of a neuron typically receive signals from hundreds or thousands of other neurons. **(b)** When action potentials arrive close together in time from the same axon or from different axons with synapses close to one another, the postsynaptic potentials sum. In this example, the first excitatory signal is insufficient to generate an action potential. Two excitatory signals arriving close together cause summation but do not reach the threshold for generating an action potential. Three excitatory signals arriving closely spaced sum to exceed the threshold. If excitatory postsynaptic potentials depolarize the axon hillock past threshold, enough Na^+ channels open to trigger an action potential.

What effect do EPSPs and IPSPs have on the postsynaptic cell? As **Figure 45.17a** shows, the dendrites and cell body of a neuron typically make hundreds or thousands of synapses with other cells. At any instant, the EPSPs and IPSPs that occur at each of these synapses lead to short-lived surges of charge in the dendrites and cell body of the postsynaptic cell. If an IPSP and EPSP occur close together in space or time, the changes in membrane potential tend to cancel each other out (see Figure 45.16b). But if several EPSPs occur close together in space or time, they sum and make the neuron likely to fire an action potential (**Figure 45.17b**). The additive nature of postsynaptic potentials is termed **summation**.

The sodium channels that trigger action potentials in the postsynaptic cell are located near the start of the axon at a site called the **axon hillock** (see Figure 45.17a). As IPSPs and EPSPs interact throughout the dendrites and cell body, charge spreads to the axon hillock. If the membrane at the axon hillock depolarizes past a point termed the **threshold**, enough sodium channels open to trigger positive feedback and an action potential. Once an action potential starts at the axon hillock, it propagates down the axon to the next synapse.

The summation that occurs prior to the axon hillock is a crucial phenomenon, because it determines whether an action potential begins in the postsynaptic cell. It also explains why the chemical transmission of electrical signals and synapses exists in the first place. If the plasma membranes and cytoplasm of adjacent neurons were continuous, then action potentials would simply propagate from one cell to another without changing. But because neurons receive input from many synapses and because IPSPs and EPSPs sum, information in the form of electrical signals is modified at the synapse before being passed along. In addition, it turns out that synapses are dynamic—their behavior can change through time in response to events. This "plasticity" in the structure and function of synapses underlies many types of processes, including learning and memory.

The general message is that the postsynaptic cell integrates information from hundreds or thousands of other neurons. The information arrives in the form of action potentials that produce EPSPs and IPSPs. If the combination of postsynaptic potentials depolarizes the membrane at the axon hillock sufficiently, the postsynaptic cell fires an action potential in response. But if the events lead to a depolarization that is below threshold, the rate of firing decreases.

What Happens when Ligand-Gated Channels Are Defective?

Recent research on synaptic transmission has concentrated on whether certain neurotransmitters and certain receptors are involved in drug addiction or mental illness. As an example of these studies, let's focus on the hypothesis that defects in a particular ligand-gated channel contribute to the development of schizophrenia.

Schizophrenia is a mental illness that is highly variable in nature. In most cases, the disease includes what physicians call positive and negative symptoms. A *positive symptom* is something that a patient does; a *negative symptom* is something that a patient cannot do. The positive symptoms of schizophrenia are often episodic and may include delusions, hallucinations, and paranoia. The negative symptoms tend to be chronic and may include social withdrawal, cognitive deficits, impaired attention, and a restricted range of emotions or a lack of emotional response. In men, the first symptoms usually appear between the ages of 18 and 25 years; in women, disease onset usually occurs between 26 and 45 years of age.

In the 1950s researchers noticed that people who were intoxicated with the street drug phencyclidine (PCP, also called angel dust) exhibited some of the symptoms of schizophrenia. Based on this observation, investigators began using PCP in lab animals to induce the symptoms of schizophrenia and document the neurological changes that occurred in response. Several of these studies indicated that PCP induced significant changes in the number or function of ligand-gated channels called *N*-methyl-D-aspartic acid (NMDA)-sensitive glutamate receptors. Glutamate is an amino acid that functions as an excitatory neurotransmitter in certain neurons. This NMDA receptor is one of three types of ligand-gated channels that respond to glutamate.

Are defects in the NMDA receptor at least partially responsible for the onset of schizophrenia? Data from several recent studies argue that the answer is yes. For example, biologists have succeeded in generating mice that produce just 5 percent of the normal amount of a key subunit of the NMDA receptor. (The technical details involved in creating these mutant mice are not important here.) These mice should exhibit some of the negative symptoms of schizophrenia, because, with fewer NMDA receptors, neurons in their brains should have a diminished ability to generate EPSPs and thus generate fewer action potentials. Researchers tested the behavior of the mutant mice and found that their actions were similar to those of mice treated with PCP. Compared with normal mice, the mutant mice perform a large number of repetitive, stereotyped movements and are socially withdrawn (**Figure 45.18a**). These mice appear to show some of the negative symptoms associated with schizophrenia.

In related work, a different team of biologists observed differences between the NMDA receptors present in brain slices preserved from the bodies of normal humans and those of schizophrenic humans. Even though the number of NMDA receptors per neuron did not differ in the two groups, brain cells from schizophrenics had abnormal levels of mRNA for two of the subunits that make up the receptor. The same subunit that had been experimentally knocked out in mice was present in lower than normal concentrations, while a second subunit was present in higher than normal concentrations. To make sense of these data, the researchers suggested that normal amounts of NMDA receptor are present but that many of the receptors do not work properly because they lack the appropriate concentration of a key subunit (**Figure 45.18b**). These data suggest that the disease may involve a problem in synthesizing the correct retio of NMDA-receptor subunits.

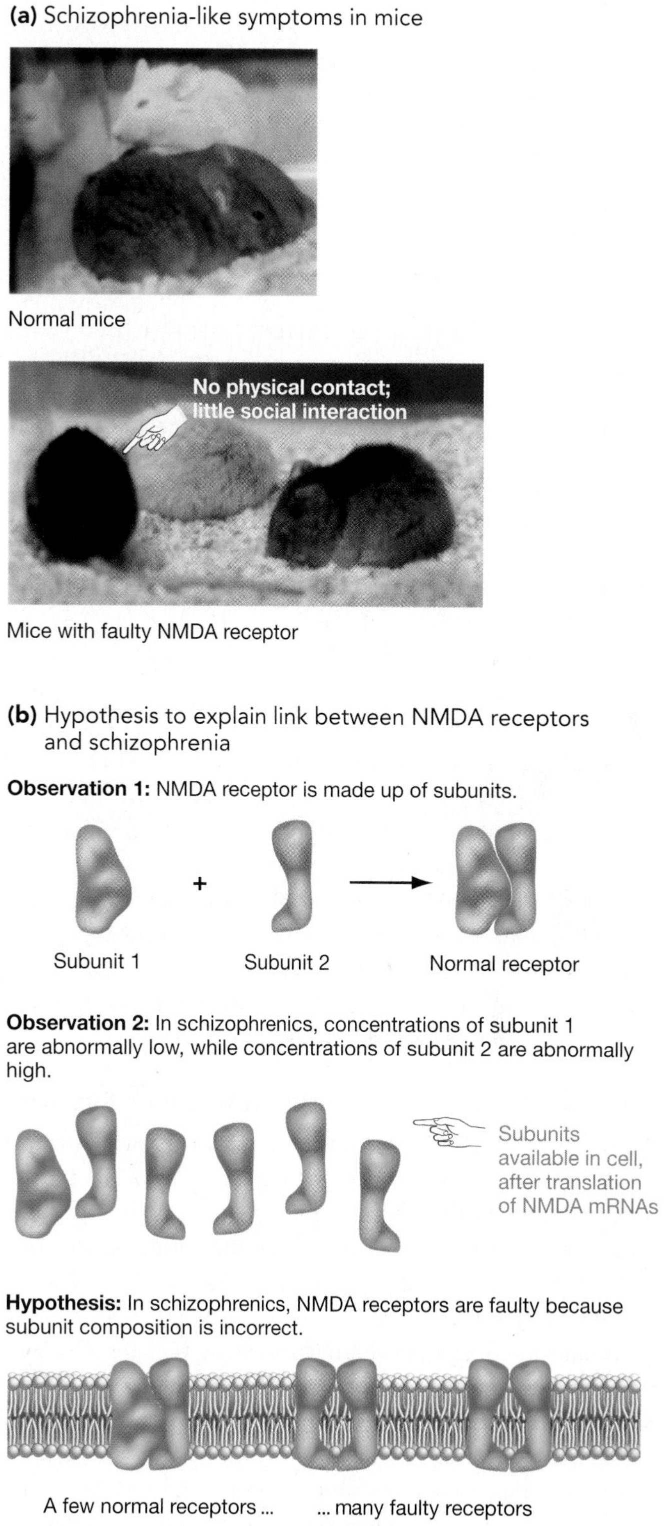
(a) Schizophrenia-like symptoms in mice
Normal mice
No physical contact; little social interaction
Mice with faulty NMDA receptor
(b) Hypothesis to explain link between NMDA receptors and schizophrenia
Observation 1: NMDA receptor is made up of subunits.
Subunit 1
Subunit 2
Normal receptor
Observation 2: In schizophrenics, concentrations of subunit 1 are abnormally low, while concentrations of subunit 2 are abnormally high.
Subunits available in cell, after translation of NMDA mRNAs
Hypothesis: In schizophrenics, NMDA receptors are faulty because subunit composition is incorrect.
A few normal receptors ...
... many faulty receptors

FIGURE 45.18 Is There a Link between the NMDA Receptor and Schizophrenia?
(a) Mice that have a faulty NMDA subunit show symptoms reminiscent of schizophrenia. These photos show typical behavior of normal and experimental mice. **(b)** Researchers who compared brain tissue of normal people with that of schizophrenics observed that mRNAs for subunit 1 of the NMDA receptor were rare while mRNAs for subunit 2 were abnormally common.

Although these data are suggestive, they are far from conclusive, because they do not establish whether faulty NMDA receptors are the cause of schizophrenia or an effect of disease. Although the mouse data argue for a causal link, it remains to be shown decisively that schizophrenia is associated with defects in a subunit of the NMDA receptor in humans. Research on the molecular basis of schizophrenia is continuing in labs around the world.

✓CHECK YOUR UNDERSTANDING

At a synapse, electrical information in the form of changes in membrane voltage is transduced to chemical information in the form of the release of neurotransmitter. The arrival of a neurotransmitter causes a change in the membrane potential of the postsynaptic cell. You should be able to explain why the presence of synapses is important to a neuron's ability to integrate and process information from many different neurons.

45.4 The Vertebrate Nervous System

The first three sections of this chapter examined electrical signaling at the level of molecules, membranes, and individual cells. The goal of this section is to discuss electrical signaling at the level of tissues, organs, and systems.

To begin, let's consider the overall anatomy of the vertebrate nervous system. Then we can ask how researchers explore the function of the most complex organ known: the human brain. The chapter concludes by returning to the molecular level and introducing recent work on learning and memory.

A Closer Look at the Peripheral Nervous System

Biologists routinely distinguish the central nervous system, or CNS, from the peripheral nervous system, or PNS. Recall from Section 45.1 that the CNS is made up of the brain and spinal cord and is concerned primarily with integrating information, and that the PNS is made up of neurons outside the CNS.

What specific functions do the cells of the PNS control? Anatomical and functional studies indicate that the PNS consists of two distinct systems: (1) an **afferent division**, which transmits sensory information to the CNS, and (2) an **efferent division**, which carries commands from the CNS to the body. Neurons in the afferent division monitor conditions inside and outside the body. Once information from afferent neurons has been processed in the CNS, neurons in the efferent division carry signals that allow the body to respond to changed conditions in an appropriate way. The afferent and efferent divisions carry out sensory and motor functions, respectively.

As **Figure 45.19** shows, the afferent and efferent divisions are part of a larger hierarchy of PNS functions. The efferent division is further divided into a **somatic nervous system**, which controls the skeletal muscles, and an **autonomic nervous system**, which controls internal processes such as digestion and heart rate. In effect, the somatic system responds to external stimuli and results in movement, while the autonomic system responds to internal stimuli and controls the activity of internal organs and glands. Many organs and glands are served by two functionally distinct types of autonomic nerves. Typically, a signal from a **parasympathetic nerve** promotes relaxation and digestion. For example, the parasympathetic nerves that innervate the heart slow it down, while those that serve the digestive tract stimulate its activity. **Sympathetic nerves**, in contrast, typically prepare organs for stressful situations. Sympathetic nerves speed up the heart rate, stimulate the release of glucose from the liver, and inhibit action by digestive organs. **Figure 45.20** summarizes the effects of signals from parasympathetic and sympathetic nerves.

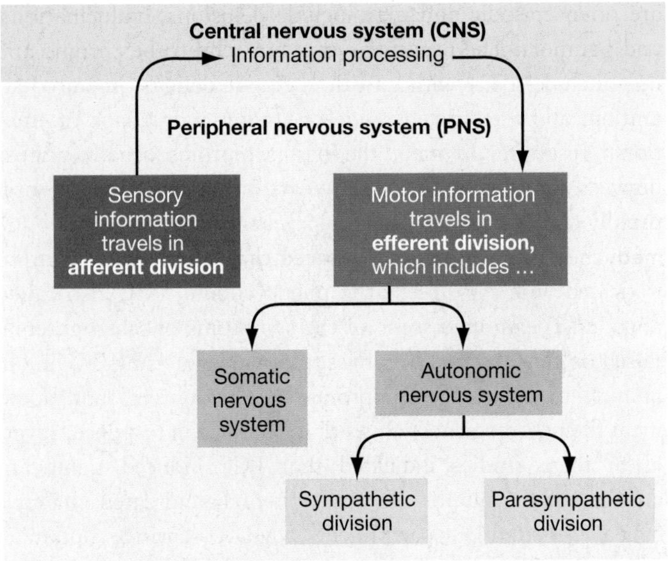

FIGURE 45.19 The Functions of the PNS Form a Hierarchy

Functional Anatomy of the CNS

Parasympathetic nerves originate at the base of the brain or the base of the spinal cord. Most sympathetic nerves also originate in the spinal cord, but they emerge along the middle of its length. Similarly, most sensory and motor neurons in the somatic nervous system project to or from the spinal cord. In effect, then, the spinal cord serves as an information conduit. It collects and transmits information from throughout the body. With a few exceptions, such as the spinal reflexes illustrated in Figure 45.1, virtually all the information that travels to or from the spinal cord is sent to the brain for processing.

The brain is far and away the most complex organ found in animals. Researchers estimate that the human brain has 100 billion neurons, each making thousands of synaptic connections with other neurons. How do biologists even begin to study how such a fantastically complex structure functions? The answer is that researchers begin with general anatomy.

Nineteenth-century anatomists established that the brain is made up of the four structures: the cerebrum, the cerebellum, the pons, and the medulla, labeled in **Figure 45.21a** (page 1046). The largest of these areas, the **cerebrum**, makes up the bulk of the brain. The cerebrum is divided into left and right hemispheres and is involved in conscious thought and memory. The **cerebellum** is a paired structure that coordinates complex motor patterns. The **pons** is a small region that relays sensory information to the cerebellum. The **medulla**, which connects the brain to the spinal cord, is the autonomic center for regulating the heart, lungs, and digestive system.

Each cerebral hemisphere has four major areas, or lobes: the **frontal lobe**, the **parietal lobe**, the **occipital lobe**, and the **temporal lobe** (**Figure 45.21b**). The two hemispheres are connected by a thick band of neurons called the **corpus callosum.** What tools do researchers use to explore the function of each area within the cerebrum?

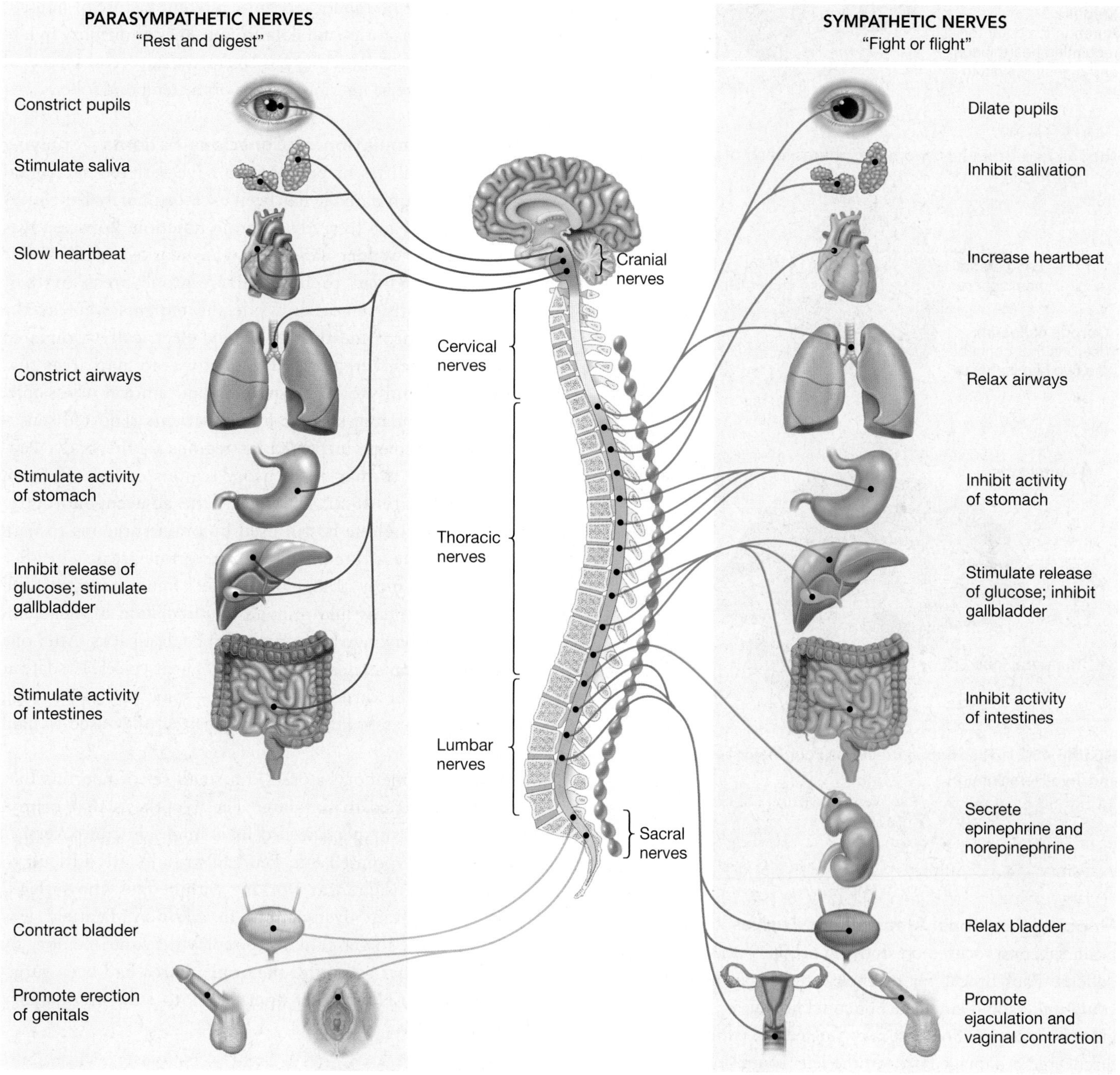

FIGURE 45.20 The Autonomic Nervous System Controls Internal Processes

QUESTION Many parasympathetic nerves promote digestion and lack of movement; many sympathetic nerves support energy utilization and muscle activity. How do each of the responses listed here support these functions?

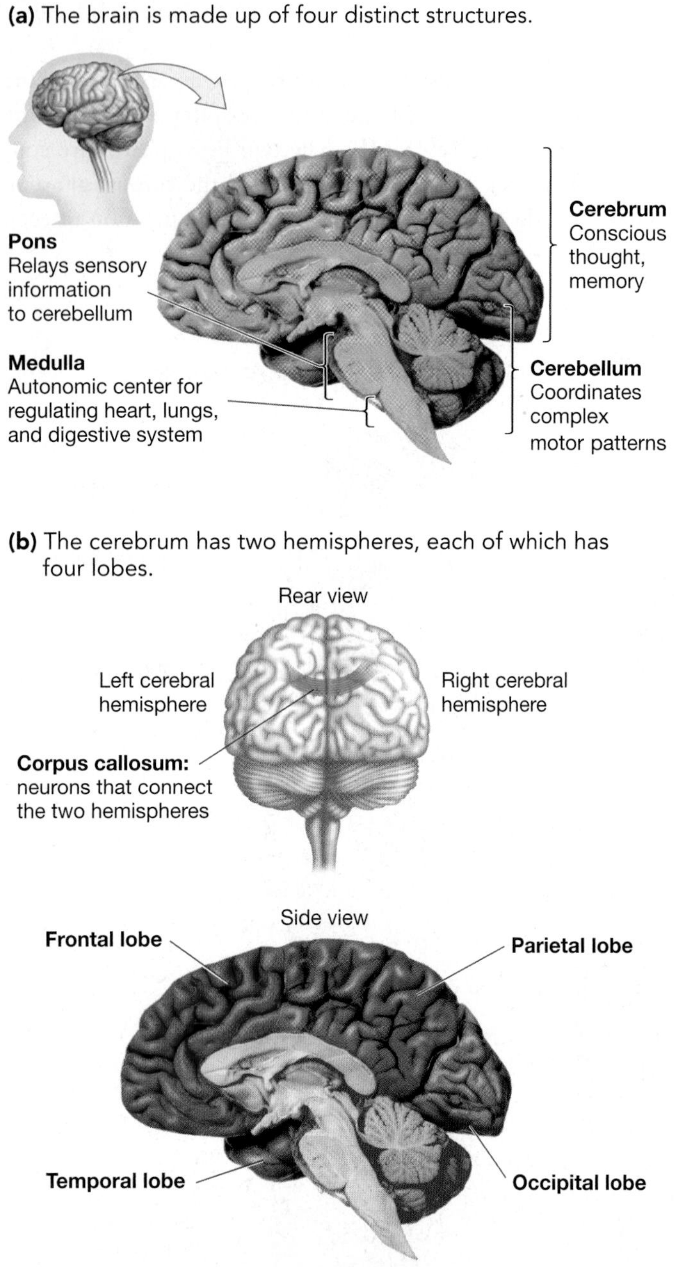
(a) The brain is made up of four distinct structures.
Cerebrum
Conscious thought, memory
Pons
Relays sensory information to cerebellum
Medulla
Autonomic center for regulating heart, lungs, and digestive system
Cerebellum
Coordinates complex motor patterns
(b) The cerebrum has two hemispheres, each of which has four lobes.
Rear view
Left cerebral hemisphere
Right cerebral hemisphere
Corpus callosum: neurons that connect the two hemispheres
Side view
Frontal lobe
Parietal lobe
Temporal lobe
Occipital lobe

FIGURE 45.21 The Human Brain Has Four Major Lobes and Two Hemispheres

EXERCISE On your own skull, point to each of the labeled areas.

Mapping Functional Areas: Lesion Studies Early work on brain function focused on studying people with specific mental deficits. Paul Broca, for example, studied an individual who could understand language but could not speak. After the individual's death in 1861, Broca examined the patient's brain and discovered a damaged area in the left frontal lobe of the cerebrum. Based on this observation, Broca hypothesized that this region is responsible for speech. More generally, he formulated the hypothesis that specific regions of the brain are specialized for coordinating particular functions.

Broca's claim that functions are localized to specific brain areas has been verified through extensive efforts to map the cerebrum. In some cases, advances were made by studying people who had to have portions of their brains removed. In 1953, for example, surgeons treated a 27-year-old man for life-threatening seizures by removing a small portion of his temporal lobe. The man recovered (he is still alive, in fact), has normal intelligence, and vividly remembers his childhood, but he has no short-term memory. Brenda Milner, who has studied this individual for over 40 years, must introduce herself to him every time they meet; he cannot recognize a recent picture of himself. Based on case histories and data from studies of memory in laboratory animals, a consensus has emerged that several aspects of memory localize to interior sections of the temporal lobe.

Electrical Stimulation of Conscious Patients Studying the mental abilities of people who have suffered accidental or surgical brain damage has been extraordinarily fruitful. A different approach to studying brain function, however, was pioneered by Wilder Penfield, who worked with severe epileptics scheduled to have seizure-prone areas of their brains surgically removed. While the patients were awake and under a local anesthetic, Penfield electrically stimulated portions of their cerebrums. His goal was to map areas that are essential and should be spared from removal if possible. When he stimulated specific areas, patients reported sensations or movement in particular regions of the body. Penfield was able to map the sensory regions of the cerebrum shown in **Figure 45.22**, as well as the adjacent motor regions. The technique is still used by brain surgeons to map critical areas near tumors and seizure-prone areas.

Perhaps the most striking of Penfield's findings was that, on occasion, patients would respond to stimulation of their temporal lobe by having what appeared to be flashbacks. After one region was stimulated, a woman said, "I hear voices. It is late at night around the carnival somewhere—some sort of traveling circus. . . . I just saw lots of big wagons that they used to haul animals in."

Was this a memory, stored in a small set of neurons that Penfield happened to simulate? The hypothesis that memories are stored in specialized cells is intensely controversial. As critics have pointed out, Penfield's results are difficult to interpret, because he was working with people who suffered from severe brain dysfunction. In addition, Penfield was sometimes able to get patients to replay the same memory by stimulating other cells after the original area had been surgically removed. Have other approaches to studying memory been productive?

How Does Memory Work?

Learning is an enduring, usually adaptive change in behavior that results from a specific experience in an individual's life. **Memory** is the retention of learned information. Learning

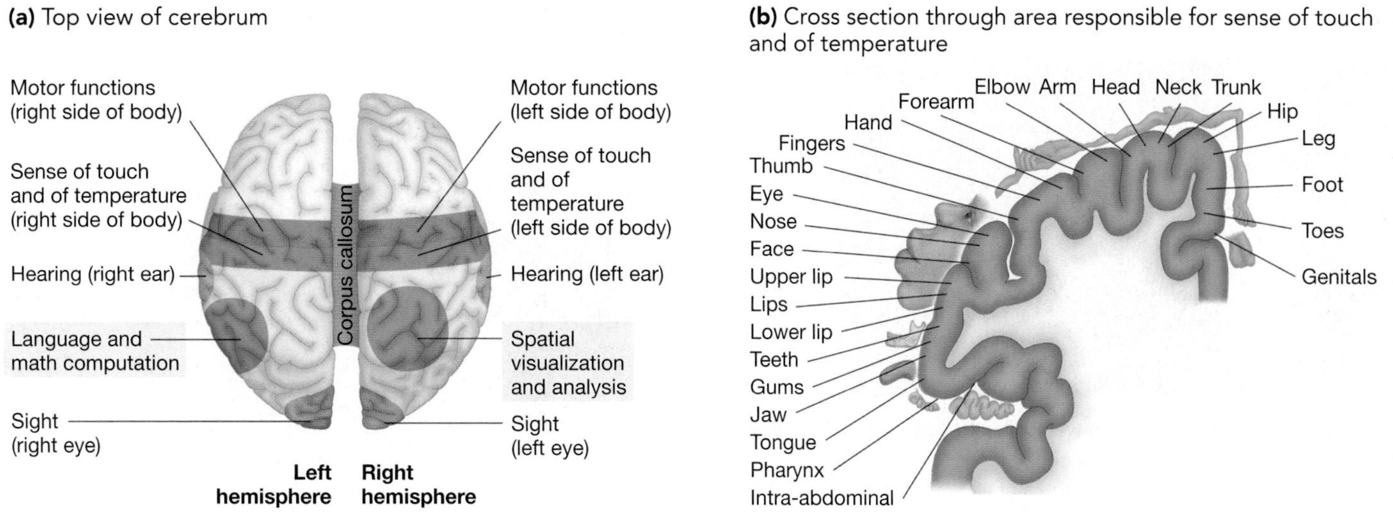

(a) Top view of cerebrum
Motor functions (right side of body)
Sense of touch and of temperature (right side of body)
Hearing (right ear)
Language and math computation
Sight (right eye)
Corpus callosum
Left hemisphere
Right hemisphere
Motor functions (left side of body)
Sense of touch and of temperature (left side of body)
Hearing (left ear)
Spatial visualization and analysis
Sight (left eye)
(b) Cross section through area responsible for sense of touch and of temperature
Elbow
Arm
Head
Neck
Trunk
Forearm
Hip
Hand
Leg
Fingers
Thumb
Eye
Foot
Nose
Toes
Face
Genitals
Upper lip
Lips
Lower lip
Teeth
Gums
Jaw
Tongue
Pharynx
Intra-abdominal

FIGURE 45.22 Specific Brain Areas Have Specific Functions
(a) Map of the brain, in top view (as if the person is looking at the top of the page), showing the functions of some major regions. The map was compiled from studies of people with damaged brain areas or with brain regions that were removed surgically. **(b)** Researchers mapped the area responsible for the sense of touch and of temperature by stimulating neurons in the brains of patients who were awake. QUESTION Is there a correlation between the size of the brain area devoted to sensing a particular body part and that body part's size? Explain.

and memory are thus closely related and are often studied in tandem. Research on these phenomena has been extensive, and progress has been particularly rapid recently. As an introduction to how researchers explore learning and memory, let's first explore work that is focused at the level of whole neurons and then review research at the molecular level.

Recording from Single Neurons during Memory Tasks

One approach to studying learning and memory is to record from individual neurons during learning and memory tasks. How do the action potentials generated by a cell change as learning and memory take place? Researchers have attempted to answer this question by recording from individual neurons in the temporal lobes of humans. Before operating on patients who were still awake and about to undergo surgery to remove seizure-prone areas of their brains, physicians have projected words or names on a screen and asked the individuals to read them silently, read them aloud, and/or remember them and repeat them later. The data shown in **Figure 45.23** are informative. In this case, the neuron being recorded was relatively quiet while the patient named objects but extremely active when the individual was remembering the objects and repeating their names aloud.

What do such data mean? Neurons in the temporal lobe are most active during memory tasks. So how could action potentials from particular cells make memory possible?

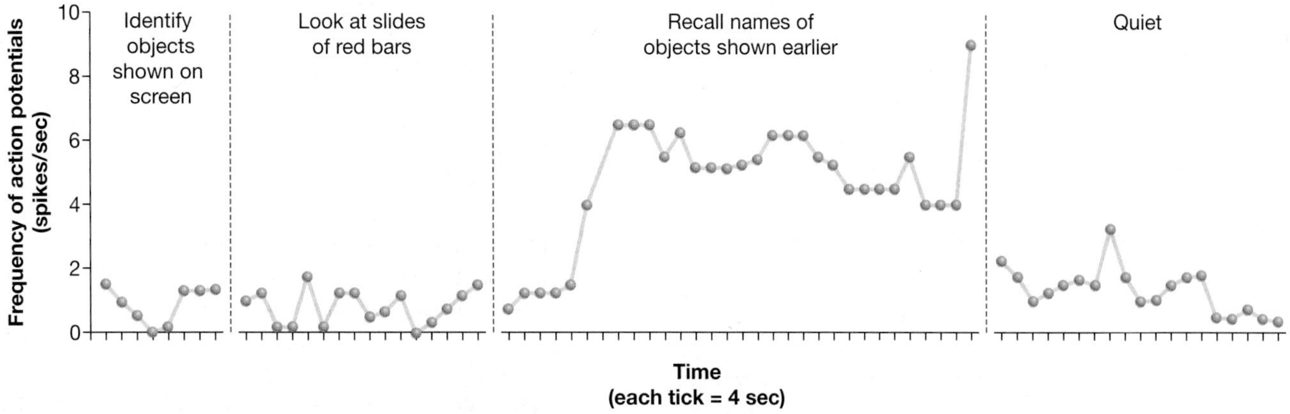

Frequency of action potentials (spikes/sec)
10
8
6
4
2
0
Identify objects shown on screen
Look at slides of red bars
Recall names of objects shown earlier
Quiet
Time (each tick = 4 sec)

FIGURE 45.23 The Frequency of Action Potentials from Brain Neurons Varies with Activity
Data showing the frequency of action potentials from a single brain neuron from the temporal lobe as a patient identified a series of objects shown on slides, looked at a series of red bars shown on slides, recalled the names of the objects shown previously, and looked at a blank screen while doing nothing.

(a) Sea slug *Aplysia californica*

(b) Gill-withdrawal reflex protects the gills during an attack.

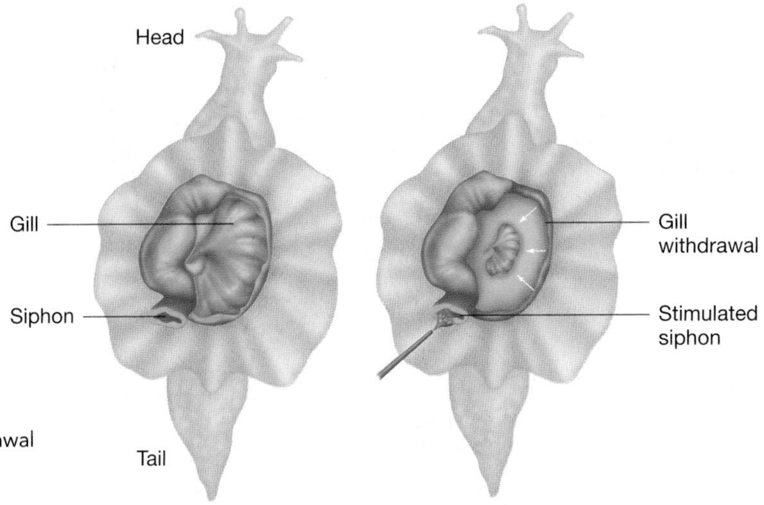

FIGURE 45.24 The Gill-Withdrawal Reflex in *Aplysia* Is a Model System in Learning and Memory
(a) *Aplysia californica* can reach lengths of 40 cm. **(b)** The gill-withdrawal reflex protects the sea slug's siphon from damage by predators.

Documenting Changes in Synapses Research on the molecular basis of memory is based on two fundamental ideas. First, learning and memory must involve some type of short-term or long-term change in the neurons responsible for these processes. This change could be structural or chemical in nature. Structural changes might include modifications in the number of synapses that a particular neuron makes. Chemical (molecular) changes might involve alterations in the amount of neurotransmitter released at certain synapses and changes in the number of receptors present in postsynaptic cells. Second, it will be much easier to understand what these changes are if an extremely simple system of neurons can be studied.

To explore the molecular basis of learning and memory, Eric Kandel's group has focused on the sea slug *Aplysia californica* (**Figure 45.24a**). Much of their work has explored the reflex diagrammed in **Figure 45.24b**. When a structure in *Aplysia* called the siphon is touched, the individual responds by withdrawing its gill. The reflex involves a sensory neuron that is activated by touch and a motor neuron that projects to a gill muscle. Retracting the gill protects it from predators.

Early work established that this simple reflex is modified by learning. For example, *Aplysia* also withdraw their siphons when their tails are given an electrical shock. If shocks to the tail are combined with a very light touch to the siphon—too light to get a response on its own—an *Aplysia* will learn to withdraw its gills in response to a light siphon touch alone.

Follow-up studies on this reflex showed that the neurons involved in learning release the neurotransmitter **serotonin** and that serotonin causes an EPSP in the motor neuron to the gill. Repeated application of serotonin, which mimics what happens at the synapse during learning, changes the serotonin-secreting cell (**Figure 45.25a**). Specifically, more serotonin is released after learning has taken place. As a result, EPSPs are higher and the motor neuron is more likely to generate action potentials. These results implied that, in *Aplysia*, changes in the nature of the synapse form the molecular basis of learning and memory. This process is termed **synaptic plasticity**.

Kandel's team was recently able to replicate these results with sensory and motor neurons growing in culture. **Figure 45.25b** shows two *Aplysia* motor neurons on a culture plate. Each motor neuron received synapses from a sensory neuron. To mimic the learning process, the investigators applied serotonin to one synapse five times over a short period. When they stimulated the sensory neuron a day later, they found a huge increase in EPSPs (**Figure 45.25c**). The cells had also established additional synapses. Neurons that had not received the repeated stimulation with serotonin had normal postsynaptic responses and numbers of synapses.

Such results reinforce a growing consensus that learning and memory involve both molecular and structural changes in synapses. Further, most researchers now agree that at least some aspects of long-term memory involve changes in gene expression. Chapter 47 explores how the chemical messengers called hormones cause changes in gene expression in target cells. But before investigating how hormones work, let's focus on the electrical signals involved in vision, hearing, taste, and movement—the subject of Chapter 46.

✓CHECK YOUR UNDERSTANDING

The CNS and PNS work together to gather information about the external and internal environments, process that information, and signal muscles, glands, and other tissues to make appropriate responses to the information. You should be able to (1) diagram the hierarchical structure of the PNS, (2) describe the types of research strategies that allowed biologists to localize particular functions to specific regions in the brain, and (3) predict whether the degree of synaptic plasticity is greater in the sensory neurons of sea slugs or in the brain neurons of humans, and why.

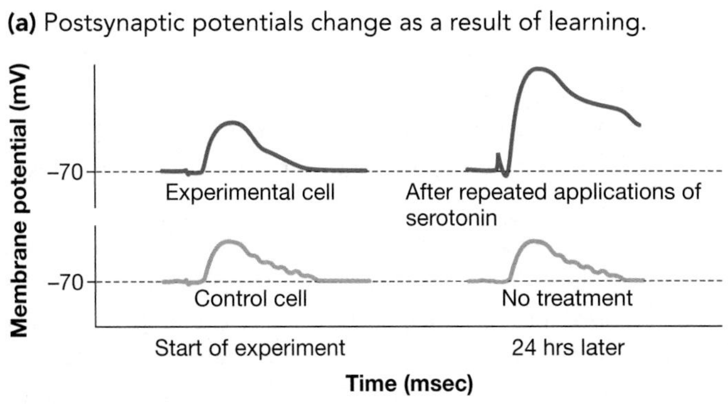

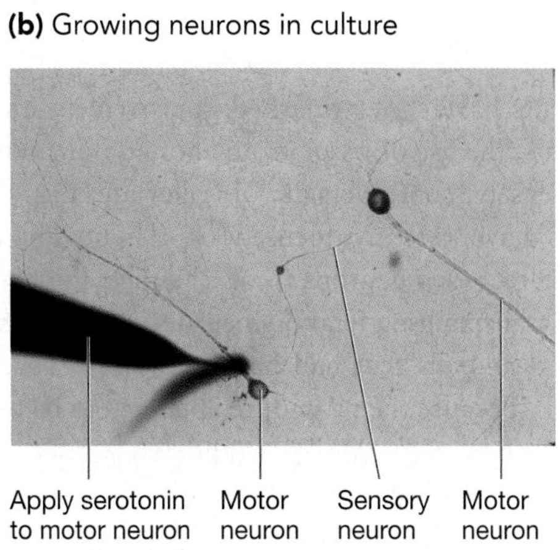

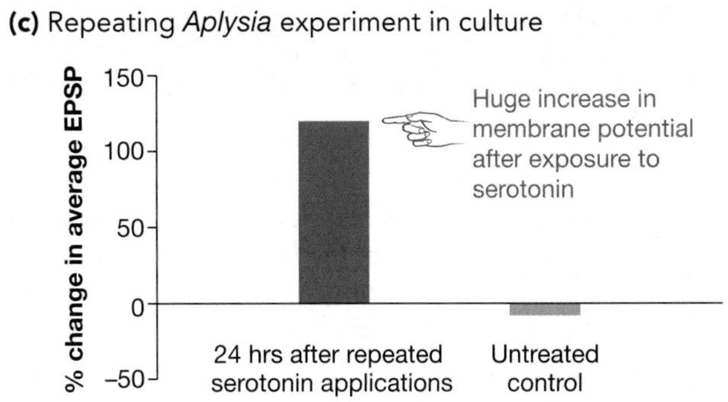

FIGURE 45.25 Learning and Memory Involve Changes in Synapses
(a) Experiment showing that repeated application of serotonin changes the behavior of postsynaptic neurons. **(b)** An in-culture study of interactions between *Aplysia* motor neurons and sensory neurons. **(c)** Histograms documenting the percentage increase in postsynaptic potentials that occurs after repeated application of serotonin, as in part (a).

ESSAY Can Brain Tissue Transplants Help People with Parkinson's Disease?

Parkinson's disease (PD) is a disorder that causes progressive deterioration of motor function in humans. The cellular basis of the disease is the inactivation or destruction of neurons in a region called the *substantia nigra*, which is located at the base of the brain. The neurons in this area secrete the neurotransmitter **dopamine** and are part of a neural circuit that inhibits action potentials in motor neurons that travel down the spinal cord. If these dopamine-secreting neurons do not function properly, motor function is impaired.

As dopamine-secreting neurons in the substantia nigra begin to inactivate, motor neurons become more excitable. As a result, muscles throughout the body become more likely to contract. As Chapter 46 will explain in detail, muscles work in pairs. For a movement to occur, one muscle must contract and another must relax. People with PD have a difficult time relaxing their muscles. They experience tremors as paired muscles alternately twitch. They also have a great deal of difficulty making smooth, coordinated movements and have a masklike expression, because they are unable to move facial muscles normally. Millions of people have become familiar with these symptoms by following the careers of former heavyweight champion boxer Muhammad Ali and actor Michael J. Fox, both of whom have PD.

In some cases, physicians have been able to alleviate these symptoms by administering dopamine-like molecules orally. Researchers continue to look for alternative therapeutic approaches, however, because many of the existing drug protocols stop working after several years of use. As the essay in Chapter 20 notes, PD is one of the diseases being targeted by researchers who are working with stem cells—undifferentiated cells that might be transplanted into tissue where cells have been lost. Another promising effort involves the direct transplantation of dopamine-secreting neurons into the brains of PD patients.

> *Another promising effort involves the transplantation of direct dopamine-secreting neurons. . . .*

Initially, the cells used in these transplants were harvested from aborted human fetuses. Although a significant number of

(Continued on next page)

(Essay continued)

PD patients have shown marked improvement after receiving transplants, the use of tissue from aborted human embryos has been intensely controversial. To alleviate this concern, researchers have recently focused on the possibility of using dopamine-secreting neurons from fetal pigs. Initial experiments involved transplanting fetal pig neurons into rats and other experimental animals that had been induced to develop PD-like symptoms. Because the procedure appeared to be both safe and helpful in alleviating PD-like symptoms, a research team was given approval to transplant these cells into 10 humans with PD. One year after the operation, most of these patients were exhibiting small but important improvements in their condition and had no measurable side effects.

The medical use of tissues from nonhuman animals, which is called **xenografting** or **xenotransplantation** (*xeno* is a Greek root meaning "stranger" or "guest"), is still in its infancy. Time will tell whether the techniques improve enough to make the approach a mainline therapy for a serious degenerative disease.

CHAPTER REVIEW

Summary of Key Concepts

- **Neurons are cells that transmit electrical signals. They have a membrane potential, which is an electric potential, or voltage, due to differences in the concentrations of ions on the two sides of the plasma membrane.**

Although neurons are highly variable in size and shape, all have a cell body and multiple short dendrites that receive electrical signals from other cells. Most neurons have axons that transmit electrical signals to other neurons or to effector cells in glands or muscles.

Studies of the squid giant axon neurons established that neurons have a resting potential due to differences in the concentrations of ions on the two sides of the membrane and the selective permeability of the membrane to ions. The difference in concentration of ions is created, in part, by the sodium-potassium pump. When Na^+/K^+-ATPase hydrolyzes ATP, it transports 3 Na^+ out of the cell and 2 K^+ in.

Web Tutorial 45.1 Membrane Potentials

- **Electrical signals are transmitted as all-or-none changes in membrane voltage called action potentials. During an action potential, an inflow of sodium ions is followed by an outflow of potassium ions.**

Studies of the squid axon established that the action potential is a rapid and stereotyped change in membrane potential. An action potential begins with an inflow of Na^+ that depolarizes the membrane. An outflow of K^+ follows and repolarizes the membrane. As charge spreads from the site of an action potential, the nearby membrane is depolarized enough to trigger additional Na^+ inflows and propagate the signal. Propagation takes place most rapidly in large axons or myelinated axons.

Voltage clamping allows researchers to hold membranes at specific potentials and record the current flows that result. This technique confirmed that both Na^+ and K^+ flow through voltage-gated channels. Patch clamping allows researchers to record the currents that flow through individual Na^+ and K^+ channels. These studies showed that the opening of K^+ channels is delayed relative to the opening of Na^+ channels. Experiments with neurotoxins allowed researchers to block these channels and observe the effect on the resting potential and action potential.

Web Tutorial 45.2 Action Potentials

- **At synapses, an electrical signal from one neuron is converted to a chemical signal in the form of a neurotransmitter. When it arrives at an adjacent neuron, the neurotransmitter causes a change in that cell's membrane potential.**

When action potentials arrive at a synapse, synaptic vesicles fuse with the axon's membrane and deliver neurotransmitters that bind to receptors on the membrane of a postsynaptic cell. One class of receptors functions as ligand-gated channels. In response to binding by a neurotransmitter, the channels open and admit ions that depolarize or hyperpolarize the postsynaptic cell's membrane. Postsynaptic potentials from nearby synapses sum. If the membrane at the axon hillock depolarizes to a threshold value, then an action potential is triggered. Defects in specific ligand-gated channels have been linked to the development of schizophrenia and other illnesses.

- **Vertebrates have a central nervous system (CNS) and a peripheral nervous system (PNS). PNS neurons receive sensory information and transmit it to the CNS for processing. The CNS then sends signals to muscles, glands, or other target tissues via other PNS neurons.**

The CNS consists of the brain and spinal cord (in vertebrates); the PNS consists of all nervous system components outside the CNS. In vertebrates, the PNS contains somatic and autonomic components. The somatic PNS is responsible for sensing external stimuli and effecting movement; the autonomic system monitors internal conditions and effects changes in the activity of organs.

Although the CNS is enormously complex, researchers have succeeded in mapping the functions of brain structures. Early mapping studies depended on individuals with lesions in certain areas or on the direct stimulation of certain regions of the cerebral cortex. Efforts to understand higher brain functions such as learning and memory form the current focus of research on the CNS. To date, research has established that learning and memory are based on modifications in synapses. After learning takes place, certain neurons release more or less neurotransmitter in response to stimulation. In the case of long-term memory, these changes depend on changes in gene expression.

Questions

Content Review

1. Why does the resting potential exist?
   a. because cells contain several types of ions
   b. because ion concentrations differ on the two sides of the membrane and because the membrane is selectively permeable to K^+
   c. because ion concentrations differ on the two sides of the membrane and because the membrane is selectively permeable to Na^+
   d. because ion concentrations differ on the two sides of the membrane and because the membrane is selectively permeable to Cl^-

2. Why did the squid axon become a model system for studying electrical signaling in animals?
   a. Its action potentials are particularly large and frequent.
   b. It is the tissue from which researchers initially isolated Na^+/K^+-ATPase.
   c. Squids are abundant and easy to obtain.
   d. It was large enough to support intracellular recording by the first microelectrodes.

3. How does myelination affect the propagation of an action potential?
   a. It speeds propagation by increasing the density of voltage-gated channels.
   b. It speeds propagation by increasing electrochemical gradients favoring Na^+ entry.
   c. It speeds propagation because charge does not leak out of the membrane as it spreads down the axon.
   d. It slows down propagation because Na^+ channels exist only at unmyelinated nodes (nodes of Ranvier).

4. In a neuron, what creates the electrochemical gradient favoring the outflow of K^+?
   a. Na^+/K^+-ATPase
   b. voltage-gated K^+ channels
   c. voltage-gated Na^+ channels
   d. ligand-gated Na^+/K^+ channels

5. Why do biologists say that positive feedback occurs during an action potential?
   a. The action potential is an all-or-none event, meaning that once it starts, it goes to completion.
   b. The opening of potassium channels repolarizes the membrane, making it less likely that sodium channels will open and depolarize the membrane.
   c. Once sodium channels open and begin to depolarize the membrane, they become more likely to open and cause further depolarization.
   d. Sodium channels are refractory—once they have opened, they are less likely to open again for a few milliseconds.

6. Why is memory thought to involve changes in particular synapses?
   a. In some systems, an increased release of neurotransmitters occurs after learning takes place.
   b. In some systems, the type of neurotransmitter released at the synapse changes after learning takes place.
   c. When researchers stimulated certain neurons electrically, individuals replayed memories.
   d. People who lack short-term memory have specific deficits in synapses within the brain regions responsible for memory.

Conceptual Review

1. Explain why the resting potential exists. Be sure to differentiate between the roles of K^+ channels and that of Na^+/K^+-ATPase in the membrane's selective permeability and ion concentration gradients.
2. Draw a graph of an action potential and label the axes. Label the parts of the graph and explain which ion flow or flows are responsible for each part.
3. Draw a diagram of a synapse. Label the parts. Then make a series of diagrams showing what happens when an action potential arrives at a synapse. Explain the events that occur at both the presynaptic cell and the postsynaptic cell.
4. Why do summation and integration occur in postsynaptic cells? Why is it important that synapses exist, instead of neurons having a continuous plasma membrane and direct cytoplasmic connections?
5. Compare and contrast the somatic and autonomic components of the PNS.
6. Compare and contrast the sympathetic and parasympathetic components of the autonomic nervous system.

Group Discussion Problems

1. Study the circuitry in the nervous system diagrammed in Figure 45.1 and the roles of the PNS described in Figure 45.19. Discuss how the nervous system functions as a mechanism for achieving homeostasis.
2. Discuss the pros and cons of lesion studies in determining the functions of particular brain structures.
3. During an epileptic seizure, muscles convulse spasmodically and the person, although awake, is unable to think, see, or hear. The cause of epilepsy is unknown. Speculate on what is happening at the level of the synapse in seizure-prone areas of the brain. How would you test your hypothesis?
4. Suppose that research establishes a causative relationship between defects in an NMDA-sensitive glutamate receptor and the development of schizophrenia. What new therapeutic strategies would this finding suggest for the treatment of this disease?

Answers to Multiple-Choice Questions 1. b; 2. d; 3. c; 4. a; 5. c; 6. a

www.prenhall.com/freeman is your resource for the following: Web Tutorials; Online Quizzes and other Online Study Guide materials; Answers to Conceptual Review Questions; Solutions to Group Discussion Problems; Answers to Figure Caption Questions and Exercises; and Additional Readings and Research.

46 Animal Sensory Systems and Movement

KEY CONCEPTS

- Sensory cells transduce sound, light, or other stimuli to a change in membrane potential. Sensory cells send action potentials to the brain, where the signals are processed and integrated.
- Hearing is based on sensory cells that move in response to sound waves of a particular frequency. The movement opens ion channels in the sensory cells and results in a change in membrane potential.
- Vision is based on sensory cells that contain the light-absorbing pigment rhodopsin. Rhodopsin changes conformation when it absorbs light and triggers events that change the cell's membrane potential.
- Taste and smell are based on membrane proteins that act as receptors for particular molecules and that trigger changes in membrane potential.
- In many cases, animals respond to sensory stimuli by moving. Movement is based on antagonistic muscle groups that act on a stiff skeleton and that flex or extend in response to electrical signals.

In many species of moth, males have much larger antennae than females do. Receptor cells on the males' feathery antennae detect airborne chemical signals that are produced by sexually mature females. As a result, males can locate females in complete darkness.

Adult moths are active at night when it is difficult or impossible to see. It is logical, then, to observe that sexually mature female and male moths do not go looking for each other. Instead, females release a chemical attractant called a pheromone into the air. A **pheromone** is a small molecule that acts as a signal between individuals and that usually causes a change in physiology or behavior. Male moths of the same species that are flying in the area can detect even a single molecule of the pheromone, due to the receptor cells located on their large, feathery antennae. In response to an airborne gradient of pheromone molecules, males fly toward a female.

As they patrol in search of these airborne pheromones, however, male moths are hunted by bats. Like moths, bats are active almost exclusively at night. Instead of hunting by sight, like a falcon or a cheetah, bats hunt with the aid of sonar. Bats emit a train of high-pitched sounds as they fly and then listen for echoes that indicate the direction and shape of objects in their path. If the object is a moth, the bat flies toward it, catches the individual in its mouth, and eats it (**Figure 46.1**). Some moth species can hear bat calls, however. When moths detect sounds from an onrushing bat, they tumble out of the sky in chaotic escape flights.

If you were out at night as these dramas unfolded, at best you might be dimly aware that bats and moths were flying about. Humans cannot smell moth pheromones or hear the sounds that bats emit when flying. It took decades of careful experimentation for biologists to understand how moths and

FIGURE 46.1 Bats Hunt Moths via Sonar
The moth (near the center) is about to be captured by an onrushing bat.

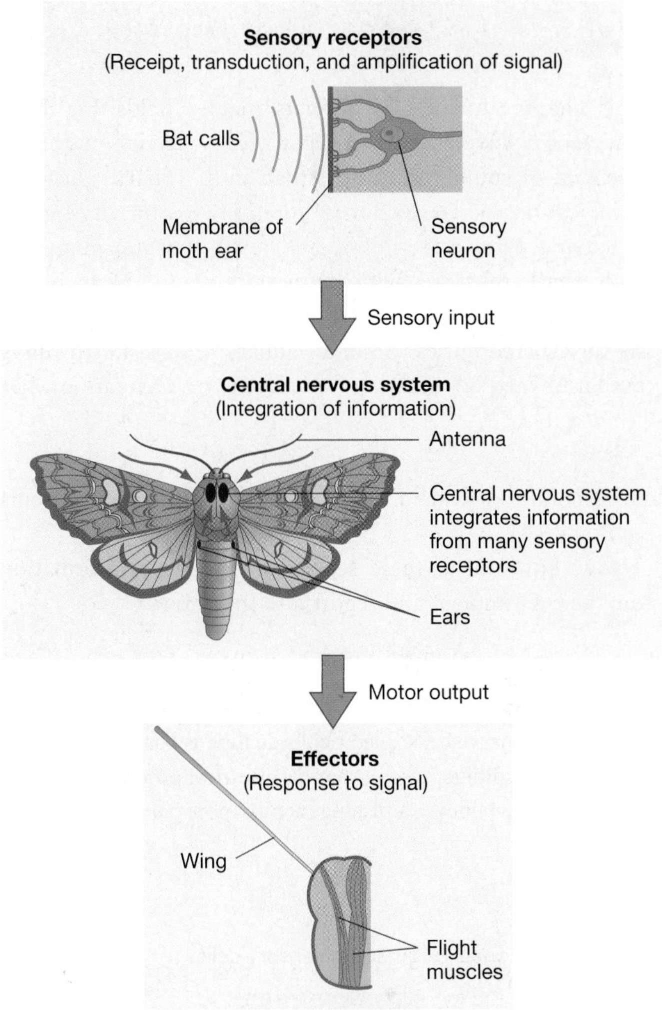

FIGURE 46.2 Sensory Systems, the CNS, and Effectors Such as Muscles Are Linked
Sensory neurons relay information about conditions inside and outside an animal to the central nervous system. After integrating information from many sensory neurons, the CNS sends signals to muscles through motor neurons.

bats sense the world around them and how they move in response to the information they receive.

Movement is fundamental to how animals work. But without accurate sensory information, animals cannot move effectively. To explore the mechanisms of sensation and movement, Section 46.1 introduces how animals receive information from the environment and respond to it. The next three sections delve into the molecular processes that are involved in hearing, vision, and taste and smell. The chapter concludes by investigating the cellular and molecular mechanisms responsible for movement.

46.1 How Do Sensory Organs Convey Information to the Brain?

As a moth flies through the night, its brain receives streams of signals from an array of sensory organs. Antennae provide information about the concentration of pheromones; "ears" located on various parts of the body send data on the presence of high-pitched sounds; detectors for balance and gravity transmit signals about the body's orientation in space. Each type of sensory information is detected by a sensory neuron or by a specialized receptor cell that makes a synapse with a sensory neuron. As **Figure 46.2** shows, the moth's nervous system integrates the sensory input—information from sensory neurons—and responds with motor output, via electrical signals, to specific muscle groups (effectors).

The ability to sense a change in the environment depends on four processes: (1) **transduction**, or the conversion of an external stimulus to an internal signal in the form of an action potential; (2) amplification of the signal; (3) transmission to the central nervous system (CNS); and (4) **integration**, or processing with other incoming signals.

The first step in the sequence requires a sensory receptor cell to convert light, sound, tension, or some other signal into an electrical signal. Sensory receptors are located throughout the body and are categorized by the type of stimulus: **Nociceptors** sense pain; **thermoreceptors** detect changes in temperature; **mechanoreceptors** respond to distortion caused by touch or pressure; **chemoreceptors** perceive specific molecules or classes of molecules; **photoreceptors** respond to particular wavelengths of light within the visible spectrum; and **electroreceptors**, if present, detect electric fields. With such a broad range of sensory detectors available, it is no wonder that animals can monitor

BOX 46.1 Senses That Humans Don't Have

This chapter introduces animals that can sense wavelengths of light, frequencies of sound, and odors that humans cannot perceive. But in addition to having eyes, ears, noses, and taste buds that are more acute than ours, many species have completely different sensory capabilities. Some aquatic predators are so sensitive to electric fields that they can detect electrical activity in the muscles of passing prey. Sharks become badly disoriented in captivity by electric fields generated by water pumps near their tanks. Similarly, many birds, sea turtles, and other animals can detect magnetic fields and use Earth's magnetic field as an aid in navigation. Homing pigeons tend to become disoriented if small magnets are glued to their heads. Some species of birds can also sense changes in barometric pressure. Biologists who work on species such as these are faced with the challenge of studying stimuli that they themselves cannot sense. In addition, it is possible that animals have senses that are yet to be discovered.

and respond to a wide array of changes in their environments (see **Box 46.1**).

Now, how do all these sensory cells receive information from the environment and report it to the brain?

Sensory Transduction

Chapter 45 introduced how electrical signals are generated and propagated in nerve cells. Electrical signaling is based on changes in membrane voltage. Recall that an **electrical potential**, or voltage, is created whenever a plasma membrane separates charges in the form of different concentrations of ions. When a sensory cell is in the resting state, the inside of its plasma membrane is more negative than its exterior.

As **Figure 46.3a** shows, researchers can record changes in a cell's membrane potential with the aid of microelectrodes. Recall from Chapter 45 that if ion flows cause the interior to become more positive (less negative), the membrane is said to be **depolarized.** If changes in ion channels cause the cell interior to become more negative than the resting potential, the membrane is said to be **hyperpolarized.**

(a) Electrical signals originate in sensory cells.

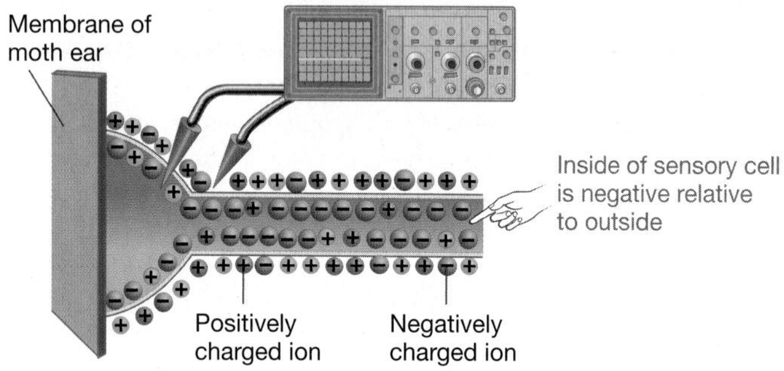

(b) Sound-receptor cells depolarize in response to sound.

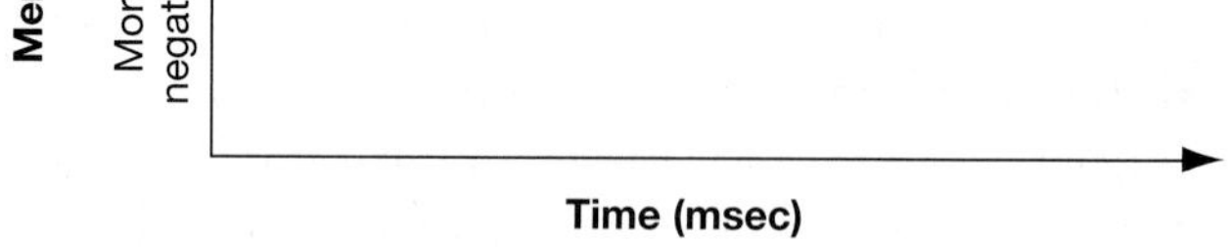

(c) Sound-receptor cells respond more strongly to louder sounds.

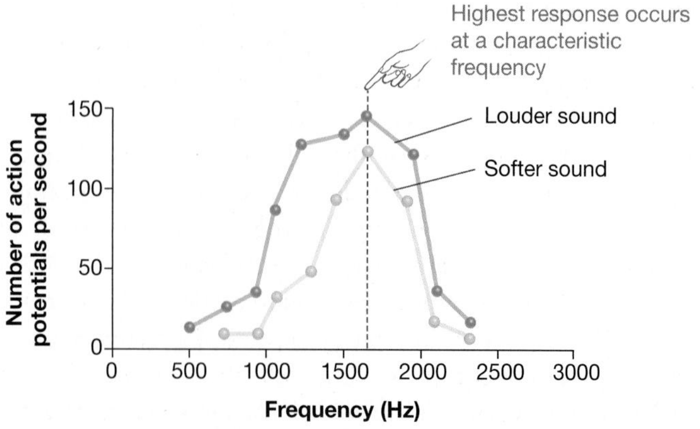

FIGURE 46.3 Sensory Inputs Change the Membrane Potential of Receptor Cells

(a) Sensory cells have a membrane potential that is measured in millivolts (mV). **(b)** In response to sensory stimuli, ions flow across the membranes of receptor cells and either depolarize or hyperpolarize the membrane. **(c)** The frequency of action potentials from a receptor transmits information about the nature and intensity of the sensory stimulus.

What do these principles have to do with sensory systems? To answer this question, examine the voltage recording from a human sound-receptor cell in **Figure 46.3b**. Notice that when the experimenter played a sound, the sound-receptor cell depolarized in response.

Other sensory cells work in a similar way. Although animals have sensory receptors that detect a remarkable variety of stimuli, they all transduce sensory input—including light, sounds, touch, and odors—to a change in membrane potential. In this way, different types of information are transduced to a common type of signal.

If a sensory stimulus induces a large change in a sensory receptor's membrane potential, action potentials are sent to the brain. For example, the amount of depolarization that occurs in a sound-receptor cell is proportional to the loudness of the sound. If the depolarization passes threshold, enough voltage-gated sodium channels open to trigger action potentials that are sent to the brain. Recall from Chapter 45 that all action potentials are identical in size and shape. As **Figure 46.3c** shows, however, louder sounds induce a higher frequency of action potentials than do softer sounds. In this way, receptor cells provide information about the intensity of a stimulus.

The universal nature of sensory transduction raises an important question: If all types of external stimuli are converted to electrical signals in the form of action potentials, and if all action potentials are alike, then how is it possible for the brain to interpret the information properly?

Transmitting Information to the Brain

There are two keys to understanding how the brain interprets sensory information. First, receptor cells tend to be highly specific. For example, each receptor cell in a human ear responds best to certain frequencies of sound. Some receptors are more sensitive to low-pitched sounds at a frequency of 1000 Hz (hertz, or cycles per second); others respond best to high-pitched sounds at 8000 Hz. The receptor-cell response depicted in Figure 46.3c is strongest for sounds at about 1650 Hz. At this frequency, the maximum number of action potentials per second emerge from the receptor. In this way, the train of action potentials from a cell contains information about the frequency of sound that is being received, its intensity, and how long the stimulus lasts.

Second, each type of sensory neuron sends its signal to a specific portion of the brain. Axons from sensory neurons in the human ear project to a particular area at the side of the brain, but axons from sensory receptors in the eye deliver action potentials to a specific area at the back of the brain. Different regions of the brain are specialized for interpreting different types of stimuli.

With this introduction to the basic principles of sensory reception and transduction, let's delve into the details of four sensory systems that are particularly well understood: hearing, vision, and chemical reception.

46.2 Hearing

Hearing is the ability to sense the wavelike changes in air pressure called sound. A **sound** consists of waves of pressure in air or in water. The number of pressure waves that occur in one second is called the **frequency** of the sound. We perceive different sound frequencies as different **pitches**.

Animals actually have a wide variety of mechanisms for sensing changes in pressure in addition to hearing airborne or waterborne sound waves. One particularly important mechanism monitors changes in body position relative to the force of gravity, via mechanical pressure changes within the sensory organ. Crabs, for example, have a fluid-filled organ that is lined with pressure-receptor cells. The organ, known as a **statocyst**, contains a small calcium-rich structure that normally rests on the bottom of the organ. If the crab is tipped or flipped over, this "stone" presses against receptors that are not on the bottom of the organ. When the brain receives action potentials from these cells, it responds by activating muscles that restore the animal to its normal posture. In addition, animals may have pressure-receptor cells that are responsible for detecting direct physical pressure on skin, as well as pressure-receptor cells that monitor how far muscles or blood vessels are stretched.

Virtually all the pressure-sensing systems found in animals are based on the same mechanism, however. Before investigating the specific structures that are involved in hearing, let's examine the general nature of a mechanoreceptor cell that responds to pressure.

How Do Sensory Cells Respond to Sound Waves and Other Forms of Pressure?

The mechanoreceptors responsible for sensing sound and vibrations in the environment are relatively simple in design. In every case, direct physical pressure on a plasma membrane or distortion by bending changes the conformation of ion channels in the membrane and causes the channels to open or close. In response to a change in ion flow, the membrane depolarizes or hyperpolarizes. The result is a new pattern of action potentials from a sensory neuron.

In vertebrate ears and several other pressure-sensing organs, the receptors are similar to the hair cell shown in **Figure 46.4a**, page 1056. **Hair cells** are pressure-receptor cells named for their appearance, owing to a set of stiff outgrowths called **stereocilia** (singular: **stereocilium**). The "hairy-looking" stereocilia are microvilli that are reinforced by actin filiments. Many hair cells also have a single **kinocilium**; the kinocilium is a true cilium that contains a 9 + 2 arrangement of microtubules (see Chapter 7).

As Figure 46.4a shows, the stereocilia in hair cells are arranged in order of increasing height; if a kinocilium is present, it is the tallest of all the projections. All these structures extend into a fluid-filled chamber. If the stereocilia are bent in the direction of the kinocilium in response to pressure

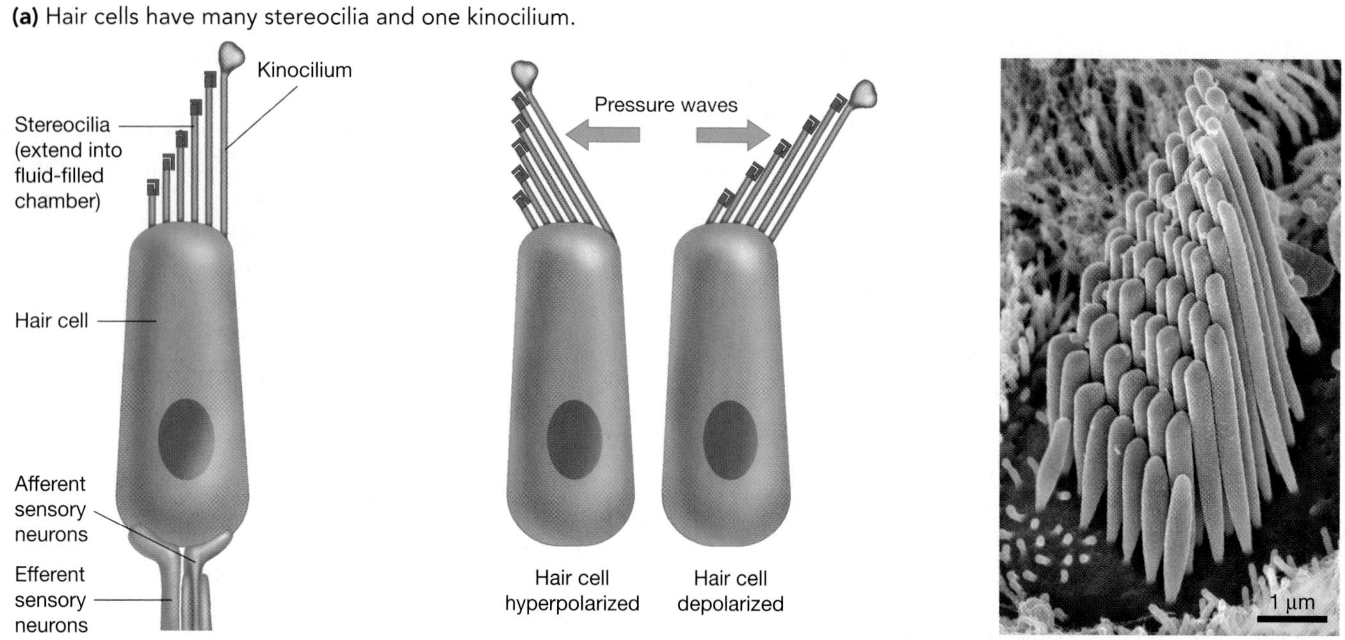
(a) Hair cells have many stereocilia and one kinocilium.
Kinocilium
Stereocilia (extend into fluid-filled chamber)
Hair cell
Afferent sensory neurons
Efferent sensory neurons
Pressure waves
Hair cell hyperpolarized
Hair cell depolarized
1 µm

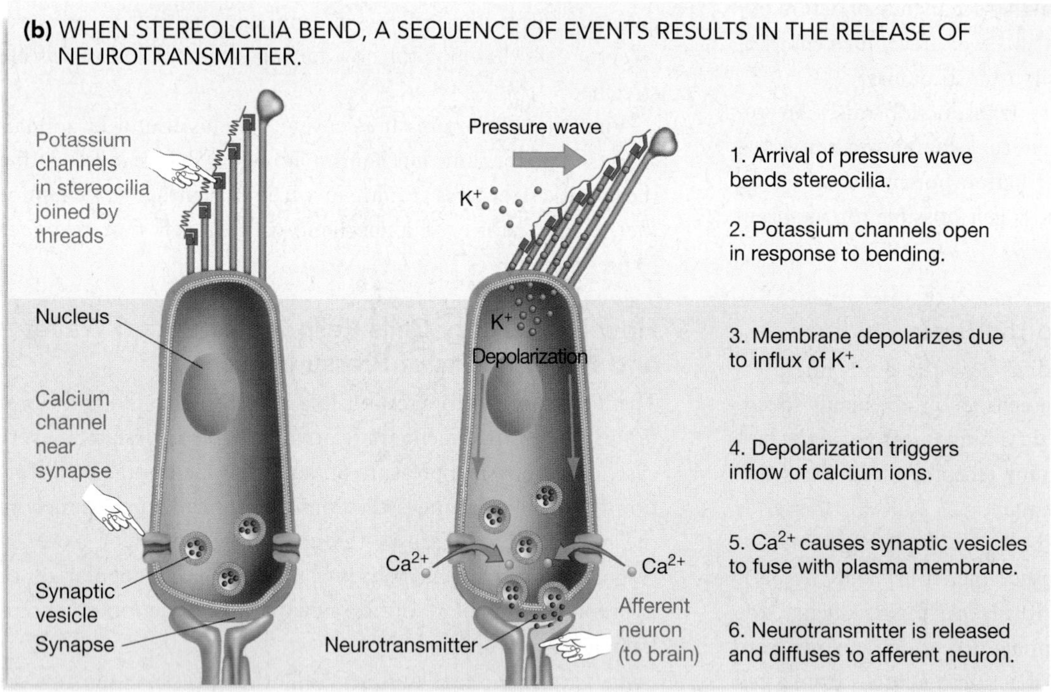
(b) WHEN STEREOLCILIA BEND, A SEQUENCE OF EVENTS RESULTS IN THE RELEASE OF NEUROTRANSMITTER.
Potassium channels in stereocilia joined by threads
Nucleus
Calcium channel near synapse
Synaptic vesicle
Synapse
Pressure wave
K+
K+
Depolarization
Ca2+
Ca2+
Neurotransmitter
Afferent neuron (to brain)
1. Arrival of pressure wave bends stereocilia.
2. Potassium channels open in response to bending.
3. Membrane depolarizes due to influx of K+.
4. Depolarization triggers inflow of calcium ions.
5. Ca2+ causes synaptic vesicles to fuse with plasma membrane.
6. Neurotransmitter is released and diffuses to afferent neuron.

FIGURE 46.4 Hair Cells Transduce Sound Waves to Electrical Signals
(a) Hair cells are named for the hairlike stereocilia that project from one end. **(b)** If potassium channels open when stereocilia bend, then the hair cell's membrane depolarizes—resulting in an influx of calcium ions and the release of neurotransmitter to an afferent neuron.

(as illustrated in **Figure 46.4b**), the membranes of the structure are distorted. The distortion causes potassium ion (K^+) channels in the stereocilia to open, and the cell depolarizes by approximately 20 mV.* The depolarization causes an inflow of calcium ions, leading to an increase in the amount of neurotransmitter released at the synapse between the hair cell and a sensory neuron. The end result is excitation of the postsynaptic cell. But if sound pressure waves bend stereocilia the other way, then the K^+ channels close and the cell hyperpolarizes by 5 mV. Hyperpolarization of the hair cell results in a decrease in neurotransmitter released at the synapse and the inhibition of the sensory neuron. Pressure that is perpendicular to the stereocilia does not change the activity of the ion channels.

How can bending affect ion channels? Electron micrographs show that tiny threads connect the tips of stereocilia to each other. One hypothesis contends that when the stereocilia are bent, the threads somehow pull open ion channels in the wall of the next tallest stereocilium—like tiny trapdoors (see Figure 46.4b, step 2). This hypothesis remains to be con-

*Recall from Chapter 45 that the opening of K^+ channels hyperpolarizes neurons. Hair-cell plasma membranes respond differently, because they are bathed by extracellular fluid with an extraordinarily high K^+ concentration. As a result, the equilibrium potential for K^+ in hair cells is 0 mV instead of the −80 mV in a typical neuron. The resting potential of the hair-cell plasma membrane is −70 mV, so K^+ rushes in and causes a depolarization.

firmed, however. Researchers still do not fully understand how the ion channels involved in pressure reception work.

The Mammalian Ear

To understand how changes in the membrane potential of hair cells result in hearing, let's focus on the human ear as a case study. **Figure 46.5** shows that the structure has three general sections: the **outer ear, middle ear,** and **inner ear.** Each of these three sections is separated from the others by a membrane.

To trace the path of sound through the three sections of the ear, begin with sound waves that are just reaching your head. Pressure waves are collected by the outer ear, which projects from the head, and are funneled into a tube known as the **ear canal.** At the end of the ear canal, the waves strike the **tympanic membrane,** or *eardrum,* which separates the outer ear from the middle ear. The repeated cycles of air compression cause the tympanic membrane to vibrate back and forth with the same frequency as the sound wave. The vibrations are passed to three tiny bones called the **ear ossicles.** In response, these bones vibrate against one another. The last ossicle, called the **stapes** (pronounced *STAY-peez*), vibrates against a membrane. That membrane, called the **oval window,** separates the middle ear from the inner ear. The oval window oscillates in response. It generates waves in the fluid inside a chamber known as the **cochlea** (pronounced *KOK-lee-ah*), where the hair cells that respond to sounds are located. These pressure waves are sensed by hair cells in the cochlea.

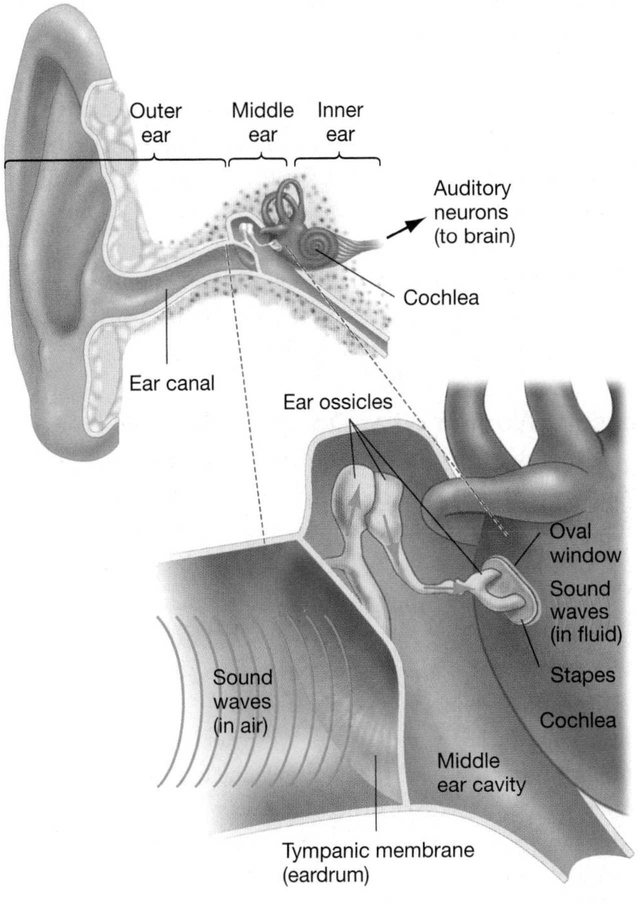

FIGURE 46.5 Mammals Have an Outer Ear, a Middle Ear, and an Inner Ear
The middle ear starts with the tympanic membrane and ends in the oval window of the cochlea.

In effect, the ear translates airborne waves to waterborne waves. The system seems extraordinarily complex, though, for such a simple result. Why doesn't the outer ear canal lead directly to the oval window? Why have a middle ear at all?

The Middle Ear Amplifies Sounds Biologists began to understand the function of the middle ear when they recognized two key aspects of its structure. First, the size difference between the tympanic membrane and the oval window is important. As the lower portion of Figure 46.5 shows, the ossicles in the middle ear receive vibrations from the tympanic membrane and transfer them to the much smaller oval window. Because the tympanic membrane is about 15 times larger than the oval window, the amount of vibration induced by sound waves is increased by a factor of 15. This phenomenon is similar to using the same amount of force to bang on a very large door versus a very small door. In addition, the three ossicles act as levers that further amplify the vibrations at the tympanic membrane. In reptiles and birds, there is just one ear ossicle instead of three, and this levering action is not possible. The overall effect in the mammalian middle ear is to amplify sound by a factor of 22—enough that the hair cells lining the cochlea are stimulated. Biologists interpret the mammalian middle ear as an adaptation for increasing sensitivity to sound.

To summarize, the outer ear transmits sound waves from the environment to the inside of a mammal; the middle ear amplifies these waves enough to stimulate the hair cells within the cochlea. How can hair cells distinguish different frequencies of sound? If all hair cells responded equally to all frequencies of sound, we would be able to perceive only one pitch. Everyone's voice—indeed, every sound—would sound the same.

The Cochlea Detects the Frequency of Sounds Careful anatomical studies and experimental work have revealed why different hair cells respond to specific frequencies of sound. As the cross section in **Figure 46.6a** (page 1058) shows, the cochlea has a set of internal membranes that divide it into three chambers. Hair cells form rows in the middle chamber. As **Figure 46.6b** indicates, the bottom of each hair cell connects to a structure called the **basilar membrane.** In addition, the hair cells' stereocilia touch yet another, smaller surface called the **tectorial membrane.** (The kinocilium is not present in a mature cochlear hair cell.) In effect, hair cells are sandwiched between membranes.

Researchers struggled for decades to understand how these membranes affect hair-cell function. It is virtually impossible to study cochleas in living organisms, because the cochleas are tiny, complex, coiled, and buried deep inside the skull. During the 1920s and 1930s, however, Georg von Békésy was able to

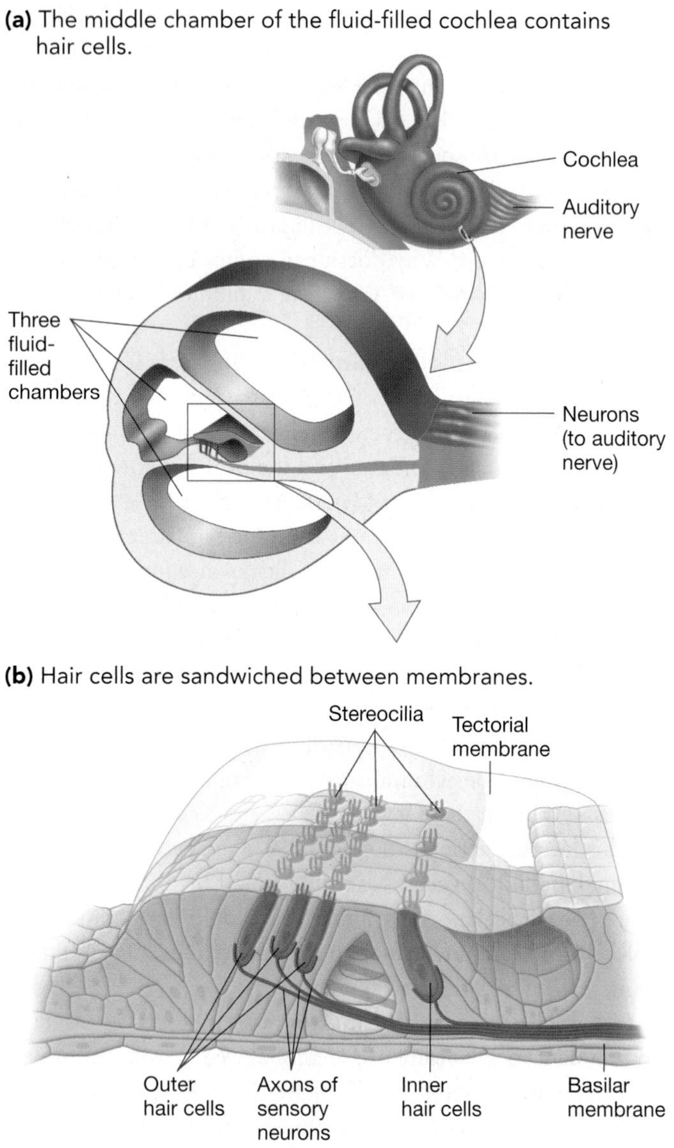
(a) The middle chamber of the fluid-filled cochlea contains hair cells.
Cochlea
Auditory nerve
Three fluid-filled chambers
Neurons (to auditory nerve)
(b) Hair cells are sandwiched between membranes.
Stereocilia
Tectorial membrane
Outer hair cells
Axons of sensory neurons
Inner hair cells
Basilar membrane

FIGURE 46.6 In the Human Cochlea, Hair Cells Are Sandwiched between Membranes
(a) The cochlea contains fluid-filled chambers separated by membranes. Hair cells are located in the middle chamber and **(b)** are sandwiched between the basilar membrane and the tectorial membrane. EXERCISE Label the tectorial membrane in (a).

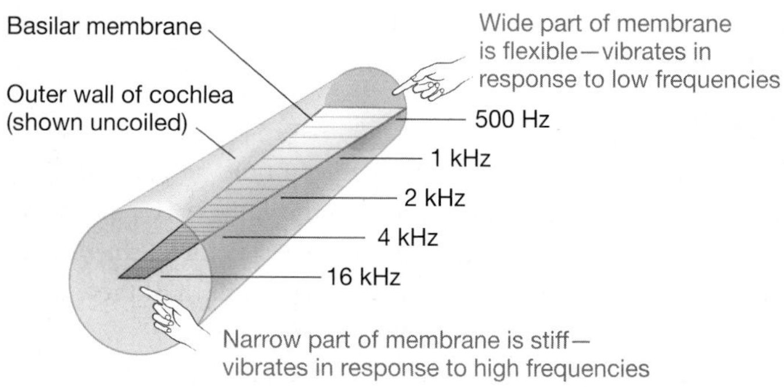
Basilar membrane
Outer wall of cochlea (shown uncoiled)
Wide part of membrane is flexible—vibrates in response to low frequencies
500 Hz
1 kHz
2 kHz
4 kHz
16 kHz
Narrow part of membrane is stiff—vibrates in response to high frequencies

FIGURE 46.7 The Basilar Membrane Varies in Stiffness
High-frequency sounds are detected at the narrow, stiff base of the cochlea; low-frequency sounds are detected at the wider end.

perform experiments on cochleas that he had dissected from human cadavers. Once a cochlea was isolated, von Békésy was able to vibrate the oval window and record how the cochlea's internal membranes moved in response. He found that when a pressure wave traveled down the fluid in the upper and lower chambers, the basilar membrane vibrated in response. His key finding, though, was that sounds of different frequencies caused the basilar membrane to vibrate in specific spots along its length (**Figure 46.7**). When the basilar membrane vibrated in a particular location, the stereocilia of the hair cells there were bent one way and then the other by the tectorial membrane.

Another observation allowed von Békésy to understand cochlea function: He noted that the basilar membrane varies in stiffness. It is very stiff near the oval window and very flexible at the other end. Thus each segment of the membrane vibrates in response to a different frequency of sound. Just as a stiff drumhead produces a high-pitched sound and a loose drumhead yields a low-pitched sound, high-frequency sounds cause the stiff part of the basilar membrane to vibrate, and low-frequency sounds cause the more flexible part to vibrate.

To summarize, certain portions of the basilar membrane vibrate in response to specific frequencies and result in the bending of hair-cell stereocilia. In this way, hair cells in a particular place on the membrane respond to sounds of a certain frequency. The brain, in turn, receives action potentials from the neurons associated with each hair cell and interprets them as a particular pitch—meaning a specific frequency of sound. Complex sounds contain a wide variety of frequencies and trigger particular combinations of hair cells. Through experience, the brain learns which combinations of frequencies represent music, a fire alarm, or a parent's voice.

Sensory Worlds: What Do Other Animals Hear?

Compared with the hearing of many mammals, human hearing is not particularly acute. This point was driven home to Katherine Payne as she was observing elephants at the Washington Park Zoo in Portland, Oregon. She noticed a throbbing sensation in the air, coming from the elephants. Payne knew that **infrasound**, or sound frequencies that are too low for humans to hear, can produce such sensations. To test the hypothesis that the elephants were producing infrasonic vocalizations, she returned to the zoo with microphones that were capable of picking up sounds at extremely low frequency. At normal speed, the tape Payne made was silent. But when she raised the pitch of the sounds by speeding the tape up, she heard a chorus of cow-like noises. The elephants were calling to each other, using low-frequency sounds.

According to follow-up research, elephants have the best infrasonic hearing of any land mammal. Biologists note that infrasound can travel exceptionally long distances, and they hypothesize that infrasonic calls allow wild elephants to coordinate their movements when they are miles apart. Recent research suggests that in addition to detecting infrasound via their large ears, elephants use their feet to detect a by-product of infrasound: seismic vibrations traveling in the ground.

The bats introduced at the start of the chapter furnish another example of an animal perceiving sounds that humans cannot. But in this case, the sounds are **ultrasonic**—that is, they have frequencies *above* the range of human hearing. Humans can hear sounds between 20 Hz and 20,000 Hz (20,000 Hz is equal to 20 kHz, or kilohertz). Ultrasonic hearing in bats was discovered in the late 1930s, when Donald Griffin borrowed the only ultrasonic apparatus then in existence from Robert Galambos, a fellow graduate student. Griffin used the machine to demonstrate that flying bats constantly emit ultrasounds. In follow-up experiments, he documented that a bat with cotton in its ears, or with its mouth taped shut, crashed into walls when released in a room. Blindfolded bats, in contrast, never crashed.

Griffin and Galambos concluded that bats use sound echoes (*sonar*) to navigate. This concept, termed **echolocation**, was an outlandish idea at the time. When Galambos described the use of sonar by bats at a meeting in 1940, another scientist shook him by the shoulders and said, "You can't really mean that!" But continued research showed that dolphins, shrews, and certain other animals besides bats use sonar. In fact, it is likely that at least some of these species perceive shapes with their ears better than they do with their eyes.

Although the use of sonar is much more extensive than originally believed, most animals rely on vision to detect the size, shape, and location of the objects around them. In fact, it turned out that Griffin and Galambos were fortunate to have started out studying echolation in the lineage of bats called Microchiroptera—also known as whispering bats. Virtually all members of the other major bat lineage, the Megachiroptera, or Old World flying foxes, do not echolocate. Instead of hunting insects at night by echolocation, flying foxes feed on fruit or nectar and rely on vision to find food. How do megachiropterans and other animals see the world?

✓CHECK YOUR UNDERSTANDING

Hearing is a type of pressure detection, and it occurs when the stereocilia on hair cells bend in response to changes in pressure. The bending movement opens ion channels and results in a change in membrane potential. You should be able to (1) make a labeled diagram of the mammalian ear and describe the function of each part; and (2) predict the type of hearing loss that results from each of the following: a punctured eardrum, a severe mutation in a gene affecting stereocilium length, and an age-related loss of flexibility in the basilar membrane.

46.3 Vision

Most animals have a way to sense light. The organs involved range from simple light-sensitive eyespots in flatworms to the sophisticated, image-forming eyes of vertebrates, cephalopod mollusks, and arthropods. Insects, for example, have a **compound eye**, which is composed of hundreds or thousands of light-sensing columns called **ommatidia** (**Figure 46.8**). Each ommatidium contributes information about one small piece of the visual field—not unlike a single pixel on a computer monitor. Thus, the more ommatidia in a compound eye, the better the resolution. In addition, the presence of many light-sensing columns makes species with compound eyes particularly good at detecting movement. Vertebrates and cephalopods, in contrast, have a **camera eye**. This is a structure that focuses incoming light onto a layer of receptor cells. Let's examine this structure more closely.

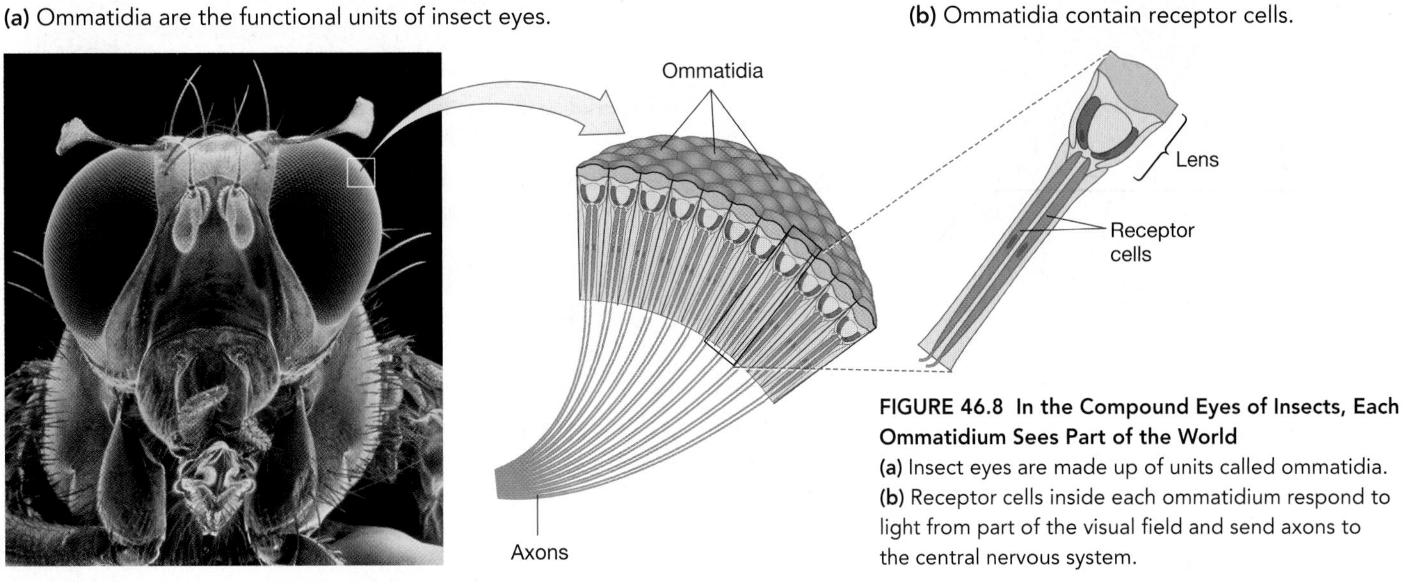
(a) Ommatidia are the functional units of insect eyes.
(b) Ommatidia contain receptor cells.
Ommatidia
Lens
Receptor cells
Axons

FIGURE 46.8 In the Compound Eyes of Insects, Each Ommatidium Sees Part of the World
(a) Insect eyes are made up of units called ommatidia. **(b)** Receptor cells inside each ommatidium respond to light from part of the visual field and send axons to the central nervous system.

WEB TUTORIAL 46.1 The Vertebrate Eye

The Vertebrate Eye

Figure 46.9a shows a typical vertebrate eye. The outermost layer of the structure is a tough rind of white tissue called the sclera. This is the "white of the eye." The front of the sclera forms the **cornea**, a transparent sheet of connective tissue. Just inside the cornea is a colored, round muscle called the **iris**. The iris can contract or expand to control the amount of light entering the eye. The hole in the center of the iris is the **pupil**. Light enters the eye through the cornea, passes through the pupil, and strikes a curved, clear **lens**. Together, the cornea and lens focus incoming light onto the retina in the back of the eye. The **retina** contains a thin layer of light-sensitive cells and several layers of neurons.

Figure 46.9b provides a closer look at the retina. The sensory cells that respond to light are called **photoreceptors**; they form a layer at the back of the retina. Photoreceptors synapse with an intermediate layer of neurons. Cells in the intermediate layer connect with one another and with the neurons called **ganglion cells**, which form the innermost layer of the retina. The axons of the ganglion cells project to the brain via the **optic nerve**. This is an unusual arrangement of neurons; **Box 46.2** discusses the consequences. Here, though, let's explore how photoreceptors work.

What Do Photoreceptor Cells Do? Early anatomists established that the photoreceptors in vertebrate eyes are small rod-shaped or cone-shaped cells called **rods** and **cones** (**Figure 46.10a**). When technical advances allowed changes in the membrane potentials of these cells to be recorded, it became clear that rods are very sensitive to dim light but not to color. Cones, in contrast, are much less sensitive to faint light but are stimulated by different wavelengths (that is, colors). These discoveries explained why night vision is largely black and white—at night, the rods do most of the work.

Early descriptions of the human retina also established that rods dominate most of the structure. There is one small spot in the center of the retina, however, that has only cones. This is the **fovea** (see Figure 46.9). Researchers noticed that when people focus on an object, their eyes move so that the image falls on the fovea of each eye. Based on these observations, biologists concluded that the high density of cones in the fovea maximizes the resolution of the image.

How Do Rods and Cones Detect Light? The effort to understand how rods and cones work required a combination of careful anatomical and biochemical studies. As Figure 46.10a shows, rods and cones have segments that are packed with membrane-rich disks. The membranes of these disks contain large quantities of a transmembrane protein called **opsin**. Each opsin molecule is associated with one much smaller molecule, the pigment **retinal**. In rod cells, the two-molecule complex is called **rhodopsin** (**Figure 46.10b**).

Experiments with isolated retinal, opsin, and rhodopsin molecules confirmed that retinal changes shape when it absorbs light. More specifically, the number-11 carbon in the retinal molecule changes from the *cis* conformation to the *trans* conformation (**Figure 46.10c**). Documenting this conformational change helped confirm that rhodopsin is the light receptor.

(a) The structure of the vertebrate eye

(b) In the retina, cells are arranged in layers.

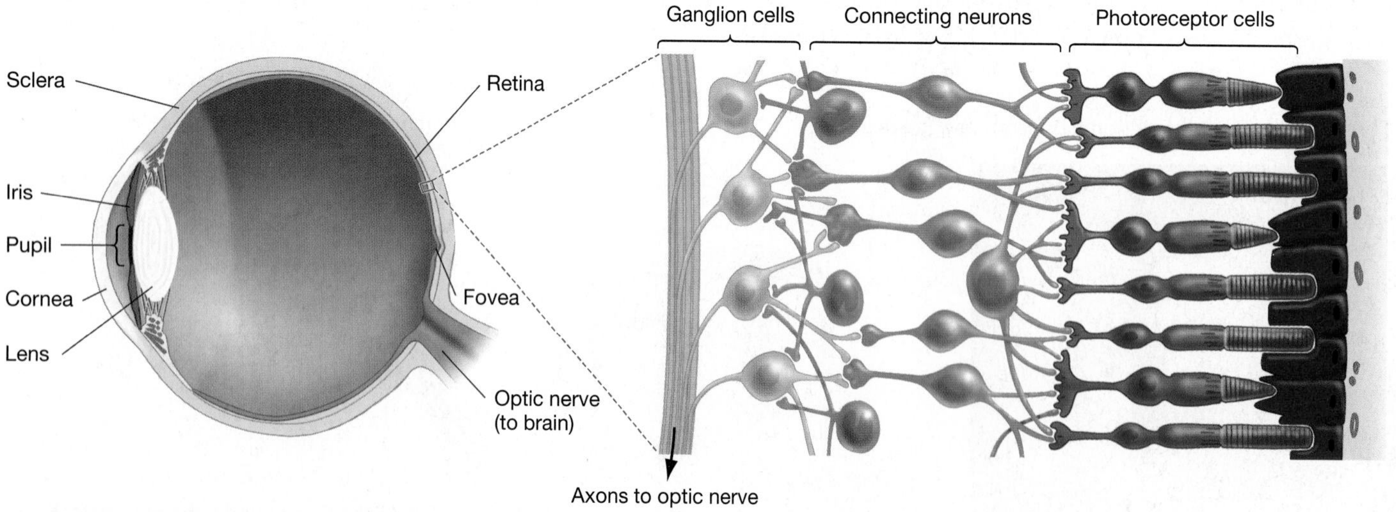

FIGURE 46.9 Camera Eyes Have a Single Lens That Focuses Incoming Light on Receptor Cells
(a) Light passes through the pupil of the eye and is focused onto the retina by the cornea and lens. **(b)** The photoreceptor cells that respond to light are in the outermost layer of the retina. QUESTION Are any photoreceptor cells located in the spot where the optic nerve leaves the retina? EXERCISE In part (b), add one arrow showing the path of light from outside the eye to a photoreceptor cell, and another arrow indicating the path of an electrical signal from this photoreceptor to the optic nerve.

BOX 46.2 Vertebrate versus Cephalopod Eyes

The photoreceptor cells in the vertebrate eye are located in the outermost layer of the retina—farthest from the incoming light. As a result, light must pass through several layers of neurons before reaching the photoreceptors. This arrangement has an important consequence: The axons of the ganglion cells in the inner layer must disrupt the photoreceptor cells in the outermost layer in order to exit the eye at the optic nerve. This disruption creates a *blind spot*—that is, a portion of the retina where there are no photoreceptors.

To convince yourself that this blind spot exists, point a finger up and hold it out at arm's length. Close your left eye, and focus your right eye on a location just beyond the tip of your finger. Now slowly move your pointed finger to the right. Can you find a place where the tip of your finger seems to vanish? Light from this region is falling on the blind spot of your right eye. If an arrow or other projectile approached from this direction, you would not detect the danger.

Squid, octopus, and other cephalopods have camera eyes analogous to those found in vertebrates, but their photoreceptors are located in the innermost layer of the retina. As a result, light strikes the photoreceptors directly, and the eye has no blind spot. Why do vertebrates have what appears to be an inferior eye design? Recall from Chapter 23 that adaptations are not perfect. To explain the existence of blind spots, biologists hypothesize that the ancestor of vertebrates had an eyelike organ with photoreceptors located in the outermost cell layer. As a camera eye evolved in the descendants of this species, the ancestral arrangement was retained. If this hypothesis is correct, our blind spots furnish another example of historical constraints on adaptation.

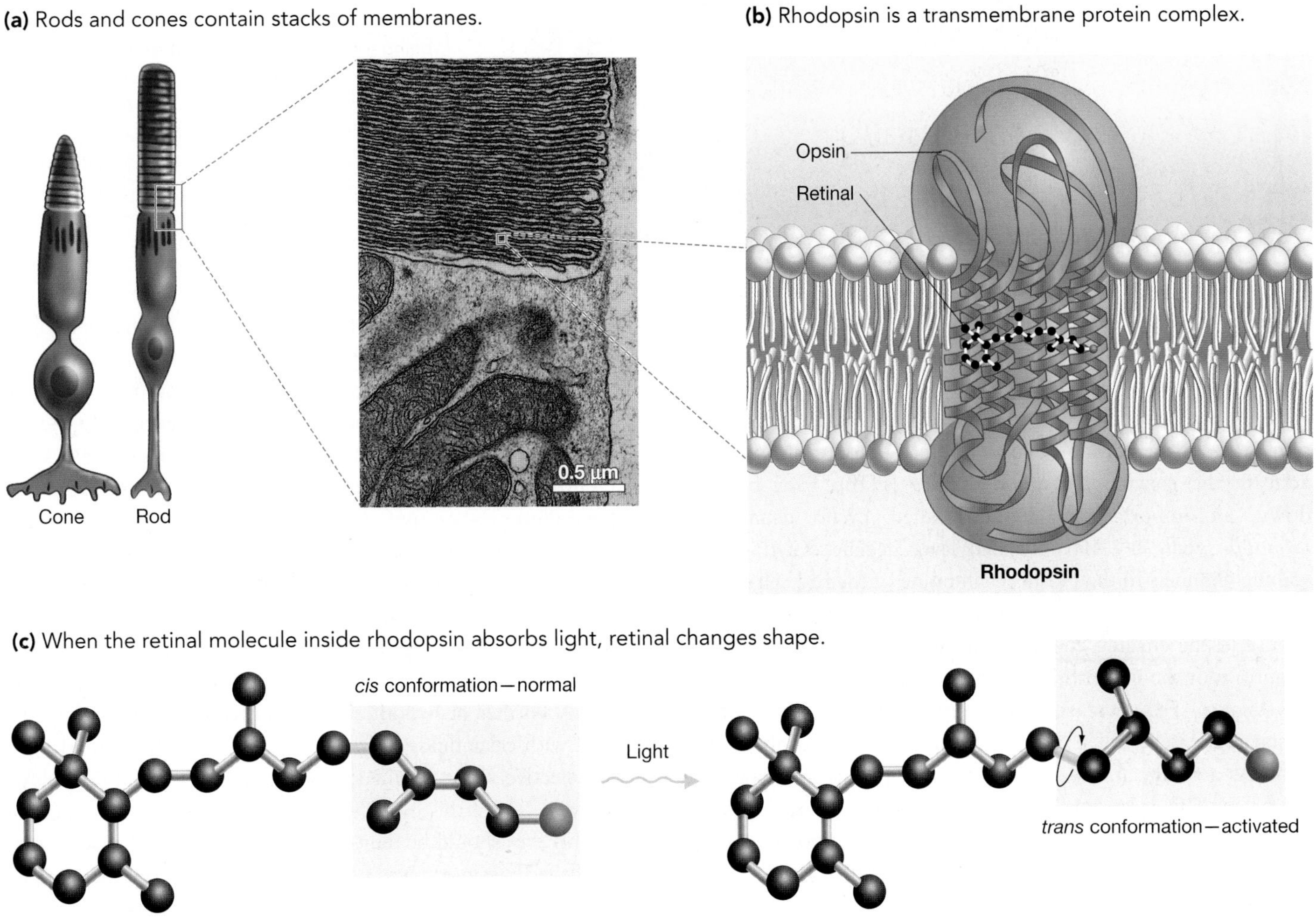
(a) Rods and cones contain stacks of membranes.
Cone
Rod
0.5 µm
(b) Rhodopsin is a transmembrane protein complex.
Opsin
Retinal
Rhodopsin
(c) When the retinal molecule inside rhodopsin absorbs light, retinal changes shape.
cis conformation—normal
Light
trans conformation—activated

FIGURE 46.10 Rods and Cones Are Packed with Transmembrane Proteins That Contain the Pigment Retinal
(a) Rods and cones have membranous disks containing thousands of opsin molecules. **(b)** Each opsin holds one retinal molecule. Retinal changes shape when it absorbs light. **(c)** In response, opsin also changes conformation.

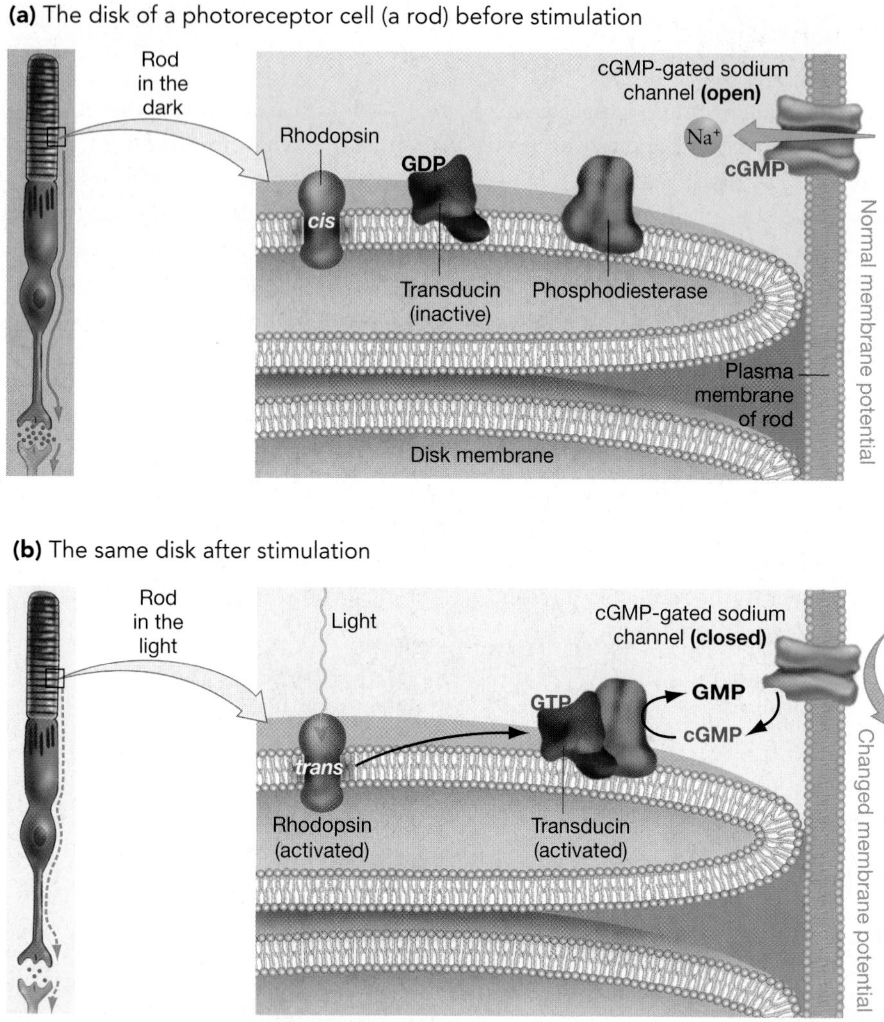

FIGURE 46.11 A Signal Transduction Pathway Connects Light Absorption with Changes in Membrane Potential
(a) An unstimulated photoreceptor. Notice that sodium ions flow into the cell when light is *not* being received. **(b)** A stimulated photoreceptor. When rhodopsin is activated, it leads to a reduction in cGMP concentration in the photoreceptor. With less cGMP available, cGMP-gated sodium channels close and the membrane hyperpolarizes. QUESTION The inflow of sodium ions into a photoreceptor cell is called "the dark current." Why?

As **Figure 46.11** shows, the shape change that occurs in retinal triggers a series of events that culminates in a change in the cell's membrane potential. The sequence begins with the activation of the opsin molecule surrounding retinal. Opsin activation causes a membrane-bound molecule called *transducin* to activate the enzyme *phosphodiesterase* (PDE). PDE breaks down an important nucleotide called *cyclic guanosine monophosphate* (cGMP). As cGMP levels decline, cGMP-gated sodium channels in the plasma membrane of the rod cell close, Na^+ entry decreases, and the membrane hyperpolarizes. In response to the ensuing change in membrane potential, smaller quantities of the neurotransmitter *glutamate* are discharged at the synapse. The decrease in neurotransmitter indicates to the postsynaptic cell, called a **bipolar cell**, that the rod absorbed light. As a result, a new pattern of action potentials is sent to the brain. This system is exquisitely sensitive: Biologists have recorded a measurable change in the membrane potentials of rod cells in response to a single photon of light.

Color Vision: The Puzzle of Dalton's Eye Studies of rhodopsin revealed the molecular mechanism responsible for light perception. But it remained to be determined how humans and other animals can respond to specific wavelengths of light. How do we perceive color?

To answer this question, consider the research program initiated by John Dalton in the late eighteenth century. At the age of 26, Dalton realized that he and his brother saw colors differently than did other people. To them, red sealing wax and green laurel leaves appeared to be the same color, and a rainbow exhibited only two hues. Dalton and his brother could not differentiate the colors red and green. This condition is called *red-green color blindness*.

In a lecture delivered in 1794, Dalton explained his perceptions by hypothesizing that red wavelengths failed to reach his retinas. Further, he hypothesized that, because a normal eyeball is filled with clear fluid, and because blue fluids absorb red light, his defective vision resulted from the presence of bluish fluid in his eyes. To test this hypothesis, Dalton left explicit instructions that his eyes should be removed after his death and examined to see if the fluid inside was blue. When he died 50 years later, an assistant dutifully removed the eyes from Dalton's corpse and examined them. The fluid inside the eye was not blue at all, however, but slightly yellow—the normal color for an older person. Further, when the back was cut off of one eye and colored objects were viewed through the lens, the objects looked perfectly normal. Dalton's hypothesis was incorrect.

What caused Dalton's color blindness? The key to answering this question was the hypothesis that the human retina contains just three types of color-sensitive photoreceptors: blue, green, and red cones, named for the colors that they best perceive. Researchers who analyzed the activity of these three types of cone cells found that each absorbed short, medium, or long wavelengths of light, respectively (**Figure 46.12**). To follow up on this result, biologists analyzed opsin molecules from the three cell types and found that each had a distinct amino acid sequence. The three proteins are called the blue, green, and red opsins (or S, M, and L, for short, medium, and long wavelengths, respectively). Although retinal is the light-absorbing molecule in all photoreceptor cells, various types of opsins respond only to distinct wavelengths of light.

Based on these results, biologists hypothesized that the brain distinguishes colors by combining signals initiated by the three classes of opsins. For example, notice in Figure 46.12 that a wavelength of 560 nm stimulates L cones strongly, M cones to an intermediate degree, and S cones hardly at all. In response to the corresponding signals from these cells, the brain perceives the color yellow.

Does this hypothesis explain Dalton's color blindness? According to the data in Figure 46.12, wavelengths from green to red do not stimulate S opsin at all. It is thus unlikely that S opsin is involved in red-green color blindness. Did Dalton fail to distinguish red and green because his M or his L cones were defective? Research has shown that color-blind

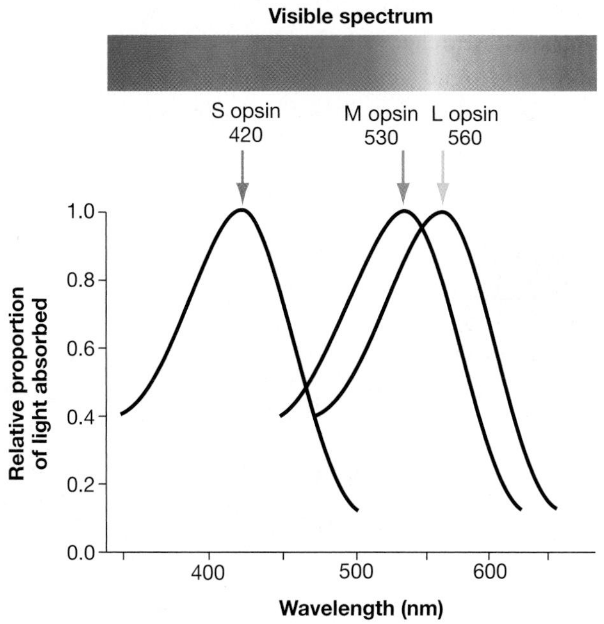

FIGURE 46.12 Color Vision Is Possible because Different Opsins Absorb Different Wavelengths of Light
Human cone cells contain three different types of opsin. Each of the opsins has a different range of absorbed wavelengths.

people lack either functional M or L cones, or both. Was the same true of Dalton? Answering this question became possible in the 1990s, when the genes for the M and L opsins were identified and sequenced. Researchers realized that if they could get a sample of Dalton's DNA, they could analyze his M opsin genes and L opsin genes directly. Remarkably, Dalton's eyes had been preserved. The research group managed to extract DNA from the 150-year-old tissue and found that Dalton had a normal *L* allele but lacked a functional *M* allele. As a result, he did not have green-sensitive cones. The puzzle of Dalton's color vision was solved.

Sensory Worlds: Do Other Animals See Color?

What about other animals—do they see color the way we do? The short answer to this question is probably not. Animals that are active at night have relatively few cone cells and many rods, giving them high sensitivity to light but poor color vision. And recent research has shown that different species with good color vision have opsins with different peak light sensitivities. A marine fish called the coelacanth, which lives in water 200 meters deep, has two opsins that respond to the blue region of the spectrum (wavelengths of 478 nm and 485 nm). Based on this observation, it is likely that coelocanths perceive several distinct hues of blue that we would perceive as a single color. The existence of these opsins is logical, because wavelengths in the yellow and red parts of the spectrum do not penetrate well into deep water. Because only blue light exists in the coelacanth's habitat, biologists interpret the opsins of this fish as an adaptation to life in deep water.

Correlations between opsin structure and visual function are extensive. For example, humans and other primates that eat fruit have opsins that are sensitive to wavelengths around 550 nm. The presence of these opsins allows individuals to make fine distinctions between the greens, yellows, and reds of unripe and ripe fruits.

In addition, numerous vertebrate and invertebrate species have four or more types of opsins and probably perceive a world of colors that is much richer than ours. Birds, insects, and a few mammals can also see **ultraviolet light**, which has shorter wavelengths than humans can see. This ability is adaptive. As explained in Chapter 40, certain flowers have ultraviolet patterns that serve as signals for insect pollinators. In addition, many birds have strong ultraviolet patterns in their plumage that are invisible to us. Recent experiments have confirmed that, in some species, females use these ultraviolet patches as a criterion for selecting mates. At the other end of the visible spectrum, rattlesnakes and other pit vipers can sense **infrared** wavelengths by using specialized infrared-sensing organs on their snouts (**Figure 46.13a**, page 1064). Infrared light has longer wavelengths than humans can see. Because endothermic animals radiate heat in the form of infrared radiation, pit vipers can detect and strike at prey in complete darkness (**Figure 46.13b**).

(a) Pit vipers can detect infrared radiation. **(b)** Warm animals give off infrared radiation.

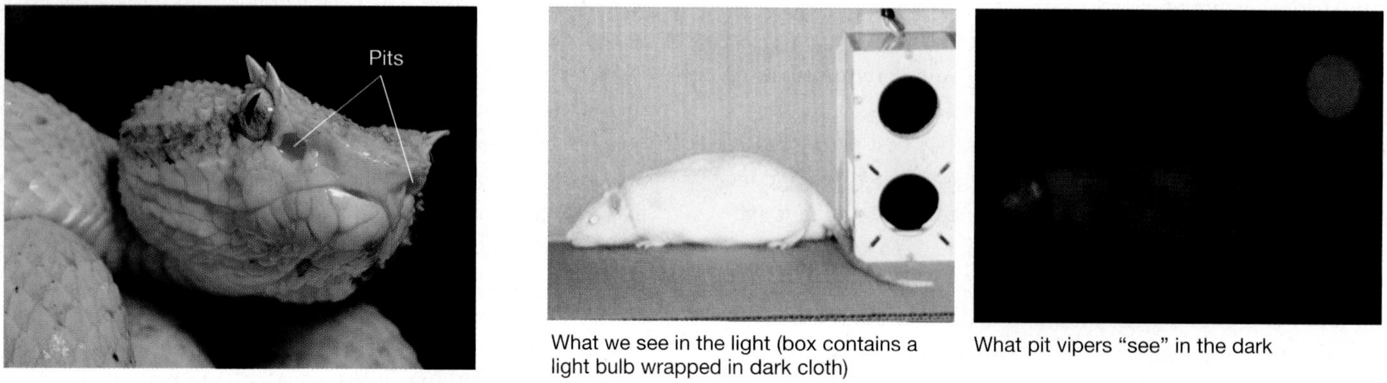
Pits
What we see in the light (box contains a light bulb wrapped in dark cloth)
What pit vipers "see" in the dark

FIGURE 46.13 Pit Vipers Detect Infrared Radiation
(a) Pit vipers are named for the "pits" on their heads. These pits contain receptor cells that detect infrared radiation. **(b)** Pit vipers are able to "see" a fuzzy image of warm objects. As a result, a rattlesnake can strike accurately at a mouse even in darkness. (They will even strike at warm, covered light bulbs, such as those hidden in the white box.) The snake loses this ability if the pits are covered experimentally.

✓CHECK YOUR UNDERSTANDING

In vertebrates, light detection begins when retinal changes shape after absorbing light. If opsin changes conformation in response, a series of events results in the closing of sodium channels in the membrane of a photoreceptor cell and thus a change in membrane voltage. You should be able to (1) make a labeled diagram of the vertebrate eye and describe the function of each part; and (2) predict the type of vision loss that results from each of the following: a tear in the fovea, a severe mutation in the gene for S opsin, and an age-related clouding and stiffening of the lens.

46.4 Taste and Smell

The senses of taste (**gustation**) and smell (**olfaction**) originate in chemoreceptors. These cells detect the presence of particular molecules. Until recently, taste and smell were poorly understood compared with vision and hearing. It is easy to see why. Eyes and ears respond to relatively simple stimuli—light and sound waves—that can be reproduced in the lab. The tongue and nose, in contrast, respond to thousands of different chemicals. Consequently, taste and smell were difficult to study until techniques became available for identifying how particular molecules bind to certain receptors. But in the last 10 years, research on the chemosenses has exploded.

Understanding taste and smell is important because animals use these senses to find food, assess mates, and avoid danger. For researchers in the food and cosmetic industries, understanding particular aspects of chemoreception, such as the sensation of sweetness and the activity of perfumes, is critical to developing new products.

Taste: Detecting Molecules in the Mouth

The chemoreceptor cells that sense taste are clustered in structures known as **taste buds**. Although humans have taste buds scattered around the mouth and throat, most taste buds are located on the tongue. As **Figure 46.14** shows, a taste bud contains about 100 spindle-shaped **taste cells**, which are taste receptors that synapse onto sensory neurons. How do these receptors work on a molecular level, and how do they produce the sensation of taste?

Early taste research focused almost exclusively on the hypothesis that four "basic tastes" existed: salty, sour, sweet, and bitter. Researchers who analyzed the membrane proteins in taste cells found strong evidence that salt and sour sensations result from the activity of ion channels. Specifically, the

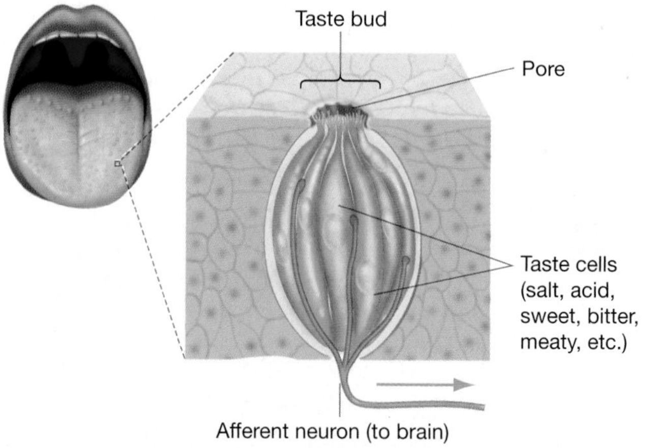
Taste bud
Pore
Taste cells (salt, acid, sweet, bitter, meaty, etc.)
Afferent neuron (to brain)

FIGURE 46.14 Taste Buds Contain Many Types of Receptor Cells
The human tongue has a high density of taste cells located in structures called taste buds.

sensation of saltiness is due primarily to sodium ions (Na^+) dissolved in food. These ions flow into certain taste cells through open Na^+ channels and depolarize the cells' membranes. Similarly, sourness is due to the presence of protons (H^+), which flow directly into certain taste cells through H^+ channels. The sour taste of grapefruit and other citrus fruits, for example, results from the release of protons by citric acid. In general, the lower the pH of a food, the more it depolarizes a taste cell's plasma membrane and the more sour the food tastes.

The molecular mechanisms responsible for the sensations of bitterness and sweetness have been much more difficult to identify, however. Researchers have only recently been able to document that certain food molecules actually bind to specific receptors on taste cells to cause bitter and sweet tastes.

Why Do Many Different Foods Taste Bitter? Bitterness has been a particularly difficult taste to understand. The problem is that molecules with very different structures are perceived as bitter. How is this possible? An answer began to emerge after researchers confirmed that some humans genetically lack the ability to taste certain bitter substances. This phenomenon came to light in 1931, when Arthur Fox was synthesizing phenylthiocarbamide (PTC) and accidentally blew some of it into the air. A nearby colleague complained of a bitter taste in his mouth. Fox could not taste anything, however. Follow-up research confirmed that the ability to taste PTC is inherited and polymorphic. About 25 percent of United States citizens cannot sense the molecule. To find the gene responsible for this trait, biologists compared the distribution of genetic markers observed in "tasters" and "nontasters." The mapping effort narrowed down this gene's location to several candidate chromosomal segments. (Chapter 19 introduced how this type of gene hunt is done.)

At about the same time, researchers who were working on taste reception in mice found that some individuals are better than others at perceiving bitter substances and avoiding them. Gene mapping narrowed the location of the bitter receptor gene down to a specific region of a mouse chromosome.

These independent lines of research converged when different research teams searched a database of known genes to look for sequences similar to those from the candidate regions of the human and mouse chromosomes. Independently, the two teams identified a family of 40 to 80 genes that encode transmembrane receptor proteins. Follow-up work documented that each protein in the family binds to just one particular type of bitter molecule. A taste cell, however, can have many different receptor proteins from this family. As a result, many kinds of molecules can depolarize the same cell and cause the taste of bitterness. These results explained why so many different molecules give rise to the same sensation.

Why do animals have such extensive molecular machinery to detect bitterness? To answer this question, biologists note that many of the molecules that bind to bitter receptors are found in toxic plants. Also, most animals react to bitter foods by spitting them out and avoiding them in the future. Based on these observations, researchers hypothesize that bitter receptors evolved in response to selection for avoiding toxic molecules. In essence, bitterness means "this food might be dangerous; don't swallow it." The proliferation of genes responsible for detecting bitterness hints that the trait is extremely important—meaning that individuals with the capacity to detect a wide array of bitter compounds produce more offspring than do individuals that lack this ability.

What Is the Molecular Basis of Sweetness and Other Tastes? Inspired by the rapid recent progress on bitter receptors, research teams are using the same techniques to search for sweetness and other types of receptors. Several proteins have been identified as possible sweetness detectors, some of which are closely related to recently discovered membrane receptors for glutamate and other amino acids. The glutamate receptor is responsible for the sensation called **umami**, which is the meaty taste of the molecule monosodium glutamate (MSG). Researchers are still struggling to understand how the sweetness receptors work, however, and exactly how many different membrane proteins might be involved in detecting the presence of sugars in food.

In addition to searching for other types of taste receptors, several teams are studying how the different taste sensations are conveyed to the brain and interpreted. Although taste is beginning to reveal its secrets, the complete story will probably not be known for many more years.

Olfaction: Detecting Molecules in the Air

Taste allows animals to assess the quality of their food before swallowing it. Smell, in contrast, allows animals to monitor airborne molecules that convey information. Wolves and domestic dogs, for example, can distinguish millions of different odors at vanishingly small concentrations. The molecules that cause odor contain information about the movements and activities of prey and other members of their own species.

How does the sense of smell work? When odor molecules, or *odorants*, reach the nose, they diffuse into a mucus layer in the roof of the nose. There, they activate olfactory receptor neurons via membrane-bound receptor proteins (**Figure 46.15**, page 1066). Axons from these neurons project up to the olfactory bulb. The **olfactory bulb** is the part of the brain where olfactory signals are processed and interpreted.

Understanding the anatomy of the odor recognition system was a relatively simple task. Understanding how receptor neurons distinguish one molecule from another turned out to be much more difficult. Initially, investigators hypothesized that receptors respond to a small set of "basic odors," such as musky, floral, minty, and so on. The idea was that each basic odor would be detected by its own type of receptor.

In 1991, however, Linda Buck and Richard Axel demolished the basic-odor hypothesis with an astonishing result: Their

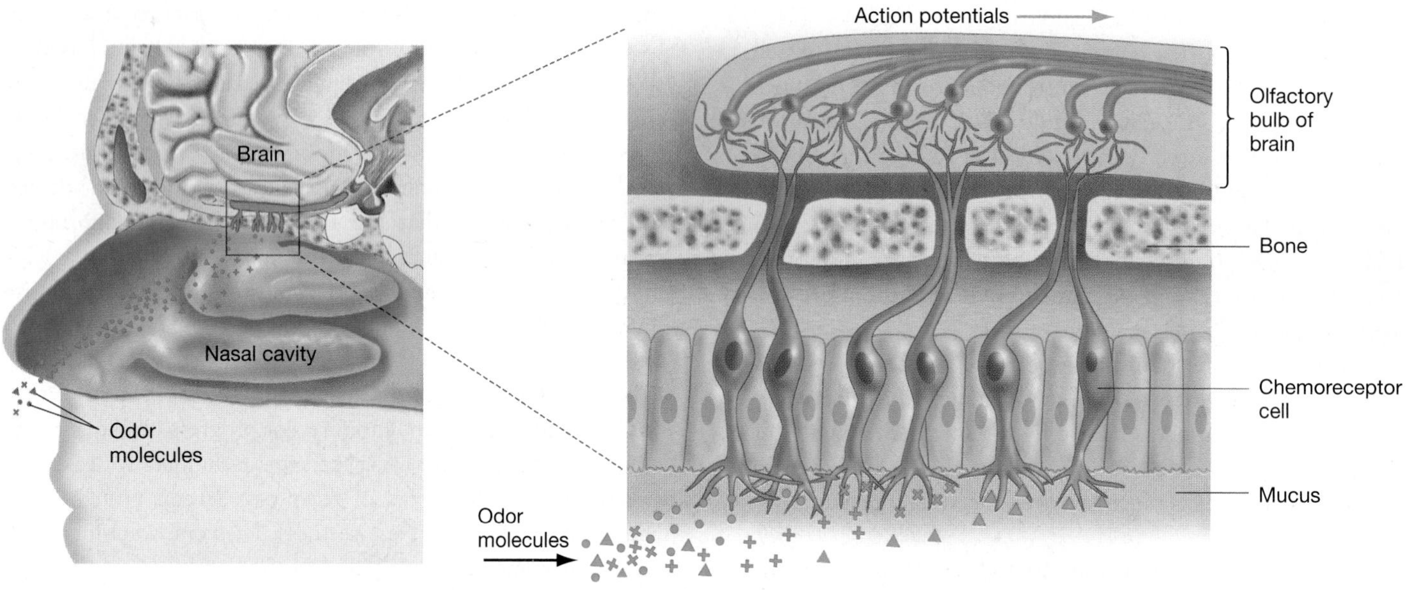

FIGURE 46.15 In Mammals, Receptor Cells in the Nose Respond to Specific Odorants
Odor molecules are detected in the nose by chemoreceptor cells. Each of these sensory neurons carries one type of odor-receptor protein on its dendrites. Sensory neurons with the same receptor project to the same glomerulus, or section within the olfactory bulb of the brain.

studies of odor recognition in mice culminated in the discovery of a gene family—containing hundreds of distinct coding regions—that encodes receptor proteins on the surface of odor-receptor neurons. Follow-up experiments confirmed that each receptor protein binds to a small set of molecules. Further work established that most, if not all, vertebrates have this family of genes. Mammals typically have between 500 and 1000 of these genes, meaning mammals can produce 500 to 1000 different odor-receptor proteins. In humans, however, about half of these genes have mutations that probably render them nonfunctional. This observation may explain why the sense of smell is so poor in humans compared with that of other mammals.

Buck and Axel's announcement inspired a series of questions. How many different receptors occur in the membrane of each neuron involved in odor reception? How does the brain make sense of the input from so many different receptors? Recently, Buck determined that each olfactory neuron has only one type of receptor and that neurons with the same type of receptor project to distinct regions in the olfactory bulb of the brain. These regions are called **glomeruli** (meaning "little balls"). Follow-up work from Axel's lab indicated that particular smells in mice are associated with the activation of a certain subset of the 2000 glomeruli in the brain. For example, the activation of clumps 130, 256, and 1502 might be perceived as the smell "cinnamon." In essence, then, the sense of smell is similar to the visual system's use of three cones to perceive many colors, but on a much larger scale. Research on this complex and impressive sense continues at a furious pace.

✓CHECK YOUR UNDERSTANDING

Chemoreception occurs when a specific molecule in air or food binds to a specific receptor in the nose or mouth and the binding event is transduced to a change in membrane voltage. You should be able to explain (1) why a loss in chemosensory ability occurs when you burn your tongue with extremely hot food, (2) why dogs have a better sense of smell than people do, and (3) why some people cannot smell the compound PTC.

46.5 Movement

The first four sections of this chapter focused on how animals sense different aspects of their environment. Acquiring information is only a first step, however. Information about changes in the environment is useless to an animal unless the individual can respond to it in an appropriate way—usually by moving. Bats and moths not only hear ultrasonic frequencies but also alter their flight paths and activities in response—in a way that allows them to catch food or avoid being eaten. Flight and other types of motion are based on muscle contractions in conjunction with a skeleton. Recall from Chapters 31–33 that skeletons, muscle tissue, and sophisticated movements are considered signature adaptations of animals.

It is important to remember that **locomotion**, or the movement of an entire animal, is only one type of movement. Sessile organisms, such as barnacles and sea anemones, may not undergo locomotion, but they rely on movement of the heart muscle or gills to accomplish circulation and respiration.

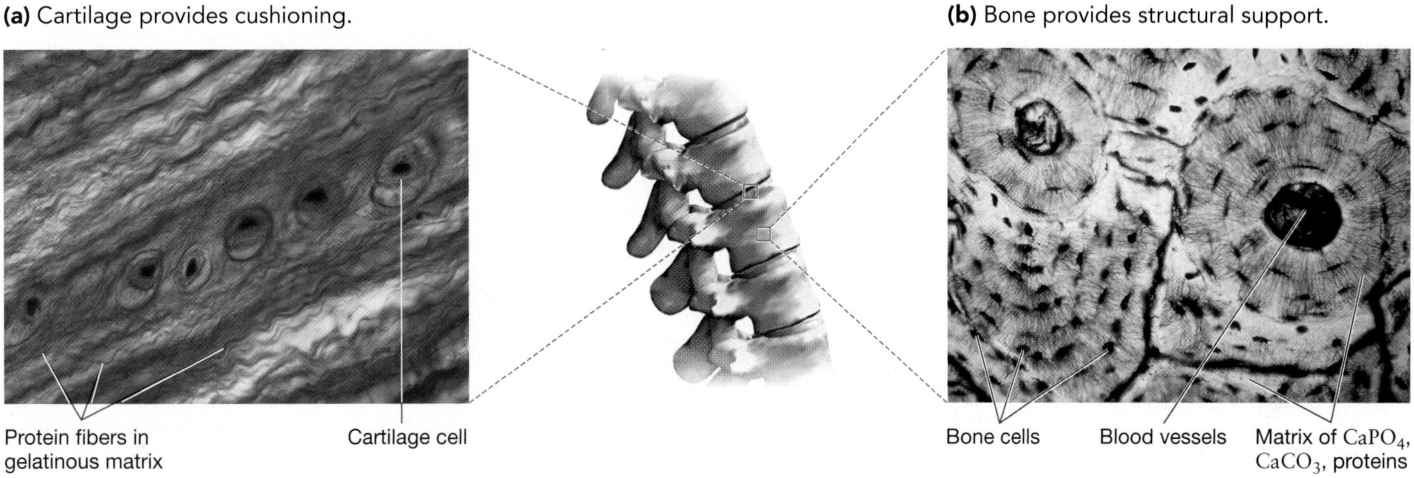
(a) Cartilage provides cushioning.
(b) Bone provides structural support.
Protein fibers in gelatinous matrix
Cartilage cell
Bone cells
Blood vessels
Matrix of $CaPO_4$, $CaCO_3$, proteins

FIGURE 46.16 Endoskeletons Contain Two Major Tissue Types
The vertebral column shown is made up of bones separated by disks of cartilage.

To explore locomotion in detail, let's briefly examine the skeleton's role as a scaffold for muscle attachment. The analysis of animal skeletons is followed by a description of the diversity of muscle cells and tissues found in vertebrates and how they work.

Skeletons

Skeletons provide attachment sites for muscles and a support system for the body's soft tissues. As indicated in Chapters 31–33, three types of skeletons are found in animals. **Exoskeletons** are hard, hollow structures that envelop the body; **hydrostatic skeletons** use the pressure of internal body fluids to support the body; and **endoskeletons** are hard structures inside the body. Because those chapters introduced the design of exoskeletons and hydrostatic skeletons in some detail, our focus here is on the structure and function of endoskeletons.

Endoskeletons are composed of connective tissues called cartilage and bone. **Cartilage** is made up of cells scattered in a gelatinous matrix of polysaccharides and protein fibers (**Figure 46.16a**). In the skeletal system, cartilage provides padding between bones. **Bone** is made up of cells in a hard extracellular matrix of calcium phosphate ($CaPO_4$) with small amounts of calcium carbonate ($CaCO_3$) and protein fibers (**Figure 14.16b**). Bones meet and interact at locations called **articulations**, or **joints**. Bones articulate in ways that allow limbs to swivel, hinge, or pivot (**Figure 46.17**).

How do skeletal muscles attach to and move elements of an endoskeleton? In vertebrates, most skeletal muscles have ends that are attached to different bones by tendons. **Tendons** are bands of tough, fibrous connective tissue. Muscles can exert force only by contracting, so pairs of muscles must work together to move a bone back and forth. In the case of a limb, one muscle pulls the limb in one direction; the other muscle pulls it in the opposite direction. This pairing of muscles is called an **antagonistic muscle group.** The muscle that swings two long bones in an arc toward each other is called a **flexor**; the muscle that straightens them out is called an **extensor** (**Figure 46.18a**, page 1068). For example, the hamstring muscle in the back of your thigh flexes your lower leg; the quadriceps muscle in the front of your thigh extends it. When the hamstring contracts, the quadriceps relaxes; when the quadriceps contracts, the hamstring relaxes. The hamstring and quadriceps are part of an antagonistic muscle group.

Locomotion in animals with exoskeletons and hydrostatic skeletons is also based on antagonistic muscle groups. Animals with exoskeletons have paired flexor-extensor muscles inside their hollow leg joints (**Figure 46.18b**). Animals with hydrostatic skeletons have both circumferential muscles and longitudinal

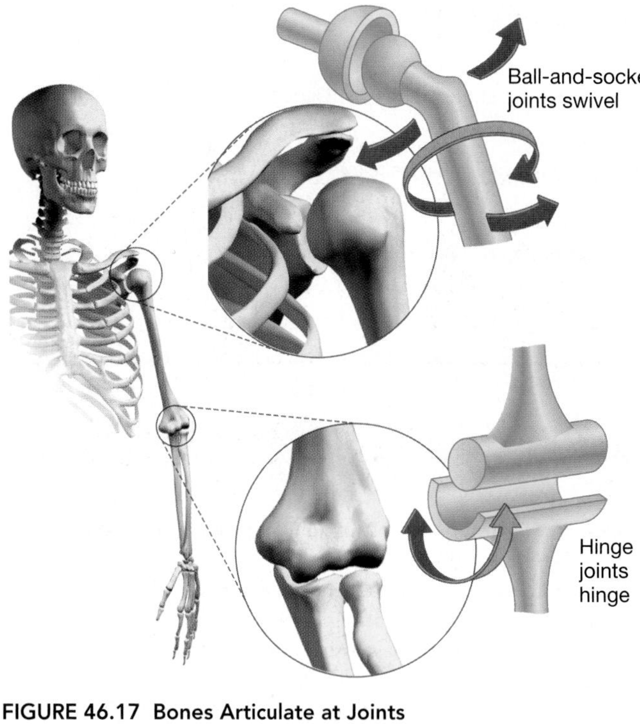
Ball-and-socket joints swivel
Hinge joints hinge

FIGURE 46.17 Bones Articulate at Joints
Bones articulate in ways that make specific types of movement possible.

muscles, and these muscles work in concert to shorten and lengthen body segments (**Figure 46.18c**).

In all vertebrate animals, the movements of paired muscles are coordinated by motor neurons that originate in the brain or spinal cord. These motor neurons project from processing centers that receive input from sensory systems. Motor neuron activity changes in response to information about balance, smells, sights, and sounds. In this way, an animal's movements are directly tied to the functioning of its sensory systems. The interplay of sensory input and motor output results in the coordinated behaviors we call running, eating, swimming, and flying.

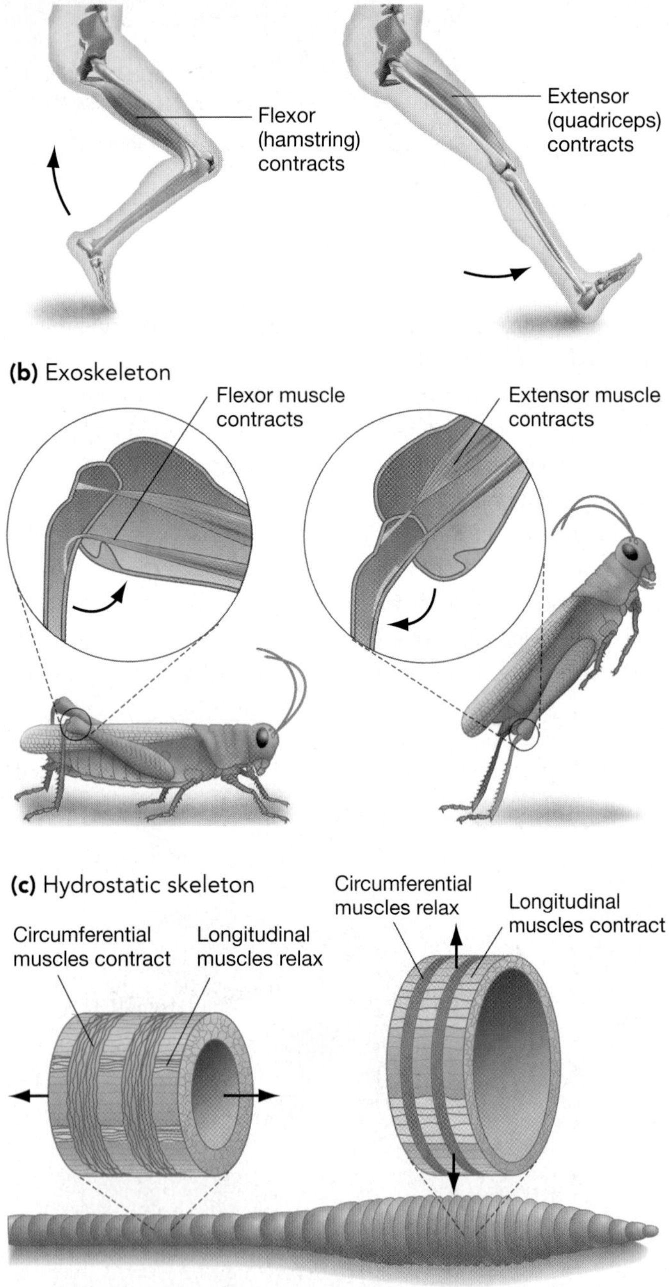
(a) Endoskeleton
Flexor (hamstring) contracts
Extensor (quadriceps) contracts
(b) Exoskeleton
Flexor muscle contracts
Extensor muscle contracts
(c) Hydrostatic skeleton
Circumferential muscles contract
Longitudinal muscles relax
Circumferential muscles relax
Longitudinal muscles contract

FIGURE 46.18 Antagonistic Muscle Groups Flex or Extend Skeletal Elements

(a) In vertebrates, which have an endoskeleton, muscles are arranged in pairs of flexors and extensors. Together, a flexor and an extensor can control a joint. **(b)** Animals with an exoskeleton also have flexors and extensors, but inside hollow joints. **(c)** In animals with a hydrostatic skeleton, the coordinated movements of longitudinal and circular muscles shorten and lengthen body segments.

How Do Muscles Contract?

Early microscopists established that the muscle tissue found in vertebrate limbs and hearts is composed of slender fibers. A **muscle fiber** is a long, thin muscle cell. Within each cell are many small strands called **myofibrils.** But when workers examined the sheets or bands of muscle that surround blood vessels and most organs, they found that myofibrils were absent. Because they lack myofibrils, these muscle cells are called **smooth muscle.**

Further anatomical studies confirmed that vertebrates have three distinct types of muscle tissue: (1) skeletal muscle, (2) cardiac muscle, and (3) smooth muscle. These muscle types were introduced briefly in Chapter 41; **Figure 46.19** provides a more detailed comparison. The three differ in overall morphology as follows:

1. Skeletal muscle consists of unbranched *multinucleate* cells—they contain multiple nuclei. This muscle is also known as *striated muscle*, because it appears to be striated, or striped.
2. Cardiac muscle contains branched cells whose ends are connected via specialized regions called intercalated discs. In an **intercalated disc**, protein-lined openings called gap junctions provide a direct cytoplasmic connection between adjacent muscle cells. As noted in Chapter 44, intercalated discs are critical to the flow of electrical signals from cell to cell and to the coordination of the heartbeat.
3. Smooth muscle is unbranched, lacks myofibrils, and is often organized into thin sheets.

Although smooth muscle is essential to the function of the lungs, blood vessels, digestive system, urinary bladder, and reproductive system, most research on muscle function has focused on how skeletal and cardiac muscles work. Let's consider the structure and function of skeletal muscle in more detail, with the goal of understanding how these cells contract to produce movement.

The Sliding-Filament Model As **Figure 46.20** shows, the myofibrils inside skeletal muscle cells look striped when viewed with the electron microscope. The pattern is caused by alternating light-dark units called **sarcomeres,** which repeat down the length of a myofibril. The sarcomeres in a myofibril lengthen when a muscle fiber is relaxed and shorten when the cell is contracted.

Based on these observations, it became clear that the question of how muscles contract simplifies to the question of how sarcomeres contract. Hugh Huxley and Jean Hanson proposed an answer to this question in 1954. These biologists had noticed that the relationship between different parts of the sarcomere—

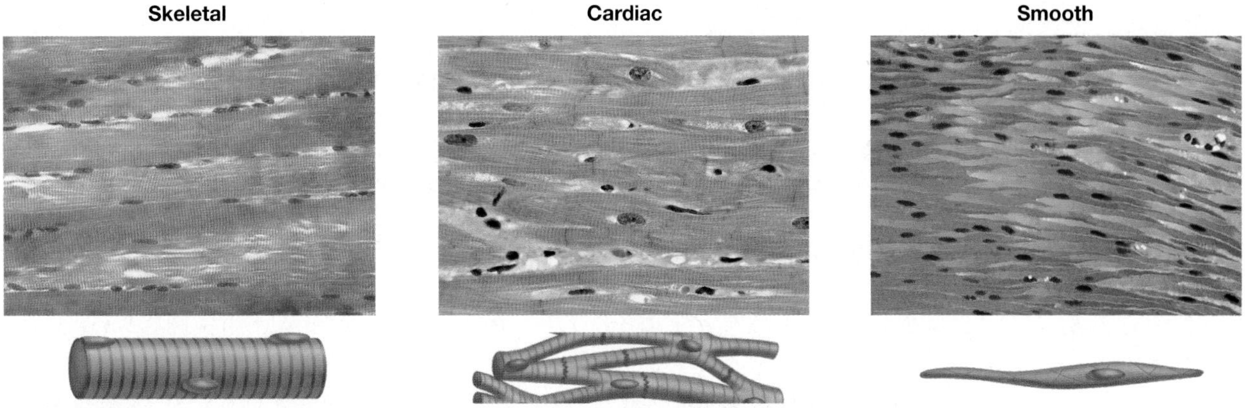

	Skeletal	Cardiac	Smooth
Location:	Attached to bones	Heart	Intestines, arteries, other
Function:	Move skeleton	Pump blood	Move food, help regulate blood pressure, etc.
Characteristics of cells:	• Multinucleate	• 1 or 2 nuclei	• Single nucleus
	• Unbranched	• Branched; intercalated discs form direct cytoplasmic connection end to end	• Unbranched
	• Activity is "voluntary," meaning that signal from motor neuron is required	• Activity is "non-voluntary," meaning that signal from motor neuron is not required	• Activity is "non-voluntary," meaning that signal from motor neuron is not required

FIGURE 46.19 Vertebrates Have Three Types of Muscle

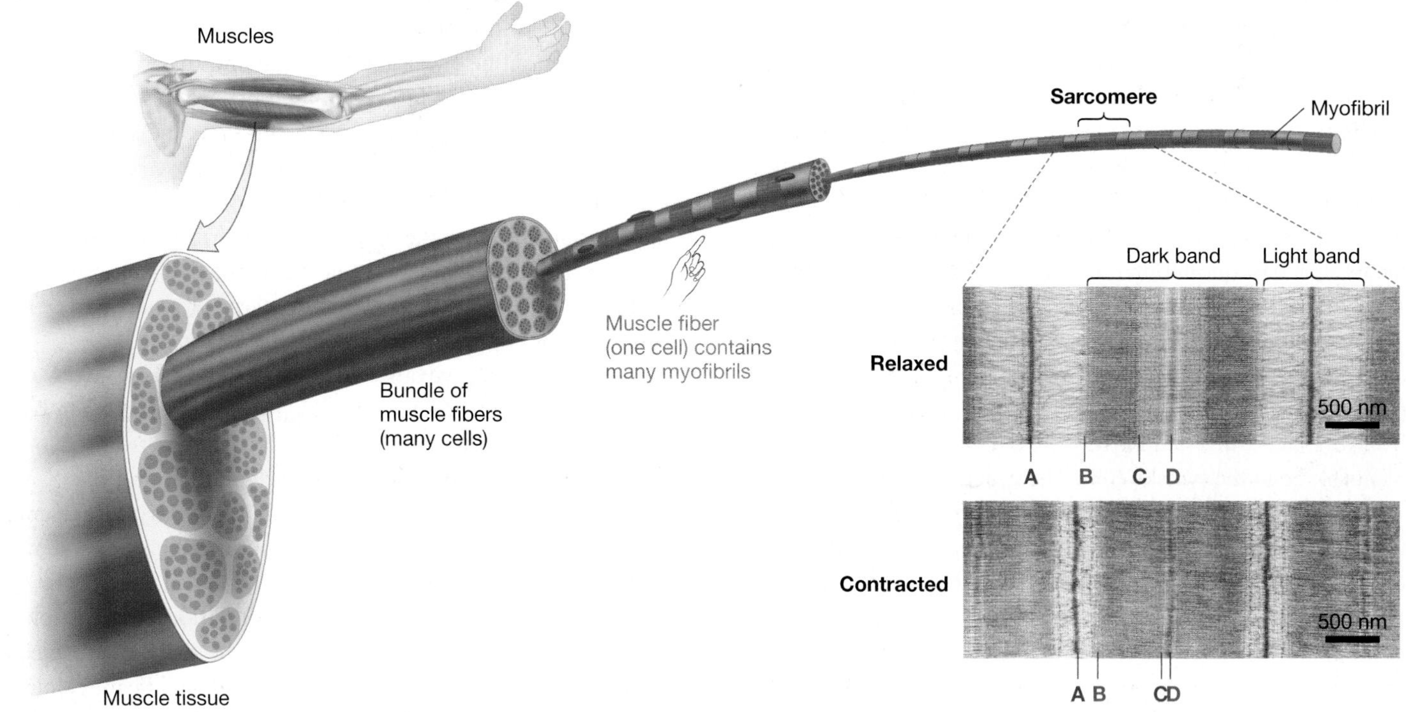

FIGURE 46.20 Muscle Cells Contain Many Myofibrils, Which Contain Many Sarcomeres
Skeletal muscle cells (fibers) have a striped appearance due to repeating sarcomeres, which are units of alternating light-dark bands. When a sarcomere contracts, the distances between the points labeled A, B, C, and D change. QUESTION Measure the distance from A to B, A to C, A to D, B to C, B to D, and C to D. Which points get closer together during contraction? Which points stay the same distance from each other?

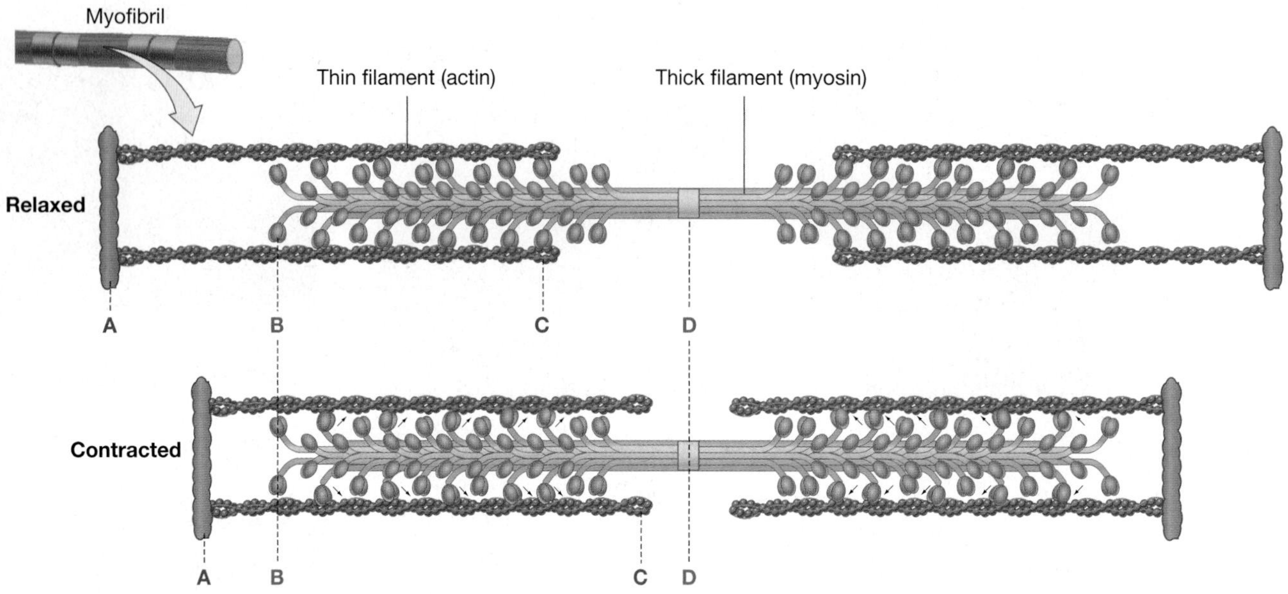

FIGURE 46.21 The Sliding-Filament Model of Sarcomere Contraction
Points A, B, C, and D are the same as those labeled on the sarcomere photographs in Figure 46.20.
QUESTION Does this model explain the changes in distances that you observed in Figure 46.20? Explain.

specifically, the points labeled A–D in the photographs in Figure 46.20—change during contraction. Points A and B move closer to each other during contraction, as do points C and D. But the distance from point A to point C does not change, and the distance from point B to point D does not change.

To explain these observations, Huxley and Hanson hypothesized that the banding patterns in the sarcomere are actually caused by two types of long filaments—**thick filaments** and **thin filaments**—and that these filaments slide past one another during contraction. **Figure 46.21** illustrates the model. Notice that thin filaments extend from A to C and thick filaments extend from B to D. According to Huxley and Hanson, the thick and thin filaments slide past one another during a contraction. This explanation became known as the **sliding-filament model.**

Follow-up research has shown that the Huxley-Hanson model is correct in almost every detail. Thin filaments are composed of two coiled chains of the globular protein **actin.** (Recall from Chapter 7 that actin is the building block of the microfilaments found in the cytoskeleton. The structure of microfilaments is similar to the structure of thin filaments.) One end of each thin filament is bound to the wall of the sarcomere; the other end is free to interact with thick filaments. Thick filaments are composed of multiple strands of a long protein called **myosin** (see Figure 46.21). The thick strands are anchored to the middle of the sarcomere; at both ends they are free to interact with thin filaments. Indeed, researchers who isolated actin filaments and myosin molecules and mixed them together on a glass slide in the presence of ATP then observed actin sliding past myosin. As predicted, thick and thin filaments interact in a way that allows them to slide past one another.

WEB TUTORIAL 46.2 Muscle Contraction

How Do Actin and Myosin Interact? Once the sliding-filament model was confirmed, researchers focused on the question of how the sliding action occurs at the molecular level. Early work on the three-dimensional structure of myosin revealed that each molecule has a head region that projects from the main body of the thick filament. Follow-up experiments established that the myosin head can bind to actin and that it can catalyze the hydrolysis of ATP into ADP and a phosphate ion. These results suggested that myosin—not actin—was the site of active movement. In addition, electron microscopy revealed that myosin and actin are locked together shortly after an animal dies and its muscles enter the stiff state known as rigor mortis. Because ATP is unavailable in dead tissue, the data suggested that ATP is involved in getting myosin to release from actin once the two molecules have bound to each other.

The next major advance in understanding how actin and myosin interact to produce movement was achieved when Ivan Rayment and colleagues solved the detailed three-dimensional structure of the myosin head (**Figure 46.22**), using the X-ray crystallographic techniques introduced in Chapter 3. Rayment's group determined the location of the actin biding site and also examined how the protein's structure changed when ATP or ADP was bound to it. Because phosphorylation or dephosphorylation can lead to dramatic changes in the shape of a protein, the investigators were not surprised to find significant changes in the protein's conformation when bound to ATP versus ADP.

Based on these data, Rayment and co-workers proposed the model for actin-myosin interaction illustrated in **Figure 46.23.** When relaxed, the myosin head of a thick filament is attached to actin in a thin filament. As step 1 shows, a molecule of

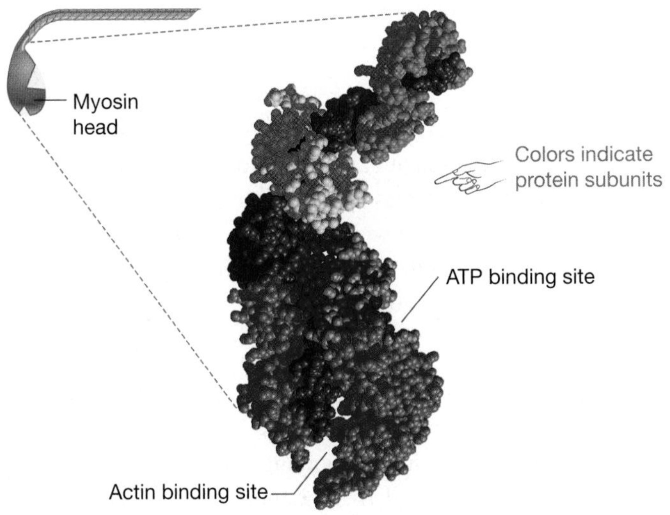
Myosin head
Colors indicate protein subunits
ATP binding site
Actin binding site

FIGURE 46.22 Myosin's "Head" Binds ATP and Actin
The head region of the myosin protein contains both an ATP binding site and an actin-binding site.

ATP then binds to the myosin head, and the myosin releases from actin. (This is the step that cannot occur after death. Rigor mortis sets in as ATP supplies run out.) When ATP is subsequently hydrolyzed to ADP and inorganic phosphate in step 2, the conformation of the protein changes. Specifically, the neck of the myosin straightens out, and the head pivots. The myosin head then binds to actin in a new location farther down the thin filament. The third step occurs when inorganic phosphate is released and the protein's conformation changes again. Specifically, the neck bends back into its original position. This bending is called the **power stroke**, because it moves the entire thin filament. In step 4, ADP is released. A new ATP molecule then binds to the myosin, causing myosin to release from actin, and the cycle starts anew.

As ATP binding, hydrolysis, and release continue, the two ends of the sarcomere are pulled closer together. The same basic mechanism is observed in cardiac and smooth muscle cells. Most animal movement, in fact, is powered by similar interactions between actin and myosin. These two proteins are also responsible for the amoeboid movement observed in amoebae and slime molds (see Chapter 28) and the streaming of cytoplasm observed in algae and land plants. The message of these observations is that actin and myosin are evolutionarily ancient proteins. Because they make movement possible in the absence of cilia and flagella, they have played a critical role in the diversification of eukaryotes.

How Does Relaxation Occur? The results summarized in Figures 46.21 and 46.23 were obtained by studying myosin and actin in isolation. Although elegant and powerful, experiments on isolated fibers indicate only what is happening in individual fibers. In vitro, thick filaments from muscle continue to crawl along thin filaments as long as ATP and a length of thin filament are available. ATP is almost always available in living muscles. How do our muscles ever stop contracting and relax?

Researchers were able to answer this question when they confirmed that, in addition to containing actin and myosin, thin filaments contain two key proteins called **tropomyosin** and **troponin**. Tropomyosin and troponin work together to block the myosin binding sites on actin. When tropomyosin and troponin are in a particular conformation, the myosin-actin interaction cannot occur, and thick and thin filaments cannot slide

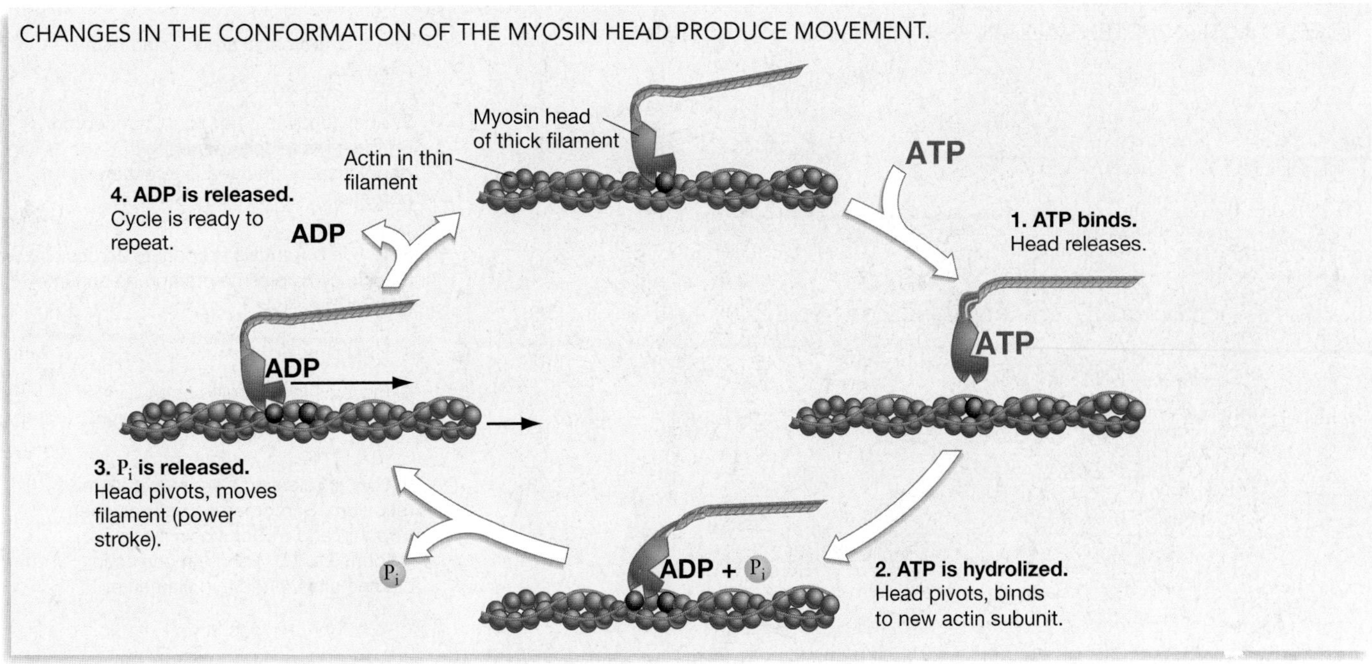
CHANGES IN THE CONFORMATION OF THE MYOSIN HEAD PRODUCE MOVEMENT.
Myosin head of thick filament
Actin in thin filament
ATP
1. ATP binds. Head releases.
ATP
2. ATP is hydrolized. Head pivots, binds to new actin subunit.
ADP + P$_i$
3. P$_i$ is released. Head pivots, moves filament (power stroke).
P$_i$
ADP
4. ADP is released. Cycle is ready to repeat.
ADP

FIGURE 46.23 Myosin and Actin Interact during Muscle Contraction
Summary of the current model of how myosin and actin interact as a sarcomere contracts. The four steps repeat rapidly.

past each other. In vivo, how are tropomyosin and troponin moved out of the way so that contraction can begin? The answer turned out to have several steps, beginning with the arrival of an electrical signal from a motor neuron.

An Overview of Events at the Neuromuscular Junction

Figure 46.24 summarizes what happens when an action potential from a motor neuron arrives at a muscle cell and initiates contraction:

1. Action potentials trigger the release of the neurotransmitter called **acetylcholine (ACh)** from the motor neuron into the synaptic cleft between the motor neuron and the muscle cell. Experiments documented this event in cardiac muscle; follow-up research confirmed that ACh is also released by motor neurons in skeletal muscle.
2. Acetylcholine diffuses across the synaptic cleft and binds to ACh receptors on the plasma membrane of the muscle cell. By recording voltage changes in muscle cells, biologists showed that a membrane depolarization occurs in response to ACh release. If enough ACh is applied to a muscle cell, depolarization triggers an action potential in the fiber itself.
3. The action potentials propagate along the length of the muscle fiber and spread into the interior of the fiber via intracellular structures called **T tubules**. (The *T* stands for *transverse*, meaning "extending across.")
4. T tubules intersect with extensive sheets of smooth endoplasmic reticulum called the **sarcoplasmic reticulum**. When an action potential passes down a T tubule and reaches one of these intersections, a protein in the T tubule membrane changes conformation and opens a calcium channel in the sarcoplasmic reticulum.
5. Calcium ions (Ca^{2+}) are released from the sarcoplasmic reticulum. The outward flow occurs because a tremendous concentration gradient exists—the concentration of calcium ions inside the sarcoplasmic reticulum is about 40,000 times that outside. Contact with Ca^{2+} causes a conformational change in troponin. The conformational change moves tropomyosin away from the myosin binding sites on actin.

The grand result is that myosin binds to actin and contraction begins. Even grander results are that an arm or a leg or a wing moves; an animal grabs or runs or flies; and a piece of ripe fruit is picked or a bat with its jaws gaping open gets farther away. Understanding movement is fundamental to understanding how animals work.

✓CHECK YOUR UNDERSTANDING

Antagonistic muscle groups work in conjunction with skeletons to produce movement. Many types of movement are based on interactions between actin and myosin. You should be able to (1) make a labeled diagram of a skeletal muscle fiber, showing the relationship between the multinucleate cell, myofibrils, and sarcomeres; (2) describe the sliding filament model; (3) explain how binding by ATP and hydrolysis of ATP affect the interaction between actin and myosin, and (4) predict the effect on muscle function of drugs that have the following actions: increase ACh release at the neuromuscular junction, prevent conformational changes in troponin, and block uptake of calcium ions into the sarcoplasmic reticulum.

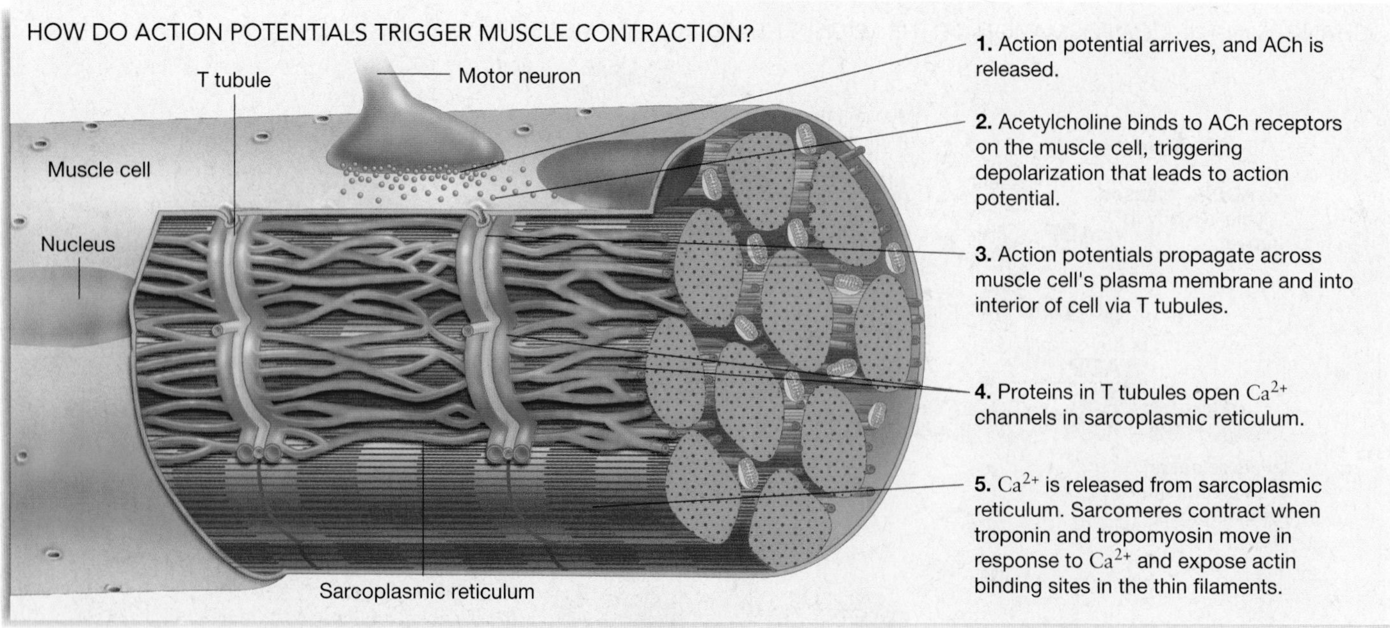
HOW DO ACTION POTENTIALS TRIGGER MUSCLE CONTRACTION?
T tubule
Motor neuron
Muscle cell
Nucleus
Sarcoplasmic reticulum
1. Action potential arrives, and ACh is released.
2. Acetylcholine binds to ACh receptors on the muscle cell, triggering depolarization that leads to action potential.
3. Action potentials propagate across muscle cell's plasma membrane and into interior of cell via T tubules.
4. Proteins in T tubules open Ca^{2+} channels in sarcoplasmic reticulum.
5. Ca^{2+} is released from sarcoplasmic reticulum. Sarcomeres contract when troponin and tropomyosin move in response to Ca^{2+} and expose actin binding sites in the thin filaments.

FIGURE 46.24 Action Potentials at the Neuromuscular Junction Trigger the Release of Ca^{2+}, Which Binds to Troponin-Tropomyosin and Allows Myosin to Walk along Actin
The arrival of action potentials from a motor neuron triggers a series of events in a muscle cell. The sequence ends with changes in troponin-tropomyosin proteins that allow contraction to begin.

ESSAY Sprinters and Marathoners—Born or Made?

Vertebrate muscle fibers come in two basic varieties: slow twitch and fast twitch. As **Table 46.1** shows, **slow-twitch fibers** excel at sustained activity, such as distance running. Slow-twitch fibers generate most of their ATP by aerobic respiration and have large stores of a protein called *myoglobin*, which holds oxygen. Because myoglobin is red and capillary density is high in these fibers, slow-twitch muscle is reddish. In contrast, fast-twitch muscle is paler. **Fast-twitch fibers** contract rapidly and excel at quick bursts of motion, such as sprinting. Because they generate most of their ATP by anaerobic metabolism (see Chapter 9), fast-twitch fibers fatigue rapidly.

Why does contraction speed vary so dramatically between these cell types? The answer is that each type of fiber contains a different type of myosin. Slow-twitch fibers make type I myosin; fast-twitch fibers make either type IIa or IIb myosin, with type IIb being faster. Type II myosin supports faster contraction because it hydrolyzes ATP more rapidly than does type I myosin.

Researchers who compare the muscle fibers of sprinters with those of endurance athletes find strong contrasts between myosin types. The leg muscles of elite sprinters are dominated by type IIb myosin, while the leg muscles of endurance athletes have mostly type I myosin. Is this difference due to training, or are people born with a certain type of myosin?

To answer this question, biologists have surveyed myosin types in people who do not train their muscles. Untrained individuals have substantial differences in their percentages of I, IIa, and IIb fibers. This result suggests that at least some of the observed phenotypic variation is due to genetic variation. But because everyone has all three myosin genes in all of their muscle fibers, it is clear that genetic variation occurs at the level of gene regulation—meaning variation in which genes are actually expressed as muscles develop.

It would be incorrect, however, to conclude that all differences in fiber types are genetic. Studies of sprinters and endurance athletes have shown that the proportion of muscle cross-sectional area devoted to each myosin type changes with level of training. For example, sprint training causes a significant increase in the percentage of fast-twitch fibers after just two months. Long-distance running takes more time to have an effect, but after 2–6 months the proportion of slow-twitch fibers increases. How do these changes occur? Although new muscle cells are never created, training can increase the number of myofibrils per cell. The proportion of cross-sectional area of muscle devoted to each myosin type changes as the cell size changes.

Although new muscle cells are never created, training can increase the number of myofibrils per cell.

Besides adding to the number of myofibrils in muscle fibers, training may cause changes in myosin types within muscle fibers. Recent data suggest that training with weights can lead to a conversion of type IIb fibers to IIa. In addition, some data imply that type I fibers can be induced to switch to type II by means of a vigorous sprinting regimen. But according to the 2- to 3-month-long studies that have been conducted to date, type II fibers do not convert to type I.

Given these data, how do elite endurance athletes end up with most of their muscle fibers being type I? The answer is not known. It is likely that marathoners have a strong genetic predisposition for expressing the type I myosin gene. If so, what molecular mechanisms trigger the expression of certain myosin genes? Alternatively, it is possible that type II fibers gradually shift to type I over many years of training. Research on the nature and extent of myosin type conversion continues.

TABLE 46.1 Muscle-Fiber Types in Vertebrates

Type	Contraction Speed	Fatigue Rate	ATP Generation	Myosin Type
Fast twitch	rapid	fast—within minutes	primarily anaerobic	type II (IIb faster than IIa)
Slow twitch	relatively slow	slow—active for hours	primarily aerobic	type I

CHAPTER REVIEW

Summary of Key Concepts

- **Sensory cells transduce sound, light, or other stimuli to a change in membrane potential. Sensory cells send action potentials to the brain, where the signals are processed and integrated.**

  If the membrane potential of a sensory cell is altered substantially enough in response to a stimulus, the pattern of action potentials that it sends to the brain changes. In this way, sensory stimuli as different as sound and light are transduced to electrical signals. The brain is able to distinguish different types of stimuli because axons from different types of sensory neurons project to different regions of the brain.

- **Hearing is based on sensory cells that move in response to sound waves of a particular frequency. The movement opens ion channels in the sensory cells and results in a change in membrane potential.**

  Pressure receptors detect direct physical stimulation, including stimulation from sound. Hair cells, the major sensory detectors in the vertebrate ear, undergo a change in membrane potential in response to bending of their stereocilia. Hair cells contact a basilar membrane, and experiments with isolated cochleas showed that sound waves of a certain frequency cause a certain part of the basilar membrane to vibrate. Because of this specificity in hair-cell response, mammals can discriminate different pitches.

- **Vision is based on sensory cells that contain the light-absorbing pigment rhodopsin. Rhodopsin changes conformation when it absorbs light and triggers events that change the cell's membrane potential.**

  In the vertebrate eye, photoreceptors are located in rods and cones. Although these two cell types differ in structure and function, both contain rhodopsin molecules that consist of retinal paired with opsin. The rhodopsin found in rods is stimulated by faint light. Color vision is possible because cones contain opsins that respond to specific wavelengths of light absorbed by retinal. Humans distinguish colors based on the pattern of stimulation of three types of opsins found in cones. People who lack one of these cone opsins are color blind, meaning they cannot distinguish as many colors as people with all three opsins can.

  Web Tutorial 46.1 The Vertebrate Eye

- **Taste and smell are based on membrane proteins that act as receptors for particular molecules and that trigger changes in membrane potential.**

  Chemoreceptors detect the presence of certain foodborne or airborne molecules. Taste buds, for example, contain taste cells with membrane proteins that respond to toxins, salt, acid, and other types of molecules in food. Sodium ions and protons enter taste cells via channels and depolarize the membrane directly. Toxic compounds bind to membrane receptors and trigger action potentials that are interpreted by the brain as bitterness. Smell, or olfaction, is used to scan molecules from the outside environment. Airborne chemicals are detected by hundreds of different odor-receptor proteins located in the membranes of receptor cells in the nose.

- **In many cases, animals respond to sensory stimuli by moving. Movement is based on antagonistic muscle groups that act on a stiff skeleton and that flex or extend in response to electrical signals.**

  All animal muscles use the same basic mechanism for contraction. Muscles shorten when thick filaments slide past thin filaments in a series of binding events mediated by the hydrolysis of ATP. Calcium ions play an essential role in making the actin in thin filaments available for binding. In animals with exoskeletons or endoskeletons, muscles are usually arranged in opposing pairs of flexors and extensors. Animals with hydrostatic skeletons have muscles arranged in opposing pairs of circular and longitudinal bands. In all cases, muscles work in concert with skeletons.

  Web Tutorial 46.2 Muscle Contraction

Questions

Content Review

1. What is the major function of the middle ear?
   a. It amplifies the sound energy transmitted to the cochlea.
   b. It transmits sound vibrations to the tympanic membrane.
   c. It circulates fluid through the cochlea.
   d. It contains hair cells that detect specific frequencies of sound.

2. In the human ear, why do different hair cells respond to different frequencies of sound?
   a. Waves of pressure move through the fluid in the cochlea.
   b. Hair cells are "sandwiched" between membranes.
   c. Receptors in the stereocilia of each hair cell are different; each receptor protein responds to a certain range of frequencies.
   d. Because the basilar membrane varies in stiffness, it vibrates in certain places in response to certain frequencies.

3. Which of the following comparisons of rods and cones is *false*?
   a. The eyes of most humans have one type of rod and three types of cones.
   b. Rods are more sensitive to dim light than cones are.
   c. There are more rods than cones in the fovea.
   d. Both rods and cones use retinal and opsins to detect light.

4. Which of the following statements about taste is *true*?
   a. Sweetness is a measure of the concentration of hydrogen ions in food.
   b. Sodium ions from foods can directly depolarize certain taste cells.
   c. All bitter-tasting compounds have a similar chemical structure.
   d. Membrane receptors are involved in detecting acids.

5. In muscle cells, myosin molecules continue moving along actin molecules as long as
   a. ATP is present and troponin is not bound to Ca^{2+}.
   b. ADP is present and tropomyosin is released from intracellular stores.
   c. ADP is present and intracellular ACh is high.
   d. ATP is present and intracellular Ca^{2+} is high.

6. Which of the following is critical to the function of exoskeletons, endoskeletons, and hydrostatic skeletons?
   a. Muscles interact with the skeleton in antagonistic groups.
   b. Muscles attach to each of these types of skeleton via tendons.
   c. Muscles extend joints by pushing them.
   d. Segments of the body or limbs are extended when paired muscles relax in unison.

Conceptual Review

1. Describe how a sound, a light, and an odor are transduced into a change in the pattern of action potentials from a sensory cell. How does the brain know which sense is which when the action potentials reach the brain?
2. Give three examples of how the sensory abilities of an animal correlate with its habitat or method of finding food and mates.
3. How did the discovery of odor-receptor genes affect our understanding of how the sense of smell works?
4. Compare and contrast the structure and function of slow-twitch and fast-twitch muscle fibers. In making the comparison, use the race between the hare and the tortoise of Aesop's fable as a metaphor.
5. Scientists generally think that a "good hypothesis" is one that is reasonable, is testable, and inspires further research into the field. Using these criteria, was Dalton's hypothesis about color vision a good hypothesis? Was it correct?
6. How did observations of sarcomere structure inspire the sliding-filament model of muscle contraction? Explain why the observation that muscle cells contain a large number of mitochondria and extensive smooth endoplasmic reticulum turned out to be logical once the molecular mechanism of muscular contraction was understood.

Group Discussion Problems

1. Myasthenia gravis is a disease that develops in humans when the immune system produces proteins that bind to the acetylcholine (ACh) receptors in muscles. The primary symptom of myasthenia gravis is muscle weakness. Why?
2. Houseflies have about 800 ommatidia in each of their compound eyes. Dragonflies, in contrast, have up to 10,000 ommatidia per eye. Houseflies feed by lapping up watery material from piles of excrement or rotting carcasses, which they locate by scent. Dragonflies are aerial predators and hunt by sight. Suggest a hypothesis to explain the difference in the structure of their eyes. How would you test your hypothesis?
3. Muscles that are dominated by slow-twitch fibers are reddish because they have a high concentration of myoglobin, while fast-twitch muscles are light in color. Chickens, turkeys, and other ground-dwelling birds can run long distances but escape from predators by flying in short, extremely fast bursts. Based on these observations, explain the distribution of "dark meat" and "white meat" in these birds. Most other birds do not have white meat. Why?
4. When looking at faint stars through a telescope, astronomers will focus their eyes just to the side of the object. Instead of landing on the fovea, then, the star's image falls next to the fovea. With this technique, faint objects pop into view. They vanish if looked at directly, however. Explain what is going on.

Answers to Multiple-Choice Questions 1. a; 2. d; 3. c; 4. b; 5. d; 6. a

www.prenhall.com/freeman is your resource for the following: Web Tutorials; Online Quizzes and other Online Study Guide materials; Answers to Conceptual Review Questions; Solutions to Group Discussion Problems; Answers to Figure Caption Questions and Exercises; and Additional Readings and Research.

47 Chemical Signals in Animals

KEY CONCEPTS

- Animals use at least six major types of chemical signals. Hormones are chemical signals that are present in tiny concentrations and travel throughout the body to affect target cells.
- The information in hormonal signals helps animals respond to environmental change, develop as embryos, undergo sexual maturation, and achieve homeostasis.
- The production of a hormone is tightly regulated by input from the nervous system and by other hormones.
- Some hormones bind to receptors inside target cells and change gene expression. Other hormones bind to receptors at the cell surface and lead to changes in protein activation.

Two mouse siblings from the same litter. As an embryo, the mouse on the right was given a copy of the gene for human growth hormone. (The structure of this recombinant gene is diagrammed in the loop around the mice.) Overexpression of this gene leads to abnormally high concentrations of a chemical signal that supports growth.

In response to sights, sounds, and other sensory stimuli, an animal's nervous system sends rapid messages to precise locations in the body. Frequently these messages result in muscle contractions and movement.

As conditions inside and outside an animal change, however, cells in the central nervous system (CNS) or cells in a specialized group of organs called the **endocrine system** may trigger the release of molecules that produce longer-term responses in a broad range of tissues and organs. A chemical signal that circulates through body fluids and affects distant target cells is called a **hormone**. As you read this, a large suite of hormones is coursing through your circulatory system. These hormones are regulating the maturation of sperm or eggs by your reproductive system, changing the composition of the urine forming in your kidneys, and controlling the release of digestive enzymes in your gastrointestinal tract. Earlier in your life, changes in hormone concentrations led to the dramatic physical changes associated with puberty.

Animal hormones are present in tiny concentrations but have huge effects on their target cells. Unlike action potentials, which are electrical impulses that have a short-term effect on a single cell or on a small population of adjacent cells, the messages that hormones carry are received by many different cells and have a relatively long-lasting effect. In combination, electrical and chemical signals allow animals to coordinate the activities of cells throughout the body. They are the mechanism responsible for maintaining trillions of cells as the integrated unit we call an individual.

The goal of this chapter is to explore how chemical signals work in animals. Let's begin with an overview of the structural and functional diversity observed in chemical signals, and then analyze how hormones regulate the activity of target cells.

47.1 Cataloging Hormone Structure and Function

For over a century, biologists have been exploring how chemical messengers function in animals. One of the most basic conclusions from this work is that the chemical signals found in animals have diverse structures and functions. **Figure 47.1** summarizes how biologists go about organizing the diversity of chemical signals, based on where the molecules originate and where they act. Notice that the names for most of the categories use the Greek word root *crin*, meaning "separated." Its use captures something essential about how chemical signals act: They are released from cells and thus are separated from them. There are six major categories of chemical signals:

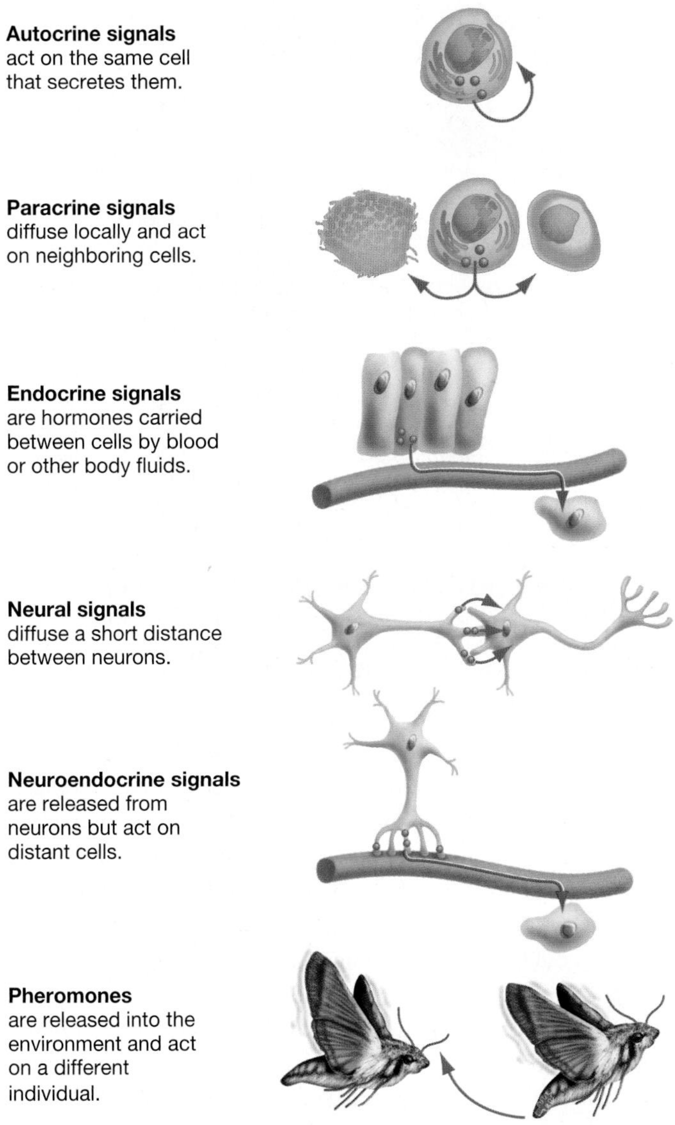

FIGURE 47.1 There Are Six Categories of Chemical Signals in Animals
Chemical signals can be categorized by how they move from the cell that produces them to the target cell(s) that respond to them.

1. **Autocrine** ("same-separated") signals act on the same cell that secretes them. Perhaps the best-studied autocrine signals are **cytokines** ("cell-movers"). The function of most cytokines is to amplify the response of a cell to a stimulus. An example is interleukin 2, which is synthesized and released by a type of white blood cell called a T cell in the course of fighting an infection. Interleukin 2 activates T cells to help eliminate the infection. It also causes the cells to divide repeatedly, producing more activated T cells for host defense.
2. **Paracrine** ("beside-separated") signals diffuse locally and act on neighboring cells. Cytokines may act as paracrine signals as well as autocrine signals, because they can trigger responses by other cells of the immune system. It is common, in fact, to observe that a single chemical messenger acts in a variety of ways. As an example, consider molecules that are produced by three distinct populations of cells within a region of the pancreas called the islets of Langerhans. The molecules named **insulin**, **glucagon**, and **somatostatin** act as hormones—meaning they target distant cells—in controlling the concentration of glucose in the blood, but they also have paracrine effects. When insulin, glucagon, and somatostatin act on nearby pancreatic cells, they ensure a smooth, steady response to changing blood-glucose levels.
3. **Endocrine** ("inside-separated") signals are hormones—molecules that are carried between cells by blood or other body fluids. The cells that produce endocrine signals may be organized into discrete organs called **glands** or may be interspersed among the cells of other organs.
4. **Neural** signals, introduced in Chapter 45, are the chemical messengers called **neurotransmitters.** Neural signaling can be very fast, because neurotransmitters have to diffuse only a short distance—the tiny gap between two neurons, called the synaptic cleft.
5. **Neuroendocrine** ("nerve-inside-separated") signals are released from neurons but act on distant cells instead of acting at the adjacent synapse. Antidiuretic hormone (ADH; also called vasopressin), introduced in Chapter 42, is a good example. ADH is produced by neurons that originate in the hypothalamus of the brain, but the molecule acts on cells in the collecting duct of the kidney to help regulate water excretion.
6. **Pheromones** are released into the environment and act on a different individual. In sea urchins, insects, and many other species, pheromones help coordinate reproduction in males and females or function in attracting mates.

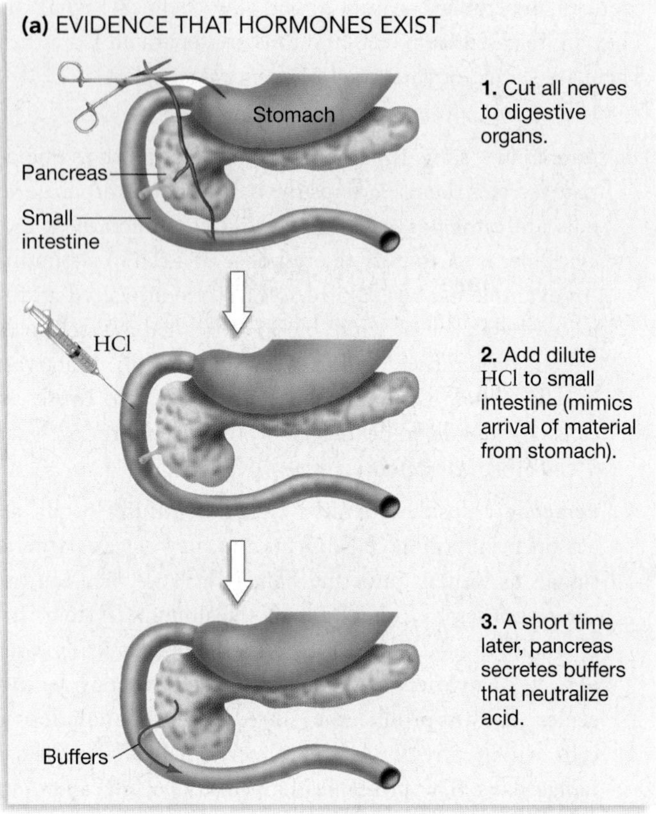
(a) EVIDENCE THAT HORMONES EXIST
Stomach
Pancreas
Small intestine
1. Cut all nerves to digestive organs.
HCl
2. Add dilute HCl to small intestine (mimics arrival of material from stomach).
3. A short time later, pancreas secretes buffers that neutralize acid.
Buffers

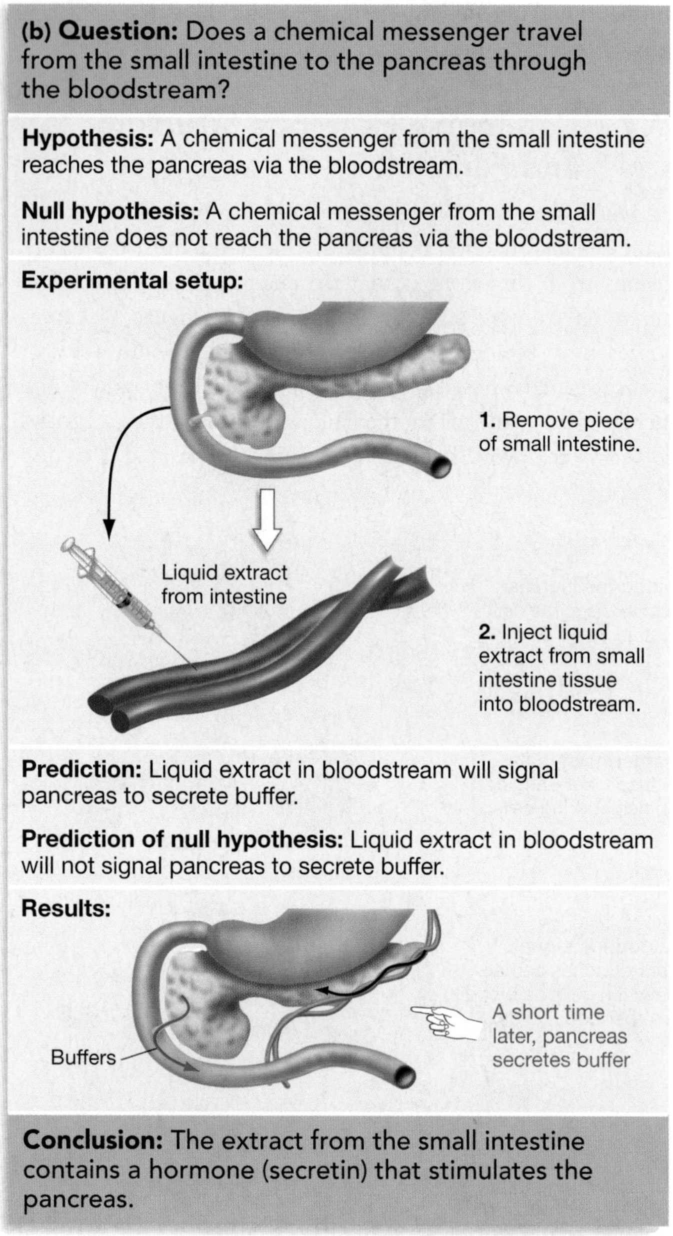
(b) Question: Does a chemical messenger travel from the small intestine to the pancreas through the bloodstream?
Hypothesis: A chemical messenger from the small intestine reaches the pancreas via the bloodstream.
Null hypothesis: A chemical messenger from the small intestine does not reach the pancreas via the bloodstream.
Experimental setup:
1. Remove piece of small intestine.
Liquid extract from intestine
2. Inject liquid extract from small intestine tissue into bloodstream.
Prediction: Liquid extract in bloodstream will signal pancreas to secrete buffer.
Prediction of null hypothesis: Liquid extract in bloodstream will not signal pancreas to secrete buffer.
Results:
Buffers
A short time later, pancreas secretes buffer
Conclusion: The extract from the small intestine contains a hormone (secretin) that stimulates the pancreas.

FIGURE 47.2 Experimental Evidence That a Hormone from the Small Intestine Signals Cells in the Pancreas, via Blood
(a) The discovery of the first animal hormone. **(b)** The molecule secretin was eventually isolated from a soluble intestinal extract.

It's important to note that these six classes of chemical messenger do not coincide with six structurally distinct classes of molecules. For example, the endocrine signals found in a particular organism routinely belong to several families of chemical compounds, ranging from amino acid derivatives to glycoproteins. And a particular family of molecules—say, peptides or the lipids called steroids—may function as endocrine, autocrine, and paracrine signals in the same individual.

The diversity of chemical signal types observed in animals is impressive. But because endocrine signals are particularly well studied, and because they help regulate such a wide array of basic physiological processes, the remainder of this chapter focuses on endocrine and neuroendocrine signals. As it turns out, most principles of hormone action are shared by the other categories of chemical signals—along with the strategies that biologists use to find and study them.

How Do Researchers Identify a Hormone?

Research on animal hormones began in earnest in the early 1900s, with the discovery of a molecule called **secretin**. Recall from Chapter 43 that researchers who were investigating how secretions from the pancreas are regulated had cut all the nerves connected to the upper part of the small intestine and pancreas of a dog. The biologists then added a small amount of dilute hydrochloric acid (HCl) to the small intestine, mimicking the arrival of acidic material from the stomach. They found that the pancreas secreted compounds that *buffered*, or neutralized, the acid in the small intestine (**Figure 47.2a**).

How could stimulating the small intestine lead to a response by the pancreas? The response could not have been mediated by the nervous system, since all the nerves to the pancreas had been cut. Instead, the biologists proposed that a chemical had traveled from the small intestine to the pancreas by way of the bloodstream to signal the arrival of acid. To test this hypothesis, the researchers did the experiment diagrammed in **Figure 47.2b**. The key step was to inject an extract from the small intestine into blood vessels in a dog's neck. A short time later, the biologists observed that the pancreas secreted an alkaline solution. This was strong evidence that the extract from the small intestine contained a hormone. The molecule was later purified and named secretin.

In addition to using liquid extracts to identify hormones, biologists have focused on analyzing the results of experimental or natural destruction of organs. For example, the experimental removal of the adrenal glands in animals or damage to the adrenal glands in humans by hemorrhage, tumor, or infection rapidly leads to death associated with very low levels of blood sugar and blood pressure. The injection of adrenal extracts into the blood of animals lacking adrenal glands corrects these abnormalities.

Experiments that followed up on these observations identified cortisol and epinephrine as the hormones that are produced by the adrenal glands. These molecules are critical to the maintenance of blood glucose and blood pressure. **Cortisol** and the related hormone **corticosterone** are referred to as glucocorticoids to emphasize their importance in regulating blood glucose. **Glucocorticoids** are hormones produced in the outer section of the adrenal glands, the adrenal cortex. **Epinephrine** is a hormone produced in the central part of the adrenal gland, the adrenal medulla. (Epinephrine is also known as **adrenaline.** The Greek word roots *epi* and *nephron* mean "top-kidney"; the Latin word roots *ad* and *renal* also mean "top-kidney.") Although cortisol and epinephrine have complementary functions in maintaining blood glucose levels and blood pressure, the two molecules are very different structurally. On a molecular level, what do hormones look like?

Chemical Characteristics of Hormones

Secretin, epinephrine, and glucocorticoids represent the three major classes of chemicals that can act as hormones in animals (**Figure 47.3**): (1) polypeptides, which are chains of amino acids linked by peptide bonds (see Chapter 3); (2) amino acid derivatives; and (3) steroids, which are a family of lipids distinguished by a four-ring structure (see Chapter 6). Secretin is a polypeptide, epinephrine is derived from the amino acid tyrosine, and cortisol is synthesized from the steroid cholesterol. Other groups of lipid-derived chemical messengers exist as well. The *prostaglandins* and *leukotrienes*, for example, serve primarily in autocrine and paracrine regulation of inflammation and blood-vessel function.

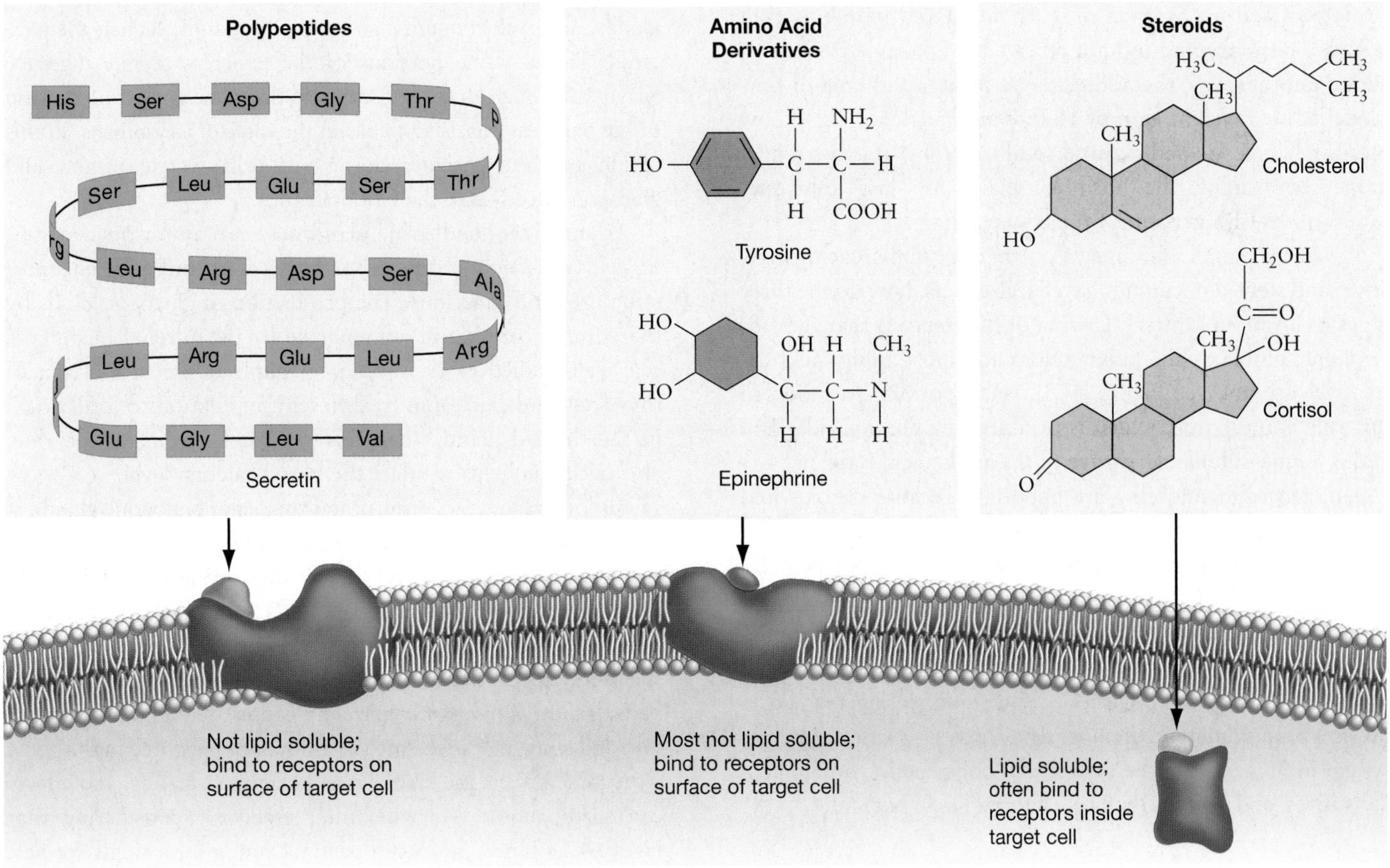

FIGURE 47.3 Most Animal Hormones Belong to One of Three Chemical Families

QUESTION Secretin and epinephrine are not lipid soluble and do not cross plasma membranes readily. Cortisol, in contrast, is lipid soluble and diffuses through plasma membranes efficiently. How do the size and charge of these molecules explain the difference in lipid solubility?

Polypeptide, amino acid derivative, and steroid hormones have important similarities, as well as important differences. All hormones share several defining characteristics:

- All hormones are organic compounds.
- All hormones are secreted from the cells where they are synthesized.
- All hormones act on target cells remote from their point of origin.
- All hormones are present in extremely small concentrations yet have large effects.

To drive this last point home, consider work on **growth hormone (GH)**. Several researchers had noted that rats and other laboratory animals stopped growing when their pituitary glands were removed. Based on this observation, it was widely suspected that the pituitary produces a chemical signal that promotes cell division and other aspects of growth. A research group was able to purify a polypeptide from cow pituitary glands, inject the polypeptide into lab rats, and document rapidly accelerated growth. When the researchers injected 0.01 mg of the molecule each day for nine days into rats that lacked pituitary glands, the width of the growth plates in the leg bones increased by 50 percent, and the individuals gained an average of 10 g compared with rats that lacked a pituitary and did not receive the hormone treatment. Stated another way, the addition of a total of 0.09 mg of hormone led to a weight gain of 10,000 mg. Further, 1 kg of cow pituitary tissue yielded a grand total of 0.04 g of growth hormone. By weight, the hormone makes up just four one-thousandths of 1 percent of the cow pituitary.

Given that small amounts of polypeptide, amino acid derivative, and steroid hormones have large effects, how do the three types of hormones differ? The major difference is that steroids are lipid soluble, while polypeptides and most amino acid derivatives are not (see Figure 47.3). An important exception to this rule is thyroxine, which is produced by the thyroid gland and is a lipid-soluble derivative of the amino acid tyrosine.

WEB TUTORIAL 47.1
Endocrine System Anatomy

Differences in solubility are important because steroids and thyroxine cross plasma membranes much more readily than do other types of hormones. To affect a target cell, all polypeptides and most amino acid derivatives bind to a receptor on the cell surface. Lipid-soluble hormones, in contrast, can diffuse through the plasma membrane and bind to receptors inside the cell as well as to receptors at the cell surface. Before exploring the consequences of this difference in more detail, let's survey the endocrine system in humans and the diversity of polypeptides, amino acid derivatives, and steroids that act as hormones in humans.

The Human Endocrine System—an Overview

Figure 47.4 illustrates some components of the human endocrine system. The endocrine system is the collection of cells, tissues, and organs responsible for hormone production and secretion. Organs that secrete a hormone into the bloodstream are called **endocrine glands**. The major glands that have endocrine functions in humans are:

- The **hypothalamus**, which is a region deep within the brain;
- The **pituitary gland**, which sits just below the hypothalamus and has distinct anterior and posterior regions;
- The **thyroid gland**, which is situated in the neck;
- The four **parathyroid glands**, which are embedded in the thyroid gland;
- The two **kidneys**, which lie in the posterior part of the abdominal cavity;
- The two **adrenal glands**, which sit atop the kidneys and have an outer cortex and a central medulla;
- The endocrine component of the **pancreas**, which is located in the anterior part of the abdominal cavity; and
- The paired **ovaries** (in females) or **testes** (in males), which are in or suspended below the pelvic cavity, respectively.

Exocrine glands, in contrast to endocrine glands, deliver their secretions through outlets called **ducts** into a space other than the circulatory system. Most of the digestive glands introduced in Chapter 43 were either exocrine glands, such as the salivary glands, or mixed endocrine and exocrine glands, such as the pancreas. The exocrine portions of the pancreas secrete digestive enzymes through ducts into the intestine. The endocrine portion of the pancreas consists of cells in the **islets of Langerhans**. In this region of the pancreas, clusters of cells secrete insulin and glucagon directly into the bloodstream.

An individual endocrine gland may contain distinct regions or cell types that produce hormones with differing chemical structures and functions. The production of glucocorticoids by the adrenal cortex and epinephrine by the adrenal medulla is one well-studied case. Another example is the production of thyroxine and calcitonin by different populations of cells within the thyroid gland. **Thyroxine** regulates the metabolic rate, and **calcitonin** helps regulate the blood calcium level.

Although Figure 47.4 illustrates the major endocrine glands, it is important to recognize that cells in many other organs and tissues secrete hormones even though they are not organized into discrete glands. For example, the intestine produces secretin, the heart produces atrial natriuretic hormone, and the cells of fat tissue produce leptin. Secretin stimulates the exocrine portion of the pancreas, **atrial natriuretic hormone** causes the kidney to excrete salt, and **leptin** helps regulate the amount of fat stored in the body.

At first glance, the diversity of hormones, glands, and effects can seem almost overwhelming, especially considering that Figure 47.4 represents just a partial catalog for a single species. Several unifying themes emerge from the diversity of molecules and functions, however.

First, recall that most animal hormones belong to one of just three major structural families: polypeptides, amino acid deriva-

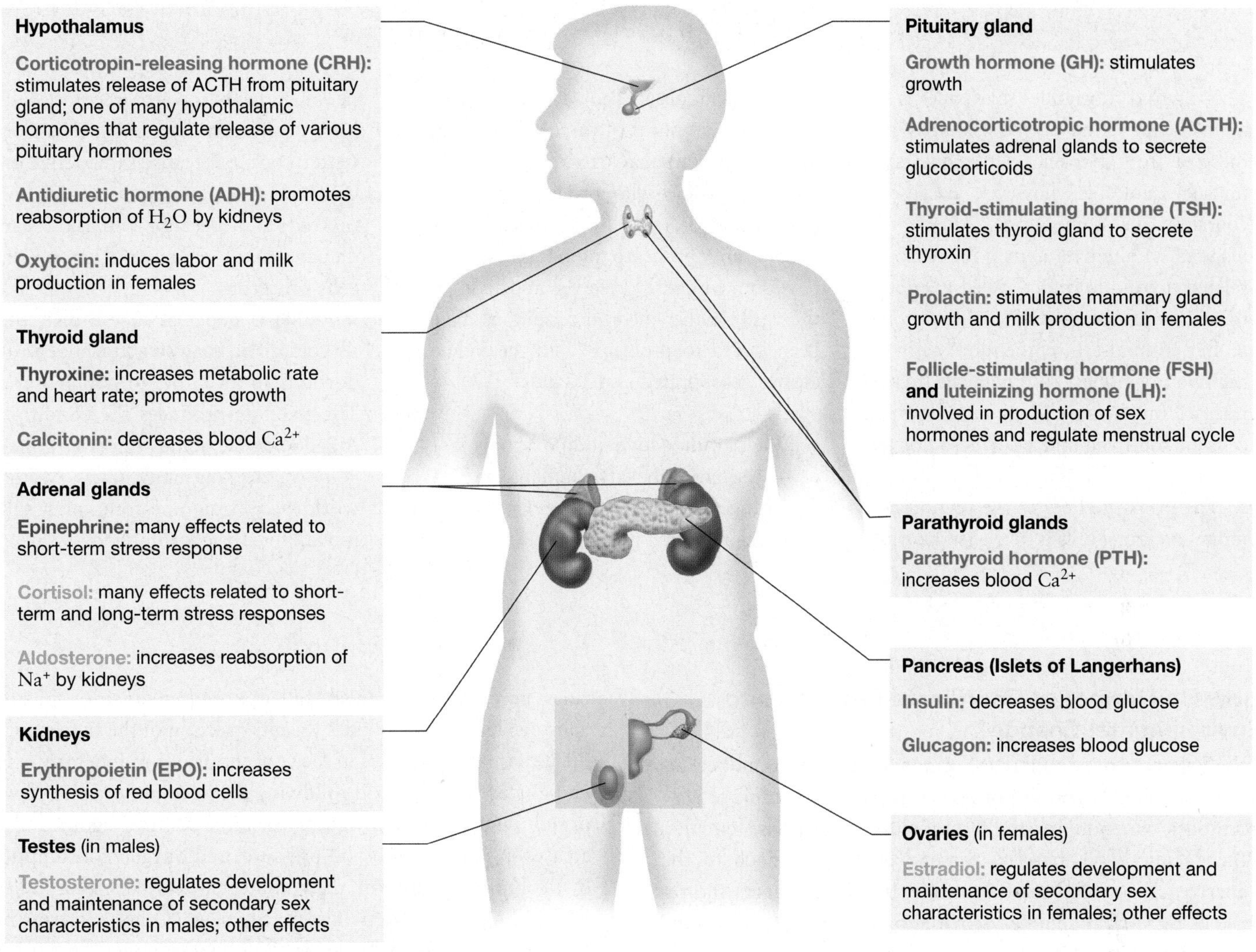
Hypothalamus
Corticotropin-releasing hormone (CRH): stimulates release of ACTH from pituitary gland; one of many hypothalamic hormones that regulate release of various pituitary hormones
Antidiuretic hormone (ADH): promotes reabsorption of H_2O by kidneys
Oxytocin: induces labor and milk production in females
Thyroid gland
Thyroxine: increases metabolic rate and heart rate; promotes growth
Calcitonin: decreases blood Ca^{2+}
Adrenal glands
Epinephrine: many effects related to short-term stress response
Cortisol: many effects related to short-term and long-term stress responses
Aldosterone: increases reabsorption of Na^+ by kidneys
Kidneys
Erythropoietin (EPO): increases synthesis of red blood cells
Testes (in males)
Testosterone: regulates development and maintenance of secondary sex characteristics in males; other effects
Pituitary gland
Growth hormone (GH): stimulates growth
Adrenocorticotropic hormone (ACTH): stimulates adrenal glands to secrete glucocorticoids
Thyroid-stimulating hormone (TSH): stimulates thyroid gland to secrete thyroxin
Prolactin: stimulates mammary gland growth and milk production in females
Follicle-stimulating hormone (FSH) and luteinizing hormone (LH): involved in production of sex hormones and regulate menstrual cycle
Parathyroid glands
Parathyroid hormone (PTH): increases blood Ca^{2+}
Pancreas (Islets of Langerhans)
Insulin: decreases blood glucose
Glucagon: increases blood glucose
Ovaries (in females)
Estradiol: regulates development and maintenance of secondary sex characteristics in females; other effects

FIGURE 47.4 An Overview of the Human Endocrine System
A partial list of the endocrine glands and hormones found in humans. Polypeptides are shown in green, amino acid derivatives in purple, and steroids in gold. The heart, gastrointestinal tract, adipose tissue, and many other hormone-producing organs and cells could be added.

tives, or steroids. Second, all hormones share a basic functional characteristic: They are released into the bloodstream or other fluids and act on target cells that are distant from the source gland. Finally, all hormones fulfill the same general role: to coordinate the activities of diverse groups of cells in response to changes in the internal or external environment. More specifically, the functions of animal hormones fall into a handful of broad categories. What are these? In general, how does the action of hormones help an individual survive and reproduce?

47.2 What Do Hormones Do?

At the beginning of this chapter, you read that hormones are chemical messengers. If so, what do hormones "say"? It is clear from Figure 47.4 that even a single type of hormone can exert a variety of effects. Consider thyroxine, for example. In humans, thyroxine stimulates metabolism and thus oxygen consumption throughout the body. But it also promotes growth, increases the heart rate, and stimulates the synthesis of many important macromolecules. (**Box 47.1**, page 1082, explores the medical implications of abnormalities in the production of thyroxine.) In other cases, several different types of hormones may affect the same aspect of physiology. For example, the blood-glucose concentration is regulated by insulin, glucagon, epinephrine, and cortisol.

Some hormones have extremely diverse effects; other hormones have functions that appear to overlap. These observations begin to make sense when hormone action is viewed in the context of the whole organism. Hormones coordinate the activities of cells in response to three general situations: (1) environmental challenges; (2) growth, development, and reproduction; and (3) homeostasis. Let's analyze each in turn.

BOX 47.1 A Closer Look at Thyroxine and the Thyroid Gland

The thyroid hormone thyroxine is an amino-acid-derived hormone that is synthesized from tyrosine. Thyroxine is an unusual molecule, however, because it contains four iodine atoms. This feature inspired its alternate name, T_4. A closely related thyroid hormone called **triiodothyronine**, or T_3, contains three iodine atoms.

In mammals, T_3 is the more active of the two hormones. Although the thyroid releases much more T_4 than it does T_3, cells in the liver and elsewhere convert T_4 to T_3.

The primary effect of the thyroid hormones on target cells is to trigger increased cellular metabolism. People who produce inadequate amounts of thyroid hormones are lethargic and unable to tolerate cold. In contrast, people who produce excessive amounts of T_3 and T_4 are restless, excitable, and prone to mood swings but have limited energy reserves and tend to tire easily. Other medical conditions have been traced to problems with the iodine atoms incorporated into T_3 and T_4:

- When radioactive iodine atoms are released into the environment, such as during the accident at the Chernobyl nuclear power plant in the Ukraine, people living in the area become highly susceptible to thyroid cancers. This pattern occurs because radioactive iodine molecules are sequestered in the thyroid and because radiation can damage DNA in a way that makes cells cancerous.
- If iodine is deficient in the diet, the thyroid gland may swell to huge proportions in an effort to compensate. The swelling produces the condition called **endemic goiter**. Endemic goiter is now rare in many parts of the world because iodine-containing salt is available for consumption.

How Do Hormones Coordinate Responses to Environmental Change?

The challenges or stimuli that hormones respond to can be simple or complex. Digestive hormones are a good example of how hormones function in simple stimulus-and-response circuits. When acidic food material passes from the stomach to the upper reaches of the small intestine, the food triggers the release of secretin and cholecystokinin by intestinal cells into the bloodstream. Recall that secretin induces the pancreas to secrete an alkaline solution that neutralizes acid. **Cholecystokinin** is a hormone that causes the pancreas to secrete digestive enzymes into the small intestine. Cholecystokinin also causes the gallbladder to eject bile salts into the intestine to emulsify fats. In this way, digestive hormones signal the arrival of food and regulate the release of molecules that aid digestion. But what about more complex environmental stimuli?

Short-Term Responses to Stress When a person is thrust into a dangerous or unpredictable situation, hormones are involved in both the short-term and long-term responses. The short-term reaction, called the **fight-or-flight response**, occurs in conjunction with the activation of the sympathetic nervous system, which was described in Chapter 45. Sympathetic nerves prepare organs for stressful situations. If you were being chased by a grizzly bear, action potentials from your sympathetic nerves would stimulate your adrenal medulla and lead to the release of epinephrine.

To understand the responses that occur, consider the experiment reported in **Figure 47.5**. To determine how epinephrine affects the body, human volunteers received injections of a saline solution, epinephrine, or the chemically related molecules **norepinephrine** (also called *noradrenaline*) or phenylephrine. The graphs in the "Results" section of the figure document dramatic increases in the concentrations of free fatty acids and glucose in the blood following the injection of epinephrine or norepinephrine. The data in the table confirm significant increases in pulse rate, blood pressure, and oxygen consumption by the brain. In addition, the volunteers reported strong subjective feelings of anxiety and excitement. Other experiments showed that epinephrine leads to dramatic changes in the distribution of blood. Specifically, the fight-or-flight response includes a redirection of blood toward the heart, brain, and muscles and away from the skin and digestive system, as discussed in Chapter 44. Epinephrine also relaxes smooth muscle and thereby opens blood vessels—increasing blood delivery to targeted tissues.

Taken together, the responses to epinephrine lead to a state of heightened alertness and increased energy use that prepares the body for rapid, intense action such as fighting or fleeing. By coordinating the activities of cells in many organs and systems throughout the body, epinephrine prepares an individual to cope with a life-threatening situation.

Long-Term Responses to Stress If you have ever been in an acutely stressful situation and experienced the fight-or-flight response, you may recall that the state is short lived. Once an epinephrine "rush" has worn off, most people feel exhausted and want to rest and eat.

What happens if the stress continues and turns into a long-term condition? In the course of a lifetime, it is not unusual for a person to experience periods of starvation or fasting, prolonged emotional distress, or chronic illness. How do hor-

Question: How does epinephrine affect the body?

Hypothesis: Epinephrine causes changes involved in the fight-or-flight response.

Null hypothesis: Epinephrine does not cause changes involved in the fight-or-flight response.

Experimental setup:

1. Inject human volunteers with saline solution, norepinephrine, epinephrine, or phenylephrine.

Control **Norepinephrine** **Epinephrine** **Phenylephrine**

(Saline solution)

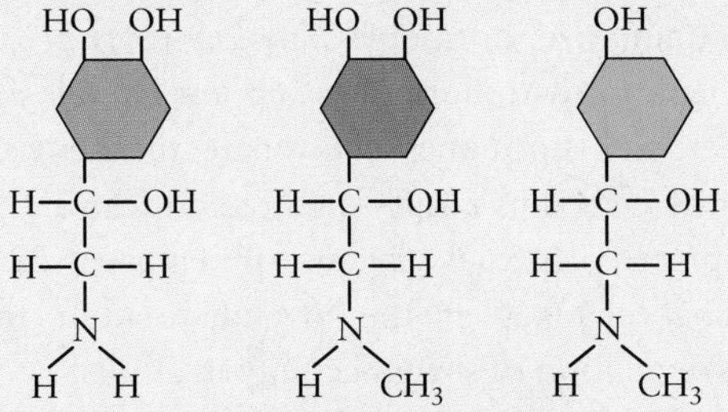

2. Document changes in fatty-acid and glucose concentrations in blood, pulse rate, blood pressure, and oxygen consumption in brain.

Prediction: Epinephrine (and possibly norepinephrine and phenylephrine) increases fatty-acid and glucose concentrations in blood, pulse rate, blood pressure, and brain oxygen consumption relative to controls.

Prediction of null hypothesis: No differences in physiological state of individuals based on molecule injected.

Results:

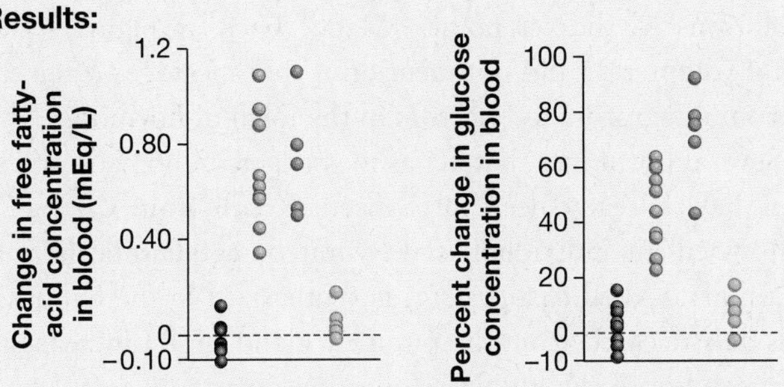

	Control (injected with saline solution)	Experimental (injected with epinephrine)
Pulse rate (beats/min)	78.3	89.6
Blood pressure (average, mm Hg)	90.9	108.7
O_2 consumption in brain (cc O_2/100 gm blood/min)	3.41	4.16

Conclusion: Epinephrine, as well as norepinephrine, causes an array of changes associated with the fight-or-flight response. Phenylephrine does not.

FIGURE 47.5 What Effect Does Epinephrine Have on the Body? Each data point on the graphs represents a human volunteer. The data in the table are average values from seven volunteers.
QUESTION Examine the chemical structures of norepinephrine, epinephrine, and phenylephrine. How do they differ? How do the effects of these molecules differ?

mones help humans and other animals cope with extended stress?

Early studies of long-term stress in human subjects suggested a role for the glucocorticoids produced in the adrenal cortex. Increased levels of cortisol or corticosterone were found in airplane pilots and crew members during long flights, athletes who were training for intense contests, the parents of children undergoing treatment for cancer, and college students who were preparing for final exams. What do these glucocorticoids do?

The primary role of glucocorticoids during prolonged stress in humans is to ensure the continued availability of fuel molecules to support important body functions. Glucocorticoids maintain glucose production by inducing the synthesis of the liver enzymes that make glucose from amino acids and other chemical precursors. Glucocorticoids also cause *adipose tissue*—that is, fat tissue—and resting muscles to become resistant to the effects of insulin. When the body is in an unstressed state, insulin stimulates **adipocytes**, or fat cells, and resting muscle cells to remove glucose from the bloodstream. When glucocorticoids make these cells resistant to insulin, the effect is to reserve glucose for use by the brain and exercising muscles, neither of which depend on insulin to stimulate glucose uptake. Glucocorticoids also promote the release of the body's major fuel molecules, fatty acids, from adipose tissue for use by the heart and muscles.

Although it may be life saving, the long-term stress response comes at a high price, as any victim of a serious injury or illness knows. Glucocorticoids make amino acids available for glucose synthesis by promoting the degradation of contractile proteins in muscle. The resulting loss of muscle mass may cause severe weakness—or even death, if respiratory muscle function is severely diminished. Glucocorticoids also impair wound healing and suppress immune and inflammatory responses that are costly in terms of energy use. The overall concept here is that the long-term stress response is a compromise. The fuel requirements of the brain are met at the expense of other tissues and organs.

It is instructive to note that glucocorticoids may mediate the response to stressful environmental challenges in species other than humans. Recall from Chapter 42 that, when moving from freshwater to saltwater, a salmon gains sodium ions via diffusion and loses water across its gills by osmosis. The salmon counteracts these stresses by actively pumping sodium and chloride ions out of its gills. The chloride cells that perform this pumping are induced to proliferate by a burst of cortisol. Cortisol release occurs as the salmon moves downstream toward the ocean. If this release fails to occur, salmon die shortly after reaching salt water.

In summary, the hormones produced by the human adrenal glands coordinate an individual's response to short- and long-term stresses. As with digestive hormones, these molecules alert cells and tissues to a particular environmental stimulus or challenge. Other hormones have dramatically different effects, however.

How Do Hormones Direct Developmental Processes?

Growth hormones and sex hormones play crucial roles in promoting cell division, increasing overall body size, and promoting sexual differentiation as an individual matures. In addition, certain hormones direct the development of particular cells and tissues at critical junctures in an individual's life. Which hormones are involved in regulating development, and what specific effects do they have? The major hormonal effects on developmental processes are as follows:

- *Early embryonic development* Events early in development dictate whether the sex organs, or **gonads**, of an embryo become male (testes) or female (ovaries). This process is called **primary sex determination.** In mammals it is not dependent on hormone action. Once testes or ovaries develop, though, they begin producing male-specific or female-specific hormones. In humans, the early testes produce two hormones: A steroid hormone called **testosterone** induces early development of the male reproductive tract, and a polypeptide hormone called **Müllerian inhibitory substance** inhibits development of the female reproductive tract. Development of the female reproductive tract depends on the action of the steroid hormone **estradiol**, which is produced in the ovaries and is in the family of molecules called **estrogens.**
- *Juvenile-to-adult transition* When humans reach early adolescence, surges of sex hormones lead to the physical and emotional changes associated with **puberty**. These developmental changes create the adult phenotype and the ability to produce offspring. In boys, surges of sex hormones lead to changes that include enlargement of the penis and testes and growth of facial and body hair. In girls, increased concentrations of estradiol lead to the enlargement of breasts, the onset of menstruation, and other changes. Hormonal control over sexual maturation is not limited to vertebrates, however. Insect metamorphosis is a well-studied process controlled by hormones. If the steroid called **juvenile hormone** (JH) is present at a high concentration, then surges of the steroid hormone **ecdysone** induce the growth of a juvenile insect via molting. But if JH levels are low, ecdysone triggers metamorphosis and the transition to adulthood and sexual maturity.
- *Attaining full adult stature* In humans and other mammals, the long bones in the limbs and the vertebrae of the spinal column must grow for full adult height to be achieved. This process is mediated by *growth factors*, which are signals that control the cell cycle (see Chapter 11) and are in turn regulated by growth hormone produced in the pituitary gland. Because the effect of growth hormone on the skeleton is enhanced by the action of sex hormones that are produced in increased amounts during adolescence, puberty is associated with a growth spurt. Sex hormones also cause skeletal maturation. As a result, bone growth stops in early adulthood, even though growth hormone is produced throughout life.
- *Seasonal or cyclical sexual activity* Most long-lived animals reproduce seasonally. In many species, environmental cues such as increasing day length, warmth, or the onset of seasonal rains trigger the release of sex hormones. Chapter 51 details how this flush of testosterone or estrogen induces the development of seasonal traits, such as sexual receptivity in female lizards. Even though humans do not breed seasonally, sex hormones are instrumental in regulating sperm production and the menstrual cycle. Chapter 48 explores these processes in detail.

To summarize, hormones play key roles in growth and development. Growth hormones support overall size increases in juveniles. Sex hormones coordinate the activities of diverse groups of cells and tissues at three critical life stages: (1) the development of sexual structures in embryos, (2) physical and emotional changes at the juvenile-to-adult transition, and (3) the regulation of sexual activity in adults.

How Are Hormones Involved in Homeostasis?

Chapter 41 introduced the concept of *homeostasis*, the maintenance of relatively constant physical and chemical conditions inside the body. Recall that homeostatic systems depend on (1) a sensory receptor that monitors conditions relative to a preferred value, or set point; (2) an integrator that processes information from the sensor; and (3) effector cells that return conditions to the set point. (Glance back at Figure 41.14.) What Chapter 41 did not mention is that messages often travel from integrators to effectors in the form of hormones.

Several hormones that act as messengers in homeostatic systems have already been introduced. Recall from Chapter 42 that when an individual is dehydrated, **antidiuretic hormone** (**ADH**, also called *vasopressin*) is synthesized in the hypothalamus and released from the pituitary gland. ADH increases the permeability of the kidney's collecting ducts to water. As a result, water is reabsorbed from the urine and saved. In this way, ADH is instrumental in achieving homeostasis with respect to water balance. If ADH's action is inhibited, homeostasis fails and illness can occur. For example, the ethanol in alcoholic beverages inhibits the release of ADH from the pituitary. People who imbibe large quantities of these beverages produce large quantities of dilute urine. The resulting water loss can lead to dehydration and nausea—symptoms associated with an alcoholic hangover.

Chapter 42 also notes that aldosterone is released from the adrenal cortex when ion concentrations in body fluids are low. Aldosterone increases the reabsorption of Na^+ in the distal tubules of the kidney. Because aldosterone results in an increase in Na^+ in blood and interstitial fluids, this hormone plays a key role in homeostasis with respect to electrolyte concentrations and the overall volume of body fluids.

Several of the hormones introduced in Figure 47.4 are involved in homeostatic systems. For example, **erythropoietin (EPO)**

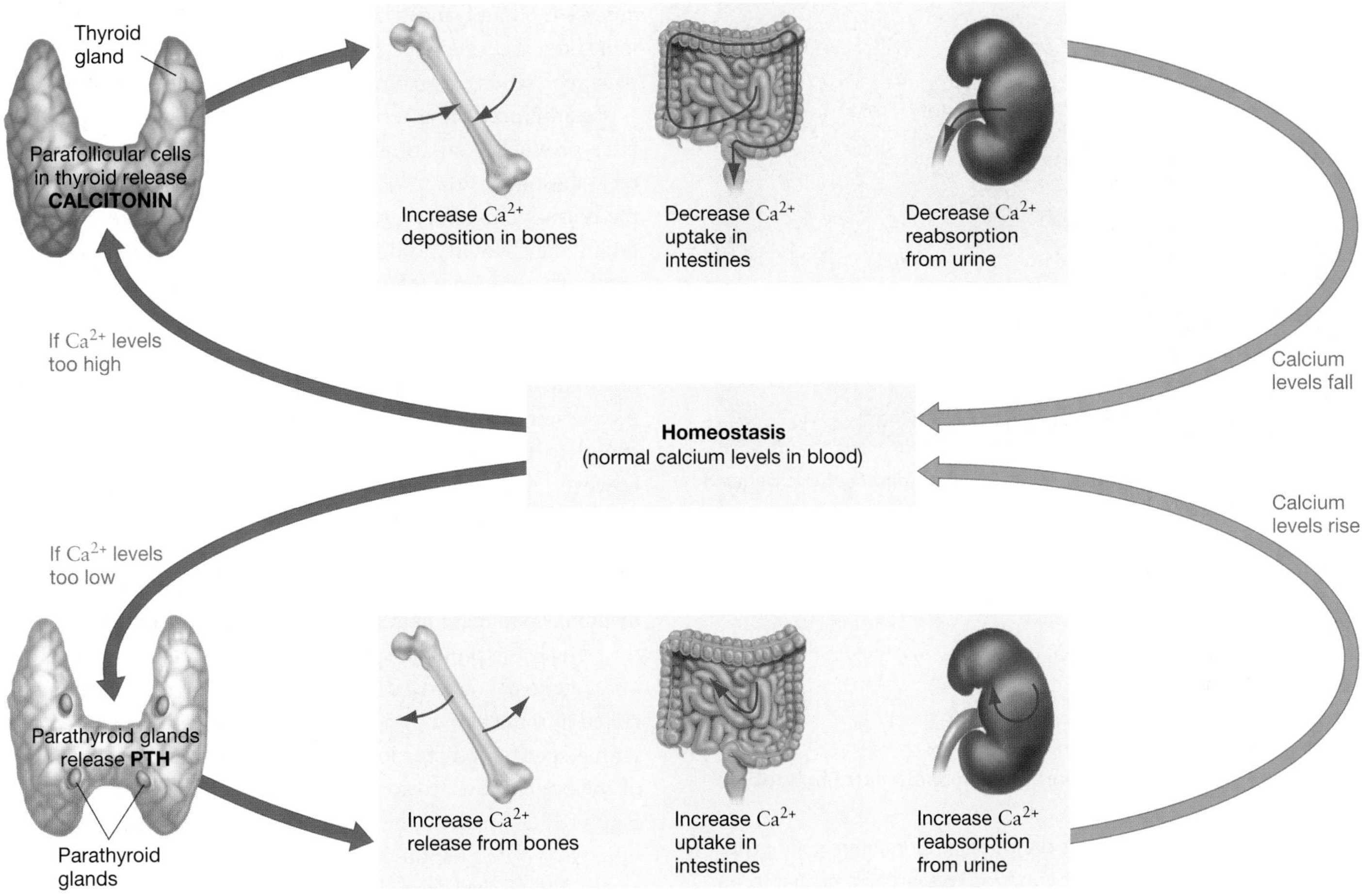

FIGURE 47.6 Hormones Act to Maintain Homeostasis
Parafollicular cells of the thyroid gland and the cells of the four adjacent parathyroid glands sense Ca^{2+} levels and secrete hormones in response to changes. Notice that this homeostatic system does not involve the CNS. **EXERCISE** Draw a similar diagram to illustrate the action of EPO, ADH, aldosterone, or glucagon and insulin. (Hint: Look back at Figure 47.4.)

is released from the kidneys and other tissues when blood oxygen levels are low. Recall from Chapter 44 that EPO acts to achieve homeostasis with respect to oxygen by stimulating the production of red blood cells. The more red blood cells, the higher the oxygen-carrying capacity of blood. In this way, EPO is a crucial element in the homeostatic system for blood oxygen levels.

Figure 47.6 diagrams how the homeostatic system for blood calcium concentration works. Calcitonin is secreted by **parafollicular cells** in the thyroid gland, and **parathyroid hormone (PTH)** is produced by the parathryroid glands. These two hormones work in tandem to keep Ca^{2+} levels in the blood close to the set point, much as insulin and glucagon interact to maintain blood glucose concentrations at preferred levels (see Chapter 43).

To drive home how important hormones are in homeostatic processes, let's consider recent work on homeostasis in energy reserves. This work has not only led to the discovery of new hormones, but also inspired research on the causes of obesity in humans.

Homeostasis of the Body's Energy Reserve For an individual to reproduce successfully, it may have to keep energy in reserve during periods of decreased food availability. This energy reserve is typically in the form of the lipid called triglyceride. **Triglyceride** is an effective energy-storage molecule, because large amounts of ATP can be generated when its three fatty-acid subunits are oxidized (see Chapter 6). Although some triglyceride is present in muscle cells, most is stored in adipocytes. These cells make up the fat bodies found in insects and other species and the adipose tissue of mammals. Even a lean 70-kg human male stores enough energy in adipose tissue to survive for over 30 days without eating.

For decades researchers thought that total adipose tissue mass varied only in response to food intake—meaning it was not subject to homeostatic regulation. This view started to change in the 1970s, when biologists began studying mutations in mouse genes called *obese* and *diabetic*. Homozygous *obese* (*ob*/*ob*) and *diabetic* (*db*/*db*) mice eat large quantities of food and move much less than heterozygous or wild-type

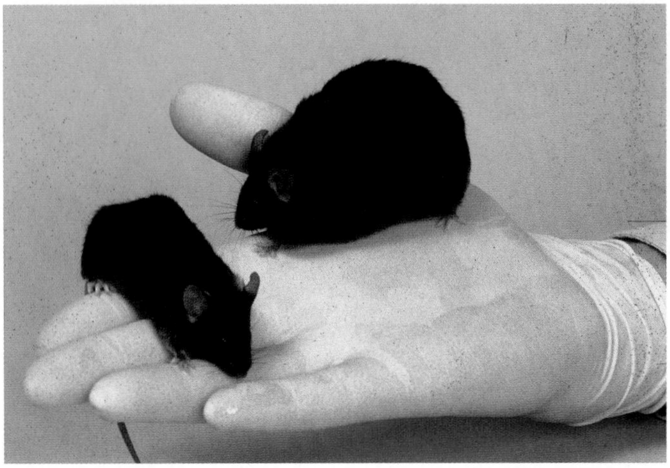

FIGURE 47.7 In Mice, Mutations in the *Obese* and *Diabetic* Genes Can Cause Obesity
Siblings from the same litter. At the *ob* gene, the lean mouse is heterozygous and the obese mouse is homozygous. The phenotypes of *ob/ob* and *db/db* mice are indistinguishable. QUESTION Suppose a mouse was doubly homozygous (*ob/ob db/db*). Would its phenotype be the same as or different from singly homozygous individuals?

siblings do. The homozygotes also become extremely obese (**Figure 47.7**).

To test the hypothesis that the *obese* and *diabetic* gene products are involved in endocrine function, researchers turned to an experimental technique called parabiosis. **Parabiosis** has been used for more than 100 years to determine whether physiological phenomena are hormone mediated. In this technique, two closely related animals are surgically united by suturing the pelvis, shoulders, and abdominal walls together. The skin of the two animals is then joined over the surgical connection, and the newly created twin animal is allowed to recover from the procedure. Within a short time, capillaries form between the two parabiotic partners, and they develop a shared circulatory system. This circulatory system permits the passage of long-lived molecules, such as certain hormones, between the two partners. In contrast, it does not allow the exchange of molecules that are rapidly metabolized, such as glucose and fatty acids. Another key observation is that no new nerves grow between the two animals. As a result, they can influence one another through endocrine signals but not through electrical signals. If an experimental manipulation of one partner produces an effect in the other partner, biologists conclude that the effect is mediated by a hormone.

Researchers performed parabiosis between *db/db*, *ob/ob*, and lean mice as shown in **Figure 47.8**. When a *db/db* animal was joined to either a lean animal or an *ob/ob* animal, the *db/db* mouse continued to eat and grow normally, but its partner stopped eating, lost weight, and eventually died of apparent starvation. But when two *db/db* animals were joined, both partners ate and grew as expected, and there was no difference between the two in survival. When *ob/ob* animals were joined to lean animals, the *ob/ob* mice ate less food and gained weight less rapidly than did *ob/ob* mice joined to other *ob/ob* mice. There was no difference in survival of pairs representing any combination of *ob/ob* and lean mice.

To interpret these results, the biologists hypothesized that mice produce a satiation, or "stop-eating," hormone. The interpretation was that *db/db* mice lack the receptor required for the hormone to affect target cells. As *db/db* mice got fatter and fatter, they would produce more and more hormone—to no avail. This model explained why the signal from *db/db* mice greatly reduced food intake in their parabiotic partners but continued to have no effect on the *db/db* mouse. The satiation hormone itself was postulated to be encoded by the *obese* gene. As a result, *ob/ob* mice do not produce the hormone. They respond to the signal if it is available from a normal partner, however. A key idea here is that the two genotypes produce the same phenotype because they disable different parts of the same hormone-signaling system.

This model was a dramatic break with traditional views of fat metabolism in mammals. It was not confirmed until 1994, when the *obese* gene product, now called leptin, was isolated and the *obese* gene characterized. Leptin is a polypeptide hormone secreted into the blood by adipocytes. Leptin signals by interacting with a specific receptor located in many tissues, including areas of the brain known to control feeding behavior. As the model in Figure 47.8 predicts, the administration of leptin corrects the obesity of *ob/ob* (leptin-deficient) mice but not of *db/db* (leptin receptor–deficient) mice. But as **Box 47.2** explains, leptin injections are not helpful for the vast majority of obese humans.

Follow-up work has shown that leptin levels in the blood vary in proportion to total adipose tissue mass. When adipose mass falls below a set point established by a variety of gene products, the leptin level in the blood also falls. The decrease in leptin level is sensed by the brain, which generates both an increase in appetite and a decrease in energy expenditure. These responses promote eating and restore energy balance. When sufficient food intake has occurred to restore triglyceride stores, leptin levels rise. The result is diminished appetite, increased energy expenditure, and the stabilization of adipose tissue mass. This is a striking example of homeostasis achieved by the process called feedback inhibition. Let's take a closer look next.

✓CHECK YOUR UNDERSTANDING

Hormones usually function in preparing an individual for environmental change, directing development and sexual maturation, or achieving homeostasis. You should be able to (1) give an example of hormones that function in each of those capacities and, for each, (2) describe how the single molecule in your example acts on a variety of cells of tissues, how the various responses that occur result in a coordinated set of changes to the same challenge, and how the overall effect of the hormone's action increases the individual's chances of surviving and reproducing.

Question: Are the *diabetic* and *obese* gene products involved in endocrine function?

Hypothesis: The gene products are involved in hormonal signaling that affects appetite and activity level.

Null hypothesis: The gene products are not involved in hormonal signaling.

Experimental setup:

db / db Lean *db / db* *ob / ob* *ob / ob* Lean

Perform parabiosis surgery on closely related mice with different genotypes, so that blood-borne products will pass between the mice in each pair.

Prediction: Phenotypes of at least some of the parabiotic mice will change.

Prediction of null hypothesis: No changes in phenotypes of parabiotic mice will occur.

Results:

db / db	Lean	*db / db*	*ob / ob*	*ob / ob*	Lean
↑ Body weight	↓ Body weight	↑ Body weight	↓ Body weight	↓ Body weight	↑ Body weight
↑ Adipose tissue mass	↓ Adipose tissue mass	↑ Adipose tissue mass	↓ Adipose tissue mass	↓ Adipose tissue mass	↑ Adipose tissue mass
	Death by starvation		Death by starvation		

Conclusion: The *diabetic* and *obese* gene products are involved in hormonal signaling that affects appetite and activity level.

Interpretation: The *obese* gene codes for a hormone that reduces appetite and increases activity in response to increases in fatness. The *diabetic* gene codes for the receptor for this hormone.

FIGURE 47.8 **Parabiosis of Genetically Obese and Lean Mice Provides Evidence of a "Satiation Hormone"**

BOX 47.2 Is Human Obesity Caused by Leptin Deficiency?

Immediately after leptin was discovered, researchers began testing the hypothesis that it could be used as a drug to treat human obesity. Unfortunately, it quickly became apparent that almost all obese individuals already have high leptin levels—consistent with their high adipose tissue mass. Administering more leptin to these individuals did not cause meaningful weight loss.

Worldwide, fewer than 20 humans with true genetic leptin deficiency have been identified. Just like *ob/ob* mice, these individuals eat vast quantities of food and are severely obese. Because they lack the hormone, obesity in these rare individuals can be cured by administering leptin.

To explain why the vast majority of obese humans do not respond to their high leptin levels, researchers hypothesize that the hormone's effect on the brain is diminished for some reason. Identifying the factor or factors involved in damping the normal response to leptin is currently the subject of intense research. Does the response drop when specific types of food are ingested, or when meal portions get large, or when exercise is inadequate? One intriguing finding is that mice placed on a high-fat diet respond much less strongly to experimental leptin administration than do mice fed a normal diet. This result supports the hypothesis that the increased ingestion of fatty foods disrupts the normal human homeostatic system in some way. If so, then increased fat content in the diet not only increases energy intake but also blocks the normal appetite-diminishing response, leading to overeating and obesity.

47.3 How Is the Production of Hormones Regulated?

Most hormones are released in response to either an environmental cue or a message from an integrator in a homeostatic system. In both cases, the nervous system may be closely involved. For example, environmental cues that signal the onset of the breeding season or the presence of a predator are received by sensory receptors and interpreted by the brain. Similarly, the integrators in most homeostatic systems consist of groups of neurons in the central nervous system (CNS)—that is, the brain and spinal cord (see Chapter 45).

Based on these observations, the short answer to the question posed in the title of this section is simple: In many cases, hormone production is directly or indirectly controlled by the nervous system. How, then, does the nervous system direct the endocrine system? To explore this question, let's focus on how cortisol and epinephrine are regulated by the hypothalamus and pituitary gland.

The Hypothalamus and Pituitary Gland

Researchers who were interested in how hormone production is controlled quickly focused on the pituitary gland. As **Figure 47.9** shows, the pituitary is located at the base of the brain and is directly connected to the brain region called the hypothalamus. The pituitary has two distinct segments: the **anterior pituitary** and the **posterior pituitary.** In 1930 a biologist showed that laboratory rats exhibit a variety of debilitating symptoms when their pituitary is removed. The animals stop growing and cannot maintain a normal body temperature. In addition, their genitals, thyroid glands, and adrenal cortexes atrophy (shrink). Not surprisingly, their life span is also dramatically shortened.

These experiments suggested that, in addition to secreting growth hormone, the pituitary secretes substances affecting the gonads, thyroid, and adrenals. Based on this observation, the pituitary took on the nickname "master gland." Follow-up work confirmed that the nickname is apt. The pituitary secretes hormones that regulate the production of a wide array of other hormones. As a case study, let's look at the pituitary hormone that acts on the adrenal glands.

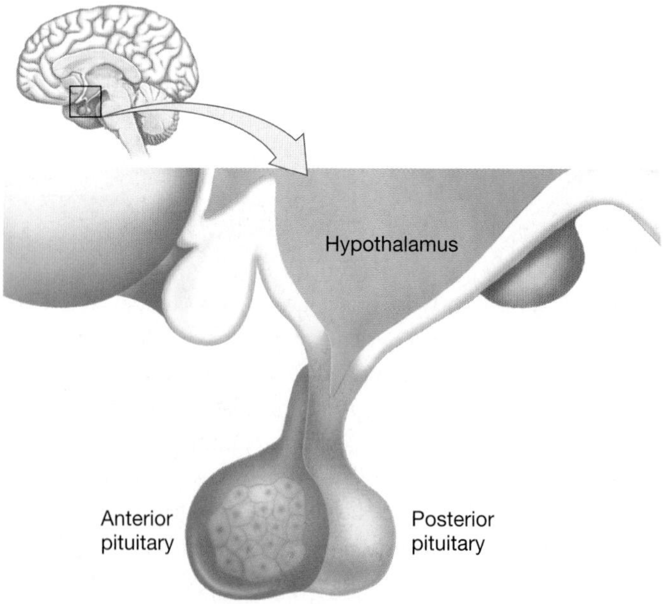

FIGURE 47.9 The Hypothalamus and the Pituitary Gland Are Physically Connected

The pituitary gland is directly connected to the hypothalamus region of the brain. The pituitary has two distinct segments, designated the anterior and posterior pituitary.

Controlling the Release of Glucocorticoids Early work on rats suggested that a molecule from the pituitary gland affects the adrenal gland. This molecule came to be called **adrenocorticotropic hormone**, or **ACTH**. (*Adreno* refers to the adrenal glands; *cortico* refers to the outer portion, or cortex, of the gland; and *tropic* means "affecting the activity of.") In 1943 ACTH was purified and characterized. The biologists who did the work confirmed that glucocorticoids are secreted from the adrenal cortex in response to ACTH released from the pituitary. As **Figure 47.10a** shows, when human volunteers are injected with ACTH, cortisol levels in their blood rise. Thus, ACTH is a regulatory hormone.

Not long after ACTH was isolated, biologists from two laboratories simultaneously and independently showed that it is released in response to a molecule produced by the hypothalamus. Researchers initially called this hormone *corticotropin-releasing factor*, or *CRF*. After years of effort, a different team of researchers finally succeeded in purifying CRF. It is a peptide—just 41 amino acids long—that has been renamed **corticotropin-releasing hormone (CRH).** When CRH is released from the hypothalamus, cells in the anterior pituitary are stimulated to secrete ACTH into the bloodstream.

Why is glucocorticoid secretion controlled by the two hormones CRH and ACTH? The key to answering this question is to recognize that glucocorticoids themselves are able to suppress ACTH production by the pituitary gland. As the data in **Figure 47.10b** show, when human volunteers are injected with cortisol, ACTH levels in their bloodstream drop dramatically. Thus, if glucocorticoid levels become too high, then ACTH levels fall. The result is that glucocorticoid production decreases and levels of glucocorticoid in the blood return to normal. But if glucocorticoid levels become too low, then ACTH levels rise and drive a compensatory increase in glucocorticoid production.

Figure 47.10c summarizes the relationships among CRH, ACTH, and the glucocorticoid cortisol. Recall from earlier chapters that when the presence of a molecule such as cortisol inhibits its own production, **negative feedback**, or **feedback inhibition**, is said to occur. Feedback inhibition in the endocrine system is analogous to temperature control by a heat-sensitive thermostat. If the temperature is too high, the furnace turns off; if the temperature is too low, the furnace turns on. The feedback control system results in a constant air temperature.

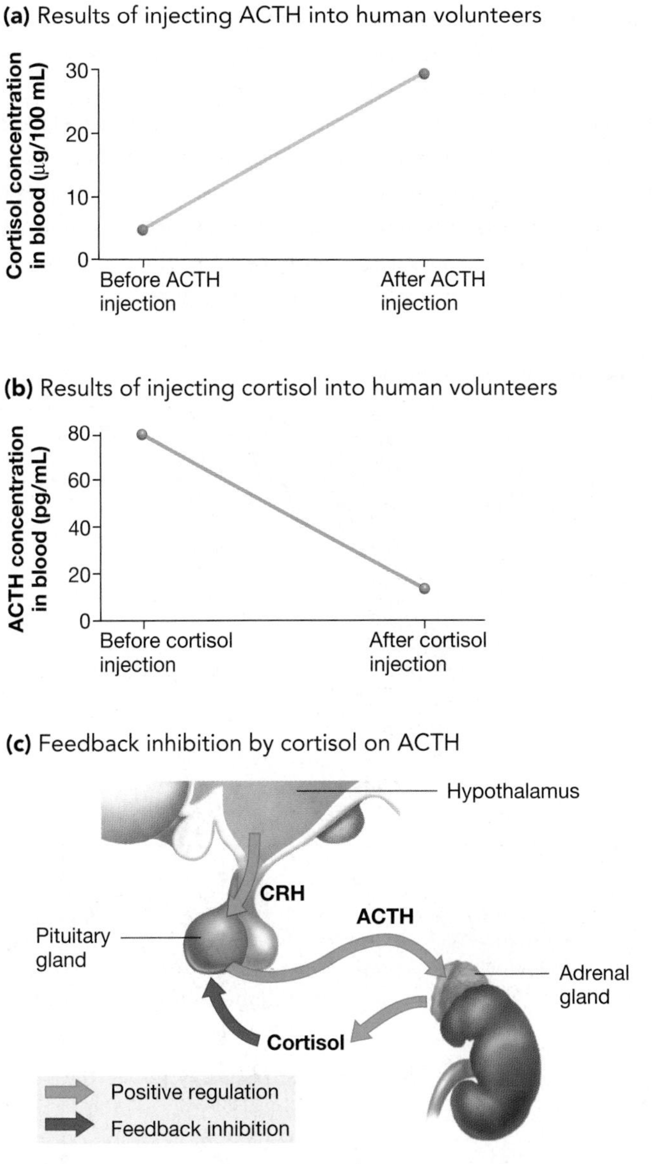

FIGURE 47.10 Feedback Inhibition by Cortisol on ACTH Release **(a)** Concentration of cortisol in the blood after human volunteers are injected with ACTH. **(b)** Concentration of ACTH in the blood after human volunteers are injected with cortisol. This response is characteristic of endocrine systems that display feedback inhibition. **(c)** The release of CRH from the hypothalamus triggers the release of ACTH from the pituitary, which induces the release of cortisol from the adrenal glands. Cortisol diminishes the production of ACTH by the pituitary. EXERCISE How would you use the data in part (b) to devise a test for adrenal failure in humans?

The importance of feedback inhibition in cortisol production can best be appreciated by observing what happens when it fails. Certain pituitary tumors diminish the ability of cortisol to suppress ACTH production. This results in a medical condition called **Cushing's disease**, which is characterized by persistently high blood ACTH and cortisol levels. In essence, Cushing's disease is an unrelenting stress response that results in progressive depletion of the body's protein reserves. It is fatal if not treated.

If the feedback relationship between cortisol and ACTH stabilizes blood cortisol levels, what is the role of CRH? In essence, CRH changes cortisol's ability to suppress ACTH production in the pituitary. Increasing the production of CRH by the hypothalamus is equivalent to raising the temperature setting on a thermostat—it makes things hotter. Under ordinary circumstances, the production of CRH by the hypothalamus displays a daily rhythm, with the highest level in the early morning hours. This drives a corresponding daily rhythm in ACTH production and blood cortisol level, as shown in **Figure 47.11**. The peak in blood cortisol levels that occurs in the morning ordinarily coincides with arousal and initiation of the day's activities—saving glucose for use by the brain. As an aside, the unpleasant symptoms of jet lag are due to the daily cortisol rhythm being out of synchrony with local time for a period of several days after you arrive in a new time zone.

Based on this observation, it should make sense to learn that the long-term stress response is initiated when stimuli that produce pain or anxiety are processed by the brain and result in a sustained increase in CRH production. Increased CRH production causes blood ACTH and cortisol levels to remain higher throughout the day than they are in the unstressed state (see Figure 47.11).

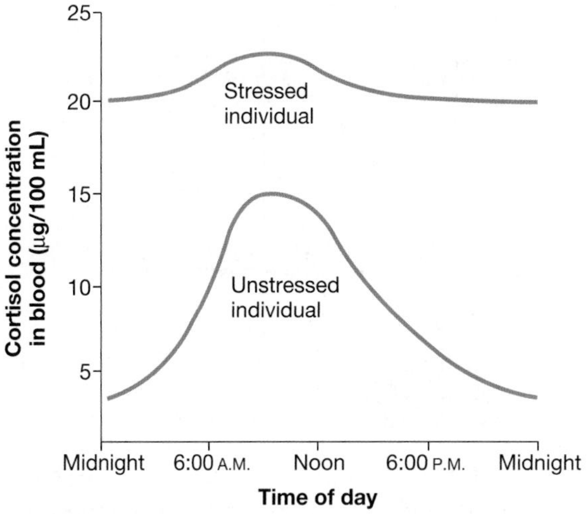

FIGURE 47.11 Blood Cortisol Levels Have a Daily Rhythm Blood cortisol levels are usually highest in the early morning and drop in the late afternoon and evening. Under stressful circumstances, blood cortisol levels are elevated throughout the day.

The Hypothalamic-Pituitary Axis—an Overview Six decades of research have revealed that the CRH-ACTH-glucocorticoid relationship is just one example in an array of hormone systems based on interactions among the hypothalamus, pituitary, and target glands or cells. As **Figure 47.12** illustrates, the **hypothalamic-pituitary axis** actually forms two anatomically distinct systems. The posterior and anterior sections of the pituitary gland are each influenced by different populations of neurons in the hypothalamus. These populations of hypothalamic neurons are called **neurosecretory cells**, because they are neurons that synthesize and release hormones. The release of hormones by cells in the hypothalamus is under the control of brain regions responsible for integrating information about the external or internal environment. For example, information about the approach of an important exam or athletic contest might trigger action potentials that lead to the release of CRH from neurosecretory cells in the hypothalamus.

It's important to recognize that the anterior pituitary and posterior pituitary function in different ways. As Figure 47.12a indicates, the posterior portion of the pituitary is actually an extension of the hypothalamus itself. Neurosecretory cells that project from the hypothalamus produce the hormones ADH and oxytocin, which are then stored in the posterior pituitary. From there, ADH and oxytocin are released into the bloodstream. Recall from Section 47.1 that ADH aids in the reabsorption of water by the kidneys. **Oxytocin** helps induce labor and milk production in females.

Neurosecretory cells in the hypothalamus do not project into the anterior portion of the pituitary, however. As Figure 47.12b illustrates, the hypothalamus and anterior pituitary are connected indirectly, by blood vessels. What happens when neurosecretory cells from the hypothalamus secrete hormones into these blood vessels? In response to the arrival of stimulatory **releasing hormones** from the hypothalamus, the anterior pituitary secretes hormones that enter the bloodstream and act on target tissues or glands. These hormones include ACTH; **follicle-stimulating hormone (FSH)** and **luteinizing hormone (LH)**, which are involved in the production of sex hormones and which regulate the men-

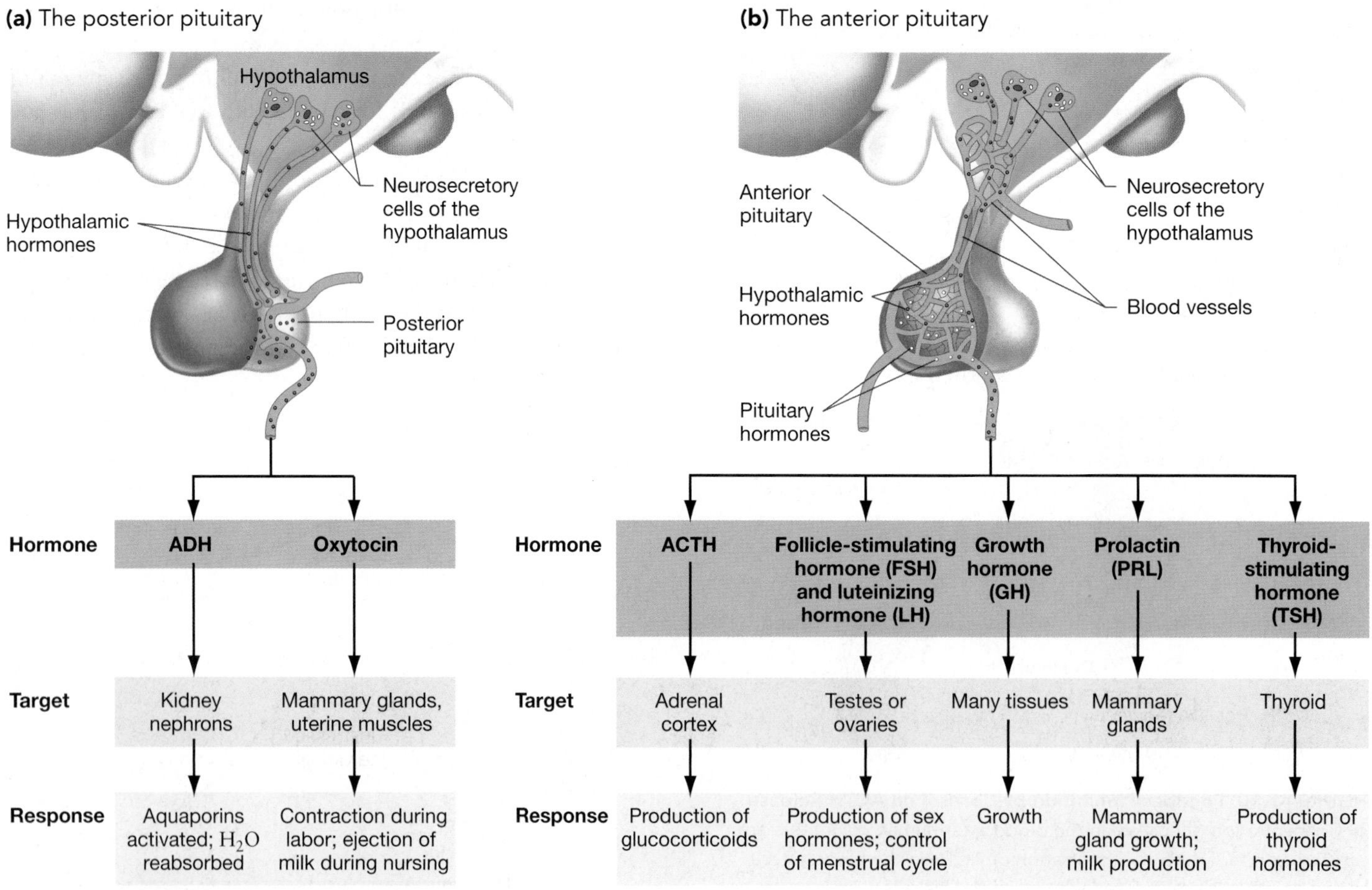

FIGURE 47.12 The Hypothalamus and Pituitary Interact Closely
(a) Developmentally and anatomically, the posterior pituitary is an extension of the hypothalamus. Neurosecretory cells in the hypothalamus extend directly into the posterior pituitary and secrete ADH (vasopressin) and oxytocin. **(b)** The hypothalamus and the anterior pituitary communicate indirectly, via blood vessels. Hormones produced by neurosecretory cells in the hypothalamus travel to the anterior pituitary in blood, where they control the release of pituitary hormones.

strual cycle; GH; **prolactin**, which stimulates mammary gland growth and milk production; and **thyroid-stimulating hormone (TSH)**, which triggers the production of thyroid hormones. Hypothalamic **inhibitory hormones**, in contrast, stop the secretion of certain pituitary hormones. These hormones include **somatostatin**, or **growth-hormone-inhibiting hormone (GHIH)**, which inhibits GH production, and **dopamine**, or **prolactin-inhibiting hormone (PIH)**, which limits prolactin production.

In several cases, the production of a pituitary hormone is controlled by both a stimulatory and an inhibitory hypothalamic hormone. The realization that many of the hormones produced by the anterior pituitary stimulate the production of other hormones justified the structure's designation as the master gland.

To summarize, interactions between the hypothalamus and pituitary gland furnish a key mechanism for the control of hormone action by the CNS. The hypothalamic-pituitary axis consists of two anatomically and functionally distinct systems. The posterior pituitary serves as a storage and release site for hormones produced by neurosecretory cells in the hypothalamus; the anterior pituitary serves as a hormone production and release site that is controlled by hypothalamic hormones.

Control of Epinephrine by Sympathetic Nerves

When biologists analyze how the nervous system and endocrine system interact to control the release of epinephrine, the distinction between the two systems begins to blur. Section 47.2 introduced how epinephrine acts as a hormone. During the fight-or-flight response, sympathetic nerves trigger the release of epinephrine from the adrenal medulla into the bloodstream. But in addition, some sympathetic nerves release the related molecule, norepinephrine, directly onto target cells. In effect, the nervous system delivers a chemical messenger directly to particular cells while the endocrine system broadcasts a similar messenger by secreting it into the bloodstream. Epinephrine and norepinephrine, which differ from one another only by the presence of an additional methyl group on epinephrine (see Figure 47.5), are members of the family of molecules called **catecholamines.**

Catecholamines function as neurotransmitters as well as hormones. The close functional similarity between hormones and neurotransmitters does not end there, however. In some cases, the mode of action of hormones and neurotransmitters is similar. To drive this point home, recall from Chapter 45 that neurotransmitters have two major types of effects on target cells. All neurotransmitters trigger a postsynaptic potential, which can make the postsynaptic neuron more or less likely to deliver an action potential. In addition, certain neurotransmitters may initiate changes in gene expression in neurons. Chapter 45 illustrated the effect of neurotransmitters on transcription by reviewing experiments on the sea slug *Aplysia*. Repeated application of serotonin led to gene activation and changes in the behavior of the synapse. By altering synapses in this way, neurotransmitters play a central role in learning and memory. Similarly, many hormones exert their effects by activating particular genes in target cells. Understanding how gene activation occurs in response to neurotransmitters and hormones is currently the subject of intense research at laboratories around the world.

> ✓CHECK YOUR UNDERSTANDING
>
> Hormone concentrations are tightly regulated. You should be able to (1) diagram interactions between the hypothalamus and pituitary during the long-term response to stress, (2) explain how feedback inhibition occurs in production of ACTH, and (3) explain how the nervous system and endocrine system interact when norepinephrine and epinephrine are produced during the short-term response to stress.

47.4 How Do Hormones Act on Target Cells?

The key to understanding how hormones act on target cells is recognizing that some animal hormones are lipid soluble and cross plasma membranes readily, while others are not (see Figure 47.3). More specifically, steroid hormones are small lipids that enter cells without difficulty. But the peptide and polypeptide hormones and most amino acid derivatives do not cross plasma membranes easily, because of their large size and electrical charge.

Differences in lipid solubility are important because they influence where a target cell receives the chemical message. Steroids often act inside the cell, while most amino acid derivatives and all polypeptides act at the cell surface. To explore these two distinct paths of hormone action, let's consider how estrogens and epinephrine affect target cells. As a steroid and a nonsteroid, they serve as model systems for hormone action.

WEB TUTORIAL 47.2 Hormone Actions on Target Cells

Steroid Hormones and Intracellular Receptors

Estrogens are steroids that direct the development of female secondary sex characteristics in many animal species. In humans and other mammals, the most important estrogen is the molecule estradiol (formally, 17β-estradiol). Because of its importance in reproduction by humans and domesticated animals, estradiol's mode of action has been the topic of intense investigation for over 50 years. How do target cells receive the signal carried by estradiol?

Identifying the Estrogen Receptor A key advance in research on how hormones affect target cells occurred in 1964, when biologists succeeded in isolating the estradiol receptor in laboratory rats. As **Figure 47.13** indicates, the researchers began by labeling purified estradiol with a radioactive atom and injecting a small amount into five adult female rats. Twenty minutes later, the biologists removed the uterus from each female and homogenized the tissue. This step was logical because the uterus, a muscular sac where embryonic development takes place, is one of the organs that changes in response to changes in the amounts of estradiol secreted from the ovaries.

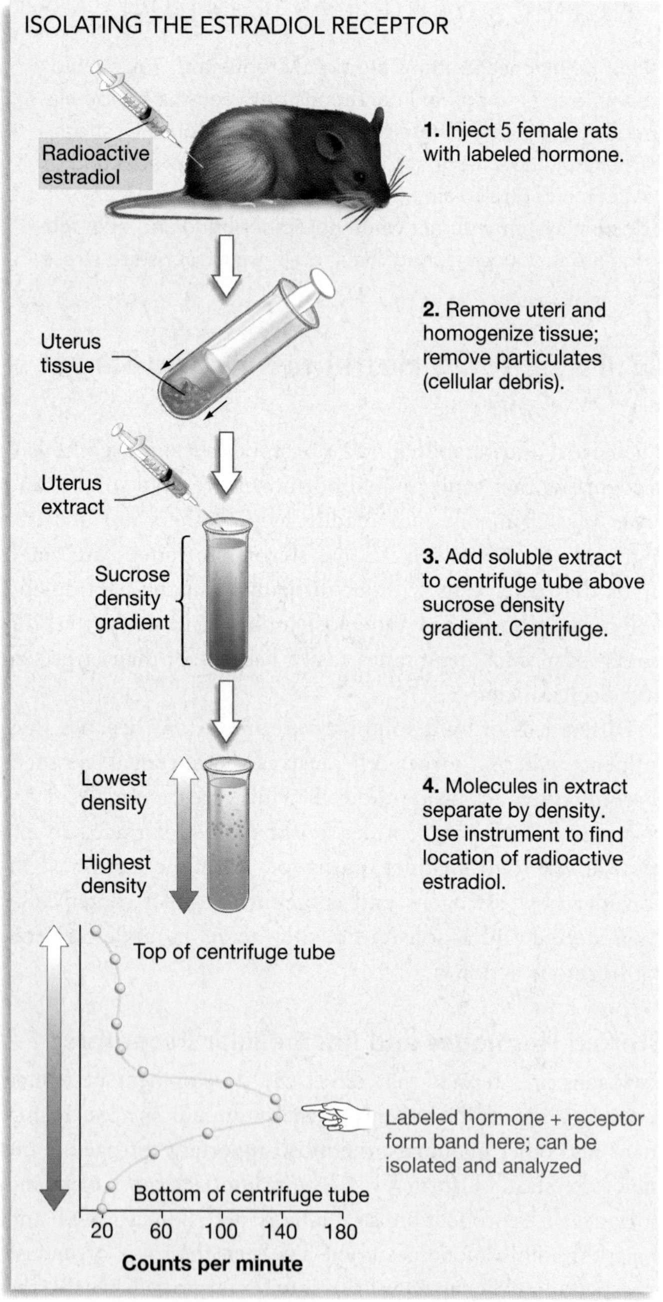
ISOLATING THE ESTRADIOL RECEPTOR
Radioactive estradiol
1. Inject 5 female rats with labeled hormone.
Uterus tissue
2. Remove uteri and homogenize tissue; remove particulates (cellular debris).
Uterus extract
Sucrose density gradient
3. Add soluble extract to centrifuge tube above sucrose density gradient. Centrifuge.
Lowest density
Highest density
4. Molecules in extract separate by density. Use instrument to find location of radioactive estradiol.
Top of centrifuge tube
Labeled hormone + receptor form band here; can be isolated and analyzed
Bottom of centrifuge tube
20
60
100
140
180
Counts per minute

FIGURE 47.13 The Discovery of the Estradiol Receptor
Researchers injected female rats with radioactive estradiol in hopes that the labeled molecule would bind to its receptor in the uterus. After molecules from uterine cells were separated by centrifugation, investigators documented the amount of radioactivity at each level in the centrifuge tube.

After removing the particulate matter from the cell homogenate, the researchers added the soluble fraction to a centrifuge tube that contained a sucrose density gradient. As explained in Chapter 7, sucrose density centrifugation allows researchers to separate molecules by density. When centrifugation was complete, the biologists found that the radioactivity was concentrated in a narrow band in the tube. The band contained radioactive estradiol bound to the estradiol receptor. When the investigators purified the receptor molecule, they found that it could readily be destroyed by treatment with proteinase enzymes. As a result, they inferred that the estradiol receptor was a protein.

Follow-up experiments confirmed that the estradiol receptor is located in the nucleus but is not associated with the nuclear envelope. Further, the receptor is found only in estradiol target tissues, including the uterus, hypothalamus, and mammary glands. The latter finding was particularly exciting, because it clarified how hormones act in a tissue-specific way. This is a crucial point: Hormones are broadcast throughout the body via the bloodstream, but they act only on cells that express the appropriate receptor. Target cells respond to particular hormones because they contain a receptor for that hormone.

Later, a different team of biologists found that the gene for the estradiol receptor has a sequence and structure that are similar to the receptors for the glucocorticoids, testosterone, and other steroids. This result suggested that all steroid receptors are related by descent from an ancestral receptor molecule and that the binding of any steroid hormone to its receptor might affect the target cell in a similar way.

Biologists understood the nature and location of the estradiol receptor. What happens once the hormone-receptor complex forms inside the nucleus of a target cell?

Documenting Changes in Gene Expression During the 1970s and 1980s, work in several laboratories suggested that estradiol and other steroid hormones affect gene transcription after they bind to their receptors. For example, researchers were able to inject laboratory animals with estradiol or other steroid hormones and document changes in the mRNAs and proteins produced in target cells. These data showed that steroid hormones can cause dramatic changes in the amount or timing of mRNA production by a large number of genes.

An important insight into the molecular mechanism of hormone action came when researchers found a distinctive DNA-binding region in the steroid-hormone receptor. The region codes for the DNA-binding domain called a *zinc finger*. Follow-up work showed that zinc fingers are found in all proteins in the steroid-hormone receptor family. To understand why this observation is significant, recall from Chapter 18 that DNA-binding domains are protein regions that make physical contact with DNA. The presence of a zinc finger in the estradiol receptor suggested that, once estradiol had bound to it, it might affect gene expression by binding directly to DNA.

Follow-up work confirmed that steroid hormone-receptor complexes bind to specific sites in DNA, which came to be called **hormone-response elements.** Hormone-response elements are located just "upstream" (in the 5′ direction) from the start of target genes. Gene expression begins when a regulatory protein such as a steroid hormone-receptor complex binds to the hormone-response element for that gene.

The model in **Figure 47.14** summarizes the current state of thinking about how steroid hormones affect target cells. When estradiol or another steroid hormone enters a target cell, the hormone binds to its receptor (steps 1 and 2 of the figure). The binding event causes a conformational change in the receptor. In response, the hormone-receptor complex binds to DNA along with other proteins required for transcription initiation (step 3), and the expression of specific genes begins. Because each hormone-receptor complex leads to the production of many copies of the gene product, the signal from the hormone is amplified (steps 4 and 5). In this way, a small number of hormone molecules produces a large change in the activity of target cells and tissues.

Currently, research on steroid hormone action is focused on the details of how the hormone-receptor protein interacts with the other proteins required to initiate transcription and how interactions between the hormone-receptor complex and other regulatory proteins alter the timing and degree of expression of target genes. Researchers are also following up on the relatively recent discovery that at least some mammals, including humans, have two distinct receptors for estradiol. As experiments continue, it will be interesting to learn whether these two receptors trigger different responses to the same hormone.

Hormones That Bind to Cell-Surface Receptors

Epinephrine, norepinephrine, and the polypeptide hormones are not lipid soluble. For these molecules to affect a cell, they have to bind to receptors on the cell surface. The messenger never enters the target cell, so its message must be *transduced*—that is, changed into a form that is active inside the cell. Recall from Chapter 8 that this phenomenon is known as **signal transduction.**

To explore how signal transduction occurs, let's first examine the nature of the hormone receptors residing in the plasma membrane. Then we can explore research on the molecules that are responsible for processing the message inside the cell. In both cases, we'll focus on epinephrine as a model system.

Identifying the Epinephrine Receptor In 1948 a biologist published the results of an exhaustive set of studies on how epinephrine and five related molecules affect dogs, cats, rats, and rabbits. The responses observed fell into two distinct categories, depending on the tissue being considered. To explain this observation, the researcher suggested that epinephrine and the related molecules bind two distinct types of receptor. He called these hypothetical proteins the alpha receptor and the beta receptor. Further, he proposed that the five epinephrine-like molecules were acting as **agonists,** meaning they bind to the same receptors as the hormone itself.

Follow-up work with agonists and with molecules that block epinephrine receptors confirmed this hypothesis. In fact, experiments documented that there are two types of alpha receptor and two types of beta receptor. In short, there are four distinct epinephrine receptors. Each receptor type is found on a distinct tissue type, and each induces a different response from the cell.

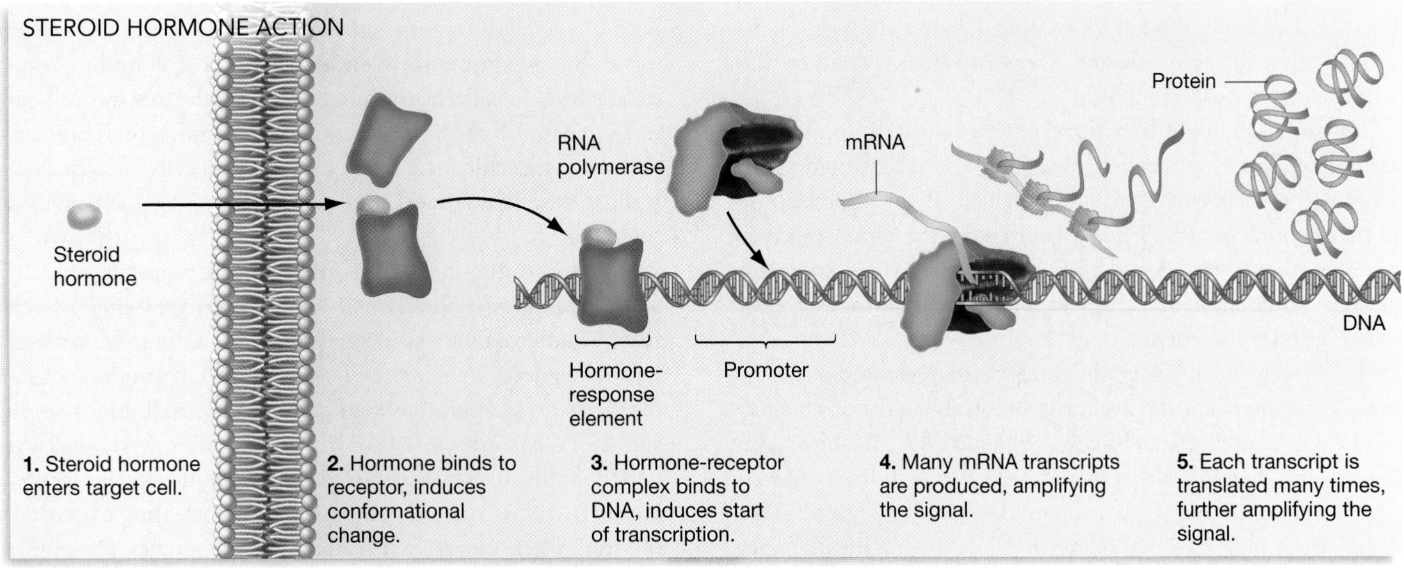
STEROID HORMONE ACTION
Protein
RNA polymerase
mRNA
Steroid hormone
DNA
Hormone-response element
Promoter
1. Steroid hormone enters target cell.
2. Hormone binds to receptor, induces conformational change.
3. Hormone-receptor complex binds to DNA, induces start of transcription.
4. Many mRNA transcripts are produced, amplifying the signal.
5. Each transcript is translated many times, further amplifying the signal.

FIGURE 47.14 Steroid Hormones Bind to Receptors inside Cells and Change Gene Expression
The current model holds that most steroid hormones bind to receptor proteins inside a cell and that hormone-receptor complexes trigger the transcription of certain genes by binding to regulatory DNA segments called hormone-response elements.

The discovery of four epinephrine receptors reinforces the concept of tissue specificity observed in experiments on the estradiol receptor. Hormones are transmitted throughout the body, not unlike a radio or television signal that is broadcast through the atmosphere. Their message is received only by cells with the appropriate receptor, however—just as a radio or TV signal is received only by equipment with an appropriate antenna. In the case of epinephrine, there are four distinct receptors instead of just one. As a result, the same hormone can trigger different effects in different cells. Now the question is, What happens once epinephrine binds to one of these receptors?

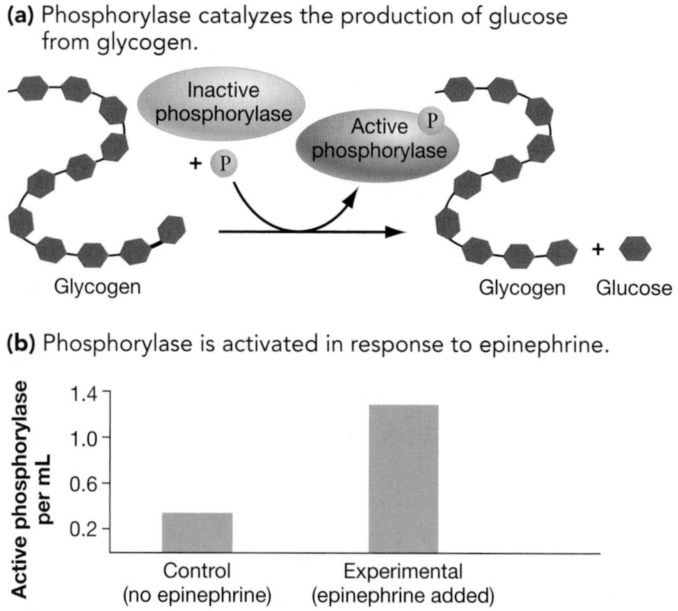

FIGURE 47.15 Epinephrine Activates the Enzyme That Catalyzes the Formation of Glucose from Glycogen
(a) Phosphorylase is activated when an enzyme adds a phosphate group to it. The activated protein catalyzes the breakdown of glycogen. **(b)** When epinephrine is added to cell-free extracts from liver tissue, the amount of activated phosphorylase increases dramatically.

Signal Transduction and the Role of Second Messengers

Signal transduction occurs when a chemical message at the cell surface elicits a response inside the cell. In effect, receptors "read" the message that is carried by a hormone and, through a sequence of events, initiate an appropriate response by the target cell. Signal transduction is currently an exceptionally active area of research, and research on epinephrine has produced fundamental insights into how the process works at a molecular level.

Recall from Section 47.2 that one of epinephrine's major effects is to increase glucose levels in the blood. To understand how epinephrine triggers the release of glucose from target cells, biologists focused on phosphorylase. **Phosphorylase** is an enzyme that was discovered in the mid-1940s by biologists who were studying how cells cleave glucose molecules off glycogen. As **Figure 47.15a** shows, phosphorylase catalyzes this reaction. The researchers found that phosphorylase exists in active and inactive forms, and that the enzyme switches between these states when it is phosphorylated or dephosphorylated by another enzyme. Phosphorylase was the first of many enzymes shown to be activated by the addition of a phosphate group.

Investigators in a different laboratory followed up on these results by exploring whether phosphorylase is activated in response to epinephrine. During the fight-or-flight response, the primary source for the glucose that enters the bloodstream is glycogen that is stored in the liver. Does epinephrine stimulate glucose production by activating phosphorylase in liver cells where glycogen is stored?

To answer this question, the research team worked out techniques for homogenizing liver cells and studying the production of glucose from glycogen in vitro. As **Figure 47.15b** shows, the team found that, when they added epinephrine to their cell-free system, large amounts of phosphorylase were activated. This observation suggested that there was something in the homogenized cells that activated phosphorylase in response to the presence of epinephrine. By purifying the components of the homogenized cells and testing them in the cell-free system, the researchers eventually found the ingredient that activated phosphorylase. It was a molecule called **cyclic adenosine monophosphate**, or **cAMP**.

The role of cAMP in signal transduction was confirmed when researchers studied the effects of epinephrine on the activity of the rat heart. During the fight-or-flight response, both the heart rate and the contractile force of the heart rise dramatically—increasing cardiac output. Recall from Chapter 44 that cardiac output is a measure of the efficiency of the heart's pumping. As the graphs in **Figure 47.16** show, there is a striking increase in cAMP levels inside heart-muscle cells soon after the cells are treated with epinephrine (at time 0). Investigators confirmed that the contractile force of the heart-muscle cells increased and, a short time later, observed a rise in phosphorylase activity in these cells.

To capture the importance of cAMP in triggering these effects, biologists referred to it as a second messenger. Recall from Chapter 8 that a **second messenger** is a nonprotein signaling molecule that increases in concentration inside a cell in response to a received signal—a molecule that binds at the surface. What does cAMP do? Follow-up work showed that cAMP binds to an enzyme called cAMP-dependent protein kinase A. The enzyme responds by phosphorylating the enzyme phosphorylase kinase, which then phosphorylates phosphorylase. This chain of events, called a **signal transduction cascade,** is initially triggered by the synthesis of cAMP, which is produced from ATP in a reaction that is catalyzed by the enzyme adenylyl cyclase.

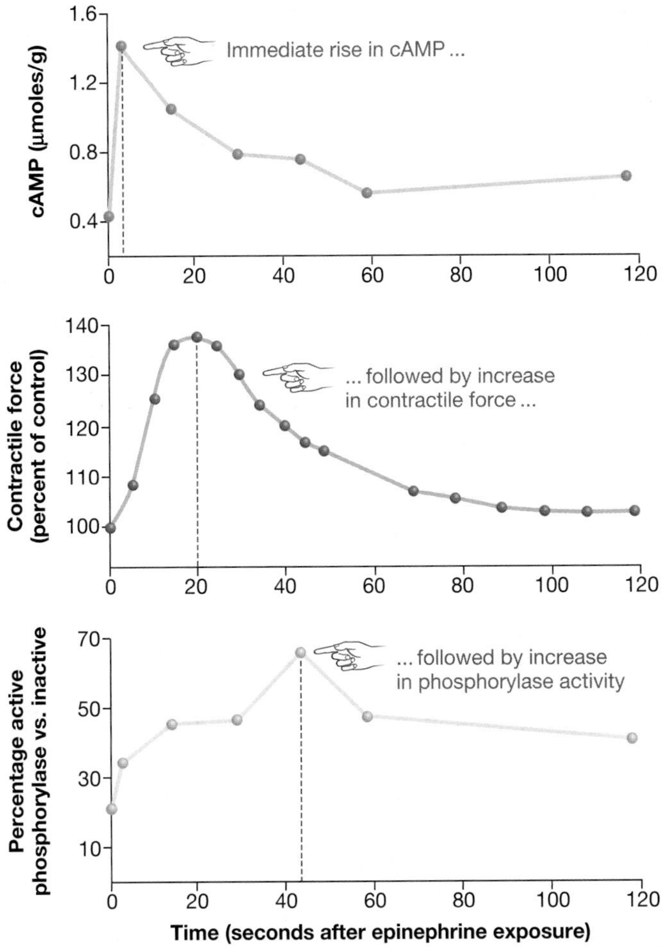

FIGURE 47.16 Chemistry and Activity of Heart-Muscle Cells Change in Response to Epinephrine
When epinephrine is added to heart-muscle cells, there is a rapid and dramatic rise in intracellular cAMP concentration. The contractile force of muscle fibers shows a sharp increase soon thereafter; slightly later, phosphorylase activity increases significantly.

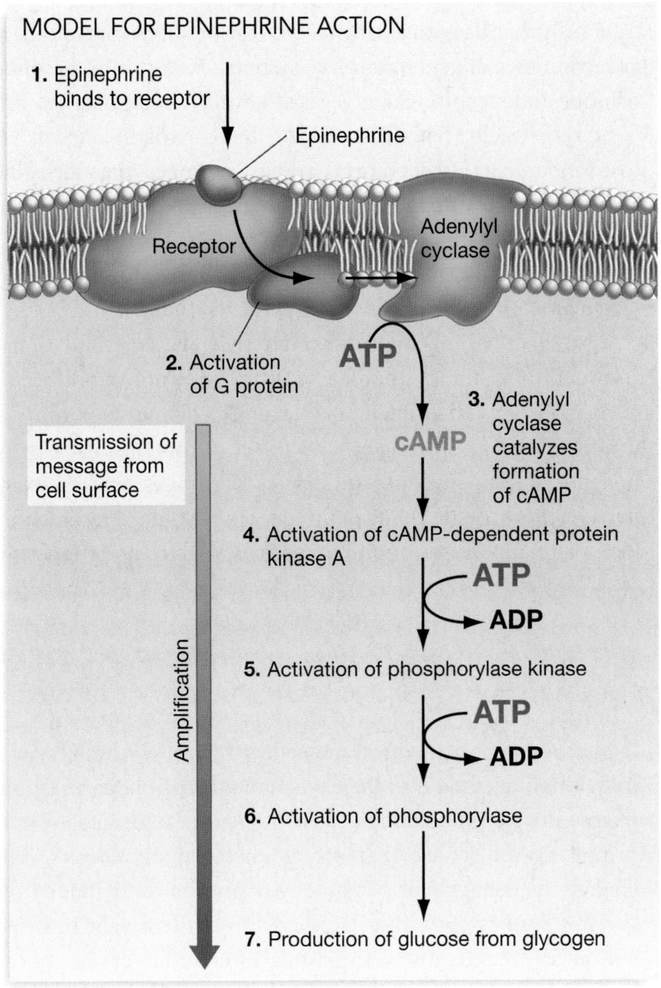

FIGURE 47.17 Epinephrine Triggers a Signal Transduction Cascade
Sequence of events after epinephrine binds to a receptor on the plasma membrane. Several steps involve phosphorylation events that activate proteins. Because each molecule in the cascade acts on many molecules downstream, the effect of the initial binding event is greatly amplified. EXERCISE Using information in the text, label the approximate amount of amplification that occurs in each of steps 3–7.

To pull all these details and reactions together, researchers proposed the model for epinephrine action diagrammed in **Figure 47.17**. In studying this model, it is crucial to recognize two points: First, the second messenger cAMP transmits the signal from the cell surface to the signaling cascade. Second, the other events in the sequence amplify that signal. To drive the latter point home, consider that, in response to stimulation by the hormone-receptor complex, adenylyl cyclase is thought to catalyze the formation of at least 100 molecules of cAMP. In turn, each of these cAMP molecules then activates many molecules of cAMP-dependent protein kinase A. Subsequently, each protein kinase molecule activates many molecules of phosphorylase kinase, and so on. In this way, the binding of just a single molecule of epinephrine may trigger the release of millions or even billions of glucose molecules. Amplification through a signal transduction cascade explains why tiny amounts of hormones can have such huge effects on an individual.

The model in Figure 47.17 was inspired by experiments on the epinephrine receptor called the beta-1 receptor. But other groups showed that, when epinephrine binds to an alpha-1 receptor, a completely different signal transduction event occurs. In this and many other receptor systems, calcium ions (Ca^{2+}) serve as the second messenger in conjunction with a second-messenger molecule called IP3. Diacylglycerol (DAG) and 3′, 5′ cyclic GMP (cGMP) are also common second messengers in hormone response systems.

Researchers are increasingly impressed with the diversity and complexity of signal transduction cascades. For example, target cells that have the same receptor protein may have different second messengers or enzyme systems. As a result, the same hormone and receptor can give rise to different responses in different target cells. This observation helps explain one of the most fundamental observations about hormones: that the same chemical messenger can trigger different responses in cells from different organs or in cells at different developmental stages. The reason is that the cells contain different receptors, second messengers, protein kinases, or amplification steps.

Although they are highly diverse, the receptors and signal transduction cascades that are affected by polypeptide and amino-acid-derivative hormones lead to a common response: the activation of a specific protein or set of proteins, usually by phosphorylation. In contrast, steroid hormones tend to exert their effects through changes in gene expression. Steroid hormones activate transcription factors that lead to the production of new proteins; nonsteroid hormones trigger signal transduction cascades that activate existing proteins.

✓CHECK YOUR UNDERSTANDING

Hormones act on target cells by binding to receptors. In response to binding, hormone-receptor complexes either bind to DNA and induce changes in gene expression or induce the production of second messengers and the activation of signal transduction cascades that result in the phosphorylation of existing proteins. You should be able to (1) draw diagrams that summarize both sequences of events and (2) predict how the systems you diagrammed would respond to a change in the DNA sequence of the hormone-response element or the enzyme responsible for production of the second messenger.

ESSAY Do Humans Produce Pheromones?

A pheromone is a chemical messenger that is synthesized by an individual, released into the environment, and elicits a response from a different individual. Pheromones are odorants, or odor-producing molecules, that act as chemical messengers. They change the behavior or physiology of another individual.

Pheromones have been studied most intensively in insects and mice. To date, over a thousand different insect pheromones have been isolated and characterized. Most of these molecules are produced by females and serve as airborne, species-specific sex attractants. In mice, pheromones have been isolated from the urine of males and females and from glands that secrete to the skin. These molecules have a variety of effects, ranging from the synchronization of ovulation and sexual receptivity in females that are living together to delaying or accelerating the onset of puberty in females.

Biologists have also collected data suggesting that pheromones exist in humans. The researchers involved had studied the role of pheromones in triggering ovulation in rodents and also documented a phenomenon in humans known as **menstrual synchrony**. When human females live in close proximity, as in college dormitories or other communal situations, their menstrual cycles tend to synchronize over time. (Chapter 48 explains the phases of the menstrual cycle and details how hormones control the timing and duration of each stage.) The researchers wanted to know whether a pheromone might be responsible for changing these phases in a way that leads to their synchronization among individuals.

To test the hypothesis that a pheromone is responsible for menstrual synchrony, the biologists recruited female volunteers at a university. For several months the research team had nine donor individuals wear cotton pads under their armpits. The researchers collected these pads daily and wiped either those pads or fresh cotton pads under the noses of recipient individuals. Then the team monitored whether exposure to armpit secretions led to a change in the length of menstrual phases in the recipients. The data showed that recipients experienced significant changes in their menstrual cycle compared with individuals who were exposed only to clean cotton pads. Because these changes corresponded to the menstrual stage being experienced by donors, the researchers concluded that recipients had responded to a pheromone.

When human females live in close proximity . . . their menstrual cycles tend to synchronize over time.

Is there any anatomical or molecular evidence that supports this result? In mice, the organ, neurons, and receptor molecules responsible for sensing pheromones are well characterized. Near the main olfactory (nasal) epithelium, mice have a specialized structure called the *vomeronasal organ*, or VNO. This organ contains sensory neurons that project to a portion of the hypothalamus that is independent of the brain area responsible for the sense of smell. Although most adult humans lack a VNO, researchers have recently found a gene in humans that is extremely similar to the genes for receptor proteins in the VNO of mice. Based on this result, the investigators claim that humans have at least one functioning VNO-like receptor. It is unknown where this receptor might be located, however.

Does a VNO-like receptor respond to the pheromone that induces menstrual synchrony? Although research continues at a rapid pace, this question remains unanswered.

CHAPTER REVIEW

Summary of Key Concepts

Animals use at least six major types of chemical signals. Hormones are chemical signals that are present in tiny concentrations and travel throughout the body to affect target cells.

Hormones are chemical messengers that are released from cells of the endocrine system, circulate in the blood or other body fluids, and trigger a response in distant target cells containing an appropriate receptor. Many hormones were discovered when researchers removed a gland from a laboratory animal and were able to correct the resulting symptoms by injecting the individual with a solution extracted from that gland. By purifying molecules in these extracts and testing their activity, biologists have been able to isolate hormones with a variety of chemical structures. Most animal hormones are polypeptides, amino acid derivatives, or steroids. Although they are produced in tiny concentrations, hormones have large effects because they trigger gene expression or because their message is amplified through a signal transduction cascade.

Web Tutorial 47.1 Endocrine System Anatomy

The information in hormonal signals helps animals respond to environmental change, develop as embryos, undergo sexual maturation, and achieve homeostasis.

With the nervous system, hormones coordinate the activities of diverse cells and tissues. A single hormone may affect a wide array of cells and tissues and induce a variety of responses, all of which help the individual accomplish one specific task.

Epinephrine and cortisol are examples of hormones that help individuals cope with environmental changes. These hormones activate the short-term and long-term responses to stress, triggering the fight-or-flight response or inducing changes that conserve glucose for use by the brain.

Estradiol is an example of a hormone that regulates development and sexual maturation. Estradiol stimulates the formation of female sex characteristics in human embryos and the maturation of these tissues in adolescence.

Hormones are involved in a wide array of homeostatic interactions. For example, hormones are involved in directing cells that modify the concentrations of glucose, Na^+, Ca^{2+}, O_2, water, and other molecules in the blood and interstitial fluid. Homeostasis with respect to triglyceride stores is also under endocrine control.

The production of a hormone is tightly regulated by input from the nervous system and by other hormones.

In many cases, the release of a hormone is regulated by chemical messengers from the anterior pituitary. The pituitary gland is located near the base of the brain in humans and other mammals. Hormone-secreting cells in the anterior pituitary are regulated by hormones released by the hypothalamus region of the brain. For example, the brain responds to long-term stress by triggering the release of the hypothalamic hormone CRH. CRH activates the release of ACTH by the pituitary gland, which stimulates the production of cortisol by cells in the adrenal cortex. This chain of events is dampened by the ability of cortisol to inhibit the production of ACTH.

Some hormones bind to receptors inside target cells and change gene expression. Other hormones bind to receptors at the cell surface and lead to changes in protein activation.

Animal hormones have two basic modes of action. Steroid hormones are lipid soluble, cross plasma membranes readily, and often bind to receptors inside cells. Most amino acid derivatives and polypeptide hormones are not lipid soluble and bind to receptors located in the membranes of target cells. In both cases, the response to a hormone is tissue specific because only certain cells express certain receptors.

Work on estradiol and ecdysone indicates that most steroid hormones act by inducing a change in gene expression. Steroid-hormone receptors are closely related structurally and have a distinctive DNA-binding domain. The hormone-receptor complex binds to hormone-response elements near the starting point of one or more genes. In response, the transcription of specific genes is activated or repressed.

Amino acid derivatives and polypeptide hormones trigger a complex sequence of events when they bind to a receptor on the plasma membrane. These signal transduction cascades induce a change in the concentration of a second messenger, such as cAMP, inside the target cell. In many cases, the endpoint of the process is the activation of a target protein by phosphorylation and a dramatic change in the cell's activity.

Web Tutorial 47.2 Hormone Actions on Target Cells

Questions

Content Review

1. Both epinephrine and cortisol are involved in the body's response to stress. How do the two molecules differ?
   a. Cortisol is an amino acid derivative; epinephrine is a steroid.
   b. Cortisol binds to receptors on the plasma membranes of target cells; epinephrine binds to receptors in the interior of target cells.
   c. Epinephrine mediates the short-term response; cortisol mediates the short- and long-term responses.
   d. Cortisol controls the release of epinephrine from the adrenal glands.

2. Why are the structural differences among amino-acid-derivative, polypeptide, and steroid hormones significant?
   a. Steroids are lipid soluble and cross plasma membranes readily.
   b. Amino-acid-derivative and polypeptide hormones are longer lived in the bloodstream and thus amplify signals more highly.
   c. Polypeptide hormones are the most structurally complex and induce permanent changes in target cells.
   d. Only polypeptide and steroid hormones bind to receptors in the plasma membrane.

3. What is signal transduction?
   a. the binding of a steroid hormone-receptor complex to DNA
   b. the release of a hormone from the anterior pituitary in response to a hypothalamic hormone
   c. the release of a hormone from the posterior pituitary in response to action potentials from the hypothalamus
   d. the production of a second chemical messenger inside a cell in response to the binding of a hormone at the cell surface
4. Which of the following developmental processes is *not* controlled by hormones?
   a. the initial development of male and female gonads, soon after fertilization
   b. overall growth
   c. molting in insects and other invertebrate animals
   d. metamorphosis in insects and other invertebrate animals
5. What is a hormone-response element?
   a. a receptor for a steroid hormone
   b. a receptor for a polypeptide hormone
   c. a segment of DNA where a hormone-receptor complex binds
   d. an enzyme that is activated in response to hormone binding and produces a second messenger
6. In hormone systems, when does feedback inhibition occur?
   a. when the presence of a hormone inhibits its synthesis
   b. when the presence of a hormone stimulates its synthesis
   c. when a second messenger triggers the phosphorylation of an inhibitory protein
   d. when a hormone from the hypothalamus inhibits the release of a hormone from the anterior pituitary

Conceptual Review

1. Compare and contrast the functions of electrical signaling and chemical signaling in animals. How do the nervous system and endocrine system interact during the fight-or-flight response? What role does the nervous system play during hormonal control of homeostasis?
2. Compare and contrast the modes of action of steroid and nonsteroid hormones.
3. The pituitary is often referred to as the "master gland." Why?
4. Why is the observation that one hormone may bind to more than one type of receptor important?
5. Hormones are present in tiny concentrations yet have large effects on target cells and on the individual as a whole. How is this possible?
6. Compare and contrast the short-term and long-term responses to stress in humans. Which hormones are involved? Why are their effects logical, in terms of maximizing an individual's ability to survive a potentially life-threatening situation?

Group Discussion Problems

1. Suppose that, during a detailed anatomical study of a marine invertebrate, you found a small, previously undescribed structure. How would you test the hypothesis that the structure is a gland that releases one or more hormones?
2. Cortisone is a glucocorticoid that suppresses inflammation and other aspects of wound healing. Cortisone was once widely used to treat athletes with joint injuries and people with arthritis. In the short term, cortisone is extremely effective in reducing swelling and pain. Over time, however, physicians found that repeated large doses of cortisone had damaging side effects. Predict what these side effects are. Explain your logic.
3. You are a physician supervising a patient's recovery from the surgical removal of the posterior pituitary. Name a hormone that you will have to administer to this patient artificially. Which symptom(s) will you monitor to assess whether the dosage and timing of your injections are having the desired effect?
4. Suppose that a researcher announces the discovery of a hormone that affects the metabolism of fats in lab rats. Preliminary data indicate that the hormone is a polypeptide, about 50 amino acids long. How would you go about isolating the receptor for this hormone? Using fat cells growing in vitro, how could you test the hypothesis that hormone binding causes a change in a second messenger?

Answers to Multiple-Choice Questions 1. c; 2. a; 3. d; 4. a; 5. c; 6. a

www.prenhall.com/freeman is your resource for the following: Web Tutorials; Online Quizzes and other Online Study Guide materials; Answers to Conceptual Review Questions; Solutions to Group Discussion Problems; Answers to Figure Caption Questions and Exercises; and Additional Readings and Research.

Animal Reproduction

48

The swollen, red rump of this female Hamadryas baboon indicates that she is about to produce an egg and will accept courtship and possibly copulation from males. She will probably mate with several males before the egg is fertilized. This chapter discusses an array of questions about animal reproduction, including how hormones control the female reproductive cycle and the consequences of multiple mating.

KEY CONCEPTS

- The reproductive systems of animals are highly variable. Many animals can switch between asexual and sexual reproduction. When sexual reproduction occurs, fertilization may be external or internal and egg development may take place inside or outside the mother's body, depending on the species.
- In humans, the male reproductive system is composed of structures specialized for the production and storage of sperm, the synthesis of fluids found in semen, or the transport and delivery of semen. The female reproductive system is composed of structures specialized for producing eggs and nourishing offspring during early development.
- In humans, the menstrual cycle is regulated by hormones from the pituitary gland and female reproductive organs. These hormones may interact with each other via positive or negative feedback. Pregnancy is maintained by hormonal signals from the embryo and from the mother's reproductive organs.

All the cells, tissues, organs, and systems introduced in Unit 8 exist for one reason: They allow animals to survive long enough and gather enough resources to reproduce. The shark rectal gland, the fish gill, the squid giant axon, the infrared sensors of pit vipers, and the human adrenal glands are all adaptations that help individuals survive in a particular environment, acquire energy, and produce offspring. Stated another way, producing offspring is the reason that adaptations exist. Reproduction is the unconscious goal of virtually everything that an animal does.

Evolution by natural selection explains why animals reproduce; the goal of this chapter is to explore *how* reproduction occurs. Section 48.1 begins by introducing research on animals that cycle between asexual and sexual modes of reproduction and then explains how both modes occur at the cellular level. Section 48.2 surveys the diversity of ways in which animals accomplish fertilization once gametes have formed. The final three sections focus on mammalian reproduction, using humans as the primary model organism. The discussion begins by analyzing male and female reproductive organs, explores how hormones control changes in those organs, and concludes with a description of human pregnancy and birth.

In the course of investigating research on the molecular and cellular mechanisms of reproduction, we'll also consider topics of urgent practical interest. These topics include the dangers of steroid hormone use by athletes and bodybuilders, the recent and dramatic declines in the death rate of mothers during childbirth, and the various available methods of contraception.

Understanding and manipulating animal reproductive systems is an important issue in the working life of physicians, veterinarians, farmers, zookeepers, and many others in biology-related professions.

48.1 Asexual and Sexual Reproduction

Several earlier chapters have explored how asexual and sexual reproduction differ. Recall that asexual reproduction is based on mitosis and that offspring that are produced asexually are genetically identical to their parent. Sexual reproduction, in contrast, is based on meiosis. Due to genetic recombination during meiosis and the fusion of haploid gametes from different parents during fertilization, offspring that are produced sexually are genetically different from each other and from their parents. Let's first consider how animals make genetically identical copies of themselves, then review the mechanisms responsible for sexual reproduction.

Mechanisms of Asexual Reproduction

In thousands of animal species, individuals can **clone** themselves—that is, produce large numbers of identical copies of themselves asexually. Which animals produce offspring asexually, and how is asexual reproduction accomplished? **Figure 48.1** addresses both questions. The hydra in Figure 48.1a is generating offspring by budding. When **budding** occurs, an offspring begins to form within or on a parent. Budding is completed when the offspring—a miniature version of the parent—breaks free and begins to grow on its own. Figure 48.1b illustrates a flatworm undergoing **fission**, in which an individual simply splits into two or more descendants. In some populations of the westerrn whiptail lizard (Figure 48.1c), reproduction occurs solely by **parthenogenesis** ("virgin-origin"), in which female offspring develop from unfertilized eggs.

The general point here is that asexual reproduction occurs in a variety of ways and in a wide variety of animal groups. To reinforce this point, consider the following observations about parthenogenesis:

- Some groups, including selected species of fishes and lizards as well as the bdelloid rotifers, reproduce exclusively by parthenogenesis, never sexually.
- In some parthenogenetic species of fish, eggs do not develop unless they are fertilized by a sperm. None of the father's genes are incorporated into the offspring, however.
- In ants, honeybees, and related insects, males are produced by parthenogenesis, but females in those same insect groups develop from fertilized eggs produced through sexual reproduction.

In addition, many animal species regularly switch between reproducing asexually and reproducing sexually. Why?

(a) Budding in hydra
(b) Fission in flatworms
(c) Parthenogenesis in lizards
Eggs
produced
via mitosis

FIGURE 48.1 Mechanisms of Asexual Reproduction in Animals Are Diverse

Switching Reproductive Modes: A Case History

Daphnia are crustaceans that live in freshwater habitats throughout the world. In a typical year, *Daphnia* produce only diploid female offspring throughout the spring and summer, via parthenogenesis. As **Figure 48.2a** shows, the eggs produced by parthenogenesis develop in a structure called a brood pouch. They are released when the female molts her exoskeleton.

In late summer or early fall, however, many *Daphnia* females begin producing male offspring parthenogenetically. Then sexual reproduction ensues: Haploid **sperm** (male gametes) produced by meiosis in these males fertilize haploid **eggs** (female gametes) that females produce via meiosis. **Fertilization** is the fusion of sperm and egg. The fertilized eggs are released into a durable case that falls to the bottom of the pond or lake for the winter. In spring, the sexually produced offspring hatch and begin reproducing asexually.

What cues trigger the change between modes of reproduction? For decades, most researchers contended that day length triggered the asexual-sexual switch. The idea was that the shortening days of late summer or fall induced the production

(a) *Daphnia* produce diploid eggs asexually.

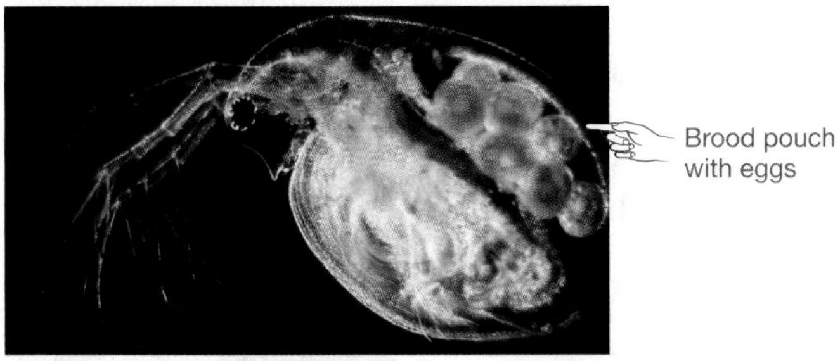

(b) Sexual reproduction is more common in crowded populations than in sparse populations.

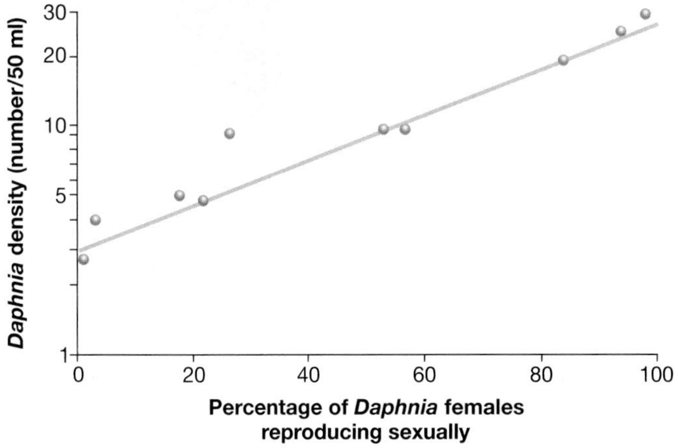

(c) Which environmental cues trigger the switch to sexual reproduction?

Water quality	Food concentration	Day length	% Sexual broods
Clean	Low	Short	0
Crowded	Low	Short	44
Clean	Low	Long	0
Crowded	Low	Long	0
Clean	High	Long	0
Crowded	High	Short	0
Clean	High	Short	0
Crowded	High	Long	0

FIGURE 48.2 In *Daphnia*, Environmental Cues Signal the Switch from Asexual to Sexual Reproduction
(a) The brood pouch of this *Daphnia* female is filled with asexually produced embryos. **(b)** Graph plotting the percentage of *D. pulex* females that reproduce sexually versus *Daphnia* density (log scale). **(c)** Environmental conditions were varied experimentally for *D. magna*. "Crowded" water was taken from tanks containing dense populations. QUESTION How would you go about determining which molecule or molecules in crowded water serve as a signal that triggers sexual reproduction?

of males and haploid eggs. Then in 1965 biologists showed that high population densities are also a factor in causing the switch. Researchers who brought *Daphnia* populations into the lab and kept day length constant found a strong correlation between population density and the percentage of females reproducing sexually (**Figure 48.2b**).

In 1992 another group of investigators extended this result by pinpointing the specific aspects of crowding that were affecting the animals. To do this work, the biologists brought *Daphnia* into the laboratory and altered day length, the amount of food available to individuals, and the quality of the water they occupied. (To vary water quality, the investigators used either clean water or water taken from tanks where *Daphnia* were being maintained at high density.) As **Figure 48.2c** shows, individuals in the study population switched to sexual reproduction only if they were exposed to water from crowded populations, low food availability, and short day lengths. In short, *Daphnia* need three different cues from the environment to switch to sexual reproduction.

These experiments don't address the question of why *Daphnia* make the switch, however. Why is it adaptive to reproduce sexually under the adverse conditions dictated by high population density, nutrient stress, and shortened days? The leading hypothesis is that, because sexually produced offspring are genetically diverse, they are more capable of thriving when they hatch the following spring than are asexually produced offspring, which are genetically identical to each other and to their parents. To date, however, this hypothesis has not been tested rigorously. Research on the adaptive significance of sex in *Daphnia* and other organisms continues.

Mechanisms of Sexual Reproduction: Gametogenesis

WEB TUTORIAL 48.1 Human Gametogenesis

The mitotic cell divisions, meiotic cell divisions, and developmental events that result in the production of male and female gametes, or sperm and eggs, are collectively called **gametogenesis** (see Chapter 21). **Spermatogenesis** is the formation of sperm, and **oogenesis** is the formation of eggs. In the vast majority of animals, gametogenesis occurs in a sex organ, or **gonad.** Male gonads are called **testes**; female gonads are called **ovaries.** Early in development, reproductive cells known as *germ cells* enter the testes and ovaries and give rise to diploid cells that will undergo gametogenesis.

Figure 48.3 (page 1102) summarizes the events that take place during gametogenesis in humans. It's important to note that in the male and female gonads, diploid cells called **spermatogonia** (singular: *spermatogonium*) and **oogonia** (singular: *oogonium*), respectively, divide by mitosis to generate the cells that undergo meiosis. In males, the **primary spermatocytes** that are produced when a spermatogonium divides mitotically undergo meiosis I and produce two **secondary spermatocytes,** which then undergo meiosis II. The result is four haploid cells called **spermatids.** Each haploid spermatid matures into a

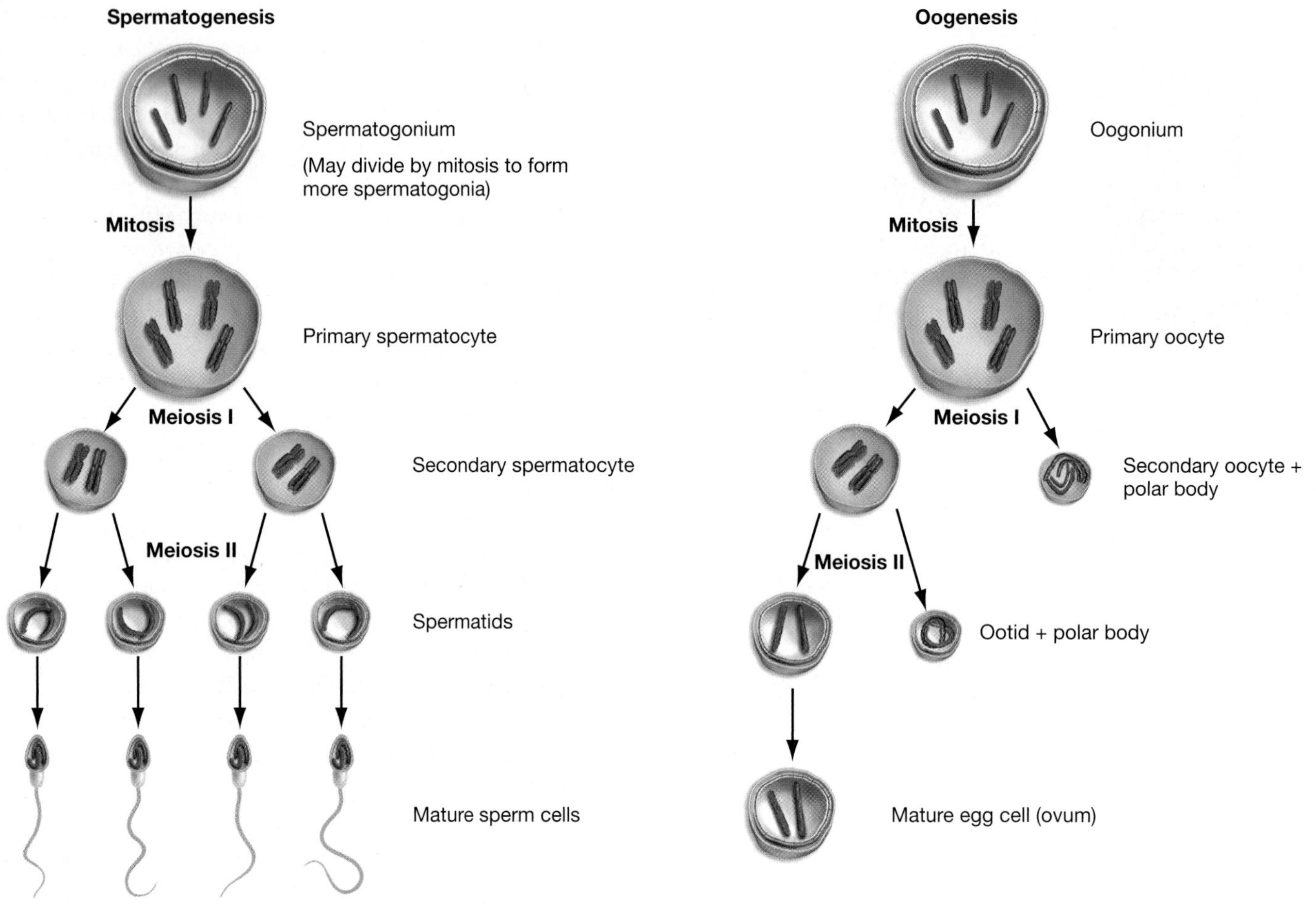

FIGURE 48.3 Gametogenesis in Humans
During gametogenesis, spermatogonia produce sperm and oogonia produce eggs. **EXERCISE** In each diagram, label the haploid cells and the diploid cells.

sperm. The production of spermatogonia and primary spermatocytes occurs continuously throughout adult life.

Gametogenesis is markedly different in human females. For example, an oogonium divides mitotically to form a **primary oocyte,** which then undergoes meiosis. Of the four haploid products, however, only one cell, known as an **ovum,** matures into an egg. The other cells produced by meiosis in females have a tiny amount of cytoplasm and do not mature into eggs. Because the distribution of cytoplasm is so unequal during each meiotic division in females, the smaller cells are called **polar bodies.** (Recall that *polar* refers to inequality or opposites.) Further, the production of oogonia stops early in development in many species; in humans, it stops before birth. And in humans and many other mammals, primary oocytes begin the first meiotic division but then stop developing for a period of months or years.

48.2 Fertilization and Egg Development

Fertilization is the joining of a sperm and an egg to form a diploid **zygote.** Chapter 21 introduced this process by describing how a sperm makes contact with an egg and penetrates the egg's membrane. That discussion focused on the molecular mechanisms of fertilization, using the sea urchin as a model organism. The objectives of this section are to introduce the variety of fertilization mechanisms observed among animals and to explore what happens to fertilized eggs. Why are eggs laid into the environment in some species but retained to develop inside the mother's body in other species? An equally striking contrast occurs in modes of fertilization. In many species, individuals release their gametes into their environment and external fertilization occurs. In other animals, males deposit sperm into the reproductive tracts of females and internal fertilization occurs. Let's take a closer look.

External Fertilization

Most animals that rely on external fertilization live in aquatic environments. The correlation between external fertilization and aquatic environments is logical, because gametes and embryos must be protected from drying. If external fertilization occurred in a terrestrial environment, either the gametes or the resulting zygote would likely die of desiccation.

Another general observation is that species with external fertilization tend to produce exceptionally large numbers of gametes. For example, a female sea star *Asterias amurensis* typically releases 100,000,000 eggs into the surrounding seawater during spawning. Males release many times that number of sperm. The leading hypothesis to explain this pattern is that the probability of a sperm and egg meeting in an ocean or lake is extremely small unless huge numbers of gametes are present.

If sperm and eggs from different individuals must be released into the environment synchronously for external fertilization to work, how is gamete release coordinated? The answer has two parts. Most research to date indicates that gametogenesis occurs in response to environmental cues such as lengthening days and warmer water temperatures. These results are logical, because the cues indicate a favorable season for breeding. But what triggers the actual release of gametes? In fishes and other aquatic animals with well-developed eyes and external fertilization, spawning is often the culmination of an elaborate courtship ritual between a male and female. In contrast, courtship behavior appears to be much less important—or even absent—in species such as clams, sea urchins, and sea cucumbers. How do animals that cannot see their mates time the release of their gametes?

Researchers have long hypothesized that the chemical messengers called **pheromones** might be involved in synchronizing gamete release. The idea is that when it is ready to mate, a sea star will release a pheromone, which then triggers gamete release in other individuals. Data to back this claim have begun to emerge only recently, however. For example, in one recent test of the pheromone hypothesis for spawning synchronization, biologists maintained two groups of sea cucumbers under natural conditions of light and temperature for 15 months. In one treatment, individuals were kept in isolated tanks; in another, they were kept in tanks with other sea cucumbers. The researchers found that all the individuals that were maintained in groups released gametes during the normal spawning period. In contrast, only about 10 percent of the individuals maintained in isolation released gametes.

Although these data suggest that pheromones might be involved in coordinating external fertilization in sea stars, they are not definitive. Most biologists will not be convinced that the pheromone hypothesis is valid until a chemical messenger is purified and can be shown to induce spawning in isolated individuals.

Internal Fertilization and Sperm Competition

Internal fertilization occurs in the vast majority of terrestrial animals as well as in a significant number of aquatic animals. To accomplish internal fertilization, males carry out **copulation**—that is, they deposit sperm directly into the female reproductive tract with the aid of a copulatory organ, usually called a **penis.** Alternatively, males may package their sperm into a structure called a **spermatophore**, which is then placed into the female's reproductive tract by the male or female. In some salamander species, for example, the male places the spermatophore on the ground within its territory. Later the female picks it up with her *cloaca*, a chamber used by both the reproductive and excretory systems that opens to the environment. In this case, the result is internal fertilization without any direct physical contact between the sexes.

Although internal fertilization appears to be straightforward, research on the topic has produced some of the most striking or even bizarre observations in all of biology. (See **Box 48.1** for some examples.) Perhaps the most important insight about

BOX 48.1 Unusual Aspects of Fertilization

Studies on animal sexual reproduction have documented some of the most remarkable behaviors and structures observed in nature. The list that follows is by no means exhaustive. It is a sample, offered simply because the diversity of mating arrangements in animals is fascinating to biologists and laypeople alike:

- *Suicidal males* When Australian redback spiders mate, the male does a somersault after inserting his penis-like organ. The somersault behavior places his dorsal surface in front of the female's mouthparts. The female then eats the male. Experiments have shown that suicidal males copulate longer and fertilize more eggs than do males that survive copulation, probably because sperm transfer continues until the meal is over.
- *Giant sperm* In the fruit fly *Drosophila bifurca*, males average 1.5 mm in total body length. Their sperm, however, are each 6 cm long. Recent research suggests that the coiled sperm fill the female's sperm storage area and prevent competition.
- *Having an affair* Researchers who use genetic markers to assess paternity in birds that appear to be monogamous find that up to 60 percent of nests contain at least one offspring fathered by a male that is not mated to the resident female. In most cases, "infidelity" occurs because females actively solicit copulations from males holding nearby territories.
- *Love darts* Many snails and slugs are **hermaphroditic**, meaning an individual has both male and female gonads. During mating, two individuals simultaneously receive and deposit sperm to fertilize each other's eggs. In some species, individuals fire "love darts" of unknown function from their genitalia into the mating partner. In other species, the penis frequently becomes stuck in the female reproductive tract and is bitten off by one partner.
- *Hypodermic insemination* Some bedbugs have hypodermic penises. Males force the organ through the female's abdominal wall and deposit sperm directly into her body cavity.

Did the process of sexual selection, introduced in Chapter 24, lead to the evolution of these structures and behaviors? If so, how and why? In many or even most cases, the answers to these questions are not known.

internal fertilization, though, originated with Geoff Parker. Parker realized that, in many animal species, females mate with more than one male before fertilization occurs. As a result, sperm from different males should compete with each other to fertilize the eggs.

In 1970 Parker published the results of experiments on dung flies that confirmed the existence of **sperm competition**—competition between sperm from different males to fertilize the eggs of the same female. Parker's experiments consisted of a series of carefully controlled matings between one female and two males. In each experiment, the two males were selected in such a way that Parker could distinguish their offspring. As predicted by the sperm competition hypothesis, the proportion of offspring fathered by each male was not 50:50. Instead, whichever male was last to copulate fathered an average of 85 percent of the offspring produced.

Follow-up research has confirmed that this **second-male advantage** is widespread, although not quite universal, in insects and some other animal groups. How does it occur? To answer this question, biologists turned to the fruit fly *Drosophila melanogaster*. A research group was able to introduce a gene into male fruit flies that resulted in the production of sperm with green tails. When a mating by a green-spermed male was followed with a mating by a male having normal-colored sperm, most of the green-tailed sperm disappeared from the female's sperm-storage area. To interpret this finding, the biologists suggest that the sperm of the second male physically dislodged the first male's gametes and inserted themselves in their place (**Figure 48.4**). The researchers also showed that the fluid that accompanies sperm during fertilization is able to poison stored sperm from competing males. These two mechanisms resulted in the second male's sperm fertilizing most of the eggs laid in the next clutch.

In addition to documenting the mechanisms responsible for second-male advantage, research on sperm competition has recently contributed another major result. In species in which multiple mating is common, males have extraordinarily large testes for their size and produce proportionately larger numbers of sperm. Although this pattern was first documented in primates, it has now been observed in a wide variety of animal groups. For example, males from species of fruit bats that live in large social groups produce more sperm, on average, than do male fruit bats from species that live in small groups. The leading hypothesis to explain this observation is that fertilization is similar to a lottery in which each sperm represents a ticket. According to the lottery hypothesis, the more tickets a male enters in the competition, the higher his chance of "winning" fertilizations and passing his alleles on to the next generation. As a result, it is logical to observe exceptionally large testes in species in which sperm competition occurs.

The lottery model has recently been challenged, however, by evidence that, at least in some species, females exert control over which sperm are successful in fertilization. In other words,

Question: How does the "second-male advantage" occur in sperm competition?

Hypothesis: Sperm from the second male displaces sperm from the first male.

Null hypothesis: Sperm from the second male does not displace sperm from the first male.

Experimental setup:

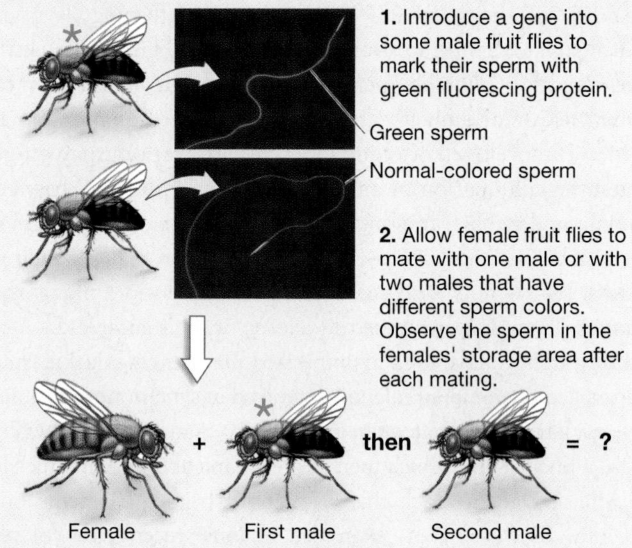
1. Introduce a gene into some male fruit flies to mark their sperm with green fluorescing protein.
Green sperm
Normal-colored sperm
2. Allow female fruit flies to mate with one male or with two males that have different sperm colors. Observe the sperm in the females' storage area after each mating.
+
then
= ?
Female
First male
Second male

Prediction: When females mate twice, the number of sperm from the first male will be much less than the amount he originally deposited—meaning the number of sperm present when only one male mates.

Prediction of null hypothesis: When females mate twice, all of the sperm deposited by the first male are still present.

Results:

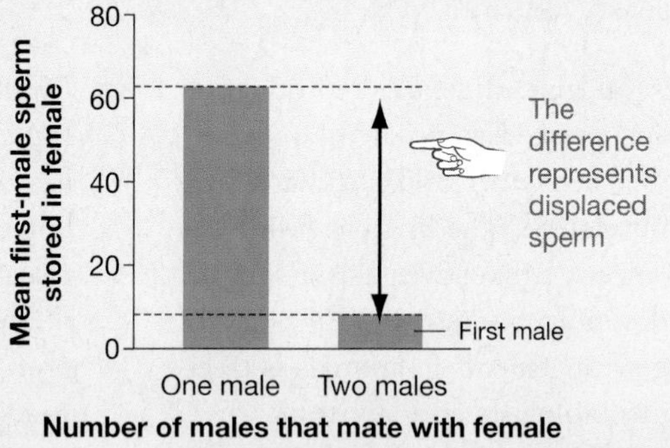
Mean first-male sperm stored in female
80
60
40
20
0
The difference represents displaced sperm
First male
One male
Two males
Number of males that mate with female

Conclusion: The sperm from the second male displaces the sperm from the first male.

FIGURE 48.4 Experimental Evidence for Second-Male Advantage during Sperm Competition in *Drosophila*

Researchers created fruit-fly males with green-tailed sperm to explore why the second male to mate tends to father more offspring than does the first male in many species. The graph shows the results when males with green sperm were the only male to mate or the second male to mate. When sperm competition occurred, the first male to mate fathered a tiny percentage of offspring, probably because most of its sperm disappeared from the sperm-storage organ in females.

females do not always accept the results of sperm competition passively. For example, females of some species actively choose which male performs the last copulation before fertilization takes place. In other species, females physically eject sperm from undesirable males. This phenomenon has been dubbed *cryptic female choice*. The name is appropriate because the selection of sperm by females is hidden from males.

Oviparity and Viviparity

Once fertilization is accomplished, the embryo is either laid as an egg outside the mother's body or retained so that it can develop inside (see Chapter 31). In **oviparous** ("egg-bearing") animals, the embryo develops in the external environment. In oviparous species such as sea stars, sea urchins, and most insects, neither the male nor the female provides any further care; the eggs and embryos are left to fend for themselves. But a substantial number of oviparous species continue to care for their young after eggs have emerged from the mother's body. For example, birds incubate their eggs and feed the young after hatching; fish may guard their eggs from predators and fan the clutches to oxygenate them.

In **viviparous** ("live-bearing") species, embryonic development takes place within the mother's body. Viviparity is possible because the embryo attaches to the reproductive tract of the mother and receives nutrition directly from the mother's circulatory system.

Why does oviparity exist in some groups and viviparity in others? One team of biologists tackled this question by studying several species from the lizard genus *Sceloporus* (**Figure 48.5a**). Some of the populations they studied are oviparous while others are viviparous; all live in the highlands of the southwestern United States and central Mexico. To understand how oviparity and viviparity evolved, the biologists analyzed a phylogenetic tree—based on molecular and morphological data—of many *Sceloporus* species to estimate the evolutionary history of each population (**Figure 48.5b**). One of the researchers' first conclusions was that, because most *Sceleporus* populations are oviparous, egg laying probably represents the original or ancestral condition. As the red branches on the tree show, however, viviparity evolved independently in two groups. Using a similar research strategy, biologists have found similar patterns in sea stars. Viviparous populations of sea stars have evolved from oviparous populations on several occasions.

Why did natural selection favor these changes between egg laying and live birth? Viviparity should evolve when it leads to higher numbers of surviving young. Researchers have long hypothesized that natural selection favoring viviparity should be especially strong in cold habitats. Low temperatures slow the development of embryos, so in cold habitats, it might be advantageous for females to retain eggs inside their bodies so that offspring can develop at a more favorable temperature. In support of this hypothesis, a researcher has shown that, in at least one oviparous species of lizard, lowering nest temperatures leads to a higher percentage of deformities in offspring. In addition, offspring that developed in colder nests run more slowly than do individuals of the same age that developed at higher temperatures. Research into why viviparity has evolved in certain populations continues, however.

(a) Some *Sceloporus* lizards lay eggs; other members of the same genus bear live young.

(b) Phylogeny of *Sceloporus* from central Mexico

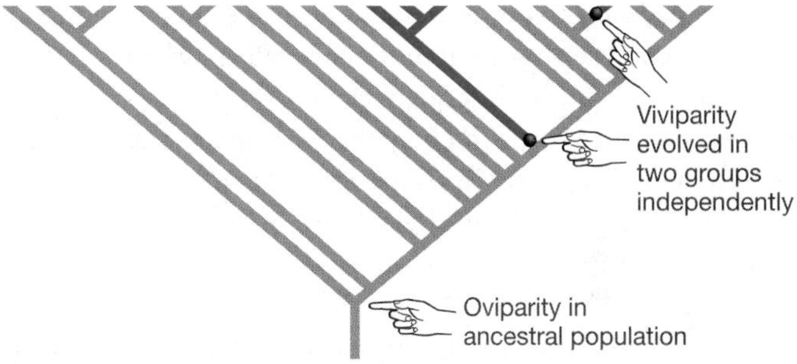
Viviparity evolved in two groups independently
Oviparity in ancestral population

FIGURE 48.5 Viviparity Has Evolved Multiple Times in *Sceloporus* Lizards
(a) *Sceloporus* is a genus of lizards from desert areas in North America. **(b)** Each twig on this phylogenetic tree represents a *Sceloporus* species or population from North America. According to these data, viviparity evolved twice as the lineage diversified.

✓CHECK YOUR UNDERSTANDING

Some of the most basic aspects of reproduction are variable among animal populations or species: whether reproduction is sexual or asexual, whether fertilization is external or internal, and whether fertilized eggs develop in the environment or in the mother's body. You should be able to (1) describe the advantages and disadvantages of each of these modes of reproduction; and (2) design experiments to test the following hypotheses: that a particular molecule is a gamete-releasing pheromone in sea stars, that a particular molecule is a toxin active in sperm competition, and that, in viviparous and oviparous populations of the same species, viviparous populations have higher reproductive success in cold habitats.

WEB TUTORIAL 48.2 Human Reproduction

48.3 Reproductive Structures and Their Functions

The first two sections of this chapter considered the broad contrast between asexual and sexual reproduction and examined an overview of general patterns in fertilization and egg care. Now let's narrow the focus and explore the mechanics of sexual reproduction in more detail. The first task is to understand the anatomy of the male and female reproductive systems.

The Male Reproductive System

In humans, the external anatomy of the male reproductive system is simple: It consists of the scrotum and the penis. The saclike **scrotum** holds the testes; the penis, functions as the organ of copulation prior to internal fertilization.

Biologists who have compared these structures among animal species have been struck by their variability and complexity. For example, a scrotum occurs only in certain mammal species. Whales, elephants, hedgehogs, moles, and many other mammal groups lack the structure entirely and instead have testes that are located well within the abdominal cavity. Among species that do have a scrotum, the structure's size and shape vary. In numerous species of primates, the scrotum is brightly colored and appears to function in sexual display behaviors.

Variation in External Anatomy Is Important in Sperm Competition Diversity in scrotal morphology pales in comparison with variation in the structure of the penis or other types of male **genitalia**. (*Genitalia* is the term used when the copulatory organ has distinct parts.) In many groups of insects and spiders, examining the animals' genitalia is the only way that biologists can distinguish species. **Figure 48.6a** shows one elaborate structure typically observed in spiders. Even among closely related species, the sizes and shapes of these organs tend to be strikingly different. Why are insect genitalia so diverse?

Researchers recently offered data that might help answer this question, based on the hypothesis that certain genitalia shapes are advantageous when sperm competition occurs. To test this idea, biologists studied how much second-male advantage varied among males with different-shaped genitalia when water-strider females mated with two males. As the data in **Figure 48.6b** show, there is substantial variation in the size and shape of two important parts of the male genitalia in this species. When sperm competition occurred, the researchers found that males with certain types of genitalia—those with genitalia shapes plotted at the lower left of Figure 48.6b—fathered a higher percentage of offspring than did others. In this and perhaps many other species, variation in the size and shape of male genitalia has a strong impact on male reproductive success. The exact mechanism responsible for this pattern is not known.

To summarize, some data suggest that, in insects and perhaps other animal groups, variation in the anatomy of genitalia is an important component of ability to produce offspring. Sexual

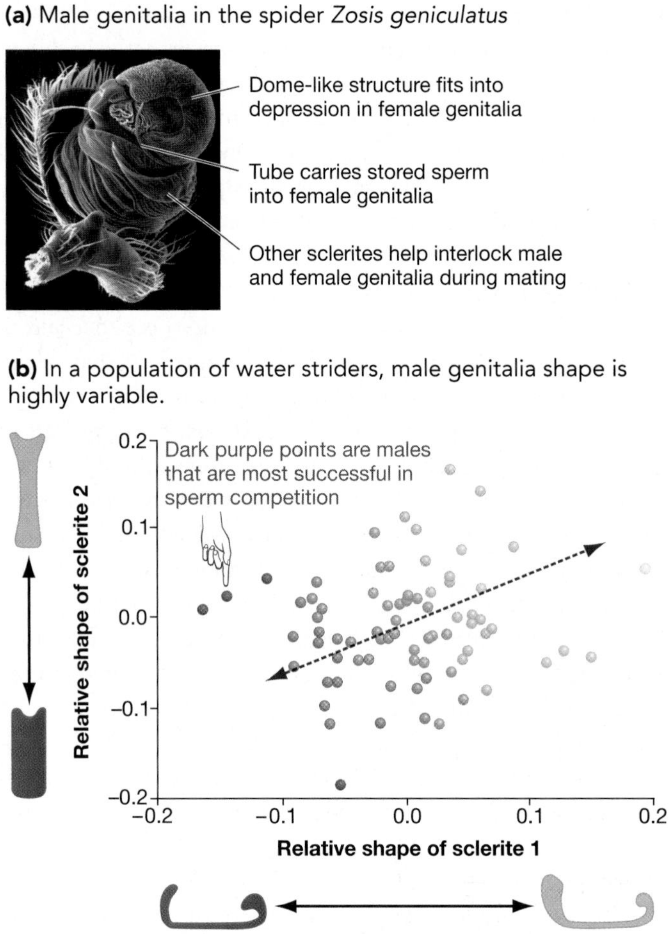
(a) Male genitalia in the spider *Zosis geniculatus*
Dome-like structure fits into depression in female genitalia
Tube carries stored sperm into female genitalia
Other sclerites help interlock male and female genitalia during mating
(b) In a population of water striders, male genitalia shape is highly variable.
Dark purple points are males that are most successful in sperm competition
Relative shape of sclerite 2
0.2
0.1
0.0
−0.1
−0.2
−0.2
−0.1
0.0
0.1
0.2
Relative shape of sclerite 1

FIGURE 48.6 In Insects, Male Genitalia May Vary among Species and Individuals and May Affect Reproductive Success
(a) An elaborate male reproductive structure, used to transfer sperm to the female, in a species of spider. In insects and spiders, species that look identical often have distinctive genitalia. **(b)** Three-dimensional plot of the relationship between the size and shape of two parts of male genitalia in water striders. Males whose genitalia plot at the lower left experience the greatest success in sperm competition.

selection may be responsible for the rapid diversification of male genitalia among species. Does the same type of variation exist in the internal anatomy of the male reproductive system?

Internal Anatomy Figure 48.7 furnishes side and front views of the internal reproductive structures in a human male. Although numerous structures are involved, notice that the system has just three basic functional components:

1. *Spermatogenesis and sperm storage* Sperm are produced in the testes and stored in the **epididymis** nearby.
2. *Production of additional fluids* Complex solutions form in the **seminal vesicles**, **prostate gland**, and **bulbourethral gland**. These solutions, called **accessory fluids**, are added to sperm prior to **ejaculation**, or expulsion from the body. Table 48.1 lists some components of these accessory fluids. **Semen** is the combination of sperm and accessory fluids.

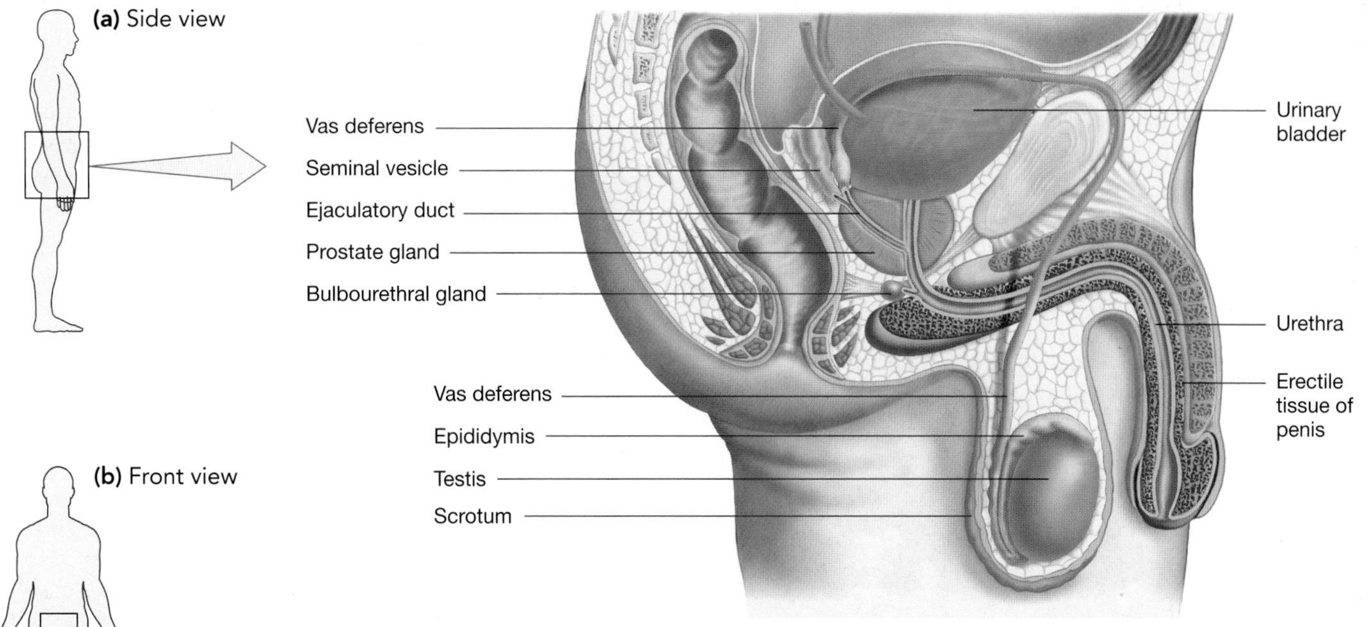
(a) Side view
Vas deferens
Seminal vesicle
Ejaculatory duct
Prostate gland
Bulbourethral gland
Vas deferens
Epididymis
Testis
Scrotum
Urinary bladder
Urethra
Erectile tissue of penis

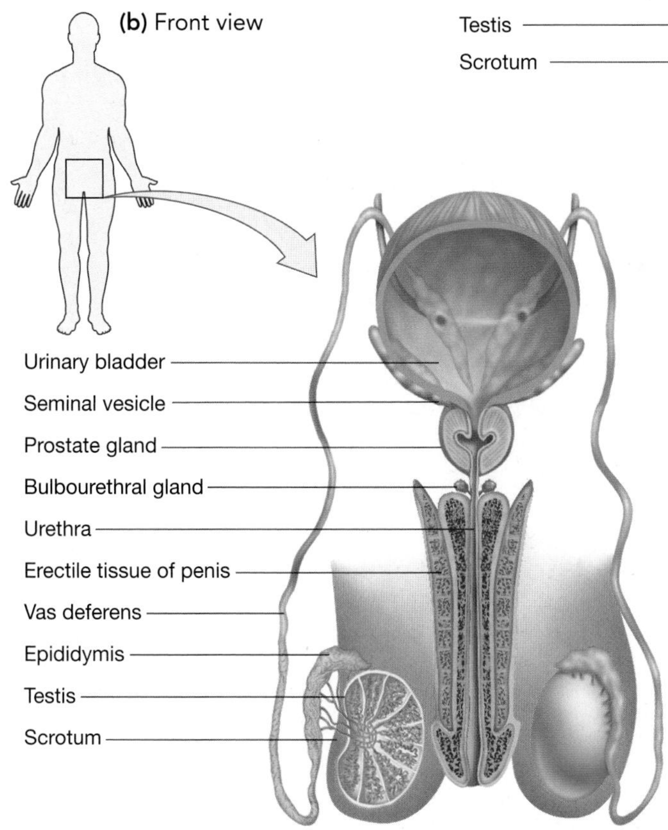
(b) Front view
Urinary bladder
Seminal vesicle
Prostate gland
Bulbourethral gland
Urethra
Erectile tissue of penis
Vas deferens
Epididymis
Testis
Scrotum

FIGURE 48.7 Anatomy of the Reproductive Tract in a Human Male

EXERCISE Label which structures function in the following capacities: (1) sperm production and storage, (2) production of accessory fluids, and (3) transport and delivery of semen.

TABLE 48.1 Accessory Fluids in Human Semen

Source	Content	Function
Seminal vesicles	Fructose (a sugar)	Source of chemical energy for sperm movement
	Prostaglandins	Stimulate smooth muscle contractions in uterus
Prostate gland	Antibiotic compound	Prevent urinary tract infections in males?
	Citric acid	Nutrient used by sperm
Bulbourethral gland	Alkaline mucus	Lubricates tip of penis; neutralizes acids in urethra

3. *Transport and delivery* The **vas deferens** are a pair of muscular tubes that store sperm and transport semen from the epididymis to the short **ejaculatory duct**. The semen then enters the **urethra**, a longer tube that passes through the penis and services both the reproductive and urinary systems in males. The semen is expelled during ejaculation.

How do these structures vary among animals? Variation in relative testes size among mammal species has already been discussed. In addition, the composition of the accessory fluids varies widely. In many insects, spiders, and vertebrates, molecules in the accessory fluids congeal after they arrive in the female reproductive tract and plug it. Experiments have shown that these copulatory plugs can serve as an effective deterrent to future matings; in some species, though, females or second males actively remove them.

Another diverse aspect of male internal anatomy is a bone inside the penis called the **baculum**. Some mammal species, including humans, lack this feature. But among rodents, the shape of the baculum is so variable that it can be used as a characteristic that distinguishes species. And in seals, baculum size appears to correlate with mating system. In seal species in which females routinely mate with several males before becoming pregnant, males have not only large testes for their size but a large baculum as well. In all species in which it appears, the baculum helps stiffen the penis during copulation.

The Female Reproductive System

Figure 48.8 shows side and front views of the anatomy of the human female reproductive system. The external anatomy features the **labia minora** (singular: labium minus) and the **labia majora** (singular: labium majus), the opening of the urethra, and the opening of the vagina. The labia are folds of skin that cover the clitoris. The **clitoris** is a sensitive organ that develops from the same population of embryonic cells that gives rise to the penis in males and likewise becomes erect during sexual stimulation. In females, however, the urethral opening, where urine is expelled, is separate from the reproductive structures. The **vagina**, or birth canal, is the chamber where semen is deposited during sexual intercourse and where the embryo is delivered during childbirth.

The internal anatomy of the female reproductive system is dominated by structures with two functions:

1. *Production and transport of eggs* Eggs are produced in the paired ovaries. During **ovulation**, a developing oocyte (egg) is expelled from the ovary and enters the **oviduct**, the tube where fertilization may take place. As Section 48.5 will detail, fertilized eggs are then transported from the oviduct to the muscular sac called the **uterus**.
2. *Development of offspring* The uterus is where embryonic development takes place. During childbirth, the developed embryo (now called a *fetus*) passes through the necklike opening of the uterus, known as the **cervix**, and into the vagina.

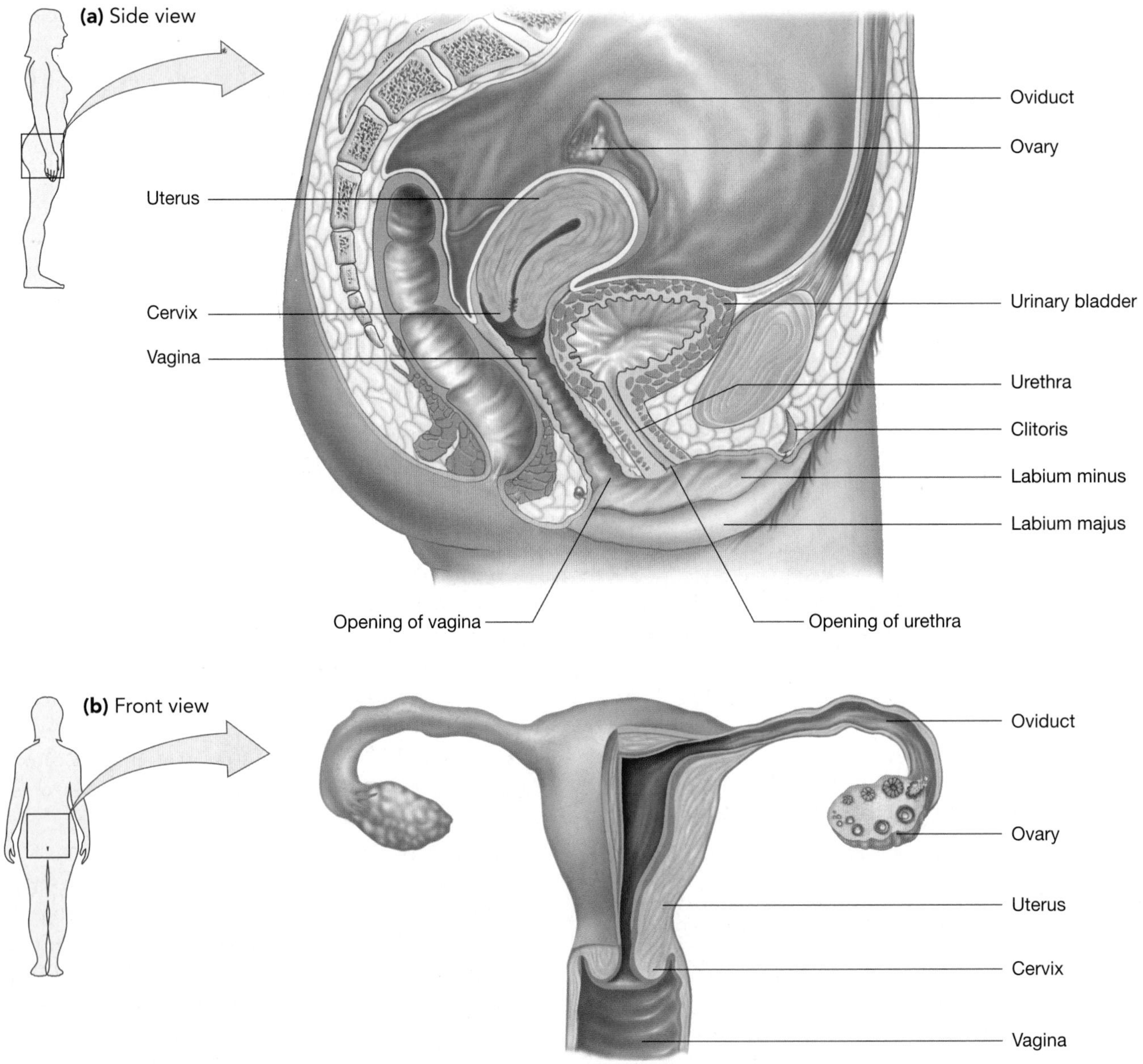
(a) Side view
Oviduct
Ovary
Uterus
Cervix
Vagina
Urinary bladder
Urethra
Clitoris
Labium minus
Labium majus
Opening of vagina
Opening of urethra
(b) Front view
Oviduct
Ovary
Uterus
Cervix
Vagina

FIGURE 48.8 Anatomy of the Reproductive Tract in a Human Female

QUESTION Generate a hypothesis to explain why male and female gonads are paired in mammals. How would you test your hypothesis?

To appreciate how the ovaries and uterus work in humans, it's essential to explore the monthly menstrual cycle and pregnancy in much more detail. It is not possible to understand either event, though, without investigating the hormones that regulate them. Let's turn now to research on the sex hormones of mammals.

48.4 The Role of Sex Hormones in Mammalian Reproduction

Chapter 47 introduced the male sex hormone **testosterone** and the female sex hormone **estradiol**, which belongs to a class of hormones known as **estrogens**. Recall that both testosterone and estradiol are steroids, that both bind to receptors inside the nucleus of target cells, and that the resulting hormone-receptor complexes bind to DNA and trigger changes in gene expression. Testosterone is synthesized in specialized cells inside the testes; estradiol and other estrogens are synthesized in the ovaries. More specifically, the female sex hormones are produced by cells that surround each developing egg, forming a structure called a **follicle** around the egg.

According to Chapter 47, the human sex hormones play a key role in three events: (1) the development of the reproductive tract in embryos, (2) the maturation of the reproductive tract during the transition from childhood to adulthood, and (3) the regulation of spermatogenesis and oogenesis in adults. To begin exploring the action of sex hormones in more detail, let's take a closer look at their role in the transition from juveniles to adults.

Puberty

Table 48.2 lists some of the changes that occur in boys and girls as they undergo **puberty**—the process that leads to sexual maturation in humans. Physicians and researchers have long known that these changes are triggered by increased levels of testosterone and estradiol in boys and girls, respectively. A group of physicians recently offered dramatic evidence for testosterone's role. They were presented with a 2-year-old boy who showed signs of entering puberty. His symptoms included an increased penis size, the development of pubic hair, and facial acne. Through careful interviewing, the doctors determined the cause of his condition. The child's father was a bodybuilder who was smearing a testosterone cream on his own shoulders and arms in an attempt to build muscle mass. The child was apparently absorbing enough testosterone from being carried by his father to trigger puberty-like symptoms. Fortunately for the 2-year-old, all his symptoms except for penile enlargement subsided once his father discontinued the use of testosterone creams. **Box 48.2** (page 1110) introduces other unfortunate side effects of steroid use by athletes and bodybuilders.

Although the roles of testosterone and estradiol in puberty are well documented, understanding how the process is initiated was a stubborn problem. Some researchers hypothesized

TABLE 48.2 Changes That Take Place during Puberty

Pediatricians use the five stages of puberty listed here to diagnose the developmental stage of their patients.

	Boys	Girls
Stage 1 (Prepubertal)	No sexual development	No sexual development
Stage 2	Testes enlarge Body odor	Breast budding First pubic hair Body odor Height spurt
Stage 3	Penis enlarges First pubic hair Ejaculation (wet dreams)	Breasts enlarge Pubic hair darkens, becomes curlier Vaginal discharge
Stage 4	Continued enlargement of testes and penis Penis and scrotum deepen in color Pubic hair curlier and coarser Height spurt Male breast development (temporary)	Onset of menstruation Nipple is distinct from surrounding areola Pelvis begins to widen
Stage 5 (Fully mature adult)	Pubic hair extends to inner thighs Increases in height slow, then stop Increased muscle mass	Pubic hair extends to inner thighs Increases in height slow, then stop Deposition of fat in hips, buttocks, thighs, and breasts

BOX 48.2 Abuse of Synthetic Steroids

Among bodybuilders and athletes, the use of synthetic testosterone-like molecules, commonly called **androgens,** has recently exploded in popularity. Both men and women use these drugs because they support the development of increased muscle mass during weight training, just as testosterone supports increased muscle mass during puberty.

Research on bodybuilders has confirmed a wide variety of damaging side effects, however. Women who use synthetic androgens routinely experience voice deepening and growth of facial hair. Both male and female users may develop acne, the emotional changes dubbed 'roid rage (for steroid rage), fluid retention, and liver ailments. There is some evidence that males may also experience fertility problems. The leading hypothesis to explain infertility is that the synthetic androgens mimic the negative feedback that testosterone exerts on LH production from the pituitary. Because LH is required to support spermatogenesis, high doses of androgens may reduce LH concentrations to a level that precludes sperm production.

Unfortunately, opinion polls have shown that most bodybuilders and athletes are willing to tolerate these side effects in hopes of increasing their performance in competitions. In response, some regulatory agencies actively test for synthetic androgens in the urine of competitors and ban individuals who are using them.

that the sex hormones are regulated by the hypothalamic-pituitary axis, introduced in Chapter 47. Recall that chemical signals from the hypothalamus region of the brain lead to the release of regulatory hormones from the pituitary gland, which then cause the release of hormones from other glands. Are the hypothalamus and pituitary involved in regulating the release of sex hormones?

Two advances made it possible to answer this question definitively and affirmatively. First, researchers isolated a hormone called **gonadotropin-releasing hormone (GnRH)** from the hypothalamus. Second, investigators noted that boys and girls who were entering puberty experienced distinctive pulses in the concentration of two pituitary hormones, called **luteinizing hormone (LH)** and **follicle-stimulating hormone (FSH).** Based on these observations, researchers hypothesized that pulses of LH and FSH occurred in response to the release of GnRH and were responsible for increases in testosterone and estradiol. To test the hypothesis of hypothalamus and pituitary control that is diagrammed in **Figure 48.9**, researchers administered pulses of GnRH to boys and girls who had deficits in their hypothalamus and were experiencing delays in the onset of puberty. As predicted, the GnRH treatment induced the developmental events that are listed in Table 48.2. This result provided strong support for the hypothesis that a hormone from the hypothalamus is responsible for initiating puberty.

What triggers GnRH increases at the appropriate age? This is the most pressing question facing researchers and physicians interested in hormonal control of sexual development. Although the question remains unanswered, there is some evidence that nutritional state is involved. For example, girls with large fat stores tend to enter puberty earlier than do girls who are thin. In addition, the current average age for the onset of menstruation in females in the United States is slightly over 12 years. This is much earlier than the average of 17 years of age that had been

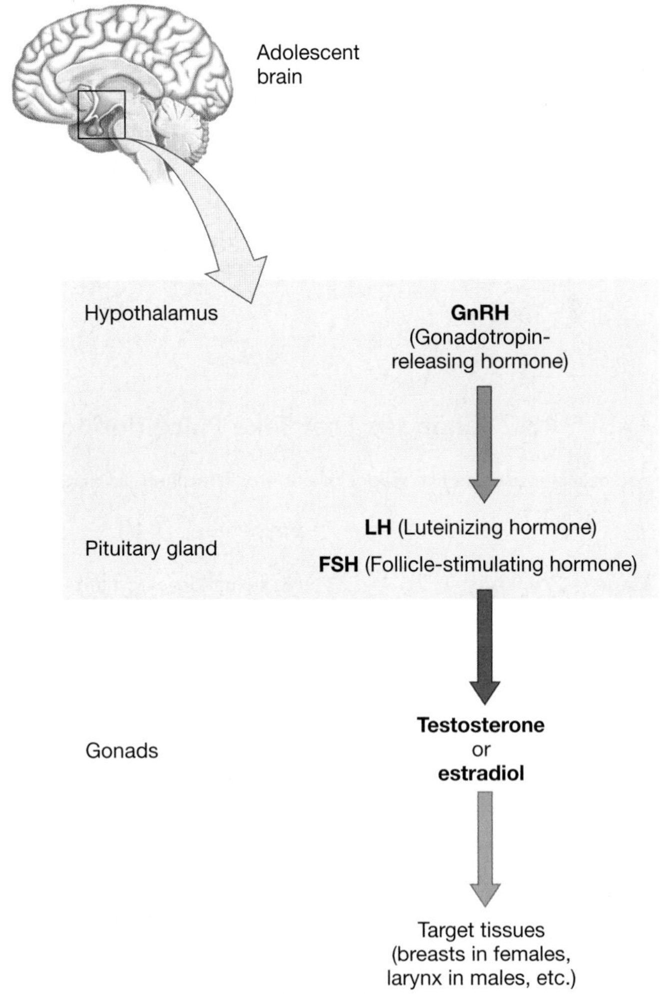

FIGURE 48.9 In Humans, Puberty Is Triggered by Hormones from the Hypothalamus and Pituitary

QUESTION How does control of testosterone and estradiol production compare with control of cortisol release by the adrenal gland? (See Figure 47.10a.)

observed in the United States during the eighteenth and nineteenth centuries, when the general nutritional state of the population was poorer. Research continues on how age, nutritional condition, and perhaps other factors interact to promote the release of GnRH.

If you recall the discussion in Chapter 47 of how the adrenal hormone cortisol is controlled, however, you might suspect that the model of sex-hormone regulation diagrammed in Figure 48.9 is probably simplified. Chapter 47 emphasized that many hormones participate in negative feedback, meaning that the presence of a hormone inhibits the factor that triggers its release. Do sex hormones participate in negative feedback? The short answer to this question is yes. To appreciate the details, let's investigate hormonal control of the human menstrual cycle.

Female Sex Hormones and the Menstrual Cycle

Figure 48.10 illustrates the sequence of events in the human **menstrual cycle**, a monthly reproductive cycle that occurs in the ovaries. Although the length of this cycle varies among women, 28 days is about average. In conjunction with changes in the ovary illustrated in the figure, the lining of the uterus undergoes a dramatic thickening and regression; ultimately, part of the lining sloughs off and is expelled through the vagina. By convention, the onset of the combined ovarian and uterine cycles is taken to be the start of **menstruation**—meaning the expulsion of the uterine lining. This event is designated as day 0 in the menstrual cycle.

The cycle itself has two distinct phases. A follicle matures during the **follicular phase**, which lasts an average of 14 days. Ovulation occurs when the follicle is mature and releases its secondary oocyte into the oviduct. (Researchers are still unsure what controls which ovary—right or left—will release an egg during ovulation.) Although the timing of ovulation is highly variable, the subsequent **luteal phase** also averages 14 days in length. Its name was inspired by the formation and subsequent degeneration of a structure called the **corpus luteum** ("yellowish-body") from the ruptured follicle.

How Do Pituitary and Gonadal Hormones Change during a Menstrual Cycle? Signals from hormones are responsible for the changes that occur in the uterus and ovaries during a menstrual cycle. For biologists who wanted to probe how hypothalamic, pituitary, and ovarian hormones interact to regulate the menstrual cycle, the first task was to document how each hormone involved changed over time. By monitoring hormone concentrations in the blood or urine of a large number of women, researchers were able to document dramatic changes in the concentrations of estradiol and several other hormones. LH and FSH are produced in the pituitary gland during each cycle in response to GnRH; progesterone is produced along with estrogens, including estradiol, in the ovaries.

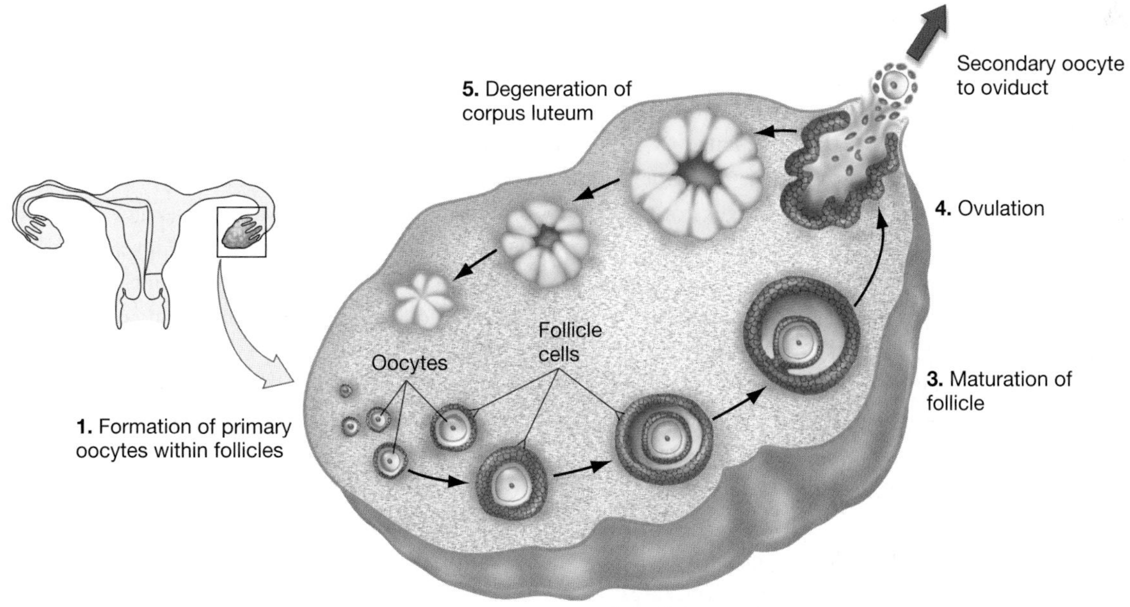

FIGURE 48.10 The Ovary Changes over the Course of a Menstrual Cycle
As a follicle matures, primary oocytes complete meiosis I. The secondary oocyte that results is expelled from the ovary and then travels down the oviduct. (Follicles don't actually move in a circle during the menstrual cycle; they are shown that way for illustrative purposes only.) QUESTION Which events illustrated here occur during the follicular phase of the menstrual cycle? Which occur during the luteal phase?

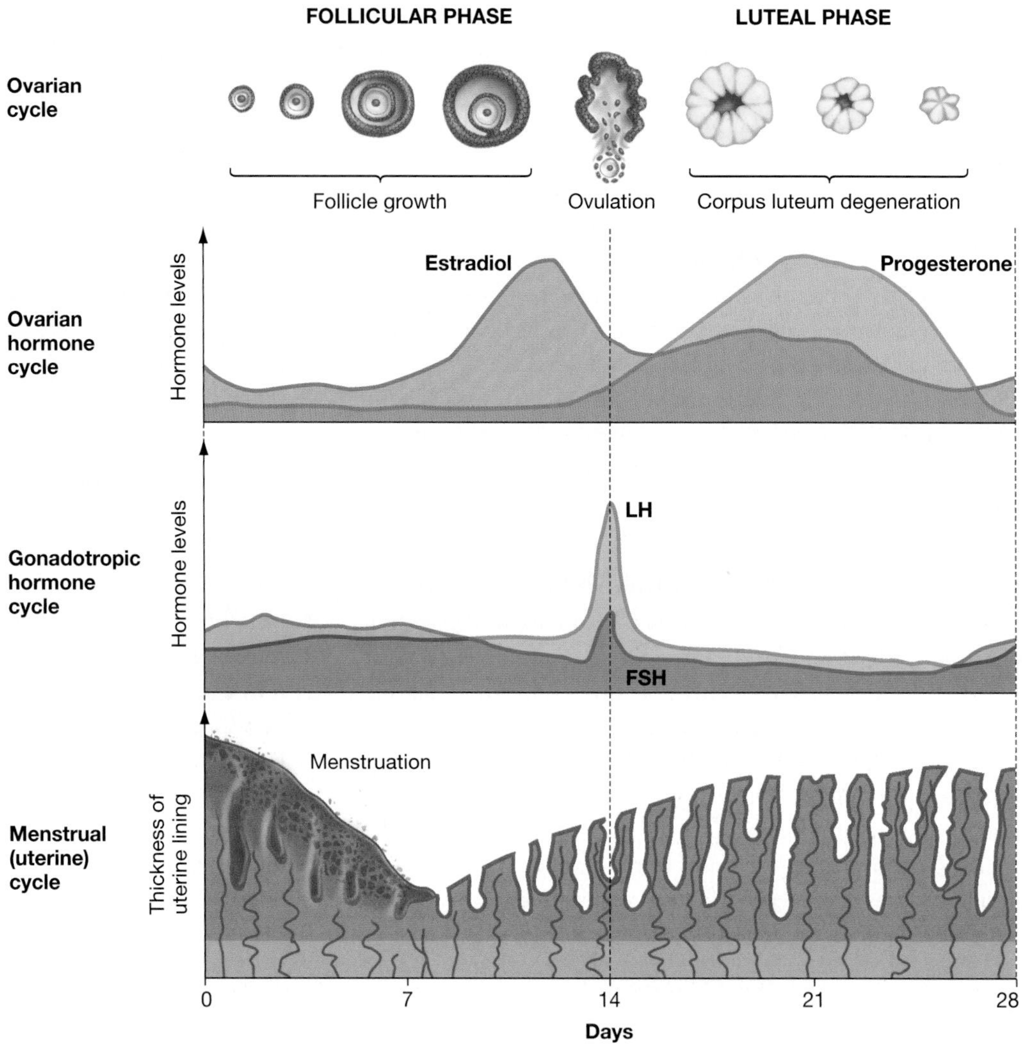
FOLLICULAR PHASE
LUTEAL PHASE
Ovarian cycle
Follicle growth
Ovulation
Corpus luteum degeneration
Ovarian hormone cycle
Hormone levels
Estradiol
Progesterone
Gonadotropic hormone cycle
Hormone levels
LH
FSH
Menstrual (uterine) cycle
Thickness of uterine lining
Menstruation
0
7
14
21
28
Days

FIGURE 48.11 Changes That Occur during a Menstrual Cycle in a Human Female
During a menstrual cycle, changes occur in the concentrations of the pituitary hormones LH and FSH, the ovary, concentrations of the ovarian hormones estradiol and progesterone, and the lining of the uterus. At the beginning of the cycle, the thickened lining of the uterus sloughs off and is expelled through the vagina.

Several observations jump out of the data in **Figure 48.11.** LH levels are fairly constant except for a spike just prior to ovulation, suggesting that LH might be the trigger for this event. FSH concentrations, in contrast, are relatively high during the follicular phase and low during the luteal phase, though they also make a small spike prior to ovulation. Progesterone is present at very low levels during the follicular phase but high levels during the luteal phase. This observation suggested that progesterone might support the maturation of the thickened uterine lining. Estradiol concentrations change in a much more complex way. What's the effect of all these changes?

How Do the Pituitary and Gonadal Hormones Interact?

Experiments helped establish that changes in the concentration of estradiol and progesterone affect the release of the pituitary hormones LH and FSH. Researchers worked with three volunteers whose ovaries had been removed because of cancerous growths or other problems. The women were receiving low, maintenance-level doses of estradiol, which appeared to exert negative feedback on LH and FSH. But when the investigators injected the women with larger doses of estradiol or with progesterone, the situation changed dramatically. For example, a large increase in estradiol stimulated a dramatic spike in LH levels. This suggested that positive feedback was occurring—that is, high levels of estradiol increased the release of its regulatory hormone, even though low doses of estradiol suppressed it. Injections of progesterone, in contrast, appeared to inhibit both FSH and LH. Progesterone exerted only negative feedback on the pituitary hormones.

Thanks to dozens of experiments like these, conducted in a wide variety of labs over the past several decades, researchers now have a fairly good grasp of how the various hormones interact. To summarize the interplay between LH, FSH, estradiol, and progesterone, let's start at day 0 and follow key events as the cycle progresses.

As a cycle begins, the uterus is shedding much of its lining during menstruation. In the ovary, a follicle is being stimulated to develop under the influence of FSH. As the follicle grows, its production of estradiol gradually begins to increase. The increase in estradiol stimulates mitosis and an increase in cell number in the uterine lining. The follicle also produces a small amount of progesterone. While estradiol levels are still rela-

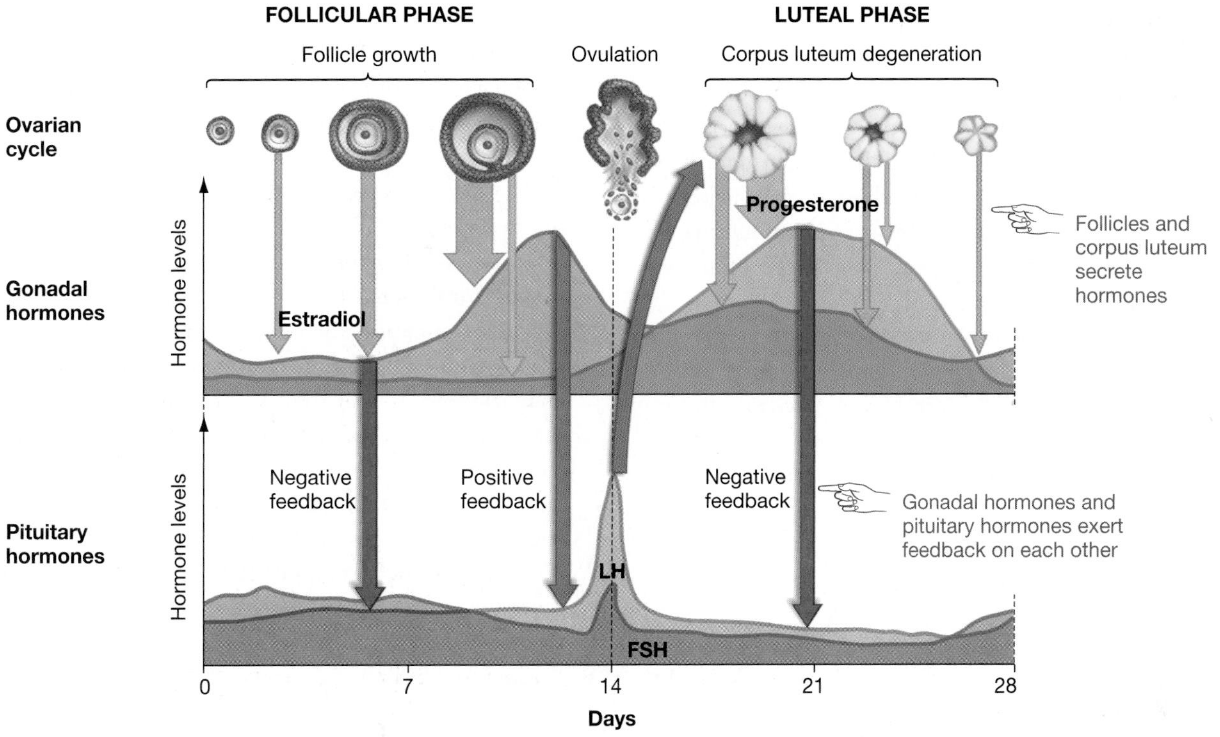
FOLLICULAR PHASE
LUTEAL PHASE
Follicle growth
Ovulation
Corpus luteum degeneration
Ovarian cycle
Gonadal hormones
Pituitary hormones
Hormone levels
Hormone levels
Progesterone
Estradiol
Follicles and corpus luteum secrete hormones
Negative feedback
Positive feedback
Negative feedback
Gonadal hormones and pituitary hormones exert feedback on each other
LH
FSH
0
7
14
21
28
Days

FIGURE 48.12 The Menstrual Cycle Is Regulated by Complex Interactions among Hormones
The pituitary hormones FSH and LH control the release of estradiol and progesterone from reproductive tissues. Estradiol and progesterone may exert negative feedback, positive feedback, or both—depending on their concentration—on the pituitary hormones.

tively low, that hormone suppresses LH secretion through negative feedback inhibition. Once the follicle has grown enough to produce large quantities of estradiol, however, the hormone begins to exert positive feedback. As Figure 48.11 shows, the positive feedback loop between LH and estradiol produces a spike in both hormones. The LH surge triggers ovulation and ends the follicular phase.

As the corpus luteum develops from the remains of the ruptured follicle, it secretes large amounts of progesterone and small quantities of estradiol in response to LH. This rise in progesterone activates the thickened uterine lining, creating an actively secreting, spongy tissue with a well-developed blood supply. In this way, progesterone creates an environment that supports embryonic development if fertilization occurs.

If fertilization does not occur, however, the corpus luteum degenerates approximately 12 days after ovulation. Progesterone levels fall as a result, which causes the thickened lining of the uterus to degenerate. Once progesterone levels have declined, the hypothalamus and pituitary are released from inhibitory control. LH and FSH levels rise, and a new cycle begins. The interactions between FSH, LH, estradiol, and progesterone are summarized in **Figure 48.12**.

If fertilization occurs, however, the corpus luteum does not degenerate, and progesterone and estradiol levels stay high. How does the corpus luteum "know" if fertilization has taken place? What other hormonal and developmental changes occur during pregnancy? To answer these questions, let's investigate the major events during pregnancy and birth in humans.

✓CHECK YOUR UNDERSTANDING

An array of hormones interact in complex ways to regulate the human menstrual cycle. You should be able to (1) describe the function of FSH, LH, estradiol, and progesterone during one cycle; (2) explain the consequences of positive feedback between LH and estradiol and negative feedback between progesterone and LH and FSH; and (3) predict the consequences of a drug that inhibits the release of any one of these four hormones during a cycle.

48.5 Human Pregnancy and Birth

Once it is released from the human ovary, a secondary oocyte is viable for less than 24 hours. Human sperm, in contrast, remain capable of fertilizing an egg for up to five days. Therefore, sexual intercourse has to occur less than five days prior to ovulation for pregnancy to result.

Although an ejaculate may contain hundreds of millions of sperm, most die as they travel through the uterus. Only 100 to 300 actually reach the oviduct, where fertilization takes place.

Chapter 21 detailed how sperm and oocyte (egg) interact when they meet. Recall that contact between sperm and oocyte triggers the **acrosomal reaction**, in which enzymes released from the head of the sperm chew a path through the material surrounding the oocyte membrane. Once the membranes of the oocyte and sperm have fused, the oocyte nucleus completes meiosis II. The two nuclei then unite to form a diploid zygote.

Major Events during Pregnancy

As smooth muscle contractions in the oviduct gradually move the zygote toward the uterus, the cell begins to divide by mitosis. By the time it reaches the lining of the uterus, the embryo consists of a hollow ball of cells. It then undergoes **implantation**—that is, it becomes embedded in the thickened, vascularized wall of the uterus. It will stay in the uterus for the full period of development, or **gestation**, of approximately 270 days (9 months) in humans.

Once the embryo has become implanted in the uterine lining, its cells begin synthesizing and secreting a hormone called **human chorionic gonadotropin (hCG)**. hCG is a chemical messenger that prevents the corpus luteum from degenerating. When hCG is present, the ovary continues secreting progesterone and the menstrual cycle is arrested. Enough hCG is produced by the embryo and excreted in the mother's urine to be used as an indicator in pregnancy tests.

The First Trimester In humans, gestation is divided into 3-month stages called **trimesters**. The first trimester is particularly eventful. Not long after implantation is complete, mass movements of cells result in the formation of the three major embryonic tissues called *ectoderm, endoderm*, and *mesoderm* (see Chapter 21). By 8 weeks of age, these tissues have differentiated into the various organs and systems of the body. Also by this time, the heart has begun pumping blood through a circulatory system. The embryo at this stage is called a **fetus** (**Figure 48.13a**).

The embryonic ectoderm also contributes to several important membranes. One of these membranes, called the **amnion**, completely surrounds the embryo. The amnion eventually fills with **amniotic fluid**, which provides the embryo with a protective cushion.

The other key event in the first trimester is the formation of the **placenta**. This organ starts to form on the uterine wall a few weeks after implantation. The placenta is composed of tissues from both mother and embryo and is the primary source of nutrition for the growing fetus. Arteries transport blood from the circulatory system of the fetus, through the **umbilical cord**, to an extensive capillary bed in the placenta. In this way, the placenta provides a large surface area for the exchange of gases, nutrients, and wastes between maternal and fetal blood, even though maternal and fetal blood do not commingle. The placenta also secretes a variety of hormones. For example, by the end of the first trimester, placental cells synthesize and secrete enough progesterone and estrogens to maintain the pregnancy, even though the corpus luteum has degenerated by this time.

The Second and Third Trimesters After the organs and placenta form during the first trimester, the remainder of development focuses on growth (**Figure 48.13b,c**). During the last weeks of pregnancy, the brain and lungs undergo particularly dramatic growth and development. If a baby is born prematurely, the primary challenge to caregivers is to maintain an adequate oxygen supply. In many cases, dramatic intervention is

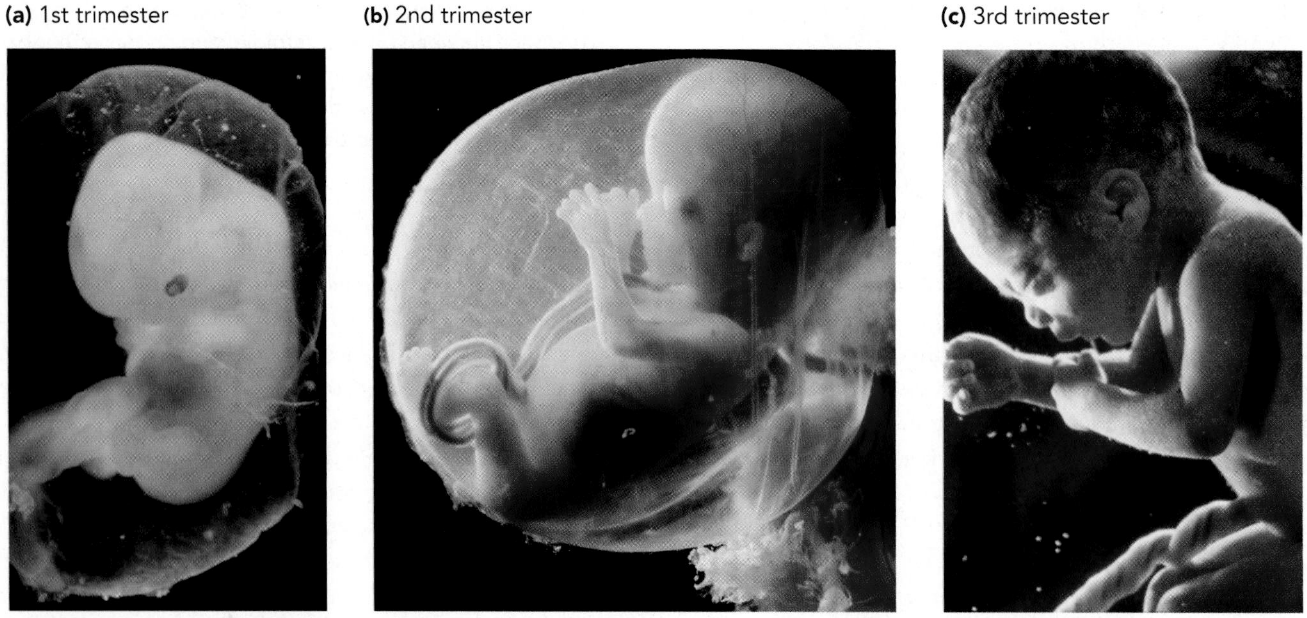
(a) 1st trimester
(b) 2nd trimester
(c) 3rd trimester

FIGURE 48.13 Development of the Human Fetus
EXERCISE On each photograph, label the hands, eye, and umbilical cord. Estimate the ratio of head length to total body length in each photograph.

required to keep the baby alive until the lungs can complete their development. (See Box 44.1.)

The machinery and level of hospital care required by premature infants emphasizes just how superbly adapted mothers are to nourishing a growing fetus in the uterus. Let's take a closer look at this critical aspect of pregnancy.

How Does the Mother Nourish the Fetus?

In viviparous species, the developing embryo is completely dependent on the mother for oxygen, chemical energy in the form of sugars, amino acids and other raw materials for growth, and waste removal. What physiological changes occur in human mothers to accommodate these demands?

To answer this question, researchers have focused on the body's basic transport and delivery systems: the heart, lungs, and blood vessels. For example, a woman's total blood volume expands by as much as 50 percent during pregnancy. To accommodate this increase, maternal blood vessels dilate (widen) and blood pressure drops. In addition, dramatic increases occur in the pumping capacity, or stroke volume, of the heart as well as in the heart rate (see Chapter 44). The mother's heart enlarges and begins to beat faster. The result is that a mother's total cardiac output increases by almost 50 percent over the course of a pregnancy.

The mother's breathing rate and breathing volume also increase to accommodate the fetus's demand for oxygen and its production of carbon dioxide. Important adaptations heighten the efficiency of gas exchange between the mother and the embryo as well. As **Figure 48.14a** shows, maternal and fetal blood flow in the placenta occurs in a countercurrent fashion in many species. As explained in Chapter 41, countercurrent flows maintain a concentration gradient that increases the efficiency of diffusion or other types of exchange. The data for partial pressures of oxygen given in Figure 48.14a are from experiments on sheep, but similar differences between the maternal and fetal circulations have been observed in humans.

Unlike arteries in some mammals, maternal arteries in humans empty into a space at the junction of the maternal and fetal portions of the placenta. This space is packed with small projections called *villi*, which contain the fetal blood vessels. Thus, a large surface area from the fetus is bathed with highly oxygenated maternal blood. The fetal villi are analogous to the alveoli of the lungs (Chapter 44), which provide a large surface area for gas exchange.

Figure 48.14b illustrates another adaptation that increases the rate of oxygen delivery to the fetus. Notice that the axes on this graph give the partial pressure of oxygen in blood or tissue versus the percentage of hemoglobin that holds oxygen at that pressure. (Recall from Chapter 44 that this type of graph is called an *oxygen-hemoglobin dissociation curve*.) The key point here is that the data for fetal hemoglobin are shifted to the left of the data for adult hemoglobin. As a result, the fetus's blood always has a higher affinity for oxygen than does the mother's blood. Biologists interpret this pattern as an adaptation. The high oxygen affinity of fetal hemoglobin ensures that the fetus is always able to acquire oxygen from the mother.

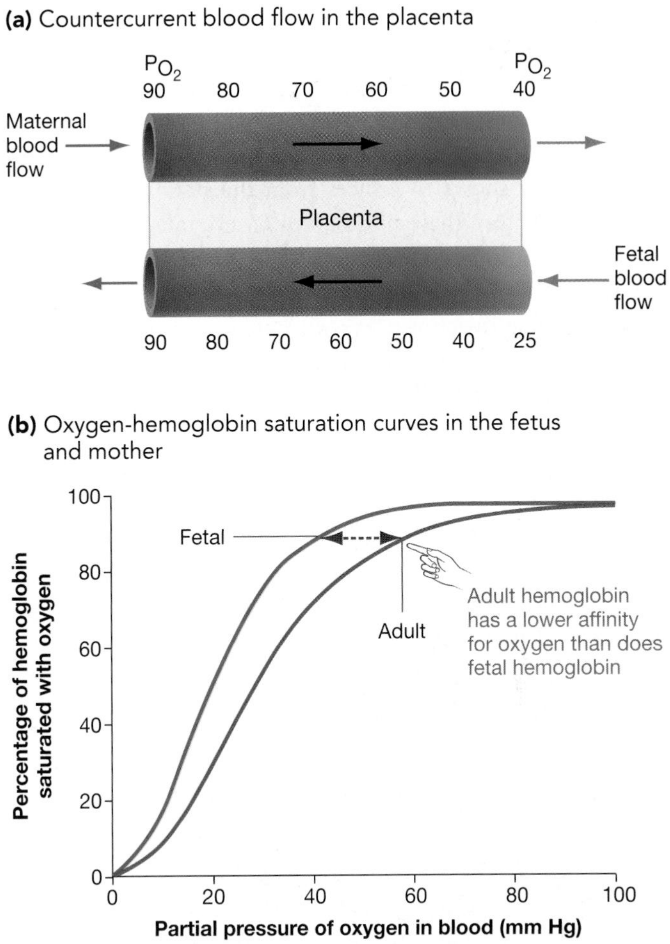

FIGURE 48.14 Adaptations That Increase Delivery of Oxygen to the Fetus
(a) Countercurrent blood flow maximizes the efficiency of oxygen diffusion from mother to fetus. **(b)** Graphs showing that fetal hemoglobin has a much higher affinity for oxygen than does adult hemoglobin.

Before leaving the topic of pregnancy, it is important to acknowledge that mothers and embryos exchange more than nutrients and wastes. For example, consider the thalidomide tragedy introduced in Chapter 3. Recall that, during the 1950s, hundreds of children were affected by their mothers' consumption of the tranquilizer thalidomide, which diffused into the fetal bloodstream and caused birth defects. Although thalidomide is now banned for use by pregnant women, alcohol use continues to take a similar toll. Children of mothers who drink alcohol are at high risk of experiencing hyperactivity, severe learning disabilities, and depression. Collectively, these symptoms are termed **fetal alcohol syndrome (FAS)**.

To understand why FAS exists, recall that the brain undergoes especially rapid growth in the third trimester of development, and then consider the data shown in **Figure 48.15**. Figure 48.15a shows two slices from the brains of newborn rats. A day before these photographs were taken, one group of rats was given two injections of a harmless salt solution at a dosage of 2.5 g per kg body weight; the other group was given the same injections, except that the salt solution contained 20 percent ethanol. Notice the contrast in brain size and in the number of black specks, which indicate degenerating neurons. The histograms in Figure 48.15b confirm the visual impression in the photographs. The data indicate marked differences in the average weights of the brain and of certain brain tissues in groups of rats subjected to each treatment. The messages of these data are that ethanol destroys growing neurons and that even a single large dose of alcohol can have a devastating effect.

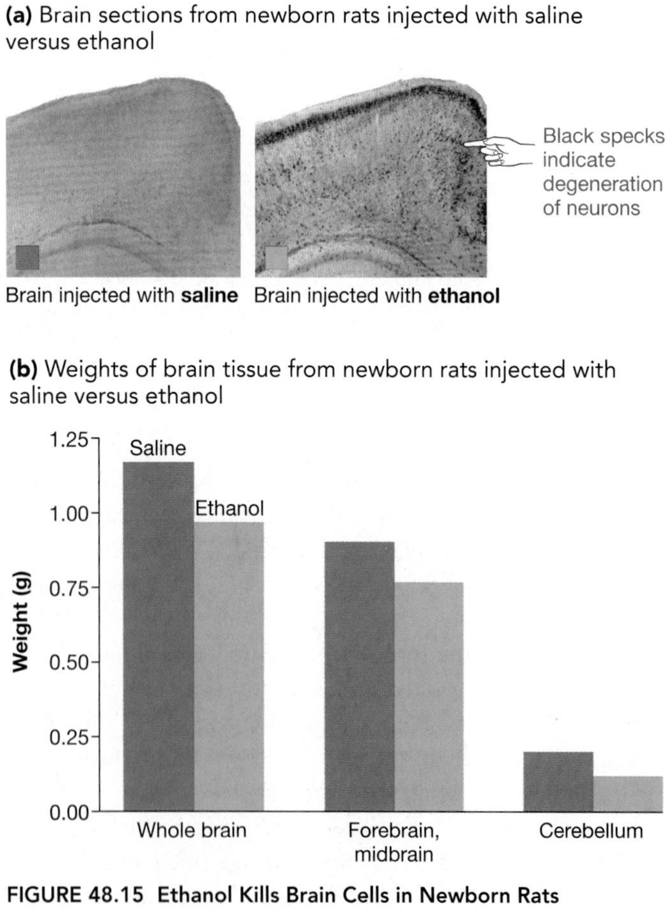
(a) Brain sections from newborn rats injected with saline versus ethanol
Black specks indicate degeneration of neurons
Brain injected with **saline**
Brain injected with **ethanol**
(b) Weights of brain tissue from newborn rats injected with saline versus ethanol
Weight (g)
1.25
1.00
0.75
0.50
0.25
0.00
Saline
Ethanol
Whole brain
Forebrain, midbrain
Cerebellum

FIGURE 48.15 Ethanol Kills Brain Cells in Newborn Rats
(a) Photographs and **(b)** histograms document the degeneration of neurons in the brains of newborn rats that were injected with ethanol.
QUESTION This study was designed to test the effects of a single "binge drinking" event. How would you modify the study to test the effects of long-term exposure to alcohol during prenancy?

Birth

Although the mechanisms responsible for triggering the birthing process are not completely understood, the pituitary hormone **oxytocin** is clearly important in stimulating smooth-muscle cells in the uterine wall to begin contractions. The contractions that expel the fetus from the uterus and out of the mother's body constitute **labor**.

During the first stage of birth, the uterus contracts at relatively low frequency. As labor progresses, the cervix at the base of the uterus begins to open, or *dilate* (**Figure 48.16**). Once the cervix is fully dilated, uterine contractions become more force-

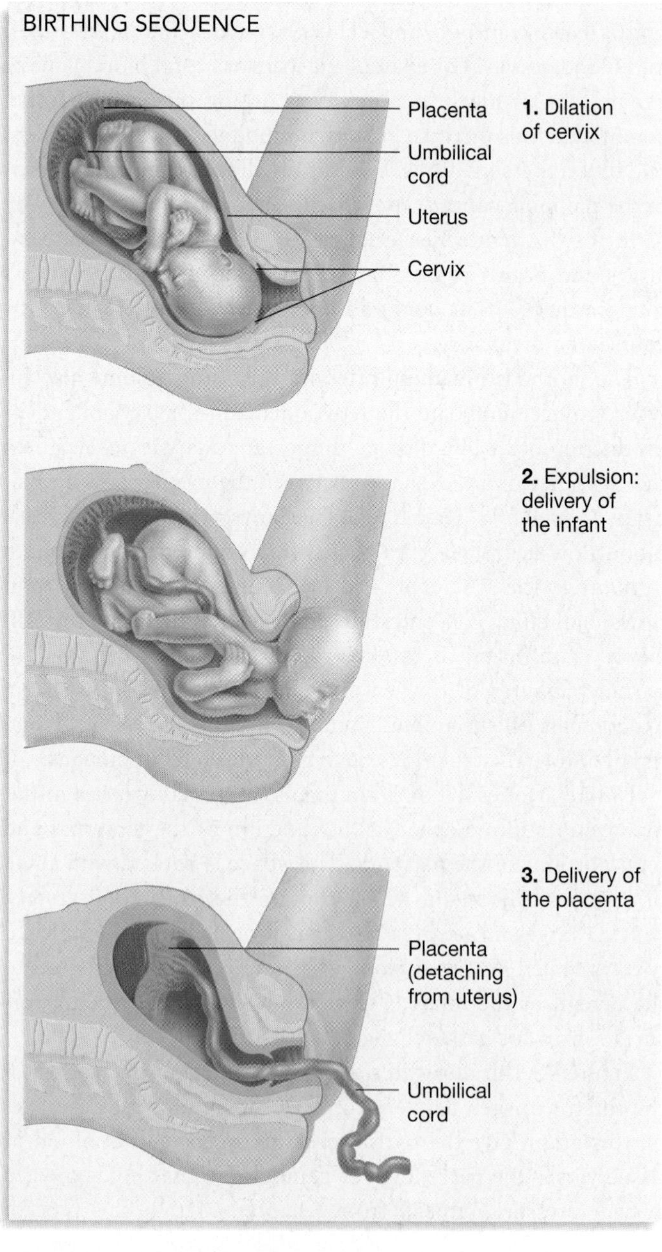
BIRTHING SEQUENCE
Placenta
Umbilical cord
Uterus
Cervix
1. Dilation of cervix
2. Expulsion: delivery of the infant
3. Delivery of the placenta
Placenta (detaching from uterus)
Umbilical cord

FIGURE 48.16 The Birthing Sequence in Humans
QUESTION Human babies are normally born face down, as shown. Predict the consequences of a baby positioned face up, or feet first, in the birth canal.

ful, longer, and more frequent. Eventually, the fetus is expelled through the cervix and into the vagina.

After the baby is delivered, the placenta remains attached to the uterine wall. At this point, caregivers clamp and then cut the umbilical cord that connects the child and the placenta. By gently tugging on the cord, the caregiver helps the mother deliver the placenta and accompanying membranes, and birth is complete.

Although this description sounds straightforward, in reality a large number of complications are possible. To drive this point home, consider the data in **Figure 48.17**. The graph shows the number of Swedish mothers who died, per 100,000 live births, in five-year intervals between 1750 and 1980. In 1750, 1.1 mothers died for every 100 infants successfully delivered. In most cases, the cause of death was blood loss or infection following delivery. As a result of sterile techniques, antibiotics, and blood transfusion technology, Sweden's mortality rate has now declined to less than 0.007 percent. The advent of cesarean sections, better care for premature infants, and improved treatments for Rh disease have also resulted in steep declines in infant mortaility rates in many countries. Although it is generally unheralded, this huge reduction in the rate of death associated with childbirth qualifies as one of the great triumphs of modern medicine. Unfortunately, because many developing nations lack sterile facilities and appropriate antibiotics, the mortality of mothers and infants remains unacceptably high in those countries.

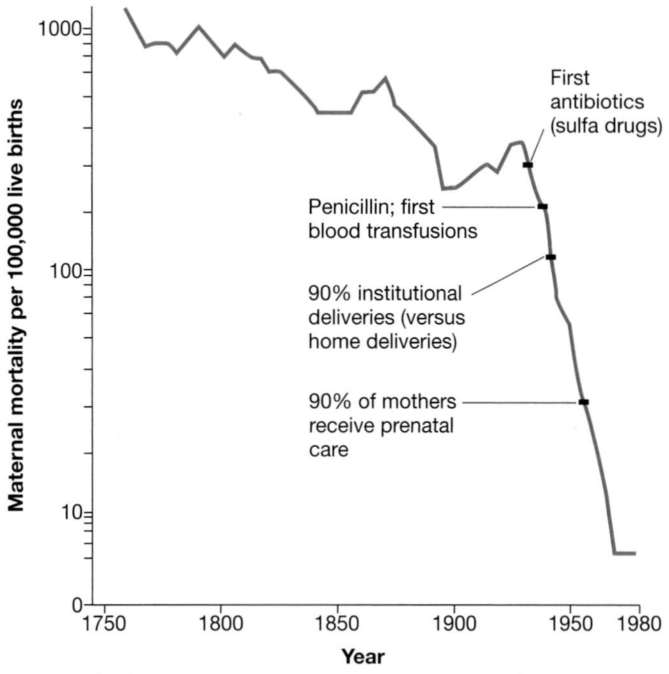

FIGURE 48.17 Maternal Mortality in Childbirth Has Decreased Dramatically since 1940

Graph documenting dramatic recent declines in the rate of maternal mortality due to childbirth in Sweden. The scale on the vertical axis is logarithmic. QUESTION If a woman living in Sweden in 1750 had nine children over the course of her lifetime, what was her overall chance of dying in childbirth?

ESSAY Contraception

This chapter details the remarkable adaptations that make fertilization, pregnancy, and birth possible. The purpose of **contraception** is to circumvent these events and prevent pregnancy. In general, birth control methods work by (1) inhibiting ovulation, (2) preventing sperm from contacting the oocyte, or (3) interfering with implantation of the embryo.

Birth control pills inhibit ovulation because they contain synthetic versions of estradiol and progesterone. These hormones suppress the release of GnRH and FSH through negative feedback inhibition and thus prevent the maturation of a follicle and ovulation. A woman uses these hormones in a cyclic fashion, taking the pills for three weeks and then stopping for one week to allow menstruation to occur. A newer method of inhibiting ovulation involves inserting a small implant of synthetic hormone under the skin; these implants are extremely effective and last for up to five years. In the United States, birth control pills remain the most widely used contraceptive method.

A variety of *barrier methods* prevent sperm from reaching the oocyte. A condom is a sheath of latex or rubber. The male version of the condom covers the penis and retains the ejaculate; the female condom covers the cervix, vagina, and the labia. In contrast, a diaphragm or cervical cap covers just the cervix of the uterus. Surgical sterilization methods can also be thought of as barrier methods for birth control. When a vasectomy is performed, the vas deferens is cut to prevent sperm from moving into the urethra along with the rest of the ejaculate. In a tubal ligation, both oviducts are cut to prevent the passage of secondary oocytes to the uterus.

Is there a completely effective method of birth control?

Three methods of preventing the implantation of the embryo are currently in widespread use. Intrauterine devices (IUDs) are inserted through the cervix and into the uterus, where they can remain for three to four years. Although it is not completely understood how IUDs work, the presence of one in the uterus effectively prevents the implantation of the embryo. The drug RU-486, in contrast, acts by blocking the action of progesterone. As a result, menstruation occurs despite the presence of a zygote. Emergency contraceptive pills (sometimes called morning-after pills) contain a high dose of progesterone and are thought to prevent implantation of the embryo.

(Continued on next page)

(Essay continued)

Although many of these methods are effective at preventing pregnancy, only abstinence offers a 100-percent guarantee. Further, only abstinence is certain to prevent the transmission of HIV, genital warts, genital herpes, and other sexually transmitted diseases.

Is there a completely effective method of birth control? Each of the methods reviewed here entails trade-offs in convenience, expense, and side effects. In the case of the hormone-based treatments, the extent and severity of side effects may vary among individuals. As a result, there is no completely effective method of birth control that is right for everyone. Each approach has advantages and disadvantages. The choice of a method, in consultation with a health-care professional, is an important decision for couples to make.

CHAPTER REVIEW

Summary of Key Concepts

The reproductive systems of animals are highly variable. Many animals can switch between asexual and sexual reproduction. When sexual reproduction occurs, fertilization may be external or internal and egg development may take place inside or outside the mother's body, depending on the species.

Only a few animal species reproduce exclusively by asexual means, but many alternate between asexual and sexual reproduction. Asexual reproduction results in offspring that are genetically identical to the parent. In contrast, every sexually produced offspring is genetically unique because of recombination during meiosis and the fusion of haploid gametes from different parents during fertilization.

In humans, the production of sperm and eggs differs in several respects. Spermatogenesis is continuous throughout the adult life of males, but in females all of the cells that divide mitotically to begin oogenesis are formed very early in development. Meiosis in females is also arrested for long periods of time. Finally, cell division during meiosis is so unequal in females that just one egg (instead of four) is produced from each primary oocyte.

Fertilization is external in many aquatic animals but internal in almost all terrestrial species. Release of gametes during external fertilization may be synchronized by courtship displays or possibly by pheromones. Research on internal fertilization has documented that multiple mating and sperm competition are widespread. Two important generalizations have emerged from studies of sperm competition. First, the last male to mate usually fathers a disproportionately large number of offspring. Second, males of species that engage in sperm competition have relatively large testes for their overall size.

Once eggs are fertilized, females may lay the eggs or retain them and give birth to live offspring. By estimating the evolutionary relationships among closely related species with viviparity and oviparity, researchers have determined that viviparity has evolved many times independently. In some groups at least, viviparity may be an adaptation that increases the survival of young in cold habitats.

Web Tutorial 48.1 Human Gametogenesis

In humans, the male reproductive system is composed of structures specialized for the production and storage of sperm, the synthesis of fluids found in semen, or the transport and delivery of semen. The female reproductive system is composed of structures specialized for producing eggs and nourishing offspring during early development.

In humans, the external anatomy of the male reproductive tract consists of the penis and scrotum. Biologists are unsure why the scrotum evolved in mammals but have documented that sexual selection has produced striking variation in the genital morphology of male insects and spiders. The internal anatomy of the human male reproductive tract includes structures devoted to sperm production and storage, synthesis of accessory fluids, and transport and delivery of semen. The external anatomy of the human female reproductive tract consists of the vaginal opening and clitoris. The internal anatomy of the human female reproductive tract is devoted to egg production and transport or to the development of offspring.

Web Tutorial 48.2 Human Reproduction

In humans, the menstrual cycle is regulated by hormones from the pituitary gland and female reproductive organs. These hormones may interact with each other via positive or negative feedback. Pregnancy is maintained by hormonal signals from the embryo and from the mother's reproductive organs.

The development and activity of male and female reproductive structures are under hormonal control. In mammals, experiments confirmed that GnRH from the hypothalamus triggers the release of FSH and LH from the pituitary gland. The pituitary hormones, in turn, regulate the production of testosterone and estradiol in the testes and ovaries, respectively. Research on the human menstrual cycle has confirmed that estradiol and progesterone exert negative and positive feedback on the production of FSH and LH. Interactions between the pituitary and ovarian hormones are responsible for regulating cyclical changes in the ovaries and uterus.

If fertilization occurs, the developing embryo secretes the hormone hCG, which arrests the menstrual cycle and allows pregnancy to continue. During the first trimester, the embryo becomes implanted in the thickened uterine wall, organs develop, and the nutritive organ called the placenta forms. To make rapid growth possible during the second and third trimesters, the mother's heart rate and pumping volume increase. Nutrients and gases are exchanged efficiently in the placenta.

Questions

Content Review

1. What term describes the mode of asexual reproduction in which offspring develop from unfertilized eggs?
   a. parthenogenesis
   b. budding
   c. regeneration
   d. fission
2. In sperm competition, what is "second-male advantage"?
   a. the observation that, when females mate with two males, each male fertilizes the same number of eggs
   b. the observation that, when females mate with two males, the second male fertilizes most of the eggs
   c. the observation that females routinely mate with at least two males before laying eggs or becoming pregnant
   d. the observation that accessory fluids form copulatory plugs and prevent matings by second males
3. How are the human penis and clitoris similar?
   a. They develop from the same population of embryonic cells.
   b. Both develop during the earliest stages of puberty.
   c. Both contain the urethra.
   d. Both produce accessory fluids required during sexual intercourse.
4. In hormone production, how do the follicle and corpus luteum compare?
   a. Both produce primarily estradiol.
   b. Both produce primarily progesterone.
   c. The follicle produces more estradiol than progesterone; the corpus luteum produces more progesterone than estradiol.
   d. The follicle produces mostly progesterone; the corpus luteum produces estradiol and progesterone.
5. What pituitary hormones are involved in regulating the human menstrual cycle?
   a. GnRH and LH
   b. estradiol and progesterone
   c. oxytocin and LH
   d. FSH and LH
6. The corpus luteum is retained upon fertilization due to the presence of which hormone?
   a. LH
   b. estradiol
   c. progesterone
   d. human chorionic gonadotropin (hCG)

Conceptual Review

1. Summarize the experimental evidence that *Daphnia* require three cues to trigger sexual reproduction. Discuss what these cues indicate about the environment. Generate a hypothesis for why sexual reproduction is adaptive for these animals.
2. Compare and contrast spermatogenesis with oogenesis. How do these two processes differ with respect to numbers of daughter cells produced, gamete size, and timing of the second meiotic division?
3. The drug RU-486 blocks progesterone receptors in the uterus. Why does this drug end pregnancy in humans, even if implantation has occurred?
4. Explain the difference between negative feedback and positive feedback in hormonal control of the human menstrual cycle. How is it possible that estrogen inhibits LH release at one point of the reproductive cycle but stimulates its release at another point?
5. Why are pregnant mothers advised to refrain from smoking and from drinking alcoholic or caffeinated beverages?
6. Suppose that females of a certain insect species routinely mate with two males and choose their mates based on courtship displays or other traits. Predict whether females should be choosier about the first male or the second male that they mate with, or equally choosy about both males. Explain the logic behind your prediction.

Group Discussion Problems

1. Researchers have recently developed methods for cloning mammals. In effect, biologists can now induce asexual reproduction in species that do not normally reproduce asexually. Suppose this practice becomes so widespread in the future that most of the sheep in the world become genetically identical. Discuss some possible consequences of this development.
2. The text claims that species with external fertilization produce extraordinarily large numbers of gametes. How would you test this hypothesis rigorously? In answering, assume that you have data on the average number of gametes produced by different species, along with data on their average body size.
3. Suppose that you've been given a chance to study populations of *Sceloporus* lizards—some of which are viviparous, and others of which are oviparous. Which populations would you expect to produce especially large eggs, relative to their overall body size? How would you go about testing this prediction?
4. HIV, hepatitis B, genital herpes, and other sexually transmitted diseases (STDs) are transferred from one person to another through semen, blood, and other body fluids that come into contact during sexual intercourse. Only one of the contraceptive methods described in the essay is effective in preventing the transmission of STDs. Which method is it, and why is it effective?

Answers to Multiple-Choice Questions 1. a; 2. b; 3. a; 4. c; 5. d; 6. d

www.prenhall.com/freeman is your resource for the following: Web Tutorials; Online Quizzes and other Online Study Guide materials; Answers to Conceptual Review Questions; Solutions to Group Discussion Problems; Answers to Figure Caption Questions and Exercises; and Additional Readings and Research.

49 The Immune System in Animals

KEY CONCEPTS

- The innate immune response to infection is mounted by cells called leukocytes, which circulate in the blood and respond in the same way to all pathogens. The acquired immune response is mounted by cells called lymphocytes, which circulate in the lymphatic system and respond to only a specific pathogen.
- The acquired immune response begins when cell-surface proteins on lymphocytes bind to a particular location on a foreign protein—usually one from a disease-causing agent.
- The humoral response occurs when pathogens are tagged for destruction with antibodies that are produced by lymphocytes called B cells.
- The cell-mediated response occurs when infected cells are destroyed by lymphocytes called T cells.

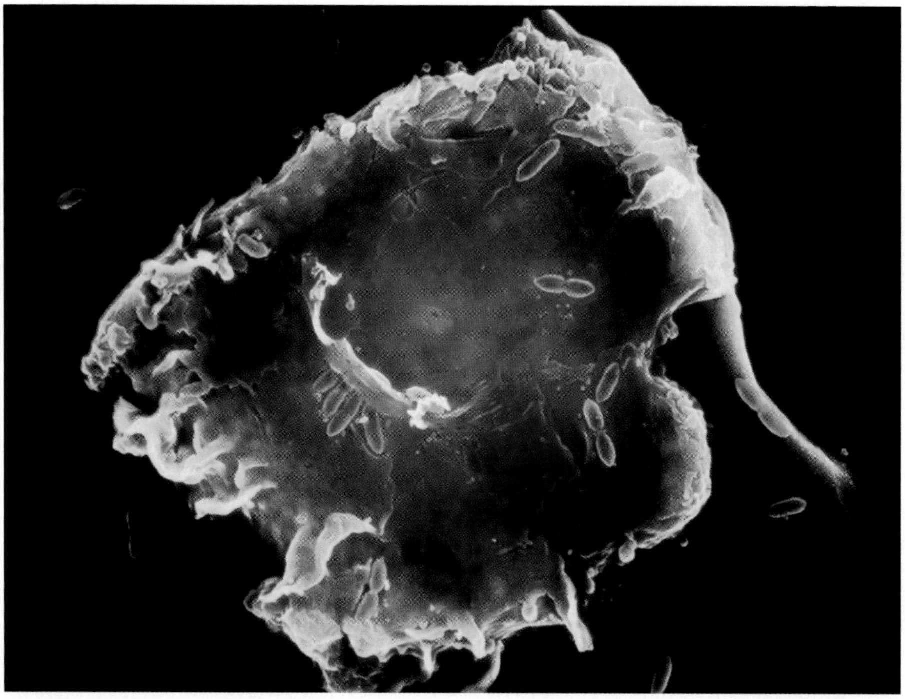

Scanning electron micrograph of an immune system cell engulfing and destroying bacteria (the small green cells). This chapter explores how immune system cells are able to recognize bacteria and viruses as foreign and eliminate them.

Disease threatens every animal. Humans alone are victimized by hundreds of different disease-causing bacteria and viruses and a wide array of parasitic worms, fungi, and protists. Given the ability of these pathogens to cause illness and death, it is remarkable that so many animals stay healthy for most of their lives.

To understand how humans and other animals are protected against disease, biologists have focused on explaining three observations. First, wounds usually heal, even if they become infected. Second, most people who contract a bacterial or viral illness eventually recover, even without help from medications. Third, people who acquire bacterial or viral infections and recover are frequently **immune** ("safe") to that disease—that is, they do not contract the same disease in the future. **Immunity** is a resistance to or protection from disease-causing pathogens.

This last observation is particularly intriguing. Why do people who get measles or chicken pox fail to get these diseases again later, during a second exposure to the same disease-causing pathogen? Thucydides commented on this pattern in 430 B.C. He wrote that when plague struck Athens, only people who had recovered from the illness could nurse the sick, because they would not become ill a second time. In the middle ages, Chinese and Turkish practitioners protected people from smallpox by intentionally exposing individuals to the dried crusts of smallpox pustules taken from infected people. These are examples of **immunization**—the conferring of immunity to a particular disease.

In the late 1700s Edward Jenner had a key insight that allowed him to refine this immunization technique. In Jenner's day, milkmaids were considered pretty because their faces were not pockmarked with scars from smallpox infections. Jenner knew that cows suffered from a smallpox-like disease called cowpox, and he reasoned that milkmaids were immune to smallpox because they had been exposed to cowpox while milking cows. To test this hypothesis, he injected a boy with fluid taken from a cowpox pustule, then later injected the same child with fluid from a smallpox pustule. That is, he **inoculated** the boy with cowpox pathogens and then with smallpox pathogens. As predicted, the boy did not contract smallpox. Jenner's technique was quickly adopted throughout Europe and was later named **vaccination.** (The Latin root *vacca* means "cow.")

Why did Jenner's technique work? More precisely, what is the molecular basis of a person's ability to fight off a bacterial or viral infection and achieve immunity? This chapter answers that question by exploring how the human immune system recognizes and eliminates pathogens. We'll consider the body's response to two events: a bacterial infection and an assault by a virus. Both infections could begin on any given day. For example, suppose that several days ago you were on a crowded bus next to someone with a persistent cough and runny nose, and 30 minutes later you stumbled while sprinting across campus and skinned your elbow. Now you realize that the wound on your arm has gotten sore and red. Worse yet, you are developing a sore throat, runny nose, and fever. Your elbow has become infected with bacteria, and your upper respiratory tract is supporting a growing population of influenza, or "flu," virus. Let's start by analyzing what is happening at the wound at the elbow.

49.1 Innate Immunity

When biologists first began analyzing the immune system, they realized that certain immune system cells are ready to respond to foreign invaders at all times, while other components must be activated first. Cells that are always ready confer **innate immunity;** cells that require activation contribute **acquired immunity.** The key to understanding the two types of immunity is to recognize that the cells involved provide different responses to antigens. An **antigen** is any foreign molecule that can initiate an immune system response. Most antigens are proteins or glycoproteins from bacteria or viruses or other invaders, but foreign carbohydrates and lipids can also function as antigens.

The cells involved in innate immunity are nonspecific in their response to antigens. Stated another way, the innate immune system responds in the same way to all antigens. The innate immune response to a wound in your elbow, for example, is the same no matter what species of bacteria or viruses or fungi enter the tissue. In contrast, cells involved in acquired immunity respond in an extremely specific way to the particular strains of bacteria or viruses or fungi involved.

To launch this investigation into immune system function, let's focus on how the body prevents entry by foreign invaders. Then we can consider what the innate response does once some do get in.

Barriers to Entry

The most effective way to avoid getting sick is to avoid contact with pathogens. In humans and other animals, the most important barrier to pathogen entry is the skin. In addition to providing a physical barrier, human skin offers a chemical deterrent. Skin cells secrete lactic acid and fatty acids, which lower the pH of the surface to between 3 and 5. The dry, acidic environment prevents the growth of most bacterial species that might be a threat to the body. Many bacteria do live on the skin, however, but most of them are adapted to the acidic conditions and are **commensal**—meaning they do not harm the host organism. The bodies of insects and other animals with exoskeletons are difficult for pathogens to enter because they are covered with the tough layer called cuticle, introduced in Chapter 42.

An animal's body has gaps in these barriers, however, where the digestive tract, reproductive tract, gas-exchange surfaces, and sensory organs make contact with the environment. As **Figure 49.1** shows, these gaps have a protective physical barrier in the form of mucus or other features that discourage pathogen entry. **Mucus** is a solution secreted by epithelial cells. It is rich in glycoproteins and **proteoglycans**—large polysaccharides bonded to small proteins. The bodies of slugs, snails, earthworms, and other soft-bodied animals are covered with a layer of mucus that shields their epithelial cells from pathogens in the environment. Mucus is equally important in vertebrates. For example, many of the pathogens that you breathe in or

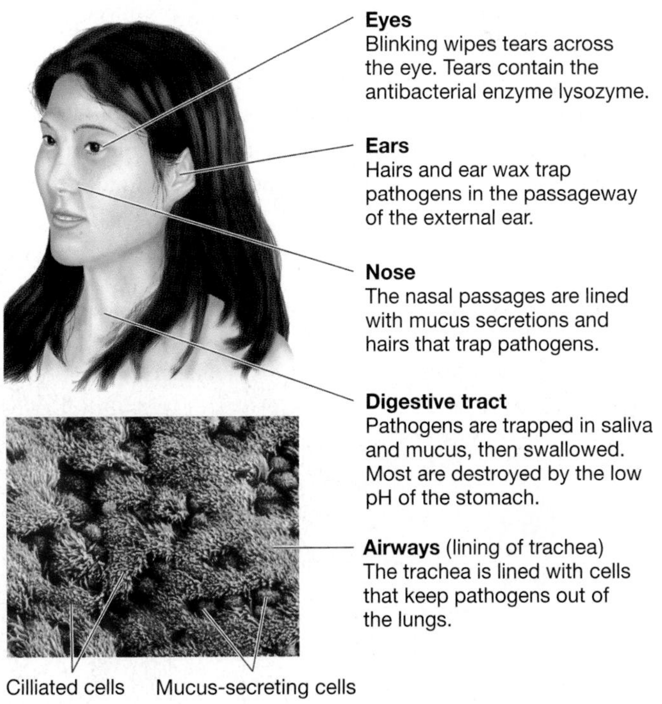
Eyes
Blinking wipes tears across the eye. Tears contain the antibacterial enzyme lysozyme.
Ears
Hairs and ear wax trap pathogens in the passageway of the external ear.
Nose
The nasal passages are lined with mucus secretions and hairs that trap pathogens.
Digestive tract
Pathogens are trapped in saliva and mucus, then swallowed. Most are destroyed by the low pH of the stomach.
Airways (lining of trachea)
The trachea is lined with cells that keep pathogens out of the lungs.
Cilliated cells
Mucus-secreting cells

FIGURE 49.1 How Does the Body Keep Pathogens Out?
QUESTION Why is a combination of hair or cilia and sticky secretions such as mucus effective at trapping pathogens?

ingest while eating or drinking stick to the mucus that lines your airways and gastrointestinal tract. Pathogens that are stuck in mucus cannot come in contact with the plasma membranes of epithelial tissues. In mammals, these pathogens are coughed out or swallowed and killed in the acidic environment of the stomach. Gaps in the body that are not covered with mucus layers, such as the eyes, are often protected by other types of secretions. Your ears are protected by waxy secretions, and your eyes by tears that contain the enzyme **lysozyme**. Lysozyme acts as an antibiotic by digesting bacterial cell walls.

With regularity, though, preventive measures fail and some pathogens gain entry to tissues beneath the skin. Flu viruses, for example, have an enzyme on their surface that disrupts the mucous lining of the respiratory tract. When the outer surface of the virus makes contact with a host cell beneath the mucous layer, the virus is able to enter the cell and begin an infection. When a fall or other physical trauma breaks your skin, bacteria and other pathogens enter and gain direct access to the tissues inside. To viruses, bacteria, and fungi, your body is a tropical paradise, brimming with resources. Within minutes, they can begin growing. What happens then?

The Innate Immune Response

When bacteria enter your body at the site of a wound, the cells pictured in **Figure 49.2** implement the **innate immune response**—the body's nonspecific response to pathogens. As a group, these cells are called **leukocytes** ("white-cells"); all of them reside in the blood. Leukocytes include mast cells, neutrophils, and macrophages. Like the red blood cells that carry oxygen in the bloodstream, leukocytes are produced in the bone marrow.

The cells involved in the innate response are alerted to the presence of foreign invaders by certain molecules. All bacteria, for example, have proteins that begin with the amino acid *N*-formylmethionine instead of the methionine found in other organisms. Gram-negative bacteria also express a distinct type of lipopolysaccharide on their surface, and certain cell-wall components in pathogens terminate with the sugar mannose. Cells in the innate system have proteins on their plasma membranes that bind to these bacteria-specific compounds. These proteins are known as **pattern-recognition receptors**. When they are activated by antigens, the leukocytes respond.

Figure 49.3 summarizes the major steps in the **inflammatory** ("in-flames") **response**, which is a type of innate immune response observed in mammals. When skin breaks, blood components called **platelets** release proteins that form clots and lessen bleeding. Other clotting proteins in the blood form cross-linked structures that help wall off the wound and reduce blood loss. In addition, leukocytes called **mast cells** release chemical messengers that cause blood vessels near the wound to constrict. As a result, blood flow and thus blood loss at the wound decreases. **Histamine** and other molecules released by mast cells induce blood vessels slightly farther from the wound to dilate, which increases the delivery of leukocytes and other blood components to the general region. Thus, signals from mast cells stop blood loss at the wound site itself but increase blood flow to surrounding tissues. Cells in nearby tissues then release chemical signals called **chemokines** ("chemical-movers").Chemokines create a chemical gradient that neutrophils and other leukocytes use in migrating toward the site.

Neutrophils are phagocytic leukocytes. **Phagocytosis** is the engulfing and digesting of foreign particles. (Translated literally, *phagocytic* means "eat-cell.") Neutrophils also secrete lysozyme—which degrades bacterial cell walls—along with free radicals, nitric oxide (NO), and molecules called **reactive oxygen intermediates (ROIs)**, including hydrogen peroxide (H_2O_2). These molecules attack bacteria and fungi. In this way, neutrophils begin to kill and degrade the foreign material present in the wound.

(a) Mast cells secrete signals that increase blood flow. **(b)** Neutrophils ingest and kill bacteria. **(c)** Macrophages ingest and kill bacteria.

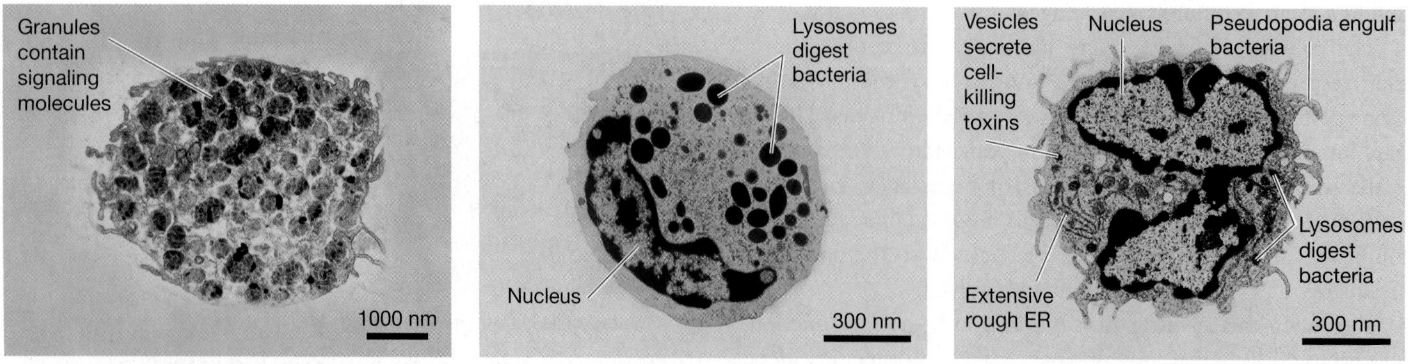
Granules contain signaling molecules
1000 nm
Lysosomes digest bacteria
Nucleus
300 nm
Vesicles secrete cell-killing toxins
Nucleus
Pseudopodia engulf bacteria
Lysosomes digest bacteria
Extensive rough ER
300 nm

FIGURE 49.2 Leukocytes Circulate in the Blood and Mount the Innate Immune Response
Transmission electron micrographs illustrating just a few of the types of leukocytes in the innate immune system. QUESTION How does the structure of these cells correlate with their function?

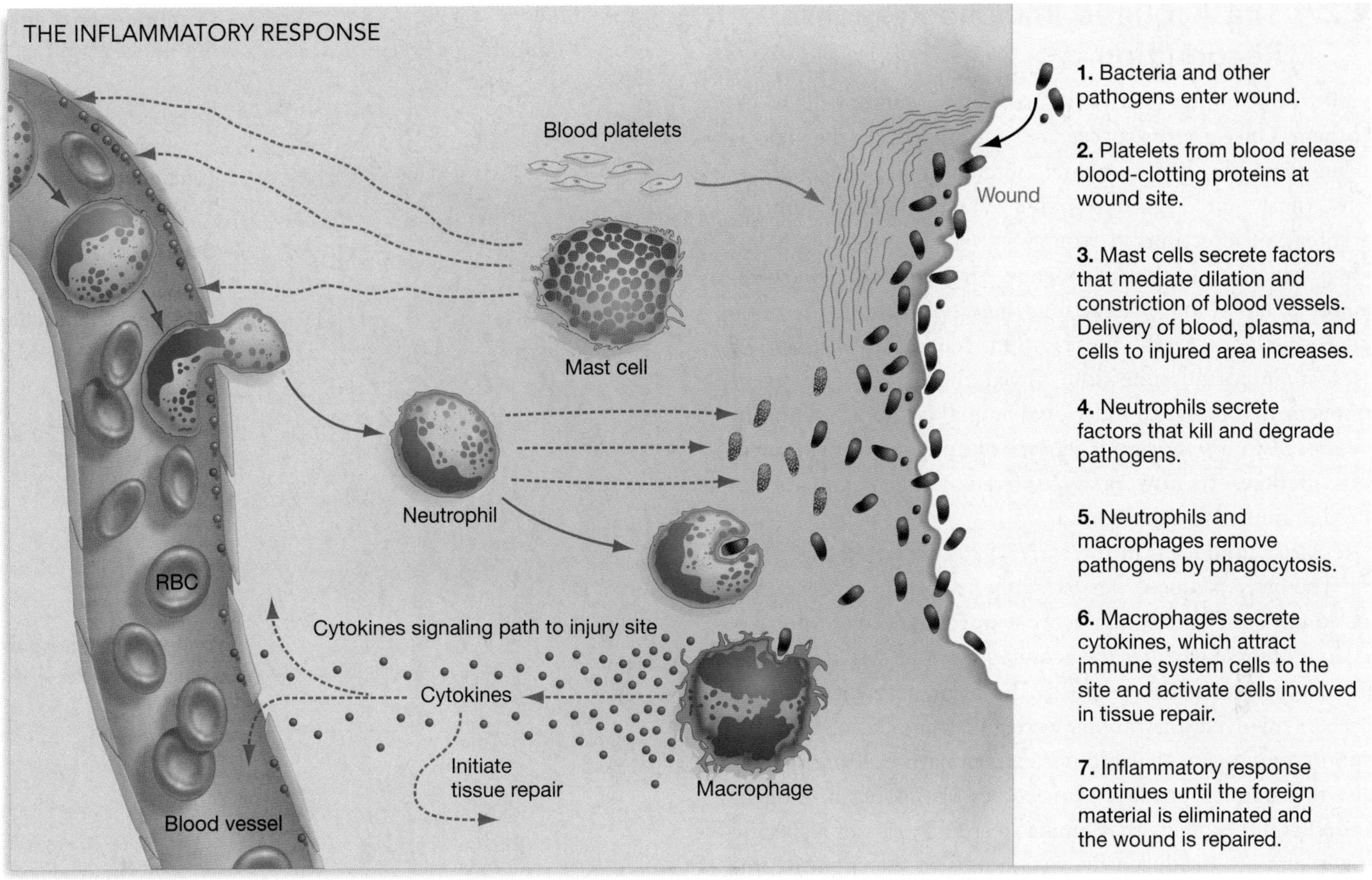
THE INFLAMMATORY RESPONSE
Blood platelets
Wound
Mast cell
Neutrophil
RBC
Cytokines signaling path to injury site
Cytokines
Initiate
tissue repair
Macrophage
Blood vessel
1. Bacteria and other pathogens enter wound.
2. Platelets from blood release blood-clotting proteins at wound site.
3. Mast cells secrete factors that mediate dilation and constriction of blood vessels. Delivery of blood, plasma, and cells to injured area increases.
4. Neutrophils secrete factors that kill and degrade pathogens.
5. Neutrophils and macrophages remove pathogens by phagocytosis.
6. Macrophages secrete cytokines, which attract immune system cells to the site and activate cells involved in tissue repair.
7. Inflammatory response continues until the foreign material is eliminated and the wound is repaired.

FIGURE 49.3 The Inflammatory Response of Innate Immunity Has Many Elements
The inflammatory response occurs at sites of tissue injury.

Leukocytes called **macrophages** arrive after neutrophils do and perform many of the same functions. In addition, macrophages secrete chemical messengers called **cytokines** ("cell-movers"). Cytokines have an array of effects. Cytokines from macrophages stimulate bone marrow to make and release additional neutrophils and macrophages. These chemical messengers also attract other immune system cells to the site, activate cells involved in tissue repair and wound healing, and induce **fever**—elevated body temperature that aids in healing (see Chapter 41).

The site of inflammation often becomes swollen due to increased numbers of cells and fluids in the area, red and warm due to increased blood flow, and painful due to signals from pain receptors. The inflammatory response continues until all foreign material is eliminated and the wound is repaired.

Humans and other vertebrates are not the only animals that have sophisticated innate immune responses, however. Innate responses make up the entire immune system observed in millions of species of invertebrates. The spectacular abundance and diversity of invertebrate animals supports the hypothesis that their innate immune systems provide efficient protection against invading bacteria, viruses, and fungi. For example, if a pathogen succeeds in entering the main body cavity of an insect through a wound or other break in the barriers to entry, cells respond by synthesizing and secreting peptides with potent antibacterial or antifungal properties. Sea stars have specialized cells that are similar in function to neutrophils and macrophages—they secrete cytokines or engulf and destroy pathogens by phagocytosis. But biologists have a great deal to learn about defense systems in invertebrates, compared with research on the immune systems of humans and other vertebrates.

What happens when the innate immune system of a vertebrate fails to contain and eliminate an infection that starts at a wound? And what happened to the virus that blasted through the mucous membrane of the human respiratory tract and entered the cells below?

✓CHECK YOUR UNDERSTANDING

The innate immune response occurs when leukocytes that circulate in the blood react in a nonspecific way to signals from invading pathogens. You should be able to describe how specific events in the innate response compare with the following steps, which are implemented by first-aid workers who are treating a wound: (1) applying direct pressure to close blood vessels, (2) applying bandages impregnated with compounds that recruit platelets to the site, (3) cleaning the wound once bleeding has stopped, and (4) applying antibiotics.

49.2 The Acquired Immune Response: Recognition

Recall from Chapter 34 that viruses actually enter cells to infect them. Once a virus succeeds in entering a cell, the virus is hidden from the host's innate immune response. In vertebrates, the acquired immune system is responsible for limiting the spread of infections at wound sites and responding to invading viruses and other pathogens. The **acquired immune response** is based on interactions between specific immune system cells and specific pieces of antigens. This observation raises an important question: Given the array of different pathogens that exist, an individual animal is almost certain to be exposed to an enormous variety of antigens in the course of its lifetime. To how many different antigens can its acquired immune system respond?

Research conducted in the early 1920s answered this question. The basic protocol was to synthesize organic compounds that do not exist in nature, inject the novel molecules into rabbits, and observe whether the acquired immune system in these animals was activated in response. To the amazement of the researchers involved, the rabbits were able to mount an immune response to antigens that did not exist in nature. More specifically, the animals produced antibodies against each antigen. **Antibodies** are proteins that bind to a specific part of a specific antigen and are produced by certain cells in the acquired immune system. The take-home message of these early injection experiments was that the immune system can produce an almost limitless array of antibodies.

How can the immune system produce antibodies to antigens that had never before existed? It took over 60 years of experimentation and debate to answer that question. To begin exploring this work, let's review the cells and organs involved in the acquired immune system.

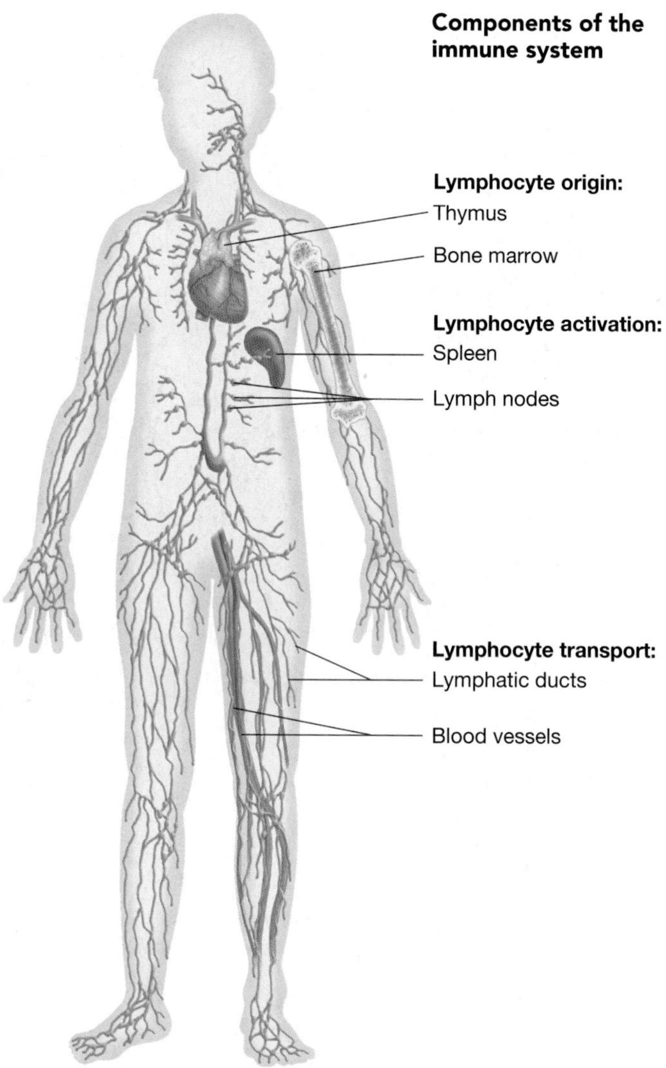

FIGURE 49.4 Lymphocytes Are Created, Activated, and Transported in the Immune System
The immune system has three functional components: lymphocyte origin, activation, and transport. The thymus and bone marrow are primary organs where lymphocytes form and mature. Lymphocytes recognize antigens and become activated in the lymph nodes and spleen. The lymphatic ducts and blood vessels transport lymphocytes throughout the body.

An Introduction to Lymphocytes and the Immune System

The cells that carry out the acquired immune response are called **lymphocytes**. Lymphocytes are created, activated, and transported in distinct components of the immune system (**Figure 49.4**):

1. *Lymphocyte origin* Lymphocytes form and mature in the primary organs of the immune system: the bone marrow and thymus. **Bone marrow** is tissue that fills the internal cavities in bones. The **thymus** is an organ located in the throat or neck of vertebrates (just behind the breastbone in humans).
2. *Lymphocyte activation* Lymphocytes recognize antigens and become activated in the spleen and lymph nodes. The **spleen** is a lymphatic organ located in the abdominal cavity. In addition to its function in the immune response, the spleen filters blood and stores iron from red blood cells. In contrast, **lymph nodes** are small, oval organs that are located all around the body. Lymph nodes filter the lymph passing through them. Recall from Chapter 44 that **lymph** is a mixture of fluid and lymphocytes. The liquid portion of lymph originates in fluid that is forced out of capillaries by blood pressure.
3. *Lymphocyte transport* In contrast to leukocytes, which circulate between the blood and tissues, lymphocytes circulate through the blood and the secondary organs of the immune system—lymph nodes, spleen, and lymphatic ducts. **Lymphatic ducts** are thin-walled, branching tubules that transport lymph throughout the body.

Lymphocytes are normally in a resting, or inactivate state as they circulate through the bloodstream and lymphatic system. As **Figure 49.5a** shows, circulating lymphocytes have a large nucleus, little cytoplasm, few mitochondria, and a ruffled membrane. Over the course of a day, an inactive lymphocyte might migrate through the spleen, enter the blood, cross over into the lymphatic vessels, migrate through a lymph node, return to the blood, and so on.

If an inactive lymphocyte does not encounter the antigen to which it is programmed to respond, then the cell eventually dies. But a lymphocyte that encounters an appropriate antigen becomes activated. As **Figure 49.5b** shows, an activated lymphocyte has a large population of mitochondria and a massive amount of rough ER. Recall from Chapter 7 that many of the proteins synthesized in the rough ER are inserted into the plasma membrane or secreted from the cell. The increased rough ER in activated lymphocytes suggests that these cells are manufacturing and possibly secreting proteins.

The contrast in morphology between inactive and active lymphocytes is remarkable. How does activation take place?

(a) Inactive lymphocyte

(b) Activated lymphocyte

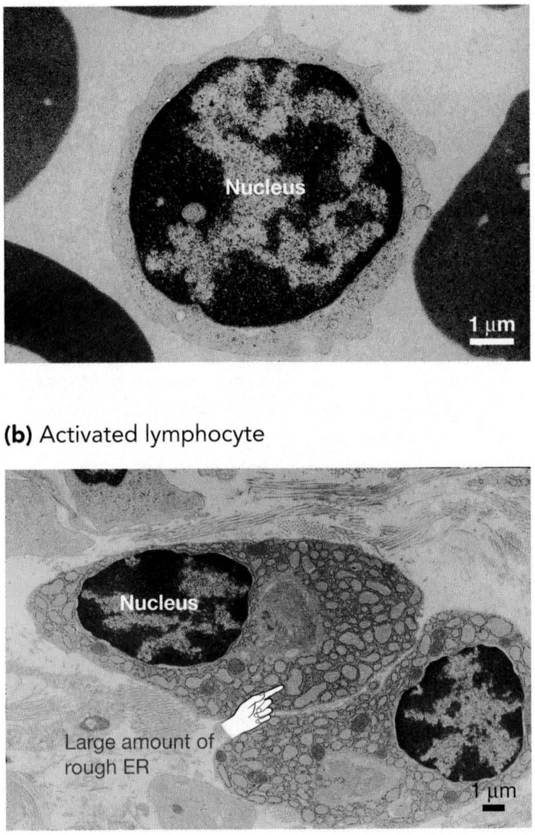

FIGURE 49.5 Lymphocytes Exist in Two States: Inactive and Activated
(a) Inactive lymphocytes have a small amount of cytoplasm and few organelles. **(b)** Activated lymphocytes have extensive rough endoplasmic reticulum. This observation suggests that a great deal of protein synthesis is taking place.

The Discovery of B Cells and T Cells

In 1956 a group of researchers provided an important insight into the acquired immune system, quite by accident. The biologists were investigating the immune system's response to the bacterium *Salmonella typhimurium*, which is a common cause of food poisoning in humans. To do this work, they needed to produce and isolate antibodies to a particular antigen from *S. typhimurium*—an antigen that is toxic. Their plan was to inject a large number of chickens with the antigen and collect the antibodies that the treated individuals produced in response.

In addition to injecting many normal chickens, though, the biologists happened to include some chickens that had undergone experimental removal of an organ called the **bursa.** Six of the nine chickens that lacked a bursa and that were injected with the antigen died. The other three birds without a bursa survived but failed to produce antibodies to the antigen. In contrast, chickens with the bursa intact produced large quantities of antibodies and survived. To make sense of these results, the researchers proposed that the bursa is critical for antibody production and that antibodies are important in neutralizing antigens.

Not long after this observation was published, three groups of scientists independently conducted a related experiment. To explore the function of the thymus in mammals, these groups removed the organ from newborn mice. Mice lacking a thymus developed pronounced defects in their immune systems. For example, when pieces of skin from other mice were grafted onto the experimental individuals, their immune systems did not recognize the tissue as foreign. In contrast, individuals with an intact thymus quickly mounted an immune response to the foreign skin cells and killed them.

The results of these and follow-up experiments showed that lymphocytes from the bursa and thymus have different functions. The two types of lymphocytes became known as bursa-dependent and thymus-dependent lymphocytes, or B cells and T cells, respectively. **B cells** produce antibodies. **T cells** are involved in an array of functions, including recognizing and killing host cells that are being infected with a virus. Later work showed that, in humans and other species that lack a bursa, B cells mature in bone marrow.

More recent research has shown that important subtypes of B cells and T cells exist. For example, T cells come in two major types that are distinguished by distinctive proteins on their plasma membranes. These membrane proteins are called **CD4** and **CD8.** T cells with the CD4 protein on their surface are known as $CD4^+$ cells. As we'll see in Section 49.3, $CD4^+$ cells have different functions than $CD8^+$ T cells. Right now, however, let's turn to the question of how B cells and T cells are able to recognize so many different antigens.

Antigen Recognition and Clonal Selection

By the 1950s biologists had made a series of important general observations about the immune system. It was clear that antibodies to a seemingly limitless number of antigens could be produced and that each antibody is specific to an antigen. By studying the time course of antibody production in rabbits and other experimental animals, researchers established that the immune response intensified through time after an infection had begun. In addition, the response is "remembered"—meaning individuals do not get sick at all, or recover extremely quickly, if they are exposed to the same pathogen again in the future.

To explain these patterns, researchers developed the **clonal-selection theory** of immune system function. This theory made three central claims about how the acquired immune system works:

1. Each lymphocyte formed in the bone marrow or thymus has a unique receptor on its surface. The receptor is a membrane protein and recognizes one antigen. Stated another way, there is a one-to-one correspondence between the receptors found on B cells and T cells and the antigens to which they respond.
2. When the receptor on a lymphocyte binds to an antigen, the lymphocyte is activated. An activated lymphocyte divides and makes many identical copies of itself. In this way, specific cells are selected and cloned in response to an infection.
3. Some of the cloned cells descended from an activated lymphocyte persist after the pathogen is eliminated. As a result, the cloned cells are able to respond quickly and effectively if the infection recurs.

The clonal-selection theory provided a coherent explanation for most major attributes of the acquired immune system. Just as important, it made a number of specific predictions that could be tested. Which aspects of the theory have been supported, and which rejected?

The Discovery of B-Cell Receptors and T-Cell Receptors

To test the prediction that lymphocytes have unique receptors on their surfaces, researchers injected experimental animals with radioactively labeled antigens. The presence of these labeled antigens allowed researchers to isolate and purify a protein on the surface of B cells that bound to the antigens. Chemical analysis of this **B-cell receptor (BCR)** showed that the protein has the same structure as the antibodies produced by B cells. This was a crucial result and led to the realization that both the BCR and the antibodies belong to a family of proteins called the *gamma globulins*. Any one gamma globulin is referred to as an **immunoglobulin (Ig)**. These proteins are the key to the acquired immune response.

As **Figure 49.6a** shows, the BCR has three distinct components. The first is a protein with a molecular weight of about 25,000 dal-

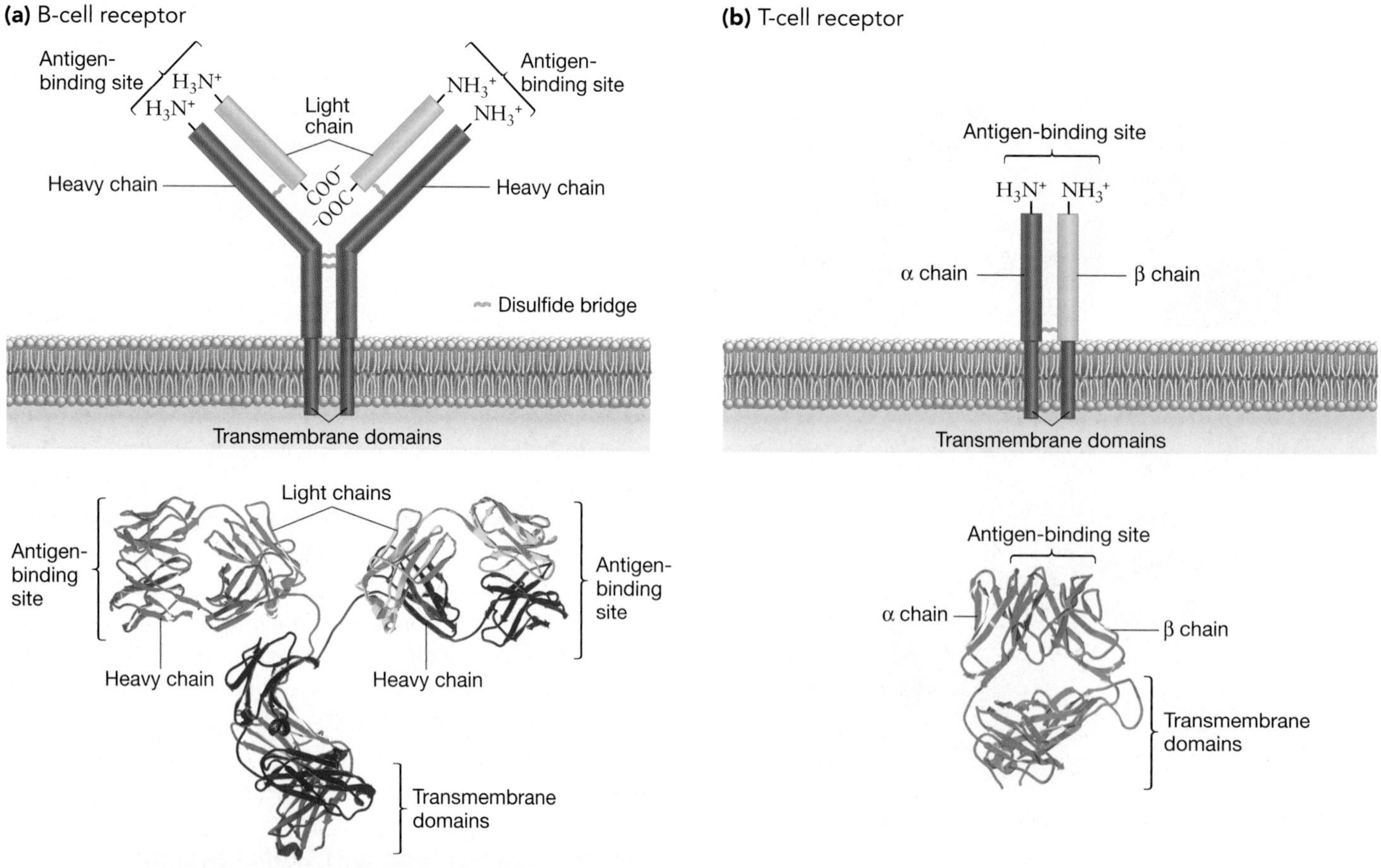

FIGURE 49.6 B-Cell Receptors and T-Cell Receptors Have Transmembrane Domains and Antigen-Binding Sites
(a) Schematic and ribbon models of the B-cell receptor, which is shaped like a Y. The antibodies and receptors produced by each B cell are identical, except that antibodies lack the transmembrane domain and are secreted.
(b) The shape of the T-cell receptor resembles one "arm" of the Y-shaped B-cell receptor.

tons, called the **light chain.** The second component is roughly twice the size of the light chain and is called the **heavy chain.** Each BCR has two copies of the light chain and two copies of the heavy chain. The third component is a **transmembrane domain,** a structure that anchors the protein in the plasma membrane of the B cell. Antibodies are identical in structure to the BCR, except that they lack this transmembrane domain. Instead of being inserted into the plasma membrane, antibodies are secreted from the cell.

Table 49.1 shows the five classes of immunoglobulin proteins that act as receptors or antibodies. The five types are symbolized IgG, IgD, IgE, IgA, and IgM. Each class is distinguished by unique amino acid sequences in the heavy-chain region, and each has a distinct function in the immune response.

It took much longer for researchers to isolate and characterize the **T-cell receptor (TCR).** It turns out that the TCR binds only to antigens after the antigens have been modified by other immune system cells and then **presented,** or displayed, on the plasma membranes of these cells. This is a fundamentally important distinction: B cells can bind to antigens directly; T cells bind only to antigens that are displayed by other immune system cells. Investigators who finally succeeded in isolating the TCR found that it is composed of two chains: a single **alpha (α) chain** and a single **beta (β) chain.** As **Figure 49.6b** shows, the overall shape of the TCR is very similar to the "arm" of an antibody or BCR molecule.

TABLE 49.1 Five Classes of Immunoglobulins

Name	Structure	Function
IgG	Monomer 	The most abundant type of secreted antibody. Circulates in blood and interstitial fluid. Protects against bacteria, viruses, and toxins.
IgD	Monomer 	Present on membranes of mature B cells; probably involved in activation of B cells.
IgE	Monomer 	Rarest type of antibody. Involved in hypersensitive reaction that produces allergies. (See essay at end of chapter.)
IgA	Dimer 	Most common antibody in breast milk, tears, saliva, and mucus that lines the respiratory and digestive tracts. Prevents bacteria and viruses from attaching to mucous membranes; helps immunize breastfed newborns.
IgM	Pentamer 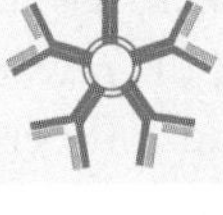	First type of secreted antibody to appear during an infection. Binds many antigens at once; effective at clumping viruses and bacteria so that they can be killed.

Antibodies, BCRs, and TCRs do not bind to the entire antigen but to a selected region of the antigen called an **epitope.** To understand the relationship between an antigen and an epitope, consider that every bacterium, virus, fungus, and protist is made up of a large number of different molecules. Each of these molecules is an antigen, because each is foreign to your cells. Each antigen may have many different epitopes, however, where binding by antibodies and lymphocyte receptors actually takes place. **Figure 49.7** illustrates a protein called **hemagglutinin,** which is found on the surface of the influenza virus. This antigen has six epitopes. Each of these epitopes is recognized by a particular antibody, B-cell receptor, or T-cell receptor. It is not unusual for an antigen to have between 10 and 100 different epitopes.

How are the immunoglobulins able to recognize specific epitopes? The answer to this question came through detailed studies of the BCR's heavy and light chains.

What Is the Molecular Basis of Antigen Specificity? In the 1950s biologists developed an important model system for studying the BCR and antibody production. The cells involved were B-cell tumors, or **myelomas,** that could be grown in laboratory culture. Like other cancers, myelomas grow in an uncontrolled fashion. Further, each type of myeloma produces a single type of antibody. (**Box 49.1**, page 1128, explains how myelomas are involved in producing the antibodies used in research.) By isolating and comparing the light chains produced by different myelomas, researchers were able to (1) document that each light chain is different and (2) infer why different antibodies bind to different epitopes on antigens.

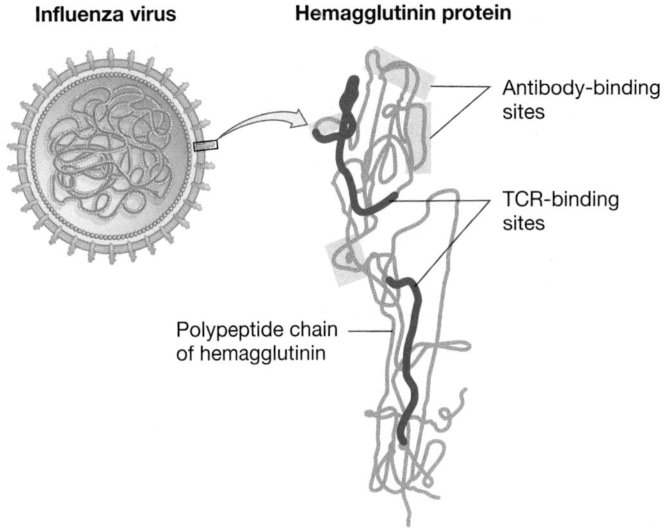

FIGURE 49.7 Most Antigens Have Multiple Binding Sites for B-Cell Receptors, Antibodies, and T-Cell Receptors
The envelope of the influenza virus includes the protein hemagglutinin. Hemagglutinin has four different sites where antibodies bind and two distinct places where T-cell receptors bind. Each binding site is called an epitope.

BOX 49.1 How Are Monoclonal Antibodies Produced?

When an animal is injected with an antigen, many different B cells respond. Some of these B cells produce antibodies to different epitopes on the antigen, and some produce slightly different antibodies that bind to the same epitope. As a result, biologists say that the immune response is **polyclonal** ("many-clones").

In many situations, though, it is extremely valuable for investigators to have antibodies from a single B cell that bind to the same epitope of a single antigen in the same way. For example, having an antibody that binds to a specific site on the insulin receptor would allow you to analyze cells from different parts of the body, or in different stages of development, and understand where and when the insulin receptor protein is expressed. Such antibodies are termed **monoclonal** ("one-clone") **antibodies.**

How can monoclonal antibodies be produced? In 1975 two researchers announced that they had developed an approach for producing monoclonal antibodies. The approach relied on the use of myelomas, which can grow continuously in culture. **Figure 49.8** outlines the protocol these researchers used to create myelomas that produced large quantities of a specific, desired antibody.

The process begins when a mouse is injected with the antigen of interest (step 1 of Figure 49.8). In the example used here, the antigen is hemagglutinin, found on the surface of influenza virus. In response, B cells in the animal's spleen activate and produce antibodies to hemagglutinin (step 2). Spleen cells are then removed from the animal. Next, the spleen cells are added to a test tube containing myeloma cells. A compound that encourages cells to fuse is added to the mixture (step 3). The resulting solution is cultured under conditions that allow growth only by cells that have fused. The hybrid of a myeloma cell and a B cell is called a **hybridoma.**

To complete the process, each hybridoma is isolated, grown separately, and tested for its ability to produce antibodies to the antigen of interest (step 4). Once the most effective hybridomas are identified, they can be grown in quantity and provide researchers with a virtually limitless supply of a monoclonal antibody (step 5).

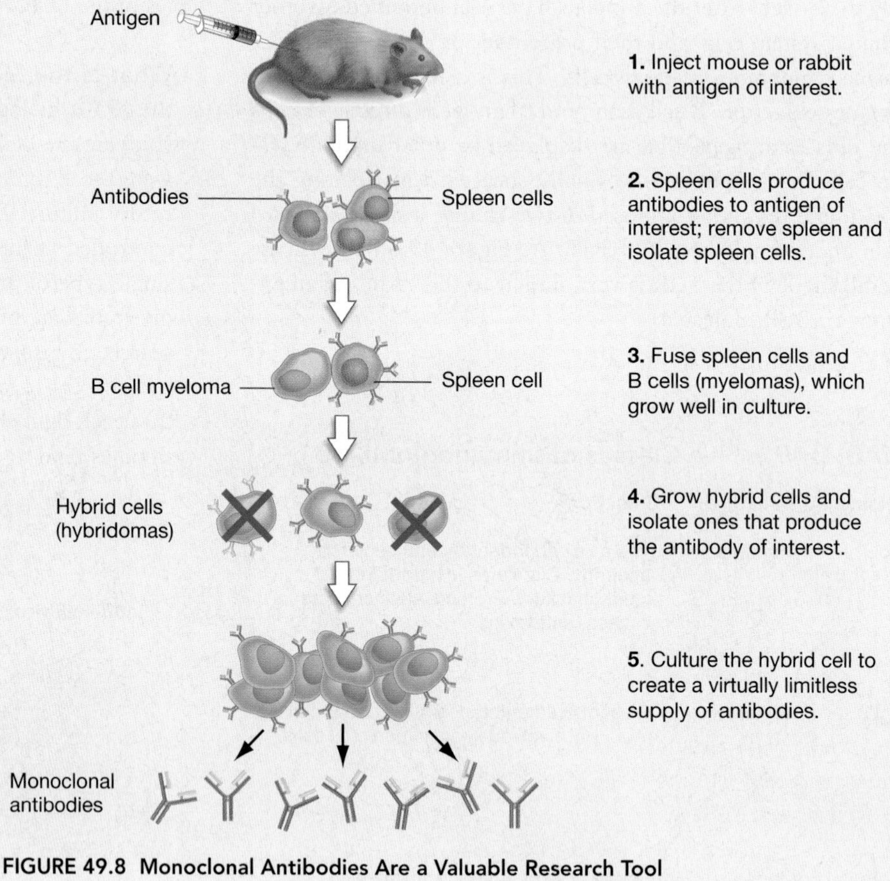
PRODUCING MONOCLONAL ANTIBODIES
Antigen
1. Inject mouse or rabbit with antigen of interest.
Antibodies
Spleen cells
2. Spleen cells produce antibodies to antigen of interest; remove spleen and isolate spleen cells.
B cell myeloma
Spleen cell
3. Fuse spleen cells and B cells (myelomas), which grow well in culture.
Hybrid cells (hybridomas)
4. Grow hybrid cells and isolate ones that produce the antibody of interest.
5. Culture the hybrid cell to create a virtually limitless supply of antibodies.
Monoclonal antibodies

FIGURE 49.8 Monoclonal Antibodies Are a Valuable Research Tool

The initial studies that explored the composition of BCRs and antibodies used enzymes to chop the light chains from different B cells into pieces. Researchers used the technique called electrophoresis to separate the pieces and noticed that light chains from different B cells appeared to have one segment in common. In addition, there seemed to be light-chain regions that were unique to each B cell. These light-chain segments came to be known as the **constant (C) regions** and **variable (V) regions**, respectively.

When techniques to determine the amino acid sequence of proteins became available, the data that resulted confirmed that both the light chain and the heavy chain consist of a variable region at the amino terminus and a constant region at the carboxy terminus. Consider, for example, the data in **Figure 49.9a.** The biologists who collected these data compared the amino acid sequences of 77 different light chains. To identify exactly which points in the chain varied the most, the researchers first counted how many of the 77 proteins had a different amino

(a) The amino acid sequences of antibodies vary.

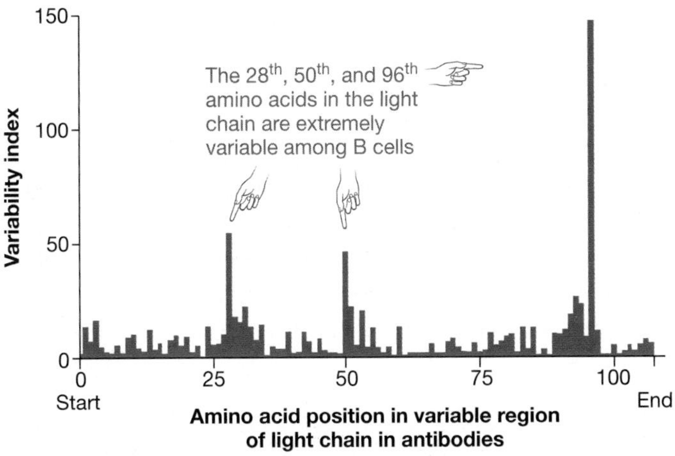

(b) B-cell receptors and T-cell receptors have variable regions and constant regions.

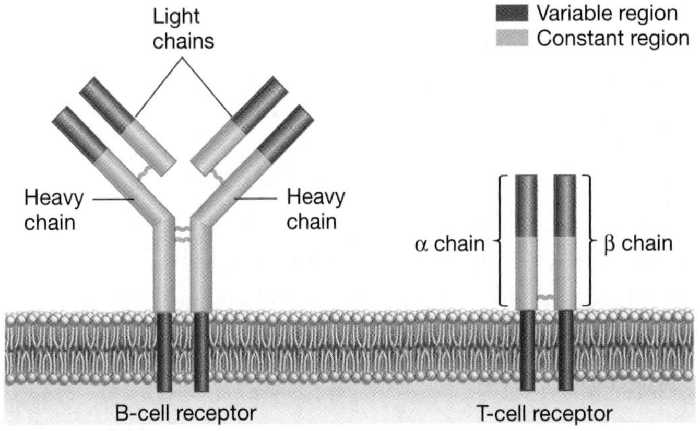

FIGURE 49.9 In B-Cell Receptors and T-Cell Receptors, the Amino Acid Sequence of the Light Chain Is Highly Variable
(a) Data indicating which amino acids in the immunoglobulin light chain are variable when 77 different antibody molecules are compared. "Variability index" is the number of antibodies that had a different amino acid at each chain position, divided by the frequency of the most common amino acid at that position.
(b) The variable regions of a B-cell receptor and T-cell receptor face away from the plasma membrane.
EXERCISE Label the transmembrane domains of the receptors shown in part (b). Circle the antigen-binding sites on both molecules.

acid at each position in the chain and then divided this number by the frequency of the most common amino acid at that position. As the resulting histogram shows, certain amino acids in the light chain are extremely variable among B cells. These portions of the light chain are within the V region. Researchers who performed the same analysis for the C region from the same group of molecules observed almost no variation. Similar patterns are found in the V and C regions of the heavy chain.

Why is this observation important? First, the presence of unique amino acids at certain positions in the light chain and heavy chain explains why the BCR and antibodies bind to unique epitopes: Receptors with different amino acid sequences bind to different epitopes. Second, the binding of epitopes occurs at the tips of the BCR molecule's and antibody's arms (**Figure 49.9b**). Analogous studies showed that the TCR also has a variable region near its amino terminus of both the α and β chains.

This observation solved one of the most fundamental questions about the acquired immune system: That system can respond to an almost limitless number of antigens because there is a virtually limitless number of different immunoglobulins and TCRs.

Analyses of the variable region raised a paradox, however. The acquired immune system can produce an almost limitless number of antibodies and receptors, but the genome of animals is not limitless. How does the genome code for so many different immune system proteins?

The Discovery of Gene Recombination In 1965 W. J. Dryer and J. Claude Bennett proposed a controversial explanation for how immunoglobulin genes code for so many different variable regions and thus so many different proteins. Dryer and Bennett suggested that a single gene codes for the C region of the light chain and that a separate gene codes for the V region. In addition, they hypothesized that, early in the life of a lymphocyte, sequences from the V gene are combined with sequences from the C gene to form a new gene for the light-chain protein. In effect, Dryer and Bennett hypothesized that, as a B cell or T cell is maturing, the two immunoglobulin genes are recombined in a way that results in the production of a sequence unique to that particular cell. The type of DNA shuffling they proposed had never been observed, however. (The introns and alternative splicing introduced in Chapter 16 had not been discovered yet.) As a result, most researchers considered the hypothesis wildly implausible.

In 1976, however, Nobumichi Hozumi and Susumu Tonegawa published data supporting the recombination hypothesis. Hozumi and Tonegawa reasoned that if light-chain diversity was produced by DNA recombination, then the variable and constant regions of mature B cells—meaning B cells that were producing BCR and antibodies—should be shorter than the same region in immature cells that are not producing antibodies. Stated another way, the V + C region should be shorter in mature cells than it is in embryonic cells.

To test this prediction, the researchers began by isolating DNA fragments from a mature mouse B-cell myeloma and from embryonic mouse tissue (step 1 of **Figure 49.10**). After isolating the DNA and using restriction endonucleases to break it up into smaller pieces, Hozumi and Tonegawa used gel electrophoresis to separate the fragments by size (steps 2–4). Then they probed the DNA fragments from each type of cell with radioactive mRNA from either the entire light chain (V and C regions) or from just the C region (step 5). (For more on this technique, see Chapter 19.)

The data are plotted in the "Results" section of Figure 49.10. Notice that two large DNA fragments from the embryonic cell hybridized with the entire light chain and that the larger of these two fragments also hybridized with the C-region mRNA. To interpret this result, the biologists proposed that the smaller fragment contained the V region and that the larger fragment included the C region. In contrast, only one small fragment from the mature B-cell DNA hybridized with the entire light chain. This fragment also hybridized with the C-region mRNA.

The researchers viewed this result as strong evidence that DNA recombination brings V and C genes closer together as B cells mature. Subsequent studies confirmed their result and showed that the heavy chain of the immunoglobulins and the α and β chain of the TCR are also produced by DNA recombination.

Hozumi and Tonegawa's paper inspired a flurry of studies on the mechanism of DNA recombination. One central finding is that

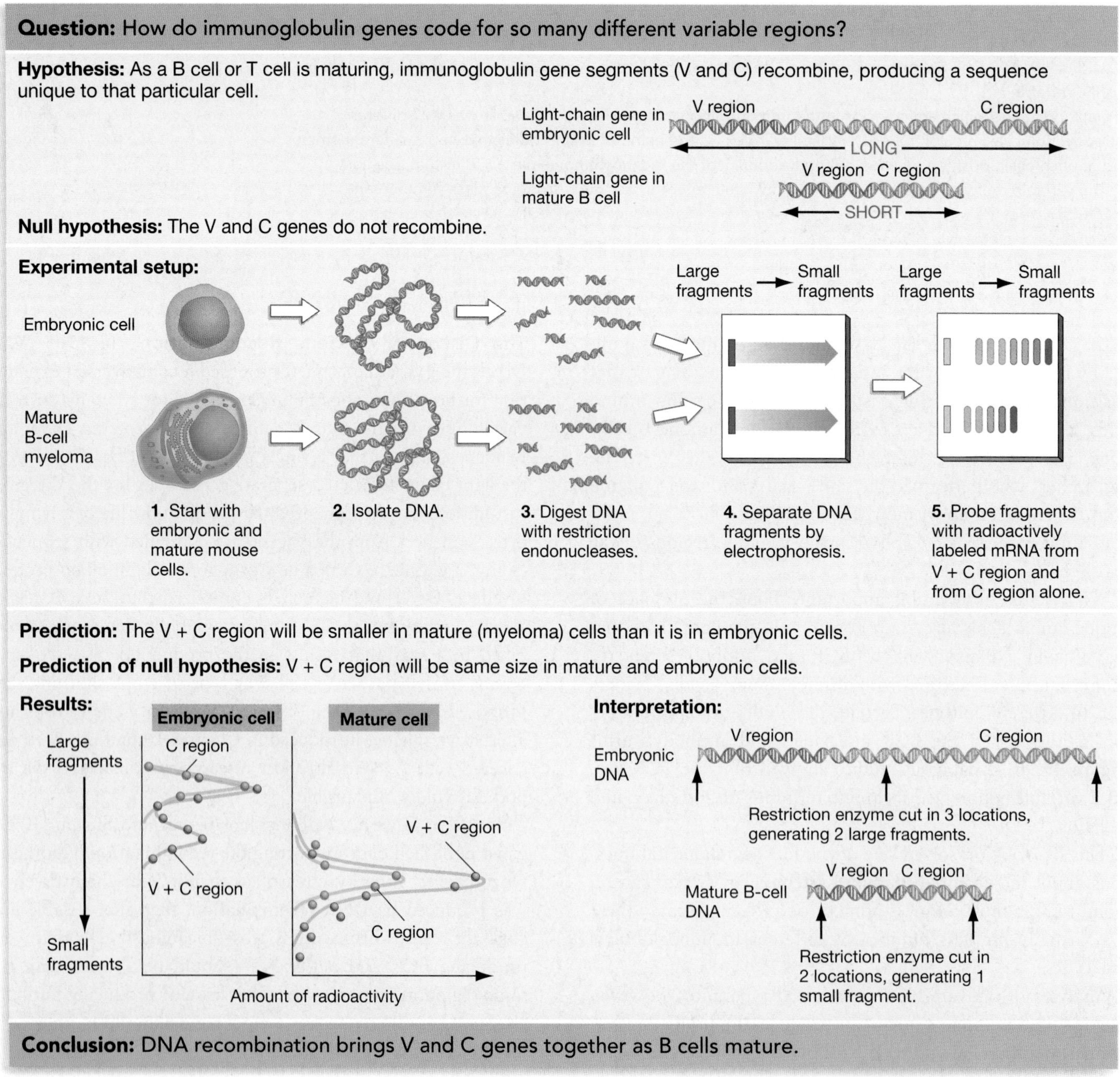
Question: How do immunoglobulin genes code for so many different variable regions?
Hypothesis: As a B cell or T cell is maturing, immunoglobulin gene segments (V and C) recombine, producing a sequence unique to that particular cell.
Light-chain gene in embryonic cell
V region
C region
LONG
Light-chain gene in mature B cell
V region C region
SHORT
Null hypothesis: The V and C genes do not recombine.
Experimental setup:
Embryonic cell
Mature B-cell myeloma
Large fragments
Small fragments
Large fragments
Small fragments
1. Start with embryonic and mature mouse cells.
2. Isolate DNA.
3. Digest DNA with restriction endonucleases.
4. Separate DNA fragments by electrophoresis.
5. Probe fragments with radioactively labeled mRNA from V + C region and from C region alone.
Prediction: The V + C region will be smaller in mature (myeloma) cells than it is in embryonic cells.
Prediction of null hypothesis: V + C region will be same size in mature and embryonic cells.
Results:
Embryonic cell
Mature cell
Large fragments
Small fragments
C region
V + C region
V + C region
C region
Amount of radioactivity
Interpretation:
Embryonic DNA
V region
C region
Restriction enzyme cut in 3 locations, generating 2 large fragments.
Mature B-cell DNA
V region C region
Restriction enzyme cut in 2 locations, generating 1 small fragment.
Conclusion: DNA recombination brings V and C genes together as B cells mature.

FIGURE 49.10 Experimental Evidence for DNA Recombination in B Cells

QUESTION If DNA recombination had not occurred, what would the data reported in part (b) look like?

mammals have two genes that code for the light chain and one for the heavy chain. Each of these genes contains distinctive segments, however. For example, one of the light-chain genes in humans includes about 40 different variable (V) segments, 5 joining (J) segments, and a single constant (C) segment. The heavy-chain locus includes about 51 V segments, 27 diversity (D) segments, and 6 joining (J) segments. A light-chain protein is made up of portions encoded by a V, a J, and a C segment; a heavy-chain protein is made up of portions encoded by a V, a D, and a J segment. The genes that encode the α and β chains of the TCR have a similar arrangement of distinct segments, each with multiple versions.

Figure 49.11 summarizes the current model for how these segments are processed as a B cell matures, using one of the immunoglobulin light-chain genes as an example (step 1). Early in lymphocyte development, the gene segments are brought closer together by DNA recombination to create a functional gene (step 2). For example, in B cells any one of the 40 V light-chain segments can recombine with any one of the 5 J segments. Thus, $40 \times 5 = 200$ different light chains can be produced by recombination. In the heavy chain, any one of the 51 V segments, 27 D segments, and 6 J segments can recombine, giving a total of $51 \times 27 \times 6 = 8262$ possible H chains. Because light-chain and heavy-chain rearrangements occur independently, DNA recombination can produce $200 \times 8262 = 1{,}652{,}400 = 1.65 \times 10^6$ different antigen-specific BCRs. And this number does not include BCRs that result from recombination in the other light-chain gene.

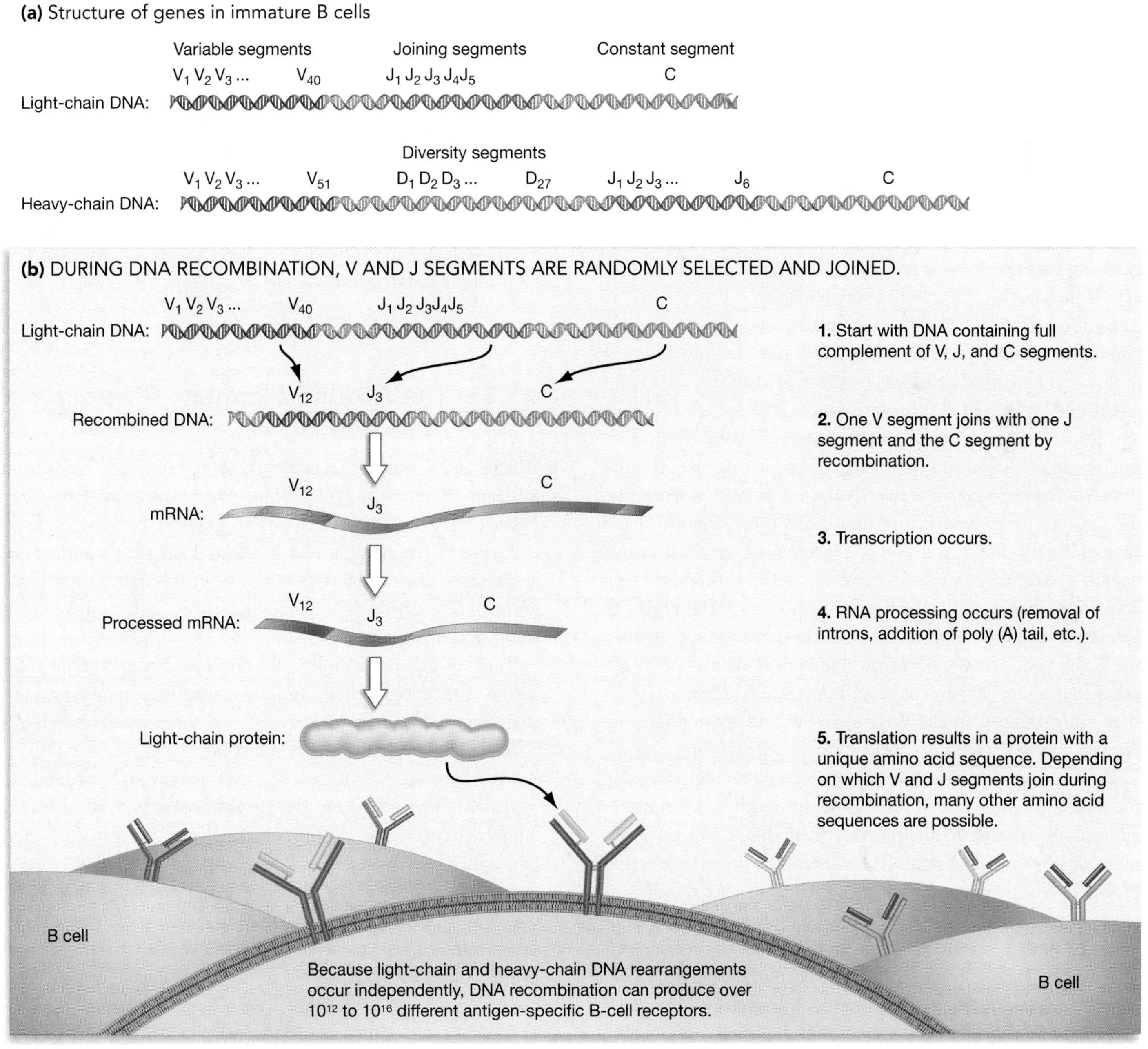
(a) Structure of genes in immature B cells
Variable segments
Joining segments
Constant segment
V1 V2 V3 ... V40
J1 J2 J3 J4 J5
C
Light-chain DNA:
Diversity segments
V1 V2 V3 ... V51
D1 D2 D3 ... D27
J1 J2 J3 ... J6
C
Heavy-chain DNA:
(b) DURING DNA RECOMBINATION, V AND J SEGMENTS ARE RANDOMLY SELECTED AND JOINED.
V1 V2 V3 ... V40
J1 J2 J3 J4 J5
C
Light-chain DNA:
1. Start with DNA containing full complement of V, J, and C segments.
V12
J3
C
Recombined DNA:
2. One V segment joins with one J segment and the C segment by recombination.
V12
J3
C
mRNA:
3. Transcription occurs.
V12
J3
C
Processed mRNA:
4. RNA processing occurs (removal of introns, addition of poly (A) tail, etc.).
Light-chain protein:
5. Translation results in a protein with a unique amino acid sequence. Depending on which V and J segments join during recombination, many other amino acid sequences are possible.
B cell
B cell
Because light-chain and heavy-chain DNA rearrangements occur independently, DNA recombination can produce over 10^12 to 10^16 different antigen-specific B-cell receptors.

FIGURE 49.11 DNA Recombination Results in Antibodies with Unique Amino Acid Sequences
Diagram summarizing current data on how light-chain proteins are produced as a B cell matures.

The functional gene is then transcribed, and the resulting mRNA is processed and translated into a light-chain protein (steps 3–5).

The story does not end here. Gene segments are not always joined precisely during DNA recombination. Some variation occurs where the V and D segments join and where the D and J segments join. This **junctional diversity** results in many additional novel sequences. Thus, by recombining small DNA segments in the two light-chain genes, an estimated 10^{12} to 10^{16} different BCRs can be created. A similar number pertains for the TCRs. These results revealed why each B cell and T cell in the body can have a different receptor on its surface. Due to gene recombination, each receptor has a unique amino acid sequence. As a result, it has the ability to bind a unique epitope on an antigen.

How Does the Immune System Distinguish Self from Nonself?

As B cells and T cells mature and begin producing receptors for specific antigens, an important problem arises: How does the immune system ensure that the receptors on these cells don't respond to molecules that are part of normal host cells? If a B-cell receptor or T-cell receptor responded to a **self** molecule—that is, a molecule belonging to the host—the receptor would trigger an immune response. Immune system cells would turn on the host and begin destroying parts of its own body. An anti-self reaction such as this is known as **autoimmunity**.

Biologists predicted that there must be some mechanism for eliminating young B cells and T cells that have receptors for self molecules. This prediction was confirmed when researchers introduced B cells and T cells with anti-self receptors into mice and found that the experimental lymphocytes were eliminated. Follow-up work showed that if B cells and T cells that are maturing in the bone marrow and thymus have anti-self receptors, the cells are destroyed before they leave these organs. In the case of T cells, it appears that the mechanism involved in this negative selection is virtually identical to the process of T-cell activation, introduced in Section 49.3. But if a T cell begins to activate in response to a self molecule presented in the thymus, the T cell is destroyed. Based on the work done to date, it appears that about 90 percent of all immature B cells and T cells that are produced in the bone marrow and thymus are destroyed before they mature, because they are anti-self.

A great deal of research is currently focused on understanding the molecular mechanisms responsible for negative selection. In addition to its intrinsic interest, this research has important implications for human health. If the process of negative selection breaks down, lymphocytes are produced that react to self proteins, and illness results. For example, multiple sclerosis (MS), introduced in Chapter 46, results from the production of anti-self T cells. The T cells attack the myelin sheath of nerve fibers. Because damage to myelin reduces the efficiency of nerve signaling, muscular and coordination problems result. Other autoimmune diseases include rheumatoid arthritis, type-1 diabetes mellitus, myasthenia gravis, and lupus (systemic lupus erythematosus).

In the vast majority of cases, however, cells in the bone marrow and thymus correctly eliminate maturing B cells and T cells that have receptors for self proteins. To summarize, each B cell and T cell has a unique receptor. These receptors are checked to make sure that they do not respond to self molecules. Now the question is, How do mature B cells and T cells interact with the antigens from invading bacteria and viruses that flood into your blood and lymph from an infected elbow or a blossoming case of the flu?

✓CHECK YOUR UNDERSTANDING

The acquired immune response is performed by lymphocytes that circulate in the lymph, blood, and organs of the immune system. The acquired response is specific because it is initiated by receptors on the surface of lymphocytes that bind to epitopes on antigens. Lymphocytes can respond to a wide array of antigens because their receptors are produced when immunoglobulin gene segments recombine. This process randomly generates unique DNA sequences prior to transcription. You should be able to (1) describe the three major claims of the clonal selection theory and (2) predict the consequences of mutations that affect one of the light-chain genes and that cause the recombination of V and J segments to fail or the joining of C segments to fail.

49.3 The Acquired Immune Response: Activation

When the cell-surface receptor of a B cell or T cell binds to an antigen, the event is like pushing the button that launches a "smart bomb"—a missile that is programmed to destroy a specific target. The binding of a B cell or T cell to an antigen unleashes a series of events that results in the destruction of the antigen. The lethal power of activated B cells and T cells is nothing short of awesome.

Perhaps because lymphocytes are so destructive, their interaction with antigens is a carefully controlled, stepwise process that eventually leads to activation of the relevant B cell and T cells. The mechanism is reminiscent of the precautions that nations with powerful missiles take to avoid accidental deployment. For the most dangerous weapons, the signal to launch is checked and cross-checked, using a series of codes and signals.

For lymphocytes, the sequence that leads to activation begins when antigens are taken up by a specific type of leukocyte or an infected cell, cut into pieces, packaged with specific cell proteins, and then transferred to the cell surface. Once antigens are presented in this way, T cells can bind to them via the TCR and begin their transformation from an inactive to an activated state. The antigen-presentation step is crucial to the activation of the acquired immune response.

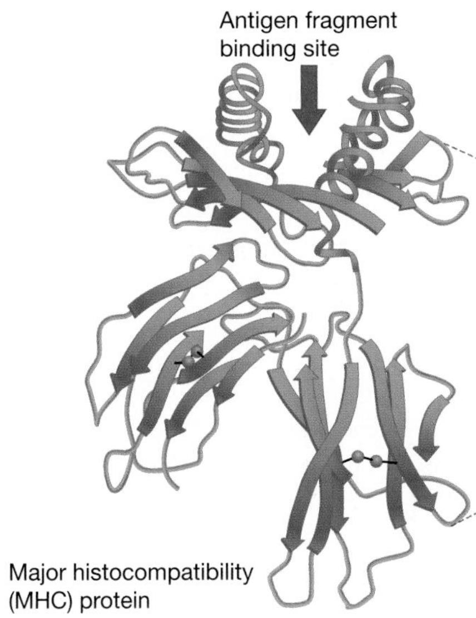

Antigen fragment
binding site
Major histocompatibility
(MHC) protein

Antigen Presentation by MHC Proteins: Activating T Cells

If bacteria begin to multiply rapidly at a wound site, leukocytes called dendritic ("tree-like") cells are recruited to the area. **Dendritic cells** ingest some of the antigens present and migrate to the nearest lymph node.

Figure 49.12 details how antigens and $CD4^+$ T cells interact in the lymph node. As Figure 49.12a shows, dendritic cells process the antigens they have ingested (step 1). Enzymes inside the cell break the antigens into small pieces (step 2). These pieces then become bound to a **major histocompatibility (MHC) protein** inside the endoplasmic reticulum (step 3). As the figure shows, MHC proteins have a groove where small antigen fragments, typically from 8 to 20 amino acids in length, bind. MHC proteins come in two types, called **Class I** and **Class II MHC proteins.** In dendritic cells, antigen fragments are attached to both Class I and Class II proteins. Once an MHC-antigen complex forms inside a dendritic cell, the molecules are transported to the cell surface via the Golgi apparatus (step 4). When the MHC protein is displayed on the cell surface, **antigen presentation** is complete (step 5).

Figure 49.12b illustrates what happens next. $CD4^+$ T cells move through the lymph node in search of antigens. T cells with receptors that are complementary to an antigen presented on a dendritic cell interact with the corresponding antigen's epitope. To a T cell, the combination of an MHC protein and a bound antigen carries the message "self cell has found antigen—response required." The T cell is activated if the following two events occur:

1. Its receptor binds to the antigen held by the MHC molecule. The receptor-antigen complex initiates a sequence of events that biologists call **signal 1.**
2. Additional receptors on the same T cell bind to additional MHC-antigen complexes on the same dendritic cell. This creates **signal 2.**

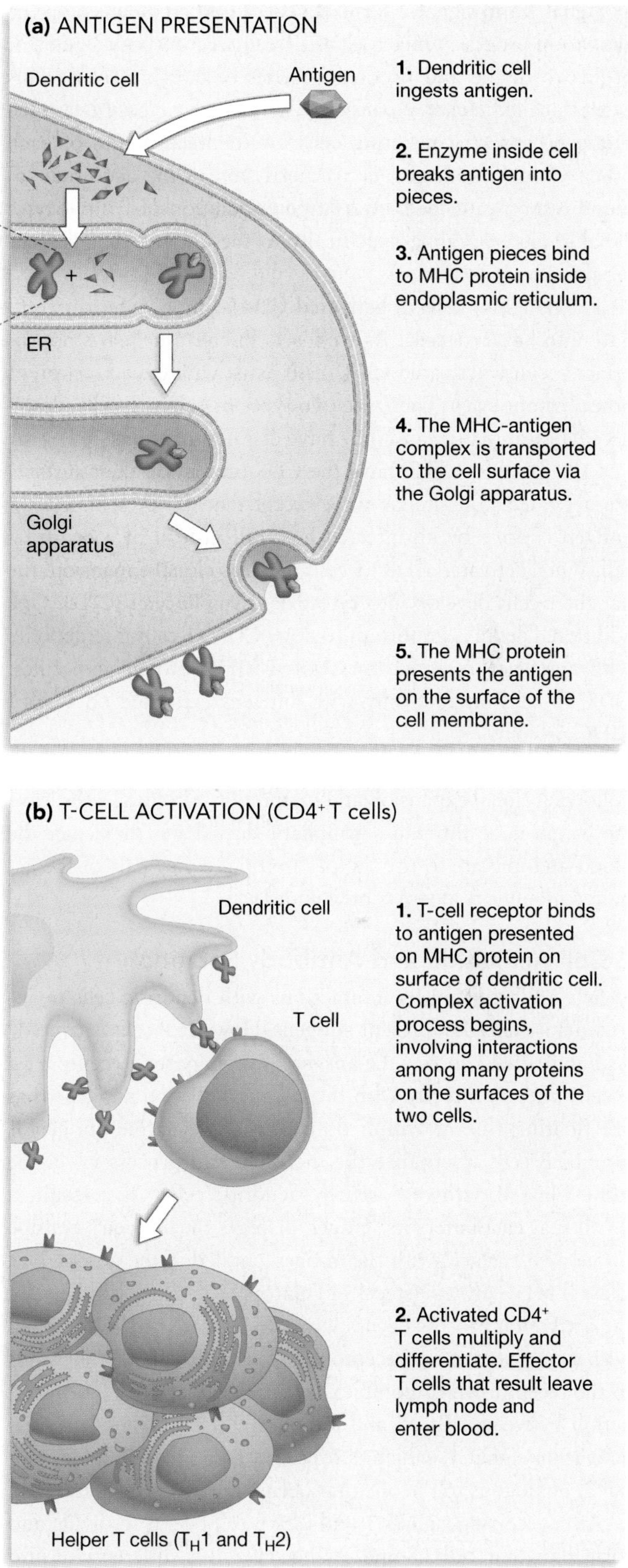

(a) ANTIGEN PRESENTATION
Dendritic cell
Antigen
1. Dendritic cell ingests antigen.
2. Enzyme inside cell breaks antigen into pieces.
3. Antigen pieces bind to MHC protein inside endoplasmic reticulum.
ER
4. The MHC-antigen complex is transported to the cell surface via the Golgi apparatus.
Golgi apparatus
5. The MHC protein presents the antigen on the surface of the cell membrane.
(b) T-CELL ACTIVATION (CD4+ T cells)
Dendritic cell
T cell
1. T-cell receptor binds to antigen presented on MHC protein on surface of dendritic cell. Complex activation process begins, involving interactions among many proteins on the surfaces of the two cells.
2. Activated CD4+ T cells multiply and differentiate. Effector T cells that result leave lymph node and enter blood.
Helper T cells (TH1 and TH2)

FIGURE 49.12 T Cells Are Activated after Interacting with Antigens Presented by Dendritic Cells
(a) Dendritic cells take in antigens, break them into pieces, and then present the fragments in the groove of an MHC protein. **(b)** If a $CD4^+$ T cell has a receptor that is complementary to the antigen presented on the surface of a dendritic cell, the T cell will begin activation.

Signal 1 and signal 2 furnish a check and cross-check on the activation process. Once a $CD4^+$ T cell receives both signals, it begins to divide and produce a series of daughter cells. This event is called **clonal expansion.** The adjective *clonal* is appropriate because the resulting cells are identical copies of each other. Clonal expansion is a crucial step in the acquired immune response. It leads to a large population of lymphocytes capable of responding specifically to the antigen that has entered the body.

The daughter cells of activated $CD4^+$ lymphocytes differentiate into **helper T cells.** As we'll see, the term *helper* is appropriate because activated $CD4^+$ cells assist with the activation of other lymphocytes. There are two types of helper T cells, designated T_H1 and T_H2, and they have distinct functions.

$CD8^+$ T cells, which have the CD8 protein on their surface, are activated in a similar way, except that the presentation of antigen is done by an infected host cell instead of a dendritic cell. When activated $CD8^+$ T cells undergo clonal expansion, the daughter cells develop into **cytotoxic T lymphocytes (CTLs).** One additional detail is important to note: $CD4^+$ T cells interact only with antigens presented on Class II MHC protein molecules; $CD8^+$ T cells interact only with antigens presented on Class I MHC proteins.

Helper T cells and CTLs are often referred to as **effector T cells.** Following clonal expansion and maturation, effector T cells leave the lymph node through a lymphatic duct. Later they enter the blood and migrate to the site of infection. A key part of the acquired immune response is now under way.

B-Cell Activation and Antibody Secretion

T cells are activated by interactions with dendritic cells or infected host cells that present antigens. How are B cells activated? As **Figure 49.13** shows, the answer has two parts. Receptors on B cells interact directly with the bacterial or viral antigens that are floating free in lymph or blood. Once a free antigen is bound, B cells internalize the molecule and process it via the same Class II pathway used by dendritic cells. As a result, a B cell that encounters its antigen displays the antigen's epitope on its own surface, with the antigen cradled in the groove of a Class II MHC protein (step 1 of Figure 49.13). The second part of the activation process occurs when an activated $CD4^+$ T_H2 cell with a complementary receptor arrives. This helper T cell binds to the receptor-MHC complex on the B cell (step 2). The interaction between a B cell and a helper T cell supplies an initial activation signal 1, which is followed by a co-stimulatory signal 2, analogous to the one that occurs during T-cell activation.

After receiving signals 1 and 2, a B cell begins to divide and form daughter cells. Some of the daughters differentiate into activated B lymphocytes called **plasma cells.** Plasma cells produce and secrete large quantities of antibodies (step 3).

Once B cells are activated, then, the acquired immune response gains an important dimension. Antibodies specific to the invading bacterium or virus begin to circulate in the blood. When an antibody encounters an antigen, it binds the antigen tightly. Antigens that are coated with antibody are marked for destruction.

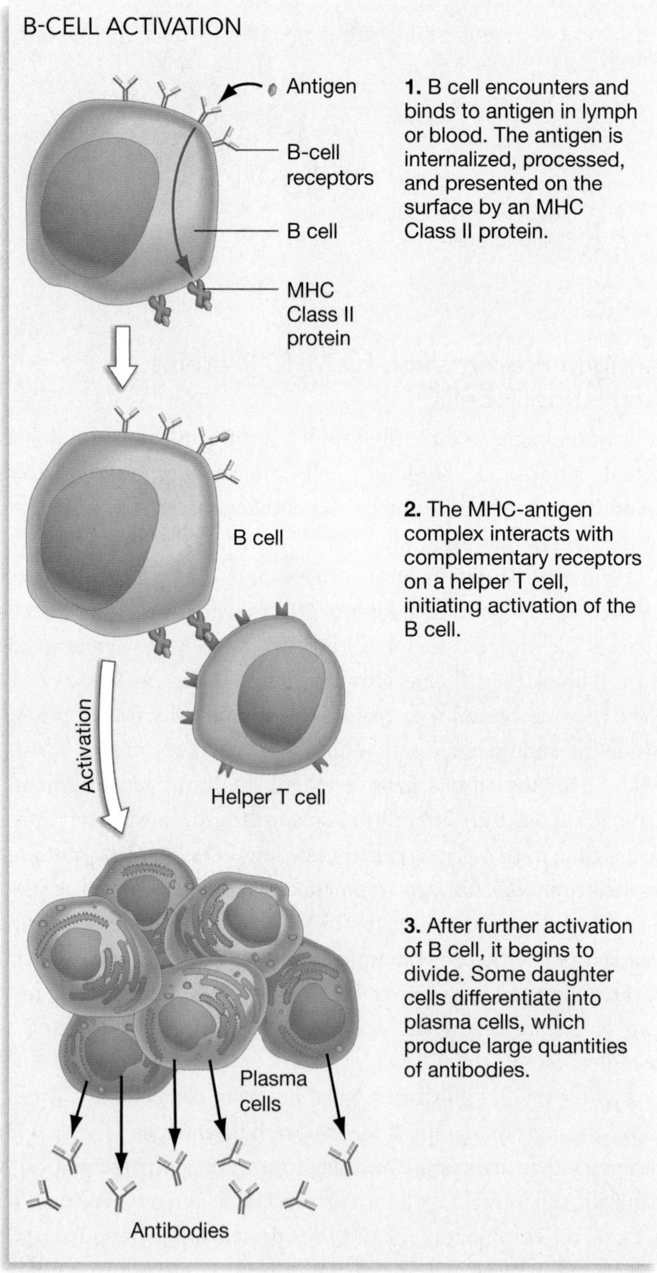

FIGURE 49.13 B Cells Are Activated after Interacting with Helper T Cells

A series of steps is involved in activating B cells. In addition to interactions between many MHC-antigen complexes on a B cell and complementary receptors on a helper T cell, other proteins on the surface of a B cell and T cell must interact. Activation eventually leads to the production of antibodies by plasma cells.

Antigen Presentation by Infected Cells: A Signal for Action by $CD8^+$ T Cells

Once T cells and B cells have been activated, most of the major elements of the acquired immune response are in place. But another fundamental event in the acquired response occurs as the receptors on $CD8^+$ T cells interact with antigens on the surface of infected host cells.

Infected cells respond to the arrival of a virus by processing antigens from the invader. MHC Class I proteins attach to the viral antigens and present the antigens on the surface of the infected cell. Every nucleated cell in the body expresses MHC Class I proteins and has the ability to signal that it is infected. Cells that display viral antigens bound to Class I MHC molecules are effectively waving a flag that says, "I'm infected. Kill me."

How are infected cells disposed of before the viruses inside replicate and spread the infection? What happens to viruses and bacteria that become coated with antibodies? And finally, what events follow a successful immune response and provide protection against future infections by the same pathogen?

✓CHECK YOUR UNDERSTANDING

T cells are activated when their receptor binds to an antigen presented by the MHC proteins on a dendritic cell or an infected cell. B cells are activated when their receptor binds to an antigen, the antigen is presented by an MHC protein on their surface, and they contact a helper T cell activated by the same antigen. Activated lymphocytes undergo dramatic changes in morphology and divide rapidly. You should be able to generate a hypothesis to explain the observation that humans who are heterozygous for the genes encoding MHC proteins tend to be healthier than individuals who are homozygous for these genes.

49.4 The Acquired Immune Response: Culmination

Activated B cells, helper T cells, and CTLs present a formidable response to invading pathogens. In combination with the leukocytes involved in the innate immune system, the cells of the acquired immune system are almost always successful in eliminating threats from bacteria, parasites, fungi, and viruses. To drive this point home, consider what happens when the acquired immune system does *not* work. For example, children who are born with severe combined immunodeficiency (SCID) have a genetic defect in the enzyme responsible for DNA recombination in maturing lymphocytes. As a result, they are unable to generate normal T-cell and B-cell receptors and lack a normal immune system (see Chapter 19). In particular, it is their acquired immune response that is badly impaired. Afflicted individuals suffer debilitating illness from infections that other children fight off easily, and they die before they are two years old. Similarly, people who are infected with the **human immunodeficiency virus (HIV)** suffer a progressive loss of $CD4^+$ T cells, because HIV parasitizes $CD4^+$ T cells and macrophages. Eventually, HIV-infected people develop **AIDS—acquired immune deficiency syndrome.** They succumb to illnesses that physicians almost never see in people with healthy immune systems.

To understand how the acquired response kills pathogens, let's again return to a bacterial infection in a wounded elbow and an upper respiratory tract infection caused by the flu virus. How do activated B cells and T cells eliminate these invaders?

Killing Bacteria

During the innate immune response to bacteria that enter a wound, macrophages at the site phagocytose some of the invaders. In addition to killing the foreign cells, these leukocytes process and present antigens via the MHC Class II pathway. As a result, macrophages at the site of infection display epitopes on their surfaces that can be recognized by helper T cells. If an activated T_H1 cell binds to these antigen-laden macrophages, two things happen: First, the phagocytic activity of the macrophages is enhanced. Second, the T_H1 cells secrete cytokines that kill bacteria and viruses, recruit additional phagocytic cells to the site, and increase the inflammatory response.

Bacteria at the infection site also begin to be coated with antibodies from plasma cells. In many cases, antibody action causes **agglutination**, or clumping of bacteria. Each antibody has at least two binding sites, so a single antibody can bind epitopes on two bacterial cells and link them, forming a clump. Bacteria or clumps of bacteria that are tagged with antibodies are readily destroyed by macrophages, either by phagocytosis or by a stream of reactive oxygen intermediates released onto their cell walls. Antibodies that are bound to antigens also stimulate a group of proteins called the **complement system.** Complement proteins circulate in the bloodstream and assemble at antigen-antibody complexes. When complement proteins activate, they punch lethal holes in the plasma membranes of bacteria. Within a few days, this combination of killing mechanisms usually eliminates all invading bacteria.

Destroying Viruses

Just as the immune system has an array of mechanisms to dispose of bacteria, it has two major ways to eliminate viruses. One route involves CTLs (activated $CD8^+$ cells) and is called the **cell-mediated response**; the other route involves antibodies and is called the **humoral response.** (The Latin root *humor* means "fluid." The humoral response takes place in blood and lymph; the cell-mediated response takes place at the surface of cells.) Let's take a closer look at both.

Cell-Mediated Immunity To understand how CTLs act, recall from Section 49.3 that influenza-infected cells present viral antigens on their cell surface in conjunction with MHC Class I proteins. As $CD8^+$ cells migrate into the area, they recognize and bind to the antigen's epitopes. Once binding occurs, a circle of molecules forms that holds the CTL tightly to the infected cell. A hole in the infected cell's plasma membrane forms inside this seal, and signaling molecules from the CTL enter the cell and activate a self-destruct response (**Figure 49.14a**). The CTL then releases and seeks out another infected cell to kill. Over time, all virus-infected cells are eliminated. In this way, the cell-mediated response limits the spread of the infection by preventing new generations of virus particles from maturing.

Humoral Immunity Even as CTLs swing into action, plasma cells are producing antibodies to viral proteins. In most cases, the most effective antibodies are those that bind to epitopes on the surface of the virus. As **Figure 49.14b** shows, two things happen when antibodies bind to the outside of a virus: First, the virus is blocked from making contact with plasma membranes and entering new host cells. Second, macrophages recognize the antibody-coated particles and phagocytose them. Because of this humoral response, the number of virus particles floating free in lymph and blood is eventually reduced to zero.

Why Does the Immune System Reject Foreign Tissues and Organs?

Antibodies and CTLs have devastating effects on invading bacteria and viruses. Unfortunately, they are equally deadly in response to tissues or organs that are introduced into a patient in the hopes of healing a wound or disease. To understand why, consider the problems that can arise with blood transfusions. Certain individuals have red blood cells with membrane proteins called A and B. These proteins act as antigens if they are introduced into a person whose own blood cells lack those proteins. For example, if you have type A blood, it means that your blood cells have the A antigen. If your blood is introduced to a person who lacks the A antigen—meaning a person who has type B or type O blood—the recipient's immune system will recognize the A antigen as foreign and mount a devastating response against it. For a blood transfusion to be successful, a recipient has to receive blood that lacks the A and B proteins entirely or that contains the same antigens found in his or her own blood.

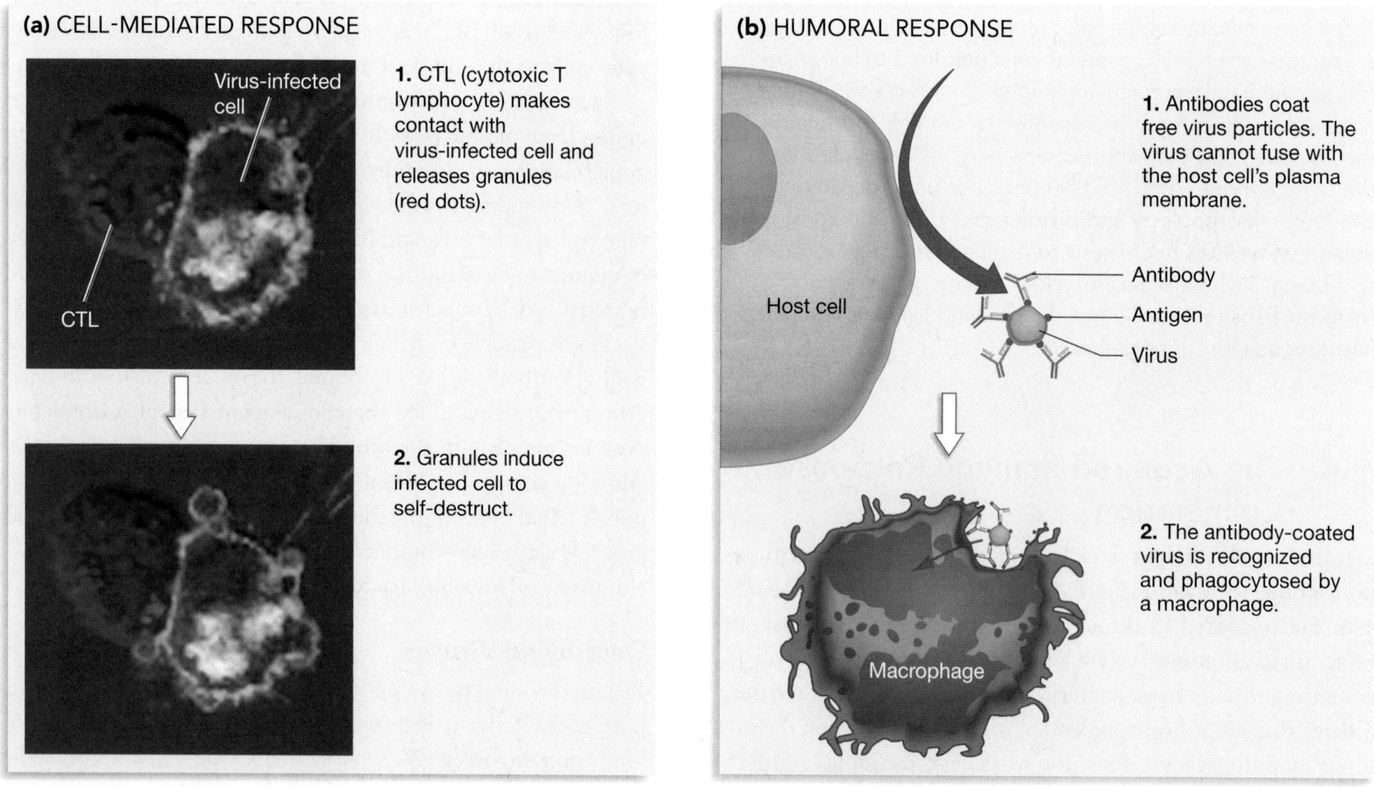
(a) CELL-MEDIATED RESPONSE
Virus-infected cell
CTL
1. CTL (cytotoxic T lymphocyte) makes contact with virus-infected cell and releases granules (red dots).
2. Granules induce infected cell to self-destruct.
(b) HUMORAL RESPONSE
1. Antibodies coat free virus particles. The virus cannot fuse with the host cell's plasma membrane.
Host cell
Antibody
Antigen
Virus
2. The antibody-coated virus is recognized and phagocytosed by a macrophage.
Macrophage

FIGURE 49.14 Lymphocytes Kill Viruses via the Cell-Mediated Response and the Humoral Response
(a) Micrographs showing CTLs killing a virus-infected cell. The red-stained granules are loaded with molecules that activate a self-destruct program in the infected cell. **(b)** After antibodies bind to intact virus particles, the particles are unable to attach to cells and begin an infection. Antibody-coated virus particles are readily phagocytosed by macrophages.

TABLE 49.2 A Summary of the Acquired Immune System

Type of Lymphocyte	Method of Activation	Cells That Result from Activation and Clonal Expansion	Function of Resulting Cells
B cell	Receptor binds to free antigen, then interacts with T_H2 cell.	Plasma cells	Secrete antibodies
$CD4^+$ T cell	Receptor binds to antigen-MHC Class II protein complex on dendritic cell.	T_H1 (helper T cell)	Activates CTLs, regulates inflammatory response
		T_H2 (helper T cell)	Activates B cells
		Memory cells	Participate in secondary response
$CD8^+$ T cell	Receptor binds to antigen-MHC Class I protein complex on infected cell. May also interact with T_H1 cell.	Cytotoxic T lymphocytes (CTLs)	Kill infected host cells
		Memory cells	Participate in secondary response

Similar problems arise in organ transplants, except that the antigenic molecules in foreign organs are the MHC proteins found on the surfaces of their cells. To prevent strong immune reactions to a transplanted kidney, heart, or liver, physicians do two things: (1) obtain the organ to be transplanted from a sibling or other donor whose MHC proteins are extremely similar in structure to those of the recipient and (2) treat the recipient with drugs that suppress the immune response. Thanks to steady improvements in drug development and in systems for matching MHC types between donors and recipients, the success rate for organ transplants has improved dramatically in recent years.

As the blood transfusion and organ transplant examples show, the immune system rejects foreign tissues because they contain non-self proteins. To your T cells and B cells, a blood transfusion or an organ transplant is indistinguishable from a massive influx of bacteria, viruses, or other foreign invaders.

Responding to Future Infections: Immunological Memory

When B cells and T cells are activated and undergo clonal expansion, their daughter cells differentiate into plasma cells, helper T cells, and cytotoxic T lymphocytes. **Table 49.2** summarizes how activation occurs and what the resulting cells do. But in addition to producing the cells that implement the humoral and cell-mediated responses, activated B cells and T cells produce specialized daughter cells called memory cells. **Memory cells** do not participate in the initial acquired response, or **primary immune response**. Instead, they provide a surveillance service after the original infection has been cleared. Memory cells circulate through blood and tissue for years or decades, ready to provide an extremely rapid response should an infection with the same antigen recur. The production of memory cells is a hallmark of the vertebrate immune response. It occurs to a limited degree, if at all, in invertebrates.

If the same antigen enters the body a second time, memory cells bind the antigen and trigger a second acquired response, or **secondary immune response**. The launching of a secondary immune response by means of memory cells is known as **immunological memory**. The secondary response is faster and more efficient than the primary response. It is faster because circulating memory T and B cells increase the likelihood that lymphocytes with the correct antigen-specific receptors will find the antigen and activate quickly. It is more efficient because some of the memory B cells that respond migrate to a specialized area in the lymph node called the *germinal center*. There the DNA sequences that code for the variable region of the immunoglobulin gene begin to undergo rapid mutation. As **Figure 49.15** shows, the mutations in the DNA of the variable region modify the receptors and antibodies produced by the memory cell. Memory B cells with receptors that end up

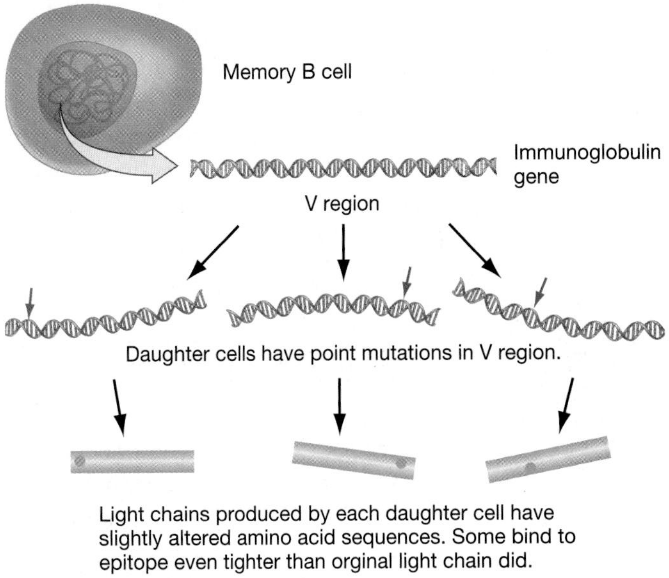
Memory B cell
Immunoglobulin gene
V region
Daughter cells have point mutations in V region.
Light chains produced by each daughter cell have slightly altered amino acid sequences. Some bind to epitope even tighter than orginal light chain did.

FIGURE 49.15 Somatic Hypermutation Results in Antibodies with Higher Affinity for the Antigen
Somatic hypermutation occurs as memory B cells divide in response to an influx of antigen.

binding best to the antigen's epitope live and produce daughter cells; those that bind to antigens less well die.

In effect, this process of **somatic hypermutation** leads to a fine-tuning of the immune response. The B cells that result from somatic hypermutation produce antibodies that bind to the antigen more tightly than their ancestor cells did during the primary immune response. As the secondary immune response proceeds and somatic hypermutation continues, better-fitting antibodies are produced.

Figure 49.16 underlines the effectiveness of the secondary immune response by comparing the rate of antibody production during the first and second exposures to a virus. The results are similar to data that researchers collect by inoculating a laboratory mouse with influenza virus, collecting blood from the mouse every other day, and then measuring the amount of antiviral antibodies present in the fluid phase of the blood. (**Box 49.2** explains how researchers measure antibody concentrations in blood.)

The data in Figure 49.16 also explain why vaccination is an effective defense against certain viruses. A **vaccine** contains epitopes from a pathogen or a killed or weakened version of the pathogen itself. After vaccination (inoculation with a vaccine), the body mounts a primary immune response that results in the production of memory cells. If an actual infection occurs later, these memory cells respond quickly and eliminate the threat before illness occurs. For example, Edward Jenner's vaccination strategy worked because the antigens presented by cowpox are extremely similar to the antigens presented by smallpox. As a result, exposing people to cowpox virus elicited the production of memory cells that effectively thwarted future infections by smallpox virus.

Unfortunately, viruses such as influenza, HIV, and the rhinoviruses that cause colds mutate so rapidly that they present the immune system with a constantly changing array of epitopes. Memory cells that were effective during a previous infection by these viruses are unlikely to respond to a later infection. As a result, it is extremely difficult for biologists to design an effective vaccine against these agents. Currently, the only "cure" for HIV infection is prevention; the most effective treatment for the common cold is rest and nourishment. To keep up with rapidly mutating flu populations, flu vaccines must be redesigned and administered every year.

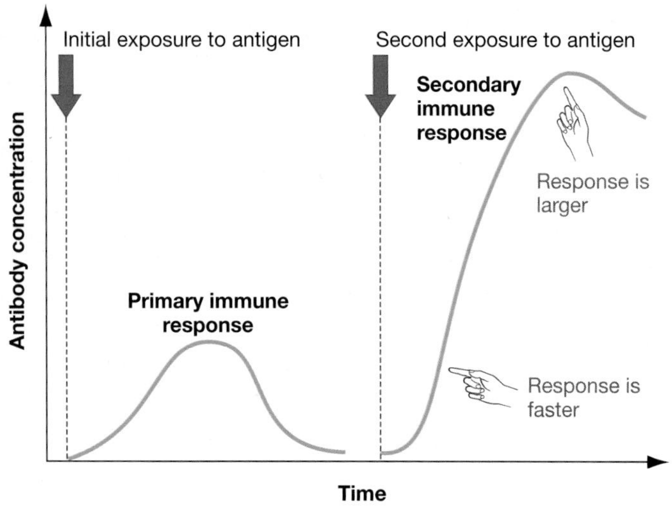

FIGURE 49.16 The Secondary Immune Response Is Faster and Stronger than the Primary Response

Data showing the types of results that biologists get when they inject a mouse with an antigen, document changes in antibody concentration over time, and then inject the same individual with the same antigen later. QUESTION In light of these data, why is vaccination effective?

✓CHECK YOUR UNDERSTANDING

The antibodies produced by activated B cells mark invading pathogens for destruction. Activated CTLs recognize epitopes displayed by infected cells and kill them before the pathogens inside can replicate. Memory cells produced during B-cell and T-cell activation remain in the body and provide secondary immunity against future infections. You should be able to (1) explain how vaccination works and (2) propose a strategy for developing an HIV vaccine.

BOX 49.2 The ELISA Test

One of the common strategies used to measure antibody concentrations in blood is called an indirect or sandwich **ELISA** (enzyme-linked immunosorbent assay). The standard screening test for HIV infection, for example, is an ELISA for antibodies that the immune system produces in response to this virus. **Figure 49.17** illustrates how an ELISA might be used to determine the amount of antibody present in a mouse that has been infected with influenza.

To begin an ELISA test, each well in a cell-culture plate containing many wells is coated with antigen (step 1 of Figure 49.17). In our example, the antigen is purified hemagglutinin protein from the influenza virus. Next, a defined amount of plasma from the patient is added to the plate. (Recall from Chapter 44 that plasma is the extracellular matrix—the cell-free fraction of the blood.) If antibodies are present, they bind to the antigen (step 2). To continue the assay, a secondary antibody is added to the mixture of antigen and plasma (step 3). A **secondary antibody** attaches to another antibody. In ELISA tests, the secondary antibody contains an enzyme such as alkaline phosphatase (shown in the figure as AP). Alkaline phosphatase is useful because it catalyzes a reaction with a substrate and because the product of the reaction changes the color of the solution. As a result, the presence of color in a well indicates a positive reaction—meaning the antibody to hemagglutinin is present (step 4). A positive result confirms that the patient is infected.

To determine the concentration of antibody, researchers make a series of dilutions of the plasma. The most diluted solution that gives a positive response is then recorded as the antibody titer, meaning the relative amount of antibody produced.

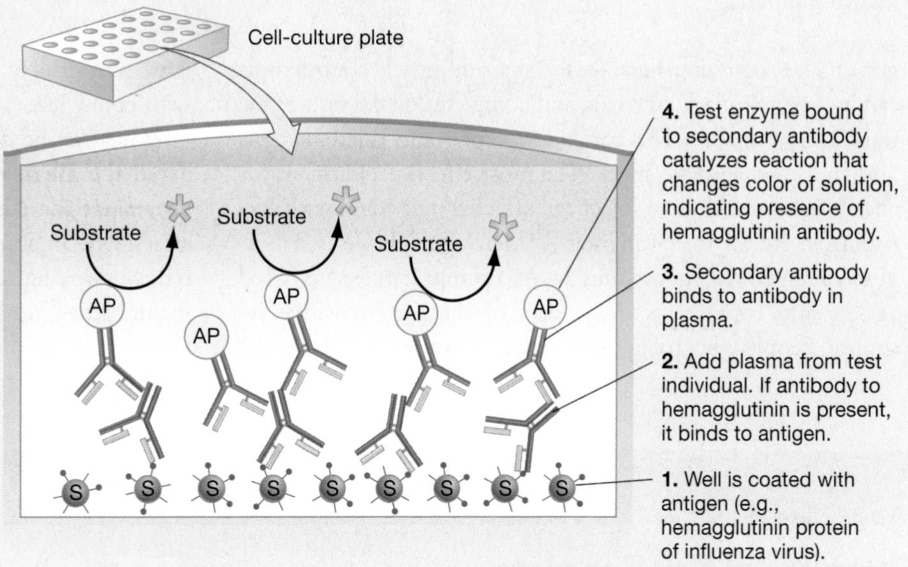
Cell-culture plate
Substrate
Substrate
Substrate
AP
S
4. Test enzyme bound to secondary antibody catalyzes reaction that changes color of solution, indicating presence of hemagglutinin antibody.
3. Secondary antibody binds to antibody in plasma.
2. Add plasma from test individual. If antibody to hemagglutinin is present, it binds to antigen.
1. Well is coated with antigen (e.g., hemagglutinin protein of influenza virus).

FIGURE 49.17 The ELISA Test Measures Antibody Concentrations in Blood
The antigen, antibody, and secondary antibody-test enzyme complex used in an ELISA test are added sequentially. The sample is washed after the addition of mouse antibody to remove any protein that is not bound to the antigen. After the secondary antibody and test enzyme are added, the sample is washed again to remove any free-floating secondary antibody. Then the substrate is added and the resulting color observed. The amount of color produced is an index of the amount of antibody present.

ESSAY Allergies

For some people, walking through a field of ragweed or petting a cat is a prescription for developing a runny nose, itchy eyes, and labored breathing. For other people, being stung by a bee or eating a peanut is a life-threatening experience. Why are some people allergic to certain molecules?

An **allergy**, or allergic reaction, is an abnormal response to an antigen. What happens during an allergic reaction? For reasons that are still not clear, certain people produce the IgE class of antibodies in response to specific molecules found in cat dander, nuts, plant pollen, or other products. Molecules that trigger this response are called **allergens** instead of antigens. In people who have no allergies, IgE antibodies are produced only in response to infections by worms.

The presence of IgE antibodies is important, because it triggers a series of events known as the **hypersensitive reaction**. When a person is first exposed to an allergen, receptors on leukocytes called mast cells and other **basophils** bind to the IgE antibodies that are produced. Once this binding occurs, the cells (and the person) are said to be sensitized. If the person is later exposed to the same allergen, the sensitized cells rapidly produce large quantities of histamine, cytokines, chemokines, and other compounds. In response to these molecules, blood vessels dilate and become more permeable, smooth-muscle cells contract, and mucus-producing cells secrete.

What happens during an allergic reaction?

If the immune response is relatively mild, symptoms such as a runny nose, watery eyes, and mild wheezing develop. The syndrome known as *hay fever*, for example, is a common allergic response to the allergens in grass or ragweed or birch-tree pollen. In other cases, tissues may respond to an allergen by reddening, swelling, and itching. (These symptoms are called *hives*.) If the smooth-muscle constrictions triggered by the hypersensitive response are localized to the respiratory passages and lead to airway constriction, the condition **asthma** develops.

If the immune response is more severe, however, blood vessels can dilate enough that blood pressure plummets and the

(Continued on next page)

(Essay continued)

person loses consciousness. Severe smooth-muscle contractions can induce vomiting, diarrhea, and complete constriction of the respiratory passages. This combination of events is lethal and is known as **anaphylactic shock**. The most effective treatment for anaphylactic shock is to inject the affected individual with epinephrine. Recall from Chapter 47 that epinephrine is a hormone that relaxes smooth muscle. As a result, it opens up blood vessels and counteracts the effects of the hypersensitive response. People who are hypersensitive to bee stings or other allergens routinely carry small syringes, called Epipens, filled with epinephrine.

What can be done about allergies? The best approach is to avoid the allergen. If exposure does occur, drugs called **antihistamines** sometimes reduce histamine release from mast cells effectively. In some cases, corticosteroids applied to skin can reduce swelling. Longer-term solutions are still in the experimental stage, however. Developing effective treatments for allergies is an important research frontier in biomedicine.

CHAPTER REVIEW

Summary of Key Concepts

- **The innate immune response to infection is mounted by cells called leukocytes, which circulate in the blood and respond in the same way to all pathogens. The acquired immune response is mounted by cells called lymphocytes, which circulate in the lymphatic system and respond to only a specific pathogen.**

  The immune system employs two major responses to invading pathogens: an innate immune response and an acquired immune response. During the innate immune system's inflammatory response, mast cells release chemical messengers that result in increased blood flow to wounds or areas of tissue damage. After neutrophils migrate to the site, they respond to bacteria that stimulate their pattern-recognition receptors by releasing toxic molecules and phagocytosing the invading cells. Macrophages also phagocytose pathogens and release chemical messengers that raise body temperature.

  Web Tutorial 49.1 The Inflammatory Response

- **The acquired immune response begins when cell-surface proteins on lymphocytes bind to a particular location on a foreign protein—usually one from a disease-causing agent.**

  Lymphocytes are activated when T-cell receptors and B-cell receptors recognize an epitope on an antigen. The receptor protein on the surface of a T cell or B cell is generated through DNA recombination, a rearrangement of gene segments. Because every receptor that results from this process is slightly different, the immune system is able to recognize and respond to an almost limitless array of antigens.

- **The humoral response occurs when pathogens are tagged for destruction with antibodies that are produced by lymphocytes called B cells.**

  During the humoral response, activated B cells produce antibodies to specific epitopes on the pathogen that has invaded the body. Viruses that become coated with antibodies are unable to enter host cells and are destroyed by macrophages. Bacteria that are tagged with antibodies are destroyed by complement proteins or by macrophages.

- **The cell-mediated response occurs when infected cells are destroyed by lymphocytes called T cells.**

  During the cell-mediated response, host cells that are infected with a virus display antigens on their surface and are destroyed by cytotoxic T lymphocytes (CTLs). As a result, the virus particles inside are destroyed and cannot contribute further to the infection.

  Because memory cells are created as activated B cells and T cells undergo clonal expansion, the immune system is able to respond rapidly and effectively to future infections by the same antigen. A vaccine triggers the production of memory cells because it contains epitopes from antigens.

  Web Tutorial 49.2 The Acquired Immune Response

Questions

Content Review

1. *Innate* means "inborn" or "untrained." What contrast with the acquired immune system inspired the name *innate immune system*?
   a. It is present in newborns and occurs throughout life.
   b. It depends primarily on barriers to pathogen entry, such as skin and mucous linings in tissues.
   c. It is not based on "trained" changes in immune system cells; there is no change through time.
   d. Leukocytes mount the same response to any invading pathogen (the innate response is not specific).

2. Which of the following events is *not* involved in the inflammatory response?
   a. CTLs kill infected host cells.
   b. Neutrophils phagocytose pathogens that stimulate their pattern-specific receptors.
   c. Mast cells secrete chemical messengers that lead to increased blood flow.
   d. Macrophages release chemical messengers that lead to increased body temperature.

3. What is the difference between an epitope and an antigen?
   a. An epitope is any foreign substance; an antigen is a foreign protein.
   b. An epitope is the part of an antigen where an antibody or lymphocyte receptor binds.
   c. An antigen is the part of an epitope where an antibody or lymphocyte receptor binds.
   d. Antigens are recognized by B cells and antibodies; epitopes are recognized by T cells.
4. How do B-cell receptors and antibodies differ?
   a. B-cell receptors are made up of heavy chains; antibodies are made up of light chains.
   b. B-cell receptors are made up of light chains; antibodies are made up of heavy chains.
   c. Only antibodies include a variable region.
   d. Antibodies lack a transmembrane domain and are secreted.
5. How do memory cells become activated?
   a. They undergo somatic hypermutation.
   b. They encounter an antigen.
   c. They undergo clonal expansion.
   d. They are stimulated by cytokines and chemokines.
6. In terms of function, T cells come in two major types. How are they distinguished?
   a. They have a CD4 or CD8 protein on their surfaces.
   b. They have extensive ER and many mitochondria.
   c. Only one type has pattern-recognition receptors.
   d. One type functions in the primary response; the other in the secondary response.

Conceptual Review

1. To a physician, the classical signs of the inflammatory response are ruber (reddening), calor (heat), dolor (pain), and tumor (swelling). Explain why each symptom occurs. Mention at least two of the leukocytes and two of the chemical messengers that are involved in the inflammatory response.
2. Sketch the general structure of a B-cell receptor. Explain how this protein interacts with an antigen.
3. Why does vaccination work? Why don't we have vaccines for HIV or cold viruses? Why do people need to get a new flu shot every year?
4. Summarize the clonal-selection theory of acquired immune system function. How was the theory tested and verified?
5. Explain why helper T cells and cytotoxic T cells are aptly named.
6. Explain how DNA recombination leads to the production of almost limitless numbers of different B-cell receptors, T-cell receptors, and antibodies.

Group Discussion Problems

1. If you were being treated for a badly skinned knee from a bicycle accident, health care workers would irrigate the wound with large volumes of sterile water and scrub it as thoroughly as possible with warm, soapy water. Explain why these measures are effective, in the context of how the innate and acquired immune responses operate.
2. The signals that are required to activate B cells and T cells are exceedingly complex. Some researchers maintain that the requirement for multiple signals precludes lymphocytes from activating in response to antigens on the surface of self cells. For example, suppose that both a B cell and a T cell in your body have a receptor that recognizes a protein found in your heart and that these cells are mistakenly released from the bone marrow and thymus. Summarize the signals that are needed to activate the B cell and T cell. Explain why the lymphocytes are likely or unlikely to attack your heart.
3. It seems astonishing that the immune system can produce antibodies to compounds that were recently synthesized for the first time in the lab. But because viruses and other pathogens are constantly mutating and producing new antigens over time, this observation should not be surprising at all. Explain.
4. During World War II, physicians discovered that if they removed undamaged skin from a patient and grafted it onto the site of a burn on the same patient, the tissue healed well. But if the grafted tissue came from a different individual, the skin graft was rejected. Explain why nonself tissues are killed by the immune system.

Answers to Multiple-Choice Questions 1. c; 2. a; 3. b; 4. d; 5. b; 6. a

www.prenhall.com/freeman is your resource for the following: Web Tutorials; Online Quizzes and other Online Study Guide materials; Answers to Conceptual Review Questions; Solutions to Group Discussion Problems; Answers to Figure Caption Questions and Exercises; and Additional Readings and Research.

UNIT 9

Ecology

Bald eagles make their living by preying on fish and small mammals and by scavenging carcasses. Although bald eagles were once in danger of going extinct, most populations have made a strong recovery.

50 An Introduction to Ecology

KEY CONCEPTS

- Ecology focuses on how organisms interact with their environment. Because its goal is to understand the distribution and abundance of organisms, ecology provides a scientific foundation for the conservation of species and natural areas.
- An organism's environment consists of both abiotic, or physical, factors and biotic factors, meaning other organisms.
- On land, environments are defined by climate—specifically, both the average value and annual variation in temperature and in moisture. In aquatic habitats, environments are defined by physical structure—particularly water depth.
- A species' distribution is constrained by a combination of historical, abiotic, and biotic factors.

In addition to providing habitat for a frog, the puddle of water inside this plant provides an environment for an array of microscopic species. The plant itself is growing in the branch of a rain forest tree and is just one of hundreds of plant and insect species that live in association with the tree. Ecology explores how all of these organisms interact with each other and with their environment.

Environments are currently changing faster than they have changed in the past 3.5 billion years—except for the episode 65 million years ago when a mountain-sized asteroid struck Earth (see Chapter 26). As human impacts on the environment have accelerated, ecology has become an increasingly important field in biological science. **Ecology** is the study of how organisms interact with their environment. Efforts to maintain human health and welfare depend on our ability to understand and predict the consequences of the rapid environmental changes that are occurring now.

One of ecology's central goals is to understand the distribution and abundance of organisms. In many cases an ecologist's job is to identify factors that dictate why certain species live where they do and how many individuals live there. Some biologists ask why orangutans are restricted to forests in Borneo and why their numbers are declining rapidly. Other biologists create mathematical models to predict how quickly the human immunodeficiency virus will increase in India and how long it will take for the current human population of 6.35 billion to double. Ecologists seek to recognize and explain patterns in nature—patterns that involve where particular organisms live and in what numbers.

This chapter has two goals: to explore how biologists go about studying ecology and to introduce the major types of environments that organisms occupy. It serves as a springboard to subsequent chapters, which analyze types of interactions between organisms and their envorinment and ask how environmental changes affect humans and other species. In terms of understanding the biological problems that humans are facing now, this may be the most important unit in the book.

50.1 Areas of Ecological Study

To understand why organisms live where they do and in what numbers, biologists break ecology into several levels of analysis. This is a common strategy in biological science. For example, cell biologists study how cells work at increasingly complex levels of organization—from individual molecules to multimolecular machines, organelles, whole cells, and cell-to-cell interactions. Physiologists may analyze processes at the level of ions and molecules as well as what is happening to whole cells, tissues and organs, or a complete system. In ecology, researchers work at four main levels: (1) organisms, (2) populations, (3) communities, and (4) ecosystems. Let's examine each.

Organismal Ecology

At the finest level of organization, ecology focuses on how individuals interact with their environment. Researchers who study organismal ecology explore the morphological, physiological, and behavioral adaptations that allow individuals to live successfully in a particular area. The study of behavior is an important aspect of organismal ecology; it focuses on how organisms respond to particular stimuli from their environment (see Chapter 51). The environmental stimuli that trigger behavior may consist of changes in temperature or moisture, an escape response from prey, or a rival challenging for a mate. Organismal ecology also focuses on the physiological adaptations that allow individuals to thrive in heat, drought, cold, or other demanding physical conditions.

As an example, consider the sockeye salmon (**Figure 50.1a**). After spending four or five years feeding and growing in the ocean, these salmon travel hundreds or thousands of kilometers to return to the stream where they hatched. Females create nests in the gravel stream bottom and lay eggs. Nearby males compete for the chance to fertilize eggs as they are laid. When breeding is finished, all of the adults die.

At the level of organismal ecology, biologists want to know how these individuals interact with their physical surroundings and with other organisms in and around the stream. Which females get the best nesting sites and lay the most eggs, and which males are most successful in fertilizing eggs? How do individuals cope with the transition from living in salt water to living in freshwater?

Population Ecology

A **population** is a group of individuals of the same species that lives in the same area at the same time. When biologists study population ecology, they focus on how the numbers of individuals in a population change over time. Some of the theory that population ecologists use to analyze and predict changes in population size is now being used to evaluate the fate of endangered species. For example, mathematical models of population growth have been used to predict the future of particular salmon populations (**Figure 50.1b**). This work is important. Many

(a) Organismal ecology

How do individuals interact with each other and their physical environment?

Male salmon fight over females during the breeding season

(b) Population biology

How and why does population size change over time?

Each female salmon produces thousands of eggs. Only a few will survive to adulthood. On average, only two will return to the stream of their birth to breed

(c) Community ecology

How do species interact, and what are the consequences?

Salmon are prey as well as predators

(d) Ecosystem ecology

How do energy and nutrients cycle through the environment?

When salmon die and decompose, the nutrients that are released are used by bacteria, archaea, plants, protists, young salmon, and other organisms

FIGURE 50.1 Biologists Study Ecology at Four Main Levels

salmon populations have declined precipitously as their habitats have become dammed or polluted, and salmon are an important source of food for both people and wildlife. If the factors that affect population size can be described accurately enough, mathematical models can assess the impact of proposed dams, changes in weather patterns, altered harvest levels, or specific types of protection efforts.

Community Ecology

A biological **community** consists of the species that interact with each other within a particular area. Researchers who study community ecology ask questions about the nature of the interactions between species and the consequences of those interactions. The work might concentrate on predation, parasitism, competition, or other types of contact and impact. At this level of organization in ecology, biologists also analyze how groups of species respond to disturbances such as fires, floods, and volcanic eruptions. Because human activities are driving species to extinction and disturbing communities on a massive scale, community ecologists are being called on to generate hypotheses and data on how human impacts can be lessened.

As an example of the types of questions asked at the level of community ecology, consider the interactions among salmon and other species in the marine and stream communities where they live. When they are at sea, salmon eat smaller fish and are themselves hunted and eaten by orcas, sea lions, humans, and other mammals. When the salmon return to freshwater to breed, they are preyed on by bears and bald eagles (**Figure 50.1c**). In both marine and freshwater habitats, salmon are subject to parasitism and disease. They are also heavily affected by disturbances—particularly changes in their food supply due to overfishing and the damming and degradation of breeding streams.

Ecosystem Ecology

Ecosystem ecology is an extension of community ecology. An **ecosystem** consists of all the organisms in a particular region along with nonliving components. These nonliving, or **abiotic** ("not-living"), components include air, water, and soil. At the ecosystem level, biologists analyze how chemical elements that act as nutrients cycle through ecosystems and how energy flows through the organisms in a region. More specifically, biologists study how nutrients and energy move between organisms and the surrounding atmosphere and soil or water. Because humans are adding massive amounts of nutrients to ecosystems all over the world and affecting energy flows and climate through global warming, this work has direct public policy implications. Ecosystem ecologists are responsible for assessing the impact of pollution and increased temperature on the distribution and abundance of species—in fact, on Earth's ability to support life.

Salmon are interesting to study at the ecosystem level, because they form a link between marine and freshwater ecosystems. They harvest nutrients in the ocean and then, when they die and decompose, transport those molecules to streams (**Figure 50.1d**). In this way, salmon transport chemical energy and nutrients from one habitat to another. Because salmon are sensitive to pollution and to changes in water temperature, human-induced changes in marine and freshwater ecosystems may have a large impact on their populations.

How Do Ecology and Conservation Efforts Interact?

The four levels of ecological study are synthesized and applied in conservation biology. **Conservation biology** is the effort to study, preserve, and restore threatened populations, communities, and ecosystems. Ecologists study how interactions between organisms and their environments result in a particular species being found in a particular area and population size; conservation biologists collect the same types of data and apply it by advising efforts to preserve species and restore environments. This is a common theme in biological science. Physicians, dentists, and veterinarians use basic results from molecular biology and physiology to improve the health of people and domesticated animals. Agricultural scientists and farmers apply information from genetics and physiology to improve the quality and quantity of the food we eat. Conservation biologists do the same for natural systems. Ecologists provide theory and data—the knowledge base for understanding how ecosystems work. Conservation biologists prescribe remedies for threatened species and manage land to produce a diversity of species, clean air, pure water, and healthy soils.

50.2 The Nature of the Environment

If ecology is the study of how species interact with their environment, what makes up the environment? The short answer is that an organism's environment has both physical and biological components. The abiotic components include temperature, precipitation, wind, and sunlight. In addition, terrestrial and aquatic environments have unique abiotic components—the chemical content and particle size of soil in terrestrial environments and the chemical composition of water in aquatic habitats. The **biotic** ("living") components, in contrast, consist of other members of the organism's own species as well as individuals of other species. Every organism that interacts with an individual is part of its biotic environment. Biotic factors include interactions with offspring, mates, competitors, predators, prey, and parasites.

It is important to understand the nature of the environments that organisms face, because humans are changing those environments rapidly. People alter the abiotic environment through air and water pollution, global warming, and soil erosion. Human impacts on the biotic environment are also nothing short of massive, through reductions in populations of other species and even outright extinction of species, removal of predators, and the introduction of **exotics**—organisms that are introduced into a region where they do not naturally live.

Because all environments have both abiotic and biotic components, it is not possible to understand ecology without considering both nonliving and living factors. But because subsequent chapters focus on biological interactions, the emphasis here is on the nature of the physical environments that organisms face. What components of the physical world have the most impact on organisms?

Climate

Climate is defined as the prevailing long-term weather conditions in a particular region. **Weather** consists of the specific short-term atmospheric conditions of temperature, moisture, sunlight, and wind. All of these factors are important to organisms.

Temperature is critical because the enzymes that make life possible work at optimal efficiency only in a narrow range of temperatures (see Chapter 3). Most organisms cannot regulate their own body temperatures, so their ability to maintain high levels of enzyme activity, metabolism, and growth or movement depends on the surrounding temperature. Vertebrate animals are an exception to this rule; many vertebrates can maintain a fairly constant body temperature by manufacturing their own heat or changing their position or activity to gain or lose heat. Temperature is also important, because it affects the availability of moisture: Water is frozen at low temperatures and evaporates rapidly at high temperatures.

Moisture is significant to terrestrial organisms because they constantly lose water to the environment through evaporation or transpiration. As a result, they must not only reduce water loss but also replace lost water. Aquatic organisms are surrounded by water and are exquisitely sensitive to changes in its ionic composition (see Chapter 42).

Sunlight is essential to organisms because it is required for photosynthesis and because photosynthesis provides the chemical energy that virtually all organisms depend on to stay alive. In terrestrial habitats the amount of sunlight available to any one organism depends on the extent of shading; in aquatic habitats the critical factor is depth below the water surface—because water absorbs and scatters light. In most habitats the availability of sunlight also varies with the time of year.

Wind is important because it exacerbates the effects of temperature and moisture. Wind increases heat loss due to evaporation and convection, and it increases water loss due to evaporation and transpiration. It can also have a direct physical impact on organisms—particularly birds, flying insects, and plants.

Of the four components of climate, temperature and moisture are far and away the most important to plants. Photosynthesis and plant growth are maximized when temperatures are warm and conditions are wet; conversely, photosynthesis cannot occur efficiently in cold temperatures or under drought stress. Because terrestrial animals depend on plants for food, it follows that temperature and moisture are the most important climatic elements for them as well.

To get a better idea of how ecologists analyze the impact of climate, let's consider a series of specific questions about why climate varies around the globe. Why are some parts of the world warmer and wetter than others, and why do seasons exist?

WEB TUTORIAL 50.1 Tropical Atmospheric Circulation

Why Are the Tropics Warm and the Poles Cold? In general, areas of the world are warm if they receive a large amount of sunlight per unit area; they are cold if they receive a small amount of sunlight per unit area. The shape of Earth dictates that regions at or near the equator receive a great deal of sunlight per unit area relative to regions that are closer to the poles. At the equator, the Sun is often directly overhead. As a result, sunlight strikes Earth at or close to a 90° angle. At these angles, Earth receives a maximum amount of solar radiation per unit area (**Figure 50.2**). But because Earth's surface slopes away from the equator, the Sun strikes the surface at lower and lower angles toward the poles. When sunlight arrives at a low angle, much less energy is received per unit area. The pattern of decreasing average temperature with increasing **latitude**, or distance from the equator, is a result of Earth's largely spherical shape.

Why Are the Tropics Wet? One of the most striking weather patterns on Earth involves precipitation. When average annual rainfall is mapped for regions around the globe, it is clear that areas along the equator receive the most moisture, while locations about 30 degrees latitude north and south of the

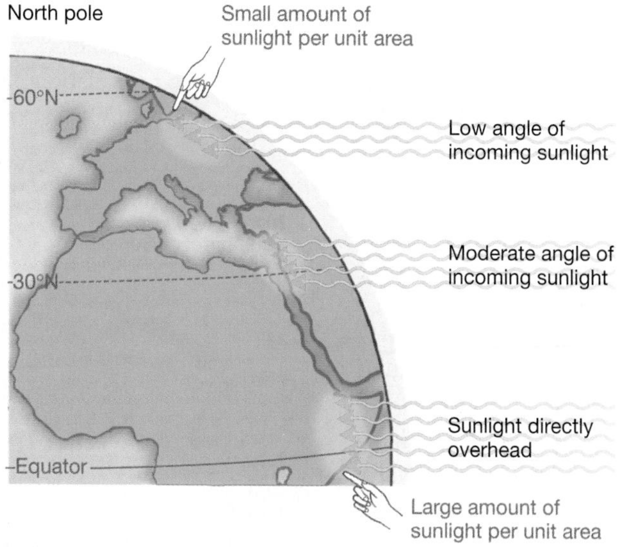

FIGURE 50.2 Solar Radiation per Unit Area Declines with Increasing Latitude
Over the course of a year, the Sun is frequently almost directly overhead at the equator. As a result, equatorial regions receive a large amount of solar radiation per unit area. At latitudes greater than 23.5°, the Sun is never directly overhead. As a result, high-latitude regions receive less solar radiation per unit area than do low-latitude regions.

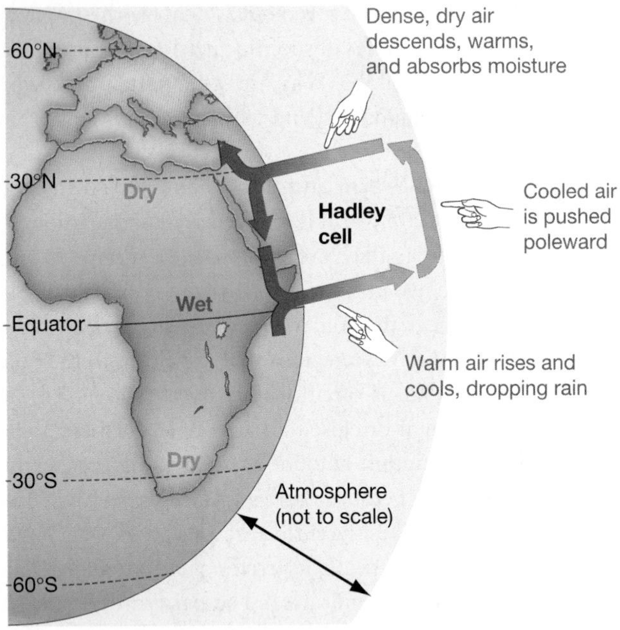

FIGURE 50.3 Global Air Circulation Patterns Affect Rainfall
Hadley cells explain why the tropics are much wetter than regions near 30° latitude north and south. Regions with rising air tend to be wetter than regions with descending air. There are three major air circulation cells on each side of the equator. (Only one of the six is shown.)

equator are among the driest on the planet. Most of the world's great deserts—including the Sahara, Gobi, Sonoran, and Australian outback—lie on or about 30°N or 30°S latitude. Why does this pattern occur?

A major cycle in global air circulation, called a **Hadley cell**, is responsible for making the Amazon River basin wet and the Sahara dry. Hadley cells are named after George Hadley, who in 1735 conceived of the idea of enormous air circulation patterns. As **Figure 50.3** indicates, air that is heated by the strong sunlight along the equator expands and rises. Warm air can hold a great deal of moisture, because warm water molecules tend to stay in vapor form instead of condensing into droplets. As the air rises, however, it radiates heat to space and begins to cool. As it cools, its ability to hold water declines. When water vapor cools, water condenses. The result? High levels of precipitation occur along the equator.

As more air is heated along the equator, the cooler, "older" air above Earth's surface is pushed poleward. When the air mass has cooled enough that its density increases, it begins to sink. As it sinks, it absorbs more and more solar radiation reflected from Earth's surface and begins to warm. As the air warms, it also gains water-holding capacity. Thus, little rain occurs where that air returns to Earth's surface, and the area is bathed in warm, dry air. The result is a band of deserts in the vicinity 30° latitude north and south.

What Causes Seasonality in Weather? The striking global patterns in temperature and moisture that we've just reviewed are complicated by the phenomenon of **seasons:** regular, annual fluctuations in temperature, precipitation, or both. As **Figure 50.4** shows, seasonality occurs because Earth is tilted on its axis by 23.5 degrees. As a result of this incline, the Northern Hemisphere is tilted toward the Sun in June and is the region that faces the Sun most directly. Thus the Northern Hemisphere receives the largest amount of solar radiation per unit area in June. But in December, the Southern Hemisphere is tilted toward the Sun, faces the Sun most directly, and receives the largest quantity of solar radiation per unit area. As a result, it is summer in the Northern Hemisphere in June but summer in the Southern Hemisphere in December. In March and September the equator faces the Sun most directly, so the tropics receive the most solar radiation then. If Earth did not tilt on its axis, there would be no seasons.

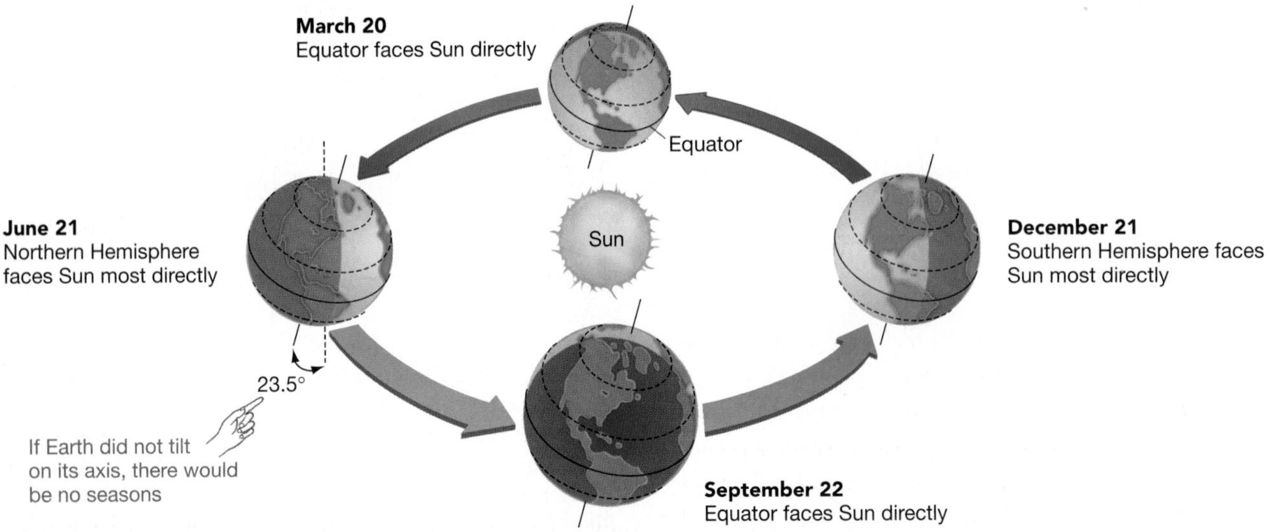

FIGURE 50.4 Earth's Orbit and Tilt Create Seasons at High Latitudes
Seasons occur due to annual variation in the amount of solar radiation that different parts of the Earth receive.

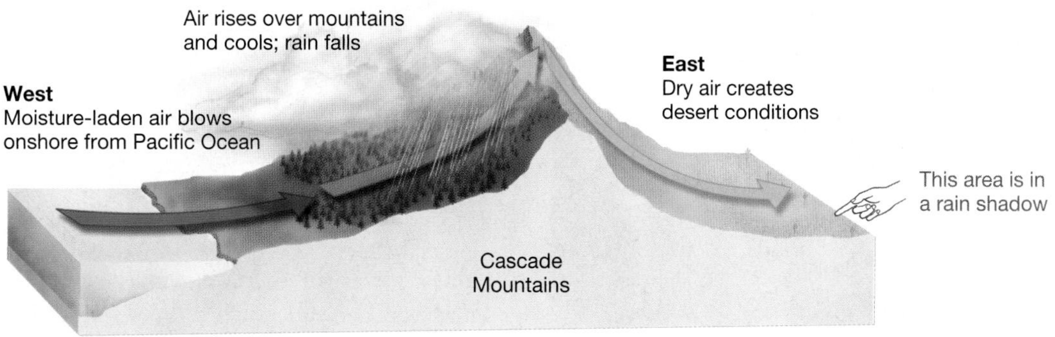

FIGURE 50.5 Mountain Ranges Create Rain Shadows
When moisture-laden air rises over a mountain range, the air cools enough to condense water vapor. As a result, moisture falls to the ground as precipitation. On the other side of the mountain, little water is left to provide precipitation.

Mountains and Oceans: Regional Effects on Climate The broad patterns of climate that are dictated by global heating patterns, Hadley cells, and seasonality are overlain by regional effects. The most important of these are due to the presence of mountain ranges and proximity to an ocean.

Figure 50.5 shows how mountain ranges affect regional climate, using the Cascade Mountain range along the Pacific Coast of North America as an example. In this area of the world, the prevailing winds are from the west. These winds bring moisture-laden air from the Pacific Ocean onto the continent. As the air masses begin to rise over the mountains, the air cools and releases large volumes of water as rain. As a result, the area along the Pacific Coast from northern California to southeast Alaska hosts some of the only high-latitude rain forests in the world. One area along the Pacific Coast of Washington State gets an average of 340 cm (134 in.) of precipitation each year. (Mountain ranges also occur along the Pacific Coast of Southern California and Mexico; but these areas are arid because they are dominated by dry air that is part of the Hadley circulation.)

Once cooled air has passed the crest of a mountain range, the air is relatively dry, because much of its moisture content has already been released. Areas that receive this dry air are said to be in a **rain shadow**. One of the only high-latitude deserts in the world exists to the east of the Cascade Mountains. It is created by a rain shadow and averages less than 25 cm (10 in.) of moisture annually. Similar rain shadows create the Great Basin desert of western North America and the dry conditions of the Tibetan Plateau.

While the presence of mountain ranges tends to produce extremes in precipitation, the presence of an ocean has a moderating influence on temperature. To understand why this is so, recall from Chapter 2 that water has an extremely high **specific heat**, meaning that it has a large capacity for storing heat energy. Because water molecules form hydrogen bonds with each other, it takes a great deal of heat energy to boil water or melt ice—in fact, to change the temperature of water at all. The hydrogen bonds have to be broken before the water molecules themselves can begin to absorb heat and move faster, which we measure as increased temperature.

Because water has a high specific heat, it can absorb a great deal of heat from the atmosphere in summer, when the water temperature is cooler than the air temperature. As a result, the ocean moderates summer temperatures on nearby landmasses. Similarly, the ocean releases heat to the atmosphere in winter, when the water temperature is warmer than the air temperature. As a result, islands and coastal areas have much more moderate climates than do nearby areas inland.

50.3 Types of Terrestrial and Aquatic Ecosystems

Climate varies dramatically around the globe, and the type of vegetation present varies just as dramatically. If you could walk from the equator in South America to the North Pole, you would notice startling changes in the organisms around you. Lush tropical forests would give way to seasonally dry forests and then to deserts. The deserts would yield to the vast grasslands of central North America, which end at the boreal forests of the subarctic. If you pressed on, you would reach the end of the trees and the beginning of the most northerly community—the arctic tundra.

Biologists have documented that similar types of changes occur with increased elevation, because of similar changes in climate. In the Siskiyou Mountains of Oregon, for example, researchers have documented that Douglas fir and other drought-tolerant tree species dominate the forests at low elevations, where the climate is warm and dry. At higher elevations, where the climate is cooler and moister, a few white fir trees join the Douglas fir. Douglas fir is absent on higher and cooler slopes, leaving just white fir. The highest-elevation forests, where the climate is cold and wet, are dominated by noble fir and mountain hemlock. Above timberline—meaning elevations higher than trees can grow—the landscape is covered with alpine tundra.

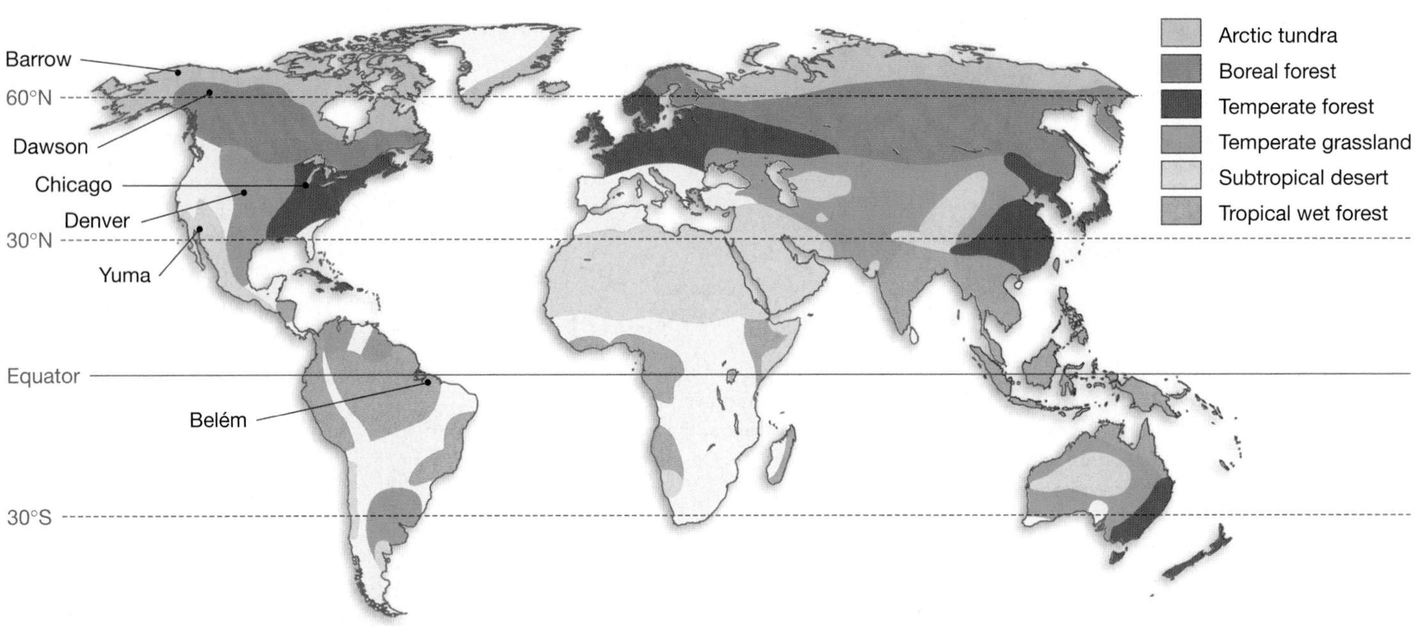

FIGURE 50.6 Distinct Biomes Are Found throughout the World
The distribution of the community types discussed in Section 50.2, and the locations of the specific sites discussed.

Terrestrial Biomes

A journey from the equator to the poles or from sea level to a mountaintop has a simple message: As climate changes, so does the predominant type of vegetation present. To capture this point, biologists identify and analyze ecological units called biomes. A **biome** is a type of terrestrial ecosystem that is unique to a given region and is characterized by a distinct type of vegetation. Grasslands, evergreen forests, and deserts are all types of biomes. The key observation is that each biome is associated with a distinctive combination of temperature and moisture conditions. Biomes represent the major types of environments available to terrestrial organisms.

How do biologists identify climatic regimes and characterize biomes? The most common approach categorizes climate types according to two major features: (1) average annual temperature and precipitation, and (2) annual variation in temperature and precipitation. The idea is that biomes correspond to a particular set of abiotic factors and that temperature and moisture are the most important of these factors—specifically their average values and their annual variations.

Classification schemes that are based on temperature and precipitation data have identified about 24 distinct climate regions around the world. Here we survey six of the most extensive terrestrial biomes, ranging from the wet tropics to the arctic. **Figure 50.6** shows the global distribution of these biomes, while **Figure 50.7** graphs the average temperature and precipitation of regions where these biomes occur in North America.

Tropical Wet Forests Tropical wet forests, or rain forests, are found in equatorial regions where temperatures and rainfall are high, variation in temperature is low, and enough rainfall is available to support year-round growth. Although plants shed older leaves throughout the year, there is no complete, seasonal loss of leaves as in temperate climates and many other regions. In tropical wet forests, plants are broadleaved evergreens.

As an example of the tropical wet forest biome, consider Belém, Brazil, which is near the equator. Even in the driest month of the year, November, this region receives over 5 cm

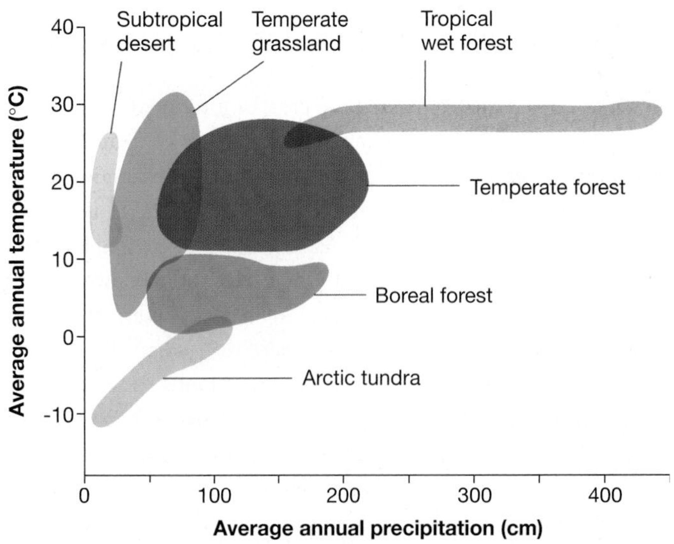

FIGURE 50.7 Biomes Occupy Regions with Distinct Average Annual Temperature and Precipitation
Data showing the distinct combination of temperature and moisture in which North American biomes are found.

(2 in.) of rainfall (**Figure 50.8a**), considerably more than the *annual* rainfall of many deserts. Mean monthly temperatures never drop below 25°C and never exceed 30°C. Tropical wet forests show much less seasonal variation than other biomes do.

As **Figure 50.8b** shows, the favorable year-round growing conditions in tropical wet forests produce riotous growth, leading to extremely high productivity and aboveground biomass. **Productivity** is the total amount of carbon fixed by photosynthesis per unit area per year; **aboveground biomass** is the total mass of living plants, excluding roots. Aboveground biomass reflects **net primary productivity**, which is the total amount of carbon that is fixed per year minus the amount of fixed carbon oxidized during cellular respiration. Fixed carbon that is consumed in cellular respiration provides energy for the organism but is not used for growth—that is, production of biomass.

(a) Climate characteristics of tropical wet forest

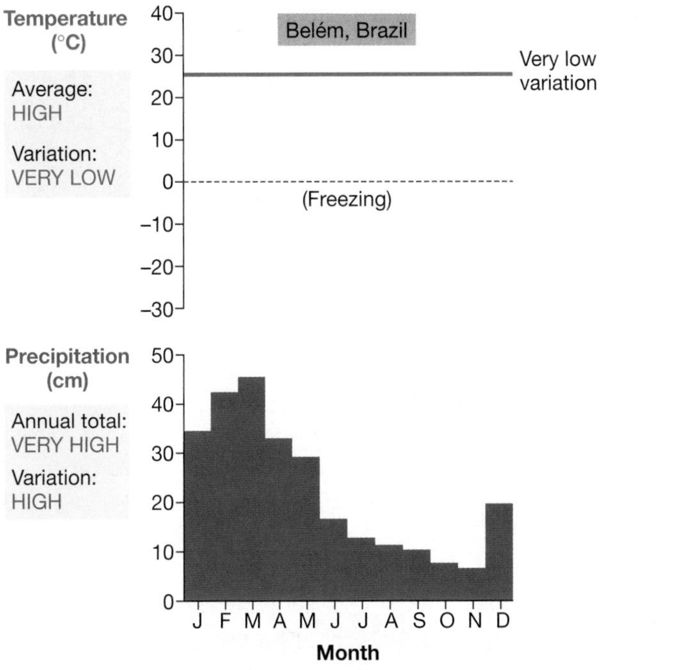
Temperature (°C)
Average: HIGH
Variation: VERY LOW
Belém, Brazil
Very low variation
(Freezing)
40
30
20
10
0
−10
−20
−30
Precipitation (cm)
Annual total: VERY HIGH
Variation: HIGH
50
40
30
20
10
0
J F M A M J J A S O N D
Month

(b) Appearance of tropical wet forest

FIGURE 50.8 Tropical Wet Forest
(a) Graph illustrating the climate of coastal Brazil, typical of a wet tropical climate. Climate graphs have months on the horizontal axis and precipitation and temperature on the vertical axes. The graph shows the average monthly temperature and precipitation for an area. **(b)** Tropical wet forests are extremely rich in species.

Tropical wet forests are also renowned for their species diversity. In this biome, researchers routinely find over 200 tree species in a single 10 m × 100 m study plot. Based on counts of the insects and spiders collected from single trees, some biologists contend that the world's tropical wet forests may hold up to 30 million species of arthropods alone.

Further, the diversity of plant sizes and growth forms in wet forest communities produces extraordinary structural diversity. In tropical wet forests, a **canopy** of trees (the uppermost layers of branches) is intermingled with vines, **epiphytes** (plants that grow entirely on other plants), shrubs, and herbs. The diversity of growth forms presents a wide array of habitat types for animals, which in turn pursue an impressive array of feeding strategies. In tropical wet forests, animals that eat fruit, insects, plant tissues, small animals, and dead material all coexist.

Subtropical Deserts Tropical wet forests are found along or near the equator. Because the Hadley cells generated at the equator deliver warm, dry air to areas about 30° latitude north and south of the equator, there is a corresponding band of subtropical deserts in these regions.

To appreciate the climatic regime in subtropical deserts, look at **Figure 50.9a** (page 1152), which provides temperature and precipitation data for Yuma, Arizona, in the Sonoran Desert of southwestern North America. Mean monthly temperatures vary more than in tropical wet climates, but in Yuma temperatures never fall below freezing. (Freezing nighttime temperatures are common elsewhere in the Sonora and in other subtropical deserts.) The most striking feature of the climate of these regions, though, is the low precipitation. The average annual precipitation in Yuma is just 7.5 cm (3 in.).

The scarcity of water in deserts has profound implications. Because conditions are rarely good enough to support photosynthesis, the productivity of desert communities is a tiny fraction of average values for tropical forest communities. Further, as **Figure 50.9b** shows, individual plants are widely spaced—a pattern that may reflect intense competition for water.

Desert species adapt to the extreme temperatures and aridity in one of two ways: They grow at a low rate year-round or break dormancy and grow rapidly in response to any rainfall. Cacti are examples of desert plants that grow year-round. They have small leaves, a waxy coating on the leaves, and the CAM pathway for photosynthesis, introduced in Chapter 10. These adaptations allow cacti to tolerate hot, dry conditions and grow at an extremely low rate throughout the year. Other desert species, though, have traits that allow them to escape drought.

(a) Climate characteristics of subtropical desert

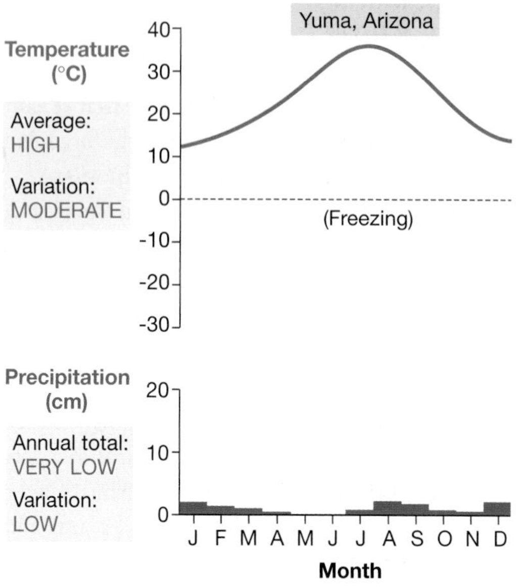
Yuma, Arizona
Temperature (°C)
Average: HIGH
Variation: MODERATE
40
30
20
10
0
-10
-20
-30
(Freezing)
Precipitation (cm)
Annual total: VERY LOW
Variation: LOW
20
10
0
J F M A M J J A S O N D
Month

(b) Appearance of subtropical desert

FIGURE 50.9 Subtropical Desert
(a) The climate of Yuma, Arizona, is typical of a subtropical desert. **(b)** Saguaro cacti are a prominent feature of the Sonoran Desert in the southwestern United States and northern Mexico.

Some spend much of the year in a dormant state, resuming growth only after rain falls. The shrub called ocotillo, for example, sprouts leaves within days of a rainfall, then drops them a week or two later when soils dry out again. Each ocotillo individual produces and sheds leaves in this way several times a year. Similarly, desert amphibians such as the spadefoot toad spend much of the year dormant in underground burrows. They respond to rain by emerging for a brief time to feed and mate.

In short, deserts provide a stark contrast to tropical forests. Although both biomes feature high temperatures that rarely or never drop below freezing, the absence of water in subtropical deserts leads to low species diversity and productivity.

Temperate Grasslands Figure 50.10a presents climate data for a typical grassland community—in this case, Denver, Colorado. Precipitation in Denver is over four times greater than that in Yuma, Arizona. Nonetheless, conditions are still quite dry; no month has more than 5 cm (2 in.) of precipitation. In comparison with regions to the south and north, temperatures in the area are moderate. Biologists use the term **temperate** to describe regions with temperatures that are moderate relative to the tropics and polar regions, and where summers are typically long and warm and winters short and cold. Grasses are the dominant life-form in temperate regions with relatively low rainfall. Temperate grassland communities

(a) Climate characteristics of temperate grassland

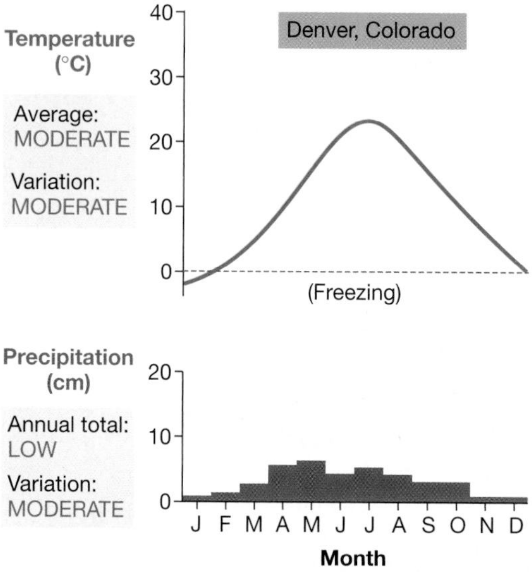
Denver, Colorado
Temperature (°C)
Average: MODERATE
Variation: MODERATE
40
30
20
10
0
(Freezing)
Precipitation (cm)
Annual total: LOW
Variation: MODERATE
20
10
0
J F M A M J J A S O N D
Month

(b) Appearance of temperate grassland

FIGURE 50.10 Temperate Grassland
(a) Denver, Colorado, has a climate typical of a midlatitude steppe. **(b)** Grasses are the dominant life-form in prairies and steppes.

are found throughout central North America and the heartland of Eurasia (see Figure 50.6).

Temperature profiles in these regions are also highly seasonal. In Denver, the mean monthly temperature exceeds 20°C in the summer but drops below freezing in the winter. Temperature variation is important because it dictates a well-defined growing season. In the temperate zone, plant growth is possible only in spring, summer, and fall months when moisture and warmth are adequate.

Temperate grasslands, also called **prairies** or **steppes**, usually exist because conditions are too dry to support tree growth but too cold, wet, and seasonal for drought-adapted desert species (**Figure 50.10b**). In temperate areas where rainfall is high enough to support forests, grasslands may develop if recurring fires burn out encroaching trees . Prairie fires, which begin naturally as a result of lightning strikes, were also set intentionally by native people. The plants that dominate prairie communities tolerate fire, because, unlike many plants, they have meristematic tissue, which produces new growth, at the base of their stems. As a result, they resprout quickly after burning.

Although the productivity of temperate grasslands is generally lower than that of forest communities, grassland soils are often highly fertile. The subsurface is packed with roots, which add organic material to the soil as they die and decay. Further, grassland soils retain nutrients, because rainfall is low enough to keep key ions from dissolving and leaching out of the soil. It is no accident, then, that the grasslands of North America and Eurasia are the breadbaskets of those continents. The conditions that give rise to natural grasslands are ideal for growing wheat, corn, and other cultivated grasses.

Temperate Forests In temperate areas with relatively high precipitation, grasslands give way to forests. A typical climate graph for an area near a grassland-forest border is shown in **Figure 50.11a**. Most of the area around Chicago, Illinois, was forested prior to the arrival of European settlers. Temperate forests are found in eastern North America, most of Europe, Chile, and New Zealand (see Figure 50.6).

Temperate forests experience a period in which mean monthly temperatures fall below freezing and plant growth stops. Unlike grassland climates, however, precipitation is moderately high and relatively constant throughout the year. Chicago, for example, has an annual precipitation of 85 cm (34 in.); most months record more than 5 cm (2 in.).

The abundance of moisture allows trees to dominate the landscape (**Figure 50.11b**). Unlike plants in some regions in the tropics, however, plants in temperate forests experience a dormant period. In North America and Europe, temperate forests are dominated by deciduous species, which drop their leaves in the autumn and grow new leaves in the spring. Evergreens, which also enter dormancy each fall, are common as well. In the temperate forests of New Zealand and Chile, however, broad-leaved evergreens predominate.

(a) Climate characteristics of temperate forest

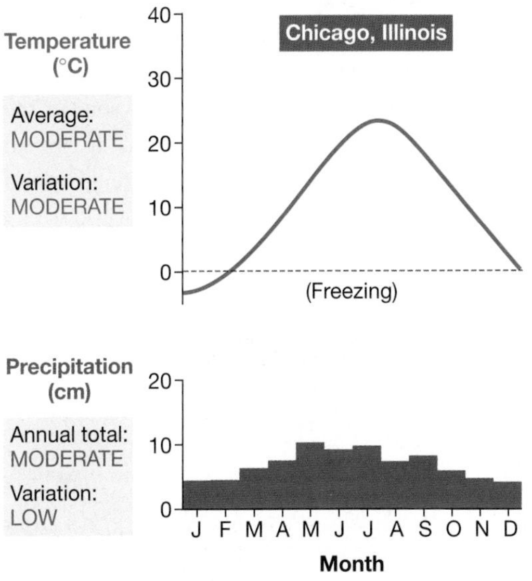
Chicago, Illinois
Temperature (°C)
Average: MODERATE
Variation: MODERATE
40
30
20
10
0
(Freezing)
Precipitation (cm)
Annual total: MODERATE
Variation: LOW
20
10
0
J F M A M J J A S O N D
Month

(b) Appearance of temperate forest

FIGURE 50.11 Temperate Forest
(a) Climate graph for Chicago, Illinois, typical of a midlatitude cold and humid climate. **(b)** Temperate forests are dominated by broad-leaved deciduous trees.

Although some of the warmer temperate forests rival or exceed the productivity of tropical forests, most have productivity levels that are lower than those of tropical forests yet higher than those of deserts or grasslands. The level of diversity is also moderate. A temperate forest in southeastern North America may have more than 20 tree species; similar forests to the north may have fewer than 10.

Boreal Forests The boreal forest, or **taiga**, stretches across most of Canada, Alaska, Russia, and northern Europe. Because these regions are just south of the Arctic Circle, they are referred to as subarctic. Their climate is characterized by very cold winters; cool, short summers; and extraordinarily high annual variation in temperature. In the course of a year, subarctic areas may be subject to temperature ranges of more than 70°C.

Figure 50.12a shows a climate graph from Dawson, in the Yukon Territory of Canada. Annual precipitation, at 36 cm (14 in.), is virtually identical to that of the temperate grassland climate of western North America. Regions occupied by boreal forests are so cold, however, that evaporation is minimal; moisture is usually abundant enough to support tree growth as a result. The climate graph also documents the extreme temperature swings typical of these areas. Summer temperatures hover around 10°C, but hot spells in which temperatures exceed 20°C are not uncommon. Winters are cold, with mean monthly temperatures dropping well below freezing for nearly half the year.

(a) Climate characteristics of boreal forest

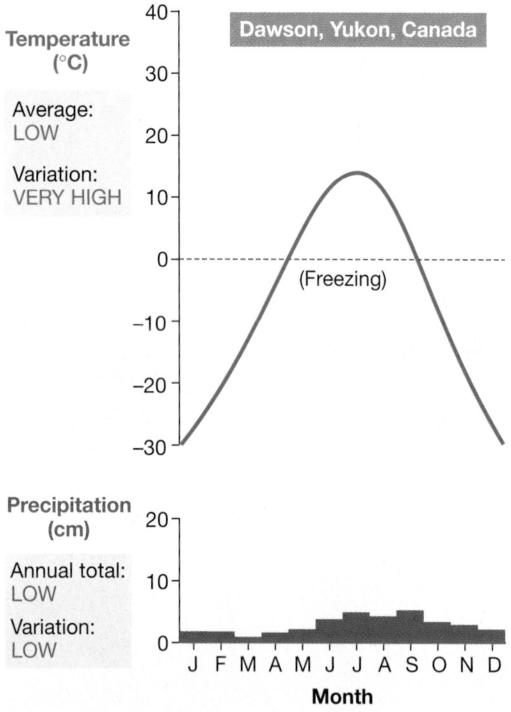

Dawson, Yukon, Canada
Temperature (°C)
Average: LOW
Variation: VERY HIGH
40
30
20
10
0
−10
−20
−30
(Freezing)
Precipitation (cm)
Annual total: LOW
Variation: LOW
20
10
0
J F M A M J J A S O N D
Month

(b) Appearance of boreal forest

FIGURE 50.12 Boreal Forest
(a) Climate graph for Dawson, Yukon Territory, Canada, in the heart of the boreal forest. **(b)** The boreal forest is dominated by needle-leaved evergreens, such as spruce and fir.

As **Figure 50.12b** shows, boreal forests are dominated by highly cold-tolerant conifers, including spruce, pine, fir, and larch trees. With the exception of larch, these species are evergreen. Two hypotheses have been offered to explain why evergreens predominate in cold environments, even though they do not photosynthesize in winter. The first is that they can begin photosynthesizing early in the spring, even before the snow melts, when sunshine is intense enough to warm their needles. The second hypothesis is based on the observation that soils tend to be acidic and contain little available nitrogen. Because leaves are nitrogen rich, species that must produce an entirely new set of leaves each year might be at a disadvantage. To date, however, these hypotheses have not been tested rigorously.

Based on these observations, it is not surprising that the productivity of boreal forests is low. Aboveground biomass is high, however, because slow-growing tree species may be long-lived and gradually accumulate large standing biomass. Boreal forests also have exceptionally low species diversity. The boreal forests of Alaska, for example, typically contain seven or fewer tree species.

Arctic Tundra Lying poleward from the subarctic is the arctic tundra. This region has climatic conditions similar to those shown in **Figure 50.13a** for Barrow, on the northern coast of Alaska. The growing season is 6–8 weeks long at most; otherwise, temperatures are below freezing. Precipitation is also low. The annual precipitation in Barrow is actually less than that in the Sonoran Desert of southwestern North America. Because of the extremely low evaporation rates, however, arctic soils are saturated year-round.

As **Figure 50.13b** shows, the arctic tundra is treeless. The leading hypothesis to explain the lack of trees is that the growing season is too short and cool to support the production of large amounts of non-photosynthetic tissue. Also, tall plants that poke above the snow in winter might experience substantial damage from wind-driven snow and ice crystals. Woody shrubs, such as willows, birch, and blueberries, are common but rarely exceed the height of a child. Most arctic tundra species hug the ground.

Arctic tundra has low species diversity, low productivity, and low aboveground biomass. Most tundra soils are in the perennially frozen state known as **permafrost**, and the cold temperatures inhibit both the release of nutrients from decaying organic matter and the uptake of nutrients into live roots. Unlike plants in the desert, however, plants in many tundra communities may completely cover the ground. Animal diversity also tends to be low in the arctic tundra, although insect abundance—particularly of biting flies—can be staggeringly high.

(a) Climate characteristics of artic tundra

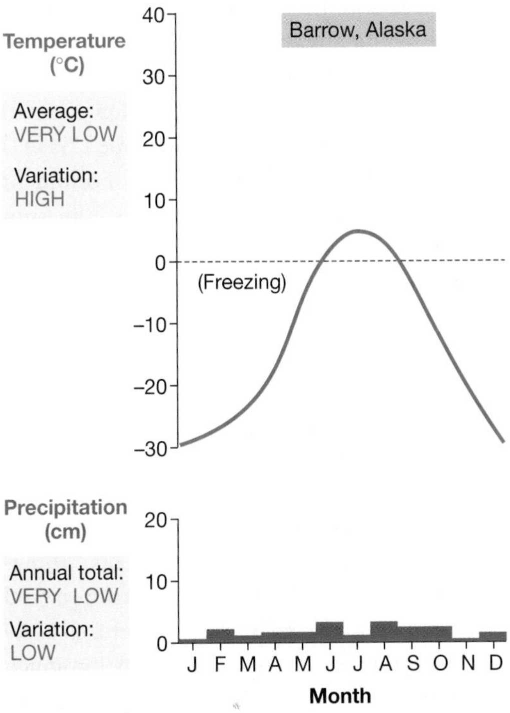

(b) Appearance of artic tundra

FIGURE 50.13 Arctic Tundra
(a) Climate graph for Barrow, Alaska, on the north slope of Alaska. **(b)** Arctic tundra is treeless. Very cold-tolerant shrubs and herbaceous plants dominate plant communities.

Aquatic Environments

Different types of aquatic habitats, like different types of terrestrial biomes, are distinguished by abiotic factors. But instead of being defined by different types of moisture and temperature regimes, aquatic environments are distinguished by depth of water and the rate of water movement. Depth of water is particularly important, because it dictates how much light reaches the organisms that live in a particular region. Light, in turn, is required for photosynthesis and the production of organic matter that other organisms can eat.

The types of wavelengths and the amount of light available to organisms changes dramatically as water depth increases. Water absorbs and scatters light. As **Figure 50.14a** shows, ocean water near the coast specifically removes light in the red and blue regions of the visible spectrum. This is an important observation, because wavelengths in the red and blue regions are required for photosynthesis (see Chapter 10). The total amount of light available to organisms also diminishes rapidly with increasing depth. In pure seawater, the total amount of light available at a depth of 10 m is less than 40 percent of what it is at the surface. Virtually no light reaches depths greater than 40 m in pure seawater (**Figure 50.14b**). In seawater that contains organisms or debris, light penetration is dramatically less than that observed in pure seawater.

(a) Only certain wavelengths of light are available under water.

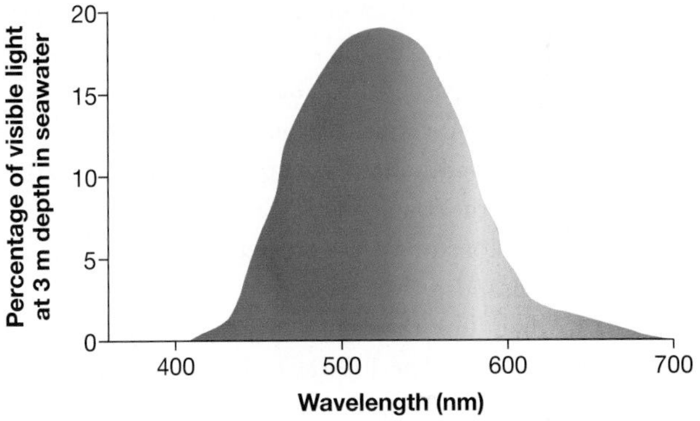

(b) Intensity of light declines with water depth.

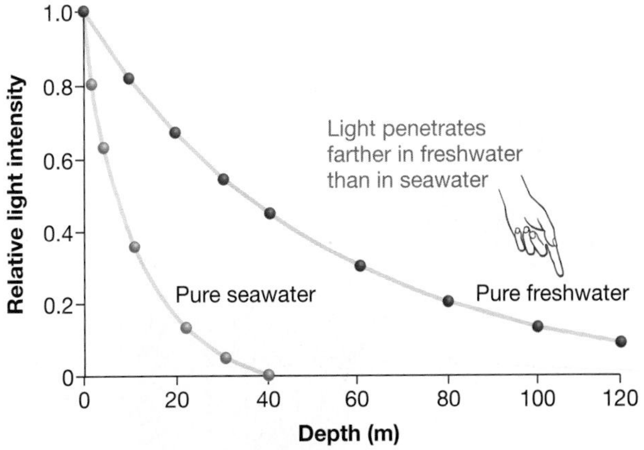

FIGURE 50.14 Availability of Light Changes Dramatically with Increasing Water Depth
(a) Graph showing the wavelengths of light available at a depth of 3 meters in seawater near an ocean coastline. **(b)** Graphs showing how rapidly the amount of light declines with depth of pure freshwater and pure seawater. When water contains organisms and organic debris, light availability declines even faster. EXERCISE Photosynthesis is driven most efficiently by wavelengths of about 425 nm and 680 nm. Mark and label these wavelengths in part (a).

In addition to water depth, the type and amount of water movement or flow are major influences in aquatic environments. For example, organisms that live in fast-flowing streams have to cope with the physical force of the water, which constantly threatens to move them downstream. Marine organisms that live in intertidal regions are exposed to the air periodically each day as well as to violent wave action during storms.

Different water depths and types of water movement define the array of aquatic environments that are available to organisms, just as different types of climate define the terrestrial biomes. Let's examine the various types of freshwater environments that organisms occupy, and then analyze the array of marine habitats and the challenges that they pose for organisms.

Lakes, Ponds, and Wetlands Bodies of standing freshwater are classified as lakes, ponds, or wetlands. Lakes and ponds are distinguished from each other by size—ponds are small, while lakes are large enough that the water in them can be mixed by wind and wave action—and from wetlands by water depth. **Wetlands** are shallow-water habitats where the soil is saturated with water for at least part of the year. Wetlands are distinct from lakes and ponds for two reasons: They have shallow water and **emergent vegetation**—meaning plants that grow above the surface of the water. Most lakes, ponds, and wetlands form in depressions that were created by the scouring action of glaciers thousands of years ago.

Figure 50.15 diagrams the major regions or zones found within lakes and ponds. The **littoral** ("seashore") **zone** consists of the shallow waters along the shore, where flowering plants are rooted. The littoral zone is shallow enough for sunlight to penetrate and support photosynthesis. As a result, it hosts abundant **planktonic** organisms, which float or swim near the surface and for the most part are microscopic. In deeper waters, rooted plants are absent and the upper, sunlit portions of the water are dominated by cyanobacteria and unicellular algae. Open water that receives enough light to support photosynthesis is referred to as the **limnetic** ("lake") **zone**. Although too little light penetrates to the bottom of most lakes and ponds to support photosynthesis, the substrate provides habitat for burrowing animals and bacteria and is called the **benthic** ("depths") **zone**. Animals that consume dead organic matter, or **detritus**, are common in the benthic zone, because dead organisms rain down from the littoral and limnetic zones to form sediments. Regions of the littoral, limnetic, and benthic zones that receive sunlight are part of the **photic zone**; regions that do not receive light are termed the **aphotic zone**.

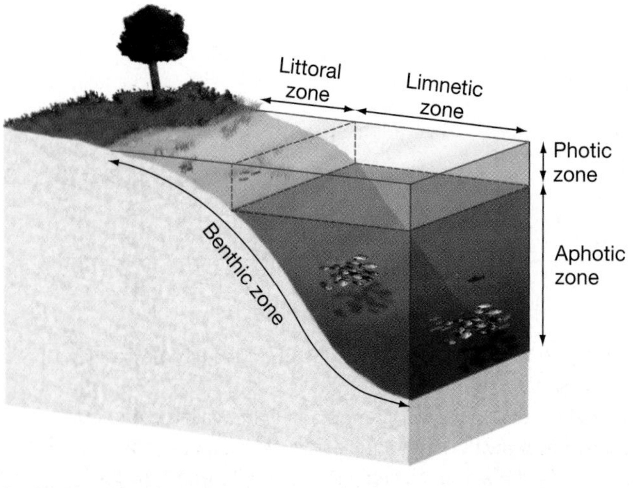

FIGURE 50.15 Lakes Have Distinctive Zones Defined by Water Depth and Distance from Shore
The littoral, limnetic, and benthic zones in a lake present discrete environments to organisms.

The major message of Figure 50.14 is that water depth is the key abiotic factor in lake and pond environments, primarily because it dictates the extent of light penetration and photosynthesis. Water depth also affects temperature, however. The littoral and limnetic zones are typically much warmer than the benthic zone, simply because they receive so much more solar radiation. **Box 50.1** details how differences in temperature at various depths in a lake, along with changes in the temperature of the surface water throughout the year, promote the mixing of water from different depths. The mixing effect of wind and temperature changes allows both well-oxygenated water from the surface to reach the benthic zone and nutrient-rich water from the benthic zone to enter the littoral and limnetic zones.

Biologists identify three major types of wetlands, as shown in **Figure 50.17**: marshes, swamps and bogs. **Marshes** lack trees and typically exhibit a slow but steady rate of water flow. Marshes are typically connected to lake or stream systems. **Swamps** are similar to marshes but are dominated by trees and shrubs. Because marshes and swamps offer ample supplies of water and sunlight, they are extraordinarily productive habitats. Bogs, in contrast, are remarkably unproductive. **Bogs** develop in depressions where water flow is low or nonexistent. The water in bogs is largely stagnant, so oxygen is used up during the decomposition of dead organic matter faster than it enters via diffusion from the atmosphere. As a result, bog water is oxygen poor or even anoxic (see Chapter 37). Once the oxygen in the water is depleted, decomposition slows. Organic acids and other acids build up, lowering the pH of the water. At low pH, nitrogen becomes unavailable to plants. The combination of acidity, lack of available nitrogen, and anoxic conditions makes bogs a demanding environment. As a consequence, productivity in bogs is low.

Bogs are an exception to the rule that productivity in wetlands is limited by light, but they conform to the rule that the productivity of aquatic ecosystems is limited by nutrients. What features of streams affect productivity?

BOX 50.1 Thermoclines and Lake Turnover

The northern and temperate regions of the world host many lakes. Each year, these bodies of water undergo remarkable changes known as the spring and fall **turnovers.**

The spring and fall lake turnovers occur in response to changes in air temperature. To see how this happens, examine **Figure 50.16**, which shows the seasonal temperature profiles of a lake. In winter, water at the surface is locked up in ice at a temperature of 0°C. The water just under the ice is slightly warmer, and the water at the bottom of the lake is at 4–5°C. A gradient in temperature such as this is called a **thermocline** ("heat-slope"); thermal stratification is said to occur.

When spring arrives, the ice begins to melt. The temperature of the water rises until it reaches 4°C. This is important, because the density of liquid water is highest at 4°C. The water at the surface of the lake is now heavier than the water below it. As a result, it sinks. The water at the bottom of the lake is displaced and comes to the surface, completing the spring turnover. During the spring turnover, water at the bottom of the lake carries sediments and nutrients from the benthic zone up to the limnetic zone. This flush of nutrients triggers a rapid increase in the growth of algae and bacteria that biologists call the *spring bloom.*

When temperatures cool in the fall, the water at the surface reaches a temperature of 4°C and sinks, displacing water at the bottom and creating the fall turnover. The fall turnover is important because it brings oxygen-rich water from the surface down to the benthic layer, and because it again brings nutrients from the benthic zone up to the limnetic zone.

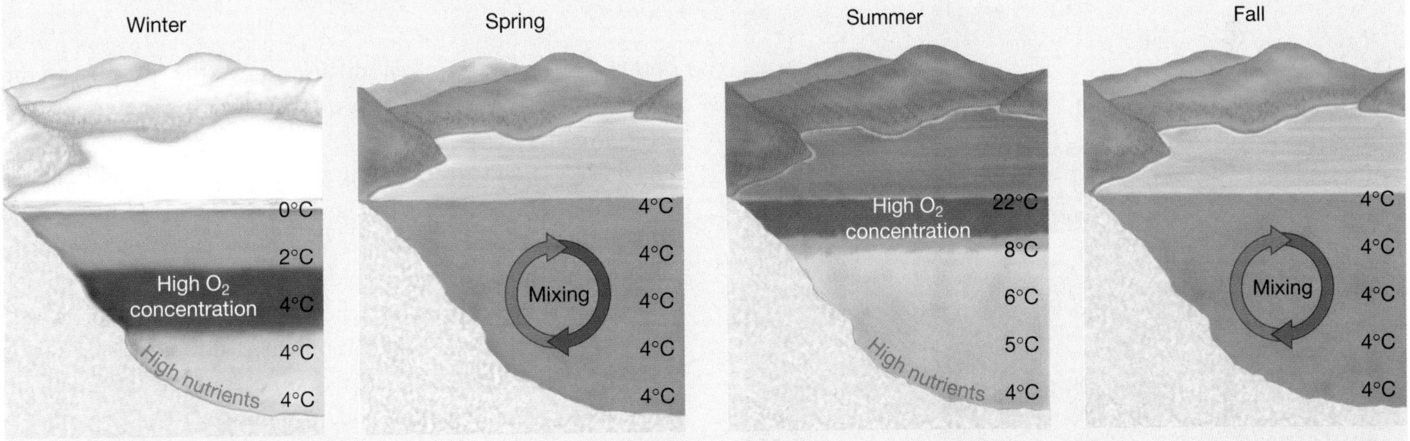

FIGURE 50.16 In Temperature Regions, Lakes Turn Over Each Spring and Fall

(a) Marshes have non-woody plants.

(b) Swamps have trees and shrubs.

(c) Bogs are stagnant and acidic.

FIGURE 50.17 Wetland Types Are Distinguished by Water Flow and Vegetation
Water flows through **(a)** marshes and **(b)** swamps, but the water in **(c)** bogs is stagnant.

Streams Streams are bodies of water that move constantly in one direction. Creeks are small streams, and rivers are large. As **Figure 50.18** shows, the structure of a stream usually varies along its length. Where it originates at a mountain glacier or a spring, a stream tends to be cold, narrow, and fast. As it descends toward a lake or an ocean, a stream accepts water from tributaries and becomes larger, warmer, and slower.

In terms of the environment that is available to organisms, the major physical variables in streams are the speed of the current and the availability of oxygen and nutrients. In small, fast-moving streams, it is rare to find photosynthetic organisms; most of the organic matter present consists of leaves, branches, and other materials that have fallen into the water from outside the stream. Although nutrient levels tend to be low in fast-moving streams, oxygen levels tend to be high because water droplets are exposed to the atmosphere when rapidly moving water splashes over rocks or other obstacles. Oxygen diffuses into the droplets, contributing to the oxygen-rich quality of fast-moving streams (see Chapter 44).

FIGURE 50.19 Estuaries Are Highly Productive Environments
Estuaries are productive because they are sunlit and nutrient rich. The salinity of estuary water is intermediate between that of freshwater and that of seawater, so organisms that thrive in estuaries may not do well in either freshwater or marine environments.

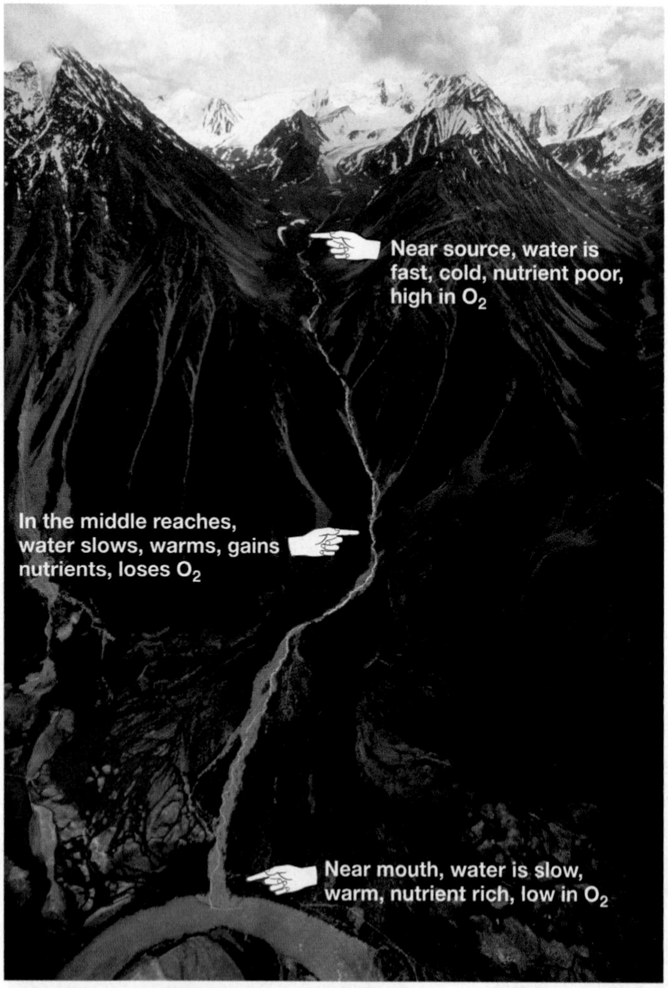
Near source, water is fast, cold, nutrient poor, high in O_2
In the middle reaches, water slows, warms, gains nutrients, loses O_2
Near mouth, water is slow, warm, nutrient rich, low in O_2

FIGURE 50.18 Distinctive Environments Appear Along a Stream's Length
The speed, temperature, nutrient level, and oxygen level of stream water usually vary along the stream's length.

As creeks and rivers widen and slow down, conditions become more favorable for the growth of algae and plants, and the amount of organic matter and nutrients in the water increases. Without riffles or rapids to oxygenate the water, however, slow-moving streams tend to become relatively oxygen poor. Based on these observations, it is not surprising that a stream often contains completely different types of organisms near its source and near its end, or mouth.

An **estuary** is the environment that forms where rivers meet the ocean (**Figure 50.19**). In these habitats, freshwater mixes with salt water. The salinity of the water in an estuary varies with proximity to the ocean. Salinity and water depth also fluctuate in estuaries in response to tides, storms, and floods. Because the water is shallow and sunlit, and because nutrients are constantly replenished by incoming river water, estuaries are among the most productive environments on Earth.

Marine Environments As in lakes and ponds, the major physical factor that distinguishes marine environments is water depth. Water depth is a major determining factor in the amount of available light, as Figure 50.14 shows. Like lakes and ponds, oceans have a **photic zone**, which is the area near the surface that receives enough sunlight to support photosynthesis. In addition, oceans have a **benthic zone**, which is the area at the bottom. There may also be an **aphotic zone**, a middle area that is too dark to support photosynthesis (**Figure 50.20**).

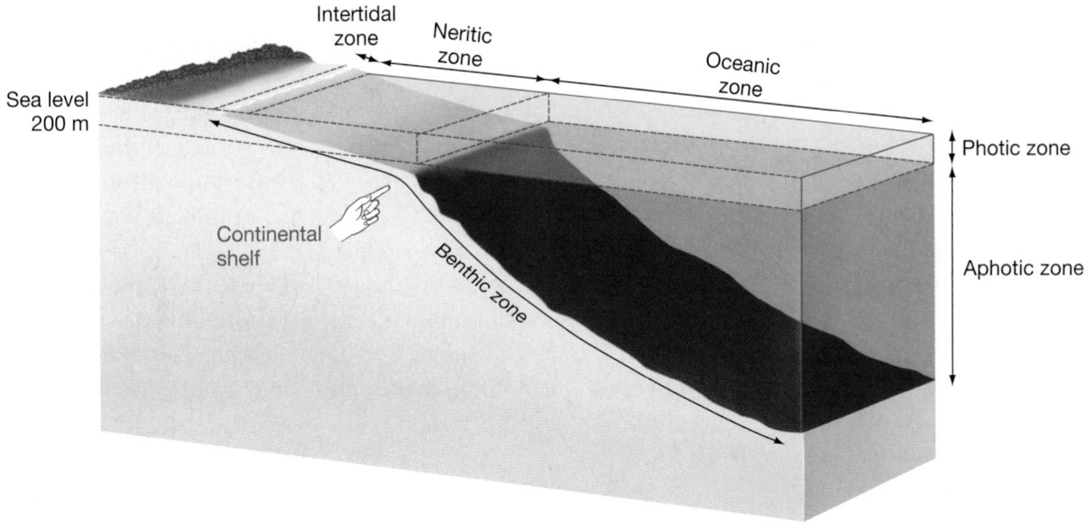

FIGURE 50.20 Oceans Have Distinctive Zones Defined by Water Depth and Distance from Shore
The intertidal, neritic, and oceanic zones present distinctive environments to organisms, as do the photic, aphotic, and benthic zones.

Water depth itself and distance from shore are used to distinguish the three major types of environment found in the ocean: (1) intertidal, (2) neritic, and (3) oceanic zones (Figure 50.20). Within each of these environments are photic and benthic zones and perhaps also an aphotic zone.

The **intertidal** ("between tides") **zone** is along the shore. The intertidal zone consists of a rocky, sandy, or muddy beach that is exposed to the air at low tide but submerged at high tide. Although organisms that live in the intertidal must be able to withstand physical pounding from waves and desiccation at low tide, productivity is high due to the ready availability of nutrients and sunlight. Nutrient levels are relatively high in the intertidal, because estuaries and rivers provide a fairly constant supply and because currents tend to sweep nutrient-laden sediments from offshore areas into the shallower waters near shore.

The **neritic zone** extends from the intertidal zone out toward the open ocean to depths of about 200 m. The edge of the neritic zone is defined by the end of the **continental shelf**—the gently sloping, submerged portion of a continental plate. Beyond the continental shelf, the bottom of the ocean changes from rock of continental origin to the rock of an oceanic plate. Instead of continuing to slope gently, the oceanic plate drops off rapidly in depth. Although it is not as productive as the intertidal zone, the neritic zone is still relatively prolific, because nutrients are available from currents that well up from the deep ocean off the continental shelf. Almost all of the world's major marine fisheries exploit organisms that live in the neritic zone.

In the tropics, shallow portions of the neritic zone may host **coral reefs.** Because the water is warm and sunlight penetrates to the ocean floor in these habitats, coral reefs are among the most productive environments in the world. Recall from Chapter 31 that coral reefs are built by species in the phylum Cnidaria that host symbiotic, photosynthetic dinoflagellates.

In contrast to the neritic zone in general and coral reefs in particular, the **oceanic zone**, or open ocean, ranks as the most unproductive environment on Earth. Although sunlight is abundant in the photic zone of the open ocean, where light penetrates, nutrients are extremely scarce. Iron, which is required for enzymes involved in cellular respiration, is particularly rare. Recall from Chapter 28 that when iron is added to the open ocean experimentally, the number of photosynthetic cells present and the overall productivity of planktonic organisms explode. This zone is nutrient poor simply because there is no mechanism for nutrients to be replenished. When the photosynthetic organisms and the animals that feed on them die, their bodies drift downward out of the photic zone and are lost. In the open ocean, there is no mechanism for bringing nutrients back up from the bottom, as there is in most lakes (see Box 50.1). The aphotic zone of the open ocean is also unproductive, because light is absent—meaning that photosynthesis is impossible. Most organisms present in the aphotic zone survive on the rain of dead bodies from the photic zone. If coral reefs are the rain forests of the ocean, then the oceanic zone is the desert. As **Box 50.2** (page 1160) shows, though, the life-forms that are present include some of the most bizarre organisms known.

The bottom of the ocean comprises the **benthic zone.** Although the benthic zone may be sunlit in the intertidal zone and parts of the neritic zone, it is more usual for light to be completely absent. In most cases, life-forms depend on a

BOX 50.2 Deep-Sea Life

The deep sea has been called the "last frontier" on Earth, with some justification. Using specialized submarines and remote-controlled video cameras that are tethered to research ships, biologists have begun to explore the ocean's coldest, darkest regions, where water is under enormous pressure. The organisms the investigators have found are nothing short of remarkable.

In the vast volume of ocean that lies between the open ocean's photic zone and benthic zone, an array of small fish, shrimp, and other organisms feed on the dead material drifting down from the surface. These organisms are preyed on by small but ferocious-looking predators such as anglerfish and fangtooth (**Figure 50.21a**). Most predators that live in the deep sea have enormous mouths and teeth relative to their overall body size, and many have tissues that emit light—possibly as a lure for unsuspecting prey.

Animals that live at the very bottom of the deep ocean may not look as bizarre as the deep-sea predators, but their adaptations to life in the abyss are just as remarkable. **Figure 50.21b**, for example, shows tubeworms that can be found 2.5 km (1.5 miles) under the surface. Tubeworms, certain clams, and other organisms live near the hot springs called black smokers (see Chapter 2). Bacteria and archaea grow there in profusion, using hydrogen gas, hydrogen sulfide, and other inorganic molecules as a source of energy (see Chapter 27). The tubeworms and clams filter these cells out of the water and digest them or protect them inside their bodies and then absorb food from them.

Although the deep ocean still qualifies as the least well-studied habitat on Earth, research has made promising strides in understanding how organisms adapt to an extraordinarily demanding physical environment.

(a) Predators from the aphotic zone of the deep sea

(b) Animals that live near black smokers

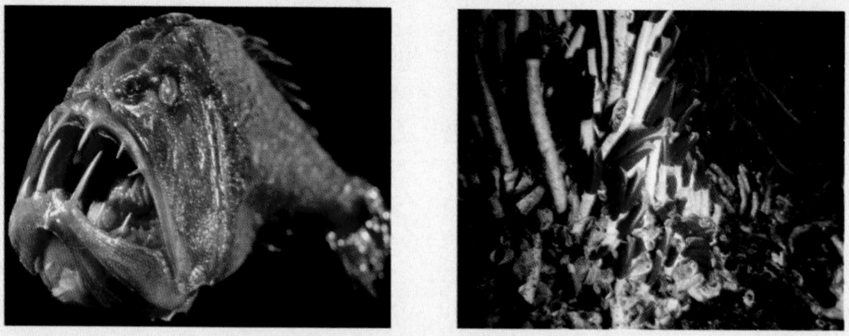

FIGURE 50.21 Deep-Sea Habitats Are Being Explored Intensively
(a) Predatory fish found in the aphotic zone of the open ocean tend to have extremely large mouths and teeth for their size. **(b)** At hot springs within deep-ocean habitats, tubeworms occur with clams and a variety of other animals.

rain of organic debris from above, which settles into the benthos and forms sediments. The exceptions to this rule are organisms that live in and around the hot springs called black smokers—organisms that are also featured in Box 50.2.

✓CHECK YOUR UNDERSTANDING

Global climate patterns are dictated by differential heating of Earth's surface, Hadley cells, and seasonality and are modified at a regional scale by the presence of mountains and oceans. Distinct patterns of temperature and moisture levels and variation are associated with regions of distinct vegetation types, or biomes. Aquatic environments are distinguished by the depth of water and the rate at which it moves. You should be able to explain how variation in the availability of light, moisture, and nutrients causes productivity to vary dramatically among terrestrial biomes and among aquatic environments.

50.4 Biogeography

The diversity of terrestrial and aquatic environments on Earth is impressive. Now the question is, how does the array of abiotic conditions that exist—along with biotic interactions—affect the distributions of organisms? Understanding why organisms are found where they are is one of ecology's most fundamental tasks, and it is a requirement for designing nature preserves in a rational and effective way. The study of how organisms are distributed geographically is called **biogeography.** In Chapter 25 we explored how continental drift has contributed to the current distributions of species; here, let's focus on how interactions with the abiotic and biotic environments affect where a particular species is able to live.

The most fundamental observation about the *range*, or geographic distribution, of species is simple: No one species can survive the full range of environmental conditions present on Earth. *Thermus aquaticus* cells thrive in hot springs with temperatures above 90°C. These archaea can live in this environment because they have enzymes that do not denature at near-boiling temperatures. The same cells would die instantly, however, if they were transplanted to the frigid waters near polar ice. But vast numbers of other archaea are present in cold seawater. These cells have enzymes that can work efficiently at near-freezing temperatures. No organism, however, lives in both hot springs and cold ocean water.

The point here is that organisms tend to be adapted to a limited set of physical conditions. Although some species have much broader geographic ranges than other species do—and humans and the bacteria that live with them may have the widest distribution of any organisms at present—no organism can live everywhere.

To understand a species' distribution fully, however, it is essential to examine historical and biotic factors in addition to the physical conditions present. For example, the temperature and moisture regimes of large areas of western North America are almost identical to those found in the Mediterranean region of Europe and in central Eurasia. Even though the physical conditions at these sites are extremely similar, the species present are almost completely different. Why?

The Role of History

To explain why certain species occur in a particular region while others don't, the first and most obvious factor to consider is history—specifically, the history of dispersal. **Dispersal** refers to the movement of an individual from the place of its birth, hatching, or origin to the location where it lives and breeds as an adult. If a particular species is missing from an area, a physical barrier to dispersal—such as a river, a glacier, a mountain range, or an ocean—may be present. For example, the obvious cause of differences in the plant species found in the grasslands and shrublands of western North America versus those of Eurasia is that the Pacific and Atlantic Oceans are effective barriers against the dispersal of seeds.

In many cases, human activity is circumventing physical barriers to dispersal. For example, public health officials closely monitor annual flu outbreaks in south China because they recognize that new strains disperse all over the planet in a matter of weeks or even days, in the respiratory passages of airplane passengers. Prior to the invention of air travel, disease epidemics in humans were much more likely to be confined to relatively small geographic areas.

Similarly, humans have transported thousands of seeds, birds, insects, and other species across physical barriers to new locations—sometimes purposefully and sometimes by accident. One accidental introduction has had disastrous consequences for the arid shrublands and grasslands of the western United States. In 1889 a native of Eurasia called cheatgrass was accidentally introduced to western North America in a shipment of crop seeds. Cheatgrass is an annual plant with seeds that germinate over winter; plants grow and set seed in early spring. Within 30 years, cheatgrass had taken over tens of thousands of square miles of habitat (**Figure 50.22**).

FIGURE 50.22 Cheatgrass Has Invaded Extensive Areas of North America
Cheatgrass is an annual plant with seeds that germinate over winter. Plants grow and set seed in early spring.

If an exotic organism is introduced into a new area, spreads rapidly, and eliminates native species, it is said to be an **invasive species.** In North America alone, dozens of invasive species have had devastating effects on native plants and animals. North American invasives include kudzu, purple loosestrife, reed canary grass, garlic mustard, Russian thistle (also known as tumbleweed), European starlings, African honeybees, and zebra mussels. Similar lists could be compiled for other regions of the world.

Regarding the distribution of organisms, the story of invasive species is simple: They and other introduced species did not exist in certain parts of the world until recently, simply because they had never been dispersed there. But data show that only about 10 percent of the species introduced to an area actually become common, and a smaller percentage than that increases in population enough to be considered invasive. If a species is introduced into an area but does not thrive, what is the cause? Conversely, why are invasives so successful?

Biotic Factors

The distribution of a species is often limited by biotic factors—meaning interactions with other organisms. As an example, recall the data presented in Chapter 25 on the current and former distributions of hermit warblers and Townsend's warblers along the Pacific coast of North America. Both hermit warblers and Townsend's warblers live in evergreen forests, but their ranges do not overlap. Experiments have shown that male Townsend's warblers directly attack male hermit warblers and evict them from breeding territories, and historical data indicate that the geographic range of

(a) Distribution of cattle is limited by distribution of tsetse flies.

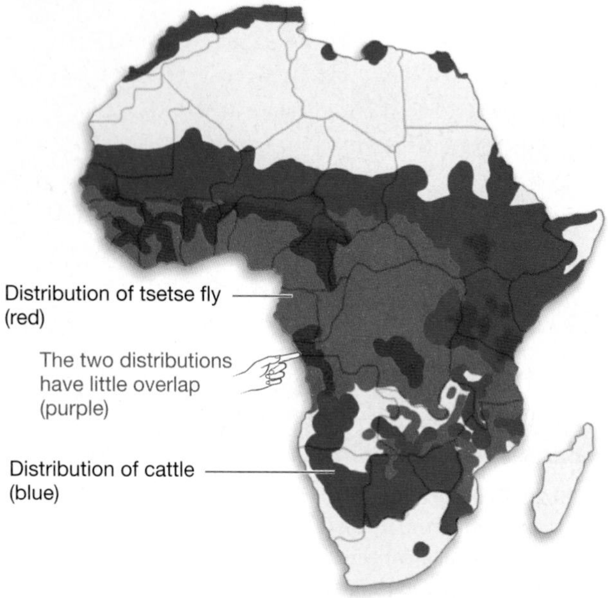
Distribution of tsetse fly (red)
The two distributions have little overlap (purple)
Distribution of cattle (blue)

(b) Yucca moth distribution is limited by yucca plant distribution.

FIGURE 50.23 Distributions Are Limited by Biotic Factors
(a) Because tsetse flies carry a disease that is fatal to cattle, tsetse flies and cattle rarely coexist. **(b)** Because yucca moths lay their eggs only in the flowers of yucca plants, the moths cannot live where yucca plants are absent.

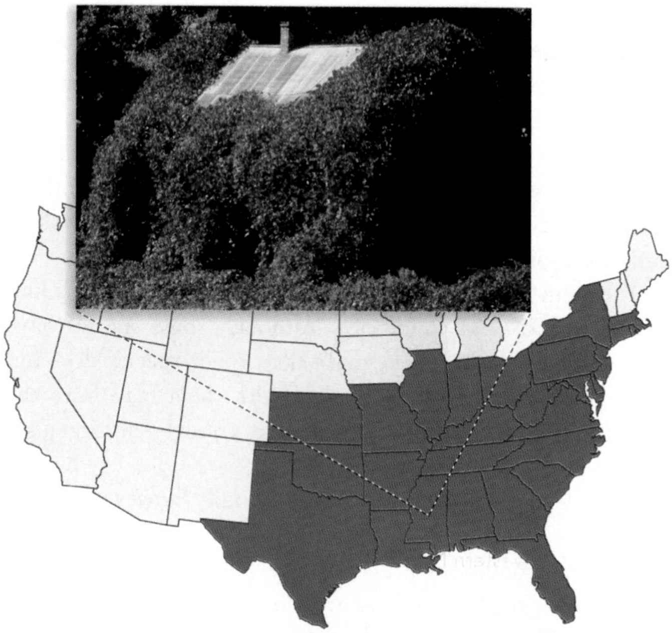

FIGURE 50.24 Distributions Are Limited by Abiotic Factors
Kudzu is an invasive plant that has spread throughout the southeastern United States. This plant's distribution is limited by its inability to grow in soils that freeze to more than a foot in depth.

Townsend's warblers has been expanding steadily at the expense of hermit warblers. The results support the hypothesis that a biotic factor—competition with another species—is limiting the range of hermit warblers.

Competition is not the only biotic factor that can limit a species' distribution, however. In Africa, the range of domestic cattle is limited by the distribution of tsetse flies, because tsetse flies transmit a parasite that causes the disease trypanosomiasis, which is fatal in cattle (**Figure 50.23a**). Similarly, most of the songbirds native to Hawaii are limited to alpine habitats—above the range of the mosquitoes that transmit avian malaria. And the moth pictured in **Figure 50.23b** lays its eggs only in the flowers of yucca plants. As a result, this species does not exist outside the range of yucca plants.

Abiotic Factors

In many cases, introduced species may not become established in a region because of abiotic factors—particularly temperature and moisture. An area may simply be too cold or too dry for an organism to survive. For example, even though the invasive plant kudzu grows prolifically throughout the southeastern United States (**Figure 50.24**), it is killed when the soil freezes to a depth of a foot or more. As a result, it has not expanded its range farther north and west.

It is often difficult to separate the effects of biotic and abiotic factors on a species' range, however. To drive this point home, consider how cheatgrass became established in North America. Cheatgrass does not grow in wet grasslands, because it does not compete well against the tall species that thrive there. But it has been able to invade two important types of biomes in North America: dry temperate grasslands and the arid, shrub-dominated habitats known as **sage-steppe.** In each case, the reasons for the success of cheatgrass were different.

Cheatgrass has been able to invade sage-steppe habitats because it is not affected by fire. Fire is an abiotic factor that can kill the shrubs that dominate sage-steppe (**Figure 50.25**). In fact, because cheatgrass grows in dense beds and dies back in early spring, it actually leads to an increase in the frequency and intensity of wildfires. Normally, fires are rare in sage-

FIGURE 50.25 Sage-Steppe Is an Arid Biome Dominated by Shrubs
Large areas of western North America are covered with sage-steppe vegetation.

steppe biomes, because plants are widely spaced. This distribution usually prevents fires from spreading and affecting a large area. But when this biome begins to dry in late summer, beds of dead cheatgrass offer a continuous, extremely flammable source of fuel. If a fire does come through, it does not affect the cheatgrass for two reasons: (1) Cheatgrass is an annual, so there is no aboveground tissue once the growing season is over, and (2) cheatgrass seeds sprout readily in soils that have been depleted of organic matter by fire. Because sagebrush and other shrubs are perennials, they are killed by hot fires. By increasing the frequency and intensity of fire in sage-steppe biomes, the presence of cheatgrass creates an environment that favors the spread of cheatgrass.

How has this species been able to invade dry grassland biomes? Prior to the arrival of cheatgrass, grasslands in arid areas of the American west were dominated by the types of bunchgrasses shown in **Figure 50.26**. Bunchgrasses do not form sod; instead, they grow in compact bunches. The spaces between the bunches are occupied by what biologists call "black-soil crust." This material is a community of soil-dwelling cyanobacteria, archaea, and other microorganisms that grow when moisture is available. Many of these species also fix nitrogen (see Chapter 37). The nitrogen becomes available to the surrounding plants when the microbial cells die and decompose.

When European settlers arrived and began grazing cattle in this biome, however, both the black soils and bunchgrasses were affected. Cattle grazed on the bunchgrasses until the grasses died, and their hooves compacted and disrupted the black soils. Neither component of the ecosystem was allowed to recover, in part because intensive grazing continued and in part because cheatgrass arrived. Cheatgrass shades out black-soil crust organisms. When black-soil crust is gone, the amount of nitrogen available to bunchgrasses is reduced. Cheatgrass also competes successfully with bunchgrasses for water and nutrients, because it germinates during the winter and completes its growth early in the spring—before the native bunchgrasses have broken dormancy and begun to grow. And as in sage-steppe habitats, the expansion of cheatgrass increased the frequency and severity of fires, which further degraded the integrity of black-soil crust communities.

To summarize, changes in abiotic and biotic factors allowed cheatgrass to invade native biomes in North America and thrive as an invasive species. Currently, managing cheatgrass is the major problem facing conservation biologists and rangeland managers in the western United States.

Although the story of cheatgrass is particularly dramatic, it is important to remember that the range of every species on Earth is limited by a combination of historical, abiotic, and biotic factors. The range of a particular species depends on dispersal ability, the capacity to survive climatic conditions, and the ability to find food and avoid being eaten. Climate and other abiotic factors were the focus of this chapter; the next several chapters focus on the biotic aspects of ecology. How do interactions among organisms of the same species and interactions among other species affect geographic range and population size?

FIGURE 50.26 Arid Grasslands Feature Bunchgrasses and Black-Soil Crust
As their name implies, bunchgrasses grow in compact bunches instead of forming a sod. The black-soil crust that forms between bunchgrasses is composed of cyanobacteria, archaea, and other microscopic organisms that add organic matter and nutrients to the soil.

ESSAY Battling Invasive Species

Conservation groups now rank invasive species as the second most serious danger to threatened and endangered species in the United States—second only to direct destruction of habitats. Globally, invasive organisms represent one of the gravest threats to native species and the integrity of biomes. Some introduced plants, such as cheatgrass and purple loosestrife—which have taken over extensive areas of marshland in North America—are outcompeting native species. The brown tree snake increased to densities of up to 13,000 individuals per square mile after being introduced to the island of Guam and has virtually wiped out all of the native birds, leading to at least 12 species extinctions. Introduced diseases have dramatically reduced populations of American chestnut and American elm as well as other forest trees. What can be done?

In conservation biology, as in medicine, an ounce of prevention is worth a pound of cure. Although species introductions were once encouraged, most governments now actively inspect imported plants, produce, and other living materials to prevent the accidental introduction of unwanted organisms. Government agencies and private conservation groups have also designed rapid response networks, so that quick action can be taken to eliminate invasive species if they are detected in a new area.

In areas where invasive species are already established, researchers are trying to devise strategies to manage their spread or reduce their numbers. Management strategies to battle invasives range from manually pulling purple loosestrife plants from marshes to lighting early spring fires in stands of cheatgrass, so that these plants will be burned when they are growing quickly and are most vulnerable.

In several cases, conservation biologists are using plant-eating insects or parasites as control agents. The hypothesis here is that invasive organisms are successful in exotic habitats because their natural predators and diseases are absent. By introducing these predators and diseases, researchers hope to control the spread of the invasive species. For example, weevils that feed on the roots or flower buds of purple loosestrife and beetles that eat purple loosestrife leaves were field tested to determine whether they control purple loosestrife effectively and have little to no impact on native vegetation. The results were promising enough that larger-scale cultivation and release of these insects are now being implemented. Similarly, researchers are actively testing the ability of a parasitic fungus to infect and control cheatgrass.

Globally, invasive organisms represent one of the gravest threats to native species. . . .

Although promising strides have been made in the management of invasive species recently, continued and expanded management and research programs are critical. As transcontinental travel and commerce continue to grow, the threat of accidental or malicious introductions continues. Keeping invasive species at bay—and native biomes intact—demands constant vigilance.

CHAPTER REVIEW

Summary of Key Concepts

- **Ecology focuses on how organisms interact with their environment. Because its goal is to understand the distribution and abundance of organisms, ecology provides a scientific foundation for the conservation of species and natural areas.**

Biologists study ecology at four main levels: (1) organisms, (2) populations, (3) communities, and (4) ecosystems. The goal of organismal biology is to understand how individuals respond to stimuli from the environment, including members of their own species and other species. Population ecology focuses on how and why populations grow or decline. Community ecology is the study of how two or more species interact, and ecosystem ecology analyzes interactions between communities of organisms and their abiotic environment—particularly the flow of energy and nutrients. In efforts to preserve endangered species and communities, conservation biologists apply analytical methods and results from all four levels.

- **An organism's environment consists of both abiotic, or physical, factors and biotic factors, meaning other organisms.**

Abiotic factors that affect organisms include the chemistry of water in aquatic habitats, the nature of soils in terrestrial habitats, and climate, which is a combination of temperature, moisture, sunlight, and wind. Biotic factors include interactions with offspring, mates, competitors, predators, prey, and parasites.

Climate varies around the globe because sunlight is distributed asymmetrically, with equatorial regions receiving more solar radiation than do regions toward the poles on average. Hadley cells create bands of wet and dry habitats, and the tilt of Earth's axis causes seasonality in the amount of sunlight that non-equatorial regions receive. The presence of mountains can create local areas of wet or dry habitats, and proximity to an ocean moderates temperatures in nearby terrestrial habitats.

Web Tutorial 50.1 Tropical Atmospheric Circulation

On land, environments are defined by climate—specifically, both the average value and annual variation in temperature and in moisture. In aquatic habitats, environments are defined by physical structure—particularly water depth.

Distinct climatic regimes are associated with distinct types of terrestrial vegetation, or biomes. Biomes represent the major types of environments available to terrestrial organisms. Because photosynthesis is most efficient when temperatures are warm and water supplies are ample, the productivity and degree of seasonality in biomes varies with temperature and moisture.

Water depth is a crucial factor in aquatic ecosystems because the wavelengths of sunlight that are required for photosynthesis do not penetrate water efficiently. Thus, only certain zones in a lake or ocean can support photosynthesis, and organisms in other aquatic zones must depend on food that rains out of the sunlit areas near the surface.

A species' distribution is constrained by a combination of historical, abiotic, and biotic factors.

The range, or geographic distribution, of a particular species depends on dispersal ability, the capacity to survive climatic conditions, and the ability to find food and avoid being eaten. For each species, a unique combination of historical, abiotic factors, and biotic factors dictates where individuals live and the size of the population.

Questions

Content Review

1. What is a rain shadow?
   a. the part of a mountain that receives prevailing winds and heavy rain
   b. the region beyond a mountain range that receives dry air masses
   c. the region along the equator where precipitation is abundant
   d. the region near 30°N and 30°S latitude that receives hot, dry air masses
2. A region receives less than 5 cm (2 in.) of precipitation annually and has temperatures that never drop below freezing. Which type of biome is present?
   a. subtropical desert b. temperate grassland
   c. boreal forest d. arctic tundra
3. What is the predominant type of vegetation in a tropical wet forest?
   a. shrubs and bunchgrasses
   b. herbs, grasses, and vines
   c. broad-leaved deciduous trees
   d. broad-leaved evergreen trees
4. The littoral zone of a lake is most similar to which of the following environments?
   a. oceanic zone b. intertidal zone
   c. neritic zone d. marsh
5. Typically, where are oxygen levels highest and nutrient levels lowest in a stream?
   a. near its source
   b. near its mouth, or end
   c. where it flows through a swamp or marsh
   d. where it forms an estuary
6. Which of the following is most important in limiting the distribution of an invasive species?
   a. historical factors
   b. abiotic factors
   c. biotic factors
   d. historical, abiotic, and biotic factors are equally important for invasive species

Conceptual Review

1. Name the four main levels of study in ecology. Write a question about the ecology of humans, salmon, or oak trees that is relevant to each level.
2. In June, does the Northern Hemisphere receive more or less solar radiation than the equator does? Explain your answer.
3. Diagram the Hadley cell that exists between the equator and 30° latitude north and south, indicating the direction of air flow. Similar air circulation patterns exist between 30° and 60° and between 60° and the poles. Diagram these cells as well.
4. Contrast the productivities of the intertidal, neritic, and oceanic zones of marine environments. Explain why large differences in productivity exist.
5. Compare and contrast the biomes found at increasing elevation on a mountain with the biomes found at increasing latitude in Figure 50.6.
6. Explain why productivity is much lower in bogs than it is in marshes or lakes that receive the same amount of solar radiation.

Group Discussion Problems

1. Mars has an even more pronounced tilt than Earth does. Explain why Mars does or does not have seasons.
2. Mountaintops are closer to the Sun than the surrounding area is. Why are mountaintops cold?
3. The southwest corners of some Caribbean islands are extremely dry, even though other areas of the islands receive substantial rainfall. Explain this observation. Base your answer on a hypothesis (one that you propose) for the direction of the prevailing winds in this region of the world.
4. Edinburgh, Scotland, and Moscow, Russia, are at the same latitude, yet these two cities have very different climates—Moscow has much colder winters. Explain why.

Answers to Multiple-Choice Questions 1. b; 2. a; 3. d; 4. b; 5. a; 6. c

www.prenhall.com/freeman is your resource for the following: Web Tutorials; Online Quizzes and other Online Study Guide materials; Answers to Conceptual Review Questions; Solutions to Group Discussion Problems; Answers to Figure Caption Questions and Exercises; and Additional Readings and Research.

51 Behavior

KEY CONCEPTS

- After describing a behavior, biologists seek to explain both its proximate and ultimate causes—meaning, how it happens at the genetic and physiological levels and how it affects the individual's fitness.
- In a single species, behavior may range from highly stereotyped, invariable responses to highly flexible, conditional responses and from unlearned to learned responses.
- The types of learning that individuals do, the way that they communicate, and the way that they orient and navigate correlate closely with their habitat and with the challenges they face in trying to survive and reproduce.
- When individuals behave altruistically, they are usually helping close relatives or individuals that help them in return.

Peregrine falcons hunt ducks and other large birds. When diving in pursuit of prey, peregrines can reach speeds of up to 320 km/hr (200 mph).

To biologists, **behavior** is action—specifically, the response to a stimulus. Pond-dwelling bacteria swim toward drops of blood that fall into the water. Time-lapse movies of sunflowers growing in a field document the steady movement of their flowering heads throughout the day, as they turn to face the shifting Sun. Filaments of the fungus *Arthrobotrys* form loops, then release a molecule that attracts roundworms. When a roundworm touches one of these loops, the fungal filaments swell—ensnaring the worm. *Arthrobotrys* cells then proceed to grow into the worm's body and digest it.

Among animals, action may be frequent and even spectacular. Peregrine falcons dive at breathtaking speeds in pursuit of ducks or other flying prey. Some Arctic terns migrate over 32,000 km (20,000 miles) each year. Deep inside a hive, in complete darkness, honeybees can communicate the position and nature of a food source a mile away.

Biologists who study behavior ask questions about what organisms do in response to short-term changes in the environment, how they do it, and why. Although all organisms respond in some way to signals from their environment, most behavioral research is performed on vertebrates, arthropods, or mollusks. With their sophisticated nervous systems and skeletal-muscular systems, these animals can sense, process, and respond rapidly to a wide array of environmental stimuli.

Although most behavioral studies start by describing what animals do in response to a particular stimulus, the eventual goal is to explain the proximate and ultimate causes of behavior. **Proximate**, or mechanistic, **causation** explains *how* actions occur in terms of the neurological, hormonal, and skeletal-muscular mechanisms involved. **Ultimate**, or evolutionary, **causation** explains *why* actions occur—based on their evolutionary consequences and history. Is a particular behavior adaptive, meaning that it increases an individual's fitness? If so, how does it help individuals produce more offspring in a particular environment?

To illustrate the proximate and ultimate levels of causation, consider the spiny lobster shown in **Figure 51.1**. These animals spend the day hiding in cracks or holes in coral reefs. At night they emerge and wander away from the reef in search of clams, mussels, crabs, or other sources of food. Before dawn, they return to one of several dens they use on a regular basis. How do they find their way back? This question addresses how an individual does what it does—representing the proximate level of causation. Although research is continuing, recent experiments have provided strong support for the hypothesis that spiny lobsters can detect Earth's magnetic field and that they navigate by using information on how the orientation and strength of the field changes as they move about. But at the ultimate level of causation, biologists want to know why the behavior occurs. The leading hypothesis to explain homing behavior is that spiny lobsters with the ability to navigate home can search for food over a wide area under cover of darkness, then return to a safe refuge before sharks and other large predators that hunt by sight can find them.

It is important to recognize that analyzing behavior at the proximate and ultimate levels is complementary. To understand what an organism is doing, biologists want to know

FIGURE 51.1 Spiny Lobsters Can Find Their Way Back to Their Home Burrow
Spiny lobsters wander widely in search of food at night, then return to their hiding places in coral reefs before dawn.

both how the behavior happens and why. Accomplishing this task requires them to draw on techniques and results from genetics, physiology, and evolutionary biology.

Because behavioral biologists work at the interface of so many fields, behavior is among the most dynamic of all disciplines in biological science. Even so, behavioral biology is essentially ecological in nature. Ecology is the study of how organisms interact with their physical and biological environments, and behavioral biology is the study of how organisms respond to particular stimuli from those environments. What does an *Anolis* lizard do when it gets too hot? How should that same individual respond when a member of the same species approaches it? Or when a house cat approaches?

The goals of this chapter are to explore some key topics in behavioral biology and illustrate the spirit of the field. To begin, let's consider one way to categorize and organize the enormous variety of behavioral responses that is observed in animals.

51.1 Types of Behavior—An Overview

Some types of behavior are performed nearly the same way every time. When people yawn, the action is so similar that every instance takes about 6 seconds. Other types of behavior are highly flexible. When people smile, the action is variable. The same person may smile in dozens of ways, and different people tend to have their own individual style and frequency of smiling. In addition, some types of behavior are readily modified by learning, while other types of behavior are not. Formally, **learning** is defined as a change in behavior that results from a specific experience in the life of an individual. Human infants learn their native language through imitation and trial and error, but they begin to yawn and smile in a normal way before they have a chance to learn these actions from their parents.

Figure 51.2 (page 1168), offers one way of summarizing these general observations about behavior. Notice that the graph has two axes. The horizontal axis maps how flexible a behavior is in terms of its expression within and between individuals of the same species. This axis forms a continuum ranging from fixed, invariant, or highly stereotyped behaviors to highly flexible and variable responses. The vertical axis plots how much learning is involved in the nature of the response. This axis ranges from behaviors that require little or no learning for normal expression to responses that are highly dependent on learning. In humans, yawning would map in the upper left-hand corner of this graph while language acquisition would be plotted in the lower right-hand corner.

When researchers first began to explore animal behavior experimentally, they focused on responses that are highly stereotyped and that are largely unlearned. This general research

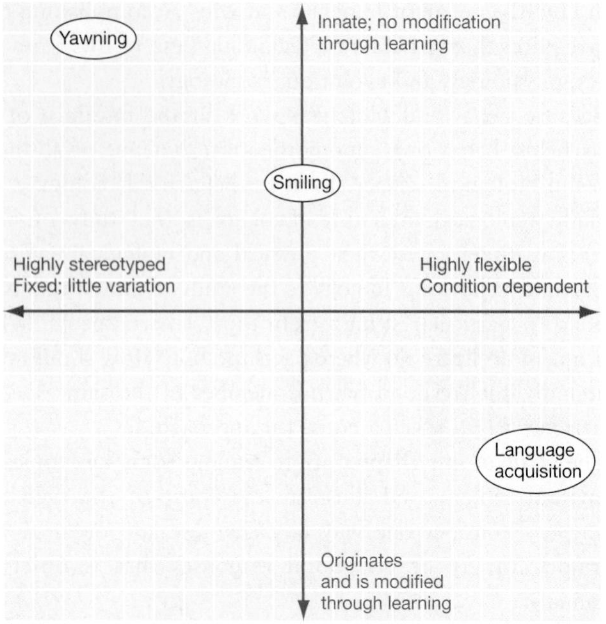
Innate; no modification through learning
Yawning
Smiling
Highly stereotyped
Fixed; little variation
Highly flexible
Condition dependent
Language acquisition
Originates and is modified through learning

FIGURE 51.2 Behavioral Traits Vary in Flexibility and Dependence on Learning
These axes represent one way to organize the diversity of behavioral traits observed within and between species. Each type of behavior observed in a species could be plotted at a particular point on the graph.

strategy is common in biological science. When biologists start to analyze a question—ranging from how DNA is transcribed to how plasma membranes work—they start with the simplest possible system. Once simple systems are understood, investigators can delve into more complex situations and questions. Behavioral biology started the same way. Early work focused on behavioral responses that map to the upper left-hand corner of the graph in Figure 51.2. Let's look at some conclusions from these pioneering studies of stereotyped responses, then consider examples of work on more flexible types of behavior.

Fixed Action Patterns

Highly stereotypical behavior patterns are called fixed action patterns, or FAPs. FAPs have three key characteristics:

1. There is almost no variation in how they are performed. For example, every "headthrow" display given by a male common goldeneye duck involves moving the head back to the tail and then swinging it forward (**Figure 51.3a**). The sequence of movements is so stereotyped that the display takes almost exactly 1.29 seconds every time.
2. They are species specific. Each species of spider builds a distinct type of web (**Figure 51.3b**). Even if a spider never sees a web built by a member of its own species, it still builds a web of its species-specific type.
3. Once the sequence of actions begins, it typically continues to completion. If you put an egg outside a goose's nest, the goose will extend its neck until the egg is just behind its bill. Then the goose retrieves the egg with a side-to-side movement of the head (**Figure 51.3c**). If researchers attach a string to an egg and pull it away after the goose has begun to retrieve it, the goose completes the retrieval movement even without the egg there.

FAPs are examples of what biologists call **innate behavior**—types of behavior that are inherited and show little variation based on learning or the individual's condition. (**Box 51.1** explores the role of genetics in behavior in more detail.) FAPs are

FIGURE 51.3 A Fixed Action Pattern Is a Stereotyped, Inflexible Sequence of Movements
(a) The headthrow display of the male common goldeneye. **(b)** Spiders of the same species make the same type of web. **(c)** Geese will continue the motion of retrieving eggs even if the egg has been pulled away.

BOX 51.1 A Closer Look at Behavior Genetics

Biologists explore the genetic bases of behavior to understand the proximate mechanisms responsible for specific behavioral traits. Are certain alleles associated with certain types of behavior? If so, how do the products of these alleles change what an individual does in response to certain stimuli?

As an introduction to the field of behavior genetics, let's consider two strategies that biologists use to explore the genetic basis of behavior. The first involves efforts to identify single genes associated with particular behavioral traits; the second focuses on the more common situation in which many genes influence a behavioral response.

The *foraging* (*for*) gene in fruit flies, discovered by Marla Sokolowski, is a particularly good example of a single gene that influences behavior. As an undergraduate research assistant, Sokolowski noticed that some of the fruit-fly larvae she was studying tended to move after feeding at a particular location, while others tended to stay put. By rearing these "rovers" and "sitters" to adulthood, breeding them, and studying the offspring, she was able to determine that the trait is inherited and that the gene responsible is located at the same location as a previously identified gene called *dg2*. (Recall from Chapter 20 that the entire fruit-fly genome has been sequenced; Chapter 19 introduced the techniques that Sokolowski used to map the *for* gene.) The protein product of *dg2* is involved in the types of signal transduction cascades introduced in Chapters 8 and 47—meaning it is involved in the response to a signal. Rovers and sitters behave differently because they have different alleles of the *for/dg2* gene.

Follow-up experiments in the lab showed that the rover allele is favored when population density is high but that the sitter allele reaches high frequency in low-density populations. These results are considered exciting because they link a proximate mechanism with an ultimate outcome—in this case, the presence of certain alleles with a difference in fitness in specific types of habitats. Sokolowski's current goals are to find out which signaling pathway uses the *for/dg2* gene product and why different alleles lead to different types of feeding behavior.

Although the *for/dg2* gene presents a compelling story, biologists recognize that it is rare for behavioral traits to be so heavily affected by a single gene product. In reality, most aspects of behavior are influenced by the action of many genes. (Think of the number of genes that are likely to be involved in the formation of the human brain and thus in aspects of human personality and decision making.) To use a term that was introduced in Chapter 13, most behavioral traits are *polygenic*. Instead of representing discrete states such as roving and sitting, polygenic traits exhibit quantitative variation—meaning that individuals differ by degree and that many different phenotypes are observed when the population is considered as a whole.

Unfortunately, it is extremely difficult to identify the specific genes and alleles responsible for polygenic traits. To organize their thinking about the genetics of polygenic traits, biologists measure how much the trait varies among individuals in a population and then recognize that every trait in an individual results from the alleles it has and the environment that it has experienced. More formally, biologists distinguish **phenotypic variation** (V_P), **genetic variation** (V_G), and **environmental variation** (V_E). These three quantities have a simple relationship: $V_P = V_G + V_E$. This equation says that the observed variation of a behavioral trait within a population is the sum of variation among individuals in the alleles they possess and the variation in the environments that they have experienced.

This simple formulation helps explain why some types of behavior show little variation among individuals while other types show a great deal of variation. For example, traits show little variability among individuals—meaning that V_P is small—because there is low genetic variation (V_G is small) and/or little environmental influence (V_E small). At the other end of a continuum of behavior, some traits are highly variable among individuals in a population, meaning that V_P is large. In traits such as these, there is a high degree of genetic variation (V_G is large) and/or a great deal of environmental influence on the trait (V_E is large).

The equation also gives biologists a key tool for exploring behavior genetics, even when the specific genes involved are not known. For example, by rearing a large group of individuals in an identical environment (so that V_e is low or zero) and then measuring variation in a particular behavioral trait, researchers can estimate the proportion of the observed variation in the population that is due to genetic differences among individuals. If most phenotypic variation is due to genetic variation, then the trait is likely to respond quickly to natural selection.

Finally, advances in behavior genetics promise to accelerate quickly due to the availability of genome sequences from an array of animals. For example, recent work on the genomes of dogs and chimpanzees is motivated in part by the goal of identifying at least some of the genes involved in complex, polygenic behavioral traits such as aggression and learning. Behavior genetics is an exciting frontier in behavioral biology.

analogous to the types of muscular reflexes introduced in Chapter 46. Recall that if a person's knee is struck with a rubber hammer, the lower leg jerks forward. Like a FAP, a muscular reflex is similar each time it occurs and goes to completion once it begins.

One of the first conclusions made about these simple, highly stereotyped behaviors is that they are responses to very simple stimuli. The stimuli that evoke FAPs are called **releasers** or **sign stimuli**. The presence of a female common goldeneye can "release" the headthrow display of the male common goldeneye. Seeing an egg outside the nest releases the egg-retrieval FAP in geese.

Some of the most classic experiments in the history of behavioral research have focused on identifying releasers. As an example of this work, consider a study that David Lack did on territorial behavior in European robins. Breeding pairs of male and female European robins establish a **territory**—an area of habitat that they defend against other members of their own species. If another robin enters the territory, the owners call, chase the intruder, or even assault it. To explore how this territory defense behavior works, Lack placed a taxidermic mount, or "stuffed" model, of a robin on an established territory. Both territory owners attacked the "intruder." This observation showed that the territorial defense response can be released by the color or shape of an intruder, in the absence of movement or vocalizations.

Lack repeated the experiment to make sure that the same response occurred in other breeding pairs. As he did, he got lucky. One particularly feisty female knocked the head off the stuffed specimen. To Lack's surprise, her attacks continued unabated. Eventually the unfortunate mount was reduced to a puff of feathers. Lack noticed that as long as the mount retained the orange breast feathers found on European robins, the attacks continued.

Based on these observations, Lack proposed a hypothesis: Territorial defense in European robins is a FAP that is released by the sight of orange feathers (**Figure 51.4**). To test the null hypothesis that orange feathers have nothing to do with the territorial response, Lack took a normal mounted specimen of a European robin and painted the orange breast feathers brown, to match the individual's back. When he presented this altered mount to breeding pairs—at the same time of day and year and under the same conditions as he had when presenting a normal mount—the territory owners barely responded. This result supported his hypothesis that orange feathers are required to release territory defense. The result also supports a more general point that releasers consist of small amounts of simple information.

Question: What characteristic of a robin intruder releases territorial behavior in breeding European robins?

Hypothesis: Orange feathers act as a releaser for territorial behavior in European robins.

Null hypothesis: Orange feathers have no effect on territorial behavior in European robins.

Experimental setup:

In territory of a breeding pair of European robins, place a full robin mount with normally orange breast feathers painted brown.

In territory of a breeding pair of European robins, place a normal robin mount.

Prediction: The bird mount lacking orange feathers will be ignored. The bird mount with orange feathers will be attacked.

Prediction of null hypothesis: Orange feathers will be ignored.

Results:

Mount that lacks orange feathers is not attacked.

Mount with orange feathers is attacked.

Conclusion: Orange feathers release territorial behavior in European robins.

FIGURE 51.4 Experimental Evidence of a Releaser
FAPs are triggered by simple stimuli. A stimulus is considered a releaser if it alone can trigger a FAP. **EXERCISE** Design an experiment to test the hypothesis that observing someone yawn acts as a releaser for yawning in humans.

What Is the Adaptive Significance of FAPs? Geese that retrieve an egg that isn't there and robins that attack puffs of orange feathers appear to be acting in a nonadaptive way. How can alleles that contribute to behavior like this help individuals produce more offspring and thus increase in frequency in a population? The answer has two parts.

First, natural selection cannot produce adaptations to unusual or "trick" situations, such as an egg being pulled away from a goose or a robin being presented with a puff of orange feathers. Second, even though the sign stimuli that release FAPs are extremely simple, they are exceptionally important for the individual's fitness—its ability to survive and produce offspring. To drive this point home, consider what happens in a darkened environment when a kangaroo rat hears the sound of a rattlesnake's rattle or when you hear a piercing scream nearby. In both cases, a simple sign stimulus releases a FAP—a rapid jumping movement backward, away from the direction of the stimulus. In both cases, the FAP may result in survival. In this light, it is logical to observe that FAPs show little variation among individuals and continue to completion once they have started. Yet the "jump-back" FAP appears nonadaptive if the stimulus is actually a tape recording of a rattlesnake or a friend who is hiding behind a door.

Conditional Strategies

Research on FAPs and releasers dominated behavioral biology through the 1950s. But today investigators focus more on behavioral traits that are triggered by complex information and that are variable and flexible in their expression. As an example of flexible behavior, let's examine a particularly startling phenomenon: female fish that change their sex and become male.

Many species of coral-reef fish have a distinctive mating system (**Figure 51.5**). Males defend territories that contain nesting sites and feeding areas. A group of females lives inside the boundaries of this territory. When these females lay eggs, the male fertilizes the eggs. Thus, a single male monopolizes all the matings in that territory. Invariably, this male is the largest individual in the group. This is logical because the male guards the territory, and because fights between fish are usually won by the biggest contestant.

FIGURE 51.5 Many Species of Coral-Reef Fish Can Change Sex, Depending on Conditions
Bluehead wrasses: a group of females (yellow) surround a male (with the blue head). Bluehead wrasses start life as females but can change sex and become male.

When the territory-holding male dies, however, something unusual happens. The largest female in the group changes sex. Her reproductive organs change from egg production to sperm production. She becomes the dominant male and begins fertilizing all of the eggs laid in the territory.

Why does she do this? What does the sex-changing female gain by making this switch? These questions are at the ultimate level of explanation and have been addressed by an idea called the size-advantage hypothesis. This hypothesis states that if a group of fish are living in a territory dominated by a single male, females should switch from female to male when they become very large. The hypothesis is based on the observation that fish have *indeterminate growth*—meaning they continue growing throughout their lives.

To understand the logic behind the size-advantage hypothesis, suppose that a small female fish can lay 10 eggs a year but a large female of the same species can lay 20 eggs a year. If six small females and two large females live in a harem, the male that owns the territory fertilizes 100 eggs each year. If the male dies, the largest female can increase the number of offspring she produces each year from 20 to 80 by changing sex and taking over the role of dominant male. Alleles that allow females to do this will increase rapidly in the population. There is no fitness advantage for smaller females to switch sex, though, because they would be defeated in fights and have 0 offspring per year instead of 10.

Biologists refer to this type of behavior as a conditional strategy. In contrast to a FAP, a **conditional strategy** is a behavioral response that depends on conditions—in this case, on the size and sex of other individuals in the same social group. The name is apt because the individuals involved are capable of behaving in several ways. It's important to note, though, that the word *strategy* does not denote any sort of conscious choice. The critical point is that individuals are observed to adopt the behavior that allows them to produce the most surviving offspring.

What other types of factors lead to flexible behavior that varies among individuals in a population? In many species, learning is crucial. Chapter 45 indicated that, at the proximate level, learning occurs because individual neurons and connections between neurons are modified in response to experience. For example, recall that in the sea slug *Aplysia*, the production of the neurotransmitter serotonin changes in response to experience. Changes in serotonin release lead to changes in the behavior of sea slugs in response to attacks by predators. Learning is usually based on changes in neurons.

Learning is a fundamentally important phenomenon in animal behavior. Let's take a closer look at learning from both proximate and ultimate perspectives. How does learning occur, and why?

✓ CHECK YOUR UNDERSTANDING

Within a given species, behavioral traits may range from highly stereotyped to highly flexible and from innate to learned. Fixed action patterns (FAPs) are highly stereotyped, innate responses released by simple stimuli. In contrast, sex change in fish is an example of a highly flexible response stimulated by complex environmental changes. You should be able to (1) think of an example of a FAP and a condition-dependent behavior in a species that you are familiar with and (2) pose hypotheses to explain why each type of behavior is adaptive.

51.2 Learning

Learning occurs when behavior changes in response to specific life experiences. Learning is particularly important in species that have large brains and a lifestyle that is dominated by complex social interactions—including humans, dolphins, chimps, and crows. In species such as these, FAPs and other types of inflexible, stereotyped behaviors are relatively rare. Instead, each individual is capable of a wide range of behavior. What an individual actually does is condition dependent or learned.

Research on learning is both extensive and fascinating. Here it is possible to touch on only a few highlights, to provide examples of what is known about learning and how biologists go about studying the phenomenon.

Simple Types of Learning: Classical Conditioning and Imprinting

Learning can be quite simple. In response to the stimulus of food, for example, dogs respond by salivating. Ivan Pavlov discovered that if he rang a bell each time food was presented to a dog, the dog would learn to associate the sound of the bell with food. When stimulated by a bell in the absence of food, the dogs Pavlov was training would begin to salivate. This type of learning is called classical conditioning. In **classical conditioning**, individuals are trained by experience to give the same response to more than one stimulus—even a stimulus that has nothing to do with the normal response.

Another type of simple learning takes place in newly hatched ducks, geese, and certain other species of birds. Upon hatching, ducklings and goslings adopt as their mother the first moving thing they see. This type of learning is called **imprinting**. In nature, the "thing" that creates the imprint is the mother duck or goose. But Konrad Lorenz found that if he incubated eggs artificially until they hatched, he could be the first thing that the offspring saw. In response, young greylag geese would imprint on whatever boots he was wearing at the time (**Figure 51.6a**). In contrast, mallard ducklings would imprint on him only if he crawled on all fours and quacked continuously.

Why does such a highly constrained type of learning exist, in response to such simple stimuli? At the ultimate level of causation, the leading hypothesis to explain imprinting is that offspring must quickly learn to recognize and respond to their mother in order to survive. This hypothesis is particularly plausible in species such as ducks and geese, which nest on the ground. Ducklings and goslings hatch synchronously—meaning all of the eggs hatch within a few hours of each other—and are relatively easy for foxes, raccoons, and other predators to find. As a result, mother ducks and geese lead the offspring away from the nest to the safety of nearby water almost immediately. In these bird species, the ability to recognize a moving object and follow it is thought to increase the fitness of offspring.

Lorenz was also able to establish that imprinting occurs only in the early life of the animal, during a short period called the **critical period**, or **sensitive period**. In addition, his research showed that imprinting lasts for life and may establish not only the identity of the offspring's mother but also its species identity. As a result, imprinting has caused problems for conservation biologists who are attempting to raise certain species of endangered birds in captivity. For example, whooping cranes that have been raised by humans not only follow them as chicks but also respond exclusively to humans as sexually mature adults. To prevent captive-reared birds from imprinting on humans, biologists are now careful to raise offspring in the absence of visual or auditory contact with humans and instead to provide recorded sounds of a natural mother in the presence of bird-like puppets or other types of appropriate visual stimulation (**Figure 51.6b**). Offspring raised in this manner imprint normally, behave normally as both chicks and adults, and can be released into the wild.

The key characteristics of imprinting—that it is fast and irreversible and occurs during a critical period—are not typical of most types of learning. Yet imprinting may be a more general type of learning than was originally thought. For example, recent research has shown that the ability of humans to learn a particular language shares at least some of the characteristics of imprinting. You may have personal experience with how quickly children learn a foreign language relative to adults. This observation suggests that there might be something like a critical period for language acquisition. The commonplace observation that babies learn the language of their parents is also important. The babbling noises that babies make are thought to be imitations of words and sounds from their parents. Children who are born deaf cannot hear these sounds. Learning a spoken language may be almost impossible for these individuals, even with intensive training later in life. In contrast, people who become deaf as adults experience relatively slight changes in their speech. These data suggest that auditory feedback is important during the learning process and that language acquisition should shift to the left in Figure 51.2. Let's explore the topic of language learning further by considering research on how birds learn to sing.

(a) Goslings imprint on the first moving object they see.

(b) With care, normal imprinting can occur in captive-reared birds.

FIGURE 51.6 Imprinting Occurs in Some Species of Birds and Lasts for Life
(a) Konrad Lorenz discovered imprinting when captive-reared geese imprinted on him. **(b)** Endangered species raised in captivity are reared with puppets to avoid imprinting on people.

More Complex Types of Learning: Birdsong

Birds sing to attract mates and to mark territories where they find food and raise offspring. In most cases, the songs that a songbird sings are unique to its species. Do young birds learn their species-specific song or suite of songs from their parents, or are songs innate? To answer these questions, biologists perform experiments based on controlling the sound stimuli that birds hear over the course of their lives. The experimental protocols involve taking eggs from the nest, hatching them, and rearing individuals in the laboratory in a controlled auditory environment.

The results of these experiments show that, depending on the species, song-learning behavior falls at various locations on the learning continuum in Figure 51.2. For example, if young chickens or phoebes are raised in isolation from other members of their species and never hear their species-specific song, they still sing a normal song as adults. Song-learning behavior is innate in certain species and may be highly stereotyped.

In contrast, white-crowned sparrows that are raised in isolation do not sing a normal song unless they hear a tape recording of their species-specific song during the early months of life. In this species, singing is learned during a critical period. Yet if researchers play the song of a song sparrow or other closely related species during the critical period for language acquisition, white-crowned sparrow chicks do not learn it. Thus, in this species the ability to learn songs is restricted to songs of their own species (**Figure 51.7**).

In addition, the critical period for song learning in white-crowned sparrows is augmented by practice that occurs when the individual is older. White-crowned sparrows don't begin to sing until they are almost one year old, during the spring. Beginning singers produce a disorganized warbling called *sub-song*.

(a) Young bird reared in silence
No song
(b) Young bird reared with white-crowned sparrow song
Normal song
(c) Young bird reared with song sparrow song
Abnormal song

FIGURE 51.7 White-Crowned Sparrows Have a Critical Period for Song Learning Early in Life
To sing a normal song, white-crowned sparrows have to hear the song of their own species while they are chicks. They also must hear themselves practice their own song when they begin singing at age 1 year.

With practice, this sub-song becomes more and more organized and progressively closer to adult song. Eventually an adult song develops and is "crystallized," meaning that it remains unchanged for the rest of the individual's life. For this improvement to occur, though, the individual must hear itself sing. Individuals that are prevented from hearing their own practice never sing a normal song. Yet if the same individual is deafened after it has developed adult song, it continues to sing a perfectly normal song. These data suggest that individuals have two critical periods for song learning during their lives: one as a nestling, when they must hear the song of their own species, and a second as a 1-year-old, when they must hear themselves practice.

To summarize, singing in white-crowned sparrows is heavily influenced by learning. But in this species, learning is constrained to certain periods of life and occurs only in response to certain types of stimuli.

If chickens are at one end of the learning continuum and if white-crowned sparrows are in the middle, then species such as mynahs, mockingbirds, lyrebirds, and some types of parrots occupy the other end. Individuals of these species continue to learn new songs throughout their lives by imitating sounds in their environment. They have relatively few age-related or species-specific limitations on song learning. "Polly want a cracker" is not a species-specific call in parrots, and the ability to use this call does not depend on hearing it during a critical period early in life.

One of the great challenges that remains for research on song learning is to explain why so much variation exists in song-learning ability and in the size of song repertoires. Why is it adaptive for chickens to sing a few innate songs while mynahs use dozens of learned songs? Research continues.

Can Animals Think?

Research on classical conditioning, imprinting, and song learning has revealed that several types of learning exist and that the ability to learn varies widely among species. It is clear that simple types of learning are widespread in animals and that at least some species can learn a great deal by imitation. But can organisms other than humans think?

Cognition can be defined as the recognition and manipulation of facts about the world, and as the ability to form concepts and gain insights. Unlike the results of classical conditioning, imprinting, and song learning, the results of cognition cannot be observed directly. Instead, to infer that animals are thinking, researchers must design experimental situations that require animals to manipulate facts or information. The experiment must be set up in a way that allows the animal to demonstrate its ability to form novel associations or insights. As an example of work on animal cognition, consider recent experiments on tool use by New Caledonian crows.

New Caledonian crows routinely make and use tools to find food in the wild. Each of these tools is held in the beak and used to remove insect larvae (grubs) or other types of food from plant stems or from crevices. Biologists have documented that the crows manufacture the following general types of tools:

- Straight or curved sticks are broken off and cleaned before use (**Figure 51.8a**). Several other bird species—including one of the Galápagos finches introduced in Chapter 41—are known to modify sticks or plant stems to "fish" for insects. Chimpanzees also make these types of tools (see Chapter 33). Recent laboratory experiments have shown that when New Caledonian crows are presented with an array of straight sticks along with a piece of food in a clear plastic tube, the crows select a stick whose width or length is appropriate for the tube's width and

(a) Straight sticks are used to fish or pry food from crevices.

(b) Hooks are made by breaking parts off leaves.

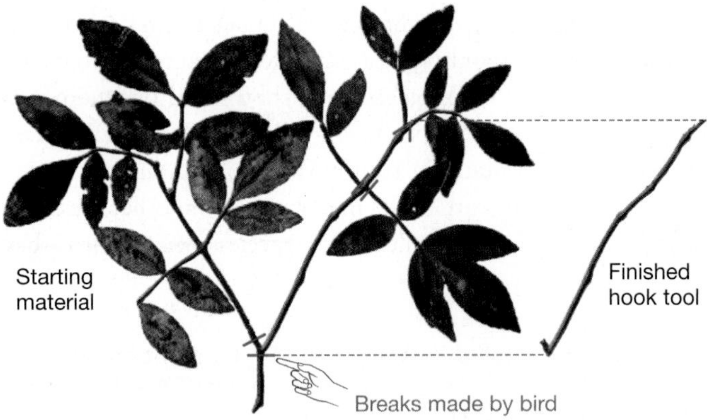
Starting material
Finished hook tool
Breaks made by bird

(c) Several cuts and tears are used to make spearing tools.

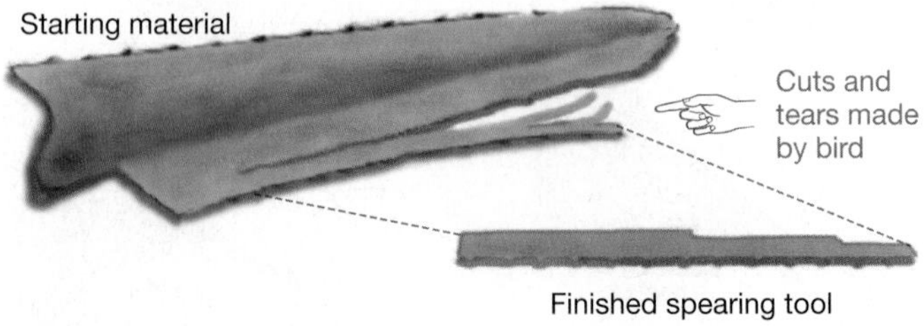
Starting material
Cuts and tears made by bird
Finished spearing tool

FIGURE 51.8 New Caledonian Crows Make and Use Several Types of Tools
(a) Straight sticks are cleaned and broken off before being used as tools. **(b)** Hooks are made from compound leaves. **(c)** Spears are made from broad, stiff leaves.

the food's distance from the end of the tube. The crows then insert the stick into the tube to fish out the food.

- Hooks are made by breaking complex leaves into pieces (**Figure 51.8b**).
- Spearing tools are constructed from a complex series of cuts and rips, made with the beak, along the edge of leaves from pandanus plants (**Figure 51.8c**).

These observations support the hypothesis that crows can think. More specifically, crows appear to understand facts about the size and shape of raw materials and the location of food. Crows recognize that if they choose or manipulate materials in a certain way, they can use the resulting structure as a tool to acquire the food. The idea is that natural selection has favored the evolution of cognition in this species, because crows with the ability to manufacture and manipulate tools have higher fitness than do crows that lack this ability.

As **Figure 51.9** shows, the crows-can-think hypothesis was supported by recent experiments with two New Caledonian crows in captivity. The experimental setup consisted of a small bucket, containing food, in the bottom of a plexiglas tube that was open at the top, with the handle of the bucket pointing up. In an initial experiment, a male crow and a female crow were given a choice of straight or bent wires to use in retrieving the food from the inside of the tube. The goal of the experiment was to test the hypothesis that the crows can distinguish the two shapes and associate the bent shape with the ability to lift the bucket and acquire the food. During the fifth trial with the apparatus and the choice of two wires, the male happened to remove the hooked wire from the area. Apparently in response, the female picked up the straight wire in her beak, bent it, and successfully used it to lift the bucket and obtain the food.

To confirm the validity of this observation, the researchers gave the pair a straight wire and let them use it until they either successfully retrieved the food or dropped the wire into the tube and were unable to retrieve it. In a total of 17 trials, the male or female dropped the wire into the tube and were unable to retrieve it 7 times. Once the male was able to lift the bucket out and obtain the food using only the unmodified, straight wire. But 9 times, the female bent the straight wire and used it successfully to obtain the food.

These observations are nothing short of astonishing. They imply not only that the female was able to associate the shape of a wire with the ability to lift the bucket successfully, but also that she had the insight to understand that exerting a force on the wire would bend it into the shape required to be successful. In more colloquial terms, she appeared to understand what sort of tool she needed and how to make one. She had no model to imitate when she made her hooks, however, and she had had no prior experience with bending wires. Instead, she appeared to be thinking. Research on animal cognition is an exciting frontier in behavioral research.

Question: Can crows use tools to solve problems?

Hypothesis: Crows can use tools to solve problems.

Null hypothesis: Crows cannot use tools to solve problems.

Experimental setup:

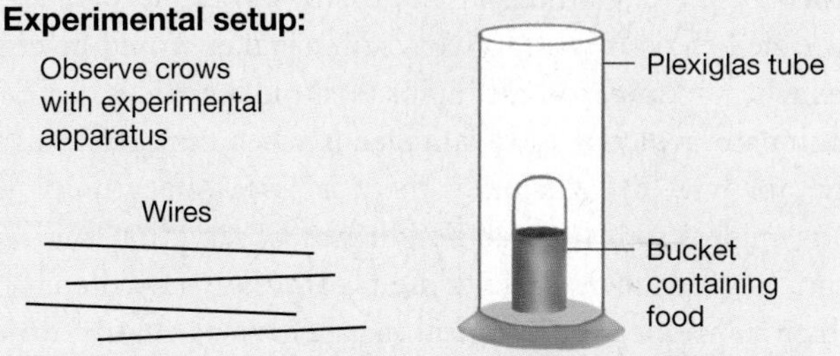
Observe crows with experimental apparatus
Wires
Plexiglas tube
Bucket containing food

Prediction: Crows will figure out how to use the wire as a tool to remove the bucket from the tube.

Prediction of null hypothesis: Crows will not figure out how to remove the bucket from the tube.

Results:

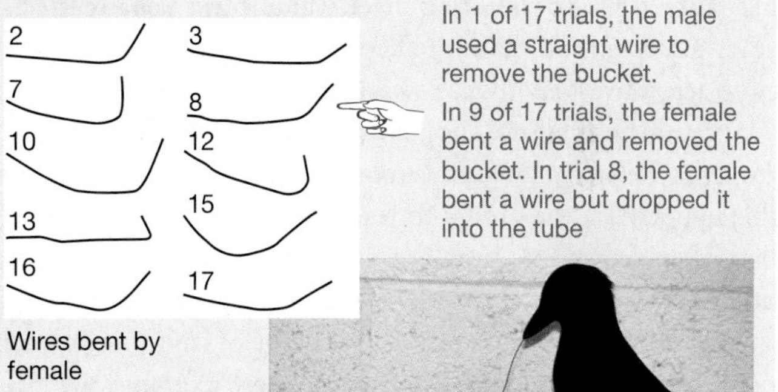
2
3
7
8
10
12
13
15
16
17
Wires bent by female
In 1 of 17 trials, the male used a straight wire to remove the bucket.
In 9 of 17 trials, the female bent a wire and removed the bucket. In trial 8, the female bent a wire but dropped it into the tube

Conclusion: Crows can use tools to solve problems. Crows can also make tools from unfamiliar materials to solve problems.

FIGURE 51.9 Experimental Evidence That New Caledonian Crows Understand How to Make Tools
Prior to this experiment, the crows involved had never seen the experimental apparatus or wires of the type available to them. Bending of materials to form hooks has not been observed in the wild.

What Is the Adaptive Significance of Learning?

Research on classical conditioning, imprinting, song learning, and cognition has revealed that several types of learning exist and that the ability to learn varies widely among species. Even within the same species, some types of behavior are innate and inflexible while other types are condition dependent, or learned, and thus highly flexible. Even in people who have devoted their lives to learning, certain simple stimuli act as releasers that trigger FAPs such as yawning and the jump-back response.

These results raise an important question: At the ultimate level of explanation, why does the ability to learn vary among species and among different types of behavior?

To answer this question in the case of FAPs and other non-learned behavior, biologists emphasize two concepts. First, inflexible behavior is adaptive when mistakes would be costly, as when a kangaroo rat jumps to avoid a rattlesnake. Second, innate behavior is advantageous when opportunities to learn are few. For example, there is little opportunity for young spiders to learn web building by observation and imitation, because their parents die before the young hatch. In such circumstances, it is logical that behavior be non-learned and inflexible.

In contrast, learning tends to have an important influence on behavior if individuals have the opportunity to make mistakes without dying and if they have parents or other sources of information to learn from. Organisms that are capable of learning also tend to be long lived, and they usually live in environments that are unpredictable. To explain why the ability to learn evolves, then, researchers hypothesize that it is an adaptation that allows individuals to change their behavior in response to a changing and unpredictable environment. Under this hypothesis, the ability to learn varies because some species live in environments that are much more unpredictable than others. In addition, the type of learning that occurs in a given species should be correlated with the type of environmental unpredictability encountered by that species. Examples will help illustrate these concepts:

- Norway rats are extremely adept at learning to navigate mazes. These animals live in sewers, the walls of homes, or other locations where the ability to learn travel routes is essential for survival (**Figure 51.10a**). Norway rats are also exceptionally good at learning to avoid foods that contain poisons. In nature they feed on a wide variety of fruits, seeds, and insects. The availability of these foods changes with the season, and many potential food items are actually toxic. Thus, it is logical to observe that rats will taste only small quantities of novel foods and that they quickly learn to avoid any foods that induce illness.
- Scrub jays cache seeds and other types of food in storage areas and are proficient at remembering where they have stored food (**Figure 51.10b**). In this species, spatial learning and a good memory are essential for survival. In addition, scrub jays learn to modify their storage behavior based on experience. This species lives in social groups, and it is not unusual for some members of the group to steal food that was cached by other jays in their group. Experiments in the laboratory have shown that if an individual that has stolen food in the past caches food while being watched by other jays, the individual will come back later by itself and move the stored food to a new location. In this species, the ability to learn and modify behavior is interpreted as an adaptation for coping with unpredictable social situations.

To summarize, the ability to learn and the type of learning that occurs vary within and among species. Learning is an adaptation that helps organisms cope with challenges from their environment.

(a) Lab rats and Norway rats learn how to navigate mazes.

(b) Scrub jays remember where they've stored food.

FIGURE 51.10 Different Species Are Adept at Different Types of Learning
There is often a strong correlation between the type of learning that occurs in a particular species and the problems that the species must solve to survive and reproduce. QUESTION Domestic dogs have been trained by people for thousands of years. Domestic dogs are adept at learning via classical conditioning. What is the relationship between these observations?

✓CHECK YOUR UNDERSTANDING

The same species of animal may learn in a variety of ways, ranging from simple forms such as imprinting and classical conditioning to imitation to cognition. It is common to observe that an individual is capable of learning only certain types of things—such as its species-specific song—only at certain times in its life. You should be able to (1) define learning and give examples of at least two distinct types of learning, (2) discuss the adaptive value of cognition, and (3) offer a hypothesis to explain why cognition appears to be rare among animals.

51.3 How Animals Act: Hormonal Control

Most behavior is modified at least slightly by some type of learning, and most animals have behavioral responses that range from highly stereotyped to highly flexible. Previous examples in the chapter have highlighted the adaptive value of learning and various types of behavior. But we have yet to explore a question about proximate causation in detail. To understand how biologists go about addressing questions at the proximate level, consider the following: When behavior is flexible and condition dependent, it means that an individual has the potential to behave in a variety of ways. How are changes in behavior implemented and controlled? To answer this question, let's explore sexual behavior in the lizard *Anolis carolinensis*.

Anolis carolinensis (**Figure 51.11a**) lives in the woodlands of the southeastern United States. After spending the winter under a log or rock, males emerge in January and establish breeding territories. Females become active a month later, and the breeding season begins in April. By May, females are laying an egg every 10–14 days. By the time the breeding season is complete three months later, the eggs produced by a female will total twice her body mass. **Figure 51.11b** maps this series of events, along with the corresponding changes observed in the male and female reproductive systems.

What causes these dramatic seasonal changes in behavior? The proximate answer is sex hormones—testosterone in males and estradiol in females. Testosterone is produced in the testes of males, and estradiol in the ovaries of females. The evidence for these statements is direct. Testosterone injections induce sexual behavior in castrated males, while estradiol injections induce sexual activity in females whose ovaries have been removed.

This result leaves some key questions unanswered, however. *Anolis* lizards are most successful if they reproduce early in the spring. In springtime, their food supplies are increasing and snakes and other predators are not yet hunting lizards to feed their own young. If testosterone and estradiol levels induced sexual activity in *Anolis* lizards at the wrong time of year, it would be a disaster for the individuals involved. In addition, it is critical for all *Anolis* individuals in a population to start their sexual activity at the same time, so that females can find males that are ready to breed, and vice versa. What environmental cues trigger the production of sex hormones in early spring? How do male and female *Anolis* synchronize their sexual behavior so that they are ready to produce gametes at the same time?

To answer these questions, a biologist brought a large group of sexually inactive adult lizards into the laboratory during the winter and divided them into five treatment groups. The physical environment was exactly the same in all treatments. Each individual received identical food, and in all treatments high "daytime" temperatures were followed by slightly lower "nighttime"

(a) Courtship display of a male *Anolis* lizard

(b) Changes in sexual organs through the year

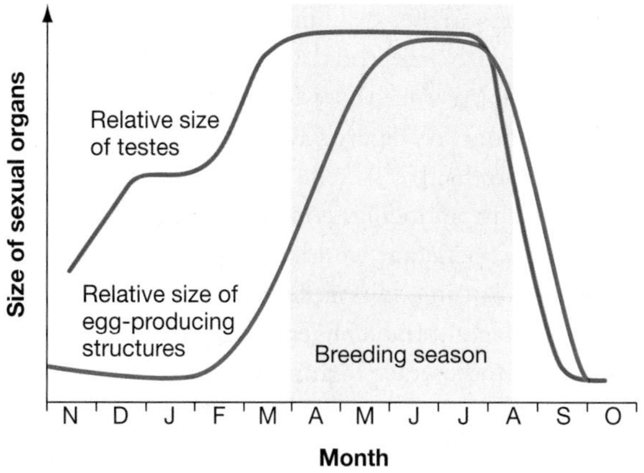

FIGURE 51.11 Sexual Behavior in *Anolis* Lizards
(a) A male *Anolis* lizard courting a nearby female. The male extends a flap of skin called a dewlap and bobs up and down. **(b)** Changes in the sexual organs of *Anolis* lizards through the year: (red curve) the relative size of the follicles inside a female's ovary, which produce eggs; (blue curve) the relative size of a male's testicles.

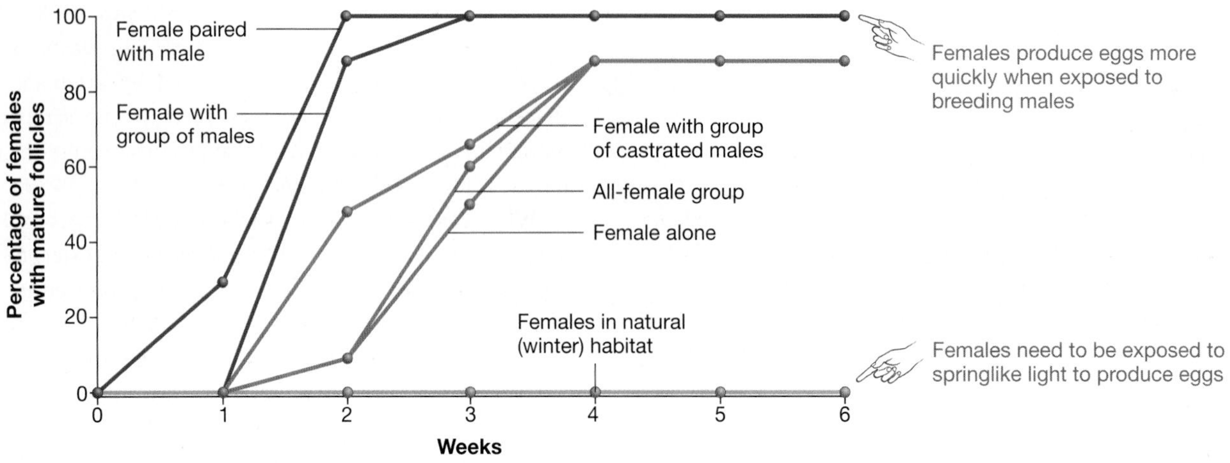

FIGURE 51.12 Exposure to Springlike Conditions and Breeding Males Stimulates Female Lizards to Produce Eggs

The percentage of female lizards with mature follicles, plotted over time. Five treatment groups were exposed to springlike light and temperature; data are also plotted for females left in a natural (winter) habitat. Each data point represents the average value for a group of 6–10 females. QUESTION Some critics contend that the females that remained outside do not represent a legitimate control in this experiment. What conditions would represent a better control?

settings. The biologist also continued to monitor the condition of lizards that remained in natural habitats nearby.

To test the hypothesis that changes in day length signal the arrival of spring and trigger initial changes in sex hormones, the biologist exposed the five treatment groups in the laboratory to artificial lighting that simulated the long days and short nights of spring. To test the hypothesis that social interactions among individuals are responsible for synchronizing sexual behavior, the social setting was varied among treatment groups: (1) single isolated females; (2) groups of females only; (3) single females, each with a single male; (4) single females, each with a group of castrated (nonbreeding) males; or (5) single females, each with a group of uncastrated (breeding) males.

Each week, the researcher examined the ovaries of females in each group. He also monitored the ovaries of females in nearby natural habitats, since those females were not exposed to springlike conditions. As **Figure 51.12** shows, the differences in the animals' reproductive systems were dramatic. Females that were exposed to springlike conditions began producing eggs; females in the field that were not exposed to springlike conditions did not. But in addition, females that were exposed to breeding males began producing eggs much earlier than did the females placed in the other treatment groups. These results support the hypothesis that two types of stimulation are necessary to produce the hormonal changes that lead to sexual behavior. Females need to experience springlike light and temperatures *and* exposure to breeding males.

What aspect of male behavior causes the difference in female egg production? The biologist suspected that visual stimulation from the male's courtship display was important. Specifically, he hypothesized that a flap of skin, called the dewlap, played a role in stimulating females to produce eggs. To court females, male *Anolis* lizards bob up and down and extend their dewlaps (see the yellow neck pouch in Figure 51.11a). To test this idea, the investigator repeated the previous experiment but added a twist: He placed some females with males that had intact dewlaps, and other females with males whose dewlaps had been surgically removed. The result? Females grouped with dewlap-less males were slow to produce eggs—just as slow, in fact, as the females in the first experiment that had been grouped with castrated males. These latter females had not been courted at all. The result suggests that the dewlap is a key visual signal. The experiments succeeded in identifying the environmental cues that trigger hormone production and the onset of sexual behavior.

To summarize, sexual activity in *Anolis* lizards is a condition-dependent behavior controlled by hormones. This research and similar experiments carry a general message: When hormone levels change in response to environmental changes, they trigger behavior that allows individuals to maximize their fitness in the changed environment. In many cases, flexible behavior is possible because hormone levels are flexible.

51.4 Communication

The bobbing dewlap of a male *Anolis carolinensis*, the song of a bird, and the sentences in this text all have the same overall goal: communication. In biology, **communication** is defined as any process in which a signal from one individual modifies the behavior of a recipient individual. A **signal** is any information-containing behavior. Communication is a crucial component of animal behavior, because it creates a stimulus that elicits a response.

By definition, communication is a social process. For communication to occur, it is not enough that a signal is sent; the signal must be received and acted on. A lizard's bobbing dewlap does not qualify as communication unless another individual sees it and responds to the message. In addition, some biologists maintain that for an event to qualify as communication, the signal that is sent must be intended as a signal. According to these investigators, predicting a person's behavior by watching their body language does not qualify as communication. Similarly, if students unintentionally fall asleep in class and the instructor changes his behavior to wake them up, the exchange does not qualify as communication.

Modes of Communication

The information that organisms communicate may be encoded and delivered in a variety of ways. Communication can be acoustic, as in the song of crickets or birds. It can be visual, as in the color patterns of birds or fish. It can be olfactory, as in scent marking by mice and wolves. It can be tactile, as in communication in some spiders.

One of the most general observations about communication is that the type of signal used by an organism correlates with its habitat. For example, light is quickly absorbed in aquatic habitats, but sound is not. Based on this observation, it is logical to observe that humpback whales rely on songs for long-distance communication. In some cases, humpback whale songs can travel hundreds of kilometers. Groups of whales use acoustic communication to keep together as they move from summer feeding areas to winter breeding grounds, and individual males sing to attract mates and warn rivals. Correlations between habitat and mode of communication are also observed in visual, olfactory, and tactile communication. Animals that are active during the day and that live in open or treeless habitats tend to rely on visual communication during courtship and territorial displays. Wolves and other animals that are active at night communicate via scent or sound; ants and termites that live underground rely on olfactory and tactile communication.

It is also common to observe several modes of communication being used in conjunction. For example, male red-winged blackbirds establish a breeding territory in the spring by giving a display that combines auditory and visual signals. The display is based on revealing bright red patches of feathers near their shoulders and giving a loud call. To test the hypothesis that both components of the display are important, researchers have manipulated both acoustic and visual elements. In one experiment, biologists used a speaker to play recordings of red-winged blackbird songs in existing territories (**Figure 51.13a**). In response, the territory-owning males flew toward the speaker and appeared to hunt for it. In a second experiment, researchers made a crude model of a male red-winged blackbird by sewing red patches on a black sock stuffed with rags. When the model was placed in existing territories, territory owners approached the model and inspected it (**Figure 51.13b**). But if the model was presented along with a recorded blackbird song, the resident males attacked the model (**Figure 51.13c**). These experiments support the hypothesis that both auditory and visual information are important in territorial displays of red-winged blackbirds. More specifically, the data indicate that while each type of stimulus

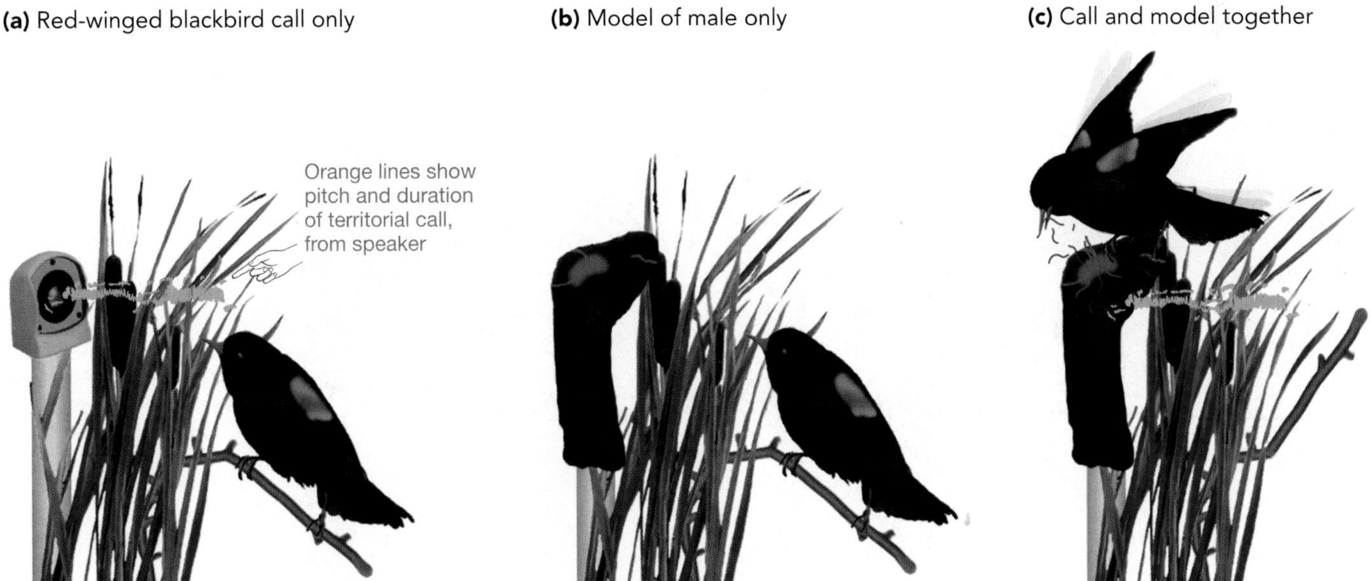
(a) Red-winged blackbird call only
Orange lines show pitch and duration of territorial call, from speaker
(b) Model of male only
(c) Call and model together

FIGURE 51.13 Experimental Evidence That Visual and Auditory Stimuli Are Important in Territorial Display
Territory-owning red-winged blackbirds react differently to **(a)** auditory stimuli, **(b)** visual stimuli, and **(c)** auditory plus visual stimuli. EXERCISE Researchers have clipped the red feathers from male red-winged blackbirds and released them into territories. Predict how territory owners respond to these individuals.

alone induces a response, the combination of both types of stimuli provokes a much more powerful change in behavior.

Each mode of communication has advantages and disadvantages. Although acoustic communication can be extremely effective in some habitats, songs and calls are short lived. A red-winged blackbird's red feathers remain in place for at least six months, but each call lasts less than 3 seconds. Thus, calls and songs have to be repeated to be effective. Frequent repetition of calls and songs requires a large expenditure of time and energy. In addition, acoustic communication has been shown to attract predators. It is no surprise that when a hawk or falcon approaches a marsh inhabited by red-winged blackbirds, things get very quiet.

Research on animal communication has gone far beyond describing types of communication and documenting correlations between modes of communication and habitat characteristics, however. To appreciate the array of questions that biologists ask about communication and how they go about answering them, let's consider two research programs on animal signaling—one classical and one more recent. The classical work revealed a remarkable form of communication in honeybees; the more recent work has focused on the question of how and why signalers lie.

A Case History: The Honeybee Dance

Honeybees are highly social animals that live in hives. Inside the hive, a queen bee lays eggs that are cared for by individuals called *workers*. In addition to caring for young and building and maintaining the hive, workers obtain food for themselves and other members of the colony by gathering nectar and pollen from flowering plants.

Biologists noticed that bees appear to recruit to food sources, meaning that if a new source is discovered by one or a few individuals, many more bees begin showing up over time. This observation inspired the hypothesis that successful food-finders have some way of communicating the location of food to other individuals. Karl von Frisch suspected that successful food-finders communicate information about food sources when they interact with other workers inside the hive. Because beehives are completely dark, von Frisch hypothesized that communication was tactile in nature.

Von Frisch began studying bee communication by observing bees that built hives inside glass-walled chambers that he constructed. He found that if he placed a feeder containing sugar water near one of these observation hives, a few of the workers began moving in a circular pattern on the vertical, interior walls of the hive. Von Frisch called these movements the "round dance" (**Figure 51.14a**). Other bee workers appeared to follow the progress of the dance by touching the displaying individual as it danced and to respond by flying away from the hive in search of the food source.

To investigate the function of these movements further, von Frisch placed feeders containing sugar water at progressively greater distances from the hive. Using this technique, he was able to get bees to visit feeders at a distance of several kilometers from the hive. By catching bees at the feeders and dabbing them with paint, he could individually mark successful food-finders. Follow-up observations at the hive confirmed that marked foragers danced when they returned to the hive. Follow-up observations at the feeders confirmed that marked foragers returned with greater and greater numbers of unmarked bees. To explain these data, von Frisch proposed that the round dance contained information about the location of food and that workers got information from the dance by touching the dancer and following the dancer's movements.

(a) The round dance

(b) The waggle dance

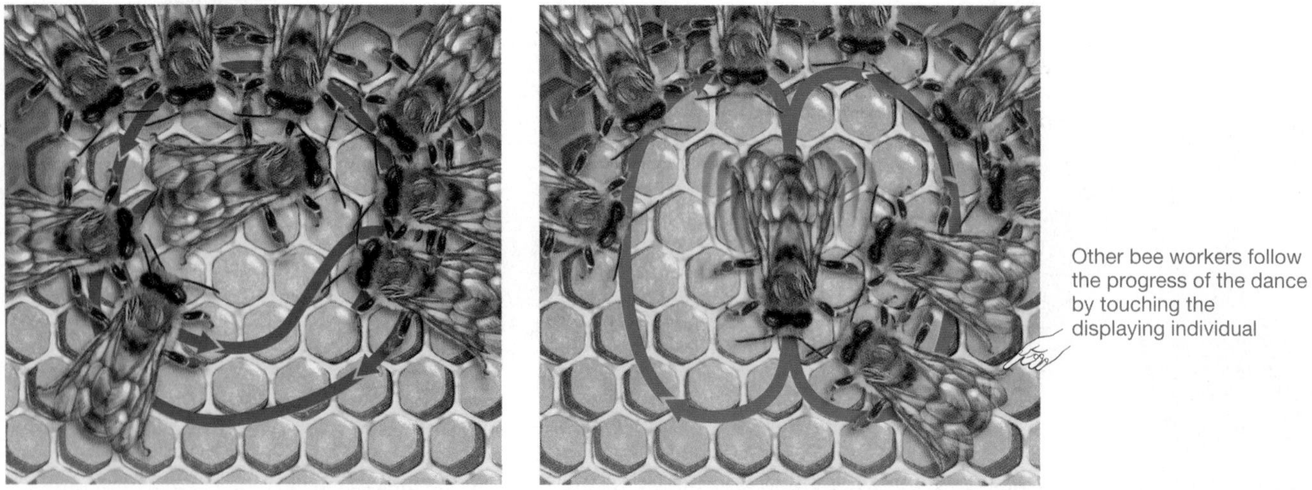

FIGURE 51.14 Honeybees Perform Two Types of Dances
(a) During the round dance, successful food-finders move in a circle. **(b)** During the waggle dance, successful foragers move in a circle but then make straight runs through the circle. During the straight part of the dance, the dancer waggles her abdomen.

When von Frisch placed feeders at longer distances from the hive, however, he found that successful food-finders no longer did the round dance. Instead, they performed a new type of display. He named these movements the "waggle dance," because they combined circular movements like those of the round dance with short, straight runs (**Figure 51.14b**). During these runs, the dancer would vigorously move her abdomen from side to side.

These observations supported the hypothesis that both the round dance and waggle dance communicated information about food sources. But a key observation allowed von Frisch to push this result further. He noticed that the orientation of the waggle part of the dance varied and that the variation correlated with the direction of the food source from the hive. In addition, he observed that the length of the straight, "waggling" run was proportional to the distance the foragers had to fly to reach the feeder. By varying the location of the food source and observing the orientation of the waggle dance given by marked workers, von Frisch was able to confirm that dancing bees were communicating the position of the food relative to the current position of the Sun. For example, if he placed food directly away from the Sun's current position, marked bees would give the waggle portion of their dance directly downward. But if the food was 90 degrees to the right of the Sun, bees would waggle 90 degrees to the right of vertical (**Figure 51.15**).

Based on these experiments, von Frisch concluded that the difference between the round dance and the waggle dance was a matter of distance. The round dance is used to indicate the presence of food within 80 to 100 m of the hive. The waggle dance is used to indicate the direction and distance to food that is over 100 m from the hive.

These results are nothing short of astonishing. Honeybees do not have large brains, yet they are capable of symbolic language. In addition, they are able to interpret the angle of the waggle dance performed on a vertical surface and to respond by flying horizontally along the corresponding angle. Further work has confirmed that the dance language of bees includes several modes of communication. In addition to the tactile information in the movements themselves, bees make sounds during the dance and give off scents that indicate the nature of the food source.

Honesty and Deceit in Communication

As far as is known, honeybees are always honest. Stated another way, honeybee dances consistently "tell the truth" about the location of food sources. This observation is logical, because the honeybees that occupy a hive are closely related and cooperate closely in the rearing of offspring. As a result, it is advantageous for an individual to convey information accurately. If a food-finder provided inaccurate or misleading information to

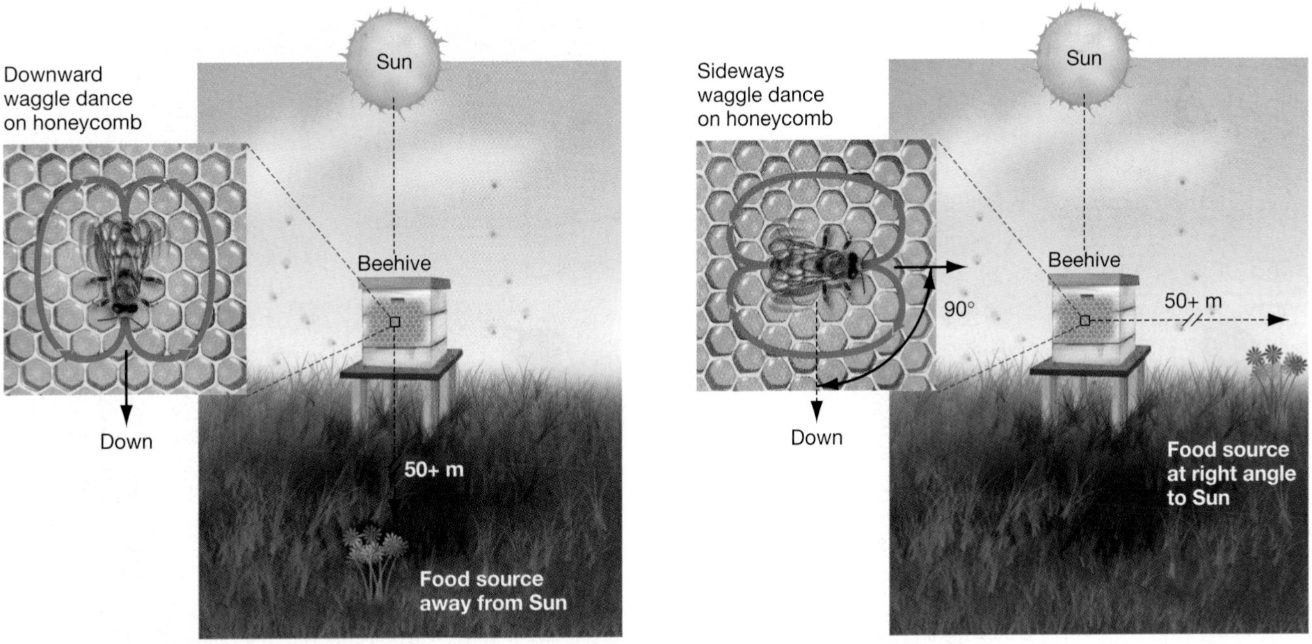

FIGURE 51.15 The Direction of the Waggle Dance Indicates the Location of Food Relative to the Sun
(a) In the language of the honeybee waggle dance, straight runs down the wall of the hive indicate that food is opposite the direction of the Sun. **(b)** Straight runs to the right indicate that food is 90° to the right of the Sun. Waggle dances are done only when the food is at least 50 m from the hive. **EXERCISE** The length of the straight run in the waggle dance indicates the relative distance of the food. Diagram a waggle dance that indicates food in the direction of the Sun and twice as far away as the food sources indicated by the dances drawn here.

its hivemates, fewer offspring would be reared and thus the food-finder's fitness would be reduced.

In many cases, however, natural selection has favored the evolution of deceitful communication, or what is commonly called lying. Recent research on deceitful communication has highlighted just how complex interactions between signalers and receivers can be. A few examples will help drive this point home.

Deceiving Individuals of Another Species Figure 51.16 illustrates a few of the hundreds of examples of deceitful communication that have been documented among members of different species. Many of the best-studied instances involve predation:

- The anglerfish in Figure 51.16a has an appendage that looks remarkably like a minnow and that dangles near its mouth. If another fish approaches this "lure" and attempts to eat it, the anglerfish attacks.
- Male and female fireflies flash a species-specific signal to each other during courtship. Predatory *Photinus* fireflies can mimic the pattern of flashes given by females of several other species, however. If a *Photinus* female responds to the courtship flashes made by a male of a different species with the appropriate set of flashes, he waits as she approaches. Instead of copulating with her, however, he is attacked and eaten (Figure 51.16b).
- The butterfly shown in Figure 51.16c is highly palatable to birds and other predators. It is rarely attacked, however, because it strongly resembles butterfly species that contain toxins and are highly unpalatable to predators. Such types of mimicry are discussed in more detail in Chapters 31 and 53.

In each of these examples, individuals increase their fitness by providing inaccurate or misleading information to members of a different species. Deceitful communication is also known in plants, where it functions in pollination. As the essay in Chapter 40 pointed out, several orchid species have flowers that look and smell like female wasps and that accomplish pollination by "fooling" male wasps into attempting copulation.

Deceiving Individuals of the Same Species In some cases, natural selection has also favored the evolution of lying to members of the same species. When mantis shrimps molt, for example, they lack any external covering and cannot use their large claws. Thus, they are unable to defend the cavities where they live. But if another mantis shrimp approaches with the intent of evicting the individual and taking over the cavity, the molting individual bluffs. It raises its claws in the normal aggressive display and may even lunge at the intruder.

(a) Anglerfish use a "lure" to attract prey.

(b) *Photuris* fireflies flash the courtship signal of another species, then eat males that respond.

(c) This butterfly looks like a bad-tasting species but actually tastes good.

FIGURE 51.16 Deceitful Communication Is Common in Nature Three examples of species that communicate misleading information to other species.

Perhaps the best-studied type of deceit in nonhuman animals, however, involves the mating system of bluegill sunfish. Male bluegills set up nesting territories in the shallow water along lake edges, fertilize the eggs laid in their nests, care for the developing embryos by fanning them with oxygen-rich water, and protect newly hatched offspring from large predators. Some males cheat on this system, however, by mimicking females. To understand how this happens, examine the female, normal male, and female-mimic male in **Figure 51.17**. Female mimic males look like females but have well-developed testes and produce large volumes of sperm. Mimics also act like females during courtship movements with territory-owning males. They even adopt the usual egg-laying posture. Then when normal females approach the nest and begin courtship, the female-mimic males join in. The territory-owning male tolerates the mimic, apparently thinking that he is successfully courting two females at the same time. But when the actual female begins to lay eggs, the mimic responds by releasing sperm and fertilizing some of them. In this way, the female-mimic male fathers offspring but does not help care for them.

When Does Deception Work? In analyzing deceitful communication in bluegill sunfish and other species, researchers point out that in most cases, lying works only when it is relatively rare. The logic behind this hypothesis is that if deceit becomes extremely common, then natural selection will strongly favor individuals that can detect and avoid or punish liars. But if liars are rare, then natural selection will favor individuals that are occasionally fooled but are more commonly rewarded by responding to signals in a normal way. In mantis shrimps, for example, molting occurs only once or twice a year. Thus, the vast majority of contests over hiding places involve individuals that have an intact exoskeleton and claws that qualify as a deadly weapon. As a result, intruders usually benefit by interpreting threat displays as honest and responding to the stimulus by retreating. But if molting is frequent and bluffing behavior common, then intruders would usually benefit by challenging territory owners, which are likely to be defenseless. Research on deceit and honesty in communication has been extraordinarily productive and continues at a rapid pace.

Territorial male
Female
Female-mimic male

FIGURE 51.17 In Bluegill Sunfish, Some Males Look and Act Like Females
Female-mimic males have fully developed, fully functional testes and produce large numbers of sperm.

✓CHECK YOUR UNDERSTANDING

Communication is an exchange of information between individuals. In most instances, the mode of communication that animals use maximizes the probability that the information will be transferred efficiently. The amount of information that is communicated can be very simple, as in the territorial display of a red-winged blackbird, or extremely complex, as in the dance language of honeybees. Communication can also be honest or deceitful, depending on the intent of the signaler. You should be able to (1) define communication; (2) explain the adaptive significance of different modes of communication and give examples of each; and (3) discuss the mechanisms that humans use to detect when they are receiving misleading or incorrect information from another human, as well as the types of behavior that humans use in response to receiving deceitful communication.

51.5 Orientation, Navigation, and Migration

In response to many stimuli, organisms move. Ants follow odor trails; bees fly toward food sources. A movement that results in a change of position is called **orientation**. The simplest type of orientation is termed **taxis**, and it involves positioning the body, or part of the body, toward or away from a stimulus. A moth attracted to a porch light is an example of a positive **phototaxis**—orientation toward light. A female cricket approaching a calling male is an example of a positive **phonotaxis**—orientation toward sound. A person who retreats from the sound of a siren blaring is an example of negative phonotaxis.

Orientation movements can be extremely sophisticated. Consider bats, which use sound to locate their prey. Hunting bats emit high-pitched sounds and listen for the echo from a flying insect. This process, called **echolocation**, is remarkably acute. If you toss a small stone into the air, a foraging bat will pursue it but then turn away at the last instant. This simple observation suggests that bats are able to tell the

Bats emit high-pitched sounds (shown in blue) and direct their flight toward objects that reflect this sound (shown in purple)

FIGURE 51.18 Bats Hunt by Echolocation
Bats detect the echo from high-pitched sounds that they emit and orient their hunting flights in response.

difference between the echo from an insect and that from a stone (**Figure 51.18**).

Not surprisingly, moths that are preyed on by bats have ears that can detect the sounds emitted by an approaching bat. When these moths hear a bat in the distance, they are able to turn and fly away. But if the bat is very close, moths make a power dive toward the ground. You can make this observation yourself if you ever see moths flying at night during the summer. If you jingle keys at them, the high-frequency sound that results may imitate a bat's hunting call closely enough to elicit a response—moths dropping from the air like stones.

Migration: Why Do Animals Move with a Change of Seasons?

Orienting movements—such as those of a hunting bat or an escaping moth—cover fairly short distances. But in some cases, bats and other species travel thousands of miles in search of places to feed and breed. In ecology, **migration** is defined as the long-distance movement of a population associated with a change of seasons. A few examples will bring home the simple point that migratory movements can be spectacular in their extent and the navigation challenge they pose:

- Arctic terns nest along the Atlantic coast of North America, fly south along the coast of Africa to wintering grounds off Antarctica, and then fly back north along the eastern coast of South America. Chicks follow their parents south but make the return trip on their own. The journey totals over 32,000 km (20,000 miles).
- Many of the monarch butterflies native to North America spend the winter in the mountains of central Mexico or southwest California. Tagged individuals are known to have flown over 3000 km (1870 miles) to get to a wintering area. In spring monarchs begin the return trip north. They do not live to complete the trip, however. Instead, mating takes place along the way, and females lay eggs en route. Although adults that lived through the winter die, offspring continue to head north. After several generations of breeding at the original habitats in northern North America, individuals again migrate south to overwinter. Instead of a single generation making the entire migratory cycle, then, the round trip takes several generations.
- Salmon that hatch in rivers along the Pacific Coast of North America and northern Asia migrate to the ocean when they are a few months to several years old, depending on the species. After spending several years feeding and growing in the North Pacific Ocean, they return to the stream where they hatched. There they mate and die.

In most cases, it is relatively straightforward to generate hypotheses for why migration exists at the ultimate level. Arctic terns feed on fish that are available in different parts of the world at different seasons. Thus, individuals that migrate achieve higher reproductive success than do individuals that do not migrate. Monarch butterflies may achieve higher reproductive success by migrating to wintering areas and new breeding areas than by trying to overwinter in northern North America.

At the proximate level, however, explaining migratory movements is often extremely difficult. How do all of these animals find their way? What cues guide them on these immense journeys? The surprising answer is that in many or even most cases, biologists don't really know.

Navigation: How Do Animals Find Their Way?

To organize research on how animals find their way during migratory movements, biologists distinguish three categories of navigation: (1) **Piloting** is the use of familiar landmarks, (2) **compass orientation** is movement that is oriented in a specific direction, and (3) **true navigation** is the ability to locate a specific place on Earth's surface. A thought experiment will help you understand these categories. Suppose you grew up on the shores of Lake Superior and were transported to the middle of this enormous lake. Because you were blindfolded and made to spin in place en route, you are disoriented and have no idea where you are. But you are given a magnetic compass. The compass tells you where north, south, east, and west are. Unfortunately this information will not help you find home, because you have no idea where you are in relation to home. You are stuck because you have no mechanism of true navigation. Now suppose that a helicopter flying overhead drops you a map. The map has an X marking your current position and an H marking your home. The map also has a scale and a compass symbol. From this information, you determine that home is 100 miles to the west. The map has solved your navigational problem. The compass has solved your orientation problem. Now that you know you live to the west, you can use the compass to find west and paddle home. As you approach the shore, you recognize familiar landmarks. Piloting gets you home.

Organisms other than humans do not have magnetic compasses or printed maps. How do they navigate? Let's answer this question for each of the three categories of navigation.

Piloting A substantial amount of data suggests that at least some migratory animals use piloting to find their way. In some species of migratory birds and mammals, offspring follow their parents south in the fall and north in the spring. Young appear to memorize the route (**Figure 51.19**). Piloting even plays a role in species with more sophisticated navigational abilities. Homing pigeons, for example, can find their way home even when they are released in strange terrain a long distance away. But if these birds are equipped with frosted spectacles so that they see the world as a foggy haze, they return to within a mile or so of their home but do not actually find it. Based on these experiments, researchers have concluded that homing pigeons use piloting to navigate the final stages of a journey home.

Compass Orientation How do animals perform compass orientation? To date, most research on compass orientation has been done on migratory birds. To determine where north is, these animals appear to use the Sun during the day and the stars at night. Let's consider each type of compass orientation separately, because they work differently.

The Sun is difficult to use as a compass reference, because its position changes during the day. It rises in the east, is due south at noon (in the Northern Hemisphere), and sets in the west. To use the Sun as a compass reference, then, an animal must have an internal clock that defines morning, noon, and evening.

FIGURE 51.19 In Some Species, Offspring Learn Migration Routes from Their Parents
In some species of birds and mammals, offspring learn migration routes by following their parents. Here two snow goose parents lead their offspring.

Fortunately, most animals, have such a clock. The **circadian clock** that exists in organisms maintains a 24-hour rhythm of chemical activity. The clock is set by the light–dark transitions of day and night. It tells individuals enough about the time of day that they can use the Sun's position to find magnetic north.

The situation is actually simpler on clear nights, because migratory birds in the Northern Hemisphere can use the North Star to find magnetic north and select a direction for migration. But what if the weather is cloudy and neither the Sun nor stars are visible? Under these conditions, migratory birds appear to orient using Earth's magnetic field. Exactly how they detect magnetism is under debate. One hypothesis contends that the birds can detect magnetism by their visual system, through an unknown molecular mechanism. An alternative hypothesis maintains that individuals have small particles of magnetic iron—the mineral called magnetite—in their bodies. Changes in the positions of magnetic particles, in response to Earth's magnetic field, could then be detected and provide reliable information for compass orientation.

Although research on mechanisms of compass orientation continues, one important point is clear: Birds and perhaps other organisms have multiple mechanisms of finding a compass direction. At least some species can use a Sun compass, a star compass, and a magnetic compass. Which system they use depends on the weather and other circumstances.

WEB TUTORIAL 51.1
Homing Behavior in Digger Wasps

51.6 The Evolution of Self-Sacrificing Behavior

The types of behavior reviewed thus far—including FAPs, learned behaviors, honest and deceitful communication, and migration—all have a key common element: They help individuals respond to environmental stimuli in a way that increases their fitness. There is a type of behavior that appears to contradict this pattern, however: altruism. **Altruism** is behavior that has a fitness cost to the individual exhibiting the behavior and a fitness benefit to the recipient of the behavior. It is the formal term for self-sacrificing behavior. Altruism decreases an individual's ability to produce offspring but helps others produce more offspring.

The existence of altruistic behavior appears to be paradoxical, because if certain alleles make an individual more likely to be altruistic, those alleles should be selected against. For example, consider the claim that rodents called lemmings sacrifice themselves for the good of the species. A widely circulated story contends that when lemming populations are high, overgrazing occurs and the habitat is damaged—enough so that most individuals in the population will die of starvation. In response, some individuals throw themselves into the sea and drown. The hypothesis is that suicide is favored because it lowers population density enough to maintain high-quality habitats for the good of the species. Even though individuals suffer, the good-of-the-species hypothesis maintains that the behavior exists because the group benefits.

FIGURE 51.20 Altruism Cannot Evolve If "Cheater" Alleles Exist Most individuals in this population have alleles that lead to self-sacrificing behavior and result in death; the lemming with the inner tube has alleles that prevent self-sacrificing behavior. [*The Far Side*® by Gary Larson ©1980 FarWorks, Inc. All Rights Reserved. Used with permission.]

QUESTION Which alleles will increase in frequency in this population, and why?

Do Organisms Act for the Good of the Species?

The lemming suicide story is false. Although lemmings do disperse from areas of high population density in order to find habitats with higher food availability, they do not throw themselves into the sea. The individual wearing the inner tube in **Figure 51.20** explains why lemmings do not kill themselves for the good of the species. The inner tube represents what biologists call a "cheater" or "selfish" allele. Most of the individuals in the cartoon's population have altruistic alleles that lead them to commit suicide. But some individuals have cheater alleles that lead them to avoid committing suicide. Individuals with self-sacrificing alleles die and do not produce offspring. But individuals with selfish, cheater alleles survive and produce offspring. As a result, selfish alleles increase in frequency while altruistic alleles decrease in frequency. Thus, it is not possible for individuals to do things for the good of the species.

Kin Selection

Even though it is not possible for behavior to evolve for the good of the species, self-sacrificing behavior does occur in nature. For example, black-tailed prairie dogs perform a behavior called alarm calling (**Figure 51.21**). These burrowing mammals live in large communities, called *towns*, throughout the Great Plains region of North America. When a badger, coyote, hawk, or other predator approaches a town, some prairie dogs give alarm calls that alert other prairie dogs to run to mounds and scan for the threat. Giving these calls is risky. In several species of ground squirrels and prairie dogs, researchers have shown that alarm-callers draw attention to themselves by calling and are in much greater danger of being attacked than non-callers are.

FIGURE 51.21 Black-Tailed Prairie Dogs Are Highly Social Prairie dogs live with their immediate and extended family within large groups called towns. The individual with the upright posture has spotted an intruder and may give an alarm call.

How can natural selection favor the evolution of self-sacrificing behavior? William Hamilton answered this question by creating a mathematical model to assess how an allele that contributes to altruistic behavior could increase in frequency in a population. To model the fate of altruistic alleles, Hamilton represented the fitness cost of the altruistic act to the actor as C and the fitness benefit to the recipient as B. Both C and B are measured in units of offspring produced. His model showed that the allele could spread if

$$Br > C$$

where r is the coefficient of relatedness. The **coefficient of relatedness** is a measure of how closely the actor and beneficiary are related. Specifically, r measures the fraction of alleles in the actor and beneficiary that are identical by descent—that is, inherited from the same ancestor (see **Box 51.2**).

This result is called **Hamilton's rule.** It is important because it shows that individuals can pass their alleles on to the next generation not only by having their own offspring but also by helping close relatives produce more offspring. Stated another way, an altruist might produce slightly fewer offspring because of the sacrifices the individual makes to help others. But if those sacrifices help close relatives produce many more offspring, then copies of the altruist's alleles will increase in frequency. According to Hamilton's rule, if the fitness benefits of altruistic behavior are high for the recipients, if the recipients are close relatives, and if the fitness costs to the altruist are low, then alleles associated with altruistic behavior will be favored by nat-

BOX 51.2 Calculating the Coefficient of Relatedness

The coefficient of relatedness, r, varies between 0.0 and 1.0. If two individuals have no identical alleles as a result of being inherited from the same ancestor, then their r value is 0.0. Because every allele in pairs of identical twins is identical by descent, their coefficient of relatedness is 1.0.

What about other relationships? **Figure 51.22a** shows how r is calculated between half-siblings. (To review what the boxes, circles, and lines in a pedigree mean, see Figure 13.21.) Half-siblings share one parent. Thus, r represents the probability that half-siblings share alleles as a result of inheriting alleles from their common parent. It is critical to realize that, in each parent-to-offspring link of descent, the probability of any particular allele being transmitted is 1/2. This is so because meiosis distributes alleles from the parent's diploid genome to their haploid gametes randomly. Thus, half the gametes produced by a parent get one of the alleles present at each gene, and half the gametes produced get the other allele. Half-siblings are connected by two such parent-to-offspring links. The overall probability of two half-siblings sharing the same allele by descent is $1/2 \times 1/2 = 1/4$.

To think about this calculation in another way, focus on the red arrows in Figure 51.22a.The left arrow represents the probability that the mother transmits a particular allele to her son. The right arrow represents the probability that the mother transmits the same allele to her daughter. Both probabilities are 1/2. Thus, the probability that the mother transmitted the same allele to both her son and daughter is $1/2 \times 1/2 = 1/4$.

Figure 51.22b shows how r is calculated between full siblings. The challenge here is to calculate the probability of two individuals sharing the same allele as a result of inheriting it through their mother or through their father. The probability that full siblings share alleles as a result of inheriting them from one parent is 1/4. Thus, the probability that full siblings share alleles inherited from either their mother or their father is $1/4 + 1/4 = 1/2$.

(a) What is the r between half-siblings?

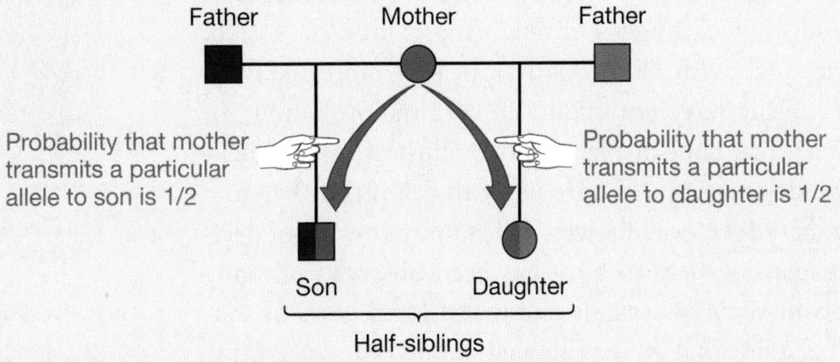

What is the probability that half-siblings inherit the same allele from their common parent?
Answer: r between half-siblings = $1/2 \times 1/2 = 1/4$

(b) What is the r between full siblings?

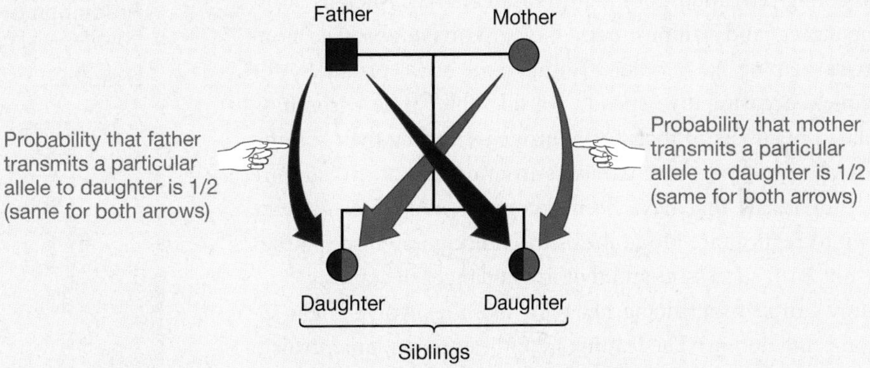

What is the probability that full siblings inherit the same allele from their father or their mother?
Answer: Probability that they inherit same allele from father = $1/2 \times 1/2 = 1/4$
Probability that they inherit same allele from mother = $1/2 \times 1/2 = 1/4$
Overall probability that they share alleles by descent = $1/4 + 1/4 = 1/2$
r between full siblings = 1/2

FIGURE 51.22 The Coefficient of Relatedness Is Calculated from Information in Pedigrees
QUESTION According to Hamilton's rule, who should be nicer to each other: half-siblings or full siblings? QUESTION What is r between first cousins?

ural selection and will spread throughout the population. The key is that close relatives are very likely to have copies of the altruistic allele. Biologists use the term **kin selection** to refer to natural selection that acts through benefits to relatives.

Does Hamilton's rule work? Do animals really favor relatives when they act altruistically? To test the kin-selection hypothesis, a reseacher studied which of the inhabitants of a black-tailed prairie dog town were most likely to give alarm calls. Within a large prairie dog town, individuals live in small groups called *coteries* that share the same underground burrow. Members of each coterie defend a territory inside the town. By tagging offspring that were born over several generations, the researcher identified the genetic relationships among individuals in the town. More specifically, the researcher knew, within each coterie, to which of the following three categories each prairie dog belonged: (1) an individual with no close genetic relatives in its coterie; (2) an individual with no offspring in the coterie but at least one sibling, cousin, uncle, aunt, niece, or nephew: and (3) an individual with at least one offspring or grandoffspring in the coterie.

The kin-selection hypothesis predicts that individuals who do not have close genetic relatives nearby will rarely give an alarm call. To evaluate this prediction, the biologist recorded the identity of callers during 698 experiments. In these studies, a stuffed badger was dragged through the colony on a sled. Were prairie dogs with close relatives nearby more likely to call, or did kinship have nothing to do with the probability of alarm calling? The data in **Figure 51.23** illustrate that black-tailed prairie dogs are much more likely to call if they live in a coterie that includes close relatives. This same pattern—of preferentially dispensing help to kin—has been observed in many other species of social mammals and birds. Most cases of self-sacrificing behavior that have been analyzed to date are consistent with Hamilton's rule and are hypothesized to be the result of kin selection.

Question: Do black-tailed prairie dogs prefer to help relatives when they give an alarm call?

Kin-selection hypothesis: Individuals give an alarm call only when close relatives are near.

Null hypothesis: The presence of relatives has no influence on the probability of alarm calling.

Experimental setup:

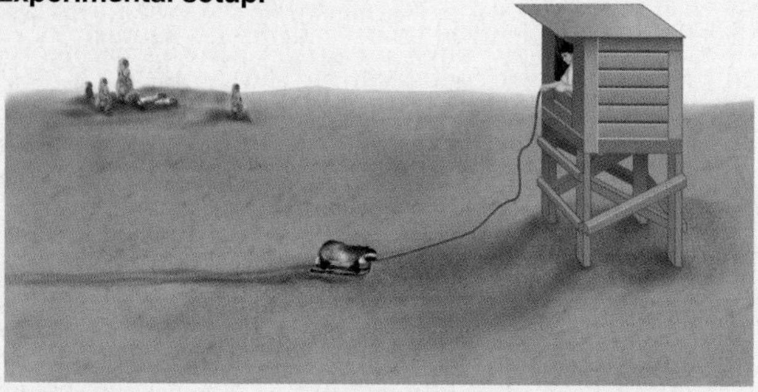

1. Determine relationships among individuals in prairie dog coterie.

2. Drag stuffed badger across territory of coterie.

3. From observation tower, record which members of coterie give an alarm call.

4. Repeat experiment 698 times. Each individual prairie dog coterie is tested 6-9 times over 3-year period.

Prediction of kin-selection hypothesis: Individuals in coteries that contain a close genetic relative are more likely to give an alarm call than are individuals in coteries that do not contain a close genetic relative.

Prediction of null hypothesis: The presence of relatives in coteries will not influence the probability of alarm calling.

Results:

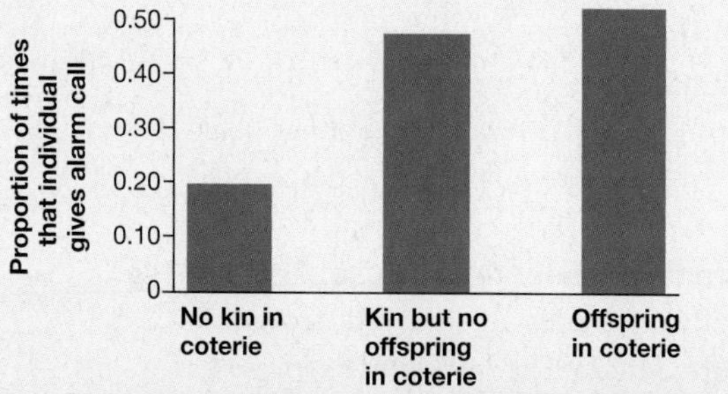

Conclusion: Alarm calling usually benefits relatives.

FIGURE 51.23 Experimental Evidence That Black-Tailed Prairie Dogs Are More Likely to Give Alarm Calls If Relatives Are Nearby

Reciprocal Altruism

During long-term studies of highly social animals, such as lions, chimpanzees, and vampire bats, biologists have observed non-relatives helping each other. Chimps and other primates, for example, occasionally spend considerable time grooming unrelated members of their social group—cleaning their fur and removing ticks and other parasites from their skin. In vampire bats, individuals that have been successful in finding food are known to regurgitate blood meals to both kin and non-kin that have not been successful and that are in danger of starving.

How can self-sacrificing behavior like this evolve if kin selection is not acting? The leading hypothesis to explain altruism among nonrelatives is called **reciprocal altruism**, which is an exchange of fitness benefits that are separated in time. Data that have been collected so far support the reciprocal-altruism hypothesis in at least some instances. Among vervet monkeys, for example, individuals are most likely to groom unrelated individuals that have groomed or helped them in the past. Similarly, vampire bats are most likely to donate blood meals to non-kin that have previously shared food with them. Reciprocal altruism is also widely invoked as an explanation for the helpful and cooperative behavior commonly observed among unrelated humans.

To summarize, altruistic behavior increases the fitness of individuals by favoring kin or by increasing the likelihood of receiving help in the futrue from non-kin. Altruism is a flexible, condition-dependent behavior that occurs in a surprisingly wide array of species that live in social groups.

✓CHECK YOUR UNDERSTANDING

Hamilton's rule states that alleles for altruistic behavior increase in frequency if the fitness cost of the behavior for the actor is low, the fitness benefit to the recipient is high, and the actor and recipient are closely related. Using Hamilton's rule, you should be able to (1) describe situations in which a black-tailed prairie dog is likely and unlikely to give an alarm call and (2) explain why reciprocal altruism has been observed only in species where individuals have good memories and live in small, long-lived social groups.

ESSAY Children at Risk

Some of the concepts discussed in this chapter—including altruism, Hamilton's rule, kin selection, and natural selection—may seem a little abstract at first. Yet Martin Daly and Margo Wilson have shown that they can be enormously important tools for understanding certain aspects of human behavior.

Daly and Wilson study the "dark side" of human behavior—homicide, domestic violence, and adultery—from a biological perspective. They try to determine whether some of these events are consistent with the types of actions favored by natural selection.

For example, let's consider data that Daly and Wilson have collected on child abuse. These researchers hypothesized that natural selection should favor parents who invest resources in biological children but not in stepchildren—with whom the parents have no genetic relationship. Based on this hypothesis, the researchers predicted that child abuse should be much more common in households containing a stepparent than it is in homes containing only biological parents. Further, Daly and Wilson predicted that, if stepparents choose to abuse their stepchildren, infants should be most at risk. The logic behind this prediction is that very young children demand the most time and resources and are least able to defend themselves.

Are the data consistent with these predictions? Daly and Wilson have analyzed the most extreme form of child abuse, which is the killing of children by parents. Using a database on all homicides reported in Canada between 1974 and 1983, the investigators found 341 cases in which a child was killed by a biological parent and 67 in which the perpetrator was a stepparent. Because households containing only biological parents were much more common, however, Daly and Wilson realized that they needed to compare the *rate* of violence in the two types of households, rather than the absolute number of incidents.

Daly and Wilson expressed the data as a rate—the number of children killed per million "child-years" by their parents. (A *child-year* is a year that a child of a particular age lived in a particular family situation.) The histograms in **Figure 51.24** resulted. The pattern is striking. Children who live with a stepparent are at much higher risk of abuse than are children who live with biological parents. Kids who are less than 2 years old are, in fact, 70 times more likely to be killed in a household with a stepparent than in a household with only biological parents.

... the "dark side" of human behavior... from a biological perspective.

The study points up the value of applying "selection thinking" to human behavior. It also has an urgent practical message: Society should be especially alert to indications of violence in households with stepparents where very young stepchildren are present. Several points about the data are worth noting, however. It is *extremely* important to recognize that violence occurs among biological kin as well as non-kin and that the vast majority of stepparents are solicitous and generous with their stepchildren. Further, the behavior is undoubtedly pathological rather than adaptive. Human perpetrators are jailed and treated as abhorrent. It is extremely unlikely, then, that the act of killing a stepchild improves the reproductive success of the murderer.

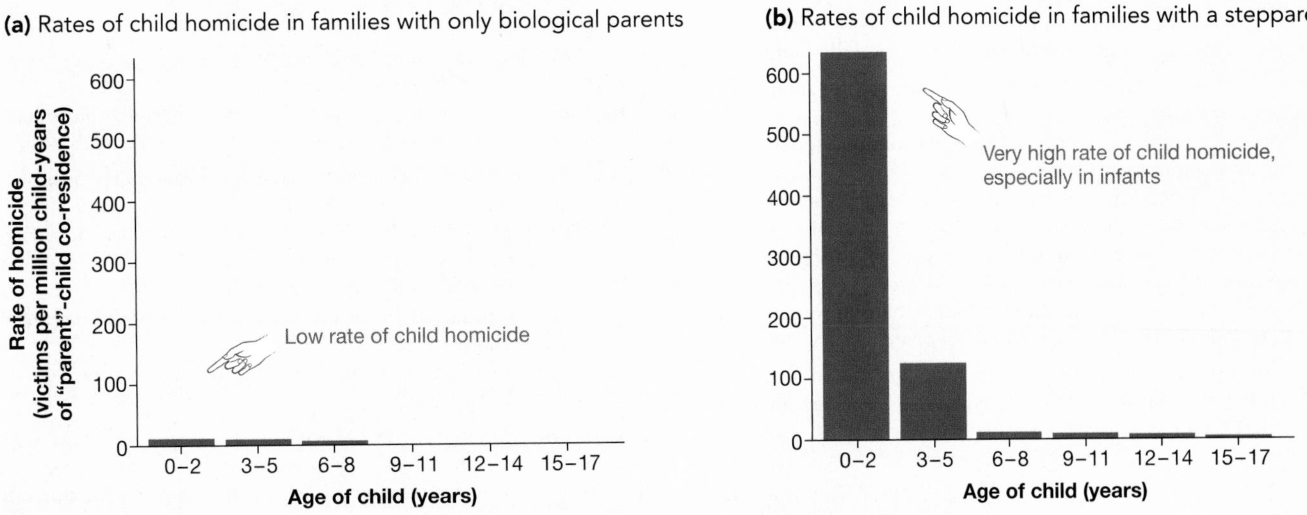

FIGURE 51.24 Rates of Child Homicide Are Higher in Families Containing a Stepparent
Rates of child homicide expressed as victims per "million child-years" in Canadian households containing only **(a)** a biological parent versus **(b)** a stepparent.

CHAPTER REVIEW

Summary of Key Concepts

- **After describing a behavior, biologists seek to explain both its proximate and ultimate causes—meaning how it happens at the genetic and physiological levels and how it affects the individual's fitness.**

  At the proximate level, experiments and observations focus on understanding how specific gene products, neuron activity, and hormonal signals cause behavior. At the ultimate level, researchers seek to understand the adaptive significance of behavior, or how it enables individuals to survive and reproduce. By combining proximate and ultimate viewpoints and studying the genetic basis of behavior, biologists can seek a comprehensive understanding of how and why animals do what they do.

  Web Tutorial 51.1 Homing Behavior in Digger Wasps

- **In a single species, behavior may range from highly stereotyped, invariable responses to highly flexible, conditional responses and from unlearned to learned responses.**

  FAPs are examples of highly stereotyped, unlearned behaviors that are released by extremely simple stimuli. Language acquisition and tool making are examples of behavior that are highly dependent on learning, cognition, and the environmental conditions experienced by an individual. Most of the behavior observed in animals is condition dependent, meaning an individual has the genetic, neuronal, and hormonal mechanisms in place for behaving in a variety of ways, and the actual behavior observed depends on environmental conditions. Sex change in coral-reef fish and the onset of sexual behavior in *Anolis* lizards are examples of condition-dependent behaviors triggered by complex environmental stimuli.

- **The types of learning that individuals do, the way that they communicate, and the way that they orient and navigate correlate closely with their habitat and with the challenges they face in trying to survive and reproduce.**

  Animals usually behave in a way that increases their ability to survive and reproduce in their current environment. Deceitful communication can be adaptive—as when anglerfish fool prey into attacking a "lure" attached to the anglerfish's face or when molting mantis shrimps try to bluff their way into retaining ownership of a hiding place.

- **When individuals behave altruistically, they are usually helping close relatives or individuals that help them in return.**

  Although it is common to observe self-sacrificing behavior in animals, individuals do not do things for the good of the species. Instead, most self-sacrificing behavior is directed toward close relatives. When this is the case, alleles that lead to self-sacrificing can increase in frequency due to kin selection. In addition, some animals that live in close-knit social groups engage in reciprocal altruism—meaning they exchange help over time.

Questions

Content Review

1. What do proximate explanations of behavior focus on?
   a. how displays and other types of behavior have changed through time, or evolved
   b. the functional aspect of a behavior, or its "adaptive significance"
   c. genetic, neurological, and hormonal mechanisms of behavior
   d. psychological interpretations of behavior—especially motivation

2. What evidence suggests that there is a critical period for song learning in some bird species?
   a. In mynahs and mockingbirds, individuals continue to sing new songs throughout their lives.
   b. Individuals do not learn to sing normally unless they are allowed to hear themselves practice.
   c. Individuals that never hear their species-specific song can still sing normally as adults.
   d. Birds that are deafened soon after hatching never sing normally, but birds that are deafened several months after hatching sing normally.

3. What is the difference between orienting behavior and piloting?
   a. Orienting movements are fast responses to visual or auditory stimuli; piloting is way-finding using landmarks.
   b. Orienting is the ability to follow a compass during navigation; piloting is the ability to use map information during navigation.
   c. Orienting is used during long-distance movements such as migration; piloting is used in short-distance movements involved in homing.
   d. Orienting is the ability to use a Sun or star compass; piloting is the ability to use a magnetic compass.

4. Which of the following statements about the waggle dance of the honeybee is not correct?
   a. The length of a waggling run is proportional to the distance from the hive to a food source.
   b. Sounds and scents produced by the dancer provide information about the nature of the food source.
   c. No elements of the round dance are used.
   d. The orientation of the waggling run provides information about the direction of the food from the hive, relative to the Sun's position.

5. Why are biologists convinced that the sex hormone testosterone is required for normal sexual activity in male *Anolis* lizards?
   a. Male *Anolis* lizards with larger testes court females more vigorously than do males with smaller testes.
   b. The molecule is not found in female *Anolis* lizards.
   c. Male *Anolis* lizards whose gonads had been removed did not develop dewlaps.
   d. Male *Anolis* lizards whose gonads had been removed did not court females.

6. What does Hamilton's rule specify?
   a. why animals do things "for the good of the species"
   b. why reciprocal altruism can lead to fitness gains for unrelated individuals
   c. when alleles that favor self-sacrificing acts increase in frequency via kin selection
   d. the conditions under which more complex behaviors evolve from simpler behaviors

Conceptual Review

1. Make a graph like that in Figure 51.2, with axes that plot the degree to which a particular behavior is modified by learning and the degree to which behavior is stereotyped and inflexible. Plot the following behaviors on the graph: tool use in New Caledonian crows; exploration of mazes by rats; singing by chickens, by white-crowned sparrows, and by mynahs; and blushing in humans.
2. Discuss the proximate and ultimate causes of the following behaviors: (a) homing behavior by spiny lobsters, (b) sexual behavior by *Anolis* lizards, and (c) the "jump-back" response of kangaroo rats.
3. Propose a hypothesis to explain the adaptive significance of tool use in New Caledonian crows. Recall that a female crow bent an unfamiliar material into a hooked tool. Explain why this observation supports the hypothesis that individuals of this species can think.
4. What environmental stimuli cause changes in hormone levels that lead to egg laying in *Anolis* lizards? Based on the data presented on sexual behavior in *Anolis*, propose a hypothesis to explain the proximate basis of sex switching in coral-reef fish.
5. For an animal to navigate, it must have a map and a compass. Explain why both types of information are needed. Describe three types of compasses that have been identified in migratory or homing species.
6. Compare and contrast kin selection and reciprocal altruism. Be sure to identify the conditions under which self-sacrificing behavior is expected to evolve under kin selection versus reciprocal altruism.

Group Discussion Problems

1. Mated pairs of cranes give territorial displays called unison calls. Cranes that are raised in isolation from other cranes give unison calls normally, and the display is performed the same way every time. Your friend argues that the unison call of cranes is a FAP, because it is stereotyped and not influenced by learning. Another friend argues that it is a condition-dependent behavior, because only mated pairs do it and because the tendency to give unison calls varies among pairs and with time of year. (Some pairs are more aggressive than others and give unison calls more frequently in territorial contexts.) Who's right?
2. Most tropical habitats are highly seasonal. But instead of alternating warm and cold seasons, there are alternating wet and dry seasons. Most animal species breed during the wet months. If you were studying a species of *Anolis* native to the tropics, what environmental cue would you simulate in the lab to bring them into breeding condition? How would you simulate this cue? Further, think about the sensory organs that lizards use to receive this cue. Are they the same as or different than the receptors that *Anolis carolinensis* uses to recognize that spring has arrived in the southeastern United States?
3. What is the significance of the observation that the product of the *for/dg2* gene in fruit flies is involved in a signal transduction pathway? Propose a hypothesis to explain why fruit-fly larvae with the rover genotype and phenotype have higher fitness when population density is high—meaning individuals are crowded.
4. J. B. S. Haldane once remarked that he'd be willing to lay down his life to save two brothers or eight cousins. Explain what he meant. Based on the theory of kin selection and reciprocal altruism, predict the conditions under which people are expected to be nice to one another.

Answers to Multiple-Choice Questions 1. c; 2. d; 3. a; 4. c; 5. d; 6. c

www.prenhall.com/freeman is your resource for the following: Web Tutorials; Online Quizzes and other Online Study Guide materials; Answers to Conceptual Review Questions; Solutions to Group Discussion Problems; Answers to Figure Caption Questions and Exercises; and Additional Readings and Research.

52 Population Ecology

KEY CONCEPTS

- Life tables summarize how likely it is that individuals of each age class in a population will survive and reproduce.
- The growth rate of a population can be calculated from life-table data or from the direct observation of changes in population size over time.
- A wide variety of patterns is observed when researchers track changes in population size over time, ranging from no growth, to regular cycles, to continued growth independent of population size.
- Data from population ecology studies help biologists evaluate prospects for endangered species and design effective management strategies.

A flock of white pelicans in a channel of the Mississippi River. This chapter explores how and why growth rates in populations change through time.

A **population** is a group of individuals from the same species that live in the same area at the same time. In both ecology and evolutionary biology, populations are the basic unit of analysis. Evolutionary biology is the study of how the characteristics of populations change through time; ecology is the study of how populations interact with their environment.

Population ecology is the study of how and why the number of individuals in a population changes over time. Biologists also analyze changes in the ages of individuals in a population, the proportion of males to females, and geographic distribution. With the explosion of human populations across the globe, the massive destruction of natural habitats, and the resulting threats to species throughout the tree of life, population ecology has become an extraordinarily vital field in biological science. The mathematical and analytical tools introduced in this chapter help biologists predict changes in population size and design management strategies to save threatened species.

Biologists ask a wide array of questions in population ecology. How are individuals distributed in space? How old are they, and how likely are they to reproduce or die? Is the population growing or declining or staying the same through time, and why?

To answer these questions, let's review some of the basic tools that biologists use to study populations, then follow up with examples of how biologists study changes in population size over time. The chapter concludes by asking how all of these elements fit together in efforts to save endangered species.

52.1 Demography

The number of individuals that are present in a population depends on four processes: birth, death, immigration, and emigration. Populations grow due to births—here meaning any form of reproduction—and **immigration**, which occurs when individuals enter a population by moving from another population. Populations decline due to deaths and **emigration**, which occurs when individuals leave a population to join another population. Analyzing birth rates, death rates, immigration rates, and emigration rates is fundamental to **demography**: the study of factors that determine the size and structure of populations through time.

To make detailed predictions about the future of a population, however, biologists need to understand the makeup of the population in more detail. If a population consists primarily of young individuals with a high survival rate and reproductive rate, the population size should increase. But if a population is comprised chiefly of old individuals with low reproductive rates and low survival rates, then it is almost certain to decline. To predict the future of a population, biologists need to know how many individuals of each age are alive, how likely individuals of different ages are to survive to the following year, how many offspring are produced by females of different ages, and how many individuals of different ages immigrate and emigrate each **generation**—the average time between a mother's first offspring and her daughter's first offspring.

These types of demographic data provide an important tool for biologists charged with designing management programs for endangered species. To understand the nature of these data and how they are used in conservation programs, let's examine a classical tool for describing the demography of a population.

Life Tables

Formal demographic analyses of populations are based on a type of data set called a life table. A **life table** summarizes the probability that an individual will survive and reproduce in any given year over the course of its lifetime. Life tables were invented almost 2000 years ago; in ancient Rome they were used to predict food needs. In modern times, life tables have been the domain of life-insurance companies, which have a strong financial interest in predicting the likelihood of a person dying at a given age. More recently, biologists have employed life tables to study the demographics of endangered species and other nonhuman populations.

To understand how researchers use life tables, consider the lizard *Lacerta vivipara* (**Figure 52.1**). *Lacerta vivipara* is a common resident of open, grassy habitats in Western Europe. Researchers set out to estimate the life table of a low-elevation population in the Netherlands, with the goal of comparing the results to data that other researchers had collected from *L. vivipara* populations in the mountains of Austria and France and in lowland habitats in Britain and Belgium. The researchers wanted to know if populations that interact with different environments vary in basic demographic features.

FIGURE 52.1 ***Lacerta vivipara*** **Are Native to Europe**
Translated literally, *vivipara* means "live birth." Females in *L. vivipara* populations from northern Spain and southwest France lay eggs; in other populations, females bear live young.

To complete the study, the researchers visited their study site daily during the seven months that these lizards are active during the year. Each day the researchers captured and marked as many individuals as possible. Because this program of daily monitoring continued for seven years, the biologists were able to document the number of young produced by each female in each year of its life. If a marked individual was not recaptured in a subsequent year, they assumed that it had died sometime during the previous year. These data allowed them to calculate the number of individuals that survived each year in each particular age group and how many offspring each female produced. What did the data reveal?

Survivorship **Survivorship** is a key component of a life table and is defined as the proportion of offspring produced that survive, on average, to a particular age. For example, suppose 1000 *L. vivipara* are born in a particular year. These individuals represent a **cohort**—a group of the same age that can be followed through time. How many individuals would survive to age 1, age 2, age 3, and so on?

As **Table 52.1** (page 1194) shows, the biologists calculated that, in the population living in the Netherlands, survivorship from birth to age 1 as 0.424. If 1000 females were born in a particular year in this population, on average 424 would still be alive one year later. Survivorship from birth to age 2 was 0.308, meaning an average of 308 female lizards would survive for two years.

TABLE 52.1 Life Table for *Lacerta vivipara*

Year	Number alive	Survivorship	Fecundity	Survivorship × fecundity = average number of offspring produced per female born
0	1000	1.000	0.00	0.00
1	424	0.424	0.08	0.03
2	308	0.308	2.94	0.91
3	158	0.158	4.13	0.65
4	57	0.057	4.88	0.28
5	10	0.010	6.50	0.07
6	7	0.007	6.50	0.05
7	2	0.002	6.50	0.01

Data are from Strijbosch and Creemers, 1988.

To recognize general patterns in survivorship and make comparisons among populations or species, biologists plot the logarithm of the number of survivors versus age. The resulting graph is called a **survivorship curve**. Studies on a wide variety of species indicate that three general types of survivorship curves exist (**Figure 52.2a**). Humans have what biologists call a type I survivorship curve. In this pattern, survivorship throughout life is high, and most individuals approach the species' maximum life span. Songbirds, in contrast, experience relatively constant mortality throughout their lives once they have left the nest, resulting in a type II survivorship curve. Many plants have type III curves due to extremely high death rates for seeds and seedlings. **Figure 52.2b** provides a graph for you to plot survivorship of *L. vivipara*.

Survivorship curves are important in conservation work, because they pinpoint the stage of life when endangered species have particularly low survivorship. Biologists can then put measures into place to protect individuals during that period. For example, trees or herbaceous plants can be protected from drought when they are very young by germinating them in a nursery and from predation by putting a wire cage over seedlings.

Fecundity The number of female offspring produced by each female in the population is termed **fecundity.** (In most cases, biologists only keep track of females when calculating life-table data, because only females produce offspring.) Researchers documented the reproductive output of the same *L. vivipara* lizard females year after year and thus were able to calculate a quanti-

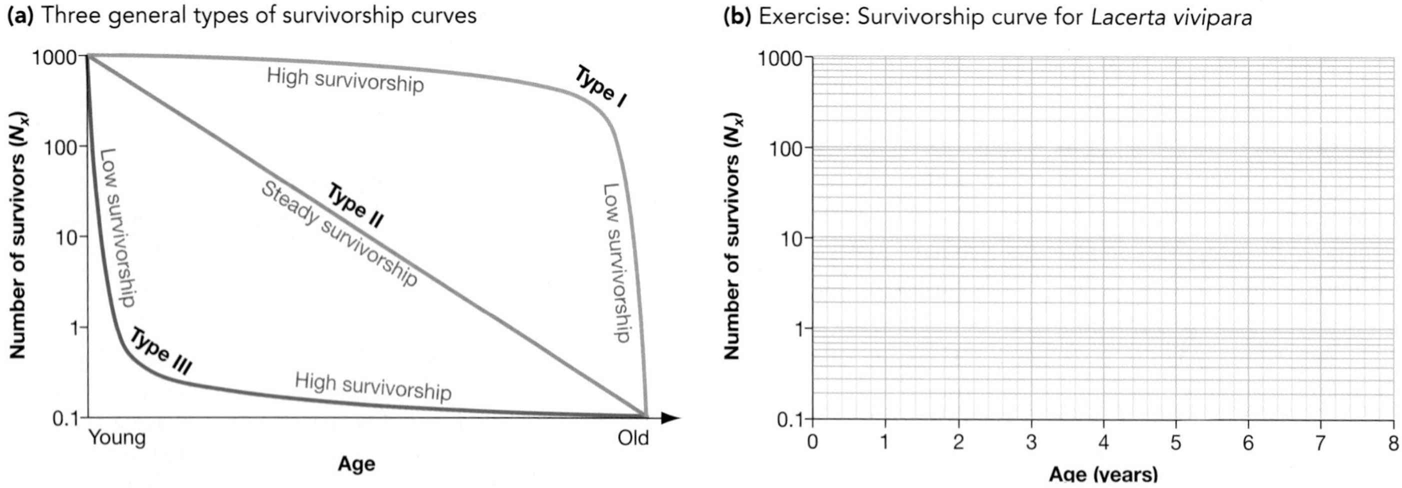

FIGURE 52.2 Three General Types of Survivorship Curves
(a) Generalized graphs showing the number of survivors, on a logarithmic scale, plotted against age.
EXERCISE Fill in the graph in part **(b)** with data on survivorship of *Lacerta vivipara* given in Table 52.1. Compare the shape of the curve to the generalized graphs in part (a).

ty called **age-specific fecundity**, which is defined as the average number of female offspring produced by a female in age class *x*. An **age class** is a group of individuals of a specific age—for instance, all female lizards between 4 and 5 years old. As **Box 52.1** (page 1196) shows, data on survivorship and fecundity allow researchers to calculate the growth rate of a population. How do data on survivorship and fecundity in the Netherlands populations compare with populations in different types of habitats?

The Role of Life History

The life-table data in Table 52.1 are interesting because they contrast with results from other populations of *L. vivipara*. For example, notice that almost no 1-year-old female *L. vivipara* lizards in the Netherlands breed. But in Brittany, France, 50 percent of 1-year-old female lizards breed. Both situations contrast sharply with data on females in the mountains of Austria. In populations that live at high elevation, females do not begin breeding until they are 4 years old. In addition, most females in the Austrian population live much longer than do individuals in either lowland population—in the Netherlands or France. When the three sites are compared, it is clear that fecundity is high but survivorship is low in Brittany. In contrast, fecundity is low but survivorship is high in Austria. The population in the Netherlands is intermediate in both characteristics.

To make sense of this type of variation in fecundity and survivorship, biologists point out that every individual has a restricted amount of energy and resources at its disposal. Resources are limited. As a result, there is a fundamental trade-off between the size of eggs that a female can produce and the number of eggs that she can produce. No female can produce a very large number of very large eggs. Similarly, individuals may allocate resources to reproduction and achieve high fecundity, or they may allocate resources to increasing survivorship. But as the data in **Figure 52.3** indicate, it is not possible to achieve both high fecundity and high survivorship. To capture this point, biologists say that a trade-off exists between survival and reproduction.

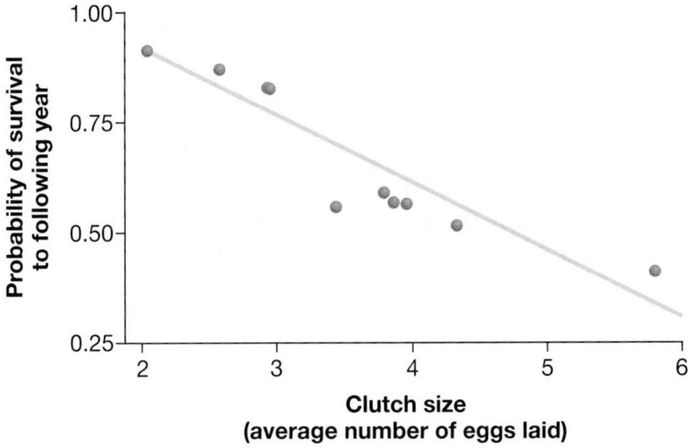

FIGURE 52.3 There Is a Trade-off between Survival and Reproduction
Graph of survivorship versus fecundity. Each data point represents a different species of bird. A statistical analysis called regression was used to create a "line of best fit" to these data points.

Organisms with high fecundity tend to grow quickly, reach sexual maturity at a young age, and produce many small eggs or seeds. The mustard plant *Arabidopsis thaliana*, for example, germinates and grows to sexual maturity in just four to six weeks. In this species, individuals usually live only a few months but may produce as many as 10,000 tiny seeds. In contrast, organisms with high survivorship tend to grow slowly and invest resources in traits that reduce damage from enemies and increase their own ability to compete for water, sunlight, or food. An oak tree, for example, may take decades to reach sexual maturity but can live for several hundred years and produce offspring each year. Oaks invest resources in making wood, an extensive root system, large seeds (acorns), and bitter-tasting molecules (tannins) that deter predators. These traits increase the survivorship of both seedlings and adults but decrease fecundity.

An organism's **life history** consists of how an individual allocates resources to growth, reproduction, and activities or structures that are related to survival. An *Arabidopsis thaliana* plant and an oak tree represent two ends of a broad continuum that exists in life-history characteristics (**Figure 52.4**). Even within the species *Lacerta vivipara*, analyses of life tables have shown that there is considerable variation in life-history traits.

Research has also documented that, within populations, life-history traits can change if conditions change. For example, suppose that a series of wet years produced an abundance of food for lizards. Females might survive better in response, or they might begin breeding at an earlier age or produce more offspring at a particular age. Such changes might directly affect a population's growth rate. How do biologists analyze growth rates in populations? And what factors make growth rates increase or decrease?

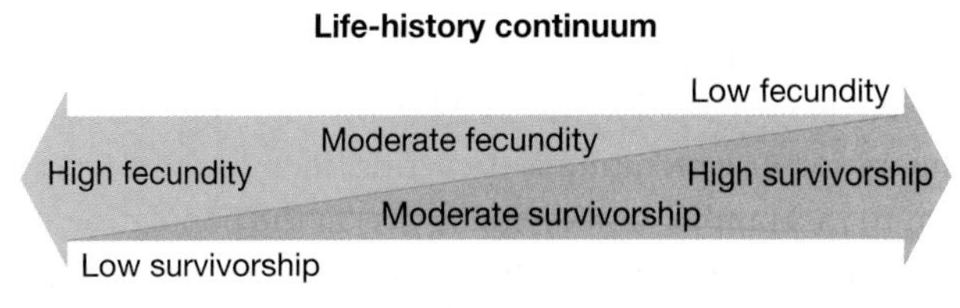

FIGURE 52.4 Life-History Traits Form a Continuum
Every organism can be placed somewhere on this life-history continuum. The placement of a species is most meaningful when it is considered relative to closely related species. EXERCISE Place *Arabidopsis thaliana* and an oak tree on the continuuum. For plants, list the types of traits you would expect to see on the left, middle, and right of the continuum. Include growth habit (herbaceous, shrub, or tree) as well as relative disease- and predator-fighting ability, seed size, seed number, body size, and investment in roots versus shoots.

BOX 52.1 Using Life Tables to Calculate Population Growth Rates

If immigration and emigration are ignored, the data in a life table can be used to calculate a population's growth rate. This is logical, because survivorship and fecundity are ways to express death rates and birth rates—the other two factors that influence population size. To see how a population's growth rate can be estimated from life-table data, let's look at each component in a life table more carefully.

Survivorship is symbolized as l_x, where x represents the age class being considered. Survivorship for age class x is calculated by dividing the number of individuals in that age class (N_x) by the number of individuals that existed as offspring (N_0):

$$l_x = \frac{N_x}{N_0} \qquad \text{(Eq. 52.1)}$$

Age-specific fecundity is symbolized m_x, where x again represents the age class being considered. Age-specific fecundity is calculated as the total number of offspring produced by females of a particular age, divided by the total number of females of that age class present. It represents the average number of offspring produced by a female of age x.

Documenting age-specific survivorship and fecundity allows researchers to calculate the net reproductive rate, R_0, of a population:

$$R_0 = \sum_{i=0}^{x} l_x m_x \qquad \text{(Eq. 52.2)}$$

The **net reproductive rate** represents the growth rate of a population per generation. The logic behind the equation for R_0 is that the growth rate of a population per generation is identical to the average number of female offspring that each female produces over the course of her lifetime. A female's average lifetime reproduction, in turn, is a function of survivorship to each age class and fecundity at each age class. In Table 52.1, R_0 is the sum of the survivorship × fecundity values in the rightmost column.

If R_0 is greater than 1, then the population is increasing in size. If R_0 is less than 1, then the population is declining. According to the data in Table 52.1, R_0 in the Netherlands population of *L. vivipara* is 2.0—in other words, an average *L. vivipara* female produces 2.00 female offspring over the course of her lifetime. When this study was being conducted, the population in the Netherlands was growing rapidly.

52.2 Population Growth

The most fundamental questions that biologists ask about populations involve growth or decline in numbers of individuals. For conservationists, the ability to analyze and predict changes in population size is fundamental in efforts to manage threatened species.

Recall that four processes affect a population's size: Births and immigration add individuals to the population; deaths and emigration remove them. It follows that a population's overall growth rate is a function of birth rates, death rates, immigration rates, and emigration rates. In this section we consider only the impact of births and deaths on population growth; Section 52.4 includes an analysis of the movement of individuals into and out of populations.

A population's growth rate is the change in the number of individuals in the population (ΔN) per unit time (Δt). If no immigration or emigration is occurring, then the growth rate is equal to the number of individuals (N) in the population times the difference between the birth rate per individual (b) and death rate per individual (d). The difference between the birth rate and death rate per individual is called the **per-capita rate of increase** and is symbolized r. (*Per capita* means "per individual.") If the per-capita birth rate is greater than the per-capita death rate, then r is positive and the population is growing. But if the per-capita death rate begins to exceed the per-capita birth rate, then r becomes negative and the population declines. Within populations, r varies through time and can be positive, negative, or 0.

When conditions are optimal for a particular species—meaning birth rates per individual are as high as possible and death rates per individual are as low as possible—then r reaches a maximal value called the **intrinsic rate of increase,** r_{max}. When this occurs, a population's growth rate is expressed as

$$\Delta N/\Delta t = r_{max} N$$

Exponential Growth

The graphs in **Figure 52.5** plot changes in population size for various values of the intrinsic rate of increase. In species such as *Arabidopsis* and fruit flies, which breed at a young age and produce many offspring each year, r_{max} is high. In contrast, r_{max} is low in species such as giant pandas and oak trees, which take years to mature and produce few offspring each year. Stated another way, r_{max} is a function of a species' life-history traits.

The sweeping curves in Figure 52.5 illustrate exponential growth. **Exponential growth** occurs when r does not change over time. The key point about exponential growth is that the growth rate does not depend on the number of individuals in the population. When increases in the size of the population do not affect r, biologists say that growth is **density independent.**

It's important to emphasize that exponential growth adds an increasing number of individuals as the total number of individuals N gets larger. As an extreme example, an r of 0.02 per year in a population of 1 billion adds over 20 million individuals per year. The same growth rate in a population of 100 adds just over 2 individuals per year. Even if r is constant, the number of individuals added to a population is a function of N.

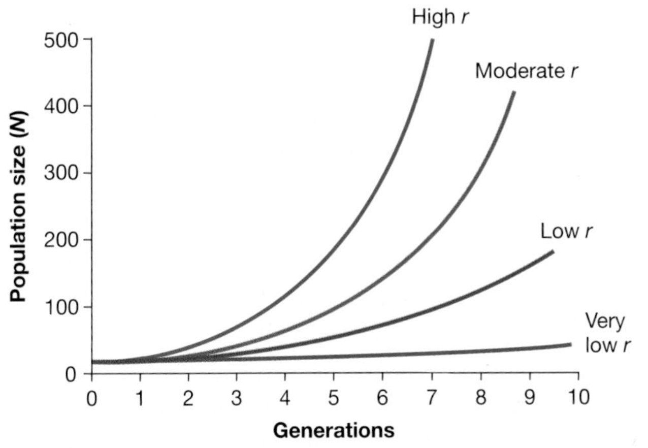

FIGURE 52.5 Exponential Growth Is Independent of Population Size

When the per-capita growth rate r does not change over time, exponential growth occurs. Population size may increase slowly or rapidly, depending on the size of r. Exponential growth occurs when growth is not affected by population density. QUESTION Consider a species on the left-hand side of the life-history continuum in Figure 52.4 (high fecundity, low survivorship) versus a species on the right-hand side (low fecundity, high survivorship). Under ideal conditions, which species has higher r? Explain your reasoning.

In nature, exponential growth is observed when a few individuals found a new population in a new habitat or when a population has been devastated by a storm or some other type of catastrophe and then begins to recover, starting with a few surviving individuals. But it is not possible for exponential growth to continue indefinitely. If *Arabidopsis* populations grew exponentially for a long period of time, they would eventually fill all available breeding habitat. When **population density**—the number of individuals per unit area—gets very high, we would expect the population's per-capita birth rate to decrease and the per-capita death rate to increase, causing r to decline. Stated another way, growth would be **density dependent.**

Logistic Growth

To analyze what happens when growth is density dependent, biologists use a parameter called carrying capacity. **Carrying capacity, K,** is defined as the maximum number of individuals in a population that can be supported in a particular habitat over a sustained period of time. The carrying capacity of a habitat depends on a large number of factors: food, space, water, soil quality, resting or nesting sites, and the intensity of disease and predation. Carrying capacity can change from year to year, depending on conditions.

If a population of size N is below the carrying capacity K, then the population should continue to grow. More specifically, a population's growth rate should be proportional to $(K - N)/K$:

$$\frac{\Delta N}{\Delta t} = r_{max} N \left(\frac{K - N}{K} \right)$$

This expression is called the **logistic growth equation.** In the expression $(K - N)/K$, the numerator defines the number of additional individuals that can accommodated in a habitat with carrying capacity K; dividing $K - N$ by K turns this number of individuals into a proportion. Thus, $(K - N)/K$ describes the proportion of "unused resources and space" in the habitat. It can be thought of as the environment's resistance to growth. When N is small, then $(K - N)/K$ is large and the growth rate should be high. But as N gets larger, $(K - N)/K$ gets smaller. When N is at carrying capacity, then $(K - N)/K$ is equal to 0 and growth should stop. The key idea here is that as a population approaches a habitat's carrying capacity, its growth rate should slow.

The logistic growth equation describes **logistic population growth,** or changes in growth rate that occur as a function of population size. Just as exponential growth is density independent, logistic growth is density dependent. **Box 52.2** explores population growth models and how they are applied in more detail.

WEB TUTORIAL 52.1 Modeling Population Growth

BOX 52.2 Developing and Applying Population Growth Equations: A Closer Look

To explore how biologists model changes in population size in more detail, let's consider data on whooping cranes, which are large, wetland-breeding birds native to North America. Whooping cranes may live more than 20 years in the wild and over 40 years in captivity, but it is common for females to be six or seven years old before they breed for the first time. Female cranes lay just two eggs per year and usually rear just one chick. It is logical to expect, based on these life-history characteristics, that the growth rate of whooping crane populations will be extremely low.

Conservationists monitor this species closely, because hunting and habitat destruction reduced the total number of individuals in the world to about 20 in the mid-1940s. Since then, intensive conservation efforts have resulted in a current total population of about 425. That number includes a group of 189 individuals that breeds in Wood Buffalo National Park in the Northwest Territories of Canada.

In addition to the Wood Buffalo population, two new populations of cranes have been established recently by releasing offspring raised in captivity. One of

(Continued on next page)

(Box 52.2 continued)

these groups lives near the Atlantic coast of Florida year round; the other group migrates from breeding areas in northern Wisconsin to wintering areas along Florida's Gulf coast.

According to the biologists who are in charge of managing whooping crane recovery, the species will no longer be considered endangered when the two newly established populations have at least 25 breeding pairs—meaning there would be about 125 individuals in each—and when neither population needs to be supplemented with captive-bred young in order to be self-sustaining. How long will this take?

Discrete Growth

Whooping cranes breed once per year, so the simplest way to express a crane population's growth rate is to compare the number of individuals at the start of one breeding season to the number at the start of the following year's breeding season. The current total at Wood Buffalo is 189. Suppose that biologists count 198 cranes on the breeding grounds next year. The population growth rate could be calculated as 198/189 = 1.048. Stated another way, the population will have grown at the rate of 4.8 percent per year.

When populations breed during discrete seasons, their growth rate can be calculated as for whooping cranes. To create a general expression for how these populations grow, biologists use N to symbolize population size. N_0 is the population size at time zero (the starting point), and N_1 the population size one breeding interval later. In equation form, the growth rate is given as

$$N_1/N_0 = \lambda \qquad \textbf{(Eq. 52.3)}$$

The parameter λ (lambda) is called the **finite rate of increase.** (In mathematics, a *finite rate* refers to an observed rate over a given period of time. A *parameter* is a variable or constant term that affects the shape of a function but does not affect its general nature.) In the whooping crane example, λ was 1.048. Rearranging the expression in Equation 52.3 gives

$$N_1 = N_0 \lambda \qquad \textbf{(Eq. 52.4)}$$

Stated more generally, the size of the population at the end of time t will be given by

$$N_t = N_0 \lambda^t \qquad \textbf{(Eq. 52.5)}$$

This equation summarizes how populations grow when breeding takes place seasonally. The size of the population at time t is equal to the starting size, times the finite rate of increase multiplied by itself t times. In a sense, λ works like the interest rate at a bank. For species that breed once per year, the "interest" on the population is compounded annually. A savings account with a 5 percent annual interest rate increases by a factor of 1.05 per year. If a population is growing, then its λ is greater than one. The population is stable when λ is 1.0 and declining when λ is less than 1.0.

Continuous Growth

The parameters λ and r—the finite rate of increase and the per-capita growth rate—have a simple mathematical relationship. The best way to understand their relationship is to recall that λ expresses a population's growth rate over a discrete interval of time. In contrast, r gives the population's per-capita growth rate at any particular instant. As a result, r is also called the *instantaneous rate of increase*. The relationship between the two parameters is given by

$$\lambda = e^r \qquad \textbf{(Eq. 52.6)}$$

where e is the base of the natural logarithm, or about 2.72. (In general, the relationship between any finite rate and any instantaneous rate is given by finite rate $= e^{\text{instantaneous rate}}$.)

Substituting Equation 52.6 into Equation 52.5 gives

$$N_t = N_0 e^{rt} \qquad \textbf{(Eq. 52.7)}$$

This expression summarizes how populations grow when they breed continuously, as do humans and bacteria, instead of at defined intervals. For species that breed continuously, the "interest" on the population is compounded continuously. When the growth rates λ and r are equivalent, however, the differences between discrete and continuous growth are negligible.

Because r represents the growth rate at any given time, and because r and λ are so closely related, biologists routinely calculate r for species that breed seasonally. In the whooping crane example, $\lambda = 1.048 = e^r$. To solve for r, take the natural logarithm of both sides. In this case, $r = 0.0469$.

The instantaneous rate of increase, r, is also directly related to the net reproductive rate, R_0, introduced in Box 52.1. In most cases, r is calculated as $\ln R_0/g$, where g is the generation time. Thus, r can be calculated from life-table data. It is a more useful measure of growth rate than R_0, because r is independent of generation time.

To summarize, biologists have developed several ways of calculating and expressing a population's growth rate. λ has the advantage of being easy to understand, and R_0 has the advantage of being calculated directly from life-table data. Although r is slightly more difficult conceptually, it is the most useful expression for growth rate, because it is independent of generation time and is relevant for species that breed either seasonally or continuously.

Applying the Models

To get a better feel for r and for Equation 52.7, consider the following series of questions about whooping cranes. The key to answering these questions is to realize that Equation 52.7 has just four parameters. Given three of these parameters, you can calculate the fourth:

1. If 20 individuals were alive in 1941 and 425 existed in 2004, what is r? Here $N_t = 425$, $N_0 = 20$, and $t = 63$ years. Substitute these values into Equation 52.7 and solve for r. Then check your answer at the end of this box.[1]

2. In the most recent report issued by the biologists working on the Wood Buffalo crane recovery program, it is estimated that the flock should be able to sustain an r of 0.046 for the foreseeable future. If the flock currently contains 189 individuals, how long will it take that population to double? Here $N_0 = 189$ and $N_t = 2 \times 189 = 378$. In this case, you solve Equation 52.7 for t. Then check your answer.[2]

3. In 2002 a pair of birds in the flock that lives in Florida year-round successfully raised offspring (nine years after the first cranes were introduced there). In 2003 another pair of cranes in this population bred successfully. If the number of breeding pairs continues to double each year, how long will it take to reach the goal of 25 breeding pairs? (Note that $\lambda = 2.0$ if a population is doubling each year.) Given that $N_0 = 2$ and $N_t = 25$, solve for t again, and check your answer.[3]

4. The whooping crane flock that migrates between Wisconsin and Florida was founded in 2001. If its development is like that of the resident flock in Florida, the first successful breeding attempt will occur in 2010—nine years after the initial introduction. Suppose that the instantaneous growth rate for the number of breeding pairs in this population will be 0.05. In what year will the breeding population reach 25 pairs? This is the year that whooping cranes should come off the endangered species list. Here $N_0 = 1$, $N_t = 25$, and $r = 0.05$. Solve for t, and add this number of years to 2010; then check your answer.[4]

[1] $r = 0.05$
[2] $t = 15$ years
[3] $t = 3.64$ years
[4] 2074

Figure 52.6a illustrates density-dependent growth for a hypothetical population. Notice that the graph has three sections. Initially, growth is rapid. With time, N increases to the point at which competition for resources or other density-dependent factors begins to occur. As a result, the growth rate begins to decline and eventually reaches 0. When the population is at the habitat's carrying capacity, the graph of population size versus time is flat.

Figure 52.6b shows data on logistic growth from laboratory populations of two species of protists in the group called ciliates (see Chapter 28): *Paramecium aurelia* and *P. caudatum*. In this experiment, an investigator placed 20 individuals from one of the *Paramecium* species into 5 mL of a solution. He set up made many replicates of these 5-mL environments for each species separately. He kept conditions as constant as possible in each replicate by adding the same number of bacterial cells

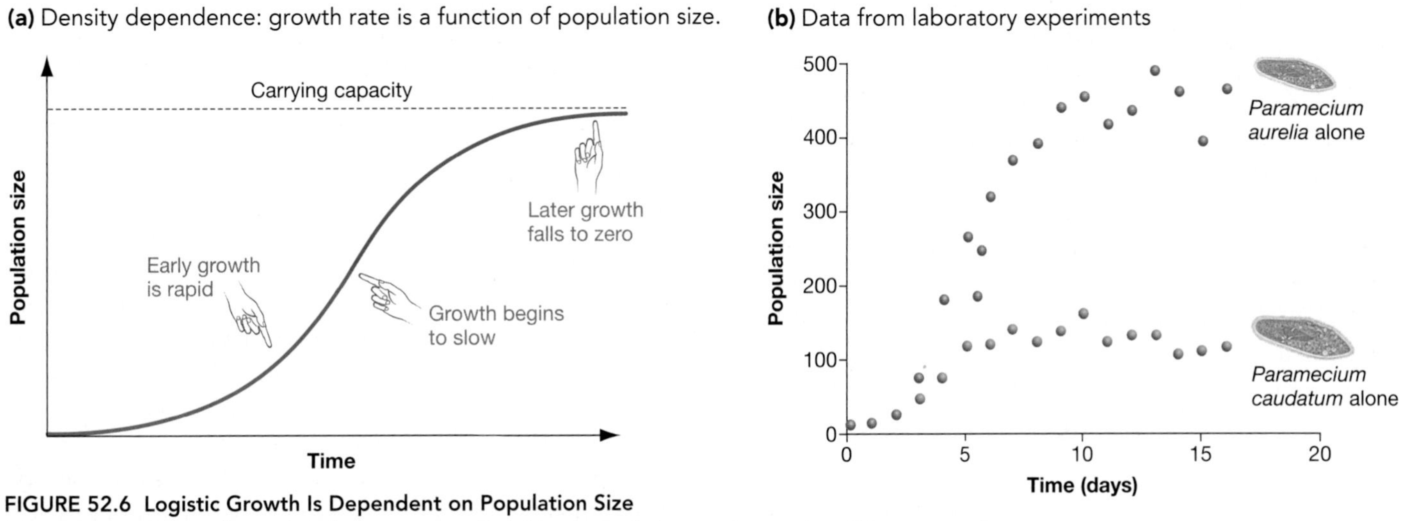

FIGURE 52.6 Logistic Growth Is Dependent on Population Size
(a) Hypothetical curve illustrating the pattern predicted by the logistic growth equation. This pattern often occurs when a small number of individuals colonizes an unoccupied habitat. Initially, r is high because competition for resources is low to nonexistent. Carrying capacity depends on the quality of the habitat.
(b) Data from laboratory experiments with two species of *Paramecium*. EXERCISE In part (b), draw dashed horizontal lines indicating the carrying capacity for each species.

every day as food, washing the solution every second day to remove wastes, and maintaining the pH at 8.0. As the graphs show, both species exhibited logistic growth in this environment. The carrying capacity differed in the two species, however. The maximum density of *P. aurelia* averaged 448 individuals per mL, but that for *P. caudatum* averaged just 128 individuals per mL.

To summarize, exponential growth can occur only for a relatively short time. Eventually a population's growth rate has to decline. Why? What factors cause growth rates to change?

What Limits Growth Rates and Population Sizes?

Population sizes change as a result of two general types of factors: (1) density-independent and (2) density-dependent factors. **Density-independent factors** change birth rates and death rates irrespective of the number of individuals in the population. Density-independent changes are usually triggered by changes in the abiotic environment—variation in weather patterns or catastrophic events such as cold snaps, hurricanes, volcanic eruptions, or drought. In contrast, **density-dependent factors** are usually biotic and change in intensity as a function of population size. For example, predation rates on deer may increase when population sizes are high, or competition for food and starvation may become more common. When trees are crowded, they have less water, nutrients, and sunlight at their disposal and make fewer seeds. Death rates may increase and birth rates decrease when populations are at high density.

To get a better understanding of how biologists study factors that affect population size, consider the data in **Figure 52.7**:

- The graph in Figure 52.7a is from an experimental study of a coral-reef fish called the bridled goby. Each data point represents an identical artificial reef, constructed by a researcher from the rubble of real coral reefs. The initial density, plotted along the horizontal axis, is the number of marked bridled gobies that were released on each artificial reef at the start of the experiment. The proportion surviving, plotted on the vertical axis, is the number of these introduced individuals that were still living at that reef 2.5 months later. This graph shows a strong density-dependent relationship in survivorship.
- The graph in Figure 52.7b is from a long-term study of song sparrows on Mandarte Island, British Columbia. Each data point represents a different year. The density of females, plotted along the horizontal axis, is the number of females that bred on the island; the clutch size, plotted on the vertical axis, is the average number of eggs laid by each female. This graph indicates a strong density-dependent relationship in fecundity.

Density-dependent changes in on survivorship and fecundity cause logistic population growth. In this way, density-dependent factors define a particular habitat's carrying capacity. If gobies get crowded, they die or emigrate. If song sparrows are crowded, they cannot produce a normal number of eggs and offspring.

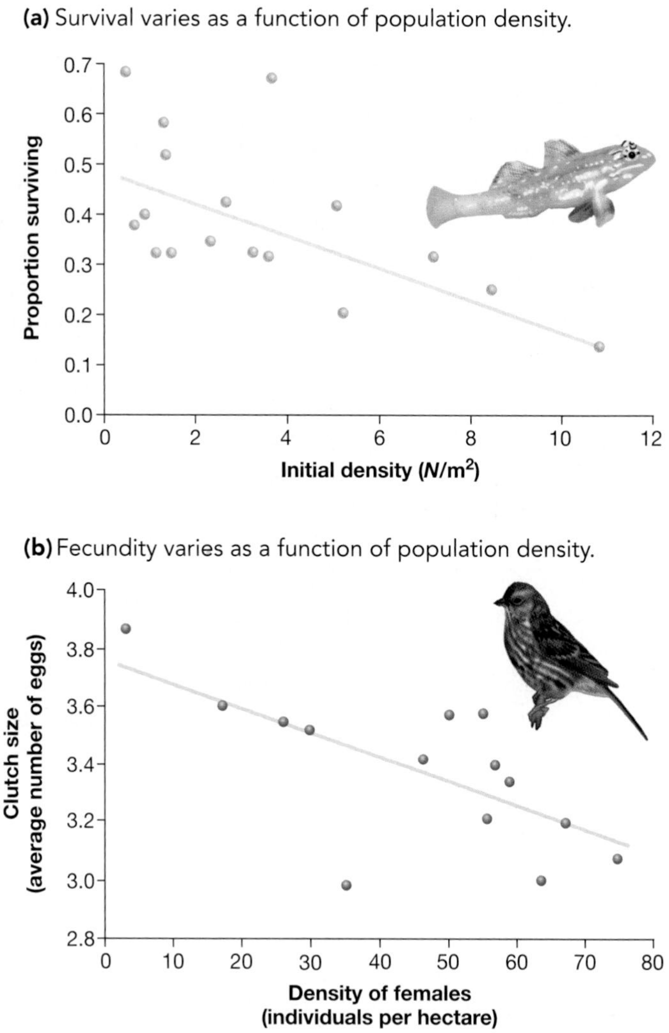

FIGURE 52.7 Density-Dependent Growth Results from Changes in Survivorship and Fecundity
(a) When bridled gobies are introduced at different population densities to artificial reefs, survivorship is highest at low density. **(b)** In the song sparrow population on Mandarte Island, there is a strong negative correlation between population density in a particular year and the average number of eggs (clutch size) produced by females. QUESTION When song sparrow populations are at high density and extra food is provided to females experimentally, average clutch size is much higher than expected from the data in part (b). Based on these data, state a hypothesis to explain why average clutch size declines at high density.

It's important to recognize that K varies among species and populations, however, and that this variation affects both growth rates and population sizes. K varies because for any particular species, some habitats are better than other habitats due to differences in food availability, predator abundance, and other density-dependent factors. Stated another way, K varies in space. It also varies with time, as conditions in some years are better than others. In addition, the same region may have a very different carrying capacity for different species. The same area will tend to support many more individuals of a small-bodied species than of a large-bodied species, simply because large individuals demand more space and resources. These simple observations help explain the variation in total population size that exists among species and among populations of the same species.

When populations of the same species are considered, r_{max} may vary as well. To drive this point home, recall that life-table data vary in low-elevation and high-elevation populations of *Lacerta vivipara* lizards. Because female lizards in low-elevation areas mature quickly and begin breeding early in life, they have higher lifetime fecundity than do females in alpine regions, which start breeding much later. Thus, r_{max} is higher in lowland populations. In general, species or populations that have relatively long life spans and large body size tend to have much lower r_{max} than do small-bodied, short-lived species.

✓CHECK YOUR UNDERSTANDING

Populations grow exponentially unless slowed by density dependence. Density-dependent factors that influence population growth rates include predation, disease, and competition for resources such as food and sunlight. You should be able to (1) propose a density-dependent factor that limits growth in *Paramecium aurelia* and design an experiment to test this hypothesis, and (2) propose a hypothesis to explain why the carrying capacity for *P. caudatum* in Figure 52.6b is so much lower than the carrying capacity for *P. aurelia* and design an experiment to test this hypothesis.

52.3 Population Dynamics

The theory introduced in the previous two sections provides a foundation for exploring how biologists study **population dynamics**—changes in populations through time. As an initial example, consider data on the longest-running experiment in the history of biological science: the Park Grass study in Rothamsted, United Kingdom. In 1856 researchers established a series of 0.1–0.2 ha (about 1/3-acre) plots in a hay meadow that had uniform soil characteristics and vegetation. Since then, some of the plots have received regular fertilizer treatments or lime (a mix of calcium and magnesium oxides), and some have been left untreated as controls. Each year researchers recorded which plant species were present in each plot, as an index of population size.

The weather was the same in all plots, the fertilizing and liming regimes were constant in treated plots, and presumably all plots were exposed to the same predators and diseases. Even so, several dramatic changes in population size have taken place. When the data from 1920–1979 were analyzed, four major patterns emerged among the 43 species present. Ten of the species showed a clear spike in abundance, followed by a decline over the course of the 60-year period (**Figure 52.8a**). Seven species increased in abundance and then reached a plateau (**Figure 52.8b**). Five species declined steadily, and 21 showed no discernible change (**Figure 52.8c**).

In several cases, patterns of change over time appeared to correlate with life-history traits. For example, species that maintained a constant population size throughout the 60-year interval have longer life spans and a lower r_{max} than do other species at the site. To interpret this observation, biologists suggest that these species allocate more energy and resources to competitive ability than to reproduction. As a result, these species are able to maintain abundance over a longer period. In contrast, species that spiked and declined or that increased and maintained high abundance had a higher r_{max}, an earlier age at first reproduction, and shorter life spans than did species that maintained a constant population size. Researchers are still working on the question of why some of these species remained constant over time and why other species showed gradual declines over the course of the study.

Population Cycles: The Case of the Red Grouse

One of the most striking patterns in population dynamics has also been one of the most difficult to understand. **Population cycles** are regular fluctuations in size that some animal populations exhibit. For example, red grouse are native to heather moorlands in Britain. Over three-fourths of red grouse populations in Britain

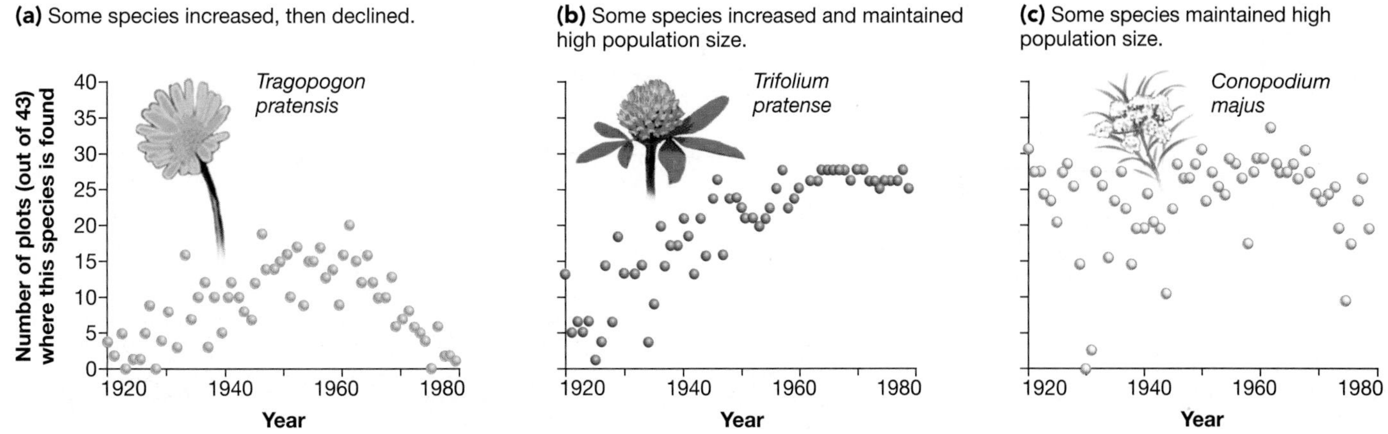

FIGURE 52.8 Distinct Patterns of Long-Term Changes in the Populations of Meadow Species
Graphs showing some of the distinct patterns in population dynamics from 1920 to 1979 among species in the Park Grass experiment, which began in 1856.

rise and fall in regular cycles. Grouse cycles in Britain average between four and eight years in length.

Most hypotheses to explain population cycles hinge on some sort of density-dependent factor. The idea is that predation, disease, or food shortages intensify dramatically at high population density and cause population numbers to crash. It has been extremely difficult, however, to document a causal relationship between population cycles and factors such as predation intensity, disease outbreaks, or food availability.

What factor causes red grouse populations to cycle? A group of biologists suspected that a parasitic roundworm called *Trichostrongylus tenuis* is responsible. They contended that infection rates skyrocket when grouse densities are high, causing populations to drop rapidly. The biologists also hypothesized that infection rates lessen when grouse densities are low, allowing populations to recover. The idea here is that the parasite is transmitted readily at high population densities, and that individuals in dense populations are less well fed and thus less able to fend off an infection.

To test these propositions experimentally, the researchers caught from 1000 to 3000 red grouse in each of several populations during the year when a crash was expected. They treated the individuals with a drug that kills roundworms, released the grouse, and later monitored the number of birds shot by hunters in the ensuing years in each of the treatment populations. Finally, they compared these data to hunting success in two nearby populations where hunting intensity was comparable, and where birds were not treated with the anti-roundworm drug.

Some results of the experiment are shown in **Figure 52.9**. The control populations showed a typical dramatic four-year

Question: What factors cause red grouse populations to cycle?

Hypothesis: Density-dependent infections by a parasitic roundworm cause red grouse populations to cycle.

Null hypothesis: Parasitic roundworm infections do not influence red grouse population cycles.

Experiment setup:

Experimental populations

1. Treat red grouse with anti-roundworm drug during a year when a population crash is expected.

2. Monitor the number of grouse shot by hunters in ensuing years.

Control populations

1. No drug treatment.

2. Monitor the number of grouse shot by hunters in ensuing years.

Prediction: The experimental populations will not crash following the treatments, but the control populations will.

Prediction of null hypothesis: Both the experimental and control populations will crash.

Results:

Two experimental populations (anti-roundworm drug treatments)

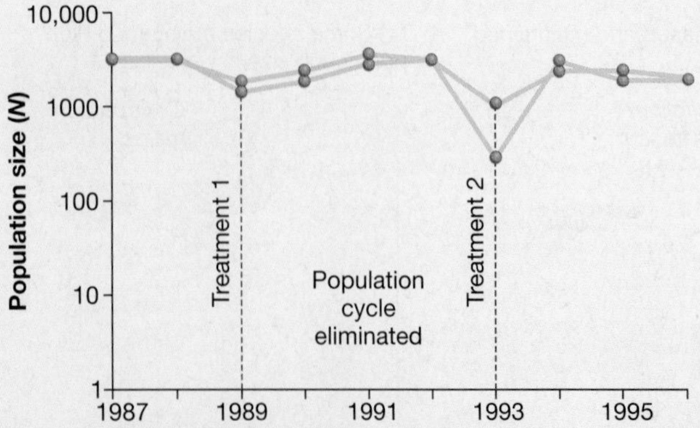

Two control populations (no drug treatment)

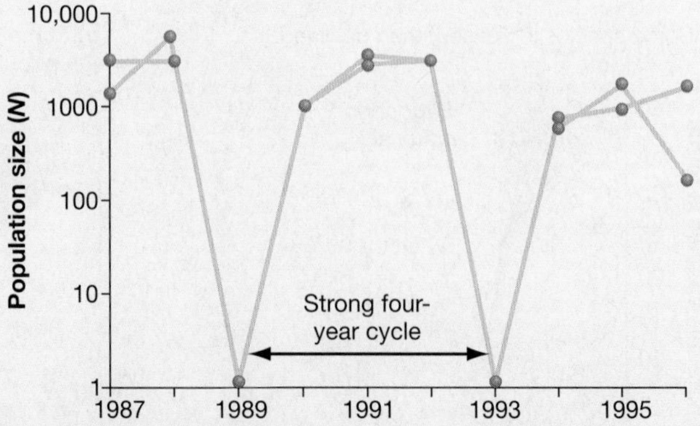

Conclusion: Density-dependent factors such as parasite infection cause red grouse populations to cycle.

FIGURE 52.9 Experimental Evidence That Disease Outbreaks Cause Population Cycles in Red Grouse
The data points on the graphs indicate the number of birds shot by hunters each year out of experimental and control populations. The number killed is interpreted as an index of total population size.

cycle in numbers, but the treated populations maintained high densities. This is strong evidence that red grouse cycles are driven by density-dependent changes in disease rates. The study also illustrates the power of a well-designed experiment to solve long-standing controversies in biology.

Age Structure

As the life-table data presented in Section 52.1 showed, age has a dramatic effect on the probability that an individual will reproduce and survive to the following year. But the previous analysis of life-table data left out a critical point about the ages of individuals in a population: A population's **age structure**—meaning the proportion of individuals that are at each possible age—has a dramatic influence on the population's growth over time. To see how changes in age structure can affect population dynamics, let's consider two case histories for which biologists have documented changes in age structure—one example concerns a flowering plant, and the other focuses on our own species.

Age Structure in a Woodland Herb The common primrose, pictured in **Figure 52.10a**, grows in the woodlands of Western Europe. Primroses grow on the forest floor and can germinate, mature, and produce offspring only in the relatively high-light environment created when a tall tree falls and opens a gap in the forest canopy. These sunny spaces are short lived, however, because mature trees and saplings in and around the gap grow and fill the space. As they do, light levels in the gap decline and the environment becomes increasingly unsuitable for herbs such as primrose.

Biologists recently set out to document how primrose populations respond to this ephemeral woodland environment. The researchers hypothesized that populations in new gaps and populations in large gaps—meaning those with the highest light availability—would experience high rates of reproduction and an age structure characterized by large numbers of juveniles. The researchers also hypothesized that as light levels declined in a particular gap due to the growth of surrounding trees, primrose reproduction would decline. As a result, the age structure of populations in older gaps should be characterized by a large proportion of individuals in adult stages, with fewer juveniles.

To test these hypotheses, the investigators selected eight common primrose populations growing in a range of light levels, from high to low. The team established 1-square-meter study plots inside each gap and studied approximately 350 individuals per population. Each individual was assigned to one of five ages: seedlings, juveniles, or adults in one of three size classes. As **Figure 52.10b** shows, the data provided strong support for their hypothesis that in gaps with high light levels, populations have a larger proportion of juveniles than do gaps with low light levels. In addition, the proportion of juvenile individuals declined over time in all gaps as light levels declined. These data have three important messages about the dynamics of primrose populations:

1. Populations that are dominated by juveniles should experience rapid growth. Population size should then decline with time, however, due to a density-independent factor: shading by trees.
2. The long-term trajectory of the overall primrose population in an area may depend primarily on the frequency and severity of windstorms that knock down trees and create sunlit gaps. If so, it means that the dynamics of primrose populations are governed by an abiotic, density-independent factor.
3. In a large tract of forest, the primrose population will consist of a large number of subpopulations, each found in

(a) Common primroses live in sunlit gaps in forests.

(b) Age structure of primrose population varies with age of gap.

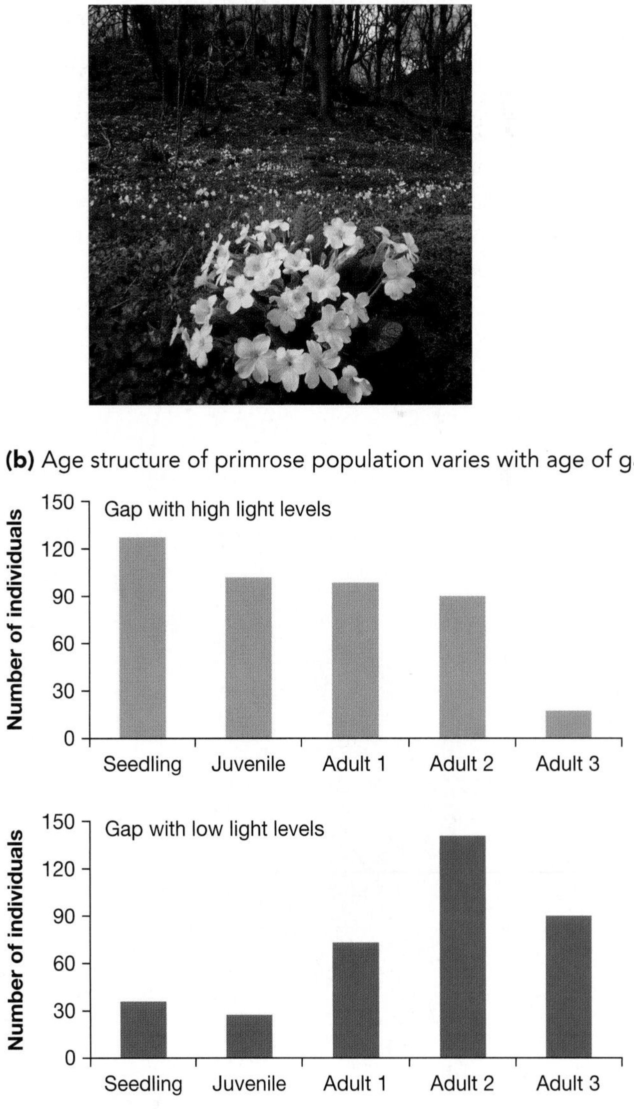

FIGURE 52.10 The Age Structure of Common Primrose Populations Changes over Time
(a) Common primroses are forest herbs. **(b)** "Adult 1," "Adult 2," and "Adult 3" individuals are of increasingly large size and thus age.

canopy gaps. Some of these subpopulations are likely to be very small, and each subpopulation will arise, grow, and then go extinct over time. Biologists say that a group of small, isolated populations such as this represents a **metapopulation**—a population of populations. Metapopulation structure turns out to be extremely common. The history of a metapopulation is driven by the birth and death of populations, just as the dynamics of a single population are driven by the birth and death of individuals. As Section 52.4 will show, this aspect of population structure has important implications for conservation biology.

Age Structure in Human Populations As another example of how age structure affects population dynamics, consider humans. In countries where industrial and technological development occurred many decades ago, survivorship has been high and fecundity low for several decades. As a result, these populations are not growing quickly. When researchers create a graph with horizontal bars representing the numbers of males and females of each age group for such a population, an **age pyramid** like that in **Figure 52.11a** results. The most striking pattern in the data is that there are similar numbers of people in most age classes. The evenness occurs because the same number of infants are being born each year and because most survive to old age. In this way, analyzing an age pyramid can give biologists important information about a population's history.

Studying age distributions can also help researchers predict a population's future. For example, the populations of developed nations are not expected to grow especially quickly, because only modest numbers of individuals will reach reproductive age in the near future. The red and blue lines in Figure 52.11a show the projected age structure of a developed nation (Sweden) in 2050. Modest changes in the age distribution have occurred since 2000 because survivorship has increased while fecundity has remained the same. The projections highlight a major public policy concern in countries such as this: how to care for an increasingly aged population.

In contrast, the age distribution is bottom-heavy in the less-developed nations. As **Figure 52.11b** shows, these populations are dominated by the very young. This type of age distribution occurs when populations have undergone rapid growth—with more children being born than exist in older age classes. Due to dramatic improvements in health care, most people in the younger age classes now survive to reproductive age.

For Figure 52.11b, the projected age distribution in 2050 is based on continued high survivorship and a pattern that has occurred repeatedly around the world: When infant mortality drops in a country, fecundity rates begin to decline within a generation or two. The red and blue lines in Figure 52.11b are relatively even for ages 0–55—meaning that rapid growth has stopped—because fecundity is expected to be relatively low and survivorship high. The red and blue lines also illustrate a major public policy concern in less-developed countries—providing

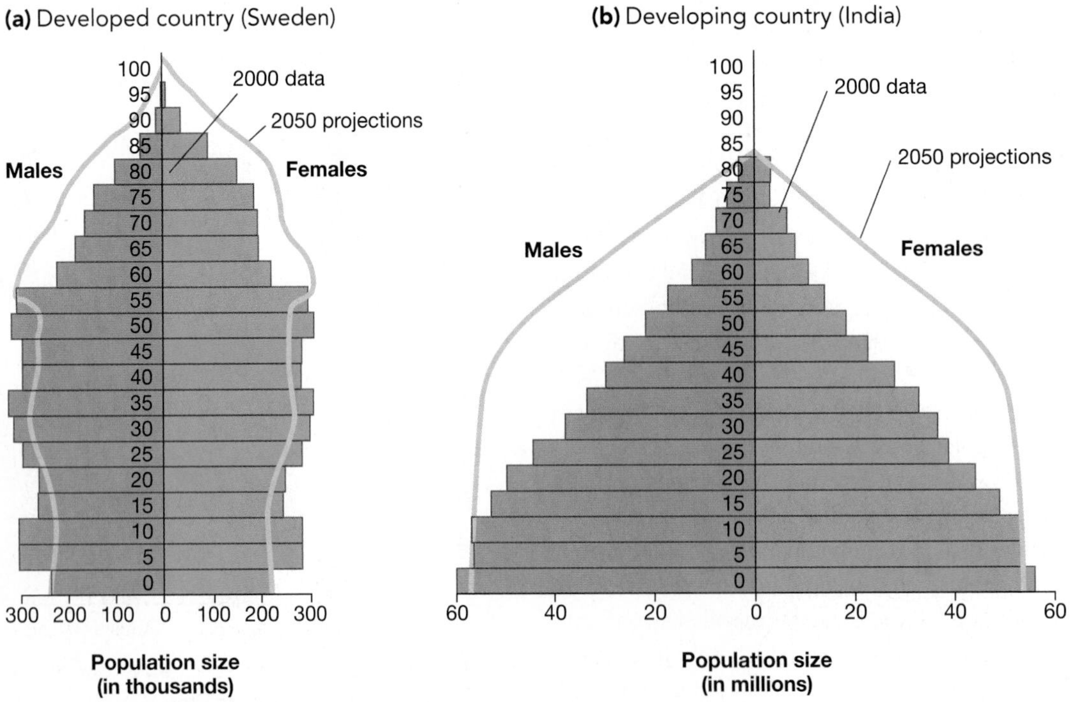

FIGURE 52.11 The Age Structure of Human Populations Varies Dramatically
Age distributions showing the number of males and females in 5-year age increments from 0 to 100.

education and jobs for an enormous group of young people who will be reaching adulthood. As these examples show, understanding age structure is a key component of analyzing population dynamics in humans and other species.

Analyzing Change in the Growth Rate of Human Populations

Studies on changes in whooping crane and Park Grass populations over 60-year periods have been extraordinarily valuable. But six decades turns out to be short in comparison to data available on human population dynamics.

Archeological and anthropological data have been used to estimate the size of human populations over the past 250 years (**Figure 52.12**). Although the shape of the curve is similar to the examples of exponential growth in Figure 52.5, the growth rate for humans has increased over time since 1750, leading to a very steeply rising curve over the past few decades. The highest growth rates occurred between 1965 and 1970, when populations increased at an average of 2.04 percent per year.

Since 1970, however, the growth rate in human populations has been dropping. Between 1990 and 1995 the overall growth rate in human populations averaged 1.46 percent per year; currently the rate is 1.2 percent annually. In humans, r may be undergoing the first long-term decline in history. What will human population size be at its peak?

To answer this question, the UN Population Division makes regular projections for how population will change between now and 2050. The UN projections are based on three scenarios. These scenarios hinge on different values for fertility rates—the average number of children that each woman has during her lifetime. Currently, the worldwide average is 2.7. (This is a huge reduction from the fertility rates during the 1950s, which averaged 5.0 children per woman.) **Figure 52.13a** shows how total population size is expected to change by 2050 if average fertility continues to decline and averages 2.5, 2.1, or 1.7 children per woman. The middle number, 2.1, is particularly important because it represents the replacement rate. When fertility is at the **replacement rate**, each woman produces exactly enough offspring to replace herself and her offspring's father. When this fertility rate is sustained for a generation, $r = 0$—meaning there is **zero population growth (ZPG)**.

A glance at Figure 52.13a should convince you that the three scenarios are starkly different. The high projection predicts that world population in 2050 will reach nearly 11 billion—a 75 percent increase over today's population, which is about 6.4 billion—with no signs of peaking. The low projection, in contrast, predicts that the total human population in 2050 will be at 7.4 billion and will already have peaked.

Figure 52.13b makes another important point. The UN had to alter its population projections dramatically between 1992 and 1998, primarily to account for the impact of AIDS. Growth rates are dropping dramatically in areas of the world that are being hit hard by the epidemic.

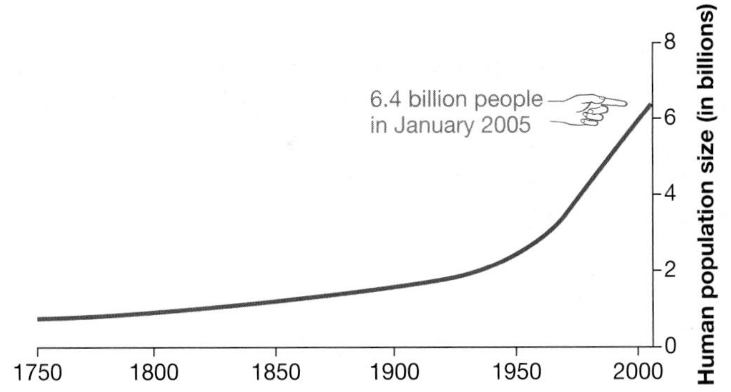

FIGURE 52.12 The Human Population Has Been Growing Rapidly
Graph based on estimates from historical and census data.

(a) 2002 Projections based on three average fertility rates

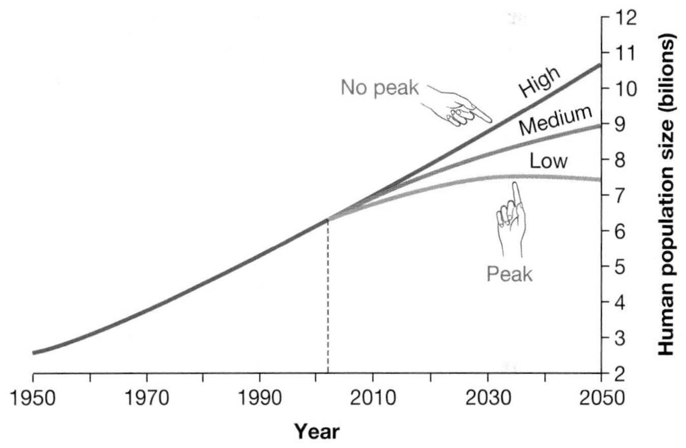

(b) 1992 Projections

Fertility rate	Projected population in 2050
High	12.5 billion
Medium	10.15 billion
Low	7.8 billion

The 1992 projections for 2050 are higher than those from 2002, primarily because the earlier projections did not account for the impact of AIDS.

FIGURE 52.13 Projections for Human Population Growth Depend on Fertility Rates
(a) The UN Population Division's population growth database includes a 50-year projection based on high, medium, and low average fertility rates over the interval. These projections, released in 2002, were based on an average of 2.5 (high), 2.1 (medium), or 1.7 (low) children born per woman. **(b)** Summary of projections made in 1992.

To summarize, humans may be approaching the end of a period of rapid growth that lasted well over 500 years. How quickly growth rates decline and how large the population can eventually become will be decided primarily by changes in fertility rates, and to a lesser extent by the future of the AIDS epidemic.

Are density-dependent processes such as disease and resource shortages responsible for these changes? In other words, are humans approaching Earth's carrying capacity? This question is controversial and is explored in more depth in the essay at the end of this chapter. For now, let's turn to questions about how conservation biologists use the concepts and data presented in this chapter.

✓ CHECK YOUR UNDERSTANDING

To explain changes in the size of populations over time, biologists test hypotheses using observational data—such as the life history of species in the Park Grass experiment—or experiments, such as treating red grouse populations with an antiparasite drug. In many cases, understanding population dynamics depends on understanding changes in a population's age structure. You should be able to explain (1) why the age structure of human populations differs between developed and developing nations and (2) why changes in fertility rates have such a dramatic impact on projections for the total human population in 2050.

52.4 How Can Population Ecology Help Endangered Species?

It should come as no surprise to learn that conservationists draw heavily on concepts and techniques from population ecology when designing programs to save species threatened with extinction. As habitat destruction and invasive species push other species from throughout the world into decline, the study of population growth rates and population dynamics has taken on an increasingly practical tone.

To illustrate how biologists use the theory and results introduced in this chapter, let's analyze how an understanding of geographic structure and life-table data can help direct conservation action.

Preserving Metapopulations

If you browse through an identification guide for trees or birds or butterflies, you'll likely find range maps indicating that most species occupy a broad area. In reality, however, the habitat preferences of many species are restricted, and individuals occupy only isolated patches within that broad area. The sunlit patches occupied by common primroses are an example of this pattern, and so are the meadows occupied by Glanville fritillaries—an endangered species of butterfly native to the Åland islands off the coast of Finland (**Figure 52.14a**).

Recall that if species occupy many small patches of habitat, such as the canopy gaps occupied by primroses, they are said to represent a metapopulation, or population of populations. Glanville fritillaries and common primroses naturally exist as metapopulations. But because humans are reducing large, contiguous areas of forest and grasslands to isolated patches or reserves, more and more species are being forced into a metapopulation structure.

Recent research on Glanville fritillaries by Ilkka Hanski and colleagues illustrates the consequences of metapopulation structure for endangered species. The work began with a survey of meadow habitats on the islands. Because fritillary caterpillars feed on just two types of host plant, *Plantago lanceolata* and *Veronica spicata*, the team was able to pinpoint potential butterfly habitats. They estimated the population size within each patch by counting the number of larval webs in each. Of the 1502 meadows that contained the host plants, 536 had Glanville fritillaries. The patches ranged from 6 m^2 to 3 ha in area. Most had only a single breeding pair of adults and one larval group, but the largest population contained hundreds of pairs of breeding adults and 3450 larvae.

Figure 52.14b illustrates how a metapopulation like this is expected to change over the years. Given enough time, each population within the larger metapopulation is expected to go extinct. The cause could be catastrophic, such as a storm or an oil spill; it could also be a disease outbreak or a sudden influx of predators. Migration from nearby populations can reestablish populations in these empty habitat fragments, however. In this way, there is a balance between extinction and recolonization. Even though subpopulations blink on and off over time, the overall population is maintained at a stable number of individuals.

Does this metapopulation model correctly describe the population structure of fritillaries? To answer this question, Hanski's group conducted a *mark-recapture study*. (**Box 52.3**, page 1208, provides details on how a mark-recapture study is done.) Of the 1731 butterflies that the biologists marked and released, 741 were recaptured over the course of the summer study period. Of the recaptured individuals, 9 percent were found in a previously unoccupied patch. This migration rate is high enough to suggest that patches where a population has gone extinct will eventually be recolonized.

Hanski and co-workers repeated the survey two years after their initial census. Just as the metapopulation model predicts, some populations had gone extinct and others had been created. On average, butterflies were lost from 200 patches each year; 114 unoccupied patches were colonized and newly occupied each year. In addition to confirming that the metapopulation model is valid, the work has important implications for endangered species. A small, isolated population—even one within a nature preserve—is unlikely to survive over the long term. For Glanville fritillaries, the data indicate that subpopu-

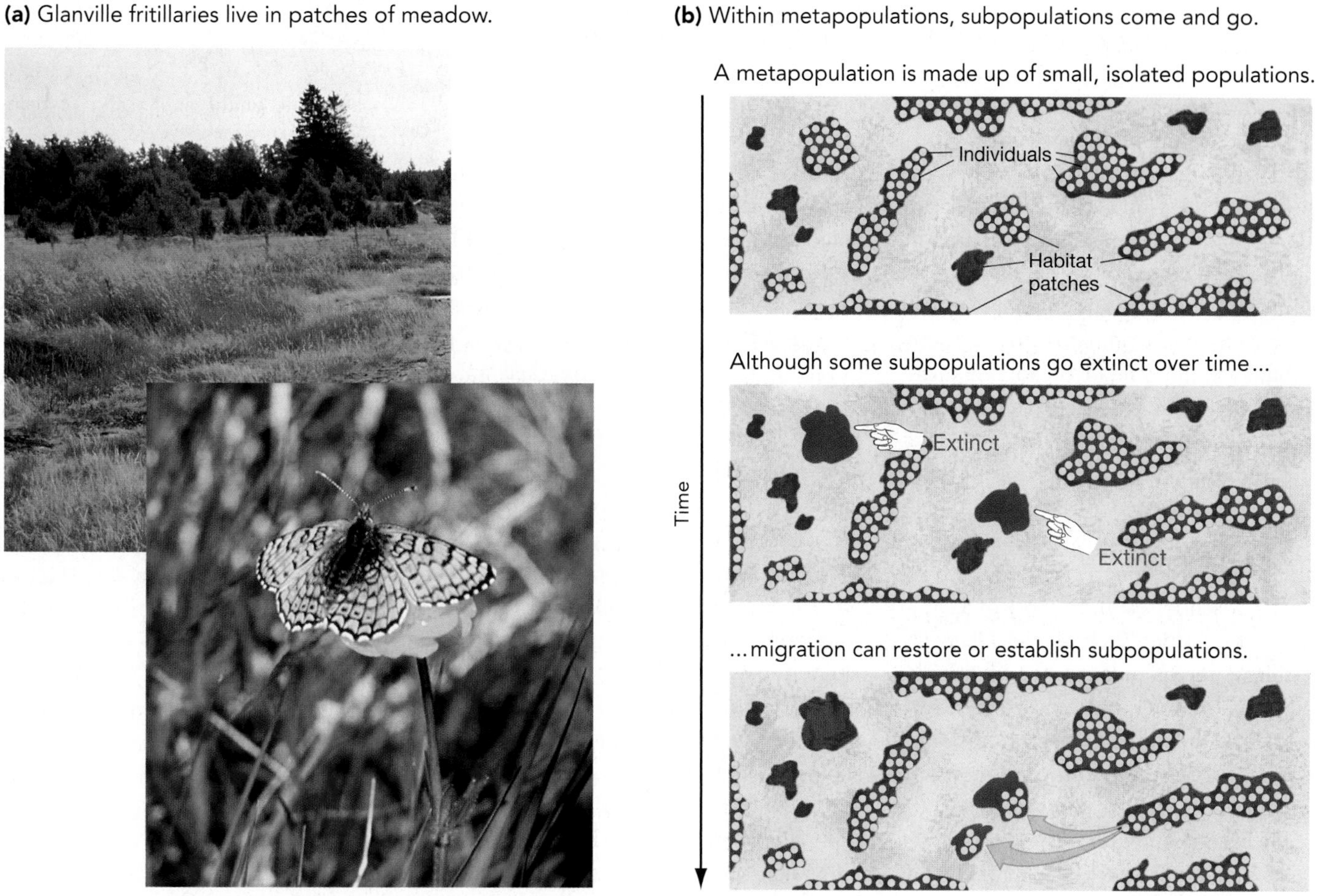
(a) Glanville fritillaries live in patches of meadow.
(b) Within metapopulations, subpopulations come and go.
A metapopulation is made up of small, isolated populations.
Individuals
Habitat patches
Although some subpopulations go extinct over time...
Extinct
Extinct
...migration can restore or establish subpopulations.
Time

FIGURE 52.14 Metapopulation Dynamics Depend on Extinction and Recolonization
(a) The Glanville fritillary is extinct in much of its former range; it is now found only on the Åland islands near Finland. **(b)** The overall population size of a metapopulation stays relatively stable even if subpopulations go extinct. These populations may be restored by migration, or unoccupied habitats might be colonized.

lations that were large, that occupied larger areas, and that were closer to neighboring populations (and hence more likely to be colonized) were less likely go extinct. The data also show that fritillary populations with low genetic diversity—probably due to inbreeding—were more likely to go extinct than were populations of the same size that had higher genetic diversity.

Results like these have important messages for conservation biologists:

- Areas that are being protected for threatened species should be substantial enough in area to maintain large populations that are unlikely to go extinct in the near future.
- When it is not possible to preserve large tracts of land, an alternative is to establish systems of smaller tracts that are connected by corridors of habitat, so that migration between patches is possible.
- If the species that is threatened exists as a metapopulation, it is crucial to preserve at least some patches of unoccupied habitat to provide future homes for immigrants.

These results also have a more general message: Although traditional population growth models such as the expressions for exponential and logistic growth are simple and elegant, the factors they ignore—immigration and emigration—are crucial to understanding the dynamics of most populations. Because metapopulation structure is common, biologists must use more sophisticated models to predict the fates of populations.

Using Life-Table Data to Make Population Projections

Demographic data such as age-specific survivorship and fecundity are important for saving endangered species and for other applied problems. To understand why, suppose that you were in charge of reintroducing a population of lizards to a nature reserve, and that research conducted when this species occupied the site previously documented the survivorship and fecundity data. Your initial plan is to take 1000 newly hatched females from a captive breeding center and release them into

BOX 52.3 Mark-Recapture Studies

To estimate the population size of sedentary (largely immobile) organisms, researchers sample appropriate habitats by counting the individuals that occur along transects or inside rectangular plots—called **quadrats**—set up at random locations in the habitat. These counts can be extrapolated to the entire habitat area to estimate the total population size. In addition, they can be compared to later censuses to document trends in population size.

In contrast, estimating the total population size of organisms that are mobile and that do not congregate into herds or flocks or schools is much more of a challenge. In species such as the Glanville fritillary, the population inside sample quadrats or along transects changes constantly as individuals move in and out. Further, it can be difficult to track whether a particular individual has already been counted.

If individuals can be captured and then tagged in some way, however, the total population size of a mobile species can be estimated by using an approach called mark-recapture. To begin a **mark-recapture study**, researchers catch individuals in live traps and mark them with leg bands, ear tags, or some other methods of identification. After the marked individuals are released, they are allowed to mix with the unmarked animals in the same population for a period of time. Then a second trapping effort is conducted, and the percentage of marked individuals that were captured is recorded.

To estimate the total population from these data, researchers make a key assumption: The percentage of marked and recaptured individuals is equal to the percentage of marked individuals in the entire population. This assumption should be valid when no bias exists regarding which individuals are caught in each sample attempt. It is important that individuals do not learn to avoid traps after being caught once and that they do not emigrate or die as a result of being trapped.

The relationship between marked and unmarked individuals can be concisely expressed algebraically:

$$\frac{m_2}{n_2} = \frac{n_1}{N}$$

In this equation, m_2 is the number of marked animals in the second sample (the recapture), n_2 is the total number of animals (marked and unmarked) in the second sample, n_1 is the number of marked animals in the first sample, and N is the total population size. Having measured m_2, n_2, and n_1, the researcher can estimate N. For example, suppose researchers marked 255 animals and later were able to trap a total of 162 individuals in the population, of which 78 were marked. What is the estimate for total population size? Solve for N and check your answer.[5]

[5] 530

the habitat. Is this population likely to become established and grow, or will you need to keep introducing offspring that have been raised in captivity?

To answer this question, you need to use the life table data in **Figure 52.15a** and calculate (1) how many adults will survive to each age class each year and (2) how many juveniles will be produced by each adult age class each year over the course of several years. **Figure 52.15b** starts these calculations for you. The fate of the original 1000 females is indicated in red. In the second year, just 330 of these individuals have survived; 40 are left as 3-year-olds.

How many offspring did this generation produce? After one year, 330 of the original 1000 females remain. As the first purple number in Figure 52.16b indicates, these survivors have an average of 3.0 female offspring apiece, so they contribute 990 new juveniles to the population. In the second year, the 200 females that are left have an average of 4.0 female offspring each and contribute 800 juveniles. In their third year, the 40 surviving females average 5.0 offspring and contribute 200 juveniles.

Figure 52.15c extends the calculations by showing what happens as the offspring of the original females begin to breed. Their offspring are shown in green. By adding subsequent generations and continuing the analysis, you could predict whether the population will stay the same, decline, or increase over time. In this way, life-table data can be used to predict the future of populations.

Part of the value of a population projection like the one begun in Figure 52.15 is that it allows biologists to alter values for survivorship and fecundity at particular ages and assess the consequences. For example, suppose that a predatory snake began preying on juvenile lizards. According to the model in Figure 52.15, what would be the impact of a change in juvenile mortality rate? Analyses like this allow biologists to determine which aspects of survivorship and fecundity are particularly sensitive for particular species. The studies done to date support some general conclusions:

- Whooping cranes, sea turtles, spotted owls, and many other endangered species have high juvenile mortality, low adult mortality, and low fecundity. In these species, the fate of a population is extremely sensitive to increases in adult mortality. Based on this insight, conservationists have recently begun an intensive campaign to reduce the loss of adult

(a) Life table

Age (x)	Survivorship (l_x)	Fecundity (m_x)
0 (birth)	----	0.0
1	0.33	3.0
2	0.2	4.0
3	0.04	5.0

(b) Fate of first-generation females

Year	0 (newborns)	1-year-olds	2-year-olds	3-year-olds	Total population size (N)
1st	1000 (just introduced)				1000
2nd	990 (= 330 × 3.0)	330 (= 1000 × 0.33)			1320 (= 990 + 330)
3rd	800 (= 200 × 4.0)		200 (= 1000 × 0.20)		
4th	200 (= 40 × 5.0)			40 (= 1000 × 0.04)	
5th					

(c) Fate of first- and second-generation females

Year	0 (newborns)	1-year-olds	2-year-olds	3-year-olds	Total population size (sum across all rows)
1st	1000				1000
2nd	990	330			1320
3rd	800 + 981 (981 = 327 × 3.0)	327 (= 990 × 0.33)	200		2308 (= 800 + 981 + 327 + 200)
4th	200 + 792 (792 = 198 × 4.0)		198 (= 990 × 0.20)	40	
5th	195 (195 = 39 × 5.0)			39 (= 990 × 0.04)	

FIGURE 52.15 Life-Table Data Can Be Used to Project the Future of a Population
(a) Life table providing age-specific survivorship and fecundity for a hypothetical population. **(b)** Predicted fate of 1000 1-year-old females introduced into a habitat just before the breeding season. The number of individuals in this cohort is shown in red; the offspring they produce each year is indicated in purple. **(c)** Extension of the data in part (b), indicating how many of the offspring produced by the original females in their first year survived in subsequent years, shown in purple, and how many offspring they produced in each subsequent year, shown in green. **EXERCISE** Assume that all 4-year-old females die after producing three young. Fill in the fourth and fifth years in the table.

female sea turtles in fishing nets. Previously, most conservation action had focused on protecting eggs and nesting sites.

- In humans and other species with high survivorship in most age classes, rates of population growth are extremely sensitive to changes in age-specific fecundity. Based on this result, programs to control human population growth focus on two issues: lowering fertility rates through the use of birth control, and delaying the age of first reproduction by improving women's access to education.

In some or even most cases, however, the population projections made from life-table data may be too simplistic to be useful. For example, conservationists may need to expand the basic demographic models to account for occasional disturbances such as fires or storms or disease outbreaks. If the population exists in the fragmented habitats typical of a metapopulation, planners also need to assess the impact of emigration and immigration from nearby populations. Let's close this chapter with a look at how that assessment is done.

Population Viability Analysis

A **population viability analysis**, or PVA, is a model that estimates the likelihood that a population will avoid extinction for a given time period. In most cases, PVA attempts to combine basic demographic models for a species with data on geographic structure and the rate and severity of habitat disturbance. Typically, a population is considered viable if the analysis predicts that it has a 95 percent probability of surviving for at least 100 years. Natural resource managers are currently using PVA to assess the effects of logging, suburbanization, and other land management practices on sensitive species and to evaluate the merits of alternative recovery plans for endangered species.

A recent study—conducted on Leadbeater's possum, an endangered marsupial—illustrates the value of a carefully constructed PVA. Leadbeater's possum (**Figure 52.16a**) inhabits old-growth forests in southeastern Australia and relies on dead trees for nest sites. The goal of the PVA was to assess the effects that logging these habitats might have on the viability of the species. The question was, if logging reduced the ability of possums to migrate to new habitats, what would the impact on the population be?

Life-table data are difficult to obtain for Leadbeater's possum, because this species is rare, lives in trees, and is active at night. Enough data were available from field studies, however, to allow for an estimate of age-specific survival and average fecundity per female per year. Biologists used these data to make population projections while varying the spatial configuration of habitat patches. The analysis allowed the researchers to assess the impact of migration and to simulate the effects of logging, fires, storms, and other disturbances.

Figure 52.16b illustrates the researchers' results. The four curves on the graph describe the changes in population size predicted to occur over time, based on different assumptions about the migration rates between groups of possums in patches of old-growth forest. When migration is high, the overall population size is predicted to stabilize at approximately 65 individuals. When no migration among fragments occurs, the final population size is predicted to be fewer than 20 individuals. Based on these results, the biologists concluded that extensive timber harvesting would pose a serious threat to the species. The PVA showed that reducing the size and number of remaining old-growth fragments would reduce the possibility for migration and lead to rapid population declines.

Like all analyses based on simulations, though, a PVA makes many assumptions about future events and is only as accurate as the data entered into it. In particular, the usefulness of PVAs has been challenged because data on age-specific fecundity and survivorship and other basic demographic information are lacking or poorly documented in many endangered species.

A group of biologists recently defended the approach, however, by analyzing 21 long-term studies that have been completed on threatened populations of animals. To do their analysis, the researchers split each of the 21 data sets in two. Data from the first half of each study was used to run a PVA on each of the 21 species. After comparing the predictions of the PVAs to the actual data from the second half of each study, the researchers found that the correspondence was extremely close. This result strengthens confidence in PVA as a good predictive method for land managers.

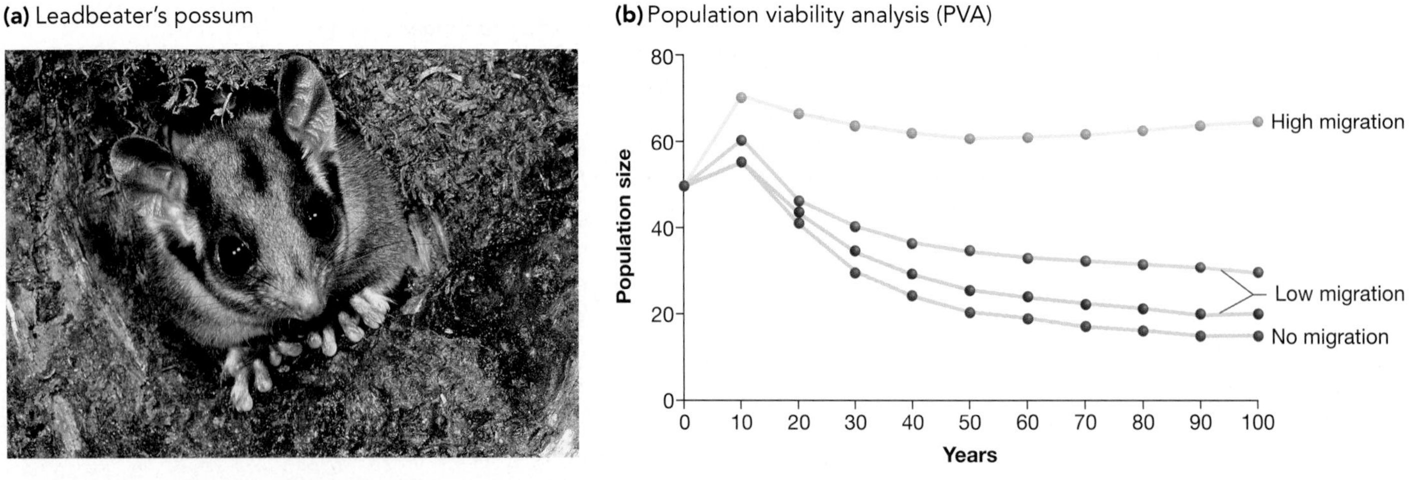

FIGURE 52.16 A PVA Can Predict the Consequences of Habitat Loss and Fragmentation
(a) Leadbeater's possum, a marsupial that breeds in old-growth forests of southeast Australia. **(b)** Results of a PVA: the fate of possum populations for four different rates of migration among habitat patches. The rates range from a case in which patches are isolated and no migration occurs to a case in which 5 percent of the individuals migrate each year. The no-migration scenario simulates what might happen if extensive deforestation occurred and created isolated patches of old-growth forest.

ESSAY What Limits Human Population Growth?

On October 12, 1999, the United Nations Population Division estimated that the human population exceeded six billion (6,000,000,000) for the first time. Why has the growth rate of the human population increased over time? Is our species at or near Earth's carrying capacity?

Population biologists point to several key events in human history that have contributed to the rising growth rate. The first event occurred about 10,000 years ago, when groups of humans began to develop agriculture. The transition from a nomadic existence to a settled, agricultural one was associated with dramatic increases in fecundity. More recently, improvements in medicine and hygiene have led to dramatic declines in death rates—particularly infant mortality. These technological advances are responsible for the dramatic rise in human populations over the past 100 years.

Until recently, most people considered population growth to be a good thing for religious, economic, and political reasons. As the twentieth century progressed and human population size continued to explode, however, a few people expressed concern about the ecological consequences of human population growth. Paul Ehrlich, who began his career studying butterflies, has been one of the most prominent voices arguing that human population growth must cease. His concern is that humans are exceeding Earth's carrying capacity and are causing an ecological collapse that might lead to our own downfall. At the root of Ehrlich's arguments is the concern that exponential growth is shooting us past Earth's carrying capacity for humans and is causing irreparable harm to the environment. What is Earth's carrying capacity, and how do we know if we have surpassed it?

Population demographers who have attempted to calculate Earth's carrying capacity for humans have produced estimates that range from 2 billion people to over 1 trillion people. The low estimate is much less than the current population; the high estimate assumes that people would live in floating cities above the planet's surface. The estimates are variable because carrying capacity depends on technological advancements and on the standard of living that people are willing to accept. Ehrilich's dire predictions have not proven correct to date, because the green revolution of the 1970s and 1980s—dramatic improvements in crop plant productivity, introduced in Chapter 29—produced enough food to keep pace with human population growth. Now, however, the availability of groundwater is declining in areas that depend on irrigated agriculture, and the amount of productive farmland available is beginning to peak. Do these observations mean that Earth's carrying capacity for humans has already been exceeded?

Is our species at or near Earth's carrying capacity?

Joel Cohen argues that the classical concept of carrying capacity has little meaning for humans. Although he acknowledges that density-dependent factors such as competition and disease are already affecting human populations, Cohen emphasizes that our future hinges much more directly on how many children the 6.4 billion of us decide to have and how many resources each of us demands. His point is that human choice is the most important element in defining carrying capacity for our species.

From this point of view, Earth's carrying capacity is what we decide it will be. The human population is already too large for all 6.4 billion of us to live with the same standard of living as that of Americans and Western Europeans. Increased demand for this level of resource use, or fertility rates that continue to exceed the replacement rate, would almost certainly bring about the type of ecological collapse Ehrlich envisioned. Over the course of your lifetime, the decisions made by 6.4 billion people about the number of children they have and the amount of resources they use will determine the future of the human population—and by extension, the future of life on Earth.

CHAPTER REVIEW

Summary of Key Concepts

A population is a group of individuals of the same species that occupy the same area at the same time. Population ecology is the study of how and why the number of individuals in a population change over time. Population size changes in response to changes in birth, death, immigration, and emigration rates. Biologists use a variety of mathematical and analytical tools to study population ecology.

- **Life tables summarize how likely it is that individuals of each age class in a population will survive and reproduce.**

Life tables are a basic tool in demography—the study of patterns in births and deaths in populations. When life tables of different species or populations are compared, it is clear that individuals may have distinct ways of allocating energy and resources to activities that promote survival versus activities that promote reproduction. Because the resources available to an individual are always limited, any increase in allocation of resources to survival and competitive ability necessitates a decrease in the resources allocated to reproduction. This trade-off between survival and reproduction is the most fundamental aspect of a species' life history.

The growth rate of a population can be calculated from life-table data or from the direct observation of changes in population size over time.

One of the most basic characteristics of a population is its growth rate. Exponential growth occurs when the per-capita growth rate, r, does not change over time. Eventually, however, growing populations approach the carrying capacity of their environment—resulting in logistic growth.

Laboratory and field studies have confirmed the existence of exponential and logistic growth in species ranging from whooping cranes to *Paramecium*. Field research has documented that as population density increases, survivorship and fecundity may decrease—leading to increased death rates and lowered birth rates and thus a reduction in growth rate. Density dependence in population growth is due to competition for resources, disease, predation, or other factors that increase in intensity when population size is high.

Web Tutorial 52.1 Modeling Population Growth

A wide variety of patterns is observed when researchers track changes in population size over time, ranging from no growth, to regular cycles, to continued growth independent of population size.

In the Park Grass experiment, species with high fecundity and low survivorship tended to increase over time, although some later declined. Experiments have shown that density-dependent factors drive the regular population cycles observed in certain species. In the case of red grouse, roundworm infections spread rapidly when population density is high and then cause the population to crash.

Changes in population size through time may occur as a result of changes in the age structure of populations—specifically the number of individuals in various age classes. A population with few juveniles and many adults past reproductive age, like the human populations of developed nations, may be declining or stable in size. In contrast, a population with a large proportion of juveniles is likely to increase rapidly in size. Human populations in the developing nations currently have this type of age distribution.

The total human population has been increasing rapidly since about 1750 and is currently about 6.4 billion. In countries where survivorship has increased due to medical advances, fecundity has decreased. Based on various scenarios for average female fertility, the total human population in 2050 is expected to be between 7 and 11 billion.

Web Tutorial 52.2 Human Population Growth and Regulation

Data from population ecology studies help biologists evaluate prospects for endangered species and design effective management strategies.

Human activities are isolating populations into metapopulations occupying small, fragmented habitats. The history of a metapopulation is driven by the birth and death of populations, just as the dynamics of a population are driven by the birth and death of individuals. Migration among habitat patches is essential for the stability of a metapopulation, so conservationists are trying to preserve unoccupied patches of habitat and establish corridors that link habitat fragments.

Demographic data are the basis of population projections that are fundamental to population viability analysis, or PVA. Most PVAs attempt to model the effects that different management strategies might have on populations of endangered species. A PVA estimates the probability that a population will persist for a certain number of years under a prescribed set of demographic and habitat conditions.

Questions

Content Review

1. What is the defining feature of exponential growth?
   a. It lasts indefinitely.
   b. The growth rate is constant.
   c. The growth rate increases rapidly over time.
   d. The growth rate is very high.
2. What four factors define population growth?
   a. age-specific birth rates, age-specific death rates, age structure, and metapopulation structure
   b. survivorship, age-specific mortality, fecundity, death rate
   c. mark-recapture, census, quadrat sampling, transects
   d. births, deaths, immigration, emigration
3. In what populations does exponential growth tend to occur?
   a. in populations that colonize new habitats
   b. in populations that experience intense competition
   c. in populations that experience high rates of predation
   d. in metapopulations
4. Which of the following is not a reason why population growth declines as population size approaches the carrying capacity?
   a. Climate becomes unfavorable.
   b. Competition for resources increases.
   c. Predation rates increase.
   d. Disease rates increase.
5. If most individuals in a population are young, why is the population likely to grow rapidly in the future?
   a. Death rates will be low.
   b. The population has a skewed age distribution.
   c. Immigration and emigration can be ignored.
   d. Many individuals will begin to reproduce soon.
6. Why have population biologists become particularly interested in the dynamics of metapopulations?
   a. because humans exist as a metapopulation
   b. because whooping cranes exist as a metapopulation
   c. because many populations are becoming restricted to small islands of habitat
   d. because metapopulations explain why populations occupying large, contiguous areas are vulnerable to extinction

Conceptual Review

1. Explain Equations 52.3 and 52.5 in words.
2. Draw type I, II, and III survivorship curves on a graph with labeled axes. Explain why the growth rate of species with a type I survivorship curve depends primarily on fertility rates. Explain why the growth rate of species with a type III survivorship curve is extremely sensitive to changes in adult survivorship.
3. Offer a hypothesis to explain why humans have undergone near-exponential growth for over 500 years. Why can't exponential growth continue indefinitely? Give two examples of density-dependent factors that influence population growth in natural populations.
4. Compare and contrast the dynamics of a population that resides in a large contiguous habitat to the dynamics of a metapopulation. Be sure to consider density-dependent and density-independent factors on birth rates and death rates, and the importance of immigration and emigration. Assume that the total amount of area occupied is the same in both populations.
5. Make a rough sketch of the age distribution in developing versus developed countries, and explain the significance of the differences. How is AIDS, which is a sexually transmitted disease, affecting the age distribution in countries hard hit by the epidemic?
6. Compare the pros and cons of using R_0, λ, and r to express growth rates. What is the difference between r (the per-capita rate of increase) and r_{max} (the maximum or intrinsic growth rate)?

Group Discussion Problems

1. When wild plant and animal populations are logged, fished, or hunted, only the oldest or largest individuals tend to be taken. What impact does harvesting have on a population's age structure? How might harvesting affect the population's life table and growth rate?
2. Design a system of nature preserves for an endangered species of beetle whose larvae feed on only one species of sunflower. The sunflowers tend to be found in small patches that are scattered throughout dry grassland habitats. Explain the rationale behind your proposal.
3. Snowshoe hares are preyed on by lynx. Both species show pronounced population cycles. The cycles are roughly synchronized, although the rise and fall of lynx populations (the predator) slightly lags the rise and fall of hare populations (the prey). Design an experiment to test the hypotheses that these cycles are driven by (1) predation of hares by lynx, (2) food shortages when hares are at high density, or (3) a combination of food shortages and predation. Assume that you are able to exclude lynx from certain study areas and add food for hares in other areas.
4. In most species the sex ratio is at or near 1.00, meaning that there is an approximately equal number of males and females. In China, however, there is a strong preference for male children. According to the 2000 census there, the sex ratio of newborns is almost 1.17, meaning that close to 117 boys are born for every 100 girls. Based on these data, researchers project that there will soon be about 50 million more men than women of marriageable age in China. Discuss how this skewed sex ratio might affect the population growth rate in China.

Answers to Multiple-Choice Questions 1. b; 2. d; 3. a; 4. a; 5. d; 6. c

www.prenhall.com/freeman is your resource for the following: Web Tutorials; Online Quizzes and other Online Study Guide materials; Answers to Conceptual Review Questions; Solutions to Group Discussion Problems; Answers to Figure Caption Questions and Exercises; and Additional Readings and Research.

53 Community Ecology

KEY CONCEPTS

- Interactions among species, such as competition and consumption, have two main outcomes: (1) They affect the distribution and abundance of the interacting species, and (2) they are agents of natural selection and thus affect the evolution of the interacting species.
- The assemblage of species found in a biological community changes over time and is primarily a function of climate and chance historical events.
- Species diversity is high in the tropics and lower toward the poles. The mechanism responsible for this pattern is still being investigated.
- Within at least some biomes, there is a strong positive relationship between the number of species present and the productivity and stability of the community.

Like many regions of the world, the area near Australia's Great Barrier Reef contains an array of biological communities.

Chapter 52 focused on the dynamics of populations—how and why they grow or decline and how they change over time and through space. That chapter considered each population as an isolated entity. But in reality, populations of different species form complex assemblages called communities. A biological **community** consists of interacting species, usually living within a defined area.

The concept of a biological community was introduced in Chapter 50. Each of the biomes analyzed in that chapter represents a broad type or category of plant community. The goal of that chapter was to describe the general characteristics of selected biomes and analyze broad correlations between a region's climate and the type of biome present.

This chapter focuses on a spatial scale between biomes and populations. The task is to analyze the dynamics of biological communities—how they develop and change over time. Biologists want to know how communities work. How do species inside communities interact with each other, and what are the long-term consequences? Why is the number of species higher in some areas than others, and why is species diversity important? Let's delve in.

53.1 Species Interactions

The species in a community interact almost constantly. Members of different species eat one another, pollinate each other, exchange nutrients, compete for resources, and provide habitats for each other. As a result, the fate of a particular population may be tightly linked to the other species that share its habitat.

Biologists organize all the possible interactions among species by considering the interactions' effects on the fitness of the individuals involved. Recall from Chapter 23 that fitness is defined as the ability to survive and produce offspring. Does the relationship between two species provide a fitness benefit to members of one species (a "+" interaction) but hurt members of the other species (a "−" interaction)? Or does the association have no effect on the fitness of a participant (a "0" interaction)? Three broad categories of interaction are analyzed in this chapter: the −/− relationship known as *competition*, the +/− interactions called *consumption* and *parasitism*, and the +/+ association termed *mutualism*. A fourth category of interaction, **commensalism**, is defined as a +/0 association. An example is the birds that follow moving army ants in the tropics. As the ants march along the forest floor, they hunt insects and small vertebrates. As they do, birds follow and pick off prey species that fly or jump up out of the way of the ants (**Figure 53.1**). The birds are **commensals** that benefit from the association (+) but have no impact on the ants (0).

In addition to introducing the array of tools that biologists use to study competition, consumption, and mutualism, it is important to focus on three key themes:

1. Species interactions may affect the distribution and abundance of a particular species. Recall from Chapter 52 that most of the density-dependent factors that produce logistic growth are based in species interactions—predation, disease, or competition for space and resources. Changes in species interactions often explain short-term changes in population size and distribution.
2. Species act as agents of natural selection when they interact. Deer are fast and agile in response to natural selection exerted by their major predators, wolves and cougars. The speed and agility of deer, in turn, promote natural selection that favors wolves and cougars that are fast and that have superior eyesight and senses of smell. To capture this point, biologists say that species interactions resemble an arms race. In humans, an arms race is said to occur when one nation develops a new weapon, which prompts a rival country to develop a defensive weapon, which pushes the original country to manufacture a more powerful weapon, and so on. In biology, a **coevolutionary arms race** occurs between predators and prey, parasites and hosts, and other types of interacting species. The adjective *coevolutionary* is appropriate because the pairs of species influence each others' evolution. In this way, changes in species interactions lead to long-term changes in the characteristics of populations—in addition to short-term impacts on population size.
3. The outcome of interactions among species is dynamic and conditional. Consider the relationship between army ants and birds that follow them. If attacks by birds force other insects into the path of the ants, then both ants and birds benefit and the relationship is mutualistic. But if birds begin to steal prey from ants, then the relationship becomes parasitic. The outcome of the interaction may depend on the number and types of prey, birds, and ants present and may change over time.

FIGURE 53.1 Commensals Gain a Fitness Advantage but Don't Affect the Species They Depend On

Birds that associate with army ants are commensals. They have no measurable fitness effect on the ants but gain from the association by capturing insects that try to fly or climb out of the way of the ants.

Competition

Competition is a −/− interaction that occurs when different individuals use the same resources and when those resources are limited—meaning they are in short supply relative to demand. Ever since ecological studies began, researchers have focused on competition as an important interaction within and between species. The attention is justified in part by the central place that competition holds in Darwin's theory of evolution by natural selection. Darwin pointed out that individuals within a population compete for the resources that are required for growth and reproduction. Further, some individuals are more successful in this competition and leave more offspring than others do. If the traits that lead to success are heritable, then the frequency of alleles in the population changes and evolution by natural selection occurs.

Competition that occurs between members of the same species is called **intraspecific** ("within species") **competition.** In addition to creating opportunities for natural selection to occur, intraspecific competition for space, sunlight, food, and other resources intensifies as a population's density increases. As a result, intraspecific competition is a major cause of density-dependent growth (see Chapter 52).

How does competition affect members of different species? Stated another way, what happens when **interspecific** ("between species") **competition** occurs? To answer this question, it is important to recognize that every species has a unique **niche**, or set of habitat requirements. A niche can be thought of as the range of resources that the species is able to use or the range of

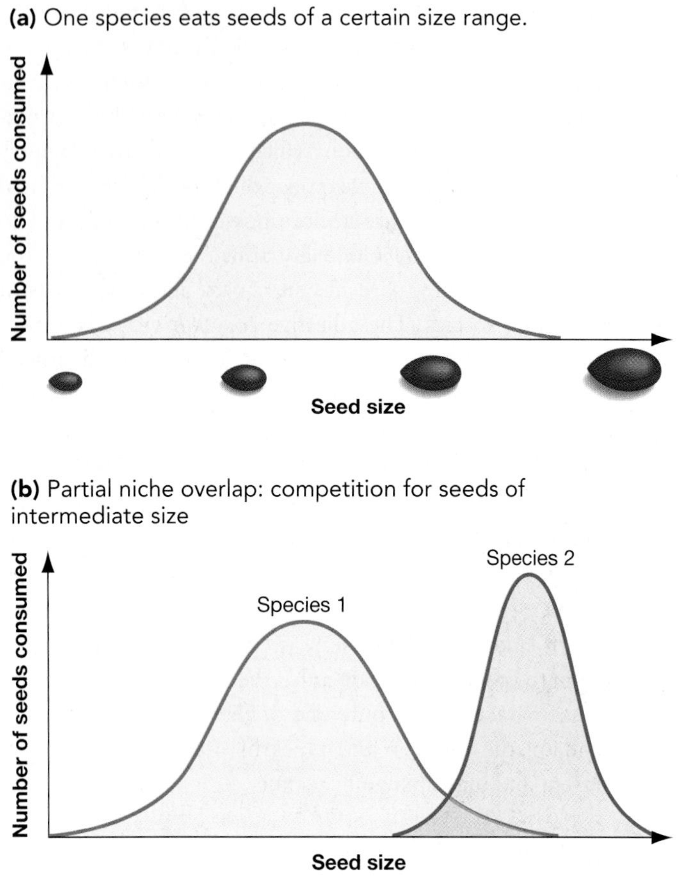

FIGURE 53.2 Niche Overlap Leads to Competition
(a) Graph describing one aspect of a species' fundamental niche, meaning the range of resources that it can use or range of conditions it can tolerate. **(b)** Competition occurs when the niches of different species overlap. In this case, both species use seeds of intermediate size. **EXERCISE** Mark the horizontal axis to indicate the range of seed sizes where competition occurs.

conditions it can tolerate. G. Evelyn Hutchinson proposed that a species' niche could be envisioned by plotting these habitat requirements along a series of axes. **Figure 53.2a**, for example, represents one niche axis for a hypothetical species. In this case, the habitat requirement plotted is the size of seeds eaten by members of this population, which might be a function of mouth or tooth size. Other niche axes could represent other types of foods used or the temperatures, humidity, and other environmental conditions tolerated by the species.

Interspecific competition occurs when the niches of two species overlap. The two species plotted in **Figure 53.2b**, for example, compete for seeds of intermediate size. When competition occurs, each individual will get fewer seeds on average. In this way, competitive interactions reduce the amount of resources available to each species. This is why competition is considered a –/– interaction.

The Consequences of Competition—Theory According to G. F. Gause, it is not possible for species with the same niche to coexist. This hypothesis is called the **competitive exclusion principle**, and it was inspired by a series of competition experiments Gause did with similar species of the unicellular pond-dweller *Paramecium*. When Gause placed small populations of *Paramecium caudatum* and *P. aurelia* in separate laboratory cultures, both species exhibited logistic growth (see Chapter 52). But Gause also showed that when the two species are put in the same culture together, only the *P. aurelia* population exhibits a logistic growth pattern. *Paramecium caudatum*, in contrast, is driven to extinction (**Figure 53.3a**).

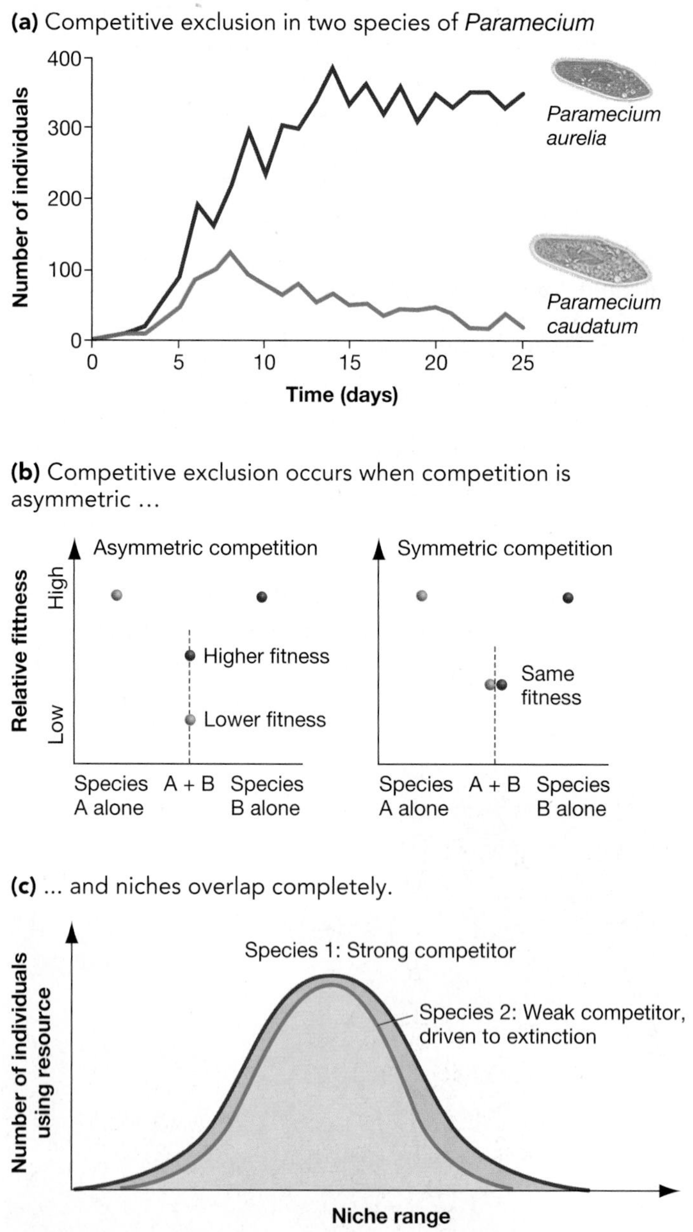

FIGURE 53.3 Competitive Exclusion Occurs when One Species Is a Superior Competitor
(a) In laboratory culture, *Paramecium aurelia* outcompetes *P. caudatum*. **(b)** When competition is asymmetric, one of the interacting species suffers a larger fitness loss than the other does. **(c)** If two species have completely overlapping niches, there is no refuge for the weaker competitor. It may be driven to extinction.

A result such as Gause's is produced by what contemporary investigators call asymmetric competition. When **asymmetric competition** occurs, one species suffers a much greater fitness decline than the other species does (**Figure 53.3b**). Under **symmetric competition**, however, each species experiences a roughly equal decrease in fitness. In Gause's laboratory cultures, competition between *P. aurelia* and *P. caudatum* was asymmetric. Subsequent research has shown that most competition is asymmetric.

Gause's work highlights a central question in the study of competition: If competition is a common interaction, why haven't the superior competitors outcompeted all other species? Is coexistence possible despite competition?

Before considering experimental work on this question, let's consider the predictions made by theory. Imagine a scenario in which asymmetric competition is occurring. If the two species have completely overlapping niches, as diagrammed in **Figure 53.3c**, then the stronger competitor is likely to drive the weaker competitor to extinction, just as Gause predicted.

If the niches do not overlap completely, then the weaker species should be able to retreat to an area of non-overlap. In cases like this, an important distinction arises between a species' **fundamental niche**, which is the combination of resources or areas used or conditions tolerated in the absence of competitors, and its **realized niche**, which is the portion of resources or areas used or conditions tolerated when competition occurs (**Figure 53.4a**). When competition does not result in the complete exclusion of one species, it may result in **niche differentiation**, in which competing species evolve traits that allow them to exploit different resources or live in different areas (**Figure 53.4b**).

It's important to recognize that some species may be stronger competitors in some habitats or environments than others and that the types of competition that occur vary widely. The actual mechanism of competition depends on the species and resource involved. As **Figure 53.5** (page 1218) shows, mechanisms of competition include the following:

1. *Consumptive competition*, in which two species consume the same resources (Figure 53.5a).
2. *Preemptive competition*, in which one species makes space unavailable to other species (Figure 53.5b).
3. *Overgrowth competition*, in which one species grows above another (Figure 53.5c).
4. *Chemical competition*, in which one species produces toxins that negatively affect another species (Figure 53.5d).
5. *Territorial competition*, in which a mobile species protects its feeding or breeding territory against other species (Figure 53.5e).
6. *Encounter competition*, in which two species interfere directly for access to specific resources (Figure 53.5f).

To summarize, competition occurs when niches overlap, and it can result in extinction, changes in niche use, and niche differentiation over time. Competition theory makes predictions that are clear, logical, and compelling. Is there evidence that niche differentiation actually occurs in nature? What types of experiments can reveal the existence of competition and identify its consequences?

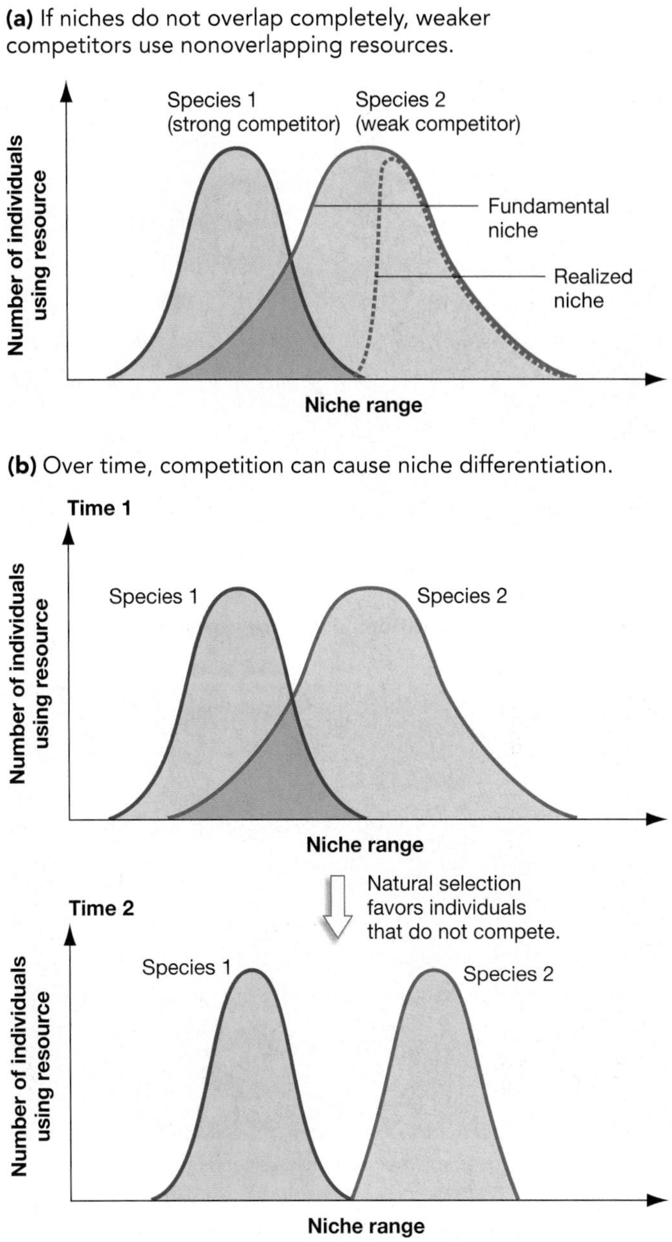

FIGURE 53.4 Partial Niche Overlap Has Several Consequences
(a) When the niches of competing species overlap, the weaker competitor uses only a subset of its fundamental niche. **(b)** Over time, evolution can lead to changes in niche breadth—a process called niche differentiation.

The Consequences of Competition—Experimental Studies

Joseph Connell initiated a classic study of competition after observing an interesting pattern in an intertidal rocky shore in Scotland. He noticed that there were two species of barnacles with

(a) Consumptive competition: Organisms consume the same resources.

These trees are competing for water and nutrients.

(b) Preemptive competition: Individuals occupy space, preventing access to resources.

Space preempted by these barnacles is unavailable to competitors.

(c) Overgrowth competition: One organism grows over another.

The large fern has overgrown other individuals and is shading them.

(d) Chemical competition: One species produces toxins that negatively affect another.

Few plants are growing under these *Salvia* shrubs.

(e) Territorial competition: Mobile organisms protect a feeding or breeding territory.

Grizzly bears drive off black bears.

(f) Encounter competition: Organisms interfere directly for access to specific resources.

Spotted hyenas and vultures fight over a kill.

FIGURE 53.5 Many Types of Competition Exist
A few of the major types of competition.

(a) Barnacle species are distributed in distinct zones.

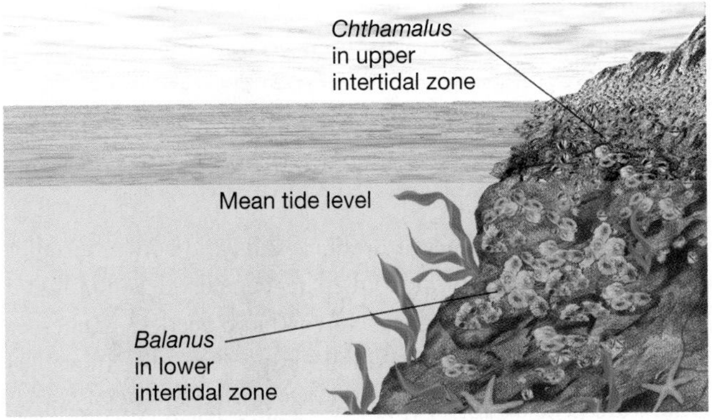
Chthamalus
in upper
intertidal zone
Mean tide level
Balanus
in lower
intertidal zone

(b) Testing the hypothesis that competition occurs

Question: Why is the distribution of adult *Chthamalus* restricted to the upper intertidal zone?

Hypothesis: Adult *Chthamalus* are competitively excluded from the lower intertidal zone.

Alternative hypothesis: Adult *Chthamalus* do not thrive in the physical conditions of the lower intertidal zone.

Experimental setup:

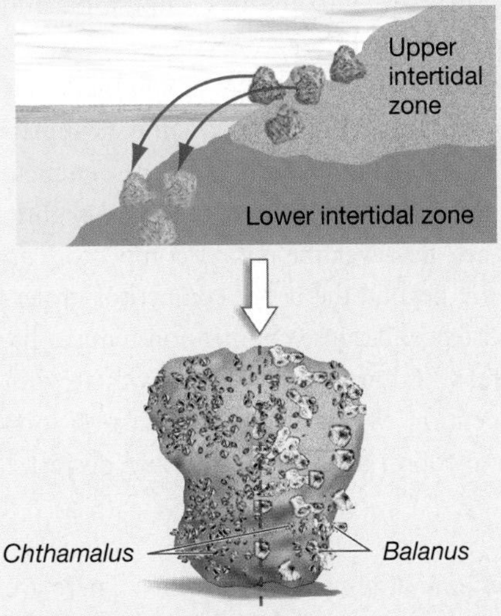
Upper
intertidal
zone
Lower intertidal zone
Chthamalus
Balanus

1. Transplant rocks containing young *Chthamalus* to lower intertidal zone.

2. Let *Balanus* colonize the rocks.

3. Remove *Balanus* from half of each rock. Monitor survival of *Chthamalus* on both sides.

Prediction: *Chthamalus* will survive better in the absence of *Balanus*.

Prediction of alternative hypothesis: *Chthamalus* survival will be low and the same in the presence or absence of *Balanus*.

Results:

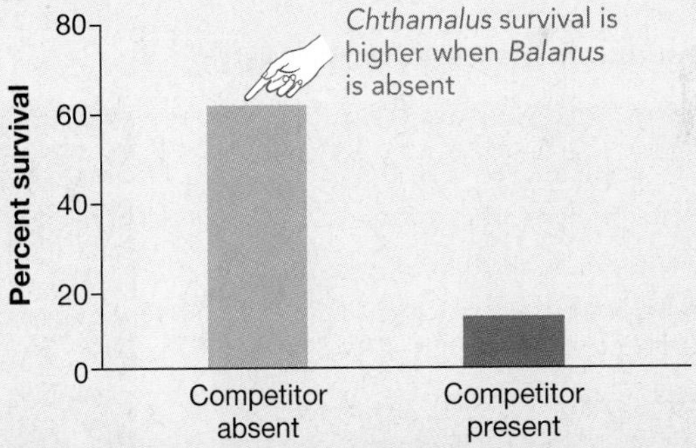
Chthamalus survival is
higher when Balanus
is absent
Percent survival
80
60
40
20
0
Competitor
absent
Competitor
present

Conclusion: *Balanus* is competitively excluding *Chthamalus* from the lower intertidal zone.

FIGURE 53.6 Experimental Evidence for Competitive Exclusion **(a)** In natural habitats, adult *Chthamalus* and *Balanus* barnacles do not coexist. **(b)** Testing the hypothesis that *Chthamalus* is competitively excluded from the lower intertidal zone by *Balanus*.
EXERCISE In part (b), label the experimental and control treatments on each rock. Why was it important to carry out the experimental and control treatments on the same rock? Why not use separate rocks?

distinctive distributions. Barnacle larvae are mobile, but adults live attached to rocks. The adults of one species, *Chthamalus stellatus*, occurred in an upper intertidal zone, while the adults of the other species, *Balanus balanoides*, were restricted to a lower intertidal zone (**Figure 53.6a**). The upper intertidal zone is a more severe environment for barnacles, because it is exposed to the air for longer periods each day. The young of both species were found together in the lower intertidal zone, however.

To explain these observations, Connell hypothesized that adult *C. stellatus* were competitively excluded from the lower intertidal zone. His view was that *Chthamalus* larvae occupy the species' fundamental niche but that adults are constrained by competition for attachment space. The alternative hypothesis is that adult *Chthamalus* are absent from the lower intertidal zone because they do not thrive in the physical conditions there.

To test these hypotheses, Connell performed the experiment shown in **Figure 53.6b**. He began by removing a number of rocks from the upper intertidal zone that had been colonized by *Chthamalus* and transplanting them into the lower intertidal zone. He screwed the rocks into place and allowed *Balanus* larvae to colonize them. Once the spring colonization period was over, Connell divided each rock into two treatments. In one half, he removed all *Balanus* that were in contact with or next to a *Chthamalus*.

This experimental design allowed Connell to document *Chthamalus* survival in the absence of competition with *Balanus* and compare it with survival during competition. This is a common experimental strategy in competition studies: One of the competitors is removed, and the response by the remaining species is observed.

Connell's results support the hypothesis of competitive exclusion. In the unmodified areas, *Balanus* killed many of the young *Chthamalus* by growing against them and lifting them off the substrate. As the graph in Figure 53.6b shows, *Chthamalus* survival was much higher when all of the *Balanus* were removed.

Connell's experiment was one of the first examples of how competition can modify the distribution of a species. In this case, the poorer competitor (*Chthamalus*) is restricted to the upper intertidal zone, which is a more severe habitat because it is prone to drying and high temperatures. *Chthamalus* is able to grow there, however, because it is more resistant to drying than *Balanus* is. *Chthamalus* has a much broader fundamental niche than *Balanus* has, so there is a region where the niches of the two species do not overlap. As a result, the overall outcome of competition between these species is coexistence by habitat partitioning. If both species had identical fundamental niches, *Chthamalus* would probably be eliminated entirely from the area.

In the area where Connell worked, the fundamental and realized niches of *Balanus* are identical, but the fundamental and realized niches of *Chthamalus* are different. The situation is dynamic, however. If an environmental change were to bring a new parasite into the area and *Chthamalus* were more resistant to it than *Balanus* was, then the outcome of competition and the realized niches might be quite different.

To summarize, competition is a common interaction that has a wide range of ecological outcomes. Most competition is asymmetrical, and in extreme cases competition can lead to the complete exclusion of one species. When the two competitors have slightly different fundamental niches, however, the poorer competitor can take refuge in areas that are beyond the better competitor's tolerance, thus achieving coexistence. Because competition reduces fitness in both species, natural selection should favor individuals with traits that minimize competition. When species interact over long periods of time, then, competition for resources should decrease.

Consumption

Consumption occurs when one organism eats another. There are three major types of consumption:

1. **Herbivory** takes place when **herbivores** ("plant-eaters") consume plant tissues. Bark beetles mine cambium; aphids suck sap; caterpillars chew leaves.
2. **Parasitism** occurs when a **parasite** consumes relatively small amounts of tissue or nutrients from another individual, called the **host**. Parasitism often occurs over a long period of time. It is not necessarily fatal, and parasites are usually small relative to their host. An array of worms and unicellular protists parasitizes humans; ticks parasitize cattle and other large mammals.
3. **Predation** occurs when a **predator** kills and consumes all or most of another individual. The consumed individual is called the **prey**. Woodpeckers eat bark beetles; finches eat seeds; ladybird beetles devour aphids; wasps kill caterpillars.

To illustrate how biologists analyze the impact of consumption, let's consider a series of questions about predators, herbivores, parasites, and their victims.

How Do Prey Defend Themselves? With respect to fitness, predation is costly for the prey species and beneficial for the predator. Prey individuals do not passively give up their lives to increase the fitness of their predators, however. Instead, prey may hide or run, fly, or swim away when they sense the presence of a predator. Analyses have shown that many species find safety in numbers—schooling and flocking behavior is an effective way to reduce the risk of predation, in part because predators become confused when they dive or swim into a mass of prey. Other prey species sequester or spray toxins or employ weaponry such as spines or kicking hooves. Traits such as these are called **standing** or **constitutive defenses**, because they are always present (**Figure 53.7a**).

(a) Constitutive defenses of animals vary.

Camouflage: blending into the background

Schooling: safety in numbers

Weaponry: fighting back

(b) Mimicry can protect both dangerous and harmless species.

Müllerian mimics
Batesian mimics
Common wasp
Bumblebee
Honeybee
Hornet moth
Wasp beetle
Hoverfly

FIGURE 53.7 Constitutive Defenses Are Always Present
(a) Prey have an array of adaptations to reduce the likelihood of predation. **(b)** Müllerian mimicry occurs among dangerous prey species; Batesian mimicry occurs between a dangerous prey species and harmless prey species.

Some of the best-studied constitutive defenses involve the phenomenon called mimicry. **Mimicry** occurs when one species closely resembles another species. **Figure 53.7b** illustrates two major types of mimicry. The wasp on the far left is brightly colored and dangerous, because of its stinger. It looks like other dangerous insects—particularly other wasps and bees. When harmful prey species resemble each other, **Müllerian mimicry** is said to occur. To explain the existence of Müllerian mimics, biologists propose that the existence of similar-looking dangerous prey in the same habitat increases the likelihood that predators will learn to avoid them. In this way, Müllerian mimicry should reduce the likelihood of dangerous individuals being attacked. But wasps also act as a model for harmless species of moth, beetle, and fly that resemble wasps. To explain these mimics, biologists propose that predators avoid the harmless mimics because they mistake them for a dangerous wasp. This resemblance is known as **Batesian mimicry.**

The central point here is that prey have adaptations that reduce the likelihood that they will become victims. These adaptations are responses to natural selection exerted by predators. Constitutive defenses are expensive, however, in terms of the energy and resources that must be devoted to producing and maintaining them. Based on this observation, it should not be surprising to learn that many prey species have **inducible defenses**—meaning defensive traits that are produced only in response to the presence of a predator. Induced defenses then decline if predators leave.

To see how inducible defenses work, consider recent research on blue mussels that live in an estuary along the coast of Maine (**Figure 53.8a**). Biologists had documented that predation on mussels by crabs was high in an area of the estuary with relatively slow tidal currents (a "low-flow" area) but low in an area of the estuary with relatively rapid tidal currents (a "high-flow" area). The researchers hypothesized that if blue mussels possess inducible defenses, then heavily defended prey individuals should occur in the low-flow area, where predation pressure is higher, but not in the high-flow area, where water movement reduces the number of crabs present.

To evaluate this hypothesis, the biologists measured mussel shell characteristics in the two areas. They found that mussels in the high-predation area were more strongly attached to their substrate and had thicker shells than did mussels in the low-predation area (**Figure 53.8b**). These traits make the mussels more difficult to remove from the substrate and harder to crush and so function as effective antipredator defenses.

Like any correlational study, however, the data in Figure 53.7b are open to interpretation. A critic of the inducible defense explanation could offer a reasonable alternative hypothesis—for example, that only constitutive defenses exist and that crabs have eliminated weakly attached mussels with thin shells from the low-flow areas. The observed differences could also be due to differences in sunlight, water temperature, or other abiotic factors and have nothing to do with predation.

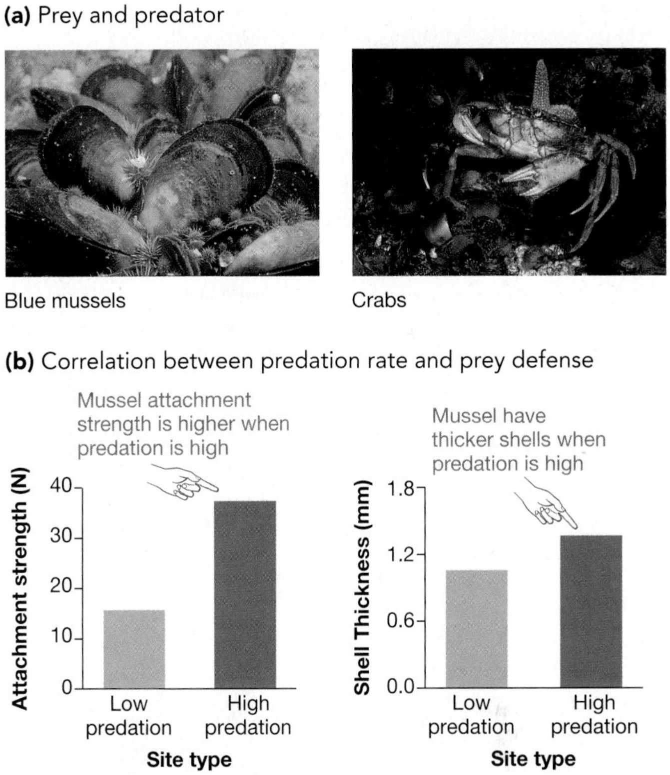

FIGURE 53.8 Inducible Defenses Are Produced Only when Prey Are Threatened

(a) In shallow-water environments in Maine, blue mussels (left) are preyed on by crabs (right). **(b)** To measure how strongly mussels are attached to the substrate, researchers drilled a hole in their shells and measured the force, in newtons (N), required to pull them off. They also measured how large mussel shells were relative to an individual's size by dividing shell weight by soft-tissue weight.

To test the inducible-defense hypothesis more rigorously, biologists did the experiment diagrammed in **Figure 53.9** (page 1222). The tank on the left allowed the researchers to measure shell growth in mussels that were "downstream" from crabs that were fed fish. As predicted by the induced-defense hypothesis, the mussels exposed to a crab in this way developed significantly tougher shells than did mussels that were not exposed to a crab. These results suggest that, even without direct contact, mussels can sense the presence of crabs and increase their investment in defenses. In a similar experiment, the investigators compared mussels that were exposed to water running through broken mussel shells versus intact but empty mussel shells. A significant increase in shell thickness in the tank downstream from the broken shells was measured. This result supports the hypothesis that mussels can detect the presence of predators from molecules released by broken shells.

Results like these underscore several themes of this section: Species interactions are dynamic and result in coevolution—in this case, adaptations and counteradaptations in an evolutionary arms race between predators and prey.

Question: Are mussel defenses induced by the presence of crabs?

Hypothesis: Mussels increase investment in defense in the presence of crabs.

Null hypothesis: Mussels do not increase investment in defense in the presence of crabs.

Experimental setup:

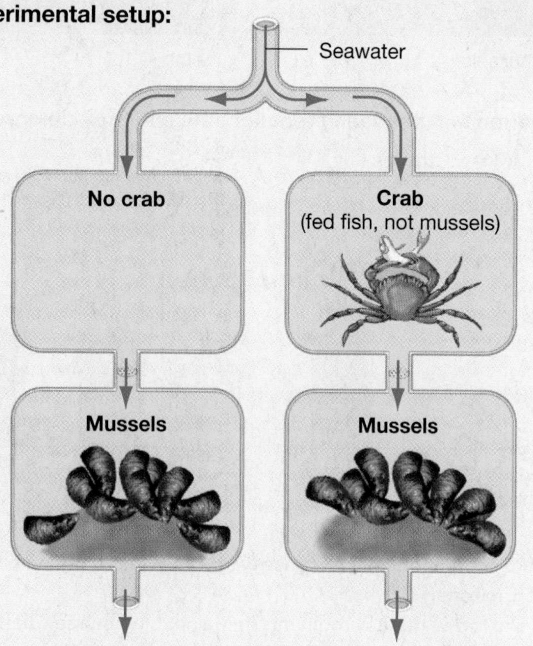
Seawater
No crab
Crab
(fed fish, not mussels)
Mussels
Mussels

Prediction: Mussels downstream of the crab tank will have thicker shells than mussels downstream of the empty tank.

Prediction of null hypothesis: Mussels in the two tanks will have shells of equal thickness.

Results:

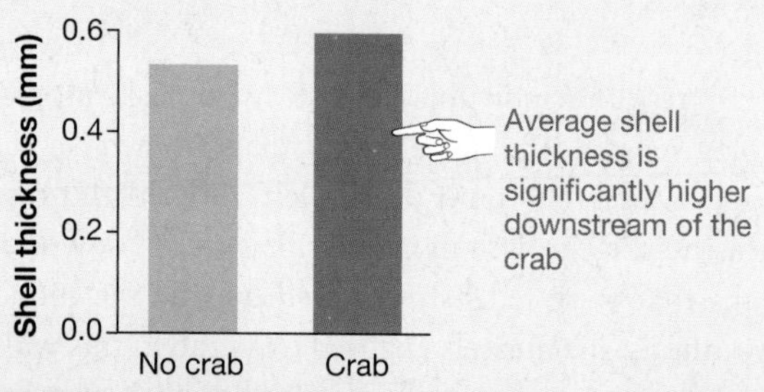
Shell thickness (mm)
0.6
0.4
0.2
0.0
No crab
Crab
Average shell thickness is significantly higher downstream of the crab

Conclusion: Mussels increase investment in defense when they detect crabs. Shell thickness is an inducible defense.

FIGURE 53.9 Experimental Evidence for Inducible Defenses
Experiment conducted in tanks in a laboratory.

Are Animal Predators Efficient Enough to Reduce Prey Populations? Research on animal defense systems supports the hypothesis that species interactions have a strong impact on the evolution of predator and prey populations. Can predators also affect the short-term distribution and abundance of prey populations? Prey are typically smaller than predators, have larger litter or clutch sizes, and tend to begin reproduction at a younger age. As a result, they have a much larger *intrinsic growth rate*, r_{max}—the maximum growth rate that a population can achieve under ideal conditions (see Chapter 52). If prey reproduce rapidly and are also well defended, it is not clear whether predators should be able to kill enough of them to reduce the prey population significantly—particularly if predators tend to take old or sick members of a population.

In several cases, data from predator-removal programs—in which wolves, cougars, coyotes, or other predators are actively killed by human hunters—support the hypothesis that predators actually do reduce the size of prey populations. For example, a wolf control program in Alaska during the 1970s decreased predator abundance to 55–80 percent below precontrol density. Concurrently, the population of moose, on which wolves prey, tripled. This observation suggests that wolf predators had reduced this moose population far below the number that could be supported by the available space and food.

Other types of experiments have also supported the hypothesis that predators play a role in density-dependent growth of prey populations. Recall from Chapter 52 that some populations go through regular cycles and that the regular population cycle of red grouse in England appears to be driven, at least in part, by density-dependent increases in parasitism. **Figure 53.10** shows another classic case of population cycling: snowshoe hare and lynx populations in northern Canada. Hares are herbivores, and lynx prey primarily on hares. Two hypotheses have been proposed to explain the cycles these populations exhibit: (1) Hares use up all their food when their populations reach high density and starve; in response, lynx also starve. (2) Lynx populations reach high density in response to increases in hare density. At high density, lynx eat so many hares that the prey population crashes. Stated another way, either hares control lynx population size or lynx control hare population size.

To test these hypotheses, researchers set up a series of 1-km^2 study plots in boreal forest habitats that were as identical as possible (**Figure 53.11**). Three plots were left as unmanipulated controls. One plot was ringed with an electrified fence with a mesh that excluded lynx but allowed hares to pass freely. Two plots received additional food for hares year-round. One plot had a predator-exclusion fence and was also supplemented with food for hares year-round. The biologists then monitored the size of the hare and lynx populations over an 11-year period, or enough time for a complete cycle in the two populations. As the data in Figure 53.11 show, plots with predators excluded showed higher hare populations at the peak of the cycle than did control plots. This result supports the hypothesis that predation by lynx reduces hare populations. Plots with supplemental food had even higher peak populations, however, and the plot with supplemental food and predators excluded had a hare population over 15 times as large as the average in control plots. These data support the hypotheses that hare populations are limited by availability of

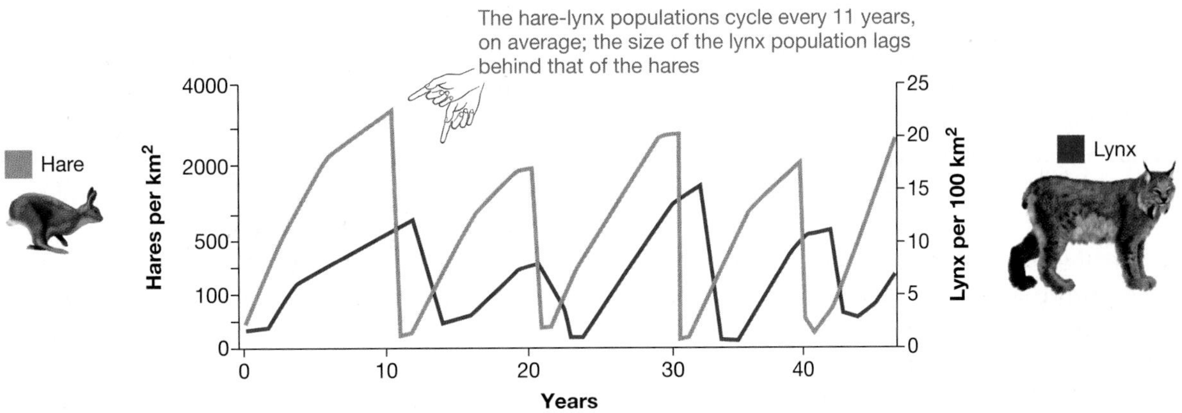

FIGURE 53.10 **Hare and Lynx Populations Cycle in Synchrony**
To make it easier to read, the left-hand vertical axis, which indicates hare density, is not linear.

Question: What factors control the hare-lynx population cycle?

Hypothesis: Predation, food availability, or a combination of those two factors controls the hare-lynx cycle.

Null hypothesis: Predation, food availability, or a combination of those two factors does *not* control the hare-lynx cycle.

Experimental setup:

Document hare population in seven study plots (similar boreal forest habitats, each 1 km^2) over 11 years (duration of one cycle).

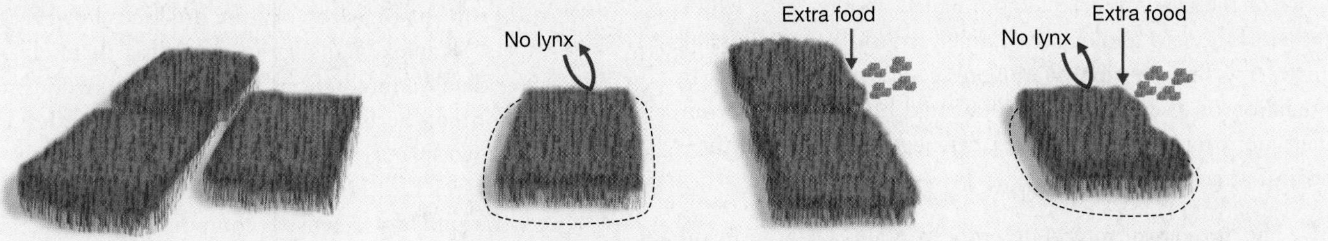

3 plots: Unmanipulated controls

1 plot: Electrified fence excludes lynx but allows free access by hares

2 plots: Supply extra food for hares

1 plot: Electrified fence excludes lynx but allows free access by hares; supply extra food for hares

Prediction: Hare populations in at least one type of manipulated plot will be higher than the average population in control plots.

Prediction of null hypothesis: Hare populations in all of the plots will be the same.

Results:

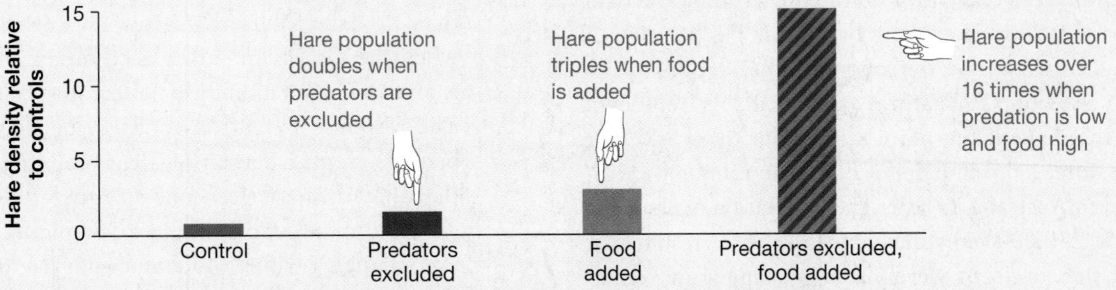

Conclusion: Hare populations are limited by both predation and food availability. When predation and food limitation occur together, they have a greater effect than either factor does independently.

FIGURE 53.11 **Experimental Evidence That Predation and Food Availability Drive the Hare-Lynx Cycle**
QUESTION State a hypothesis to explain why lack of predation and increased food, in combination, have such a large effect on hare population size.

food as well as by predation, and that food availability and predation intensity interact—meaning the combined effect of food and predation is much larger than their impact in isolation. Research continues to figure out why this is so. In the meantime, it is increasingly clear that, in many instances, predators are efficient enough to reduce prey populations.

Why Don't Herbivores Eat Everything—Why Is the World Green? If predators affect the size of populations in prey that can run or fly or swim away, then consumers should have a devastating impact on plants and on mussels, anemones, sponges, and other sessile (nonmoving) animals. In some cases, this prediction turns out to be correct. For example, consider the results of a recent **meta-analysis**—a study of studies, meaning an analysis of a large number of data sets on a particular question. Biologists who compiled the results of more than 100 studies on herbivory found that the median percentage of mass removed from aquatic algae by herbivores was 79 percent. This figure supports the hypothesis that herbivores eat the vast majority of algal food available in aquatic biomes. The figure dropped to just 30 percent for aquatic plants, however, and only 18 percent for terrestrial plants.

These data raise the question of why herbivores don't eat more of the available plant food. Stated another way, why is the world green? Biologists routinely consider three possible answers to this question. Herbivores could be kept in check by predation or disease; plant tissues could offer poor or incomplete nutrition; or plants could defend themselves effectively against attack:

1. The **top-down control** hypothesis states that herbivore populations are limited by predation or disease. The term *top-down* is inspired by the concept of a food chain, which is explained in detail in Chapter 54; see also Chapter 28. As an example, consider the boreal forest biome of northern Canada. Lynx are top predators, meaning they are not subject to predation. Lynx eat hares, which in turn eat vegetation on the forest floor. The data in Figure 53.11 showed that lynx limit the number of hares and thus increase the amount of vegetation. Biologists can also point to "natural experiments" in top-down control that have taken place throughout the world. For example, after a dozen pairs of European rabbits were introduced to Australia in 1859, the population exploded—exceeding 250,000 individuals by 1865. As a result, vast areas of the continent were virtually stripped of vegetation. Similar consequences have occurred when goats and pigs have escaped from captivity and established large, wild populations in areas that lack their natural predators and diseases.
2. The **poor-nutrition hypothesis** contends that plants are a poor food source in terms of the nutrients they provide for herbivores. More specifically, plant tissues have less than 10 percent of the nitrogen found in animal tissues, by weight. If the growth and reproduction of herbivores is limited by the availability of nitrogen, then populations will be kept too low to consume more than a small fraction of the available food. But if the nitrogen concentration of plants is increased via fertilization, then herbivores should grow faster and reproduce more. To test this prediction, a recent meta-analysis examined 185 studies of how insect herbivores had responded to plant fertilization. In over half of the cases, herbivores showed a significant increase in growth rate or reproduction when the plants that they feed on were fertilized. Based on this result, the researchers concluded that nitrogen limitation is an important factor in a large fraction of plant-herbivore interactions.
3. The **plant-defense hypothesis** holds that plants defend themselves effectively enough to limit herbivory. Most plant tissues are defended by weapons such as thorns, prickles, or hairs or by potent poisons such as nicotine, caffeine, and cocaine. In addition, no animal species can, without help from protists or bacteria, digest cellulose or lignin, which are components of wood. Manufacturing defensive compounds seems like the perfect solution to the problem of herbivory. In practice, however, plants face a complex challenge in defending themselves. Consider data on interactions between cottonwood trees and two of their herbivores: beavers and leaf beetles. Cottonwoods resprout after they are cut down by a beaver (**Figure 53.12a**), and resprouted trees contain high concentrations of a defensive compound that deters further attack by beavers (**Figure 53.12b**). This is an induced defense. But caterpillars of a particular leaf beetle species eat this defensive compound readily. In fact, the caterpillars store enough of it in their bodies to act as a defensive compound against their own major predator—ants. The data in **Figure 53.12c** show that caterpillars that grow on resprouted cottonwood trees survive longer than do caterpillars that grow on normal cottonwoods when the caterpillars are placed on an ant mound. This is an example of an indirect effect in species interactions. The response by cottonwoods to herbivory by beavers benefits another herbivore—leaf beetles. The net result is that there is no perfect, one-size-fits-all defensive strategy. Natural selection should favor plants that evolve an ever-changing suite of compounds to deter the ever-changing array of herbivores they face.

Taken together, the data reviewed here suggest that there is no single answer to the question of why herbivores don't eat a greater fraction of the available plant food. All three of the hypotheses we have examined are correct. Top-down control, nitrogen limitation, and effective defense are all important factors

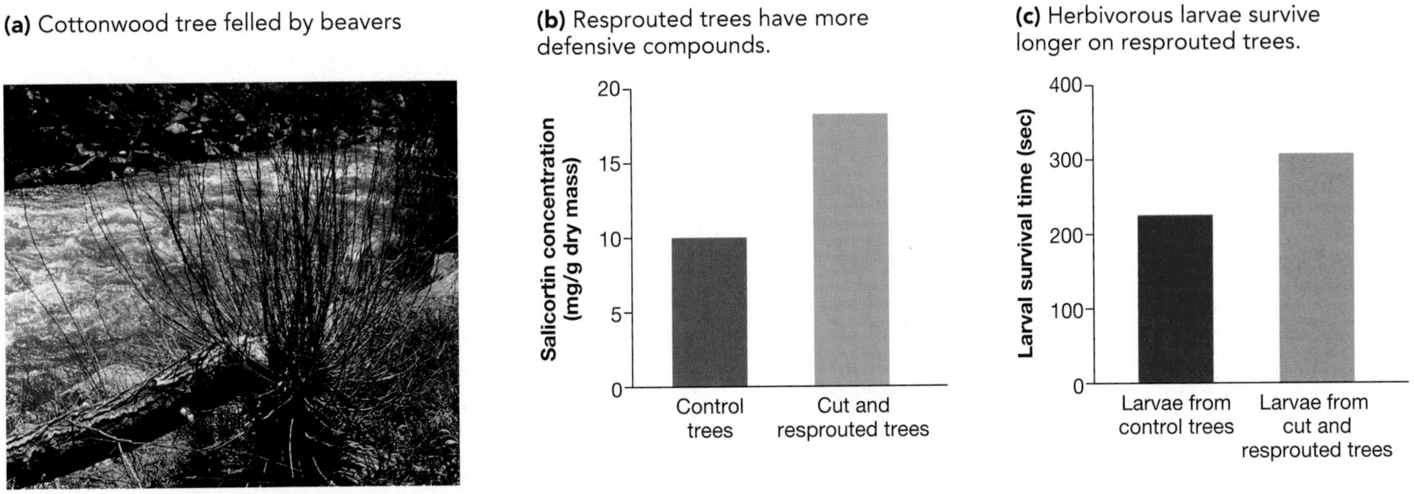

FIGURE 53.12 Defensive Compounds in Cottonwood Trees Discourage Beavers but Favor Beetles
(a) If felled by a beaver, most cottonwood trees respond by sending up new shoots.
(b) Cottonwoods that have resprouted following attack by beavers have significantly higher concentrations of defensive compounds than do cottonwoods that were never attacked by beavers. **(c)** Leaf beetle larvae placed on the mounds of ants (their primary predator) survived significantly longer if the larvae had been reared on cottonwood leaves with high levels of defensive compounds.

in limiting the impact of herbivory, although the particular mix of factors will vary from plant species to plant species and from habitat to habitat.

Adaptation and Arms Races Over the long term, how do species that interact via consumption affect each other's evolution? To answer this question, let's focus on interactions between humans and the most serious of all human parasites—species in the genus *Plasmodium*. Recall from Chapter 28 that *Plasmodium* are unicellular protists that cause malaria in a wide array of vertebrates. Malaria kills at least a million people a year, most of them children of preschool age. Recent data suggest that humans and the causative agent of malaria—four species in the genus *Plasmodium*—are locked in a coevolutionary arms race.

Plasmodium has a complex life cycle. It is a unicellular organism, but it has several distinct cell types. Each cell type corresponds to the host and host tissue being infected. For example, one of these cell types lives in the salivary glands of infected mosquitoes and is transferred to a human host when the infected mosquito bites a person. The *Plasmodium* cells travel to the human's liver, reproduce asexually, and form a large population of a new cell type that is later released into the host's bloodstream. This generation of parasite cells invades the host's red blood cells, reproduces asexually, and eventually causes the infected blood cells to rupture. Loss of red blood cells causes severe anemia in the human host; the infected person's immune system responds so strongly to the release of the parasite cells that debilitating fevers result. The parasite's life cycle continues when a mosquito bites an infected person and ingests infected red blood cells.

Cells in the human immune system do not sit passively by as *Plasmodium* destroys liver cells and red blood cells. Instead, certain cells (introduced in Chapter 49) seek out and destroy *Plasmodium* cells. In West Africa, for example, there is a strong association between an allele called *HLA-B53* and protection against malaria. In liver cells that are infected by *Plasmodium*, HLA-B53 proteins on the surface of the liver cell display a parasite protein called cp26 (**Figure 53.13a**, page 1226). The display is a signal that immune system cells can read. It means "I'm an infected cell. Kill me before they kill all of us." Immune system cells destroy the liver cell before the parasite cells inside can multiply. In this way, people who have at least one copy of the *HLA-B53* allele are better able to beat back malaria infections before the infection progresses. People who have the *HLA-B53* allele appear to be winning the arms race against malaria.

Unfortunately, follow-up research has shown that the arms race is far from won. *Plasmodium* populations in West Africa now have a variety of alleles for the protein recognized by HLA-B53. Some of these variants bind to HLA-B53 and trigger an immune response in the host, but others escape detection. Furthermore, many people in West Africa are infected with several different strains of *Plasmodium*. In some cases, the recognition step by HLA-B53 breaks down when certain strains are found together (**Figure 53.13b**). To make sense of these observations, researchers suggest that natural selection has favored the evolution of *Plasmodium* strains with weapons that counter HLA-B53.

WEB TUTORIAL 53.1
Life Cycle of a Malaria Parasite

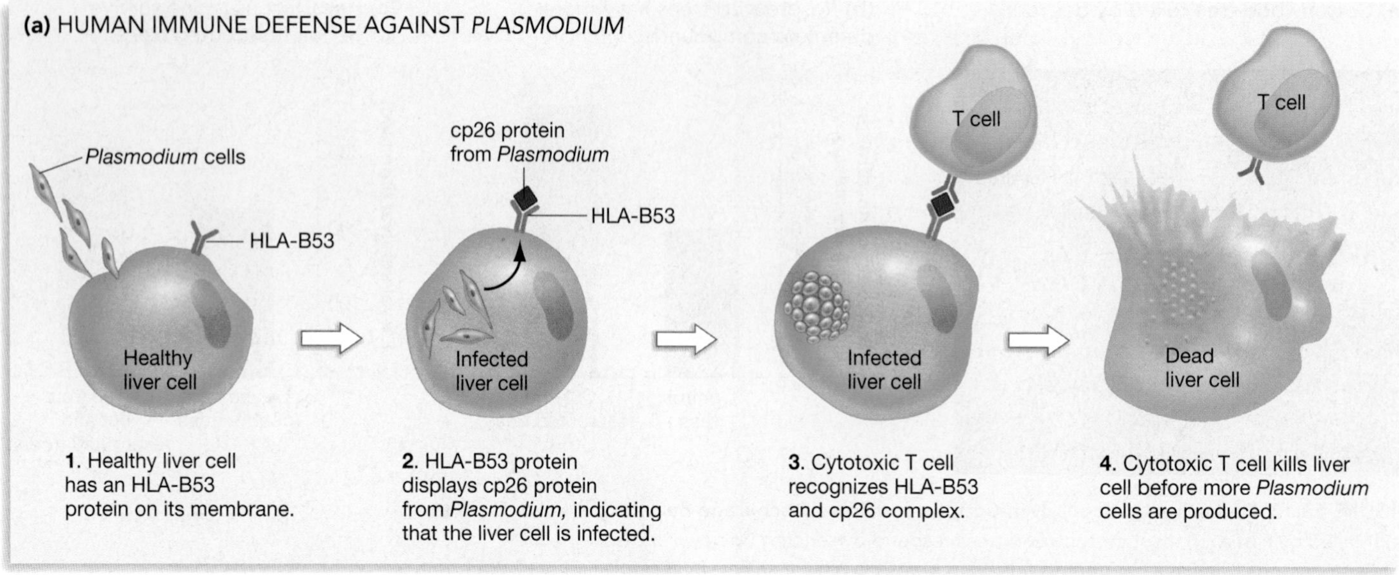
(a) HUMAN IMMUNE DEFENSE AGAINST PLASMODIUM
Plasmodium cells
HLA-B53
Healthy liver cell
cp26 protein from Plasmodium
HLA-B53
Infected liver cell
T cell
Infected liver cell
T cell
Dead liver cell
1. Healthy liver cell has an HLA-B53 protein on its membrane.
2. HLA-B53 protein displays cp26 protein from Plasmodium, indicating that the liver cell is infected.
3. Cytotoxic T cell recognizes HLA-B53 and cp26 complex.
4. Cytotoxic T cell kills liver cell before more Plasmodium cells are produced.

(b) *Plasmodium* strains have different versions of the cp protein. How successful are these strains at infecting people?

Plasmodium strain	Infection rate	Interpretation
cp26	Low	HLA-B53 binds to these proteins. Immune response is effective.
cp29	Low	
cp26 and cp29 strains together	High	Immune response fails when these strains infect the same person.
cp27	High	HLA-B53 does not bind to these proteins. Immune response is not as effective.
cp28	Average	

FIGURE 53.13 Interactions between the Human Immune System and *Plasmodium*
(a) If HLA-B53 binds to a particular *Plasmodium* protein, then infected cells are recognized and destroyed. **(b)** Some strains of *Plasmodium* appear to avoid detection by the immune system better than others do. Also, certain strains defeat the immune response if they infect the same person at the same time.

To summarize, an arms race continues to rage between *Plasmodium* and humans. Certain human proteins act as antimalarial weapons. But as predicted by coevolutionary theory, *Plasmodium* has evolved effective responses and continues to evolve.

Can Parasites Manipulate Their Hosts? Parasites do not just have to invade tissues and grow while evading defensive responses by their host. They also have to be transmitted to new hosts. To a parasite, an uninfected host represents uncolonized habitat, teeming with resources. What have biologists learned about how parasites are transmitted to new hosts?

To answer this question, consider species of land snails that are parasitized by flatworms—specifically, flukes in the genus *Leucochloridium*. Researchers who studied this association discovered something unusual. When the flukes have matured and are ready to be transmitted to their next host, a bird, they burrow into the snail's tentacles and wriggle. Infected snails also become attracted to light, even though uninfected snails avoid sunlit areas and prefer dark, shady environments (**Figure 53.14**). When infected snails move out of the shade into the open and glide about with wriggling tentacles, they are more easily spot-

Birds that prey on snails are the next host for the parasite
Infected snails move to open sunny areas; tentacles wriggle
Uninfected snails stay in shaded areas; tentacles do not wriggle

FIGURE 53.14 A Parasite That Manipulates Host Behavior
The behavior of snails that are infected with flukes in the genus *Leucochloridium* is dramatically different from the behavior of uninfected snails. EXERCISE Design an experiment to test the hypothesis that infected snails are more likely than uninfected snails to be eaten by birds.

ted and consumed by birds. To interpret these observations, biologists suggest that the worms manipulate the behavior of the snail, and that the change in snail behavior makes the parasite more likely to be transmitted to a new host.

Studies of how parasites are transmitted to new hosts reinforce a general message: Extensive coevolution occurs among species that interact via consumption. Experiments on mutualistic interactions carry the same message.

Mutualism

Mutualisms are +/+ interactions that involve a wide variety of organisms and rewards. Many species of bees, for example, visit flowers to harvest nectar and pollen. Bees benefit because nectar is used as a food source for adult bees while the pollen is fed to larvae. Flowering plants also benefit because, in the process of visiting flowers, foraging bees carry pollen from one plant to another and accomplish pollination. Chapters 29 and 40 detailed some of the adaptations found in flowering plants that increase the efficiency of pollination. Other chapters highlighted an array of other mutualisms:

- One of the most important of all mutualisms occurs between fungi and plant roots. Chapter 30 reviewed experimental evidence indicating that fungi receive sugars and other carbon-containing compounds in exchange for nitrogen or phosphorus needed by the plant partner.
- Arguably the most critical of all mutualisms involves bacteria that fix nitrogen and certain species of plants. As pointed out in Chapters 27 and 37, the partnership is based on host plants providing sugars and protection and the bacteria supplying nitrogen in return.
- Chapter 31 cited interesting mutualisms between rancher ants that protect aphids in exchange for sugar-rich honeydew and between farmer ants that cultivate fungi for food.

Figure 53.15 illustrates some other interesting mutualisms. Figure 53.15a shows ants in the genus *Crematogaster*, which live in acacia trees native to Africa and the New World tropics. The ants live in bulbs at the base of acacia thorns and feed on small structures that grow from tree branches. These ants protect the tree by attacking and biting herbivores and by cutting vegetation from the ground below the host tree. Figure 53.15b illustrates cleaner shrimp in action. These shrimp pick external parasites from the jaws and gills of fish. In this mutualism, one species receives dinner while the other obtains medical attention.

As these examples show, the rewards from mutualistic interactions range from the transportation of gametes to food, housing, medical help, and protection. It is important to note, however, that even though mutualisms benefit both species, the interaction does not involve individuals from different species being altruistic or "nice" to each other. Expanding on a point that Charles Darwin introduced in 1862, Judith Bronstein recently described mutualisms as "a kind of reciprocal parasitism; that is, each partner is out to do the best it can by obtaining what it needs from its mutualist at the lowest possible cost to itself." Her point is that the benefits received in a mutualism are a by-product of each individual pursuing its own self-interest, by maximizing its ability to survive and reproduce.

In this light, it is not surprising that some species that are closely related to mutualists "cheat" on the system. For

(a) Mutualism between ants and acacia trees

(b) Mutualism between cleaner shrimp and fish

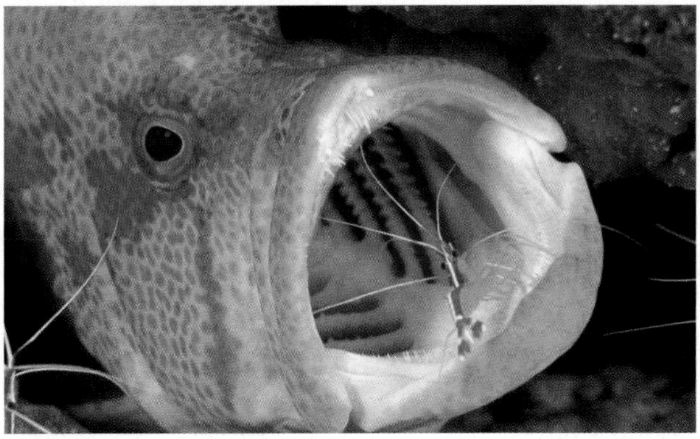

FIGURE 53.15 Mutualisms Take Many Forms
(a) In certain species of acacia tree, ants in the genus *Crematogaster* live in large bulbs at the base of spines and attack herbivores that threaten the tree. The ants eat nutrient-rich tissue produced at the tips of leaves or on branches. **(b)** Cleaner shrimp remove and eat parasites that take up residence on the gills of fish. EXERCISE In part (a), label the entrance to the ant colony and the nutrient-rich tissue on which the ants are feeding.

example, Chapter 40 introduced deceit pollination, in which certain species of plants produce a showy flower but no nectar reward. Pollinators have to be deceived to make a visit and carry out pollination. Evolutionary studies show that deceit pollinators evolved from ancestral species that did provide a reward. Over time, a +/+ interaction evolved into a +/− interaction.

A recent experimental study of mutualism provides another good example of the dynamic nature of these interactions. This study focused on ants and treehoppers. Ants are insects that live in colonies; treehoppers are small, herbivorous insects that feed by sucking sugar out of the phloem of plants. Treehoppers excrete the sugary solution honeydew from their posteriors. The honeydew, in turn, is harvested for food by ants.

It is clear that ants benefit from this association. But do the treehoppers? Biologists hypothesized that the ants might protect the treehoppers from their major predator, jumping spiders. These spiders feed heavily on juvenile treehoppers.

To test the hypothesis that ants protect treehoppers, the researchers studied ant-treehopper interactions over a three-year period. As **Figure 53.16** shows, the researchers marked out a 1000-m^2 study plot. Each year they randomly assigned the treehopper host plants inside the plot to one of two groups. The biologists removed the ants from one group but left the others alone to serve as a control. Then they compared the growth and survival of treehoppers on plants with and without ants. Recall that this is a common research strategy for studying species interactions. To assess the fitness costs or benefits of the interaction, researchers remove one of the participants experimentally and document the effect on the other participant's survival and reproduction, compared with the survival and reproduction of control individuals that experience a normal interaction.

In both the first and third years of the study, the number of treehopper young on host plants increased in the treatment with ants but showed a significant decline in the treatment with no ants (Figure 53.16). This result supports the hypothesis that treehoppers benefit from the interaction with ants because the ants protect the treehoppers from predation by jumping spiders.

In the second year of the study, however, the researchers found a very different pattern. There was no difference in offspring survival, adult survival, or overall population size between treehopper populations with ants and those without ants. Why? The researchers were able to answer this question because they also measured the abundance of spiders in each of the three years. Their census data showed that in the second year of the study, spider populations were very low.

Based on these results, the investigators concluded that the benefits of the ant-treehopper interaction depend entirely on predator abundance. Treehoppers benefit from their interaction with ants in years when predators are abundant but are unaffected in years when predators are scarce. If producing honeydew is costly to treehoppers, then the +/+ mutualism changes to a +/− interaction when spiders are rare.

To summarize, mutualism is like parasitism, competition, and other types of species interactions in an important respect: The outcome of the interaction depends on current conditions. Because the costs and benefits of species interactions are fluid, an interaction between the same two species may vary from par-

Question: Is the relationship between ants and treehoppers mutualistic?

Hypothesis: Ants harvest food from treehoppers and protect treehoppers from jumping spiders.

Null hypothesis: Ants harvest food from treehoppers but are not beneficial to treehopper survival.

Experimental setup:

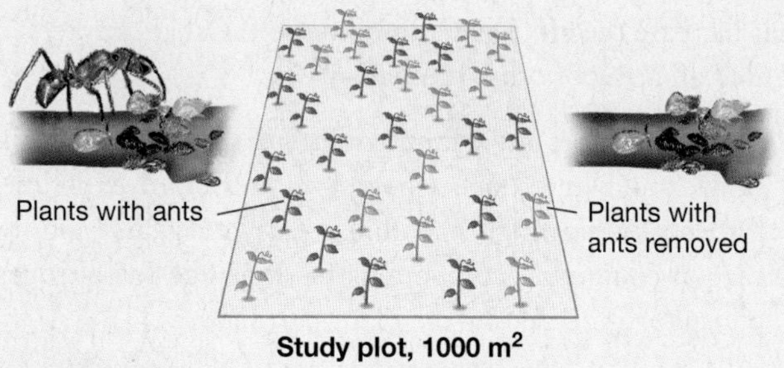

Prediction: More young treehoppers will be found when ants are present than when ants are absent.

Prediction of null hypothesis: There will be no difference in the number of young treehoppers on the plants.

Results (Year 1):

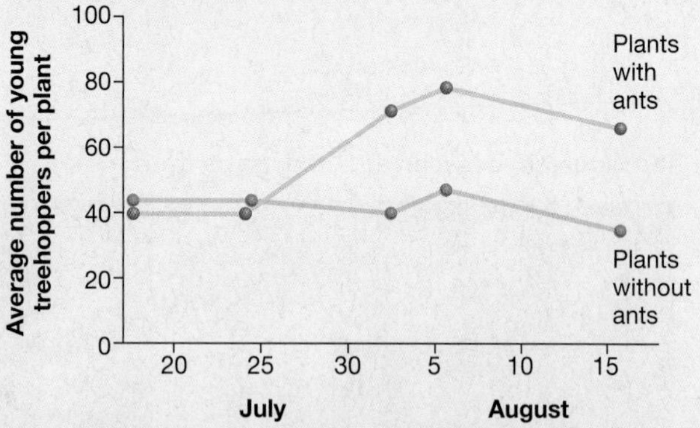

Conclusion: Treehoppers benefit from the interaction with ants, which protect treehoppers from predation by jumping spiders.

FIGURE 53.16 Experimental Evidence That the Treehopper-Ant Interaction Is Mutualistic
An experiment designed to test whether the presence of ants is beneficial for treehoppers. Treehoppers are small herbivorous insects that suck sugar out of the phloem of plants and secrete honeydew. In some years, the number of young treehoppers increased over time when ants were present and declined over time when ants were absent.

TABLE 53.1 Summary of Species Interactions

Type of Interaction	Fitness Effects	Short-Term Impact: Distribution and Abundance	Long-Term Impact: Coevolution
Competition	−/−	Reduces population size of both species; if competition is asymmetric, competitive exclusion reduces range of one species.	Niche differentiation via selection to reduce competition
Consumption	+/−	Impact on prey population depends on prey density and effectiveness of defenses.	Strong selection on prey for effective defense; strong selection on consumer for traits that overcome defenses
Parasitism	+/−	Impact on host population depends on parasite density and effectiveness of defenses.	Strong selection on host for effective defense; strong selection on parasite for traits that overcome defenses
Mutualism	+/+	Population size and range of both species are dependent on each other.	Strong selection on both species to maximize fitness benefits and minimize fitness costs of relationship
Commensalism	+/0	Population size and range of commensal may depend on size and distribution of host.	Strong selection on commensal to increase fitness benefits in relationship; no selection on host

asitism to mutualism to competition. **Table 53.1** summarizes the fitness effects, short-term impacts on population size, and long-term evolutionary aspects of species interactions.

✓CHECK YOUR UNDERSTANDING

Species interactions affect the short-term distribution and abundance of the populations involved and result in coevolution. For example, natural selection favors the evolution of traits that decrease competition, resulting in changes in a species' niche over time. Consumers may reduce the population size of the species on which they feed and often exert strong natural selection for effective defense mechanisms. Mutualisms benefit the species involved and can lead to highly coevolved associations, such as mycorrhizal fungi and symbiotic nitrogen-fixing bacteria. You should be able to (1) explain what a coevolutionary arms race is and (2) give a specific example of how competition, consumption, and mutualism affect population size and long-term evolution.

53.2 Community Structure

Biologists have made important progress in understanding the nature of species interactions and their consequences. In terms of understanding the structure and function of biological communities, however, research on species interactions has a limitation: It usually focuses on just two species at a time. But biological communities contain many thousands of species. To understand how communities work, biologists broaden the scope of research and explore how combinations of many species interact.

The question that biologists initially asked about communities concerned structure: Do biological communities have a tightly prescribed organization and composition, or are they merely loose assemblages of species? If communities are highly structured entities, then their makeup should be predictable. For example, if a community is destroyed by a disturbance and then allowed to recover, the diversity and abundance of species at that site should be identical when recovery is complete. But if communities can be made up of many different combinations of species, depending on which arrive earlier or later, then community composition will be difficult to predict. The diversity and abundance of species found in a region would vary substantially before a disturbance and after recovery.

Clements and Gleason: Two Views of Community Dynamics

Beginning with a paper published in 1936, Frederick Clements promoted the view that biological communities are stable, integrated, and orderly entities with a highly predictable composition. His hypothesis was that species interactions are so extensive and coevolution is so important that the groups of species called communities have become highly integrated and interdependent units in nature. Stated another way, the species within a community cannot live without each other.

To drive his idea home, Clements likened the development of a plant community to the development of an individual organism. He argued that communities develop by passing through a series of predictable stages dictated by extensive interactions among species and that this development culminates in a stable final stage known as a **climax community**. According to Clements, the nature of the climax community is determined by the area's climate and does not change over time. Further, he held that if a fire or other disturbance destroys the climax community, it will reconstitute itself by repeating its predictable developmental stages.

Henry Gleason, in contrast, contended that the community found in a particular area is neither stable nor predictable. He claimed that plant and animal communities are ephemeral associations of species that just happen to share similar climatic requirements. According to Gleason, it is largely a matter of chance whether a similar community develops in the same area after a disturbance occurs. Gleason downplayed the role of biotic factors, such as species interactions, in structuring communities. To him, abiotic factors and history—for example, which seeds and juvenile animals happened to arrive after a disturbance—were the key elements in determining which species are found at a particular location.

Which viewpoint is more accurate? To answer this question, let's consider a series of data sets that are relevant to the issue, then synthesize the results into a general picture of how communities are structured.

Keystone Species According to Clements, the composition of a community is dictated by climate and structured by biotic interactions. Some of the best data on the role of biotic interactions in communities contradict his view of communities as static, predictable entities, however. Experiments have shown that if a single species of predator or herbivore is removed from a community, the structure of an entire community can change dramatically.

As an example of this work, consider a series of classical experiments that Robert Paine did in intertidal habitats of the Pacific Northwest of North America. In this environment, the sea star *Pisaster ochraceous* is an important predator. When Paine removed *Pisaster* from experimental areas, what had been diverse communities of algae and invertebrates became overgrown with solid stands of the California mussel *Mytilus californianus* (**Figure 53.17**). Although *M. californianus* is a dominant competitor, its populations had been held in check by sea-star predation. When this predator was gone, the species richness and structural complexity of the habitat changed radically.

To capture the effect that a predator such as *Pisaster* can have on a community, Paine coined the term *keystone species.* A **keystone species** is a species that has a much greater impact on the surrounding species than its abundance would suggest. In grassland biomes of central North America, bison act as a keystone species even though they represent a tiny fraction of the total mass of organisms present. Because they graze primarily on grasses, bison herds decrease the abundance of grasses and increase the importance of **forbs**—herbaceous, flowering plants such as daisies and members of the pea family.

Sea stars, bison, and other keystone species may vary in abundance in space or over time. As a result, the composition and structure of their communities will vary. These observations suggest that communities are dynamic entities whose composition is difficult to predict.

Historical Data on Community Structure Data on the historical composition of plant communities began to accumulate during the 1970s and 1980s. Studies of fossil pollen documented that the distribution of plant species and communities at specific locations throughout North America has changed radically since the end of the last ice age about 11,000 years ago. An important pattern emerged from these data: Contrary to what Clements would predict, groups of species do not come and go in the fossil record in tightly integrated units. Instead, the ranges of individual species tended to change independently of one another. For example, at a study site in Mirror Lake, New Hampshire, the percentage of pollen present from pines and oaks rose and fell in tandem over the past 14,000 years—just as Clements would predict for a tightly integrated, pine-oak forest community. In contrast, the percentage of pollen from spruce, fir, cedar, beech, and other tree species changed independently of each other. In general, studies of fossil pollen suggest that the composition of most plant communities has been dynamic and contingent on historical events rather than static.

Experimental Tests As an example of experimental work on how communities are structured, consider a recent study of

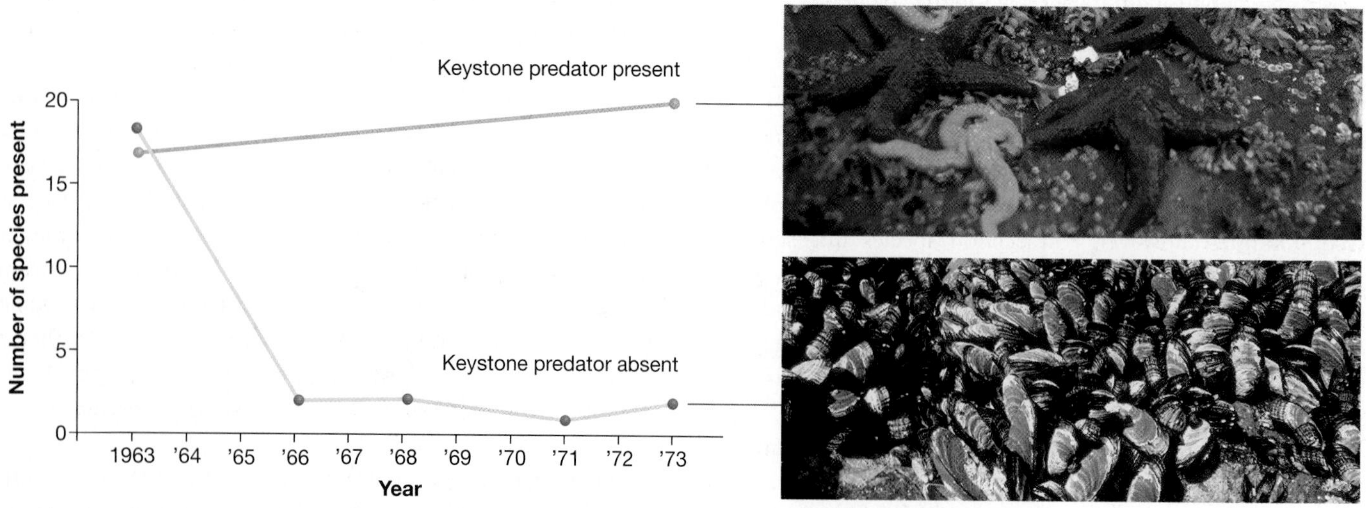

FIGURE 53.17 Keystone Predation Alters Community Structure in a Rocky Intertidal Habitat
When the sea star *Pisaster ochraceous* is present, rocky intertidal habitats have a large diversity of species. When *Pisaster* is absent, the same habitats are dominated by beds of California mussels.

the planktonic organisms in ponds. To explore how predictable community structure is, biologists constructed 12 identical ponds (**Figure 53.18**). They filled the ponds at the same time with water that contained enough chlorine to kill any preexisting organisms. At the outset, then, the ponds were sterile. If Clements's view of community structure is correct, each pond should develop the same community of species once the chlorine vaporized and made the water habitable. If Gleason's view of community structure is correct, then each pond should develop a different community.

To test this prediction, the researchers sampled water from the ponds repeatedly for one year. They measured temperature, chemical makeup, and other physical characteristics of the water and recorded the diversity and abundance of each species of plankton by examining the samples under the microscope. They found a total of 61 species in all of the ponds but discovered that individual ponds had only 31 to 39 species. This observation is important. Each pond contained just half to two-thirds of the total number of species that lived in the experimental area and that were available for colonization. Most species occurred in most or all of the 12 ponds, but each pond had a unique species assemblage. Why?

To explain their results, the researchers contended that some species are particularly good at dispersing and are likely to colonize all or most of the available habitats. Other species disperse more slowly and tend to reach only one or a few of the available habitats. Further, the investigators proposed that the arrival of certain competitors or predators early in the colonization process greatly affects which species are able to invade successfully later. As a result, the specific details of community assembly and composition are somewhat contingent and difficult to predict. At least to some degree, communities are a product of chance and history.

The overall message of research on keystone species, the history of plant communities, and experiments is that Clements's position was clearly too extreme and that Gleason's view is closer to being correct. Although both biotic interactions and climate are important in determining which species exist at a certain site, chance and history also play a large role.

Question: Do identical communities develop in identical environments?

"Clementsian" hypothesis: Identical communities will always develop in identical environments.

"Gleasonian" hypothesis: Identical communities will not always develop in identical environments.

Experimental setup:

1. Construct 12 identical ponds. Fill at the same time and sterilize water so that there are no preexisting organisms.

2. Examine water samples from each pond. Identify each plankton species present in each sample.

"Clementsian" prediction: Identical plankton communities will develop in all 12 ponds.

"Gleasonian" prediction: Different plankton communities will develop in different ponds.

Results (after 1 year):

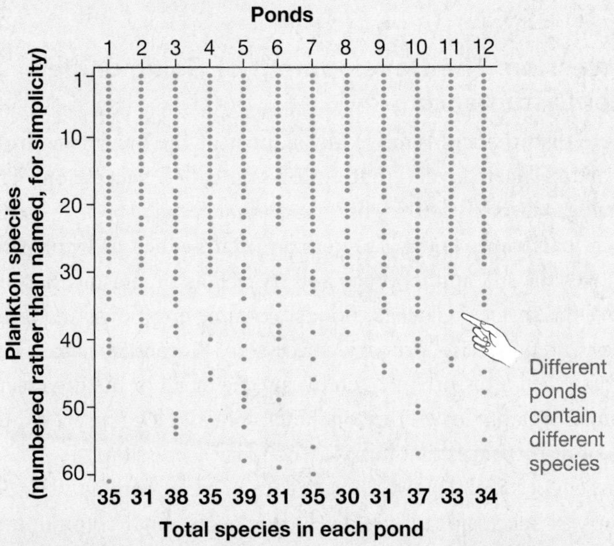

Conclusion: Although about half of the species present appear in all or most ponds, each pond has a unique composition. Both hypotheses are partially correct.

FIGURE 53.18 Experimental Evidence That Identical Communities Do Not Develop in Identical Habitats

QUESTION How many species are found in all 12 ponds?

Disturbance and Change in Ecological Communities

Perhaps the most dramatic contrast between the views held by Clements and contemporary biologists concerns how communities change over time. Instead of viewing communities as the static entities that Clements envisioned, researchers now recognize that communities are dynamic. Community composition and structure may change radically in response to changes in abiotic and biotic conditions. Biologists have become particularly interested in how communities respond to disturbance. A **disturbance** is any event that removes some individuals or biomass from a community. (Recall from Chapter 50 that biomass is the total mass of living organisms.)

Forest fires, windstorms, floods, the fall of a large canopy tree, disease epidemics, and herbivore outbreaks all qualify as disturbances. These events are important because they alter light levels, nutrients, unoccupied space, or some other aspect of resource availability. What biologists have come to realize is that the impact of disturbance is a function of three factors: (1) the type of disturbance, (2) its frequency, and (3) its severity—for example, the speed and duration of a flood, or the intensity of heat during a fire. Let's take a closer look.

What Disturbances Occur, and How Frequent and Severe Are They?

Most communities experience a characteristic type of disturbance. In most cases, disturbances occur with a predictable frequency and severity. To capture this point, biologists refer to a community's **disturbance regime**. For example, fires kill all or most of the existing trees in a boreal forest every 100 to 300 years, on average. In contrast, numerous small-scale tree falls, usually caused by windstorms, occur in temperate and tropical forests every few years.

How Do Researchers Determine a Community's Disturbance Regime? Ecologists use two strategies to determine a community's natural pattern of disturbance. The first approach is based on inferring long-term patterns from data obtained in a short-term analysis. For example, an observational study might document that 1 percent of all boreal forest on Earth burns in a given year. Assuming that fires occur randomly, researchers project that any particular piece of boreal forest has a 1 in 100 chance of burning each year. According to this reasoning, fires will recur in that particular area every 100 years, on average.

This extrapolation approach is straightforward to implement, but it has important drawbacks. In boreal forests, for example, fires do not occur randomly in either space or time. They are more likely in some areas than in others, and they tend to occur in particularly dry years. Unless sampling is extensive, it is difficult to avoid errors caused by extrapolating from particularly disturbance-prone or disturbance-free areas or years.

The second approach to determining disturbance regimes is based on reconstructing the history of a particular site. Flooding frequency, for example, can be estimated by analyzing sediments, because floods deposit distinctive groups of sediment particles. Researchers estimate the frequency and impact of storms by finding wind-killed trees and determining their date of death. (Investigators do so by comparing patterns in the growth rings of the dead trees with those of living individuals nearby; see Chapter 35.) The disturbances that have been most extensively studied using historical techniques, however, are forest fires.

Forest fires often leave a layer of burned organic matter and charcoal on the surface of the ground. As a result, researchers can dig a soil pit, find charcoal layers, and use radioisotope dating to establish when the fires occurred. It is also possible to date the death of trees killed by fire by comparing their growth rings with those of living trees. Further, trees that are not killed by fire are often scarred. When a fire burns close enough to kill a patch of cambium tissue, a scar forms. These fire scars occur most often at the tree's base, where dead leaves and twigs accumulate and furnish fuel. Fire scars can be dated by analyzing growth rings.

Why Is It Important to Understand Disturbance Regimes? To appreciate why biologists are so interested in understanding disturbance regimes, consider a recent study on the fire history of giant sequoia groves in California (**Figure 53.19a**). Giant sequoias grow in small, isolated groves on the western side of the Sierra Nevada range. Individuals live more than a thousand years, and many have been scarred repeatedly by fires. A biologist obtained samples of cross sections through the bases of 90 giant sequoias in five different groves. As **Figure 53.19b** shows, the cross sections contained numerous rings that had been scarred by fire. To determine the date of each disturbance, the researcher counted tree rings back from the present. He found that, in most of the groves, 10 to 53 fires had occurred each century for the past 1530 to 2000 years (**Figure 53.19c**). The data indicated that each tree had been burned an average of 64 times.

This study established that fires are extremely frequent in the community examined. Because not enough time would pass for large amounts of fuel to accumulate between fires, they were probably of low severity. Partly because of this work, the biologists responsible for managing sequoia groves now set controlled fires or let low-intensity natural fires burn instead of suppressing them immediately. Similarly, studies of disturbance regimes along the Colorado River in southwestern North America inspired land managers to release a huge pulse of water from the reservoirs behind dams on the waterway recently. The flood that resulted was designed to mimic a natural disturbance. According to follow-up studies, the artificial flood appears to have benefited the plant and animal communities downstream.

Succession: The Development of Communities after Disturbance

Severe disturbances remove all or most of the organisms from an area. The recovery that follows is called **succession**. The name was inspired by the observation that certain species succeed others over time. **Primary succession** occurs when a disturbance removes the soil and its organisms as well as organisms that live above the surface. Glaciers, floods, volcanic eruptions, and landslides often initiate primary succession. **Secondary succession** occurs when a disturbance removes some or all of the organisms from an area but leaves the soil intact. Fire and logging are examples of disturbances that initiate secondary succession.

As **Figure 53.20** shows, a sequence of plant communities develops as succession proceeds. Early successional communities are dominated by species that are short lived and small and that disperse their seeds over long distances. Late successional communities are dominated by species that tend to be long lived, large, and good competitors for resources such as light and nutrients. The specific sequence of species that appears over time is called a **successional pathway**. What determines the pattern and rate of species replacement during succession at a particular time and place?

(a) Giant sequoias after a fire

(b) Fire scars in the growth rings

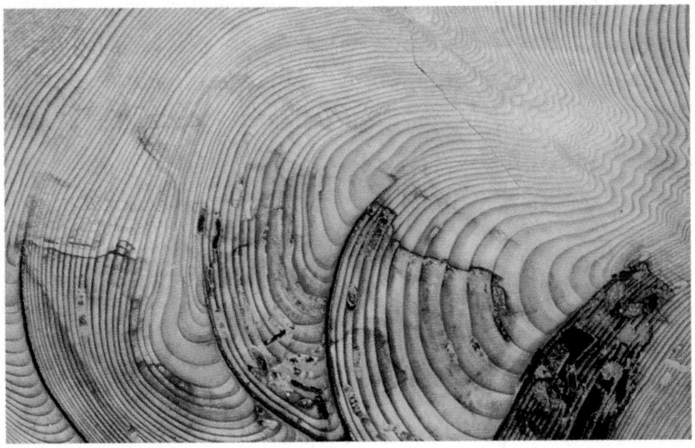

(c) Reconstructing history from fire scars

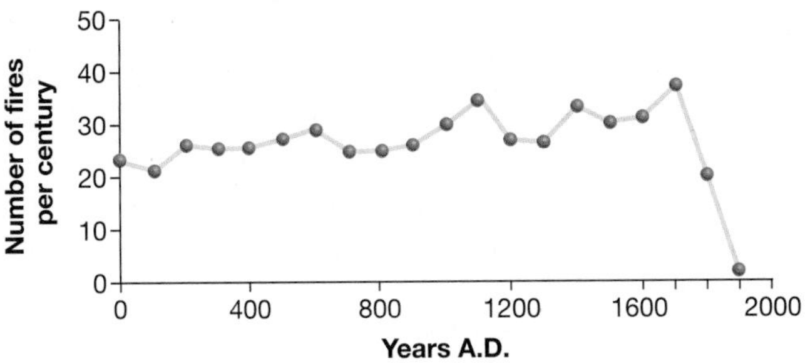
Number of fires per century
0
10
20
30
40
50
0
400
800
1200
1600
2000
Years A.D.

FIGURE 53.19 History of Disturbance in a Fire-Prone Community (a) A giant sequoia grove. (b) Trunk cross section of a giant sequoia; the thick black arcs are fire scars. (c) Because trees form one ring (light band/dark band) every year, researchers can count the rings to determine how often fires have occurred during the last 2000 years.

Old field
Disturbance (plowing) ends, site is invaded by short-lived weedy species.
Pioneering species
Weedy species are replaced by longer-lived herbaceous species and grasses.
Early successional community
Shrubs and short-lived trees begin to invade.
Mid-successional community
Short-lived tree species mature; long-lived trees begin to invade.
Late-successional community
Long-lived tree species mature.
Climax community

FIGURE 53.20 Succession in Midlatitude Temperate Forests Sequence of photos showing how succession leads to the development of a temperate forest from a disturbed state (in this case, an abandoned agricultural field).

Theoretical Considerations Biologists focus on three factors to predict the outcome of succession in a community: (1) the particular traits of the species involved, (2) how species interact, and (3) historical and environmental circumstances such as the size of the area involved and weather conditions. Before going on to consider a detailed case history, we need to explore these factors in more detail.

Species traits, such as dispersal capability and the ability to withstand extreme dryness, are particularly important early in succession. As common sense would predict, recently disturbed sites tend to be colonized by plants and animals with good dispersal ability. When these organisms arrive, however, they often have to endure harsh environmental conditions. These **pioneering species** tend to have "weedy" life histories. (A **weed** is a plant that is adapted for growth in disturbed soils.) Early successional species devote most of their energy to reproduction and little to competitive ability. They have small seeds, rapid growth, and a short life span, and they begin reproducing at an early age. As a result, they have a high reproductive rate. But in addition, they can tolerate severe abiotic conditions, such as high light levels, poor nutrient availability, and drying.

Once colonization is under way, the course of succession tends to depend less on how species cope with aspects of the abiotic environment and more on how they interact with other species. This change occurs because plants that grow early in succession change abiotic conditions in a way that makes the conditions less severe. Because plants provide shade, they reduce temperatures and increase humidity. Their dead bodies also add organic material and nutrients to the soil. As abiotic conditions improve, biotic interactions become more important.

During succession, existing species can have one of three effects on subsequent species: (1) facilitation, (2) tolerance, or (3) inhibition. **Facilitation** takes place when the presence of an early-arriving species makes conditions more favorable for the arrival of certain later species by providing shade or nutrients. **Tolerance** means that existing species do not affect the probability that subsequent species will become established. **Inhibition** occurs when the presence of one species inhibits the establishment of another, as when a plant species that requires high light levels to germinate is inhibited late in succession.

In addition to species traits and species interactions, the pattern and rate of succession depend on the historical and environmental context in which they occur. For example, researchers found that the communities that developed after forest fires disturbed Yellowstone National Park in 1988 depended on the size of the burned patch and how hot the fire had been at that location. Succession is also affected by the particular weather or climate conditions that occur during the process. Variation in weather and climate causes different successional pathways to occur in the same place at different times.

In summary, analyzing species traits, species interactions, and the historical/environmental context provides a useful structure for understanding why particular successional pathways occur. To see this theoretical framework in action, let's examine data on the course of primary succession that has occurred in Glacier Bay, Alaska.

A Case History: Glacier Bay, Alaska An extraordinarily rapid and extensive glacial recession is occurring at Glacier Bay (**Figure 53.21a**). In just 200 years, glaciers that once filled the bay have retreated approximately 100 km, exposing extensive tracts of barren glacial sediments to colonization. Because of this event, Glacier Bay has become an important site for studying succession.

Figure 53.21a shows the plant communities found in the area. The oldest sites are dense forests of Sitka spruce and western hemlock. Areas that have been deglaciated for about 100 years are inhabited by scattered spruce trees and dense thickets of a shrub called Sitka alder. Sites that have been deglaciated for 45 to 80 years are also covered with dense alder thickets, but the emergent trees are primarily cottonwood. Locations that have been ice free for 20 years or less do not have a continuous plant cover. Instead, they host scattered individuals of willow and a small shrub called *Dryas*.

These observations inspired a hypothesis for the pattern of succession in Glacier Bay: With time, the youngest communities of *Dryas* and willow succeed to alder thickets, which subsequently become dense spruce-hemlock forests. A recent study has challenged this hypothesis, however. Researchers who reconstructed the history of each community by studying tree rings found that three distinct successional pathways have occurred (**Figure 53.21b**): (1) In the lower part of the bay, soon after the ice retreated, Sitka spruce began growing and quickly formed dense forests. Western hemlock arrived after spruce and is now common in the understory. (2) At middle-aged sites in the upper part of the bay, alder thickets were dominant for several decades, and spruce is just beginning to become common. These forests will probably never be as dense as the ones in the lower bay, however, and there is no sign that western hemlock has begun to establish itself. (3) In contrast, the youngest sites in the uppermost part of the bay may be following a third pathway. Alder thickets became dominant fairly early, but spruce trees are scarce. Instead, cottonwood trees are abundant.

These data present a challenge: How do species traits, species interactions, and dispersal patterns interact to generate the three observed successional pathways?

Species traits may be especially important in explaining certain details about the successional pathways. Western hemlock, for example, is abundant at older sites but largely absent from young ones. This is logical, because its seeds germinate and grow only in soils containing a substantial amount of organic matter and because the young trees can tolerate deep shade but not bright sunlight. Western hemlock's intolerance of early-successional conditions explains why none of the three pathways began with colonization by this species.

(a) Hypothesis 1: Only one successional pathway occurs in Glacier Bay.

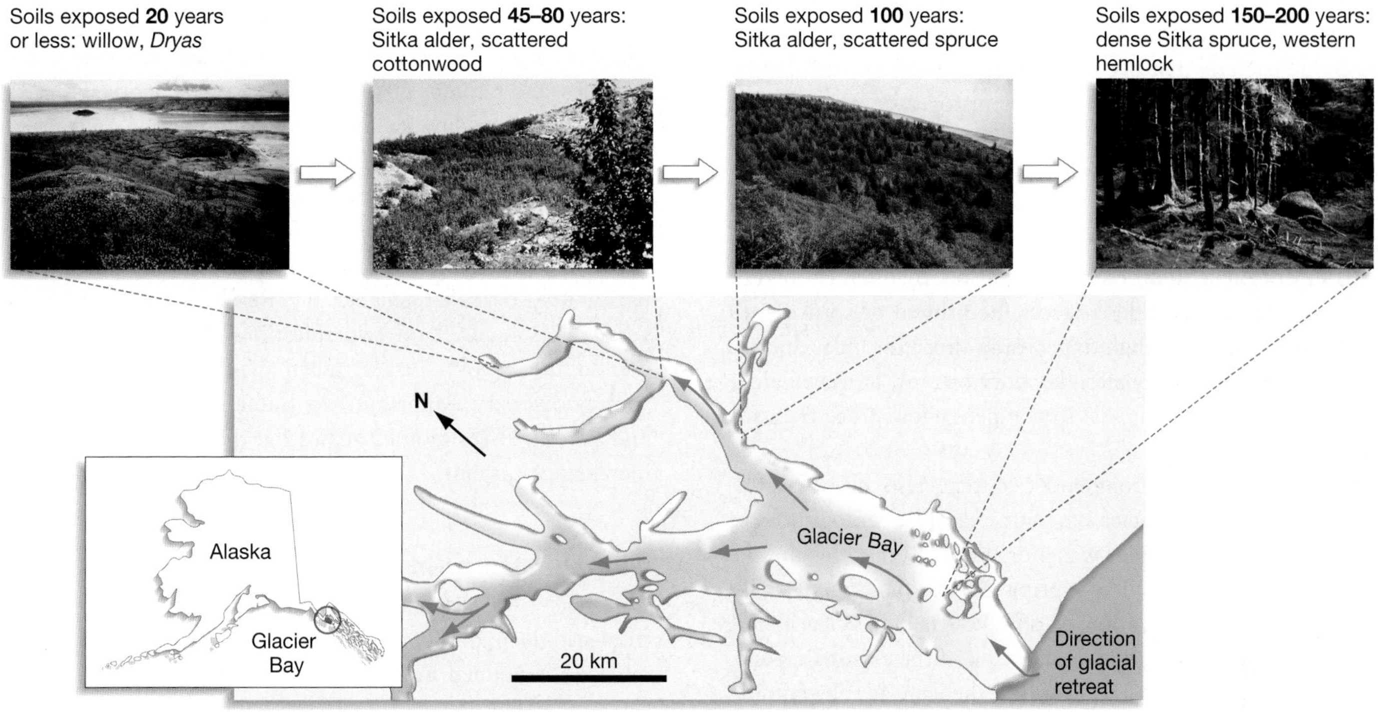
Soils exposed **20** years or less: willow, *Dryas*
Soils exposed **45–80** years: Sitka alder, scattered cottonwood
Soils exposed **100** years: Sitka alder, scattered spruce
Soils exposed **150–200** years: dense Sitka spruce, western hemlock
N
Alaska
Glacier Bay
Glacier Bay
20 km
Direction of glacial retreat

(b) Hypothesis 2: Three distinct successional pathways occur in Glacier Bay.

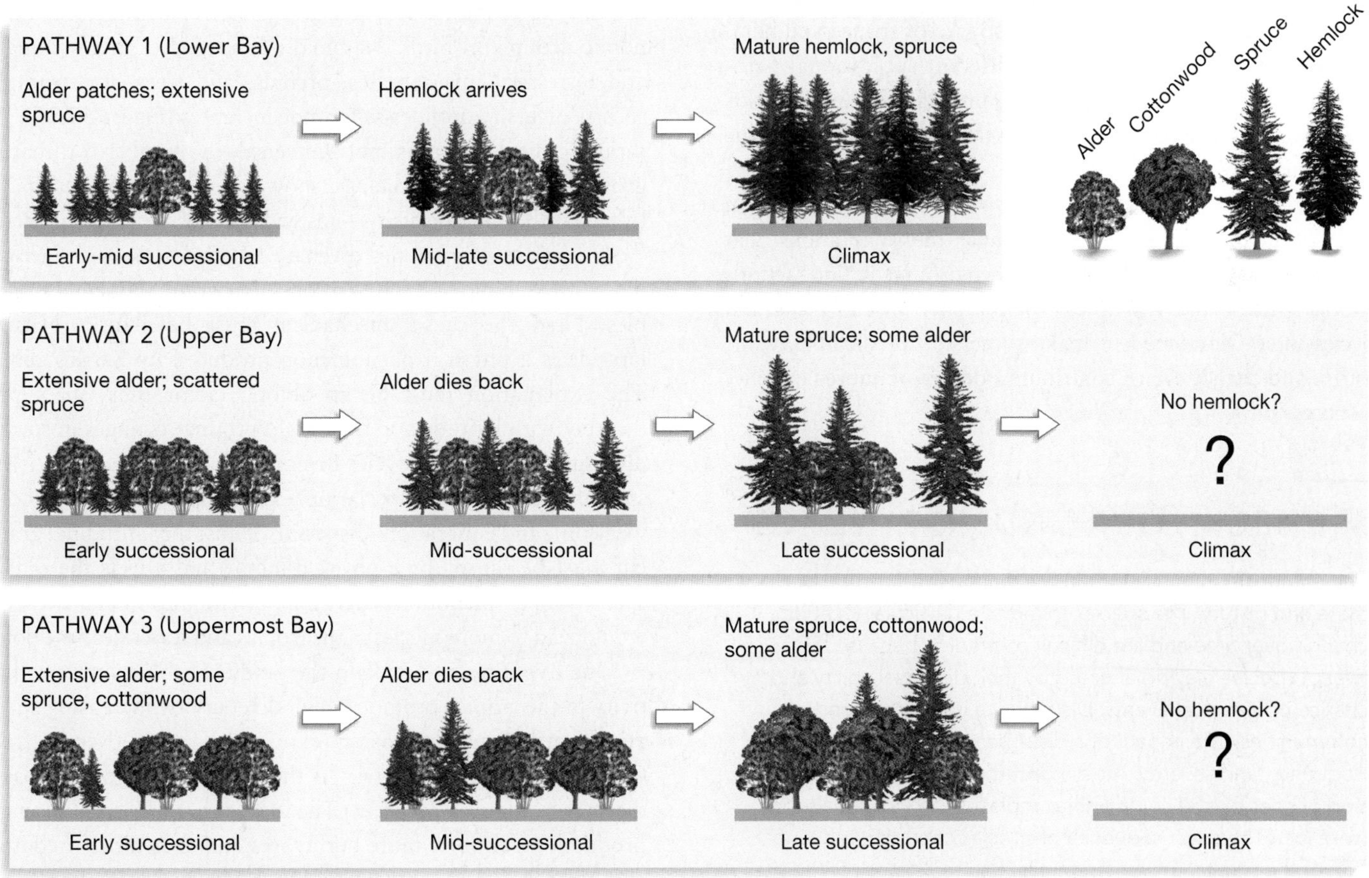
PATHWAY 1 (Lower Bay)
Alder patches; extensive spruce
Hemlock arrives
Mature hemlock, spruce
Early-mid successional
Mid-late successional
Climax
Alder
Cottonwood
Spruce
Hemlock
PATHWAY 2 (Upper Bay)
Extensive alder; scattered spruce
Alder dies back
Mature spruce; some alder
No hemlock?
?
Early successional
Mid-successional
Late successional
Climax
PATHWAY 3 (Uppermost Bay)
Extensive alder; some spruce, cottonwood
Alder dies back
Mature spruce, cottonwood; some alder
No hemlock?
?
Early successional
Mid-successional
Late successional
Climax

FIGURE 53.21 Evidence for Multiple Successional Pathways at Glacier Bay
(a) Several distinct plant communities are found in Glacier Bay, Alaska. The first hypothesis about succession at Glacier Bay proposed that these communities represent stages in the same successional pathway. **(b)** Recent data indicate that three successional pathways exist.

Species interactions have been important in all three pathways. For example, research has shown that alder facilitates the growth of Sitka spruce. The facilitating effect occurs because symbiotic bacteria that live inside nodules on the roots of alder convert atmospheric nitrogen (N_2) to nitrogen-containing molecules that alder can use to build proteins and nucleic acids. Although spruce trees are capable of invading and growing without the presence of alder, they grow faster when alder stands have added nitrogen to the soil.

Competition is another important species interaction. For example, shading by alder reduces the growth of spruce until spruce trees are tall enough to protrude above the alder thicket. Once the spruce trees breach the alder canopy, however, alder dies out, because it is unable to compete with spruce trees for light.

Historical and environmental context also clearly influences succession at Glacier Bay. For example, geologists have found evidence that the ice was more than 1100 m thick in the upper part of Glacier Bay during the mid-1700s. Because forests grow to an elevation of only 700 m or 800 m in this part of Alaska, the glacier eliminated all of the existing forests. In the lower part of the bay, however, the ice was substantially thinner. As a result, some forests remained on the mountain slopes beyond the ice. As the glacier retreated from these areas, the forests on the slopes provided a source of spruce and hemlock seeds and set a dramatically different successional pathway in motion. In this way, environmental context—in this case, distance to existing forests—helped determine how the community developed.

To summarize, successional pathways are determined by an array of factors. These factors include the adaptations that certain species have to their abiotic environment, interactions among species, and the history of the site. Species traits and species interactions tend to make succession predictable, while history and chance events contribute a degree of unpredictability to succession.

✓CHECK YOUR UNDERSTANDING

The composition of a community is primarily a function of climate and history. The species found at a particular location change over time and are difficult to predict, because the arrival of species may be affected by their dispersal ability and by chance or historical events. Disturbance is a normal part of communities. The impact of a disturbance depends on its type, frequency, and severity. After a disturbance occurs, a succession of species and communities replaces the individuals that were lost. The exact sequence of species observed is a function of their traits, their interactions, and the history of the site. You should be able to cite specific observational or experimental evidence that supports each of these statements.

53.3 Species Diversity in Ecological Communities

The diversity of species present is a key feature of biological communities, and it can be quantified in two ways. **Species richness** is a simple count of how many species are present in a given community. **Species diversity**, in contrast, is a weighted measure that incorporates a species' relative abundance as well as its presence or absence (see **Box 53.1**). At a global scale, species abundance data are often not available. In such cases, ecologists sometimes use the two terms interchangeably. Research on richness and diversity has focused on two questions: First, why are some communities more species rich than others? Second, why is species diversity important? As human populations increase and species losses mount, answering these questions becomes increasingly urgent.

Global Patterns in Species Diversity

Biologists who began cataloguing the flora and fauna of the tropics in the mid-1800s immediately recognized that these communities contained many more species than did temperate or subarctic environments. Data compiled in the intervening years have confirmed the existence of a latitudinal gradient in species diversity for communities as a whole as well as for many taxonomic groups. In birds, mammals, fish, reptiles, many aquatic and terrestrial invertebrates, orchids, and trees, for example, species diversity declines as latitude increases (**Figure 53.23**). Although this pattern is not universal—a number of marine groups, as well as shorebirds, show a positive relationship between latitude and diversity—it is widespread. Why does it occur?

To explain why species diversity might decline with increasing latitude, biologists have to consider two fundamental principles. First, the causal mechanism must be abiotic, because latitude is a physical phenomenon produced by Earth's shape. The explanation must be an abiotic factor that varies predictably with latitude and that could produce changes in species diversity. Second, the species diversity of a particular area is the sum of four processes: speciation, extinction, immigration (colonization), and emigration (dispersal). Thus, the latitudinal gradient must be caused by a physical factor that affects the rate of speciation, extinction, immigration, or emigration in a way that would lead to more species in the tropics and fewer near the poles.

One hypothesis to explain the gradient is that high productivity in the tropics promotes high diversity by increasing speciation rates and decreasing extinction rates. (Recall from Chapter 50 that productivity is the total amount of photosynthesis per unit area per year.) The idea is that increased biomass production supports more herbivores and thus more predators and parasites and scavengers. In addition, speciation rates should increase when niche differentiation occurs within populations of herbivores, predators, parasites, and scavengers.

BOX 53.1 Measuring Species Diversity

To measure species diversity, a biologist could simply count the number of species present in a community. The problem is that simple counts provide an incomplete picture of diversity. The relative abundance of species is also an important component of diversity. To grasp this point, consider the composition of the three hypothetical communities shown in **Figure 53.22**. These communities are nearly identical in species composition but differ greatly in the relative abundance of each species.

These data can be used to compare two measures of species diversity. Species richness is simply the number of species found in a community. In this case, communities 1 and 2 have equal species richness and community 3 is lower in richness by one species. It is important to note, however, that communities 2 and 3 have similar relative abundances of each species, or what biologists call high *evenness*. Community 1, in contrast, is highly uneven. Fifty-five percent of the individuals in community 1 belong to species A, and other species are relatively rare. An uneven community has lower effective diversity than its species richness would indicate.

To take evenness into account, other diversity indices have been developed. A simple example is the **Shannon-Weaver index**, given by the following equation:

$$H' = -\Sigma p_i \log(p_i)$$

In this equation, p_i is the proportion of individuals in the community that belong to species i. The index is summed over the i species in the study. The Shannon-Weaver index for the three hypothetical communities is shown at the bottom of Figure 53.22. Notice that while communities 1 and 2 have the same species richness, community 2 has higher diversity because of its greater evenness. Community 3 has lower species richness than community 1 but higher diversity.

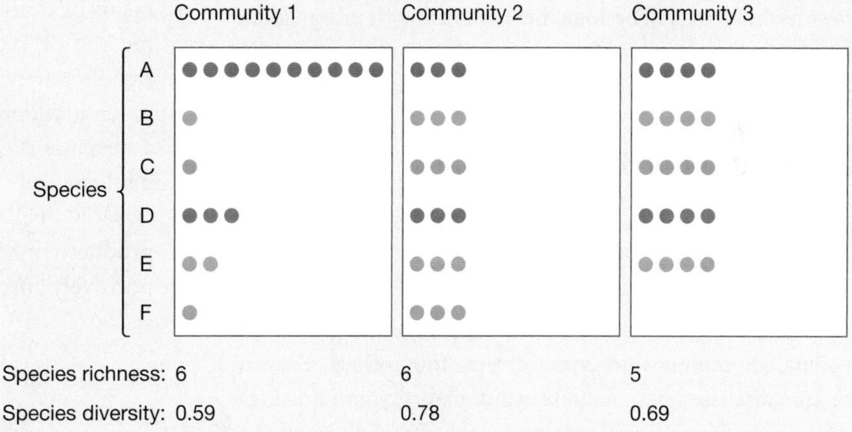

FIGURE 53.22 Species Diversity Can Be Quantified

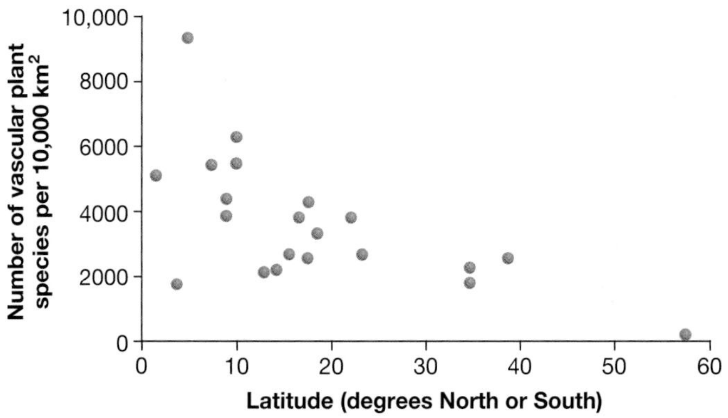

FIGURE 53.23 There Is a Strong Latitudinal Gradient in Species Diversity
One example of a latitudinal gradient in species diversity. The number of plant species in a 10,000-km^2 area declines from up to 10,000 in equatorial regions to less than 1000 at very high latitudes.

Although this **high-productivity hypothesis** is supported by the global-scale correlation between productivity and diversity, experimental studies challenge it. For example, researchers who add fertilizer to aquatic or terrestrial communities routinely observe significant increases in productivity but decreases in diversity.

Based on results such as these, researchers have concluded that productivity alone is probably not a sufficient explanation for the higher species diversity in the tropics. Research continues, however, and investigators have recently begun to focus on an abiotic factor that is correlated with productivity: temperature. Biologists who analyzed data on gastropods and other marine invertebrates have documented a strong correlation between the temperature of marine waters and species diversity. This observation has inspired the *energy hypothesis*: Temperature causes changes in species diversity by influencing productivity *and* the ability of organisms to tolerate the physical conditions in a region.

A third hypothesis is that tropical regions have had more time for speciation to occur than other regions have. Temperate and arctic latitudes were repeatedly scoured by ice sheets over the last 2 million years, but tropical regions were not. Recent data suggest, however, that tropical forests were dramatically reduced in size by widespread drying trends during the ice ages. Existing forests may be much younger than originally thought. If so, then the contrast in the age of northern and southern habitats may not be enough to explain the dramatic difference in species diversity.

A fourth hypothesis was inspired by the observation that species diversity is much higher in mid-successional communities than in pioneer or mature communities. The **intermediate disturbance hypothesis** holds that regions with a moderate type, frequency, and severity of disturbance should have high species richness and diversity. The logic here is that, with intermediate levels of disturbance, communities will contain pioneering species as well as species better adapted to late-successional conditions. For example, recent studies have confirmed that tree falls and canopy gaps occur regularly in tropical forests and that fires occur in these biomes occasionally. As yet, however, there are no convincing data showing that intermediate levels of disturbance are more likely to occur in the tropics than they are at higher latitudes.

To summarize, there is no simple answer to the question of why some communities are more diverse than others. Each of the factors discussed here may influence diversity, but no single one offers a convincing explanation for the global diversity gradient. According to the data available at present, the presence of high energy inputs seems most responsible for communities with high species diversity.

The Role of Diversity in Ecological Communities

Research on species richness and species diversity has recently taken a practical turn: Does the number of species present affect the productivity and stability of communities? If so, then the rapid loss of diversity that is occurring now is almost certain to have serious consequences for biological communities all over the planet.

Does Species Richness Affect Productivity?

Primary productivity is the total amount of photosynthesis per unit area per year. **Net primary productivity (NPP)** is the amount of primary productivity that ends up in biomass. NPP is a key component of communities, because it represents the amount of food that is available to consumers and **detritivores**—species that eat dead material.

One of the more robust conclusions emerging from recent work in community ecology is that, at least in grassland biomes, productivity increases with increasing species richness. Consider an experiment in which researchers grew seedlings from 24 species native to North American prairies and planted them into a total of 147 study plots. Each plot received a similar number of individuals but a different number of species: 1, 2, 4, 6, 8, 12, or 24. The research team allowed the plants to grow for a year and weeded the plots to remove any invading species. In the second year, the team measured productivity by quantifying the area of the plot covered by plants. As **Figure 53.24** shows, plant cover was higher in the more species-rich plots. Similar results have been obtained in other types of grassland habitats and when the investigators measure aboveground biomass instead of total cover. The more species in a grassland, the more biomass is produced.

Does Species Richness Affect Community Stability?

When biologists refer to the stability of a community, they mean its ability to (1) withstand disturbance without changing and (2) recover to former levels of productivity or species richness after a disturbance. **Resistance** is a measure of how much a community is affected by a disturbance. **Resilience** is a measure of how quickly a community recovers following a disturbance.

How does species richness affect resistance, resilience, and productivity? Several biologists have proposed that diversity is positively correlated with stability, meaning that highly diverse

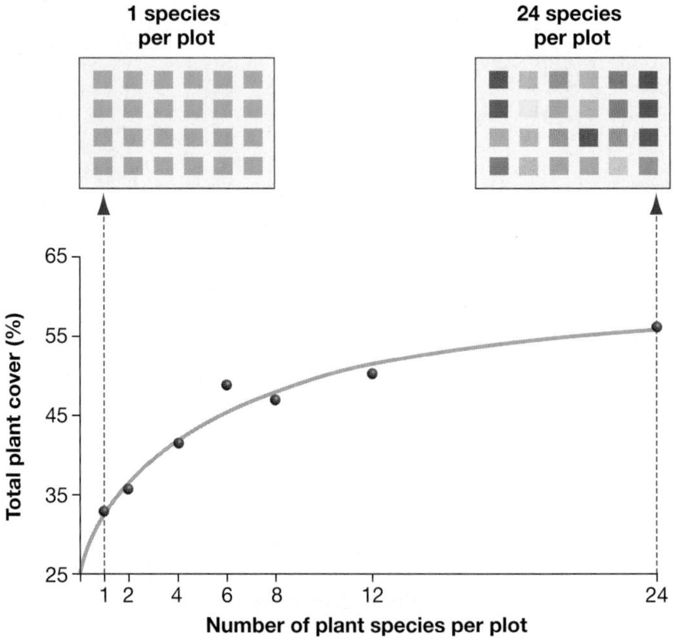

FIGURE 53.24 Evidence That Species-Rich Communities Are More Productive

Total plant cover is defined as the percentage of a study plot's area that is covered by plants. It is interpreted as an index of productivity.

QUESTION Extrapolating from these data, what would total plant cover be in a plot that contained 53 plant species?

communities are more resistant and more resilient than are less diverse communities. The logic behind this hypothesis is that when a large number of species are present, some are likely to be redundant in their role in the community. For example, there are likely to be many types of herbivores, carnivores, pollinators, and so on. Because of this redundancy, wiping out one or a few species is unlikely to affect the community as a whole profoundly.

To test the hypothesis that diversity leads to high resistance and resilience, biologists followed up on a natural experiment. In 1987–1988 a severe drought hit sites in Minnesota where a research team had been measuring species richness and other characteristics in a series of permanent study plots. When the drought ended, the team was able to ask whether resistance and resilience correlated with the number of species that existed before the disturbance. Some of their results are shown in **Figure 53.25**. The graph documents the change in total biomass that occurred from the year prior to the drought to the height of the drought. This quantity reflects how resistant the community is to disturbance. A completely resistant community would show no change in biomass. As predicted, drought resistance appeared to be higher in more diverse communities than in less diverse ones. In addition, the group analyzed the change in biomass in each plot four years after the drought versus biomass prior to the drought. This analysis focused on how resilient the community is. A completely resilient community would recover quickly from the disturbance and have the same biomass at both times. The data indicated that most plots containing five or fewer species showed a significant lowering of biomass after the disturbance, indicating that they had not recovered. But in all of the plots that contained more than five species, biomass after the drought was the same as biomass prior to the disturbance.

These experiments support the hypothesis that species richness affects how communities function. In North American grasslands, at least, communities that are more diverse appear to be more productive, more resistant, and more resilient than communities that are less diverse. Why this pattern occurs is not clear, however. More experiments are needed to confirm these results and determine the mechanisms responsible for these patterns.

Question: Are more species-rich communities more stable than less-diverse communities?

Hypothesis: Resistance to disturbance should increase with increasing species richness.

Null hypothesis: No relationship exists between resistance and species richness.

Experimental setup:

Compare biomass of experimental plots before drought and at peak of drought. (This was a natural experiment—severe drought just happened to occur during study.)

Prediction: Plots that were more species rich before the drought will be more resistant to change.

Prediction of null hypothesis: All the plots will have similar resistance regardless of species richness before the drought.

Results:

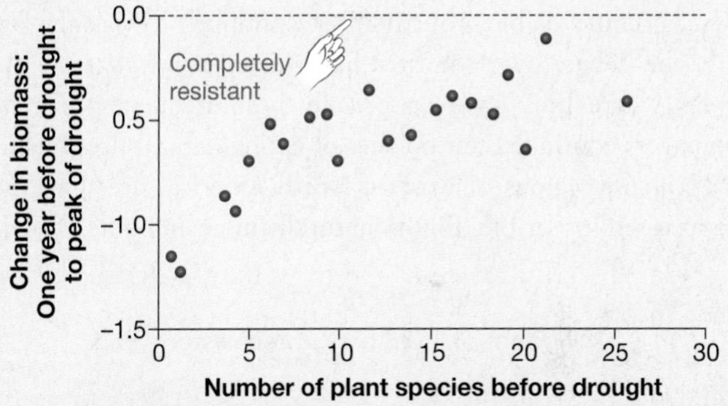

Conclusion: Resistance to disturbance increases with increasing species richness.

FIGURE 53.25 Evidence That Species-Rich Communities Resist Disturbance

Data support the hypothesis that diverse communities are more stable, meaning they change less during a disturbance, than less-diverse communities are. QUESTION The 0.0 values on the vertical graph axes are labeled "Completely resistant." Why?

ESSAY Let-It-Burn Policies

The summer of 1988 was extremely dry in the western United States. In the Greater Yellowstone Area (GYA), which includes Yellowstone National Park, almost no rain fell from mid-July through August. Thunderstorms rolled through with ample lightning but no rain. As a result of the severe conditions, fires ignited. A great deal of fuel (in the form of dry brush) was available in forested areas of the GYA, because park managers had actively suppressed forest fires since the park's founding in 1872. But in 1988 park personnel allowed the fires to burn. Research had shown that the GYA experienced frequent fires

(Continued on next page)

(Essay continued)

before the arrival of European settlers, and that many of the region's biological communities depend on fires to thin out dense stands of trees and allow regeneration. By late July, however, the blazes had grown so large that the National Park Service decided to suppress them. The conflagration continued to grow, however. By September 1 the fire perimeter enclosed approximately 353,800 ha—an area substantially larger than the state of Rhode Island (**Figure 53.26**).

The onset of winter, combined with a $120-million firefighting effort, eventually extinguished the fires. Cold, wet weather did not extinguish a firestorm of controversy, however. Critics of the let-it-burn policy decried the destruction of the National Park Service's crown jewel.

But was the park destroyed? The species traits of trees that are particularly abundant in the GYA suggest that the answer is no. For example, both aspen and lodgepole pine require fire in order to reproduce. Aspen cannot tolerate shade and does not sprout beneath its own canopy. Lodgepole pine is also shade intolerant and has cones that are glued shut with resin. The cones open and shed their seeds only in the heat of a fire.

Since 1988 the regeneration promised by biologists has started to become visible. Tourists have continued to visit the park, and the debate over the fires has quieted somewhat. In other areas where fires have historically been frequent, most land managers continue their policies of letting natural fires burn.

In many regions of western North America, fires have been suppressed for so long that unnaturally large amounts of fuel, in the form of dead and drying trees, have built up. Once fires start in an area like this, they can quickly reach the proportions of the Yellowstone fires. In some of the sequoia preserves introduced earlier in this chapter, for example, enough fuel has built up that an uncontrolled fire would threaten to kill even the largest, oldest trees. To remedy the situation, biologists are attempting to reintroduce fire through controlled burnings. The idea is that setting and managing a series of fires when conditions are moist will gradually reduce the amount of fuel and reduce the danger of a catastrophic fire. In many cases, land managers are also being urged to use logging as a way to reduce the amount of fuel in fire-prone forests quickly. This tactic has been disputed, however. Logging opponents maintain that, for logging operations to be economical, valuable old trees will have to be removed and smaller trees—which have little economic value but pose high fire danger—will be left.

Controversy over how to manage natural disturbances is almost certain to continue.

Management of these forests is also complicated by the dramatic increase in the number of homes that are being built in fire-prone areas. In addition, efforts to fight forest fires aggressively take an enormous cost in money and lives, with sometimes little gain in reducing the actual number of hectares burned. Controversy over how to manage natural disturbances is almost certain to continue.

(a) Extent of 1988 fires in Yellowstone Park

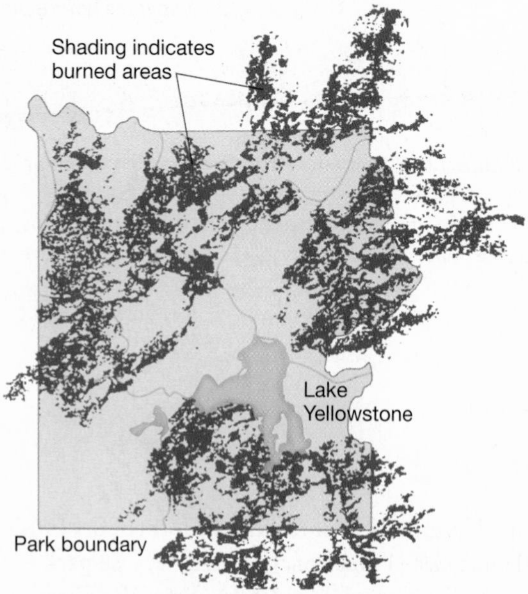
Shading indicates burned areas
Lake Yellowstone
Park boundary

(b) If fuel is abundant, fires burn canopy trees as well as ground cover.

FIGURE 53.26 Extensive Forest Fires in and around Yellowstone Park, 1998
(a) Map of the Greater Yellowstone Area, showing the regions that burned in the 1988 fires. **(b)** One of the blazes in progress.

CHAPTER REVIEW

Summary of Key Concepts

- **Interactions among species, such as competition and consumption, have two main outcomes: (1) They affect the distribution and abundance of the interacting species, and (2) they are agents of natural selection and thus affect the evolution of the interacting species.**

A community is an assemblage of interacting species.

To categorize the different types of interactions that occur among species, biologists consider whether each participant experiences a net fitness cost or benefit from the interaction. These costs and benefits depend on the conditions that prevail at a particular time and place and may change through time.

Competition is a −/− interaction that occurs when the niches of two species overlap—meaning they use the same resources. Competition may result in the complete exclusion of one species. It may also result in niche differentiation, in which competing species evolve traits that allow them to exploit different resources or live in different areas.

Consumption is a +/− interaction that occurs when consumers eat prey, which resist through standing defenses or inducible defenses. Predators are efficient enough to reduce the size of many prey populations. Levels of herbivory are relatively low in terrestrial ecosystems, however, because predation and disease limit herbivore populations, because plants provide little nitrogen, and because many plants contain toxic compounds or other types of defenses. Parasites generally spend all or part of their life cycle in or on their host (or hosts) and usually have traits that allow them to escape host defenses. In turn, hosts have evolved counteradaptations that help fight off parasites.

Mutualism is a +/+ interaction that provides participating individuals with food, shelter, transport of gametes, or defense against predators. For each species involved, the costs and benefits of a mutualism may vary over time and from place to place.

Web Tutorial 53.1 Life Cycle of a Malaria Parasite

- **The assemblage of species found in a biological community changes over time and is primarily a function of climate and chance historical events.**

Within at least some biomes, there is a strong positive relationship between the number of species present and the productivity and stability of the community.

Ecologists have debated whether communities are fixed, predictable entities or simply places where the distributions of various species overlap. Historical and experimental evidence support the view that communities are dynamic rather than static and that their composition is neither entirely predictable nor stable over time.

In addition to climate, the composition of a community is influenced by disturbance. In extreme cases, disturbance may remove all organisms and all soil from a large area. Each community has a characteristic disturbance regime—meaning a type, severity, and frequency of disturbance that it experiences. Three types of factors influence the pattern of succession that occurs after a disturbance. First, the historical and environmental context of the site affects which species are available to join the resulting communities. The dispersal ability of different species also affects their availability. Second, a species' physiological traits influence the kinds of abiotic environmental conditions it can tolerate and dictate when it can successfully join a community. Third, interactions among species influence if and when a species appears during succession.

Web Tutorial 53.2 Primary Succession

- **Species diversity is high in the tropics and lower toward the poles. The mechanism responsible for this pattern is still being investigated.**

One of the most widely studied patterns in community ecology is the latitudinal gradient in species diversity. Within most specific taxonomic groups and within communities as a whole, species diversity declines from the equator to the poles. Recent research suggests that the high species diversity observed in the tropics results from a combination of higher rates of speciation and lower rates of extinction.

- **Within at least some biomes, there is a strong positive relationship between the number of species present and the productivity and stability of the community.**

Experimental evidence supports the hypothesis that, at least in grassland biomes, the productivity and stability of communities increase as a function of species richness. If further research confirms this conclusion, the rapid loss of species occurring now will have profound impacts on the integrity of communities worldwide.

Questions

Content Review

1. What is competitive exclusion?
   a. the evolution of traits that reduce niche overlap and competition
   b. interactions that allow species to occupy their fundamental niche
   c. the degree to which the niches of two species overlap
   d. the claim that species with the same niche cannot coexist

2. What is niche differentiation?
   a. the evolution of traits that reduce niche overlap and competition
   b. interactions that allow species to occupy their fundamental niche
   c. the degree to which the niches of two species overlap
   d. the claim that species with the same niche cannot coexist

3. Why is the phrase "coevolutionary arms race" an appropriate way to characterize the long-term effects of species interactions?
   a. Both plants and animals have evolved weapons for defense that are so effective that many plants are not eaten and predators cannot reduce prey populations to extinction.
   b. Adaptations that give one species a fitness advantage in an interaction are likely to be countered by adaptations in the other species that eliminate this advantage.
   c. In all species interactions except for mutualism, at least one species loses (suffers decreased fitness).
   d. Even mutualistic interactions can become parasitic if conditions change. As a result, interacting species are always "at war."
4. Why are inducible defenses advantageous?
   a. They are always present—thus, an individual is always able to defend itself.
   b. They make it impossible for a consumer to launch surprise attacks.
   c. They result from a coevolutionary arms race.
   d. They make efficient use of resources, because they are produced only when needed.
5. Which of the following is *not* correlated with species diversity?
   a. latitude
   b. productivity
   c. longitude
   d. resilience
6. What is net primary productivity?
   a. an individual's lifetime reproductive success (lifetime fitness)
   b. an individual's average annual reproductive success
   c. the total amount of photosynthesis that occurs in an area of a given size per year
   d. the amount of energy that is stored in standing biomass per year

Conceptual Review

1. The text claims that species interactions are conditional and dynamic. Do you agree with this statement? Why or why not? Cite specific examples to support your answer.
2. Species interactions affect the distribution and abundance of populations. Summarize experimental evidence that population size for snowshoe hares depends on both predation rates by lynx and competition for food among hares.
3. State three hypotheses that have been proposed to explain the low level of herbivory in terrestrial plant communities. Are these hypotheses mutually exclusive? (In other words, can more than one be correct?) Explain why or why not.
4. State the hypotheses that Clements and Gleason proposed to explain community composition and structure. State the predictions that each of these hypotheses makes with respect to (a) the presence and impact of keystone species, (b) changes in the distribution of the species in a particular community over time, and (c) the communities that should develop at sites where abiotic conditions are identical. Which hypothesis appears to be more accurate?
5. What is a disturbance? List five examples of disturbance. Compare and contrast their effects. For each type of disturbance, compare and contrast the consequences of high-frequency and low-frequency disturbances and high and low severity of disturbances.
6. Describe the latitudinal gradient in species diversity that exists for most taxonomic groups. Discuss the pros and cons of one hypothesis to explain this pattern.

Group Discussion Problems

1. Some insects harvest nectar by chewing through the wall of the structure that holds the nectar As a result, they obtain a nectar reward, but pollination does not occur. Suppose that you observed a certain bee species obtaining nectar in this way from a particular orchid species. Over time, how would you expect the characteristics of the orchid population to change in response to this bee behavior?
2. Using this chapter's information on fire regimes in giant sequoia groves, propose a management plan for Sequoia National Park. Explain the logic behind your plan.
3. Suppose that a two-acre lawn on your college's campus is allowed to undergo succession. Describe how species traits, species interactions, and the site's history might affect the community that develops.
4. Design an experiment to test the hypothesis that increasing species richness increases a community's productivity and ability to resist disturbance and recover from disturbance. Use a biome other than grasslands.

Answers to Multiple-Choice Questions 1. b; 2. d; 3. a; 4. d; 5. c; 6. d

www.prenhall.com/freeman is your resource for the following: Web Tutorials; Online Quizzes and other Online Study Guide materials; Answers to Conceptual Review Questions; Solutions to Group Discussion Problems; Answers to Figure Caption Questions and Exercises; and Additional Readings and Research.

Ecosystems

54

The Qori Kalis glacier in the Andes Mountains of Peru in the year 2000. In 1978 the glacier extended past the edge of the lake in the lower left-hand corner of this photograph. Global warming is causing dramatic reductions in the size of glaciers and ice fields in many locations throughout the world.

KEY CONCEPTS

- An ecosystem has four components: (1) the abiotic environment, (2) primary producers, (3) consumers, and (4) decomposers. These components are linked by the movement of energy and nutrients.
- The productivity of terrestrial ecosystems is limited by warmth and moisture, while nutrient availability is the key constraint in aquatic ecosystems. As energy flows from producers to consumers and decomposers, much of it is lost.
- To analyze nutrient cycles, biologists focus on the nature of the reservoirs where elements reside and on how quickly elements move between reservoirs.
- Humans are causing large, global changes in the abiotic environment. The burning of fossil fuels has led to rapid global warming. Extensive fertilization is increasing productivity and causing pollution.

An **ecosystem** consists of the organisms that live in an area together with their physical, or abiotic, environment. With the explosive growth of human populations, dramatic changes are occurring in both the biotic and abiotic components of ecosystems around the globe. Extinctions and other effects on the biotic elements of ecosystems are explored in Chapter 55; this chapter analyzes changes to the chemical and physical characteristics of the environment. Human impacts on the abiotic environment include global warming, acid rain, the development of a hole in the ozone layer over Antarctica, the formation of anaerobic "dead zones" in the oceans due to nitrate pollution, and the development of algal blooms and eutrophication in lakes as a result of phosphate pollution.

This chapter has three goals: to introduce (1) what ecosystem ecology is, (2) how biologists study ecosystems, and (3) how human activities are affecting ecosystems. One of its most fundamental tasks is to explore how organisms interact with the physical environment to produce an ecosystem. The relevant abiotic components include energy and nutrients. Biologists analyze how the energy in sunlight dissipates as it flows through an ecosystem, and how atoms and molecules cycle between the abiotic and biotic worlds. Because human activities are adding massive amounts of energy to ecosystems, disrupting nutrient cycles, and changing the chemistry of lakes, oceans, and the atmosphere, ecosystem ecology has taken on an increasingly applied role in biological science.

How does an ecosystem differ from a community? In most cases, ecosystems are composed of multiple communities along with their chemical and physical environments. For example, biologists who study lakes recognize a number of communities within a lake. There is a distinct community of interacting species along the lake bottom, another community near the surface of the water, and others in other locations (see Chapter 50). Those different communities are studied as a unit, called the lake ecosystem, because energy and matter flow among them.

This chapter introduces ecosystem studies with a look at how energy flows among the components of an ecosystem; it continues by exploring how carbon, nitrogen, and other key elements cycle through organisms, sediments, the oceans, and the atmosphere; and it ends by analyzing how humans are affecting the abiotic environment. Farming, logging, suburbanization, and other human activities are having a massive impact on energy flows and nutrient cycles. Understanding ecosystem ecology is fundamental to managing the future of our planet.

54.1 Energy Flow and Trophic Structure

If an ecosystem is thought of as an economy, then energy is its currency. As **Figure 54.1** shows, ecosystems have four components that are linked by a flow of energy: (1) the abiotic environment, (2) primary producers, (3) consumers, and (4) decomposers.

A **primary producer** is an **autotroph** ("self-feeder"), which is any organism that can synthesize its own food—from inorganic sources (Chapter 10). In most ecosystems, primary producers use solar energy to manufacture their own food via photosynthesis. But in some ecosystems, such as the deep-sea hydrothermal vents (introduced in Chapter 2) and the iron-rich rocks deep below Earth's surface (explored in Chapter 27), bacteria use the chemical energy contained in inorganic compounds such as hydrogen (H_2), methane (CH_4), or hydrogen sulfide (H_2S) to make food.

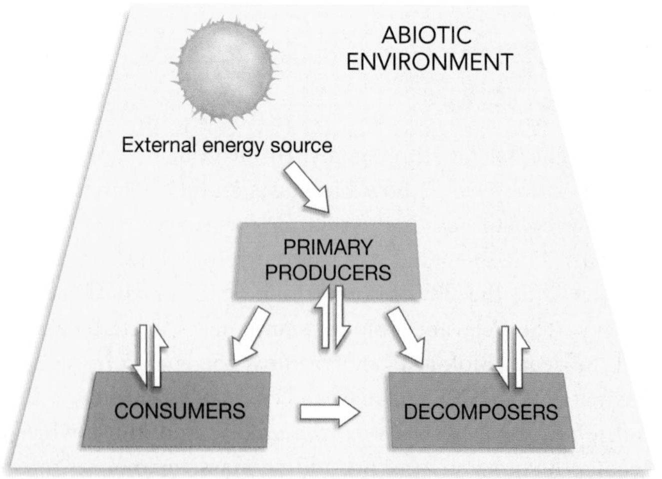

FIGURE 54.1 The Four Components of an Ecosystem Interact
The arrows represent energy. A similar diagram could be drawn to represent the flow of nutrients among the abiotic environment, primary producers, consumers, and decomposers.

Primary producers form the basis of ecosystems by transforming the energy in sunlight or inorganic compounds into the chemical energy stored in sugars. Primary producers use the chemical energy at their disposal in two ways: Most is used for maintenance or respiratory costs, but some supports growth and reproduction. The energy that is invested in new tissue is called **net primary productivity (NPP).**

Net primary productivity represents the amount of energy that is available to the second component of an ecosystem: consumers. **Consumers** eat other organisms. **Herbivores** are consumers that eat plants; **carnivores** are consumers that eat animals. **Decomposers**, or *detritivores*, are consumers that obtain energy by feeding on the dead remains of other organisms or waste products. Decomposers form the third component of ecosystems. The fourth and final component is the abiotic environment, which includes the soil, the climate, the atmosphere, and the particulate matter and solutes in water. Energy moves from the Sun or inorganic compounds to consumers, decomposers, and the abiotic environment.

Because the amount of energy available to consumers and decomposers depends on NPP, biologists have focused on that as a key topic of research. Let's consider two of the most fundamental questions about net primary productivity: How does it vary among the world's ecosystems, and what happens to it?

Global Patterns in Productivity

Figure 54.2 summarizes data on NPP from around the globe. To interpret this map, begin by focusing on terrestrial ecosystems. A quick look at the color key should convince you that the terrestrial ecosystems with highest productivity are located in the wet tropics. Notice that, with the exception of the world's major deserts, NPP on land declines from the equator toward the poles. Productivity patterns in marine ecosystems are somewhat different, however. Marine productivity is highest along coastlines, and it can be as high near the poles as it is in tropics. The oceanic zones, introduced in Chapter 50, in contrast, have extremely low NPP. Typically, a square meter of open ocean produces a maximum of 35 gr (1.2 oz) of organic matter each year. In terms of productivity, the open ocean is a desert.

Figure 54.3 presents the same NPP data a different way—organized by biome instead of by geography. Figure 54.3a provides data on average NPP per square meter per year for each biome; Figure 54.3b documents the total area that is covered by each type of ecosystem; and Figure 54.3c presents the percentage of the world's total productivity—the result of multiplying the data in part (a) by the data in part (b). It's

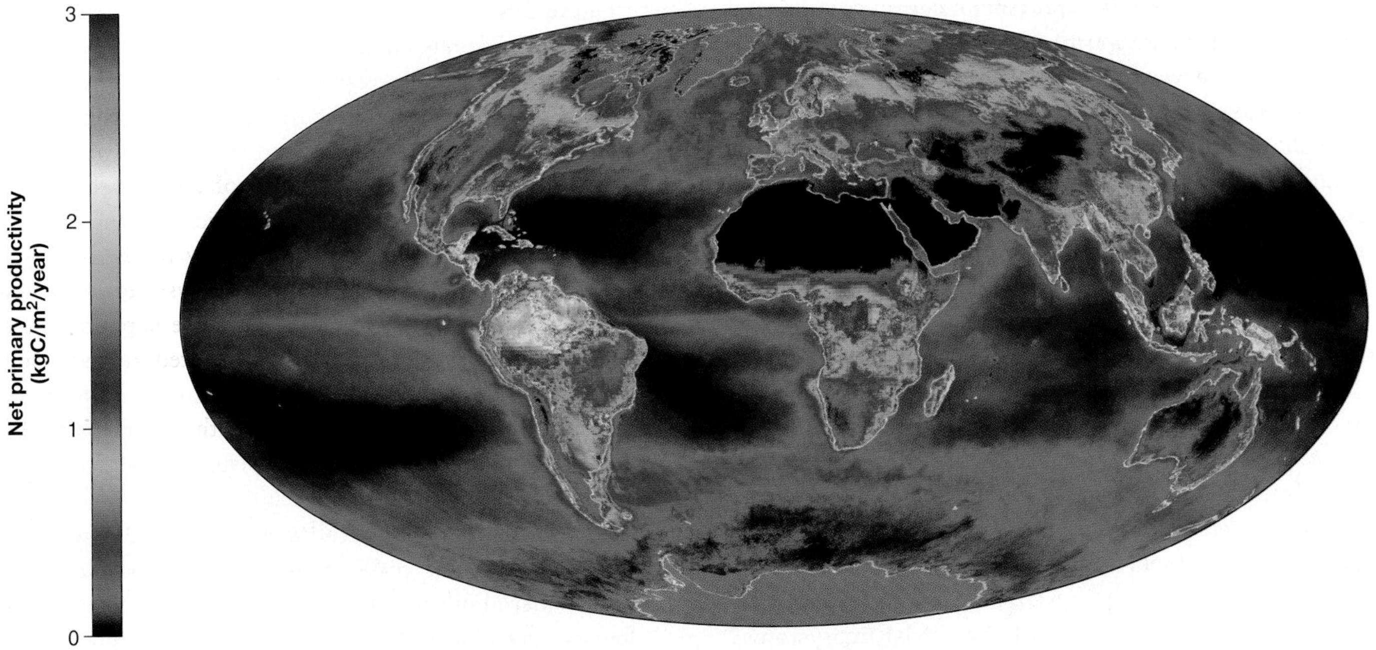

FIGURE 54.2 Net Primary Productivity Varies among Regions
The terrestrial ecosystems with the highest primary productivity are found in the tropics, where warm temperatures and high moisture encourage high rates of photosynthesis. Tundras and deserts have the lowest productivity. The highest productivity in the oceans occurs in nutrient-rich coastal areas.

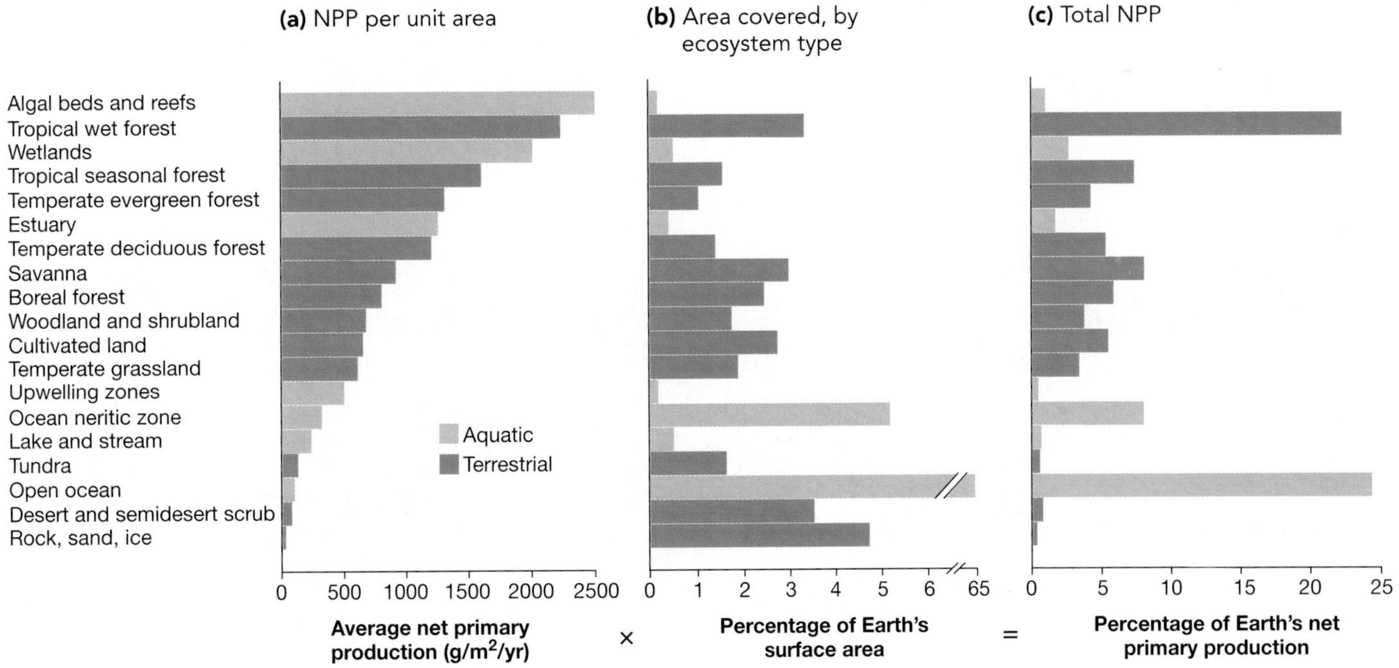

FIGURE 54.3 Net Primary Productivity Varies among Biomes
(a) Among biomes, average annual NPP per square meter varies by over three orders of magnitude. **(b)** Most of Earth's surface is covered by open ocean. The most common terrestrial habitat consists of unvegetated rock, sand, or ice. **(c)** Even though NPP per square meter is low, the open ocean is so vast that it is responsible for over 25 percent of Earth's total NPP.

important to note several important patterns. Tropical seasonal forests, which are warm year-round but have a dry season, and tropical wet forests cover less than 5 percent of Earth's surface but together account for over 30 percent of total NPP. Among aquatic ecosystems, the most productive habitats are marshes, estuaries, and algal beds and coral reefs. Most of the total NPP provided by aquatic ecosystems derives from the open ocean, however. Even though NPP per square meter is low in these regions, the open ocean is so extensive that its total production is high.

These patterns raise an interesting question: What limits NPP in terrestrial and marine ecosystems? The short answer is that productivity is limited by any factor that limits the rate of photosynthesis—specifically, temperature and the availabilities of water, sunlight, and nutrients. Different limiting factors prevail in different environments, however. Let's take a closer look.

What Limits the Productivity of Terrestrial Ecosystems?

The data in Figure 54.2 and Figure 54.3 document that terrestrial productivity is lowest in deserts and arctic regions. This observation suggests that the overall productivity of terrestrial ecosystems is limited by a combination of temperature and availability of water and sunlight. The same conclusion has been reinforced by a recent analysis of changes in the productivity of terrestrial ecosystems between 1982 and 1999. When temperatures are averaged over the globe, the interval covered by this study included the warmest years since data collection began in 1861.

Figure 54.4 shows how average annual NPP changed in terrestrial environments during the 1982–1999 warm interval. Notice that NPP increased dramatically in most tropical wet forests, particularly in the Amazon River basin of South America. The leading hypothesis to explain this observation is that increased temperatures led to decreased cloud cover in these regions, and thus higher incident sunlight and higher productivity. Regions that decreased in productivity, in contrast, are thought to have experienced increased cloud cover or drier conditions in response to the elevated temperatures. Even though global terrestrial NPP increased by over 6 percent during the 17-year study interval, variation among regions was high because of complex interactions between temperature and the availability of sunlight and moisture.

What Limits the Productivity of Marine Ecosystems?

To explain why the productivity of marine habitats is so much higher along coastlines than in deepwater regions, biologists focus on nutrient limitation. As pointed out in Chapter 50, the neritic and intertidal zones along coasts receive nutrients from two major sources: (1) rivers that carry and deposit nutrients from terrestrial ecosystems and (2) near-shore ocean currents that bring nutrients that have rained down into the cold, deep water of the oceanic zone back up to the surface. Both of these sources are absent in the surface waters of the open ocean. In addition, nutrients found in organisms near the surface of the open ocean—where sunlight is abundant—constantly rain down to dark, deeper waters in the form of dead cells and are lost.

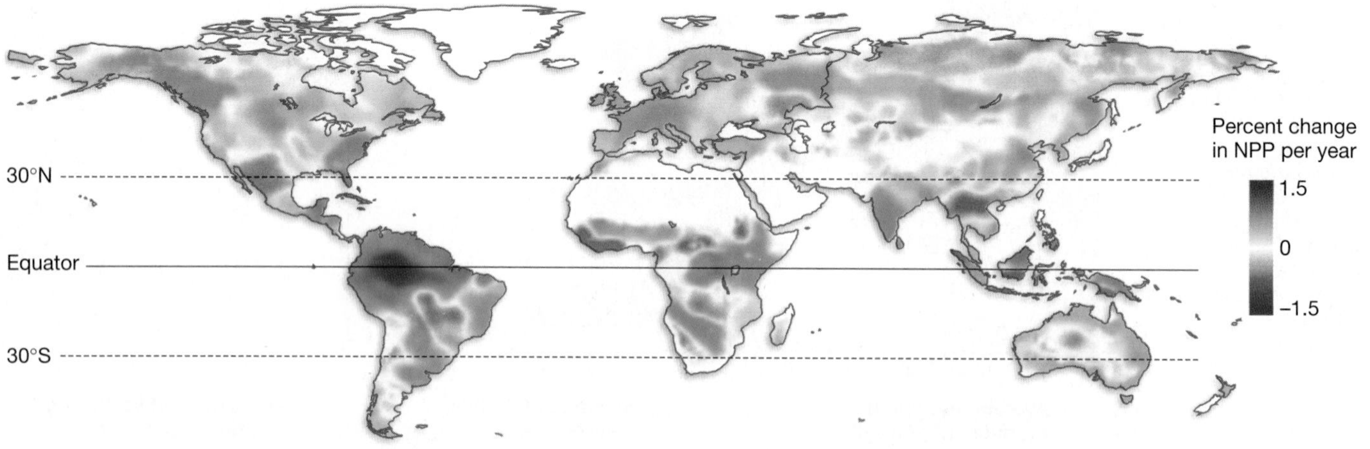

FIGURE 54.4 Changes in NPP during a Warming Trend, 1982–1999
The 1980s and 1990s had the highest average global temperature on record. During this time interval, NPP increased in the green regions and decreased in the red areas, compared with average NPP before this interval.

Analyses of waters that support populations of planktonic organisms have confirmed that trace elements such as zinc, iron, and magnesium are particularly rare in the open ocean. These atoms are important because they are required as enzyme cofactors. Iron, for example, is essential to the proteins that are involved in electron transport chains (see Chapter 9). On the basis of these observations, biologists have proposed that the productivity of open-ocean ecosystems could be increased dramatically by fertilizing them with iron. The results of one recent iron-fertilization experiment are shown in **Figure 54.5**. The data show large increases in the concentration of chlorophyll *a* in surface waters over a two-week interval. Results like these provide strong support for the hypothesis that NPP in marine ecosystems is limited primarily by the availability of nutrients.

How Does Energy Flow through an Ecosystem?

What happens to NPP? Net primary productivity results in biomass—organic material that non-photosynthetic organisms can eat. In every terrestrial and marine environment in the world, the chemical energy in primary producers eventually moves to one of two types of organisms: primary consumers or decomposers.

A **primary consumer** is an herbivore—an organism that eats plants or algae or other photosynthetic cells. Primary consumers are a key link in what biologists call the **grazing food web**: the collection of organisms that eat plants, along with the organisms that eat the herbivores. Consumers that eat herbivores are called **secondary consumers.**

Not all plant tissue is consumed by herbivores, however. Tissues that are not consumed eventually die. When they do, they enter a **decomposer food web**, which is composed of species that eat the dead remains of organisms. In forest ecosystems, for example, dead animals and the dead tissues that accumulate, creating plant *litter*, are collectively known as **detritus.** Detritus is consumed by a variety of **primary decomposers:** bacteria, archaea, fungi, roundworms, earthworms, and millipedes (**Figure 54.6**, page 1248). These primary decomposers are in turn consumed by secondary consumers, such as centipedes, spiders, salamanders, and shrews.

Trophic Levels, Food Chains, and Food Webs To describe how energy flows from primary producers to primary consumers and secondary consumers, biologists identify distinct feeding levels in an ecosystem. Organisms that obtain their energy from the same type of source are said to occupy the same **trophic** ("feeding") **level.** As an example, consider the organisms that eat the living or dead plants illustrated in Figure 54.6. These primary consumers occupy the same trophic

Question: Is net primary productivity (NPP) in the open ocean limited by nutrients?

Hypothesis: NPP in the open ocean is limited by availability of iron.

Null hypothesis: NPP in the open ocean is not limited by availability of iron.

Experimental setup:

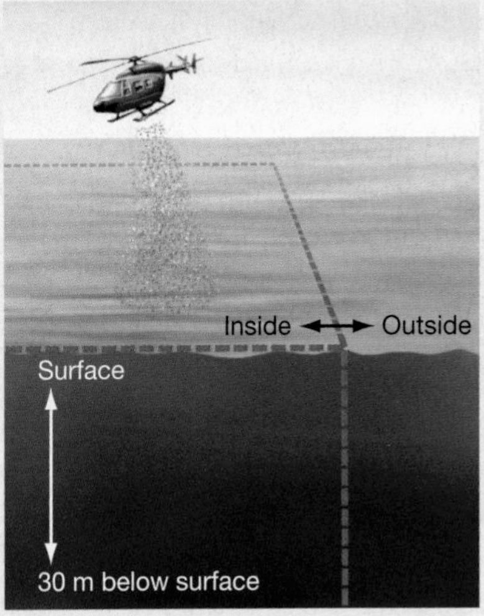

1. Add 350 kg iron (as $FeSO_4$) to a patch of ocean 8 km × 10 km.

2. Take water samples for a two-week period outside and inside the patch, at surface and a depth of 30 m, and record amount of chlorophyll *a* present (as indicator of NPP).

Prediction: Amount of chlorophyll *a* near the surface inside the patch will increase relative to amounts outside the patch or at 30 m below the surface.

Prediction of null hypothesis: Amount of chlorophyll *a* will be the same in all measurements.

Results:

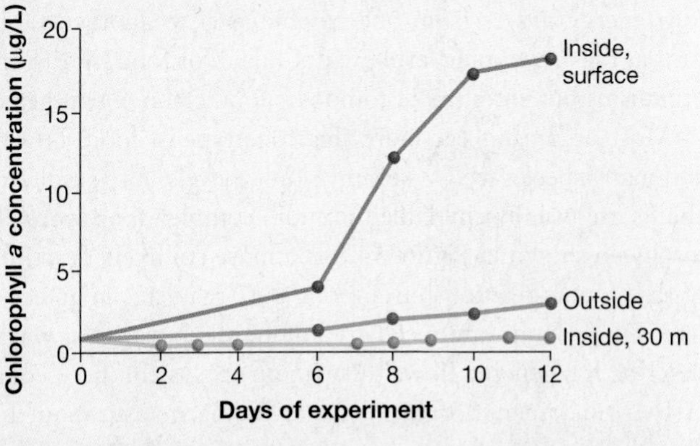

Conclusion: NPP in the open ocean is limited by the scarcity of nutrients—specifically, iron.

FIGURE 54.5 Fertilization with Iron Increases NPP in the Open Ocean

QUESTION Most of the increased chlorophyll *a* was present in a single species of diatom. Recall from Chapter 28 that diatoms have glassy cases made of silicon. What happens to the biomass present in diatoms after they die?

FIGURE 54.6 The Decomposer Food Web
Dead leaves, sticks, dead animals, and other types of detritus are eaten by an enormous variety of organisms. These decomposers, in turn, are eaten by salamanders, shrews, spiders, centipedes, and other predators.

level, as illustrated in **Figure 54.7**. A **food chain** connects the trophic levels in a particular ecosystem. In doing so, it describes how energy moves from one trophic level to another. Organisms at the top trophic level are not killed for food by any other organisms but enter the decomposer food chain when they die.

Most organisms eat more than one type of food, however, and many species feed at several trophic levels. As a result, food chains are usually embedded in more complex **food webs**. The food web shown in **Figure 54.8** is a more complete description of the trophic relationships among the organisms in an ecosystem. Food chains and webs are among the most basic ways to describe how energy flows through an ecosystem. Let's consider two fundamental questions that biologists ask about food chains and webs.

Why Is Energy Lost at Each Trophic Level? Ecosystems share a characteristic pattern: Total biomass present declines from one trophic level up to the next. Although the biomass and thus the chemical energy present at each trophic level varies widely among ecosystems, the data in **Figure 54.9** are representative of many terrestrial communities. The most striking observation is that as much as 90 percent of the energy present in a given trophic level is lost at the next highest trophic level.

Trophic level	Feeding strategy	Decomposer food chain	Grazing food chain
4	Tertiary consumer	Fox 	Cooper's hawk 
3	Secondary consumer	Salamander 	Robin 
2	Primary consumer	Earthworm 	Cricket 
1	Producer	Dead maple leaves 	Maple tree leaves 

FIGURE 54.7 Organisms That Share a Trophic Level Obtain Their Energy from the Same Source
Examples from a temperate-forest ecosystem are shown; many other species exist at each level in this ecosystem.

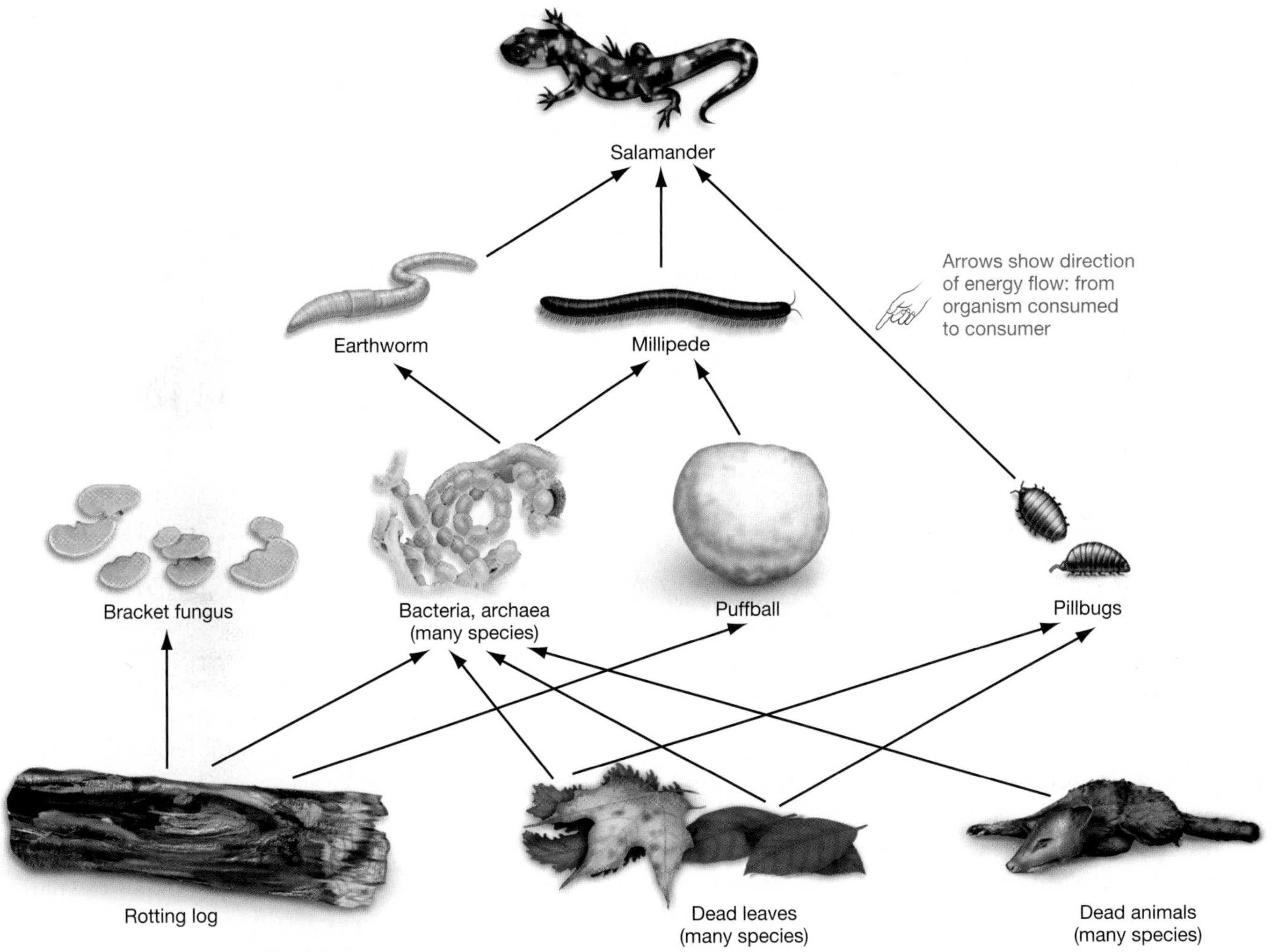

FIGURE 54.8 Food Webs Describe Trophic Relationships in an Ecosystem
Food webs offer a more comprehensive analysis of feeding relationships than food chains do. This food web shows only a fraction of the total feeding relationships and species in the decomposer food web of a temperate deciduous forest.

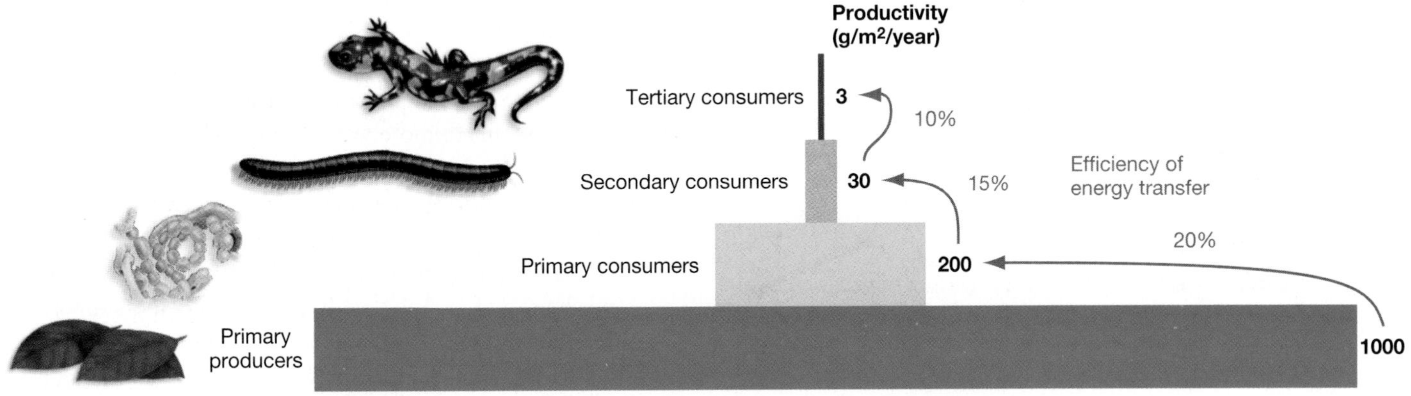

FIGURE 54.9 Productivity Declines at Higher Trophic Levels
In all ecosystems, productivity is highest at the first (bottom) trophic level and declines at higher levels. This pattern is the pyramid of productivity.

To understand why this pattern occurs, consider the interface between primary producers and primary consumers. Much of the net primary productivity that exists is unavailable to herbivores, because it resides in indigestible substances such as the lignin in wood or because it is protected by noxious defensive compounds. NPP that is not eaten by an herbivore enters the decomposer food web. Even if the material is ingested by an herbivore, most of the energy stored in the chemical bonds of carbon compounds is lost as heat as it is metabolized or used to keep the consumer alive—not to grow or reproduce. At the next higher trophic level, many herbivores are never consumed by secondary consumers, because these herbivores defend themselves effectively and die of other causes. Of the energy that is successfully ingested by secondary consumers, some is again lost as heat or used up in staying alive and trying to capture prey.

The general point here is simple: The amount of biomass present at the second trophic level must be less than productivity at the first trophic level; productivity at the third trophic level must be less than that at the second. This pattern holds true for the entire food chain and produces a **pyramid of productivity.** Production of biomass is highest at the lowest trophic level.

What Limits the Length of Food Chains? It is interesting to note that none of the food chains, food webs, or trophic structures presented in this chapter thus far have more than four trophic levels. When a biologist reviewed the characteristics of food chains that had been documented by researchers working in a wide variety of ecosystems, he found that the maximum number of links ranged from 1 to 6 (**Figure 54.10**). Each link joins two trophic levels. In this study, terrestrial and lake ecosystems had 3.7 links per food chain, on average, while streams had an average of 3.2 links. Why don't ecosystems have 8 or 9 or 10 trophic levels? Why does the overall average number of levels seem to be about 3.5? Several competing hypotheses have been offered to explain this observation. Let's consider three:

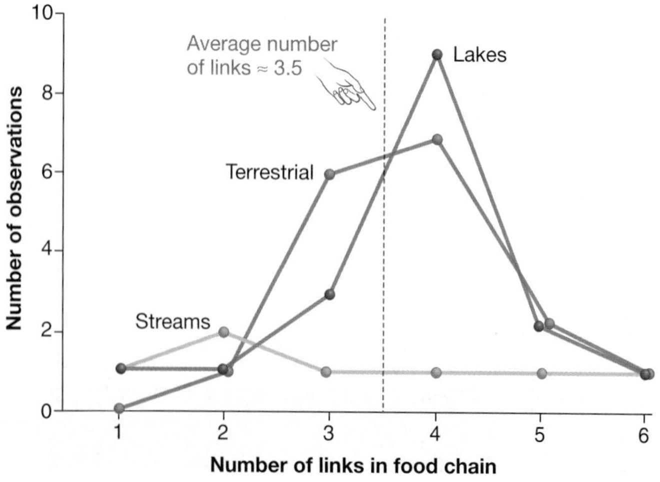

FIGURE 54.10 Most Food Chains Are Short
The vertical axis plots the number of research studies that described food chains with 1 to 6 links in stream, lake, and terrestrial habitats.

- *Hypothesis 1: Energy Transfer* As energy is transferred up a food chain, a large fraction of that energy is lost. By the time energy reaches the top trophic level, there may not be enough left to support an additional suite of consumers. Suppose that the efficiency of energy transfer between each pair of trophic levels is 10 percent. If the initial trophic level in a hypothetical ecosystem produces 10,000 kcal per day, then the second, third, and fourth levels will produce 1000, 100, and 10 kcal per day, respectively. Can any organism obtain enough energy to survive at a fifth trophic level?

  The hypothesis that food-chain length is limited by productivity leads to a strong prediction: There should be more trophic levels in ecosystems with higher productivity or higher efficiency of energy transfer than in ecosystems with lower productivity or lower efficiency of energy transfer. When Stuart Pimm analyzed data on the productivity of four aquatic and four terrestrial ecosystems, however, he found that low-productivity ecosystems were as likely to contain four trophic levels as high-productivity ones were. To date, research has not supported the prediction that the most productive ecosystems have the longest food chains.
- *Hypothesis 2: Stability* Pimm proposed an alternative hypothesis to the energy-transfer explanation for why food chains are short. His idea was that long food chains are easily disrupted by environmental perturbations and thus tend to be eliminated. Using mathematical models, he demonstrated that long food chains took longer to return to their previous state following a disturbance than did short food chains. He then proposed that long food chains are unlikely to persist in a variable environment.

  Pimm's hypothesis also predicts that the length of food chains should increase with the stability of the environment. A biologist tested this prediction by comparing the animal communities that develop inside tree holes in Australia and Great Britain. After water accumulates in depressions or in the holes in tree branches or roots, leaf litter from the trees falls in and forms the basis of a food web. The researcher found that tree-hole communities in Australia and Great Britain were similar in many respects, except that annual leaf fall was much more variable in the British communities than in the Australian ones. As predicted by the stability hypothesis, the British ecosystems had only two trophic levels while Australian habitats supported three.

  Other researchers have challenged the assumptions behind Pimm's theoretical analysis, however, and support for the stability hypothesis remains very tentative. Experimental and observational tests of this hypothesis are continuing.
- *Hypothesis 3: Environmental Complexity* Biologists have also hypothesized that food-chain length is a function of an ecosystem's physical structure. Specifically, researchers have

proposed that tundra, grasslands, lake and sea bottoms, streambeds, and intertidal zones offer a largely flat, or two-dimensional, surface to the organisms living there. In contrast, forests and open-water environments in lakes and rivers offer three-dimensional volumes. This hypothesis predicts that three-dimensional ecosystems should have longer food chains than two-dimensional sites do.

To test this hypothesis, researchers examined 113 publications that described food chains in a wide variety of aquatic and terrestrial environments. These data supported the prediction that food webs are significantly longer in three-dimensional ecosystems. As a result, the investigators concluded that dimensionality does influence food-web structure. The mechanism for this pattern remains to be determined, however. It remains to be seen why three-dimensional ecosystems allow longer food webs to develop.

To summarize, there is unlikely to be a single, simple answer to the question of what limits food-chain length. Inefficiencies of energy transfer, environmental stability, and environmental complexity may all influence the number of trophic levels that can be supported in a given ecosystem. Multiple causation is not unusual in biological science. In genetics, for example, gene expression is rarely triggered by a single regulatory site in DNA and just one regulatory protein. Instead, the amount and timing of gene expression result from interactions among dozens of regulatory DNA sequences and proteins. Recognizing multiple causation is productive, because it can suggest exciting new experiments and analyses. For example, if food-chain length does have multiple causes, then habitats that are particularly productive, stable, and highly complex in structure should have the longest food chains. Research continues on this fundamental question in ecosystem ecology.

Analyzing Energy Flow: A Case History

Research on energy flow in ecosystems has uncovered a series of important general principle. In both terrestrial and marine ecosystems, NPP varies with region and biome. On land, the most important constraints on productivity are warmth and moisture; in aquatic habitats, nutrient availability is key. The chemical energy in biomass flows into consumer or decomposer food webs, where it is passed up a series of trophic levels. Productivity declines with each trophic level, however.

To see these general principles in action, let's analyze energy flow in a specific ecosystem. For almost five decades, a team of researchers has been studying how energy and nutrients flow through a temperate forest ecosystem in the northeastern United States: the Hubbard Brook Experimental Forest in New Hampshire.

As **Figure 54.11** shows, energy flow in this ecosystem begins when plants capture the energy in solar radiation via photosynthesis. At Hubbard Brook, the amount of energy entering the ecosystem from sunlight varies throughout the year and, to a lesser extent, from year to year. From June 1, 1969, to May 31, 1970, which the research team describes as a typical year, 1,254,000 kilocalories (kcal) of solar radiation per square meter (m^2) reached the forest. If this amount of energy were available in the form of electricity, it would easily power two 75-watt lightbulbs that burned continuously for one year.

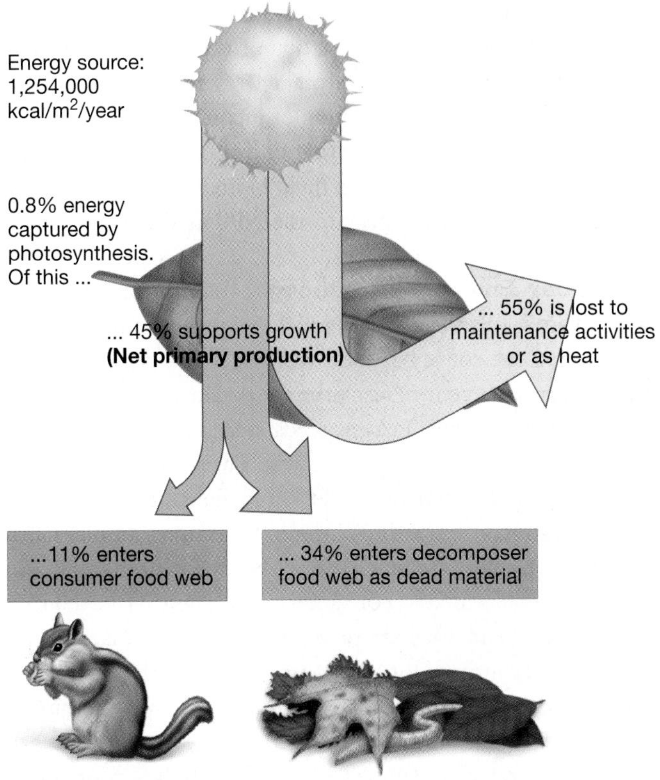

FIGURE 54.11 Energy Flow through the Hubbard Brook Forest Ecosystem
Energy from sunlight is transformed to chemical energy by photosynthesis. Photosynthetic products fuel new plant growth, which eventually enters either the consumer or decomposer food web. **EXERCISE** Calculate gross photosynthetic efficiency and NPP in kcal/m^2/year. Calculate the amount of energy that enters the consumer and decomposer food webs each year, in kcal/m^2.

By documenting rates of photosynthesis in a large sample of forest plants, the biologists calculated that the plants used 10,400 kcal/m^2 of energy in photosynthesis. This value represents **gross primary productivity**, which is the total amount of photosynthesis in a given area and time period. The team also calculated **gross photosynthetic efficiency**, or the efficiency with which plants use the total amount of energy available to them, as the ratio of gross photosynthesis to solar radiation in kcal/m^2. At Hubbard Brook, efficiency was 10,400 ÷ 1,254,000 = 0.8 percent. This value is typical of other ecosystems as well.

Why is efficiency so low? The answer has several components. Plants cannot photosynthesize in winter or during dry periods of a day or year, although sunlight is available. Even when conditions are ideal, the pigments that drive photosynthesis can respond to only a fraction of the wavelengths available and thus a fraction of the total energy received.

Given that only a tiny fraction of incoming sunlight is converted to chemical energy and that only a fraction of this gross primary productivity is used to build biomass, let's consider the next question: What happens to the NPP at Hubbard Brook?

How Does Energy Flow through the Hubbard Brook Ecosystem? Between 1969 and 1971 the amount of energy that entered the consumer food web at Hubbard Brook varied from about 1 percent of net primary productivity per year to more than 40 percent. Leaves, seeds, and fruits were the most commonly consumed plant parts; wood was rarely eaten.

Figure 54.12 illustrates what happened to the energy obtained by consumers, using chipmunks as an example. Chipmunks are small rodents that are seed predators. On average, these primary consumers harvested 31 kcal/m^2 of energy each year. Of that total, 17.7 percent was excreted and 80.7 percent was lost to respiration and other processes. Just 1.6 percent went into the production of new chipmunk tissue by growth and reproduction. The production of new tissue by primary consumers is called **secondary production.** Secondary production was much higher in ectothermic consumers such as caterpillars, which transformed about 5.4 percent of the energy they ingested into new tissue. Because ectotherms rely principally on heat gained from the environment and do not oxidize sugars to keep warm, they devote much less energy to cellular respiration than endotherms do. Even so, it is clear that only a tiny fraction of the available solar radiation is involved in secondary production. These data reflect the general principles that productivity drops sharply at each trophic level.

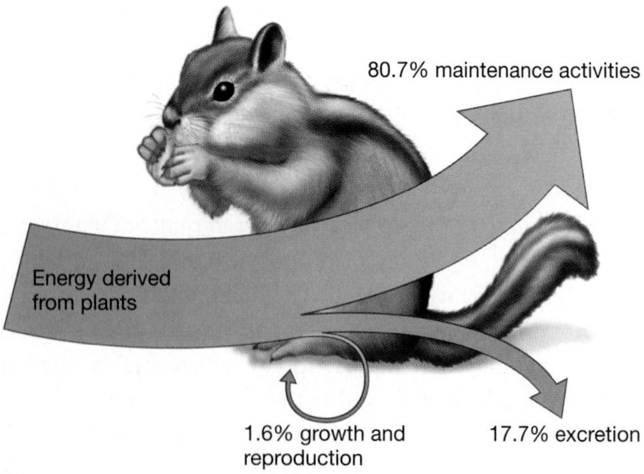
80.7% maintenance activities
Energy derived from plants
1.6% growth and reproduction
17.7% excretion

FIGURE 54.12 How Do Consumers Use Primary Production?
A small fraction of the energy consumed by chipmunks, which are primary consumers (herbivores), is used for secondary production. Most of the energy is used for cellular respiration.

What about biomass that enters the decomposer food web? At Hubbard Brook, about 75 percent of net primary productivity is not eaten while alive. This energy enters the decomposer food web and is passed up subsequent trophic levels, just as it is in the consumer food web. But in addition, the researchers at Hubbard Brook found that large amounts of energy leave the decomposer food web in the form of detritus that washes into streams. This transfer of energy into the aquatic ecosystem is important, because it is a major source of energy for aquatic organisms. At Hubbard Brook, photosynthesis by aquatic plants introduced only about 10 kcal/m^2/yr of energy. In contrast, each year 6039 kilocalories washed into each square meter of streambed from the surrounding forest.

To summarize, long-term studies have documented the flow of energy into, through, and out of a temperate-forest ecosystem. Although the specific numbers found at Hubbard Brook are unique to that site and the time interval of that study, the general patterns have turned out to be typical of ecosystems around the globe.

✓CHECK YOUR UNDERSTANDING

In an ecosystem, energy flows from sunlight or inorganic compounds with high potential energy to producers, and from there to consumers and decomposers. In most ecosystems, NPP is limited by conditions that limit the rate of photosynthesis: temperature and/or the availability of sunlight, water, and nutrients. Productivity diminishes at each subsequent trophic level in the ecosystem. You should be able to (1) summarize and provide a brief explanation of variation in NPP among ecosystems, (2) explain why productivity diminishes with increasing trophic levels, and (3) evaluate at least one hypothesis to explain why most food chains have only two or three links.

54.2 Biogeochemical Cycles

Energy is not the only quantity that is transferred when one organism eats another. The organisms that are eaten also contain carbon (C), nitrogen (N), phosphorus (P), calcium (Ca), and other elements that act as nutrients. Atoms are constantly reused as they cycle through trophic levels and air, water, and soil. The path that an element takes as it moves from abiotic systems through organisms and back again is referred to as its **biogeochemical** ("life-Earth-chemical") **cycle.**

Because humans are now disturbing biogeochemical cycles on a massive scale, research on this aspect of ecosystem ecology is exploding. To get a basic understanding of how a particular biogeochemical cycle works, researchers focus on three fundamental questions:

1. What are the nature and size of the **reservoirs**, or pools where elements are stored for a period of time? In the case of carbon, the biomass of living organisms is an important reservoir, along with sediments and soils. Another significant carbon reservoir is buried in the form of coal and oil.
2. How fast does the element move between reservoirs, and what factors influence these rates? The global photosynthetic rate, for example, measures the rate of carbon flow from CO_2 in the atmosphere to living biomass. When fossil fuels burn, carbon that was buried in coal or petroleum moves into the atmosphere as carbon dioxide (CO_2).
3. How does this biogeochemical cycle interact with other cycles? For example, researchers are trying to understand how changes in the nitrogen cycle affect carbon.

Before taking up these issues, though, let's begin with a very general overview of what nutrient cycling is.

Biogeochemical Cycles in Ecosystems

Figure 54.13 shows a simplified version of a generalized terrestrial nutrient cycle. In this case, the cycle starts when nutrients are taken up from the soil by plants and incorporated into plant tissue. If the plant is eaten, the nutrients pass to the animal members of the ecosystem's consumer food web; if the plant dies, the nutrients enter the decomposer food web. Once ingested by an animal, nutrients are excreted in fecal matter or urine, taken up by a parasite or predator, or added to the dead biomass reservoir when the animal dies.

The nutrients in plant litter, animal excretions, and dead animal bodies are broken down by bacteria, archaea, roundworms, fungi, and other organisms. The combination of breakdown products and microscopic decomposers forms the soil organic matter. **Soil organic matter** is a complex mixture of partially and completely decomposed detritus. Completely decayed organic material is called **humus**, because it is rich in a family of carbon-containing molecules called *humic acids*. (Chapter 37 describes other components of the soil.) Eventually, the nutrients in soil organic matter are converted to an inorganic form. For example, cellular respiration by soil-dwelling bacteria and archaea converts the nitrogen present in amino acids that are found in detritus to ammonium (NH_4^+) or nitrate (NO_3^-) ions. Once this step is accomplished, the nutrients are available for uptake by plants.

A key feature of this process is that nutrients are reused. Reuse is not total, however. Nutrients leave the ecosystem whenever plant or animal biomass is removed. For example, plant biomass is removed when an herbivore enters the ecosystem, eats a plant, and leaves the ecosystem before excreting nutrients or dying. Nutrients also leave ecosystems when flowing water or wind removes particles or inorganic ions. Soil erosion has a huge impact on nutrient cycles: It removes nutrients rapidly and in potentially large quantities.

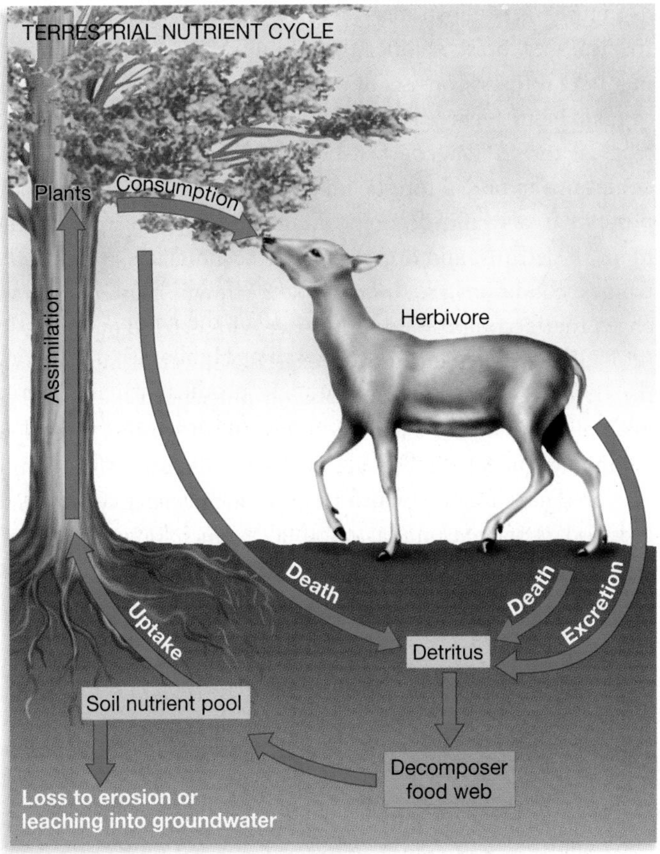

FIGURE 54.13 Generalized Terrestrial Nutrient Cycle
Nutrients cycle from organism to organism in an ecosystem via assimilation, consumption, and decomposition. Nutrients are exported from ecosystems through the migration of organisms out of the area or in flowing water or groundwater.

What Factors Control the Rate of Nutrient Cycling in Ecosystems?

Of the many links in a nutrient cycle, the decomposition of detritus most often limits the overall rate at which nutrients move through an ecosystem. The decomposition rate, in turn, is influenced by two types of factors: abiotic conditions such as temperature and precipitation, and the quality of the detritus as a nutrient source for the fungi, bacteria, and archaea that accomplish decomposition.

To appreciate the importance of abiotic conditions on the decomposition rate, consider the difference in detritus accumulation between boreal forests and tropical wet forests. Chapter 50 indicated that boreal forests occur in areas where temperature is low. As a result, soils in these ecosystems are cold and wet. Tropical wet forests, in contrast, occur in areas where temperatures and rainfall are high. Soils tend to remain moist and warm all year long.

Figure 54.14 illustrates typical soils from boreal forests and tropical wet forests. Notice that the uppermost part of the soil in a boreal forest consists of partially decomposed detritus and organic matter. There is no such layer at the top of the soil in a tropical forest. The contrast occurs because the cold and wet conditions in boreal forests limit the metabolic rates of decomposers. As a result, decomposition fails to keep up with the input of detritus, and organic matter accumulates there. In the tropics, conditions are so favorable for fungi, bacteria, and archaea that decomposition keeps pace with detrital inputs.

The quality of detritus also exerts a powerful influence on the decomposition rate and thus on nutrient availability. For example, decomposers can be limited by the presence of large compounds in detritus that are difficult to digest, such as lignin. The presence of lignin, which is one of the primary constituents of wood, is one reason that wood takes much longer to decompose than leaves do. The growth of decomposers is also inhibited if detritus is low in nitrogen.

(a) Boreal forest: Extensive litter accumulation (dead wood and leaves)

(b) Tropical wet forest: Almost no litter accumulation

FIGURE 54.14 Temperature and Moisture Affect Decomposition Rates

(a) In boreal forests, decomposition rates are limited by cold soil temperatures. Organic matter builds up, because the input of detritus into the soil exceeds the decomposition rate. **(b)** In tropical wet forests, warm temperatures allow decomposition to proceed rapidly, so organic matter does not build up.

What Factors Influence the Rate at Which Nutrients Are Exported from Ecosystems? Nutrient availability has a profound effect on productivity. The rate of nutrient loss is thus an important characteristic of an ecosystem. Several of the major impacts that humans have on ecosystems—such as farming, logging, burning, and soil erosion—accelerate nutrient loss.

To test the effect of vegetation removal on nutrient cycling, researchers initiated a large-scale experiment at the Hubbard Brook Experimental Forest. They chose two similar **watersheds**—areas drained by a single stream—for study (**Figure 54.15**). They then removed all vegetation, including the trees, from the forests in one of the two watersheds. In the following two years, this *clear-cut* area was treated with an herbicide to prevent vegetation from regrowing. As a result, the experimental watershed was devegetated. The untreated watershed served as a control.

Before removing the vegetation, the researchers had documented that 90 percent of the nutrients in the ecosystems were in soil organic matter, and an additional 9.5 percent was in plant biomass. After the vegetation was removed, the team monitored the concentrations of nutrients in the streams exiting the two watersheds. The graphs in Figure 54.15 document the amount of dissolved substances that subsequently washed out of the stream in each watershed over the course of four years. Losses were typically 10 times as high in the clear-cut site as in the control site.

Based on these data, the researchers concluded that devegetation has a huge impact on nutrient export. Instead of being held in the ecosystem and recycled, nutrients wash out. The loss occurs because nutrients in the soil are either dissolved in water or attached to small particles of sand or clay. If plant roots no longer hold the soil particles in place and if they no longer take up and recycle nutrients, then the molecules and ions wash out of the soil and are lost to the ecosystem. Consequently, the productivity that an area can support may decline over time if the area is kept in a devegetated state. Long-term devegetation of this type has occurred in formerly forested areas of the Middle East, North Africa, and elsewhere due to intensive farming and grazing.

Global Biogeochemical Cycles

When nutrients leave one ecosystem, they enter another. In this way, the movement of ions and molecules among ecosystems links local biogeochemical cycles into one massive global system. Local and global cycles interact when water, organisms, or wind move nutrients. As an introduction to how these global biogeochemical cycles work, let's consider the global water, carbon, and nitrogen cycles. These cycles have recently been heavily modified by human activities—with serious ecological consequences.

The Global Water Cycle A simplified version of the **global water cycle** appears in **Figure 54.16**. The diagram shows the estimated amount of water that moves between major components of the cycle over the course of a year.

Question: How does the presence of vegetation affect the rate of nutrient export in an ecosystem?

Hypothesis: Presence of vegetation lowers the rate of nutrient export because it increases soil stability and recycling of nutrients.

Null hypothesis: Presence of vegetation has no effect on the rate of nutrient export.

Experimental setup:

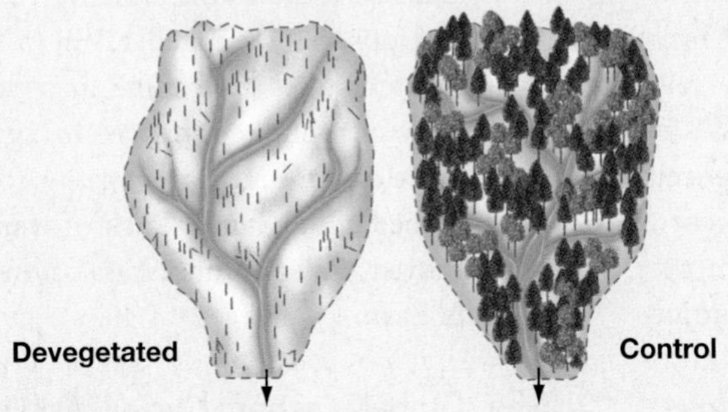
Devegetated
Control

1. Choose two similar watersheds. Document nutrient levels in soil organic matter, plants, and streams.

2. Devegetate one watershed, and leave the other intact.

3. Monitor the amount of dissolved substances in streams.

Prediction: Amount of dissolved substances in stream in devegetated watershed will be much higher than amount of dissolved substances in stream in control watershed.

Prediction of null hypothesis: No difference will be observed in amount of dissolved substances in the two streams.

Results:

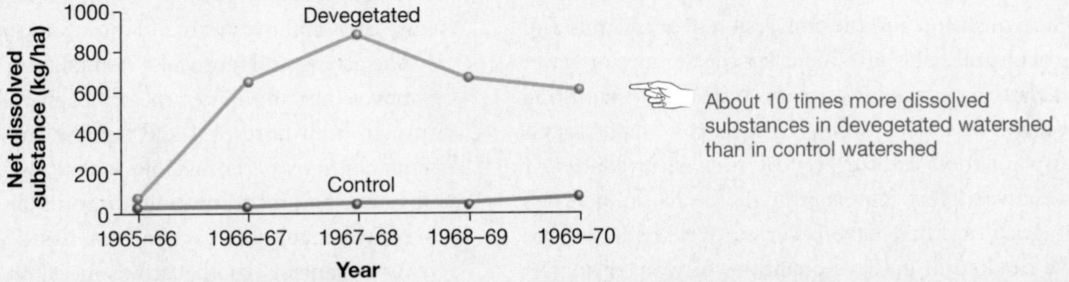
Net dissolved substance (kg/ha)
1000
800
600
400
200
0
Devegetated
Control
1965–66
1966–67
1967–68
1968–69
1969–70
Year
About 10 times more dissolved substances in devegetated watershed than in control watershed

Conclusion: Presence of vegetation limits nutrient loss. Removing vegetation leads to large increases in nutrient export.

FIGURE 54.15 Land-Use Changes Affect the Rate of Nutrient Loss from Ecosystems
QUESTION In effect, this experiment removed one of the arrows in Figure 54.13. Which one?

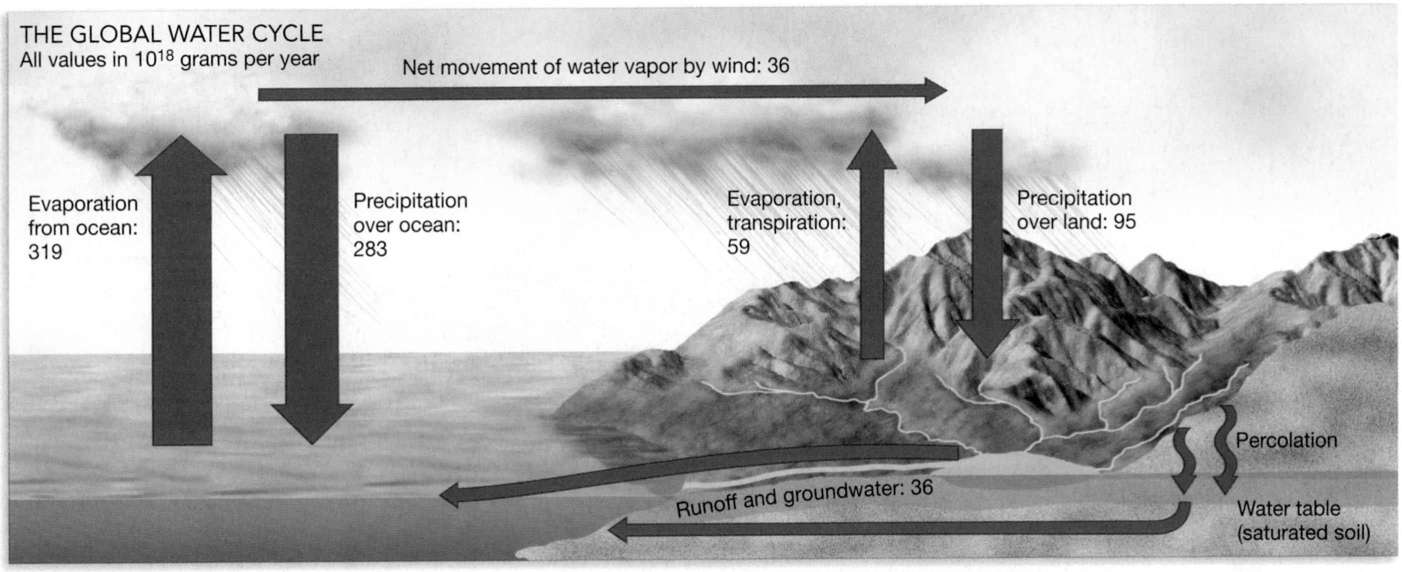
THE GLOBAL WATER CYCLE
All values in 10^{18} grams per year
Net movement of water vapor by wind: 36
Evaporation from ocean: 319
Precipitation over ocean: 283
Evaporation, transpiration: 59
Precipitation over land: 95
Percolation
Runoff and groundwater: 36
Water table (saturated soil)

FIGURE 54.16 The Global Water Cycle
QUESTION Predict how the amount of water evaporated from the ocean is changing in response to global warming. Discuss one possible consequence of this change.

To analyze this cycle, begin with evaporation of water out of the ocean and precipitation back into it. The key observation is that, for the marine component of the cycle, evaporation exceeds precipitation—meaning that, over the oceans, there is a net gain of water to the atmosphere. When this water vapor moves over the continents, it is augmented by a small amount of water that evaporates from lakes and streams and a large volume of water transpired by plants. The total volume of water in the atmosphere over land is balanced by the amount of precipitation that occurs on the continents. The cycle is completed by both water that moves from the land to the oceans via streams and **groundwater**—meaning water that is found in soil.

Humans are affecting the water cycle in complex ways. Perhaps the simplest and most direct impacts concern rates of groundwater replenishment. Asphalt and concrete surfaces from suburbanization reduce the amount of precipitation that percolates through the soil from the surface and enters deep soil layers. Surbanization and the conversion of grasslands and forests to agricultural fields also increases the amount of water that runs off Earth's surface into streams instead of penetrating to groundwater layers. This occurs because soft, shaded soils packed with plant roots are replaced by soils with reduced or no vegetative cover. Most importantly, the dramatic increases in irrigated agriculture that have occurred over the past three decades have resulted in massive quantities of water being removed from groundwater storage and brought to the surface.

In combination, these impacts have resulted in alarming drops in groundwater storage. The **water table** is the upper limit below the surface that the ground is saturated with stored water, and it is dropping on every continent. Between 1991 and 1996 the water table north of Beijing, China, experienced drops that averaged 1.5 m (5 ft) per year. Throughout India, water tables are falling at a rate of between 1 m and 3 m per year. Similar rates are being documented in Yemen, parts of Mexico, and states in the southern Great Plains region of the United States. Lack of water is exacerbating political tensions in several areas of the world, including the Middle East.

The Global Carbon Cycle As **Figure 54.17** shows, the **global carbon cycle** documents the movement of carbon among terrestrial ecosystems, the oceans, and the atmosphere. Of these three reservoirs, the ocean is by far the largest. The atmospheric reservoir is also important despite its relatively small size, because carbon moves into and out of it rapidly.

The arrows in Figure 54.17 emphasize that carbon frequently moves into and out of the atmospheric pool through organisms. In both terrestrial and aquatic ecosystems, for example, photosynthesis is responsible for taking carbon out of the atmosphere and incorporating it into tissue. Cellular respiration, in contrast, releases carbon that has been incorporated into living organisms to the atmosphere, in the form of carbon dioxide.

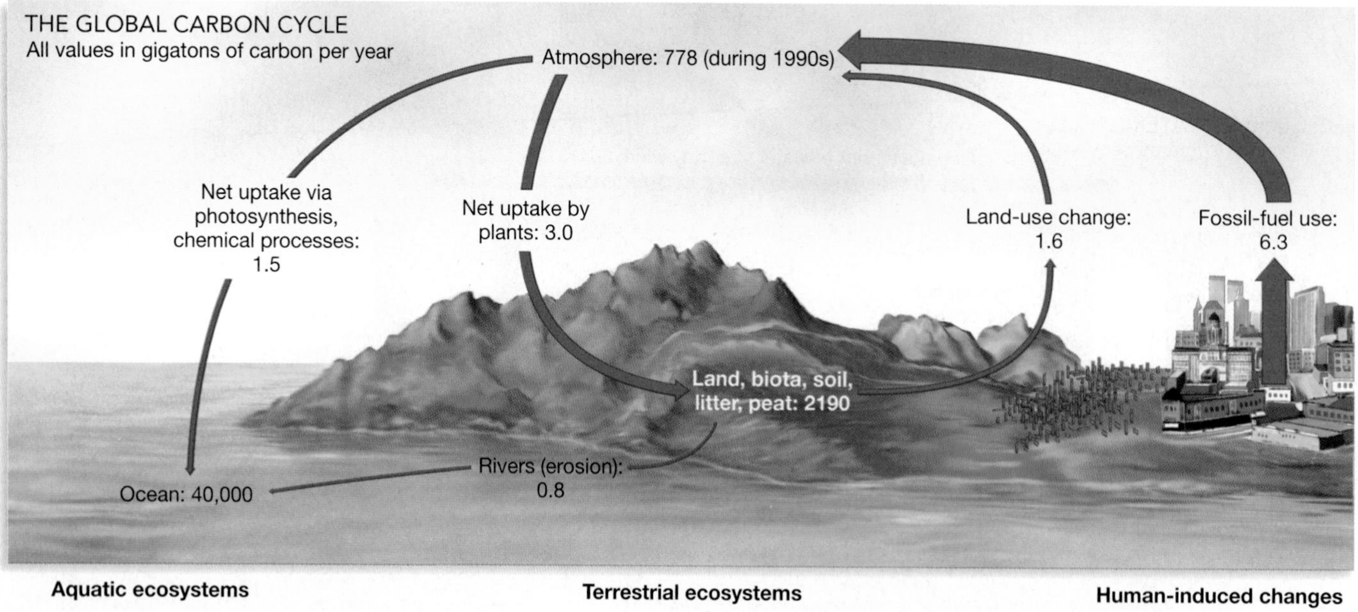

FIGURE 54.17 The Global Carbon Cycle
The arrows indicate how carbon moves into and out of ecosystems. Deforestation and the use of fossil fuels are adding 7.9 gigatons of carbon to the atmosphere each year. Of that amount, 2 gigatons are fixed by photosynthesis in terrestrial ecosystems and 2 gigatons are fixed by physical and chemical processes in the oceans. The remainder—3.9 gigatons—is added to the atmosphere.

How have humans changed the carbon cycle? The fossil fuels found on Earth, which are derived from carbon-rich sediments, are estimated to contain a total of 5000–10,000 gigatons of carbon. (A gigaton, symbolized Gt, is a billion tons). This is one-eighth to one-fourth the size of the oceanic reservoir and 2.5 to 5 times the size of the terrestrial reservoir. In effect, burning fossil fuels moves carbon from an inactive geological reservoir to an active reservoir—the atmosphere.

Land-use changes have also altered the global carbon cycle. Deforestation, for example, reduces an area's net primary productivity. It also releases CO_2 when fire is used for clear-cutting or when dead limbs, twigs, and stumps are left to decompose. Researchers who used data on the amount of land cleared for agriculture and forestry to estimate the amount of carbon released found that, at a global scale, a net movement of carbon to the atmosphere has been occurring from terrestrial ecosystems for at least the past 100 years.

Figure 54.18a shows the amount of carbon released from fossil-fuel burning and land-use changes over the past century; **Figure 54.18b** shows the consequences. In just 45 years, CO_2 in the atmosphere at Mauna Loa Observatory on the island of Hawaii has increased from about 315 to over 375 parts per million (ppm)—meaning milligrams of CO_2 per kilogram of air. The same trend has been observed at sites around the globe. As the data in **Figure 54.18c** show, CO_2 concentrations in the atmosphere have now risen to levels far above the 280 ppm that were typical prior to 1860.

These changes in the global carbon cycle are important because carbon dioxide functions as a **greenhouse gas**: It traps heat that has been radiated from Earth and keeps it from being lost to space, similar to the way the glass of a greenhouse traps heat. More specifically, carbon dioxide is one of several gases in the atmosphere that absorb and reflect the infrared wavelengths radiating from Earth's surface. Increases in amounts of greenhouse gases have the potential to warm Earth's climate by increasing the atmosphere's heat-trapping potential.

The Global Nitrogen Cycle **Figure 54.19** (page 1258) illustrates the **global nitrogen cycle**. A key aspect of this biogeochemical cycle is that plants are able to use nitrogen only in the form of ammonium or nitrate ions (NH_4^+ or NO_3^-). As a result, the vast pool of molecular nitrogen (N_2) that is in the air blanketing Earth—N_2 makes up 78 percent of the atmosphere—is unavailable to plants. Nitrogen is added to ecosystems in a usable form only when it is reduced, or "fixed," meaning when it is converted from N_2 to NH_3. Nitrogen fixation results from lightning-driven reactions in the atmosphere and from enzyme-catalyzed reactions in bacteria that live in the soil and oceans. (Chapter 37 explained how bacteria fix nitrogen.)

The nitrogen cycle has been profoundly altered by human activities. The amount of nitrogen fixation from human sources is now approximately equal to the amount of nitrogen

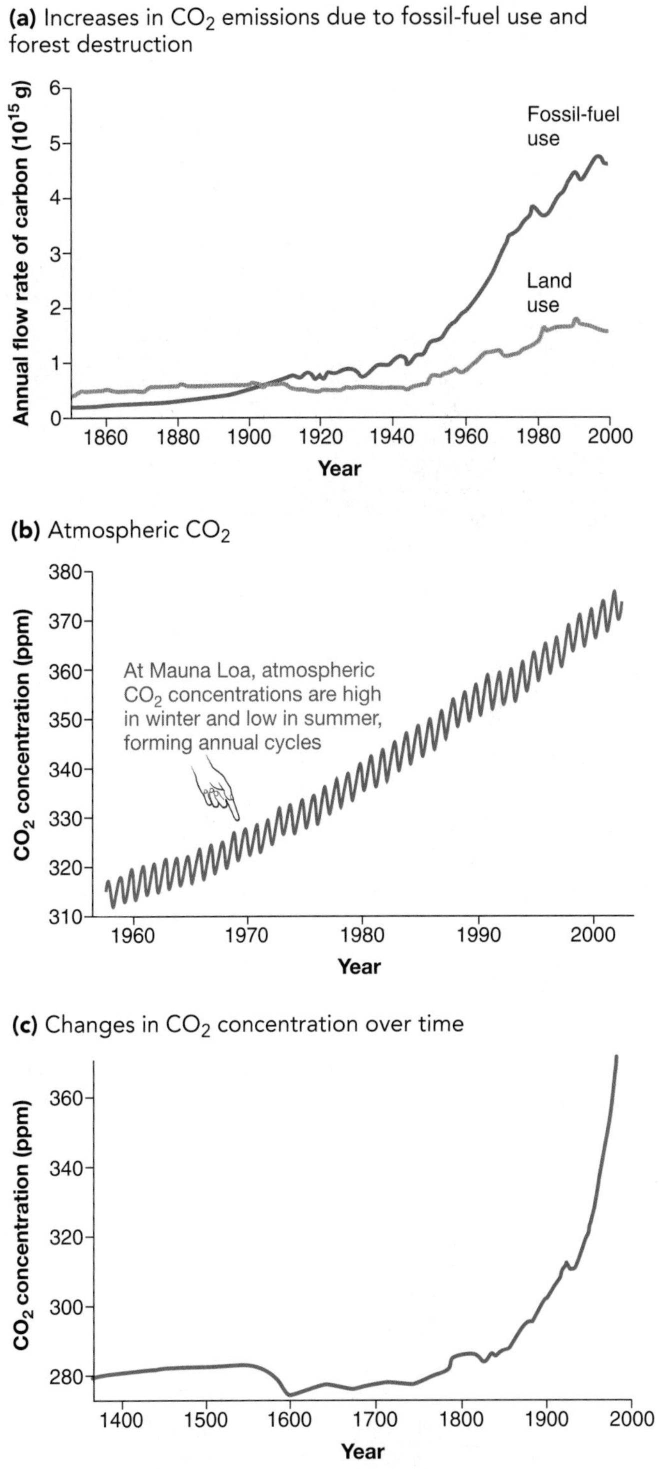

FIGURE 54.18 Humans Are Causing Increases in Atmospheric Carbon Dioxide
(a) Rates of carbon flow from fossil-fuel burning and land-use changes have been increasing since 1860. **(b)** Because the Mauna Loa Observatory in Hawaii is far from large-scale human influences, it should accurately represent the average condition of the atmosphere in the Northern Hemisphere. **(c)** For centuries, average CO_2 concentrations in the atmosphere were about 280 parts per million (ppm). **QUESTION** Why are atmospheric CO_2 concentrations low in the Northern Hemisphere in summer and high in winter? What pattern would you expect in the Southern Hemisphere?

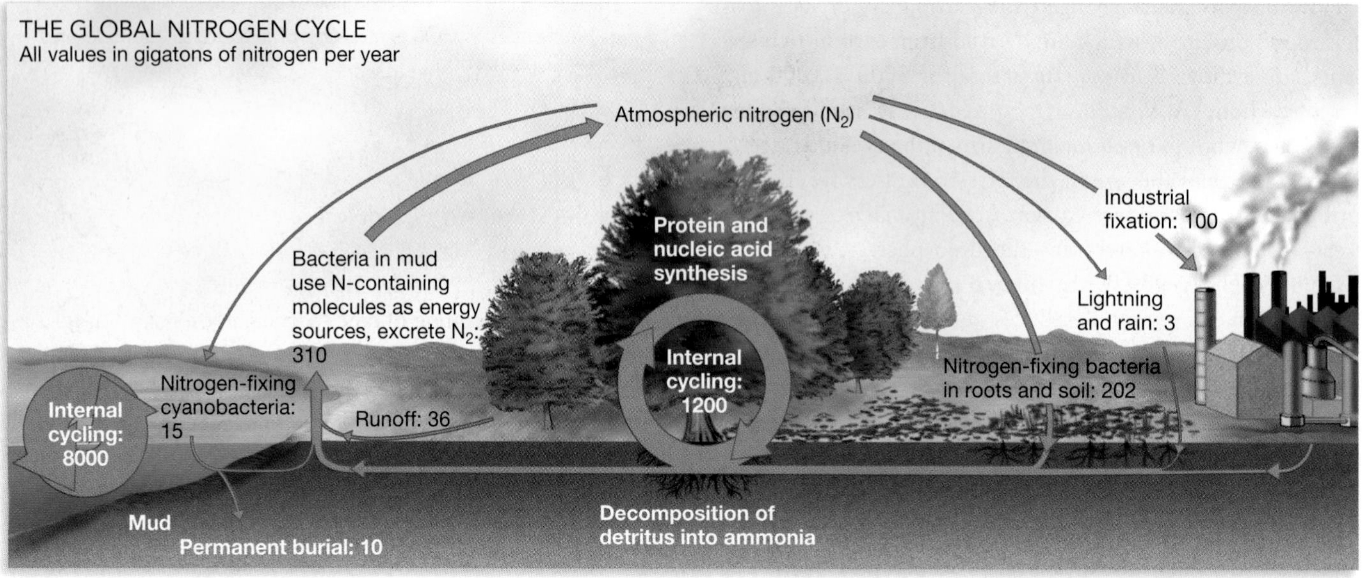
THE GLOBAL NITROGEN CYCLE
All values in gigatons of nitrogen per year
Atmospheric nitrogen (N_2)
Industrial fixation: 100
Protein and nucleic acid synthesis
Bacteria in mud use N-containing molecules as energy sources, excrete N_2: 310
Lightning and rain: 3
Internal cycling: 1200
Nitrogen-fixing bacteria in roots and soil: 202
Nitrogen-fixing cyanobacteria: 15
Internal cycling: 8000
Runoff: 36
Mud
Permanent burial: 10
Decomposition of detritus into ammonia

FIGURE 54.19 The Global Nitrogen Cycle
Nitrogen enters ecosystems as ammonia or nitrate via fixation from atmospheric nitrogen. It is exported in runoff and as nitrogen gas given off by bacteria that use nitrogen-containing compounds as an electron acceptor.

fixation from natural sources, as **Figure 54.20** shows. There are three major sources of this human-fixed nitrogen: (1) industrially produced fertilizers, (2) the cultivation of crops, such as soybeans and peas that harbor nitrogen-fixing bacteria, and (3) the release of nitric oxide during the combustion of fossil fuels.

Adding nitrogen to terrestrial ecosystems usually increases productivity. In some cases, then, a massive increase in nitrogen availability is beneficial. But in other situations, it is not.

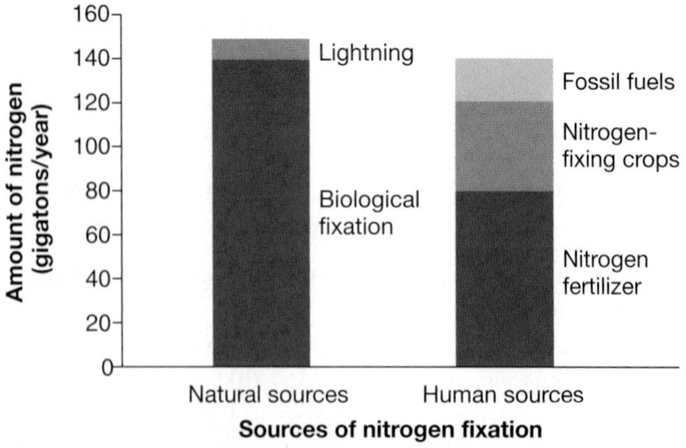
160
140
120
100
80
60
40
20
0
Amount of nitrogen (gigatons/year)
Lightning
Biological fixation
Fossil fuels
Nitrogen-fixing crops
Nitrogen fertilizer
Natural sources
Human sources
Sources of nitrogen fixation

FIGURE 54.20 Humans Are Adding Large Amounts of Nitrogen to Ecosystems
Human activities now fix almost as much nitrogen each year as natural sources do. Thus, human activities have almost doubled the total amount of nitrogen available to organisms.

Chapter 27, for example, detailed how excessive applications of nitrogen-containing fertilizers on farmlands have produced nitrogen-laced runoff, which has had disastrous impacts on numerous aquatic ecosystems. Researchers have also documented that nitrogen inputs can lead to a significant loss of biodiversity in terrestrial ecosystems. For example, when biologists added nitrogen to plots of native grasslands in midwestern North America, a few competitively dominant species tended to take over. As those species grew rapidly, they displaced other species that did not respond to nitrogen inputs as strongly. In this ecosystem, increased nitrogen boosted productivity but decreased species diversity, by altering the balance of competitive interactions. The same result has occurred in study plots within the Park Grass experiment, introduced in Chapter 52. The plots that have been fertilized with nitrogen since 1856 contain many fewer species than unfertilized plots do.

To summarize, several of the most pressing environmental problems facing our species result from recent and massive alterations in biogeochemical cycles. Even local changes in biogeochemical cycles tend to have large-scale consequences, because nutrients are transported among ecosystems. If you are in your late teens or early twenties as you read this text, you are part of a generation that is projected to experience the most traumatic episode of environmental change in human history. The trauma has two sources: the massive loss of species documented in Chapter 26 and Chapter 55 and the profound changes in global biogeochemistry recorded here. Next we'll take a closer look at the consequences of human activities.

✓CHECK YOUR UNDERSTANDING

Biogeochemical cycles trace the movement of water or elements among components of an ecosystem. Biologists analyze three aspects of these cycles: (1) the nature and size of pools where elements are stored or captured, (2) the rate of movement between reservoirs, and (3) interactions among cycles. You should be able to evaluate these three aspects of the water, carbon, and nitrogen cycles.

54.3 Human Impacts on Ecosystems

Two factors are responsible for the human impacts on ecosystems currently being documented by biologists. The first is the rapid increase in human population size, analyzed in Chapter 52. The second is the rapid increase in resources used by people. Figure 54.18a introduced this second factor by showing increases in fossil-fuel consumption and land use that have occurred over about the past 150 years. **Figure 54.21** follows up by providing data on the variation in average annual energy consumption per person around the world. The key point is that residents of industrialized countries, though relatively few in number, have a disproportionately large impact on biogeochemical cycles because they use so much energy, water, food, and other resources. Let's look at two of the consequences of population growth and resource use: global warming and productivity increases.

Global Warming

Recall that carbon dioxide concentration in the atmosphere has been increasing throughout the twentieth century and that most analyses indicate that the increase is due to the burning of fossil fuels and clear-cutting of land, particularly forests, for agriculture. Whether the increase in CO_2 is producing **global warming**—an increase in Earth's surface temperature, averaged over the globe, has been intensely controversial until recently, however.

In 1988 an international group of scientists, called the Intergovernmental Panel on Climate Change (IPCC), was formed to evaluate the consequences of rising CO_2. The group has since produced a series of reports summarizing the state of scientific knowledge on the issue. In 1998 the report concluded that current evidence suggests a "discernible human influence on climate." This was the IPCC's first statement supporting the hypothesis that rising CO_2 concentrations due to human activities are having a measurable impact on climate. The IPCC's most recent report, released in 2001, stated that "There is new

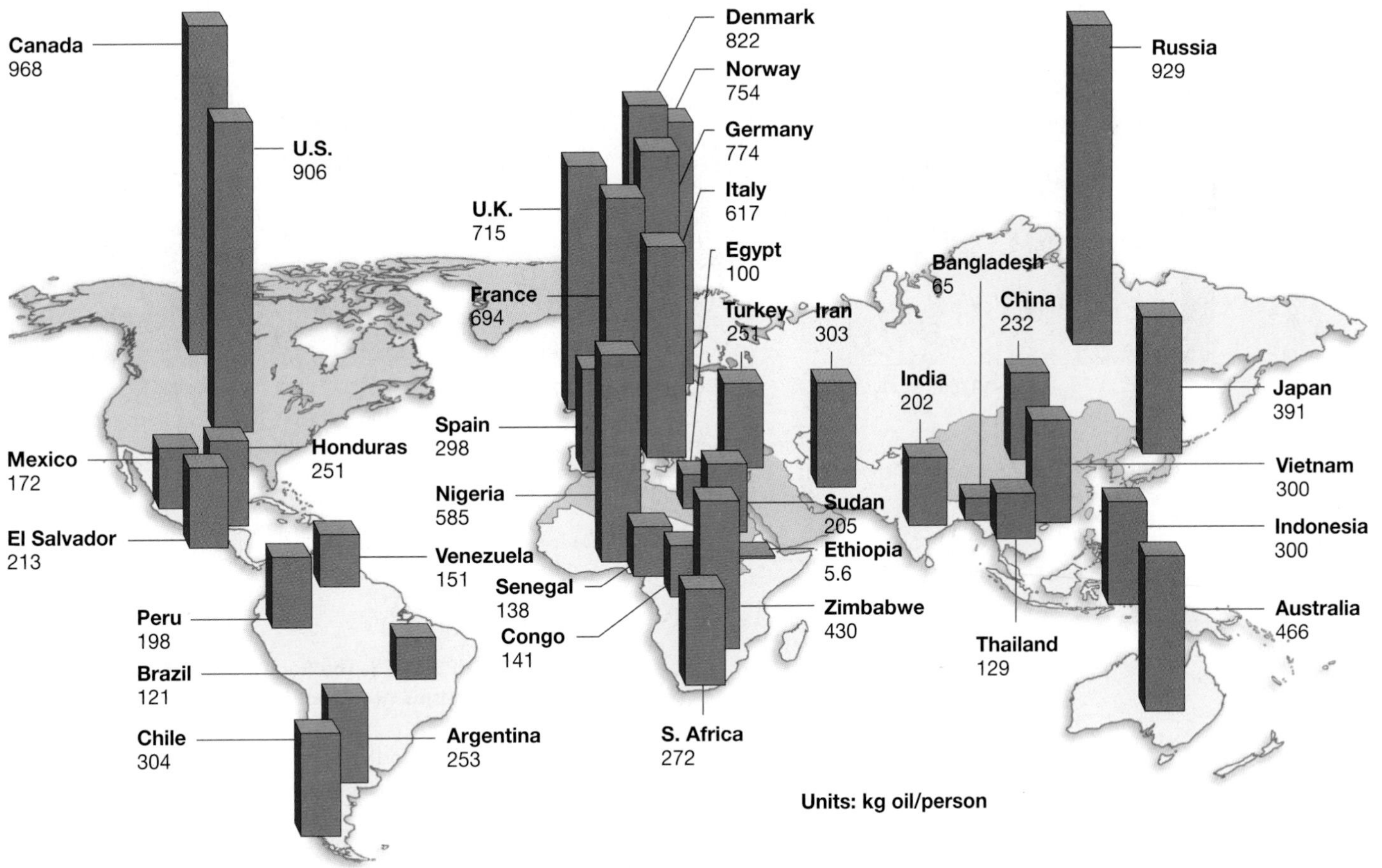

FIGURE 54.21 Energy Consumption Varies among Countries
Per-capita energy use in 1999.

and stronger evidence that most of the warming observed over the last 50 years is attributable to human activities." The panel has now taken the position that rising concentrations of greenhouse gases are at least partially, and perhaps primarily, responsible for global warming.

How much will average temperatures rise in our lifetimes? Predicting the future state of a system as complex and variable as Earth's climate is extremely difficult. Global climate models are the primary tools that scientists use to make these projections. A global climate model is based on a large series of equations that describe how the concentrations of various gases in the atmosphere, solar radiation, transpiration rates, and other parameters interact to affect climate. The models currently being used suggest that average global temperature will undergo additional increases of up to 3.5°C over the next 50 years and 1.4–5.8°C (2.2–10°F) by the year 2100.

How will ecosystems respond to this increase? Answering this question is equally difficult, because ecosystems respond to warming in ways that increase or decrease CO_2 concentrations and thus exacerbate or mitigate warming. Positive feedbacks, for example, occur when warmer and drier climate conditions lead to more fires, which in turn release more CO_2, which leads to even more warming. Researchers recently documented that a form of positive feedback is already occurring in arctic tundras. Traditionally, tundras sequester carbon in the form of soil organic matter, because decomposition rates are extremely low there. During a series of warm summers in the 1980s, however, researchers found that decomposition rates increased sufficiently to release carbon from stored soil organic matter and to produce a net flow of carbon to the atmosphere.

Negative feedbacks, in contrast, arise when warmer conditions lead to increased photosynthetic rates and hence to an increase in the uptake of CO_2. In addition, the growth rates of several tree species and some agricultural crops have been shown to increase in direct response to increasing atmospheric CO_2. Because CO_2 is required for photosynthesis, it can act as a fertilizer.

Will ecosystem responses increase or decrease global warming? Currently, it is not clear whether positive or negative feedbacks will predominate. Researchers are working to answer this question by using computer models, experiments, and analyses of how ecosystems responded to past climate changes.

Biologists have also documented dramatic impacts that warming temperatures are already having on organisms. **Figure 54.22a** shows recent data on the ranges of copepods in the North Atlantic. Copepods are small crustaceans that are often the dominant predator in marine plankton. The maps

(a) Copepods that live in cold water are declining in the North Atlantic.

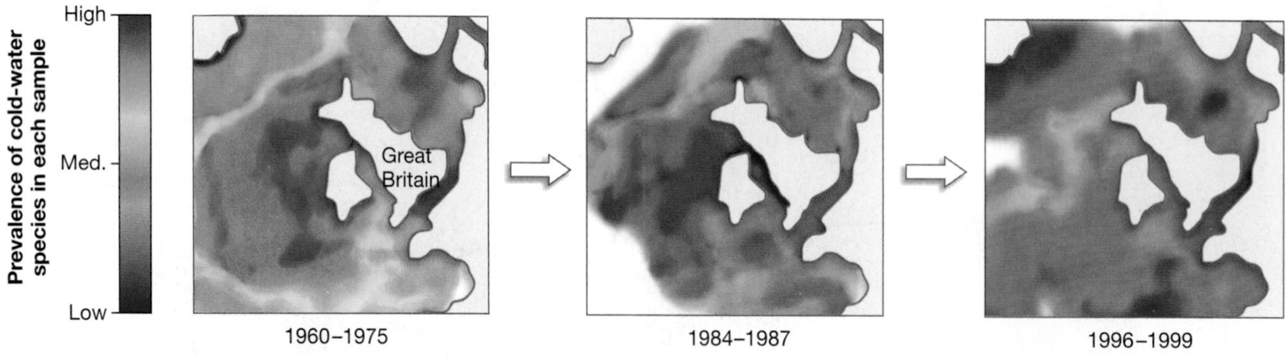

(b) Flowering times for some species in midwestern North America are earlier in the year.

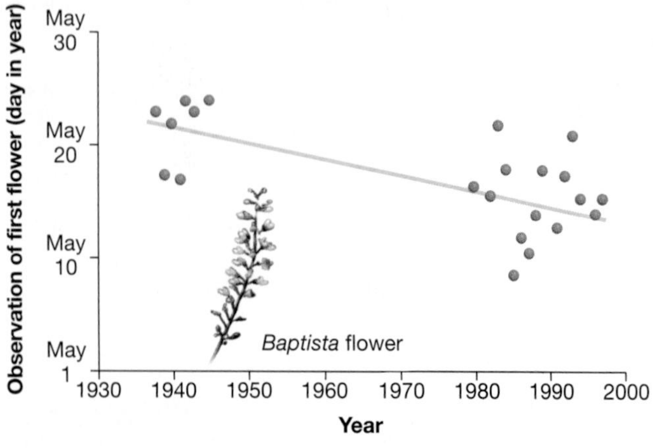

FIGURE 54.22 The Ranges and Behavior of Some Organisms Are Changing in Response to Global Warming
(a) In the North Atlantic, species that live in cold water are declining in frequency; species that live in warm water are increasingly common. **(b)** In a data set from Wisconsin, 19 of the 55 events tracked over a 61-year period showed a significant increase in earliness. Only one event declined in earliness; the rest showed no change.

show changes that occurred in a 40-year period, beginning in 1960, in the number of copepod species typical of marine waters off southern Europe versus the number of copepod species typical of northern European waters. The key observation is that southern species have moved north, while the range of cold-water species has declined. The graph in **Figure 54.22b**, which shows changes in the date of first flowering for a plant native to the midwestern United States, has a similar message. Over the past 60 years, the first day of the year when flowers are produced by certain species has moved up by almost 20 days. To interpret this observation, biologists hypothesize that the climate has warmed enough to promote growth and flowering earlier in the year. Global warming has caused significant changes in the geographic ranges and behavior of organisms, and it promises to cause more.

Productivity Increases

Several of the changes that humans are inducing in biogeochemical cycles have the same effect: They increase productivity. Warming temperatures, the addition of nitrogen and other nutrients, and rising CO_2 levels have all been shown to increase NPP, at least in certain ecosystems. The question is, What are the short-term and long-term consequences of large-scale increases in productivity?

In previous chapters, increased productivity has been viewed as beneficial for plant communities. Chapter 53 pointed out that high productivity resulting from high temperatures is one of the hypotheses to explain why tropical wet forests are so species rich. But several other lines of evidence suggest that increased productivity may have negative effects, at least in certain ecosystems:

- Data introduced in Section 54.2 suggest that if grassland habitats are fertilized with nitrogen, productivity increases but species richness declines over time.
- Nitrate ions derived from fertilizers applied to cornfields in the midwestern United States are washing into the Gulf of Mexico. Increased nitrate concentrations have stimulated the growth of planktonic organisms, whose subsequent decomposition has used up available oxygen and triggered the formation of anaerobic "dead zones" in the gulf (see Chapter 27). In this way, excessive fertilization has caused a severe pollution problem.
- Biologists hypothesize that recent increases in levels of nitrate and other fertilizers of human origin have contributed to the frequency or intensity of harmful algal blooms—large populations of dinoflagellates that release toxins into the surrounding water (see Chapter 28).
- During the 1960s and 1970s it became popular in industrialized nations to use phosphate-based detergents for washing clothes. Much of this phosphate ended up in lakes and streams, where it triggered a massive growth of cyanobacteria—which normally are phosphorus limited in these ecosystems. The response was **eutrophication**: the conversion of a lake to a highly productive ecosystem with rapid decomposition, low oxygen levels, and rapid filling with decomposing organic matter.

Overall, it is not clear whether increased productivity will be beneficial or detrimental to ecosystems. The answer should be forthcoming, however. Because large-scale additions of nitrogen, phosphorus, and carbon dioxide are continuing, humans are implementing a global-scale fertilization experiment.

ESSAY Phosphate Pollution, Acid Rain, and the Ozone Hole: Hope for Ecosystem Recovery?

A central message of this chapter is that energy flows and nutrient cycling are undergoing extraordinary changes in ecosystems throughout the world. Although the outcome of these changes is uncertain, it is important to recognize that humans have already identified and acted on several recent changes in the abiotic environment that clearly had negative consequences. The events and responses took place at the local, regional, and global levels.

The first example of an effective response to an abiotic change took place at a local level and involved a global nutrient cycle. Like many other elements and molecules, phosphorus cycles through ecosystems. As the close of this chapter notes, the use of phosphate-based detergents in the industrialized countries led to a large increase in the concentrations of phosphate in lakes and streams, triggering rapid and widespread eutrophication—particularly in shallow lakes that received outflow from municipal sewage systems. In response, governments in North America and Western Europe encouraged or required the use of phosphate-free detergents, and sewage plants were upgraded to remove more phosphate during treatment. Although

Effective responses have occurred ... when biologists documented serious problems in ecosystem ecology.

(Continued on next page)

(Essay continued)

phosphate pollution from farm fertilizers remains a serious problem, the crisis conditions of the 1960s and 1970s (**Figure 54.23a**) have largely been alleviated.

The second example involves changes in the pH of rainwater at a regional level. The problem began with sulfur oxides and nitrogen oxides that are pumped into the atmosphere by coal-burning electrical power plants and vehicles that lack catalytic converters. When exposed to sunlight and water vapor, the molecules react to form sulfuric acid (H_2SO_4) and nitric acid (HNO_3). Normal rainwater has a slightly acidic pH, about 5.6. But in areas affected by **acid rain**, precipitation can have a pH as low as 4.2 or 4.4. During the 1980s and 1990s, biologists documented that forests and lakes in eastern North America and northern Europe were being affected by acid rain (**Figure 54.23b**). Tree growth slowed in response to the acidification of soils, and lakes became less productive and less diverse. Once biologists had documented the problem, governments instituted stricter controls on the amounts of sulfur oxides and nitrogen oxides that could be emitted from power plants, cars, and trucks. Over the past decade, the intensity of acid rain has diminished and some of the ecosystems that were being affected have begun to recover.

A third example involves changes in atmospheric chemistry that were global in scale. Widespread use of the compounds called *chlorofluorocarbons* (CFCs) in refrigeration and aerosol products resulted in the release of thousands of tons of CFCs into the atmosphere. When CFCs accumulated in the upper atmosphere, they participated in chemical reactions that released chlorine atoms. These chlorine atoms subsequently reacted with ozone (O_3) molecules, which also accumulate in the upper atmosphere. In some years, the loss of ozone due to these reactions was so severe that an **ozone hole** opened over Antarctica (**Figure 54.23c**). This issue concerned scientists from around the world, because ozone absorbs large amounts of ultraviolet (UV) radiation. When the ozone layer that surrounds Earth thins or is wiped out, an excess of UV radiation can reach Earth's surface and act as a mutagen and carcinogen (see Chapter 18). Fortunately, soon after the problem was documented in the early 1990s, international treaties scheduled and enforced the rapid phasing out of CFC production and use. Scientists have recently been able to document the first signs that the size and duration of the ozone hole may be moderating.

The message of these examples is clear: Effective responses have occurred at the local, regional, and global levels when biologists documented serious problems in ecosystem ecology. It remains to be seen whether the same success can be achieved in response to global warming, nitrate pollution, and other current problems in the abiotic environment.

(a) Phosphate pollution: lake eutrophication

(b) Sulfur oxide and nitrogen oxide pollution: acid rain

(c) CFC pollution: ozone hole (blue region) over Antarctica

FIGURE 54.23 Some Pollution Problems Are Getting Better
(a) Although eutrophication reached a crisis situation in some lakes in the 1960s, declining phosphate release from sewage systems has alleviated the problem in many areas. **(b)** Acid rain was an increasing problem in some regions of the world until clean air legislation resulted in air-quality improvements in the 1990s. **(c)** The ozone hole has started to show its first signs of shrinking, in response to lowered CFC emissions.

CHAPTER REVIEW

Summary of Key Concepts

An ecosystem has four components: (1) the abiotic environment, (2) primary producers, (3) consumers, and (4) decomposers. These components are linked by the movement of energy and nutrients.

An ecosystem consists of one or more communities of interacting species and their abiotic environment. As energy flows through ecosystems and as nutrients cycle through them, energy and nutrients are exchanged between biotic and abiotic components of the ecosystem. Energy flows into ecosystems as a result of photosynthesis or the respiration of inorganic molecules with high potential energy. Chemical energy from producers enters the consumer food web or the decomposer food web.

The productivity of terrestrial ecosystems is limited by warmth and moisture, while nutrient availability is the key constraint in aquatic ecosystems. As energy flows from producers to consumers and decomposers, much of it is lost.

Among terrestrial ecosystems, net primary productivity (NPP) is highest in tropical wet forests and tropical dry forests. Among aquatic ecosystems, productivity is highest in coral reefs, wetlands, and estuaries. Although productivity is extremely low in the oceanic zone, the area covered by this ecosystem is so extensive that the open ocean accounts for the highest percentage of Earth's overall productivity.

Organisms that acquire energy from the same type of source are said to occupy the same trophic level. Most ecosystems have at least three trophic levels: primary producers, herbivores or primary decomposers, and carnivores. Because energy transfer from one trophic level to the next is inefficient, ecosystems have a pyramid of productivity: Biomass is highest at the lowest trophic level and lower at each higher trophic level.

The feeding relationships among species in a particular ecosystem are described by a food chain or food web. Food chains rarely exceed five or six trophic levels. The energy-flow, dynamic-stability, and environmental-complexity hypotheses have been proposed to explain why food webs and food chains are not longer.

To analyze nutrient cycles, biologists focus on the nature of the reservoirs where elements reside and on how quickly elements move between reservoirs.

Nutrients move through ecosystems in biogeochemical cycles. The rate of nutrient cycling is strongly affected by the rate of decomposition of detritus. The decomposition rate, in turn, is affected by abiotic environmental conditions such as temperature and by the quality of the detritus. Nutrients are also lost from ecosystems. Experiments have shown that the loss of vegetation greatly increases the rate of nutrient loss.

Web Tutorial 54.1 The Global Carbon Cycle

Humans are causing large, global changes in the abiotic environment. The burning of fossil fuels has led to rapid global warming. Extensive fertilization is increasing productivity and causing pollution.

Average global temperatures are increasing rapidly because land-use changes and burning of fossil fuels have increased the flow of carbon in the form of CO_2 into the atmosphere, and because carbon dioxide acts as a greenhouse gas. Nitrogen fixation from fertilizer production and from the planting of nitrogen-fixing crop species is now approximately equal to the amount of nitrogen fixation from natural sources. These increases have led to increased productivity, but also to pollution and to loss of biodiversity.

Questions

Content Review

1. What is the difference between a community or group of communities and an ecosystem?
   a. A community and the abiotic environment comprise an ecosystem.
   b. An ecosystem is a type of community.
   c. A biome includes only the plant community or communities present in an environment.
   d. An ecosystem includes only the abiotic aspects of a particular environment.
2. Which of the following ecosystems would you expect to have the highest primary production?
   a. subtropical desert
   b. temperate grassland
   c. boreal forest
   d. tropical dry forest
3. Most of the net primary productivity in an ecosystem is used for what purpose?
   a. respiration by primary consumers
   b. respiration by secondary consumers
   c. growth by primary consumers
   d. growth by secondary consumers
4. According to the dynamic-stability hypothesis for food-chain length, food chains will be shorter in which type of environment?
   a. cold
   b. constant
   c. variable
   d. low in nutrient availability

5. Which of the following is normally the longest-lived reservoir for carbon?
   a. atmosphere (CO_2)
   b. marine plankton (primary producers *and* consumers)
   c. petroleum
   d. wood

6. Devegetation has what effect on ecosystem dynamics?
   a. It increases belowground biomass.
   b. It increases nutrient export.
   c. It increases rates of groundwater recharge (penetration of precipitation to the water table).
   d. It increases the pool of soil organic matter.

Conceptual Review

1. Draw a pyramid of productivity for a temperate forest ecosystem, and explain its shape.
2. Explain the difference between gross primary productivity and net primary productivity. Which is larger, and why? Why is total primary production always greater than total secondary production?
3. Explain why decomposition rates are higher in some ecosystems than in others, and give examples. How does the decomposer food web regulate nutrient availability in an ecosystem?
4. Compare and contrast the energy-flow, dynamic-stability, and environmental-complexity hypotheses for food-chain length.
5. Draw a diagram of the global water, carbon, or nitrogen cycle. Label major reservoirs and flows. Compare and contrast the life span of water or an element in each reservoir, and evaluate factors that affect the rate of movement between reservoirs.
6. Why are the open oceans nutrient poor? Why are neritic zones and intertidal habitats relatively nutrient rich?

Group Discussion Problems

1. Suppose you had a small set of experimental ponds at your disposal and an array of pond-dwelling algae, plants, and animals. How could you use radioactive isotopes of carbon or phosphorus to study energy flows or nutrient cycling in these experimental ecosystems?
2. Some researchers are concerned that fertilizing the open oceans with iron would lead to overfertilization of neritic zones, because larger amounts of nutrients would be carried into coastal regions by ocean currents. Outline an experiment and a computer simulation study to test this hypothesis. Evaluate which approach—the experiment or the computer simulation—would likely be more effective in addressing the hypothesis.
3. Suppose that herbivores were removed from a temperate deciduous forest ecosystem. Predict what would happen to the rate of nitrogen cycling. Explain the logic behind your prediction.
4. Explain why human-caused changes to the global carbon cycle are affecting Earth's climate. State why you think that these changes are beneficial or detrimental. List three things that you, local institutions, and state and national governments can do to either augment or mitigate these changes.

Answers to Multiple-Choice Questions 1. a; 2. d; 3. a; 4. c; 5. c; 6. b

www.prenhall.com/freeman is your resource for the following: Web Tutorials; Online Quizzes and other Online Study Guide materials; Answers to Conceptual Review Questions; Solutions to Group Discussion Problems; Answers to Figure Caption Questions and Exercises; and Additional Readings and Research.

Biodiversity and Conservation

55

Lowland wet forest along the Segama River in Borneo, Indonesia. Most of the world's biodiversity is found in tropical wet forests.

KEY CONCEPTS

- Biodiversity can be analyzed at the genetic, species, and ecosystem levels.
- Humans depend on biodiversity for the products that wild species provide and for ecosystem services that protect the quality of the abiotic environment.
- If present trends in human population growth and habitat destruction continue, a mass extinction event is under way.
- Effective conservation programs start with sound biological research and involve economists, politicians, and local residents.

In a technologically advanced world filled with dizzying amounts of information, it is easy to assume that biologists have already answered most of the important questions about the natural world. But this assumption could not be further from the truth. No one knows how many species live on Earth. Of the approximately 1.4 million species that have been described to date, only a tiny percentage could be described as well studied. Biologists are only beginning to understand the details of how species interact with one another; we have even more to learn about how groups of species interact with the abiotic world.

One thing is certain, however: Researchers who study biological diversity are doing their work with a newfound urgency. The world population of humans will surpass 6.5 billion people by the time you read this chapter and promises to exceed 9 billion during your lifetime. Human activities are changing ecosystems all over Earth at an ever-increasing rate. For existing species, most of these changes are negative. The global loss of biodiversity and degradation of ecosystems are arguably the greatest challenge our species has ever faced.

Meeting this challenge will require an expanded global community dedicated to the conservation of biodiversity. The role of the biologist in this community is central. While conservation is a complex issue requiring the expertise and cooperation of economists, conservationists, government administrators, educators, business professionals, and community leaders, any effective course of action must be based on accurate scientific information. To design and implement successful conservation measures, we need to understand more about how ecosystems work—particularly the roles of their component species. The data that biologists collect are the starting point for conservation action.

Let's delve into the science behind conservation work by examining what is known about biodiversity, beginning with its often unrecognized significance to our own lives. Then we can review the principal human activities that cause environmental degradation and loss of biodiversity, and close by exploring some of the local and global efforts to conserve Earth's biodiversity.

55.1 What Is Biodiversity?

Perhaps the simplest way to think about biodiversity is in terms of the tree of life that was introduced in Chapter 1. Recall that biologists have used DNA sequence data and morphological traits to estimate the evolutionary relationships among all forms of life. These analyses have supported Charles Darwin's hypothesis that all organisms are related to each other by descent from a common ancestor. The exact relationships among lineages and species are also coming into clearer and clearer focus as biologists compare DNA sequences from more species, measure and interpret additional morphological traits, and discover and analyze new fossil remains. Stated another way, biologists are learning more and more about which species are closely or distantly related, and thus about the size and shape of the tree of life (see Chapters 27 through 34).

Biodiversity Can Be Measured and Analyzed at Several Levels

In a fundamental sense, **biodiversity** can be thought of as the tips of the branches on the tree of life—meaning all distinctive populations and species living today. More formally, biologists recognize and analyze biodiversity on three levels:

1. **Genetic diversity** is the total genetic information contained within all individuals of a species and is measured as the number and relative frequency of all alleles present in a species. Because no two members of the same species are genetically identical, each species is the repository of an immense array of alleles. At this level, biodiversity is everywhere around you, from the variety of apples at the grocery store to the variation in coloration and singing ability of sparrows in your backyard.
2. **Species diversity** represents the variety of life-forms on Earth—the twigs on the tree of life. Recall from Chapter 53 that, in practice, species diversity is measured as the number and relative frequency of species in a particular region. There is an additional aspect of species diversity, however, that biologists call *taxonomic diversity*. When biologists assess taxonomic diversity, they attempt to incorporate the evolutionary relationships among the species present. For example, some lineages on the tree of life are extremely species rich, while others may be represented by only a single living species. African cichlid fish, introduced in Chapter 26, and the 35,000 species of orchids are examples of species-rich lineages. In contrast, the red panda (**Figure 55.1a**) and the Indian river dolphin (**Figure 55.1b**) have few close relatives. They are examples of species-poor lineages. Some biologists argue that it is particularly important to preserve populations from species-poor lineages, because they are the last living representatives of their lineages. While measures of species diversity focus on the number and relative abundance of species present, measures of taxonomic diversity incorporate how distinct each species is in its evolutionary history.
3. **Ecosystem diversity** is the variety of biotic communities in a region along with abiotic components, such as soil and water and nutrients. Ecosystem diversity is more difficult to define and measure than genetic diversity or species diversity, because ecosystems do not have sharp boundaries. Recall from Chapter 54 that ecosystems are complex and

(a) Red panda

(b) Indian river dolphin

FIGURE 55.1 Phylogenetically Distinct Species May Be High-Priority Targets for Conservation
(a) The red panda and **(b)** the Indian river dolphin have few close relatives and represent distinct branches on the tree of life. If conservation measures attempt to preserve taxonomic diversity, these species would be a high priority for preservation.

dynamic assemblages of organisms that interact with each another and their nonliving environment. Attempts to measure ecosystem diversity focus on capturing the array of biotic communities in a region, along with variation in the physical conditions present.

In addition to recognizing that biodiversity can be recognized and quantified on several distinct levels, it is important to note that biodiversity is dynamic. Mutations that create new alleles increase genetic diversity; natural selection, genetic drift, and gene flow may eliminate certain alleles or change their frequency, leading to an increase or decrease in overall genetic diversity. Speciation increases species diversity, while extinction decreases it (see Chapters 25 and 26). Changes in climate or other physical conditions can result in the formation of new ecosystems, as can the evolution of new species that interact in novel ways. Disturbances such as volcanic eruptions, human activities, and glaciation can destroy ecosystems.

Biodiversity is not static. It has been changing since life on Earth began.

Why Is Biodiversity Important?

If you live in a climate-controlled building, buy your food at a grocery store, and get your medicine from a pharmacy, it can be difficult to recogize the extent to which humans depend on biodiversity. But human health and well-being have always been closely tied to biodiversity.

When someone asks a biologist why biodiversity is important, the answers make up a long list that can be broken into two general categories: (1) ways that the presence of many species benefits humans directly and (2) ways that biodiversity aids humans indirectly by contributing to a healthy overall environment. Although most of these benefits are explained in detail in other chapters, it will be helpful to pull them together into a compact list.

Direct Benefits of Biodiversity: Providing Goods and Services The first large-scale, highly organized human societies began to form about 10,000 years ago, when people at several locations around the world began to domesticate wild plants and farm their food. Since that time, wild species have provided the raw material to fuel the continued development of human societies. Vast tracts of forest have been felled for building materials and fuel; wild plants have been selectively bred to yield food and fibers; animals have been domesticated to provide food, labor, and material goods; the oceans have been harvested for protein; and plants, animals, and fungi have been processed as sources of medicines. For thousands of years, humans have relied on a diversity of wild species to survive.

The direct use of biodiversity continues today:

- Research programs collectively known as *bioprospecting* focus on assessing bacteria, archaea, plants, and fungi as novel sources of drugs or ingredients in consumer products (see Chapters 27, 29, and 30). Bioprospecting has benefited from the recent explosion of genetic and phylogenetic information, because biologists can now search genomes from a wide array of species to find alleles with desired functions. In addition, recent advances in biotechnology facilitated the development of a new painkiller from the paralyzing sting of tropical cone snails and a blood anticoagulant from the saliva of vampire bats.
- Agricultural scientists are preserving diverse strains of crop plants in seed banks and continue to use wild relatives of domesticated species in breeding programs aimed at improving crop traits (see Chapter 29). In addition, genetic engineering techniques are being used to transfer alleles from a diverse array of species into crop plants. In some cases these efforts have reduced dependence on pesticides, increased resistance to diseases and drought, and improved nutritional value and overall crop yields (see Chapter 19).
- The production of almonds, apples, cherries, chocolate, alfalfa, and an array of other crops depends on the presence of wild pollinators. In the United States alone, insect-pollinated crops produce $40 billion worth of products annually.
- Strategies for cleaning up oil spills, abandoned mines, and contaminated industrial sites are incorporating *bioremediation*—the use of bacteria, archaea, and plants to metabolize pollutants and render them harmless (see Chapters 27 and 29).
- Recreation based on visiting wild places, or *ecotourism*, is a major industry internationally; it is growing rapidly. In South Africa, for example, the number of tourists visiting wildlife preserves increased from 454,428 in 1986 to almost 6 million in 1997. In developing nations with well-designed nature preserves, it is not unusual for ecotourism to represent the first or second most important source of foreign currency earnings.
- High ecosystem diversity—particularly the presence of forests or grasslands on steep slopes and the presence of wetlands in low-lying areas—dramatically reduces flood damage.

Indirect Benefits of Biodiversity: Ecosystem Services The benefits of biodiversity extend beyond the direct use of diverse genes and species by humans to include **ecosystem services**—processes that increase the quality of the abiotic environment. Recall from Chapter 10 that green plants and other photosynthetic organisms produce the oxygen we breathe, and from Chapter 29 that the presence of land plants builds soil, reduces soil erosion, moderates local temperature and wind conditions, and increases the volume of water retained in lakes, streams, and soils. Species from throughout the tree of life are involved in cycling nitrogen, carbon, and other nutrients through ecosystems (see Chapters 27 through 30 and 54).

Current research on ecosystem services is focused on two issues: Economists are attempting to quantify the dollar value of ecosystem services as a way of justifying the cost of preserving natural areas, and biologists continue to document how the loss of biodiversity is affecting the quality of the abiotic environment.

Loss of biodiversity also affects biotic interactions in ecosystems. Recall from Chapter 53 that the loss of keystone predators or herbivores has had a dramatic effect on species diversity and the species present in several types of habitats. Similarly, the food chains, food webs, and mutualistic and parasitic interactions introduced in Chapter 53 change markedly when species are lost. With respect to the biotic aspects of ecosystem functioning, eliminating a species can be like breaking a thread in a piece of cloth and pulling on it. The species itself is lost, and its interactions with other organisms and the physical environment are lost. When the broken thread is an organism, the possibility that it will continue to evolve is lost also.

55.2 How Do Biologists Study Biodiversity?

Our current understanding of Earth's biodiversity is poor. Scientists have a more accurate estimate of how many stars there are in the known universe than of how many species coexist with us on this planet. It is not even clear how many species have been formally described to date. There is no central registry or database for the estimated 1.4 to 1.75 million described species. Work is now under way to use the World Wide Web to build such an encyclopedia of life.

The knowledge base of biodiversity is also heavily biased. The most intensively studied lineages on the tree of life are the vertebrates, flowering plants, conifers, and other groups that are large, attractive, and relatively easy to study. Bacteria, archaea, insects, mites, and other groups are acknowledged to be species rich but are much less well known. In addition, species that inhabit the temperate regions of the world have been studied much more extensively than tropical species.

Even among the larger plants and animals, however, new species are described each year. The number of recognized amphibian species increased by 36 percent in the last 20 years. Since 2000, 30 new species of reptiles have been documented. Biodiversity surveys in less well studied parts of the world, especially the tropics and the oceans, regularly find new species of flowering plants, crustaceans, and mollusks.

Understanding the extent and importance of biodiversity is one of the most massive and exciting challenges facing biological science today. To get a feel for how biologists address this challenge, let's explore four questions about biodiversity: (1) How can biologists estimate levels of genetic diversity? (2) How many species are alive today? (3) What does a thorough understanding of species diversity entail? and (4) How does species diversity affect the function of ecosystems?

Quantifying Genetic Diversity

Recall that, at the genetic level, biodiversity means allelic diversity. When the first techniques for analyzing genes and proteins from many individuals became available in the 1960s, biologists immediately began using those techniques to quantify the level of genetic diversity within species. The basic approach was to collect tissue samples from many individuals within a population or species, isolate proteins or DNA from the samples, and document how many different alleles existed at a particular gene or suite of genes. With some notable exceptions, the overall message of this work is that most species contain a remarkably high level of allelic diversity. This was an important result, because it meant that changes in natural selection or other evolutionary processes could cause rapid changes in the frequencies of existing alleles and thus a rapid evolutionary response to environmental change. It also meant that losing a species to extinction signifies losing a large number of unique alleles.

Research on the genetic aspects of biodiversity continues and has recently taken on a new goal: quantifying levels of genetic diversity in entire communities or ecosystems rather than in individual species. To appreciate how this work is done and why it is important, consider recent research on genetic diversity among bacteria that live in the Sargasso Sea. The Sargasso Sea is located in the Caribbean region near the islands of the Bahamas and is exceptionally nutrient poor and species poor. It is considered one of the ocean's great deserts in terms of the abundance and diversity of life present. Because it represents such a simple system, the Sargasso Sea was an attractive choice for one of the first studies of genetic diversity in an ecosystem.

To inventory the complete array of bacterial genes present, a research team of investigators collected cells from different water depths and locations. The team isolated DNA from the samples and used the "shotgun sequencing" approach, introduced in Chapter 20, to sequence the complete genome of each species present. They did the gene sequencing in the absence of any information about the nature of the species present—the researchers did not try to culture or name the cells but instead simply sequenced all the bacterial DNA present in the Sargasso Sea samples. After analyzing over 1 billion base pairs of distinct sequences and grouping them by gene and by species, the team concluded that at least 1800 bacterial species were present, of which 148 were previously undiscovered. The researchers identified more than 1.2 million alleles that had never before been characterized. This result suggests that an enormous amount of genetic diversity exists even in Earth's simplest ecosystems.

The research team also claims that this genetic diversity is important. For example, they found more than 780 new bacterial alleles that code for proteins similar to the rhodopsin molecules that are found in your eyes. Do these proteins function as light receptors, like rhodopsin does? If so, does the presence of so many distinct rhodopsin-like genes mean that different bacterial species have different ways of responding to light? Answers to questions like these will come only with continued research. Stay tuned.

Estimating the Total Number of Species Living Today

Biologists acknowledge that only a fraction of the number of species living on Earth has been discovered and described. Consider recent work by Philippe Bouchet and colleagues. This group conducted a massive survey of marine mollusks in coral-reef habitats along the west coast of New Caledonia, a tropical island in the southwest Pacific Ocean. The team spent more than a year collecting mollusks, using an array of techniques, at 42 sites over a total area of almost 300 square kilometers. The survey represented the most thorough sampling effort ever made to determine the species diversity of mollusks.

The survey produced more than 127,000 mollusks representing 2738 species. These numbers far exceed the mollusk diversity recorded for any comparable-sized area. The species total was 2 to 3 times the total number of mollusk species that had been reported for similar habitats in the region.

In reporting their findings, the biologists emphasized that 20 percent of the species found were represented by a single specimen. This observation suggests that many species are exceedingly rare and thus likely to be missed by less-intensive sampling efforts. And when the investigators compared their data with a survey in progress at a second site, different in reef structure but only 200 kilometers away, they found that only 36 percent of species were shared between the two sites. The results from the New Caledonia survey support the hypothesis that the global biodiversity of mollusks—currently thought to be about 93,000 species—is an underestimate.

Given that only a fraction of the organisms alive have been discovered to date, how can biologists go about estimating the total number of species on Earth? Two general approaches have been used. One is based on intensive surveys of species-rich groups at very small sites, and a second is based on attempts to identify all taxa in a particular region.

Taxon-Specific Surveys To estimate the number of insect species that live in the canopy of tropical trees, Terry Erwin and J. C. Scott used an insecticidal fog to knock down insects from the top of a *Luehea seemannii* tree. The researchers identified over 900 species of beetles among the individuals that fell. Most of these species were new to science. To use these data as an indicator of total arthropod species diversity, the researchers used the following train of logic: Based on earlier work with insects on this tree species, they estimated that 160 of the 900 beetle species live only on *L. seemannii*. Worldwide, beetles represent about 40 percent of all known arthropods. Thus, it was reasonable to suggest that 400 species of arthropods live only in the canopy of *L. seemannii*. By adding an estimate of arthropods specializing on the trunk and roots of this tree, Erwin projected that it is host to 600 specialist arthropods. If each of the 50,000 species of tropical tree harbors the same number of arthropod specialists, then the world total of arthropod species exceeds 30 million species. Based on such studies, biologists estimate that at least 10 million, and possibly as many as 100 million, species of all types exist today.

(a) Great Smoky Mountains National Park

(b) All-taxon survey

FIGURE 55.2 The First All-Taxon Survey Is Now Under Way
The first attempt to document every species living in a prescribed area is under way at Great Smoky Mountains National Park, along the border of Tennessee and North Carolina.

All-Taxon Surveys To obtain a more direct estimate of total species numbers, the first effort to find and catalog all of the eukaryotic species present in a large area is now under way. The location is the Great Smoky Mountains National Park in the southeastern United States (**Figure 55.2**). A consortium of biologists, volunteers, and research organizations initiated this all-taxon survey in 1999. To date, the survey has discovered over 230 species that are new to science and over 1750 species that had never before been found in the park. When the inventory is complete, in 2015, biologists will have a much better database to use in estimating the extent of global biodiversity.

Steps in Understanding Biodiversity

A thorough understanding of biodiversity involves several levels of research. The first step is to name a species. When a new species is discovered, researchers write a paper describing its morphology or other distinguishing traits and assign it a genus and species name. Usually this written description is associated with a preserved physical specimen that is deposited in a collection at an herbarium or natural history museum. These repositories serve as the world's biodiversity libraries. Museum collections and published descriptions provide the baseline of biodiversity knowledge.

Once a species is described, the second step is to ask what other species are related to it. Using techniques introduced in Chapter 25, researchers use molecular or morphological data to locate the organism on the tree of life. Understanding a species' evolutionary relationships is the single most powerful tool available for predicting its characteristics. For example, physicians in Australia administer antivenom to snakebite victims based on the phylogenetic position of the snake in question. Each type of antivenom is effective against poisons from closely related snakes (**Figure 55.3**). Phylogenetic analyses also alerted researchers to a threat from an alga that recently began growing at several sites in coastal California. Because the species is closely related to an invasive algal species that has taken over large areas of the Mediterranean Sea, officials from California immediately put an eradication program in place.

A third step in biodiversity studies is to assess the species' geographic range. Biogeographic studies that compare the distributions of a large number of species have been extremely useful in the design of conservation programs. For example, the observation that species richness increases from the poles to the equator, introduced in Chapter 53, has helped focus attention on the importance of preserving habitats in the tropics. In addition, biologists have been able to map specific locations that have a high proportion of **endemic species**—taxa that are found nowhere else—or a high proportion of endangered species. Conservationists have used these biogeographic analyses to set priorities for programs designed to protect the highest diversity of species possible.

A fourth level of biodiversity studies involves understanding a species' ecology, or how it interacts with other organisms and the physical environment. Ecological studies establish what an organism eats and what eats it, whether it is a parasite or a mutualist, and which species it competes with. In addition to deepening our understanding of what a particular species is, ecological studies are required to probe how the diversity of species present at a particular location affects how the ecosystem functions. Let's take a closer look.

Biodiversity and Ecosystem Function

Intact ecosystems control flooding, provide oxygen and clean water, build soil, reduce soil erosion, and maintain nutrient levels. It is legitimate to ask, however, whether the services that ecosystems provide for humans depend on a diversity of species being present. Do ecosystems function "better" in some sense when biodiversity is high?

The answer is not necessarily yes. For example, the *redundancy hypothesis* of ecosystem function holds that many species perform similar roles in an ecosystem, such as primary producer or decomposer. If so, then a minimal level of diversity

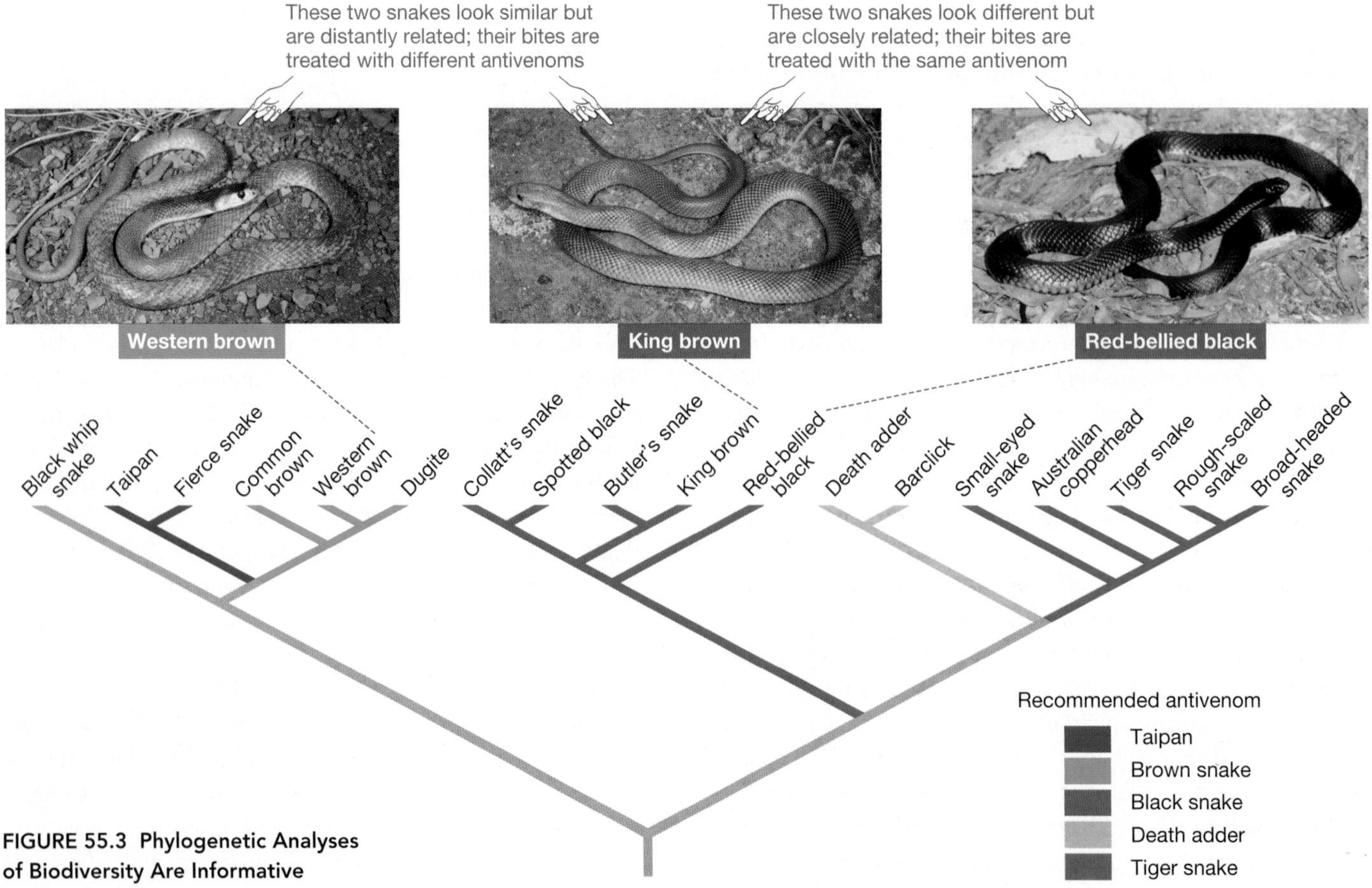

FIGURE 55.3 Phylogenetic Analyses of Biodiversity Are Informative

is required for an ecosystem to function properly. Additional species would likely be redundant and therefore not contribute to overall ecosystem products and services. In contrast, according to the *rivet hypothesis*, an ecosystem is like an airplane wing held together by rivets. If so, then an ecosystem's function may be compromised if certain species are lost, just as an airplane wing may function worse if particular rivets are missing.

To assess how biodiversity affects ecosystem function, David Tilman and colleagues classified 32 grassland plant species into the five types, or functional categories, described in **Figure 55.4**. Notice that some of the types differ by the timing of their growing season, while others are defined by whether they allocate most of their resources to manufacturing woody stems or seeds. To test the effect that these ecological differences have on productivity, the researchers planted plots measuring 13 m × 13 m with a mixture of between 0 and 32 randomly chosen species, representing from 0 to 5 functional categories. After 2 years of plant growth, the researchers harvested and weighed the aboveground tissues from the plots, as an index of net primary productivity. Recall

Question: Do high species richness and high functional diversity of species increase aspects of ecosystem function such as net primary productivity (NPP)?

Hypothesis: NPP increases with increasing species richness and with increasing functional diversity of species.

Null hypothesis:

Experimental setup:

Plant a total of 289 experimental plots, each with up to 32 species and up to 5 functional groups:

Cool-season grasses: Grow in spring
Warm-season grasses: Grow in summer
Legumes: Fix nitrogen
Woody plants: Trees, shrubs
Forbs: Lots of seeds

Examples of experimental plots:

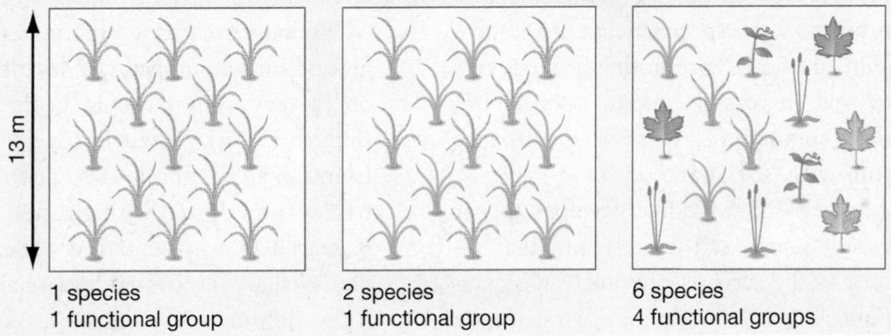

13 m
1 species
1 functional group
2 species
1 functional group
6 species
4 functional groups

Prediction:

Prediction of null hypothesis:

Results:

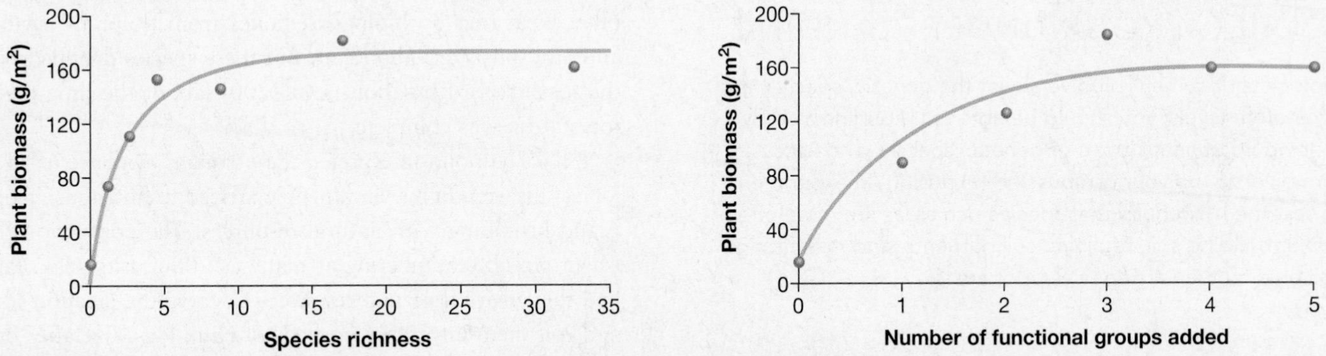

Plant biomass (g/m²)
200
160
120
80
40
0
0 5 10 15 20 25 30 35
Species richness
Plant biomass (g/m²)
200
160
120
80
40
0
0 1 2 3 4 5
Number of functional groups added

Conclusion: In this plant community, NPP increases with increasing species richness and increasing functional diversity of plants, at least up to a point.

FIGURE 55.4 Evidence That the Productivity of Ecosystems Depends on the Number and Types of Species Present

EXERCISE Write in the null hypothesis and both sets of predictions.

from Chapter 53 that **net primary productivity (NPP)** is the total amount of photosynthesis per unit area per year that ends up in biomass, the total mass of living organisms.

Tilman's group compared the productivity of plots with different numbers of species and functional groups, and found that both the number and type of species present had important effects. Plots with more species and with a wider diversity of functional groups were more productive. Total biomass leveled off as species richness and functional diversity increased, however. This observation suggests that increasing species diversity improves ecosystem function only up to a point. The data support the hypothesis that at least some species in ecosystems are redundant.

Other experiments in grassland ecosystems are consistent with the conclusion that species richness causes an increase in productivity and thus in ecosystem function. One hypothesis to explain this pattern states that diverse assemblages of plant species make more efficient use of the sunlight, water, and other resources available and thus lead to greater overall productivity. A second hypothesis states that productivity increases with species richness because certain species or functional groups facilitate the growth of other species by providing them with nutrients, partial shade, or other benefits. These hypotheses are not mutually exclusive, however. Both could be occurring, and research on the relationship between species richness and productivity continues. And in addition to promoting productivity, some research indicates that increasing species richness improves the ability of plant communities to resist disturbance and to recover from disturbance (see Chapter 53).

The message of this research is that biodiversity does appear to provide benefits beyond human self-interest and that increased biodiversity increases the services provided by ecosystems. By implication, biologists can infer that if ecosystems are simplified by extinctions, then productivity and other attributes might decrease.

✓CHECK YOUR UNDERSTANDING

Biologists document biodiversity at the genetic, species, and ecosystem levels. You should be able to (1) outline a study that would quantify levels of genetic, species, and ecosystem diversity on your campus and (2) design an experiment to test the hypothesis that species-rich areas are more effective at building soil, retaining soil nutrients, and minimizing soil erosion than are species-poor areas.

55.3 Threats to Biodiversity

No species lasts forever. Extinction, climate change, competition from newly arrived species, and habitat alteration are all natural processes. They have been happening since life began. Extinction, like death, is a fact of life.

But today, species are vanishing faster than at virtually any other time in Earth's history. Modern rates of extinction are 100 to 1000 times greater than the average, or "background," rate recorded in the fossil record over the past 550 million years. Either directly or indirectly, the extinctions are being caused by the demands of a rapidly growing human population, which is currently increasing by about 76 million people per year. If present trends in human population growth and species extinctions continue, a mass exinction on the scale of events described in Chapter 26 is almost certain to occur. Human impacts on the planet promise to equal or exceed those of the gigantic asteroid that smashed into Earth 65 million years ago. It's important to note, however, that human impacts on biodiversity are not a new phenomenon.

Humans Have Affected Biodiversity throughout History

Historically, humans have a poor record of conserving resources and protecting species. For example, recent research on fossil birds found on islands in the South Pacific suggests that about 2000 bird species were wiped out as people colonized this area between about A.D. 400 and 1600. Many of these extinctions occurred due to predation by humans or rats, pigs, and other animals introduced by humans.

Humans even have a poor record of preserving species they depend on for survival. To drive this point home, consider data on the easternmost island in the South Pacific—Easter Island (**Figure 55.5a**). When European explorers arrived on Easter Island in 1722, about 1000 people lived there. The island was treeless and dotted with gigantic stone statues (**Figure 55.5b**). Researchers who analyzed pollen cores taken from swamps on the island discovered that it had once been covered with lush forest dominated by palm trees. A steady decline in tree pollen coincided with the arrival of the first human settlers around A.D. 400.

Fossil digs confirmed that the fauna of the island underwent drastic changes at the same time. In the oldest human garbage piles excavated by biologists, bones from dolphins, seabirds, and land birds are abundant. But these species dropped out of the fossil record by about A.D. 1200—about the time that deforestation was complete.

Jared Diamond interpreted these data as evidence of an ecological disaster. In his view, people arrived to find a lush tropical island brimming with natural resources. The population flourished, possibly numbering as many as 7000 at its peak. People had the leisure time and resources to carve the gigantic statues and roll them into place on beds of palm logs. But after deforestation was complete and local extinctions had begun, the system collapsed. Without palm trunks to make canoes, Easter Island natives did less dolphin hunting and less fishing. Soil erosion may have cut into the productivity of banana and sweet potato plantations. The population crashed, and the great statues fell.

(a) Easter Island is the easternmost island in the South Pacific.

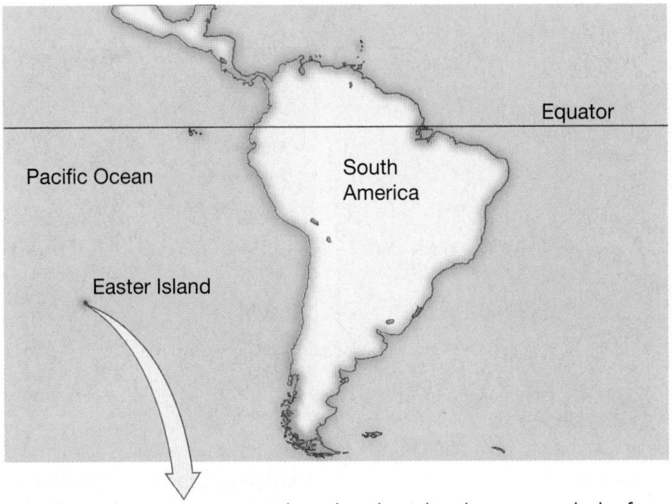

(b) Gigantic stone statues that dot the island were made before an ecological collapse occurred.

FIGURE 55.5 On Easter Island, Social Decline Coincided with the Exhaustion of Natural Resources
(a) As humans spread across the South Pacific, Easter Island was one of the last islands to be colonized. **(b)** The prosperity that supported the carving of these statues was based on resources that were overexploited.

Current Threats to Biodiversity

Over the past 1000 years, most recorded extinctions have occurred on islands and have been associated with overhunting or human-assisted introductions of competitors, diseases, or predators that were not native to the area. Starting in the twentieth century, however, this situation began to change. Human impacts now include entire continents as well as islands and they go far beyond overhunting and species introductions. As a growing human population consumes Earth's resources to fulfill its needs and desires, biodiversity is being degraded or destroyed at the genetic, species, and ecosystem levels.

The direct exploitation of plants and animals for food, medicines, trophies, or pets continues to be a problem for some species in some regions (**Figure 55.6a**, page 1274). Overexploitation is a particular concern for the world's major marine fisheries, two-thirds of which are currently fully exploited or depleted. A recent study of the global fishing industry over the last 50 years concluded that 90 percent of large ocean fish—such as tuna, swordfish, and marlin—and the same percentage of important bottom-dwelling fish—such as cod, halibut, and flounder—have been removed from the world's oceans.

In both aquatic and terrestrial ecosystems, human introductions of *exotic*, or nonnative, species are still causing severe problems (**Figure 55.6b**). **Invasive species**—exotic species that are introduced to a new area, spread rapidly, and eliminate native species (see Chapter 50)—are implicated in the demise of one-third to one-half of all species currently listed as endangered or threatened in the United States. These exotic species are largely free of their normal diseases, competitors, and parasites, and either directly consume native species or outcompete them.

Figure 55.6c highlights the problems caused by pollution such as sewage, toxic chemicals in fertilizers, pesticides and industrial wastes, oil spills, and acid rain. Pollution can kill organisms directly or cause reproductive abnormalities and failure. Some toxic compounds are highly persistent and can be transported long distances from their point of origin. Pollutants found in polar ice and in tissues from oceangoing whales reveal that chemical contaminants have reached every corner of the globe.

By burning forests and fossil fuels, humans have caused carbon dioxide levels in the atmosphere to rise enough to cause dramatic global warming (see Chapter 54). While there is no scientific consensus about the exact effects resulting from higher average temperatures worldwide, global climate models suggest that sea levels will increase and regional climate patterns will change (**Figure 55.6d**). Coral reefs, alpine tundras, and low-lying islands are among the most vulnerable ecosystems. The impact of climate change on biodiversity is unknown and represents a major research frontier in conservation biology.

With respect to the numbers of species and ecosystems affected, human alteration of natural habitat is now the dominant cause of biodiversity decline. Humans cause **habitat destruction** by logging and burning forests, damming rivers, dredging and trawling the oceans, plowing prairies, grazing livestock, filling in wetlands, excavating and extracting minerals, and building housing developments, golf courses, shopping centers, office complexes, airports, and roads (**Figure 55.6e**).

In addition to destroying natural areas outright, human activities fragment large, contiguous areas of natural habitats into small, isolated fragments. **Habitat fragmentation** concerns biologists for several reasons. First, it can reduce habitats to a size that is too small to support some species (**Figure 55.6f**). This is especially true for predators such as mountain lions, grizzly bears, and bluefin tuna that need vast natural spaces in which

WEB TUTORIAL 55.1
Habitat Fragmentation

(a) Overexploitation

Overhunting led to the extinction of the passenger pigeon.

(b) Introduced species

When species that are not native are introduced to an area, a number of problems can occur.

Competition: In North American marshes, purple loosestrife is crowding out native organisms.

Disease: An introduced fungus has virtually wiped out the American chestnut.

Predation: The brown tree snake has extinguished dozens of bird species on Guam.

(c) Pollution

In the late 1940s, DDT was widely used to control mosquitoes in marshy areas. By the mid-1950s, fish were accumulating large quantities of DDT in their bodies. Fish-eaters such as bald eagles accumulated even larger quantities of DDT, which turned out to be toxic.

(d) Global warming

Due to increased levels of CO_2 in the atmosphere, average temperatures are expected to increase by 5.5°C over the next 100 years. Species that disperse slowly and take a long time to grow to maturity, such as beech trees, may not be able to shift their ranges quickly enough to survive.

(e) Habitat destruction

Logging in the Pacific Northwest of North America has removed so many large nesting trees that the northern spotted owl is now endangered.

(f) Habitat fragmentation

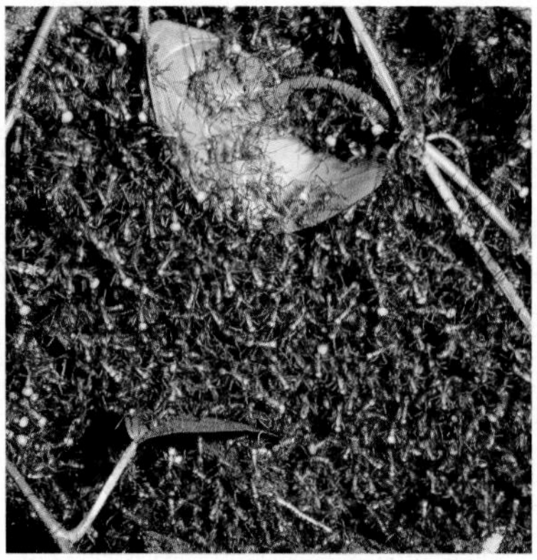

Army ants range widely in search of suitable food. Small forest plots are not large enough to support them.

(g) Domino effects

Antbirds feed on the insects stirred up by army ants. When army ants disappear from forest fragments, so do antbirds.

FIGURE 55.6 Seven Deadly Sins That Affect Biodiversity

[Photo (b) middle ©Gary Braasch; (g) ©Doug Wechsler/VIREO.]

to feed, find mates, and reproduce successfully. Second, habitat fragmentation reduces the ability of individuals to disperse from one habitat to another by creating islands of habitat in a sea of human-dominated landscapes. Small, isolated populations can suffer from inbreeding depression and random loss of alleles due to genetic drift (see Chapter 24). Third, fragmentation creates large amounts of "edge" habitat that is subject to invasion by weedy species and exposed to more intense sunlight and wind.

The decline in habitat quality caused by fragmentation was documented in a recent experiment in a tropical wet forest. In an area near Manaus, Brazil, that was slated for clear-cutting, a research group set up 66 square, 1-hectare experimental plots that remained uncut. (A hectare, abbreviated ha, is 100 meters by 100 meters, approximately the size of two football fields.) Thirty-nine of these study plots were located in fragments designed to contain 1, 10, or 100 hectares of intact forest. Twenty-seven of the plots were set up nearby, in continuous wet forest. As the "Experimental setup" section of **Figure 55.7** shows, the distribution of the study plots allowed the research team to monitor changes inside forest fragments of different sizes and to compare these changes with conditions in unfragmented forest.

When the research group surveyed the plots 10–17 years after the initial cut, they recorded two predominant effects: a rapid loss of species diversity, especially from the smaller fragments, and a startling drop in **biomass**, or the total amount of fixed carbon, in the study plots located near the edges of logged fragments. In tropical wet forests, most of the biomass is concentrated in large trees. The edges of the experimental plots contained many downed and dying trees. Based on this observation, the researchers inferred that the decrease in biomass occurred because large trees near the edges of fragments died from exposure to high winds and dry conditions. This edge effect is important because it eliminates treetop, or canopy, habitats. As a result, the quality as well as the quantity of habitat in the study area declined drastically.

Researchers are also beginning to document population reductions due to **domino effects**—that is, impacts on a given species as a result of the loss of a different species (**Figure 55.6g**). When one species goes extinct, any other species that had specialized in parasitizing or preying on that species will also go extinct.

Now, how many extinctions are the seven problems highlighted in Figure 55.6 projected to cause?

How Can Biologists Predict Future Extinction Rates?

One strategy for predicting the future of Earth's biodiversity focuses on the consequences of the most urgent problem: habitat destruction. Given reasonable projections of habitat loss, biologists can estimate rates of extinction based on well-documented **species–area relationships**.

Question: How does fragmentation affect the quality of tropical wet forest habitats?

Hypothesis: Fragmentation reduces the quality of wet forest habitats.

Null hypothesis: Fragmentation has no effect on the quality of wet forest habitats.

Experimental setup:

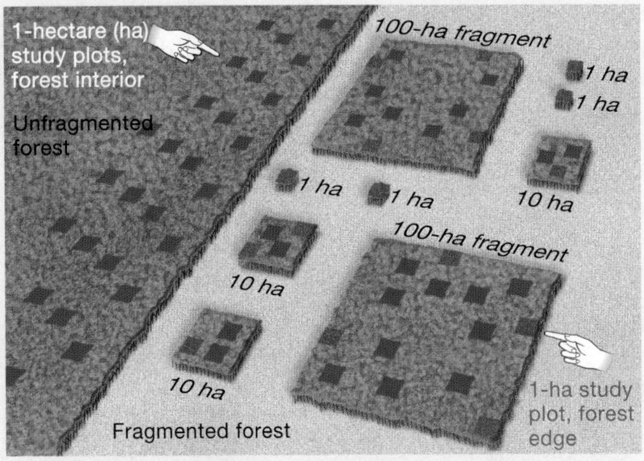

Prediction: Species diversity and biomass will decline in forest fragments compared with those of the forest interior, particularly along edges of fragments.

Prediction of null hypothesis: Species diversity and biomass will be the same inside forest fragments and along edges as in the forest interior.

Results:

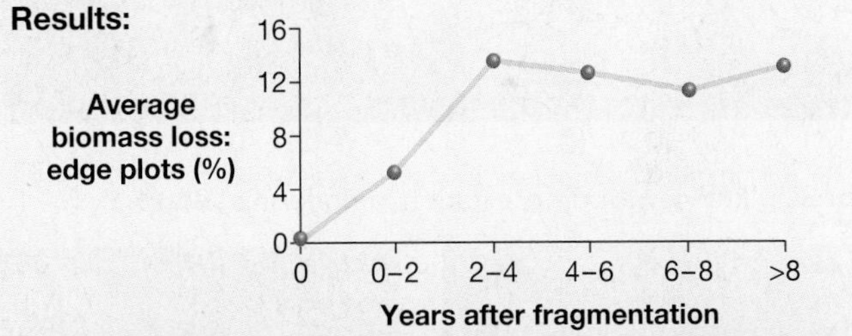

Conclusion: Biomass declines sharply along edges of forest fragments. To interpret this result, hypothesize that large trees near forest edges tend to die.

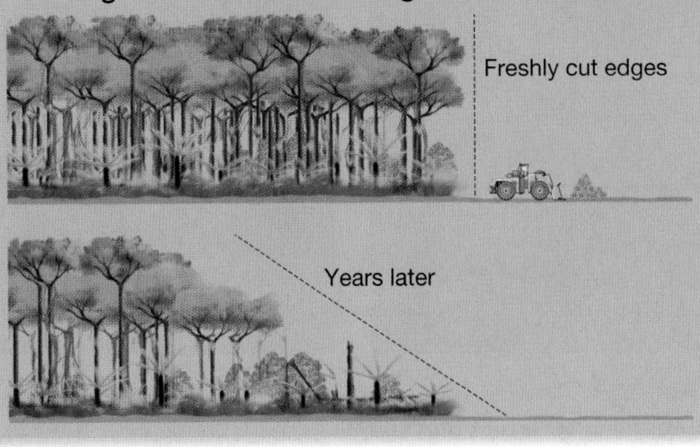

FIGURE 55.7 Experimental Evidence for Edge Effects in Fragmented Forests
Researchers tracked 66 study plots among four 1-hectare fragments, three 10-hectare fragments, and two 100-hectare fragments.

To understand how this work is done, consider recent estimates of habitat destruction in the world's largest continuous wet forest and most species-rich region: the Amazon River basin (**Figure 55.8a**). In the tropics, deforestation is the result of logging, the conversion of forest to pasture or cropland, and wildfires set by humans. To appreciate the extent of deforestation in some areas, consider the satellite images of Rondônia, Brazil, that were made in 1975 and in 2001 (**Figure 55.8b**). Analyses of satellite photos such as these have shown that an average of 5000 km^2, an area roughly the size of Connecticut, was deforested in the Amazon each year during this decade. In the 1990s losses of Amazonian wet forest increased to as much as 30,000 km^2 annually. On the basis of such studies, conservationists estimate that about 10 percent of the original area covered by tropical forests worldwide is being lost each decade. It is not unreasonable, then, to project that 90 percent of the world's tropical forest habitat will be lost over the next century.

To predict how many species will be driven to extinction as a result, biologists use species–area curves like the one shown in **Figure 55.9**. This graph was generated by a biologist who analyzed the number of bird species found on islands in the Bismarck Archipelago near New Guinea, in the South Pacific. The graph shows the number of species on islands of various sizes. Notice that both axes on the graph are logarithmic. The solid line drawn through the points is described by the function $S = (18.9)A^{0.15}$, where S is the number of species and A is the area. In this island group, the function indicates that each tenfold increase in island area increases the number of bird species present by about 40 percent; a tenfold decline in area reduces bird species by about 30 percent.

These data turn out to be typical for other habitats and taxonomic groups as well. When biologists have plotted species–area relationships for plants, butterflies, mammals, or birds from islands or continental habitats around the globe, the relationship is consistently described by a function of the form $S = cA^z$. The c term is a constant that scales the data. Its value is high in species-rich areas, such as coral reefs or tropical wet forests, and low in species-poor areas, such as arctic tundras.

(a) The devastation of deforestation

(b) Satellite view of deforestation in Rondônia, Brazil

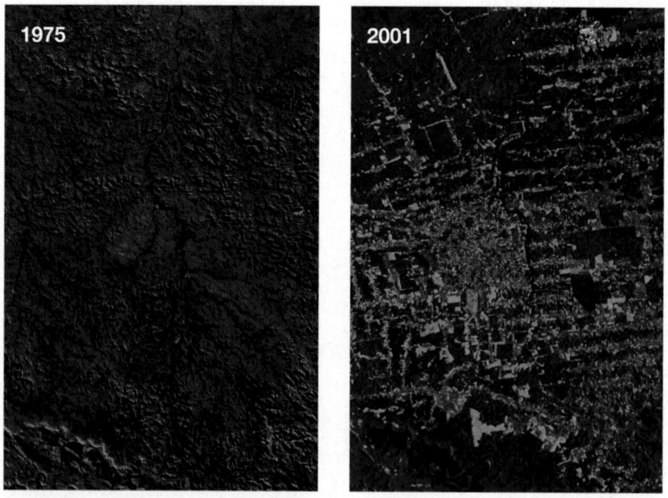

FIGURE 55.8 Rates of Deforestation in Tropical Wet Forests Are Approaching 1 Percent per Year
(a) Deforestation is occurring in the tropics due to logging, the conversion of forest to pasture or cropland, and wildfires set by humans. **(b)** The satellite photos show the extent of deforestation in Rondônia, Brazil, in less than three decades.

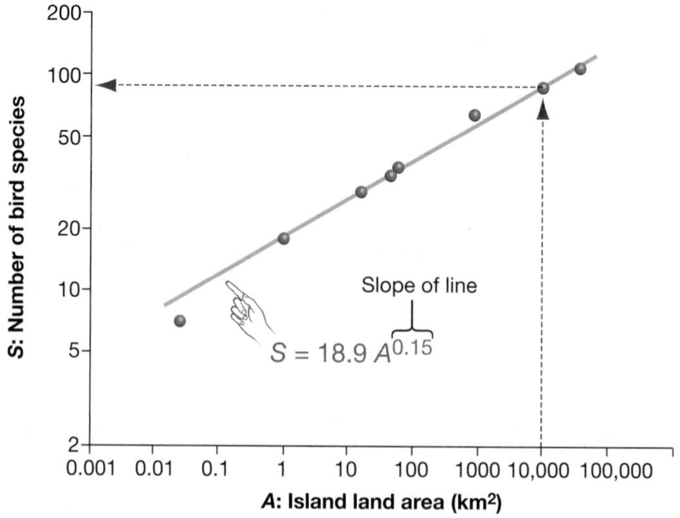

FIGURE 55.9 Species–Area Plots Quantify the Relationship between Species Richness and Habitat Area
Graph plotting the sizes of islands in the South Pacific (horizontal axis) versus the number of bird species present on each island (vertical axis). Notice the log-log scale. The dashed lines show the number of bird species that are expected to live on an island with an area of 10,000 km^2.

The exponent z represents the slope of the line on a log-log plot. Thus, z describes how rapidly species numbers change with area. In analyses of species–area relationships in island groups, z is typically 0.25 to 0.30. For continental habitats, z usually ranges from 0.10 to 0.20.

To understand how biologists use these analyses to project extinction rates, study Figure 55.9 again. Ask yourself, If 90 percent of the habitats were destroyed—for example, if A were reduced from 10,000 km^2 to 1000 km^2—what percentage of species would disappear? According to the graph, the number of bird species in the Bismarck Archipelago would drop from about 75 to about 53. Thus, the answer is roughly 30 percent. This is an important conclusion. If z is about 0.15 for taxa that inhabit tropical wet forests and other threatened habitats, the prediction is that 30 percent of all species will be wiped out in the next 100 years.

This prediction is consistent with projections based on analyses of species currently threatened with extinction. For example, each year the International Union for the Conservation of Nature (IUCN) publishes a book summarizing data on the status of virtually all species whose conservation status has been studied. If the total population in the wild is declining rapidly enough to threaten a species with extinction, then the species is "red-listed." By studying how fast these red lists are growing over time, biologists can project extinction rates. Consider that IUCN red-listed 5205 mammals in 1996 but 5435 in 2000. If this rate of growth in the number of threatened species continues, and if all these threatened species go extinct, then 60 percent of the mammal species living today will be wiped out in less than 500 years. Similar trends are occurring in listings of threatened birds, mollusks, butterflies, conifers, and flowering plants.

Recall from Chapter 26 that a mass extinction event is defined as the loss of 60 percent of all species in less than 1 million years. Based on analyses of habitat destruction and red lists, biologists have concluded that a mass extinction event is now well under way.

✓CHECK YOUR UNDERSTANDING

Until recently, most human-induced extinctions have occurred on islands and were due to overhunting or the impact of introduced species. Although invasive species remain a major threat to biodiversity, the majority of current problems are on continents and are due to habitat destruction and fragmentation. You should be able to (1) explain why fragmentation of habitats reduces the habitats' ability to support biodiversity and provide ecosystem services, (2) respond to individuals who argue that the biodiversity crisis began with the emergence of technologically sophisticated industrial societies, and (3) cite data relevant to the hypothesis that the biodiversity crisis is overblown and may already have peaked in intensity.

55.4 Conserving Biodiversity: Biology, Sociology, Economics, and Politics

The conservation of biodiversity is among the most urgent but least-recognized issues facing us today. From subsistence fishermen in Pacific islands to Wall Street CEOs, humans depend on the intricate web of life that is biodiversity. Unlike ozone holes, acid rain, phosphorus pollution, and other environmental problems that humans have addressed recently, extinction is irreversible. Preventing the loss of alleles, species, and ecosystems is the only conceivable solution to the biodiversity crisis.

Solving any global problem requires a common goal to be defined. In the case of preserving biodiversity, the objective is to sustain a diversity of alleles, populations, species, habitats, and ecosystems in natural landscapes while supporting the extraction of resources and environmental services required to maintain the health and well-being of an increasingly large human population.

Fundamentally, the conservation of biodiversity boils down to maintaining viable populations of all species. Because this is a biological issue, biologists are the anchors of any sound conservation effort. One of the most important biological goals in conservation work is to establish a **baseline**, or reference point, that provides a standard for evaluating changes in the diversity of alleles, species, and ecosystems. In establishing baselines, it is crucial to recognize that ecosystems around the world have already deteriorated significantly. For example, if you are a fisheries biologist charged with maintaining salmon diversity in the Pacific Northwest of North America, you might think you are doing an excellent job if salmon numbers rise to double what they were in the 1930s. But by the 1930s salmon in this region had already been reduced by 90 percent of the populations present in the 1800s. By using the 1930s as a baseline, you have accepted a highly degraded state for measuring the success of conservation programs.

Ideally, the baseline for all conservation efforts would be the state of genetic, species, and ecosystem diversity prior to significant human impacts. But with each passing generation, this baseline becomes less and less realistic. Baselines are shifting. As a result, biologists often have to accept impaired ecosystems as reference points. In addition, biologists are realizing that there may be no such thing as a state prior to significant human impacts. Around the world, people have been changing ecosystems for thousands of years. These impacts range from the intentional burning of prairies and woodlands by Native Americans, to grazing of domestic livestock by aboriginal peoples in Africa, to the slash-and-burn agricultural systems of native peoples in the Amazonian forest.

Finally, biologists realize that they do not work in isolation when they are establishing baselines and working on management plans for threatened species and ecosystems. In almost every case, the underlying causes of the biodiversity crisis are socioeconomic factors that encourage the short-term overexploitation of land and other resources and discourage the long-term sustainability of resources. **Sustainability** is the planned use of resources at a rate only as fast as the rate at which they are replaced. For this reason, successful conservation practice requires a complex mixture of economic, social, and political action based on robust scientific knowledge. Let's take a closer look at some of the organizations involved in conservation action.

The Role of Governmental and Private Agencies

Because most species live in more than one country and because changes to the atmosphere, surface water, groundwater, and soils tend to affect large geographic areas, efforts to preserve biodiversity are often international in scope. At the same time, conservation programs tend to have a disproportionate impact on specific individuals, even though the general public may benefit. For example, it is possible for restrictions on fishing quotas or logging contracts to threaten families and businesses with bankruptcy if these programs are not managed carefully. Elected officials like to say that "All politics is local." The same might be said of conservation programs.

Implementing conservation programs effectively requires a combination of global support and local support. As a result, a blend of international, regional, and local agencies are often involved.

International Agreements The biodiversity crisis is recognized as a global problem, and many international programs have emerged to help solve it. By far the most important is an international agreement called the *Convention on Biological Diversity* (CBD). The CBD was proposed in Brazil in 1992 at the United Nations Conference on Environment and Development, which still ranks as the largest meeting of world leaders ever held. The CBD was the first global agreement on the conservation and sustainable use of biological diversity and has now been formally ratified by over 175 countries.

The CBD has three main goals: (1) to conserve biodiversity, (2) to promote sustainable human use of biodiversity, and (3) to share the commercial benefits of biodiversity equitably, especially with respect to genetic diversity. The CBD was hailed as a landmark international law, because it recognized that the conservation of biological diversity is a "common concern of mankind" and because it laid the groundwork for a new philosophy of environmentally sound economic development.

Even though the CBD established goals to which all signatories agreed, the responsibility for achieving those goals rests with each nation. Under the CBD, each government agreed to undertake the conservation and sustainable use of biodiversity and to develop and implement national biodiversity action plans. Countries vary widely in the resources, expertise, and political will they devote to carrying out their obligations, however. Although much has been achieved, successful implementation of the CBD will require significant and sustained efforts by world governments.

National Agencies Most countries have an array of offices or programs designed to preserve biodiversity. In the United States, for example, biodiversity is managed by a broad range of government agencies at the federal, state, and local levels. The federal government is by far the country's largest landholder, however. The U.S. Department of the Interior has several agencies that are charged with managing biological resources, including the Bureau of Land Management, the Fish & Wildlife Service, and the National Park Service. The U.S. Forest Service, under the Department of Agriculture, has primary responsibility for managing huge tracts of forested land, especially in the Southeast and Northwest.

In the United States, most federal land is managed for multiple uses, including extractive industries, such as timber logging and mining, gas and oil development, cattle grazing, and hunting and fishing. The major exception to this rule is wilderness areas, which are large tracts of land—usually greater then 5000 acres—left in their natural state. Currently, the United States has 662 wilderness areas that total over 100,000,000 acres; more than half of that total acreage is in Alaska. These wilderness areas are considered critical for research focused on establishing baselines for conservation programs.

In the ocean realm, the National Marine Fisheries Service, under the U.S. Department of Commerce, is charged with managing and conserving marine organisms. The ocean territory of the United States encompasses more than 4 million square miles of coastal waters and open seas, but less than 0.1 percent of this total is protected from extractive uses, such as fishing, oil drilling, dredging, or sea-floor mining. A large body of recent research points to negative trends in marine biodiversity, from declining biomass in many commercial fisheries, to massive coral-bleaching events, growing anoxic dead zones, and outbreaks of invasive species. Biologists know far less about all aspects of marine biodiversity than about its terrestrial counterpart, and governments are doing much less to conserve it.

Nongovernmental Organizations *Nongovernmental organizations* (NGOs) play a vital role in drawing public attention to specific conservation issues and generating grassroots awareness and support. Many of these organizations have staffs made up of scientists, economists, and attorneys who work together to create comprehensive solutions to biodiversity problems. Some of the largest and most familiar NGOs involved in biodiversity conservation are Conservation International, The Nature Conservancy, and the World Wildlife Fund. In addition to sponsoring research and coordinating efforts to set up nature preserves or even purchasing land outright, NGOs may monitor legislation that affects conservation and lobby governmental authorities to fulfill their commitment to the CBD and other biodiversity agreements.

Another important role for NGOs is to organize collaborations among different levels of government, university-based scientists, local businesses and institutions, and individuals. The IUCN, for example, is a unique union of government agencies, individual scientists, and conservation organizations working together on many aspects of biodiversity conservation, including the production of the global red list of endangered species. The World Conservation Monitoring Center acts as a centralized information resource for policy and action to conserve biodiversity.

Conservation Strategies

Biologists are documenting the extent of biodiversity and identifying threats to species and ecosystems; an array of international and national public and private organizations are focused on solving the biodiversity crisis. What general strategies are being used to preserve biodiversity, and what is the outlook for the future?

In Situ Conservation: Protected Areas Preserving a significant portion of the world's biodiversity requires in situ ("in place") conservation programs that maintain organisms in their wild state and within their existing range. Conserving species where they already exist is more practical and less expensive than ex situ ("out of place") conservation approaches, in which organisms are maintained in zoos, aquaria, or botanical gardens. In situ conservation is usually achieved by government protection of designated areas from development. In theory, it allows wild populations to continue to evolve and to be used by indigenous people who depend on them for food, fuel, or medicine.

The IUCN monitors the success of in situ conservation programs by sponsoring the World Commission on Protected Areas, which holds the World Parks Congress every decade. In 1992 the Congress met in Caracas, Venezuela, and established a goal of setting aside 10 percent of Earth's land surface in protected areas. In 2003 the Congress met in Durban, South Africa, and announced that this goal had been surpassed: Protected areas covered 11.5 percent of Earth's terrestrial surface.

How efficiently is the existing network of protected areas protecting biodiversity? Researchers are attempting to answer this question via a geographic approach called the **Gap Analysis Program (GAP)**. A GAP analysis tries to identify gaps between geographic areas that are particularly rich in biodiversity and areas that are actually managed for conservation. One recent GAP analysis combined data sets on the distribution of mammals, birds, amphibians, and freshwater turtles with a map of world protected areas. The analysis revealed that many species' ranges occur completely outside any protected areas. It also pinpointed regions in Mexico, Madagascar, and elsewhere where the gap between species range and protected areas is particularly high. To date, most GAP analyses suggest that, because relatively few "biodiversity hotspots" are included in existing protected areas, the 11.5 percent of Earth's surface area that is now being managed for biodiversity will not be enough to conserve many species.

Ex Situ Conservation: Zoos, Aquaria, and Botanical Gardens In some cases, it is not possible to save species in the wild, and only intensive management efforts can save them from extinction. By propagating species in captivity, zoos, aquaria, and botanical gardens function like the emergency room of a hospital. To date, over 20 animal species that were exterminated in the wild, including the California condor and Pere David's deer, have been reintroduced into their native habitats after successful captive-breeding programs. Many aquaria are now implementing similar efforts. Because plants are easier to cultivate in captivity than animals are, the world's approximately 1500 botanic gardens can maintain a large percentage of existing plant species in captivity. One of the largest—the Royal Botanic Gardens, Kew, in England—grows representatives of one in eight of Earth's known plants.

In addition to pursuing research and captive propagation, zoos, aquaria, and botanic gardens play a key role in educating the general public about the importance of biodiversity and the need to conserve it. These institutions host tens of millions of visitors every year and in many cases offer visitors their only direct experience of biodiversity, its importance, and its threats. Experience has shown that people have to care about biodiversity before they will work to protect it.

Looking to the Future

For many years, the conservation community's common goal was **sustainable development**, or economic progress for local communities based on using certain species or resources carefully enough that they could regenerate and not decline over time. This concept has recently been expanded to a new community-based conservation paradigm called **adaptive management**, which recognizes that the design, management, and monitoring of conservation programs are inseparable. Under adaptive management, information from monitoring programs is used to review and modify the project's original goals and implementation continuously.

Under both sustainable development and adaptive management, biologists and conservation officials emphasize that, for biodiversity protection programs to be successful, local people must be involved and must benefit from them. If you are a biologist charged with reintroducing the black-footed ferret to the Montana prairies, you are much more likely to be successful if ranchers and other local residents are informed and committed to the project. If you were working in Myanmar to increase the amount of area protected for one of the last remaining Asian tiger populations, it would be essential that you get approval of and participation from local villagers who hunt and collect firewood in the forest, and that local people benefit from any increases in ecotourism based on viewing tigers.

Another cornerstone of sustainable development and adaptive management is active involvement by economists, politicians, and educators, as well as by biologists. Worldwide, there is an urgent need for economists who are trained in biology and who can (1) design profitable but sustainable ways to harvest resources and (2) estimate the dollar value of ecosystem services such as flood control, soil erosion control, and groundwater recharge. In the political arena, elected officials with a background in biology are needed to create new legislation aimed at promoting the sustainable use of biodiversity and to support the enforcement of existing regulations and international conventions. Environmental educators are needed at all levels, from kindergarten to adult education, to increase awareness of biodiversity and commitment to its preservation.

Biology will continue to be the scientific discipline that is central to our understanding of biological diversity and to our wise use of biodiversity. Millions of species are yet to be discovered, named, and studied, and the organized effort to survey genetic diversity within species and ecosystems has only just begun. Much more research is required for us to understand how species interact within ecosystems and how we can best preserve the ecosystem services and functions that support life. In addition to studying life, biologists are now charged with helping save it.

ESSAY Models for the Future: The Malpai Borderlands and Guanacaste

Emotionally, studying biodiversity and implementing conservation programs can be a tough business. The historical record is discouraging, and predictions for the future are not optimistic. It would be misleading, though, to conclude that the future of biodiversity is hopeless.

Recall from the essay at the end of Chapter 54 that governments, corporations, and private individuals banded together to take effective action after researchers identified serious problems with phosphorus pollution, acid rain, and depletion of ozone in the atmosphere. In at least some small pockets around the world, the same thing is happening in response to the biodiversity crisis.

In the United States, one of the greatest threats to biodiversity is suburban sprawl. In response to threats from development in southeast Arizona and southwest New Mexico, a group of ranchers banded together to form an association they called the Malpai Borderlands group. To preserve biodiversity and their livelihood, the ranchers (1) protected 42,000 acres of their land with conservation easements, which are legal agreements that prevent land from being developed; (2) set up cooperative "grassbanks," or pastureland that is made available to ranchers whose own grazing areas are suffering from short-term drought and need relief from continued grazing; and (3) reintroduced fire to the area via prescribed (intentional) burns that remove encroaching woody shrubs and encourage the growth of native grasses (**Figure 55.10**). The ranchers have also been cooperating with each other and with biologists to monitor or

In addition to studying ecosystems, biologists have begun to restore them.

(a) Before prescribed burn

(b) After prescribed burn: 2 days after ... and 2 months later

FIGURE 55.10 Using Fire to Restore Grassland Ecosystems in the Malpai Borderlands
The same region, in the Malpai Borderlands of the southwestern United States **(a)** before and **(b)** after a prescribed burn. The burn occurred in a wet year and resulted in a dramatic increase in native grasses.

reintroduce native vegetation and animals such as the thick-billed parrot, bighorn sheep, Mexican jaguar, Yaqui chub, and Chiricahua leopard frog into the region.

Today, the Malpai Borderlands group is considered a model of innovative action by private individuals, in cooperation with governmental agencies and NGOs, that benefits local people as well as biodiversity. One aspect of the Malpai Borderlands program—restoration of native vegetation through active management—is turning out to be an important theme in efforts to preserve biodiversity. In many areas of the world, ecosystems are already heavily degraded or lost. The state of Illinois, for example, was estimated to contain 22 million acres of prairie when European settlers arrived; but only about 2000 acres of prairie remain there today. It is much easier to save intact ecosystems than to restore damaged ones. But when natural areas are badly degraded or gone, biologists turn to restoration.

Hundreds of large-scale and small-scale restoration and reforestation projects are now occurring around the globe. In the tropics, one of the most successful is the restoration of dry forest at the Area de Conservación Guanacaste in northwestern Costa Rica. The primary task facing conservation area staff there was to stop human-caused fires. In just 15 years, their efforts have succeeded in transforming a vast swath of marginal ranching land into an increasingly popular tourist destination and a water source for neighboring farms and ranches (**Figure 55.11**).

Conservationists contend that, to preserve biodiversity over the long term, it will not be enough to save patches of undisturbed habitat in parks and preserves; extensive work to restore damaged ecosystems is also crucial. In addition to studying ecosystems, biologists have begun to restore them. From small-scale restorations of woodland and prairie habitats in North America to efforts the size of the Malpai Borderlands and Guanacaste projects, the field of "restoration ecology" is burgeoning.

(a) Start of restoration project

(b) 17 years later

FIGURE 55.11 Using Fire Prevention to Restore Forest Ecosystems in the Guanacaste Reserve
(a) Before and **(b)** after views of the same region, 17 years apart, in the Area de Conservación Guanacaste, Costa Rica. To appreciate how quickly the trees grew, notice the person standing near the middle of the photo in (b).

CHAPTER REVIEW

Summary of Key Concepts

- **Biodiversity can be analyzed at the genetic, species, and ecosystem levels.**

Genetic diversity is well characterized within some species, but biologists are only beginning to use genome sequencing techniques to explore the extent of genetic diversity in ecosystems. Research on the total number of species alive today is also at a preliminary stage, with estimates ranging from 10 to 100 million. At the ecosystem level, early studies support the hypothesis that high species diversity increases aspects of ecosystem function such as productivity and resistance to disturbance. To date, the message from efforts to characterize biodiversity is that it is much more extensive than expected and that a great deal remains to be learned.

- **Humans depend on biodiversity for the products that wild species provide and for ecosystem services that protect the quality of the abiotic environment.**

Humans gain direct economic benefits from fishing, forestry, agriculture, tourism, and other activities that depend on biodiversity. Species diversity is also important for maintaining the productivity of natural ecosystems and their ability to build and hold soil, moderate local climates, retain and cycle nutrients, retain surface water and recharge groundwater, prevent flooding, and produce oxygen.

If present trends in human population growth and habitat destruction continue, a mass extinction event is under way.

Historically, most human-caused extinctions have occurred on islands because of direct exploitation or the introduction of exotic herbivores and predators. Habitat destruction is currently the leading cause of extinctions, however. Experiments in the Brazilian Amazon have shown that habitat loss leads not only to a rapid decline in biodiversity but also to a decline in the quality of the remaining habitats due to fragmentation.

To estimate how many species will go extinct in the near future, biologists combine data on current rates of habitat loss—usually estimated from satellite images taken over time—with data on the average number of species found in habitats of a given size. These species–area analyses suggest that if 90 percent of habitats are destroyed as expected during the next century, then over 30 percent of all species will become extinct. According to data on the rate at which lists of threatened mammals and other species are growing, it is likely that 60 percent of all species will be wiped out within 500 years.

Web Tutorial 55.1 Habitat Fragmentation

Effective conservation programs start with sound biological research and involve economists, politicians, and local residents.

Preserving biodiversity may well be the most important—and difficult—enterprise our species has ever undertaken. Although 11.5 percent of Earth's total land area has been officially designated as protected from development, many areas with extraordinarily high biodiversity remain vulnerable and relatively few marine reserves exist.

The Convention on Biological Diversity has set broad goals for the preservation of biodiversity and has been endorsed by most countries. Various governmental agencies and nongovernmental organizations have been created to address the biodiversity crisis, through programs that protect species and ecosystems while encouraging the sustainable use of resources and promoting the welfare of local people.

Biologists play a key role in conservation programs by collecting the data required to make sound political and economic decisions. In addition to documenting the extent and benefits of biodiversity, biologists alert officials and the general public to biodiversity problems, establish baselines for biodiversity, evaluate priorities for conservation action, and monitor the effectiveness of conservation measures.

Questions

Content Review

1. What does a species–area plot show?
   a. The overall distribution, or area, occupied by a species.
   b. The relationship between the body size of a species and the amount of territory or home range it requires.
   c. The number of species found, on average, in tropical versus northern areas.
   d. The number of species found, on average, in a habitat of a given size.
2. What does a GAP analysis do?
   a. It compares the current distributions of species with the locations of preserved habitats.
   b. It quantifies gross *a*boveground *p*roductivity.
   c. It uses data on the rates at which lists of threatened species are growing to project the rate of future extinctions.
   d. It uses genome sequencing techniques to quantify genetic (allelic) diversity in an ecosystem.
3. Why do most existing data tend to support the redundancy hypothesis of ecosystem function?
   a. It is common to observe that many species in an ecosystem belong to the same functional group.
   b. Productivity and resistance to disturbance increase with increasing species richness but quickly level off.
   c. Species function like rivets in an airplane wing, because the loss of even one species alters the function of the entire ecosystem.
   d. Ecosystems with low species richness still provide high levels of ecosystem services.
4. What types of data suggest that Easter Island once had high biodiversity and productivity?
   a. comparisons with islands of similar size, latitude, and soil type
   b. historical records from the first Europeans to visit the island
   c. oral traditions of the indigenous people
   d. fossil pollen and animal bones
5. Why is the Convention on Biological Diversity considered a landmark achievement in efforts to preserve biodiversity?
   a. It halted international trade in wildlife pets and products from endangered species.
   b. It established the goal of designating 10 percent of Earth's land surface as protected area.
   c. It is a formal commitment by the signatory nations to preserve biodiversity.
   d. It set up the first international system of marine reserves devoted to preservation of biodiversity.
6. What are biologists doing when they establish a baseline for preservation of biodiversity?
   a. providing documentation for a conservation goal in the management or restoration of a species or an ecosystem
   b. estimating the current number of species alive today, as a basis for measuring change in the future
   c. creating a coalition of international, national, and regional governmental agencies and NGOs focused on a particular conservation goal
   d. documenting the direct and indirect benefits of biodiversity, as a justification for conservation programs

Conceptual Review

1. List two direct benefits to humans for preserving biodiversity and two indirect benefits from ecosystem services. Next to each entry in your list, state which members of society receive this benefit when biodiversity is preserved. Then state who, in your opinion, should be responsible for paying the costs associated with preserving biodiversity in return for this benefit.
2. Compare and contrast biodiversity at the genetic, species, and ecosystem levels. How do biologists analyze the extent of biodiversity at each of these levels?
3. Biologists claim that the all-taxa survey now under way at the Great Smoky Mountains National Park in the United States will improve their ability to estimate the total number of species living today. Discuss the benefits and limitations that this data set will provide in understanding the extent of global biodiversity.
4. How are species–area curves used to relate rates of habitat destruction to projected extinction rates?
5. Discuss evidence to support the hypothesis that species richness increases ecosystem functions such as productivity and resistance to disturbance, and thus the ecosystem services that benefit humans.
6. Explain why the fragmentation of habitats reduces their ability to support biodiversity.

Group Discussion Problems

1. Projections for an impending mass extinction are contingent on the continuation of present trends in human population growth and in habitat destruction. Do you believe that these trends will continue? In general, are you optimistic or pessimistic about the future of biodiversity? Explain your logic.
2. You are helping design a series of reserves in a tropical country. List the steps you would recommend for gathering data and creating a plan that would protect a large number of species in a small amount of land.
3. The maps that follow chronicle the loss of old-growth forest (>200 years old) that occurred in Warwickshire, England, and in the United States. In your opinion, under what conditions is it ethical for conservationists who live in these countries to lobby government officials in Brazil, Indonesia, and other tropical countries to slow the rate of loss of old-growth forest?

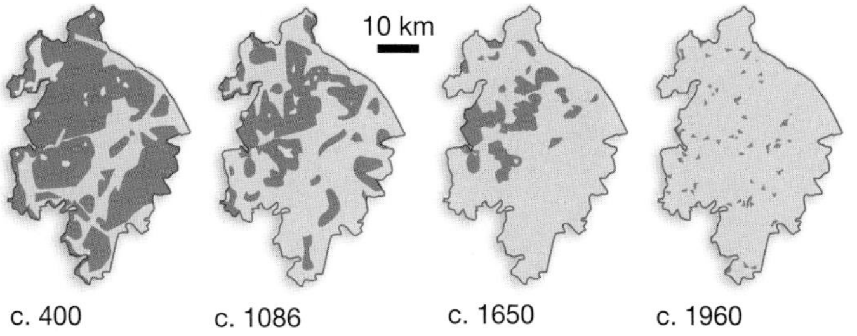
10 km
c. 400
c. 1086
c. 1650
c. 1960

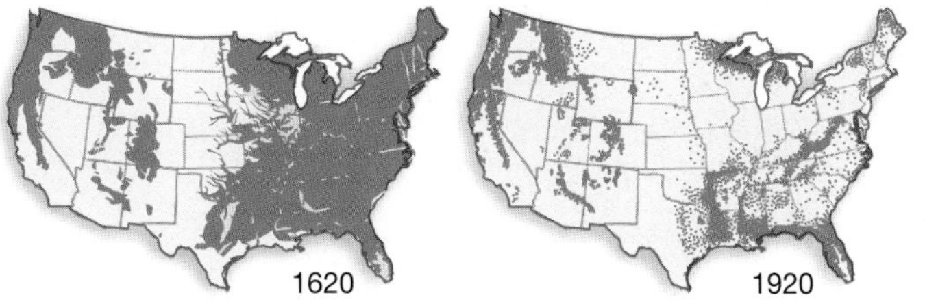
1620
1920

4. Make a list of characteristics that would render a species particularly vulnerable to extinction by humans. Make a list of characteristics that would render a species particularly resistant to pressure from humans. Try to think of an example of each type of species.

Answers to Multiple-Choice Questions **1.** d; **2.** a; **3.** b; **4.** d; **5.** c; **6.** a

www.prenhall.com/freeman is your resource for the following: Web Tutorials; Online Quizzes and other Online Study Guide materials; Answers to Conceptual Review Questions; Solutions to Group Discussion Problems; Answers to Figure Caption Questions and Exercises; and Additional Readings and Research.

Appendix

TABLE A.1 The Metric System

Measurement	Unit of Measurement and Abbreviation	Metric System Equivalent	Converting Metric Units to English Units
Length	kilometer (km)	1 km = 1000 m = 10^3 m	1 km = 0.62 miles
	meter (m)	1 m = 100 cm	1 m = 1.09 yards = 3.28 feet = 39.37 inches
	centimeter (cm)	1 cm = 0.01 m = 10^{-2} m	1 cm = 0.3937 inch
	millimeter (mm)	1 mm = 1000 μm = 10^{-3} m	1 mm = 0.039 inches
	micrometer (μm)	1 μm = 1000 nm = 10^{-6} m	
	nanometer (nm)	1 nm = 10^{-9} m	
	angstrom (Å)	1 Å = 0.1 nm = 10^{-10} m	
Area	hectare (ha)	1 ha = 10,000 m^2	1 ha = 2.47 acres
	square meter (m^2)	1 m^2 = 10,000 cm^2	1 m^2 = 1.196 square yards
	square centimeter (cm^2)	1 cm^2 = 100 mm^2 = 10^{-4} m^2	1 cm^2 = 0.155 square inches
Mass	kilogram (kg)	1 kg = 1000 g	1 kg = 2.20 pounds
	gram (g)	1 g = 1000 mg	1 g = 0.035 ounces
	milligram (mg)	1 mg = 1000 μg = 10^{-3} g	
	microgram (μg)	1 μg = 10^{-6} g	
Volume	liter (L)	1 L = 1000 mL	1 L = 1.06 quarts
	milliliter (mL)	1 mL = 1000 μL = 10^{-3} L	1 mL = 0.034 fluid ounces
	microliter (μL)	1 μL = 10^{-6} L	
Temperature	†Kelvin (K)		K = °C + 273.15
	Degrees Celsius (°C)		$°F = \frac{9}{5}°C + 32$
	Degrees Fahrenheit (°F)		$°C = \frac{5}{9}(°F - 32)$

†Absolute zero is −273.15°C = 0 K

TABLE A.2 Prefixes Used in the Metric System

Prefix	Abbreviation	Definition
micro-	μ	0.000001 = 10^{-6}
milli-	m	0.001 = 10^{-3}
centi-	c	0.01 = 10^{-2}
deci-	d	0.1 = 10^{-1}
		1 = 10^{0}
kilo-	k	1000 = 10^{3}

Questions

(1) Some friends of yours just competed in a 5-kilometer run. How many miles did they run?

(2) An American football field is 100 yards long, while rugby fields are 140 meters long. In yards, how much longer is a rugby field than a football field?

(3) What is your normal body temperature in degrees Celsius? (Normal body temperature is 98.6°F.)

(4) What is your current weight in kilograms?

(5) A friend asks you to buy a gallon of milk. How many liters would you buy to get approximately the same volume?

Glossary

–10 box The six-pair base sequence (usually TATAAT) found in most prokaryotic promoters 10 bases upstream from the start of mRNA transcription. Also called Pribnow box.

30-nanometer fiber A fiber of DNA and protein found in eukaryotic nuclei and consisting of many nucleosomes wound together and tightly packed.

–35 box The six-pair base sequence (usually TTGACA) in most prokaryotic promoters 35 bases upstream from the start of mRNA transcription.

ABC model A model of pattern formation of flowers. Proposes that three genes (A, B, and C) are necessary to build a normal flower containing sepals, petals, stamens, and carpels.

abdomen The region of the body that houses most of the digestive tract. In insects, the third and hindmost of the three major body regions.

abiotic Not alive (e.g., air, water, and soil).

aboveground biomass The total mass of living plants in an area, excluding roots.

abscisic acid (ABA) A plant hormone that inhibits cell elongation and stimulates leaf shedding and dormancy.

abscission The shedding of leaves from a plant.

abscission zone The region of a petiole that thins and breaks during dropping of leaves.

absorption The uptake of the ions and small molecules from food across the lining of the intestine and into the bloodstream.

absorption spectrum A graph depicting how well a pigment absorbs various wavelengths of light.

absorptive feeding The absorption of nutrients directly from the environment, across a cell's plasma membrane.

accessory fluids The fluid portion of semen, containing nutrients, enzymes, and other substances. Added to spermatozoa by male reproductive glands.

accessory pigments Pigments, such as carotenoids, that absorb light in plant cells and pass the energy on to chlorophyll to drive photosynthesis.

acclimatization Gradual physiological adjustment of an organism to new environmental conditions.

acetylation Addition of an acetyl group (CH_3COO^-) to a molecule.

acetylcholine (ACh) A neurotransmitter that is released by nerve cells onto muscle cells to trigger muscle contraction. Also used as a neurotransmitter in the nervous system.

acetyl CoA The final product of glycolysis, produced by oxidation of pyruvate and binding to CoA. Can enter the Krebs cycle.

acid Any compound that gives up protons or accepts electrons during a chemical reaction or that releases hydrogen ions when dissolved in water.

acid-base reaction A chemical reaction that involves a transfer of protons.

acid-growth hypothesis The hypothesis that auxin's effect on plant cell elongation occurs via installation of proton pumps that make the cell wall more acidic, causing the cell wall to expand.

acid rain Rain that is substantially more acidic than normal (usually with a pH below 5), enough to adversely affect plants and other species.

acoelomate Not having a coelom (internal body cavity). Diploblasts (sponges, cnidarians, and ctenophores), flatworms, and the Acoelomorpha are all acoelomate.

acquired immune deficiency syndrome (AIDS) A human disease characterized by death of immune system cells and subsequent vulnerability to other infections. Caused by the human immunodeficiency virus (HIV).

acquired immunity Immunity to a particular pathogen conferred by antibodies produced by previous exposure to the pathogen or to a vaccine or acquired passively through the placenta or breast milk.

acrosomal process A protrusion that forms on the head of a sperm cell during fertilization in some species, such as some echinoderms.

acrosomal reaction A set of events occurring in a sperm cell upon encountering an egg cell, including release of acrosomal enzymes and formation of microfilaments that help sperm reach the egg.

acrosome A packet of enzymes found on the head of a sperm cell that help dissolve the zona pellucida or jelly layer.

actin A globular protein that can be polymerized to form actin filaments, which are involved in cell movement.

actin filament A long, thin fiber composed of two intertwined strands of polymerized actin. Involved in cell movement. Also called a microfilament.

action potential A rapid, temporary change in electrical potential across a membrane, from negative to positive and back to negative.

action spectrum The range of wavelengths of light that can drive photosynthesis.

activation energy The amount of energy required to cause a chemical reaction; specifically, the energy required to reach the transition state.

active site The location on an enzyme where substrates (reactant molecules) bind and react.

active transport The movement of ions or molecules across a plasma membrane against a concentration gradient or an electrochemical gradient. Requires energy, such as ATP.

adaptation A heritable trait that increases the fitness of an individual with that trait compared with individuals without it, in a certain environment.

adaptive management A conservation strategy in which information from environmental monitoring is continuously used to review and modify ongoing conservation programs.

adaptive radiation Rapid evolutionary diversification within one lineage, producing numerous descendant species with a wide range of adaptive forms.

addiction Compulsive use of a substance despite recognition of its undesirable consequences, coupled with increasing resistance to the substance (tolerance) and withdrawal symptoms upon cessation of use.

adenosine diphosphate (ADP) A molecule consisting of adenine, a sugar, and two phosphate groups. Addition of a third phosphate group produces adenosine triphosphate (ATP).

adenosine triphosphate (ATP) A molecule consisting of adenine, a sugar, and three phosphate groups that can be hydrolyzed to release energy. Universally used by cells to store and transfer energy.

adenylyl cyclase An enzyme that can catalyze the formation of cyclic AMP (cAMP) from ATP. Involved in control of the gene transcription of various operons in prokaryotes.

adhesion The tendency of certain dissimilar molecules to cling together due to attractive forces.

adipocyte A fat cell.

adipose tissue A connective tissue whose cells store fats.

ADP See **adenosine diphosphate (ADP)**.

adrenal glands Small glands that sit above each kidney. The outer portion (cortex) secretes several steroid hormones; the inner portion (medulla) secretes epinephrine and norepinephrine.

adrenaline A catecholamine hormone from the adrenal medulla. Triggers rapid responses of the fight-or-flight response. Also called epinephrine.

adrenocorticotropic hormone (ACTH) A peptide hormone from the anterior pituitary gland that stimulates release of cortisol, corticosterone, and aldosterone from the adrenal cortex.

adventitious root A root that develops from a plant's shoot system instead of from the plant's root system.

aerobe An organism that uses aerobic respiration (i.e., uses oxygen as a final electron acceptor).

aerobic respiration Cellular respiration using oxygen as the electron acceptor, typically using the Krebs cycle and the electron transport chain.

afferent division The part of the nervous system that transmits sensory information to the central nervous system. Consists mainly of sensory neurons.

agar A gelatinous mix of polysaccharides that is commonly used in labs to grow cell cultures.

age class A group of individuals of a specific age.

agent A nonliving infectious entity, such as a virus. Also called a particle.

age pyramid A graph of a population's age structure, with horizontal bars representing the numbers of males and females, stacked against age on the vertical axis.

age-specific fecundity The average number of female offspring produced by a female in a certain age class.

age structure The proportion of individuals in a population that are of each possible age.

agglutination Clumping together of cells, typically caused by antibodies.

agonist A compound that can bind to and activate a receptor such as a hormone receptor or neurotransmitter receptor.

AIDS See **acquired immune deficiency syndrome (AIDS)**.

air sacs Thin-walled sacs of air that extend throughout most of a bird's body and connect to the lungs. Function as bellows that continuously feed air to the lungs.

albumen A solution of water and protein (particularly albumins), found in amniotic eggs, that nourishes the growing embryo. Also called egg white.

albumin A class of large proteins found in plants and animals, particularly in the albumen of eggs and in blood plasma.

alcohol fermentation Fermentation (production of ATP without oxygen) in which the end product of glycolysis is converted to ethanol.

aldosterone A hormone from the adrenal cortex that stimulates the kidney to conserve salt and water and promotes retention of sodium.

aleurone layer In a seed, a layer that releases the starch-digesting enzyme α-amylase during germination.

alkaptonuria A medical condition characterized by an accumulation of homogentisic acid in the body, caused by a defect in the enzyme that metabolizes homogentisic acid.

allantois In an amniotic egg, the membrane-bound sac that holds waste materials.

allele A particular version of a gene.

allergen Any molecule that can trigger an allergic response.

allergy An abnormal immune response to a non-pathogenic substance, usually involving production of the IgE class of antibodies.

allometry A disproportionate relationship of body size to another feature, such as limb size or heart rate. Also, the general study of the effects of body size on biological processes.

allopatric speciation The divergence of populations into different species by physical isolation of populations in different geographic areas.

allopatry Living in an area different from the area in which some other population lives.

allopolyploid Polyploid (having multiple copies of chromosomes) due to hybridization between different species.

allosteric regulation Regulation of enzymatic catalysis via a change in the enzyme's shape, usually caused by a regulatory molecule binding at a specific site.

α-amylase A plant enzyme that digests the starch in a seed so that the plant embryo can use the resulting sugars for growth.

alpha chain (α chain) One of the two polypeptide chains that make up a T-cell receptor protein, enabling the T cells of the immune system to bind to antigens on other cells.

alternation of generations A life cycle involving alternation of a multicellular haploid stage (gametophyte) with a multicellular diploid stage (sporophyte). Occurs in most plants and some protists.

alternative splicing In eukaryotes, the splicing of the same primary RNA transcript in different ways to produce different mature mRNAs and thus different proteins.

altruism Any behavior that has a cost to the individual (such as lowered survival or reproduction) and a benefit to the recipient.

alveolus (plural: alveoli) One of the tiny air-filled sacs of a mammalian lung.

ambisense virus A virus whose genome contains both positive-sense and negative-sense sequences.

amino acid A small organic molecule with a central carbon atom bonded to an amino group ($-NH_3$), a carboxyl group ($-COOH$), a hydrogen atom, and a side group. Proteins are polymers of 20 common amino acids.

aminoacyl tRNA A transfer RNA molecule that is covalently bound to an amino acid.

aminoacyl tRNA synthetases Enzymes responsible for catalyzing the addition of amino acids to tRNA molecules.

ammonia The molecule NH_3, a toxin produced by the breakdown of proteins and nucleic acids.

ammonium ion The ion NH_4^+.

amnion The membrane in an amniotic egg that surrounds the embryo and encloses it in a protective pool of fluid (amniotic fluid).

Amniota A major lineage of vertebrates that reproduce with amniotic eggs. Includes all reptiles, birds, and mammals.

amniotic egg An egg containing a membrane-bound supply of water (the amnion), a membrane-bound supply of food (yolk sac), and a waste sac (allantois), all encased in a leathery or hard shell.

amniotic fluid The fluid inside the amnion of a developing mammalian embryo.

amoeba Any unicellular protist that lacks a cell wall and is extremely flexible in shape.

amphipathic Containing hydrophilic and hydrophobic elements.

amylase Any enzyme that can break down starch by catalyzing hydrolysis of the glycosidic linkages between the glucose monomers.

amyloplast Dense, starch-storing organelles that settle to the bottom of plant cells and that may be used as gravity detectors.

anabolic pathway A set of chemical reactions that synthesizes larger molecules from smaller ones.

anadromous Having a life cycle in which adults live in the ocean (or large lakes) but migrate up freshwater streams to breed and lay eggs.

anaerobic respiration Cellular respiration using an electron acceptor other than oxygen, such as nitrate or sulfate.

analogous trait Similarity between different species due to adaptation to a similar way of life rather than inheritance from a common ancestor. Often the result of convergent evolution.

anal pore An organelle in some protists used to expel wastes from the cell.

anaphase A stage in nuclear division (mitosis or meiosis) during which chromosomes are moved to opposite ends of the cell.

anaphylactic shock A life-threatening condition in which an allergic response constricts respiratory passages and dilates blood vessels, reducing blood pressure to dangerously low levels.

anatomy The study of the physical structure of organisms.

androgens A class of steroid hormones that generally promote male-like traits (although females have some androgens, too). Secreted primarily by the gonads and adrenal glands.

aneuploidy The state of having an abnormal number of copies of a certain chromosome.

angiosperm A flowering plant; a member of the lineage of plants that produces seeds within mature ovaries (fruits).

animal hemisphere The upper half of an amphibian egg cell, containing little of the yolk. Gives rise to most of the animal's body.

animal model Any disease that occurs in a non-human animal and has many parallels to a similar disease of humans. Studied by medical researchers in hope that findings may apply to human disease.

animals A major lineage of eukaryotes. Typically have a complex, large, multicellular body, eat other organisms, and are mobile.

anion A negatively charged ion.

annual plant A plant whose life cycle normally lasts only one growing season—less than one year.

anoxic For an environment, lacking oxygen.

anoxygenic photosynthesis Photosynthesis that does not include photosystem II and hence does not split water or produce oxygen.

antagonistic muscle groups Sets of muscles whose actions oppose each other and whose control must be coordinated. Often a pair of muscles that move a body part back and forth.

antenna (plural: antennae) A long appendage that is used to touch or smell.

antenna complex An array of chlorophyll molecules that receives energy from light and directs the energy to a reaction center. Part of a photosystem of the light-dependent reactions of photosynthesis.

anterior Toward an animal's head and away from its tail. The opposite of posterior.

anterior pituitary The anterior part of the pituitary gland, containing endocrine cells that release a variety of peptide hormones in response to other hormones from the hypothalamus.

anther The pollen-producing structure at the end of a flower stamen of an angiosperm plant.

antheridium (plural: antheridia) The sperm-producing structure in most land plants except angiosperms.

anthropoids One of the two major lineages of primates, including apes, humans, and all monkeys, but not prosimians.

antibiotic Any substance, such as penicillin, that can kill or inhibit the growth of bacteria.

antibiotic resistance The ability of a microorganism to grow and reproduce effectively in the presence of a certain antibiotic.

antibody A Y-shaped immunoglobulin protein, secreted from B cells, that can bind to a specific part of an antigen, tagging it for attack by the immune system.

anticodon The three bases of a transfer RNA molecule that can bind to a messenger RNA codon with a complementary sequence.

antidiuretic hormone (ADH) A peptide hormone from the posterior pituitary gland that stimulates water retention by the kidney.

antigen Any foreign molecule, often a protein, that can stimulate a specific response by the immune system.

antigen presentation Display of an antigen on the surface of an immune system cell to recruit other immune system cells.

antihistamines Drugs that oppose histamine release from mast cells, reducing inflammatory and allergic symptoms such as swelling and itching.

antiporter A membrane protein that allows an ion to diffuse down a concentration gradient, using the energy of that process to transport some other substance against its concentration gradient and in the opposite direction.

anus In a multicellular animal, the end of the digestive tract where wastes are expelled.

aorta In terrestrial vertebrates, the major artery carrying oxygenated blood away from the heart.

aphotic zone Deep water receiving no sunlight.

apical Toward the top. In plants, at the tip of a branch. In animals, on the side of an epithelial layer that faces the environment and not other body tissues.

apical bud A bud at the tip of a stem, where growth occurs to lengthen the stem.

apical dominance Inhibition of lateral bud growth by the apical meristem at the tip of the plant branch.

apical meristem A group of undifferentiated cells at the tip of a stem or root of a vascular plant. Responsible for primary growth.

apoplast A transport pathway in plant roots through the porous cell walls of adjacent cells.

apoptosis Programmed cell death. Occurs frequently during embryonic development; may occur later in response to infections or cell damage.

aquaporin A membrane protein through which water can move by osmosis.

arbuscular mycorrhizal fungi (AMF) Fungi whose hyphae enter the root cells of their host plants.

Archaea A domain consisting of unicellular prokaryotes distinguished by cell walls made of certain polysaccharides, unique plasma membranes but ribosomes and RNA polymerase similar to those of eukaryotes.

archegonium (plural: archegonia) The egg-producing structure in most land plants except angiosperms.

arteriole Tiny arteries that deliver blood to capillaries. Has muscular walls.

artery Any blood vessel that carries blood from the heart to capillaries. Has thick muscular walls that withstand and control blood pressure.

articulation A movable joint of a skeleton.

artificial selection Deliberate manipulation by humans, as in animal and plant breeding, of the hereditary traits of a population by allowing only certain individuals to reproduce.

asci (singular: ascus) Specialized spore-producing cells found at the ends of hyphae in sac fungi.

ascocarp A large, cup-shaped reproductive structure produced by some ascomycete fungi. Contains many microscopic asci, which produce spores.

asexual reproduction Any form of reproduction that results in offspring that are genetically identical to the parent.

A site The site in a ribosome where an aminoacyl transfer RNA pairs with a messenger RNA codon, in preparation for adding an amino acid to the growing peptide chain.

assimilation hypothesis The hypothesis that modern humans (*Homo sapiens*) evolved in Africa but acquired modern traits after interbreeding with other *Homo* species in Europe and Asia.

asthma A chronic lung condition characterized by periodic episodes of difficulty in breathing, due to smooth-muscle contractions in the respiratory passages.

astrobiology The study of biological molecules that occur in space (e.g., on meteors) and of possible extraterrestrial life.

asymmetric competition Ecological competition between two species in which one species suffers a much greater fitness decline than the other.

atherosclerosis A disease in which lipids accumulate on artery walls, eventually resulting in blockage of the artery.

atom The smallest unit of a chemical element that retains the characteristics of that element.

atomic mass unit (amu) A unit of mass equal to 1/12 the mass of one carbon-12 atom; about the mass of 1 proton or 1 neutron. Also called a dalton.

atomic number The number of protons in the nucleus of an atom, giving the atom its identity as a certain chemical element.

ATP See **adenosine triphosphate (ATP)**.

ATP synthase A membrane-bound enzyme in chloroplasts and mitochondria that uses the energy of protons flowing through it to synthesize ATP.

atrial natriuretic hormone A hormone from the heart that stimulates the kidney to excrete sodium, lowering blood pressure.

atrioventricular (AV) node A location in the heart between the atria and ventricles where electrical signals from the atria are slowed before spreading to the ventricles, allowing them to fill with blood before contracting.

atrium (plural: atria) A thin-walled chamber of the heart that receives blood from veins and pumps it to a neighboring chamber (the ventricle).

attenuated virus A virus that is functional but that has been rendered nonvirulent for a certain species, usually by culturing in cells of a different species. Used for vaccines.

attenuation Gradual reduction in strength. In bacterial genetics, fine-tuning of gene expression such that genes produce only as much of their proteins as required.

australopithecines Species of early hominins that appear in the fossil record shortly after the split from chimpanzees. Generally bipedal, but retained a chimpanzee-size brain.

autocrine signal A chemical signal that affects the same cell that released it.

autoimmunity A pathological condition in which the immune system attacks part of the body.

autonomic nervous system The part of the peripheral nervous system that controls internal organs and involuntary processes, such as stomach contraction, hormone release, and heart rate. Includes parasympathetic and sympathetic nerves.

autophagy The process by which damaged organelles are surrounded by a membrane and delivered to a lysosome to be destroyed.

autopolyploid Polyploid due to a mutation that doubled the chromosome number.

autosomal inheritance The inheritance patterns that occur when genes are located on autosomes rather than on sex chromosomes.

autosome One of any pair of chromosomes that do not carry the gene(s) that determine gender.

autotroph Any organism that can synthesize its own food—complex organic compounds—from simple inorganic sources such as CO_2 or CH_4, including most plants and some bacteria and archaea.

auxin Indoleacetic acid, a plant hormone that stimulates phototropism and some other responses.

avirulence (*avr*) loci Genes in pathogens that trigger a defense response in plants.

axon A long projection of a neuron that can propagate an action potential.

axoneme An arrangement of two central microtubules surrounded by nine doublet microtubules. Found in eukaryotic cilia and flagella and responsible for their motion.

axon hillock The location at which an axon joins the cell body of a neuron; the site at which action potentials are first triggered.

BAC (bacterial artificial chromosome) A lab-created loop of DNA that can be inserted into a living bacterial cell, which can then be grown in a cell culture to create many copies of the BAC.

background extinction The average rate of extinction that occurs normally, as opposed to the higher rate of extinction that occurs during mass extinction events.

BAC library A set of bacterial colonies containing BACs (bacterial artificial chromosomes) that each have a section of the DNA being studied. Commonly used in genome sequencing.

Bacteria A taxonomic domain that includes all bacteria—prokaryotes that have cell walls made of peptidoglycan, distinct forms of ribosomes and RNA polymerase, and certain other traits.

bacteriophage Any virus that infects bacteria.

baculum A bone inside the penis. Usually present in mammals that do not have erectile tissue.

bark The protective outer layer around woody plants, containing cork cells, cork cambium, and secondary phloem.

baroreceptors Specialized nerve cells in the walls of the heart and certain major arteries that detect changes in blood pressure and trigger appropriate responses by the brain.

basal body A structure of nine pairs of microtubules arranged in a circle at the base of eukaryotic cilia and flagella where they attach to the cell.

basal cell The lower cell produced by an angiosperm zygote. Gives rise to a structure that transfers nutrients from the parent plant to the developing embryo.

basal lamina A thick, collagen-rich extracellular matrix found in animal skin.

basal meristem A group of undifferentiated plant cells located below ground, from which the plant can regenerate the shoot system if necessary.

basal metabolic rate (BMR) The total energy consumption by an organism at rest in a comfortable environment. For aerobes, often measured as the amount of oxygen consumed per hour.

basal transcription complex A multi-protein structure that initiates transcription of eukaryotic genes.

basal transcription factors Proteins that bind to eukaryotic promoters and help initiate transcription.

base Any compound that acquires protons or gives up electrons during a chemical reaction or accepts hydrogen ions when dissolved in water.

baseline A measurement that represents normal values before a change or disturbance and that serves as a reference point for future comparisons.

basidia (singular: basidium) Specialized spore-producing cells at the ends of hyphae in club fungi.

basilar membrane The membrane in the vertebrate cochlea on which the hair cells sit.

basolateral Toward the bottom and sides. In animals, the side of an epithelial layer that faces other body tissues and not the environment.

basophil A circulating leukocyte involved in the inflammatory response and in allergies.

Batesian mimicry A type of mimicry in which a harmless species resembles a harmful species.

B cells A class of leukocytes that can produce antibodies. Produced by the bursa of birds and the bone marrow of mammals. Also called B lymphocytes.

B-cell receptor (BCR) A Y-shaped immunoglobulin protein found on the surfaces of B cells and to which antigens bind.

behavior Any action by an organism.

benign tumor A mass of abnormal tissue that grows slowly or not at all, does not disrupt surrounding tissues, and does not metastasize to other organs. Benign tumors are not cancers.

benthic Living at the bottom of an aquatic environment.

benthic zone The area along the bottom of an aquatic environment.

beta chain (β chain) One of the two polypeptide chains that make up a T-cell receptor protein, enabling the T cells of the immune system to bind to antigens on other cells.

bilateral symmetry An animal body pattern in which there is one plane of symmetry dividing the body into a left side and a right side. Typically, the body is long and narrow, with a distinct head end and tail end.

Bilateria A major lineage of animals that are bilaterally symmetrical at some point in their life cycle, have three embryonic germ layers, and have coeloms. Includes protostomes and deuterostomes.

bile A fluid produced by the liver, stored in the gall bladder, and secreted into the intestine, where it emulsifies fats during digestion.

bile salts Small lipids in bile that can emulsify fats.

binary fission Division of a bacterial cell to produce two daughter cells. Similar to mitosis of eukaryotic cells.

binomial nomenclature A system of naming species by using two-part Latinized names with a genus name and a species name. Always italicized, with genus name capitalized.

biodiversity The variety and relative abundance of species present in a certain area.

biofilm A tough, polysaccharide-rich substance secreted by certain prokaryotes. Encases and protects cells and attaches them to the organism's surface.

biogeochemical cycle The pattern of circulation of an element or molecule among living organisms and the environment.

biogeography The study of how species and populations are distributed geographically.

bioinformatics The field of study concerned with managing, analyzing, and interpreting biological information, particularly DNA sequences.

biological species concept The concept that species are best identified as groups that are reproductively isolated from each other; thus, different species cannot crossbreed in nature to produce viable and fertile hybrid offspring.

bioluminescence The emission of light by a living organism.

biomass The total mass of all organisms in a given population or geographical area; usually expressed as total dry weight.

biome A major category of ecosystem, characterized by a distinct type of vegetation and climate.

bioprospecting The search for naturally occurring compounds that can be useful to humans (e.g., as drugs, fragrances, or insecticides). Also called chemical prospecting.

bioremediation The use of living organisms, usually bacteria and archaea, to degrade environmental pollutants.

biotechnology The application of biological techniques and discoveries to medicine, industry, and agriculture.

biotic Living, or produced by a living organism.

bipedal Walking primarily on two legs.

bipolar cell Cells in the vertebrate retina that receive information from one or more photoreceptors and pass it to other bipolar cells and ganglion cells.

bivalves A lineage of mollusks that have two shells, such as clams and mussels.

bladder A mammalian organ that holds urine until it can be excreted.

blade The wide, flat part of a plant leaf.

blastomere The small cells created by cleavage divisions in early animal embryos.

blastopore A small pore in the surface of an early vertebrate embryo, through which cells move during gastrulation.

blastula An early stage of embryonic development in vertebrates, consisting of a ball of cells (a blastomere) enclosing a fluid-filled space. Immediately precedes gastrulation.

blood A type of liquid connective tissue consisting of red blood cells and leukocytes suspended in the fluid plasma.

blood doping The practice of giving an athlete a transfusion of his or her own red blood cells (removed and stored earlier) to increase aerobic capacity for a competition.

body mass index A mathematical relationship of weight and height used to assess obesity in humans. Calculated as weight (kg) divided by the square of height (m^2).

body plan The basic architecture of an animal's body, including the number and arrangement of limbs, body segments, and major tissue layers.

bog A wetland that has no or almost no water flow, resulting in very low oxygen levels and acidic conditions.

Bohr shift The rightward shift of the oxygen-hemoglobin dissociation curve that occurs with decreasing pH. Results in hemoglobin being more likely to release oxygen in the acidic environment of exercising muscle.

bone A vertebrate tissue consisting of living cells and blood vessels within a hard extracellular matrix of calcium phosphate ($CaPO_4$) with small amounts of calcium carbonate ($CaCO_3$) and protein fibers.

bone marrow The soft tissue filling the inside of long bones. Produces red blood cells and leukocytes and also contains fat.

Bowman's capsule The hollow, sphere-like end of a kidney nephron that surrounds a glomerulus.

braincase A bony, cartilaginous, or fibrous case that encloses and protects the brain of vertebrates. Forms part of the skull. Also called the cranium.

branch (1) A part of a phylogenetic tree that represents populations. (2) Any extension of a plant's shoot system.

bronchi (singular: bronchus) In mammals, the large tubes that lead from the trachea to each lung.

bronchioles The small tubes in mammalian lungs that carry air from the bronchi to the alveoli.

brown adipose tissue A specialized form of fat tissue whose cells have a high density of mitochondria as well as stored fats and that can produce extra body heat. Found in some mammals.

bryophytes Several phyla of green plants that lack vascular tissue. Includes liverworts, hornworts, and mosses. Also called nonvascular plants.

budding Asexual reproduction via growth of a small individual from part of the parent, eventually breaking free as an independent individual.

buffer A substance that, in solution, acts to minimize changes in the pH of that solution.

bulbourethral glands Small, paired glands at the base of the urethra in male mammals that secrete pre-ejaculatory fluid for lubrication during copulation. In humans, also called Cowper's glands.

bulk flow The directional movement of a substantial volume of fluid due to pressure differences, such as movement of water through plant phloem and movement of blood in animals.

bundle-sheath cell A type of cell that is located around the vascular tissue in the interior of plant leaves.

Burgess shale fauna A characteristic set of soft-bodied animal fossils found in several early Cambrian rock formations (525–515 million years old), particularly the Burgess Shale in Canada and the Chengjiang deposits in China.

bursa An immune system organ of birds that produces B cells.

C_3 photosynthesis A common form of photosynthesis in which atmospheric carbon is used to form 3-phosphoglycerate, a three-carbon compound.

C_4 photosynthesis A variant of photosynthesis in which carbon from CO_2 is first fixed into four-carbon compounds. Enhances photosynthetic efficiency in hot, dry environments by reducing loss of oxygen due to photorespiration.

cadherins Cell-surface proteins involved in cell adhesion and important for coordinating movements of cells during embryological development.

calcitonin A hormone from the thyroid gland that lowers blood calcium by preventing calcium and phosphorus withdrawal from bone.

callus A mass of undifferentiated plant cells that can generate roots and other tissues necessary to create a mature plant.

Calorie A unit of energy, often used to measure energy content of food. Also called a kilocalorie.

Calvin cycle In photosynthesis, the set of reactions that use the NADPH and ATP formed earlier (in the light-dependent reactions) to drive the reduction of atmospheric CO_2, ultimately producing sugars. Also called light-independent reactions.

CAM See **crassulacean acid metabolism (CAM)**.

cambium (plural: cambia) A layer of undifferentiated plant cells found in woody plants. Responsible for secondary growth. Also called secondary meristem or lateral meristem.

Cambrian explosion The rapid diversification of animal body types that began about 543 million years ago, during approximately 40 million years.

camera eye A type of eye in vertebrates and cephalopods, consisting of a hollow chamber with a hole at one end (through which light enters) and a sheet of light-sensitive cells against the opposite wall.

cAMP See **cyclic AMP (cAMP)**.

canopy The uppermost layers of branches in a forest (i.e., those fully exposed to the Sun).

5′ cap A structure added to the 5′ end of newly transcribed messenger RNA molecules. Consists of 7-methylguanylate and three phosphate groups.

CAP binding site A DNA sequence upstream of certain prokaryotic operons, to which catabolite activator protein can bind, increasing gene transcription.

capillarity The tendency of water to move up a narrow tube due to surface tension, adhesion, and cohesion.

capillary The smallest blood vessel, where gases and other molecules are exchanged between blood and tissues.

capillary bed A thick network of capillaries.

capsid A shell of protein enclosing the genome of a virus particle.

carapace In crustaceans, a large platelike section of the exoskeleton that covers and protects the cephalothorax (e.g., a crab's "shell").

carbohydrates A class of molecules, including sugars, starches, glycogen, and cellulose, that have a carbonyl group, several hydroxyl groups, and several to many carbon-hydrogen bonds.

carbonic anhydrase An enzyme that catalyzes the formation of carbonic acid (H_2CO_3) from carbon dioxide and water.

carbon sink A reservoir that holds carbon for very long time periods; e.g., fossil fuels, sedimentary rocks.

carcinogen Any cancer-causing agent.

cardiac cycle One complete heartbeat cycle, including systole and diastole.

cardiac muscle The muscle tissue of the vertebrate heart. Consists of long branched fibers that are electrically connected and that initiate their own contractions; not under voluntary control.

cardiac output The total volume of blood leaving the left ventricle per minute. Calculated as stroke volume times heart rate.

carnivore (1) Any animal whose diet consists predominantly of other animals. (2) Any member of the mammalian taxon Carnivora. (Most members of the Carnivora are meat eaters.)

carotenoids A class of plant pigments that absorb wavelengths not absorbed by chlorophyll. Typically appear yellow, orange, or red. Includes carotenes and xanthophylls.

carpel In a flower, the reproductive structure that produces female gametophytes; consists of the stigma, the style, and the ovary.

carrier (1) A heterozygous individual carrying a normal allele and a recessive allele for an inherited condition; does not display the phenotype of the condition but can pass the recessive gene to offspring. (2) A transmembrane protein that assists with facilitated diffusion by binding a specific ion or molecule and transporting it across the membrane.

carrying capacity The maximum population size of a certain species that a given habitat can support. Symbolized by *K*.

cartilage A type of connective tissue in the skeletons of vertebrates, consisting of relatively few cells scattered in a stiff matrix of polysaccharides and protein fibers.

Casparian strip A waxy layer that prevents movement of water through the walls in plant roots.

catabolic pathway Any set of chemical reactions that breaks down larger, complex molecules into smaller ones, releasing energy in the process.

catabolite The end product of a catabolic pathway; the final breakdown product of a substance.

catabolite activitor protein (CAP) A protein that can bind to the CAP binding site upstream of certain prokaryotic operons, facilitating binding of RNA polymerase and stimulating gene expression.

catabolite repression A type of inhibition of gene transcription in which a gene codes for an enzyme in a catabolic pathway and the end product of that pathway inhibits further transcription of that gene.

catalysis Acceleration of the rate of a chemical reaction, produced by lowering the potential energy of the transition state.

catalyst Any substance that increases the rate of a chemical reaction without itself undergoing any permanent chemical change. Functions by lowering the activation energy of the reaction.

catecholamines A class of small compounds, derived from the amino acid tyrosine, that are used as hormones or neurotransmitters. Includes epinephrine, norepinephrine, and dopamine.

cation A positively charged ion.

cation exchange The release of cations such as magnesium and calcium from soil particles, due to displacement by protons in acidic soil water.

CD4 A membrane protein on the surface of some T lymphocytes of the human immune system; these $CD4^+$ T cells can give rise to helper T cells.

CD8 A membrane protein on the surface of some T lymphocytes of the human immune system; these $CD8^+$ T cells can give rise to cytotoxic T cells.

Cdk See **cyclin-dependent kinase (Cdk)**.

cDNA (complementary DNA) DNA created in a lab from an RNA transcript, using the enzyme reverse transcriptase. Corresponds to a certain gene but lacks introns. Is also created naturally by some viruses (retroviruses).

cDNA library A set of cDNAs from a particular organism or cell type. Usually exists as a set of bacteria colonies, each with a plasmid containing a particular cDNA sequence.

cecum A blind sac between the small intestine and the colon. Used in some species as a fermentation vat for digestion of cellulose.

cell A highly organized living entity that is bounded by a thin, flexible structure called a plasma membrane and that contains concentrated chemicals in an aqueous (watery) solution.

cell body The part of a neuron that contains the nucleus and where incoming signals are integated. Also called the soma.

cell-cell signal A molecule that is secreted from one cell and affects the activity of a nearby cell.

cell crawling A form of cellular movement in which the cell produces bulges (pseudopodia) that stick to the substrate and pull the cell forward.

cell cycle The sequence of stages that a dividing eukaryotic cell goes through from the time it is created (by division of a parent cell) to the time it undergoes mitosis.

cell-cycle checkpoint A regulated point at which the cell cycle can be stopped.

cell division Creation of new cells by division of pre-existing cells (i.e., by mitosis or meiosis).

cell enlargement Growth in plants in which individual cells expand in size in a certain direction.

cell extract or cell homogenate A solution created by breaking cells apart. Cell extracts contain organelles, free macromolecules, and small membrane-bound vesicles.

cell-mediated response Defense mounted by cytotoxic T lymphocytes against infections.

cell plate A double layer of new plasma membrane that appears in the middle of a dividing plant cell; ultimately divides the cytoplasm into two separate cells.

cell sap An aqueous solution found in the vacuoles of plant cells.

cell theory The theory that all organisms are made of cells and that all cells come from preexisting cells.

cellular respiration A common pathway for production of ATP, involving transfer of electrons from compounds with high potential energy (often NADH and $FADH_2$) to an electron transport chain and ultimately to an electron acceptor (often oxygen).

cellulases Enzymes that can digest cellulose.

cellulose A polysaccharide that is the major component of plant cell walls. Consists of β-glucose monomers joined with β-1,4-glycosidic linkages in long, straight chains.

cell wall A protective layer located outside the plasma membrane and usually composed of polysaccharides. Found in algae, plants, bacteria, fungi, and some other groups.

central dogma The long-accepted hypothesis that information in cells flows in one direction: DNA codes for RNA, which codes for proteins. Exceptions are now known (e.g., retroviruses).

central nervous system (CNS) The brain and spinal cord of vertebrate animals.

centrifugation A lab technique for separating substances by density and size by spinning them rapidly in a circle, so that larger or denser particles are flung to the outside of the sample container.

centriole One of two small cylindrical structures found together near the nucleus of a eukaryotic cell. Collectively called the centrosome, they serve as a microtubule hub for the cell's cytoskeleton.

centromere The structure that joins two sister chromatids during meiosis.

centrosome Structure in animal and fungal cells, consisting of two centrioles together near the nucleus. Serves as a microtubule organizing center for the cell's cytoskeleton and for the mitotic spindle during cell division.

cephalization The formation of a distinct head—an anterior region where sense organs and a mouth are clustered.

cephalochordates A lineage of small, mobile chordates also called lancelets or amphioxi.

cephalopods A lineage of mollusks including the squid, octopuses, and nautiluses. Distinguished by large brains, excellent vision, tentacles, and a reduced or absent shell.

cerebellum A posterior section of the vertebrate brain, involved in coordination of complex muscle movements, such as in locomotion and maintaining balance.

cerebrum The anteriormost section of the vertebrate brain. Divided into left and right hemispheres and involved in memory, interpretation of information, decision making, and (in humans) conscious thought.

cervix The narrow passageway between the vagina and the uterus of female mammals.

channel protein, or channel A membrane protein that forms a pore that selectively allows a specific ion or molecule to cross the membrane via diffusion. Channel proteins provide passive transport (facilitated diffusion) and do not require energy.

chelicerae Appendages found around the mouth of the chelicerates (spiders, mites, and allies).

chemical bond A strong attractive force binding two atoms together. Covalent bonds, ionic bonds, and hydrogen bonds are types of chemical bonds.

chemical carcinogen Any chemical that can cause cancer.

chemical energy The potential energy stored in covalent bonds between atoms.

chemical equilibrium A dynamic but stable state of a reversible chemical reaction in which the forward reaction and reverse reactions proceed at the same rate, so that the concentrations of reactants and products remain constant.

chemical evolution The hypothesis that simple chemical compounds in the ancient atmosphere and ocean combined by spontaneous chemical reactions to form larger, more complex substances, eventually leading to the origin of life and the start of biological evolution.

chemical reaction An event in which one compound or element is combined with others or is broken down.

chemiosmosis The production of ATP via proton movement, through ATP synthase, across a membrane, driven by a proton gradient.

chemiosmotic hypothesis The hypothesis that ATP synthesis in mitochondria and chloroplasts occurs indirectly via proton movement across a membrane.

chemokine A chemical signal that attracts leukocytes to a site of tissue injury or infection.

chemoreceptor A sensory cell or organ specialized for detection of specific molecules or classes of molecules.

chemotaxis Movement toward or away from a certain chemical.

chemotherapy Treatment with anticancer chemicals or drugs. Most chemotherapy drugs kill dividing cells or stop the cell cycle.

chiasma (plural: chiasmata) The X-shaped structure formed during meiosis by crossing over between adjacent chromatids of a pair of homologous chromosomes.

chitin A polysaccharide consisting of monomers of *N*-acetylglucosamine joined end to end in long, straight chains. Found in cell walls of fungi and many algae, and in external skeletons of insects and crustaceans.

chitons Marine mollusks that have a protective shell formed of eight calcium carbonate plates.

chloride cells In the gills of marine fish with bony skeletons, cells that excrete excess salt and maintain electrolyte balance.

chlorophyll A green pigment molecule that absorbs light energy to power photosynthesis. Found in plant cells and in photosynthetic protists.

chloroplast In plants, a chlorophyll-containing organelle in which photosynthesis occurs. Also, the location of amino-acid, fatty-acid, purine, and pyrimidine synthesis.

chloroplast DNA DNA found within a chloroplast of a eukaryotic plant cell. Consists of a single circular chromosome.

choanocytes The specialized feeding cells found in sponges.

choanoflagellates A phylum of protists thought to be the closest living relatives of animals.

cholecystokinin A peptide hormone from the small intestine that stimulates the secretion of digestive enzymes from the pancreas and bile from the liver and gallbladder.

cholera A human infectious disease characterized by watery diarrhea that can lead to dehydration and death. Caused by the bacterium *Vibrio cholerae*.

cholesterol A steroid that is a major component of plasma membranes. Required for synthesis of steroid hormones. In excess, can contribute to atherosclerosis.

chordate An animal of the phylum Chordata, distinguished by such traits as a dorsal hollow nerve cord, pharyngeal gill slits, and a post-anal tail. Includes vertebrates.

chorion In an amniotic egg, a highly vascularized membrane across which gas exchange occurs with the environment.

chromatid One of the daughter strands of a chromosome that has recently been copied (prior to mitosis or meiosis) and that is still connected to the other daughter strand.

chromatin The material that makes up eukaryotic chromosomes: a DNA molecule complexed with histone proteins. Can be highly compact (heterochromatin) or filamentous (euchromatin).

chromatin remodeling The process by which the DNA in chromatin is unwound from its associated proteins to allow transcription or replication.

chromatin-remodeling complex A protein involved in restructuring chromatin (DNA packed with proteins) in eukaryotic cells. The remodeling requires ATP.

chromosome A single long molecule of DNA and any associated proteins (e.g., histones).

chromosome inversion A mutation in which a segment of a chromosome breaks from the rest of the chromosome, flips, and rejoins with the opposite orientation as before.

chromosome painting A technique for producing high-resolution karyotypes by "painting" chromosomes with fluorescent tags that bind to particular regions of certain chromosomes. Also called spectral karyotyping (SKY).

chromosome theory of inheritance The theory that Mendel's rules of inheritance can be explained by the independent segregation of homologous chromosomes during meiosis.

chylomicron A ball of protein-coated lipids, used to transport the lipids through the bloodstream.

cilia (singular: cilium) Short, numerous, filamentous projections of some eukaryotic cells, used to move the cell and/or to move fluid or particles along a stationary cell.

circadian Lasting approximately one day.

circadian clock An internal mechanism found in most organisms that regulates many body processes (sleep-wake cycles, hormonal patterns, etc.) in a roughly 24-hour cycle.

circadian rhythm Any biological process that occurs on an approximately 24-hour cycle and that is controlled by an internal circadian clock.

circulation In physiology, mass movement of blood throughout the body.

cisterna (plural: cisternae) A flattened, membrane-bound compartments of the Golgi apparatus.

clade An evolutionary unit that includes an ancestral population, all of its descendants, and only its descendants. Also called a monophyletic group or a lineage.

cladistic approach A method for constructing a phylogenetic tree that is based on identifying the unique traits of each monophyletic group.

class A classic taxonomic rank above the order level and below the phylum level.

Class I MHC protein A type of MHC protein that is present on the plasma membrane of every body cell, marking the cells as "self" to the immune system.

Class II MHC protein A type of MHC protein, present only on the plasma membranes of macrophages, B cells, and T cells, that helps them join together during B-cell activation.

classical conditioning A type of learning in which an animal learns to associate two stimuli, so that a response originally given to just one stimulus can be evoked by the second stimulus as well.

cleavage Rapid cell division without production of new cytoplasm. Seen only in early embryonic development in animals. Cleavage transforms a zygote into a blastula.

cleavage furrow A pinching-in of the plasma membrane that occurs as the cytoplasm of an animal cell begins to divide; one of the final events of cell division in animals, fungi, and slime molds.

climate The prevailing long-term weather conditions in a particular region.

climax community The stable, final stage of an ecological community that develops from ecological succession.

clitoris A small rod of erectile tissue in the external genitalia of female mammals. Forms from the same embryonic tissue as the male penis and has a similar function in sexual arousal.

clonal expansion Rapid cell division by a particular T cell or B cell of the immune system in response to a particular antigen. Produces a large population of descendant cells that can attack the antigen.

clonal-selection theory The dominant theory of the development of acquired immunity in vertebrates, proposing that the immune system retains a vast pool of inactive lymphocytes, each with a unique receptor for a unique antigen. Lymphocytes that encounter their antigens are stimulated to divide (selected and cloned), producing daughter cells that combat infection and confer immunity.

clone An individual that is genetically identical to another individual.

closed circulatory system A circulatory system in which the circulating fluid (blood) is confined to blood vessels and flows in a continuous circuit.

clot A mass of red blood cells, platelets, and protein fibers that forms to plug a hole in the circulatory system and stop blood loss.

clutch size The number of eggs laid by a female in a single nest or a single breeding effort.

cnidocyte A specialized stinging cell found in cnidarians.

coactivators A class of regulatory proteins that help initiate transcription by bringing together the necessary transcription factor proteins.

cochlea A coiled, fluid-filled tube in the inner ear of mammals, birds, and crocodilians. Contains nerve cells that detect sounds of different pitches.

codominance An inheritance pattern in which heterozygotes exhibit both of the traits seen in either kind of homozygous individual.

codon A sequence of three nucleotides of DNA or RNA that codes for a certain amino acid or that initiates or terminates protein synthesis.

coefficient of relatedness A measurement of how closely two individuals are related. Calculated as the probability that an allele in two individuals is inherited from the same ancestor. Symbolized as *r*.

coelom An internal, usually fluid-filled, body cavity that forms within the mesoderm.

coelomate Possessing a coelom.

coenocytic Containing many nuclei and a continuous cytoplasm through a filamentous body, without the body being divided into distinct cells. Some fungi are coenocytic.

coenzyme Any non-protein molecule or ion that is a required cofactor for an enzyme-catalyzed reaction. Often transfers or receives electrons or functional groups.

coenzyme A (CoA) A coenzyme that is required for many cellular reactions involving transfer of acetyl groups (–$COCH_3$). In glycolysis, pyruvate reacts with CoA to produce acetyl CoA, which can then enter the Krebs cycle.

coenzyme Q A molecule that transfers electrons in the electron transport chain of cellular respiration. Also called ubiquinone or Q.

coevolutionary arms race A pattern of evolution in which one species evolves a defense against another species (e.g., a predator or parasite), which then evolves a counterdefense, and so on.

cofactor A non-protein molecule or ion that is required for an enzyme to function normally. May be a metal ion or a coenzyme.

cognition The mental processes involved in recognition and manipulation of facts about the world, particularly to form novel associations or insights.

cohesion A phenomenon seen in some liquids, such as water, in which attractive forces between molecules cause the liquid molecules to cling together and resist disruption.

cohesion-tension theory The theory that water movement upward through plant vascular tissues is due to transpiration (loss of water from leaves), which pulls a cohesive column of water upward.

cohort A group of individuals that are the same age and can be followed through time.

coleoptile A modified leaf that covers and protects the stems and leaves of young grasses.

collagen A fibrous, cable-like protein secreted by animal cells into the extracellular matrix. Forms a pliable, strong substance in which the cells sit.

collenchyma cell An elongated type of plant cell with cell walls thickened at the corners that provides support to growing plant parts; usually found in strands along leaf veins and stalks.

colonial growth Growth in which unicellular or multicellular individuals join to form a structure in which the multiple individuals are not specialized and each retains the ability to reproduce.

colony An assemblage of individuals. May refer to an assemblage of semi-independent cells or to a breeding population of multicellular organisms.

combination therapy Medical therapy that involves dosing an infected patient with several drugs simultaneously, to lessen the chances of the pathogen evolving resistance.

commensal A species that is dependent on another species but does not harm that species.

commensalism A species interaction in which one species benefits and the other is not harmed.

communication Any process in which a signal from one individual modifies the behavior of another individual.

community All of the species that interact with each other in a certain area.

companion cell Cell in phloem tissue that provides materials and nutrients to sieve-tube members.

compass orientation Motion in a specific compass direction.

competition The effect of two species or two individuals trying to use the same limited resource.

competitive exclusion principle The principle that two species cannot coexist in the same ecological niche in the same area because one species will out-compete the other.

competitive inhibition Inhibition of an enzyme's ability to catalyze a chemical reaction via a nonreactant molecule that competes with the substrate(s) for access to the active site.

complementary base pairing The specific pairing that occurs between nitrogenous bases of nucleic acids: Adenine pairs only with thymine (in DNA) or uracil (in RNA), and guanine pairs only with cytosine. Allows accurate replication of DNA and RNA sequences.

complementary DNA (cDNA) DNA created in the lab from an RNA transcript, using reverse transcriptase; corresponds to a particular gene but lacks introns. Created naturally by retroviruses.

complementary strand A new strand of RNA or DNA that has a base sequence complementary to that of the template strand.

complement system A class of proteins that circulate in the bloodstream and attack plasma membranes of bacteria.

compound eye An arthropod eye formed of many independent light-sensing columns (ommatidia).

concentration gradient Variation across space in the concentration of a dissolved substance, from a region of high concentration to a region of low concentration.

condensation reaction A type of chemical reaction involving the bonding together of two molecules by removal of an –OH from one and an –H from another to form water. Also called a dehydration reaction. The reverse of hydrolysis.

conditional strategy A behavioral response that varies with the current environmental and social conditions.

conduction Direct transfer of heat between two objects that are in physical contact.

cone (1) A photoreceptor cell with a cone-shaped outer portion that is particularly sensitive to bright light of a certain color. Found in eyes of vertebrates and some other animals. (2) The reproductive structure found in conifers.

confocal microscopy A technique for obtaining a focused image of a certain plane within a live cell.

conformational homeostasis Homeostasis (steady internal body conditions) that is achieved by the body's passively matching the conditions of a stable external environment.

conjugation The process by which DNA is exchanged between unicellular individuals. Occurs in bacteria, archaea, and some protists.

conjugation tube A connection between two prokaryotes that are in the process of transferring plasmids to each other.

connective tissue A class of animal tissue consisting of scattered cells in a liquid, jellylike, or solid extracellular matrix. Includes bone, cartilage, tendons, ligaments, and blood.

conservation biology The effort to study, preserve, and restore threated populations, communities, and ecosystems.

conservative replication A now-disproven hypothesis of DNA replication that proposed that one of the daughter molecules retains both original strands of DNA, while the other daughter molecule is built with two new strands.

constant (C) region A section of the light chains of antibodies that has the same amino acid sequence in every B cell of an individual.

constitutive Always occurring; always present (as in enzymes that are synthesized continuously).

constitutive defense A defensive trait that is always present, regardless of need. Also called standing defense.

constitutive mutant A mutant in which certain genetic loci are constantly transcribed due to flaws in gene regulation.

consumer An organism that consumes food created by other organisms; not a primary producer.

continental drift The motion of continents over large periods of time due to plate tectonics.

continental shelf The shallow, gently sloping portion of the ocean floor near continents.

continuous strand In DNA replication, the strand of new DNA synthesized in one continuous piece, with nucleotides added to the 3′ end of the growing molecule. Also called leading strand.

contraception Any method to prevent pregnancy.

control In an experiment, a group of organisms or samples that do not receive the experimental treatment but are otherwise identical to the group that does.

convection Transfer of heat by movement of large volumes of a gas or liquid.

convergent evolution Evolution of similar traits in distantly related organisms due to adaptation to similar environments and a similar way of life. Often produces analogous traits.

convergent trait Similarity between different species that is due to adaptation to a similar way of life rather than inheritance from a common ancestor. Often the result of convergent evolution. Also called an analogous trait.

cooperative binding The tendency of the protein subunits of hemoglobin to affect each other's oxygen binding such that each bound oxygen molecule increases the likelihood of further oxygen binding.

copulation The act of transferring sperm from a male directly into a female's reproductive tract.

coral reef A large assemblage of colonial marine corals that usually serves as shallow-water, sunlit habitat for many other species as well.

co-receptor Any membrane protein that acts with some other membrane protein in a cell interaction or cell response.

core enzyme The part of a holoenzyme that contains the active site for catalysis.

cork cambium A ring of undifferentiated plant cells found just under the cork layer of woody plants. Produces new cork cells on its outer side.

cork cells The waxy cells in the protective outermost layer of a woody plant.

corm A rounded, thick underground stem that can produce new plants via asexual reproduction.

cornea The transparent sheet of connective tissue at the very front of the eye in vertebrates and some other animals. Protects the eye and helps focus light.

corolla All of the petals of a flower.

corona The cluster of cilia at the anterior end of a rotifer.

coronary heart disease Progressive weakening of the heart muscle due to chronic oxygen deprivation caused by blocked coronary arteries.

corpus callosum A thick band of neurons that connects the two cerebral hemispheres of the mammalian brain.

corpus luteum A yellowish structure in an ovary, formed from a follicle that has recently ovulated. Secretes progesterone.

cortex (1) The outermost region of an organ, such as the kidney or adrenal gland. (2) In plants, a layer of ground tissue found outside the vascular bundles and pith of a plant stem.

cortical granules Small enzyme-filled vesicles in the cortex of an egg cell. Involved in formation of the fertilization envelope after fertilization.

corticosterone The major glucocorticoid hormone released by the cortex of the adrenal gland in most reptiles, birds, and many mammals. Increases blood glucose and prepares the body for stress.

corticotropin-releasing hormone (CRH) A peptide hormone from the hypothalamus that stimulates the anterior pituitary gland to release ACTH.

cortisol The major glucocorticoid hormone released by the cortex of the adrenal gland in some mammals. Increases blood glucose and prepares the body for stress. Also called hydrocortisone.

cotransporter A membrane protein that allows an ion to diffuse down a previously established concentration gradient and uses the energy of that process to transport some other substance in the same direction. Also called a symporter.

cotyledon The first leaf, or seed leaf, of a plant embryo. Used for storing and digesting nutrients and/or for early photosynthesis.

countercurrent exchange A particularly efficient mechanism for the exchange of heat or a soluble substance, based on transfer between parallel tubes carrying fluids in opposite directions.

countercurrent heat exchanger A specialized network of blood vessels that recirculates body heat within a certain part of the body. A type of countercurrent exchanger.

covalent bond A type of molecular bond in which two atoms share one pair of electrons.

cranium A bony, cartilaginous, or fibrous case that encloses and protects the brain of vertebrates. Forms part of the skull. Also called braincase.

crassulacean acid metabolism (CAM) A variant of photosynthesis in which CO_2 is stored in organic compounds at night when stomata are open and released to enter the Calvin cycle during the day when stomata are closed. Helps reduce loss of water and oxygen by photorespiration in hot, dry environments.

crista (plural: cristae) Membranous sacs that articulate with the inner membrane of a mitochondrion. Location of the electron transport chain and ATP synthase.

critical period A short time span in a young animal's life during which the animal can learn certain things, such as song, language, or imprinting. Also called the sensitive period.

Cro-Magnon A prehistoric European population of modern humans (*Homo sapiens*) known from fossils, paintings, sculptures, and other artifacts.

crossing over The exchange of segments of nonsister chromatids between a pair of homologous chromosomes that occurs during meiosis.

cross-pollination Pollination of a flower by pollen from another individual, rather than by self-fertilization.

cryptochromes A class of plant photoreceptors that detect blue light and affect stem growth and flowering in shady conditions.

cud A partially digested package of food and symbiotic bacteria from the rumen that ruminants regurgitate for further chewing.

culture A collection of cells growing under controlled conditions in a lab, usually in suspension or on the surface of a dish on solid growth medium.

cup fungi A monophyletic lineage of fungi that produce large, often cup-shaped reproductive structures that contain spore-producing asci. Also called sac fungi.

current A flow of electrical charge past a point. Also called electric current.

Cushing's disease A human endocrine disorder caused by loss of feedback inhibition of cortisol on ACTH secretion. Characterized by high ACTH and cortisol levels and wasting of body protein reserves.

cuticle A protective coating secreted by the outermost layer of cells of an animal or a plant.

cyanobacteria A lineage of photosynthetic bacteria formerly known as blue-green algae. Likely the first life-forms to evolve by oxygenic photosynthesis.

cyclic AMP (cAMP) Cyclic adenosine monophosphate; a small molecule, derived from ATP, that is widely used in cells for signaling and regulation (e.g., in gene transcription, enzyme control, and hormone signal transduction).

cyclic photophosphorylation An alternative pathway in the light-dependent reactions of photosynthesis, in which excited electrons from photosystem I are transferred back to the electron transport chain of photosystem II to increase ATP generation.

cyclin-dependent kinase (Cdk) A protein kinase that is active only when bound to a cyclin. Involved in control of the cell cycle.

cyclins A class of proteins whose concentrations fluctuate cyclically, following the cell cycle.

cyst In some species, a protective structure containing a diploid cell in a resting state.

cystic fibrosis A human disease caused by a defective chloride channel. Causes thickened mucus in the respiratory tract and deterioration of the gastrointestinal and reproductive tracts.

cytochrome C A protein that helps transfer electrons between the parts of the electron transport chain in mitochondria.

cytokine Generally, any substance that stimulates cell division. Many cytokines are secreted by macrophages and helper T cells during an immune response, stimulating leukocyte production, tissue repair, and fever.

cytokinesis Division of the cytoplasm to form two daughter cells. Typically occurs immediately after division of the nucleus by mitosis or meiosis.

cytokinins A group of plant hormones that stimulate cell division.

cytoplasm All of the contents of a cell, excluding the nucleus.

cytoplasmic determinant A regulatory molecule that is distributed unevenly in the cytoplasm of the egg cells of many animals and that directs the differentiation of embryonic cells.

cytoplasmic streaming The directed flow of cytosol and organelles around the interior of a plant or fungal cell. Occurs along actin filaments and is powered by myosin.

cytoskeleton A network of protein fibers embedded in the cytoplasm of eukaryotic cells. Involved in cell shape, support, locomotion, and transport of materials within the cell. Includes microtubules, intermediate filaments, and actin filaments.

cytosol The fluid portion of the cytoplasm in a cell (i.e., not including the organelles).

cytotoxic T lymphocyte (CTL) A T cell that destroys infected cells and cancer cells. Cytotoxic T lymphocytes are descendants of an activated $CD8^+$ T cell. Also called killer T cell.

Darwinian fitness The ability of an organism to produce surviving fertile offspring. Also called fitness.

day-neutral plant A plant whose flowering time is not affected by photoperiod (relative length of day and night).

dead space Portions of the air passages that are not involved in gas exchange with the blood, such as the trachea and bronchi.

deciduous Shedding leaves annually.

decomposer A species that feeds on the dead bodies of other organisms. Decomposers include various bacteria, fungi, and protists.

decomposer food web An ecological network of detritus, decomposers that eat detritus, and predators and parasites of the decomposers.

definitive host The host species in which a parasite reproduces sexually.

dehydration reaction A type of chemical reaction involving the bonding of two molecules by removal of an –OH from one subunit and an –H from another to form water. Also called a condensation reaction. The reverse reaction of hydrolysis.

deleterious allele Any allele that reduces an individual's fitness.

deleterious mutation Any mutation that reduces an individual's Darwinian fitness.

demography The study of factors that determine the size and structure of populations through time.

denatured For a protein's three-dimensional structure, unfolded; usually due to breakage of hydrogen bonds and disulfide bonds.

dendrite A short extension from a neuron's cell body that receives neurotransmitters from other neurons.

dendritic cell A type of leukocyte that ingests foreign antigens, moves to a lymph node, and presents the antigens on its membrane to other immune system cells.

density-dependent factors Factors that affect birth rates or death rates differently, depending on population size.

density-dependent growth Population growth that is limited by increasing population size.

density-independent factors Factors that change birthrates and death rates irrespective of population size.

density-independent growth Population growth that is not affected by the population size.

dental plaque A biofilm of bacteria that can form on mammalian teeth.

deoxynucleoside triphosphate (dNTP) A monomer that can be polymerized to form DNA. Consists of deoxyribose, a base (A, T, G, or C), and three phosphate groups; similar to a nucleotide, but with two more phosphate groups.

deoxyribonucleic acid (DNA) A polymer consisting of two strings of deoxyribonucleotides, wound together in a double helix. Contains the genetic information of a cell.

deoxyribonucleotide A nucleotide consisting of the sugar deoxyribose, a phosphate group, and one of four nitrogen-containing bases (adenine, cytosine, guanine, or thymine). A subunit of deoxyribonucleic acid (DNA).

dephosphorylation Removal of a phosphate group from a molecule. A common mechanism of controlling protein shape or function.

depolarization A change in membrane potential from its resting negative state to a less negative or a positive state.

deposit feeder An organism that eats its way through a substrate.

dermal tissue Tissue forming the outer layer of an organism. In plants, also called epidermis; in animals, forms two distinct layers: dermis and epidermis.

descent with modification The phrase used by Darwin to describe his hypothesis of evolution by natural selection.

desmosome A complex physical connection between two animal cells, consisting of proteins that bind the cells' cytoskeletons together. Found where cells are strongly attached to each other.

detergent An amphipathic molecule that forms micelles in water and that can cleanse by suspending hydrophobic molecules (such as oily dirt) in water.

determination The irreversible commitment during development of a cell to becoming a particular cell type (e.g., liver cell, brain cell).

detritivore An organism whose diet consists mainly of detritus.

detritus A layer of dead organic matter that accumulates at ground level or on seafloors and lake bottoms.

deuterostomes A major lineage of animals that share a pattern of embryological development, including radial cleavage, formation of the anus earlier than the mouth, and formation of the coelom by pinching off of layers of mesoderm from the gut. Includes echinoderms and chordates.

developmental homology A similarity in embryonic form, or in the fate of embryonic tissues, that is due to inheritance from a common ancestor.

diabetes insipidus A human disease caused by defects in the kidney's system for conserving water. Characterized by production of large amounts of dilute urine.

diabetes mellitus A human disease caused by defects in insulin production or response. Characterized by abnormally high blood glucose levels and large amounts of glucose-containing urine.

diaphragm An elastic, sheetlike structure. In mammalian ventilation, the muscular sheet that separates the chest from the abdominal cavities and that can expand the chest cavity by moving downward.

diastole The portion of the heartbeat cycle during which the atria or ventricles of the heart are relaxed.

diastolic blood pressure Blood pressure in arteries during relaxation of the heart's left ventricle.

dicot A dicotyledonous plant (i.e., any plant that has two cotyledons upon germination).

dideoxy sequencing A lab technique for determining the exact nucleotide sequence of DNA. Relies on the use of dideoxynucleotide triphosphates (ddNTPs), which terminate DNA replication.

differential centrifugation Separation of cellular components by spinning a cell homogenate in a series of centrifuge runs at progressively higher velocities.

differential gene expression Expression of different genes in different cell types.

differentiation The process by which a cell becomes a particular cell type (e.g., liver cell, brain cell) by differential gene expression.

diffusion Spontaneous movement of molecules and ions along a concentration gradient, from an area of high concentration to one of low concentration.

digestion The breakdown of food into pieces that are small enough to be absorbed.

digestive tract The long tube that begins at the mouth and ends at the anus. Also called alimentary canal, gastrointestinal tract, or the gut.

dihybrid cross A mating between two parents that are heterozygous for both of the two genes being studied.

dimer An association of two identical molecules.

dioecious Having either male flowers or female flowers but not both.

diploblast An animal whose body develops from two basic embryonic cell layers—ectoderm and endoderm; includes cnidarians and ctenophores.

diploid With two sets of chromosomes. Most animals and many plants are diploid.

directional selection A pattern of natural selection in which individuals with a particular extreme phenotype have higher fitness than individuals with average phenotypes or with the other extreme of the phenotype.

direct sequencing A lab technique for discovery and study of unknown microscopic organisms that will not grow easily in the lab. Relies on detecting and amplifying copies of their DNA.

disaccharide A carbohydrate consisting of two monosaccharides (sugar subunits) linked together.

discontinuous replication In DNA replication, the process by which the lagging strand is synthesized in separate pieces (Okazaki fragments) that are joined together later.

discontinuous strand In DNA replication, the strand of new DNA that is synthesized discontinuously in a series of short pieces that are later joined together. Also called lagging strand.

discrete trait A phenotypic trait that exhibits distinct forms rather than the continuous variation seen in quantitative traits such as body height.

dispersal The movement of individuals from their place of origin (birth, hatching) to a new location.

dispersive replication A now-disproven hypothesis of DNA replication, in which each strand in the daughter molecules was proposed to consist of a mixture of old and new segments of DNA.

disruptive selection A pattern of natural selection in which individuals with extreme phenotypes (at either end of the range of phenotypic variation) have higher fitness than do individuals with an average phenotype.

dissociation curve The graphical relationship that depicts the percentage of hemoglobin in the blood that will bind to oxygen at various partial pressures of oxygen.

distal Away from the center of the body; toward the furthest tip of an appendage. The opposite of proximal.

disturbance Any event that removes some individuals or biomass from a community.

disturbance regime The characteristic disturbances that affect a given ecological community.

disulfide bond A covalent bond between two sulfur atoms, typically in the side groups of some amino acids (e.g., cysteine). Often contributes to tertiary structure of proteins.

DNA See **deoxyribonucleic acid (DNA)**.

DNA fingerprinting Any of several methods for identifying individuals by unique features of their genomes. Commonly involves using PCR to create many copies of certain simple sequence repeats and then analyzing their lengths.

DNA footprinting A technique used to find and sequence stretches of DNA that are bound by particular regulatory proteins.

DNA ligase An enzyme that can connect pieces of DNA by catalyzing formation of a phosphodiester bond between the different pieces.

DNA microarray A lab tool involving the use of cDNAs to investigate whether any of several thousand genes are expressed in a certain cell or tissue.

DNA polymerase Any enzyme that catalyzes synthesis of DNA from deoxyribonucleotides.

domain (1) A section of a protein that has a distinctive tertiary structure and function. (2) A fundamental taxonomic group of organisms sharing similarities in basic cellular biochemistry, such as Bacteria, Archaea, and Eukarya.

dominant An allele that determines the phenotype of a heterozygous individual (i.e., one that can hide the presence of a recessive allele).

domino effect Progressive loss of species diversity in a fragmented habitat, in which the loss of one species causes the loss of further species.

dopamine A catecholamine neurotransmitter that functions mainly in a part of the mammal brain involved with muscle control.

dormancy A temporary state of greatly reduced, or no, metabolic activity.

dorsal Toward an animal's back and away from its belly. The opposite of ventral.

dorsal hollow nerve cord A bundle of nerves extending from the brain along the dorsal (back) side of a chordate animal, with cerebrospinal fluid inside a hollow central channel. One of the defining features of chordates.

double fertilization An unusual form of reproduction seen in flowering plants, in which one sperm nucleus fuses with an egg to form a zygote and the other sperm nucleus fuses with two polar nuclei to form the triploid endosperm.

double helix The three-dimensional shape of a molecule of DNA, consisting of two antiparallel DNA strands wound around each other.

Doushantuo fossils A characteristic assemblage of microscopic fossils found in pre-Cambrian rocks in China from about 580 million years ago, including evidence of sponges, different cell types, and multicellular embryos.

downstream In genetics, the direction in which RNA polymerase moves along a DNA strand.

Down syndrome A human developmental disorder caused by trisomy of chromosome 21.

duct A thin tube through which an exocrine gland secretes some substance.

dyad symmetry A type of symmetry in which an object can be superimposed on itself if rotated 180°. Occurs in some regulatory sequences of DNA. Also called two-fold rotational symmetry.

dynein A motor protein that produces movement of cilia and flagella. Dynein bridges use the chemical energy of ATP to "walk" along the adjacent microtubule doublets.

early endosome A membrane-bound vesicle, formed by endocytosis, that is an early stage in the process of becoming a lysosome.

ecdysone An insect hormone that triggers either molting (to a larger larval form) or metamorphosis (to the adult form), depending on the level of juvenile hormone.

Ecdysozoa A lineage of protostomes that grow by shedding their external skeletons and expanding their bodies, rather than by increasing the length of their skeletons. Includes arthropods and nematodes.

echolocation The use of echoes from vocalizations to obtain information about locations of objects in the environment.

ecology The study of how organisms interact with each other and with their surrounding environment.

ecosystem All organisms that live in a geographic area, together with abiotic components that affect or exchange materials with the organisms.

ecosystem diversity The variety of biotic components in a region along with abiotic components, such as soil, water, and nutrients.

ecosystem services Alteration of the physical components of an ecosystem by living organisms, especially beneficial changes in the quality of the atmosphere, soil, water, etc.

ectoderm One of the three basic embryonic cell layers of a triploblast animal. Forms the outer covering and nervous system.

ectomycorrhizal fungi (EMF) Fungi whose hyphae form a dense network that covers their host plant's roots but do not enter the root cells.

ectoparasite A parasite that lives on the outer surface of the host's body.

ectotherm An animal that does not use internally generated heat to regulate its body temperature.

Ediacaran fauna A characteristic set of animal fossils found in various pre-Cambrian rocks around *565–544* million years old, containing sponges, jellyfish, comb jellies, and other filter-feeding, shallow-water marine animals.

effector Any structure, cell, or organ with which an animal can respond to external or internal stimuli. Usually under control of the nervous system.

effector T cells T lymphocytes—descendants of activated T cells—that are actively involved in combating an infection. Includes helper T cells and cytotoxic T lymphocytes.

efferent division The part of the nervous system that carries commands from the central nervous system to the body. Consists primarily of motor neurons.

egg A mature female gamete and any associated external layers (such as a shell). Larger and less mobile than the male gamete. In animals, also called an ovum.

ejaculation The release of semen from the copulatory organ of a male animal.

ejaculatory duct A short duct connecting the vas deferens to the urethra, through which sperm move during ejaculation.

elasticity The ability to stretch and then spring back to the original shape.

electrical potential Potential energy created by a separation of electric charges between two points. Also called voltage.

electric current A flow of electrical charge past a point. Also called current.

electrocardiogram (EKG) A recording of the electrical activity of the heart, as measured through electrodes on the skin.

electrochemical gradient The combined effect of a concentration gradient and an electrical gradient. Affects the movement of ions across plasma membranes.

electrolyte Any compound that dissociates into ions when dissolved in water. In nutrition, refers to the major ions necessary for normal cell function.

electromagnetic spectrum The full range of wavelengths of electromagnetic radiation (light).

electron An extremely tiny, negatively charged particle that usually occupies an orbital around an atomic nucleus. Exhibits wave as well as particle characteristics.

electron acceptor A reactant that gains an electron and is reduced in a reduction-oxidation reaction.

electron carrier A molecule that readily donates electrons to other molecules.

electron donor A reactant that loses an electron and is oxidized in a reduction-oxidation reaction.

electronegativity The tendency of an atom to attract electrons toward itself.

electron microscope A microscope that uses beams of electrons instead of light to produce images of specimens. Can magnify hundreds of thousands of times.

electron shell Atomic orbitals of similar energies. Arranged in roughly concentric layers around the nucleus of an atom, with electrons in outer shells having more energy than those in inner shells.

electron transport chain (ETC) A set of molecules involved in a coordinated series of redox reactions in which the potential energy lost by electrons is used to pump protons from one side of a membrane to the other.

electroreceptor A sensory cell or organ specialized to detect electric fields.

element A fundamental chemical entity consisting of atoms with a specific number of protons. Elements preserve their identity in chemical changes.

elemental ion An ion that is a single element, such as K^+ or Cl^-, and not a molecule.

ELISA (enzyme-linked immunosorbent assay) A lab technique that can measure the concentration of a substances present in very small amounts.

elongation (1) The process by which messenger RNA lengthens during transcription. (2) The process by which polypeptide chains lengthen during translation.

elongation factors Proteins involved in the elongation phase of translation, assisting ribosomes in the synthesis of the growing peptide chain.

elongation phase The phase of DNA transcription in which RNA polymerase moves along the DNA molecule and synthesizes messenger RNA.

embryo A young developing organism; the stage after fertilization and zygote formation.

embryogenesis The process by which a single-celled zygote becomes a multicellular embryo.

embryophyte A plant, including a land plant, that nourishes its embryos inside its own body.

emergent vegetation Any plants in an aquatic habitat that extend above the surface of the water.

emerging disease Any infectious disease that suddenly afflicts significant numbers of individuals, often due to changes in host species or host population movements.

emerging virus Any of several pathogenic viruses that that suddenly afflict significant numbers of individuals, often due to changes in host species or host population movements.

emigration The movement of individuals from one population into another.

emphysema A lung disease caused by breakdown of alveoli and loss of elasticity of the lungs.

emulsified Broken up; the result of mixing of fat into an aqueous solution, usually with the aid of an emulsifying agent (an amphipathic substance such as a detergent) that can break large fat globules into microscopic fat droplets.

endemic species A species that occurs only in one limited area.

endergonic Said of a chemical reaction that will not occur spontaneously but requires an input of energy. For such a reaction, ΔG (Gibbs free-energy change) > 0.

endocrine Relating to hormones.

endocrine gland Any organ that secretes hormones into the blood.

endocrine system All of the glands that produce and secrete hormones into the bloodstream.

endocytosis Uptake of extracellular material by engulfing and pinching-off the plasma membrane to form a small membrane-bound vesicle in the cell.

endoderm One of the three basic embryonic cell layers of a triploblast animal. Forms the digestive tract and organs that connect to it (liver, lungs, etc.).

endodermis In plant roots, a cylindrical layer of cells that separates the cortex from the vascular tissue.

endomembrane system A system of organelles in eukaryotic cells that performs most protein and lipid synthesis. Includes the endoplasmic reticulum (ER), Golgi apparatus, and lysosomes.

endoparasite Any parasite that lives inside the host's body.

endophytic Living inside of a plant.

endoplasmic reticulum (ER) A network of interconnected membranous sacs and tubules found inside eukaryotic cells. Either rough ER (with ribosomes attached) or smooth ER.

endoskeleton An internal skeleton, such as that found in vertebrates.

endosome A membrane-bound vesicle created by endocytosis. Gradually transformed into a lysosome.

endosperm A triploid tissue in the seed of a flowering plant. Serves as food for the plant embryo.

endosymbiont An organism that lives in a symbiotic relationship inside the body of its host.

endosymbiosis A type of symbiotic relationship in which individuals of one species live inside the bodies of individuals of another species.

endosymbiosis theory The theory that mitochondria and chloroplasts evolved from prokaryotes that were engulfed by host cells and took up a symbiotic existence within those cells.

endotherm An animal whose primary source of body heat is internally generated heat.

endothermic (1) Said of a chemical reaction that absorbs heat. (2) Able to maintain a high body temperature using internally generated heat.

energy The capacity to do work or to supply heat. May be stored (potential energy) or available in the form of motion (kinetic energy).

enhancer In eukaryotes, a regulatory DNA sequence to which certain proteins can bind, enhancing the transcription of certain genes.

enrichment culture A method of growing cells in the lab that involves providing cells from the environment with a very specific set of conditions and isolating those that grow rapidly in response.

enterokinase An intestinal enzyme that converts trypsinogen (from the pancreas) to active trypsin, which then activates protein-digesting enzymes.

entomology The study of insects.

entropy The amount of disorder in any system, such as a group of molecules. Commonly symbolized by S.

envelope A membrane-like covering that encloses some viruses and their capsid coats, shielding them from attack by the host's immune system.

enveloped In a virus, having an envelope surrounding its capsid coat.

environmental variation The proportion of phenotypic variation in a trait that is due to environmental influences rather than genetic influences, in a certain population in a certain environment.

enzyme A protein catalyst used by living organisms to speed up and control biological reactions.

epicotyl In some embryonic plants, a portion of the embryonic stem that extends above the cotyledons.

epidemic The spread of an infectious disease throughout a large population in a short time.

epidermis The outermost layer of cells of any multicellular organism.

epididymis A coiled tube wrapped around the testis in reptiles, birds, and mammals. The site of the final stages of sperm maturation and storage.

epinephrine A catecholamine hormone from the adrenal medulla. Triggers rapid responses relating to the fight-or-flight response. Also called adrenaline.

epiphyte A plant that grows on trees or other solid objects and is not rooted in soil. Not parasitic—they do not harm the host.

epistasis An interaction of independently inherited genes, such that alleles at one locus alter the phenotypic effect of alleles at another locus.

epithelial tissue, or epithelium A class of animal tissues consisting of layers of tightly packed cells that line an organ, a duct, or a body surface. Also called epithelium (plural: epithelia).

epitope The unique region of a particular antigen to which antibodies or lymphocytes bind.

equilibrium A state of balance between forward and reverse processes, such as forward and reverse chemical reactions, or between diffusion rates from one side to the other of a selectively permeable membrane.

equilibrium potential The membrane potential at which there is no net movement of a particular ion into or out of a cell.

ER signal sequence A specific sequence of 20 amino acids, found at the beginning of certain newly synthesized proteins, that marks them for transport to the endoplasmic reticulum.

erythropoietin (EPO) A peptide hormone, released by the kidney in response to low blood oxygen levels, that stimulates the bone marrow to produce more red blood cells.

E site The site in a ribosome where an unattached transfer RNA is shunted after its amino acid is added to the growing polypeptide chain.

esophagus The muscular tube that connects the mouth to the stomach.

essential amino acid An amino acid that an animal cannot synthesize and must obtain from the diet. May refer specifically to one of the eight essential amino acids of adult humans: isoleucine, leucine, lysine, methionine, phenylalanine, threonine, tryptophan, and valine.

essential element A chemical element that an organism must obtain from its environment.

essential nutrient Any chemical element or compound required for normal growth, reproduction, and maintenance of a living organism.

ester linkage The covalent bond that joins a fatty acid to glycerol to form a fat or phospholipid.

estradiol The major estrogen produced by the ovaries of female mammals. Stimulates development of the female reproductive tract, growth of ovarian follicles, and growth of breast tissue.

estrogens A class of steroid hormones, including estradiol, estrone, estriol, and others, that generally promote female-like traits. Secreted by the gonads, fat tissue, and some other organs.

estuary An environment of brackish (partly salty) water where a river meets the ocean.

ethnobotanists Plant biologists who study how humans use plants, often focusing on indigenous cultures' knowledge of medical uses for plants.

ethylene A gaseous plant hormone that induces fruit to ripen, flowers to fade, and leaves to drop.

euchromatin Chromatin (a eukaryotic chromosome and its histone proteins) that is unwound in a long, filamentous structure.

eudicot A member of the lineage of angiosperms that includes complex flowering plants and trees. Eudicots include some of the plants classically called dicots.

Eukarya The taxonomic domain that includes all eukaryotes (protists, fungi, plants, animals, etc.), which share a cell nucleus, a cytoskeleton, and other features.

eukaryote A member of the domain Eukarya; an organism with complex cells with distinctive traits such as a nucleus, membrane-bound organelles, a cytoskeleton, and the presence of introns in genes. May be unicellular or multicellular.

eumetazoans Animals whose cells are organized into distinct tissues. All animals except sponges.

eutherians A lineage of mammals whose young develop in the uterus and are not housed in an abdominal pouch. Also called placental mammals.

eutrophication Deoxygenation of an aquatic ecosytem due to a bloom of photosynthetic algae that produces a bloom of decomposers, which use up all the oxygen.

evaporation The energy-absorbing phase change from a liquid state to a gaseous state. Many organisms evaporate water as a means of heat loss.

evo-devo The study of the developmental and molecular causes of major evolutionary changes such as novel body parts.

evolution (1) The theory that all organisms on Earth are related by common ancestry and that they have changed over time, predominantly via natural selection. (2) Any change in the genetic characteristics of a population over time; especially, a change in allele frequencies.

excision repair system A set of enzymes that identify and repair damaged sections of DNA.

excitable membrane A plasma membrane that is capable of generating an action potential.

excitatory postsynaptic potential (EPSP) A change in membrane potential at a neuron dendrite that makes an action potential more likely. Usually a depolarization.

exergonic Said of a chemical reaction that will occur spontaneously, releasing heat and/or increasing entropy. For such reactions, ΔG (Gibbs free-energy change) < 0.

exocrine gland A gland that secretes some substance through a duct into a space other than the circulatory system, such as the digestive tract or the skin sufrace.

exocytosis Secretion of cellular contents to the outside of the cell by fusion of vesicles to the plasma membrane.

exon A region of a eukaryotic gene that is translated into a peptide or protein.

exonuclease Any enzyme that can remove nucleotides from the end of a strand of DNA or RNA.

exoskeleton A hard covering secreted on the outside of the body, used for body support, protection, and muscle attachment.

exothermic Said of a chemical reaction that releases heat.

exotic From a different area. May refer specifically to exotic species introduced into a new area.

expansins A class of plant proteins that actively increase the length of the cell wall when the pH of the wall falls below 4.5.

exponential growth A constantly accelerating increase in population size that occurs when growth rate is constant (not affected by population size).

exportins A class of intracellular proteins whose function is to transport certain large molecules out of the nucleus.

extensor A muscle that pulls two bones further apart from each other, as in the extension of a limb or the spine.

extinct Said of a species that has died out.

extracellular digestion Digestion that takes place outside of an organism, as occurs in many fungi.

extracellular matrix (ECM) The substance that animal cells secrete and in which they are embedded; often has a fiber-composite structure with protein fibers (e.g., collagen, elastin).

extremophiles Any of several groups of bacteria and archaea that thrive in "extreme" (e.g., high-salt, high-temperature, low-temperature, or low-pressure) environments.

F_1 generation First filial generation. The first generation of offspring produced from a mating (i.e., the offspring of the parental generation).

facilitated diffusion Diffusion of a substance across a plasma membrane down a concentration gradient with the assistance of carrier proteins.

facilitation Ecological succession in which early-arriving species make conditions more favorable for later-arriving species.

facultative aerobe Any organism that can perform aerobic respiration when oxygen is available to serve as an electron acceptor but can switch to fermentation when it is not.

fallopian tube A narrow tube connecting the uterus to the ovary in humans, through which the egg travels after ovulation. Site of fertilization and cleavage. In nonhuman animals, called oviduct.

family A classic taxonomic rank above genus and below order. In animals, usually ends in the suffix -idae.

fast-twitch fibers A class of muscle fibers that contract rapidly and powerfully but fatigue quickly.

fate The likely developmental path that a certain embryonic cell will follow (i.e., which tissue types it will give rise to).

fate map A diagram of an embryo, showing the eventual fate of cells in that embryo—the ultimate location and the tissues each cell will give rise to.

fats A class of lipids consisting of three fatty acids joined to a glycerol molecule. Also called triacylglycerols or triglycerides.

fatty acid A type of lipid consisting of a hydrocarbon chain bonded to a carboxyl group (–COOH) at one end. Used by many organisms to store chemical energy; a major component of animal and plant fats.

fatty-acid binding protein A membrane protein of intestinal cells that binds to lipids from food and takes them into the cell.

feather A specialized skin outgrowth in all birds and only in birds. Composed of β-keratin. Used for flight, insulation, display, and other purposes.

feces The waste products of digestion.

fecundity The average number of female offspring produced by a single female per unit time.

feedback inhibition Inhibition of a process by high concentrations of the product of that process.

female gametophyte A multicellular haploid structure that can produce haploid eggs, in a species that exhibits alternation of generations. In flowering plants, consists of a sac (the embryo sac) containing haploid nuclei.

fermentation Any of several metabolic pathways that allow continued production of ATP via glycolysis by transferring electrons from a reduced compound such as glucose to an electron acceptor other than oxygen.

ferredoxin A molecule involved in the electron transport chain of photosystem I of photosynthesis. Passes electrons to the enzyme $NADP^+$ reductase, which catalyzes formation of NADPH.

fertilization Fusion of the nuclei of two haploid gametes to form a zygote with a diploid nucleus.

fertilization envelope A physical barrier that forms around a fertilized egg in amphibians and some other animals. Formed by an influx of water under the vitelline membrane.

fetal alcohol syndrome A condition thought to be caused by exposure to high blood alcohol concentrations during embryonic development.

fetus A later developmental stage of an embryo, usually developed sufficiently to be recognizable as a certain species.

fever A sustained, regulated elevation of body temperature in an endotherm, typically as part of a response to infection.

fiber In botany, a type of elongated sclerenchyma cell that provides support to vascular tissue.

fibronectins A class of proteins in the extracellular matrix to which cells bind to stay in place.

Fick's law of diffusion A law that gives the rate of diffusion of a gas into a liquid as a function of gas solubility, temperature, surface area, difference in partial pressures, and the thickness of the diffusion barrier.

fight-or-flight response Rapid physiological changes that prepare the body for emergencies. Includes increased heart rate, increased blood pressure, and decreased digestion. Triggered by catecholamines.

filament Any thin, threadlike structure, particularly (1) the threadlike extensions of a fish's gills or (2) the slender stalk that bears the anthers in a flower.

filter feeder Any organism that obtains food by filtering small particles or small organisms out of water or air. Also called suspension feeder.

filtrate Any fluid produced by filtration, such as the fluid in kidney nephrons.

filtration A process of removing large components from a fluid by forcing it through a filter, such as that in a renal corpuscle of the kidney.

finite rate of increase The rate of increase of a population over a given period of time. Calculated as the ending population size divided by the starting population size. Symbolized by lambda (λ).

first law of thermodynamics A fundamental principle of physics stating that energy is conserved; it cannot be created or destroyed.

fission The splitting of an organism into two daughter organisms.

fitness The ability of an organism to produce surviving fertile offspring. Also called Darwinian fitness.

fixed action patterns (FAPs) Highly stereotyped behavior patterns that occur in a certain invariant way in a certain species. A form of innate behavior.

flaccid Limp, as a result of low internal pressure; e.g., a wilted plant leaf.

flagellum (plural: flagella) A long, cellular projection that undulates (in eukaryotes) or rotates (in prokaryotes) to move the cell through an aqueous environment.

flavonoids A class of accessory pigments, found in many plants, that can absorb ultraviolet (UV) radiation and hence protect cells from damage by UV light.

flexor A muscle that pulls two bones closer together, as in the flexing of a limb or the spine.

floral meristem A group of undifferentiated plant cells that can produce flowers.

florigen A hypothesized hormone that may stimulate flowering but that has not yet been isolated.

flower The reproductive structure of angiosperm plants, containing microsporangia and/or megasporangia. Typically includes a calyx, a corolla, and one or more stamens or carpels.

flowering plant An angiosperm; a member of the lineage of plants that produces seeds within mature ovaries (fruits).

fluid-mosaic model The widely accepted hypothesis that plasma membranes consist of proteins embedded in a phospholipid bilayer and that these components move fluidly around the membrane.

fluorescence The spontaneous emission of light from an excited electron falling back to its normal (ground) state.

follicle An egg cell and its surrounding ring of supportive cells in a mammalian ovary.

follicle-stimulating hormone (FSH) A peptide hormone from the anterior pituitary that stimulates (in females) growth of eggs and follicles in the ovaries or (in males) sperm production in the testes.

follicular phase The first major phase of a menstrual cycle, when follicles are growing and estrogen levels are increasing. Ends with ovulation.

food A digestible material that contains nutrients.

food chain A simple pathway of energy through a few species in an ecosystem; e.g., a primary producer, a primary consumer, a secondary consumer, and a decomposer.

food vacuole A membrane-bound organelle containing food engulfed by the cell.

food web Any complex pathway along which energy moves among many species at different trophic levels of an ecosystem.

foot A muscular appendage of mollusks, used for movement and/or burrowing into sediment.

forb Herbaceous flowering plants such as daisies and coneflowers.

forensic science The use of scientific analyses to help solve crimes.

formed elements Cells and cell fragments found in blood, including red blood cells, leukocytes and platelets.

fossil Any trace of an organism that existed in the past. Includes tracks, burrows, fossilized bones, casts, etc.

fossil record All of the fossils that have been found anywhere on Earth and that have been formally described in the scientific literature.

founder effect A change in allele frequences that often occurs when a new population is established from a small group of individuals, due to chance variations in gene frequency in the small group.

founder event The establishment of a new population from a small group of individuals.

fovea The small region of the vertebrate retina in which the photoreceptors are very tightly packed, producing the most acute vision.

F-plasmid A particular type of bacterial plasmid that gives the bacterium the ability to initiate conjugation with another bacterium.

fragile-X syndrome The most common form of inherited mental retardation in humans. Caused by an increase in the number of copies of a CGG codon at a certain location on the X chromosome.

free radical An atom with an unpaired electron. Unstable and highly reactive.

freeze-fracture electron microscopy A technique in which plasma membranes are frozen and split to obtain a highly magnified view of both the outer surface and the interior of the bilayer.

frequency The number of wave crests per second traveling past a stationary point. Determines the pitch of sound and the color of light.

fronds The large leaves of ferns.

frontal lobe The anteriormost region of each cerebral hemisphere of the mammal brain. In humans, involved in complex decision making.

fruit A mature, ripened plant ovary (or group of ovaries), along with the seeds it contains and any adjacent fused parts.

fruiting body A structure formed in some fungi and prokaryotes for spore dispersal, usually consisting of a base, a stalk, and a mass of spores at the top.

functional genomics The study of how a genome works (how the genes identified in genome sequencing interact to produce a functional organism).

functional group A small group of atoms bonded together in a precise configuration. Each group has particular chemical properties that it imparts to any organic molecule in which it occurs.

fundamental asymmetry of sex The fact that eggs are larger and more expensive to produce than sperm, and the resulting evolutionary consequences for males and females.

fundamental niche The ecological space that a species occupies in its habitat in the absence of competitors.

fungi A major lineage of eukaryotes that typically have a filamentous body (mycelium) and obtain nutrients by absorption.

fungicide Any substance that can kill fungi or slow their growth.

fur An insulative layer that covers most mammals' bodies, consisting of longer rain-shedding hairs covering a shorter layer of soft underfur.

G_1 phase The phase of a cell cycle that constitutes the first part of interphase before DNA synthesis (S phase).

G_2 phase The phase of a cell cycle between synthesis of DNA (S phase) and mitosis (M phase); the last part of interphase.

gall A tumorlike growth that forms on plants that are infected with certain bacteria or parasites.

gallbladder A small pouch that stores bile from the liver and releases it to the small intestine during digestion for emulsification of fats.

gametangium (plural: gametangia) (1) The gamete-forming structure found in all land plants except angiosperms. Contains a sperm-producing antheridium and an egg-producing archegonium. (2) The gamete-forming structure of chytrid fungi.

gamete A haploid reproductive cell that can fuse with another haploid cell to form a zygote. Most multicellular eukaryotes have two distinct forms of gametes: egg cells (ova) and sperm cells.

gametogenesis The production of gametes (eggs or sperm).

gametophyte The multicellular haploid phase in a species that exhibits alternation of generations. Arises from a single spore and produces gametes.

ganglion cells Neurons in the vertebrate retina that collect visual information from one or several bipolar cells and send it to the brain via the optic nerve.

Gap Analysis Program (GAP) An analysis aimed at identifying the gaps between geographic areas that are rich in biodiversity and areas that are being protected or managed for conservation.

gap genes A class of fruit-fly segmentation genes that organize embryonic cells into major groups of segments. Active in broad regions of the embryo.

gap junction A direct physical connection between two animal cells, consisting of a gap in the ECM and plasma membranes lined by specialized proteins that allow passage of water, ions and small molecules between the cells.

gastrin A hormone produced by the stomach in response to the arrival of food or to a signal via nerves from the brain. Stimulates other stomach cells to release hydrochloric acid.

gastropods The slugs and snails; mollusks distinguished by a large muscular foot and a unique feeding structure, the radula.

gastrula A vertebrate embryo just after gastrulation, containing the three embryonic germ layers (ectoderm, mesoderm, and endoderm) but with no nerve cord yet.

gastrulation The process by which some cells on the outside of a young embryo move to the interior of the embryo, resulting in the three distinct cell layers of endoderm, mesoderm, and ectoderm.

gated channel A channel protein that opens and closes in response to a certain stimulus, such as the binding of a particular molecule or a change in the electrical charge on the outside of the membrane.

gel electrophoresis A technique for separating molecules on the basis of size and electric charge, which affect their differing rates of movement through a gelatinous substance in an electric field.

gemma (plural: gemmae) A small reproductive structure produced in some liverworts by a gametophyte and can grow into another gametophyte.

gene A section of DNA (or RNA, for some viruses) that encodes information for building a polypeptide or a functional molecule of RNA.

gene duplication The creation of an additional copy of a gene, typically by misalignment of chromosomes during crossing over.

gene expression The transcription and translation of a gene, producing a protein.

gene family A set of genetic loci whose DNA sequences are extremely similar. Thought to have arisen by duplication of a single ancestral gene.

gene flow The movement of alleles between populations.

gene-for-gene hypothesis The hypothesis that there is a one-to-one correspondence between the resistance (*R*) loci of plants and the avirulence (*avr*) loci of pathogenic fungi; particularly, that *R* genes produce receptors and *avr* genes produce molecules that bind to those receptors.

gene pool All of the alleles of all of the genes in a certain population.

generation The average time between a mother's first offspring and her daughter's first offspring.

generative cell A small haploid cell within the male gametophyte of a flowering plant. Gives rise to two haploid sperm cells.

gene therapy The treatment of an inherited disease by introducing normal alleles.

genetic bottleneck A reduction in allelic diversity via genetic drift due to a population bottleneck.

genetic cloning A technique for producing many identical copies of a certain gene, usually by inserting a cDNA version of the gene into a bacterial plasmid and growing the bacteria.

genetic code The set of all 64 codons of DNA and the particular amino acids that each specifies.

genetic correlation A type of evolutionary constraint in which selection on one trait causes a change in another trait as well, because the same gene(s) affect both traits.

genetic diversity The diversity of alleles in a population, species, or group of species.

genetic drift Any change in allele frequencies due to random events. Causes allele frequencies to drift up and down randomly over time, and eventually can lead to the fixation or loss of alleles.

genetic engineering The field of study of the manipulation of DNA sequences in living organisms.

genetic homology Similarities among certain organisms in DNA sequences or amino acid sequences of proteins that are due to inheritance from a common ancestor.

genetic map A map of the relative locations of specific genes on a certain chromosome. Also called a linkage map or meiotic map.

genetic marker A genetic locus that can be identified and traced in populations by lab techniques or by a distinctive phenotype. Includes reporter genes.

genetic model A set of hypotheses that explain how a certain trait is inherited.

genetic recombination A change in the combination of genes or alleles on a given chromosome or in a given individual. Also called recombination.

genetics The field of study concerned with the inheritance of traits.

genetic screen Any of several techniques for identifying individuals with a particular type of mutation. Also called a screen.

genetic variation (1) The number and relative frequency of alleles present in a particular population. (2) The proportion of phenotypic variation in a trait that is due to genetic rather than environmental influences in a certain population in a certain environment.

genitalia External copulatory organs.

genome All of the hereditary information in an organism, including not only genes but also other non-gene stretches of DNA.

genomic library A set of all the DNA sequences in a particular genome, split into small segments, each inserted into a vector for further study.

genomics The field of study concerned with sequencing, interpreting, and comparing whole genomes.

genotype All of the alleles of every gene present in a given individual. May refer specifically to the alleles of a particular set of genes under study.

genus (plural: genera) A taxonomic category of closely related species. Always italicized and capitalized to indicate that it is a recognized scientific genus.

geologic time scale The sequence of eons, epochs, and periods used to describe the geologic history of Earth.

geometric isomer A molecule that shares the same molecular formula as another molecule but differs in the arrangement of atoms or groups on either side of a double bond or ring structure.

germination The process by which a seed becomes a young plant.

germ layer In animals, one of the three basic types of tissues of early embryonic development that give rise to all other tissues: endoderm, mesoderm, or ectoderm.

germ theory of disease The theory that infectious diseases are caused by bacteria, viruses, and other microorganisms.

gestation The duration of embryonic development from fertilization to birth, in those species that have live birth.

gibberellic acid (GA) A plant hormone that stimulates elongation of shoots; a gibberellin.

gibberellins A class of plant hormones that stimulate growth.

Gibbs free-energy change A measure of the change in potential energy and entropy that occurs in a given chemical reaction. Calculated as $\Delta G = \Delta H - T\,\Delta S$, where ΔH is the change in potential energy, T is the temperature in kelvins, and ΔS is the change in entropy. Determines whether a reaction will be spontaneous.

gill Any organ in aquatic animals that exchanges gases and other dissolved substances between the blood and the surrounding water. Typically, a filamentous outgrowth of a body surface.

gill arch In aquatic vertebrates, curved regions of tissue between the gills. Gills are suspended from the gill arches.

gill filament The thin, pink strands of fish gills that extend from the gill arches into the water and across which gas exchange occurs.

gill lamellae (singular: lamella) Tiny, crescent-shaped flaps on the gill filaments of fish that serve to increase surface area for gas exchange.

gland An organ whose primary function is to secrete some substance, either into the blood (endocrine gland) or into some other space such as the gut or skin (exocrine gland).

glia Several types of cells in nervous tissue that are not neurons and do not conduct electrical signals but provide support, nourishment, and electrical insulation and perform other functions.

global carbon cycle The movement of carbon among terrestrial ecosystems, the oceans, and the atmosphere.

global nitrogen cycle The movement of nitrogen among terrestrial ecosystems, the oceans, and the atmosphere.

global warming A sustained increase in Earth's average surface temperature.

global water cycle The movement of water among terrestrial ecosystems, the oceans, and the atmosphere.

glomerulus (1) A ball-like cluster of capillaries at the beginning of a kidney nephron. Surrounded by Bowman's capsule. (2) The ball-shaped clusters of neurons in the olfactory bulb of the brain.

glucagon A peptide hormone produced by the pancreas in response to low blood glucose. Raises blood glucose by triggering breakdown of glycogen and stimulating gluconeogenesis.

glucocorticoids A class of steroid hormones released from the adrenal cortex that increase blood glucose and prepare the body for stress.

gluconeogenesis Synthesis of new glucose from non-carbohydrate sources, such as proteins and fatty acids. Occurs in the liver in response to low insulin levels and high glucagon levels.

glyceraldehyde-3-phosphate (G3P) A phosphorylated sugar produced during the Calvin cycle of photosynthesis.

glycerol A three-carbon molecule that forms the "backbone" of phospholipids and most fats.

glycogen A polysaccharide that is the major form of stored carbohydrate in animals. Consists of α-glucose monomers joined end to end in highly branched chains.

glycolipid Any lipid molecule that is covalently bonded to a carbohydrate group.

glycolysis A series of 10 chemical reactions that oxidize glucose to produce pyruvate and ATP. Used by all organisms as part of fermentation or cellular respiration.

glycoprotein Any protein with one or more covalently bonded carbohydrate groups.

glycosidic linkage The covalent bond between two sugar subunits of a polysaccharide.

glycosylation Addition of a carbohydrate group to a molecule.

glyoxisomes Specialized peroxisomes found in plant cells and that contain enzymes for processing the products of photosynthesis.

goblet cells The cells in the stomach lining that secrete mucus.

goiter A pronounced swelling of the thyroid gland in the neck, usually caused by a deficiency of iodine in the diet.

Golgi apparatus A stack of flattened membranous sacs (cisternae) in eukaryotic cells; processes proteins and lipids that will be secreted or directed to other organelles.

gonad An organ that produces reproductive cells; e.g., a testis or an ovary.

gonadotropin-releasing hormone (GnRH) A peptide hormone from the hypothalamus that stimulates release of FSH and LH from the anterior pituitary.

G proteins A class of peripheral membrane proteins that are important in signal transduction, typically by binding to GTP and activating a second messenger.

gracile australopithecines Several species of slender, lightly built hominins that appear in the fossil record shortly after the split from chimpanzees.

grade A group of species that share a position in an inferred evolutionary sequence of lineages but that are not a monophyletic group.

gram-negative bacteria Bacteria that look pink when treated with a Gram stain. Have a cell wall composed of a thin layer of peptidoglycan and an outer membrane.

gram-positive bacteria Bacteria that look purple when treated with a Gram stain. Have thick cell walls containing peptidoglycan.

Gram stain A dye that stains different types of bacteria different colors.

granum (plural: grana) A stack of the flattened, membrane-bound thylakoid disks inside plant chloroplasts.

gravitropism The growth or movement of a plant in a particular direction in response to gravity.

gray crescent A region of an amphibian zygote that becomes visible shortly after fertilization, opposite the point of sperm entry. Eventually gives rise to the blastopore and the organizer, which determines major body axes.

grazing food web The ecological network of herbivores and the predators and parasites that consume them.

great apes The hominids; members of the family Hominidae, including humans and extinct related forms, chimpanzees, gorillas, and orangutans. Distinguished by large body size, no tail, an exceptionally large brain, and a tendency toward bipedalism.

green fluorescent protein (GFP) A jellyfish protein that spontaneously emits green light after stimulation. Widely used in lab research to mark the location of certain molecules or cells.

greenhouse gas An atmospheric gas that absorbs and reflects infrared radiation, so that heat radiated from Earth is retained in the atmosphere instead of being lost to space.

green plants A lineage of eukaryotes that includes green algae and land plants.

gross photosynthetic efficiency The efficiency with which all the plants in a given area use the light energy available to them to produce sugars.

gross primary productivity The total amount of carbon fixed by photosynthesis, including that used for cellular respiration, in a given area over a given time period.

ground meristem The middle layer of a young plant embryo. Gives rise to the ground tissue.

ground tissue A plant tissue consisting of all cells beneath the outer protective layers of epidermis and cork, except for vascular tissue.

groundwater Any water below the land surface.

growth factor Any of several compounds that are secreted by certain cells and that stimulate other cells to divide or to differentiate.

growth hormone A peptide hormone produced by the mammalian pituitary gland. Involved in lengthening the long bones during childhood and in muscle growth, tissue repair, and lactation in adults.

growth-hormone-inhibiting-hormone (GHIH) A peptide hormone from the hypothalamus that inhibits release of growth hormone from the anterior pituitary.

guanosine triphosphate (GTP) A molecule consisting of guanine, a sugar, and three phosphate groups. Can be hydrolyzed to release free energy. Commonly used in many cellular reactions.

guard cell A specialized, crescent-shaped cell forming the border of a plant stoma. Changes shape to open or close the stoma.

gustation The perception of taste.

guttation Excretion of water droplets from plant leaves in the early morning, due to root pressure.

gymnosperms Four lineages of green plants that have vascular tissue and make seeds but that do not produce flowers. Includes cycads, ginkgoes, conifers, and gnetophytes.

H1 The histone protein associated with DNA in the "linker" stretches between the nucleosomes.

habitat destruction Human-caused destruction of a natural habitat with replacement by an urban, suburban, or agricultural landscape.

habitat fragmentation The breakup of a large region of a habitat into many smaller regions, separated from others by a different type of habitat.

Hadley cell An atmospheric cycle of large-scale air movement in which warm equatorial air rises, moves north or south, and then descends at approximately 30°N or 30°S latitude.

hair Mammalian fur that lacks an insulative layer of underfur; or a single strand of hair or fur consisting of keratin, dead cells, and pigments.

hair cell A pressure-detecting sensory cell that has tiny "hairs" (stereocilia) jutting from its surface.

hairpin A stable loop formed in an RNA molecule by hydrogen bonding between purine and pyrimidine bases on the same strand.

half-life The characteristic time taken for half of any amount of a particular radioactive isotope to decay.

halophile A bacterium or archaean that thrives in high-salt environments.

halophyte A plant that thrives in salty habitats.

Hamilton's rule The proposition that an allele for altruistic behavior will be favored by natural selection only if $Br > C$, where B = the fitness benefit to the recipient, C = the fitness cost to the actor, and r = the coefficient of relatedness between recipient and actor.

haploid Having one set of chromosomes. Bacteria, archaea, animal gametes, plant gametophytes, and many algae are haploid.

haploid number The number of different types of chromosomes in a cell. Symbolized as n.

Hardy-Weinberg principle A principle of population genetics stating that, if mutation, migration, genetic drift, random mating, and selection do not occur, then genotype frequencies will not occur in predictable ratios.

harmful algal bloom An extreme increase in the abundance of a toxin-producing protist in a particular aquatic environment.

HDL See **high-density lipoprotein (HDL)**.

head A distinct anterior region of an organism's body, usually bearing a cluster of sensory organs and/or a mouth.

heart A muscular pump that circulates blood throughout the body.

heart attack An episode of cramping or death of heart muscle due to oxygen deprivation, usually caused by a blood clot in one or more partially blocked coronary arteries. Also called a myocardial infarction (MI).

heart murmur A distinctive sound caused by backflow of blood through a defective heart valve.

heartwood The older xylem in the center of an older stem or root, containing protective compounds and no longer functioning in water transport.

heat Thermal energy that is transferred between two objects.

heat of vaporization The energy required to vaporize 1 gram of a liquid into a gas.

heavy chain One of two identical polypeptides that together form the base of immunoglobulin proteins. Differences in heavy chains determine the different classes of immunoglobulins (IgA, IgE, etc.).

helicase An enzyme that catalyzes the breaking of hydrogen bonds between nucleotides of DNA, "unzipping" a double-stranded DNA molecule.

helix-turn-helix motif A motif seen in many repressor proteins in prokaryotes, consisting of two α-helices connected by a short stretch of amino acids that form a turn.

helper T cell A T cell that assists with the activation of other lymphocytes. Helper T cells are descendants of an activated $CD4^+$ T cell.

hemagglutinin A protein that juts out from the surface of influenza viruses, enabling them to bind to host cells.

heme group A small molecule containing an iron atom that can bind to oxygen. Myoglobin and hemoglobin contain heme groups held within large specialized heme-carrying proteins (globins).

hemimetabolous metamorphosis A type of metamorphosis in which the animal increases in size from one stage to the next, but does not dramatically change its body form.

hemocoel A body cavity of arthropods and some mollusks, containing a pool of circulatory fluid (hemolymph) bathing the internal organs.

hemoglobin An oxygen-binding protein consisting of four polypeptide subunits, each containing a heme group. The major oxygen carrier of mammalian blood.

hemolymph The circulatory fluid of animals with open circulatory systems (e.g., insects) in which the fluid is not confined to blood vessels.

hemophilia A human disease characterized by defects in the blood-clotting system. Caused by an X-linked recessive allele.

herb A seed plant that lacks wood and has a relatively short-lived stem.

herbaceous Said of a plant that is not woody.

herbivore An animal that eats primarily plants and rarely or never eats meat.

herbivory The practice of eating plant tissues.

hereditary nonpolyposis colorectal cancer (HNPCC) A type of colon cancer that is associated with a mutated version of a mismatch repair gene on chromosome 2.

heredity The transmission of traits from parents to offspring via genetic information.

heritable Refers to traits that are influenced by hereditary genetic material (DNA, or RNA for some viruses).

hermaphroditic Producing both eggs and sperm.

heterochromatin Chromatin (a eukaryotic chromosome and its histone proteins) that is highly compact and supercoiled.

heterokaryotic Containing nuclei that are genetically distinct. Occurs naturally in many fungi.

heterospory The production of two distinct types of spore-producing structures and thus two distinct types of spores. Occurs in seed plants, which produce both microspores (which become the male gametophyte) and megaspores (which become the female gametophyte).

heterotherm An animal whose body temperature varies markedly with environmental conditions.

heterotroph Any organism that cannot synthesize reduced organic compounds from inorganic sources and that must obtain them by eating other organisms. Some bacteria, some archaea, and virtually all fungi and animals are heterotrophs.

heterozygote advantage A pattern of natural selection in which heterozygotes have greater fitness than either parental homozygote. Also called heterozygote superiority.

heterozygous Having two different alleles of a certain gene.

hexose A monosaccharide (simple sugar) with six carbons.

Hfr strain Any strain of bacteria in which an F-plasmid has been incorporated into the main chromosome. These bacteria have a high frequency of recombination.

hibernation A period of torpor (including a decreased body temperature) that continues for weeks or months.

high-density lipoprotein (HDL) Balls of protein and fat that transport cholesterol from body tissues to the liver for breakdown. Associated with decreased risk of heart disease.

hindgut The posterior portion of the digestive tract of an animal. Often functions to reabsorb water from wastes.

hinge helix A motif found in many repressor proteins in bacteria. Involved in locking the helix-turn-helix motif onto DNA.

histone A globular protein that is tightly associated with DNA in eukaryotic cells.

histone acetyl transferases In eukaryotes, a class of enzymes that loosens chromatin structure by adding acetyl groups to histone proteins.

histone deacetylases In eukaryotes, a class of enzymes that recondense chromatin by removing acetyl groups from histone proteins.

HIV See **human immunodeficiency virus (HIV)**.

holoenzyme A multipart enzyme consisting of a core enzyme (containing the active site for catalysis) along with other required proteins.

holometabolous metamorphosis A type of metamorphosis in which the animal completely changes its form.

homeobox A 180-base-pair sequence present in the *HOM*/*Hox* genes of animals. Codes for a helix-turn-helix motif in the proteins produced by these genes.

homeosis The occurrence of extra, fully formed segments or appendages normally found elsewhere in the body, replacing the structures usually formed at that location.

homeostasis The maintenance of a relatively constant physical and chemical environment within an organism.

homeotherm An animal that has a constant or relatively constant body temperature.

homeotic complex (*HOM-C*) A set of eight fruit-fly genes, closely linked on the same chromosome, that are active in different regions of the fly embryo, specifying the identity of body segments.

homeotic genes A class of genes that specify a location within an embryo, leading to the development of structures appropriate for that location.

homeotic mutant An individual with a mutation in a homeotic gene, causing the development of extra body parts or body parts in the wrong places.

hominids The great apes; members of the family Hominidae, which includes humans and extinct related forms; chimpanzees, gorillas, and orangutans. Distinguished by large body size, no tail, and an exceptionally large brain.

hominins Humans and extinct related forms; species in the lineage that branched off from chimpanzees and eventually led to humans.

homologous chromosomes In a diploid organism, chromosomes that are similar in size, shape, and gene content. Also called homologs.

homologous trait Any trait showing marked similarity between different species, due to inheritance from a common ancestor.

homology Similarity between organisms or DNA sequences that is due to inheritance from a common ancestor.

homospory The production of just one type of spore. Occurs in the seedless vascular plants, in contrast to the heterospory of seed plants.

homozygous Having two identical alleles of a certain gene.

hormone A signaling molecule that circulates throughout the body in blood or other body fluids; can trigger pronounced responses in distant target cells at very low concentrations.

hormone-response elements Sites on DNA to which a steroid hormone-receptor complex can bind and affect gene transcription.

host An individual or a species in or on which a parasite lives.

host cell A cell that has been invaded by an organism such as a parasite or a virus.

***Hox* genes** A class of homeotic genes found in several animal phyla, including vertebrates. Controls pattern formation in early embryos.

human Any member of the genus *Homo*. Includes modern humans (*Homo sapiens*) and several extinct species.

human chorionic gonadotropin (hCG) A glycoprotein hormone produced by the human placenta from about week 3 to week 14 of pregnancy. Maintains the corpus luteum, which produces hormones that preserve the uterine lining.

Human Genome Project The multinational research project that sequenced the human genome.

human immunodeficiency virus (HIV) A retrovirus that causes AIDS (acquired immune deficiency syndrome) in humans.

humoral response Defense against infections and cancers via antibodies produced by B cells.

humus The completely decayed organic matter in soils.

Huntington's disease A degenerative brain disease of humans, caused by an autosomal dominant allele.

hybrid The offspring of parents from two different strains, populations, or species.

hybridoma A mass of cells produced in a lab from a myeloma cell that fused with an antibody-producing B cell. Used to produce large amounts of a monoclonal antibody.

hybrid zone A geographic area where interbreeding occurs between two species, sometimes producing fertile hybrid offspring.

hydrocarbon A molecule that contains only hydrogen and carbon, usually with the carbon atoms bonded covalently to form chains or rings.

hydrogen bond A weak interaction between two molecules, due to attraction between a hydrogen atom with a partial positive charge on one molecule and another atom (usually O or N) with a partial negative charge on the other molecule.

hydrogen ion A proton (H^+); typically, one that has dissolved in solution or that is being transferred from one atom to another in a chemical reaction.

hydrolysis A type of chemical reaction in which a compound reacts with water to break down into smaller molecules. In biology, most hydrolysis reactions involve polymers breaking down into monomers. The reverse of a condensation, or dehydration, reaction.

hydrophilic Mixing readily with water. Hydrophilic compounds are typically polar compounds with charged or electronegative atoms.

hydrophobic Not mixing readily with water. Hydrophobic compounds are typically nonpolar compounds without charged or electronegative atoms and often contain many C–C and C–H bonds.

hydroponic growth Growth of plants in liquid cultures instead of soil.

hydrostatic skeleton A system of body support involving fluid-filled compartments that can change in shape but cannot easily be compressed.

hydroxide ion An oxygen atom bonded to a hydrogen atom and carrying a negative charge (OH^-).

hyperpolarization A change in membrane potential from its resting negative state to an even more negative state.

hypersensitive reaction An allergic immune response in which previously sensitized mast cells and basophils produce large quantities of histamine, cytokines, etc., in response to a small amount of allergen. Triggers symptoms of allergies.

hypersensitive response The rapid death of a plant cell that has been infected by a pathogen. Thought to be a strategic mechanism that protects the rest of the plant.

hypertension Abnormally high blood pressure.

hypertonic Having a greater solute concentration, and therefore a lower water concentration, relative to another solution.

hypha (plural: hyphae) One of the strands of a fungal mycelium (the meshlike body of a fungus). Also found in some protists.

hypocotyl The stem of a very young plant; the region between the cotyledon (embryonic leaf) and the radicle (embryonic root).

hypothalamic-pituitary axis The combination of the hypothalamus and the pituitary gland, which together regulate most of the other endocrine glands in the body.

hypothalamus A part of the brain that regulates the body's internal physiological state, such as the autonomic nervous system and endocrine system.

hypothesis A proposed explanation for a phenomenon or for a set of observations.

hypotonic Having a lower solute concentration, and therefore a higher water concentration, relative to another solution.

immigration The movement of individuals into a certain population from some other population.

immune system In vertebrates, the system of tissues and organs whose primary function is to defend the body against pathogens. Includes lymphocytes, lymph nodes, and other small organs.

immunity Protection against infection by disease-causing pathogens.

immunization The conferring of immunity to a particular disease.

immunoglobulins (Ig) A class of Y-shaped proteins capable of binding to specific antigens. Responsible for acquired immunity.

immunological memory The ability of the immune system to "remember" an antigen—i.e., to mount a rapid, effective response to a pathogen encountered years or decades earlier.

impact hypothesis The hypothesis that the mass extinction at the end of the Cretaceous period, in which dinosaurs died out, was due to an asteroid impact.

imperfect For a flower, containing male parts (stamens) or female parts (carpels) but not both.

implantation The process by which an embryo buries itself in the uterine wall and forms a placenta. Occurs in mammals and a few other vertebrates.

importins A class of intracellular proteins whose function is to transport certain large molecules into the nucleus.

imprinting A type of rapid, irreversible learning in which young animals learn the distinctive appearance of the individual caring for them.

inactivated virus A virus that has been deliberately damaged to render it incapable of causing infection, for use in vaccines.

inbreeding Mating between closely related individuals or within a genetically homozygous strain.

inbreeding depression A loss of fitness that occurs when homozygosity increases in a population.

incomplete dominance An inheritance pattern in which the heterozygote phenotype is a blend or combination of both homozygote phenotypes.

independent assortment The inheritance of the alleles of one gene independently of other genes. True only for genes on different chromosomes.

indeterminate growth A pattern of growth in which an organism continues to increase its overall body size throughout its life.

indicator plate A laboratory technique for detecting mutant cells by observing color changes due to cells cleaving (or not cleaving) the bonds in a pigmented substance.

induced defense A defensive structure or compound produced by plants only in direct response to attack by pathogens or herbivores.

induced fit The phenomenon whereby initial weak binding of a substrate to the active site of an enzyme causes the enzyme to change shape so as to bind the substrate more tightly.

inducer A molecule that triggers transcription of a specific gene.

inducible defense A defensive trait that is produced only when needed—i.e., in response to the presence of a predator or pathogen.

induction The process by which one embryonic cell, or group of cells, alters the differentiation of neighboring cells.

infection thread An invagination of a root hair membrane, through which beneficial nitrogen-fixing bacteria enter the roots of their host plants, legumes.

infectious disease Any disease that can be transmitted from infected individuals to uninfected individuals.

infertile Unable to produce offspring.

inflammatory response An aspect of the innate immune response, seen in most cases of infection or tissue injury, in which the affected tissue becomes swollen, red, warm, and painful.

infrared light Light with a wavelength longer than visible red light.

infrasound Sound frequencies that are too low for humans to hear—lower than about 20 hertz.

ingestion Taking nutrients or other substances into the body; feeding.

inheritance of acquired characters The now-disproven theory that traits acquired during the lifetime of an organism due to noninherited causes will be passed genetically to its offspring.

inhibition Ecological succession in which early-arriving species make conditions less favorable for the establishment of certain other species.

inhibitory hormone A hormone that inhibits the release of some other hormone.

inhibitory postsynaptic potential (IPSP) A change in membrane potential at a neuron dendrite that makes an action potential less likely. Usually a hyperpolarization.

initiation The first stage of a molecular reaction, in which the necessary molecules are brought together. (1) In an enzyme-catalyzed reaction, the stage during which enzymes orient reactants precisely as they bind at specific locations within the enzyme's active site. (2) In DNA transcription, the binding of RNA polymerase to the promotor sequence. (3) In RNA translation, the binding of the ribosome to the mRNA molecule.

initiation factors A class of proteins that assist ribosomes in binding to a messenger RNA molecule to initiate translation. Occur both in prokaryotes and in eukaryotes.

innate behavior Behavior that is inherited genetically and does not have to be learned.

innate immune response The body's nonspecific response to pathogens; the action of leukocytes.

innate immunity A set of nonspecific defenses against pathogens that occurs even without previous exposure to the pathogen and that does not involve antibodies. Includes responses by mast cells, neutrophils, and macrophages; typically results in an inflammatory response.

inner ear The innermost portion of the mammalian ear, consisting of a fluid-filled system of tubes that includes the cochlea (which receives sound vibrations from the middle ear) and the semicircular canals (which function in balance).

inoculation Introduction of some substance into the body to increase immunity to disease. Usually involves vaccination.

in situ hybridization A lab technique for revealing the locations of specific mRNAs in organisms (i.e., revealing which tissues transcribe certain genes).

insulin A peptide hormone produced by the pancreas in response to high levels of glucose (or amino acids) in blood. Enables cells to absorb glucose and coordinates synthesis of fats, proteins, and glycogen.

integral membrane protein Any membrane protein that spans the entire membrane. Also called transmembrane protein.

integration Processing of information from many sources.

integrator A component of an animal's nervous system that evaluates sensory information and triggers appropriate responses.

integrins A class of transmembrane proteins that binds to fibronectins in the extracellular matrix, thus holding cells in place.

intercalated discs The physical connections between adjacent heart muscle cells. Contain gap junctions to allow electrical signals to pass between the cells.

intermediate disturbance hypothesis The hypothesis that moderate ecological disturbance causes high species diversity.

intermediate filament Any of various long, thin cellular fibers composed of thin filaments of various protein polymers wound into thicker cables. Forms part of a cell's cytoskeleton.

intermediate host The host species in which a parasite reproduces asexually.

interneuron A neuron that passes information between two other neurons.

internode The section of a plant stem between two nodes (sites where leaves attach).

interphase The part of the cell cycle during which no cell division occurs. Includes the G_1 phase, the S phase, and the G_2 phase.

interspecific competition Competition between members of different species for the same limited resource.

interstitial fluid The plasma-like fluid found in the spaces between cells.

interstitial space Any area between two cells. Usually filled with fluid called interstitial fluid.

intertidal zone The region between the low-tide and high-tide marks on a seashore.

intracytoplasmic sperm injection (ICSI) A technique used to help infertile couples bear children, by injecting a single sperm directly into the cytoplasm of the egg.

intraspecific competition Competition between members of the same species for the same limited resource.

intrinsic rate of increase The rate at which a population will grow under optimal conditions (i.e., when birthrates are as high as possible and death rates are as low as possible). Symbolized by r_{max}.

intron A region of a eukaryotic gene that is transcribed into RNA but is later excised from the mRNA transcript before translation into protein.

invasive species An exotic species that, upon introduction to a new area, spreads rapidly and competes with native species.

invertebrates All multicellular animals that are not vertebrates. This is a paraphyletic group.

in vitro Outside the living body. Refers to a process or experiment done "in glass" (i.e., in a dish or test tube rather than in an intact living body).

in vitro fertilization (IVF) A technique for fertilizing mammalian eggs with sperm in a lab dish.

in vivo In a living body. Refers to a process or experiment done "in life."

ion An atom or a molecule that has gained or lost electrons and carries an electric charge.

ion channels A class of membrane proteins that allow certain ions to diffuse across a plasma membrane, via passive transport.

ionic bond A bond between atoms that is formed when an electron is completely transferred from one atom to another so that the atoms remain associated due to their opposite electric charges.

ionophore Any compound that increases membrane permeability for a certain ion. Many ionophores are membrane proteins that bind to the ion and carry it through the membrane.

iris A ring of pigmented muscle just inside the vertebrate eye that contracts or expands to control the amount of light entering the eye.

islets of Langerhans Clusters of cells in the pancreas that secrete insulin and glucagon directly into the blood.

isomer A molecule that has the same molecular formula as another molecule but differs from it in three-dimensional structure.

isometric (1) Characterized by scaling that is proportionate to some other feature, such as body size. (2) A type of muscle contraction in which the muscle exerts force without shortening.

isotonic Having the same solute concentration and water concentration relative to another solution.

isotope Any of several forms of an element that have the same number of protons but differ in the number of neutrons.

jelly layer A gelatinous layer that encloses the egg cells of some vertebrates.

jet propulsion Movement accomplished by forcibly ejecting water in the opposite direction. Used by various aquatic invertebrates.

joint A place where two pieces (bones, cartilages, etc.) of a skeleton meet. May be movable (an articulated joint) or immovable (e.g., skull sutures).

junctional diversity Genetic diversity created by variations in the joining of gene segments; occurs during gene recombination in lymphocytes of the immune system.

juvenile hormone An insect hormone that prevents larvae from metamorphosing into adults.

karyogamy Fusion of two haploid nuclei to form a diploid nucleus. Occurs in many fungi, and in animals and plants during fertilization of gametes.

karyotype The distinctive appearance of all of the chromosomes in an individual, including the number of chromosomes and their length and banding patterns (after staining with dyes).

keystone species A species that has an exceptionally great impact on the other species in its ecosystem relative to its abundance.

kidneys Paired organs situated at the back of the abdominal cavity that filter the blood, produce urine, and secrete several hormones.

kilocalorie (kcal) A unit of energy, often used to measure energy content of food. Also called a Calorie.

kinesin A motor protein that uses the chemical energy of ATP to transport vesicles, particles, or chromosomes along microtubules.

kinetic energy The energy of motion, as opposed to potential energy (stored in position or shape).

kinetochore The structure on chromatids where spindle fibers attach. Contains motor proteins that move the chromosome along the microtubule.

kingdom A classic taxonomic rank that is above the phylum level and below the domain level.

kinocilium A single cilium that juts from the surface of many hair cells and functions in detection of sound or pressure.

kin selection A form of natural selection that favors traits that increase survival or reproduction of an individual's kin at the expense of the individual.

Klinefelter syndrome A syndrome seen in humans who have an XXY karyotype. People with this syndrome have male sex organs, may have some female traits, and are sterile.

knock-out mutant A mutant allele that does not function at all, or an organism homozygous for such a mutation. Also called null mutant or loss-of-function mutant.

Koch's postulates Four criteria used to determine whether a putative infectious agent causes a particular disease.

Krebs cycle A series of chemical reactions, during which acetyl CoA (from glycolysis) is broken down to CO_2, producing ATP and reduced compounds for the electron transport chain. Occurs in the mitochondria.

labia majora (plural: labium majus) One of two outer folds of skin that protect the labia minora, clitoris, and vaginal opening of female mammals.

labia minora (plural: labium minus) One of two inner folds of skin that protect the opening of the urethra and vagina.

labor The strong muscular contractions of the uterus that expel the fetus during birth.

***lac* operon** The operon in *E. coli* that includes genes responsible for metabolism of lactose.

lactation Production of milk from mammary glands of mammals.

lacteal Lymphatic vessels in the center of the villi of the small intestine. Receive chylomicrons containing fat absorbed from food and send them into the lymph system.

lactic acid fermentation Production of ATP without an electron transport chain, and the end product of glycolysis (pyruvate) is converted to lactic acid.

lagging strand In DNA replication, the strand of new DNA that is synthesized discontinuously in a series of short pieces that are later joined together. Also called discontinuous strand.

lamellae (singular: lamella) Any set of parallel platelike structures (e.g., the crescent-shaped flaps on the gill filaments of fish gills that serve to increase surface area for gas exchange).

large intestine The posterior part of the intestine, between the small intestine and the rectum.

larva (plural: larvae) An immature stage of a species in which the immature and adult stages have different body forms.

late endosome A membrane-bound vesicle, formed by endocytosis, that is a late stage in the process of becoming a lysosome.

lateral bud A bud on a plant stem that is capable of producing a new side branch.

lateral gene transfer Transfer of DNA between two different species, especially distantly related species. Commonly occurs among bacteria and archaea via plasmid exchange; also can occur in eukaryotes via viruses and some other mechanisms.

lateral meristem A layer of undifferentiated plant cells found in older stems and roots. Responsible for secondary growth. Also called cambium or secondary meristem.

lateral root A plant root extending from another, older root.

LDL See **low-density lipoprotein (LDL)**.

leaching Loss of nutrients from soil via percolating water.

leading strand In DNA replication, the strand of new DNA that is synthesized in one continuous piece, with nucleotides added to the 3′ end of the growing molecule. Also called continuous strand.

leaf primordium (plural: leaf primordia) Small protuberances that form around an apical meristem of a plant shoot and that will develop into leaves.

leak channels The potassium channels in the membranes of nerve cells, which allow potassium ion to leak out of the cell in its resting state.

learning An enduring change in an individual's behavior that results from specific experience(s).

leghemoglobin An iron-containing protein similar to hemoglobin. Found in root nodules of legume plants where it binds oxygen, preventing it from poisoning a bacterial enzyme needed for nitrogen fixation.

legumes Members of the pea plant family. Form symbiotic associations with nitrogen-fixing bacteria in their roots.

lens A transparent, crystalline structure that focuses incoming light onto a retina or other light-sensing apparatus of an eye.

leptin A hormone produced by fat cells (adipocytes) that signals how much body fat is stored. Inhibits appetite.

leukocytes Immune system cells, including neutrophils, macrophages, B cells, and T cells, that circulate in blood or lymph and function in defense against disease. Also called white blood cells.

lichen A symbiotic association of a fungus and a photosynthetic alga.

life cycle The sequence of developmental events and phases that occurs during the life span of an organism, from fertilization to offspring production.

life history The sequence of events in an individual's life, from birth to reproduction to death. Also, the study of how the organism allocates resources and energy to these different activities.

life table A data set that summarizes the probability that an individual in a certain population will survive and reproduce in any given year over the course of its lifetime.

ligand Any molecule that binds to a specific site on a receptor molecule.

ligand-gated channel An ion channel that opens or closes in response to binding by a certain molecule.

light chain One of two identical short polypeptides forming part of an immunoglobulin protein (antibody, B-cell receptor, etc.). The light chains determine which antigen the molecule will bind to.

light-dependent reactions In photosynthesis, the set of reactions that use the energy of sunlight to split water, producing ATP, NADPH, and oxygen.

light-independent reactions In photosynthesis, the set of reactions that use the NADPH and ATP formed earlier (in the light-dependent reactions) to drive the reduction of atmospheric CO_2, ultimately producing sugars. Also called the Calvin cycle.

lignin An extremely strong polymer in the secondary cell walls of plant cells in woody plant parts. Composed of six-carbon rings joined.

lignin peroxidase A fungal enzyme that can digest lignin.

limiting nutrient An essential nutrient whose scarcity in the environment significantly affects growth and reproduction.

limnetic zone Open water (not near shore) that receives enough light to support photosynthesis.

lineage An evolutionary unit that includes an ancestral population, all of its descendants, and only its descendants. Also called a monophyletic group or a clade.

linkage A physical association between two genes because they are on the same chromosome; the inheritance patterns resulting from this association.

linkage map A map of the relative locations of specific genes on a certain chromosome. Also called a genetic map or meiotic map.

lipase Any enzyme that can digest fats.

lipid A carbon-containing subtance that is hydrophobic and thus does not dissolve in water, but dissolves well in nonpolar organic solvents. Lipids include fats, oils, phospholipids, and waxes.

lipid bilayer A double layer of phospholipid molecules, with hydrophobic tails oriented toward the inside and hydrophilic heads toward the outside. The fundamental component of membranes.

liposome An artificially formed tiny membrane-bound structure composed of a phospholipid bilayer membrane.

lithotroph An organism that produces ATP by oxidizing inorganic molecules with high potential energy, such as ammonia (NH_3) or methane (CH_4).

littoral zone Shallow waters near shore that receive enough sunlight to support photosynthesis. May be marine or freshwater.

liver An abdominal organ of vertebrates that performs many biochemical processes, including storage of glycogen, processing and conversion of food and wastes, and production of bile.

live virus vaccine A vaccine containing intact viruses that have been rendered nonvirulent for a certain species. Used for vaccines. Also called an attenuated virus vaccine.

lobe-finned fishes A lineage of fishes with fins supported by an arrangement of bones and muscles similar to that seen in tetrapod limbs. Includes two living groups: coelacanths and lungfishes.

lock-and-key model The hypothesis that enzymes have a precise three-dimensional structure into which substrates (reactant molecules) fit. Modified to reflect knowledge that enzymes are not rigid and may change shape.

locomotion Movement of an organism under its own power.

locus (plural: loci) A gene's physical location on a chromosome.

logistic growth equation The mathematical equation that defines how fast a population will grow over time, given a certain carrying capacity.

logistic population growth Changes in growth rate that occur as a function of population size.

long-day plant A plant that blooms in midsummer, in response to short nights.

long interspersed nuclear element (LINE) A type of parasitic DNA sequence commonly found in eukaryotic genomes. Contains genes for reverse transcriptase, integrase, and a promoter and can create copies of itself, inserted elsewhere in the genome.

loop of Henle A long loop of the nephrons in mammalian kidneys. Functions to set up a concentration gradient that allows reabsorption of water from a subsequent section of the nephron.

loose connective tissue A type of connective tissue consisting of fibrous proteins in a soft matrix. Often functions as padding for organs.

lophophore A specialized feeding structure found in three phyla of the Lophotrochozoa and used in filter feeding.

Lophotrochozoa The lineage of protostomes that includes mollusks and annelids. Many phyla have lophophore feeding structures, have trochophore larvae, and grow by extending the size of their skeletons rather than by molting.

loss-of-function mutant A mutant allele that does not function at all, or an organism homozygous for such a mutation. Also called knock-out mutant or null mutant.

low-density lipoprotein (LDL) Balls of protein and fat that transport cholesterol from the liver to the rest of the body for storage and use. Associated with increased risk of heart disease.

lumen The interior space of any hollow structure (e.g., the rough ER) or organ (e.g., the stomach).

lung Any respiratory organ used for gas exchange between blood and air.

luteal phase The second major phase of a menstrual cycle, after ovulation, when the progesterone levels are high and the body is preparing for a possible pregnancy.

luteinizing hormone (LH) A peptide hormone from the anterior pituitary that stimulates estrogen production, ovulation, and formation of the corpus luteum in females and testosterone production in males.

lymph The mixture of fluid and lymphocytes that circulates through the ducts and lymph nodes of the lymphatic system in vertebrates.

lymphatic duct One of the thin-walled tubes that collect excess fluid from body tissues and return it to the circulatory system, passing through lymph nodes along the way.

lymphatic system In vertebrates, a network of thin-walled tubes that collects excess fluid from body tissues and returns it to the veins of the circulatory system. Includes lymph nodes.

lymph nodes Small oval structures through which lymph ducts run. Filter the lymph and screen it for infection.

lymphocyte A type of leukocyte that circulates through the bloodstream and lymphatic system and that is responsible for the development of acquired immunity. Includes B cells and T cells.

lysogeny A type of viral replication in which the viral DNA is inserted into the host's chromosome, remaining there indefinitely and passively replicating whenever the host cell divides.

lysosome A small organelle in an animal cell containing acids and enzymes that catalyze hydrolysis reactions and can digest large molecules. In plants, fungi and some other groups, may be called vacuoles.

lysozyme An enzyme that acts as an antibiotic by digesting bacterial cell walls. Occurs in saliva, tears, mucus, and egg white.

lytic replication cycle A type of viral replication in which new virus particles are made inside a host cell and eventually burst out of the cell, killing it.

macromolecule Any very large molecule, usually made up of smaller molecules joined together. Include proteins, nucleic acids, and polysaccharides.

macronutrient An essential nutrient required in large quantities. Usually a major component of many organic molecules.

macrophage A type of leukocyte, capable of amoeboid movement through body tissues, that engulfs and digests pathogens and other foreign particles. Also plays roles in secreting cytokines and presenting foreign antigens to other immune system cells.

MADS box A DNA sequence found in some genes of fungi, animals, and plants (genes that control pattern formation in flowers). Codes for a stretch of 58 amino acids that can bind to DNA.

major histocompatibility proteins (MHC proteins) Mammalian cell-surface glycoproteins involved in immunity and in marking cells as "self" to the immune system.

maladaptive For a trait, reducing the fitness of individuals.

malaria A human disease caused by four species of the protist *Plasmodium*, which is passed to humans by mosquitoes.

male gametophyte A multicellular haploid structure that can produce haploid sperm cells in a species that exhibits alternation of generations. In flowering plants, consists of two sperm cells within a larger tube cell.

malignant tumor A tumor that is actively growing and disrupting local tissues and/or is spreading to other organs. A cancer consists of one or more malignant tumors.

Malpighian tubules A major excretory organ of insects, consisting of blind-ended tubes that extend from the gut into the hemocoel. Filter hemolymph to form pre-urine and then send it to the hindgut for further processing.

mammary glands Specialized skin glands that can produce milk for nursing offspring. A diagnostic feature of mammals.

mandibles Any chewing mouthparts. In vertebrates, the lower jaw. In insects, crustaceans, and myriapods, the first pair of mouthparts.

mantle In mollusks, the thick outer tissue that protects the visceral mass and that may secrete a calcium carbonate shell.

Marfan syndrome A human syndrome involving increased height, long limbs and fingers, an abnormally shaped chest, and heart disorders.

mark-recapture study A method of estimating population size involving release of marked individuals into a population and an assessment of how many individuals captured later are marked vs. unmarked.

marsh A wetland that lacks trees and usually has a slow but steady rate of water flow.

marsupial A member of the taxon Marsupiala, a lineage of mammals that nourish their young in an abdominal pouch after a very short period of development in the uterus.

mass extinction Rapid extinction of an unusually large number of diverse evolutionary groups across a wide geographic area. May occur due to sudden and extraordinary environmental changes.

mass number The total number of protons and neutrons in an atom.

mast cell A type of leukocyte that is stationary (embedded in tissue) and that helps trigger the inflammatory response to infection or injury, including secretion of histamine. Particularly important in allergic responses and defense against parasites.

master plate In replica plating, the original plate of bacteria that is used as the source for producing replica plates for further study.

maternal chromosome A chromosome inherited from the mother.

maternal effect inheritance A pattern of inheritance in which an individual's phenotype is determined by its mother's genotype. Common in embryological development, during which egg components made by the mother can influence development of the offspring.

mating type A form of fungal hyphae that carries certain alleles for the genes involved in mating and mates only with hyphae of a different mating type.

matrix A general term for any liquid or semisolid that fills or surrounds some structure and that has some role in maintaining the shape or function of organelles, cells, or tissues.

medulla The innermost part of an organ (e.g., kidney or adrenal gland). In the brain, the posteriormost portion, responsible for rhythmic body functions such as heart rate, respiration, and digestion.

medullary respiratory center An area at the base of the brain (in the region known as the medulla oblongata) that stimulates breathing.

medusa (plural: medusae) The free-floating stage of a cnidarian life cycle.

megapascal (MPa) A unit of pressure (force per unit area), equivalent to 1 million pascals (Pa).

megaphyll A leaf of a fern, horsetail, or vascular plant.

megasporangium (plural: megasporangia) In seed plants, a spore-producing structure that produces megaspores, which can grow to become the female gametophytes.

megaspore A type of small spore produced by the megasporangia of seed plants and that can grow to become a female gametophyte.

megasporocyte A diploid cell contained in the megasporangium of a flower ovule. Undergoes meiois to produce four megaspores.

meiosis A type of cell division in which one diploid parent cell produces four haploid reproductive cells (gametes). In meiosis, chromosome pairs synapse and can exchange genes via crossing over.

meiosis I The first cell division of meiosis, in which synapsis and crossing over occur, and homologous chromosomes are separated from each other, producing daughter cells with half as many chromosomes as the parent cell.

meiosis II The second cell division of meiosis, in which sister chromatids are separated from each other. Similar to mitosis.

meiotic map A map of the relative locations of specific genes on a certain chromosome. Also called a genetic map or linkage map.

membrane potential A difference in electric charge across a cell membrane; a form of potential energy. Also called membrane voltage.

membrane protein Any protein found in a cell membrane. May span the entire membrane (transmembrane or integral membrane proteins) or may be found on only one side of the membrane (peripheral membrane proteins).

memory Retention of learned information.

memory cell A type of lymphocyte responsible for maintenance of immunity for years or decades after an infection. Descendants of B or T cells activated during a previous infection.

meniscus (plural: menisci) The concave boundary layer formed at most air-water interfaces, due to surface tension.

menstrual cycle A female reproductive cycle seen in Old World monkeys and apes (including humans), consisting of a follicular phase, ovulation, a luteal phase, and then (if no pregnancy occurs) menstruation.

menstrual synchrony The phenomenon in which human women living in close proximity experience synchronization of their menstrual cycles.

menstruation The periodic shedding of the uterine lining through the vagina that occurs in females of Old World monkeys and apes, including humans.

meristem A group of undifferentiated plant cells that can produce cells that differentiate into specific adult tissues.

mesoderm One of three basic embryonic cell layers of a triploblast animal. Forms the middle tissues between skin and gut: muscles, bones, blood, and some internal organs (kidney, spleen, etc.).

mesoglea A gelatinous material with scattered ectodermal cells, found in cnidarians between ectoderm and endoderm.

mesophyll cell A type of cell found near the surfaces of plant leaves and where most photosynthesis occurs.

messenger RNA (mRNA) An RNA molecule that carries encoded information, transcribed from DNA, and that can be used for the synthesis of one or more proteins.

meta-analysis An analysis that combines and compares the results of many smaller, previously published studies.

metabolic pathway A series of distinct chemical reactions that build up or break down a particular molecule. Often, each reaction is catalyzed by a different enzyme.

metabolic rate The total energy consumption of all the cells of an individual. For aerobic organisms, often measured as the amount of oxygen consumed per hour.

metabolic water The water that is produced as a by-product of cellular respiration.

metabolism All the chemical reactions occurring in a living cell or organism.

metallothioneins Small plant proteins that bind to and prevent excess metal ions from acting as toxins.

metamorphosis A dramatic change from the larval to the adult form of an animal.

metaphase A stage in cell division (mitosis or meiosis) during which chromosomes line up in the middle of the cell.

metaphase plate The plane along which chromosomes line up during metaphase of cell division (mitosis or meiosis).

metapopulation A population of a single species that is divided into many smaller populations.

metastasis The process by which cancerous cells leave the primary tumor and establish additional tumors elsewhere in the body.

methanogen A group of archaea that produce methane (CH_4; natural gas) as a by-product of cellular respiration.

methanotroph An organism that uses methane (CH_4; natural gas) as its primary electron donor and source of carbon.

methylation The addition of a methyl ($-CH_3$) group to a molecule.

MHC See **major histocompatibility complex proteins (MHC proteins)**.

micelle A small droplet of similar molecules clumped together in a solution.

microbe Any microscopic organism. Includes bacteria, archaea, and various tiny eukaryotes.

microfilament A long, thin fiber composed of two intertwined strands of polymerized actin. Involved in cell movement. Also called actin filament.

micrograph A photograph of an image produced by a microscope.

micronutrient An essential nutrient required in very small quantities—usually an enzyme cofactor.

microphyll A type of small leaf on lycopods.

micropyle The tiny pore in a plant ovule through which the pollen tube reaches the embryo sac.

microsatellite A noncoding stretch of eukaryotic DNA that contains a repeating sequence one to five base pairs long. A type of simple sequence repeat.

microsporangium (plural: microsporangia) In seed plants, a spore-producing structure that produces microspores, which can grow to become the male gametophytes.

microspore A small haploid spore produced by the microsporangia of seed plants and that can grow to become a male gametophyte.

microsporidians A lineage of single-celled, parasitic eukaryotes that are closely related to fungi.

microsporocytes Diploid cells contained within the microsporangium of a flower anther. Undergo meiosis to produce microspores.

microtubule A long, tubular polymer of protein subunits α-tubulin and β-tubulin. Involved in cell movement and transport of materials within the cell.

microtubule organizing center Any structure that organizes microtubules.

microvilli (singular: microvillus) Tiny protrusions from the surface of an epithelial cell that increase the surface area for absorption of substances.

middle ear The air-filled middle portion of the mammal ear, connecting to the throat via the Eustachian tube. Transmits and amplifies sound from the tympanic membrane to the inner ear.

middle lamella A central layer of gelatinous pectins between the primary cell walls of adjacent plant cells. Helps hold the cells together.

migration (1) A cyclical movement of large numbers of organisms from one geographic location or habitat to another. (2) In population genetics, movement of individuals from one population to another.

millivolt (mV) A unit of voltage equal to 1/1000 of a volt.

mimicry A phenomenon in which one species has evolved (or learns) to look or sound like another species.

minisatellite A noncoding stretch of eukaryotic DNA that contains a repeating sequence 6 to 500 base pairs long. A type of simple sequence repeat.

mismatch repair The process by which mismatched base pairs in DNA are fixed.

missense mutation A point mutation (change in a single base pair) that causes a change in the amino acid sequence of a protein. Also called replacement mutation.

mitochondrial DNA DNA found inside the mitochondria of eukaryotic cells.

mitochondrial matrix The solution inside the inner membrane of a mitochondrion. Contains the enzymes of the Krebs cycle.

mitochondrion (plural: mitochondria) A eukaryotic organelle that is the site of aerobic respiration.

mitosis Nuclear division of a eukaryotic cell producing two daughter nuclei that are genetically identical to the parent.

mitosis-promoting factor (MPF) A complex of two proteins (cyclin and cyclin-dependent kinase) that causes eukaryotic cells to initiate mitosis.

mitotic phase (M phase) The phase of the cell cycle during which cell division occurs. Includes mitosis and cytokinesis.

mitotic spindle An array of microtubules that moves chromosomes to opposite sides of the cell during cell division (mitosis or meiosis).

model organism An organism selected for intensive scientific study based on features that make it easy to work with (e.g., body size, life span), in hope that the findings will apply to other species.

molarity The number of moles of a dissolved solute in 1 liter of solution; a unit of concentration.

mole An amount of any substance containing 6.022×10^{23} atoms, ions, or molecules. This number of molecules of a compound will have a mass equal to the molecular weight of that compound expressed in grams.

molecular chaperones A class of proteins that facilitate the three-dimensional folding of newly synthesized proteins.

molecular clock The hypothesis that certain types of mutations tend to reach fixation in populations at a steady rate over large spans of time. As a result, comparisons of DNA sequences can be used to infer the timing of evolutionary divergences.

molecular formula A notation that indicates the numbers and types of atoms in a molecule, such as H_2O for the water molecule.

molecular ion An ion consisting of a group of several different atoms (rather than one element).

molecular weight The sum of the mass numbers of all of the atoms in a molecule; roughly, the total number of protons and neutrons in the molecule.

molecule Two or more atoms held together by covalent bonds.

molting A method of body growth, used by ecdysozoans, that involves the shedding of an external protective cuticle or skeleton, expansion of the soft body, and growth of a new external layer.

monoclonal antibody An antibody produced in the lab from a hybridoma derived from a single B cell. Such an antibody has a unique amino acid sequence and thus the ability to bind to a specific site of a particular antigen.

monocot A plant that has a single cotyledon (embryonic leaf) upon germination.

monoecious Having both male and female flowers.

monohybrid cross A mating between two parents that are both heterozygous for a given gene.

monomer A small molecular subunit that can bond to other subunits to form long macromolecules, or polymers.

monophyletic group An evolutionary unit that includes an ancestral population, all of its descendants, and only its descendants. Also called a clade or a lineage.

monosaccharide A single sugar monomer, such as glucose. Formally, a small carbohydrate of the chemical formula $(CH_2O)_n$ that cannot be hydrolyzed to form any smaller carbohydrates.

monosomy Having only one copy of a particular type of chromosome.

monotremes A member of the Monotremata, a lineage of mammals that lay eggs and then nourish the young with milk. Includes just three living species: the platypus and two species of echidna.

morphogenesis A process of embryologic development during which cells become organized into recognizable tissues, organs, and other structures.

morphology The shape and appearance of an organism's body and its component parts.

morphospecies concept The concept that species are best identified as groups that have measurably different anatomical features.

motif A repeating theme. In molecular biology, a domain (a section of a protein with a distinctive tertiary structure) in many different proteins.

motile Not sessile (not permanently attached to a substrate); capable of moving to another location.

motor neuron A nerve cell that carries signals from the central nervous system (brain and spinal cord) to an effector, such as a muscle or gland.

motor protein A protein whose major function is to convert the chemical energy of ATP into motion.

mRNA See **messenger RNA (RNA)**.

mucigel A slimy substance secreted by plant root caps to ease passage of the growing root through the soil.

mucins Glycoproteins, produced by salivary glands, that form mucus when mixed with water.

mucus A slimy mixture of glycoproteins and water, secreted by many organs for lubrication.

Müllerian inhibitory substance A peptide hormone secreted by the embryonic testis that causes regression (withering away) of the female reproductive ducts.

Müllerian mimicry A type of mimicry in which two (or more) harmful species resemble each other.

multicellular growth Growth in which individual cells join together to form a multicelled body and cells differentiate to perform specialized roles.

multicellularity The condition whereby an organism's body contains more than one cell and only certain cells pass genes to the next generation.

multiple allelism The occurrence of more than two alleles of a gene in a given population.

multiple sclerosis (MS) A human autoimmune disease caused by the immune system attacking the myelin sheaths that insulate nerve axons.

muscle fiber A single muscle cell.

muscle tissue A class of animal tissue consisting of bundles of long, thin contractile cells (muscle fibers).

mutagen Anything that can increase the rate of mutation.

mutant An individual that carries a mutation, particularly a new or rare mutation.

mutation Any change in the hereditary material of an organism (DNA in most organisms, RNA in some viruses).

mutualism A relationship between two species that benefits both species.

mutualist An organism that lives in a close relationship with a host and that benefits its host.

Myb proteins DNA-binding proteins that turn gene transcription on or off, acting as transcription activators or transcription repressors, respectively.

mycelium (plural: mycelia) A mass of underground filaments (hyphae) that form the body of a fungus. Also found in some protists and bacteria.

mycorrhiza (plural: mycorrhizae) A mutualistic association between certain fungi and most vascular plants, sometimes visible as nodules or nets in or around plant roots.

myelin sheath Multiple layers of myelin, a lipid, that are wrapped around the axons of neurons to provide electrical insulation.

myeloma A cancer or tumor of the cells of bone marrow. Myelomas of B cells are used in lab production of monoclonal antibodies.

MyoD A regulatory protein involved in differentiation of muscle cells during embryological development. Enhances transcription of muscle-specific genes.

myofibril A bundle of strands of contractile proteins organized into repeating units (sarcomeres) in vertebrate heart muscle and striated muscle.

myoglobin An oxygen-binding muscle protein consisting of a single globin and one heme group.

myosin A eukaryotic protein that can be polymerized to form thick filaments that are used in muscle contraction and intracellular movement.

natural history The branch of biology concerned with describing what exists in nature (i.e., primarily observational and not experimental).

natural selection The process by which individuals with certain heritable traits tend to produce more surviving offspring than do individuals without those traits, resulting in a change in the genetic makeup of the population. A major mechanism of evolution.

nauplius A distinct planktonic larval stage seen in many crustaceans.

Neanderthal A recently extinct European species of hominid, *Homo neanderthalensis*, closely related to but distinct from modern humans.

nectar The sugary fluid produced by flowers to attract and reward pollinating animals.

nectary The nectar-producing gland at the base of a flower.

negative control A type of gene regulation in which a transcription can occur only when a certain substance—typically a repressor protein that binds to a control sequence in the DNA and prevents transcription—is removed.

negative feedback A self-limiting, corrective response in which a deviation in some variable (e.g., body temperature, blood pH) triggers responses aimed at returning the variable to normal.

negative pressure ventilation Ventilation of the lungs that is accomplished by "pulling" air into the lungs by expansion of the rib cage.

negative result An experimental result that fails to show a predicted difference between two groups or conditions.

negative-sense virus A virus whose genome contains sequences complementary to those in the mRNA required to produce viral proteins.

nephron One of the tiny tubes within the vertebrate kidney that filter blood and concentrate salts to produce urine.

neritic zone Shallow marine waters beyond the intertidal zone, extending down to about 200 m, where the continental shelf ends.

Nernst equation A formula that converts the energy of a concentration gradient to the energy of an electrical potential, for a particular ion or ions.

nerve A long, tough strand of nervous tissue typically containing thousands of axons that carry information to or from the central nervous system.

nervous tissue A class of animal tissue consisting of nerve cells (neurons) and various supporting cells, and functioning in rapid transmission of complex information.

net primary productivity (NPP) In ecology, the amount of primary productivity that is stored in new biomass (not used for cellular respiration).

net reproductive rate The growth rate of a population per generation; equivalent to the average number of female offspring that each female produces over her lifetime. Symbolized by R_0.

neural Relating to nerve cells.

neural tube A folded tube of ectoderm that forms along the dorsal side of a young vertebrate embryo and that will give rise to the brain and spinal cord.

neuroendocrine Refers to nerve cells that release hormones into the blood.

neuron A nerve cell; a cell that is specialized for the transmission of nerve impulses. Typically has dendrites, a cell body, and a long axon that forms synapses with other neurons.

neurosecretory cell A neuron that secretes hormones into the blood; a neuroendocrine cell.

neurotoxin Any poison that specifically affects neuron function.

neurotransmitter A molecule that conveys information from one neuron to another or from a neuron to a muscle or gland. Released from the end of an axon and diffuse a very short distance to the next cell, where they can trigger an action potential.

neutral Any mutation that has no effect on an individual's fitness.

neutron An uncharged particle found in atomic nuclei. Variations in neutron number (with no change in proton number) produce different isotopes of the same element.

neutrophil A type of leukocyte, capable of amoeboid movement through body tissues, that engulfs and digests pathogens and other foreign particles and secretes various compounds that attack bacteria and fungi.

niche The particular set of habitat requirements of a certain species and the role that species plays in its ecosystem.

niche differentiation The tendency of competing species to use different ecological niches because of competition.

nitrogen fixation The incorporation of atmospheric nitrogen (N_2) into forms such as ammonia (NH_3) or nitrate (NO_3^-), which can be used to make many organic compounds. Occurs in only a few lineages of bacteria and archaea.

nociceptor A sensory cell or organ specialized to detect tissue damage, usually producing the sensation of pain.

node (1) Any small thickening (e.g., a lymph node); (2) The part of a stem where leaves or leaf buds are attached. (3) The point on a phylogenetic tree where two branches diverge, representing the point in time when an ancestral group split into two or more descendant groups.

node of Ranvier A point on a neuron's axon between sections of myelin sheath, where an action potential can be regenerated.

Nod factor Molecules produced by nitrogen-fixing bacteria that help them recognize and bind to legume roots.

nodules The lumplike structures in roots of the pea family, containing symbiotic nitrogen-fixing bacteria.

nondisjunction An error that can occur during meiosis or mitosis in which both homologous chromosomes of a pair move to the same side of the dividing cell. One daughter cell receives two copies of the chromosome, and the other daughter cell receives none.

nonenveloped For a virus, lacking an envelope surrounding its capsid coat.

nonpolar covalent bond A symmetrical covalent bond (i.e., one in which electrons are equally shared between the two atomic nuclei).

non-self Property of a molecule or cell whereby immune system cells will recognize it as "different" and attack it.

nonsense mutation A point mutation (change in a single base pair) that results in an early stop codon, resulting in a truncated polypeptide.

non-sister chromatids The chromosome copies of homologous chromosomes. Crossing over occurs between non-sister chromatids.

non-template strand The strand of DNA that is not transcribed to create RNA.

nonvascular plants Several phyla of green plants that lack vascular tissue. Includes liverworts, hornworts, and mosses. Also called bryophytes.

norepinephrine A catecholamine used as a neurotransmitter in the sympathetic nervous system and also released as a hormone from the adrenal medulla. Stimulates increased heart rate, increases blood pressure, decreases digestion, and produces other effects.

norm of reaction The range of phenotypes that are possible for an individual of a given genotype.

Northern blotting A lab technique for identifying the RNA produced by a particular gene. Involves separating RNAs by gel electrophoresis, transferring to a filter paper, and hybridizing to a labeled DNA probe.

notochord A long, gelatinous, supportive rod down the back of a chordate embryo, below the developing spinal cord. Replaced by vertebrae in adult vertebrates. A defining feature of chordates.

nuclear envelope The double-layered membrane enclosing the nucleus of a eukaryotic cell.

nuclear lamina A lattice-like sheet of fibrous nuclear lamin proteins that line the inner membrane of the nuclear envelope. Stiffens the envelope and helps organize the chromosomes.

nuclear lamins A class of fibers that form a dense mesh (the nuclear lamina) just inside the nuclear envelope. A form of intermediate filament.

nuclear localization signal (NLS) A certain sequence of amino acids that tags a protein for delivery to the nucleus by importins.

nuclear pore An opening in the nuclear envelope that connects the inside of the nucleus with the cytoplasm and through which molecules such as mRNA and some enzymes can pass.

nuclear pore complex A large complex of dozens of proteins lining a nuclear pore, defining its shape and transporting substances through the pore.

nuclease Any enzyme that can break down RNA and DNA molecules.

nucleic acids Polymers consisting of a chain of nucleotides. Generally used by cells to store or transmit hereditary information. Includes ribonucleic acid and deoxyribonucleic acid.

nucleoid The region of a prokaryotic cell that contains chromosomes.

nucleolus The structure in a eukaryotic nucleus where ribosomal RNA processing occurs and ribosomal subunits are assembled.

nucleoside A purine or pyrimidine base attached to a five-carbon sugar (ribose or deoxyribose).

nucleosome A repeating, bead-like structure of a eukaryotic chromosome, consisting of about 200 nucleotides of DNA wrapped twice around eight histone proteins.

nucleotide A monomer that can be polymerized to form the nucleic acid DNA or RNA. Consists of a five-carbon sugar (ribose or deoxyribose), a phosphate group, and one of several nitrogen-containing bases. Equivalent to a nucleoside plus one phosphate group.

nucleus (1) The center of an atom, containing protons and neutrons. (2) In eukaryotic cells, the membrane-bound organelle that contains DNA. (3) A discrete clump of neuron cell bodies in the brain, usually sharing a distinct function.

null hypothesis A hypothesis that specifies what the results of an experiment will be if the main hypothesis being tested is wrong. Often states that there will be no difference between experimental groups.

null mutant A mutant allele that does not function at all; or an organism homozygous for such a mutation. Also called knock-out mutant or loss-of-function mutant.

nutrient A substance that an organism requires for normal growth, maintenance, or reproduction.

nutritional balance A state in which an organism is taking in enough nutrients to maintain normal health and activity.

nymph An immature stage of an insect species in which the immature form looks like a miniature adult, such as that in dragonflies.

obesity The condition of having extremely and abnormally high reserves of body fat.

occipital lobe In the mammal brain, the posteriormost region of each cerebral hemisphere. Receives and interprets visual information.

oceanic zone The waters of the open ocean beyond the continental shelf.

oil A lipid that is liquid at room temperature.

Okazaki fragments Short fragments of DNA produced during DNA replication. Now known to be pieces of the lagging strand.

olfaction The perception of odors.

olfactory bulb A bulb-shaped projection of the brain just above the nose. Receives and interprets odor information from the nose.

oligopeptide A polypeptide with fewer than 50 amino acids. May also be referred to simply as "peptides."

ommatidium (plural: ommatidia) A light-sensing column of an arthropod's compound eye.

omnivore An animal whose diet regularly includes both meat and plants.

oncogene An allele that has mutated so as to stimulate cell growth at all times and thus promotes cancer development.

one-gene, one-enzyme hypothesis The hypothesis that each gene is responsible for making one (and only one) particular protein, in most cases an enzyme that catalyzes a specific reaction. Many exceptions to this hypothesis are now known.

oocyte A cell in the ovary that can undergo meiosis to produce an ovum.

oogenesis The production of egg cells (ova).

oogonia (singular: oogonium) The diploid cells in an ovary that can divide by mitosis to create more oogonia and primary oocytes, which can undergo meiosis.

open circulatory system A circulatory system in which the circulating fluid (hemolymph) is not confined to blood vessels.

open reading frame (ORF) Any DNA sequence that is suspected to be a functional gene because it has a start codon and a stop codon separated by a long stretch of DNA (and sometimes has other characteristic features, such as promoters).

operator The DNA binding site for a repressor protein in a prokaryotic operon.

operculum The stiff flap of tissue that covers the gills of teleost fishes.

operon A region of bacterial DNA that codes for a series of functionally related genes.

opsin One of several proteins involved in animal vision. An opsin joins with retinal to form the light-detecting pigment rhodopsin in rod cells.

optical isomer A molecule that shares the same molecular formula as another molecule but differs in the arrangement of atoms or groups around a carbon atom; left-handed or right-handed form of a molecule.

optic nerve A nerve that carries information from the eye to the brain. Vertebrates have two.

orbital The region around an atomic nucleus in which an electron orbits.

order A classic taxonomic rank above the family level and below the class level.

organ A group of tissues organized into a functional and structural unit.

organelle Any discrete, membrane-bound structure in the cytoplasm of a cell (e.g., mitochondrion).

organic For a compound, containing carbon and hydrogen and usually containing carbon-carbon bonds. Organic compounds are widely used by living organisms.

organism Any living entity that contains one or more cells.

organizer A region of an amphibian embryo (around the upper side of the blastopore) that can organize the development of the entire embryo.

organogenesis A stage of embryonic development just after gastrulation in vertebrate embryos, during which major organs develop from the three embryonic cell layers.

organotroph An organism that produces ATP, oxidizing organic molecules with high potential energy, such as sugars.

orientation A deliberate movement that results in a change in position relative to some external cue, such as toward the Sun or away from a sound.

origin of replication The place on a chromosome at which DNA replication begins.

osmoconformer An animal that does not actively regulate the osmolarity of its tissues but conforms to the osmolarity of the surrounding environment.

osmolarity The concentration of dissolved substances in a solution, measured in moles per liter.

osmoregulation The process by which a living organism controls the concentration of water and salts in its body.

osmoregulator An animal that actively regulates the osmolarity of its tissues.

osmosis Diffusion of water across a selectively permeable membrane from areas of high water concentration (low solute concentration) to areas of low water concentration (high solute concentration).

osmotic stress A condition in which there are abnormal concentrations of water and salts in an organism's cells or tissues.

ouabain A plant toxin that poisons the sodium-potassium pumps of animals.

outcrossing Reproduction by fusion of the gametes of different individuals, rather than self-fertilization. Typically refers to plants.

outer ear The outermost portion of the mammal ear, consisting of the pinna (ear flap) and the ear canal. Funnels sound to the tympanic membrane.

outgroup A taxon known to have diverged earliest from all the other taxa under study. Used to determine the root (most ancient node) of a phylogenetic tree.

out-of-Africa hypothesis The hypothesis that modern humans (*Homo sapiens*) evolved in Africa and spread to other continents, replacing other *Homo* species without interbreeding with them.

oval window A membrane separating the fluid-filled cochlea from the air-filled middle ear. The stapes, a middle ear bone, transmits sound vibrations to the cochlea by vibrating on the oval window.

ovary The egg-producing organ of a female animal, or the seed-producing structure of the female part of a flower.

oviduct A narrow tube that connects the uterus to the ovary, and through which the egg travels after ovulation. Fertilization and cleavage occur in the oviduct. In humans, called the fallopian tube.

oviparous Reproducing by laying eggs, rather than giving live birth.

ovoviviparous Reproducing by retaining eggs inside the body until they are ready to hatch. The embryos are nourished by egg yolk, not via a placenta.

ovulation The process by which an ovum is released from the ovary of a female vertebrate.

ovule A structure inside a flower ovary that produces the female gametophyte and eventually (if fertilized) becomes a seed.

ovum (plural: ova) An egg cell; a mature female gamete and any associated external layers. Larger and less mobile than the male gamete.

oxidation The loss of electrons from an atom during a redox reaction, either by donation of an electron to another atom or by the shared electrons in covalent bonds moving farther from the atomic nucleus.

oxidative phosphorylation Production of ATP molecules from the redox reactions of an electron transport chain, starting with reduced compounds (such as NADH or $FADH_2$) and ending with oxygen as the final electron acceptor.

oxygenic photosynthesis Photosynthesis that involves photosystem II, which catalyzes the splitting of water and produces oxygen. Occurs in cyanobacteria, algae, and plants.

oxytocin A peptide hormone from the posterior pituitary that triggers labor and milk production in females and that stimulates pair bonding, parental care, and affiliative behavior in both sexes.

p53 A tumor-suppressor protein that responds to DNA damage by stopping the cell cycle and/or triggering apoptosis. Codes for a protein with a molecular weight of 53 kilodaltons.

pacemaker cells Cells with an inherent rhythm that can set a rhythm for other cells.

pair-rule genes A class of fruit-fly segmentation genes that organize embryonic cells into particular segments. Active in alternating segments.

paleontology The study of organisms that lived in the distant past.

palisade mesophyll Elongated parenchyma cells found in the ground tissue of leaves. Contain many chloroplasts and perform most photosynthesis.

pancreas A gland attached to the small intestine that secretes digestive enzymes into the intestine and several digestion-related hormones (notably, insulin and glucagon) into the bloodstream.

pancreatic amylase A carbohydrate-digesting enzyme secreted by the pancreas into the small intestine.

pancreatic lipase A fat-digesting enzyme secreted by the pancreas into the small intestine.

paper chromatography A technique for separating molecules on the basis of their size and solubility, by wicking them through a piece of filter paper using a certain solvent.

parabiosis An experimental technique for determining whether a certain physiological phenomenon is regulated by a hormone, by surgically uniting two individuals so that hormones can pass between them.

parabronchi (singular: parabronchus) The tiny parallel air tubes that run through a bird's lung.

paracrine signal A chemical signal that is released by one cell and affects neighboring cells.

parafollicular cells The cells of the thyroid gland that release calcitonin.

paraphyletic group A group of organisms that is not monophyletic; i.e, a group that includes some but not all descendants of the last common ancestor of the group. Paraphyletic groups are not meaningful evolutionary groups.

parasite An organism that lives on or in a host species and that damages its host.

parasitism A long-term relationship between two organisms that is beneficial to one organism (the parasite) but detrimental to the other (the host).

parasitoid An organism that has a parasitic larval stage and a free-living adult stage. Most parasitoids are insects that lay eggs in the bodies of other insects.

parasympathetic nervous system The part of the autonomic nervous system that stimulates activities of relaxation, repair, and rebuilding, such as reduced heart rate and increased digestion.

parathyroid glands Four small glands that are near or embedded in the thyroid gland of vertebrates. Secrete parathyroid hormone, which increases blood calcium.

parathyroid hormone (PTH) A peptide hormone from the parathyroid glands that increases blood calcium.

parazoans Animals whose cells are not organized into distinct tissues. Contains one living group: the sponges.

parenchyma cell A thin-walled type of plant cell found in leaves, the centers of stems and roots, and fruits. Involved in photosynthesis, starch storage, and new growth.

parental care Any action by which an animal expends energy or assumes risks to benefit its offspring (e.g., nest-building, feeding of young, defense).

parental generation The adult organisms used in the first experimental cross in a formal breeding experiment.

parietal cells The cells in the stomach lining that secrete hydrochloric acid.

parietal lobe In the mammal brain, the region of each cerebral hemisphere that is behind and above the frontal lobe. In humans, involved in integrating sensory and motor control.

Parkinson's disease A human neurological disorder that causes progressive deterioration of motor function. Due to the inactivation or destruction of dopamine-secreting neurons at the base of the brain.

parsimony The principle that the phylogenetic tree most likely to be correct is the one that requires the fewest evolutionary changes.

parthenogenesis Development of offspring from unfertilized eggs. A form of asexual reproduction.

partial pressure The pressure of one particular gas in a mixture; the contribution of that gas to the overall pressure.

particle A nonliving infectious entity, such as a virus. Also called an agent.

pascal (Pa) A unit of pressure (force per unit area).

passive transport Diffusion of a substance across a cell membrane down a concentration gradient. When this occurs with the assistance of carrier proteins, it is also called facilitated diffusion.

patch clamping A lab technique for studying the electrical currents that flow through individual ion channels, by sucking a tiny patch of membrane to the hollow tip of a microelectrode.

paternal chromosome A chromosome inherited from the father.

pathogen Any entity capable of causing disease, such as a microbe, virus, or prion.

pathogenic Capable of causing disease.

pattern formation The series of events that determines the spatial organization of an embryo, including alignment of the major body axes and orientation of the limbs.

pattern-recognition receptor Leukocyte membrane proteins that bind to molecules in many bacteria. Part of the innate immune response.

PCR See **polymerase chain reaction (PCR)**.

peat Semidecayed organic matter that accumulates in moist, low-oxygen environments such as bogs.

pectins A class of gelatinous polysaccharides found in the primary cell walls of plant cells. Attract and hold water, forming a gel that helps keep the cell wall moist.

pedigree A family tree of parents and offspring, showing inheritance of particular traits of interest.

pellet Any solid material that collects at the bottom of a test tube below a layer of liquid (the supernatant) during centrifugation.

penis The copulatory organ of male mammals, used to insert sperm into a female.

pentaradial symmetry A form of radial symmetry, found in adult echinoderms, in which the body has exactly five planes of symmetry. Typically, five (or multiples of five) body parts radiate from a central hub.

pentose A monosaccharide (simple sugar) with five carbons.

PEP carboxylase An enzyme that catalyze addition of CO_2 to three-carbon compounds, forming four-carbon compounds. Occurs in mesophyll cells of plants that perform C_4 photosynthesis.

pepsin A protein-digesting enzyme produced in the stomach.

pepsinogen The precursor of the digestive enzyme pepsin. Converted to pepsin by the acidic environment of the stomach.

peptide bond The C–N bond between two amino acid residues in a peptide or protein.

peptide hormone A hormone that is a chain of two or more amino acids.

peptidoglycan A polysaccharide found in bacterial cell walls.

per-capita rate of increase The growth rate of a population, expressed per individual. Calculated as the per-capita birthrate minus the per-capita death rate and symbolized *r*. Also called per-capita growth rate.

perennial plant A plant that normally lives for more than one year.

perfect For a flower, containing both male parts (stamens) and female parts (carpels).

perforations In plants, small holes in the primary and secondary cell walls of vessel elements that allow passage of water.

pericarp The part of a fruit that surrounds the seeds, formed from the ovary wall. The flesh of most edible fruits; the hard shells of most nuts.

pericycle In plant roots, a layer of cells that give rise to lateral roots.

peripheral membrane protein Any membrane protein that is found only on one side of the membrane, rather than spanning the entire membrane.

peripheral nervous system (PNS) All the components of the nervous system that are outside the central nervous system (the brain and spinal cord). Includes the somatic nervous system and the autonomic nervous system.

peristalsis Rhythmic waves of muscular contraction that push food along the digestive tract.

permafrost A permanently frozen layer of icy soil found in most tundra and some taiga.

permeability The tendency of a structure, such as a membrane, to allow a given substance to diffuse across it.

peroxisome A eukaryotic organelle that contains oxidative enzymes, usually for degrading fatty acids and amino acids and the resulting hydrogen peroxide.

petal One of the modified leaves arranged around the reproductive structures of a flower. Often colored to attract pollinators.

petiole The stalk of a leaf.

phagocytosis The engulfment and uptake of a small particle or cell by an extension of another cell's plasma membrane.

pharyngeal gill slits A set of parallel openings from the throat through the neck to the outside. One of the diagnostic traits of chordates.

pharyngeal jaw A secondary jaw in the back of the mouth, found in some fishes. Derived from modified gill arches.

phenetic approach A method for constructing a phylogenetic tree by computing a statistic that summarizes the overall similarity among populations, based on the available data.

phenotype Any of the observable traits of an individual. Commonly includes physical, physiological, and behavioral traits.

phenotypic variation The total observable variation in a particular trait in a certain population in a certain environment.

phenylketonuria (PKU) A human genetic disease caused by lack of the enzyme that converts the amino acid phenylalanine to tyrosine.

pheophytin A molecule that acts as an electron acceptor in the light-dependent reactions of photosynthesis, accepting excited electrons from chlorophyll and passing them to an electron transport chain.

pheromone A chemical signal, released by one individual into the external environment, that can trigger responses in a different individual.

phloem A plant vascular tissue that conducts sugars. Contains sieve-tube members and companion cells.

phloem loading The movement of sugars into plant phloem.

phloem sap The sugary fluid found in phloem tissue of plants.

phloem unloading The movement of sugars out of plant phloem.

phonotaxis Orientation toward or away from sound.

phosphate The functional group $–OPO_3^{2-}$. Breaking the O–P bonds between adjacent phosphate groups releases large amounts of energy.

phosphodiester bond The type of bond that links the nucleotides in DNA or RNA. Joins the phosphate group of one nucleotide to the hydroxyl group on the sugar of another nucleotide.

phosphofructokinase The enzyme that catalyzes the synthesis of fructose-1,6-bisphosphate from fructose-6-phosphate, a key reaction (step 3) in glycolysis.

phospholipid A type of lipid having a hydrophilic head (a phosphate group) and a hydrophobic tail (one or more fatty acids), often linked by a glycerol molecule. Major components of plasma membranes.

phosphorylase An enzyme that breaks down glycogen, by catalyzing hydrolysis of the α-glycosidic linkages between the glucose monomers.

phosphorylation The addition of a phosphate group to a molecule. Commonly, refers to phosphorylation of proteins to control protein shape or function.

phosphorylation cascade A sequence of events in which one enzyme phosphorylates other enzymes, which in turn phosphorylate many more, leading to phosphorylation of thousands of proteins. Commonly used in signal transduction of hormone messages; also called signal transduction cascade.

photic zone In an aquatic habitat, water that is shallow enough to receive some sunlight (whether or not it is enough to support photosynthesis).

photon A discrete packet of light energy; a particle of light.

photoperiodism Any response by an organism to photoperiod—the relative lengths of day and night.

photophosphorylation Production of ATP molecules by using the energy of light to excite electrons, which are then passed down an electron transport chain. Occurs during photosynthesis.

photoreceptor A molecule, a cell, or an organ that is specialized to detect light.

photorespiration A series of light-driven chemical reactions that consumes oxygen and releases carbon dioxide, basically reversing photosynthesis. Usually occurs when there are high O_2 and low CO_2 concentrations inside plant cells, often in bright, hot, dry environments when stomata must be kept closed.

photoreversibility A change in conformation that occurs in certain plant pigments when they are exposed to their preferred wavelengths of light and that triggers responses by the plant.

photosynthesis A series of chemical reactions and electron transfer events that converts the energy of light into chemical energy stored in glucose.

photosynthesis-transpiration compromise The balance that plants must strike between maximizing photosynthesis and conserving water.

photosystem A system of 200–300 chlorophyll molecules, accessory pigments, and proteins, found in plant chloroplasts and involved in the light-dependent reactions of photosynthesis.

photosystem I A system of molecules and enzymes in chloroplasts that absorbs light energy, using it to produce NADPH.

photosystem II A system of molecules and enzymes in plant chloroplasts that absorbs light energy, using it to produce ATP and to split water into protons and oxygen.

phototaxis Orientation toward or away from light.

phototroph An organism that produces ATP through photosynthesis.

phototropins A class of plant photoreceptors that detect blue light and initiate phototropic responses.

phototropism Growth or movement in a particular direction in response to light.

pH scale A measure of the concentration of protons in a solution and thus of acidity or alkalinity. Defined as the negative of the base-10 logarithm of the proton concentration: $pH = -\log[H^+]$.

phylogenetic species concept The concept that species are best identified as the smallest monophyletic group in a phylogenetic tree.

phylogenetic tree A diagram that depicts the evolutionary history of a group of organisms.

phylogeny The evolutionary history of a group of organisms.

phylum (plural: phyla) A classic taxonomic rank above the class level and below the kingdom level. (In plants, sometimes called a division.)

physical map A map of a chromosome that shows the number of base pairs between various genetic markers.

physiology The study of how an organism's body functions.

phytoalexin Any small plant compound produced to combat an infection (usually a fungal infection).

phytochromes A class of light-sensitive plant proteins involved in detecting light and timing certain physiological processes, such as flowering and germination.

phytoplankton Plankton (small drifting aquatic organisms) that are photosynthetic.

phytoremediation The use of plants to clean contaminated soils.

pigment Any molecule that absorbs only certain wavelengths of visible light and reflects or transmits other wavelengths.

piloting Finding one's way by using familiar landmarks (i.e., without a specific compass direction).

pinocytosis Uptake of extracellular fluid by endocytosis (i.e., by pinching off the plasma membrane to form small membrane-bound vesicles).

pioneering species Species that are often the first to appear in a recently disturbed area.

pitch The sensation produced by a particular frequency of sound. Low frequencies are perceived as low pitches, high frequencies as high pitches.

pith The center of a plant stem.

pits In plants, small holes in the secondary cell walls of tracheids that allow passage of water.

pituitary dwarfism Dwarfism (abnormally small body size) in mammals that is caused by defects in production of growth hormone by the pituitary gland.

pituitary gland A small gland directly under the brain, close to the hypothalamus. Releases hormones that affect many other glands and organs.

placenta An organ formed by a union of maternal and fetal tissues. Exchanges nutrients and wastes between mother and fetus, anchors the fetus to the uterine wall, and produces some hormones. Occurs in most mammals and in a few other vertebrates.

placental mammals Members of the Eutheria, a major lineage of mammals whose young develop in the uterus and are not housed in an abdominal pouch. Also called eutherians.

planar bilayer A lipid bilayer (double-layered membrane) constructed across a hole in a glass or plastic wall separating two aqueous solutions.

plankton Any small organism that drifts near the surface of oceans or lakes and swims little if at all.

plant A red alga, green alga, glaucophyte alga, or green plant.

plant-defense hypothesis The hypothesis that rates of herbivory are limited by plant defenses such as toxins and spines.

plantlet A small plant, particularly one that forms on a parent plant via asexual reproduction and drops, becoming an independent individual.

plasma The fluid portion of blood that remains when red blood cells, leukocytes, and platelets are removed. (Equivalent to serum plus clotting factors.)

plasma cell A type of leukocyte that produces large quantities of antibodies to combat an ongoing infection. A descendant of an activated B cell.

plasma membrane A membrane that surrounds a cell, separating it from the external environment and selectively regulating passage of molecules and ions into and out of the cell.

plasmid A small, usually circular, supercoiled DNA molecule independent of the cell's main chromosome(s) in prokaryotes and some eukaryotes.

plasmodesmata (singular: plasmodesma) Physical connections between two plant cells, consisting of gaps in the cell walls through which the two cells' plasma membranes, cytoplasm, and smooth ER can connect directly.

plasmogamy Fusion of the cytoplasm of two individuals. Occurs in many fungi.

plastocyanin A small protein that shuttles electrons from photosystem II to photosystem I during photosynthesis.

plastoquinone The molecule involved in the light-dependent reactions of photosynthesis that receives excited electrons from pheophytin and passes them to more electronegative molecules in the chain. Also carries protons to the lumen side of the thylakoid membrane.

platelet A small membrane-bound cell fragment in vertebrate blood, important in blood clotting.

platelet-derived growth factor (PDGF) A protein secreted by platelets and some other cells at the site of an injury. Promotes wound healing.

plate tectonics The theory that Earth's crust is made up of separate plates that have moved throughout geologic history.

pleiotropy A pattern of genetic expression in which one gene affects more than one phenotypic trait.

ploidy The number of each type of chromosome present. Haploid cells have a ploidy of 1; diploid cells have a ploidy of 2.

pneumonia Inflammation of the lungs.

podium (plural: podia) The part of the tube foot of an echinoderm that extends outside of the body and makes contact with the substrate.

point mutation A mutation that results in a change in a single nucleotide pair of DNA.

polar (1) Asymmetrical or unidirectional. (2) Carrying a partial positive charge on one side of a molecule and a partial negative charge on the other. Polar molecules are generally hydrophilic.

polar bodies The tiny, nonfunctional cells produced during meiosis of a primary oocyte, due to most of the cytoplasm going to the ovum.

polar covalent bond A covalent bond that is asymmetrical, such that the electrons spend more time near one atomic nucleus than the other. Often results in the molecule being polar.

polar nuclei (singular: polar nucleus) The nuclei in the female gametophyte of a flowering plant that will fuse with one sperm nucleus to produce the endosperm. Most species have two.

pollen grain In flowering plants, a male gametophyte enclosed within a protective coat.

pollen tube In flowering plants, a tube that grows out of a pollen grain and toward the ovule and through which the two sperm nuclei move.

pollination The process by which pollen reaches the carpel of a flower (in flowering plants) or reaches the ovule directly (in conifers and their relatives).

poly (A) tail In eukaryotes, a long sequence of 100–250 adenine nucleotides added to the 3′ end of newly transcribed messenger RNA molecules.

polycistronic For an mRNA molecule, containing more than one protein-coding segment, each with its own start and stop codons and each coding for a different protein. Common in prokaryotes.

polyclonal antibody A mix of several different antibodies that all bind to the same antigen, typically produced by a living animal after injection with the antigen.

polygenic inheritance The inheritance patterns that result when many genes influence one trait.

polymer Any long molecule composed of small repeating subunits (monomers) bonded together.

polymerase chain reaction (PCR) A lab technique for rapidly generating millions of identical copies of a specific stretch of DNA. Involves incubating the original DNA sequence with primers, free nucleotides, and DNA polymerase.

polymerization The process by which one monomer (a small subunit molecule) is bound to others to form a polymer (a long chain molecule).

polymorphism (1) The occurrence of more than one allele at a certain genetic locus in a population. (2) The occurrence of more than two distinct phenotypes of a trait in a population.

polyp The sessile stage of a cnidarian life cycle.

polypeptide A peptide of three or more amino acids linked together in a chain. Very large polypeptides may be called proteins.

polyploid Having more than two chromosome sets.

polyribosome A structure consisting of one messenger RNA molecule along with many attached ribosomes and their growing peptide strands. Occurs in prokaryotes and eukaryotes.

polysaccharide A large carbohydrate polymer consisting of many monosaccharides linked together in a chain.

polyspermy Fertilization of an egg by multiple sperm. This is usually an abnormal situation.

polytomy A multibranched node on a phylogenetic tree. Represents a time when an ancestral population split into descendant populations.

pons A small region of the brain that relays information to the cerebellum and is also involved in control of breathing.

poor-nutrition hypothesis The hypothesis that herbivore populations are limited by the poor nutritional content of plants, especially low nitrogen.

population A group of individuals of the same species living in the same area at the same time.

population bottleneck An extreme reduction in population size, followed by re-expansion to a larger population size. May cause genetic bottleneck.

population cycles Regular fluctuations in size exhibited by certain populations.

population density The number of individuals of a population per unit area.

population dynamics Changes in population size through time.

population ecology The study of how and why the number of individuals in a population changes over time.

population viability analysis (PVA) A method of estimating the likelihood that a population will avoid extinction for a given time period.

pore Any small opening, such as the small opening in the stoma of a plant leaf or in the septa of fungal filaments.

positive control A type of gene regulation in which transcription can occur only when a certain substance—typically an activator protein that binds to a control sequence in the DNA and promotes transcription—is present.

positive feedback Stimulation of a reaction or process by the end result of that process. Tends to accelerate processes rapidly.

positive pressure ventilation Ventilation of the lungs that is accomplished by "pushing" air into the lungs by positive pressure in the mouth.

positive-sense virus A virus whose genome contains the same sequences as the mRNA required to produce viral proteins.

posterior Toward an animal's tail and away from its head. The opposite of anterior.

posterior pituitary The posterior part of the pituitary gland, consisting of the ends of neurosecretory cells from the hypothalamus, which secrete oxytocin and antidiuretic hormone.

postsynaptic neuron A neuron that receives neurotransmitters from another neuron at a particular synapse.

post-translational control Regulation of gene expression by modification of proteins after translation.

post-translational modification Any chemical alteration of a protein after it has been synthesized; includes cleavage of side chains, phosphorylation.

postzygotic isolation Result of mechanisms that prevent gene flow between different species even if mating occurs between them, typically due to death of hybrid embryos or reduced fitness of hybrids.

potential energy Energy stored in matter through its position or its shape, as opposed to kinetic energy (the energy of motion).

power stroke During muscle contraction, the motion of a myosin "head" that pulls myosin and actin filaments further along each other.

prairie An extensive grassland. Typically found in the dry interiors of continents in the temperate latitudes. Also called a steppe.

prebiotic soup A hypothetical solution of sugars, amino acids, nitrogenous bases, and other building blocks of larger molecules that may have formed in shallow waters or deep-ocean vents of ancient Earth and given rise to larger biological molecules.

predation The killing and eating of one organism by another.

predator Any organism that kills other organisms for food.

presentation Display of an antigen on the surface of an immune system cell to recruit other immune system cells.

pressure-flow hypothesis The hypothesis that sugar movement through phloem tissue is due to differences in the turgor pressure of phloem sap.

pressure potential Potential energy of water caused by pressure differences. Equals the sum of all the types of pressure that affect water, such as atmospheric pressure, wall pressure, and tension.

presynaptic neuron A neuron that releases neurotransmitters to another neuron at a synapse.

pre-urine The fluid in kidney nephrons that is formed by filtration of blood. Upon further processing, becomes urine. Also called filtrate.

prey A species that is commonly attacked and eaten by a predator species.

prezygotic isolation Anything that prevents individuals of two different species from mating.

primary cell wall The outermost layer of a plant cell wall, made of cellulose fibers with gelatinous polysaccharides, that defines the shape of the plant cell and withstands the turgor pressure of the plasma membrane.

primary consumer An herbivore; an organism that eats plants, algae, or other primary producers.

primary decomposer A decomposer that consumes detritus from plants.

primary growth Plant growth that results in an increase in length of stems and roots. Produced by apical meristems.

primary immune response An immune response to a pathogen that the immune system has not encountered before.

primary oocyte The large diploid cell in an ovarian follicle that can initiate meiosis to produce a haploid ovum.

primary producer An autotroph; a species that creates its own food through photosynthesis or from reduced inorganic compounds and that is a source of food for other species in its ecosystem.

primary productivity The total amount of carbon fixed by photosynthesis per unit area per year, including that used for cellular respiration.

primary RNA transcript In eukaryotes, a newly transcribed messenger RNA molecule that has not yet been processed—i.e., that still contains introns and has not received a 5′ cap or a poly (A) tail.

primary sex determination The process by which an embryonic gonad becomes either a testis or an ovary.

primary spermatocyte A diploid cell in the testis that can initiate meiosis I to produce two secondary spermatocytes.

primary structure The sequence of amino acids in a peptide or protein; also the sequence of nucleotides in a nucleic acid.

primary succession The gradual colonization of a habitat of bare rock or gravel, usually after an environmental disturbance that removes all soil and previous organisms.

primase An enzyme that synthesizes a short stretch of RNA to use as a primer during DNA replication.

primates The lineage of mammals that includes prosimians (lemurs, lorises, etc.), monkeys, and apes (including humans).

primer A short, single-stranded sequence of RNA that enables the start of replication of a DNA sequence.

principle of independent assortment The concept that each pair of hereditary elements (alleles of the same gene) behaves independently of other genes during meiosis. One of Mendel's two principles of genetics.

principle of segregation The concept that each pair of hereditary elements (alleles of the same gene) separate from each other during the formation of offspring (i.e., during meiosis). One of Mendel's two principles of genetics.

prion An infectious protein that is thought to cause disease by inducing other proteins to assume an abnormal three-dimentional structure. Likely cause of spongiform encephalopathies, such as mad cow disease.

probe A single-stranded fragment of a labeled, known DNA sequence that will bind to a complementary sequence in the sample being analyzed.

proboscis A long, narrow feeding appendage through which food can be obtained.

procambium A group of cells in the center of a young plant embryo that will give rise to the vascular tissue.

product The final atoms or molecules after a chemical reaction has occurred.

productivity The amount of energy used by a certain component of an ecosystem (e.g., a community or a trophic level) in a given area over a given time period.

progesterone A steroid hormone produced, along with estrogens, in the ovaries. Secreted by the corpus luteum after ovulation; causes the uterine lining to thicken.

prokaryote A member of the domain Bacteria or Archaea; a unicellular organism lacking certain complex cell features, such as a membrane-bound nucleus and gene introns.

prolactin A peptide hormone from the pituitary gland that promotes milk production in female mammals and that has a variety of effects on parental behavior and seasonal reproduction in other vertebrates.

prolactin-inhibiting hormone (PIH) A peptide hormone from the hypothalamus that inhibits release of prolactin from the anterior pituitary.

prometaphase A stage of cell division (mitosis or meiosis), during which the nuclear envelope breaks down and spindle fibers attach to chromatids.

promoter A short sequence of DNA that facilitates binding of RNA polymerase to enable transcription of downstream genes.

promoter-proximal elements In eukaryotes, regulatory sequences that are close to a promoter and that can bind regulatory transcription factors.

prophase The first stage of cell division (mitosis or meiosis) during which chromosomes become visible and the mitotic spindle forms. Synapsis and crossing over occur during prophase of meiosis I.

proplastid A colorless organelle found in undifferentiated plant cells that matures to become a plastid.

prosimians One of the two major lineages of primates, including lemurs, tarsiers, pottos, and lorises but not monkeys or apes.

protease An enzyme that can break apart proteins, by cleaving the peptide bonds between amino acids.

protein A long chain of 50 or more amino acids linked together; a large polypeptide.

proteinase inhibitors Defensive compounds produced by plants that block the enzymes in animal digestive tracts responsible for digesting proteins.

protein kinase An enzyme that catalyzes the addition of a phosphate group to another protein, typically activating or inactivating the other protein.

proteoglycans A type of glycoprotein commonly found in the extracellular matrix. Proteoglycans have a greater carbohydrate content and are larger than most other glycoproteins.

proteomics The study of the three-dimensional structure and function of proteins.

protist Any microscopic eukaryote that is not a green plant, animal, or fungus.

protoderm The exterior layer of a young plant embryo. Gives rise to the epidermis.

proton A small, positively charged particle in atomic nuclei. The number of protons in an atom gives that atom its characteristics as an element.

proton-motive force The combined effect of a proton gradient and an electric potential gradient across a membrane, which can drive protons across the membrane. Used by mitochondria and chloroplasts to power ATP synthesis.

proton pump A membrane protein that uses the energy of ATP to transport protons across the membrane against an electrochemical gradient.

proto-oncogene Any gene that encourages cell growth, typically by triggering specific phases in the cell cycle.

protostomes A major lineage of animals that share a pattern of embryological development, including spiral cleavage, formation of the mouth earlier than the anus, and formation of the coelom by splitting of a block of mesoderm. Includes arthropods, mollusks, and annelids.

proximal Toward or from the center of the body; away from the furthest tip of an appendage. The opposite of distal.

proximal tubule The convoluted section of a kidney nephron into which filtrate moves from Bowman's capsule. Involved in active reabsorption of certain solutes and water.

proximate causation In biology, the immediate, mechanistic cause of a phenomenon (how it happens), as opposed to the ultimate cause (why it evolved). Also called proximate explanation.

pseudocoelom A body cavity that forms between the endoderm and mesoderm layers. Occurs in roundworms (nematodes) and rotifers.

pseudogene A DNA sequence that closely resembles a working gene but is not transcribed. Thought to have arisen by duplication of the working gene followed by accidental inactivation due to a mutation.

pseudopodium (plural: pseudopodia) A mobile, outward bulging of a cell's plasma membrane, used in cell crawling or ingestion of food.

P site The site in a ribosome where peptide bonds are formed between amino acids.

puberty The process by which an immature animal attains reproductive maturity.

pulmonary artery A short, thick-walled artery that carries oxygen-poor blood from the heart to the lungs.

pulmonary circulation The part of the circulatory system that sends oxygen-poor blood to the lungs. In many vertebrates, the pulmonary circulation is separate from the rest of the circulatory system (the systemic circulation).

pulmonary vein A short, thin-walled vein that carries oxygen-rich blood from the lungs to the heart.

pulse-chase experiment A lab technique for marking a population of cells or molecules at a particular moment in time by means of a labeled molelcule and then following their fate over time.

Punnett square A diagram that depicts the genotypes and phenotypes that should appear in offspring of a certain cross.

pupa A metamorphosing insect that is enclosed in a protective case.

pupation A developmental stage of many insects, in which the body metamorphoses from the larval form to the adult form while enclosed in a protective case.

pupil The hole in the center of the iris through which light enters a vertebrate or cephalopod eye.

pure line In animal or plant breeding, a strain of individuals that produce offspring identical to themselves when self-pollinated or crossed to another member of the same population. Pure lines are homozygous for most, if not all, genetic loci.

purines A class of small, nitrogen-containing, double-ringed bases (guanine, adenine) found in nucleotides.

pyramid of productivity The characteristic pattern of productivity in ecosystems, in which productivity declines with each higher trophic level.

pyrimidines A class of small, nitrogen-containing, single-ringed bases (cytosine, uracil, thymine) found in nucleotides.

pyruvate dehydrogenase A large enzyme complex, located in the inner mitochondrial membrane, that is responsible for conversion of pyruvate to acetyl CoA during glycolysis.

Q A molecule in mitochondria that helps transfer electrons between the complexes of the electron transport chain during cellular respiration. Also called coenzyme Q or ubiquinone.

quadrat A small rectangular plot set up in a habitat to mark an area under intensive study.

quantitative trait A phenotypic trait that exhibits variation along a smooth, continuous scale of measurement (for example, human height), rather than the distinct forms seen in discrete traits.

quantitative variation Variation that exhibits differences in degree across a smooth, continuous scale of measurement.

quaternary structure The three-dimensional shape of several polypeptide chains and sometimes other small functional groups, arranged together to form a large, multiunit protein.

race A population that has different characteristics from another population of the same species, whether or not there are significant genetic differences between them.

radial cleavage The pattern of embryonic cleavage seen in protostomes, in which cells divide at right angles to each other to form tiers.

radial symmetry An animal body pattern in which there are least two planes of symmetry. Typically, the body is in the form of a cylinder or disk, with body parts radiating from a central hub.

radiation Production of electromagnetic energy (light). One of the mechanisms of heat transfer between organisms and the environment.

radicle The root of a plant embryo.

radioactive decay A spontaneous change in the mass number of an atomic nucleus via emission of radiation or a particle.

radioactive isotope An isotope that spontaneously emits radiation (gamma rays) and/or subatomic particles. In the latter case, the isotope decays to form a different isotope or element.

radiometric dating A technique for determining the age of a rock by measuring the amount of radioactive decay that has occurred since the rock solidified, usually by measuring the amount of new daughter element that has formed.

radula A rasping feeding appendage in gastropods.

rain shadow The dry region on the side of a mountain range away from the prevailing wind.

rays Lines of parenchyma cells that extend laterally through the xylem of plant wood.

reactant The atoms or molecules in a chemical reaction, in their starting states.

reaction center A central area in a photosystem of the light-dependent reactions of photosynthesis. Surrounded by chlorophyll molecules of the antenna complex and receives energy from them.

reactive oxygen intermediates (ROIs) Oxygen-containing compounds that are highly reactive and that are used in plant and animal cells to kill infected cells and for other purposes.

reading frame The division of a sequence of DNA or RNA into a particular series of three-nucleotide codons. There are three possible reading frames for any sequence.

realized niche The ecological niche that a species occupies in the presence of competitors.

receptor (1) A molecule, cell, or group of cells specialized for detecting environmental signals. (2) A molecule that binds to a particular chemical (e.g., hormone, sperm protein) and triggers a cellular response. (3) A cell-surface molecule necessary for a virus to gain entry to a cell.

receptor-mediated endocytosis Endocytosis triggered by the binding of certain macromolecules outside the cell to membrane proteins.

receptor tyrosine kinases Transmembrane proteins involved in signal transduction. Typically bind to a signalling molecule, triggering a phosphorylation cascade of other proteins inside the cell.

recessive Property of an allele whereby its influence on phenotype can be entirely hidden by the presence of another, dominant allele.

reciprocal altruism Altruistic behavior that is exchanged between a pair of individuals at different points in time (i.e., sometimes individual A helps individual B, and sometimes B helps A).

reciprocal cross A breeding experiment in which the mother's and father's phenotypes are the reverse of that examined in a previous breeding experiment.

recognition sequence A sequence of amino acids that binds to a specific sequence of DNA and that is found within the helix-turn-helix motif of many repressor proteins.

recognition site The specific sequence of DNA bases cut by a certain restriction endonuclease.

recombinant Possessing a new combination of alleles. May refer to a single chromosome or an entire organism.

recombinant DNA Any DNA altered by exchange with, or inclusion of, foreign DNA. May be produced via meiosis, viruses, or lab manipulation.

recombinant DNA technology A variety of lab techniques for isolating specific DNA fragments and introducing them into different regions of DNA and/or a different host organism. Also called biotechnology or genetic engineering.

recombination A change in the combination of genes or alleles on a given chromosome or in a given individual. Also called genetic recombination.

rectal gland A salt-excreting gland in the digestive system of sharks, skates, and rays.

rectum The last part of the digestive tract; a short tube that holds feces until they are expelled.

red blood cells Hemoglobin-containing cells that circulate in the blood and deliver oxygen from the lungs to the tissues.

redox reactions A class of chemical reactions that involve the loss and gain of electrons. Also called reduction-oxidation reactions.

reduction An atom's gain of electrons during a redox reaction, either by acceptance of an electron from another atom or by the electrons in covalent bonds moving closer to the atomic nucleus.

reduction-oxidation reactions A class of chemical reactions that involve the loss and gain of electrons. Also called redox reactions.

redundant code A code in which different sequences can represent the same information. The genetic code is redundant because some amino acids are coded for by two or three different codons.

reflex An involuntary response to environmental stimulation. May involve the brain (e.g., conditioned reflex) or not (e.g., spinal reflex).

refractory No longer responding to stimuli that previously elicited a response. For example, the tendency of voltage-gated sodium channels to remain closed immediately after an action potential.

regeneration Growth of a new body part to replace a lost body part.

regulatory cascade In embryonic development, a progressive series of interactions among genes and/or cytoplastmic determinants that organizes the body plan of the embryo.

regulatory evolution The evolution of new body parts and patterns via changes in the regulatory genes that affect pattern formation in embryos.

regulatory homeostasis Steady internal body conditions achieved by active physiological processes.

regulatory protein Any protein that affects gene transcription by binding to specific enhancers, silencers, or other sites in DNA.

regulatory sequence Any section of DNA that is involved in controlling the activity of other genes.

regulatory site A site on an enzyme to which a regulatory molecule can bind and affect the enzyme's activity, separate from the active site where catalysis occurs.

regulatory transcription factors Proteins that bind to eukaryotic enhancers, silencers, and promoter-proximal elements, but not to the promoter itself.

reinforcement Natural selection for traits that prevent interbreeding between recently diverged species.

release factors Proteins that can trigger termination of RNA translation when a ribosome reaches a stop codon.

releaser A simple stimulus that elicits an invariant, stereotyped behavioral response (fixed action pattern) from an animal. Also called a sign stimulus.

releasing hormone A hormone that stimulates release of some other hormone.

renal corpuscle The ball-like structure at the beginning of a kidney nephron, consisting of Bowman's capsule surrounding a glomerulus.

replacement mutation A point mutation (change in a single base pair) that causes a change in the amino acid sequence of a protein. Also called missense mutation.

replacement rate The reproductive rate at which each female produces two surviving offspring over her entire life—enough to exactly replace herself and her mate, resulting in zero population growth.

replica plate In replica plating, a copy of the master plate produced by transferring bacteria from it with a velvet-covered block.

replica plating A method of identifying bacterial colonies that have certain mutations by observing their growth on a plate that is exposed to different conditions.

replicated chromosome A chromosome that has been copied; consists of two chromatids joined at the centromere.

replication The exact copying of something—e.g., DNA replication.

replication fork The Y-shaped site at which a double-stranded molecule of DNA is separated into two single strands for replication.

repolarization A return to a normal membrane potential after a depolarization.

repressor Any regulatory protein that inhibits transcription of certain genes, typically by binding to a silencer upstream of the promoter.

reservoir In biogeochemical cycles, a location in the environment where elements are stored for a time.

resilience A measure of how quickly a community recovers following a disturbance.

resistance (1) The ability of an organism to defend itself against drugs, pathogens, or parasites. (2) A measure of how much a community is affected by a disturbance.

resistance (*R*) loci Genes associated with disease resistance in plants.

respiration The biochemical pathways that produce ATP from compounds with high potential energy, via an electron transport chain and using an inorganic final electron acceptor. Also called cellular respiration.

respiratory distress syndrome A syndrome in which premature infants can suffocate due to insufficient surfactant in their lungs.

resting potential The membrane potential of a cell in its resting, or normal, state.

restriction endonucleases Bacterial endonucleases that cut DNA at a specific base-pair sequence (recognition site). Also called restriction enzymes.

restriction fragment length polymorphisms (RFLPs) Variations in the size of DNA fragments that are produced by restriction endonucleases, due to differences in the DNA sequences at recognition sites.

retina A thin layer of light-sensitive cells (rods or cones) and neurons at the back of a camera-type eye, such as that of cephalopods or vertebrates.

retinal A carotenoid pigment derived from vitamin A that, with opsin, forms rhodopsin, the light-detecting pigment in rods and cones of animal eyes.

retrovirus A virus with an RNA genome that reproduces by transcribing its RNA into a DNA sequence and then inserting that DNA into the host's genome for replication.

reverse transcriptase A enzyme of retroviruses (RNA viruses) that can synthesize double-stranded DNA from a single-stranded RNA template.

rhizobia Members of the bacterial genus *Rhizobia*, nitrogen-fixing bacteria that live in root nodules of members of the pea family.

rhizoid The hairlike structure that anchors a bryophyte (nonvascular plant) to the substrate.

rhizome A plant stem extending horizontally underground.

rhodopsin A combination of two molecules (retinal and one of various opsins) instrumental in detection of light by rods and cones of vertebrate eyes.

ribonucleic acid (RNA) A polymer of ribonucleotides, usually single stranded. RNAs function as structural components of ribosomes (rRNA), transporters of amino acids (tRNA), and translaters of the message of the DNA code (mRNA).

ribonucleotide A nucleotide consisting of the five-carbon sugar ribose, a phosphate group, and one of several nitrogen-containing bases (adenine, guanine, thymine, or uracil). Can be polymerized to form ribonucleic acid (RNA).

ribosomal RNAs (rRNAs) A class of RNA molecules that form part of the structure of a ribosome.

ribosome A molecular machine that synthesizes proteins by using the genetic information encoded in messenger RNA strands. Consists of two subunits, each composed of ribosomal RNA and proteins.

ribosome binding site In bacteria, the sequence at the beginning of an mRNA molecule to which a ribosome binds to initiate translation. Also called the Shine-Dalgarno sequence.

ribozyme Any RNA molecule that can act as a catalyst for a chemical reaction.

ribulose bisphosphate (RuBP) A five-carbon compound that is the initial reactant in the Calvin cycle of photosynthesis.

rickets A human disorder characterized by malformed, soft bones. Usually caused by environmental factors (e.g., inadequate vitamin D intake), but also may be caused by an X-linked dominant allele.

RNA See **ribonucleic acid (RNA)**.

RNA polymerases A class of enzymes that catalyze synthesis of RNA from ribonucleotides, using a DNA template. Also called RNA pols.

RNA processing In eukaryotes, the changes that a primary RNA transcript undergoes to become a mature mRNA molecule. Includes splicing and the addition of 5′ caps and poly (A) tails.

RNA replicase A viral enzyme that can synthesize RNA from an RNA template.

RNA world hypothesis The hypothesis that life on Earth began as a polymer of ribonucleic acid.

robust australopithecines Several species of comparatively large, strong hominins that appear in the fossil record shortly after the split from chimpanzees.

rod A type of photoreceptor cell with a rod-shaped outer portion. Found in vertebrate retinas. Particularly sensitive to dim light, but not used to distinguish colors.

root (1) An underground appendage of a plant that anchors the plant and absorbs water and nutrients. (2) In a phylogenetic tree, the bottom, most ancient node.

root apical meristem A group of undifferentiated plant cells at the tip of a plant root.

root cap A small group of cells that covers and protects the tip of a plant root. Senses gravity and determines the direction of root growth.

rooted For a phylogenetic tree, oriented so that its bottom-most node is the most ancient node. Usually, the root is identified through the use of an outgroup.

root hair A long, thin outgrowth of the epidermal cells of plant roots, providing increased surface area for absorption of water and nutrients.

root pressure Positive pressure that is generated in plant roots during the night, due to accumulation of ions from the soil and subsequent movement of water into the root.

root system The belowground part of a plant.

rosette In plants, a compact growth form in which leaves pile up on each other in whorls.

rough ER (rough endoplasmic reticulum) A type of endoplasmic reticulum dotted with ribosomes. Involved in processing of membrane proteins and secretory proteins.

rRNAs See **ribosomal RNAs (rRNAs)**.

rubisco The enzyme that catalyzes the first step of the Calvin cycle of photosynthesis: the addition of a molecule of CO_2 to ribulose bisphosphate. Also called ribulose 1,5-bisphosphate carboxylase/oxygenase.

rumen The largest chamber of a ruminant's stomach, containing a large vat of symbiotic cellulose-digesting bacteria. Creates cud, which is sent back to the mouth for further chewing.

ruminants A group of hoofed mammals that have a four-chambered stomach specialized for digestion of plant cellulose, with one chamber containing symbiotic cellulose-digesting bacteria.

sac fungi A monophyletic lineage of fungi that produce large, often cup-shaped reproductive structures that contain asci. Also called cup fungi.

sage-steppe An arid, shrub-dominated habitat with some characteristics of deserts and grasslands.

salicylic acid A compound thought to play a role in the systemic acquired resistance (SAR) defensive mechanism of plants; a component of aspirin.

salivary glands Mammalian glands that secrete saliva (a mixture of water, mucins, and digestive enzymes) into the mouth.

sampling error The accidental selection of a nonrepresentative sample from some larger population, due to chance.

saprophyte An organism that feeds primarily on dead plant material. Usually, a fungus or plant.

sapwood The younger xylem in the outer layer of wood of a stem or root, functioning primarily in water transport.

sarcomere A single contractile unit of a skeletal muscle cell.

sarcoplasmic reticulum Sheets of smooth endoplasmic reticulum in a muscle cell. Contains high concentrations of calcium, which can be released into the cytoplasm to trigger contraction.

SARS See **severe acute respiratory syndrome (SARS)**.

saturated fat A fat that contains the maximum number of hydrogen atoms because all the carbon atoms in its fatty acid chains are joined by single bonds. Such fats have relatively high melting points.

scanning electron microscope (SEM) A microscope that produces images of the surfaces of objects by reflecting electrons from a specimen coated with a layer of metal atoms.

scarify Scraping, rasping, or other damage to the coat of a seed. Necessary in some species to trigger germination.

schizophrenia A human psychological disorder characterized by delusions, hallucinations, social withdrawal, and other symptoms.

Schwann cells Specialized cells that wrap around axons of neurons outside the brain and spinal cord, providing electrical insulation.

sclereid A type of plant sclerenchyma cell that usually functions in protection, such as in seed coats and nutshells.

sclerenchyma cells A thick-walled class of plant cells that provide support and typically contain the tough structural polymer lignin. Usually dead at maturity.

screen Any of several techniques for identifying individuals with a particular type of mutation. Also called genetic screen.

scrotum A sac of skin, containing the testes, suspended just outside the abdominal body cavity of many male mammals.

secondary antibody An antibody that will bind to another antibody. Used in ELISAs and in some other lab tests.

secondary cell wall An inner layer of the cell wall, formed by certain types of plant cells as they mature.

secondary consumer A carnivore; an organism that eats herbivores.

secondary endosymbiosis The presence of an organelle that originated with ingestion of a cell containing that organelle, which in turn was originally derived from ingestion and symbiosis.

secondary growth Plant growth that results in an increase in width of stems and roots.

secondary immune response The immune response, using memory cells, to an infection that the immune system has encountered before.

secondary meristem A layer of undifferentiated plant cells found in older stems and roots. Responsible for secondary growth (increase in width). Also called cambium or lateral meristem.

secondary metabolite Any poison produced by a plant that is synthesized by a variation of a biosynthetic pathway used for other purposes.

secondary phloem Phloem tissue produced by a lateral meristem rather than by an apical meristem.

secondary production The total amount of new body tissue produced by animals that eat plants. May involve growth and/or reproduction.

secondary spermatocyte A cell produced by meiosis I of a primary spermatocyte in the testis. Can undergo meiosis II to produce spermatids.

secondary structure A type of protein structure created by hydrogen bonding between C=O and N–H groups of the polypeptide backbone; most notably, the α-helix and β-pleated sheet structures.

secondary succession Gradual colonization of a habitat after an environmental disturbance (e.g., fire, windstorm, logging) that removes some or all previous organisms but leaves the soil intact.

secondary xylem Xylem tissue produced by a lateral meristem, rather than by an apical meristem.

second law of thermodynamics A fundamental principle of physics stating that, in an isolated system, entropy always increases during any chemical reaction.

second-male advantage The reproductive advantage of a male who mates with a female last, after other males have mated with her.

second messenger A nonprotein signaling molecule produced or activated inside a cell in response to stimulation at the cell surface. Commonly used to relay the message of a protein hormone.

secretin A peptide hormone produced by the small intestine in response to the arrival of food from the stomach. Stimulates secretion of bicarbonate (HCO_3^-) from the pancreas.

sedimentary rock A type of rock formed by gradual accumulation of sediment, as in riverbeds and on the ocean floor. Most fossils are found in sedimentary rocks.

seed A plant embryo with nutritive tissue (endosperm) to fuel its early growth, surrounded by an outer protective layer (seed coat). In angiosperms, forms from the fertilized ovule of a flower.

seed coat A protective layer around a seed that encases both the embryo and the endosperm.

seed dormancy A state of suspended development of a plant seed. Can be terminated by cues indicating that favorable environmental conditions have arrived or that unfavorable ones have passed.

seedless vascular plants Several phyla of green plants that have vascular tissue but do not make seeds. Include horsetails, ferns, lycophytes, and whisk ferns.

seedling A young plant that has emerged from a seed.

seed plants A group of several phyla of green plants that have vascular tissue and make seeds. Includes gymnosperms and angiosperms.

segment A well-defined region of the body along the anterior-posterior body axis, containing similar structures as other, nearby segments.

segmentation A body plan involving division of the body into many similar segments that bear similar or identical structures.

segmentation genes Genes that affect body segmentation in embryonic development. Includes gap genes, pair-rule genes, and segment polarity genes.

segment polarity genes A class of fruit-fly segmentation genes that establish the anterior-posterior orientation of each embryonic segment. Active in particular regions of each segment.

selective adhesion The tendency of cells of one tissue type to adhere to other cells of the same type.

selectively permeable membrane A membrane that some solutes can cross more readily than other solutes can.

selective permeability The property of a structure, such as a membrane, that allows some substances to diffuse across it much more readily than other substances.

self Property of a molecule or cell such that immune system cells do not attack it, due to certain molecular similarities to other body cells.

self-fertilization The fusion of two gametes from the same individual to form a diploid offspring. Also called selfing.

self-incompatible Incapable of self-fertilization.

selfing The fusion of two gametes from the same individual to form a diploid offspring. Also called self-fertilization.

selfish genes DNA sequences that survive and reproduce but reduce the fitness of the host genome.

self-pollination Pollination in which pollen from a certain individual lands on a flower stigma of that same individual.

semen The combination of sperm and accessory fluids that is released by male mammals and reptiles during ejaculation.

semiconservative replication The type of replication used by cells to copy DNA, in which each daughter DNA molecule is composed of one old strand and one new strand.

seminal vesicles Paired reproductive glands that, in mammals, secrete an alkaline fluid into semen to counteract the acidic environment of the vagina. In other vertebrates and invertebrates, often stores sperm.

senescence The process of aging.

sensitive period A short time span in a young animal's life during which the animal can learn certain things, such as song, language, or imprinting. Also called the critical period.

sensor A cell, organ, or structure that senses some aspect of the external or internal environment.

sensory neuron A nerve cell that carries sensory information to the central nervous system.

sepal One of the protective leaflike structures enclosing a flower bud and later supporting the blooming flower.

septa (singular: septum) Any wall-like structure. In fungi, cross-walls that divide fungul filaments into cell-like compartments.

serotonin A neurotransmitter involved in many brain functions, including sleep, pleasure, and mood.

serum The liquid that remains when clotted cells are removed from blood; plasma without clotting factors. Contains water, dissolved gases, hormones, food molecules, and other soluble substances.

sessile Permanently attached to a substrate; not capable of moving to another location.

set point A normal or target value for a regulated internal factor, such as body heat or blood pH.

severe acute respiratory syndrome (SARS) A human disease characterized by sudden and intense flu-like symptoms.

severe combined immune deficiency (SCID) A human disease characterized by an extremely high vulnerability to infectious disease, due to a genetic defect in the immune system.

sex chromosome One of the pair of chromosomes carrying the gene(s) that determine sex.

sex-linked inheritance Inheritance patterns observed in genes carried on sex chromosomes, so females and males have different numbers of alleles of a gene and may pass its trait only to one sex of offspring. Also called sex-linkage.

sexual dimorphism Any trait that differs between males and females.

sexual reproduction Reproduction in which genes from two parents are combined via fusion of gametes, producing offspring that are genetically distinct from both parents.

sexual selection The process by which individuals with certain heritable traits leave more offspring than other individuals specifically due to superiority in competing for mating opportunities.

Shannon-Weaver index A common measurement of species diversity, calculated as $H' = -\Sigma p_i \log(p_i)$, where p_i is the proportion of individuals in the community that belong to species *i*.

shell A hard protective outer structure. In protists, also called a test.

Shine-Dalgarno sequence In bacteria, the sequence at the beginning of an mRNA molecule to which a ribosome binds to initiate translation. Also called the ribosome binding sequence.

shoot The aboveground portion of a young plant, including stem and leaves.

shoot apical meristem A group of undifferentiated plant cells at the tip of a plant stem.

shoot system The aboveground part of a plant.

short-day plant A plant that blooms in response to long nights.

shotgun sequencing A method of sequencing genomes that is based on breaking the genome into small pieces, sequencing each piece separately, and then figuring out how the pieces are connected.

sieve plates The pore-containing structure at one end of a sieve-tube member in plant phloem tissue.

sieve-tube member An elongated sugar-conducting cell in phloem. Has sieve plates at both ends, allowing sap to flow to adjacent cells.

sigma A detachable protein subunit of RNA polymerase that binds to the –35 and –10 boxes to initiate transcription of DNA in prokaryotes.

signal Any information-containing behavior.

signal 1 The first step in activation of a T cell, in which the T cell engulfs an antigen presented by a dendritic cell.

signal 2 The second step in the activation of a T cell, in which additional receptors on the T cell bind to additional MHC-antigen complexes on a dendritic cell.

signal hypothesis The hypothesis that proteins destined for secretion are directed to the rough ER by a certain amino acid sequence at the beginning of the proteins.

signal receptor Any cellular protein that binds to a particular signaling molecule (such as a hormone or neurotransmitter) and triggers a response by the cell, usually by changing conformation or activity upon binding.

signal recognition particle (SRP) A complex of RNA and protein that transports certain newly synthesized proteins to the endoplasmic reticulum.

signal transducers and activators of transcription (STATs) In mammals, a group of regulatory transcription factors that, upon phosphorylation, can activate transcription of certain genes.

signal transduction The process by which a stimulus (e.g., a hormone, a neurotransmitter, or sensory information) outside a cell is translated into a response by the cell.

signal transduction cascade A sequence of events in which one enzyme phosphorylates other enzymes, which in turn phosphorylate many more, ultimately leading to phosphorylation of thousands of proteins. Commonly used in signal transduction of hormone messages. Also called phosphorylation cascade.

signal transduction pathway The exact sequence of molecular events by which signal transduction occurs.

sign stimulus A simple stimulus that elicits an invariant, stereotyped behavioral response (fixed action pattern) from an animal. Also called a releaser.

silencer A regulatory DNA sequence in eukaryotes to which repressor proteins can bind, inhibiting transcription of certain genes.

silent mutation A mutation that does not detectably affect the phenotype of the organism. Typically, a point mutation in the third position of certain codons that does not alter the amino acid coded for.

simian immunodeficiency viruses (SIVs) A family of lentiviruses that infect monkeys and apes and that are thought to be closely related to human immunodeficiency virus (HIV).

simple eye An eye with only one light-collecting apparatus (e.g., one lens), as in vertebrates.

simple sequence repeat A stretch of eukaryotic DNA consisting of repeats of a short, simple sequence that does not code for any protein or RNA.

single nucleotide polymorphism (SNP) A genetic locus where a single base pair often varies between individuals of a certain species. Used as a genetic marker to help track the inheritance of nearby genes.

single-strand DNA-binding proteins A class of proteins that attach to separated strands of DNA during replication or transcription, preventing them from re-forming a double helix.

sink A location where an element or a molecule is consumed or taken out of circulation.

sinoatrial node (SA node) A cluster of heart muscle cells that initiates the heartbeat and determines the heart rate. In the wall of the right atrium.

siphon A tubelike appendage of many mollusks. Often used for feeding or propulsion.

sister chromatids The paired strands of a recently replicated chromosome that has not yet divided.

sister groups or sister taxa Closely related taxa.

skeletal muscle The muscle tissue attached to the bones of the vertebrate skeleton. Consists of long, unbranched muscle fibers with a characteristic striped (striated) appearance; controlled voluntarily. Also called striated muscle.

sliding clamp A doughnut-shaped structure that holds the enzyme DNA polymerase in place during DNA replication.

sliding-filament model The hypothesis that the contraction of muscle cells is caused by filaments of actin and myosin sliding past each other.

slow-twitch fibers A class of muscle fibers that contract relatively slowly, but do not fatigue easily.

slug (1) A member of a certain lineage of terrestrial gastropods, closely related to snails but lacking a shell. (2) A mobile aggregation of cells of a cellular slime mold.

small intestine The first section of the intestine, immediately after the stomach. The site of the final stages of digestion and of most nutrient absorption.

smooth ER (endoplasmic reticulum) Endoplasmic reticulum that does not have ribosomes attached to it. Involved in synthesis and secretion of lipids.

smooth muscle The unstriated muscle tissue that lines the intestine, blood vessels, and some other organs. Consists of tapered, unbranched cells that can sustain long contractions. Not voluntarily controlled.

snRNPs (small nuclear ribonucleoproteins) A complex of proteins and small RNA molecules that catalyze splicing (removal of introns from mRNA). Components of spliceosomes.

sodium-potassium pump A membrane protein that uses the energy of ATP to move sodium ions out of the cell and potassium ions in. Formally known as Na^+/K^+-ATPase.

soil erosion The removal of soil from an area by wind or water.

soil organic matter A mixture of partially and completely decomposed detritus.

solute Any substance that is dissolved in a liquid.

solute potential The potential energy of water caused by a difference in solute concentrations at two locations. Also called osmotic potential.

solution A liquid containing one or more dissolved solids or gases in a homogeneous mixture.

solvent Any liquid in which some substance will dissolve.

soma The part of a neuron that contains the nucleus and where incoming signals are integated. Also called the cell body.

somatic cell Any type of cell that does not pass its genes on to the next generation. In a multicellular organism, all cells except eggs, sperm, and their parent cells (oogonia and spermatogonia) are somatic cells.

somatic hypermutation Rapid DNA mutation that occurs in a somatic cell. Occurs in certain cells of the immune system, such as memory cells that are fine-tuning antibody performance.

somatic nervous system The part of the peripheral nervous system (outside the brain and spinal cord) that controls skeletal muscles and is under voluntary control.

somatostatin A hormone secreted by many organs including the pancreas and hypothalamus, with a wide variety of effects.

somite A block of mesoderm on both sides of the developing spinal cord in a vertebrate embryo. Gives rise to muscle tissue, vertebrae, ribs, limbs, etc.

soredia (singular: soredium) Small reproductive structures produced by lichen. Contain both symbionts of the lichen (the fungus and green alga).

sorus (plural: sori) One of the small dots on the underside of fern fronds. Consists of many sporangia, each of which contains spores.

source A location where a substance is produced or enters circulation (e.g., in plants, the tissue where sugar enters the phloem).

Southern blotting A lab technique for the isolation and analysis of pieces of DNA. Involves cleavage with restriction enzymes, separation with gel electrophoresis, hybridization to a labeled probe, and visualization with autoradiography.

speciation The evolution of two or more distinct species from a single ancestral species.

species A distinct, identifiable group of populations that is thought to be evolutionarily independent of other populations. Generally distinct from other species in appearance, behavior, habitat, ecology, genetic characteristics, etc.

species-area relationship The mathematical relationship between the area of a certain habitat and the number of species that it can support.

species diversity The variety and relative abundance of the species present in a given ecological community.

species richness The number of species present in a given ecological community.

specific heat The amount of energy required to raise the temperature of 1 gram of a substance by 1°C; a measure of the capacity of a substance to absorb energy.

spectral karyotyping (SKY) A technique for producing high-resolution karyotypes by "painting" chromosomes with fluorescent tags that bind to particular regions of certain chromosomes. Also called chromosome painting.

spectrophotometer A lab instrument used to measure the wavelengths of light that are absorbed by a particular pigment.

sperm A mature male gamete, smaller and more mobile than the female gamete.

spermatid An immature sperm cell.

spermatogenesis The production of sperm.

spermatogonia (singular: spermatogonium) The diploid cells in a testis that can give rise to primary spermatocytes.

spermatophore A gelatinous package of sperm cells that is produced by males of species that have internal fertilization without copulation.

sperm competition Competition between sperm from different males to fertilize the eggs of the same female.

sperm nuclei The two haploid nuclei within the pollen grain of a flowering plant. One sperm nucleus fuses with the egg cell to form a zygote; the other fuses with two other nuclei to form triploid endosperm.

sphincter A muscular valve that can close off a tube, as in a blood vessel or a part of the digestive tract.

spicules Stiff spikes of silica or calcium carbonate found in the bodies of sponges.

spindle fibers Groups of microtubules that attach to chromosomes and pull them to opposite sides of the cell during cell division.

spine A modified plant leaf that functions as a sharp protective structure.

spiracle In insects, the small openings that connect air-filled tracheae to the external environment.

spiral cleavage The pattern of embryonic cleavage seen in deuterostomes, in which cells divide at oblique angles to form a spiral coil of cells.

spleen A dark red organ, found near the stomach of most vertebrates, that filters blood, stores extra red blood cells in case of emergency, and plays a role in immunity.

spliceosome In eukaryotes, an organized complex of snRNPs (small nuclear ribonucleoproteins) that catalyzes removal of introns from primary RNA transcripts.

splicing The process by which introns are removed from messenger RNA molecules and the remaining exons are connected together.

spongy mesophyll Rounded parenchyma cells found in the ground tissue of leaves near stomata. The site of most gas exchange.

spontaneous For a chemical reaction, occurring on its own, without any continuous external influences such as added energy.

spontaneous generation The disproven hypothesis that living organisms can develop spontaneously and rapidly from nonliving, noncellular materials under certain conditions.

sporangium (plural: sporangia) A spore-producing structure found in some plants, such as liverworts, and in some fungi, such as chytrids.

spore A single cell produced by mitosis or meiosis (not by cell fusion) that is capable of developing into an adult organism.

sporophyte The multicellular diploid phase of a species that exhibits alternation of generations; arises from two fused gametes; produces spores.

sporopollenin A watertight material that encases spores and pollen of modern land plants.

stabilizing selection A pattern of natural selection in which individuals with an average phenotype have higher fitness than those with extreme phenotypes.

stamen The male reproductive structure of a flower. Consists of an anther, which produces pollen grains, and a filament, which supports the anther.

standing defense Any defensive mechanism that is always present, regardless of need. Also called constitutive defense.

stapes A stirrup-shaped bone in the middle ear of vertebrates. Receives vibrations from the tympanic membrane and passes them to the cochlea.

starch A mixture of polysaccharides amylose and amylopectin; used primarily for food storage in plants. Both are helical polymers of α-glucose subunits, branched in amylopectin.

start codon The mRNA sequence AUG, which induces the beginning of protein synthesis and codes for the amino acid methionine.

statocyst A sensory organ of many arthropods that detects the animal's orientation in space (i.e., whether the animal is flipped upside down).

statolith A tiny stone or dense particle that can be used to sense gravity.

statolith hypothesis The hypothesis that amyloplasts (dense, starch-storing plant organelles) serve as statoliths (i.e., gravity detectors).

STATs See **signal transducers and activators of transcription (STATs)**.

stem A vertical aboveground part of a plant, usually bearing leaves, fruit, or flowers.

stem cell Undifferentiated cells that have the potential to give rise to any tissue type.

steppe An extensive grassland. Typically found in the dry interiors of continents in the temperate latitudes. Also called a prairie.

stereocilia (singular: stereocilium) Stiff outgrowths from the surface of a hair cell that is involved in detection of sound by terrestrial vertebrates or of waterborne vibrations by fishes.

steroids A class of lipids with a characteristic four-ring structure.

sticky ends The short, single-stranded ends of a DNA molecule cut by a restriction endonuclease. Tend to form hydrogen bonds with other sticky ends that have complementary sequences.

stigma The moist tip at the end of a flower carpel, to which pollen grains adhere.

stolon Modified stems that run horizontally over the soil surface.

stomata (singular: stoma) Generally, pores or openings. In plants, microscopic pores on the surface of a leaf or stem, through which gas exchange occurs for photosynthesis.

stomach A tough, muscular pouch in the digestive tract that breaks up food and delivers it to the intestine.

stop codon One of three messenger RNA triplets (UAG, UGA, or UAA) that cause termination of protein synthesis. Also called a termination codon.

strain A population of genetically similar or identical individuals.

striated muscle The muscle tissue attached to the bones of the vertebrate skeleton. Consists of long, unbranched muscle fibers with a characteristic striped (striated) appearance; controlled voluntarily. Also called skeletal muscle.

stroke An episode of oxygen deprivation to part of the brain that results in death of some neurons; usually due to a blood clot that blocks blood vessels of the brain.

stroke volume The volume of blood ejected from the left ventricle of the heart in one contraction.

stroma The fluid between a chloroplast's outer membrane and its thylakoid disks.

structural formula A graphical representation of the atoms and bonds in a molecule, often with covalent bonds represented by straight lines.

structural gene A stretch of DNA that codes for a functional protein or functional RNA molecule—i.e., not a promoter, enhancer, etc.

structural homology Similarities in organismal structures (e.g., limbs, shells, flowers) that are due to inheritance from a common ancestor.

structural isomer A molecule that shares the same molecular formula as another molecule but differs in the order in which covalently bonded atoms are attached.

style The slender stalk of a flower carpel, connecting the stigma and the ovary.

suberin A water-repellent compound that is a major component of the waxy Casparian strip in plant roots.

subspecies A population that has distinctive traits and some genetic differences relative to other populations of the same species but that is not distinct enough to be called a separate species.

substrate (1) A reactant that interacts with an enzyme in a chemical reaction. (2) A surface on which a cell or organism sits.

substrate-level phosphorylation Production of ATP molecules via transfer of a phosphate group from an intermediate substrate directly to ADP, unmediated by an electron transport chain. Occurs in glycolysis and in the Krebs cycle.

subzonal injection (SUZI) A technique used to help infertile couples bear children by selecting a few healthy sperm and injecting them directly into the zona pellucida of the woman's egg.

succession Gradual colonization of a habitat after an environmental disturbance (e.g., fire, flood), usually by a series of species assemblages.

successional pathway The specific sequence of species that appears over a period of time in an environment undergoing primary succession or secondary succession.

sugars A class of small, water-soluble organic compounds containing a carbonyl (–C=O) group and several hydroxyl (–OH) groups. Include monosaccharides and disaccharides.

sulfate-reducers A group of archaea that produce hydrogen sulfide (H_2S) as a by-product of cellular respiration.

summation The additive effect of different postsynaptic potentials at a nerve or muscle cell, such that several subthreshold stimulations can cause an action potential.

supernatant The liquid above a layer of solid particles (the pellet) in a tube after centrifugation.

surface tension The attractive force between liquid molecules that causes the liquid to form a rounded surface at an air-liquid interface.

surfactant A mixture of phospholipids and proteins produced by lung cells that reduces surface tension, allowing the lungs to expand more.

survivorship The average proportion of offspring produced that survive to a particular age in a certain population.

survivorship curve A graph depicting the percentage of a population that survives to different ages.

suspension culture A population of cells grown in the lab in a liquid flask of nutrients that is rotated continuously to keep the cells suspended and the nutrients mixed.

suspension feeder Any organism that obtains food by filtering small particles or small organisms out of water or air. Also called filter feeder.

sustainability The planned use of environmental resources at a rate no faster than the rate at which they are naturally replaced.

sustainable agriculture Agricultural techniques that are designed to maintain long-term soil quality and productivity.

sustainable development Economic development based on sustainability (i.e., that uses natural resources no faster than they are naturally replaced).

swamp A wetland that has a steady rate of water flow and is dominated by trees and shrubs.

swim bladder A gas-filled organ of many ray-finned fishes that regulates buoyancy.

symbiosis Any close and prolonged physical relationship between individuals of two different species. May be mutualistic, parasitic, or commensal.

symmetric competition Ecological competition between two species in which both suffer similar declines in fitness.

sympathetic nervous system The part of the autonomic nervous system that stimulates fight-or-flight responses, such as increased heart rate, increased blood pressure, and decreased digestion.

sympatric speciation Speciation that occurs while the two diverging species are living in the same area.

sympatry Living in the same geographic area as some other population.

symplast A pathway for water transport in plants roots through the cytoplasm of adjacent cells that are connected by plasmodesmata.

synapomorphy A shared, derived (altered from the ancestral state) character that can be used to infer evolutionary relationships.

synapse The connection between two neurons, or between a neuron and a muscle cell: a tiny space into which neurotransmitters are released.

synapsis The physical pairing of two homologous chromosomes during prophase I of meiosis. Crossing over occurs during synapsis.

synaptic cleft The gap of a synapse; the space between two communicating nerve cells, across which neurotransmitters diffuse.

synaptic plasticity Long-term changes in the responsiveness or physical structure of a synapse that can occur after particular stimulation patterns. Thought to be the basis of learning and memory.

synaptic vesicles Tiny neurotransmitter-containing vesicles at the end of an axon. Can fuse with the axon membrane to release a neurotransmitter into a synapse, stimulating the next neuron or effector cell.

synaptonemal complex A network of proteins that hold non-sister chromatids together during synapsis in meiosis.

syndrome A group of medical symptoms that often occur together and that are suspected to have the same underlying cause.

syngamy Fusion of two gametes to create a diploid zygote. In multicellular species, also called fertilization.

synthesis phase (S phase) The phase of the cell cycle during which DNA is synthesized and chromosomes are duplicated.

system A set of interacting elements, such as (1) a set of reactants and products at chemical equilibrium and (2) a group of organs that work together to perform a function.

systemic acquired resistance (SAR) A set of events through which plant tissues prepare to combat a possible infection.

systemic circulation The part of the circulatory system that sends oxygen-rich blood from the lungs out to the rest of the body. In mammals and birds, separate from the pulmonary circulation.

systemin A peptide hormone, produced by plant cells damaged by herbivores, that initiates a protective response in undamaged cells.

systole The portion of the heartbeat cycle during which the heart muscles are contracting.

systolic blood pressure Blood pressure in arteries during ventricular systole (heart contraction).

taiga A vast forest biome throughout subarctic regions, consisting primarily of short conifer trees. Characterized by intensely cold winters, short summers, and high annual variation in temperature.

taproot A large vertical main root of a plant.

***Taq* polymerase** A DNA polymerase commonly used in PCR due to its stability at high temperatures. Derived from *Thermus aquaticus*, a bacterium found in hot springs.

taste buds Sensory structures, found chiefly in the mammalian tongue, that are responsible for the sense of taste.

taste cells Spindle-shaped cells found in a taste bud of the mammalian tongue. Respond to certain chemical stimuli.

TATA box A DNA sequence in many eukaryotic promoters 30 base pairs upstream from the transcription start site. Recognized by RNA polymerase II.

taxis Movement toward or away from some external cue.

taxon (plural: taxa) Any named group of species at any level of a classification system.

taxonomy The branch of biology concerned with the classification and naming of organisms.

TBP (TATA-binding protein) A protein that binds to eukaryotic promoters and helps initiate transcription.

T-cell receptor (TCR) A T-cell membrane protein that can bind to antigens displayed on the surfaces of other cells.

T cells Lymphocytes that mature in the thymus and are involved in acquired immunity in vertebrates.

May be involved in activation of other lymphocytes (helper T cells) or destruction of infected cells (cytotoxic T cells). Also called T lymphocytes.

tectorial membrane A membrane in the vertebrate cochlea that takes part in the transduction of sound by bending the stereocilia of hair cells in response to sonic vibrations.

telomerase An enzyme that replicates telomeres by catalyzing DNA synthesis from an RNA template.

telomere The region at the end of a linear chromosome.

telophase The final stage in cell division (mitosis or meiosis), during which chromosomes finish moving and new nuclear envelopes begin to form around each set of daughter chromosomes.

temperate Having a climate with pronounced annual fluctuations in temperature (i.e., warm summers and cold winters) but typically neither as hot as the tropics nor as cold as the poles.

template strand (1) The strand of DNA that is transcribed by RNA polymerase to create RNA. (2) An original strand of RNA used to make a complementary strand of RNA.

temporal lobe In the mammalian brain, the region of each cerebral hemisphere near the ears. Functions in memory, interpretation of information from the ears, and, in humans, language.

tendon A band of tough, fibrous connective tissue that connects a muscle to a bone.

tension Any pulling force; the opposite of compression.

tentacle A long, thin, muscular appendage of gastropod mollusks.

terminal cell The upper cell produced by an angiosperm zygote. Gives rise to the plant embryo.

termination (1) In enzyme-catalyzed reactions, the final stage in which the enzyme returns to its original conformation and products are released. (2) In DNA transcription, the dissociation of RNA polymerase from DNA. (3) In RNA translation, the dissociation of ribosomes from mRNA.

territory An area that is actively defended by an animal from others of its species.

tertiary structure The overall three-dimensional shape of a single polypeptide chain, created by a variety of interactions among R-groups and the peptide backbone.

test A hard protective outer structure seen in some protists. Also called a shell.

testcross The breeding of an individual of unknown genotype with an individual having only recessive alleles for the traits of interest, in order to infer the unknown genotype from the phenotypic ratios seen in offspring.

testis (plural: testes) The sperm-producing organ of a male animal.

testosterone A steroid hormone from the testes that stimulates sperm production and various male traits and reproductive behaviors.

tetrad A pair of synapsed homologous chromosomes, each containing two chromatids.

tetraploid With four sets of chromosomes ($n = 4$).

tetrapod Any member of the taxon Tetrapoda, including all descendants of the first four-footed animals to move onto the land.

texture In soil biology, the proportions of different-sized particles present in soil.

theory A proposed explanation for a very general class of phenomena or observations.

thermal energy The kinetic energy of molecular motion.

thermocline A gradient (cline) in environmental temperature across a large geographic area.

thermophiles Bacteria or archaea that thrive in very hot environments.

thermoreceptor A sensory cell or an organ specialized for detection of changes in temperature.

thermoregulation Regulation of body temperature.

thick filaments Strands of myosin found in the middle of a muscle sarcomere. Bind to thin filaments (actin) and cause muscle contraction.

thin filaments Strands of actin found at the two ends of a muscle sarcomere. Bind to thick filaments (myosin) during muscle contraction.

thorax In mammals, the anterior region of the torso, containing the lungs. In insects, the middle of the three major body regions.

thorn A modified plant stem shaped as a sharp protective structure.

threshold The membrane potential at which a neuron's voltage-gated sodium channels will trigger an action potential.

thylakoid A flattened, membrane-bound disk inside a plant chloroplast. A stack of thylakoids is a granum.

thymus An organ, located in the anterior chest or neck of vertebrates, that, in young animals, processes T cells and sends them to lymph nodes.

thyroid gland A gland in the neck that releases thyroid hormone (which increases metabolic rate) and calcitonin (which lowers blood calcium).

thyroid-stimulating hormone (TSH) A peptide hormone from the pituitary gland that stimulates release of thyroid hormones from the thyroid gland.

thyroxine An iodine-containing peptide hormone from the thyroid gland that increases metabolic rate, both directly and via conversion to the more active hormone triiodothyronine.

tight junction A physical attachment between two adjacent animal cells, consisting of proteins that "stitch" the cells' plasma membranes together.

tip The end of a branch on a phylogenetic tree. Represents a specific population or species that has not (yet) produced descendants—either a group living today or a group that ended in extinction.

Ti plasmid A plasmid carried by *Agrobacterium* (a bacterium that infects plants) that can integrate into the plant cell's chromosomes and induce formation of a gall.

tissue A group of similar cells that function as a unit, such as muscle tissue or epithelial tissue.

tissue culture A collection of cells of a certain tissue type grown in a lab, typically in liquid suspension or in a petri dish on a solid food medium.

tolerance Ecological succession in which early-arriving species do not affect the probability that subsequent species will become established.

tonoplast The membrane surrounding a plant vacuole.

top-down control The limitation of herbivore population size by predation or disease rather than by limited or toxic nutritional resources.

topoisomerase An enzyme that cuts and rejoins DNA downstream of the replication fork, to ease the twisting that would otherwise occur as the DNA "unzips."

torpor A regulated, "deliberate" decrease in metabolic rate and body temperature in an endotherm.

totipotent Capable of dividing and developing to form a complete, mature organism.

tracheae (singular: trachea) A system of small air-filled tubes that extends throughout an insect's body and functions in gas exchange.

tracheid An elongated water-conducting plant cell that has gaps (pits) in its secondary cell wall to allow water movement from one cell to the next.

trade-off An inescapable compromise between two traits that cannot be optimized simultaneously.

trait Any characteristic of an individual.

transcription The process by which messenger RNA is made from a DNA template.

transcriptional control Regulation of gene expression via changes in the rate at which genes are transcribed to form messenger RNA.

transcription termination signal A DNA sequence that can terminate messenger RNA synthesis, usually because the RNA transcribed from this sequence forms a hairpin that prevents further transcription.

transduction Conversion of information from one mode to another. For example, the process by which a stimulus outside a cell is translated into a response by the cell.

transfer cells Cells that transfer nutrients from a parent plant to a developing plant seed. Occur in land plants.

transfer RNAs (tRNAs) A class of RNA molecules with an anticodon at one end and an amino acid binding site at the other. Match amino acids to messenger RNA codons during translation.

transformation Incorporation of DNA obtained directly from the environment into the genome. Occurs naturally in some bacteria; can be induced in the lab by certain processes.

transgenic Containing DNA that has been modified by genetic engineering.

transitional form A fossil species or population with traits that are intermediate between older and younger species.

transition state A high-energy intermediate state occurring during a chemical reaction that determines the activation energy necessary for the reaction to proceed.

translation The process by which proteins and peptides are synthesized from messenger RNA.

translational control The regulation of gene expression by altering the life span of messenger RNA or the efficiency of translation. A type of post-transcriptional regulation.

translocation (1) The movement of sugars through a plant by bulk flow. (2) A type of mutation in which a piece of a chromosome moves to a nonhomologous chromosome. (3) The process by which a ribosome moves down a messenger RNA molecule during translation.

transmembrane domain A structure that anchors a B-cell receptor molecule to the plasma membrane of the B cell.

transmembrane protein Any membrane protein that spans the entire membrane. Also called integral membrane protein.

transmission electron microscope (TEM) A microscope that forms an image from electrons that pass through a specimen.

transmission genetics The study of the genotypic and phenotypic patterns that occur as genes pass from one generation to the next.

transpiration Water loss from aboveground plant parts. Occurs primarily through the stomata.

transport protein Any membrane protein that enables specific molecules to cross plasma membranes, sometimes by causing a conformational change in the protein. Also called transporter.

transposable elements Any of several kinds of parasitic DNA sequences that are capable of moving themselves, or copies of themselves, to other locations in the genome.

tree of life A diagram depicting the genealogical relationships of all living organisms on Earth, with a single ancestral species at the base.

triacylglycerols Lipids consisting of three fatty acids joined to a glycerol molecule. Also called triglycerides or fats.

trichomes Protective hairlike appendages of the epidermal cells of some plants.

triglycerides Lipids consisting of three fatty acids joined to a glycerol molecule. Also called triacylglycerols or fats.

triiodothyronine An iodine-containing peptide hormone, secreted by the thyroid gland, that increases metabolic rate. Has a stronger effect than does the related hormone thyroxine.

trilobite A member of an extinct lineage of arthropods that were abundant from 550–440 million years ago.

trimester In humans, one of three major stages of gestation, each three months long.

triose A monosaccharide (simple sugar) that has three carbons.

triplet code A code in which a "word" of three letters encodes one piece of information. The genetic code is a triplet code because a codon is three nucleotides long and encodes one amino acid.

triploblast An animal whose body develops from three basic embryonic cell layers: ectoderm, mesoderm, and endoderm. All animals except sponges, cnidarians, and ctenophores are triploblasts.

triploid With three sets of chromosomes ($n = 3$).

trisomy The state of having three copies of one particular type of chromosome.

tRNAs See **transfer RNAs (tRNAs)**.

trochophore A type of larva that has a ring of cilia around its middle. Occurs in several phyla of the Lophotrochozoa lineage of protostomes.

trophic level A feeding level in an ecosystem.

tropomyosin A muscle protein that blocks the myosin-binding sites on actin filaments, preventing muscle contraction. Can be moved out of the way by troponin when intracellular calcium is high.

troponin A muscle protein that can trigger muscle contraction by moving tropomyosin off the myosin-binding sites on actin filaments. Activated by high intracellular calcium.

***trp* operon** An *E. coli* operon that includes five cotranscribed genes involved in the synthesis of the amino acid tryptophan.

true navigation Navigation by which an animal can reach a specific point on Earth's surface.

trypsin A protein-digesting enzyme produced by the pancreas, secreted into the intestine, and activated by enterokinase. Trypsin in turn activates several other protein-digesting enzymes.

trypsinogen The precursor of protein-digesting enzyme trypsin. Secreted by the pancreas and activated by the intestinal enzyme enterokinase.

T tubules Membranous tubes that extend into the interior of muscle fibers. Propagate action potentials throughout the muscle cell and trigger release of calcium from the sarcoplasmic reticulum.

tube cell A large cell in a male gametophyte that will give rise to the pollen tube.

tube feet Small, mobile, fluid-filled extensions of the water vascular system of echinoderms. Used in locomotion.

tuber Plant rhizomes that are modified to function as carbohydrate-storage organs.

tuberculosis (TB) A highly contagious human disease caused by the bacterium *Mycobacterium tuberculosis*.

tube-within-a-tube Describing the basic body plan of all triploblast animals (i.e., an inner tube of endoderm within an outer tube of ectoderm, with mesoderm between the two tubes).

tumor A mass of cells formed by uncontrolled cell division. Can be benign or malignant.

tumor suppressor A gene that prevents cell division, particularly when the cell has DNA damage. Mutated forms are associated with cancer. Also, the protein produced by such a gene.

tundra The treeless biome in polar and alpine regions, characterized by short, slow-growing vegetation, permafrost, and a climate of long, intensely cold winters and very short summers.

turgid Refers to a plant cell that is firm (i.e., containing enough water for the cell cytoplasm to press against the cell wall).

turgor pressure The outward pressure exerted by the fluid contents of a plant cell against its cell wall.

Turner syndrome A human syndrome caused by the presence of only one X chromosome and no Y chromosome ("XO"). Individuals with this condition are female but sterile.

turnover In lake ecology, the complete mixing of upper and lower layers of water that occurs each spring and fall in temperate-zone lakes.

two-fold rotational symmetry A type of symmetry in which an object can be superimposed on itself if rotated 180°. Occurs in some regulatory sequences of DNA. Also called dyad symmetry.

tympanic membrane The eardrum; a membrane separating the middle ear from the outer ear in terrestrial vertebrates, or similar structures in insects.

type I diabetes mellitus Diabetes mellitus that is caused by insufficient secretion of insulin from the pancreas.

type II diabetes mellitus Diabetes mellitus that is caused primarily by insufficient responsiveness of tissues to insulin, despite normal secretion of insulin from the pancreas.

typhoid fever A human disease caused by the bacterium *Salmonella typhi*.

ubiquinone A molecule that transfers electrons in the electron transport chain of cellular respiration in mitochondria. Also called coenzyme Q or Q.

ulcer A hole or thin spot in the stomach or intestinal wall.

ultimate causation In biology, the reason that a trait or phenomenon is thought to have evolved; the adaptive advantage of that trait. Also called ultimate explanation.

ultrasound Sound frequencies that are too high for humans to hear—higher than about 20,000 Hz (20 kHz).

ultraviolet light Light with a wavelength shorter than visible blue light.

umami The taste of glutamate, responsible for the "meaty" taste of most proteins and of monosodium glutamate.

umbilical cord The cord that connects a developing mammal embryo or fetus to the placenta and through which the embryo or fetus receives oxygen and nutrients.

undershoot The brief phase after an action potential when a cell's membrane potential temporarily becomes more negative than the resting potential.

unequal crossover An error in crossing over during meiosis I in which the two chromatids match up at different sites. Results in gene duplication and gene loss in the two resulting chromatids.

uniformitarianism The concept that the laws of the physical universe are constant throughout time and space.

unsaturated fat A fat with one or more carbon-carbon double bonds in its fatty acid chains and thus fewer than the maximum number of hydrogen atoms. Double bonds produce kinks in the fatty acid chains and decrease the compound's melting point.

upstream Opposite to the direction in which RNA polymerase moves along a DNA strand.

urea A water-soluble excretory product of mammals and sharks, used to excrete excess nitrogen from amino acids.

ureter In mammals, a tube that transports urine from one kidney to the bladder.

urethra The tube that drains urine from the bladder to the outside environment. In male vertebrates, also used for passage of sperm during ejaculation.

uric acid A whitish excretory product of birds, reptiles, and terrestrial arthropods, used to get rid of excess nitrogen derived from amino acids.

urochordates The tunicates or sea squirts. A lineage of sessile, filter-feeding chordates.

uterus The organ in which developing embryos are housed in those vertebrates that give live birth. Common in most mammals and in some lizards, sharks, and other vertebrates.

vaccination Injection with weakened, killed, or altered pathogens to stimulate development of immunity against those pathogens.

vaccine Preparation containing pieces of a pathogen, or entire killed or weakened pathogens, administered to stimulate immunity without causing illness.

vacuole An organelle usually used for bulk storage of substances such as pigments, oils, carbohydrates, water, or toxins. In animal cells, these organelles are smaller and are called lysosomes.

vagina The birth canal of female mammals; a muscular tube that extends from the uterus through the pelvis to the exterior.

valence The number of electrons in the valence (outermost) electron shell of an atom; determines how many covalent bonds the atom can form.

valence electron An electron in the valence (outermost) electron shell of an atom. Valence electrons tend to be involved in chemical bonding.

valence shell The outermost electron shell of an atom.

valves In circulatory systems, flaps of tissue that prevent backward flow of blood, particularly in veins and between the chambers of the heart.

van der Waals interaction A weak electrical attraction between two hydrophobic side chains. Often contributes to tertiary structure in proteins.

variable (V) region A section of the light chains of antibodies that has a highly variable amino acid sequence, unique to each B cell.

vasa recta A network of blood vessels that runs alongside the loop of Henle of a kidney nephron, reabsorbing water and solutes.

vascular bundle A cluster of xylem and phloem strands in a plant stem.

vascular cambium A ring of undifferentiated plant cells inside the cork cambium of woody plants. Produces secondary xylem and secondary phloem.

vascular tissue In plants, a tissue that is involved in conducting water or solutes from one part of a plant to another, or that gives rise to such tissues.

vas deferens A pair of muscular tubes that store and transport semen from the epididymus to the ejaculatory duct. In nonhuman animals, called the ductus deferens.

vector (1) A plasmid or other vehicle used to transfer recombinant genes to a new host. (2) A biting insect or other organism that transfers pathogens between two other species.

vegetal hemisphere The lower half of an amphibian egg cell, containing most of the yolk. This portion eventually becomes part of the gut.

vein (1) Any blood vessel that carries blood from capillaries to the heart (oxygenated or not). Has thinner walls and lower blood pressure than an artery. (2) A strip of vascular tissue in a plant leaf. (3) A supporting filament of an insect wing.

veliger A unique form of larva in bivalves.

venae cavae (singular: vena cava) Large veins that return oxygen-poor blood to the heart.

ventilation Movement of air or water through the lungs or gills.

ventral Toward an animal's belly and away from its back. The opposite of dorsal.

ventricle (1) A thick-walled chamber of the heart that receives blood from an atrium and pumps it to the body or to the lungs. (2) One of several small fluid-filled chambers in the vertebrate brain.

vertebra (plural: vertebrae) One of the cartilaginous or bony elements that form the spine of vertebrate animals.

vertebrate An animal in the lineage Vertebrata, characterized by a skull, usually a spinal column, usually an endoskeleton of bone, and other traits.

vessel element An elongated water-conducting plant cell found in the xylem of certain advanced plants. Has gaps through both the primary and secondary cell walls, allowing unimpeded passage of water from one cell to the next.

vestigial trait Any rudimentary structure of unknown or minimal function that is homologous to functioning structures in other species. Vestigial traits are thought to reflect evolutionary history.

vicariance The physical splitting of a population into smaller, isolated populations by a geographic barrier.

villi (singular: villus) Small, fingerlike projections of the lining of the digestive tract. Function to increase surface area for absorption.

virion A single mature virus particle.

virulence The ability of a pathogen or parasite to cause disease and death.

virulent Tending to cause severe disease rather than mild disease.

virus A tiny infectious parasitic entity consisting of DNA or RNA enclosed in a protective covering (a capsid and sometimes an envelope). The DNA or RNA contain the necessary instructions to make more viruses but must use the machinery of a host cell to do so.

visceral mass The part of a mollusk containing most of the internal organs and external gill.

visible light The range of wavelengths of electromagnetic radiation that humans can see, from about 400 to 700 nanometers.

vitamin Any micronutrient that is a carbon-containing compound rather than a single chemical element. Usually functions as a coenzyme.

vitelline envelope A fibrous sheet of glycoproteins that surrounds a mature egg cell. Found in many vertebrates.

viviparous For animals, reproducing by live birth rather than by laying eggs.

volt A unit of voltage (electrical potential).

voltage Potential energy created by a separation of electric charges between two points. Also called electrical potential.

voltage clamping A lab technique for imposing a certain constant membrane potential on a cell. Widely used to investigate ion channels.

voltage-gated channel An ion channel that opens or closes in response to changes in membrane voltage.

wall pressure The inward pressure exerted by a cell wall against the fluid contents of a plant cell.

water potential The potential energy of water in a certain environment. In living organisms, the sum of solute potential and pressure potential. Determines movement of water into or out of cells.

water potential gradient A difference in water potential in one location compared with another. Determines the movement of water through plant tissues.

watershed The area drained by a single stream or river.

water table The upper limit of the underground layer of soil that is saturated with water.

water vascular system A system of fluid-filled tubes and chambers in echinoderms. Functions as a hydrostatic skeleton.

wavelength The distance between two successive wave crests in any regular wave, such as light waves, sound waves, or waves in water.

wax A type of lipid with long hydrocarbon tails, usually combinations of long-chain alcohols with fatty acids. Harder and less greasy than fats.

weather The specific short-term atmospheric conditions of temperature, moisture, sunlight, and wind in a certain area.

weathering The gradual wearing down of large rocks by rain, running water, and wind; one of the processes that transform rocks into soil.

weed A plant that is adapted for growth in disturbed soils.

Western blotting A lab technique for identifying proteins that will bind to a certain antibody. Involves separation of proteins with gel electrophoresis, transfer to a filter, and probing with a labeled antibody.

wetland A shallow-water habitat where the soil is saturated with water for at least part of the year.

white blood cells Immune system cells that circulate in the blood or lymph and function in the defense against disease. Also called leukocytes.

wild type The most common phenotype seen in a population; especially the most common phenotype in wild populations compared with inbred lab strains of the same species.

wilt To lose turgor, in a plant tissue.

wobble hypothesis The hypothesis that some tRNA molecules can pair with more than one codon of mRNA, tolerating some variation in the third base, as long as the first and second bases are correctly matched.

wood The secondary xylem of older woody stems and roots.

xanthophylls Carotenoid pigments found in many algae and some plants, typically appearing yellow.

xenotransplantation or xenografting The transplantation of tissues from one species to another.

xeroderma pigmentosum (XP) A human disease characterized by extreme sensitivity to ultraviolet light. Caused by an autosomal recessive allele that results in a defective DNA repair system.

X-linked inheritance Inheritance resulting from a gene being located on the mammalian X chromosome. Also called X-linkage.

X-ray crystallography A lab technique used to infer the three-dimensional structure of a molecule, on the basis of the diffraction patterns produced by X-rays beamed at the crystallized molecule.

xylem A plant vascular tissue that conducts water and ions; contains tracheids and/or vessel elements.

xylem sap The watery fluid found in xylem tissue of plants.

yeast Any fungus growing as a single-celled form. Also, a specific lineage of ascomycetes.

Y-linked inheritance Inheritance resulting from a gene being located on the mammalian Y chromosome. Also called Y-linkage.

yolk The nutrient-rich cytoplasm inside an egg cell, used as food for the growing embryo.

yolk sac In an amniotic egg, the membrane-bound sac that contains the yolk.

zeaxanthin A carotenoid pigment of plants that initiates the opening of stomata in response to blue light so that carbon dioxide can diffuse into photosynthesizing cells.

zero population growth (ZPG) A state of stable population size due to fertility staying at the replacement rate for at least a generation.

zona pellucida The gelatinous layer around the egg cell of a mammal.

zone of cellular division A group of plant cells just behind the root cap. Contains the apical meristem, where cells are actively dividing.

zone of cellular elongation A group of plant cells behind the apical meristem in plant roots. Consists of young cells that are increasing in length.

zone of cellular maturation A group of plant cells several centimeters behind the root cap that are differentiating into mature tissues.

Z scheme A widely accepted model for the passage of electrons from photosystem II to photosystem I during photosynthesis. Electrons on a graph of energy level versus location trace a Z shape.

zygosporangium (plural: zygosporangia) The spore-producing structure in fungi that are members of the Zygomycota.

zygote The diploid cell formed by the union of two haploid gametes. Capable of undergoing embryological development to form an adult.

Image Credits

Frontmatter

xv Jacob Halaska/Index Stock Imagery, Inc. **xvii** ©E.H. Newcomb & W.P. Wergin/Biological Photo Service **xviii** Dr. Gopal Murti/Science Photo Library/Photo Researchers, Inc. **xxi** Kent & Donna Dannen/Photo Researchers, Inc. **xxii** ©Beth Donidow/Visuals Unlimited **xxv** David Nunuk/Science Photo Library/Photo Researchers, Inc. **xxvi** ©Rob Nunnington/Foto Natura/Minden Pictures **xxix** ©Thomas Mangelsen/Minden Pictures **xxx** ©J.P. Ferrero/Peter Arnold **53.1** ©Christian Ziegler

Chapter 1

Opener ©Frans Lanting/Minden Pictures**1.1a** Burndy Library/Omikron/Photo Researchers, Inc. **1.1b** M.I. Walker/Photo Researchers, Inc. **1.3a** Photo by Kelly Buono/Courtesy of Dr. Richard Amasino, University of Wisconsin **1.3c** Bruce Forster/Getty Images Inc. - Stone Allstock **1.5a** Samuel F. Conti and Thomas D. Brock **1.5b** Kwangshin Kim/Photo Researchers, Inc. **1.6/1** ©Dr. David Phillips/Visuals Unlimited **1.6/2** Dennis Kunkel/Dennis Kunkel Microscopy, Inc. **1.6/3** Kolar, Richard/Animals Animals/Earth Scenes **1.6/4** Biophoto Associates/Photo Researchers, Inc. **1.6/5** Darwin Dale/Photo Researchers, Inc. **1.9b** Michael Hughes/Aurora & Quanta Productions Inc. **1.10a** Joshua J. Tewksbury **1.10b** William Weber/Visuals Unlimited **1.10c** Robert Dobbs **1.11** Data from Tewksbury and Nabhan. 2001. *Nature* 412: 403–404. Fig. 1a. **1.12** Data from Tewksbury and Nabhan. 2001. *Nature* 412: 403–404. Fig. 1b.

Unit 1

©Eric Meola/Image Bank/Getty Images

Chapter 2

Opener Science Photo Library/Photo Researchers, Inc. **2.1b** Mitchell Layton/Duomo Photography Incorporated **2.2** US Department of Energy/SPL/Photo Researchers, Inc. **2.4b** Minik Rosing, Geological Museum, Copenhagen **2.9c** Albert Copley/Visuals Unlimited **2.14a** P.W. Lipman/U.S. Geological Survey/U.S. Department of the Interior **2.14b** Richard Megna/Fundamental Photographs **2.17** Don Farrall/Getty Images Inc. - PhotoDisc **2.24** Robert and Beth Plowes Photography **2.27c** Geostock/Getty Images Inc. - PhotoDisc **2.29a** TSADO/NASA/Tom Stack & Associates, Inc. **2.29b** Jet Propulsion Laboratory/NASA Headquarters **T2.2** Source: Table 8.1, p. 312, in John McMurry and Robert C. Fay, *Chemistry*, 4th Edition, ©2004. Reprinted by permission of Pearson Education, Inc., Upper Saddle River, NJ

Chapter 3

Opener Jacob Halaska/Index Stock Imagery, Inc. **3.2** NOAA Vents Program **3.3** Photo courtesy R. Kempton, New England Meteoritical Services **3.6** Martin Bough/Fundamental Photographs **3.7b** Clare Sansom, Birkbeck College, University of London, London, England **3.12** Clare Sansom, Birkbeck College, University of London, London, England **3.13a** ©Microworks/Phototake **3.13b** ©Walter Reinhart/Phototake **3.15b** Clare Sansom, Birkbeck College, University of London, London, England **3.16** Clare Sansom, Birkbeck College, University of London, London, England **3.22** T.A. Steitz, Yale University **3.25** T.A. Steitz, Yale University **3.27a** Data from Nawani and Kapadnis. 2001. *Journal of Applied Microbiology* 90: 803–808. Fig. 3. Data also from Nawani et al. 2002. *Journal of Applied Microbiology* 93: 965–975. Fig. 7. **3.27b** Data from Hansen et al. 2002. *FEMS Microbiology Letters* 216: 249–253. Fig. 1. **3.29** ©Paul Fievez/Hulton Archive/Getty Images

Chapter 4

Opener Micheal Simpson/Getty Images, Inc. - Taxi **4.6** Reproduced by permission from J.P. Ferris et al., Synthesis of long prebiotic oligomers on mineral surfaces. Nature 381:59–61 (1996), Fig. 2. Copyright ©1996 Macmillan Magazines Limited. Image courtesy of James P. Ferris, Rensselaer Polytechnic Institute. **4.7** Omikron/Photo Researchers, Inc. **4.8** A. Barrington Brown/Science Source/Photo Researchers, Inc. **4.15** Reprinted with permission from *Science* 292: 1319–1325 Fig. 4B (2001) by Wendy K. Johnston, Peter J. Unrau, Michael S. Lawrence, Margaret E. Glasner, David P. Bartel "RNA-Catalyzed RNA Polymerization: Accurate and General RNA-Templated Primer Extension."

Chapter 5

Opener Dr. Jeremy Burgess/Photo Researchers, Inc. **5.8a** ©Biophoto Associates/ Photo Researchers, Inc. **5.8b** Dr. Jacob S. Ishay, and Dr. Eyal Rosenzweig. ©2000 E. Rosenzweig **5.8c** M. Jericho, Dept of Physics, Dalhousie University, Halifax, Nova Scotia and T.J. Beveridge, Dept of Microbiology, University of Guelph, Guelph, Ontario

Chapter 6

Opener Kit Pogliano and Marc Sharp, University of California at San Diego **6.1aL** JEOL USA Inc. **6.1aR** Photo courtesy of Peter M. O'Day, Juan Bacigalupo, Joan E. Haab, and Cecilia Vergara. The Journal of Neuroscience, October 1, 2000. 20(19): 7193–7198, Fig. 1C. ©2004 by the Society for Neuroscience. **6.1bL** SEM photo courtesy of JEOL-USA, Peabody, MA **6.1bR** ©Dennis Kunkel/Phototake **6.2a** Alec D. Bangham, M.D., F.R.S. **6.2b** Fred Hossler/Visuals Unlimited **6.8aL** James J. Cheetham, Carleton University **6.12a** ©Dorling Kindersley **6.12b** Clive Streeter ©Dorling Kindersley **6.12c** Phil Degginger/Color-Pic, Inc. **6.20** Don W. Fawcett/Photo Researchers, Inc. **6.25b** Kovacs, F., Quine, J. & Cross, T.A. (1999) "Validation of the Single-Stranded Channel Formation of Gramicidin A by Solid-State NMR" *Proc. Natl. Acad. Sci. U.S.A.* 96:7910–7915. **6.26L** Andrew Syred/Getty Images Inc. - Stone Allstock **6.26M** ©Dr. David Phillips/Visuals Unlimited **6.26R** Joseph F. Hoffman, Yale University School of Medicine **6.32** ©Science Photo Library/Photo Researchers, Inc.

Unit 2

©Yorgos Nikas/Stone/Getty Images

Chapter 7

Opener ©Albert Tousson/Phototake **7.1** ©Dr. T.J. Beveridge/Visuals Unlimited **7.2** ©Stanley C. Holt/Biological Photo Service **7.3** ©Dr. Gopal Murti/Visuals Unlimited **7.4** COVER (upper left image), from *Science* vol 285, no. 5436, September 1999. Image: J.H. Cate, M.M. Yusupov, G. Zh. Yusupova, T.N. Earnest, H.F. Noller. Copyright 1999 American Association for the Advancement of Science. **7.5** Wanner/Eye of Science/Photo Researchers, Inc. **7.7** ©Fawcett/Photo Researchers, Inc. **7.8** Photo Researchers, Inc. **7.9** ©Dr. Don Fawcett/Photo Researchers, Inc. **7.10** ©Biophoto Associates/Photo Researchers, Inc. **7.11** ©Don W. Fawcett/Photo Researchers, Inc. **7.12** ©Dr. Don Fawcett, Daniel Friend, and Richard Wood/Photo Researchers, Inc. **7.13** ©Dr. Gopal Murti/Visuals Unlimited **7.15** E.H. Newcomb & W.P. Wergin/Biological Photo Service/Getty Images Inc. - Stone Allstock **7.16** K.R. Porter/Photo Researchers, Inc. **7.17** W.P. Wergin/Biological Photo Service **7.18** ©E. H. Newcomb/Biological Photo Service **7.19a** ©Dr. Don Fawcett/S. Ito & A. Like/Photo Researchers, Inc. **7.19b** ©Dr. Don Fawcett/Photo Researchers, Inc. **7.19c** Courtesy of Dr. Julian R. Thorpe, Electron Microscope Division, The Sussex Centre for Advanced Microscopy, School of Life Sciences, University of Sussex. **7.19d** ©Jan Robert Factor/Photo Researchers, Inc. **7.21** ©Michael W. Davidson, Florida State University/Molecular Expressions (http://www.microscopy.fsu.edu) **7.22b** Don W. Fawcett/Photo Researchers, Inc. **7.27** Jamieson, J.D. and Palade, G.E. (1967). Intracellular transport of secretory proteins in the pancreatic exocrine cell. II. Transport to condensing vacuoles and zymogen granules. Reproduced from *The Journal of Cell Biology*, 1967, v. 34 , pp. 597–615. **7.33a** K.G. Murti/Visuals Unlimited **7.33b** Peter Dawson/Science Photo Library/Photo Researchers, Inc. **7.33c** ©Dr. Gopal Murti/Science Photo Library/Photo Researchers, Inc. **7.35** ©Dr. Conly L. Rieder/Biological Photo Service **7.36** Reproduced by permission of the American Society for Cell Biology from *Molecular Biology of the Cell* 9(12), December 1998, cover. Copyright ©1999 by the American Society for Cell Biology. Image courtesy of Bruce J. Schnapp, Oregon Health Sciences. **7.37a** John E. Heuser, M.D., Washington University School of Medicine, St. Louis, Missouri **7.38** Dennis Kunkel/Phototake NYC **7.39a** Dr. Gopal Murti/Science Photo Library/Photo Researchers, Inc. **7.40** Reproduced from C.J. Brokaw, Microtubule sliding in swimming sperm flagella: direct and indirect measurements on sea urchin and tunicate spermatozoa. *Journal of Cell Biology* 114:1201–1215 (1991), cover. Reproduced by copyright permission of The Rockefeller University Press.

Chapter 8

Opener ©E.H. Newcomb & W.P. Wergin/Biological Photo Service **8.2a** ©Spencer Grant, Photo Researcher, Inc. **8.2b** ©Biophoto Associates/Photo Researchers, Inc. **8.3a** Ken Eward/Science Source/Photo Researchers, Inc. **8.3b** Barry King, University of California, Davis/Biological Photo Service **8.5** C..T. Huang, Karen Xu, Gordon McFeters, and Philip S. Stewart **8.6** ©K.R. Porter/Photo Researchers, Inc. **8.8** Biophoto Associates/Science Source/Photo Researchers, Inc. **8.9a** ©Dr. Don Fawcett/Photo Researchers **8.10a** ©Dr. Don Fawcett/Gida Matoltsy/Photo Researchers, Inc. **8.13a** ©E. H. Newcomb & W.P. Wergin/Biological Photo Service **8.13b** ©Dr. Don Fawcett/Photo Researchers, Inc.

Chapter 9

Opener ©2004 Richard Megna/Fundamental Photographs **9.1/1** Peter Anderson ©Dorling Kindersley **9.1/2** ©Dorling Kindersley **9.1/3** ©Darren McCollester/CORBIS **9.2c** Clare Sansom, Birkbeck College, University of London, London, England **9.9** Clare Sansom, Birkbeck College, University of London, London, England. **9.19b** Modified from Wang and Oster. 1998. *Nature* 396: 279. Fig. 1.

Chapter 10

Opener J.P. Nacivet/Photo Researchers, Inc. **10.2a** John Durham/Science Photo Library/Photo Researchers, Inc. **10.2b** ©Dr. J. Burgess, Science Photo Library/Photo Researchers, Inc. **10.3L** Biophoto Associates/Photo Researchers, Inc. **10.3M** ©Dr. George Chapman/Visuals Unlimited **10.3R** Richard Green/Photo Researchers, Inc. **10.5b** Sinclair Stammers/Science Photo Library/Photo Researchers, Inc. **10.10** David Newman/Visuals Unlimited **10.19** James A. Bassham, Lawrence Berkeley Laboratory, UCB (retired) **10.21** Andersson I. Journal of Molecular Biology, May 1996, vol. 259, iss. 1, pp. 160–174(15). Courtesy of Rolf Bergmann, University of Hamburg **10.23a** Dr. Jeremy Burgess/Science Photo Library/Photo Researchers, Inc. **10.26a** Adam Hart-Davis, Science Photo Library/Photo Researchers, Inc. **10.26b** ©David Muench/CORBIS **10.28** ©David Woodfall/DRK Photo

Chapter 11

Opener Professor G. Schaten/Science Photo Library/Photo Researchers, Inc. **11.2a** Originally published in Walter Flemming, Zellsubstanz, Kern, und Zelltheilung. Leipzig: Verlag von F.C.W. Vogel, 1882. Image courtesy of Conly Rieder, Wadsworth Center, New York State Department of Health. **11.2b** ©Photo Researchers, Inc. **11.3** Mark E. Warchol, Central Institute for the Deaf, Washington University, and Jeffrey T. Corwin, University of Virginia School of Medicine **11.10a** ©Dr. Richard Kessel/Visuals Unlimited **11.10b** R. Calentine/Visuals Unlimited **11.11** Micrographs by Conly L. Rieder, Division of Molecular Medicine, Wadsworth Center, Albany, New York 12201–0509. **11.17** Source: American Cancer Society's *Cancer Facts and Figures – 2003*. Reprinted with permission. **11.18** Dr. E. Walker/Science Photo Library/Photo Researchers, Inc.

Unit 3

©Eye of Science/Photo Researchers, Inc.

Chapter 12

Opener David Phillips/The Population Council/Photo Researchers, Inc. **12.2a** Applied Image Corp., San Jose, CA **12.2b** ©Addenbrookes Hospital/Photo Researchers, Inc. **12.2c** ©Wessex Reg. Genetics Centre /Wellcome Photo Library **12.7** David A. Jones **12.10** Doug Sokell/Visuals Unlimited **12.16** ©Robert and Beth Plowes Photography

Chapter 13

Opener Margaret H. Peaslee, Vice President for Academic Affairs and Professor of Biology, University of Pittsburgh at Titusville. **13.7** Dr. Madan K. Bhattacharyya **13.10a** Robert Calentine/Visuals Unlimited **13.10b** Carolina Biological Supply Company/Phototake NYC **13.11a** ©Addenbrookes Hospital/Photo Researchers, Inc. **13.17a** Robert Calentine/Visuals Unlimited **13.18a** David Cavagnaro/Peter Arnold, Inc. **13.19a** Albert F. Blakeslee, *Journal of Heredity*, 1 Fig., 1914. Reproduced by permission of Oxford University Press. **13.23** ©Bettmann/CORBIS

Chapter 14
Opener Dr. Gopal Murti/Science Photo Library/Photo Researchers, Inc. **14.1** Reproduced from O.T. Avery, C.M. MacLeod, and M. McCarty, Studies on the chemical nature of the substance inducing transformation of pneumcoccal types, *The Journal of Experimental Medicine*, 1944, 79:137–157, plate 1, by copyright permission of The Rockefeller University Press. **14.4b** Oliver Meckes/Max Planck Institut-Tubingen/Photo Researchers, Inc. **14.11a** Dr. Gopal Murti/Science Photo Library/Photo Researchers, Inc. **14.20** Vesalius Studios

Chapter 15
Opener Halaska, Jacob/Index Stock Imagery, Inc. **15.11** Omikron/Photo Researchers, Inc.

Chapter 16
Opener Oscar Miller/Science Photo Library/Photo Researchers, Inc. **16.2** Reproduced with permission from Structural basis of transcription initiation: an RNA polymerase holoenzyme-DNA complex. Murakami KS, Masuda S, Campbell EA, Muzzin O, Darst SA. *Science*. 2002 May 17;296(5571):1285–90. Fig 2a pg 1287. Copyright ©2004 **16.4** Hans Reinhard/Bruce Coleman Inc. **16.5** Bert W. O'Malley, M.D., Baylor College of Medicine **16.13** Reprinted from *Cell Press*, "Ribosome Structure and the Mechanism of Translation" by V. Ramakrishnan from Vol. 108, 557–572, Fig 1b, February 22, 2002, with permission from Elsevier. **16.16T** ©E.V. Kiseleva and Donald Fawcett/Visuals Unlimited **16.19b** Bill Longcore/Photo Researchers, Inc. **16.21a** "Structural basis for the interaction of antibiotics with peptidyl transferase centre in eubacteria: by Schluenzen et al." Figure 5 in *Nature* (2001) 413: 814–821 by Schluenzen et al.

Chapter 17
Opener EM Unit, VLA/Science Photo Library/Photo Researchers, Inc. **17.11** Reproduced by permission from A. Schmitz and D.J. Galas, The interaction of RNA polymerase and lac repressor with the lac control region. *Nucleic Acids Research* 6:111–137 (1979), fig. 2b. Copyright ©1979 by Oxford University Press. Image courtesy of David J. Galas, Keck Graduate Institute. **17.15** ©Dr. Dennis Kunkel/Visuals Unlimited **p. 381** Michael Gabridge/Visuals Unlimited

Chapter 18
Opener Prof. Oscar L. Miller/Science Photo Library/Photo Researchers, Inc. **18.2a** Ada Olins/Don Fawcett/Photo Researchers, Inc. **18.4b** Copyright ©2002 from *Molecular Biology of the Cell* 4/e fig. 4.23 by Bruce Alberts et al. Reproduced by Permission of Garland Science Taylor & Francis Books, Inc., and Dr. Barbara Hamkalo. **18.4c** Copyright ©2002 from *Molecular Biology of the Cell* 4/e fig. 4.23 by Bruce Alberts et al. Reproduced by Permission of Dr. Victoria E. Foe, and Garland Science/Taylor & Francis Books, Inc. **18.16** Walter J. Gehring

Chapter 19
Opener Work of Atsushi Miyawaki, Qing Xiong, Varda Lev-Ram, Paul Steinbach, and Roger Y. Tsien at the University of California, San Diego. **19.2** ©Dennis Kunkel/ Phototake **19.12** Baylor College of Medicine/Peter Arnold, Inc. **19.14** ©Wayne P. Armstrong, Palomar College **19.16a** Brad Mogen/Visuals Unlimited **19.17b** Peter Beyer, University of Freiburg, Germany

Chapter 20
Opener ©Sanger Institute/Wellcome Photo Library **20.7** ©David Parker/Photo Researchers **20.8** AP Wide World Photos **20.10** Camilla M. Kao and Patrick O. Brown, Stanford University.

Unit 4
©Ted Levin/Animals Animals

Chapter 21
Opener Photo courtesy Dr. Andrew Ewald **21.2a** Microworks/Phototake NYC **21.2b** Cabisco/Visuals Unlimited **21.4a** Gregory Ochocki/Photo Researchers, Inc. **21.6a** Michael Whitaker/Science Photo Library/Photo Researchers, Inc. **21.6b** Victor D. Vacquier, Scripps Institution of Oceanography, University of California at San Diego **21.7b** Biodisc/Visuals Unlimited **21.8c** David M. Phillips/Visuals Unlimited **21.10a** David S. Addison/Visuals Unlimited **21.11a** Michael V. Danilchik, Oregon Health Sciences University **21.11b** Douglas A. Melton, Harvard University **21.12** Douglas A. Melton, Harvard University **21.15** Richard Hutchings/Photo Researchers, Inc. **21.18a** Holt Studios International/Photo Researchers, Inc. **21.24** ©Jochen Tack, Das Fotoarchiv/Peter Arnold

Chapter 22
Opener Gary C. Schoenwolf, University of Utah School of Medicine **22.1** F. Rudolf Turner, Indiana University **22.3a** Christiane Nusslein.Volhard, Max Planck Institute of Developmental Biology, Tubingen, Germany **22.3b** Wolfgang Driever, University of Freiburg, Germany **22.4a** Jim Langeland, Stephen Paddock, and Sean Carroll, University of Wisconsin at Madison **22.4b** Stephen J. Small, New York University **22.4c** Jim Langeland, Stephen Paddock, and Sean Carroll, University of Wisconsin at Madison **22.6** F. Rudolf Turner, Indiana University **22.7a/1** ©Oliver Meckes/Photo Researchers, Inc. **22.7a/2** Copyright ©Carolina Biological Supply Company/ Phototake **22.7b/1** ©Oliver Meckes/Photo Researchers, Inc. **22.7b/2** Edward B. Lewis, California Institute of Technology **22.11** John L. Bowman, University of California at Davis **22.12** Reprinted by permission from *Nature* (Vol 424, July 24, 2003, pg 439, by Kim et al) copyright (2004) Macmillan Publishers Ltd. Photograph supplied by Dr. Neelima Sinah. **22.13b** Roslin Institute/PA Photos Ltd **22.15** Reproduced by permission of Elsevier Science from K. Kuida et al., Reduced apoptosis and cytochrome *c*.mediated caspase activation in mice lacking caspase 9. *Cell* 94:325–337 (1998), figs. 2E and 2F. Copyright ©1998 by Elsevier Science Ltd. Image courtesy of Keisuke Kuida, Vertex Pharmaceuticals. **22.16a** Kathryn W. Tosney, University of Michigan

Unit 5
©Gary Meszaros/Photo Researchers

Chapter 23
Opener Kent & Donna Dannen/Photo Researchers, Inc. **23.1a** ©Bettmann/Corbis **23.1b** Alfred Russel Wallace by unknown artist, after a photograph by Thomas Sims, fl. 1860s. Reg. No.: 1765. National Portrait Gallery, London. **23.2a** ©Ken Lucas/Visuals Unlimited **23.2b** ©T. A. Wiewandt/DRK PHOTO **23.2c** Courtesy photographer, Knut Finstermeier, MPI.EVA Leipzig **23.3aL** Based on Gingerich et al. 2001. *Science* 293: 2239–2242 and Thewissen et al. 2001. *Nature* 413: 277–281. **23.3aR** Robert Lubeck/Animals Animals/Earth Scenes **23.4aL** Vincent Zuber/Custom Medical Stock Photo, Inc. **23.4aR** CMCD/Getty Images Inc./PhotoDisc **23.4bL** Custom Medical Stock Photo, Inc. **23.4bR** Mary Beth Angelo/Photo Researchers, Inc. **23.5aTL** Marie Read/Animals Animals/Earth Scenes **23.5aTR** Tui De Roy/Bruce Coleman Inc. **23.5aBL** George D. Lepp/Photo Researchers, Inc. **23.5aBR** Mickey Gibson/Animals Animals/Earth Scenes **23.7L** Photo by Michael K. Richardson, reproduced by permission from *Anatomy and Embryology* 305, Fig. 7. Copyright ©Springer-Verlag GmbH & Co KG, Heidelberg, Germany. **23.7M** Photo k by Professor R. O'Rahilly. Photos a–j by Dr. Michael K. Richardson. Reproduced by permission *from Anatomy and Embryology* 305, Fig. 8. Copyright ©Springer-Verlag GmbH & Co KG, Heidelberg, Germany. **23.7R** From: Richardson, M. K., et al. *Science*. Vol 280: Pg 983c, Issue # 5366, May 15, 1998. Embryo from Professor R. O'Rahilly. National Museum of Health and Medicine/Armed Forces Institute of Pathology **23.10** Arthur C. Aufderheide, M.D., University of Minnesota School of Medicine, Duluth **23.12a** Frans Lanting/Minden Pictures **23.11** Modified from Grant and Grant. 2001. *Science* 296: 707–711. Fig. 1.

Chapter 24
Opener ©Wayne Lynch/DRK Photo **24.5** Otorohanga Kiwi House, New Zealand **24.6** Reprinted with permission from Fig. 1B, C from B Faivre et al., *Science* 300:103 (2003). Copyright 2004 American Association for the Advancement of Science. Photo courtesy of Bruno Faivre. **24.7a** Cyril Laubscher ©Dorling Kindersley **24.7b** Data from Blount et al. 2003. *Science* 300: 125–127. Fig. 1a. **24.8** David H. Funk **24.9a** Marc Moritsch/National Geographic Image Collection **24.10aT&B** Robert & Linda Mitchell/Robert & Linda Mitchell Photography **24.10bT** ©B. Schorre/VIREO **24.10bB** ©R. & A. Simpson/VIREO **24.10cT** Jeremy Woodhouse/Getty Images, Inc./Photodisc. **24.10cB** Van Os, Joseph/Getty Images Inc. - Image Bank **24.11a** ©James Hughes/Visuals Unlimited **24.14** Data from Johnston, M. 1992. *Evolution* 46: 688–702.

Chapter 25
Opener ©Wayne Lankinen/DRK PHOTO **25.2** Credit is to Don McGranaghan in Tattersall, I. 1995. *The Fossil Trail.* (Oxford: Oxford University Press) **25.4** Diane Pierce/NGS Image Collection **25.7a** Joseph T. Collins/Photo Researchers, Inc. **25.7b** Alvin E. Staffan/National Audubon Society/Photo Researchers, Inc. **25.12** H. Douglas Pratt/NGS Image Collection **25.13** Jason Rick/Loren H. Rieseberg **25.15a** ©Anthony Bannister; Gallo Images/CORBIS **25.15b** ©Barbara Cushman Rowell/DRK PHOTO **25.15c** ©Bob Krist/CORBIS **25.15d** ©Nik Wheeler/CORBIS **25.15e** ©Matthew McKee; Eye Ubiquitous/CORBIS **25.15f** ©Richard Cummins/CORBIS

Chapter 26
Opener ©1985 David L. Brill **26.7a** Reproduced by permission from P.S. Herendeen, W.L. Crepet, and K.C. Nixon, Chloranthus like stamens from the Upper Cretaceous of New Jersey, *American Journal of Botany* 80(8):865–871. ©Botanical Society of America. Micrograph courtesy of William L. Crepet, Cornell University. **26.7b** Martin Land/Science Photo Library/Photo Researchers, Inc. **26.7c** Monte Hieb & Harrison Hieb/Geocraft/www.geocraft.com **26.7d** John Gerlach/DRK Photo **26.8B** Neg. no. 325097, shot from Sanford Bird Hall. Courtesy Department of Library Services, American Museum of Natural History. **26.12b** Reproduced by permission from S. Xiao, Y. Zhang, and A.H. Knoll, Three-dimensional preservation of algae and animal embryos in a Neoproterozoic phosphorite. *Nature* 391:553–558 (February 5, 1998), fig. 5. Copyright ©1998 Macmillan Magazines Limited. **26.12cT** Simon Conway Morris, University of Cambridge, Cambridge, United Kingdom **26.12cB** Ken Lucas/Visuals Unlimited **26.12dT** B. Miller/Biological Photo Service **26.12dB** ©Ed Reschke/Peter Arnold, Inc. **26.14** Denis Duboule, University of Geneva, Geneva, Switzerland **26.16a** Jonathan B. Losos, Washington University in St. Louis **26.17a** Colin Keates ©Dorling Kindersley, Courtesy of the Natural History Museum, London **26.17b** ©Tim Fitzharris/Minden Pictures **26.17d** ©Joe Tucciarone/SPL/Photo Researchers, Inc. **26.19b** Glen A. Izett/U.S. Geological Survey, Denver **26.19c** Peter H. Schultz, Brown University; Steven D'Hondt, University of Rhode Island Graduate School of Oceanography

Unit 6
©John Gerlach /DRK Photo

Chapter 27
Opener ©Beth Donidow/Visuals Unlimited **27.2** ©Science VU/Visuals Unlimited **27.3** J. Robert Waaland/Biological Photo Service **27.4** ©Richard L. Carlton/Visuals Unlimited **27.6a** ©T. Bannor/Custom Medical Stock Photo **27.6b** ©Carolina Biological/Visuals Unlimited **27.7** Reprinted with permission of the American Association for the Advancement of Science from S.V. Liu et al., Thermophilic Fe(III)-reducing bacteria from the deep subsurface. *Science* 277:1106–1109 (1997), fig. 2. Copyright©1997 American Association for the Advancement of Science. Image courtesy of Yul Roh, Oak Ridge National Laboratory. **27.11aL** CNRI/Science Photo Library/Photo Researchers, Inc. **27.11aR** Reprinted with permission from H.N. Schulz, et al., Dense populations of a giant sulfur bacterium in Namibian shelf sediments. *Science* 284:493–495, Fig. 1b, August 4, 1999. Copyright 1999 American Association for the Advancement of Science. Image courtesy of Dr. Heide Schulz, Max-Planck-Institute for Marine Microbiology, Bremen, Germany. **27.11bL** ©Gary Gaugler/Visuals Unlimited **27.11bR** David M. Phillips/Visuals Unlimited **27.11cL** Linda Stannard, University of Cape Town/Science Photo Library/Photo Researchers, Inc. **27.11cR** Richard W. Castenholz, University of Oregon **27.12c** ©Jack Bostrack/Visuals Unlimited **27.16** ©Eric Grave/Phototake **27.17** ©National Cancer Institute/SPL/Photo Researchers, Inc. **27.18** ©Michael Abbey/Visuals Unlimited **27.19** ©R. Calentine/Visuals Unlimited **27.20** ©J. C. Revy/Phototake **27.21a** Photo by Yves B. Brun **27.21b** Hans Reichenbach, Gesellschaft fur Biotechnologische Forschung mbH, Braunschweig, Germany **27.23** Photograph courtesy Dr. Kenneth M. Stedman/Portland State University **27.23 (inset)** ©Corale Brierley/Visuals Unlimited **27.24** Marli Miller **27.24 (inset)** Photograph courtesy Dr. Bonnie K. Baxter/Westminster College

Chapter 28
Opener Dr. Ann Smith/Photo Researchers, Inc. **28.1a** Norman T. Nicoll **28.1b** Gregory Ochocki/Photo Researchers, Inc. **28.1c** Tom and Therisa Stack/Tom Stack & Associates, Inc. **28.1R** John Anderson/Animals Animals/Earth Scenes **28.2** Dr. Gopal Murti/Photo Researchers, Inc. **28.3** Sanford Berry/Visuals Unlimited **28.3 (inset)** J. Robert Waaland/Biological Photo Service **28.6** Biophoto Associates/Photo Researchers, Inc. **28.12L** M.I. Walker/Photo Researchers, Inc. **28.13** David L. Kirk, Washington University **28.14a–b** Andrew Syred/Science Photo Library/Photo Researchers, Inc. **28.14c** David M. Phillips/Visuals Unlimited **28.15a** Biophoto Associates/Photo Researchers, Inc. **28.15b** Bruce Coleman Collection **28.16** Eric Grave/Photo Researchers,

Inc. **28.17a** Astrid and Hanns-Frieder Michler/Science Photo Library/Photo Researchers, Inc. **28.17b** Cabisco/Visuals Unlimited **28.17c** Sherman Thomson/Visuals Unlimited **28.17d** ©OSF/D. Fleetham/Animals Animals Earth Scenes **28.23** M. Abbey/Photo Researchers, Inc. **28.24** ©Carolina Biological/Visuals Unlimited **28.25** ©Michael Abbey/Visuals Unlimited **28.26** Roland Birke/Photo Researchers, Inc. **28.27** ©Wim van Egmond/Visuals Unlimited **28.28** Courtesy of Philip J. Spagnuolo, et al., *CID*:14:1084–1088 (1992) **28.29** ©Andrew Syred/SPL/Photo Researchers **28.30** ©Phototake **28.31** Joyce Photographics/Photo Researchers, Inc. **28.32** ©Peter Parks/imagequestmarine.com **28.33** ©Robert De Goursey/Visuals Unlimited **28.34** Scott Camazine/Photo Researchers, Inc.

Chapter 29

Opener Darrell Gulin/DRK Photo **29.1a** ©Tom Myers/Photo Researchers, Inc. **29.1b** ©D. P. Burnside/Photo Researchers, Inc. **29.3a1** Runk/Schoenberger/Grant Heilman Photography, Inc. **29.3a2** D. Cavagnaro/Visuals Unlimited **29.3a3** Ed Reschke/Peter Arnold, Inc. **29.3aR** Robert and Beth Plowes Photography **29.3R** Courtesy Jason P. Londo **29.5** ©D. Cavagnaro/DRK Photo **29.6aR** Lee W. Wilcox **29.6bBR** Lee W. Wilcox **29.6bT** Lee W. Wilcox **29.6bWL** K.G. Vock/Okapia/Photo Researchers, Inc. **29.6c** ©Biophoto Associates/Photo Researchers, Inc. **29.6cUL** Milton Rand/Tom Stack & Associates, Inc. **29.6cUR** Rod Planck/Photo Researchers, Inc. **29.6cWL** Alvin E. Staffan/Photo Researchers, Inc. **29.6d** ©Tom Willock/Photo Researchers, Inc. **29.6dML** Degginger, E R/Animals Animals/Earth Scenes **29.6dMR** Art Wolfe/Getty Images Inc. - Stone Allstock **29.6dU** Stephen J. Krasemann/DRK Photo **29.6dWR** Walter H. Hodge/Peter Arnold, Inc. **29.7** ©Holt Studios Int./Photo Researchers, Inc. **29.8a** Dr. James L. Castner **29.8aL** Linda Graham, University of Wisconsin-Madison **29.8T1** Paul K. Strother, Boston College **29.8T2** David L. Dilcher and Ge Sun **29.8T3** Thomas A. Wiewandt/DRK Photo **29.8T4** Robert & Linda Mitchell/Robert & Linda Mitchell Photography **29.10** Lee W. Wilcox **29.16** Lee W. Wilcox **29.17** David T. Webb, Botany, University of Hawaii **29.18a** ©Carolina Biological Supply Company/ Phototake **29.18b** ©Bill Banaszewski/Visuals Unlimited **29.18bM** C.C. Lockwood/Animals Animals/Earth Scenes **29.18bT** Runk/Schoenberger/Grant Heilman Photography, Inc. **29.19a** ©Holt Studios Int. (Nigel Cattlin)/Photo Researchers, Inc. **29.19b1** Gerald C. KelleyPhoto Researchers, Inc. **29.19b2** ©John Cancalosi/Peter Arnold **29.20a** Steven D. Johnson, University of Natal, Pietermaritzburg, South Africa **29.20b** Steven D. Johnson, University of Natal, Pietermaritzburg, South Africa **29.20c** Gerry Ellis/GLOBIO **29.23.2** Robert & Linda Mitchell/Robert & Linda Mitchell Photography **29.23.4** Don Farrall/Getty Images, Inc./Photodisc. **29.23.5** Ed Reschke/Peter Arnold, Inc. **29.23.6** Runk/Schoenberger/Grant Heilman Photography, Inc. **29.23.7** Robert & Linda Mitchell/Robert & Linda Mitchell Photography **29.23.8** Kim Heacox Photography/DRK Photo **29.24a** ©Holt Studios Int./Photo Researchers, Inc. **29.24TL** Lee W. Wilcox **29.24TLb** ©Pat O'Hara/CORBIS **29.26a** Lee W. Wilcox **29.26aL** David T. Webb, Botany, University of Hawaii **29.26b** Lee W. Wilcox **29.26c** Lee W. Wilcox **29.27** ©Andrew J. Martinez/Photo Researchers, Inc. **29.28** ©Wim van Egmond **29.29** Photograph courtesy of Dr. John Clayton/ NIWA (National Institute of Water and Atmospheric Research) **29.30** ©Carolina Biological/Visuals Unlimited **29.31** ©Henry Robinson/Visuals Unlimited **29.32** Robert and Beth Plowes Photography **29.33** Lee W. Wilcox **29.34** Photo courtesy Karen S. Renzaglia **29.35** ©Michael P. Gadomski/Photo Researchers, Inc. **29.36aR** ©David Cavagnaro/DRK Photo **29.36b** Lee W. Wilcox **29.37** Nick Green/Photolibrary.Com **29.38a** ©Ken Lucas/Visuals Unlimited **29.38b** David T. Webb, Botany, University of Hawaii **29.39a** ©Martin Land/Science Photo Library/Photo Researchers, Inc. **29.39b** Biophoto Associates/Photo Researchers, Inc. **29.40** Lee W. Wilcox **29.41** Lee W. Wilcox

Chapter 30

Opener ©Steve Austin; Papilio/CORBIS **30.1** Photography courtesy Mychorrizal Applications, Inc./www.mycorrhizae.com **30.4a** David M. Dennis/Tom Stack & Associates, Inc. **30.4b** Poehlmann, Mauritius, GMBH/Phototake **30.5L** George Musil/Visuals Unlimited **30.5R** Tony Brain/Science Photo Library/Photo Researchers, Inc. **30.6b** Biophoto Associates/Photo Researchers, Inc. **30.11a** Stan Flegler/Visuals Unlimited **30.11b** David T. Webb, Botany, University of Hawaii **30.12** Bruce M. Herman/Photo Researchers, Inc. **30.14** Photograph courtesy of Dr. Daniel A. Wubah, James Madison University **30.15** ©Biophoto Associates/Photo Researchers **30.16** Courtesy of Hans Kerp and Hagen Hass, Palaeobotanical Research Group, University of Munster, Germany. **30.17** Jeff Lepore/Photo Researchers, Inc. **30.18b** Pat O'Hara/DRK Photo **30.19** ©Stephen Sharnoff/Visuals Unlimited **30.20** Dr. Lee Berger, James Cook University

Chapter 31

Opener ©David B. Fleetham/SeaPics.com **31.1a** ©Nuridseny et Perennou/Photo Researchers, Inc. **31.1b** Mark W. Moffett/Minden Pictures **31.2a** ©Harry Rogers/Photo Researchers **31.2b** ©Glenn Oliver/Visuals Unlimited **31.2c** ©C & M Fallows/SeaPics.com **31.3a** ©Walker England/Photo Researchers **31.3b** ©Ken Lucas/Visuals Unlimited **31.3c** David J. Wrobel/Visuals Unlimited **31.4** ©Roland Birke/Okapia/Photo Researchers **31.9bT** ©Robert Steene/imagequestmarine.com **31.9bB** ©L. Newman & A. Flowers/Photo Researchers **31.12a** Reprinted with permission from Chia.Wei Li et al. Fig 1A in *Science* Vol 279, pg. 879 (1998). ©1998 AAAS. **31.12b** ©South Australian Museum **31.12c** Chip Clark **31.15a** J.P. Ferrero/Jacana/Photo Researchers, Inc. **31.15b** Robert & Linda Mitchell/Robert & Linda Mitchell Photography **31.17a** ©Satoshi Kuribayashi/OSF/DRK Photo **31.17b** Jim Brandenburg/Minden Pictures **31.19a** ©Peter Parks/imagequestmarine.com **31.19b** ©Roger Steene/imagequestmarine.com **31.20a** ©Roland Seitre/Peter Arnold **31.20b** ©Jeff Foott/DRK Photo **31.20c** ©Chamberlain, MC/DRK Photo **31.20d** ©Jim Greenfield/imagequestmarine.com **31.21** Reproduced by permission from Grace Panganiban et al., The origin and evolution of animal appendages. *Proceedings of the National Academy of Sciences* 94:5162–5166 (May 1997), p. 5163, Fig. 1a. Copyright ©1997 National Academy of Sciences, U.S.A. **31.22** Photograph courtesy Dr. David Mark Welch. **31.23a** ©Gregory G. Dimijian/Photo Researchers **31.23b** ©Doug Perrine/SeaPics.com **31.24a** ©Doug Perrine/DRK Photo **31.24b** ©Dan Suzio/Photo Researchers **31.25a–b** Phototake/Carolina Biological Supply Company **31.25c** Sinclair Stammers/Science Photo Library/Photo Researchers, Inc. **31.28** Photograph courtesy of Dr. Kalle Olli **31.29** ©Andrew J. Martinez **31.30a** ©William P. Leonard/DRK Photo **31.30b** ©Mark Conlin/SeaPics.com **31.31** ©Andrew J. Martinez/Photo Researchers **31.32** Photography courtesy Matthew D. Hooge, University of Maine **31.33a** ©Steven Wolper/DRK Photo **31.33b** ©Denise Tackett/DRK Photo

Chapter 32

Opener Edward S. Ross, California Academy of Sciences **32.2a** ©Jean Claude Revy - ISM /Phototake **32.2b** ©Sinclair Stammers/Science Photo Library/Photo Researchers **32.5a** ©Erling Svensen/UWPhoto ANS **32.5b** ©Robert Calentine/Visuals Unlimited **32.7** ©Roger Steene/Imagequestmarine.com **32.8a** Sinclair Stammers/Science Photo Library/Photo Researchers, Inc. **32.8b** Photograph courtesy Dr. Ed Jarzembowski **32.10a** ©Fred Bavendam/Minden Pictures **32.10b** ©Norbert Wu/ Peter Arnold, Inc. **32.10c** Mark W. Moffett/Minden Pictures **32.10d** Nigel Cattlin/Photo Researchers, Inc. **32.10e** ©Harry Rogers/Photo Researchers **32.11a** ©Stephan Dalton/Photo Researchers **32.11b** ©M.C. Chamberlain/DRK Photo **32.11c** ©Stephen Dalton/Photo Researchers **32.12a** ©Peter Parks/Imagequestmarine.com **32.13** ©Jan Van Arkel/Foto Natura/Minden Pictures **32.14a** ©Newman & Flowers/Photo Researchers **32.14b** ©Manfred Kage/Peter Arnold, Inc. **32.14c** ©Oliver Meckes/Photo Researchers **32.15a** ©Peter Batson/Imagequestmarine.com **32.15b** Photograph courtesy of Alexa Bely/University of Maryland **32.15c** ©Robert Calentine/Visuals Unlimited **32.16** Degginger, E R/Animals Animals/Earth Scenes **32.17** ©Norbert Wu/Peter Arnold, Inc. **32.18** ©Kjell B. Sandved/Photo Researchers **32.19** ©Fred Bavendam/Peter Arnold, Inc. **32.20** ©Bill Beatty/Visuals Unlimited **32.21a** ©Michael Fogden/DRK Photo **32.21b** ©Diane Nelson/Visuals Unlimited **32.22** ©Luis M. de la Maza, Ph.D. M.D./Phototake **32.23** ©Mark Smith/Photo Researchers **32.24a** ©Gerard Blondeau/ISM/Phototake **32.24b** ©Dr. David M. Phillips/Visuals Unlimited **32.25a** ©WHM Bildarchiv/Peter Arnold, Inc. **32.25b** ©Stephan Dalton/Photo Researchers **32.26a** ©E. R. Degginger/Photo Researchers **32.26b** ©M C Chamberlain/DRK Photo

Chapter 33

Opener ©Konrad Wothe/Minden Pictures **33.2a** University of Saskatchewan Archives, T.C. Lacalli fonds **33.2b** ©Jett Britnell/DRK Photo **33.3b** John D. Cunningham/Visuals Unlimited **33.4** ©Ria Tan/www.wildsingapore.com **33.9a** Brian Parker/Tom Stack & Associates, Inc. **33.9b** ©Mark Smith/Photo Researchers **33.9c** AP Wide World Photos **33.10a** Visuals Unlimited **33.10b** ©MC Chamberlain/DRK Photo **33.12** ©Dr. John D. Cunningham/Visuals Unlimited **33.14** Boban Filipovic **33.19a** ©James D. Watt/SeaPics.com **33.19b** ©S. Boon Fu/UNEP/Peter Arnold **33.20a** ©Fred Bavendam/Minden Pictures **33.20b** ©Marty Snyderman/Visuals Unlimited **33.20c** ©Fred Bavendam/Minden Pictures **33.21** ©James D. Watt/imagequestmarine.com **33.22a** ©Al Lowry/Photo Researchers **33.22b** ©David Wrobel/Visuals Unlimited **33.23a** Tom McHugh/Photo Researchers, Inc. **33.23b** Tom Stack/Tom Stack & Associates, Inc. **33.24a** ©Fred McConnaughey/Photo Researchers **33.24b** ©Jeff Jaskolski/SeaPics.com **33.25** ©Richard Herrmann Photography **33.26** Peter Scoones/TCL/Getty Images, Inc. **33.27a** ©Michael Fogden/DRK Photo **33.27b** ©R. Andrew Odum/Peter Arnold, Inc. **33.28** ©Tom McHugh/Photo Researchers, Inc. **33.29** ©Tom Vezo/Peter Arnold **33.30** ©Gerry Ellis/Minden Pictures **33.31** ©Gerald and Buff Corsi/Visuals Unlimited **33.32** George Grall/NGS Image Collection **33.33** ©Doug Perrine/DRK Photo **33.34** ©Gerald and Buff Corsi/Visuals Unlimited **33.35a** Dr. M. S. Mayilvahnan/TROPICAL TRANNIES **33.35b** ©Tom McHugh/Photo Researchers **33.40** Gunter Ziesler/Peter Arnold, Inc.

Chapter 34

Opener Jed Fuhrman, University of Southern California **34.2** National Institute for Biological Standards and Control (U.K.)/Science Photo Library/Photo Researchers, Inc. **34.5L** Biophoto Associates/Photo Researchers, Inc. **34.5M** Dennis Kunkel/Phototake NYC **34.5R** David M. Phillips/Visuals Unlimited **34.6a** Omikron/Photo Researchers, Inc. **34.6b** Biophoto Associates/Photo Researchers, Inc. **34.6c** K.G. Murti/Visuals Unlimited **34.6d** Oliver Meckes/Eye of Science/Max Planck Institut Tubingen/Photo Researchers, Inc. **34.12** Photos provided by Abbott Laboratories. **34.13aR** National Institute for Biological Standards and Control (U.K.)/Science Photo Library/Photo Researchers, Inc. **34.13bR** James L. Van Etten, University of Nebraska. Originally published in R.H. Meints, J.L. Van Etten, D. Kuczmarski, K. Lee, and B. Ang, Viral infection of the symbiotic Chlorella-like alga present in *Hydra viridis*. *Virology* 113:698–703 (1981), Fig. C. **34.15** Hans Gelderblom/Eye of Science/Photo Researchers, Inc. **34.16** Photograph courtesy of Dr. Philip Leder, Dept. of Genetics, Harvard Medical School **34.17** ©Nigel Cattlin/Holt Studios/Photo Researchers, Inc. **34.18** ©Lowell Georgia/Science Source/Photo Researchers, Inc. **34.19** ©David Parker/SPL/Photo Researchers

Unit 7

©2004 Susan Middleton & David Liittschwager

Chapter 35

Opener ©Tom McHugh/Photo Researchers **35.4a** Tim Hauf Photography/Visuals Unlimited **35.4b** Reproduced from John E. Weaver, *North American Prairie*, fig. 32. Lincoln, Nebraska: Johnsen Publishing Co. (1954). J.E. Weaver, Papers, Archives and Special Collections, University of Nebraska-Lincoln Libraries. **35.5a** ©Wally Eberhart/Visuals Unlimited **35.5b** ©Rick Poley/Visuals Unlimited **35.7a–b** Gerald D. Carr **35.7c** ©Frans Lanting/Minden Pictures **35.8a** ©Ken Wagner/Phototake **35.8b** Kenneth W. Fink/Photo Researchers, Inc. **35.8c** Lee W. Wilcox **35.8d** ©Bill Beatty/Visuals Unlimited **35.8e** Richard Shiell/Animals Animals/Earth Scenes **35.9a–b** Lee W. Wilcox **35.9c–d** David T. Webb, Botany, University of Hawaii **35.10a** Lee W. Wilcox **35.10b** ©M. Harvey/DRK Photo **35.10c–f** Lee W. Wilcox **35.12** ©Wally Eberhart/Visuals Unlimited **35.14a** Photograph courtesy of Elisabeth S. Pierson, Radboud University, Nijmegen/www.vcbio.sci.kun.nl/ **35.14b** Lee W. Wilcox **35.15** Lee W. Wilcox **35.16a** Biophoto Associates/Science SourcePhoto Researchers, Inc. **35.16b** Lee W. Wilcox **35.17** Lee W. Wilcox **35.18** Andrew Syred/Science Photo Library/Photo Researchers, Inc. **35.19a** Bruce Iverson/Bruce Iverson, Photomicrography **35.19b** G. Shih and R. Kessel/Visuals Unlimited **35.19c** ©Dr. Richared Kessel & Dr. Gene Shih/Visuals Unlimited **35.20** Lee W. Wilcox **35.22b** Walker/Photo Researchers, Inc. **35.23** Ed Reschke/Peter Arnold, Inc. **35.24a–b** Lee W. Wilcox **35.24c** Ed Reschke/Peter Arnold, Inc. **35.25R** Ken Wagner/Phototake NYC **35.26a** ©Ed Reschke/Peter Arnold **35.26b** David T. Webb, Botany, University of Hawaii **35.28a** James W. Richardson/Visuals Unlimited **35.28b** Lee W. Wilcox **35.29a** Andrea H. Lloyd, Middlebury College **35.29b** Henri D. Grissino.Mayer, Department of Geography, University of Tennessee **35.30a** Gerry Ellis/GLOBIO **35.30b** Richard Phelps Frieman/Photo Researchers, Inc.

Chapter 36

Opener David Nunuk/Science Photo Library/Photo Researchers, Inc. **36.5** Photograph courtesy of Dr. Arthur C. Gibson **36.6** David T. Webb, Botany, University of Hawaii **36.7** Lee W. Wilcox **36.9** ©Larry Lefever/Grant Heilman Photography **36.11 (leaf cross section)** Ken Wagner/Phototake NYC **36.15a** John D. Cunningham/Visuals Unlimited

36.15a (inset) Lee W. Wilcox 36.15b Lee W. Wilcox 36.18L Jean Claude Revy/Phototake NYC 36.18R ©Biophoto Associates/Photo Researchers Inc. 36.20 Photograph courtesy of Dr. Art J.E. an Bel. 36.23 Reproduced by permission from N.D. DeWitt and M.R. Sussman, Immunocytological localization of an epitope-tagged plasma membrane proton pump (H(+)-ATPase) in phloem companion cells. *Plant Cell* 7:2053–2067 (1995), fig. 6. Copyright ©1995 American Society of Plant Biologists. Image courtesy of Michael R. Sussman, University of Wisconsin.

Chapter 37

Opener John Eastcott/Yva Momatiuk/Animals Animals/Earth Scenes **37.2** Diane Schiumo/Fundamental Photographs, NYC **37.3b** Emanuel Epstein, University of California at Davis **37.6a** The UT Institute of Texan Cultures, No. 87–286. Collection of Flaine Moczygemba; Courtesy of Emilia Dziuk Moczygemba. **37.6b** Thony Belizaire/Agence France Presse/Getty Images **37.8** ©Dennis Drenner/Visuals Unlimited **37.10b** Photograph courtesy of Dr. Eduardo Blumwald. Reprinted with permission from Apse et al., *Science* 285:1256–1258, Fig 3 (1999). Copyright 2004 AAAS. **37.11** Hugh Spencer/Photo Researchers, Inc. **37.12** E.H. Newcomb and S.R.Tandon, Science 212:1394 (1981)/Biological Photo Service **37.13a** ©Richard Thom/Visuals Unlimited **37.13b** Gerry Ellis/GLOBIO **37.14** Frank Greenaway ©Dorling Kindersley **37.14 (inset)** Photograph courtesy of Carol A. Wilson and Clyde L. Calvin **37.16** ©Carlos Munoz-Yague/Eurelios **37.17** Ken W. Davis/Tom Stack & Associates, Inc.

Chapter 38

Opener Lee W. Wilcox **38.1** Malcolm B. Wilkins, University of Glasgow, Glasgow, Scotland, U.K. **38.5** Malcolm B. Wilkins, University of Glasgow, Glasgow, Scotland, U.K. **38.6a3** Reprinted with permission from John M. Christie et al., Arabidopsis NPH1: A Flavoprotein with the Properties of a Photoreceptor for Phototropism. *Science* 282:1698–1701, Fig. 1c (1998). Copyright 1998 American Association for the Advancement of Science. **38.8** Runk/Schoenberger/Grant Heilman Photography, Inc. **38.9a** Reproduced by permission from E.B. Blancaflor et al., Mapping the functional roles of cap cells in the response of Arabidopsis primary roots to gravity. *Plant Physiology* 116:213–222 (1998), Fig. 1A. Copyright ©1998 American Society of Plant Biologists. Image courtesy of Simon Gilroy, Penn State University. **38.11** Thomas Bjorkman, Cornell University, from the experiments published in L.C. Garner and T. Bjorkman, Mechanical conditioning for controlling excessive elongation in tomato transplants: sensitivity to dose, frequency, and timing of brushing. *Journal of the American Society for Horticultural Science* 121:894-900 (1996). **38.14L** Runk/Schoenberger/Grant Heilman Photography, Inc. **38.14R** ©David M. Dennis/Animals Animals/Earth Scenes **p. 887** David Sieren/Visuals Unlimited

Chapter 39

Opener Photograph courtesy of Professor Adel A. Kader, University of California at Davis **39.5** Malcolm B. Wilkins, University of Glasgow, Glasgow, Scotland, U.K. **39.8** Malcolm B. Wilkins, University of Glasgow, Glasgow, Scotland, U.K. **39.10** Reproduced by permission from F. Gubler et al., Gibberellin-regulated expression of a *myb* gene in barley aleurone cells: evidence for Myb transactivation of a high-p1 alpha-amylase gene promoter. *Plant Cell* 7:1879–1891, p. 1883, fig. 5A (1995). Copyright ©1995 American Society of Plant Biologists. Image courtesy of Frank Gubler, CSIRO Division of Plant Industry, Canberra, ACT, Australia. **39.13** ©Ed Reschke/Peter Arnold **39.16** Nigel Cattlin/Holt Studios International/Photo Researchers, Inc.

Chapter 40

Opener Walter H. Hodge/Peter Arnold, Inc. **40.1aL** Wayne P. Armstrong, Palomar College **40.1aR** ©Mitsuhiko Imamori/Nature Production **40.1bL** Jerome Wexler/Photo Researchers, Inc. **40.1bR** ©Danny Ellinger/Foto Natura/Minden Pictures **40.2a** ©D. Cavagnaro/DRK Photo **40.2b** Lee W. Wilcox **40.4a** Dan Suzio/Photo Researchers, Inc. **40.4b** ©Jerome Wexler/Photo Researchers **40.4c** Jerome Wexler/Photo Researchers, Inc. **40.6bTL** Tom & Therisa Stack/Tom Stack & Associates, Inc. **40.6bTR** Rod Planck/Photo Researchers, Inc. **40.6bB** John Gerlach/DRK Photo **40.7** Leonard Lessin/Photo Researchers, Inc. **40.13** ©Biophoto Associates/Science Source/ Photo Researchers **40.17** Lee W. Wilcox **40.18** Wayne P. Armstrong, Palomar College **40.19TL** R.J. Erwin/Photo Researchers, Inc. **40.19TM** Mark Stouffer/Animals Animals/Earth Scenes **40.19TR** Gerard Lacz/Animals Animals/Earth Scenes **40.19ML** Tom Edwards/Visuals Unlimited **40.19MM** David Stuckel/Visuals Unlimited **40.19MR** Gregory K. Scott/Photo Researchers, Inc. **40.19BL** Fritz Prenzel/Peter Arnold, Inc. **40.19BM** David M. Schleser/Nature's Images, Inc./Photo Researchers, Inc. **40.19BR** Sylvan Wittwer/Visuals Unlimited **40.22** Dr. Rod Peakall

Unit 8

©1996 Susan Middleton & David Liittschwager

Chapter 41

Opener ©Rob Nunnington/Foto Natura/Minden Pictures **41.2** Reproduced by permission from K. Petren et al., A phylogeny of Darwin's finches based on microsatellite DNA length variation. *Proceedings of the Royal Society of London*, Series B, 266:321–329. Image courtesy of Kenneth Petren, University of Cincinnati. **41.6** ©Stewart Cohen/Index Stock Imagery **41.9** Natalie Fobes/Getty Images Inc. - Stone Allstock **41.11a** Fred Hossler/Visuals Unlimited **41.11b** Oliver Meckes & Nicole Ottawa/Photo Researchers, Inc. **41.11c** P.M. Motta, A. Caggiati, G. Macchiarelli/Science Photo Library/Photo Researchers, Inc. **41.16a** Kent, Breck P./Animals Animals/Earth Scenes **41.16b** Mendez, Raymond A./Animals Animals/Earth Scenes **41.16c** ©Wayne Lynch/DRK PHOTO **41.17L** M.I. Walker/Photo Researchers, Inc. **41.17R** Reproduced by permission of Editrice Kurtis S.r.l. from Saverio Cinti, *The Adipose Organ*. Milan: Editrice Kurtis S.r.l., 1999. Copyright ©1999 by Editrice Kurtis S.r.l., Milan, Italy.

Chapter 42

Opener Frans Lanting/Minden Pictures **42.6a** ©Mark Conlin/SeaPics.com **42.7** Reproduced by permission from K. Uchida et al., Morphometrical analysis of chloride cell activity in the gill filaments and lamellae and changes in Na^+,K^+-ATPase activity during seawater adaptation in chum salmon fry. *Journal of Experimental Zoology* 276:193–200 (1996), p. 196, fig. 4A. Copyright ©1996 by Wiley-Liss, Inc., a subsidiary ofJohn Wiley & Sons, Inc. Image courtesy of Toyoji Kaneko, University of Tokyo. **42.8** Stephen Dalton/Photo Researchers, Inc. **42.14a** Fred Hossler/Visuals Unlimited **42.19** Richard R. Hansen/Photo Researchers, Inc.

Chapter 43

Opener ©Jonathan Blair/CORBIS **43.1** Michael Newman/PhotoEdit **43.3a** Craig K. Lorenz/Photo Researchers, Inc. **43.3bL** ©Phototake **43. 3bR** ©Dennis Kunkel/Phototake **43.3cL** ©S.J. Krasemann/Peter Arnold **43.3cM** Wayne Lynch/DRK Photo **43.3cR** David Bassett/Photo Researchers, Inc. **43.4** Mark Smith/Photo Researchers, Inc. **43.5b** Reproduced by permission from K.F. Liem, Evolutionary strategies and morphological innovations: cichlid pharyngeal jaws. *Systematic Zoology* 22:425–441 (1974), p. 436, fig. 15B. Copyright ©1974 by the Society of Systematic Biologists. Image courtesy of Karel F. Liem, Harvard University. **43.10aL** John D. Cunningham/Visuals Unlimited **43.10aR** ©Carolina Biological Supply Company/Phototake **43.14a** Ed Reschke/Peter Arnold, Inc. **43.19** ©Science Photo Library/Photo Researchers, Inc.

Chapter 44

Opener Tony Freeman/PhotoEdit **44.3a** Russell Illig/Getty Images, Inc. - Photodisc **44.3b** ©Michael P. Gadomski/Photo Researchers **44.5a** Walter E. Harvey/Photo Researchers, Inc. **44.5b** John D. Cunningham/Visuals Unlimited **44.15a** Photo Lennart Nilsson/Albert Bonniers Forlag **44.15b** Carolina Biological Supply/Visuals Unlimited

Chapter 45

Opener Courtesy Marcus E. Raichle, M.D., Washington University Medical Center, from research based on S.E. Petersen et al., Positron emission tomographic studies of the cortical anatomy of single.word processing. *Nature* 331:585–589 (1988). **45.2a** ©Dr. Dennis Kunkel/Phototake **45.12a** C. Raines/Visuals Unlimited **45.14** ©Dennis Kunkel/Visuals Unlimited **45.16a** Oliver Meckes & Nicole Ottawa/Photo Researchers, Inc. **45.18a** Reproduced by permission of Elsevier Science from A.R. Mohn et al., Mice with reduced NMDA receptor expression display behaviors related to schizophrenia. *Cell* 98:427–436 (August 20, 1999). Copyright ©1999 Cell Press. Image courtesy of Amy R. Mohn, Duke University. **45.24a** Norbert Wu/Peter Arnold, Inc. **45.25b** Reproduced by permission of Elsevier Science from K.C. Martin et al., Synapse-specific, long-term facilitation of Aplysia sensory to motor synapses: a function for local protein synthesis in memory storage. *Cell* 91:927–938 (1997), fig. 2a. Copyright ©1998 by Elsevier Science Ltd. Image courtesy of Kelsey C. Martin, University of California, Los Angeles.

Chapter 46

Opener Stanley Breeden/DRK Photo **46.1** Stephen Dalton/Photo Researchers, Inc. **46.4** Photograph courtesy of Carole M. Hackney, Mackay Institute of Communication and Neuroscience, School of Life Sciences, Keele University **46.8a** ©David Scharf/Peter Arnold, Inc. **46.10aR** ©Dr. Don Fawcett/T. Kwwabara/Photo Researchers **46.13a** Michael and Patricia Fogden/Fogden Natural History Photographs **46.13b** John A. Pearce, University of Texas at Austin. Research support from Air Force Office of Scientific Research. **46.16L** Ed Reschke/Peter Arnold, Inc. **46.16R** Fred Hossler/Visuals Unlimited **46.19L** Brian Eyden/Science Photo Library/Photo Researchers, Inc. **46.19M** Innerspace Imaging/Science Photo Library/Photo Researchers, Inc. **46.19R** Brian Eyden/Science Photo Library/Photo Researchers, Inc. **46.20** James E. Dennis/Phototake NYC

Chapter 47

Opener Ralph L. Brinster and Robert E. Hammer, School of Veterinary Medicine, University of Pennsylvania **47.7** Photograph courtesy of D. Scott Weigle, M.D.

Chapter 48

Opener Leszczynski, Zig/Animals Animals/Earth Scenes **48.1a** Oxford Scientific Films/Animals Animals/Earth Scenes **48.1b** Tom Adams/Visuals Unlimited **48.1c** Photograph courtesy of Charles J. Cole **48.2a** Oxford Scientific Films/Animals Animals/Earth Scenes **48.4** Reprinted by permission from Catherine S.C. Price et al., Sperm competition between *Drosophila* males involves both development and incapacitation. *Nature* 400:449–452 (1999), figs. 2 and 3. Copyright ©1999 Macmillan Magazines Limited. Image courtesy of Jerry A. Coyne, University of Chicago. **48.5a** G. & C. Merker/B. Tomberlin/Visuals Unlimited **48.6a** Brent D. Opell, Virginia Polytechnic Institute and State University. **48.13a** Claude Edelmann/Photo Researchers, Inc. **48.13b** Photo Lennart Nilsson/Bonnier Alba AB, A CHILD IS BORN, Dell Publishing Company. **48.13c** Petit Format/Nestle/Science Source/Photo Researchers, Inc. **48.15aT** Reprinted with permission from C. Ikonomidou et al., Ethanol-induced apoptotic neurodegeneration and fetal alcohol syndrome. *Science* 287:1056–1060 (February 11, 2000), p. 1057, fig. 1. Copyright ©2000 by the American Association for the Advancement of Science. Image courtesy of Chrysanthy Ikonomidou, Humboldt University, Berlin, Germany.

Chapter 49

Opener Dennis Kunkel/CNRI/Phototake NYC **49.1R** Fred Hossler/Visuals Unlimited **49.2aL** Rosalind King/Science Photo Library/Photo Researchers, Inc. **49.2aM** Chapman-Williams-Blake/Visuals Unlimited **49.2aR** Don Fawcett/Visuals Unlimited **49.5** David M. Phillips/Visuals Unlimited **49.14a** Giovanna Bossi and Gillian Griffiths, University of Oxford, Oxford, U.K.

Unit 9

©Fritz Polking/Peter Arnold, Inc.

Chapter 50

Opener Stephen J. Krasemann/DRK Photo **50.1** ©Natalie B. Fobes Photography **50.8b** Hans Reinhard/OKAPIA/Photo Researchers, Inc. **50.9b** C.K. Lorenz/Photo Researchers, Inc. **50.10b** Tom Bean/DRK Photo **50.11b** Tom Edwards/Animals Animals/Earth Scenes **50.12b** Francis Lepine/Animals Animals/Earth Scenes **50.13b** Tom Bean/DRK Photo **50.17a** ©Gerry Ellis/Minden Pictures **50.17b** ©Tim Fitzharris/Minden Pictures **50.17c** ©Gerry Ellis/Minden Pictures **50.18** ©Georg Gerster/Photo Researchers **50.19** ©Michael Collier **50.21a** ©Gregory Ochocki/SeaPics.com **50.21b** J. Frederick Grassle, Institute of Marine and Coastal Sciences, Rutgers University **50.22** ©Raymond Gehman/CORBIS **50.22 (inset)** ©William H. Mullins/Photo Researchers **50.23b** Michael Fogden/DRK Photo **50.24** Chuck Pratt/Bruce Coleman Inc. **50.25** ©Phil Schermeister/Peter Arnold, Inc. **50.26** ©Scott T. Smith/CORBIS **50.26 (inset)** ©Mark Newman/Photo Researchers

Chapter 51

Opener ©Thomas Mangelsen/Minden Pictures **51.1** ©A. Flowers & L. Newman/Photo Researchers **51.3a** ©Stephen J. Krasemann/DRK Photo **51.3b** ©Gerard Fuehrer/DRK Photo **51.3c** ©Stephen J. Krasemann/DRK Photo **51.5** ©Fred McConnaughey/Photo Researchers **51.6a** Nina Leen/Time Life Pictures/Getty Images **51.6b** ©Frans Lanting/Minden Pictures **51.8a** Photograph courtesy of Dr. Gavin Hunt **51.9** Modified from Weir et al. *Science* 297: 981. Fig. 1. Photo courtesy of the Behavioural Ecology Research Group, Department of Zoology, Oxford University **51.10a** ©Matt Meadows/Peter Arnold **51.10b** McDonald, Joe/Animals Animals/Earth Scenes **51.11a**

Manuel Leal, Department of Biological Sciences, Union College **51.16a** ©Ken Lucas/Visuals Unlimited **51.16b** ©Dr. Paul A. Zahl/Photo Researchers **51.16c** ©Ken Preston.Mafham/Premaphotos Wildlife **51.17** ©Bryan D. Neff **51.18** Stephen Dalton/Photo Researchers, Inc. **51.19** ©Arthur Morris/VIREO **51.20** The Far Side® by Gary Larson ©1980 FarWorks, Inc. All Rights Reserved. Used with permission. **51.21** ©Tom Vezo/Peter Arnold **51.23** Data from Hoogland, J.L. 1983. *Anim. Behav.* 31: 472–479.

Chapter 52

Opener National Geographic Image Collection **52.1** Robert Maier/Animals Animals/Earth Scenes **52.3** Modified from Ghalambor and Martin. 2001. *Science* 292: 494–496. Fig. 1c. **52.6b** Data from Gause, G.F. 1934. *The Struggle for Existence.* (New York: Hafner Press) **52.8** Data from Dodd, M. et al. 1995. *Journal of Ecology* 83: 277–285. **52.10a** ©Paal Hermansen/NHPA/Photo Researchers **52.10b** Data from Valverde and Silvertown. 1998. *J. Ecology* 86: 545–562. **52.14** Marko Nieminen, University of Helsinki, Helsinki, Finland **52.16a** BIOS (F. Mercay)/Peter Arnold, Inc.

Chapter 53

Opener ©J.P. Ferrero/Peter Arnold **53.1** ©Christian Ziegler **53.3** Data from Gause, G.F. 1934. *The Struggle for Existence.* (New York: Hafner Press) **53.5a** G.R. "Dick" Roberts Photo Library/The Natural Sciences Image Library (NSIL) **53.5b** Fred Bavendam/Minden Pictures **53.5c** Tom Till/DRK Photo **53.5d** D. Cavagnaro/Visuals Unlimited **53.5e** ©Stephen J. Krasemann/DRK Photo **53.5e** Lior Rubin/Peter Arnold, Inc. **53.5f** Hal Beral/Visuals Unlimited **53.7L** Dave B. Fleetham/Visuals Unlimited **53.7M** ©Chris Newbert/Minden Pictures **53.7R** ©J. Sneesby/B. Wilkins/Stone **53.8aL** Andrew J. Martinez/Photo Researchers, Inc. **53.8aR** Zig Leszczynski/Animals Animals/Earth Scenes **53.12a** Thomas G. Whitham, Northern Arizona University **53.15a** Mark Moffett/Minden Pictures **53.15b** ©Doug Perrine/SeaPics.com **53.17 (graph)** Data from Paine, R.T. 1974. Intertidal community structure. *Oecologia* 15: 93–120. **53.17T** Nancy Sefton/Photo Researchers, Inc. **53.17B** Thomas Kitchin/Tom Stack & Associates, Inc. **53.19a** Larry Ulrich/DRK Photo **53.19b** Tony C. Caprio **53.20/1** Breck P. Kent/Animals Animals/Earth Scenes **53.20/2** G. Carleton Ray/Photo Researchers, Inc. **53.20/3** Michael P. Gadomski/Animals Animals/Earth Scenes **53.20/4** James P. Jackson/Photo Researchers, Inc. **53.20/5** Bruce Heinemann/Getty Images, Inc./Photodisc **53.20/6** Michael P. Gadomski/Photo Researchers, Inc. **53.20a1–3** Glacier Bay National Park Photo. **53.20a4** Christopher L. Fastie, Middlebury College **53.26b** Jeff Henry/Peter Arnold, Inc.

Chapter 54

Opener Photograph courtesy of Dr. Ellen Mosley-Thompson **54.4** Modified from Nemani et al. 2003. *Science* 300: 1560–1563. **54.5** Data from Tsuda et al. 2003. *Science* 300: 958–961. **54.6BL** Jeremy Burgess/Science Photo Library/Photo Researchers, Inc. **54.6BR** R.F. Ashley/Visuals Unlimited **54.14a** Stephen J. Krasemann/DRK Photo **54.14b** James P. Rowan/DRK Photo **54.22a** Data from Beaugrand et al. 1999. *Science* 296: 1692–1694. **54.22b** Data from Bradley et al. 1999. *PNAS* 96: 9701–9704. **54.23a** Scott Camazine/Photo Researchers, Inc. **54.23b** Oliver Strewe/Getty Images Inc. - Stone Allstock **54.23c** NASA/Goddard Space Flight Center

Chapter 55

Opener Frans Lanting/Minden Pictures **55.1a** ©Tom & Pat Lesson/DRK Photo **55.1b** ©Roland Seitre/Peter Arnold **55.2a** Carr Clifton/Minden Pictures **55.2b** ©Kevin W. Fitz Patrick/All Species Photography & Sound **55.3L** ©Greg Harold/Auscape International **55.3M** ©Greg Harold/Auscape International **55.3R** ©Peter Robinson/Viridans Biological Databases **55.5b** Fred Bruemmer/DRK Photo **55.6a** Stephen J. Krasemann/DRK Photo **55.6bL** John Mitchell/Photo Researchers, Inc. **55.6bM** ©Gary Braasch **55.6bR** David Dennis/Animals Animals/Earth Scenes **55.6c** Tom & Pat Leeson/Photo Researchers, Inc. **55.6d** N. Benvie/Animals Animals/Earth Scenes **55.6e** Gerry Ellis/GLOBIO **55.6f** Norbert Wu/Peter Arnold, Inc. **55.6g** ©Doug Wechsler/VIREO **55.8a** Jacques Jangoux/Peter Arnold, Inc. **55.8b** NASA/Goddard Space Flight Center **55.10** Charles Curtain/Arid Lands Project **55.11** Daniel H. Janzen, University of Pennsylvania

Index

BIOLOGICAL SCIENCE

ABOUT THE PHOTO:

Great tiger moth, *Arctia caja*, as photographed by Joseph Scheer. Scheer, of New York's Alfred University, is the author of *Night Visions: The Secret Designs of Moths* (Prestel, 2003).